Cell Physiology Sourcebook

Cell Physiology Sourcebook

Fifth Edition

Edited by

Francisco Javier Alvarez-Leefmans
Department of Pharmacology and Toxicology, Boonshoft School of Medicine, Wright State University, Dayton, OH, United States

Eric Delpire
Department of Anesthesiology, Vanderbilt University School of Medicine, Nashville, TN, United States

Associate Editor

Edna S. Kaneshiro
Department of Biological Sciences, University of Cincinnati, Cincinnati, OH, United States

ELSEVIER

ACADEMIC PRESS
An imprint of Elsevier

Academic Press is an imprint of Elsevier
125 London Wall, London EC2Y 5AS, United Kingdom
50 Hampshire Street, 5th Floor, Cambridge, MA 02139, United States

For accessibility purposes, images in electronic versions of this book are accompanied by alt text descriptions provided by Elsevier. For more information, see https://www.elsevier.com/about/accessibility.

Books and Journals published by Elsevier comply with applicable product safety requirements. For any product safety concerns or queries, please contact our authorised representative, Elsevier B.V., at productsafety@elsevier.com.

Publisher's note: Elsevier takes a neutral position with respect to territorial disputes or jurisdictional claims in its published content, including in maps and institutional affiliations.

Notices
Knowledge and best practice in this field are constantly changing. As new research and experience broaden our understanding, changes in research methods, professional practices, or medical treatment may become necessary.

Practitioners and researchers must always rely on their own experience and knowledge in evaluating and using any information, methods, compounds, or experiments described herein. In using such information or methods they should be mindful of their own safety and the safety of others, including parties for whom they have a professional responsibility.

To the fullest extent of the law, neither the Publisher nor the authors, contributors, or editors, assume any liability for any injury and/or damage to persons or property as a matter of products liability, negligence or otherwise, or from any use or operation of any methods, products, instructions, or ideas contained in the material herein.

ISBN: 978-0-12-811114-7

Publisher: Megan Ball
Acquisitions Editor: Ali Afzal-Khan
Editorial Project Manager: Harshita Bathla
Production Project Manager: Kiruthigadevi
Cover Designer: Greg Harris

Typeset by TNQ Tech

Dedication

This book is dedicated to the late Nicholas ("Nick") Sperelakis, PhD (1930–2013), renowned pioneer and leader in the field of membrane biophysics and cardiac electrophysiology. In a career that spanned over 5 decades, Nick authored over 400 original scientific papers, and innumerable abstracts, presentations, and book chapters. Nick held faculty positions at Case Western Reserve University, the University of Virginia, and served as the Chairman of the Department of Physiology and Biophysics at the University of Cincinnati College of Medicine. Nick's contributions to the field were recognized by multiple organizations including the American Heart Association's Research Merit Award (1996) and the Samuel Kaplan Visionary Award for Cardiac Research (1996).

Nick was also a committed educator and mentor to generations of students and junior faculty. Research was his life but teaching was his passion! Nick loved to quote the Greek philosopher Diogenes (c 350 bc): "The foundation for every state is the education of its youth." At the University of Cincinnati, his students honored his dedication with the Professor of the Year award, and the Outstanding Research, Mentoring and Teaching Award, among many accolades. Nick edited the first through fourth editions of *Cell Physiology Sourcebook* and authored multiple chapters. His insights and rigorous biophysical explanations of membrane excitation are now considered classics.

Nick was a proud veteran of the US Marine Corps, a devoted patron of over 60 organizations concerned with the environment, animal welfare, and human rights, and a devoted humanitarian who volunteered for the Wood's Hole Marine Biology Research facility and Project HOPE in Peru and Mexico. Nick's seemingly boundless energy impressed all those who had the privilege of interacting with him as a colleague, collaborator, teacher, and mentor.

In Memoriam

This book represents the culmination of a long-standing collaboration among its authors. It is with profound sadness that we note the sudden and unexpected passing away of our esteemed colleague Professor Juan Vicente Sánchez-Andrés on September 4, 2023. Juan Vicente was the director of the Department of Medicine at Jaume I University in Castellón de la Plana, Spain. He achieved great recognition as a scientist in the field of pancreatic beta-cell physiology. Among his key contributions was the first in vivo intracellular recording of the electrical activity of beta cells, demonstrating glucose-dependent membrane potential oscillations that occurred within the physiological range. Beta cells were not his only passion; he was a tireless promoter of physiology teaching and research in Spain and editor of an important Spanish medical journal (*Revista de Neurología*). Juan Vicente was a contributing author to this book. In collaboration with Drs Yukari Takeda and Colin A. Leech, he wrote Chapter 26, "*Pancreas and Insulin Secretions*," which was revised and updated by his coauthors and the editors. His scholarly contributions are woven throughout these pages. He will be greatly missed by everyone who had the privilege of knowing him.

We were also deeply saddened by the passing of Professor Erich Sackmann in May 2024. He was a pioneer of biophysics in Europe and an influential researcher and educator in his native Germany. He was a professor of physics and head of the biophysics department at the Physics Department of the Technical University of Munich, where he became an emeritus professor. His work contributed to foundational knowledge of biological membrane mechanics, including elasticity and response to external forces. Among his many contributions was the refinement and application of magnetic tweezers capable of exerting very small pulling forces, of tens of piconewtons, to study the dynamics of single actin filaments, actin networks, cell adhesion, and cell migration in living cells, the subject of Chapter 33, "*Cell locomotion driven by pulsed solitary actin gelation waves and actin-microtubule crosstalk.*" His contribution was revised and updated by the editors. Any errors or omissions that remain are our own.

Francisco Javier Alvarez-Leefmans
Eric Delpire

Contents

Contributors

Francisco J. Alvarez-Leefmans, Department of Pharmacology and Toxicology, Boonshoft School of Medicine, Wright State University, Dayton, OH, United States

Chinwendu Amazu, Department of Obstetrics and Gynecology, School of Medicine, Washington University, Saint Louis, MO, United States

Clive M. Baumgarten, Department of Physiology and Biophysics, Virginia Commonwealth University School of Medicine, Richmond, VA, United States

Paolo Bernardi, Dipartimento di Scienze Biomediche, Università degli Studi di Padova, Padova, Italy

Albertino Bigiani, Dipartimento di Scienze Biomediche, Metaboliche e Neuroscienze, Università di Modena e Reggio Emilia, Modena, Italy

Matthew R. Bowers, Colorado State University, Department of Biomedical Sciences, Fort Collins, CO, United States; Colorado State University, Molecular, Cellular and Integrative Neurosciences Program, Fort Collins, CO, United States

David P. Clark, Department of Microbiology, Southern Illinois University, Carbondale, IL, United States

Charles D. Cox, Mechanosensory Biophysics Laboratory, Victor Chang Cardiac Research Institute & St Vincent's Clinical School, Darlinghurst, NSW, Australia

Eric Delpire, Department of Anesthesiology, Vanderbilt University School of Medicine, Nashville, TN, United States

Neil D. Detweiler, Department of Biology, Goshen College, Goshen, IN, United States

Roberto Docampo, Center for Tropical and Emerging Global Diseases and Department of Cellular Biology, University of Georgia, Athens, GA, United States

Sarah K. England, Department of Obstetrics and Gynecology, School of Medicine, Washington University, Saint Louis, MO, United States

Joseph J. Feher, Department of Physiology and Biophysics, Virginia Commonwealth University School of Medicine, Richmond, VA, United States

Federico Fogolari, Dipartimento di Scienze Matematiche, Informatiche e Fisiche, Università degli Studi di Udine, Udine, Italy

Kenneth W. Foster, Physics Department, Syracuse University, Syracuse, NY, United States

Tamara Frank, Halmos College of Natural Sciences and Oceanography, Nova Southeastern University, Dania Beach, FL, United States

Jeffrey C. Freedman, Department of Neuroscience and Physiology, SUNY Upstate Medical University, Syracuse, NY, United States

Kamil Godula, Department of Chemistry and Biochemistry and the Glycobiology Research and Training Center, University of California San Diego, La Jolla, CA, United States

Dennis W. Grogan, Department of Biological Sciences, University of Cincinnati, Cincinnati, OH, United States

Dorothee Günzel, Clinical Physiology/Nutritional Medicine, Charité - Universitätsmedizin Berlin, Berlin, Germany

Dan R. Halm, Department of Neuroscience, Cell Biology, and Physiology, Wright State University Boonshoft School of Medicine, Dayton, OH, United States

Taviare L. Hawkins, Department of Physics, Wagner College, Staten Island, NY, United States

Judith A. Heiny, Department of Pharmacology and Systems Physiology, University of Cincinnati, College of Medicine, Cincinnati, OH, United States

Aldebaran M. Hofer, VA Boston Healthcare System and the Department of Surgery, Brigham and Women's Hospital and Harvard Medical School, West Roxbury, MA, United States

Daniel Honigfort, Department of Chemistry and Biochemistry and the Glycobiology Research and Training Center, University of California San Diego, La Jolla, CA, United States

Mia L. Huang, Department of Chemistry and Biochemistry and the Glycobiology Research and Training Center, University of California San Diego, La Jolla, CA, United States

Tom Iwanicki, The Earth Commons Institute, Georgetown University, Washington, DC, United States; Smithsonian Institution, National Museum of Natural History, Department of Invertebrate Zoology, Washington, DC, United States; Department of Biology, University of Hawai'i at Mānoa, Honolulu, HI, United States

Colin A. Leech, Department of Medicine, State University of New York (SUNY) Upstate Medical University, Syracuse, NY, United States

Simon Rock Levinson, Department of Pharmacology, University of Colorado School of Medicine, Aurora, CO, United States; Department of Physiology and Biophysics, University of Colorado School of Medicine, Aurora, CO, United States

Giovanna Lippe, Dipartimento di Area Medica, Università degli Studi di Udine, Udine, Italy

David Lloyd, Microbiology, School of Biosciences, Cardiff University, Wales, United Kingdom

Frédéric H. Login, Department of Clinical Medicine, Aarhus University, Aarhus, Denmark

Boris Martinac, Mechanosensory Biophysics Laboratory, Victor Chang Cardiac Research Institute & St Vincent's Clinical School, Darlinghurst, NSW, Australia

Marisa McDonald, School of Life Science, University of Hawai'i at Mānoa, Honolulu, HI, United States

David R. Mitchell, Cell and Developmental Biology, SUNY Upstate Medical University, Syracuse, NY, United States

Santhosh Kumar Nagarajan, Clinical Physiology/Nutritional Medicine, Charité - Universitätsmedizin Berlin, Berlin, Germany

Lene N. Nejsum, Department of Clinical Medicine, Aarhus University, Aarhus, Denmark

Michal K. Oklinski, Department of Clinical Medicine, Aarhus University, Aarhus, Denmark

Mark D. Parker, The Jacobs School of Medicine and Biomedical Sciences, The State University of New York, University at Buffalo, Buffalo, NY, United States

Matthew R. Pincus, Department of Pathology, SUNY Health Sciences University, Brooklyn, NY, United States

Jörg Piontek, Clinical Physiology/Nutritional Medicine, Charité - Universitätsmedizin Berlin, Berlin, Germany

Megan L. Porter, School of Life Science, University of Hawai'i at Mānoa, Honolulu, HI, United States; Department of Biology, University of Hawai'i at Mānoa, Honolulu, HI, United States

Srinivasan Ramakrishnan, Center for Tropical and Emerging Global Diseases and Department of Cellular Biology, University of Georgia, Athens, GA, United States

Noreen E. Reist, Colorado State University, Department of Biomedical Sciences, Fort Collins, CO, United States; Colorado State University, Molecular, Cellular and Integrative Neurosciences Program, Fort Collins, CO, United States

Nancy J. Rusch, Department of Pharmacology and Toxicology, College of Medicine, University of Arkansas for Medical Sciences, Little Rock, AR, United States; Department of Pharmaceutical Sciences, College of Pharmacy, University of Arkansas for Medical Sciences, Little Rock, AR, United States

Erich Sackmann, Physics Department, Technical University Munich, Garching, Germany

Juan V. Sanchez-Andres, Department of Medicine, University Jaume I-UJI, Castellon, Spain

Jorge A. Sánchez, Department of Pharmacology, Centro de Investigación y de Estudios Avanzados del Insitututo Politécnico Nacional, Mexico City, Mexico

William A. Sather, Department of Physiology and Biophysics and Department of Pharmacology, University of Colorado School of Medicine, Aurora, CO, United States

Mallory C. Shields, Colorado State University, Department of Biomedical Sciences, Fort Collins, CO, United States; Colorado State University, Molecular, Cellular and Integrative Neurosciences Program, Fort Collins, CO, United States

Nicholas Sperelakis[†], Department of Physiology and Biophysics, University of Colorado School of Medicine, Aurora, CO, United States

Amanda J. Stolarz, Department of Pharmacology and Toxicology, College of Medicine, University of Arkansas for Medical Sciences, Little Rock, AR, United States; Department of Pharmaceutical Sciences, College of Pharmacy, University of Arkansas for Medical Sciences, Little Rock, AR, United States

[†] Deceased.

Shiro Suetsugu, Nara Institute of Science and Technology, Ikoma, Japan

Ildikó Szabó, Dipartimento di Biologia, Università degli Studi di Padova, Padova, Italy

Yukari Takeda, Department of Bioinformatics, Ritsumeikan University, Shiga, Japan

Kenneth R. Tovar, University of California Santa Barbara, Santa Barbara, CA, United States

Megan Valentine, State University of New York at Plattsburgh, Department of Biological Sciences, Plattsburgh, NY, United States

Judith Van Houten, Department of Biology, University of Vermont, Burlington, VT, United States

Richard D. Veenstra, Department of Pharmacology, SUNY Upstate Medical University, Syracuse, NY, United States

Gary L. Westbrook, Vollum Institute, Oregon Health & Science University, Portland, OR, United States

Preface

The history of this book goes back to its influential predecessor, *A Textbook of General Physiology,* written by a single author: the noted British physiologist Hugh Davson (1909–96). His book reached its fourth and last edition in 1970. The successor of Davson's work was *Cell Physiology Sourcebook*, first published in 1995 and edited by Nicholas Sperelakis (1930–2013) with a forward written by Hugh Davson himself. At that time, molecular and cell physiology knowledge had already become so vast that a single author's work, like its forerunner, was materially impossible, thus requiring the participation of many contributors. The highly successful *Cell Physiology Sourcebook* was later updated in the second (1998), third (2001), and fourth (2012) editions. Twelve years have passed since its last edition, and we are delighted to bring the book's fifth edition to light. The success of *Cell Physiology Sourcebook* and the gap it fills motivated the present editors to continue this project, updating it entirely to reflect new developments. This was a challenging task, given the rapid succession of new discoveries in various fields of molecular and cell physiology. Significant technological advances in molecular biology, computing, and microscopy in the past decade have propelled this new knowledge explosion. Not less challenging was the fact that we began this project in the midst of a devastating pandemic that impacted all realms of life and society. We strived to incorporate new molecular insights in the present edition without losing the integrative perspective of cell physiology and biophysics and its foundational concepts. We have added or significantly expanded classic topics such as diffusion in intracellular and extracellular environments and across biological membranes, the biophysical properties of tight junctions, osmotic balance, and cell volume regulation, and added new chapters such as water channels, F-ATPases, transepithelial transport mechanisms, bacteria migration, taxis, kinesis, quorum sensing, and tropism, among others.

Without the dedication of all authors involved, to whom we are forever grateful, this book would not exist. The editors are also particularly thankful to André G. Wolff, Samuel Young, and Selvaraj Raviraj, our publisher, editorial project manager, and project manager, respectively, at Elsevier, for their unfailing support and encouragement throughout all the steps of this project, from conception to production. We would also like to thank the collaboration in the initial steps of this project to our associate editor, Professor Edna S. Kaneshiro. We gratefully acknowledge our respective institutions, Wright State University, Boonshoft School of Medicine, and Vanderbilt University School of Medicine, for their continued support. Finally, we cannot stress enough the foresight of our predecessor, Professor Nicholas Sperelakis, who conceived the concept of a *Cell Physiology Sourcebook* and to whom we dedicate this edition.

Francisco Javier Alvarez-Leefmans
Eric Delpire

Chapter 1

Biophysical chemistry of physiological solutions

Jeffrey C. Freedman
Department of Neuroscience and Physiology, SUNY Upstate Medical University, Syracuse, NY, United States

1. Summary

This initial chapter begins by describing the hydrogen-bonded structure of liquid water and its dipolar and dielectric properties. In salt solutions, water exhibits three regions of structure as follows: (1) oriented dipoles near ions; (2) an intermediate structure-breaking region; and (3) flickering clusters with short-range order characteristic of ice. Electrostatic interactions between ions and water based on Coulomb's law account for the enthalpy of hydration, as well as for the sequence of ionic mobilities of the hydrated alkali cations. Salts and nonelectrolytes dissolve in intracellular water with nearly the same solubility and activity as in extracellular water; even the viscosities are similar. Protons in solution are hydrogen-bonded to water forming H_3O^+, which is further hydrated due to electrostatic forces. Protons migrate through water by a Grotthus chain mechanism; in contrast, other hydrated ions migrate through water as hard spheres that interact with water dipoles according to their ionic radius and charge. Nonspecific ion−ion interactions reduce the activity coefficients of ions, as described by the Debye−Hückel theory, which conceptualizes a central ion surrounded by a cloud of counterions with smeared charge. Selective interactions of ions with sites on enzymes or on channel proteins occur in certain predictable specific patterns (Eisenman sequences) that depend on the energy of interaction between the ions and water dipoles, relative to that between ions and their binding sites.

All of the solutes in a cell contribute to the free energy of the intracellular solution. The Gibbs equation, which is derived in the Appendix by combining the first and second laws of thermodynamics, enables estimation of the changes in free energy when solutes traverse cell membranes. The change in free energy is zero at equilibrium and negative for spontaneous processes. Solutes will redistribute until the electrochemical potential is the same in every compartment to which that solute has access. The Nernst equation (see Appendix), which follows from the Gibbs equation, relates the ratio of intracellular and extracellular ion concentrations at equilibrium to the membrane potential, and can be used to test whether a solute is in electrochemical equilibrium across a cell membrane.

2. Introduction

All living cells contain proteins, salts and water enclosed in membrane-bounded compartments. These biochemical and ionic cellular constituents, along with a set of genes, enzymes, substrates, and metabolic intermediates, function to maintain cellular homeostasis and enable cells to replicate and to perform chemical, mechanical and electrical work. *Homeostasis*, a term introduced by the physiologist Walter Cannon in *The Wisdom of the Body* (Cannon, 1932), means that certain parameters, including cellular volume, intracellular pH, the transmembrane electrical potential and intracellular concentrations of salts are maintained relatively constant in resting cells. Cellular homeostasis depends on a relative constancy of the extracellular fluids that bathe cells. The extracellular fluid compartment was termed the *milieu intérieur*, or internal environment, by Claude Bernard, who recognized around 1865 that "*La fixité du milieu intérieur est la condition de la vie libre*", that the constancy of the internal environment is the condition for independent life. Thus, in the 1949 translation of *An Introduction to the Study of Experimental Medicine* (Bernard, 1949) we find that "only in the physico-chemical

Cell Physiology Sourcebook. https://doi.org/10.1016/B978-0-12-811114-7.00018-9

conditions of the inner environment can we find the causation of the external phenomena of life". The development of cell physiology has been greatly influenced by the Bernard—Cannon theory of physical-chemical homeostasis.

Biophysical chemistry concerns the application of the concepts and methods of physical chemistry to the study of biological systems. Physical chemistry includes such physiologically relevant subjects as thermodynamics, chemical equilibria and reaction kinetics, solutions and electrochemistry, properties and the kinetic theory of gases, transport processes, surface phenomena and molecular structure and spectroscopy. Throughout this book, it is seen that many cellular physiological phenomena are best understood with a rigorous and comprehensive understanding of physical chemistry. During the past 80 years, a succession of outstanding monographs and reviews on biophysical chemistry include the following: Höber (1945); Edsall and Wyman (1958); Tanford (1961); Eisenberg and Kauzmann (1969); Cooke and Kuntz (1975); Cantor and Schimmel (1980); Silver (1985); Bergethon and Simons (1990); Weiss (1996). To help the reader understand how cellular homeostasis is achieved, this chapter will introduce some of the conceptual underpinnings of cell physiology by describing certain physicochemical properties of water and electrolytes that are important for understanding living cells.

3. Structure and properties of water

Biological cells contain a large amount of water, ranging from 0.66 g H_2O/g cells in human red blood cells to around 0.8 g H_2O/g tissue in skeletal muscle. The amount of water in cells is determined by osmosis. Liquid water is a highly polar solvent, with a structure stabilized by extensive intermolecular *hydrogen bonds* (Fig. 1.1). The hydrogen bonds between adjacent water molecules are linear (O—H O) but are able to bend by about 10 degrees. From X-ray diffraction studies of ice crystals, it is known that each of the two covalent O—H bonds in water is 1 Å in length, that each of the two hydrogen bonds is about 1.8 Å in length and that these occur around each oxygen atom at angles of 104.5 degrees in a tetrahedral array.

The energy required to break hydrogen bonds in liquid water is 5—7 kcal/mol, much less than the 109.7 kcal/mol required to break the covalent O—H bond. A calorie is a unit of work equal to 4.18 Joule (J) in the SI system of units, where 1 J equals 1 Newton meter. The newton ($N = k\ gm/s^2$) is the unit of force in the International System of Units (SI); the dyne ($dyn = g\ cm/s^2$) is the corresponding unit of force in the older cgs system ($1\ N = 10^5$ dyn). Recall that, according to Newton's first law, the force (F) is mass times acceleration and that work is force times distance. Because of the strength of hydrogen bonds, liquid water has unique physicochemical properties, including a high boiling point (100°C), a high *molar heat capacity* (18 cal/mol°K), a high *molar heat of vaporization* (9.7 kcal/mol at 1 atm) and a high *surface tension*

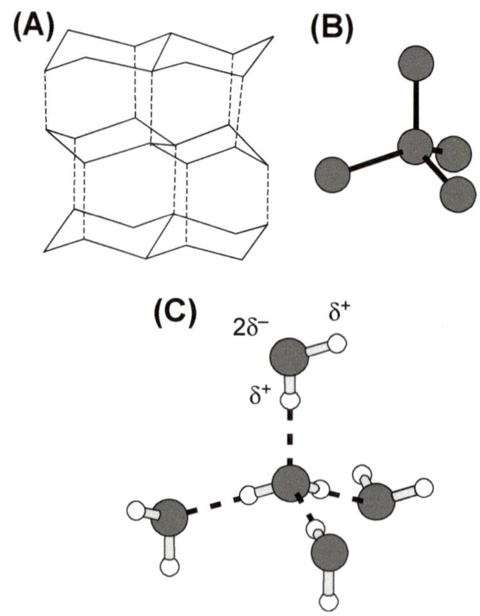

FIGURE 1.1 Hydrogen bonds in ice and liquid water. (A) Crystal structure of ice, showing the puckered hexagonal rings of oxygen atoms. (B) Each oxygen atom is connected to four others in a tetrahedron. (C) The linear hydrogen bonds are indicated by dotted lines between adjacent water molecules (O—H O). Two covalent bonds (O—H), indicated by solid lines, and two hydrogen bonds occur around each oxygen atom in a tetrahedral array. The partial charge separation of the O—H dipoles is indicated on the upper water molecule.

(72.75 dyn/cm). As discussed by Bergethon and Simons (1990), all of these thermodynamic parameters are considerably higher for H_2O than for the analogous compound H_2S, which boils at −59.6°C. The outer electrons of the S atom shield its nuclear charge and reduce its electronegativity, making the S—H bond weaker, longer and less polar than the O—H bond. Also, the bond angle of H_2S is only 92.1 degrees, an angle that does not form a tetrahedral array and thus H_2S does not form a hydrogen-bonded network such as water.

The extensive intermolecular association in liquid water is due to the geometry of the water molecule and also to its strong permanent *dipole moment* (see Debye, 1929). Water molecules have no net charge, yet possess permanent charge separation along each of the two O—H bonds. Since oxygen is more *electronegative* than hydrogen, its electron density is preferentially greater. Oxygen thus acquires a partial negative charge (δ^-), leaving the hydrogens with a partial positive charge (δ^+) (see Fig. 1.1C). The magnitude of the dipole moment (μ) of a chemical bond is computed as the product of the separated charge (q) at either end, and the distance (d) between the centers of separated charge ($\mu = qd$). The dipole moment of a molecule is the vector sum of the dipole moments of each bond. For water in the gaseous phase, the dipole moment is 1.85 debye, where one debye $= 10^{-18}$ esu·cm. One esu (electrostatic unit), or stat-coulomb, equals 3.336×10^{-10} C, the coulomb (C) being a unit of electric charge. In liquid water, the dipole moment becomes even larger (2.5 debye) due to association with other water molecules. In comparison, the linear molecule carbon dioxide (O=C=O) also has a separation of charge, with a partial negative charge on each oxygen atom and a partial positive charge on the carbon atom. In this case, the oppositely directed dipole moments of the 2 C=O bonds sum to a zero dipole moment for the CO_2 molecule. In H_2S, the S—H bond has a smaller dipole moment (1.1 debye) than the O—H bond in water because sulfur is less electronegative than oxygen. Electrostatic attractions between opposite charges in adjacent water dipoles stabilize the hydrogen-bonded structure of liquid water.

Water strongly affects the forces between ions in solution by virtue of its high *dielectric constant*. The dielectric constant (ε) of a medium known as a *dielectric* is defined as the ratio of the coulombic force (F_{coul}) between two charges in a vacuum to the actual force (F) between the same two charges in the dielectric medium.

$$\varepsilon = \frac{F_{coul}}{F} \tag{1.1}$$

In a vacuum, the coulombic force, F_{coul} (newtons, N), between two ions with charges q^+ and q^- (coulombs, C) separated by a distance d (meters, m) is given by Coulomb's law:

$$F_{coul} = \frac{q^+q^-}{4\pi\varepsilon_0 d^2} \tag{1.2}$$

where ε_0 is the permittivity constant ($8.854 \times 10^{-12} C^2/N\ m^2$). Coulomb's law states that the force between two point charges is directly proportional to the magnitude of each charge and inversely proportional to the square of the distance between the charges. Substituting F_{coul} into the above expression and rearranging yields the following:

$$F = \frac{q^+q^-}{4\pi\varepsilon_0\varepsilon d^2} \tag{1.3}$$

The dielectric constant of a vacuum is unity and that of air at 25°C is close to unity ($\varepsilon_{air} = 1.00059$). Increasing the dielectric constant of the solvent decreases the attractive force between oppositely charged ions. The force felt by a distant ion is reduced in a dielectric medium, such as water, as compared with a vacuum, because part of the interaction energy is spent aligning the intervening water dipoles and distorting their polarizable electron clouds. The dielectric constant of pure water at 25°C is 78.4, much greater than methanol ($\varepsilon = 32.7$), ethanol (24.3), or methane (1.7). In liquid water, the high dielectric constant weakens the coulombic attractive forces between oppositely charged particles and thus promotes dissociation and ionization of salts.

Although the thermodynamic properties of liquid water are explicable in terms of extensive hydrogen-bonding, the actual structure of water is complex. In the flickering cluster model, groups of 50—70 water molecules, resembling a slightly expanded broken piece of the ice lattice (icebergs), are continuously associating and dissociating on a picosecond time scale. This dynamic model contrasts with the extended order of solid ice. At a given instant, some water molecules are unattached to the clusters and are located in the interstitial regions of the network, but may attach and detach as the clusters continuously form and break down. Other theories treat water as a mixture of distinct states, or as a continuum of states, with considerable short-range order, characteristic of the crystalline lattice of ice.

Despite these uncertainties regarding the structure of liquid water and the influence of macromolecules on its properties in cells, the ability of water to act as a solvent inside cells closely resembles that of extracellular water, as discussed in detail in Chapter 9. Thus, a variety of permeant, hydrophilic, nonmetabolized nonelectrolytes distribute at equilibrium across human red blood cell membranes with ratios of intracellular to extracellular concentrations that deviate from unity by less than 10% (Gary-Bobo, 1967). In mouse diaphragm muscle, Miller (1974) found that several alcohols, diols and monosaccharides exhibit distribution ratios within 2% of unity, whereas certain other sugars appear to be excluded from membrane-bounded intracellular compartments. The amount of solute that dissolves at equilibrium is also nearly normal in water that is constrained in gels, which are cross-linked networks of fibrous macromolecules. The diffusion of solutes within gels, however, may be hindered by collisions and interactions with the macromolecules and by the tortuosity of the diffusion paths, as discussed in detail in Chapter 9 (Biophysics of intracellular and extracellular diffusion).

Viscosity, another property of water, contributes to the resistance to flow. A fluid with a greater viscosity exhibits less flow under the influence of a given pressure gradient than a fluid with a lesser viscosity. For example, molasses, a concentrated sugar solution, is much more viscous than pure water. During *laminar flow*, a frictional force develops between adjacent layers (laminae) in the fluid and this force impedes the sliding of one lamina past its neighbor. In a *Newtonian fluid*, the frictional force per unit area, or shear stress (τ, dyn/cm^2), is proportional to the velocity gradient, or rate of strain (dv/dy, s^{-1}) between laminae,

$$\tau = \eta \cdot (dv / dy) \tag{1.4}$$

where the viscosity (η) is the proportionality constant with units of poise (=1 dyn s/cm^2), named after Poiseuille. The viscosity of H_2O at 20.3°C is 0.01 P, or one cP (cp). In cultured fibroblasts, the fluid-phase cytoplasmic viscosity, as determined from rotational motions of fluorescent probes on a picosecond time scale, is only 1.2–1.4 times that of pure water (Fushimi & Verkman, 1991). Fluorescence studies also show that the viscosity is the same in the cytoplasm and nucleoplasm and is unaffected by large decreases in cell volume or by disruption of the cytoskeleton with cytochalasin B. The fluid-phase viscosity, as determined from fluorophore rotational motions, is not nearly as affected by macromolecules as is the bulk viscosity and it thus provides a more accurate view of the physical state of the aqueous domain of the cytoplasm. These and other studies (e.g., Horowitz & Miller, 1984; Schwan & Foster, 1977) set limits on the extent to which the physicochemical properties of intracellular water differ from those of extracellular water. (See Chapter 9, Section 7.2.4 for further discussion).

4. Interactions between water and ions

Ions in solution behave as charged, hard spheres that interact with and orient water dipoles. When crystals of sodium chloride are dissolved in water, the electrostatic attractive forces between water dipoles and ions in the crystal lattice overcome the interionic attractive forces between oppositely charged ions in the crystal. The dissociated ions then acquire the freedom of translational motion as they diffuse into the solution accompanied by a layer of tightly associated *hydration water* (Fig. 1.2). *Ion* is the Greek word for "wanderer". The strength of the attraction between ions and water dipoles

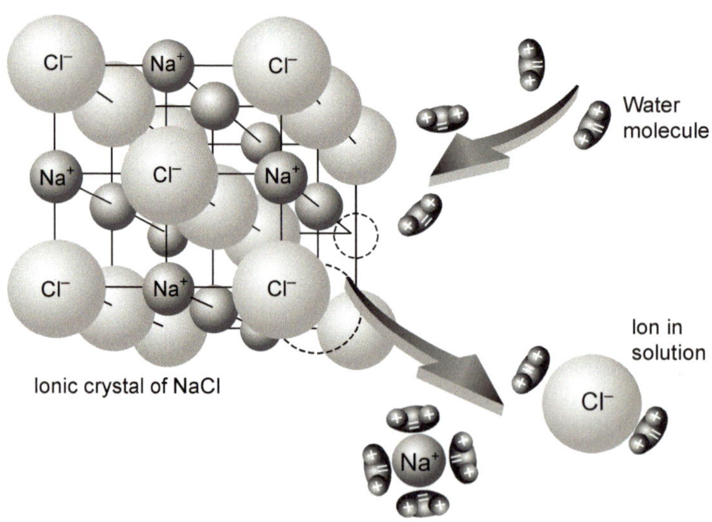

FIGURE 1.2 Dissociation of NaCl in water into hydrated Na^+ and Cl^- ions.

depends to a large extent on the ionic charge and radius. For the alkali metal cations, the force of attraction and the energy of interaction of ions with water, decreases according to the following series: $Li^+ > Na^+ > K^+ > Rb^+ > Cs^+$.

Li^+, being the smallest of the alkali metal cations, has the strongest interaction with water because its positively charged nucleus can approach most closely to the negative side of neighboring water dipoles. As the ionic radius increases with increasing atomic number in the alkali metal series, the filled outer shells of electrons effectively shield the cationic charge and reduce the distance of closest approach to water molecules. The smallest ion thus acquires the greatest degree of hydration and has the largest hydrated ionic radius (see Chapter 9 for further details).

According to Frank and Wen (1957), the orienting influence of ions on water dipoles results in three regions of water structure (Fig. 1.3). The electric field of an ion is sufficiently strong to remove water dipoles from the bulk water clusters and to attract to itself the oppositely charged ends of from one to five water dipoles. A certain number of these water dipoles, called the *hydration number*, then become trapped and oriented in the ion's electric field. The inner hydration shell includes water molecules that are aligned by the force field and in direct contact with the ion, 5 ± 1 for Li^+, 4 ± 1 for Na^+, 3 ± 2 for K^+ and Rb^+, 4 ± 1 for F^-, 2 ± 1 for Cl^- and Br^- and 1 ± 1 for I^-. Thus, in physiological saline at 0.15 M NaCl in water (55.5 M), about 1.6% ($=100 \times 0.15 \times 6/55.5$) of the water is located in the inner hydration shells of Na^+ and of Cl^-. Water in the inner hydration shells of Na^+, K^+ and Ca^{2+} rapidly exchanges with bulk water on a nanosecond time scale but, in contrast, the smaller divalent cation Mg^{2+}, with its high charge density, is some four orders of magnitude slower in exchanging its inner hydration water. The inner hydration sheath of water molecules moves together with the ion as a distinct and single kinetic entity.

The mobilities of the alkali cations in water decrease as the nonhydrated ionic radius decreases, but as the hydrated ionic radius increases (Table 1.1; for discussion, see Hille, 2001). The ionic *mobility* (u) is defined as the proportionality constant that relates the velocity (v) of ionic migration to the force exerted by an external electric field E ($v = u \cdot E$); in other words, mobility is the velocity per unit electric field. Farther away from the ion, where the ion's electric field falls towards zero, water retains the structure of bulk water. In the region between the inner hydration sheath and the bulk water, the orienting influences of the ion and the bulk water network tend to compete and to disrupt the structure of water. In this intermediate structure-breaking region, the water dipoles are partially oriented toward the central ion, yet do not migrate with the ion, and they join only infrequently with the hydrogen-bonded clusters of the bulk water. If the net effect of an ion is to disorganize more water in the intermediate region than is found in the primary hydration shell, then the ion is termed a *structure-breaker*. Conversely, water may form a variety of hydrogen-bonded clathrate structures around apolar protein side chains, whose action resembles that of the class of solutes termed *structure-makers* (see Klotz, 1970, for review).

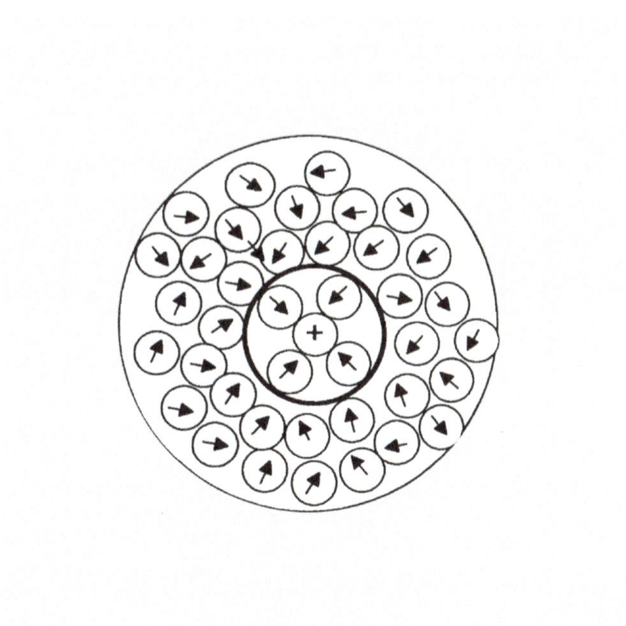

FIGURE 1.3 Three regions of water structure near an ion. A central cation is shown with a primary hydration shell of four oriented water dipoles, surrounded by a partially oriented secondary sheath. The square *gray* region represents the hydrogen-bonded structure of the bulk water.

TABLE 1.1 Radii, enthalpies of hydration and mobilities of Selected ions.

Ion	Nonhydrated Radius[a] (Å)	$\Delta H^0_{hydration}$[b] (kcal/mol)	Mobility[c] $10^{-4}\frac{cm/s}{V/cm}$
H^+	—	−269	36.25
Li^+	0.60	−131	4.01
Na^+	0.95	−105	5.19
K^+	1.33	−85	7.62
Rb^+	1.48	−79	8.06
Cs^+	1.69	−71	8.01
Mg^{2+}	0.65	−476	2.75
Ca^{2+}	0.99	−397	3.08
Sr^{2+}	1.13	−362	3.08
Ba^{2+}	1.35	−328	3.30
Cl−	1.81	−82	7.92

[a]*Radii are from Pauling (1960).*
[b]*Standard enthalpies of hydration at 25°C are from Edsall and McKenzie (1978).*
[c]*Mobilities in water at 25°C are from Hille (2001, p. 317).*

The *enthalpy of hydration* of an ion is a measure of the strength of the interaction between ions and water and is defined as the increase in enthalpy when 1 mole of free ion in a vacuum is dissolved in a large quantity of water. The enthalpy of hydration may be estimated from the heat released upon dissolving salts in water, usually less than 10 kcal/mol, taking into account the energy needed to dissociate the salt crystal and then to hydrate the ions. The enthalpy of hydration may be calculated using electrostatic theory. Accurate results are obtained if water is considered to be an electric quadrupole with four centers of charge, two partial positive charges near the hydrogen nuclei and two partial negative charges on the nonbonded electron orbitals near the oxygen nucleus. A further improvement in the calculated enthalpy of hydration is obtained when the polarizability (α) of the water dipole by the ion is taken into account. The electric field of the ion distorts the electron cloud of the hydration water along its permanent dipole axis, thus inducing an additional increment of charge separation and increasing the dipole moment. For small fields, this *induced dipole moment* (μ_{ind}) is proportional to the electric field strength (E) and the constant of proportionality is the *polarizability* ($\mu_{ind} = \alpha \cdot E$). In Fig. 1.4, the measured

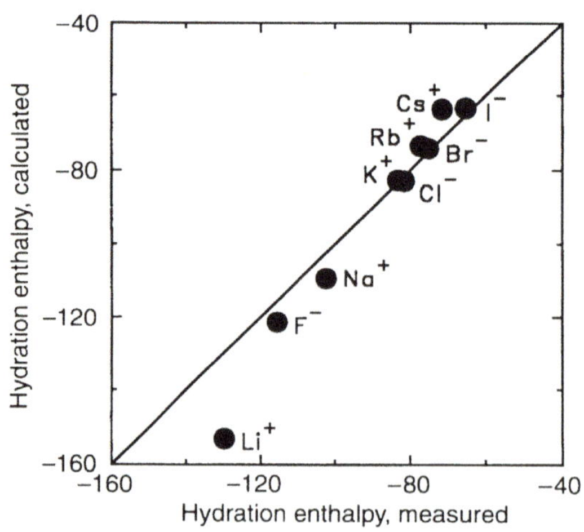

FIGURE 1.4 Comparison of measured enthalpies of hydration of alkali metal cations and halides in water (abscissa), with those calculated (ordinate) taking into account the Born charging energy, ion-dipole interactions, ion-quadrupole interactions and ion-induced dipole interactions. *Data from Bockris and Reddy (1970), p. 107.*

enthalpies of hydration for alkali metal cations and halides are compared with calculated values; the impressive agreement demonstrates the primary importance of electrostatic forces in determining the solvation of ions in water. The enthalpy of hydration for the smallest alkali metal cation Li^+ is quite large at −131 kcal/mol. With increasing ionic radius, the enthalpy falls progressively for Na^+, K^+ and Rb^+, reaching −71 kcal/mol for Cs^+, the largest nonhydrated but smallest hydrated ion in the series (see Table 1.1). For the divalent, alkaline earth metal ions Mg^{2+} and Ca^{2+}, the enthalpies of hydration also follow the ionic radius but are considerably larger at −476 kcal/mol and −397 kcal/mol, respectively. As pointed out by Hille (2001), the magnitude of ionic hydration enthalpies approximates the cohesive strength of the ionic bonds in a crystalline salt lattice.

Physical chemical analyses show that, in small spaces that restrict the formation of tetrahedral water clusters, the dielectric constant will be reduced and electrostatic forces between ions will then be correspondingly increased. In the primary hydration shell of ions, the oriented water is polarizable but cannot be reoriented by applied fields and so the dielectric constant is reduced from its bulk value of 78 to a value of about 6, with intermediate values in the partially oriented structure-breaking region between the primary hydration shell and the bulk water. For this reason, the dielectric constant of a solution decreases with increasing salt concentration but, in protein-free solutions at physiological salt concentrations, the extent of the decrease is small due to the small fraction of hydration water.

5. Protons in solution

The hydrogen-bonded structure of liquid water also contributes to high proton mobility by a mechanism that differs fundamentally from the migration of other hydrated ions. The proton is a highly reactive, positively-charged hydrogen nucleus, devoid of electrons. Whereas ions with electron shells typically have diameters in angstroms, the diameter of the proton is only about 10^{-5} Å. With such a small size, the proton has a strong attraction to electrons, as indicated by the high *ionization energy* of 323 kcal/mol needed to remove an electron from a hydrogen atom to form a proton. By comparison, the ionization energies for the alkali cations decrease with increasing atomic number in the series as follows:

$$\begin{aligned} Li^+(124\ \text{kcal/mol}) &> Na^+(118\ \text{kcal/mol}) \\ &> K^+(100\ \text{kcal/mol}) \\ &> Rb^+(96\ \text{kcal/mol}) \\ &> Cs^+(90\ \text{kcal/mol}) \end{aligned}$$

As more filled electron shells separate and shield the outer shell from the positively charged atomic nucleus, it becomes easier to form a cation by removing an electron. The high affinity of the proton for electrons, and its small size, explain its tendency to form hydrogen bonds with the unshared electrons of oxygen in water. Infrared spectra indicate that protons in solution exist predominantly in the form of hydronium ions H_3O^+. Nuclear magnetic resonance (NMR) data are consistent with a flattened trigonal pyramidal structure, with O—H bond lengths of 1.02 Å and H—O—H bond angles of 115 degrees, a structure resembling that of NH_3 but with a radius similar to that of K^+. The free energy of formation of H_3O^+ from H_2O and H^+ is about −170 kcal/mol, which corresponds to a hypothetical concentration of free protons in solution at room temperature of about 10^{-150} M — "about as zero as one can get" say Bockris and Reddy (1970). The heat of hydration of H_3O^+ is an additional −90 kcal/mol. The change in the density of water with temperature is consistent with the hydration of H_3O^+ with an additional three water molecules, forming the tetrahedral cluster $H_9O_4^+$ (Bockris & Reddy, 1970). The proton mobility in water is 36 × 10^{-4} cm^2/s V, much slower than expected on the basis of hydrodynamic theory if free protons were to carry the current, but curiously about seven times as fast as expected if H_3O^+ migrates like K^+. The abnormally high proton mobility in water (and also in ice) is consistent with a *Grotthus chain mechanism* (Fig. 1.5) in which the successive breakage and formation of hydrogen bonds, accompanied by proton jumps between neighboring water molecules, effectively results in the passage of H_3O^+ along a chain. This transport process is rate-limited by the time needed for each successive acceptor water molecule to rotate and reorient its nonbonded orbital into a suitable position to accept the donated proton. The

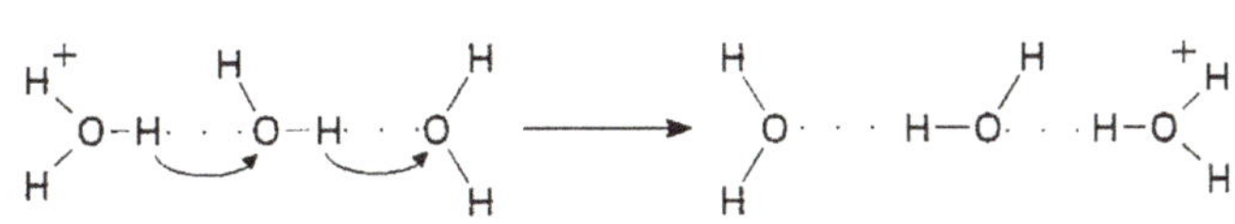

FIGURE 1.5 Grotthus chain mechanism for proton mobility in water.

Grotthus mechanism accounts for about 80% of proton mobility, with the remaining 20% due to H_3O^+ itself, as a single kinetic entity, undergoing a translational migratory movement through the solvent like other ions.

6. Interactions between ions

Interactions between ions may be weak or highly selective. Ions in solution attract ions of the opposite charge, in accordance with Coulomb's law. Since the long-range attractive forces are inversely proportional to the square of the distance between charges, the interaction energies are greater in more concentrated solutions. In an uni-univalent salt solution, defined as having monovalent cations and anions, of concentration *c* (mol/L), the average distance (*d*, angstroms or Å) between any two ions is given by:

$$d = \left(10^{27}/2N_{A}c\right)^{1/3} \tag{1.5}$$

where N_A is Avogadro's number (6.023×10^{23}). The factor 2 is included because the anions and cations are both counted and the factor 10^{27} is Å^3/L. For 0.15 M NaCl, the average distance between Na^+ and the nearest Cl^- is 17.7 Å, at which range coulombic forces are highly significant. At pH 7.4, where the concentration of "protons" (really hydronium ions) is only 40 nM, the average distance between "protons" is 0.28 μm.

Attractive ion–ion interactions are stabilizing and they lower the chemical potential of an ion from its value in an ideal, infinitely-dilute solution. The *activity* (a) of an ion is defined in terms of its *chemical potential* (μ) as follows:

$$\mu = \mu^0(T, P) + RT \text{ In } a \tag{1.6}$$

where $\mu^0(T, P)$ is the *standard-state chemical potential*, or the chemical potential when the activity is 1. Increases in chemical activity increase the chemical potential. The chemical potential of a solute in an ideal dilute solution, where the activity equals the chemical concentration (c), is:

$$\mu_{\text{ideal}} = \mu^0(T, P) + RT \text{ In } c \tag{1.7}$$

In order to describe the properties of more concentrated nonideal solutions, Gilbert N. Lewis (1875–1946) introduced the *activity coefficient* (γ), such that the activity is the product of the concentration (c) and the activity coefficient ($a = \gamma \cdot c$). The chemical potential of an ion in a nonideal solution is:

$$\mu = \mu^0(T, P) + RT \text{ In } c + RT \text{ In } \gamma \tag{1.8}$$

where the term $RT \ln \gamma$ includes the effect of ion–ion interactions on the chemical potential.

When a salt ($c_+ c_-$) is added to water, the cations and anions, at concentrations c_+ and c_- respectively, both contribute to the free energy of the solution. The chemical potentials of the cations and anions (μ_+ and μ_-, respectively) are given by:

$$\mu_+ = \mu_+^0 + RT \text{ In } c_+ + RT \text{ In } \gamma_+ \tag{1.9}$$

$$\mu_- = \mu_-^0 + RT \text{ In } c_- + RT \text{ In } \gamma_- \tag{1.10}$$

Adding these two expressions and taking the average, gives:

$$\frac{\mu_+ + \mu_-}{2} = \frac{\mu_+^0 + \mu_-^0}{2} + RT \text{ In } (c_+ c_-)^{1/2} + RT \text{ In } (\gamma_+ \gamma_-)^{1/2} \tag{1.11}$$

The mean ionic activity coefficient $\gamma\pm$ is defined as:

$$\gamma_{\pm}(\gamma_+ \gamma_-)^{1/2} \tag{1.12}$$

Mean ionic activity coefficients of salts have been estimated theoretically by the *Debye–Hückel theory*, which supposes that central ions attract oppositely-charged ions (or *counterions*) in a diffuse and structureless ion cloud (or atmosphere). The forces of coulombic attraction between the central ion and its cloud of counterions are opposed by the randomizing influence of the thermal motion of the ions. In dilute solutions, in which the theory treats central ions as point charges relative to the size of the ion cloud, the mean ionic activity coefficient ($\gamma\pm$) for salt ions with charges z_+ and z_-, is given by the limiting law as follows:

$$\log \gamma_{\pm} = -A|z_+ \quad z_-|I^{1/2} \tag{1.13}$$

where I is the *ionic strength*, defined as:

$$I = \frac{1}{2}\sum_i c_i z_i^2 \tag{1.14}$$

The constant A, which equals 0.5108 $kg^{1/2}/mol^{1/2}$ is given in terms of B, which equals 0.3287×10^8 $kg^{1/2}/mol^{1/2}/cm$, both at 25°C, as follows:

$$A = \frac{1}{2.303}\frac{N_A e_0^2}{2\varepsilon RT}B \tag{1.15}$$

$$B = \sqrt{\frac{8\pi N_A e_0^2}{1000\varepsilon K_B T}} \tag{1.16}$$

where N_A is Avogadro's number (6.023×10^{23} ions/mol), e_0 is the charge on an electron (4.80×10^{-10} statcoul $= 1.60 \times 10^{-19}$ C), ε is the dielectric constant, R is the gas constant (8.32 J/mol·K), T is the absolute temperature (°K) and K_B ($=R/N_A$) is Boltzmann's constant (1.38×10^{-23} J/K).

The derivation of the Debye–Hückel limiting law (see Bockris & Reddy, 1970) computes the spherically symmetric electric field around an ion, leading to an expression relating the charge density (ρ_r) of the ionic cloud to the electrostatic potential (ψ_r) at a distance r from the central point charge. The counterions, at concentration n_i for ionic species i at a distance r from the central ion, are considered to distribute in the electric field according to a *Boltzmann distribution*:

$$n_i = n_i^0 e^{-z_i e_0 \psi_r / k_B T} \tag{1.17}$$

where n_i^0 is the bulk ion concentration (ions/L). By linearizing the Boltzmann equation, using a Taylor series expansion and retaining only the first two terms, which assumes that $z_i e_0 \psi_r \ll k_B T$, a resulting expression shows how the electrostatic potential ψ decreases as a function of distance (r) from the central ion (Fig. 1.6A):

$$\psi_r = \frac{z_i e_0}{\varepsilon}\frac{e^{-\kappa_B r}}{r} \tag{1.18}$$

where κ is given by:

$$\kappa = \sqrt{\frac{4\pi}{\varepsilon k_B T}\sum_i n_i^0 z_i^2 e_0^2} \tag{1.19}$$

The charge density (ρ_r) of the ionic cloud also decreases with increasing distance from the central ion (see Fig. 1.6B) according to:

$$\rho_r = -\frac{r_i e_0}{4\pi}\kappa^2 \frac{e^{-\kappa r}}{r} \tag{1.20}$$

The amount of charge dq contained in a concentric spherical shell of thickness dr located at a distance r from the central ion (see Fig. 1.6C) is given by:

$$dq = -z_i e_0 e^{-\kappa r}\kappa^2 r dr \tag{1.21}$$

Furthermore, the maximum amount of charge contained in such a spherical shell occurs at a distance κ^{-1}, known as the *Debye length*, which defines the effective radius of the counterion atmosphere. It can also be shown that the Debye length is the distance away from the central ion where an ion of equal and opposite charge would contribute to the electrostatic field by an amount equivalent to that of the dispersed cloud of counterions (Bockris & Reddy, 1970). The Debye length decreases as the ionic concentration increases (see Fig. 1.6D).

The Debye–Hückel limiting law predicts activity coefficients accurately only to concentrations of about 0.01 M. In more concentrated solutions, the finite radius (a, in cm) of the ion is taken into account in the *extended Debye–Hückel equation*, given by:

$$\log \gamma_\pm = \frac{A|z_+ z_-|I^{1/2}}{1 + Ba\ I^{1/2}} \tag{1.22}$$

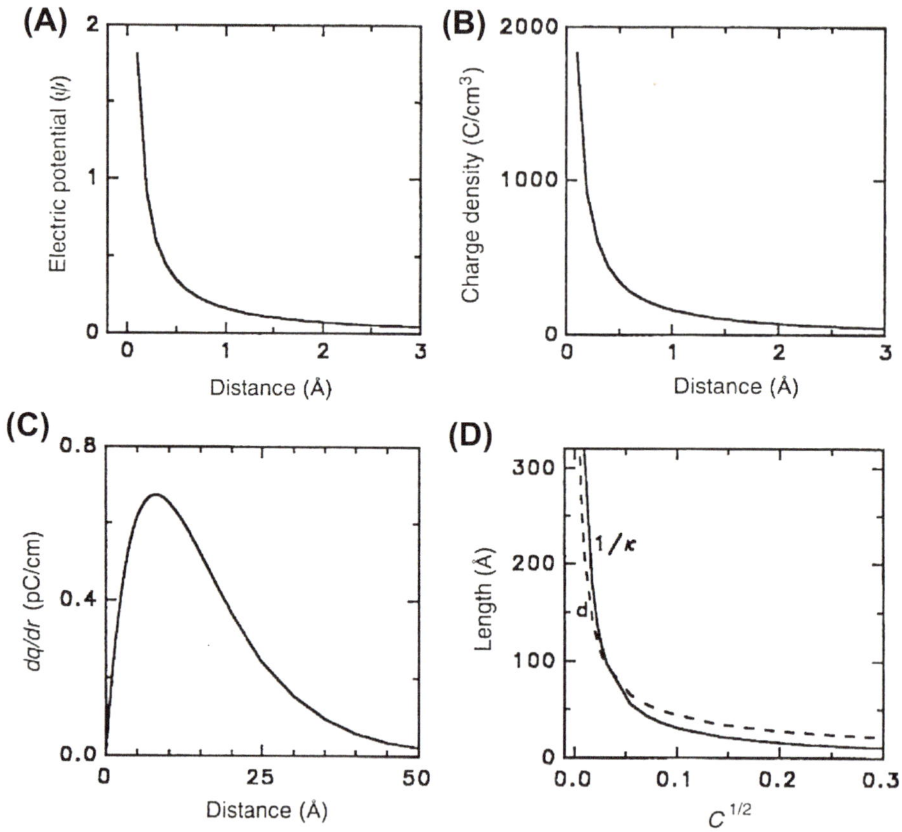

FIGURE 1.6 Debye–Hückel theory. (A) Electrostatic potential (ψ), (B) charge density (ρ) of the counterion cloud and (C) the charge (dq) enclosed in a spherical shell of thickness (dr), each calculated and plotted versus distance d from the central ion. (D) Debye length κ^{-1} (*solid line*) and average distance d between ions (*dashed line*) versus square root of ion concentration.

where the constants A and B are still 0.5108 $kg^{1/2}/mol^{1/2}$ and 0.3287×10^8 $kg^{1/2}/mol^{1/2}/cm$, respectively, for water at 25°C. The ion-size parameter (a) is adjustable, but reasonable values extend the range of concentrations for which the Debye–Hückel theory accurately predicts ionic activity coefficients (Fig. 1.7).

At still higher salt concentrations above about 0.7 M, the activity coefficient stops decreasing and instead begins to increase with increasing salt concentration. This behavior has been explained by the increasing fraction of hydration water, which both reduces the amount of free water in the solution and raises the effective concentration and, therefore, the

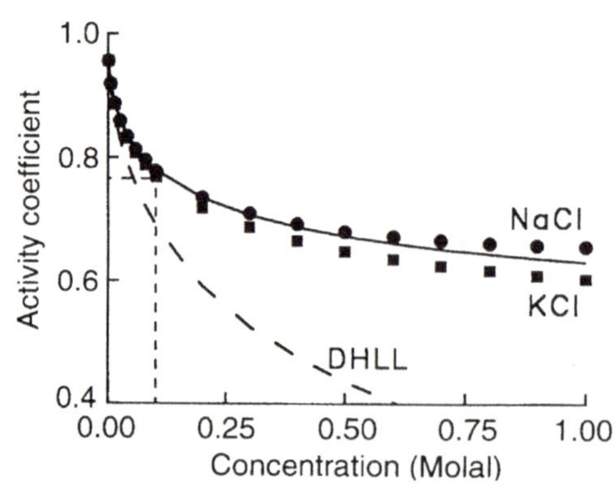

FIGURE 1.7 Activity coefficients of NaCl (circles) and KCl (squares). The dashed curve represents the prediction of the debye–Hückel limiting law. The *solid* line represents the prediction of the extended debye–Hückel equation with an ion size parameter a of 4.7 Å. The dashed vertical line indicates that at 0.1 M, the activity coefficients of NaCl and KCl are 0.77 and 0.76, respectively. *Data are from Robinson and Stokes (1959).*

activity, of the dissolved ions. Corrections for hydration enable prediction of the activity coefficient as a function of salt concentration over the full range of salt concentrations reaching to several molar. A related effect is that, in concentrated protein solutions, a certain shell of volume known as *excluded volume* around each protein is unavailable for solutes because the center of the solutes with their own finite size can only approach to within one solute radius of the protein surface. The fraction of excluded volume increases with increasing protein concentration and with increasing radius of the solutes which themselves may be proteins with enzymatic or regulatory activities. Such *macromolecular crowding* can dramatically increase activities of proteins in concentrated solutions (Kuznetsova et al., 2015).

The activity coefficients calculated by the Debye–Hückel theory indicate that the forces between ions in dilute solutions are weak and nonselective. The predicted activity coefficients depend on the ionic charge and the ionic strength of the solution and they are largely independent of the specific ion within, say, the alkali metal series. A very small degree of selectivity is introduced with the ion size parameter. The mean ionic activity coefficients for NaCl and KCl as a function of salt concentration are shown in Fig. 1.7. At 0.1 M concentrations, the activity coefficient of Na^+ is 0.77, whereas those of K^+ and Cl^- are both 0.76. These relatively high activity coefficients correspond to ion–ion interaction energies of only about −0.3 kcal/mol (see Hille, 2001).

Agreement between theory and data, such as seen in Fig. 1.7 for the Debye–Hückel theory, does not necessarily imply the correctness of the theory, or of the model on which the theory is based. A fundamental criticism of the Debye–Hückel theory concerns the smeared charge model for the cloud of counterions (Frank & Thompson, 1960). Below a concentration of about 0.001 M, the Debye length κ^{-1} is greater than the average distance between ions in the solution (see Fig. 1.6D), in which case the counterions could reasonably appear as a cloud of smeared charge from the vantage point of the central ion. Above 0.001 M, however, the Debye length incongruously falls below the average distance between ions in the solution. Moreover, at only 0.01 M, just one ion is needed to account for 50% of the effect of the counterion cloud on the central ion, yet this ion must be smeared around the central ion in a cloud of spherically symmetric charge distribution located only 25 Å away from the central ion (Bockris & Reddy, 1970). Thus, well below physiological salt concentrations, the assumption of smeared charge in the Debye–Hückel theory is not in accord with the coarse-grained structure of salt solutions, despite the impressive ability of the theory to predict activity coefficients. A quasi-lattice approach to salt solutions minimizing the free energy using modern computational power would seem to be a preferable alternative (see Horvath, 1985, for references).

Intracellular solutions may be quite concentrated. The red-cell membrane and associated cytoskeleton enclose a viscous cytoplasmic solution of the oxygen-binding pigment *hemoglobin* at a concentration of 34 g/100 mL cells, corresponding to 5.2 mM, 7.3 millimolal, or 44 g Hb/100 g cell water. This concentration is near the threshold for gelation, but hemoglobin is one of the most soluble of all proteins and constitutes more than 98% of red-cell protein by mass. The hemoglobin $\alpha_2\beta_2$ tetrameters have a molecular weight of 64,373 Da and are approximately spheroidal with dimensions of 65 × 55 Å × 50 Å. In the interior of red blood cells, hemoglobin is nearly close-packed, with the distance between the surfaces of neighboring proteins averaging only about 20 Å. With the 86 carboxylates of aspartic and glutamic acids and with the 98 basic amino and amine groups of arginine, lysine and histidine per hemoglobin tetramer, the total cellular concentration of the titratable amino acids of human hemoglobin is around 0.9 M. Thus, hemoglobin contributes significantly, depending on intracellular pH, to a high intracellular ionic strength. Since the charged amino acids are located on the surface of hemoglobin and taking the radius of hemoglobin to be 28 Å, the average distance between charged sites is only about 7 Å. About 7600 water molecules per hemoglobin tetramer occupy the narrow interstices between the protein molecules. The hydration of hemoglobin in dilute solution is 0.2–0.3 g water/g Hb, representing about 15% of the intracellular water. When the tortuosity of the surface of soluble proteins is taken into account, as much as 30% of red-cell water could reside in the first monolayer around the protein surface.

To determine the *mean ionic activities* of the intracellular KCl and NaCl in the concentrated charged environment of human red blood cells, studies were conducted in which the red-cell membrane was rendered permeable to cations by exposure of the cells to the channel-forming antibiotic nystatin, thus allowing K^+, Na^+ and Cl^- to reach *Gibbs–Donnan equilibrium*. In these experiments, sufficient extracellular sucrose was added to prevent cell swelling by balancing the colloid osmotic pressure of hemoglobin and other impermeant cell solutes. The *mean ionic activity coefficient* of KCl and NaCl in the concentrated intracellular hemoglobin solution was found to be within 2% of that in the extracellular solution (Freedman & Hoffman, 1979). In view of the high intracellular ionic strength and the high volume fraction of cell water in direct contact with hemoglobin, it is both curious and remarkable that intracellular salts and nonelectrolytes appear to behave as if in dilute solution. Either the intracellular solution is indeed like a dilute solution or, alternatively, the expected effect of protein–solvent interactions in altering the activity of intracellular solutes is offset by the effect of interactions between proteins and the solutes themselves.

TABLE 1.2 Eisenman's selectivity sequences for binding of alkali cations to negatively charged sites.

Highest field strength of site	
Li > Na > K > Rb > Cs	Sequence XI
Na > Li > K > Rb > Cs	Sequence X
Na > K > Li > Rb > Cs	Sequence IX
Na > K > Rb > Li > Cs	Sequence VIII
Na > K > Rb > Cs > Li	Sequence VII
K > Na > Rb > Cs > Li	Sequence VI
K > Rb > Na > Cs > Li	Sequence V
K > Rb > Cs > Na > Li	Sequence IV
Rb > K > Cs > Na > Li	Sequence III
Rb > Cs > K > Na > Li	Sequence II
Cs > Rb > K > Na > Li	Sequence I
Lowest field strength of site	

Specific effects of small anions on the solubility, aggregation or denaturation of proteins often follow the *Hofmeister (lyotropic) series* in the following order of effectiveness:

$$F^- < PO_4^{3-} < SO_4^{2-} < CH_3COO^- < Cl^- < Br^- < NO_3^- < I^- < SCN^-$$

The ions to the right of Cl^- are referred to as *chaotropic*, since they tend to destabilize proteins (for review, see Collins & Washabaugh, 1985). Ions may also associate with sites on proteins, as exemplified by the binding to albumin of Ca^{2+} (Katz & Klotz, 1953), and of Cl^- (Scatchard et al., 1957; see Chapter 8 in Tanford, 1961). Certain enzymes, ion channels and membrane transport proteins interact with the alkali cations with a high degree of *ionic selectivity*. Considering the five alkali metal cations, there are 5! (=120) possible orders of selectivity that might arise. However, only 11 cationic selectivity orders commonly occur in chemical and biological systems, as listed in Table 1.2. Eisenman (1967) predicted these selectivity orders by calculating the ion-site interaction energies. If the field strength of a negatively-charged site is weak, then an associated ion remains hydrated. In this case, Cs^+, having the smallest hydrated ionic radius, is favored (Sequence I) because it can approach most closely to the site and thus has the strongest coulombic force of attraction. If the field strength of the site is strong and the interaction energy between the ion and the site is stronger than the energy between the ion and water dipoles, then the ion is stripped of its associated water and becomes dehydrated. In this case, Li^+, having the smallest nonhydrated ionic radius, becomes favored (sequence XI). With intermediate field strengths, ions partially dehydrate, giving rise to the intervening selectivity sequences. For example, K^+ functions as a cofactor for the glycolytic enzyme pyruvate kinase, with the maximal catalytic velocity following Eisenman sequence IV. During in vitro protein synthesis, K^+ maintains an active conformation of 50S ribosomal subunits that catalyze peptide bond formation following Eisenman sequence III. Other sequences are possible when the sites are assumed to be polarizable (for review, see Eisenman & Horn, 1983).

7. Solute transport: Basic definitions

In the intracellular or extracellular solutions, a solute is said to be at *equilibrium* when its concentration is constant in time without requiring the continuous input of energy from metabolism or other sources. In human red blood cells, for example, Cl^-, HCO_3^- and H^+ are at thermodynamic equilibrium, i.e., they are passively distributed. A solute is said to be at *steady-state* when its concentration is constant in time, but is dependent on the continuous input of energy from metabolism or other sources. *Active transport* utilizes energy and results in steady-state distributions that represent a deviation from equilibrium. In biological cells, the concentrations of Na^+, K^+ and Ca^{2+} are at a steady-state. Their concentrations are constant in time, but are dependent on the continuous hydrolysis of ATP by the Na^+-K^+ pump (for a classic review, see Glynn, 1985) and the Ca^{2+} pump. With isolated red cell membranes ("ghosts"), active transport of Na^+ and K^+ against their respective concentration gradients has been observed directly by measurements of net fluxes (Freedman, 1976). The active production and maintenance of ion concentration gradients by membrane pumps, such as the Na^+, K^+-ATPase and the Ca^{2+}-ATPase, represent *chemical work* and *electrical work* done by the cell. The energy required for continuous pumping of Na^+ by resting frog sartorius muscles has been estimated to represent some 14%–20% of the energy available from the hydrolysis of ATP and the value is similar in red blood cells.

Passive transport is the movement of solutes toward a state of equilibrium. Passive transport of *hydrophobic* substances across cell membranes usually occurs directly by diffusion across the lipid bilayer. Passive transport of *hydrophilic* substances is usually mediated by specific membrane proteins through a process known as *facilitated diffusion*. Passive ion transport may occur through pores or ion channels under the influence of concentration gradients and electrical forces by a process known as *electrodiffusion*. The passive flow of ionic currents down their concentration gradients through specific ion channel proteins constitutes negative work done by the cell. The changes in free energy associated with ion transport and the Nernst equilibrium equation are described in the Appendix (See also Chapter 10, Membrane diffusion and permeability).

8. Measurement of electrolytes and membrane potential

Ion concentrations in biological fluids may be expressed as millimoles per liter of solution (millimolar, or mM), or as millimoles per kilogram of water (millimolal). When a solution containing metallic ions is aspirated into a flame, each type of ion burns with a characteristic color, Na^+ giving a yellow flame, K^+ giving a violet flame and Ca^{2+} giving a red flame. In the technique of *flame photometry*, the intensity of the emitted light, in comparison with that produced by solutions containing known concentrations of ions, provides a convenient measure of ion concentration in extracellular fluids and in acid extracts of cells. With a uniform rate of aspiration, a flame photometer accurately measures the intensity of the emitted light, which is related linearly to the cation concentrations in suitably diluted standards and unknowns (see e.g., Funder & Wieth, 1966). *Atomic absorption spectroscopy* is an alternative technique that measures the light absorbed by ions during electronic excitation in a flame. Flame photometry and atomic absorption spectroscopy both measure total ionic concentrations in cell extracts irrespective of any intracellular compartmentation and they are sensitive in the millimolar range of cellular concentrations. K^+, Na^+, Ca^{2+} and other elements in single cells, or even in single cell organelles, may be measured by *electron probe microanalysis*, a technique that utilizes an electron beam to excite the emission of X-rays with energies characteristic of the various elements in cells. A limitation of these methods is that they are invasive and destructive and yield total concentrations that do not discriminate between bound, free ions and ion activities (Tsien, 1983).

The development of *ion-specific glass microelectrodes* by Eisenman (1967) and then of selective *liquid ion exchange microelectrodes*, made possible the direct determination of intracellular cation activities. Palmer and Civan (1977) found that for Na^+, K^+ and Cl^- of *Chironomus* salivary gland cells, the ion activities are the same in the nuclear and cytoplasmic compartments. In a related study, Palmer et al. (1978) found that during development of frog oocytes, the ratio of the cytoplasmic concentration of Na^+ to K^+ increases, while the corresponding ratio of ion activities decreases. This observation could reflect the development of yolk platelets and intracellular vesicles that contain ions at differing concentrations and activities than does the bulk cytoplasm. In frog skeletal muscle, the sarcoplasmic reticulum contains a solution enriched in calcium, whereas the ionic composition of the solution in the t-tubules is extracellular. The extracellular space, however, consists of the interstitial space between the muscle fibers as well as the vascular space; solutes leave these two extracellular compartments with differing rate constants, thus considerably complicating the interpretation of experiments that assess the rate of membrane transport with radioactive isotopes of Na^+, K^+ and other solutes (Neville, 1979; Neville & White, 1979).

To measure transient changes of intracellular Ca^{2+} in the micromolar and submicromolar range, *fluorescent*—indicator *dyes* such as Quin-2, Fura-2, Indo-1 and Fluo-3 were developed by Roger Y. Tsien (Tsien, 1988). Quin-2, Fura-2, and Indo-1 are fluorescent analogs of ethylenediaminetetraacetic acid (EDTA), which contains four carboxylate groups that specifically bind two divalent cations. EGTA is a nonfluorescent analog with a higher binding affinity for Ca^{2+} as compared with its affinity for Mg^{2+} and it is thus quite useful in experiments where the extracellular concentration of Ca^{2+} is systematically varied. Fluo-3 is a tetracarboxylate fluorescein analog that exhibits a shift in the emission spectrum upon binding Ca^{2+}; in contrast, Fura-2 undergoes a shift in its excitation spectrum. By measuring the ratio of Fura-2 fluorescence upon excitation at two exciting wavelengths, changes in the concentration of Ca^{2+} may be monitored. The cells are incubated with a permeant ester form of the dye to enable the dye to permeate into cells; intracellular esterases then release the Ca^{2+}-sensitive chromophore. With video microscopy of cells stained with fluorescent Ca^{2+} indicators, it is also possible to obtain time-resolved and spatially-resolved light microscopic images of the changes in intracellular Ca^{2+}. Confocal laser scanning microscopy, and multiphoton microscopy, allows the precise spatial and temporal analysis of intracellular Ca^{2+} activity at the subcellular level in addition to measurement of its concentration (Takahashi et al., 1999). Another fluorescent probe (SPQ), developed by Helsey and Verkman (1987), has been used to measure intracellular Cl^- and to study its transport across cell membranes. Current methods to measure intracellular Cl^- are discussed in detail in Chapter 15, (Intracellular chloride regulation)

The transmembrane electrical potential is usually measured by means of open-tipped microelectrodes such as those developed and used by Ling and Gerard (1949) to obtain accurate and stable measurements of the membrane potential of frog skeletal muscle (see Chapter 12, Origin of the resting membrane potential). With human red blood cells, stable potentials have not been achieved with microelectrodes. As an alternative technique, Hoffman and Laris (1974) utilized fluorescent cyanine dyes to monitor and measure red cell membrane potentials. Fluorescent cyanines, merocyanines, oxonols, sytryls, rhodamines and other dyes have since been used in numerous electrophysiological studies of red blood cells, neutrophils, platelets and other nonexcitable cells and organelles that are too small for the use of microelectrodes (for review, see Freedman & Novak, 1989a, 1989b; Balach et al., 2019). The equilibrium distribution of permeant, lipophilic, radioactively labeled ions such as triphenylmethylphosphonium ($TPMP^+$) may also be used to assess the membrane potential using the Nernst equilibrium equation (see Freedman & Novak, 1989a, 1989b).

9. Appendix: Thermodynamics of membrane transport

9.1 Free energy

When ions cross cell membranes, changes in free energy are involved. Consider a membrane permeable only to cations separating two solutions of KCl of differing concentrations, c_i and c_o, where the subscripts i and o represent the intracellular and extracellular compartments, respectively (Fig. A.1). If the intracellular concentration (c_i) is greater than the extracellular concentration (c_o), then potassium will tend to diffuse out of the cell down its concentration gradient. When only a very slight amount of potassium crosses the membrane (which for this example is assumed to be impermeable to chloride), the separation of charge creates a transmembrane electrical potential (E_m) that is negative inside, relative to outside. The resultant electrical force retards further efflux of potassium. An equilibrium is reached when the diffusion force favoring efflux of potassium exactly balances the electrical force, preventing efflux of potassium (see Chapter 12, Origin of the resting membrane potential).

Prior to the attainment of equilibrium, the process of moving dn moles of K^+ ions out of the cell from compartment i to compartment o involves a change in *free energy* (dG) of the system. According to the laws of thermodynamics, the process will occur spontaneously only if $dG < 0$ and the system will be at equilibrium if $dG = 0$. Thus, in order to compute the change in free energy (dG), we need to review certain basic thermodynamic principles.

The *first law of thermodynamics*, also known as the law of conservation of energy, states that for any system, the increase in *energy* (dE) is the gain in *heat* (dQ) minus the *work* (dW) done by the system:

$$dE = dQ - dW \tag{1.23}$$

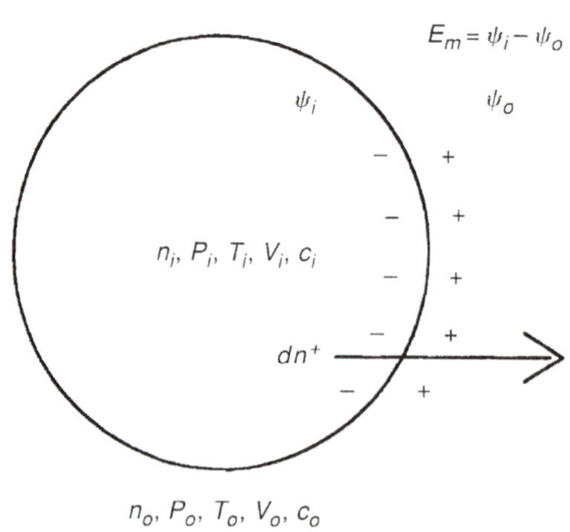

FIGURE A.1 Free energy of ion transport. A cell membrane separates the intracellular (subscript i) and the extracellular (subscript o). Each compartment is at pressure P and temperature T, has volume V and electrical potential ψ and contains solutes at concentration c. A free energy change is involved when dn^+ moles of cation leave the cell. The membrane potential E_m is $\psi_i - \psi_o$.

For reversible processes, the *second law of thermodynamics* defines the change in entropy (dS) of a system in terms of the heat gained (dQ) and the absolute temperature (T) as follows:

$$dS = \frac{dQ}{T} \tag{1.24}$$

Combining the first and second laws gives:

$$dE = TdS - dW \tag{1.25}$$

Since the difference in pressure (dP) between the internal and external solutions of animal cells is negligible, the work (dW) done by the system includes *pressure—volume work*, PdV, and electrical work, $zF E_m dn$, summed as follows:

$$dW = PdV + zFE_m dn \tag{1.26}$$

where P is the pressure, V is the total volume of the system, z is the ionic valence (eq/mol), F is the Faraday constant (=96,490 C/eq) and dn is the number of moles of solute crossing the membrane and leaving the cell. The transmembrane electrical potential (E_m) is the difference in electrical potential (ψ, in volts or J/C), between the two compartments:

$$E_m = \psi_i - \psi_0 \tag{1.27}$$

The electrical potential (ψ) represents the work done in moving a unit positive test charge from infinity to a point in the solution. Note that when the transmembrane electrical potential (E_m) is negative inside, the electrical work done by the system is negative whenever Na^+ or K^+ leave the cell, but positive when K^+ or Na^+ enter the cell, irrespective of the mechanism of transport.

According to the first and second laws, the change in energy (dE) of the system is:

$$dE = TdS = PdV - zFE_m dn \tag{1.28}$$

For systems of variable chemical composition, the free energy (G) is by definition:

$$G = H - TS + \sum \mu_j n_j \tag{1.29}$$

where the *enthalpy* (H) is defined as;

$$H = E + PV \tag{1.30}$$

The *chemical potential* μ_j of the jth solute in the system at constant T, P, and n_k is defined by:

$$\mu_j = \left(\partial G/\partial n_j\right)_{T,P,n_k} \tag{1.31}$$

where n_j and n_k are the number of moles of the jth and kth solutes, respectively. The chemical potential, also called the *partial molar free energy*, represents the incremental addition of free energy to the system upon incremental addition of a solute. All solutes contribute to the free energy of a solution. While the free energy itself is a parameter of state for the whole system, the chemical potential refers to a particular solute. The free energy is thus given by:

$$G = E + PV - TS + \sum \mu_j n_j \tag{1.32}$$

Differentiating yields:

$$dG = dE + PdV + Vdp - TdS - SdT + \sum \mu_j dn_j \tag{1.33}$$

Substituting the expression for dE given previously and simplifying yields a form of the *Gibbs equation*,

$$dG = -SdT + VdP - zFE_m dn + \sum \mu_j dn_j \tag{1.34}$$

which states that the free energy of a system of variable chemical composition is a function of the temperature, the pressure and the number of moles of each component in the mixture, or $G = G(T, P, n_j)$. For processes that occur at constant temperature and pressure, where $dT = dP = 0$, the Gibbs equation simplifies to:

$$dG = -zFE_m dn + \sum \mu_j dn_j \tag{1.35}$$

which states that the increase in free energy of a system is equal to the sum of the electrical work done on the system plus the total change in free energy due to changes in chemical composition. Furthermore, when the system is at equilibrium, *dG* must equal zero. The second law of thermodynamics also implies that the change in free energy (*dG*) is negative for all spontaneous processes. Thus, the first and second laws of thermodynamics, when combined with the definitions of free energy and enthalpy, result in the Gibbs equation, which is the fundamental equation for the estimation of free energy changes when water, ions, or other solutes cross cell membranes.

9.2 Nernst equilibrium

The *Nernst equation* describes the relationship between voltage across a semipermeable membrane and the ion concentrations at equilibrium in the compartments adjacent to the membrane. The Nernst equation provides a simple method of testing whether or not a particular solute is at equilibrium.

Considering the example described in the previous section and using the Gibbs equation, the change in the free energy of the system that occurs when *dn* moles of K^+ ions move from compartment *i* containing K^+ at activity a_i to compartment *o* containing K^+ at activity a_o is given by:

$$dG = -zFE_m dn + \mu_i dn_i + \mu_o dn_o \tag{1.36}$$

where the change in free energy has been summed for the two compartments. For this process, the decrease in the number of moles of solute inside the cell ($-dn_i$) equals the increase outside the cell (dn_o), so that both may be represented by *dn*:

$$-dn_i = dn_o = dn \tag{1.37}$$

and therefore

$$dG = [-zFE_m - (\mu_i - \mu_o)]dn \tag{1.38}$$

The chemical potential in each solution is:

$$\mu_i = \mu_i^0 + RT \ln a_i \tag{1.39}$$

$$\mu_o = \mu_o^0 + RT \ln a_o \tag{1.40}$$

For the system under consideration, the intracellular and extracellular standard-state chemical potentials are assumed to be identical:

$$\mu_i^0 = \mu_o^0 \tag{1.41}$$

and thus:

$$dG = [-zFE_m - RT\ In\ (a_i/a_o)]dn \tag{1.42}$$

At equilibrium, $dG = 0$ and, since, $dn \neq 0$ therefore

$$-zFE_m - RT \ln (a_i/a_o) = 0 \tag{1.43}$$

Rearranging yields the Nernst equation for a cationic concentration cell,

$$E_m = -\frac{RT}{zF}\ln\frac{a_i}{a_o} \tag{1.44}$$

Converting the natural logarithm to base 10 yields:

$$E_m = -2.303\frac{RT}{zF}\log\frac{a_i}{a_o} \tag{1.45}$$

The value of $2.303RT/F$ is 58.7 mV at 23°C and 61.5 mV at 37°C.

For charged solutes, the *electrochemical potential* ($\mu_{j;} = dG/dn_j$) of the *jth* solute is defined as the sum of a chemical and an electrical component. The electrical contribution to the electrochemical potential is $zF\psi$ and the chemical contribution is $RT \ln a$.

$$\mu_j = \mu_j^0(T,P) + RT \ln a_j + zF\psi \tag{1.46}$$

Note that a fundamental condition of equilibrium is that the difference in electrochemical potential between compartments i and o is zero. The electrochemical potential of the solute is the same in each compartment to which that solute has access.

The Nernst equation is independent of the mechanism of transport and is often used to ascertain whether or not an intracellular ion is at electrochemical equilibrium. If so, then:

$$a_i = a_o e^{-zFEm/RT} \tag{1.47}$$

Another way of understanding the Nernst equation is that, at equilibrium, ions distribute across the membrane electric field in accordance with a Boltzmann distribution.

If $z = 0$, as for nonelectrolytes, then $a_i = a_o$, and the activities of the solutes will be the same at equilibrium on both sides of the membrane, as was found to be nearly the case for the distribution of nonelectrolytes across the membranes of human red blood cells (Gary-Bobo, 1967). If $z \neq 0$, as for electrolytes then, at equilibrium, each permeant monovalent ion will reach the same ratio of intracellular to extracellular activity. Such is the case for the apparent passive distribution of Cl^-, HCO_3^- and H^+ in human red blood cells.

$$r = \frac{[Cl^-]_i}{[Cl^-]_o} = \frac{[HCO_3^-]_i}{[HCO_3^-]_o} = \frac{[H^+]_o}{[H^+]_i} = e^{FEm/RT}$$

In red blood cells, Na^+, K^+ and Ca^{2+} deviate from this ratio due to the action of the Na^+-K^+ pump and Ca^{2+} pump. In skeletal muscle, K^+ and Cl^- have Nernst equilibrium potentials that are close to the actual measured resting potential, while Na^+ is far from equilibrium. In squid axons, the Nernst equilibrium potential for K^+ is closest to the resting potential, Cl^- is somewhat removed and, as in muscle and red blood cells, Na^+ is far from equilibrium.

References

Balach, Melisa M., Casale, Cesar H., & Campetelli, Alexis N. (2019). Erythrocyte plasma membrane potential: past and current methods for its measurement. *Biophysical Reviews, 11*, 995–1005. https://doi.org/10.1007/s12551-019-00603-5

Bergethon, P. R., & Simons, E. R. (1990). *Biophysical chemistry. Molecules to membranes*. New York: Springer-Verlag.

Bernard, C. (1949). *An introduction to the study of experimental medicine*. New York: H.C. Green, translator.

Bockris, J. O.'M., & Reddy, A. K. N. (1970). *Modern electrochemistry* (Vol. 1). New York: Plenum Press.

Cannon, W. B. (1932). *The wisdom of the Body*. New York: Norton.

Cantor, C. R., & Schimmel, P. R. (1980). *Biophysical chemistry. , parts I, II, III*. San Francisco: W.H. Freeman.

Collins, K. D., & Washabaugh, M. W. (1985). The Hofmeister effect and the behavior of water at interfaces. *Quarterly Reviews of Biophysics, 18*, 323–422.

Cooke, R., & Kuntz, I. D. (1975). The properties of water in biological systems. *Annual Review of Biophysics and Bioengineering, 3*, 95–126.

Debye, P. (1929). *Polar molecules*. New York: Dover Publications.

Edsall, J. T., & McKenzie, H. A. (1978). Water and proteins. I. The significance and structure of water; its interaction with electrolytes and nonelectrolytes. *Advances in Biophysics, 10*, 137–207.

Edsall, J. T., & Wyman, J. (1958). *Physical biochemistry*. New York: Academic Press.

Eisenberg, D., & Kauzmann, W. (1969). *The structure and properties of water*. Oxford: Oxford University Press.

Eisenman, G. (1967). *Glass electrodes for hydrogen and other cations*. New York: M. Dekker.

Eisenman, G., & Horn, R. (1983). Ionic selectivity revisited: The role of kinetic and equilibrium processes in ion permeation through channels. *Journal of Membrane Biology, 76*, 197–225.

Frank, H. S., & Thompson, P. T. (1960). A point of view on ion clouds. In W. J. Hamer (Ed.), *The structure of electrolyte solutions* (pp. 113–134). New York: John Wiley & Sons.

Frank, H. S., & Wen, W.-Y. (1957). Structural aspects of ion-solvent interaction in aqueous solutions: A suggested picture of water structure. *Discussions of the Faraday Society, 24*, 133–140.

Freedman, J. C., & Hoffman, J. F. (1979). Ionic and osmotic equilibria of human red blood cells treated with nystatin. *The Journal of General Physiology, 74*, 157–185.

Freedman, J. C., & Novak, T. S. (1989). Optical measurement of membrane potentials of cells, organelles, and vesicles. *Methods in Enzymology, 172*, 102–122.

Freedman, J. C., & Novak, T. S. (1989). Use of triphenylmethylphosphonium to measure membrane potentials in red blood cells. *Methods in Enzymology, 173*, 94–100.

Freedman, J. C. (1976). Partial restoration of sodium and potassium gradients by human erythrocyte membranes. *Biochimica et Biophysica Acta, 455*, 989–992.

Funder, J., & Wieth, J. O. (1966). Determination of sodium, potassium, and water in human red blood cells. Elimination of sources of error in the development of a flame photometric method. *Scandinavian Journal of Clinical and Laboratory Investigation, 18*, 151–166.

Fushimi, K., & Verkman, A. S. (1991). Low viscosity in the aqueous domain of cell cytoplasm measured by picosecond polarization microfluorimetry. *The Journal of Cell Biology, 112*, 719–725.

Gary-Bobo, C. M. (1967). Nonsolvent water in human erythrocytes and hemoglobin solutions. *The Journal of General Physiology, 50*, 2547–2564.

Glynn, I. M. (1985). The Na+, K+-transporting adenosine triphosphatase. In A. N. Martonosi (Ed.), *2nd ednThe enzymes of biological membranes* (Vol. 3, pp. 35–114). New York: Plenum Press.

Helsey, N. P., & Verkman, A. S. (1987). Membrane chloride transport measured using a chloride-sensitive fluorescent probe. *Biochemistry, 26*, 1215–1219.

Hille, B. (2001). *Ion channels of excitable membranes* (3rd ed.). Sunderland, MA: Sinauer Associates.

Höber, R. (1945). *Physical chemistry of cells and tissues*. Philadelphia: The Blakiston Company.

Hoffman, J. F., & Laris, P. C. (1974). Determination of membrane potentials in human and Amphiuma red blood cells by means of a fluorescent probe. *The Journal of Physiology, 239*, 519–552.

Horowitz, S. B., & Miller, D. S. (1984). Solvent properties of ground substance studied by cryomicrodissection and intracellular reference-phase techniques. *The Journal of Cell Biology, 99*, 172s–179s.

Horvath, A. L. (1985). *Handbook of aqueous electrolyte solutions. Physical properties, estimation and correlation methods*. New York: John Wiley and Sons.

Katz, S., & Klotz, I. M. (1953). Interactions of calcium with serum albumin. *Archives of Biochemistry, 44*, 351–361.

Klotz, I. M. (1970). Water: Its fitness as a molecular environment. In E. E. Bittar (Ed.), *Membranes and ion transport* (Vol. 1, pp. 93–122). New York: Wiley.

Kuznetsova, I. M., Zaslavsky, B. Y., Breydo, L., Turoverov, K. K., & Uversky, V. N. (2015). Beyond the excluded volume effects: Mechanistic complexity of the crowded milieu. *Molecules, 20*, 1377–1409. https://doi.org/10.3390/molecules20011377

Ling, G. N., & Gerard, R. W. (1949). The normal membrane potential of frog sartorius fibers. *Journal of Cellular and Comparative Physiology, 34*, 383–396.

Miller, C. (1974). Nonelectrolyte distribution in mouse diaphragm muscle. I. The pattern of nonelectrolyte distribution and reversal of the insulin effect. *Biochimica et Biophysica Acta, 339*, 71–84.

Neville, M. C. (1979). The extracellular compartments of frog skeletal muscle. *The Journal of Physiology, 288*, 45–70.

Neville, M. C., & White, S. (1979). Extracellular space of frog skeletal muscle in vivo and in vitro: Relation to proton magnetic resonance relaxation times. *The Journal of Physiology, 288*, 71–83.

Palmer, L. G., Century, T. J., & Civan, M. M. (1978). Activity coefficients of intracellular Na+ and K+ during development of frog oocytes. *Journal of Membrane Biology, 40*, 25–38.

Palmer, L. G., & Civan, M. M. (1977). Distribution of Na+, K+, and Cl− between nucleus and cytoplasm in Chironomus salivary gland cells. *Journal of Membrane Biology, 33*, 41–61.

Pauling, L. (1960). *The nature of the chemical bond and the structure of molecules and crystals; an introduction to modern structural chemistry* (3rd ed.). Ithaca, NY: Cornell University Press.

Robinson, R. A., & Stokes, R. H. (1959). *Electrolyte solutions* (2nd ed.). London: Butterworths.

Scatchard, G., Coleman, J. S., & Shen, A. L. (1957). Physical chemistry of protein solutions. VII. The binding of some small anions to serum albumin. *Journal of the American Chemical Society, 79*, 12–20.

Schwan, H. P., & Foster, K. R. (1977). Microwave dielectric properties of tissue. Some comments on the rotational mobility of tissue water. *Biophysical Journal, 17*, 193–197.

Silver, B. L. (1985). *The physical chemistry of membranes*. Boston: Allen and Unwin.

Takahashi, A., Camacho, P., Lechleiter, J. D., & Herman, B. (1999). Measurement of intracellular calcium. *Physiological Reviews, 79*, 1089–1125.

Tanford, C. (1961). *Physical chemistry of macromolecules*. New York: Wiley.

Tsien, R. (1983). Intracellular measurements of ion activities. *Annual review of biophysics and bioengineering, 12*, 91–116.

Tsien, R. Y. (1988). Fluorescence measurement and photochemical manipulation of cytosolic free calcium. *Trends in Neurosciences, 11*, 419–424.

Weiss, T. F. (1996). *Cellular biophysics, vol. 1: Transport* (Vol. 2). Cambridge, MA: The MIT Press. Electrical Properties.

Chapter 2

Structure and function of proteins

Matthew R. Pincus
Department of Pathology, SUNY Health Sciences University, Brooklyn, NY, United States

1. Molecular structure of proteins

Proteins are biopolymers that are essential for all plant and animal life. All antibodies, enzymes, channels, transporters, and cell receptors are proteins. The basis for all connective tissue is the fibrous proteins such as elastin and collagen. The entire genetic apparatus of every cell, regardless of how simple or complex its functions, is dedicated to the synthesis of proteins. The building blocks for proteins are α-amino acids that are linked to one another by peptide (or amide) bonds. The 20 naturally occurring amino acids (Fig. 2.1) all have the same **backbone** structure, the $N{-}C^{\alpha}{-}C{=}O$ unit, but differ from one another in that they have different side chains attached to the α-carbon, as shown in Fig. 2.1. Also, all of these amino acids are chiral—optically active and in the L-configuration—except for the simplest amino acid, glycine, which has two H atoms attached to the α-carbon. Note that cysteine has a sulfhydryl group, giving it the capacity to form disulfide bonds (Fig. 2.1). Disulfide bonds are important in stabilizing the three-dimensional structures of proteins, as discussed in Section 1.3.

Topographically, proteins are divided into the **fibrous proteins,** which are long repeating helical proteins, and the **globular proteins,** which tend to fold up into spherical or ellipsoidal shapes. As we will show, how proteins fold is governed uniquely by their linear sequences of amino acids.

1.1 Primary amino acid structure

All proteins are composed of linear sequences of amino acids that are unique to each protein and are covalently bound to one another by peptide bonds (Fig. 2.2). This peptide bond is almost always arranged so that the C^{α} atoms of two successive residues are *trans* to one another and the atoms attached to the C' and N of the peptide all lie in the same plane since the $C'{-}N$ bond has double-bond character, namely, the *trans*-planar peptide bond described long ago by Linus Pauling. The linear sequence of amino acids is called the **primary structure.**

Note that the amino acids in Fig. 2.1 are classified as **polar, nonpolar,** and **neutral** depending on the nature of the side chain. Thus, for example, leucine and valine are nonpolar amino acids with aliphatic side chains that are hydrophobic, or nonmiscible with water and other polar solvents. Amino acids like lysine and glutamic acid, on the other hand, contain charged side chains that are highly soluble in water and are completely miscible in polar solvents. Amino acids such as glycine and alanine do not contain groups that predispose them to be classified as either polar or nonpolar, so they are referred to as neutral.

1.2 Regular or secondary structure in proteins

The three-dimensional structures of over 300 proteins have been determined by x-ray crystallography. One striking feature in all of these proteins is the presence of recurrent regular structures, in particular, α-helices, β-sheets, and reverse turns (Fig. 2.3). The existence of the first two of these structural types was predicted by Linus Pauling in the early 1950s on the basis of x-ray diffraction patterns of model poly-α-amino acids.

Note that in the **α-helix** in Fig. 2.3A, the residues form a spiral such that at every 3.6 residues, the spiral or helix makes one complete turn. The repeat distance of this helix is 5.4 Å. This structural motif is observed in virtually all proteins, and some proteins like myoglobin are virtually completely α-helical. One very important interaction for stabilizing helices is

Cell Physiology Sourcebook. https://doi.org/10.1016/B978-0-12-811114-7.00037-2

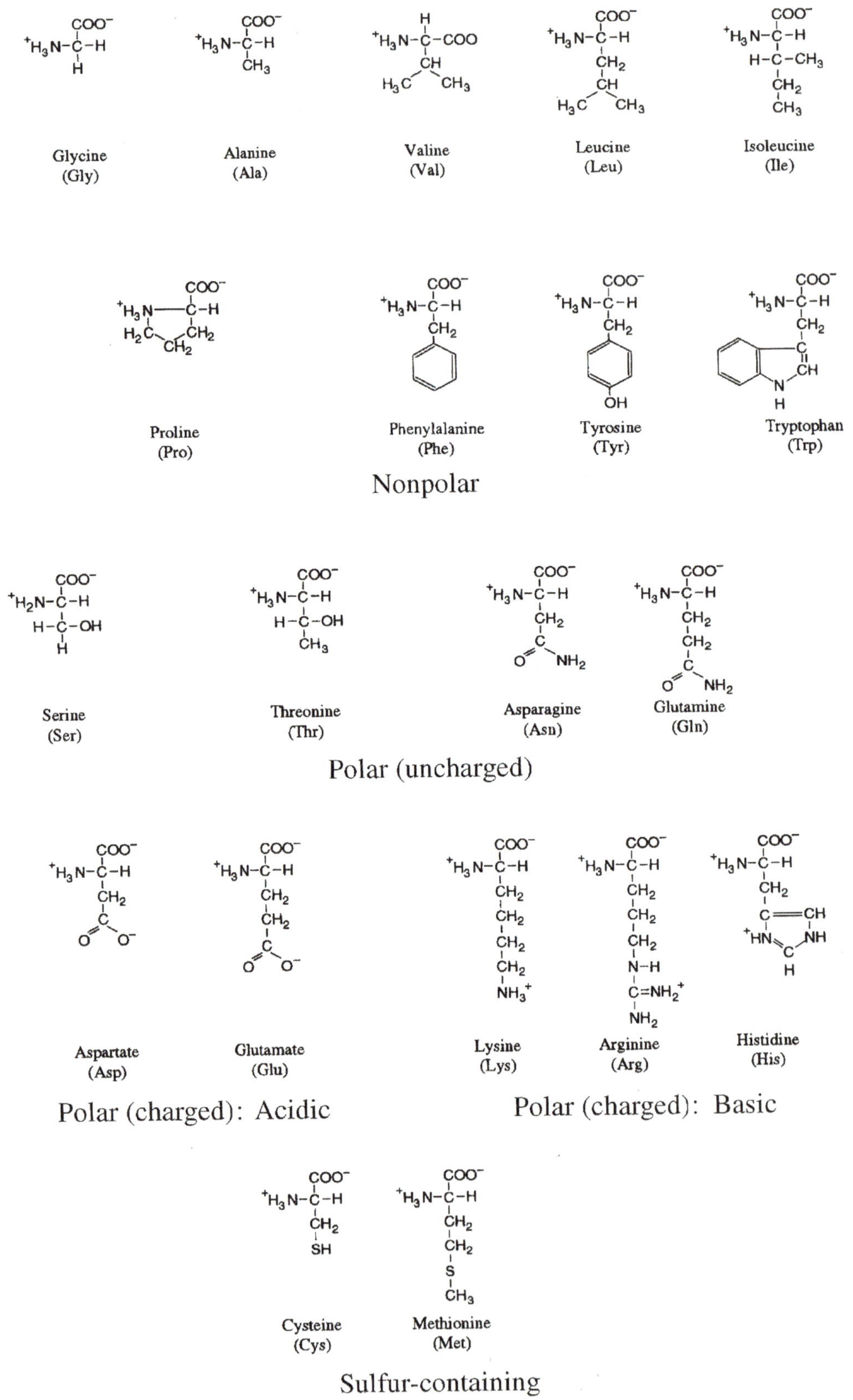

FIGURE 2.1 The 20 naturally occurring amino acids grouped by type. Note that Cys residues can form disulfide bonds in the following reaction: $R{-}SH + R{-}SH \rightarrow R{-}S{-}S{-}R$, where R is the backbone atoms of Cys + the side chain CH_2 group.

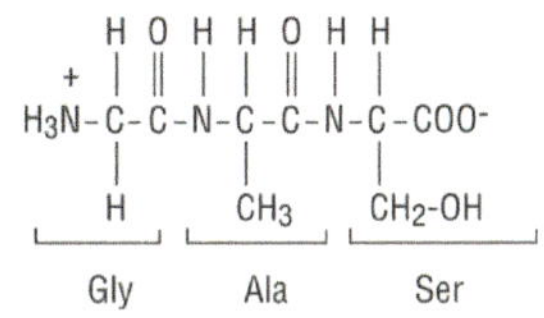

FIGURE 2.2 A typical tripeptide, Gly-Ala-Ser, showing the CO—NH peptide bond linkage.

the i-to-i+4 hydrogen-bonding scheme shown in Fig. 2.3A. The C=O of the *ith* residue accepts a hydrogen bond from the N—H of the i+4th residue. The hydrogen bond can provide up to 3 kcal/mol in stabilization energy. For an amino acid residue in the middle of an α-helix, two hydrogen bonds form to its N—H and C=O groups. This occurrence tends to cause helices to propagate because the more such double hydrogen bonds form, the greater the stabilization energy provided. Two other long-range factors stabilize α-helices. The first is the presence of negatively charged amino acid residues on the amino terminal end of a helix and positively charged amino acids on the carboxyl terminal end of the helix. This stabilization results from the backbone dipole moment of an α-helix, where the center of positive charge is on the amino terminus of the helix and the center of negative charge is on the carboxyl terminus of the helix. The presence of the oppositely charged amino acids stabilizes the helical dipole.

The second stabilizing factor is **amphipathicity.** This refers to the i-to-i+4 side chain—side chain interactions such that nonpolar or hydrophobic amino acids contact one another on one face of the helix, while other polar i-to-i+4 interactions occur on the opposite side of the helix. This segregation of polar and nonpolar faces allows for hydrophobic clustering to occur on one face and hydrogen bonding and charge neutralization to occur on the opposite face. This motif is vital in stabilizing α-helices in membrane proteins such as ion channel proteins.

α-helices are extremely important in membrane proteins, in the transmembrane domains of cellular receptors, and in ion channel proteins, as we discuss in Section 4.

In contrast, the **β-sheet** is composed of alternating residues of amino acids that are "flipped" 180° with respect to their nearest neighbors, as shown in Fig. 2.3B. We may regard the structure of single strands in these sheets as a flat helix whose repeat is every two residues. Note that within each strand of a β-sheet is minimal interactions between the backbone atoms of different amino acid residues. The hydrogen-bonding stabilization comes from other strands that lie in proximity to one another. As shown in Fig. 2.3B, the arrangement of the strands is either **parallel** or **antiparallel.** Regardless of the arrangement of the two interacting strands, it should be noted from Fig. 2.3B that two hydrogen bonds between the backbone atoms form for **alternate** residues; every other residue forms no hydrogen bond to the neighboring strand. Thus there are twice the number of hydrogen bonds in long α-helices as in long β-sheets. Evidently, other interactions are important in stabilizing β-sheets besides backbone-backbone hydrogen bonding.

Another type of regular structure is the **reverse turn,** the prototype of which is shown in Fig. 2.3C. In this structure, the polypeptide chain reverses its direction. A minimum of two amino acids is needed to form a reverse or hairpin turn, which is also called a **β-bend.** Usually three amino acids are involved, however, because the C=O of the first residue can form an i-to-i+3 hydrogen bond with the N—H of the i+3rd residue, as shown in Fig. 2.3.

Of the three regular structures described, α-helices tend to propagate most strongly due to the double-hydrogen bonding of the NH and C=O groups of central residues in the helix discussed above and illustrated in Fig. 2.3D. This type of "medium-range" interaction is not present for the two other types of regular structures.

The specific sequences of amino acids in proteins will adopt specific structures such as the three basic regular structures just described. Thus, the amino acid sequences of proteins determine their respective three-dimensional structures. This conclusion has been verified as we now describe.

In 2009, a Vanderbilt University Medical Center investigator discovered a new chemical bond in proteins: the sulfilimine bound. This chemical bound occurs between sulfur and nitrogen atoms (-S=N-). It was found linking collagen IV chains at the cell basement membrane (Vanacore et al., 2009).

1.3 Tertiary or three-dimensional structure of proteins

1.3.1 The amino acid sequences of proteins determine their three-dimensional structures

In a classic series of experiments, Christian Anfinsen and his coworkers at the National Institutes of Health showed in the early 1960s that the primary structure (linear sequence of amino acids) of a protein determines its unique three-dimensional structure (Anfinsen et al., 1961). These investigators took the protein ribonuclease A, which hydrolyzes RNA and whose amino acid sequence had just been determined and was known to contain four disulfide bonds, and denatured it in the

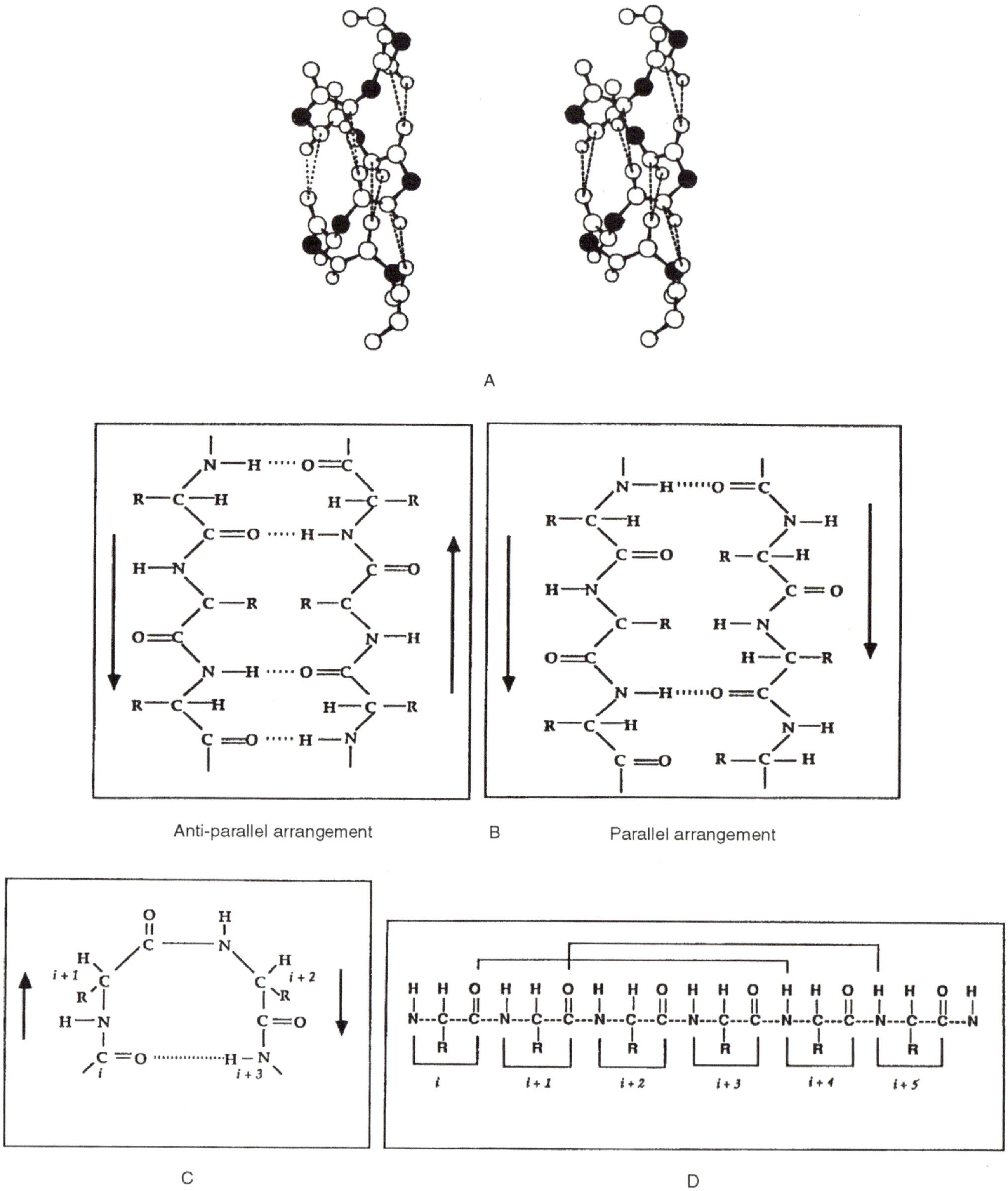

FIGURE 2.3 Three types of regular (secondary) structure. (A) Stereo view of an α-helix for a polypeptide backbone showing the i-to-i+4 hydrogen bond. The filled circles are α-carbons; double dashed lines are drawn from C=O to NH (shorter dashed line) and from C=O to N (longer dashed line). (B) Two types of β-sheets. The left side shows the antiparallel arrangement while the right side shows the parallel arrangement, as indicated by the arrows in both figures. (C) The arrangement of backbone atoms in a β-bend showing how an i-to-i+3 hydrogen bond stabilizes the reverse turn structure. (D) Illustration of an extended α-helical chain showing how internal residues have both their NH and CO groups involved in hydrogen bonds.

denaturing agent 6 M guanidine hydrochloride. This agent and 8 M urea are both known to disrupt the hydrogen bonds in proteins and destroy their three-dimensional structures. In addition, using β-mercaptoethanol, they reduced the four disulfides to eight sulfhydryl (SH) groups, thereby destroying all of the major determinants of the three-dimensional structure. The denatured protein was completely inactive toward RNA hydrolysis and possessed spectral properties that differed greatly from those of the native protein. They then allowed the sulfhydryl groups to reoxidize to disulfides and dialyzed the 6 M guanidine hydrochloride. Within 2 hours, virtually all of the native enzymatic activity returned, and the spectral properties of the refolded protein were identical to those of the undenatured protein. In addition, all of the disulfides were paired exactly as they were in the native protein.

This experiment illustrated that the primary sequence of amino acids in a protein determines its three-dimensional structure. Further, it demonstrated that the interactions that govern the correct folding of a protein must be highly specific and strong because, of the vast number of possible structures that the polypeptide could have adopted, only one such structure actually ultimately formed. This conclusion is reinforced by the following consideration. Once the protein is reduced to the eight-sulfhydryl state, there are many ways in which these eight sulfhydryls can pair to give four disulfides. In particular, there are seven ways in which the first disulfide can form, five for the second, three for the third, and one for the fourth, or $7 \times 5 \times 3 \times 1 = 105$ possible ways. Of this total, only one pairing scheme, the native pairing, ultimately was found to form in this experiment.

Similar results have been obtained for a number of different proteins. In every case, the native structure is regenerated. Such dramatic results imply that, given the amino acid sequence, it should be possible to infer the three-dimensional structure of the protein.

Indeed, methods have been and are continuing to be developed that allow us to **predict** the three-dimensional structure of a polypeptide or a protein from its amino acid sequence. At present, it is possible to compute the three-dimensional structures of polypeptides from their amino acid sequences for polypeptides and proteins containing more than 100 amino acid residues. As might be imagined, the field of structure prediction is a burgeoning one that is developing rapidly. All of the methods used to predict protein structure are based on the physicochemical principle that the observed structure of a molecule is the one with the lowest free energy. If we can compute the energies for the possible conformations of a protein, we can pick out the structure of lowest energy, which is the structure that the protein is most likely to adopt. In order to perform this task, we must understand the types of energetic interactions within a polypeptide chain that must be optimized for the protein to fold correctly.

1.4 Possible interactions between amino acids in a protein chain

As might be expected from the differing nature of the side chains of the naturally occurring amino acids, a vast number of possible interactions between different side chains and backbone are possible for given amino acid sequences. These types of interactions include **electrostatic interactions** (given by the Coulombic potential q_1q_2/DR_{12}, where the qs are the charges on two different interacting atoms, R is the distance between them, and D is the dielectric constant of the medium) between positive and negative charges of oppositely charged side chains, such as the $-COO^-$ group of glutamic acid and the $-NH_3^+$ group of lysine as well as the interactions between the partial charges on all of the atoms of the protein; **nonbonded** interactions between the individual atoms, explained below; **hydrogen-bonding interactions** between polar atoms with H atoms and other polar atoms; and a **solvation energy** for most proteins.

This solvation energy is such that the nonpolar side chains of nonpolar amino acids tend to avoid water and therefore "bury" themselves in the core of the protein, while the side chains of polar amino acids tend to interact strongly with water and become solvated. This favorable solvation energy is due to water dipole interactions with charged residues such that the charges on the side-chain atoms are reduced by favorable interactions with the water dipoles. It is also due to hydrogen bonding between water molecules and the polar atoms of the exposed polar side chains.

1.4.1 Specific interactions: Hydrogen bonding, nonbonded, and hydrophobic interactions

In the preceding paragraph, we noted that hydrogen-bonding interactions between polar atoms in a protein can greatly stabilize the structure of a protein. These interactions are due to the sharing of a hydrogen atom between the donor atom to which the H atom is covalently linked and the receptor polar atom as in Fig. 2.4. This interaction stabilizes the distribution of charges over the three atoms involved in the hydrogen-bonding system (Fig. 2.4). For a given protein sequence, hydrogen-bonding patterns can be arranged in many possible ways. However, in every protein whose three-dimensional structure is known, only one or a few hydrogen-bonding schemes are observed, many of these occurring between backbone atoms (NH—O=C) and others occurring between the polar atoms of side chains.

```
δ-δ+   δ-
R-N-H---O=C-R'
  |
  R
```

FIGURE 2.4 Illustration of how hydrogen bonds are stabilized by partial sharing of a polar H atom between two polar heavy atoms (N and O in this case).

In addition to the hydrogen-bonding interactions between the atoms of a protein, all pairs of atoms in a protein interact with one another so as to attract each other weakly until they become sufficiently close that their electron clouds repel one another. The weak attractive interactions between a pair of atoms are due to the polarization of charges in each atom between the negatively charged electrons and the positively charged nucleus that is induced by the other interacting atom. This dipole—induced-dipole interaction is attractive because the positive end of the dipole of the first atom interacts attractively with the negative end of the dipole of the other atom of the pair and vice versa. The overall attractive energy for this interaction can be simply expressed as $-B_{ij}/R_{ij}^6$, where B_{ij} represents a constant that characterizes the magnitude of the attractive energy between any two interacting atoms and R_{ij} is the distance between the two interacting atoms. As shown in Fig. 2.5, as two interacting atoms approach one another, the attractive energy becomes progressively stronger until the electrons in the outer shells repel one another. This repulsive energy increases rapidly as the internuclear distance between the atoms decreases.

Most proteins fold in an aqueous environment. Changing the solvent, such as by adding nonpolar solvents to water, is known to denature proteins so that they cannot function properly. Evidently, water provides important interactions with proteins, such that they fold correctly to their native conformations.

The three-dimensional structures of over 300 proteins have been determined mostly by the technique of x-ray crystallography (see Section 2.2). In all of these proteins, the nonpolar amino acid residues tend to "pack" in the interior of the protein, while the polar residues tend to interact with water on the surface of the protein. Avoidance of interaction of the nonpolar residues with water is due to the **hydrophobic (water-avoiding) effect.** This behavior is caused predominantly by an entropy effect. Water molecules tend to become highly ordered around the side chains of nonpolar or hydrophobic amino acid residues. On the other hand, water molecules can interact in a large number of low-energy complexes with polar side chains and so are less constricted structurally. This ordering of water molecules by nonpolar residues causes the latter to "pack" into the interior of the protein, allowing the water to be more disordered on the outside of the protein.

If one measures the free energy of transfer, ΔG, of nonpolar compounds (such as benzene or *n*-hexane) from nonpolar to polar solvents, it is found that the enthalpy of transfer, ΔH, is small, whereas the entropy of transfer, ΔS, is a large negative value. From the Gibbs free energy expression,

$$\Delta G = \Delta H - T\ \Delta S \tag{2.1}$$

it can be seen that a large negative entropy makes the overall process unfavorable; in other words ΔG becomes more positive. This effect can be attributed to the water structure effect (see Scheraga, 1984).

All of the above interactions have been taken into account in a variety of computer programs that generate the conformations of polypeptides and proteins and calculate their conformational energies. One such program is ECEPP (Empirical Conformational Energies of Peptides Program) developed in the laboratory of Professor Harold A. Scheraga of Cornell University. The energy parameters, such as the constants for the nonbonded interaction energies, have been determined experimentally. This program has been used to compute the structures of many polypeptides and proteins with

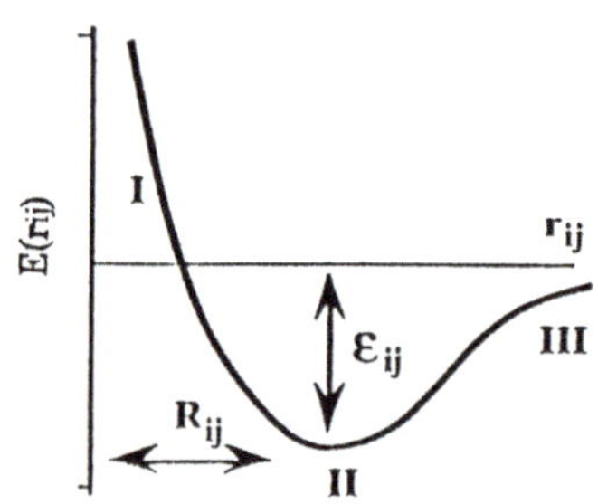

FIGURE 2.5 Interatomic van der Waals energy as a function of distance. Notice that two atoms can approach one another with decreasing energy (regions II and III) until the energy reaches a minimum ($\in_{ij}$) at a distance of R_{ij}. If the atoms become closer, they then repel one another so that the energy rises steeply (region I).

excellent agreement between predicted and experimentally determined structures (Scheraga, 1984). We give examples below of how these methods have been used to predict protein structure and provide insight into the relationship between the structure of a protein and its function. First, we describe the properties of the structures of proteins.

1.5 Properties of the structures of proteins

1.5.1 Geometry and dihedral angles

Among the many protein structures that have been determined by crystallography, each of the amino acid residues that compose the protein have geometries that are remarkably constant throughout the protein and when compared between any two proteins. Remarkably, this geometry is essentially the same as that found in single crystals of the individual amino acids. Geometry refers to the **bond lengths** and **bond angles** of the individual amino acid. An example of the geometry of an amino acid is illustrated for glycine in Fig. 2.6. The basic variables that allow for changes in chain conformation are the **dihedral angles,** defined as the angle made between two overlapping planes. For example, in Fig. 2.7, the angle between the planes determined by the N—C^{α}—C′ atom and the C^{α}—C′—N atoms of the two amino acid residues shown is a dihedral angle. These angles are generated by rotation around single bonds. As shown in Fig. 2.7, for the backbone of a single amino acid residue, three dihedral angles exist: Φ, Ψ, and ω.

The dihedral angle Φ is determined by the angle between the planes of C′—N—C^{α} and N—C^{α}—C′, whereas the dihedral angle ψ is determined by the angle between the planes of N—C^{α}—C′ and C^{α}—C′—N. To determine Φ, one merely has to look down the N—C^{α} axis using the C′-N bond as a reference and determine the angle made between the C′-N bond and the C^{α}—C′ bond as shown in Fig. 2.7. To determine Ψ, one uses the N—C^{α} bond as a reference and measures its angle with the C′—N bond. The dihedral angle ω is the interpeptide dihedral angle, which is determined by rotation around the C′-N bond shown in Fig. 2.7. It can be measured by determining the angle made between the C^{α}—C′ bond and the N—C^{α} bond when sighting down the C′-N bond. This angle is almost always close to 180°, the *trans*-planar peptide conformation.

Given the geometry and a complete set of dihedral angles for a given sequence of amino acids in a polypeptide, a unique three-dimensional structure can be generated, called the **conformation** of the protein. All regular structures described above have repeating values for Φ and ψ from residue to residue. All α-helices have values for Φ and ψ that are close to −60° while, for the individual extended chains of β-sheets, the values for both of these dihedral angles lie close to 180°.

In fact, the α-helical or A and extended or E conformations are minimum-energy conformations for all of the naturally occurring amino acids. Using conformational energy calculations based on ECEPP, all of the low-energy minimum conformations for the 20 naturally occurring amino acids have been computed as a function of the dihedral angles Φ and ψ.

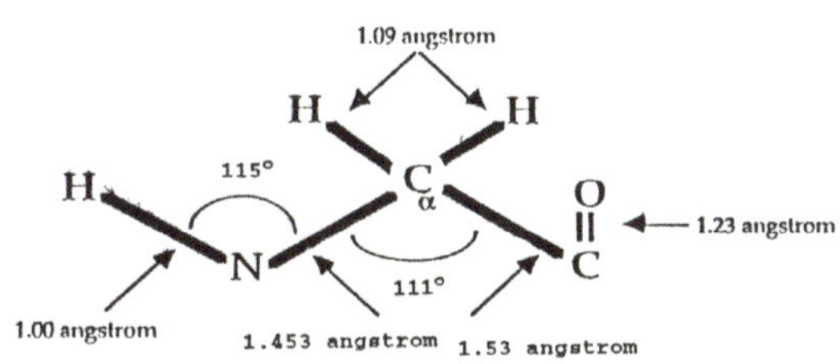

FIGURE 2.6 The geometry of a glycine residue showing typical bond lengths and bond angles.

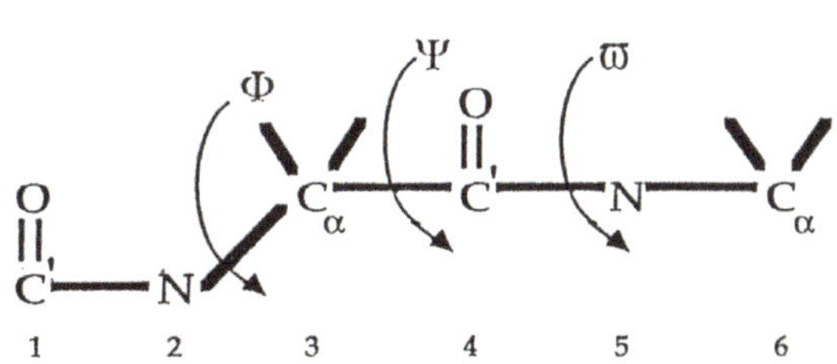

FIGURE 2.7 Dihedral angles, Φ, ψ, and ω, for the backbone of an amino acid. Φ is determined by atoms 1, 2, 3, and 4. Clockwise rotations of the 3—4 bond relative to the 1—2 bond are considered positive while counterclockwise rotations are considered negative. 0° is the conformation in which the 3—4 bond is eclipsed relative to the 1—2 bond, while 180° is the conformation in which the two bonds are *trans* relative to one another. ψ is determined by the 4—5 bond relative to the 2—3 bond. The same considerations apply to this dihedral angle as to those for Φ. ω is determined by the angle between the 5—6 bond and the 3—4 bond. This dihedral angle is almost always very close to 180° (i.e., the 5—6 bond is *trans* to the 3—4 bond).

In Fig. 2.8, these energies are plotted as isoenergetic contours in the same way that different elevations are plotted on contour maps. All of the amino acids have at least seven basic low-energy minima as indicated by the dots in Fig. 2.8.

Note in this figure that large regions of the map are energetically forbidden and that the allowed (low-energy) regions of the map are relatively restricted (Fig. 2.8). Analysis of the conformations of the individual amino acids in proteins whose three-dimensional structures are known reveals that virtually all of these amino acids adopt one of these eight basic conformational states, consistent with the results of the energy calculations.

1.5.2 Computation of the structures of polypeptides and proteins

Using programs such as ECEPP, systematic methods, reviewed extensively in Vasquez et al. (1994), have been developed that generate representative sets of the low-energy structures for a given polypeptide chain. These methods include the chain build-up procedure (Pincus, 1988), molecular dynamics (Karplus & McCammon, 1986), Monte Carlo procedures, and more specialized methods that allow for "jumping" potential energy barriers between different conformations of the given polypeptide chain (Scheraga, 1989). All of these methods take advantage of the energy map shown in Fig. 2.8, since the backbone of each amino acid is constricted to adopting one of the low-energy states shown in this figure. Once the low-energy conformations for a given polypeptide are generated and their energies computed, the lowest energy structure is selected and this structure is the one that should be observed experimentally.

Successful ECEPP computations of the three-dimensional structures of a number of polypeptides and proteins have been performed. Examples are shown in Fig. 2.9 for the cyclic decapeptide gramicidin A, collagen, and melittin (Pincus & Scheraga, 1985). There is close agreement between these computed structures and the experimentally determined ones. Excellent agreement between theory and experiment has also been achieved for other proteins such as avian pancreatic polypeptide and bovine pancreatic trypsin inhibitor (Scheraga, 1989), and a 100-residue segment of human leukocyte interferon (Gibson et al., 1986). The experimental structures of these polypeptides and proteins have been determined by specific methods, namely, x-ray crystallography, two-dimensional nuclear magnetic resonance spectroscopy (NMR), or circular dichroism.

More recently, methods have been developed that allow for the computation of the lowest energy structures of larger proteins. In one of these, called UNRES (Scheraga et al., 2004), a virtual protein chain represents a given sequence. This chain consists of C^{α} atoms attached to centroids inside ellipsoids representing side chains of the individual amino acids. Peptide centroids along each C^{α}- C^{α} bond are inserted. These centroids represent the backbone peptide (amide) groups. The conformational energies are then computed as the sum of backbone centroid-backbone centroid, side-chain-side chain,

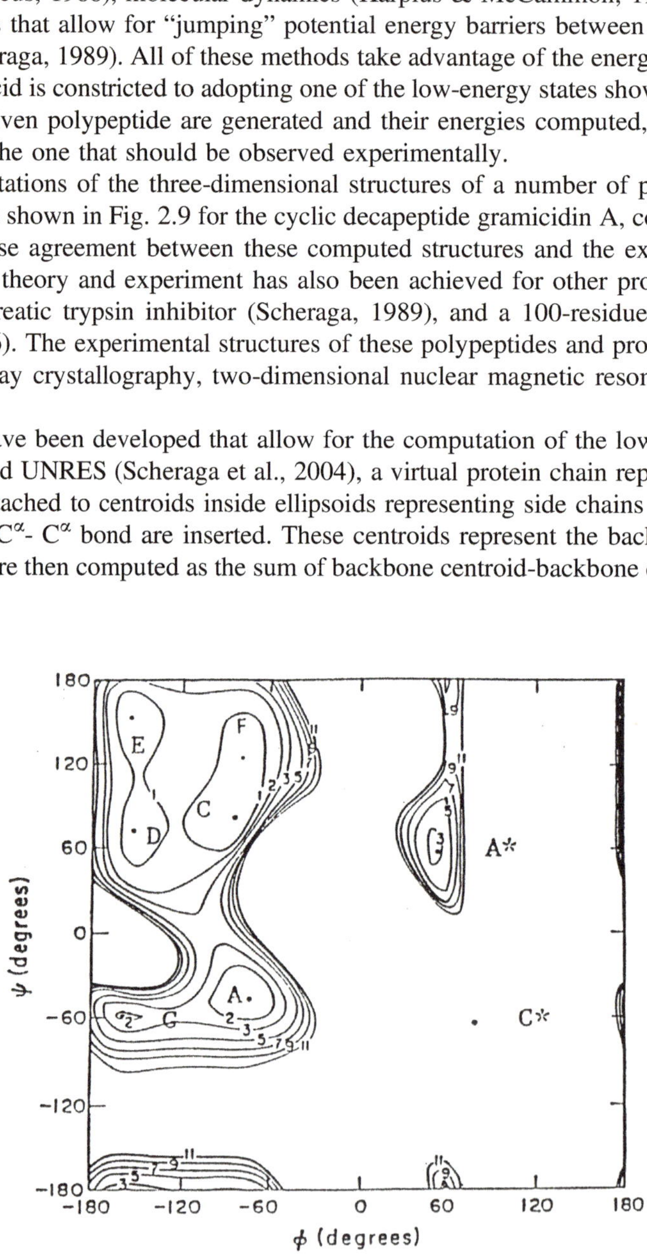

FIGURE 2.8 Contour map for Φ and ψ for N-acetyl-alanine-N′-methyl amide as computed by ECEPP (Scheraga, 1984). Note that the actual energy minima are denoted by dots on the energy contour map. Seven low-energy minima are shown: (A) (α-helical), (C and D) (both of which are seen in bends), (G–F) (both of which are seen in β-sheets), A* (left-handed α-helix), and C*, which is energetically forbidden for all L-amino acids except glycine. Note also that some of these regions are "split" (i.e., more than one minimum can exist for a given conformational region).

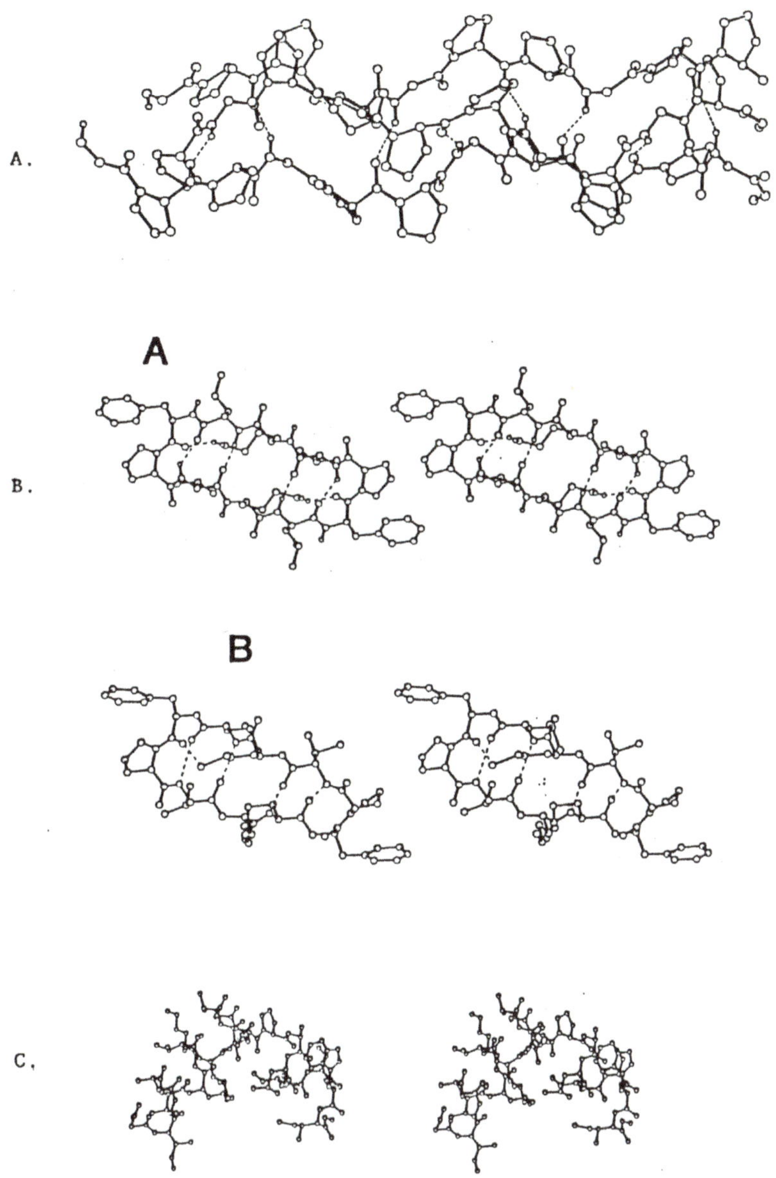

FIGURE 2.9 Views of computed structures. (A) The structure of a section of collagen model polypeptide, (Gly-pro-Pro)$_{20}$ (Miller & Scheraga, 1976), which was directly confirmed by single-crystal x-ray crystallographic studies subsequent to the computations of the structure of this molecule. (B) The computed structure for the cyclic decapeptide gramicidin A (upper part) and the x-ray structure of the same polypeptide (lower part) show the close agreement between the predicted structure and the x-ray crystal structure (Dygert et al., 1975). (C) The computed structure for the membrane-active protein melittin, which consists of two α-helical rods, separated by a bend in the middle of the chain (Pincus et al., 1982). These features have been described in the x-ray crystal structure of this protein. Each polypeptide represents a type of structure seen in many proteins, viz., the collagen helix (A) in fibrous proteins, β-sheets with reverse turns (B), and α-helices (C). *See Pincus and Scheraga (1985), for a further description of these structures.*

backbone centroid-side chain, and torsion terms for rotations around the virtual bonds and the side chain centroids. This simplified polypeptide chain greatly reduces the complexity of the calculation of conformational energies.

Search procedures for low energy conformations are then carried out on this virtual chain. These include molecular dynamics, genetic algorithms, and a technique called the electrostatically-Driven-Monte Carlo method (EDMC) technique. In this latter approach, the electric field for a given energy minimum structure is searched for the backbone dipole moment(s) that is(are) the least optimally aligned with the field. The backbone dihedral angles are changed such that the alignment is optimized. The resulting structure is then subjected to energy minimization, and the process is repeated.

If the same energy minima are repeated frequently in the search, the chain is randomly perturbed via the Monte Carlo procedure, and the new chain is subjected to energy minimization and then the procedure described above. Once low energy conformations are computed, they are classified into different structure types. Then for each structure, the full atom representation of the chain is generated, and the search is repeated for each of these structures. The lowest energy structures form this search are then determined. This approach has been used to compute the low energy structures for proteins of 140 amino acids or more using supercomputers that can perform the computations in a little over an hour. The lowest energy structures are in good agreement with the experimentally determined ones. A good case in point is the agreement between the computed structure for the first 75 residues of bovine calbindin and the x-ray crystal structure of this protein, the root-mean square deviation being 3.9 Å (Scheraga et al., 2004).

2. Techniques for the determination of the structures of proteins

2.1 Regular (secondary) structure: Circular dichroism

One of the most widely used techniques for determining regular structure is **circular dichroism,** which is based on the principle that molecules with asymmetric structures absorb light asymmetrically. Light is electromagnetic radiation, which can be plane-polarized to the right or to the left. Normally light consists of both types of plane-polarized waves. When a molecule with an asymmetric structure absorbs light, it preferentially absorbs either the left- or right-polarized light wave. The amount of light absorbed, A, is equal to $E \times C$, where E is the molar extinction coefficient and C is the concentration of the molecules. For molecules with asymmetric structures, the E values for left-and right-plane-polarized light differ. The difference $E_L - E_R$ depends on the wavelength of the incident light. A plot of $E_L - E_R$ (called the **molar ellipticity**) versus wavelength is referred to as the circular dichroism or CD spectrum. Regular structures in proteins are asymmetric both because the amino acids in these structures are themselves asymmetric and because these structures have a **handedness** or a twist-sense as in the case of the α-helix, which is right-handed (see van Holde, 1985; for a complete treatment of circular dichroism.) Since the peptide bond has strong absorption in the far-UV wavelengths from 230 nm down to about 190 nm, CD spectra for each different type of regular structure should be unique in this range of wavelengths. The patterns in the CD spectra found for each structure type are shown in Fig. 2.10. In this figure, the term ***random coil*** is a misnomer: the term refers collectively to all structures that are not α-helices or β-sheets, but these other structures are unique, not random.

For proteins of unknown three-dimensional structures, a knowledge of the contents of regular structures (α-helices and β-pleated sheets) can be obtained by taking their CD spectra. Since patterns for α-helix, β-sheet, and random coil structures are known, these patterns may be combined to reproduce the observed CD spectrum of a particular protein by curve-fitting. The combination of regular structure that best fits the observed CD curve is the one most likely to exist in the protein. This methodology has been tested on proteins of known three-dimensional structures (so that the percentage of regular structures can be determined) and found to reproduce the regular structure quite satisfactorily.

2.2 Three-dimensional structure of proteins: X-ray crystallography

This technique is based on the principle that the electron clouds around atoms diffract incident light in a predictable way. If the atoms are parts of an oriented molecule, in other words, the protein molecules are all oriented in a **crystal,** the same

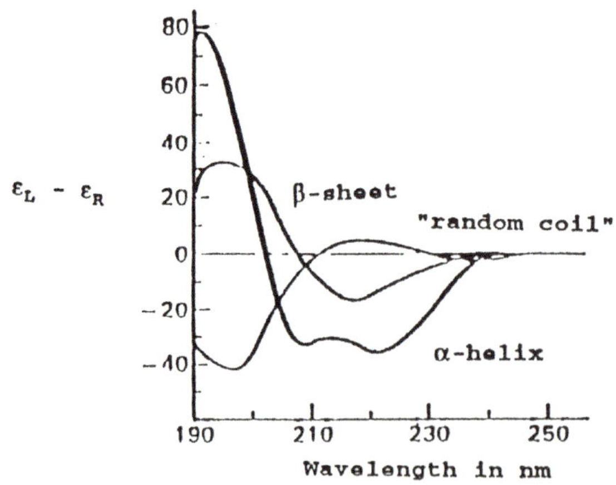

FIGURE 2.10 CD spectra for regular structures: α-helix, β-sheet, and so-called random coil. Note that α-helices have a characteristic "dip" in the region around 210 nm. For a given protein, one can compute the amount of each such structure from its CD spectrum by curve-fitting, using different amounts of each structure type and assuming that the resulting spectrum is the sum of the individual spectra for each structure type.

atoms of different molecules diffract the light in the same way. The closer two atoms are in the molecule, the farther apart the diffraction pattern will be. Conversely, the farther away two atoms are in the molecule, the closer the diffraction patterns from each atom will be. Thus, the diffraction pattern is the reciprocal of the atomic pattern. The diffraction pattern can be analyzed with respect to the amplitude of diffraction and the relative positions of the scattering pattern. The light diffracted must be x-rays because the wavelength of light used must be of the order of single bond lengths, around 1.5–2.0 Å. Only x-rays have this range of wavelengths.

Once a diffraction pattern has been obtained, the problem is to relate the positions of the atoms to one another from the diffraction pattern. The relative positions of two atoms can be computed from their (diffraction) amplitudes and a phase factor that is related to the angular displacement of one atom relative to the other. This phase factor cannot be determined from the diffraction pattern. Since the position of at least one atom is required, in the x-ray diffraction of proteins, a heavy atom bound to the protein is used in the so-called heavy atom replacement method. Once the phase factor is determined, the individual atomic positions can be computed from the diffraction pattern as a Fourier transform that converts from reciprocal (diffraction) space to real (three-dimensional) space.

2.3 Two-dimensional high-resolution nuclear magnetic resonance spectroscopy (NMR)

In one of the most exciting developments in protein chemistry, this technique is being applied to solving the structure of proteins **in solution** without requiring them to be oriented in a crystal. NMR is based on the principle that most nuclei of atoms have spins. If a magnetic field is applied to the atom, the nuclei tend to align their spins with the field. The energy difference between the spins in the absence and the presence of the field is proportional to the magnetic field strength and is given as

$$\Delta E = h\nu = g\beta H \tag{2.2}$$

where ΔE is the energy difference, h is the Planck constant, ν is the resonance frequency corresponding to the energy absorbed by the spinning nuclei, $g\beta$ is a constant, and H is the magnetic field strength. Nuclei like hydrogen exist in a protein in different environments so that some are more shielded by electrons than others, thereby requiring higher ν values or field strengths to excite them. The frequencies corresponding to the transitions of the spins of individual nuclei to align themselves with the field is in the radiofrequency range. Because different H atoms absorb at different field strengths or values, one can obtain an absorption spectrum for a given compound or protein, where, for a given field strength, H, different radio frequencies excite the nuclei to change their spins to orient with the field. When two H atoms from different parts of a protein approach one another, their spins either add to or subtract from one another and affect the intensity of each other's absorption in a phenomenon known as the **nuclear Overhauser effect** or NOE. This change in intensity is inversely proportional to the sixth power of the distance between the two H atoms. If the individual resonant frequencies for the two interacting H atoms are known, the distance between them can be calculated.

The specific H atoms that give rise to different NOE values in a protein can be identified by an elegant technique called two-dimensional NMR (2D-NMR) (Wuthrich, 1986). By irradiating the nuclei at two different frequencies, it is possible to construct a correlation map between the different interacting atoms. A large number of inter-H distances allows us to fit well-defined structures that have these distances so that a class of structures can be directly determined for the given polypeptide or protein.

2.4 Cryo-Electron microscon microscopy

The crystalization of membrane proteins has been a challenge, and with a few exceptions mostly crystals made of bacterial proteins have been successffully studied. As a result, a limited number of unique crystal structures of membrane proteins are found in the Protein Data Bank (Finkelstein, 2014).

In contrast, the structure of thousands of membrane proteins has been obtained by cryo-electromicroscopy (cryo-EM). Major advances in cryo-EM technology have enabled researchers to resolve structure at atomic resolution, rivaling those obtained by X-ray crystallography. While the resolution of apoferritin was achieved at 1.25 angstoms (Yip et al., 2020), most protein structures get resolved at a resolution ranging from 2.5 to 4.0 angstroms. At this level, some details like individual atoms can still be difficult to resolve, especially in flexible and disordered regions. Individual atoms like inorganic ions are also difficult to identify and in many occasions, the identification is only possible through the analysis of the surrounding environment. That said, the field of cryo-EM is in an exponential phase with several hundreds of paper published since 2020 and the resolution of structures gets better and better.

3. Folding of proteins

Synthesis of proteins occurs intracellularly on ribosomes in the rough endoplasmic reticulum (RER) wherein the specific amino acids that compose the protein are added one-at-a-time from the amino to the carboxyl terminal end of the protein. After the protein is synthesized, it is subjected to posttranslational modification involving cleavage of the leader sequence from its amino terminal end and, in some cases, addition of carbohydrates to specific residues and, if the protein is synthesized as a proprotein, cleavage of the amino terminal propeptide. At this point, the protein undergoes folding to its final three-dimensional structure. This folding, which often takes place in the presence of chaperone proteins that guide folding, discussed below in this section, occurs in minutes. Many proteins are very large and can contain more than 1000 amino acids in a single polypeptide chain. For "small" proteins, say with 150—200 amino acids, to compute, using the techniques described in Section 1.5.2, the low energy states for these proteins would take weeks to months. Evidently, there is some guiding principle that allows the chains to fold rapidly even in the absence of chaperones. Thus, there is great interest in discovering how proteins fold. To accomplish this task, it is desirable to understand the pathway of protein folding, i.e., to determine what intermediates lie on the pathway from the fully denatured state to the fully folded protein and/or vice versa.

Currently, it has not proved possible to crystallize any of the intermediates on the folding pathway so that x-ray diffraction can be performed on them. Apparently, because of the multiple conformations that exist in the denatured forms of proteins, it has proved likewise not possible thus far to analyze the structures that constitute denatured forms by multidimensional NMR. Thus, it is necessary to fall back on less direct methods that allow inference at least of what domains of folding proteins fold first and which domains fold up sequentially after the initial folding events. This is an active area of investigation and involves multiple approaches. Here we describe several of these to give some idea as to how the problem is being approached.

In one approach (Chavez & Scheraga, 1980), in the denaturation and then renaturation of ribonuclease A (RNase A), the 124-residue protein discussed above that was used in the original Anfinsen experiment, antibodies were made against specific segments of this protein, viz, residues 1—13, 31—79, and 80—124 in the native protein. These antibodies do not react with RNase A when the protein is reduced and denatured. Upon reoxidation and renaturation of this protein, it was found that the first domain that reacted with the antibodies was 80—124, followed by the 31—79 segment. Importantly, if residues 121—124 were omitted, the kinetics of refolding were strongly affected. These results implied that the first domain to fold in RNase A is the 105—124 segment, which is consistent with the conclusions of a theoretical treatment that predicts where nucleation of folding is likely to occur in folding proteins and predicted that folding would occur in a hydrophobic pocket involving residues 106—118. It is also consistent with the results of peptide inhibition of protein refolding. Peptide segments of RNase A, including residues 105—124, were prepared and were incubated with fully reduced, denatured RNase A. The 105—124 peptide uniquely completely blocked refolding from the beginning of the incubation suggesting that the isolated segment folds and then competes with its counterpart in the protein chain.

A second approach involves trapping intermediates of refolding proteins that contain disulfide bonds by reacting their free —SH groups with suitable reagents such as 2-aminoethyl methanethiosulfonate and then separating these species using high-performance liquid chromatography (HPLC) (Rothwarf et al., 1998). When this was performed on RNase A, which has four disulfide bonds 26—84, 40—95, 58—110, and 65—72 in the native state, it was found that there were actually two intermediates on the folding pathways, both containing three disulfide bonds, that were formed in rate determining steps, des 40—95 and des 65—72. Both of these species then form the fourth disulfide bond after the rate-determining steps (Rothwarf et al., 1998). These two species form on separate pathways. Therefore, there are two productive pathways in the folding of RNase A, each giving rise to the respective three-disulfide species. On the folding pathway at the 1- and 2-disulfide stages, there were multiple species none of which had native or native-like structure. These studies indicated that protein refolding is an ordered process and contains key structures on the pathway that are vital for correct folding.

Chaperone Proteins. For large proteins containing 500—1000 amino acid residues (such as those that occur in ion channel proteins and in proteins with multiple functions such as the MDM-2 protein), the folding process becomes complex due to the large number of possible interactions between folding domains making folding "errors" more probable on their folding pathways. Also, as proteins are synthesized on ribosomes in the rough endoplasmic reticulum (RER), the possibility of misfolding of the elongating nascent chain exists. To reduce the occurrence of misfolding, proteins called chaperones exist that appear to guide newly synthesized proteins to fold correctly.

These chaperone proteins themselves have multiple functions and include such members as heat-shock protein (Hsp), bacterial Dnaj protein, the chaperonins including chaperonin 60 (cpn60) and TriC (Hendrick and Hartl, 2017). These proteins bind to proteins that are being synthesized on ribosomes and/or to newly synthesized proteins that are denatured in the intracellular environment. Often, the process of guiding the correct folding of newly synthesized proteins by chaperones requires hydrolysis of ATP. The structures of a number of chaperone proteins have been determined. Tri C actually is composed of multimeric subunits that are arranged as two stacked rings that bind to nascent polypeptide chains in the RER to prevent aggregation of folding intermediates. Release of the correctly folded chains from complexes with TriC requires ATP hydrolysis (Hendrick and Hartl, 2017).

It is of great interest to understand how chaperone proteins function because this would shed further light on how proteins fold. Recently, a technique has been developed that allows inferences to be made as to what the structures are of folding proteins bound to a chaperone, by combining x-ray crystallography with computation of structure (Horowitz et al., 2016). The method has been applied to the folding of the *E. coli* protein Im7 (an immunity protein), which interacts with the chaperone protein called Spy. The structure of Spy has been determined by x-ray crystallography, and the x-ray diffraction pattern for Spy complexed with the known Spy-binding domain of Im7, residues 6–45 has also been determined. The structure of Spy can be determined from this diffraction pattern, leaving what is termed the "residual" electron density (In the complex, Spy undergoes a change in structure such that two alpha-helices that make up of the structure of this protein change their angle of orientation by about 9°).

In addition, eight iodine-labeled forms of the Im7 6–45 sequence were cocrystallized with Spy, and the x-ray diffraction pattern for each of these eight Im7 derivatives was determined. The structure of Spy in each of these complexes was found to be the same as that of the original Spy structure. The heavy iodine atom allowed for the positions of the residue containing it in the Im7 6–45 peptide to be determined in the x-ray structure.

Now, the low-energy conformations for Im7 were computed using molecular dynamics as mentioned in Section 1.5.2. These low-energy conformations constituted a "pool" of structures that were then tested for their abilities to reproduce the residual electron density and would also reproduce the positions of the specific substituted residues. The final structures that met these criteria were regarded as the possible intermediates on the folding pathway. These structures were superimposed on the structure of Im7 and were then further classified by families whose structures were found to have a given root mean square (rms) deviation. The structures whose rms deviations were the lowest were considered to be folded or almost folded while those that had deviations that were larger were considered less folded or unfolded.

The less folded and unfolded structures in the higher rms range were found to have a large preponderance of hydrophobic interactions with Spy and few if any polar interactions with Spy. In direct contrast, the more or completely folded (low rms deviation) structures were found to have almost exclusively polar interactions with Spy residues. This result suggests that the mechanism of chaperone-induced folding is to sequester nonpolar amino acid residues in the incipiently folded chain; when these hydrophobic residues aggregate correctly in the folding process, the folding protein then interacts with the chaperone protein via polar interactions and is finally released.

Once a protein has adopted its final folded form, it possesses a unique size and shape that confer on it certain bulk properties, among the most important of which are electrical properties, which we now discuss.

4. Bulk properties of proteins: Proteins as polyelectrolytes

Folded proteins will, in general, have different shapes and sizes depending on their specific sequences of amino acids. Virtually, all folded proteins have their nonpolar groups buried in the interior of the protein as a result of hydrophobic interactions and their polar and charged groups on the surface of the protein interacting with the solvent. Extensive work in the early 1980 has addressed the interaction proteins with water, salts, and organic osmolytes and measure the effects of these comopnents on protein stabilization (Arakawa and Timasheff, 1982a, 1982b, 1983, 1985). The effect of pH is discussed in the next section. This distribution of charged groups on the surfaces of proteins strongly affects the manner in which these proteins will be oriented and migrate in electric fields that are either applied in electrophoresis experiments or that result from the transmembrane potential in cells. Proteins with net charges also contribute to the fixed charge inside cells. How proteins migrate in electric fields gives much information about their sizes, shapes, and molecular masses. We must therefore consider the properties of proteins as polyelectrolytes dissolved in solution.

4.1 Acid-base properties of amino acids

All amino acids have free α-amino (NH_2) groups and carboxyl (COOH) groups. Amino groups are **basic,** that is, **take up** hydrogen ions (H^+), while carboxyl groups are **acidic,** that is, **give up** hydrogen ions. Using glycine as an example, at neutral pH (7.0), the only form to exist is the dipolar, or zwitterionic, form, $^+H_3N{-}CH_2{-}COO^-$. Fig. 2.11 shows a prototypical titration curve for a dibasic acid such as glycine over the pH range of 1–10. In this figure, note that there are two regions of the titration curve where the change in pH is minimal for added base: at pH of around 2.0 and 9.0, the pK_a values for the COOH and NH_2 groups, respectively. At pH 2.0, the buffering capacity of the COOH group is maximal,

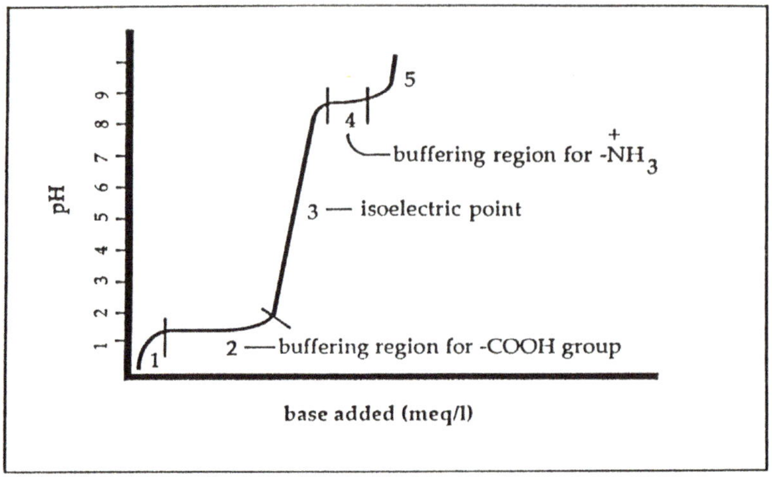

1) $H_3\overset{+}{N}-CH_2-COOH$

2) $H_3\overset{+}{N}-CH_2-COOH \leftrightarrow H_3\overset{+}{N}-CH_2-COO^-$

3) $H_3\overset{+}{N}-CH_2-COO^-$

4) $H_3\overset{+}{N}-CH_2-COO^- \leftrightarrow H_2N-CH_2-COO^-$

5) $H_2N-CH_2-COO^-$

FIGURE 2.11 Titration curve for a dibasic acid like glycine to show the two buffering regions for the COOH and NH_3^+ groups and the isoelectric point that can be computed as $(pK_{a1} + pK_{a2})/2$. A protein has multiple buffering regions, characteristic of each residue type that is involved in a prototropic dissociation.

while at pH 9.0, the buffering of the NH_2 is maximal. The buffering capacity is governed by the Henderson–Hasselbalch equation,

$$pH = pK_a + \log\frac{[\text{conjugate base form}]}{[\text{acid form}]} \tag{2.3}$$

where the pK_a is the negative logarithm of the proton dissociation constant for the acid. For the acid segment of the titration curve in Fig. 2.11, these two forms are $H_3N^+{-}CH_2{-}COO^-$ and $H_3N^+{-}CH_2{-}COOH$, respectively; for the high-pH portion of the curve, these forms are $H_2N{-}CH_2{-}COO^-$ and $H_3N^+{-}CH_2{-}COO^-$, respectively. Note also that the following equilibria exist over this range:

$$\begin{aligned} &H_3N^+ - CH_2 - COOH \rightleftharpoons H_3N^+ - CH_2 - COO^- + H^+ \\ &H_3N^+ - CH_2 - COO^- \rightleftharpoons H_2N - CH_2 - COO^- + H^+ \end{aligned} \tag{2.4}$$

We can write two separate sets of equilibrium conditions for the two reactions in Eq. (2.4) as follows:

$$K_1 = \frac{[H_3N^+ - CH_2 - COO^-]\,[H^+]}{[H_3N^+ - CH_2 - COOH]}$$

$$K_2 = \frac{[H_2N - CH_2 - COO^-]\,[H^+]}{[H_3N^+ - CH_2 - COO^-]} \tag{2.5}$$

Solving both of these equations for the common zwitterionic species, $H_3N^+{-}CH_2{-}COO^-$, we obtain the expression

$$K_1K_2 = \frac{[H_3N^+ - CH_2 - COOH]}{[H_2N - CH_2 - COO^-]} \times [H^+]^2 \tag{2.6}$$

Taking the negative logarithms of both sides of Eq. (2.6), we obtain the expression

$$pH = \frac{pK_1 + pK_2}{2} + \frac{1}{2}\log\frac{[H_2N - CH_2 - COO^-]}{[H_3N^+ - CH_2 - COOH]} \tag{2.7}$$

where pK_1 and pK_2 are the negative logarithms of K_1 and K_2. This equation is the same as the Henderson–Hasselbalch equation for any acid–base equilibrium except, in this case, it is written for two equilibria. The **isoelectric point,** pI, is defined as the pH at which the total negative charge of the molecule is equal to its total positive charge (i.e., where only the zwitterionic species exists). In terms of Eq. (2.7), this condition is met if the numerator and denominator are equal. In this case,

$$pH = \frac{pK_1 + pK_2}{2} \tag{2.8}$$

For amino acids with charged side chains, the isoelectric point is easily shown to be that pH equal to $(pK_1 + pK_2 + pK_3)/3$. In general, for multiple prototropic dissociations for a whole protein, the isoelectric point of the protein is simply that point on the titration curve for the whole protein where

$$pH = \sum pK_i/N \tag{2.9}$$

where the sum is taken over all proton-dissociating groups, and N is the number of these groups in the protein.

4.2 Protein charge and solubility

Since all proteins are charged because of charged side chains and the α-NH_3^+ group and the $-COO^-$ carboxyl terminal group, there is generally a **net** charge on the protein. These net charges tend to be solvated and interact with counterions in solution. The solvation of these charges makes the protein soluble in H_2O. However, interactions of the charged side chains of the protein with surrounding ions and aqueous solvent become minimal at the isoelectric point, where the charges on the protein exactly balance one another. At its isoelectric point, the solubility of the protein in water is minimal, and it will tend to precipitate from solution. Different proteins have different sequences and different charges and isoelectric points so that their tendencies to precipitate will differ from one another and depend on the pH of the solution. Thus, by changing the pH of protein solutions, it is possible to effect differential precipitation of these proteins. A more sophisticated version of isolation of proteins based on their isoelectric points is **isoelectric focusing,** an electrophoretic method discussed in Section 4.4.

4.2.1 Isoionic point

This term is used when proteins are dissolved in aqueous solutions where the only counterions are H^+ and OH^-; in other words, the counterion for the anionic groups is H^+, and the counterion for the cationic groups is OH^-. Since the total number of positive charges for the system must equal the total number of negative charges for the system,

$$(H^+) + (\text{total positive charges on protein}) = (OH^-) + (\text{total negative charges on protein}) \tag{2.10}$$

From Eq. (2.10), if the isoionic point of a protein is at pH 7 (i.e., $H^+ = OH^-$), **the isoelectric point is the same as the isoionic point** since from Eq. (2.10), if $H^+ = OH^-$, then total positive charges on the protein equal total negative charges on the protein.

4.3 Titration of proteins

Because of their large number of exposed acidic and basic groups, proteins behave as buffers. Generally, the pH of the buffering region will depend on the relative numbers of acidic and basic groups on the protein surface. Binding of protons to proteins may be considered to be formally the same as the binding of ligands (small molecules) to proteins. To understand the behavior of proteins as buffers, it is important to understand ligand binding theory.

4.3.1 Ligand binding theory

Suppose a protein has n equivalent sites for binding to a ligand and each site is completely independent of any other site. The equilibrium for the binding of a ligand, A, to any one site of the protein can be written as

$$\mathrm{A} + \mathrm{P} \rightleftharpoons \mathrm{AP} \tag{2.11}$$

where A is the free ligand, P is the unbound protein, and AP is the ligand bound to protein. The equilibrium association constant, K, for this process is

$$K = \frac{[\mathrm{A}][\mathrm{P}]}{[\mathrm{AP}]} \tag{2.12}$$

where [A] is the free ligand concentration, [P] is the concentration of unbound protein, and [AP] is the concentration of ligand bound to protein. If there are n equivalent binding sites per protein molecule, then the total concentration of sites is n $[\mathrm{P}]_o$, where $[\mathrm{P}]_o$ is the concentration of total protein and is equal to the sum of the concentrations of free and bound protein:

$$[\mathrm{P}]_o = [\mathrm{P}] + [\mathrm{AP}] \tag{2.13}$$

Combining Eqs. (2.12) and (2.13), we find that

$$[\mathrm{P}]_o = [\mathrm{P}] + K[\mathrm{A}][\mathrm{P}] \tag{2.14}$$

Solving for [P], we obtain

$$[\mathrm{P}] = \frac{[\mathrm{P}]_o}{1 + K[\mathrm{A}]} \tag{2.15}$$

Since from Eq. (2.12), $[\mathrm{AP}] = K[\mathrm{A}][\mathrm{P}]$, the ratio of bound protein to total protein is

$$\frac{[\mathrm{AP}]}{[\mathrm{P}]_o} = \frac{K[\mathrm{A}]}{1 + K[\mathrm{A}]} \tag{2.16}$$

If there are n equivalent binding sites per protein molecule, the fraction of sites bound from Eq. (2.15) is

$$\frac{[\mathrm{AP}]}{n[\mathrm{P}]_o} = \frac{K[\mathrm{A}]}{1 + K[\mathrm{A}]} \tag{2.17}$$

This equation can be linearized. Defining $R = [\mathrm{AP}]/[\mathrm{P}]_o$,

$$\frac{R}{[\mathrm{A}]} = nK - KR \tag{2.18}$$

Plots of $R/[\mathrm{A}]$ versus R (called **Scatchard plots**) should give a straight line whose slope is K and whose intercept is nK. Thus both K and n are readily determined.

For n equivalent proton binding sites on a protein, we can write Eq. (2.16) as

$$R = \frac{K[\mathrm{H}^+]}{1 + K[\mathrm{H}^+]} \tag{2.19}$$

In this case, $R = [\mathrm{PH}^+]/n\ [\mathrm{P}]_o$. If we use the dissociation constant $K' = 1/K$, Eq. (2.19) may be recast as

$$[\mathrm{PH}^+]K' = (n\ [\mathrm{P}]_o - [\mathrm{PH}^+])\ [\mathrm{H}^+] \tag{2.20}$$

Rearranging and taking logs of both sides of Eq. (2.20), we obtain the equation

$$\mathrm{pH} = \mathrm{p}K_a + \log\frac{n[\mathrm{P}]_o - [\mathrm{PH}^+]}{[\mathrm{PH}^+]} \tag{2.21}$$

where $\mathrm{p}K_a$ is $-\log(K')$. Note that Eq. (2.21) is the Henderson–Hasselbalch equation for multiple dissociations. The denominator in the logarithmic term in Eq. (2.21) represents the concentration of H^+-bound protein while the numerator represents nonproton-bound protein. This equation is useful for titration of proteins provided that the $\mathrm{p}K_a$ of each group is independent of that of any other group for a given set of groups, such as the carboxyl groups of Glu and Asp.

For proteins, it is not generally correct to assume that the above condition holds because, as each carboxyl group dissociates from a protein, the protein becomes progressively negatively charged, so that it becomes progressively difficult for the next proton to dissociate. In fact, the protein behaves as a polyelectrolyte wherein proton dissociations are strongly dependent on the electric field of the protein.

4.3.2 The electrostatic field effect (Bull, 1943)

To account for the electrostatic field effect of the charged protein on the dissociation process, we note that the free energy of dissociation per mole of acid group, ΔG, can be written as

$$\Delta G = \Delta G_i + \Delta G_e \tag{2.22}$$

where ΔG_i is the intrinsic dissociation free energy for the group and ΔG_e is the electrostatic free energy for dissociating a proton from the acid group in the presence of the field of the protein. From electrostatic theory,

$$\Delta G_e = eU \tag{2.23}$$

where $e = 1$ electrostatic charge unit and U is the electrical potential from the protein. Since the standard free change for dissociation, ΔG, equals $-2.303 \times RT \log K$, and ΔG_i equals $-2.303 \times RT \log K_i$ (where R is the gas constant and T is the temperature in K) from Eq. (2.22), we obtain the relation

$$\mathrm{p}K = \mathrm{p}K_i + eU / 2.303RT \tag{2.24}$$

From Eq. (2.21), we have

$$\mathrm{pH} - \log\frac{n[\mathrm{P}]_o - [\mathrm{PH}^+]}{[\mathrm{PH}^+]} = \mathrm{p}K_i + eU / 2.303RT \tag{2.25}$$

It is of obvious importance to obtain an explicit expression for U since it strongly influences the dissociation of protons from proteins.

4.3.3 Computation of the electrical potential

To compute the electrical potential U, it is necessary to use a model for the interaction of protein charges with those of small ions. The simplest model is to assume that the protein is a charged sphere and that the counterions surrounding it, like protons, form a spherical shell around the central charged protein sphere as represented in Fig. 2.12. The overall

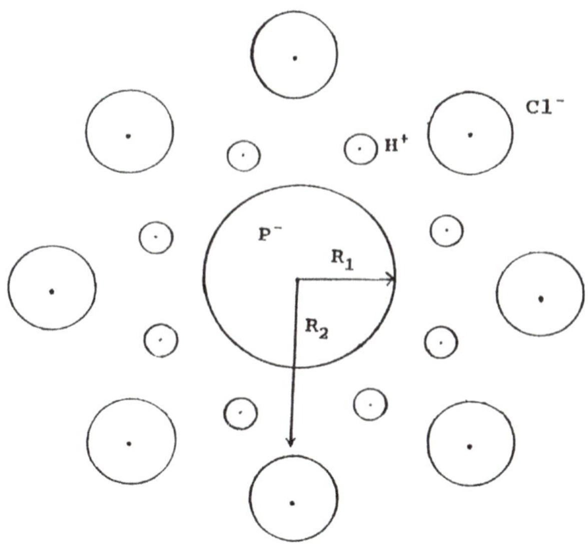

FIGURE 2.12 The model for the effect of the charge on a protein (p^-) on the dissociation of protons from groups on the protein involved in prototropic dissociations. R_1 is the radius of the protein, assuming that it has a spherical shape. R_2 is the distance from the center of the protein sphere to the mean center of the ions surrounding the protein. In this figure, H^+ ions are shown as the positive ions and Cl^- as the negative ions. Their distributions around the protein will depend upon the charge on the protein and the Boltzmann factors for each surrounding ion type. R_2 is therefore a mean distance from the center of the protein to the mean center of the surrounding ions.

arrangement is that of two concentric spheres of radii R_1 and R_2, the first term being the distance from the center of the protein to its surface, and the second term being the distance from the center of the protein to the center of mass of the ions surrounding the protein. It is clear from this figure that the interactions of protons, and/or any ions, with the central, charged protein molecule depend on the distance of the centers of these ions from the center of the protein ion. The distribution of ions around the central charged protein ion is given by the Boltzmann distribution,

$$\psi = \sum en_i Z_i \exp - (en_i Z_i U / RT) \tag{2.26}$$

where Z_i is the valence of the ith ion. We can expand the exponential term in a Taylor series as follows:

$$\begin{aligned}\psi &= \sum n_i Z_i e - \sum n_i Z_i e(eZ_i U/RT) \\ &\quad + \left(\sum n_i Z_i e/2\right)(Z_i eU/RT)^2 + \cdots\end{aligned} \tag{2.27}$$

We make the assumption that the eZU term is much less in value than RT, and that electrical neutrality must be preserved (i.e., the first sum on the right of Eq. (2.27) is 0). Then,

$$\psi \approx -\sum n_i Z_i^2 e^2 U/RT \tag{2.28}$$

This charge distribution must satisfy the Poisson charge distribution equation,

$$\nabla^2 U = -4\pi\psi/D \tag{2.29}$$

where $\nabla^2 U$ is the second derivative of the potential with respect to the sum of the coordinates of the system, ψ is the charge distribution, and D is the dielectric constant of the medium, water. Combining Eqs. (2.28) and (2.29), we obtain,

$$\nabla^2 U = (4\pi e^2) / (DRT) \times \sum n_i Z_i^2 U \tag{2.30}$$

Solution of this equation for U as a function of the coordinates of the system constitutes the Debye–Hückel theory, and it depends on the assumed configuration of small ions around the central large protein ion, assumed here to be spherical in shape as in Fig. 2.12.

Solution of Eq. (2.30) for this circumstance at R_1 gives

$$U = Ze\,(R_2 - R_1) \;/\; DR_1 R_2$$

where $Z = \sum n_i Z_i$. Letting $1/\chi = R_2 - R_1$,

$$U = \frac{Ze}{DR_1(\chi R_1 + 1)} \tag{2.31}$$

Combining Eqs. (2.24) and (2.31), we obtain

$$\mathrm{p}K = \mathrm{p}K_i + \frac{Ze^2}{2.303RTD} \times \frac{1}{R_1(\chi R_1 + 1)} \tag{2.32}$$

Setting

$$W = \frac{e^2}{2.303RTD} \times \frac{1}{R_1(\chi R_1 + 1)} \tag{2.33}$$

Eq. (2.25) may now be written as

$$\mathrm{pH} - \log \frac{n[\mathrm{P}]_o - [\mathrm{PH}^+]}{[\mathrm{PH}^+]} = \mathrm{p}K_i + 0.868\ WZ \tag{2.34}$$

This equation now takes into account the field effect of protein and surrounding ion charges provided that the system conforms to the assumptions made for this model: that $U << RT$, that the protein is spherical in shape, and that the ions around the protein form a spherical shell around it. Plots of the left side of Eq. (2.34) versus Z, the overall protein charge, should yield a straight line whose slope is $0.868W$. This term may be regarded as a structure term that depends on the radius of the protein and the position of the center of mass of the surrounding ion cloud.

4.3.4 More explicit models based on protein structures

The above theoretical expression is based on several critical assumptions that do not always apply to proteins, namely, that the charge distribution around the protein is uniform and spherical and that its electrical potential is such that the *ZU* term << RT. We know from the many x-ray crystallographically determined protein structures that protein surfaces are far from spherical and that the charge distributions are not uniform. Thus, to understand the effects of protein charge and the effect of the medium on proton dissociations from the protein, more explicit and detailed models are needed. One treatment is based on conformational energy calculations on proteins surrounded explicitly by solvent molecules. These calculations have thus far been aimed at determination of the intrinsic pK_as of specific groups on protein surfaces.

In this approach (Russell & Warshel, 1985), the energy of dissociation of a particular acid group, AH, is computed for the whole protein surrounded by water molecules placed on a grid around the protein. The energy of interaction of the atoms of the protein with one another is computed using energy functions similar to those discussed in Section 1.4. Included in this energy is the effect of the permanent dipoles of the atoms of the protein that induce dipoles on the other atoms of the protein. The magnitude of this energy depends on the polarizability of the interacting atoms. Further account is taken of the effect of the interaction of the dipoles of the atoms of the protein with water molecules which themselves are represented as dipoles. The calculations are then applied to the dissociation of a proton from a particular group, for example, Asp 3 and Glu 7 in bovine pancreatic trypsin inhibitor (BPTI) whose structure has been determined at high resolution by x-ray crystallography (see Section 2.2).

In this procedure, a thermodynamic cycle is employed in which the free energy difference of the system is computed for the acid group in the AH form in the protein and for the AH group dissolved in water (process 1). The group is then allowed to dissociate in water at a given pH (process 2). The free energy difference between AH and A^- in water is known (for Asp residues, this would be the free energy of the dissociation of acetic acid in water at a given pH) and is equal to $2.303RT(\text{p}K_a - \text{pH})$. Finally, the free energy difference between solvated A^- in water and A^- in the protein (process 3) is computed. The overall free energy for dissociation is then computed as the difference between the free energy for process 1 and the free energy for process 3 plus a constant (free energy for dissociation of AH in water).

These calculations have been carried out on a supercomputer using molecular dynamics and iterative procedures (for the polarization energies) and have reproduced ionization constants satisfactorily. The method, which is computer intensive, includes the effects of both solvent and the atoms of the whole protein on the ionization process. Thus the approach appears to be a fruitful one in the investigation of prototropic dissociations in proteins.

4.4 Protein charge and electrophoresis

Different proteins have different net charges as discussed above. If these proteins are placed in an electric field, positively charged proteins migrate toward the negatively charged pole, the cathode, while negatively charged proteins migrate to the positively charged pole, the anode. If a protein is present at a pH equal to its isoelectric point, it will generally not move toward either pole. The electrostatic force on a protein present in an electric field is equal to ZeU, where U is the electric field, e is the charge of an electron (or a proton), and Z is the number of charges present. There is a drag force on the proteins that is proportional to the velocity at which they move toward either pole, fv, where v is the velocity of the protein moving in the electric field and f is a proportionality constant. This force balances the electrostatic force so that

$$ZeU = fv \tag{2.35}$$

and

$$v = ZeU/f \tag{2.36}$$

The value of f generally increases with increasing size and molecular weight so that, for proteins of similar charge but different molecular weights, the higher molecular weight proteins move more slowly than the lower molecular weight proteins.

There are several types of electrophoresis, all of which are extremely effective not only in separating proteins but in directly determining their molecular weights.

4.4.1 Slab gel electrophoresis

Proteins are separated on gels made up of such polymers as starch, polyacrylamide, and agarose. The separation is based mainly on charge, and, to an extent, on size.

4.4.2 Sodium dodecyl sulfate (SDS) electrophoresis

Sodium dodecyl sulfate (SDS) is a detergent that contains a long aliphatic chain and a sulfate group. This detergent interacts with denatured proteins to form a strongly negatively charged complex (the negative charge arising from the SO_4^{-2} groups of SDS). The proteins are first denatured by heat, and then the SDS is added in large excess. The SDS−protein complexes all contain about the same negative charge because the SDS swamps out all of the protein charges. Since the charges are all the same, the proteins all separate from one another strictly on the basis of their sizes. From Eq. (2.36), the larger polypeptide chains of higher molecular weight migrate the most slowly while the lower molecular weight proteins migrate more rapidly toward the anode. Proteins of known molecular weight can be subjected to this procedure and used as "markers." A plot of the molecular weights versus the log of the distance traveled from the point of application yields a straight line. The log of the distance of migration of a protein of unknown molecular weight can then be plotted on this line and the molecular weight directly determined.

4.4.3 Isoelectric focusing

In this elegant technique, a polymer is used in which acidic and basic groups change in density from one end of the polymer to the other. When this polymer is placed in solution, a continuous pH gradient is established along the polymer. A mixture of proteins is then applied to the polymer in a weak buffer. When the proteins migrate in the electric field, they experience local differences in pH on the polymer and eventually reach a local pH equal to their individual pIs. At the pI, the protein no longer migrates in the electric field. This method of separation depends only on the presence of proteins with different pI values.

4.4.4 Two-dimensional gel electrophoresis

Proteins can be separated from one another in one lane as described in the preceding three sections; however, these approaches may not meet the conditions for separating all of the proteins. For example, on agarose gel electrophoresis, two proteins may not separate well at the pH of the buffer used. A buffer of different pH may then be added and the electrophoresis carried out at right angles to the original direction of migration to allow further separation of the proteins.

4.4.5 Western blots (immunoblots)

All proteins are antigens and can provoke the production of antibodies against them when injected into animals. The antibodies are immunoglobulins which are proteins that bind very specifically to and with high affinity for specific antigens. Antibodies and, in fact, all proteins can be conjugated covalently to fluorescent dyes or to enzymes. If we wish to identify a particular protein on an electrophoretogram, we can use a "tagged" antibody to the protein and reveal its presence on the electrophoretic gel. The antibody is "tagged" with a covalently labeled enzyme that catalyzes a reaction that produces a chromophoric (colored) reaction product. For example, alkaline phosphatase catalyzes the hydrolysis of *p*−nitrophenol phosphate, which is colorless, to *p*−nitrophenol (more accurately, p-nitrophenoxide anion), which is bright yellow. Since antibody cannot be added directly to the gel from electrophoresis, the gel itself is blotted onto nitrocellulose or another suitable membrane that contains all of the separated bands of proteins as they were on the original electrophoretic gel. This nitrocellulose strip is then incubated in a solution containing the antibody. The antibody itself may be conjugated to an enzyme-like alkaline phosphatase, and the band can be identified, or antibodies to the primary antibodies, such as goat antirabbit IgG, are conjugated to the enzyme, and the blot is incubated with these secondary antibodies after being treated with the primary antibody. The secondary antibodies are then markers for the desired protein band.

5. Relationship of protein structure to function

In this section, we discuss some of the relationships of the conformations of proteins to their functions. This is a vast field, and so we concentrate on one aspect of fundamental importance to cell physiology, membrane polypeptides, and proteins.

5.1 Membrane polypeptides and proteins

Membrane proteins have exceptional physiological importance in a number of different ways. Virtually all extracellular receptors have a transmembrane domain that is essential to the functioning of the receptor protein. Other membrane polypeptides are vital in the transport of secreted proteins across cell membranes. Still other membrane proteins, such as the components of the complement system, are involved in intercalating into the cell membrane, causing cell lysis. Yet other membrane proteins form ion channels, which allow for selective entry or exit of specific ions to and from the cell. In this section, we discuss the relationship of the structures of some of these membrane proteins to their functions. It should be emphasized that this entire field of structure–function relationships of membrane proteins is quite new. Most of our knowledge in this field comes from experimental cell biological data on specific proteins and from the results of computations on membrane and membrane-associated protein structure.

5.1.1 Structure and function of leader peptides

Each protein, after it is synthesized on the ribosome, must be transported across the rough endoplasmic reticulum (RER) membranes. If the protein is to be secreted from the cell, or to intercalate into the cell membrane, it must undergo transport across and/or into the cell membrane. For virtually every protein that is synthesized in the cell, there is a polypeptide segment beginning at the N-terminal methionine residue that consists of about 20–30 amino acid residues. Typically, these sequences consist of several hydrophilic amino acids followed by a long stretch of hydrophobic amino acids. When the protein is secreted across the RER membrane, this **leader sequence** is cleaved off, presumably by intracellular proteases. Absence of these leader sequences results in the inability of the protein to traverse the RER membrane, resulting in ultimate intracellular degradation of the protein. Thus, these leader sequences are of vital importance to cellular protein function in general.

The regulation of transmembrane transport of newly synthesized proteins is delicate. It has been shown in in vitro systems that newly synthesized proteins that are secreted across reticulocyte membranes do not undergo leader sequence cleavage. These proteins are nonfunctional because they do not fold correctly. Thus, leader sequences must be attached to the polypeptide chain to enable protein secretion, but must be cleaved off to allow correct protein folding.

To explain how proteins are secreted across membranes, Engelman and Steitz (1981) proposed the so-called **helical hairpin hypothesis** in which the leader sequence adopts an α-helical conformation. At the end of this segment, there is a hairpin turn followed by another α-helix involving 30–40 residues of the protein itself. Both the amino terminal end of the leader sequence and the carboxyl terminal end of the growing polypeptide chain lie on the same side of the membrane while the hairpin turn lies on the opposite side of the membrane. Because the leader sequence helix and the succeeding helical sequence are both independently stable, they interact minimally with one another so that the growing polypeptide chain can slide past the hydrophobic leader sequence in the membrane. As the nascent polypeptide chain pushes through the membrane, anchored by the leader sequence, it begins to fold in the aqueous environment of the cytoplasm on the opposite side of the RER membrane. The leader sequence, however, remains in the membrane because of its hydrophobic character. The hairpin connection between the two helices is a signal for intracellular proteases that ultimately results in cleavage of the leader sequence.

Pincus and Klausner (1982), using ECEPP, have computed the low energy structures for the leader sequence of the κ-light chain immunoglobulin. This sequence, which contains 16 amino acids, is Asp-Thr-Glu-Thr-Leu-Leu-Leu-Trp-Val-Leu-Leu-Leu-Trp-Val-Pro-Gly. These investigators found that the first four polar residues tend to adopt an extended conformation followed by a long α-helix that ends in a hairpin turn at the carboxyl terminal Val-Pro-Gly sequence. This structure was lower in energy than any other competing structure by at least 10 kcal/mol. Thus, these calculations support the essential features of the Engelman–Steitz hypothesis, and have been further corroborated experimentally. For example, the CD spectrum of synthetically prepared leader sequences shows high α-helical content when the sequence is dissolved in hexafluoroisopropanol, a nonpolar solvent that simulates the low dielectric medium of the membrane (Rosenblatt et al., 1980).

In site-specific mutagenesis experiments, Thomas Silhavy and his coworkers then at the National Cancer Institute have placed proline residues in the middle of the hydrophobic transmembrane domain of a bacterial leader sequence (Emr & Silhavy, 1982). Proline residues tend to disrupt α-helices. These investigators found that if they introduced two proline residues close together in the sequence, the protein to which the leader sequence was attached was no longer secreted. This result strongly corroborates the conclusion that leader sequences must adopt α-helices in the membrane.

In a series of elegant experiments, Blobel et al. (1979) have shown that specific sequences in proteins will insert into membranes but will not lead to cleavage of the inserted sequence from the rest of the protein. These specific sequences contain what are referred to as “stop signals” that do not permit cleavage. Because the transmembrane domain is hydrophobic, it remains inserted in the membrane and there is no impetus for it to move through the membrane, resulting in secretion. This mechanism is extremely important for the function of receptor proteins, as will be discussed in Section 5.1.3.

5.1.2 Membrane-active peptides: Melittin and magainin

These are peptides that insert into cell membranes and induce lysis of these membranes, by forming pores.

5.1.2.1 Melittin

Melittin is one of the smallest proteins known to fold spontaneously. Unlike most proteins, it folds in nonpolar media and is denatured in aqueous solutions and consists of 26 amino acids, i.e., $^{+}$H-Gly-Ile-Gly-Ala-Val-Leu-Lys-Val-Leu-Thr-Thr-Gly-Leu-Pro-Ala-Leu-Ile-Ser-Trp-Ile-Lys-Arg-Lys-Arg-Gln-Gln-NH_2. As the major component by mass of bee venom, it specifically induces lysis of red blood cell membranes. Note that this peptide contains six positive charges and **no** negative charges. Also note that most of the positively charged residues occur at the carboxyl terminal end of the molecule. As discussed in Section 1.2, alpha-helices are stabilized by the presence of positively charged residues on their carboxyl terminal ends as is present in this sequence.

As noted in Section 1.5, the structure of melittin has been computed (Fig. 2.9) and determined by x-ray crystallography (Terwilliger et al., 1982). Overall, the structure of this protein may be thought of as a bent α-helical rod with α-helices from Gly 1 to Thr 10, a reverse turn at Thr 11 to Gly 12, followed by another α-helix from Pro 14 to Gln 26. Leu 13 can adopt an energetically favorable conformation, a D state (see Fig. 2.8) that results in a compact structure for the monomeric state. In the x-ray structure of tetrameric melittin, this residue adopts an energetically less favorable α-helical conformation, causing it to adopt a more open, "straight" structure. In both structures, melittin forms an amphipathic helix in which the nonpolar groups contact one another on the inside of the structure while the polar groups protrude out toward the solvent.

Inspection of the structure for the tetrameric protein reveals that packing between the monomeric units is such that the positive charges of each lie as far away from one another as possible while the hydrophobic cores pack as tightly as possible with one another. This can be accomplished if the monomeric units become slightly less compact by allowing Leu 13 to adopt the α-helical conformation. The sacrifice in conformational energy is about 3 kcal/mol (Kempf et al., 1982).

5.1.2.2 Magainin

This 23 residue peptide intercalates into the cell membranes of bacteria and some fungi and induces lysis of these membranes, but has low activity in lysing red blood cell membranes. The sequence of magainin is H - Gly - Ile - Gly - Lys — Phe - Leu - His - Ser — Ala - Lys - Lys - Phe - Gly - Lys - Ala - Phe - Val - Gly - Glu — Ile- Met - Asn - Ser — OH. As with melittin, this sequence contains a preponderance of positive charges (6 Lys, 1 His [positive at pH's below 7] and one amino terminal positive charge vs two negative charges, one Glu and carboxyl terminal carboxylate). Unlike melittin, magainin has the positively charged residues closer to the amino terminal domain of the peptide. This peptide has been found to adopt an alpha-helical structure.

Multiple theories about the mechanism by which melittin and magainin cause cell lysis have been proposed. Most recently, studies with artificial membranes using 1,2- dioleoyl-*sn*-glycero-3-phosphatidylcholine (DOPC) and 1,2-dioleoyl-*sn*-glycero-3-phosphatidylglycerol (DOPG) indicate that melittin interacts strongly with DOPG, which is negatively charged in two ways: flat, i.e., in the plane of the membrane and perpendicular to the plane of the membrane. In the latter manner, melittin can form channels that disrupt the membrane while molecules in the former mode of binding block channel formation (van Den Bogaart et al., 2008). Deductions about the arrangement of melittin molecules in artificial membranes using fluorescence quenching of the Trp 19 and use of Raman spectroscopy suggest that melittin forms terameric pores through the membrane (Vogel & Jahnig, 1986). These pores would allow extrusion of the intracellular contents resulting in cell death.

The structure of magainin in artificial membranes has been determined using neutron diffraction of magainin in membranes containing 1,2-Dimyristoyl-sn-glycero-3-phosphatidylcholine (DMPC) and 1,2-dimyristoyl-sn-glycero-3-phosphatidylglycerol (DMPG). It appears to form hexagonal pores that are stacked across the membrane that can explain membrane lysis (Yang et al., 2000). The pores appear to have variable sizes over time that are also dependent on the concentration of magainin (Tamba et al., 2010). Thus far it is not clear what causes the selectivity of each peptide for specific cell membranes, i.e., melittin for red blood cell membranes and magainin for bacterial and fungal membranes.

Other theories include the possibility that melittin and magainin have act detergent actions because they are both amphipathic, containing both polar positively charged groups and hydrophobic groups, which allows them to interact favorably with water and the hydrophobic membrane lipid bilayer, respectively, resulting in solubilization of the membrane causing its breakdown. This theory does not account for the fact that virtually all biological membranes contain lyso-lecithins and like fatty acid derivatives that contain nonpolar and positively charged headgroups that actually stabilize membrane structure.

A third theory (Kempf et al., 1982), based on evidence that formation of melittin channels depends on the transmembrane potential, proposes that melittin undergoes a conformational change in the presence of a transmembrane electric field such that it adopts a less compact structure and becomes a more rigid rod-like structure that spans the whole membrane and disrupts the lipids in the lipid bilayer. In this theory, monomeric melittin would form pores or channels. Evidence for this theory comes from experimental work in which melittin was added to artificial bilayers. Proteolytic enzymes (pronase) were added to the aqueous phase under the bilayer. If the transmembrane potential was 0 or such that the outside was negative relative to the inside of the bilayer, no proteolysis of melittin occurred. However, if the inside of the membrane was made negative with respect to the outside such that the transmembrane potential was −60 mv or less, proteolysis of melittin occurred. A potential of −60 mv, found in most resting cells, corresponds to an energy of 2−3 kcal/mol, sufficient to convert compact monomeric melittin to a straightened structure that then spans the membrane and encounters the proteolytic enzymes. This example illustrates the great biological relevance of the relationship of charges on the protein surface to their interactions with electric fields, as in electrophoresis, and to the effects of an electric field on the conformations of proteins. This relationship is vital in understanding the functioning of ion channel proteins as described in Section 5.1.4.

5.1.2.3 New membrane-active anticancer peptides kill cancer cells (Sarafraz-Yazdi et al., 2010)

Recently, peptides derived from the p53 protein have been developed that kill cancer cells but not normal cells. p53 is an antioncogenic protein that slows down or blocks cell division and induces apoptosis of cancer cells. Its lifetime is shortened when it binds to another important protein in the cell, the double-minute-binding protein of human (H) or mouse (M) origin, and called H(M)DM-2 (called HDM-2 in this section). HDM-2 induces the addition of a 76 amino acid protein, called ubiquitin, onto p53, targeting p53 for proteolysis. P53 binds to HDM-2 using its amino terminal amino acid residues 12−26 in the human sequence.

Originally, this p53 peptide was to be introduced into cancer cells to block the binding of p53 to ubiquitin, resulting in increasing the half-life of p53. In order to get the p53 12−26 peptide into cancer cells, it had to be attached to a transmembrane carrier. There is a group of peptides, called penetratins, that are highly positively charged and include the tat sequence of hiv, that cause the transport of proteins across cell membranes, much in the same manner as leader sequences, discussed in Section 5.1.1. Since the structure of the HDM-2-binding domain of p53 is largely alpha-helical when bound to HDM-2 (Kussie et al., 1996), the positively charged transmembrane-penetrating sequence (or membrane residency peptide or MRP) was added to the carboxyl terminal end of the p53 12−26 peptide using the principal that alpha-helices are stabilized by positive charges on their carboxyl terminal ends. When the p53 12−26-MRP, called PNC-27, was incubated with cancer cells, it killed them within 1 h. Surprisingly, it killed p53-defective cells just as rapidly. On the other hand, it had no effect on the growth or viability of normal cells. Electron micrographs of cancer cells killed by PNC-27 and a shorter analogue consisting of p53 residues 17-26-MRP, called PNC-28, showed that transmembrane pores formed. The two-dimensional NMR structure of PNC-27 showed that it had the classical alpha-helix-loop-alpha-helix structure as shown in Fig. 2.13 (Rosal et al., 2004). This structure was highly amphipathic. Thus PNC-27 and PNC-28 kill cancer cells selectively by forming pores in their membranes as membrane-active peptides.

The question, as with melittin and magainin, was why these peptides were selective for 1 cell type, in this case, cancer cells. Recent work has shown that cancer cells express HDM-2 in their cell membranes, while normal cells do not. The reason is thought to be that HDM-2 in the cell membrane binds to cadherin, a cell-to-cell adhesion protein. When it binds to cadherin, it targets its catabolism so that the cells do not stick together and therefore do undergo contact-inhibited cell growth. Thus PNC-27 and PNC-28, which adopt a membrane-active amphipathic helix-loop-helix structure, bind to and are held by HDM-2 in the cell membrane. In this conformation, they form transmembrane pores, resulting in cell lysis. Thus, it is possible that membrane-active peptides can become effective agents in the treatment of cancer.

5.1.3 The function of transmembrane proteins

Currently, there appear to be three classes of transmembrane proteins: intercellular adhesion molecules, receptor proteins, and ion channel proteins. Studies on the first category of these proteins are incipient, whereas the second and third categories of proteins have been better studied. Generally, receptor proteins contain three specific domains: an extracellular domain (ECD), an intramembrane domain, and an intracytoplasmic domain, which often contains a tyrosine kinase. This motif has been found to exist for several growth factor receptors and for the T-cell receptor. The X-ray crystal structures of several growth factor receptors (mostly their extracellular domains) have been determined (see, for example Cho et al., 2003). Mutagenesis experiments and these X-ray structures have elucidated critical features concerning their functioning. A seemingly constant feature of growth factor receptor proteins is that, when the growth factor, which is often a

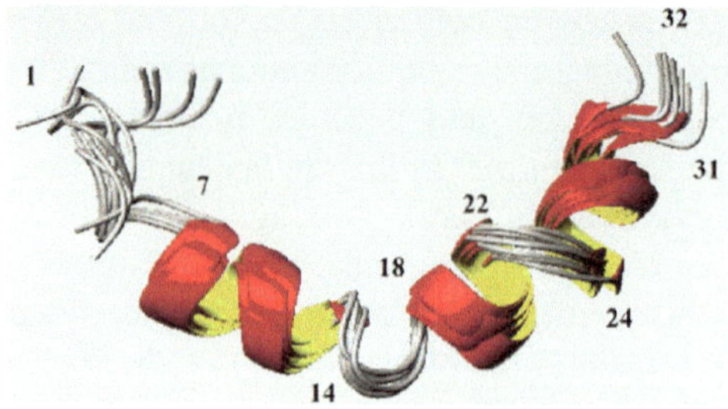

FIGURE 2.13 Ribbon structure for PNC-27, an anticancer, amphipathic, helix-loop-helix peptide structure that binds to membrane-bound HDM-2 that holds it in the membrane allowing it for form transmembrane pores. The two alpha-helical domains are shown as the spiral ribbons.

polypeptide, binds to its transmembrane receptor, the receptor dimerizes. This dimerization induces the activation of the intracellular tyrosine kinase that results in activation of critical cellular processes in a so-called signal transduction pathway.

One of the critical growth factor receptors involved in mitogenic signaling is the epidermal growth factor receptor (EGFR). The x-ray crystal structure of the ECD of EGFR has been determined. In it unbound state, the ECD is folded such that it cannot dimerize spontaneously. When EGF binds to the ECD, however, the internal interactions in the native state are disrupted such that the dimerization domains are exposed and able to interact with one another (Ferguson et al., 2003). Thus binding of the growth factor EGF to EGFR ECD results in critical changes in the structure of EGFR allowing for dimerization and activation of this protein.

As another example, the sequence for the growth factor receptor called *neu* or *her-2,* which is a protein of molecular weight 185 kDa, or p185, is known (Padhy et al., 1982). This growth factor receptor is extremely important in the control of mitogenesis in epithelial cells (Padhy et al., 1982). Defects in this receptor protein can result in unregulated cell division and have been found to be associated highly with breast cancer (King et al., 1985; Slamon et al., 1987). It is known that activation of this protein requires that its growth factor bind simultaneously to two receptors (Ben-Levy et al., 1992), thus crosslinking them. As a result of this crosslinking process, tyrosine kinases become activated, as noted above, resulting in phosphorylation of critical target proteins in a mitogenic signal transduction pathway (Bargmann & Weinberg, 1988).

It is known that substitution of single amino acids within the transmembrane domain of this protein activates the protein, presumably by facilitating dimerization in the membrane (Bargmann & Weinberg, 1988). The essential segment of the transmembrane domain contains the sequence (residues 650–683) -Glu-Gln-Arg-Ala-Ser-Pro-Val-Thr-Phe-Ile-Ile-Ala-Thr-Val-XXX-Gly-Val-Leu-Leu-Phe-Leu-Ile-Leu-Val-Val-Val-Val-Gly-Ile-Leu-Ile-Lys-Arg-Arg-. The XXX amino acid residue can be Val, His, Tyr, Lys, or Gly for normal proteins and Gln or Glu for transforming proteins.

Calculations (Brandt-Rauf et al., 1989) on the structure of the transmembrane domain of the *neu* protein indicate that the entire transmembrane domain, most of which is hydrophobic, can exist in two states, as shown schematically in Fig. 2.14. The normal nonmutated transmembrane domain is a bent α-helix in which two helices are separated by a β-turn at residues 664–665 (these conformations are defined in Section 1). In the "off" state for the normal protein, not bound to ligand, the α-helices are bent and are not favorably disposed to interact with one another. Amino acid substitutions at critical positions around where the bend occurs that obliterate the bend between the helices result in the formation of two regular α-helices that can easily associate with one another.

A complete analysis (Brandt-Rauf et al., 1990) of the Boltzmann distribution of the low-energy conformations of the p185 protein shows that 90% of the protein is in the bent helical state while 10% is in the all-helical state for all normal proteins. For the transforming proteins with Gln or Glu at position 664, this distribution reverses so that 90% of the protein is in the all-helical conformation while 10% is in the bent-helical conformation. These results predict therefore that if the normal protein is overexpressed by about a factor of 10, then sufficient all-helical forms would be present in the cell to cause cell transformation. This computed result has been confirmed in experiments where overexpression of the normal protein by factors of one to four causes no effect on the cell (Hudziac et al., 1987), while ten-fold overexpression of the normal protein has been found to cause cell transformation (Di Fiore et al., 1987). They have been further directly confirmed in ligand-binding experiments using Scatchard plots (see Section 1.3.1) (Ben-Levy et al., 1992), in which it was found that, for the normal cellular protein with Val at position 664, 90% of the protein was in a low-affinity state for the crosslinking ligand and only about 10% of the protein was in a high-affinity state. However, with Glu at position 664, a transforming substitution, the reverse was found: over 90% of the receptors were in the high-affinity state, or readily crosslinked.

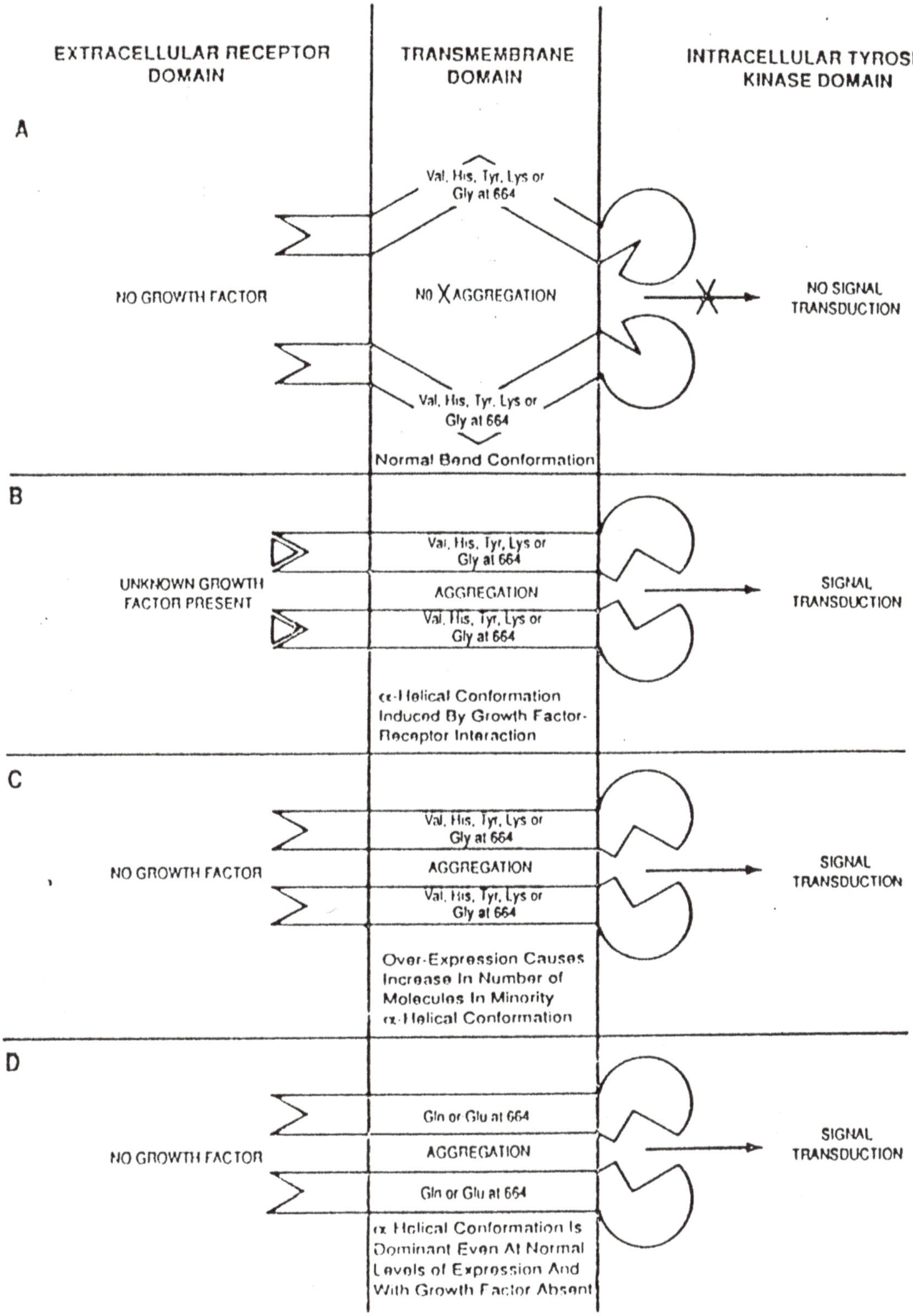

FIGURE 2.14 The effects of transmembrane domain structure of transmembrane proteins on their function. The *neu*-oncogene-encoded p185 protein is a transmembrane protein that has three distinct domains: An extracellular ligand-binding domain (left), a transmembrane domain (middle), and an intracellular signaling domain (right). To get signal transduction of a mitogenic signal to the nucleus, two p185 molecules must associate by the binding of their transmembrane domains to one another. This can be accomplished in a variety of ways as shown in this figure. (A) The normal protein has Val, His, Tyr, Lys, or Gly at position 664 in the amino acid sequence for the normal protein. The structure of the transmembrane domain is a bent α-helix, preventing the two domains from associating with one another. (B) When a growth factor binds to the extracellular domain of the protein, the bend is removed from the middle of the helix, allowing two straight helices to associate. This dimerization allows for activation of mitogenic signaling elements via the intracellular domain in the pathway ending in the nucleus. (C) From the calculations of the distribution of conformations for the normal transmembrane domain (Brandt-Rauf et al., 1990) of p185, 90% of the molecules exist in the bent helix state, while 10% are straight helices. Overexpression of the normal protein by a factor of 10 was therefore predicted to produce sufficient straight helices to cause mitogenic cell signaling, leading to cell transformation. (D) Finally, substitution of Glu or Gln at position 664 reverses the distribution of bent and straight helices directly so that 90% of the molecules are straight helices while only 10% are bent helices. This results in significant levels of dimerization, resulting in permanent mitogenic cell signaling and oncogenic transformation of the cells. *Modified from Brandt-Rauf et al. (1990).*

Thus a change in conformation from a bent to a straight helix may cause major changes in the functioning of a transmembrane protein, much like the proposed change in structure of melittin in the presence of an electric field. It is clear from the above examples that to understand cell function it is necessary to understand the conformational properties of proteins, the basis of which have been presented in this chapter.

5.1.4 Ion channels (Caterall, 1995)

As discussed in the preceding section on melittin and as explained in detail in several later chapters of this textbook, cell membranes are polarized, resulting in a voltage change across them as a result of the fact that the inside of the cell is negative with respect to the outside. Furthermore, in the resting state of most cells, the extracellular sodium ion concentration is much higher than the intracellular sodium, while the reverse is true for potassium ions. During action potentials in electrically excitable cells, such as neurons and myocytes, the membrane becomes depolarized. During this period, there is a large influx of sodium ions, followed during repolarization by an almost equivalent efflux of potassium ions. These fluxes are then rapidly reversed, during which time the cell becomes refractory to further excitation.

The change in the permeability of the cell to sodium and potassium ions is caused by changes in voltage across the cell membrane. Thus, the motion of these ions across the membrane is said to be **voltage-dependent.** During action potentials, there is a **voltage-dependent** increase in the conductance of the membrane to sodium, followed by a rapid and then a slower inactivation of sodium conductance.

All of these changes in conductance and permeability-selectivity of the membrane are mediated by sodium, potassium, and calcium channel proteins, all of which have been cloned, purified, and expressed in different cell lines and in artificial lipid bilayers (Caterall, 1995). More recently, x-ray crystallographic structures of the potassium channel protein have been determined (see, for example, Long et al., 2005). All of these channel proteins have remarkably similar linear amino acid sequences with differences that cause each to bind selectively to different ions.

All of the channel proteins are composed of an α-subunit of molecular mass between 240 and 250 kDa that is heavily glycosylated. Sequencing of the gene for this protein reveals four domains that contain six repeating sequences (called S1–S6). These repeats are characterized by regular spacing in the sequence between hydrophobic amino acid residues and positively charged residues. As with leader sequences, melittin, and the transmembrane domain of the *neu/her*-2 protein, the six repeat sequences are highly hydrophobic. Therefore, they are thought to be transmembrane domains and have been postulated to be α-helical. The basic arrangement of the four domains is shown schematically in Fig. 2.15. In three dimensions, these four domains would be arranged in a cylindrical fashion and would surround a pore. The α-subunit in the

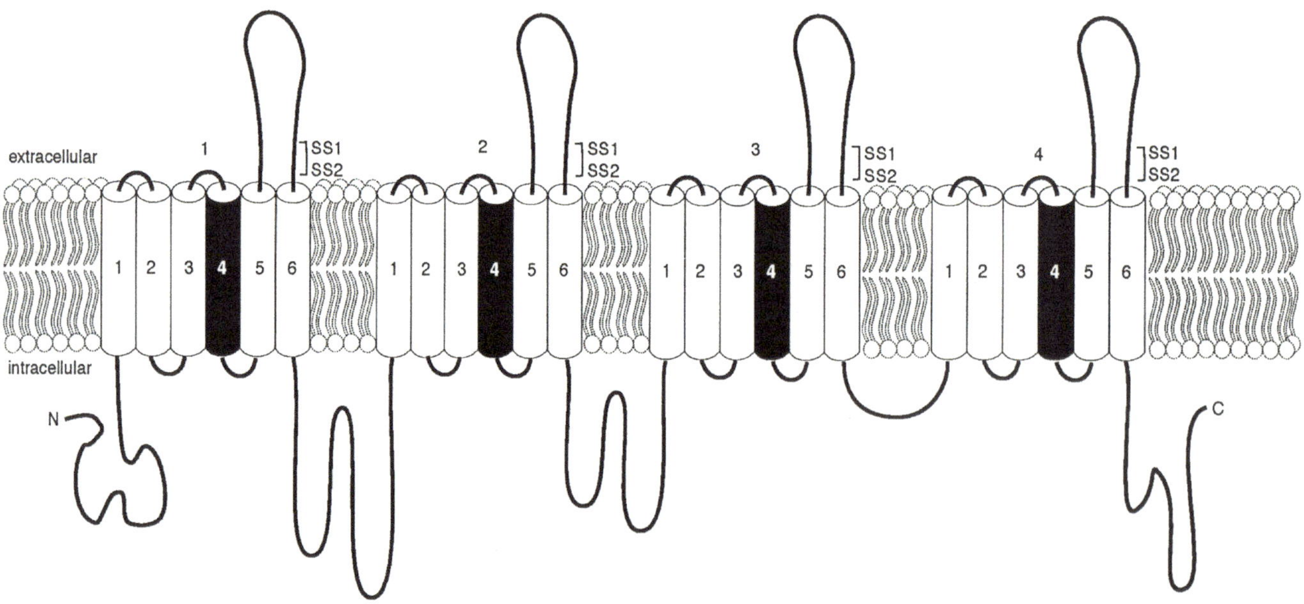

FIGURE 2.15 Model for the arrangement of the domains in the α-subunit of the sodium ion channel protein. The four domains are labeled 1–4. Each domain is shown to contain six transmembrane helical segments labeled 1–6. Subdomain four is highlighted in each domain because it contains the critical residues for voltage-dependent gating. The short segments one and two that are critical for ion selectivity are labeled SS1 and SS2, respectively. This region contains the critical residues Glu 387 and Phe/Tyr or Cys 385. *This model is modified from Hall (1992), p. 109.*

sodium channel protein is noncovalently linked to a heavily glycosylated β-1 subunit of molecular mass 36 kDa and by a disulfide link to a β-2 subunit.

5.1.4.1 Voltage activation

The actual voltage at which activation of the sodium channel occurs has recently been shown to depend on positively charged and hydrophobic amino acid residues in the S4 amphipathic helical segment of each transmembrane domain. As shown schematically in Fig. 2.16, seven critical positively charged Arg and Lys residues are interspersed among many hydrophobic residues. Each of these positive charges must interact with a negative charge from an Asp or Glu residue from another subunit.

In site-specific mutagenesis experiments, each of the positively charged residues has been replaced with neutral amino acid residues. Neutralization of positively charged residues 1, 3, 5, and 7 results in shifting the activation transmembrane potential to more positive values so that the influx of sodium ions begins to occur at more depolarized states. Neutralization of charged residues 2 and 4 produces the opposite effect. Thus residues 1, 3, 5, and 7 are involved with activation of sodium influx while residues 2 and 4 are involved with pore closing and inactivation of sodium influx.

To explain how the S4 segment causes activation of sodium influx, it has been hypothesized that at a certain level of depolarization, the S4 helix rotates such that positive charges on the helix change negatively charged partners in an upward spiral from the inner to the outer membrane, resulting in effective transfer of a positive charge to the outside of the membrane. Since, as indicated in the above discussion, not all positive charges perform the same function, this hypothesis may be oversimplified.

Another explanation for charge transfer across the membrane is that, at a critical transmembrane voltage, the S4 helix unfolds into a fully extended conformation, allowing for the shifting of positive charges toward the outer membrane surface. One problem with this explanation is that the energy required to convert an all-helical protein into an all-extended form requires the disruption of a large number of hydrogen bonds, each about 1–2 kcal/mol in energy. Unless the fully extended form can make hydrogen bonds with other segments in a β-pleated sheet arrangement, the energy barriers for this conformational transition would be very high.

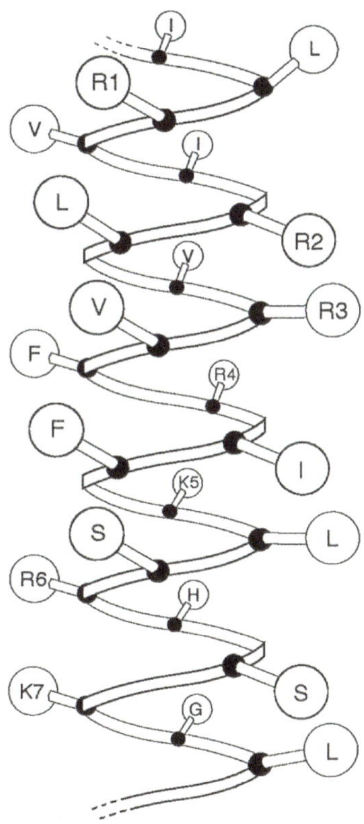

FIGURE 2.16 Helical model for the critical segment of the S4 subdomain of the sodium channel. The critical positively charged residues are labeled in numerical order. *Modified from Caterall (1995), Fig. 2.3, p. 515.*

Nonetheless, it is clear that the change in the transmembrane potential causes a change in the conformation of the S4 helix, resulting in the opening or closing of the sodium channel. This effect is strongly analogous to the proposed effect of the transmembrane potential on the conformation and function of melittin discussed above. Pore formation is clearly influenced by transmembrane voltage and results from a conformational change in the transmembrane protein.

5.1.4.2 Permeability-selectivity

Neural toxins, such as from snake venom, have long been known to bind with high affinity for and selectivity to specific ion channels. The toxin tetrodotoxin is known to bind to the sodium channel pore at its outermost surface. Localization of the tetrodotoxin (and sodium channel pore) binding site has recently been accomplished. Between the S5 and S6 helical segments are two short segments, named short segments 1 and 2 (SS1 and SS2), shown in Fig. 2.15. In SS2, neutralization of Glu 387 by site-specific mutagenesis results in a ten-thousand-fold decrease in the binding of tetrodotoxin to the mutated sodium channel. In similar experiments, it has further been found that additional residues are vital to perm-selectivity of the sodium ion channel. These include negatively charged residues in domains 1 and 2, a positively charged amino acid residue in domain 3, and a neutral amino acid in domain 4.

Position 385 is also a vital amino acid residue in SS2. In brain and skeletal muscle sodium channels, this residue is Phe or Tyr; in cardiac muscle, it contains the nonconservative substitution, Cys. The cardiac muscle sodium channel has a several hundred-fold lower affinity for tetrodotoxin than does brain or skeletal muscle channel protein. If the Cys residue of cardiac muscle channel protein is changed to Tyr or Phe, the high affinity for tetrodotoxin is restored. Thus Tyr/Phe 385 appears to be vital in tetrodotoxin binding and ion selectively.

Sodium and calcium channel structures are similar to one another. However, calcium ions bind with only about one-tenth the affinity to the sodium channel protein as that for sodium ions. However, mutation of Lys 1422 and Ala 1714 to Glu residues completely reverses the order of affinity of these two ions to the mutated sodium channel protein. These two residues are therefore implicated as being present at the mouth of the pore of the sodium channel.

5.1.4.3 Inactivation

Normally, after the voltage-gated increase in sodium ion permeability, there is a rapid inactivation of the sodium conductance. This inactivation process can be blocked by treatment of the intracellular domain of the sodium channel protein with proteases and by a monoclonal antibody directed against a segment of the protein connecting domains 3 and 4. Further studies involving site-specific mutagenesis have identified three critical amino acid residues as being vital to the inactivation process: residues 1488–1490, the sequence of which is Ile-Phe-Met, all hydrophobic amino acid residues. Mutation of Phe 1489 to Gln completely blocks inactivation of sodium channel conductance.

To explain the inactivation process, the three-residue segment has been proposed to constitute a "hinged lid," which moves to form a block in the sodium conductance channel after the conformational changes occur that increase sodium ion conductance. Phe 1489 would bind to another as-yet unidentified pocket in the protein, blocking further movement of sodium ions.

Similar experiments on the potassium channel protein have identified a completely different segment of this protein involved in the inactivation of potassium conductance. This constitutes the amino terminal domain consisting of hydrophobic (especially Leu 7) and hydrophilic amino acid residues.

The model for inactivation for the potassium channel protein is the "ball and chain" model in which the amino terminal arm of the protein swings into the pore, blocking further potassium conductance. Convincing evidence for this model has been provided by experiments in which native amino terminal peptides hybridized with mutant potassium channel proteins, which do not inactivate potassium ion conductance, completely restore their abilities to inactivate potassium ion conductance. These results compared with those for the sodium channel suggest that inactivation of sodium channels occurs on the intracellular surface, whereas that for the potassium channels occurs on the extracellular surface.

Our knowledge of the functioning of ion channels is incipient but has been greatly advanced by the cloning of the sodium, potassium, and calcium channel proteins and the site-specific mutagenesis work. The x-ray structures of the potassium channel are also providing more insights. These studies all suggest that the functioning of all of these proteins is based on the principles discussed earlier in this chapter. Selective binding of ions is caused by specific ionic (electrostatic) interactions; changes in conductance of the channel proteins are caused by changes in the three-dimensional structure of the proteins induced by changes in the electric field across the cell membrane as proposed for the simpler melittin protein. The basic structure of the channel protein consists of a series of amphipathic α-helices containing large numbers of hydrophobic amino acid residues. As found for the transmembrane domain of the *neu/her*-2 receptor protein, the interaction of the helices with one another is responsible for the functioning of the channel protein.

5.1.5 Effects of amino acid substitutions on the ras-p21 protein

One of the most striking discoveries over the past 3 decades has been the finding that a single base change at codon 12 in the human *ras* gene, which encodes a protein of Mr 21 kDa called the p21 protein (with 189 amino acids in its sequence), results in the malignant transformation of quiescent (NIH 3T3) cells in culture. The encoded p21 protein contains a Val in place of the normally occurring Gly at position 12 in its amino acid sequence. Transfection of the oncogene, but not its normal counterpart proto-oncogene, into NIH 3T3 cells in culture results in their malignant transformation (Barbacid, 1987).

Microinjection of cloned, purified oncogene-encoded (Val 12-containing) p21 protein, but not its normal counterpart protein, into NIH 3T3 cells also causes these cells to undergo cell transformation. Thus the transforming agent is the *ras*-gene-encoded p21 protein itself. Similarly, oncogenic but not normal p21 induces maturation (meiotic division) of frog *(Xenopus laevis)* oocytes. The biochemical changes induced by oncogenic p21 in all of these cells are very similar, suggesting that *ras*-p21 induces stereotypical events in each cell line. Incredibly, oncogenic *ras* genes have been identified overall in about one out of three common human cancers, in over 90% of human pancreatic and 70% of human colon cancers (Almoguerra et al., 1988). Oncogenic p21 protein has also been found to occur at elevated concentrations in the sera of patients with specific types of malignant neoplasms (Pincus et al., 2006).

There are other critical positions in the polypeptide chain of p21, such as at Gly 13, Ala 59, and Gln 61, where single amino acid substitutions result in a transforming protein. By far the most common site for oncogenic substitution is Gly 12. Since 18 out of 20 amino acids that occur at position 12 cause the protein to become oncogenic, it seems reasonable to conclude that these substitutions induce changes in the three-dimensional structure of the protein, resulting in its being permanently activated.

5.1.5.1 Activation *of ras*-p21 (Pincus et al., 2000)

ras-p21 is a **G protein** that becomes activated when GDP, bound to the protein in its inactive state, is exchanged for GTP. This process of activation is set in motion when a specific growth factor such as epidermal growth factor (EGF), itself a polypeptide, binds to its receptor. As discussed under the *neu/her*-2 growth factor receptor section (Section 5.1.3), the transmembrane receptor contains an extracellular growth factor binding domain, a transmembrane domain, and an intracytoplasmic domain which, for *neu* and EGF receptors, contains a tyrosine kinase domain. This kinase becomes activated when the growth factor receptor dimerizes as a consequence of the binding of the growth factor to it.

As a result, the intracytoplasmic domain binds to an "adapter" molecule, called *grb*-2, which simultaneously binds to a protein called SOS that, in turn, binds to p21, promoting the exchange of GDP for GTP. Modulating the effect of GTP on activation of p21 is the GTPase activating protein (GAP), which binds to activated p21 and induces the hydrolysis of GTP to GDP, thereby resulting in the inactivation of *ras*-p21.

All of these events occur in or near the inner cell membrane. *ras*-p21 is attached to the inner cell membrane by a covalent link at Cys 186 (Willumsen et al., 1984) in which its sulfur atom is in thioether link to the farnesyl moiety, a lipophilic hydrocarbon that intercalates in the lipid medium of the cell membrane. Thus, like melittin, *ras*-p21 is active as a membrane-associated protein.

Thioether synthesis is catalyzed by the enzyme farnesyl transferase. In site-specific mutagenesis experiments, if Cys 186 is replaced by a Ser residue, ras-p21 not only is inactive, even if it contains oncogenic amino acid substitutions, but it competes with its cell−membrane associated counterpart protein so as to **inhibit** mitogenesis. The double-inactive mutant of p21 with the Val-for-Gly 12 and Ser-for-Cys 186 substitutions is therefore called a dominant negative mutant of *ras*-p21. A number of farnesyl transferase inhibitors have been synthesized that block insertion of p21 into the membrane and allow the cytosolic p21 to compete with its membrane-bound counterpart protein, resulting in the slowing down or halting of mitogenesis in *ras*-induced cancers (Downward, 2003; James et al., 1993).

Signal transduction induced by G proteins is a generalized phenomenon that explains the action of many hormones and external stimuli. For example, polypeptide hormones such as ACTH, growth hormone, calcitonin, cholecystokinin, and insulin all activate G proteins that stimulate phosphorylation cascades. Light induces electrical signals on the retina; conversion of light to electrical signals traveling to the brain is performed by the G protein transducin in the retina. Much pharmacotherapy is currently aimed at interfering with the signal transduction events that occur after the hormone has bound to its receptor.

5.1.5.2 Function of ras-p21

A natural question that arises from the foregoing discussion is that, if *ras*-p21 is membrane bound, what is the mechanism by which it induces mitosis, when the latter process occurs in the nucleus? The answer to this question appears to be that activated *ras*-p21 induces a number of phosphorylation cascades in the cytosol in which successive proteins become activated until one or more of these proteins directly activates nuclear proteins that are intimately involved in DNA synthesis. These successive activation cascades are referred to as signal transduction pathways.

Activated *ras*-p21 binds directly to a protein called *raf*-p74 (Mr, 74 kDa). When activated by *ras* or other intracellular proteins, *raf* binds to and phosphorylates an extracellular **mitogen-stimulated kinase** called MEK, which, in turn, binds to and phosphorylates a vital protein, MAP kinase (mitogen-activated protein kinase) or ERK. This protein is critical in catalyzing cytoskeletal rearrangements in mitosis and in activating the highly important nuclear transcriptional activating factor, *fos*. To be active, *fos* forms a heterodimeric complex with another protein, *jun*. Together these two activated proteins bind to promoter regions of the genome and induce the transcription of mRNA for a number of mitogenesis-promoting proteins such as the cyclins. The factor *jun* is activated by another pathway in which it is activated by the critical kinase, *jun* kinase (JNK). Further, oncogenic (Val 12-containing) *ras*-p21 interacts preferentially with both JNK and *jun* proteins, providing a "short circuit" pathway for activating direct nuclear transcription.

Remarkably, the above-described cascades occur in a wide variety of different cells with differing effects depending on the cell type. Thus, for example, in frog oocytes, oncogenic *ras*-p21 induces meiosis and cell **maturation,** not malignant transformation. In cardiac muscle that has suffered injury, infarction, or hypertrophy, the above cascades also become activated.

5.1.5.3 Relationship of structure to function of *ras*-p21

The fact that arbitrary amino acid substitutions at critical positions in the polypeptide chain cause the *ras*-p21 protein to become oncogenic suggests that critical changes in its three-dimensional structure occur in response to these substitutions. Another point of view is that these substitutions affect the ability of GAP to induce an increase in the GTPase activity of p21. There is an excellent correlation of the abilities of some oncogenic p21 proteins to induce cell transformation with their diminished levels of GAP-induced GTPase activities and with their binding affinities to GAP. On the other hand, there are striking exceptions to this correlation. For example, a Glu-for-Asp 38−substituted p21 protein binds tightly to GAP and has low GTPase activity but does **not** transform cells. Some substituted proteins efficiently hydrolyze GTP to GDP but transform cells. Also there are triply substituted forms of p21 that do not bind at all to any nucleotide but are permanently activated and strongly induce cell transformation. Thus it appears that for many substituted oncogenic p21 proteins, permanent activation is related to structural changes induced by the oncogenic amino acid substitutions themselves.

The x-ray structures of wild-type and oncogenic forms of p21 bound to GDP and GTP and of wild-type p21 bound to the *ras*-binding domain of *raf*, to SOS, and to GAP proteins have all been determined. Comparison of oncogenic and normal unbound p21 structures reveals some conformational changes in an effector loop from residues Tyr 32-Asp 47, which is involved in the binding of p21 to GAP, *raf*, and SOS. Because the protein occurs in a crystal lattice, wherein interactions between protein molecules limit conformational flexibility, the x-ray structures may not identify other regions that undergo conformational changes.

5.1.5.4 Use of conformational energy calculations to identify effector domains of p21

To identify other regions of the protein that may change conformation when the oncogenic form is compared with that of the wild-type protein, calculations of the low-energy conformations of these proteins have been carried out. These calculations are based on the principle, discussed in Section 1.5.1, that the observed structure of a protein is the one of lowest energy; there may be a group of lowest energy structures for a protein that have the same overall chain fold but which differ from one another in the conformation of local regions.

In these calculations, using sampling methods such as molecular dynamics (discussed in Section 1.5.1), the x-ray crystal structure is subjected to energy minimization and then allowed to move so that all of the low-energy structures that lie close to this structure, but which may differ from it regionally, are sampled. The average structure is the average of these low-energy structures. If oncogenic amino acid substitutions induce permanent, regional conformational changes in the three-dimensional structure of the protein, these will be reflected in differences in the average structures for the oncogenic proteins compared with that of the normal (wild-type) protein. We might expect that synthetic segments of p21 corresponding to these regions may inhibit the functioning of the oncogenic protein.

Results of these calculations on the wild-type protein bound to GDP and GTP and of the Val 12− and Leu 61−substituted proteins bound to GTP (Monaco et al., 1995; Pincus et al., 2007) show that the overall fold of the proteins is virtually identical, but local regions differ in conformation between oncogenic and the inactive (GDP-bound) protein.

There are six such regions: residues 4–20, 35–47, 55–71, 81–93, 96–110, and 115–126. Peptides corresponding to each of these regions have been synthesized and tested for their abilities to alter the ability of *ras*-p21 to induce mitogenesis.

5.1.5.5 Test cell system for identification of functional domains of *ras*-p21

As it happens, a very convenient cell system for measuring the abilities of agents to inhibit *ras*-p21 is oocytes. Oocyte maturation induced by a particular agent, such as oncogenic *ras*-p21, is easy to evaluate because of the large structural changes, like disappearance of the brown-colored animal pole, that each matured oocyte undergoes. Maturation of all oocytes in a given experiment occurs in a relatively short time, within 1–2 days after injection or treatment of the oocytes with a maturation-promoting agent. Finally, oocytes contain an abundance of insulin receptors; insulin is a growth factor that induces oocyte maturation by activating normal cellular *ras*-p21.

Proof that insulin acts through *ras*-p21 was provided by the finding that if oocytes are first injected with an inactivating antibody against *ras*-p21 and then incubated in insulin, they do not undergo maturation. Prior injection of other, unrelated antibodies has no effect; insulin induces maturation in these oocytes. Thus oocytes are ideal for studying the specificity of anti–*ras*-p21 agents to study whether they have a preferential effect on the oncogenic form versus the normal form of p21 by either co-injecting the agent with oncogenic p21 or by injecting the agent into oocytes that are then incubated with insulin.

5.1.5.6 Effects of p21 peptides from effector domains identified by conformational analysis (Pincus et al., 2000, 2007)

Microinjection into oocytes of each of the synthesized peptides from the *ras*-p21 domains identified from the conformational energy calculations mentioned above, either together with oncogenic *ras*-p21 or into oocytes subsequently incubated with insulin, has been performed. Three peptides, 35–47 (PNC-7) from the previously identified effector domain, 96–110 (PNC-2), and 115–126 (PNC-1), all strongly block the ability of oncogenic p21 to induce oocyte maturation. Significantly, these peptides, especially peptides corresponding to residues 96–110, and 115–126, only minimally affected the ability of insulin, hence activated normal p21, to induce maturation. This finding suggests that oncogenic and normal p21 induce mitogenesis by differing signal transduction pathways. As it happens, the latter two peptides have been found to interfere in the interaction of oncogenic p21 with JNK and *jun* proteins; their abilities to inhibit oncogenic p21-induced oocyte maturation strongly correlate with their abilities to inhibit this interaction.

The above results suggest that activated normal and oncogenic *ras*-p21 stimulate different pathways; the oncogenic form can preferentially activate JNK and *jun* directly, leading to unregulated nuclear transcriptional processes. By breaking this "short circuit" with effector peptides, selective inhibition of oncogenic *ras*-p21 can be achieved. This result has important implications for the design of new selective chemotherapeutic agents.

5.1.5.7 The two *ras* peptides, PNC-2 and PNC-7, block cancer cell but not normal cell growth (Pincus et al., 2007)

Two *ras*-p21 peptides, PNC-2 and PNC-7, attached to the transmembrane-penetrating sequence discussed in Section 4.3, were used to treat ras-transformed cancer cells. It was found that a ras-transformed rodent pancreatic cancer cell line called TUC-3, when incubated with either peptide but not a control peptide, are induced to revert to the untransformed phenotype. These cells, unlike their untreated or control peptide-treated cancer cells, do not grow when transplanted into nude mice. Neither PNC-2 nor PNC-7 has any effect on the growth of normal cells such as the untransformed counterpart cells to TUC-3, called BMRPA1. Both peptides likewise exert the same effect on ras-transformed HT1080 human fibrosarcoma cells and have cytotoxic effects on other cancer cell lines. It seems, therefore, that these *ras*-p21 peptides have potential for treating cancers.

Thus, use of the methods of conformational analysis based on the principles of protein structure described in this chapter has resulted in our being able to identify critical residues of an oncogenic protein that are vital to their function, to design peptides that selectively inhibit oncogenic *ras*-p21, and to detect differences in signal transduction pathways induced by oncogenic and normal p21 protein.

6. Summary

We have seen in this chapter that the three-dimensional structures of proteins are dictated by their linear sequences of amino acids. A protein's structure is the one of lowest free energy for the given polypeptide chain, the detailed conformation of which is determined by the dihedral angles of the backbone and side chains. Proteins are structured in such a way that their polar amino acids tend to be directed toward the aqueous solvent while the nonpolar or hydrophobic residues tend to point toward the interior of the protein. The three-dimensional structures of proteins can be determined by a variety of techniques, which include circular dichroism (for regular structure), x-ray crystallography, two-dimensional NMR, and theoretical techniques.

The three-dimensional structures of proteins determine bulk properties of these proteins such as their behavior toward titration with acids or bases, their isoionic points, and their migration in electric fields, as in electrophoresis. The ionization of charged groups on the surfaces of proteins can be treated, and thus their titration curves can be predicted using ligand-binding theory combined with electrostatic field theory (using the Debye—Hückel formulation) or using more explicit models based on the actual three-dimensional structure of the protein surrounded by water dipoles.

The structure of a protein and its functioning have a well-defined relationship. Membrane proteins, for example, tend to fold into α-helices. This structure is of critical importance to the functioning of leader peptides that allow for proteins to be secreted both within the cell and outside of the cell. The ability of melittin to intercalate into cell membranes and to cause cell lysis depends on its ability to adopt an α-helical conformation and to change the angle of these helices. Membrane active peptides, like PNC-27, fold into membrane-active helix-loop-helix amphipathic structures, and bind to unique protein targets in the membranes of cancer cells allowing them to kill the cancer cells but not affect normal cells making them potentially effective anticancer agents.

Certain large proteins, like *neu*/HER-2, contain major transmembrane domains, which can adopt several different conformations, some of which allow them to dimerize and thus initiate cellular signal transduction. This process is especially critical for oncogene-encoded proteins that are involved in mitogenic signal transduction to the nucleus, causing cell division. We have begun to understand the structural basis for this process using conformational analysis as described in this chapter. In this regard, conformational analysis has now been extended to the identification of the regions of oncogenic proteins involved in activation of downstream target proteins and has opened new possibilities in the field of drug design. Voltage-dependent conformational changes in the transmembrane domains of ion channel proteins are vital to their regulated functioning. The basic functioning of cells is therefore controlled by the proper folding and functioning of cellular proteins, making vital the understanding of protein structure.

References

Almoguerra, C., Shibata, D., Forrester, K., Martin, J., Arnheim, M., & Perucho, M. (1988). Most human carcinomas of the endocrine pancreas contain mutant c-K-*ras* genes. *Cell, 53*, 813—815.

Anfinsen, C. B., Haber, E., Sela, M., & White, F. H., Jr. (1961). The Kinetics of formation of native ribonuclease during oxidation of the reduced polypeptide chain. *Proceedings of the National Academy of Sciences, 47*, 1309—1314.

Barbacid, M. (1987). *Ras* genes. *Annual Review of Biochemistry, 56*, 779—827.

Bargmann, C. I., & Weinberg, R. A. (1988). Oncogenic activation of the *neu*-encoded receptor protein by point mutation and deletion. *The EMBO Journal, 7*, 2043—2052.

Ben-Levy, R., Peles, E., Goldman-Michael, R., & Yarden, Y. (1992). An oncogenic point mutation confers high affinity ligand binding to the *neu* receptor. *Journal of Biological Chemistry, 267*, 17304—17313.

Blobel, G., Walter, P., Chang, C. N., Goldman, B. M., Erickson, A. H., & Lingappa, R. (1979). Translocation of proteins across membranes: The signal hypothesis and beyond. *Symposia of the Society for Experimental Biology, 33*, 9—37.

Brandt-Rauf, P. W., Pincus, M. R., & Chen, J. M. (1989). Conformational changes induced by the transforming amino acid substitution in the transmembrane domain of the *neu*-oncogene-encoded p185 protein. *Journal of Protein Chemistry, 8*, 749—755.

Brandt-Rauf, P. W., Rackovsky, S., & Pincus, M. R. (1990). Correlation of the transmembrane domain of the *neu*-oncogene-encoded p185 protein with its function. *Proceedings of the National Academy of Sciences, 87*, 8660—8664.

Bull, H. B. (1943). *Physical biochemistry*. London: John Wiley and Sons.

Caterall, W. A. (1995). Structure and function of voltage-gated ion channels. *Annual Review of Biochemistry, 64*, 493—531.

Chavez, L., & Scheraga, H. A. (1980). Folding of ribonuclease, S-protein, and des (121-124)-ribonuclease during glutathione oxidation of the reduced proteins. *Biochemistry, 19*, 996—1004.

Cho, H. S., Mason, K., Ramyar, K. X., Stanley, A. M., Gabelli, S. B., Denney, D. W., Jr., & Leahy, D. J. (2003). Structure of the extracellular region of HER2 alone and in complex with the Herceptin Fab. *Nature, 421*, 756—760.

Di Fiore, P. P., Pierce, J. H., Kraus, M. H., Segatto, O. S., King, R., & Aaronson, S. A. (1987). *erb*B-2 is a potent oncogene when overex-pressed in NIH/3T3 cells. *Science, 237*, 178—182.

Downward, J. (2003). Targeting the Ras signalling pathway in cancer therapy. *Nature Reviews Cancer, 3*, 11—22.

Dygert, M., Go, N., & Scheraga, H. A. (1975). Use of a symmetry condition to compute the conformation of gramicidin S. *Macro-molecules, 8*, 750—761.

Emr, S. D., & Silhavy, T. J. (1982). Molecular components of the signal sequence that function in the initiation of protein export. *The Journal of Cell Biology, 95*, 689—696.

Engelman, A. M., & Steitz, T. A. (1981). The spontaneous insertion of proteins into and across membranes: The helical hairpin hypothesis. *Cell, 23*, 411–422.

Ferguson, K. M., Berger, M. B., Mendrola, J. M., Cho, H., Leahy, D. J., & Lemmon, M. A. (2003). EGF activates its receptor by removing interactions that autoinhibit ectodomain dimerization. *Molecular Cell, 11*, 507–517.

Gibson, K., Chin, S., Pincus, M. R., Clementi, E., & Scheraga, H. A. (1986). Parallelism in conformational energy calculations on proteins: Partial structure of interferon. In M. Dupuis (Ed.), *Montreal symposium on supercomputer simulation in chemistry. Lecture notes in chemistry* (44, pp. 198–213).

Hall, Z. W. (1992). Ion channels. In Z. W. Hall (Ed.), *An introduction to molecular neurobiology* (pp. 81–118). Sunderland, MA: Sinauer Associates, Inc.

Hendrick, J. P., & Hartl, F.-U. (2017). The role of molecular chaperones in protein folding. *The FASEB Journal, 9*, 1559–1569.

Horowitz, S., Salmon, L., Koldewey, P., Ahlstrom, L. S., Martin, R., Quan, S., Alfine, P. V., van der Bedem, H., Wang, L., Xu, Q., Trievel, R. C., Brooks, C. L., & Bardwell, J. C. A. (2016). Visualizing chaperone-assisted protein folding. *Nature Structural & Molecular Biology, 23*, 691–697.

Hudziak, R. M., Schlessinger, J., & Ullrich, A. (1987). Increased expression of the putative growth factor reception p185^{HER2} causes transformation and tumorigenesis of NIH/3T3 cells. *Proceedings of the National Academy of Sciences of the United States of America, 84*, 7159–7163.

James, G. L., Goldstein, G. L., Brown, M. S., Rawson, T. E., Somers, T. C., McDowell, R. S., Crowley, C. W., Lucas, B. K., Levinson, A. D., & Marsters, J. C., Jr. (1993). Benzodiazepine peptidomimetics: Potent inhibitors of *ras* farnesylation in animal cells. *Science, 260*, 1937–1942.

Karplus, M., & McCammon. (1986). The dynamics of proteins. *Scientific American, 254*, 42–52.

Kempf, C., Klausner, R. D., Weinstein, J. N., van Renswoude, J., Pincus, M. R., & Blumenthal, R. (1982). Voltage-dependent transbilayer orientation of melittin. *Journal of Biological Chemistry, 257*, 2469–2475.

King, C. F., Kraus, M. H., & Aaronson, S. A. (1985). Amplification of a novel v-*erbB*-related gene in a human mammary carcinoma. *Science, 229*, 974–976.

Kussie, P. H., Gorina, S., Marechal, V., Elenbaas, B., Moreau, J., Levine, A. J., & Pavletich, N. P. (1996). Structure of the MDM2 oncoprotein bound to the p53 tumor suppressor transactivation domain. *Science, 274*, 921–922.

Long, S. B., Campbell, E. B., & MacKinnon, R. (2005). Crystal structure of a mammalian voltage-dependent shaker family K^+ channel. *Science, 309*, 897–903.

Miller, M. H., & Scheraga, H. A. (1976). Calculation of the structure of collagen models. Role of interchain interactions in determining the triple helical coiled coil conformation. *Journal of Polymer Science. Polymer Symposium, 54*, 171–200.

Monaco, R., Chen, J. M., Friedman, F. K., Brandt-Rauf, P. W., & Pincus, M. R. (1995). Structural effects of the binding of GTP to the wild-type and oncogenic forms of the *ras*-gene-encoded p21 proteins. *Journal of Protein Chemistry, 14*, 721–730.

Padhy, L. C., Shih, C., Cowing, D., Finkelstein, R., & Weinberg, R. A. (1982). Identification of a phosphoprotein specifically induced by the transforming DNA of rat neuroblastomas. *Cell, 28*, 865–871.

Pincus, M. R., Brandt-Rauf, P. W., & Bluth, M. (2006). Oncoproteins and early tumor detection (Chapter 75). In R. McPherson, & M. R. Pincus (Eds.), *Henry's clinical diagnosis and management by laboratory methods* (21st ed., pp. 1367–1380). Philadelphia: Elsevier.

Pincus, M. R., & Klausner, R. D. (1982). Prediction of the three-dimensional structure of the leader sequence of murine pre-kappa light chain, a hexadecapeptide. *Proceedings of the National Academy of Sciences, 79*, 3413–3417.

Pincus, M. R., Klausner, R. D., & Scheraga, H. A. (1982). Calculation of the three-dimensional structure of the membrane-bound portion of melittin from its amino acid sequence. *Proceedings of the National Academy of Sciences, 79*, 5107–5110.

Pincus, M. R., Michl, J., Bowne, W., & Zenilman, M. (2007). Anti-Cancer Peptides from the ras- p21 and p53 Proteins. In R. M. Mohan (Ed.), *Research advances in cancer* (pp. 65–90). Kerala, India: Global Research Network, Publishers.

Pincus, M. R., & Scheraga, H. A. (1985). Conformational analysis of biologically active polypeptides, with application to oncogenesis. *Accounts of Chemical Research, 18*, 372–379.

Pincus, M. R. (1988). The chain build-up procedure in computing the structures of biologically active polypeptides and proteins. *International Journal of Quantum Chemistry: Quantum Biology Symposium, 15*, 209–220.

Pincus, M. R., Brandt-Rauf, P. W., Michl, J., & Friedman, F. K. (2000). *ras*-p21-induced cell transformation: unique signal transduction pathways and implications for the design of new chemotherapeutic agents. *Cancer Investigation, 18*, 39–50.

Rosal, R., Pincus, M. R., Brandt-Rauf, P. W., Fine, R. L., Michl, J., & Wang, H. (2004). NMR Solution structure of a peptide from the mdm-2 binding domain of the p53 protein that is selectively cytotoxic to cancer cells. *Biochemistry, 43*, 1754–1861.

Rosenblatt, M., Beaudette, N. V., & Fasman, G. D. (1980). Conformational studies of the synthetic precursor-specific region of preproparathyroid hormone. *Proceedings of the National Academy of Sciences, 77*, 3983–3987.

Rothwarf, D. M., Li, Y.-J., & Scheraga, H. A. (1998). Regeneration of bovine pancreatic ribonuclease A: Detailed kinetic analysis of two independent folding pathways. *Biochemistry, 37*, 3767–3776.

Rothwarf, D. M., Li, Y. J., & Scheraga, H. A. (1998). Regeneration of bovine pancreatic ribonuclease A: Identification of two native-like three disulfide intermediates involved in separate pathways. *Biochemistry, 37*, 3760–3766.

Russell, S. T., & Warshel, A. (1985). Calculations of electrostatic energies in proteins. The energetics of ionized groups in bovine pancreatic trypsin inhibitor. *Journal of Molecular Biology, 185*, 389–404.

Sarafraz-Yazdi, E., Bowne, W. B., Adler, V., Sookra, K. A., Wud, V., Shteylerd, V., Patel, H., Oxbury, W., Brandt-Rauf, P. W., Zenilman, M. E., Michl, J., & Pincus, M. R. (2010). Anticancer peptide PNC-27 adopts an HDM-2-binding conformation and kills cancer cells by binding to HDM-2 in their membranes. *The Proceedings of the National Academy of Sciences, 107*, 1918–1923.

Scheraga, H. A., Liwo, A., Oldziej, S., Czaplewski, C., Pillardy, J., Ripol, D. R., Vila, J. A., Kazmierkiewicz, R., Saunders, J. A., Arnatova, Y. A., Jagielska, A., Chinchio, M., & Nanias, M. (2004). The protein folding problem: Global optimization of force fields. *Frontiers in Bioscience, 9*, 3296–3323.

Scheraga, H. A. (1989). Calculations of stable conformations of polypeptides, proteins, and protein complexes. *Chimica Scripta, 29A*, 3–13.

Scheraga, H. A. (1984). Protein structure and function, from a colloidal to a molecular point of view. *Carlsberg Research Communications, 49*, 1–55.

Slamon, D. J., Clark, G. M., Wong, S. G., Levin, W. J., Ullrich, A., & McGuire, W. L. (1987). Human breast cancer: Correlation of relapse and survival with amplification of the *her-2/neu* oncogene. *Science, 235*, 177–182.

Tamba, Y., Arivama, H., Levadny, V., & Yamazaki, M. (2010). Kinetic pathway of antimicrobial peptide magainin 2-induced pore formation in lipid membranes. *Journal of Physical Chemistry B, 114*, 2018–2026.

Terwilliger, T. C., Weissman, L., & Eisenberg, D. (1982). The structure of melittin in the form I crystals and its implication for melittin's lytic and surface activities. *Biophysical Journal, 37*, 353–361.

van Den Bogaart, G., Guzman, J. V., Mica, J. T., & Poolman, B. (2008). On the mechanism of pore formation by melittin. *Journal of Biological Chemistry, 283*, 33854–33857.

van Holde, K. E. (1985). *Physical biochemistry*. Englewood Cliffs, NJ: Prentice Hall, Inc.

Vasquez, M., Nemethy, G., & Scheraga, H. A. (1994). Conformational energy calculations on polypeptides and proteins. *Chemistry Review, 94*, 2183–2239.

Vogel, H., & Jahnig, F. (1986). The structure of melittin in membranes. *Biophysical Journal, 50*, 573–582.

Willumsen, B. M., Christensen, A., Humbert, N. L., Papageorge, A. G., & Lowy, D. R. (1984). The p21 *ras* C-terminus is required for transformation and membrane association. *Nature, 310*, 583–586.

Wuthrich, K. (1986). *NMR of proteins and nucleic acids*. New York: John Wiley and Sons.

Yang, L., Weiss, T. M., Lehrer, R. I., & Huang, H. W. (2000). Crystallization of antimicrobial pores in membranes: Magainin and Protegrin. *Biophysical Journal, 79*, 2002–2009.

Chapter 3

Carbohydrates—The glycocalyx and its biological roles

Mia L. Huang, Daniel Honigfort and Kamil Godula
Department of Chemistry and Biochemistry and the Glycobiology Research and Training Center, University of California San Diego, La Jolla, CA, United States

1. An introduction to glycans

Carbohydrates, or glycans, are a large class of essential biopolymers intimately linked to cellular physiology. Although the public mostly knows carbohydrates as macronutrient energy sources, they are also intriguing cellular messengers, molecules that control and report on a cell's physiological state. Their central importance to life is reflected by the fact that evolution has produced no single organism devoid of glycans. Glycans are prominently found conjugated to lipids or proteins populating the outer leaflet of cell membranes or integrated as components of bacterial cell walls (Fig. 3.1). The entire collection of cell surface glycoconjugates is typically referred to as the glycocalyx. The chemical diversity of glycan structures found in the glycocalyx is rather vast, with additional complexity arising from the spatial organization of glycans on glycoconjugate scaffolds. For instance, glycans can be presented in close proximity to the cell membrane when attached to small glycolipids and glycopeptides, or they can extend hundreds of nanometers away from the cell surface when displayed on large proteins, such as mucins and proteoglycans. This chapter focuses on the biology of the glycocalyx, with a special emphasis on their macromolecular mucin glycoprotein and proteoglycan components.

1.1 Structural complexity of glycans

Glycans are composed of monosaccharide building blocks with the general formula $C_n(H_2O)_n$ (Fig. 3.2). One must also take note of the number of building blocks for glycans, in contrast to other types of biopolymer molecules. Only 22 amino acid building blocks give rise to proteins, whereas DNA and RNA are made from five nucleobases. On the other hand, glycans can be made from upwards of 50 different monosaccharides, a number that greatly outnumbers the English alphabet; thus significant recent efforts have been made to standardize how glycans should be cataloged and represented graphically. The glycoscience community is coalescing around an agreement that each monosaccharide should be represented by a specific shape and color. For example, glucose is represented as a blue circle, whereas a yellow circle indicates galactose. The study of glycans is complex; it begins with determining the chemical structures of each monosaccharide building block and extends toward the understanding of the biosynthesis and function of large macromolecular glycoconjugates. The entirety of glycans, free or in the form of glycoconjugates, within an organism is referred to as the Glycome. The Glycome is involved in virtually every aspect of cellular biology and it is fascinating to think of how it works in concert with the Proteome and the Genome to control the physiological state of each cell.

Before we turn our attention to some of the complex glycan structures dominating the cellular glycocalyx, we should consider perhaps one of the simpler monosaccharides, glucose. Glucose is the most widely used energy source, which animals derive by the hydrolysis of carbohydrates such as lactose from milk or sucrose from sugar cane. The released glucose can be directly metabolized or it can be stored in the form of glycogen, a protein decorated with densely branched glucose polysaccharides. In addition to serving as an energy molecule, glucose is also a key building block used for the construction of more complex glycans. It is a hexose with six carbon atoms on its backbone, and four out of the six are chiral centers (Fig. 3.2). Because the chemical groups around each chiral center can be arranged in two different

Cell Physiology Sourcebook. https://doi.org/10.1016/B978-0-12-811114-7.00031-1

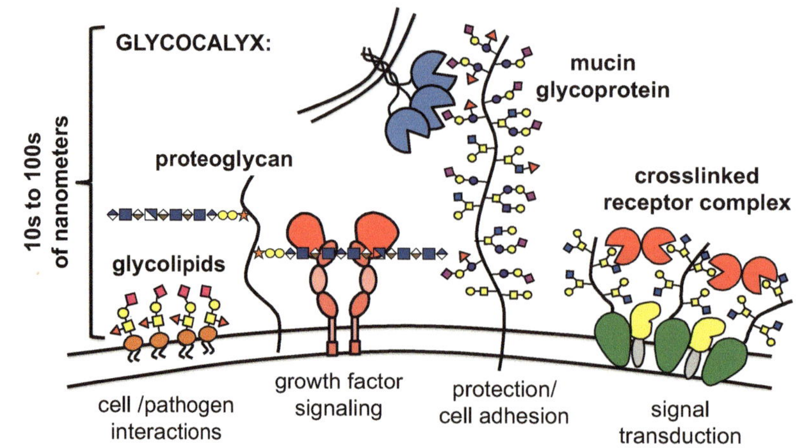

FIGURE 3.1 **The nanoscale architecture of the glycocalyx.** Cell surface glycans and glycoconjugates that decorate cell membranes and bacterial cell walls serve important physiological functions. The glycocalyx can extend tens to hundreds of nanometers away from the cellular membrane. *Reprinted with permission from Huang, M. L. and Godula, K. (2016). Nanoscale materials for probing the biological functions of the glycocalyx. Glycobiology. 26, 797–803. Copyright 2016 Oxford University Press.*

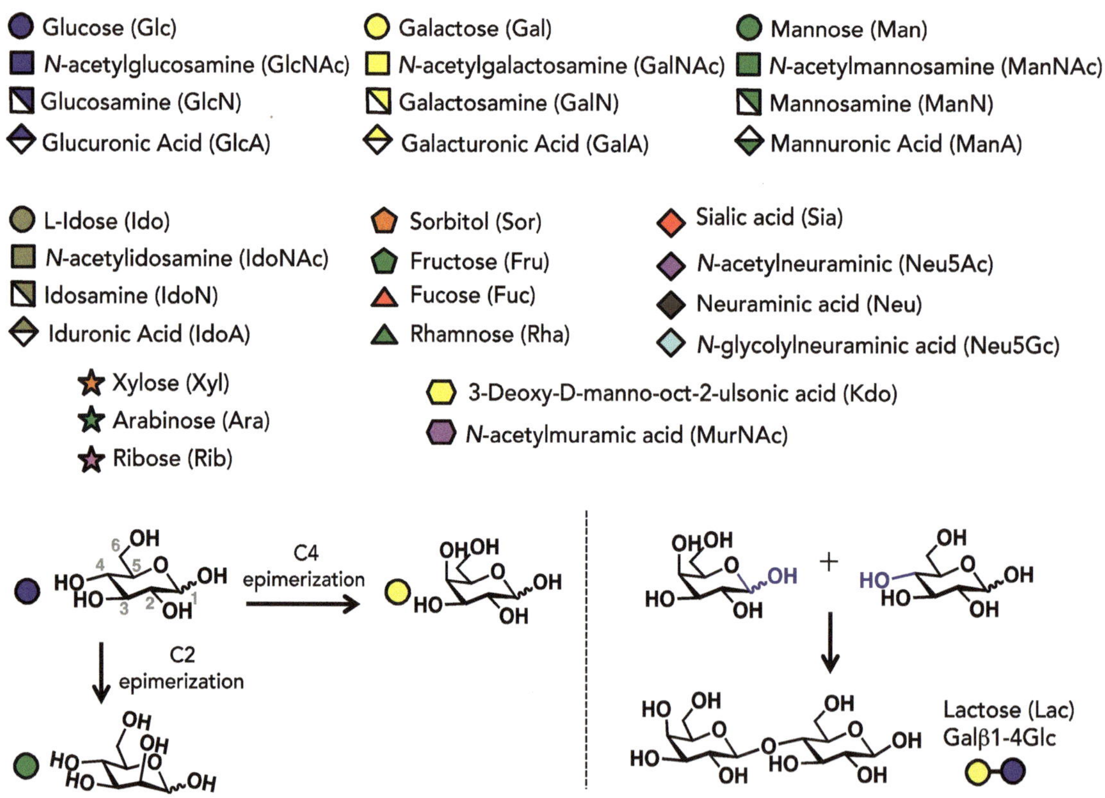

FIGURE 3.2 **Common monosaccharide building blocks and their symbol representation.** The simple monosaccharide glucose can generate epimers at carbon-2 to yield mannose (Man) or at carbon-4 to generate galactose (Gal). Glycans contain carbonyl functionality at the anomeric carbon-1 position to form covalent glycosidic linkages with hydroxyl groups of another monosaccharide, to yield a disaccharide.

configurations, glucose can exist as (4^2) 16 aldohexose stereoisomers. Interestingly, D-glucose is a naturally occurring isomer, whereas L-glucose is not (but other L-monosaccharides can be). Epimerization, a change in the stereochemical configuration of a single carbon atom of a sugar molecule, also leads to a completely different monosaccharide. Glucose can epimerize into two other glycans, epimerization of carbon 2 yields mannose (Man), whereas epimerization of carbon 4 generates galactose (Gal). In water, a monosaccharide exists in equilibrium as the open and closed form.

The individual monosaccharides containing carbonyl functionality at the anomeric carbon-1 position and several hydroxyl groups are strung together into larger structures through covalent glycosidic linkages formed between a hydroxyl group on one monosaccharide and the carbonyl group of another. Since most monosaccharides contain multiple hydroxyl groups, branched structures can be obtained, which distinguishes glycans from other biopolymers, such as DNA, RNA, or proteins, which only exist in the form of linear chains. Each glycosidic linkage can also exist in α or β configuration. Thus, the disaccharide lactose (a milk glycan) is properly notated as β-D-galactopyranosyl-(1−>4)-D-glucopyranose or Galβ1-4Glc.

1.2 Glycosylation and glycan biosynthesis

The process of stringing glycans together or attaching them to other biomolecules is called "glycosylation." In eukaryotes, glycans are biosynthesized in the endoplasmic reticulum (ER) and Golgi apparatus as posttranslational modifications on peptides, proteins, or lipids to form glycoconjugates. Various glycosyltransferases and glycosidase enzymes work in concert to add or remove individual monosaccharides to a growing chain.

At any given point in time, the abundance and diversity of glycans is a tightly regulated process (Fig. 3.3). Glycosylation has been known to be a highly dynamic process that changes with respect to cell cycle, developmental stage, and disease states. A number of factors can dictate glycan levels, including the (1) metabolic flux of monosaccharide building

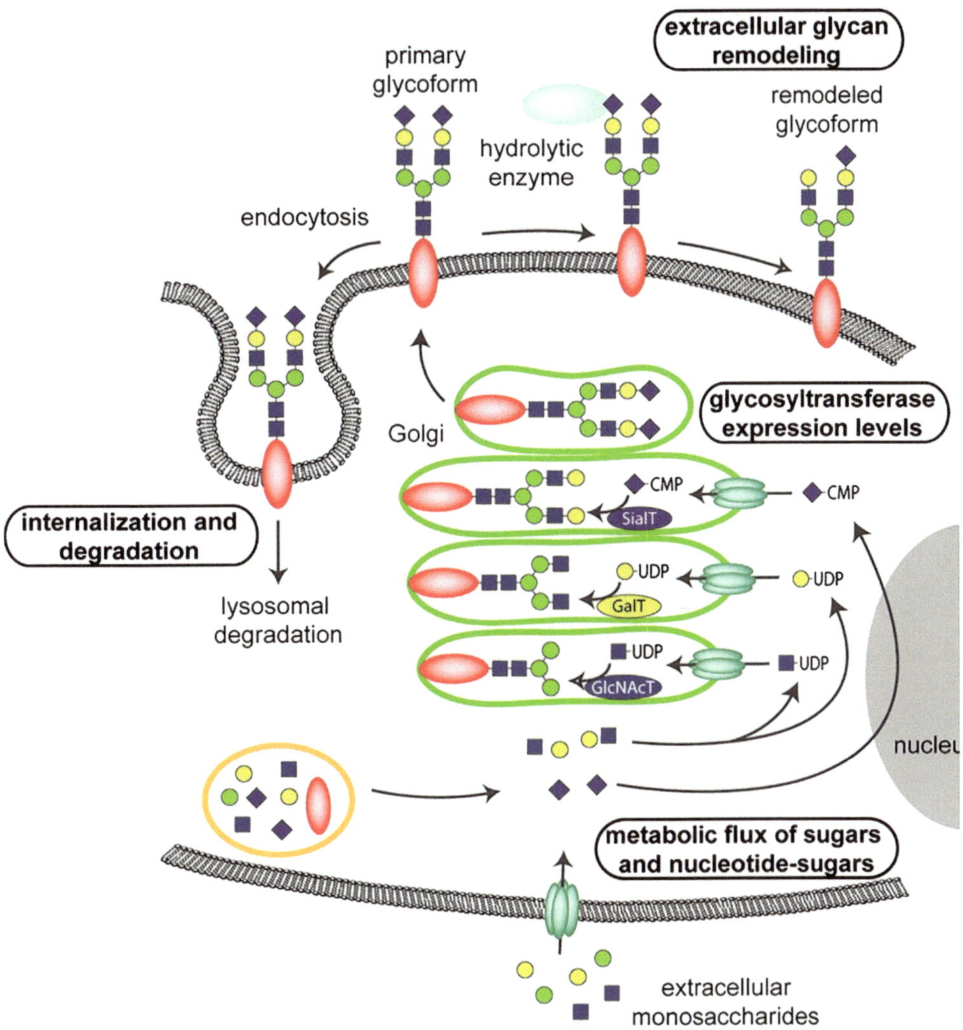

FIGURE 3.3 **Glycan abundance is dictated by multiple steps in its biosynthesis.** Intracellular concentrations of monosaccharides and nucleotide sugar precursors can influence the extent of biosynthesis, as well as glycosyltransferases expression levels. For cell surface glycans that are exported to the cell surface upon completion of biosynthesis, extracellular glycan remodeling enzymes, such as sialidases or sulfatases, can further alter the presented glycans. Internalization and degradation mechanisms also reduce the expression of cell surface glycans.

blocks and nucleotide sugars, as well as the (2) expression levels of glycosyltransferases and glycosidases. For glycans present on the glycocalyx, (3) extracellular glycan remodeling, as well as (4) internalization and degradation can also affect glycan abundance at the cell surface. Glycosylation is extremely sensitive to intracellular pools of monosaccharide building blocks and nucleotide sugars (e.g., UDP-GlcNAc, CMP-sialic acid), the activated forms of monosaccharides that act as donors in glycosylation reactions. Some monosaccharides can be directly transported into the cellular membrane by transporters, whereas others can be biosynthesized from other precursors. For example, the transmembrane glucose transporter brings extracellular glucose and mannose into the cell, and sialic acid can be biosynthesized from N-acetylmannosamine. Catabolism of glycoproteins can also feed monosaccharides into the cellular pool. Upon synthesis in the cytoplasm or nucleus, the nucleotide sugars are transferred to the Golgi or ER lumen by nucleotide sugar transporters. There, the levels of glycosyltransferases and glycosidases affect the extent of glycan biosynthesis. Some cancers exhibit aberrant glycosylation due to altered levels of glycosylation. Upon completion of biosynthesis, cell surface glycans are exported to the cell surface, where primary glycoforms can be altered by extracellular glycan remodeling enzymes, such as glycosidases and sulfatases. Sialidase enzymes can cleave terminal sialic acids off glycan chains to expose underlying glycans, whereas sulfatases can remove sulfate groups already present on the glycan chain. Cell surface glycans can be removed from the glycocalyx by clathrin-mediated or endocytic mechanisms. These can also be recycled back to the cell surface or targeted for degradation into the lysosomes. Although the enzymes responsible for elongating and decorating the glycan chains are encoded by the genome, there is no template for the exact manner, pattern, and length, in which an individual glycan is constructed. Moreover, a single glycosylation site in a protein is not always associated with a unique glycan structure, providing considerable heterogeneity in glycoconjugates and their glycoforms.

Glycans can be classified based on their connectivity. Glycans attached to proteins via a nitrogen atom are referred to as N-glycans, whereas those that are attached through an oxygen atom are called O-linked glycans. The O-linked glycans can be further divided into two distinct classes: branched mucin type glycans and linear glycosaminoglycan polysaccharides of proteoglycans. While the two families of mucin and proteoglycan-type O-glycans will be discussed in considerable detail later in this chapter, a short introduction and comparison of the two uniquely distinct biosynthetic pathways to N-linked and O-linked glycans is warranted.

1.3 N-linked glycans

The assembly of N-linked glycans can occur cotranslationally, or simultaneous with protein translation (Fig. 3.4). This timing creates a logistical challenge, where the growing peptide chain emerging from a ribosome must meet the ER- or Golgi-resident enzymes responsible for attaching glycans. N-glycosylation begins in the cytoplasmic side of the ER where a dolichol phosphate lipid-linked GlcNac2-Man5 oligosaccharide is generated. This lipid-linked precursor is flipped into the ER lumen, where glycosyltransferases further elaborate the glycan chain into GlcNac2Man9Glc3. This completed glycan is then transferred to an asparagine residue of a nascent polypeptide chain by an oligosaccharyltransferase (OST) enzyme. The resulting glycoconjugate then undergoes extensive processing, where the three terminal glucose residues are removed by α-glucosidases as part of a quality control check for proper folding of the protein. Correctly folded proteins are exported to the Golgi apparatus, whereas improperly folded proteins bind to the chaperone calnexin/calreticulin protein to be refolded. At the Golgi, mannosidases edit some of the Man residues and glycosyltransferases further add various monosaccharides. It is at this step that three major types of N-glycans are generated: high-mannose, hybrid, or complex.

N-glycans play important roles in cellular physiology and immunology. Inhibition of N-glycan biosynthesis is costly for cellular function, and molecules that disrupt N-glycan biosynthesis have been implicated for chemotherapeutic purposes. For example, the natural product tunicamycin, which inhibits GlcNAc phosphotransferase to prevent the formation of GlcNAc-PP-dolichol, has been shown to induce the unfolded protein response and arrest cells in the G1 phase. N-linked glycans also play important roles in regulating the immune system, and N-glycosylation of recombinant therapeutic proteins, such as antibodies, have recently come into focus as problems in biotechnology. In order to produce glycoprotein antibodies, eukaryotic expression systems must be used, as prokaryotes cannot efficiently generate posttranslational modifications. However, these eukaryotic expression systems, such as the yeast *Saccharomyces cerevisiae*, generate different glycosylation patterns compared to humans (e.g., high mannose) that can cause severe undesired immunogenic reactions.

1.4 O-linked glycosylation

In this short section, we will focus on O-linked glycans comprising only one or a few monosaccharides. The first attachment of an O-linked monosaccharide to serine/threonine protein residues occurs in the Golgi through the action of specific glycosyltransferases. GalNAc is transferred by UDP-GalNAc:polypeptide N-acetylgalactosaminyltransferase

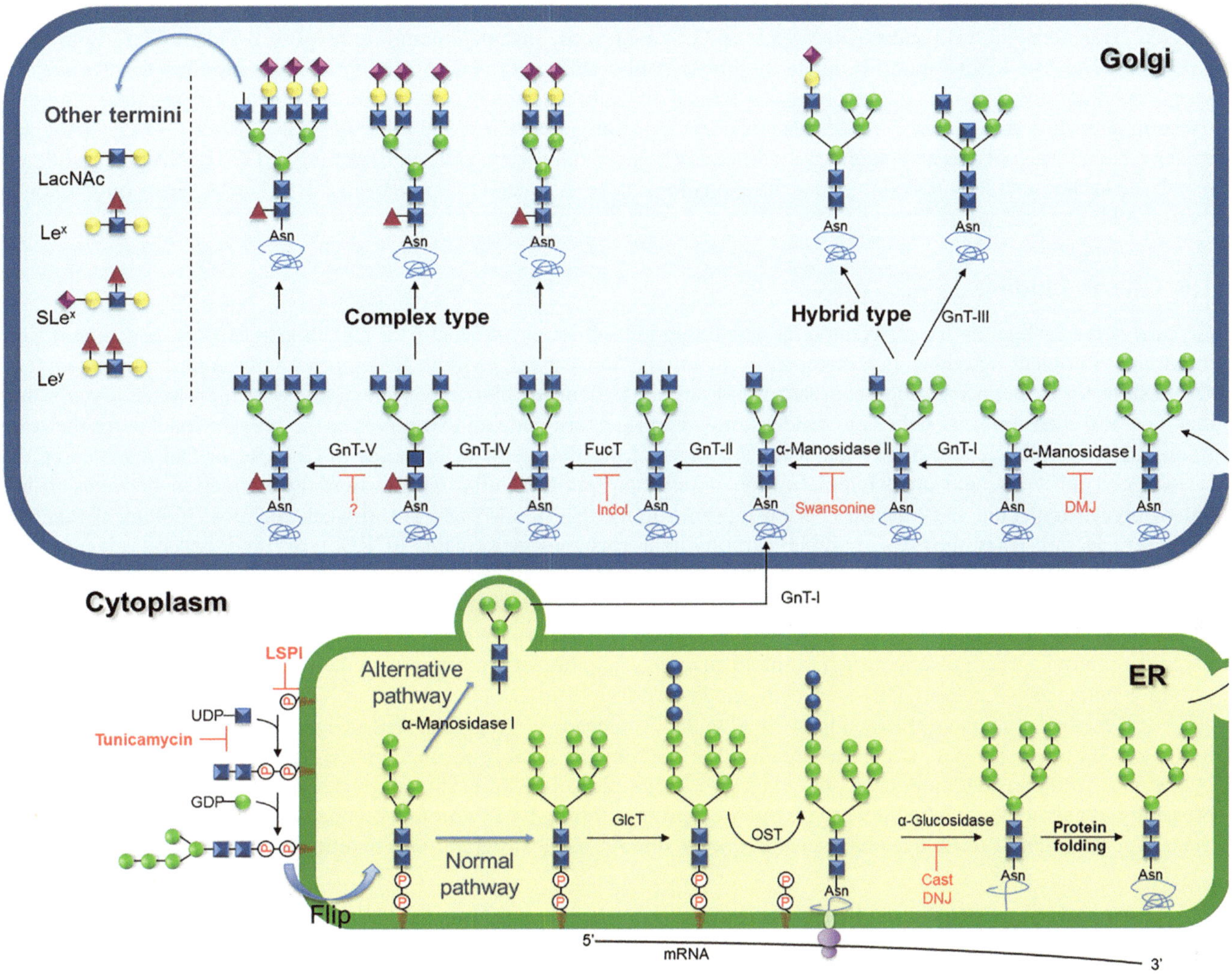

FIGURE 3.4 **Biosynthesis of N-linked glycans and known chemical inhibitors.** A dolichol phosphate lipid is initially glycosylated at the cytoplasmic side of the ER (*bottom left*). Upon flipping and assembly of the Glc2Man9Glc3 precursor in the ER lumen, the glycan is transferred to the side chain of an asparagine residue. Glucosidases trim the terminal Glc residues, and after checking for proper protein folding, the glycoconjugate is transported to the Golgi, where further processing occurs. This process results in complex type and hybrid-type glycans. Molecules that disrupt this pathway can exert anticancer activity. *Used under the Creative Commons Attribution License. Vasconcelos-dos-Santos, A.; Oliveira I. A.; Lucena, M. C.; Mantuano, N. R.; Whelan, S. A.; Dias, W. B.; Todeschini, A. R. Biosynthetic machinery involved in aberrant glycosylation: promising targets for developing of drugs against cancer. Front. Oncol. (2015) 5:138.*

(ppGalNAcT), followed by the attachment of other sugars, for instance fucose, which is introduced by the action of fucosyltransferases. Further extension of these monosaccharides can occur to generate antigenic markers. For example, the simple GalNAcαSer/Thr modification is known as the Tn antigen and is commonly found on the cancer cell glycocalyx. Subsequent sialylation or galactosylation generates other types of antigens, but further elongation generates O-glycan core structures found on the surface of healthy cells. O-linked glycans also carry the basis for ABO blood group determinants. The O blood group is characterized by a Fucα1-2Gal modification, the A group by a GalNAcα1-3(Fucα1-2)Gal, and the B blood group by a Galα1-3(Fucα1-2)Gal modification.

1.5 O-GlcNAc

Whereas most mucin and proteoglycan-type O-linked glycans are heterogeneous and complex, even very simple glycosylation modifications can produce significant functional effects. For example, the simple GlcNAc modification of proteins serves an important regulatory function in signaling and gene expression, similar to protein phosphorylation. Indeed, some competitive phosphorylation or O-GlcNAc modification of serine/threonine residues has been observed. Two enzymes, O-GlcNAc Transferase (OGT) and O-GlcNAcase (OGA) control the installation and removal of β-linked GlcNAc to serine/threonine. The O-GlcNAc modification is highly dynamic and is known to modify histone proteins to regulate transcription, ubiquitination, and other processes critical to cellular homeostasis. O-GlcNAc can also regulate autophagy (lysosomal protein degradation), subcellular localization of target proteins, and other nuclear functions. O-GlcNAcylation has also been implicated as an epigenetic modification that regulates stem cell reprograming. UDP-GlcNAc is considered a metabolic sensor, and its intracellular concentrations have been estimated to range from 0.1 to 1 mM, levels only second to that of ATP.

1.6 Glycan binding

Glycans serve as ligands for many proteins, including cell adhesion molecules or signaling receptors, and thus regulate important biological functions. For example, in mammals, the Ashwell–Morell Receptor (AMR), found in hepatocytes, is responsible for detecting asialoglycoproteins and targeting them for clearance. These asialoglycoproteins, glycoproteins that have lost their terminal sialic acid residues, now expose a terminal Gal that serves as the recognition site for the AMR. In another example, ricin, a highly lethal lectin isolated from the castor bean, binds to GalNAc or Gal residues on cell-surface glycoproteins, and upon internalization inhibits protein biosynthesis. The binding interaction between an individual glycan molecule and a lectin is often low in affinity (K_D in mM) and is too weak for most biological functions. However, lectins often possess multiple carbohydrate recognition domains (CRDs) to bind several glycans. This "multivalent" interaction is a hallmark of glycan–protein interactions and results in avidity enhancements (K_D in μM to nM), stronger binding and more selective recognition events. Thermodynamic studies on glycan–lectin interactions have proposed a few binding models to describe that ability of lectins to cross-link glycoconjugates (Fig. 3.5). Because glycan–protein interactions often trigger biological events, significant effort has been devoted to developing tools to study glycan binding and function. This area of research is referred to as Functional Glycomics. Among the most powerful analytical methods that have emerged in the field in the last decade is the glycan array. Similar to the DNA or protein array, it presents libraries of glycan structures organized in micrometer-sized spots on a chip. Taking the multivalency of glycan interactions into consideration, each spot consists of multiple copies of a unique glycan attached to the surface in close proximity to each other, where they can be interrogated for binding by any protein receptor of interest or even whole cells. Such high-throughput techniques have enabled rapid survey of the structural and functional complexity of the Glycome.

1.7 Consequences of glycosylation defects

There is increasing evidence that proves glycans can play critical roles for proper biological function. Much of this knowledge results from the observation of phenotypes resulting from mutations in glycosyltransferases and glycan-binding receptors. Many of these mutations lead to abnormalities that span from embryonic development to aging and disease. Some of these mutations in humans can cause severe diseases, classified collectively as Congenital Disorders of Glycosylation (CDGs). CDGs are autosomal recessive, inborn diseases that cause serious (sometimes lethal) malfunctions of the nervous, muscular, and gastrointestinal systems. The incidence of CDG worldwide is still unclear, although minimal estimates report 1 to every 5000 births. To date, there are over 1000 patients, but many still remain undiagnosed. Historically described as underglycosylation of transferrin, a plasma protein that is responsible for binding

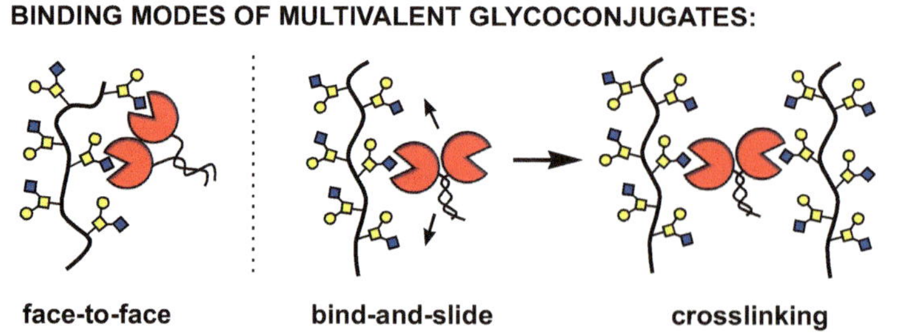

FIGURE 3.5 **Lectins, or carbohydrate-binding proteins engage in multivalent interactions with glycans to generate high avidity interactions.** Lectins may engage in a face-to-face, or a bind-and-slide interaction. The latter provides an explanation for the ability of lectins to oligomerize cell surface glycoconjugates. *Reprinted with permission from Huang, M. L. and Godula, K. (2016). Nanoscale materials for probing the biological functions of the glycocalyx. Glycobiology. 26, 797–803. Copyright 2016 Oxford University Press.*

free iron to maintain homeostasis, significant progress has now been made in understanding the molecular basis of CDGs. There are now over 50 genes known to cause CDGs. In the most common subtype of CDG, CDG-1a, a genetic mutation that affects phosphomannomutase, an isomerase enzyme that converts mannose-6-phosphate, a marker for lysosomal degradation, to mannose-1-phosphate, leads to a functional loss. Patients with CDG-1a can exhibit different severities in abnormalities, but affected infants exhibit developmental delays and underdeveloped cerebellums. There are currently no treatments available for CDGs, although dietary glycan supplementation (e.g., fucose, mannose) exhibits partial effects on some forms of CDG.

While the functional contributions of individual glycan structures within the cellular glycocalyx are coming into focus, their collective properties are much less understood. The cellular glycocalyx (Fig. 3.1) at biological barriers is often dominated by large glycoconjugates, including mucins and proteoglycans, performing unique functions in directing the passage of molecular nutrients or cargoes, while delineating self from nonself and keeping opportunistic organisms at bay. For instance, mucosal barriers are composed predominantly of mucin glycoproteins, whereas the endothelial tissues are populated by proteoglycans. The following sections will discuss these two glycocalyx-dominant classes of glycoproteins in detail and highlight how the unique structural features of their glycans give rise to the distinct properties and functions of these glycoconjugates.

2. Mucins

Mucins are a family of large, heavily O-glycosylated proteins present primarily on epithelial luminal surfaces, where they serve lubricating, hydrating, and protective functions. This section will describe the structure, biosynthesis, and properties of mucins, as well as their roles in host–pathogen interactions, cancer, and cell surface biophysics.

2.1 Mucin structure

Mucins are distinct from other glycoconjugates, because of their large size, dense glycosylation, extended structure, and ability to form cross-linked gels. The characteristic "mucin domain" is composed of a Variable Number of Tandem Repeat (VNTR) region (Fig. 3.6), which provides the structural features that give mucins their unique functions. The VNTR is composed of repeating sequences rich in serine and threonine sites for glycan attachment and is also enriched for proline residues to prevent protein folding and facilitate glycosylation. Depending on the mucin, these VNTR regions can contain >100 repeat sequences, each of which can harbor potential glycosylation sites. The composition of the VNTR and the number of repeats are mucin-specific. For example, the VNTR of human MUC1 (Mucin1) contains 20–100 repeats of the sequence GSTAPPAHGVTSAPDTRPAP. This 20 amino acid polypeptide contains five serine (S) or threonine (T) residues, and roughly 50%–100% of these are glycosylated, creating a glycan-rich domain. In addition to the VNTR module, some mucins have Irregular Repeat (IR) domains. IR domains, like the VNTR modules, are densely O-glycosylated through S/T residues, but lack a specific repeating amino acid sequence. In some cases (e.g., in MUC 16), this domain can be larger than the VNTR.

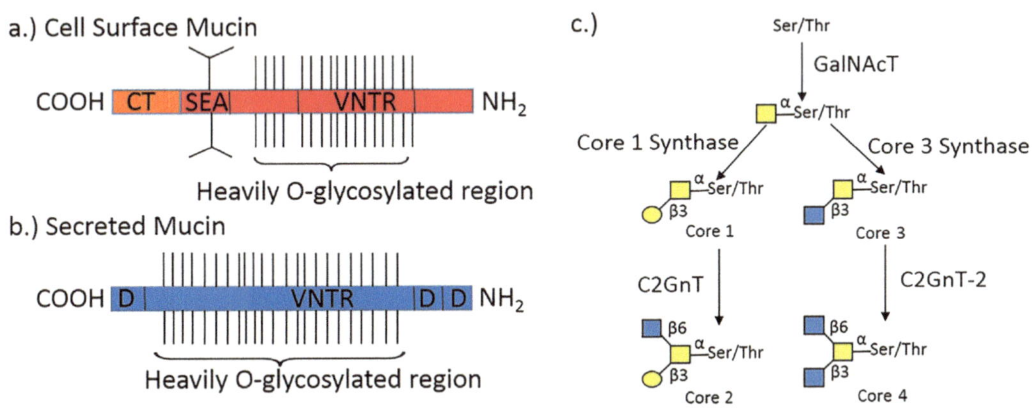

FIGURE 3.6 **Structure of mucin glycoproteins.** (A) Cell surface–bound mucins contain a cytosolic tail (CT) at their C-terminus, a SEA domain to facilitate cleavage and reassociation, and a heavily glycosylated region comprised of VNTR regions as well as irregular repeats (B) Secreted mucins contain D domains rich in cysteine at both the N and C terminus to facilitate crosslinking and mucus gel formation in addition to their heavily o-glycosylated region (C) Biosynthetic pathway of the four common core structures present in O-glycans.

O-glycosylation of the VNTR and IR domains occurs in the Golgi. The enzyme ppGalNacT initiates this process by transferring an α-O-GalNAc to the mucin chain (see Section 1.4). Subsequent elaboration of this core structure is achieved with mucin-specific glycosyltransferases. Just like other glycans, the structure and chemical composition of these complex O-glycans is highly variable. In general, there are four common core structures of O-glycans (and an additional four cores have been identified) (Fig. 3.6), each of which can be extended to produce a broad range of structures. These glycans typically comprise 50%–80% of the glycoprotein by mass.

Such dense glycosylation is often necessary to exert the protective function of mucins. The steric constraints of packing so many glycans into this region force the protein to adopt a more linear secondary structure extended hundreds of nanometers in length above the cell surface. An additional consequence of such heavy glycosylation is that the VNTR and IR regions of mucins are often quite resistant to proteolysis, because the presence of the glycan chains can shield the protein backbone from enzymatic degradation. In order for these domains to be proteolyzed, glycosidases must first remove or reduce the glycan chains to allow access to the core protein. Although all mucins contain the characteristic highly glycosylated VNTR and IR regions, differences are often observed in the sparsely glycosylated regions outside this domain between secreted and cell surface bound mucins.

2.1.1 Cell surface mucins

Cell surface mucins are confined to the apical surfaces of cells via a noncovalent linkage between its core protein and a separate transmembrane domain. This class of mucins includes MUC1, MUC3A, MUC3B, MUC4, MUC12, MUC13, MUC15, MUC16, MUC17, MUC20, and MUC21, each of which vary in size, composition, distribution, and function. These mucins are variably expressed throughout the body, where they fill a range of functions. The single-pass transmembrane domains contain a cytoplasmic tail for anchoring onto the cell surface. Although the extracellular and cytoplasmic domains of the protein are translated, they quickly undergo autoproteolysis into a heterodimer within the ER to yield a large extracellular region that contains a short cell surface linker (Fig. 3.7). These cleaved subunits then associate noncovalently, often through a SEA (Sea urchin sperm protein, Enterokinase, and Agrin) module. Additionally in the ER, the protein is N-glycosylated before it is transported to the Golgi, where O-glycosylation takes place before transport to the cell membrane. On the cell surface, the extracellular portion of these mucins can be cleaved through a variety of mechanisms, ranging from proteolysis of the protein backbone near the surface, to decoupling of the two subunits. This release is especially significant in mucins function as a barrier to pathogens (Section 2.4).

2.1.2 Secreted mucins

Secreted mucins produced by specialized goblet cells in the epithelium lack the transmembrane and cytoplasmic domains of their surface bound counterparts and are instead outfitted with domains tailored to extracellular function and mucus formation. This comes primarily in the form of "D" domains on either side of the VNTR/IR region, which are cysteine rich and allow for oligomerization and generation of viscous gels through formation of disulfide bonds. The family of secreted mucins includes MUC2, MUC5AC, MUC5B, MUC6, MUC7, MUC8, MUC9, and MUC19. Similar to cell surface mucins, secreted mucins first receive N-glycan modifications in the ER. However, rather than cleavage of the mucin into discretely functional subunits, dimerization occurs via their C-terminal D domains. This dimer is then transported to the Golgi for O-glycosylation and is packed into secretory granules. Within these granules, oligomerization of the dimers into networks occurs through the D-domains on their N-terminus. Upon secretion, hydration of the mucin layer occurs, swelling in size by 100–1000 fold into a gel-like mucus layer.

2.2 Mucins and mucosal barriers

Epithelial cell surfaces are the physical barriers protecting underlying tissues from exposure to pathogens, and must often present a selectively permeable barrier to allow transport of beneficial materials while excluding harmful particles. Mucosal barriers are found in tissues that directly interface with the outside world and include the gastrointestinal (GI), respiratory, and reproductive tracts, as well on the ocular surface or in the middle ear. The chemical and physical properties of mucins provide key contributions to the defense mechanism of mucosal barriers. In this section, we will examine the structure and function of mucins as they relate to physiological function as well as barrier formation against pathogenic advancement. The opposite side of the same proverbial coin is of course the continuous evolution of pathogens that take advantage of host glycans to circumvent these defense mechanisms.

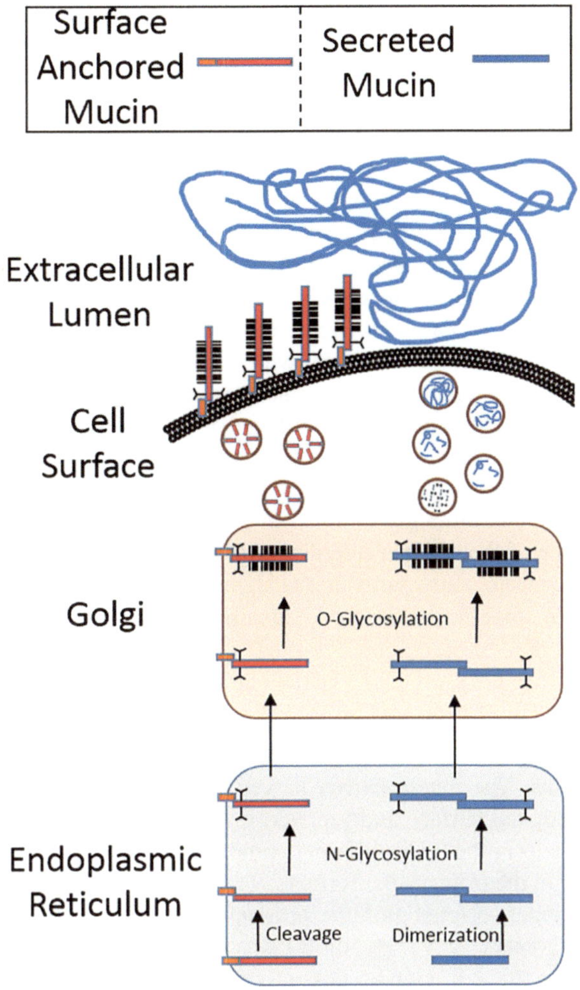

FIGURE 3.7 **Biosynthesis of mucins.** Transmembrane mucins are transcribed in the ER, where they are N-glycosylated and cleaved at their SEA domain. The two subunits reassociate before transport to the Golgi for O-glycosylation prior to surface presentation. Secreted mucins are transcribed in the ER, where they are N-glycosylated and dimerize before transport to the Golgi for O-glycosylation. The mucins are packaged into vesicles where they oligomerize and are secreted into the extracellular space.

2.2.1 Gastrointestinal tract

From the moment food reaches the stomach to its excretion from the colon, a single layer of columnar epithelial cells acts as a barrier between the underlying host tissue and the intestinal lumen. This barrier balances a variety of competing protective and harmful processes. For instance, the mucus layer must serve as a harbor for commensal bacteria (microorganisms that have evolved a mutually beneficial relationship with their human host) and must allow food to be digested, absorbed, and processed, but cannot allow pernicious pathogens to penetrate the cell surface. The mucosal barrier is composed of a secreted and a cell surface–bound layer. Each layer has its own mucin composition and provides specialized barrier functions to the underlying epithelium. Goblet cells are the primary constituent of this mucosal lining.

MUC5AC is the most highly expressed secreted mucin in the stomach, whereas MUC2 is the primary component of mucus in the small and large intestines. Although mucins are the main contributors, the extracellular barrier also contains a wide variety of proteins, most of which do not yet have defined roles. It is thought that many of these proteins, such as trefoil peptides, interact noncovalently with mucins to fine-tune the rheological properties of the mucus. Antimicrobial molecules and secretory antibodies are also present in the mucosal barrier to provide defense against invasive pathogens. These molecules are contained within the mucin matrix, where they can be better retained and localized at increased concentrations. Without the mucus barrier, these secreted molecules would be lost quickly to diffusion in the lumen. The mucus matrix also allows for these components to be present in a gradient, which promotes a robust pathogen defense system close to the cell, while still enabling colonization of the outer layer by beneficial commensal bacteria.

The mucosal barrier is composed of two continuous layers of mucin matrix, termed the inner (closer to the epithelial surface) and outer layer (directly in contact with the GI tract). The inner mucus layer is formed from mucins of the secretory vesicles released by goblet cells and is primarily defined by its viscoelastic and antibacterial properties. It is extremely viscous as a result of the cross-linked high molecular weight mucins. The compounded effects of viscosity and the presence of antimicrobial molecules provide resistance to colonization of the inner mucus layer of healthy epithelium by both commensal and pathogens. The outer layer of mucus is formed from the partial enzymatic degradation of the inner mucus, making it less viscous and allowing for greater diffusion of its contents. This decrease in antimicrobial peptide concentration allows commensal bacteria to colonize the outer mucus. This microenvironment provides protective conditions and nutrients for commensals. The continuous turnover of the mucosal barrier generates a constant flux of secreted mucins passing from the inner mucus layer to the outer, which is in turn degraded and removed through mucociliary motion, a cycle which takes ~24 h to complete under normal conditions.

A variety of cell surface–bound mucins are also produced throughout the GI tract, including MUC1, MUC3A, MUC3B, MUC4, and MUCs 12–17. MUC1 is the most common and well characterized. In the gut epithelium, transmembrane mucins (Section 2.2) localize only to the apical (outward facing, into the GI tract) surface of epithelia cells. Due to the increased size of mucins relative to other glycocalyx components (>100 nm compared to ~10 nm), localization of mucins to the lateral or basal surface of the cell can interfere with adhesion of the cell to the surrounding tissue. As such, the primary functions of these glycoprotein structures are localized on the apical surface, where they maintain protective functions (see Section 2.3), and can serve as "outside-in" signaling molecules.

2.2.2 Respiratory tract

Similar to the GI tract, the respiratory tract (RT) serves as an interface with the outside environment to facilitate nutrient exchange (in this case, gases like oxygen). As with the GI tract, the air passing through the RT also contains a variety of physical and pathogenic components. The mucin barrier serves as a critical first line of defense against these airborne particulates and pathogens, and both cell surface and secreted mucins are present to provide two layers of protection.

The composition of secreted mucus in the RT is different compared to the GI tract and is formed by a variety of mucins, of which MUC5B and MUC5AC are the primary gel formers. The secretion and crosslinking of these mucins provide the scaffold, which then absorbs additional secreted globular proteins that further modulate the viscoelastic properties of the mucus gel. Secreted mucus also contains a variety of RT-specific defensive molecules (e.g., antimicrobial peptides, reactive oxygen species) that directly attack foreign bodies and pathogens. Particulates caught in the secreted mucus matrix can also be mechanically removed through mucociliary action or coughing. The common over-the counter cough drug, Guaifenesin (sold as Robitussin or Mucinex in the US), assists in relieving the symptoms of a respiratory infection by facilitating the clearance of pathogens from the mucus. In experiments involving differentiated human airway epithelial cells, researchers have shown that guaifenesin can reduce the expression of mucins and decrease mucus viscosity.

Transmembrane mucins form a final protective barrier on the epithelial cells lining the RT. Like in the GI tract, MUC1, MUC4, and MUC16 have been identified as the primary mucins present in the airway epithelium, with MUC1 again being the best characterized. As in the gut, MUC1 is capable of providing outside-in feedback from the extracellular environment, although the mechanisms are unclear. Overexpression of MUC1 in response to the proinflammatory factor TNF-α has been shown to decrease inflammation during infection.

2.3 Mucins in innate immune defense against pathogens

Because pathogens often contain glycan-binding proteins on their surface, various mucin glycans can serve as decoy ligands for these receptors, trapping pathogens in the mucus before they can reach underlying cells. Pathogen binding to decoy cell surface mucins often triggers shedding of the extracellular region of the glycoprotein. Ideally, the pathogen remains bound to the glycans as the mucin diffuses away, removing it from the cell surface. Research suggests that cleavage of the extracellular subunit could alert the cell to the pathogen's presence and stimulates signaling cascades to mount a defense. As part of this mechanism, the detection of pathogenic infection results in increased flux of mucins to the cell membrane, allowing for the untethered decoy to be carried away into the secreted mucus layer for clearance. However, more evidence is still needed to fully corroborate this hypothesis.

Aside from direct binding interactions, the viscous, cross-linked mucin network can mechanically exclude larger particles, such as viruses and bacteria, from penetrating. Cell surface mucins, because they tower over the rest of the glycocalyx, can also sterically hinder incoming pathogens from reaching their receptors closer to the cell surface. Even if the relevant mucin does not contain the appropriate ligand to act as a decoy, its sheer steric bulk can act as a shield to prevent the pathogen from engaging the cell surface.

These defense mechanisms have been beautifully demonstrated by the interactions of the bacterium, *H. pylori*. In recent in vitro experiments, gastric epithelial cells exhibited resistance to the bacterium depending on the presence of the mucin-binding adhesins SabA (Sialic acid–binding adhesin) and BabA (blood group–binding adhesin). In the case that the bacterium contains these adhesins and can bind to mucins, infection is prevented through shedding of the mucins. However, infection is still not observed for bacterium without the adhesins. In this case, the mucin is not shed, but may be physically blocking *H. pylori* from binding to the cell surface, preventing infection.

2.4 Pathogen subversion of mucosal barriers

Despite the innate defenses presented by the mucus layer, pathogens are at times able to breach this barrier and infect the host cells (Fig. 3.8). To overcome its protective features, pathogens have evolved strategies to circumvent or destroy the cell's natural defenses. Broadly speaking, pathogen countermeasures are designed to either degrade mucins on their way to the cell, or secrete factors that interfere with their production or release.

The degradation of mucins is a challenging task, as the densely glycosylated VNTR/IR regions are inaccessible to peptidases. Therefore, bacterial enzymes must act on the nonglycosylated regions (such as the cysteine-rich crosslinking domains) or be able to digest the glycans prior to deploying proteolytic enzymes. Indeed, bacteria often express glycosidases that trim decoy glycans to expose the protein backbone. The breakdown of the crosslinked mucin network results in reduced viscosity, which disrupts the steric exclusionary effects of the mucus barrier, and release of antimicrobial agents. Some pathogens can also secrete small molecule toxins that pass through the mucus layer and cell membrane to disrupt normal cellular protective functions, such as mucin biosynthesis. The mucin barrier is constantly being degraded even during normal cell function, so without the secretion of new mucus layers from goblet cells, the barrier will lose its protective functions rather quickly.

A classic example of pathogenic invasion comes from the influenza A virus, which is notable for its interaction with the mucin barrier. In a healthy mucosal barrier, influenza A virions can be sterically excluded from the cell. Influenza has two sialic acid–recognizing proteins that allow it to cut its way through the mucus: the hemagglutinin (HA) binding receptor and neuraminidase (NA), which cleaves sialic acids. There are hundreds of copies of HA on the viral envelope acting as receptors for sialic acids present on mucins. Although the affinity of each HA receptor to individual sialic acids is relatively low, multivalent binding of the collective HA on the viral envelope to ubiquitous sialic acids present on mucin glycoproteins leads to high avidity binding. NA then cleaves the sialic acids off of the glycoprotein, removing the carbohydrate ligand for HA to promote viral release and passage through the mucus layer.

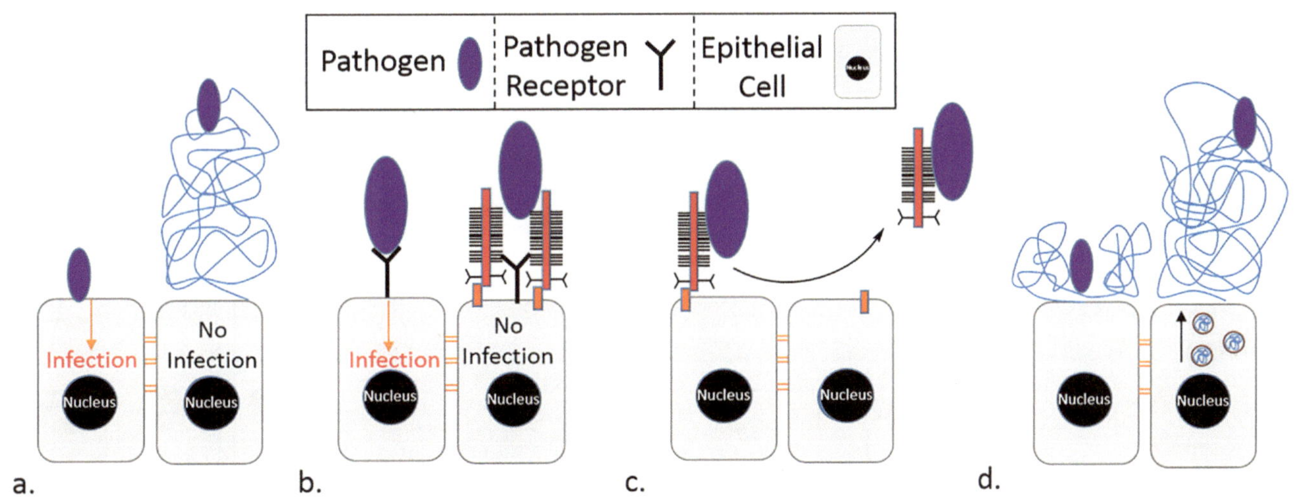

FIGURE 3.8 **Mucin defenses against pathogen invasion.** (A) Secreted mucins can protect the cell from pathogens through steric hindrance or through glycan binding interactions. (B) Transmembrane mucins can sterically prevent pathogens from reaching receptors on the cell surface. (C) Upon binding of pathogens to transmembrane mucins, the extracellular domain can be released from the cell surface. (D) During pathogen invasion mucin production is upregulated, increasing the flux of mucus and amplifying its protective effects.

2.5 Mucins in cancer

Although the primary role of mucins is to protect cells, they can also be coopted by cancer cells to evade immune responses and, by altering cell adhesion, to aid in metastasis and tumor spreading. Mucins are aberrantly expressed in roughly 65% of all cancers and more than 90% of epithelial cell adenocarcinomas. Both overproduction and anomalous glycosylation of mucins are common in cancer cells. Altered glycosylation stems from differential expression of glycotransferases in cancer cells and the resulting cancer-specific glycans can be used as biomarkers for cancer detection and prognosis. While the transmembrane MUC1 was the first mucin recognized as a cancer biomarker, other mucins have been implicated in various carcinomas, including MUC2, MUC4 (pancreatic, ovary, breast, lung, etc.), MUC13 (colon), MUC16 (ovarian), and MUC17 (pancreatic). In this section, we will review some of the mechanisms through which mucins can aid in cancer cell survival and tumorigenesis.

2.5.1 Disruption of cell–cell adhesion

Often in cancerous states, the normal basal-apical polarity of the cell is lost, allowing mucins to localize to nonapical surfaces. Because these mucins extend hundreds of nanometers from the cell surface, they cover a much larger distance than the gap between cells in the epithelium, which are sealed with tight junctions (to maintain polarity) and adherens junctions (to maintain epithelial morphogenesis). In cancer cell progression, mucins can promote detachment and metastasis through disruption of these connections (Fig. 3.9).

Tight junctions serve as a partition between the lateral and apical surfaces of the cell, a function maintained through PAR complexes with atypical protein kinase C (aPKC). Formation of the PAR complex and establishment of polarity is mediated by ERBB2 receptor tyrosine kinase, which disrupts the PAR complex association and leads to loss of polarity. Mucins (specifically MUC1 and MUC4) can bind to ERBB2, which leads to promotion of ERBB2 signaling and disruption of the tight junctions. The resulting loss of polarity allows mucins to localize to the basal and lateral surfaces of the cell and paves the way for loss of cell–cell adhesion through disruption of adherens junctions.

Adherens junctions maintain the spatial integrity of cells by forming intercellular homotypic cadherin interactions between adjacent cells. In addition to cell–cell binding, the intracellular domain of cadherins directs the actin cytoskeleton network through interactions with cytosolic β-catenin. MUC1 is able to doubly disrupt this interaction on both the extracellular and cytosolic fronts. When MUC1 is overexpressed on the lateral region of the cell, the steric bulk of the extended mucins can physically separate cadherins on adjacent cells. In addition, the glycans on the mucins are capable of creating cis-interactions with adhesive receptors on the cell surface, preventing them from forming trans-interactions with other cells. Additionally, the intracellular CT domain of MUC1 is capable of disrupting the β-catenin/cadherin complex. Upon phosphorylation, the CT domain competes with cadherin for binding of β-catenin, sequestering it, and interfering with the

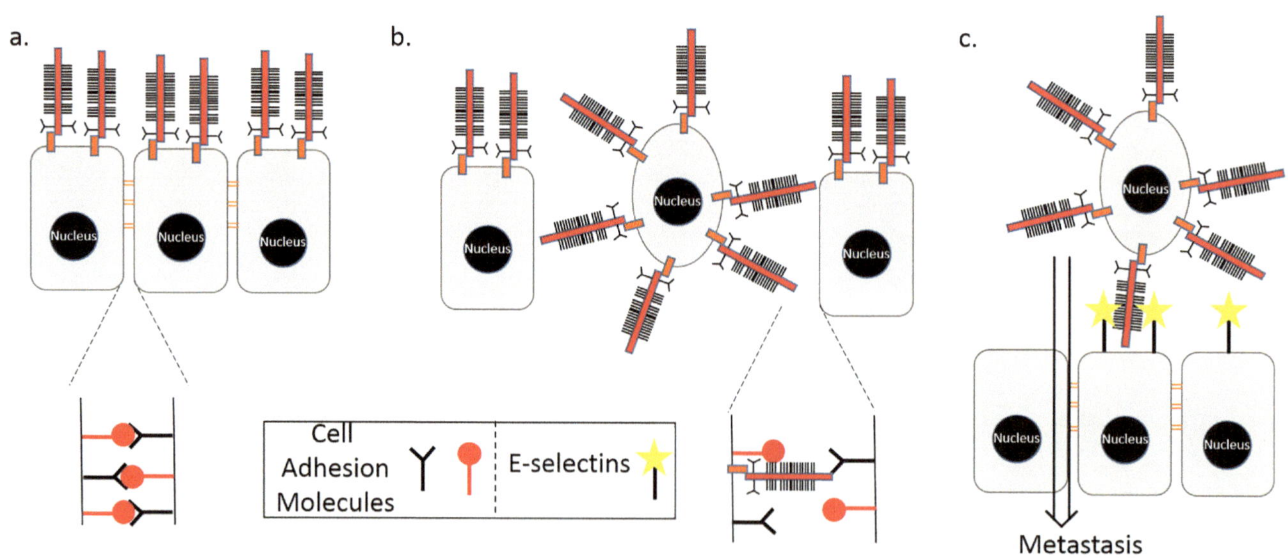

FIGURE 3.9 **Mucins in metastasis.** (A) In normal epithelial tissue, tight junctions are maintained between cells. (B) Loss of polarity results in loss of tight junctions through steric exclusion and cis-binding to receptors on the cell surface. (C) Glycans on mucins can bind to receptors on new tissues and can trigger invasion and metastasis.

cytosolic components of the adherens junctions. Additionally, the β-catenin/MUC1 complex can traffic to the nucleus and influence signaling and can modulate Wnt genes involved in proliferation and differentiation. This disruption of the adherens junctions is one part of the complex process of detachment and metastasis of cancer cells. In short, mucins appear to be crucial in the loss of polarity and intercellular adhesion in cancer and have emerged as key players in motility and metastasis.

2.5.2 Contributions to adhesion

Paradoxically, while MUC1 is heavily involved in detachment of cancerous cells, it is also implicated in promoting their subsequent adhesion in their target tissues. To fully metastasize, circulating cancer cells must readhere and recolonize to create a new tumor. To do so, cancer cells often borrow a strategy from the immune system. During inflammation, epithelial cells are known to express E-selectins, which are sialic acid–binding proteins. This binding event pulls the circulating immune cells into the tissue so they can perform their function. Mucins in cancer cells also contain sialic acid residues that serve as E-selectin ligands on epithelial cells, in addition to other adhesion markers, such as CD-44 (which itself is a nonmucin glycoprotein). Through this interaction, rolling of the tumor cells along the metastatic site is initiated, but full adhesion to the tissue is dependent on secondary anchoring mechanisms to prevent release back into circulation. Integrin-mediated interactions with ECM components (such as ICAM-1) then occur to complete metastasis. Mucins are likely involved in this process as well, as certain MUC1 isoforms are involved with binding to ICAM-1.

Recent studies also point to the biophysical properties of mucins, which may play a role in tumor cell adhesion through cell-surface integrin organization. In experiments utilizing synthetic mucin-mimetic glycopolymers, researchers discovered that the presence of sterically demanding glycopolymers causes integrins to cluster in "kinetic traps", which resulted in enlarged focal adhesions and integrin-dependent activation of proliferation and cell survival.

2.5.3 Inhibition of immune response

Mucins also play a role in cancer cell survival by protecting it from immune surveillance. Similar to the shielding action of epithelial mucins toward pathogens, cancer cells also utilize mucins to protect themselves from the body's natural defense mechanisms. For example, when healthy epithelial cells lose their adhesive properties and detach from their local environment, a specialized apoptotic process called anoikis is initiated to remove the cell from circulation. In the case of metastasizing cancer cells, MUC1 shields the associated death receptors on the cell surface to inhibit anoikis. More generally, when cells become cancerous, they can present factors that are recognized by various immune responses to target for cell death and removal. Although research in the field is ongoing, it is plausible that through their roles in barrier function, secreted or cell surface mucins could prevent the approach of antitumor molecules designed to detect and remove cancerous cells. Significantly, the glycosylated domains of mucins are capable of interacting with leukocytes, and at high levels of soluble mucin secretion, may trap immune cells before they reach cancer cells. Similar mechanisms are possible for sequestering other antitumor immune effectors, such as natural killer cells.

In summary, mucins are formidable biomolecules that provide robust protective barriers at the interface of organisms and hostile outside environments. Their structure and glycosylation provides mucins with unique physical and biological properties to control the passage of nutrients into cells while restricting access from particulates and pathogens. Mucins not only provide passive protection but also report on the cell's physiological state, modulate its interactions with the surrounding matrix, and mediate cellular communication. Mucins can also contribute to the pathophysiology of cells, such as carcinogenesis, by facilitating detachment and adhesion in metastasis or by protecting aberrant cells from innate defenses and immune surveillance. Although the exact mechanisms that cells use to control these seemingly contradictory functions are not yet fully known, mucins are increasingly recognized as biomarkers for cancer and other diseases and will continue to be a focus of intensive biomedical research.

3. Proteoglycans

Proteoglycans are glycoconjugates modified with extended O-linked glycosaminoglycan (GAG) polysaccharide chains. Proteoglycans are a major component of the endothelial glycocalyx and are abundantly present in the ECM. Like mucins, proteoglycans serve both as a structural element and as a relay for biological information. Structural PGs can serve as scaffolds for tissues and are found in joints and cartilage. As information molecules, PGs can function as coreceptors for adhesion molecules, growth factors, and cytokines at the cell surface and facilitate cellular signaling. At the ECM, proteoglycans can sequester growth factors (GFs) away from their receptors to inhibit stimulation, or they can transport and stabilize gradients for GFs and morphogens. A plethora of proteoglycans are present in various tissue to regulate cellular physiology (Table 3.1).

TABLE 3.1 Membrane-bound and secreted proteoglycans, their locations, and functions.

Proteoglycan	Number/type GAG chains	Presence	Functions
Membrane-bound proteoglycans			
Syndecans 1-4	1–3 HS, 0–2 CS	Nucleated cells	Regulates cell adhesion, migration, GF signaling, ligand clearance
Betaglycan	0–1 HS, 0–1 CS	Fibroblasts	Regulates TGFβ signaling
Glypicans 1-6	1–3 HS	Epithelial & mesenchymal cells	GF coreceptors to regulate signaling
Phosphacan	2–5 CS/DS	Brain	Regulates migration of cortical neurons
Thrombomodulin	1 CS/DS	Endothelial cells	Anticoagulant
CD44	0–2 CS/DS/HS	Lymphocytes, among many others	Receptor for hyaluronan
NG2/CSPG4	2–3 CS/DS	Stem cells, glial progenitors,	Modulates cell-ECM interactions, GF interaction, integrin activation
CD74	1 CS	Antigen-processing cells	Chaperones MHC II; enhances T-cell activation
SV2	1–3 KSI	Synaptic vesicles	Regulates exocytosis of neurotransmitters
Secreted proteoglycans			
Aggrecan	~100 CS, ~20 KSII	Cartilage, brain	Hydration barrier
Versican	0–15 CS/DS	ECM, blood vessels, brain, leukocytes	Regulates inflammation, cell migration
Neurocan	1–2 CS/DS	Brain	Regulates neurite outgrowth
Brevican	0–4 CS/DS	Brain	Regulates synaptic plasticity
Decorin	1 CS/DS	Connective tissue	Regulates collagen fibrillogenesis
Biglycan	0–2 CS/DS	Macrophages	Activates innate immunity via TLR2/4
Lumican	3–4 KS I	Widely distributed	Collagen matrix assembly
Keratocan	3–4 KS I	Cornea	Corneal transparency
Fibromodulin	2–4 KS I	Widely distributed	Collagen matrix assembly
Osteoglycan	2–3 KS I	Cornea	Corneal transparency
Perlecan	1–3 HS, 0–2 CS	ECM	Regulates cell migration
Agrin	1–3 HS	Brain, NMJ	Regulates NMJ maturation
Collagen type IX	1 CS/DS	Cartilage	Stabilizes cartilage-collagen fibrils
Collagen type XVIII	2–3 HS	Basement membranes	Regulates stability of basement membranes

3.1 Proteoglycan biosynthesis, structure, and the "sulfation code"

The biosynthesis of proteoglycans begins with the genome-templated expression of the core protein in the ribosomes (Fig. 3.10). The protein transcript is then processed in the Golgi apparatus, where the attachment of glycosaminoglycan (GAG) chains occurs within Ser-Gly consensus sequences. This process begins with the priming of protein Ser/Thr residues with a tetrasaccharide (GlcA-Gal-Gal-Xyl-) linker. During this process, xylosyltransferase enzymes (XYLT1, XYL2) attach a xylose monosaccharide to the Ser residue of the core protein. The attached xylose residue then serves as a substrate

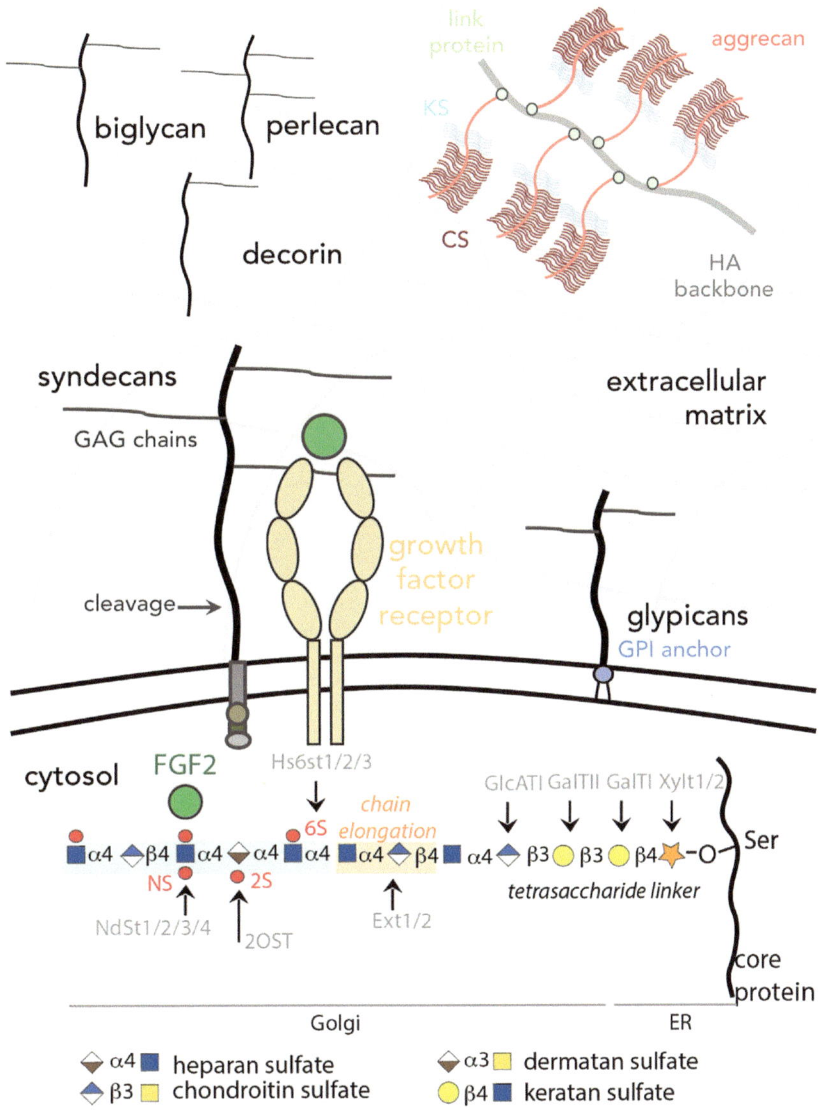

FIGURE 3.10 Proteoglycan examples, structure, and biosynthesis. Proteoglycans can be present as transmembrane or soluble forms. Proteoglycan biosynthesis occurs in the ER and Golgi, where a tetrasaccharide GlcA-Gal-Gal-Xyl linker is first attached to Ser/Thr residues of the core protein. Subsequent elongation and modification by exostosin and sulfotransferase enzymes, respectively, generate linear long-chain sulfated glycosaminoglycan chains that encode binding sites for growth factors. At the glycocalyx, proteoglycans can serve as coreceptors for growth factors and their receptors.

for galactosyltransferases (B4GALT7, B4GALT6, and B4GALT3, respectively) that successively attach the Gal, Gal, and GlcA residues of the linker. Following linker synthesis, elongation of the GAG chain occurs. GAGs are composed of a repeating disaccharide motif, composed by an uronic acid linked to an amino sugar. For example, heparan sulfate (HS) chains are composed of GlcAβ(1−4)GlcNAc disaccharides, whereas chondroitin sulfate (CS) GAGs are composed of GlcAβ(1−3)GalNAc. The resulting long, linear, and highly negatively charged GAG chains are considered the most abundant heteropolysaccharides in the body. Individual GAG chains can be as long as 100 disaccharide units long. Given the large size of GAGs, they tend to dominate the chemical and physical properties of proteoglycans.

Sulfate groups can decorate the GAG chains, and proteoglycans are known to possess domains of high, medium, and low sulfation. These modifications are the result of sulfotransferases that transfer sulfate groups from activated PAPS (3′-phosphoadenosine-5′-phosphosulfate) substrates during chain polymerization. A suite of sulfotransferases work in concert to transfer sulfate groups to various positions of the sugar backbone. For example, NDST1/2 (N-deacetylase/N-sulfotransferase) is a bifunctional enzyme that can add or remove sulfate groups to or from the amino group of the glucosamine residues of HS. There are also 6-O-sulfotransferases and 2-O-sulfotransferases that catalyze the transfer of sulfate groups to the 6- and 2- hydroxyl position groups of HS, respectively. For CS, chondroitin-4-O-sulfotransferase (C4ST) and chondroitin-6-O-sulfotransferase (C6ST) catalyze the sulfation of sulfate groups at 4-O or 6-O positions of the CS GalNAc residue, respectively, whereas 2-O sulfation of the GlcA residue is achieved with uronyl 2-O-sulfotransferase (UST). A few naming schemes have emerged throughout the years for classifying the disaccharide repeating units present in GAGs. Table 3.2 outlines the structures and naming schemes for the major naturally occurring disaccharide motifs.

In contrast to the other major biopolymers (e.g., DNA/RNA, proteins), the acknowledgment that GAGs (or even glycans in general) possess some type of specificity and that it harbors information in a sequence-specific manner has historically been slower to come. Given the complexity of glycans, and the heterogeneity of sulfation in GAGs, it is not surprising that the "sulfation code" has taken years to decipher and still remains an ongoing enterprise. Through painstaking work, we now know that specific sulfation patterns and three-dimensional structures of HS and CS chains can encode for growth factor–binding sequences to affect various biological processes.

Proteoglycans can be present as membrane-anchored or secreted proteins (Fig. 3.10). In some cases, alternative mRNA splicing heavily dictates these two forms. Proteoglycans can be anchored onto the plasma membrane via transmembrane protein domains, such as those found in syndecans. On the other hand, smaller proteoglycans, such as glypicans, use the glycophosphatidylinositol (GPI) glycolipid anchor. Transmembrane proteoglycans, such as syndecans, are known to also exist in the soluble form via an event called "ectodomain shedding," in which the extracellular portions can be clipped off

TABLE 3.2 GAG nomenclature. Disaccharide structure code (DSC) of HS and CS repeating units.

DSC	Roman numeral	Structure
Heparan sulfate disaccharides		
D0H0	IV-H	ΔUA-GlcNH2
D0H6	II-H	ΔUA-GlcNH2-6S
D2H0	III-H	ΔUA-GlcNH2
D2H6	I-H	ΔUA2S-GlcNH2-6S
D0A0	IV-A	ΔUA-GlcNAc
D0A6	II-A	ΔUA-GlcNAc6S
D2A0	III-A	ΔUA2S-GlcNAc
D2A6	I-A	ΔUA2S-GlcNAc6S
D0S0	IV-S	ΔUA-GlcNS
D0S6	II-S	ΔUA-GlcNS6S
D0S3		ΔUA-GlcNS3S
D2S0	III-S	ΔUA2S-GlcNS
D2S6	I-S	ΔUA2S-GlcNS6S
D2S3		ΔUA2S-GlcNS3S
D0S9		ΔUA-GlcNS3S6S
D2S9		ΔUA2S-GlcNS3S6S
Chondroitin sulfate disaccharides		
D0a0		ΔUA-GalNAc
D0a4		ΔUA-GalNAc4S
D0a6		ΔUA-GalNAc6S
D2a0		ΔUA2S-GalNAc
D2a4		ΔUA2S-GalNAc4S
D2a6		ΔUA2S-GalNAc6S
D0a10		ΔUA-GalNAc4S6S
D2a10		ΔUA2S-GalNA4Sc6S

the membrane by "sheddases" that recognize a region of the core protein proximal to the cell membrane. Ectodomain shedding is a highly regulated event, and it can rapidly alter the resulting phenotype of affected cells by reducing the amount of ligands on the cell surface or competing with membrane-bound ligand. On the other hand, glypicans can be released from the cell surface by the action of phospholipases that cleave the ester bonds of the GPI anchor.

3.2 Other major classes of GAGs

3.2.1 Hyaluronic acid

In contrast to other GAGs that are sulfated and anchored onto a core protein, hyaluronic acid (HA) or hyaluronan is nonsulfated and soluble. Named after the Greek word for vitreous, hyaloid, this glass-like substance, was first isolated in the early 1930s. This linear polysaccharide is larger than other types of GAGs (10^7 Da) and can be composed of up to 10^4 repeating GlcA-GlcNAc disaccharide units. Rather than being produced in the Golgi, HA is formed in the plasma membrane. Hyaluronan synthases are responsible for the biosynthesis of HA. These integral membrane proteins extend HA by the repeated addition of GlcA and GlcNAc to a growing polysaccharide. The resulting HA is extruded via the trans-membrane ABC (ATP-binding cassette) transporter into the extracellular space. The carboxylic acid groups on the GlcA residues of HA are anionic at physiological pH, and its high anionicity and large size confers its high hydration volume. HA is a major component of the ECM, as well as the connective, epithelial, and neural tissues. HA has been known to exist in multiple sizes, low (LMWHA, <500 kDa), and high-molecular weight HA (>500 kDa). Peculiarly, there is feedback inhibition that occurs among different HA sizes. For example, HA synthesis can be inhibited by LMWHA, but is stimulated by HMWHA in human synovial fibroblasts. Moreover, it is thought that fragments of HA that occur as a result of injury can be proinflammatory, whereas the native HMWHA are not. Due to its ubiquitous presence in the ECM, HA is a popular scaffold for tissue engineering and biomaterials applications.

3.2.2 Keratan sulfate (KS)

KS is a type of linear sulfated GAG molecule found enriched in the corneal and skeletal tissue. KSPGs include lumican, aggrecan, mimecan, among others. KS is composed of a Galβ1-4GlcNAc repeating units that can be sulfated at carbon 6 of either or both of the Gal or GlcNAc residues. Interestingly, KS can be mannosylated at their linkage region, and some regions along the chain can be fucosylated, and a large majority of KS chains are terminally sialylated. There are two major classes of KS, originally classified based on where they were isolated, but have now been assigned based on their linkage to the core protein. KSI is N-linked to Asn via GlcNAc, whereas KSII is O-linked to Ser/Thr via GalNAc residues. At the cornea, KS is produced by specialized fibroblasts called keratocytes. KS retains the hydration content of the cornea. Lumican specifically is known to organize collagen fibrils in the cornea to achieve optical transparency.

3.3 Glycocalyx proteoglycans and glycosaminoglycans in the vasculature

GAGs are prominent players in the vasculature. HS alone comprises 50%–90% of the total endothelial GAGs. Heparin is perhaps the most well-studied GAG in the context of the circulatory system. In contrast to the easily interchangeable "heparan sulfate," heparin is a soluble form of HS. Whereas domains of sulfation occur in HS, heparin is isolated mostly as a trisulfated product and is often considered an oversulfated variant of HS, with ~2.3 sulfate groups per disaccharide versus 0.8 for HS. In its natural form, heparin has a molecular weight ranging from 3 to 30 kDa. Heparin has significant anticoagulant activity and is often used in the clinic as a blood thinner in coronary syndromes and kidney dialysis. The anticoagulant activity of heparin lies in its unique pentasaccharide sequence that includes a 3-O-sulfated glucosamine residue that is critical for binding to the enzyme inhibitor antithrombin III (AT). This rare modification occurs in 1 out of 20 residues in heparin, and 1 out of 200 in HS. Seven mammalian 3-O-sulfotransferase (Hs3st) enzymes that install these sulfate groups on HS constitute the largest group of sulfotransferases involved in HS biosynthesis. Despite this large number, there is still a limited array of proteins known to be modulated by 3-O-sulfation.

Until recently, pharmaceutical heparin was isolated from porcine mucosa or bovine lungs. There is substantial chemical heterogeneity in these samples, and improperly purified and characterized sources can lead to the presence of deleterious contaminants. In 2008, oversulfated CS was found to be a major contaminant of heparin that led to a number of adverse reactions in patients. To circumvent problems associated with insufficiently pure heparin, low-molecular-weight heparins (LMWHs) that have a molecular mass of <8 kDa are now being used in the clinic. LMWHs such as Arixtra (Fondaparinux) are now being used in the clinic for deep vein thrombosis and pulmonary embolisms.

3.4 Glycocalyx proteoglycans in development

During development, GFs and morphogens provide the cues for directing cell fate. Proteoglycans can interact with a number of GFs through their GAG chains. The secreted morphogens, FGF, Wnt, Hedgehog, and BMP emanate from cells and can trigger various signaling events in a concentration-dependent manner. The ability of proteoglycans and morphogens to work in concert and create these concentration gradients has been a key issue in understanding mammalian development.

3.4.1 Embryonic development

Studies involving genetic knockouts of the HS elongating enzyme, exostosin-1 (Ext1), have revealed the importance of proteoglycans in embryonic development. Ext1 deficiency in mouse embryos is considered embryonic lethal, as these embryos fail to gastrulate or develop. The mouse embryonic stem cells isolated from these embryos revealed that HS at the stem cell glycocalyx functions as a coreceptor to organize FGF2 and FGFR ternary complexes. The formation of this complex triggers the activation of signaling pathways that provide cues for the stem cell to exit from pluripotency and differentiate. Void of Ext1, these mutant stem cells cannot be different and remain fully pluripotent.

Random mutagenesis experiments have also identified that mutations in the lazy mesoderm gene in mice (sugarless in Drosophila), involved in proteoglycan synthesis causes improper mesoderm migration during gastrulation, and is also embryonic lethal. Lazy mesoderm encodes for UDP-glucose dehydrogenase, the enzyme that oxidizes UDP-Glc into UDP-GlcA. These defects in cell migration resulted from the loss of FGF signaling. Further confirming the importance of proteoglycans in FGF signaling, mutations in the HS domains required for FGF binding have also been explored. Genetic knockouts of 2-O-sulfotransferases and glucuronyl C5-epimerase leading to the loss of 2-O-sulfate or GlcA residues of HS, respectively, also result in embryonic lethality due to abnormal development, but such mice progress further than those with complete loss of GAGs. In addition to HS, mutation of Hyal2 has also been shown to be embryonic lethal. It is worthwhile to note, however, that not all single mutations or knockouts lead to embryonic lethality, as the function of some enzymes can be compensated by others.

In murine embryonic stem cell differentiation, the sulfation patterns on HS are known to be dynamic and can change depending on the day of differentiation. In studies highlighting the importance of sulfation on differentiation, pluripotent embryonic stem cells synthetically engineered to display high degrees of sulfated HS displayed accelerated exit from self-renewal and differentiation through increased activation of the FGF/Erk mediated signaling pathways.

3.4.2 Myogenesis

Lipid rafts are important constituents of the plasma membrane. Through these localized structures, signals arising from cell surface events can be efficiently localized to increase binding and enhance cell sensitivity to propagate to downstream events. Glypican-1 has been found to be an important component of lipid rafts in order to allow the development of mature myofibers from precursor myoblast cells. Here, glypican-1 is localized into membrane microdomains, where they sequester FGF2 away from their receptors, which are diffused on the membrane. Through this sequestration, FGF2 signaling is inhibited and myogenic differentiation occurs. Other proteoglycans have also been known to modulate FGF2 signaling by congregating into lipid rafts. For example, syndecan-4, have been known to recluster into lipid rafts upon activation by FGF2, although the significance of this event is less known.

3.4.3 Neuromuscular junction

The neuromuscular junction (NMJ) is the synapse connecting a motoneuron and a muscle fiber. The development of the NMJ is protracted and extends beyond the embryonic stage and continues after birth. The NMJ harbors clusters of acetylcholine receptors that allow the proper propagation of the action potential. To form the NMJ, motoneurons must secrete an important proteoglycan called agrin. Agrin is an "extracellular synaptic organizing molecule," the connecting piece between the motoneuron and the myofiber, and agrin deficiency halts NMJ development. Agrin expression is highly regulated by alternative mRNA splicing. Through its various forms, agrin is present in the NMJ, in neuronal synapses, as well as in immune synapses. Agrin at the NMJ is a secreted multidomain 600 kDa proteoglycan, composed of a 200 kDa core protein, and GAG chains.

3.4.4 Neuronal development and regeneration

CS proteoglycans (CSPGs) are critical regulators of the developing and regenerating nervous system. CSPGs have been known to serve dual—both enhancing and inhibitory roles in neuronal growth. CSPGs can enhance neurite outgrowth through interactions with bone-derived neurotrophic factor (BDNF), midkine (MK), or pleiotrophin (PTN). During embryonic development, CS glycans can guide the path of axons as they elongate and form synapses with neurons. Interestingly, growing axons avoid CSPGs in the developing growth cone, indicating that CSPGs act as negative cues to guide axonal growth. The expression of CSPGs, such as neurocan, brevican, phosphacan, and versican, is significantly upregulated during adult injury, and CSPGs are also known to inhibit axon regeneration after an injury to the spinal cord. CSPGs also contribute to glial scarring postinjury, and this scar acts as a barrier for new axons trying to reach the injury site. CSPGs also play crucial roles for the inability of the spinal cord for regeneration after injury, and some research has shown that chondroitinase enzymes that digest CS chains can restore some of this regeneration function. The phosphacan or receptor-type-protein-tyrosine phosphatase (PTP/RPTP) family of proteins is expressed in neurons and astrocytes and is known to interact with CS chains to exert its inhibitory functions. Some sulfation patterns are highly preferred for binding, such as the disulfated CS-D motif, which strongly interact with PTNs.

3.5 Glycocalyx proteoglycans and glycosaminoglycans in infection and immunity

Similar to mucins, proteoglycans at the glycocalyx can also serve dual functions in the progression of an infection. The GAG chains at the cell surface can serve as a point of entry or interaction with foreign agents to promote infection. On the other hand, soluble GAGs at the ECM can also serve a protective role, inhibiting the interaction of the foreign agents with their receptors.

3.5.1 Sexually transmitted infections

HS has a prime role in interactions with sexually transmitted infections. For instance, the herpes simplex virus type 1 (HSV1), or herpes, which commonly causes "cold sores," but can also cause genital herpes, interacts with mammalian cells via glycocalyx HS chains. Interestingly, an additional interaction of the HSV-1 glycoprotein gD with the glycocalyx 3-O-sulfate HS chains is required for viral entry. In HIV-1, treatment of T-cells with heparinase enzymes or a metabolic sulfation inhibitor (sodium chlorate) has been shown to inhibit HIV-1 binding and infection. These results indicate that glycocalyx HS chains participate in HIV-1 infection by facilitating the binding of HIV-1 virions to T cells, and that sulfation, in general, is important for this binding event. Subsequent studies revealed that it is the HIV-1 viral envelope glycoprotein gp120 that binds GAGs, and that although this binding event is dependent on general sulfation, gp120 prefers to bind to heparin and HS, compared to CS and dermatan. Similar relations with other surface receptor proteins have also been revealed in HPV, Chlamydia, Gonorrhea, Syphilis, and yeast infections.

3.5.2 Other pathogens

HA is a component of some pathogenic bacteria, including the extracellular capsule of group A Streptococcus. Capsular HA can promote the adhesion and colonization to the host tissue. Similar to other capsular polysaccharides, it can also aid in pathogen virulence by mimicking host HA, and thus shielding the pathogens from the action of host defenses (e.g., inhibiting phagocytosis and complement). Some bacteria (e.g., *Staphylococcus aureus, Clostridium perfringens*) can also synthesize hyaluronidase enzymes to utilize the host HA as a carbon nutrient source.

3.5.3 The immune response

The HSPG biglycan has recently been implicated as an endogenous key mediator of inflammation. During an inflammatory challenge, circulating macrophages have been found to engage in the de novo synthesis of biglycan. Biglycan acts as an agonist of the innate immune Toll-like receptors TLR2 and 4 (TLR2/4), where it promotes receptor clustering with the purinergic P2X7/P2X4 receptors to rapidly activate the NLRP3 inflammasome. Thus, biglycan is a new DAMP (Damage-associated molecular pattern) that links the soluble matrix and the innate system. Biglycan and its analog, decorin, are master regulators of tumorigenesis, perhaps by binding other receptors such as LRP6. Several cancers are known to overexpress biglycan. This increased expression of GAG chains at the glycocalyx could point to enhanced metastasis and drug resistance.

In the lungs, GAGs are distributed along the bronchial walls and airway secretions. In this role, they maintain the hydrated state and structure of the epithelial ECM. GAGs also modulate interactions with respiratory pathogens, as well as influence the resulting inflammatory response. Mast cells that release heparin are heavily enriched in the lungs. Heparin mostly resides within secretory granules in mast cells, but is released into the vasculature during injury. At these sites, heparin serves a defensive rather than an anticoagulant function, inhibiting the entrance of invading foreign pathogens.

3.6 Proteoglycan disorders

Given the importance of proteoglycans and GAGs in physiology, it is not surprising that abnormalities and pathophysiological conditions arise from mutations in genes involved in the biosynthesis of breakdown of proteoglycans. As described above, many mutations that result in a lack of GAGs are embryonic lethal. On the other hand, mutations involved in the catabolism of GAGs most often result in lysosomal storage disorders (LSDs), a group of 50 or so rare inherited disorders, mostly recessive, that are characterized by dysfunctional lysosomal breakdown. Mucopolysaccharidoses (MPS) are a group of lysosomal storage disorders characterized by a lack of one of more eight GAG-degrading enzymes, resulting in an accumulation of pathological GAGs in the lysosome. The effects of MPS are devastating. Patients suffering from MPS exhibit brain and development abnormalities. Hunter syndrome is caused by mutations in MPS II, and MPSIIIA Sanfilippo A is specific for its lack of sulfamidase enzyme. Mutations in HYAL1 encoding for a lysosomal hyaluronidase also result in mucopolysaccharidosis IX. There are currently no therapeutic options for this group of diseases, but one promising method lies in enzyme replacement therapy (ERT), the delivery of enzymes that aid in the breakdown of GAGs to treat somatic symptoms.

While this chapter focuses only on the two main families of large protein glycoconjugates of the cellular glycocalyx, mucins and proteoglycans, the structural and compositional complexity and functional diversity of this biological interface are immense. Although still poorly understood, the contributions of the glycocalyx to cellular physiology are well appreciated. We are currently focusing on improving tools to study glycans and we anticipate discoveries of new functional roles of glycans in human disease as well as the emergence of biomedical technologies that target glycans for therapeutic gain.

Further Reading

Bond, M. R., & Hanover, J. A. (2015). A little sugar goes a long way: The cell biology of O-GlcNAc. *Journal of Cell Biology, 208*, 869.

Corfield, A. P. (2015). Mucins: A biologically relevant glycan barrier in mucosal protection. *Biochim. Biophys. Acta BBA - Gen. Subj., 1850*, 236–252.

Derrien, M., van Passel, M. W. J., van de Bovenkamp, J. H. B., Schipper, R. G., de Vos, W. M., & Dekker, J. (2010). Mucin-bacterial interactions in the human oral cavity and digestive tract. *Gut Microbes, 1*, 254–268.

Gama, C. I., Tully, S. E., Sotogaku, N., Clark, P. M., Rawat, M., Vaidehi, N., Goddard, W. A., III, Nishi, A., & Hsieh-Wilson, L. C. (2006). Sulfation patterns of glycosaminoglycans encode molecular recognition and activity. *Nature Chemical Biology, 2*, 467–473.

Guerrini, M., Beccati, D., Shriver, Z., Naggi, A., Viswanathan, K., Basio, A., Capila, I., Lansing, J. C., Guglieri, S., Fraser, B., Al-Hakim, A., Gunay, N. S., Zhang, Z., Robinson, L., Buhse, L., Nasr, M., Woodcock, J., Langer, R., Venkataraman, G., ... Sasisekharan, R. (2008). Oversulfated chondroitin sulfate is a contaminant in heparin associated with adverse clinical events. *Nature Biotechnology, 26*, 669–675.

Hattrup, C. L., & Gendler, S. J. (2008). Structure and function of the cell surface (Tethered) mucins. *Annual Review of Physiology, 70*, 431–457.

Hollingsworth, M. A., & Swanson, B. J. (2004). Mucins in cancer: Protection and control of the cell surface. *Nature Reviews Cancer, 4*, 45–60.

Kim, K. C. (2012). Role of epithelial mucins during airway infection. *Pulmonary Pharmacology Therapeutics, 25*, 415–419.

Kufe, D. W. (2009). Mucins in cancer: Function, prognosis and therapy. *Nature Reviews Cancer, 9*, 874–885.

Lawrence, R., Lu, H., Rosenberg, R. D., Esko, J. D., & Zhang, L. (2008). Disaccharide structure code for the easy representation of constituent oligosaccharides from glycosaminoglycans. *Nature Methods, 5*, 291–292.

McGuckin, M. A., Lindén, S. K., Sutton, P., & Florin, T. H. (2011). Mucin dynamics and enteric pathogens. *Nature Reviews Microbiology, 9*, 265–278.

Miller, G. M., & Hsieh-Wilson, L. C. (2015). Sugar-dependent modulation of neuronal development, regeneration, and plasticity by chondroitin sulfate proteoglycans. *Experimental Neurology, 274*, 115–125.

Neill, T., Schaefer, L., & Iozzo, R. V. (2015). Decoding the matrix: Instructive roles of proteoglycan receptors. *Biochemistry, 54*, 4583.

Paszek, M. J., DuFort, C. C., Rossier, O., Bainer, R., Mouw, J. K., Godula, K., Hudak, J. E., Laking, J. N., Wijekoon, A. C., Cassereau, L., Rubashkin, M. G., Magbanua, M. J., Thorn, K. S., Davidson, M. W., Rugo, H. S., Park, J. W., Hammer, D. A., Giannone, G., Bertozzi, C. R., & Weaver, V. M. (2014). The cancer glycocalyx mechanically primes integrin-mediated growth and survival. *Nature, 511*, 319–325.

Shukla, D., Liu, J., Blaiklock, P., Shworak, N. W., Bai, X., Esko, J. D., Cohen, G. H., Eisernberg, R. J., Rosenberg, R. D., & Spear, P. G. (1999). A novel role for 3-O-sulfated heparan sulfate in herpes simplex virus 1 entry. *Cell, 99*, 13–22.

Taylor, M. E., & Drickamer, K. (2011). *Introduction to glycobiology* (3rd ed.). New York: Oxford University Press.

Tiwari, V., Maus, E., Sigar, I. M., Ramsey, K. H., Shukla, D. Role of heparan sulfate in sexually transmitted infection.

Thacker, B. E., Xu, D., Lawrence, R., & Esko, J. D. (2014). Heparan sulfate 3-O-sulfation: A rare modification in search of a function. *Matrix Biology, 35*, 60–72.

Yan, D., & Lin, X. (2009). Shaping morphogen gradients by proteoglycans. *Cold Spring Harb Perspect Biol, 1*, Article a002493.

Chapter 4

Cell membranes and their shaping

Shiro Suetsugu
Nara Institute of Science and Technology, Ikoma, Japan

1. Summary

The cellular membrane is composed of lipids. In this chapter, the characteristics of cellular membrane lipids, including lipid composition, are reviewed. The lipid composition contributes to the fluidity of the membrane. The outermost cellular membrane, the plasma membrane, is supported by an actin cytoskeleton. Membrane-binding soluble proteins (peripheral membrane proteins) are considered to be responsible for the remodeling and shaping of the membrane.

2. The cell membrane and lipids

Lipids that constitute the cell membrane have amphiphilic properties. Amphiphilic lipids have a high structural diversity and are composed of a head, backbone (typically glycerol), and fatty tails (Robertson, 2018; van Meer & de Kroon, 2011; van Meer et al., 2008). The amphipathic properties of lipids result in a bilayer membrane, where the fatty acid tails face each other, forming a hydrophobic core. Lipids that form biological membranes include phospholipids (glycerophospholipids) and sphingolipids. Phospholipids are found on the both cytoplasmic and extracellular surfaces, while sphingolipids are present on the extracellular surface.

In phospholipids, fatty acids are linked to two of the three hydroxyl groups of glycerol, while the head groups are linked to the remaining group (Fig. 4.1). In sphingolipids, a fatty acid is attached to a sphingoid, which is a backbone having a fatty acid and a hydrophilic polar group. In addition to phospholipids and sphingolipids, biological membranes contain a large number of sterols, including cholesterol. Sterols are present between the fatty acids of amphiphilic lipids and regulate the density and fluidity of the lipid membranes (Antonny et al., 2015; Vanni et al., 2014) (Fig. 4.1).

Surfactants (or detergents) can be used to understand the properties of biological membranes. The surfactant is also composed of a fatty acid portion and a hydrophilic portion. However, surfactants cannot adopt the bilayer structure, forming micelles because of the smaller volume of the hydrophobic fatty acid than the hydrophilic head group. In contrast, the lipids that make up the lipid membrane have a larger volume by possessing two fatty acids. This characteristic is the reason phospholipids assemble into membranes in an aqueous solution rather than into micelles.

3. Cell membrane lipid composition

Phospholipid classification is determined by the polar and hydrophilic groups. The polar groups of major phospholipids are ethanolamine, serine, choline, and inositol, resulting in phosphatidylethanolamine (PE), phosphatidylserine (PS), phosphatidylcholine (PC), and phosphatidylinositol (PI), respectively (Marszalek & Lodish, 2005). Inositol, choline, and serine are larger than ethanolamine. In addition, serine is characterized by having a negative charge. Inositol can also be phosphorylated to become phosphoinositide, which presents a large negative charge. Without a polar group, a proton can be at the place of the polar group, resulting in phosphatidic acid (PA). Charged PS is abundant in plasma membranes and endosome systems formed via cellular endocytosis (Bigay & Antonny, 2012).

Depending on the size of the polar group and the bulk height of the fatty acid, the shape of the phospholipid varies and is classified as cylindric, inverted-cone-shaped, or cone-shaped. The typical fatty acids include oleic and linoleic acids. These fatty acids are largely present in food. The carbon-to-carbon bonds of fatty acids are often saturated; however, unsaturated bonds, namely double bonds, are also observed. Fatty acids in which hydrocarbons are bonded only to

Cell Physiology Sourcebook. https://doi.org/10.1016/B978-0-12-811114-7.00005-0

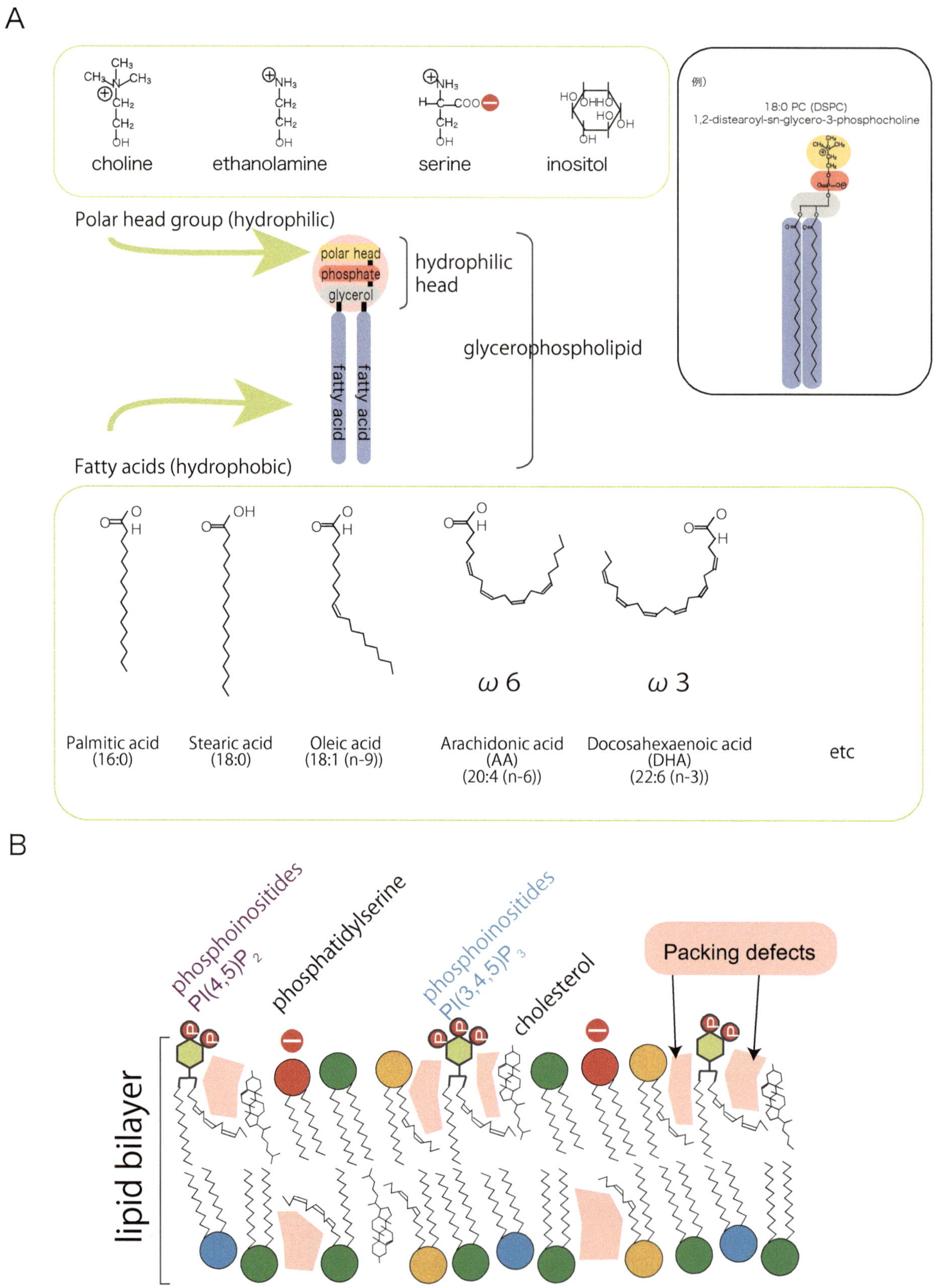

FIGURE 4.1 **Packing defect between phospholipid fatty acids and phospholipids in lipid membranes.** (A) Phospholipid composition. A hydrophilic polar group binds to the third position of glycerol via phosphoric acid, and two fatty acids bind to the first and second positions. The type of fatty acids varies according to the number of hydrocarbons and the number and position of unsaturated bonds. Combined with the diversity of polar groups, phospholipids can form a vast number of combinations. (B) Packing defects between lipid molecules in lipid membranes. A gap can form between the phospholipid molecules depending on the nature of the fatty acid or depending on the curvature of the membrane. These gaps are known as packing defects. Cholesterol can interact with this gap. Also, charged lipids (phosphoinositide and phosphatidylserine) determine the charge of the membrane surface. The gap between lipid molecules and the charge are two major factors that determine the nature of the membrane.

saturated bonds are called "saturated." Saturated fatty acids include palmitic acid (16:0; number of carbons: number of double bonds) and stearic acid (18:0) (Fig. 4.1) (Marszalek & Lodish, 2005). The main unsaturated fatty acids are oleic acid (18:1), arachidonic acid (20:4), and docosahexaenoic acid (22:6) (Fig. 4.1) (Marszalek & Lodish, 2005). Arachidonic acid, docosahexaenoic acid, and other fatty acids with multiple unsaturated bonds are called "poly-unsaturated fatty acids" (PUFAs), because they have two or more double bonds. PUFAs are classified into ω3 (such as docosahexaenoic acid) and ω6 (such as arachidonic acid) depending on the position of the double bond in the hydrocarbon chain. Because phospholipids are composed of a combination of two fatty acids and a hydrophilic head, the types of phospholipids are numerous (Subra et al., 2007).

Interestingly, the ω3 and ω6 types are essential fatty acids, meaning that they cannot be synthesized by animal cells and must be ingested from diet. Because these fatty acids are taken up and metabolized into phospholipids, they are considered to have a significant impact on the physical properties of biological membranes.

To date, phospholipid classification based on polar groups has been widely used. This classification is determined through analytical techniques, such as thin-layer chromatography. However, because the fatty acids of phospholipids cannot be easily analyzed using thin-layer chromatography, their diversity had not been widely taken into consideration. This limitation was overcome with the development of mass spectrometry techniques (Aoyagi et al., 2017; Tsugawa et al., 2017).

4. The fluid mosaic model and the packing gaps—of the membrane

As described above, the membrane consists of a phospholipid bilayer, but it also contains carbohydrates and proteins. Proteins can be either membrane-embedded, if presenting a transmembrane region, or peripheral, i.e., the surface binding. Phospholipids do not have covalent bonds between them; therefore, the bilayer lipid membrane is considered to move freely. The same movement should be applicable to transmembrane proteins if they do not have any anchorage. This "fluid mosaic model" was first proposed by Singer and Nicholson (Singer & Nicolson, 1972) (Fig. 4.2). This model was supported by an experiment that showed the mixing of membrane components after fusion (Frye and Edidin, 1970).

The free movement of lipids in the bilayer induces spontaneous gaps or small pits in the membrane (Fig. 4.1). These stochastic gaps between phospholipid molecules are called "packing gaps" or "packing defects." This gap is related to the shape of the lipid itself as well as the membrane curvature. When the lipid membrane is deformed, the outer layer needs to increase its area compared to the inner layer. Therefore, the curvature of the lipid membrane increases the packing defect in one of the layers (Bigay & Antonny, 2012; McMahon & Gallop, 2005).

Phospholipids constituting the lipid membrane move randomly, as shown by the fluid mosaic model. Therefore, packing gaps can occur randomly, even in a steady state. The gap between lipid molecules is considered to occur between phospholipid fatty acids and can increase according to the degree of unsaturation. An unsaturation of fatty acid induces a kink that causes a fatty acid to occupy a larger space.

Cholesterol is inserted between fatty acid portion of the membrane and can regulate the size of the packing gap. Cholesterol forms a so-called raft in the lipid membrane by increasing membrane density and inducing phase separation (Korade & Kenworthy, 2008; Simons & Ehehalt, 2002). Because it is often difficult to directly visualize the one-molecule kinetics of lipid, the estimates of the packing gap are often achieved by molecular dynamics calculations (Berkowitz, 2009; Risselada & Marrink, 2008).

5. The erythrocyte membrane

The plasma membrane of erythrocytes is one of the well-studied membranes with their associated proteins. Because erythrocytes lack a nuclear membrane and most intracellular organelles, their plasma membrane proteins are intensively studied using the "ghost," obtained by removing intracellular components via osmotic rupture. The proteins found in the membrane of erythrocytes provided a model of the shaping of the membrane by the proteins. Erythrocyte membrane proteins include transporters, receptors, adhesion proteins, and membrane-associated peripheral proteins (Steck & Yu, 1973). The proteins of α-spectrin, β-spectrin, band 4.1, band 4.2, ankyrin, actin, adducin, tropomyosin, and tropomodulin are components of the cytoskeletal matrix responsible for the mechanical stability of erythrocytes. α and β spectrin are two subunits of spectrin proteins assembled in an antiparallel arrangement to produce a rod-like structure. These α/β spectrin dimers form tetramers that are 200 nm in length (Roberts et al., 2002). The spectrin tetramer interacts with the actin filament to form a junction complex that contributes to the actin cytoskeleton matrix beneath the membrane (Djinovic-Carugo et al., 2002; Machnicka et al., 2014). Moreover, ankyrin interacts with band 3, and protein 4.1 binds to glycophorin. These cytoskeletal protein networks are bound to membrane bilayer (Bennett, 2013; Hoock et al., 1997).

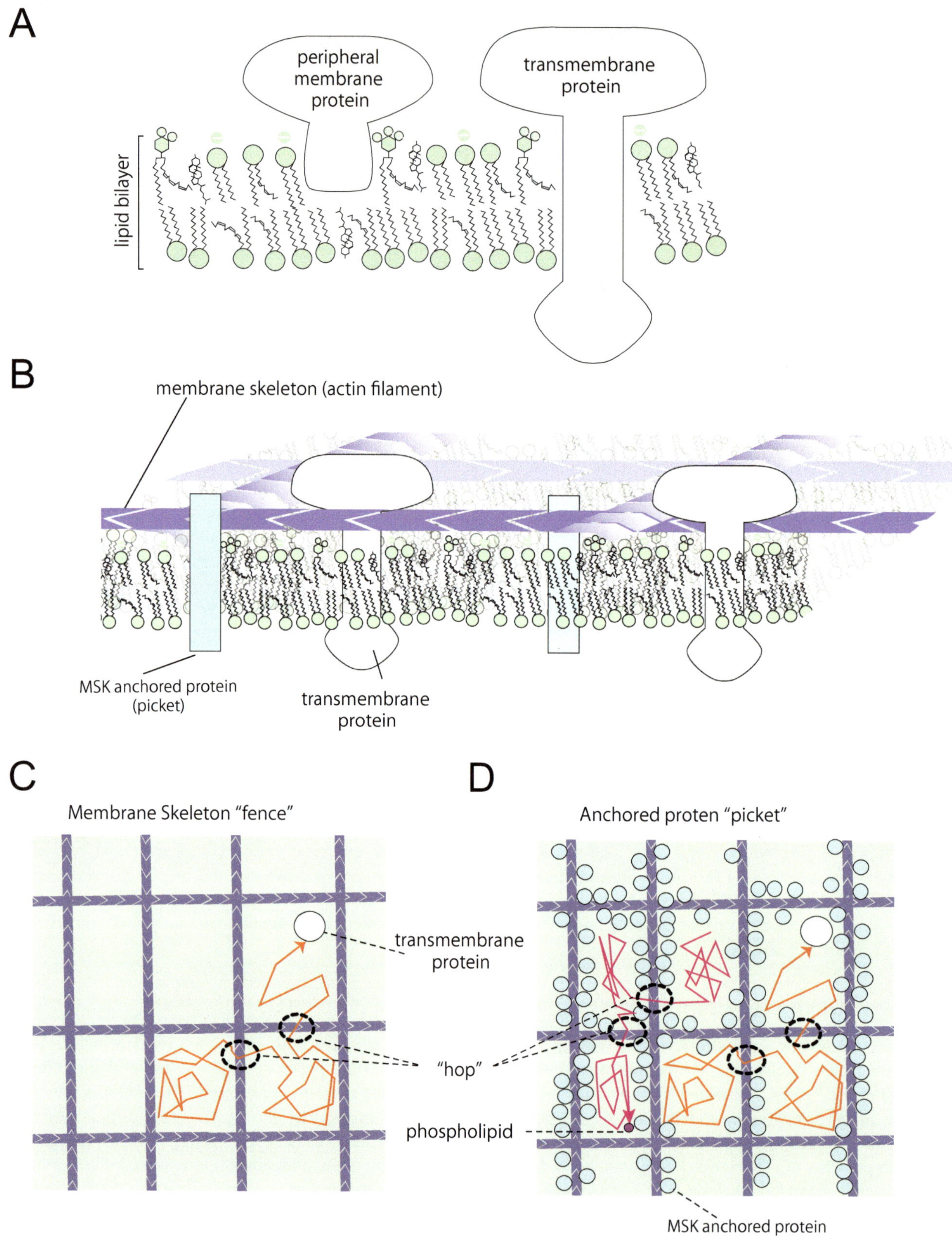

FIGURE 4.2 **The fluid mosaic and anchored picket fence models.** (A) The fluid mosaic model. The membrane cross-section with transmembrane proteins in the phospholipid bilayer is illustrated. The ionic and polar portions of the lipid heads are indicated with + and − symbols. (B−D) the anchored picket fence model. (B) An illustration of the plasma membrane showing the actin cytoskeleton, the diffusing transmembrane protein, and a transmembrane protein anchored to the actin cytoskeleton that acts as a picket (*blue*). (C) The actin cytoskeleton "fence" model (bottom view of the plasma membrane, i.e., the view from inside the cell) and (D) the anchored transmembrane protein "picket" model. In (C) and (D), trajectories of a transmembrane protein and a phospholipid are shown. In the fence and picket, these molecules diffuse freely, exactly as in the fluid mosaic model. *(A) Modified from (Singer & Nicolson, 1972), (B−D) Modified from Kusumi et al. (2012)*

The most abundant erythrocyte transmembrane proteins are band 3 (anion exchanger), aquaporin-1 (water channel), and glycophorin (Poole, 2000). Glycophorins contain several asparagine-N-linked oligosaccharides and some O-linked oligosaccharides that contain sialic acid attached to the serine or threonine residues (Tomita & Marchesi, 1975). The presence of high concentrations of sialic acid in the extracellular domain of glycophorin provides a substantial negative charge, which is important in limiting the interaction between erythrocytes and other cells in circulation (Varki et al., 2012).

Fluorescence recovery after photobleaching (FRAP) is a technique used to study the lateral mobility of lipids and proteins. FRAP studies in human erythrocytes revealed that a fraction of band 3 protein moves, while another is static if anchored to the cytoskeletal matrix proximal to the membrane (Steck, 1978).

6. The anchored picket fence model

The detailed organization of the cytoskeleton–membrane structure was first described in erythrocytes as described above. However, similar structures have been observed in eukaryotic cells, where the actin cytoskeleton is found beneath the plasma membrane. This architecture is key to the behavior of lipids and proteins in the membrane. Singer and Nicolson's two-dimensional fluid mosaic model is the most well-accepted model describing cell membranes (Singer & Nicolson, 1972). However, the diffusion coefficient of membrane proteins and lipids on cell membranes is very small, approximately 1/5–1/100, compared to that of liposome artificial membranes. For example, the diffusion coefficient of a protein complex is slower; the diffusion of a dimerized receptor upon ligand binding is greatly reduced compared to that of a monomeric receptor (Jacobson et al., 2019).

This issue was resolved with single-molecule tracking, attaching colloidal gold to lipids and proteins. Lipids and proteins diffuse freely, as in liposomes, in the small compartments defined by actin filaments and spectrin, as in erythrocytes. Actin filaments and spectrin are membrane-associated and, therefore, are supposed to function as a "membrane-skeleton fence" to be hopped by membrane-embedded proteins, such as receptors (Morone et al., 2006). Therefore, the need for hopping is thought to be the reason for the slow diffusion of these proteins within the membrane.

Lipids also diffuse slowly within the cell membrane as they are required to hop between compartments, but in small compartments, they can move freely, as predicted by the fluid mosaic model. In most cells, the actin cytoskeleton is associated with many transmembrane proteins, as in erythrocytes. These transmembrane proteins function as pickets that hinder the free diffusion of lipid molecules. This revised model, including transmembrane proteins, is called "the anchored picket fence model." A simulation study showed that if about 20%–30% of the compartment boundary, namely the actin and spectrin fence, is covered by membrane proteins, this can explain the behavior of lipids (Kusumi et al., 2012) (Fig. 4.2).

7. Proteins and sugars in plasma membrane shaping

Proteins bound to the lipid membrane are of two types: transmembrane and peripheral. Peripheral membrane-binding proteins do not have transmembrane domains and are localized on the membrane surface. These proteins can be present in both the cytoplasm and the lipid membrane surface in the cell, and they bind dynamically to lipid membranes in response to certain circumstances, such as changes in charge, local appearance of packing gaps, and morphological changes. It is likely that proteins bind differently to membranes with diverse lipid composition.

Peripheral membrane-binding proteins bind to membranes via their lipid binding domains and motifs; these may either recognize the polar group of the phospholipid, a packing defect in the lipid membrane, or the membrane curvature.

The biological membrane contains many negatively charged lipids, such as phosphatidylserine and phosphoinositide, allowing the binding of peptides or motifs containing many basic (positive) charged amino acids. For example, actin-remodeling WASP family proteins and small G proteins, especially K-Ras, have polybasic amino acid sequences for membrane binding (Bigay & Antonny, 2012; Suetsugu et al., 2014).

The typical protein domain that recognizes lipid polar groups is the pleckstrin homology (PH) domain, which binds specifically to the hydrophilic polar groups of phosphoinositides. Proteins with a PH domain are known to regulate protein localization and activity by binding to phosphoinositide (Suetsugu et al., 2014). Because of its binding to a lipid through a domain, the PH domain has not been considered to shape the membrane.

Two mechanisms are involved in the recognition of packing defects caused by membrane curvature: the half-insertion of an amphiphilic helix into the lipid membrane, and the Bin-amphiphysin-Rvs (BAR) domain, which generates and/or sense membrane curvature. The proteins having BAR domain often bind to WASP family proteins, cooperating in the remodeling of actin cytoskeleton for the membrane shaping.

8. Amphiphilichelix

Amphiphilic helices are α-helices with hydrophobic and hydrophilic surfaces. Some amphiphilic helices are not structured in solution but form α-helices upon membrane binding. The alpha helix forms one turn at approximately four amino acids but acquires amphiphilicity when hydrophilic and hydrophobic amino acids appear at a cycle of approximately four amino acids (Fig. 4.3). In solution, the amphiphilic helix does not always form a helical structure, but when it interacts with the lipid membrane it is inserted so that the α-helix lies on one side of the lipid bilayer (Fig. 4.3). When the amphiphilic helix is inserted into a single layer of the membrane bilayer, an enlargement in the area induces membrane deformation leading to either vesicle or tubule formation (Bigay & Antonny, 2012; Campelo et al., 2008; McMahon & Gallop, 2005).

Insertion is thought to depend on the packing defects of lipid molecules within the membrane (Boucrot et al., 2012; Drin & Antonny, 2010). In general, an increase in packing defects owing to polyunsaturated fatty tails promotes the binding of amphiphilic helices (Brummel et al., 2017; Hofbauer et al., 2018). However, their binding to densely packed membranes rich in saturated fatty acids is challenging (Long et al., 2010; Manneville et al., 2008) (Fig. 4.3). Amphiphilic helices are found in a variety of proteins, but a typical one is the ArfGAP1 lipid-packing sensor (ALPS) motif.

ArfGAP is involved in membrane transport in the endoplasmic reticulum and Golgi apparatus. These inner membrane systems are low in charge and cholesterol, but contain relatively high levels of phospholipids with unsaturated fatty acids (Bigay & Antonny, 2012). The ALPS motif has the property of an amphiphilic helix, but its hydrophilic surface contains

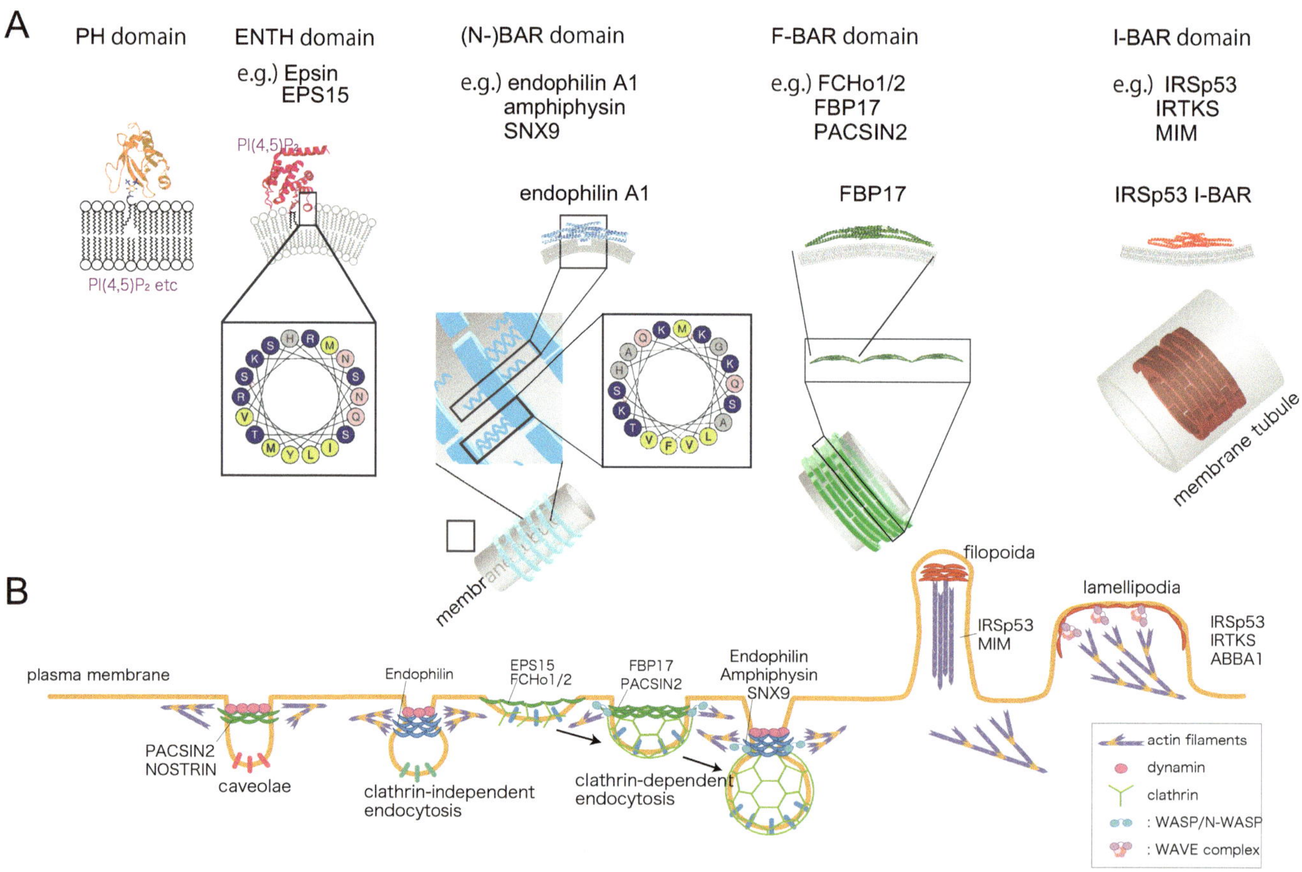

FIGURE 4.3 **Membrane-binding domains and the plasma membrane shape.** (A) The lipid-binding domain of superficial membrane-binding proteins that function primarily in cell membranes. The PH domain binds to a hydrophilic polar group, such as phosphoinositide. The ENTH domain has an amphiphilic helix and also binds to PI(4,5) P_2, which is a phosphoinositide. The BAR domain forms an α-helix bundle and binds to a lipid membrane having a curvature matching its conformation, or induces a membrane with that curvature. Some N-BAR domains also have amphiphilic helixes. The membrane curvature to which the I-BAR domain binds is convex, and the (N-)BAR or F-BAR domain is often concave. The amphiphilic helix becomes an α-helix when bound to the lipid membrane and is partially inserted into the lipid bilayer. This insertion depends on the packing defect between lipid molecules and thus binds in a curvature-dependent manner to the lipid membrane. A wheel diagram (helical wheel) of the amphiphilic helices of the ENTH domain and N-BAR domain were designed using HeliQuest (http://heliquest.ipmc.cnrs.fr). (B) Localization of membrane organelles of cell membranes and BAR domains, actin cytoskeletal regulators WASP, WAVE, and dynamin that cleaves lipid membranes.

large amounts of serine and threonine. In the absence of a lipid membrane, the helical structure is not formed; the ALPS motif forms an amphiphilic helix when bound and inserted into the lipid membrane (Vanni et al., 2013). It has been reported that the smaller the diameter of the lipid membrane (liposome), the stronger the binding of the ALPS motif. If the size of the lipid membrane is small, the membrane bending degree is higher and the gap in the packing of phospholipid molecules increases. The gap depends on the population of unsaturated fatty acid tails among the membrane lipids. In cells, ArfGAP1 has been reported to be localized in the Golgi apparatus, with an increase in oleic acid, an unsaturated fatty acid, in the phospholipid tail (Vanni et al., 2014).

Epsin is involved in the formation of clathrin-coated vesicles in cells (Itoh et al., 2001) and has an amphiphilic helix at the N-terminus of its ENTH domain. This helix is disordered in solution but contains many basic amino acid residues on its hydrophilic surface for the binding to the negatively charged lipids, including phosphatidylinositol-4,5-diphosphate (PI(4,5)P2) (Chen et al., 1998; Ford et al., 2002; Itoh et al., 2001). This interaction induces the cleavage of the lipid membrane, namely membrane vesicle formation (Boucrot et al., 2012), as the amphipathic helix wedges into the membrane. Moreover, epsin is involved in the formation of clathrin-coated vesicles from clathrin-coated pits (Ford et al., 2002).

Arf1 and Sar1p are low molecular weight GTPases. In the GTP-bound form, their amphiphilic helices are exposed (Lee et al., 2005; Paczkowski & Fromme, 2014) and inserted into the membrane, contributing to the formation of tubules and vesicles. In cells, Arf1 is involved in the formation of coat protein I (COPI) vesicles from the Golgi apparatus, and Sar1p is involved in the formation of coat protein II (COPII) from the ER (Krauss et al., 2008; Lee et al., 2005). Arf1 binds to phosphatidic acid (PA) at C18:0—C22, deforms liposomes containing PA (stearoyl-docosahexaenoyl: SDPA), and forms tubules with constrictions (Park et al., 2019; Randazzo, 1997). However, the fatty acid dependence of Sar1p is currently unknown.

9. The BAR domains

The Bin-Amphiphysin-Rvs (BAR) domain was identified for the first time as a protein with a three-dimensional structure matching the curvature of the lipid membrane (Suetsugu et al., 2014). The three-dimensional structure of each BAR domain is different each other, allowing the various lipid membrane curvatures to be accommodated by these diverse domains. The BAR domains are classified into three types: canonical, F-BAR, and I-BAR. Some canonical BAR domains also have N-terminal amphiphilic helices and are called "N-BAR domains" as a subtype of BAR domains (Masuda et al., 2006; Peter et al., 2004). The amphiphilic helix of the N-BAR domain is inserted into the lipid membrane (Mim et al. 2012) (Fig. 4.3). Hence, this domain is believed to cooperate with the amphiphilic helices to form the tube structure of the lipid membrane.

It is believed that all BAR domains form homodimers and that the three-dimensional homodimer structure matches the curvature of the membrane. The representative proteins with an N-BAR domain are amphiphysin and endophilin, those with an F-BAR domain are formin-binding protein17 (FBP17) and Cdc42-interacting protein 4 (CIP4), and those with an I-BAR domain include insulin receptor substrate of 53 kDa (IRSp53) and Missing in metastasis (MIM). The representative cellular functions of the BAR and F-BAR domain proteins include clathrin-mediated endocytosis and other invaginations, such as caveolae, whereas that of I-BAR domain proteins is the formation of filopodia or other protrusive structures (Suetsugu et al., 2014). Binding to membrane scission GTPase dynamin and actin regulatory proteins of the WASP family is a common feature of BAR domain-containing proteins (Suetsugu et al., 2014).

Most of the BAR and F-BAR domains have a crescent-shaped or banana-like structure and a concentration of positively charged amino acids on the concave surface (Peter et al., 2004; Shimada et al., 2007). Phospholipid molecules that make up the lipid membrane contain negatively charged phosphatidylserine and phosphoinositides. The BAR domain binds to negatively charged lipids including PS and phosphoinositides, at one surface of a three-dimensional structure rich in basic amino acid residues with positive charge. Therefore, the BAR domain can curve the negatively charged lipid membrane.

Another important property of the BAR domain is its multimer formation. Since the size of one dimer in the domain is smaller than that of lipid membrane vesicles with diameters of several hundred nanometers, it is considered insufficient to form a membrane structure. The BAR domain is thought to curve the membrane via ordered assembly into multimers. When purified proteins in the BAR or F-BAR domains are incubated with a reconstituted lipid membrane (liposome), they deform the liposomes into tubules on which the BAR and F-BAR domains form a spiral (Frost et al., 2008; Shimada et al., 2007) (Fig. 4.3). Furthermore, the shape of the protein surface of the BAR domain roughly coincides with the curvature of tubules. Therefore, there is a correlation between the shape of the membrane and structure of the BAR domains (Suetsugu et al., 2010).

Some BAR domains contain an amphiphilic helix at the N-terminus and are referred to as N-BAR domains (Peter et al., 2004). In addition to the tubular formation of lipid membranes, some N-BAR domains, such as amphiphysin and endophilin, form membrane vesicles in vitro (Boucrot et al., 2012; Gallop et al., 2005; Isas et al., 2015; Masuda et al., 2006; Peter et al., 2004). In particular, endophilin is thought to be involved in high-speed endophilin-mediated endocytosis (Boucrot et al., 2015). During this cleavage, endophilin binds and cooperates with dynamin, a GTPase for membrane scission (Kaksonen & Roux, 2018). In vitro membrane cleavage via the endophilin–dynamin complex is dependent on fatty acid tails (Pinot et al., 2014). In terms of membrane lipid composition, cholesterol is involved in PACSIN2-mediated membrane deformation for caveolae (Amir et al., 2020), but the role of cholesterol in other BAR domains is still unclear.

Some BAR domain structures have protruding wedges, which have been reported in the F-BAR domains of PACSIN2 and GAS7. The wedge of PACSIN2 is hydrophobic and is essential for binding to the membrane (Shimada et al., 2010). In contrast, the wedge in the F-BAR domain of GAS7 is hydrophilic, so it is not inserted into the membrane, and the loop of GAS7 contributes to oligomeric formation during phagocytosis (Hanawa-Suetsugu et al., 2019).

10. Ankyrin repeat domain (ARD)

Most ARDs are involved in protein–protein interactions, but some can modify lipid membrane conformation through their curved shape and protein three-dimensional structure (Li et al., 2006; Mosavi et al., 2004). ARDs with amphiphilic helices transform lipid membranes into tubules and vesicles (Kitamata et al., 2019; Wolf et al., 2019). The C-terminus of ANKHD1 contains ARD and an amphiphilic helix, and the N-terminal of Ankycorbin contains ARD and an amphiphilic helix (Kitamata et al., 2019; Wolf et al., 2019). ANKHD1 vesiculates the membrane of the early endosomes, and ankycorbin forms the dendritic branch protrusions of neurons (Kitamata et al., 2019; Wolf et al., 2019). The relatively small TRPV4 channel ARD binds to the membrane without causing deformation (Takahashi et al., 2014). However, canonical ARD-containing proteins, including ANK1, ANK2, ANK3, Ankyrin B, Ankyrin G, and Ankyrin R, have not been reported to bind to membranes.

11. Lectin-mediated lipid membrane deformation

Cell membrane deformation is also induced extracellularly. Lectins are proteins that bind to glycans of sugar-modified glycoproteins and glycolipids. Common glycolipids are sphingosaccharides, which include cerebrosides, gangliosides, and glycosylated ceramides. Polyvalent binding of lectins to sphingosaccharide lipids induces membrane curvature through lipid clustering. Such phenomena are observed in secretory and pathogen proteins.

Pathogenic *Escherichia coli*, in addition to the pathogenic bacteria *Shigella* and O157, causes diarrhea along with several other symptoms. These pathogens produce Shiga toxin B subunits (STxB) that bind to globotriaosylceramide (Gb3), a sphingosaccharide lipid on the extracellular membrane surface. STxB form a pentamer, creating clusters of Gb3 and deforming a cell membrane into invaginations (Romer et al., 2007). The unsaturated acyl chain of Gb3 is necessary for STxB infiltration (Watkins et al., 2014).

Simian virus 40 (SV40) is a DNA virus with a capsid and a cancer-inducing polyomavirus. The capsid contains VP1, a pentameric lectin that recognizes sphingolipids. VP1 binds specifically to GM1 (monosialotetrahexosylganglioside) and can induce the invagination of reconstituted liposome membranes depending on the fatty acid tails (Ewers et al., 2010).

Vibrio cholerae produces the cholera toxin that causes the cholera disease. The cholera toxin B subunit (CTxB) is another example of GM1-bound lectin. CTxB has five GM1 binding sites and presents a similarity to the SV40 VP1 structure. CTxB also induces liposomal membrane invagination (Ewers et al., 2010). Interestingly, CTxB cannot invade hard cell membranes that contain saturated fatty acids (Ewers et al., 2010), as invagination of the cell membrane is an essential step in host cell invasion.

Galectins are produced by mammalian cells and are secreted extracellularly (Johannes et al., 2018). They are involved in many physiological functions including immune responses, inflammation, cell migration, and autophagy. Galectins are contained in various glycolipids, including GM1, and have a high affinity for β-galactosides, namely glycosides containing galactose. Among galectins, galectin-3 invades liposomes containing sphingosaccharide lipids and is involved in clathrin-independent endocytosis of β1 integrin and CD44 (Lakshminarayan et al., 2014). Galectin-8 also induces clathrin-independent endocytosis of CD166 (Renard et al., 2020).

12. Concluding remarks

The structure of the biological membrane and the peripheral membrane binding proteins responsible for its shaping are described in this chapter. In recent years, the importance of fatty acids in metabolic research, such as cancer research, has been recognized. Fatty acids are directly metabolized into the phospholipids that make up membranes; however, the role of the variation of fatty acid composition in cellular membrane is still largely unknown. This could be revealed by clarifying how certain proteins recognize fatty acids specifically. At present, such proteins are unidentified. Instead, there are proteins that can recognize the gap between phospholipids through fatty acid tail variations. Such lipid variations are not easily visible, and probably many of these are still undiscovered.

References

Amir, A. G., Takemura, K., Hanawa-Suetsugu, K., Oono-Yakura, K., Yasuhara, K., Kitao, A., & Suetsugu, S. (2020). Regulation of caveolae through cholesterol-depletion dependent tubulation by PACSIN2/Syndapin II. *bioRxiv*. https://doi.org/10.1101/2020.03.25.008854, 2020.2003.2025.008854.

Antonny, B., Vanni, S., Shindou, H., & Ferreira, T. (2015). From zero to six double bonds: Phospholipid unsaturation and organelle function. *Trends in Cell Biology, 25*, 427–436. https://doi.org/10.1016/j.tcb.2015.03.004

Aoyagi, R., Ikeda, K., Isobe, Y., & Arita, M. (2017). Comprehensive analyses of oxidized phospholipids using a measured MS/MS spectra library. *Journal of Lipid Research, 58*, 2229–2237. https://doi.org/10.1194/jlr.D077123

Bennett, V. (2013). *Functional organization of vertebrate plasma membrane*. Elsevier.

Berkowitz, M. L. (2009). Detailed molecular dynamics simulations of model biological membranes containing cholesterol. *Biochimica et Biophysica Acta, 1788*, 86–96. https://doi.org/10.1016/j.bbamem.2008.09.009

Bigay, J., & Antonny, B. (2012). Curvature, lipid packing, and electrostatics of membrane organelles: Defining cellular territories in determining specificity. *Developmental Cell, 23*, 886–895. https://doi.org/10.1016/j.devcel.2012.10.009

Boucrot, E., Ferreira, A. P., Almeida-Souza, L., Debard, S., Vallis, Y., Howard, G., Bertot, L., Sauvonnet, N., & McMahon, H. T. (2015). Endophilin marks and controls a clathrin-independent endocytic pathway. *Nature, 517*, 460–465. https://doi.org/10.1038/nature14067

Boucrot, E., Pick, A., Çamdere, G., Liska, N., Evergren, E., McMahon, H. T., & Kozlov, M. M. (2012). Membrane fission is promoted by insertion of amphipathic helices and is restricted by crescent BAR domains. *Cell, 149*, 124–136. https://doi.org/10.1016/j.cell.2012.01.047

Brummel, B. E., Braun, A. R., & Sachs, J. N. (2017). Polyunsaturated chains in asymmetric lipids disorder raft mixtures and preferentially associate with alpha-Synuclein. *Biochimica et Biophysica Acta (BBA) - Biomembranes, 1859*, 529–536. https://doi.org/10.1016/j.bbamem.2016.10.006

Campelo, F., McMahon, H. T., & Kozlov, M. M. (2008). The hydrophobic insertion mechanism of membrane curvature generation by proteins. *Biophysical Journal, 95*, 2325–2339. https://doi.org/10.1529/biophysj.108.133173

Chen, H., Fre, S., Slepnev, V. I., Capua, M. R., Takei, K., Butler, M. H., Di Fiore, P. P., & De Camilli, P. (1998). Epsin is an EH-domain-binding protein implicated in clathrin-mediated endocytosis. *Nature, 394*, 793–797. https://doi.org/10.1038/29555

Djinovic-Carugo, K., Gautel, M., Ylanne, J., & Young, P. (2002). The spectrin repeat: A structural platform for cytoskeletal protein assemblies. *FEBS Letters, 513*, 119–123. https://doi.org/10.1016/s0014-5793(01)03304-x. Pii s0014-5793(01)03304-x.

Drin, G., & Antonny, B. (2010). Amphipathic helices and membrane curvature. *FEBS Letters, 584*, 1840–1847. https://doi.org/10.1016/j.febslet.2009.10.022

Ewers, H., Romer, W., Smith, A. E., Bacia, K., Dmitrieff, S., Chai, W., Mancini, R., Kartenbeck, J., Chambon, V., Berland, L., et al. (2010). GM1 structure determines SV40-induced membrane invagination and infection. *Nature Cell Biology, 12*, 11–18. https://doi.org/10.1038/ncb1999

Ford, M. G., Mills, I. G., Peter, B. J., Vallis, Y., Praefcke, G. J., Evans, P. R., & McMahon, H. T. (2002). Curvature of clathrin-coated pits driven by epsin. *Nature, 419*, 361–366. https://doi.org/10.1038/nature01020

Frost, A., Perera, R., Roux, A., Spasov, K., Destaing, O., Egelman, E. H., De Camilli, P., & Unger, V. M. (2008). Structural basis of membrane invagination by F-BAR domains. *Cell, 132*, 807–817.

Frye, L. D., & Edidin, M. (1970). The rapid intermixing of cell surface antigens after formation of mouse-human heterokaryons. *Journal of Cell Science, 7*, 319–335.

Gallop, J. L., Butler, P. J., & McMahon, H. T. (2005). Endophilin and CtBP/BARS are not acyl transferases in endocytosis or Golgi fission. *Nature, 438*, 675–678. https://doi.org/10.1038/nature04136

Hanawa-Suetsugu, K., Itoh, Y., Ab Fatah, M., Nishimura, T., Takemura, K., Takeshita, K., Kubota, S., Miyazaki, N., Wan Mohamad Noor, W. N. I., Inaba, T., et al. (2019). Phagocytosis is mediated by two-dimensional assemblies of the F-BAR protein GAS7. *Nature Communications, 10*, 4763. https://doi.org/10.1038/s41467-019-12738-w

Hofbauer, H. F., Gecht, M., Fischer, S. C., Seybert, A., Frangakis, A. S., Stelzer, E. H. K., Covino, R., Hummer, G., & Ernst, R. (2018). The molecular recognition of phosphatidic acid by an amphipathic helix in Opi1. *The Journal of Cell Biology, 217*, 3109–3126. https://doi.org/10.1083/jcb.201802027

Hoock, T. C., Peters, L. L., & Lux, S. E. (1997). Isoforms of ankyrin-3 that lack the NH2-terminal repeats associate with mouse macrophage lysosomes. *Journal of Cell Biology, 136*, 1059–1070. https://doi.org/10.1083/jcb.136.5.1059

Isas, J. M., Ambroso, M. R., Hegde, P. B., Langen, J., & Langen, R. (2015). Tubulation by amphiphysin requires concentration-dependent switching from wedging to scaffolding. *Structure, 23*, 873–881. https://doi.org/10.1016/j.str.2015.02.014

Itoh, T., Koshiba, S., Kigawa, T., Kikuchi, A., Yokoyama, S., & Takenawa, T. (2001). Role of the ENTH domain in phosphatidylinositol-4,5-bisphosphate binding and endocytosis. *Science, 291*, 1047–1051. https://doi.org/10.1126/science.291.5506.1047

Jacobson, K., Liu, P., & Lagerholm, B. C. (2019). The lateral organization and mobility of plasma membrane components. *Cell, 177*, 806–819. https://doi.org/10.1016/j.cell.2019.04.018

Johannes, L., Jacob, R., & Leffler, H. (2018). Galectins at a glance. *Journal of Cell Science, 131*. https://doi.org/10.1242/jcs.208884

Kaksonen, M., & Roux, A. (2018). Mechanisms of clathrin-mediated endocytosis. *Nature Reviews Molecular Cell Biology, 19*, 313–326. https://doi.org/10.1038/nrm.2017.132

Kitamata, M., Hanawa-Suetsugu, K., Maruyama, K., & Suetsugu, S. (2019). Membrane-deformation Ability of ANKHD1 is involved in the early endosome enlargement. *iScience, 17*, 101–118. https://doi.org/10.1016/j.isci.2019.06.020

Korade, Z., & Kenworthy, A. K. (2008). Lipid rafts, cholesterol, and the brain. *Neuropharmacology, 55*, 1265–1273. https://doi.org/10.1016/j.neuropharm.2008.02.019

Krauss, M., Jia, J. Y., Roux, A., Beck, R., Wieland, F. T., De Camilli, P., & Haucke, V. (2008). Arf1-GTP-induced tubule formation suggests a function of Arf family proteins in curvature acquisition at sites of vesicle budding. *Journal of Biological Chemistry, 283*, 27717–27723. https://doi.org/10.1074/jbc.M804528200

Kusumi, A., Fujiwara, T. K., Chadda, R., Xie, M., Tsunoyama, T. A., Kalay, Z., Kasai, R. S., & Suzuki, K. G. (2012). Dynamic organizing principles of the plasma membrane that regulate signal transduction: Commemorating the fortieth anniversary of singer and nicolson's fluid-mosaic model. *Annual Review of Cell and Developmental Biology, 28*, 215–250. https://doi.org/10.1146/annurev-cellbio-100809-151736

Lakshminarayan, R., Wunder, C., Becken, U., Howes, M. T., Benzing, C., Arumugam, S., Sales, S., Ariotti, N., Chambon, V., Lamaze, C., et al. (2014). Galectin-3 drives glycosphingolipid-dependent biogenesis of clathrin-independent carriers. *Nature Cell Biology, 16*, 595–606. https://doi.org/10.1038/ncb2970

Lee, M. C., Orci, L., Hamamoto, S., Futai, E., Ravazzola, M., & Schekman, R. (2005). Sar1p N-terminal helix initiates membrane curvature and completes the fission of a COPII vesicle. *Cell, 122*, 605–617. https://doi.org/10.1016/j.cell.2005.07.025

Li, J., Mahajan, A., & Tsai, M. D. (2006). Ankyrin repeat: A unique motif mediating protein-protein interactions. *Biochemistry, 45*, 15168–15178. https://doi.org/10.1021/bi062188q

Long, K. R., Yamamoto, Y., Baker, A. L., Watkins, S. C., Coyne, C. B., Conway, J. F., & Aridor, M. (2010). Sar1 assembly regulates membrane constriction and ER export. *The Journal of Cell Biology, 190*, 115–128. https://doi.org/10.1083/jcb.201004132

Machnicka, B., Czogalla, A., Hryniewicz-Jankowska, A., Bogusławska, D. M., Grochowalska, R., Heger, E., & Sikorski, A. F. (2014). Spectrins: A structural platform for stabilization and activation of membrane channels, receptors and transporters. *Biochimica et Biophysica Acta, 1838*, 620–634. https://doi.org/10.1016/j.bbamem.2013.05.002

Manneville, J. B., Casella, J. F., Ambroggio, E., Gounon, P., Bertherat, J., Bassereau, P., Cartaud, J., Antonny, B., & Goud, B. (2008). COPI coat assembly occurs on liquid-disordered domains and the associated membrane deformations are limited by membrane tension. *Proceedings of the National Academy of Sciences of the United States of America, 105*, 16946–16951. https://doi.org/10.1073/pnas.0807102105

Marszalek, J. R., & Lodish, H. F. (2005). Docosahexaenoic acid, fatty acid-interacting proteins, and neuronal function: Breastmilk and fish are good for you. *Annual Review of Cell and Developmental Biology, 21*, 633–657. https://doi.org/10.1146/annurev.cellbio.21.122303.120624

Masuda, M., Takeda, S., Sone, M., Ohki, T., Mori, H., Kamioka, Y., & Mochizuki, N. (2006). Endophilin BAR domain drives membrane curvature by two newly identified structure-based mechanisms. *EMBO Journal, 25*, 2889–2897.

McMahon, H. T., & Gallop, J. L. (2005). Membrane curvature and mechanisms of dynamic cell membrane remodelling. *Nature, 438*, 590–596. https://doi.org/10.1038/nature04396

Mim, C., Cui, H., Gawronski-Salerno, J. A., Frost, A., Lyman, E., Voth, G. A., & Unger, V. M. (2012). Structural basis of membrane bending by the N-BAR protein endophilin. *Cell, 149*, 137–145. https://doi.org/10.1016/j.cell.2012.01.048

Morone, N., Fujiwara, T., Murase, K., Kasai, R. S., Ike, H., Yuasa, S., Usukura, J., & Kusumi, A. (2006). Three-dimensional reconstruction of the membrane skeleton at the plasma membrane interface by electron tomography. *The Journal of Cell Biology, 174*, 851–862. https://doi.org/10.1083/jcb.200606007

Mosavi, L. K., Cammett, T. J., Desrosiers, D. C., & Peng, Z. Y. (2004). The ankyrin repeat as molecular architecture for protein recognition. *Protein Science: A Publication of the Protein Society, 13*, 1435–1448. https://doi.org/10.1110/ps.03554604

Paczkowski, J. E., & Fromme, J. C. (2014). Structural basis for membrane binding and remodeling by the exomer secretory vesicle cargo adaptor. *Developmental Cell, 30*, 610–624. https://doi.org/10.1016/j.devcel.2014.07.014

Park, S. Y., Yang, J. S., Li, Z., Deng, P., Zhu, X., Young, D., Ericsson, M., Andringa, R. L. H., Minnaard, A. J., Zhu, C., et al. (2019). The late stage of COPI vesicle fission requires shorter forms of phosphatidic acid and diacylglycerol. *Nature Communications, 10*, 3409. https://doi.org/10.1038/s41467-019-11324-4

Peter, B. J., Kent, H. M., Mills, I. G., Vallis, Y., Butler, P. J., Evans, P. R., & McMahon, H. T. (2004). BAR domains as sensors of membrane curvature: The amphiphysin BAR structure. *Science, 303*, 495–499. https://doi.org/10.1126/science.1092586

Pinot, M., Vanni, S., Pagnotta, S., Lacas-Gervais, S., Payet, L. A., Ferreira, T., Gautier, R., Goud, B., Antonny, B., & Barelli, H. (2014). Lipid cell biology. Polyunsaturated phospholipids facilitate membrane deformation and fission by endocytic proteins. *Science, 345*, 693–697. https://doi.org/10.1126/science.1255288

Poole, J. (2000). Red cell antigens on band 3 and glycophorin A. *Blood Reviews, 14*, 31–43. https://doi.org/10.1054/blre.1999.0124

Randazzo, P. A. (1997). Functional interaction of ADP-ribosylation factor 1 with phosphatidylinositol 4,5-bisphosphate. *Journal of Biological Chemistry, 272*, 7688–7692.

Renard, H. F., Tyckaert, F., Lo Giudice, C., Hirsch, T., Valades-Cruz, C. A., Lemaigre, C., Shafaq-Zadah, M., Wunder, C., Wattiez, R., Johannes, L., et al. (2020). Endophilin-A3 and Galectin-8 control the clathrin-independent endocytosis of CD166. *Nature Communications, 11*, 1457. https://doi.org/10.1038/s41467-020-15303-y

Risselada, H. J., & Marrink, S. J. (2008). The molecular face of lipid rafts in model membranes. *Proceedings of the National Academy of Sciences of the United States of America, 105*, 17367–17372. https://doi.org/10.1073/pnas.0807527105

Roberts, K., Alberts, B., Johnson, A., Walter, P., & Hunt, T. (2002). *Molecular biology of the cell* (Vol 32). New York: Garland Science.

Robertson, J. L. (2018). The lipid bilayer membrane and its protein constituents. *Journal of General Physiology, 150*, 1472–1483. https://doi.org/10.1085/jgp.201812153

Romer, W., Berland, L., Chambon, V., Gaus, K., Windschiegl, B., Tenza, D., Aly, M. R., Fraisier, V., Florent, J. C., Perrais, D., et al. (2007). Shiga toxin induces tubular membrane invaginations for its uptake into cells. *Nature, 450*, 670–675. https://doi.org/10.1038/nature05996

Shimada, A., Niwa, H., Tsujita, K., Suetsugu, S., Nitta, K., Hanawa-Suetsugu, K., Akasaka, R., Nishino, Y., Toyama, M., Chen, L., et al. (2007). Curved EFC/F-BAR-domain dimers are joined end to end into a filament for membrane invagination in endocytosis. *Cell, 129*, 761–772. https://doi.org/10.1016/j.cell.2007.03.040

Shimada, A., Takano, K., Shirouzu, M., Hanawa-Suetsugu, K., Terada, T., Toyooka, K., Umehara, T., Yamamoto, M., Yokoyama, S., & Suetsugu, S. (2010). Mapping of the basic amino-acid residues responsible for tubulation and cellular protrusion by the EFC/F-BAR domain of pacsin2/Syndapin II. *FEBS Letters, 584*, 1111–1118. https://doi.org/10.1016/j.febslet.2010.02.058

Simons, K., & Ehehalt, R. (2002). Cholesterol, lipid rafts, and disease. *The Journal of Clinical Investigation, 110*, 597–603. https://doi.org/10.1172/jci16390

Singer, S. J., & Nicolson, G. L. (1972). The fluid mosaic model of the structure of cell membranes. *Science, 175*, 720–731.

Steck, T. L. (1978). The band 3 protein of the human red cell membrane: A review. *Journal of Supramolecular Structure, 8*, 311–324.

Steck, T. L., & Yu, J. (1973). Selective solubilization of proteins from red blood cell membranes by protein perturbants. *Journal of Supramolecular Structure, 1*, 220–232.

Subra, C., Laulagnier, K., Perret, B., & Record, M. (2007). Exosome lipidomics unravels lipid sorting at the level of multivesicular bodies. *Biochimie, 89*, 205–212. https://doi.org/10.1016/j.biochi.2006.10.014

Suetsugu, S., Kurisu, S., & Takenawa, T. (2014). Dynamic shaping of cellular membranes by phospholipids and membrane-deforming proteins. *Physiological Reviews, 94*, 1219–1248. https://doi.org/10.1152/physrev.00040.2013

Suetsugu, S., Toyooka, K., & Senju, Y. (2010). Subcellular membrane curvature mediated by the BAR domain superfamily proteins. *Seminars in Cell & Developmental Biology, 21*, 340–349. https://doi.org/10.1016/j.semcdb.2009.12.002. S1084-9521(09)00247-X [pii].

Takahashi, N., Hamada-Nakahara, S., Itoh, Y., Takemura, K., Shimada, A., Ueda, Y., Kitamata, M., Matsuoka, R., Hanawa-Suetsugu, K., Senju, Y., et al. (2014). TRPV4 channel activity is modulated by direct interaction of the ankyrin domain to PI(4,5)P_2. *Nature Communications, 5*, 4994. https://doi.org/10.1038/ncomms5994

Tomita, M., & Marchesi, V. T. (1975). AMINO-ACID sequence and oligosaccharide attachment sites of human erythrocyte glycophorin. *Proceedings of the National Academy of Sciences of the United States of America, 72*, 2964–2968. https://doi.org/10.1073/pnas.72.8.2964

Tsugawa, H., Ikeda, K., Tanaka, W., Senoo, Y., Arita, M., & Arita, M. (2017). Comprehensive identification of sphingolipid species by in silico retention time and tandem mass spectral library. *Journal of Cheminformatics, 9*, 19. https://doi.org/10.1186/s13321-017-0205-3

van Meer, G., & de Kroon, A. I. (2011). Lipid map of the mammalian cell. *Journal of Cell Science, 124*, 5–8. https://doi.org/10.1242/jcs.071233

van Meer, G., Voelker, D. R., & Feigenson, G. W. (2008). Membrane lipids: Where they are and how they behave. *Nature Reviews Molecular Cell Biology, 9*, 112–124. https://doi.org/10.1038/nrm2330

Vanni, S., Hirose, H., Barelli, H., Antonny, B., & Gautier, R. (2014). A sub-nanometre view of how membrane curvature and composition modulate lipid packing and protein recruitment. *Nature Communications, 5*, 4916. https://doi.org/10.1038/ncomms5916

Vanni, S., Vamparys, L., Gautier, R., Drin, G., Etchebest, C., Fuchs, P. F., & Antonny, B. (2013). Amphipathic lipid packing sensor motifs: Probing bilayer defects with hydrophobic residues. *Biophysical Journal, 104*, 575–584. https://doi.org/10.1016/j.bpj.2012.11.3837

Varki, A., Gagneux, P., & Annals, N. Y. A. S. (2012). Multifarious roles of sialic acids in immunity. *Glycobiology of the Immune Response, 1253*, 16–36. https://doi.org/10.1111/j.1749-6632.2012.06517.x

Watkins, E. B., Gao, H., Dennison, A. J. C., Chopin, N., Struth, B., Arnold, T., Florent, J. C., & Johannes, L. (2014). Carbohydrate conformation and lipid condensation in monolayers containing glycosphingolipid Gb3: Influence of acyl chain structure. *Biophysical Journal, 107*, 1146–1155. https://doi.org/10.1016/j.bpj.2014.07.023

Wolf, D., Hofbrucker-MacKenzie, S. A., Izadi, M., Seemann, E., Steiniger, F., Schwintzer, L., Koch, D., Kessels, M. M., & Qualmann, B. (2019). Ankyrin repeat-containing N-Ank proteins shape cellular membranes. *Nature Cell Biology, 21*, 1191–1205. https://doi.org/10.1038/s41556-019-0381-7

Chapter 5

Bacterial and archaeal cells

Dennis W. Grogan
Department of Biological Sciences, University of Cincinnati, Cincinnati, OH, United States

1. The diversity of unicellular life

The cells of an animal or plant are genetically equivalent but differentiate during development to perform distinct functions. This functional specialization results in cell types distinguished from each other by various prominent physiological properties (contraction, sensing, signal propagation, photosynthesis, etc.). The organism, in turn, supports each type of cell with the resources and environmental conditions it needs. Cells of most microbial species, in contrast, remain essentially undifferentiated, functioning as populations of largely autonomous individuals. Each microbial cell must survive in an environment that is typically heterogeneous, stressful, and changing. From this environment, it must extract all the nutrients and energy needed for growth and reproduction while defending itself against various threats.

The autonomy of microbial cells and the challenges posed by their environments demand metabolic versatility and innovation, yet microbial body plans, genomes, and reproduction are remarkably simple compared to those of plants and animals. This simplicity and other lines of evidence support the idea that the first organisms were unicellular, yet extant microorganisms cannot be considered primitive. On the contrary, considering that unicellular organisms predate multicellular life by billions of years, maintain much larger populations, and reproduce much more quickly, it follows that extant microorganisms are more extensively and thoroughly evolved at the cellular and molecular levels than plant and animals are. Diversification within the unicellular context results in species which seem superficially similar (in being small and simple), but which span a tremendous spectrum of physiological properties, reflecting a long history of functional specialization and optimization.

The functional divergence of unicellular life can be seen qualitatively in the wide range of habitats colonized by microorganisms, and the metabolic diversity that this requires. In addition to populating all the environments that host plants and animals, microorganisms thrive under conditions that exclude plant and animal life. This ecological diversity reflects a corresponding diversity of metabolic capabilities. These include CO_2 fixation, anaerobic respiration, nitrogen fixation, phototrophy, and chemolithotrophy, and many of the underlying biochemical pathways are not represented among multicellular organisms. Because they help define the cellular energetics and ecology of microorganisms, metabolic processes are considered to be important aspects of microbial physiology.

Microbial diversity can also be quantified, as illustrated by comparative analysis of conserved macromolecules. Analyzing the sequences of small-subunit ribosomal RNAs, for example, shows that all cellular life is represented by three distinct lines of descent termed Domains: Eukarya, Bacteria, and Archaea (Woese et al., 1990). Within the Domain Eukarya, representing all eukaryotes, most of the major lineages are unicellular, and these unicellular lineages encompass greater molecular diversity than that of fungi, plants, and animals combined (Fig. 5.1). The other two Domains, Bacteria and Archaea (Woese et al., 1990), consist solely of unicellular organisms. This pattern indicates that the earliest organisms were unicellular and that the lineages which remained unicellular have continued to diversify at high rates.

Although unicellular organisms generally reproduce clonally, exchange of DNA contributes to their diversification. Phylogenetic analysis applied to complete microbial genomes shows that genes can be transferred rather indiscriminately among lineages that often are unrelated. This process, commonly called horizontal gene transfer, is erratic, but has been found to have contributed to essentially every microbial species (Ochman et al. 2000). Comparative analysis of other cellular components has also revealed that important components of eukaryotic cells, including organelles, have bacterial origins. In particular, the **mitochondrion** and the **chloroplast** each represent endosymbiotic acquisitions of bacteria by

Cell Physiology Sourcebook. https://doi.org/10.1016/B978-0-12-811114-7.00004-9

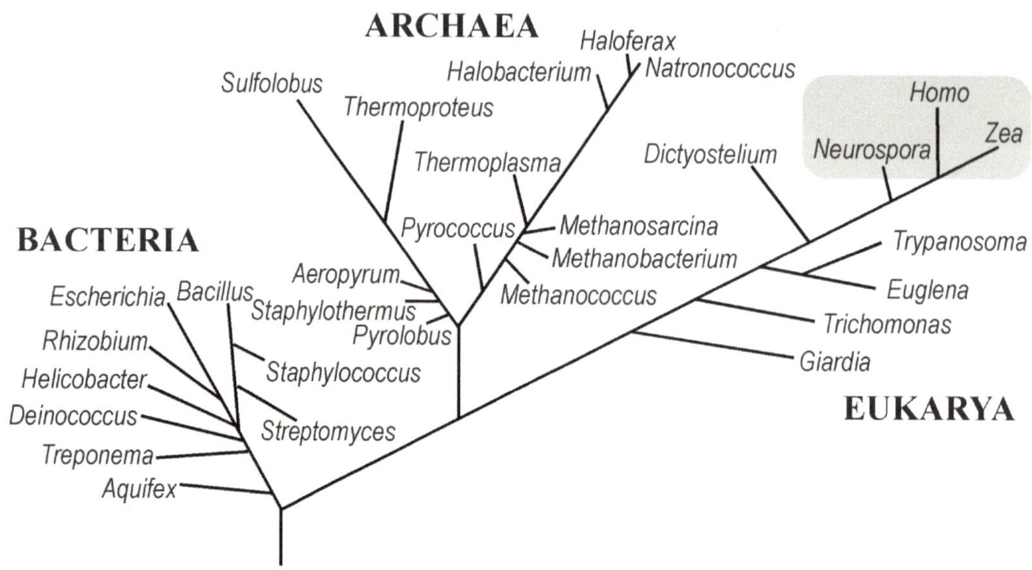

FIGURE 5.1 Phylogenetic relatedness of cellular organisms. Major lineages are indicated by genus names. The lengths of branches indicate the molecular divergence between modern genera, as defined by the sequences of small-subunit ribosomal RNAs (16S rRNAs of bacteria and archaea, 18S rRNAs of eukarya). The *gray box* illustrates the corresponding molecular diversity encompassed by multicellular organisms. *See Pace, 1997.*

progenitors of modern eukaryotic cells (Archibald, 2015). This relationship provides a context for understanding molecular structure and function of both the eukaryotic organelle and bacterial cells.

2. Cytology of bacteria and archaea

2.1 Basic cellular structure

Bacteria and archaea (singular: archaeon, or less commonly, archaeum) have the smallest and simplest cells known. These cells usually measure about 1 μm in diameter, and thus have only about 0.1% of the volume and mass of typical eukaryotic cells. In particular, bacterial and archaeal cells lack the nuclei and other membrane-enclosed organelles found in eukaryotic cells, and historically have been designated "prokaryotic", although the term "akaryotic" is arguably more accurate (Forterre, 2013) and is used in this chapter. This extremely small size limits the structural complexity of these cells and makes them difficult to visualize by conventional light microscopy (Fig. 5.2). Historiclly, most of the information regarding the internal organization of akaryotic cells, which can be represented by four structural types, as summarized in Fig. 5.3, has come from transmission electron microscopy.

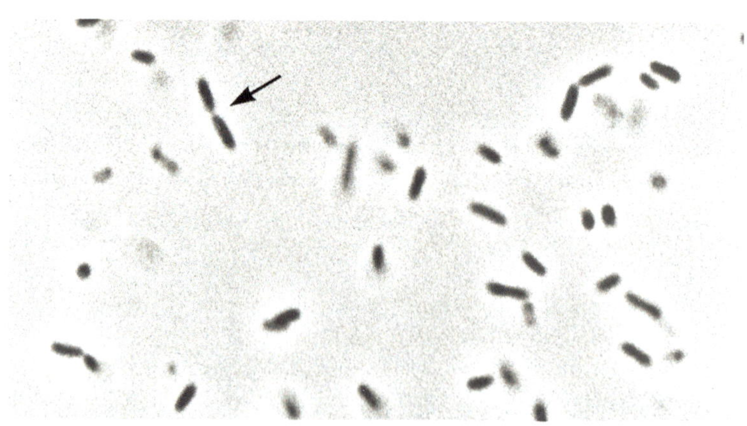

FIGURE 5.2 Light microscopy of bacteria. Live *Escherichia coli* cells suspended in growth medium are shown under a phase-contrast microscope at high magnification. The average width of the cells is about 0.5 m; arrowheads show the site of septum formation and cell separation at division. The micrograph illustrates the limited structural detail revealed by light microscopy of akaryotic cells.

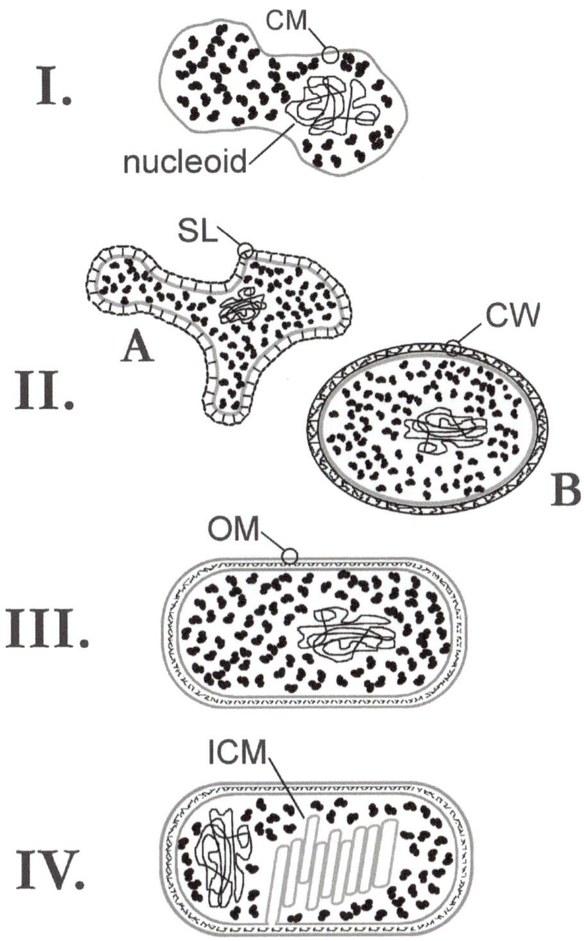

FIGURE 5.3 Common structural types (body plans) of akaryotic cells. I: The minimal cell includes only a nucleoid, cytoplasm, and cytoplasmic membrane (CM). II. Most archaea (A) have an additional S-layer (SL), composed of glycoprotein subunits, providing external structural support for the cell membrane, whereas Gram-positive bacteria (B) have a thick cell wall (CW) composed of peptidoglycan. III. A Gram-negative bacterium has a thin cell wall sandwiched between the CM and a second diffusional barrier, the outer membrane (OM). IV. Many Gram-negative photosynthetic bacteria produce a system of intracytoplasmic membranes (ICM). These contain specialized light-harvesting pigments and may be topologically continuous with each other and the CM.

The simplest type of akaryotic cell (Fig. 5.3, panel I) has only a **nucleoid**, a **cytoplasm**, and a **cytoplasmic membrane**, each of which is described in more detail below. This minimal cell plan is relatively rare in nature; it is represented by bacteria of the genus *Mycoplasma*, some of which are pathogens, and archaea of the genus *Thermoplasma*, which colonize geothermally heated acidic soils (Kletzin, 2007). Lack of external walls makes these cells relatively vulnerable to osmotic stress (see Section 2.5). However, they have molecular features not common in other akaryotic cells which provide some compensating structural reinforcement. These include sterols and lipoglycans in the cytoplasmic membrane and some form of cytoskeleton (Kornspan & Rottem, 2012; Nakane & Miyata, 2009; Smith, 1984).

2.2 Nucleoid

Most bacteria and archaea have a single circular DNA of several million base-pairs that serves as a chromosome. Although some species maintain multiple copies of this circular DNA per cell, the copies are identical, so that akaryotic cells are genetically haploid. In cytological terms, the typical akaryotic chromosome is neither contained within a nuclear membrane nor evenly dispersed throughout the cytoplasm, but is instead localized to a convoluted region near the center of the cell, called the **nuclear region** or **nucleoid** (Fig. 5.3, panel I). Cations, primarily Mg^{++}, spermidine, and similar polyamines, and small, basic proteins contribute to this condensation, which the nucleoid maintains despite ongoing replication, transcription, and repair. In *Escherichia coli*, longer-range organization of the circular chromosome takes the form of 50–100 topologically constrained domains or "loops", which appear to be constrained by protein complexes. Lysis of cells under appropriate conditions (nonionic detergent and high salt concentration) releases nucleoids in a relatively intact form and allows the loops to be visualized by electron microscopy (Pettijohn, 1996).

2.3 Cytoplasm

The remaining components of the cell interior are collectively termed the **cytoplasm,** or **cytosol**, which represents a very concentrated mixture of soluble proteins, RNAs, and metabolites. The cytosol of an *E. coli* cell adapted to growth in simple glucose medium at 37°C contains about 19,000 ribosomes, which, together with transfer RNA and various translation factors, account for nearly half of the cell's dry weight. This large investment of cellular resources in the machinery of protein synthesis enables the *E. coli* cell to reproduce itself every 40 min under these conditions (Neidhardt et al., 1990). Supplementing the simple medium with additional nutrients, such as peptides, leads to an increase in growth rate and the number of ribosomes per cell.

Small size and a lack of compartmentalization make diffusion an effective and rapid way to distribute molecules within bacterial and archaeal cells. Absence of a nuclear membrane also allows transcription and translation to be coupled, both temporally and spatially. Thus, bacterial and archaeal ribosomes begin translating the mRNA before it has been completed by the transcription complex. In addition to supporting rapid growth, this coupling provides for some uniquely akaryotic modes of genetic regulation, as demonstrated by the phenomenon called **attenuation**. In this regulatory mechanism, adequate levels of particular amino acids cause RNA polymerase to terminate transcription before it reaches the biosynthetic genes, thereby avoiding wasteful production of unneeded enzymes; conversely, if the amino acid is limiting, gene expression proceeds at full efficiency. The molecular basis of this regulation hinges on the coupling between ribosome and RNA polymerase. When it is tight, biosynthetic genes are poorly expressed, but when it is loose, i.e., when the ribosome fails to keep pace with RNA polymerase, transcription increases (Yanofsky, 2000).

2.4 Cytoplasmic membrane

The third component of the minimal cell depicted in Fig. 5.3 panel I is a unit membrane called the **cytoplasmic** (or **cell) membrane** (CM), which chemically separates the cytosol plus nucleoid from the external environment. A bacterial CM generally conforms to the fluid mosaic model of biological membranes (see Chapter 4, Cell membranes). Its matrix is a phospholipid bilayer intrinsically permeable to gases, water, and other small, uncharged molecules, but impermeable to ions or large polar solutes. Impermeability to the latter types of molecules is essential for viability, because the cell must retain metabolites and transmembrane ion potentials in order to survive. However, metabolism also requires nutrient uptake and exchange of other chemical species with the surroundings. Controlled access to the environment is provided by many different integral and peripheral proteins that breach the phospholipid bilayer with solute-specific pumps and gates. In bacteria, these proteins comprise about 70% of the CM by weight, and they actively control the transmembrane distribution of critical solutes. Archaeal membranes have analogous function, although the polar lipids that provide the diffusion barrier differ radically from the phospholipids of other cells. Archaea contain sn-(2,3 di-O-alkyl)-glycerol membrane lipids in which the hydrocarbon chains are C_{20} or C_{40} saturated isoprenoids attached to the glycerol backbone via ether linkages (Boucher, 2007).

2.5 Cell walls

Most archaea and bacteria have some form of **cell wall** immediately outside the cytoplasmic membrane (Fig. 5.3, IIA, IIB). This structure solves a specific threat created by the fact that the cytoplasm has, by necessity, a relatively low concentration of water relative to most microbial environments, whereas the CM is intrinsically permeable to water. Therefore, without appropriate countermeasures, the hypotonic environments of akaryotic cells would lead to uncontrolled influx of water, causing the cell to swell and rupture, i.e., **lyse**. Bacterial cell walls prevent lysis by enclosing the cell in a container that resists expansion. In contrast to the cellulosic cell walls of plants, however, bacterial cell walls are composed of a covalently cross-linked heteropolymer called **peptidoglycan,** or **murein**, The repeating unit (two modified amino sugars with an attached pentapeptide) is first polymerized into linear glycan chains, which are then joined laterally at frequent intervals by cross-linking of the peptide (Fig. 5.4). This results in one bag-like macromolecule, the **sacculus**, that completely surrounds the cell membrane. The tensile strength of the murein sacculus allows the bacterial cell to maintain an osmotic (turgor) pressure as high as 20 atm. The high and constant turgor pressure in most bacterial cells has the consequence that the bacterial cell wall determines cell shape, and thus plays a major role in cell division. This can be demonstrated by the effects on cell morphology of antibiotics that specifically disrupt the normal synthesis of

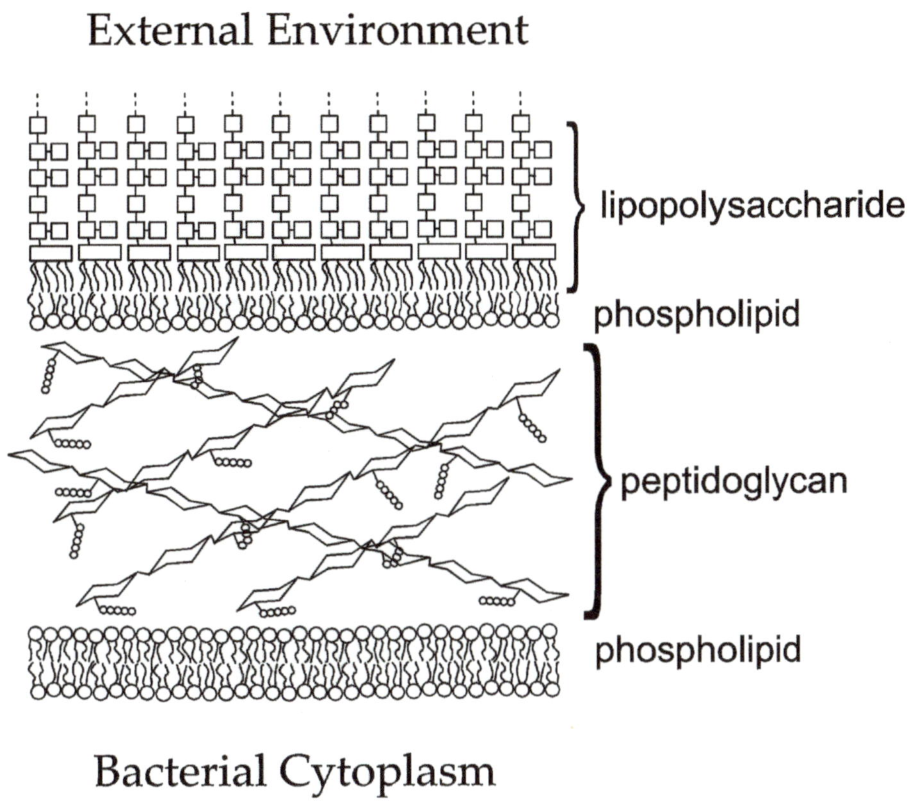

FIGURE 5.4 **The cell envelope of Gram-negative bacteria.** The cytoplasm of Gram-negative bacteria is bounded by three distinct layers. The central layer is the bacterial cell wall (CW) composed of peptidoglycan, whose repeating unit consists of a dimer of two modified amino sugars: N-acetyl muramic acid and N-acetyl glucosamine. Each repeating unit has a short peptide (circles) attached to the N-acetyl muramic acid; the peptides of adjacent glycan chains become joined by enzymes in the periplasm to form the periodic cross-links found in the mature peptidoglycan polymer. The outer membrane (OM) consists of one leaflet of lipopolysaccharide (LPS) and one of phospholipid. Each LPS molecule has a large polysaccharide chain (linked squares) whose precise structure depends on the bacterial species. The cytoplasmic membrane (CM), in contrast, has two leaflets of phospholipid. In addition to lipid, both the OM and CM contain protein species, which are not depicted in the figure.

peptidoglycan. In rod-shaped bacteria, sublethal concentrations of these inhibitors do not lead to cell lysis, but instead induce altered cell shapes (Schwarz et al., 1969).

Although archaea appear to use analogous strategies of structural reinforcement, they can differ in important ways from the bacterial strategy. Some archaea have cell walls composed of polysaccharides similar to peptidoglycan, but most have only a layer of protein or glycoprotein for exterior mechanical support for the CM (Rodrigues-Oliveira et al. 2017). These archaeal **surface- (S-) layers** (Fig. 5.3, panel IIa; Fig. 5.5) form by noncovalent association of individual subunits, but they

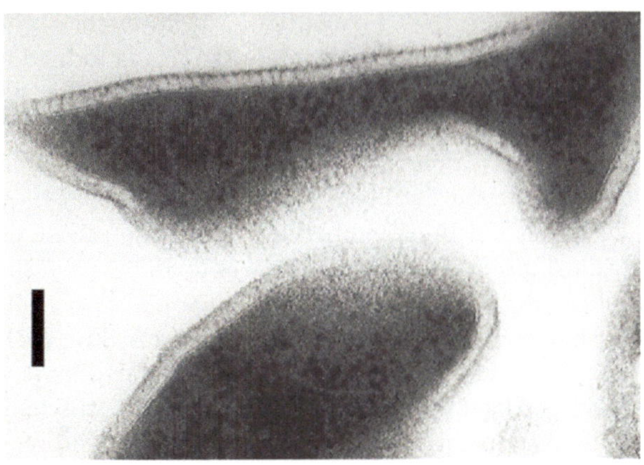

FIGURE 5.5 Cells of *Sulfolobus acidocaldarius*, an archaeon from geothermal environments (Section 7). The ultrathin sections show the irregular shape of the cell and the regular repeating structure of the glycoprotein S-layer. The S-layer is held at a fixed distance from the cytoplasmic membrane by a spacer protein (Baumeister et al. 1989); this results in a "picket fence" boundary visible at points around the cell periphery (Grogan, 1996a).

can be mechanically strong. For example, the hot SDS-extraction technique used to purify bacterial cell walls yields correspondingly pure and virtually intact S-layer sacculi when applied to *Sulfolobus* cells under mildly acidic conditions; the sacculi can dissociate into monomers, however, at elevated pH (Grogan, 1996b). Some archaeal S-layer sacculi appear to be relatively rigid; like bacterial cell walls, they retain the contours of the cell after isolation, whereas sacculi of other archaea appear to be very flexible. Proteinaceous S-layers also occur on many bacterial cells, but the bacterial structures do not function as cell walls, and most are nonessential (Fagan & Fairweather, 2014).

2.6 Cytoskeletal elements

The fact that naturally wall-less bacteria can maintain and alter their characteristic irregular shapes in hypotonic media suggests that their cells have some functionally analogous feature to a cytoskeleton. In the case of *Mycoplasma*, electron microscopy of cells treated with nonionic detergent reveals a network of fibrous structures, some of which appear to play a role in the organism's characteristic motility (Hegermann et al., 2002; Nakane & Miyata, 2009). Other methods, primarily genetic analysis and high-resolution fluorescence microscopy, suggest that even the bacteria and archaea that make cell walls have filament-forming proteins, which play important roles in determining cell shape. Although the cell wall defines the shape of most bacterial cells in the immediate, i.e., mechanical, sense, cellular growth and reproduction require the wall to expand in a way that is controlled and coordinated with protein synthesis and DNA replication. Some of this control involves dynamic structures, formed by cytoplasmic proteins related to actin and tubulin, that appear to regulate the topology of peptidoglycan synthesis.

For example, MreB proteins, which resemble actin, occur only in rod-shaped bacteria and are required for their characteristic shape. Inactivating MreB through mutation or by treating cells with specific MreB inhibitors converts rod-shaped cells into spheres. MreB molecules polymerize to form filamentous helices of constant diameter. Transient contact of these filaments with the interior of the CM appears to stimulate localized peptidoglycan synthesis on the opposite (outer) face of the CM. This controlled topology of peptidoglycan synthesis promotes extension of the cylindrical portion of the rod-shaped cell wall without a corresponding increase in diameter (Typas et al., 2011), thus leading to cell elongation without widening.

Another morphologically important process in bacteria, that of determining the site of cell division, involves the tubulin-like FtsZ protein. FtsZ determines the timing and location of septum formation, which then drives the physical separation of two daughter cells at cell division. The key event in this process is formation of a ring of polymerized FtsZ subunits at the inner surface of the CM at the midpoint of the fully grown cell. Peptidoglycan synthesis occurs opposite this ring, and the ring itself shrinks as new cell wall is deposited externally. The resulting inward cell wall synthesis can thus act as a radial wedge, driving the cytoplasms of the two daughter cells apart in the final stages of cell division (Typas et al., 2011).

2.7 Outer membrane

The cellular architecture found in most bacteria incorporates a second lipid bilayer, called the **outer membrane (OM)** (Fig. 5.3 panel III), outside of the cell wall. This second diffusional barrier creates a protected compartment around the cell called the **periplasm** (Fig. 5.4). In bacteria with this dual-membrane architecture, the peptidoglycan cell wall is much thinner than in bacterial cells that lack an OM (Fig. 5.3 panel IIB). This difference provides the basis of a classic bacterial identification technique, the Gram stain, in which a dark blue dye complex is formed in bacterial peptidoglycan. After the initial staining, bacterial cells are briefly rinsed with alcohol. Bacteria with a thin wall and an OM (Fig. 5.3 panel III) decolorize readily and are designated Gram-negative, whereas those with a thick peptidoglycan layer (Fig. 5.3 panel IIB) retain the dye complex and are classified as Gram-positive. Gram-negative and Gram-positive bacteria also differ with respect to other cellular features, as summarized in Table 5.1.

The OM of Gram-negative bacteria has functional properties distinct from those of the CM, and these differences are reflected in corresponding compositional differences. With respect to lipid, the OM is a hybrid membrane. Its inner leaflet, which faces the periplasm, incorporates the same phospholipids as the CM does, whereas the outer leaflet consists of a much higher molecular-weight lipid known as **lipopolysaccharide**, or **LPS** (Fig. 5.4). Similarly, with respect to protein, the OM has a large quantity of only a few protein species, in contrast to the diversity of proteins found in the CM.

These two biochemical characteristics of the OM (asymmetric bilayer and limited protein diversity) relate to its physiological functions, which include shielding the Gram-negative cell from hydrophobic compounds in the environment. The outer leaflet of LPS impedes diffusion of soluble hydrophobic (i.e., lipophilic) molecules across the OM. This apparently reflects the permeability properties of the dense array of cooperatively interacting charged and polar sugar

TABLE 5.1 Properties of Gram-positive and Gram-negative bacteria.

Property	Gram-positive bacteria	Gram-negative bacteria
Cell wall:		
Chemical constituents	Peptidoglycan	Peptidoglycan
	Teichoic, lipoteichoic and teichuronic acids	
Typical thickness	20–80 nm	2–3 nm
Outer membrane	(Absent)	Composed of LPS and phospholipid
Intrinsic resistance to detergents, dyes, and certain antibiotics	Low	High
Desiccation resistance	High	Low

residues at the membrane surface, through which nonpolar solutes must pass (Fig. 5.4). Experimental evidence of the importance of this screening function can be seen in the fact that Gram-positive bacteria and LPS-depleted Gram-negative cells are relatively sensitive to toxic dyes, detergents, and hydrophobic antibiotics (Nikaido, 2003). As in the case of the CM, the OM must nevertheless allow passage of nutrients, and this function is provided by relatively nonspecific solute channels, or **pores**. These structures form as trimers of corresponding proteins, called **porins**, which account for most of the protein in the OM. In *E. coli*, OM pores exclude neutral solutes with molecular weights higher than about 700 Da and discriminate additionally against anionic solutes (Nikaido, 2003).

2.8 Intracellular structures

Although bacteria and archaea generally lack true membrane-bound organelles, a few species contain specialized intracellular structures. Most photosynthetic bacteria, for example, have photosynthetic membranes located in the cytoplasm (Fig. 5.3 panel IV). In some photosynthetic bacteria, such as cyanobacteria, these membranes are flattened (lamellar) in form, and are called **thylakoids**, like the corresponding membranes of plant chloroplasts (see Chapter 30, Microtubules, dynein and eukaryotic cilia). Stacked intracytoplasmic membranes morphologically similar to thylakoids also occur in various bacteria that oxidize methane or ammonia as a source of energy (Hanson & Hanson, 1996). In other photosynthetic bacteria, the photosynthetic membranes are rounded vesicles, called **intracytoplasmic membranes** (ICM), that appear to be continuous with the CM (Dierstein et al., 1981).

Some nonphotosynthetic bacteria have internal membranes with no confirmed metabolic function. These bacteria, typified by a group called planctomycetes, all lack the peptidoglycan CW of most bacteria, and all have at least one large, internal membrane vesicle that divides the cytoplasm into two compartments of comparable volumes (Fuerst, 2005). One of these vesicles encloses the nucleoid, which tends to be highly condensed, forming a well-defined structure clearly visible in EM thin sections. In at least one species, a double-membrane structure surrounds the nucleoid, forming an analog of the eukaryotic nuclear membrane. Other species have additional membranes that create a series of nested cellular compartments (Fuerst, 2005).

The cytoplasms of certain bacteria and archaea contain small, regular vesicles of protein, which self-assemble by a process that excludes water, so that the resulting structures contain only gas. These **gas vesicles** (also termed gas vacuoles) function as buoyancy organs, allowing the cell to stay near the surface in natural waters. Consistent with this ecological role, synthesis of gas vesicles is often regulated by environmental conditions, such a light intensity and oxygen concentration (Walsby, 1994). The small size of these structures, and their ability to self-assemble in vivo, has created interest in biophysical applications, including development of tissue-specific contrast agents for ultrasonic medical imaging (Lakshmanan et al., 2017).

2.9 Extracellular structures

Bacteria and archaea have various types of filaments which are anchored in the CM and extend out into the surrounding medium. The **flagella** of bacteria and archaea are protein filaments that have a helical shape and mediate rapid swimming motility. Depending on the species, akaryotic flagella may be single or numerous, clustered or dispersed over the cell, or absent. Each flagellum extends several cell lengths (2–20 m) and is rotated by a protein complex at its base; the rotating helix acts like a propeller to push the cell through its liquid medium. Thus, bacterial and archaeal flagella have a cellular function (i.e., motility) analogous to that of eukaryotic cilia and flagella (see Chapter 30: Microtubules, Dynein, and Eukaryotic Cilia), but no structural, molecular, or mechanistic homology to the eukaryotic structures. Archaeal flagella and bacterial flagella also differ from each other in certain structural details.

Pili and **fimbriae** are structurally similar to each other, although pili are generally longer than fimbriae and not as numerous. Both structures consist of straight protein fibers that protrude from the cell, giving it a bristled or hairy appearance in the electron microscope. Both structures mediate various forms of attachment; pili attach akaryotic cells to each other as a first step in the transfer of DNA from donor to recipient, and some pili mediate transient interactions with surfaces to produce a type of locomotion called "twitching motility". Fimbriae often attach the bacterium to a specific surface and are used by certain pathogenic bacteria to colonize epithelia.

Flagella, pili, and fimbriae all form via self-assembly of small protein subunits. Bacterial flagella grow by adding protein subunits to the distal tip, which requires the subunits (monomers) to travel from the cytoplasm to the growing tip of the flagellum, several cell lengths away. This is accomplished by using the hollow core of the flagellum itself as a conduit. In contrast, pili, fimbriae, and archaeal flagella appear to polymerize at the base of the fiber, where it attaches to the CM (Neidhardt et al., 1990; Ng et al., 2006).

In addition to these discrete protein structures, many bacteria secrete hydrophilic polymers (usually polysaccharides) that form a gelatinous matrix around the cell. If this matrix remains attached to the cell and forms a defined zone around it, it is called a **capsule**. Capsules inhibit the ingestion of bacteria by human phagocytes or protozoa and may also help free-living bacteria to survive temporary desiccation or nutrient deprivation. Even if it does not form a defined capsule, secreted polysaccharide aids the nonspecific adherence of bacteria to wetted solid surfaces as "**biofilms**". Biofilms represent dense communities of bacteria that develop spontaneously on wetted surfaces through a relatively well-defined process (Monds & O'Toole, 2009). Bacteria in biofilms exhibit characteristics that differ from those of the same bacteria living in suspension, including the ability to withstand higher concentrations of antimicrobial compounds (Monds & O'Toole, 2009). Because of the resilience of bacteria in biofilms, biofilm formation on surfaces of implanted medical devices, such as heart valves, poses a common and significant health risk.

3. Energetics and metabolic strategies

Much of the metabolic diversity of microorganisms involves ways in which chemicals provided by the environment can be used to provide material and energy for the cell. These various chemical conversions contribute to the global cycling of many ecologically essential elements, and several are unique to archaea and bacteria.

3.1 Substrate-level phosphorylation vs. chemiosmotic coupling

As in all cells, most endergonic cellular processes of bacteria and archaea are coupled enzymatically to the hydrolysis of ATP, which requires the cell to continually resupply this central "energy currency". Catabolism of organic compounds provides one approach to meeting this need, and the range of compounds that support the growth of bacteria and archaea can be remarkably broad, particularly under aerobic conditions. As illustrated by an *E. coli* cell growing aerobically using glucose, ATP can be produced from organic compounds in two fundamentally different ways: **substrate-level phosphorylation** and **chemiosmotic coupling**. These two basic energy-capture mechanisms combine with other pathways to create various metabolic schemes which enable bacteria and archaea to utilize diverse carbon compounds and colonize a wide spectrum of ecological niches.

In the example of *E. coli* growing aerobically, glucose is first metabolized through a common pathway of glycolysis, the **Embden–Meyerhof–Parnas (EMP)** pathway. This pathway uses soluble cytoplasmic enzymes to (i) convert 1 mole of glucose to 2 moles of pyruvic acid, (ii) phosphorylate 2 moles of ADP to form ATP, and (iii) reduce 2 moles NAD^+ to form NADH. By phosphorylating certain hydroxyl groups before oxidizing adjacent C–C bonds, the pathway creates relatively reactive phosphoanhydride or phospho-enol bonds, which can be used to phosphorylate ADP. Other glycolytic pathways found in bacteria and archaea follow similar strategies: an organic compound (usually a sugar) is phosphorylated, rearranged, and used to phosphorylate ADP.

From the perspective of the microbial cell, substrate-level phosphorylation has the advantage of requiring only a few soluble enzymes and only one chemical resource from the environment (i.e., the sugar). Its disadvantages are significant, however, the yield of ATP per glucose molecule is quite low, and the NADH produced during the process must be reoxidized. An *E. coli* cell metabolizing glucose aerobically uses oxygen to solve both of these problems. Under aerobic conditions, the EMP pathway still operates, but it does not stop with substrate-level phosphorylation. Instead, an oxido-reductase transfers the electrons from NADH to a series of two- and one-electron carriers of increasingly positive midpoint potentials, ending with oxygen. Most of the electron transfers occur in the CM and are organized such that uptake of protons occurs in the cytoplasm and release of protons occurs at the cell exterior (in this case, the periplasm). This creates a **proton potential,** or **protonmotive force** (PMF), across the CM (Fig. 5.6). Analogous to mitochondria (see Chapter 13, ATP synthase: Structure, function, and channel formation), the *E. coli* cell then converts the PMF into ATP, using an $\mathbf{F_1F_0}$ **proton-translocating ATPase,** which couples the entry of three protons from outside the cell to phosphorylation of an ADP molecule (Harold & Maloney, 1996). This approach thus yields additional energy; furthermore, it is not restricted to electrons removed during glycolysis. *E. coli* and other cells use additional pathways (the Krebs or tricarboxylic acid cycle, for example) to strip electrons from the pyruvic acid and feed them into the same respiratory electron transport chain. As a result of these additional electrons, the overall yield of ATP per mole of glucose under aerobic conditions is high (about 34).

3.2 Protonmotive force

In the situation above, a proton potential across the CM, i.e., PMF, serves as an energetic intermediate between the oxidation of carbon compounds and the phosphorylation of ADP. For bacteria and archaea, however, the PMF per se represents an essential form of cellular energy, as fundamental processes use it directly as the energy source. *E. coli*, for example, uses direct coupling to H^+ influx, rather than ATP hydrolysis, to transport several nutrients (see below) and to drive flagellar rotation. Even bacteria that cannot respire maintain a PMF, as illustrated by **lactic acid bacteria**. These bacteria have no electron transport pathway, but they do have an F_1F_0 ATPase, and use it in the reverse sense of respiring bacteria, i.e., they maintain a PMF at the expense of ATP hydrolysis.

As discussed in Chapter 13 (ATP synthase: Structure, function, and channel formation), the PMF is the combination of two transmembrane potentials: the electrical potential ($\Delta\psi$) and the chemical potential (ΔpH). Although the small size of bacterial and archaeal cells precludes direct measurement by electrodes, these membrane potentials can be estimated using the distribution of chemical probes and other cellular properties (Chen & Lo, 2016). In bacteria which live at pH values near 7, both $\Delta\psi$ and the ΔpH contribute significantly to the PMF. For example, a typical *E. coli* cell in medium at pH 6.5 has a $\Delta\psi$ of about 100 mV (inside negative) and a ΔpH (insider higher) that corresponds to an additional 100 mV (Harold & Maloney, 1996). Extreme acidophiles, in contrast, may maintain an electrical potential of opposite polarity, i.e., inside-positive. This helps counteract the very large ΔpH across the cytoplasmic membrane, which can be more than 4 pH units in

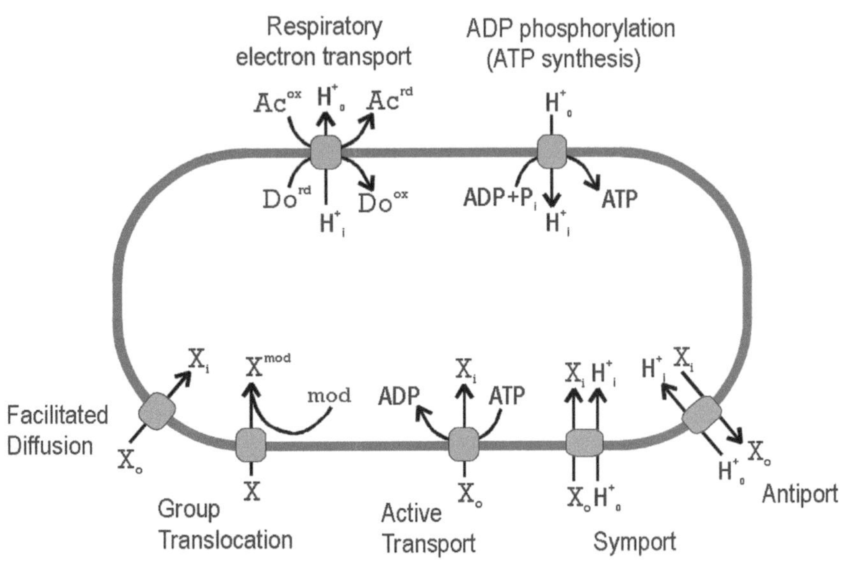

FIGURE 5.6 Energetics of solute transport across the CM. The diagram summarizes the salient features of major classes of energetic coupling to solute transport in akaryotic cells. For example, the net result of respiratory electron transport is transfer of electrons from a donor (Do) to an acceptor (Ac) coupled to the extrusion of protons. Similarly, controlled entry of a proton through a membrane-bound ATPase drives the phosphorylation of ADP. The remaining processes are those which support growth by promoting the uptake of some necessary solute (X) from the environment. Cellular location is indicated by subscripts (o, outside; i, inside), and "mod" indicates a chemical modification.

these organisms. Conversely, extreme alkaliphiles, which grow at external pH values of 10−12, have a negative ΔpH across the CM, for which they must compensate by maintaining a correspondingly large $\Delta\psi$ (White, 2000). Perhaps, the most serious physiological challenge faced by these organisms is the extremely low concentration of free protons in the extracellular environment. Several alkaliphiles solve this problem by using the environmentally more abundant Na^+ ion to drive ATP synthesis via Na^+-coupled ATP synthases, and certain neutrophilic marine bacteria use this mechanism, as well (Skulachev, 1994).

3.3 Fermentation and anaerobic respiration

The *E. coli* cell described in section 3.1 used glycolysis as the prelude to a series of electron transfers that led ultimately to oxygen reduction, thereby solving several metabolic problems. Not all environments provide oxygen, however, and a remarkable diversity of metabolic processes allows certain bacteria and archaea to derive energy from carbon compounds under anaerobic conditions. These various pathways can generally be classified as either **fermentation** or **anaerobic respiration**, and either strategy can be observed in *E. coli* cells catabolizing glucose anaerobically.

After an *E. coli* cell has converted glucose to pyruvate under anoxic conditions, it must somehow regenerate NAD^+ from NADH. The simplest option is simply to reduce the pyruvate, thereby regenerating NAD^+ and producing lactate, which the cell excretes; this approach is used by lactic acid bacteria (and also by the muscle cells of animals during intense exertion). Alternatively, *E. coli* and related bacteria can first split the pyruvate into acetaldehyde and formic acid, reducing the acetaldehyde, then regenerates NAD^+ and produces ethanol. Other options convert pyruvate to CO_2, H_2, acetate, or succinate, all of which are released as waste products. These alternatives are all examples of **fermentation**, and their common feature is that they combine with glycolysis to make it sustainable. By using NADH to reduce compounds derived from the original glucose molecule, these pathways recycle the electron carrier that glycolysis requires. In biochemical terms, therefore, fermentation is defined as catabolism of an organic compound via substrate-level phosphorylation in which the reduced electron carrier is reoxidized by a product of the pathway (White, 2000).

If *E. coli* cells fermenting glucose anaerobically are given nitrate, however, they adopt a different metabolic strategy, which results in increased growth, cessation of mixed acid production, and formation of nitrite. The cells under these conditions use nitrate as a terminal electron acceptor; the glucose is being oxidized completely to CO_2 and all the electrons are being sent down the electron-transport chain, ending in the reduction of nitrate to nitrite. This form of oxidative phosphorylation, called **anaerobic respiration**, thus substitutes another chemical for oxygen as the terminal electron acceptor. *E. coli* can use at least four compounds in this role (Table 5.2), and each requires a corresponding oxidoreductase at the oxidizing end of the electron-transport chain. *E. coli* expresses the different oxidoreductases according to a regulatory hierarchy that ensures that only the energetically most favorable electron acceptor (including oxygen) is used, should more than one be available (Gunsalus, 1992).

Anaerobic respiration illustrates the diverse ways in which bacteria and archaea can couple oxidation/reduction processes to proton extrusion as a catabolic (energy-yielding) strategy. In the example above, the exogenous electron donor is an organic compound (glucose), and the exogenous acceptor (nitrate) is inorganic, but other elements and inorganic compounds in the environment can play these roles. For example, some bacteria derive their cellular energy by transferring electrons from H_2 to sulfate, producing H_2S. The ability to use only inorganic compounds as the source of energy is termed

TABLE 5.2 Respiratory strategies of akaryotes.

Electron acceptor	Reduced product	Organism
Oxygen	Water	*E. coli*, other aerobes
Nitrate	Nitrite	*E. coli*, other bacteria
Fumarate	Succinate	*E. coli*, other bacteria
Dimethyl sulfoxide	Dimethyl sulfide	*E. coli*, other bacteria
Trimethylamine oxide	Trimethylamine	*E. coli*, other bacteria
Sulfate	Hydrogen sulfide	*Desulfovibrio, Archaeoglobus*[a]
Sulfur	Hydrogen sulfide	*Thermoproteus*[a]
Carbon dioxide	Methane	*Methanobacterium*[a]

[a]denotes archaea; all others are bacteria.

chemolithotrophy, and it enables certain bacteria and archaea to colonize environments that receive no other source of energy, such as that provided by organic matter or sunlight.

3.4 Carbon and nitrogen fixation

Whereas chemolithotrophy exemplifies the metabolic diversity of bacteria and archaea, it also illustrates the fact that some environments may provide chemical energy for microbial metabolism but offer no form of carbon except CO_2. The ability to synthesize carbonaceous cell material from CO_2 (i.e., to "fix" carbon) is termed **autotrophy**. In addition to a metabolic pathway of CO_2 fixation, autotrophy requires environmental sources of energy and reducing equivalents (i.e., electrons). **Photoautotrophy** is well known in eukaryotic organisms, i.e., green plants and algae. In the chloroplasts of these organisms, light provides the energy for CO_2 fixation, whereas water provides the electrons; as a result, this form of photoautotrophy generates oxygen as a waste product of electron harvesting. This oxygenic type of photosynthesis also occurs in **cyanobacteria**, which are evolutionarily related to chloroplasts (Archibald, 2015). Like chloroplasts, cyanobacteria contain **chlorophyll *b*** and have two connected **photosystems**. Photosystem II (PS2) takes electrons from water and passes them to photosystem I (PS1), which generates the strongly reducing ferredoxins needed for CO_2 fixation and other biosynthetic processes (see Chapter 30, Microtubules, dynein and eukaryotic cilia).

Other bacteria use variations on the chloroplast/cyanobacterial strategy of harnessing light energy; each variation uses a single photosystem, and operates in the absence of oxygen. These anoxygenic photoautotrophs use **bacteriochlorophylls** to absorb light energy; these pigments resemble cyanobacterial and plant chlorophylls, but have intense absorbance maxima in the infrared region (White, 2000). The single photosystem in each of these bacteria can operate in a cyclic manner to produce a proton potential. Alternatively, for the reduction of CO_2, these organisms draw electron equivalents from H_2S, H_2, or dissolved organic compounds, depending on the species and the resources available in the environment. In many of these organisms, photoautotrophic growth represents only one of several metabolic options, and production of photosynthetic pigments occurs only in the presence of light and absence of oxygen. A fundamentally different approach to harvesting light energy is found in the archaeal species *Halobacterium salinarum*. This extreme halophile uses a type of rhodopsin to generate a PMF directly from absorbed light, but produces this protein only under anaerobic conditions (see Section 7).

The various forms of chemolithotrophy and phototrophy described above represent alternatives for harvesting the electrons and energy necessary for CO_2 assimilation. In addition to this diversity of energy generation, bacteria and archaea have diverse pathways for CO_2 fixation itself. Some use (i) the **Calvin–Benson cycle** that also operates in green plants and algae: ribulose bisphosphate is carboxylated to yield two molecules of 3-phosphoglycerate, followed by a complex series of reductions and rearrangements, which produce sugars and regenerate ribulose bisphosphate (see Chapter 33, Microtubules, dynein and eukaryotic cilia). Other autotrophic bacteria and archaea use (ii) the **Acetyl CoA pathway**, (iii) the **Reductive Citric Acid cycle,** or (iv) the **Hydroxypropionate pathway** (White, 2000; Zarzycki et al., 2009). None of the latter three autotrophic pathways (described below) are found in plants or algae.

The Acetyl-CoA pathway (ii) is found in methanogenic archaea and sulfate-reducing bacteria; it uses H_2 to reduce two CO_2 molecules; one is converted into a methyl group and the other is reduced to form a carbonyl unit. Condensing these two units yields an acetyl group that is transferred to coenzyme A for assimilation into cell material. The reductive citric acid cycle (iii) begins with carboxylation of succinyl CoA to yield 2-oxoglutarate, which is then carboxylated to yield isocitrate. The transformations of the citric acid cycle then continue in reverse, leading to the cleavage of citrate by an ATP-dependent citrate lyase to yield acetyl CoA and oxaloacetate (White, 2000). The hydroxypropionate pathway (iv) begins by carboxylating the acyl group of acetyl CoA and then of the resulting propionyl CoA; after rearrangement of the C_4 acyl group, glyoxylate is cleaved off to regenerate acetyl CoA (Zarzycki et al., 2009).

The ability to fix nitrogen, i.e., convert N_2 into ammonia, is found only in akaryotes. It is distributed widely but sporadically among aerobic, anaerobic, and facultatively aerobic bacteria (i.e., those able to grow either aerobically or anaerobically) and methanogenic archaea. The N_2 reduction reactions are catalyzed by a large enzyme complex, **nitrogenase**, which consists of the **dinitrogenase,** which reduces N_2, and an iron-containing **dinitrogenase reductase**. The dinitrogenase active site has a covalently bound molybdenum- and iron-containing cofactor, commonly abbreviated "FeMoCo". The N atoms bind to this cofactor and are reduced in multiple steps. The dinitrogenase reductase uses a low-potential ferredoxin or flavoprotein as the source of electrons and further increases their reducing potential through ATP hydrolysis.

Dinitrogenase reductases are intrinsically O_2-sensitive, and N_2-fixing akaryotes use a variety of strategies to protect these enzymes from oxidative inactivation. For example, facultative anaerobes such as *Klebsiella* spp. express nitrogenase genes only under anaerobic conditions. Filamentous cyanobacteria, which generate O_2 during photosynthesis, sequester nitrogenase in specialized cells called **heterocysts**, in which the oxygenic photosystem II (PS2) does not operate. *Rhizobium* spp. fix nitrogen only when they are protected from oxygen within the specialized nodules which they colonize on the roots of legumes. *Azotobacter* spp. represent the rare examples of obligate aerobes that can fix nitrogen; they appear to scavenge intracellular O_2 by maintaining high respiration rates.

4. Solute transport

Cells of bacteria and archaea cannot phagocytose, and so they must take in nutrients and other chemical resources from the environment as small molecules. Fig. 5.6 summarizes the major types of protein-mediated transport mechanisms that have been identified in bacterial cells.

4.1 Facilitated diffusion

Facilitated diffusion uses a solute-specific channel to allow the solute to cross the CM down a concentration gradient (Fig. 5.6). It is mechanistically simple and requires little energy, but also has limited effectiveness as a form of nutrient uptake. A well-studied example of facilitated diffusion in akaryotes is glycerol transport in *E. coli*. In this process, extracellular glycerol enters passively through a polyol-specific membrane channel (encoded by the *glpF* gene) as long as a favorable concentration difference exists (see Chapter 10, Membrane diffusion and permeability)

4.2 Group translocation

In an elaboration of facilitated diffusion, the solute becomes chemically modified upon entering the cytoplasm. This modification, termed "group translocation," is typically phosphorylation or phosphoribosylation, which converts an uncharged solute molecule into an ion that cannot diffuse out through the same transmembrane carrier. This allows the uncharged species actually transported by the channel to remain at a lower steady-state concentration inside the cell than outside. The chemical modification thus serves as a "trapping reaction" and represents the input of chemical energy required to accumulate the nutrient. In many cases, it also doubles as the first step in metabolizing the solute.

4.3 Active transport

In bacterial physiology, the term “active transport” has been reserved historically for processes with the following characteristics (Neidhardt et al., 1990): (i) specific steric recognition between the solute and a membrane-bound transport protein, (ii) release of the solute into the cytoplasm in its unmodified form, (iii) net accumulation against a solute concentration gradient, and iv) expenditure of energy. Two types of active transport are distinguished from each other by the immediate source of energy used (Fig. 5.6). The first type requires the hydrolysis of ATP or an equivalent high-energy phosphoryl bond. In these transport systems, the membrane-bound carriers typically consist of several subunits, one of which contains an ATP-binding site. This site can be detected in the protein sequence as an ATP-binding “cassette,” yielding the designation “ABC protein” (Holland & Blight, 1999). Transporters of this type are also found in eukaryotic cells, but the eukaryotic proteins generally mediate solute efflux, not uptake (Holland & Blight, 1999).

In Gram-negative bacteria, an ABC transporter often uses an accessory soluble protein located externally, i.e., in the periplasm, to initially bind the solute and deliver it to the membrane proteins. Periplasmic solute-binding proteins are integral parts of the corresponding uptake systems, as demonstrated by the fact that cold hypotonic shock treatment of bacterial cells, which releases only periplasmic proteins, also destroys transport capability. Studies of bacterial mutants and of cytoplasmic membrane vesicles confirm that the membrane-bound transporter interacts specifically, not with external solute in its free form, but with the solute complexed to its cognate binding protein (Holland & Blight, 1999).

4.4 Secondary (ion-coupled) transport

A distinct type of transport that can accumulate solutes against a concentration difference utilizes the energy of the PMF or another ionic transmembrane potential. Typified by the lactose permease of *E. coli*, many of these catalyze **symport** (same-direction transfer) of an organic compound and a proton (Fig. 5.6), although some systems utilize a Na^+ ion. Conversely, proton-coupled **antiport** systems provide the energy to pump certain solutes, such as Ca^{++}, out of bacterial cells (Harold & Maloney, 1996).

4.5 Efflux systems

Whereas uptake systems are abundant and diverse in akaryotes, **efflux transporters** are comparatively rare. Most that have been identified pump toxic substances, such as heavy metals or antibiotics, out of bacterial cells. Although this function enhances microbial survival in nature, it is a concern in clinical settings because of its ability to compromise the effectiveness of antibacterial therapy. Several distinct families of multidrug resistance (MDR) systems have been found in pathogenic bacteria and studied intensively (Nikaido, 2009). One family uses ABC-type transporters to export antibiotics, whereas the others use proton-antiporters (Piddock, 2006). Bacterial MDR systems tend to exhibit relatively low substrate specificity, exporting broad classes of relatively complex organic compounds. This suggests that the natural physiological function of MDR efflux pumps may relate to protecting bacteria against a wide spectrum of potentially threatening organic compounds encountered in natural communities.

5. Sensing the environment and responding

Unicellular organisms cannot insulate themselves from their environments, and therefore must mount cellular responses to stress and external change in order to optimize their growth or simply to survive. Rapid responses to environmental change typically involve modifying the activity of existing enzymes, whereas longer-term responses typically involve changing the abundance of particular proteins by changing the transcription rates of their genes.

5.1 Osmotic stress

Hypotonic conditions are normal for most akaryotes, and evidence of this is seen in the homeostatic responses of bacterial and archaeal cells to a sharp increase in the concentration of extracellular solute. In *E. coli*, the initial response is net K^+ uptake by transporters in the CM, which respond to some parameter of the osmotically stressed cell (Wood, 2011). If the increase in external ionic strength is very large, however, K^+ accumulation alone does not fully alleviate the osmotic stress, and the cells begin to build up high concentrations of **compatible solutes** in the cytoplasm. Responses of this type are widely represented among bacteria and archaea (Roesser & Muller, 2001).

Most compatible solutes are small organic compounds; two of the best-studied examples are L-proline and glycine betaine. Accumulation of these solutes in the cytoplasm, whether by scavenging from the environment or synthesis de novo, allows microbial cells to grow at otherwise inhibitory external salt concentrations (Wood, 2011). Conversely, impairment of either K^+ or compatible-solute accumulation systems limits the concentration of external solutes that *E. coli* can tolerate. These observation suggest that turgor pressure is important for normal function of bacterial cells. Turgor provides the physical force that expands the bacterial cell, for example, allowing it to keep pace with biosynthesis of cytoplasmic components during cellular growth. Similarly, the importance of compatible solutes in hyperosmotic stress responses suggests that high intracellular concentrations of these organic compounds are physiologically preferable to high concentrations of the salts which are generally more abundant in the environment.

Although akaryotic cells generally require hypotonic conditions, it is possible for some environmental changes to induce hypotonic stress. For example, if bacterial cells have adapted physiologically to high-solute medium, sudden dilution of the medium can drive turgor pressure to damaging levels. Under such conditions, mechanosensitive channels in the bacterial CM open, releasing water and low molecular weight solutes transiently but nonspecifically (Wilson et al., 2013). In the absence of these osmotic "safety valves", hypoosmotic shock can induce death or even lysis of bacterial cells (Hoffmann et al., 2008).

5.2 Acid stress

When the extracellular pH decreases sharply, *E. coli*, *Salmonella enterica*, and other bacteria increase the synthesis rate of about 50 proteins, which act in various ways to help the cell adapt to acid stress. For example, arginine, ornithine, and lysine decarboxylases induced by acid stress produce the weakly acidic CO_2 and basic diamines. Other acid-induced proteins alter pathways of sugar fermentation in bacteria; some remove cytoplasmic protons by reducing them to H_2, for example. By diverting the flow of metabolic intermediates into production of ethanol, CO_2, and H_2, these enzymes decrease the amount of acid produced by the bacteria (Kanjee & Houry, 2013; Slonczewski & Foster, 1996).

The bacterial response to acid stress exemplifies many other microbial responses to environmental change, in that the expression levels of various genes are changed dramatically, causing shifts among alternative pathways. Such "metabolic retooling" can be rapid in akaryotic cells, reflecting their relatively large biosynthetic capacity, simple organization, and rapid turnover of messenger RNA (in *E. coli* under standard conditions, for example, mRNA half-life averages 1.3 min). Studies of specific responses in bacteria have revealed an array of elegant and diverse molecular mechanisms for controlling gene expression, and a comprehensive summary of these lies outside the scope of this chapter. One example will be described, however, which involves transmembrane signaling and mediates many environmental responses in bacteria.

5.3 Two-component regulatory systems

When Gram-negative bacteria begin to starve for inorganic phosphate, they activate the transcription of several families of genes whose products help the cell scavenge additional phosphate from the environment (Wanner, 1993). The process involves the specific interaction and phosphorylation of two proteins, PhoR and PhoB (named after their respective genes in *E. coli),* and it typifies a type of signal transduction system characteristic of akaryotic cells.

The phosphorylation state of the PhoB protein determines expression of the specialized set of genes that mediate the phosphate-starvation response. Only the phosphorylated form of PhoB (here designated "*PhoB") activates transcription of these genes; as a result, the relative abundance of *PhoB determines the level of the cell's phosphate-starvation response. The phosphorylation state of PhoB is determined, however, by the conformational state of a CM protein, called PhoR (Fig. 5.7). In its "inactive" conformation, PhoR acts as a *PhoB phosphatase, whereas in its "active" conformation ("*PhoR"), it is a PhoB kinase. The conformational state of PhoR responds to the availability of inorganic phosphate in the external medium, as sensed by phosphate-specific periplasmic binding protein. If the phosphate-bound form of the binding protein is low, PhoR assumes its active conformation. The resulting *PhoR phosphorylates PhoB, and the resulting *PhoB stimulates the transcription of *phoA* and other genes for phosphate-scavenging pathways (Fig. 5.7) (Wanner, 1993).

Pairs of proteins resembling PhoR and PhoB mediate a wide variety of bacterial responses to environmental change and are collectively termed **"two-component" regulatory systems** (Table 5.3). The PhoR protein represents the corresponding **component I**, or the **sensor-kinase,** of these systems. With few exceptions, bacterial sensor-kinases are CM proteins that

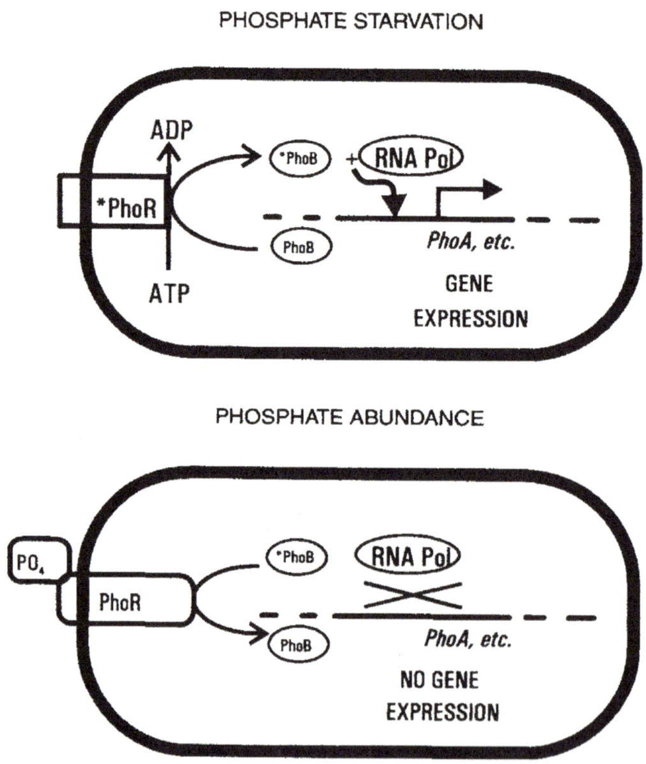

FIGURE 5.7 Example of a two-component regulatory system. The diagram shows a simplified version of phosphate-regulated gene expression. The PhoR protein senses the extracellular availability of phosphate, either directly through an externally exposed domain or indirectly via interaction with the high-affinity phosphate transporter (Neidhardt et al., 1990; Wanner, 1993). When the external phosphate concentration is low, PhoR becomes "activated"; it autophosphorylates at the expense of intracellular ATP and then phosphorylates the PhoB protein. When the external phosphate concentration is high, PhoR becomes "deactivated" and removes phosphate groups from phosphorylated PhoB protein. PhoB is a soluble transcription factor for a specific set of genes, but stimulates transcription only when phosphorylated.

TABLE 5.3 Examples of two-component regulatory systems.

Mnemonic	Environmental stimulus	Cellular response	Component I	Component II
***Pho*sphate**	Low external [PO_4]	Induction of various PO_4-metabolizing enzymes	PhoR	PhoB
Nitrogen regulation	Low external [NH_4^+]	Induction of glutamine synthetase	NR_I	NR_{II}
Outer membrane protein	External osmolarity	Regulation of alternate porin synthesis	EnvZ	OmpR
Aerobic respiratory control	Oxygen concentration	Regulation of central metabolic enzymes	ArcB	ArcA
Nitrate reductase	External NO_3^- present, Oxygen absent	Induction of nitrate reductase	NarX	NarL
***Che*motaxis**	Gradients of attractants and repellents	Change in swimming interval	CheA	CheY

have two conformations and sense the status of some environmental parameter. In its "active" conformation, the component I has two characteristic phosphotransferase activities: (i) it phosphorylates one of its own histidine residues, using ATP and (ii) it transfers the phosphate to an aspartyl residue on a cognate **component II** or **response regulator**, corresponding to PhoB in the phosphate-starvation response. This results in the net transfer of a phosphate group from ATP to the specific component II, which is usually a transcriptional regulator. Phosphorylation of the response regulator either increases or decreases the transcription of relevant genes, depending on the regulatory logic of the system in question.

Although they respond to diverse signals, all components I have two highly conserved sequence motifs at their carboxy termini. Similarly, all components II have sequence homology near their amino termini, and various subfamilies may share regions of additional homology elsewhere in the amino acid sequence, despite the fact that they target diverse regulatory sequences at various genes. The conserved regions of components I recognize and interact with the conserved regions of components II as an essential part of the signal-transmission process. Conversely, the nonconserved regions among the components I and II include domains defining the specificity for the external stimulus detected, and the genetic regulatory response, respectively.

5.4 Modes of taxis

Many bacteria and archaea swim through liquid and respond to various environmental parameters that include nutrients, light, and oxygen, resulting in the simple behaviors of chemotaxis, phototaxis, and aerotaxis, respectively. These tactic responses might be presumed to follow a scheme in which a microbial cell first detects a spatial gradient and then orients its swimming in the direction of maximal increase (or decrease). Interestingly, however, at the micron/submicron scale of akaryotic cells, neither element of this scheme is feasible. Biologically relevant gradients do not produce concentration differences large enough to be detected across even the most elongated akaryotic cell, for example. In addition, Brownian motion reorients akaryotic cells on a subsecond time scale, as can be seen easily by phase-contrast microscopy of nonmotile bacterial cells suspended in water. This randomization prevents any given swimming direction from being maintained for more than a few cell lengths, which accordingly limits the effectiveness of cell steering.

These fundamental problems have been solved in various ways by archaea and bacteria, but the example of chemotaxis in *E. coli* has been particularly well studied. *E. coli* cells have multiple helical flagella distributed around the cell, which are coordinated so that all rotate in the same direction at any given time. When rotating counterclockwise, the flagella coalesce into one helical bundle that pushes the cell in the opposite direction from the bundle. When the flagella reverse and rotate clockwise, they separate and cause the cell to rotate erratically without net locomotion in a behavior called "tumbling". The cell reverses direction of flagellar rotation every few seconds, creating short intervals of swimming that alternate with tumbling. This results in a three-dimensional "random walk" in which short "runs" in randomly determined directions are punctuated by reorientation.

Spatial gradients of attractants or repellents influence, or bias, this random walk. The *E. coli* cell prolongs those runs which lead toward an attractant (or away from a repellent) and shortens those leading in the opposite direction. Over time, this yields a net migration along the spatial gradient, despite the cell's inability to control the direction of any given run. In order to implement this strategy, *E. coli* uses receptors in the CM to monitor the relative change in attractant or repellent concentration over time. If the temporal trend is favorable, the reversal of flagellar motors is delayed; if unfavorable, the flagella are reversed sooner. Although this strategy is conceptually simple, it nevertheless involves a complex network of associated regulatory proteins that monitors the environment as a function of time and controls and coordinates the multiple flagellar motors (Armitage, 1999).

The difficulties of directing the motility of micron-sized cells have been solved in a very different way by certain bacteria that swim along the lines of an externally applied magnet field. The guidance mechanism of bacterial **magnetotaxis** seems to be passive, in the sense that the cell aligns itself with the magnetic field, while rotation of the helical flagellum drives it forward (Lefevre et al., 2009). This requires a strong magnetic dipole that is fixed within the cell on the main axis of the flagellum and can maintain the orientation of the bacterial cell within the earth's magnetic field (Goswami et al., 2022). In magnetotactic bacteria, the dipole is provided by a short chain of single-domain magnetic crystals, called **magnetosomes**, each of which is enclosed in a phospholipid membrane. Mechanistic details of magnetosome synthesis, including what determines the magnetic polarity of the chain and how the parallel orientation of internal magnet and flagellum is maintained, have been the focus of active research (Jogler & Schüler, 2009).

6. Physiological aspects of pathogenesis

Bacteria capable of causing disease, termed **pathogens,** represent a tiny minority of bacterial species. Pathogens differ from the majority of bacteria in being able to invade an animal or plant host, elude multiple defense mechanisms (including the immune system), reproduce within host fluids and tissues, and damage the host in some way. These processes, and the defenses mounted by the host, often involve corresponding specialized functions of pathogen and host cells.

6.1 Iron acquisition

Iron is critical to many metabolic processes and makes up a significant proportion of the metal content of most microbial cells; accordingly, pathogens must acquire it in corresponding amounts from the host. Ferric ion (Fe^{+++}) predominates under aerobic conditions, but has extremely low solubility at neutral pH. In addition, iron is sequestered in various ways within the human body, decreasing its availability even more. As a result, sophisticated iron acquisition systems are critical for most bacterial pathogens to establish infections and cause disease (Becker & Skaar, 2014).

Most bacteria, including nonpathogenic species, scavenge ferric iron using specialized iron chelators called **siderophores** and transport systems specific for the iron–siderophore complexes. Under low-iron conditions, bacteria synthesize siderophores and release them into the environment, where they bind ferric iron specifically and with extremely high affinity; for example, the dissociation constant of ferric ion complexed with the siderophore enterobactin is 10^{-52} M. The bacteria then use specialized, high-affinity transport systems to recover the resulting complexes. In the case of Gram-negative bacteria, transport is complicated by the fact that the extracellular iron–siderophore complexes are too large to pass through the OM pores. Gram-negative bacteria therefore have specialized OM receptors, which appear to function as gated channels that allow the iron–siderophore complexes to enter the periplasm (Earhart, 1996). This initial transport step involves a specific protein (TonB) that energizes transport of these and other solutes across the OM.

In addition to iron–siderophore uptake systems, some pathogenic bacteria produce surface proteins that bind transferrin or lactoferrin and allow the bacterium to use these complexes as sources of iron (Becker & Skaar, 2014). In addition, some pathogenic bacteria can use hemoglobin as a source of iron. This involves the production of hemolysins, which are small proteins secreted by the bacterium that bind to the plasma membrane of red blood cells and either degrade or puncture it to release the hemoglobin. Enzymes on the surface of the bacterial cell can then bind the hemoglobin, extract the heme, and transport it into the bacterial cytoplasm, where the iron is released by heme-degrading enzymes (Becker & Skaar, 2014).

6.2 Surviving phagocytosis

If a bacterium succeeds in invading the body, it is usually attacked by some form of **phagocyte**, i.e., one of the specialized motile cells that circulate in the blood and consume bacteria. After a phagocyte engulfs a bacterial cell, the resulting cytoplasmic vesicle fuses with a lysosome to form a **phagolysosome** (or phagosome) (Fig. 5.8). In addition to the low pH and digestive enzymes typical of lysosomes, phagolysosomes generate high concentrations of nitrous oxide, hypochlorous

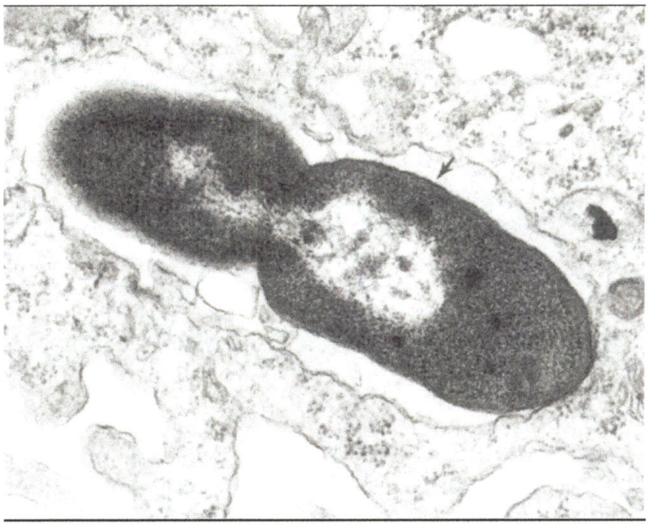

FIGURE 5.8 Phagocytosis of bacterial cells. The transmission electron micrograph shows an *E. coli* cell in the late stages of cell division that has been engulfed by a macrophage. The nucleoid is visible as a clear zone in each of the nascent daughter cells, surrounded by darker-staining, granular cytoplasm. The arrowhead shows a region in which the three-layer structure of the Gram-negative cell envelope (CM-CW-OM) can be seen. The wider clear space around the dividing cell is the interior of the phagolysosome. *Photograph courtesy of A. J. Mukkada.*

acid, and reactive oxygen species such as hydrogen peroxide, singlet oxygen, superoxide, and the hydroxyl radical. Synthesis of these reactive oxidants is triggered when a phagocyte is activated and is accompanied by a sudden demand for oxygen, called the **respiratory burst**. These reactive oxidants, combined with the low pH inside the phagolysosome, normally kill the ingested bacterium.

Certain pathogenic bacteria can survive in the phagolysosome, however, and escape it to multiply in the cytoplasm of the phagocyte. Some accomplish this by secreting a toxin termed a **leukocidin**, which kills the phagocyte. Other bacterial countermeasures against phagocytosis include synthesis of an external capsule (section 2.1), which inhibits the initial engulfment of the bacterial cell, and synthesis of cell-associated carotenoid pigments, which quench singlet oxygen.

6.3 Toxins

The leukocidins used by some bacteria to kill phagocytes are an example of bacterial **exotoxins**, which are soluble proteins released by pathogenic bacteria that damage particular host tissues or otherwise cause symptoms of disease. Among pathogenic bacteria that cause intestinal disease, the production of exotoxins that induce diarrhea is a common theme. The disease cholera, for example, occurs when cells of the bacterium *Vibrio cholerae* are ingested and colonize the intestines, where they produce a toxin that binds to specific gangliosides in the plasma membranes of intestinal epithelial cells (O'Brien & Holmes, 1996). This cholera toxin then covalently modifies and inactivates a GTP-binding protein that normally regulates an adenylate cyclase in the plasma membrane. Loss of this regulation causes overproduction of cAMP in the epithelial cell, which, in turn, activates a number of cAMP-stimulated cellular functions, including export of chloride ions. As large amounts of chloride are pumped into the lumen of the intestine, water follows by passive diffusion across the plasma membrane (O'Brien & Holmes, 1996). The massive secretion of salts and water into the intestine produces a severe diarrhea, which may be lethal if water and salts are not replaced at an adequate rate.

Other pathogenic bacteria secrete exotoxins that affect neuron function, as illustrated by two anaerobic, spore-forming species: *Clostridium tetani* and *Clostridium botulinum*. *Cl. tetani* infects host tissue, typically through deep puncture wounds, and produces **tetanus toxin**. This protein binds at the ends of inhibitory spinal neurons, blocking the release of GABA and glycine, both of which are inhibitory neurotransmitters that prevent muscle contraction. This inhibition of inhibitory spinal neurons leads to uncontrolled muscle contractions, resulting in generalized spastic paralysis, which can cause death by asphyxia. In contrast, the neurotoxin of *Cl. botulinum*, **botulinum toxin,** is normally produced during growth outside the host, but can be lethal if ingested, even in small quantities. Botulinum toxin prevents the release ofacetylcholine at the neuromuscular junction, blocking muscle contraction leading to flaccid paralysis.

7. Bacteria and archaea in extreme environments

Some of the most striking examples of the physiological diversity of microorganisms are provided by certain bacteria and archaea which not only tolerate harsh conditions but actually require them for normal cellular function.

Extreme halophiles thrive in concentrated brines, such as those that occur in natural salt lakes or artificial evaporation ponds. Most of these are archaea, and well-studied species include the rod-shaped *H. salinarum* and the irregularly shaped *Haloferax volcanii. H. salinarum* grows best in media containing 3–4 M NaCl and will not grow at salt concentrations less than 1.5 M. The cell maintains an internal K^+ concentration of up to 5M and contains Cl^- as the major counterion. Consistent with this, most *H. salinarum* enzymes function best in extremely high salt concentrations and denature at salt concentrations below about 1 M NaCl (Madern et al., 2000). The unusual salt-stability and salt-dependence of these enzymes correlate with an unusually high number of acidic amino acid residues on the protein surface, which are thought to help the protein compete successfully with salt ions to retain a hydration layer.

Other notable features of the *H. salinarum* cell include gas vesicles in the cytoplasm and patches of a special retinal protein, **bacteriorhodopsin**, in the cytoplasmic membrane. The gas vesicles make the cells buoyant, which helps keep them in contact with two sources of energy: O_2 and light. The former allows oxidation of various organic compounds that occur in hypersaline environments. If oxygen becomes limiting, however, light energy can be harvested by bacteriorhodopsin, which is a membrane-bound retinal protein that functions as a light-driven ion pump. When the cell is illuminated, bacteriorhodopsin pumps protons from the cytoplasm; this contributes to the cell's PMF and thus, to the production of ATP. Related proteins, collectively termed **proteorhodopsins,** have been identified in some other archaea and also bacteria (Fuhrman et al., 2008).

The Domains Bacteria and Archaea include a number of extreme **acidophiles,** i.e., organisms that require low pH for optimal cell growth. *Thiobacillus* spp. are Gram-negative bacteria that derive energy from the oxidation of sulfur compounds. Since the oxidized end-product is usually H_2SO_4, these bacteria acidify their environment in the normal course of their metabolism. Some *Thiobacillus* cells grow optimally at about pH 2 but maintain a cytoplasmic pH near 6.5. At least one species, *Tb. ferrooxidans,* can also oxidize Fe^{++}; this metabolic strategy is aided by the fact that autooxidation of Fe^{++}, which would compete with the biologically mediated process, occurs very slowly at low pH values.

The most extreme acidophile known to date is a thermophilic archaeon, *Picrophilus oshimae*. This organism, isolated from the soil of a Japanese geothermal field, grows optimally in dilute sulfuric acid, at about pH 0.9. *P. oshimae* cells can grow at pH 0 but not at pH values above 3.5, and at pH values greater than 5, the cells lyse (Schleper et al., 1995).

Bacteria that require low temperatures are termed **psychrophiles**, and typically occur in the polar oceans. For some, the minimum temperature for growth lies below about $-10°C$. The proteins of psychrophiles generally have structural flexibility, which allows the enzymes to catalyze reactions at low ambient temperature. This property also makes psychrophilic enzymes prone to denature, however, which limits the maximal growth temperature of these organisms. Similarly, the membrane phospholipids of psychrophiles usually contain high proportions of unsaturated and short-chain fatty acids, ensuring membrane fluidity at physiological temperatures (Russell, 1990).

At the other extreme of temperature, certain akaryotes, termed **extreme thermophiles** or "**hyperthermophiles**", grow optimally at 80°C or higher temperatures (Stetter et al., 1990). Most hyperthermophiles that have been cultured are archaea isolated from geothermal habitats such as terrestrial hot springs or submarine thermal vents. Aerobic representatives include *Sulfolobus* spp., which require acidic conditions as well as high temperatures. The combination of high temperature and low pH required for optimal growth (about 80°C and pH 3) is extremely effective at denaturing proteins. The *Sulfolobus* cell helps protect its cytoplasmic proteins, which appear to be generally heat-stable but not acid-stable (Grogan, 1996b), by maintaining its cytoplasmic pH at about 6 (Schäfer et al., 1990). Physical properties of the unique lipids of *Sulfolobus* spp. probably help maintain the resulting large ΔpH. The ether-linked phytanyl chains of these lipids span the entire membrane, forming a stable monolayer. Such membranes exhibit very low rates of proton leakage at high temperature compared to ester-linked phospholipid bilayers (Van De Vossenberg et al., 1995), which represents an important prerequisite for efficient function of these cells.

The most thermophilic organisms known are anaerobic archaea isolated from areas of ocean volcanism, where hydrostatic pressure permits liquid water to be superheated. *Pyrolobus fumarii*, for example, was cultured from a submarine hydrothermal vent and grows optimally at about 106°C (Blöchl et al., 1997). The very existence of cells requiring such temperatures raises fundamental questions regarding how nucleic acids, proteins, reactive metabolites, and other molecules necessary for all life can be stabilized. The enzymes of most hyperthermophiles are intrinsically stable, reflecting a structural rigidity to which several features of the native conformation of these enzymes contribute (Jaenicke & Böhm, 1998). In contrast to proteins, the DNA of hyperthermophiles relies on extrinsic factors, primarily polyamines and small basic proteins, to ensure that its duplex form is thermodynamically favored in vivo. Whether these organisms have special strategies to deal with the threat to the primary structure of their DNA, posed by accelerated chemical decomposition, remains less clear, however (White & Grogan, 2008).

8. Summary

Though structurally simple, the akaryotic cells of bacteria and archaea carry out all the basic functions necessary for life, including nutrient uptake, energy conversion and conservation, growth, genome replication, cell division, and regulation of gene expression in response to environmental change. Several alternative mechanisms of phototrophy and of CO_2 fixation are represented among bacteria and archaea in addition to the Calvin–Benson cycle adopted by green plants. Some bacteria and archaea can use electron acceptors other than O_2 for oxidative phosphorylation; some species can grow chemolithotrophically, and some can fix N_2.

All bacteria and archaea maintain a transmembrane proton potential and depend on solute-specific transport proteins located in the CM to accumulate nutrients and to regulate ion flux. A common type of active transport in Gram-negative bacteria utilizes soluble proteins that bind the solute in the periplasm and deliver it to the corresponding membrane-bound transporter. Homeostatic responses to environmental stress use various regulatory mechanisms to alter the solute composition of the cytoplasm or relevant pathways of metabolism.

Specialized mechanisms enable a small minority of bacteria to invade multicellular organisms, persist in their tissues, and cause damage. Another minority of bacteria, and many archaea, require one or more environmental extremes (pH, salt concentrations, or temperature, for example) for optimal growth. Many of these "extremophiles" have enzymes that resist denaturation because of various structural features encoded in the amino-acid sequence.

References

Archibald, J. M. (2015). Endosymbiosis and eukaryotic cell evolution. *Current Biology, 25*, R911–R921. https://doi.org/10.1016/j.cub.2015.07.055

Armitage, J. P. (1999). Bacterial tactic responses. *Advances in Microbial Physiology, 41*, 229–289.

Baumeister, W., Wildhaber, I., & Phipps, B. (1989). Principles of organization in eubacterial and archaebacterial surface proteins. *Canadian Journal of Microbiology, 35*, 215–227.

Becker, K. W., & Skaar, E. P. (2014). Metal limitation and toxicity at the interface between host and pathogen. *FEMS Microbiology Reviews, 38*, 1235–1249. https://doi.org/10.1111/1574-6976.12087

Blöchl, E., Rachel, R., Burggraf, S., Hafenbrandl, D., Jannasch, H., & Stetter, K. O. (1997). *Pyrolobus fumarii*, gen. and sp. nov., represents a novel group of archaea, extending the upper temperature limit for life to 113 °C. *Extremophiles, 1*, 14–21.

Boucher, Y. (2007). Lipids: Biosynthesis, function, and evolution. In R. Cavicchioli (Ed.), *Archaea: Molecular and cellular biology* (pp. 341–353). Washington, D.C: American Society for Microbiology.

Chen, M.-T., & Lo, C.-J. (2016). Chapter 6. Using biophysics to monitor the essential protonmotive force in bacteria. In M. C. Leake (Ed.), *Biophysics of infection, advances in experimental medicine and biology* (79, p. 69). Switzerland: Springer International Publishing.

Dierstein, R., Schumacher, A., & Drews, G. (1981). On insertion of pigment-associated polypeptides during membrane biogenesis in *Rhodopseudomonas capsulata*. *Archives of Microbiology, 128*, 376–383.

Earhart, C. F. (1996). Uptake and metabolism of iron and molybdenum. In F. C. Neidhardt (Ed.), Escherichia coli *and* Salmonella*: Cellular and molecular biology* (2nd ed., pp. 1079–1090). Washington, D.C: American Society for Microbiology.

Fagan, R. P., & Fairweather, N. F. (2014). Biogenesis and functions of bacterial S-layers. *Nature Reviews Microbiology, 12*, 211–222.

Forterre, P. (2013). The common ancestor of archaea and eukarya was not an archaeon. *Archaea, 2013*, Article 372396. https://doi.org/10.1155/2013/372396

Fuerst, J. A. (2005). Intracellular compartmentation in planctomycetes. *Annual Reviews in Microbiology, 59*, 299–328.

Fuhrman, J. A., Schwalbach, M. S., & Stingl, U. (2008). Proteorhodopsins: An array of physiological roles? *Nature Reviews Microbiology, 6*, 488–494.

Goswami, P., Kuang, H., Jinhua, L., et al. (2022). Magnetotactic bacteria and magnetofossils: ecology, evolution and environmental implications. *Npj Biofilms and Microbiomes, 8*(43). https://doi.org/10.1038/s41522-022-00304-0

Grogan, D. W. (1996a). Isolation and fractionation of cell envelope from the extreme thermo-acidophile *Sulfolobus acidocaldarius*. *Journal of Microbiological Methods, 26*, 35–43.

Grogan, D. W. (1996b). Organization and interactions of cell envelope proteins of the extreme thermoacidophile *Sulfolobus acidocaldarius*. *Canadian Journal of Microbiology, 42*, 1163–1171.

Gunsalus, R. P. (1992). Control of electron flow in *Escherichia coli*: Co-Ordinated transcription of respiratory pathway genes. *Journal of Bacteriology, 174*, 7069–7074.

Hanson, R. S., & Hanson, T. E. (1996). Methanotrophic bacteria. Microbiol. *In Review, 60*, 439–471.

Harold, F. M., & Maloney, P. C. (1996). Energy transduction by ion currents. In F. C. Neidhardt (Ed.), Escherichia coli *and* Salmonella*: Cellular and molecular biology* (2nd ed., pp. 283–306). Washington, D.C: American Society for Microbiology.

Hegermann, J., Herrmann, R., & Mayer, F. (2002). Cytoskeletal elements in the bacterium *Mycoplasma pneumoniae*. *Naturwissenschaften, 89*, 453–458. https://doi.org/10.1007/s00114-002-0359-2

Hoffmann, T., Boiangiu, C., Moses, S., & Bremer, E. (2008). Responses of *Bacillus subtilis* to hypotonic challenges: Physiological contributions of mechanosensitive channels to cellular survival. *Applied and Environmental Microbiology, 74*, 2454–2460. https://doi.org/10.1128/AEM.01573-07

Holland, I. B., & Blight, M. A. (1999). ABC-ATPases, adaptable energy generators fuelling transmembrane movement of a variety of molecules in organisms from bacteria to humans. *Journal of Molecular Biology, 293*, 381–399. https://doi.org/10.1006/jmbi.1999.2993

Jaenicke, R., & Böhm, G. (1998). The stability of proteins in extreme environments. *Current Opinion in Structural Biology, 8*, 738–748.

Jogler, C., & Schüler, D. (2009). Genomics, genetics, and cell biology of magnetosome formation. *Annual Review of Microbiology, 63*, 501–521.

Kanjee, U., & Houry, W. (2013). Mechanisms of acid resistance in *Escherichia coli*. *Annual Reviews in Microbiology, 67*, 65–81.

Kletzin, A. (2007). General characteristics and important model organisms. In R. Cavicchioli (Ed.), *Archaea: Molecular and cellular biology* (pp. 14–92). Washington, D.C: American Society for Microbiology.

Kornspan, J. D., & Rottem, S. (2012). The phospholipid profile of mycoplasmas. *J. Lipids., 2012*, Article 640762. https://doi.org/10.1155/2012/640762

Lakshmanan, A., Lu, G. J., Farhadi, A., Nety, S. P., Kunth, M., Lee-Gosselin, A., Maresca, D., Bourdeau, R. W., Yin, M., Yan, J., Witte, C., Malounda, D., Foster, F. S., Schröder, L., & Shapiro, M. G. (2017). Preparation of biogenic gas vesicle nanostructures for use as contrast agents for ultrasound and MRI. *Nature Protocols*, 2050–2080.

Lefèvre, C. T., Song, T., Yonnet, J. P., & Wu, L. F. (2009). Characterization of bacterial magnetotactic behaviors by using a magnetospectrophotometry assay. *Applied and Environmental Microbiology, 75*, 3835–3841.

Madern, D., Ebel, C., & Zaccai, G. (2000). Halophilic adaptation of enzymes. *Extremophiles, 4*, 91–98.

Monds, R. D., & O'Toole, G. A. (2009). The developmental model of microbial biofilms: Ten years of a paradigm up for review. *Trends in Microbiology, 17*, 73–87. https://doi.org/10.1016/j.tim.2008.11.001

Nakane, D., & Miyata, M. (2009). Cytoskeletal asymmetrical dumbbell structure of a gliding mycoplasma, *Mycoplasma gallisepticum*, revealed by negative-staining electron microscopy. *Journal of Bacteriology, 191*, 3256–3264. https://doi.org/10.1128/JB.01823-08

Neidhardt, F. C., Ingraham, J. L., & Schaechter, M. (1990). *Physiology of the bacterial cell*. Sunderland, MA: Sinauer Associates, Inc.

Ng, S. Y., Chaban, B., & Jarrell, K. F. (2006). Archaeal flagella, bacterial flagella and type IV pili: A comparison of genes and posttranslational modifications. *Journal of Molecular Microbiology and Biotechnology, 11*, 167–191. doi:94053 [pii].

Nikaido, H. (2003). Molecular basis of bacterial outer membrane permeability revisited. *Microbiology and Molecular Biology Reviews, 67*, 593–656.

Nikaido, H. (2009). Multidrug resistance in bacteria. *Annual Review of Biochemistry, 78*, 119–146.

O'Brien, A. D., & Holmes, R. K. (1996). Protein toxins of *Escherichia coli* and *Salmonella*. In F. C. Neidhardt (Ed.), Escherichia coli *and* Salmonella*: Cellular and molecular biology* (2nd ed., pp. 2788–2802). Washington, D.C: American Society for Microbiology.

Ochman, H., Lawrence, J. G., & Groisman, E. A. (2000). Lateral gene transfer and the nature of bacterial innovation. *Nature, 405*, 299–304.

Pace, N. R. (1997). A molecular view of microbial diversity and the biosphere. *Science, 276*, 734–740.

Pettijohn, D. E. (1996). The nucleoid. In F. C. Neidhardt (Ed.), Escherichia coli *and* Salmonella*: Cellular and molecular biology* (2nd ed., pp. 158–166). Washington, D.C: American Society for Microbiology.

Piddock, L. J. (2006). Multidrug-resistance efflux pumps - not just for resistance. *Nature Reviews Microbiology, 4*, 629–636. doi:nrmicro1464 [pii].

Rodrigues-Oliveira, T., Belmok, A., Vasconcellos, D., Schuster, B., & Kyaw, C. M. (2017). Archaeal S-layers: Overview and current state of the art. *Frontiers in Microbiology, 8*, 2597.

Roesser, M., & Muller, V. (2001). Osmoadaptation in bacteria and archaea: Common principles and differences. *Environmental Microbiology, 3*, 743–754.

Russell, N. J. (1990). Cold adaptation of microorganisms. *Philosophical Transactions of the Royal Society of London B Biological Sciences, 326*, 595–608. discussion 608-11.

Schäfer, G., Anemüller, S., Moll, R., Meyer, W., & Lübben, M. (1990). Electron transport and energy conservation in the archaebacterium *Sulfolobus acidocaldarius*. *FEMS Microbiology Reviews, 75*, 335–348.

Schleper, C., Pühler, G., Holz, I., Gambacorta, A., Janekovic, D., Santarius, U., Klenk, H.-P., & Zillig, W. (1995). *Picrophilus* gen. Nov., fam. Nov: A novel heterotrophic, thermoacidophilic genus and family comprising archaea capable of growth around pH 0. *Journal of Bacteriology, 177*, 7050–7059.

Schwarz, U., Asmus, A., & Frank, H. (1969). Autolytic enzymes and cell division of *Escherichia coli*. *Journal of Molecular Biology, 41*, 419–429.

Skulachev, V. P. (1994). The latest news from the sodium world. *Biochimica et Biophysica Acta, 1187*, 216–221.

Slonczewski, J. L., & Foster, J. W. (1996). Ch. 96. pH-regulated genes and survival at extreme pH. In F. C. Neidhardt (Ed.), Escherichia coli *and* Salmonella*: Cellular and molecular biology* (2nd ed., pp. 1539–1549). Washington, D.C: American Society for Microbiology.

Smith, P. F. (1984). Lipoglycans from mycoplasmas. *Critical Reviews in Microbiology, 11*, 157–186. https://doi.org/10.3109/10408418409105476

Stetter, K. O., Fiala, G., Huber, G., Huber, R., & Segerer, A. (1990). Hyperthermophilic microorganisms. *FEMS Microbiology Reviews, 75*, 117–124.

Typas, A., Banzhaf, M., Gross, C. A., & Vollmer, W. (2011). From the regulation of peptidoglycan synthesis to bacterial growth and morphology. *Nature Reviews Microbiology, 10*, 123–136. https://doi.org/10.1038/nrmicro2677

Van De Vossenberg, J., Ubbink-Kok, T., Elferink, M., Driessen, A., & Konings, W. N. (1995). Ion permeability of the cytoplasmic membrane limits the maximum growth temperature of bacteria and archaea. *Molecular Microbiology, 18*, 925–932.

Walsby, A. E. (1994). Gas vesicles. Microbiol. *In Review, 58*, 94–144.

Wanner, B. L. (1993). Gene regulation by phosphate in enteric bacteria. *Journal of Cellular Biochemistry, 51*, 47–54.

White, D. (2000). *The physiology and biochemistry of prokaryotes* (2nd ed.). New York, N.Y: Oxford Univ. Press.

White, M. F., & Grogan, D. W. (2008). DNA stability and repair. In F. Robb, G. Antranikian, D. Grogan, & A. Driessen (Eds.), *Thermophiles: Biology and technology at high temperatures* (pp. 179–188). Boca Raton, FL: CRC Press.

Wilson, M. E., Maksaev, G., & Haswell, E. S. (2013). MscS-like mechanosensitive channels in plants and microbes. *Biochemistry, 52*, 5708–5722.

Woese, C. R., Kandler, O., & Wheelis, M. (1990). Towards a natural system of organisms: Proposal for the domains archaea, bacteria, and eucarya. *Proceedings of the National Academy of Sciences of the United States of America, 87*, 4576–4579.

Wood, J. M. (2011). Bacterial osmoregulation: A paradigm for the study of cellular homeostasis. *Annual Review of Microbiology, 65*, 215–238. https://doi.org/10.1146/annurev-micro-090110-102815

Yanofsky, C. (2000). Transcription attenuation: Once viewed as a novel regulatory strategy. *Journal of Bacteriology, 182*, 1–8.

Zarzycki, J., Brecht, V., Muller, M., & Fuchs, G. (2009). Identifying the missing steps of the autotrophic 3-hydroxypropionate CO2 fixation cycle in *Chloroflexus aurantiacus. Proceedings of the National Academy of Sciences of the United States of America, 106*, 21317–21322. https://doi.org/10.1073/pnas.0908356106

Further reading

Baumeister, W., & Lembcke, G. (1992). Structural features of archaebacterial cell envelopes. *Journal of Bioenergetics and Biomembranes, 24*, 567–575.

Celler, K., Koning, R. I., Koster, A. J., & van Wezel, G. P. (2013). Multidimensional view of the bacterial cytoskeleton. *Journal of Bacteriology, 195*, 1627–1636. https://doi.org/10.1128/JB.02194-12

Grogan, D. W. (1998). Hyperthermophiles and the problem of DNA instability. *Molecular Microbiology, 28*, 1043–1050.

Chapter 6

Nucleic acids and the cell nucleus

David P. Clark
Department of Microbiology, Southern Illinois University, Carbondale, IL, United States

1. Introduction

The cells of higher organisms, i.e., eukaryotes, contain a variety of organelles, some of which are surrounded by membranes. The nucleus is the most important organelle in the eukaryotic cell and is surrounded by a double membrane, the nuclear envelope. Pores in the nuclear envelope allow entry and exit of macromolecules in a strictly controlled manner.

The nucleus is generally ovoid, typically around 5 μ diameter, and often centrally located. The nucleus may vary considerably in size and shape in certain tissues, and some cell types contain multiple nuclei. Cancer cells often have nuclei that are aberrant in both size and shape (Jevtić et al., 2014).

The nucleus contains a set of chromosomes that carry most of the cell's genetic information in the form of DNA. Transcription of DNA to RNA also occurs in the nucleus. In many cases, this RNA is extensively modified before exiting the nucleus and entering the cytoplasm. A dense intranuclear body, the nucleolus is specialized in the synthesis of ribosomal RNA.

During cell division, the nuclear envelope dissolves. After the chromosomes have divided and segregated, two new nuclear envelopes are formed around the duplicate sets of chromosomes, creating two new nuclei. Chromosome segregation requires the formation of the spindle, which consists of microtubules. The organization of microtubules is controlled by the centrosomes, which are associated with the nucleus, although they are located outside the nuclear envelope.

2. Cell compartmentalization

A metabolically active membrane, the cell membrane, encloses all living cells (Chapter 4). This membrane controls the entrance and exit of molecules into and out of the cell. In addition, it may play a role in energy generation, cell motility, and a variety of other functions. Internally, the cell may be further subdivided into compartments by additional membranes. Depending on their plan of compartmentalization, biological cells may be divided into two major groups, the prokaryotes and eukaryotes.

2.1 Cell layout: Prokaryotes versus eukaryotes

The two major structural alternatives for cells are prokaryotes and eukaryotes. The prokaryotes (Bacteria and Archaea) show only minimal subdivision by membranes; in particular the genome, the machinery of gene expression, and the ribosomes, the sites of protein synthesis, all share the same major cell compartment, the cytoplasm (Fig. 6.1). Most important, the prokaryotes lack the nucleus characteristic of higher organisms—the eukaryotes. For this reason, they are also referred to as akariotics, as discussed in Chapter 5.

Prokaryotes consist of a single major compartment that contains the cytoplasm, nucleus, and ribosomes. The cytoplasmic membrane surrounds the cytoplasm, and outside this, there is usually a cell wall, as discussed in Chapter 5. Many bacteria possess further outer layers and a variety of appendages (not shown).

In contrast, the eukaryotes typically possess many internal membranes that subdivide the cell into a variety of compartments and organelles. Most prominent is the nucleus, which is surrounded by a double membrane, the nuclear envelope, and which contains the genome (Fig. 6.2). The eukaryotic genome consists of several chromosomes—long molecules of DNA that also carry many bound proteins. Most higher organisms are diploid and so contain pairs of homologous chromosomes in their nuclei. Transcription of the genomic DNA to give RNA also occurs within the nucleus and

Cell Physiology Sourcebook. https://doi.org/10.1016/B978-0-12-811114-7.00028-1

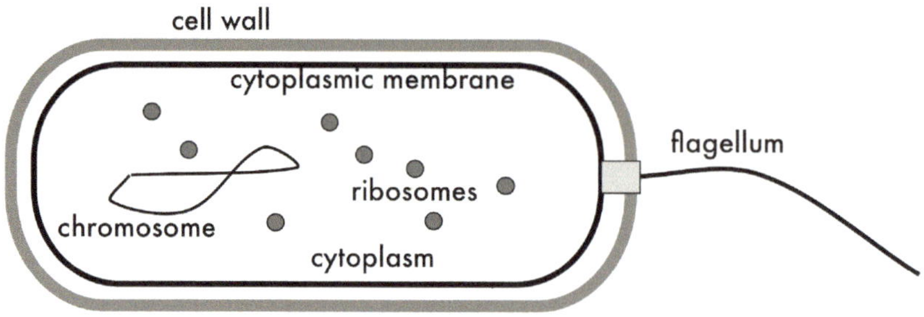

FIGURE 6.1 Typical prokaryotic cell layout.

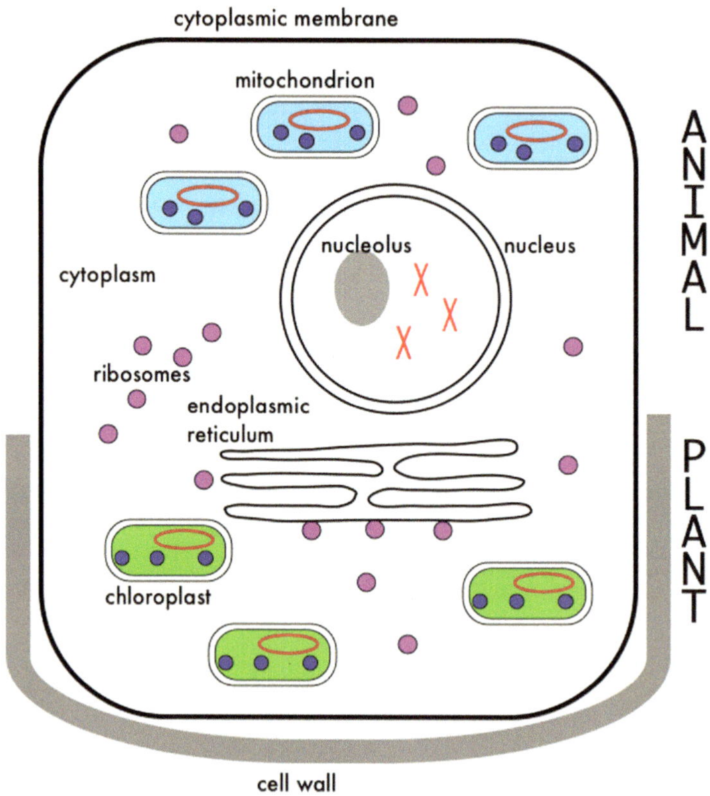

FIGURE 6.2 Stylized eukaryotic cell compartments.

most RNA molecules, including messenger RNA, are further processed before being permitted to exit through the nuclear pores into the cytoplasm. In eukaryotes, the ribosomes are located outside the nucleus in the cytoplasm and messenger RNA must therefore exit the nucleus in order to be translated into proteins.

Eukaryotic cells contain a variety of other membrane-bound compartments, although these are mostly peripheral to the subject of this chapter. The mitochondria and chloroplasts are notable because they both contain their own genomes and are surrounded by double membranes. As for the nucleus, entry into and exit from the mitochondria and chloroplasts is strictly controlled. Other membrane-bound organelles (that lack their own DNA) include the Golgi apparatus, lysosomes, and a variety of vesicles.

Eukaryotes are subdivided into several compartments by internal membranes. The nucleus contains the genome carried on several chromosomes. The endoplasmic reticulum is an internal membrane system that connects to the nucleus and forms a variety of vesicles in the cell. Mitochondria generate energy by respiration and contain their own ribosomes and a small genome. Plant cells also contain chloroplasts, which also contain their own ribosomes and genome and generate energy by photosynthesis. A variety of other vesicles and granules are often present (not shown). The cytoplasmic membrane surrounds the cell as a whole, and in plants, there is a cell wall further outside.

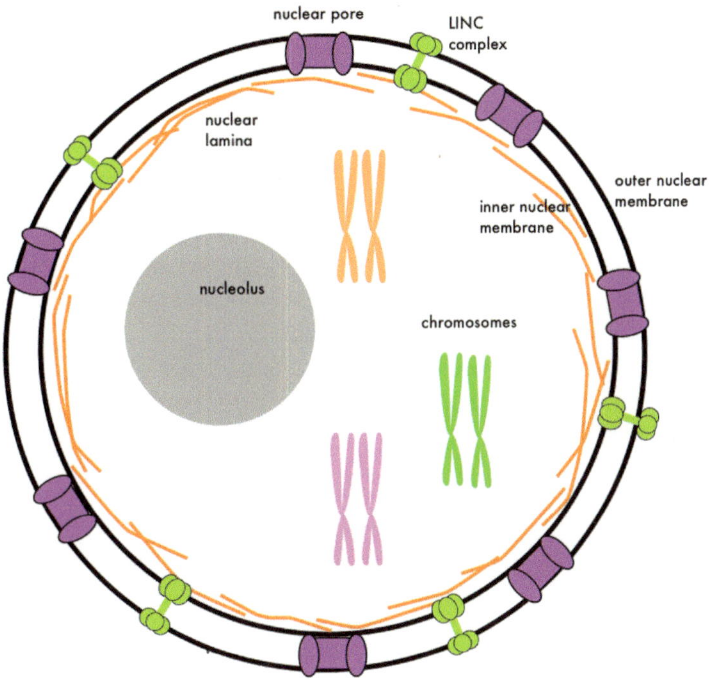

FIGURE 6.3 Schematic view of eukaryotic nucleus.

2.2 Overview of the nucleus

The nucleus is the most important of the organelles in the eukaryotic cell (Fig. 6.3). Firstly, it contains the genome, except for the small contributions from the mitochondrial and chloroplast genomes. Secondly, it is the location where DNA is transcribed into RNA and many RNA molecules remain in the nucleus where they are involved in processing and gene regulation. Thirdly, about a third of the cell's proteins belong in the nucleus. From an evolutionary perspective, the nucleus seems to have appeared as an extension of the endoplasmic reticulum (ER). Indeed, the nucleus is disassembled and reassembled from the ER every time the cell divides.

Although we tend to think of the nucleus as a storage facility for the genome, it is actually highly active in many respects. RNA synthesis, DNA repair, entry and exit of macromolecules, and positioning of chromosomes all occur routinely even in the "resting" nucleus between cell divisions (Rout & Karpen, 2014).

The nucleus contains a variety of substructures (Fig. 6.3). Around the outside is a double membrane—the nuclear envelope. Nuclear pores that permit the entry and exit of macromolecules cross the envelope. The LINC complex, which links the nuclear interior to the cytoskeleton outside, also crosses the envelope. Just inside the envelope is the nuclear lamina, composed of lamins, proteins that play several roles, including anchoring the chromosomes. The nucleolus is the most prominent of several nuclear bodies with specialized roles (discussed below). The nucleolus itself is devoted to synthesizing ribosomal RNA and assembling ribosomal subunits.

An envelope comprising inner and outer nuclear membranes surrounds the nucleus. On the inside of the envelope is the nuclear lamina. Nuclear pores penetrate the nuclear envelope to allow entry and exit of macromolecules. The LINC complex connects the nuclear interior to the cytoskeleton. The nucleolus is the site of ribosomal RNA synthesis and, although distinct, is not surrounded by a membrane.

2.3 Advantages of having a nucleus

The presence of the nuclear envelope, in addition to the cell membrane, provides the genome with an extra layer of protection from external dangers. Furthermore, the structure of the nucleus accommodates the presence and arrangement of multiple chromosomes. This allows the genome to be much larger than in prokaryotes.

Dividing the cell into subcompartments has advantages, irrespective of the nucleus itself. Macromolecules are extremely crowded in cells and occupy around 20%–30% of the available volume. As a consequence, the gap between macromolecules is less than their diameters (see Chapter 9). This situation favors the association of macromolecules into

TABLE 6.1 Entry of viruses into nucleus.

Virus	DNA or RNA	Where replicates	Parts entering nucleus	Comments
Herpes	DNA	Nucleus	Genome	
Adenovirus	DNA	Nucleus	Genome	
Orthomyxovirus, e.g., influenza	RNA	Nucleus	Genome plus proteins	
Lentivirus, e.g., HIV	RNA in virion DNA in nucleus	Nucleus	dsDNA plus proteins	Reverse transcription in cytoplasm
Parvovirus	RNA	Nucleus	Whole virion	
Polyomavirus	RNA	Nucleus	Genome	Disassemble in ER
Poxvirus	DNA	Cytoplasm	None	Cytoplasmic factories
Mimivirus	DNA	Cytoplasm	None	Cytoplasmic factories
Most RNA viruses	RNA	Cytoplasm	None	Carry own RNA replicase

complexes with related roles (Ellis, 2001), as discussed in detail in Chapter 9. It also favors division of the internal space of the cell into compartments that specialize in different functions. Thus, each compartment has its own complement of proteins and nucleic acid molecules that need to interact. Other molecules that might interfere are excluded. Strictly speaking, division does not require membranes (consider the nucleolus, for example); however, this is obviously an effective approach. From this perspective, the nucleus is just one of several subdivisions of the eukaryotic cell—the one that specializes in DNA replication and transcription.

An advantage that is specific to the nucleus is protection of the genome and the replication and RNA synthesis machinery from viruses. The majority of viruses fail to enter the nucleus and must replicate in the cytoplasm. RNA viruses that cannot enter the nucleus must carry their own RNA polymerases instead of stealing the host enzymes. Further, some DNA viruses replicate in the eukaryotic cytoplasm; these include the poxviruses and mimiviruses, which build cytoplasmic factories where they carry out both DNA replication and transcription using virus-coded components. A few viruses, such as retroviruses, avoid the issue by infecting dividing cells and waiting for the nuclear envelope to be disassembled. This introduces the disadvantage of waiting for host cell division. Table 6.1 summarizes the entry and replication of selected viruses relative to the cell nucleus (Kobiler et al., 2012).

Those viruses that do penetrate the nucleus possess a variety of strategies. Most go through the nuclear pore. However, some viruses have capsids too large to transit the pore, and so these viruses must disassemble first, and only the virus genome enters the nucleus (e.g., herpesvirus, adenovirus, influenza). Only a few tiny viruses can get through the pore before disassembling inside (e.g., hepadnavirus, parvovirus). Even those viruses that possess the ability to enter the nucleus pay a price. Often, only a minority of the invading virus particles succeed in getting across the nuclear envelope, and the majority fail and are trapped outside.

One possible outcome of the presence of a nucleus may be that eukaryotes tend to be mostly infected by RNA viruses. Most of the viruses that infect bacteria have DNA genomes and relatively few contain RNA. In higher organisms, the converse is true. When DNA viruses inject their DNA into bacteria, the viral DNA ends up in the same compartment as the bacterial DNA. The virus can then easily take advantage of host DNA synthesis enzymes and precursors to replicate its own DNA genome. In contrast, if viral DNA enters a eukaryotic cell, it finds itself in the cytoplasm, a compartment quite separate from the nucleus where the genome resides. No DNA polymerases or nucleotide precursors are available for the virus to commandeer, as they too are segregated in the nucleus.

2.4 Problems with having a nucleus

The presence of the nuclear envelope cuts off the genetic control center of the cell from the cytoplasm, where protein synthesis and most metabolic reactions occur. As a consequence, a complicated scheme is needed to send the mRNA carrying genetic information to the ribosomes. After the primary transcript is synthesized, complex processing is required to convert it into the mature mRNA. Only then is it allowed out of the nucleus via the nuclear pores. This, in turn, requires an array of protein factors and consumes energy.

Similarly, signals entering the cell from outside must also cross the cytoplasm and then the nuclear envelope in order to affect gene expression. This often involves relays of signaling proteins, although certain signal molecules, such as lipophilic steroid hormones, can diffuse directly to the nucleus via the membrane system.

Possession of a nucleus is clearly incompatible with the rapidly dividing lifestyle and emphasis on quantity over quality of bacteria. In contrast, for complex, relatively slow-growing cells with large genomes, the benefits of having a nucleus (and other subcompartments) outweigh the hindrances and extra resources needed.

3. Packaging the genome

3.1 DNA packaging in prokaryotes

In both prokaryotes and eukaryotes, the DNA molecules that make up the genome are extremely long and must be coiled and compacted in order to fit into the cell. If the DNA of a typical prokaryotic cell were spread out in a linear manner, it would be about 1 mm, roughly a thousand times longer than the cell it came from. Eukaryotes contain far more DNA than prokaryotes, so their situation is even more extreme. In order to fit into the cell, or in the case of eukaryotes, into the nucleus, the DNA must be folded up—but in such a way that the genes may be expressed as needed.

The prokaryotes and eukaryotes take totally different approaches to DNA packaging. In bacteria, the DNA, which is already a double helix, is supercoiled to compact it. The original double helix has a right-handed twist, but the supercoils are left-handed or "negative" supercoils. There is roughly one supercoil for every 200 nucleotides in typical bacterial DNA. A second level of compaction consists of arranging the supercoiled DNA in giant loops around a protein scaffold (Fig. 6.4) (Badrinarayanan et al., 2015). In *E. coli* with around 4000 genes and close to 4 megabases of DNA, there are approximately 50 such giant loops.

Supercoiling of bacterial DNA results in around 50 giant loops of supercoiled DNA extending from a central scaffold. As the inset indicates, each single line in this diagram represents the DNA double helix.

Supercoiling in bacteria is carried out by DNA gyrase, which inserts negative supercoils into circular DNA molecules such as bacterial chromosomes or plasmids. Gyrase cuts both strands of DNA, rotates the strands relative to each other, and then reseals the break. Each twist is energized by ATP hydrolysis. DNA gyrase is only found in prokaryotes. Consequently, antibiotics that inhibit DNA gyrase are harmless to animals, although lethal to bacteria. These antibiotics include fluoroquinolones such as norfloxacin and ciprofloxacin.

The compacted bacterial DNA, together with its associated proteins, is known as the nucleoid. The nucleoid stains densely when visualized under the electron microscope. Although it is distinct from the rest of the cytoplasm, the nucleoid differs from the nucleus of eukaryotes in that it is not surrounded by a membrane. A variety of DNA binding proteins help to compact the nucleoid. Some of these, e.g., the HU and H-NS proteins, are sometimes referred to as "histone-like". However, they do not have any sequence or structural homology to the true histones of eukaryotes. [Note that the Archaea do have genuine histone proteins that are smaller but homologous to those of eukaryotes.]

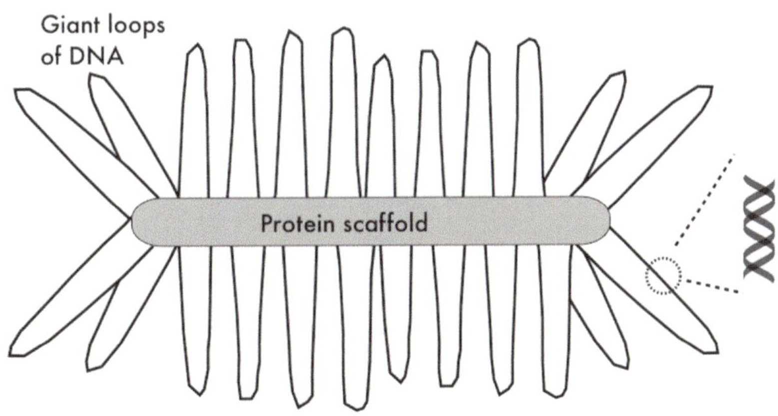

FIGURE 6.4 Arrangement of bacterial chromosome.

Despite their very different overall approaches to DNA condensation, prokaryotes and eukaryotes show similarity at the molecular level. They both share the condensin protein complex that binds and compacts DNA. As expected, there are minor variations in condensin substructure between prokaryotes and eukaryotes. In bacteria, there are two pairs of condensin subunits, whereas in eukaryotes all the subunits are different, although closely related.

Condensin forms a cage structure that can hold DNA strands together. The condensin cage uses energy from ATP hydrolysis to open and shut. This allows it to enclose strands of DNA. In bacteria, the condensins diffuse along the DNA and hold the various domains in the correct location within the nucleoid. Although it has no surrounding membrane, nucleoid organization by condensins is essential for bacteria to survive.

3.2 DNA packaging in eukaryotes

Eukaryotes possess multiple linear chromosomes, which contrasts with the single circular chromosome characteristic of bacteria. Furthermore, these chromosomes are located inside the nucleus, separated from the rest of the cell by the nuclear membrane. Eukaryotic chromosomes may be as much as a centimeter long and must be folded to fit into the cell nucleus, which is typically around five microns across, although there is considerable variation in size between organisms. This means that the DNA must be compacted about 2000-fold. Instead of supercoiling using DNA gyrase, the packaging mechanism in eukaryotes involves several levels of protein interaction, which are summarized in Table 6.2 (Corless & Gilbert, 2016).

The term chromatin refers to DNA plus the associated proteins that are responsible for DNA folding and chromosome structure. Depending on the level of condensation, chromatin varies considerably in composition and structure. At the lowest level of folding, eukaryotic DNA is covered with proteins known as histones. These are absent from bacteria, although shorter histone proteins are found in most members of the Archaea. Histones are extremely conserved, due to their critical role in protecting and compacting the genome. At the first level of folding, eukaryotic DNA, together with the histone proteins, forms structures known as nucleosomes. Each nucleosome consists of a core plus a linker (Fig. 6.5). The core comprises eight histones: two each of H2A, H2B, H3, and H4. The linker contains just one molecule of H1.

TABLE 6.2 Summary of chromosome folding.

Level of folding	Consists of	Base pairs per turn	Width (nm)
DNA double helix	Nucleotides	10	2
Nucleosomes	200 base pairs each	100	11
30 nm fiber	6 nucleosomes per turn	1200	30
Loops	50 helical turns per loop	60,000	300
Chromatid	2000 loops	n/a	800

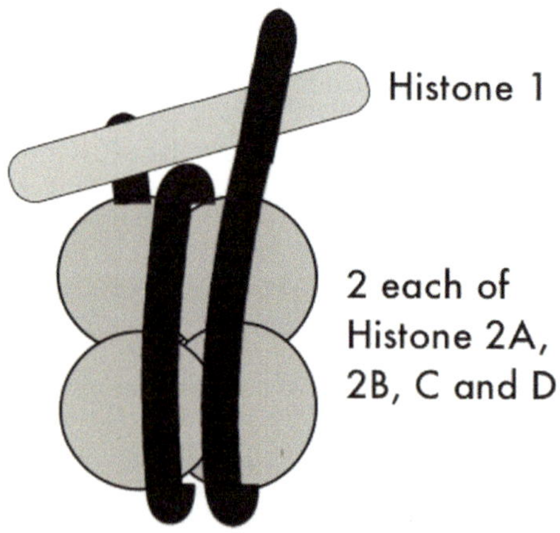

FIGURE 6.5 Nucleosome with histones.

The nucleosome is the basic unit in folding eukaryotic DNA. The DNA is wrapped around a core of eight histones. A separate linker histone (H1) is located where the wrapped DNA emerges. The core contains two copies each of histones 2A, 2B, 3, and 4.

About 200 bp of DNA is bound to each nucleosome. Most of the DNA (around 160 bp) is wrapped twice around the core histones , and the other 40 bp are bound to H1, forming the linker region that lies between successive nucleosomes along a strand of DNA. The linker DNA is relatively exposed and may be cut with DNases. This releases the core and linker components of the nucleosomes. A DNA strand covered in nucleosomes is sometimes referred to as "beads on a string".

The linker histone, H1, is longer than the core histones and has about 220 amino acid residues. H1 has two arms that project on either side and bind to the two neighboring nucleosomes. The central, more spherical domain of H1 binds to its own nucleosome.

The core histones are small, consisting of 100–135 amino acid residues, and more or less spherical. They have N-terminal tails that protrude from the nucleosome core and may be chemically modified in a variety of ways including methylation, acetylation, and phosphorylation. Histone modification is one factor that controls the state of DNA compaction. This affects the access of RNA polymerase to the DNA and hence the transcription of genes. When histones are highly methylated, access tends to be restricted, whereas highly acetylated histones are typically found in DNA that is transcriptionally active.

3.3 Higher levels of DNA condensation in eukaryotes

As noted in Table 6.2, the next level of compaction consists of a giant helical structure with six nucleosomes per turn known as the 30 nm fiber. The third level of compaction involves the looping of the 30 nm fiber with the ends attached to the chromosomal protein scaffold—in some ways reminiscent of the structure of the bacterial chromosome shown in Fig. 6.4. The details of these levels of folding are less well defined than the nucleosome, in particular, the structure of the protein scaffold is still vague in both bacteria and eukaryotes (Moindrot et al., 2013).

Part of the problem is that chromosome structure is dynamic and changes to accommodate gene expression by transcription. In particular, eukaryotic chromosomes are subdivided into domains. These unfold and refold individually and their structure is controlled partly by condensin and the related protein complex, cohesin.

When cell division is near, the chromosomes condense even further than normal. The precise details are uncertain, but condensed chromosomes at the start of mitosis are 50,000 times shorter than fully extended DNA. The oft-depicted chunky X-shaped chromosomes are those characteristic of metaphase. These actually consist of two identical double-helical DNA molecules that are still held together at the centromere. Each DNA molecule is a chromatid (Fig. 6.6). In nondividing cells, each chromosome consists of a single chromatid.

Between cell division, the interphase chromosomes consist of a single DNA molecule shown in an extended conformation. Before the next cell division, the DNA is replicated and each chromosome consists of two DNA molecules or chromatids linked at the centromere. Just before mitosis, condensation occurs, giving the metaphase chromosomes that are visible under the light microscope. Each daughter cell will acquire one of the chromatids (the anaphase chromosome).

3.4 Arrangement of chromosomes in the nucleus

The chromosomes do not merely fit inside the nucleus; they are divided into structural domains and arranged into chromosomal territories that are oriented specifically with respect to the nuclear envelope and nuclear interior.

Starting from the chromatin loops mentioned in Table 6.2, chromosomes are further organized into larger domains known as TADs (topologically associated domains). Typically, there may be 5–10 TADs for an average sized chromosome. Furthermore, each chromosome tends to occupy a specific region of the nucleus, known as a chromosomal territory (Fig. 6.7) (Nguyen & Bosco, 2015). Relative to the chromosomal territory, active genes tend to be found at the periphery. Genes at the periphery probably have better access to the machinery of transcription. A complication to this rule is that when a chromosome is located at the edge of the nucleus, those genes that lie close to the nuclear lamina tend to be inactive even though they are obviously on the periphery of the chromosomal territory.

Interphase chromosomes are arranged into large-scale domains and occupy specific regions of the nucleus, known as chromosomal territories. Architectural proteins maintain the structure of the TADs.

4. Nuclear synthesis and processing of RNA

The nucleus is not only the location where the genome is stored, but it is also a site of intense molecular activity. In particular, both the synthesis of RNA and much of the processing and modification of RNA occur in the nucleus.

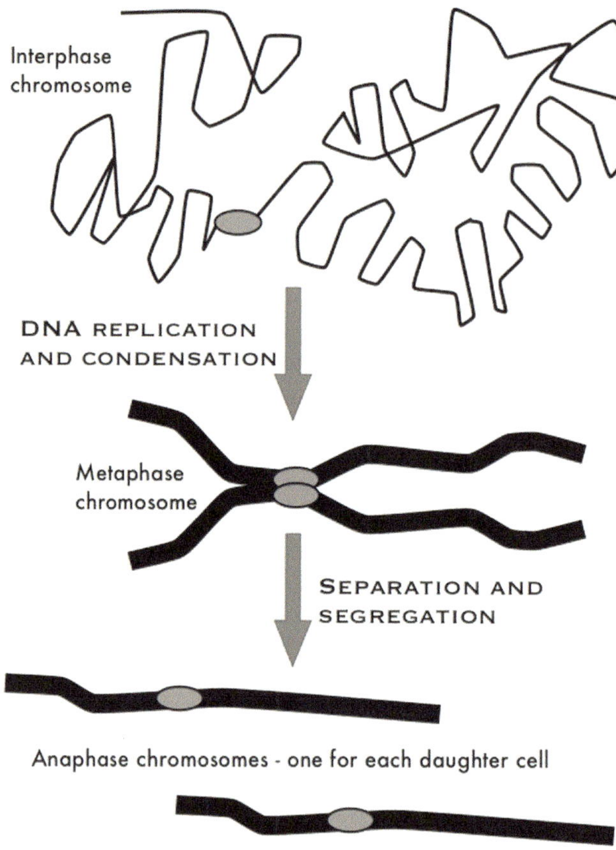

FIGURE 6.6 Interphase and metaphase chromosomes.

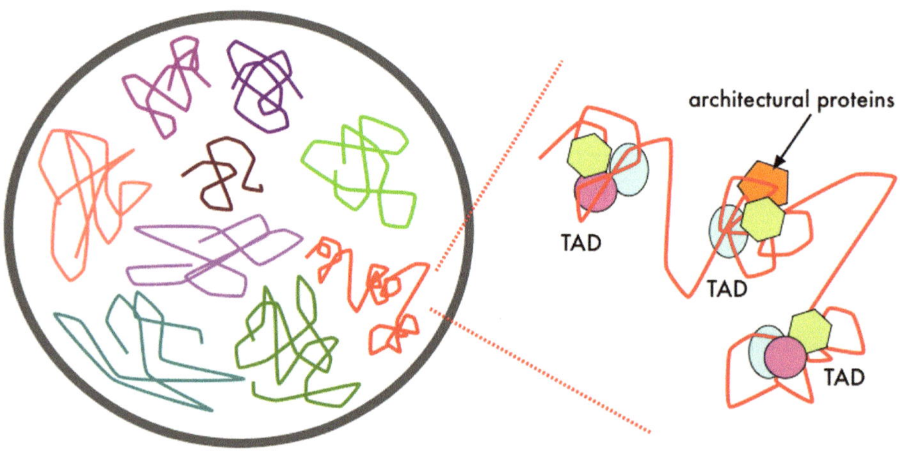

FIGURE 6.7 Chromosome TADs and territories.

4.1 RNA synthesis in the nucleus

The DNA genome resides in the nucleus, whereas synthesis of protein occurs on the ribosomes in the cytoplasm. The intermediary is messenger RNA, which is made in the nucleus and then exits through the nuclear pores to the cytoplasm. In addition to mRNA, protein synthesis also requires ribosomal RNA and transfer RNA. These are also made in the nucleus and exported. In addition, many other kinds of RNA are transcribed from the DNA in the eukaryotic nucleus. Most of these remain in the nucleus, where they take part in maintaining genomic structure, regulating gene expression, and processing RNA. A few exit to the cytoplasm to carry out their biological roles (e.g., the 7SL RNA of the signal recognition particle that is involved in protein export across membranes).

The list of noncoding RNAs has expanded greatly over the last few years, and a detailed discussion of all of these is way beyond the scope of this chapter (Cech & Steitz, 2014). However, Table 6.3 summarizes the main classes of RNA found in eukaryotes and their major roles. Note that many of these classes of RNA have individual members whose role is different from the majority.

The nucleus contains three distinct RNA polymerases that transcribe RNA from the DNA that encodes it. RNA polymerase I transcribes the genes for 5.8S rRNA, 18S rRNA, and 28S rRNA and RNA polymerase III transcribes the genes for tRNA, 5S rRNA, and many other small RNA molecules. RNA polymerase II transcribes most eukaryotic genes that encode proteins as well as some larger noncoding RNA molecules. Since all types of cells need rRNA and tRNA all the time, RNA polymerases I and III usually operate constitutively, whereas RNA polymerase II is highly regulated. Although rRNA only consists of four different molecules, in terms of bulk it makes up a great majority of the cellular RNA content. Perhaps because of this, a special zone of the nucleus, the nucleolus, is dedicated to the synthesis of 5.8S, 18S, and 28S rRNA by RNA polymerase I.

4.2 RNA processing in the nucleus

Before leaving the nucleus, most RNA molecules are processed in a variety of ways (Fig. 6.8). In some cases, further processing occurs in the cytoplasm. RNA may be cut and joined, bases may be altered, inserted or removed and, most common of all, bases may be chemically modified.

RNA processing can be classified as (A) cutting/joining RNA segments or (B) alterations to the individual bases.

Many, perhaps most, RNA molecules are modified by base alterations. The most heavily modified are rRNA and tRNA, although even mRNA has a few modified bases. Many of these modifications occur in the nucleus, although some are carried out in the cytoplasm. Base modification of tRNA depends largely on individual enzymes for each kind of base alteration. This also applies to bacterial rRNA, which is modified in only a handful of locations. The processing of eukaryotic mRNA and rRNA is more complex and discussed in more detail in the sections below.

4.3 Production of messenger RNA

Protein encoding genes are transcribed in the nucleus by RNA polymerase II. The primary transcript for eukaryotic mRNA must be processed in the nucleus before it is allowed to exit via the nuclear pores. It will then move to the cytoplasm where it will be translated by the ribosomes.

Processing of the primary transcript for mRNA involves removing the introns (splicing), adding a cap at the 5′ end and adding a tail to the 3′ end (Fig. 6.9). The 5′ cap is added first. This consists of a guanosine triphosphate (GTP) residue added in reverse orientation relative to the other nucleotides in the RNA. The guanine is methylated at position 7 to give a "cap0" structure (Cowling, 2009). Lower eukaryotes and plants retain the cap0 but higher eukaryotes add methyl groups to the ribose sugars of the first one or two nucleosides of the original mRNA to give cap1 and cap2 structures. In practice, human mRNA is a mixture with about 50:50 cap1 and cap2. The significance of this is obscure.

Next, the polyA tail consisting of 100–200 adenine residues is added to the 3′ end (Norbury, 2013). The mRNA primary transcript has a tail recognition sequence (AAUAAA) near its 3′ end. The RNA is cut a little downstream of this and the tail is added by poly(A) polymerase.

Transcription of the DNA generates the RNA primary transcript, which still contains the introns and exons plus a tail signal. Processing the RNA involves adding a cap and a poly A tail before splicing out the introns.

Finally, the introns are spliced out (Lee & Rio, 2015). This requires a complex known as the spliceosome, which consists of several enzymes and short RNA molecules known as snRNA (short nuclear RNA). Each snRNA plus its protein partners forms a small nuclear ribonucleoprotein (snRNP) and five of these comprise the spliceosome. The snRNAs recognize the splicing signals on the pre-mRNA and their protein partners carry out the cutting and rejoining reactions.

4.4 The nucleolus and other nuclear bodies

The nucleus contains several substructures known as nuclear bodies. These were originally visualized by electron microscopy but many can now be detected or analyzed by modern light microscopy techniques (Nunes & Moretti, 2017). Table 6.4 summarizes several nuclear bodies, although only the most important are considered further. The largest by far of the nuclear bodies is the nucleolus, which is discussed separately below.

Some nuclear bodies are built around DNA and function as specialized transcription factories. The nucleolus is the site where rRNA is made and is formed by the aggregation of those chromosomal regions where rRNA genes are clustered.

TABLE 6.3 Major classes of RNA.

	Class	Abbreviation	Size in nucleotides	Role
Genomic Integrity	Piwi interacting RNA	piRNA	25–32	Transposon silencing in germ line cells
	Small interfering RNA	siRNA	22	Defense against foreign RNA
	Telomerase RNA	TERC	451	Synthesis of telomeres
	Xist RNA	—	17,000	X chromosome inactivation
Regulation of transcription	Micro-RNA	miRNA	22	Regulation of mRNA expression
	Enhancer RNAs	eRNAs	200–500	Genetic regulation
	Circular RNA	circRNA	1000 or more	Regulation of miRNA abundance
	Long noncoding RNA	lncRNA	Wide range	Various regulatory roles
	tRNA fragments	tRF	20–40	Various regulatory roles
RNA processing	Guide RNA	gRNA		Editing of mRNA
	Small nuclear RNA	snRNA	100–300	Splicing of RNA
	Small nucleolar RNA	snoRNA	60–300	Nucleotide modification in RNA
Protein translation	Messenger RNA	mRNA	Wide range	Protein synthesis
	Transfer RNA	tRNA	70–90	Protein synthesis
	Ribosomal RNA	rRNA	120, 160, 1868, 5025	Protein synthesis
	Dual function RNA	—		Protein coding plus regulatory roles
Enzymatic function	Ribozymes	–	>250	Function as enzymes
	Signal recognition particle RNA	7SL RNA or SRP RNA	300	Membrane insertion of proteins

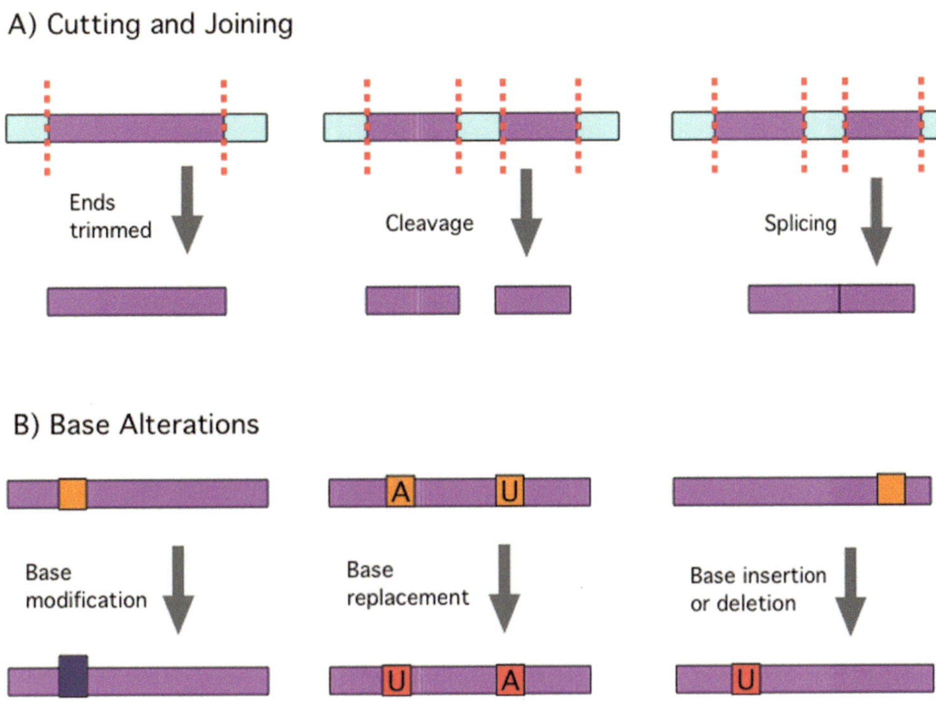

FIGURE 6.8 RNA processing overview.

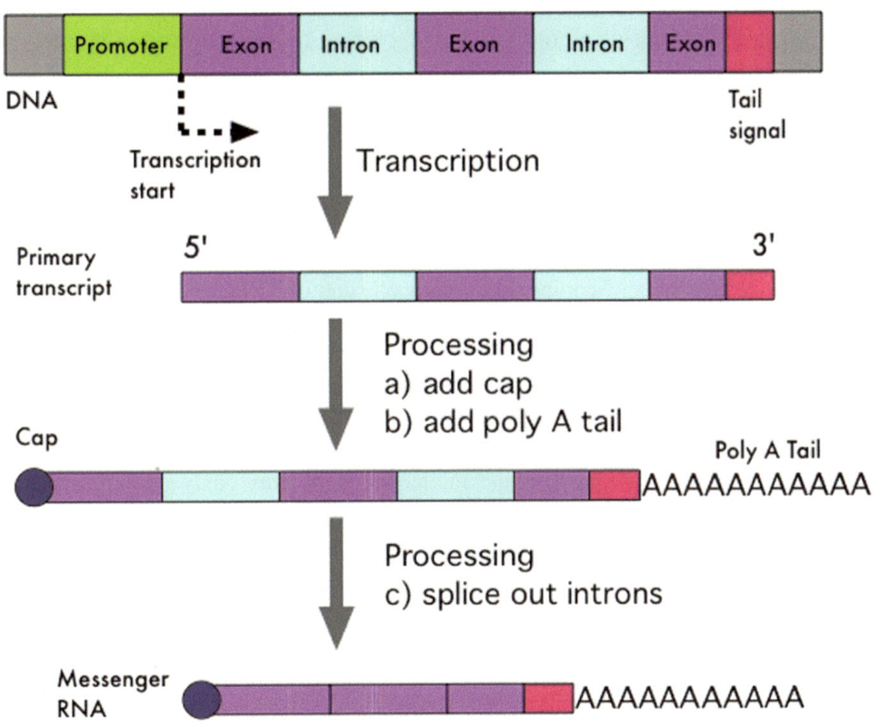

FIGURE 6.9 Overview of eukaryotic mRNA processing.

Similarly, the Cajal bodies are sites for the synthesis and further processing of snRNA and snoRNA. These two classes of small nuclear RNA are themselves involved in RNA processing: snRNA in splicing of mRNA and snoRNA in guiding base modifications in rRNA and other RNAs. Coilin is a structural protein characteristic of Cajal bodies and also found in histone bodies (or histone locus bodies).

TABLE 6.4 Major nuclear bodies.

Name	Size (μM)	Typical proteins and RNAs	Major function
Nucleolus	0.5–8	RNA polymerase 1	Synthesis of rRNA Assembly of ribosomal subunits
Cajal bodies	0.1–2	Coilin	Synthesis and processing of snRNA and snoRNA
Speckles	0.8–2	SRSF1, SRSF2 Malat1 RNA	Reservoir for splicing factors
Paraspeckles	~0.5	PSP1, PSF NEAT1 RNA	mRNA regulation; RNA editing
Histone body	0.2–1	NPAT, FLASH	Histone RNA synthesis
Stress body	0.3–3	HSF1, HAP SatIII RNA	Regulation of transcription and splicing during stress
Clastosome	0.1–1	Proteasome	Protein degradation
PcG bodies	0.3–1	Polycomb proteins	Repressing gene expression

Other nuclear bodies do not carry out transcription and are therefore not built around DNA. Instead they are built using long noncoding RNA molecules as scaffolds. Speckles rely on Malat1 RNA and paraspeckles rely on NEAT1 RNA (Caudron-Herger & Rippe, 2012). Speckles are mainly storage compartments for components of splicing machinery not in active service. However, they do appear to play an active role in the assembly of some RNA–protein complexes involved in splicing. Paraspeckles are less well defined but play a role in editing and regulating mRNA.

None of these intranuclear compartments is surrounded by a membrane. Instead, their macromolecular components crowd together, excluding other molecules, and this generates a form of phase separation that maintains them as separate entities (Ulianov et al., 2015).

4.5 The nucleolus and ribosomal RNA

Although not surrounded by a membrane, the nucleolus is a distinct area within the nucleus that specializes in manufacturing ribosomal RNA and also oversees ribosomal subunit assembly. When the protein synthesis rate is high, rRNA and ribosome production increases and the nucleolus may expand to occupy up to 25% of the volume of the nucleus.

The genes for rRNA exist in hundreds of copies. In humans clusters of tandem repeats are found on five chromosomes. The genes for 18S, 5.8S, and 28S rRNA form a single transcription unit that is transcribed by RNA polymerase I to give a 45S pre-rRNA (Henras et al., 2015). The 5S rRNA is encoded separately and is transcribed by RNA polymerase III. The 45S pre-rRNA is then cleaved and processed to generate the three final 18S, 5.8S, and 28S rRNAs (Fig. 6.10).

The genes for rRNA are clustered at multiple sites on the DNA. Each transcription unit of DNA produces an initial 45S pre-RNA molecule that is processed to yield the final 18S, 5.8S, and 28S ribosomal RNA molecules.

The chromosome segments that congregated in the nucleolus were previously called "nucleolar organizers." These correspond to the location of clusters of rRNA genes. The new rRNA molecules bind proteins forming ribonucleoprotein (RNP) particles that may be visualized as a dense granular region under the microscope. These RNP particles will ultimately become part of the ribosomes.

Eukaryotic rRNA is also heavily modified by base alteration. The two major modifications are the conversion of uridine to pseudouridine and the methylation of the 2′-OH group of the ribose backbone. Both processes depend on short guide RNA molecules to locate the correct sites for modification. Both processes are carried out in the nucleolus and the guide RNAs are known as small nucleolar RNAs (snoRNAs). Human rRNA is modified at around 200 positions and the base sequences around each site are almost never related. Consequently, there is a different snoRNA with a specific recognition sequence for each site. The snoRNAs are 70–100 bases long and bind both to the rRNA and to the modification enzymes.

4.6 The nucleolus and ribosomal assembly

Although ribosomal proteins are manufactured in the cytoplasm, the two subunits of the ribosome are mostly assembled in the nucleolus. Thus, ribosomal proteins must enter the nucleus and associate with rRNA in the nucleolus. This results in

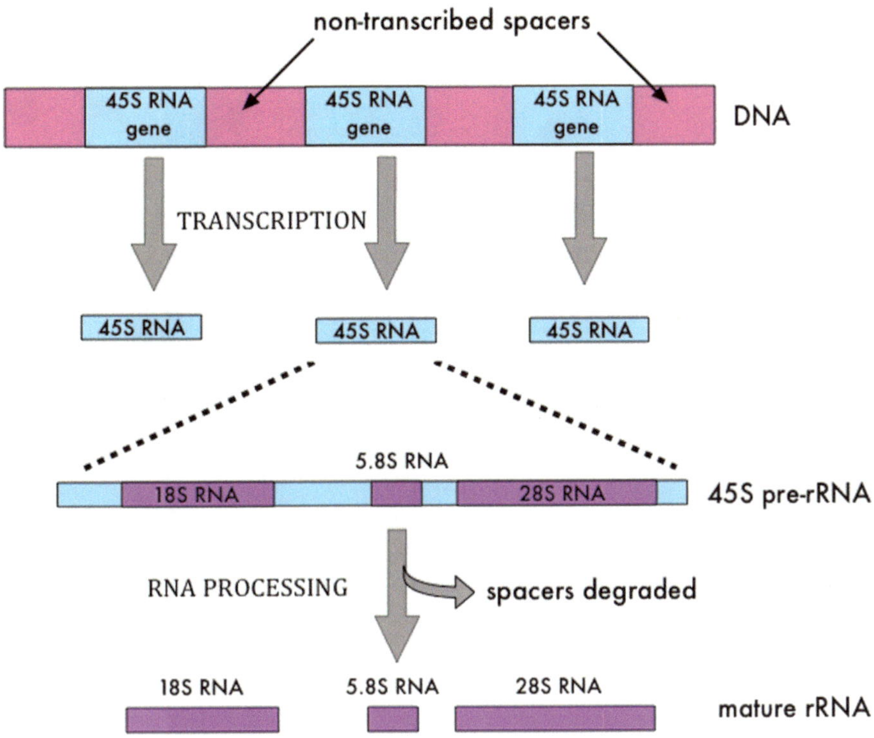

FIGURE 6.10 Transcription and processing of rRNA genes.

creation of pre-40S and pre-60S ribosomal subunits. These are exported through the nuclear pores into the cytoplasm where final processing occurs (Fig. 6.11). Although the pre-60S subunit leaves the nucleus with mature 5S rRNA, 5.8S rRNA, and 28S rRNA, the pre-40S subunit contains a 20S precursor to 18S rRNA that is finally processed in the cytoplasm.

Assembly and export of the ribosome subunits is extremely complex. In addition to the 4 rRNA and 139 ribosomal proteins that comprise the final ribosome more than 350 additional protein factors are involved in processing and export (Nerurkar et al., 2015). Not surprisingly, the individual roles of many of these factors are still vague.

Indeed, before assembly can begin the nucleus must import the ribosomal proteins. In yeast, a typical cell must import around 15 million ribosomal proteins per generation. Specialized proteins, known as importins, are needed for ribosomal proteins to transit the nuclear pore. Since many ribosomal proteins bind strongly to RNA, they cannot be allowed to diffuse freely around the nucleus, which is full of RNA molecules. Instead, another set of special carriers termed escortins transfer the ribosomal proteins from the nuclear pore to the nucleolus (Fig. 6.11).

Transcription of rRNA occurs in the nucleosome but protein synthesis occurs in the cytoplasm. Ribosomal proteins must therefore be imported into the nucleus by importins and carried to the nucleosome by escortins in order to assemble onto the rRNA. Some ribosomal proteins plus the 45S pre-rRNA and the small 5S rRNA aggregate into the 90S complex. After processing, this splits into precursor complexes for the 40S and 60S ribosomal subunits. These travel to the cytoplasm where they form ribosomes. At each stage of transit (grey arrows in Fig. 6.11), a variety of protein factors arrive and depart, and more ribosomal proteins are added (details not shown). The rRNA is also processed along the route to the outside.

5. The nuclear envelope

5.1 Structure of the nuclear envelope

The envelope surrounding the nucleus consists of a double membrane that is penetrated by nuclear pores (see below). In addition to merely sealing off the genome from the rest of the cell, the nuclear envelope plays an active role in several processes such as DNA replication, regulating transcription, and establishing chromosome territories.

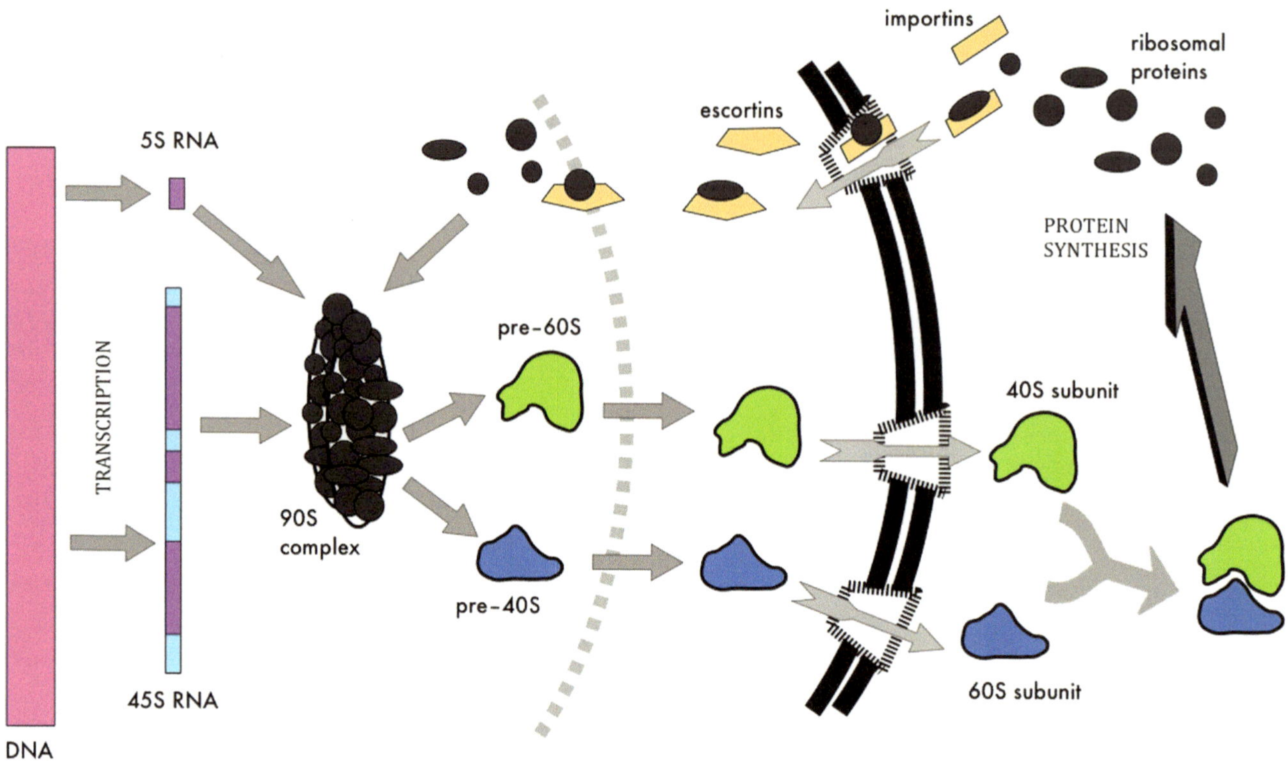

FIGURE 6.11 Assembly and export of ribosomal subunits.

The outer nuclear membrane is often covered with ribosomes and is continuous with the endoplasmic reticulum. In contrast, the inner nuclear membrane is isolated from the endoplasmic reticulum and has a unique protein composition. Inside the inner nuclear membrane is a layer of protein, the nuclear lamina, composed of proteins named lamins (see Fig. 6.3, above). The lamins form an interconnected meshwork of fibers (Yáñez-Cuna & van Steensel, 2017).

The lamina aids in the folding and location of chromosomes within the nucleus. Many chromosomes possess segments that bind to the lamina—the lamina-associated domains (LADs). The mammal genome contains 1000 or so LADs. Some of these LADs are functional in all cell types (i.e., constitutive LADs), whereas others only function in some cell types. The constitutive LADs are highly conserved among different mammals suggesting that they form stable anchor points for chromosomes. Regions of chromosomes located at the periphery of the nucleus due to LAD binding tend to show low gene expression. Conversely, genes on those regions of chromosomes that are further interior tend to be more highly expressed (Gruenbaum & Foisner, 2015).

A relatively recent finding is that the cytoskeleton in the cytoplasm is connected to the interior of the nucleus. LINC complexes span the nuclear envelope and link to the cytoskeleton via connector proteins known as KASH proteins (see Fig. 6.3). Different KASH proteins can link to microtubules, actin, myosin, intermediate filaments, and the centrosomes (Luxton & Starr, 2014). Internally, the LINC complexes bind to the nuclear lamina.

5.2 Nuclear pores control entry and exit

The nuclear envelope allows a variety of macromolecules, including proteins and RNA, to enter or exit via the nuclear pores. This process is strictly controlled by the nuclear pore complex, which surrounds each pore and is the largest protein complex in the cell (about 65 MDa in yeast and double that in higher animals). The pore complex shows eight-fold symmetry and consists of 30 or more different proteins, the nucleoporins or Nups (Fig. 6.12).

The nuclear pore has a basket structure on the inside and cytosolic filaments projecting outward into the cytoplasm. The central pore structure is roughly cylindrical but with eightfold symmetry.

Exit or entry or through the nuclear pores is under the control of proteins known respectively as exportins and importins (Cautain et al., 2015). Separate exportins and importins are specific for different classes of molecules. For example,

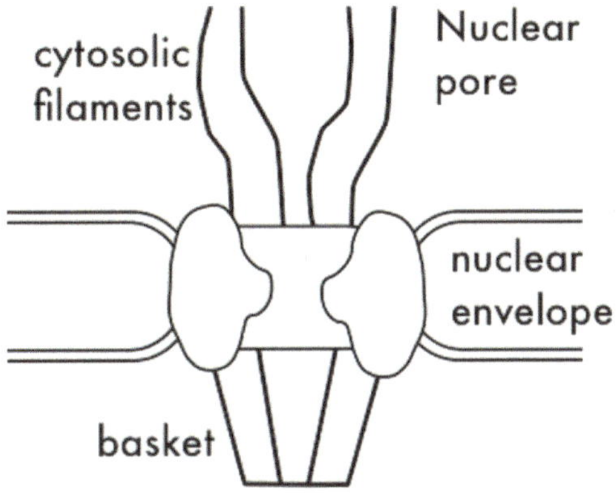

FIGURE 6.12 Nuclear pore cross section.

exportin-t recognizes tRNA and exportin 5 is specific for microRNA precursors. Small molecules can diffuse freely, but large molecules need energy for transit through the pores. RAN protein couples the hydrolysis of GTP to transport for the majority of exportins and importins (Okamura et al., 2015).

The exit of messenger RNA is a special case and requires ATP as an energy source. Moreover, mRNA must be fully processed, that is, the mRNA must have the introns spliced out and the cap and tail added before exiting the nucleus. The mRNA is bound to the spliceosome while being processed, and this prevents exit until splicing is finished. Exit occurs in three stages: docking onto the nuclear pore basket, transiting the central channel, and then being released by the cytoplasmic filaments. Export of mRNA typically involves binding to the Nxf1/Nxt1 complex. Upon energy release, this alternates between two conformations one of which promotes release. Oddly, the mRNA fails to exit successfully about two thirds of the time and must return to the nucleus and try again!

5.3 Division of the nucleus

When the eukaryotic cell divides, it faces the problem that its genome is sealed inside the nuclear envelope. During the familiar process of mitosis, the DNA replicates and the chromosomes divide while the nuclear envelope is still intact (prophase). The nuclear envelope then disassembles. The mitotic spindle assembles and binds to the condensed chromosomes (see Fig. 6.6), which then line up at the cell equator (metaphase). Eventually, the spindle pulls the separated chromatids toward the cell poles (anaphase). Finally, the spindle disaggregates and nuclear membranes are reformed around each of the daughter nuclei (telophase) (McIntosh, 2016, pii).

Thus, each time the cell divides, the nuclear envelope must be disassembled and, later, two new envelopes are assembled around the two new sets of chromosomes. In fact, the nuclear envelope does not really vanish. It merges temporarily into the endoplasmic reticulum, and its components are then reused when the two new nuclear envelopes are assembled. When mitosis begins, many proteins of the nuclear envelope are released into the endoplasmic reticulum, whereas most of the proteins from the nuclear pores are released into the cytoplasm. Disassembly of the nuclear pore complex is due to phosphorylation of its key components (Champion et al., 2017). Disassembly of the nuclear lamina is also due to phosphorylation of the lamin proteins.

Membrane reassembly is at least partly driven by the chromosomes themselves. Changes in histone modification are involved, including removal of phosphate groups from histone threonine and serine and of methyl groups from lysine on histone side chains. The altered nucleosomes then attract the nucleoporins responsible for nuclear pore formation, the lamins required to form the nuclear lamina, and other proteins characteristic of the nuclear envelope. The two nuclear membranes then assemble around these core proteins (LaJoie & Ullman, 2017). Envelope reassembly involves the removal of the phosphate groups from those components of the nuclear lamina and the nuclear pores that were phosphorylated during membrane breakdown. Some spindle components must also be dephosphorylated to promote spindle breakdown.

5.4 The centriole and mitotic spindle

The segregation of chromosomes during mitosis and meiosis is carried out by a structure known as the spindle from its appearance under the microscope. The spindle fibers are in fact microtubules that are constantly growing or contracting due to the addition or removal of tubulin protein subunits (Fig. 6.13). One outstanding puzzle is how the spindle can exert the

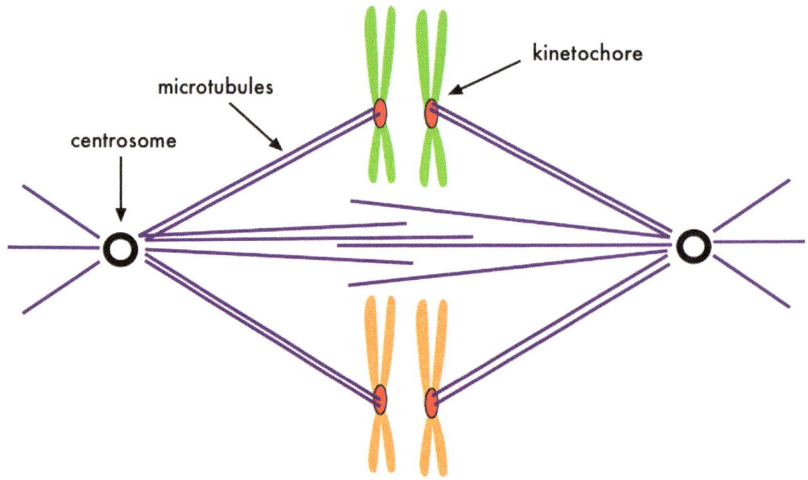

FIGURE 6.13 Stylized spindle organization.

force needed to pull chromosomes apart. There appears to be no solid evidence for any obvious anchor points or scaffold to provide support for this activity (Schweizer et al., 2014).

The spindle is made up of microtubules with several different roles. Some attach to the kinetochores of the chromosomes and function directly in chromosome segregation. Other microtubules overlap between the chromosomes and maintain spindle structure and the astral microtubules project outward from the centrosomes.

A variety of other proteins are bound to the microtubules, including the molecular motors dynein and kinesin. These motors bind to the kinetochores and use energy to move chromosomes along the microtubules. This is important for lining up the chromosomes at the cell equator during metaphase.

In most animals, the spindle forms around the centrosome, which consists of a pair of centrioles lying at right angles to each other. These organelles organize not only the spindle but also the cilia and flagella. Thus, they act as microtubule-organizing centers as all these structures are constructed largely from microtubules (Meunier & Spassky, 2016). The centrioles themselves consist of nine bundles of three microtubules each arranged in a cylinder with a surrounding sheath and protein appendages. Preceding cell division, each centriole acts as a template for the assembly of a new centriole (Fırat-Karalar & Stearns, 2014, pii). The new centrioles are built attached sideways onto the old centriole, rather like a bud (Fig. 6.14). Thus, centrioles do not actually divide and, indeed, in some organisms they can be built from scratch. After duplication, the two centrosomes move to opposite poles of the cell and the spindle forms between them.

The centrosome consists of two centrioles that are connected by a protein tether. Just before cell division, each centriole begins building a new centriole that is attached sideways via a stalk. Eventually, two new centrioles are completed, and there are now two centrosomes. The tether is then broken and one centrosome migrates to the opposite side of the nucleus (not shown).

6. Macro- and micronuclei in ciliates

Even in familiar higher animals, there is considerable variation in the size and shape of the nucleus from one cell type to another. However, one of the most bizarre examples of nuclear behavior is found in the ciliates, which are unique in possessing two distinct types of nucleus—the macronucleus and the micronucleus (Bracht et al., 2013).

The micronucleus may be regarded as equivalent in some ways to the germline cells of typical animals and the macronucleus to the somatic cells. Thus, gene expression takes place only in the macronucleus, whereas the micronucleus preserves the genome from one generation to another. During vegetative (i.e., nonsexual) reproduction, both types of nucleus divide in synchrony with cell division.

Conversely, during sexual reproduction no cell division occurs, instead two ciliates conjugate and exchange haploid micronuclei. In each partner, these fuse to form a diploid micronucleus. The previous macronuclei are degraded and must be replaced to allow future gene expression. To achieve this, the diploid micronucleus divides and one of the copies morphs into a new macronucleus by a bizarre program of DNA rearrangement. The details of the *Paramecium* sexual cycle are shown in Fig. 6.15.

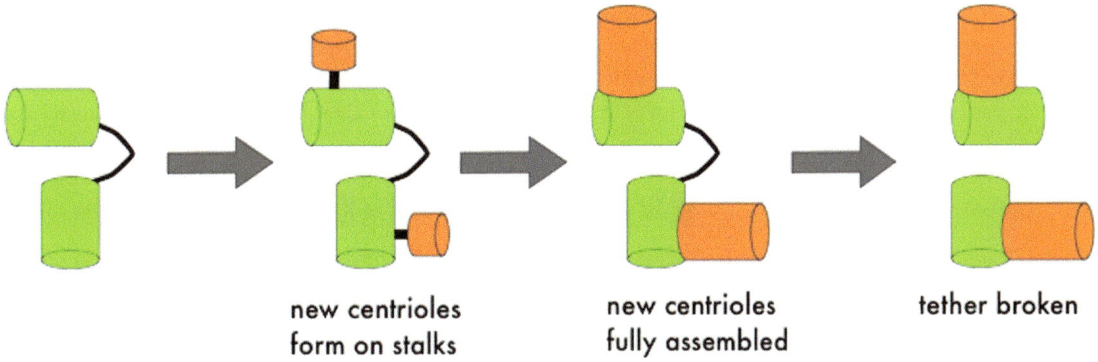

FIGURE 6.14 Overview of centriole duplication.

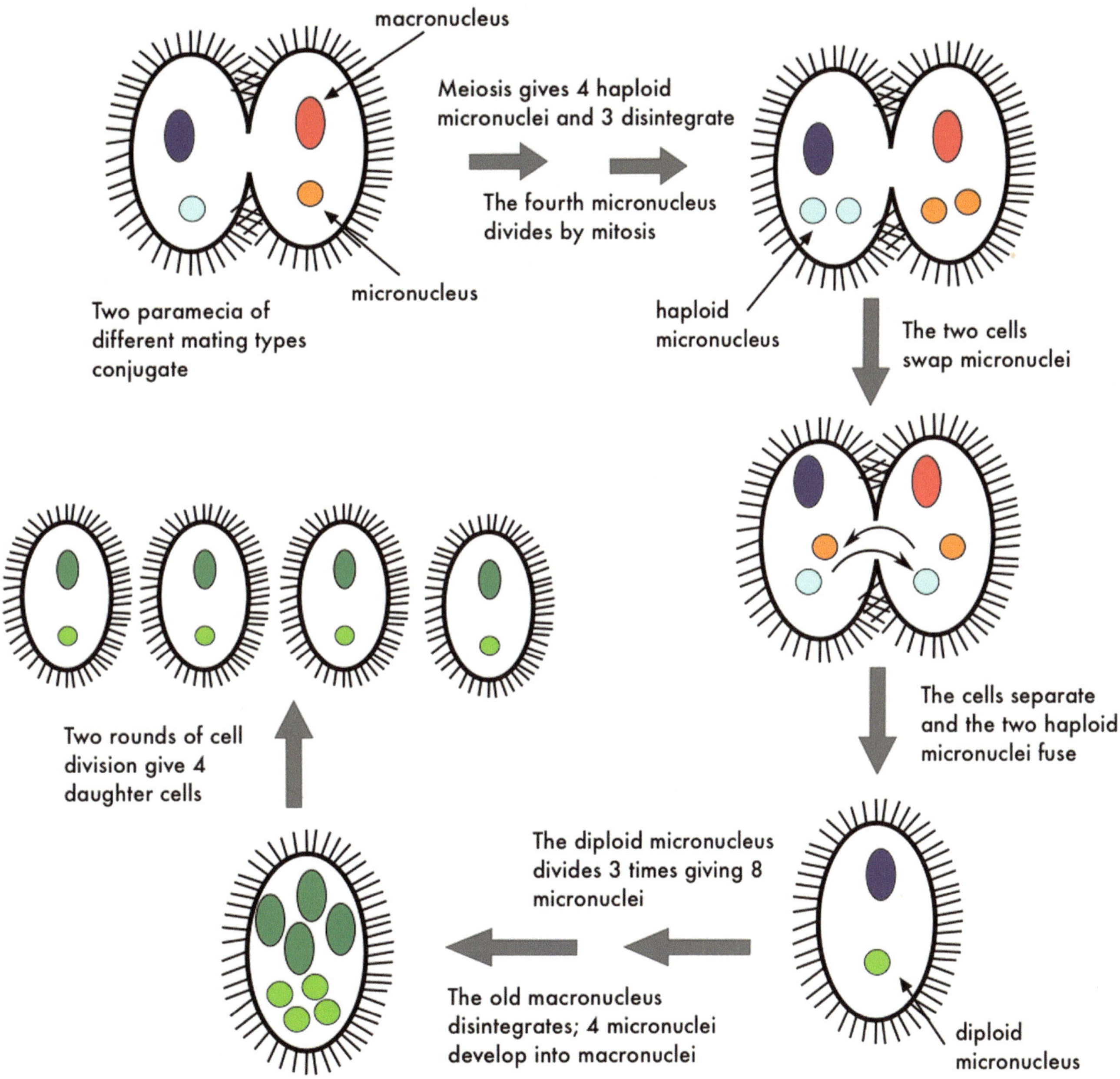

FIGURE 6.15 The paramecium sexual cycle.

Between cell division, the interphase chromosomes consist of a single DNA molecule shown in an extended conformation. Before the next cell division, the DNA is replicated and each chromosome consists of two DNA molecules or chromatids linked at the centromere. Just before mitosis, condensation occurs, giving the metaphase chromosomes that are visible under the light microscope. Each daughter cell will acquire one of the chromatids (the anaphase chromosome).

The macronucleus is thus reconstructed from a micronucleus after every sexual cycle. During this process, the chromosomes are both amplified around 400-fold to give multiple copies and also fragmented. The final macronucleus contains thousands of small mini-chromosomes that range in size from around 50 Kbp to 1 Mbp some, thus some are so short they only carry a single gene (Betermier & Duharcourt, 2014).

This restructuring involves a bizarre program of DNA rearrangement. During this process, a considerable amount of DNA is spliced out, some segments from the interior of genes, others from intergenic regions. Nearly half of the 40,000 estimated genes of *Paramecium* are interrupted by such Internal Eliminated Sequences (IESs) while in the micronucleus and are therefore nonfunctional. Overall, up to 30% of the DNA sequence of the micronucleus may be discarded in this manner.

The generation of the macronucleus is extremely complex and only partly understood. Nonetheless, it is clear that histone modifications and small guide RNA molecules are critical. One truly strange aspect of this is that excision of the IES sequences is carried out by a tamed transposase that derives from a transposon of the widely distributed piggyBac

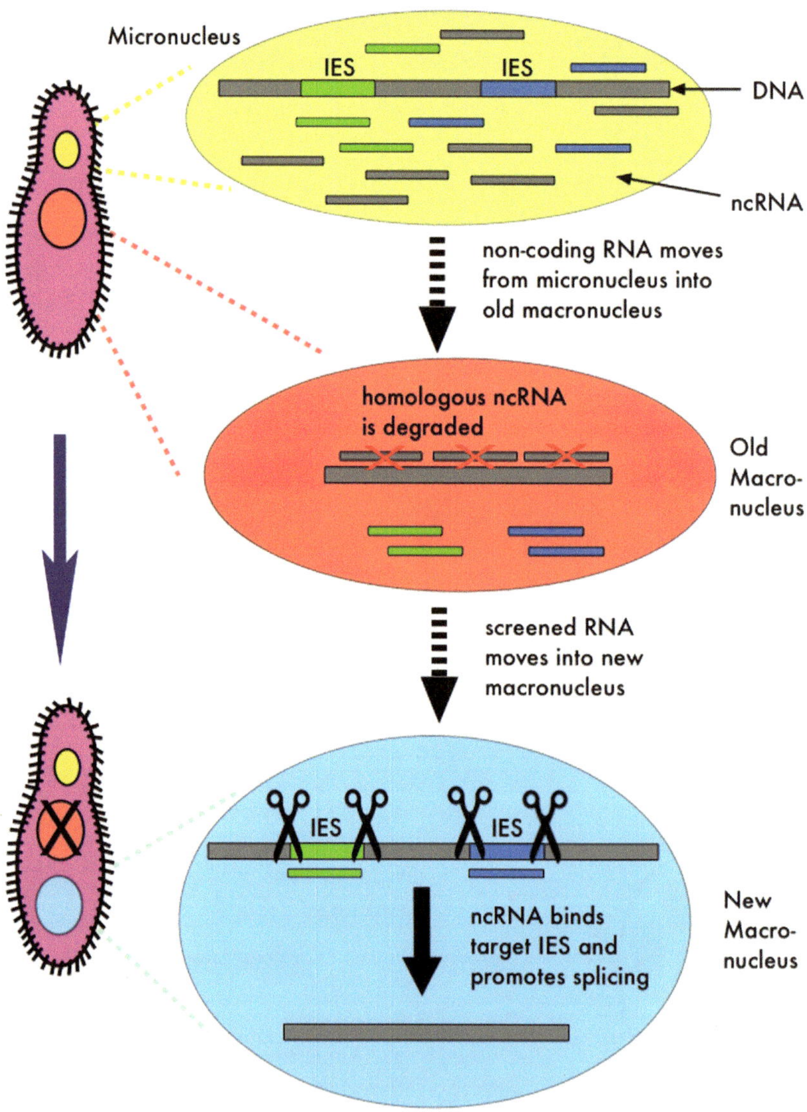

FIGURE 6.16 Model for IES excision in *Paramecium*.

family. Some IES possess inherent recognition sites for this enzyme, but many do not and instead rely on being identified by noncoding RNA. The precise mode of operation of these noncoding RNAs is still speculative and several schemes have been proposed, one of which is shown in Fig. 6.16. Whatever the precise mechanism, it seems clear that noncoding RNAs bind to sequences within their target IES (Chalker et al., 2013).

Noncoding RNA (ncRNA) supervises the excision of IES sequences. Transcription in the micronucleus produces a large number of ncRNA molecules that originally recognize both those DNA sequences to be retained and those to be excised. The ncRNA passes through the old macronucleus, where those molecules that are homologous to retained DNA are eliminated. The remainder of the ncRNA molecules (i.e., those that match IES sequences) then move to the developing macronucleus and target the corresponding IES sequences for excision.

References

Badrinarayanan, A., Le, T. B., & Laub, M. T. (2015). Bacterial chromosome organization and segregation. *Annual Review of Cell and Developmental Biology, 31*, 171–199.

Betermier, M., & Duharcourt, S. (2014). Programmed rearrangement in ciliates: Paramecium. *Microbiology Spectrum, 2*(6).

Bracht, J. R., Fang, W., Goldman, A. D., Dolzhenko, E., Stein, E. M., & Landweber, L. F. (2013). Genomes on the edge: Programmed genome instability in ciliates. *Cell, 152*(3), 406–416.

Caudron-Herger, M., & Rippe, K. (2012). Nuclear architecture by RNA. *Current Opinion in Genetics & Development, 22*(2), 179–187.

Cautain, B., Hill, R., de Pedro, N., & Link, W. (2015). Components and regulation of nuclear transport processes. *FEBS Journal, 282*(3), 445–462.

Cech, T. R., & Steitz, J. A. (2014). The noncoding RNA revolution-trashing old rules to forge new ones. *Cell, 157*(1), 77–94.

Chalker, D. L., Meyer, E., & Mochizuki, K. (2013). Epigenetics of ciliates. *Cold Spring Harbor Perspectives in Biology, 5*(12), Article a017764.

Champion, L., Linder, M. I., & Kutay, U. (2017). Cellular reorganization during mitotic entry. *Trends in Cell Biology, 27*(1), 26–41.

Corless, S., & Gilbert, N. (2016). Effects of DNA supercoiling on chromatin architecture. *Biophysical Reviews, 8*(Suppl. 1), 51–64.

Cowling, V. H. (2009). Regulation of mRNA cap methylation. *Biochemical Journal, 425*(2), 295–302.

Ellis, R. J. (2001). Macromolecular crowding: An important but neglected aspect of the intracellular environment. *Current Opinion in Structural Biology, 11*(1), 114–119.

Fırat-Karalar, E. N., & Stearns, T. (2014). The centriole duplication cycle. *Philosophical Transactions of the Royal Society of London B Biological Sciences, 369*(1650).

Gruenbaum, Y., & Foisner, R. (2015). Lamins: Nuclear intermediate filament proteins with fundamental functions in nuclear mechanics and genome regulation. *Annual Review of Biochemistry, 84*, 131–164.

Henras, A. K., Plisson-Chastang, C., O'Donohue, M. F., Chakraborty, A., & Gleizes, P. E. (2015). An overview of pre-ribosomal RNA processing in eukaryotes. *Wiley Interdisciplinary Reviews: RNA, 6*(2), 225–242.

Jevtić, P., Edens, L. J., Vuković, L. D., & Levy, D. L. (2014). Sizing and shaping the nucleus: Mechanisms and significance. *Current Opinion in Cell Biology, 28*, 16–27.

Kobiler, O., Drayman, N., Butin-Israeli, V., & Oppenheim, A. (2012). Virus strategies for passing the nuclear envelope barrier. *Nucleus, 3*(6), 526–539.

LaJoie, D., & Ullman, K. S. (2017). Coordinated events of nuclear assembly. *Current Opinion in Cell Biology, 46*, 39–45.

Lee, Y., & Rio, D. C. (2015). Mechanisms and regulation of alternative pre-mRNA splicing. *Annual Review of Biochemistry, 84*, 291–323.

Luxton, G. W., & Starr, D. A. (2014). KASHing up with the nucleus: Novel functional roles of KASH proteins at the cytoplasmic surface of the nucleus. *Current Opinion in Cell Biology, 28*, 69–75.

McIntosh, J. R. (2016). Mitosis. *Cold Spring Harbor Perspectives in Biology, 8*(9).

Meunier, A., & Spassky, N. (2016). Centriole continuity: Out with the new, in with the old. *Current Opinion in Cell Biology, 38*, 60–67.

Moindrot, B., Bouvet, P., & Mongelard, F. (2013). Chromatin structure and organization: The relation with gene expression during development and disease. *Subcellular Biochemistry, 61*, 373–396.

Nerurkar, P., Altvater, M., Gerhardy, S., Schütz, S., Fischer, U., Weirich, C., & Panse, V. G. (2015). Eukaryotic ribosome assembly and nuclear export. *International Review of Cell and Molecular Biology, 319*, 107–140.

Nguyen, H. Q., & Bosco, G. (2015). Gene positioning effects on expression in eukaryotes. *Annual Review of Genetics, 49*, 627–646.

Norbury, C. J. (October 2013). Cytoplasmic RNA: A case of the tail wagging the dog. *Nature Reviews Molecular Cell Biology, 14*(10), 643–653.

Nunes, V. S., & Moretti, N. S. (January 2017). Nuclear subcompartments: An overview. *Cell Biology International, 41*(1), 2–7.

Okamura, M., Inose, H., & Masuda, S. (2015). RNA export through the NPC in eukaryotes. *Genes, 6*(1), 124–149.

Rout, M. P., & Karpen, G. H. (2014). Editorial overview: Cell nucleus: The nucleus: A dynamic organelle. *Current Opinion in Cell Biology, 28*, iv–vii.

Schweizer, N., Weiss, M., & Maiato, H. (2014). The dynamic spindle matrix. *Current Opinion in Cell Biology, 28*, 1–7.

Ulianov, S. V., Gavrilov, A. A., & Razin, S. V. (2015). Nuclear compartments, genome folding, and enhancer-promoter communication. *International review of cell and molecular biology, 315*, 183–244.

Yáñez-Cuna, J. O., & van Steensel, B. (2017). Genome-nuclear lamina interactions: From cell populations to single cells. *Current Opinion in Genetics & Development, 43*, 67–72.

Chapter 7

Structural and biophysical properties of tight junctions

Santhosh Kumar Nagarajan, Jörg Piontek and Dorothee Günzel
Clinical Physiology/Nutritional Medicine, Charité - Universitätsmedizin Berlin, Berlin, Germany

1. Introduction: The era before tight junction protein discovery

Trans-epithelial and trans-endothelial solute and fluid transport occur via two pathways: transcellular and paracellular. *Transcellular transport* is the movement of solutes and fluid through the cells, across both apical and basolateral membrane domains, and is the main subject of Chapter 18. *Paracellular transport* occurs between cells through tight junctions (TJs), which are the focus of the present chapter. TJs are intercellular adhesion complexes that control paracellular permeability. These paracellular diffusion barriers are semipermeable and size- and charge-selective, and are formed by several junctional proteins which will be analized in detail in this chapter.

"Terminal bars" (Bonnet, 1895; Cohn, 1897; Cohn, 1895; Zimmermann, 1911) have been described in epithelial histology since the second half of the 19th century. Initially, the expression appears to have described the lateral space that could be filled with dye and that was assumed to contain an elastic sealant ("Kittsubstanz") providing a mechanical connection between neighboring cells (Arnold, 1875, 1876; Thoma, 1875). Cohn (1895, 1897) systematically stained different epithelia and endothelia with Heidenhain's iron-hematoxylin (Heidenhain, 1892) and observed terminal bars in the paracellular cleft, close to the apical pole, and thus truly tight junction-like. He interpreted these structures as sealants against the intrusion of pathogens (Fig. 7.1A).

A role of the terminal bars as diffusion barriers against macro(cmecules or smaller solutes was assumed to be unlikely by Cohn (1895, 1897) and was only proven in electron microscopy studies more than half a century later, when employing electron dense tracers such as hemoglobin, ferritin, or lanthanum (Miller, 1960; Revel & Karnovsky, 1967; Thoenes & Langer, 1969).

The expressions "tight junction" (TJ) and "zonula occludens" were coined in 1963, in the seminal, systematic electron microscopic study on "Junctional complexes in various epithelia" by Marilyn G. Farquhar (1928–2019) and George E. Palade (1912–2008): "The epithelia of a number of glands and cavitary organs of the rat and guinea pig have been surveyed, and in all cases investigated, a characteristic tripartite junctional complex has been found between adjacent cells. [...] The elements of the complex, identified as zonula occludens (tight junction), zonula adherens (intermediary junction), and macula adherens (desmosome), occupy a juxtaluminal position and succeed each other in the order given in an apical-basal direction."

The zonula occludens (tight junction) is characterized by the fusion of the adjacent cell membranes, resulting in obliteration of the intercellular space over variable distances. Within the obliterated zone, the dense outer leaflets of the adjoining cell membranes converge to form a single intermediate line. [...] The tight junction is impervious to concentrated protein solutions and appears to function as a diffusion barrier or "seal" (Farquhar & Palade, 1963).

During the following years, numerous studies were carried out that proved the sealing function of TJs by employing lanthanum as tracer substance (Fig. 7.1B–D, see e.g., Brightman & Reese, 1969; Goodenough & Revel, 1970; Friend & Gilula, 1972; Neaves, 1973; Giacomelli & Wiener, 1976). In intact TJs, lanthanum penetration stopped at what was dubbed the "kisses" or "kissing points" of the TJs, the points at which the membranes of the neighboring cells touched and the outer leaflets of the plasma membranes appeared to merge (Claude & Goodenough, 1973; Diamond, 1974). This hypothesis for the (cmecular structure of TJs will be referred to as the lipid model.

Cell Physiology Sourcebook. https://doi.org/10.1016/B978-0-12-811114-7.00027-X

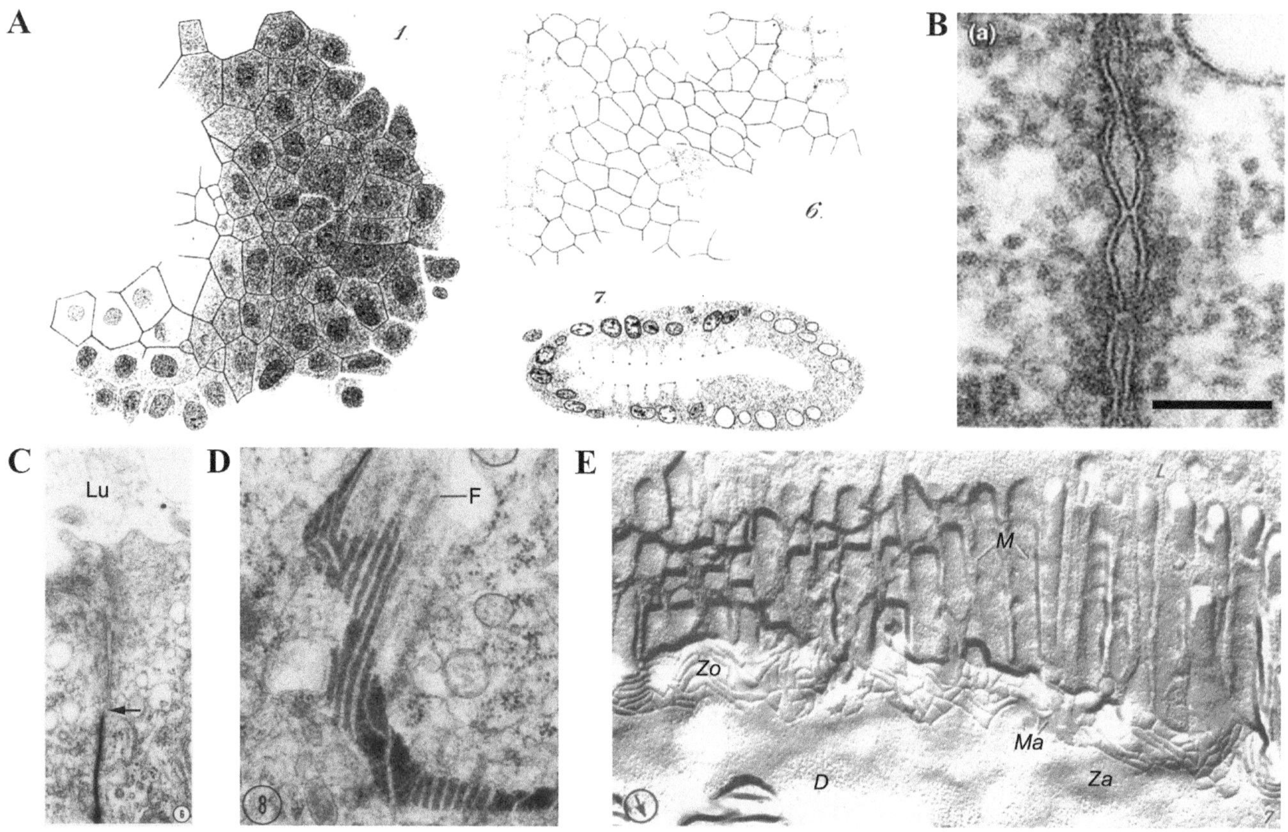

FIGURE 7.1 **Tight junction microscopy.** (A) Light microscopy. Various preparations stained with iron hematoxylin to visualize terminal bars. (1) uppermost layer of axolotl epidermis. (6) Tangential section of the surface epithelium of a cat stomach. (7) Oblique section through a cat gastric gland. (B–D) Transmission electron microscopy of ultrathin sections. (B) Mouse epididymis. Tight junction kissing points, transversely cut. Scale bar 100 nm. (C) Rat sublingual acinar cells. Lanthanum penetrates the intercellular space and stops at a kissing point. Lu, lumen. ×35,000. (D) Sertoli cells from rat testis 1 week after efferent ductule ligation. Tight junctions are permeated with lanthanum and sectioned in the plane of the participating membranes. Tight junction strands appear as electron-translucent lines in lakes of electron-dense tracer. Cytoplasmic filaments (F) associated with the tight junction appear in a longitudinal array. ×43,000. (E) Freeze fracture image of mouse intestine. Surface view of a junctional complex split along the intercellular space. The meshwork of tight junction strands (zonula occludens, Zo) is localized at the base of the microvilli (M). Strands (diameter 65 ± 10 Å) are described as being embedded in a matrix (Ma), which has been nearly completely broken away in this micrograph. Adhering zonule (Za), desmosomes (D), lumen (L). ×45,000. *(A) Reprinted by permission from RightLinks: Springer Nature, Cohn (1895). (B) Reprinted from Furuse and Tsukita (2006), with permission from Elsevier. (C) Reprinted from Shimono et al. (1980), with permission from Elsevier. (D) Used with permission of Rockefeller University Press, from Neaves (1973); permission conveyed through Copyright Clearance Center, Inc. (E) Reprinted by permission from RightLinks: Springer Nature, Staehelin et al. (1969b).*

In 1969, Staehelin et al. presented the first freeze-fracture images of TJ strand networks (Staehelin et al., 1969a, 1969b), which they described as being "comparable to a concertina, which can be stretched or compressed and twisted" (Fig. 7.1E). They observed that "After glutaraldehyde fixation the fibrils of the meshwork are usually less fragmented […] than in material without prefixation" and concluded "Since glutaraldehyde is known to cross-link proteins but not polysaccharides […] it is suggested that the fibrils may contain protein." (Staehelin et al., 1969b)—a first emergence of the protein model of the TJ.

Staehelin et al. (1969b) also observed the tricellular junction, i.e., the place where three epithelial cells meet, as being structurally distinct from the bicellular junction: "The fibrillar meshwork does not seem to be organized according to any specific and regular pattern […]. Only where three epithelial cells form a joint may some degree of orderly arrangement be detected. In such circumstances, the meshwork is extended along the triple junction in a ladder-like arrangement with a central common fibril connected to parallel fibrils laid down between the adjacent cells." In a follow-up study, Staehelin (1973) identified three sealing elements linked by cross bridges that make up this central common fibril. He speculated that, together, they may form a central tube and estimated the dimensions to about 1 μm in length and about 10 nm in diameter.

That the fibrils in the meshworks observed in freeze-fracture images are indeed the same structures as the "kissing points" was further demonstrated by sectioning lanthanum-filled TJs in parallel to the lateral plasma membrane. These images show the TJ strand network "as pale lines in lakes of electron-opaque tracer" (Fig. 7.1D; Neaves, 1973; see also Giacomelli & Wiener, 1976). Electron micrographs of cross sections through tricellular junction from lanthanum-incubated epithelial cells, in which the center of the tricellular junction was lanthanum-filled, further demonstrated that Staehelin's assumption of the central fibril being a tube may, at least under certain conditions, be correct (Staehelin, 1973; Walker et al., 1985).

The transepithelial electrical resistance (TER) of different epithelia varies widely, from 5 $\Omega\cdot cm^2$ in leaky epithelia, such as the proximal tubule, to 300 $k\Omega\cdot cm^2$ in the extremely tight epithelia of mammalian urinary bladder (reviewed by Diamond, 1974, 1977). Claude and Goodenough (1973) and Claude (1978), established a correlation between the number of parallel TJ strands and the TER. Interestingly, the relationship between paracellular resistance and the number of TJ strands was found not to be linear (as expected for resistors added in series), but exponential. Claude (1978) interpreted these results in terms of strand break probability. She proposed that the paracellular resistance ($R^{para)}$ could be calculated by the equation:

$$R^{para} = R_{min}\, p^{-n} \qquad 7.1$$

where R_{min} = paracellular resistance in the absence of strands,
p = probability for a strand to be "open" (i.e., strand break probability), and
n = number of horizontal strands.

Estimating p for different epithelia, she derived a *P* value of 0.4, suggesting that TJs are "remarkably labile" (Claude, 1978).

The hypothesis that the number of horizontal strands determines the paracellular resistance was contested by Møllgård et al. (1976). As already discussed by Claude and Goodenough (1973), many additional factors may affect epithelial resistance, such as cell size (smaller cells resulting in more junctional length per unit epithelial area) or the straightness of the cell borders (with interdigitation of cells causing junctional elongation). Yet, Møllgård et al. (1976) conclude that "some other feature not revealed by electron microscopy could account for some of these changes in permeability. Both the (cmecular structure of the junctional membrane and the biochemical composition of the material enclosed by the junction fibrils might be important for passive ion transport."

The paracellular ion permeability (P_{ion}) properties of the rabbit gallbladder, used as a model of a leaky epithelium, were investigated in great detail. Initial observations showed that $P_K > P_{Na} > P_{Cl}$ (Wright & Diamond, 1968). These observations were extended to a full analysis of the Eisenman sequences (Barry et al., 1971). The authors found cation permeabilities corresponding to Eisenman sequence VIa ($P_K > P_{Rb} > P_{Na} > P_{Li} > P_{Cs}$) in freshly prepared gall bladders that, within a short time, changed to Eisenman sequence V ($P_K > P_{Rb} > P_{Na} > P_{Cs} > P_{Li}$). To explain their data, the authors presented the model depicted in Fig. 7.2, which includes an unspecific shunt pathway in parallel to a cation preferring channel. The pore of the channel is lined with dipolar groups, the negatively charged end of which points into the channel lumen. The shunt is likely due to tissue damage that develops over time.

In 1977, Diamond (Diamond, 1977) ascribed three functions to epithelial junctions: bridge, gate, and fence. Bridge referred to the coupling of neighboring cells through gap junctions, but gate and fence described a dual diffusion barrier formed by TJs. Gate envisioned a size selectivity filter within the TJ that keeps (cmecules above a certain diameter from passing. Fence postulates a role of the TJ in preventing proteins within the apical plasma membrane from diffusing into the basolateral plasma membrane compartment and vice versa, and is thus the prerequisite of cell polarity.

A question requiring special attention is whether TJs are permeable to water. Such a permeability is the basis of concepts such as *solvent drag*—the idea that in leaky epithelia, the flow of water takes ions and nutrients along. In the intestine, Pappenheimer and Reiss (1987) postulated that the presence of luminal glucose widens the paracellular pathway, triggering a bulk uptake of small organic substances such as creatinine, oligopeptides as well as glucose itself by solvent drag. Pappenheimer et al. (1994) estimated that under these conditions, more than 70% of the ingested glucose is absorbed along the paracellular pathway.

The three decades after the 1963 publication of Farquhar and Palade's epoch-making study were characterized by intense electrophysiological investigations of TJ properties, which, in many aspects, still shape our current understanding of TJ physiology. Yet, while in parallel, the first membrane ion channels were discovered and analyzed in detail, the (cmecular identity of TJ components remained elusive.

The pre-TJ protein era ended with the discovery of the TJ-associated protein zonula occludens-1 (ZO-1, Stevenson et al., 1986), followed by the first transmembrane TJ proteins, occludin in 1993 (Furuse et al., 1993), the junctional adhesion (cmecule (JAM) in 1998 (Martin-Padura et al., 1998), and especially, the characterizations of the first members of

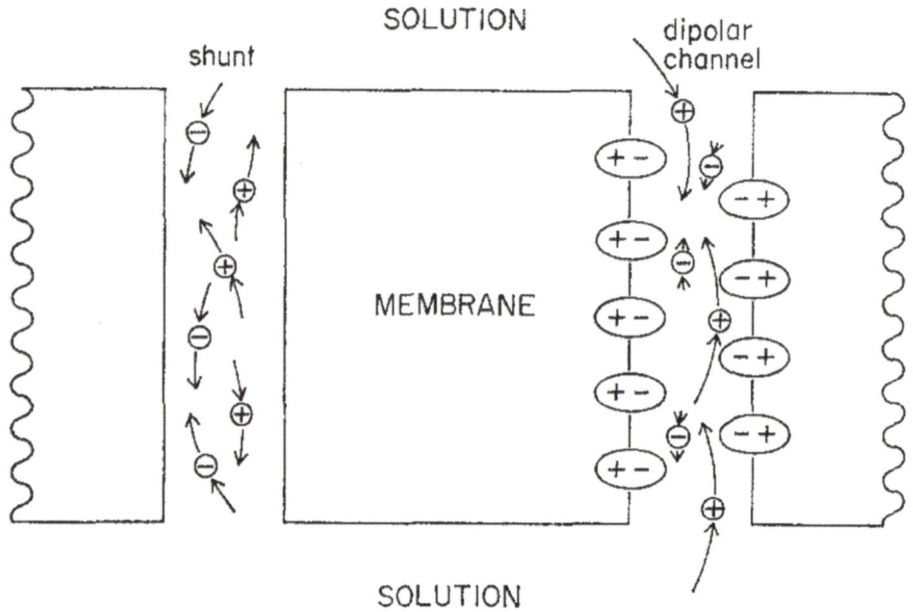

FIGURE 7.2 **Epithelial ion permeation mechanism suggested by Barry et al. (1971).** Tight junctional channels are lined with dipolar groups (e.g., δ+ >C=O δ-). For cation selective channels, negative ends point into the channels. The number of cations (symbolized by ⊕) and anions (symbolized by ⊖) in the channels are assumed to be equal, but their mobilities (indicated by lengths of arrows) differ. In a parallel shunt, cations and anions move with free-solution mobility ratios. *Reprinted by permission from RightLinks: Springer Nature, Barry et al. (1971).*

the claudin protein family by Tsukita and his group in 1998 (Furuse et al., 1998). It soon turned out that several members of this family had been discovered previously, but their function as TJ components had not been recognized, e.g., rat ventral prostate 1 (RVP.1/claudin-3, Briehl & Miesfeld, 1991), oligodendrocyte-specific protein (OSP/claudin-11, Bronstein et al., 1996), and *Clostridium perfringens* enterotoxin receptor (CPE-R/claudin-4, Katahira, Inoue, et al., 1997).

2. Tight junction proteins

Phylogenetically, TJs multiprotein complexes (recently reviewed by Citi et al., 2024) appear to be limited to tunicates and vertebrates (Georges, 1979; Green & Bergquist, 1982; Lane et al., 1986), and absent in nonchordates and cephalopods (see however, Lane & Chandler, 1980) which, instead, possess septate junctions as paracellular barriers (Furuse & Tsukita, 2006; Green & Bergquist, 1982; Lane et al., 1987; Simske & Hardin, 2011; Staehelin, 1974; Welsch, 1983).

TJs are made up of at least three major families of transmembrane proteins: the numerous members of the claudin family composed of about 25 members in tetrapods, (for review see Günzel & Yu, 2013) and almost 60 members in teleosts (Baltzegar et al., 2013; Loh et al., 2004; Sun et al., 2015), the three proteins, occludin, tricellulin, and MarvelD3 making up the TJ associated MARVEL protein (TAMP) family, and the junctional adhesion protein (JAM) family which, depending on definition, contains either three members (JAM-A, -B, —C) or in addition several more remote members such as JAM4, JAM-like (JAML), coxsackie- and adenovirus receptor (CAR), CAR-like membrane protein (CLMP), and endothelial cell-selective adhesion (cmecule (ESAM) (for review, see e.g., Zihni et al., 2016; Hartmann et al., 2020). Further transmembrane TJ proteins include blood vessel epicardial substance (BVES), Crumbs protein ho(cmog 3 (Crb3), and the tricellular angulins (or lipolysis-stimulated lipoprotein receptors, LSR) (Zihni et al., 2016). Of all these transmembrane proteins, only certain members of the claudin family are able to directly polymerize into TJ strand-like structures when overexpressed in cells devoid of TJs, such as fibroblasts, indicating their ability to interact both in *cis* (with claudins within the same cell membrane) and in *trans* (with claudins within the membrane of a neighboring cell).

Most transmembrane TJ proteins attach to—and are regulated by—cytoplasmic scaffolding proteins, such as the zonula occludens proteins (ZO-1, -2, -3), MAGI proteins, MUPP1, PATJ, cingulin, and/or JACOP/paracingulin, to mention a few (Rouaud et al., 2020; Zihni et al., 2016). In the current context, these scaffolding proteins contribute to the dynamics and stability of the junctions and the fine-tuning of the paracellular properties. The main building blocks that span the paracellular cleft and directly regulate paracellular permeation are members of the claudin family, occludin, and JAMs for the bicellular junction, and tricellulin together with the angulins for the tricellular junction. These proteins will therefore be the main focus of the present chapter.

3. Barrier, channel, and leak

3.1 Permeation of ions—The concept of paracellular ion channels

As mentioned above, the tightness of the paracellular pathway differs greatly between different epithelia. One straightforward way to assess tightness is to measure transepithelial resistance (TER). It must be kept in mind that both the transcellular and the paracellular pathways contribute toward this resistance. Given that the transcellular and paracellular resistances (R^{trans}, R^{para}) act in parallel, the inverse of their individual values must be added to obtain the inverse of the total resistance (R^{total}):

$$\frac{1}{R^{total}} = \frac{1}{R^{trans}} + \frac{1}{R^{para}};\ R^{total} = \frac{R^{trans} \cdot R^{para}}{R^{trans} + R^{para}} \qquad 7.2$$

This implies that (i) the total resistance is always lower than the partial resistances and that (ii) the changes in the total resistance only reflect changes in the paracellular pathway, if $R^{trans} >> R^{para}$. In epithelia, in which R^{trans} and R^{para} are similar, alteration in membrane resistance, e.g., through ion channel activation, is directly visible as reduction of R^{total}.

Such considerations resulted in definitions of epithelial leakiness or tightness, based on the ratio R^{trans}/R^{para}: If this ratio was <1, an epithelium was considered to be leaky, at a ratio between 1 and 100 as medium tight (or tight), at a somewhat arbitrarily chosen ratio of >100 as tight (or impermeable; see e.g., (Boulpaep & Sackin, 1980; Piontek et al., 2020; Schultz, 1977). Thus, in a leaky epithelium, it is easier to drive ions through the TJ as compared to through the cells whereas in tight epithelia, this is the other way round. As a consequence, leaky epithelia cannot maintain large transepithelial ion gradients, at least not if they are equally leaky for cations and anions, whereas tight epithelia can generate and maintain large electrochemical gradients (Schultz, 1977). The definition has two major disadvantages. (i) It is difficult to separately assess transcellular and paracellular resistances (see below). (ii) As the definition is based on electrical parameters, it cannot be applied to uncharged solutes or to water. Thus, an epithelium that is leaky with respect to certain ions may indeed be very tight with respect to other solutes and water, such as the thick ascending limb of the nephron.

Originally, the term "leaky" was strictly applied to ion permeability, as proposed byDiamond (1978) who already predicted the presence of paracellular ion channels in leaky epithelia: "Epithelia may be classified as "tight" or "leaky," depending on whether there is a significant pathway for transepithelial ion permeation via the junctions and bypassing the cells. The resistance of this paracellular channel may depend partly on structures visible in the electron microscope, partly on wall charge. Permeability determinations in the leaky junctions of gallbladder epithelium, using many different organic cations, suggest that the critical barriers to ion permeation are 5–8 Å in radius and bind cations by up to four strongly proton-accepting oxygens."

The discovery of occludin and the first members of the claudin protein family (Furuse et al., 1993, 1998) marked a turning point in the understanding of TJ biology. The names chosen for these proteins imply their tightening role and contribution toward epithelial barrier function. However, it immediately became clear that these proteins played a role not only in the tightness but also in the ion permeability of TJs, i.e., they determined their "leakiness". Thus, transepithelial resistance (TER) increased upon overexpression of claudin-1 in MDCK (Madin–Darby canine kidney) cells, as expected for a barrier-forming TJ protein (Inai et al., 1999). In contrast, mutations in claudin-16 (then called paracellin-1) caused renal Mg^{2+} and Ca^{2+} wasting, suggesting its role as a paracellular channel for divalent cations (Simon et al., 1999; Wong & Goodenough, 1999). Overexpression of claudin-2 in MDCK C7 cells caused a dramatic decrease in TER without affecting the permeability of the small uncharged solute mannitol. However, when applying a transepithelial NaCl gradient and measuring the resulting diffusion potentials, claudin-2 transfected cell layers showed a strong paracellular preference for monovalent cations versus anions, and thus acted as a paracellular cation channel while forming a barrier against the permeation of anions (Amasheh et al., 2002).

Using diffusion potential measurements, several additional claudins were identified that act as paracellular channels for monovalent ions. It was found that, when overexpressed in high resistance cell lines: claudin-10a and -17 act as anion channels (Krug et al., 2012; Van Itallie et al., 2006) and claudin-10b, -15, -21[25] as cation channels (Colegio et al., 2002; Günzel et al., 2009; Tanaka et al., 2016). It is important to note that there is some confusion about the nomenclature of mammalian claudins 21 to 25 as discussed by Mineta et al. (2011). Tanaka et al. (2016) give accession number NG_020884 as a source for the claudin-21 they analyzed, which, according to the NCBI database, is "*Mus musculus claudin* 25, pseudogene (Cldn25-ps) on chromosome 9". However, the identical sequence can be found at NM_001384265, where it is named *Mus musculus* claudin 25 (Cldn25). Therefore, in the present chapter, this claudin will be referred to as claudin-21[25].

Taken together, the variety of properties that different claudins conveyed to TJs demonstrated that claudins play a dual role; they form the TJ strands and thus the paracellular barrier. However, some claudins form channels for certain ions within these strands, while they still fulfil their tightening function against counterions and uncharged solutes. In contrast to ion channels embedded within and conducting through a plasma membrane, claudins are thus both the channels as well as the matrix in which these channels are embedded.

This can give rise to confusion in the interpretation of data: For example, Yu et al. (2003) overexpressed claudin-8 in low-resistance MDCK II cells and interpreted the observed effects as an augmentation of the barrier against the passage of cations. They later could attribute this augmentation to the replacement of the endogenous cation channel-forming claudin-2 by the overexpressed claudin-8 (Angelow et al., 2007). Such "dilution" effects of endogenous claudins in overexpression experiments explain contradicting results when different cell lines are employed as overexpression systems. Examples of these discrepancies include the initial differences observed in the ion permeability characteristics of the two claudin-10 isoforms, the anion channel-forming claudin-10a, and the cation channel-forming claudin-10b (Günzel et al., 2009; Van Itallie et al., 2006); discrepancies for claudin-4 which was either considered as barrier-forming (Van Itallie et al., 2001) or as forming an anion channel itself (Hou et al., 2006) or together with claudin-8 (Hou et al., 2010); for claudin-7, which was considered to be a cation channel and anion barrier (Alexandre et al., 2005) or a cation barrier and anion channel (Hou et al., 2006); for the channel formed by claudin-16 alone (channel for divalent cations, Ikari et al., 2004; Kausalya et al., 2006) or when interacting with claudin-19 (claudin-16 as mono- and divalent cation channel together with claudin-19 as anion barrier, Hou et al., 2008; Hou et al., 2009).

In contrast, claudin-14 appears to physiologically inhibit the channel for (divalent) cations formed by the interaction of claudin-16 and -19 in the thick ascending limb of the nephron (Gong et al., 2012). At high serum Ca^{2+} levels, the basolateral calcium-sensing receptor is activated, which leads to an upregulation of claudin-14. Claudin-14 can bind to claudin-16 and thus disrupts its interaction with claudin-19. This reduces Mg^{2+} and Ca^{2+} permeability and, by increasing renal excretion of divalent cations, helps to correct their serum concentrations (Gong et al., 2012, 2015a).

3.2 Permeation of uncharged solutes (leak pathway)

Permeation of uncharged solutes from a donor compartment across an epithelium into an acceptor compartment requires specific and highly sensitive methods for detecting the investigated solute on the acceptor side. Frequently applied methods include radioactive labeling, especially for small solutes (e.g., ^{14}C-urea, ^{14}C-creatinine, ^{14}C-erythitol, ^{14}C- or ^{3}H-mannitol) but also for larger (cmecules, such as polyethylene glycols (^{14}C-PEG), labeling with fluorophores for solutes that are considerably larger than the employed fluorophores (i.e., FITC-labeled dextrans), or enzymatic activities, if the solutes under investigation are enzymes (such as horseradish peroxidase). Alternatively, the appearance of the solutes in the acceptor compartment can be analyzed by high-performance liquid chromatography (HPLC), such as in the case of PEGs. Care must be taken to ensure that the solutes are not additionally transported along alternative transcellular routes (e.g., by specific carriers or by transcytosis). When using fluorophore-labeled macro(cmecules, dialysis of the stock solution is paramount to avoid detection of free fluorophore, rather than the labeled solute.

Several systematic investigations employing small radioactively labeled compounds (Artursson et al., 1993), PEGs (Chadwick et al., 1977a, 1977b; Fanning et al., 2012; Linnankoski et al., 2010; Van Itallie & Anderson, 2011; Van Itallie et al., 2008, 2009; Watson et al., 2001, 2005) or dextrans (Buschmann et al., 2013; Neuhaus et al., 2006a, 2006b, 2008) of various sizes were carried out on different tissues and cell layers. These studies demonstrated the presence of at least two types of paracellular pathways for small organic compounds: one with a cut-off radius of about 4 to 4.5 Å, and another with much lower transport capacity, which did not appear to have a well-defined upper size limit (see Fig. 7.3). Interestingly, both pathways could be modulated independently. Transport capacity of the larger diameter pathway increased upon ZO-1 knockdown (Van Itallie et al., 2009), an effect that was further enhanced by the additional knockdown of ZO-2 (Fanning et al., 2012). Independent of ZO-1, this pathway was also affected by knockdown of occludin (Al-Sadi et al., 2011; Buschmann et al., 2013).

In contrast to Al-Sadi et al. (2011) and Buschmann et al. (2013), Yu and collaborators (2005) found a specific increase in the transport capacity of the smaller diameter pathway (<7 Å diameter) upon occludin knockdown. Furthermore, this smaller diameter pathway was observed to be increased when claudin-2 was overexpressed, but did not decrease when claudin-2 was knocked out (Van Itallie et al., 2008). The authors hypothesize that claudin-2 may be able to modulate strand stability (Van Itallie & Anderson, 2006; Van Itallie et al., 2008). However, this hypothesis is somewhat at variance with their own observations as well as with the findings of Amasheh et al. (2002), showing that claudin-2 overexpression did not cause any alterations in mannitol and lactulose permeability, substances that are of lower (cmecular weight than the small PEG used by Van Itallie et al. (2008). Whereas Amasheh et al. (2002) found that MDCK C11 cells (expressing very high

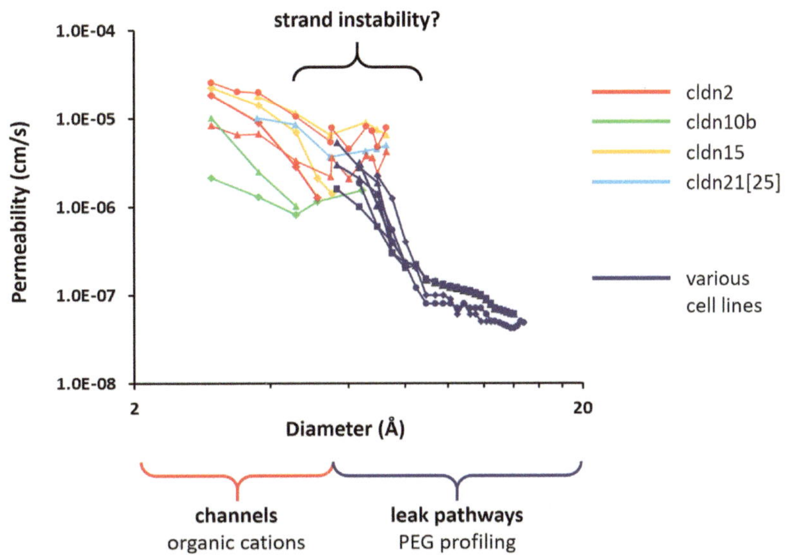

FIGURE 7.3 **Size restriction of paracellular channel and leak pathways.** Permeabilities of MDCK C7 or MDCK I cell layers stably transfected with channel-forming claudins are plotted against solute diameter. Red, Cldn2 (Rosenthal et al., 2020; Yu et al., 2003); green, Cldn10b (Hempel et al., 2022; Günzel, unpublished data); yellow, Cldn15 (Rosenthal et al., 2010; Tanaka et al., 2016); light blue Cldn21[25] (Tanaka et al., 2016). Dark blue lines, permeabilities of various cell lines determined by PEG profiling. *Data from Watson et al. (2001), Van Itallie et al. (2008), (2009).*

levels of endogenous claudin-2) had a higher permeability for mannitol than MDCK C7 cells (not expressing any endogenous claudin-2), a comparison of two similar MDCK cell subclones, MDCK II (high levels of endogenous claudin-2) with MDCK I (no endogenous claudin-2), and the high-resistance T84 cells, showed, that mannitol flux did not differ between these cell lines (Tang & Goodenough, 2003). Differences between mannitol and PEG permeabilities are also reviewed and discussed by Linnankoski et al. (2010). They conclude that the discrepancies may be due to an over-estimation of PEG diameters, as the shape of these (cmecules is nonspherical, therefore, cannot be directly estimated from their (cmecular weight.

Similar to claudin-2, claudin-15 and -21 have also been reported to increase permeability for uncharged solutes (Tanaka et al., 2016). Conversely, Tanaka et al. (2015) suggest that claudin-7 prevents the permeation of small organic solutes. They report that deficiency in claudin-7 increases flux for organic solutes with a molecular weight of about 500 Da (N-formyl- L-methionyl- L-leucyl- L-phenylalanine, 438 Da; Lucifer Yellow, 457 Da), but not for the larger macromolecules (FITC-Dextran, 4400 Da). Similarly, loss of claudin-5 increases permeability of molecules $\leq \sim$800 Da across the endothelial blood–brain barrier but not for larger macromolecules (Nitta et al., 2003). It thus remains to be investigated whether there is a separate pathway for small uncharged solutes, e.g., through minor strand instabilities or fluctuations (Van Itallie & Anderson, 2006) and whether certain claudins (or other TJ proteins, such as occludin) are able to differentially affect strand stability and thus affect permeability of small, uncharged organic compounds (Fig. 7.3).

As at least pore populations permeable to larger organic compounds appear to be distinct from the pores of the paracellular ion channels formed by claudins, Shen et al. (2011) suggested to use the term "channel pathway" for the structures conveying charge- and size selective ion permeability and the term "leak pathway" for the pathways for larger solutes (see also Fig. 7.3). Thus, the shunt pathway postulated by Barry et al. (1971), (Fig. 7.2) would be part of the leak pathway.

TJ proteins that are involved in the regulation of macromolecular permeability are TAMPs and JAMs. In a very elegant and thorough study, Otani et al. (2019) used MDCK II cells to knock out various TJ proteins and thus to generate five different cell lines: (1) unaltered control cells; (2) cells deficient in both scaffolding proteins, ZO-1 and ZO-2; (3) cells deficient in the five major claudins of this cell line (claudin quin KO; claudin-1, -2, -3, -4 and -7); (4) cells deficient in JAM-A; and (5) claudin quin KO cells deficient in JAM-A (Otani et al., 2019). In these cell layers, they investigated TER, macromolecular permeability, presence of tight junctional kissing points, and the width of the apical paracellular space in ultrathin section/transmission electron microscopy, presence of TJ strands in freeze-fracture electron microscopy, and cell polarity as localization/separation of various apical and basolateral membrane proteins. Their findings are summarized in Table 7.1.

TABLE 7.1 Summary of the results by Otani et al. (2019).

	Control	ZO-1/-2 KO	Claudin quin KO	JAM-A KO	Claudin quin KO with JAM-A KO
TER	50–60 $\Omega \cdot cm^2$	10–15 $\Omega \cdot cm^2$	10–20 $\Omega \cdot cm^2$	50–60 $\Omega \cdot cm^2$	5–10 $\Omega \cdot cm^2$
Fluorescein (332 Da, anion)	Low	High	High	low	High
Macromolecular permeability	Low	Very high 4 kDa ≈ 40 kDa ≈ 150 kDa	Increased: 4 kDa >> 40 kDa >> 150 kDa	Low	High 4 kDa > 40 kDa > 150 kDa
Kissing points	Yes	No (occasional remnants)	No (some particles present)	Yes	No
Apical para-cellular cleft	Narrow	Wide	Increased	Narrow	Increased
TJ strands	Yes	No	No	Yes	(Not shown)
Cell polarity	Intact	Disturbed	Intact	Intact	Disturbed

The authors conclude that claudins form the paracellular barrier against macromolecules (>~10 kDa, 20 Å diameter), and, depending on the specific claudin composition, against specific inorganic and organic ions. Except for fluorescein, no typical small molecules (180 Da–4 kDa; 8 to 14 Å diameter) were tested. In the absence of claudins, JAM-A was able to restrict macromolecular permeation and thus appears to present a second line of defense against intrusion/loss of macromolecules.

How do these findings relate to those shown in Fig. 7.3? The size of dextrans can be estimated from their molecular weight (MW) by using the following equation (Haller, 1977):

$$\text{diameter}\left(\text{in Å}\right) = 0.215\left(\text{MW}^{0.587}\right) \tag{7.3}$$

The radius of a 4 kDa dextran would thus be on the order of 14 Å and thus already almost beyond the range investigated by PEG profiling. However, even for the larger dextrans (40 and 150 kDa with radii of about 55 and 120 Å, respectively), permeability in control MDCK II cells is > 0.

A hypothesis on how uncharged small-/medium-sized solutes and even macromolecules may pass the TJ without compromising the barrier for small inorganic ions (i.e., while maintaining high TER) is the strand break hypothesis put forward by Anderson et al. (2004). This hypothesis is based on live cell imaging observations by Sasaki et al. (2003) that show that TJ-like strands formed in fibroblasts transfected with claudin-1 are flexible and detach/reattach from other strands, resulting in a network that permanently reshapes. Anderson et al. (2004) highlight the fact that as long as the same average number of intact horizontal strands is present, TER is constant. Molecules enclosed within the meshes of the TJ strands, however, may gradually traverse the TJ. Recent computer models of this mechanism indeed demonstrate that such a model can explain many features of TJ permeation (see below).

A completely independent pathway for macromolecular passage would be permeation along the tricellular pathway. Krug, Amasheh, et al. (2009) were able to demonstrate that increased amounts of tricellulin within the tricellular TJ restricted the permeability of macromolecules with diameters between about 10 and 100 Å. Conversely, Sugawara et al. (2021) found that knockdown of angulin-1 disrupted the integrity of the tricellular TJ and caused increased flux of fluorescein. In claudin quin KO MDCK II cells, angulin-1 retained its tricellular localization. In these cells, the barrier against the passage of 150 kDa dextran was still intact, but was lost if angulin-1 was additionally knocked out (Sugawara et al., 2021). In this study, knockout of tricellulin in MDCK II cells had no effect on fluorescein permeability. In contrast, in an earlier study on tight (high TER) Eph4 cell layers, knockdown of tricellulin greatly reduced TER and increased permeability of a 4 kDa dextran (Ikenouchi et al., 2005). This may mean that an additional increase in fluorescein or 4 kDa dextran permeability upon tricellulin knockout may not be visible in MDCK II cells due to an already high permeability background for these solutes, especially in light of the bicellular leak pathway.

Pathophysiologically relevant alterations of the tricellular leak pathway have been demonstrated in inflammatory bowel disease (Hu et al., 2021; Krug et al., 2018). In ulcerative colitis, but not in Crohn's disease, tricellulin is downregulated

(Krug et al., 2018), whereas levels of tricellulin as well as angulin-1 reach control levels in remission (Hu et al., 2021). In mouse models for multiple sclerosis (experimental autoimmune encephalomyelitis model) and stroke (middle cerebral artery occlusion model), angulin-1 is downregulated (Sohet et al., 2015). Thus, loss of tricellular tightness can contribute to intestinal barrier loss and pathological blood–brain barrier leakage.

Further leak pathways for macromolecules as well as for entire pathogens will be provided by loss of TJ integrity, e.g., at apoptotic sites, and are observed under many pathophysiological conditions (see e.g., Epple et al., 2010; Gitter et al., 2000; Heller et al., 2005; Nielsen et al., 2011).

3.3 Permeation of water

Before the discovery of aquaporins, the main pathway for water transport, especially across leaky epithelia, such as the proximal tubule of the nephron or the small intestine, was assumed by many authors to be the paracellular pathway (for review and critical discussion see Diamond, 1979). Coupling between ion and water transport was inferred from "streaming potentials", transepithelial potential differences that were induced by osmotic or pressure gradients across epithelia (Diamond, 1979; Imai et al., 1986). The concept of solvent drag is based on this assumption: the stream of water is sweeping along ions and small uncharged solutes.

The presence of aquaporins entirely changed this picture: Schnermann et al. (1998) concluded from studies in aquaporin-1-deficient mice that "the high water permeability in proximal tubule of wild-type mice is primarily transcellular, mediated by AQP1 water channels, and required for efficient near-isosmolar fluid absorption". In 2010, Rosenthal et al. discovered that claudin-2 not only affects paracellular cation permeability but also water permeability (Rosenthal et al., 2010). They were able to demonstrate an interdependence of cation and water movement: in the absence of an osmotic gradient, a sodium gradient was able to drive water flux, and conversely, in the absence of a sodium gradient, an osmotic gradient resulted in concomitant $^{22}Na^{+}$ transport. It was estimated that the coupling ratio of water and sodium was about 1000:1, and thus considerably more than the amount of water within the hydration shell of one sodium ion. As claudin-2 is strongly expressed within the proximal tubule, Schnermann et al. (2013) reinvestigated proximal fluid reabsorption (PFR) and were able to demonstrate that both paracellular and transcellular pathways contribute to PFR (Schnermann et al., 2013). They conclude that the deletion of Cldn2, AQP1, or both Cldn2 and AQP1 reduces PFR by 22.7%, 19.6%, and 26%, respectively, and that their data are consistent with up to 25% paracellular contribution to PFR. They explain the absence of additive effects upon the combined deletion of Cldn2 and AQP1 by an increase in luminal hypotonicity caused by the reduced osmotic water permeability in the absence of AQP1 and by a decrease in filtration rate. As a consequence, in the absence of AQP1 and claudin-2, the capacity of alternative pathways is able to maintain PFR at 75% of control. (Schnermann et al., 2013).

In contrast to claudin-2, the cation-permeable claudin-10b, which is strongly expressed in the thick ascending limb of the nephron, was found to be water tight, as expected for this nephron segment (Breiderhoff et al., 2012; Rosenthal et al., 2010). Similarly, both anion channel-forming claudins, claudin-10a and claudin-17, are impermeable to water (Krug et al., 2012; Rosenthal et al., 2017). The third cation channel-forming claudin, claudin-15, is permeable to water, however, its properties differ greatly from claudin-2 (Rosenthal et al., 2020). Pore size of claudin-15 is larger than that of claudin-2 (Fig. 7.3), and sodium ions are only partially dehydrated when passing through these claudin-15-based channels. The more bulky, hydrated sodium ions appear to prevent additional water movement along the channel. Consequently, water movement increases in solutions, in which sodium is iso-osmotically replaced by the impermeant cation N-methyl D-glucamine. Conversely, unidirectional movement of $^{22}Na^{+}$ in apical to basolateral as well as in basolateral to apical direction was inhibited by an osmotic gradient induced by the addition of 200 mM mannitol to the apical compartment.

Another potential pathway for water transport is the tricellular junction. Knockdown of tricellulin as well as knockout of angulin-1 in the tight epithelial cell line MDCK C7 caused an increase in osmotically driven water flux (Ayala-Torres et al., 2019, 2021). In contrast, in the less tight intestinal cell line HT-29/B6, no effects were observed, neither through tricellulin knockdown nor through angulin-1 knockout (Ayala-Torres et al., 2021). It is speculated that in this cell line, any additional water transport may be masked by water transport along other pathways (aquaporins, claudin-2).

The tricellular pathway was also found to be permeable to water in the distal nephron of angulin-2 (also called Ig-like domain containing receptor 1, ILDR1)—deficient mice. As angulin-2 did not affect the localization of tricellulin, it is concluded that tricellulin is not directly involved in the tricellular water permeability of the distal nephron (Gong et al., 2017). Conversely, MDCK II cells without detectable expression of endogenous angulin-2 were transfected with angulin-2. This did neither affect sodium nor chloride permeability but resulted in a marked decrease (by 73%) in fluorescein permeability and a decrease in water permeability (by 35%, Gong et al., 2017). Together, these results indicate that the tricellular pathway may be a relevant pathway not only for solutes but also for water.

In summary, recent advances in the study of paracellular water transport suggest that the concept of solvent drag should be reconsidered. Although claudin-2 allows the coupling of cation and water transport, it is impermeable to other solutes such as glucose or amino acids. The tricellular pathway may provide coupling of water and solute transport, however, under physiological conditions (i.e., in the presence of angulins and tricellulin), this pathway appears to be closed and even upon knockout of tricellular proteins, the total transport capacity of this pathway appears to be low and therefore only detectable in otherwise tight epithelia. Thus, the tricellular pore can be ruled out as a basis for solvent drag in leaky epithelia.

4. Methods to determine paracellular transport properties

Paracellular transport is passive, i.e., it follows external gradients, such as transepithelial pressure, concentration, or electrical potential differences. Consequently, paracellular transport, according to current knowledge, is symmetrical, i.e., it works equally well in both directions if driving forces are reversed experimentally. Currently, there is no indication for rectifying properties of the paracellular channels. In contrast, due to the asymmetric localization of transporter proteins in epithelia, transcellular transport usually has a preferred direction. Furthermore, the paracellular pathway is not associated with a significant capacitance, as it does not result from crossing a lipid membrane.

These specific properties of paracellular transport are the basis of several electrophysiological methods that have been developed to evaluate transcellular and paracellular transport characteristics separately. Many of these techniques were developed decades ago but are still in use.

A classical technique used for epithelial transport studies is the Ussing chamber. It is named after its inventor, the Danish physiologist Hans Henriksen Ussing (1911–2000), and was developed in the mid-20th century (Ussing, 1949; Ussing & Zerahn, 1951). It is a device that consists of two electrolyte-filled hemichambers separated by the epithelium under investigation. The initial aim was to isolate and quantify transcellular active transport by eliminating all gradients across an epithelial layer that might drive paracellular transport, i.e., pressure, concentration, or electrical potential. While pressure and concentration differences between the apical and basolateral side can be easily eliminated, transepithelial voltage has to be assessed by two electrodes and, through a feedback system, current is applied through two electrodes to clamp the transepithelial voltage to 0 mV. The current necessary to achieve this is the so-called short circuit current, and its absolute value equals the net current produced by the active, transcellular transport through the epithelium.

There are ways to repurpose Ussing chambers to assess paracellular rather than transcellular transport properties: applying defined gradients of tracer substances for flux measurements, ion gradients for diffusion (dilution, biionic) potential measurements, or applying current pulses or AC currents of different frequencies for impedance spectroscopic analyses. To avoid contamination of transcellular transport processes, specific transporters may be inhibited pharmacologically. Alternatively, one may use the fact that paracellular transport is symmetrical, i.e., it works equally well from the apical to the basolateral compartment or the other way round. Given that this is not the case for transcellular transport it is possible to experimentally choose a transport direction that is least affected by transcellular transport.

4.1 Diffusion potential measurements: Determining charge selectivity

Diffusion potential measurements have been applied since the early 1960s to characterize paracellular ion selectivity (Diamond, 1962a). The basic idea behind these experiments is to apply an ion gradient across an epithelium and calculate the relative ion permeabilities from the resulting potential difference, using the Goldman–Hodgkin–Huxley equation (Chapter 12, Origin of Resting Membrane Potential). Dilution potential measurements usually determine the ratio of sodium and chloride permeability (P_{Na}/P_{Cl}) by "diluting" the NaCl concentration on one side of an epithelium, by iso-osmotically replacing it with an uncharged, impermeant substance (e.g., mannitol or sucrose). More generally speaking, dilution potential measurements allow for determining the P_{cation}/P_{anion} ratio of the predominant cation—anion pair within the bath solution used in the experiment. Although such experiments appear to be relatively straightforward, many confounders have to be taken into consideration, such as liquid junction potentials (Barry & Diamond, 1970), unstirred layer effects (Diamond, 1962a), active transcellular transport (Diamond, 1962a), or leakage/shunts (Barry et al., 1971).

Further difficulties may be encountered if charge selectivity of channel-forming claudins is quantified from overexpression experiments, as the resulting values additionally depend on the properties of the endogenous claudins present in the cell line used, and also strongly on the amount of channel-forming claudin within the TJ. This can be visualized by plotting relative permeabilities versus transepithelial resistance. In these plots, P_{Na}/P_{Cl} for cation channel-forming claudins (P_{Cl}/P_{Na} for anion channel-forming claudins) increases with decreasing transepithelial resistance (see e.g., Alzahrani et al., 2021; Klar et al., 2017; Sewerin et al., 2021).

When strongly overexpressed in MDCK C7 cells, claudin-2, -10b and -15 roughly display similar P_{Na}/P_{Cl} values of about 10:1 (Alzahrani et al., 2021; Amasheh et al., 2002; Hempel et al., 2022; Klar et al., 2017; Rosenthal et al., 2010, 2020; Sewerin et al., 2021), however, individual cell layers may display considerably higher P_{Na}/P_{Cl} (up to 25:1 for claudin-10b, Alzahrani et al., 2021; Sewerin et al., 2021). P_{Na}/P_{Cl} of claudin-21 (previously known as claudin 25) appears to be somewhat lower, but this may reflect lower expression levels (Tanaka et al., 2016). Similarly, P_{Cl}/P_{Na} for claudin-10a on average is on the order of three, but may reach values of five to six in highly expressing cells (Alzahrani et al., 2021; Sewerin et al., 2021), whereas P_{Cl}/P_{Na} for claudin-17 is about 2 (Krug et al., 2012). Function as ion channels has been confirmed in mouse models deficient for claudin-2 (Muto et al., 2010; Wada et al., 2013), claudin-10a (Breiderhoff et al., 2022), claudin-10b (Breiderhoff et al., 2012), claudin-15 (Tamura et al., 2008), and claudin-16/19 (Milatz, Himmerkus, et al., 2017).

Another type of diffusion potential determination is the biionic potential measurements. These refer to experimental conditions under which the potential is evoked by replacing the predominant cation (or anion) on one side of the epithelium isoosmotically by another cation (or anion, respectively). Thus, permeability ratios, e.g., for all alkali metal ions (Li^+, K^+, Rb^+, Cs^+) can be determined relative to Na^+ (P_x/P_{Na}). In 1962, George Eisenman (1962) described selectivities of ion exchanger binding sites for alkali metal ions depending on electrostatic interactions and ion dehydration. Depending on the properties of the ion exchangers, different alkali metal ions were preferred by the binding site. However, Eisenman's calculations suggested that of all the 120 possible permutations for the five ion species, only 11 different orders of binding strengths for the different ions can be obtained. These so-called Eisenman sequences are also often observed in biological systems such as ion channels. As already mentioned above, Eisenman sequences for TJs have already been extensively investigated in the pre-TJ protein era and interpreted as properties of paracellular channels (e.g., Barry et al., 1971; Salas & Moreno, 1982; Tang & Goodenough, 2003; Wright & Diamond, 1968).

For all four claudins forming monovalent cation channels, detailed studies of Eisenman sequences have been carried out, using biionic potential measurements (Günzel et al., 2009; Hempel et al., 2022; Rosenthal et al., 2020; Tanaka et al., 2016; Yu et al., 2009). In these studies

- claudin-10b and −21[25] follow Eisenman's sequence X ($Na^+ > Li^+ > K^+ > Rb^+ > Cs^+$),
- claudin-2 follows Eisenman's sequence IX ($Na^+ \geq K^+ \geq Li^+ > Rb^+ > Cs^+$), whereas
- claudin-15 follows Eisenman's sequence I to IV ($Cs^+ > Rb^+ = K^+ = Na^+ > Li^+$).

Furthermore, claudin-15 is more permeable to NH_4^+ than to Na^+ (P_{NH4}/P_{Na} ~ 1.5), whereas the opposite applies for claudin-10b (P_{NH4}/P_{Na} ~ 0.8; Hempel et al., 2022).

In summary, to cross claudin-2, -10b, and -21[25] channels, monovalent inorganic cations have to be more extensively dehydrated, but for claudin-15 channels ions retain their hydration shell. In terms of absolute values, claudin-2 and claudin-21[25] are most similar (see Table 7.2).

Overall, the properties of all claudin-based cation channels are more comparable to nonselective cation channels in the plasma membrane, such as the nicotinic acetylcholine receptor (nACh-R) or members of the transient receptor potential (Trp) channel family, rather than to highly selective Na^+ channels such as the voltage-gated Na^+ channel (Na_v) or the epithelial Na^+ channel (ENaC) or to potassium channels (see Table 7.2).

Pore diameters were estimated from biionic potential measurements employing organic cations such as methylammonium, dimethylammonium, ethylammonium, trimethylammonium, tetramethylammonium, tetraethylammonium (see Fig. 7.3):

- Claudin-15, 6.0 Å; claudin-2, 5.7 Å; claudin-10b, 5.2 Å (Rosenthal et al., 2020),
- Claudin-15, ~5.9 − 6.4 Å, claudin-10b, ~4.6 Å (Hempel et al., 2022),
- Claudin-15, ~5.6 Å, claudin-21[25], ~5.6 (Tanaka et al., 2016).

Together, these studies indicate that the pore diameters of these claudins are:

Claudin-10b < claudin-2 < claudin-15 = claudin-21[25].

Permeation properties of anion channel-forming claudins are less well characterized.

Claudin-10a and claudin-17 appear to differ in pore size. Claudin-10a increases the permeability for $NO_3^- > Cl^-$ but neither for HCO_3^- nor for the organic anion pyruvate (Breiderhoff et al., 2022; Günzel et al., 2009). Claudin-17-formed channels are permeable to $Cl^- \approx Br^- \approx I^- \approx HCO_3^- \approx$ pyruvate $> F^- \approx NO_3^- >$ fluorescein (Krug et al., 2012).

4.2 Flux measurements: Determining size selectivity

Flux is defined as the amount of substance moving across a membrane (or across an epithelium) per unit time and area. Its units are (mol/(cm^2·s)). It's essentially the rate of transport. As absolute flux values depend on the concentration in the donor compartment, it is advisable to convert flux values into permeabilities for better comparability between different

TABLE 7.2 Cation selectivity of paracellular and plasma membrane ion channels.

	P_{NH4}/P_{Na}	P_{Li}/P_{Na}	P_K/P_{Na}	P_{Rb}/P_{Na}	P_{Cs}/P_{Na}	Eisenman sequence
Cldn2[a]	1.45	0.93	0.98	0.8	0.67	IX
Cldn10b[b]	0.58	0.86	0.45	0.29	0.21	X
Cldn15[c]	1.6	0.83	1.02	1.01	1.06	I – IV
Cldn21[d]	n.d.	0.94	0.8	0.76	0.74	X
nACh-R[e]	2.09	0.75	1.23	1.11	1.27	I – IV
TrpV4[f]	n.d.	0.78	1.35	1.27	1.32	IV
Na_v[g]	0.16	0.93	0.086	n.d.	n.d.	X-XI
ENaC[h]	0	1.5	<0.01	0	0	XI
K^+ channels[i,j]	2.14	1.29	14.3	10.6	2.57	—[i]
	13	1.8	100	91	7.70	—[j]

n.d., not determined.
[a]*Rosenthal et al. (2010).*
[b]*Hempel et al. (2022).*
[c]*Rosenthal et al. (2020).*
[d]*Estimated from Tanaka et al. (2016).*
[e]*Nutter and Adams (1995).*
[f]*Voets et al. (2002).*
[g]*Hille (1972).*
[h]*Kellenberger et al. (1999), (2001).*
[i]*Delayed K^+ channels from snail neurons, Reuter and Stevens (1980).*
[j]*K^+ channel from frog myelinated nerve, Hille (1973).*

studies. Permeability is defined as flux/donor concentration, (cm/s). In epithelial studies, the donor and acceptor compartments refer to the two sides of the epithelial cell layer involved in transport. The donor compartment is the side from which substances are entering or being released, while the acceptor compartment is the side where the substance is being transferred or accumulating.

Flux measurements can be carried out under zero-*trans* conditions, i.e., the tracer substance is only added to the donor compartment and the acceptor volume is sufficiently large that back flux is negligible. Alternatively, flux measurements can also be carried out under steady-state conditions, i.e., using equal concentrations on both sides of the epithelium. This is often done when radioactive tracers are employed; these tracers are used at very low concentrations over a symmetrical background of the "cold" substance. For example, this is important for Ca^{2+} flux measurements. Withdrawal of Ca^{2+} from the bath solution may affect TJ integrity. However, when employing $^{45}Ca^{2+}$ as a tracer, normal Ca^{2+} concentrations can be maintained on both sides and $^{45}Ca^{2+}$ added to the donor compartment at trace concentrations.

Both experimental conditions, zero-*trans* and steady-state, may yield different results and may explain the discrepancies observed in the literature, especially with respect to results from mannitol versus PEG flux experiments mentioned above.

Using Ussing chambers for flux measurements, rather than simply measuring flux, e.g., in filter insert in 12-well plates, has several advantages. If the tracer substance is electrically charged (as e.g., in the case of $^{22}Na^+$, $^{36}Cl^-$, $^{45}Ca^{2+}$, or fluorescein), then the transepithelial potential can be clamped in order to avoid adverse effects of potential changes on the observed flux. Furthermore, the constant perfusion of an Ussing chamber should reduce unstirred layer effects, which have long been known to limit diffusion processes (Diamond, 1962a; Kidder et al., 1964). Unstirred layer effects are discussed in Chapter 10 (Membrane diffusion and permeability).

4.3 Temperature dependence of the channel and leak pathways

Transport along both channels and leak pathways is temperature dependent. In the channel pathway, this manifests itself as an increase in transepithelial resistance as temperature decreases; in the leak pathway, it results in a decrease of tracer flux (Blume et al., 2010; González-Mariscal et al., 1984). Yu et al. (2003) made use of the temperature dependence of paracellular conductance and Na^+ permeation in MDCK II to estimate the necessary activation energy for ion permeation through claudin-2 (the predominant channel-forming claudin in this cell line) from Arrhenius plots. They found that it was similar to the activation energy described for membrane channels. They were also able to demonstrate that overexpression of claudin-8 in these cells decreased paracellular Na^+ permeability but did not alter the activation energy for Na^+ permeation, indicating that claudin-8 acted by replacing claudin-2 rather than by increasing the activation energy necessary for Na^+ to enter the channel.

As demonstrated by Sasaki et al. (2003), Matsuda et al. (2004), and Shen et al. (2008), TJs are permanently remodeled, indicating flexibility of TJ proteins within the strands. This remodeling can be visualized in FRAP (fluorescence recovery after photobleaching) experiments that demonstrate the presence of mobile and immobile fractions of TJ proteins within the membrane (Shen et al., 2008), likely corresponding to nonpolymerized and polymerized TJ proteins, respectively (Piontek et al., 2011). As expected, movement of TJ proteins within the membrane, like all biological processes, was temperature dependent (Shen et al., 2008). As temperatures were decreased from 37°C to about 20°C, $t_{1/2}$ of occludin and ZO-1 recovery increased linearly and in parallel with an increase in transepithelial resistance. Changes in $t_{1/2}$ for ZO-1 continued with the same slope as temperature decreased further, whereas there was an abrupt change in slope for occludin and in the size of its mobile fraction, indicating that its mobility was affected by a transition of the membrane lipids (Shen et al., 2008). The reduced mobility of TJ proteins at low temperatures should result in a decreased probability of strand breaks and may thus explain the decrease in macromolecular permeability.

Tanaka et al. (2016) compared the permeability of claudin-15 and -21[25] overexpressing cell layers at 37 and 4°C. They found that both the permeability for ions and small organic solutes was decreased at low temperature. In the case of cell layers overexpressing claudin-21[25], the permeability for organic solute ions was completely abolished. Interestingly, the estimated pore sizes for cation permeability remained constant for both claudins.

4.4 Determining water and proton permeability

As already discussed by Diamond (1979), the major challenge when measuring paracellular water permeability is to separate it from transcellular permeability. There are no suitable inhibitors for water transport through aquaporins to isolate paracellular transport, and there are no tracers other than water.

The other challenge is the separate assessment of water and proton/hydronium flux. Because of the Grotthus effect, i.e., the rapid exchange of protons within a network of water molecules, adding deuterated or tritiated water as a tracer will not yield meaningful results.

Methods for measuring paracellular water transport always involve comparison of control epithelia with epithelia that either overexpress or are deficient in the TJ protein under investigation. Water transport can be assessed by weighing (e.g., in intestinal sack preparations, gall bladder, Diamond, 1962b), by assessing volume changes by fitting small-diameter tubes either directly to an epithelial tissue (Diamond, 1962c) or by Ussing-chamber-like devices to determine increase/decrease in water levels (Ayala-Torres et al., 2019, 2021; Rosenthal et al., 2010, 2020), or by adding an impermeable dye to one compartment and measuring changes in dye concentration (Gong et al., 2017).

Because of the Grotthus effect, transport of water and acid or base equivalents is difficult to separate. It is thus not clear whether the two water permeable claudins also provide a pathway for protons/hydronium ions. Measuring proton permeability is difficult because the application of a proton gradient may alter TJ properties. Furthermore, physiological solutions are usually well buffered against pH changes, so that samples from the acceptor side need to be back-titrated to the original pH value in order to determine the amount of transported protons. One such experiment was carried out by Angelow et al. (2006) in claudin-8 overexpressing MDCK II cells. Claudin-8 overexpression decreased proton permeability. As claudin-8 has been demonstrated to replace the cation (and water) channel-forming claudin-2 (Angelow et al., 2007), it may be speculated that claudin-2 provides a paracellular pathway for protons. However, first, the percentage of claudin-8-induced reduction in NH_4^+ and alkali metal cation permeability amounted to about 55%, whereas the change in H^+ permeability was only 25%–30%. Second, unlike the activation energy for alkali metal cation permeation, activation energy for H^+ permeation changed considerably upon claudin-8 overexpression, indicating a direct blocking effect of claudin-8 on the H^+ permeation pathway (Angelow et al., 2006). Both results argue against the involvement of claudin-2 in proton permeability. It has to be kept in mind, however, that the paracellular gradient applied by Angelow et al. (2006) involved acidifying the donor compartment to a pH-value of 3–4. It has previously been reported that paracellular charge selectivity is highly pH-dependent, especially at pH values < 4.5 (Moreno & Diamond, 1974), so that the permeability determined by Angelow et al. (2006) may not reflect physiological conditions. Thus, the exact mechanism for paracellular H^+ permeation remains elusive.

Another claudin that is able to block unwanted paracellular proton permeation is the stomach-specific isoform of claudin-18. Similar to claudin-10, there are two major claudin-18 isoforms that differ in the N-terminal part of the protein that is encoded by one of two alternative exons 1. Claudin-18 isoform one is expressed in the lung, whereas isoform two is exclusively expressed in the stomach. As for claudin-8 in MDCK II cells, neither the exact mechanism of proton permeation nor of blocking this permeation through claudin-18 is known (Tamura et al., 2012).

So far, little is known about the paracellular transport of hydroxide ions. However, in biological systems, base exchange often occurs through the transport of HCO_3^-. Of the two anion channel-forming claudins known to date, only claudin-17 is permeable to HCO_3^- (Krug et al., 2012), whereas claudin-10a is not (Breiderhoff et al., 2022; Günzel et al., 2009). There may be other, yet unidentified paracellular transport pathways for HCO_3^-, as Breiderhoff et al. (2022) found considerable HCO_3^- permeability in the thick ascending limb of the kidney, which does not express claudin-17.

4.5 Impedance spectroscopy: Exploiting capacitive properties of the transcellular pathway

Impedance spectroscopy makes use of the fact that a lipid membrane that separates two salt solutions can be described as an electric capacitor. Under alternating current conditions, a capacitor causes a -90° phase shift between current and voltage (voltage lagging behind the current). As a biological membrane also contains ion channels, which act as ohmic resistors (0° phase shift between current and voltage), the plasma membrane can be described as an RC-element, i.e., as an ohmic resistor, R, and a capacitor, C, in parallel, acting as a current divider. If AC currents of a wide range of frequencies (typically between 1 Hz and 100 kHz) are applied, the phase angle will vary between 0 and -90 degrees, depending on the proportion of current flowing through each electric component. Impedance, Z, i.e., the frequency-dependent resistance to AC current, can be described by amplitude (voltage divided by current) and phase angle, and be depicted in a polar coordinate system. Alternatively, complex numbers can be used to describe the impedance, Z, and depict it in a Cartesian coordinate system.

If the real part of Z is plotted against the imaginary part (Nyquist diagram), the resulting graph for an RC element is a semicircle with the diameter R, whereas C can be calculated from the angular frequency, ω', at which Z reaches a minimum $\omega' = \frac{1}{R \bullet C}$.

In contrast to a cell membrane, the TJ does not appear to be associated with any relevant capacitance. Thus, it can be described by a pure ohmic resistor. A simple epithelium, consisting of a monolayer of cells, can thus be represented by two RC elements in series, one each for the apical and the basolateral membrane compartment, and a paracellular resistor that is in parallel to (and thus bypassing) these two RC elements.

A further component that has to be added to this equivalent circuit is the (ohmic) resistance of the subepithelium (or filter support) on which the epithelial cells grow.

The electrical equivalent circuit of a simple epithelium thus consists of four resistived components (subepithelial resistance R^s, apical membrane resistance R^{ap}, basolateral membrane resistance R^{bl}, paracellular resistance R^{para}), and the apical and basolateral capacitances (C^{ap}, C^{bl}). This equivalent circuit was dubbed the "lumped model" by Clausen et al. (1979), as it neglects to consider other resistances, e.g., of the fluid within the paracellular cleft. Despite being a simplification, this model explains many features of simple epithelia and can be used to extract the individual components.

Although impedance measurements on biological materials have been conducted since the beginning of the 20th century to investigate plasma membrane properties (see e.g., Höber, 1910), one of the first descriptions of an epithelium by impedance spectroscopy is probably a Nyquist diagram obtained from frog skin by Cole (1932). There are many studies employing impedance spectroscopy aimed at distinguishing between the subepithelial and the epithelial resistance (e.g., Bürgel et al., 2002; Fromm et al., 1985; Gitter et al., 1998; Kroesen et al., 2008; Schifferdecker & Frömter, 1978; Teorell & Wersäll, 1945; Zeissig et al., 2007).

Using impedance spectroscopy to directly distinguish between transcellular and paracellular pathways is not possible, as there are too many variables. It can be achieved, however, by modeling pairs or sequences of spectra obtained under different experimental conditions, e.g., by affecting the paracellular resistance (sodium caprate and sodium caprylate, Sawada et al., 1991; Ca^{2+} chelation with EGTA, Van Driessche et al., 2007; Krug, Fromm, & Günzel, 2009; Rosenthal et al., 2012; Schmid et al., 2013), by activating or inhibiting transcellular transport components (Clausen et al., 1979; Gögelein & Van Driessche, 1981; Kottra & Frömter, 1990, 1993; Păunescu & Helman, 2001a, 2001b; Schmid et al., 2013; Singh et al., 2002), or by changing the membrane area and thus the capacitance by activating exocytosis- or endocytosis (Bertrand et al., 1998; Bertrand & Hopfer, 2002; Păunescu & Helman, 2001a, 2001b). Under these experimental conditions, not only can the different resistance components, R^{ap}, R^{bl}, R^{para}, be determined, but also the capacitances, C^{ap} and C^{bl}, which can be used to estimate apical and basolateral membrane areas (Bertrand et al., 1998; Bertrand & Hopfer, 2002; Clausen et al., 1979).

Instead of altering epithelial parameters, individual parameters of the equivalent circuit can also be measured separately and then combined with impedance spectroscopy to estimate all the parameters of the equivalent circuit. Such approaches include quantification of the apical and basolateral membrane resistance by intracellular microelectrodes measurements (Boulpaep, 1972; Clausen et al., 1979; Kottra & Frömter, 1990, 1993; Lewis & Diamond, 1976; Suzuki et al., 1982); use of extracellular microelectrodes to estimate the paracellular conductance (i.e., conductance or voltage scanning, see also below for a more detailed description; Frömter, 1972; Frömter & Diamond, 1972; Schifferdecker & Frömter, 1978; Gitter et al., 2000), or measuring flux of paracellular tracer substances (Amasheh et al., 2009; Krug et al., 2009a, 2009b; Milatz et al., 2010; Savolainen et al., 2011; Schulzke et al., 2011).

Many of the interventions mentioned above induce only brief or very rapid changes in individual parameters of the equivalent circuit. Measuring full impedance spectra, on the other hand, takes 10s of a second or up to 1 minute (Singh et al., 2002). During this time period, all parameters should be stable, as changes would distort the recorded impedance spectra. Different approaches have therefore been developed to speed up measurements, e.g., by applying either rectangular pulses or at least several frequencies at the same time and employing Fourier transform to analyze the resulting responses (Bertrand et al., 1998; Bertrand & Hopfer, 2002; Suzuki et al., 1982; Teorell, 1946; for review see also Van Driessche et al., 2007).

Impedance spectroscopy has not only been applied to simple but also to stratified epithelia (Kiesewetter et al., 2019; Mannweiler et al., 2021; Tunggal et al., 2005). In these epithelia, TJs are only found in the uppermost layers of the stratum granulosum. In addition, the stratum corneum supplies a further RC element that has to be taken into consideration. Therefore, circuit models have to be adapted to different situations. Nonetheless, estimates of TJ contribution toward epidermal barrier function are possible (Mannweiler et al., 2021).

4.6 Determining local conductance inhomogeneities

Junctional properties may vary greatly in different regions of one epithelial cell layer. Apoptosis will happen sporadically, cells may move or divide, and therefore need to remodel their TJs, etc. The disadvantage of all methods described above is that they integrate measured signals over the whole area of the cell layer or tissue. Yet, leaks may, e.g., be rare and localized events, but their effects may still be large enough to mask specific junctional transport properties. Thus, imaging methods are needed to identify local alterations of junctional properties. Various electrical and optical methods have been developed to enable the detection of inhomogeneities. The basic principle of electrical methods is the application of an electrical field across an epithelium or cell layer. Glass microelectrodes are then used to scan parallel to the surface of the cells and detect either voltage differences between different vertical positions in a vibrating probe approach (electrode oscillating between a position close to the cells and a position at a defined vertical distance) or using a pair of electrodes with a fixed vertical distance and record the voltage deflection between the two electrode tips. In the vicinity of a wounded/apoptotic cell or junctional disturbance, the amplitude of the microelectrode signals would increase. Depending on the distance between the electrode tip and the cell surface, larger leaks can be identified, or even differences between the cell body and TJ detected (Cereijido et al., 1980; Chen et al., 2013; Frömter, 1972; Gitter et al., 1997; Günzel et al., 2006; Yu et al., 2009; Zeissig et al., 2007; Zhou et al., 2013).

Alternatively, regions of increased paracellular transport can be detected optically. Richter et al. (2016) and (2019), developed a "sandwich assay" in which the basolateral side of an epithelial cell layer is labeled with membrane-adhesive streptavidin. Adding fluorescently labeled biotin to the apical side allows for the detection of sites of permeation, as biotin reaches the basolateral side and is bound by the streptavidin. Using this method Krug et al. (2017) and (2018) were able to demonstrate that the removal of angulin-1 from the tricellular junction by using the angulin binder angubindin-1 or a reduction in tricellulin expression, as observed in intestinal tissue from ulcerative colitis patients, locally increases macromolecular permeation at the tricellular junction. This latter effect could be mimicked by IL-13 treatment of a cell layer of the colon carcinoma cell line HT-29/B6 (Krug et al., 2018). In an alternative live cell imaging approach, Stephenson et al. (2019) injected the zinc-sensitive fluorescence dye FluoZin-3 (FZ3) into the blastocoel of an early frog embryo and applied zinc to the outside solution to visualize local, TJ breaches, and study their repair.

4.7 Patch-clamp technique

Many experimental approaches based on glass microelectrodes were developed during the second half of the 20th century. In epithelial research, they were used to determine membrane resistances directly from intracellular recordings, and thus to indirectly assess paracellular resistance. Microelectrodes were also employed in extracellular scanning techniques to separate transcellular and paracellular pathways as mentioned above. Although patch-clamp techniques boosted research on membrane-ion channels since the 1970s; to date, there is only one published attempt to assess paracellular channel properties by patch-clamping (Weber et al., 2015, Fig. 7.4).

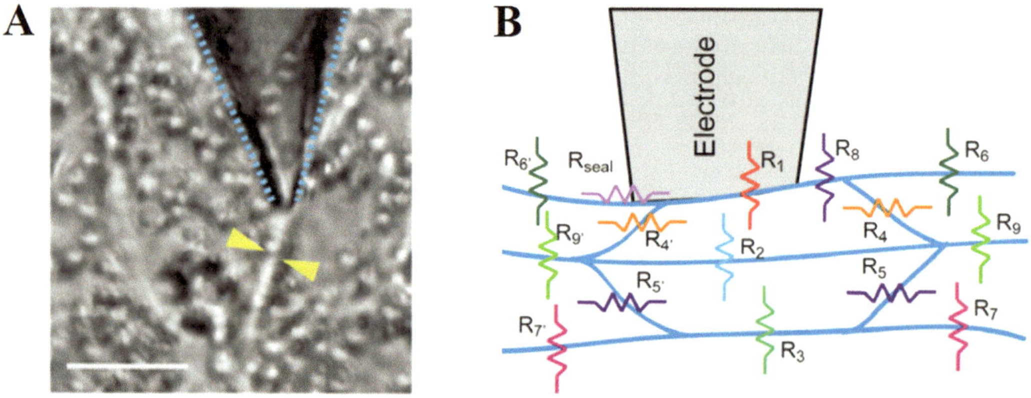

FIGURE 7.4 **Patch-clamping tight junctions.** (A) *Trans*-tight junction patch clamp placement. The intercellular junction is marked by yellow arrowheads, the patch pipette by blue lines. Bar = 10 μm. (B) The tight junction meshwork can be modeled as a complex resistor array and here considers the strands in the vicinity of the recording electrode. Each resistor within the array represents multiple parallel claudin-2 channels. *(A) Reprinted from Weber et al. (2015). (B) Reprinted by permission from John Wiley and Sons and the New York Academy of Sciences, Weber and Turner (2017).*

Weber et al. (2015) chose MDCK cells with claudin-2 overexpression controlled by a tet-off regulated system to investigate channel properties of claudin-2. As the tet-off system was "leaky", it resulted in cells expressing only small amounts (in the presence of doxycycline) or large amounts (in the absence of doxycycline) of claudin-2. They adapted the patch-clamp technique to patch small stretches of TJ from the apical side, with cell apical plasma membranes of the two neighboring cells providing the seal (Fig. 7.4A). This approach implies that the current pathway is perpendicular to the whole TJ network, i.e., crossing about three—four horizontal TJ strands.

Weber et al. (2015) were able to demonstrate that the observed currents strictly depended on the presence of functional claudin-2. Current-Voltage (I—V) relationships suggested channels with a conductance of about 90 pS and a reversal potential of about 0 mV, which, under the given experimental conditions, would not be expected for any membrane channels. The I—V relationships were linear over a wide voltage range (−100 to + 100 mV), i.e., there was no rectification. Channels were opening in bursts, with a time constant $\tau_{open} < 1$ ms. In contrast, there were two time constants for channel closing, one between two bursts ($\tau_{closed(stable)} > 1$ s) and one within bursts ($\tau_{closed(transient)} < 2$ ms). Current amplitude as well as opening and closing kinetics were independent of the amount of claudin-2 expressed. P_{Na}/P_{Cl} was 7.4 ± 3.7 and thus similar to P_{Na}/P_{Cl} calculated from dilution potentials in claudin-2 overexpressing cell layers. Furthermore, methylammonium permeability was similar to P_{Na}, whereas relative permeabilities of the large diameter cations tetraethylammonium (TEA) and N-methyl-D-glucamine were close to zero, indicating a pore diameter <6 Å. Finally, mutant claudin-2 I66C, but not wildtype (WT) claudin-2, could be inhibited by the thiol-reactive 2-(trimethylammonium) ethyl methanethiosulfonate (MTSET; Angelow & Yu, 2009). This inhibition could also be demonstrated in the patch-clamp study, when cell layers expressing claudin-2 I66C were investigated (Weber et al., 2015). WT Claudin-2 channels were inhibited by La^{3+}, but not by other membrane channel blockers investigated (4-aminopyridine, TEA, charybdotoxin, apamin, amiloride, CFTR inhibitor-172, 4,4′-Diisothio-cyano-2,2′-stilbenedisulfonic acid (DIDS)).

In addition to the claudin-dependent size- and charge-selective pore pathway, a charge non-selective leak pathway that allows paracellular flux of small and large molecules is also known to be present. However, Weber et al. (2015) did not detect this relatively large diameter (62.5 Å) leak pathway. They speculate that such events may have been interpreted as a loss of seal and therefore discarded. Detecting strand-break events, as well as further interpretation of the single channel data, especially deriving an open probability for single claudin-2-based channels, was complicated by the fact that within the pathway under the pipette, there were likely several channels both in series and in parallel. Molecular models of the channel geometry are only just emerging (see below), and molecular correlates of the channel gate are not yet known. It is expected that in the near future the structure—function characteristics of paracellular ion channels will be elucidated.

5. In silico models for paracellular permeation

In a follow-up to their patch-clamp study, Weber and Turner (2017) presented a computer model for claudin-2 channels within a multistrand TJ network (see Fig. 7.4B). The model suggested a claudin-2 channel density of about 36 channels per μm TJ length in highly expressing claudin-2 cells versus 6 channels per μm TJ length in cells expressing little claudin-2. The open probabilities, expressed as N Po = Σ(open time × number of open channels)/(total time of recording), were similar to those directly estimated from patch-clamp recordings, i.e., N Po_{high} of 0.021 and N Po_{low} of 0.0045 for high and low expressing cells. The ratio N Po_{high}/N Po_{low} amounts to 4.6, i.e., close to the ratio 36/6 of the number of channels per μm TJ length assumed in the present model. The difference in N Po under the two conditions thus reflects the differences in the number of open channels, N, rather than real differences in single channel open probabilities Po.

The arrangement of channels within a strand network, i.e., channels being in parallel as well as in series, explained variations in single channel conductance observed in the patch-clamp recordings, when two channels opened at the same time. Again, simulated and measured values were in good agreement. Finally, the authors speculate that the meshwork resulting from strand branching shielded the recordings from detecting larger strand breaks. The authors predict that, due to strand-break effects, it would be far more difficult to patch-clamp TJs with few branching points. Considering the success rate for obtaining a gigaseal of about 5% within the current study (Weber et al., 2015), it is unlikely that this hypothesis will be tested in the near future.

Recently, the computer model presented by Weber and Turner (2017) was adapted to simulate the leak pathway, i.e., paracellular permeation of uncharged solutes through strand breaks and tricellular pores (Saito et al., 2021; Tervonen et al., 2019).

Tervonen et al. (2019) used a model of TJ networks with four horizontal strands and mesh sizes of 100 × 100 × 8 nm. The network is bypassed by a tricellular tube with a diameter of 10 nm and a length of 1 μm. 20 nm. Strand breaks occur randomly and allow permeation of macromolecules and ions. Therefore, breaks reduce strand resistance (calculated from the bulk resistivity of the electrolyte within the gap).

TER and macromolecular permeability were calculated in parallel, the two variables being the probability of strand breaks and the resistance of a single TJ strand. Using published data for different MDCK subclones and for Caco-2 cells, Tervonen et al. (2019) estimated strand break permeability to be about 0.04 $\mu m^{-1}s^{-1}$ for MDCK II and Caco-2 cells but only about 0.005 $\mu m^{-1}s^{-1}$ for MDCK C7 cells. On the other hand, strand resistances for Caco-2 and MDCK C7 cells are high (10.62 and 7.65 GΩ μm, respectively) compared to MDCK II cells with about 0.4 GΩ μm. Thus, when comparing MDCK subclones to Caco-2 cells, MDCK C7 cells display higher strand stability, whereas MDCK II cells have a lower strand resistance. Tervonen et al. (2019) used these data to simulate the proposed effects of ZO-1 knockdown and ZO-1/ZO-2 double knockdown on strand stability (Fanning et al., 2012; Van Itallie et al., 2009). They found that results from these studies could be explained by an increase in strand break probability, which was, however, accompanied by an increase in strand resistance.

An increase in the number of horizontal strands caused an increase in TER and a decrease in macromolecular permeability. For both parameters, the relative effect of the tricellular junction increased with increasing strand numbers as it bypasses the bicellular TJ network. However, the model does not reproduce the exponential relationship between strand number and TER inferred by Claude (1978), so other parameters not included in the model must play a role when comparing different cell types or tissues.

One of these parameters may be the gating behavior of claudin-based channels, as already predicted by Liang and Weber (2014), before the claudin-2 single channels study had been published (Weber et al., 2015). Other parameters may be the complexity of the TJ network. Saito et al. (2021) extended the model by Tervonen et al. (2019) to include the number of branching points per unit length of TJ. They found that this parameter had a considerable effect on paracellular permeability. Interestingly, they found that a double knockdown of occludin and tricellulin in MDCK II cells, in which claudin-2 had previously been knocked out, greatly reduced the number of branching points. This caused a decrease in TER and an increase in permeability for 4 and 10 kDa FITC-dextrans, whereas the 250 kDa FITC dextran permeability was unaffected. Simulating the combined effects of the number of branching points and the number of horizontal strands, Saito et al. (2021) were able to demonstrate that increasing the number of branching points is a highly efficient strategy for increasing the tightness of an epithelial barrier. They estimate that in the absence of strand branching, 12 to 200 horizontal strands are necessary to obtain a barrier strength that is similar to a crosslinked network with three to seven horizontal strands (Saito et al., 2021). The authors conclude that vertebrate TJs and their invertebrate septate junctions, which are the counterpart of TJs, follow different strategies to achieve the tightness of the paracellular barrirer. Invertebrates increased the number of unbranched sptate junction strands to achieve a tight barrier, while vertebrates express crosslinked TJ strands increasing network complexity in improving the tightness of the barrier.

6. Fence function of tight junctions

While an overwhelming number of publications deal with the "gate" function of TJs, there are comparatively few studies directly addressing their "fence" function, i.e., the role of TJs in separating the apical from the basolateral membrane compartment and thus keeping up cell polarity (Diamond, 1977). Fence function does not only refer to the separation of apical and basolateral membrane proteins but also to lipids and is thus to some extent intertwined with the question, to what extent the plasma membrane itself contributes to the TJ. One hypothesis proposes that the TJ purely consists of merged outer plasma membrane leaflets and a series of cylindrical inverted lipid micelles (e.g., Kachar & Reese, 1982; Meyer, 1983), and has been adapted into a protein lipid hybrid model in which the inverted micelles are stabilized by TJ proteins (Lee et al., 2008), see also reviews by Lingaraju et al. (2015), Zihni et al. (2016), and Otani and Furuse (2020). The the lipid model implies (i) that lipids within the inner leaflet of the membrane can freely diffuse between the cell poles and (ii) that lipids within the outer membrane can move from one cell to the neighboring cell. There is evidence for and against lipid diffusion, and thus, the involvement of lipids in TJ formation has not yet been proven (or disproven) conclusively.

With respect to membrane proteins, fence function would imply that, in the absence of TJ strands, apical membrane proteins should diffuse to the basolateral compartment and vice versa. Evidence for and against this hypothesis has been published (for review, see Vasquez et al., 2021). Furthermore, a strict association of TJ gate and fence function is not necessarily obligatory: Compartmentation of membrane domains in nonepithelial cells, e.g., the change in ion channel expression of neuron soma and axon, at the axon hillock, is stable without any TJ-like structure (Winckler et al., 1999). Recent work by Otani et al. (2019) indicates that claudins, and thus gate function, are not necessary to maintain the polarity of MDCK II cells. However, when ZO-1 and ZO-2 are both knocked out, polarity is lost.

7. Molecular structure of claudins and tight junction strands

7.1 Introduction to claudin structure

Francis Crick, in his autobiography "What Mad Pursuit", says, *"If you want to understand function, study structure"* (Crick, 1990). The structure of a biological macromolecule is a framework to understand its functioning mechanism. In the case of TJs, information about the 3D structure of claudins is paramount since they form the backbone of TJ strands. However, due to the multispanning integral membrane protein nature and the ability to oligomerize, it is experimentally challenging to express, purify, and crystallize claudins to obtain structural information. Nevertheless, methodological advances in the purification and crystallization of transmembrane proteins have improved the situation enormously (Li et al., 2021).

A major milestone was made in 2014, when Suzuki et al. reported the crystal structure of mouse claudin-15 at 2.4 Å resolution (PDB ID: 4P79) (Suzuki et al., 2014). The structure (Fig. 7.5) revealed the conformation of a major portion of

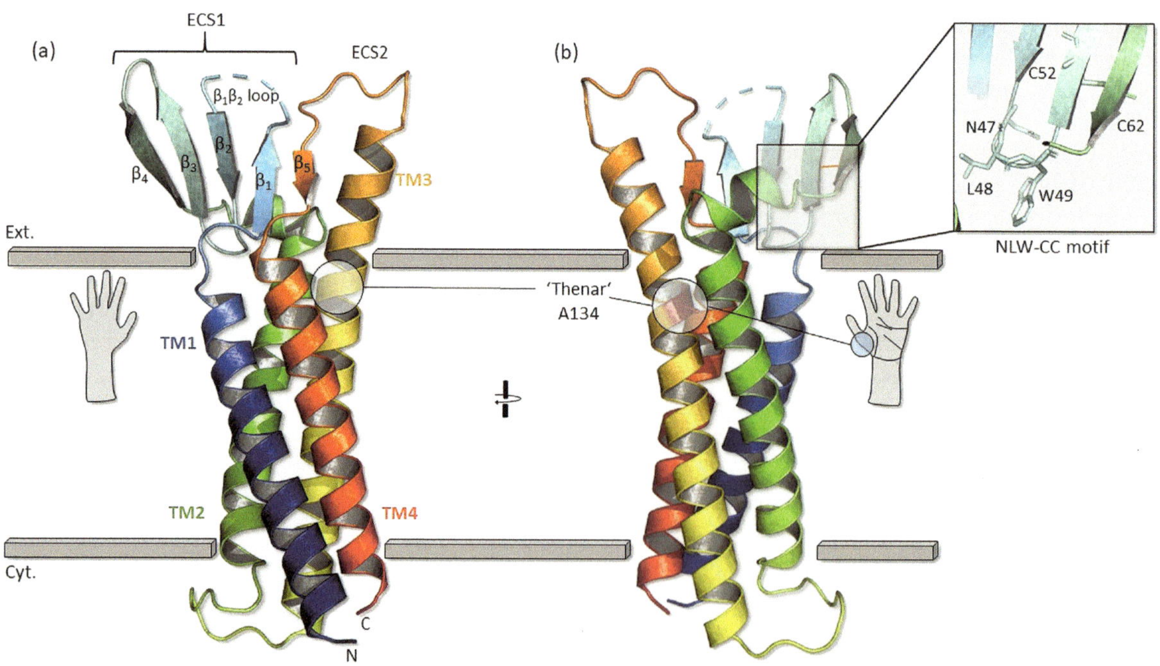

FIGURE 7.5 **Mouse Claudin15 crystal structure.** Structure of mCldn15 (PDB ID: 4P79) is shown. The protein is represented as cartoon, with the color changing gradually from the N-terminus (blue) to the C-terminus (red). The monomer is shown at two different angles (A and B), which corresponds to the back and front of a left hand, respectively. The NLW-C-C motif with LW forming a "hydrophobic anchor" and C—C stabilizing the β-sheet is shown in an inset, with the residues represented as sticks. The position of A134, which is linked to TM3 bending at the "thenar" is marked with circles. The presumable unstructured $\beta_1\beta_2$ loop in ECS1 and intracellular C-terminal region with the PDZ-binding motif at the C-terminal end were not resolved.

the two extracellular segments (ECS1 and ECS2) in addition to the typical left-handed helical bundles formed by four transmembrane (TM) segments. The length of the helices was consistent with the membrane thickness, except for TM3, which was longer than the membrane and had an extracellular segment. The extracellular segments formed an antiparallel β-sheet with five β-strands and gradually twisted along the membrane axis. One portion of the ECS1 formed four strands, whereas ECS2formed the remaining one. The overall conformation is like an extended left-hand, with the TM helices as the forearm, the long TM3 as the thumb, and the β-sheet as the other four fingers (Fig. 7.5). A disulfide bridge between the conserved cysteines present at β3 and β4 strengthens the β-sheet formation and is crucial for claudin functioning (Li et al., 2013; Van Itallie & Anderson, 2006). These cysteines are part of the consensus W−N/GLW-C-C motif conserved in all claudin family members. Residues W29, L48, and W49 of claudin-15, which are part of this motif, reside in the groove at the top of the helical bundles near the extramembrane surface. These residues serve as a "hydrophobic anchor" to the β-sheet at the transition between the membrane and the extracellular surface. A limitation of this structural study is that the loop connecting the β_1 and β_2 strands (residues 34 to 41, '$\beta_1\beta_2$ loop') was not resolved, most likely due to a disordered/ flexible structure (conformation).

The ECS1 connects with the TM2 through a short extracellular helix (ECH), whereas the ECS2 connects with the β5 strand via a more flexible turn region. The rather poorly conserved sequences in the ECS1 $\beta_1\beta_2$ loop and the ECS2 turn were defined as variable regions, V1 and V2, respectively (Suzuki et al., 2014). However, highly conserved hydrophobic residues in part of the V1 region were proposed to be critically involved in the ability of classic claudins to form the TJ strand (Krause et al., 2008; Piontek et al., 2020). In addition, the V1 and V2 regions are considered to be involved in homo- and hetero-typic *trans*-interactions of claudins. In the crystal lattice reported by Suzuki et al. (2014), the mouse claudin-15 (mCldn15) protomers are arranged as a linear polymer, mediated by interactions between two neighboring protomers. M68, one of the conserved hydrophobic residues, sticks out from the ECH region in one protomer and nestles inside a complementary hydrophobic pocket formed by the conserved aromatic residues in TM3 (F146 and F147) and L158 in ECS2 of the adjacent protomer (linear-*cis* interface). Suzuki et al. (2014) studied the Cldn15 strand formation using freeze-fracture electron microscopy and observed no TJ strands when these residues were mutated with small or charged amino acids. This suggested that TJ strands consist of claudin polymers in which protomers are connected at least partly by the linear-*cis* interface (Suzuki et al., 2014).

7.2 Crystal structures of cCPE−claudin complexes

Clostridium perfringens is one of the most common causal agents of foodborne illnesses in Europe and the United States. Its virulence is mediated by the virulence factor *Clostridium perfringens* enterotoxin (CPE). The C-terminal domain of CPE (cCPE) binds to the ECSs of certain claudin subtypes. Even though cCPE is not cytotoxic, it binds to claudins with high affinity and influences the formation of TJ strands by claudins. Except for the mCldn15 structure, all other crystal structures of claudins reported to date are crystallized in complex with the cCPE. Therefore, in addition to the structural information about different claudin subtypes, they provide vital information regarding claudin−CPE interaction. In 2015, Saitoh et al. reported a 3.7 Å resolution crystal structure of mammalian Cldn19 in complex with cCPE (PDB ID: 3X29, Saitoh et al., 2015). The structure appeared similar to that of mCldn15. For example, the structure formed an almost identical four-helix bundle, a similar β-sheet, and the W-LW consensus motif formed a similar "hydrophobic anchor" for the β-sheet dipping into the four-helix bundle. However, there are also structural differences, for which it is at least partly unclear if they are due to the presence of cCPE or the sequence difference between Cldn15 and -19. The most notable difference in Cldn19 structure is that the $\beta_1\beta_2$ loop, especially the consensus motif (A/N/S)−I-(I/L/V)-(T/V) which was disordered in Cldn15 structure. The motif was seen interacting with the surface of cCPE. Another significant difference is that the $\beta_3\beta_4$ loop and the ECH region seen in Cldn15 are disordered in Cldn19.

Even though the Cldn19 crystal structure provides much information about claudin−cCPE binding, it should be taken with a grain of salt, considering the fact that Cldn19 is a low-affinity CPE receptor. Hence, expectedly, there were noticeable differences when Shinoda et al. (2016) reported a 3.5 Å resolution structure of cCPE-bound Cldn4, a high affinity receptor of CPE (PDB ID: 5B2G). Like Cldn19, the ECS1 and ECS2 interact with the cCPE in Cldn4. Shinoda et al. reported that the Cldn4-cCPE conformation looked like "a left-hand grasping the ellipsoidal C-CPE". Certain differences between the two cCPE-bound structures include: the orientation of the extracellular part of TM3, which is bent by ~10 degrees outwards, details of the interactions between ECS2 and cCPE, and weaker hydrophobic interactions in Cldn19. For both Cldn19 and Cldn4 complex structures, the ability for a *cis* interaction as proposed by Suzuki et al. (2014) was appraised. According to Shinoda et al. (2016), if two cCPE bind to two adjacent Cldn4 in *cis*, it would generate an unavoidable steric hindrance. The competition of space created by cCPE was proposed as a potential mechanism for the disruption of the TJ.

In 2019, Nakamura et al. (2019) published a 3.6 Å crystal structure of mCldn3 wt in complex with cCPE wt (PDB ID: 6AKE)). Cldn3 is one of the most widely studied CPE receptors (Briehl & Miesfeld, 1991; Katahira, Sugiyama, et al., 1997; Morita, Furuse, et al., 1999; Veshnyakova et al., 2012). Except for the ECH residues (residues 66–70), the electron density of the structure was effectively determined. Even though the cCPE bound with ECS1 and ECS2, it was positioned distant to the segments in comparison with the same in Cldn19 and Cldn4. Nakamura et al. (2019) explain this with the following analogy: "Explaining with left-hand model, the hands of mCldn19 and hCldn4 grasp and enwrap C-CPE with the palms, whereas the hand of mCldn3 pinches C-CPE with the thumb and the index and middle fingers."

Vecchio and Stroud (2019) reported crystal structures of cCPE-bound human Cldn9 at resolutions 3.2 and 3.3 Å (PDB IDs: 6OV2, closed form; 6OV3, open form). Except for the ECH, other regions are well conserved, along with those of other crystal structures. In 2021, Vecchio et al. (2021) reported a structure of human Cldn4 with cCPE at 3.37 Å (PDB IDs: 7KP4), which was similar to PDB ID 5B2G but differed in the detailed conformation of the ECH region and the extracellular extension of TM3 as well as some aspects cCPE–claudin interaction.

7.3 The different claudin crystal structures differ in bending of TM3

All the mentioned crystal structures have almost identical TM four-helix bundles. However, in the structure of Cldn3, the extracellular part of TM3 (C-terminal of Pro134) is bent, probably due to the kink caused by Pro134. This conserved residue is either a proline, a glycine, or an alanine, depending on the claudin subtype. This residue, independent of cCPE binding, heavily influences the bending of TM3. For example, in Cldn15 and Cldn19, where proline is replaced by alanine, the bending of the TM3 extracellular part is not seen. In contrast, in Cldn3 where proline is present, the TM3 is clearly bent outward (Fig. 7.6). Due to the prominence of the residue Pro134, Nakamura et al. (2019) studied different point mutations of the residue and the region where the TM3 bends as "thenar" (the mound at the palm that connects the thumb to the wrist, Fig. 7.5). Based on crystal structures of the Cldn3 mutants (PDB IDs: 6AKF, 6AKG), the proline at the thenar bent TM3

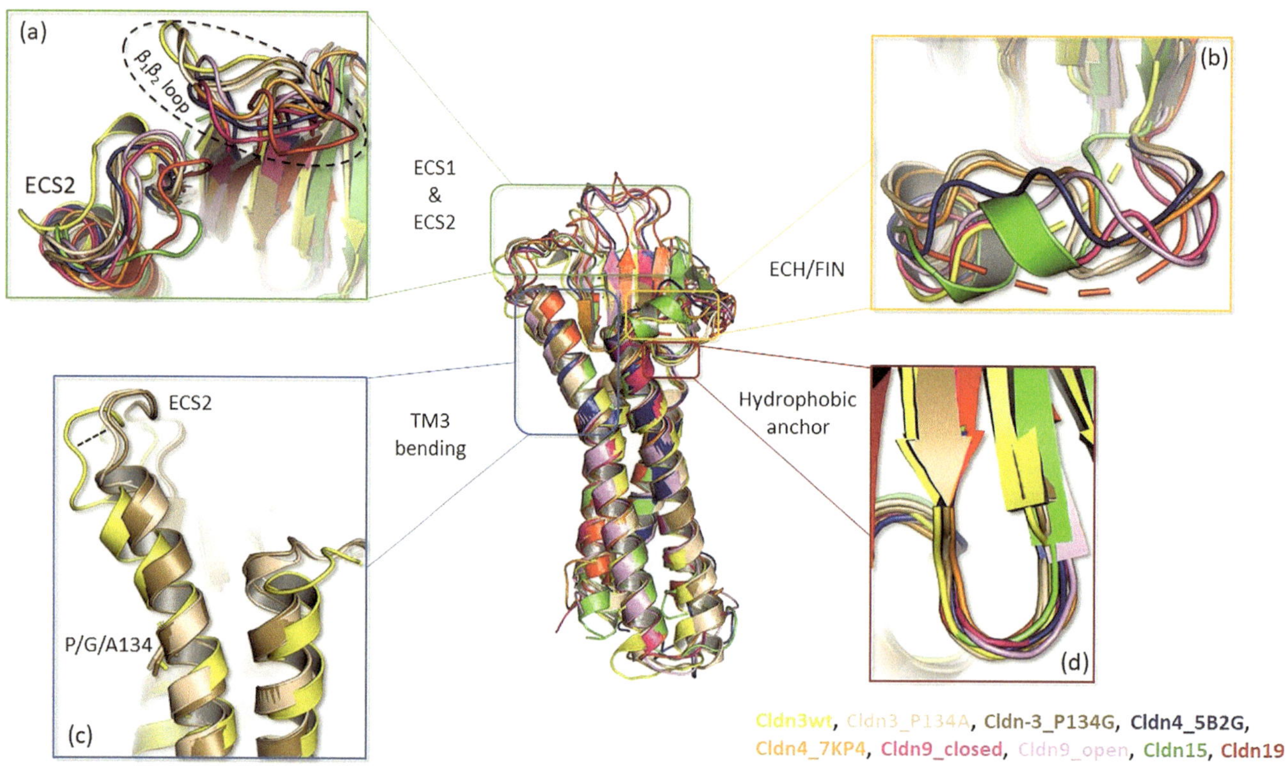

FIGURE 7.6 **Superimposition of the crystal structures of claudins.** (A) The difference in structural conformation of the extracellular segments, ECS1 and ECS2 is shown. Most notably, the conformation of β1β2 loop is different among the different crystal structures. (B) The extracellular helix/flexible interaction nub, an important player in the linear-*cis* interface, differs among the crystal structures. In most structures (in presence of cCPE), the ECH region is not α-helical. (C) Superimposition of the structures of Cldn3 wt and P134 mutants reported by Nakamura et al. (2019) that differ in TM3 bending. TM3 of Cldn3 wt with Pro134 bent more than the mutants with alanine/glycine at the position. (D) The structural conformation of the hydrophobic anchor formed by the conserved N/GLW motif is highly similar among claudin structures.

the most. Glycine and alanine at this position did not bend the TM3 but made the helix wobbly. All the mutations showed no difference in terms of cCPE affinity. Hence, the bending of TM3 is determined by proline rather than the cCPE binding (Nakamura et al., 2019). Also, other studies indicated TM3 bending (Irudayanathan et al., 2017; Rossa, Protze, et al., 2014; Shinoda et al., 2016). Its importance for claudin polymerization is discussed below.

8. Models of strand and channel formation by claudins

Based on the crystal structure of Cldn15, Suzuki et al. (2015) suggested a model for the basic molecular architecture of claudin polymers responsible for the size-selective barrier function and charge-selective ion channel function of TJ strands (Suzuki et al., 2015, joined double rows; below named *Suzuki-JDR* model, Fig. 7.7). This model proposes that claudin monomers polymerize into antiparallel double rows within one membrane, via two different *cis* alignments with neighboring protomers. One interface, named side-by-side or *linear-cis* (Hempel et al., 2020; Piontek et al., 2020), is seen in the crystal packing of Cldn15 and is formed between the ECH and ECS2 of adjacent claudin protomers within the same row (Suzuki et al., 2014). The other *cis*-interface, called as the *"face-to-face"* interface (Suzuki et al., 2015), is formed between two mCldn15 monomers of the opposing antiparallel rows by a series of H-bonds between β4 strands that face each other. This results in an extended antiparallel β-sheet structure that is energetically favorable and stable (Suzuki et al., 2015). Different crosslinking studies (Angelow & Yu, 2009; Hempel et al., 2020; Suzuki et al., 2015) support the *"face-to-face"* interface.

Moreover, in the Suzuki-JDR model, two claudin *cis*-double rows from two opposing cell membranes are joined in trans, resulting in the formation of β-barrel-like pores, which form the paracellular pathway along the membrane plane. These tetrameric pore or channel scaffold tightly interact with neighboring tetrameric pores and form a continuous polymeric barrier for solutes that are excluded from the pore due to size or charge.

This model represents the steady state of the claudin polymer. Concerning the polymerization process, it was suggested that claudins first form *cis*-oligomers (e.g., dimers or tetramers) and that *trans*-interaction at cell–cell contacts triggers *cis-/trans*-polymerization (Hempel et al., 2020).

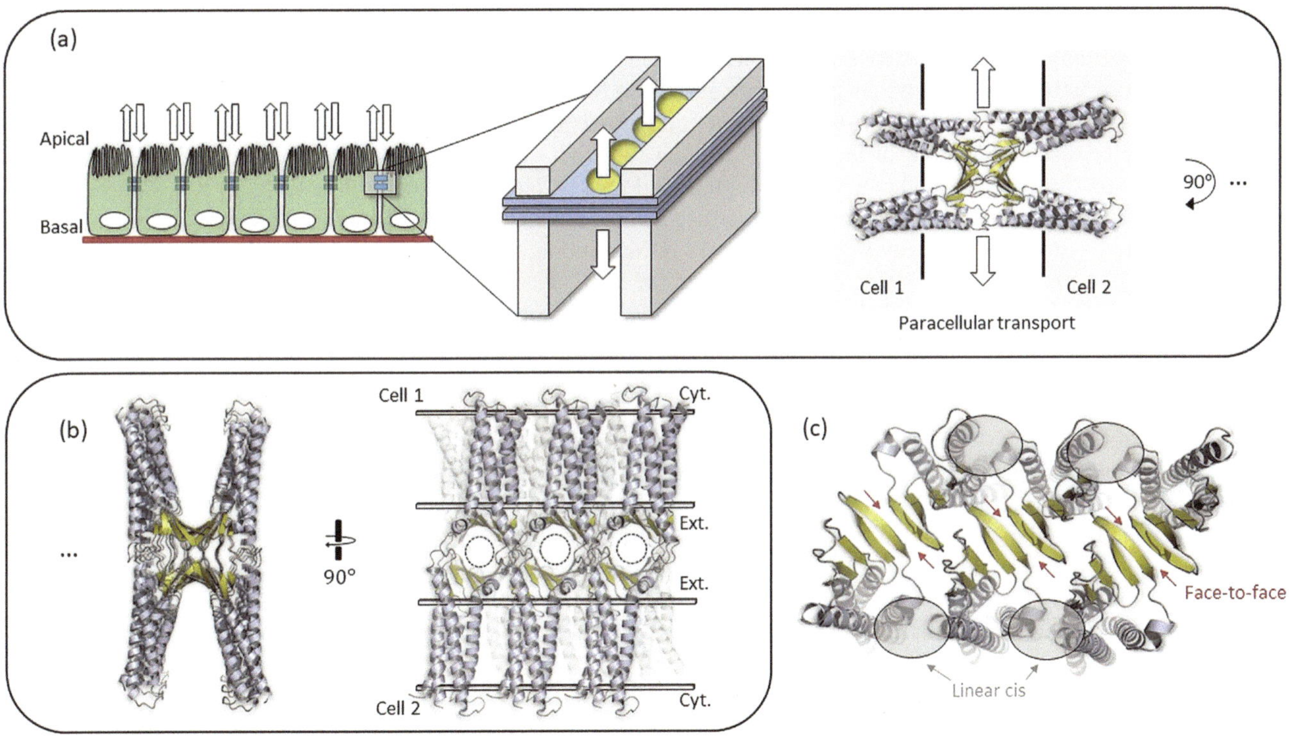

FIGURE 7.7 **Tight junctions strand assembly and paracellular channel architecture.** (A) Representation of the paracellular ion channels formed by TJ strands in epithelia. Claudin protomers (shown in cartoon representation) form ion permeation pores in the paracellular space between two cells. White double arrows represent the orientation of flow of the paracellular transport. (B) Cldn15 joined double rows (JDR) model of paracellular TJ channels as proposed by Suzuki et al. (2015). The paracellular pores formed by the β-sheets are represented as dotted circles. (C) Two important interfaces of Suzuki Cldn15 JDR model are shown. Linear-*cis* interfaces, formed between ECH and ECS2 of two neighboring protomers, are marked with gray circles. Red arrows represent the face-to-face interface formed between β_4 strands of two adjacent protomers.

The inside of β-barrel-like pores is lined by the extracellular domains ("palms") of mCldn15 protomers that contain a negatively charged surface, which can serve as a route for the paracellular pathway. The resultant pore diameter of the paracellular pore would be less than 10 Å in diameter, which is consistent with studies probing the pore diameter (Hempel et al., 2022; Rosenthal et al., 2020; Samanta et al., 2018; Van Itallie & Anderson, 2006; Yu et al., 2009). Channels involved in the transcellular pathway allow for the perpendicular (vectorial) movement of ions or water across the membranes. In contrast, along the paracellular pathway in TJs, the solutes travel in between and parallel to the membranes (Fig. 7.7). Despite the difference in orientation and protein structure, the underlying mechanism of ion conduction through the pores should, to a certain extent, be similar for both channel types (Burnashev et al., 1996; Villarroel et al., 1995).

Various studies support the linear-*cis* polymer model of TJ strands proposed by Suzuki et al. It was shown that the ECS2 residues in Cldn3 and Cldn5 corresponding to interfacial residues in the Cldn15 strands model play a critical role in formation of TJ strands (Piontek et al., 2008, 2011) and of paracellular barriers (Piehl et al., 2010; Zhang et al., 2010). Conserved ECS2 and ECH residues, form the linear-*cis* interface, which is the "most-critical-cog-in-the-wheel" of the linear-*cis* polymerization model.

Based on sequence homology, claudins can be classified into classic (Cldn1 to −10, −14, −15, −17, −19) and nonclassic (Cldn11, -12, -13, -16, -18, -20 to -27) (Krause et al., 2015), where the former display higher sequence homology than the latter. The classic claudins are proposed to follow similar TJ assembly mechanism (Hempel et al., 2020; Krause et al., 2015; Suzuki et al., 2015) mediated by common sequence motifs: I (in ECS2: [F, F/Y, N/D/S, P/s, x(5−6), K/R,Y/F/R, E, L/I/M/F], II in ECH: [S/T, L/M/I/V, L/F, X, L/V], and motif III: in $\beta_1\beta_2$-loop: [V/I, V/I/L, V/T/I]), which are missing or incomplete in nonclassic claudins (Piontek et al., 2020). These structural motifs are not present, or partly present, in nonclassic claudins, and hence the structural conformation and interaction properties of these regions are likely to be significantly different between classic and nonclassic claudins. Many studies have reported the importance of these regions in TJ assembly (Daugherty et al., 2007; Hempel et al., 2020; Piontek et al., 2008, 2011; Rossa, Ploeger, et al., 2014; Suzuki et al., 2014; Zhao et al., 2018).

Based on the claudin subtype that forms TJ strands, the TJs are either a barrier for paracellular solute permeation or a charge- and size-selective paracellular ion channel. Barrier-forming claudins such as Cldn1, -3, -5 and channel-forming claudins such as Cldn2, -10a, −10b, −15, −17 are proposed to share a similar overall molecular strand architecture (Hempel et al., 2020). Interactions between the extracellular domains of the barrier forming ones seem to close the paracellular space completely, whereas the channel-forming claudins are modeled to form the above-mentioned β-barrel-like pores. These pore-lining claudin subtype-specific residues have been shown to determine the ion selectivity of the channel. In a single protomer, this proposed pore-lining region corresponds to the "palm" of the "hand" model shown in Fig. 7.5. Based on the electrostatic potential of the "palm," the ion selectivity of the pore would be determined. Anion-selective claudins like Cldn10a and Cldn17 have a positive palm and cation-selective claudins like Cldn2, -10b, -15, or -21 have a negative palm (Conrad et al., 2016; Suzuki et al., 2014; Tanaka et al., 2016). Key residues affecting ion selectivity are in (or close to) the β4-strand in the palm region (Anderson & Van Itallie, 2009; Angelow & Yu, 2009; Colegio et al., 2002; Conrad et al., 2016; Krause et al., 2015; Tanaka et al., 2016; Yu et al., 2009). For example, in mCldn15 crystal structure, negatively charged residues like D55 extend outwardly, so much that, if an ion-permeable pore is formed, they will be facing the pore. Colegio et al. (2002) observed that mutation of D55 resulted in a reversal of ion selectivity. Similarly, in Cldn2, D65 lines the putative pore center region and was shown to determine charge-selectivity (Yu et al., 2009).

Claudins assemble into TJ strands through homophilic and heterophilic *cis*- and *trans*-interactions (homo- or heteromeric or -typic), where the two ECS serve distinct but cooperative functions. In most epithelial and endothelial cells, mainly heteropolymeric strands are formed in a claudin subtype-specific fashion (Daugherty et al., 2007; Furuse et al., 1999; Piontek et al., 2011). Nearly all classic claudins are able to form homopolymeric strands (Furuse et al., 1999; Krause et al., 2008; Milatz, Piontek, et al., 2017; Piontek et al., 2020; Yamazaki et al., 2011; Gonschior et al., 2022). In addition, heteropolymerization occurs for a subset of compatible claudins (Daugherty et al., 2007; Furuse et al., 1999; Krause et al., 2008; Milatz, Himmerkus, et al., 2017; Piontek et al., 2020; Gonschior et al., 2022). Nonclassic claudins appear to form only heteropolymeric TJ strands mainly with classic claudins (Hou et al., 2009; Milatz, Himmerkus, et al., 2017; Piontek et al., 2020). The nonclassic Cldn11 is an exception that forms also homopolymeric strands (Berndt et al., 2019; Krystofiak et al., 2019; Morita, Sasaki, et al., 1999).

8.1 Refinement of the Suzuki-JDR Cldn15 model by molecular dynamics simulation studies

In the Cldn15 Suzuki-JDR model, nine residues in the ECS1 and two residues in the ECS2 are missing. Suzuki et al. (2015) stated that the space between the two "half-pipe" structures is big enough to accommodate the disordered ECS1 residues (β_1-β_2 loop). However, insertion of these residues led to substantial steric clashes as reported in different studies (Alberini et al., 2017; Krause et al., 2015; Milatz, Piontek, et al., 2017; Piontek et al., 2017; Samanta et al., 2018). At the same time,

Miltaz et al. (2015) reported that the molecular TJ strand architecture is consistent with the antiparallel double-row *cis*-arrangement, observed via a FRET analysis of Cldn1 and Cldn3 protomers in a TJ strand. Thus, the Suzuki-JDR model provides a starting point for understanding the molecular architecture of TJ strands made up of claudins. Based on the Cldn15 Suzuki-JDR model template (Suzuki et al., 2015), similar models for other claudin subtypes, like Cldn3 (Piontek et al., 2017) and Cldn10b (Milatz, Piontek, et al., 2017), were proposed. Subsequently, molecular dynamics (MD) studies on different claudin subtypes based on the Cldn15 Suzuki-JDR model were reported. As the model proposed by Suzuki et al. is a static representation of how claudins could interact, with important residues missing, and expected steric clashes, the observations from the MD studies help in refining the Suzuki-JDR model with atomic precision.

Based on the Suzuki-JDR model, Alberini et al. (2017) published atomic models of paracellular Cldn15 single and double pore architectures and observations from the all-atom MD simulations of the membrane-embedded claudin-15 oligomer models. One of the significant observations from the study is that the disordered segments in the Cldn15 model could be arranged without clashes in different conformations, which confirms the possibility of pore formation within claudin polymers based on the Suzuki-JDR model. The single, tetrameric pore, and the octameric double pore were constructed using replicas of a well-equilibrated Cldn15 monomer and simulated in a double-lipid bilayer and water environment, respectively. The average minimal pore radius over the simulated time is ~2.5 Å and 3 Å, for a single and a double pore, respectively. This agrees with the narrowest point of Cldn2 pores with a diameter of ~6.5 Å, as suggested by electrophysiological studies (Yu et al., 2009). It is important to note that the maximal constriction of the pore occurred in correspondence of the D55 residues. During MD simulations, the models were stable with a well-defined β-barrel-like pore, and the interactions in the linear-*cis* and face-to-face interfaces were maintained. *Trans*-interacting residues showed differences in interactions between the two models, which could be due to the different conformations of the ECS segments used in the two models. ECS2 region, especially, showed a set of interactions in the double pore model, which was missing in the single pore model. For example, hydrophobic residues in the ECS2, like F146, F147, P149, L150, A152, and G153, formed a dense hydrophobic cluster. This clustering suggests their importance in TJ strand formation and is in agreement with the corresponding hypothesis suggested by Suzuki et al. (2015) and Krause et al. (2015).

Alberini et al. (2018) further validated their single-pore model by extending the simulations and estimated the cation permeation capability of the pore. The calculation of the permeation timescale of Na^+, K^+, and Cl^- ions and their corresponding potential of mean force (PMF) profiles revealed that the channel allowed the passage of Na^+ nearly 30 times more than Cl^-. This fits to experimentally determined charge-selectivity of Cldn15 channels (Colegio et al., 2002; Rosenthal et al., 2020). Strikingly, the overall architecture was stable with a minimum pore radius of ~2.5–3 Å, over a period of ~400 ns. Partial dehydration of a Na^+ ion in the central pore region, and the ion binding to the side chains of D55 residues in a partially dehydrated state, was also observed.

Samanta et al. studied the architecture and channel selectivity of Cldn15 channels using all-atom MD simulations and in vitro analyses in Cldn15-overexpressing MDCK I cells (Samanta et al., 2018). Their model indicated that Cldn15 protomers form well-defined channels that allowed the transport of water and small cations. A dodecamer model was constructed with six monomers present in two 1-Palmitoyl-2-oleoyl-sn-glycero-3-phosphocholine (POPC) bilayers each, resulting in a system with three pores. Trajectory analysis showed that the overall structure was stable over the long simulation period (~250 ns), supported by stable linear-*cis*, *face-to-face*, and *trans*-hydrophobic interfaces. Importantly, E157, a highly conserved residue in barrier- and channel-forming claudins (Krause et al., 2015; Piehl et al., 2010; Piontek et al., 2008, 2017), formed an H-bond with S67 present in the ECH. P149 and L150 of two protomers in *trans* interacted with each other steadily over time. Hence, Samanta et al. suggested that a partial tilting of TM3 with a slight adjustment of the ECS2 loop residues conformation could remove potential clashes in *trans* alignment of opposing claudin monomers.

Like in Alberini et al. (2017), the highly flexible $\beta_1\beta_2$ loops in two opposing protomers were involved in *trans*-interactions. However, the mode of these interactions differs between these two studies, suggesting different conformations of the $\beta_1\beta_2$ loop that should be tested in further studies. The analysis of water density confirmed that the flow of water was restricted to the pores and the space between the pores was sealed by *cis*- and *trans*-interfaces. The overall length of the simulated pore is 49 Å with a mean radius of 8.1 Å and 4.2 Å in the widest and narrowest regions, respectively. The size of the pore is large enough to selectively accommodate the flow of partially hydrated small ions like Na^+ and Cl^-, but restricting the larger ions, which agrees with the in vitro cell models of Cldn15 channels (Colegio et al., 2002; Samanta et al., 2018). Samanta et al. reported negatively charged residues like E46, D55, and D64 to line the wall of the permeation pathway. While E46 and D64 were present at the narrowest region, D55, an important residue that regulates the charge-selectivity, resides at the widest region of the pore model. Further mutation studies of these residues confirmed the key role played by D55 in charge-selectivity of Cldn15 channels, with D64 and E46 having only small effects. However, in the Cldn2 cation-selective channels, the residue critical for charge-selectivity is D65, corresponding to W63 in Cldn15, suggesting different selectivity filters for Cldn15 and Cldn2 (Li et al., 2014; Yu et al., 2009).

9. Alternative models of claudin polymerization

Concerns were raised against the Suzuki-JDR-like architecture of TJ strands: (i) The pore-lining residues did not include many of the residues that correspond to putative pore-lining residues of Cldn2 cation channels as indicated by their accessibility toward thiol-sensitive probes (Li et al., 2014; Krause et al., 2015). (ii) Lack of direct interaction between the four TM-helix bundles of the claudin protomers, which was assumed to stabilize oligomerization of the transmembrane proteins (Conrad et al., 2016; Krause et al., 2015) and was suggested by in vitro data (Gong, Renigunta, et al., 2015; Milatz, Piontek, et al., 2017; Rossa, Ploeger, et al., 2014; Van Itallie et al., 2011).

Alternative oligomeric interfaces were suggested by comprehensive coarse-grained MD-based screens for homo *cis*-dimers of different classic claudins (Irudayanathan et al., 2016, 2018). These self-assembly studies resulted in four different dimeric interfaces (named dimer A–D). Dimer A corresponds to the linear-*cis* dimer proposed in the Suzuki-JDR model. In Dimer B, TM2-TM3 from two adjacent protomers interact directly. In dimer C, TM3-TM4 from two adjacent protomers, along with ECS2, form the *cis* interface. Dimer D is analogous to the *face-to-face* interface. Rajagopal et al. further expanded the dimer simulations by developing a novel approach named Protein Association Energy Landscape (PANEL) method and applying it to Cldn-2, -4, -5, and -15 (Rajagopal et al., 2020).

Two of the above-mentioned dimers contained *cis*-interfaces between TM helices. The interface in dimer B (also called *back-to-back* interface) was notable since it was supported by experimental data (Rossa, Ploeger, et al., 2014; Van Itallie et al., 2011). Mutations in Cldn5 (F139S and I142T), shown to affect oligomerization (Rossa, Ploeger, et al., 2014), inhibited the formation of the back-to-back dimer (dimer B) in MD simulations (Irudayanathan et al., 2017).

Irudayanathan et al. (2017) also analyzed the *cis*- and *trans*-assembly of Cldn3 and Cldn5 oligomers through *trans*-docking of either the face-to-face or the back-back dimer, leading to tetramers containing either the Suzuki-JDR-like arrangement "pore one" (Pore I) or pore II, which has similar overall *trans* interfaces like pore I, but containing a *back-to-back* interface instead of a *face-to-face*. The highly conserved cysteine 64 is at the center of pore I, whereas it is present at the mouth of pore II. Given the fact that Cldn3 and Cldn5 are accepted as barrier-forming claudins (Amasheh et al., 2005; Colegio et al., 2003; Milatz et al., 2010; Nitta et al., 2003; Piehl et al., 2010; Wen et al., 2004), it is interesting that the two pore models had average diameters that would allow the passage of water. However, due to insufficient experimental evidence, it is debatable that Cldn5 could form a functional pore. As an extension of this study, Irudayanathan et al. (2018) performed self-assembly simulations of classic claudins including Cldn-1, -2, -3, -4, -15, and −19 and molecular docking to further test the tetrameric pore I, which is narrow at the center and wider at the entrance. In terms of the pore minimal diameters, there was no clear difference among the different claudins. On the one hand, this suggests that both barrier- and channel-forming claudins could follow a similar oligomerization mechanism. These results indicate that the simulation studies must be substantialy refined and substantiated by functional in vitro data.

Recently, Hempel et al. studied the architecture of channel-forming Cldn10b and barrier-forming Cldn3. In their study, they compared models with two different linear-*cis* interfaces, which correspond to *face-to-face* and, *back-to-back* —interfaces (Hempel et al., 2020). Cysteine-scanning mutagenesis and Förster resonance energy transfer (FRET) assays supported the face-to-face but not the back-to-back interface. In sum, this study supported the Suzuki-JDR-like architecture for the prototypic channel-forming Cldn10b as well as for the prototypic barrier-forming Cldn3. Hempel et al. suggested that classic claudins follow a multistep polymerization. First, claudins *cis*-oligomerize, then *trans* interactions between claudins at the cell–cell contacts stabilize further *cis* interactions and lead to *cis-/trans*-polymerization. The oligomeric building blocks are linearly joined into polymers, which form pores only in channel-forming claudins. $\beta_1\beta_2$ loops clustered together in *cis* and *trans* that allowed the stabilization of the polymers via steric, hydrophobic, and hydrophilic fitting. In addition, the *trans* interactions between ECS2 is important in forming stable *cis/trans* interactions between claudin tetramers. It was suggested that electrostatic repulsion between four D56 residues from four Cldn10b protomers contributes to pore formation. In contrast to this, electrostatic attraction between the corresponding Q56 residues from Cldn3 residues helps in the barrier formation.

10. Structural base for lateral flexibility of TJ strands

Using live-cell imaging microscopy, Zhao et al. extensively studied the inherent flexibility of TJ strands within the membrane. Additionally, they evaluated different *cis* axial interfaces for Cldn15 using in silico methods, including MD, protein–protein docking, and coevolutionary-coupled mutation (Zhao et al., 2018). Their results indicated that the *cis* interactions could be structurally variable and could result in a wide-scale lateral bending of TJ strands. These authors suggest that the bending would create tensile and compressive forces on opposing rows in the case of the Suzuki-JDR model. This would require local complementary rotations of the protomers. A new *cis* interface, named *cis-1*, was suggested to be important for strand formation. *Cis-1* interface is like the linear-*cis* interface, but has a ~17° lateral rotation between two protomers. Despite the rotation, hydrogen bonds were present between ECH (S67, M68) and ECS2 (E157, L158) residues in both *Cis-1* and linear-*cis* interfaces. Direct interaction between these residues was supported by the observation that single mutations, including S67A and E157A, disturbed Cldn15 strand formation, but the swapped double mutation S67E/E157S retained the strand formation. In contrast to previously reported results (Suzuki et al., 2014), the M68E mutation did not block the formation of TJ strands by Cldn15. M68 forms an integral part of the proposed linear-*cis* interface, and the mutation of the residue is expected to prevent the hydrophobic fit that leads to the linear-*cis* interface. In the *cis-1* model, R79 forms an intermolecular salt-bridge with E157, which is missing in the linear-*cis* interface (Zhao et al., 2018). In brief, Zhao et al. suggested that *cis*-1 and linear-*cis* interfaces contribute to the structural variability and inherent flexibility of tight junction strands.

Nakamura et al. (2019) observed that mutation P134A or P134G in TM3 decreased the bending of this TM segment in Cldn3 crystal structures, and increased the curvature of reconstituted Cldn3 TJ strands as detected by Freeze-Fracture Transmission Electron Microscopy. The study suggested that TM3 bending influences the linear-*cis* interface and thereby the lateral flexibility of TJ strands.

Recently, Fuladi et al. (2022) analyzed the stability and mechanical flexibility of Cldn15 strands using extensive MD simulations. The authors used a previously refined model of Cldn15 pores (Samanta et al., 2018) to model and simulate claudin strands with 36–300 protomers in two membranes. The results suggested that the claudin strands are dynamic and flexible within the plane of the membrane, as suggested by Zhao et al. (2018). They also suggested that three different interactions control the claudin polymerization into TJ strands and their functional flexibility: (i) linear-*cis* interactions, (ii) the *face-to-face* interactions, and (iii) *trans* interactions. The linear-*cis* interactions were reported to be the most flexible of the three interactions, which is facilitated by the rotational pivot point at the ECH. The other two interactions help in stabilizing the overall tetrameric structure as they are least affected by strand curvature. Another interesting suggestion by the study of Fuladi et al. (2022) is that the tetrameric pores are able to slide with respect to each other within the strand. This is facilitated by the structural flexibility of the linear-*cis* interactions and enables the lateral flexibility of TJ strands. In addition, their study supports the stability and thus the validity of the Suzuki-JDR-like architecture of TJ strands.

11. Conclusions and outlook

Over the past decades, research to elucidate tight junction structure and biophysical properties went through distinct phases. Novel staining techniques for light microscopy sparked a vivid interest in the "terminal bars" visible in epithelial preparations in the late 19th and early 20th century. Research on tight junctions intensified again in the 1960s and 1970s, triggered by the electron-microscopic work of Farquhar and Palade (1963) and Staehelin et al. (1969b), which was paralleled by extremely fruitful electrophysiological approaches, many of which were driven by the ingenious conceptions of Jared M. Diamond and his collaborators. During the 1980s and early 1990s, interest in tight junctions ebbed again, but was rekindled by the discovery of the major tight junction protein families by Shoichiro Tsukita and his group. The latest milestone was the presentation of the claudin-15 protomer structure by Suzuki et al. (2014).

Thus, we finally have the tools to reevaluate the structural and functional data from the 20th century and explain them at molecular level. The discovery of distinct paracellular—and thus purely extracellular—anion, cation, and water channels is only a first step. Understanding of tight junction strand formation and regulation, as well as the interplay of the different molecular components, is only just emerging.

TJ strand models reported in recent times suggest that the assembly and architecture of TJ strands are more complex and flexible than previously thought. One main limitation of current models is that they are largely based on a single crystal structure of a channel-forming claudin (Cldn15) that is missing some important regions involved in the assembly of TJ strands. The other resolved crystal structures, of largely barrier-forming claudins, are affected by cCPE binding, which makes the scenario and comparison more complex. Hence, there is a preeminent need for more cCPE-free crystal structure studies of claudins with important regions like ECS1, ECS2, and ECH resolved. Furthermore, a structure with quaternary assembly of claudins, e.g., resolved by cryogenic electron microscopy, that reveals details of the interaction sites between neighboring protomers could be most beneficial. However, molecular dynamics simulations of claudin oligomers and polymers are also expected to provide further structural information about channel and strand assembly and function.

Thus, future TJ-related research that combines biophysicalcell biological and computational methods will open up new avenues for functional, structural, and mechanistic understanding of the tight junction.

References

Al-Sadi, R., Khatib, K., Guo, S., Ye, D., Youssef, M., & Ma, T. (2011). Occludin regulates macromolecule flux across the intestinal epithelial tight junction barrier. *American Journal of Physiology - Gastrointestinal and Liver Physiology, 300*, G1054–G1064.

Alberini, G., Benfenati, F., & Maragliano, L. (2017). A refined model of claudin-15 tight junction paracellular architecture by molecular dynamics simulations. *PLoS One, 12*, Article e0184190.

Alberini, G., Benfenati, F., & Maragliano, L. (2018). Molecular dynamics simulations of ion selectivity in a claudin-15 paracellular channel. *Journal of Physical Chemistry B*, 10783–10792.

Alexandre, M. D., Lu, Q., & Chen, Y. H. (2005). Overexpression of claudin-7 decreases the paracellular Cl- conductance and increases the paracellular Na+ conductance in LLC-PK1 cells. *Journal of Cell Science, 118*, 2683–2693.

Alzahrani, A. S., Hussein, M., Alswailem, M., Mouna, A., Albalawi, L., Moria, Y., Jabbar, M. A., Shi, Y., Günzel, D., & Dasouki, M. (2021). A novel claudin-10 mutation with a unique mechanism in two unrelated families with HELIX syndrome. *Kidney International, 100*, 415–429.

Amasheh, S., Meiri, N., Gitter, A. H., Schöneberg, T., Mankertz, J., Schulzke, J. D., & Fromm, M. (2002). Claudin-2 expression induces cation-selective channels in tight junctions of epithelial cells. *Journal of Cell Science, 115*, 4969–4976.

Amasheh, S., Milatz, S., Krug, S. M., Bergs, M., Amasheh, M., Schulzke, J. D., & Fromm, M. (2009). Na+ absorption defends from paracellular back-leakage by claudin-8 upregulation. *Biochemical and Biophysical Research Communications, 378*, 45–50.

Amasheh, S., Schmidt, T., Mahn, M., Florian, P., Mankertz, J., Tavalali, S., Gitter, A. H., Schulzke, J. D., & Fromm, M. (2005). Contribution of claudin-5 to barrier properties in tight junctions of epithelial cells. *Cell and Tissue Research, 321*, 89–96.

Anderson, J. M., Van Itallie, C. M., & Fanning, A. S. (2004). Setting up a selective barrier at the apical junction complex. *Current Opinion in Cell Biology, 16*, 140–145.

Anderson, J. M., & Van Itallie, C. M. (2009). Physiology and function of the tight junction. *Cold Spring Harbor Perspectives in Biology, 1*, Article a002584.

Angelow, S., Kim, K. J., & Yu, A. S. (2006). Claudin-8 modulates paracellular permeability to acidic and basic ions in MDCK II cells. *The Journal of Physiology, 571*, 15–26.

Angelow, S., Schneeberger, E. E., & Yu, A. S. (2007). Claudin-8 expression in renal epithelial cells augments the paracellular barrier by replacing endogenous claudin-2. *Journal of Membrane Biology, 215*, 147–159.

Angelow, S., & Yu, A. S. L. (2009). Structure-function studies of claudin extracellular domains by cysteine-scanning mutagenesis. *Journal of Biological Chemistry, 284*, 29205–29217.

Arnold, J. (1876). Ueber die Kittsubstanz der Endothelien. *Archiv für Pathologische Anatomie und Physiologie und für Klinische Medicin, 66*, 77–109.

Arnold, J. (1875). Ueber die Kittsubstanz der Epithelien. *Archiv für Pathologische Anatomie und Physiologie und für Klinische Medicin, 64*, 203–243.

Artursson, P., Ungell, A. L., & Lofroth, J. E. (1993). Selective paracellular permeability in two models of intestinal absorption: Cultured monolayers of human intestinal epithelial cells and rat intestinal segments. *Pharmaceutical Research, 10*, 1123–1129.

Ayala-Torres, C., Krug, S. M., Rosenthal, R., & Fromm, M. (2021). Angulin-1 (LSR) affects paracellular water transport, however only in tight epithelial cells. *International Journal of Molecular Sciences, 22*, 7827.

Ayala-Torres, C., Krug, S. M., Schulzke, J. D., Rosenthal, R., & Fromm, M. (2019). Tricellulin effect on paracellular water transport. *International Journal of Molecular Sciences, 20*, 5700.

Bürgel, N., Bojarski, C., Mankertz, J., Zeitz, M., Fromm, M., & Schulzke, J. D. (2002). Mechanisms of diarrhea in collagenous colitis. *Gastroenterology, 123*, 433–443.

Baltzegar, D. A., Reading, B. J., Brune, E. S., & Borski, R. J. (2013). Phylogenetic revision of the claudin gene family. *Marine Genomics, 11*, 17–26.

Barry, P. H., & Diamond, J. M. (1970). Junction potentials, electrode standard potentials, and other problems in interpreting electrical properties of membranes. *Journal of Membrane Biology, 3*, 93–122.

Barry, P. H., Diamond, J. M., & Wright, E. M. (1971). The mechanism of cation permeation in rabbit gallbladder: Dilution potentials and biionic potentials. *Journal of Membrane Biology, 4*, 358–394.

Berndt, P., Winkler, L., Cording, J., Breitkreuz-Korff, O., Rex, A., Dithmer, S., Rausch, V., Blasig, R., Richter, M., Sporbert, A., Wolburg, H., Blasig, I. E., & Haseloff, R. F. (2019). Tight junction proteins at the blood-brain barrier: Far more than claudin-5. *Cellular and Molecular Life Sciences: CMLS, 76*, 1987–2002.

Bertrand, C. A., Durand, D. M., Saidel, G. M., Laboisse, C., & Hopfer, U. (1998). System for dynamic measurements of membrane capacitance in intact epithelial monolayers. *Biophysical Journal, 75*, 2743–2756.

Bertrand, C. A., & Hopfer, U. (2002). Measurement of membrane capacitance in epithelial monolayers. *Methods in Molecular Biology, 188*, 315–327.

Blume, L. F., Denker, M., Gieseler, F., & Kunze, T. (2010). Temperature corrected transepithelial electrical resistance (TEER) measurement to quantify rapid changes in paracellular permeability. *Pharmazie, 65*, 19–24.

Bonnet, R. (1895). Über die Schlußleisten der Epithelien. *Deutsche Medizinische Wochenschrift, 21*.

Boulpaep, E. L., & Sackin, H. (1980). Electrical analysis of intraepithelial barriers. *Current Topics in Membranes and Transport, 13*, 169–197.

Boulpaep, E. L. (1972). Permeability changes of the proximal tubule of Necturus during saline loading. *American Journal of Physiology, 222*, 517–531.

Breiderhoff, T., Himmerkus, N., Meoli, L., Fromm, A., Sewerin, S., Kriuchkova, N., Nagel, O., Ladilov, Y., Krug, S. M., Quintanova, C., Stumpp, M., Garbe-Schönberg, D., Westernstroer, U., Merkel, C., Brinkhus, M. A., Altmüller, J., Schweiger, M. R., Müller, D., Mutig, K., … Günzel, D. (2022). Claudin-10a deficiency shifts proximal tubular Cl(-) permeability to cation selectivity via claudin-2 redistribution. *Journal of the American Society of Nephrology, 33*, 699–717.

Breiderhoff, T., Himmerkus, N., Stuiver, M., Mutig, K., Will, C., Meij, I. C., Bachmann, S., Bleich, M., Willnow, T. E., & Müller, D. (2012). Deletion of claudin-10 (Cldn10) in the thick ascending limb impairs paracellular sodium permeability and leads to hypermagnesemia and nephrocalcinosis. *Proceedings of the National Academy of Sciences of the U S A, 109*, 14241–14246.

Briehl, M. M., & Miesfeld, R. L. (1991). Isolation and characterization of transcripts induced by androgen withdrawal and apoptotic cell death in the rat ventral prostate. *Molecular Endocrinology, 5*, 1381–1388.

Brightman, M. W., & Reese, T. S. (1969). Junctions between intimately apposed cell membranes in the vertebrate brain. *The Journal of Cell Biology, 40*, 648–677.

Bronstein, J. M., Popper, P., Micevych, P. E., & Farber, D. B. (1996). Isolation and characterization of a novel oligodendrocyte-specific protein. *Neurology, 47*, 772–778.

Burnashev, N., Villarroel, A., & Sakmann, B. (1996). Dimensions and ion selectivity of recombinant AMPA and kainate receptor channels and their dependence on Q/R site residues. *The Journal of Physiology, 496*, 165–173.

Buschmann, M. M., Shen, L., Rajapakse, H., Raleigh, D. R., Wang, Y., Wang, Y., Lingaraju, A., Zha, J., Abbott, E., McAuley, E. M., Breskin, L. A., Wu, L., Anderson, K., Turner, J. R., & Weber, C. R. (2013). Occludin OCEL-domain interactions are required for maintenance and regulation of the tight junction barrier to macromolecular flux. *Molecular Biology of the Cell, 24*, 3056–3068.

Cereijido, M., Stefani, E., & Palomo, A. M. (1980). Occluding junctions in a cultured transporting epithelium: Structural and functional heterogeneity. *Journal of Membrane Biology, 53*, 19–32.

Chadwick, V. S., Phillips, S. F., & Hofmann, A. F. (1977a). Measurements of intestinal permeability using low molecular weight polyethylene glycols (PEG 400). I. Chemical analysis and biological properties of PEG 400. *Gastroenterology, 73*, 241–246.

Chadwick, V. S., Phillips, S. F., & Hofmann, A. F. (1977b). Measurements of intestinal permeability using low molecular weight polyethylene glycols (PEG 400). II. Application to normal and abnormal permeability states in man and animals. *Gastroenterology, 73*, 247–251.

Chen, C. C., Zhou, Y., Morris, C. A., Hou, J., & Baker, L. A. (2013). Scanning ion conductance microscopy measurement of paracellular channel conductance in tight junctions. *Analytical Chemistry, 85*, 3621–3628.

Claude, P., & Goodenough, D. A. (1973). Fracture faces of zonulae occludentes from "tight" and "leaky" epithelia. *The Journal of Cell Biology, 58*, 390–400.

Citi, S., Fromm, M., Furuse, M., González-Mariscal, L., Nusrat, A., Tsukita, S., & Turner, J. R. (2024). A short guide to the tight junction. *Journal of Cell Science, 137*, Article jcs261776.

Claude, P. (1978). Morphological factors influencing transepithelial permeability: A model for the resistance of the *zonula occludens*. *Journal of Membrane Biology, 39*, 219–232.

Clausen, C., Lewis, S. A., & Diamond, J. M. (1979). Impedance analysis of a tight epithelium using a distributed resistance model. *Biophysical Journal, 26*, 291–317.

Cohn, T. (1897). Über epitheliale Schlußleisten an embryonalen und ausgebildeten Geweben. *Verhandl d Physik-med Ges, 21*, 171–200.

Cohn, T. (1895). Über Intercellularlücken und Kittsubstanz. *Anatomische Hefte, 5*, 293–333.

Cole, K. S. (1932). Electric phase angle of cell membranes. *The Journal of General Physiology, 15*, 641–649.

Colegio, O. R., Van Itallie, C., Rahner, C., & Anderson, J. M. (2003). Claudin extracellular domains determine paracellular charge selectivity and resistance but not tight junction fibril architecture. *American Journal of Physiology - Cell Physiology, 284*, 1346–1354.

Colegio, O. R., Van Itallie, C. M., McCrea, H. J., Rahner, C., & Anderson, J. M. (2002). Claudins create charge-selective channels in the paracellular pathway between epithelial cells. *American Journal of Physiology - Cell Physiology, 283*, 142–147.

Conrad, M. P., Piontek, J., Günzel, D., Fromm, M., & Krug, S. M. (2016). Molecular basis of claudin-17 anion selectivity. *Cellular and Molecular Life Sciences: CMLS, 73*, 185–200.

Crick, F. (1990). *What mad pursuit. A personal view of scientific discovery*. Harmondsworth: Penguin.

Daugherty, B. L., Ward, C., Smith, T., Ritzenthaler, J. D., & Koval, M. (2007). Regulation of heterotypic claudin compatibility. *Journal of Biological Chemistry, 282*, 30005–30013.

Diamond, J. M. (1978). Channels in epithelial cell membranes and junctions. *Federation Proceedings, 37*, 2639–2643.

Diamond, J. M. (1979). Osmotic water flow in leaky epithelia. *Journal of Membrane Biology, 51*, 195–216.

Diamond, J. M. (1962a). The mechanism of solute transport by the gall-bladder. *The Journal of Physiology, 161*, 474–502.

Diamond, J. M. (1962c). The mechanism of water transport by the gall-bladder. *The Journal of Physiology, 161*, 503–527.

Diamond, J. M. (1962b). The reabsorptive function of the gall-bladder. *The Journal of Physiology, 161*, 442–473.

Diamond, J. M. (1974). Tight and leaky junctions of epithelia: A perspective on kisses in the dark. *Federation Proceedings, 33*, 2220–2224.

Diamond, J. M. (1977). Twenty-first bowditch lecture. The epithelial junction: Bridge, gate, and fence. *Physiologist, 20*, 10–18.

Eisenman, G. (1962). Cation selective glass electrodes and their mode of operation. *Biophysical Journal, 2*, 259–323.

Epple, H. J., Allers, K., Tröger, H., Kuhl, A., Erben, U., Fromm, M., Zeitz, M., Loddenkemper, C., Schulzke, J. D., & Schneider, T. (2010). Acute HIV infection induces mucosal infiltration with CD4+ and CD8+ T cells, epithelial apoptosis, and a mucosal barrier defect. *Gastroenterology, 139*, 1289–1300.

Fanning, A. S., Van Itallie, C. M., & Anderson, J. M. (2012). Zonula occludens-1 and -2 regulate apical cell structure and the zonula adherens cytoskeleton in polarized epithelia. *Molecular Biology of the Cell, 23*, 577–590.

Farquhar, M. G., & Palade, G. E. (1963). Junctional complexes in various epithelia. *The Journal of Cell Biology, 17*, 375–412.

Frömter, E., & Diamond, J. (1972). Route of passive ion permeation in epithelia. *Nature - New Biology, 235*, 9–13.

Frömter, E. (1972). The route of passive ion movement through the epithelium of Necturus gallbladder. *Journal of Membrane Biology, 8*, 259–301.

Friend, D. S., & Gilula, N. B. (1972). Variations in tight and gap junctions in mammalian tissues. *The Journal of Cell Biology, 53*, 758–776.

Fromm, M., Schulzke, J. D., & Hegel, U. (1985). Epithelial and subepithelial contributions to transmural electrical resistance of intact rat jejunum, in vitro. *Pflügers Archiv, 405*, 400–402.

Fuladi, S., McGuinness, S., Shen, L., Weber, C. R., & Khalili-Araghi, F. (2022). Molecular mechanism of claudin-15 strand flexibility: A computational study. *Journal of General Physiology, 154*(12), Article e202213116. https://doi.org/10.1085/jgp.202213116

Furuse, M., Fujita, K., Hiiragi, T., Fujimoto, K., & Tsukita, S. (1998). Claudin-1 and -2: Novel integral membrane proteins localizing at tight junctions with no sequence similarity to occludin. *The Journal of Cell Biology, 141*, 1539−1550.

Furuse, M., Hirase, T., Itoh, M., Nagafuchi, A., Yonemura, S., & Tsukita, S. (1993). Occludin: A novel integral membrane protein localizing at tight junctions. *The Journal of Cell Biology, 123*, 1777−1788.

Furuse, M., Sasaki, H., & Tsukita, S. (1999). Manner of interaction of heterogeneous claudin species within and between tight junction strands. *The Journal of Cell Biology, 147*, 891−903.

Furuse, M., & Tsukita, S. (2006). Claudins in occluding junctions of humans and flies. *Trends in Cell Biology, 16*, 181−188.

Gögelein, H., & Van Driessche, W. (1981). The effect of electrical gradients on current fluctuations and impedance recorded from Necturus gallbladder. *Journal of Membrane Biology, 60*, 199−209.

Günzel, D., Florian, P., Richter, J. F., Troeger, H., Schulzke, J. D., Fromm, M., & Gitter, A. H. (2006). Restitution of single-cell defects in the mouse colon epithelium differs from that of cultured cells. *American Journal of Physiology - Regulatory, Integrative and Comparative Physiology, 290*, R1496−R1507.

Günzel, D., Stuiver, M., Kausalya, P. J., Haisch, L., Krug, S. M., Rosenthal, R., Meij, I. C., Hunziker, W., Fromm, M., & Müller, D. (2009). Claudin-10 exists in six alternatively spliced isoforms that exhibit distinct localization and function. *Journal of Cell Science, 122*, 1507−1517.

Günzel, D., & Yu, A. S. (2013). Claudins and the modulation of tight junction permeability. *Physiological Reviews, 93*, 525−569.

Georges, D. (1979). Gap and tight junctions in tunicates. Study in conventional and freeze-fracture techniques. *Tissue and Cell, 11*, 781−792.

Giacomelli, F., & Wiener, J. (1976). Specialized junction in the distal convoluted tubule of rat kidney. *The Anatomical Record, 185*, 197−207.

Gitter, A. H., Bendfeldt, K., Schulzke, J. D., & Fromm, M. (2000). Trans/paracellular, surface/crypt, and epithelial/subepithelial resistances of mammalian colonic epithelia. *Pflügers Archiv, 439*, 477−482.

Gitter, A. H., Bertog, M., Schulzke, J., & Fromm, M. (1997). Measurement of paracellular epithelial conductivity by conductance scanning. *Pflügers Archiv, 434*, 830−840.

Gitter, A. H., Fromm, M., & Schulzke, J. D. (1998). Impedance analysis for the determination of epithelial and subepithelial resistance in intestinal tissues. *Journal of Biochemical and Biophysical Methods, 37*, 35−46.

Gong, Y., Himmerkus, N., Plain, A., Bleich, M., & Hou, J. (2015). Epigenetic regulation of microRNAs controlling CLDN14 expression as a mechanism for renal calcium handling. *Journal of the American Society of Nephrology, 26*, 663−676.

Gong, Y., Himmerkus, N., Sunq, A., Milatz, S., Merkel, C., Bleich, M., & Hou, J. (2017). ILDR1 is important for paracellular water transport and urine concentration mechanism. *Proceedings of the National Academy of Sciences of the U S A, 114*, 5271−5276.

Gong, Y., Renigunta, V., Himmerkus, N., Zhang, J., Renigunta, A., Bleich, M., & Hou, J. (2012). Claudin-14 regulates renal Ca(+)(+) transport in response to CaSR signalling via a novel microRNA pathway. *The EMBO Journal, 31*, 1999−2012.

Gong, Y., Renigunta, V., Zhou, Y., Sunq, A., Wang, J., Yang, J., Renigunta, A., Baker, L. A., & Hou, J. (2015). Biochemical and biophysical analyses of tight junction permeability made of claudin-16 and claudin-19 dimerization. *Molecular Biology of the Cell, 26*, 4333−4346.

Gonschior, H., Schmied, C., Van der Veen, R. E., Eichhorst, J., Himmerkus, N., Piontek, J., Günzel, D., Bleich, M., Furuse, M., Haucke, V., & Lehmann, M. (2022). Nanoscale segregation of channel and barrier claudins enables paracellular ion flux. *Nature Communications, 13*, Article 4985.

González-Mariscal, L., Chávez de Ramírez, B., & Cereijido, M. (1984). Effect of temperature on the occluding junctions of monolayers of epithelioid cells (MDCK). *Journal of Membrane Biology, 79*, 175−184.

Goodenough, D. A., & Revel, J. P. (1970). A fine structural analysis of intercellular junctions in the mouse liver. *The Journal of Cell Biology, 45*, 272−290.

Green, C. R., & Bergquist, P. R. (1982). Phylogenetic-relationships within the invertebrata in relation to the structure of septate junctions and the development of occluding junctional types. *Journal of Cell Science, 53*, 279−305.

Höber, R. (1910). Eine Methode, die elektrische Leitfähigkeit im Innern von Zellen zu messen. *Pflüger's Archiv für die gesamte Physiologie des Menschen und der Tiere, 133*, 237−253.

Haller, W. (1977). Critical permeation size of dextran molecules. *Macromolecules, 10*, 83−86.

Hartmann, C., Schwietzer, Y. A., Otani, T., Furuse, M., & Ebnet, K. (2020). Physiological functions of junctional adhesion molecules (JAMs) in tight junctions. *Biochimica et Biophysica Acta (BBA) - Biomembranes, 1862*, Article 183299.

Heidenhain, M. (1892). *Über Kern und Protoplasma. Festschrift. Herrn Geheimrat Albert von Koelliker zur Feier seines fünfzigjährigen medicinischen Doktorjubiläums gewidmet. Verlag von Wilhelm Engelmann, Leipzig.*

Heller, F., Florian, P., Bojarski, C., Richter, J., Christ, M., Hillenbrand, B., Mankertz, J., Gitter, A. H., Burgel, N., Fromm, M., Zeitz, M., Fuss, I., Strober, W., & Schulzke, J. D. (2005). Interleukin-13 is the key effector Th2 cytokine in ulcerative colitis that affects epithelial tight junctions, apoptosis, and cell restitution. *Gastroenterology, 129*, 550−564.

Hempel, C., Protze, J., Altun, E., Riebe, B., Piontek, A., Fromm, A., Lee, I. M., Saleh, T., Günzel, D., Krause, G., & Piontek, J. (2020). Assembly of tight junction strands: Claudin-10b and claudin-3 form homo-tetrameric building blocks that polymerise in a channel-independent manner. *Journal of Molecular Biology, 432*, 2405−2427.

Hempel, C., Rosenthal, R., Fromm, A., Krug, S. M., Fromm, M., Günzel, D., & Piontek, J. (2022). Tight junction channels claudin-10b and claudin-15: Functional mapping of pore-lining residues. *Annals of the New York Academy of Sciences.*

Hille, B. (1973). Potassium channels in myelinated nerve. Selective permeability to small cations. *The Journal of General Physiology, 61*, 669−686.

Hille, B. (1972). The permeability of the sodium channel to metal cations in myelinated nerve. *The Journal of General Physiology, 59*, 637−658.

Hou, J., Gomes, A. S., Paul, D. L., & Goodenough, D. A. (2006). Study of claudin function by RNA interference. *Journal of Biological Chemistry, 281*, 36117–36123.

Hou, J., Renigunta, A., Konrad, M., Gomes, A. S., Schneeberger, E. E., Paul, D. L., Waldegger, S., & Goodenough, D. A. (2008). Claudin-16 and claudin-19 interact and form a cation-selective tight junction complex. *Journal of Clinical Investigation, 118*, 619–628.

Hou, J., Renigunta, A., Yang, J., & Waldegger, S. (2010). Claudin-4 forms paracellular chloride channel in the kidney and requires claudin-8 for tight junction localization. *Proceedings of the National Academy of Sciences of the U S A, 107*, 18010–18015.

Hou, J. H., Renigunta, A., Gomes, A. S., Hou, M. L., Paul, D. L., Waldegger, S., & Goodenough, D. A. (2009). Claudin-16 and claudin-19 interaction is required for their assembly into tight junctions and for renal reabsorption of magnesium. *Proceedings of the National Academy of Sciences of the United States of America, 106*, 15350–15355.

Hu, J. E., Weiss, F., Bojarski, C., Branchi, F., Schulzke, J. D., Fromm, M., & Krug, S. M. (2021). Expression of tricellular tight junction proteins and the paracellular macromolecule barrier are recovered in remission of ulcerative colitis. *BMC Gastroenterology, 21*, 141.

Ikari, A., Hirai, N., Shiroma, M., Harada, H., Sakai, H., Hayashi, H., Suzuki, Y., Degawa, M., & Takagi, K. (2004). Association of paracellin-1 with ZO-1 augments the reabsorption of divalent cations in renal epithelial cells. *Journal of Biological Chemistry, 279*, 54826–54832.

Ikenouchi, J., Furuse, M., Furuse, K., Sasaki, H., Tsukita, S., & Tsukita, S. (2005). Tricellulin constitutes a novel barrier at tricellular contacts of epithelial cells. *The Journal of Cell Biology, 171*, 939–945.

Imai, Y., Miyamoto, M., Nakahari, T., Murakami, M., & Yoshida, H. (1986). Observation of electro-kinetic phenomena by imposing oscillating pressure and voltage gradients across some epithelial membranes. *The Japanese Journal of Physiology, 36*, 397–402.

Inai, T., Kobayashi, J., & Shibata, Y. (1999). Claudin-1 contributes to the epithelial barrier function in MDCK cells. *European Journal of Cell Biology, 78*, 849–855.

Irudayanathan, F. J., Trasatti, J. P., Karande, P., & Nangia, S. (2016). Molecular architecture of the blood brain barrier tight junction proteins–A synergistic computational and in vitro approach. *Journal of Physical Chemistry B, 120*, 77–88.

Irudayanathan, F. J., Wang, N., Wang, X., & Nangia, S. (2017). Architecture of the paracellular channels formed by claudins of the blood-brain barrier tight junctions. *Annals of the New York Academy of Sciences, 1405*, 131–146.

Irudayanathan, F. J., Wang, X., Wang, N., Willsey, S. R., Seddon, I. A., & Nangia, S. (2018). Self-assembly simulations of classic claudins-insights into the pore structure, selectivity, and higher order complexes. *Journal of Physical Chemistry B, 122*, 7463–7474.

Kachar, B., & Reese, T. S. (1982). Evidence for the lipidic nature of tight junction strands. *Nature, 296*, 464–466.

Katahira, J., Inoue, N., Horiguchi, Y., Matsuda, M., & Sugimoto, N. (1997). Molecular cloning and functional characterization of the receptor for *Clostridium perfringens* enterotoxin. *The Journal of Cell Biology, 136*, 1239–1247.

Katahira, J., Sugiyama, H., Inoue, N., Horiguchi, Y., Matsuda, M., & Sugimoto, N. (1997). *Clostridium perfringens* enterotoxin utilizes two structurally related membrane proteins as functional receptors in vivo. *Journal of Biological Chemistry, 272*, 26652–26658.

Kausalya, P. J., Amasheh, S., Günzel, D., Wurps, H., Müller, D., Fromm, M., & Hunziker, W. (2006). Disease-associated mutations affect intracellular traffic and paracellular Mg2+ transport function of Claudin-16. *Journal of Clinical Investigation, 116*, 878–891.

Kellenberger, S., Auberson, M., Gautschi, I., Schneeberger, E., & Schild, L. (2001). Permeability properties of ENaC selectivity filter mutants. *The Journal of General Physiology, 118*, 679–692.

Kellenberger, S., Hoffmann-Pochon, N., Gautschi, I., Schneeberger, E., & Schild, L. (1999). On the molecular basis of ion permeation in the epithelial Na+ channel. *The Journal of General Physiology, 114*, 13–30.

Kidder, G. W., 3rd, Cereijido, M., & Curran, P. F. (1964). Transient changes in electrical potential differences across frog skin. *American Journal of Physiology, 207*, 935–940.

Kiesewetter, L., Littau, L., Walles, H., Boccaccini, A. R., & Groeber-Becker, F. (2019). Reepithelialization in focus: Non-invasive monitoring of epidermal wound healing in vitro. *Biosensors and Bioelectronics, 142*, Article 111555.

Klar, J., Piontek, J., Milatz, S., Tariq, M., Jameel, M., Breiderhoff, T., Schuster, J., Fatima, A., Asif, M., Sher, M., Mäbert, K., Fromm, A., Baig, S. M., Günzel, D., & Dahl, N. (2017). Altered paracellular cation permeability due to a rare CLDN10B variant causes anhidrosis and kidney damage. *PLoS Genetics, 13*, Article e1006897.

Kottra, G., & Frömter, E. (1990). Barium blocks cell membrane and tight junction conductances in Necturus gallbladder epithelium. Experiments with an extended impedance analysis technique. *Pflügers Archiv, 415*, 718–725.

Kottra, G., & Frömter, E. (1993). Tight-junction tightness of Necturus gall bladder epithelium is not regulated by cAMP or intracellular Ca2+. II. Impedance measurements. *Pflügers Archiv, 425*, 535–545.

Krause, G., Protze, J., & Piontek, J. (2015). Assembly and function of claudins: Structure-function relationships based on homology models and crystal structures. *Seminars in Cell & Developmental Biology, 42*, 3–12.

Krause, G., Winkler, L., Müller, S. L., Haseloff, R. F., Piontek, J., & Blasig, I. E. (2008). Structure and function of claudins. *Biochimica et Biophysica Acta, 1778*, 631–645.

Kroesen, A. J., Dullat, S., Schulzke, J. D., Fromm, M., & Buhr, H. J. (2008). Permanently increased mucosal permeability in patients with backwash ileitis after ileoanal pouch for ulcerative colitis. *Scandinavian Journal of Gastroenterology, 43*, 704–711.

Krug, S. M., Amasheh, S., Richter, J. F., Milatz, S., Günzel, D., Westphal, J. K., Huber, O., Schulzke, J. D., & Fromm, M. (2009). Tricellulin forms a barrier to macromolecules in tricellular tight junctions without affecting ion permeability. *Molecular Biology of the Cell, 20*, 3713–3724.

Krug, S. M., Bojarski, C., Fromm, A., Lee, I. M., Dames, P., Richter, J. F., Turner, J. R., Fromm, M., & Schulzke, J. D. (2018). Tricellulin is regulated via interleukin-13-receptor alpha2, affects macromolecule uptake, and is decreased in ulcerative colitis. *Mucosal Immunology, 11*, 345–356.

Krug, S. M., Fromm, M., & Günzel, D. (2009). Two-path impedance spectroscopy for measuring paracellular and transcellular epithelial resistance. *Biophysical Journal, 97*, 2202–2211.

Krug, S. M., Günzel, D., Conrad, M. P., Rosenthal, R., Fromm, A., Amasheh, S., Schulzke, J. D., & Fromm, M. (2012). Claudin-17 forms tight junction channels with distinct anion selectivity. *Cellular and Molecular Life Sciences: CMLS, 69*, 2765–2778.

Krug, S. M., Hayaishi, T., Iguchi, D., Watari, A., Takahashi, A., Fromm, M., Nagahama, M., Takeda, H., Okada, Y., Sawasaki, T., Doi, T., Yagi, K., & Kondoh, M. (2017). Angubindin-1, a novel paracellular absorption enhancer acting at the tricellular tight junction. *Journal of Controlled Release, 260*, 1–11.

Krystofiak, E. S., Heymann, J. B., & Kachar, B. (2019). Carbon replicas reveal double stranded structure of tight junctions in phase-contrast electron microscopy. *Communications Biology, 2*, 98.

Lane, N. J., & Chandler, H. J. (1980). Definitive evidence for the existence of tight junctions in invertebrates. *The Journal of Cell Biology, 86*, 765–774.

Lane, N. J., Dallai, R., Burighel, P., & Martinucci, G. B. (1986). Tight and gap junctions in the intestinal tract of tunicates (urochordata): A freeze-fracture study. *Journal of Cell Science, 84*, 1–17.

Lane, N. J., Dallai, R., Martinucci, G. B., & Burighel, P. (1987). Cell junctions in amphioxus (cephalochordata): A thin section and freeze-fracture study. *Tissue and Cell, 19*, 399–411.

Lee, D. B., Jamgotchian, N., Allen, S. G., Abeles, M. B., & Ward, H. J. (2008). A lipid-protein hybrid model for tight junction. *American Journal of Physiology - Renal Physiology, 295*, F1601–F1612.

Lewis, S. A., & Diamond, J. M. (1976). Na+ transport by rabbit urinary bladder, a tight epithelium. *Journal of Membrane Biology, 28*, 1–40.

Li, F., Egea, P. F., Vecchio, A. J., Asial, I., Gupta, M., Paulino, J., Bajaj, R., Dickinson, M. S., Ferguson-Miller, S., Monk, B. C., & Stroud, R. M. (2021). Highlighting membrane protein structure and function: A celebration of the protein data bank. *Journal of Biological Chemistry, 296*, Article 100557.

Li, J., Angelow, S., Linge, A., Zhuo, M., & Yu, A. S. (2013). Claudin-2 pore function requires an intramolecular disulfide bond between two conserved extracellular cysteines. *American Journal of Physiology - Cell Physiology, 305*, C190–C196.

Li, J., Zhuo, M., Pei, L., Rajagopal, M., & Yu, A. S. (2014). Comprehensive cysteine-scanning mutagenesis reveals Claudin-2 pore-lining residues with different intrapore locations. *Journal of Biological Chemistry, 289*, 6475–6484.

Liang, G. H., & Weber, C. R. (2014). Molecular aspects of tight junction barrier function. *Current Opinion in Pharmacology, 19*, 84–89.

Lingaraju, A., Long, T. M., Wang, Y., Austin, J. R., 2nd, & Turner, J. R. (2015). Conceptual barriers to understanding physical barriers. *Seminars in Cell & Developmental Biology, 42*, 13–21.

Linnankoski, J., Makela, J., Palmgren, J., Mauriala, T., Vedin, C., Ungell, A. L., Lazorova, L., Artursson, P., Urtti, A., & Yliperttula, M. (2010). Paracellular porosity and pore size of the human intestinal epithelium in tissue and cell culture models. *Journal of Pharmaceutical Sciences, 99*, 2166–2175.

Loh, Y. H., Christoffels, A., Brenner, S., Hunziker, W., & Venkatesh, B. (2004). Extensive expansion of the claudin gene family in the teleost fish, Fugu rubripes. *Genome Research, 14*, 1248–1257.

Møllgård, K., Milinowska, D. H., & Saunders, N. R. (1976). Lack of correlation between tight junction morphology and permeability properties in developing choroid plexus. *Nature, 264*, 293–294.

Mannweiler, R., Bergmann, S., Vidal, Y. S. S., Brandner, J. M., & Günzel, D. (2021). Direct assessment of individual skin barrier components by electrical impedance spectroscopy. *Allergy, 76*, 3094–3106.

Martin-Padura, I., Lostaglio, S., Schneemann, M., Williams, L., Romano, M., Fruscella, P., Panzeri, C., Stoppacciaro, A., Ruco, L., Villa, A., Simmons, D., & Dejana, E. (1998). Junctional adhesion molecule, a novel member of the immunoglobulin superfamily that distributes at intercellular junctions and modulates monocyte transmigration. *The Journal of Cell Biology, 142*, 117–127.

Matsuda, M., Kubo, A., Furuse, M., & Tsukita, S. (2004). A peculiar internalization of claudins, tight junction-specific adhesion molecules, during the intercellular movement of epithelial cells. *Journal of Cell Science, 117*, 1247–1257.

Meyer, H. W. (1983). Tight junction strands are lipidic cylinders. *Naturwissenschaften, 70*, 251–252.

Milatz, S., Himmerkus, N., Wulfmeyer, V. C., Drewell, H., Mutig, K., Hou, J., Breiderhoff, T., Müller, D., Fromm, M., Bleich, M., & Günzel, D. (2017). Mosaic expression of claudins in thick ascending limbs of Henle results in spatial separation of paracellular Na+ and Mg2+ transport. *Proceedings of the National Academy of Sciences of the U S A, 114*, E219–e227.

Milatz, S., Krug, S. M., Rosenthal, R., Günzel, D., Muller, D., Schulzke, J. D., Amasheh, S., & Fromm, M. (2010). Claudin-3 acts as a sealing component of the tight junction for ions of either charge and uncharged solutes. *Biochimica et Biophysica Acta, 1798*, 2048–2057.

Milatz, S., Piontek, J., Hempel, C., Meoli, L., Grohe, C., Fromm, A., Lee, I. M., El-Athman, R., & Günzel, D. (2017). Tight junction strand formation by claudin-10 isoforms and claudin-10a/-10b chimeras. *Annals of the New York Academy of Sciences, 1405*, 102–115.

Milatz, S., Piontek, J., Schulzke, J. D., Blasig, I. E., Fromm, M., & Günzel, D. (2015). Probing the *cis*-arrangement of prototype tight junction proteins claudin-1 and claudin-3. *Biochemical Journal, 468*, 449–458.

Miller, F. (1960). Hemoglobin absorption by the cells of the proximal convoluted tubule in mouse kidney. *The Journal of Biophysical and Biochemical Cytology, 8*, 689–718.

Mineta, K., Yamamoto, Y., Yamazaki, Y., Tanaka, H., Tada, Y., Saito, K., Tamura, A., Igarashi, M., Endo, T., Takeuchi, K., & Tsukita, S. (2011). Predicted expansion of the claudin multigene family. *FEBS Letters, 585*, 606–612.

Moreno, J. H., & Diamond, J. M. (1974). Discrimination of monovalent inorganic cations by "tight" junctions of gallbladder epithelium. *Journal of Membrane Biology, 15*, 277–318.

Morita, K., Furuse, M., Fujimoto, K., & Tsukita, S. (1999). Claudin multigene family encoding four-transmembrane domain protein components of tight junction strands. *Proceedings of the National Academy of Sciences of the United States of America, 96*, 511–516.

Morita, K., Sasaki, H., Fujimoto, K., Furuse, M., & Tsukita, S. (1999). Claudin-11/OSP-based tight junctions of myelin sheaths in brain and Sertoli cells in testis. *The Journal of Cell Biology, 145*, 579–588.

Muto, S., Hata, M., Taniguchi, J., Tsuruoka, S., Moriwaki, K., Saitou, M., Furuse, K., Sasaki, H., Fujimura, A., Imai, M., Kusano, E., Tsukita, S., & Furuse, M. (2010). Claudin-2-deficient mice are defective in the leaky and cation-selective paracellular permeability properties of renal proximal tubules. *Proceedings of the National Academy of Sciences of the U S A, 107*, 8011–8016.

Nakamura, S., Irie, K., Tanaka, H., Nishikawa, K., Suzuki, H., Saitoh, Y., Tamura, A., Tsukita, S., & Fujiyoshi, Y. (2019). Morphologic determinant of tight junctions revealed by claudin-3 structures. *Nature Communications, 10*, 816.

Neaves, W. B. (1973). Permeability of Sertoli cell tight junctions to lanthanum after ligation of ductus deferens and ductuli efferentes. *The Journal of Cell Biology, 59*, 559–572.

Neuhaus, W., Bogner, E., Wirth, M., Trzeciak, J., Lachmann, B., Gabor, F., & Noe, C. R. (2006). A novel tool to characterize paracellular transport: The APTS-dextran ladder. *Pharmaceutical Research, 23*, 1491–1501.

Neuhaus, W., Plattner, V. E., Wirth, M., Germann, B., Lachmann, B., Gabor, F., & Noe, C. R. (2008). Validation of in vitro cell culture models of the blood-brain barrier: Tightness characterization of two promising cell lines. *Journal of Pharmaceutical Sciences, 97*, 5158–5175.

Neuhaus, W., Trzeciak, J., Lauer, R., Lachmann, B., & Noe, C. R. (2006). APTS-Labeled dextran ladder: A novel tool to characterize cell layer tightness. *Journal of Pharmacy Biomedicine Analytical, 40*, 1035–1039.

Nielsen, H. L., Nielsen, H., Ejlertsen, T., Engberg, J., Günzel, D., Zeitz, M., Hering, N. A., Fromm, M., Schulzke, J. D., & Bücker, R. (2011). Oral and fecal Campylobacter concisus strains perturb barrier function by apoptosis induction in HT-29/B6 intestinal epithelial cells. *PLoS One, 6*, Article e23858.

Nitta, T., Hata, M., Gotoh, S., Seo, Y., Sasaki, H., Hashimoto, N., Furuse, M., & Tsukita, S. (2003). Size-selective loosening of the blood-brain barrier in claudin-5-deficient mice. *The Journal of Cell Biology, 161*, 653–660.

Nutter, T. J., & Adams, D. J. (1995). Monovalent and divalent cation permeability and block of neuronal nicotinic receptor channels in rat parasympathetic ganglia. *The Journal of General Physiology, 105*, 701–723.

Otani, T., & Furuse, M. (2020). Tight junction structure and function revisited. *Trends in Cell Biology, 30*, 805–817.

Otani, T., Nguyen, T. P., Tokuda, S., Sugihara, K., Sugawara, T., Furuse, K., Miura, T., Ebnet, K., & Furuse, M. (2019). Claudins and JAM-A coordinately regulate tight junction formation and epithelial polarity. *The Journal of Cell Biology, 218*, 3372–3396.

Păunescu, T. G., & Helman, S. I. (2001a). cAmp activation of apical membrane Cl(-) channels: theoretical considerations for impedance analysis. *Biophysical Journal, 81*, 838–851.

Păunescu, T. G., & Helman, S. I. (2001b). PGE(2) activation of apical membrane Cl(-) channels in A6 epithelia: Impedance analysis. *Biophysical Journal, 81*, 852–866.

Pappenheimer, J. R., Dahl, C. E., Karnovsky, M. L., & Maggio, J. E. (1994). Intestinal absorption and excretion of octapeptides composed of D amino acids. *Proceedings of the National Academy of Sciences of the U S A, 91*, 1942–1945.

Pappenheimer, J. R., & Reiss, K. Z. (1987). Contribution of solvent drag through intercellular junctions to absorption of nutrients by the small intestine of the rat. *Journal of Membrane Biology, 100*, 123–136.

Piehl, C., Piontek, J., Cording, J., Wolburg, H., & Blasig, I. E. (2010). Participation of the second extracellular loop of claudin-5 in paracellular tightening against ions, small and large molecules. *Cellular and Molecular Life Sciences: CMLS, 67*, 2131–2140.

Piontek, A., Rossa, J., Protze, J., Wolburg, H., Hempel, C., Günzel, D., Krause, G., & Piontek, J. (2017). Polar and charged extracellular residues conserved among barrier-forming claudins contribute to tight junction strand formation. *Annals of the New York Academy of Sciences, 1397*, 143–156.

Piontek, J., Fritzsche, S., Cording, J., Richter, S., Hartwig, J., Walter, M., Yu, D., Turner, J. R., Gehring, C., Rahn, H. P., Wolburg, H., & Blasig, I. E. (2011). Elucidating the principles of the molecular organization of heteropolymeric tight junction strands. *Cellular and Molecular Life Sciences: CMLS, 68*, 3903–3918.

Piontek, J., Krug, S. M., Protze, J., Krause, G., & Fromm, M. (2020). Molecular architecture and assembly of the tight junction backbone. *Biochimica et Biophysica Acta (BBA) - Biomembranes, 1862*, Article 183279.

Piontek, J., Winkler, L., Wolburg, H., Müller, S. L., Zuleger, N., Piehl, C., Wiesner, B., Krause, G., & Blasig, I. E. (2008). Formation of tight junction: Determinants of homophilic interaction between classic claudins. *The FASEB Journal, 22*, 146–158.

Rajagopal, N., Durand, A. J., & Nangia, S. (2020). Predicting selectivity of paracellular pores for biomimetic applications. *Molecular Systems Design & Engineering, 5*, 686–696.

Reuter, H., & Stevens, C. F. (1980). Ion conductance and ion selectivity of potassium channels in snail neurones. *Journal of Membrane Biology, 57*, 103–118.

Revel, J. P., & Karnovsky, M. J. (1967). Hexagonal array of subunits in intercellular junctions of the mouse heart and liver. *The Journal of Cell Biology, 33*, C7–C12.

Richter, J. F., Hildner, M., Schmauder, R., Turner, J. R., Schumann, M., & Reiche, J. (2019). Occludin knockdown is not sufficient to induce trans-epithelial macromolecule passage. *Tissue Barriers, 7*, Article 1612661.

Richter, J. F., Schmauder, R., Krug, S. M., Gebert, A., & Schumann, M. (2016). A novel method for imaging sites of paracellular passage of macromolecules in epithelial sheets. *Journal of Controlled Release: Official Journal of the Controlled Release Society, 229*, 70–79.

Rosenthal, R., Günzel, D., Piontek, J., Krug, S. M., Ayala-Torres, C., Hempel, C., Theune, D., & Fromm, M. (2020). Claudin-15 forms a water channel through the tight junction with distinct function compared to claudin-2. *Acta Physiologica, 228*, Article e13334.

Rosenthal, R., Günzel, D., Theune, D., Czichos, C., Schulzke, J. D., & Fromm, M. (2017). Water channels and barriers formed by claudins. *Annals of the New York Academy of Sciences, 1397*, 100–109.

Rosenthal, R., Heydt, M. S., Amasheh, M., Stein, C., Fromm, M., & Amasheh, S. (2012). Analysis of absorption enhancers in epithelial cell models. *Annals of the New York Academy of Sciences, 1258*, 86–92.

Rosenthal, R., Milatz, S., Krug, S. M., Oelrich, B., Schulzke, J. D., Amasheh, S., Günzel, D., & Fromm, M. (2010). Claudin-2, a component of the tight junction, forms a paracellular water channel. *Journal of Cell Science, 123*, 1913–1921.

Rossa, J., Ploeger, C., Vorreiter, F., Saleh, T., Protze, J., Günzel, D., Wolburg, H., Krause, G., & Piontek, J. (2014). Claudin-3 and claudin-5 protein folding and assembly into the tight junction are controlled by non-conserved residues in the transmembrane 3 (TM3) and extracellular loop 2 (ECL2) segments. *Journal of Biological Chemistry, 289*, 7641–7653.

Rossa, J., Protze, J., Kern, C., Piontek, A., Günzel, D., Krause, G., & Piontek, J. (2014). Molecular and structural transmembrane determinants critical for embedding claudin-5 into tight junctions reveal a distinct four-helix bundle arrangement. *Biochemical Journal, 464*, 49–60.

Rouaud, F., Sluysmans, S., Flinois, A., Shah, J., Vasileva, E., & Citi, S. (2020). Scaffolding proteins of vertebrate apical junctions: Structure, functions and biophysics. *Biochimica et Biophysica Acta (BBA) - Biomembranes, 1862*, Article 183399.

Saito, A. C., Higashi, T., Fukazawa, Y., Otani, T., Tauchi, M., Higashi, A. Y., Furuse, M., & Chiba, H. (2021). Occludin and tricellulin facilitate formation of anastomosing tight-junction strand network to improve barrier function. *Molecular Biology of the Cell, 32*, 722–738.

Saitoh, Y., Suzuki, H., Tani, K., Nishikawa, K., Irie, K., Ogura, Y., Tamura, A., Tsukita, S., & Fujiyoshi, Y. (2015). Tight junctions. Structural insight into tight junction disassembly by *Clostridium perfringens* enterotoxin. *Science, 347*, 775–778.

Salas, P. J., & Moreno, J. H. (1982). Single-file diffusion multi-ion mechanism of permeation in paracellular epithelial channels. *Journal of Membrane Biology, 64*, 103–112.

Samanta, P., Wang, Y., Fuladi, S., Zou, J., Li, Y., Shen, L., Weber, C., & Khalili-Araghi, F. (2018). Molecular determination of claudin-15 organization and channel selectivity. *The Journal of General Physiology, 150*, 949–968.

Sasaki, H., Matsui, C., Furuse, K., Mimori-Kiyosue, Y., Furuse, M., & Tsukita, S. (2003). Dynamic behavior of paired claudin strands within apposing plasma membranes. *Proceedings of the National Academy of Sciences of the U S A, 100*, 3971–3976.

Savolainen, V., Juuti-Uusitalo, K., Onnela, N., Vaajasaari, H., Narkilahti, S., Suuronen, R., Skottman, H., & Hyttinen, J. (2011). Impedance spectroscopy in monitoring the maturation of stem cell-derived retinal pigment epithelium. *Annals of Biomedical Engineering, 39*, 3055–3069.

Sawada, T., Ogawa, T., Tomita, M., Hayashi, M., & Awazu, S. (1991). Role of paracellular pathway in nonelectrolyte permeation across rat colon epithelium enhanced by sodium caprate and sodium caprylate. *Pharmaceutical Research, 8*, 1365–1371.

Schifferdecker, E., & Frömter, E. (1978). The AC impedance of Necturus gallbladder epithelium. *Pflügers Archiv, 377*, 125–133.

Schmid, T., Bogdan, M., & Günzel, D. (2013). Discerning apical and basolateral properties of HT-29/B6 and IPEC-J2 cell layers by impedance spectroscopy, mathematical modeling and machine learning. *PLoS One, 8*, Article e62913.

Schnermann, J., Chou, C. L., Ma, T. H., Traynor, T., Knepper, M. A., & Verkman, A. S. (1998). Defective proximal tubular fluid reabsorption in transgenic aquaporin-1 null mice. *Proceedings of the National Academy of Sciences of the United States of America, 95*, 9660–9664.

Schnermann, J., Huang, Y., & Mizel, D. (2013). Fluid reabsorption in proximal convoluted tubules of mice with gene deletions of claudin-2 and/or aquaporin1. *American Journal of Physiology - Renal Physiology, 305*, F1352–F1364.

Schultz, S. G. (1977). The role of paracellular pathways in isotonic fluid transport. *Yale Journal of Biology and Medicine, 50*, 99–113.

Schulzke, J. D., Andres, S., Amasheh, M., Fromm, A., & Günzel, D. (2011). Anti-diarrheal mechanism of the traditional remedy Uzara via reduction of active chloride secretion. *PLoS One, 6*, Article e18107.

Sewerin, S., Piontek, J., Schönauer, R., Grunewald, S., Rauch, A., Neuber, S., Bergmann, C., Günzel, D., & Halbritter, J. (2021). Defective claudin-10 causes a novel variation of HELIX syndrome through compromised tight junction strand assembly. *Genes & Disease, 9*, 1301–1314.

Shen, L., Weber, C. R., Raleigh, D. R., Yu, D., & Turner, J. R. (2011). Tight junction pore and leak pathways: A dynamic duo. *Annual Review of Physiology, 73*, 283–309.

Shen, L., Weber, C. R., & Turner, J. R. (2008). The tight junction protein complex undergoes rapid and continuous molecular remodeling at steady state. *The Journal of Cell Biology, 181*, 683–695.

Shimono, M., Yamamura, T., & Fumagalli, G. (1980). Intercellular junctions in salivary glands: Freeze-fracture and tracer studies of normal rat sublingual gland. *Journal of Ultrastructure Research, 72*, 286–299.

Shinoda, T., Shinya, N., Ito, K., Ohsawa, N., Terada, T., Hirata, K., Kawano, Y., Yamamoto, M., Kimura-Someya, T., Yokoyama, S., & Shirouzu, M. (2016). Structural basis for disruption of claudin assembly in tight junctions by an enterotoxin. *Scientific Reports, 6*, Article 33632.

Simon, D. B., Lu, Y., Choate, K. A., Velazquez, H., Al-Sabban, E., Praga, M., Casari, G., Bettinelli, A., Colussi, G., Rodriguez-Soriano, J., McCredie, D., Milford, D., Sanjad, S., & Lifton, R. P. (1999). Paracellin-1, a renal tight junction protein required for paracellular Mg2+ resorption. *Science, 285*, 103–106.

Simske, J. S., & Hardin, J. (2011). Claudin family proteins in *Caenorhabditis elegans*. *Methods in Molecular Biology, 762*, 147–169.

Singh, A. K., Singh, S., Devor, D. C., Frizzell, R. A., van Driessche, W., & Bridges, R. J. (2002). Transepithelial impedance analysis of chloride secretion. *Methods in Molecular Medicine, 70*, 129–142.

Sohet, F., Lin, C., Munji, R. N., Lee, S. Y., Ruderisch, N., Soung, A., Arnold, T. D., Derugin, N., Vexler, Z. S., Yen, F. T., & Daneman, R. (2015). LSR/angulin-1 is a tricellular tight junction protein involved in blood-brain barrier formation. *The Journal of Cell Biology, 208*, 703–711.

Staehelin, L. A., Mukherjee, T. M., & Williams, A. W. (1969a). Fine structure of frozen-etched tight junctions. *Naturwissenschaften, 56*, 142.

Staehelin, L. A., Mukherjee, T. M., & Williams, A. W. (1969b). Freeze-etch appearance of the tight junctions in the epithelium of small and large intestine of mice. *Protoplasma, 67*, 165–184.

Staehelin, L. A. (1973). Further observations on the fine structure of freeze-cleaved tight junctions. *Journal of Cell Science, 13*, 763–786.

Staehelin, L. A. (1974). Structure and function of intercellular junctions. *International Review of Cytology, 39*, 191–283.

Stephenson, R. E., Higashi, T., Erofeev, I. S., Arnold, T. R., Leda, M., Goryachev, A. B., & Miller, A. L. (2019). Rho flares repair local tight junction leaks. *Developmental Cell, 48*, 445–459.

Stevenson, B. R., Siliciano, J. D., Mooseker, M. S., & Goodenough, D. A. (1986). Identification of ZO-1: A high molecular weight polypeptide associated with the tight junction (zonula occludens) in a variety of epithelia. *The Journal of Cell Biology, 103*, 755–766.

Sugawara, T., Furuse, K., Otani, T., Wakayama, T., & Furuse, M. (2021). Angulin-1 seals tricellular contacts independently of tricellulin and claudins. *The Journal of Cell Biology, 220*, Article e202005062.

Sun, L. Y., Liu, S. K., Bao, L. S., Li, Y., Feng, J. B., & Liu, Z. J. (2015). Claudin multigene family in channel catfish and their expression profiles in response to bacterial infection and hypoxia as revealed by meta-analysis of RNA-Seq datasets. *Comparative Biochemistry & Physiology D, 13*, 60–69.

Suzuki, H., Nishizawa, T., Tani, K., Yamazaki, Y., Tamura, A., Ishitani, R., Dohmae, N., Tsukita, S., Nureki, O., & Fujiyoshi, Y. (2014). Crystal structure of a claudin provides insight into the architecture of tight junctions. *Science, 344*, 304–307.

Suzuki, H., Tani, K., Tamura, A., Tsukita, S., & Fujiyoshi, Y. (2015). Model for the architecture of claudin-based paracellular ion channels through tight junctions. *Journal of Molecular Biology, 427*, 291–297.

Suzuki, K., Kottra, G., Kampmann, L., & Fromter, E. (1982). Square wave pulse analysis of cellular and paracellular conductance pathways in Necturus gallbladder epithelium. *Pflügers Archiv, 394*, 302–312.

Tamura, A., Kitano, Y., Hata, M., Katsuno, T., Moriwaki, K., Sasaki, H., Hayashi, H., Suzuki, Y., Noda, T., Furuse, M., & Tsukita, S. (2008). Megaintestine in claudin-15-deficient mice. *Gastroenterology, 134*, 523–534.

Tamura, A., Yamazaki, Y., Hayashi, D., Suzuki, K., Sentani, K., Yasui, W., & Tsukita, S. (2012). Claudin-based paracellular proton barrier in the stomach. *Annals of the New York Academy of Sciences, 1258*, 108–114.

Tanaka, H., Takechi, M., Kiyonari, H., Shioi, G., Tamura, A., & Tsukita, S. (2015). Intestinal deletion of Claudin-7 enhances paracellular organic solute flux and initiates colonic inflammation in mice. *Gut, 64*, 1529–1538.

Tanaka, H., Yamamoto, Y., Kashihara, H., Yamazaki, Y., Tani, K., Fujiyoshi, Y., Mineta, K., Takeuchi, K., Tamura, A., & Tsukita, S. (2016). Claudin-21 has a paracellular channel role at tight junctions. *Molecular and Cellular Biology, 36*, 954–964.

Tang, V. W., & Goodenough, D. A. (2003). Paracellular ion channel at the tight junction. *Biophysical Journal, 84*, 1660–1673.

Teorell, T., & Wersäll, R. (1945). Electrical impedance properties of surviving gastric mucosa of the frog. *Acta Physiologica Scandinavica, 10*, 243–257.

Teorell, T. (1946). Application of "square wave analysis" to bioelectric studies. *Acta Physiologica Scandinavica, 12*, 235–254.

Tervonen, A., Ihalainen, T. O., Nymark, S., & Hyttinen, J. (2019). Structural dynamics of tight junctions modulate the properties of the epithelial barrier. *PLoS One, 14*, Article e0214876.

Thoenes, W., & Langer, K. H. (1969). Die endocytose-phase der eiweißresorption im proximalen nierentubulus. Untersuchungen am ferritin-resorbierenden einzeltubulus der rattenniere. *Virchows Archiv - B: Cell Pathology, 2*, 361–379.

Thoma, R. (1875). Ueber die Kittsubstanz der Epithelien. *Archiv für Pathologische Anatomie und Physiologie und für Klinische Medicin, 64*, 394–422.

Tunggal, J. A., Helfrich, I., Schmitz, A., Schwarz, H., Günzel, D., Fromm, M., Kemler, R., Krieg, T., & Niessen, C. M. (2005). E-cadherin is essential for in vivo epidermal barrier function by regulating tight junctions. *The EMBO Journal, 24*, 1146–1156.

Ussing, H. H., & Zerahn, K. (1951). Active transport of sodium as the source of electric current in the short-circuited isolated frog skin. *Acta Physiologica Scandinavica, 23*, 110–127.

Ussing, H. H. (1949). The active ion transport through the isolated frog skin in the light of tracer studies. *Acta Physiologica Scandinavica, 17*, 1–37.

Van Driessche, W., Kreindler, J. L., Malik, A. B., Margulies, S., Lewis, S. A., & Kim, K. J. (2007). Interrelations/cross talk between transcellular transport function and paracellular tight junctional properties in lung epithelial and endothelial barriers. *American Journal of Physiology Lung Cellular and Molecular Physiology, 293*, L520–L524.

Van Itallie, C., Rahner, C., & Anderson, J. M. (2001). Regulated expression of claudin-4 decreases paracellular conductance through a selective decrease in sodium permeability. *Journal of Clinical Investigation, 107*, 1319–1327.

Van Itallie, C. M., & Anderson, J. M. (2006). Claudins and epithelial paracellular transport. *Annual Review of Physiology, 68*, 403–429.

Van Itallie, C. M., & Anderson, J. M. (2011). Measuring size-dependent permeability of the tight junction using PEG profiling. *Methods in Molecular Biology, 762*, 1–11.

Van Itallie, C. M., Fanning, A. S., Bridges, A., & Anderson, J. M. (2009). ZO-1 stabilizes the tight junction solute barrier through coupling to the perijunctional cytoskeleton. *Molecular Biology of the Cell, 20*, 3930–3940.

Van Itallie, C. M., Holmes, J., Bridges, A., Gookin, J. L., Coccaro, M. R., Proctor, W., Colegio, O. R., & Anderson, J. M. (2008). The density of small tight junction pores varies among cell types and is increased by expression of claudin-2. *Journal of Cell Science, 121*, 298–305.

Van Itallie, C. M., Mitic, L. L., & Anderson, J. M. (2011). Claudin-2 forms homodimers and is a component of a high molecular weight protein complex. *Journal of Biological Chemistry*, 3442–3450.

Van Itallie, C. M., Rogan, S., Yu, A., Vidal, L. S., Holmes, J., & Anderson, J. M. (2006). Two splice variants of claudin-10 in the kidney create paracellular pores with different ion selectivities. *American Journal of Physiology - Renal Physiology, 291*, 1288–1299.

Vasquez, C. G., de la Serna, E. L., & Dunn, A. R. (2021). How cells tell up from down and stick together to construct multicellular tissues - interplay between apicobasal polarity and cell-cell adhesion. *Journal of Cell Science, 134*, Article jcs248757.

Vecchio, A. J., Rathnayake, S. S., & Stroud, R. M. (2021). Structural basis for *Clostridium perfringens* enterotoxin targeting of claudins at tight junctions in mammalian gut. *Proceedings of the National Academy of Sciences of the U S A, 118*, Article e2024651118.

Vecchio, A. J., & Stroud, R. M. (2019). Claudin-9 structures reveal mechanism for toxin-induced gut barrier breakdown. *Proceedings of the National Academy of Sciences of the U S A, 116*, 17817–17824.

Veshnyakova, A., Piontek, J., Protze, J., Waziri, N., Heise, I., & Krause, G. (2012). Mechanism of *Clostridium perfringens* enterotoxin interaction with claudin-3/-4 protein suggests structural modifications of the toxin to target specific claudins. *Journal of Biological Chemistry, 287*, 1698–1708.

Villarroel, A., Burnashev, N., & Sakmann, B. (1995). Dimensions of the narrow portion of a recombinant NMDA receptor channel. *Biophysical Journal, 68*, 866–875.

Voets, T., Prenen, J., Vriens, J., Watanabe, H., Janssens, A., Wissenbach, U., Bodding, M., Droogmans, G., & Nilius, B. (2002). Molecular determinants of permeation through the cation channel TRPV4. *Journal of Biological Chemistry, 277*, 33704–33710.

Wada, M., Tamura, A., Takahashi, N., & Tsukita, S. (2013). Loss of claudins 2 and 15 from mice causes defects in paracellular Na+ flow and nutrient transport in gut and leads to death from malnutrition. *Gastroenterology, 144*, 369–380.

Walker, D. C., MacKenzie, A., Hulbert, W. C., & Hogg, J. C. (1985). A re-assessment of the tricellular region of epithelial cell tight junctions in trachea of Guinea pig. *Acta Anatomica, 122*, 35–38.

Watson, C. J., Hoare, C. J., Garrod, D. R., Carlson, G. L., & Warhurst, G. (2005). Interferon-gamma selectively increases epithelial permeability to large molecules by activating different populations of paracellular pores. *Journal of Cell Science, 118*, 5221–5230.

Watson, C. J., Rowland, M., & Warhurst, G. (2001). Functional modeling of tight junctions in intestinal cell monolayers using polyethylene glycol oligomers. *American Journal of Physiology - Cell Physiology, 281*, 388–397.

Weber, C. R., Liang, G. H., Wang, Y., Das, S., Shen, L., Yu, A. S., Nelson, D. J., & Turner, J. R. (2015). Claudin-2-dependent paracellular channels are dynamically gated. *Elife, 4*, Article e09906.

Weber, C. R., & Turner, J. R. (2017). Dynamic modeling of the tight junction pore pathway. *Annals of the New York Academy of Sciences, 1397*, 209–218.

Welsch, U. (1983). Septate junctions with paired septa in the cephalochordate *Branchiostoma lanceolatum. Cell and Tissue Research, 229*, 175–181.

Wen, H., Watry, D. D., Marcondes, M. C., & Fox, H. S. (2004). Selective decrease in paracellular conductance of tight junctions: Role of the first extracellular domain of claudin-5. *Molecular and Cellular Biology, 24*, 8408–8417.

Winckler, B., Forscher, P., & Mellman, I. (1999). A diffusion barrier maintains distribution of membrane proteins in polarized neurons. *Nature, 397*, 698–701.

Wong, V., & Goodenough, D. A. (1999). Paracellular channels. *Science, 285*, 62.

Wright, E. M., & Diamond, J. M. (1968). Effects of pH and polyvalent cations on the selective permeability of gall-bladder epithelium to monovalent ions. *Biochimica et Biophysica Acta, 163*, 57–74.

Yamazaki, Y., Tokumasu, R., Kimura, H., & Tsukita, S. (2011). Role of claudin species-specific dynamics in reconstitution and remodeling of the zonula occludens. *Molecular Biology of the Cell, 22*, 1495–1504.

Yu, A. S., Cheng, M. H., Angelow, S., Günzel, D., Kanzawa, S. A., Schneeberger, E. E., Fromm, M., & Coalson, R. D. (2009). Molecular basis for cation selectivity in claudin-2-based paracellular pores: Identification of an electrostatic interaction site. *The Journal of General Physiology, 133*, 111–127.

Yu, A. S., Enck, A. H., Lencer, W. I., & Schneeberger, E. E. (2003). Claudin-8 expression in Madin-Darby canine kidney cells augments the paracellular barrier to cation permeation. *Journal of Biological Chemistry, 278*, 17350–17359.

Yu, A. S., McCarthy, K. M., Francis, S. A., McCormack, J. M., Lai, J., Rogers, R. A., Lynch, R. D., & Schneeberger, E. E. (2005). Knockdown of occludin expression leads to diverse phenotypic alterations in epithelial cells. *American Journal of Physiology - Cell Physiology, 288*, C1231–C1241.

Zeissig, S., Bürgel, N., Günzel, D., Richter, J., Mankertz, J., Wahnschaffe, U., Kroesen, A. J., Zeitz, M., Fromm, M., & Schulzke, J. D. (2007). Changes in expression and distribution of claudin 2, 5 and 8 lead to discontinuous tight junctions and barrier dysfunction in active Crohn's disease. *Gut, 56*, 61–72.

Zhang, J., Piontek, J., Wolburg, H., Piehl, C., Liss, M., Otten, C., Christ, A., Willnow, T. E., Blasig, I. E., & Abdelilah-Seyfried, S. (2010). Establishment of a neuroepithelial barrier by Claudin5a is essential for zebrafish brain ventricular lumen expansion. *Proceedings of the National Academy of Sciences of the U S A, 107*, 1425–1430.

Zhao, J., Krystofiak, E. S., Ballesteros, A., Cui, R., Van Itallie, C. M., Anderson, J. M., Fenollar-Ferrer, C., & Kachar, B. (2018). Multiple claudin-claudin *cis* interfaces are required for tight junction strand formation and inherent flexibility. *Communications Biology, 1*, 50.

Zhou, Y., Chen, C. C., Weber, A. E., Zhou, L., Baker, L. A., & Hou, J. (2013). Potentiometric-scanning ion conductance microscopy for measurement at tight junctions. *Tissue Barriers, 1*, Article e25585.

Zihni, C., Mills, C., Matter, K., & Balda, M. S. (2016). Tight junctions: From simple barriers to multifunctional molecular gates. *Nature Reviews Molecular Cell Biology, 17*, 564–580.

Zimmermann, K. W. (1911). Zur Morphologie der Epithelzellen der Saugetierniere. *Archiv für Mikroskopische Anatomie, 87*, 199–231.

Chapter 8

Biology of gap junctions

Richard D. Veenstra
Department of Pharmacology, SUNY Upstate Medical University, Syracuse, NY, United States

1. Introduction

1.1 Early evidence for intercellular communication

The first evidence for intercellular communication between like cells in a tissue was electrically based. Current injection by microelectrodes into cardiac Purkinje fibers, septated giant axons, and teleost Mauthner cells revealed that electrical signals (voltage responses) traveled beyond the cellular boundary of the injection site (see Fig. 8.1; Weidmann, 1952; Furshpan & Potter, 1959; Furshpan & Furukawa, 1962). First thought to be a property of excitable cells like nerve and muscle, this phenomenon of electrical communication passed on to the nonexcitable world with the observation of electrical coupling between secretory epithelial cells (Kanno & Loewenstein, 1964a). Often referred to as the electrical synapse, the apparent site of electrotonic coupling was described as the fusion of adjacent cell membranes termed the "nexus" (Dewey & Barr, 1962). Experimental proof that this nexal structure was the site of electrotonic interaction responsible for action potential propagation in cardiac muscle was provided when disruption of the nexal membrane structure by hypertonic solutions also blocked action potential transmission across a sucrose gap (Barr et al., 1965). The term "gap junction" did not appear until 1967 when a narrow 20 Å gap was observed between apposed cell membranes from mouse liver and heart, owing to improved fixation and higher resolution transmission electron microscopy (TEM) methods (Revel & Karnovsky, 1967; Uehara & Burnstock, 1970).

1.2 From ions to molecules

The early observations of ionic electrical coupling between adjacent cells were amended when intercellular transfer of fluorescent dyes like fluorescein and Procion Yellow was demonstrated after intracellular dye injection (see Fig. 8.1; Kanno & Loewenstein, 1964b; Payton et al., 1969). The demonstration of passage of larger molecules led to additional characterization of gap junctions as pores with a limiting diameter of 10−14 Å, molecular size limit of 1200 Da (1 kD for mammalian gap junctions), and fixed negative charge (Brink & Dewey, 1980; Flagg-Newton et al., 1979; Simpson et al., 1977). Thus, gap junctions not only couple cells electrically via the passage of ions like Na^+ and K^+ but also biochemically via the passage of metabolites and second messengers like glucose, cyclic AMP (cAMP), Ca^{2+} ions, inositol-3,4,5-trisphosphate (IP_3), and even small RNAs (De Mello, 2015; Gilula et al., 1972; Sáez et al., 1989; Tsien & Weingart, 1976; Valiunas et al., 2005). So gap junctions have a vital role to play in tissue development, homeostasis, and functional integration in addition to rapid electrical impulse propagation in nerve and muscle.

2. Gap junction structure

2.1 Anatomical and structural models of gap junctions

The anatomical interpretation of the nexus as a fusion of appositional membranes changed with the observations of Robertson and Revel and Karnovsky of hexagonal protein arrays containing a central electron opaque core within a 2 nM extracellular gap (Revel & Karnovsky, 1967; Robertson, 1963). In transmission electron microscope (TEM) images, uranium-stained gap junction membranes resemble parallel railroad tracks and the extracellular gap is visualized by adding electron dense lanthanum (La^{3+}) ions to the extracellular solution (Fig. 8.1A−D). In scanning electron (SEM)

Cell Physiology Sourcebook. https://doi.org/10.1016/B978-0-12-811114-7.00015-3

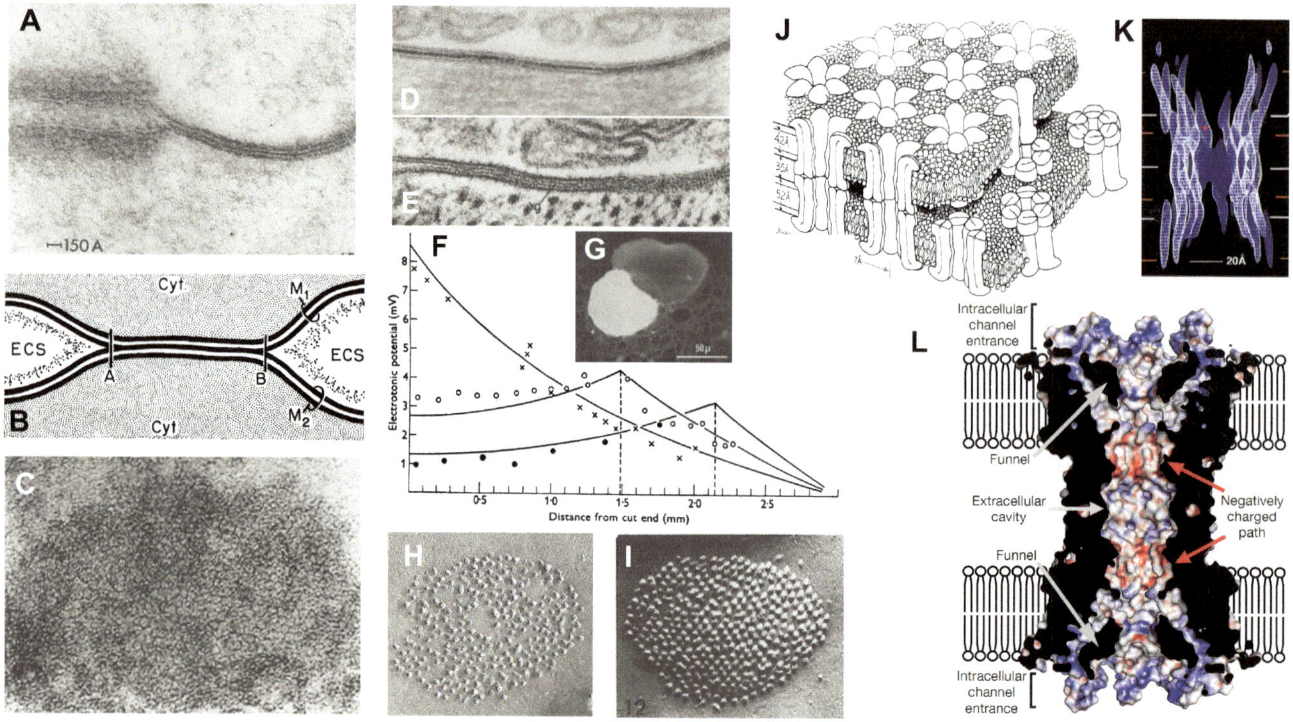

FIGURE 8.1 Gap junction structure. (A) Transmission electron (TEM) micrograph of a desmosome and nexus from a Guinea pig papillary muscle. (B) Early drawing of a nexus from dog intestinal smooth muscle depicting the fusion of the outer membrane leaflet of adjoining cell membranes. (C) First publication of hexagonal (protein) subunit arrays appearing within the electrical synapse between goldfish brain Mauthner cells. (D and E) TEM images of mouse heart gap junctions with (D) and without (E) lanthanum (La^{3+}) ions filling the narrow 2 nm gap between adjacent cell membranes, thus prompting the term "gap" junctions. (F) Classical figure of the electrotonic transmission of membrane voltage deflections in response to intracellular current injection near the left end or center of a cardiac Purkinje fiber (<30 mm in length), providing the first evidence of electrical coupling between heart cells ($\approx$100 μm in length). (G) Transfer of Procion yellow dye between lateral giant axons from a crayfish illustrating the ability of larger fluorescent molecules to pass through electrical synapses, i.e., gap junctions. (H and I) Freeze fracture (E-face) scanning electron (SEM) micrograph of gap junction plaques between crayfish septate lateral giant axons under conditions of electrical coupling (H) and uncoupled (I) illustrating the shift from a random order of irregularly spaced gap junction particles to a condensed hexagonal array under nonfunctional conditions. (J) Summary illustration of the molecular structure of gap junctions based on the X-ray diffraction patterns from mouse liver gap junctions. Each 60 Å diameter connexon is formed by a hexamer of protein subunits that dock end-to-end to form a gap junction channel. (K) Three-dimensional structure of the CT-truncated Cx43 gap junction channel at 18 Å resolution with half of the density projection removed to show the side view of the channel interior. The red asterisk indicates the narrowest region of the pore. (L) Similar view of the atomic structure of the Cx26 gap junction channel at 3.5 Å resolution illustrating an electronegative (red) narrow pore region at the extracellular membrane interface with a wider central extracellular compartment and pore funnel extending into the membrane from a wide intracellular pore opening. The transmembrane pore is formed by the NT and M1 domains. *(A) Barr et al. (1965). (B) From Dewey and Barr (1962). Reprinted with permission from AAAS. (C) Robertson (1963). (D and E) Revel and Karnovsky (1967). (F) From Weidmann (1952). Reprinted with permission from John Wiley and Sons. (G) From Payton et al. (1969). Reprinted with permission from AAAS. (H and I) Peracchia and Dulhunty (1976). (J) Makowski et al. (1977). (K) From Unger et al. (1999). Reprinted with permission from AAAS. (L) Reprinted by permission from Maeda et al. (2009).*

micrographs, gap junctions appear as clusters, or "plaques", of randomly distributed integral membrane particles (E-face particles and P-face pits) that crystalize into a tightly packed hexagonal array during uncoupling procedures (Peracchia & Dulhunty, 1976). A three-dimensional structural model of the gap junction with 18 Å resolution was developed based on X-ray diffraction studies of mouse liver gap junction membranes during the mid-1970s that defined a hexamer of protein subunits 60 Å in diameter and 80 Å long with a 20 Å wide central pore as a "connexon" (Makowski et al., 1977). Two connexons dock end-to-end to form a gap junction channel with a hexagonal lattice center-to-center spacing of 82–87 Å (Fig. 8.1J–L). The salient features of this model remain preserved in higher resolution (7.5 Å, 3.5 Å) gap junction structures published in 1999 and 2009 (Maeda et al., 2009; Unger et al., 1999).

2.2 From structure to molecular identity—The connexins

The earliest identification of the gap junction protein composition was performed by isolation of detergent-insoluble gap junction enriched membranes from mouse liver followed by SDS-PAGE electrophoresis of the constituent components

(Revel et al., 1985). The mouse liver gap junction protein migrated as two bands of 34,000 and 18,000 Da that could be further reduced to a 10 kDa doublet named connexins A and B (Goodenough, 1974). Similar procedures performed on mouse and rabbit myocardial gap junctions revealed multiple protein bands ranging in apparent molecular weight from 28.5 to 47 kDa (Kensler & Goodenough, 1980; Manjunath et al., 1982). Questions about the existence of distinct or common gap junction proteins in liver and heart were answered when partial amino acid sequence analysis of the two source proteins revealed 43% sequence identity, 25% conservative substitutions, and conservation of a hydrophobic domain within the first amino terminal 32 amino acids of both proteins (Nicholson et al., 1985). The distinct molecular identity of liver and heart gap junction proteins was confirmed shortly thereafter by the molecular cloning of the liver and heart gap junction proteins, connexin32 (Cx32), and connexin43 (Cx43), respectively (Beyer et al., 1987; Kumar & Gilula, 1986; Paul, 1986). Cardiac and hepatic gap junctions were initially thought to be composed of only one gap junction protein in each tissue, but this perspective quickly changed with the cloning of additional connexins from each tissue (Beyer, 1990; Kanter et al., 1992; Zhang & Nicholson, 1989).

2.3 Connexin nomenclature

The discovery of multiple connexin proteins necessitated the need for a method of classifying the rapidly growing family of gap junction proteins. Initially, the newly cloned connexins were named according to their predicted molecular weights, though an alternative nomenclature of alpha/beta-connexins was also considered based on primary sequence homology (Beyer et al., 1990; Kumar & Gilula, 1992). However, what began as two has culminated in 20 mouse and 21 human gap junction genes and proteins, resulting in additional sequence diversities beyond the α/β-connexin classification and necessitating the need to use decimal points for the connexin molecular weight classification scheme (Fig. 8.2A). A final nomenclature system for the mouse/human connexin genes and proteins was adopted where the proteins retain their connexin predicted molecular weight names and the corresponding gene names adopted a modified α/β/γδε classification system as shown in Table 8.1 (Söhl & Willecke, 2003; Beyer & Berthoud, 2009). The entire coding sequence is found in a single exon for all of the connexin genes except Cx36, which has two exons (Condorelli et al., 1998).

2.4 Connexin topology

The connexin proteins share a common membrane topology commencing with a cytoplasmic amino-terminus (NT domain), four transmembrane (M1-4) domains separated by two extracellular (E1 and E2) loops, and one cytoplasmic loop (CL) domain, culminating with a cytoplasmic carboxyl-terminal (CT) domain (Fig. 8.2B). The CL and CT domains exhibit the greatest diversity in amino acid sequence and length among the connexins, accounting for the observed differences in molecular weight (Fig. 8.2C). An essential feature of the connexin extracellular loops is the strict conservation of three cysteine residues in each loop that form E1-E2 intramolecular disulfide bonds in an antiparallel manner that are essential for functional docking of the two connexons (Foote et al., 1998). The membrane topology was predicted by hydropathy plots of the cloned connexin primary sequences and verified by proteolytic digests, calmodulin (CaM) binding, and site-directed antibody localization of the rat and human liver Cx32 sequences (Milks et al., 1988; Zimmer et al., 1987).

3. Are gap junctions static or dynamic?

3.1 Gap junction closure

The anatomical features of the nexal or gap junction membrane may lead one to believe that these structures are quite stable and long-lived. This, however, is not the case. The first evidence that gap junctions are dynamic structures was derived from freeze-fracture SEM images of crayfish septate axon and rat liver and stomach gap junctions that illustrated a transition from randomly packed gap junction particles with a center-to-center spacing of 10 nM to hexagonally packed arrays with an 8.5 nm center-to-center spacing and condensation of the gap junction particle diameter from 8.6 to 7.7 nm (Peracchia, 1977; Peracchia & Dulhunty, 1976). These structural changes were correlated with electrical uncoupling and could be induced by metabolic uncoupling with 2,4-dinitrophenol (DNP), near μM Ca^{2+} or mM Mg^{2+} ion concentrations, or pH 6.5 or lower (Peracchia, 1977; Peracchia & Peracchia, 1980a, 1980b). These structural and functional correlates resemble the "healing-over" process of cardiac myocardium whereby lesioned myocardial tissue recovers its resting membrane potential and electrical excitability within minutes after being cut when bathed in normal Ca^{2+}-containing saline solutions (De Mello et al., 1969; Délèze, 1970). These experimental observations led to the first structural model for the closing of a gap junction channel based on X-ray diffraction studies of rat liver gap junction membranes in the presence or

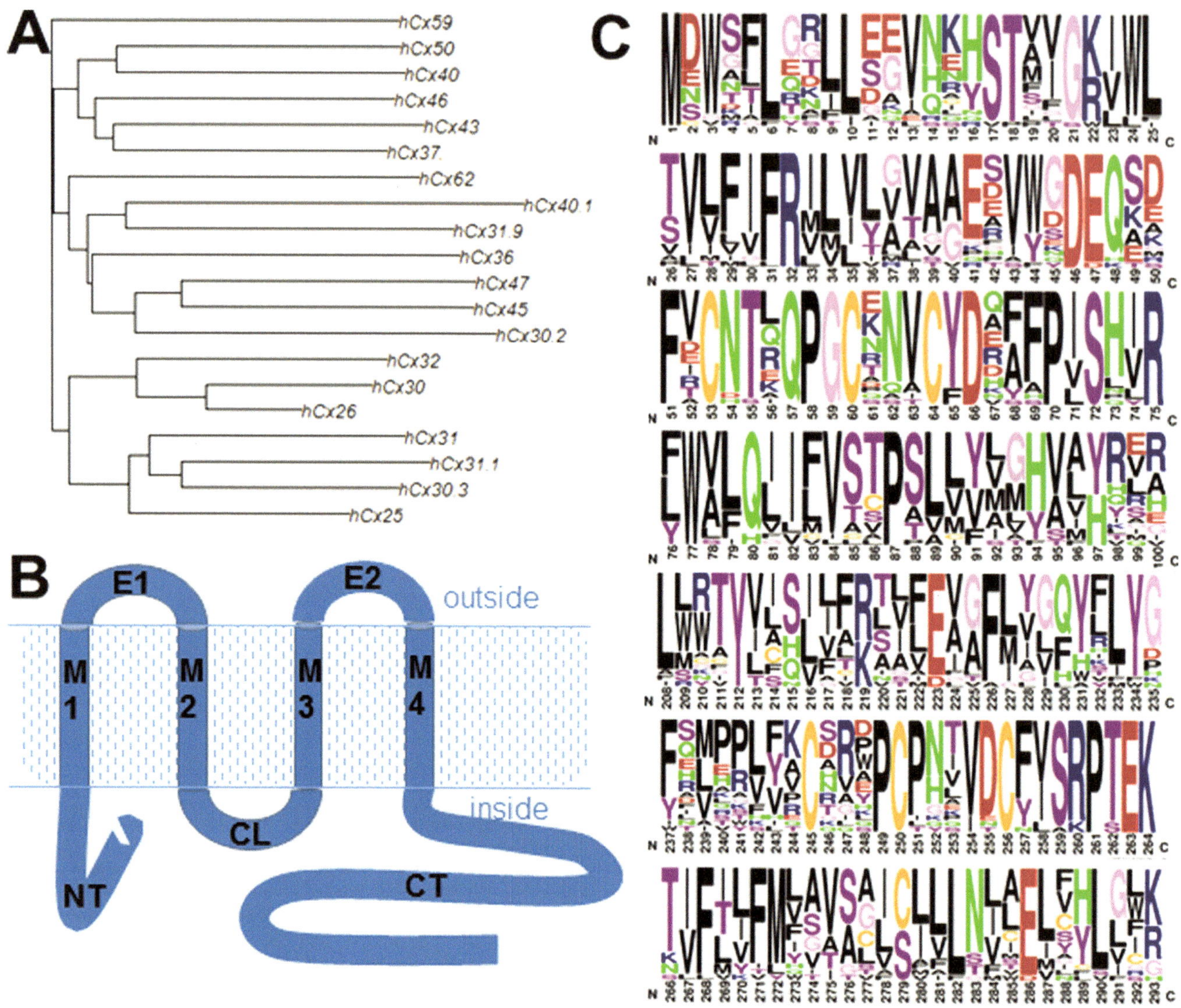

FIGURE 8.2 Connexin homology and topology. (A) Dendrogram of 20 human connexin protein sequences after alignment using the CLUSTALW software (http://embnet.vital-it.ch/software/ClustalW.html) with the α-group connexins clustered at the top, β-group connexins clustered at the bottom, and the miscellaneous δ/γ/ε-connexins in the middle. (B) Illustration of the conserved connexin membrane topology. *NT*, amino-terminal domain; *M1-M4*, first through fourth transmembrane domains; *E1 and E2*, first and second extracellular loop domains; *CL*, cytoplasmic loop domain; *CT*, carboxyl terminal domain. (C) Frequency plot logos (http://weblogo.berkeley.edu/logo.cgi) of the aligned primary sequences for 20 human connexin proteins beginning with the NT domain in 25 amino acid blocks through the M2 domain and resuming with the M3-E2-M4 domains. The three conserved cysteine residues in the E1 and E2 domains appear in the third and sixth logos from the top. There is negligible sequence conservation among the CL and CT domains of the 20 connexins. The maximal sequence alignment was achieved by deleting the G2 residue from Cx31.9, Cx36, Cx37, Cx40, Cx43, Cx46, Cx50, Cx59, Cx62, and Cx40.1 (G3); Cx36 A14; G236 from Cx25, Cx26, Cx30, Cx31, Cx32, Cx30.3 (D236), Cx31.1 (K236), and N247 from Cx31.

absence of 50 μM Ca^{2+} (Unwin & Ennis, 1983, 1984). Ca^{2+} ions, but not 0.16 mM Mg^{2+} or pH 6.0, caused a 2% decrease in the lattice dimension (center-to-center spacing) but a 4% increase in the dimension perpendicular to the lattice (length). These results were interpreted as a 5-degree reduction in the tilt and 6 Å radial displacement of the connexin hexameric protein subunits, resulting in closure of the central pore in a manner analogous to the iris of a camera (Unwin & Ennis, 1984). Thus, Ca^{2+} ions can reversibly induce the closure of gap junctions within minutes of exposure. These early indications suggest that gap junction channels are not static, but gated structures that are modulated by a variety of factors presented in Section 4.

TABLE 8.1 Human and mouse connexin nomenclature.

Human		Mouse	
Gene	**Protein**	**Gene**	**Protein**
GJE1	hCx23	Gje1	mCx23
GJB7	hCx25	–	–
GJB2	hCx26	Gjb2	mCx26
GJC3	hCx30.2	Gjc3	mCx29
GJB6	hCx30	Gjb6	mCx30
GJB4	hCx30.3	Gjb4	mCx30.3
GJB3	hCx31	Gjb3	mCx31
GJB5	hCx31.1	Gjb5	mCx31.1
GJD3	hCx31.9	Gjd3	mCx30.2
GJB1	hCx32	Gjb1	mCx32
–	–	Gja6	mCx33
GJD2	hCx36	Gjd2	mCx36
GJA4	hCx37	Gja4	mCx37
GJA5	hCx40	Gja5	mCx40
GJD4	hCx40.1	Gjd4	mCx39
GJA1	hCx43	Gja1	mCx43
GJC1	hCx45	Gjc1	mCx45
GJA3	hCx46	Gja3	mCx46
GJC2	hCx47	Gjc2	mCx47
GJA8	hCx50	Gja8	mCx50
GJA9	hCx59	–	–
GJA10	hCx62	Gja10	mCx57

a = α; b = β; d = δ; c = γ; e = ε based on the nomenclature proposed by Kumar & Gilula, 1992. This table is modified from the classification system that appeared in Sohl & Willecke, 2003 and reviewed in Beyer & Berthoud, 2009.

3.2 Life cycle of the connexins

The gap junction proteins also turnover with time. In vivo ^{14}C-labeling of de novo protein synthesis in mice was utilized to follow the time course of liver gap junction protein turnover (Fallon & Goodenough, 1981). Using this biochemical approach, a half-life ($T_{1/2}$) of 5 h was measured for the 21 kD protein band from isolated liver gap junction membranes. Since the cloning of the constituent connexins and development of connexin-specific antibodies, ^{35}S-methionine labeling of protein synthesis and immunoprecipitation of Cx43 with Cx43-specific antibodies have revealed a surprising short $T_{1/2}$ of only 1.3–2 h for Cx43 in cultured rat cardiomyocytes or ex vivo perfused adult hearts (Beardslee et al., 1998; Laird et al., 1991). Lysosomal inhibition with leupeptin or alkaline conditions and proteasome inhibition with N-acetyl-leucyl-leucylnorleucine (ALLN) increased the level of phosphorylated and nonphosphorylated Cx43, respectively. Thus, Cx43, the most abundantly expressed mammalian connexin, is rapidly synthesized and degraded by lysosomal and proteosomal pathways every couple of hours. So, how are the gap junctions degraded once they are formed, since gap junction membranes are notoriously resistant to detergent solubilization and require strong hyperosmotic conditions (e.g., 630 mOsm sucrose:saline solution, Barr et al., 1965) to separate the double membrane structure? The first clue was provided by the observation of internalized gap junction-like vesicles in TEM images from the late 1970s (Larsen & Tung, 1978; Marquart, 1977). These "annular" gap junctions were thought to be formed by endocytosis of gap junction membranes. The connexin composition of these annular gap junctions was later confirmed using green fluorescent protein (GFP)-tagged

Cx43- and Cx43-specific antibodies (Jordan et al., 2001). In an excellent review of the connexin lifecycle, the term "connexosome" was proposed to affirm the connexin composition of these internalized gap junction vesicles (Fig. 8.3A, Laird, 2006). Thus, intact gap junction membranes are internalized and degraded by an endocytotic mechanism and lysosomal and proteosomal protein degradation pathways.

So, how are the connexins assembled into connexons and subsequently docked to form a gap junction? Like typical proteins, the connexins are synthesized and inserted into the microsomal endoplasmic reticulum (ER) membrane, where many of the connexins oligomerize to form a connexon (Falk et al., 1997). The assembled connexons may be homomeric, composed of only one connexin type, or heteromeric, resulting from the coexpression of two or more types in the same cell (Brink et al., 1997; Jiang & Goodenough, 1996; Stauffer, 1995). Cx43, however, does not appear to oligomerize into connexons until reaching the trans-Golgi compartment, and this assembly into Triton X-100 insoluble connexons apparently is dependent upon phosphorylation (Musil & Goodenough, 1991, 1993). Prior to Cx43 being identified as a phospho-protein, cAMP was known to diffuse through cardiac gap junctions, increase myometrial gap junction area, and increase myocardial gap junctional conductance, all of which are now known to contain Cx43 (Estapé-Wainwright & De Mello, 1983; Garfield et al., 1980; Tsien & Weingart, 1976). cAMP and its membrane permeable derivatives (dibutyryl-cAMP (db-cAMP) and 8-bromo-cAMP (br-cAMP)) similarly increased the junctional conductance between rat hepatocytes (Sáez et al., 1986). Cx32 and Cx43, but not Cx26, were subsequently shown to be phospho-proteins (Crow et al., 1990; Sáez et al., 1990). Cx43 is now known to be phosphorylated by a myriad of serine/threonine and tyrosine kinases, but selective inhibition of casein kinase 1 (CK1) reportedly decreases the phosphorylated amount of Cx43 assembled into Triton X-100 insoluble gap junctions while increasing surface expression of nonphosphorylated Cx43 (Cooper & Lampe, 2002; Solan & Lampe, 2009). The current model for how phosphorylation regulates Cx43 gap junction assembly and degradation is that CK1 promotes Cx43 gap junction assembly, protein kinase A (PKA) increases Cx43 trafficking to the

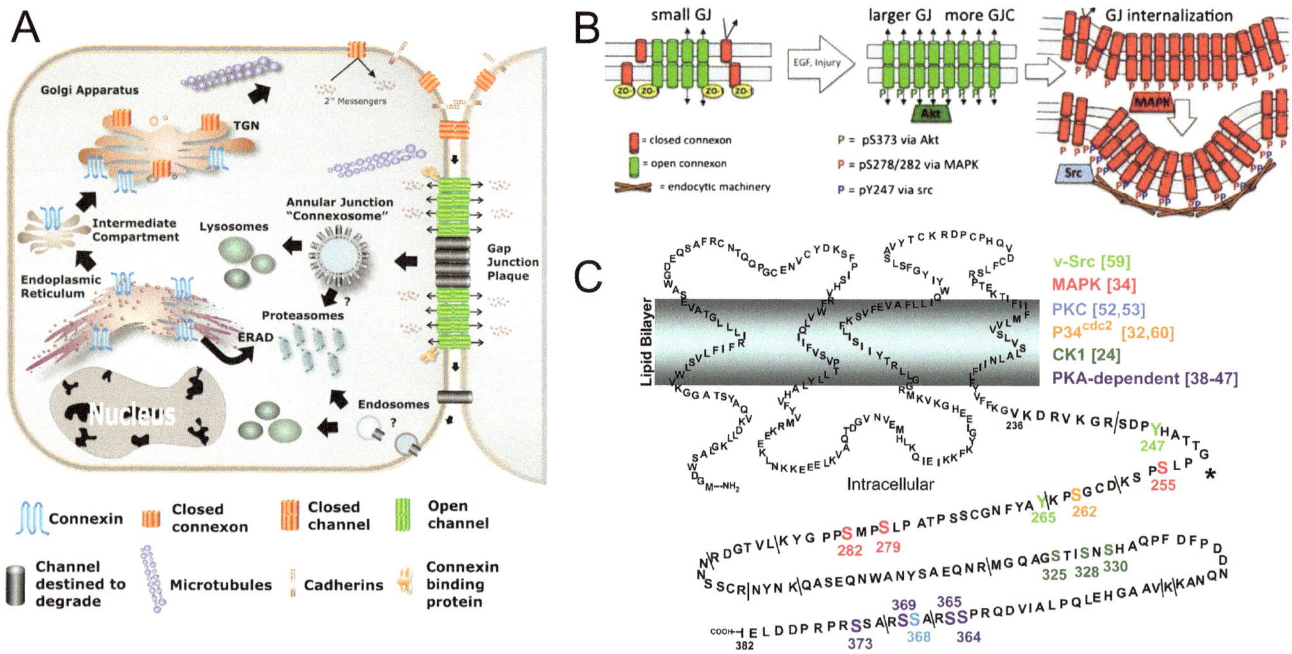

FIGURE 8.3 Life cycle of the connexins. (A) Illustration of the life cycle of a connexin gap junction protein from its synthesis in the endoplasmic reticulum, transport to and thru the Golgi apparatus and assembly into connexons, transport from the trans-Golgi network (TGN) to the plasma membrane along microtubules, docking and assembly into gap junction plaques with open channels located around the periphery of the plaque, internalization from the center of the gap junction plaque to form an annular gap junction or connexosome, and degradation by the lysosomal or proteosomal pathways. Aberrant connexins may be degraded by the endoplasmic reticulum associated with protein degradation (ERAD) pathway and connexons may be internalized by endocytosis. (B) Illustration summarizing the regulation of Cx43 gap junction plaque assembly by ZO-1 and phosphorylation wherein Akt phosphorylation of S373 inhibits the PDZ-dependent ZO-1 interaction with Cx43, thus increasing plaque size, and downregulatory phosphorylation events mediated by MAPK and Src kinases close gap junction channels and signal for these plaques to be internalized for degradation. (C) Membrane topological primary amino acid sequence for Cx43 with known CT domain protein kinase phosphorylation sites indicated (v-Src: Y247 & Y265; MAPK: S255, S279, S282; PKC: S368; p34^{cdc2}: S262; CK1: S325, S328, S330; PKA-dependent: S364, S365; and Akt: S373). *(A) This figure was originally published in Laird (2006). (B) This figure was originally published in Solan and Lampe (2014). (C) Reprinted from Solan and Lampe (2005), with permission from Elsevier.*

membrane and gap junction content, and Akt/PKB (protein kinase B) increases gap junction size by phosphorylating Cx43 S373, thus decreasing its interaction with zonula occludens-1 (ZO-1), which normally limits Cx43 gap junction size (Fig. 8.3B; Hunter et al., 2005; Dunn & Lampe, 2014). Conversely, Cx43 phosphorylation by mitogen-associated protein kinases (MAPKs) and Src tyrosine kinases downregulate Cx43-mediated gap junction communication and promote gap junction degradation (Lin et al., 2001; Warn-Cramer et al., 1996). Cx43 contains at least 19 phosphorylation sites, and how phosphorylation regulates Cx43 gap junctions is summarized in recent reviews of this extensive subject (Fig. 8.3C; Solan & Lampe, 2005, 2014). As these figures indicate, the carboxyl-tail domain of Cx43 interacts with many protein kinases and ZO-1. The Cx43 CT domain also binds tubulin via a unique highly basic K234−K243 motif near the membrane, but disruption of microtubules with nocodazole did not alter gap junction coupling (Giepmans et al., 2001). A further discussion of connexin interacting proteins appears in the final section of this chapter, Section 6, Gap Junctions are not alone.

4. Gap junction function

4.1 Channel basis of gap junction communication

Correlations between the phenomena of electrical coupling and observations of the nexal or gap junction structures led to the conceptualization of "permeable membrane elements ... permitting a central aqueous passageway insulated peripherally from the extracellular space. Many such units in parallel ... make up the entire junctional pathway ... junctional units proposed were thus discrete, continuous, and direct cell-to-cell passageways or channels" (Loewenstein, 1966). Further correlation of measured gap junction area and electrical gap junction conductances provided early estimates of the electrical resistivity of gap junction membranes to be between 8.8 and $12.8 \times 10^{-9}\ \Omega^{-1}/\mathrm{cm}^2$ (nanoSiemen/cm^2, nS/cm^2) and the observation of quantal step increases in gap junctional conductance between newly formed pairs of *Xenopus laevis* blastocysts provided evidence for the formation of stable cell-to-cell channels with a postulated conductance of 100 pS ($10 \times 10^{-9}\ \Omega^{-1}$) (Loewenstein et al., 1978; Sheridan et al., 1978). Definitive proof of the existence of discrete gap junction channels was provided by double whole cell patch-clamp recordings from isolated pairs of rat lacrimal gland cells and embryonic chick ventricular cardiomyocytes (Neyton & Trautmann, 1985; Veenstra & DeHaan, 1986). A distinguishing trait of these dual cell channel recordings is the mirror image "equal (amplitude) and opposite (polarity)" appearance of the gap junction channel currents, which exhibited unitary channel conductances of 120−165 pS, in close agreement with earlier estimates (Fig. 8.4A). Confirmation that the newly cloned connexins formed gap junction channels with these characteristic channel conductances was provided by exogenous expression of the connexins in communication-deficient cell lines like mouse Neuro2a neuroblastoma (N2a) cells and human SkHep1 hepatoma or HeLa cell lines subjected to dual whole cell patch-clamp recording methods (Eghbali et al., 1990; Traub et al., 1994; Veenstra et al., 1992). The first exogenous expression system utilized to test the function of the newly cloned connexins was paired *Xenopus* oocytes, which later became useful for the study of connexin hemichannels (connexon channel) that were first discovered during the cloning and expression of Cx46 (Dahl et al., 1987; Ebihara & Steiner, 1993; Paul et al., 1991). N2a and HeLa cells are routinely used for gap junction channel current recordings to this day, though N2a cells are especially low in endogenous connexin expression (Lin et al., 2014).

Since the cloning and functional expression of more than a dozen mammalian connexins, gap junction channel conductances have been found to vary by at least 200-fold from 10 to 15 pS for the δ-connexins Cx31.9 and Cx36 to 300 pS for Cx37 (Veenstra et al., 1994; Srinivas et al., 1999; White et al., 2002; reviewed in Veenstra, 2015, Chapter 13). Although one might think that the maximum channel conductances correlate with the relative pore diameters of the connexin-specific gap junction channels, this assumption is likely not true (Fig. 8.4B−D). Comparison of the channel conductance ratios from six different connexin-specific gap junction channels (Cx26, Cx32, Cx37, Cx40, Cx43, Cx45) in equimolar Kglutamate/KCl solutions and observation of their permeability to 6-carboxyfluorescein (6-CF) and 2,7-dichlorofluorescein (diCl-F) fluorescent dyes revealed no correlation between channel conductance, anion-to-cation conductance ratios, and fluorescent dye permeabilities (Veenstra, 1996; Veenstra et al., 1995). Thus, the divergent conductance and selective permeability properties of distinct connexin gap junction channels are not consistent with a simple aqueous pore model and are indicative of electrostatic interactions of permeant molecules with the channel pore. Based on the cation permeability of the alkali metal (Rb^+, Cs^+, K^+, Na^+) and tetraalkylammonium series (tetramethylammonium (TMA^+), tetraethylammonium (TEA^+), tetrabutylammonium (TBA^+)) of cations relative to Li^+ ions (P_X/P_{Li}) and the hydrodynamic equation, the Cx40 and Cx43 gap junction channel pore radii were estimated to be 6.6 ± 0.7 Å and 6.3 ± 0.4 Å, respectively (Beblo & Veenstra, 1997; Wang & Veenstra, 1997). These values are within the range of 10−14 Å from initial estimates and are in close agreement with the 15 Å limiting diameter of the Cx43 pore (omitting amino acid side chains) based on the crystal structure (Maeda et al., 2009; Simpson et al., 1977; Unger et al., 1999).

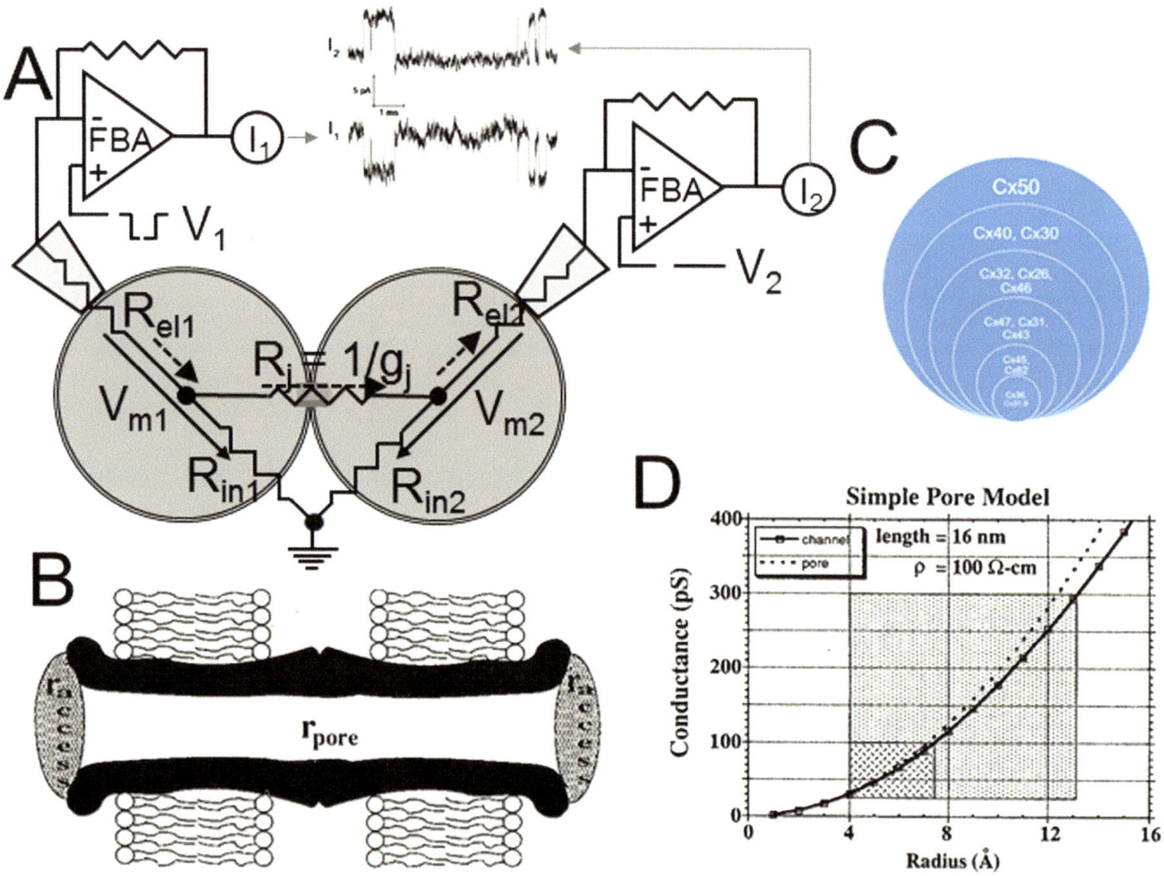

FIGURE 8.4 Connexin gap junction channel currents and conductances. (A) Illustration of the two cell (dual) whole cell patch-clamp configuration for directly recording gap junctional currents (I_j, center arrow). An example of an "equal amplitude and opposite polarity" gap junction channel current is shown in the inset between the two feedback amplifiers (FBA, patch-clamp amplifiers). The membrane voltages (V_{m1} and V_{m2}) are determined by the nonjunctional membrane current flow (solid lines, small arrowheads) across the input resistance (R_{in}) of each cell and V_m will differ from the command potentials (V_1 and V_2) for each cell by the product of the total whole cell current (I_1 and I_2 = sum of nonjunctional and junctional (dashed lines, large arrowheads) currents) flowing across the resistance of each patch electrode (R_{el1} and R_{el2}). The junctional conductance (g_j) or resistance (R_j) is directly measured by stepping V_1 to a different voltage from a common holding potential (=V_2). (B) Diagram of the simple aqueous pore model for a gap junction channel wherein the channel conductance (γ_j) is determined by the reciprocal of the sums of the cytoplasmic access (r_{access}) and pore (r_{pore}) resistances. (C) Diagrammatic illustration of the known gap junction channel conductances for homomeric, homotypic gap junctions formed by a single connexin from the smallest γ_j channels (Cx31.9, Cx36; 10–15 pS) to the largest γ_j channel (Cx50, 300 pS). (D) Graph of the predicted γ_j for gap junction channels with a limiting pore radius in Angstroms (Å) and fixed length of 16 nm and resistivity (ρ) of 100 Ω·cm assuming a right cylindrical pore $\left[\frac{1}{\gamma_j} = r_{pore} + r_{access} = \left(1 + \frac{\pi a}{2}\right) \cdot \left(\frac{\rho}{\pi a^2}\right)\right]$. An observed 15 pS channel would have a radius <3 Å and a 300 pS would require a radius of 13 Å according to this simple aqueous pore model, but observed permeabilities to fluorescent dyes did not correlate with observed γ_j values for Cx37, Cx40, Cx43, Cx45, suggesting that there are electrostatic interactions between the pore and permeant ions and molecules. *Panels B and D are from Veenstra et al. (1995).*

Despite possessing similar pore diameters of ≈13 ± 1 Å, Cx40 is selectively blocked by the polyamine spermine (valence = +4), indicative of selective interactions between large cations and gap junction channel pores (Musa & Veenstra, 2003). Molecular dynamic simulations based on the published crystal structure of the Cx26 gap junction channel suggest that the localized permeability barrier to cations and anions reside in the vicinity of K41 and E42 within the narrowest region of the pore near the M1-E1 interface plus A49 (E49 in Cx30) and D2 located on the NT domain, respectively (Maeda et al., 2009; Kwon et al., 2011; Zonta et al., 2012).

4.2 Gap junction inhibitors

We've already learned that the half-life of connexins is surprisingly short, only 2–5 h, and that gap junctions change their morphological appearance within minutes, lending credence to the concept that gap junction channels are dynamic structures and not static aqueous conduits for intercellular exchange of soluble molecules of <1 kDa in mass and 12–14 Å in diameter. Numerous chemical agents have been identified over the decades that are capable of inhibiting gap junction function (Harris, 2001). The first factors to influence electrical gap junction communication were cytoplasmic Ca^{2+} ($[Ca^{2+}]_i$), pH (-log[H^+]), and transjunctional voltage (V_j) (Rose & Loewenstein, 1975; Spray et al., 1979; Turin & Warner, 1977, 1980). This "voltage" and "chemical" sensitive gating of gap junctions is now thought to occur via distinct "fast" and "slow" gating mechanisms, though V_j gating also exhibits "slow" gating at higher voltages (Bukauskas & Peracchia, 1997; Bukauskas & Verselis, 2004). Alkanols, the alcohols of long chain alkanes like heptanol and octanol, were also discovered to reversibly uncouple electrical synapses, i.e., gap junctions, thus adding lipophilic compounds (lipophiles) to the growing list of gap junction modifiers (Johnston et al., 1980). Saturated and monounsaturated long chain fatty acids like myristic, oleic (and its amide, oleamide), and arachidonic acids also cause gap junction uncoupling (Aylsworth et al., 1987; Burt et al., 1991; Giaume et al., 1989). 18α- and 18β-glycyrrhetinic acid, the active ingredient in licorice, and the synthetic derivative carbenoxolone (CBX) also inhibit gap junction communication when applied in micromolar amounts (Davidson et al., 1986). Pharmacological inhibition of gap junction communication has also been achieved with volatile anesthetics like halothane, chloride channel blockers like niflumic and flufenamic acids, quinine, 2-aminoethoxydiphenyl borate (2-APB), cyclodextrins, protonated aminosulfonate pH buffers like MES and HEPES, large tetraalkylammonium (TAA^+) cations, and polyamines like spermine (Bai et al., 2006; Bevans & Harris, 1999; Burt & Spray, 1989; Eskandari et al., 2002; Harks et al., 2003; Locke et al., 2004; Musa et al., 2001; Musa & Veenstra, 2003; Srinivas et al., 2001). Some of these compounds (TAA^+, spermine, cyclodextrans) only block gap junction conductance when injected intracellularly and some block only certain connexin-specific gap junctions (e.g., quinine, spermine, 2-APB).

A significant limitation to the study of gap junction communication in vivo or ex vivo using intact tissues or organs has been the lack of specific blockers of connexin gap junction channels. Long considered the gold standard gap junction inhibitors, heptanol and octanol also affect the activity of voltage-gated Na^+ channels, nicotinic acetylcholine (ACh) receptor channels, and intracellular Ca^{2+} oscillations in pancreatic acini, presumably by increasing the membrane fluidity or direct channel blockade (Deutsch et al., 1995; Goldstein, 1984; Murrell & Haydon, 1991; Nelson & Makielski, 1991; Swenson & Narahashi, 1980). Carbenoxolone (CBX, 50−100 μM) may be useful as a relatively selective gap junction inhibitor in cardiac tissues, but reportedly has effects on ACh, $GABA_A$, and AMPA-induced endothelial hyperpolarizations and neuronal inhibitory or excitatory postsynaptic potentials (IPSCs or EPSCs) (de Groot et al., 2003; Tare et al., 2002; Tovar et al., 2009). CBX, glycyrrhetinic acid, and its metabolites also affect prostaglandin and cortisol metabolism by inhibiting metabolic enzymes and CBX inhibits pannexin1 channels at lower concentrations (5 μM) than Cx43 gap junctions (Bruzzone et al., 2005; Peskar, 1980; Stewart et al., 1987). Another approach toward the development of gap junction inhibitors was the design of connexin-mimetic peptides that initially targeted extracellular loop (E1 and E2) sequences of Cx43 (Boitano & Evans, 2000; Evans & Boitano, 2001). Gap 26 and Gap 27 peptides reversibly inhibited gap junction formation when applied extracellularly and are marketed to this day as gap junction inhibitors. However, application of the E1 Cx43 Gap 26 peptide to a chimeric $Cx32E_143$ connexin hemichannel (Cx32 connexon with Cx43 E1 domain inserted) in *Xenopus* oocytes failed to inhibit channel activity and has no effect on existing gap junctions (Wang et al., 2007). Complicating matters further, Gap 27 partially inhibits pannexin1 (Panx1) channels and a Panx1-mimetic peptide (^{10}panx1) partially inhibits both Panx1 channels and Cx46 hemichannels, suggesting that these connexin- and pannexin-mimetic peptides are not as specific in their action as originally reported (Dahl, 2007). Gap 19, a recently developed Cx43-mimetic peptide directed against a CL sequence, reportedly inhibits Cx43 hemichannel activity without affecting gap junction coupling or Panx1 channel activity (Wang et al., 2013). A different approach was taken to identify Cx43 CT domain-binding peptides containing an RxP motif that prevented acidosis-induced and octanol-induced uncoupling of full-length, but not the M257-truncated, Cx43 (Shibayama et al., 2006; Verma et al., 2009). Since the RXP-E peptides prevent Cx43 gap junction channel closure, they could be considered Cx43 gap junction agonists. Rotigaptide (ZP123), a synthetic peptide based on a naturally occurring atrial antiarrhythmic peptide (AAP10), preserves myocardial conduction during and after ischemia by enhancing cardiac ventricular (Cx43) gap junction conductance by phosphorylation-dependent mechanisms (Dhein et al., 2010; Haugan et al., 2005; Kjølbye et al., 2003; Lin et al., 2008).

In summary, a universal and specific chemical inhibitor of gap junction communication does not exist, though 50−100 μM CBX may be the best alternative at this time. Synthetic (partial) gap junction agonists have been identified, but none of them have developed into clinical antiarrhythmic therapeutics. Numerous physiological factors such as intracellular Ca^{2+}, pH, voltage, and posttranslational modifications also regulate gap junction communication.

4.3 Gating of gap junctions by pH and Ca^{2+}

There remains a controversy about the relative ability of Ca^{2+} ions and H^+ (protons) to close gap junction channels. Elevations in $[Ca^{2+}]_i$ and cytoplasmic acidosis both cause gap junction uncoupling, but whether they do so independently or synergistically is still an unanswered question (Rose & Loewenstein, 1975; Turin & Warner, 1980; Spray et al., 1981;

Peracchia et al., 1996; reviewed in Peracchia, 2004). Since Ca^{2+} and H^+ influence each other's binding to acidic proteins, it is difficult to assess the action of one of these factors without influencing the other (Vaughan-Jones et al., 1983). Calmodulin (CaM), a Ca^{2+}-binding protein that modulates the activity of numerous enzymes and ion channels, binds to connexin gap junctions and reportedly mediates the gating of gap junctions by $[Ca^{2+}]_i$ (Kovalevskaya et al., 2013; Simon et al., 2015; Török et al., 1997; Unwin & Ennis, 1984; Van Eldik et al., 1985; Welsh et al., 1982; Zhou et al., 2007; Zou et al., 2014). CaM inhibition by calmidazolium (CDZ), W-7, or antisense oligonucleotides prevents Ca^{2+}-induced uncoupling and deletion of a Cx32 CaM binding site increased intracellular retention of Cx32 and limited connexon formation and plasmalemmal membrane targeting, suggesting a role of CaM in connexin oligomerization and trafficking in addition to gap junction channel gating (Ahmad et al., 2001; Cole & Garfield, 1988; Peracchia, 2004; Peracchia et al., 1996; Xu et al., 2012). The Ca^{2+}/CaM gating of gap junction channels, eluded to by earlier structural studies under functional and uncoupling conditions, results from a reduction in open channel probability (P_o) that occurs within seconds (Chen et al., 2011; Xu et al., 2012). The precise molecular mechanism by which Ca^{2+}-bound CaM closes gap junction channels still isn't fully understood and it could result from large-scale connexin/connexon conformational changes like those proposed by Unwin and Ennis, connexin transdomain interactions, or direct pore blockade by CaM binding to the cytoplasmic mouth of the gap junction channel pore (i.e., steric hindrance, "cork" model; Unwin & Ennis, 1984; Dodd et al., 2008; Zou et al., 2014). The Cx43 pH gating mechanism apparently involves an intramolecular interaction between a CL (L2) domain and the distal CT domain, as deduced from a pH-insensitive Cx43 residues 258–382 (M257) deletion mutant that regained its pH-dependent gating when the CT domain peptide was coexpressed with the Cx43 M257 mutant protein (Morley et al., 1996). The proposed "ball-and-chain" particle-receptor Cx43 pH gating model, based on the N-type inactivation of voltage-gated K^+ channels, describes a pH-induced increase in α-helicity of CL (receptor) residues 119–144, which increases the binding of the CT domain (particle) to induce the pH-dependent closure of the Cx43 gap junction channel (Duffy et al., 2002; Hoshi et al., 1990). The key pH-dependent helical residues of 122–127 and 139–143 are juxtaposed with the basic-x-basic-1-5-10 (BxB1xxx5xBxx10) primary Cx43 CaM binding motif residues 144–156, again lending controversy to the independent or synergistic actions of these two intracellular cationic gating factors on gap junctions (Duffy et al., 2002; Zhou et al., 2007). Furthermore, the location of the Cx43 pH gating and CaM binding motifs don't directly translate to Cx32, suggesting that a universal model for the gating of connexin-specific gap junctions by intracellular pH or Ca^{2+}/CaM doesn't necessarily exist (Dodd et al., 2008; Wang & Peracchia, 1997).

4.4 Modulation of gap junctions by posttranslational modifications

Connexin phosphorylation by serine/threonine and tyrosine protein kinases is implicated in the assembly and degradation of connexin gap junctions, especially Cx43 (Dunn & Lampe, 2014; Lin et al., 2001; Musil & Goodenough, 1991, 1993; Solan & Lampe, 2005, 2009; Warn-Cramer et al., 1996). Nearly all connexins possess phosphorylation sites with Cx43 being the most studied connexin (Solan & Lampe, 2005). The CT domain of Cx43 contains at least 19 phosphorylation sites, so the ability of phosphorylation to regulate Cx43-mediated gap junction communication is extensive. In addition to the role of CK1 phosphorylation of S325/328/330 in promoting the assembly of Cx43 gap junctions, phosphorylation of S364/365 is associated with the cAMP-dependent upregulation of gap junction communication and S365 phosphorylation is proposed to serve as a "gate-keeper" by preventing downregulating dephosphorylating/phosphorylating events at these and other Cx43 phosphorylation sites from occurring (Solan et al., 2007; Solan & Lampe, 2005). Hypoxia and ischemia promote dephosphorylation of Cx43 at these CK1 and presumed PKA sites, leading to the downregulatory phosphorylation of S368 by PKC and S255/279/282 by MAPK (Beardslee et al., 1998; Lampe et al., 2000; Solan et al., 2007; Warn-Cramer et al., 1996). PKC-mediated phosphorylation also decreases the Cx43 single gap junction channel conductance (γ_j) (Moreno et al., 1994; Kwak & Jongsma, 1996). Cx43 gap junction communication is also downregulated during mitosis, a process mediated by p34(cdc2)/cyclin B kinase phosphorylation of S255 and/or S262 (Kanemitsu et al., 1998; Lampe et al., 1998). Tyrosine phosphorylation of Cx43 Y247 and Y265 by Src kinases also downregulates gap junction communication (Lin et al., 2001). While not affecting gating per se, Akt/PKB phosphorylation of S373 increases Cx43 gap junction size by affecting ZO-1 binding to Cx43 (Dunn & Lampe, 2014). To summarize, Cx43 gap junction function is favorably affected by CK1, PKA, and Akt/PKB serine phosphorylation, while cdc2(p34)/cyclin B, PKC, MAPK, and Src kinases downregulate Cx43 gap junction communication.

For every kinase, there is a phosphatase and protein phosphatase 1 (PP1), PP2A, and PP2B (calcineurin) have all been implicated in the dephosphorylation of Cx43 by various and sometimes conflicting reports. One report implicates the Ca^{2+}-dependent phosphatase calcineurin (PP2B) in the hypoxia-induced dephosphorylation of Cx43, while another states that calcineurin acts indirectly via activation of PP1 (Jabr et al., 2016; Li & Nagy, 2000). Another study indicates that PP2A is associated with Cx43 dephosphorylation in heart failure, whereas no effect of PP2A inhibition was detected in the other two studies (Ai & Pogwizd, 2005). Perhaps different phosphatases are activated under different conditions, i.e., Ca^{2+} overload, hypoxia, and heart failure, and affect different Cx43 serine phosphorylation sites. PTEN (phosphatase and tensin homolog), a tyrosine phosphatase, binds to the CT of Cx43 and counteracts the oncogenic activity of cellular Src kinase (c-Src), which may help explain the antioncogenic activity of Cx43 (González-Sánchez et al., 2016; see Connexin-associated Diseases).

Within the last decade, a second form of protein posttranslational modification has become recognized as another major modulator of protein function: N^{ε}-lysine acetylation (Kouzarides, 2000). At least 1750 acetylated proteins affecting diverse cellular processes (e.g., cell signaling and metabolism) have been identified (Choudhary et al., 2009; Menzies et al., 2016). Cx26 was the first acetylated connexin to be identified and Cx43 is now also known to be acetylated (Locke et al., 2009). Acetylated Cx43 acetylation was detected in the hearts of muscular dystrophic (mdx) mice and was associated with the dissociation (or "lateralization") of Cx43 from the cardiac intercalated disc and a presumed loss of gap junction coupling (Colussi et al., 2011). The mdx mouse exhibits increased activity of histone acetyltransferases (HATs) like p300/CBP-associated factor (PCAF) which leads to a hyperacetylated state. Inhibition of histone deacetylases (HDACs), the enzymes that deacetylate proteins, was later found to reduce Cx43 expression, gap junction area, and functional coupling in normal cardiac ventricular myocytes, providing definitive proof that acetylation of Cx43 downregulates function by transcriptional and possibly posttranslational mechanisms (Xu et al., 2013). The precise acetylation sites on Cx43 and the functional consequences thereof remain to be determined. Nitrosylation of Cx43 CT cysteine (C) residues, C271 in particular, increases the activity of hemichannels and gap junction channels, providing a link between the cellular reduction—oxidation (redox) state and plasmalemmal hemichannel and gap junction channel permeability (Retamal, 2014; Retamal et al., 2006; Straub et al., 2011).

5. Connexin-associated diseases

5.1 Connexin-linked diseases

One way to gain insight into the functional importance of connexin-specific gap junctions is to determine the consequences of their loss-of-function either by genetic deletion (knockout, KO) of a particular connexin gene (see Table 8.1) or naturally occurring mutations that affect their function, either gain-of-function (GOF) or loss-of-function (LOF). One of the earliest links of gap junctions to a human disease was the observation of a loss of electrical communication and paucity of nexuses between cancer cells (Loewenstein & Kanno, 1967; McNutt et al., 1971). Thus far, Cx26, Cx30, Cx32, Cx37, Cx43, and Cx46 have been linked to various forms of cancer, including breast, cervical, gastric, lung, prostate, gliomas, and neuroblastomas (Jiang & Penuela, 2016). Germline deletion of Cx26, Cx32, and Cx43 (murine *Gjb2*, *Gjb1*, and *Gja1* genes, respectively) increases breast, liver, and lung tumor formation and breast cancer metastasis (Aasen et al., 2016; Temme et al., 1997). In general, the loss of gap junction intercellular communication (GJIC) promotes cell proliferation and metastasis, while reestablishment of GJIC suppresses tumor growth, invasiveness, and promotes apoptosis (programmed cell death) via a "bystander effect" wherein cytotoxic signals are transferred to cancerous cells via gap junctions (Defamie et al., 2014; Denning & Pitts, 1997). Cx43 also serves as a tumor suppressor by promoting the binding of PTEN and Csk (c-terminal Src kinase) to c-Src, which inactivates c-Src by dephosphorylation of Y416 and phosphorylation of Y527, respectively (González-Sánchez et al., 2016).

The bystander effect of GJIC may also be detrimental to tissue survival by propagating cell death and increasing infarct size during ischemia and ischemia/reperfusion injury in the brain and heart (Lin et al., 1998; Garcia-Dorado et al., 1997). The myocardial "healing-over" process is essential to preserving cardiac function by uncoupling the damaged and dying cells from the adjacent healthy cells, thus preventing their cell death via the bystander effect. Another form of cardio- and neuro-protection exists in the form of ischemic preconditioning (IPC), wherein short intervals of ischemia "precondition" the cardiac and neural tissues to limit infarct size in response to subsequent prolonged ischemic episodes (Murray et al., 1986; Liu et al., 1992; Schulz et al., 2015). Cx43 is required for IPC, as evidenced by the absence of IPC in Cx43 heterozygous mice (Schwanke et al., 2002). IPC delays gap junction uncoupling, increases Cx43 phosphorylation and lateralization, prevents hypoxia-induced Cx43 dephosphorylation, and reduces gap junction permeability by a process involving PKCε and ERK1/2 MAPK phosphorylation of Cx43 (Ping et al., 1997; Jain et al., 2003; Naitoh et al., 2006; Hund et al., 2007; Srisakuldee et al., 2009; Miura et al., 2010; Schulz et al., 2015). PKC phosphorylation of S368 is known to decrease Cx43 gap junction channel conductance and permeability, but there is conflicting evidence about the effect of IPC on Cx43 phospho-S368 content (Ek-Vitorin et al., 2006; Hund et al., 2007; Kwak & Jongsma, 1996; Moreno et al., 1994; Naitoh et al., 2009). The induction of IPC also requires the activity of the mitochondrial K_{ATP} (mK_{ATP}) channel, and working models for IPC induction indicate that activity of mK_{ATP} and/or Cx43 hemichannels located in the inner mitochondrial membrane (IMM) is required for K^+-flux-dependent activation of a reactive oxygen species (ROS) signaling cascade that is required for induction of IPC (Miura et al., 2010; Schulz et al., 2015). Cx43 is imported into mitochondria via a heat shock protein 90 (Hsp90)-dependent translocase of the outer membrane 20 (Tom20) pathway, and Hsp90 inhibition by geldanamycin prevented drug-induced IPC (Rodriguez-Sinovas et al., 2006). IPC also increases the nitrosylation of subsarcolemmal mitochondrial Cx43, resulting in increased IMM permeability and ROS presumably by increased Cx43 hemichannel activity (Soetkamp et al., 2014). Thus, mitochondrial Cx43 appears to be an effector of IPC induction, whereas junctional Cx43 is a target of IPC, thereby limiting gap junctional molecular permeability and passage of apoptotic or Ca^{2+}-overload cell death signals and ultimately limiting the infarct size after prolonged ischemia and reperfusion.

X-linked Charcot-Marie-Tooth disease (CMTX) was the first human disease linked to connexin mutations, specifically Cx32 (Bergoffen et al., 1993). Since this initial discovery, mutations in at least 11 of the 20 human connexin proteins (Cx23 protein has not been detected) have been linked to human diseases (Fig. 8.5). Today more than 300 *GJB1* (Cx32) mutations have been identified in association with CMTX1 and a summary of these mutations with references can be found at the Inherited Peripheral Neuropathies (IPN) Mutation Database (http://www.molgen.vib-ua.be/CMTMutations/Mutations/MutByGene.cfm; GJB1). The loss of Cx32 function results in peripheral nerve demyelination in Cx32-null mice since Cx32 is expressed in the myelin-producing Schwann cells (Scherer et al., 1998). The demyelination is thought to occur because of the loss of "reflexive" gap junctions that form a radial pathway for the transport of nutrients between the layers of the myelin sheath (Bone et al., 1997; Herr, 1976). Cx32 is also expressed in some central nervous system (CNS) oligodendrocytes, yet no CNS demyelination was found, at least not until the *GJC2* (formerly *GJA12*, Cx47) gene was also deleted (Menichella et al., 2003). Mutations in Cx47 were subsequently found in association with Pelizaeus–Merzbacher-like disease (PMLD), an inherited early onset form of hypomyelinating leukoencephalopathy (Uhlenberg et al., 2004). Cx47 mutations are now also linked to spastic paraplegia 44 and hereditary lymphedema type 1C (Orthmann-Murphy et al., 2009; Ferrell et al., 2010; see http://www.ncbi.nlm.nih.gov/omim/608803). There is one report of a Cx31 (*GJB3*) mutation causing a peripheral neuropathy in addition to mild hearing impairment (López-Bigas et al., 2001).

Autosomal dominant forms of nonsyndromic hereditary deafness (DFNA#) are associated with mutations in the *GJB2* (Cx26; DFNA3A), *GJB6* (Cx30; DFNA3B), and *GJB3* (Cx31; DFNA2B) genes (Grifa et al., 1999; Kelsell et al., 1997; Xia et al., 1998). Autosomal recessive forms of nonsyndromic hereditary deafness (DFNB#) include mutations in the *GJB2* (DFNB1A), *GJB6* (DFNB1B), and digenic dominant forms of DFN1A involving the *GJB2/GJB3* and *GJB2/GJB6* genes (Kelsell et al., 1997; Zelante et al., 1997; Liu et al., 2000; del Castillo et al., 2002). Two mutations in the *GJA1* (Cx43) gene have also been associated with autosomal recessive deafness and hearing impairment is sometimes associated with CMTX (Bone et al., 1997; Liu et al., 2001). Nonsyndromic deafness means that the hearing loss is not associated with any

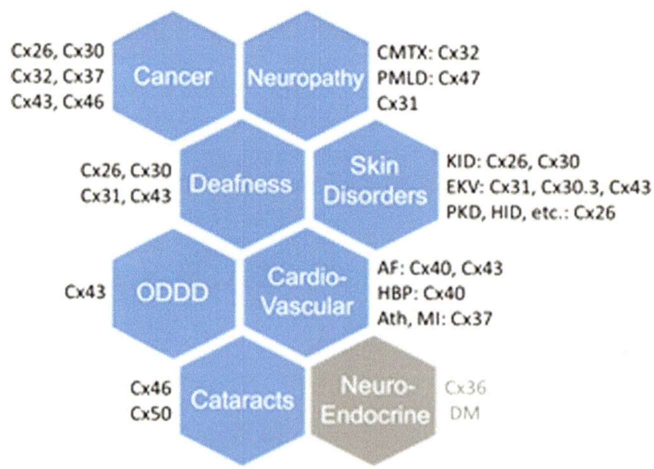

FIGURE 8.5 Connexin-associated diseases. The downregulation of connexin gap junction-mediated intercellular communication is associated with numerous forms of cancer. Connexin mutations (connexinopathies) resulting in loss-of-function or aberrant hemichannel activity have been linked to neuropathies like X-linked Charcot–Marie–Tooth disease (CMTX) and Pelizaeus–Merzbacher-like disease (PMLD); autosomal dominant (DFNA#) and recessive (DFNB#) forms of nonsyndromic and syndromic hereditary deafness; skin disorders like Keratitis-ichthyosis-deafness syndrome (KID), erythrokeratodermia variabilis (EKV), palmoplantar keratoderma with deafness (PKD), Vohwinkel syndrome, Hystrix-like ichthyosis with deafness (HID), and Bart-Pumphrey syndrome; oculodentodigital dysplasia (ODDD), cardiovascular disorders like atrial fibrillation (AF), hypertension (high blood pressure, HBP), myocardial infarction (MI), and atherosclerosis (Ath); cataracts; and neuroendocrine disorders like diabetes mellitus (DM).

other symptoms and syndromic deafness means that the hearing loss is associated with other symptoms like peripheral neuropathies or skin diseases. A summary of connexin deafness mutations can be found on the Connexin-deafness homepage (http://davinci.crg.es/deafness/). Cx26 mutations are known to cause skin disorders including Keratitis-ichthyosis-deafness syndrome (KID), palmoplantar keratoderma with deafness (PKD), Vohwinkel syndrome, Hystrix-like ichthyosis with deafness (HID), and Bart-Pumphrey syndrome (Maestrini et al., 1999; Heathcote et al., 2000; Richard et al., 2002, 2004; van Geel et al., 2002). A Cx30 (*GJB6*) mutation previously associated with Clouston syndrome (nail and hair dysplasia with palmar hyperkeratosis) was found in a patient with KID syndrome (Jan et al., 2004; Smith et al., 2002). Another hyperkeratotic skin disease, erythrokeratodermia variabilis (EKV), is caused by mutations in the *GJB3* (Cx31), *GJB4* (Cx30.3), and *GJA1* (Cx43) genes (Boyden et al., 2015; Macari et al., 2000; Richard et al., 1998).

Cx43 (*GJA1*) mutations were only rarely associated with the sensorineural hearing loss and skin disorders predominantly associated with many of the β-group connexins, but is strongly associated with a different multisymptomatic disorder called oculodentodigital dysplasia (ODDD, Paznekas et al., 2003). Characterized by microphthalmia (small eyes), small nose, microdontia and enamel hypoplasia, fourth and fifth finger syndactyly (webbed fingers), and numerous other skeletal-facial and neurological phenotypes, at least 62 *GJA1*/Cx43 mutations are known to occur in patients with ODDD (Paznekas et al., 2009). The vast majority of Cx43 mutations in ODDD are autosomal dominant, with only a couple of known autosomal recessive Cx43 ODDD mutations. Although Cx43 is the predominant cardiac gap junction protein, cardiac defects are rare, occurring in only 3% of human ODDD patients, 10−20 times less frequent than the neurological and skeletal-facial disorders observed in these patients with Cx43 mutations. However, three murine models of Cx43 ODDD mutations, G60S, I130T, an G138R, exhibit an increased susceptibility to cardiac arrhythmias (Dobrowolski et al., 2008; Kalcheva et al., 2007; Tuomi et al., 2011). Perhaps the incidence of cardiac abnormalities in patients with ODDD isn't higher because severe defects likely result in embryonic or neonatal lethality or sudden cardiac death. Deletion of the *Gja1* gene in mice results in right ventricular cardiac outflow tract defects and neonatal lethality (Reaume et al., 1995). Homozygous knock-in of a truncated Cx43K258stop *Gja1* gene (Cx43-M257 protein) produces an epidermal barrier defect that usually results in neonatal lethality and sterility of any surviving females (Maass et al., 2004). Thus, expression of Cx43 is linked to heart development and folliculogenesis. Consistent with a role in cardiogenesis, mutations in Cx43 were identified in patients with hypoplastic left heart syndrome and other lateral heart defects (Britz-Cunningham et al., 1995; Dasgupta et al., 2001). Cx43, being the most abundantly expressed connexin in the human body, is linked to a wide variety of cellular functions too numerous to mention (see http://www.uniprot.org/uniprot/P17302), but relatively few inheritable genetic diseases, primarily limited to those already mentioned (see https://www.omim.org/entry/121014).

Cardiac arrhythmias are associated with promoter region single nucleotide polymorphisms (SNPs) and mutations in the *GJA5* (Cx40) gene. Cx40 mutations are found in patients with familial atrial fibrillation type 11 (ATFB11) and digenic atrial standstill involving the cardiac sodium channel (*SCN5A*, $Na_V1.5$), respectively (Gollob et al., 2006; Groenewegen et al., 2003; Shi et al., 2013; Wirka et al., 2011; Yang et al., 2010). There is one report of a somatic *GJA1* mutation occurring in a patient with lone atrial fibrillation that resulted in an aberrant form of Cx43 (Thibodeau et al., 2010). Recently, a novel Cx40 mutation was found in a patient with progressive familial heart block type 1, indicative of an essential role for Cx40 in cardiac conduction and consistent with findings obtained in Cx40-null mice (Kirchoff et al., 1998; Simon et al., 1998; Makita et al., 2012). Later it was determined that Cx40-null mice also exhibit hypertension owing to ectopic expression of renin secreting cells and disruption of the regulatory feedback mechanism of blood pressure on renin secretion (Krattinger et al., 2007; Kurtz et al., 2007; Wagner et al., 2007). *GJA5* promoter region SNPs are also associated with increased susceptibility to hypertension (Firouzi et al., 2006). Cx40 and Cx43 are expressed in the vasculature along with Cx37, yet only polymorphisms in the *GJA4* gene have been associated with increased risk for atherosclerotic coronary artery disease and myocardial infarction (Boerma et al., 1999; Van Rijen et al., 1997; Yamada et al., 2002). Mutations in and genetic deletion of the Cx46 (*GJA3*) and Cx50 (*GJA8*) genes, two α-group connexins expressed in the lens of the eye, cause autosomal dominant cataracts (Gong et al., 1997; Mackay et al., 1999; Shiels et al., 1998; White et al., 1998). The functional loss of Cx50 also causes microphthalmia, a function that is unique to Cx50 since knock-in of Cx46 prevents cataract formation but does not reverse the adverse effects of Cx50 deletion on lens growth (White, 2002).

These "connexinopathies", human diseases linked to connexin mutations, exemplify the vital functions that gap junctions perform in the development, growth, functional integration, and homeostasis of the tissues in which they are expressed throughout the human body (García et al., 2016; Laird, 2008).[1]

1. Srinivas, M., Verselis, V.K., & White, T.W. (2018 Jan). Human diseases associated with connexin mutations. *Biochimica et Biophysica Acta - Biomembranes*, *1860*(1), 192−201. https://doi.org/10.1016/j.bbamem.2017.04.024. Epub 2017 Apr 27. PMID: 28457858; PMCID: PMC5659969.

5.2 Disease mechanisms

As we have just learned, connexin mutations and models of connexin-null (knockout) mice provide valuable insight into the essential roles that gap junctions play in their respective tissues and organs. But what are the mechanisms by which these numerous connexin mutations cause the pathophysiological condition of the disease they are associated with and how can scientists discern these mechanisms? The vast majority of disease-linked connexin mutations result in a loss-of-function, as determined by exogenous expression of the mutated connexins in *Xenopus* oocytes, N2a, or HeLa cells and electrophysiological analysis of their inability to induce intercellular electrical coupling (García et al., 2016; Laird, 2008; Paznekas et al., 2009). The resulting loss of gap junction coupling impairs the cellular/tissue homeostasis, eventually causing cell death and the phenotypic symptoms of the disease based on the organ wherein the specific connexin is expressed. The mechanistic basis for the lack of functional coupling caused by a mutated connexin may vary from abnormal protein trafficking and impaired ability to form gap junction plaques, dominant-negative functional effects resulting in nonfunctional gap junction channels, reduced connexin expression, or accelerated connexin degradation (Bone et al., 1997; Deschênes et al., 1997; Gemel et al., 2014; Kabzińska et al., 2011; Laird, 2008; Omori et al., 1996; Paznekas et al., 2009). Those mutant disease connexins that are capable of forming functional gap junctions typically result in reduced coupling and molecular permeability that negatively impact cellular homeostasis (Beltramello et al., 2005; Oh et al., 1997).

There is one type of gain-of-function connexin disease mutation, aberrant hemichannel formation. As previously defined, a connexin hemichannel refers to the opening of an unopposed (undocked) connexon hemi-gap junction channel located within the plasma membrane. The first evidence that connexins could form hemichannels was inadvertently discovered while expressing the newly cloned Cx46 in *Xenopus* oocytes (Paul et al., 1991). Unlike the previously cloned Cx32 and Cx43 that readily formed gap junctions when exogenously expressed in paired *Xenopus* oocytes, expression of Cx46 caused oocyte lysis, a phenomenon resulting from the opening of unpaired Cx46 connexons or hemichannels (Ebihara & Steiner, 1993). Connexin hemichannel activity is enhanced by membrane depolarization and low extracellular $[Ca^{2+}]$ ($[Ca^{2+}]_o$) levels, and several connexin disease mutations are known to increase the opening of these mutant hemichannels by interfering with the voltage-gating or Ca^{2+}-gating mechanisms (Abrams et al., 2002; Gerido et al., 2007; Sánchez et al., 2010). For instance, the Cx26 A40V KID mutation exhibits impaired regulation of hemichannel activity by $[Ca^{2+}]_o$, whereas the $[Ca^{2+}]_o$ modulation of G45E hemichannel activity is only mildly affected (Sánchez et al., 2010). However, the Cx26 G45E KID mutant hemichannel exhibits increased permeability to $[Ca^{2+}]_o$, a likely cause of cell death, and the selective expression of Cx26-G45E in keratinocytes in an inducible mouse model of this KID mutation recapitulated the symptoms of the disease (hyperkeratosis, scales, hair loss, etc.), confirming that the aberrant hemichannel permeability manifests the disease phenotype (Mese et al., 2011). The plethora of human diseases caused by loss-of-function or aberrant hemichannel mutations in more than half of the known connexins provides evidence of the vital role that these intercellular channels play in the routine maintenance and function of the specific tissues where each connexin is expressed.

6. Gap junctions are not alone

This section began with the discovery of electrical coupling within tissues being associated with the presence of an intercellular structure called the gap junction, formerly referred to as the nexus. What wasn't previously mentioned is that gap junctions are typically found in an intercellular junctional complex consisting of adhesion junctions termed intermediary junctions (*zonula* or *fascia adherens*) and desmosomes (*macula adherens*), and sometimes with occluding junctions called tight junctions or *zonula occludens* (Farquhar & Palade, 1963) (see Chapter 7).

6.1 Tight junctions—Zonula occludens

In polarized (barrier and transport) epithelia, the tight junctions occupy the most apical (luminal) localization and are characterized by the occlusion of the extracellular space by the formation of a continuous circumferential adhesive belt (intermediate line) that occludes the paracellular (between the cells) transport route, thus dividing the epithelial cell into apical and basolateral compartments. The intermediate (fusion) line forms down the middle of an approximately 90 Å intercellular gap between the cell membranes and early on there were opposing hypotheses about the protein or lipid composition of these tight junction adhesion belts (reviewed in Tsukita et al., 2001). Barrier "tight" epithelia (e.g., urinary bladder, blood–brain barrier) contain numerous anastomosing junctional strands (belts), whereas "leaky" epithelia (e.g., renal proximal convoluted tubule) may contain only one junctional strand separating the apical and basolateral compartments of the epithelium (Brightman & Reese, 1969; Claude & Goodenough, 1973). In TEM images, the tight junction appears relatively devoid of a cytoplasmic matrix of proteins, unlike the *zonula adherens* and *macula adherens* junctions (Fig. 8.6). Proving the protein composition of the tight junction (zipper-like) adhesion belt proved difficult as numerous tight junction-associated cytoplasmic peripheral membrane proteins, beginning with zonula occludens 1 (ZO-1) were identified prior to the identification of any integral membrane proteins (Stevenson et al., 1986; Anderson et al., 1988, reviewed in Schneeberger & Lynch, 2004). The first cloned tight junction associated integral membrane protein, occludin-1, was thought to be the principal tight junction protein until its germline deletion failed to fully disrupt the tight junction structure in occludin-1 null mice (Furuse et al., 1993; Saitou et al., 2000; Schneeberger & Lynch, 2004). This prompted further studies and the same group of investigators later cloned a second class of tight junction integral membrane proteins known as the claudins (Furuse et al., 1998). Though the claudins and occludins share no sequence identity, their membrane topologies are similar, consisting of four transmembrane domains with two extracellular loops, one cytoplasmic loop, and cytoplasmic amino- and carboxyl-termini, just like the connexins (Schneeberger & Lynch, 2004). Today, tight junctions are known to contain three distinct integral membrane proteins: occludins (2 isoforms), claudins (24 isoforms), and a single

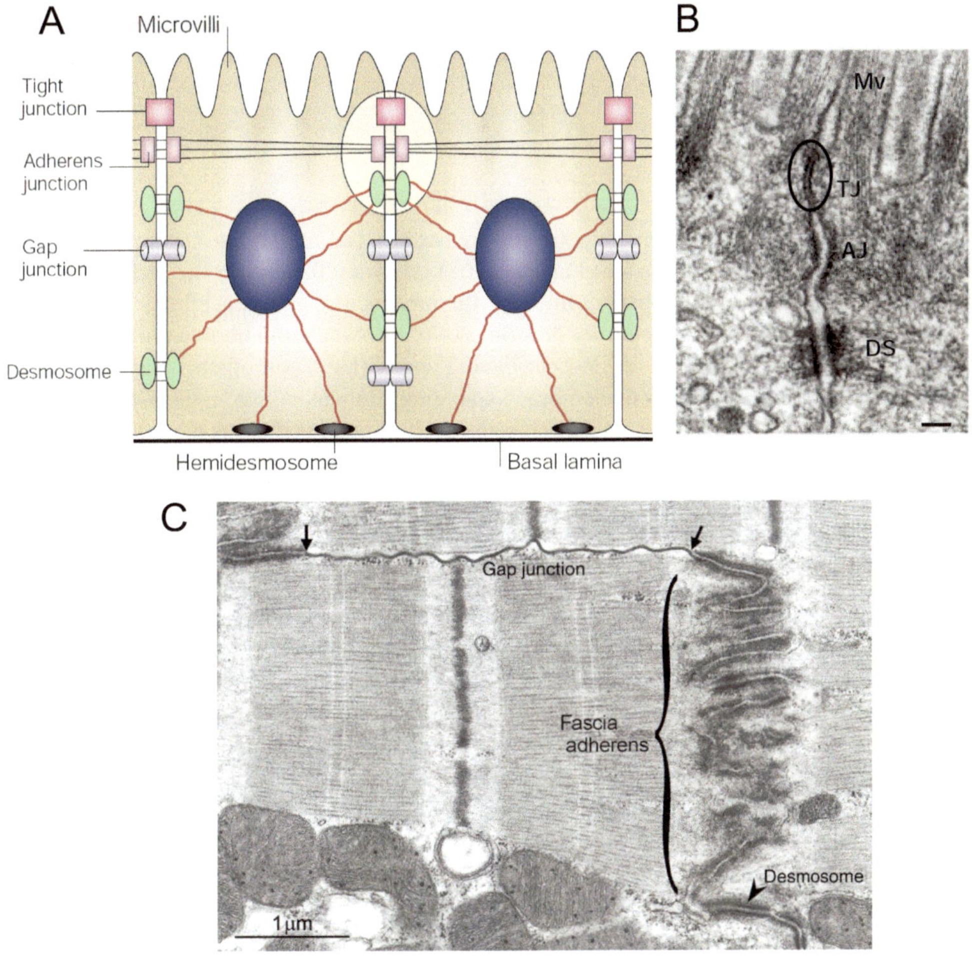

FIGURE 8.6 Intercellular junctional complexes. (A) Diagram illustrating the organization of the intercellular junctional complex of a polarized epithelium, intestinal epithelium, with the tight junctions separating the cell into apical and basolateral membrane compartments. The adhesive adherens junctions and desmosomes are located on the basolateral surface below the tight junctions with gap junctions located at a further distance from the apical junctional complex in the basolateral membrane. (B) TEM image of the apical junctional complex from mouse intestinal epithelium with the tight junctions (TJ, circled) forming just below the microvilli (Mv) and the adherens junctions (AJ) and desmosomes (DS) located basal to the TJs in the basolateral membrane. (C) TEM image of a cardiac intercalated disc illustrating the localization of the adhesive fascia adherens junctions and desmosomes to the plicate zone and the large gap junctions to the adjacent interplicate zone of the intercalated disc. Cardiac muscle does not contain tight junctions. *(B) Panels A and B were reprinted by permission from Tsukita et al. (2001). (C) Reprinted from Severs et al. (2008), with permission from European Society of Cardiology.*

transmembrane spanning junction adhesion molecule (JAM, four isoforms); more than a dozen different forms of PDZ (PSD-95, Discs-large, ZO-1)-containing proteins; and a half a dozen nonPDZ-containing cytoplasmic proteins (e.g., cingulin; Citi et al., 1988; Kennedy, 1995; Schneeberger & Lynch, 2004). The claudins play a more integral role than the occludins in the formation of the tight junction as evidenced by the severity of the phenotypes associated with their genetic deletion, the variety of human diseases associated with claudin mutations, and their genetic diversity and tissue-selective expression patterns, like those observed for the connexins (reviewed in Van Itallie & Anderson, 2006). The paracellular route through tight junctions is not absolutely impermeable. Though the 100,000 fold difference in trans-epithelial electrical resistance (TEER) between "leaky" and "tight" epithelia does correlate with the number tight junction strands between epithelial cells, the concept that gated pores exist within these strands is decades old and substantiated by the ability of lanthanum (La^{3+}) ions to penetrate all layers of a tight junction (Diamond, 1977; Machen et al., 1972; Van Itallie & Anderson, 2006). These tight junction pores are typically cation selective with an upper limit of 30:1, have an estimated diameter of 8−9 Å, are thought to be formed by the claudins, and are modulated by intracellular kinase signaling pathways (Van Itallie & Anderson, 2006; Zihni et al., 2016). A much slower macromolecular paracellular diffusion pathway for solutes of 30−60 Å in diameter is thought to occur by a dynamic strand mechanism wherein transient remodeling of short segments creates brief openings within the anastomosing junctional strands (Zihni et al., 2016). In conclusion, tight junctions are a molecularly diverse and dynamic intercellular junctional structure integrally composed of claudins and occludins that bear basic similarities to the connexins in their membrane topology, molecular diversity, and association of genetic mutations to human diseases based on the tissue-specific expression pattern of the affected claudin. The PDZ domain-binding protein ZO-1 is also known to bind to multiple connexins in addition to the claudins and occludins (Giepmans, 2004).

6.2 Zonula adherens junctions

The *zonula adherens* (adhesion) junctions are characterized by parallel cell membranes separated by a 200 Å extracellular space and a dense cytoplasmic plaque (Farquhar & Palade, 1963). They are formed by the cadherin superfamily of proteins consisting of >100 structurally diverse members defined by the presence of an extracellular cadherin (EC) repeat domain (Gooding et al., 2004; Gumbiner, 2005; Halbleib & Nelson, 2006; http://calcium.uhnres.utoronto.ca/cadherin). This brief discussion will focus on the Type I (E−, N-, P-, R-, etc.) and Type II (VE-, and others) classical cadherins that form the *zonula adherens* junctions. They are single transmembrane spanning proteins with a large N-terminal extracellular domain consisting of five EC domain repeats. The first four EC repeats of the Type I cadherins contain a His-Ala-Val (HAV) motif involved in strong binding that is lacking in the Type II cadherins. All of the classical cadherins contain a carboxyl-terminal cytoplasmic (α-, β-, p120-) catenin binding domain that in turn binds to α-actinin, vinculin, and formin-1 to form a linkage to the actin cytoskeleton, which is one of the key functions of the cadherin-catenin-based adhesion junctions. The cadherins are also essential to tissue morphogenesis, cancer metastasis, and cell sorting by virtue of their cellular adhesive and homotypic binding properties. When not bound to the cadherin−catenin complex, β-catenin also serves as a transcriptional activator in the Wnt signaling pathway, and dysregulation of β-catenin/Wnt signaling, perhaps by β-catenin activating mutations, is linked to cancer (Hecht & Kemler, 2000).

Formation of the adherens junctions is thought to facilitate the formation of gap junctions. Two different studies expressed either chicken E-cadherin (E = epithelial, aka uvomorulin, L-CAM) or N-cadherin (N = neuronal, aka A-CAM) in cadherin-deficient (mouse sarcoma S180 or P3/22) cell lines and observed the induction of endogenous Cx43 gap junctions (Jongen et al., 1991; Mege et al., 1988). Two studies using antiN-cadherin antibody Fab fragments to block endogenous N-cadherin binding concluded that adherens junctions need to form to permit gap junction formation (Keane et al., 1988; Meyer et al., 1992). This cadherin-dependent gap junction formation is sometimes referred to as the precedence hypothesis. ZO-1, the PDZ domain cytosolic tight junction associated protein that also interacts with Cx43, also binds to α-catenin and actin filaments within the adherens junctions complex, providing another molecular and functional link between the adherens, tight, and gap junctions (Itoh et al., 1997). The association between cadherins and connexins does not end there. N-cadherin, the predominant cadherin expressed in the myocardium, and Cx43, the major myocardial connexin, are both essential to normal heart development via modulation of cardiac neural crest cell motility (Xu et al., 2001). Furthermore, conditional cardiac-specific knockout of the *Cdh2* (N-cad) gene disrupts the intercalated disc structure of cardiomyocytes consisting of adherens junctions, gap junctions, and desmosomes (Kostetskii et al., 2005). Cx43 protein expression levels were concomitantly decreased along with N-cad and these mice often died from sudden cardiac death resulting from spontaneous ventricular tachycardias that degraded into ventricular fibrillation. N-cad and Cx43 and also closely associated with the cardiac sodium ($Na_V1.5$) channel in cardiac intercalated discs, suggesting the existence of a macromolecular protein complex that is essential to transmission of the cardiac action potential (Malhotra et al., 2004). In general, reductions in N-cad expression are thought to negatively influence Cx43 expression and gap junction formation, though genetic deletion of the muscle LIM protein (MLP) resulting in the upregulation of numerous cardiac intercalated disc intercellular junction proteins (e.g., cadherins, catenins, plakoglobin) except Cx43, which was downregulated (Ehler et al., 2001).

6.3 Desmosomes—Macula adherens

Desmosomes (*macula adherens*) are another form of adhesive or anchoring intercellular junction that are shorter in length than the *zonula adherens* with a disc-like appearance, a 240 Å intercellular gap between the adjacent parallel cell membranes, and also stain densely for cytoplasmic proteins (Farquhar & Palade, 1963). The desmosomes are formed by the desmosomal cadherins, namely the desmoglein (Dsg1-4) and desmocollin (Dsc1-3), and associated cytoplasmic linker proteins such as plakoglobin (Jup; aka γ-catenin), desmoplakin (Dsp), and plakophilins1-3 (Pkp; Gooding et al., 2004; Gumbiner, 2005; Kowalczyk & Green, 2013). The general molecular membrane organization of the desmosomal proteins resembles the cadherin/catenin adherens junction complex in that the integral membrane proteins possess a large extracellular (EC1-4) binding domain, a single transmembrane domain, and an intracellular cadherin-like sequence (ICS) that binds plakoglobin (Kowalczyk & Green, 2013). The desmocollin and desmoglein also contain unique extracellular and intracellular anchoring domains not found in the cadherins, and the desmoglein possesses unique intracellular proline-rich linker (IPL), repeat unit domains (RUD), and desmoglein terminal domains (DTD). Plakoglobin binds to the ICS and links the desmosomal cadherins to the classical cadherins, desmoplakin, and the plakophilins, which belong to the p120catenin subfamily. Dsp is an obligatory component of the desmosomes that links the desmosomal complex to the intermediate filaments like keratin and desmin. Thus, mutations in the desmosomal protein genes are linked to numerous skin diseases like pemphigus vulgaris, palmoplantar keratoderma including Carvajal syndrome and Naxos disease. Surprisingly, several desmosomal mutations also affect the heart and result in cardiomyopathies. Mutations in Dsp, Jup, and Dsc2 affect both heart and skin while mutations in Pkp2 and Dsg2 only affect the heart. Mutations resulting in the loss of PKP2 also cause reductions in Cx43 and $Na_V1.5$ in the cardiac intercalated discs (Sato et al., 2011). One particularly lethal cardiomyopathy linked to desmosomal protein mutations is the arrhythmic right ventricular cardiomyopathy (ARVC) or arrhythmogenic cardiomyopathy (ACM) (Saffitz, 2017). ACM is caused by mutations in Dsc2, Dsg2, Pkp2, Pkg, and Jup and is characterized by progressive ventricular myocardial remodeling and degeneration with fibrofatty infiltration and inflammation. First diagnosed in the 1980s, ACM is a leading cause of sudden death in young athletes that commences in the right ventricle and involves both ventricles as the pathology progresses with age. Patients with ACM exhibit a loss Cx43 expression and gap junctions and a concomitant loss of $Na_V1.5$ protein and cardiac sodium current density, again demonstrating the interrelationship between the adherens junctions and desmosomes of the cardiac intercalated disc (cardiac myocytes do not form tight junctions), the gap junctions formed by Cx43, and $Na_V1.5$-mediated sodium currents (Noorman et al., 2013).

To summarize, in polarized epithelia, gap junctions are part of an intercellular junctional complex consisting of tight junctions, adherens junctions, desmosomes, and gap junctions, whereas in nonpolarized epithelia and other tissues like the myocardium, tight junctions are lacking but gap junctions remain localized to membrane regions where adherens junctions and desmosomes are present. Inhibition of cadherin-mediated adhesion junction formation typically results in the reduction of other intercellular junctions, suggesting that adherens junctions must form in order to bring the adjacent cell membranes into close enough proximity (i.e., 200 Å) for the other junctional proteins (claudins, occludins, connexins, desmosomal cadherins) to bind and form their respective intercellular junctions. Formation of the tight junctions, adherens junctions, and desmosomes is Ca^{2+}-dependent owing to the presence of Ca^{2+}-binding sites on the respective integral membrane proteins, but not the connexins, whose hemichannel function is inhibited by extracellular Ca^{2+} ions, but do not explicitly bind Ca^{2+} with high affinity. Removal of extracellular Ca^{2+} also causes the tight junctions, adherens junctions, and desmosomes to split apart and become susceptible to degradation from proteases like trypsin, but not the gap junctions, which remain bound and are degraded by internalization of the intact gap junction membrane known as annular gap junctions or connexosomes. There are numerous protein interactions among the various intercellular junctional structures, many of them mediated by the PDZ domain protein ZO-1, which associates not only with multiple tight junctions proteins from which ZO-1 was first identified but also with approximately 10 connexins, α-catenin, actin, α-actinin, and many more proteins, thus forming a link between the tight, adhesion, and communicating gap junctions (Fanning & Anderson, 2009; Hervé et al., 2014). Some of these junctional protein interactions, N-cadherin and plakophilin-2 for example, influence Cx43 expression and gap junction function in unexpected ways that promote arrhythmogenesis in the heart that are critical to the pathophysiology of known diseases.

References

Aasen, T., Mesnil, M., Naus, C. C., Lampe, P. D., & Laird, D. W. (2016). Gap junctions and cancer: Communicating for 50 years. *Nature Reviews Cancer, 16*, 775–788.

Abrams, C. K., Bennett, M. V., Verselis, V. K., & Bargiello, T. A. (2002). Voltage opens unopposed gap junction hemichannels formed by a connexin 32 mutant associated with X-linked Charcot-Marie-Tooth disease. *Proceedings of the National Academy of Sciences of the U S A, 99*, 3980–3984.

Ahmad, S., Martin, P. E., & Evans, W. H. (2001). Assembly of gap junction channels: Mechanism, effects of calmodulin antagonists and identification of connexin oligomerization determinants. *European Journal of Biochemistry, 268*, 4544–4552.

Ai, X., & Pogwizd, S. M. (2005). Connexin 43 downregulation and dephosphorylation in nonischemic heart failure is associated with enhanced colocalized protein phosphatase type 2A. *Circulation Research, 96*, 54–63.

Anderson, J. M., Stevenson, B. R., Jesaitis, L. A., Goodenough, D. A., & Mooseker, M. S. (1988). Characterization of ZO-1, a protein component of the tight junction from mouse liver and Madin-Darby canine kidney cells. *The Journal of Cell Biology, 106*, 1141–1149.

Aylsworth, C. F., Welsch, C. W., Kabara, J. J., & Trosko, J. E. (1987). Effects of fatty acids on gap junctional communication: Possible role in tumor promotion by dietary fat. *Lipids, 22*, 445–454.

Bai, D., del Corsso, C., Srinivas, M., & Spray, D. C. (2006). Block of specific gap junction channel subtypes by 2-aminoethoxydiphenyl borate (2-APB). *Journal of Pharmacology and Experimental Therapeutics, 319*, 1452–1458.

Barr, L., Dewey, M. M., & Berger, W. (1965). Propagation of action potentials and the structure of the nexus in cardiac muscle. *The Journal of General Physiology, 48*, 797–823.

Beardslee, M. A., Laing, J. G., Beyer, E. C., & Saffitz, J. E. (1998). Rapid turnover of connexin43 in the adult rat heart. *Circulation Research, 83*, 629–635.

Beblo, D.,A., & Veenstra, R. D. (1997). Monovalent cation permeation through the connexin40 gap junction channel. Cs, Rb, K, Na, Li, TEA, TMA, TBA, and effects of anions Br, Cl, F, acetate, aspartate, glutamate, and NO3. *The Journal of General Physiology, 109*, 509–522.

Beltramello, M., Piazza, V., Bukauskas, F. F., Pozzan, T., & Mammano, F. (2005). Impaired permeability to Ins(1,4,5)P3 in a mutant connexin underlies recessive hereditary deafness. *Nature Cell Biology, 7*, 63–69.

Bergoffen, J., Scherer, S. S., Wang, S., et al. (1993). Connexin mutations in X-linked Charcot-Marie-Tooth disease. *Science, 262*, 2039–2042.

Bevans, C. G., & Harris, A. L. (1999). Regulation of connexin channels by pH. Direct action of the protonated form of taurine and other aminosulfonates. *Journal of Biological Chemistry, 274*, 3711–3719.

Beyer, E. C., Paul, D. L., & Goodenough, D. A. (1987). Connexin43: A protein from rat heart homologous to a gap junction protein from liver. *The Journal of Cell Biology, 105*, 2621–2629.

Beyer, E. C., & Berthoud, V. M. (2009). The family of connexin genes. In A. L. Harris, & D. Locke (Eds.), *Connexins, A guide* (pp. 3–26). New York: Humana Press.

Beyer, E. C. (1990). Molecular cloning and developmental expression of two chick embryo gap junction proteins. *Journal of Biological Chemistry, 265*, 14439–14443.

Beyer, E. C., Paul, D. L., & Goodenough, D. A. (1990). Connexin family of gap junction proteins. *Journal of Membrane Biology, 116*, 187–194.

Boerma, M., Forsberg, L., Van Zeijl, L., et al. (1999). A genetic polymorphism in connexin 37 as a prognostic marker for atherosclerotic plaque development. *Journal of Internal Medicine, 246*, 211–218.

Boitano, S., & Evans, W. H. (2000). Connexin mimetic peptides reversibly inhibit Ca(2+) signaling through gap junctions in airway cells. *American Journal of Physiology - Lung Cellular and Molecular Physiology, 279*, L623–L630.

Bone, L. J., Deschênes, S. M., Balice-Gordon, R. J., Fischbeck, K. H., & Scherer, S. S. (1997). Connexin32 and X-linked Charcot-Marie-tooth disease. *Neurobiology of Disease, 4*, 221–230.

Boyden, L. M., Craiglow, B. G., Zhou, J., et al. (2015). Dominant de novo mutations in GJA1 cause erythrokeratodermia variabilis et progressiva, without features of oculodentodigital dysplasia. *Journal of Investigative Dermatology, 135*, 1540–1547.

Brightman, M. W., & Reese, T. S. (1969). Junctions between intimately apposed cell membranes in the vertebrate brain. *The Journal of Cell Biology, 40*, 648–677.

Brink, P. R., Cronin, K., Banach, K., et al. (1997). Evidence for heteromeric gap junction channels formed from rat connexin43 and human connexin37. *American Journal of Physiology, 273*, C1386–C1396.

Brink, P. R., & Dewey, M. M. (1980). Evidence for fixed charge in the nexus. *Nature, 285*, 101–102.

Britz-Cunningham, S. H., Shah, M. M., Zuppan, C. W., & Fletcher, W. H. (1995). Mutations of the Connexin43 gap-junction gene in patients with heart malformations and defects of laterality. *New England Journal of Medicine, 332*, 1323–1329.

Bruzzone, R., Barbe, M. T., Jakob, N. J., & Monyer, H. (2005). Pharmacological properties of homomeric and heteromeric pannexin hemichannels expressed in Xenopus oocytes. *Journal of Neurochemistry, 92*, 1033–1043.

Bukauskas, F. F., & Peracchia, C. (1997). Two distinct gating mechanisms in gap junction channels: CO2-sensitive and voltage-sensitive. *Biophysical Journal, 72*, 2137–2142.

Bukauskas, F. F., & Verselis, V. K. (2004). Gap junction channel gating. *Biochimica et Biophysica Acta, 1662*, 42–60.

Burt, J. M., Massey, K. D., & Minnich, B. N. (1991). Uncoupling of cardiac cells by fatty acids: Structure-activity relationships. *American Journal of Physiology, 260*, C439–C448.

Burt, J. M., & Spray, D. C. (1989). Volatile anesthetics block intercellular communication between neonatal rat myocardial cells. *Circulation Research, 65*, 829–837.

Chen, Y., Zhou, Y., Lin, X., et al. (2011). Molecular interaction and functional regulation of connexin50 gap junctions by calmodulin. *Biochemical Journal, 435*, 711–722.

Choudhary, C., Kumar, C., Gnad, F., et al. (2009). Lysine acetylation targets protein complexes and co-regulates major cellular functions. *Science, 325*, 834–840.

Citi, S., Sabanay, H., Jakes, R., Geiger, B., & Kendrick-Jones, J. (1988). Cingulin, a new peripheral component of tight junctions. *Nature, 333*, 272–276.

Claude, P., & Goodenough, D. A. (1973). Fracture faces of zonulae occludentes from "tight" and "leaky" epithelia. *The Journal of Cell Biology, 58*, 390–400.

Cole, W. C., & Garfield, R. E. (1988). Effects of calcium ionophore, A23187 and calmodulin inhibitors on intercellular communication in the rat myometrium. *Biology of Reproduction, 38*, 55–62.

Colussi, C., Rosati, J., Straino, S., et al. (2011). N^{ε}-lysine acetylation determines dissociation from gap junctions and lateralization of connexin 43 in normal and dystrophic heart. *Proceedings of the National Academy of Sciences of the United States of America, 108*, 2795–2800.

Condorelli, D. F., Parenti, R., Spinella, F., et al. (1998). Cloning of a new gap junction gene (Cx36) highly expressed in mammalian brain neurons. *European Journal of Neuroscience, 10*, 1202–1208.

Cooper, C. D., & Lampe, P. D. (2002). Casein kinase 1 regulates connexin-43 gap junction assembly. *Journal of Biological Chemistry, 277*, 44962–44968.

Crow, D. S., Beyer, E. C., Paul, D. L., Kobe, S. S., & Lau, A. F. (1990). Phosphorylation of connexin43 gap junction protein in uninfected and Rous sarcoma virus-transformed mammalian fibroblasts. *Molecular and Cellular Biology, 10*, 1754–1763.

Délèze, J. (1970). The recovery of resting potential and input resistance in sheep heart injured by knife or laser. *The Journal of Physiology, 208*, 547–562.

Dahl, G. (2007). Gap junction-mimetic peptides do work, but in unexpected ways. *Cell Communication and Adhesion, 14*(6), 259–264.

Dahl, G., Miller, T., Paul, D., Voellmy, R., & Werner, R. (1987). Expression of functional cell-cell channels from cloned rat liver gap junction complementary DNA. *Science, 236*, 1290–1293.

Dasgupta, C., Martinez, A. M., Zuppan, C. W., Shah, M. M., Bailey, L. L., & Fletcher, W. H. (2001). Identification of connexin43 (alpha1) gap junction gene mutations in patients with hypoplastic left heart syndrome by denaturing gradient gel electrophoresis (DGGE). *Mutation Research, 479*, 173–186.

Davidson, J. S., Baumgarten, I. M., & Harley, E. H. (1986). Reversible inhibition of intercellular junctional communication by glycyrrhetinic acid. *Biochemical and Biophysical Research Communications, 134*, 29–36.

de Groot, J. R., Veenstra, T., Verkerk, A. O., et al. (2003). Conduction slowing by the gap junctional uncoupler carbenoxolone. *Cardiovascular Research, 60*, 288–297.

De Mello, W. C. (2015). Cell-to-cell diffusion of glucose in the mammalian heart is disrupted by high glucose. Implications for the diabetic heart. *Experimental Cell Research, 334*, 239–245.

De Mello, W. C., Motta, G. E., & Chapeau, M. (1969). A study on the healing-over of myocardial cells of toads. *Circulation Research, 24*, 475–487.

Defamie, N., Chepied, A., & Mesnil, M. (2014). Connexins, gap junctions and tissue invasion. *FEBS Letters, 588*, 1331–1338.

del Castillo, I., Villamar, M., Moreno-Pelayo, M. A., et al. (2002). A deletion involving the connexin 30 gene in nonsyndromic hearing impairment. *New England Journal of Medicine, 346*, 243–249.

Denning, C., & Pitts, J. D. (1997). Bystander effects of different enzyme-prodrug systems for cancer gene therapy depend on different pathways for intercellular transfer of toxic metabolites, a factor that will govern clinical choice of appropriate regimes. *Human Gene Therapy, 8*, 1825–1835.

Deschênes, S. M., Walcott, J. L., Wexler, T. L., Scherer, S. S., & Fischbeck, K. H. (1997). Altered trafficking of mutant connexin32. *Journal of Neuroscience, 17*, 9077–9084.

Deutsch, D. E., Williams, J. A., & Yule, D. I. (1995). Halothane and octanol block Ca2+ oscillations in pancreatic acini by multiple mechanisms. *American Journal of Physiology, 269*, G779–G788.

Dewey, M. M., & Barr, L. (1962). Intercellular connection between smooth muscle cells: The nexus. *Science, 137*, 670–672.

Dhein, S., Hagen, A., Jozwiak, J., et al. (2010). Improving cardiac gap junction communication as a new antiarrhythmic mechanism: The action of antiarrhythmic peptides. *Naunyn-Schmiedeberg's Archives of Pharmacology, 381*, 221–234.

Diamond, J. M. (1977). Twenty-first bowditch lecture. The epithelial junction: Bridge, gate, and fence. *Physiologist, 20*, 10–18.

Dobrowolski, R., Sasse, P., Schrickel, J. W., et al. (2008). The conditional connexin43G138R mouse mutant represents a new model of hereditary oculodentodigital dysplasia in humans. *Human Molecular Genetics, 17*, 539–554.

Dodd, R., Peracchia, C., Stolady, D., & Török, K. (2008). Calmodulin association with connexin32-derived peptides suggests trans-domain interaction in chemical gating of gap junction channels. *Journal of Biological Chemistry, 283*, 26911–26920.

Duffy, H. S., Sorgen, P. L., Girvin, M. E., et al. (2002). pH-dependent intramolecular binding and structure involving Cx43 cytoplasmic domains. *Journal of Biological Chemistry, 277*, 36706–37614.

Dunn, C. A., & Lampe, P. D. (2014). Injury-triggered Akt phosphorylation of Cx43: A ZO-1-driven molecular switch that regulates gap junction size. *Journal of Cell Science, 127*, 455–464.

Ebihara, L., & Steiner, E. (1993). Properties of a nonjunctional current expressed from a rat connexin46 cDNA in Xenopus oocytes. *The Journal of General Physiology, 102*, 59–74.

Eghbali, B., Kessler, J. A., & Spray, D. C. (1990). Expression of gap junction channels in communication-incompetent cells after stable transfection with cDNA encoding connexin 32. *Proceedings of the National Academy of Sciences of the United States of America, 87*, 1328–1331.

Ehler, E., Horowits, R., Zuppinger, C., et al. (2001). Alterations at the intercalated disk associated with the absence of muscle LIM protein. *The Journal of Cell Biology, 153*, 763–772.

Ek-Vitorin, J. F., King, T. J., Heyman, N. S., Lampe, P. D., & Burt, J. M. (2006). Selectivity of connexin 43 channels is regulated through protein kinase C-dependent phosphorylation. *Circulation Research, 98*, 1498–1505.

Eskandari, S., Zampighi, G. A., Leung, D. W., Wright, E. M., & Loo, D. D. (2002). Inhibition of gap junction hemichannels by chloride channel blockers. *Journal of Membrane Biology, 185*, 93–102.

Estapé-Wainwright, E., & De Mello, W. C. (1983). Cyclic nucleotides and calcium: Their role in the control of cell communication in the heart. *Cell Biology International Reports, 7*, 91–97.

Evans, W. H., & Boitano, S. (2001). Connexin mimetic peptides: Specific inhibitors of gap-junctional intercellular communication. *Biochemical Society Transactions, 29*, 606–612.

Falk, M. M., Buehler, L. K., Kumar, N. M., & Gilula, N. B. (1997). Cell-free synthesis and assembly of connexins into functional gap junction membrane channels. *The EMBO Journal, 16*, 2703–2716.

Fallon, R. F., & Goodenough, D. A. (1981). Five-hour half-life of mouse liver gap-junction protein. *The Journal of Cell Biology, 90*, 521–526.

Fanning, A. S., & Anderson, J. M. (2009). Zonula occludens-1 and -2 are cytosolic scaffolds that regulate the assembly of cellular junctions. *Annals of the New York Academy of Sciences, 1165*, 113–120.

Farquhar, M. G., & Palade, G. E. (1963). Junctional complexes in various epithelia. *The Journal of Cell Biology, 17*, 375–412.

Ferrell, R. E., Baty, C. J., Kimak, M. A., et al. (2010). GJC2 missense mutations cause human lymphedema. *The American Journal of Human Genetics, 86*, 943–948.

Firouzi, M., Kok, B., Spiering, W., et al. (2006). Polymorphisms in human connexin40 gene promoter are associated with increased risk of hypertension in men. *Journal of Hypertension, 24*, 325–330.

Flagg-Newton, J., Simpson, I., & Loewenstein, W. R. (1979). Permeability of the cell-to-cell membrane channels in mammalian cell junction. *Science, 205*, 404–407.

Foote, C. I., Zhou, L., Zhu, X., & Nicholson, B. J. (1998). The pattern of disulfide linkages in the extracellular loop regions of connexin 32 suggests a model for the docking interface of gap junctions. *The Journal of Cell Biology, 140*, 1187–1197.

Furshpan, E. J., & Furukawa, T. (1962). Intracellular and extracellular responses of the several regions of the Mauthner cell of the goldfish. *Journal of Neurophysiology, 25*, 732–771.

Furshpan, E. J., & Potter, D. D. (1959). Transmission at the giant motor synapses of the crayfish. *The Journal of Physiology, 145*, 289–325.

Furuse, M., Fujita, K., Hiiragi, T., Fujimoto, K., & Tsukita, S. (1998). Claudin-1 and -2: Novel integral membrane proteins localizing at tight junctions with no sequence similarity to occludin. *The Journal of Cell Biology, 141*, 1539–1550.

Furuse, M., Hirase, T., Itoh, M., et al. (1993). Occludin: A novel integral membrane protein localizing at tight junctions. *The Journal of Cell Biology, 123*, 1777–1788.

García, I. E., Prado, P., Pupo, A., et al. (2016). Connexinopathies: A structural and functional glimpse. *BMC Cell Biology, 17*(Suppl. 1), 17.

Garcia-Dorado, D., Inserte, J., Ruiz-Meana, M., et al. (1997). Gap junction uncoupler heptanol prevents cell-to-cell progression of hypercontracture and limits necrosis during myocardial reperfusion. *Circulation, 96*, 3579–3586.

Garfield, R. E., Merrett, D., & Grover, A. K. (1980). Gap junction formation and regulation in myometrium. *American Journal of Physiology, 239*, C217–C228.

Gemel, J., Simon, A. R., Patel, D., et al. (2014). Degradation of a connexin40 mutant linked to atrial fibrillation is accelerated. *Journal of Molecular and Cellular Cardiology, 74*, 330–339.

Gerido, D. A., DeRosa, A. M., Richard, G., & White, T. W. (2007). Aberrant hemichannel properties of Cx26 mutations causing skin disease and deafness. *American Journal of Physiology - Cell Physiology, 293*, C337–C345.

Giaume, C., Randriamampita, C., & Trautmann, A. (1989). Arachidonic acid closes gap junction channels in rat lacrimal glands. *Pflügers Archiv, 413*, 273–279.

Giepmans, B. N. (2004). Gap junctions and connexin-interacting proteins. *Cardiovascular Research, 62*, 233–245.

Giepmans, B. N., Verlaan, I., Hengeveld, T., et al. (2001). Gap junction protein connexin-43 interacts directly with microtubules. *Current Biology, 11*, 1364–1368.

Gilula, N. B., Reeves, O. R., & Steinbach, A. (1972). Metabolic coupling, ionic coupling and cell contacts. *Nature, 235*, 262–265.

Goldstein, D. B. (1984). The effects of drugs on membrane fluidity. *Annual Review of Pharmacology and Toxicology, 24*, 43–64.

Gollob, M. H., Jones, D. L., Krahn, A. D., et al. (2006). Somatic mutations in the connexin 40 gene (GJA5) in atrial fibrillation. *New England Journal of Medicine, 354*, 2677–2688.

Gong, X., Li, E., Klier, G., et al. (1997). Disruption of alpha3 connexin gene leads to proteolysis and cataractogenesis in mice. *Cell, 91*, 833–843.

González-Sánchez, A., Jaraíz-Rodríguez, M., Domínguez-Prieto, M., Herrero-González, S., Medina, J. M., & Tabernero, A. (2016). Connexin43 recruits PTEN and Csk to inhibit c-Src activity in glioma cells and astrocytes. *Oncotarget, 7*, 49819–49833.

Goodenough, D. A. (1974). Bulk isolation of mouse hepatocyte gap junctions. Characterization of the principal protein, connexin. *The Journal of Cell Biology, 61*, 557–563.

Gooding, J. M., Yap, K. L., & Ikura, M. (2004). The cadherin-catenin complex as a focal point of cell adhesion and signalling: New insights from three-dimensional structures. *BioEssays, 26*, 497–511.

Grifa, A., Wagner, C. A., D'Ambrosio, L., et al. (1999). Mutations in GJB6 cause nonsyndromic autosomal dominant deafness at DFNA3 locus. *Nature Genetics, 23*, 16–18.

Groenewegen, W. A., Firouzi, M., Bezzina, C. R., et al. (2003). A cardiac sodium channel mutation cosegregates with a rare connexin40 genotype in familial atrial standstill. *Circulation Research, 92*, 14–22.

Gumbiner, B. M. (2005). Regulation of cadherin-mediated adhesion in morphogenesis. *Nature Reviews Molecular Cell Biology, 6*, 622–634.

Halbleib, J. M., & Nelson, W. J. (2006). Cadherins in development: Cell adhesion, sorting, and tissue morphogenesis. *Genes & Development, 20*, 3199–3214.

Harks, E. G., Camiña, J. P., Peters, P. H., et al. (2003). Besides affecting intracellular calcium signaling, 2-APB reversibly blocks gap junctional coupling in confluent monolayers, thereby allowing measurement of single-cell membrane currents in undissociated cells. *The FASEB Journal, 17*, 941–943.

Harris, A. L. (2001). Emerging issues of connexin channels: Biophysics fills the gap. *Quarterly Reviews of Biophysics, 34*, 325–472.

Haugan, K., Olsen, K. B., Hartvig, L., et al. (2005). The antiarrhythmic peptide analog ZP123 prevents atrial conduction slowing during metabolic stress. *Journal of Cardiovascular Electrophysiology, 16*, 537–545.

Heathcote, K., Syrris, P., Carter, N. D., & Patton, M. A. (2000). A connexin 26 mutation causes a syndrome of sensorineural hearing loss and palmoplantar hyperkeratosis (MIM 148350). *Journal of Medical Genetics, 37*, 50–51.

Hecht, A., & Kemler, R. (2000). Curbing the nuclear activities of beta-catenin. Control over Wnt target gene expression. *EMBO Reports, 1*, 24–28.

Herr, J. C. (1976). Reflexive gap junctions. Gap junctions between processing arising from the same ovarian decidual cell. *The Journal of Cell Biology, 69*, 495–501.

Hervé, J. C., Derangeon, M., Sarrouilhe, D., & Bourmeyster, N. (2014). Influence of the scaffolding protein Zonula Occludens (ZOs) on membrane channels. *Biochimica et Biophysica Acta, 1838*, 595–604.

Hoshi, T., Zagotta, W. N., & Aldrich, R. W. (1990). Biophysical and molecular mechanisms of Shaker potassium channel inactivation. *Science, 250*, 533–538.

Hund, T. J., Lerner, D. L., Yamada, K. A., Schuessler, R. B., & Saffitz, J. E. (2007). Protein kinase cepsilon mediates salutary effects on electrical coupling induced by ischemic preconditioning. *Heart Rhythm, 4*, 1183–1193.

Hunter, A. W., Barker, R. J., Zhu, C., & Gourdie, R. G. (2005). Zonula occludens-1 alters connexin43 gap junction size and organization by influencing channel accretion. *Molecular Biology of the Cell, 16*, 5686–5698.

Itoh, M., Nagafuchi, A., Moroi, S., & Tsukita, S. (1997). Involvement of ZO-1 in cadherin-based cell adhesion through its direct binding to alpha catenin and actin filaments. *The Journal of Cell Biology, 138*, 181–192.

Jabr, R. I., Hatch, F. S., Salvage, S. C., Orlowski, A., Lampe, P. D., & Fry, C. H. (2016). Regulation of gap junction conductance by calcineurin through Cx43 phosphorylation: Implications for action potential conduction. *Pflügers Archiv, 468*, 1945–1955.

Jain, S. K., Schuessler, R. B., & Saffitz, J. E. (2003). Mechanisms of delayed electrical uncoupling induced by ischemic preconditioning. *Circulation Research, 92*, 1138–1144.

Jan, A. Y., Amin, S., Ratajczak, P., Richard, G., & Sybert, V. P. (2004). Genetic heterogeneity of KID syndrome: Identification of a Cx30 gene (GJB6) mutation in a patient with KID syndrome and congenital atrichia. *Journal of Investigative Dermatology, 122*, 1108–1113.

Jiang, J. X., & Goodenough, D. A. (1996). Heteromeric connexons in lens gap junction channels. *Proceedings of the National Academy of Sciences of the United States of America, 93*, 1287–1291.

Jiang, J. X., & Penuela, S. (2016). Connexin and pannexin channels in cancer. *BMC Cell Biology, 17*(Suppl. 1), 12.

Johnston, M. F., Simon, S. A., & Ramón, F. (1980). Interaction of anesthetics with electrical synapses. *Nature, 286*, 498–500.

Jongen, W. M., Fitzgerald, D. J., Asamoto, M., et al. (1991). Regulation of connexin 43-mediated gap junctional intercellular communication by Ca^{2+} in mouse epidermal cells is controlled by E-cadherin. *The Journal of Cell Biology, 114*, 545–555.

Jordan, K., Chodock, R., Hand, A. R., & Laird, D. W. (2001). The origin of annular junctions: A mechanism of gap junction internalization. *Journal of Cell Science, 114*, 763–773.

Kabzińska, D., Kotruchow, K., Ryniewicz, B., & Kochański, A. (2011). Two pathogenic mutations located within the 5'-regulatory sequence of the GJB1 gene affecting initiation of transcription and translation. *Acta Biochimica Polonica, 58*, 359–363.

Kalcheva, N., Qu, J., Sandeep, N., et al. (2007). Gap junction remodeling and cardiac arrhythmogenesis in a murine model of oculodentodigital dysplasia. *Proceedings of the National Academy of Sciences of the United States of America, 104*, 20512–20516.

Kanemitsu, M. Y., Jiang, W., & Eckhart, W. (1998). Cdc2-mediated phosphorylation of the gap junction protein, connexin43, during mitosis. *Cell Growth & Differentiation, 9*, 13–21.

Kanno, Y., & Loewenstein, W. R. (1964b). Intercellular diffusion. *Science, 143*, 959–960.

Kanno, Y., & Loewenstein, W. R. (1964a). Low-resistance coupling between gland cells. Some observations on intercellular contact membranes and intercellular space. *Nature, 201*, 194–195.

Kanter, H. L., Saffitz, J. E., & Beyer, E. C. (1992). Cardiac myocytes express multiple gap junction proteins. *Circulation Research, 70*, 438–444.

Keane, R. W., Mehta, P. P., Rose, B., Honig, L. S., Loewenstein, W. R., & Rutishauser, U. (1988). Neural differentiation, NCAM-mediated adhesion, and gap junctional communication in neuroectoderm. A study in vitro. *The Journal of Cell Biology, 106*, 1307–1319.

Kelsell, D. P., Dunlop, J., Stevens, H. P., et al. (1997). Connexin 26 mutations in hereditary non-syndromic sensorineural deafness. *Nature, 387*, 80–83.

Kennedy, M. B. (1995). Origin of PDZ (DHR, GLGF) domains. *Trends in Biochemical Sciences, 20*, 350.

Kensler, R. W., & Goodenough, D. A. (1980). Isolation of mouse myocardial gap junctions. *The Journal of Cell Biology, 86*, 755–764.

Kirchhoff, S., Nelles, E., Hagendorff, A., Krüger, O., Traub, O., & Willecke, K. (1998). Reduced cardiac conduction velocity and predisposition to arrhythmias in connexin40-deficient mice. *Current Biology, 8*, 299–302.

Kjølbye, A. L., Knudsen, C. B., Jepsen, T., Larsen, B. D., & Petersen, J. S. (2003). Pharmacological characterization of the new stable antiarrhythmic peptide analog Ac-D-Tyr-D-Pro-D-Hyp-Gly-D-Ala-Gly-NH2 (ZP123): In vivo and in vitro studies. *Journal of Pharmacology and Experimental Therapeutics, 306*, 1191–1199.

Kostetskii, I., Li, J., Xiong, Y., et al. (2005). Induced deletion of the N-cadherin gene in the heart leads to dissolution of the intercalated disc structure. *Circulation Research, 96*, 346–354.

Kouzarides, T. (2000). Acetylation: A regulatory modification to rival phosphorylation? *The EMBO Journal, 19*, 1176–1179.

Kovalevskaya, N. V., van de Waterbeemd, M., Bokhovchuk, F. M., et al. (2013). Structural analysis of calmodulin binding to ion channels demonstrates the role of its plasticity in regulation. *Pflügers Archiv, 465*, 1507–1519.

Kowalczyk, A. P., & Green, K. J. (2013). Structure, function, and regulation of desmosomes. *Progress in Molecular Biology and Translational Science, 116*, 95–118.

Krattinger, N., Capponi, A., Mazzolai, L., et al. (2007). Connexin40 regulates renin production and blood pressure. *Kidney International, 72*, 814–822.

Kumar, N. M., & Gilula, N. B. (1986). Cloning and characterization of human and rat liver cDNAs coding for a gap junction protein. *The Journal of Cell Biology, 103*, 767–776.

Kumar, N. M., & Gilula, N. B. (1992). Molecular biology and genetics of gap junction channels. *Seminars in Cell Biology, 3*, 3–16.

Kurtz, L., Schweda, F., de Wit, C., et al. (2007). Lack of connexin 40 causes displacement of renin-producing cells from afferent arterioles to the extraglomerular mesangium. *Journal of the American Society of Nephrology, 18*, 1103–1111.

Kwak, B. R., & Jongsma, H. J. (1996). Regulation of cardiac gap junction channel permeability and conductance by several phosphorylating conditions. *Molecular and Cellular Biochemistry, 157*, 93–99.

Kwon, T., Harris, A. L., Rossi, A., & Bargiello, T. A. (2011). Molecular dynamics simulations of the Cx26 hemichannel: Evaluation of structural models with brownian dynamics. *The Journal of General Physiology, 138*, 475–493.

López-Bigas, N., Olivé, M., Rabionet, R., et al. (2001). Connexin 31 (GJB3) is expressed in the peripheral and auditory nerves and causes neuropathy and hearing impairment. *Human Molecular Genetics, 10*, 947–952.

Laird, D. W. (2008). Closing the gap on autosomal dominant connexin-26 and connexin-43 mutants linked to human disease. *Journal of Biological Chemistry, 283*, 2997–3001.

Laird, D. W. (2006). Life cycle of connexins in health and disease. *Biochemical Journal, 394*, 527–543.

Laird, D. W., Puranam, K. L., & Revel, J. P. (1991). Turnover and phosphorylation dynamics of connexin43 gap junction in cultured cardiac myocytes. *Biochemical Journal, 273*, 67–72.

Lampe, P. D., Kurata, W. E., Warn-Cramer, B. J., & Lau, A. F. (1998). Formation of a distinct connexin43 phosphoisoform in mitotic cells is dependent upon p34cdc2 kinase. *Journal of Cell Science, 111*, 833–841.

Lampe, P. D., TenBroek, E. M., Burt, J. M., Kurata, W. E., Johnson, R. G., & Lau, A. F. (2000). Phosphorylation of connexin43 on serine368 by protein kinase C regulates gap junctional communication. *The Journal of Cell Biology, 149*, 1503–1512.

Larsen, W. J., & Tung, H. N. (1978). Origin and fate of cytoplasmic gap junctional vesicles in rabbit granulosa cells. *Tissue and Cell, 10*, 585–598.

Li, W. E., & Nagy, J. I. (2000). Connexin43 phosphorylation state and intercellular communication in cultured astrocytes following hypoxia and protein phosphatase inhibition. *European Journal of Neuroscience, 12*, 2644–2650.

Lin, J. H., Weigel, H., Cotrina, M. L., et al. (1998). Gap-junction-mediated propagation and amplification of cell injury. *Nature Neuroscience, 1*, 494–500.

Lin, R., Warn-Cramer, B. J., Kurata, W. E., & Lau, A. F. (2001). v-Src phosphorylation of connexin 43 on Tyr247 and Tyr265 disrupts gap junctional communication. *The Journal of Cell Biology, 154*, 815–827.

Lin, X., Xu, Q., & Veenstra, R. D. (2014). Functional formation of heterotypic gap junction channels by connexins-40 and -43. *Channels, 8*, 433–443.

Lin, X., Zemlin, C., Hennan, J. K., Petersen, J. S., & Veenstra, R. D. (2008). Enhancement of ventricular gap-junction coupling by rotigaptide. *Cardiovascular Research, 79*, 416–426.

Liu, X. Z., Xia, X. J., Adams, J., et al. (2001). Mutations in GJA1 (connexin 43) are associated with non-syndromic autosomal recessive deafness. *Human Molecular Genetics, 10*(25), 2945–2951.

Liu, X. Z., Xia, X. J., Xu, L. R., et al. (2000). Mutations in connexin31 underlie recessive as well as dominant non-syndromic hearing loss. *Human Molecular Genetics, 9*, 63–67.

Liu, Y., Kato, H., Nakata, N., & Kogure, K. (1992). Protection of rat hippocampus against ischemic neuronal damage by pretreatment with sublethal ischemia. *Brain Research, 586*, 121–124.

Locke, D., Bian, S., Li, H., & Harris, A. L. (2009). Post-translational modifications of connexin26 revealed by mass spectrometry. *Biochemical Journal, 424*, 385–398.

Locke, D., Koreen, I. V., Liu, J. Y., & Harris, A. L. (2004). Reversible pore block of connexin channels by cyclodextrins. *Journal of Biological Chemistry, 279*, 22883–22892.

Loewenstein, W. R., & Kanno, Y. (1967). Intercellular communication and tissue growth. I. Cancerous growth. *The Journal of Cell Biology, 33*, 225–234.

Loewenstein, W. R., Kanno, Y., & Socolar, S. J. (1978). Quantum jumps of conductance during formation of membrane channels at cell-cell junction. *Nature, 274*, 133–136.

Loewenstein, W. R. (1966). Permeability of membrane junctions. *Annals of the New York Academy of Sciences, 137*, 441–472.

Maass, K., Ghanem, A., Kim, J. S., et al. (2004). Defective epidermal barrier in neonatal mice lacking the C-terminal region of connexin43. *Molecular Biology of the Cell, 15*, 4597–4608.

Macari, F., Landau, M., Cousin, P., et al. (2000). Mutation in the gene for connexin 30.3 in a family with erythrokeratodermia variabilis. *The American Journal of Human Genetics, 67*, 1296–1301.

Machen, T. E., Erlij, D., & Wooding, F. B. (1972). Permeable junctional complexes. The movement of lanthanum across rabbit gallbladder and intestine. *The Journal of Cell Biology, 54*, 302–312.

Mackay, D., Ionides, A., Kibar, Z., et al. (1999). Connexin46 mutations in autosomal dominant congenital cataract. *The American Journal of Human Genetics, 64*, 1357–1364.

Maeda, S., Nakagawa, S., Suga, M., et al. (2009). Structure of the connexin 26 gap junction channel at 3.5 Å resolution. *Nature, 458*, 597–602.

Maestrini, E., Korge, B. P., Ocana-Sierra, J., et al. (1999). A missense mutation in connexin26, D66H, causes mutilating keratoderma with sensorineural deafness (Vohwinkel's syndrome) in three unrelated families. *Human Molecular Genetics, 8*, 1237–1243.

Makita, N., Seki, A., Sumitomo, N., et al. (2012). A connexin40 mutation associated with a malignant variant of progressive familial heart block type I. *Circulation: Arrhythmia and Electrophysiology, 5*, 163–172.

Makowski, L., Caspar, D. L., Phillips, W. C., & Goodenough, D. A. (1977). Gap junction structures. II. Analysis of the x-ray diffraction data. *The Journal of Cell Biology, 74*, 629–645.

Malhotra, J. D., Thyagarajan, V., Chen, C., & Isom, L. L. (2004). Tyrosine-phosphorylated and nonphosphorylated sodium channel beta1 subunits are differentially localized in cardiac myocytes. *Journal of Biological Chemistry, 279*, 40748–40754.

Manjunath, C. K., Goings, G. E., & Page, E. (1982). Isolation and protein composition of gap junctions from rabbit hearts. *Biochemical Journal, 205*, 189–194.

Marquart, K. H. (1977). So-called annular gap junctions in bone cells of normal mice. *Experientia, 33*, 270–272.

McNutt, N. S., Hershberg, R. A., & Weinstein, R. S. (1971). Further observations on the occurrence of nexuses in benign and malignant human cervical epithelium. *The Journal of Cell Biology, 51*, 805–825.

Mege, R. M., Matsuzaki, F., Gallin, W. J., Goldberg, J. I., Cunningham, B. A., & Edelman, G. M. (1988). Construction of epithelioid sheets by transfection of mouse sarcoma cells with cDNAs for chicken cell adhesion molecules. *Proceedings of the National Academy of Sciences of the United States of America, 85*, 7274–7278.

Menichella, D. M., Goodenough, D. A., Sirkowski, E., Scherer, S. S., & Paul, D. L. (2003). Connexins are critical for normal myelination in the CNS. *Journal of Neuroscience, 23*, 5963–5973.

Menzies, K. J., Zhang, H., Katsyuba, E., & Auwerx, J. (2016). Protein acetylation in metabolism - metabolites and cofactors. *Nature Reviews Endocrinology, 12*, 43–60.

Mese, G., Sellitto, C., Li, L., et al. (2011). The Cx26-G45E mutation displays increased hemichannel activity in a mouse model of the lethal form of keratitis-ichthyosis-deafness syndrome. *Molecular Biology of the Cell, 22*, 4776–4786.

Meyer, R. A., Laird, D. W., Revel, J. P., & Johnson, R. G. (1992). Inhibition of gap junction and adherens junction assembly by connexin and A-CAM antibodies. *The Journal of Cell Biology, 119*, 179–189.

Milks, L. C., Kumar, N. M., Houghten, R., Unwin, N., & Gilula, N. B. (1988). Topology of the 32-kd liver gap junction protein determined by site-directed antibody localizations. *The EMBO Journal, 7*, 2967–2975.

Miura, T., Miki, T., & Yano, T. (2010). Role of the gap junction in ischemic preconditioning in the heart. *American Journal of Physiology - Heart and Circulatory Physiology, 298*, H1115–H1125.

Moreno, A. P., Sáez, J. C., Fishman, G. I., & Spray, D. C. (1994). Human connexin43 gap junction channels. Regulation of unitary conductances by phosphorylation. *Circulation Research, 74*, 1050–1057.

Morley, G. E., Taffet, S. M., & Delmar, M. (1996). Intramolecular interactions mediate pH regulation of connexin43 channels. *Biophysical Journal, 70*, 1294–1302.

Murrell, R. D., & Haydon, D. A. (1991). Actions of n-alcohols on nicotinic acetylcholine receptor ion channels in cultured rat muscle cells. *Annals of the New York Academy of Sciences, 625*, 365–374.

Murry, C. E., Jennings, R. B., & Reimer, K. A. (1986). Preconditioning with ischemia: A delay of lethal cell injury in ischemic myocardium. *Circulation, 74*, 1124–1136.

Musa, H., Gough, J. D., Lees, W. J., & Veenstra, R. D. (2001). Ionic blockade of the rat connexin40 gap junction channel by large tetraalkylammonium ions. *Biophysical Journal, 81*, 3253–3274.

Musa, H., & Veenstra, R. D. (2003). Voltage-dependent blockade of connexin40 gap junctions by spermine. *Biophysical Journal, 84*, 205–219.

Musil, L. S., & Goodenough, D. A. (1991). Biochemical analysis of connexin43 intracellular transport, phosphorylation, and assembly into gap junctional plaques. *The Journal of Cell Biology, 115*, 1357–1374.

Musil, L. S., & Goodenough, D. A. (1993). Multisubunit assembly of an integral plasma membrane channel protein, gap junction connexin43, occurs after exit from the ER. *Cell, 74*, 1065–1077.

Naitoh, K., Ichikawa, Y., Miura, T., et al. (2006). MitoKATP channel activation suppresses gap junction permeability in the ischemic myocardium by an ERK-dependent mechanism. *Cardiovascular Research, 70*, 374–383.

Naitoh, K., Yano, T., Miura, T., et al. (2009). Roles of Cx43-associated protein kinases in suppression of gap junction-mediated chemical coupling by ischemic preconditioning. *American Journal of Physiology - Heart and Circulatory Physiology, 296*, H396–H403.

Nelson, W. L., & Makielski, J. C. (1991). Block of sodium current by heptanol in voltage-clamped canine cardiac Purkinje cells. *Circulation Research, 68*, 977–983.

Neyton, J., & Trautmann, A. (1985). Single-channel currents of an intercellular junction. *Nature, 317*, 331–335.

Nicholson, B. J., Gros, D. B., Kent, S. B., Hood, L. E., & Revel, J. P. (1985). The Mr 28,000 gap junction proteins from rat heart and liver are different but related. *Journal of Biological Chemistry, 260*, 6514–6517.

Noorman, M., Hakin, S., Kessler, E., et al. (2013). Remodeling of the cardiac sodium channel, connexin43, and plakoglobin at the intercalated disk in patients with arrhythmogenic cardiomyopathy. *Heart Rhythm, 10*, 412–419.

Oh, S., Ri, Y., Bennett, M. V., Trexler, E. B., Verselis, V. K., & Bargiello, T. A. (1997). Changes in permeability caused by connexin 32 mutations underlie X-linked Charcot-Marie-Tooth disease. *Neuron, 19*, 927–938.

Omori, Y., Mesnil, M., & Yamasaki, H. (1996). Connexin 32 mutations from X-linked Charcot-Marie-tooth disease patients: Functional defects and dominant negative effects. *Molecular Biology of the Cell, 7*, 907–916.

Orthmann-Murphy, J. L., Salsano, E., Abrams, C. K., et al. (2009). Hereditary spastic paraplegia is a novel phenotype for GJA12/GJC2 mutations. *Brain, 132*, 426–438.

Paul, D. L., Ebihara, L., Takemoto, L. J., Swenson, K. I., & Goodenough, D. A. (1991). Connexin46, a novel lens gap junction protein, induces voltage-gated currents in nonjunctional plasma membrane of Xenopus oocytes. *The Journal of Cell Biology, 115*, 1077–1089.

Paul, D. L. (1986). Molecular cloning of cDNA for rat liver gap junction protein. *The Journal of Cell Biology, 103*, 123–134.

Payton, B. W., Bennett, M. V., & Pappas, G. D. (1969). Permeability and structure of junctional membranes at an electrotonic synapse. *Science, 166*, 1641–1643.

Paznekas, W. A., Boyadjiev, S. A., Shapiro, R. E., et al. (2003). Connexin 43 (GJA1) mutations cause the pleiotropic phenotype of oculodentodigital dysplasia. *The American Journal of Human Genetics, 72*, 408–418.

Paznekas, W. A., Karczeski, B., Vermeer, S., et al. (2009). GJA1 mutations, variants, and connexin 43 dysfunction as it relates to the oculodentodigital dysplasia phenotype. *Human Mutation, 30*, 724–733.

Peracchia, C. (2004). Chemical gating of gap junction channels; roles of calcium, pH and calmodulin. *Biochimica et Biophysica Acta, 1662*, 61–80.

Peracchia, C., & Dulhunty, A. F. (1976). Low resistance junctions in crayfish. Structural changes with functional uncoupling. *The Journal of Cell Biology, 70*, 419–439.

Peracchia, C. (1977). Gap junctions. Structural changes after uncoupling procedures. *The Journal of Cell Biology, 72*, 628–641.

Peracchia, C., & Peracchia, L. L. (1980a). Gap junction dynamics: Reversible effects of divalent cations. *The Journal of Cell Biology, 87*, 708–718.

Peracchia, C., & Peracchia, L. L. (1980b). Gap junction dynamics: Reversible effects of hydrogen ions. *The Journal of Cell Biology, 87*, 719–727.

Peracchia, C., Wang, X., Li, L., & Peracchia, L. L. (1996). Inhibition of calmodulin expression prevents low-pH-induced gap junction uncoupling in Xenopus oocytes. *Pflügers Archiv, 431*, 379–387.

Peskar, B. M. (1980). Effect of carbenoxolone on prostaglandin synthesizing and metabolizing enzymes and correlation with gastric mucosal carbenoxolone concentrations. *Scandinavian Journal of Gastroenterology - Supplement, 65*, 109–114.

Ping, P., Zhang, J., Qiu, Y., et al. (1997). Ischemic preconditioning induces selective translocation of protein kinase C isoforms epsilon and eta in the heart of conscious rabbits without subcellular redistribution of total protein kinase C activity. *Circulation Research, 81*, 404–414.

Reaume, A. G., de Sousa, P. A., Kulkarni, S., et al. (1995). Cardiac malformation in neonatal mice lacking connexin43. *Science, 267*, 1831–1834.

Retamal, M. A. (2014). Connexin and pannexin hemichannels are regulated by redox potential. *Frontiers in Physiology, 5*, 80.

Retamal, M. A., Cortés, C. J., Reuss, L., Bennett, M. V., & Sáez, J. C. (2006). S-Nitrosylation and permeation through connexin 43 hemichannels in astrocytes: Induction by oxidant stress and reversal by reducing agents. *Proceedings of the National Academy of Sciences of the United States of America, 103*, 4475–4480.

Revel, J. P., & Karnovsky, M. J. (1967). Hexagonal array of subunits in intercellular junctions of the mouse heart and liver. *The Journal of Cell Biology, 33*, C7–C12.

Revel, J. P., Nicholson, B. J., & Yancey, S. B. (1985). Chemistry of gap junctions. *Annual Review of Physiology, 47*, 263–279.

Richard, G., Brown, N., Ishida-Yamamoto, A., & Krol, A. (2004). Expanding the phenotypic spectrum of Cx26 disorders: Bart-Pumphrey syndrome is caused by a novel missense mutation in GJB2. *Journal of Investigative Dermatology, 123*, 856–863.

Richard, G., Rouan, F., Willoughby, C. E., et al. (2002). Missense mutations in GJB2 encoding connexin-26 cause the ectodermal dysplasia keratitis-ichthyosis-deafness syndrome. *The American Journal of Human Genetics, 70*, 1341–1348.

Richard, G., Smith, L. E., Bailey, R. A., et al. (1998). Mutations in the human connexin gene GJB3 cause erythrokeratodermia variabilis. *Nature Genetics, 20*, 366–369.

Robertson, J. D. (1963). The occurrence of a subunit pattern in the unit membranes of club endings in Mauthner cell synapses in goldfish brains. *The Journal of Cell Biology, 19*, 201–221.

Rodriguez-Sinovas, A., Boengler, K., Cabestrero, A., et al. (2006). Translocation of connexin 43 to the inner mitochondrial membrane of cardiomyocytes through the heat shock protein 90-dependent TOM pathway and its importance for cardioprotection. *Circulation Research, 99*, 93–101.

Rose, B., & Loewenstein, W. R. (1975). Permeability of cell junction depends on local cytoplasmic calcium activity. *Nature, 254*, 250–252.

Sáez, J. C., Connor, J. A., Spray, D. C., & Bennett, M. V. (1989). Hepatocyte gap junctions are permeable to the second messenger, inositol 1,4,5-trisphosphate, and to calcium ions. *Proceedings of the National Academy of Sciences of the United States of America, 86*, 2708–2712.

Sáez, J. C., Nairn, A. C., Czernik, A. J., et al. (1990). Phosphorylation of connexin 32, a hepatocyte gap-junction protein, by cAMP-dependent protein kinase, protein kinase C and Ca2+/calmodulin-dependent protein kinase II. *European Journal of Biochemistry, 192*, 263–273.

Sáez, J. C., Spray, D. C., Nairn, A. C., Hertzberg, E., Greengard, P., & Bennett, M. V. (1986). cAMP increases junctional conductance and stimulates phosphorylation of the 27-kDa principal gap junction polypeptide. *Proceedings of the National Academy of Sciences of the United States of America, 83*, 2473–2477.

Sánchez, H. A., Mese, G., Srinivas, M., White, T. W., & Verselis, V. K. (2010). Differentially altered Ca^{2+} regulation and Ca^{2+} permeability in Cx26 hemichannels formed by the A40V and G45E mutations that cause keratitis ichthyosis deafness syndrome. *The Journal of General Physiology, 136*, 47–62.

Söhl, G., & Willecke, K. (2003). An update on connexin genes and their nomenclature in mouse and man. *Cell Communication and Adhesion, 10*, 173–180.

Saffitz, J. E. (2017). Molecular mechanisms in the pathogenesis of arrhythmogenic cardiomyopathy. *Cardiovascular Pathology, 28*, 51–58.

Saitou, M., Furuse, M., Sasaki, H., et al. (2000). Complex phenotype of mice lacking occludin, a component of tight junction strands. *Molecular Biology of the Cell, 11*, 4131–4142.

Sato, P. Y., Coombs, W., Lin, X., et al. (2011). Interactions between ankyrin-G, Plakophilin-2, and Connexin43 at the cardiac intercalated disc. *Circulation Research, 109*, 193–201.

Scherer, S. S., Xu, Y. T., Nelles, E., Fischbeck, K., Willecke, K., & Bone, L. J. (1998). Connexin32-null mice develop demyelinating peripheral neuropathy. *Glia, 24*, 8–20.

Schneeberger, E. E., & Lynch, R. D. (2004). The tight junction: A multifunctional complex. *American Journal of Physiology - Cell Physiology, 286*, C1213–C1228.

Schulz, R., Görge, P. M., Görbe, A., Ferdinandy, P., Lampe, P. D., & Leybaert, L. (2015). Connexin 43 is an emerging therapeutic target in ischemia/reperfusion injury, cardioprotection and neuroprotection. *Pharmacology & Therapeutics, 153*, 90–106.

Schwanke, U., Konietzka, I., Duschin, A., Li, X., Schulz, R., & Heusch, G. (2002). No ischemic preconditioning in heterozygous connexin43-deficient mice. *American Journal of Physiology - Heart and Circulatory Physiology, 283*, H1740–H1742.

Severs, N. J., Bruce, A. F., Dupont, E., & Rothery, S. (2008). Remodelling of gap junctions and connexin expression in diseased myocardium. *Cardiovascular Research, 80*, 9–19.

Sheridan, J. D., Hammer-Wilson, M., Preus, D., & Johnson, R. G. (1978). Quantitative analysis of low-resistance junctions between cultured cells and correlation with gap-junctional areas. *The Journal of Cell Biology, 76*, 532–544.

Shi, H. F., Yang, J. F., Wang, Q., et al. (2013). Prevalence and spectrum of GJA5 mutations associated with lone atrial fibrillation. *Molecular Medicine Reports, 7*, 767–774.

Shibayama, J., Lewandowski, R., Kieken, F., et al. (2006). Identification of a novel peptide that interferes with the chemical regulation of connexin43. *Circulation Research, 98*, 1365–1372.

Shiels, A., Mackay, D., Ionides, A., Berry, V., Moore, A., & Bhattacharya, S. (1998). A missense mutation in the human connexin50 gene (GJA8) underlies autosomal dominant "zonular pulverulent" cataract, on chromosome 1q. *The American Journal of Human Genetics, 62*, 526–532.

Simon, A. M., Goodenough, D. A., & Paul, D. L. (1998). Mice lacking connexin40 have cardiac conduction abnormalities characteristic of atrioventricular block and bundle branch block. *Current Biology, 8*, 295–298.

Simon, B., Huart, A. S., & Wilmanns, M. (2015). Molecular mechanisms of protein kinase regulation by calcium/calmodulin. *Bioorganic & Medicinal Chemistry, 23*, 2749–2760.

Simpson, I., Rose, B., & Loewenstein, W. R. (1977). Size limit of molecules permeating the junctional membrane channels. *Science, 195*, 294–296.

Smith, F. J., Morley, S. M., & McLean, W. H. (2002). A novel connexin 30 mutation in Clouston syndrome. *Journal of Investigative Dermatology, 118*, 530–532.

Soetkamp, D., Nguyen, T. T., Menazza, S., et al. (2014). S-nitrosation of mitochondrial connexin 43 regulates mitochondrial function. *Basic Research in Cardiology, 109*, 433.

Solan, J. L., & Lampe, P. D. (2005). Connexin phosphorylation as a regulatory event linked to gap junction channel assembly. *Biochimica et Biophysica Acta, 1711*, 154–163.

Solan, J. L., & Lampe, P. D. (2009). Connexin43 phosphorylation: Structural changes and biological effects. *Biochemical Journal, 419*, 261–272.

Solan, J. L., & Lampe, P. D. (2014). Specific Cx43 phosphorylation events regulate gap junction turnover in vivo. *FEBS Letters, 588*, 1423–1429.

Solan, J. L., Marquez-Rosado, L., Sorgen, P. L., Thornton, P. J., Gafken, P. R., & Lampe, P. D. (2007). Phosphorylation at S365 is a gatekeeper event that changes the structure of Cx43 and prevents down-regulation by PKC. *The Journal of Cell Biology, 179*, 1301–1309.

Spray, D. C., Harris, A. L., & Bennett, M. V. (1981). Gap junctional conductance is a simple and sensitive function of intracellular pH. *Science, 211*, 712–715.

Spray, D. C., Harris, A. L., & Bennett, M. V. (1979). Voltage dependence of junctional conductance in early amphibian embryos. *Science, 204*, 432–434.

Srinivas, M., Hopperstad, M. G., & Spray, D. C. (2001). Quinine blocks specific gap junction channel subtypes. *Proceedings of the National Academy of Sciences of the United States of America, 98*, 10942–10947.

Srinivas, M., Rozental, R., Kojima, T., et al. (1999). Functional properties of channels formed by the neuronal gap junction protein connexin36. *Journal of Neuroscience, 19*, 9848–9855.

Srisakuldee, W., Jeyaraman, M. M., Nickel, B. E., Tanguy, S., Jiang, Z. S., & Kardami, E. (2009). Phosphorylation of connexin-43 at serine 262 promotes a cardiac injury-resistant state. *Cardiovascular Research, 83*, 672–681.

Stauffer, K. A. (1995). The gap junction proteins beta 1-connexin (connexin-32) and beta 2-connexin (connexin-26) can form heteromeric hemichannels. *Journal of Biological Chemistry, 270*, 6768–6772.

Stevenson, B. R., Siliciano, J. D., Mooseker, M. S., & Goodenough, D. A. (1986). Identification of ZO-1: A high molecular weight polypeptide associated with the tight junction (zonula occludens) in a variety of epithelia. *The Journal of Cell Biology, 103*, 755–766.

Stewart, P. M., Wallace, A. M., Valentino, R., Burt, D., Shackleton, C. H., & Edwards, C. R. (1987). Mineralocorticoid activity of liquorice: 11-beta-hydroxysteroid dehydrogenase deficiency comes of age. *Lancet, 2*, 821–824.

Straub, A. C., Billaud, M., Johnstone, S. R., et al. (2011). Compartmentalized connexin 43 s-nitrosylation/denitrosylation regulates heterocellular communication in the vessel wall. *Arteriosclerosis, Thrombosis, and Vascular Biology, 31*, 399–407.

Swenson, R. P., & Narahashi, T. (1980). Block of sodium conductance by n-octanol in crayfish giant axons. *Biochimica et Biophysica Acta, 603*, 228–236.

Török, K., Stauffer, K., & Evans, W. H. (1997). Connexin 32 of gap junctions contains two cytoplasmic calmodulin-binding domains. *Biochemical Journal, 326*, 479–483.

Tare, M., Coleman, H. A., & Parkington, H. C. (2002). Glycyrrhetinic derivatives inhibit hyperpolarization in endothelial cells of Guinea pig and rat arteries. *American Journal of Physiology - Heart and Circulatory Physiology, 282*, H335–H341.

Temme, A., Buchmann, A., Gabriel, H. D., Nelles, E., Schwarz, M., & Willecke, K. (1997). High incidence of spontaneous and chemically induced liver tumors in mice deficient for connexin32. *Current Biology, 7*, 713–716.

Thibodeau, I. L., Xu, J., Li, Q., et al. (2010). Paradigm of genetic mosaicism and lone atrial fibrillation: Physiological characterization of a connexin 43-deletion mutant identified from atrial tissue. *Circulation, 122*, 236–244.

Tovar, K. R., Maher, B. J., & Westbrook, G. L. (2009). Direct actions of carbenoxolone on synaptic transmission and neuronal membrane properties. *Journal of Neurophysiology, 102*, 974–978.

Traub, O., Eckert, R., Lichtenberg-Fraté, H., et al. (1994). Immunochemical and electrophysiological characterization of murine connexin40 and -43 in mouse tissues and transfected human cells. *European Journal of Cell Biology, 64*, 101–112.

Tsien, R. W., & Weingart, R. (1976). Inotropic effect of cyclic AMP in calf ventricular muscle studied by a cut end method. *The Journal of Physiology, 260*, 117–141.

Tsukita, S., Furuse, M., & Itoh, M. (2001). Multifunctional strands in tight junctions. *Nature Reviews Molecular Cell Biology, 2*, 285–293.

Tuomi, J. M., Tyml, K., & Jones, D. L. (2011). Atrial tachycardia/fibrillation in the connexin 43 G60S mutant (Oculodentodigital dysplasia) mouse. *American Journal of Physiology - Heart and Circulatory Physiology, 300*, H1402–H1411.

Turin, L., & Warner, A. (1977). Carbon dioxide reversibly abolishes ionic communication between cells of early amphibian embryo. *Nature, 270*, 56–57.

Turin, L., & Warner, A. E. (1980). Intracellular pH in early Xenopus embryos: Its effect on current flow between blastomeres. *The Journal of Physiology, 300*, 489–504.

Uehara, Y., & Burnstock, G. (1970). Demonstration of "gap junctions" between smooth muscle cells. *The Journal of Cell Biology, 44*, 215–217.

Uhlenberg, B., Schuelke, M., Ruschendorf, F., et al. (2004). Mutations in the gene encoding gap junction protein alpha-12 (connexin 46.6) cause Pelizaeus-Merzbacher-like disease. *The American Journal of Human Genetics, 75*, 251–260.

Unger, V. M., Kumar, N. M., Gilula, N. B., & Yeager, M. (1999). Three-dimensional structure of a recombinant gap junction membrane channel. *Science, 283*, 1176–1180.

Unwin, P. N., & Ennis, P. D. (1983). Calcium-mediated changes in gap junction structure: Evidence from the low angle X-ray pattern. *The Journal of Cell Biology, 97*, 1459–1466.

Unwin, P. N., & Ennis, P. D. (1984). Two configurations of a channel-forming membrane protein. *Nature, 307*, 609–613.

Valiunas, V., Polosina, Y. Y., Miller, H., et al. (2005). Connexin-specific cell-to-cell transfer of short interfering RNA by gap junctions. *The Journal of Physiology, 568*, 459–468.

Van Eldik, L. J., Hertzberg, E. L., Berdan, R. C., & Gilula, N. B. (1985). Interaction of calmodulin and other calcium-modulated proteins with mammalian and arthropod junctional membrane proteins. *Biochemical and Biophysical Research Communications, 126*, 825–832.

van Geel, M., van Steensel, M. A. M., Kuster, W., et al. (2002). HID and KID syndromes are associated with the same connexin 26 mutation. *British Journal of Dermatology, 146*, 938–942.

Van Itallie, C. M., & Anderson, J. M. (2006). Claudins and epithelial paracellular transport. *Annual Review of Physiology, 68*, 403–429.

Van Rijen, H., van Kempen, M. J., Analbers, L. J., et al. (1997). Gap junctions in human umbilical cord endothelial cells contain multiple connexins. *American Journal of Physiology, 272*, C117–C130.

Vaughan-Jones, R. D., Lederer, W. J., & Eisner, D. A. (1983). Ca^{2+} ions can affect intracellular pH in mammalian cardiac muscle. *Nature, 301*, 522–524.

Veenstra, R. D., & DeHaan, R. L. (1986). Measurement of single channel currents from cardiac gap junctions. *Science, 233*, 972–974.

Veenstra, R. D. (2015). Gap junction channels: The electrical conduit of the intercellular world. In A. H. Delcour (Ed.), *Springer series in biophysics: Vol. 18. Electrophysiology of unconventional channels and pores* (pp. 313–341). Switzerland: Springer International Publishing.

Veenstra, R. D. (1996). Size and selectivity of gap junction channels formed from different connexins. *Journal of Bioenergetics and Biomembranes, 28*, 327–337.

Veenstra, R. D., Wang, H. Z., Beblo, D. A., et al. (1995). Selectivity of connexin-specific gap junctions does not correlate with channel conductance. *Circulation Research, 77*, 1156–1165.

Veenstra, R. D., Wang, H. Z., Beyer, E. C., Ramanan, S. V., & Brink, P. R. (1994). Connexin37 forms high conductance gap junction channels with subconductance state activity and selective dye and ionic permeabilities. *Biophysical Journal, 66*, 1915–1928.

Veenstra, R. D., Wang, H. Z., Westphale, E. M., & Beyer, E. C. (1992). Multiple connexins confer distinct regulatory and conductance properties of gap junctions in developing heart. *Circulation Research, 71*, 1277–1283.

Verma, V., Larsen, B. D., Coombs, W., et al. (2009). Novel pharmacophores of connexin43 based on the "RXP" series of Cx43-binding peptides. *Circulation Research, 105*, 176–184.

Wagner, C., de Wit, C., Kurtz, L., Grünberger, C., Kurtz, A., & Schweda, F. (2007). Connexin40 is essential for the pressure control of renin synthesis and secretion. *Circulation Research, 100*, 556–563.

Wang, J., Ma, M., Locovei, S., Keane, R. W., & Dahl, G. (2007). Modulation of membrane channel currents by gap junction protein mimetic peptides: Size matters. *American Journal of Physiology - Cell Physiology, 293*, C1112–C1119.

Wang, N., De Vuyst, E., Ponsaerts, R., et al. (2013). Selective inhibition of Cx43 hemichannels by Gap19 and its impact on myocardial ischemia/reperfusion injury. *Basic Research in Cardiology, 108*, 309.

Wang, H. Z., & Veenstra, R. D. (1997). Monovalent ion selectivity sequences of the rat connexin43 gap junction channel. *The Journal of General Physiology, 109*, 491–507.

Wang, X. G., & Peracchia, C. (1997). Positive charges of the initial C-terminus domain of Cx32 inhibit gap junction gating sensitivity to CO2. *Biophysical Journal, 73*, 798–806.

Warn-Cramer, B. J., Lampe, P. D., Kurata, W. E., et al. (1996). Characterization of the mitogen-activated protein kinase phosphorylation sites on the connexin-43 gap junction protein. *Journal of Biological Chemistry, 271*, 3779–3786.

Weidmann, S. (1952). The electrical constant of Purkinje fibers. *The Journal of Physiology, 118*, 348–360.

Welsh, M. J., Aster, J. C., Ireland, M., Alcala, J., & Maisel, H. (1982). Calmodulin binds to chick lens gap junction protein in a calcium-independent manner. *Science, 216*, 642–644.

White, T. W., Goodenough, D. A., & Paul, D. L. (1998). Targeted ablation of connexin50 in mice results in microphthalmia and zonular pulverulent cataracts. *The Journal of Cell Biology, 143*, 815–825.

White, T. W., Srinivas, M., Ripps, H., Trovato-Salinaro, A., Condorelli, D. F., & Bruzzone, R. (2002). Virtual cloning, functional expression, and gating analysis of human connexin31.9. *American Journal of Physiology - Cell Physiology, 283*, C960–C970.

White, T. W. (2002). Unique and redundant connexin contributions to lens development. *Science, 295*, 319–320.

Wirka, R. C., Gore, S., Van Wagoner, D. R., et al. (2011). A common connexin-40 gene promoter variant affects connexin-40 expression in human atria and is associated with atrial fibrillation. *Circulation: Arrhythmia and Electrophysiology, 4*, 87–93.

Xia, J., Liu, C., Tang, B., et al. (1998). Mutations in the gene encoding gap junction protein beta-3 associated with autosomal dominant hearing impairment. *Nature Genetics, 20*, 370–373.

Xu, Q., Kopp, R. F., Chen, Y., Yang, J. J., Roe, M. W., & Veenstra, R. D. (2012). Gating of connexin 43 gap junctions by a cytoplasmic loop calmodulin binding domain. *American Journal of Physiology - Cell Physiology, 302*, C1548–C1556.

Xu, Q., Lin, X., Andrews, L., Patel, D., Lampe, P. D., & Veenstra, R. D. (2013). Histone deacetylase inhibition reduces cardiac connexin43 expression and gap junction communication. *Frontiers in Pharmacology, 4*, 44.

Xu, X., Li, W. E., Huang, G. Y., et al. (2001). N-cadherin and Cx43alpha1 gap junctions modulates mouse neural crest cell motility via distinct pathways. *Cell Communication and Adhesion, 8*, 321–324.

Yamada, Y., Izawa, H., Ichihara, S., et al. (2002). Prediction of the risk of myocardial infarction from polymorphisms in candidate genes. *New England Journal of Medicine, 347*, 1916–1923.

Yang, Y. Q., Liu, X., Zhang, X. L., et al. (2010). Novel connexin40 missense mutations in patients with familial atrial fibrillation. *Europace, 12*, 1421–1427.

Zelante, L., Gasparini, P., Estivill, X., et al. (1997). Connexin26 mutations associated with the most common form of non-syndromic neurosensory autosomal recessive deafness (DFNB1) in Mediterraneans. *Human Molecular Genetics, 6*, 1605–1609.

Zhang, J. T., & Nicholson, B. J. (1989). Sequence and tissue distribution of a second protein of hepatic gap junctions, Cx26, as deduced from its cDNA. *The Journal of Cell Biology, 109*, 3391–3401.

Zhou, Y., Yang, W., Lurtz, M. M., et al. (2007). Identification of the calmodulin binding domain of connexin 43. *Journal of Biological Chemistry, 282*, 35005–35017.

Zihni, C., Mills, C., Matter, K., & Balda, M. S. (2016). Tight junctions: From simple barriers to multifunctional molecular gates. *Nature Reviews Molecular Cell Biology, 17*, 564–580.

Zimmer, D. B., Green, C. R., Evans, W. H., & Gilula, N. B. (1987). Topological analysis of the major protein in isolated intact rat liver gap junctions and gap junction-derived single membrane structures. *Journal of Biological Chemistry, 262*, 7751–7763.

Zonta, F., Polles, G., Zanotti, G., & Mammano, F. (2012). Permeation pathway of homomeric connexin 26 and connexin 30 channels investigated by molecular dynamics. *Journal of Biomolecular Structure and Dynamics, 29*, 985–998.

Zou, J., Salarian, M., Chen, Y., Veenstra, R., Louis, C. F., & Yang, J. J. (2014). Gap junction regulation by calmodulin. *FEBS Letters, 588*, 1430–1438.

Chapter 9

Biophysics of intracellular and extracellular diffusion

Francisco J. Alvarez-Leefmans
Department of Pharmacology and Toxicology, Boonshoft School of Medicine, Wright State University, Dayton, OH, United States

1. Summary

This chapter discusses the physical principles of diffusion, that is, the net passive displacement (transport) of "matter" through a "medium," arising from the random motion of ions or molecules due to thermal energy. The "matter" considered here are uncharged or charged particles (e.g., ions), and the "medium" in which matter moves is the aqueous phase of the extracellular space, or the crowded cytoplasmic environment. The factors determining the diffusion of biomolecules in the extracellular and intracellular spaces are discussed. Core concepts are built from classic observations of Brownian motion and Fick's laws, the fundamental constants of the microscopic theory of diffusion derived by Einstein and experimentally tested by Perrin, to contemporary methods applying physical principles to describe quantitatively cytoplasmic and interstitial space diffusion phenomena.

2. Molecular dynamics of diffusion in gases and solutions

Diffusion underlies basic biological phenomena such as solute and water transport across biological membranes and net solute movement in the extracellular and intracellular space. In the extracellular space, diffusion is fundamental for spreading signaling molecules, ions, hormones, or neurotransmitters, whereas in the intracellular space, diffusion of biomolecules determines the rate of metabolic reactions.

2.1 What is diffusion?

Diffusion is the net displacement (transport) of matter through a medium, from one region to another, due to random molecular motions caused by thermal agitation. It is a *passive transport process* since particles (atoms or molecules) in a liquid or a gas move from regions of higher concentration to regions of lower concentration due to random thermal motion of both the particles that diffuse and those composing the medium. Thus, diffusion is an irreversible process by which a difference in concentration is reduced by the spontaneous flow of matter. This process reaches *equilibrium* when the particles are uniformly distributed. The stochastic molecular motions continue at equilibrium, but the *net movement* of particles is zero. In the absence of *advection*, which refers to the transport of mass by bulk flow of fluid, such as that caused by stirring, the movement of the molecules is by diffusion only.

Diffusion can readily be observed by placing a concentrated aqueous solution of a dye, e.g., methylene blue, at the bottom of a glass cylinder, carefully layering pure water above the colored solution. Initially, there is a sharp boundary between the two solutions. Still, as time progresses, the upper solution becomes increasingly colored, whereas the lower solution becomes paler until the entire column looks uniform. At this stage, diffusion of the dye ceases; the system has achieved equilibrium. The individual molecules of both the dye (solute) and the solvent (water) are in constant motion with velocities dependent upon temperature; the higher the temperature, the faster they move, hence the terms "thermal agitation" or "thermal movement." Thus, relatively small solute molecules, like methylene blue (Mr 319.85 g/mol) in the example considered, move on their own because of their thermal energy and also move by being bombarded by solvent molecules which, in this

Cell Physiology Sourcebook. https://doi.org/10.1016/B978-0-12-811114-7.00002-5

example, are water molecules. Large molecules, such as proteins, lipids, or particles in suspension or emulsion, move mainly by solvent bombardment. This random movement of molecules or particles in a fluid is called *Brownian motion or Brownian movement.* A broader definition of Brownian motion is "any of various physical phenomena in which some quantity is constantly undergoing small, random fluctuations" (Gregersen, 2017). As discussed below, Brownian motion was named after the Scottish botanist Robert Brown, who first studied it systematically (Brown, 1828).

2.2 Diffusion is a macroscopic manifestation of Brownian motion at the microscopic level

The macroscopic laws of diffusion were first derived in the 19th century by the German-born physiologist Adolf Fick (1829–1921) while working at the University of Zurich. Fick's first law relates the diffusive flux of an electrically neutral solute. It postulates that the net solute flux direction goes from regions of high concentration to regions of low concentration of that solute. Fick's second law describes how the concentration of a solute changes with position and time as a result of diffusion (Fick, 1855). *Diffusion is a macroscopic manifestation of Brownian motion.* Thus, to better understand diffusion, it is convenient to start by considering the behavior of Brownian particles, thereby building from the microscopic to the macroscopic level. Following this scheme, we first discuss the theory of Brownian movement followed by Fick's equations. Returning to the example considered above (2.1), if it were possible to observe the individual molecules of methylene blue, it would be found that the movement of each molecule is random. In a dilute solution, each dye molecule behaves independently of the others, and each is constantly undergoing collisions with solvent molecules (water in this case); as a result of these collisions, dye molecules sometimes move toward a region of higher concentration, other times toward a region of lower concentration, without preferred direction. The motion of a single molecule can be described as a "random walk; " it is possible to calculate the mean square distance traveled in a given time interval, but the direction in which a given molecule will move at that time is random. Further, the ceaseless motion of each molecule is unrelated to past motion, i.e., the system has no memory, and the tortuous molecular path is totally random (Berg, 1993).

A similar situation to that described for particles in liquid solutions occurs in gasses, where particles (molecules or atoms) are also in constant random motion, moving rapidly in all directions, frequently colliding with each other, and tending to distribute evenly. As in liquids, gas particles are bombarded by the other moving particles in the medium in which they are suspended, e.g., air. The process in which gas particles intermingle due to random thermal motion is diffusion.

2.3 Kinetic energy and molecular movement

Given that gas particles are in constant motion, the kinetic energy (E_K) of an individual particle can be calculated by the following equation where m is the particle's mass and v is its velocity.

$$E_K = \frac{1}{2}mv^2 \tag{9.1}$$

When considering an ensemble of similar particles, the mean kinetic energy will be

$$\bar{E}_K = \frac{1}{2}\overline{mv^2} \tag{9.2}$$

The ideal gas law equation relates pressure (P), molar volume (V), and temperature (T), as $PV = nRT$, where n is the number of moles of the gas, R is the molar gas constant. The temperature T is related to the mean translational kinetic energy of the gas molecules, $\bar{E}_K$. Thus, according to the kinetic molecular theory, the $\bar{E}_K$ of gas particles is proportional to the absolute temperature of the gas (T). This is expressed in the following equation, where K_B is the Boltzmann Constant.

$$\bar{E}_K = \frac{3}{2}K_BT \tag{9.3}$$

The Boltzmann constant (K_B) is simply the molar gas constant (R), divided by Avogadro's number (N), and its value is $1.380{,}649 \times 10^{-23}\ J\,K^{-1}$, where J stands for Joules and K for Kelvin degrees. Thus, the Boltzmann constant K_B relates temperature with energy, bridging a macroscopic property (temperature) with a microscopic quantity (energy). Note that the translational motion velocity vector v has three degrees of freedom (one for each dimension), and thus, the average kinetic energy is $3/2K_BT$ (Eq. 9.3). Hence, the average kinetic energy of translational motion for one dimension (e.g., in the *x direction*) is $1/2\ K_BT$ (see below). Since $\bar{E}_K$ is related to both absolute temperature and average molecular velocity, we can combine Eqs. (9.2) and (9.3) and define the root-mean-square velocity as follows:

$$\bar{E}_K = \frac{1}{2}\overline{mv^2} = \frac{3}{2}K_BT \tag{9.4}$$

It follows that,

$$\sqrt{\overline{v^2}} = \sqrt{\frac{3k_BT}{m}} \tag{9.5}$$

Eq. (9.5) shows that the root mean square velocity of gas particles is directly proportional to the square root of temperature and inversely proportional to the square root of the gas molar mass. This explains why light molecules move faster than heavy molecules at the same temperature.

In sum, gas particles (atoms or molecules) move rapidly and randomly, in all directions, but without preferred direction. They frequently collide with each other and tend to spread evenly. The ambient temperature and the mass of the particles affect this movement. The average kinetic energy of gas particles is proportional to the absolute temperature of the gas, and all gases at the same temperature have the same average kinetic energy. The root mean square (rms) speed of gas molecules is also related to the molar mass of the substance. Comparing gases of different molar mass at the same temperature, we see that despite having the same average kinetic energy, the gas molecules with the smaller molar mass will have a higher rms speed.

The above insights were a prominent part of the kinetic theory of gases developed in the third quarter of the 19th century by physicists Rudolf Clausius (1822–88), James Clerk Maxwell (1831–79), and Ludwig Boltzmann (1844–1906). As we have seen, according to the theory, the temperature of a gas is proportional to the average kinetic energy with which the molecules are moving or vibrating. A reasonable guess from the kinetic theory is that the molecular motion might be imparted to larger particles that could be observed under a microscope. This line of reasoning lead to the mathematical theory of Brownian motion first developed by Albert Einstein in a series of epoch-making papers published between 1905 and 1908 (Einstein, 1956). The theoretical predictions from these papers were verified by quantitative microscopy during the first decade of the 20th century, notably by the work of the French physicist Jean-Baptiste Perrin (1870–1942), who in 1926 received the Nobel Prize in physics. Perrin's work was a turning point in our understanding of the elementary constitution of matter. This is because, despite the success of the kinetic-molecular theory of gases in the 19th century, there was no consensus between chemists and physicists about the reality of atoms and molecules. After Perrin's work, atoms and molecules became a physical reality and not just a conceptual construct (Brush, 1968; Nye, 1972).

3. Brownian motion and random walks

Recapitulating, Brownian motion is the ceaseless random movement of minute particles of matter suspended in a fluid medium caused by the continuous stochastic bombardment of the particles by atoms or molecules of the suspending medium. As mentioned above, Brownian motion takes its name after the Scottish botanist Robert Brown (1773–1858), who made the first systematic observation of this phenomenon (Brown, 1828). Initially, he used a single-lens microscope with a focal length of "about 1/32 of an inch" and an estimated magnification of about 370X (Pearle et al., 2010). In his first paper, Brown described the continuous motion of *minute particles ejected by pollen grains* of the plant *Clarkia pulchella* immersed in water (Brown, 1828). He found that the movement "arose neither from currents in the fluid, nor from its gradual evaporation, but belonged to the particle itself," an early demonstration that *convection* was not the cause of the motion. Further, the motion could be observed in inorganic particles from a wide range of minerals, suggesting that it was a general property of particulate matter.

In a second paper, Brown confirmed and extended his conclusions by observing oil-immersed water droplets containing the particles. "... in all the drops thus formed and protected, the motion of the particles takes place with undiminished activity, while the principal causes assigned for that motion, namely, evaporation, and their mutual attraction and repulsion, are either materially reduced or absolutely null." He emphasized that the "active molecules" were not animated, as proposed by others that had observed a similar motion before him (Brown, 1829). The word "molecules" used by Brown was in common usage in his time to denote *minute particles*. Although Brown might not have been the first to observe the motion that now carries his name, as he himself acknowledged, his great contribution was to find that almost any kind of matter, organic or inorganic, broken into fine particles exhibit the same jiggly motion. Thus, he extended the subject from the realm of biology into the realm of physics (Brush, 1968).

A common misconception found in the literature is that Brown observed the motion of the pollen grains themselves, while in fact, what he observed was the motion of smaller particles that reside *within* the pollen (Cadée, 1991; Pearle et al., 2010). This is evident just by reading the title of his 1828 paper: "*A brief account of microscopical observations made in the months of June, July and August 1827, on the particles contained in the pollen of plants; and on the general existence of active molecules in organic and inorganic bodies.*" This is an important point because Brownian motion is sensitive to the size of the particles in suspension.

Brownian motion can be readily observed and quantified using current imaging microscopy techniques (e.g., a CCD camera interfaced with a frame grabber and a computer) to visualize polystyrene microspheres of known diameter, suspended in liquids of known viscosity in temperature-controlled chambers. At constant temperature and viscosity of the medium, Brownian motion becomes less evident when the particle diameter is ≥ 2 μm (Jia et al., 2007; Newburgh et al., 2006). *Clarkia pulchella* pollen grains have a tetrahedral shape, with the longest dimension of around 100 μm, too large for their Brownian motion to be detected by the eye, as discussed in depth in an excellent paper titled "What Brown saw and You Can Too" (Pearle et al., 2010), which also shows beautiful micrographs of *Clarkia* pollen grains and their content. To understand why pollen contents, but not the pollen grains, can be seen to undergo Brownian motion, let us first consider what determines the mean square distance $\overline{x^2}$ traveled by a spherical particle suspended in water, at time *t*, in one dimension:

$$\overline{x^2} = \frac{RT}{3\pi N\eta r}t \tag{9.6}$$

This is Einstein's fundamental equation for Brownian motion, where *R*, *T*, and *N* were defined above, η is the dynamic viscosity of the medium, and *r* is the sphere radius (Einstein, 1905). In its usual notation, this equation was derived by the physicist Paul Langevin in 1908 (Langevin, 1908; Lemons & Gythiel, 1997). Note that the measure of the motion is not the mean velocity of the particle in suspension but the mean square of its displacement $\Delta x = x - x_o$ from its initial position x_0, that is $\overline{\Delta x^2}$. Thus, Eq. (9.6) shows that the mean square distance traveled in a given direction *x* by a spherical particle suspended in water, as a result of Brownian motion, at time *t*, is inversely proportional to the particle size, in this case, a sphere of radius *r*.

Brown mentioned that the particles ejected from *Clarkia* pollen were of two shapes: some were oblong and others were circular. The latter being smaller than the oblong ones. Both types of particles move randomly in water. We now know that the oblong particles are amyloplasts, i.e., starched-containing organelles, whereas the spherical ones are spherosomes. i.e., lipid-filled organelles. The average radii (half the long axis length) of *Clarkia* amyloplasts is about 2 μm, whereas that of spherosomes is about 0.5 μm (Pearle et al., 2010). Brown estimates of the size of these particles were slightly larger than those verified with modern microscopes, probably because of misinterpretation of the Airy-Disk, as discussed by Pearle et al. (2010). This, however, does not compromise Brown's conclusions. In this context, it is useful to have an expression for the average mean distance traveled $|\overline{x}|$ for a particle undergoing Brownian motion (Pearle et al., 2010). It can be shown that

$$|\overline{x}| \approx 0.80 \sqrt{\overline{x^2}} \tag{9.7}$$

Thus,

$$|\overline{x}| \simeq 0.80 \sqrt{\frac{K_B T\, t}{3\, \pi\, \eta r}} \tag{9.8}$$

This result holds for various particle shapes, where *r* is replaced with the effective radius r_{eff} for an oblong object. Using Eq. (9.8), Pearle and coworkers calculated the time dependence (in seconds) of the average mean distance $|\overline{x}|$ traveled (in μm) in water at 20°C, for particles with r_{eff} between 0.5 and 50 μm. Note that 50 μm is the r_{eff} of a pollen grain 100 μm across. In 1 s, a particle of 0.5 μm r_{eff}, travels 0.74 μm, whereas a particle of $r_{eff} = 50$ μm travels only 0.07 μm. An explanation of why the pollen contents, but not the pollen, can be seen to undergo Brownian motion has been put forward by Pearle et al. (2010). Their argument goes as follows. The human eye has an angular resolution of less than 1 arc-minute (0.02° or 0.0003 radians). Thus, at a distance of about 25 cm (10 inches), the eye cannot see a displacement <73 μm. Considering the optical properties of Brown's microscope (focal length of about 1/32 of an inch and 370X), a displacement <73/370 ≈ 0.2 μm cannot be detected by the eye with a 370X lens. Based on the Rayleigh criterion, they conclude that if the eye cannot resolve two separate points, it cannot detect an object's motion traveling between them. Using Eq. (9.8) to estimate $|\overline{x}|$ for particles having various effective radii, they conclude that particles with $r_{eff} < 4$ μm could be seen to move in 1 s, but not a pollen grain with $r_{eff} \approx 50$ μm.

4. The microscopic theory of diffusion

Brownian motion remained unexplained until 1905, when Einstein produced the first mathematical theory that could be tested experimentally (Maiocchi, 1990; Smith & Seth, 2020). Einstein's explanation of Brownian motion based on the kinetic-molecular conception of matter is considered one of the fundamental pillars supporting atomism and the basis for a microscopic theory of diffusion, for which there were only macroscopic phenomenological descriptions, i.e., Fick's laws. Further, Brownian motion theory laid the foundation of our understanding of stochastic phenomena (Andrews et al., 2009; Chandrasekhar, 1943; Cohen, 2005). Today, nobody doubts the existence of atoms and molecules, but at the beginning of the 20th Century, not everyone accepted the discontinuous nature of matter. Several emblematic physicists at that time maintained that, "it would be dangerous for the credibility of science, to base a complete theory of matter on some

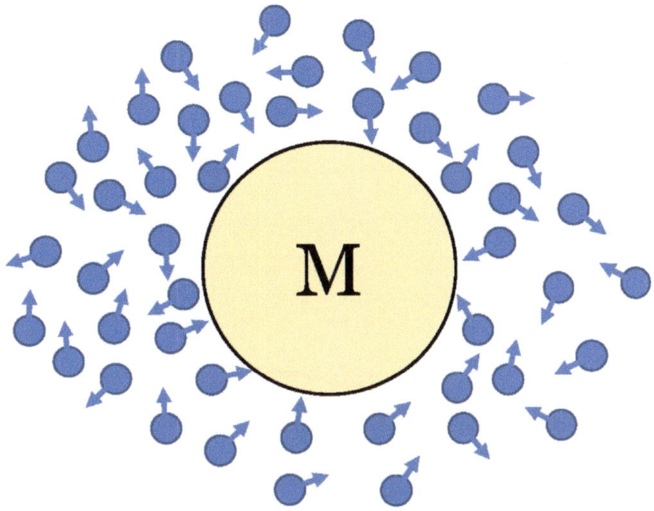

FIGURE 9.1 Representation of a large Brownian particle of mass M immersed in a fluid of much smaller and lighter particles (water molecules). The large Brownian particle is represented as a large yellow-filled circle, and the lighter particles (water molecules) are represented as blue-filled circles with arrows. The agitated motion of the large particle results from random and rapid collisions with the smaller molecules of the fluid.

hypothetical object that has never been seen" (Haw, 2005). In fact, Einstein's main motivation for a theory of Brownian motion was not to explain this phenomenon per se but to find measurable and observable manifestations of atoms to infer their physical reality. In the opening sentence of his 1905 paper (Einstein, 1905), he wrote:

> *It will be shown in this paper that, according to the molecular-kinetic theory of heat, bodies of microscopically visible size suspended in liquids must, as a result of thermal molecular motions, perform motions of such magnitude that these motions can easily be detected by a microscope.*

Einstein's work was founded on three basic ideas (Maiocchi, 1990): first, Van't Hoff's theory, according to which gas laws can be extended to dilute solutions, provided that osmotic pressure is considered instead of mechanical pressure, i.e., gas pressure and osmotic pressure are identical (Van't Hoff, 1887, 1901). This allowed the application of principles of the kinetic theory of gases to molecules in solutions. Thus, a liquid with *suspended particles* (like *Clarkia* spherosomes) could, for this purpose, be treated just as a liquid with *dissolved particles* (e.g., glucose molecules). Thus, Einstein pictured a relatively large particle (a Brownian particle) immersed (suspended) in a fluid of much smaller particles (atoms). The large particle is seen as a big atom pushed around by the "real" atoms (Fig. 9.1). According to energy equipartition, the statistical properties of the large particle are the same as the invisible, much smaller particles (atoms or molecules). That is, the mean kinetic energy $\bar{E}_K$ of the Brownian particle (e.g., *Clarkia* spherosomes) is the same as the $\bar{E}_K$ of the surrounding atoms or molecules (e.g., water molecules). Therefore, one can use the big particle as a probe for the ones we cannot see (Cohen, 2005; Rigden, 2005).

The second basic assumption was the extension of Stoke's law to Brownian motion. The key ideas here are *viscosity* and *diffusion*. Accordingly, *diffusion is thermal agitation opposed by a frictional force*. The assumption was that the frictional coefficient ξ for the diffusion of a large spherical particle should be the same as the classical frictional coefficient of a ball falling through a viscous fluid, which is given by Stoke's law:

$$\xi = 6\pi\eta r \tag{9.9}$$

where η is the dynamic viscosity and r is the radius of the sphere. Implicit in this assumption is that all the suspended particles have the same radius. It is important to note that the value of ξ is a minimal value; the asymmetric shape of a particle (molecule) or its interaction with the solvent (e.g., hydration) will increase ξ. Combining the kinetic theory of gases and Stokes' law, Einstein derived the *diffusion coefficient D* in terms of the fundamental constants,

$$D = \frac{RT}{N}\frac{1}{6\pi\eta r} \tag{9.10}$$

Thus, the diffusion coefficient D depends on the size and shape of the molecule, its interaction with the solvent, the viscosity of the solvent, as well as the temperature T. Given that R/N is K_B, from Eqs. (9.9) and (9.10), we obtain:

$$D = \frac{1}{\xi}K_BT \tag{9.11}$$

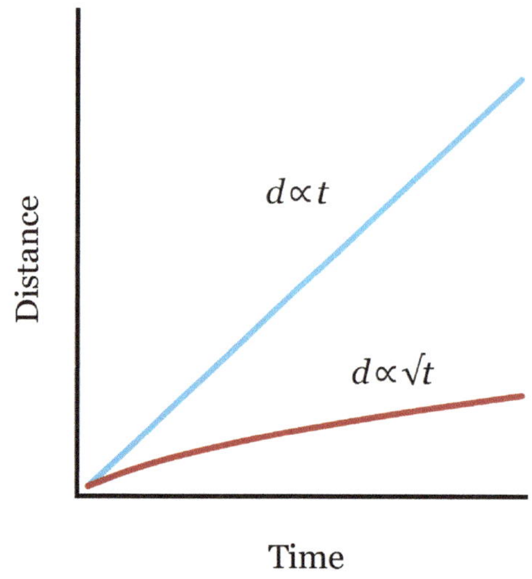

FIGURE 9.2 Comparison of linear and square-root dependence of diffusion distance over time. With a linear time dependence (blue line), equal increments of time (t) yield equal increments of distance traveled (d), as in linear motion. With a square-root time dependence (red line), as the distance to be traveled becomes greater, the time required to cover the distance becomes disproportionately longer. Thus, diffusion is a relatively inefficient process over large distances.

The derivation of the diffusion coefficient *D,* in terms of its determining constants, was a breakthrough in our understanding of the microscopic dynamics of diffusion. Before 1905, the diffusion coefficient or *diffusivity* was a constant that made Fick's laws work (Fick, 1855). Fick called it *k*, "a constant dependent upon the nature of the substances," but the essential parameters on which this constant depends were unknown. Eq. (9.10) is known as the *Stokes–Einstein relation.* It applies to diffusion in liquids when spherical solute molecules are much larger than the solvent molecules. It holds, however, even when the solute particles are smaller than water (Hille, 2001). According to this relation, the diffusion coefficient *D* of a molecule increases in inverse proportion to its radius. A larger molecular weight is related to a larger radius, so the diffusion coefficient is smaller. Likewise, the smaller the radius, the larger the mobility of the particle.

The third Einstein's idea was the revolutionary insight that the root-mean-square displacement of the Brownian particle with respect to its starting point (the origin), increases with the square root of time, independently of the path followed (Fig. 9.2). The path is the random walk of the particle (Figs. 9.3 and 9.6A–C). The central idea is that the proper measure of the motion is not, as many physicists thought at the time, the mean velocity of the particle in suspension, but the mean square of its displacement $\Delta x = x\text{-}x_o$, from the initial position x_0, that is $\overline{\Delta x^2}$, or simply $\overline{x^2}$. Einstein found that the mean square displacement $\overline{x^2}$ from the origin (x_o, at t_o) in the direction of the *x*-axis at time *t* is given by

$$\overline{x^2} = 2Dt \tag{9.12}$$

Thus, the sums of random displacements grow as the square root of time rather than in direct proportion to time, as in linear motion (Fig. 9.2). Note that in two dimensions, the right side of Eq. (9.12) would be $4Dt$, and in three dimensions, $6Dt$. From this equation, we can calculate the diffusion time *t* of an ion or a molecule over a given distance. For one dimension this is given by

$$t \approx \frac{\overline{x^2}}{2D} \tag{9.13}$$

where *D* is the diffusion coefficient of the solute in free solution. *D* is a measure of the rapidity with which a substance diffuses, and has the units of area/time (typically cm^2/s or $\mu m^2/s$). As mentioned above, Fick realized that the value of *D* is unique for each solute. From Eq. (9.10), it is evident that *D* is a function of both the physicochemical properties of the solute (shape, size, and molecular mass) and those of the medium (viscosity, temperature). In sum, diffusion time increases with the square of diffusion distance, and it is inversely proportional to *D*. Eq. (9.13) is often referred to as the "Einstein's approximation equation." For example, using this equation, one may ask, how long it takes for a glucose molecule to diffuse a given distance in free solution? The diffusion coefficient of glucose is 5×10^{-6} cm^2/s (in water at 25°C). Using Eq. (9.13), we can calculate that, in one dimension, glucose diffuses an average of 1 μm in 1 ms, but it will take 100 ms to diffuse 10 μm, 10 s to diffuse 100 μm, 16.7 min to diffuse 1 mm, and 28 h to diffuse 1 cm!

From the above discussion, it is clear that diffusion is adequate for the movement of ions/molecules over short distances (10–1000 nm), but for movement over long distances (e.g., = 100 μm), it is a relatively inefficient process. As an example of diffusion within short distances relevant to cell physiology, consider the diffusion of acetylcholine across the neuromuscular junction where the synaptic cleft is approximately 50 nm. The effective diffusion coefficient of acetylcholine in the extracellular space of the neuromuscular junction has been estimated to be 4×10^{-6} cm^2/s (Land et al., 1984). It can be calculated that it takes acetylcholine approximately 3.1 μs to diffuse from its presynaptic release site to reach the postsynaptic nicotinic acetylcholine receptors. This diffusion time is a minor fraction of the total synaptic delay of approximately 500 μs, measured at the neuromuscular junction (Katz & Miledi, 1965). In the Physiology Web, the reader can find an online calculator for the time required for diffusion over a given distance for other physiologically relevant solutes (https://www.physiologyweb.com/calculators/diffusion_time_calculator.html).

4.1 Experimental demonstration of the microscopic theory of diffusion: The endeavor of Jean Perrin

The theory of Brownian motion and diffusion proposed by Einstein was first verified by Jean Perrin, in a series of painstaking experiments (Perrin, 1910, 1990). For this purpose, Perrin and his students, Chaudesaigues and Dabrowski, developed techniques to produce small, spherical particles of uniform radius. These particles were prepared either from gum resin ("gamboge") or from mastic. Gamboge is a latex extracted from the tree *Garcinia morella* (Indian Gamboge), whereas mastic, used in the production of varnishes, is extracted from the bark of *Pistacia lentiscus* (mastic tree). Having prepared the particles with known uniform diameter, Perrin and Chaudesaigues verified the law of Brownian displacement by direct observation, sequenced every 30 s, using gamboge grains of radius 0.212 μm. Similar measurements were carried out by Perrin and Dabrowski using mastic particles having a radius of 0.52 μm. These measurements yielded the famous diagrams of recorded random walk trajectories (Fig. 9.3). The experimental approach of Perrin and Dabrowski was to use a camera lucida attached to a microscope so they were able to observe the particle and its projection on a piece of graph paper. Using this setup, they observed and recorded the Brownian motion of gamboge particles suspended in a liquid of known viscosity at constant temperature. They marked the particle's position on the graph paper at fixed time intervals (30 s). Three of these Brownian particle trajectories from Perrin's original work are reproduced in Fig. 9.3, in which successive dots were joined by line segments. For instance, in the leftmost trace, the particle was followed for 24 min (48 segments of 30 s each). This was the first recording of random walk trajectories of a Brownian particle!

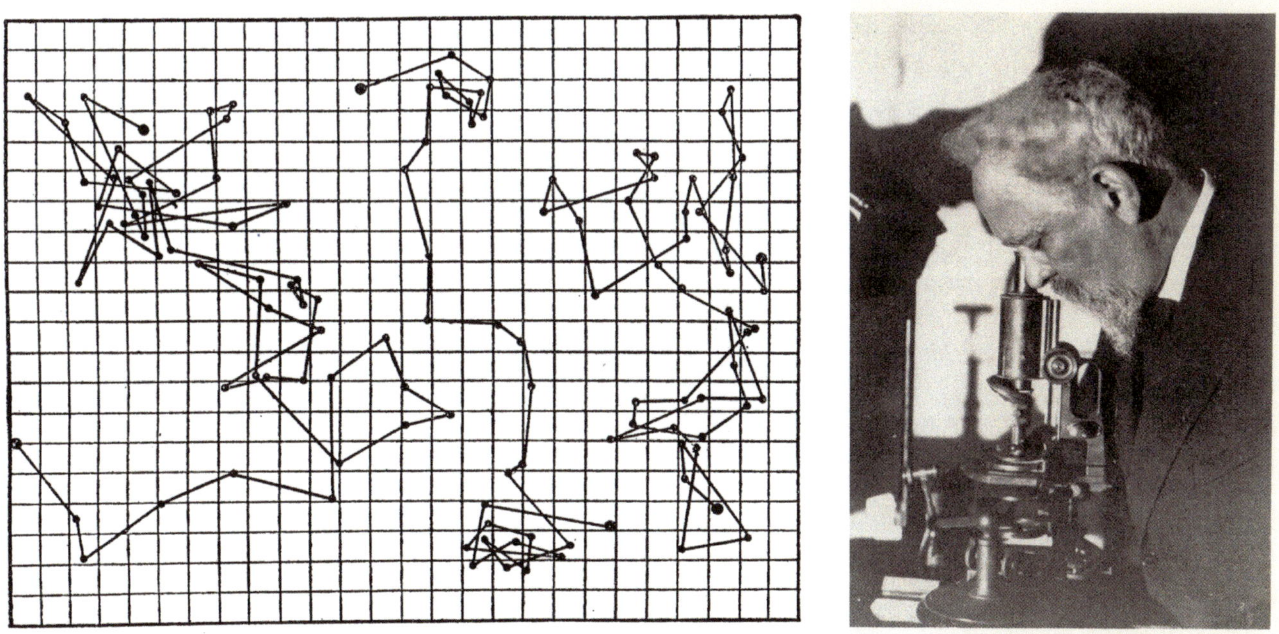

FIGURE 9.3 The first recording of random walk trajectories of Brownian particles. *Left:* Tracks of three Brownian particles recorded by Jean Perrin. Using a camera lucida he marked the successive positions of each particle at regular intervals of time. In this case, 30 s. Thus, the dots show the particle positions at 30 s intervals with straight lines joining successive points. Scale: 1 division = 3.125 μm. The radius of the mastic particles was 0.52 μm. *Reproduced from Fig. 6 of Perrin (1910).* Right: *Jean Perrin in his laboratory (1924). The photograph is reproduced from plate III of Nye (1972), by kind permission of Professor Mary Jo Nye.*

Using this particle tracking method, Perrin applied Einstein's Brownian motion equation (Eq. 9.6) to experimentally derive the value of N, the Avogadro's constant. From Eq. (9.6), it follows that:

$$N = \frac{t}{\overline{x^2}}\left(\frac{RT}{3\pi\eta r}\right) \tag{9.14}$$

Perrin showed that the mean square displacement $\overline{x^2}$ of the Brownian particles in the x direction for time t, behaved as Einstein predicted and led to an experimentally measured value for the Avogadro's constant N of 70.5×10^{22} mol^{-1}, close enough to the current value of 60.2×10^{22} mol^{-1}. Perrin tested the constancy of N using particles of various sizes and materials suspended in solutions of varying viscosity, demonstrating that Brownian motion depends on the viscosity and size of the particles, as predicted by Einstein's theory.

Another experimental insight we owe to Perrin was the demonstration that the random motion of the Brownian particles followed a Gaussian statistical distribution. Following a suggestion by his colleague Paul Langevin (1872–1946), Perrin plotted the motion of the particles giving them all a common origin. "The extremities of the vectors [i.e., the endpoints] so obtained should distribute themselves around this origin as bullets fired at a target distribute themselves around the bull's-eye" (Perrin, 1910). This was the first ever empirically confirmed instance of random motion in nature, i.e., motion exhibiting a Gaussian and hence Maxwellian distribution (Fig. 9.4).

4.2 The random walk: Deconstructing Einstein and Perrin

A deeper understanding of the microscopic theory of diffusion can be achieved by computational modeling of the spatial location of diffusing particles as a function of time. This computational model is the "random walk", a term introduced in 1905 by the English mathematician and biostatistician Karl Pearson (1857–1936), in a short note published in *Nature* (Pearson, 1905):

"A man starts from a point *0* and walks ℓ yards in a straight line; he then turns through any angle whatever and walks another ℓ yards in a second straight line. He repeats this process n times. I require the probability that after these n stretches he is at a distance between r and $r + \delta r$ from his starting point, *0*"

Let us start by considering a two-dimensional (2D) random walk, as the one represented in Fig. 9.5, where the motion of the Brownian particle occurs randomly in predefined discrete points in a Cartesian grid called a *lattice,* where the

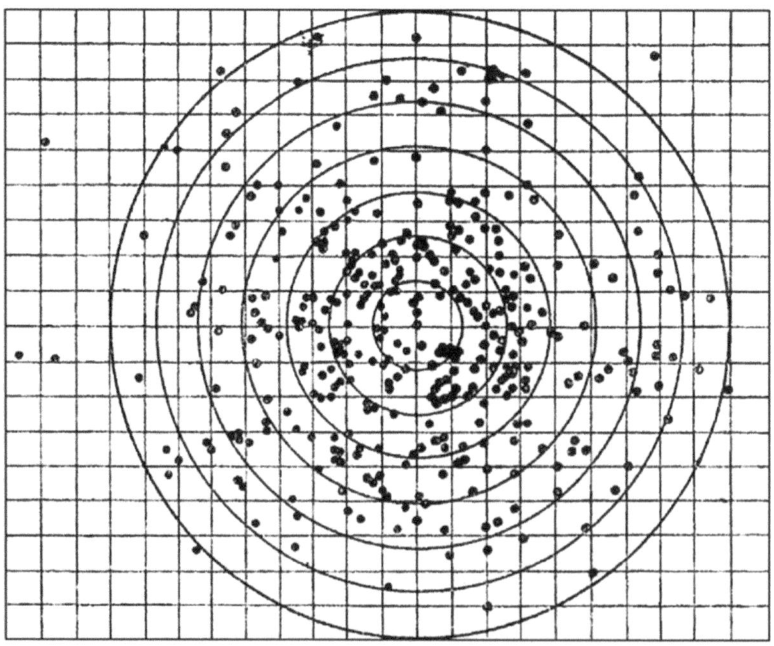

FIGURE 9.4 Graphical display of the observed displacements of particles, shifting the starting point of each motion to the origin. This yields a plot that looks like bullets fired at a target. To generate this graph, Perrin plotted 365 observations of granules of mastic (r = 0.52 μm). This illustrated the Gaussian distribution visually. *Reproduced from Fig. 7 of Perrin (1910).*

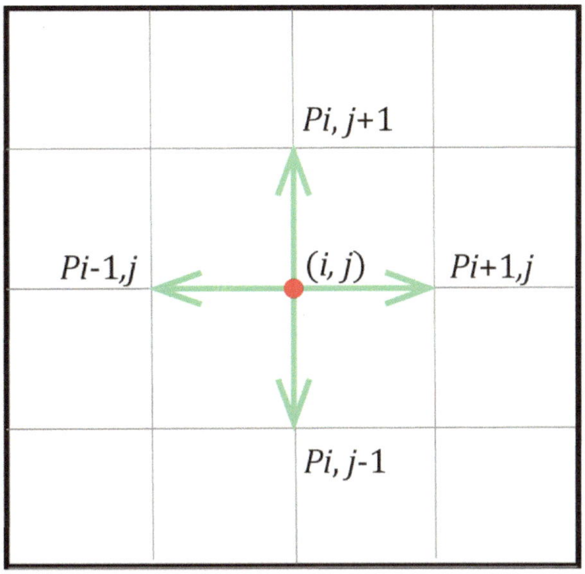

FIGURE 9.5 The random walk in 2D. A particle starting at location (*i, j*) at time *t* on a Cartesian grid of points can move to one of the four neighboring grid points with probability $Pi\pm1, j\pm1$ where $i \neq j$. *Modified from Frank (2023).*

locations of the points are named *vertices* (Frank, 2023). At time $t = 0$, the particle starting point in the lattice is (i,j). At time $t+1$, the particle jumps to any of the four neighboring grid points. For the sake of argument, let us say that the particle jumps to location *i, j+1,* with a probability *P i, j+1.* Since the sum of the probabilities to jump to any of the four points must be 1, the probability to jump to any of the four points is $P = ¼$. Note that if we were considering this problem in 1D, there would be only two points in the grid to jump to, whereas in 3D, there would be six points in the grid to jump to. Now, if the particle starts at the origin and we let this process run for *n* discrete time steps in a 2D computer simulation, we observe the random walk patterns shown in Fig. 9.6A–C. These patterns are similar to those that Perrin observed using real particles (Fig. 9.3). As predicted, the distribution of final particle locations obtained at the end of 1000 runs (Fig. 9.6D) results in a Gaussian distribution similar to that shown by Perrin in his graph of bullets fired at a target (Fig. 9.4).

The above concepts can be more clearly visualized in Fig. 9.7, showing the distribution of endpoints and the corresponding histogram along the *x*-axis (blue double-headed arrow line through the origin). Drawing the blue line in the x-axis in Fig. 9.7A at any angle, as long as it crosses the origin, will yield the same Gaussian distribution because the particles spread symmetrically about the origin. This is called *isotropic* Gaussian distribution, and the process that generates it, is *Gaussian diffusion* (Frank, 2023); it implies that the mean and the variance of the Gaussian distribution are the same irrespective of the angle of the blueline in Fig. 9.7A, as long as the line passes through the origin. When the variance is independent of the direction, the phenomenon is referred to as *isotropic diffusion.* Using this model, it is possible to find an expression for the diffusion coefficient consistent with that in Eq. (9.12). An excellent quantitative argument leading to the derivation of the diffusion coefficient from the random walk and on how the random walk becomes Gaussian is found elsewhere (Berg, 1993; Frank, 2023).

The one-dimensional (1D) Gaussian distribution is:

$$p\ (x) = \frac{1}{\sqrt{2\pi\sigma^2}} e^{-x^2/2\sigma^2} \tag{9.15}$$

where the variance is $\sigma^2 = 2\ Dt$. It is possible to extend this to a 2D random walk. An assumption in the 2D model is that the particles move independently, with the same probability, in each direction—that is, the probability of moving in the *x* direction is independent of the probability of moving in the *y* direction. Thus, $p\ (x,y) = p(x)p(y)$. Extending Eq. 9.15 to 2D, the final distribution of particles is given by the following expression:

$$p\ (x, y) = \frac{1}{2\pi\sigma^2} e^{-(x^2+y^2)/2\sigma^2} \tag{9.16}$$

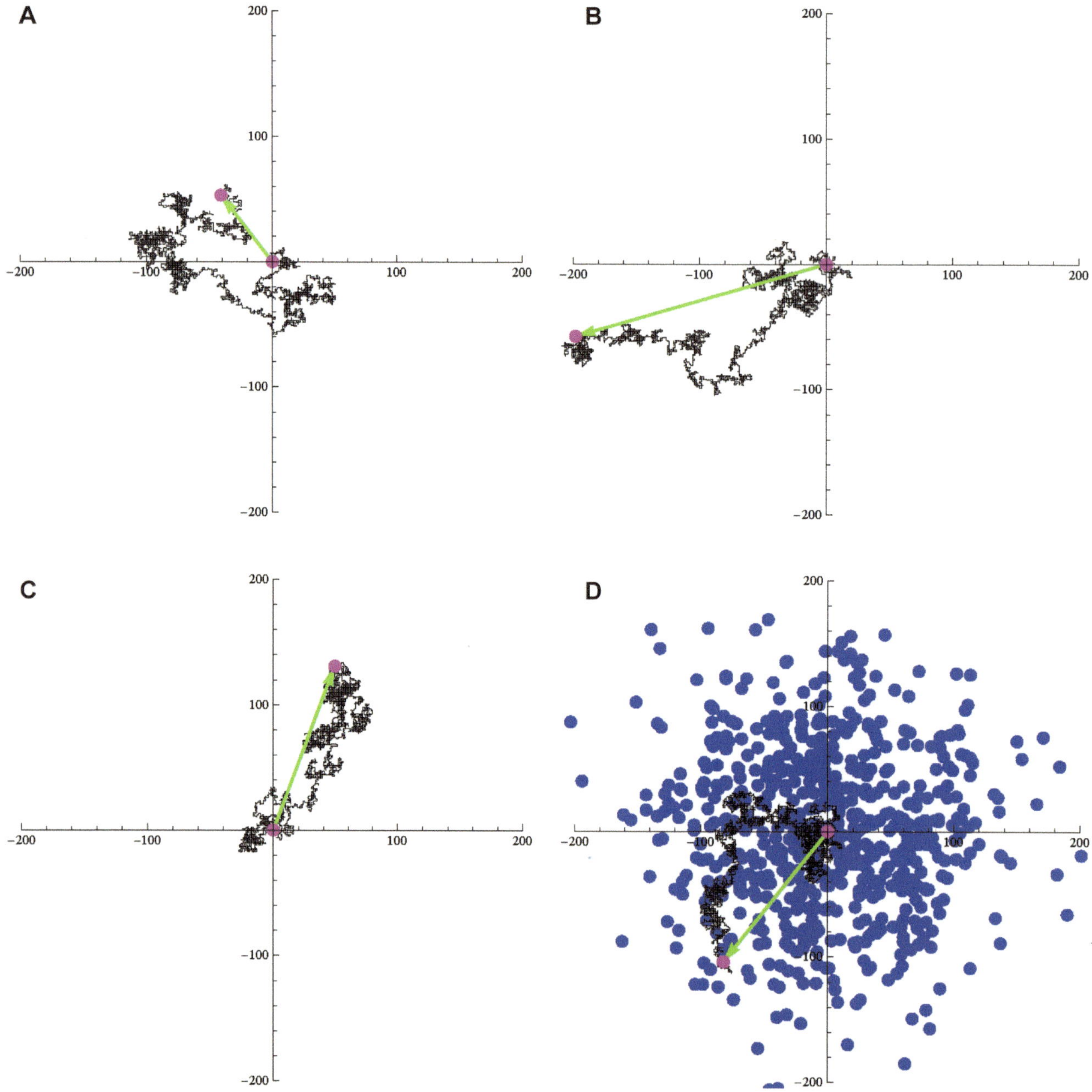

FIGURE 9.6 Two-dimensional random walk for $n = 10{,}000$ time steps for equal probabilities of jumping between any two grid points. Three realizations are shown: (A) Run 1, (B) Run 2, and (C) Run 3. Repeating this many times results in the distribution of final particle locations shown in D. The simulations were kindly provided by Professor Lawrence R. Frank, UCSD Center for functional MRI.

This can be readily extended to three dimensions, as shown and illustrated by Frank (2023).

5. Fick's laws of diffusion

As mentioned in Section 2 B, the macroscopic laws of diffusion were empirically derived by Adolf Fick. He published his paper on diffusion in 1855 when he was just 26 years old (Fick, 1855). The paper was first published in German, but in the same year, Fick published his own English translation, which can be readily accessed in current sources (Fick, 1995).

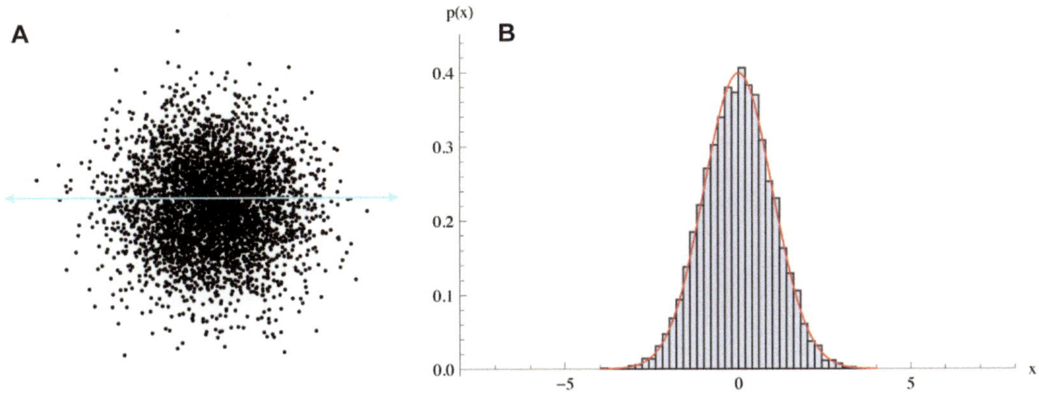

FIGURE 9.7 Distribution of final points (A) of random walk of Brownian particles and the corresponding histogram along the x-axis (B). The blue line in A demarcates the x-axis. The histogram of final points along the x-axis is Gaussian (B). *This figure was kindly provided by Professor Lawrence R. Frank, UCSD Center for Functional MRI.*

Although Fick's original work pertained to diffusion in liquids, the laws also apply to diffusion in gases and solids. Further, he recognized the similarity between solute diffusion in liquids, conduction of heat in solids (Fourier's law), and Ohm's law.

"It was quite natural to suppose that this law for the diffusion of a salt in its solvent must be identical with that, according to which the diffusion of heat in a conducting body takes place; upon this law, Fourier founded his celebrated theory of heat, and it is the same which Ohm applied with such extraordinary success, to the diffusion of electricity in a conductor".

We have seen (2A) that any diffusing substance tends to move from regions of higher concentration to regions of lower concentration until the substance is uniformly distributed at *equilibrium.* The particles continue to move at equilibrium, but the *net movement* is zero. In the absence of *advection*, which refers to bulk flow of fluid such as that produced by stirring or a pressure gradient, the movement of particles (atoms, molecules, ions) is by diffusion only. The phenomenon of diffusion reflects the behavior of a population of particles, not the behavior of any single particle, as no particle is prevented from moving in the opposite direction. It is a game of numbers; as more particles exist in regions of high concentration, more particles will move toward regions of low concentration.

To grasp these concepts more clearly, consider the scenario represented in Fig. 9.8, showing a horizontal square tube containing a solution of some solute in water. The concentration of this solute decreases from left to right, as depicted in the superimposed graph. The vertical square in the middle represents a very thin layer, having thickness *dx, through* a region of the solution at which there is a concentration gradient, *dc/dx*. The thin layer offers no hindrance to the diffusion of solute. By random walk, some fraction of the solute molecules just to the left of the layer will meander across into the

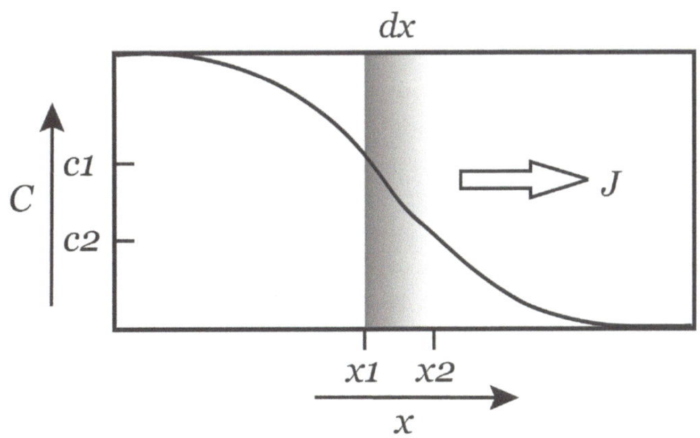

FIGURE 9.8 A concentration gradient along a square tube drawn in two dimensions. The gradient is negative in the x-direction; therefore, the flux (J) is positive in the x-direction. C is the concentration of a solute in water. The solute concentration decreases from left to right. The vertical shadowed square represents a very thin layer of thickness dx.

region just to the right. In this same time period, the same fraction of solute just to the right of the layer will meander across into the region just to the left. Since there are more solute molecules just to the left than there are just to the right, more molecules pass to the right per unit time than pass to the left; that is, there is *net* diffusion to the right. As solute diffuses to the right, an equal volume of solvent (water) diffuses to the left. The process continues until equilibrium is reached, at which time the concentration is the same everywhere. At equilibrium, solute and water molecules swiftly and randomly walk in both directions but at equal rates; *unidirectional fluxes* continue, but the *net flux* stops. We use the term *flux,* symbolized as J, to describe quantitatively the intensity of the net transport of matter. It is the amount of matter passing through a unit area normal to the direction of the flow per unit time (Fig. 9.8). Its units are mol cm^{-2} s^{-1}. Note that the *flux density* is a vector analogous to the current density in electricity, i.e., the amount of charge per unit time that flows through a cross-sectional unit area in a conductor. In SI base units, the current density is measured in amperes per square meter or Coulomb per square meter per second.

5.1 Fick's first law of diffusion

For diffusion across a plane in a one-dimensional system

$$\frac{dm}{dt} = -AD\frac{dc}{dx} \tag{9.17}$$

Where m is the net amount (moles) of solute crossing the plane in time t (seconds), c is the solute concentration (moles cm^{-3}), x is the distance (cm) perpendicular to the plane, A is the area of the plane (cm^2), D is the diffusion coefficient (cm^2 sec^{-1}), dm/dt is the instantaneous rate of net diffusion in the x direction, and dc/dx is the concentration gradient in the x direction, at the plane, i.e., referring to Fig. 9.8, (*c*2-*c*1/*x*2-*x*1). Eq. (9.17) is *Fick's first law of diffusion*; it states that the instantaneous rate of diffusion across a plane is proportional to both the concentration gradient at the plane and the plane's area. The concentration gradient, *dc/dx*, is the slope of the curve relating concentration to distance. The negative sign in Eq. (9.17) stems from the fact that a negative concentration gradient causes net movement of solute in a positive direction. Note that the dimensions of D are area per unit time, usually expressed as $cm^2 \cdot sec^{-1}$. Dividing both sides of Eq. (9.17) by A, we obtain:

$$\frac{dm}{dtA} = -D\frac{dc}{dx} \tag{9.18}$$

where $\frac{dm}{dtA}$ is the *flux density, J.* The units for *flux density* are moles$\cdot$$sec^{-1}$ cm^{-2}. Thus, we obtained the typical form in which Fick's first law is expressed:

$$J = -D\frac{dc}{dx} \tag{9.19}$$

$$\frac{\text{mol}}{\text{s}\cdot\text{cm}^2} = \frac{\text{cm}^2}{\text{s}}\,\frac{\text{mol/cm}^3}{\text{cm}}$$

Fick's first law expresses the instantaneous rate of diffusion, which can be measured only if it stays constant. In other words, Fick's first law can be directly applied only in a *steady-state* condition, where the composition does not change with time. In addition, there should not be any external driving forces present other than the concentration gradient. A *steady state* is a situation in which the *net rate* of a process and the *concentrations of the reactants* remain constant with time. In other words, a system or a process is in a steady state if the "state variables" defining the behavior of the system or the process do not change with time. A typical example of a steady state is a bathtub with the drain opened and water being added. If water leaves the tub at the same rate at which it is added, the state variable "volume" will be constant over time. The difference between *steady state* and *equilibrium state* is that in the latter, the *net* rate is zero.

The derivation of Fick's law includes the following assumptions: (1) Statistical laws apply, (2) the average duration of a collision is short compared to the average time between collisions, a condition that is valid in dilute solutions, (3) the particles move independently, (4) classical mechanics can be used to describe molecular collisions, (5) energy, momentum, and mass are conserved in every collision, and (6) the diffusing solute particles are much larger than the solvent molecules of the liquid (Sperelakis & Freedman, 2012).

If the imaginary thin layer we have been considering were replaced with a thin membrane permeable to molecules with unhindered diffusion, the concentration gradient in the steady-state is equal to the difference in the concentration ($\Delta = c2-c1$) of the solute on both sides of the membrane divided by the membrane thickness (Δx), so that:

$$J = -D\frac{\Delta c}{\Delta x} \tag{9.20}$$

The concentration gradient through the thin membrane is assumed to be linear over its thickness (Δx). This diffusion across the membrane in which the solute is not partitioned in the membrane is known as "diffusion without partitioning" (see Chapter 10 in this volume).

Besides the concentration gradient and the thickness of the layer (or the membrane), the other critical parameter determining diffusion is the *diffusion coefficient*, *D*, or *diffusivity*. As discussed in Section 4, the diffusion coefficient *D* is the proportionality factor in Fick's laws. However, *D* is not a constant; it can vary with concentration, the nature of the diffusing particles, and the environment in which the particles diffuse (e.g., viscosity and temperature). Further, the diffusion coefficient of substances in free solution is dependent upon molecular size (and shape for large molecules). Although the relation between diffusion coefficients and molecular weight can be complex in a solution, molecules with higher molecular weights tend to have lower diffusion coefficients. That is, the smaller the particle, the more mobile it tends to be. However, as noted by Hille (2001), the observed mobilities of very small nonelectrolyte particles rise faster than the Stokes–Einstein relation predicts. Nonetheless, the theory holds with relatively small errors; the deviation is less than a factor of 2 at a radius of 1 Å, i.e., 0.1 nm (Hille, 2001).

So far, we have considered the diffusion of uncharged particles. For the case of ions, the diffusion coefficient decreases as the size of the hydrated ion increases. The effective or hydrated radius of an ion in water is larger than its crystal lattice radius (the unhydrated radius). When the ion interacts with water, it becomes surrounded by a hydration shell. The larger water shell causes diffusion to be slower. For example, unhydrated Na^+ is smaller than unhydrated K^+; the crystal radius of Na^+ is 0.95 Å, and that of K^+ is 1.33 Å (Pauling, 1960). In contrast, the hydrated radius of Na^+ is 3.67 Å, whereas that of K^+ is 3.3 Å (Israelachvili, 2011). The diffusion coefficients of Na^+ and K^+ (infinite dilution, 25°C) are $1.33 \times 10^{-5}\ cm^2\ s^{-1}$ and $1.96 \times 10^{-5}\ cm^2 s^{-1}$, respectively (Vanysek, 2005). Thus, although Na^+ has a lower atomic weight than K^+, Na^+ diffuses in water slower than K^+ because it holds a larger hydration shell.

The diffusion coefficient is inversely related to the resistance to free diffusion and, therefore, is much lower in water than in a gas. In liquids, the frictional coefficient is the resistance to free diffusion (Eqs. 9.9 and 9.11). According to **Graham's law** (named after Thomas Graham, 1805–1869), the diffusion coefficient of a gas is inversely proportional to the square root of its molecular weight (*M*):

$$D\ \alpha \frac{1}{\sqrt{M}} \tag{9.21}$$

Graham's law was originally discovered from experimental observations and was later derived theoretically by Maxwell. The random molecular motions due to thermal energy are opposed by intermolecular attractive forces, but in dilute gases, these attractive forces are negligible because the molecules are relatively far apart and do not interact with each other. Thus, a dilute gas behaves ideally.

Another factor that critically affects the diffusion rate is the temperature (*T*), as discussed in Section 4. The relative increase in the diffusion coefficient when the temperature is raised by 10°C is known as *Q*10 (temperature coefficient). For dilute solutions, the *Q*10 value should be the same for all molecules and is 1.03 (between 25 and 35°C). A *Q*10 of 2 or 3 indicates that a process other than diffusion is likely responsible for transport (Sperelakis & Freedman, 2012).

5.2 Fick's second law of diffusion

Fick's second law quantitatively describes diffusion in space and time. For instance, if a substance is released in one location, how long will it take before we can detect that substance at various distances? Discussion on these phenomena implies the need to determine the dependence of concentration *c* on both time and distance. Diffusion Fick's first law (Eq. 9.19) relates the diffusive flux (flux density) to the concentration gradient. Given that the change in concentration in the *x*-direction is defined in Cartesian coordinates at some moment in time (constant *t*) and for specified values of *y* and *z*, a partial differential equation is necessary to formally express Fick's first law:

$$J = -D\frac{\partial c}{\partial x} \tag{9.22}$$

In most cases, c varies with time and distance. We will see that Eq. (9.22) can be converted into a second-order partial differential equation connecting c, x, and time t (Robinson & Stokes, 2002). Consider again Fig. 9.8, in which a square tube of uniform unit cross-section is intersected by two planes $x1$ and $x2$ of unit area A, normal to the x-axis. Thus, the two planes, $x1$ and $x2$, are situated at x and $x + dx$, respectively. How will the concentration change in this volume element in time dt? The net change in the amount of matter (mass of solute) in the volume element between these planes is given by the flux density at x (Jx) minus the flux density *out* at $x + dx$ $(Jx + dx)$, yielding the continuity equation (Nobel, 2020; Van Holde, 1985):

$$\frac{\partial c}{\partial t} = -\frac{\partial J}{\partial x} \tag{9.23}$$

The continuity equation is a mathematical expression of the law of conservation of mass, i.e., matter cannot be created or destroyed under ordinary conditions. Substituting Fick's first law (Eq. 9.22) into the continuity equation (Eq. 9.23), we obtain the **diffusion equation**, also known as **Fick's second law**

$$\frac{\partial c}{\partial t} = -\frac{\partial}{\partial x}\left(-D\frac{\partial c}{\partial x}\right) \tag{9.24}$$

Thus,

$$\frac{\partial c}{\partial t} = D\frac{\partial^2 c}{\partial x^2} \tag{9.25}$$

The diffusion equation may be solved for specific cases and boundary conditions (Crank, 1979). It describes how the concentration of a substance changes with position and time as a result of diffusion. The diffusion equation is isomorphic to the heat (diffusion) equation (Carslaw & Jaeger, 1986) and the cable equation (Davis & Lorente de Nó, 1947; Hodgkin & Rushton, 1946; Koch, 1999). For instance, replacing c with n in Eq. (9.25), where n (x, t) is the number of particles per unit volume around x at time t, the diffusion equation is written as follows:

$$\frac{\partial n}{\partial t} = D\frac{\partial^2 n}{\partial x^2} \tag{9.25.1}$$

The solution of Eq. (9.25.1) corresponding to a condition in which at time $t = 0$ all particles are at the origin is:

$$n(x,t) = \frac{n}{\sqrt{4\pi Dt}}\exp\left(-\frac{x^2}{4Dt}\right) \tag{9.26}$$

where $n = \int n(x)dx$. Then, the mean square displacement $\overline{x^2}$ from the origin is given by

$$\overline{x^2} = \frac{1}{n}\int x^2\, n(x,t)dx = 2Dt \tag{9.27}$$

From Eqs. (9.25.1) and (9.10), we arrive at Eq. (9.6) (Section 3), which is Einstein's fundamental equation for Brownian motion (Pais, 2005).

6. Diffusion in the extracellular space

The extracellular space (ECS), also known as "interstitial space," is the relatively narrow space between cells, containing interstitial fluid and extracellular matrix (ECM). This complex microenvironment, surrounding every cell in solid tissues, is the medium through which diffusion of ions, gases, metabolites, neurotransmitters, signaling molecules, and therapeutic agents designed to target tumors takes place. Further, the ECS acts as a reservoir for ions essential for the function of channels and solute transport proteins embedded in the plasma membrane of cells. Four groups of proteins form the ECM: elastin, structural glycoproteins, proteoglycans, and collagens (Truscott, 2021; Wiig & Swartz, 2012). These proteins form labyrinthic networks that confer the ECS not only stability and elasticity but also porous-like properties. Diffusion of particles in the ECS depends on their size, charge, and shape, as well as the geometry of this space and the physicochemical properties of both the interstitial fluid (e.g., viscosity) and those of the ECM proteins, as discussed below. In sum, diffusion in the ECS is determined by local geometry, frictional forces such as the viscosity of the interstitial fluid, and obstructive and binding properties of the ECM. Not surprisingly, diffusion in the ECS is hindered.

Quantitative characterization of extracellular diffusion in solid tissues not only gives information about diffusion coefficients in the ECS but also provides insights into the structure of this complex environment. The focus here is on local diffusion in interstitial spaces at nano-to micrometer scale. Considerable knowledge about the biophysics of macroscopic diffusion in ECS has been gathered over more than 5 decades. Recently, with the advent of more powerful computers and superresolution microscopy, it has been possible to apply mathematical models such as Monte Carlo simulations, as well as optical imaging and single particle tracking of carbon nanotubes or nanoparticles, to explore the microscopic behavior of diffusing particles in the ECS, and the interstitial fluid properties, at nanoscale (Soria et al., 2020; Tønnesen et al., 2023).

6.1 Extracellular space volume fraction and tortuosity

Much of what we know about local diffusion in parenchymal extracellular space comes from biophysical studies in brain tissue and solid tumors (Syková & Nicholson, 2008; Verkman, 2013; Tønnesen et al., 2023). The motivation for this focused interest is not difficult to grasp; the ECS of brain nervous tissue is a labyrinthic corridor through which diffusion of neurotransmitters at synapses, extrasynaptic neurotransmitter's spillover (volume transmission), neuromodulators, metabolic products, ions, and other solutes critical for nervous system function occurs. Further, brain ECS is a dynamic compartment; it changes under physiological and pathological conditions (Syková et al., 1994; Xie et al., 2013). In turn, these changes in ECS impact brain function and dysfunction. In tumors, understanding the factors that determine extracellular diffusion is key to therapeutic drug delivery (Bulthuis et al., 2023).

Two key physical parameters constraining diffusion in the ECS are its *volume fraction*, α, and its *tortuosity*, λ (Patlak & Fenstermacher, 1975; Nicholson & Phillips, 1981). Thus, a full description of macroscopic diffusion in the ECS incorporating these parameters requires a modification of the diffusion equation (Eq. 9.25), as discussed below. The *extracellular volume fraction* (α) determines what percentage of the total tissue volume is accessible to the diffusing molecules, and it is defined as:

$$\alpha = V_{ECS}/V_{Tissue} \tag{9.28}$$

where V_{ECS} and V_{Tissue} are the volumes of ECS and the whole tissue, respectively. V_{Tissue} is often referred to as "sample tissue volume" or "unit volume of tissue." The extracellular volume fraction α is an important determinant of small solute and macromolecule diffusion between cells and between tissue microvasculature and cells. In an unhindered or free medium, such as aqueous physiological solutions or a very dilute agarose gel (0.3%), $\alpha = 1$. The latter type of gels is used as a control in diffusion studies because they allow measurement of free diffusion coefficients circumventing thermal convection currents that can compromise the accuracy of measurements in water or physiological saline solutions (Nicholson, 2001). When measured in solid tissues, α is $\ll 1$. In brain gray matter, where it has been measured with various methods, on average, $\alpha \approx 0.2$ (Levin et al., 1970; Nicholson & Phillips, 1981; Syková & Nicholson, 2008). This means that about 20% of the brain is ECS, despite having a width that ranges between 38 and 64 nm (Thorne & Nicholson, 2006).

Alan Verkman and collaborators developed a method to measure α by simultaneous microfiberoptic detection of two fluorescent dyes, calcein, at self-quenching concentrations, i.e., > 5 mM (Hamann et al., 2002), and sulforhodamine 101 (SR), which does not undergo self-quenching. The dyes were delivered by iontophoresis into solid organ tissues (brain cortex, kidney) through a dual-lumen micropipette positioned 20 μm from the tip of the microfiber optic (Magzoub et al., 2009; Verkman, 2013). They showed that the maximum increase in calcein fluorescence after iontophoresis is linearly related to α, whereas the SR fluorescence is independent of it. The kinetic of SR fluorescence decay provides a direct measure of its diffusion coefficient. They validated the method using cell-embedded gels consisting of myeloma SP 2/0 Ag14 spherical cells suspended in 0.3% agarose, to simulate the cellular and extracellular compartments of solid tissues with varying values of α and predefined ECS viscosity. Calibration measurements were done for α of 1.0 (0.3% agarose gels without cells), 0.42, 0.21, and 0.12. To this end, α was computed from cell density and average cell volume. As expected, α decreased as the cell number and density increased (Fig. 9.9). They applied the method to in vivo mice solid organs, which included brain neocortex, where abundant quantitative information on the ECS exists using independent methods (Levin et al., 1970; Syková & Nicholson, 2008; Tønnesen et al., 2023), and kidney and skeletal muscle, where relatively little information about α using modern methods is available. Values of α measured in various mammalian tissues using more traditional methods (e.g., equilibrium concentration of radiolabeled extracellular markers) can be found elsewhere (Law, 1982). Verkman and collaborators confirmed that α in the neocortex is 0.2 ± 0.01 and found that in skeletal muscle (gluteus maximus) $\alpha = 0.074 \pm 0.01$, whereas in kidney cortex $\alpha = 0.13 \pm 0.02$. The values in skeletal muscle are consistent with those obtained using quantitative MRI in muscles from several body regions (Marty et al., 2023), and those found with methods in which an ECS marker molecules like C14-labeled inulin is introduced into the ECS and its equilibrium concentration measured (Poole-Wilson & Cameron, 1975). The values obtained in the kidney

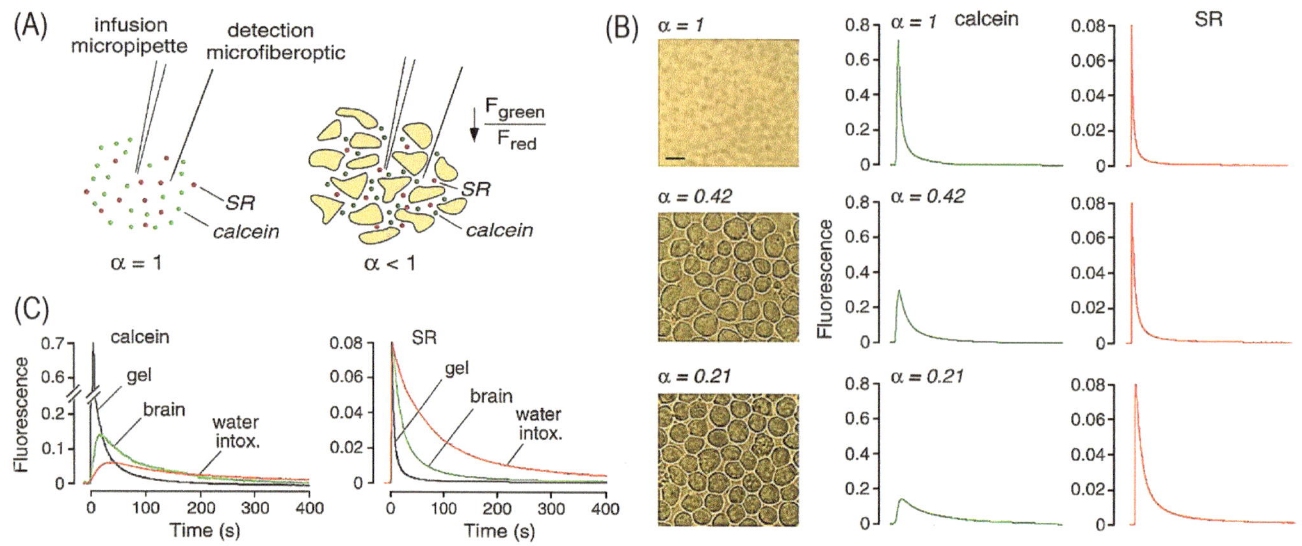

FIGURE 9.9 Extracellular volume fraction (a) measurement by simultaneous microfiberoptic detection of two membrane-impermeant fluorescent dyes, calcein free-acid, at self-quenching concentrations, and sulforhodamine-dextran (SR), which does not undergo self-quenching and serves as a reference. (A) Basic principle of the method. The dyes are delivered in the tissue of interest by a brief (1 s) iontophoretic pulse through a micropipette (infusion micropipette) positioned in close proximity (20 μm) to the tip of a microfiberoptic inserted through the lumen of another glass micropipette (detection microfiber optic). Reduced extracellular space volume fraction (α) results in a relative increase in extracellular calcein concentration and consequent decrease in calcein fluorescence due to self-quenching. SR fluorescence provides an α-independent reference. (B) Measurements in agarose gels (bright field images) without cells (α = i) and in gels embedded with live SP2/0 cells at various densities. As the density of cells increases, α is reduced (α = 0.42 and α = 0.21). The traces on the right are recordings of calcein (green) and SR (red) fluorescence following iontophoretic pulsed delivery of the dyes. Note that, as expected, calcein fluorescence decreases with α. (C) Measurements were made at a depth of 400 μm in mouse brain cortex before and after inducing brain swelling by intraperitoneal water administration. As expected cell swelling produces a decrease in α. *Reproduced with permission from Verkman (2013).*

cortex (0.13) are slightly lower than the ~0.19–0.26 measured using C14-labeled inulin as ECS marker (Macknight, 1968; Rosenberg et al., 1962). In this context, it is worth mentioning that an ideal ECS volume marker should fulfill the following characteristics; it should: (a) readily and uniformly distribute throughout the entire anatomical ECS, (b) be excluded from cells, i.e., remain in the ECS without entering cells, (c) not alter the size of the ECS, (d) not be metabolized, i.e., produced or consumed, (e) be of uniform molecular size and diffusibility, and (f) be easily and accurately estimated at low concentrations (Law, 1982). Among these ECS volume markers, the uncharged polysaccharide inulin (~5 kDa) has been favored because it behaves close to an ideal marker. Although inulin is considered a reliable ECS marker, data reported varies considerably depending on temperature and metabolic conditions of the tissues studied (Law, 1982). Further, as pointed out by McKnight "... measured inulin space may not necessarily provide a measure of the true extracellular water. There may for instance be extracellular compartments which are relatively inaccessible to a molecule the size of inulin. Alternatively, some inulin may always be diffusing across cell membranes or entering damaged cells, and the extracellular space may always be overestimated" (Macknight, 1968).

The other key determinant of diffusion in the ECS is its *tortuosity, λ*, which represents the hindrance posed to diffusion by a complex medium in comparison to an environment free of any obstacles (Patlak & Fenstermacher, 1975; Hrabe et al., 2004; Safford & Bassingthwaighte, 1977). *Tortuosity* can be formally defined as:

$$\lambda = \sqrt{\frac{D}{D_{eff}}} \tag{9.29}$$

where D is the diffusion coefficient in water (or in a dilute agarose gel) of the substance, as defined in Eq. (9.10), and D_{eff} is the effective (or apparent) diffusion coefficient of that substance measured in the ECS of the tissue of interest (Syková & Nicholson, 2008). From Eq. (9.29), it follows that $Deff = D/\lambda^2$. Both diffusion coefficients ($Deff$ and D, in cm^2/s) can be determined experimentally. D is typically measured in diluted (0.3%) agarose gels prepared with 154 mM NaCl solution or PBS. As mentioned above, agar gels are used as a reference medium to prevent possible convection, which would hamper diffusion measurements in a free solution. It has been shown that the obstruction effect and charge on the agar gel have an insignificant influence on diffusion (Magzoub et al., 2009; Nicholson, 2001). In a free medium, such as dilute agarose gel,

$\alpha = 1 = \lambda$. Average tortuosity in isotropic brain regions, like most of the cerebral cortex, is ~1.6 for small molecules, meaning that D_{eff} is about 40% of its value in free medium (Syková & Nicholson, 2008). As expected, in nerve fiber tracks, like the corpus callosum, diffusion is anisotropic; molecules diffuse more readily along the fibers (x-axis), $\lambda x = 1.46$ than across the fibers (y- and z-axes), in which case $\lambda y = 1.70$ and $\lambda z = 1.72$ (Vořišek & Syková, 1997). Also, as expected, λ increases with the hydrodynamic diameter of the molecules estimated from the Stokes–Einstein equation (Eq. 9.10).

In brief, hindrance to ECS diffusion in brain tissue is determined by several factors, which are schematized in Fig. 9.10. They include (1) geometric and structural complexity such as width, convoluted structure, and dead-space microdomains (spaces surrounded by astrocytic wrappings, cell invaginations, or voids between cells); (2) viscosity of the ISF; (3) binding interactions and obstructions with ECM components, e.g., negatively charged proteoglycans or hyaluronan (Nicholson & Hrabětová, 2017), (4) Size of the diffusing molecules (Syková & Nicholson, 2008; Tønnesen et al., 2023). It is likely that the above factors, in various degrees, hinder diffusion in tissues other than the brain. However, further research using the techniques that have been successfully applied to brain tissue is needed in other tissues (Soria et al., 2020).

The pioneering work of Joseph D. Fenstermacher, Clifford Patlak, and colleagues in the 70′S of last century, using radiolabeled extracellular space markers and applying mathematical principles of diffusion to brain tissue, was a turning point in our understanding of macroscopic diffusion in the ECS of the brain and other tissues (Patlak & Fenstermacher, 1975). They measured the diffusion of radiolabeled sucrose, inulin, or mannitol from the brain ventricular surface into the caudate nucleus of anesthetized dogs. They found that sucrose (0.342 kDa) was the optimal extracellular probe because it did not penetrate cells, thus remaining in the ECS. They estimated the effective diffusion coefficient (D_{eff}) for sucrose based on the depth profile of concentration through the caudate nucleus and applied a one-dimensional solution to the diffusion equation. They found that D_{eff} for sucrose was about 45% of its value in water. Specifically, the diffusion coefficient of sucrose in water (37°C), D, was 7×10^{-6} cm^2/s, whereas D_{eff} was 3.1×10^{-6} cm/s. Thus, λ calculated from Eq. (9.29) was 1.5. Similar values were found for inulin. Their estimated value of α using sucrose was 0.17. These values of α and λ fall within the range reported later with a variety of methods and probes with increased resolution and sensitivity (Syková & Nicholson, 2008; Tønnesen et al., 2023), as discussed below.

Another significant advance in our understanding of the biophysics of extracellular local diffusion at the scale of tens of micrometers was the development of theory and measurement of the spread of ions or molecules in tissues released from a point source. Charles Nicholson and coworkers introduced this "point-source paradigm" to study extracellular diffusion in brain tissue (Nicholson & Phillips, 1981). The method uses small monovalent ions, typically tetramethylammonium ions

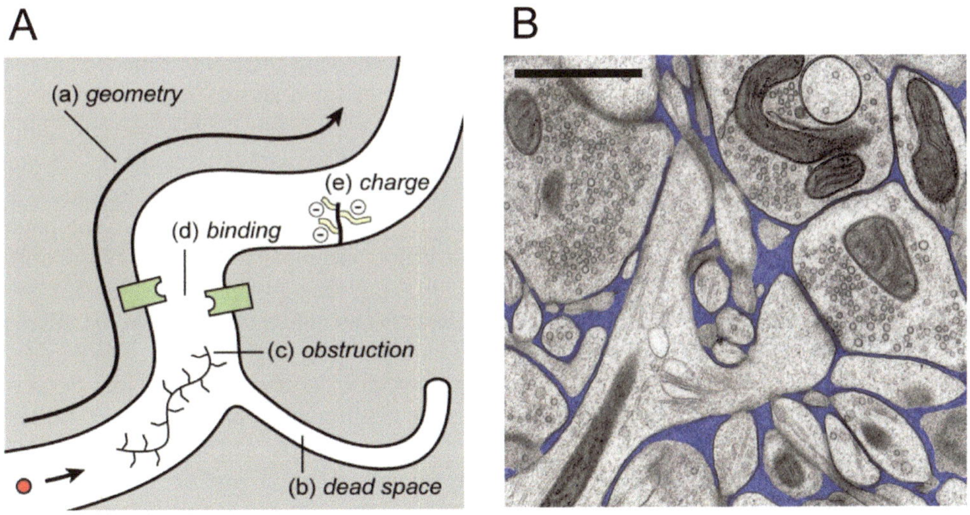

FIGURE 9.10 Properties and structure of brain extracellular space. (A) Factors affecting the diffusion of molecules in brain ECS. These include (a) geometry of ECS, which imposes a longer path length on a diffusing molecule (red-filled circle) compared to a free medium; (b) dead-space microdomains, where molecules lose time exploring a dead-end which may be an invagination, as shown, or a glial wrapping, or a local expansion of the ECS; (c) obstruction by extracellular matrix molecules such as hyaluronan; (d) binding sites for the diffusing molecule either on cell membranes or extracellular matrix; and (e) fixed negative charges on the extracellular matrix that may affect the diffusion of charged molecules. (B) Electron micrograph of cryofixed brain tissue from adult mouse cerebral cortex. The ECS is pseudocolored in blue. Cryofixation preserves the volume of the ECS. The calibration bar represents 1 μm. *(A) Reproduced with permission from Syková and Nicholson (2008). (B) Reproduced with permission from Korogod et al. (2015).*

(TMA^+), as ECS space markers or macromolecular probes labeled with a fluorophore whose dispersion is recorded with fluoresce imaging microscopy (Nicholson & Tao, 1993). TMA^+ are released by real-time iontophoresis from the tip of a glass micropipette, whereas macromolecular probes are released by pressure injection. The tip of the releasing pipette effectively behaves as a point source in a region of interest in brain tissue. The change in probe concentration (mM) in space (cm) and time (s) is recorded with an ion-sensitive microelectrode whose tip is positioned at a distance of 100 μm from the source or with fluorescence imaging microscopy methods (Soria et al., 2020). The typical average value of λ measured in brain tissue for small molecules and ions is ~1.6, as Syková and Nicholson (2008) discussed in detail. As predicted, λ increases with the hydrodynamic diameter of the diffusing molecule, indicating that the extracellular space has local constrictions that hinder the diffusion of molecules above a critical size (Hrabetova et al., 2018; Nicholson & Tao, 1993).

Like in any other medium, diffusion in the ECS can be described at microscopic and macroscopic levels. As discussed in Section 4, microscopic diffusion of a particle (e.g., a molecule) consists of the sum of the random walks that the particle performs as it collides with water molecules and occasionally with obstructions such as plasma membranes or ECS matrix components. This motion can be visualized with computer-based Monte Carlo simulations of particles performing random walks (Syková & Nicholson, 2008) or with state-of-the-art single-particle tracking methods (Godin et al., 2017; Paviolo et al., 2022; Tønnesen et al., 2023). As discussed in Section 4, a fundamental result of the microscopic theory of diffusion is that, for a population of particles released from the origin at $t = 0$, the mean square of their displacements $\overline{x^2}$ from the release point at time t is given by

$$\overline{x^2} = 2dDt \tag{9.30}$$

where d is the dimension of the space (1, 2, or 3) and D is the diffusion coefficient. This equation is a general expression of Eq. (9.12). Also, as discussed above, $\overline{x^2}$ from the origin is described by a Gaussian distribution.

As we have seen throughout this chapter, the behavior of a large ensemble of random walks can be described by the equations of classical macroscopic diffusion, i.e., Fick's laws. The diffusion equation (Fick's second law) has been applied to describe molecular migration in brain ECS. However, given that the diffusion equation applies to particles moving in a free medium and not in a porous-tortuous-viscous medium like brain ECS, a modification of this equation applicable to local macroscopic diffusion over distances of about 100 μm was proposed and verified experimentally by Nicholson and colleagues (Syková & Nicholson, 2008; Nicholson & Hrabětová, 2017). Thus, incorporating α and λ, the diffusion equation (Fick's second law in x, y, and z) for ECS, the following expression is obtained:

$$\frac{\partial C}{\partial t} = D_{eff}\,\nabla^2 C + \frac{Q\,(x,y,z,t)}{\alpha} - k'C - \frac{f(C)}{\alpha} + \mathbf{v}\cdot\nabla C \tag{9.31}$$

where C is the change in concentration of the diffusing molecule (mM) with respect to time t, at a given location in the interstitial space (volume), represented by the vector $\mathbf{x} = (x,y,z)$, D_{eff} (defined above) is the effective diffusion coefficient ($D_{eff} = D/\lambda^2$); and ∇^2 symbolizes the second spatial derivatives in three-dimensional Cartesian coordinates. A "source term" $Q\,(x,y,z,t)$ accounts for the release of ions or molecules into the ECS and has units mMs^{-1}. Note that Q is divided by α, reflecting the fact that substances released into the ECS are restricted to a small tissue volume. As discussed below, the point source can be the tip of a microelectrode from which the substance is released by iontophoresis or a pressure injection pulse (Nicholson, 1993, 2001; Nicholson & Hrabětová, 2017; Nicholson & Phillips, 1981), or it can be the release of a neurotransmitter from a presynaptic terminal (Zheng et al., 2017). If the substance is released by iontophoresis from a micropipette, $Q = In_t/zF$, where I is the current applied to the electrolyte solution contained in the micropipette, and n_t, the transport number, i.e., the fraction of the current that expels the ion of interest (e.g., TMA^+), F is Faraday's equivalent that relates mass to charge, and z is the valence. The first-order rate constant k', accounts for any loss or clearance of the probe into cells or blood vessels and has units s^{-1}; and $f(C)$ represents the binding of the probe to the ECM, transporters, or other sites, including uptake of the extracellular probe by cells, and its units are mMs^{-1}. The fourth term of Eq. (9.31) includes the bulk flow velocity vector $\mathbf{v}$ and forms a scalar product with the concentration gradient if bulk flow (advection) occurs. Thus, if extracellular local transport is by diffusion only, the fourth term of the equation is zero. In the brain, local extracellular transport of substances in the nano-to micrometers length-scale appears to be mainly, if not exclusively, by diffusion (Tønnesen et al., 2023). However, there is growing evidence supporting the notion that interstitial fluid transport is not only driven by diffusion but also by advection, particularly over relatively long distances. Advection is a fast and directional flow driven by a pressure gradient that can efficiently cover large distances in a brief time (Rasmussen et al., 2022).

Eq. (9.31) can be solved in a spherically symmetric coordinate system so C is a function of radial distance r from the source (Nicholson & Hrabětová, 2017). If $f\,(C)$ is negligible and there is no bulk flow, $f\,(C)$ and the bulk flow term are neglected, and the source is an instantaneous point source (δ-function source density) at the origin with magnitude Q, the solution of Eq. (9.31) is:

$$C = \frac{Q}{\alpha}\frac{1}{\left(4D_{eff}\,t\pi\right)^{3/2}}\exp\left(-\frac{r^2}{4D_{eff}\,t}\right)\exp\left(-k't\right) \tag{9.32}$$

Eq. (9.32) is a Gaussian curve multiplied by an exponential loss factor. The shape of the curve yields D_{eff}, and thus λ can be estimated. The amplitude is a function of $Q/\alpha\,(D_{eff})^{3/2}$ and therefore, knowing Q it is possible to find α (Nicholson & Hrabětová, 2017).

The fact that the experimental measurements using the point source paradigm can be fitted to the diffusion equation implies that local diffusion in the ECS is a Gaussian process reflecting Brownian motion. In this type of diffusion, also known as "normal" diffusion, the mean squared displacement ($\overline{x^2}$) of solute particles increases linearly with time. However, as discussed above, the ECS is a labyrinthine environment of complex geometry containing EMS proteins (Fig. 9.10). This complex environment can potentially form a crowded media in which diffusion is not only hindered but can also become anomalous, i.e., nonlinear. In three-dimensional space, the non-Brownian behavior is described by the following equation:

$$\overline{x^2} = 6D\,t^a \tag{9.33}$$

If $a < 1$, the diffusion is called anomalous subdiffusion, if $a > 1$, it is called anomalous superdiffusion, and when $a = 1$ it is called normal diffusion (Fig. 9.11). Note that when $a = 1$ Eq. (9.33) becomes Eq. (9.30) in a 3D space. Anomalous diffusion in the ECS appears to be the exception rather than the rule; it has only been experimentally demonstrated in one instance, namely, the granular layer of the cerebellum, a complex anatomical structure in the brain where $a = 0.4$. Anomalous subdiffusion, in this instance, has been attributed to the presence of dead-space microdomains (Xiao et al., 2015). The biophysics of anomalous diffusion is discussed in detailed elsewhere (Metzler et al., 2014; Sposini et al., 2022)

We have discussed the transport of substances in the extracellular space by diffusion only, and as an example, we have examined the brain's extracellular space, which has been thoroughly studied with multiple techniques. Although diffusion in the brain's extracellular space is effective over short distances, early evidence showed that it is insufficient to explain the clearance of metabolites in the brain. The pioneering work of Helen Cserr showed that when markers of different molecular weights were injected into the brain parenchyma, they dispersed away from the injection site at similar rates, suggesting convective flow rather than diffusion, which would have been dependent on the molecular weight of the probes (Cserr et al., 1981). More recently, the "glymphatic system" has been proposed as a key mechanism for solute and fluid transport over relatively long distances in brain tissue. The discussion of this important system and its functional implications is beyond the scope of the present chapter. The interested reader should consult the landmark studies of Maiken Nedergaard and collaborators, which have been recently reviewed (Rasmussen et al., 2022).

7. Intracellular diffusion

The cell interior is a microscopic universe whose complexity is far beyond a membrane-bound bag containing organelles suspended in a homogeneous aqueous solution in which ions, signaling molecules, and enzymes freely diffuse. In fact, the

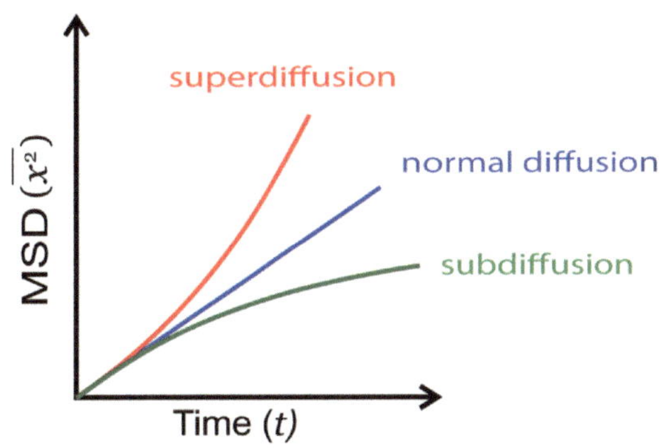

FIGURE 9.11 Anomalous and normal diffusion. Mean squared displacement (MSD) as a function of time for different values of exponent a in Eq. (9.33).

cell interior is a complex, heterogenous, and crowded environment in which diffusive transport is restricted by several factors that include macromolecular crowding, geometry, steric hindrance, binding interactions, fluid phase micro- and nanoviscosity, and effective hydrodynamic radii of the diffusing particles (Luby-Phelps, 2000, 2013; van Tartwijk & Kaminski, 2022). Luby-Phelps (2013) imagined the cytoplasmic space as a crowded party in a house full of furniture, presenting obstacles that retard the movement of guests across the room. Extending this metaphor, the translational and rotational mobility of the guests will also depend on their body mass index.

Intracellular solute transport is accomplished not only by diffusion but also by advection. Diffusion, as discussed in this chapter, is passive thermal Brownian motion, whereas advective cytoplasmic flow (bulk flow) is chemically driven, enhanced, and directed motion caused by the activity of molecular motors like microtubules and other intracellular cytoskeletal elements (Mogilner & Manhart, 2018). Active (energy-consuming) cytoplasmic motion sometimes has the appearance of diffusion; it plays key functional roles for cargo delivery for relatively long distances in large cells (e.g., neurons, oocytes, plant cells) where intracellular transport of matter exclusively by diffusion would be neither sufficient nor efficient (Bellotto et al., 2022). The intracellular transport of substances involving molecular motors is not discussed here. This section focuses on intracellular diffusion, specifically cytoplasmic diffusion, a transport mechanism crucial for local intracellular signaling, metabolic reactions, and nanoparticle-based drug delivery. Diffusion is a rate-limiting factor of metabolic reactions in living cells. This is because most biochemical reactions occur when the reactant molecules approach each other. Changes in the diffusion rate of the reactants affect the probability of molecular encounters leading to biochemical reactions.

7.1 Structure and architecture of the intracellular space

Before going deeper into the subject of intracellular diffusion, it is important to clarify the terminology used here. In eukaryotic cells, the *cytoplasm* is the gelatinous, viscoelastic material enclosed by the plasma membrane, excluding the nucleus and the membrane-bound organelles. The so-called membraneless organelles (e.g., RNA–protein granules) not contained within the nucleus are also part of the cytoplasm. These membraneless structures are biomolecular condensates resulting from biological liquid–liquid phase separation (Jalihal et al., 2021; Hirose et al., 2023; Lyon et al., 2021). The cytoskeletal elements (actin filaments, microtubules, and intermediate filaments) form the malleable cytoplasmic *solid matrix phase*, whereas the *cytosol* is the *fluid phase*. The membrane-bound organelles suspended in the cytoplasm include the nucleus, mitochondria, endoplasmic reticulum, Golgi apparatus, vacuoles, lysosomes, and, in plant cells, chloroplasts. The material contained within the nuclear membranes is the *nucleoplasm*.

Cytoplasmic diffusion refers to the diffusion process occurring in the cytoplasmic matrix and the cytosol and thus excludes the nucleus and the membrane-bound organelles. In prokaryotic cells, which lack a nucleus and other membrane-bound organelles, the cytoplasm is everything inside the plasma membrane. The cytoplasm of prokaryotic cells is also a crowded environment that contains ribosomes, DNA, cytoskeletal filaments, enzymes, inorganic molecules, and ions (Bellotto et al., 2022). It is important to highlight that in the literature on intracellular diffusion, the "cytosol" is often used as a synonym of "cytoplasm," although, as discussed, they are separate entities within a single system; the cytosol is an integral part of the cytoplasm.

The tortuous, malleable, *solid matrix phase* structure formed by cytoskeletal filaments, macromolecules (proteins, nucleic acids, lipids), and organelle envelopes form a porous, viscoelastic material that allows fast diffusion of small molecules but hinders the diffusion of larger molecules like proteins. Plasma membrane deformation using atomic force microscopy cantilevers reveals that the cytoplasm behaves as a sponge with interconnected nanopores ($<<30-60$ nm depending on cell type and cell region) filled with aqueous intracellular fluid (the cytosol) that has <2 to 3 times higher microviscosity than water (Mitchison et al., 2008; Moeendarbary et al., 2013; Sachs & Sivaselvan, 2015). Other approaches to studying cytoplasmic nanostructure and properties include various optical methods such as fluorescence correlation spectroscopy (FCS) and its variants (Baum et al., 2014; Kwapiszewska et al., 2020; Yamamoto et al., 2021) and superresolution mapping of single-molecule displacement/diffusivity (SMdM), as discussed below (Choi et al., 2023; Xiang et al., 2020). In these methods, biologically inert tracers (dye molecules, fluorescent proteins, fluorescently labeled polymers, or nanoparticles) of various hydrodynamic radii are introduced into cells, and their diffusion coefficients are measured by fluorescence correlation spectroscopy (FCS) or SMdM. Based on measurements of *effective* nanoviscosity using probes with hydrodynamic radii ranging from 1 to 150 nm, it has been possible to explore the cytoplasmic structure and properties at a nanoscopic length scale in various human cell lines (Kwapiszewska et al., 2020). These studies conclude that the cytoplasm behaves as a liquid for length scales smaller than 100 nm and as a physical gel for larger length scales. Nanoparticles of diameters bigger than 100 nm do not diffuse freely through the cytoplasm, suggesting a critical length scale crossover to gel-like structure in the cytoplasm. Kwapiszewska's and collaborators propose that large cytoplasmic structures like cytoskeleton or endoplasmic reticulum create a gel-like structure of a mesh size of ~100 nm. The size of the mesh differs in different cells or cell regions, so it is topologically heterogeneous as nanoparticles of hydrodynamic radius $r = 68$ nm exhibit free diffusion in some cases while in others, nanoparticles $r = 68-81$ nm are trapped. They estimate that the radii of the effective intercrowder gaps is 4.6 nm. Small molecules having hydrodynamic radii smaller than the intercrowder gaps experience the viscosity of the solvent. Whereas large molecules or probes having hydrodynamic radii much larger than the hydrodynamic radius of the main crowders experience the effective viscosity measured by microscopic methods. These observations are consistent with the model of the cytoplasm having a gel-like structure with a liquid phase, derived from atomic force microscopy measurements (Moeendarbary et al., 2013). It is also consistent with the data showing that diffusion of proteins is affected by large crowders in the cytoplasm, while smaller solutes experience the viscosity of the solvent. Another important conclusion from this study is that the *effective* nano-viscosity is length-scale dependent. Probes with a radius of <1 nm exhibit cytoplasmic water viscosity, while for probes >20 nm, the *effective* viscosity can reach values of 10 times the viscosity of water.

Fig. 9.12 illustrates schematically the organization of the solid and liquid phases of the cytoplasm of mammalian (eukaryotic) cells, derived from combined methodological approaches including rheological analysis using atomic force microscopy (AFM), diffusion measurements using fluorescence recovery after photobleaching (FRAP), and quantum dots diffusion (Moeendarbary et al., 2013). Independent experiments on the osmotic properties of mammalian cytoplasm, as determined in permeabilized cells, also conclude that the cytoplasm behaves as a porous hydrogel, i.e., a tangled network of cross-linked molecules immersed in a liquid medium (Fels et al., 2009; Orlov et al., 2018). It is worth mentioning that the view that cytoplasm behaves as a hydrogel is not new (Lechene, 1985). For instance, working on the giant squid axon, an unmyelinated nerve fiber (0.5–1.5 mm diameter) that has been an invaluable model for biophysics of neuronal excitability, membrane transport, and intracellular diffusion, Alan Hodgkin and Bernard Katz already noticed that the axoplasm "is normally in a gelatinized state and behaves as a solid" (Hodgkin & Katz, 1949). Squid cytoplasm can be extruded as toothpaste does by squeezing a toothpaste tube (Baker et al., 1962).

The spaces inside the nucleus and organelles are also aqueous compartments with their own properties. For instance, the mitochondrial matrix exhibits barriers to diffusion due to the presence of cristae (folds of the inner membrane) and macromolecular crowding. The microviscosity of the fluid phase of the mitochondrial matrix is 3.69–4.57 cP, which is 3–5-fold higher than water, although higher values have been reported (Bulthuis et al., 2023). The mitochondrial matrix's physicochemical and geometric properties impact diffusion-dependent biomolecular reactions occurring inside these organelles and matrix Ca^{2+} diffusion, which regulates mitochondrial function.

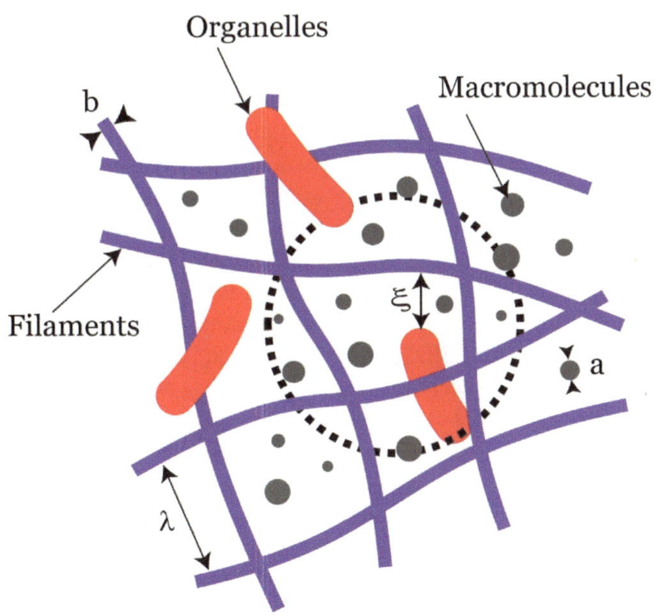

FIGURE 9.12 Schematic representation of the cytoplasm derived from rheological analysis using atomic force microscopy, FRAP, and quantum dots diffusion measurements. The cytoskeleton and the macromolecular crowders participate in setting the hydraulic pore size through which water and solutes can diffuse. The length scales involved in setting cellular rheology are the average filament diameter "b," the size "a" of particles in the cytosol, the hydraulic pore size ξ, and the entanglement length "λ" of the cytoskeleton. The estimated hydraulic pore radius ξ is ~15 nm, comparable to the hydrodynamic radius of 400-Dalton polyethylene glycol-passivated quantum dots. *Reproduced with permission from Moeendarbary et al. (2013).*

7.2 Determinants of cytoplasmic diffusion of small solutes and macromolecules

The translational diffusion of small and macromolecular solutes in the cytoplasm is determined by solute properties (size, shape, and charge) and by the properties and geometric constraints of the cytoplasm, including viscosity, collisional forces, and crowding. Translational diffusion of small solutes is relevant for second messenger signaling, whereas diffusion of macromolecules, like enzymes or transcription factors, is fundamental for cell metabolism and gene regulation.

As mentioned, several fluorescence microscopy methods exist to study translational diffusion in the cytoplasm and plasma membranes. In these methods, fluorescently labeled molecules are introduced into cells by microinjection, incubation, membrane permeabilization, or targeted expression of green fluorescent protein (GFP). These methods include fluorescence recovery after photobleaching (FRAP) (Verkman, 2002), fluorescence correlation spectroscopy (FCS) (Kwapiszewska et al., 2020; Yamamoto et al., 2021), single-molecule localization microscopy (Lelek et al., 2021), and single molecule displacement mapping (Choi et al., 2023; Xiang et al., 2020). These methods are often used in combination, and they continue to be refined with the advent of superresolution microscopy. A detailed description of these methods is beyond the scope of this chapter. The references cited above will provide basic information for the interested reader.

7.2.1 Diffusion of proteins in the cytoplasm

Which factors affect the diffusion of macromolecules in the cytoplasm? From the Stokes–Einstein relation (Eqs. 9.10 and 9.11), the translational diffusion coefficient is:

$$D_T = \frac{K_B T}{6\pi\eta r} \tag{9.34}$$

where k_B, T, η, and r are the Boltzmann constant, absolute temperature, viscosity, and *hydrodynamic radius* of the diffusing particle, respectively. Measurements of translational diffusion of various proteins show that it is significantly slowed down in the cytoplasm compared to free diffusion (in water/or saline). For instance, the cytoplasmic translational diffusion coefficient (D_{cyto}) of creatine kinase, a dimeric protein with 83 kDa molecular weight and 3.3 nm hydrodynamic radius, is 4.5 $\mu m^2/s$, whereas its diffusion coefficient in water (D_{water}) is 65 $\mu m^2/s$. Thus, $D_{cyto}/D_{water} \approx 0.07$, which means that diffusion of this enzyme in the cytoplasm is about 7% of that in water (Arrio-Dupont et al., 1997). The green fluorescent protein (GFP) has long been used as an inert probe to quantify both translational and rotational diffusion in the cytoplasm (Swaminathan et al., 1997; Lippincott-Schwartz & Patterson, 2003; Yamamoto et al., 2021). GFP is a ~27 kDa monomeric protein of cylindrical shape and a hydrodynamic radius of 2.82 nm. Its diffusion in the cytoplasm is ~3 to 4-fold slower than in water (Verkman, 2002; Yamamoto et al., 2021). In contrast to macromolecules, diffusion of small molecules (<~1 kDa) is modestly slowed down to 60%–70% with respect to that in phosphate-buffered saline solutions, which have a viscosity nearly identical to water (Choi et al., 2023).

What are the underlying factors of translational diffusion retardation of macromolecules in the cytoplasm? We have seen that there are factors that depend on the biophysical properties of the diffusing molecules and those that depend on the properties of cytoplasmic components, such as the viscosity of the fluid phase and molecular crowding. Among the factors that depend on the properties of the diffusing molecules are their *size and shape*; proteins vary widely in size, and size is inversely proportional to the rate of diffusion. Proteins are not chemically inert spheres; they exhibit complex chemical surfaces and often deviate from simple spheres in shape. However, their hydrodynamic radii in solution (Stokes' radius) account for nonspherical shapes and can be used as an accurate parameter in calculating a diffusion coefficient applying the Stokes–Einstein relationship (Parker & Lollar, 2021). Other factors impacting cytoplasmic protein diffusion are *binding interactions* resulting from their chemical surface properties (hydrophobic domains and surface charge) and *collisions* with other molecules (*crowding*) or with *physical barriers*. Regarding binding interactions, sites for binding to mobile or immobile elements have the potential to decrease the apparent mobility of proteins significantly. For instance, the nonspecific binding of proteins to cytoskeletal elements largely accounts for the very low and often size-independent diffusion coefficients measured for some proteins inside intact cells (Jacobson & Wojcieszyn, 1984).

Regarding physical barriers, single molecule displacement mapping (SMdM) imaging microscopy shows that cytoplasmic diffusion of an average-size protein (mEos3.2-C1; ~28 kDa) is spatially heterogeneous and anisotropic at the nanoscale and that the variations in local diffusivity are correlated with the actin cytoskeleton architecture (Xiang et al., 2020). These authors also explored the effect of altering the *net charge* of mEos3.2 to −14, −7, 0 + 7, and +14 on the cytoplasmic translational diffusion coefficient of this protein by adding short consecutive Asp/Glu and Arg/Lys to the carboxy terminus. They found that the two negatively charged (−14, −7) species yielded D_{cyto} comparable to that of the neutral species but slightly higher than that of the original mEos3.2-C1 (+2 charges). For the +7-charged protein, however, a ~50% reduction in D_{cyto} was found across all cytoplasmic mapped locations, whereas extremely slow diffusion was found for the +14-charged protein. These findings highlight the importance of macromolecular charge on diffusion, probably contributing to the heterogeneity of intracellular diffusivities at the nanoscale.

Other factors that depend on the cytoplasmic medium in which diffusion occurs are *viscosity,* discussed below in Section 7.2.4, and *hydrodynamic interactions* (Ando & Skolnick, 2010). The latter refers to the observation that particles moving through a fluid affect each other's velocity through the flow their motions cause. This phenomenon has been described as the effects of moving objects in a fluid on other objects, similar to the effect of boats on each other via their wake (Milo & Phillips, 2016). Hydrodynamic interactions lead approaching bodies to repel each other, whereas two bodies moving away are attracted.

7.2.2 Diffusion of small molecules in the cytoplasm

Diffusion of small solutes (<1.5 kDa) has also been studied using SMdM microscopy (Choi et al., 2023). This methodological approach enables the nanoscale mapping of intracellular diffusivity through local statistics of the instantaneous displacements of diffusing single molecules (Xiang et al., 2020). The method has been used to quantify and spatially map the fast-translational diffusion of small solutes in the cytoplasm of mammalian cells loaded with water-soluble dyes, e.g., Cy3B (560 Da) and dye tag nucleotides (ATP and cAMP). The D_{cyto}/D_{water} of the solutes tested was ~ 60%–70%, suggesting that the intracellular diffusion of small solutes is only modestly reduced by the ~20%–40% higher viscosity of the cytosol over water but otherwise not further hindered by macromolecular crowding.

7.2.3 Water in the intracellular crowded environment

Water mobility and fluid viscosity are two key physicochemical parameters of aqueous cytoplasmic fluid (cytosol) that determine intracellular diffusion. As discussed, the cytoplasmic viscoelastic porous gel allows diffusion of ions and small molecules but hinders the movement of larger particles, depending on the gel's pore/mesh size, the size and shape of the diffusing particles, their charge, and the physicochemical properties of the aqueous solution that moves through the pores (Baum et al., 2014; Kwapiszewska et al., 2020).

The liquid medium of cells is water; typically, 70% of the total cell mass is water, whereas proteins constitute 55% of the total dry weight (Milo & Phillips, 2016). Estimates of the total concentration of macromolecules (proteins, nucleic acids, polysaccharides) in a cell range from 200 to over 400 g/L (Rivas & Minton, 2016). Water around proteins is classified into two major categories: "bulk water," i.e., water that has the properties of pure liquid water, and "hydration water," which is water tightly interacting with protein surfaces and ions. Water molecules not only lie on protein surfaces but are also found in protein clefts and cavities (Purkiss et al., 2001; Chen et al., 2008). Altogether, we will refer to hydration water and water sequestered in protein clefts and cavities as "bound water" or "slow dynamic water." Other names, such as "structured water," "interfacial water," "solvation water," and "nonsolvent water" are also encountered in the literature in this field.

The average distance between macromolecules in the cytoplasm is around 1 nm, corresponding to three to four molecular layers of water, which, based on solvation theory, cannot be considered bulk-like (Ball, 2017). Thus, given this crowded environment, with a high concentration of enzymes and other macromolecules, and the extensive surface area presented by intracellular membranes in eukaryotic cells, intracellular water's mobility and solvent properties are thought to be different from those of bulk water. However, the extent these properties differ from bulk water has been the subject of debate (Clegg, 1984; Pollack, 2001; Ball, 2008, 2017; Luby-Phelps, 2013; Jungwirth, 2015; Tros et al., 2017; Piazza et al., 2018; You et al., 2021). Thus, a central, recurrent, and unsettled question in biophysics and cell biology is how the crowded intracellular environment affects the properties of intracellular water, specifically its viscosity, rotational and translational mobility, and solvent properties, compared with bulk water. If the properties of intracellular water differ from those of pure water, what are those differences, and how do they impact diffusion? To what extent does intracellular water behave differently than bulk water? These are crucial questions not only because water is the fluid medium in which diffusion occurs in the cytosol and other intracellular aqueous compartments but also because it is a major determinant of biochemical interactions, enzyme reactions, and protein folding and stability (Purkiss et al., 2001; Bellissent-Funel et al., 2016).

The emerging view, based on in vivo experiments using quasi-elastic neutron scattering (QENS) and relaxation dispersion NMR (nuclear magnetic resonance) spectroscopy in various cell types, is that 85%–90% of intracellular water behaves like bulk water, and the remaining 10%–15% is hydration water interacting with protein surfaces, confined to clefts and cavities, and solvating ions (Persson & Halle, 2008; Tros et al., 2017; Piazza et al., 2018). QENS spectroscopy is a particularly valuable technique for the measurement of intracellular water dynamics in the Angstroms/picosecond space/

time scale (Zaccai, 2020). The translational and rotational diffusion coefficients, D_T and D_R, of bulk water at room temperature (25°C) determined by QENS are ~2.3 10^{-5} cm^2/s and 0.17 ps^{-1}, respectively (Piazza et al., 2018). Applying this technique to *E. coli* cytoplasm, Piazza et al. (2018) found two populations of intracellular water: the largest one (85%) having dynamic properties such as D_T and D_R similar to bulk water, and a small one (15%) having slow dynamics, which corresponds to "bound" water. The D_T and D_R of the bulk-like (fast-dynamics) intracellular water at 25°C were 2.13×10^{-5} cm^2/s and 0.2 ps^{-1}, respectively. These values are similar to those of bulk water, whereas the D_T and D_R for the slow-dynamic water population were 2.1×10^{-6} cm^2/s and 0.043 ps^{-1}, respectively. The latter indicated that translational and rotational diffusion of the "slow-water" was retarded by a factor of 10 and 5, compared with bulk water. The fraction of bulk-like water (fast) and "slow" intracellular water found in *E. coli* using QENS was consistent with that measured with relaxation dispersion NMR spectroscopy (Persson & Halle, 2008). Another study in *E. coli* using ultrafast vibrational spectroscopy and dielectric-relaxation spectroscopy to observe the random orientational motion of intracellular water molecules found similar results, namely that the fraction of bulk-like intracellular water was 80%, whereas the "slow" fraction was 20% (Tros et al., 2017). Comparison with the water dynamics observed in solutions mimicking the chemical composition of the cytosol shows that the "slow water" is mostly in the hydration shells of proteins and, to a lesser extent, in other biomolecules and ions (Tros et al., 2017). Earlier NMR studies also concluded that the fraction of water buried in proteins is relatively small (Persson & Halle, 2008).

Most studies on intracellular water dynamics come from prokaryotes like *E. coli*, as discussed above, and a few from eukaryotic tumor cells (Likhtenshtein, 2021). The question arises as to what the state of intracellular water is in normal eukaryotic cells. This question has been addressed using QENS in human red blood cells (RBCs) by Giuseppe Zaccai and coworkers (Jasnin et al., 2008). The QENS analysis revealed two populations of water in RBCs: a major fraction of ~90%, which has translational properties similar to those of bulk water, and a fraction of ~10% of immobile water, which could be identified as dynamically bound water in the hydration layer of hemoglobin. However, RBCs are not typical eukaryotic cells; they lack nuclei and membrane-bound organelles. Hemoglobin makes up 92% of their dry weight. The confined water in these cells corresponds to a monolayer of hydration around each hemoglobin molecule (Zaccai, 2020). Further studies are needed in typical eukaryotic cells to determine the extent of cytoplasmic water having bulk-like properties. In conclusion, the existing evidence so far suggests that more than 80% of intracellular water behaves as bulk-free water.

7.2.4 Viscosity of the cytoplasmic aqueous phase

The properties of the cytosolic water can be probed by measuring the rotational mobility of labeled molecules. The *rotational mobility* largely reflects the *local viscosity* of the solvent (cytosolic water) because it is insensitive to obstruction and confinement (Fig. 9.13). Viscosity is a fundamental parameter necessary for understanding the intrinsic rheological properties of the cytosol, the translational mobility of solutes, and the nanoscopic architecture of the cytoplasm.

The concept that the rotation time (τ_{rot}) of a molecule that describes the relaxation of its orientation is related to the solvent microviscosity was introduced by Debye (1929). The equation relating rotational correlation (relaxation) time, τ_{rot}, for a spherical particle in a medium of viscosity η is given by

$$\tau_{rot} = \frac{1}{6D_R} \tag{9.35}$$

In this equation, D_R is the rotational diffusion coefficient. Thus, from the Stoke—Einstein equation,

$$D_R = \frac{K_B T}{8\pi\eta r^3} \tag{9.36}$$

This mathematical expression is often referred to as the Stokes—Einstein—Debye equation.

A crucial point when considering rotational mobility measurements in living cells is the verification that the probe does not interact directly with the cytoplasmic biopolymers.

The question arises as to whether the microviscosity of the cytosol differs from that of water. NMR of labeled or endogenous proteins, spin probes, and fluorescently labeled molecules have been used to determine the viscosity of the cytosolic fluid. In all mammalian cell types studied in which probe rotation was not affected by binding, or free and bound fractions of the probe could be discerned unequivocally, the apparent solvent viscosity of the cytosol was approximately 1.3—3 times the viscosity of water (Luby-Phelps, 2000; Verkman, 2002; Baum et al., 2014; Yamamoto et al., 2021). For reference, the dynamic viscosity of water is 1 cP at 20°C, 0.890 cP at 25°C, and 0.75 cP at 37°C. These values of the viscosity of pure water are not different from the viscosity of an aqueous solution of 0.15 M KCl under the same conditions. The picture that emerges is that intrinsic cytosolic viscosity has only a modest impact on cytoplasmic diffusion.

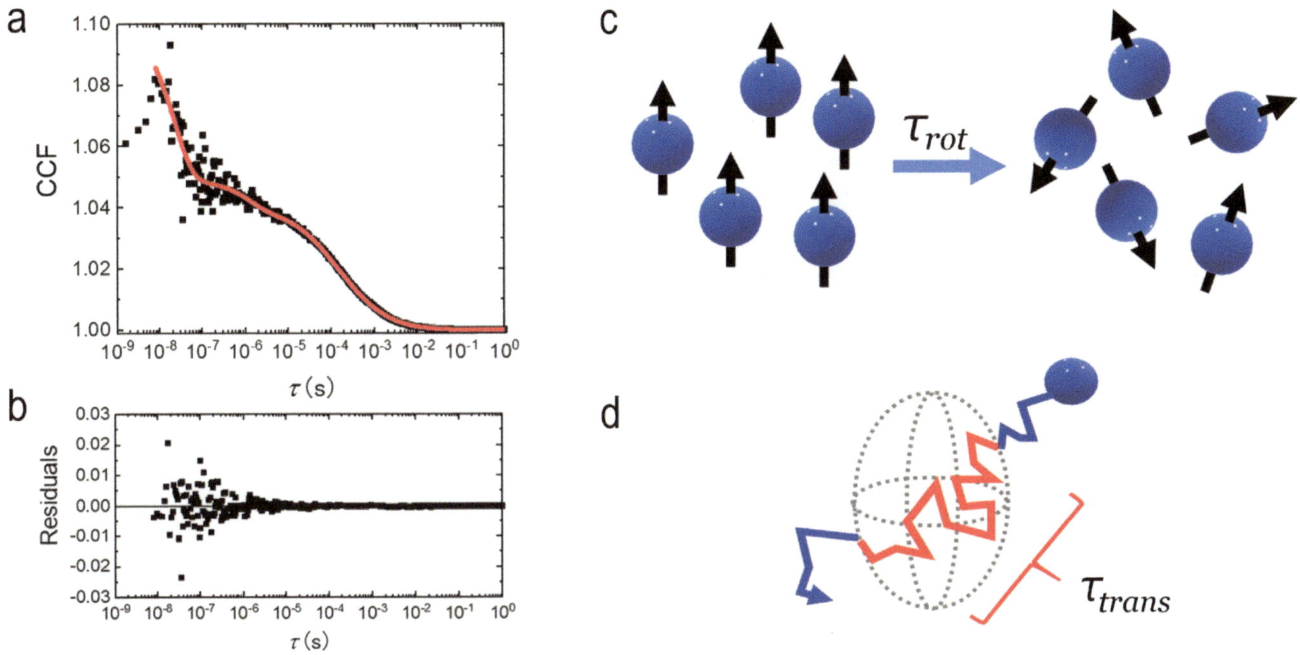

FIGURE 9.13 Rotational and translational diffusion times of green fluorescent protein (GFP) in phosphate-buffered saline (PBS) measured with polarization fluorescence correlation spectroscopy (Pol-FCS). (a) Representative cross-correlation function (CCF) of 100 nM GFP in PBS. The rapidly decaying component around 10^{-8}–10^{-7} s is a rotational diffusion component, and its characteristic decay time (rotational diffusion time) τ_{rot} was ~22.5 ns. The decay component around 10^{-5}–10^{-3} s is the translational diffusion component, and its characteristic decay time (translational diffusion time) τ_{trans} was ~ 180 μs. (b) Residuals from the fitted model. (c) Schematic diagram of rotational diffusion time. (d) Schematic diagram of translational diffusion time. *Reproduced with permission from Yamamoto et al. (2021).*

Thus, translational diffusion retardation of proteins cannot be explained in terms of the cytosolic viscosity. Factors other than viscosity, such as the mesh size of the cytoplasmic matrix and binding interactions, underlie the translational diffusion retardation of proteins, as discussed above (Section 7.2.1).

Swaminathan et al. (1997) compared the translational diffusion of GFP measured by FRAP and the rotational diffusion measured by time-resolved fluorescence in the cytoplasm of CHO cells, a mammalian cell line. They found that the rotation of GFP in the cytosol was only 1.5-fold slower than that in saline solution, whereas the translation of GFP was ~3.2-fold slower than that in water. The rapid GFP rotational rate in cytoplasm (36 ns) is consistent with the notion that particle rotation is relatively unhindered when particle size is smaller than the spacings between obstacles. These concepts can be clearly visualized in experiments in which the D_R and D_T of GFP are measured in gels using polarization-dependent fluorescence correlation spectroscopy (Yamamoto et al., 2021). The idea behind these experiments is to mimic the structure of the cytoplasmic matrix, using chemically cross-linked hydrogels as a model system to study the effect of an immobile environment, like that resulting from cytoskeletal elements, on the diffusion of GFP. Fig. 9.14 shows a comparison of the translational and rotational diffusion times of GFP between PBS and gels soaked in PBS. The translational diffusion time increases with the decrease in the mesh size above a threshold value while the mesh size ξ hardly influences the rotational diffusion (The mesh size ξ should not be confused with ξ the frictional coefficient, although they are inversely related; dynamic viscosity decreases as mesh size increases). The threshold mesh size around ~4 nm is consistent with the GFP size (~2.8 nm), indicating that only when the mesh size of the gels is larger than the probe size it can diffuse inside the gels, as illustrated in Fig. 9.14B.

Is viscosity of cytoplasm uniform? High-resolution intracellular viscosity measurements at submicron lateral- and axial-resolution using time-dependent fluorescence anisotropy in J774 macrophage cells show that cytoplasmic viscosity is spatially heterogeneous, varying in different cytoplasmic locations, ranging from 1.00 ± 0.03 cP to 2.21 ± 0.05 cP (Parker et al., 2010). The implication is that even though the local cytosolic microenvironment is heterogeneous, the range of intracellular viscosities is close to that of water. Further exploration of the cytoplasmic environment has been made by diffusivity measurements using high-resolution multifocal fluorescence imaging microscopy and fluorescence correlation spectroscopy (FCS) in single cells with inserted inert monomers and multimers of GFP covering the size range of most enzymes, and measuring their translational diffusion kinetics (Baum et al., 2014). The probes used were GFP1, a GFP

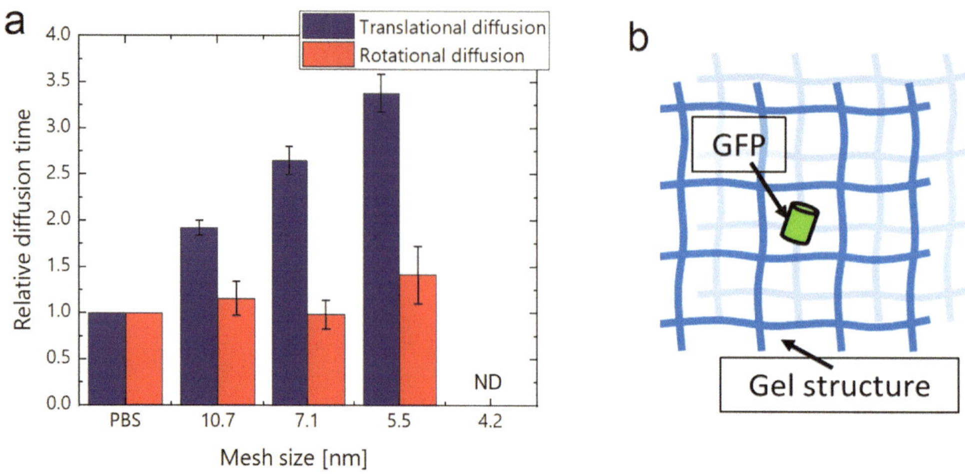

FIGURE 9.14 Translational and rotational diffusion times of GFP in gels of various mesh sizes (nm) measured with Pol-FCS. (a) Relative rotational diffusion times (red) and relative translational diffusion times (blue). The sample with 4.2 nm mesh size could not be measured (ND) because GFP did not penetrate the gel. (b) Schematic representation of GFP in a gel. *Reproduced with permission from Yamamoto et al. (2021).*

monomer with a hydrodynamic radius of ~2.8 nm, the trimer GFP3 (r ~ 5.5 nm), and the pentamer GFP5 (r ~ 7.9 nm). These authors obtained molecular mobility maps, linking protein mobility and intracellular architecture in a time- and length-scale-dependent manner. Protein translocations were measured with microsecond time resolution on multiple length scales from 0.2 to ~3 μm. Using this approach, they confirmed that the cellular interior behaves as a porous medium formed by random obstacles with characteristic size and density and that the intrinsic cellular viscosity is very similar to that of an aqueous solution. The local viscosity of the cytosolic fluid was only 1.3-fold larger than that of water for GFP1, consistent with earlier measurements of its rotational diffusion.

Acknowledgments

The author wishes to thank Professor Lawrence R. Frank for kindly providing Figs. 9.5–9.7, Professor Mary Jo Nye, for permission to reproduce the photograph of Jean Perrin shown in Fig. 9.3 and Professor Eric Delpire for his comments in an earlier version of the manuscript.

References

Ando, T., & Skolnick, J. (2010). Crowding and hydrodynamic interactions likely dominate in vivo macromolecular motion. *Proceedings of the National Academy of Sciences of the U S A, 107*, 18457–18462.

Andrews, S. S., Dinh, T., & Arkin, A. P. (2009). Stochastic models of biological processes. In R. A. Meyers (Ed.), *Encyclopedia of complexity and systems science* (pp. 8730–8749). New York, New York, NY: Springer.

Arrio-Dupont, M., Foucault, G., Vacher, M., Douhou, A., & Cribier, S. (1997). Mobility of creatine phosphokinase and beta-enolase in cultured muscle cells. *Biophysical Journal, 73*, 2667–2673.

Baker, P. F., Hodgkin, A. L., & Shaw, T. I. (1962). Replacement of the axoplasm of giant nerve fibres with artificial solutions. *The Journal of Physiology, 164*, 330–354.

Ball, P. (2008). Water as an active constituent in cell biology. *Chemistry Review, 108*, 74–108.

Ball, P. (2017). Water is an active matrix of life for cell and molecular biology. *Proceedings of the National Academy of Sciences of the U S A, 114*, 13327–13335.

Baum, M., Erdel, F., Wachsmuth, M., & Rippe, K. (2014). Retrieving the intracellular topology from multi-scale protein mobility mapping in living cells. *Nature Communications, 5*, 4494.

Bellissent-Funel, M. C., Hassanali, A., Havenith, M., Henchman, R., Pohl, P., Sterpone, F., van der Spoel, D., Xu, Y., & Garcia, A. E. (2016). Water determines the structure and dynamics of proteins. *Chemistry Review, 116*, 7673–7697.

Bellotto, N., Agudo-Canalejo, J., Colin, R., Golestanian, R., Malengo, G., & Sourjik, V. (2022). Dependence of diffusion in *Escherichia coli* cytoplasm on protein size, environmental conditions, and cell growth. *Elife, 11*, Article e82654.

Berg, H. C. (1993). *Random walks in biology*. Princeton, N.J: Princeton University Press.

Brown, R. (1828). A brief account of microscopical observations made in the months of June, July and August 1827, on the particles contained in the pollen of plants; and on the general existence of active molecules in organic and inorganic bodies. *The Philosophical Magazine, 4*, 161–173.

Brown, R. (1829). XXIV. Additional remarks on active molecules. *The Philosophical Magazine, 6*, 161–166.

Brush, S. G. (1968). A history of random processes: I. Brownian movement from Brown to Perrin. *Archive for History of Exact Sciences, 5*, 1–36.

Bulthuis, E. P., Dieteren, C. E. J., Bergmans, J., Berkhout, J., Wagenaars, J. A., van de Westerlo, E. M. A., Podhumljak, E., Hink, M. A., Hesp, L. F. B., Rosa, H. S., Malik, A. N., Lindert, M. K., Willems, P., Gardeniers, H., den Otter, W. K., Adjobo-Hermans, M. J. W., & Koopman, W. J. H. (2023). Stress-dependent macromolecular crowding in the mitochondrial matrix. *The EMBO Journal, 42*, Article e108533.

Cadée, G. C. (1991). Brownian emotion. *Nature, 354*, 180.

Carslaw, H. S., & Jaeger, J. C. (1986). *Conduction of heat in solids*. Oxford Oxfordshire & New York: Clarendon Press ; Oxford University Press.

Chandrasekhar, S. (1943). Stochastic problems in physics and astronomy. *Reviews of Modern Physics, 15*, 1–89.

Chen, X., Weber, I., & Harrison, R. W. (2008). Hydration water and bulk water in proteins have distinct properties in radial distributions calculated from 105 atomic resolution crystal structures. *Journal of Physical Chemistry B, 112*, 12073–12080.

Choi, A. A., Xiang, L., Li, W., & Xu, K. (2023). Single-molecule displacement mapping indicates unhindered intracellular diffusion of small ($\lesssim$1 kDa) solutes. *Journal of the American Chemical Society, 145*, 8510–8516.

Clegg, J. S. (1984). Properties and metabolism of the aqueous cytoplasm and its boundaries. *American Journal of Physiology, 246*, R133–R151.

Cohen, L. (2005). The history of noise [on the 100th anniversary of its birth]. *IEEE Signal Processing Magazine, 22*, 20–45.

Crank, J. (1979). *The mathematics of diffusion*. Oxford, Eng: Clarendon Press.

Cserr, H. F., Cooper, D. N., Suri, P. K., & Patlak, C. S. (1981). Efflux of radiolabeled polyethylene glycols and albumin from rat brain. *American Journal of Physiology, 240*, F319–F328.

Davis, L., Jr., & Lorente de Nó, R. (1947). Contribution to the mathematical theory of the electrotonus. *Studies from the Rockefeller Institute for Medical Research. Reprints. Rockefeller Institute for Medical Research, 131*, 442–496.

Debye, P. (1929). *Polar molecules*. New York: Dover.

Einstein, A. (1956). *Investigations on the theory of Brownian movement*. New York, N.Y: Dover Publications.

Einstein, A. (1905). On the motion of small particles suspended in liquids at rest required by the molecular-kinetic theory of heat. In J. Stachel (Ed.), *Einstein's miraculous year five papers that changed the face of physics* (pp. 85–98). Princeton, New Jersey: Princeton University Press.

Fels, J., Orlov, S. N., & Grygorczyk, R. (2009). The hydrogel nature of mammalian cytoplasm contributes to osmosensing and extracellular pH sensing. *Biophysical Journal, 96*, 4276–4285.

Fick, A. (1855). On liquid diffusion. *Philosophical Magazine, 10*, 30–39.

Fick, A. (1995). On liquid diffusion. *Journal of Membrane Science, 100*, 33–38.

Frank, L. (2023). *The microscopic theory of diffusion*. http://csci.ucsd.edu/courses/cogs260w2018/files/lecture-03/Lecture3_Readings_Chap23.pdf.

Godin, A. G., Varela, J. A., Gao, Z., Danné, N., Dupuis, J. P., Lounis, B., Groc, L., & Cognet, L. (2017). Single-nanotube tracking reveals the nanoscale organization of the extracellular space in the live brain. *Nature Nanotechnology, 12*, 238–243.

Gregersen, E. (2017). Brownian motion. In Encyclopaedia britannica. *Encyclopaedia britannica*.

Hamann, S., Kiilgaard, J. F., Litman, T., Alvarez-Leefmans, F. J., Winther, B. R., & Zeuthen, T. (2002). Measurement of cell volume changes by fluorescence self-quenching. *Journal of Fluorescence, 12*, 139–145.

Haw, M. (2005). Einstein's random walk. *Physics World, 18*, 19.

Hille, B. (2001). *Ion channels of excitable membranes*. Sunderland, Massachusetts: Sinauer Associates.

Hirose, T., Ninomiya, K., Nakagawa, S., & Yamazaki, T. (2023). A guide to membraneless organelles and their various roles in gene regulation. *Nature Reviews Molecular Cell Biology, 24*, 288–304.

Hodgkin, A. L., & Katz, B. (1949). The effect of calcium on the axoplasm of giant nerve fibers. *Journal of Experimental Biology, 26*, 292–294.

Hodgkin, A. L., & Rushton, W. A. (1946). The electrical constants of a crustacean nerve fibre. *Proceedings of the Royal Society of Medicine, 134*, 444–479.

Hrabe, J., Hrabĕtová, S., & Segeth, K. (2004). A model of effective diffusion and tortuosity in the extracellular space of the brain. *Biophysical Journal, 87*, 1606–1617.

Hrabetova, S., Cognet, L., Rusakov, D. A., & Nägerl, U. V. (2018). Unveiling the extracellular space of the brain: From super-resolved microstructure to in vivo function. *Journal of Neuroscience, 38*, 9355–9363.

Israelachvili, J. N. (2011). *Interactions involving polar molecules in* Intermolecular and surface forces (3rd ed., pp. 71–90). San Diego, California: Academic Press -Elsevier.

Jacobson, K., & Wojcieszyn, J. (1984). The translational mobility of substances within the cytoplasmic matrix. *Proceedings of the National Academy of Sciences of the United States of America, 81*, 6747–6751.

Jalihal, A. P., Schmidt, A., Gao, G., Little, S. R., Pitchiaya, S., & Walter, N. G. (2021). Hyperosmotic phase separation: Condensates beyond inclusions, granules and organelles. *Journal of Biological Chemistry, 296*, Article 100044.

Jasnin, M., Moulin, M., Haertlein, M., Zaccai, G., & Tehei, M. (2008). Down to atomic-scale intracellular water dynamics. *EMBO Reports, 9*, 543–547.

Jia, D., Hamilton, J., Zaman, L. M., & Goonewardene, A. (2007). The time, size, viscosity, and temperature dependence of the Brownian motion of polystyrene microspheres. *American Journal of Physics, 75*, 111–115.

Jungwirth, P. (2015). Biological water or rather water in biology? *Journal of Physical Chemistry Letters, 6*, 2449–2451.

Katz, B., & Miledi, R. (1965). The measurement of synaptic delay, and the time course of acetylcholine release at the neuromuscular junction. *Proceedings of the Royal Society of London B Biological Sciences, 161*, 483–495.

Koch, C. (1999). *Biophysics of computation: Information processing in single neurons*. New York: Oxford University Press.

Korogod, N., Petersen, C. C., & Knott, G. W. (2015). Ultrastructural analysis of adult mouse neocortex comparing aldehyde perfusion with cryo fixation. *Elife, 4*, Article e05793. https://doi.org/10.7554/eLife.05793

Kwapiszewska, K., Szczepański, K., Kalwarczyk, T., Michalska, B., Patalas-Krawczyk, P., Szymański, J., Andryszewski, T., Iwan, M., Duszyński, J., & Hołyst, R. (2020). Nanoscale viscosity of cytoplasm is conserved in human cell lines. *Journal of Physical Chemistry Letters, 11*, 6914–6920.

Land, B. R., Harris, W. V., Salpeter, E. E., & Salpeter, M. M. (1984). Diffusion and binding constants for acetylcholine derived from the falling phase of miniature endplate currents. *Proceedings of the National Academy of Sciences of the U S A, 81*, 1594–1598.

Langevin, P. (1908). Sur la théorie du mouvement brownien. *Comptes Rendus, 146*, 530–533.

Law, R. O. (1982). Techniques and applications of extracellular space determination in mammalian tissues. *Experientia, 38*, 411–421.

Lechene, C. (1985). Cellular volume and cytoplasmic gel. *Biologie Cellulaire, 55*, 177–180.

Lelek, M., Gyparaki, M. T., Beliu, G., Schueder, F., Griffié, J., Manley, S., Jungmann, R., Sauer, M., Lakadamyali, M., & Zimmer, C. (2021). Single-molecule localization microscopy. *Nature Reviews Methods Primers, 1*(39). https://doi.org/10.1038/s43586-021-00038-x

Lemons, D. S., & Gythiel, A. (1997). Paul langevin's 1908 paper "on the theory of Brownian motion" ["sur la théorie du mouvement brownien," cr acad. sci.(Paris) 146, 530–533 (1908)]. *American Journal of Physics, 65*, 1079–1081.

Levin, V. A., Fenstermacher, J. D., & Patlak, C. S. (1970). Sucrose and inulin space measurements of cerebral cortex in four mammalian species. *American Journal of Physiology, 219*, 1528–1533.

Likhtenshtein, G. I. (2021). Hydration of cells and tissues. In *Biological water: Physicochemical aspects* (pp. 439–479). Cham: Springer International Publishing.

Lippincott-Schwartz, J., & Patterson, G. H. (2003). Development and use of fluorescent protein markers in living cells. *Science, 300*, 87–91.

Luby-Phelps, K. (2000). Cytoarchitecture and physical properties of cytoplasm: Volume, viscosity, diffusion, intracellular surface area. *International Review of Cytology, 192*, 189–221.

Luby-Phelps, K. (2013). The physical chemistry of cytoplasm and its influence on cell function: An update. *Molecular Biology of the Cell, 24*, 2593–2596.

Lyon, A. S., Peeples, W. B., & Rosen, M. K. (2021). A framework for understanding the functions of biomolecular condensates across scales. *Nature Reviews Molecular Cell Biology, 22*, 215–235.

Macknight, A. D. (1968). The extracellular space in rat renal cortical slices incubated at 0.5 degrees and 25 degrees. *Biochimica et Biophysica Acta, 163*, 85–92.

Magzoub, M., Zhang, H., Dix, J. A., & Verkman, A. S. (2009). Extracellular space volume measured by two-color pulsed dye infusion with microfiberoptic fluorescence photodetection. *Biophysical Journal, 96*, 2382–2390.

Maiocchi, R. (1990). The case of Brownian motion. *The British Journal for the History of Science, 23*, 257–283.

Marty, B., Baudin, P. Y., Caldas de Almeida Araujo, E., Fromes, Y., Wahbi, K., & Reyngoudt, H. (2023). Assessment of extracellular volume fraction in becker muscular dystrophy by using MR fingerprinting. *Radiology, 307*, Article e221115.

Metzler, R., Jeon, J.-H., Cherstvy, A. G., & Barkai, E. (2014). Anomalous diffusion models and their properties: Non-stationarity, non-ergodicity, and ageing at the centenary of single particle tracking. *Physical Chemistry Chemical Physics, 16*, 24128–24164.

Milo, R., & Phillips, R. (2016). In *Cell biology by the numbers*. New York: Garland Science.

Mitchison, T. J., Charras, G. T., & Mahadevan, L. (2008). Implications of a poroelastic cytoplasm for the dynamics of animal cell shape. *Seminars in Cell & Developmental Biology, 19*, 215–223.

Moeendarbary, E., Valon, L., Fritzsche, M., Harris, A. R., Moulding, D. A., Thrasher, A. J., Stride, E., Mahadevan, L., & Charras, G. T. (2013). The cytoplasm of living cells behaves as a poroelastic material. *Nature Materials, 12*, 253–261.

Mogilner, A., & Manhart, A. (2018). Intracellular fluid mechanics: Coupling cytoplasmic flow with active cytoskeletal gel. *Annual Review of Fluid Mechanics, 50*, 347–370.

Newburgh, R., Peidle, J., & Rueckner, W. (2006). Einstein, Perrin, and the reality of atoms: 1905 revisited. *American Journal of Physics, 74*, 478–481.

Nicholson, C., & Hrabětová, S. (2017). Brain extracellular space: The final frontier of neuroscience. *Biophysical Journal, 113*, 2133–2142.

Nicholson, C., & Phillips, J. M. (1981). Ion diffusion modified by tortuosity and volume fraction in the extracellular microenvironment of the rat cerebellum. *The Journal of Physiology, 321*, 225–257.

Nicholson, C., & Tao, L. (1993). Hindered diffusion of high molecular weight compounds in brain extracellular microenvironment measured with integrative optical imaging. *Biophysical Journal, 65*, 2277–2290.

Nicholson, C. (2001). Diffusion and related transport mechanisms in brain tissue. *Reports on Progress in Physics, 64*, 815.

Nicholson, C. (1993). Ion-selective microelectrodes and diffusion measurements as tools to explore the brain cell microenvironment. *Journal of Neuroscience Methods, 48*, 199–213.

Nobel, P. S. (2020). Cells and diffusion. In P. S. Nobel (Ed.), *Physicochemical and environmental plant physiology* (5th ed., pp. 3–50). San Diego: Academic Press.

Nye, M. J. (1972). *Molecular reality; a perspective on the scientific work of Jean Perrin*. London, New York: Macdonald; American Elsevier.

Orlov, S. N., Shiyan, A., Boudreault, F., Ponomarchuk, O., & Grygorczyk, R. (2018). Search for upstream cell volume sensors: The role of plasma membrane and cytoplasmic hydrogel. *Current Topics in Membranes, 81*, 53–82.

Pais, A. (2005). The reality of molecules. In *'Subtle is the lord': The science and the life of Albert Einstein* (New ed ed., pp. 79–107). Oxford: Oxford University Press.

Parker, E. T., & Lollar, P. (2021). Measurement of the translational diffusion coefficient and hydrodynamic radius of proteins by dynamic light scattering. *Bio-protocol, 11*, Article e4195.

Parker, W. C., Chakraborty, N., Vrikkis, R., Elliott, G., Smith, S., & Moyer, P. J. (2010). High-resolution intracellular viscosity measurement using time-dependent fluorescence anisotropy. *Optics Express, 18*, 16607–16617.

Patlak, C. S., & Fenstermacher, J. D. (1975). Measurements of dog blood-brain transfer constants by ventriculocisternal perfusion. *American Journal of Physiology, 229*, 877–884.

Pauling, L. (1960). *The nature of the chemical bond and the structure of molecules and crystals* (3rd ed). Ithaca, New York: Cornell University Press ionic.

Paviolo, C., Ferreira, J. S., Lee, A., Hunter, D., Calaresu, I., Nandi, S., Groc, L., & Cognet, L. (2022). Near-infrared carbon nanotube tracking reveals the nanoscale extracellular space around synapses. *Nano Letters, 22*, 6849—6856.

Pearle, P., Collett, B., Bart, K., Bilderback, D., Newman, D., & Samuels, S. (2010). What Brown saw and you can too. *American Journal of Physics, 78*, 1278—1289.

Pearson, K. (1905). The problem of the random walk. *Nature, 72*, 294.

Perrin, J. (1990). *Atoms*. Woodbridge, CT: Ox Bow Press.

Perrin, J. (1910). *Brownian movement and molecular reality*. London: Taylor and Francis.

Persson, E., & Halle, B. (2008). Cell water dynamics on multiple time scales. *Proceedings of the National Academy of Sciences of the U S A, 105*, 6266—6271.

Piazza, I., Cupane, A., Barbier, E. L., Rome, C., Collomb, N., Ollivier, J., Gonzalez, M. A., & Natali, F. (2018). Dynamical properties of water in living cells. *Frontiers of Physics, 13*, Article 138301.

Pollack, G. H. (2001). *Cells, gels and the engines of life: A new, unifying approach to cell function*. Seattle, WA: Ebner & Sons.

Poole-Wilson, P. A., & Cameron, I. R. (1975). ECS, intracellular pH, and electrolytes of cardiac and skeletal muscle. *American Journal of Physiology, 229*, 1299—1304.

Purkiss, A., Skoulakis, S., & Goodfellow, J. M. (2001). The protein—solvent interface: A big splash. *Philosophical Transactions of the Royal Society A, 359*, 1515—1527.

Rasmussen, M. K., Mestre, H., & Nedergaard, M. (2022). Fluid transport in the brain. *Physiological Reviews, 102*, 1025—1151.

Rigden, J. S. (2005). *Einstein 1905: The standard of greatness*. Cambridge, Massachusetts: Harvard University Press.

Rivas, G., & Minton, A. P. (2016). Macromolecular crowding in vitro, in vivo, and in between. *Trends in Biochemical Sciences, 41*, 970—981.

Robinson, R. A., & Stokes, R. H. (2002). *Electrolyte solutions*. Mineola, New York: Dover Publications, Inc.

Rosenberg, L. E., Downing, S. J., & Segal, S. (1962). Extracellular space estimation in rat kidney slices using C saccharides and phlorizin. *American Journal of Physiology, 202*, 800—804.

Sachs, F., & Sivaselvan, M. V. (2015). Cell volume control in three dimensions: Water movement without solute movement. *The Journal of General Physiology, 145*, 373—380.

Safford, R. E., & Bassingthwaighte, J. B. (1977). Calcium diffusion in transient and steady states in muscle. *Biophysical Journal, 20*, 113—136.

Smith, G. E., & Seth, R. (2020). Brownian motion and molecular reality : A study in theory-mediated measurement. In *Oxford studies in philosophy of science* (p. 1). New York, NY: Oxford University Press.

Soria, F. N., Miguelez, C., Peñagarikano, O., & Tønnesen, J. (2020). Current techniques for investigating the brain extracellular space. *Frontiers in Neuroscience, 14*, Article 570750.

Sperelakis, N., & Freedman, J. C. (2012). Diffusion and permeability. In N. Sperelakis (Ed.), *Cell physiology sourcebook. Essentials of membrane biophysics* (4th ed., pp. 113—120). San Diego: Academic Press-Elsevier.

Sposini, V., Krapf, D., Marinari, E., Sunyer, R., Ritort, F., Taheri, F., Selhuber-Unkel, C., Benelli, R., Weiss, M., Metzler, R., & Oshanin, G. (2022). Towards a robust criterion of anomalous diffusion. *Communications Physics, 5*, 305.

Swaminathan, R., Hoang, C. P., & Verkman, A. S. (1997). Photobleaching recovery and anisotropy decay of green fluorescent protein GFP-S65T in solution and cells: Cytoplasmic viscosity probed by green fluorescent protein translational and rotational diffusion. *Biophysical Journal, 72*, 1900—1907.

Syková, E., & Nicholson, C. (2008). Diffusion in brain extracellular space. *Physiological Reviews, 88*, 1277—1340.

Syková, E., Svoboda, J., Polák, J., & Chvátal, A. (1994). Extracellular volume fraction and diffusion characteristics during progressive ischemia and terminal anoxia in the spinal cord of the rat. *Journal of Cerebral Blood Flow and Metabolism, 14*, 301—311.

Tønnesen, J., Hrabĕtová, S., & Soria, F. N. (2023). Local diffusion in the extracellular space of the brain. *Neurobiology of Disease, 177*, Article 105981.

Thorne, R. G., & Nicholson, C. (2006). In vivo diffusion analysis with quantum dots and dextrans predicts the width of brain extracellular space. *Proceedings of the National Academy of Sciences of the U S A, 103*, 5567—5572.

Tros, M., Zheng, L., Hunger, J., Bonn, M., Bonn, D., Smits, G. J., & Woutersen, S. (2017). Picosecond orientational dynamics of water in living cells. *Nature Communications, 8*, 904.

Truscott, R. J. W. (2021). Long-lived cells and long-lived proteins in the human body. In R. J. W. Truscott (Ed.), *Long-lived proteins in human aging and disease* (1st ed., pp. 5—41). Weinheim, Germany: Wiley.

Van Holde, K. E. (1985). *Physical biochemistry*. Englewood Cliffs, NJ: Prentice-Hall.

van Tartwijk, F. W., & Kaminski, C. F. (2022). Protein condensation, cellular organization, and spatiotemporal regulation of cytoplasmic properties. *Advanced biology, 6*, Article e2101328.

Vanysek, P. (2005). Ionic conductivity and diffusion at infinite dilution. In D. R. Lide (Ed.), *CRC handbook of chemistry and physics* (pp. 93—95). Boca Raton, Florida: CRC Press.

Verkman, A. S. (2013). Diffusion in the extracellular space in brain and tumors. *Physical Biology, 10*, Article 045003.

Verkman, A. S. (2002). Solute and macromolecule diffusion in cellular aqueous compartments. *Trends in Biochemical Sciences, 27*, 27—33.

Van't Hoff, J. H. (1901). Osmotic pressure and chemical equilibrium. In *Nobel lectures, chemistry 1901-1921* (pp. 5—10). Amsterdam: Elsevier Publishing Company. Available at: https://www.nobelprize.org/prizes/chemistry/1901/hoff/lecture/.

Van't Hoff, J. H. (1887). The role of osmotic pressure in the analogy between solutions and gases-translated from "Die rolle des osmotischen druckes in der analogie zwischen lösungen und gasen," Z. Physik. Chemie 1:481-493, 1887). In G. R. Kepner (Ed.), *Cell membrane permeability and transport* (pp. 20–28). Dowden, Stroudsburg, Pennsylvania: Hutchinson & Ross, Inc.

Voříšek, I., & Syková, E. (1997). Evolution of anisotropic diffusion in the developing rat corpus callosum. *Journal of Neurophysiology, 78*, 912–919.

Wiig, H., & Swartz, M. A. (2012). Interstitial fluid and lymph formation and transport: Physiological regulation and roles in inflammation and cancer. *Physiological Reviews, 92*, 1005–1060.

Xiang, L., Chen, K., Yan, R., Li, W., & Xu, K. (2020). Single-molecule displacement mapping unveils nanoscale heterogeneities in intracellular diffusivity. *Nature Methods, 17*, 524–530.

Xiao, F., Hrabe, J., & Hrabetova, S. (2015). Anomalous extracellular diffusion in rat cerebellum. *Biophysical Journal, 108*, 2384–2395.

Xie, L., Kang, H., Xu, Q., Chen, M. J., Liao, Y., Thiyagarajan, M., O'Donnell, J., Christensen, D. J., Nicholson, C., Iliff, J. J., Takano, T., Deane, R., & Nedergaard, M. (2013). Sleep drives metabolite clearance from the adult brain. *Science, 342*, 373–377.

Yamamoto, J., Matsui, A., Gan, F., Oura, M., Ando, R., Matsuda, T., Gong, J. P., & Kinjo, M. (2021). Quantitative evaluation of macromolecular crowding environment based on translational and rotational diffusion using polarization dependent fluorescence correlation spectroscopy. *Scientific Reports, 11*, Article 10594.

You, X., Shirley, J. C., Lee, E., & Baiz, C. R. (2021). Short- and long-range crowding effects on water's hydrogen bond networks. *Cell Reports Physical Science, 2*, Article 100419.

Zaccai, G. (2020). Molecular dynamics in cells: A neutron view. *Biochimica et Biophysica Acta (BBA) - General Subjects, 1864*(3), Article 129475.

Zheng, K., Jensen, T. P., Savtchenko, L. P., Levitt, J. A., Suhling, K., & Rusakov, D. A. (2017). Nanoscale diffusion in the synaptic cleft and beyond measured with time-resolved fluorescence anisotropy imaging. *Scientific Reports, 7*, Article 42022.

Chapter 10

Membrane diffusion and permeability

Eric Delpire[1] and Francisco J. Alvarez-Leefmans[2,a]

[1]*Department of Anesthesiology, Vanderbilt University School of Medicine, Nashville, TN, United States;* [2]*Department of Pharmacology and Toxicology, Boonshoft School of Medicine, Wright State University, Dayton, OH, United States*

1. Summary

The biophysical principles of diffusion and their application to confined compartments of biological relevance, such as the extracellular space and the cytoplasm, are the subject of Chapter 9. The focus of the present chapter is the diffusion of uncharged particles (nonelectrolytes) and ions across cell membranes. The simplest transport process of a solute across membranes is passive diffusion. It occurs when the solute's electrochemical potentials on the two sides of a permeable barrier are different. The basic equations governing this process are presented. The relationship between the *diffusion coefficient* (D) and the *permeability coefficient* (P) is discussed, including the factors that determine these coefficients. The dependence of P on the mobility of an ion through the membrane under a voltage gradient is also succinctly described. Electrochemical potential is defined, and the interconversion between flux and current is discussed. Finally, the *Ussing flux ratio* equation is discussed, and examples of its significance are provided.

2. Cell membrane diffusion

The cell membrane, a complex and fascinating structure composed mainly of lipids and proteins (Chapter 4), constitutes a barrier to the rapid and free reciprocal exchange of solutes between the extracellular compartment and the cytosol. However, cell membranes are not just simple barriers but are semipermeable, since some molecules can diffuse across the lipid bilayer, but others cannot. Thus, membranes are selectively permeable. This permselectivity results from the intricate arrangement of lipids and proteins in the membrane. Cell membranes restrict the diffusion of highly charged molecules, such as ions, and relatively large molecules, such as sugars and amino acids. These large molecules and ions cross the membrane through selective channels and transporters (carriers), which are proteins embedded in the lipid bilayer. Because of Donnan forces and active transport, cells maintain a cytosolic concentration of ions and molecules distinct from the extracellular milieu, creating the perfect conditions for these particles to diffuse back across the membrane down their concentration and/or electric gradients.

We have seen (Chapter 9) that diffusion occurs because of the random thermal motion of particles, which flow from a region of high concentration to one of low concentration. If a solution of high solute concentration is adjacent to one of low concentration but separated by an imaginary plane, as illustrated in Fig. 9.8, Chapter 9 in this volume, it is probable that more molecules per unit of time will be crossing the plane from the side of higher concentration to the side of lower concentration than in the opposite direction. As discussed also in Chapter 9, there are *fluxes* (transport of matter, e.g., molecules) in both directions (the unidirectional fluxes), but the *net flux* is from the side of higher concentration to the side of lower concentration. If the imaginary plane is replaced with a thin membrane permeable to a certain type of solute molecules, the same situation applies; the solute molecules diffuse across the membrane from the side of higher concentration to the side of lower concentration, which in the examples here considered would be from the outside (out) fluid compartment, where the bulk solute concentration is C_{1b}, to the inside (in) fluid compartment, where the bulk solute

[a] The authors contributed equally to this work.

Cell Physiology Sourcebook. https://doi.org/10.1016/B978-0-12-811114-7.00001-3

concentration is C_{2b}. The solute molecules are continuously colliding with the membrane. Thus, at any moment, the number of molecules crossing the membrane from the outside (out) compartment to the internal compartment (in) will be proportional to the number impinging on the membrane, multiplied by the membrane unit area, A. Similarly, the number of solute molecules crossing from the inside to the outside (in→out) will be proportional to the solute concentration inside (C_{2b}). The net movement from the outside toward the inside (out→in) will be determined by the difference in the solute concentration between the two fluid compartments ($C_{2b}-C_{1b}$). The other factor determining the rate of transmembrane solute movements is the *resistance* that the membrane offers to the movement of the solute across it (not to be confused with electrical resistance), as discussed below.

In living cells, it is usually assumed that the bulk solutions are relatively well mixed, and diffusion of most substances through the plasma membrane is much slower than that through a free solution. Therefore, diffusion through the membrane is a rate-limiting step. For simplicity, we first assume that the solutions on either side of the membrane are well stirred. However, this is a simplification given that unstirred layers (ULs) in both the external and internal leaflets of the membrane occur, as discussed below. We focus here on the diffusion of small (nonelectrolytes) molecules and ions across the cell membrane. For thin membranes, we first consider the hypothetical case of *diffusion without partitioning*, as illustrated in Fig. 10.1 (left), in which the diffusing molecule can freely enter the membrane. This simplified view is first discussed as a base case to explain more complex transport mechanisms across cell membranes. We will then introduce the effect of partitioning into a lipid membrane. This process involves a change of chemical potential upon entrance into the membrane phase, as illustrated in Fig. 10.1 (right). For this discussion, we disregard possible obstacles, such as potential energy barriers within the membrane (Danielli, 1943; Hille, 2001).

2.1 Diffusion without partitioning

Let us consider the hypothetical case in which the diffusing molecule can freely enter a homogeneous thin membrane from the aqueous solution on each side of the membrane (Fig. 10.1, left); that is, the molecules do not partition between the aqueous phase and the membrane phase. Further, the bulk solutions are well stirred, and thus, there are no local concentration gradients in the immediate vicinity of the membrane surfaces. The concentration of the diffusible substance outside this membrane (i.e., in the external aqueous bulk solution) is $C_{1b} = C_o$, and that in the internal aqueous bulk solution is $C_{2b} = C_i$. If $C_{1b} > C_{2b}$, there will be a concentration gradient across the membrane, i.e., a chemical potential difference across it, and thus, the net diffusion will be out → in. The concentration profile *inside* this membrane shows that there is no change in the concentration of the diffusing molecules within the membrane in the outer boundary adjacent to

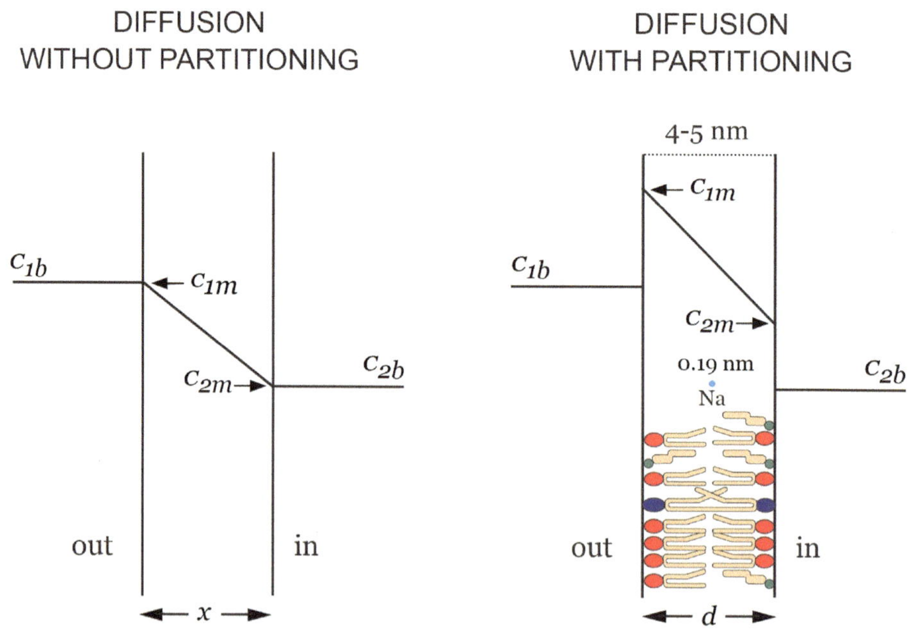

FIGURE 10.1 **Diffusion across a membrane without and with partitioning.** *Left*: *Diffusion across a thin membrane without partitioning*. The concentrations of the diffusible solute in the external (out) and the internal (in) aqueous bulk solutions are C_{1b} and C_{2b}, respectively, and the thickness of the membrane is x. The corresponding solute concentrations within the membrane are C_{1m} and C_{2m}. Since the solute does not partition in the membrane and there are no unstirred layers, $C_{1b} = C_{1m}$ and $C_{2m} = C_{2b}$. The concentration profile is depicted as a line with a negative slope within the membrane. *Right*: *Diffusion across a lipid bilayer with partitioning*. The membrane is composed of two layers of lipids with their heads (colored) on the outside and tails on the inside. Phosphatidylserine (*red*), sphingolipids (*blue*), and cholesterol (*green*) are represented. The membrane thickness (d) is 4–5 mm, which is ~20–25 times the size of a sodium atom (Na) represented by a *light blue* dot. The solute at concentrations in the bulk aqueous phases on each side are C_{1b} (outside) and C_{2b} (inside), and the corresponding concentrations inside the membrane are C_{1m} and C_{2m}. The concentration profile inside the membrane in this diagram is for a lipophilic compound with a partition coefficient $K_p > 1$.

the external solution (C_{1m}), i.e., $C1_b = C1_m$; that is, there is no chemical potential difference between these two adjacent points. Inside the membrane itself, there is a concentration gradient represented by the negative slope line (the concentration profile), and the concentration at the inner boundary between the membrane and the internal bulk C_{2m} is the same as C_{2b}. In other words, the concentrations at the internal leaflet of the membrane adjacent to the intracellular solution and the latter solution are the same ($C_{2m} = C_{2b}$). This ideal situation in which there are no ULs on each side of the membrane and the diffusing solute molecule does not partition as it does in lipid bilayers is referred to as *diffusion without partition.* It is a simplification because, in biological membranes, ULs usually exist in the external and internal interfaces of the membrane, and since membranes are composed of lipid bilayers, solutes partition within the lipid phase (Missner & Pohl, 2009; Pohl et al., 1997), as discussed below. Nevertheless, the thin membrane model in which diffusion occurs without partition provides a "base case" that can be compared against the actual, more complex biological membranes.

Recapping: the net movement of solute from side one (out) to side two (in) of the membrane is determined by the difference in bulk solute concentration in the two bathing solutions (C_{1b}–C_{2b}) and by the *resistance* that the membrane offers to the movement of the solute across it (Stein & Litman, 2015). Among the factors that determine this resistance are the nature of the barrier (loosely vs. tightly packed components) and the strength of the physical-chemical forces of interaction between the diffusing molecules and the molecules that compose the membrane, e.g., membrane viscosity (Trimble & Grinstein, 2015). A barrier with high resistance leads to a low rate of net transbarrier solute movement (*dm/dt*). Besides the resistance, *dm/dt* will also be determined by the barrier thickness (x). Expressed in mathematical form, the rate of the net transbarrier solute movement (*dm/dt*) is inversely proportional to the resistance (ρ) of the barrier and its thickness (x) and directly proportional to the concentration gradient across the membrane and to the unit area (A), and thus,

$$\frac{dm}{dt}_{1\rightarrow2} = -A\frac{(C_{1b} - C_{2b})}{\rho\, x} \tag{10.1}$$

where m is the net amount of solute (moles) crossing the membrane plane in time t (seconds), C is the solute concentration ($\text{mol}\cdot\text{cm}^{-3}$), x is the thickness of the barrier (cm), and A is the barrier area (cm^2). The minus sign in this equation indicates that the direction of transport is down the concentration gradient (negative slope). The resistance to diffusion ρ is analogous to electrical resistance, as noted by Fick himself (Chapter 9, Section 5). In studies on transbarrier diffusion, the reciprocal of the resistance of the barrier (ρ) is defined as D, the diffusion coefficient or diffusivity of the solute ($\text{cm}^2\cdot\text{s}^{-1}$). Resistance and diffusion always have this reciprocal relationship with each other (Stein & Litman, 2015). The quantitative factors that determine D are thoroughly discussed in Chapter 9, Section 4 of this volume. Expressed in terms of the diffusion coefficient D, Eq. (10.1) becomes:

$$\frac{dm}{dt}_{1\rightarrow2} = -DA\frac{(C_{1b} - C_{2b})}{x} \tag{10.2}$$

Eq. (10.2) is a form of Fick's first law of diffusion, discussed in Chapter 9, Section 5.1. Dividing both sides of Eq. (10.2) by A, we obtain:

$$\frac{dm}{dtA}_{1\rightarrow2} = -D\frac{(C_{1b} - C_{2b})}{x} = -D\frac{dc}{dx} \tag{10.3}$$

where *dm/dtA* is the transmembrane flux density, J, which units are $\text{mol}\cdot\text{sec}^{-1}\cdot\text{cm}^{-2}$. Thus, from Eq. (10.3), we obtain the typical form in which Fick's first law is usually expressed:

$$J = -D\frac{dc}{dx} \tag{10.4}$$

Eq. (10.4) is identical to- Eq. (9.19) of Chapter 9.

2.2 Diffusion across the lipid bilayer of plasma membranes: The solubility diffusion mechanism or diffusion with partitioning

Molecular permeation through lipid membranes is a fundamental biological process that is of great importance for small neutral molecules, including several drug molecules of clinical relevance. In order for a molecule to get across the lipid bilayer, it must first leave the aqueous phase and dissolve in the bilayer; in other words, the molecule must detach itself from its surrounding water molecules and jump into the membrane phase. Then, the molecule must diffuse across the hydrocarbon interior of the bilayer, perhaps by a series of small jumps over energy barriers, and finally leave the bilayer to dissolve in the aqueous phase on the opposite side of the membrane. This passive diffusion process is called the *solubility-*

diffusion mechanism of membrane permeation. It is also known as the *Meyer–Overton rule*, after Charles Ernst Overton (1865–1933) and Hans Horst Meyer (1853–1939), who first postulated it independently (Meyer, 1901; Overton, 1901).

According to the Meyer–Overton rule, the entry of any molecule into a cell is governed by its lipid solubility. Overton found that the permeability of cells to many different solutes is proportional to the solutes' *partition coefficient-* from water into organic phases. Independently, Meyer proposed the theory for the permeation of several anesthetics. The rule does not account for transport processes mediated by membrane channels and carriers, which were unknown at the time, and ignores membrane inhomogeneities such as rafts, which exist in biological membranes (Al-Awqati, 1999; Simons & Ikonen, 1997). The rule works satisfactorily for molecules that cross the lipid matrix by simple diffusion (Missner & Pohl, 2009).

In the solubility-diffusion mechanism, the relevant diffusion coefficient to be used in Fick's first law is not that in the external and internal bulk aqueous solutions but that in the bilayer (D_m). The relevant concentrations are not those in the bulk aqueous solutions on either side of the membrane but those *within* the membrane at the interface between the aqueous and lipid phases on each side of the membrane. Let us suppose that there is a positive flux, J, of a solute across the membrane from outside to inside, that is, from side one to side two in the notation used here, and that J remains constant. Also, assume that D_m, the diffusion coefficient inside the membrane, remains constant. From Fick's first law (Eq. 10.4 above), if J and D_m are constant, dc/dx must also be constant. This means that *there must be a linear concentration profile through the membrane*, and thus,

$$\frac{dc_m}{dx} = \frac{(C_{1m} - C_{2m})}{d} \tag{10.5}$$

where d is the membrane thickness, and the concentrations of the permeating molecule inside the membrane, $C_{1m,}$ and C_{2m} are those within the inner and outer leaflets of the bilayer, respectively (Fig. 10.1, right). Substituting Eq. (10.5) into Eq. (10.4), we obtain:

$$J = -D_m\frac{(C_{1m} - C_{2m})}{d} = -D_m\frac{dc_m}{dx} \tag{10.6}$$

So far, there is no way to measure the concentration of a solute within the membrane directly. However, it is possible to measure the solute concentrations in the aqueous phases near the outer and inner surfaces of the membrane. Assuming that there are no unstirred layers (i.e., the solute distribution in water is statistically uniform), those concentrations are C_{1b} and C_{2b}, respectively. The concentrations within the membrane (C_{1m} and C_{2m}) are some constant fraction of the concentration just outside the membrane (C_{1b} and C_{2b}); this fraction is the *partition coefficient* K_p. In a homogeneous membrane, $K_p = C_{1m}/C_{1b} = C_{2m}/C_{2b}$. Thus, Eq. (10.6) can be rewritten:

$$J = \frac{D_m K_p}{d}(C_{2b} - C_{1b}) \tag{10.7}$$

The coefficients in Eq. (10.7) are lumped into a single measurable factor called the permeability coefficient, P_M thus:

$$J = P_M(C_{2b} - C_{1b}) \tag{10.8}$$

$$\frac{\text{mol}}{\text{s}\cdot\text{cm}^2} = \frac{\text{cm}}{\text{s}}\frac{\text{mol}}{\text{cm}^3}$$

where $P_M = \dfrac{D_m K_p}{d}$

$$\frac{\text{cm}}{\text{s}} = \frac{\text{cm}^2\cdot\text{s}^{-1}}{\text{cm}}$$

Eq. (10.8) is called the *permeability equation*. It is the most commonly used mathematical expression to describe the passive free diffusion of nonelectrolytes across a membrane. It shows that the determinants of the experimentally measurable permeability coefficient (P_M) are: (1) the diffusion coefficient of the solute of interest within the membrane (D_m, $\text{cm}^2\cdot\text{s}^{-1}$), (2) the partition coefficient K_p of the permeant solute from the aqueous phase into the organic membrane phase, and (3) the membrane thickness (d, cm). Thus, according to the Meyer–Overton rule (solubility-diffusion mechanism), the transmembrane flux density, J, of a membrane-permeating solute molecule can be predicted if its K_p is known. Partitioning into organic solvents is usually used as a proxy for the K_p of a solute in the membrane. A $K_p > 1$ indicates that the permeant solute dissolves better in the membrane than in the aqueous solution, the concentration gradient within the membrane is enhanced, and the flux will be proportionally increased. This is the situation illustrated in Fig. 10.1 (right). Conversely, $K_p < 1$ indicates that the permeant solute dissolves better in the aqueous phase than in the membrane; the concentration gradient within the membrane is reduced, and thus, the flux will decrease. A $K_p = 0$ indicates that the solute is insoluble in the membrane; the membrane flux J will be zero, indicating that the membrane is entirely impermeable to the solute that is not soluble in the membrane. A $K_p = 1$ indicates that the solute dissolves equally well in the membrane and in the aqueous bulk solution. If the membrane depicted in Fig. 10.1 (left) were a lipid bilayer, the concentration profile within the membrane for $K_p = 1$ would be as the one illustrated. The Meyer–Overton rule has been experimentally tested for thousands of compounds, although few molecules do not obey it (Menichetti & Bereau, 2019; Missner & Pohl, 2009; Stein & Litman, 2015).

As discussed above, a basic assumption of the Meyer–Overton rule is that given the high lipid content of biological membranes, the ease with which molecules cross the membrane by passive diffusion depends on their partition between water and a bulk nonpolar solvent (e.g., oil). In other words, the ease with which solutes can cross the bilayer strongly depends upon the degree to which they are lipid soluble. The value of K_p is determined by measuring the ratio of the equilibrium of the solute of interest in a lipid phase and an aqueous phase. This can be done in a separatory funnel or in a test tube holding a sample of the substance whose K_p is to be measured, water, and a solvent immiscible in water (Cumming & Rücker, 2017; Stein & Litman, 2015). The funnel, or the test tube, is shaken for several hours until the substance of interest reaches equilibrium between the two phases and left to stand until the two separate out again. The concentration of the solute of interest in the upper and the lower phases is then determined. Octanol-water partition coefficients are commonly used to assess a compound's lipophilicity (fat solubility) or hydrophilicity (water solubility). In this case, the immiscible phases are n-octanol and water (Cumming & Rücker, 2017). Other solvents, such as hexadecane or olive oil, have also been used to determine the K_p. Walter and Gutknecht (1986) studied the diffusion of small nonelectrolytes through planar lipid bilayer membranes (egg phosphatidylcholine-decane) and, by correlating the permeability coefficients of 22 solutes with their partition coefficients between water and four organic solvents concluded that the highest correlations were observed with hexadecane and olive oil ($r = 0.95$ and 0.93), but not octanol and ether ($r = 0.75$ and 0.74). The optimal solvent system depends on the permeant and its mechanism of permeation. More recently, coarse grain simulation models of water/octanol and water/cyclohexane have been used to predict K_ps successfully (Venable et al., 2019). As expected from Eq. (10.7), there is a very strong correlation between K_p and the ease of membrane permeation for a large number of water-soluble substances. Further, a strong correlation of permeability with the octanol-water K_p has been demonstrated in many biological membranes and in lipid bilayer membranes of varying composition (Finkelstein, 1976). More recently, the permeability of membranes has been studied using molecular dynamics simulations, allowing evaluation of Fick's first law and considering inhomogeneous solubility-diffusion (Frallicciardi et al., 2022; Venable et al., 2019).

The ease of bilayer permeation is also dependent upon the molecular size of the permeating solute; the smaller the molecule, the faster it penetrates. The effect of molecular size on P_M is due to the effect of size on D_m. The effect of molecular size on D_m is much more pronounced than the effect of molecular size on the aqueous diffusion coefficient D. The factors that determine the diffusion coefficients are discussed in Chapter 9. Other factors governing bilayer permeation are the shape and polarity of the permeant substance. Passive diffusion is not only dependent on the molecular characteristics of the solute but also on those of the lipid bilayer, such as thickness, lipid composition, structural organization (e.g., packing), and phase behavior (Frallicciardi et al., 2022; Trimble & Grinstein, 2015). The plasma membrane lipid bilayer has an actual thickness of 4–5 nm (4000–5000 pm). For comparison, the size of a sodium atom (Na) is 0.190 nm. Thus, the membrane is ~21–26 thicker than a Na atom (Fig. 10.1, right). The overall thickness of the membrane can be higher due to lipopolysaccharides and proteins sticking out of it.

2.3 Unstirred layer effects on passive and active transport processes

The governing equations for passive diffusion across the membrane, discussed in the preceding sections, assume that the aqueous solutions on both sides of the membrane are "well-stirred," meaning that there are no local concentration gradients at the membrane–solution interfaces, i.e., the solute concentrations at the membrane interfaces are the same as the bulk concentrations. We have seen that the solute concentrations that determine nonelectrolyte and water transport across a membrane are those immediately adjacent to each side of the membrane at the membrane–solution interfaces. However, the solute concentrations adjacent to each side of the membrane differ from their value in the bulk solution due to intervening *unstirred layers* (ULs), also known as "Nernst diffusion layers" (Barry & Diamond, 1984; Dainty, 1963). The ULs are diffusion barriers at the surface of all membranes, whether biological or artificial, resulting from stagnant water layers at the membrane interfaces (Fig. 10.2). Thus, ULs are not only extracellular but intracellular.

The ULs influence physiological processes such as the coupling between water transport and active solute transport in the lateral intercellular spaces of epithelia (Barry & Diamond, 1984; Larsen et al., 2009). Another example is the brain

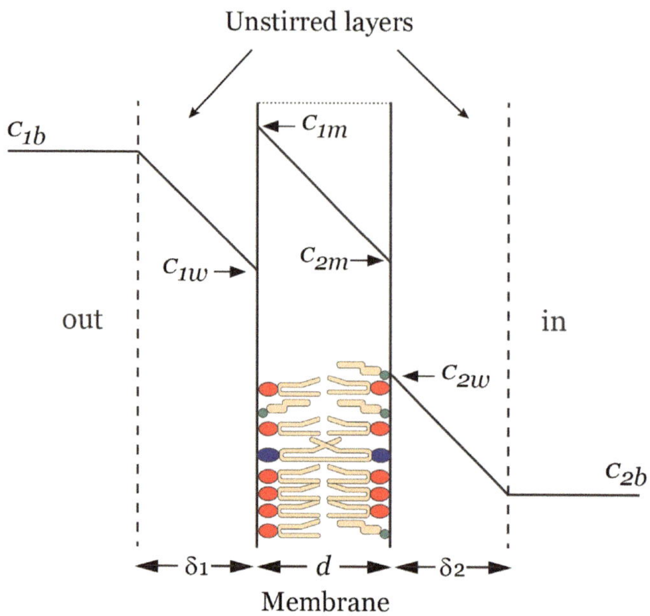

FIGURE 10.2 **Concentration profile of a lipophilic compound within a homogeneous lipid membrane in the presence of unstirred layers and in the absence of solvent flow.** C_{1b}, C_{2b}, C_{1w}, C_{2w}, C_{1b}, C_{1m}, and C_{2m} denote the solute concentration of the permeable solute in the outside bulk solution, in the inside bulk solution, in the outside aqueous phase adjacent to the bilayer, in the inside aqueous phase adjacent to the bilayer, in the lipid phase adjacent to the outside compartment, and in the lipid phase adjacent to the inside compartment. Unstirred layers on both sides of the bilayer are denoted as δ1 (outside) and δ2 (inside). Transport through the bilayer occurs solely by diffusion. The concentration profile inside the membrane is for a lipophilic compound with a partition coefficient $K_p > 1$. Note that the actual transmembrane concentration difference ($C_{1m}-C_{2m}$) is less than the overall concentration difference ($C_{1b}-C_{2b}$).

extracellular microenvironment, which behaves as an unstirred layer region in which diffusion is hindered by geometric factors such as tortuosity, as discussed in Chapter 9. ULs impact the permeation of small-molecule drugs. For instance, the intestinal solute permeation rate, which determines a drug's absorption and distribution, is significantly influenced by ULs (Kang et al., 2024). For some fast permeant molecules (e.g., ketoprofen, a nonsteroidal antiinflammatory drug, and propranolol, a commonly used Beta-blocker), the resistance to permeation due to ULs overshadows the intrinsic membrane resistance, i.e., ρ. As noted by Dainty (1963), the transport of a rapidly penetrating solute across a membrane may be wholly or partially rate-controlled by the unstirred layers and not by the membrane at all. When measuring fluxes and membrane permeability coefficients (P_M) for solutes or water, ignoring the presence of the unstirred water layers leads to artifactually low estimates of permeability coefficients for passive transport processes and overestimates of the Michaelis–Menten constant (*Km*) values for active transport processes (Wilson & Dietschy, 1974).

The concept of UL was developed originally by physical chemists Arthur A. Noyes (1866–1936) and Willis R. Whitney (1868–1958) in 1897 and later, in 1904, by Walther H. Nernst (1864–1941). Nernst envisaged it as a thin layer of static fluid at the interface of a solid body immersed in a bathing fluid (House, 1974; Nernst, 1904). ULs are regions of poorly mixed solution often referred to as "convection-free" regions, in which transport occurs solely by diffusion. In experimental conditions, convection can be introduced by stirring the bulk solutions. Vigorous stirring can reduce the thickness of ULs. However, even with vigorous stirring of the bulk solutions, stagnant water layers adjacent to the membrane persist (Missner & Pohl, 2009; Pedley, 1983). "Convection-free" is not to be taken literally; convection increases gradually; as the distance from the membrane increases, stirring becomes increasingly important, and after a certain distance, the velocity of convective fluid flow approaches bulk values (Pohl et al., 1998). Thus, the fluid throughout an unstirred layer is not stationary; instead, it is a region of slow laminar flow parallel to the membrane in which the only transport mechanism is by diffusion (Dainty, 1963). The thickness of the layer, δ, is also not the actual thickness but rather an operational thickness defined as the width of a uniform concentration gradient. Within each UL, the concentration of solute is a function of the distance to the membrane, and it is different from that in the bulk solution , as illustrated in Fig. 10.2. Thus, the thickness of the unstirred layer δ is defined in terms of the concentration gradient at the membrane–water interface:

$$\frac{dc_{\mathrm{UL}}}{dx} = \frac{(C_b - C_w)}{\delta} \tag{10.9}$$

where x is the distance from the membrane interface, dc_{UL}/dx is the concentration gradient at the interface, and C_b and C_w denote the solute concentration at the bulk solution and the interface, respectively (Fig. 10.2). For the membrane interface on side one in Fig. 10.2, to be consistent with the notation used here, $C_b = C_{1b}$ and $C_w = C_{1w}$.

An UL can be envisaged as an equivalent membrane in series with the actual membrane, with an equivalent permeability coefficient P_{UL} given by:

$$P_{\mathrm{UL}} = \frac{D}{\delta} \tag{10.10}$$

where D has its usual meaning and units. Thus, the diffusion barrier arising from the presence of stagnant water layers can be estimated from the value of the solute diffusion coefficient, D, and the thickness of the unstirred layer (δ). The solute s crosses the UL only by diffusion, and the flux Js in the UL, say on side 1 (out) in Fig. 10.2, is:

$$Js = P_{\mathrm{UL1}}(C_{1b} - C_{1w}) = \frac{D_{1b}}{\delta_1}(C_{1b} - C_{1w}) \tag{10.11}$$

From Eq. (10.10) and Fig. 10.2, it can be inferred that C_{1w} approaches C_{1b} as the thickness of the unstirred layer approaches zero; otherwise, Js would become infinite. When C_{1b} is assumed to be equal to C_{1w}, a common assumption, the bulk phase is said to be "well-stirred." As discussed, this assumption is just an approximation.

Now the question arises: how to determine δ? Several methodological approaches have been used to assess δ. One of them is estimating it from the time course for the accumulation or depletion of a solute adjacent to the membrane under defined experimental conditions (e.g., stirring rates, bulk solution mixing, chamber geometry, and the diffusion coefficient D of the probe). This approach has been applied to various types of epithelia, cell monolayers, and planar lipid bilayer membranes, and some representative examples are discussed below. It isn't easy to make meaningful comparisons between results reported by various research groups that have estimated δ because of differences in experimental conditions. Equally important is the fact that the UL depends on the solute diffusing, and thus on its D, and has no clearly defined boundary (Pedley, 1983; Pohl et al., 1998). Small molecules with a larger diffusion coefficient generally encounter an apparently thicker UL than larger molecules with a smaller diffusion coefficient. Most of the reported UL thicknesses for small solutes fall within 30—150 μm range.

A thorough study in *Necturus* gallbladder epithelial cells, using tetramethylammonium (TMA^+) and tetrabutylammonium (TBA^+) as probes, measuring the changes in probe concentration with ion-sensitive microelectrodes whose tip was positioned at the apical membrane, estimated δ to be 37 μm for TMA^+ and 34 μm for TBA^+, estimates of δ varied by <10% (Cotton & Reuss, 1989). With other methods, estimates of mucosal (apical) UL thickness in rabbit gallbladder have yielded values between 70 and 113 μm (For references and discussion, see Barry & Diamond, 1984; Cotton & Reuss, 1989). More recently, molecular dynamics simulations of intestinal permeation of a series of small molecule drugs of therapeutic interest found UL effective thickness between 36.1 μm for metoprolol and 60 μm for salicylate (Kang et al., 2024). The value of δ for other compounds tested (ketoprofen, naproxen, atenolol, and propranolol) fell within the two values mentioned.

Using lipid bilayers and double-barreled ion-sensitive microelectrodes to measure two ions simultaneously at various distances from the membrane, Pohl et al. (1998) found that δ for Ca^{2+} was 116 μm and for K^+ 153 μm, respectively, and for Na^+ and Ca^{2+}, 130 and 105 μm, respectively. The use of a double-barreled microelectrode ensured that the concentration profiles of two different ions were measured exactly under the same conditions. Pohl et al. (1998) found that measured δ values are related to the diffusion coefficients as follows:

$$\frac{\delta 1}{\delta 2} = \sqrt[3]{\frac{D1}{D2}} \tag{10.12}$$

Measuring the UL thickness for one solute, it should be possible to predict the corresponding δ of any other solute knowing the diffusion coefficients (Eq. 10.10). Pohl et al. concluded that there is a UL thickness corresponding to each solute diffusing in the solution, that is the UL thickness depends on the D for each solute, and thus there is no "one single" UL where the movement of all solutes takes place by diffusion.

Since ULs act as additional diffusion barriers in series with the membrane, the difference between the bulk concentrations of the permeant solute (C_{1b}—C_{2b}) differs significantly from the transmembrane concentration difference

(C_{1w}−C_{2w}), as can be observed in Fig. 10.2. Denoting the permeabilities of the first and second ULs as P_{UL1} and P_{UL2} the apparent permeability P_{app} obtained by flux measurements under steady-state conditions is (Missner & Pohl, 2009):

$$\frac{1}{P_{app}} = \frac{1}{P_{UL1}} + \frac{1}{P_M} + \frac{1}{P_{UL2}} \tag{10.13}$$

The preceding equation can be viewed as the sum of various barrier resistances in series. Thus, three resistances in series separate the two bulk phases. P_{UL1} and P_{UL2} are determined by the corresponding UL size (δ) and the aqueous D on each side of the membrane. Measurement with ion-sensitive microelectrodes, for instance, allows estimation of δ, C_{1w}, and C_{2w}, and thus, the transmembrane flux and P_M in steady state can be determined (Pohl et al., 1997, 1998). The solute flux is equal to the difference in concentration of the bulk solutions, divided by the sum of these resistances, each of which is inversely proportional to a permeability:

$$J_s = \frac{C_{1b} - C_{2b}}{\frac{\delta_1}{D_1} + \frac{1}{P_M} + \frac{\delta_2}{D_2}} \tag{10.14}$$

In the preceding equation, P_M (defined in Eq. 10.8) is the "true" permeability coefficient. The apparent permeability of the membrane $J_s/(C_{ib}-C_{2b})$ is the reciprocal of the denominator. D_1 and D_2 are the diffusion coefficients in the bulk solutions. Of note, in a cell, the intracellular diffusion coefficient is likely to be different than the external one for the same solute, as discussed in Chapter 9. Further, the intracellular D coefficients of small solutes are, most of the time, unknown, and therefore, it is difficult to make good estimates of the errors in the estimation of cell membrane permeabilities caused by the intracellular diffusional resistance. In the presence of ULs, the concentration difference driving the transmembrane flux (C_{1w}−C_{2w}-) is less than the difference between the bulk concentrations (C_{1b}−C_{2b}), as illustrated in Fig. 10.2. Thus, solute permeabilities estimated using the driving force of the bulk concentrations (C_{1b}−C_{2b}) might lead to significant underestimation of the true value if the resistance of the ULs is an important fraction of the total resistance, particularly if the membrane permeability for the solute probed is high (Friedman, 2008).

We have seen that when measuring fluxes across sheets of cells, like epithelia, ULs act as series resistances, reducing the driving force for passive transport across the barrier. A second type of UL effect is observed when there is *solvent flow* across the barrier. This is the situation encountered when studying water permeability by imposing an osmotic gradient across a barrier, such as a bilayer or a sheet of cells. In this situation, even with vigorous stirring, there is a solvent flow across the barrier such that water passing through the barrier dilutes the solution it enters and concentrates the solution it leaves. This phenomenon is known as *concentration polarization* (Barry & Diamond, 1984; Pohl et al., 1997, 1998). Concentration polarization generates an osmotic driving force that opposes solvent flow across the barrier, thereby decreasing the solvent flow; thus, the effective osmotic gradient across the barrier is decreased, resulting in underestimates of *Pf*, the osmotic water permeability (Cotton et al., 1989; Horner & Pohl, 2018; Verkman, 2000).

3. Flux and permeability coefficient

Assuming for the sake of argument that there are no ULs, and considering an inward gradient for the solute (out → in) as shown in Fig. 10.2, and as expressed in Eq. (10.8), the net flux density of a nonelectrolyte across a membrane (J) is equal to the permeability coefficient P_M multiplied by the difference in concentration across the membrane (C_{2b}−C_{1b}). The net flux of a solute through a membrane is the result of balancing the unidirectional influx and efflux. *The unidirectional influx* (J_{in}) *and efflux* (J_{out}) are defined by:

$$J_{in\ (out\rightarrow in)} = P_M C_{1b} \tag{10.15}$$

$$J_{out\ (in\rightarrow out)} = P_M C_{2b} \tag{10.16}$$

The *net flux* is defined as the difference between the efflux and the influx. Thus,

$$J = J_{out} - J_{in} = P_M\,(C_{2b} - C_{1b}) \tag{10.17}$$

Eqs. (10.15) and (10.16) state that the influx of a neutral solute is equal to its permeability coefficient (P_M) times the external concentration (C_{1b}), whereas the efflux (or outflux) is equal to the permeability coefficient times the internal concentration (C_{2b}). Note that Eq. (10.17) describing the net flux is basically the same as Eq. (10.8).

TABLE 10.1 Permeability values for water, H^+, K^+, Na^+, and Cl^- in various membranes.

	Water	pH^+	pK^+	pNa^+	pCl^-	References
PC[1] bilayer					1.5×10^{-11}	Toyoshima and Thompson (1975)
Membrane[2]			4.7×10^{-14}	5×10^{-14}		Papahadjopoulos et al. (1972)
Crayfish axon			1.33×10^{-5}	1.92×10^{-8}	1.99×10^{-6}	Strickholm (1981)
Rat liver			7.6×10^{-8}	4×10^{-8}	12.8×10^{-8}	Claret and Mazet (1972)
Ehrlich mouse ascites					4×10^{-8} -6×10^{-7}	Hoffmann et al. (1979)
Cardiac Purkinje fibers				$0.8–1.5 \times 10^{-8}$		Eisner et al. (1981)
Membrane[2]	3.4×10^{-3}					Walter and Gutknecht (1986)
DLPE[3]	2.1×10^{-3}					Shinoda (2016)
DOPC[4]	16×10^{-3}					Shinoda (2016)
Phospholipid vesicle		1×10^{-9}				Nozaki and Tanford (1981)
Mouse embryo		2.2×10^{-1}				Baltz et al. (1990)
Human lymphocytes	4.2×10^{-4}					Porsche et al. (1986)
Intestinal brush border	1.2×10^{-3}					Worman and Field (1985)

[1]*PC*, phosphatidylcholine; [2]artificial membrane; [3]1,2-Dilauroyl-sn-glycero-3-phosphoethanolamine; [4]dioleoyl phosphatidylcholine. Values are provided in cm/sec.

3.1 Barrier properties of biological membranes

We have seen in the preceding sections that nonelectrolyte solutes can cross the plasma membrane by a solubility-diffusion mechanism involving the partition of the solute into the lipid phase, the diffusion from one side of the bilayer to the other, and the reemergence into the aqueous phase on the other side. As for water, it can readily penetrate the phospholipids head groups but is slowed down in the vicinity of the hydrocarbon chains of the lipids. For water, and small nonelectrolytes like urea and ammonia, the length of the lipid chains and their ability to form "kinks" significantly affect their permeability (Zeidel, 1996). For charged ions, like H^+, Na^+, K^+, and Cl^-, the permeabilities do not vary much as a function of lipid composition and fluidity, indicating that ion permeation occurs through different mechanisms. As shown in Table 10.1, the permeability coefficient for ions in artificial membranes is exceedingly small. This coefficient is enhanced by several orders of magnitude in plasma membranes, likely due to the insertion of membrane proteins (e.g., channels and carriers).

Water permeability differs greatly among cells, depending upon their function in the body. Cells like red blood cells have a relatively high water permeability, 2.4 10^{-3} cm/s, allowing them to shrink rapidly when they enter the kidney's vasa recta (from Latin, "vasa": vessels; "recta": straight), whereas epithelial cells lining the urinary bladder are known to have extremely low water permeability (3–5 10^{-5} cm/s). Thus, the water permeability of the bladder is 60 times lower than that of red blood cells. As we will see, the nature of lipids affects the permeation of water through the lipid bilayer. However, it is not the lipid composition that explains the striking difference between the high-water permeability of a red blood cell compared to the urinary bladder epithelial cells, but the expression of a large number of specialized water-conducting proteins, i.e., AQP1 water channels in the red blood cell membrane that explains the difference as discussed in Chapter 19. As we have seen, the overall permeability of a membrane is defined by thickness, lipid composition, and proteins, among other factors. Aside from the standard fluorescent-based fast measurements of membrane permeability, molecular dynamics simulations are now being used to "visualize" and study the movement of particles through the plasma membrane (Frallicciardi et al., 2022).

Of interest is the difference in the composition of the external and internal leaflets of the membrane. In 2000, Warren Hill and Mark Zeidel created distinct liposomes with outer and inner leaflets lipids of a barrier epithelium, Madin–Darby canine kidney type one cells (MDCK-I), and measured their water permeability. They assumed that the apical to basolateral membrane surface area ratio was 1:4, that cholesterol and phospholipids in the apical membrane were present in a 1:1 ratio, that glycosphingolipids were distributed between apical and basolateral membranes in a 2:1 ratio, that half of the lipid moles were in each leaflet, that all glycosphingolipids, sphingomyelin, and phosphatidylcholine in the apical membrane were in the exofacial leaflet, and that all of the phosphatidylethanolamine, phosphatidylserine, and phosphatidylinositol in the MDCK-I apical membrane were in the cytoplasmic leaflet (Hill & Zeidel, 2000). The water permeability of exofacial and cytoplasmic liposomes was $2.4 \pm 0.4\ 10^{-4}$ versus $4.4 \pm 0.3\ 10^{-3}$ cm/s, respectively. Thus, the cytoplasmic side lipids were 18.1 times more permeable than exofacial lipids, revealing striking bilayer asymmetry. Using liposomes made of exofacial lipids, they individually removed glycosphingolipids, sphingomyelin, and cholesterol and demonstrated a significant increase in water permeability upon the removal of these individual lipids (Hill & Zeidel, 2000). Similar data were obtained for the permeation of NH_3 and glycerol. Similar water permeability data were observed while comparing different lipids from archaebacteria. Lipids properties allowed some bacteria to live in extremely hostile environments, such as those of extremely high temperatures, or extreme pHs. Unsurprisingly, liposomes formed from the type of lipids found in these extremophile bacteria show markedly reduced permeability to water, NH_3, urea, and glycerol (Mathai et al., 2001) (Fig. 10.3).

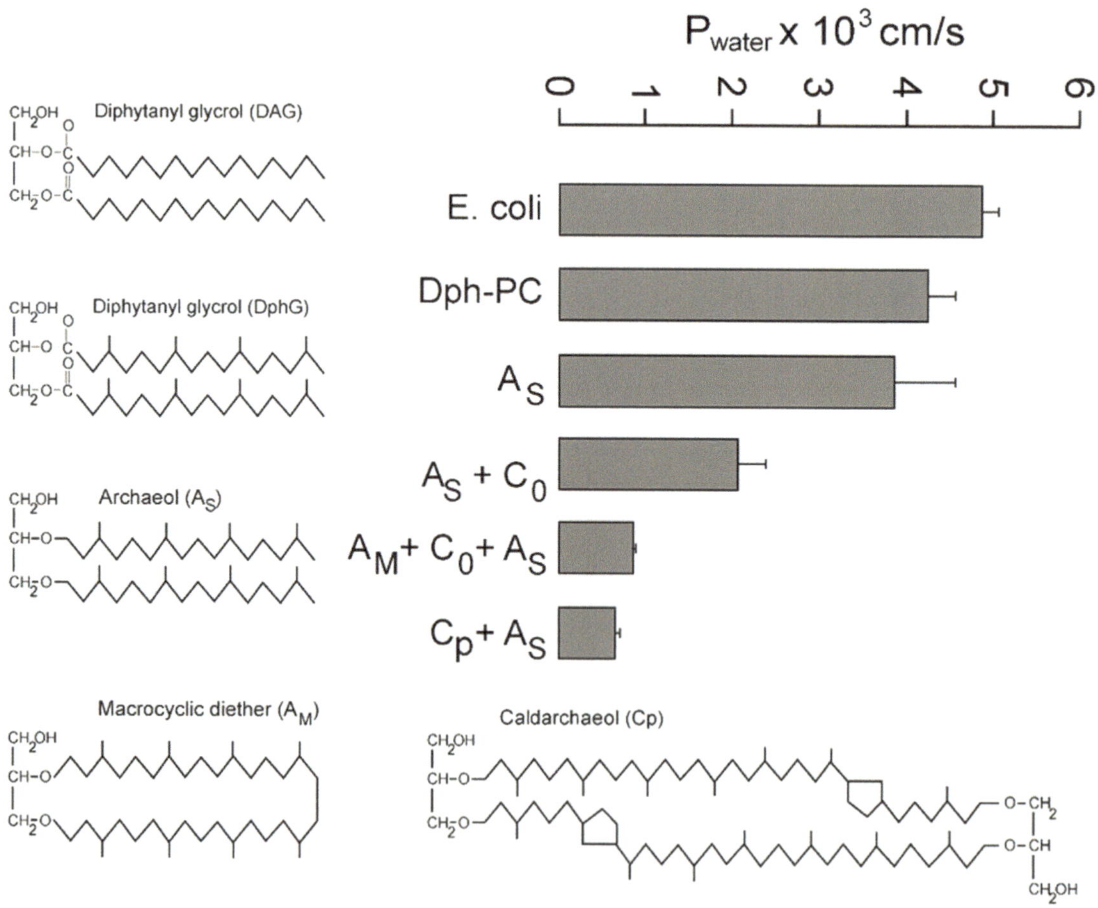

FIGURE 10.3 Decreased water permeability of membranes with increasing complexity/rigidity of carbon chains. Lipids isolated from extremophile bacteria (e.g., methanogenic archaeobacteria) have increased complexity when compared to lipids found in *Escherichia coli* membranes or membranes from higher organisms. The combination of those lipids in liposomes affects water permeability. *Modified from Mathai et al. (2001).*

3.2 Electrodiffusion

The passive diffusion of charged particles under the influence of an electric field is known as *electrodiffusion*. Eqs. (10.15)−(10.17) and the first Fick's law diffusion (Eq. 9.19 in Chapter 9) apply only to uncharged solutes (nonelectrolytes). If the solute carries a net charge (e.g., ions), the net fluxes are not only determined by the difference in concentration but also by the electric field across the membrane.

The electric current (i) carried by a charged particle across the membrane is equal to the flux (J', in mol/s) multiplied by zF (coul/mol or coul/equiv), where z is the valence, and F is the Faraday constant:

$$i = J' zF \tag{10.18}$$

Expressed in terms of current density, the preceding equation becomes:

$$I = JzF \tag{10.19}$$

$$\frac{\text{amp}}{\text{cm}^2} = \frac{\text{mol}}{\text{cm}^2 \cdot \text{sec}} \frac{\text{coul}}{\text{mol}}$$

where I is the current density in amp/cm^2 and J is the flux density in mol/cm^2 s.

The net flux of a charged solute is determined by both the concentration difference ($C_{1b} - C_{2b}$), which can also be written as ($C_0 - C_i$), and the electric field across the membrane. The equation describing the net flux of a charged solute across the membrane is (Sperelakis & Freedman, 2012):

$$J = P_M f(E_m)\left(C_{1b} - C_{2b} e^{E_m F/RT}\right) \quad \text{or} \quad J = P_M f(E_m)\left(C_o - C_i e^{E_m F/RT}\right) \tag{10.20}$$

where P_M is the permeability coefficient, $f(E_m)$ is a function of the *electrical potential difference* across the membrane, and F, R, and T are the Faraday constant, the gas constant, and the absolute temperature, respectively. Note the similarity of Eq. (10.20) to Eqs. (10.8) and (10.17), except for the membrane potential terms. The electrical potential difference across the membrane (E_m) equals $E_{in} - E_{out}$, where E_{in} is the inside potential and E_{out} is the outside potential. When $f(E_m)$ is equal to $E_m F/RT$, this term is dimensionless ($E_m F$ is the electrical energy, whereas RT is the thermal energy); $RT/F = 0.026$ V at 25°C. The term $e^{EmF/RT}$ is also dimensionless, and therefore, the units for flux are the same as those given in Eq. (10.8).

For a monovalent cation, Einstein's law of diffusion states that:

$$D = UK_B T = U\frac{RT}{F} \tag{10.21}$$

where D is the diffusion coefficient, U is the electrophoretic mobility of the ion through the membrane and has units of a velocity per unit driving voltage gradient (cm/s per V/cm), K_B is the Boltzmann constant, T, R, and F have their usual meaning. Substituting Eq. (10.21) into the definition of the permeability coefficient (Eq. 10.8), we obtain:

$$P_M = U\frac{K_p}{d}\frac{RT}{F} \tag{10.22}$$

$$\frac{\text{cm}}{\text{sec}} = \frac{\text{cm/sec}}{V/\text{cm}} \frac{1}{\text{cm}} \frac{V}{1}$$

where K_p is the partition coefficient for the ion between the bulk solution and the membrane (dimensionless), and d is the membrane thickness (cm). Eq. (10.22) indicates that the permeability coefficient of an ion is directly related to the electrophoretic mobility of the ion through the membrane.

The *Nernst−Planck equation* describes electrodiffusion. It states that the total flux density of an ion species (i), J_i (mol/cm$^2 \cdot$s), across the membrane is the sum of the flux density due to diffusion and the flux density due to the electrical potential:

$$J_i = J_{i(\text{diff})} + J_{i(\text{elec})} \tag{10.23}$$

The flux density due to diffusion is given by the first Fick's law (Eq. 9.19, Chapter 9).

$$J_{i(diff)} = -D\frac{dC_i}{dx} \tag{10.24}$$

The flux density due to the electric flux is given by

$$J_{i(elec)} = -zUC_i\frac{dE}{dx} \tag{10.25}$$

In the preceding equation, *dE/dx* is the electrical potential gradient (V/cm). This equation bears a resemblance to Fick's first law (Eq. 9.19 in Chapter 9), which describes the diffusive flux. From Eqs. (10.23–10.25), the total net flux J_i of ion i is given by:

$$J_i = J_{i(diff)} + J_{i(elec)} = -D\frac{dC_i}{dx} - zUC_i\frac{dE}{dx} \tag{10.26}$$

Eq. (10.26) is the Nernst–Planck electrodiffusion equation. At *electrochemical equilibrium*, the flux driven by the concentration gradient is balanced by the flux driven in the opposite direction by the electrical potential gradient, and thus, the *net flux is zero*.

$$J_i = J_{i(diff)} + J_{i(elec)} = -D\frac{dC_i}{dx} - zUC_i\frac{dE}{dx} = 0 \tag{10.27}$$

Therefore,

$$zUC_i\frac{dE}{dx} = -D\frac{dC_i}{dx} \tag{10.28}$$

Incorporating Einstein's Eq. (10.21) into Eq. (10.28):

$$zUC_i\frac{dE}{dx} = -U\frac{RT}{F}\frac{dC_i}{dx} \tag{10.29}$$

And thus,

$$\frac{dE}{dx} = -\frac{RT}{zF}\frac{1}{Ci}\frac{dC_i}{dx} \tag{10.30}$$

Integration of Eq. (10.30) yields the following expression:

$$E_{\text{in}} - E_{\text{out}} = -\frac{RT}{zF}\left(\ln C_{i(\text{in})} - \ln C_{i(\text{out})}\right) = \frac{RT}{zF}\ln\left(\frac{C_{i(\text{in})}}{C_{i(\text{out})}}\right) \tag{10.31}$$

where (E_{in}-E_{out}) is the membrane potential E_m. The equation indicates that if the membrane is selectively permeable to ion i, and the ion is in electrochemical equilibrium across the membrane, it is possible to calculate the membrane potential E_m that would result from known concentrations of ion i on each side of the membrane ($C_{i(\text{out})}$ and $C_{i(\text{in})}$). In other words, the E_m that results when an ion is in electrochemical equilibrium is the *equilibrium potential* for that ion, E_{eq}. From Eq. (10.31) and the preceding arguments, the equilibrium potential for ion i is:

$$E_{\text{eq}} = \frac{RT}{zF}\ln\frac{C_{i(\text{out})}}{C_{i(\text{in})}} \tag{10.32}$$

Eq. (10.32) is the *Nernst equation.*

Now, let us say that we know E_m and $C_{i(\text{out})}$, and we want to determine what the internal concentration $C_{i(\text{in})}$ would be when ion i is in electrochemical equilibrium:

$$C_{i\text{in}}^{\text{eq}} = C_{i(\text{out})}e^{E_mF/RT} \tag{10.33}$$

3.3 Ussing flux ratio equation

As pointed out in Eq. (10.17), the net flux of a solute across the membrane is the difference between the movement of the solute in one direction, influx, and its movement in the opposite direction, i.e., efflux. The Danish physiologist Hans Eriksen Ussing (1911–2000) introduced the notion of flux ratio instead of flux difference (Ussing, 1949). With passive diffusion and not flux coupling, the ratio of the unidirectional fluxes should be equal to the ratio of the ions' electrochemical activities outside and inside the membrane. Thus, the ratio of influx to efflux will be unity when the electrochemical potential across the membrane is zero, i.e., at electrochemical equilibrium, or differ from unity when a gradient of electrochemical potential exists. The relationship between flux ratios and driving forces on ions is given in the *flux ratio equation* that generally bears the eponym "the Ussing flux ratio equation":

$$\frac{J_{\text{in}}}{J_{\text{out}}} = \frac{C_{\text{out}}}{C_{\text{in}}} e^{-E_m \ zF/RT} \tag{10.34}$$

It is important to emphasize that this equation will be obeyed when the fluxes are carried by ions without their being coupled. Coupling can occur when there is more than one ion flowing through an ion channel or when fluxes use a source of energy other than that provided by the electric field across the membrane (Hille, 2001). One example is exchange diffusion, where a carrier makes a stoichiometric exchange of an equal number of ions. Other examples include cotransporters and exchangers. When such coupling occurs, Eq. (10.34) might not hold (Stein & Litman, 2015).

Taking as an example the ratio of Na^+ fluxes for a cell kept at 37°C having $E_m = -70$ mV in which the Na^+ concentrations outside and inside are 140 and 10 mM, respectively:

$$\frac{J_{\text{in}}^{\text{Na}}}{J_{\text{out}}^{\text{Na}}} = \frac{[Na^+]_{\text{out}}}{[Na^+]_{\text{in}}} e^{-E_m F/RT} \tag{10.35}$$

Thus,

$$\frac{J_{\text{in}}^{\text{Na}}}{J_{\text{out}}^{\text{Na}}} = \frac{140 \text{ mM}}{10 \text{ mM}} e^{+70 \text{ mV}/26.73 \text{ mV}} = 192 \tag{10.36}$$

This means that the *passive influx* of Na^+ is 192 times greater than the *passive efflux* of Na^+ in the example considered, in which the concentrations are close to those found in most vertebrate cells. This influx is of such magnitude because of the large inward-directed chemical gradient. For K^+ ions with an extracellular concentration of 4 mM and an intracellular concentration of 140 mM, the ratio of passive flux is 0.392, which means that the *influx* of K^+ is 2.6 times smaller than the *efflux*. Thus, what makes the substantial difference observed between Na^+ and K^+ is not the cation per se but the concentrations of each ion on each side of the membrane. If the concentration at the external side is 72.9 times greater than inside, the *influx* will be 1000× greater than the *efflux*. If the external concentration is now 7.29 times greater than inside, the *influx* will be 100× greater than the *efflux*. With an external concentration 0.729 times greater (or 1.37 times *smaller* than inside) the *influx* will still be 10× greater than the *efflux*. Note that while we have now reversed the gradient, the *influx* is still greater than the *efflux*. This is due to the electrical *pull* that the membrane potential (negative inside) exerts on an external cation. It is the membrane potential that creates this asymmetry. Also note that the ion itself does not matter, e.g., that the ion is Na + or, let's say, Tl^+. What matters are the outside and inside concentrations. If Tl^+ was placed across the membrane at the same concentration as Na^+, its *influx* would also be 192 times larger than its *efflux*, irrespective of its permeability coefficient. For example, if the permeability coefficient of Tl^+ is 20 times lower than that of Na^+, the net passive flux of Tl^+ would be 20 times lower than that of Na^+ but still with an *influx* 192 times greater than *efflux*.

A modified form of Eq. (10.34) can be obtained by substituting the ratio of ion concentrations with $e^{-\left(E_m - E_i^{\text{eq}}\right)F/RT}$ (derived from the Nernst equation), giving:

$$\frac{J_{\text{in}}}{J_{\text{out}}} = e^{-\left(E_m - E_i^{\text{eq}}\right)F/RT} \tag{10.37}$$

where E_i^{eq} is the equilibrium potential for the ion *i*. For instance, for Na^+, since E_{eq} is +70.5 mV, the term in the exponential is now −(−70 mV-(+70.5 mV))/26.73 mV, also yielding 192. This value of 192 obtained here for the flux ratio is identical to that obtained from Eq. (10.36). One advantage of Eq. (10.37) is that it is obvious that when E_i^{eq} equals E_m, the flux ratio is exactly one (*influx* = *efflux*).

Thus, if an ion is passively distributed across the membrane, i.e., $E_i^{\text{eq}} = E_m$, its flux ratio should be one; *influx* equals *efflux*. The larger the difference between E_i^{eq} and E_m, i.e., the farther the ion is from equilibrium, the greater the flux ratio.

Sometimes, Eq. (10.37) does not fit the experimental data. Those data are best fitted if the exponential term contains another factor (*n*), which is an empirical factor:

$$\frac{J_{\text{in}}}{J_{\text{out}}} = e^{-n\left(E_m - E_i^{\text{eq}}\right)F/RT} \tag{10.38}$$

The best fit to the data is when n has a value of 2.5–4.0, depending on the membrane under investigation. One interpretation given to n is that if the length of the water-filled pore that the ion must traverse to cross the membrane is much longer than the ion diameter, as is likely, then for so-called single-file diffusion, n hits on the same side are required for the ion to complete its journey across the membrane. One could consider, for example, that there are three potential energy wells, or a chain of reactive sites, along the length of the pore, and the only way for the ion to escape the well is to receive a kinetic bump from an adjacent ion in the file. Complete permeation of an ion through the pore is more likely to happen if the ion is moving in the same direction as the majority of ions, i.e., down the electrochemical gradient. Therefore, this additional factor in Eq. (10.38) makes the flux ratio much greater than would otherwise be predicted.

4. Special transport processes

Most nonpolar molecules pass directly through the lipid bilayer matrix of the membrane, i.e., not through special sites (e.g., water-filled pores or channels). Small charged ions (e.g., Na^+, K^+, Cl^-, Ca^{2+}) pass through water-filled channels, many of which have a voltage-dependent gating mechanism (Chapter 21). These channels can exhibit a high degree of selectivity for specific ions, with the selectivity orders not being based solely on the hydrated or unhydrated sizes of the ions. Another type of ion-conducting protein is the ionophores, small, lipid-soluble molecules, generally of microbial origin, that function to transport ions across membranes. Some of them are channel formers, and others are mobile carriers (Stillwell, 2016). An example of the latter is valinomycin, a naturally occurring antibiotic extracted from Streptomyces that forms a hydrophobic cage around K^+. This K^+-ionophore complex is lipid soluble and passes rapidly through the lipid bilayer matrix.

Special transport proteins (carriers) are normally present in the cell membrane for the net uptake of nutrients, such as glucose and amino acids. These carriers are for *downhill transport* mechanisms, i.e., down the chemical gradient of the transported solute. One type of protein-mediated transport in cells (e.g., for glucose) is known as *facilitated diffusion*. This carrier-mediated passive transport contrasts with *simple diffusion* across the cell membrane (Stein & Litman, 2015). They are functionally similar to enzymes, with activities described by Michaelis–Menten kinetics. They are members of the Major Facilitator Superfamily (MFS) transporters (Drew et al., 2021), being responsible for the largest portion of nutrient transport in cells. Another type of mediated transport is known as *exchange diffusion*, in which one molecule of substance (or ion) inside the cell is exchanged for one molecule outside the cell, in which case there is no net movement across the membrane. In either type of mediated transport, the rate of downhill movement of the substance across the membrane is enhanced by the transport protein.

Mediated transport systems differ from simple diffusion in that they exhibit *saturation kinetics* (Friedman, 2008; Stein & Lieb, 1986). That is, the rate of transport increases with an increase in the substrate concentration up to a maximum, after which the rate levels off due to a finite number of available transport sites. Other characteristics of mediated transport systems include *competitive inhibition*, in which an inhibitor substance competes at the same site for binding to the transporter. In *noncompetitive inhibition*, the *inhibitor* (nontransported) substance binds to the transporter at a site different from the transport site but still alters or prevents the binding of the usual substrate. Mediated transport exhibits greater specificity than simple diffusion; e.g., the rate of mediated transport of D-glucose is greater than that of L-glucose. Because mediated transport resembles an enzymatic process involving protein conformational changes, the temperature coefficient (Q10) is also higher than for simple diffusion.

Acknowledgments

This chapter is an updated and revised version of the one written by N. Sperelakis and J.C. Friedman for the previous edition of this book. Thus, some parts of the text were left intact in the present version.

References

Al-Awqati, Q. (1999). One hundred years of membrane permeability: Does Overton still rule? *Nature Cell Biology, 1*, E201–E202.

Baltz, J. M., Biggers, J. D., & Lechene, C. (1990). Apparent absence of Na^+/H^+ antiport activity in the two-cell mouse embryo. *Developmental Biology, 138*, 421–429.

Barry, P. H., & Diamond, J. M. (1984). Effects of unstirred layers on membrane phenomena. *Physiological Reviews, 64*, 763–872.

Claret, M., & Mazet, J. L. (1972). Ionic fluxes and permeabilities of cell membranes in rat liver. *The Journal of Physiology, 223*, 279–295.

Cotton, C. U., & Reuss, L. (1989). Measurement of the effective thickness of the mucosal unstirred layer in *Necturus* gallbladder epithelium. *The Journal of General Physiology, 93*, 631–647.

Cotton, C. U., Weinstein, A. M., & Reuss, L. (1989). Osmotic water permeability of *Necturus* gallbladder epithelium. *The Journal of General Physiology, 93*, 649–679.

Cumming, H., & Rücker, C. (2017). Octanol–water partition coefficient measurement by a simple 1H NMR method. *ACS Omega, 2*, 6244–6249.

Dainty, J. (1963). Water relations of plant cells. In R. D. Preston (Ed.), *Advances in botanical research* (pp. 279–326). Academic Press.

Danielli, J. F. (1943). The theory of penetration of a thin membrane: Appendix. In H. Davson, & J. F. Danielli (Eds.), *The permeability of natural membranes*. London: Cambridge University Press.

Drew, D., North, R. A., Nagarathinam, K., & Tanabe, M. (2021). Structures and general transport mechanisms by the major facilitator superfamily (MFS). *Chemical Reviews, 121*, 5289–5335.

Eisner, D. A., Lederer, W. J., & Vaughan-Jones, R. D. (1981). The effects of rubidium ions and membrane potentials on the intracellular sodium activity of sheep Purkinje fibres. *The Journal of Physiology, 317*, 189–205.

Finkelstein, A. (1976). Water and nonelectrolyte permeability of lipid bilayer membranes. *The Journal of General Physiology, 68*, 127–135.

Frallicciardi, J., Melcr, J., Siginou, P., Marrink, S. J., & Poolman, B. (2022). Membrane thickness, lipid phase and sterol type are determining factors in the permeability of membranes to small solutes. *Nature Communications, 13*, 1605.

Friedman, M. H. (2008). *Principles and models of biological transport*. New York, NY: Springer.

Hill, W. G., & Zeidel, M. L. (2000). Reconstituting the barrier properties of a water-tight epithelial membrane by design of leaflet-specific liposomes. *Journal of Biological Chemistry, 275*, 30176–30185.

Hille, B. (2001). *Ion channels of excitable membranes*. Sunderland, Massachusetts: Sinauer Associates.

Hoffmann, E. K., Simonsen, L. O., & Sjøholm, C. (1979). Membrane potential, chloride exchange, and chloride conductance in Ehrlich mouse ascites tumour cells. *The Journal of Physiology, 296*, 61–84.

Horner, A., & Pohl, P. (2018). Single-file transport of water through membrane channels. *Faraday Discussions, 209*, 9–33.

House, C. R. (1974). *Water transport in cells and tissues* (Vol. 24). London: Edward Arnold.

Kang, C., Shoji, A., Chipot, C., & Sun, R. (2024). Impact of the unstirred water layer on the permeation of small-molecule drugs. *Journal of Chemical Information and Modeling, 64*, 933–943.

Larsen, E. H., Willumsen, N. J., Møbjerg, N., & Sørensen, J. N. (2009). The lateral intercellular space as osmotic coupling compartment in isotonic transport. *Acta Physiologica, 195*, 171–186.

Mathai, J. C., Sprott, G. D., & Zeidel, M. L. (2001). Molecular mechanisms of water and solute transport across archaebacterial lipid membranes. *Journal of Biological Chemistry, 276*, 27266–27271.

Menichetti, R., & Bereau, T. (2019). Revisiting the Meyer-Overton rule for drug-membrane permeabilities. *Molecular Physics, 117*, 2900–2909.

Meyer, H. (1901). Zur Theorie der Alkoholnarkose. *Archiv für experimentelle Pathologie und Pharmakologie, 46*, 338–346.

Missner, A., & Pohl, P. (2009). 110 years of the Meyer-Overton rule: Predicting membrane permeability of gases and other small compounds. *ChemPhysChem: A European Journal of Chemical Physics and Physical Chemistry, 10*, 1405–1414.

Nernst, W. (1904). Theorie der Reaktionsgeschwindigkeit in heterogenen Systemen. *Zeitschrift für Physikalische Chemie, 47*, 52–55.

Nozaki, Y., & Tanford, C. (1981). Proton and hydroxide ion permeability of phospholipid vesicles. *Proceedings of the National Academy of Sciences of the United States of America, 78*, 4324–4328.

Overton, C. E. (1901). *Studien über die Narkose: zugleich ein Beitrag zur allgemeinen Pharmakologie*. G. Fischer.

Papahadjopoulos, D., Nir, S., & Oki, S. (1972). Permeability properties of phospholipid membranes: Effect of cholesterol and temperature. *Biochimica et Biophysica Acta, 266*, 561–583.

Pedley, T. J. (1983). Calculation of unstirred layer thickness in membrane transport experiments: A survey. *Quarterly Reviews of Biophysics, 16*, 115–150.

Pohl, P., Saparov, S. M., & Antonenko, Y. N. (1997). The effect of a transmembrane osmotic flux on the ion concentration distribution in the immediate membrane vicinity measured by microelectrodes. *Biophysical Journal, 72*, 1711–1718.

Pohl, P., Saparov, S. M., & Antonenko, Y. N. (1998). The size of the unstirred layer as a function of the solute diffusion coefficient. *Biophysical Journal, 75*, 1403–1409.

Porsche, A. M., Körber, C., Englich, S., Hartmann, U., & Rau, G. (1986). Determination of the permeability of human lymphocytes with a microscope diffusion chamber. *Cryobiology, 23*, 302–316.

Shinoda, W. (2016). Permeability across lipid membranes. *Biochimica et Biophysica Acta, 1858*, 2254–2265.

Simons, K., & Ikonen, E. (1997). Functional rafts in cell membranes. *Nature, 387*, 569–572.

Sperelakis, N., & Freedman, J. C. (2012). Diffusion and permeability. In N. Sperelakis (Ed.), *Cell physiology sourcebook: Essentials of membrane biophysics* (4th ed., pp. 113–120). San Diego: Academic Press-Elsevier.

Stein, W. D., & Lieb, W. R. (1986). *Transport and diffusion across cell membranes*. Academic Press.

Stein, W. D., & Litman, T. (2015). *Channels, carriers, and pumps: An introduction to membrane transport*. London: Academic Press, Elsevier.

Stillwell, W. (2016). *An introduction to biological membranes: composition, structure and function*. Academic Press-ELsevier.

Strickholm, A. (1981). Ionic permeability of K, Na, and Cl in potassium-depolarized nerve. Dependency on pH, cooperative effects, and action of tetrodotoxin. *Biophysical Journal, 35*, 677–697.

Toyoshima, Y., & Thompson, T. E. (1975). Chloride flux in bilayer membranes: Chloride permeability in aqueous dispersions of single-walled, bilayer vesicles. *Biochemistry, 14*, 1525–1531.

Trimble, W. S., & Grinstein, S. (2015). Barriers to the free diffusion of proteins and lipids in the plasma membrane. *The Journal of Cell Biology, 208*, 259–271.

Ussing, H. H. (1949). The distinction by means of tracers between active transport and diffusion. *Acta Physiologica Scandinavica, 19*, 43–56.

Venable, R. M., Krämer, A., & Pastor, R. W. (2019). Molecular dynamics simulations of membrane permeability. *Chemistry Review, 119*, 5954–5997.

Verkman, A. S. (2000). Water permeability measurement in living cells and complex tissues. *Journal of Membrane Biology, 173*, 73–87.

Walter, A., & Gutknecht, J. (1986). Permeability of small nonelectrolytes through lipid bilayer membranes. *Journal of Membrane Biology, 90*, 207–217.

Wilson, F. A., & Dietschy, J. M. (1974). The intestinal unstirred layer: Its surface area and effect on active transport kinetics. *Biochimica et Biophysica Acta, 363*, 112–126.

Worman, H. J., & Field, M. (1985). Osmotic water permeability of small intestinal brush-border membranes. *Journal of Membrane Biology, 87*, 233–239.

Zeidel, M. L. (1996). Low permeabilities of apical membranes of barrier epithelia: What makes watertight membranes watertight? *American Journal of Physiology, 271*, F243–F245.

Chapter 11

Osmotic pressure and water movement across cell membranes

Joseph J. Feher and Clive M. Baumgarten
Department of Physiology and Biophysics, Virginia Commonwealth University School of Medicine, Richmond, VA, United States

1. Summary

Many of the long-standing uncertainties about osmosis and water transport across cell membranes have been solved. Mathematical expressions for osmotic pressure can be derived from thermodynamic principles. The assumptions of dilute solutions and ideal behavior give rise to the classical van't Hoff equation. Corrections for nondilute solutions and nonideal behavior are available, and the perfect prediction of osmotic pressure from other colligative properties is explained. Nevertheless, these thermodynamic arguments shed no clue as to the mechanisms of osmotic flow. It is now understood that osmotic flow across biological membranes has multiple routes: diffusion through the bilayer membrane, pressure-driven flow through narrow pores provided by aquaporins, and cotransport by solute carriers. Aquaporins are widely expressed throughout the biological world. Molecular portraits of the aquaporin family are available and provide insight into the disposition of water passage through their narrow pores. In most cells, aquaporins provide the major route of water flow across their membranes. The importance of water transport via other carriers is not yet established.

2. Early studies defined osmosis and osmotic pressure

Probably the first recorded observation of osmosis and osmotic pressure was made by Jean-Antoine Nollet in 1748 (Nollet (1752) was translated and reprinted in 1995. Nollet was interested in how liquids were boiled. He degassed some ethanol in a cylindrical flask, sealed the opening with a pig bladder, and then submerged the entire apparatus in water. He noted the volume in the ethanol compartment increased, and the membrane bulged outward with great pressure. If he reversed the components so that water was in the flask, water moved out of the flask, developing a partial vacuum inside. In separate experiments, Nollet demonstrated that water could cross the bladder, whereas ethanol could not.

This is the fundamental observation of osmosis: water movement across a membrane that is impermeant to solutes, creating a pressure on the solute side. Henri Dutrochet (1826), however, is regarded as the "father" of osmosis. He investigated osmotic flow in both animal and plant tissues and coined the terms "endosmose" and "exosmose" from the Greek ωσμος meaning "thrust" or "impulse". The modern terms "osmosis" and "osmotic pressure" were introduced by Graham in 1854 (Graham, 1854). The earliest and clearest advance in osmosis, however, was provided by the plant physiologist, Wilhelm Pfeffer (1877).

Pfeffer formed a rigid precipitation membrane within the walls of unglazed porcelain cups by reacting copper salts with potassium ferricyanide. He used this artificial copper ferrocyanide membrane to separate a sucrose solution inside the cup from water on the outside. He observed that water flowed from the outside to the inside. This reprises the fundamental observation of osmosis: the movement of fluid across a semipermeable membrane in response to differing concentrations of solute on the two sides of the membrane. Importantly, Pfeffer further noted that flow was proportional to the sucrose concentration in the cup. Moreover, when water was inside the cup, he observed that a pressure applied to the inside compartment would force water out of the cup, and the flow was proportional to the pressure. When sucrose was inside a closed cup, and water outside, a pressure would develop inside the cup and at equilibrium the pressure was proportional to the sucrose concentration. He recognized this as an equilibrium in which the outward filtration of water by pressure

Cell Physiology Sourcebook. https://doi.org/10.1016/B978-0-12-811114-7.00033-5

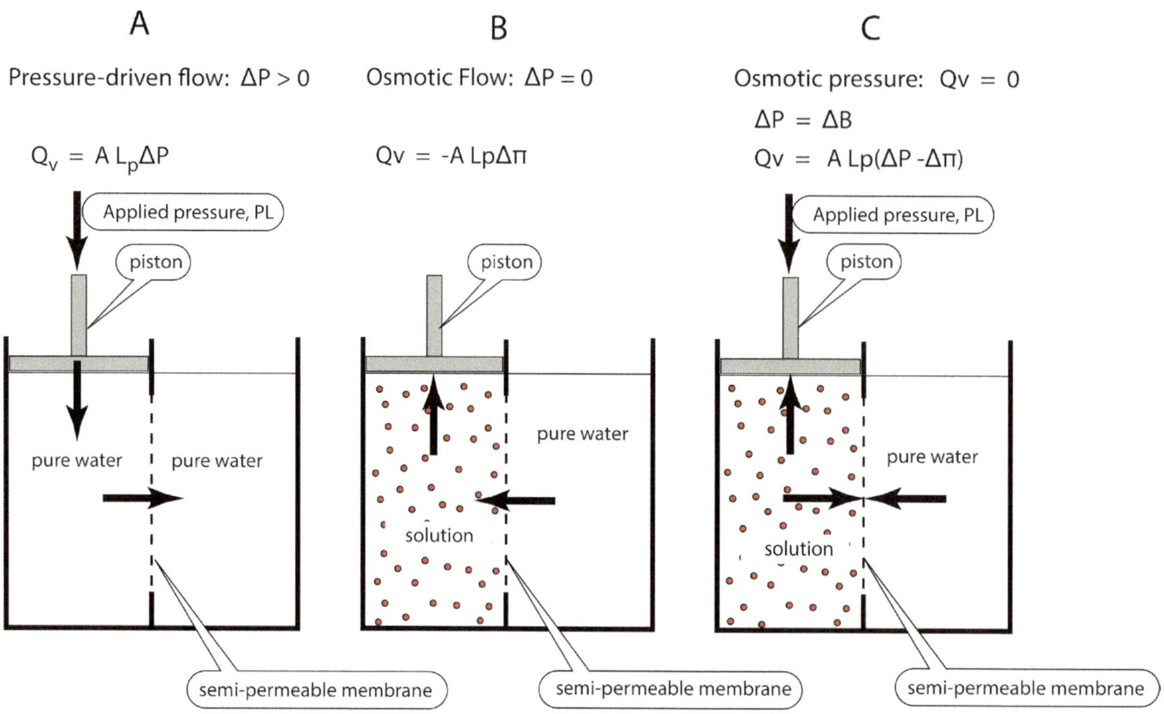

FIGURE 11.1 Equivalence of hydrostatic and osmotic pressure in driving fluid flow across a membrane. Panel A: An ideal, semipermeable membrane is permeable to water but not solute. When it separates pure water on the left from pure water on the right, application of a pressure on the left, P_L, forces water across the membrane to the right. The flow (Q_V) is linearly related to the pressure by the area of the membrane (A) and a proportionality constant, L_P, that is characteristic of the membrane. This constant is variously called the *hydraulic conductivity, hydraulic permeability, or filtration coefficient*. Positive Q_V is taken as flow to the right. Panel B: The ideal semipermeable membrane separates a solution on the left from pure water on the right, and water moves into the solution side by osmosis. The flow is linearly related to the difference in osmotic pressure, $\Delta\pi$, by the area and the hydraulic conductivity, L_P. The flow causes expansion of the left compartment and movement of the piston, which is assumed here to be weightless. Q_V is negative because fluid flow is to the left. Panel C: Application of additional pressure to the solution so that $\Delta P = \Delta\pi$ results in no net flow across the membrane. The osmotic pressure of the solution is defined as the pressure necessary to stop fluid flow when an ideal semipermeable membrane separates pure water from the solution. *From Feher (2017).*

balanced the inward movement of water by osmosis (see Fig. 11.1). Pfeffer *defined* the **osmotic pressure** as **the pressure necessary to stop osmotic flow across a barrier that is impermeable to the solute**. This definition is key. It defines osmotic pressure as the pressure needed to *stop* osmotic flow rather than the pressure that *drives* osmotic flow. Osmotic pressure is also defined only for a semipermeable membrane, one that is impermeant to solute but permeant to the solvent, water. Historically, the osmotic pressure is given the symbol π. Pfeffer's data on the linear relation of π to concentration of sucrose are shown in Fig. 11.2.

3. van't Hoff's law, $\pi = RTC$

From Pfeffer's data and thought experiments considering gases in equilibrium with water, van't Hoff (1887) argued that the osmotic pressure should be given by

$$\pi = RT\sum_i C_i \tag{11.1}$$

where π is the osmotic pressure, R is the gas constant, T is the temperature in Kelvins, and C_i is the concentration of solute i. Note that this equation is analogous to the ideal gas law, $P = RTn/V$. van't Hoff's law tells us that the osmotic pressure of a solution is the pressure the *solute* would exert if it were a gas! However, the osmotic pressure is not a positive pressure, but a pressure *deficit*. The solution acts as if its internal pressure is *reduced* by π, because water is drawn into the solution as if it were driven by the pressure π.

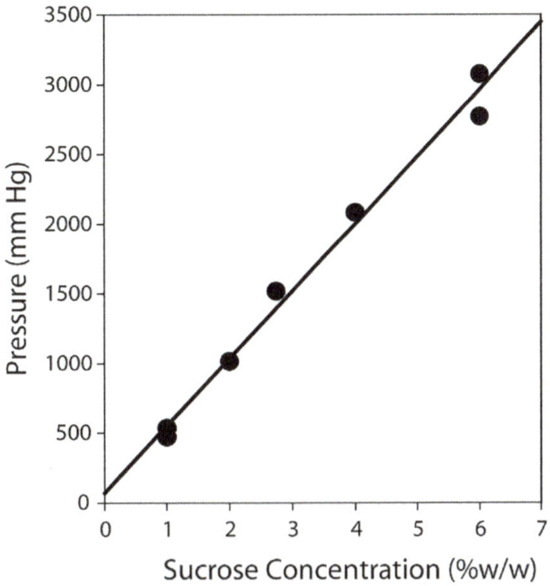

FIGURE 11.2 Plot of the data of Pfeffer (1877) for the osmotic pressure of sucrose solutions. A copper ferrocyanide precipitation membrane was formed in the walls of an unglazed porcelain cup. The membrane separated a sucrose solution in the inner chamber from water outside the cup. The inner chamber was attached to a manometer and sealed. The linear relationship between the pressure measured with this device and the sucrose concentration was the experimental impetus for deriving van't Hoff's law.

The concentration used in van't Hoff's law, $\sum C_i$, refers to the concentration of osmotically active particles, or the solution's *osmolarity*. Organic compounds such as glucose or sucrose form one osmotically active particle per molecule. For these compounds, the osmolarity is the same as its molarity. Strong salts such as NaCl or $CaCl_2$ ideally dissociate into two (Na^+ and Cl^-) or three (Ca^{2+} and two Cl^-) particles per molecule of salt, respectively, and the osmolarity is twice (NaCl) or three times ($CaCl_2$) their molarity. In fact, the notion of salt dissociation proposed by Arrhenius came about partly to fit osmotic pressure data (Smith, 1960).

The osmolarity of a solution is expressed in osmol per liter to indicate that we are referring to the number of osmotically active particles, or osmolytes, rather than the concentration of solute. An alternate scale, *osmolality,* defines $\sum C_i$ as osmoles per kg of solvent. This osmolal scale better describes the osmotic pressure in van't Hoff's equation, but the osmolar scale is more generally used in physiological studies. van't Hoff's law is a limiting law that is strictly true only for dilute and ideal solutions. In this limit, osmolal and osmolar scales both converge to the same results.

To illustrate the magnitude of osmotic pressure, consider 0.9% NaCl and 5% glucose solutions that are isotonic to red blood cells. What this means is that red blood cells that are placed in 0.9% NaCl or 5% glucose do not change their volume—there is no net movement of fluid across the erythrocyte cell membrane. The molarity is 0.154 M for 0.9% NaCl and 0.278 M for 5% glucose. The calculated osmotic pressure at 37°C for 0.154 M NaCl is $0.082\ \text{L atm/osmol/K} \times 310\ \text{K} \times 0.154\ \text{mol/L} \times 2\ \text{osmol/mol} = 7.82\ \text{atm} = 5950\ \text{mm Hg}$. For 0.278 M glucose, the osmotic pressure is $7.06\ \text{atm} = 5371\ \text{mm Hg}$. Thus, osmotic pressures are enormous in comparison to normal hydrostatic pressures in physiological systems. Table 11.1 lists common units for osmotic pressure along with the values and units of R needed to make the calculations.

TABLE 11.1 Units for the calculation of osmotic pressures.

Pressure units	1 atm equivalent	Gas constant (R)	Solute concentration
Atm	1	0.082 L atm/mol/K	mol/L
mm Hg	760	62.36 L mmHg/mol/K	mol/L
Pa = N/m^2	1.013×10^5	8.314 N m/mol/K	mol/m^3 = mol/(1000 L)
$Dyne/cm^2$	1.013×10^6	8.314×10^7 dyne cm/mol/K	mol/cm^3

4. van't Hoff's law can be derived from thermodynamics

One of the conclusions of chemistry is that all spontaneous processes are accompanied by a decrease in free energy. The total free energy of a solution can be divided among its components. This parceling out of the Gibbs Free Energy, G, is embodied in the concept of the chemical potential:

$$\mu_i = \left[\frac{\partial G}{\partial n_i}\right]_{T,P,n_k} \tag{11.2}$$

where μ_i is the chemical potential of component i, n_i is the number of moles of component i, and $i \neq k$. The chemical potential of a component of a solution consists of three terms: a standard potential, which is the chemical energy involved in the formation of the material from standard states; a compositional term, which depends on the presence of other constituents; and a work term, encompassing other work required, per mole, to bring additional material into the mixture. The work term can involve pressure—volume work or electrical work. The work term in the chemical potential of water is $\int d\overline{V_W}P$, where $\overline{V_W}$ is the volume of water per mole and P is the pressure. The electrochemical potential of ions in solution requires the inclusion of the electrical work term, $\int z_i F d\psi$ where z_i is the ion's valence, F is Faraday's constant that converts the charge into coulombs per mole, and ψ is the electrical potential.

In the case of a solution on the left separated from pure water on the right by an ideal semipermeable membrane, water movement will occur when there is a difference in the chemical potential of water on the two sides of the membrane, so that water movement will result in a decrease in free energy. When the pressure applied to the solution is equal to the osmotic pressure, equilibrium is established and the chemical potential for water is the same on both sides of the membrane, and no water movement occurs. This equality of chemical potential of water is written as

$$\mu_L^0 + \overline{V_W}P_L + RT \ln a_{W,L} = \mu_L^0 + \overline{V_W}P_R + RT \ln a_{W,R} \tag{11.3}$$

where the subscripts L and R refer to the left and right sides of the semipermeable membrane, μ^0 is the chemical potential of pure liquid water in its standard state, and a_W is the **activity** of the water. Solute enters Eq. (11.3) by altering the activity of water on its side. For an ideal solution, the activity of water is its mole fraction:

$$a_W = X_W = \frac{n_W}{n_W + n_S} \tag{11.4}$$

where X_W is the usual symbol for mole fraction of water, and n_W and n_S are the moles of water and solute, respectively, in the solution. Substituting in X_W for a_W, and canceling the μ^0 on both sides of Eq. (11.3), and rearranging, we come to

$$\overline{V_W}P_L - \overline{V_W}P_R = RT \ln X_{W,R} - RT \ln X_{W,L} \tag{11.5}$$

Since the fluid on the right is pure water, $X_{W,R} = 1$ and $\ln X_{W,R} = 0$, and we have

$$\overline{V_W}(P_L - P_R) = -RT \ln X_{W,L} \tag{11.6}$$

The mole fractions of water and solute in the solution must sum to 1.0. This is written as

$$\begin{aligned} X_{W,L} + X_{S,L} &= 1.0 \\ \ln X_{W,L} &= \ln(1 - X_{S,L}) \end{aligned} \tag{11.7}$$

In dilute solutions, $X_{S,L} \ll 1.0$, and thus, $\ln(1 - X_{S,L}) \approx -X_{S,L}$. Substitution of this approximation ($\ln X_{W,L} \approx -X_{S,L}$) in Eq. (11.6) gives

$$P_L - P_R = \frac{RT}{\overline{V_W}} X_{S,L} \tag{11.8}$$

The left-hand side of Eq. (11.8) is the osmotic pressure, π, which is the extra pressure that must be applied to the solution on the left side in order to establish equilibrium with water on the right side. Inserting the definition of mole fraction of solute, and using the assumption of dilute solutions so that $n_S \ll n_W$, we get

$$\pi = \frac{RT}{\overline{V_W}} X_{S,L} = \frac{RT}{\overline{V_W}} \frac{n_S}{n_S + n_W} \approx \frac{RT}{\overline{V_W}} \frac{n_S}{n_W} \approx RT \frac{n_S}{V} = RTC_S \tag{11.9}$$

where $n_W \overline{V_W} = V$ is the total volume of solution, and C_S is the concentration of impermeant solute on the solution side of the membrane. The last expression is the van't Hoff equation for the osmotic pressure. This thermodynamic derivation entails some assumptions: (1) the solution is ideal, (2) the solution is dilute, and (3) the solution is incompressible, so that $\int d\,\overline{V_W}P = \overline{V_W}\Delta P$. The compressibility of water is 4.4×10^{-10} /Pa = 4.46×10^{-5} /atm, which justifies its omission. It is important to recognize that **van't Hoff's law is not exact for physiological solutions.** Rather, it is an approximation that is exactly true only for ideal dilute solutions.

The thermodynamic derivation of van't Hoff's Law is abstract and does not yield any intuitive feel as to what really causes osmosis. This agrees in general with the power of thermodynamic reasoning, which lies in its independence of mechanism. Conceivably, the semipermeable membrane could be like a sieve that allows water to pass while blocking the solute. Alternatively, solvent water might dissolve in the membrane, whereas solute is insoluble. Both of these types of membranes would exhibit osmotic flow from the region of low osmotic pressure (pure water) to the region of high osmotic pressure (the solution). The mechanism by which osmosis occurs must be determined by other methods and must be determined for every membrane.

5. Osmotic pressure is a property of solutions that is related to other colligative properties

One of the astounding conclusions from van't Hoff's Law is that the nature of the solute has no consequence whatever for the magnitude of the osmotic pressure, provided that the solution is ideal and dilute and that the membrane is impermeant to the solute. All that matters is that dissolution of the solute lowers the chemical potential of water by the amount $\sim RTC_S$. Several other properties of solutions are also consequences of the reduced chemical potential of water because of dissolved solutes. These properties were named *colligative* properties by Ostwald (1891). Ostwald identified three classes of solution properties. *Additive* properties, such as mass, are the sums of the composite particles making up the solution; *constitutive* properties depend on the molecular structure of the components; and *colligative* properties depend on the concentration of solute and temperature but are independent of the nature of the solute. The colligative properties include:

- **osmotic pressure**
- **vapor pressure depression**
- **boiling point elevation**
- **freezing point depression**

These are properties of the solution in the sense that a solution in a beaker on the benchtop has an osmotic pressure, a depressed vapor pressure, an elevated boiling point, and a depressed freezing point. Colligative properties are evident when you measure them, but clearly they are not evident while the beaker just sits on the benchtop. To see the connection between vapor pressure depression and osmotic pressure, consider two open compartments enclosed in a chamber (see Fig. 11.3). One compartment contains pure water and the other contains a solution of a non-volatile solute. The vapor pressure above a solution is defined as the partial pressure of water vapor in equilibrium with the solution. Since the vapor pressure of pure water is higher than that of the solution, water vapor above the pure water will be higher than that above the solution. As a result, water will evaporate from the water compartment and condense into the solution compartment. Thus, water will move from the pure water to the solution. "Osmosis" would occur through a "semipermeable membrane" that consists of the surfaces of the two compartments and the intervening air. This flow would be very slow, much slower than what we would expect for osmosis through a semipermeable membrane because evaporation is slow and diffusion across the intervening air is expected to be very slow due to the macroscopic distances involved.

Lowering of the vapor pressure by solutes is a direct consequence of Raoult's law (Raoult, 1886), which was published shortly before van't Hoff's law. Raoult's law gives:

$$p_W = p_W^0 X_W \tag{11.10}$$

where p_W is the vapor pressure of water, p_W^0 is the vapor pressure of pure water, and X_W is the mole fraction of water in the solution. **Raoult's law is also an ideal law**. In a solution, $X_W < 1.0$ and the vapor pressure is correspondingly reduced. Raoult's law means that the mole fraction of water in an ideal solution can be obtained experimentally as

$$X_W = \frac{p_W}{p_W^0} \tag{11.11}$$

This can be used in Eq. (11.6) to calculate the osmotic pressure. For a nonideal solution, the ratio of the vapor pressures in Eq. (11.11) gives the activity of water, a_W. Thus, solutions that have the same vapor pressure have the same osmotic pressure; vapor pressure depression is a perfect predictor of the osmotic pressure because, as shown in Fig. 11.3, they are

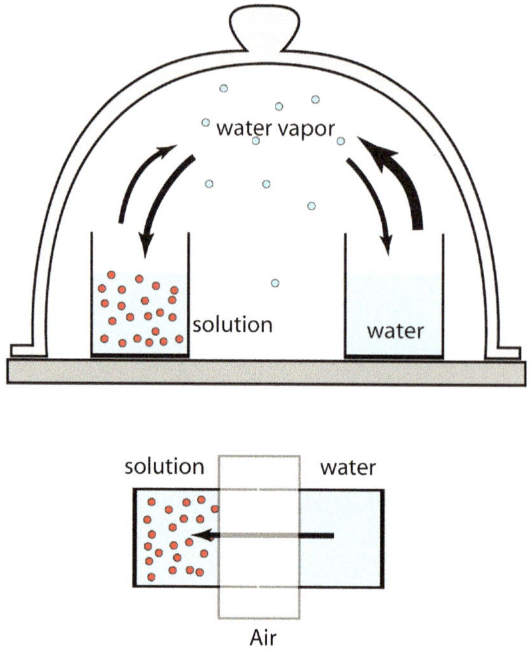

FIGURE 11.3 Equivalence of vapor pressure and osmotic pressure. Two beakers containing a solution or pure water are placed in a sealed chamber. The air above the beakers contains air molecules and water vapor. The partial pressure of water is its contribution to the total gas pressure and is proportional to the water concentration in the gas phase. The vapor pressure is defined as the partial pressure of water in equilibrium with the liquid phase. Water will leave the liquid to moisten the gas phase. At equilibrium, water will evaporate from the liquid phase and condense from the gas phase at equal rates. The vapor pressure of a solution is lower than that of pure water. This vapor pressure depression is one of the colligative properties of solutions. Because there is only one vapor pressure in the sealed container, it cannot simultaneously be in equilibrium with water and solution. Water from the pure water beaker will continue to evaporate, producing a vapor pressure higher than that of the solution, and therefore, water will condense into the solution. If we were to reduce the size of the intervening air and bring the beakers close together, the situation looks identical to the situation of osmosis, in which water moves from pure water to solution across a semipermeable membrane. In this case, the semipermeable membrane is the two surfaces and the intervening air. Thus, osmosis and vapor pressure depression are two aspects of a single cause: Solutes lower the chemical potential of water in a solution. *From Feher (2017).*

essentially the same phenomenon. This equivalence was known by earlier investigators of the mechanism of osmosis (Andrews, 1976; Hammel, 1976; Hildebrand, 1955; Hulett, 1903; Kiil, 1989; Smith, 1960).

From thermodynamic arguments, the freezing point depression can be shown to be given as

$$T - T_0 = \Delta T = -\frac{RT_0^2}{\Delta H_{fus}} X_S \tag{11.12}$$

where T is the freezing point of the solution, T_0 is the freezing point of pure water, ΔH_{fus} is the molar heat of fusion of water (the heat change upon freezing of pure water), and X_S is the mole fraction of solute. Using the definition of mole fraction and the approximation of dilute solution, Eq. (11.12) becomes

$$\Delta T \approx -\frac{RT_0^2 \overline{V_W}}{\Delta H_{fus}} C_S = -K_f C \tag{11.13}$$

The freezing point depression constant, K_f, is evaluated as

$$\Delta T = -\frac{8.314\ \text{joule/mol/K}\ (273.15\ \text{K})^2\ (0.018\ \text{L/mol})}{6014.9\ \text{joule/mol}} C = -1.86\ \text{K/M}\ C \tag{11.14}$$

This equation predicts that the freezing point depression is linearly related to the concentration of solute, just as van't Hoff's law predicts that the osmotic pressure is linearly related to concentration. Deviations from dilute solution and nonideality similarly affect osmotic pressure, vapor pressure depression, and freezing point depression. Because osmotic pressure and vapor pressure depression are more difficult to measure, freezing point depression is often used in clinical and research laboratory devices to measure the osmolarity of solutions.

6. Osmotic coefficients correct osmotic pressure for nonideality

van't Hoff's law is exactly true only in the limit of dilute and ideal solutions. Because most solutions are not ideal and not dilute, van't Hoff's law must be modified to include a correction term, the **osmotic coefficient**, ϕ_i, that depends on the nature of the solute, the temperature, the concentration of solute, and other factors:

$$\pi = RT\sum_i \phi_i C_i \tag{11.15}$$

Values for the osmotic coefficients of electrolytes can be found in Robinson and Stokes (1959) or can be calculated from the parameters tabulated by Pitzer and Mayorga (1973). Nonelectrolytes and polyelectrolytes, especially proteins, also show deviations from van't Hoff's law. The osmotic coefficients correct van't Hoff's law due to interactions for only the particular solute. When other solutes are present, interactions could occur that are not accounted for in the osmotic coefficients. Therefore, calculations of the expected osmotic pressures of mixtures of solutes, even when osmotic coefficients are used, must be regarded as approximations.

Approximate values for the osmotic coefficients of commonly encountered physiological solutes, under approximate physiological conditions, are given in Table 11.2.

The osmotic coefficient for a single solute can be calculated from Eq. (11.15) as

$$\phi = \frac{\pi_{observed}}{\pi_{calculated}} = \frac{\pi_{observed}}{RTC_S} \tag{11.16}$$

The osmotic coefficient for sucrose is plotted against the sucrose concentration in Fig. 11.4. The osmotic coefficient is nearly 1.0 in dilute solutions but approaches 3 in highly concentrated solutions. Thus, the van't Hoff law successfully predicts the osmotic pressure of dilute sucrose solutions but fails at high sucrose concentrations. This is because van't Hoff's law incorporates the assumptions of ideal behavior and dilute solution. We can eliminate the assumption of dilute solution by calculating the osmotic pressure using the mole fraction, according to Eq. (11.6):

$$\pi = -\frac{RT}{\overline{\overline{V}}_W}\ln X_W \tag{11.17}$$

This prediction of the osmotic pressure still requires the assumption of ideality—it assumes that the activity of water is equal to its mole fraction. The **rational osmotic coefficient** is defined as

$$g = \frac{\pi_{observed}}{\pi_{calculated}} = \frac{\pi_{observed}}{-\frac{RT}{\overline{\overline{V}}_W}\ln X_W} \tag{11.18}$$

TABLE 11.2 Approximate value of the osmotic coefficient for some physiological solutes.

Solute	Particles Formed	Molecular weight (g/mol)	Osmotic coefficient (ϕ)
NaCl	2	58.4	0.93
KCl	2	74.6	0.92
$CaCl_2$	3	111.0	0.85
Na_2SO_4	3	142.0	0.74
$MgCl_2$	3	95.2	0.89
$MgSO_4$	2	120.4	0.58
$NaHCO_3$	2	84.0	0.96
Alanine	1	89.1	1.00
Mannitol	1	182.2	1.00
Glucose	1	180.2	1.01
Sucrose	1	342.2	1.02

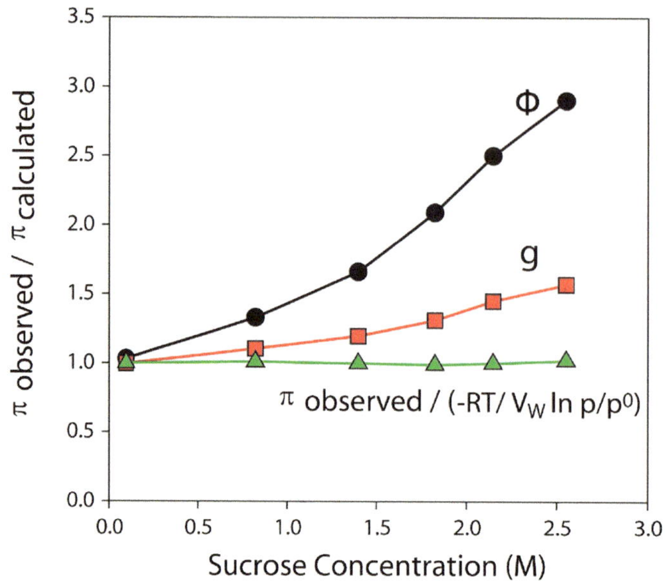

FIGURE 11.4 Osmotic coefficients as a function of sucrose concentration. The osmotic coefficient, ϕ, was calculated according to Eq. (11.16) by dividing the observed osmotic pressure by RTC (circles). The rational osmotic coefficient, g, was obtained according to Eq. (11.18) by dividing the observed osmotic pressure by $-\frac{RT}{\overline{V}_W}\ln X_W$ (squares). The correlation of vapor pressure depression to osmotic pressure was tested by dividing the osmotic pressure by $-\frac{RT}{\overline{V}_W}\ln\frac{p}{p^0}$ (triangles). *Data from Glasstone (1946) and Figure from Feher (2017).*

The rational osmotic coefficient is also shown in Fig. 11.4. It is closer to 1.0, but the predicted osmotic pressure still deviates from the observed pressure at high sucrose concentrations. This is because we have removed one assumption (dilute solutions) but not the second (ideal behavior). The expression for osmotic pressure without assuming either a dilute solution or ideality is given from Eq. (11.3) as:

$$\pi = -\frac{RT}{\overline{V}_W}\ln a_W \tag{11.19}$$

We can know the activity of water from the vapor pressure of the solution from Raoult's law, Eq. (11.10), applied to a nonideal solution:

$$a_W = \frac{p_W}{p_W^0} \tag{11.20}$$

If we calculate the osmotic pressure according to Eq. (11.19) using Eq. (11.20) and divide it into the observed osmotic pressure, we find that the ratio is close to 1.0 over the entire range of sucrose concentrations (see Fig. 11.4). Thus, the activity of water calculated from the vapor pressure measurements allows perfect prediction of the osmotic pressure. Thus, we have three ways of calculating the osmotic pressure:

$$\begin{aligned}
\pi &= RT\phi C \;\therefore\; \phi = \frac{\pi_{observed}}{RTC} \\
\pi &= -\frac{RT}{\overline{V}_W} g \ln X_W \;\therefore\; g = \frac{\pi_{observed}}{-\frac{RT}{\overline{V}_W}\ln X_W} \\
\pi &= -\frac{RT}{\overline{V}_W}\ln\frac{p_W}{p_W^0} \;\therefore\; 1.0 = \frac{\pi_{observed}}{-\frac{RT}{\overline{V}_W}\ln\frac{p_W}{p_W^0}}
\end{aligned} \tag{11.21}$$

In the top equation, van't Hoff's law requires the empirically observed osmotic coefficient, ϕ, which varies with solute concentration (see Fig. 11.4), to correct for the assumption of dilute solution and for nonideality; in the middle equation, the rational osmotic coefficient, g, uses X_W to effectively correct for the assumption of dilute solution, but g varies with solute concentration to account for nonideal behavior. The bottom equation uses the empirically determined vapor pressure depression, p_W/p_W^0, which incorporates the consequences when a solution is neither ideal nor dilute, to perfectly predict the osmotic pressure. The practical problem is that we have no theory that allows us to calculate or predict ϕ or g for specific solution conditions without some other measurement, such as the activity of water.

7. Osmotic pressure of proteins is not ideal

Because of their importance in maintaining fluid distribution in physiological systems, the non-ideality of the osmotic pressure of protein requires attention. Adair (1928) found that the observed osmotic pressure of hemoglobin solutions curved upwards when graphed against its concentration, whereas for simple physiological salts such as NaCl the same plot would curve downward. In the limit of low concentrations, Adair's plot for hemoglobin converges to van't Hoff's law, but at high concentrations, the osmotic coefficient was quite high. At the concentrations expected in red blood cells (34.4 g hemoglobin per 100 mL of solution), $\phi_{Hb} = 4.03$ (see Fig. 11.5). Part of the osmotic pressure of hemoglobin is due to the unequal distribution of ions across the semipermeable membrane caused by the electric charge on the protein molecules. This is the Gibbs–Donnan equilibrium, which is discussed in Chapters 12 and 24 of this volume. However, the Gibbs–Donnan distribution makes only a small contribution to the total osmotic pressure, and nearly all of the nonlinearity is due to the protein itself.

Plasma proteins also have osmotic coefficients that increase with protein concentration. The degree of deviation depends on the individual proteins, with plasma albumin deviating more than γ-globulins. Landis and Pannenheimer (1963) give empirical fits of the osmotic pressure as follows:

$$\begin{aligned} \pi_{albumin} &= 2.8C + 0.18C^2 + 0.012C^3 \\ \pi_{globulin} &= 1.6C + 0.15C^2 + 0.006C^3 \end{aligned} \tag{11.22}$$

where π is the osmotic pressure in mm Hg and C is the concentration of the protein in g/dL at 37°C. In each of these equations, the first term represents the limiting van't Hoff law, and the molecular weight can be determined from the coefficient of the first power of C. According to the empirical equation for albumin, a 1 g/dL (= 10 g/L) solution of albumin will have an osmotic pressure of 2.8 mm Hg, which is 2.8 mm Hg/760 mm Hg/atm = 0.00368 atm. Assuming this is dilute

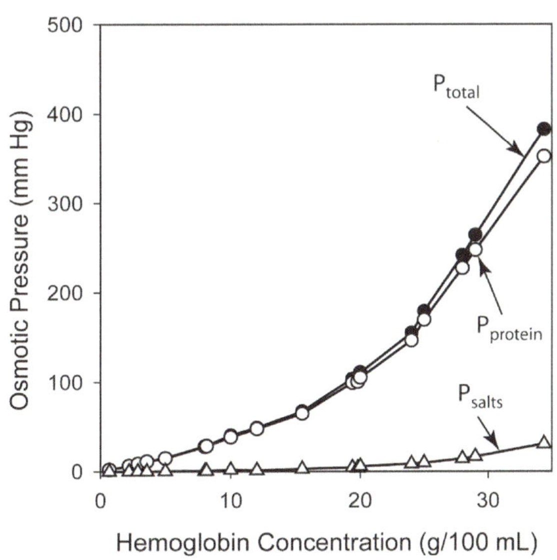

FIGURE 11.5 Dependence of the observed osmotic pressure on hemoglobin concentration. P_{total} (closed circles) is the observed osmotic pressure of the solutions and is due to both protein and salts. $P_{protein}$ (open circles) is the pressure attributed to the proteins and P_{salt} (open triangles) is the contribution of the salts to the osmotic pressure as calculated from the Gibbs–Donnan distribution and the van't Hoff equation.

enough to approximate van't Hoff's law, this pressure is RTC, where C is in mol/L. Thus, $C = 0.00368$ atm/(0.082 L atm/mol/K $\times$ 310 K) $= 1.448 \times 10^{-4}$ M. Thus, 10 g/L $= 1.448 \times 10^{-4}$ mol/L. The molecular weight is obtained by dividing the mass by the number of moles: $M_r = 10$ g/1.448 $\times$ 10^{-4} mol $=$ 69,076 g/mol. The molecular weight of albumin from its amino acid composition is 66, 400 g/mol.

The large ϕ for proteins and polymers is due in part to *excluded volume* effects. That is, proteins and polymers exclude solvent water from a larger volume than do small solutes. The lowering of the free energy of solvent water upon dissolution of solute, which gives rise to osmosis, can be calculated from the increase in entropy on mixing. This entropy of mixing is much larger if the solute occupies a much larger volume (Hildebrand, 1955). Tanford (1961) has shown that the expected osmotic pressure of proteins can be given as

$$\pi = RTC\left(1 + \frac{\overline{V_S}}{\overline{V_W}}C\right) \tag{11.23}$$

Consideration of the excluded volume explains in part the behavior of the osmotic coefficient, ϕ, for proteins. However, these excluded volume effects also affect the osmotic pressure of all other solutes in the presence of proteins. Thus, the osmotic pressure of salts is increased when protein concentration is high.

8. "Antifreeze" proteins act by binding ice

Animals and plants that live in subfreezing temperatures can survive by employing two general adaptive strategies: avoid freezing or tolerate it. "Freeze tolerant" means being able to survive freezing of body fluids. Microbes, ciliates, fungi, plants, insects, and fish have developed a variety of adaptations to achieve these ends. Marine teleost fish have body fluids hypoosmolar to sea water, with an equilibrium freeze-melting point (eqFMP) of −0.6 to −0.9°C, whereas the freezing point of the surrounding ice water is −1.9°C. Why don't they freeze? These fish have "antifreeze proteins" (AFPs), first discovered by Devries and coworkers (Devries, 1971; Devries & Wohlschlag, 1969), that prevent the fish blood and extracellular fluid from freezing. This diverse set of fish antifreeze glycoproteins, ranging in molecular weight from 2.6 to 33.7 kDa (Devries, 1986), lower the freezing point of their solutions but do not affect the melting points (Knight et al., 1984). The difference between the freezing point and melting point is called thermal hysteresis (TH). These proteins lower the freezing point by a noncolligative mechanism in which the proteins absorb into the ice and prevent further ice crystal growth, thereby lowering the nonequilibrium hysteretic freezing point (hFP) below the equilibrium freezing and melting point (eqFMP). This creates a thermal hysteresis and allows fish to maintain hypoosmolar fluids that do not freeze at the eqFMP. In arthropods, there are a variety of ice-binding proteins that can produce supercooling of tissues to −20°C and below. The reader is referred to recent reviews of these proteins and their mechanisms for additional details (Baskaran et al., 2021; Dolev et al., 2016; Duman, 2015).

9. The reflection coefficient corrects van't Hoff's law for permeable solutes

As discussed earlier, when a semipermeable membrane separates a solution from pure water, the measured osmotic pressure is perfectly predictable from the vapor pressure depression, or by van't Hoff's law that incorporates an osmotic coefficient to correct for nonideality and nondilute solutions. A semipermeable membrane is one that is impermeant to the solute. When a membrane is somewhat permeable to the solute, the measured osmotic pressure is less than that predicted from $RT\phi C$. This phenomenon has led to the definition of another parameter, the **reflection coefficient**, σ, defined as

$$\sigma = \frac{\pi_{observed}}{\pi_{theoretical}} = \frac{\pi_{observed}}{RT\phi_S C_S} \tag{11.24}$$

The reflection coefficient was first proposed and named by Staverman (1951). It refers to the fraction of collisions of the solute molecules with the membrane that are reflected back into the solution compared to the fraction of the collisions of solvent water that are reflected back. In this sense, it is an index of the ability of the membrane to distinguish solute from solvent. For a semipermeable membrane, the membrane perfectly distinguishes solute from solvent, and $\sigma = 1.0$. In this case, the observed osmotic pressure is predicted by $RT\phi C$. When the solute can permeate through the membrane, some fraction of the collisions with the membrane are not reflected, and $\sigma < 1.0$. Thus, the observed osmotic pressure will be $\sigma RT\phi C$. The values of σ depend on both the size of solutes, being larger for larger solutes (Durbin, 1960), and also characteristics of the membrane. Consequently, **the reflection coefficient is uniquely defined for every membrane–solute pair**.

10. Equivalence of osmotic and hydrostatic pressure

As noted earlier, Pfeffer found that a hydrostatic pressure applied to one side of a membrane could drive a fluid flow whose magnitude is directly proportional to the pressure. The relation is written as

$$Q_V = AL_P\Delta P \tag{11.25}$$

where Q_V is the macroscopic volume flow, in L/min, A is the area, ΔP is the pressure difference across the membrane and L_P is variously called the **hydraulic conductivity, hydraulic permeability, or filtration coefficient**.

Pfeffer found that an osmotic pressure also produces a flow that is proportional to the osmotic pressure, given as

$$Q_V = -AL_P\Delta\pi \tag{11.26}$$

where the symbols have the same meaning as before, and $\Delta\pi$ is the difference in osmotic pressure across the membrane. The negative sign in the equation is a consequence of the fact that osmotic flow is from the side of low osmotic pressure to the side of high osmotic pressure: the osmotic pressure is the apparent *reduction* in the internal pressure of solution brought about by dissolution of the solute.

Using collodion membranes, Meschia and Setnikar (1958) found that the proportionality constant relating hydrostatic pressure and filtration flow was identical to the proportionality constant relating osmotic pressure to osmotic flow. That is, the L_P in Eq. (11.25) is identical to the L_P in Eq. (11.26). Thus, the flow induced by osmotic pressure can be nulled by a hydrostatic pressure (which is how osmotic pressure is defined), and the proportionality between hydrostatic pressure and filtration flow is the same as the proportionality between osmotic pressure and osmotic flow. This allows us to combine both pressures in a single equation:

$$\begin{aligned} Q_V &= AL_P[(P_L - P_R) - (\pi_L - \pi_R)] \\ Q_V &= AL_P[\Delta P - \Delta\pi] \end{aligned} \tag{11.27}$$

This is the proper equation to use for semipermeable membranes. If the membrane is permeable to solute, we must incorporate the reflection coefficient of all of the solutes in the solution:

$$Q_V = AL_P\left[\Delta P - \sum_i \sigma_i \Delta\pi_i\right] \tag{11.28}$$

where σ_i is the reflection coefficient of solute i and $\Delta\pi_i$ is the difference in osmotic pressure for solute i. The $\Delta\pi_i$ used in this equation would be the osmotic pressure observed for a semipermeable membrane, equal to ϕRTC_i. This situation of combined hydrostatic and osmotic pressures is shown in Fig. 11.6.

11. The mechanism of osmosis

11.1 Osmosis is driven by a pressure difference

As described above, a solution separated from pure water by a semipermeable membrane gives rise to an osmotic flow that can be stopped by the application of a pressure to the solution side. The pressure that stops the osmotic flow is the osmotic pressure, and that pressure is perfectly predicted by the vapor pressure of the solution, or by any of the other colligative properties of the solution. Suppose we look at this situation the other way around. What pressure do we need to apply to the pure water side to stop the osmotic flow? This question is already answered from Eq. (11.8):

$$P_L - P_R = \frac{RT}{\overline{V}_W} X_{S,L} = \pi \tag{11.8}$$

where the subscripts indicate the left and right compartments and the solution is on the left and pure water is on the right. From this, we see that

$$P_R = P_L - \pi \tag{11.29}$$

That is, *reduction* in the pressure by π on the pure water side (the right side) would bring the solution into equilibrium with pure water. It can be shown by thermodynamic reasoning that reduction in the pressure on pure water by π reduces the vapor pressure from p_W^0 to that of the solution, p_w, with osmotic pressure π. To see this, we start with the definition of the Gibbs free energy for liquid water:

$$G = VP - TS \tag{11.30}$$

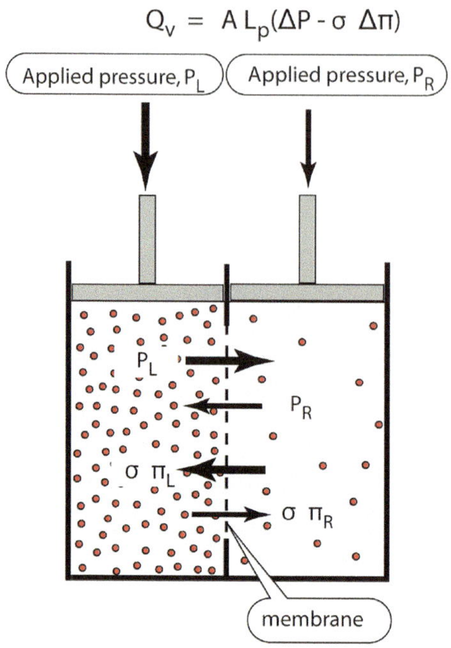

FIGURE 11.6 Net flow in the presence of osmotic and hydrostatic pressures on both sides of the membrane. For a real membrane, the effective osmotic pressure on the left, π_L, causes flow to the left while the applied hydrostatic pressure, P_L, drives flow to the right. The balance of forces on the left favoring fluid flow to the right is $P_L - \sigma\pi_L$. A similar situation occurs on the right-hand side, with the effective osmotic pressure, $\sigma\pi_R$, favoring flow to the right and the applied hydrostatic pressure, P_R, driving flow to the left. The balance of forces driving fluid flow to the left is $P_R - \sigma\pi_R$. The net force is the difference between the two net forces: $(P_L - \sigma\pi_L) - (P_R - \sigma\pi_R) = (P_L - P_R) - \sigma(\pi_L - \pi_R)$ or $\Delta P - \sigma\Delta\pi$. The resulting flow is proportional to this net driving force and the area, A, with the proportionality constant equal to the filtration coefficient, L_P.

The free energy defined in this way is an extensive variable that depends on the number of moles of material, n. Dividing both sides of Eq. (11.30) by n, we get

$$\begin{aligned} \frac{G}{n} &= \frac{V}{n}P - T\frac{S}{n} \\ \mu &= \overline{V}P - T\overline{S} \end{aligned} \tag{11.31}$$

where $\overline{V}$ is the molar volume (the volume per mole) and $\overline{S}$ is the molar entropy. Taking the partial derivative with respect to pressure at constant temperature, we find

$$\left(\frac{\partial \mu}{\partial P}\right)_T = \overline{V} \tag{11.32}$$

For the liquid phase, water is nearly incompressible. Its molar volume is given as

$$\overline{V_{liq}} = (1 + \alpha P)\overline{V_{liq}^0} \tag{11.33}$$

where $\overline{V_{liq}^0}$ is the molar volume under standard conditions (1 atm pressure) and α is the coefficient of compressibility. This coefficient varies with temperature. At ordinary temperatures, $\alpha = 4.35 \times 10^{-10}$/Pa or 4.4×10^{-5} /atm. This is so small that it is ordinarily ignored and $\overline{V_{liq}}$ in Eq. (11.33) is treated as a constant.

Inserting $\overline{V_{liq}}$ into Eq. (11.32) and integrating, we find that the difference between state 1 and state 2 is

$$\begin{aligned} \int_1^2 d\mu_{liq} &= \int_1^2 \overline{V_{liq}}\, dP \\ \mu_{liq,2} - \mu_{liq,1} &= \overline{V_{liq}}(P_2 - P_1) \end{aligned} \tag{11.34}$$

where state 1 is the standard state and state 2 is at the same temperature but different pressure. For the vapor in equilibrium with the liquid water in state 1 and state 2, we write

$$\mu_{liq,1} = \mu_{vap,1} = \mu^0_{vap} + RT \ln P_{vap,1}$$
$$\mu_{liq,2} = \mu_{vap,2} = \mu^0_{vap} + RT \ln P_{vap,2} \tag{11.35}$$

Inserting these results into Eq. (11.34), we obtain

$$RT \ln \frac{P_{vap,2}}{P_{vap,1}} = \overline{V_{liq}}(P_2 - P_1) \tag{11.36}$$

We identify state 1 with water at standard state at 1 atm and state 2 with the water on the right in Fig. 11.7 with the vapor pressure of a solution, we can rewrite Eq. (11.36) as

$$P - P^0 = \frac{RT}{\overline{\overline{V_W}}} \ln \frac{P_W}{P^0_W} \tag{11.37}$$

Here P is the pressure of pure water in equilibrium with vapor pressure p_W and P^0 is the standard 1 atmosphere pressure of pure water in equilibrium with vapor pressure P^0_W. We recognize the right-hand side of Eq. (11.37) from Eq. (11.21) as being equal to $-\pi$. Thus, Eq. (11.37) becomes

$$P = P^0 - \pi \tag{11.38}$$

That is, reduction in the hydrostatic pressure on pure water by π reduces the vapor pressure from P^0_W to P_W and allows pure water to be in equilibrium with the solution when they are separated by a semipermeable membrane. This argument demonstrates that the air–water interface is a semipermeable barrier that functions exactly like a semipermeable membrane. Besides showing again the connection between vapor pressure and osmotic pressure, this strongly suggests that both osmotic pressure and vapor pressure depression have the same origin. Both are due to an apparent reduction in the pressure on the solution side. This interpretation was well known to early investigators of osmosis and is at the root of the controversial proposal that *solvent tension* causes both osmotic pressure and flow and vapor pressure depression (Hammel, 1979; Hildebrand, 1979; Hulett, 1903; Mauro, 1979; Rawlins et al., 1967; Scholander, 1967). One problem with this view is the difficulty in actually carrying out the procedure. If the osmotic pressure is smaller than 1 atm, a negative pressure could be

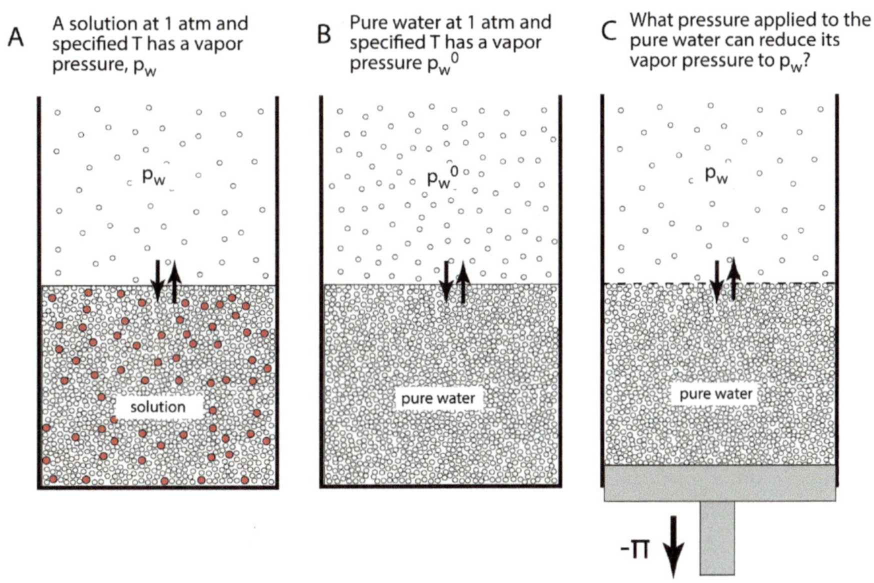

FIGURE 11.7 Hypothetical experiment to determine the pressure required to lower the partial pressure of pure water to that of a solution. The solution at 1 atm pressure has a vapor pressure, P_W, that is lower than that of pure water, P^0_W. If it were possible to apply a negative pressure only to the water phase, what would it be to lower its vapor pressure to P_W? Thermodynamic arguments indicate that the applied pressure would be $-\pi$, where π is the osmotic pressure.

applied to the system, but how can one exert a large negative pressure, greater than 1 atm, only on the water phase? How can it not be communicated to the gas phase where a negative pressure is impossible? Also, there is no evidence of a reduced pressure within a solution that produces the reduced vapor pressure of water. Finally, there seems to be no way to measure the solvent tension, as it is the tension exerted only by the solvent in a mixture with solvent and solute.

11.2 Osmosis is due to an interaction between membrane and solute

Osmosis is an apparent reduction in the pressure on the solution side because water moves toward the solution from pure water. This reduction in pressure is not evident in the bulk phase, but appears only at the membrane and only because of the membrane. It is caused by the solute because pure water on both sides of the membrane produces no net flow. Application of a mechanical pressure on the solution side can stop the osmotic flow, further suggesting that there is a pressure deficit on the solution side. The coefficient relating osmotic pressure to osmotic flow is the same as the filtration coefficient relating hydrostatic pressure to filtration flow, indicating that pressure differences drive osmotic flow. The magnitude of the osmotic pressure, or its derived flow, does not depend on the nature of the solute at all, but only on its concentration (taking into account the osmotic coefficient) and its impermeability to the membrane. Further, if the membrane is permeable to the solute, the observed osmotic pressure is reduced according to the reflection coefficient. The reflection coefficient is an index of the degree of discrimination of the membrane between solvent and solute (Meschia & Setnikar, 1958). Experiments show that the reflection coefficient differs with the size of the solute and with the kind of membrane. All of these findings show that **the origin of the osmotic pressure must be found in the interaction of the membrane with the solute** (Guell & Brenner, 1996). This conclusion is also true for the vapor pressure depression. The origin of the vapor pressure depression is found in the interaction of the solute with the air–water interface.

11.3 Is osmotic flow diffusional or convective?

Thermodynamics is powerful because it is true for all mechanisms and, therefore, is independent of them. Mechanisms must be consistent with thermodynamics, but thermodynamics cannot predict the rates at which processes occur because the rate of a process depends on the mechanism. The thermodynamic derivation of van't Hoff's law sheds no light on its mechanism, and similarly, the statement that osmotic water flow occurs because the free energy of water decreases does not help us understand how osmosis occurs. Historically, two opposing models of osmotic flow have been proposed. These are: (1) a microporous membrane, in which an impermeant substrate is pierced by pores through which material passes by bulk flow—by pressure-driven flow or convection and (2) a dissolution-diffusion model in which water penetrates by dissolving within the membrane and diffusing across. In the microporous membrane, flow may be envisioned as being hydrodynamic or nonhydrodynamic. Finkelstein (1987) argued elegantly that the two mechanisms can be distinguished by the ratio of the **diffusional permeability**, P_d, to the **osmotic permeability**, P_f. If osmotic fluid flow occurs by diffusion, the theoretical prediction is that $P_f/P_d = 1.0$. If it occurs by convection, then $P_f/P_d > 1.0$. The definitions of P_f and P_d and why their ratio is expected to be different for a diffusion mechanism and a convection mechanism are presented in Appendix 1.

11.3.1 P_f / P_d ratio for lipid bilayers is 1.0

A number of investigators determined P_f/P_d for lipid bilayers by using planar lipid bilayers formed over a narrow hole drilled into a thin Teflon partition or over a narrow orifice in a polyethylene tube (see Fig. 11.8). The bilayers so formed contain some remnant of the solvent and, therefore, are never completely comparable to the lipid bilayers found in biological membranes (Fettiplace & Haydon, 1980). P_f typically was measured by a null method by adjusting the volume of the solution on one side of the membrane to prevent bulging of the membrane as determined optically (see Fig. 11.8). P_d was determined by adding ^3H_2O water to one side of the membrane and measuring ^3H in aliquots taken from the opposite side at successive times.

Measurement of both P_f and P_d in lipid bilayer membranes showed P_f/P_d in the range of 4–8 (Hanai et al., 1966; Hanai & Haydon, 1966; Huang & Thompson, 1966). However, these investigators realized that unstirred layers adjacent to the bilayer membrane would affect the experimental determination of P_d much more than P_f, leading to an erroneously high ratio. Cass and Finkelstein (1967) showed that the experimentally determined values of P_d depended on the thickness of the sidewall of the teflon partition, implying that unstirred layers indeed were a significant experimental problem. By correcting for the unstirred layers in various ways, investigators concluded that the P_f/P_d ratio for planar lipid bilayers is indeed 1.0 (Cass & Finkelstein, 1967; Holz & Finkelstein, 1970; Andreoli & Troutman, 1971), indicating that osmotic

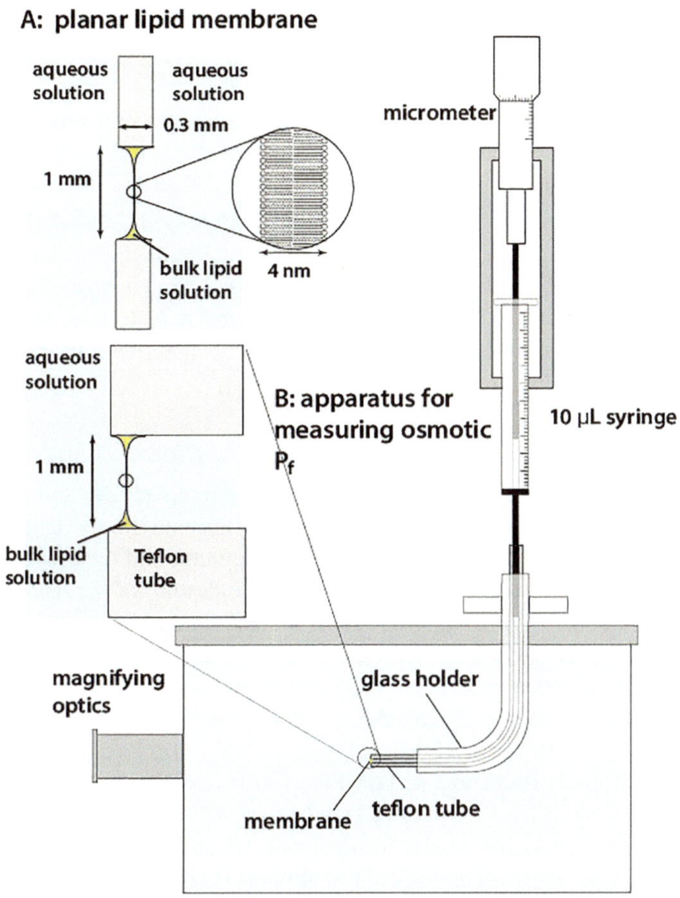

FIGURE 11.8 Method of measuring P_f in planar lipid bilayers. Panel A: Planar lipid bilayers can be made by "painting" a solution of lipids across a narrow orifice such as a hole drilled in a thin Teflon partition or the end of a Teflon tube. The solvent drains away to form an annulus around the perimeter, and a central region that is a bilayer some 3–6 nm thick, depending on the lipids used. Such membranes can be stable for hours but will not withstand large pressure differences. Panel B: The osmotic flow can be measured by addition solute to the outside solution and optically monitoring the inward bulge of the membrane as water leaves the tube compartment. Moving the microliter syringe nulls the bulge in the membrane, and the record of micrometer displacement matches the osmotic flow out of the tube. *Adapted from Fettiplace and Haydon (1980).*

water flow through lipid bilayers occurs by diffusion. This conclusion is strengthened by experiments with pore-forming antibiotics. Insertion of pore-forming antibiotics such as nystatin, amphotericin B, or gramicidin A into planar lipid bilayers should form pores that allow water to move by convection so that P_f/P_d takes on small values of N (3–5) where N is the number of water molecules in the pore (Holz & Finkelstein, 1970; Rosenberg & Finkelstein, 1978). The increase in P_f/P_d to small values of N that was observed when pore-forming antibiotics were added gives confidence in the validity of the result that $P_f/P_d = 1.0$ in lipid bilayers without the antibiotics. The conclusion that $P_f/P_d = 1$ for lipid bilayers is challenged by the report that P_f/P_d in unilammelar liposomes, hollow spheres some 10–25 nm in diameter formed by a single lipid bilayer, exceeds 1.0 (Jansen & Blume, 1995). However, the molecular arrangement of lipids in these highly curved unilamellar liposomes differs from that in biological membranes and distinct mechanisms for the flux of water may occur.

11.3.2 P_f / P_d *ratio for biological membranes exceeds 1.0*

One of the earliest comparisons of osmotic water flow and diffusional water flow in biological systems was provided by Hevesy et al. (1935), who estimated that P_f/P_d averaged 4.5 in frog skin. Prescott and Zeuthen (1953) reported P_f/P_d was 69 in frog eggs. Ussing and Anderson (1956) noted that the equivalent of P_f/P_d in frog skin is three to five, but more importantly, it is increased to as much as 16 when stimulated with antidiuretic hormone. This was early evidence not only of pores within biological membranes but also of their regulation by physiological signals. These authors were fully aware that P_f/P_d in excess of 1.0 implied bulk flow of water through pores in the membrane (Koefoed-Johnsen & Ussing, 1953). Using the frog urinary bladder, Parisi and Bourguet (1983) found that both P_f and P_d were increased with ADH, but P_f/P_d greatly exceeded 1.0, and the $\Delta P_f/\Delta P_d$ caused by ADH was about 9.

Sidel and Solomon (1957) used the time course of volume change with rapid mixing of human red blood cells with anisotonic solutions, determined by light scattering, to calculate that $P_f = 1.28 \times 10^{-2}$ cm/s at 296 K for the red cell membrane. Paganelli and Solomon (1957) used rapid filtration and ^3H_2O equilibration across human red blood cell membranes to determine that $P_d = 5.3 \times 10^{-3}$ cm/s. These values give a P_f/P_d of 2.5, from which it was concluded that erythrocytes have pores with an "equivalent pore radius" of about 0.35 nm. Refinement of the techniques have led to revisions of these numbers: the accepted P_f for human red blood cells has been revised upward to about 2.2×10^{-2} cm/s (Mlekoday et al., 1983; Moura et al., 1984; Rich et al., 1968; Terwilliger & Solomon, 1981), whereas P_d has been revised downward to 2.35×10^{-3} cm/s (Brahm, 1982; Brahm, 2013). The accepted P_f/P_d for human red blood cells is thus about 9, strongly suggesting that water traverses the membrane through a convective mechanism using some kind of pore.

11.3.3 Low activation energy of P_f *suggests a convection mechanism*

Diffusion depends on molecular motion, which in turn depends on the temperature, and so we expect diffusion to depend on temperature (Chapter 9). Diffusion can be analyzed in terms of Eyring rate theory, using jumps and energy barriers as a basis. In the bulk phase, water has only small energy barriers for movement, and therefore, the diffusion or bulk water flow should have a relatively low slope on an Arrhenius plot (a plot of the logarithm of the rate constant against $1/T$, where T is the absolute temperature, giving a slope $= E_a/RT$ where E_a is the activation energy, which is a measure of the size of the energy barriers). The self-diffusion coefficient of water has an E_a of about 4.8 kcal/mol (Wang, 1965). Diffusion through a lipid bilayer, on the other hand, encompasses larger energy barriers at both lipid—water boundaries and smaller energy barriers in the lipid phase (Issack & Peslherbe, 2015). Accordingly, the E_a for P_d or P_f through the lipid bilayer is much larger and depends on the lipids used to form the bilayer: egg lecithin (12.7 kcal/mol; Price & Thompson, 1969); egg lecithin and cholesterol (14.6 kcal/mol; Redwood & Haydon, 1969); oxidized cholesterol (6.8 kcal/mol; Tien & Ting, 1968); egg lecithin vesicles (8.3—9.8 kcal/mol; Reeves & Dowben, 1970). The activation energy for P_d for dog and human red blood cells was 4.9 and 6.0 kcal/mol, respectively; for P_f it was 3.7 and 3.3 kcal/mol, respectively (Vieira et al., 1970). Thus, the activation energy of flow across most biological membranes is consistent with bulk phase water movement rather than diffusion through the lipid bilayer.

12. Discovery of water channels

Macey and Farmer (1970) found that sulfhydryl reagents p-chlorormercuribenzoate (PCMB) and p-chloromercuribenzenesulfonate (PCMBS) reversibly inhibit water permeability of red blood cells. With saturating levels of PCMBS (about 2 mM), P_f was reduced by a factor of 10 while P_d was decreased by less than a factor of 2; the ratio P_f/P_d became 1.0, and E_a increased from 4.8 to 11.5 kcal/mol. In short, sulfhydryl reagents make the water transport in red blood cells lose all of the properties that distinguishes it from water transport across lipid bilayers. This effect could be reversed by cysteine, suggesting that some protein in the membrane was responsible for water transport. Investigators attempted to identify the protein by linking it to PCMBS made with radioactive Hg (Benga et al., 1986; Solomon et al., 1983) and then separating the proteins by sodium dodecyl sulfate polyacrylamide gel electrophoresis (SDS-PAGE). However, there are a lot of sulfhydryl groups in a variety of different proteins in the red blood cell membrane, and labeling with ^{203}Hg-PCMBS does not label only the water channel. Solomon's and Begna's groups enhanced the likelihood of success by pretreating the membranes with N-ethylmaleimide (NEM), a sulfhydryl reagent that earlier was shown to not block the channel, presumably because NEM attacks sulfhydryl groups exposed to the water phase rather than deep within the membrane (Benga et al., 1986). Both groups identified two bands, one of 95 kDa (band 3) and another of 55 kDa (band 4.5), as potential water channels (Benga et al., 1986; Solomon et al., 1983). The identity of the channel was solved when Agre and coworkers (Preston et al., 1992) injected mRNA that coded for red cell CHIP28 protein into *Xenopus laevis* oocytes and found a marked increase in osmotic water permeability. The increased P_f was inhibited by mercurial reagents, and the inhibition was reversed by a sulfhydryl reducing agent, β-mercaptoethanol.

CHIP28 previously had been isolated from red blood cells as part of an effort to identify Rh factors (Agre et al., 1987). Antibodies to the purified 28 kDa protein cross-reacted with material on SDS-PAGE with M_r between 35 and 60 kDa, which was shown to be an N-glycosylated form of the protein (Denker et al., 1988). Both the native form and glycosylated forms interact in the membrane to form a tetramer, and it has high sequence homology with MIP26, the **major intrinsic** protein of the lens (Smith & Agre, 1991). The apical membrane of renal tubules, known to have high water permeability, reacts with antibodies to the 28 kDa protein (Denker et al., 1988). Isolation of the cDNA and search of the DNA sequence data base led to the recognition that this protein was an integral protein related to a family of membrane channels. Preston and Agre (1991) named it "CHIP28" for "**ch**annel-like **i**ntegral membrane **p**rotein of **28** kDa".

Following expression of CHIP28 in Xenopus oocytes, two groups (Van Hoek & Verkman, 1992; Zeidel et al., 1992) nearly simultaneously incorporated purified CHIP28 into liposomes and found that it increased P_f up to 50 times that of control liposomes, with a much lower activation energy (3.4 kcal/mol or 3.6 kcal/mol) than for liposomes without the protein (16.0 kcal/mol). Importantly, CHIP28 did not alter the permeability of the liposomes to H^+ ions (Zeidel et al., 1992) or urea (Van Hoek & Verkman, 1992; Zeidel et al., 1992). Both groups showed reversible inhibition of P_f by $HgCl_2$. These results demonstrated conclusively that the CHIP28 protein is a water channel.

Nearly simultaneously, Sasaki and coworkers used primers constructed from the MIP family to clone complementary DNA for "WCH-CD" (for **w**ater **ch**annel of the **c**ollecting **d**uct) (Fushimi et al., 1993). The protein was localized to the apical and subapical regions of principal cells of the collecting duct, distinct from the localization of CHIP28 in kidney, and encodes a M_r 28,900 protein with 43% sequence homology to CHIP28% and 60% to MIP. Injection of WCH-CD mRNA into *Xenopus laevis* oocytes increased P_f. The increased water permeability was sensitive to $HgCl_2$, and the inhibition was reversed by β-mercaptoethanol. The activation energy of the WCH-CD mediated osmotic permeability was 3.6 kcal/mol. These biophysical characteristics are analogous to those of CHIP28 and indicate that renal collecting ducts contain a water channel distinct from CHIP28.

Chrispeel and coworkers (Maurel et al., 1993) compared the responses to expression in *Xenopus* oocytes of three homologs of MIP: a 27 kDa membrane protein named γ-TIP (tonoplast intrinsic protein) that is abundantly expressed in vacuoles of higher plants; GLpF, a 25 kDa protein that facilitates glycerol transport in *E. coli*; and CHIP28. They found that expression of γ-TIP or CHIP28 increased the osmotic water permeability of the oocytes, but expression of GlpF did not. In contrast, GlpF increased ^{14}C-glycerol uptake into the transfected oocytes, whereas CHIP28 or γ-TIP did not. Increased osmotic water permeability in oocytes expressing γ-TIP was inhibited by $HgCl_2$, and the authors concluded that γ-TIP forms a water channel similar to CHIP28. But what about bacterial GLpF? Is it fundamentally different than CHIP28 and γ-TIP? Another mammalian MIP homolog expressed in basolateral membrane of renal collecting duct, intestine, colon and other locations was identified a few years later (Ishibashi et al., 1994). Although predominantly permeant to water, this protein (later named AQP3, see below), like bacterial GLpF, also facilitates the flux of glycerol as well as urea.

In 1993 Agre, Sasaki, and Chrispeels together proposed that proteins that increase water permeability but do not conduct ions should be called *aquaporins*. At that time, there were three members of the group: CHIP28, WCH-CD, and γ-TIP. CHIP was renamed AQP1 and is found in erythrocytes, in proximal tubules, descending limb of the loop of Henle, and descending portion of the vasa recta in kidney. The collecting duct water channel, WCH-CD, was renamed AQP2 and notably is regulated by vasopressin, and γ-TIP was renamed AQP. At1, as it was identified in the vacuolar membrane (tonoplast) of *Arabidopsis thaliana*.

13. The mosaic model of water transport

13.1 Water transport has two main parallel and independent routes

The experiments that revealed that CHIP28, WCH-CD, and γ-TIP were water channels were possible because *Xenopus laevis* oocytes are relatively impermeant to water—they lack water channels. Therefore, addition of water channels to their plasma membranes produces a marked increase in the osmotic water permeability. This fact illuminates an important concept: water permeability can be regulated by the number of channels in the membrane, and membranes without water channels are generally poorly permeable to water but not completely impermeant. One can infer that there must be at least two routes and mechanisms of water transport across cell membranes: one through the lipid bilayer and one through water channels. The mechanisms appear to be different. Permeation through the lipid bilayer appears to be diffusive, with $P_f/P_d = 1.0$, $E_a \approx 10-16$ kcal/mol and insensitive to sulfhydryl reagents. Permeation through the water channels appears to be hydrodynamic, or approximately so, with $P_f/P_d >> 1.0$, $E_a \approx 4$ kcal/mol and sensitive to sulfhydryl reagents. The situation is shown in Fig. 11.9.

The two pathways are arranged in parallel, so that the total flux across the membrane is the sum of them, and thus the filtration and diffusion constants are sum of the contributions of each:

$$\begin{aligned} P_{f,membrane} &= P_{f,pores} + P_{f,lipid} \\ P_{d,membrane} &= P_{d,pores} + P_{d,lipid} \end{aligned} \tag{11.39}$$

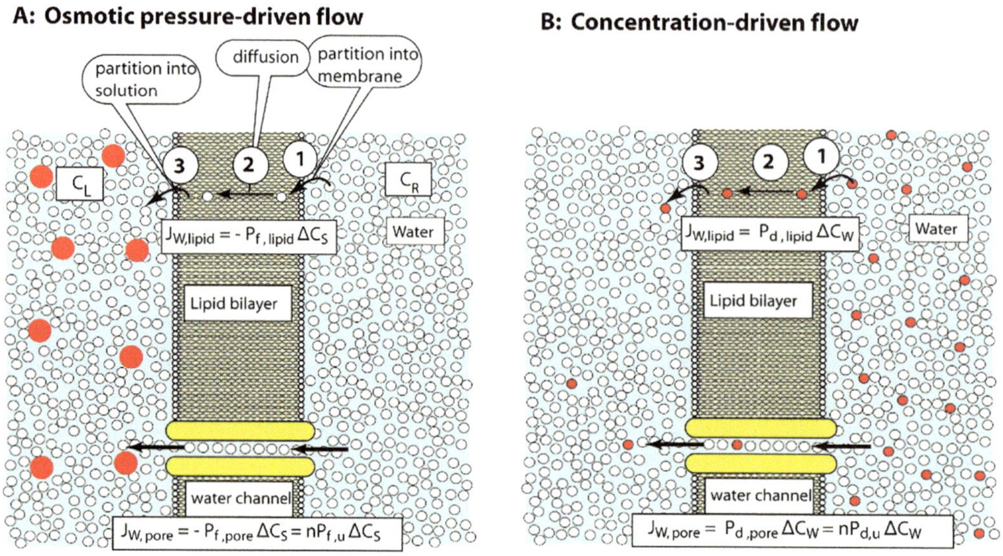

FIGURE 11.9 Osmotic and concentration-driven flow across biological membranes. Panel A: Osmotic flow and concentration flow occurs through the lipid bilayer and in parallel through aquaporins, protein pores inserted into the membrane. The flux of water is given by the equations shown in the figure. Larger filled circles on the left side indicate solute molecules. Small white circles indicate water molecules. Panel B: Concentration-driven flow is typically assessed by adding labeled water, using ^3H_2O for example, and measuring its appearance on the opposite side. Filled small circles indicate labeled water. The coefficient for osmotic and concentration-driven flow is the same for a diffusional mechanism (the lipid phase), but the osmotic flow is greater for pores in which bulk flow of water occurs. Inactivation of the pores thus reduces the osmotic flow much more than it reduces the diffusive flow.

where the subscript "membrane" indicates the entire membrane, "pore" indicates the contribution due to the pore, and "lipid" denotes the contribution of the lipid bilayer. Since PCMBS closes the pores in an all-or-none fashion and does not affect the lipid pathway, titration of the pores with PCMBS allows us to determine both contributions. The derived values for human erythrocytes are given in Table 11.3 (Moura et al., 1984). From these it can be seen that closing the channels by reacting with PCMBS should reduce P_d of the composite membrane about 50%, whereas P_f should be reduced by 90%, and the ratio P_f/P_d should be reduced from 5.8 to 1.0. The data also show that the ratio P_f/P_d for the channel alone is much higher than the ratio determined from the whole membrane.

13.2 How many channels are there in the erythrocyte membrane?

The values of P_f, P_d and P_f/P_d for the composite membrane depend on both the channel values for these parameters and the number of channels in the membrane. Solomon et al. (1983) calculated the number of pores using a variant of Eq. (A.1.14)

$$P_d = \frac{n\pi r^2 D_W}{\delta} A_{sd}/A_p \tag{11.40}$$

TABLE 11.3 Contribution of the lipid bilayer and pores to the osmotic and diffusional permeability of human red blood cells.

	Permeability (10^{-3} cm/s)		P_f/P_d
Source	Filtration	Diffusion	
Whole membrane	**21.5**	**3.67**	**5.85**
Pores	**19.7**	**1.86**	**10.6**
Lipid bilayer	**1.81**	**1.81**	**1.0**

Values from Moura et al. (1984).

where n is the number of pores per unit area of membrane, r is its radius, D_W is the diffusion coefficient for water, δ is the thickness of the membrane (i.e., the length of the diffusion path), and A_{sd}/A_p is an estimate of steric hindrance (calculated from experimental data to be 0.064). The number of pores is given by the area of the erythrocyte membrane (1.35×10^{-6} cm^2; Solomon, 1989) times n. They calculated that there are 2.7×10^5 pores in the membrane of a single erythrocyte, or about 2000 channels per μm^2. Quantitative immunoblots indicated there are about 1.4×10^5 copies of AQP1 per erythrocyte (Denker et al., 1988) or about 1040 per μm^2. Considering the uncertainties in many of the factors of Eq. (11.40), this level of agreement could be regarded as being either remarkable or fortuitous.

13.3 What is the single-channel permeability of AQP1?

The unitary osmotic permeability, or single-channel permeability, of an aquaporin can be determined as

$$P_{f,u} = \frac{P_{f,pores}}{n} \tag{11.41}$$

where $P_{f,u}$ is the single-channel osmotic permeability, in cm^3/s, $P_{f,pores}$ is the osmotic or pressure filtration coefficient of the aggregate pores, and n is the pore density (number/cm^2). Zeidel et al. (1992) estimated the number of copies of CHIP28 (i.e., AQP1) incorporated into liposomes from the total protein divided by the molecular weight (28,500 Da). From the measured P_f of the liposomes, they calculated $P_{f,u} = 1.17 \times 10^{-13}$cm^3/s at an unspecified temperature. Using $P_{f,pores} = 1.97 \times 10^{-2}$ cm/s (at 23°C) for the erythrocyte from Table 11.3, and $n = 1.04 \times 10^{11}$ cm^{-2} from Denker et al. (1988), we calculate $P_{f,u} = 1.89 \times 10^{-13}$ cm^3/s, which agrees reasonably well with the earlier calculation. Van Hoek and Verkman (1992) found $P_{f,u} = 0.68 \times 10^{-13}$ cm^3/s (at 10°C). They extrapolated this to be about 1.0×10^{-13} cm^3/s at 23 °C, close to that estimated by Zeidel et al. (1992). The values of $P_{f,u}$ for AQP4 incorporated into liposomes was in the range of the values calculated here, but varied from 0.9 to 3.5×10^{-13} cm^3/s with the kinds of lipids used in the liposomes (Tong et al., 2016).

13.4 The mosaic model of osmosis

The mosaic model proposes that there are at least two mechanisms for osmosis: osmosis occurs through the lipid bilayer part of the membrane by diffusion and through the water channels by a mechanism that is either hydrodynamic or single-file or something else. It is instructive to compare the rates of flow through these components.

The ***net*** osmotic flux of water through the lipid portion of membranes depends on its driving force. It is given from Eq. (A.1.3) as

$$J_{W,lipid} = \frac{Q_W}{A} = -P_{f,lipid}\Delta C_S \tag{11.42}$$

Suppose we investigate a relatively small osmotic difference of 10 mOsM so that $\Delta C_S = 0.01$ mol/L. Using $P_{f,lipid} = 1.81 \times 10^{-3}$ cm/s for erythrocytes (Table 11.3), we calculate that $J_{W,lipid} = 1.81 \times 10^{-8}$ mol/cm^2/s. Given that each phospholipid typically occupies an area of 0.7 nm^2, we can calculate that each phospholipid molecule on average mediates a net flux of about 76 molecules of water per s. The corresponding net flux through a single AQP1 pore in the erythrocyte membrane can be calculated by

$$\begin{aligned} J_{W,pores} &= \frac{Q_W}{A} = -P_{f,pores}\Delta C_S = -nP_{f,u}\Delta C_S \\ \frac{J_{W,pores}}{n} &= q_W = -P_{f,u}\Delta C_S \end{aligned} \tag{11.43}$$

where q_W is the flow of water through a single pore. Using $P_{f,u} = 1.17 \times 10^{-13}$cm^3/s and $\Delta C_S = 0.01$ mol/L, we can calculate that $q_W = 1.17 \times 10^{-18}$ mol/s or 700,000 water molecules per s. Flux through the water channel for a modest osmotic difference is about 10,000-times faster than around a single phospholipid. However, there are a lot more phospholipid molecules than AQP1 in the erythrocyte membrane: one leaflet contains about 1.43×10^6 phospholipid molecules per μm^2 and only about 1500 or so copies of AQP1. Thus, because there are about 1000 phospholipid molecules in each of the half bilayer for every AQP1 molecule, and AQP1 carries about 10,000 times as many water molecules, AQP1 increases the osmotic permeability of the erythrocyte membrane about 10-fold. In this case about 90% of the net water transport goes through the pores. These figures agree with the degree of inhibition of water flux across red blood cell membranes by mercurial reagents.

The numbers calculated here are for the *net* transport of water in response to a modest osmotic difference. The *unidirectional* water transport is much higher (about 10^9 molecules s^{-1} for a single pore) in both directions, and the net transport is just the difference between the two unidirectional fluxes.

14. The aquaporin family portrait

The family of membrane intrinsic proteins (MIPs), also known as aquaporins (Chapter 19 in this volume), so far has nearly 1700 members identified from publicly available genomes and are present in virtually every living organism including bacteria, plants, invertebrates, and vertebrates (Abascal et al., 2014). The human genome contains 13 of these, named AQP0 through AQP12. Based on their sequences, these have been divided into three subfamilies. Classical or orthodox aquaporins (AQP0, AQP1, AQP2, AQP4, AQP5, AQP6, and AQP8) primarily are permeable to water; the aquaglyceroporins (AQP3, AQP7, AQP9, and AQP10) are permeable to small neutral solutes such as glycerol; and unorthodox aquaporins are intracellular with as yet incompletely described permeabilities (Direito et al., 2016) (see Fig. 11.10). Aquaporins play important roles in transepithelial transport, brain edema, cell migration, cell proliferation, cell differentiation, cell adhesion, and metabolism (reviewed in Direito et al., 2016).

The structures of several of the aquaporins (Murata et al., 2000; Horsefield et al., 2008; Harries et al., 2004) have revealed a common theme: aquaporins have molecular weights of about 28 kDa and form homotetramers in the membrane, and one or more may be glycosylated, but each protomer has its own pore. Each protomer consists of six highly conserved transmembrane helices, and two helices that insert from opposite sides of the membrane form a seventh pseudo-transmembrane helix. These place the highly conserved asparagine-proline-alanine (NPA) motifs (two per protomer) near the center of the water channel. The electrostatic potential barrier that peaks at the NPA region appears to be responsible for aquaporin's impermeability to H^+ ions (De Groot and Grubmuller, 2005). Fig. 11.11 shows the structure of AQP1 and the arrangement of the helices in the protomer with respect to the primary amino acid sequence. The proposed disposition of water through the channel is shown in Fig. 11.12. The water is believed to pass through the pore with six or 7 water molecules arranged in a single file (Binesh & Kamali, 2015).

15. Water transport through other transporters

A growing body of evidence suggests that a number of transporters, unrelated to the aquaporins, can also increase the passive permeability when expressed in *Xenopus* oocytes. Examples include the urea transporter, UT-B (Yang & Verkman, 1998, 2002); the glucose transporter, GLUT1 (Fischbarg et al., 1990); the sodium-linked glucose transporter, SGLT1 (Loo et al., 1996); the neuronal glutamate transporter, EAAT1 (Vandenberg et al., 2011); the GABA re-uptake transporter, GAT1 (Santacroce et al., 2010); and the Na^+-K^+-$2Cl^-$ cotransporter type 1 (NKCC1) but not type 2 (NKCC2), as discussed in Chapter 16, and in Hamann et al. (2005). These molecules are proposed to cotransport water as part of their transport mechanism or to incorporate parallel

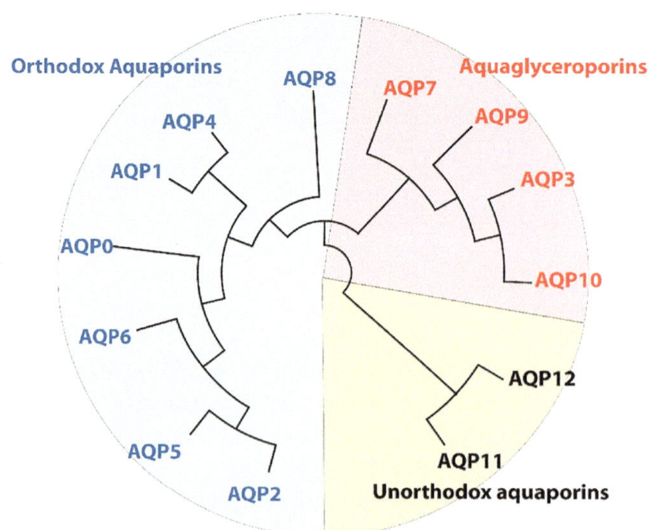

FIGURE 11.10 Dendrogram of the phylogenetic relationship of the mammalian aquaporins, showing three main types of aquaporins. *Redrawn from Direito et al. (2016).*

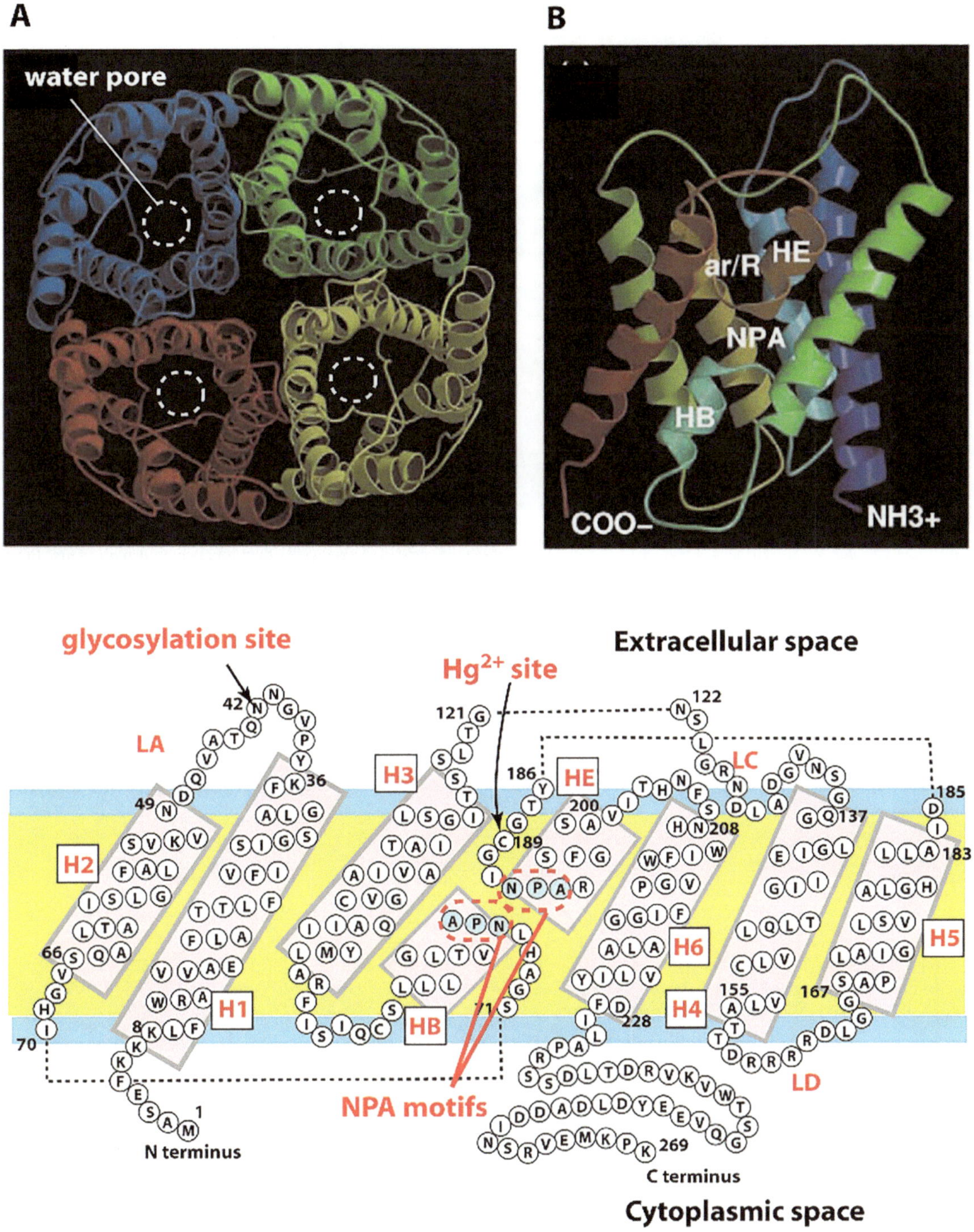

FIGURE 11.11 Panel A: Structure of AQP1 as an archetypical AQP structure. AQP1 associates as a homotetramer, and each protomer has its own water channel, as shown. Panel B: Viewed from the side, six transmembrane helical sections are apparent, with a pseudo seventh provided by half sections (HE and HB) that enter from opposite sides of the membrane. Panel C: The amino acid sequence is shown with the sites for glycosylation and inhibition with mercurial reagents. The six transmembrane helices are identified (H1–H6) along with the two half helices (HE and HB) that carry the NPA motif in each half transmembrane segment. The pair of overlapping NPA motifs create the selectivity filter. *Panels A and B are reproduced by permission from De Groot and Grubmuller (2005). Panel C is redrawn from Murata et al. (2000).*

pathways that allow passive water transport in response to local osmolyte transport (Zeuthen, 2010). Some water cotransporters are proposed to be able to pump water against an osmotic gradient (Zeuthen & MacAulay, 2012). The proposed stoichiometry of transport ranges from 0 (NKCC2) to 590 water molecules per turnover (Zeuthen, 2010). There are at least three separate possible

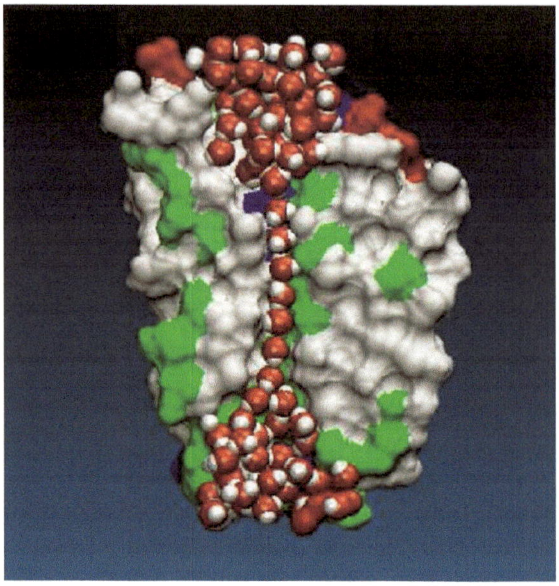

FIGURE 11.12 Single file of water molecules is shown in the center of the water pore of a monomer of AQP2. The protein and water molecules are represented by their van der Waals surfaces. Oxygen of the water molecules is shaded darker (red) than small bound hydrogens. Protein shading is according to residue type. *Reproduced by permission from Binesh and Kamali (2015).*

explanations for this phenomenon (see Fig. 11.13). First, water could be cotransported by an alternate-accessibility model in which water along with transported solute is occluded in the reaction mechanism of the transporter. Transport of the solute is accompanied by cotransport of solvent water in which one conformation of the transporter essentially squeezes out the water to follow the solute. In a second kind of process, water could follow solute by the solute's local osmotic effect within the transporter

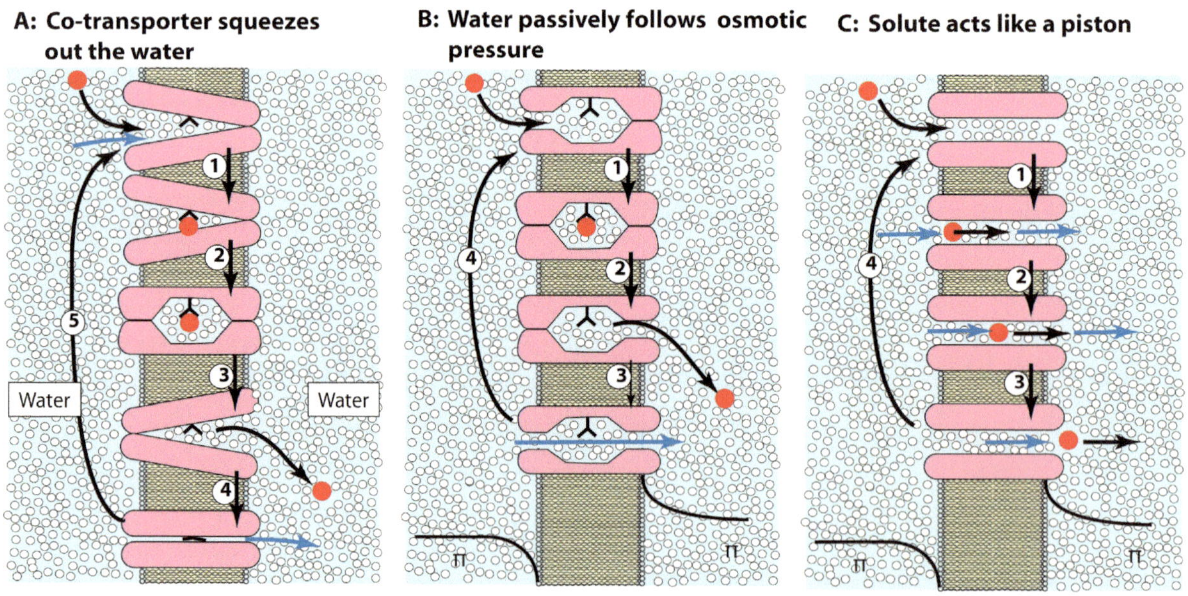

FIGURE 11.13 Possible mechanisms of water cotransport. Panel A: Solute is transported by an alternate accessibility model. In step 1, solute binds in a pocket surrounded by water molecules. In step 2, the solute and water is occluded within the transporter. In step 3, the solute is delivered to the opposite side, but water cannot follow because it would leave behind a vacuum. In step 4, transition of the transporter to the initial state squeezes the water out of the occluded space. In step 5, opening of the channel draws water in because of the vacuum transiently created. Panel B: This method has features of the model in Panel A, except a physical squeezing out of the water is not required. Instead, desorption of the solute to the opposite side creates a local osmotic gradient that water can follow due to opening of a channel to the left compartment. The transport creates a drop in the concentration of solute on the input side and an increase in the concentration on the output side and water follows passively. Panel C: The solute is transported through a channel that is just large enough to receive the solute but does not allow water to pass the solute molecule. As the solute travels down the pore, it pushes water ahead, and water collects behind the solute due to the space left when the solute moves. The solute molecule acts like a piston, driving a volume of water across the membrane.

itself. In this case, it would be expected that a passive permeability pathway might be available even in the absence of transport. Third, the solute molecule might prevent water passing by it, so that transport through a pore in the transporter pushes water ahead and makes water follow in the space left behind: the solute acts like a piston. Transport of osmolytes could also engage water transport through parallel pathways, to follow the local osmotic pressure, but this is not what is usually meant by water cotransport.

It has been difficult to distinguish among these different mechanisms, and the field remains controversial (Lapointe et al., 2002). Structural evidence suggests that there may be water cotransported with solutes but not at the high stoichiometry that has been proposed (Sasseville et al., 2011). Although investigators believe that water can cross membranes using SGLT1 in a passive manner that is inhibited by SGLT1 inhibitors, it is not clear if such routes are physiologically relevant. It is likely that the bulk of the enormous volume of water transported in the kidney (170 L/day) does not occur via cotransport mechanisms (Mollajew et al., 2010), whereas the contribution of cotransport mechanisms to water absorption in the intestine may be much more important (Zeuthen et al., 2016). The small intestine absorbs some 5 L of fluid per day, and this absorption is tightly coupled to solute absorption, particularly Na^+. Some of this Na^+ crosses the apical membrane of enterocytes by SGLT1 with cotransport of glucose (2 Na^+ per glucose). Glucose that enters the cell leaves by GLUT2-facilitated diffusion, and Na^+ leaves via the basolateral Na^+/K^+-ATPase (Augustin, 2010). The normal human diet contains about 300 g of carbohydrates daily, or nearly 2 mol of monosaccharides. Cotransport of 100 molecules of water per monosaccharide would allow the absorption of nearly 200 mol of water, or 3.6 L. This is a major fraction of water absorption by the intestine. Molecular dynamic simulations suggest that the SGLT1 has a water-filled pathway containing about 100 water molecules (Sasseville et al., 2011). The cotransport of Na^+ and glucose and their effect on water absorption from the intestine is the basis for oral rehydration therapy and the rescue of persons with life-threatening acute diarrhea (Binder et al., 2014). On the other hand, at 100 mg% plasma levels, the filtered load of glucose by the kidneys, assuming a glomerular filtration rate (GFR) of 120 mL /min, is about 1 mol per day. Consequently, water-cotransport by this mechanism would account for only about 2 L of the about 170 L reabsorbed by the kidneys each day.

Appendix 1

A.1.1 Definitions of P_f and P_d

The diffusional permeability is defined as

$$Q_W = AP_d\Delta C_W \tag{A.1.1}$$

where Q_W is the flow of water, in mole s^{-1}, A is the area of the membrane through which flow occurs; and ΔC_W is the difference in water concentration that drives diffusive flow. If A is in units of cm^2 and ΔC_W is in mole cm^{-3}, then P_d is in units of cm s^{-1}, a typical unit for permeability of solutes. The proportionality constant between bulk volume flow and pressure is given as L_P in Eqs. (11.25)–(11.27). Its units differ from P_d and cannot be directly compared. We can rewrite Eq. (11.27) in terms of ΔC:

$$\begin{aligned} Q_V &= AL_P(\Delta P - \Delta\pi) = AL_P(\Delta P - RT\Delta C_S) \\ Q_W &= \frac{Q_V}{\overline{V_W}} = \frac{AL_PRT}{\overline{V_W}}\left(\frac{\Delta P}{RT} - \Delta C_S\right) \end{aligned} \tag{A.1.2}$$

where the volume flow, Q_V, in cm^3/s, can be converted to the molar flow, Q_W, in mole s^{-1} by dividing by the partial molar volume, $\overline{V_W}$. We can rewrite Eq. A.1.2 as

$$Q_W = AP_f\left(\frac{\Delta P}{RT} - \Delta C_S\right) \tag{A.1.3}$$

Note that $\Delta P/RT$ has units of concentration. If $\Delta C_S = 0$, we can write this as

$$Q_W = AP_f\left(\frac{\Delta P}{RT}\right) \tag{A.1.4}$$

which has the same form at Eq. (A.1.1). Comparison of Eqs. (A.1.2) and (A.1.3) allows us to conclude

$$P_f = \frac{L_PRT}{\overline{V_W}} \tag{A.1.5}$$

A.1.2 P_f/P_d for microporous membranes

Here we consider a microporous membrane that is a flat thin sheet of thickness δ that is pierced by right cylindrical pores of radius r with N pores per area A, giving a density of pores $n = N/A$. For the moment, we will assume that the pores are large enough and the thickness of the membrane is great enough that laminar flow occurs. Under these conditions, flow in response to a pressure difference is given by the Poiseuille equation. The flow through a single pore will be

$$q_V = \frac{\pi r^4}{8\eta\delta}\Delta P \qquad \text{(A.1.6)}$$

where q_V is the flow through a single pore, in cm^3/s, η is the viscosity of the flowing fluid, δ is the thickness of the membrane, which is taken to be the length of the pore, and ΔP is the difference in pressure across the pore. The π in the numerator is the geometric ratio and should not be confused with the symbol for osmotic pressure. The overall macroscopic volume flow through an area, A, of the membrane would be the flow through each pore times the number of pores in the area:

$$Q_V = Nq_V = \frac{N}{A}A\frac{\pi r^4}{8\eta\delta}\Delta P = A\frac{n\pi r^4}{8\eta\delta}\Delta P \qquad \text{(A.1.7)}$$

Comparison of this result with Eq. (11.25) allows us to identify L_P as

$$L_P = \frac{n\pi r^4}{8\eta\delta} \qquad \text{(A.1.8)}$$

Thus, L_P is a parameter of the membrane that depends on characteristics of the membrane and the fluid: the density of its pores, n; the radius of the pores, r; the viscosity of the medium, η; and the length of the pores, δ. Using Eq. (A.1.5), we identify P_f as

$$P_f = \frac{n\pi r^4 RT}{8\eta\delta\overline{V_W}} \qquad \text{(A.1.9)}$$

Since the same L_P pertains to both osmotic flow and pressure-driven flow, the same P_f would also pertain to osmotic flow as pressure-driven flow. This implies that a real hydrostatic pressure within the pore drives osmotic pressure-induced flow, with the pressure being lower on the solution side (Dainty, 1965; Mauro, 1957; Ray, 1960). In this view, P_f would be the same for osmotic flow as for pressure-driven flow.

In the absence of a pressure gradient and bulk flow, solvent and solute will cross a microporous membrane by diffusing through the pores. If we assume that the membrane substrate is impermeable except at the pores, the flux of solute through a single pore should be given by Fick's first and second laws of diffusion:

$$\begin{aligned} j_S &= -D\frac{\partial C}{\partial x} \\ \frac{\partial C}{\partial t} &= D\frac{\partial^2 C}{\partial x^2} \end{aligned} \qquad \text{(A.1.10)}$$

where j_S is the flux of solute through a single pore. The lower expression describes the time dependence of the concentration profile over the distance, x, within the pore. At steady-state, the concentration profile no longer changes: $\partial C/\partial t = D\partial^2 C/\partial x^2 = 0$. This means that the concentration gradient, $\partial C/\partial x$, is constant and C is a linear function of x: $\partial C/\partial x = (C_L - C_R)/(0 - \delta) = -\Delta C/\delta$, where δ is the thickness of the membrane. Thus, the flux through a single pore is

$$j_S = D\frac{\Delta C}{\delta} \qquad \text{(A.1.11)}$$

The flow through the single pore is its flux times the area of the pore. The flow across the macroscopic membrane is just the flow through a single pore times the number of pores. We write

$$\begin{aligned} Q_S &= Nq_S = A\frac{N}{A}q_S = An\pi r^2 D\frac{\Delta C}{\delta} \\ Q_S &= AP_d\Delta C \end{aligned} \qquad \text{(A.1.12)}$$

Here A is the macroscopic area of the membrane and the area of the pore is πr^2. According to these results, the diffusional permeability, P_d, for a solute is

$$P_d = \frac{n\pi r^2 D}{\delta} \tag{A.1.13}$$

If pure water is on both sides of the membrane, $\Delta C_W = 0$ and the net diffusive flow is zero. However, we can measure unidirectional flows by tagging one side of the compartment with isotopic water. The isotopic water should obey these equations, and the diffusional permeability of water would be

$$P_d = \frac{n\pi r^2 D_W}{\delta} \tag{A.1.14}$$

where n is the number of pores per unit area, r is the radius of the right cylindrical pore, D_W is the diffusion coefficient of water, and δ is the thickness of the membrane, taken to be equal to the length of the pore. The ratio P_f/P_d from Eq. (A.1.9) and Eq. (A.1.14) is predicted to be

$$\frac{P_f}{P_d} = \frac{r^2 RT}{8\eta \overline{V_W} D_W} \tag{A.1.15}$$

Inserting approximate values for all but r ($R = 8.314$ N m/mol/K, $T = 298$ K; $\eta = 1 \times 10^{-3}$ N/m^2 s; $\overline{V_W} = 18 \times 10^{-6}$ m^3/mol and $D_W = 2.3 \times 10^{-9}$ m^2/s), we calculate from Eq. (A.1.15)

$$\frac{P_f}{P_d} = 7.48 \times 10^{18}\, m^{-2}\, r^2 \tag{A.1.16}$$

For all values of $r \geq 0.37$ nm, which is a little more than twice the size of a single water molecule, P_f/P_d exceeds 1.0. However, if we set $r = 0.15$ nm, the approximate size of a water molecule, the value of P_f/P_d is *less* than 1.0. This is impossible, because flow can never be lower than the diffusive flow. A reason for this discrepancy for small values of r is that the hydrodynamic derivation of Poiseuille's law no longer applies, and a new derivation must be sought. For pores for which Poiseuille's law holds, separate experimental determinations of P_f and P_d could be used to calculate the "equivalent pore radius" for microporous membranes. The validity of hydrodynamic flow in small pores, which are still larger than a single water molecule has been both challenged (Mikulecky, 1972) and supported (Levitt, 1973).

A.1.3 P_f / P_d for lipid bilayer membranes

One model of water penetration of a lipid bilayer membrane supposes that the water dissolves in the lipid phase on one side of the membrane, diffuses through the lipid interior, and then equilibrates out of the lipid phase into the water phase on the other side. The supposition is that equilibration between the aqueous phase and the lipid phase at the left and right boundaries is rapid compared to the diffusion of water through the lipid phase. Equilibration between the aqueous phase and lipid phase is described by equating the chemical potential of water in the two phases:

$$\mu^0_{W(solution)} + RT \ln X_{W(solution)} + P\overline{V_W}_{(solution)} = \mu^0_{W(membrane)} + RT \ln X_{W(membrane)} + P\overline{V_W}_{(membrane)} \tag{A.1.17}$$

The *partition coefficient* describes the equilibrium distribution of water between the solution and membrane phase. It is given as

$$K_W = \frac{X_{W(membrane)}}{X_{W(solution)}} = e^{\frac{\mu^0_{W(solution)} - \mu^0_{W(membane)} + \Delta P\overline{V_W}}{RT}} \tag{A.1.18}$$

Here we consider the case where only osmotic pressure drives water flow and the hydrostatic pressure difference across the membrane, ΔP, is zero. Since water partitions poorly into hydrocarbon solvents, we may assume that the mole fraction of water in the membrane phase is low. Thus, the concentration of water in the membrane phase is dilute, and we may write:

$$\begin{aligned} K_W X_{W(solution)} &= X_{W(membrane)} = \frac{n_W}{n_W + n_{lipid}} \approx \frac{n_W}{n_{lipid}} = \frac{n_W \overline{V}_{lipid}}{n_{lipid}\overline{V}_{lipid}} = \overline{V}_{lipid}\frac{n_W}{V} \\ K_W \frac{X_{W(solution)}}{\overline{V}_{lipid}} &= C_{W(membrane)} \end{aligned} \tag{A.1.19}$$

where $\overline{V}_{lipid}$ is the partial molar volume of lipid in the membrane. If the solution is dilute, we may replace its mole fraction, $X_{W(solution)}$, with $\left(1 - \overline{V_W}C_S\right)$ to obtain

$$C_{W(membrane)} = \frac{K_W}{\overline{V}_{lipid}}\left(1 - \overline{V_W}C_S\right) \tag{A.1.20}$$

This expresses the concentration of water within the lipid bilayer immediately adjacent to the solution in terms of the concentration of *solute* in the solution in equilibrium with the membrane water. If partitioning of water is fast compared to diffusion, such an equation would be valid on either side of the membrane. If the concentration of solute on the left is not the same as the concentration on the right, $C_{S,L} \neq C_{S,R}$, then we would expect a concentration difference in water in the membrane, and a net diffusive flux given by Fick's first law of diffusion:

$$J_W = -D_W^m\frac{(C_{W,L} - C_{W,R})}{(0 - \delta)} \tag{A.1.21}$$

where J_W is the flux of water in mol/cm^2/s, D_W^m is the diffusion coefficient of water within the lipid bilayer, and δ is the thickness of the membrane. Inserting the results from Eq. (A.1.20), we find

$$J_W = -D_W^m\frac{K_W\overline{V_W}(C_{S,L} - C_{S,R})}{\overline{V}_{lipid}\delta} \tag{A.1.22}$$

The water flow, in mole/s, can be obtained from J_W by multiplying by the area, A, through which flow occurs:

$$Q_W = AJ_W = -A\frac{D_W^m K_W\overline{V_W}}{\overline{V}_{lipid}\delta}\Delta C_S \tag{A.1.23}$$

Comparing this to Eq. (A.1.3), with $\Delta P = 0$, we can identify

$$P_f = \frac{D_W^m K_W\overline{V_W}}{\overline{V}_{lipid}\delta} \tag{A.1.24}$$

This expression for P_f was derived for an osmotic pressure difference ($C_{S,L} \neq C_{S,R}$) in the absence of a hydrostatic pressure difference ($\Delta P = 0$). The same expression can be derived for a hydrostatic pressure difference by returning to Eq. A.1.17 and setting the mole fractions of water to be 1.0 on both sides of the membrane, while the hydrostatic pressures differ (Finkelstein, 1987).

The diffusional permeability, P_d, of water through a lipid membrane is defined by

$$J_W = P_d\Delta C_W \tag{A.1.25}$$

If we assume as we did in the derivation of P_f that the rate-limiting step is the diffusion of water through the membrane phase, Eq. (A.1.25) can be rewritten as

$$J_W = \frac{D_W^m}{\delta}\Delta C_{W(membrane)} = \frac{D_W^m}{\delta}\frac{K_W}{\overline{V}_{lipid}}\Delta X_{W(solution)} \tag{A.1.26}$$

For dilute solutions, $\Delta X_{W(solution)} = \overline{V}_W\Delta C_W$. Insertion of this into Eq. (A.1.26) gives

$$J_W = \frac{D_W^m K_W\overline{V_W}}{\overline{V}_{lipid}\delta}\Delta C_W \tag{A.1.27}$$

Comparison of Eqn. A.1.27 with Eqn. (A.1.25) allows us to identify P_d as

$$P_d = \frac{D_W^m K_W\overline{V_W}}{\overline{V}_{lipid}\delta} \tag{A.1.28}$$

Using the results of Eqs. A.1.24 and A.1.28 we conclude that

$$\frac{P_f}{P_d} = 1 \tag{A.1.29}$$

for a lipid bilayer membrane.

A.1.4 P_f / P_d for single-file pores

The discussion above for P_f/P_d for microporous membranes suggests that this ratio approaches 1.0 as the size of the pore approaches that of a water molecule. This conclusion rests on the use of hydrodynamic equations for the pressure-driven flow, which do not pertain to narrow pores in which laminar flow is impossible. Single-file pores are a particular case of a narrow pore because in this case water molecules cannot pass one another. The narrow pore restricts free diffusion of water through the pore because diffusion of a water molecule from one position in the pore to the next requires its neighbor to move away. Thus, diffusion within the restricted geometry of the pore becomes a collective property of all of the molecules in the pore.

Suppose that there are N water molecules in a pore of length δ as shown in Fig. A.1.1. We assume that the water molecules reside in positions separated by δ/N, so that the number of water molecules in the pore is proportional to its length. Application of a pressure to one side would push molecules through the pore all at once, with a retarding force proportional to the number of molecules, N. Steady-state flow would occur when the hydrostatic force is balanced by the drag force within the pore, and so, the pressure-driven flow should be inversely proportional to N. In the case of diffusion in free water, the flux of water is proportional to $1/\delta$ or $1/N$. In the case of narrow pores, the diffusion of a vacancy through the water within the pore looks exactly like the diffusion of water through a series of vacancies in free diffusion. Thus, the flux of the vacancy is proportional to $1/N$. The movement of tracer water from one position to the next in the pore requires diffusion of the vacancy all the way through the pore. Consequently, the flux of tracer by one step also is proportional to $1/N$. In order to diffuse through the pore, the tracer must make $N+1$ such jumps. The unidirectional flux of tracer over $N+1$ steps of equal flux is given by Stein (1976):

$$J_{0\to N} = \frac{J_{ij}^{N+1}}{(N+1)J_{ij}^{N}} = \frac{1}{(N+1)}J_{ij} \tag{A.1.30}$$

where $J_{0\to N}$ is the unidirectional flux of water through N intermediate positions, involving $N+1$ steps, and J_{ij} is the unidirectional flux of tracer over one step. Since J_{ij} requires the diffusion of a vacancy all the way through the pore, it is proportional to $1/N$. Therefore, the unidirectional diffusional flux of tracer through a single-file pore should be proportional to

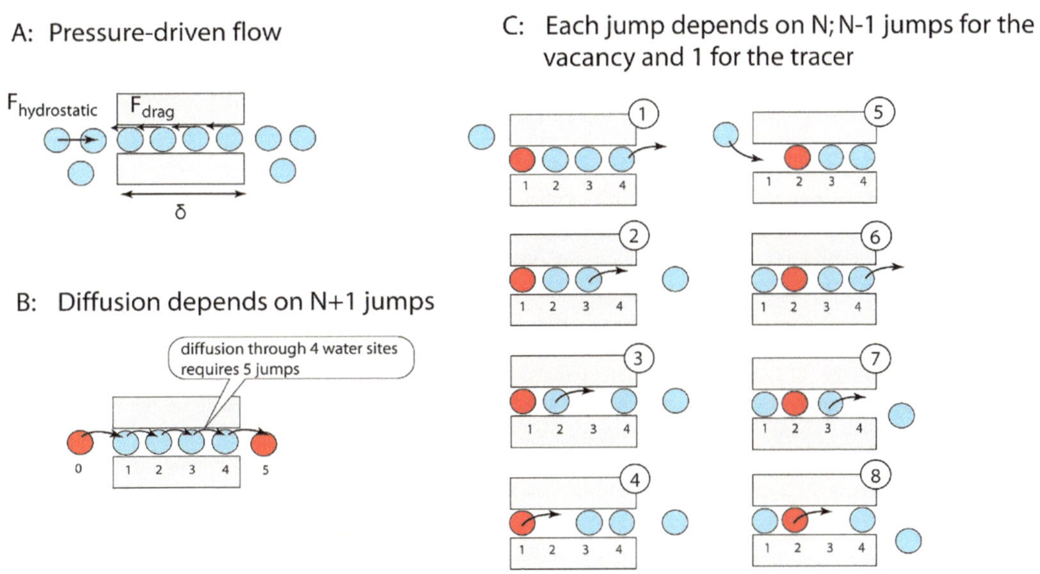

FIGURE A.1.1 Pressure-driven flow and diffusion through a single-file pore. The pore is narrow and does not allow water molecules to pass one another. Panel A: Pressure driven flow is opposed by a drag force contributed by each molecule within the pore. The total drag force depends on the number of water molecules in the pore, and their velocity relative to the pore. The flow for any given pressure is inversely proportional to the number, N, of molecules in the pore. Panel B: Diffusion of tracer (differently shaded circle) through the pore depends on making a number of "jumps" along the pore. If there are N water molecules in the pore, the tracer must make $N+1$ jumps to be detected on the other side. Panel C: Each jump of tracer depends on N, because the tracer cannot move until a vacancy is available, and the vacancy must move through the entire pore. The vacancy requires $N-1$ jumps to become available, and then the tracer must also jump, making N jumps of water for each jump of tracer water. Two jumps of tracer water (differently shaded circles) are shown. The diffusive flux is inversely proportional to the number of "jumps". Panel B and Panel C suggest that the diffusive flux should be inversely related to $(N+1)N$.

$1/N(N+1)$. Finkelstein (1987) analyzed the theory of both pressure-driven flow and diffusion through single-file pores and found that the proportionality constants relating P_f to $1/N$ and P_d to $1/N(N+1)$ are identical. The result is the prediction that, for single-file pores

$$\frac{P_f}{P_d} = N + 1 \quad \text{(A.1.31)}$$

This result was first proposed by Dick (1966) by an entirely different line of reasoning and was substantiated by the later analysis of Hernandez and Fischbarg (1992) and Zhu et al. (2004).

References

Abascal, F., Irisarri, K., & Zardoya, R. (2014). Diversity and evolution of membrane intrinsic proteins. *Biochimica et Biophysica Acta, 1840*, 1468–1481.

Adair, G. S. (1928). A Theory of partial osmotic pressures and membrane equilibria, with special reference to the application of Dalton's Law to haemoglobin solutions in the presence of salts. *Proceedings of the Royal Society of London, 120A*, 573–603.

Agre, P., Saboori, A. M., Asimos, A., & Smith, B. L. (1987). Purification and partial characterization of the Mr 30,000 integral membrane protein associated with the erythrocyte Rh(D) antigen. *Journal of Biological Chemistry, 262*, 17497–17503.

Agre, P., Sasaki, S., & Crispeels, M. J. (1993). Aquaporins: A family of water channel proteins. *American Journal of Physiology, 265*, F461.

Andreoli, T. E., & Troutman, S. L. (1971). An analysis of unstirred layers in series with "tight" and "porous" lipid bilayer membranes. *Journal of General Physiology, 72*, 341–350.

Andrews, F. C. (1976). Colligtive properties of simple solutions. *Science, 194*, 567–571.

Augustin, R. (2010). The protein family of glucose transport facilitators: it's not only about glucose after all. *Life, 62*, 315–333.

Baskaran, A., Kaari, M., Venugopal, G., Manikkam, R., Joseph, J., & Bhaskar, P. V. (2021). Anti freeze proteins (Afp): Properties, sources and applications – a review. *International Journal of Biological Macromolecules, 189*, 292–305.

Benga, G., Popescu, O., & Pop, V. I. (1986). p-(Chloromercuri)benzenesulfonate binding by membrane proteins and the inhibition of water transport in human erythrocytes. *Biochemistry, 25*, 1535–1538.

Binder, H. J., Brown, I., Ramakrishna, B. S., & Young, G. P. (2014). Oral rehydration therapy in the second decade of the twenty-first century. *Current Gastroenterology Reports, 16*, 376–384.

Binesh, A. R., & Kamali, R. (2015). Molecular dynamics insights into human aquaporin 2 water channel. *Biophysical Chemistry, 207*, 107–113.

Brahm, J. (1982). Diffusional water permeability of human erythrocytes and their ghosts. *Journal of General Physiology, 79*, 791–819.

Brahm, J. (2013). The permeability of red blood cells to chloride, urea and water. *Journal of Experimental Biology, 216*, 2236–2246.

Cass, A., & Finkelstein, A. (1967). Water permeability of thin lipid membranes. *Journal of General Physiology, 50*, 1765–1784.

Dainty, J. (1965). Osmotic flow. *Symposia of the Society for Experimental Biology, 19*, 75–85.

De Groot, B. L., & Grubmuller, H. (2005). The dynamics and energetics of water permeation and proton exclusion in aquaporins. *Current Opinion in Structural Biology, 15*, 176–183.

Denker, B. M., Smith, B. L., Kuhajda, F. P., & Agre, P. (1988). Identification, purification and partial characterization of a novel Mr 28,000 integral protein from erythrocytes and renal tubules. *Journal of Biological Chemistry, 263*, 15634–15642.

Devries, A. L. (1986). Antifreeze glycopeptides and peptides: Interactions with ice and water. *Methods in Enzymology, 127*, 293–303.

Devries, A. L. (1971). Glycoproteins as biological antifreeze agents in Antarctic fishes. *Science, 172*, 1152–1155.

Devries, A. L., & Wohlschlag, C. (1969). Freezing resistance in some Antarctic fishes. *Science, 163*, 1073–1075.

Dick, D. A. T. (1966). *Cell water*. Washington: Butterworths.

Direito, I., Madeira, A., Brito, M. A., & Soveral, G. (2016). Aquaporin-5: From structure to function and dysfunction in cancer. *Cellular and Molecular Life Sciences, 73*, 1623–1640.

Dolev, M. B., Braslavsky, I., & Davies, P. L. (2016). Ice-binding proteins and their function. *Annual Review of Biochemistry, 85*, 515–542.

Duman, J. G. (2015). Animal ice-binding (antifreeze) proteins and glycolipids: An overview with emphasis on physiological function. *Journal of Experimental Biology, 218*, 1846–1855.

Durbin, R. P. (1960). Osmotic flow of water across permeable cellulose membranes. *Journal of General Physiology, 44*, 315–326.

Dutrochet, H. (1826). *L'Agent immediate du mouvement vital devoile dans sa nature et son mode d'action chez les vegetaux et chez les animaux*. Paris.

Feher, J. J. (2017). *Quantitative Human Physiology*. Academic Press.

Fettiplace, R., & Haydon, D. A. (1980). Water permeability of lipid membranes. *Physiological Reviews, 60*, 510–550.

Finkelstein, A. (1987). *Water movement through lipid bilayers, pores and plasma membranes: Theory and reality*. New York: Wiley Interscience.

Fischbarg, J., Kuang, K., Vera, J. C., Arant, S., Silverstein, S. C., Loike, J., & Rosen, O. M. (1990). Glucose transporters serve as water channels. *Proceedings of the National Academy of Sciences of the United States of America, 87*, 3244–3247.

Fushimi, K., Uchida, S., Hara, Y., Hirata, Y., Marumo, F., & Sasaki, S. (1993). Cloning and expression of apical membrane water channel of rat kidney collecting tubule. *Nature, 361*, 549–552.

Glasstone, S. (1946). *Textbook of physical chemistry* (pp. 669–671). Princeton: Van Nostrand.

Graham, T. (1854). VII. The Bakerian lecture — On osmotic force. *Philosophical Transactions of the Royal Society of London, 144*, 177–228.

Guell, D. C., & Brenner, H. (1996). Physical mechanism of membrane osmotic phenomena. *Industrial & Engineering Chemistry Research, 35*, 3004–3014.

Hamann, S., Herrea-Perez, J. J., Bundgaard, M., Alvarez-Leefmans, J., & Zeuthen, T. (2005). Water permeability of Na-K-2Cl cotransporters in mammalian epithelial cells. *The Journal of Physiology, 568*, 123–135.

Hammel, H. T. (1976). Colligative properties of a solution. *Science, 192*, 748–756.

Hammel, H. T. (1979). Forum on osmosis I. Osmosis: Diminished solvent activity or enhanced solvent tension? *American Journal of Physiology, 237*, R95–R107.

Hanai, T., Haydon, D. A., & Redwood, W. R. (1966). The water permeability of artificial bimolecular leaflets: A comparison of radio-tracer and osmotic methods. *Annals of the New York Academy of Sciences, 137*, 731–739.

Hanai, T., & Haydon, D. A. (1966). The permeability to water of bimolecular lipid membranes. *Journal of Theoretical Biology, 11*, 370–382.

Harries, W. E. C., Akhaven, D., Miercke, L. J. W., Khademi, S., & Stroud, R. M. (2004). The channel architecture of aquaporin 0 at a 2.2 Å resolution. *Proceedings of the National Academy of Sciences, 101*, 14045–14050.

Hernandez, J. A., & Fischbarg, G. (1992). Kinetic analysis of water transport through a single-file pore. *Journal of General Physiology, 99*, 645–662.

Hevesy, G., Hofer, E., & Krogh, A. (1935). The permeability of the skin of frogs to water as determined by D_2O and H_2O. *Acta Physiologica Scandinavica, 72*, 199–214.

Hildebrand, J. H. (1979). Forum on osmosis II. A criticism of "solvent tension" in osmosis. *American Journal of Physiology, 237*, R108–R109.

Hildebrand, J. H. (1955). Osmotic pressure. *Science, 121*, 116–119.

Holz, R., & Finkelstein, A. (1970). The water and nonelectrolyte permeability induced in thin lipid membranes by the polyene antibiotics nystatin and amphotericin B. *Journal of General Physiology, 56*, 125–145.

Horsefield, R., Norden, K., Fellert, M., Backmark, A., Tornroth-Horsefield, S., Terwisscha van Schletinga, A. K., Kvassman, J., Kjellbom, P., Johanson, U., & Neutze, R. (2008). High resolution x-ray structure of human aquaporin 5. *Proceedings of the National Academy of Sciences, 105*, 13327–13332.

Huang, C., & Thompson, T. E. (1966). Properties of lipid bilayer membranes separating two aqueous phases: Water permeability. *Journal of Molecular Biology, 15*, 539–554.

Hulett, G. A. (1903). Beziehung zwischen negative Druck und osmotichem Druck. *Zeitschrift für Physikalische Chemie, 42*, 353–368.

Ishibashi, K., Sasaki, S., Fushimi, K., Uchida, S., Kuwahara, M., Saito, H., Furukawa, T., Nakajima, K., Yamaguchi, Y., Godoberi, T., & Marumo, F. (1994). Molecular cloning and expression of a member of the aquaporin family with permeability to glycerol and urea in addition to water expressed at the basolateral membrane of kidney collecting duct cells. *Proceedings of the National Academy of Sciences, 91*, 6269–6273.

Issack, B. B., & Peslherbe, G. H. (2015). Effects of cholesterol on the thermodynamics and kinetics of passive transport of water through lipid membranes. *Journal of Physical Chemistry, 119*, 9391–9400.

Jansen, M., & Blume, A. (1995). A comparative study of diffusive and osmotic water permeation across bilayers composed of phospholipids with different head groups and fatty acyl chains. *Biophysical Journal, 68*, 997–1008.

Kiil, F. (1989). Molecular mechanisms of osmosis. *American Journal of Physiology, 256*, R801–R808.

Knight, C. A., Devries, A. L., & Oolman, L. D. (1984). Fish antifreeze protein and freezing and recrystallization of ice. *Nature, 308*, 295–296.

Koefoed-Johnsen, V., & Ussing, H. H. (1953). The contributions of diffusion and flow to the passage of D_2O through living membranes. *Acta Physiologica Scandinavica, 28*, 60–76.

Landis, E. M., & Pannenheimer, J. R. (1963). Exchange of substances through the capillary walls. In *Handbook of Physiology* (Vol. 2, pp. 961–1034). Washington DC: American Physiological Society Sec.

Lapointe, J. Y., Gagnon, M. P., Gagnon, D. G., & Bissonnette, P. (2002). Controversy regarding the secondary active water transport hypothesis. *Biochemistry and Cell Biology, 80*, 525–533.

Levitt, D. G. (1973). Kinetics of diffusion and convection in 3.2 Å pores. *Biophysical Journal, 13*, 186–206.

Loo, D. F., Zeuthen, T., Chandy, G., & Wright, E. M. (1996). Cotransport of water by the Na^+/glucose transporter. *Proceedings of the National Academy of Sciences of the United States of America, 93*, 13367–13370.

Macey, R. I., & Farmer, R. E. L. (1970). Inhibition of water and solute permeability in human red cells. *Biochimica et Biophysica Acta, 211*, 104–106.

Maurel, C., Reizer, J., Schroeder, J. K., & Chrispeels, M. J. (1993). The vacuolar membrane protein γ-TIP creates water specific channels in Xenopus oocytes. *The EMBO Journal, 12*, 2241–2247.

Mauro, A. (1979). Forum on osmosis III. Comments on Hammel and Scholander's solvent tension theory ant its application to the phenomenon of osmotic flow. *American Journal of Physiology, 237*, R110–R113.

Mauro, A. (1957). Nature of solvent transfer in osmosis. *Science, 126*, 252–253.

Meschia, G., & Setnikar, I. (1958). Experimental study of osmosis through a collodion membrane. *Journal of General Physiology, 43*, 429–444.

Mikulecky, D. C. (1972). A continuum mechanical approach to the flow equations for membrane transport. *Biophysical Journal, 12*, 1642–1660.

Mlekoday, H. J., Moore, R., & Levitt, D. G. (1983). Osmotic water permeability of the human red cell. *Journal of General Physiology, 81*, 213–220.

Mollajew, R., Zocher, F., Horner, A., Wiesner, B., Klussmann, E., & Pohl, P. (2010). Routes of epithelial water flow: Aquaporins versus cotransporters. *Biophysical Journal, 99*, 3647–3656.

Moura, T. F., Macey, R. I., Chien, D. Y., Karan, D., & Santos, H. (1984). Thermodynamics of all-or-none water channel closure in red cells. *Journal of Membrane Biology, 81*, 105–111.

Murata, K., Mitsuoka, K., Hirai, T., Walz, T., Agre, P., Heymann, J. B., Engel, A., & Fujiyoshi, Y. (2000). Structural determinants of water permeation through aquaporin-1. *Nature, 407*, 599–605.

Nollett, J. A. (1995). Reserches sur les causes du Bouillonnement des Liquides. Histoire de l'Academie Royale des Sciences, Paris, 1752. [Translated and reprinted as Investigations on the causes for the ebullition of liquids]. *Journal of Membrane Science, 100*, 1–3.

Ostwald, W. (1891). *Lehrbuch der allgemeinen Chemie* (2nd ed., Vol. 2).

Paganelli, C. V., & Solomon, A. K. (1957). The rate of exchange of tritiated water across the human red cell membrane. *Journal of General Physiology, 41*, 259–277.

Parisi, M., & Bourguet, J. (1983). The single file hypothesis and the water channels induced by antidiuretic hormone. *Journal of Membrane Biology, 71*, 189–193.

Pfeffer, W. (1877). In W. Englelmann, & Leipzig (Eds.), *Osmotiche Untersuchungen. Studien zur Zellmechanik.*

Pitzer, K. S., & Mayorga, G. (1973). Thermodynamics of electrolytes II: Activity and osmotic coefficients for strong electrolytes with one or both ions univalent. *Journal of Physical Chemistry, 77*, 2300–2308.

Prescott, D. M., & Zeuthen, E. (1953). Comparison of water diffusion and water filtration across cell surfaces. *Acta Physiologica Scandinavica, 38*, 77–94.

Preston, G. M., & Agre, P. (1991). Isolation of the cDNA for erythrocyte integral membrane protein of 28 kilodaltons: Member of an ancient channel family. *Proceedings of the National Academy of Sciences, 88*, 11110–11114.

Preston, G. M., Carroll, T. P., Guggino, W. B., & Agre, P. (1992). Appearance of water channels in Xenopus oocytes expressing red cell CHIP28 protein. *Science, 256*, 385–387.

Price, H. D., & Thompson, T. E. (1969). Properties of liquid bilayer membranes separating two aqueous phases: Temperature dependence of water permeability. *Journal of Molecular Biology, 41*, 443–457.

Raoult, F.-M. (1886). Loi generale des tensions de vapeur des dissolvants. *Comptes Rendus, 104*, 1430–1433.

Rawlins, S. L., Oster, J. D., Levitt, J., Knoerr, K. R., & Scholander, P. F. (1967). Osmotic mechanism and negative pressure. *Science, 158*, 1210–1212.

Ray, P. M. (1960). On the theory of osmotic water movement. *Plant Physiology, 35*, 783–795.

Redwood, W. R., & Haydon, D. A. (1969). Influence of temperature and membrane composition on the water permeability of lipid bilayers. *Journal of Theoretical Biology, 22*, 1–8.

Reeves, J. P., & Dowben, R. M. (1970). Water permeability of phospholipid vesicles. *Journal of Membrane Biology, 3*, 123–141.

Rich, G. T., Sha'afi, R. I., Romualdez, A., & Solomon, A. K. (1968). Effect of osmolality on the hydraulic permeability coefficient of red cells. *Journal of General Physiology, 52*, 941–954.

Robinson, R. A., & Stokes, R. H. (1959). *Electrolyte solutions* (pp. 480–490). London: Butterworth.

Rosenberg, P. A., & Finkelstein, A. (1978). Water permeability of gramicidin A-treated lipid bilayer membranes. *Journal of General Physiology, 72*, 341–350.

Santacroce, M., Castagna, M., & Sacchi, V. F. (2010). Passive water permeability of some wild type and mutagenized amino acid cotransporters of the SL6/NSS family expressed in *Xenopus laevis* oocytes. *Comparative Biochemistry and Physiology - Part A, 156*, 509–517.

Sasseville, L. J., Cuervo, J. E., Lapointe, J. Y., & Noskov, S. Y. (2011). The structural pathway for water permeation through sodium-glucose cotransporters. *Biophysical Journal, 101*, 1887–1895.

Scholander, P. F. (1967). Osmotic mechanism and negative pressure. *Science, 156*, 67–69.

Sidel, V. W., & Solomon, A. K. (1957). Entrance of water into human red cells under an osmotic pressure gradient. *Journal of General Physiology, 41*, 243–257.

Smith, B. L., & Agre, P. (1991). Erythrocyte Mr 28,000 transmembrane protein exists as a multisubunit oligomer similar to channel proteins. *Journal of Biological Chemistry, 266*, 6407–6415.

Smith, H. W. (1960). Theory of solutions. *Circulation, 21*, 808–817.

Solomon, A. K., Chasan, B., Dix, J. A., Lukacovic, M. F., Toon, M. R., & Verkman, A. S. (1983). The aqueous pore in the red cell membrane: Band 3 as a channel for anions, cations, nonelectrolytes, and water. *Annals of the New York Academy of Sciences, 414*, 97–124.

Solomon, A. K. (1989). Water channels across the red blood cell and other biological membranes. *Methods in Enzymology, 173*, 192–222.

Staverman, A. J. (1951). The theory of measurement of osmotic pressure. *Recueil des Travaux Chimiques des Pays-Bas, 70*, 344–352.

Stein, W. D. (1976). An algorithm for writing down flux equations for carrier kinetics, and its application to co-transport. *Journal of Theoretical Biology, 62*, 467–478.

Tanford, C. (1961). *Physical chemistry of macromolecules.* New York: John Wiley.

Terwilliger, T. C., & Solomon, A. K. (1981). Osmotic water permeability of human red cells. *Journal of General Physiology, 77*, 549–570.

Tien, H. T., & Ting, H. P. (1968). Permeation of water through bilayer lipid membranes. *Journal of Colloid and Interface Science, 27*, 702–713.

Tong, J., Wu, Z., Briggs, M. M., Schulten, K., & McInstosh, T. J. (2016). The water permeability and pore entrance structure of aquaporin-4 depend on lipid bilayer thickness. *Biophysical Journal, 111*, 90–99.

Ussing, H. H., & Anderson, B. (1956). The relation between solvent drag and active transport of ions. In *Proceedings of the third international congress of biochemistry 1955*. New York, NY: Academic Press.

Van Hoek, A. N., & Verkman, A. S. (1992). Functional reconstitution of the isolated erythrocyte water channel CHIP28. *Journal of Biological Chemistry, 267*, 18267–18269.

Vandenberg, R. J., Handford, C. A., Campbell, E. M., Ryan, R. M., & Yool, A. J. (2011). Water and urea permeability pathways of the human excitatory amino acid transporter EAAT1. *Biochemical Journal, 439*, 333–340.

Van't Hoff, J. H. (1887). Die Rolle des osmotichen Drukes in der analogie zwischen Losungen und gasen. *Zeitschrift für Physikalische Chemie, 1*, 481–493.

Vieira, F. L., Sha'afi, R. I., & Solomon, A. K. (1970). The state of water in human and dog red cell membranes. *Journal of General Physiology, 55*, 451–466.

Wang, J. H. (1965). Self-diffusion coefficients in water. *Journal of Physical Chemistry, 69*, 4412.

Yang, B., & Verkman, A. S. (2002). Analysis of double knockout mice lacking aquaporin-1 and urea transporter UT-B. *Journal of Biological Chemistry, 277*, 36782–36786.

Yang, B., & Verkman, A. S. (1998). Urea transporter UT3 functions as an efficient water channel. *Journal of Biological Chemistry, 273*, 9369–9372.

Zeidel, M. L., Ambudkar, S. V., Smith, B. L., & Agre, P. (1992). Reconstitution of functional water channels in liposomes containing purified red cell CHIP28 protein. *Biochemistry, 31*, 7436–7440.

Zeuthen, T., Gorraitz, E., Her, K., Wright, E. M., & Loo, D. D. F. (2016). Structural and functional significance of water permeation through transporters. *Proceedings of the National Academy of Sciences, 113*, E6887–E6894.

Zeuthen, T., & MacAulay, N. (2012). Transport of water against its concentration gradient: Fact or fiction? *WIRES Membrane Transport and Signaling, 1*, 373–381.

Zeuthen, T. (2010). Water transporting proteins. *Journal of Membrane Biology, 234*, 57–73.

Zhu, F., Tajkhorshid, E., & Schulten, K. (2004). Theory and simulation of water permeation in aquaporin-1. *Biophysical Journal, 86*, 50–57.

Chapter 12

Origin of the resting membrane potential

Jorge A. Sánchez
Department of Pharmacology, Centro de Investigación y de Estudios Avanzados del Insitututo Politécnico Nacional, Mexico City, Mexico

1. Introduction

The cell membrane (Chapter 4) keeps the concentrations of ions in the intracellular medium at levels that are entirely different from those of the extracellular medium. These concentration differences between the intracellular and extracellular compartments are maintained by specialized ion transporter proteins embedded in the lipid bilayer. The membrane also contains ion channel proteins whose activity modifies the transmembrane flow of ions in the short and long term.

Fig. 12.1 shows a schematic representation of the cell membrane from two different perspectives. The upper panel shows the lipid bilayer and the membrane proteins that transport ions from the extracellular to the intracellular medium and vice versa. Cells express different types of ion channels that allow the selective flow of K^+, Na^+, Ca^{2+}, or Cl^- through their aqueous pores. Ion flow through ion channels leads to changes in the membrane potential when the channel gate opens upon specific stimuli, such as voltage or membrane ligands, among others. Other membrane proteins illustrated in Fig. 12.1 are ion transporters like the Na^+/K^+ ATPase, also known as the Na^+-K^+ pump, or the Na^+/Ca^{2+} exchanger (NCX). The activity of these transporters also changes the membrane potential because they are *electrogenic,* that is, the transport process leads to the translocation of net charge across the membrane. Two of the three membrane components shown in the upper panel of Fig. 12.1, the lipid bilayer and ion channels, can be represented as simple components of an electric circuit. The lower panel in Fig. 12.1 is an electrical representation or "equivalent circuit" of Na^+, K^+, Ca^{2+}, and Cl^- ion channels, symbolized by ohmic resistances in series with batteries representing the corresponding equilibrium potentials. Ion flow through the channels gives rise to ionic currents that obey Ohm's law. Thus, for K^+ channels:

$$I_K = g_K\,(E - E_K) \tag{12.1}$$

where I_K represents the currents through K^+ channels; g_K the conductance to K^+; E, the membrane potential, and E_K the K^+ equilibrium potential calculated with the Nernst equation. Similar equations describe the relation between current and voltage for other channels. Currents flowing through a single ion channel are quite small and reach values up to a few picoamperes.

The lipid bilayer is represented by a capacitor (Cm) and has a capacitance per unit area close to 1 $\mu F/cm^2$. Capacitors store electric charges (Q) when a voltage (V) is applied to them. Therefore, $Q = CV$, where C is called the capacitance. Assuming that C is constant, the first derivative with respect to time (rate of change) of this relation yields:

$$dQ/dt = C\,dV/dt \tag{12.2}$$

Since the electric current (I) in a circuit represents the flow of charge (dQ/dt), the capacitive current $I_c = C\,dV/dt$. This implies that the capacitive current (I_c) depends on the changes in voltage across the circuit terminals with respect to time (t) and therefore, if the voltage is constant, $I_c = 0$.

According to the equivalent circuit of the membrane of Fig. 12.1, the membrane current (I_m) is the sum of the currents through the individual branches of the circuit, and it can be written as:

$$I_m = I_c + I_{Na} + I_K + I_{Ca} + I_{Cl} \tag{12.3}$$

This equation shows that I_m is made up of two components, one associated with the membrane capacitor (the capacitive current I_c) and the other with ion channels (the ionic currents).

Cell Physiology Sourcebook. https://doi.org/10.1016/B978-0-12-811114-7.00007-4

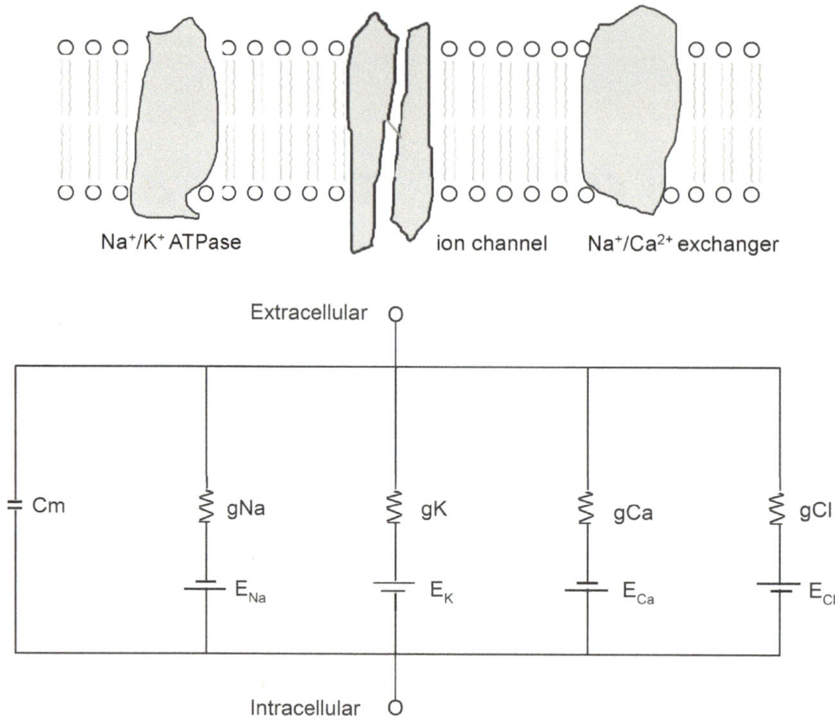

FIGURE 12.1 View of the cell membrane showing the lipid bilayer, ion channels, and ion transporters (upper part). In the equivalent circuit (lower diagram), ion channels are represented by ohmic conductive pathways to Na^+, K^+, Ca^{2+}, and Cl^- ions. The equilibrium potential for each ion is represented by batteries, and the lipid bilayer as a capacitor.

The different conductance values associated with ion channels cause the cells to become electrically polarized, or, in other words, there is normally a membrane potential. The latter is present in all types of cells and has been studied in greater detail in excitable cells such as skeletal muscle fibers and neurons. Still, other cells, such as blood cells or epithelial cells, also have a resting potential.

The relatively stable membrane potential in the absence of stimuli contrasts with the sudden changes that occur in excitable cells during their activity (action potentials). It is therefore called the *resting potential*, even when the cell invests energy in maintaining it. In many cells, the resting potential is relatively constant, and therefore, we can ignore the capacitive current and focus on the ion channels and transporters involved.

Some of the most influential studies on the origin and properties of the resting potential were carried out in skeletal muscle, and it is from this system that the most detailed information has been obtained. Therefore, I will begin by describing in some detail the origin of the resting potential in muscle fibers and then discuss it for other cell types.

The existence of resting potential in muscle fibers was proposed during the first half of the 19th century when the *injury* current was measured with a galvanometer. This current flows between an intact area of the muscle and an injured region. The lesion makes the membrane permeable by exposing the cellular interior to the external environment. In the 19th century, the German physiologist Emil du Bois-Reymond (1818–1896) showed that the injury current flows from the damaged area to the intact area and concluded that the inside of the muscle had a negative potential (Needham, 1971).

At the beginning of the 20th century, another German physiologist, Julius Bernstein (1839–1917), a disciple of du Bois-Reymond, suggested that the muscle fiber's membrane is responsible for the potential differences between the injured and intact parts. If the membrane were impermeable to an ion of an electrolyte (such as the anion of the salt KH_2PO_4), the tendency of K^+ ions to diffuse out due to their high intracellular concentration would establish an "electric double layer" with the interior negatively charged (the high cellular content of K^+ had been known since the late 19th century).

However, it was the extensive work of Boyle and Conway (1941) that laid the foundation for our current understanding of the resting potential. They confirmed that K^+ and Cl^- ions can cross the muscle membrane and quantitatively explained the origin of the resting potential by the larger permeability of the membrane to K^+ relative to other cations. The analysis of Boyle and Conway was based on the Donnan equilibrium, which will be discussed below (Boyle & Conway, 1941).

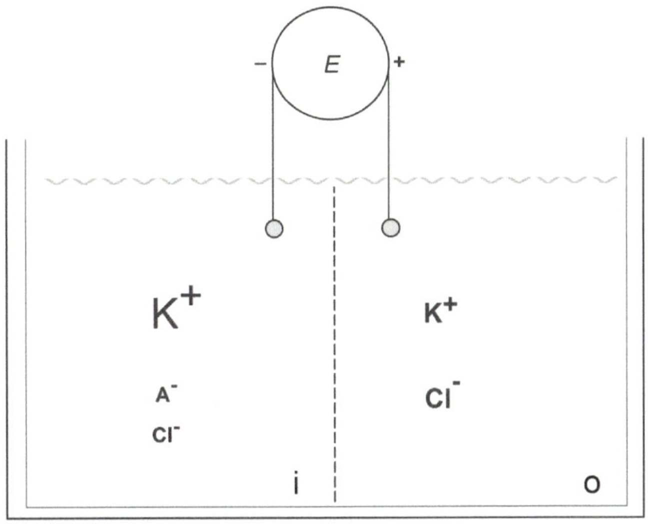

FIGURE 12.2 Donnan's balance. In this system, compartment i containing the impermeable anion A^- has a higher concentration of K^+ than compartment o. The opposite is true for the case of Cl^-. The tendency of ions to migrate according to their concentration gradient is balanced by the development of an electric potential, negative in i relative to o. The membrane that divides both compartments is assumed to be impermeable to water.

2. Donnan equilibrium

We begin by considering a simplified condition in which a KCl solution is ionized into K^+ and Cl^- ions in a container. Suppose we now separate two compartments by a membrane that is permeable to both ions. In that case, no net flow of ions takes place since they are at equilibrium, the permeable membrane plays no apparent role, and the electrical potential between both compartments is zero (the compartments will be called hereinafter "i" representing the cellular interior and "o" the exterior). Let us now assume for the sake of argument that we add a potassium salt, KA, to compartment i, and that A^- is an impermeant anion. The concentration of K^+ in compartment i will then be larger than that in o. Thus there will be a net movement of K^+ from i to o due to the difference in concentrations (Fig. 12.2). Since A^- is impermeable, the movement of K^+ will necessarily be accompanied by the only permeable anion, Cl^-, otherwise the electrical neutrality of the solutions within each compartment would not be preserved. But as the Cl^- moves from i to o, the concentration of this anion in each compartment will change (remember that initially both were the same), and thus:

$$[Cl]_i^- < [Cl]_o^- \tag{12.4}$$

The movements of K^+ and Cl^- will not continue permanently but will eventually stop when equilibrium is reached, but what is the equilibrium condition? This concept is discussed in detail in Chapters 10 and 15. Briefly, to calculate the equilibrium condition, let us first consider the case of Cl^-. Since $[Cl^-]_i < [Cl^-]_o$, the equilibrium condition for this ion is reached when a voltage difference is established between the two compartments that counteracts the effect of the higher concentration of Cl^- in o, that is, the potential in i must be negative with respect to o and can be calculated using the Nernst equation for Cl^- ions.

$$E_{Cl} = \frac{RT}{Fz}\ln\frac{[Cl^-]o}{[Cl^-]_i} \tag{12.5}$$

where R is the universal gas constant, T is the temperature in Kelvin degrees, F the Faraday constant (all of which are positive constants), and z is the valence of the ion. Since $[Cl^-]_o > [Cl^-]_i$, the ratio $[Cl^-]_o/[Cl^-]_i > 1$, so the logarithm is positive. Since $z = -1$, we obtain that E_{Cl} is negative, as our intuitive analysis suggested, and as discussed in detail in Chapter 15 (Intracellular Chloride Regulation).

It has been assumed in this idealized case that the membrane that divides the two compartments is permeable to both ions, K^+ and Cl^-, so that both will eventually reach equilibrium. The movement of K^+ and Cl^- will gradually establish a voltage (E) between the compartments until the equilibrium is reached for both ions, which implies that the potential E must be equal to the equilibrium potential of K^+ (E_K):

$$E_k = \frac{RT}{zF} \ln \frac{[K^+]_o}{[K^+]_i} \quad (12.6)$$

and E must also be equal to the equilibrium potential of Cl^- (E_{Cl}). Therefore, it follows that:

$E = E_K$ and $E = E_{Cl}$, indicating that $E_K = E_{Cl^-}$. Since E_{Cl} is negative, E_K is also negative, and using the Nernst equation for K^+ and Cl^-, we get:

$$\ln \frac{[K^+]_o}{[K^+]_i} = -\ln \frac{[Cl^-]_o}{[Cl^-]_i} \quad (12.7)$$

Therefore:

$$[K^+]_o x [Cl^-]_o = [K^+]_i \, x \, [Cl^-]_i \quad (12.8)$$

This equation means that at equilibrium, the product of the concentrations of permeant ions in one compartment is equal to their product in the other compartment (Donnan's theory).

Given that $[Cl^-]_i < [Cl^-]_o$, it follows that:

$$[K^+]_o < [K^+]_i \quad (12.9)$$

Thus, K^+ ions are distributed in the inverse form as Cl^- ions (Fig. 12.2).

Qualitatively, this idealized situation of two compartments, a single permeable cation (K^+) and two anions, one permeable (Cl^-) and one impermeable (A^-), resembles the distribution of ions in a living cell such as a muscle fiber, where K^+ ions reach a much higher concentration inside the cell, and the opposite is true for Cl^- ions. It would seem then that the resting potential could originate in the manner described above. However, as we will see below, this is not so.

Given the condition of electroneutrality within each compartment, the number of cations must be equal to the number of anions, from which it follows that:

$$[K^+]_o = [Cl^-]_o \quad (12.10)$$

And

$$[K^+]_i = [Cl^-]_i + [A^-]_i \quad (12.11)$$

If we take nd use Eqs. (12.10) and (12.11), we get that $[Cl^-]_i + [A^-]_i > [Cl^-]_o$, and we conclude that:

$$[K^+]_i + [Cl^-]_i + [A^-]_i > [K^+]_o + [Cl^-]_o \quad (12.12)$$

Examination of the preceding equation indicates that the osmolarity at i is greater than at o and therefore, the equilibrium has been reached under nonisotonic conditions. This difference in osmolarities, deduced directly from Donnan's theory, indicates that the origin of the resting potential of a cell cannot be explained in these simple terms. This is because the plasma membrane of a cell is permeable to water, which is incompatible with osmotic differences since it would flow from the outside to the inside until the concentration of the impermeable anion is negligible. Thus, the Donnan balance can only be established in a system with a membrane impermeable to water.

Boyle and Conway (1941) recognized this fact and then analyzed the case of a double Donnan equilibrium condition described below. Assume now that we have a slightly different case in which compartment i contains K^+ and Cl^- ions only, while compartment o also contains (in addition to K^+ and Cl^- ions) a cation to which the membrane is impermeable, C^+. Analyzing this new situation, with a procedure similar to that described in previous paragraphs, we conclude that the movement of Cl^- will be accompanied by the permeable cation K^+ (which was originally at the same concentration in i and o), in doing so $[K^+]_i > [K^+]_o$ and given that $E = E_K = E_{Cl}$, therefore by applying the Nernst equation E will be negative as in the previous case. Similarly, $[K^+]_i \times [Cl^-]_i = [K^+]_o \times [Cl^-]_o$.

But in this case, there is an inverse situation in terms of osmolarity:

$$[K^+]_o + [C^+]_o + [Cl^-]_o > [K^+]_i + [Cl^-]_i \quad (12.13)$$

That is, the osmolarity is larger in o than in i.

If conditions described by Eqs. (12.12) and (12.13) take place at the same time, where a nonpermeable anion is present in i and a nonpermeable cation is present in o, an isosmotic condition through a semipermeable membrane, with the development of a negative potential E in i relative to o would be expected. Since the most abundant cation in the extracellular medium is Na^+, it is natural to think that this is the impermeable cation located in the extracellular medium.

The double Donnan equilibrium can then occur in a living cell and indeed, Boyle and Conway showed in skeletal muscle cells that as established by the Donnan's theory (Eq. 12.8) the product of the concentrations of K^+ and Cl^- ions is approximately equal on both sides of the membrane, even for K^+ concentrations$^-$ in the extracellular medium as high as 300 mM. Thus, they showed that the ratio: $[K^+]_o \times [Cl^-]_o/[K^+]_i \times [Cl^-]_i$ is very close to one when $[K^+]_o$ is larger than 10 mM. For example, this ratio is equal to 1.05, 0.98, 1.06, and 1.05 when $[K^+]_o$ (mM) is 12, 18, 30, and 300, respectively. However, despite good agreement between the experimental data and Eq. (12.8) for many values of $[K^+]_o$, Boyle and Conway observed that the ratio is significantly less than one when $[K^+]_o$ is close to its physiological value (0.36 and 0.74 when $[K^+]_o$ is 3 and 6 mM, respectively). This is probably related to the fact that the membrane of skeletal muscle is slightly permeable to Na^+ ions as discussed below.

The origin of the resting potential can thus be explained by a double Donnan equilibrium, which assumes that the cell membrane contains permeability pathways for K^+ and Cl^-, and that it is impermeable to the rest of the intracellular anions and to the rest of the extracellular cations. As will be shown later, this explanation is only approximately correct.

3. Effect of K^+ and Cl^- ions on the resting potential under equilibrium conditions

The Donnan's double equilibrium proposed by Boyle and Conway (1941) postulates that both K^+ and Cl^- ions reach the equilibrium potential predicted by the Nernst equation, as if the cell membrane behaves as a selective electrode for K^+ and Cl^- ions. If this is the case, the resting potential should follow the changes predicted by the Nernst equation (Eqs. 12.5 and 12.6) for these ions, and in particular, if the intracellular concentrations of the permeant ions ($[K^+]_i$ and $[Cl^-]_i$), remain constant. If these conditions are met, the resting potential should follow a logarithmic relationship with the changes in K^+ and Cl^- concentrations in the extracellular medium.

Hodgkin and Horowicz (1959) examined this hypothesis in detail in isolated muscle fibers, by measuring the membrane potential intracellularly with glass micropipettes (microelectrodes) in the presence of extracellular solutions of varied ionic composition. Their results established the origin of the resting potential as we understand it today. Notably, in these studies, Cl^- was assumed to be passively distributed across the plasma membrane. This assumption is justified because of the relatively high permeability to Cl^- characteristic of skeletal muscle fibers. However, this does not preclude Cl^- being actively transported, as discussed in depth in Chapter 15 (Intracellular Chloride Regulation).

Initially, Hodgkin and Horowicz measured variations in the membrane potential that resulted from rapid changes in the extracellular concentrations of the permeable ions, without altering their intracellular concentrations; that is, under equilibrium conditions.

According to Eq. (12.18), the equilibrium condition is satisfied when the product of the extracellular concentration of K^+ and Cl^- is kept constant while changing the individual concentrations of both ions. This condition implies that if the extracellular concentration of K^+ is increased, the extracellular concentration of Cl^- must decrease by a similar ratio. To keep a constant osmolarity of the solutions, they replaced Cl^- with relatively membrane-impermeable sulfate (SO_4^{2-}) or sucrose.

As shown below, if the product $[K^+]_o \times [Cl^-]_o$ and the osmolarity are kept constant, the $[K^+]_i$ and $[Cl^-]_i$ will remain constant. This, in turn, allows the analysis of changes in extracellular concentrations of these ions without altering their intracellular concentrations.

Let us consider an idealized condition in which the osmolarity of the intracellular medium is only determined by K^+, Cl^-, and A^- ions, then:

$$[K^+]_i = \frac{N_i}{2V_i} \tag{12.14}$$

where N_i is the sum of the total number of moles of intracellular solute, and V_i is the intracellular volume. This follows from the fact that K^+ is the only intracellular cation considered and that electrical neutrality must be preserved.

Now, let us assume that we vary $[K^+]_o$ and $[Cl^-]_o$ in such a way that an x moles of KCl enters the cell (if, on the contrary, the movement of KCl were in the opposite direction, x would have a negative sign in our analysis). In this situation, $[K^+]_i$ and V_i will reach new values $[K^+]_i'$ and V_i'. We then conclude that:

$$[K^+]_i' = \frac{(N_i/2) + x}{V_I'} \tag{12.15}$$

To calculate $[K^+]_i'$, we now need the value of x, which is the number of moles that, under equilibrium conditions, have entered the cell. If the osmolarity in the extracellular medium has remained constant, the intracellular one will not vary either because the membrane is permeable to water and thus:

$$\frac{N_i}{V_i} = \frac{N_i + 2x}{V_i^f} \tag{12.16}$$

Solving the preceding equation for x and substituting into Eq. (12.15), we obtain:

$$[K^+]_i' = \frac{N_i}{2V_i} \tag{12.17}$$

Eqs. (12.14 and 12.17) are identical, thus:

$$[K^+]_i = [K^+]_i' \tag{12.18}$$

This means that if $[K^+]_o$ is changed under isotonic conditions (changing $[Na^+]_o$ accordingly), $[K^+]_i$, should not be altered.

However, as explained in the example of Section 10, maintaining constant osmolarity in the extracellular medium (by replacing Na^+ with K^+) is not a sufficient condition to prevent changes in intracellular concentrations of ions (of Cl^- in particular) and cell volume; Hodgkin and Horowicz (1959) concluded that it is also necessary to keep the product $[K^+]_o \times [Cl^-]_o$ constant since according to Eq. (12.18) if $[K^+]_o \times [Cl^-]_o$ does not change, the product $[K^+]_i \times [Cl^-]_i$ will remain constant and since $[K^+]_i$ does not vary under isotonic conditions, as discussed above, it is finally concluded that $[Cl^-]_i$ will also be constant. That is, we have an equilibrium condition.

To keep the product $[K^+]_o \times [Cl^-]_o$ constant, while increasing $[K^+]_o$, Hodgkin and Horowicz (1959) decreased $[Cl^-]_o$ and used SO_4^{2-} to replace it since the latter is a nondiffusible anion across the muscle membrane. Fig. 12.3 shows the values of the resting membrane potential when the product $[K^+]_o \times [Cl^-]_o$ remains constant. The symbols correspond to experimental values, and the line is the Nernst equation for K^+ and Cl^-. In this experiment, the increase in $[K^+]_o$ was accompanied by the corresponding decrease in $[Na^+]_o$. It is evident that for $[K^+]_o > 10$ mM, the agreement between the experimental values and the Nernst equation is excellent. We can also note that the equilibrium condition is preserved at all times, since the membrane potentials measured in an interval of 10–60 min after the solution change (+ in Fig. 12.3) are similar to the values obtained 20–60 s after the change (o in Fig. 12.3). In conclusion, under stationary conditions, the

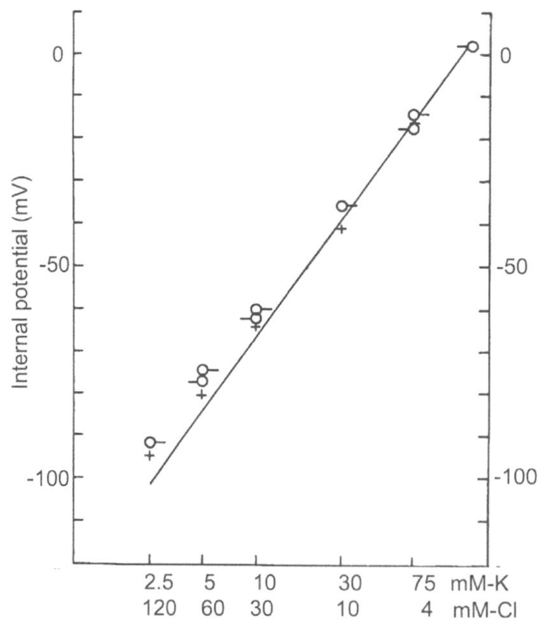

FIGURE 12.3 The relationship between the resting membrane potential and the logarithm of the external concentration of K^+ and Cl^- in a skeletal muscle fiber. The product $[K^+]_o \times [Cl^-]_o$ remained constant. Crosses are potentials measured after equilibrating for 10–60 min; circles are membrane potentials measured 20–60 s after a sudden change in concentration, -o after an increase in $[K^+]_o$, o- after a decrease in $[K^+]_o$. The line is the Nernst equation for both ions. The membrane behaves like a K^+ or Cl^- electrode. *Modified from Hodgkin and Horowicz (1959).*

resting potential in muscle can be explained by a double Donnan equilibrium, except for small values of $[K^+]_o$. These results provided direct evidence for the conclusions of Boyle and Conway (1941).

4. Effect of K^+ and Cl^- ions on the resting potential under nonequilibrium conditions

In the previous section, we considered equilibrium conditions, keeping constant both the product $[K^+]_o \times [Cl^-]_o$ and the osmolarity. Under these conditions, E_K is identical to E_{Cl}, and the intracellular ionic concentrations remain unchanged.

Suppose now that we change the concentration of a single ion of the extracellular medium, either K^+ or Cl^-, this necessarily changes the product $[K^+]_o \times [Cl^-]_o$ and leads to nonequilibrium conditions. Net flows of K^+ and Cl^- ions and water will occur until equilibrium is reached. The net movements of these ions will alter their intracellular concentrations but we can consider, as a first approximation, that they will not have a large effect on their concentrations in the extracellular medium, due to its large volume (though in some cases, the extracellular concentration of K^+ can be substantially altered in regions where diffusion is restricted). In this way, changing only one ion at a time alters the equilibrium potentials of both K^+ and Cl^-. However, there is a big difference between both ions, since the intracellular concentration of Cl^- is relatively small, approximately 2.5 mM, while that of K^+ is relatively high, of around 120 mM, so that although the flows of K^+ and Cl^- ions are equal, the intracellular concentration of Cl^- is altered to a greater extent than that of K^+. Furthermore, E_K will be altered even less by a change in $[K^+]_i$ since the relationship between E_K and $[K^+]_o$ is logarithmic. Thus, an increase of 10 mM in $[K^+]_i$ causes a change in E_K of just 2.0 mV. On the other hand, a similar change in $[Cl^-]_i$ alters E_{Cl} by 40.5 mV. E_K is determined almost entirely by the extracellular concentration, and when equilibrium is reached, $E = E_K = E_{Cl}$, a movement of Cl^- takes place until it reaches its equilibrium potential and matches that of K^+. Changes in $[Cl^-]_o$ will then have only a temporary effect that ends when $E_{Cl} = E_K$. Therefore, it is $[K^+]_o$ that ultimately determines the value of the resting potential in skeletal muscle fibers. This conclusion could be obtained by measuring the changes in the resting potential following a change in $[Cl^-]_o$ or $[K^+]_o$ independently.

Fig. 12.4 shows results from an experiment by Hodgkin and Horowicz (1959) in which they reduced the extracellular concentration of Cl^- from 120 to 30 mM, without modifying that of K^+. They observed that the reduction of $[Cl^-]_o$ produced a sudden depolarization of the resting potential, which was not sustained, and it subsequently returned to its initial value. When they raised $[Cl^-]_o$ to its original value of 120 mM, the resting potential changed in an opposite manner. Prior to changes in the extracellular solution, under normal ionic conditions, E_{Cl} and E_K had values very close to the resting

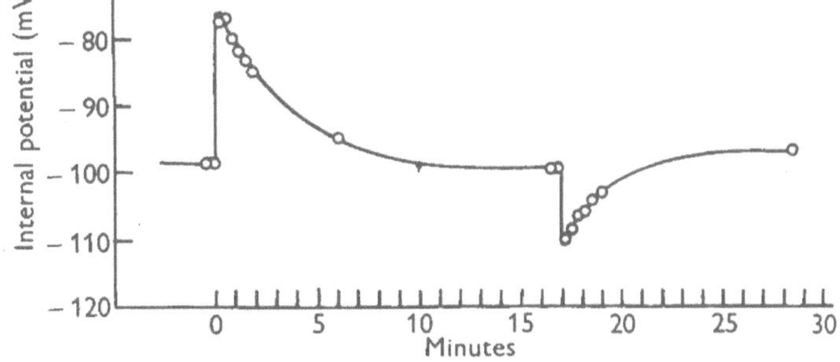

FIGURE 12.4 The effect of a rapid decrease in $[Cl^-]_o$ on the resting potential. The sudden fall of $[Cl^-]_o$ causes a transient depolarization of the muscle fiber. The resting potential recovers with a time course of several minutes as E_{Cl} returns to its original value due to the outflow of Cl^- from the fiber. E_K remains virtually constant throughout. *Modified from Hodgkin and Horowicz (1959).*

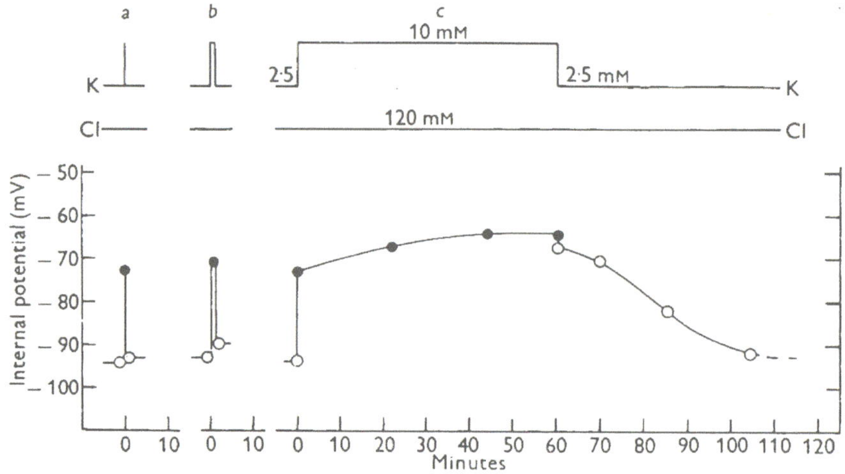

FIGURE 12.5 The effect on membrane potential of changing $[K^+]_o$ from 2.5 to 10 mM. The concentration of K^+ remained elevated for different time periods. The initial rapid change is as expected for the new E_K with a conductance ratio g_{Cl}/g_K equal to 3 (recordings *a* and *b*); while the gradual change (recording *c*) reflects the entry of K^+ and Cl^- toward the equilibrium condition that is reached when E_{Cl} matches E_K. *Modified from Hodgkin and Horowicz (1959).*

potential. When $[Cl^-]_o$ was reduced from 120 to 30 mM, E_{Cl} suddenly changed from −98 to −77 mV. This is a nonequilibrium condition. Equilibrium requires that the product of the concentrations of the K^+ and Cl^- ions be equal on each side of the membrane (Eq. 12.8). In this case, the product $[K^+]_o \times [Cl^-]_o$ was sharply reduced from 300 to 75 $(mM)^2$, so that the ions began to move through their permeability pathways, reaching equilibrium when $[K^+]_i \times [Cl^-]_i = 75\ mM^2$. This leads to the efflux of K^+, Cl^- ions, and water from the cell, reducing both E_K and E_{Cl}. Since $[Cl^-]_i$ is very small and $[K^+]_i$ very high, $[K^+]_i$ is almost unchanged, while $[Cl^-]_i$ drops substantially to 0.6 mM.

The gradual efflux of K^+ and Cl^- leads to a decrease in the resting potential with a time course of minutes from the value of −77 to −98 mV. The final values, in equilibrium, of E_K and E_{Cl} are practically the same as they are in the normal Ringer solution, so that *E* is very close to its initial value of −98 mV and the product $[K^+]_i \times [Cl^-]_i$ descends to be the same value in the extracellular medium (mainly at the expense of the reduction of intracellular Cl^-).

Let us now consider the experimental results when $[K^+]_o$ is changed. In the experiment shown in Fig. 12.5, $[K^+]_o$ was increased for different time intervals from 2.5 to 10 mM while maintaining $[Cl^-]_o$ constant. Therefore, the product $[K^+]_o \times [Cl^-]_o$ increased from 300 to 1200 $(mM)^2$. Again, this is a nonequilibrium condition. Initially, the muscle fiber was exposed to an elevated $[K^+]_o$ (10 mM) for such a short time that intracellular concentrations remain unchanged. The resting potential returned to a value close to the original one when the extracellular concentration of K^+ was restored to the initial value (2.5 mM).

Different results were observed following prolonged exposure to K^+ (Fig. 12.5c). Since the product of the extracellular concentrations of $K^+ \times Cl^-$ ions is very high, the product of their intracellular concentrations will rise to reach

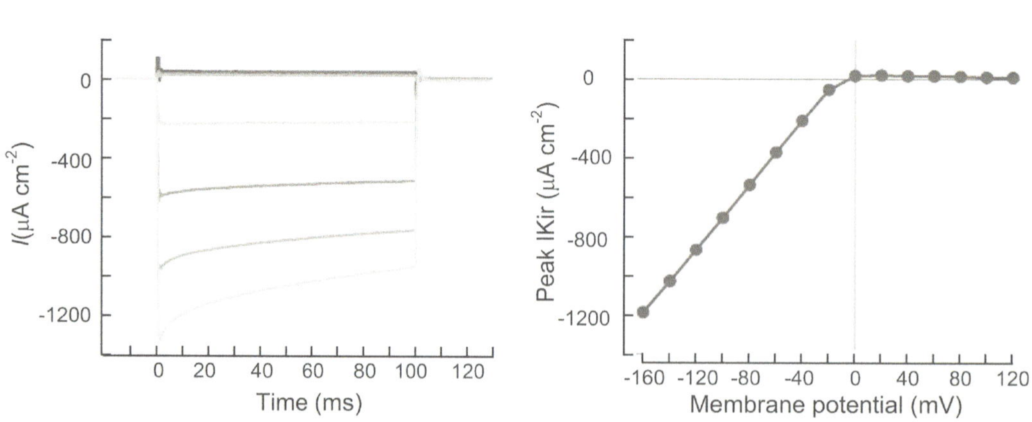

FIGURE 12.6 (A) Inward rectifier K^+ currents in an adult mammalian skeletal muscle fiber in high external K^+ in response to pulses to +40, +20, −40, −80, −120, and −160 mV. (B) The current−voltage relation of inward rectifier currents. *Modified from DiFranco et al. (2015).*

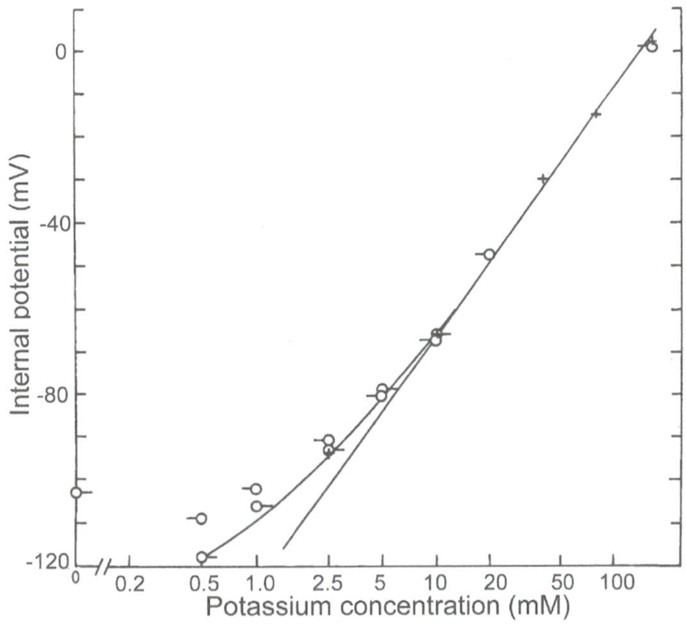

FIGURE 12.7 The relationship between the resting potential and the logarithm of the extracellular concentration of K^+ in the absence of Cl^-. In these experiments, Cl^- was replaced with SO_4^{2-}. Symbols represent single experimental values (crosses are membrane potentials measured after equilibrating for 10–60 min; circles are potentials measured 20–60 s after a sudden change in concentration, –O after increase in $[K^+]_o$, O– after decrease in $[K^+]_o$. The line is the Nernst equation for K^+ ions, and the curve is the Goldman–Hodgkin–Katz equation (Eq. 12.19), which assumes a slight membrane permeability to Na^+ ions. The resting potential is mainly determined by $[K^+]_o$, while the role of Na^+ is relatively minor. *Modified from Hodgkin and Horowicz (1959).*

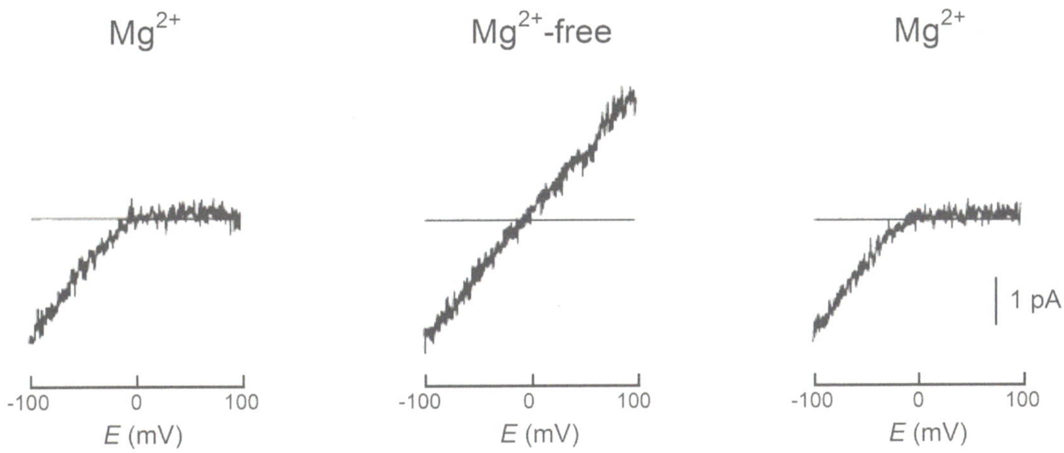

FIGURE 12.8 Single-channel currents during a voltage ramp at the indicated potentials. Currents were recorded with the inside-out configuration of the patch-clamp technique (Hamill et al., 1981) from a patch of membrane excised from a single cardiac cell. Intracellular Mg^{2+} concentration was estimated at 1.2 mM. *Modified from Vandenberg (1987).*

equilibrium; this will lead to the influx of K^+ and Cl^-. This ion movement barely affects the internal concentration of K^+, but it increases that of Cl^- until $E_{Cl} = E_K$. The resting potential gradually becomes less negative until it reaches equilibrium. When $[K^+]_o$ is reduced to initial levels, the resting potential gradually returns to its original value.

We can note, however, that the instantaneous change in the membrane potential when the concentration of K^+ was reduced from 10 to 2.5 mM is 3 mV, compared to a change of 23 mV when $[K^+]_o$ was increased from 2.5 to 10 mM. This asymmetry in the membrane potential change suggests that K^+ moves much more easily into the muscle cell interior than in the opposite direction, as if the conductance to K^+ ions in Fig. 12.5 were not constant, being larger when K^+ ions move inward than when they move outward. This means that the K^+ conductance depends on the direction of the K^+ flow. In electrical circuits, when the curve that describes the relation between the current and voltage is not a straight line but

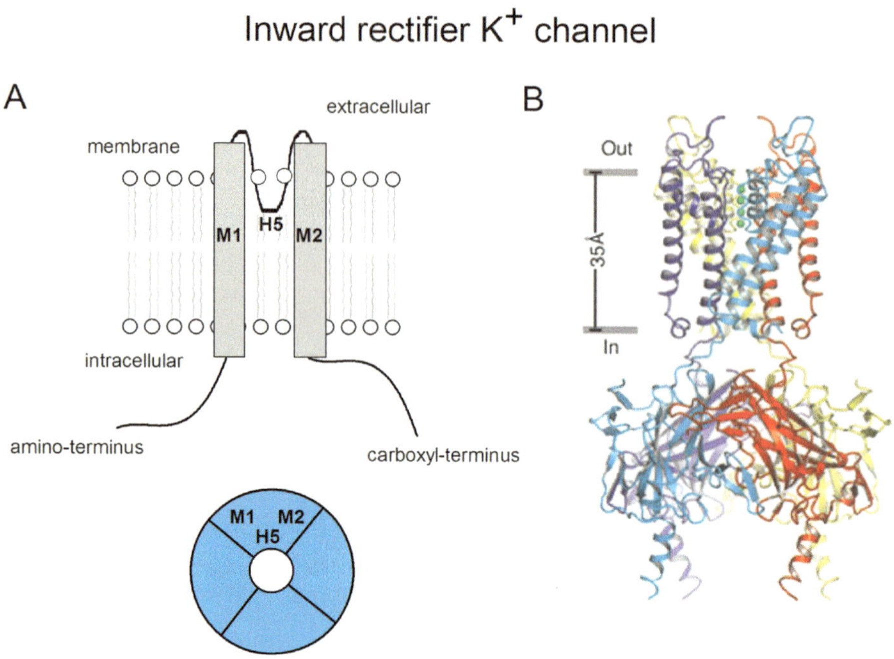

FIGURE 12.9 (A) Topology of the inward rectifier K^+ channel (IRK1). Two segments, M1 and M2, span the membrane. The channel has a long pore whose sequence (H5) extends from the extracellular surface into the membrane. The channel has a tetrameric organization. (B) Stereoview of the ribbon representation of the Kir2.2 tetrameric channel from the side. Each color represents a different subunit of the channel. *(A) Modified from Kubo et al. (1993). (B). Modified from Tao et al. (2009).*

behaves like a diode, it is said that the curve shows *rectification*. Therefore, we conclude that the K^+ current associated with the resting potential flows through *inward rectifier channels*.

4.1 Inward rectifier K^+ currents

Inward rectification for the movement of K^+ across the muscle cell membrane was first proposed by Bernard Katz who described in the late 1940s: "*les conditions électriques de chaque côté de mcmbrane ne sont pas rendues symétriques … comme si on pouvait faire plus facilement pénétrer le potassium á l'intérieur de la fibre que de l'en faire sortir*" (Katz, 1949). Once voltage-clamp methods became available for skeletal muscle fibers, currents through inward rectifier K^+ channels were first reported in the early 1970s. The gene/protein family consists of seven members, denoted as $K_{ir}1 - K_{ir}7$ and several submembers denoted as $K_{ir}2.1$, $K_{ir}2.2$, $K_{ir}2.3$, etc. (Adelman et al., 2023). Fig. 12.6 shows $K_{ir}2.1$ K^+ currents from an adult mammalian skeletal muscle fiber (DiFranco et al., 2015). Note that the current amplitude is much larger during pulses more negative than E_K than during pulses in the opposite direction (Fig. 12.6, right).

In conclusion, movements of K^+ ions are accompanied by movements of Cl^- ions, and a variation in E_K is followed by the corresponding one in E_{Cl}. Thus, $[K^+]_o$ is the main parameter that governs the resting potential. This close relation between the resting potential and the concentration of external K^+ ions will be further discussed below.

5. The effect of the extracellular K^+ concentration on the resting potential

One way to study the effect of changes in the concentration of K^+ in the extracellular medium on the membrane potential without the involvement of Cl^- is by replacing this anion in the extracellular medium with one that is impermeable to the membrane. In this case, Cl^- ions will flow toward the extracellular space until all the intracellular Cl^- content is lost.

Hodgkin and Horowicz (1959) replaced Cl^- with SO_4^{2-}, an anion that does not penetrate through Cl^- channels, and analyzed the effect of several concentrations of K^+ ions on the resting potential. Fig. 12.7 summarizes their results. If we take into consideration that $[K^+]_i$ remains constant, the relationship between the resting potential and $[K^+]_o$ should be linear on the semilogarithmic scale of Fig. 12.7. However, this is only true for values of $[K^+]_o$ larger than 10 mM. For smaller values, discrepancies with the theory, i.e., departures from the predicted Nernst potential (Eq. 12.6) are clearly observed (Eq. 12.6). The experimental values are several millivolts less negative than what the Nernstian line predicts, and the discrepancy increases as $[K^+]_o$ is further decreased.

The fact that the resting potential is less negative than expected could be explained by the influx of a cation (since there are no permeable anions moving outward) to which the membrane is not completely impermeable, Na^+, for example. Hodgkin and Horowicz (1959) hypothesized that at the resting level, the membrane potential of the muscle membrane exhibits a slight permeability to Na^+, 100 times lower than that to K^+ ions, and this explained the observed deviation from the Nernst equation. As $[K^+]_o$ decreases, $[Na^+]_o$ becomes increasingly more significant despite its low membrane permeability.

Quantitatively, this phenomenon is described by the Goldman–Hodgkin–Katz equation (Hodgkin & Katz, 1949) that, in the absence of Cl^- in the extracellular and intracellular spaces (or when the permeability to this anion is zero), takes the form:

$$E = \frac{RT}{F} \ln \frac{[K^+]_o + \alpha[Na^+]_o}{[K^+]_i + \alpha[Na^+]_i} \tag{12.19}$$

where $\alpha = P_{Na}/P_K$, the permeability ratio of Na^+ and K^+ ions. Note that this equation is identical to the Nernst equation when $\alpha = 0$, for a membrane permeability pathway that is perfectly selective to K^+. Thus, the Goldman–Hodgkin–Katz (GHK) equation represents a better fit to the experimental values than those predicted by the Nernst equation. However, the GHK equation is still unsatisfactory for very low values of $[K^+]_o$ (see the extrapolated value at "zero" $[K^+]_o$ in Fig. 12.7). The Na^+ conductance associated with resting potential values, less negative than those expected for a selective K^+ electrode at low $[K^+]_o$ has not been characterized in muscle fibers, though some experiments suggest that it may be an amiloride-sensitive cation channel (Estrada & Sanchez, 1991). In neurons, a sodium leak channel (NALCN) that contributes to the resting membrane potential and controls neuronal excitability has also been described, and it appears to be a non-selective monovalent cation channel (Xie et al., 2020).

In conclusion, the origin of the resting potential in muscle can be explained quantitatively by the Donnan's double equilibrium theory developed by Boyle and Conway, which Hodgkin and Horowicz tested under equilibrium and nonequilibrium conditions. This theory is complemented with the addition of a small modification that assumes that the muscle membrane has a relatively small permeability to Na^+ ions under resting conditions.

6. The involvement of K^+ and Cl^- ion channels on the resting potential of skeletal muscle

The identification of the K^+ and Cl^- permeability pathways involved in the generation of the resting potential required extensive electrophysiological, molecular, and structural experimental work over many years. Conway and Boyle (1939) proved that the membrane of muscle fibers is not impermeable to Cl^- ions, and they already proposed at the time that K^+ and Cl^- ions cross the membrane of skeletal muscle fibers through "pores", the hallmark of ion channels. They claimed that "the membrane … considered is one of special type, with cation and anion pores, permeable to potassium but not to larger sodium ions, and permeable at the same time to small anions of the type of chloride …"

6.1 K^+ channels

The kinetic properties of K^+ channels have been studied by several electrophysiological techniques. The direct measurement of ion currents under voltage-clamp conditions allowed the identification of several types of K^+ channels in the membrane of muscle fibers. One of them is the voltage-dependent delayed rectifier K^+ channel with similar properties to those originally described for squid giant axons. As in axons, this channel is responsible for the repolarization of the muscle action potential. As discussed above (Section 4.1), the K^+ channel associated with the resting potential is a different type of channel that exhibits inward rectification, since ion currents flowing through this channel are distinctly asymmetric, being larger when the membrane potential is clamped to values below EK than for higher values.

Inward rectification was a puzzling phenomenon when it was first discovered in muscle in the late 1940s (Katz, 1949) and was called "anomalous." It took some time to understand its physiological significance. As discussed, K^+ currents through inward rectifier K^+ channels are large at potentials more negative than E_K. This prevents muscle fibers from being hyperpolarized (and less excitable) as a result of the electrogenic activity of the Na^+–K^+ pump, and therefore the resting potential remains close to E_K. On the other hand, at potentials more positive than E_K, K^+ currents through these channels are quite small, which allows depolarization to be maintained unopposed during action potentials, especially during the long plateau phase of cardiac action potentials in ventricular cardiomyocytes.

The mechanism underlying inward rectification has been described in some detail. Rectification is not an intrinsic property of the channel; it depends on intracellular factors that block K^+ permeation, namely polyamines and intracellular Mg^{2+}. Blockade by intracellular Mg^{2+} was first demonstrated in cardiac muscle cells in the late 1980s by recording single channel activity with the patch-clamp technique developed by Erwin Neher, Bert Sakmann and coworkers (Hamill et al., 1981). Fig. 12.8 shows current–voltage relations from recordings of single channel inward rectifier K^+ currents obtained during applied voltage ramps. Strong rectification is observed in the presence of Mg^{2+} while in Mg^{2+}-free solutions, no rectification is observed, and therefore the curve describing the relation between the amplitude of the currents and the membrane potential becomes a straight line (Vandenberg, 1987).

Molecular biology and structural experiments have greatly contributed to the characterization of inward rectifier K^+ channels. IRK1 ($K_{ir}2.1$, Hibino et al., 2010), the first member of the superfamily of inward rectifier channels, was cloned and functionally expressed in the early 1990s by Lily Jan's lab (Kubo et al., 1993) from a mouse macrophage cell line. $K_{ir}2.1$ channels contain four subunits each of which has only two putative transmembrane segments that correspond to the inner core structure of voltage-gated K^+ channels (Fig. 12.9A). The crystal structure of K_{ir} channels has been described in detail down to the atomic level by Roderick MacKinnon and colleagues (Fig. 12.9B). The channels contain a transmembrane domain and a large cytoplasmic domain. The transmembrane domain is composed of two α-helices, an outer helix (M1) and a pore-lining inner helix (M2) (reviewed by Baronas & Kurata, 2014).

The structural basis for inward rectification has been revealed by a combination of site-directed mutagenesis and functional studies. The replacement of a single amino acid in the second transmembrane segment of a weakly inward rectifier K^+ channel by the corresponding amino acid of the IRK1 channel (strong inward rectifier) alters the Mg^{2+} blockade by an electrostatic mechanism, and the rectification of the current–voltage relation of the channel changes from weak to strong, leading to a complete blockade of K^+ currents in the outward direction (Lu & MacKinnon, 1994).

6.2 Cl^- channels

The Cl^- channels responsible for the high Cl^- permeability in skeletal muscle have also been characterized in detail (See Chapter 15). These channels belong to the CLC family, a large group of Cl^- transport proteins (Chen & Hwang, 2008). CLC-0 was the first member of the CC gene family to be identified; it was cloned from the electric organ of the Torpedo ray and expressed in Xenopus oocytes in the early 1990s. Several members of the same family were described in mammalian species by homology cloning, among them, CLC-1, a dimeric double-pore Cl^- channel that is predominant in skeletal muscle. This channel shares ~50%–60% amino acid identity with CLC-0 in the primary sequence.

In muscles from humans and other mammals, amphibians, reptiles, and birds, ClC-1 chloride ion channels account for 80% of the membrane conductance (reviewed by Pedersen et al., 2016). Cl^- currents have also been studied under voltage-clamp conditions in muscle cells and in heterologous expression systems. Fig. 12.10A shows macroscopic Cl^- currents from muscle CLC-1 channels recorded in *Xenopus oocytes* with the pulse protocol illustrated in the inset. The transmembrane currents are voltage and time dependent, and slowly decay during pulses, especially at large positive and negative potentials. Experiments done at the single channel level have revealed that when the ClC-1 channel opens, currents display two equidistant current levels. This observation is compatible with the view that channels have a double-barrel structure with a common gate (Saviane et al., 1999).

Fig. 12.10B–C shows the structure of the human dimeric ClC-1 channel that was unraveled by cryo-electron microscopy techniques. The channel has a narrow connecting pore and positively charged vestibules that attract Cl^- ions. The chloride conducting pathway includes a central glutamate residue that is related to the channel's voltage-dependence. The overall structure consists of 17 helices with connecting loops and two cystathionine-β-synthase (CBS) domains that interact with regulators of the channel, like pH. CBS domains are involved in slow gating of the channel and closure of the common gate. The dimer structure of the ClC-1 channel is illustrated in Fig. 12.10C, depicting the flow of Cl^- ions.

7. Origin of the resting potential in other cells

The experiments in muscle fibers described in Sections 3 and 4, led to the conclusion that the resting potential in this cell type depends on the activity of K^+ and Cl^- channels, with a minor contribution of Na^+ ions. The origin of the resting potential has also been studied in other cell types. The general conclusion is that the relative influence of these ions is different from one preparation to another. For instance, the resting potential of mammalian ventricular muscle fibers basically follows the behavior expected for a K^+ electrode when exposed to different $[K^+]_o$. No large differences have

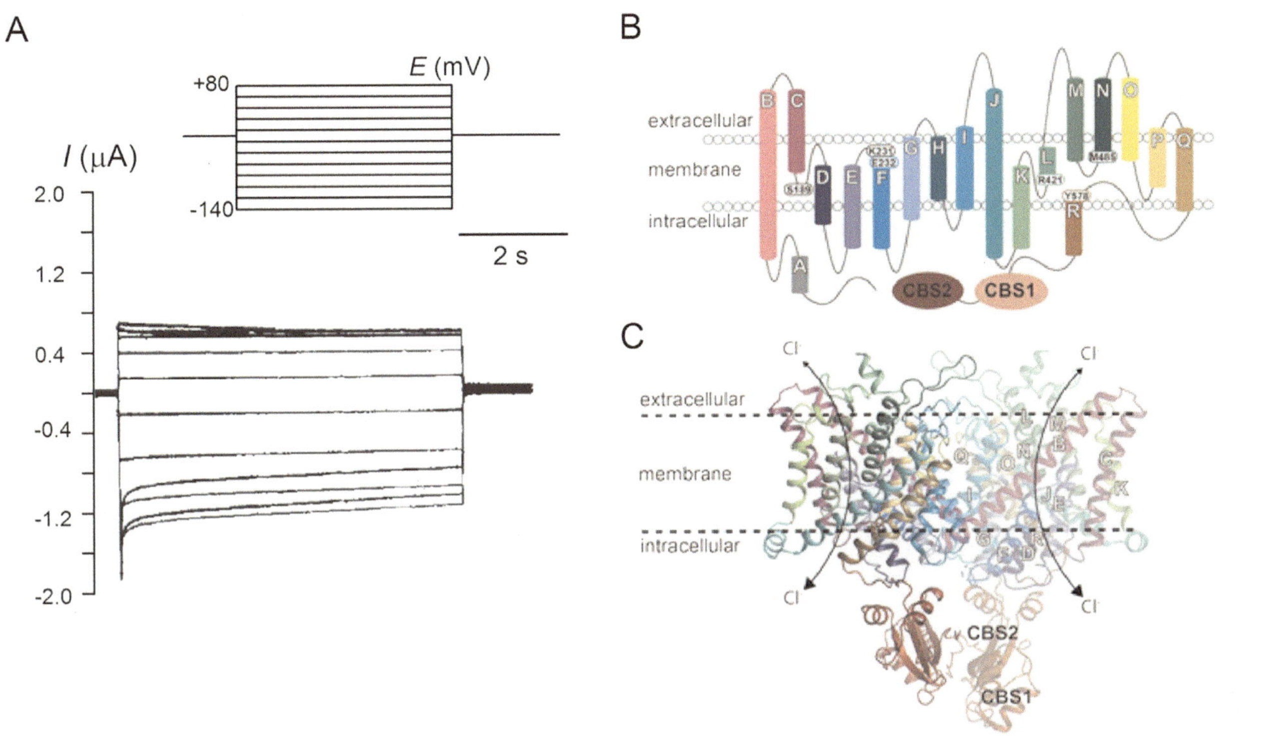

FIGURE 12.10 (A) Chloride currents through ClC-1 channels expressed in *Xenopus* oocytes recorded with the pulse protocol shown in the inset. (B) Structure of the human ClC-1 channel. A single monomer is shown. Helices are labeled with white letters. Numbers indicate amino acids associated with ion permeation through the pore. (C) Overall structure with one of the monomers in pale colors. CBS: Cystathionine-β-synthase, domains involved in channel gating. Letters correspond to helices shown in *B*. *(A) Modified from Steinmeyer et al. (1991). (B) Modified from Wang et al. (2019). (C) Modified from Wang et al. (2019).*

been found when the product $[K^+]_o \times [Cl^-]_o$ did not remain constant. The influence of Cl^- ions in this preparation is much less relevant than in skeletal muscle since the conductance ratio g_{Cl}/g_K is only 0.17 (Fozzard & Lee, 1976), whereas in skeletal muscle it is about 3.

The effect of ions on the resting potential of smooth muscle cells has been studied in many preparations with likewise different results. For example, the resting potential of the guinea pig's intestinal smooth muscle deviates significantly from the behavior predicted by the Nernst equation for a K^+ electrode when $[K^+]_o$ is less than 30 mM, leading to the conclusion that in this preparation Na^+ ions contribution is much larger than in skeletal muscle and that Cl^- ions do not appear to be passively distributed (See Chapter 15) and Kuriyama (1963). On the other hand, in arteriolar smooth muscle, the Cl^- permeability is only one tenth that of K^+, and the resting Na^+ permeability is about half the relative value obtained in skeletal muscle. Therefore, the resting potential of arteriolar smooth muscle is determined almost exclusively by K^+ ions. However, experimental results indicated that when the extracellular K^+ concentration is very low, there is also a significant deviation from the expected Nernst potential (Hirst & Van Helden, 1982).

A similar diversity of responses to external ions on membrane potential is observed in experiments conducted in neurons. For instance, the membrane potential of some marine mollusk ganglion neurons is almost completely insensitive to changes in extracellular Cl^- concentration, while they also show a significant resting permeability to Na^+ (Marmor and Gorman, 1970). A similar conclusion has been reached for experiments performed in dorsal root ganglion neurons, where Cl^- movements occur without noticeable changes in the resting potential and the input resistance of those cells does not depend on the extracellular concentration of Cl^-, supporting the idea that the resting permeability to this anion is extremely low, as discussed in Chapter 15 and Alvarez-Leefmans et al. (1988). However, different results have been reported for other neurons. In mammalian spinal motoneurons, a high K^+ permeability sets the resting potential at high extracellular concentrations. However, K^+ ions appear to have no role when $[K^+]_o$ is less than 3 mM, which is the physiological extracellular concentration. This indicates that a significant resting permeability to Na^+ is involved. In fact, the input conductance ratio of $g_{Na}/g_K = 0.13$ has been estimated. Cl^- ions also have an impact on the resting potential of these cells since the absence of these ions in the extracellular medium results in a depolarization of up to 7 mV. The input conductance ratio is $g_{Cl}/g_K = 0.25$. This value is significant, though much lower than in skeletal muscle (Forsythe & Redman, 1988).

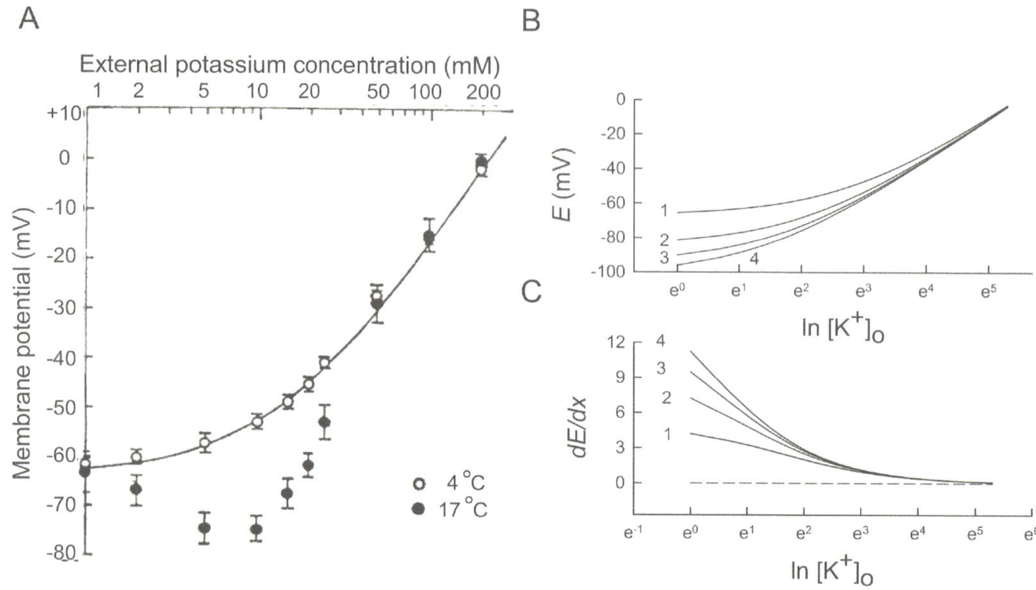

FIGURE 12.11 (A) The contribution of the Na^+–K^+ pump to the resting potential of a neuron of the marine mollusk *Anisodoris nobilis*. The empty circles correspond to the resting potential values measured at different concentrations of external K^+, at $T = 4°C$. The solid line corresponds to the Goldman–Hodgkin–Katz equation (Eq. 12.19) with $\alpha = 0.033$ and $[K^+]_i = 235$ mM. Values indicated by filled circles were obtained at $T = 17°C$. (B) Plots of the Goldman–Hodgkin–Katz equation for different values of r in Eq. (12.20), at $T = 20°C$. (C) Plots of the first derivative of traces in *B*. *(A) Modified from Marmor and Gorman (1970).*

8. Contribution of the $Na^+ - K^+$ pump to the resting potential

The asymmetry in the ionic concentrations across the cell membrane with a high content of K^+ and a low content of Na^+ in the intracellular space, and high concentration of Na^+ and low K^+ in the extracellular medium, indicates that the membrane effectively separates these two media. However, the membrane is not an impenetrable barrier for these ions, and in the long term, both the resting potential and the differences in ionic concentrations would be cancelled. The fact that this does not happen is due to a mechanism that continuously and actively transports Na^+ and K^+ ions: that is the Na^+–K^+ ATPase (NKA) also called the Na^+–K^+ pump.

The Na^+–K^+ pump is electrogenic because, for each cycle, it outwardly transports a net positive charge, thereby contributing to the resting potential in muscle and other cells. Several laboratories have found that the transport number of the pump, defined as the ratio of the number of Na^+ ions to that of K^+ ions, is close to 1.5 (3 Na^+ ions pumped out of the cell for every 2 K^+ ions pumped into the cell. Therefore, the continuous pump activity hyperpolarizes the cell.

Fig. 12.11A shows the contribution of the Na^+–K^+ pump to the resting potential of a ganglion neuron of a marine mollusk from the experiments of Marmor and Gorman (1970). The pump's activity is highly temperature-dependent, so its contribution to the resting potential becomes more evident at high temperatures. As shown in Fig. 12.11, the resting potential at 17°C (filled circles) reaches the most negative value at $[K^+]_o = 10$ mM, and at both lower and higher concentrations, the neuron is depolarized.

The depolarization with increasing $[K^+]_o$ was explained by the high membrane permeability of the neuron to these ions, while that observed at low $[K^+]_o$ was related to a decrease in the pump's activity due to the virtual absence of K^+ in the extracellular medium. A similar finding has been reported in experiments performed in skeletal muscle (Fig. 12.7). Fig. 12.7 shows that when the temperature is lowered to 4°C (•), the pump's activity decreases, and the membrane follows the behavior predicted by the Goldman–Hodgkin–Katz equation (Eq. 12.19). Ouabain, a Na^+–K^+ pump inhibitor, suppresses the hyperpolarization, all of which indicates that this transport mechanisms is electrogenic and contributes to the resting potential under physiological conditions.

Mullins and Noda (1963) examined quantitatively the pump's role in determining the resting potential in muscle and proposed a modification to the Goldman–Hodgkin–Katz equation to include the pump contribution to the membrane potential. Accordingly, the Goldman–Hodgkin–Katz equation (Eq. 12.19) leads to Eq. (12.20) by including the electrogenic contribution of the pump:

$$E = \frac{RT}{F} \ln \frac{r[K^+]_o + \alpha[Na^+]_o}{r[K^+]_i + \alpha[Na^+]_i} \tag{12.20}$$

Note that Eq. (12.20) is reduced to Eq. (12.19) when r, the stoichiometry of the pump, is 1.

However, it is interesting to note that although the Mullins–Noda equation (Eq. 12.20) takes into consideration the stoichiometry of the Na^+–K^+ pump, it is not a completely satisfactory description of the resting potential since Eqs. (12.19) and (12.20) do not predict minimum or the maximum values (peaks) of the resting potential, and experimentally, the relationship between the resting potential and $[K^+]_o$ has a distinct peak minimum, as shown in Fig. 12.11. From a mathematical point of view, Eq. (12.19) is equivalent to the Goldman–Hodgkin–Katz equation (to prove it, it is sufficient to divide the numerator and the denominator by r in the argument of the logarithm in Eq. (12.19). A simple approach to demonstrate the absence of peaks in Eq. (12.19) or Eq. (12.20) is to take the first derivative of the function since maximum and minimum values are found where the derivative changes sign. Fig. 12.11B shows the predicted values of E as a function of ln $[K^+]_o$ for different values of r, and the corresponding plots of dE/dx are shown in Fig. 12.11C. All dE/dx values are positive and decrease asymptotically to zero with no negative values compatible with a peak. The equation predicts that when $[K^+]_o$ tends to zero, E tends asymptotically to a constant value, and when $[K^+]_o$ increases, E also increases continuously. However, as mentioned above, the experimental values of the resting potential as a function of $[K^+]_o$ do not decrease asymptotically at low $[K^+]_o$.

The ion current generated by the Na^+–K^+ pump has been measured in different cell types under voltage-clamp conditions. Since more Na^+ ions are transported out of the cell than K^+ ions inwards, an outward current is generated. These currents have been characterized in isolated cardiac cells using the whole-cell patch-clamp technique (Gadsby et al., 1985). Currents associated with the pump are activated when $[Na^+]_i$ is increased and are blocked following addition of ouabain, a Na^+–K^+ pump inhibitor. The pump current shows marked voltage dependence; it declines steadily from a maximal level near 0 mV, becoming very small at −140 mV. The Na^+–K^+ pump thus contributes to the electrical activity of the heart. The pump current has been measured in other cell types using the voltage-clamp technique (Apell, 1989).

9. The sodium–calcium exchanger

The sodium–calcium exchanger (NCX) will be only briefly discussed in this chapter, since it appears to play no direct role in the resting potential. However, the NCX is an important ion transport system in excitable cells like cardiac myocytes, where it plays a significant role in regulating action potential duration (Armoundas et al., 2003). Under physiological conditions, the exchanger is involved in Ca^{2+} homeostasis, and it is the most important mechanism transporting Ca^{2+} from the intracellular to the extracellular medium in cardiac myocytes. The exchanger is electrogenic since it transports 3 Na^+ per 1 Ca^{2+}. Therefore, an ionic current associated with its activity is expected. The NCK current has indeed been measured with the whole-cell patch-clamp technique in isolated cardiac myocytes by several groups.

The equilibrium potential of the exchanger (E_{Na-Ca}) is:

$$E_{Na-Ca} = 3E_{Na} - 2E_{Ca} \tag{12.21}$$

where E_{Na} and E_{Ca} are the Na^+ and Ca^{2+} equilibrium potentials calculated with the Nernst equation, respectively. E_{Na-Ca} in cardiac myocytes is close to −20 mV. Therefore, in these cells, a depolarizing inward current at the resting potential (~−80 mV) would be expected. However, inhibition of the exchanger by specific blockers has only negligible effects on the resting potential.

10. Maintaining the product $[K^+]_o \times [Cl^-]_o$ constant preserves cell volume in skeletal muscle fibers

To illustrate the importance of maintaining constant the product $[K^+]_o \times [Cl^-]$, as discussed above in Sections 3 and 4, let us consider the following example: suppose that a skeletal muscle fiber has as main intracellular ions K^+, Cl^- and impermeable anions A^- in the following concentrations (mM): $[K^+]_i = 119.5$, $[A^-]_i = 117$, $[Cl^-]_i = 2.5$, and has a volume: $V_i = 1\ mm^3$. The extracellular medium has the following composition: $[K^+]_o = 2.5$, $[Na^+]_o = 117$, and $[Cl^-]_o = 119.5$.

Using the equations described in the preceding sections, and applying the Nernst equation for K^+ ions (Eq. 12.6), we calculate an equilibrium potential $E_K = -97.4$ mV. Similarly, we can calculate E_{Cl} (Eq. 12.5), resulting in $E_{Cl} = -97.4$ mV. Suppose now that $[K^+]_o$ is increased to 40 mM, keeping the osmolarity constant, for which the extracellular concentration of Na^+ is reduced in the same proportion. Thus, the new concentrations in the extracellular medium are (mM): $[Na^+]_o' = 79.5$, $[K^+]_o' = 40$, $[Cl^-]_o' = 119.5$.

Based on these results, we can now answer the following questions: What will be the new intracellular concentration of K^+ ions? What will be the new cell volume? What will be the value of the equilibrium potential for K^+ ions under these conditions?

According to Eq. (12.17), it becomes clear that $[K^+]_i$ does not change in this example, so $[K^+]_i' = 119.5$, so we can calculate the equilibrium potential for K^+, applying the Nernst equation: $E_K = -27.6$ mV.

However, there are changes in cell volume. To calculate these changes, we have to consider that the osmotic equilibrium is determined only by the final concentrations of the nondiffusible ions. Thus, due to the high permeability of the cell membrane to water, the concentration of these ions both intra- and extracellular is the same (if it were higher, for example, inside the cell, water would pass from the outside to equalize it). In the present case, the concentrations are (mM): $[A^-]_i = 117$ and $[Na^+]_o = 117$.

On the other hand, the number of moles of the nondiffusible anion remains constant, which implies that: $V_i \times [A^-]_i = V_i' \times [A^-]_i'$ and since $[A^-]_i = [Na^+]_o$ it follows that:

$$\frac{V_i}{V_i'} = \frac{[Na^+]_o'}{[Na^+]_o} \tag{12.22}$$

In this example, the relative cell volume change $V_i/V_i' = 0.68$ and $V_{i'} = 1.47\ mm^3$.

In short, the cell increases its volume but it maintains the same concentration of K^+. On the other hand, the concentration of Cl^- increases since the relation given by Eq. (12.8) must be fulfilled, from which $[Cl^-]_i' = 40$ mM is obtained. The new equilibrium potential E_{Cl}' is obtained from Eq. (12.5) and it turns out to be -27.6 mV.

In conclusion, when testing different concentrations of K^+, keeping the extracellular osmolarity constant by replacing Na^+ with K^+, is not a sufficient condition to prevent variations in either intracellular ion concentrations (particularly Cl^-) or in cell volume. Hodgkin and Horowicz (1959) concluded that it is essential to maintain a constant product $[K^+]_o \times [Cl^-]_o$ to preserve the cell volume.

11. Derivation of the Nernst equation

The Nernst equation is essential for understanding the direction of ion movement across the membrane when ion channels open, not only under resting conditions but also during action potentials in excitable cells. Ions can move as a result of concentration and electrical gradients. When these two forces act in opposite directions, it is not obvious whether ions would move in or out of the cell. Based on the intracellular and extracellular ionic concentrations, the Nernst equation calculates the equilibrium potential for a particular ion species, which is the voltage at which no net ion movement occurs across the membrane. The derivation of the Nernst equation is dealt with in detail in Chapter 10 (Membrane Diffusion and Permeability).

Consider the movement of ions by diffusion (flux) due to concentration gradients. As discussed in Chapter 10, Fick's first law, states that the flux occurs from a high to a low concentration region. In mathematical terms:

$$J = -D\frac{\partial C}{\partial x} \tag{12.23}$$

where J is the flux, C is the concentration of a substance (which is position-dependent), and D is the diffusion coefficient. In the physiological literature, the term "permeability" is commonly used when describing ion flow across cell membranes, as discussed in detail in Chapter 10.

D is related to the mobility by the Einstein law of diffusion that states that:

$$D = U\frac{RT}{F} \tag{12.24}$$

where U is the mobility, R is the universal gas constant, and T is the absolute temperature.

Therefore, J can be written as:

$$J = -U\frac{RT}{F}\frac{\partial C}{\partial x} \tag{12.25}$$

On the other hand, in the presence of an electric field, the flux J' is:

$$J' = -zUC\frac{\partial V}{\partial x} \tag{12.26}$$

where C is the ionic concentration at position x, and the voltage gradient (∂V/∂x) acts on ion charges.

No net movement of ions is expected to occur when $J = -J'$, and therefore, combining Eqs. (12.25) and 12.26:

$$-\frac{RT}{zF}\frac{\partial C}{\partial x} = C\frac{\partial V}{\partial x} \tag{12.27}$$

And after rearranging:

$$dV = \frac{-RT}{zF}\frac{\partial C}{C} \tag{12.28}$$

Finally, integration of both sides of Eq. (12.28) yields:

$$V_1 - V_2 = \frac{RT}{zF}\ln(C_2 - C_1) \tag{12.29}$$

Eq. (12.29) is the Nernst equation.

12. The Goldman–Hodgkin–Katz equation

The Goldman–Hodgkin–Katz (GHK) equation calculates the expected membrane potential based on the intra- and extracellular concentrations of permeant ions and the membrane permeability values to these ions. For Na^+, K^+, and Cl^- ions, it takes the form:

$$\mathrm{E} = \frac{RT}{F}\ln\frac{P_{Na}[\mathrm{Na}^+]_\mathrm{o} + P_K[\mathrm{K}^+]_\mathrm{o} + P_{Cl}[\mathrm{Cl}^-]_\mathrm{i}}{P_{Na}[\mathrm{Na}^+]_\mathrm{i} + P_K[\mathrm{K}^+]_\mathrm{i} + P_{Cl}[\mathrm{Cl}^-]_\mathrm{o}} \tag{12.30}$$

where R, T, and F are thermodynamic constants and P is the membrane permeability to each ion, which depends on their relative mobility (See Chapter 10).

This equation is reduced to Eq. (12.19) in the absence of Cl^-, or when $PCl \rightarrow 0$. The derivation of the GHK equation is described in detail in the seminal paper of Hodgkin and Katz (1949). The basic assumptions for its derivation are:

- Ions move under the influence of diffusion and electric fields in a way that is similar to that in a free solution.
- The voltage gradient through the membrane is constant.
- The membrane is homogenous.
- The ion concentrations at the edge of the membrane are directly proportional to those in the solutions bounding the membrane.

The first assumption leads to equations linking currents carried by ions with concentration and voltage gradients, following an approach similar to that used in Section 11 for the derivation of the Nernst equation. The second and third assumptions allow direct integration of the differential equations involved. Finally, using the last assumption, the boundary concentration of an ion species can be calculated by multiplying its concentration in the bulk solution times its partition coefficient (β_i). The permeability coefficient of an ion (Pi) in the GHK equation is defined as:

$$Pi = U_i\beta iRT/dF \tag{12.31}$$

where μ_i is the ion mobility, d is the membrane thickness, and R, T, and F are thermodynamic constants. Permeability coefficients are discussed in detail in Chapter 10.

The Goldman–Hodgkin–Katz equation, first derived by David Goldman, has been extremely useful in the study of ion channels and has been recently reviewed from a historical perspective (Alvarez & Latorre, 2017).

References

Adelman, J. P., Clapham, D. E., Hibino, H., Inanobe, A., et al. (2023). Inwardly rectifying potassium channels (K_{IR}) in GtoPdb v.2023.1. https://doi.org/10.2218/gtopdb/F74/2023.1.

Alvarez, O., & Latorre, R. (2017). The enduring legacy of the "constant field equation" in membrane ion transport. *J Gen Physiol, 149*, 911–920.

Alvarez-Leefmans, F. J., Gamiño, S. M., Giraldez, F., & Noguerón, I. (1988). Intracellular chloride regulation in amphibian dorsal root ganglion neurones studied with ion-selective microelectrodes. *The Journal of Physiology, 406*, 225–246. https://doi.org/10.1113/jphysiol.1988.sp017378

Apell, H. J. (1989). Electrogenic properties of the Na,K pump. *Journal of Membrane Biology, 110*(2), 103–114. https://doi.org/10.1007/BF01869466

Armoundas, A. A., Hobai, I. A., Tomaselli, G. F., Winslow, R. L., & O'Rourke, B. (2003). Role of sodium-calcium exchanger in modulating the action potential of ventricular myocytes from normal and failing hearts. *Circulation Research, 93*(1), 46–53. https://doi.org/10.1161/01.RES.0000080932.98903.D8

Baronas, V. A., & Kurata, H. T. (2014). Inward rectifiers and their regulation by endogenous polyamines. *Frontiers in Physiology, 5*, 1–4.

Boyle, P. J., & Conway, E. J. (1941). Potassium accumulation in muscle and associated changes. *The Journal of Physiology, 100*, 1–63.

Chen, T.-Y., & Hwang, T.-C. (2008). CLC-0 and CFTR: chloride channels evolved from transporters. *Physiological Reviews, 88*(2), 351–387. https://doi.org/10.1152/physrev.00058.2006. PMID: 18391167.

Conway, E. J., & Boyle, P. J. (1939). A mechanism for the concentrating of potassium by cells, with experimental verification in muscle. *Nature, 114*, 709–710.

DiFranco, M., Yu, C., Quiñonez, M., & Vergara, J. L. (2015). Inward rectifier potassium currents in mammalian skeletal muscle fibres. *The Journal of Physiology, 593*, 1213–1238.

Estrada, F., & Sanchez, J. A. (1991). The effect of amiloride on the resting potential and the electrical constants of frog skeletal muscle fibres. *The Journal of Physiology, 433*, 705–717.

Forsythe, I. D., & Redman, S. J. (1988). The dependence of motoneurone membrane potential on extracellular ion concentrations studied in isolated rat spinal cord. *Journal of Physiology, 404*, 83–99. https://doi.org/10.1113/jphysiol.1988.sp017280

Fozzard, H. A., & Lee, C. O. (1976). Influence of changes in external potassium and chloride ions on membrane potential and intracellular potassium ion activity in rabbit ventricular muscle. *The Journal of Physiology, 256*, 663–689.

Gadsby, D. C., Kimura, J., & Noma, A. (1985). Voltage dependence of Na/K pump current in isolated heart cells. *Nature, 315*, 63–65.

Hamill, O. P., Marty, A., Neher, E., Sakmann, B., & Sigworth, F. J. (1981). Improved Patch-Clamp techniques for high-resolution current recording from cells and cell-free membrane patches. *Pflügers Archiv, 391*, 85–100.

Hibino, H., Inanobe, A., Furutani, K., Murakami, S., Findlay, I., & Kurachi, Y. (2010). Inwardly rectifying potassium channels: Their structure, function, and physiological roles. *Physiological Reviews, 90*, 291–366.

Hirst, G. H. S., & Van Helden, D. F. (1982). Ionic basis of the resting potential of submucosal arterioles in the Ileum of the Guinea pig. *The Journal of Physiology, 333*, 53–67.

Hodgkin, A. L., & Horowicz, P. (1959). The influence of potassium and chloride ions on the membrane potential of single muscle fibres. *The Journal of Physiology, 148*, 127–160.

Hodgkin, A. L., & Katz, B. (1949). The effect of sodium ions on the electrical activity of the giant axon of the squid. *The Journal of Physiology, 108*, 37–77.

Katz, B. (1949). Les constantes éléctriques de la membrane du muscle. *Archives des Sciences Physiologiques, 2*, 285–299.

Kubo, Y., Baldwin, T. J., Jan, Y. N., & Jan, L. Y. (1993). Primary structure and functional expression of a mouse inward rectifier potassium channel. *Nature, 362*, 127–133.

Kuriyama, H. (1963). The influence of potassium, sodium and chloride on the membrane potential of the smooth muscle of taenia coli. *Journal of Physiology, 166*(1), 15–28. https://doi.org/10.1113/jphysiol.1963.sp007088

Lu, Z., & MacKinnon, R. (1994). Electrostatic tuning of Mg^{2+} affinity in an inward-rectifier K^{+} channel. *Nature, 371*, 243–246.

Marmor, M. F., & Gorman, A. L. F. (1970). Membrane potential as the sum of ionic and metabolic components. *Science, 67*, 65–67.

Mullins, L., & Noda, K. (1963). The influence of sodium-free solutions on the membrane potential of frog muscle fibers. *J Gen Physiol, 47*, 117–132.

Needham, D. M. (1971). The biochemistry of muscular contraction in its historical development. *Machina Carnis*. Cambridge, UK: Cambridge University Press.

Pedersen, T. H., Riisager, A., de Paoli, F. V., Chen, T. Y., & Nielsen, O. B. (2016). Role of physiological ClC-1 Cl^{-} ion channel regulation for the excitability and function of working skeletal muscle. *The Journal of General Physiology*, 147, 291–308.

Saviane, C., Conti, F., & Pusch, M. (1999). The muscle chloride channel ClC-1 has a double-barreled appeara nce that is differentially affected in dominant and recessive myotonia. *J Gen Physiol, 113*, 457–467.

Steinmeyer, K., Ortland, C., & Jentsch, T. J. (1991). Primary structure and functional expression of a developmentally regulated skeletal muscle chloride channel. *Nature, 354*, 301–304.

Tao, X., Avalos, J. L., Chen, J., & MacKinnon, R. (2009). Crystal structure of the eukaryotic strong inward-rectifier K^{+} channel Kir2.2 at 3.1 Å resolution. *Science, 326*, 1668–1674.

Vandenberg, C. A. (1987). Inward rectification of a potassium channel in cardiac ventricular cells depends on internal magnesium ions. *Proc Natl Acad Sci, 84*, 2560–2564.

Wang, K., Preisler, S. S., Zhang, L., Cui, Y., Missel, J. W., Grønberg, C., et al. (2019). Structure of the human ClC-1 chloride channel. *PLoS Biology, 17*, 1–20.

Xie, J., Ke, M., Xu, L., Lin, S., et al. (2020). Structure of the human sodium leak channel NALCN in complex with FAM155A. *Nature Communications, 11*(1), 5831. https://doi.org/10.1038/s41467-020-19667-z

Chapter 13

ATP synthase: Structure, function, and channel formation

Giovanna Lippe[1], Federico Fogolari[2], Ildikó Szabó[3] and Paolo Bernardi[4]

[1]*Dipartimento di Area Medica, Università degli Studi di Udine, Udine, Italy;* [2]*Dipartimento di Scienze Matematiche, Informatiche e Fisiche, Università degli Studi di Udine, Udine, Italy;* [3]*Dipartimento di Biologia, Università degli Studi di Padova, Padova, Italy;* [4]*Dipartimento di Scienze Biomediche, Università degli Studi di Padova, Padova, Italy*

1. Summary

This chapter presents the structure and function of ATP synthases, the ubiquitous enzymes of energy conservation. Although the focus of the chapter is on F_1F_O ATP synthase, we will first illustrate the structural features of the simpler bacterial enzyme. This should allow a better understanding of the additional features conferred to F_1F_O ATP synthases by unique subunits that are not essential for ATP synthesis. These "accessory" subunits mediate interactions between enzyme monomers in mitochondria resulting in the formation of dimers and then of long rows of oligomers located in the depth of the cristae, which they critically contribute to forming. An emerging function of mitochondrial ATP synthase is the Ca^{2+}-dependent formation of a high-conductance channel, the permeability transition pore (PTP). We will cover the features of this channel and potential mechanisms through which ATP synthase is turned from an energy-conserving into an energy-dissipating system.

2. Introduction

F_1F_O-ATP synthase (ATP synthase), one of the most ubiquitous and abundant enzymes of mitochondria, chloroplasts, and bacteria, makes ATP from ADP and Pi using the proton motive force across the membranes (Kühlbrandt, 2019), as first proposed in the chemiosmotic theory of energy conservation by Peter Mitchell over 60 years ago (see Mitchell, 2011). The gradient is created by H^+ pumping by the respiratory chain complexes coupled to electron transfer from NADH and $FADH_2$ to oxygen in bacteria and mitochondria; and by the light-driven redox activity of photosystems I and II and the *b6f* complex in chloroplasts. All ATP synthases are able to transform the energy of the H^+ gradient into ATP through a common mechanism of rotary catalysis achieving an apparently perfect energy conversion (Soga et al., 2017). In eukaryotes, the enzyme has additional functions. One is to shape the inner membrane cristae by formation of dimers that assemble into long rows of oligomers; and an emerging unexpected function is to form the permeability transition pore (PTP), a Ca^{2+}-activated, high-conductance channel that is involved in Ca^{2+} signaling and cell death.

3. Structure and function of ATP synthases

ATP synthases are ~600-kDa complexes formed by 8–17 different subunits depending on their source. They are all are organized in two principal domains, the spherical, catalytic F_1 sector protruding from the membrane and the F_O subcomplex embedded in the membrane, which are connected by a central stalk (CS) and a peripheral stalk (PS). F_1 possesses the catalytic sites where ATP is synthesized, while F_O translocates H^+ or Na^+ in some archaea living in extreme environments (Mayer et al., 2015) across the membrane. Interestingly, it has recently been proposed that in addition to H^+ the mitochondrial enzyme can also translocate K^+ (Juhaszova et al., 2021). F_1 and F_O are firmly attached to each other by connections between (i) the CS rotating within F_1 and the ring of subunits c (c-ring) rotating within F_O and (ii) the

Cell Physiology Sourcebook. https://doi.org/10.1016/B978-0-12-811114-7.00029-3

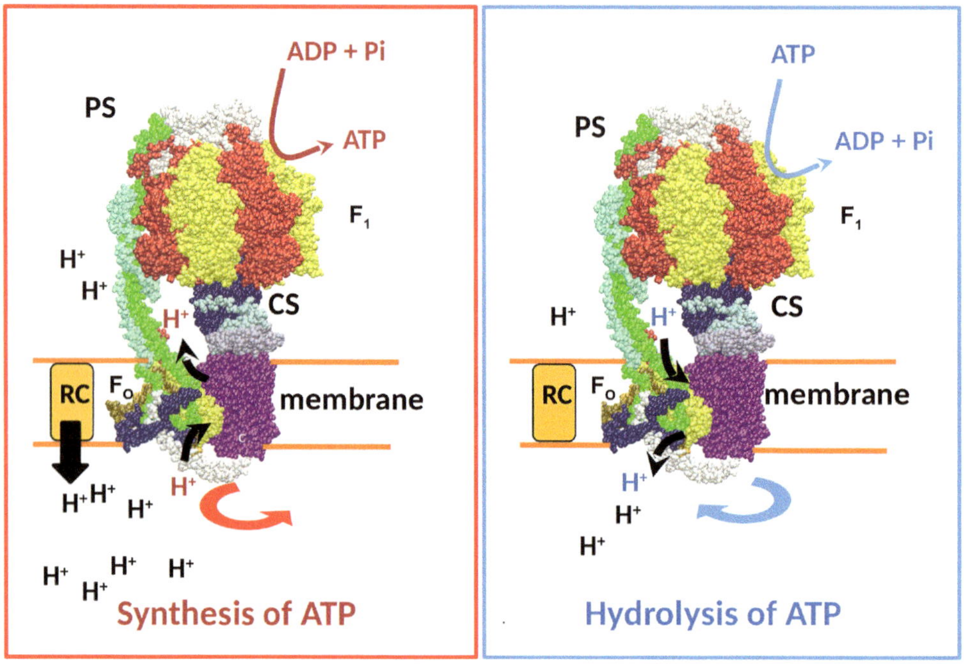

FIGURE 13.1 ATP synthase is a fully reversible rotary machine. ATP synthase can operate in the "ATP synthesis mode" in respiring mitochondria, where the H^+ gradient generated by the respiratory chain is used to power the counterclockwise rotation of the c-ring (as seen from the matrix). Conversely, in the "ATP hydrolysis mode" the energy released by ATP hydrolysis is used to spin the c-ring in the opposite direction, resulting in H^+ pumping across the membrane. *CS*, Central stalk; *PS*, Peripheral stalk; *RC*, respiratory chain.

stationary PS and F_1 (formed by three α and three β subunits) on one side of the membrane and the set of F_O subunits in contact with the c-ring from the opposite side (Gu et al., 2019; Pinke et al., 2020; Spikes et al., 2020). This arrangement allows the ATP synthase to operate in the "ATP synthesis mode" in respiring mitochondria, where the H^+ (or Na^+) gradient is used to power c-ring rotation; or in the "ATP hydrolysis mode" when the H^+ gradient is dissipated and ATP is provided e.g., by glycolysis, a condition where the energy released by ATP hydrolysis is used to spin the c-ring, resulting in H^+ pumping across the membrane (Fig. 13.1). A notable feature of the mitochondrial ATP synthase is its association into dimers that self-assemble into long rows of oligomers to develop the typical cristae, a process that is mediated by specific subunits connecting within the membrane (Kühlbrandt, 2019). Because of their central role in cellular metabolism, ATP synthases have been studied intensely since their discovery in the 1960s (Kagawa & Racker, 1966). The importance of these studies is highlighted by the award of the Nobel Prize to Peter Mitchell for his formulation of the chemiosmotic theory in 1978, and to Paul Boyer and John Walker for unraveling the structure and catalytic mechanism of ATP synthase in 1997.

3.1 Overall structural organization and functional features

Over a period of about 30 years crystallography first, and high-resolution cryo-electron microscopy (cryo-EM) next, revealed the entire structure of ATP synthases from bacteria (Guo, Suzuki, & Rubinstein, 2019; Morales-Rios, Montgomery, Leslie, & Walker, 2015; Sobti et al., 2016), chloroplasts (Hahn et al., 2018), and mitochondria from various sources (Gu et al., 2019; Guo et al., 2017; Hahn et al., 2016; Pinke et al., 2020; Spikes et al., 2020, 2021; Srivastava et al., 2018; Vinothkumar et al., 2016; Zhou et al., 2015). These studies have shown that the catalytic F_1 and CS sectors are highly conserved while the F_O and PS display considerable variation in subunit composition, sequence and structure suggesting an evolutionary divergence that may underlie additional functions for the enzyme complex (Pinke et al., 2020).

Crystallography has defined the structure of many F_1 and F_O subcomplexes, while it came short of obtaining the structure of the entire ATP synthase with the exception of the bacterial complex (Morales-Rios, Montgomery, Leslie, & Walker, 2015). The first crystal structure of bovine F_1 revealed most of its basic structural features (Abrahams et al., 1994), which were confirmed in subsequent studies of F_1 from other sources (Bowler et al., 2007; Cingolani & Duncan, 2011; Kabaleeswaran et al., 2006). The F_1 sector is always an assembly of three α and three β subunits arranged in near-three-fold symmetry constituting a ring stator around the γ subunit, which forms the CS by associating at the foot with subunit ε.

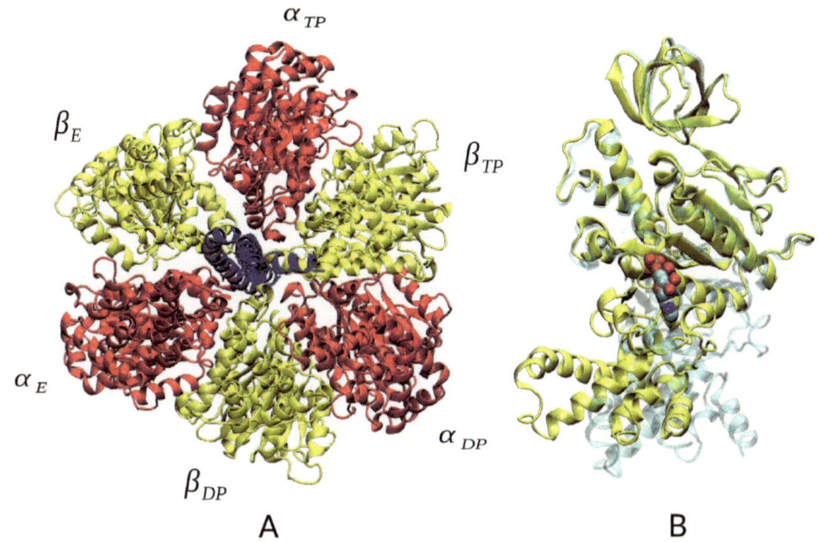

FIGURE 13.2 Crystal structure of F_1 from bovine mitochondria. In (A) F_1 is shown as a section view from the matrix side (PDB ID: 1BMF) (Abrahams et al., 1994). The catalytic sites are located at the interfaces between α (yellow chains) and β subunits (red chains). Each β subunit catalytic site at any time during the catalytic cycle is designated as β_{TP}, β_{DP}, or β_E, depending on the bound nucleotide (ATP, ADP and none) respectively. β_E takes an open conformation, and both β_{TP} and β_{DP} bind nucleotides and take a closed conformation. In (B) the N-terminal region of β_E (shown as a cyan transparent cartoon) is superimposed to the same region of β_{DP} (shown as a yellow cartoon) binding the nucleotide (shown as van der Waals spheres). The C-terminal domain of the closed β subunit appears to push the γ subunit.

In mitochondria, the foot of the CS also contains subunit δ. In chloroplasts and bacteria subunit δ is instead located on top of F_1 to connect the PS to the $\alpha_3\beta_3$ stator, a function that in mitochondria is fulfilled by subunit oligomycin sensitivity conferral protein (OSCP). The three αβ pairs contain six nucleotide binding sites, of which only three are catalytically active and mostly contributed by the β subunits. During the catalytic cycle, the β subunits undergo three sequential conformational changes into states called β_{DP} (bound ADP), which is thought to represent the catalytically active conformational state; β_{TP} (bound ATP); and β_E (empty), which has no bound nucleotides. β_{DP} and β_{TP} adopt the so-called closed conformation in which the C-terminal domain rotates inwardly to the γ subunit, while β_E adopts an open conformation, swinging the C-terminal domain outwardly (Fig. 13.2). Very importantly, these structures have provided a fundamental support to the catalytic rotary mechanism of intact ATP synthase, which was first hypothesized in the "cyclical binding change" mechanism by Paul Boyer. According to this mechanism, in the synthesis mode the rotation of γ subunit sequentially forces each of the three catalytic sites into the three major conformations (β_{DP} and β_{TP} and β_E), thereby catalyzing the synthesis of one ATP molecule per each 120° rotation of the γ subunit (Abrahams et al., 1994; Boyer, 1993). Rotation of subunit γ was first visualized in *Bacillus PS3* F_1 immobilized on a glass surface after attaching a fluorescent actin filament to subunit γ. Upon addition of ATP the γ subunit rotates clockwise when viewed from the side of F_1 (Noji et al., 1997). A further technical sophistication allowed to follow in real time the binding and release of nucleotides at the three catalytic sites during rotation of subunit γ (Adachi et al., 2007). Since then, the F_1 rotary dynamics have been intensively studied by single-molecule assays in F_1 from many sources, including the human (Suzuki et al., 2014) and bovine species (Kobayashi et al., 2020).

3.2 The bacterial enzyme

As already mentioned, the F_O part and the PS have a variable composition in different species. The extraordinary progress made in recent years by the "resolution revolution" of cryo-EM (Kühlbrandt, 2014) has made determination of the structure in many organisms possible. The bacterial and chloroplast species have the simplest F_O, which comprises one subunit a and a c-ring with a variable number of copies of subunit c (usually 10–15 with a record 17 in *Burkholderia pseudomallei*) (Nirody et al., 2020). A brief description of these structures is presented here, because it can help to better appreciate the basic mechanistic features of rotary catalysis, as well as the functional meaning of the higher complexity of the mitochondrial enzyme. In the c-ring, each c subunit folds into a hairpin of two α-helices (the N- and C-termini corresponding to the inner and outer ring surface, respectively) connected by a short loop essential for the c-ring-CS interaction (Dmitriev et al., 1999). Subunit a (the largest in F_O) is arranged in a concave four-helix horizontal bundle surrounding the c-ring, and

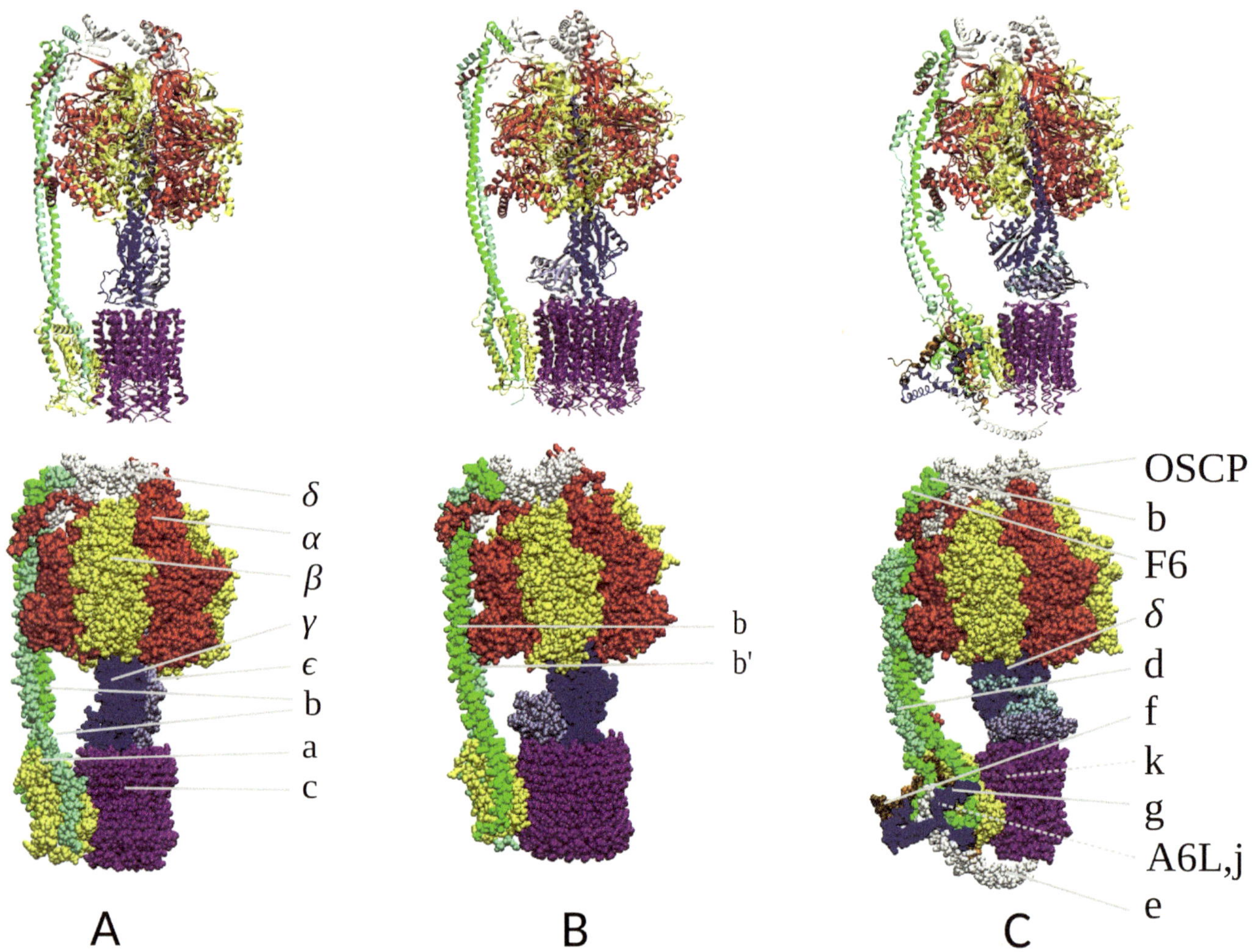

FIGURE 13.3 Subunits of ATP synthases. (A) Bacterial ATP synthase from *E. coli* (PDB id.: 6OQR) (Sobti et al., 2020) is shown in cartoon (upper) and as solvent accessible surface (lower), subunits are indicated. (B) Chloroplast ATP synthase from spinach (PDB id.: 6FKF) (Hahn et al., 2018) is shown. Subunits are indicated, except for those already indicated in (A). (C) Bovine ATP synthase (PDB id.: 6ZPO) (Spikes et al., 2020) is shown with subunits indicated. Subunit e is taken from the ovine structure (PDB id.: 6ZA9) (Pinke et al., 2020), where the chain is completely resolved. Subunits which cannot be seen because hidden by other subunits are indicated by dashed lines.

forms two aqueous half-channels separated by a conserved arginine through which H^+ ions are translocated. To ensure the chemomechanical coupling between F_O and F_1, subunit a associates with the PS. The latter is formed by a homodimer of subunits b, which within the membrane is bifurcated to clamp subunit a close to the c-ring, while in their extrinsic portions form an α-helical coiled-coil structure that is looser at the top of F_1 to allow different interactions with subunit δ (Guo, Suzuki, & Rubinstein, 2019; Sobti et al., 2020). In chloroplasts, the overall arrangement is similar but the PS is formed by subunits b and b', which both interact with subunit δ (Hahn et al., 2018) (Fig. 13.3A and B). As already mentioned bacterial δ subunits are essentially the same species as mitochondrial OSCP subunit, as also indicated by their sequence similarity that is prominent in the regions that interact with the PS and the α subunits (Kühlbrandt, 2019).

3.3 The mitochondrial enzyme

The mitochondrial enzyme, which is located in the inner membrane with the F_1 sector facing the matrix, contains an additional set of subunits—d, e, f, g, F6, A6L (also called 8), j (previously called 6.8 proteolipid), and k (previously called DAPIT) —that make the F_O and PS structures more complex than the corresponding bacterial structures. Another striking feature is the size of the c-ring, which in mammals is composed of eight c subunits, the smallest reported so far (Nirody et al., 2020) (Fig. 13.3C). Given that a 360 degrees rotation of the c-ring yields three molecules of ATP, the mammalian enzyme is the most efficient in terms of the H^+/ATP stoichiometry (2.67) (Watt et al., 2010). In line with earlier results

(Watt et al., 2010), recent cryo-EM structures of the ovine and bovine complexes have shown that the cavity of the c-ring is occupied by phospholipids (Pinke et al., 2020; Spikes et al., 2020). The lipids at the intermembrane space (IMS) and matrix end of the c-ring are functionally and chemically distinct. Only the matrix-side lipids rotate with the c-ring, thus the lipid at the IMS provides a "lubricated" lipid plug facilitating rotation.

Like in bacteria and chloroplasts, at the interface with the PS, the c-ring is surrounded by two long (up to 80 Å), membrane-intrinsic hairpins generated by helices H3 to H6 of subunit a, which form two aqueous half-channels separated by a conserved arginine (R159), providing a funnel for H^+ access to the c-ring. In the structure of the dimeric ATP synthase from bovine mitochondria, the channels are clearly defined, especially the one from the IMS, which is also contributed by elements of subunits b and f bound to specific lipids, possibly to prevent proton leakage (Fig. 13.4A). This hemichannel contains several polar and charged residues, including the conserved E203, that face inwards forming a scaffold for five water molecules (Spikes et al., 2020). The cryo-EM map of the ovine complex (Pinke et al., 2020) revealed that subunit a is tightly bound to subunits k, j and 8 at its distal part from the PS, thus forming a structure that has been denoted as "proton translocation cluster"; while it interacts with helices of subunit b (H3) and f (H2) at its proximal part to the PS, which is close to the membrane-intrinsic portion. Remarkably, helices H2 and H3 of subunit b make a "U-turn" forming a compact triple transmembrane bundle with helices of subunits e and g through GXXXG motifs, creating a "hook apparatus" that allows the C-terminus of subunit e to reach out and attach to the c-ring lipids from the IMS side (Fig. 13.4B). In its membrane-extrinsic part subunit b forms a ~150 Å long, continuous α-helix stiffened by parallel (not coiled) interactions with helices of subunits F6 and d, and in the lower region with the C-terminal region of subunit 8, reaching out to the C-terminus of OSCP subunit on top of the F_1 sector. The N-terminus of OSCP forms many tight

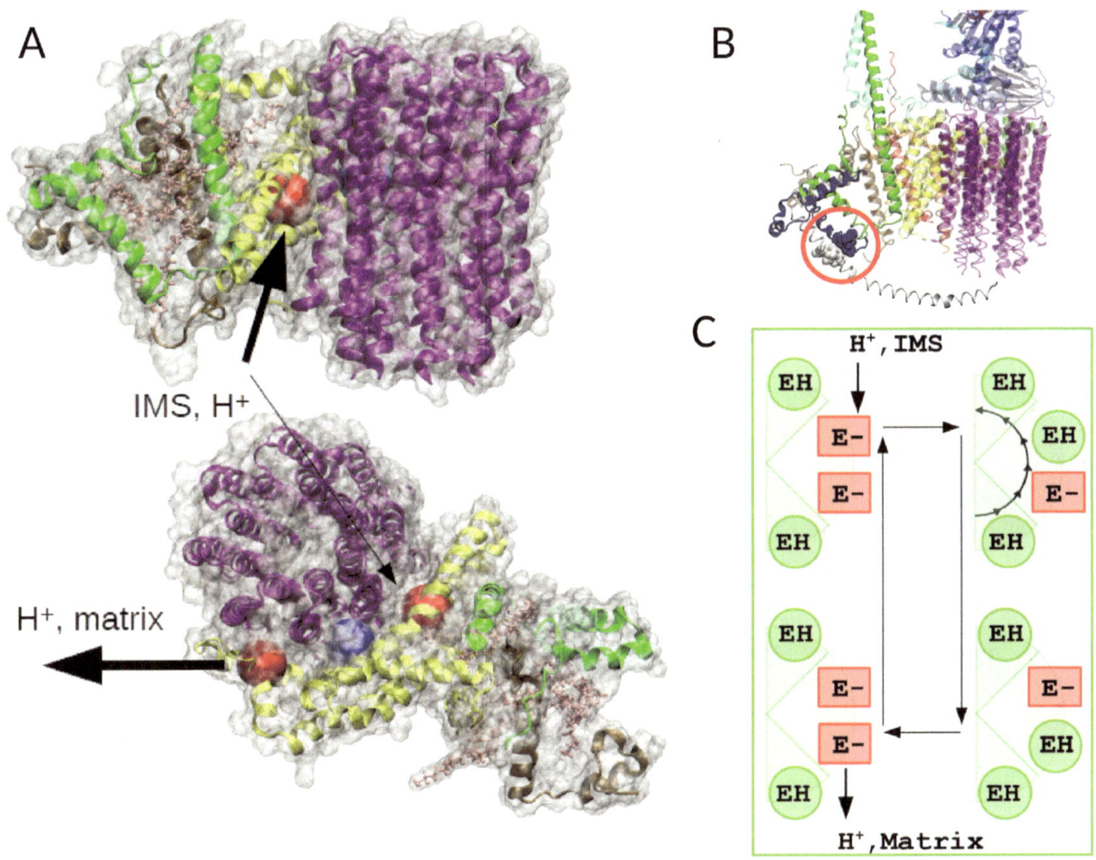

FIGURE 13.4 Conversion of concentration gradient in rotation. (A) The half-channels for protons are shown (PDB id.: 6ZPO) (Spikes et al., 2020). The C_δ atom of the two key residues of glutamic acids 145 and 203 of subunit a (ovine sequence numbering) in the half-channels are shown as red spheres. The phospholipids specifically bound to the F_O membrane-intrinsic part are colored in pink. Protons from the IMS (at lower pH) can neutralize the ionized E58 amino acid of the c-ring subunit in the half-channel open to the IMS. Upon protonation the rotation taking the neutralized amino acid in the membrane becomes more favorable and a neutral amino acid is exposed in the half-cannel open to the matrix (at higher pH) where it releases the proton. (B) The "hook apparatus" linking the peripheral stalk to the Fo region. In the red circle the interacting two GXXXG motifs on subunits e and g are shown as van der Waals spheres. (C) Sketch of the mechanism of rotation generation.

interactions with three α and two β subunits, resulting in a strong but flexible link between F_O and F_1 that resists the corotation of the three αβ pairs with subunit γ (Pinke et al., 2020; Spikes et al., 2020). Anchoring of F_O and F_1 is further strengthened by the recently described connection of the c-ring lipid plug with subunit e on the IMS side (Pinke et al., 2020).

Comparison of these intact, functionally competent ATP synthase structures highlighted the high conservation of the subunit a/c-ring interface and provided a fundamental basis for the detailed description of H^+ translocation and torque generation (Allegretti et al., 2015; Kühlbrandt, 2019; Pinke et al., 2020; Spikes et al., 2020; Srivastava et al., 2018; Zhou et al., 2015). In the synthesis mode, H^+ ions traveling down the H^+ gradient from the access channel reach an essential ionized glutamate residue of the proximal subunit c that is exposed around the midpoint of the membrane on the external surface of the ring. Once protonated, this glutamate moves to a more hydrophobic environment generating the counter-clockwise rotation of the c-ring (as seen from the matrix), which allows the ionized glutamate of the next c subunit to access the channel, and the protonated glutamate of the next c subunit to access the other half-channel where it gets deprotonated. After a rotation of (n-1) subunits (where *n* is the number of subunits in the c-ring), the first protonated glutamate in the cycle described here is exposed to the release channel and its proton ripped out by the higher pH of the mitochondrial matrix. An additional step brings this glutamate to its starting position and the cycle is completed. During its passage through the access channel, the deprotonated glutamate likely interacts with the positively charged and highly conserved R159 of subunit a located between the two half-channels (Fig. 13.4C). R159 both prevents short circuiting of the two half-channels and shifts the pKa of neighboring glutamate toward lower values (Fogolari et al., 2012), facilitating proton release. Rotation of the c-ring promotes corotation of the CS, to which it is firmly attached, thus driving the synthesis of ATP.

3.4 The role of IF1

Because ATP synthases are fully reversible, regulatory mechanisms have evolved to prevent uncontrolled ATP hydrolysis when synthesis stops. One key mechanism is inhibition of rotary hydrolysis through reversible interaction of the regulatory protein IF1 with the F_1 sector, which occurs in both mammalian and yeast ATP synthases. IF1 is a small protein that under specific conditions binds the α_{DP}/β_{DP} interface at a complex site mainly contributed by the C terminus of subunit β (Gledhill et al., 2007; Robinson et al., 2013). Interestingly, the inhibitory ζ subunit of proteobacteria occupies an equivalent position at the α_{DP}/β_{DP} interface (Morales-Rios, Montgomery, Leslie, Garcia-Trejo, & Walker, 2015; Morales-Rios, Montgomery, Leslie, & Walker, 2015) and despite its distinct structure the autoinhibitory ε subunit of *E. coli* shows a broad similarity in the way it interacts with F_1 (Cingolani & Duncan, 2011). In the case of the mammalian species, the complex binding process of IF1 to F_1, which requires the hydrolysis of two ATP molecules (i.e., two 120 degrees rotations of the γ subunit), has been defined based on the crystallographic structure of the F_1 sector inhibited by a mutated form of bovine IF1 (Bason et al., 2014) (Fig. 13.5A). It is well established that active IF1 is a dimer that forms under acidic conditions. Cryo-EM of a porcine ATP synthase tetramer revealed that, thanks to its shape, the IF1 dimer can simultaneously bind the F_1 heads of two neighboring ATP synthase dimers in a row, thus providing an excellent way to block enzyme activity (Gu et al., 2019) (Fig. 13.5B). However, based on structural considerations, formation of inhibited tetramers in vivo has been questioned (Spikes et al., 2021).

The role of IF1 in ischemic conditions has been well elucidated starting from the 1980s both in *in vitro* and ex vivo models (Green & Grover, 2000; Rouslin, 1983), in cultured cardiomyocytes (Harris & Das, 1991), after silencing and overexpressing IF1 under hypoxic conditions (Campanella et al., 2008) and in in vivo studies (Di Pancrazio et al., 2004). In these conditions, mitochondria would turn from ATP producers into ATP consumers precipitating ATP depletion. Due to enhanced glycolysis matrix pH can drop below seven leading to formation of the inhibitory IF1 dimers, which would prevent ATP hydrolysis thus limiting and/or delaying cell death and tissue damage. More recently, IF1 inhibition of ATP hydrolysis has been observed even under aerobic conditions in cells expressing a constitutively active form of IF1 (IF1−H49K), where prevention of futile ATP synthesis/hydrolysis cycles led to optimization of mitochondrial production of ATP (Rieger et al., 2021). Conversely, a growing body of evidence supports the ability of IF1 to inhibit mitochondrial ATP synthesis mediating the concurrent ROS-dependent activation of different survival pathways able to propitiate mitohormesis, thus stressing a role of IF1 in the metabolic reprogramming in cancer and in cellular differentiation (García-Aguilar & Cuezva, 2018; García-Bermúdez & Cuezva, 2016) and in tissue homeostasis, especially in neurons (Esparza-Moltó et al., 2021). A mechanism of IF1 regulation additional to pH has been described that is mediated by IF1 phosphorylation at a conserved serine residue, making the inhibitor unable to bind ATP synthase (García-Bermúdez et al., 2015). Interestingly, some studies of the mitochondrial proteome reported IF1 to be acetylated (Rardin et al., 2013) and succinylated (Park et al., 2013), but the physiological relevance of these modifications remains to be investigated (García-

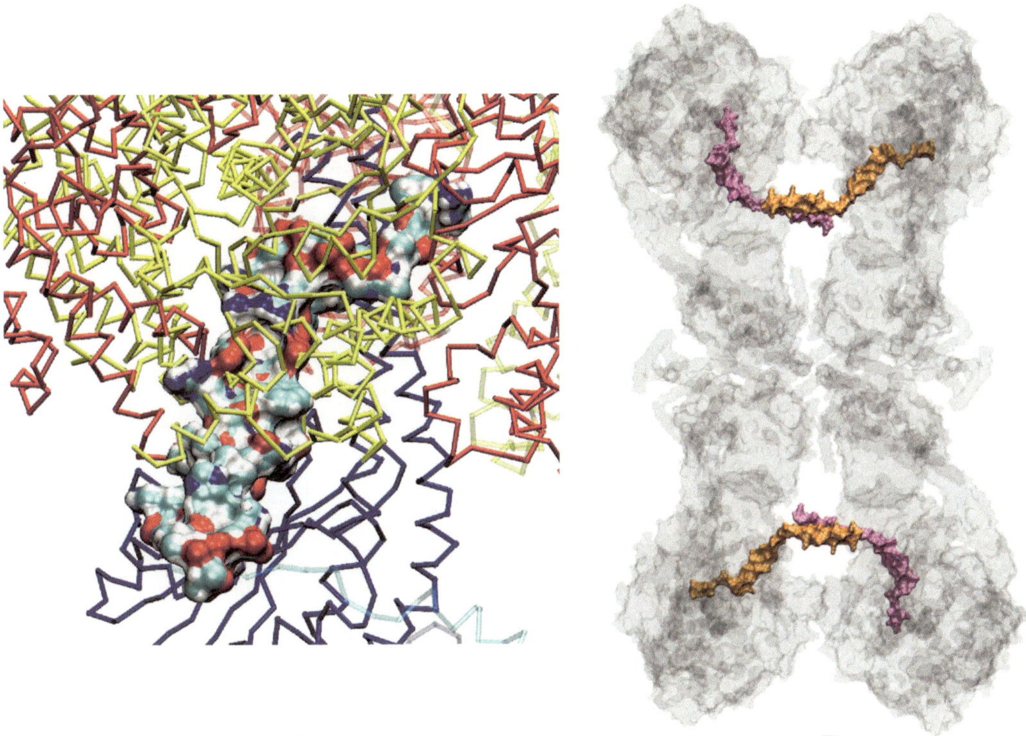

FIGURE 13.5 IF1 inhibitor. (A) IF1 inhibitor (colored by atom) interacting with ATP synthase monomer (PDB id.: 6ZPO) (Spikes et al., 2020). The flexible N-terminus is not resolved in the monomer. (B) IF1s chains interacting in the ATP synthase tetramer (PDB id.: 6J5K) (Gu et al., 2019). Interacting chains are colored in orange and mauve. ATP synthase is shown in transparency.

Bermúdez & Cuezva, 2016). Unexpectedly, a conserved BH3-like domain has been very recently identified, which overlaps with the inhibitory domain of IF1 (Juhaszova et al., 2022). This fits the proposal that IF1 can be engaged by survival-regulating proteins such as Bcl-xL via their intrinsic BH3-binding grooves. This interaction would cause IF1 displacement from F_1 and rearrangement of Bcl-xL, enabling its binding to F_1 β subunits and resulting in increased ATP synthesis (Alavian et al., 2011).

3.5 Oligomeric states of ATP synthase

As mentioned earlier, a remarkable feature of the mitochondrial ATP synthase complexes is their ubiquitous association into dimers, which in animals and fungi self-assemble into long rows of oligomers at the cristae edges, thus mediating formation of the cristae themselves (Allen, 1995; Blum et al., 2019; Davies et al., 2012; Paumard et al., 2002). Pioneering experiments in yeast by genetic approaches and cross-linking analyses established that formation of dimers depends on the dimer-specific subunits e, g, and k (Paumard et al., 2002). Contrary to expectation, these subunits only partially contribute to formation of the dimer interface, while possibly providing an oligomerization interface through homotipic interactions (Habersetzer et al., 2013; Spikes et al., 2021). Indeed, in situ cryo-EM structures revealed that dimers form at an angle of 80–90 degrees between two CS, with the PS turned away from one another, and are mainly stabilized through interactions of subunits a and j (Guo et al., 2017; Spikes et al., 2020). Interestingly, the dimerization interface is made up of different subunits in different organisms. In yeast, subunit a establishes direct dimer contacts together with subunit j (Guo et al., 2017), while in mammals the weak interaction between monomers is mainly mediated by the two j subunits and the a subunit loop near the N-terminal end of helix 1, thus accounting for the lower stability of the bovine dimer (Fig. 13.6). Such an interface appears indeed to be highly dynamic, i.e., it changes over time to accommodate motions both dependent on and independent of catalysis (Spikes et al., 2021).

The mechanism of formation of ATP synthase rows is still debated. It has been proposed that the elastic energy of the membrane is sufficient, without the need for specific protein–protein contacts (Anselmi et al., 2018; Davies et al., 2012; Kühlbrandt, 2019). Based on the dimeric structure, others have proposed formation of a "fluid interface" that mainly occurs

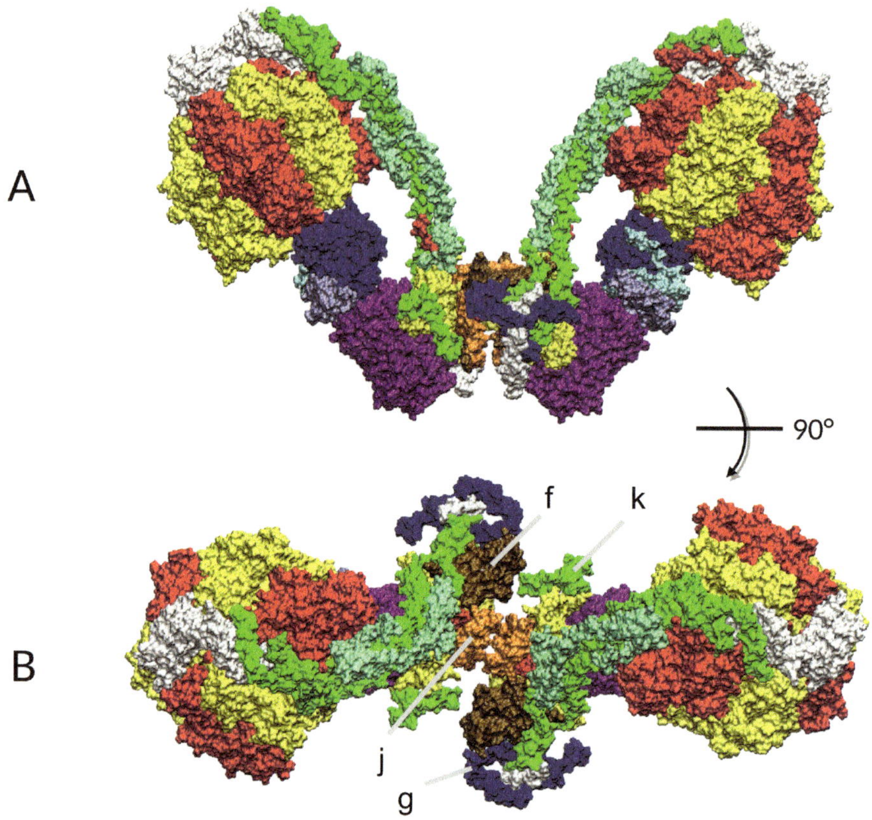

FIGURE 13.6 Mitochondrial ATP synthase dimers. ATP synthase dimers (pdb id. 7AJB) (Spikes et al., 2020) as seen from along the cristae of mitochondria (A) and from the matrix (B). The same colors of subunits as in Fig. 13.2 are used. In (A) the curvature of the membrane follows the ends of the c-rings. In (B) subunits not visible in Fig. 13.2, including those at the interface, are indicated.

through homotipic interactions between the matrix-exposed part of subunit g and subunit k, which is able to follow the cristae curvature (Spikes et al., 2021). Participation of g and k subunits to the oligomerization interface has also been observed in cryo-EM analyses of the ovine and porcine tetrameric complexes, where also two IF1 dimers link two ATP synthase dimers (Gu et al., 2019; Pinke et al., 2020). These observations may suggest that different ATP synthase oligomeric structures can coexist at the cristae edges within a mitochondrion, consistent with an a specific mechanism of oligomer formation. In addition, the presence of a subpopulation of monomeric ATP synthase complexes prevailing in the inner boundary membrane has been revealed by live cell microscopy experiments (Rieger et al., 2021; Salewskij et al., 2020). These monomeric complexes are possibly working as ATPases and are characterized by higher mobility compared to the dimeric complexes localized at the cristae edges. ATP synthase thus seems to react to metabolic changes not only by functional but also by spatial and structural reorganization, which also involves interactions with the cristae-shaping protein OPA-1 (Quintana-Cabrera et al., 2018). Intriguingly, the existence of a supramolecular structure composed of ATP synthase, adenine nucleotide translocator (ANT) and Pi carrier, the "ATP synthasome" has been proposed (Ko et al., 2003; Nusková et al., 2015), the formation of which seems to depend on the interaction with CyPD and on the bioenergetic state of the mitochondria (Beutner et al., 2017).

3.6 The two stepping motors and their cooperation

F_O and F_1 are two tethered stepping motors driven by different fuels. In energized membranes rotation of the c-ring within F_O is driven by the proton motive force (pmf) and leads to ATP synthesis; in the absence of a pmf rotation of the CS within F_1 is instead driven by ATP hydrolysis. In the whole complex, F_O and F_1 therefore exert a rotational torque against one another. When the pmf is larger than the free energy of ATP hydrolysis the rotational torque of F_O is stronger than that of F_1, leading to ATP synthesis. Vice versa, when the free energy of ATP hydrolysis predominates, F_1 reverses the rotation of F_O to build a pmf (Kobayashi et al., 2020). Single-molecule analysis has elucidated the basic features of F_1 rotation, which has a unitary step size of 120 degrees. However, variations in the rotation dynamics have been observed in the different F_1.

In thermophilic *Bacillus PS3* the 120 degrees rotation of the central stalk is divided into two substeps of 80 degrees and 40 degrees, preceded by the ATP-waiting dwell (binding dwell) or the catalysis-waiting dwell (catalytic dwell), respectively (Adachi et al., 2007). In human F_1 the 120 degrees rotation consists of three substeps of 65 degrees, 25 degrees, and 30 degrees, each with the corresponding binding dwell, Pi-release dwell, and catalytic dwell, respectively (Suzuki et al., 2014). In spite of the close sequence homology with the human enzyme, in bovine F_1 the three substeps appear to differ, with the pause between the binding and the catalytic dwell being of 10–20 degrees only (Kobayashi et al., 2020). Importantly, despite such variations in the stepping behavior determination of the rate and equilibrium constants of the steps by "manipulation experiments" with magnetic tweezers showed that the mechanism of torque generation is common among F_1 of different origin, i.e., that in all types of F_1 the ATP-binding process generates a torque larger than the hydrolysis process (Kobayashi et al., 2020). This provides further, important support to the "binding change mechanism", according to which in the ATP synthesis mode energy serves primarily to drive the release of tightly bound ATP (Boyer, 1993).

In F_O, the 360 degrees rotation of the c-ring is generated by sequential protonation and deprotonation of a conserved glutamate residue exposed on each c subunit and produces three ATP molecules. This implies that the bioenergetic cost of making ATP depends on the number of c subunits in the ring. In all vertebrates, where the c-ring has eight c subunits, only 2.67 protons are required by the ATP synthase to make each ATP molecule, while in prokarya, chloroplasts and fungi, where the c-ring contains from 10 to 17 subunits, the bioenergetic cost is higher, ranging from 3.3 to 5.7 H^+ per ATP (Watt et al., 2010).

In all organisms, the existence of two motors (F_O and F_1) differing in the number of steps during the catalytic cycle poses a challenge to their cooperation (Kühlbrandt, 2019). Thanks to numerous studies, it is now widely accepted that energy conservation is ensured by elastic power transmission and not by fine tuning of the respective steps (Junge et al., 2009). Most interestingly, the cryo-EM structures of active complexes from mammals (Zhou et al., 2015), *E. coli* (Sobti et al., 2020), and the green alga *Polytomella* sp. (Murphy et al., 2019) enabled the separation of independent rotary states, which revealed the molecular movements that accompany the F_1 and F_O rotation. At the beginning of each 120° step the $\alpha_3\beta_3$ F_1 stator rotates together with the central stalk and c-ring as a rigid body, while the single-peptide chain connecting the N-terminal and C-terminal domains of the conserved OSCP/δ subunit is flexible and undergoes hinge-like conformational changes, ensuring a flexible coupling between F_1 and F_O (Murphy et al., 2019). The ability of the OSCP subunit to accommodate the rocking motions of the $\alpha_3\beta_3$ domain during catalysis has also been observed in the dimeric ATP synthase from bovine heart (Spikes et al., 2020), thus suggesting that the hinge movement of the two-domain OSCP subunit is universal. However, work with ATP synthase subcomplexes led to propose that flexible coupling between F_1 and F_O may be mediated by the region where the "foot" contacts the c-ring (Sielaff et al., 2008) or by the CS (Martin et al., 2018). Despite the debate about the elastic coupling mechanism, measurement of the H^+:ATP ratio in a bacterial complex excluded "slips" of the rotor, i.e., a rotation occurring without H^+ translocation, revealing that ATP synthase performs a "perfect chemomechanical coupling" between proton translocation, rotary motion, and ATP synthesis/hydrolysis (Soga et al., 2017).

3.7 Biogenesis of ATP synthase

Another notable feature of the mitochondrial ATP synthase is that this enzyme complex is a "genetic hybrid" constituted of subunits that are encoded in nuclear genes, synthesized on cytoplasmic ribosomes, and imported into the mitochondrial matrix; and of subunits encoded by the mitochondrial genome (subunits a and A6L in the human species and subunits a, A6L and c in yeast). The dual genetic origin requires complex mechanisms for insuring a balanced output of mitochondrial and nuclear-encoded subunits that are still largely unknown (Franco et al., 2020). Biogenesis of the ATP synthase has been studied more extensively in yeast (Franco et al., 2020) and more recently in human cells (Carroll et al., 2021; He et al., 2018). In each case, the enzyme is assembled in a stepwise process through formation of subcomplexes, which differ depending on the organism. The terminal step is always the interaction between the a subunit with the c-ring to prevent even a transient energy-wasting leakage of protons across the membrane (Franco et al., 2020; Wittig et al., 2010). All steps take place in the cristae rather than in the boundary membrane, as revealed by superresolution microscopy (nanoscopy) and quantitative cryo-immunogold electron microscopy (Stoldt et al., 2018). The F_1 sector and the c-ring assemble independently and then form a transient F_1-c-ring complex, which has been observed both in yeast (Rak et al., 2011) and in a human near-haploid cell line (HAP1) (He et al., 2018). Based on the composition of intermediate "vestigial" ATP synthase complexes obtained by disruption of individual genes for nuclear-encoded F_O subunits in HAP1 cells, an assembly pathway has been proposed according to which the F_1-c8 complex, which is inhibited by IF1, attaches to the PS at subunits e, f, and g providing the template for insertion of subunits a and A6L, whose association is stabilized by subunit j. The monomeric complexes then dimerize via interactions between subunits a and j (He et al., 2018). In yeast an intermediate containing subunits a and 8 (A6L in the human species) and the PS except OSCP was found early in the assembly pathway, suggesting its subsequent interaction with the F_1-c-ring complex to form the whole enzyme by addition of OSCP (the presence of the subunits e, f, g, i/j, and k was not examined) (Rak et al., 2011). Another study in yeast suggests that the last step of the pathway is incorporation of the c-ring into an intermediate species consisting of F_1, the PS, and other F_O subunits (Naumenko et al., 2017).

The assembly pathway is assisted by several protein factors that partly differ between lower and higher eukaryotes (Franco et al., 2020; Hejzlarová et al., 2014). Well characterized in yeast are Atp11, Atp12 (Ackerman & Tzagoloff, 1990), and FMC1 (Lefebvre-Legendre et al., 2001), which are necessary for the proper insertion of the α and β subunits in F_1; the Ina17 and Ina22 proteins, which facilitate PS formation (Lytovchenko et al., 2014; Naumenko et al., 2017); and the Atp10 and Atp23 factors, which target subunit a and are implicated in the formation of the proton channel (Franco et al., 2020; Hejzlarová et al., 2014). In mammals ATPAF1 and ATPAF2, the homologs of the yeast Atp11 and Atp12 are essential for F_1 and ATP synthase assembly (Wang et al., 2001; Zhou et al., 2021) and both TMEM70 and TMEM242 are required for the formation of the c8 ring (Carroll et al., 2021; Kovalcíková et al., 2019). Interestingly, these latter factors also interact with the mitochondrial complex I assembly complex (MCIA) that supports assembly of the membrane arm of complex I, suggesting that they could be part of a shared regulation of the CI and ATP synthase levels in mitochondria (Carroll et al., 2021).

4. ATP synthase and the mitochondrial permeability transition

4.1 The mitochondrial permeability transition

A set of very recent findings has identified a novel function for F-ATP synthase, i.e., formation of a high conductance channel, the mitochondrial megachannel (MMC), which coincides with the permeability transition pore (PTP) and is responsible for the mitochondrial permeability transition (PT). The PT, which requires matrix Ca^{2+}, is a transition of the mitochondrial inner membrane from a highly impermeable state, which is essential for ATP synthesis, to a highly permeable state (Hunter et al., 1976). If long-lasting, openings of the PT cause dissipation of ion gradients, osmotic swelling of the matrix and outer membrane damage with release of cytochrome *c* and other proapoptotic factors, often resulting in cell death. Transient PTP openings also occur under physiological conditions and may provide mitochondria with a channel for Ca^{2+} release (Bernardi et al., 2015; Bernardi & Petronilli, 1996), as will be discussed later in the chapter.

The molecular nature of the PT has long been debated. One theory posited that the permeability increase occurred within the lipid bilayer after activation of a Ca^{2+}-dependent phospholipase A_2 (Beatrice et al., 1980); while the pore theory postulated that the PT was due to opening of a regulated channel (Hunter et al., 1976). The discoveries (i) that the PTP can be inhibited by nanomolar concentrations of cyclosporin A (CsA) without affecting phospholipase A_2 activity (Broekemeier et al., 1989); (ii) that the inner mitochondrial membrane does possess ion channels, including the high-conductance, Ca^{2+}-activated MMC (Kinnally et al., 1989; Petronilli et al., 1989); and (iii) that the MMC and PTP are manifestations of the same molecular entities (Szabó & Zoratti, 1992) tipped the balance in favor of the channel theory. The effects of CsA on the PTP are best described as desensitization and are mediated by its high-affinity binding to cyclophilin (CyP) D, a member of the CyP family of peptidyl prolyl *cis-trans* isomerases encoded by the *Ppif* gene that sensitizes the PTP to opening (Nicolli et al., 1996). The most studied is cytosolic CyPA, with which CsA creates a complex that acquires the ability to inhibit cytosolic calcineurin, a Ca^{2+}-calmodulin-dependent phosphatase that catalyzes dephosphorylation of several proteins including nuclear factor of activated T cells (NFAT). Dephosphorylated NFAT translocates to the nucleus and activates the immune response against transplants, which is inhibited by CsA through calcineurin inhibition (Clipstone & Crabtree, 1992; Liu et al., 1991). It has been established that calcineurin inhibition is not required for PTP inhibition by CsA (Nicolli et al., 1996).

CyPD has been instrumental in redirecting PTP research toward the ATP synthase. In an unbiased study aimed at identifying CyPD partners in mitochondria, we discovered its regulatory association with the PS of ATP synthase. Binding of CyPD required inorganic phosphate, which is a powerful inducer of PTP opening, while it was counteracted by CsA, which removed CyPD from ATP synthase. The association of CyPD inhibited the rate of ATP hydrolysis by about 30%, and inhibition was relieved by CsA (Giorgio et al., 2009). These intriguing findings led us and others to explore the possibility that ATP synthase contributes to PTP formation, a hypothesis that was not readily supported by the available structural features of the enzyme complex. A few years later it was shown that the binding site for CyPD is the ATP synthase OSCP subunit, the hinge connecting F_1 to subunit b (Giorgio et al., 2013), and that ATP synthase does display Ca^{2+}-dependent channel activity after partial purification from bovine (Giorgio et al., 2013), human (Alavian et al., 2014), yeast (Carraro et al., 2014), and *Drosophila* mitochondria (von Stockum et al., 2015). Consistent with a role of ATP synthase in the PT, downregulation of subunit c, which does not allow proper assembly of the enzyme, caused PTP inhibition (Bonora et al., 2013). PTP formation by ATP synthase is now strongly supported (i) by reconstitution studies with highly purified and catalytically active ATP synthase from bovine (Urbani et al., 2019) and porcine heart mitochondria (Mnatsakanyan et al., 2019) and (ii) by site-directed mutagenesis (Alavian et al., 2014; Antoniel et al., 2018; Carraro et al., 2018, 2020; Giorgio et al., 2017; Guo et al., 2018; Guo, Carraro, et al., 2019; Morciano et al., 2021). A controversy arose when it was found that genetic inactivation of several subunits of ATP synthase, which prevent its correct assembly, did not prevent the PT (Carroll et al., 2019; He, Carroll et al., 2017; He, Ford et al., 2017). It is now becoming clear that in the absence of a fully assembled ATP synthase a permeability pathway is provided by the ANT (Carrer et al., 2021; Neginskaya et al., 2019). The ANT was the first candidate for PTP formation (Hunter & Haworth, 1979) and we refer the reader to a recent review that also covers its role in the PT (Bernardi et al., 2021).

4.2 Electrophysiology

The hypothesis that the PT is mediated by a regulated pore (Hunter et al., 1976) was corroborated by the application of the electrophysiological patch-clamp technique to mitoplasts, i.e., mitochondria devoid of most of the outer membrane (Sorgato et al., 1987). This revolutionary technique (Neher & Sakmann, 1976) allows detection of ionic currents passing through the inner mitochondrial membrane in real time, by exploiting the formation of a high-resistance seal among an electrolyte-filled glass micropipette and a biological membrane. The main biophysical characteristics that can be determined using this system include channel conductance (expressed in Siemens), voltage-dependence, selectivity, and kinetic behavior. In addition, the effect of pharmacological agents on channel activity and/or biophysical characteristics can be directly detected. Inhibitors or activators either decrease or increase the probability of the channel being open, respectively. The same kind of information can be obtained also by incorporating purified channel-forming proteins into an artificial lipid bilayer and measuring the current flowing through the membrane.

Most of the ion channels measured in biological membranes are characterized by a conductance ranging up to a few hundred picosiemens (pS). Thus, the discovery of an unselective, inner membrane high-conductance channel with a preferential maximal conductance value of 1.3 nS (value ranging up to 1.5 nS) (Kinnally et al., 1989; Petronilli et al., 1989) was extremely surprising, and initially met by considerable skepticism. On the other hand, previous solute exclusion experiments suggested the presence of a Ca^{2+}-activated, nonselective pore with a diameter of 2–3 nm (Massari & Azzone, 1972). Such a large diameter fits well with the large conductance value of the observed channel. Indeed, the high-conductance MMC has been later on identified as the PTP on the basis of its pharmacological profile, in particular (i) activation by matrix-side Ca^{2+} in the sub-mM range; (ii) inhibition by CsA (Szabó & Zoratti, 1991), mildly acidic matrix pH, and matrix divalent cations other than Ca^{2+} such as Sr^{2+}, Mn^{2+}, Ba^{2+}, and Mg^{2+}, which inhibited MMC activity in a Ca^{2+}-competitive manner (Bernardi et al., 1992; Szabó et al., 1992). From an electrophysiological point of view, MMC is characterized by a binary cooperative behavior, with a "full-conductance state" and a prominent "half-conductance" fast-gating substate, in addition to numerous substates occurring with lower conductances. The half-conductance channel can also be observed as a stand-alone channel (De Marchi et al., 2008; Zoratti et al., 2010), indicating a dimeric structure. MMC is unselective and shows voltage-dependence with full opening of the channel at depolarizing membrane potential values near to zero. One interesting feature of MMC that can be appreciated in electrophysiological experiments visualizing single channel behavior is that the full-size pore appears to develop gradually from smaller pores. At very low Ca^{2+} levels, the probability of observing a transient MMC activity in a membrane patch is low (De Marchi et al., 2008). Thus, Ca^{2+} seems to be necessary for full activation and/or assembly of the pore, without the need of Pi or other coadjuvants. When Ca^{2+} is chelated, activity disappears immediately (Szabó et al., 1992), suggesting that this divalent cation is required not only for assembly but also for keeping the pore open. Independently of the tissue origin, the MMC described in various studies shows very similar biophysical and pharmacological characteristics, pointing to a wide tissue-distribution of the protein(s) forming the channel. Both F-ATP synthase and the ANT fulfill this criterion.

The above-described electrophysiological and pharmacological features of MMC are best recapitulated by the channel activity of purified bovine ATP synthase reconstituted in planar lipid bilayers (Alavian et al., 2014; Giorgio et al., 2013; Mnatsakanyan et al., 2019; Urbani et al., 2019). In the first reported link between MMC and F-ATP synthase, the dimeric form of the latter from different organisms was eluted directly from native PAGE (see Section 4.3 for details) and shown to give rise to channel activity resembling MMC (Giorgio et al., 2013; von Stockum et al., 2015; Carraro et al., 2018). The partial purity of the preparations did not allow to draw definitive conclusions on whether F-ATP synthase alone is sufficient to form MMC. In this respect, the finding that purified reconstituted c-subunit ring of the F_O of the F_1F_O ATP synthase was able to form a voltage-sensitive high-conductance channel (Alavian, Beutner et al., 2014) was of interest, although the question of purity still persisted. In more recent work, incorporation of proteoliposomes containing native gel-purified oligomers and dimers of fully active lauryl-maltose-neopentyl glycol (LMNG)-solubilized bovine F-ATP synthase, and direct incorporation of such purified F-ATP synthase into artificial planar lipid bilayers gave rise to currents matching those of the MMC (Urbani et al., 2019). This F-ATP synthase preparation was of high purity, as determined by mass spectrometry, and inserted into the membrane with its hydrophobic F_O domain, allowing to determine the site of action of channel agonists and antagonists to the F_1 side, at the same time ruling out that their effects were unspecific. The observed current, similarly to MMC, showed multiple substates and a Ca^{2+} concentration-dependent increase of the current amplitude. Ca^{2+}-elicited currents were fully reversible by chelating this ion with EGTA and were stabilized by benzodiazepine 423 (Bz-423), a ligand of the OSCP subunit of F-ATP synthase that is known to activate the MMC/PTP. While Bz-423 was not able to trigger channel opening per se, it synergistically acted with Ca^{2+} and allowed channels to reach the maximal conductance typical of MMC (Urbani et al., 2019). The purified F-ATP synthase was inhibited by Mg^{2+} and adenine nucleotides, which also block PTP activity. In contrast, channel activity was not affected by pharmacological modulators of other mitochondrial proteins previously proposed to contribute to PTP formation like ANT and outer membrane voltage-dependent anion channel (Urbani et al., 2019). In an independent study, purified monomeric ATP synthase from porcine heart mitochondria was incorporated into small unilamellar vesicles formed with lipids resembling the lipid composition of mitochondrial inner membrane. This preparation gave rise to a voltage-dependent activity with a peak conductance of up to 1.5 nS (Mnatsakanyan, Llaguno et al., 2019). Altogether these findings resolved the long-standing mystery of the molecular identity of MMC/PTP and clearly pointed at F-ATP synthase as the complex forming MMC/PTP.

While experiments using the highly purified F-ATP synthase in artificial membranes were instrumental for providing a molecular identity for MMC/PTP, it has to be kept in mind that a hallmark feature of MMC/PTP, i.e., the inhibitory action of CsA on channel activity cannot be observed in this experimental setup due to the lack of CyPD. As mentioned above, CyPD is required for desensitization of MMC/PTP by CsA (De Marchi et al., 2006). Therefore, patch-clamp experiments recording MMC in the native membrane expressing wild-type or mutated F-ATP synthase are of utmost importance. In addition, it might be of interest to exploit a series of other MMC modulators other than CsA such as ubiquinone analogues (Martinucci et al., 2000), the dopaminergic D2-receptor agonist pramipexole (Sayeed et al., 2006), 17β-estradiol (Thiede et al., 2012), as well as the calpain inhibitor calpeptin (Derksen et al., 2016).

4.3 Mechanisms of channel formation from ATP synthase

Together with site-directed mutagenesis, recent structural work has provided a convincing framework to envision plausible if hypothetical mechanisms through which the ATP synthase could generate the PTP (Gu et al., 2019; Pinke et al., 2020; Spikes et al., 2020). Initially proposed by Gerle as the "death finger" hypothesis (Gerle, 2016, 2020), a current consensus model combines elements of two previous hypotheses, i.e., (i) that the PTP forms at the dimer interface through a conformational change originating at OSCP (Giorgio et al., 2013) or (ii) that the c-ring provides a permeability pathway after dissociation of F_1 (Alavian et al., 2014; Bonora et al., 2013).

- When Mg^{2+} is bound to the catalytic site, rotation of the γ subunit is smooth, allowing dissipation of elastic force by OSCP acting as a leaf spring oscillating on its central "hinge" region. This shock absorber function would minimize transmission of mechanical force to the PS, hence to the "hook apparatus" connecting with the outer lipid plug of the c-ring through subunit e.
- When binding of Ca^{2+} occurs at the catalytic metal site, its larger van der Waals radius causes a rearrangement of the F_1 sector, which increases its overall rigidity (Giorgio et al., 2017). Through the connecting loop made up by residues 82–131 of subunit β, the rearrangement is transmitted to the "crown region", i.e., the β barrel-shaped ring at the interface of F_1 and OSCP (Rees et al., 2009).
- More mechanical stress occurs at OSCP, challenging its shock absorber function during rotation of subunit γ (Murphy et al., 2019) in a process that is enhanced by CyPD binding (Giorgio et al., 2013), which decreases OSCP compliance. Intriguingly, it has been recently shown that OSCP also interacts with IF1, which regulates PTP opening through interaction with the p53–CyPD complex (Guo, 2022). Mechanical energy is transmitted to the PS and is relayed at its point of entrance in the membrane (Hahn et al., 2016; Zhou et al., 2015). This is where subunit b forms a hairpin making the already described tight association with the C- and N-termini of subunits g and e, respectively, forming a strong "wedge" (Spikes et al., 2020) or "bundle" (Pinke et al., 2020).

- The C-terminus of subunit e extends outside the bundle region and makes contacts with the c-ring lipids (Pinke et al., 2020). Subunit e exerts a pulling action on the lipids of the c-ring, allowing formation of a channel within the c-ring itself by displacing the outer lipid plug (Pinke et al., 2020) (Fig. 13.7).
- At low levels of matrix Ca^{2+} (i.e., when very few ATP synthase units bind Ca^{2+}), the PTP flickers between the closed and open states. When matrix Ca^{2+} increases (or as the result of additional inducing events) openings become more stable leading to the displacement of the inner lipids and perhaps of the CS. It should be noted, however, that also the fully open state is reversible when Ca^{2+} is removed (Petronilli et al., 1994).

The mechanism described above fits cryo-EM structures of ATP synthase prepared in the presence of 5 mM Ca^{2+}, which yielded density classes never observed with Mg^{2+}. These densities could be arranged in a sequence that may provide snapshots of the sequential stages of PTP opening (Pinke et al., 2020). The mechanism through which the channel would reversibly form within the c-ring (which in the fully open state must also imply rearrangement or even displacement of F_1) remains a mystery. Also, it is not clear whether all the conductance substates of the PTP can be explained by this hypothesis.

As mentioned in paragraph IVB, in the native membranes the PTP exhibits substates ranging from 45 pS to >1000 pS (Kinnally et al., 1989; Petronilli et al., 1989). These substates are also observed in purified dimeric ATP synthase preparations reconstituted in lipid bilayers (Urbani et al., 2019). As discussed above, it is possible that the low-conductance flickering events reflect short-range oscillations of the lipid plug within the c-ring. A nonmutually exclusive alternative, however, is that a channel can form at the monomer–monomer interface of adjacent subunits j, which appear to form a cavity not filled by lipids in reconstructions of the dimer interface (Spikes et al., 2020, 2021). At the C-terminal end of subunits j (which are in contact), the large number of spatially close basic residues (which are mostly conserved in mammalian sequences) suggests interaction with negative moieties. However, the latter were not identified in the available cryo-EM structures. Whether the space at the dynamic dimer interface could provide a channel formed by the mobile subunits j remains to be established (Fig. 13.8). This hypothetical channel could contribute to the reversible, transient flickerings of the PTP that have been observed both in isolated mitochondria and in intact cells (Hüser & Blatter, 1999; Hüser et al., 1998; Petronilli et al., 1999; Sambri et al., 2020).

As already mentioned, the catalytic sector and the rotor (which includes the c-ring) are very conserved. Since bacteria do not have a PT, it appears logical to conclude that the c-ring may be necessary but not sufficient for PTP formation, which indeed involves the PS and its "accessory" subunits that are not found in prokaryotes. Like the PTP, subunit e

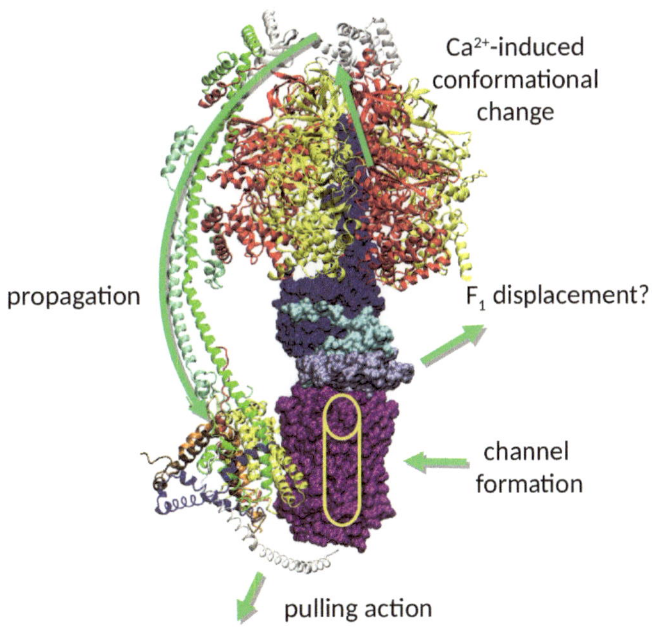

FIGURE 13.7 The "death finger" hypothesis of channel formation from ATP synthase. When Ca^{2+} binds to the catalytic metal site, it causes a rearrangement of the F_1 sector, which is transmitted to the "crown region" at the interface of F_1 and OSCP, challenging the absorber function of OSCP. Mechanical energy is transmitted to the PS and is relayed in the membrane to the "hook apparatus", distorting the C-terminus of subunit e. Subunit e exerts a pulling action on the lipids of the c-ring, allowing formation of a channel within the c-ring itself by displacing the outer lipid plug (see Section 4.3).

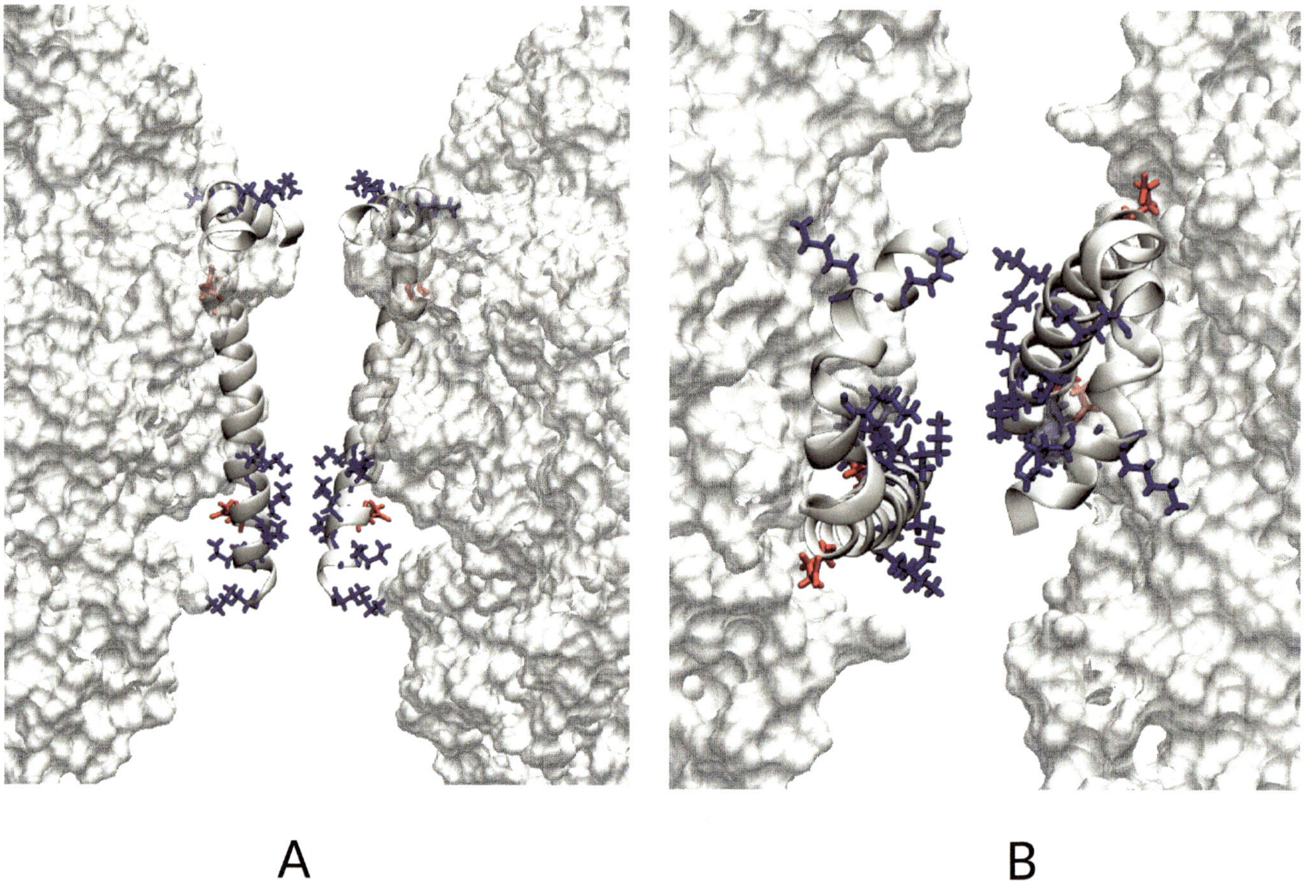

FIGURE 13.8 Dimer interface. Subunits j at the dimer interface (pdb id.: 7AJB) (Spikes et al., 2020) shown as cartoon. Acidic (red) and basic (blue) residues are shown in licorice. In (A) the view is along the membrane and in (B) from the matrix. Other subunits of ATP synthase are shown in transparency. Molecules filling the space between monomers, if any, are not resolved.

displays species-specific features. The well-characterized GXXXG motif (position 23–27 in *Homo sapiens*), which allows packing with the α helices of subunits g and b, is conserved both for primary sequence and for position within the protein, and sequence similarity is also present in the transmembrane region. Where subunits e differ considerably is at the C-terminus, which in the bovine and porcine species makes contacts with the outer face of the c-ring lipid plug. In this region, unique sequences can be found in yeast and *Drosophila*, i.e., organisms in which the orientation of subunit e also differs from that of mammals (Bernardi et al., 2021). Can these differences also explain the different features of the channels formed by ATP synthase? The PTP has distinct properties in different species (Azzolin et al., 2010) with mean conductance ranging between the 500 pS of the bovine enzyme (Giorgio et al., 2013) to the 300 pS of *S. cerevisiae* (Carraro et al., 2014) to the mere 53 pS *Drosophila* (von Stockum et al., 2015), a species where the PTP does not mediate in situ swelling owing to its high selectivity for Ca^{2+} (von Stockum et al., 2011). It is therefore tempting to speculate that the unique C-termini of subunit e confer specific features to the PTP, possibly together with the unique features of the monomer/monomer interfaces. It will be of great interest to study the structure of ATP synthase from *Artemia franciscana*, a salt- and anoxia-tolerant organism that is strikingly resistant to PTP opening (Menze et al., 2005), and which may provide a stringent test to the hypotheses discussed above.

5. The mitochondrial permeability transition in pathophysiology

Although this chapter is dedicated to the ATP synthase, we think that it may be useful to include a very short summary of the role of the PTP in pathology and physiology.

5.1 Role in disease pathogenesis

The role of the PTP in cell death has been studied thoroughly. Stable pore openings cause mitochondrial depolarization, dissipation of ion gradients, cessation of ATP synthesis, matrix swelling with rupture of the outer membrane, release of proapoptotic factors that may lead to activation of procaspase 9 and apoptosis but also to necro(pto)sis, if ATP levels are not sufficient and caspase 8 is not active (Bernardi et al., 2001).

The number of pathophysiological conditions in which a role of PTP has been documented is extremely large. A partial list includes ischemia-reperfusion injury of the heart (Baines et al., 2005; Griffiths & Halestrap, 1993; Nakagawa et al., 2005) and of the brain (Uchino et al., 1998), traumatic brain injury (Readnower et al., 2011), muscular dystrophies (Irwin et al., 2003; Millay et al., 2008), neurodegenerative diseases (Du et al., 2008; Martin et al., 2009, 2014), vascular pathology (Davidson & Duchen, 2007; Di Lisa et al., 2009; Vindis et al., 2005), hemorrhagic shock (Kim et al., 2015; Lei et al., 2015; Li et al., 2015), coagulation and platelet apoptosis (de Witt et al., 2014; Leytin et al., 2018), onset of diabetes and vascular damage including dementia (Fujimoto et al., 2010; Widlansky & Hill, 2018; Yan et al., 2016), pulmonary dysfunction and hypertension (Pardo Semo et al., 2019; Ten & Ratner, 2019), sepsis (Stanzani et al., 2019), acute pancreatitis (Mukherjee et al., 2019), kidney diseases (Priante et al., 2019; Shah et al., 2019), somatic cell reprogramming (Ying et al., 2019), aging (Angeli et al., 2021; Mather & Rottenberg, 2000, 2002; Panel et al., 2018), and the effects of a large variety of toxicants of both pharmacological and environmental concern (Wallace, 2017). Over the years, these conditions have been extensively covered (Bernardi et al., 2015) and will not be discussed further here.

5.2 Role as a Ca^{2+} release channel

The mitochondrial inner membrane has an intrinsically low permeability to H^+ and charged species in general. This is key to energy conservation because maintenance of the proton electrochemical gradient (Δp, the sum of the electrical potential and pH difference across the inner membrane of about 200 mV, negative inside) is essential for ATP synthesis, as discussed earlier in this Chapter. The existence of a large membrane potential poses the major problem of cation distribution, since equilibration with the Δp would predict a matrix concentration of 150 M for K^+ and 0.1−1 M for Ca^{2+} (Pozzan et al., 1977). Equilibrium is never achieved because electrophoretic K^+ and Ca^{2+} uptake is compensated by specific release systems, see (Bernardi, 1999) for discussion.

Ca^{2+} is taken up through the mitochondrial Ca^{2+} uniporter (MCU), a complex and finely regulated channel that in energized mitochondria drives Ca^{2+} uptake down its electrochemical gradient (De Stefani et al., 2016). Equilibrium is not achieved because of the operation of the $3Na^+$-Ca^{2+} antiporter (Palty et al., 2010) and of a Na^+-independent efflux pathway that may be related to LETM1 (Austin & Nowikovsky, 2019). As a result of the operation of this set of transport systems, steady state extramitochondrial [Ca^{2+}] can be increased (with a matching decrease of matrix [Ca^{2+}]) by increasing the activity of the efflux pathways, by decreasing the activity of the MCU, or both. And yet, mitochondria are exposed to the hazards of Ca^{2+} overload. Indeed, the rate of Ca^{2+} uptake increases steeply with extramitochondrial [Ca^{2+}], reaching values as high as 1400 nmol $Ca^{2+} \times$ mg protein$^{-1} \times$ min^{-1} while the maximal rate of the combined efflux pathways does not exceed about 20 nmol $Ca^{2+} \times$ mg protein$^{-1} \times$ min^{-1} (Bernardi, 1999). Transient PTP openings, which have been documented in intact cells and organs (Barsukova et al., 2011; Elrod et al., 2010; Lu et al., 2016; Parone et al., 2013; Petronilli et al., 1999; Sambri et al., 2020), may provide mitochondria with a convenient mechanism to release Ca^{2+} and thus prevent Ca^{2+} overload (Bernardi & Petronilli, 1996; Hüser & Blatter, 1999). It may be asked why would a high-conductance, apparently unselective channel be needed for mitochondrial Ca^{2+} release. The answer may lie in the very low permeability of the inner membrane to H^+ and charged species, which would oppose fast Ca^{2+} release even after full membrane depolarization. Indeed, Ca^{2+} efflux down its concentration gradient via a selective channel (like for example the MCU) is opposed by the buildup of a Ca^{2+} diffusion potential which, according to the Nernst equation, is −30 mV per decade resulting in a very slow rate of Ca^{2+} efflux that could only be increased by the independent increase of membrane ion permeability. The clear advantage of a pore of large size is to provide charge compensation within the channel itself, which maximizes Ca^{2+} flux as it occurs at zero potential, thus permitting effective Ca^{2+} release even for very small [Ca^{2+}] gradients. Given that no K^+ and Na^+ concentration gradients are maintained across the inner membrane, PTP opening leads to selective Ca^{2+} release without affecting K^+ and Na^+ homeostasis.

6. Conclusions

Consistent with its central role in cellular energy metabolism, ATP synthase is one of the most ubiquitous and abundant enzymes. In the presence of an adequate transmembrane H^+ gradient, the mitochondrial complex is able to synthesize ATP at a rate that can reach 400 ATP molecules per second. In deenergized membranes, the enzyme catalyzes the hydrolysis of ATP at a comparable rate, so that regulatory mechanisms have evolved to prevent excess ATP waste under these conditions. The "overall structural organization" comprising the two principal domains F_O and F_1 connected by a central and a peripheral stalk is well conserved among organisms. Such a conservation is reflected in the conservation of the mechanisms of proton translocation and ATP synthesis. Nevertheless, the evolutionary origin of ATP synthase is still an unresolved issue. The finding that these catalytic mechanisms are conserved in the mitochondrial and chloroplast enzyme has suggested that they predate the divergence of these lineages 1.5 billion years ago, leading to speculate that, along with the ribosome, ATP synthase is one of the two most ancient macromolecular machines of life (Kühlbrandt, 2019).

ATP synthases, however, do show clear differences in structure, sequence, and subunit composition among organisms, especially in the F_O and peripheral stalk domains. Moreover, only the mitochondrial complex forms dimers and oligomers at the cristae edges, thus mediating formation of the cristae themselves. Another property of the mitochondrial enzyme is its involvement in the Ca^{2+}-dependent formation of the PTP. Definition of the mechanisms underlying the transformation of the energy-conserving enzyme of life into a potential effector of death is at present an intense area of research. These novel findings also make this complex a viable drug target in a variety of diseases where the PTP is involved. Indeed, although malfunction of ATP synthase has been associated with genetic defects (Zech et al., 2022) and with a variety of pathological conditions, such as cardiovascular and neurodegenerative diseases (Long et al., 2015), aging (Angeli et al., 2021; Goldberg et al., 2018), and cancer (Esparza-Moltó & Cuezva, 2018), this enzyme has only recently been considered as a potential drug target (Lippe et al., 2019), despite the fact that more than 300 natural and synthetic molecules are known to bind and inhibit the complex (Hong & Pedersen, 2008). In addition, ATP synthase is a very promising molecular target for the development of new antimicrobial agents that selectively inhibit the bacterial enzyme, such as bedaquiline (Guo et al., 2021).

Cryo-EM and X-ray crystallography have provided extremely detailed description of ATP synthases that could now be accompanied by molecular dynamics simulations to fully understand the catalytic processes at the atomic level. Information about the nature and inventory of all the specific lipids interacting with ATP synthase is not complete, but most recent structures provided atomic coordinates of some cardiolipins and phospholipids in close contact with subunits in the membrane. With the continuing rapid increase in computer power, the demanding calculations of the whole catalytic cycle of ATP synthase and its transformation into PTP could soon be made possible.

References

Abrahams, J. P., Leslie, A. G., Lutter, R., & Walker, J. E. (1994). Structure at 2.8 Å resolution of F_1-ATPase from bovine heart mitochondria. *Nature, 370*, 621–628.

Ackerman, S. H., & Tzagoloff, A. (1990). Identification of two nuclear genes (*ATP11, ATP12*) required for assembly of the yeast F_1-ATPase. *Proceedings of the National Academy of Sciences of the United States of America, 87*, 4986–4990.

Adachi, K., Oiwa, K., Nishizaka, T., Furuike, S., Noji, H., Itoh, H., Yoshida, M., & Kinosita, K., Jr. (2007). Coupling of rotation and catalysis in F_1-ATPase revealed by single-molecule imaging and manipulation. *Cell, 130*, 309–321.

Alavian, K. N., Beutner, G., Lazrove, E., Sacchetti, S., Park, H. A., Licznerski, P., Li, H., Nabili, P., Hockensmith, K., Graham, M., Porter, G. A., Jr., & Jonas, E. A. (2014). An uncoupling channel within the c-subunit ring of the F_1F_O ATP synthase is the mitochondrial permeability transition pore. *Proceedings of the National Academy of Sciences of the United States of America, 111*, 10580–10585.

Alavian, K. N., Li, H., Collis, L., Bonanni, L., Zeng, L., Sacchetti, S., Lazrove, E., Nabili, P., Flaherty, B., Graham, M., Chen, Y., Messerli, S. M., Mariggio, M. A., Rahner, C., McNay, E., Shore, G. C., Smith, P. J., Hardwick, J. M., & Jonas, E. A. (2011). Bcl-xL regulates metabolic efficiency of neurons through interaction with the mitochondrial F1FO ATP synthase. *Nature Cell Biology, 13*, 1224–1233.

Allegretti, M., Klusch, N., Mills, D. J., Vonck, J., Kühlbrandt, W., & Davies, K. M. (2015). Horizontal membrane-intrinsic α-helices in the stator a-subunit of an F-type ATP synthase. *Nature, 521*, 237–240.

Allen, R. D. (1995). Membrane tubulation and proton pumps. *Protoplasma, 189*, 1–8.

Angeli, S., Foulger, A., Chamoli, M., Peiris, T. H., Gerencser, A., Shahmirzadi, A. A., Andersen, J., & Lithgow, G. (2021). The mitochondrial permeability transition pore activates the mitochondrial unfolded protein response and promotes aging. *eLife Sciences, 10*, e63453.

Anselmi, C., Davies, K. M., & Faraldo-Gomez, J. D. (2018). Mitochondrial ATP synthase dimers spontaneously associate due to a long-range membrane-induced force. *The Journal of General Physiology, 150*, 763–770.

Antoniel, M., Jones, K., Antonucci, S., Spolaore, B., Fogolari, F., Petronilli, V., Giorgio, V., Carraro, M., Di Lisa, F., Forte, M., Szabó, I., Lippe, G., & Bernardi, P. (2018). The unique histidine in OSCP subunit of F-ATP synthase mediates inhibition of the permeability transition pore by acidic pH. *EMBO Reports, 19*, 257–268.

Austin, S., & Nowikovsky, K. (2019). LETM1: Essential for mitochondrial biology and cation homeostasis? *Trends in Biochemical Sciences, 44*, 648–658.

Azzolin, L., von Stockum, S., Basso, E., Petronilli, V., Forte, M. A., & Bernardi, P. (2010). The mitochondrial permeability transition from yeast to mammals. *FEBS Letters, 584*, 2504–2509.

Baines, C. P., Kaiser, R. A., Purcell, N. H., Blair, N. S., Osinska, H., Hambleton, M. A., Brunskill, E. W., Sayen, M. R., Gottlieb, R. A., Dorn, G. W., Robbins, J., & Molkentin, J. D. (2005). Loss of cyclophilin D reveals a critical role for mitochondrial permeability transition in cell death. *Nature, 434*, 658–662.

Barsukova, A., Komarov, A., Hajnoczky, G., Bernardi, P., Bourdette, D., & Forte, M. (2011). Activation of the mitochondrial permeability transition pore modulates Ca^{2+} responses to physiological stimuli in adult neurons. *European Journal of Neuroscience, 33*, 831–842.

Bason, J. V., Montgomery, M. G., Leslie, A. G., & Walker, J. E. (2014). Pathway of binding of the intrinsically disordered mitochondrial inhibitor protein to F_1-ATPase. *Proceedings of the National Academy of Sciences of the United States of America, 111*, 11305–11310.

Beatrice, M. C., Palmer, J. W., & Pfeiffer, D. R. (1980). The relationship between mitochondrial membrane permeability, membrane potential, and the retention of Ca^{2+} by mitochondria. *Journal of Biological Chemistry, 255*, 8663–8671.

Bernardi, P., Carraro, M., & Lippe, G. (2021). The mitochondrial permeability transition: Recent progress and open questions. *FEBS Journal*, Article 16254.

Bernardi, P. (1999). Mitochondrial transport of cations: Channels, exchangers, and permeability transition. *Physiological Reviews, 79*, 1127–1155.

Bernardi, P., Petronilli, V., Di Lisa, F., & Forte, M. (2001). A mitochondrial perspective on cell death. *Trends in Biochemical Sciences, 26*, 112–117.

Bernardi, P., & Petronilli, V. (1996). The permeability transition pore as a mitochondrial calcium release channel: A critical appraisal. *Journal of Bioenergetics and Biomembranes, 28*, 131–138.

Bernardi, P., Rasola, A., Forte, M., & Lippe, G. (2015). The mitochondrial permeability transition pore: Channel formation by F-ATP synthase, integration in signal transduction, and role in pathophysiology. *Physiological Reviews, 95*, 1111–1155.

Bernardi, P., Vassanelli, S., Veronese, P., Colonna, R., Szabó, I., & Zoratti, M. (1992). Modulation of the mitochondrial permeability transition pore. Effect of protons and divalent cations. *Journal of Biological Chemistry, 267*, 2934–2939.

Beutner, G., Alanzalon, R. E., & Porter, G. A., Jr. (2017). Cyclophilin D regulates the dynamic assembly of mitochondrial ATP synthase into synthasomes. *Scientific Reports, 7*, Article 14488.

Blum, T. B., Hahn, A., Meier, T., Davies, K. M., & Kühlbrandt, W. (2019). Dimers of mitochondrial ATP synthase induce membrane curvature and self-assemble into rows. *Proceedings of the National Academy of Sciences of the U S A, 116*, 4250–4255.

Bonora, M., Bononi, A., De Marchi, E., Giorgi, C., Lebiedzinska, M., Marchi, S., Patergnani, S., Rimessi, A., Suski, J. M., Wojtala, A., Wieckowski, M. R., Kroemer, G., Galluzzi, L., & Pinton, P. (2013). Role of the c subunit of the F_O ATP synthase in mitochondrial permeability transition. *Cell Cycle, 12*, 674–683.

Bowler, M. W., Montgomery, M. G., Leslie, A. G., & Walker, J. E. (2007). Ground state structure of F_1-ATPase from bovine heart mitochondria at 1.9 Å resolution. *Journal of Biological Chemistry, 282*, 14238–14242.

Boyer, P. D. (1993). The binding change mechanism for ATP synthase–some probabilities and possibilities. *Biochimica et Biophysica Acta, 1140*, 215–250.

Broekemeier, K. M., Dempsey, M. E., & Pfeiffer, D. R. (1989). Cyclosporin A is a potent inhibitor of the inner membrane permeability transition in liver mitochondria. *Journal of Biological Chemistry, 264*, 7826–7830.

Campanella, M., Casswell, E., Chong, S., Farah, Z., Wieckowski, M. R., Abramov, A. Y., Tinker, A., & Duchen, M. R. (2008). Regulation of mitochondrial structure and function by the F_1F_o-ATPase inhibitor proteinIF_1. *Cell Metabolism, 8*, 13–25.

Carraro, M., Checchetto, V., Sartori, G., Kucharczyk, R., di Rago, J.-P., Minervini, G., Franchin, C., Arrigoni, G., Giorgio, V., Petronilli, V., Tosatto, S. C. E., Lippe, G., Szabò, I., & Bernardi, P. (2018). High-conductance channel formation in yeast mitochondria is mediated by F-ATP synthase e and g subunits. *Cellular Physiology and Biochemistry, 50*, 1840–1855.

Carraro, M., Giorgio, V., Šileikyte, J., Sartori, G., Forte, M., Lippe, G., Zoratti, M., Szabó, I., & Bernardi, P. (2014). Channel formation by yeast F-ATP synthase and the role of dimerization in the mitochondrial permeability transition. *Journal of Biological Chemistry, 289*, 15980–15985.

Carraro, M., Jones, K., Sartori, G., Schiavone, M., Antonucci, S., Kucharczyk, R., di Rago, J.-P., Franchin, C., Arrigoni, G., Forte, M., & Bernardi, P. (2020). The unique cysteine of F-ATP synthase OSCP subunit participates in modulation of the permeability transition pore. *Cell Reports, 32*, Article 108095.

Carrer, A., Tommasin, L., Šileikyte, J., Ciscato, F., Filadi, R., Urbani, A., Forte, E., Rasola, A., Szabò, I., Carraro, M., & Bernardi, P. (2021). Defining the molecular mechanisms of the mitochondrial permeability transition through genetic manipulation of F-ATP synthase. *Nature Communications, 12*, 4835.

Carroll, J., He, J., Ding, S., Fearnley, I. M., & Walker, J. E. (2019). Persistence of the permeability transition pore in human mitochondria devoid of an assembled ATP synthase. *Proceedings of the National Academy of Sciences of the United States of America, 116*, 12816–12821.

Carroll, J., He, J., Ding, S., Fearnley, I. M., & Walker, J. E. (2021). TMEM70 and TMEM242 help to assemble the rotor ring of human ATP synthase and interact with assembly factors for complex I. *Proceedings of the National Academy of Sciences of the United States of America, 118*, Article e2100558118.

Cingolani, G., & Duncan, T. M. (2011). Structure of the ATP synthase catalytic complex (F_1) from *Escherichia coli* in an autoinhibited conformation. *Nature Structural & Molecular Biology, 18*, 701–707.

Clipstone, N. A., & Crabtree, G. R. (1992). Identification of calcineurin as a key signalling enzyme in T-lymphocyte activation. *Nature, 357*, 695–697.

Davidson, S. M., & Duchen, M. R. (2007). Endothelial mitochondria: Contributing to vascular function and disease. *Circulation Research, 100*, 1128–1141.

Davies, K. M., Anselmi, C., Wittig, I., Faraldo-Gomez, J. D., & Kühlbrandt, W. (2012). Structure of the yeast F_1F_o-ATP synthase dimer and its role in shaping the mitochondrial cristae. *Proceedings of the National Academy of Sciences of the United States of America, 109*, 13602–13607.

De Marchi, U., Basso, E., Szabó, I., & Zoratti, M. (2006). Electrophysiological characterization of the cyclophilin D-deleted mitochondrial permeability transition pore. *Molecular Membrane Biology, 23*, 521–530.

De Marchi, U., Szabò, I., Cereghetti, G. M., Hoxha, P., Craigen, W. J., & Zoratti, M. (2008). A maxi-chloride channel in the inner membrane of mammalian mitochondria. *Biochimica et Biophysica Acta, 1777*, 1438–1448.

De Stefani, D., Rizzuto, R., & Pozzan, T. (2016). Enjoy the trip: Calcium in mitochondria back and forth. *Annual Review of Biochemistry, 85*, 161–192.

de Witt, S. M., Verdoold, R., Cosemans, J. M. E. M., & Heemskerk, J. W. M. (2014). Insights into platelet-based control of coagulation. *Thrombosis Research, 133*, S139–S148.

Derksen, M., Vorwerk, C., & Siemen, D. (2016). Calpeptin, not calpain, directly inhibits an ion channel of the inner mitochondrial membrane. *Protoplasma, 253*, 835–843.

Di Lisa, F., Kaludercic, N., Carpi, A., Menabò, R., & Giorgio, M. (2009). Mitochondria and vascular pathology. *Pharmacological Reports, 61*, 123–130.

Di Pancrazio, F., Mavelli, I., Isola, M., Losano, G., Pagliaro, P., Harris, D. A., & Lippe, G. (2004). In vitro and in vivo studies of F_0F_1ATP synthase regulation by inhibitor protein IF_1 in goat heart. *Biochimica et Biophysica Acta, 1659*, 52–62.

Dmitriev, O. Y., Jones, P. C., & Fillingame, R. H. (1999). Structure of the subunit c oligomer in the F_1F_O ATP synthase: Model derived from solution structure of the monomer and cross-linking in the native enzyme. *Proceedings of the National Academy of Sciences of the United States of America, 96*, 7785–7790.

Du, H., Guo, L., Fang, F., Chen, D., Sosunov, A. A., McKhann, G. M., Yan, Y., Wang, C., Zhang, H., Molkentin, J. D., Gunn-Moore, F. J., Vonsattel, J. P., Arancio, O., Chen, J. X., & Yan, S. D. (2008). Cyclophilin D deficiency attenuates mitochondrial and neuronal perturbation and ameliorates learning and memory in Alzheimer's disease. *Nature Medicine, 14*, 1097–1105.

Elrod, J. W., Wong, R., Mishra, S., Vagnozzi, R. J., Sakthievel, B., Goonasekera, S. A., Karch, J., Gabel, S., Farber, J., Force, T., Brown, J. H., Murphy, E., & Molkentin, J. D. (2010). Cyclophilin D controls mitochondrial pore-dependent Ca^{2+} exchange, metabolic flexibility, and propensity for heart failure in mice. *Journal of Clinical Investigation, 120*, 3680–3687.

Esparza-Moltó, P. B., & Cuezva, J. M. (2018). The role of mitochondrial H^+-ATP synthase in cancer. *Frontiers Oncology, 8*, 53.

Esparza-Moltó, P. B., Romero-Carramiñana, I., Núñez de Arenas, C., Pereira, M. P., Blanco, N., Pardo, B., Bates, G. R., Sánchez-Castillo, C., Artuch, R., Murphy, M. P., Esteban, J. A., & Cuezva, J. M. (2021). Generation of mitochondrial reactive oxygen species is controlled by ATPase inhibitory factor 1 and regulates cognition. *PLoS Biology, 19*, Article e3001252.

Fogolari, F., Corazza, A., Yarra, V., Jalaru, A., Viglino, P., & Esposito, G. (2012). Bluues: A program for the analysis of the electrostatic properties of proteins based on generalized born radii. *BMC Bioinformatics, 13*(Suppl. 4), S18.

Franco, L. V. R., Su, C. H., & Tzagoloff, A. (2020). Modular assembly of yeast mitochondrial ATP synthase and cytochrome oxidase. *Biological Chemistry, 401*, 835–853.

Fujimoto, K., Chen, Y., Polonsky, K. S., & Dorn, G. W. (2010). Targeting cyclophilin D and the mitochondrial permeability transition enhances β-cell survival and prevents diabetes in Pdx1 deficiency. *Proceedings of the National Academy of Sciences of the U S A, 107*, 10214–10219.

García-Aguilar, A., & Cuezva, J. M. (2018). A review of the inhibition of the mitochondrial ATP synthase by IF1 in vivo: Reprogramming energy metabolism and inducing mitohormesis. *Frontiers in Physiology, 9*, 1322.

García-Bermúdez, J., & Cuezva, J. M. (2016). The ATPase inhibitory factor 1 (IF1): A master regulator of energy metabolism and of cell survival. *Biochimica et Biophysica Acta, 1857*, 1167–1182.

García-Bermúdez, J., Sánchez-Aragó, M., Soldevilla, B., del Arco, A., Nuevo-Tapioles, C., & Cuezva, J. M. (2015). PKA phosphorylates the ATPase inhibitory factor 1 and inactivates its capacity to bind and inhibit the mitochondrial H^+-ATP synthase. *Cell Reports, 12*, 2143–2155.

Gerle, C. (2020). Mitochondrial F-ATP synthase as the permeability transition pore. *Pharmacological Research, 160*, Article 105081.

Gerle, C. (2016). On the structural possibility of pore-forming mitochondrial F_oF_1 ATP synthase. *Biochimica et Biophysica Acta, 1857*, 1191–1196.

Giorgio, V., Bisetto, E., Soriano, M. E., Dabbeni-Sala, F., Basso, E., Petronilli, V., Forte, M. A., Bernardi, P., & Lippe, G. (2009). Cyclophilin D modulates mitochondrial F_0F_1-ATP synthase by interacting with the lateral stalk of the complex. *Journal of Biological Chemistry, 284*, 33982–33988.

Giorgio, V., Burchell, V., Schiavone, M., Bassot, C., Minervini, G., Petronilli, V., Argenton, F., Forte, M., Tosatto, S., Lippe, G., & Bernardi, P. (2017). Ca^{2+} binding to F-ATP synthase β subunit triggers the mitochondrial permeability transition. *EMBO Reports, 18*, 1065–1076.

Giorgio, V., von Stockum, S., Antoniel, M., Fabbro, A., Fogolari, F., Forte, M., Glick, G. D., Petronilli, V., Zoratti, M., Szabó, I., Lippe, G., & Bernardi, P. (2013). Dimers of mitochondrial ATP synthase form the permeability transition pore. *Proceedings of the National Academy of Sciences of the United States of America, 110*, 5887–5892.

Gledhill, J. R., Montgomery, M. G., Leslie, A. G., & Walker, J. E. (2007). How the regulatory protein, IF_1, inhibits F_1-ATPase from bovine mitochondria. *Proceedings of the National Academy of Sciences of the United States of America, 104*, 15671–15676.

Goldberg, J., Currais, A., Prior, M., Fischer, W., Chiruta, C., Ratliff, E., Daugherty, D., Dargusch, R., Finley, K., Esparza-Molto, P. B., Cuezva, J. M., Maher, P., Petrascheck, M., & Schubert, D. (2018). The mitochondrial ATP synthase is a shared drug target for aging and dementia. *Aging Cell, 17*, Article e12715.

Green, D. W., & Grover, G. J. (2000). The IF_1 inhibitor protein of the mitochondrial F_1F_0-ATPase. *Biochimica et Biophysica Acta, 1458*, 343–355.

Griffiths, E. J., & Halestrap, A. P. (1993). Protection by Cyclosporin A of ischemia/reperfusion-induced damage in isolated rat hearts. *Journal of Molecular and Cellular Cardiology, 25*, 1461–1469.

Gu, J., Zhang, L., Zong, S., Guo, R., Liu, T., Yi, J., Wang, P., Zhuo, W., & Yang, M. (2019). Cryo-EM structure of the mammalian ATP synthase tetramer bound with inhibitory protein IF1. *Science, 364*, 1068–1075.

Guo, H., Bueler, S. A., & Rubinstein, J. L. (2017). Atomic model for the dimeric F_O region of mitochondrial ATP synthase. *Science, 358*, 936–940.

Guo, H., Courbon, G. M., Bueler, S. A., Mai, J., Liu, J., & Rubinstein, J. L. (2021). Structure of mycobacterial ATP synthase bound to the tuberculosis drug bedaquiline. *Nature, 589*, 143–147.

Guo, H., Suzuki, T., & Rubinstein, J. L. (2019). Structure of a bacterial ATP synthase. *eLife Sciences, 8*.

Guo, L., Carraro, M., Carrer, A., Minervini, G., Urbani, A., Masgras, I., Tosatto, S. C. E., Szabó, I., Bernardi, P., & Lippe, G. (2019). Arg-8 of yeast subunit e contributes to the stability of F-ATP synthase dimers and to the generation of the full-conductance mitochondrial megachannel. *Journal of Biological Chemistry, 294*, 10987–10997.

Guo, L., Carraro, M., Sartori, G., Minervini, G., Eriksson, O., Petronilli, V., & Bernardi, P. (2018). Arginine 107 of yeast ATP synthase subunit g mediates sensitivity of the mitochondrial permeability transition to phenylglyoxal. *Journal of Biological Chemistry, 293*, 14632–14645.

Guo, L. (2022). Mitochondrial ATP synthase inhibitory factor 1 interacts with the p53-cyclophilin D complex and promotes opening of the permeability transition pore. *Journal of Biological Chemistry*, Article 101858.

Habersetzer, J., Ziani, W., Larrieu, I., Stines-Chaumeil, C., Giraud, M. F., Brèthes, D., Dautant, A., & Paumard, P. (2013). ATP synthase oligomerization: From the enzyme models to the mitochondrial morphology. *The International Journal of Biochemistry & Cell Biology, 45*, 99–105.

Hahn, A., Parey, K., Bublitz, M., Mills, D. J., Zickermann, V., Vonck, J., Kühlbrandt, W., & Meier, T. (2016). Structure of a complete ATP synthase dimer reveals the molecular basis of inner mitochondrial membrane morphology. *Molecular Cell, 63*, 445–456.

Hahn, A., Vonck, J., Mills, D. J., Meier, T., & Kühlbrandt, W. (2018). Structure, mechanism, and regulation of the chloroplast ATP synthase. *Science, 360*.

Harris, D. A., & Das, A. M. (1991). Control of mitochondrial ATP synthesis in the heart. *Biochemical Journal, 280*, 561–573.

He, J., Carroll, J., Ding, S., Fearnley, I. M., & Walker, J. E. (2017). Permeability transition in human mitochondria persists in the absence of peripheral stalk subunits of ATP synthase. *Proceedings of the National Academy of Sciences of the United States of America, 114*, 9086–9091.

He, J., Ford, H. C., Carroll, J., Ding, S., Fearnley, I. M., & Walker, J. E. (2017). Persistence of the mitochondrial permeability transition in the absence of subunit c of human ATP synthase. *Proceedings of the National Academy of Sciences of the United States of America, 114*, 3409–3414.

He, J., Ford, H. C., Carroll, J., Douglas, C., Gonzales, E., Ding, S., Fearnley, I. M., & Walker, J. E. (2018). Assembly of the membrane domain of ATP synthase in human mitochondria. *Proceedings of the National Academy of Sciences of the U S A, 115*, 2988–2993.

Hejzlarová, Mrácek, T., Vrbacky, M., Kaplanová, V., Karbanová, V., Nusková, H., Pecina, P., & Houštek, J. (2014). Nuclear genetic defects of mitochondrial ATP synthase. *Physiological Research, 63*(Suppl. 1), S57–S71.

Hong, S., & Pedersen, P. L. (2008). ATP synthase and the actions of inhibitors utilized to study its roles in human health, disease, and other scientific areas. *Microbiology and Molecular Biology Reviews, 72*, 590–641.

Hunter, D. R., Haworth, R. A., & Southard, J. H. (1976). Relationship between configuration, function, and permeability in calcium-treated mitochondria. *Journal of Biological Chemistry, 251*, 5069–5077.

Hunter, D. R., & Haworth, R. A. (1979). The Ca^{2+}-induced membrane transition in mitochondria. I. The protective mechanisms. *Archives of Biochemistry and Biophysics, 195*, 453–459.

Hüser, J., & Blatter, L. A. (1999). Fluctuations in mitochondrial membrane potential caused by repetitive gating of the permeability transition pore. *Biochemical Journal, 343*, 311–317.

Hüser, J., Rechenmacher, C. E., & Blatter, L. A. (1998). Imaging the permeability pore transition in single mitochondria. *Biophysical Journal, 74*, 2129–2137.

Irwin, W. A., Bergamin, N., Sabatelli, P., Reggiani, C., Megighian, A., Merlini, L., Braghetta, P., Columbaro, M., Volpin, D., Bressan, G. M., Bernardi, P., & Bonaldo, P. (2003). Mitochondrial dysfunction and apoptosis in myopathic mice with collagen VI deficiency. *Nature Genetics, 35*, 267–271.

Juhaszova, M., Kobrinsky, E., Zorov, D. B., Nuss, H. B., Yaniv, Y., Fishbein, K. W., de Cabo, R., Montoliu, L., Gabelli, S. B., Aon, M. A., Cortassa, S., & Sollott, S. J. (2021). ATP synthase K+- and H+-fluxes drive ATP synthesis and enable mitochondrial K+-"uniporter" function: I. Characterization of ion fluxes. *Function*, Article zqab065.

Juhaszova, M., Kobrinsky, E., Zorov, D. B., Nuss, H. B., Yaniv, Y., Fishbein, K. W., de Cabo, R., Montoliu, L., Gabelli, S. B., Aon, M. A., Cortassa, S., & Sollott, S. J. (2022). ATP synthase K^+- and H^+-fluxes drive ATP synthesis and enable mitochondrial K^+-"Uniporter" function: II. Ion and ATP synthase flux regulation. *Function, 3*, Article zqac001.

Junge, W., Sielaff, H., & Engelbrecht, S. (2009). Torque generation and elastic power transmission in the rotary F_OF_1-ATPase. *Nature, 459*, 364–370.

Kabaleeswaran, V., Puri, N., Walker, J. E., Leslie, A. G., & Mueller, D. M. (2006). Novel features of the rotary catalytic mechanism revealed in the structure of yeast F_1 ATPase. *The EMBO Journal, 25*, 5433–5442.

Kagawa, Y., & Racker, E. (1966). Partial resolution of the enzymes catalyzing oxidative phosphorylation. 8. Properties of a factor conferring oligomycin sensitivity on mitochondrial adenosine triphosphatase. *Journal of Biological Chemistry, 241*, 2461–2466.

Kim, K., Shin, J. H., Lee, J. H., Jo, Y. H., Kim, M. A., Lee, K., Kang, K. W., & Hong, Y. S. (2015). The effect of cyclosporine A in hemorrhagic shock model of rats. *Journal of Trauma and Acute Care Surgery, 78*.

Kinnally, K. W., Campo, M. L., & Tedeschi, H. (1989). Mitochondrial channel activity studied by patch-clamping mitoplasts. *Journal of Bioenergetics and Biomembranes, 21*, 497–506.

Ko, Y. H., Delannoy, M., Hullihen, J., Chiu, W., & Pedersen, P. L. (2003). Mitochondrial ATP Synthasome. Cristae-enriched membranes and a multiwell detergent screening assay yield dispersed single complexes containing the ATP synthase and carriers for Pi and ADP/ATP. *Journal of Biological Chemistry, 278*, Article 12305.

Kobayashi, R., Ueno, H., Li, C. B., & Noji, H. (2020). Rotary catalysis of bovine mitochondrial F_1-ATPase studied by single-molecule experiments. *Proceedings of the National Academy of Sciences of the United States of America, 117*, 1447–1456.

Kovalcíková, J., Vrbacký, M., Pecina, P., Tauchmannová, K., Nusková, H., Kaplanová, V., Brázdová, A., Alán, L., Eliáš, J., Cunátová, K., Korínek, V., Sedlacek, R., Mrácek, T., & Houstek, J. (2019). TMEM70 facilitates biogenesis of mammalian ATP synthase by promoting subunit c incorporation into the rotor structure of the enzyme. *The FASEB Journal, 33*, 14103–14117.

Kühlbrandt, W. (2014). Biochemistry. The resolution revolution. *Science, 343*, 1443–1444.

Kühlbrandt, W. (2019). Structure and mechanisms of F-type ATP synthases. *Annual Review of Biochemistry, 88*, 515–549.

Lefebvre-Legendre, L., Vaillier, J., Benabdelhak, H., Velours, J., Slonimski, P. P., & di Rago, J. P. (2001). Identification of a nuclear gene (*FMC1*) required for the assembly/stability of yeast mitochondrial F_1-ATPase in heat stress conditions. *Journal of Biological Chemistry, 276*, 6789–6796.

Lei, Y., Peng, X., Liu, L., Dong, Z., & Li, T. (2015). Beneficial effect of cyclosporine A on traumatic hemorrhagic shock. *Journal of Surgical Research, 195*, 529–540.

Leytin, V., Gyulkhandanyan, A. V., & Freedman, J. (2018). Role of mitochondrial membrane permeabilization and depolarization in platelet apoptosis. *British Journal of Haematology, 181*, 281–285.

Li, P., Meng, X., Bian, H., Burns, N., Zhao, K.s., & Song, R. (2015). Activation of sirtuin 1/3 improves vascular hyporeactivity in severe hemorrhagic shock by alleviation of mitochondrial damage. *Oncotarget, 6*, 36998–37011.

Lippe, G., Coluccino, G., Zancani, M., Baratta, W., & Crusiz, P. (2019). Mitochondrial F-ATP synthase and its transition into an energy-dissipating molecular machine. *Oxidative Medicine and Cellular Longevity, 2019*, Article 8743257.

Liu, J., Farmer, J. D. J., Lane, W. S., Friedman, J., Weissman, I., & Schreiber, S. L. (1991). Calcineurin is a common target of cyclophilin-cyclosporin A and FKBP-FK506 complexes. *Cell, 66*, 807–815.

Long, Q., Yang, K., & Yang, Q. (2015). Regulation of mitochondrial ATP synthase in cardiac pathophysiology. *American Journal of Cardiovascular Disease, 5*, 19–32.

Lu, X., Kwong, J. Q., Molkentin, J. D., & Bers, D. M. (2016). Individual cardiac mitochondria undergo rare transient permeability transition pore openings. *Circulation Research, 118*, 834–841.

Lytovchenko, O., Naumenko, N., Oeljeklaus, S., Schmidt, B., von der Malsburg, K., Deckers, M., Warscheid, B., van der Laan, M., & Rehling, P. (2014). The INA complex facilitates assembly of the peripheral stalk of the mitochondrial F_1F_O-ATP synthase. *The EMBO Journal, 33*, 1624–1638.

Martin, J. L., Ishmukhametov, R., Spetzler, D., Hornung, T., & Frasch, W. D. (2018). Elastic coupling power stroke mechanism of the F_1-ATPase molecular motor. *Proceedings of the National Academy of Sciences of the United States of America, 115*, 5750–5755.

Martin, L. J., Gertz, B., Pan, Y., Price, A. C., Molkentin, J. D., & Chang, Q. (2009). The mitochondrial permeability transition pore in motor neurons: Involvement in the pathobiology of ALS mice. *Experimental Neurology, 218*, 333–346.

Martin, L. J., Semenkow, S., Hanaford, A., & Wong, M. (2014). Mitochondrial permeability transition pore regulates Parkinson's disease development in mutant α-synuclein transgenic mice. *Neurobiology of Aging, 35*, 1132–1152.

Martinucci, S., Szabó, I., Tombola, F., & Zoratti, M. (2000). Ca^{2+}-reversible inhibition of the mitochondrial megachannel by ubiquinone analogues. *FEBS Letters, 480*, 89–94.

Massari, S., & Azzone, G. F. (1972). The equivalent pore radius of intact and damaged mitochondria and the mechanism of active shrinkage. *Biochimica et Biophysica Acta, 283*, 23–29.

Mather, M., & Rottenberg, H. (2000). Aging enhances the activation of the permeability transition pore in mitochondria. *Biochemical and Biophysical Research Communications, 273*, 603–608.

Mather, M. W., & Rottenberg, H. (2002). The inhibition of calcium signaling in T lymphocytes from old mice results from enhanced activation of the mitochondrial permeability transition pore. *Mechanism of Ageing and Development, 123*, 707–724.

Mayer, F., Lim, J. K., Langer, J. D., Kang, S. G., & Müller, V. (2015). Na^+ transport by the A_1A_O-ATP synthase purified from *Thermococcus onnurineus* and reconstituted into liposomes. *Journal of Biological Chemistry, 290*, 6994–7002.

Menze, M. A., Hutchinson, K., Laborde, S. M., & Hand, S. C. (2005). Mitochondrial permeability transition in the crustacean *Artemia franciscana*: Absence of a calcium-regulated pore in the face of profound calcium storage. *American Journal of Physiology, 289*, R68–R76.

Millay, D. P., Sargent, M. A., Osinska, H., Baines, C. P., Barton, E. R., Vuagniaux, G., Sweeney, H. L., Robbins, J., & Molkentin, J. D. (2008). Genetic and pharmacologic inhibition of mitochondrial-dependent necrosis attenuates muscular dystrophy. *Nature Medicine, 14*, 442–447.

Mitchell, P. (2011). Chemiosmotic coupling in oxidative and photosynthetic phosphorylation. 1966. *Biochimica et Biophysica Acta, 1807*, 1507–1538.

Mnatsakanyan, N., Llaguno, M. C., Yang, Y., Yan, Y., Weber, J., Sigworth, F. J., & Jonas, E. A. (2019). A mitochondrial megachannel resides in monomeric F_1F_O ATP synthase. *Nature Communications, 10*, 5823.

Morales-Rios, E., Montgomery, M. G., Leslie, A. G., Garcia-Trejo, J. J., & Walker, J. E. (2015). Structure of a catalytic dimer of the α- and β-subunits of the F-ATPase from *Paracoccus denitrificans* at 2.3 Å resolution. *Acta Crystallographica Section F: Structural Biology Communications, 71*, 1309–1317.

Morales-Rios, E., Montgomery, M. G., Leslie, A. G., & Walker, J. E. (2015). Structure of ATP synthase from *Paracoccus denitrificans* determined by X-ray crystallography at 4.0 Å resolution. *Proceedings of the National Academy of Sciences of the United States of America, 112*, 13231–13236.

Morciano, G., Pedriali, G., Bonora, M., Pavasini, R., Mikus, E., Calvi, S., Bovolenta, M., Lebiedzinska-Arciszewska, M., Pinotti, M., Albertini, A., Wieckowski, M. R., Giorgi, C., Ferrari, R., Galluzzi, L., Campo, G., & Pinton, P. (2021). A naturally occurring mutation in ATP synthase subunit c is associated with increased damage following hypoxia/reoxygenation in STEMI patients. *Cell Reports, 35*, Article 108983.

Mukherjee, R., Nunes, Q., Huang, W., & Sutton, R. (2019). Precision medicine for acute pancreatitis: Current status and future opportunities. *Precision Clinical Medicine, 2*, 81–86.

Murphy, B. J., Klusch, N., Langer, J., Mills, D. J., Yildiz, Ö., & Kühlbrandt, W. (2019). Rotary substates of mitochondrial ATP synthase reveal the basis of flexible F_1-F_o coupling. *Science, 364*, Article eaaw9128.

Nakagawa, T., Shimizu, S., Watanabe, T., Yamaguchi, O., Otsu, K., Yamagata, H., Inohara, H., Kubo, T., & Tsujimoto, Y. (2005). Cyclophilin D-dependent mitochondrial permeability transition regulates some necrotic but not apoptotic cell death. *Nature, 434*, 652–658.

Naumenko, N., Morgenstern, M., Rucktaschel, R., Warscheid, B., & Rehling, P. (2017). INA complex liaises the F_1F_O-ATP synthase membrane motor modules. *Nature Communications, 8*, 1237.

Neginskaya, M. A., Solesio, M. E., Berezhnaya, E. V., Amodeo, G. F., Mnatsakanyan, N., Jonas, E. A., & Pavlov, E. V. (2019). ATP synthase C-Subunit-Deficient mitochondria have a small cyclosporine A-sensitive channel, but lack the permeability transition pore. *Cell Reports, 26*, 11–17.

Neher, E., & Sakmann, B. (1976). Single-channel currents recorded from membrane of denervated frog muscle fibres. *Nature, 260*, 799–802.

Nicolli, A., Basso, E., Petronilli, V., Wenger, R. M., & Bernardi, P. (1996). Interactions of cyclophilin with the mitochondrial inner membrane and regulation of the permeability transition pore, a cyclosporin A-sensitive channel. *Journal of Biological Chemistry, 271*, 2185–2192.

Nirody, J. A., Budin, I., & Rangamani, P. (2020). ATP synthase: Evolution, energetics, and membrane interactions. *The Journal of General Physiology, 152*, Article e201912475.

Noji, H., Yasuda, R., Yoshida, M., & Kinosita, K., Jr. (1997). Direct observation of the rotation of F_1-ATPase. *Nature, 386*, 299–302.

Nusková, H., Mrácek, T., Mikulová, T., Vrbacký, M., Kovárová, N., Kovalcíková, J., Pecina, P., & Houštek, J. (2015). Mitochondrial ATP synthasome: Expression and structural interaction of its components. *Biochemical and Biophysical Research Communications, 464*, 787–793.

Palty, R., Silverman, W. F., Hershfinkel, M., Caporale, T., Sensi, S. L., Parnis, J., Nolte, C., Fishman, D., Shoshan-Barmatz, V., Herrmann, S., Khananshvili, D., & Sekler, I. (2010). NCLX is an essential component of mitochondrial Na^+/Ca^{2+} exchange. *Proceedings of the National Academy of Sciences of the United States of America, 107*, 436–441.

Panel, M., Ghaleh, B., & Morin, D. (2018). Mitochondria and aging: A role for the mitochondrial transition pore? *Aging Cell, 17*, Article e12793.

Pardo Semo, A., Luis-Garcia, E. R., Becerril, C., Mendoza, C., Maldonado, M., Aparicio Trejo, E., & Selman, M. (2019). Alterations in mitochondrial permeability transition pore contribute to apoptosis resistance in IPF fibroblasts. *European Respiratory Journal, 54*, Article A584.

Park, J., Chen, Y., Tishkoff, D. X., Peng, C., Tan, M., Dai, L., Xie, Z., Zhang, Y., Zwaans, B. M., Skinner, M. E., Lombard, D. B., & Zhao, Y. (2013). SIRT5-mediated lysine desuccinylation impacts diverse metabolic pathways. *Molecular Cell, 50*, 919–930.

Parone, P. A., Da Cruz, S., Han, J. S., McAlonis-Downes, M., Vetto, A. P., Lee, S. K., Tseng, E., & Cleveland, D. W. (2013). Enhancing mitochondrial calcium buffering capacity reduces aggregation of misfolded SOD1 and motor neuron cell death without extending survival in mouse models of inherited amyotrophic lateral sclerosis. *Journal of Neuroscience, 33*, 4657–4671.

Paumard, P., Vaillier, J., Coulary, B., Schaeffer, J., Soubannier, V., Mueller, D. M., Brèthes, D., di Rago, J. P., & Velours, J. (2002). The ATP synthase is involved in generating mitochondrial cristae morphology. *The EMBO Journal, 21*, 221–230.

Petronilli, V., Miotto, G., Canton, M., Brini, M., Colonna, R., Bernardi, P., & Di Lisa, F. (1999). Transient and long-lasting openings of the mitochondrial permeability transition pore can be monitored directly in intact cells by changes in mitochondrial calcein fluorescence. *Biophysical Journal, 76*, 725–734.

Petronilli, V., Nicolli, A., Costantini, P., Colonna, R., & Bernardi, P. (1994). Regulation of the permeability transition pore, a voltage-dependent mitochondrial channel inhibited by cyclosporin A. *Biochimica et Biophysica Acta, 1187*, 255–259.

Petronilli, V., Szabó, I., & Zoratti, M. (1989). The inner mitochondrial membrane contains ion-conducting channels similar to those found in bacteria. *FEBS Letters, 259*, 137–143.

Pinke, G., Zhou, L., & Sazanov, L. A. (2020). Cryo-EM structure of the entire mammalian F-type ATP synthase. *Nature Structural & Molecular Biology, 27*, 1077–1085.

Pozzan, T., Bragadin, M., & Azzone, G. F. (1977). Disequilibrium between steady-state Ca^{2+} accumulation ratio and membrane potential in mitochondria. Pathway and role of Ca^{2+} efflux. *Biochemistry, 16*, 5618–5625.

Priante, G., Gianesello, L., Ceol, M., Del Prete, D., & Anglani, F. (2019). Cell death in the kidney. *International Journal of Molecular Sciences, 20*, 3598.

Quintana-Cabrera, R., Quirin, C., Glytsou, C., Corrado, M., Urbani, A., Pellattiero, A., Calvo, E., Vazquez, J., Enriquez, J. A., Gerle, C., Soriano, M. E., Bernardi, P., & Scorrano, L. (2018). The cristae modulator optic atrophy 1 requires mitochondrial ATP synthase oligomers to safeguard mitochondrial function. *Nature Communications, 9*, 3399.

Rak, M., Gokova, S., & Tzagoloff, A. (2011). Modular assembly of yeast mitochondrial ATP synthase. *The EMBO Journal, 30*, 920–930.

Rardin, M. J., Newman, J. C., Held, J. M., Cusack, M. P., Sorensen, D. J., Li, B., Schilling, B., Mooney, S. D., Kahn, C. R., Verdin, E., & Gibson, B. W. (2013). Label-free quantitative proteomics of the lysine acetylome in mitochondria identifies substrates of SIRT3 in metabolic pathways. *Proceedings of the National Academy of Sciences of the United States of America, 110*, 6601–6606.

Readnower, R. D., Pandya, J. D., McEwen, M. L., Pauly, J. R., Springer, J. E., & Sullivan, P. G. (2011). Post-injury administration of the mitochondrial permeability transition pore inhibitor, NIM811, is neuroprotective and improves cognition after traumatic brain injury in rats. *Journal of Neurotrauma, 28*, 1845–1853.

Rees, D. M., Leslie, A. G., & Walker, J. E. (2009). The structure of the membrane extrinsic region of bovine ATP synthase. *Proceedings of the National Academy of Sciences of the United States of America, 106*, 21597–21601.

Rieger, B., Arroum, T., Borowski, M. T., Villalta, J., & Busch, K. B. (2021). Mitochondrial F_1F_O ATP synthase determines the local proton motive force at cristae rims. *EMBO Reports, 22*, Article e52727.

Robinson, G. C., Bason, J. V., Montgomery, M. G., Fearnley, I. M., Mueller, D. M., Leslie, A. G., & Walker, J. E. (2013). The structure of F_1-ATPase from *Saccharomyces cerevisiae* inhibited by its regulatory protein IF_1. *Open Biol, 3*, Article 120164.

Rouslin, W. (1983). Protonic inhibition of the mitochondrial oligomycin-sensitive adenosine 5'-triphosphatase in ischemic and autolyzing cardiac muscle. Possible mechanism for the mitigation of ATP hydrolysis under nonenergizing conditions. *Journal of Biological Chemistry, 258*, 9657–9661.

Salewskij, K., Rieger, B., Hager, F., Arroum, T., Duwe, P., Villalta, J., Cogliati, S., Richter, C. P., Psathaki, O. E., Enriquez, J. A., Dellmann, T., & Busch, K. B. (2020). The spatio-temporal organization of mitochondrial F_1F_O ATP synthase in cristae depends on its activity mode. *Biochimica et Biophysica Acta, 1861*, Article 148091.

Sambri, I., Massa, F., Gullo, F., Meneghini, S., Cassina, L., Patanella, L., Carissimo, A., Iuliano, A., Santorelli, F., Codazzi, F., Grohovaz, F., Bernardi, P., Becchetti, A., & Casari, G. (2020). Impaired flickering of the permeability transition pore causes SPG7 spastic paraplegia. *EBioMedicine, 61*, Article 103050.

Sayeed, I., Parvez, S., Winkler-Stuck, K., Seitz, G., Trieu, I., Wallesch, C. W., Schönfeld, P., & Siemen, D. (2006). Patch clamp reveals powerful blockade of the mitochondrial permeability transition pore by the D2-receptor agonist pramipexole. *The FASEB Journal, 20*, 556–558.

Shah, S. S., Lannon, H., Dias, L., Zhang, J. Y., Alper, S. L., Pollak, M. R., & Friedman, D. J. (2019). APOL1 kidney risk variants induce cell death *via* mitochondrial translocation and opening of the mitochondrial permeability transition pore. *Journal of the American Society of Nephrology, 30*, 2355–2368.

Sielaff, H., Rennekamp, H., Wachter, A., Xie, H., Hilbers, F., Feldbauer, K., Dunn, S. D., Engelbrecht, S., & Junge, W. (2008). Domain compliance and elastic power transmission in rotary F_OF_1-ATPase. *Proceedings of the National Academy of Sciences of the United States of America, 105*, 17760–17765.

Sobti, M., Smits, C., Wong, A. S., Ishmukhametov, R., Stock, D., Sandin, S., & Stewart, A. G. (2016). Cryo-EM structures of the autoinhibited *E. coli* ATP synthase in three rotational states. *eLife Sciences, 5*.

Sobti, M., Walshe, J. L., Wu, D., Ishmukhametov, R., Zeng, Y. C., Robinson, C. V., Berry, R. M., & Stewart, A. G. (2020). Cryo-EM structures provide insight into how *E. coli* F_1F_O ATP synthase accommodates symmetry mismatch. *Nature Communications, 11*, 2615.

Soga, N., Kimura, K., Kinosita, K., Jr., Yoshida, M., & Suzuki, T. (2017). Perfect chemomechanical coupling of FoF1-ATP synthase. *Proceedings of the National Academy of Sciences of the United States of America, 114*, 4960–4965.

Sorgato, M. C., Keller, B. U., & Stühmer, W. (1987). Patch-clamping of the inner mitochondrial membrane reveals a voltage-dependent ion channel. *Nature, 330*, 498–500.

Spikes, T. E., Montgomery, M. G., & Walker, J. E. (2021). Interface mobility between monomers in dimeric bovine ATP synthase participates in the ultrastructure of inner mitochondrial membranes. *Proceedings of the National Academy of Sciences of the United States of America, 118*, Article e2021012118.

Spikes, T. E., Montgomery, M. G., & Walker, J. E. (2020). Structure of the dimeric ATP synthase from bovine mitochondria. *Proceedings of the National Academy of Sciences of the United States of America, 117*, 23519–23526.

Srivastava, A. P., Luo, M., Zhou, W., Symersky, J., Bai, D., Chambers, M. G., Faraldo-Gomez, J. D., Liao, M., & Mueller, D. M. (2018). High-resolution cryo-EM analysis of the yeast ATP synthase in a lipid membrane. *Science, 360*, Article eaas9969.

Stanzani, G., Duchen, M. R., & Singer, M. (2019). The role of mitochondria in sepsis-induced cardiomyopathy. *Biochimica et Biophysica Acta, 1865*, 759–773.

Stoldt, S., Wenzel, D., Kehrein, K., Riedel, D., Ott, M., & Jakobs, S. (2018). Spatial orchestration of mitochondrial translation and OXPHOS complex assembly. *Nature Cell Biology, 20*, 528–534.

Suzuki, T., Tanaka, K., Wakabayashi, C., Saita, E., & Yoshida, M. (2014). Chemomechanical coupling of human mitochondrial F_1-ATPase motor. *Nature Chemical Biology, 10*, 930–936.

Szabó, I., Bernardi, P., & Zoratti, M. (1992). Modulation of the mitochondrial megachannel by divalent cations and protons. *Journal of Biological Chemistry, 267*, 2940–2946.

Szabó, I., & Zoratti, M. (1991). The giant channel of the inner mitochondrial membrane is inhibited by cyclosporin A. *Journal of Biological Chemistry, 266*, 3376–3379.

Szabó, I., & Zoratti, M. (1992). The mitochondrial megachannel is the permeability transition pore. *Journal of Bioenergetics and Biomembranes, 24*, 111–117.

Ten, V. S., & Ratner, V. (2019). Mitochondrial bioenergetics and pulmonary dysfunction: Current progress and future directions. *Paediatric Respiratory Reviews, 34*, 37–45.

Thiede, A., Gellerich, F. N., Schönfeld, P., & Siemen, D. (2012). Complex effects of 17β-estradiol on mitochondrial function. *Biochimica et Biophysica Acta, 1817*, 1747–1753.

Uchino, H., Elmer, E., Uchino, K., Li, P. A., He, Q. P., Smith, M. L., & Siesjö, B. K. (1998). Amelioration by cyclosporin A of brain damage in transient forebrain ischemia in the rat. *Brain Research, 812*, 216–226.

Urbani, A., Giorgio, V., Carrer, A., Franchin, C., Arrigoni, G., Jiko, C., Abe, K., Maeda, S., Shinzawa-Itoh, K., Bogers, J, F. M., McMillan, D. G. G., Gerle, C., Szabó, I., & Bernardi, P. (2019). Purified F-ATP synthase forms a Ca^{2+}-dependent high-conductance channel matching the mitochondrial permeability transition pore. *Nature Communications, 10*, 4341.

Vindis, C., Elbaz, M., Escargueil-Blanc, I., Auge, N., Heniquez, A., Thiers, J. C., Negre-Salvayre, A., & Salvayre, R. (2005). Two distinct calcium-dependent mitochondrial pathways are involved in oxidized LDL-induced apoptosis. *Arteriosclerosis, Thrombosis, and Vascular Biology, 25*, 639–645.

Vinothkumar, K. R., Montgomery, M. G., Liu, S., & Walker, J. E. (2016). Structure of the mitochondrial ATP synthase from *Pichia angusta* determined by electron cryo-microscopy. *Proceedings of the National Academy of Sciences of the United States of America, 113*, 12709–12714.

von Stockum, S., Basso, E., Petronilli, V., Sabatelli, P., Forte, M. A., & Bernardi, P. (2011). Properties of Ca^{2+} transport in mitochondria of *Drosophila melanogaster*. *Journal of Biological Chemistry, 286*, 41163–41170.

von Stockum, S., Giorgio, V., Trevisan, E., Lippe, G., Glick, G. D., Forte, M. A., Da-Rè, C., Checchetto, V., Mazzotta, G., Costa, R., Szabò, I., & Bernardi, P. (2015). F-ATPase of *D. melanogaster* forms 53 picosiemen (53-pS) channels responsible for mitochondrial Ca^{2+}-induced Ca^{2+} release. *Journal of Biological Chemistry, 290*, 4537–4544.

Wallace, K. B. (2017). Mitochondrial toxicity. *Toxicology, 391*, 1.

Wang, Z. G., White, P. S., & Ackerman, S. H. (2001). Atp11p and Atp12p are assembly factors for the F_1-ATPase in human mitochondria. *Journal of Biological Chemistry, 276*, 30773–30778.

Watt, I. N., Montgomery, M. G., Runswick, M. J., Leslie, A. G., & Walker, J. E. (2010). Bioenergetic cost of making an adenosine triphosphate molecule in animal mitochondria. *Proceedings of the National Academy of Sciences of the United States of America, 107*, 16823–16827.

Widlansky, M. E., & Hill, R. B. (2018). Mitochondrial regulation of diabetic vascular disease: An emerging opportunity. *Translational Research, 202*, 83–98.

Wittig, I., Meyer, B., Heide, H., Steger, M., Bleier, L., Wumaier, Z., Karas, M., & Schägger, H. (2010). Assembly and oligomerization of human ATP synthase lacking mitochondrial subunits a and A6L. *Biochimica et Biophysica Acta, 1797*, 1004–1011.

Yan, S., Du, F., Wu, L., Zhang, Z., Zhong, C., Yu, Q., Wang, Y., Lue, L. F., Walker, D. G., Douglas, J. T., & Yan, S. S. (2016). F1F0 ATP synthase-cyclophilin D interaction contributes to diabetes-induced synaptic dysfunction and cognitive decline. *Diabetes, 65*, 3482–3494.

Ying, Z., Xiang, G., Zheng, L., Tang, H., Duan, L., Lin, X., Zhao, Q., Chen, K., Wu, Y., Xing, G., Lv, Y., Li, L., Yang, L., Bao, F., Long, Q., Zhou, Y., He, X., Wang, Y., Gao, M., … Liu, X. (2019). Short-term mitochondrial permeability transition pore opening modulates histone lysine methylation at the early phase of somatic cell reprogramming. *Cell Metabolism, 29*, 502.

Zech, M., Kopajtich, R., Steinbrucker, K., Bris, C., Gueguen, N., Feichtinger, R. G., Achleitner, M. T., Duzkale, N., Périvier, M., Koch, J., Engelhardt, H., Freisinger, P., Wagner, M., Brunet, T., Berutti, R., Smirnov, D., Navaratnarajah, T., Rodenburg, R. J. T., Pais, L. S., … Prokisch, H. (2022). Variants in mitochondrial ATP synthase cause variable neurologic phenotypes. *Annals of Neurology, 91*, 225–237.

Zhou, A., Rohou, A., Schep, D. G., Bason, J. V., Montgomery, M. G., Walker, J. E., Grigorieff, N., & Rubinstein, J. L. (2015). Structure and conformational states of the bovine mitochondrial ATP synthase by cryo-EM. *eLife Sciences, 4*, Article e10180.

Zhou, Z., Zhang, K., Liu, Z., Gao, X., Huang, K., Chen, C., Wang, D., Yang, Q., & Long, Q. (2021). ATPAF1 deficiency impairs ATP synthase assembly and mitochondrial respiration. *Mitochondrion, 60*, 129–141.

Zoratti, M., De Marchi, U., Biasutto, L., & Szabó, I. (2010). Electrophysiology clarifies the megariddles of the mitochondrial permeability transition pore. *FEBS Letters, 584*, 1997–2004.

Chapter 14

Calcium transport and signaling

Srinivasan Ramakrishnan and Roberto Docampo
Center for Tropical and Emerging Global Diseases and Department of Cellular Biology, University of Georgia, Athens, GA, United States

1. Chemical properties of calcium

Calcium is the fifth most abundant element and third most abundant metal in nature. Being available to cells at high concentrations, calcium was adopted as an important signaling regulator early during evolution. Another alkali metal, magnesium, is also believed to have been present at high concentrations in the primordial soup. However, the unique chemical properties and reaction chemistry of calcium allowed it to be selected over magnesium as the perfect intracellular communicator. Both calcium and magnesium have a valence of two, indicating that they can easily lose two electrons and form stable cations. However, a larger ionic radius of calcium (0.099 nm) as compared to magnesium (0.065 nm) indicates that electrons in the outermost orbit of calcium are loosely held, thereby allowing calcium to participate in reactions more readily. Additionally, the larger size confers higher polarizability to calcium, which means that the electron distribution of calcium can be easily distorted by external forces. Such flexibility is necessary for interacting with sites of irregular geometry offered by biological molecules like proteins (Carafoli & Krebs, 2016). This was especially important during evolution since calcium concentration within cells is high and the ability of proteins to bind calcium allows cells to easily buffer the calcium concentration to mediate signaling. Another peculiar property of calcium is its lower charge density (2.24) in comparison to magnesium (4.3) (Collins, 1997). This means that upon reacting with an anion like phosphate, calcium phosphate unlike magnesium phosphate would form only sparingly soluble salts. Considering phosphate-based energetics evolved early within cells, this property provided an additional advantage for calcium to be selected as the main signaling ion. Therefore, the chemical properties of calcium which allowed lower calcium concentrations within cells, efficient coordination chemistry for binding to proteins and permissibility for phosphate-based energetics were together responsible for its evolution as the main signaling molecule.

2. Ca^{2+}-binding proteins

Ca^{2+}-binding proteins play a wide variety of roles within cells, ranging from Ca^{2+} buffering for regulating Ca^{2+} concentration, transcriptional regulation, cell proliferation, and homeostasis. Identification of Ca^{2+}-binding proteins is still a developing field. While thousands of proteins are already known to bind Ca^{2+}, more proteins are being discovered with time. However, most of these proteins can be broadly classified into one of the following superfamilies.

3. EF-hand superfamily

The name EF hand originated from the Ca^{2+} binding motif of parvalbumin whose structure was first described in 1973 by Kretsinger and Nockolds (1973). The EF-hand domain is approximately 30 amino acids, which forms a Ca^{2+} binding loop and is sandwiched by two alpha helices. Parvalbumin possesses three EF hands with six alpha helices. The fifth (E) and six (F) helix, along with Ca^{2+} binding loop, look structurally similar to a hand and led to the use of the term: EF hand. To date, more than 800 proteins have been identified to contain an EF-hand domain. Calmodulin, stromal interaction molecule 1 (STIM1), and S100 proteins are the most widely recognized EF-hand containing Ca^{2+} binding proteins (Lewit-Bentley & Rety, 2000). Calmodulin is the most conserved Ca^{2+}-binding protein and is present in all eukaryotes with extremely identical amino acid sequence in vertebrates. Ca^{2+} binding allows calmodulin to change conformation and thereby bind and regulate proteins that contain a calmodulin-binding domain. STIM1 has a unique EF-hand domain with very low

Cell Physiology Sourcebook. https://doi.org/10.1016/B978-0-12-811114-7.00003-7

affinity for Ca^{2+}. Therefore, STIM1 can modulate its role in Ca^{2+} entry under low Ca^{2+} conditions. S100 proteins have acquired their name mainly due to their property of 100% solubility in ammonium sulfate. In addition to a canonical EF-hand domain S100 proteins harbor another N-terminal noncanonical 14 amino acid sequence pseudo-EF-hand domain. Therefore, S100 proteins have weak specificity for Ca^{2+} and can also bind zinc and copper.

4. C2 domain superfamily

C2 domain is a stretch of 120–130 amino acids that was first identified in the Ca^{2+}-dependent protein kinase C. Structurally, it is a pair of four beta sheets with a total of six connecting loops. Ca^{2+} is coordinated between the loops connecting sheets 2-3 and sheets 6–7. The best-studied C2 domain-containing protein is synaptotagmin with two C2 domains and which at synaptic vesicles is involved in Ca^{2+}-based exocytosis of neurotransmitters (Sutton et al., 1995). C2 domain proteins are mainly involved in signal transduction and membrane trafficking events.

5. Annexin Ca^{2+}-binding fold

Annexins play important roles in both intracellular and extracellular environments. Extracellular annexins play a role in blood coagulation and fibrinolysis. Intracellular annexins play numerous roles ranging from endocytosis, exocytosis, and as their name suggests anchoring of other proteins to membranes. In their Ca^{2+}-bound state, annexins can interact with phospholipids and bind cell membranes. All annexins have a highly diverse N-terminal head domain and a very conserved C-terminal core domain. The core domain comprises five helices. The connecting loop region of these helices binds Ca^{2+} (Carafoli & Krebs, 2016; Niki et al., 1996).

6. Second messengers that release Ca^{2+}

Second messengers are defined as intracellular molecules that are released in response to a signal from an extracellular molecule such as a hormone or a neurotransmitter and in turn, mediate an intracellular response to the extracellular signal. The three best known second messengers in Ca^{2+} signaling are (1) cyclic nucleotides, (2) inositol 1,4,5-trisphosphate, and lastly (3) Ca^{2+} itself.

Cyclic nucleotides such as cAMP are generated from ATP by the action of adenylyl cyclases, which are activated by binding of hormones to their receptors. cAMP can then act as the second messenger and mediate Ca^{2+} signaling (Bugrim, 1999; Kamp & Hell, 2000). Similarly, cGMP, which is generated from GTP, by the action of nitric oxide and guanylate cyclase can also act as a second messenger of Ca^{2+} signaling (Francis et al., 2010). Another cyclic nucleotide, cyclic ADP ribose (cADPR), is produced from nicotinamide adenine dinucleotide (NAD) by the action of ADP-ribosyl cyclases and mediates Ca^{2+}-induced Ca^{2+} release from ryanodine receptors. Another nucleotide, nicotinic acid adenine dinucleotide phosphate (NAADP), was confirmed as a second messenger only recently. NAADP is produced from NADP by the action of CD38 (Lee, 2001). It is believed that NAADP could act on either the acidic Ca^{2+} stores or the endoplasmic reticulum (Guse, 2015). Nucleotides acting as second messengers in Ca^{2+} signaling have also been reported in plants (Wu et al., 1997) and protozoan parasites (Nagamune et al., 2008). One of the most important second messengers in Ca^{2+} signaling is a product of lipid hydrolysis. Upon external stimulus, phospholipase C acts on phosphatidylinositol 4,5-bisphosphate (PIP_2) and generates inositol-1,4,5-trisphosphate (IP_3) and diacyl glycerol (DAG). IP_3 then activates the IP_3-receptor (IP_3R) present in the endoplasmic reticulum and induces Ca^{2+} release. Owing to a high Ca^{2+} concentration in the endoplasmic reticulum and also due to its main role in overall Ca^{2+} homeostasis, IP_3 has been considered as a very important second messenger in Ca^{2+} signaling pathway in animal cells (Berridge, 2009). However, the role of IP_3 in Ca^{2+} signaling is still under debate for higher plant cells, fungi, and some protozoan parasites, as an orthologue for the mammalian IP_3R has still not been identified in some of these organisms (Prole & Taylor, 2016). Another phospholipid, sphingosine-1-phosphate, can also act as a second messenger in Ca^{2+} release from the endoplasmic reticulum. It is believed that sphingosine-1-phosphate acts on the protein sphingolipid calcium release-mediating protein of the endoplasmic reticulum (SCaMPER) to mediate Ca^{2+} release from endoplasmic reticulum.

Interestingly, Ca^{2+} signaling can also be modulated by Ca^{2+} itself, acting as a messenger molecule. Ca^{2+}-induced Ca^{2+} release has been reported for several different cell types and can play a function in release of neurotransmitters, secretion of hormones, muscle contraction, and apoptosis (Endo, 2009; Finch et al., 1991; Petersen et al., 2005).

7. Ca^{2+} homeostasis and signaling

Ca^{2+} homeostasis and signaling is a complex process orchestrated by the cooperative action of Ca^{2+} binding proteins, second messengers, and a variety of transporters, all of which act together to regulate intracellular and intercellular communication.

8. Plasma membrane Ca^{2+} channels and transporters

In all cells, cytosolic Ca^{2+} concentration is very low. The plasma membrane acts as a barrier between the cytoplasm and the extracellular environment. Additionally thus, a plethora of Ca^{2+} transporters expressed at the plasma membrane contribute toward the maintenance of a low cytosolic Ca^{2+} concentration, thereby playing a central role in Ca^{2+} homeostasis.

8.1 Plasma membrane Ca^{2+} import

There are several different types of Ca^{2+} transporters that have been reported to localize to the plasma membrane and facilitate Ca^{2+} uptake into the cell. These include (see Fig. 14.1) (1) Voltage-gated Ca^{2+} channels (VGCC), (2) Ligand-gated Ca^{2+} channels (LGCC), (3) Transient receptor potential (TRP) channels, and (4) Store operated Ca^{2+} entry (SOCE) channels.

8.2 Voltage-gated Ca^{2+} channels

VGCC's open in response to membrane depolarization and are mainly found in excitable cells like neurons and muscle cells (Deliot & Constantin, 2015; Parekh & Putney, 2005). Depending on their gating characteristics and pharmacological response, VGCCs have been classified into high-voltage (HVA) L, N, P/Q, R-types or low-voltage (LVA) T-type channels (https://www.ncbi.nlm.nih.gov/books/NBK6181/). VGCC's are comprised of four subunits namely $\alpha 1$, $\alpha 2\delta$, β, and γ subunits. $\alpha 1$ is the pore forming subunit while others are ancillary subunits. Both $\alpha 2\delta$ and β subunits interact with the pore forming $\alpha 1$ subunit and ensure its membrane localization. The δ subunit of $\alpha 2\delta$ is associated with the membrane while the $\alpha 2$ subunit directly interacts with the $\alpha 1$ subunit. Similarly, the β subunit has a beta interaction domain that interacts directly with the alpha interaction domain of the $\alpha 1$ subunit. The γ subunit does not contribute toward membrane localization of $\alpha 1$ subunit but may play a role in modulating the biophysical properties of the channel (Arikkath & Campbell, 2003; Simms & Zamponi, 2012). An orthologue for the mammalian $\alpha 1$ subunit has been identified in yeast and exhibits a role in cellular Ca^{2+} uptake indicating that voltage-gated Ca^{2+} channels may exist in yeast as well (Paidhungat & Garrett, 1997). In plants, electrophysiological studies with plasma membranes demonstrated the existence of hyperpolarization or depolarization-induced Ca^{2+} channels as well (White, 2000). However, identification and characterization of such channels at molecular level is still pending.

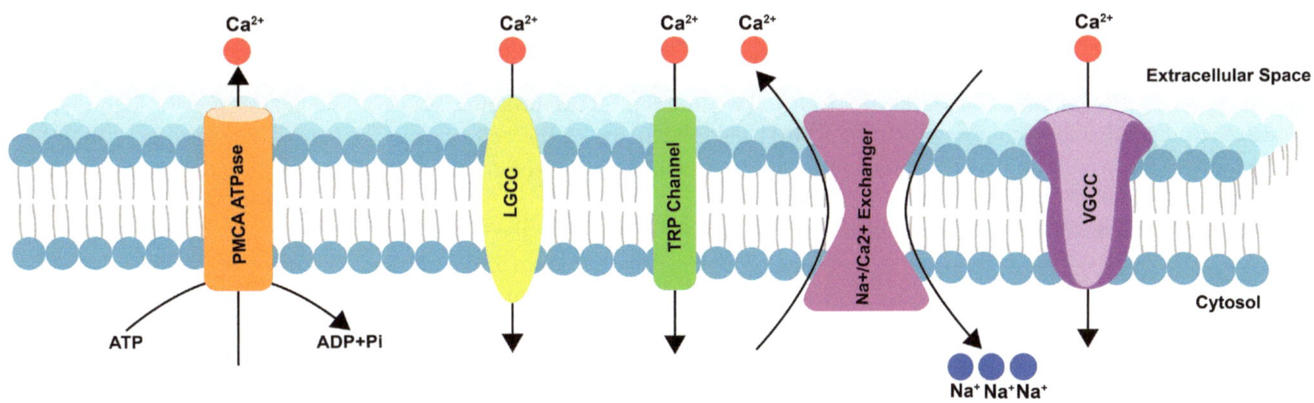

FIGURE 14.1 Calcium entry through plasma membrane can occur by several different mechanisms. Ligand-gated Ca^{2+} channels, voltage-gated Ca^{2+} channels, and TRP channels are all responsible for Ca^{2+} uptake into the cell. The Na^+/Ca^{2+} exchanger and the PMCA-ATPase, which uses the energy generated from ATP to pump Ca^{2+} out, are involved in Ca^{2+} release.

8.3 Ligand-gated Ca^{2+} channels

LGCCs are also known as ionotropic Ca^{2+} channels and they open in response to a ligand, usually a neurotransmitter. LGCCs can be receptor-operated Ca^{2+} channels or second messenger–operated Ca^{2+} channels. Second messenger–operated LGCCs are mostly located on intracellular membranes, while receptor-operated LGCCs are present on cell surface and contribute toward the role of plasma membrane in Ca^{2+} uptake (Mooren & Kinne, 1998). Receptor-operated LGCCs are mainly found in excitable cells (Parekh & Putney, 2005; Striggow & Ehrlich, 1996). The best-studied example of this is the glutamate receptor (Courtney et al., 1990; Dingledine et al., 1999), an orthologue of which has also been identified in plants (Qi et al., 2006). Another example is the ATP-activated Ca^{2+} channel present in muscle cells, neurons, and exocrine cells.

8.4 Transient receptor potential channels

TRP channels open in response to a variety of physical and chemical stimuli. In humans 28 TRP channel proteins have been identified and classified into six subfamilies according to sequence similarities. The six families are TRPC (canonical), TRPM (melastatin), TRPV (vanniloid), TRPA (ankyrin), TRPML (mucolipin), and TRPP (polycystin). All families have two common characteristics, they all harbor six transmembrane domains and they allow only cations to transit their pore. Among these, at least members from four different families TRPC, TRPA, TRPV, and TRPP are believed to participate in plasma membrane Ca^{2+} permeability (Prole & Taylor, 2011). Most members exhibit both plasma membrane and intracellular localization with a lack of specificity toward cation permeability. However, two proteins TRPV5 and TRPV6 have been localized to plasma membrane and are known to be specific for Ca^{2+} transport (Clapham, 2003; den Dekker et al., 2003). TRP channels are absent in plants (Nilius & Owsianik, 2011) but have been identified in yeast and protozoan parasites. The best-characterized TRP channel in yeast localizes to the vacuole. Another yeast TRP, which is an orthologue for the mammalian TRPP2/PKD2 (polycystic kidney disease 2), has been identified as well. While the human TRPP2/PKD localizes to ER in most cell systems (Giamarchi et al., 2006), the yeast PKD2 has been reported to also localize to the plasma membrane and is necessary for Golgi trafficking and growth (Aydar & Palmer, 2009; Kaleta & Palmer, 2011). TRP channels have not been characterized in all protozoan parasites (Prole & Taylor, 2011) but in some, they localize to intracellular compartments (Docampo & Huang, 2015; Gupta et al., 2006).

8.5 Store-operated Ca^{2+} entry channels

Store-operated Ca^{2+} entry (SOCE) channels participate in a unique mechanism of Ca^{2+} entry. This process is tightly controlled by the intracellular Ca^{2+} concentration. Components of the SOCE are absent in yeast, higher plants, and protozoan parasites but store-opreated Ca^{2+} channels exist in all eukaryotes (Collins & Meyer, 2011; Docampo & Huang, 2015; Prole & Taylor, 2011). Sequence similarities indicate that ORAI orthologues, which encode the calcium release-activated calcium channel protein 1, could be present in algae but this requires further experimental verification (Collins & Meyer, 2011). At the mammalian plasma membrane, ORAI or TRPC proteins play a role in SOCE. The protein STIM, which is present in the ER membrane, can control Ca^{2+} entry into the cell from the extracellular environment (see Fig. 14.2). Absence or reduced Ca^{2+} levels in the ER lumen trigger STIM to oligomerize at ER–plasma membrane junctions. At these junctions, STIM can communicate with the ORAI (Hogan & Rao, 2015) and TRPC (Cheng et al., 2013) channels present in the plasma membrane to allow Ca^{2+} influx into the cells. This process is termed SOCE (Hogan & Rao, 2015). An RNAi screen focused on changes in cytosolic Ca^{2+} concentration was used to uncoover the role of STIM proteins in Ca^{2+} signaling (Zhang et al., 2006). Two human STIM proteins, STIM1 and STIM2, were identified and localize to the ER with a minor amount of STIM1 also localizing to the cell surface (Dziadek & Johnstone, 2007). Both these proteins have similar overall architecture with N-terminal in the ER lumen, a single transmembrane domain in the ER membrane and the cytosolic C-terminal domain. The N-terminal portion of STIM senses the low Ca^{2+} concentration in the ER lumen. This triggers the oligomerization and movement of STIM to the ER–plasma membrane junctions (Carrasco & Meyer, 2011). ORAI proteins in humans exist as ORAI1-3 and have cytosolic N and C-termini, as well as four transmembrane domains. ORAI proteins are believed to move through the plasma membrane by diffusion and to interact with STIM upon reaching ER–plasma membrane junctions to become reversibly trapped by STIM interaction. Experimental evidence suggests that the C-terminus of ORAI channel interacts with STIM1 (Hogan & Rao, 2015). Among TRPC channels, TRPC1, TRPC3, TRPC4, and TRPC5 have been indicated to play a role in SOCE (Alkhani et al., 2014; Shin et al., 2010; Wang et al., 2004). However, the data is most consistent with TRPC1. The first evidence for involvement of TRPC1 in SOCE came from experiments with human submandibular gland (HSG) ductal cells (Liu et al., 2000). This was

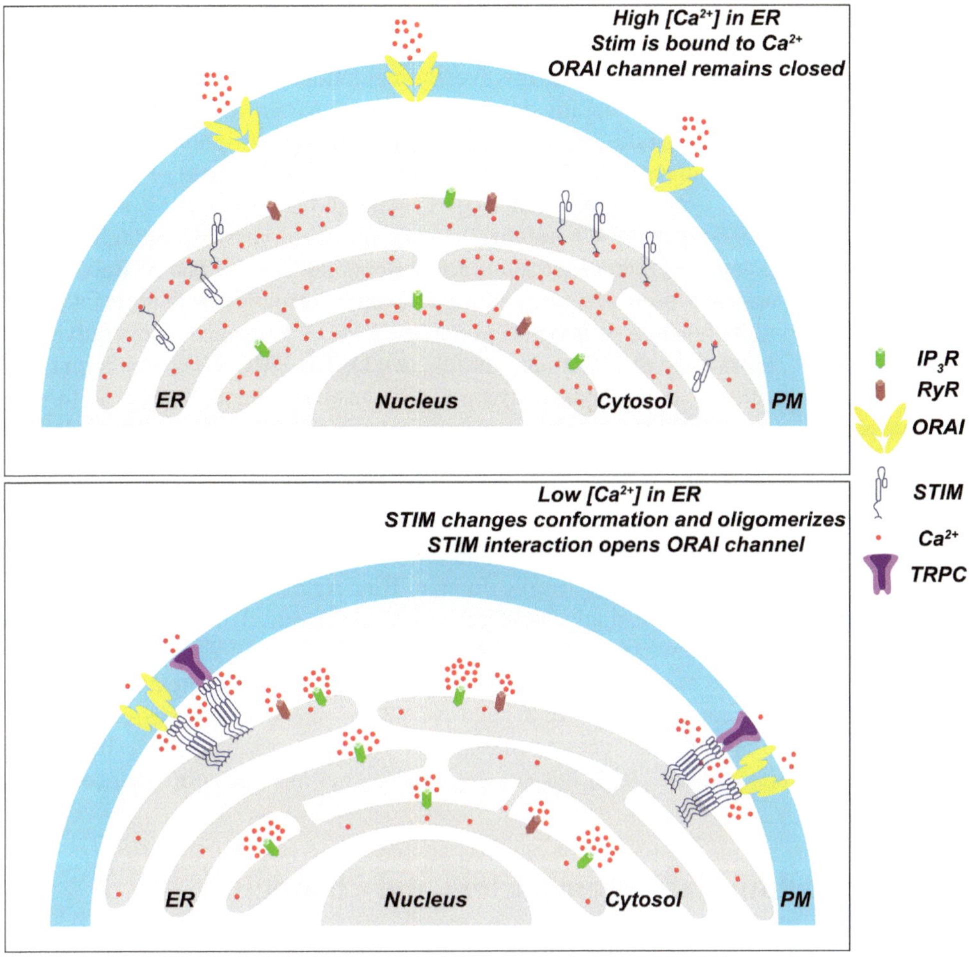

FIGURE 14.2 Store-operated calcium entry—STIM is an ER resident protein. Ca^{2+}-binding domains of STIM are located in the ER lumen. Under resting conditions (Upper panel) of high Ca^{2+} concentration in the ER lumen, STIM is bound to Ca^{2+} and therefore remains inactive. However, upon stimulation, IP_3R and RyR's release Ca^{2+} and cause the Ca^{2+} concentration in the ER lumen to drop significantly (lower panel). This results in dissociation of Ca^{2+} ions from STIM, which results in conformational change, oligomerization, and the activation of STIM. STIM then moves to ER–plasma membrane junctions and interacts with the plasma membrane localized ORAI channels. Opening of ORAI channel results in Ca^{2+} influx from the exterior of the cells. It has been suggested that the activity of ORAI channel can then cause insertion of TRPC channel into the plasma membrane. TRPC channels are also gated by STIM, allowing Ca^{2+} influx from the cell exterior. This Ca^{2+} is then uptaken by the ER-localized SERCA to refill the depleted ER Ca^{2+} stores.

later confirmed with reduced SOCE in a variety of human cell types and also with cells obtained from TRPC1 knockout mice. The ER domain in STIM1 C-terminus is responsible for binding TRPC1, however, activation of TRPC1 occurs by a different mechanism. The STIM1 C-terminal domain has positively charged lysine residues that interact with the negatively charged aspartate residues in TRPC1. This electrostatic interaction allows TRPC1 to be gated by STIM1 (Zeng et al., 2008). Since the aspartate residues are also conserved in several other TRPC channels, it is believed that STIM1 could also gate the activity of these other TRPC channels. TRPC-mediated SOCE is also dependent on the activity of ORAI1 channels. Current model suggests that Ca^{2+} entry via ORAI1 channel causes TRPC1 to be recruited to the plasma membrane (Cheng et al., 2011). Therefore, ORAI1 controls the surface expression of TRPC1, whereas STIM1 controls the gating indicating that a coordinated mechanism of action between ORAI, STIM, and TRPC is required for SOCE into cells (Cheng et al., 2013).

8.6 Plasma membrane Ca^{2+} export

PMCA is a P-type ATPase pump, which uses one ATP molecule per Ca^{2+} transported. The PMCA pumps are autoinhibited by a CaM-binding domain present at the C-terminus in human cells, but present at the N-terminus in plant cells

(Bonza et al., 2000). In humans, PMCA has four isoforms PMCA1-4. PMCA2-3 has tissue-specific roles but PMCA1 is expressed ubiquitously and has a housekeeping function. PMCA proteins have 10 transmembrane domains, two cytosolic loop regions and a long cytosolic C-terminal tail that is involved in regulation of the PMCA activity. The C-terminal tail of PMCA contains the autoinhibitory calmodulin-binding domain (CaM-BD). When CaM binds to the CaM-BD of PMCA, it reduces the $K_m[Ca^{2+}]$ of PMCA and renders it active. Protein kinases act by phosphorylating the target residues on the C-terminal tail of PMCA, whereas calpain regulates PMCA activity by cleaving off a part of the C-terminal tail and making it constitutively active. Additionally, phospholipids can also regulate PMCA activity by acting on its C-terminal tail (Lopreiato et al., 2014). An orthologue of PMCA pump has been identified in yeast (Cunningham & Fink, 1994) and its loss can be complemented by the human protein indicating a functional similarity (Ton & Rao, 2004). A PMCA pump localized to plasma membrane has also been identified in some protozoan parasites (Luo et al., 2004), indicating the versatile nature of this protein and its function. It is believed that the PMCA pump acts cooperatively with the Na^+/Ca^{2+} exchanger (NCX) to pump Ca^{2+} outside the cell (Brini & Carafoli, 2011). The PMCA has high affinity but low capacity for Ca^{2+} transport and NCX has low affinity but high capacity for Ca^{2+} transport. Therefore, PMCA functions in maintaining the cytosolic Ca^{2+} concentration but NCX activity plays an important role during large cytosolic Ca^{2+} variations (Brini & Carafoli, 2011). The NCX exports one Ca^{2+} while importing three Na^+ ions into the cell. The NCX has a cytosolic loop region that is important for regulation of its activity (Maack et al., 2005). Similar to PMCA, a number of proteins such as phospholamban, ankyrin, calcineurin, etc. can bind to the cytosolic loop region and modulate the NCX activity. Additionally, Ca^{2+} (Hilgemann, Collins, & Matsuoka, 1992) and Na^+ ions (Hilgemann, Matsuoka, et al., 1992) can also modulate the activity of NCX via binding to the cytosolic domain. Binding of Ca^{2+} to the cytosolic loop activates NCX, whereas binding of Na^+ deactivates it. Interestingly NCX activity is reversible. Under resting conditions, NCX exports Ca^{2+} in exchange for importing Na^+. However, in heart cells when the Na^+ levels rise during systole, the NCX imports Ca^{2+} while exporting Na^+ (Armoundas et al., 2003; Noble et al., 1991). Three NCX proteins NCX1-3 have been identified in mammals. NCX1 is almost ubiquitous, whereas NCX2 and NCX3 are selectively expressed in brain, skeletal muscles, and some neurons (Linck et al., 1998). Orthologues for the human Na^+/Ca^{2+} exchanger localizing to plasma membrane have been reported in plants (Wang et al., 2012) but not in yeast (Cui et al., 2009) or protozoan parasites (Moreno & Docampo, 2003).

9. Endoplasmic reticulum Ca^{2+} transporters and channels

The Ca^{2+} concentration in the ER lumen of mammalian cells was found to be in the range of 80–500 μM by using Ca^{2+} binding fluorescent probes and genetically encoded Ca^{2+} indicators (Henderson et al., 2015; Solovyova et al., 2002). This is about 5–10 times higher than the cytosolic Ca^{2+} concentration highlighting the role of endoplasmic reticulum as Ca^{2+} store in the cell. The endoplasmic reticulum is an especially important Ca^{2+} store in mammalian cells because it can regulate the overall Ca^{2+} uptake by store operated Ca^{2+} entry mechanism (SOCE) (Hogan & Rao, 2015) and can also transfer Ca^{2+} to other organelles via recently described membrane contact sites (Phillips & Voeltz, 2016).

9.1 Endoplasmic reticulum Ca^{2+} import

Ca^{2+} uptake into the ER lumen is mediated by a sarcoplasmic/endoplasmic reticulum Ca^{2+} ATPase (SERCA) which is a P-type ATPase that transports Ca^{2+} ion into the ER lumen against the gradient with the help of energy derived from ATP (Periasamy & Kalyanasundaram, 2007). The SERCA family cotains three isoforms, SERCA1-3, encoded by three genes ATP2A1-3. In cardiac cells, phospholamban, activates the SERCA pump when phosphorylated and inactivates it and when dephosphorylated (Gustavsson et al., 2013). In skeletal muscles, the SERCA activity is regulated by a similar protein called sarcolipin (Bal et al., 2012). In addition to these proteins, SERCA activity is also regulated by thyroid hormones at transcriptional level and miRNA-25 at posttranscriptional level. Furthermore, posttranslational modifications such as glutathionylation, nitration, SUMOylation, acetylation and glycosylation can also increase or decrease the SERCA activity (Stammers et al., 2015). In yeast two P-type ATPases namely Pmr1p and Cod1p/Sfp1p work together to pump Ca^{2+} into the ER lumen (Vashist et al., 2002). The Ca^{2+} concentration in yeast ER lumen is only about 10 μM (Strayle et al., 1999), which is 10–50 fold lower than in mammalian cells. However, despite this difference, an appropriate Ca^{2+} concentration in the yeast ER lumen is necessary for export of secretory proteins and protein folding and degradation (Durr et al., 1998). Human SERCA can complement Ca^{2+}, but not Mn^{2+} importing activity of yeast Pmr1p indicating a moderate functional similarity between these proteins (Ton & Rao, 2004). SERCA has also been identified in some protozoan parasites and can act as a virulence factor (Moreno & Docampo, 2003).

9.2 Endoplasmic reticulum Ca^{2+} export

Ca^{2+} release from the endoplasmic reticulum of mammalian cells is mediated by an inositol 1,4,5-triphosphate receptor (IP_3R) or Ryanodine receptor (RyR) receptor. IP_3 induces Ca^{2+} release by IP_3R, whereas cyclic-ADP ribose (cADPR) induces Ca^{2+} release by RyR. Although both IP_3R and RyR exist as three isoforms (Ivanova et al., 2014; Lanner et al., 2010) in humans, their expression level differs across different tissues. In humans, IP_3R exists as three isoforms namely IP_3R1, IP_3R2, and IP_3R3. IP_3R1 is upregulated in brain tissue, whereas IP_3R2 is upregulated in cardiac tissues. Similar to IP_3Rs, RyR family isoforms are also named RyR1-3. RyR3 is ubiquitously expressed. RyR2 is expressed in central nervous system and striated muscles, whereas RyR3 is expressed in central nervous system and cardiac tissues. IP_3R is posttranslationally modified (Vanderheyden et al., 2009) and can interact with several other proteins (Ando et al., 2014), which could modulate its activity as the ER Ca^{2+} release channel. RyR also interact with wide variety of proteins which may moderate their activity (Van Petegem, 2015). It seems IP_3R and RyR receptors are mostly specific to mammalian cells. An IP_3R has been characterized in *Paramecium* (Ladenburger et al., 2006), in the protozoan parasites *Trypanosoma brucei* and *T. cruzi* (Huang et al., 2013; Lander et al., 2016) and in some lower plants (Wheeler et al., 2008). However, in *Paramecium* and trypanosomes the IP_3R is not localized to the ER and could be very different from the mammalian orthologues. Additionally, molecular evidence for IP_3R has not been found in any other protozoan parasite (Moreno & Docampo, 2003) or higher plants (Stael et al., 2012). Homologs for both IP_3R and RyR are absent in yeast cells. However, a calcineurin regulated cls2p protein has been proposed to mediate Ca^{2+} release from the yeast ER (Tanida et al., 1996). Recently a new protein, calcium load-activated channel (TMCO1), has been identified as an ER Ca^{2+} channel. It has been suggested that at physiological Ca^{2+} concentration, TMCO1 remains as inactive monomers, but when there is Ca^{2+} overloading of the ER, above a certain threshold, TMCO1 forms active tetramers, which extrude Ca^{2+} from the ER matrix (Wang et al., 2016).

10. Mitochondrial Ca^{2+} transporters and channels

Ca^{2+} is an important modulator of mitochondrial function. Three enzymes within the mitochondria namely pyruvate dehydrogenase, α-ketoglutarate dehydrogenase, and isocitrate dehydrogenase are regulated by Ca^{2+} (Denton, 2009). A Ca^{2+}-dependent phosphatase dephosphorylates the catalytic subunit of pyruvate dehydrogenase to activate it (Denton et al., 1972). α-Ketoglutarate dehydrogenase (Mccormack & Denton, 1979) and isocitrate dehydrogenase (Mccormack & Denton, 1979) are directly activated by binding Ca^{2+}. Therefore, an increase in mitochondrial Ca^{2+} results in activation of these enzymes and an increase in ATP production by mitochondria. Mitochondrial Ca^{2+} accumulation also has specialized function of aldosterone production in glomerulosa cells (Mccormack & Denton, 1979), glutamate production in pancreatic beta cells (Maechler & Wollheim, 1999), and activation of the permeability transition pore (Crompton, 1999).

10.1 Mitochondria Ca^{2+} import

The contribution of mitochondria toward Ca^{2+} signaling was believed to be minimal until relatively recently. It was known that a voltage-dependent anion channel (VDAC1) could facilitate the transfer of Ca^{2+} through the outer mitochondrial membrane (Gincel et al., 2001). However, how Ca^{2+} could be transferred through the inner mitochondrial membrane into the mitochondrial matrix remained unclear. Over the past decade, our understanding of mitochondrial Ca^{2+} uptake has improved dramatically. With these recent developments, it is now clear that Ca^{2+} uptake into the mitochondrial matrix is mediated by a mitochondrial Ca^{2} uniporter (MCU) complex on the mitochondrial inner membrane. Interestingly MCU complex orthologues and mitochondrial Ca^{2+} uptake activity is absent in yeast (Yamamoto et al., 2016). In plants only one out of the six MCU complex orthologues have been characterized (Carraretto et al., 2016; Pittis et al., 2020). Therefore, the vast majority of information on the MCU complex described below comes from mammalian cells. The current model for the MCU complex in mammalian cells comprises five proteins (see Fig. 14.3)—MCU (mitochondrial Ca^{2+} uniporter), MCUb, mitochondrial Ca^{2+} uptake 1 (MICU1), MICU2, and EMRE (Essential MCU regulator). However, it must be noted that EMRE is absent in plants, and protozoans (Sancak et al., 2013). This is particularly interesting because, in all these organisms, mitochondrial Ca^{2+} uptake has been experimentally validated. Therefore, it seems that MCU along with MICU1 might be the only two components that are indispensable for mitochondrial Ca^{2+} uptake. Although early studies suggested the absence of EMRE in fungi, indicating a potential EMRE-independent mechanism for MCU activity, recent research has identified EMRE orthologs in certain chytrid fungi (Pittis et al., 2020). This discovery suggests that some fungi also rely on an EMRE-dependent MCU system.

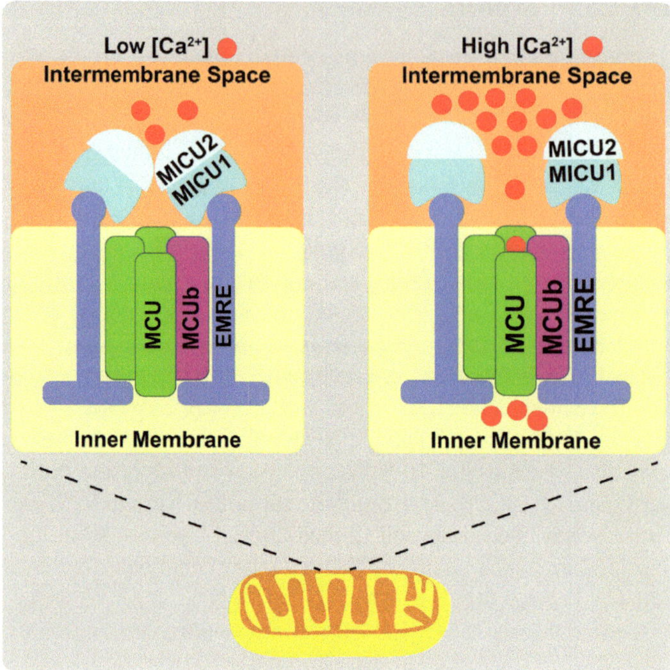

FIGURE 14.3 The MCU complex of *Homo sapiens*—MCU is the pore-forming subunit and is complexed to MICU proteins via the EMRE. Another protein, MCUb, is a dominant negative isoform of MCU. The presence of MCUb in the MCU oligomers reduces the activity of MCU complex. Therefore, the ratio of MCU and MCUb expression in any given cell type determines the activity of MCU complex in that cell. Under low Ca^{2+} concentrations (left panel) in the intermembrane space, MICU proteins inhibit the uptake of Ca^{2+} through the MCU. However, during a signaling event, Ca^{2+} concentration in the intermembrane space increases (right panel) resulting in uptake through the MCU complex.

10.2 Mitochondrial Ca^{2+} uniporter or MCU

In the MCU complex, the protein MCU forms the ion transport pore, which along with its mitochondrial-targeting signal is a 40-kDa protein (De Stefani et al., 2011). The role of MCU has been confirmed by several independent experiments. In vivo validation was obtained by testing Ca^{2+} uptake activity in tissues from MCU knockout mice (Pan et al., 2013) as well as by silencing MCU expression in rat cardiomyocytes (Drago et al., 2012), pancreatic β-cells (Alam et al., 2012; Tarasov et al., 2012), neurons (Qiu et al., 2013), and breast epithelial cells (Curry et al., 2013). In vitro evidence for role of MCU comes from in vitro synthesis of MCU or MCU with point mutations and lipid bilayer reconstitution to test Ca^{2+} uptake activity (De Stefani et al., 2011). All these experiments provide a common conclusion for a clear role of MCU as the pore-forming unit in the MCU complex (Chaudhuri et al., 2013; De Stefani et al., 2011).

10.3 MCUb

MCUb, which is present in most vertebrates but absent in other organisms, is closely related to MCU as it shares 50% sequence similarity. As compared to MCU, MCUb has an E256V substitution, which reduces its Ca^{2+} permeability. It is believed that MCUb acts by forming a heterooligomeric complex with the MCU to reduce the overall Ca^{2+} uptake efficiency of the MCU complex (Raffaello et al., 2013). Therefore, the ratio of MCU to MCUb expression could control the mitochondrial Ca^{2+} uptake in any given cell type. In support of this idea, a low MCU:MCUb ratio of 3:1 was observed in cardiomyocytes, whereas a high ratio of 40:1 was observed in the skeletal muscles (De Stefani et al., 2016). Therefore, it seems that MCUb provides a novel mechanism of regulation for mitochondrial Ca^{2+} uptake.

10.4 EMRE

EMRE is a broadly expressed 10 kDa protein that is required to keep the MCU together with the MICU1 and MICU2 in the MCU complex. In support of an essential role for EMRE in the MCU complex, either silencing or knockout of EMRE completely abolishes mitochondrial Ca^{2+} uptake (Sancak et al., 2013). However, the absence of EMRE in plants indicates that an alternative EMRE-independent mechanism for MCU activity may exist in nature. In consensus with this hypothesis, expression of EMRE in yeast is required for forming a functional channel with the mammalian MCU but is dispensable for some fungal MCU (Kovacs-Bogdan et al., 2014).

10.5 MICU1 and MICU2

MICU1 was the first protein to be identified among all the MCU complex components (Perocchi et al., 2010). Loss of MICU1 reduces mitochondrial Ca^{2+} uptake efficiency (Csordas et al., 2013). MICU1 has two isoforms MICU2 and MICU3 (Plovanich et al., 2013). It appears that MICU1 forms a heterodimeric complex with MICU2 and that absence of MICU1 causes instability of MICU2 protein specifically and not MICU2 mRNA (Patron et al., 2014). The role of MICU3 is yet unknown but is believed to be tissue specific. Current prevailing model for MICU family is that, at low Ca^{2+} concentrations, MICU2 exerts an inhibitory effect and prevents Ca^{2+} accumulation in the mitochondria, whereas at high Ca^{2+} concentrations, MICU2 inhibitory activity is reduced, while MICU1 is activated resulting in rapid Ca^{2+} uptake into the mitochondria by the MCU complex (De Stefani et al., 2016).

10.6 MCUR1 and SLC25A23

Two additional proteins, the mitochondrial Ca^{2+} regulator 1 (MCUR1) and the solute carrier protein 25 member 23 (SLC25A23), have also been suggested as regulators of the MCU complex (Kwong et al., 2014; Mallilankaraman et al., 2012). Both these proteins were identified from an RNAi screen where loss of proteins resulted in a Ca^{2+} uptake defect. However, proteomic studies resolved MCU, MCUb, EMRE, MICU1, and MICU2 in the same complex, but did not include MCUR1. Additionally, SLC25A23 belongs to the family of proteins that act as solute carriers by facilitating the transport of Mg^{2+}-ATP/Pi across the inner mitochondrial membrane. Therefore, their role is debated, considering that they may be having an indirect effect on the activity of MCU complex. It has been suggested that both proteins could affect the mitochondrial membrane potential and thereby exert an indirect effect on the functionality of the MCU complex (Paupe et al., 2015). Currently, the role of MCUR1 and SLC25A23 in the MCU complex is debated (Paupe et al., 2015; Vais et al., 2015).

10.7 Mitochondria Ca^{2+} export

Ca^{2+} release from mitochondria is mediated by the action of two Ca^{2+} exchangers namely the H^+/Ca^{2+} exchanger and the Na^+/Ca^{2+} exchanger. A protein named LETM1 (Tsai et al., 2014) is believed to act as the H^+/Ca^{2+} exchanger in the mitochondria (Doonan et al., 2014). However, some researchers have shown that LETM1 possesses a K^+/H^+ activity instead of a Ca^{2+} exchange activity (Hashimi et al., 2013; McQuibban et al., 2010). In contrast, the Na^+/Ca^{2+} exchanger activity for mitochondrial Ca^{2+} release is well documented and the molecular identity of the candidate is known (Palty et al., 2010). A protein named NCLX has been identified as the Na^+/Ca^{2+} exchanger of the mitochondria. NCLX is ubiquitously expressed with different spliced variants found in different cell types. Structurally, NCLX is similar to other proteins of the NCX family with two transmembrane domains providing four ion binding sites: three for Na^+ ions and one for a Ca^{2+} ion (Boyman et al., 2013). NCLX can also export Ca^{2+} in exchange for Li^+ which is why it has been named as NCLX for $Na^+/Ca^{2+}/Li^+$ exchanger. The role of NCLX as Na^+/Ca^{2+} exchanger has been validated in several mammalian cell types (Nita et al., 2014, 2015; Parnis et al., 2013; Takeuchi et al., 2013).

11. Acidic stores channels and transporters

11.1 Acidocalcisomes

Acidic Ca^{2+} stores, termed acidocalcisomes, have been identified in both prokaryotes and eukaryotes (Docampo et al., 2005), indicating an early role for these organelles during evolution. Acidocalcisomes, occasionally termed as volutin granules, are electron dense structures with an internal pH range of 5–5.5. A vacuolar pyrophosphatase or a vacuolar proton ATPase, or both, may drive acidification of these organelles. They play an important role in polyphosphate metabolism, storage of cations, pH homeostasis, and osmoregulation. Acidocalcisomes contain high concentration of Ca^{2+}, usually in the millimolar range and can therefore play a very crucial role in Ca^{2+} signaling events within the cell. Acidocalcisomes are known to contain a PMCA-type Ca^{2+}-ATPase and a Ca^{2+}/H^+ exchanger, which may facilitate the import of Ca^{2+} into these acidic stores (Docampo et al., 2005; Huang et al., 2014). Earlier studies have indicated that release of Ca^{2+} from acidocalcisomes may be mediated by the co-operative action of Ca^{2+}/H^+ exchanger and Na^+/H^+ exchanger (Docampo & Moreno, 2011). However, recent studies implicate a role for IP_3R in Ca^{2+} release from acidocalcisomes. While IP_3R usually localizes to ER, recent research demonstrated that in trypanosomes it could localize exclusively to acidocalcisomes and participate in Ca^{2+} release (Huang et al., 2014; Lander et al., 2016). In addition to acidocalcisomes, there are other acidic organelles, which can contribute towards Ca^{2+} storage and release in specific organisms. Vacuoles, endosomes, lysosomes, and Golgi bodies maintain an acidic environment and harbor channels and transporters involved in Ca^{2+} uptake and release.

11.2 Vacuoles

Contractile vacuoles from *D. discoideum* and *Paramecium,* as well as vacuoles from yeast and plants, contain a PMCA type Ca^{2+}-ATPase for Ca^{2+} import (Cunningham & Fink, 1994; Ferrol & Bennett, 1996; Malmstrom et al., 1997). Additionally, yeast and plant vacuoles also have a Ca^{2+}/H^{+} exchanger at their disposal for Ca^{2+} uptake (Cunningham & Fink, 1996; Hirschi et al., 1996). However, the Ca^{2+} efflux mechanisms differ greatly among the vacuoles of these organisms. In *D. discoideum* an ion channel P2XA is shown to release Ca^{2+}, resulting in Rab11a activation and vesicle fusion (Parkinson et al., 2014). Whereas, *Paramecium* harbors a variant of IP_3R, which localizes to contractile vacuole system and plays a role in Ca^{2+} release (Ladenburger et al., 2006). Ca^{2+} release from yeast vacuoles may be mediated by a TRP channel (Denis & Cyert, 2002). However, higher plants lack TRP channels (Nilius & Owsianik, 2011), but harbor two pore channels (Peiter et al., 2005) for Ca^{2+} release. It has also been shown that IP_3 and cyclic ADP-ribose can stimulate Ca^{2+} release from plant vacuoles (Allen et al., 1995; Ranjeva et al., 1988), indicating that there may be other Ca^{2+} channels for Ca^{2+} release in plant vacuoles which have not been genetically characterized yet.

11.3 Endolysosomal system

Endosomes are especially important in Ca^{2+} signaling because they can also serve as mobilizable Ca^{2+} stores. The original pH of endosomes is acidic and decreases further during maturation. Pumps responsible for import of Ca^{2+} into endosomes have yet to be been reported. However, it is likely that endosomes obtain Ca^{2+} from the extracellular environment during their formation at the plasma membrane. For Ca^{2+} release, endosomes from mammalian cells harbor TRP channels (Saito et al., 2007) and TPC channels (Brailoiu et al., 2009; Calcraft et al., 2009). Lysosomes, which are highly acidic organelles with pH ranging from four to five can also act as Ca^{2+} stores. Lysosomal Ca^{2+} is important for regulation of autophagy (Medina & Ballabio, 2015). Lysosomes from mammalian cells may contain a yet uncharacterized Ca^{2+}-ATPase for Ca^{2+} import (Ezaki et al., 1992; Lemons & Thoene, 1991). It remains to be determined if a recently identified P-type ATPase localizing to the lysosomes of mammalian cells (Dehay et al., 2012) is capable of transporting Ca^{2+} or not. Both TRP (Lange et al., 2009; Medina et al., 2015; Puertollano & Kiselyov, 2009) channels and TPC (Brailoiu et al., 2009; Calcraft et al., 2009; Zong et al., 2009) have been localized to lysosomes of mammalian cells, which may facilitate Ca^{2+} release. Lysosome-related dense granules in human platelets have a pH of about 6.1 (Rosado, 2011) and store Ca^{2+} at high concentrations which can be released by Ca^{2+} ionophores (Ruiz et al., 2004). Ca^{2+} accumulation in dense granules is believed to be mediated by SERCA-type Ca^{2+} ATPase (Lopez et al., 2005), whereas Ca^{2+} release may be controlled by NAADP (Lopez et al., 2006). Secretory granules from endocrine cells also seem to contain a SERCA-type Ca^{2+}-ATPase and a P-type Ca^{2}-ATPase (Mitchell et al., 2001) for Ca^{2+} accumulation (Santodomingo et al., 2008). Melanosomes, which are also lysosome-related organelles, harbor a Na^{+}/Ca^{2+} exchanger that is believed to work cooperatively with a Na^{+}/H^{+} exchanger to accumulate Ca^{2+} and thereby play a role in human pigmentation (Lamason et al., 2005; Salceda & Sanchez-Chavez, 2000). There is some limited evidence for c-ADPR- and NAADP-mediated Ca^{2+} release from secretory granules via a ryanodine receptor. However, recent extensive studies support an IP_3 induced mechanism of Ca^{2+} release from secretory granules. Such studies were conducted in mammalian neuroendocrine cells as well as unicellular phytoplanktons (Yoo, 2011).

11.4 Secretory system

Secretory vesicles from endocrine cells contain a SERCA-type Ca^{2+}-ATPase. Additionally, secretory vesicles may also contain an H^{+}/Ca^{2+} exchanger, and a Na^{+}/Ca^{2+} exchanger to facilitate the uptake process. Synaptic vesicles from sheep brain also contain a Ca^{2+}-ATPase of unknown type and a Ca^{2+}/H^{+} exchanger (Goncalves et al., 2000). Golgi apparatus, which form an important part of the secretory system, are mildly acidic with pH ranging from 6 to 6.5. Ca^{2+} within the Golgi bodies maintains its structure and also plays a functional role in vesicle trafficking (Yang et al., 2015). In addition to having a SERCA-type Ca^{2+}-ATPase, the ER and Golgi also have a specific SPCA-type Ca^{2+}-ATPase for Ca^{2+} uptake.

Ca^{2+} release is then likely mediated by IP_3R, which is common to the ER and Golgi (Micaroni, 2012) or via a yet unidentified ryanodine receptor (Cifuentes et al., 2001).

12. Ca^{2+} and disease

Ca^{2+} signaling is an important process that affects many cellular functions in our tissues and organs. Most cellular processes are affected by the action of Ca^{2+} signaling, Ca^{2+}-binding proteins, Ca^{2+} channels and transporters. Biological processes affected by Ca^{2+} signaling include muscle contraction and relaxation, neuronal signaling, hormone secretion, and cell proliferation and apoptosis. Therefore, defects in Ca^{2+} signaling will interfere with these key physiological processes and result in diseases.

12.1 Diseases of the reproductive system

Ca^{2+} signaling is important during fertilization. A monoclonal antibody against the IP_3R indicated the role of IP_3-mediated Ca^{2+} signaling during fertilization and embryonic development. It is now clear that during fertilization, sperm injects phospholipase C zeta (PLCζ) into the oocytes, which triggers the Ca^{2+} oscillations that activates egg development. Therefore, mutations in PLCζ can cause male infertility (Kashir et al., 2012; Nomikos et al., 2011).

12.2 Diseases of the immune system

Alterations in Ca^{2+} signaling can also have detrimental effects on the immune system. Ca^{2+} signaling plays a role in activation, gene expression, and differentiation and effector function of T-cells, B-cells, NK cells, mast cells, monocytes, and macrophages. Mutations in ORAI1 channel can result in severe combined immunodeficiency disorder (SCID) (Thompson et al., 2009). SCID also known, as the "bubble boy" disease, is a genetic disorder characterized by improper development of T and B cells leading to severe immunodeficiency. X-linked agammaglobulinemia is another inherited disease with linked defect in B-cell development. It is caused by mutations in a tyrosine kinase that enhances PLCγ activation and plays a role in IP_3-mediated Ca^{2+} signaling. Patients suffering from Wiskott–Aldrich syndrome (WAS) show reduced Ca^{2+} influx in their NK cells. Autoimmune diseases such as systemic lupus erythematosus (SLE) have been linked to defects in B-cell signaling and increased Ca^{2+} signaling responses. This was shown through experiments performed in mouse models, as well as experiments utilizing B-cells isolated from human patients (Feske, 2007). Impaired Ca^{2+} signaling is also commonly associated with chronic inflammatory diseases. such as Crohn's disease.

12.3 Diseases of the nervous system

Phospholipase C, gamma (PLCγ) plays an important role in brain development. Similarly, IP_3-mediated Ca^{2+} signaling also plays an important role in memory retention and regulation of brain rhythms, which occur during the sleep wake cycle. Therefore, defects in Ca^{2+} signaling can have severe consequences in the nervous system. Presenilin and bcl-2 proteins, which regulate the activity of IP_3R, have been found to be associated with Alzheimer's disease (Cai et al., 2015; Passer et al., 1999). Ca^{2+} signaling is also involved in the motor neuron degenerative disease called Lou-Gehrig's disease or amyotrophic lateral sclerosis. Indeed, mouse studies have shown that reduced expression of glutamate receptor or increased expression of IP_3R or PLCζ may be associated with amyotrophic lateral sclerosis (Leal & Gomes, 2015; Milanese et al., 2014; Staats et al., 2012, 2013). Other neurological diseases such as autism (Filice et al., 2016; Wohr et al., 2015), bipolar disorder (Udawela et al., 2011), Huntington's disease (Giacomello et al., 2011), and schizophrenia (Lidow, 2003) have all been linked to deregulation of Ca^{2+} signaling. Reduced expression of a Ca^{2+} buffering protein parvalbumin (Wohr et al., 2015) is associated with autism, whereas high levels of PLCβ1 are associated with bipolar disorder (Berridge, 2014). Additionally, genetic variants of oxytocin receptor and reduced IP_3-mediated Ca^{2+} signaling in tissues from autism patients also indicate an indirect role for Ca^{2+} signaling in autism (Ma et al., 2013). Another disease of the central nervous system called spinocerebellar ataxia (SCA) can also be a result of defective Ca^{2+} signaling. Symptoms of SCA include poor coordination of movement and speech. Many cases of SCA have been linked to mutations in the gene encoding for IP_3R1 in neurons (Huang et al., 2012; Tada et al., 2016). Mutations in the α1 subunit of the P/Q type voltage-gated Ca^{2+} channel also lead to SCA (Burk et al., 2014).

12.4 Diseases of the musculoskeletal system

In muscle cells, Ca^{2+} signaling is of prime importance. Increased IP_3R expression has been linked to increased myogenic tone, hypertension, and cardiac disease (Abou-Saleh et al., 2013; Althoff & Offermanns, 2015) whereas, downregulation of IP_3 induced Ca^{2+} signaling is a characteristic feature of congestive heart failure (Lou et al., 2012). Kawasaki's disease in which inflamed blood vessels affect several organs such as heart, skin, mouth, nose, and throat has been linked to a mutation in IP_3-kinase gene (Onouchi et al., 2008). Ca^{2+} signaling in muscle cells is also associated with respiratory diseases. Increased expression of muscarinic receptors causes increased IP_3-mediated Ca^{2+} signaling, excessive contraction of airway muscle cells, resulting in asthma and chronic obstructive pulmonary disease (COPD) (Gosens et al., 2006). Increased Ca^{2+} signaling in chondrocytes can cause cartilage damage and joint pain resulting in osteoarthritis or Kashin–Beck disease (Nguyen et al., 2012). RyR, which is responsible for increase in cytosolic Ca^{2+} concentration, plays an important role during the excitation-contraction-relaxation cycle of the muscle cells. Mutations in RyR have been associated with the disease named malignant hyperthermia indicated by muscle rigidity, tachycardia, hypoxia, and high blood pressure in response to muscle relaxants (MacLennan et al., 1990). Mutation in RyR has also been associated with central core disease (Zhang et al., 1993) characterized by proximal muscle weakness and fetal hypotonia. SERCA, which removes excess Ca^{2+} from the cytosol of the muscle cells, also plays an important role in muscle contraction and relaxation. Mutations in SERCA can lead to Brody disease or Brody myopathy, symptoms of which include muscle cramping and stiffening after exercise. In Brody disease, the expression of SERCA is unaffected but their activity is reduced by 50% leading to the symptoms which appear early during childhood (Hovnanian, 2007). In nemaline myopathy and mitochondrial myopathies, defective actin–myosin interactions cause Ca^{2+} hyposensitivity, which eventually results in muscle weakness (Gommans et al., 2002). Individuals that suffer from proximal myopathy and extrapyramidal movement disorder have mutations in the MICU1 protein of the MCU complex. In such patients, the mitochondrial potential is unchanged. However, there is increased mitochondrial Ca^{2+} uptake at low cytosolic Ca^{2+} concentrations (Logan et al., 2014).

12.5 Diseases of the endocrine and exocrine system

Deregulation of Ca^{2+} signaling can also affect the activity of exocrine and endocrine glands. Ca^{2+} signaling is important for insulin secretion from the pancreatic cells. Increase in plasma glucose levels results in increase in ATP/ADP ratio in pancreatic β-cells. This causes the inhibition of ATP-sensitive K^+ channels resulting in decrease in membrane potential and opening of voltage-gated Ca^{2+} channels. Ca^{2+} influx leads to insulin secretion. Therefore, decline in IP_3R expression and cholecystokinin expression has been reported in diabetes patients (Zhang et al., 2014), whereas increase IP_3R activity is believed to be associated with pancreatitis (Orabi et al., 2012). Double knockout mice for IP_3R2 an IP_3R3 exhibit defects in secretory function of the pancreas and salivary glands. Impaired Ca^{2+} signaling in pancreatic and salivary gland cells of these mice indicates a clear role for Ca^{2+} signaling in exocrine physiology (Futatsugi et al., 2005).

12.6 Ca^{2+} and cancer

Above all, Ca^{2+} signaling is especially important in cancer. Mutations in PLC encoding genes can result in several different types of cancer (Koss et al., 2014; Woyach et al., 2014). A role for IP_3R-mediated Ca^{2+} signaling is also confirmed in cancerous pancreatic duct cells (Choi et al., 2013). High expression of IP_3R has been observed in acute myeloid leukemia (Shi et al., 2015). High expression of IP_3R2 and IP_3R3 has been reported in tissue samples from patients suffering from lung cancer and gastric cancer respectively. Decreased RyR1 activity is detected in patient tissue samples from thymus cancer. Changes of expression voltage-gated channels are mostly associated with colorectal cancer. Expression changes of both high-voltage L-type channels and low-voltage T-type channels have been detected in colorectal cancer cell lines. Colorectal cancer cell lines also show changed in expression levels of SERCA2 and SERCA3. Occasionally, SERCA2 expression changes are also observed in some thyroid, oral, and lung cancer cell lines. Transient receptor potential cannels TRPM8 and TRPV6 show increased expression in breast cancer and prostate cancer tissues from patients. A similar change in expression of TRPM1 and TRPV1 is detected in patients with melanoma and bladder cancer (Monteith et al., 2012). Ca^{2+} extrusion from cells via the action of PMCA pump on the plasma membrane is also an important process and its significance is highlighted by the observation that changes in PMCA expression are observed in skin, lung, breast, and muccosal tissues from cancer patients (Monteith et al., 2007). MCU-mediated mitochondrial Ca^{2+} uptake seems to play an important role in the growth and metastasis of breast cancer cells (Tosatto et al., 2016). Proper functioning of Ca^{2+} signaling is mediated by the action of several regulatory proteins. Bcl-2 protein performs its antiapoptotic function by modulating Ca^{2+} signaling in cells (Monteith et al., 2012). Another oncogene BRCA1 during apoptosis binds to IP_3R, sensitizes it to the ligand IP_3, and stimulates Ca^{2+} release from endoplasmic reticulum (Hedgepeth et al., 2015). The oncogene K-ras modulates the isoform composition of IP_3R and can alter the Ca^{2+} content in the lumen of endoplasmic reticulum. Bcl-2 (Vervliet et al., 2016), K-ras (Pierro et al., 2014), and breast cancer 1 (BRCA1) (Hedgepeth et al., 2015) which are all known regulators of IP_3-mediated Ca^{2+} signaling have been associated with the development and the survival of cancer cells. In human colon cancers, MCU expression is reduced, whereas expression of micro-RNA miR-25 is upregulated. It appears that miR-25 can reduce MCU levels, resulting in reduced mitochondrial Ca^{2+} uptake and protection from apoptosis, possibly leading to cancer (Marchi et al., 2013).

References

Abou-Saleh, H., Pathan, A. R., Daalis, A., Hubrack, S., Abou-Jassoum, H., Al-Naeimi, H., Rusch, N. J., & Machaca, K. (2013). Inositol 1,4,5-trisphosphate (IP3) receptor up-regulation in hypertension is associated with sensitization of Ca2+ release and vascular smooth muscle contractility. *Journal of Biological Chemistry, 288*, 32941–32951.

Alam, M. R., Groschner, L. N., Parichatikanond, W., Kuo, L., Bondarenko, A. I., Rost, R., Waldeck-Weiermair, M., Malli, R., & Graier, W. F. (2012). Mitochondrial Ca2+ uptake 1 (MICU1) and mitochondrial ca2+ uniporter (MCU) contribute to metabolism-secretion coupling in clonal pancreatic beta-cells. *Journal of Biological Chemistry, 287*, 34445–34454.

Alkhani, H., Ase, A. R., Grant, R., O'donnell, D., Groschner, K., & Seguela, P. (2014). Contribution of TRPC3 to store-operated calcium entry and inflammatory transductions in primary nociceptors. *Molecular Pain, 10*, 43.

Allen, G. J., Muir, S. R., & Sanders, D. (1995). Release of Ca2+ from individual plant vacuoles by both InsP3 and cyclic ADP-ribose. *Science, 268*, 735–737.

Althoff, T. F., & Offermanns, S. (2015). G-protein-mediated signaling in vascular smooth muscle cells - implications for vascular disease. *Journal of Molecular Medicine, 93*, 973–981.

Ando, H., Kawaai, K., & Mikoshiba, K. (2014). Irbit: A regulator of ion channels and ion transporters. *Biochimica et Biophysica Acta, 1843*, 2195–2204.

Arikkath, J., & Campbell, K. P. (2003). Auxiliary subunits: Essential components of the voltage-gated calcium channel complex. *Current Opinion in Neurobiology, 13*, 298–307.

Armoundas, A. A., Hobai, I. A., Tomaselli, G. F., Winslow, R. L., & O'rourke, B. (2003). Role of sodium-calcium exchanger in modulating the action potential of ventricular myocytes from normal and failing hearts. *Circulation Research, 93*, 46–53.

Aydar, E., & Palmer, C. P. (2009). Polycystic kidney disease channel and synaptotagmin homologues play roles in schizosaccharomyces pombe cell wall synthesis/repair and membrane protein trafficking. *Journal of Membrane Biology, 229*, 141–152.

Bal, N. C., Maurya, S. K., Sopariwala, D. H., Sahoo, S. K., Gupta, S. C., Shaikh, S. A., Pant, M., Rowland, L. A., Bombardier, E., Goonasekera, S. A., Tupling, A. R., Molkentin, J. D., & Periasamy, M. (2012). Sarcolipin is a newly identified regulator of muscle-based thermogenesis in mammals. *Nature Medicine, 18*, 1575–1579.

Berridge, M. J. (2014). Calcium signalling and psychiatric disease: Bipolar disorder and schizophrenia. *Cell and Tissue Research, 357*, 477–492.

Berridge, M. J. (2009). Inositol trisphosphate and calcium signalling mechanisms. *Biochimica et Biophysica Acta, 1793*, 933–940.

Bonza, M. C., Morandini, P., Luoni, L., Geisler, M., Palmgren, M. G., & De Michelis, M. I. (2000). At-ACA8 encodes a plasma membrane-localized calcium-ATPase of Arabidopsis with a calmodulin-binding domain at the N terminus. *Plant Physiology, 123*, 1495–1506.

Boyman, L., Williams, G. S., Khananshvili, D., Sekler, I., & Lederer, W. J. (2013). NCLX: The mitochondrial sodium calcium exchanger. *Journal of Molecular and Cellular Cardiology, 59*, 205–213.

Brailoiu, E., Churamani, D., Cai, X., Schrlau, M. G., Brailoiu, G. C., Gao, X., Hooper, R., Boulware, M. J., Dun, N. J., Marchant, J. S., & Patel, S. (2009). Essential requirement for two-pore channel 1 in NAADP-mediated calcium signaling. *The Journal of Cell Biology, 186*, 201–209.

Brini, M., & Carafoli, E. (2011). The plasma membrane Ca(2)+ ATPase and the plasma membrane sodium calcium exchanger cooperate in the regulation of cell calcium. *Cold Spring Harbor Perspectives in Biology, 3*.

Bugrim, A. E. (1999). Regulation of Ca2+ release by cAMP-dependent protein kinase. A mechanism for agonist-specific calcium signaling? *Cell Calcium, 25*, 219–226.

Burk, K., Kaiser, F. J., Tennstedt, S., Schols, L., Kreuz, F. R., Wieland, T., Strom, T. M., Buttner, T., Hollstein, R., Braunholz, D., Plaschke, J., Gillessen-Kaesbach, G., & Zuhlke, C. (2014). A novel missense mutation in CACNA1A evaluated by in silico protein modeling is associated with non-episodic spinocerebellar ataxia with slow progression. *European Journal of Medical Genetics, 57*, 207–211.

Cai, Y., An, S. S., & Kim, S. (2015). Mutations in presenilin 2 and its implications in Alzheimer's disease and other dementia-associated disorders. *Clinical Interventions in Aging, 10*, 1163–1172.

Calcraft, P. J., Ruas, M., Pan, Z., Cheng, X., Arredouani, A., Hao, X., Tang, J., Rietdorf, K., Teboul, L., Chuang, K. T., Lin, P., Xiao, R., Wang, C., Zhu, Y., Lin, Y., Wyatt, C. N., Parrington, J., Ma, J., Evans, A. M., Galione, A., & Zhu, M. X. (2009). NAADP mobilizes calcium from acidic organelles through two-pore channels. *Nature, 459*, 596–600.

Carafoli, E., & Krebs, J. (2016). Why calcium? How calcium became the best communicator. *Journal of Biological Chemistry, 291*, 20849–20857.

Carraretto, L., Checchetto, V., De Bortoli, S., Formentin, E., Costa, A., Szabo, I., & Teardo, E. (2016). Calcium flux across plant mitochondrial membranes: Possible molecular players. *Frontiers of Plant Science, 7*, 354.

Carrasco, S., & Meyer, T. (2011). STIM proteins and the endoplasmic reticulum-plasma membrane junctions. *Annual Review of Biochemistry, 80*, 973–1000.

Chaudhuri, D., Sancak, Y., Mootha, V. K., & Clapham, D. E. (2013). MCU encodes the pore conducting mitochondrial calcium currents. *Elife, 2*, Article e00704.

Cheng, K. T., Liu, X., Ong, H. L., Swaim, W., & Ambudkar, I. S. (2011). Local Ca(2)+ entry via Orai1 regulates plasma membrane recruitment of TRPC1 and controls cytosolic Ca(2)+ signals required for specific cell functions. *PLoS Biology, 9*, Article e1001025.

Cheng, K. T., Ong, H. L., Liu, X., & Ambudkar, I. S. (2013). Contribution and regulation of TRPC channels in store-operated Ca2+ entry. *Current Topics in Membranes, 71*, 149–179.

Choi, J. H., Ji, Y. G., & Lee, D. H. (2013). Uridine triphosphate increases proliferation of human cancerous pancreatic duct epithelial cells by activating P2Y2 receptor. *Pancreas, 42*, 680–686.

Cifuentes, F., Gonzalez, C. E., Fiordelisio, T., Guerrero, G., Lai, F. A., & Hernandez-Cruz, A. (2001). A ryanodine fluorescent derivative reveals the presence of high-affinity ryanodine binding sites in the Golgi complex of rat sympathetic neurons, with possible functional roles in intracellular Ca(2+) signaling. *Cellular Signalling, 13*, 353–362.

Clapham, D. E. (2003). TRP channels as cellular sensors. *Nature, 426*, 517–524.

Collins, K. D. (1997). Charge density-dependent strength of hydration and biological structure. *Biophysical Journal, 72*, 65–76.

Collins, S. R., & Meyer, T. (2011). Evolutionary origins of STIM1 and STIM2 within ancient Ca2+ signaling systems. *Trends in Cell Biology, 21*, 202–211.

Courtney, M. J., Lambert, J. J., & Nicholls, D. G. (1990). The interactions between plasma membrane depolarization and glutamate receptor activation in the regulation of cytoplasmic free calcium in cultured cerebellar granule cells. *Journal of Neuroscience, 10*, 3873–3879.

Crompton, M. (1999). The mitochondrial permeability transition pore and its role in cell death. *Biochemical Journal, 341*(Pt 2), 233–249.

Csordas, G., Golenar, T., Seifert, E. L., Kamer, K. J., Sancak, Y., Perocchi, F., Moffat, C., Weaver, D., De La Fuente Perez, S., Bogorad, R., Koteliansky, V., Adijanto, J., Mootha, V. K., & Hajnoczky, G. (2013). MICU1 controls both the threshold and cooperative activation of the mitochondrial Ca(2)(+) uniporter. *Cell Metabolism, 17*, 976–987.

Cui, J., Kaandorp, J. A., Sloot, P. M., Lloyd, C. M., & Filatov, M. V. (2009). Calcium homeostasis and signaling in yeast cells and cardiac myocytes. *FEMS Yeast Research, 9*, 1137–1147.

Cunningham, K. W., & Fink, G. R. (1996). Calcineurin inhibits VCX1-dependent H+/Ca2+ exchange and induces Ca2+ ATPases in *Saccharomyces cerevisiae. Molecular and Cellular Biology, 16*, 2226–2237.

Cunningham, K. W., & Fink, G. R. (1994). Calcineurin-dependent growth control in *Saccharomyces cerevisiae* mutants lacking PMC1, a homolog of plasma membrane Ca2+ ATPases. *The Journal of Cell Biology, 124*, 351–363.

Curry, M. C., Peters, A. A., Kenny, P. A., Roberts-Thomson, S. J., & Monteith, G. R. (2013). Mitochondrial calcium uniporter silencing potentiates caspase-independent cell death in MDA-MB-231 breast cancer cells. *Biochemical and Biophysical Research Communications, 434*, 695–700.

De Stefani, D., Raffaello, A., Teardo, E., Szabo, I., & Rizzuto, R. (2011). A forty-kilodalton protein of the inner membrane is the mitochondrial calcium uniporter. *Nature, 476*, 336–340.

De Stefani, D., Rizzuto, R., & Pozzan, T. (2016). Enjoy the trip: Calcium in mitochondria back and forth. *Annual Review of Biochemistry, 85*, 161–192.

Dehay, B., Ramirez, A., Martinez-Vicente, M., Perier, C., Canron, M. H., Doudnikoff, E., Vital, A., Vila, M., Klein, C., & Bezard, E. (2012). Loss of P-type ATPase ATP13A2/PARK9 function induces general lysosomal deficiency and leads to Parkinson disease neurodegeneration. *Proceedings of the National Academy of Sciences of the U S A, 109*, 9611–9616.

Deliot, N., & Constantin, B. (2015). Plasma membrane calcium channels in cancer: Alterations and consequences for cell proliferation and migration. *Biochimica et Biophysica Acta, 1848*, 2512–2522.

Den Dekker, E., Hoenderop, J. G., Nilius, B., & Bindels, R. J. (2003). The epithelial calcium channels, TRPV5 & TRPV6: From identification towards regulation. *Cell Calcium, 33*, 497–507.

Denis, V., & Cyert, M. S. (2002). Internal Ca(2+) release in yeast is triggered by hypertonic shock and mediated by a TRP channel homologue. *The Journal of Cell Biology, 156*, 29–34.

Denton, R. M., Randle, P. J., & Martin, B. R. (1972). Stimulation by calcium ions of pyruvate dehydrogenase phosphate phosphatase. *Biochemical Journal, 128*, 161–163.

Denton, R. M. (2009). Regulation of mitochondrial dehydrogenases by calcium ions. *Biochimica et Biophysica Acta, 1787*, 1309–1316.

Dingledine, R., Borges, K., Bowie, D., & Traynelis, S. F. (1999). The glutamate receptor ion channels. *Pharmacological Reviews, 51*, 7–61.

Docampo, R., De Souza, W., Miranda, K., Rohloff, P., & Moreno, S. N. (2005). Acidocalcisomes - conserved from bacteria to man. *Nature Reviews Microbiology, 3*, 251–261.

Docampo, R., & Huang, G. (2015). Calcium signaling in trypanosomatid parasites. *Cell Calcium, 57*, 194–202.

Docampo, R., & Moreno, S. N. (2011). Acidocalcisomes. *Cell Calcium, 50*, 113–119.

Doonan, P. J., Chandramoorthy, H. C., Hoffman, N. E., Zhang, X., Cardenas, C., Shanmughapriya, S., Rajan, S., Vallem, S., Chen, X., Foskett, J. K., Cheung, J. Y., Houser, S. R., & Madesh, M. (2014). LETM1-dependent mitochondrial Ca2+ flux modulates cellular bioenergetics and proliferation. *The FASEB Journal, 28*, 4936–4949.

Drago, I., De Stefani, D., Rizzuto, R., & Pozzan, T. (2012). Mitochondrial Ca2+ uptake contributes to buffering cytoplasmic Ca2+ peaks in cardiomyocytes. *Proceedings of the National Academy of Sciences of the U S A, 109*, 12986–12991.

Durr, G., Strayle, J., Plemper, R., Elbs, S., Klee, S. K., Catty, P., Wolf, D. H., & Rudolph, H. K. (1998). The medial-Golgi ion pump Pmr1 supplies the yeast secretory pathway with Ca2+ and Mn2+ required for glycosylation, sorting, and endoplasmic reticulum-associated protein degradation. *Molecular Biology of the Cell, 9*, 1149–1162.

Dziadek, M. A., & Johnstone, L. S. (2007). Biochemical properties and cellular localisation of STIM proteins. *Cell Calcium, 42*, 123–132.

Endo, M. (2009). Calcium-induced calcium release in skeletal muscle. *Physiological Reviews, 89*, 1153–1176.

Ezaki, J., Himeno, M., & Kato, K. (1992). Purification and characterization of (Ca2+-Mg2+)-ATPase in rat liver lysosomal membranes. *Journal of Biochemistry, 112*, 33–39.

Ferrol, N., & Bennett, A. B. (1996). A single gene may encode differentially localized Ca2+-ATPases in tomato. *The Plant Cell, 8*, 1159–1169.

Feske, S. (2007). Calcium signalling in lymphocyte activation and disease. *Nature Reviews Immunology, 7*, 690–702.

Filice, F., Vorckel, K. J., Sungur, A. O., Wohr, M., & Schwaller, B. (2016). Reduction in parvalbumin expression not loss of the parvalbumin-expressing GABA interneuron subpopulation in genetic parvalbumin and shank mouse models of autism. *Molecular Brain, 9*, 10.

Finch, E. A., Turner, T. J., & Goldin, S. M. (1991). Calcium as a coagonist of inositol 1,4,5-trisphosphate-induced calcium release. *Science, 252*, 443–446.

Francis, S. H., Busch, J. L., Corbin, J. D., & Sibley, D. (2010). cGMP-dependent protein kinases and cGMP phosphodiesterases in nitric oxide and cGMP action. *Pharmacological Reviews, 62*, 525–563.

Futatsugi, A., Nakamura, T., Yamada, M. K., Ebisui, E., Nakamura, K., Uchida, K., Kitaguchi, T., Takahashi-Iwanaga, H., Noda, T., Aruga, J., & Mikoshiba, K. (2005). IP3 receptor types 2 and 3 mediate exocrine secretion underlying energy metabolism. *Science, 309*, 2232–2234.

Giacomello, M., Hudec, R., & Lopreiato, R. (2011). Huntington's disease, calcium, and mitochondria. *BioFactors, 37*, 206–218.

Giamarchi, A., Padilla, F., Coste, B., Raoux, M., Crest, M., Honore, E., & Delmas, P. (2006). The versatile nature of the calcium-permeable cation channel TRPP2. *EMBO Reports, 7*, 787–793.

Gincel, D., Zaid, H., & Shoshan-Barmatz, V. (2001). Calcium binding and translocation by the voltage-dependent anion channel: A possible regulatory mechanism in mitochondrial function. *Biochemical Journal, 358*, 147–155.

Gommans, I. M., Vlak, M. H., De Haan, A., & Van Engelen, B. G. (2002). Calcium regulation and muscle disease. *Journal of Muscle Research & Cell Motility, 23*, 59–63.

Goncalves, P. P., Meireles, S. M., Neves, P., & Vale, M. G. (2000). Distinction between Ca(2+) pump and Ca(2+)/H(+) antiport activities in synaptic vesicles of sheep brain cortex. *Neurochemistry International, 37*, 387–396.

Gosens, R., Zaagsma, J., Meurs, H., & Halayko, A. J. (2006). Muscarinic receptor signaling in the pathophysiology of asthma and COPD. *Respiratory Research, 7*, 73.

Gupta, S., Raychaudhury, B., Banerjee, S., Das, B., & Datta, S. C. (2006). An intracellular calcium store is present in Leishmania donovani glycosomes. *Experimental Parasitology, 113*, 161–167.

Guse, A. H. (2015). Calcium mobilizing second messengers derived from NAD. *Biochimica et Biophysica Acta, 1854*, 1132–1137.

Gustavsson, M., Verardi, R., Mullen, D. G., Mote, K. R., Traaseth, N. J., Gopinath, T., & Veglia, G. (2013). Allosteric regulation of SERCA by phosphorylation-mediated conformational shift of phospholamban. *Proceedings of the National Academy of Sciences of the U S A, 110*, 17338–17343.

Hashimi, H., Mcdonald, L., Stribrna, E., & Lukes, J. (2013). Trypanosome Letm1 protein is essential for mitochondrial potassium homeostasis. *Journal of Biological Chemistry, 288*, 26914–26925.

Hedgepeth, S. C., Garcia, M. I., Wagner, L. E., 2nd, Rodriguez, A. M., Chintapalli, S. V., Snyder, R. R., Hankins, G. D., Henderson, B. R., Brodie, K. M., Yule, D. I., Van Rossum, D. B., & Boehning, D. (2015). The BRCA1 tumor suppressor binds to inositol 1,4,5-trisphosphate receptors to stimulate apoptotic calcium release. *Journal of Biological Chemistry, 290*, 7304–7313.

Henderson, M. J., Baldwin, H. A., Werley, C. A., Boccardo, S., Whitaker, L. R., Yan, X., Holt, G. T., Schreiter, E. R., Looger, L. L., Cohen, A. E., Kim, D. S., & Harvey, B. K. (2015). A low affinity GCaMP3 variant (GCaMPer) for imaging the endoplasmic reticulum calcium store. *PLoS One, 10*, Article e0139273.

Hilgemann, D. W., Collins, A., & Matsuoka, S. (1992). Steady-state and dynamic properties of cardiac sodium-calcium exchange. Secondary modulation by cytoplasmic calcium and ATP. *The Journal of General Physiology, 100*, 933–961.

Hilgemann, D. W., Matsuoka, S., Nagel, G. A., & Collins, A. (1992). Steady-state and dynamic properties of cardiac sodium-calcium exchange. Sodium-dependent inactivation. *The Journal of General Physiology, 100*, 905–932.

Hirschi, K. D., Zhen, R. G., Cunningham, K. W., Rea, P. A., & Fink, G. R. (1996). CAX1, an H+/Ca2+ antiporter from Arabidopsis. *Proceedings of the National Academy of Sciences of the U S A, 93*, 8782–8786.

Hogan, P. G., & Rao, A. (2015). Store-operated calcium entry: Mechanisms and modulation. *Biochemical and Biophysical Research Communications, 460*, 40–49.

Hovnanian, A. (2007). SERCA pumps and human diseases. *Subcellular Biochemistry, 45*, 337–363.

Huang, G., Bartlett, P. J., Thomas, A. P., Moreno, S. N., & Docampo, R. (2013). Acidocalcisomes of Trypanosoma brucei have an inositol 1,4,5-trisphosphate receptor that is required for growth and infectivity. *Proceedings of the National Academy of Sciences of the U S A, 110*, 1887–1892.

Huang, G., Ulrich, P. N., Storey, M., Johnson, D., Tischer, J., Tovar, J. A., Moreno, S. N., Orlando, R., & Docampo, R. (2014). Proteomic analysis of the acidocalcisome, an organelle conserved from bacteria to human cells. *PLoS Pathogens, 10*, Article e1004555.

Huang, L., Chardon, J. W., Carter, M. T., Friend, K. L., Dudding, T. E., Schwartzentruber, J., Zou, R., Schofield, P. W., Douglas, S., Bulman, D. E., & Boycott, K. M. (2012). Missense mutations in ITPR1 cause autosomal dominant congenital nonprogressive spinocerebellar ataxia. *Orphanet Journal of Rare Diseases, 7*, 67.

Ivanova, H., Vervliet, T., Missiaen, L., Parys, J. B., De Smedt, H., & Bultynck, G. (2014). Inositol 1,4,5-trisphosphate receptor-isoform diversity in cell death and survival. *Biochimica et Biophysica Acta, 1843*, 2164–2183.

Kaleta, M., & Palmer, C. (2011). TRP channels in yeast. *Advances in Experimental Medicine & Biology, 704*, 315–321.

Kamp, T. J., & Hell, J. W. (2000). Regulation of cardiac L-type calcium channels by protein kinase A and protein kinase C. *Circulation Research, 87*, 1095–1102.

Kashir, J., Konstantinidis, M., Jones, C., Lemmon, B., Lee, H. C., Hamer, R., Heindryckx, B., Deane, C. M., De Sutter, P., Fissore, R. A., Parrington, J., Wells, D., & Coward, K. (2012). A maternally inherited autosomal point mutation in human phospholipase C zeta (PLCzeta) leads to male infertility. *Human Reproduction, 27*, 222–231.

Koss, H., Bunney, T. D., Behjati, S., & Katan, M. (2014). Dysfunction of phospholipase Cgamma in immune disorders and cancer. *Trends in Biochemical Sciences, 39*, 603–611.

Kovacs-Bogdan, E., Sancak, Y., Kamer, K. J., Plovanich, M., Jambhekar, A., Huber, R. J., Myre, M. A., Blower, M. D., & Mootha, V. K. (2014). Reconstitution of the mitochondrial calcium uniporter in yeast. *Proceedings of the National Academy of Sciences of the U S A, 111*, 8985–8990.

Kretsinger, R. H., & Nockolds, C. E. (1973). Carp muscle calcium-binding protein. II. Structure determination and general description. *Journal of Biological Chemistry, 248*, 3313–3326.

Kwong, J. Q., Davis, J., Baines, C. P., Sargent, M. A., Karch, J., Wang, X., Huang, T., & Molkentin, J. D. (2014). Genetic deletion of the mitochondrial phosphate carrier desensitizes the mitochondrial permeability transition pore and causes cardiomyopathy. *Cell Death & Differentiation, 21*, 1209–1217.

Ladenburger, E. M., Korn, I., Kasielke, N., Wassmer, T., & Plattner, H. (2006). An Ins(1,4,5)P3 receptor in Paramecium is associated with the osmoregulatory system. *Journal of Cell Science, 119*, 3705–3717.

Lamason, R. L., Mohideen, M. A., Mest, J. R., Wong, A. C., Norton, H. L., Aros, M. C., Jurynec, M. J., Mao, X., Humphreville, V. R., Humbert, J. E., Sinha, S., Moore, J. L., Jagadeeswaran, P., Zhao, W., Ning, G., Makalowska, I., Mckeigue, P. M., O'donnell, D., Kittles, R., … Cheng, K. C. (2005). SLC24A5, a putative cation exchanger, affects pigmentation in zebrafish and humans. *Science, 310*, 1782–1786.

Lander, N., Chiurillo, M. A., Storey, M., Vercesi, A. E., & Docampo, R. (2016). CRISPR/Cas9-mediated endogenous C-terminal tagging of trypanosoma cruzi genes reveals the acidocalcisome localization of the inositol 1,4,5-trisphosphate receptor. *Journal of Biological Chemistry, 291*, 25505–25515.

Lange, I., Yamamoto, S., Partida-Sanchez, S., Mori, Y., Fleig, A., & Penner, R. (2009). TRPM2 functions as a lysosomal Ca2+-release channel in beta cells. *Science Signaling, 2*, Article ra23.

Lanner, J. T., Georgiou, D. K., Joshi, A. D., & Hamilton, S. L. (2010). Ryanodine receptors: Structure, expression, molecular details, and function in calcium release. *Cold Spring Harbor Perspectives in Biology, 2*, Article a003996.

Leal, S. S., & Gomes, C. M. (2015). Calcium dysregulation links ALS defective proteins and motor neuron selective vulnerability. *Frontiers in Cellular Neuroscience, 9*, 225.

Lee, H. C. (2001). Physiological functions of cyclic ADP-ribose and NAADP as calcium messengers. *Annual Review of Pharmacology and Toxicology, 41*, 317–345.

Lemons, R. M., & Thoene, J. G. (1991). Mediated calcium transport by isolated human fibroblast lysosomes. *Journal of Biological Chemistry, 266*, 14378–14382.

Lewit-Bentley, A., & Rety, S. (2000). EF-hand calcium-binding proteins. *Current Opinion in Structural Biology, 10*, 637–643.

Lidow, M. S. (2003). Calcium signaling dysfunction in schizophrenia: A unifying approach. *Brain Research Reviews, 43*, 70–84.

Linck, B., Qiu, Z., He, Z., Tong, Q., Hilgemann, D. W., & Philipson, K. D. (1998). Functional comparison of the three isoforms of the Na+/Ca2+ exchanger (NCX1, NCX2, NCX3). *American Journal of Physiology, 274*, C415–C423.

Liu, X., Wang, W., Singh, B. B., Lockwich, T., Jadlowiec, J., O'connell, B., Wellner, R., Zhu, M. X., & Ambudkar, I. S. (2000). Trp1, a candidate protein for the store-operated Ca(2+) influx mechanism in salivary gland cells. *Journal of Biological Chemistry, 275*, 3403–3411.

Logan, C. V., Szabadkai, G., Sharpe, J. A., Parry, D. A., Torelli, S., Childs, A. M., Kriek, M., Phadke, R., Johnson, C. A., Roberts, N. Y., Bonthron, D. T., Pysden, K. A., Whyte, T., Munteanu, I., Foley, A. R., Wheway, G., Szymanska, K., Natarajan, S., Abdelhamed, Z. A., … Sheridan, E. (2014). Loss-of-function mutations in MICU1 cause a brain and muscle disorder linked to primary alterations in mitochondrial calcium signaling. *Nature Genetics, 46*, 188–193.

Lopez, J. J., Camello-Almaraz, C., Pariente, J. A., Salido, G. M., & Rosado, J. A. (2005). Ca2+ accumulation into acidic organelles mediated by Ca2+- and vacuolar H+-ATPases in human platelets. *Biochemical Journal, 390*, 243–252.

Lopez, J. J., Redondo, P. C., Salido, G. M., Pariente, J. A., & Rosado, J. A. (2006). Two distinct Ca2+ compartments show differential sensitivity to thrombin, ADP and vasopressin in human platelets. *Cellular Signalling, 18*, 373–381.

Lopreiato, R., Giacomello, M., & Carafoli, E. (2014). The plasma membrane calcium pump: New ways to look at an old enzyme. *Journal of Biological Chemistry, 289*, 10261–10268.

Lou, Q., Janardhan, A., & Efimov, I. R. (2012). Remodeling of calcium handling in human heart failure. *Advances in Experimental Medicine & Biology, 740*, 1145–1174.

Luo, S., Rohloff, P., Cox, J., Uyemura, S. A., & Docampo, R. (2004). Trypanosoma brucei plasma membrane-type Ca(2+)-ATPase 1 (TbPMC1) and 2 (TbPMC2) genes encode functional Ca(2+)-ATPases localized to the acidocalcisomes and plasma membrane, and essential for Ca(2+) homeostasis and growth. *Journal of Biological Chemistry, 279*, 14427–14439.

Ma, W. J., Hashii, M., Munesue, T., Hayashi, K., Yagi, K., Yamagishi, M., Higashida, H., & Yokoyama, S. (2013). Non-synonymous single-nucleotide variations of the human oxytocin receptor gene and autism spectrum disorders: A case-control study in a Japanese population and functional analysis. *Molecular Autism, 4*, 22.

Maack, C., Ganesan, A., Sidor, A., & O'rourke, B. (2005). Cardiac sodium-calcium exchanger is regulated by allosteric calcium and exchanger inhibitory peptide at distinct sites. *Circulation Research, 96*, 91–99.

Maclennan, D. H., Duff, C., Zorzato, F., Fujii, J., Phillips, M., Korneluk, R. G., Frodis, W., Britt, B. A., & Worton, R. G. (1990). Ryanodine receptor gene is a candidate for predisposition to malignant hyperthermia. *Nature, 343*, 559–561.

Maechler, P., & Wollheim, C. B. (1999). Mitochondrial glutamate acts as a messenger in glucose-induced insulin exocytosis. *Nature, 402*, 685–689.

Mallilankaraman, K., Cardenas, C., Doonan, P. J., Chandramoorthy, H. C., Irrinki, K. M., Golenar, T., Csordas, G., Madireddi, P., Yang, J., Muller, M., Miller, R., Kolesar, J. E., Molgo, J., Kaufman, B., Hajnoczky, G., Foskett, J. K., & Madesh, M. (2012). MCUR1 is an essential component of mitochondrial Ca2+ uptake that regulates cellular metabolism. *Nature Cell Biology, 14*, 1336–1343.

Malmstrom, S., Askerlund, P., & Palmgren, M. G. (1997). A calmodulin-stimulated Ca2+-ATPase from plant vacuolar membranes with a putative regulatory domain at its N-terminus. *FEBS Letters, 400*, 324–328.

Marchi, S., Lupini, L., Patergnani, S., Rimessi, A., Missiroli, S., Bonora, M., Bononi, A., Corra, F., Giorgi, C., De Marchi, E., Poletti, F., Gafa, R., Lanza, G., Negrini, M., Rizzuto, R., & Pinton, P. (2013). Downregulation of the mitochondrial calcium uniporter by cancer-related miR-25. *Current Biology, 23*, 58–63.

Mccormack, J. G., & Denton, R. M. (1979). The effects of calcium ions and adenine nucleotides on the activity of pig heart 2-oxoglutarate dehydrogenase complex. *Biochemical Journal, 180*, 533–544.

Mcquibban, A. G., Joza, N., Megighian, A., Scorzeto, M., Zanini, D., Reipert, S., Richter, C., Schweyen, R. J., & Nowikovsky, K. (2010). A Drosophila mutant of LETM1, a candidate gene for seizures in Wolf-Hirschhorn syndrome. *Human Molecular Genetics, 19*, 987–1000.

Medina, D. L., & Ballabio, A. (2015). Lysosomal calcium regulates autophagy. *Autophagy, 11*, 970–971.

Medina, D. L., Di Paola, S., Peluso, I., Armani, A., De Stefani, D., Venditti, R., Montefusco, S., Scotto-Rosato, A., Prezioso, C., Forrester, A., Settembre, C., Wang, W., Gao, Q., Xu, H., Sandri, M., Rizzuto, R., De Matteis, M. A., & Ballabio, A. (2015). Lysosomal calcium signalling regulates autophagy through calcineurin and TFEB. *Nature Cell Biology, 17*, 288–299.

Micaroni, M. (2012). Calcium around the Golgi apparatus: Implications for intracellular membrane trafficking. *Advances in Experimental Medicine & Biology, 740*, 439–460.

Milanese, M., Giribaldi, F., Melone, M., Bonifacino, T., Musante, I., Carminati, E., Rossi, P. I., Vergani, L., Voci, A., Conti, F., Puliti, A., & Bonanno, G. (2014). Knocking down metabotropic glutamate receptor 1 improves survival and disease progression in the SOD1(G93A) mouse model of amyotrophic lateral sclerosis. *Neurobiology of Disease, 64*, 48–59.

Mitchell, K. J., Pinton, P., Varadi, A., Tacchetti, C., Ainscow, E. K., Pozzan, T., Rizzuto, R., & Rutter, G. A. (2001). Dense core secretory vesicles revealed as a dynamic Ca(2+) store in neuroendocrine cells with a vesicle-associated membrane protein aequorin chimaera. *The Journal of Cell Biology, 155*, 41–51.

Monteith, G. R., Davis, F. M., & Roberts-Thomson, S. J. (2012). Calcium channels and pumps in cancer: Changes and consequences. *Journal of Biological Chemistry, 287*, 31666–31673.

Monteith, G. R., Mcandrew, D., Faddy, H. M., & Roberts-Thomson, S. J. (2007). Calcium and cancer: Targeting Ca2+ transport. *Nature Reviews Cancer, 7*, 519–530.

Mooren, F. C., & Kinne, R. K. (1998). Cellular calcium in health and disease. *Biochimica et Biophysica Acta, 1406*, 127–151.

Moreno, S. N., & Docampo, R. (2003). Calcium regulation in protozoan parasites. *Current Opinion in Microbiology, 6*, 359–364.

Nagamune, K., Hicks, L. M., Fux, B., Brossier, F., Chini, E. N., & Sibley, L. D. (2008). Abscisic acid controls calcium-dependent egress and development in Toxoplasma gondii. *Nature, 451*, 207–210.

Nguyen, C., Lieberherr, M., Bordat, C., Velard, F., Come, D., Liote, F., & Ea, H. K. (2012). Intracellular calcium oscillations in articular chondrocytes induced by basic calcium phosphate crystals lead to cartilage degradation. *Osteoarthritis and Cartilage, 20*, 1399–1408.

Niki, I., Yokokura, H., Sudo, T., Kato, M., & Hidaka, H. (1996). Ca2+ signaling and intracellular Ca2+ binding proteins. *Journal of Biochemistry, 120*, 685–698.

Nilius, B., & Owsianik, G. (2011). The transient receptor potential family of ion channels. *Genome Biology, 12*, 218.

Nita, I., Hershfinkel, M., Kantor, C., Rutter, G. A., Lewis, E. C., & Sekler, I. (2014). Pancreatic beta-cell Na+ channels control global Ca2+ signaling and oxidative metabolism by inducing Na+ and Ca2+ responses that are propagated into mitochondria. *The FASEB Journal, 28*, 3301–3312.

Nita, I., Hershfinkel, M., Lewis, E. C., & Sekler, I. (2015). A crosstalk between Na(+) channels, Na(+)/K(+) pump and mitochondrial Na(+) transporters controls glucose-dependent cytosolic and mitochondrial Na(+) signals. *Cell Calcium, 57*, 69–75.

Noble, D., Noble, S. J., Bett, G. C., Earm, Y. E., Ho, W. K., & So, I. K. (1991). The role of sodium-calcium exchange during the cardiac action potential. *Annals of the New York Academy of Sciences, 639*, 334–353.

Nomikos, M., Elgmati, K., Theodoridou, M., Calver, B. L., Cumbes, B., Nounesis, G., Swann, K., & Lai, F. A. (2011). Male infertility-linked point mutation disrupts the Ca2+ oscillation-inducing and PIP(2) hydrolysis activity of sperm PLCzeta. *Biochemical Journal, 434*, 211–217.

Onouchi, Y., Gunji, T., Burns, J. C., Shimizu, C., Newburger, J. W., Yashiro, M., Nakamura, Y., Yanagawa, H., Wakui, K., Fukushima, Y., Kishi, F., Hamamoto, K., Terai, M., Sato, Y., Ouchi, K., Saji, T., Nariai, A., Kaburagi, Y., Yoshikawa, T., ... Hata, A. (2008). ITPKC functional polymorphism associated with Kawasaki disease susceptibility and formation of coronary artery aneurysms. *Nature Genetics, 40*, 35–42.

Orabi, A. I., Luo, Y., Ahmad, M. U., Shah, A. U., Mannan, Z., Wang, D., Sarwar, S., Muili, K. A., Shugrue, C., Kolodecik, T. R., Singh, V. P., Lowe, M. E., Thrower, E., Chen, J., & Husain, S. Z. (2012). IP3 receptor type 2 deficiency is associated with a secretory defect in the pancreatic acinar cell and an accumulation of zymogen granules. *PLoS One, 7*, Article e48465.

Paidhungat, M., & Garrett, S. (1997). A homolog of mammalian, voltage-gated calcium channels mediates yeast pheromone-stimulated Ca2+ uptake and exacerbates the cdc1(Ts) growth defect. *Molecular and Cellular Biology, 17*, 6339–6347.

Palty, R., Silverman, W. F., Hershfinkel, M., Caporale, T., Sensi, S. L., Parnis, J., Nolte, C., Fishman, D., Shoshan-Barmatz, V., Herrmann, S., Khananshvili, D., & Sekler, I. (2010). NCLX is an essential component of mitochondrial Na+/Ca2+ exchange. *Proceedings of the National Academy of Sciences of the U S A, 107*, 436–441.

Pan, X., Liu, J., Nguyen, T., Liu, C., Sun, J., Teng, Y., Fergusson, M. M., Rovira, I., Allen, M., Springer, D. A., Aponte, A. M., Gucek, M., Balaban, R. S., Murphy, E., & Finkel, T. (2013). The physiological role of mitochondrial calcium revealed by mice lacking the mitochondrial calcium uniporter. *Nature Cell Biology, 15*, 1464–1472.

Parekh, A. B., & Putney, J. W., JR. (2005). Store-operated calcium channels. *Physiological Reviews, 85*, 757–810.

Parkinson, K., Baines, A. E., Keller, T., Gruenheit, N., Bragg, L., North, R. A., & Thompson, C. R. (2014). Calcium-dependent regulation of Rab activation and vesicle fusion by an intracellular P2X ion channel. *Nature Cell Biology, 16*, 87–98.

Parnis, J., Montana, V., Delgado-Martinez, I., Matyash, V., Parpura, V., Kettenmann, H., Sekler, I., & Nolte, C. (2013). Mitochondrial exchanger NCLX plays a major role in the intracellular Ca2+ signaling, gliotransmission, and proliferation of astrocytes. *Journal of Neuroscience, 33*, 7206–7219.

Passer, B. J., Pellegrini, L., Vito, P., Ganjei, J. K., & D'adamio, L. (1999). Interaction of Alzheimer's presenilin-1 and presenilin-2 with Bcl-X(L). A potential role in modulating the threshold of cell death. *Journal of Biological Chemistry, 274*, 24007–24013.

Patron, M., Checchetto, V., Raffaello, A., Teardo, E., Vecellio Reane, D., Mantoan, M., Granatiero, V., Szabo, I., De Stefani, D., & Rizzuto, R. (2014). MICU1 and MICU2 finely tune the mitochondrial Ca2+ uniporter by exerting opposite effects on MCU activity. *Molecular Cell, 53*, 726–737.

Paupe, V., Prudent, J., Dassa, E. P., Rendon, O. Z., & Shoubridge, E. A. (2015). CCDC90A (MCUR1) is a cytochrome c oxidase assembly factor and not a regulator of the mitochondrial calcium uniporter. *Cell Metabolism, 21*, 109–116.

Peiter, E., Maathuis, F. J., Mills, L. N., Knight, H., Pelloux, J., Hetherington, A. M., & Sanders, D. (2005). The vacuolar Ca2+-activated channel TPC1 regulates germination and stomatal movement. *Nature, 434*, 404–408.

Periasamy, M., & Kalyanasundaram, A. (2007). SERCA pump isoforms: Their role in calcium transport and disease. *Muscle & Nerve, 35*, 430–442.

Perocchi, F., Gohil, V. M., Girgis, H. S., Bao, X. R., Mccombs, J. E., Palmer, A. E., & Mootha, V. K. (2010). MICU1 encodes a mitochondrial EF hand protein required for Ca(2+) uptake. *Nature, 467*, 291–296.

Petersen, O. H., Michalak, M., & Verkhratsky, A. (2005). Calcium signalling: Past, present and future. *Cell Calcium, 38*, 161–169.

Phillips, M. J., & Voeltz, G. K. (2016). Structure and function of ER membrane contact sites with other organelles. *Nature Reviews Molecular Cell Biology, 17*, 69–82.

Pierro, C., Cook, S. J., Foets, T. C., Bootman, M. D., & Roderick, H. L. (2014). Oncogenic K-Ras suppresses IP(3)-dependent Ca(2)(+) release through remodelling of the isoform composition of IP(3)Rs and ER luminal Ca(2)(+) levels in colorectal cancer cell lines. *Journal of Cell Science, 127*, 1607–1619.

Pittis, A. A., Goh, V., Cebrian-Serrano, A., Wettmarshausen, J., Perocchi, F., & Gabaldón, T. (2020). Discovery of EMRE in fungi resolves the true evolutionary history of the mitochondrial calcium uniporter. *Nature Communications, 11*(4031), 1–10. https://doi.org/10.1038/s41467-020-17705-4

Plovanich, M., Bogorad, R. L., Sancak, Y., Kamer, K. J., Strittmatter, L., Li, A. A., Girgis, H. S., Kuchimanchi, S., De Groot, J., Speciner, L., Taneja, N., Oshea, J., Koteliansky, V., & Mootha, V. K. (2013). MICU2, a paralog of MICU1, resides within the mitochondrial uniporter complex to regulate calcium handling. *PLoS One, 8*, Article e55785.

Prole, D. L., & Taylor, C. W. (2011). Identification of intracellular and plasma membrane calcium channel homologues in pathogenic parasites. *PLoS One, 6*, Article e26218.

Prole, D. L., & Taylor, C. W. (2016). Inositol 1,4,5-trisphosphate receptors and their protein partners as signalling hubs. *The Journal of Physiology, 594*, 2849–2866.

Puertollano, R., & Kiselyov, K. (2009). TRPMLs: In sickness and in health. *American Journal of Physiology - Renal Physiology, 296*, F1245–F1254.

Qi, Z., Stephens, N. R., & Spalding, E. P. (2006). Calcium entry mediated by GLR3.3, an Arabidopsis glutamate receptor with a broad agonist profile. *Plant Physiology, 142*, 963–971.

Qiu, J., Tan, Y. W., Hagenston, A. M., Martel, M. A., Kneisel, N., Skehel, P. A., Wyllie, D. J., Bading, H., & Hardingham, G. E. (2013). Mitochondrial calcium uniporter Mcu controls excitotoxicity and is transcriptionally repressed by neuroprotective nuclear calcium signals. *Nature Communications, 4*, 2034.

Raffaello, A., De Stefani, D., Sabbadin, D., Teardo, E., Merli, G., Picard, A., Checchetto, V., Moro, S., Szabo, I., & Rizzuto, R. (2013). The mitochondrial calcium uniporter is a multimer that can include a dominant-negative pore-forming subunit. *The EMBO Journal, 32*, 2362–2376.

Ranjeva, R., Carrasco, A., & Boudet, A. M. (1988). Inositol trisphosphate stimulates the release of calcium from intact vacuoles isolated from Acer cells. *FEBS Letters, 230*, 137–141.

Rosado, J. A. (2011). Acidic Ca(2+) stores in platelets. *Cell Calcium, 50*, 168–174.

Ruiz, F. A., Lea, C. R., Oldfield, E., & Docampo, R. (2004). Human platelet dense granules contain polyphosphate and are similar to acidocalcisomes of bacteria and unicellular eukaryotes. *Journal of Biological Chemistry, 279*, 44250–44257.

Saito, M., Hanson, P. I., & Schlesinger, P. (2007). Luminal chloride-dependent activation of endosome calcium channels: Patch clamp study of enlarged endosomes. *Journal of Biological Chemistry, 282*, 27327–27333.

Salceda, R., & Sanchez-Chavez, G. (2000). Calcium uptake, release and ryanodine binding in melanosomes from retinal pigment epithelium. *Cell Calcium, 27*, 223–229.

Sancak, Y., Markhard, A. L., Kitami, T., Kovacs-Bogdan, E., Kamer, K. J., Udeshi, N. D., Carr, S. A., Chaudhuri, D., Clapham, D. E., Li, A. A., Calvo, S. E., Goldberger, O., & Mootha, V. K. (2013). EMRE is an essential component of the mitochondrial calcium uniporter complex. *Science, 342*, 1379–1382.

Santodomingo, J., Vay, L., Camacho, M., Hernandez-Sanmiguel, E., Fonteriz, R. I., Lobaton, C. D., Montero, M., Moreno, A., & Alvarez, J. (2008). Calcium dynamics in bovine adrenal medulla chromaffin cell secretory granules. *European Journal of Neuroscience, 28*, 1265–1274.

Shi, J. L., Fu, L., & Wang, W. D. (2015). High expression of inositol 1,4,5-trisphosphate receptor, type 2 (ITPR2) as a novel biomarker for worse prognosis in cytogenetically normal acute myeloid leukemia. *Oncotarget, 6*, 5299–5309.

Shin, H. Y., Hong, Y. H., Jang, S. S., Chae, H. G., Paek, S. L., Moon, H. E., Kim, D. G., Kim, J., Paek, S. H., & Kim, S. J. (2010). A role of canonical transient receptor potential 5 channel in neuronal differentiation from A2B5 neural progenitor cells. *PLoS One, 5*, Article e10359.

Simms, B. A., & Zamponi, G. W. (2012). Trafficking and stability of voltage-gated calcium channels. *Cellular and Molecular Life Sciences, 69*, 843–856.

Solovyova, N., Veselovsky, N., Toescu, E. C., & Verkhratsky, A. (2002). Ca(2+) dynamics in the lumen of the endoplasmic reticulum in sensory neurons: Direct visualization of Ca(2+)-induced Ca(2+) release triggered by physiological Ca(2+) entry. *The EMBO Journal, 21*, 622–630.

Staats, K. A., Bogaert, E., Hersmus, N., Jaspers, T., Luyten, T., Bultynck, G., Parys, J. B., Hisatsune, C., Mikoshiba, K., Van Damme, P., Robberecht, W., & Van Den Bosch, L. (2012). Neuronal overexpression of IP(3) receptor 2 is detrimental in mutant SOD1 mice. *Biochemical and Biophysical Research Communications, 429*, 210–213.

Staats, K. A., Van Helleputte, L., Jones, A. R., Bento-Abreu, A., Van Hoecke, A., Shatunov, A., Simpson, C. L., Lemmens, R., Jaspers, T., Fukami, K., Nakamura, Y., Brown, R. H., JR., Van Damme, P., Liston, A., Robberecht, W., Al-Chalabi, A., & Van Den Bosch, L. (2013). Genetic ablation of phospholipase C delta 1 increases survival in SOD1(G93A) mice. *Neurobiology of Disease, 60*, 11–17.

Stael, S., Wurzinger, B., Mair, A., Mehlmer, N., Vothknecht, U. C., & Teige, M. (2012). Plant organellar calcium signalling: An emerging field. *Journal of Experimental Botany, 63*, 1525–1542.

Stammers, A. N., Susser, S. E., Hamm, N. C., Hlynsky, M. W., Kimber, D. E., Kehler, D. S., & Duhamel, T. A. (2015). The regulation of sarco(endo) plasmic reticulum calcium-ATPases (SERCA). *Canadian Journal of Physiology and Pharmacology, 93*, 843–854.

Strayle, J., Pozzan, T., & Rudolph, H. K. (1999). Steady-state free Ca(2+) in the yeast endoplasmic reticulum reaches only 10 microM and is mainly controlled by the secretory pathway pump pmr1. *The EMBO Journal, 18*, 4733–4743.

Striggow, F., & Ehrlich, B. E. (1996). Ligand-gated calcium channels inside and out. *Current Opinion in Cell Biology, 8*, 490–495.

Sutton, R. B., Davletov, B. A., Berghuis, A. M., Sudhof, T. C., & Sprang, S. R. (1995). Structure of the first C2 domain of synaptotagmin I: A novel Ca2+/phospholipid-binding fold. *Cell, 80*, 929–938.

Tada, M., Nishizawa, M., & Onodera, O. (2016). Roles of inositol 1,4,5-trisphosphate receptors in spinocerebellar ataxias. *Neurochemistry International, 94*, 1–8.

Takeuchi, A., Kim, B., & Matsuoka, S. (2013). The mitochondrial Na+-Ca2+ exchanger, NCLX, regulates automaticity of HL-1 cardiomyocytes. *Scientific Reports, 3*, 2766.

Tanida, I., Takita, Y., Hasegawa, A., Ohya, Y., & Anraku, Y. (1996). Yeast Cls2p/Csg2p localized on the endoplasmic reticulum membrane regulates a non-exchangeable intracellular Ca2+ pool cooperatively with calcineurin. *FEBS Letters, 379*, 38–42.

Tarasov, A. I., Semplici, F., Ravier, M. A., Bellomo, E. A., Pullen, T. J., Gilon, P., Sekler, I., Rizzuto, R., & Rutter, G. A. (2012). The mitochondrial Ca2+ uniporter MCU is essential for glucose-induced ATP increases in pancreatic beta-cells. *PLoS One, 7*, Article e39722.

Thompson, J. L., Mignen, O., & Shuttleworth, T. J. (2009). The Orai1 severe combined immune deficiency mutation and calcium release-activated Ca2+ channel function in the heterozygous condition. *Journal of Biological Chemistry, 284*, 6620–6626.

Ton, V. K., & Rao, R. (2004). Functional expression of heterologous proteins in yeast: Insights into Ca2+ signaling and Ca2+-transporting ATPases. *American Journal of Physiology - Cell Physiology, 287*, C580–C589.

Tosatto, A., Sommaggio, R., Kummerow, C., Bentham, R. B., Blacker, T. S., Berecz, T., Duchen, M. R., Rosato, A., Bogeski, I., Szabadkai, G., Rizzuto, R., & Mammucari, C. (2016). The mitochondrial calcium uniporter regulates breast cancer progression via HIF-1alpha. *EMBO Molecular Medicine, 8*, 569–585.

Tsai, M. F., Jiang, D., Zhao, L., Clapham, D., & Miller, C. (2014). Functional reconstitution of the mitochondrial Ca2+/H+ antiporter Letm1. *The Journal of General Physiology, 143*, 67–73.

Udawela, M., Scarr, E., Hannan, A. J., Thomas, E. A., & Dean, B. (2011). Phospholipase C beta 1 expression in the dorsolateral prefrontal cortex from patients with schizophrenia at different stages of illness. *Australian and New Zealand Journal of Psychiatry, 45*, 140–147.

Vais, H., Tanis, J. E., Muller, M., Payne, R., Mallilankaraman, K., & Foskett, J. K. (2015). MCUR1, CCDC90a, is a regulator of the mitochondrial calcium uniporter. *Cell Metabolism, 22*, 533–535.

Van Petegem, F. (2015). Ryanodine receptors: Allosteric ion channel giants. *Journal of Molecular Biology, 427*, 31–53.

Vanderheyden, V., Devogelaere, B., Missiaen, L., De Smedt, H., Bultynck, G., & Parys, J. B. (2009). Regulation of inositol 1,4,5-trisphosphate-induced Ca2+ release by reversible phosphorylation and dephosphorylation. *Biochimica et Biophysica Acta, 1793*, 959–970.

Vashist, S., Frank, C. G., Jakob, C. A., & Ng, D. T. (2002). Two distinctly localized p-type ATPases collaborate to maintain organelle homeostasis required for glycoprotein processing and quality control. *Molecular Biology of the Cell, 13*, 3955–3966.

Vervliet, T., Parys, J. B., & Bultynck, G. (2016). Bcl-2 proteins and calcium signaling: Complexity beneath the surface. *Oncogene, 35*, 5079–5092.

Wang, P., Li, Z., Wei, J., Zhao, Z., Sun, D., & Cui, S. (2012). A Na+/Ca2+ exchanger-like protein (AtNCL) involved in salt stress in Arabidopsis. *Journal of Biological Chemistry, 287*, 44062–44070.

Wang, Q. C., Zheng, Q., Tan, H., Zhang, B., Li, X., Yang, Y., Yu, J., Liu, Y., Chai, H., Wang, X., Sun, Z., Wang, J. Q., Zhu, S., Wang, F., Yang, M., Guo, C., Wang, H., Zheng, Q., Li, Y., … Tang, T. S. (2016). TMCO1 is an ER Ca(2+) load-activated Ca(2+) channel. *Cell, 165*, 1454–1466.

Wang, X., Pluznick, J. L., Wei, P., Padanilam, B. J., & Sansom, S. C. (2004). TRPC4 forms store-operated Ca2+ channels in mouse mesangial cells. *American Journal of Physiology - Cell Physiology, 287*, C357–C364.

Wheeler, G. L., Joint, I., & Brownlee, C. (2008). Rapid spatiotemporal patterning of cytosolic Ca2+ underlies flagellar excision in Chlamydomonas reinhardtii. *The Plant Journal, 53*, 401–413.

White, P. J. (2000). Calcium channels in higher plants. *Biochimica et Biophysica Acta, 1465*, 171–189.

Wohr, M., Orduz, D., Gregory, P., Moreno, H., Khan, U., Vorckel, K. J., Wolfer, D. P., Welzl, H., Gall, D., Schiffmann, S. N., & Schwaller, B. (2015). Lack of parvalbumin in mice leads to behavioral deficits relevant to all human autism core symptoms and related neural morphofunctional abnormalities. *Translational Psychiatry, 5*, e525.

Woyach, J. A., Furman, R. R., Liu, T. M., Ozer, H. G., Zapatka, M., Ruppert, A. S., Xue, L., Li, D. H., Steggerda, S. M., Versele, M., Dave, S. S., Zhang, J., Yilmaz, A. S., Jaglowski, S. M., Blum, K. A., Lozanski, A., Lozanski, G., James, D. F., Barrientos, J. C., ... Byrd, J. C. (2014). Resistance mechanisms for the Bruton's tyrosine kinase inhibitor ibrutinib. *New England Journal of Medicine, 370*, 2286–2294.

Wu, Y., Kuzma, J., Marechal, E., Graeff, R., Lee, H. C., Foster, R., & Chua, N. H. (1997). Abscisic acid signaling through cyclic ADP-ribose in plants. *Science, 278*, 2126–2130.

Yamamoto, T., Yamagoshi, R., Harada, K., Kawano, M., Minami, N., Ido, Y., Kuwahara, K., Fujita, A., Ozono, M., Watanabe, A., Yamada, A., Terada, H., & Shinohara, Y. (2016). Analysis of the structure and function of EMRE in a yeast expression system. *Biochimica et Biophysica Acta, 1857*, 831–839.

Yang, Z., Kirton, H. M., Macdougall, D. A., Boyle, J. P., Deuchars, J., Frater, B., Ponnambalam, S., Hardy, M. E., White, E., Calaghan, S. C., Peers, C., & Steele, D. S. (2015). The Golgi apparatus is a functionally distinct Ca2+ store regulated by the PKA and Epac branches of the beta1-adrenergic signaling pathway. *Science Signaling, 8*, Article ra101.

Yoo, S. H. (2011). Role of secretory granules in inositol 1,4,5-trisphosphate-dependent Ca(2+) signaling: From phytoplankton to mammals. *Cell Calcium, 50*, 175–183.

Zeng, W., Yuan, J. P., Kim, M. S., Choi, Y. J., Huang, G. N., Worley, P. F., & Muallem, S. (2008). STIM1 gates TRPC channels, but not Orai1, by electrostatic interaction. *Molecular Cell, 32*, 439–448.

Zhang, S. L., Yeromin, A. V., Zhang, X. H., Yu, Y., Safrina, O., Penna, A., Roos, J., Stauderman, K. A., & Cahalan, M. D. (2006). Genome-wide RNAi screen of Ca(2+) influx identifies genes that regulate Ca(2+) release-activated Ca(2+) channel activity. *Proceedings of the National Academy of Sciences of the U S A, 103*, 9357–9362.

Zhang, Y., Chen, H. S., Khanna, V. K., De Leon, S., Phillips, M. S., Schappert, K., Britt, B. A., Browell, A. K., & Maclennan, D. H. (1993). A mutation in the human ryanodine receptor gene associated with central core disease. *Nature Genetics, 5*, 46–50.

Zhang, Z., Tian, J., Liao, Q., & Wang, Y. (2014). The analysis of expression of CCK and IP3 receptors in gallstones patients with type 2 diabetes mellitus. *Hepato-Gastroenterology, 61*, 2173–2176.

Zong, X., Schieder, M., Cuny, H., Fenske, S., Gruner, C., Rotzer, K., Griesbeck, O., Harz, H., Biel, M., & Wahl-Schott, C. (2009). The two-pore channel TPCN2 mediates NAADP-dependent Ca(2+)-release from lysosomal stores. *Pflügers Archiv, 458*, 891–899.

Chapter 15

Intracellular chloride regulation

Francisco J. Alvarez-Leefmans

Department of Pharmacology and Toxicology, Boonshoft School of Medicine, Wright State University, Dayton, OH, United States

1. Introduction

The importance of chloride ions in cell physiology is fully recognized. Chloride (Cl^-), together with bicarbonate (HCO_3^-), is the most abundant free anion in living animal cells, and, as will be seen in this chapter, it performs or determines fundamental biological functions in virtually all tissues. For decades, it was thought that Cl^- moved through plasma membranes by a purely electro-diffusive mechanism, a passive mechanism in which membrane permeation of a given ion is driven by both its electrical and concentration gradients. According to this view, Cl^- was distributed passively across the plasma membrane in thermodynamic equilibrium as defined by Eq. (15.2) (see Section 3 below and Fig. 15.1), and from a functional perspective, it was thought to be necessary just to maintain electroneutrality of intra- and extracellular fluids and for keeping the cell osmotic balance. The notion that Cl^- was passively distributed implied that it was not actively transported, nor could it provide energy for the transport of other substances or play any role in cell signaling functions. Research carried out during the last decades has led to a total change in this simplistic view. It is now firmly established that Cl^- plays pivotal roles in cell physiology and that many, if not all, animal cells exhibit a nonequilibrium distribution of Cl^- across their plasma membranes (Alvarez-Leefmans & Delpire, 2009b; Alvarez-Leefmans & Russell, 1990; Delpire & Gagnon, 2018a; Gamba, 2005; Hartzell, 2009; Kaila et al., 2014; Portioli et al., 2021; Russell, 2000). Far from being passively distributed across the plasma membrane following the membrane potential (E_m), we now know that Cl^- is actively transported and tightly regulated. Some cells actively extrude Cl^-, others actively accumulate it, but few cells, if at all, ignore it. The intracellular Cl^- concentration, $[Cl^-]_i$, is determined by the relative contribution of various anion-transporting proteins expressed in cell membranes. These transport proteins include several types of Cl^- channels as well as numerous transporters that work primarily as cotransporters (symporters) or as exchangers (antiporters). Among the cotransporters are the cation-coupled Cl^- cotransporters (CCCs), belonging to the solute carrier gene/protein family 12 (SLC12) of membrane transporters, which play a central role in controlling the $[Cl^-]_i$.

By virtue of being distributed out of electrochemical equilibrium, Cl^- serves as a critical player in a variety of cellular functions such as pH_i regulation, cell volume regulation, transepithelial salt and water transport, and membrane potential

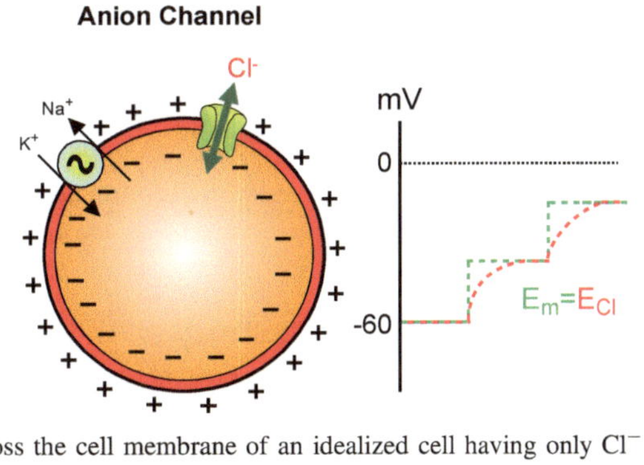

FIGURE 15.1 Passive Cl^- distribution across the cell membrane of an idealized cell having only Cl^- channels and a Na^+/K^+ ATPase. The resting membrane potential (E_m) is −60 mV. Right: E_m is changed in steps of ≈20 mV from the initial resting membrane potential. Chloride redistributes passively, following E_m. At each steady state, $E_m = E_{Cl}$. *Reproduced with permission from Alvarez-Leefmans (2012).*

Cell Physiology Sourcebook. https://doi.org/10.1016/B978-0-12-811114-7.00012-8

stabilization. Moreover, it has been recently proposed that Cl^- has an intracellular signaling function modulating the activity of various cellular proteins, including ion transporters and channels, protein kinases, and transcription factors (Valdivieso & Santa-Coloma, 2019; Lüscher et al., 2020). In addition to these functions, in the nervous system, Cl^- gradients are fundamental in synaptic signaling mediated by the neurotransmitters GABA (γ-aminobutyric acid) via $GABA_A$ receptor-channels and glycine. This signaling takes place through ligand-gated anion channels, which are primarily permeable to Cl^-, and can be in the depolarizing and hyperpolarizing directions, thereby determining mechanisms such as presynaptic and postsynaptic inhibition or the GABA-mediated depolarizing excitation observed in immature CNS neurons. In some cell types, the "ligand" that opens Cl^- channels, instead of being a neurotransmitter, is intracellular ionized calcium (Ca^{2+}). Transmembrane Cl^- gradients play key roles in neuronal growth and development, cell migration, extracellular K^+ scavenging, neurotransmitter reuptake, vesicular storage, circadian rhythms, inflammation, and sensory transduction including olfaction and nociception (Alvarez-Leefmans & Delpire, 2009a; Carlton, 2014; Frings, 2009; Hengl et al., 2010; Kaila et al., 2014; Murillo-de-Ozores et al., 2020; Takayama & Tominaga, 2024; Zeilhofer et al., 2012). This chapter presents and discusses examples illustrating how Cl^- is regulated and involved in these functions. The focus is on the basic principles of intracellular chloride regulation. Readers interested in deep diving into this ever-growing field should consult the multiple references and review papers cited throughout this chapter.

During the last decade, and certainly since the previous edition of this book in 2012, significant advances in the field have been achieved. Notably, the molecular structure and function of Cl^- transport proteins have been elucidated at the atomic level by means of cryogenic electron microscopy (cryo-EM), a technique that has complemented and, in most cases, replaced X-ray crystallography and NMR spectroscopy (Brzezinski, 2017; Chari & Stark, 2023). Using cryo-EM in the last few years, the structure and binding sites of the cation-coupled Cl^- cotransporters (CCCs) belonging to the solute carrier family SLC12 of membrane transporters have been studied, revealing at atomic resolution, structural elements that play critical roles in the function and pharmacology of these proteins (Trejo et al., 2023).

Another significant advance has been the development of various probes to measure intracellular Cl^- concentration ($[Cl^-]_i$). These techniques have grown from Cl^--sensitive microelectrodes (Alvarez-Leefmans et al., 1990; Coles, 1995), which permit simultaneous measurement of intracellular Cl^- and membrane potential (E_m) with unrivaled selectivity and sensitivity, although limited to relatively large and robust cells, to various Cl^--sensitive fluorescent dyes (Sonawane et al., 2002; Verkman, 1990, 2009). More recently, genetically encoded anion biosensors (GEABs) with adequate sensitivity and dynamic range have been developed to study Cl^- ion regulation in multiple cell types (Lodovichi et al., 2022; Zajac et al., 2020). An advantage of GEABs over other methods is that intracellular Cl^--imaging may be done in identified cell classes. An inconvenience, however, is that most GEABs are pH sensitive, which could lead to significant errors in the measurement of changes in $[Cl^-]_i$. One strategy to circumvent this problem has been developing biosensors such as ClopHensor and derivatives (e.g., LSSmClopHensor), designed to measure pH and Cl^- simultaneously, allowing correction of the Cl^- signal from pH interference. Using LSSmClopHensor, it was possible to measure in vivo the developmental shift in $[Cl^-]_i$ that has been long proposed to occur in CNS neurons (Sulis Sato et al., 2017). The most desirable outcome for designing a Cl^- probe for measuring $[Cl^-]_i$ would be the development of indicators insensitive to pH with high selectivity and dynamic range. Another feature of the ideal indicator is that it should be able to correct for cell volume changes that could alter the probe readings, leading to spurious measurements. Promising developments in this very active field are on the way; it is expected that soon, there will be Cl^- indicators with adequate sensitivity, selectivity, and response time to be able to quantitatively study intracellular chloride regulation simultaneously with other variables like cell volume and pH, that could impact the measurement of $[Cl^-]_i$.

Another breakthrough in the last 2 decades has been the identification and detailed understanding of the long sought-after Cl^--sensitive kinases that regulate the activity of CCCs via phosphorylation/dephosphorylation, the WNK kinases, an acronym for the "with-no-lysine [K]" serine-threonine kinases, which along with their downstream targets, the SPAK/OSR1 serine-threonine kinases, are crucial for the regulation of intracellular Cl^- (Goldsmith & Rodan, 2023; Murillo-de-Ozores et al., 2020; Piala et al., 2014; Shekarabi et al., 2017). WNKs regulate and are regulated by Cl^- in a feedback signaling system that has been the subject of numerous studies. But before delving further into these mechanisms controlling $[Cl^-]_i$, let us consider the foundations from which this knowledge emerged.

2. Origin of "the passive Cl^- distribution assumption"

The notion that Cl^- is nonpassively distributed and actively transported across cell membranes may sound obvious nowadays. However, it took decades to eradicate the generalization of "the passive Cl^- distribution assumption" and its implications. When dealing with the Goldman-Hodgkin-Katz equation for resting membrane potential, Cl^- is still often omitted. One of the usual arguments for this omission is that any ion that is passively distributed cannot determine the resting membrane potential (E_m); instead, E_m determines the distribution of that ion across the membrane. Another argument is that P_{Cl}, the membrane permeability to Cl^-, is negligible. However, neither of these arguments applies to all cell types. The question arises as to how and why basic facts such as the nonequilibrium distribution of Cl^- across the plasma membranes, the transport mechanisms that make it possible, and its functional significance were missed or neglected for decades. The passive Cl^- distribution assumption exemplifies how scientific knowledge progress is often slowed down due to notions that are mistakenly generalized and taken for granted.

The notion that Cl^- was passively distributed in all cells emerged from three key experimental observations made in squid axons, skeletal muscle cells, and red blood cells between the 1940s and the 1960s. First, in the early 1940s, an influential paper reported that in freshly dissected squid giant axons, one of the best animal cell models to study ion transport, the average $[Cl^-]_i$ was 36 mmol/kg axoplasm (Steinbach, 1941). Taking the water content of squid axoplasm as 880 g/kg, this corresponded to 41 mM, a value close to what would be expected if Cl^- was in electrochemical equilibrium across the axolemma of this seawater animal. However, in 1963, British physiologist Richard Keynes (1919–2010) published a landmark paper entitled "Chloride in the Squid Giant Axons," in which he reanalyzed the chloride content of extruded axoplasm from *Loligo forbesii* and Steinbach's low values for $[Cl^-]_i$ could not be confirmed (Keynes, 1963). The average $[Cl^-]_i$ measured by Keynes in squid axoplasm was 108 ± 2 mM, a concentration "... well over twice that expected for a simple passive distribution ...." Thus, E_{Cl}, the Nernst equilibrium potential for Cl^- as defined by Eq. (15.1) (see below, Section 5.3), was less than -40 mV, which was different from the resting E_m of -60 to -70 mV typically measured in squid giant axons. Based on these findings and measurements of Cl^- fluxes, Keynes suggested that Cl^- was "... being transported inwards against the electrochemical gradient by some form of active transport mechanism ...." The nature of the transport mechanism in squid axons was elucidated nearly 20 years later by the biophysicist John M. Russell. The active Cl^- uptake mechanism turned out to be an *electroneutral* Na^+-K^+-Cl^- cotransporter (Russell, 2000) similar to that described in 1980 by Peter Geck, Erich Heinz, and coworkers in Ehrlich cells, a mammalian cell line (Geck et al., 1980; Geck & Heinz, 1986). The difference between the squid and the mammalian Na^+-K^+-Cl^- is in the stoichiometry, which is $2Na^+$:$1K^+$:$3Cl^-$ in squid giant axons and $1Na^+$:$1K^+$:$2Cl^-$ in most mammalian cells. This difference in stoichiometry probably reflects some hitherto unknown differences in the protein structure (Garneau et al., 2020). An exception in mammalian cells is the stoichiometry of the Na^+-K^+-Cl^- in the membrane of ferret erythrocytes, which is $2Na^+$:$1K^+$:$3Cl^-$, as in squid axons (Hall & Ellory, 1985). Other stoichiometries have been proposed, but in all cases, they are consistent with the *electroneutral* movement of Na^+-K^+-Cl^- (Garneau et al., 2020; Lytle, 2003). Simultaneous recording of E_m and changes in intracellular Cl^- with Cl-sensitive microelectrodes finally proved that the fluxes carried by the Na^+-K^+-Cl^- cotransporter were indeed electroneutral (Alvarez-Leefmans et al., 1988), as discussed below in Section 5.2.2.

The second set of observations that gave rise to the "passive distribution assumption" came from work on frog skeletal muscle done in the late 1950s and early 1960s, which showed that the resting membrane Cl^- conductance was twice as large as that for K^+ (Hodgkin & Horowicz, 1959; Hutter & Noble, 1960). Further, measured $[Cl^-]_i$ was consistent with a distribution of Cl^- determined by a purely passive mechanism (Adrian, 1961). Thus, E_{Cl} was found to be close or equal to E_m. Direct, real-time measurements of $[Cl^-]_i$ and E_m in skeletal muscle using Cl^--sensitive microelectrodes demonstrated that, indeed, in skeletal muscle, E_{Cl} closely follows E_m (Vaughan-Jones, 1982). This is due to the relatively high P_{Cl} characteristic of the sarcolemma, as illustrated schematically in Fig. 15.1 We now know that this exceptionally high resting P_{Cl} of skeletal muscle fibers is due to the expression of ClC-1 chloride channels. ClC-1 is a member of the ClC family of chloride channels and transporter proteins (Jentsch, 2008). The functional significance of this high P_{Cl} is that it stabilizes the membrane potential of skeletal muscle fibers. Thus, for all practical purposes, Cl^- appears passively distributed across the sarcolemma. Nevertheless, the fact is that $E_m = E_{Cl}$ does not preclude the existence of carrier-mediated active transport of Cl^- in skeletal muscle cells or in any cell type. In fact, there is compelling evidence for an inward active transport of Cl^- in the skeletal muscle (Aickin, 1990). This inward transport of Cl^- is mediated primarily by one of the two isoforms of the Na^+-K^+-$2Cl^-$ cotransporters, NKCC1, a member of the electroneutral cation-coupled chloride cotransporter protein family encoded by the SLC12 gene family (Gosmanov et al., 2003; Jurkat-Rott et al., 2006; Kristensen & Juel, 2010; Mandai et al., 2017; Wong et al., 2001). However, as shown in Fig. 15.2, the relatively high P_{Cl} mediated by ClC-1 masks any effects that the operation of this inward transport system may have on $[Cl^-]_i$. Accordingly, reducing P_{Cl} by pharmacological blockage of ClC-1 with 9-anthracene carboxylic acid (9-AC) unmasks a powerful active Cl^- uptake mediated by NKCC1.

The third set of observations that led to the generalization of the "passive distribution assumption" came from red blood cells in which $[Cl^-]_i$ was also found to be close to that predicted for electrochemical equilibrium. Several Cl^- transport studies in red blood cells were consistent with the view that the Cl^- transmembrane distribution was thermodynamically passive and, in addition, showed that Cl^- crossed the membrane extremely rapidly. This latter finding,

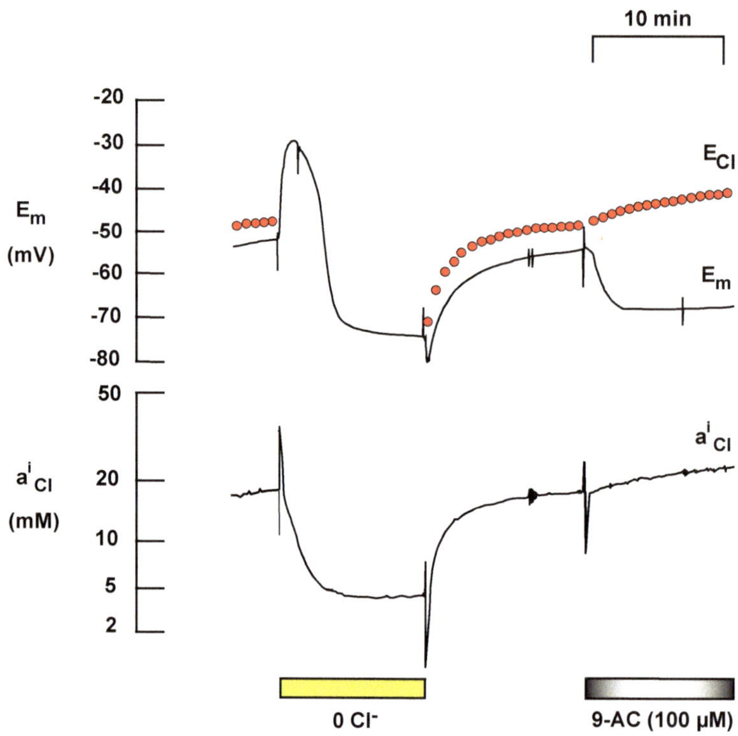

FIGURE 15.2 Active accumulation of Cl^- in rat skeletal muscle unmasked by reducing the resting Cl^- conductance (gCl^-) mediated by Cl^- channels (ClC-1). Membrane potential (E_m) and intracellular Cl^- activity a^i_{Cl} were measured simultaneously in a rat lumbrical muscle during the removal and readdition of external Cl^- and upon inhibition of the resting gCl^- by application of ClC-1 blocker 9-anthracene carboxylic acid (9-AC). The Cl^- equilibrium potential (E_{Cl}), calculated from the recorded a^i_{Cl} and a^o_{Cl}, using Eq. (15.1) has been plotted as red filled circles above the recorded E_m trace. At the beginning of the recording, E_{Cl} was already slightly positive with respect to E_m, indicating that a^i_{Cl} was maintained just above equilibrium. Recovery of a^i_{Cl} following intracellular Cl^- depletion revealed uphill Cl^- accumulation, which became more evident by inhibiting the gCl^- with 9-AC. Note the further increase in a^i_{Cl} above equilibrium during 9-AC application, unmasking uphill Cl^- accumulation by reducing the Cl^- leak (gCl^-) through ClC-1 channels. Consequently, the difference between E_m and E_{Cl} became ostensibly larger. The recordings were made with a double-barreled microelectrode, one barrel sensing E_m and the other a^i_{Cl}. The preparation was maintained in the nominal absence of CO_2 in solutions equilibrated with 100% O_2 to decrease or abolish the presence of Cl^-/HCO_3^- exchangers. It is now known that the mechanism responsible for the uphill accumulation of Cl^- is NKCC1, the Na^+-K^+-$2Cl^-$ cotransporter 1. *Modified from Aickin (1990).*

erroneously interpreted for a long time as being only the result of high passive P_{Cl}, made it quite likely that Cl^- could remain at thermodynamic equilibrium. In fact, it was later found that the high P_{Cl} of erythrocytes was primarily due to the expression of an anion exchanger transport protein, the $Cl^-/HCO3^-$ exchanger originally identified as "Band 3" because of its relative position in SDS gels. "Band 3" is now known as "anion exchanger 1" (AE1), and it is one of 10 members of the SLC4 gene and protein family of anion transporters (Alper, 2009; Romero et al., 2013), as discussed by Parker in Chapter 16.

The three sets of experimental observations discussed above were generalized so that virtually all cells were thought to have a very high P_{Cl} and a thermodynamically passive Cl^- transmembrane distribution, implying that Cl^- was not actively transported. Moreover, to make things worse for the case of Cl^-, cations claimed center stage during this period, with attention focused on the Na^+ and Ca^{2+} action potentials, the K^+ channels, and the Na^+/K^+ pump. Thus, the belief that Cl^- played no significant role in membrane physiology other than keeping electroneutrality, and the attention directed to the study of cations combined to cause research on Cl^- transport to be largely neglected for many years. This situation was eloquently described by Richard Keynes in the introductory remarks to a symposium titled "The Binding and Transport of Anions in Living Tissues" organized in 1982 (Keynes, 1982). When referring to a previous symposium that he organized in 1970 on active transport of salt and water, he wrote, "... on looking through the published proceedings, one gets the impression that all the time it was sodium that was in the limelight with the thick black arrows, while chloride was relegated to the indignity of being passively transported, and was shuffled off with dashed lines in an obscure corner of the diagram."

3. Passive and nonpassive Cl^- distribution across the plasma membrane

The Cl^- equilibrium potential, E_{Cl}, is defined by the Nernst equation:

$$E_{Cl} = \frac{RT}{zF}\ln\frac{a^i_{Cl^-}}{a^o_{Cl^-}} = \frac{RT}{zF}\ln\frac{\gamma_i[Cl^-]_i}{\gamma_o[Cl^-]_o} = \frac{RT}{zF}\ln\frac{[Cl^-]_i}{[Cl^-]_o} \tag{15.1}$$

Where a^i_{Cl} and a^i_{Cl} are the intracellular and extracellular Cl^- activities, γ_i and γ_o are the intracellular and extracellular activity coefficients, which are assumed to be equal, $[Cl^-]_i$ and $[Cl^-]_o$ are the intracellular and extracellular Cl^- concentrations, R is the gas constant, T is absolute temperature, F is the Faraday's constant, and z is the valence (=−1).

If Cl^- is *passively distributed,* E_m will have the same value as E_{Cl} and therefore:

$$E_m = \frac{RT}{zF}\ln\frac{[Cl^-]_i}{[Cl^-]_o} \tag{15.2}$$

From Eq. (15.2), the predicted $[Cl^-]_i$ at electrochemical equilibrium, $[Cl^-]_i^{eq}$, is given by:

$$[Cl^-]_i^{eq} = [Cl^-]_o e^{E_m F/RT} \tag{15.3}$$

Thus, Eq. (15.3) predicts the value that $[Cl^-]_i$ will reach when Cl^- is *passively* distributed across the plasma membrane. Obviously, this value will change with E_m, as illustrated in Fig. 15.1. For example, if $[Cl^-]_o$ is 113 mM and E_m is constant at −60 mV, at 37°C RT/F = 26.73 mV, and thus $[Cl^-]_i^{eq} \approx 12$ mM. Clearly, in this case, both E_m and E_{Cl} are −60 mV. If the cell is depolarized, say by 20 mV to −40 mV, Cl^- redistributes across the membrane until reaching equilibrium, which, for the example considered, will occur when $[Cl^-]_i^{eq} = 25.3$ mM, and thus $E_{Cl} = -40$ mV.

Another way of arriving at Eq. (15.2) and grasping its meaning is to consider once more the idealized cell shown in Fig. 15.1 that has only anion channels and Na^+/K^+pumps but no active (uphill) Cl^- transporters. In this cell, Cl^- distributes passively following E_m. The Cl^- flux through the anion channel is driven by the electrochemical potential gradient for Cl^-, $\Delta\mu_{Cl}$, which is the sum of a chemical and an electrical component. This is formally defined by Eq. (15.4):

$$\Delta\mu_{Cl} = RT\ln\frac{[Cl^-]_i}{[Cl^-]_o} + zFE_m \tag{15.4}$$

Since the cell is negative inside with respect to the outside (in the model shown $E_m = -60$ mV, and z, the Cl^- valence is −1), the electrical potential gradient will tend to drive Cl^- out of the cell. However, the chemical potential gradient is inward; $[Cl^-]_o$ is higher than $[Cl^-]_i$. Consequently, the chemical gradient will drive Cl^- toward the cell interior. The system reaches equilibrium when the two forces cancel each other (i.e., when they are equal and opposite). At that point $\Delta\mu_{Cl} = 0$. When this condition is met, we can readily derive Eq. (15.2) from Eq. (15.4). That is, when Cl^- is passively distributed, E_m will have the same value as E_{Cl}.

If Cl^- is *nonpassively distributed,* it means that it is *actively transported*, which implies energy consumption. Active Cl^- transport (i.e., transport against the Cl^- electrochemical equilibrium or "uphill" transport) requires demonstrating that $[Cl^-]_i$ is maintained at a level different from that predicted by Eq. (15.3), and this implies that E_{Cl} and E_m will have different values. In most cells, $[Cl^-]_i$ is maintained at a value different from that predicted from Eq. (15.3). However, as mentioned above, *equal values of* E_m *and* E_{Cl} *do not preclude the presence of carrier-mediated active* Cl^- *transport.* A clear example illustrating this notion is the case of skeletal muscle cells, which, as mentioned above (Section 5.2) and shown in Fig. 15.2, have a substantial channel-mediated (electrodiffusional) Cl^- permeability that masks the effects of an inwardly directed active Cl^- transport system (Aickin, 1990; Harris & Betz, 1987).

In most cell types of both vertebrates and invertebrates, in which Cl^- is not in electrochemical equilibrium, the $[Cl^-]_i$ is maintained at a *higher* value than that predicted for a passive distribution across the plasma membrane, that is, E_{Cl} has a more depolarized (less negative) value than E_m. This implies the presence of transport mechanisms that actively accumulate Cl^- often called "Cl^- loaders." Examples of mammalian cells in which $[Cl^-]_i$ is maintained above equilibrium include virtually all primary sensory neurons, sympathetic ganglion cells, immature central nervous system neurons, epithelial cells, leukocytes, and both smooth and cardiac muscle cells (Fig. 15.3). In contrast, in some cells, mature cortical neurons in particular, $[Cl^-]_i$ is maintained at a value *lower* than that predicted for a passive Cl^- distribution across the plasma membrane due to the presence of active Cl^- extrusion mechanisms (i.e., Cl^- extruders). In this case, E_{Cl} has a more hyperpolarized (more negative) value than E_m. This *lower* than passive Cl^- distribution in adult neurons in the CNS constitutes the basis for hyperpolarizing synaptic inhibition mediated by neurotransmitters such as GABA and glycine. A typical case is the hyperpolarizing inhibition produced by GABA acting via $GABA_A$ receptors,

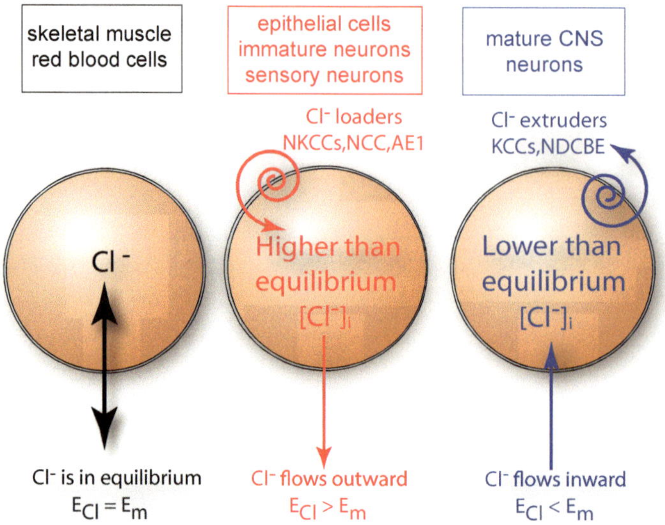

FIGURE 15.3 Control of cytosolic Cl^- in different cell types. Cl^- appears to be in electrochemical equilibrium in skeletal muscle and erythrocytes because these cells have large resting Cl^- permeabilities that mask active Cl^- transport. In epithelial cells, immature neurons, and adult sensory neurons, Cl^- loaders establish a higher than electrochemical equilibrium intracellular Cl^- concentration. In mature neurons, Cl^- extruders set a lower than electrochemical equilibrium intracellular Cl^- concentration. Examples of Cl^- loaders are: the Na^+, K^+, $2Cl^-$ cotransporters (NKCCs) like NKCC1, the Na^+-Cl^- cotransporter (NCC), and the anion exchanger 1 (AE1). Examples of Cl^- extruders are the K^+-Cl^- cotransporters, like KCC2, and the Na^+-driven Cl^-/HCO_3^- exchanger (NDCBE). *Modified from Hartzell (2009).*

resulting in the opening of Cl^- channels with consequent hyperpolarizing Cl^- influx. The ensuing hyperpolarization decreases the probability of neuronal firing of action potentials because of the transient shift in E_m away from the firing threshold. By controlling neuronal excitability, Cl^--mediated postsynaptic inhibition shapes the firing pattern of individual neurons and entire neuronal networks. Thus, this mechanism plays a central role in the processing of information in the nervous system. The carrier protein that actively extrudes Cl^- in mature CNS neurons, making possible hyperpolarizing inhibition, is one of four isoforms of the K^+-Cl^- cotransporters, KCC2, a member of the electroneutral cation-coupled chloride cotransporter gene/protein family SLC12, as discussed below (Section 6). Thus, KCC2 function (and NKCC1) *indirectly* affects neuronal excitability.

4. Active transport mechanisms for Cl^-

Intracellular Cl^- levels are determined by the relative contributions of the various Cl^- transport mechanisms expressed in the plasma membrane of a given cell type. Various mechanisms capable of transporting Cl^- across the plasma membrane have been identified in vertebrate and invertebrate cells. They include various kinds of *Cl^- channels* some of which behave as passive leak pathways, and a series of specialized *carriers that mediate* "active" Cl^- *transport*. Specifically, all Cl^- carrier proteins known in animal cells are *secondary active transport mechanisms*, and all of them are encoded by members of the solute carrier (SLC) gene superfamily (Pizzagalli et al., 2021). As we will see, secondary active transport does not directly utilize ATP hydrolysis to drive "uphill" solute transport. In animal cells there is no firm evidence for the existence of an active Cl^- pump, that is, a primary active transport mechanism *directly* energized by ATP.

Chloride channels are a functionally and structurally diverse group of anion-selective channels.

(See: https://www.guidetopharmacology.org/). They include the $GABA_A$ and glycine receptor channels referred to as ligand-gated Cl^- channels (Chapter 23), the voltage-sensitive ClC family that includes the ClC-1 expressed in skeletal muscle, the cystic fibrosis transmembrane conductance regulator (CFTR) involved in regular fluid transport across various epithelia (Chapter 17), and the volume regulated Cl^- channels which participate in cell volume regulation (Chapter 22). The carrier protein molecules that actively transport Cl^- include (1) the anion exchangers (AEs) and (2) the electroneutral cation-coupled chloride cotransporters. The AEs are dealt with in Chapter 16, whereas the electroneutral cation-coupled chloride cotransporters are the main focus of this chapter. The neurotransmitter transporters encoded by the SLC6 gene family of Na^+- and Cl^- -coupled neurotransmitter transporter family also transport Cl^- but are not involved in its regulation. An example of the latter is the GABA transporter GAT1 (SLC6A1). These transporters are located in the plasma membrane of neurons and astrocytes and couple the translocation of GABA with the dissipation of the electrochemical gradient for Na^+ and Cl^- (Nayak et al., 2023). The ion-substrate stoichiometry of GAT1 has been determined to be 2 Na^+:1 Cl^-:1 GABA, although other stoichiometries have been proposed (Goodspeed et al., 2020). The role of Cl^- in the transport process is not entirely understood, and it is unclear whether this Cl^- cotransport impacts $[Cl^-]_i$ or Cl^- is merely used as a cofactor (Wu et al., 2007). For this reason, neurotransmitter transport proteins are not considered further in the present chapter.

4.1 Anion exchangers

Anion exchangers are encoded by two evolutionarily unrelated mammalian gene families, **SLC4** and **SLC26** (Alper, 2009; Romero et al., 2009, 2013). Mark Parker discusses the function of these transport proteins in Chapter 16 of this volume. Plasmalemmal Cl^-/HCO_3^- exchangers are ubiquitously expressed in various cell types. Their primary function is the regulation of intracellular pH, but they also play key roles in regulating $[Cl^-]_i$ and cell volume. The Cl^-/HCO_3^- exchangers of polarized epithelial cells contribute to transepithelial secretion and reabsorption of acid-base equivalents and Cl^-. The mammalian SLC4 gene and protein family includes three Na^+-independent Cl^-/HCO_3^- exchangers (AE1, AE2, and AE3) that couple the electroneutral movement of 1 Cl^- in exchange for 1 HCO_3^- across plasma membranes. Under physiological conditions, most AEs of the SLC4 family behave as cell Cl^- loaders; they couple the inward movement of 1 Cl^- with the outward movement of 1 HCO_3^-. However, *reversibility is a key thermodynamic feature of carrier-mediated transport* proteins, and thus, solutes can be transported in either inward or outward direction under physiological conditions; these proteins use the chemical gradient of either one of the transported ions to move the other. The transport process of AE1, AE2, and AE3 is *electroneutral*. This means that the transport process occurs without net charge transfer across the membrane and without being affected by the membrane potential (see below). This contrasts with *electrogenic* transport, in which the driving force for the transport process is the sum of the chemical and electrical gradients of the transported ions. Thus, in this case, the transport process is affected by the membrane potential, and it generates a current across the membrane, that is, it transfers net charge across the membrane. The $Cl^-/HCO3^-$ exchanger SLC26A6 is an example of carrier electrogenicity and reversibility (Shcheynikov et al., 2006). This carrier protein mediates the coupled exchange of Cl^- and HCO_3^- with a transport stoichiometry of $1Cl^-/2HCO_3^-$ (Fig. 15.4).

One of the best examples of *transport reversibility of an electroneutral transporter* under physiological conditions is the erythroid Cl^-/HCO_3^- exchanger (AE1). This carrier is essential for removing CO_2 produced by tissue metabolism and transporting it to the lungs, where the gas can be expelled from the body. AE1 operates near equilibrium so that a relatively small variation in the chemical gradients of either Cl^- or HCO_3^- across the plasma membrane of red blood cells (RBCs)

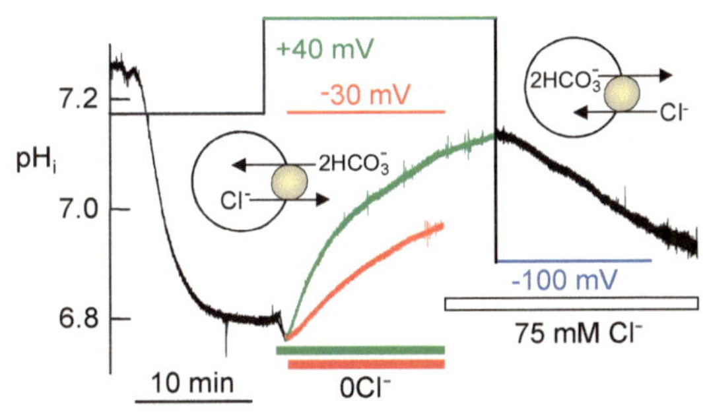

FIGURE 15.4 Electrogenicity and flux reversibility of the Cl^-/HCO_3^- exchanger SLC26A6 heterologously expressed in *Xenopus* oocytes. The membrane potential of transfected *Xenopus* oocytes was controlled by voltage clamp as indicated. Intracellular pH (pHi) changes were measured with pH-sensitive microelectrodes. Changes in $[Cl^-]_i$ were also measured (not shown) using Cl^--sensitive microelectrodes. The experiment shows the effect of membrane potential and external $[Cl^-]$ on HCO_3^- transport by SLC26A6. The HCO_3^- fluxes are inferred from the pH_i changes reflecting net H^+ fluxes. Accordingly, when HCO_3^- enters the cell in exchange for Cl^-, the cell interior is alkalinized (green and red traces), and when HCO_3^- exits the cell in exchange for Cl^-, the cell is acidified (black trace following green trace). The reversibility and direction of transport occurring during the experiment are illustrated in the inset diagrams. In this experiment, the *Xenopus* oocyte expressing SLC26A6 was bathed in HCO_3^-- buffered media. After the stabilization of pH_i, the membrane potential was clamped at −30 mV (red pH_i trace) or +40 mV (green pH_i trace) before exposing the oocyte to Cl^--free medium ($0Cl^-$). Note the effect of membrane potential on HCO_3^- influx upon removal of external Cl^- (green trace at +40 mV and red trace at −30 mV). At the time indicated by the white bar, external $[Cl^-]$ (75 mM) was restored, and after an additional period of 5 min, the membrane potential was switched from +40 mV to −100 mV. This resulted in intracellular acidification (black trace) due to HCO_3^- efflux. SLC26A6 mediates the coupled exchange of Cl^- and HCO_3^- with a transport stoichiometry of $1Cl^-/2HCO_3^-$. Accordingly, holding the membrane potential at +40 (green pH_i trace) and −100 mV (black trace) accelerated and inhibited, respectively, Cl^--mediated HCO_3^- influx, and holding the membrane potential at −100 mV increased HCO_3^--mediated Cl^- influx. *Modified from Shcheynikov et al. (2006).*

alters the direction of transport. This feature permits HCO_3^- formed from CO_2 in the systemic capillaries by the intracellular enzyme carbonic anhydrase to leave the RBCs in exchange for Cl^-, thus effectively storing CO_2 in the blood in the form of plasma HCO_3^-, greatly increasing the CO_2 carrying capacity of the blood. In the lungs, the process is reversed to regenerate CO_2, which is then exhaled. In other words, the relatively high external concentration of CO_2 generated by metabolism in tissues drives CO_2 entry into RBCs. CO_2 is then hydrated inside RBCs, a forward reaction catalyzed by the enzyme carbonic anhydrase ($CO_2 + H_2O \rightarrow H_2CO_3 \rightarrow H^+ + HCO_3^-$). The resulting increase in $[HCO_3^-]_i$ drives the Cl^-/HCO_3^- exchanger to operate in the bicarbonate-out-chloride-in mode. In the alveolar capillaries, the CO_2 concentration gradient is reversed because of the relatively low CO_2 concentration in the alveolar space, thereby driving CO_2 from the RBCs into the alveolus. This process is possible because the reactions catalyzed by carbonic anhydrase are also reversible, so this enzyme now catalyzes the net dehydration of H_2CO_3 ($H^+ + HCO_3^- \rightarrow H_2CO_3 \rightarrow H_2O + CO_2$). The consequent decrease in $[HCO_3^-]_i$ drives the Cl^-/HCO_3^- exchanger to operate in the bicarbonate-in-chloride-out mode.

Heart muscle cells are an interesting example in which $[Cl^-]_i$ is determined by the presence of electrogenic anion exchangers working in parallel with the electroneutral Cl^- loader NKCC1. Early studies using intracellular ion-selective microelectrodes demonstrated that $[Cl^-]_i$ is maintained at a *higher* value than predicted for a passive distribution across the sarcolemma of cardiac muscle cells (Vaughan-Jones, 1982). It is now known that a variety of sarcolemmal AEs, in conjunction with the Na^+-K^+-$2Cl^-$ cotransporter NKCC1, serve as uphill Cl^- accumulating mechanisms in cardiac cells. The anion exchangers predominantly expressed in adult heart muscle are AE3 (SLC4A3), SLC26A3, and particularly SLC26A6, all of which play a central role in the regulation of pH_i in cardiomyocytes (Alvarez et al., 2004). Both SLC26A3 and SLC26A6 function as dual Cl^-/OH^- and Cl^-/HCO_3^- exchangers. The transport stoichiometry of SLC26A6 is 2 HCO_3^-: 1 Cl^-, whereas that of SLC26A3 is 1 HCO_3^-:2Cl^- (Shcheynikov et al., 2006). These stoichiometries mean that both these SLC26 transporters are *electrogenic*; on each cycle, they transport net charge across the plasma membrane, thereby producing a change in membrane potential (Romero et al., 2009). Thus, their operation is also affected by the membrane potential (see Fig. 15.4). The "opposite" stoichiometry of SLC26A6 and SLC26A3 transporters means that their relative transport activity will vary with the membrane potential; SLC26A6 will operate to cause efflux of HCO_3^- at resting membrane potential (≈ -80 mV), while SLC26A3 will become active in the HCO_3^- efflux mode as the transmembrane voltage moves to more positive values during the cardiac action potential.

All these AEs (SLC4A3, SLC26A3, and SLC26A6) play key roles in the regulation of pH_i and $[Cl^-]_i$ in cardiac myocytes (Wang et al., 2014). Further, the outward Cl^- gradient of cardiac myocytes that is generated and maintained by the above-mentioned AEs in conjunction with the operation of NKCC1 is of high functional significance, as revealed by studies using double knockout mice in which both AE3 (SLC4A3) and NKCC1 (SLC12A2) are genetically deleted resulting in impaired cardiac contractility (Prasad et al., 2008). Numerous Cl^- channels are expressed in cardiac muscle (Duan, 2009, 2013). Their activation produces significant effects on cardiac action potential characteristics and pacemaker activity. The repertoire of cardiac Cl^- channels includes the cystic fibrosis transmembrane conductance regulator (CFTR), which is a member of the ATP-binding cassette (ABC) transporter superfamily and may be responsible for the Cl^- currents activated by protein kinase A (PKA), ClC-2, ClC-3, and the Ca^{2+} activated Cl^- channel TMEM16A. The resting E_m of cardiac myocytes is -80 to -90 mV, and during an action potential, the overshoot and plateau phases can be several millivolts positive. Given that E_{Cl} remains fairly constant, E_m can be either more positive or more negative than E_{Cl} during the normal cardiac action potential cycle. Thus, cardiac Cl^- channels can generate both outward and inward currents and cause both depolarization and repolarization during the action potential.

Anion exchangers, in parallel with NKCC1, also appear to play a role in intracellular Cl^- homeostasis in developing neurons in some CNS areas. For instance, in developing rat auditory brainstem neurons (Becker et al., 2003) and in embryonic chick motoneurons (Gonzalez-Islas et al., 2009), AEs contribute to the maintenance of $[Cl^-]_i$ at values higher than predicted for a passive distribution across the plasma membrane. The resulting outward Cl^- gradient underlies the *depolarizing* action of glycine and GABA, which is important in these and other central neurons in promoting neurogenesis, migration, and differentiation (Reynolds et al., 2008; Wang & Kriegstein, 2009).

The AE4 (SlC4A9) deserves special mention in the context of epithelial secretion. It is an electroneutral monovalent cation-dependent $Cl^-/HCO3^-$ exchanger that plays an essential role in exocrine gland fluid secretion by promoting, together with NKCC1, intracellular Cl^- accumulation in salivary glands secretory acinar cells (Peña-Münzenmayer et al., 2015). Both NKCC1 and AE4 are expressed in the basolateral membrane of these cells, maintaining $[Cl^-]_i$ above electrochemical equilibrium, thus making possible fluid secretion. Peña-Münzenmayer and coworkers have shown that mice salivary glands lacking the $HCO3^-$ transporter AE4 (Ae4−/−) exhibit impaired fluid secretion and a lower $[Cl^-]_i$ in their acinar cells. The same study revealed that the absence of AE2 does not impair salivation. These authors propose that the physiological role of AE4 in secretory cells is to promote Cl^- influx in exchange for K^+ (Na^+) and $HCO3^-$ ions (Peña-

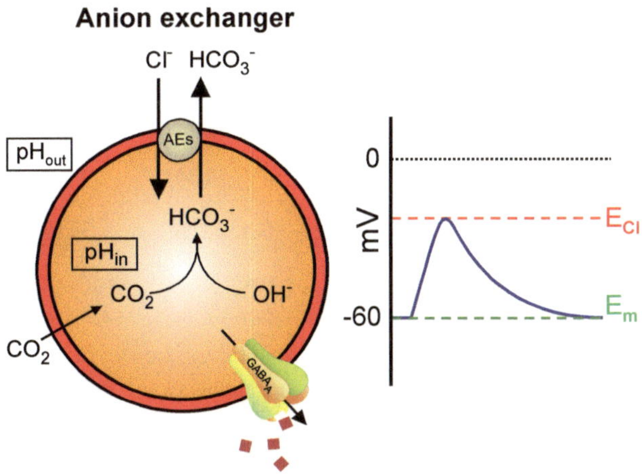

FIGURE 15.5 Model of a cell (e.g., a neuron) expressing only anion exchangers (AEs), Na^+/K^+ ATPase (not shown), and ligand-gated anion channels ($GABA_A$ receptor-channels). The ΔpH across the plasma membrane (pH_{in}-pH_{out}) is constant. Resting Em = – 60 mV. The AEs mediate the active accumulation of Cl^- in exchange for HCO_3^- (or CO_3^{2-}). Intracellular $[Cl^-]$ is maintained at higher than passive levels, i.e., higher than electrochemical equilibrium. Consequently, E_{Cl} is more positive than E_m. Opening of GABA-gated anion channels produces a depolarization mediated by electro-diffusional Cl^- efflux. The red squares represent GABA molecules acting on the $GABA_A$ receptor-channels. *Reproduced from Alvarez-Leefmans and Delpire (2009b).*

Münzenmayer et al., 2016). The cation transport pathway does not distinguish between Na^+ and K^+. Thus, AE4 is an electroneutral exchanger that transports Cl^- inwardly in exchange for monovalent cations in the same direction as $HCO3^-$.

How much Cl^- can a cell accumulate through electroneutral anion exchangers like AE1, AE2, or AE3? It depends on *the thermodynamic equilibrium conditions of the AEs*, on kinetic constraints dictated by mechanisms regulating ion translocation, and on the presence of other transporters and channels expressed in the same cell. For understanding the $[Cl^-]_i$ that would be reached when an anion exchanger attains thermodynamic equilibrium, consider the cell model shown in Fig. 15.5. Let us assume that the cell expresses only an electroneutral Cl^-/HCO_3^- exchanger with a stoichiometry of $1Cl^-:1\ HCO_3^-$ per cycle, that there are no kinetic constraints and that the only passive leak pathway is a GABA-gated Cl^- channel. Since AEs are secondary active transporters, the driving force for anion transport will be determined by the overall net free energy (ΔG) of the system, which, in this case, is the sum of the *chemical* potential gradients of the anions involved. Because the anions are transported in opposite directions, the combined chemical gradient ($\Delta\mu_{Cl/HCO_3}$) is given by the *difference* of the Cl^- and HCO_3^- chemical potential gradients $\Delta\mu_{Cl}$ and $\Delta\mu_{HCO_3}$, respectively:

$$\Delta G = \Delta\mu_{Cl/HCO_3} = \Delta\mu_{Cl} - \Delta\mu_{HCO_3} \tag{15.5}$$

Hence:

$$\Delta G = \Delta\mu_{Cl/HCO_3} = RT\ln\frac{[Cl^-]_i}{[Cl^-]_o} - RT\ln\frac{[HCO_3^-]_i}{[HCO_3^-]_o} \tag{15.6}$$

The transport process reaches thermodynamic equilibrium when $\Delta G = \Delta\mu_{Cl/HCO_3} = 0$, and Eq. (15.6) becomes:

$$\frac{[Cl^-]_i}{[Cl^-]_o} = \frac{[HCO_3^-]_i}{[HCO_3^-]_o} \tag{15.7}$$

Eq. (15.7) relates the intracellular and extracellular Cl^- and HCO_3^- concentrations when the Cl^-/HCO_3^- exchange mechanism has reached equilibrium, that is, when no further net movement of Cl^- and HCO_3^- can occur across the membrane. For the model shown in Fig. 15.5, $pH_o = 7.3$ and $pH_i = 7.1$ (i.e., $\Delta pH = 0.2$) and $[HCO_3^-]_o$ taken as 22 mM, the concentration measured in mammalian arterial plasma. In the presence of a CO_2/HCO_3^--buffered solution, the distribution of Cl^- will be determined by the distribution of HCO_3^- across the membrane. This is because CO_2 is freely permeable through the membrane, and at constant PCO_2 ($[CO_2]_i = [CO_2]_o$), the ratio of intracellular and extracellular $[HCO_3^-]$ is determined by ΔpH. Under these conditions E_H, the equilibrium potential for H^+, is equal to E_{HCO3^-}, the equilibrium potential for bicarbonate (Bevensee & Boron, 1998). In the model considered here, the cell possesses only an anion exchanger (Fig. 15.5). Thus, the $[Cl^-]_i$, when the anion exchanger reaches equilibrium, can be predicted by knowing $[Cl^-]_o$ and the ΔpH. Since we know ΔpH, we can calculate E_H, which is given by the Nernst equation:

$$E_H = \frac{RT}{F} \ln (pH_i - pH_o) \tag{15.8}$$

At 37°C, and taking $pH_o = 7.3$ and $pH_i = 7.1$, $E_H = -12.3$ mV. Since $E_H = E_{HCO_3}$, we can now calculate the $[HCO_3^-]_i$ at equilibrium using the following expression:

$$[HCO_3^-]_i = [HCO_3^-]_o \exp\left(\frac{E_{HCO3}F}{RT}\right) \tag{15.9a}$$

Hence,

$$[HCO_3^-]_i = 22 \text{ mM} \exp\left(\frac{-12.3 \text{ mVF}}{RT}\right) = 13.9 \text{ mM} \tag{15.9b}$$

We can now calculate $[Cl^-]_i$ when the Cl^-/HCO_3^- reaches equilibrium, using Eq. (15.10), which is derived from Eq. (15.7). For $[Cl^-]_o = 113$ mM,

$$[Cl^-]_i = \frac{[HCO_3^-]_i [Cl^-]_o}{[HCO_3^-]_o} = \frac{13.9 \times 113}{22} = 71.4 \text{ mM} \tag{15.10}$$

Thus, the maximal possible $[Cl^-]_i$ due to the AE under the conditions considered here would be 71.4 mM. Obviously, in a cell endowed with AEs, activation of $GABA_A$ receptors would produce a depolarizing Cl^- efflux (Fig. 15.5). For thermodynamic equilibrium conditions of other anion exchangers, the interested reader should consult the review of Romero and colleagues (Romero et al., 2009). It is important to mention that AEs may be kinetically inhibited at different set points for $[Cl^-]_i$, and thus, these "kinetic brakes" may prevent the exchanger from reaching thermodynamic equilibrium under physiological conditions (Alper et al., 2002; Kurschat et al., 2006). A key kinetic brake is pH itself (Alper, 2009; Alper et al., 2002). However, the discussion of the regulatory mechanism of AEs is beyond the scope of the present chapter.

AE1, AE2, and AE3 are often referred to as "Na^+-independent SLC4 anion exchangers" to distinguish them from the Na^+-driven Cl^-/HCO_3^- exchanger NDCBE, also an electroneutral transporter of the SLC4 family (SLC4A8) that under physiological conditions mediates the uptake of 1 Na^+ and the equivalent of 2 HCO_3^- (or one Na^+ plus one CO_3^{2-} or one $NaCO_3^-$ ion pair) in exchange for one Cl^- (Romero et al., 2013). Thus, NDCBE behaves typically as a Cl^- extruder, and because of its capacity to transport HCO_3^- inwardly, it is a major acid extruder. Although NDCBE mediates a large Cl^- efflux that can potentially affect $[Cl^-]_i$, its role in intracellular Cl^- regulation is little known (see below); instead, NDCBE appears to be a major pH_i regulator in several types of CNS neurons (Chen et al., 2008). In fact, NDCBE was the first transporter shown to be involved in pH_i regulation, a discovery made in the mid-1970s in squid giant axons by Walter Boron, Paul De Weer, and John Russell and by Roger C. Thomas working independently in snail neurons [for references see (Romero et al., 2004; Thomas, 1984):]. Human NDCBE was cloned by Boron's group (Grichtchenko et al., 2001). The cryo-EM structure of NDCBE has been recently described (Wang et al., 2021). The possible role of NDCBE in the regulation of $[Cl^-]_i$ has been addressed in certain types of mammalian neurons, the cartwheel cells of the dorsal cochlear nucleus. These neurons exhibit activity-dependent intracellular acidification that is offset by activation of NDCBE with a consequent hyperpolarizing shift in E_{Cl} due to a decrease in $[Cl^-]_i$ mediated by NDCBE (Kim & Trussell, 2009). Observations in neurons of the nematode *Caenorhabditis elegans* showed that NDCBEs function with KCCs to generate the neuronal inward Cl^- gradient necessary for the hyperpolarizing action of GABA and suggests a mechanism for the close tie between pH and excitability in the brain (Bellemer et al., 2011).

4.2 The electroneutral cation-chloride cotransporter family

The electroneutral cation-chloride cotransporter family (SLC12), which is the main focus of this chapter, comprises the major proteins responsible for intracellular Cl^- regulation (Arroyo et al., 2013; Di Fulvio & Alvarez-Leefmans, 2009; Gamba, 2005; Garbarini & Delpire, 2009; Payne, 2009; Trejo et al., 2023). The SLC12 gene/protein family encompasses nine members: SLC12A3 (the thiazide-sensitive Na^+-Cl^- cotransporter NCC), SLC12A2 (the Na^+-K^+-$2Cl^-$ cotransporter NKCC1), SLC12A1 (the Na^+-K^+-$2Cl^-$ cotransporter NKCC2), and four Na^+-independent K^+-Cl^- cotransporters, namely SLC12A4 (KCC1), SLC12A5 (KCC2), SLC123A6 (KCC3), SLC12A7 (KCC4), and two orphan members named

SLC12A9 (CIP) and SLC12A8 (CCC9) that do not appear to transport ions and whose function is little understood. Thus, only seven of the nine members of the SLC12 family have been established to function as cation-coupled Cl^- cotransporters. NKCC1, NKCC2 and NCC typically work as active Cl^- loaders, whereas the 4 KCCs (1–4) usually work as Cl^- extruders (Fig. 15.3).

As mentioned above, all of these cation-coupled Cl^- cotransporters and the Cl^- exchangers are *secondary active transport mechanisms*. In this kind of transport, the translocation of the substrate is coupled to the translocation of another substrate that uses the same carrier, whether in the opposite direction (counter-transport, antiport, or exchanger) or in the same direction (cotransport or symport). The energy consumed in this type of transport process comes secondarily or indirectly from ATP. In other words, in these transporters, the flow of the substrate that is being translocated up its electrochemical gradient is coupled to the flow of a second substrate down its electrochemical gradient; the gradient of this second substrate is maintained by a primary active transport mechanism, that is, one that consumes metabolic energy directly derived from the hydrolysis of ATP. In the case of cation-coupled Cl^- cotransporters, the primary active transport mechanism that generates and maintains the Na^+ and K^+ gradients is the Na^+/K^+ pump, also known as the Na^+/K^+ ATPase. For example, let us consider the Na^+-Cl^- cotransporter (NCC). The uphill movement of Cl^- carried out by this electroneutral cotransporter system is driven primarily by the energy stored in the Na^+ chemical potential gradient, which in turn is maintained by the ATP-requiring Na^+/K^+ pump. The Na^+ that enters the cell is pumped back out of the cell via the Na^+/K^+ pump. Hence, there is no net gain of intracellular Na^+. In essence, energy is directly invested into a primary active transport mechanism (the Na^+/K^+ pump) that is responsible for extruding Na^+ from the cell (in exchange for K^+) and thereby maintaining a low and constant intracellular Na^+ concentration that results in a steep Na^+ inward gradient. The Na^+-Cl^- cotransport mechanism can then convey the uphill movement of Cl^- energized by the downhill flow of Na^+.

5. Electroneutral Na^+-K^+-Cl^- cotransporters

5.1 Isoforms and splice variants: An overview of their functional roles

The Na^+-K^+-Cl^- cotransporters (NKCCs) are a class of membrane proteins that mediate the coupled, electrically neutral movement of Na^+, K^+, and Cl^- ions across the membrane of many animal cells, in most cases with a stoichiometry of 1Na:1K:2Cl (Russell, 2000). Under normal physiological conditions, NKCCs work as active ("uphill") transport mechanisms for accumulating Cl^- in cells. NKCCs maintain $[Cl^-]_i$ at levels above the predicted electrochemical equilibrium, generating an outward Cl^- gradient across the plasma membranes of many cell types. The higher than equilibrium $[Cl^-]_i$ is used in multifarious functions, depending on the cell type in which NKCCs are expressed. Some examples of these functions are given below and in Section 5.5.

Two distinct but highly homologous isoforms of the Na^+-K^+-Cl^- cotransporter proteins (NKCC1 and NKCC2) have been identified by cDNA cloning and functional expression (Delpire et al., 1994; Gamba et al., 1994; Payne & Forbush, 1994; Xu et al., 1994; Payne & Forbush, 1995). The two genes encoding Na^+-K^+-Cl^- cotransporter are members of the SLC12 family and are known as SLC12A1 and SLC12A2. The SLC12A1 gene encodes NKCC2, often referred to as "the renal Na^+-K^+-Cl^- cotransporter" because it is predominantly expressed in the apical membrane of the epithelial cells that form the thick ascending limb of Henle's loop (TALH) and the macula densa (Castrop & Schießl, 2014). The SLC12A2 gene encodes the ubiquitously expressed NKCC1, a cotransporter that is found in nearly all cell types, whether epithelial or not. In the vast majority of secretory epithelial cells, NKCC1 expression is restricted to the basolateral membrane, together with the Na^+/K^+-ATPase (Haas & Forbush, 2000), as illustrated in Fig. 15.14A. A significant exception is the choroid plexus epithelium, where both NKCC1 (Gregoriades et al., 2019; Piechotta et al., 2002) and the Na^+/K^+ ATPase (Baasch Christensen et al., 2021; Quinton et al., 1973) reside in the apical membrane as illustrated in Fig. 15.17B. A similar case occurs in the retinal pigment epithelium (RPE), where the Na^+/K^+-ATPase is also located in the apical membrane (Gundersen et al., 1991; Lobato-Álvarez et al., 2016) together with a Na^+, K^+, $2Cl^-$ mechanism that most likely is NKCC1 (Reichhart & Strauss, 2014; Reichhart & Strauß, 2020). However, in the RPE, the evidence comes from functional studies, but the molecular identity and exact location of this apical NKCC1 have not been established.

The apical NKCC1 in the choroid plexus epithelium functions in the inward mode, maintaining higher than equilibrium $[Cl^-]_i$ and water volume of these epithelial cells needed for CSF secretion while absorbing K^+ from the CSF toward the blood (Gregoriades et al., 2019; Wu et al., 1998). Thus, apical NKCC1 in this epithelium appears to function not only to maintain choroid plexus epithelial cell volume but also as a regulator of CSF $[K^+]$ (Alvarez-Leefmans, 2020a; Fame et al., 2023; Gregoriades et al., 2019; Xu et al., 2021). CSF $[K^+]$ must be tightly regulated and maintained within 3 mM, as in the brain interstitial fluid, to sustain normal excitability and brain cell volume. In the retinal epithelial cells, apical NKCC1 also appears to have an extracellular K^+ buffering function. Molecular mechanisms involved in buffering extracellular K^+

concentration, $[K^+]_o$, in the subretinal space (SRS) are critical for photoreceptor function. Several transporters and channels expressed in the apical membrane of retinal pigment epithelial cells are involved in extracellular K^+ buffering; they include the Na^+/K^+ pump, NKCC1, and the inward rectifier K^+ channels Kir7 (Kumar & Pattnaik, 2014; Reichhart & Strauβ, 2020). Interestingly, the other epithelium in which an NKCC is located in the apical membrane is the thick ascending limb of the loop of Henle in the kidney. This apical NKCC is NKCC2, the other isoform of the NKCCs, which also works in the inward mode, reuptaking 20%–25% of the filtered NaCl. Besides the chemical gradients of Na^+ and Cl^-, the driving force for NKCC2-dependent salt transport is maintained by the activity of the basolateral Na^+/K^+ ATPase (Fig.15.14B). Apical NKCC2 is the target of loop diuretics, and it does not appear to cotransport water as NKCC1 does (Zeuthen & Macaulay, 2012). Thus, no water is reabsorbed in this portion of the nephron due to the lack of paracellular or transcellular water permeability (Castrop & Schieβl, 2014).

The high $[Cl^-]_i$ maintained by NKCC1 is used in polarized epithelial cells to promote net salt and water transport (Haas & Forbush, 2000). A good example is the salivary glands, in which AE4, together with NKCC1, is the major Cl^- uptake mechanism across the basolateral membrane of acinar cells, a mechanism critical for driving saliva secretion in vivo. Mice with targeted disruption of the *Slc12a2* gene encoding NKCC1 exhibit a dramatic reduction (>60%) in the volume of saliva secreted in response to a muscarinic agonist, the primary in situ salivation signal (Evans et al., 2000).

Besides epithelial cells, NKCC1 is expressed in many nonpolarized cells, such as neurons (Delpire & Gagnon, 2018a). By virtue of the outward Cl^- gradient that NKCC1 generates and maintains, it makes possible the depolarizing action of neurotransmitters that open Cl^- channels, such as GABA and glycine, or second messengers, such as Ca^{2+}, that open Ca^{2+}-activated Cl^- channels. For instance, the GABA-induced depolarizations made possible by NKCC1 in the intraspinal terminals of primary sensory neurons constitute the basis for the phenomenon of presynaptic inhibition in the spinal cord (Alvarez-Leefmans, 2009; Alvarez-Leefmans et al., 1998), as illustrated in Fig. 15.15 and discussed below in more detail (Section 5.5). This GABAergic presynaptic control is a crucial mechanism in the gating of the afferent flow of information of various sensory modalities, including proprioception and pain into the spinal cord (Alvarez-Leefmans, 2009; Carlton, 2014; Guo & Hu, 2014; Rudomin & Schmidt, 1999; Willis, 1999, 2006). NKCC1 is also involved in sensory transduction. Two well-documented examples discussed below (Section 5.5) are the inner ear and the olfactory sensory neurons.

NKCCs play key roles in cell volume regulation and maintenance, as discussed below. There is evidence that in some cell types, NKCCs function to offset osmotically induced cell shrinkage by mediating the net influx of ions and water (Russell, 2009). Furthermore, NKCC1 serves to maintain cell volume under euvolemic conditions (Gregoriades et al., 2019; Rocha-Gonzalez et al., 2008), and in addition to ions, it cotransports water (Hamann et al., 2010; Ruiz Munevar et al., 2024; Zeuthen & Macaulay, 2012). NKCC1 function is also important in cell cycle and cell division (Russell, 2009). Overexpression of the NKCC1 cotransporter gene has been shown to induce cell proliferation and phenotypic transformation in mouse fibroblasts (Panet et al., 2000).

The introduction of knockout mice models in biomedical research significantly advanced our understanding of the function of cation-chloride cotransporters and many other proteins. Knockout mice are transgenic animals that contain a modification in a specific gene that makes the gene product (i.e., the protein) nonfunctional. The functional roles of various members of the SLC12 gene/protein family have been studied using knockout mice models (Garbarini & Delpire, 2009) or animals or humans with mutations in specific SLC12 genes. NKCC1-knockout mice exhibit several abnormalities. The most salient are deafness, imbalance, hypotension, and increased nociceptive threshold, suggesting NKCC1 involvement in pain perception, severe salivation impairment, and male infertility (Delpire et al., 1999; Evans et al., 2000; Flagella et al., 1999). More recently, diseases linked to NKCC1 dysfunction in humans have been studied (Delpire & Koumangoye, 2023). NKCC1 deficiency in young patients leads to clinical presentations that overlap with some of the phenotypes observed in NKCC1 knockout.

The SLC12A1 gene encodes NKCC2, which, as mentioned above, is predominantly expressed in kidney, specifically in the apical membrane and in subapical vesicles of the epithelial cells of the thick ascending limb of the loop of Henle (TALH) and the macula densa (Nielsen et al., 1998). NKCC2-dependent salt transport constitutes the major apical entry pathway for transepithelial salt reabsorption in the TALH (Castrop & Schieβl, 2014). To achieve net salt absorption, electroneutral cotransport of Na^+, K^+, and $2Cl^-$ via NKCC2 is complemented by recycling of K^+ via apical K^+ channels (the so-called ROMK), efflux of Cl^- via basolateral Cl^- channels, and basolateral extrusion of Na^+ by the Na^+/K^+-ATPase which in addition generates the inward Na^+ gradient necessary to energize the cotransport (Fig. 15.14B). Pharmacological blockade of NKCC2 by loop diuretics like bumetanide results in marked natriuresis, diuresis, and a decrease in blood pressure. This is the rationale for using loop diuretics in the treatment of high blood pressure.

Besides carrying out transcellular absorption of Na^+ and Cl^- by the medullary and cortical TALH, NKCC2 plays a secondary role in the paracellular transport of Na^+, Ca^{2+}, and Mg^{2+} by this nephron segment (Gamba & Friedman, 2009). The TALH also functions in renal acid secretion, given that NH_4^+ competes with K^+ at its binding site on the NKCC2

(Good, 1994; Weiner & Verlander, 2011). The NH_4^+ generated by the renal proximal tubule is thus reabsorbed from the tubule lumen in the TALH and excreted from the renal interstitium by more distal nephron segments.

In addition to its function in TALH salt retrieval, NKCC2-mediated transport constitutes the first step in the tubulo-vascular signaling pathway between macula densa cells in the juxtaglomerular region with the afferent arteriole. As mentioned, NKCC2 is also located in the apical membrane of macula densa cells, a specialized group of tubular cells situated at the point of apposition of renal tubules with their parent glomeruli. Macula densa cells are modified epithelial cells that sense luminal Na^+ and Cl^- concentrations and generate paracrine chemical signals to the glomerular vascular elements. Thus, these cells control two important physiological processes: tubuloglomerular feedback and tubular regulation of renin release. The tubuloglomerular feedback mechanism plays a crucial role in regulating glomerular filtration rate and blood flow. An increase in luminal Na^+ and Cl^- concentrations at the macula densa decreases glomerular filtration by inducing constriction of the afferent renal arteriole. Increases in luminal Na^+-Cl^- activity at the macula densa also inhibit renin release from juxtaglomerular cells in the afferent arteriole. Thus, NKCC2 plays a central role in the regulation of body fluid volume and osmolarity (Bell et al., 2009; Caceres & Ortiz, 2019; Peti-Peterdi & Harris, 2010).

Although NKCC2 is encoded by a single gene, differential splicing results in the generation of several alternative transcripts. Specifically, three full-length NKCC2 splice variants (NKCC2A, NKCC2B, and NKCC2F) and three variants with truncated carboxyl termini have been described. They differ in their location along the TALH and the macula densa and have different kinetic properties. Analysis of the molecular and functional properties of individual splice variants is beyond the scope of this chapter. References on this subject can be found in excellent reviews (Gamba, 2005, 2013) and in studies reported by the group of Castrop (Carota et al., 2010; Castrop & Schieβl, 2014; Castrop & Schnermann, 2008; Oppermann et al., 2006, 2007).

Inactivating homozygous mutations of the human NKCC2 gene product cause type 1 Barter's syndrome, a salt-wasting disease state characterized by hypotension, metabolic alkalosis with hypokalemia, secondary hyperaldosteronism, polyhydramnios, nephrocalcinosis, and hyposthenuria (Cunha & Heilberg, 2018). The vital function of NKCC2 proteins is also manifest in various knockout mice models. Targeted disruption of the promoter and the first three exons of mouse NKCC2 gene results in a full NKCC2 knockout with a phenotype sharing many similarities to Bartter's syndrome (Takahashi et al., 2000).

5.2 Basic functional features of the Na^+-K^+-Cl^- cotransporters

Four functional features define the NKCCs.

1. Ion translocation by the NKCCs requires the simultaneous presence of all three ions (Na^+, K^+, and Cl^-) on the same side of the membrane.
2. The cotransport process is electroneutral; in vertebrate cells the stoichiometry is $1Na^+$:$1K^+$:$2Cl^-$ in nearly all cases, as discussed above in Section 2.
3. The transporter can work in forward and backward modes, i.e., net transport may occur into or out of the cells; the magnitude and direction of this transport is determined by the sum of the chemical potential gradients of the transported ions (Lytle et al., 1998).
4. Transport is inhibited by *loop diuretics* derived from 5-sulfamoylbenzoic acid, which include furosemide, bumetanide, and benzmetanide, as shown in Fig. 15.11 (O'Grady et al., 1987b; Hannaert et al., 2002; Zhao et al., 2022a)

5.2.1 Absolute requirement for all three ions on the same side of the membrane

NKCC-mediated fluxes of all three ions require the presence of the other two ions on the side of the membrane from which the flux originates (termed the *cis*-side). This property has been definitely proved by measuring individual fluxes of Na^+, K^+, and Cl^- and looking at their interdependence. Perhaps the most stringent test for this coupling of ion fluxes has been carried out in internally dialyzed squid giant axons. This preparation has the unique advantage of permitting the simultaneous control of the composition of the intracellular as well as the extracellular milieu while measuring unidirectional fluxes. Using this preparation, Russell and his coworkers demonstrated the absolute requirement for the *cis*-side presence of all three ions for both unidirectional influx and efflux (Russell, 2000). Each of the three co-ions was systematically removed while measuring the bumetanide-sensitive influx of the other two, using a double-label flux approach. This is illustrated in Fig. 15.6A, which shows the effects on ^{36}Cl (yellow dots) and ^{24}Na (green squares) influx when an axon was sequentially superfused with external solutions without K^+ (0 K^+), with K^+ (10 mM), and finally with K^+ (10 mM) plus bumetanide (10 μM). Clearly, the presence of extracellular K^+ increased the influx of both Cl^- (by ~21 $pmol\ cm^{-2}\ s^{-1}$) and Na^+ (by ~14 $pmol\ cm^{-2}\ s^{-1}$), and these increases were reversed by blocking the cotransporter with bumetanide. A similar protocol

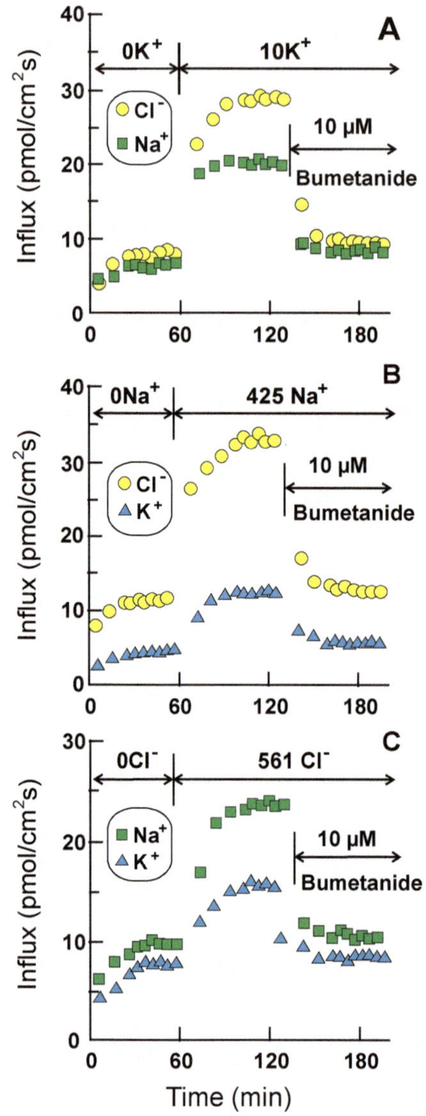

FIGURE 15.6 Dependence of cotransporter influxes of Na^+, K^+, and Cl^- on the simultaneous presence of all three ions in the external fluid. Experiments were carried out on internally dialyzed squid axons. The internal fluid in all cases contained the following (in mM): 400 K^+, 0 Na^+, 0 Cl^-,8 Mg^{2+}, and 4 ATP. Ouabain and tetrodotoxin were present to block the Na^+/K^+ pump and Na^+ channels, respectively. (A) Effect of external K^+ on $^{24}Na^+$ and $^{36}Cl^-$ influxes. After a control period (0 K^+) in which the external K^+ was replaced with N-methyl-D-glucammonium (NMDG), 10 mM K^+ was applied, and influxes of both $^{24}Na^+$ (green squares) and $^{36}Cl^-$ (yellow circles) increased. Finally, bumetanide (10 μM) was added, and both influxes decreased to levels similar to those seen in the absence of external K^+. (B) Effect of external Na^+ on $^{42}K^+$ (blue triangles) and $^{36}Cl^-$ (yellow circles) influxes. After a control period in which the external Na^+ was removed and replaced with NMDG (0 Na^+), 425 mM Na^+ was applied, leading to increases in both $^{42}K^+$ and $^{36}Cl^-$ influxes. Subsequent application of bumetanide caused influxes to decrease to values close to those observed in 0 Na^+. (C) Effect of external Cl^- on $^{42}K^+$ (blue triangles) and $^{24}Na^+$ (green squares) influxes. In the control condition (0 Cl^-), gluconate substituted for Cl^-. Note that upon admission of the external solution containing 561 mM Cl^-, both $^{42}K^+$ and $^{24}Na^+$ increased. Subsequent application of bumetanide caused influxes to decrease to levels near those observed in the absence of external Cl^-. *Adapted from Russell (2000).*

was followed to examine the requirement of external Na^+ (Fig. 15.6B) and Cl^- (Fig. 15.6C). Note that in each case, the increase in the influx of the two co-ions caused by supplying the third one is reversed by bumetanide treatment.

The absolute requirement for the *cis*-side presence of Na^+, K^+, and Cl^- for the uphill accumulation of Cl^- mediated by NKCC was first demonstrated in vertebrate cells using different methodologies. Examples include Ehrlich ascites tumor

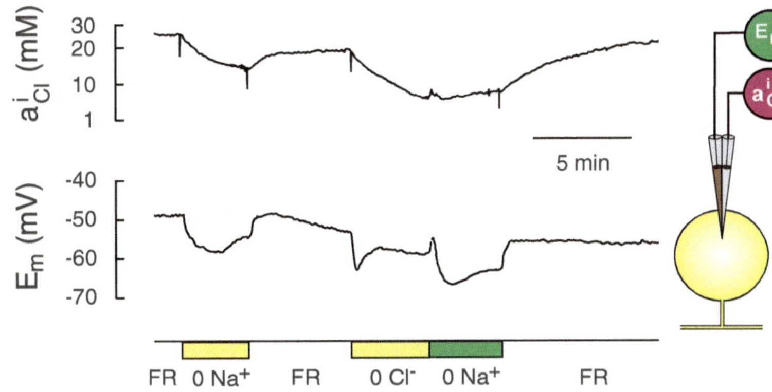

FIGURE 15.7 Effect of external Na^+ removal on intracellular Cl^- activity (a^i_{Cl}) and membrane potential (E_m) in a frog dorsal root ganglion neuron. The recording was made with a double-barreled microelectrode, one barrel sensing E_m and the other a^i_{Cl}. Inset on the right is a diagram of a double-barreled microelectrode where a^i_{Cl} (redish-purple circle) denotes the Cl^--selective barrel filled with a liquid anion exchanger, which senses a^i_{Cl}, and Em (green circle) denotes the barrel that records the membrane potential. When the microelectrode is inside the cell, the a^i_{Cl} barrel records the sum of a potential proportional to a^i_{Cl} plus E_m. To obtain a^i_{Cl}, the potential on the E_m barrel is subtracted from that recorded by the Cl^- sensitive barrel, which gives a differential voltage signal that is proportional to a^i_{Cl} (top trace). During the times indicated by the bottom bars, the bathing solution was changed from frog Ringer (FR) to a Na^+-free solution (0 Na^+), in which external Na^+ was replaced with N-methyl-D-glucammonium, and then back to FR. Note that the removal of external Na^+ led to a decrease in a^i_{Cl}. The cell was then exposed to Cl^- -free solution (0 Cl^-), which also produced a decrease in a^i_{Cl}. Then, the reaccumulation of internal Cl^- was monitored in the absence of external Na^+ (0 Na^+) and then in FR, i.e., in the presence of external Na^+. Note that in the absence of external Na^+, active Cl^- uptake was inhibited. The hyperpolarizations recorded in the 0Na^+ solutions (E_m trace) are due to the suppression of a persistent Na^+ inward current. *Modified from Alvarez-Leefmans et al. (1988).*

cells (Geck et al., 1980), Madin–Darby canine kidney (MDCK) cell line (McRoberts et al., 1982); duck red cells (Haas et al., 1982); frog and rat dorsal root ganglion (DRG) neurons (Alvarez-Leefmans et al., 1988; Rocha-Gonzalez et al., 2008); immature rat neocortical neurons (Achilles et al., 2007); and other cell types (Russell, 2000). In frog sensory neurons, measurements were made using double-barreled intracellular microelectrodes, one of which barrels contained a reference electrolyte to measure membrane potential (E_m) and the other containing a liquid anion exchanger (the Cl^--sensitive barrel) to measure intracellular Cl^- activity (a^i_{Cl}), which was then transformed into intracellular chloride concentration, $[Cl^-]_i$, by dividing it by the Cl^- activity coefficient (Alvarez-Leefmans et al., 1988, 1990). Measurements of changes in a^i_{Cl} with intracellular microelectrodes reflect *net* Cl^- *fluxes* occurring through the plasma membrane. Fig. 15.7 shows an example of one experiment in which the cell body of a single neuron was penetrated with a double-barreled microelectrode (inset). The advantages of this method are evident; it is possible to measure simultaneously intracellular a^i_{Cl} and E_m in single cells. Moreover, in contrast with methods that measure total $[Cl^-]_i$, which includes free-plus-bound or sequestered Cl^-, the microelectrodes give a measurement of free Cl^- in the cytosolic space (Alvarez-Leefmans et al., 1990). The dependence on extracellular Na^+ and K^+ of the bumetanide-sensitive active accumulation of Cl^- was explored in these cells. To this end, the effects of extracellular Na^+- or K^+-free isosmotic solutions on a^i_{Cl} were studied in the steady state and during the active (uphill) Cl^- reaccumulation following exposure to Cl^--free extracellular solutions. A typical result showing the effects of external Na^+ removal on a^i_{Cl} is shown in Fig. 15.7. The basal a^i_{Cl} (upper trace) and E_m (lower trace) were recorded in the standard frog Ringer's solution (FR) for about 3 min after impalement. The initial E_m was ~ -50 mV and a^i_{Cl} was 25 mM. The latter value was twice the a^i_{Cl} predicted for an equilibrium distribution according to Eq. (15.3). It is important to mention that the measured a^i_{Cl} is equivalent to an intracellular $[Cl^-]$ of $\sim$33 mM, considering 0.76 as the intracellular activity coefficient. Then, the cell was exposed for about 3.5 min to a solution in which Na^+ was eliminated (0 Na^+) and replaced mole by mole with the impermeant cation N-methyl-D-glucammonium, keeping a^o_{Cl} at the same level as in the FR solution. This resulted in an immediate decrease in a^i_{Cl} accompanied by a membrane hyperpolarization. The latter was due to suppression of a resting Na^+ conductance, also known as background Na^+-leak conductance (Ren, 2011). Both E_m and a^i_{Cl} returned to control levels upon readmission of FR solution to the bath. Observations like these indicate that external Na^+ is required to maintain internal Cl^- above electrochemical equilibrium in the resting state.

To explore if external Na^+ is required for the active reaccumulation of Cl^- after intracellular Cl^- depletion, external Cl^- was removed (0 Cl^-) and replaced mole by mole with the impermeant anion gluconate, keeping the other ions as in control FR. This also resulted in a decrease in a^i_{Cl}. The reaccumulation of Cl^- was then observed in the Na^+-free solution containing normal Cl^- (0 Na^+) and finally in standard FR. It was found that the rate of Cl^- reaccumulation in the Na^+-free solution was slowed to about 15% of its value in standard FR solution.

These observations indicate that extracellular Na^+ is required for active reaccumulation of Cl^- following cell Cl^- depletion. The concurrent changes in E_m and calculated values for E_{Cl} in the same experiment are shown in Fig. 15.8A, which illustrates more clearly the effects of external Na^+ on the distribution of Cl^- across the cell membrane. External Na^+ removal reduced the basal intracellular accumulation of Cl^-, as reflected by the reduction in the difference between E_m (blue filled circles) and E_{Cl} (red filled circles). In addition, Na^+ removal after Cl^- depletion impeded the reaccumulation of Cl^- even though E_{Cl} was now more negative than E_m, a situation in which the direction of the Cl^- gradient was reversed, favoring Cl^- influx. Therefore, it can be concluded that external Na^+ is required to accumulate Cl^- above equilibrium and to keep it out of equilibrium in basal conditions.

The effects of extracellular K^+ on a^i_{Cl} were similar to those observed for Na^+. Fig. 15.8B shows a plot of E_m and E_{Cl} from an experiment similar to that shown in Fig. 15.7 in which external K^+, instead of external Na^+, was removed during the periods indicated. Removal of external K^+ was accompanied by an immediate reduction in a^i_{Cl} revealed in the graph by the changes in E_{Cl} toward more negative values. In the absence of extracellular K^+, the reaccumulation of Cl^- in standard FR solution was impeded. Summarizing, these experiments show that the steady-state a^i_{Cl} levels in DRG cells depend on the simultaneous presence of extracellular Na^+ and K^+. Similarly, the active reaccumulation of Cl^- after intracellular Cl^- depletion requires the simultaneous presence of Na^+ , K^+ and Cl^- in the extracellular medium.

Similar results to those described above for frog sensory neurons have been obtained in rat DRG neurons, which are also characteristically endowed with NKCC1 (Alvarez-Leefmans et al., 2001; Plotkin et al., 1997). Mammalian neurons

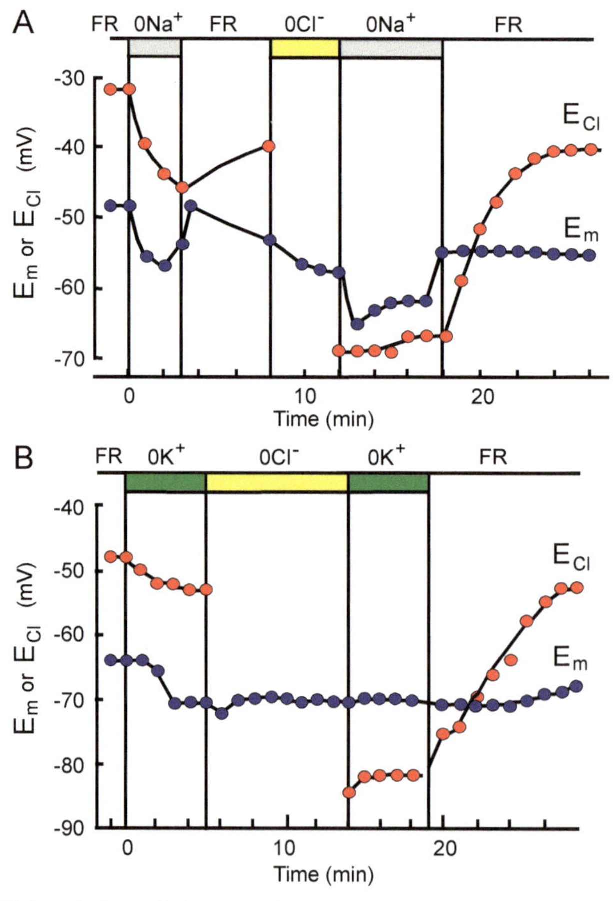

FIGURE 15.8 (A) Effects of external Na^+ on the intracellular accumulation of Cl^- in frog DRG neurons. Membrane potential (E_m, blue dots) and the calculated Cl^- equilibrium potential (E_{Cl}, red dots) from measured a^i_{Cl} and a^o_{Cl} are plotted on the same time scale using data from the experiment illustrated in Fig. 15.7. The E_m hyperpolarizations recorded in the 0Na^+ solutions are due to the suppression of a persistent Na^+ inward current. (B) Effects of external K^+ on the intracellular accumulation of Cl^-. Symbols as in (A). Data were obtained from an experiment similar to that shown in Fig. 15.7, in which external K^+ was removed instead of external Na^+. *Modified from Alvarez-Leefmans et al. (1988).*

are too delicate to stand penetration with microelectrodes, and thus, a noninvasive procedure to measure changes in $[Cl^-]_i$, such as those discussed in Section 1, was needed. The strategy used in rat DRG neurons was to study the external ion requirements of native NKCC1 by measuring changes in $[Cl^-]_i$ using the fluorescent dye MQAE (Rocha-Gonzalez et al., 2008).

5.2.2 Electroneutral cotransport process: Stoichiometry of $1Na^+:1K^+:2Cl^-$

Geck and coworkers, working with Ehrlich ascites tumor cells, provided the first comprehensive evidence for the existence of the electrically neutral cotransport of Na^+, K^+, and Cl^- ions, with a stoichiometry of $1Na^+:1K^+:2Cl^-$ (Geck et al., 1980; Geck & Heinz, 1986). A solidly documented exception is the squid axon cotransporter, which, as mentioned (Section 2), has a stoichiometry of $2Na^+:1K^+:3Cl^-$ (Russell, 2000). This different stoichiometry probably reflects differences in molecular structure with respect to the vertebrate isoforms. The $1Na^+:1K^+:2Cl^-$ stoichiometry has been found in many vertebrate cells (O'Grady et al., 1987a; Haas, 1994; Russell, 2000; Hoffmann, 2001; Lytle, 2003; Flatman, 2004; Gamba, 2005). Consistent with the 1:1:2 stoichiometry are recent structural studies using Cryo-EM, showing the detailed atomic maps of each NKCC1 protomer exhibiting one Na^+, one K^+, and two Cl^- binding sites (Chew et al., 2019; Delpire & Guo, 2020; Neumann et al., 2022). For a recent review of these structural studies, see (Trejo et al., 2023).

There is general agreement that the NKCC is electrically silent, which means that the overall transport process of Na^+, K^+, and Cl^- via the cotransporter is not driven by the transmembrane voltage, nor does the transport process directly generate a membrane current that may affect the transmembrane voltage. As pointed out by Russell (2000), the electrical silence of the NKCC is well accepted by workers in the field, yet direct evidence for the assertion is surprisingly sparse. Electrical silence is often inferred from the apparent stoichiometry of the cotransport process, but arriving at this conclusion involves a somewhat circular argument.

Conclusive proof of electrical silence requires direct and simultaneous measurement of transmembrane potential and cotransporter-mediated ion fluxes. There are only two reports in the literature fulfilling these requirements, one for the squid axon NKCC cotransporter (Russell, 1984) and one for amphibian dorsal root ganglion neurons (Alvarez-Leefmans et al., 1988). The latter remains the only existing direct evidence for the electroneutrality of Cl^- transport through NKCC in a vertebrate cell. Fig. 15.9 shows an example of one of these experiments in which a frog DRG neuron was penetrated with a double-barreled microelectrode (inset). At the time indicated by the downward orange arrow, the cell was penetrated with the microelectrode, and E_m stabilized at about −55 mV. The deflection in the differential signal (upper trace) at this time corresponded to an *apparent* intracellular Cl^- activity $(a^i_{Cl})_{app}$ of about 40 mM, which, after correction for background intracellular interference produced by other native intracellular anions on microelectrode response (Alvarez-Leefmans,

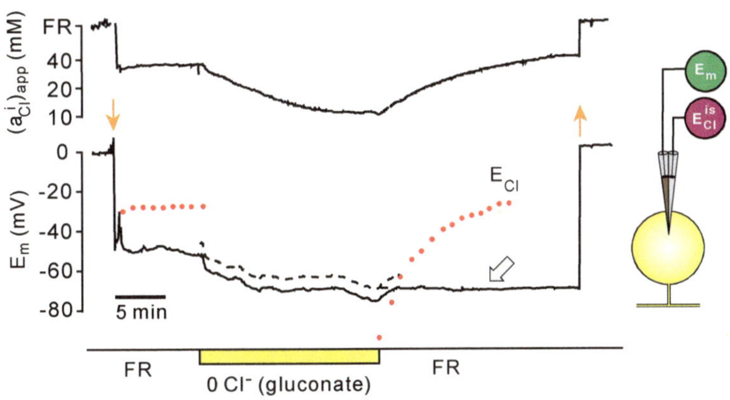

FIGURE 15.9 Electrically silent transmembrane movements of Cl^- mediated by Na^+, K^+, Cl^- cotransporter 1 (NKCC1) in a frog dorsal root ganglion (DRG) neuron. The intracellular Cl^- activity, a^i_{Cl}, was measured with a double-barrel Cl^--selective microelectrode in the neuronal cell body (diagram on the right), as explained in Fig. 15.7. After obtaining stable readings, the effects of removing and re-adding external Cl^- on a^i_{Cl} and E_m were recorded. The upper trace shows the differential signal (E_m-E^{is}_{Cl}), where E^{is}_{Cl} is the potential recorded through the ion-sensitive barrel. The differential signal indicates $(a^i_{Cl})_{app}$, the apparent intracellular Cl^- activity. The lower trace is the membrane potential (E_m). The dashed trace above a segment of E_m trace is the E_m value after the subtraction of liquid junction potentials. At the periods indicated by the yellow bar at the bottom, the bathing solution was changed from standard "frog Ringer" (FR) to a Cl^- free solution in which Cl^- was replaced with gluconate. The red-filled circles superimposed on the E_m trace are the point-by-point E_{Cl} values calculated from measured a^o_{Cl} and a^i_{Cl}. The wide gray arrow pointing toward the E_m trace indicates that the "uphill" accumulation of Cl^- following Cl^- depletion occurred without changes in E_m, as expected from electroneutral transport. The small orange arrows indicate electrode penetration and withdrawal from the cell. *Modified from Alvarez-Leefmans et al. (1988), and reproduced from Alvarez-Leefmans (2012).*

1990), gave an a^i_{Cl} of 30 mM. If Cl^- were passively distributed, a^i_{Cl} as predicted from Eq. (15.3) should be only 10.2 mM. The ion-selective electrodes measure a^i_{Cl} and not $[Cl^-]_i$. Thus, for an activity coefficient of 0.76, the basal a^i_{Cl} of 30 mM is equivalent to a basal $[Cl^-]_i$ of ~40 mM. Clearly, *intracellular Cl^- was almost three times the value expected for an equilibrium distribution.* E_{Cl} was −27 mV, and the driving force for $Cl^-(E_m - E_{Cl})$ was 28 mV. The red-filled circles superimposed on the E_m trace (lower record in Fig. 15.9) correspond to the point-by-point values of E_{Cl} calculated using Eq. (15.1). Removal of external Cl^- and its replacement with an impermeant anion (gluconate) led to a rapid decrease in $(a^i_{Cl})_{app}$, reaching a final steady-state level of ~10 mM in about 10 min. The recovery of intracellular Cl^- started immediately after external Cl^- was restored. *Note that the net Cl^- movements occurred without changes in E_m,* that is, they were electroneutral (Fig. 15.9, E_m trace, large gray arrow). The small rapid changes that can be observed in the E_m trace upon removal and readmission of external Cl^- were due to liquid junction potentials at the reference (bath) electrode. Subtraction of this liquid junction potential yielded the dashed line superimposed on the recorded E_m trace. If Cl^- movements had occurred through an electrodiffusional pathway (i.e., channel-mediated) or through an electrogenic "carrier," E_m should have changed both upon Cl^- efflux and upon Cl^- influx. Clearly, that was not the case. Since the cell membrane of DRG neurons has a relatively low resting Cl^- permeability, the movements of Cl^- should have occurred through a nonelectrodiffusional pathway, that is, an electroneutral carrier mechanism. As mentioned, the red circles superimposed on the E_m trace in Fig. 15.9 correspond to the calculated values of E_{Cl}. Note that on exposure to the Cl^- free Ringer's solution, the cell was depleted of Cl^- and, immediately after restoring the external Cl^-, E_{Cl} was transiently below E_m. However, during the recovery period in Ringer's solution, the movement of Cl^- occurred against its electrochemical potential difference across the cell membrane, that is, uphill, without changes in E_m. Since the Cl^- fluxes were bumetanide-sensitive and required the simultaneous presence of external Na^+, K^+, and Cl^-, this was direct functional evidence for the presence of an active process for accumulation of Cl^- in DRG cells through an electroneutral membrane transport mechanism, i.e., a $1Na^+$-$1K^+$-$2Cl^-$ cotransporter. The presence of NKCC protein in frog DRG cells was corroborated later on using immunolabeling and Western blot analyses (Alvarez-Leefmans et al., 2001). It turned out to be NKCC1 and was also found in DRG cells of cats, rats, and mice (Alvarez-Leefmans et al., 2001; Plotkin et al., 1997; Sung et al., 2000), not only in the cells bodies but also in intraspinal axon terminals (Javdani et al., 2020).

5.2.3 The magnitude and direction of the cotransport process are determined by the sum of the chemical potential gradients of the transported ions

We have seen that the transport process carried out by NKCC is electrically neutral, and thus net cotransport neither affects nor is affected by the membrane potential. Consequently, the magnitude and direction of the driving force for net ion movement must be calculated from the *chemical*, rather than the *electrochemical* potential gradients of all the transported ions.

The NKCC is a secondary active transport mechanism, and hence, it is reversible, that is, it mediates ion fluxes into or out of the cell, and the net direction of cotransport depends on the overall net free energy (ΔG) of the system. If ΔG is negative, it means that the direction of the cotransport will be inward, i.e., favoring uptake. If ΔG is zero, the system is at equilibrium. Since NKCC is electroneutral and has a $1Na^+$:1 K^+:$2Cl^-$ stoichiometry, ΔG is given by the following equation:

$$\Delta G = \Delta\mu_{Na,K,Cl} = \Delta\mu_{Na} + \Delta\mu_K + 2\Delta\mu_{Cl} \tag{15.11}$$

In other words, the driving force for the electroneutral $1Na^+$, $1K^+$, and $2Cl^-$ cotransport ($\Delta\mu_{Na,K,Cl}$) is the sum of the chemical potential differences of Na^+ ($\Delta\mu_{Na}$) and K^+ ($\Delta\mu_K$) plus twice the Cl^- potential difference ($\Delta\mu_{Cl}$), since two Cl^- ions are translocated per cycle. From the definition of chemical potential for each of the ions involved, it follows:

$$\Delta G = \Delta\mu_{Na,K,Cl} = RT\ln\frac{[Na^+]_i}{[Na^+]_o} + RT\ln\frac{[K^+]_i}{[K^+]_o} + 2RT\ln\frac{[Cl^-]_i}{[Cl^-]_o} \tag{15.12}$$

Hence

$$\Delta G = \Delta\mu_{Na,K,Cl} = RT\ln\frac{[Na^+]_i\,[K^+]_i\,[Cl^-]_i^2}{[Na^+]_o[K+]_o[Cl^-]_o^2} \tag{15.13}$$

In most vertebrate cells, under physiological conditions and for a stoichiometry $1Na^+$:$1K^+$:$2Cl^-$, ΔG is always *negative*, therefore favoring *net ion uptake*. Russell has computed ΔG for a variety of vertebrate cells; it ranges from −0.71 kJ/mol in bovine aortic endothelial cells up to −8.86 kJ/mol in rabbit ventricular myocytes (Russell, 2000).

ΔG has been estimated to be −1.29 kJ/mol and −2.8 kJ/mol in rat and frog DRG neurons, respectively (Rocha-Gonzalez et al., 2008). More recently, ΔG has been calculated for mouse choroid plexus epithelial cells from actual intracellular and extracellular ion measurements, and it was found to be −1.3 kJ/mol, thus favoring apical NKCC1 transport in the net uptake mode (Alvarez-Leefmans, 2020a; Gregoriades et al., 2019). As pointed out by Lytle (2003), the rare exception of a negative ΔG may be red blood cells which, as discussed above, have an unusually high $[Cl^-]_i$ (75−100 mM) reflecting their high anion permeability mediated by the anion exchanger protein AE1, which has translocation rates and characteristics similar to channels (Alper et al., 2008). The consequence of this high $[Cl^-]_i$ is that $\Delta\mu_{Na,K,Cl}$ is nearly 0. Indeed, in human, duck, and ferret red blood cells under physiological conditions, $\Delta\mu_{Na,K,Cl}$ is nearly 0, leaving little if any driving force to run the cotransporter under basal conditions (Lytle, 2003). This means that relatively small variations in the chemical gradients of any of the transported ions can alter the direction of transport in red blood cells.

The NKCC cotransport mechanism reaches thermodynamic equilibrium when $\Delta G = \Delta\mu_{Na,K,Cl} = 0$, that is, when the product of the concentrations of Na^+ and K^+ times the square of the Cl^- concentration on one side of the membrane is matched by the product of these ions in the *trans* side:

$$[Na^+]_i\,[K^+]_i\,[Cl^-]_i^2 = [Na^+]_o\,[K^+]_o\,[Cl^-]_o^2 \tag{15.14}$$

Since the cotransporter is electroneutral, thermodynamic equilibrium will not be affected by E_m. Assuming for the sake of argument that Na^+, K^+, Cl^- cotransport is the only Cl^- transport mechanism operating at steady state in a cell, that it has a $1Na^+{:}1K^+{:}2Cl^-$ stoichiometry, and that there are no kinetic constraints, the predicted $[Cl^-]_i$ when the cotransport reaches thermodynamic equilibrium is given by the following equation derived from Eq. (15.14):

$$[Cl^-]_i = \sqrt{\frac{[Na^+]_o[K^+]_o[Cl^-]_o^2}{[Na^+]_i[K^+]_i}} \tag{15.15}$$

For example, consider the model cell shown in Fig. 15.10, an idealized neuron in which NKCC is the sole mechanism transporting Cl^-. The $[Na^+]_i$ and $[K^+]_i$ are maintained at constant levels by the operation of the Na^+/K^+ pump. The cell has a resting E_m of −60 mV, and relevant physiological ion concentrations (in mM) are: $[Na^+]_o = 132$; $[K^+]_o = 3$; $[Cl^-]_o = 113$; $[Na^+]_i = 10$ and $[K^+]_i = 120$. Under these conditions, when the cotransporter reaches thermodynamic equilibrium, *without any kinetic constraints* (e.g., phosphorylation or dephosphorylation of the cotransporter), the $[Cl^-]_i$ will be ≈ 65 mM, and this would be the *maximal possible* $[Cl^-]_i$ under these conditions. Thus, we can conclude from the examples considered in Figs. 15.1 and 15.10 that if Cl^- were in electrochemical equilibrium, $[Cl^-]_i$ calculated from Eq. (15.3) would be 12 mM, but if we add NKCC to the same cell (Fig. 15.10), assuming that the resting channel-mediated permeability to Cl^- is negligible, the $[Cl^-]_i$ can increase up to about five times above its equilibrium level, and E_{Cl} would now be ≈ −12 mV. Since E_m is kept constant, at −60 mV, opening of Cl^- channels, for example, $GABA_A$-receptor-channels in the example considered, will produce an efflux of Cl^-. Since Cl^- is negatively charged, its electrodiffusional efflux produces an inward (depolarizing) current. However, most vertebrate cells endowed with NKCC do not have the high intracellular Cl^- level of the cell considered in this example (i.e., ≈65 mM). Why is that? Are there parallel mechanisms that prevent the NKCC from achieving thermodynamic equilibrium? The answer is yes. The first point to be

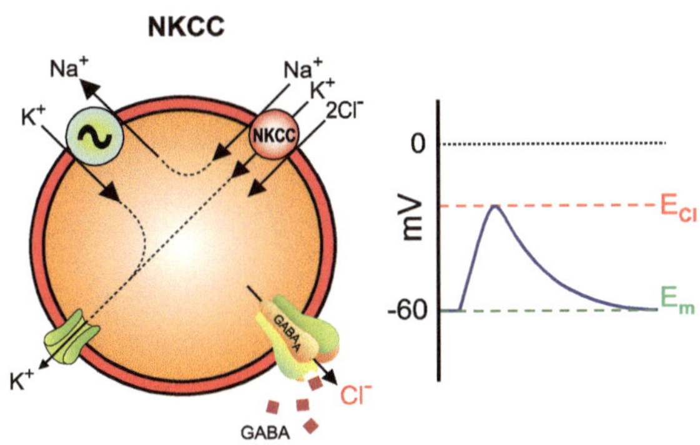

FIGURE 15.10 Model of a cell (e.g., a neuron) expressing Na^+, K^+, $2Cl^-$ cotransporters (NKCC), $GABA_A$-gated anion channels, Na^+/K^+ ATPase, and K^+ channels. The energy stored in the inward Na^+ chemical gradient generated and maintained by the Na^+/K^+ ATPase is used by NKCC for the active Cl^- uptake transport process. The Na^+/K^+ ATPase expels the Na^+ entering through NKCC, thus keeping $[Na^+]_i$ low and constant. The K^+, entering through NKCC diffuses out of the cell via K^+ channels. Cl^- is accumulated above electrochemical equilibrium, shifting E_{Cl} (red dashed line) to values more positive than E_m (green dashed line). Opening of $GABA_A$-gated anion channels produces a membrane depolarization (blue trace) due to Cl^- efflux. *Reproduced from Alvarez-Leefmans (2012).*

considered is that, besides NKCC, other Cl^- transport mechanisms such as Cl^- channels, anion exchangers (Cl^-/HCO_3^-), and KCl cotransporters (KCCs) could and in fact do coexist in the same cell with NKCC. The final $[Cl^-]_i$ will be determined by the coordinated functional interaction between the various transport mechanisms present in the cell. For instance, in skeletal muscle, active Cl^- uptake via NKCC is "shunted" by a large Cl^- conductance, as discussed (Fig. 15.2). In this case, the $[Cl^-]_i$ will be determined by a balance between the net Cl^- influx mediated by NKCC and the net Cl^- efflux occurring through Cl^- channels. Since the Cl^- leak through channels is larger than the net Cl^- influx brought up by NKCC, $[Cl^-]_i$ in skeletal muscle is very low, so low that E_{Cl} is close to $E_{m,}$ as discussed above in Section 2. In addition to Cl^- channels, in most cells, NKCC coexists with KCCs. Under physiological conditions, KCCs mediate net efflux of K^+ and Cl^- from cells, thus functioning as active Cl^- extruders (see Fig. 15.3). Hence, NKCCs and KCCs have opposing roles; the former are Cl^- loaders, and the latter are Cl^- extruders. Thus, KCCs may influence the levels of intracellular Cl^- just by counteracting Cl^- uptake mechanisms and vice versa. This interaction between two cotransporters working in opposite directions is ideally suited for achieving tight regulation of an intracellular ion, in this case Cl^- (and K^+). However, the result of these interactions is not just the simple algebraic sum of two fluxes having opposite directions; cotransporters are *kinetically* regulated by several mechanisms, notably by phosphorylation/dephosphorylation via a complex system of kinases briefly mentioned in Section 1, some of which are sensitive to $[Cl^-]i$ and thus serve as intracellular Cl^- sensors (Goldsmith & Rodan, 2023; Murillo-de-Ozores et al., 2020; Shekarabi et al., 2017).

The consequences of functional interactions between KCCs and NKCCs have been experimentally addressed using human embryonic kidney cells (HEK-293), an epithelial cell line in which the ubiquitous isoform of the KCCs, KCC1, was overexpressed by transfection (Gillen & Forbush, 1999). Overexpression of KCC1 not only led to a decrease in $[Cl^-]_i$ but also to a significant increase in endogenous NKCC activity. Indeed, the reduction of $[Cl^-]_i$ produced by KCC1 was responsible for NKCC1 activation. A steep relationship between cell Cl^- and NKCC1 activity was found over the physiological ranges, supporting a primary role for $[Cl^-]_i$ in the regulation of NKCC1. This means that *NKCC1 activity is determined not only by the thermodynamic driving force but also by kinetic constraints.* In this case, internal Cl^- acts as a "kinetic brake" on the ion translocating capacity of NKCC1. Thus, kinetic constraints can inactivate the cotransporter, even in the presence of a substantial thermodynamic driving force. The experimental results of Gillen and Forbush in transfected HEK-293 cells agree with previous observations showing that lowering $[Cl^-]_i$ activates NKCC, and increasing $[Cl^-]_i$ inhibits its activity. In other words, intracellular Cl^- acts as a negative feedback signal controlling the transport activity of NKCC. It was the Danish physiologist Hans Ussing (1911–2000) who first proposed that NKCC might be activated by a fall of $[Cl^-]_i$. Working with frog skin epithelium, Ussing realized that when the cells lose Cl^- NKCC restores it back to its initial higher than equilibrium level, but that the cotransporter "… seems to be virtually dormant unless the cells have lost chloride." (Ussing, 1982). The definitive proof came from studies in the internally dialyzed squid axon preparation, in which it was shown that elevation of $[Cl^-]_i$ inhibits bumetanide-sensitive unidirectional fluxes of all three cotransported ions in both the influx and efflux direction (Russell, 2000, 2009). The negative feedback system in which intracellular Cl^- regulates its own influx via NKCC was also demonstrated for the endogenous NKCC1 of a mammalian cell, namely rat DRG neurons (Rocha-Gonzalez et al., 2008). As mentioned above, the mechanisms by which intracellular Cl^- regulates NKCCs involve a complex system of kinases (see below). Regulation of NKCCs and other members of the SLC12 family is an extensive topic beyond the scope of the present chapter.

5.2.4 Loop diuretics derived from 5-sulfamoylbenzoic acid inhibit NKCCs

Diuretics derived from 5-sulfamoylbenzoic acid are organic anions that reversibly inhibit NKCCs in a variety of preparations by a concentration-dependent mechanism, with a half inhibitory constant (IC_{50}) ranging from ≈ 0.1 to 8.7 μM in mammalian cells (Russell, 2000). These compounds are members of the so-called *loop diuretics* because they inhibit Na^+ and Cl^- reabsorption by the thick ascending limb of the loop of Henle (TALH) by blocking NKCC2, thereby promoting diuresis (increased formation and excretion of urine). However, 5-sulfamoylbenzoic acid derivatives also block NKCC1. Native NKCC2 is a very active transport mechanism under basal conditions, permanently reabsorbing Na^+ and Cl^-. In contrast, the basal activity of native NKCC1 in extra-renal nonepithelial cells is relatively small; the cotransporter is often "dormant," which means that *net* ion fluxes in basal conditions are negligible either because the cotransporter is working near thermodynamic equilibrium, or more likely because it is inhibited by kinetic constraints, e.g., phosphorylation state. NKCC1 can be activated by either cell shrinkage produced by exposure to hypertonic solutions ("anisosmotic shrinkage") and/or by decreases in $[Cl^-]_i$. For instance, decreases in $[Cl^-]_i$ leading to activation of NKCC1 are accompanied by "isosmotic shrinkage" like that produced by experimental removal of external Cl^- in isosmotic media. Intracellular Cl^- can also be decreased by activation of Cl^- conductances, leading to a decrease in cell volume, e.g., the decrease in $[Cl^-]_i$ produced by secretagogues in airway submucosal glands and other epithelial cells (Lee et al., 2007). NKCC1 can also be activated by increases in extracellular K^+, as it occurs in the choroid plexus epithelial cells in which activation of apical NKCC1 contributes to the regulation of CSF [K^+] (Alvarez-Leefmans, 2020a; Wu et al., 1998; Xu et al., 2021). When assessing the inhibitory effect of 5-sulfamoylbenzoic acid derivatives like bumetanide, piretanide, and furosemide on the native NKCC proteins, *activated* NKCC1 has the same sensitivity to these compounds as *basal* NKCC2 (Hannaert et al., 2002). However, more recent observations using *Xenopus* oocyte heterologous expression system to compare the effects of bumetanide on two major human splice variants of NKCCs, hNKCC1a (the long NKCC1 variant) and hNKCC2A, found that half-maximal inhibitory concentrations (IC_{50}s) of bumetanide were 0.68 μM (hNKCC1a) and 4.0 μM (hNKCC2A), respectively, indicating that this compound is 6-times more potent to inhibit hNKCC1a than hNKCC2A (Lykke et al., 2015, 2016). In these studies, hNKCC1a was activated by exposure to a hyperosmolar solution that caused shrinkage of the oocytes (Lykke et al., 2016). For hNKCC2A, they used $^{86}Rb^+$ as a surrogate for K^+, and its uptake was compared between injected and uninjected (control) oocytes (Lykke et al., 2015).

The relative efficacy (IC_{50} values) and order of potency for blocking the NKCC1 were systematically investigated in the intestinal epithelium of the winter flounder (O'Grady et al., 1987b). As shown in Fig. 15.11, the order of potency of 5-sulfamoylbenzoic acid derivatives to block the cotransporter was: benzmetanide > bumetanide > piretanide > furosemide > amino-piretanide. This order of potency is similar to that determined for NKCC2 in TALH from rabbit kidney (Schlatter et al., 1983) and in other mammalian cells like human fibroblasts and dog trachea that express NKCC1 (O'Grady et al., 1987b).

Bumetanide and furosemide are among the most commonly prescribed drugs worldwide. They are clinically used to treat moderate hypertension and edema associated with congestive heart failure and hepatic and renal disease, including nephrotic syndrome (Shankar & Brater, 2003). Research during the last decades shows that NKCC1 is involved in many more functions and pathologies than previously thought, depending on the tissues in which it is expressed. Thus it is not surprising that NKCC1 has emerged as a potential therapeutic drug target for bumetanide and other loop diuretics in conditions as diverse as inflammatory pain, given its analgesic effects (Granados-Soto et al., 2005) and neonatal seizures, given that it blocks this abnormal brain electrical activity (Dzhala et al., 2010; Kahle & Staley, 2008), idiopathic intracranial hypertension (Dzhala et al., 2010), ischemic brain edema (Kahle et al., 2009), glioma cell invasion (Haas & Sontheimer, 2010), neuroinflammation (Tóth et al., 2022), and autism spectrum disorder (Ben-Ari & Cherubini, 2022), to

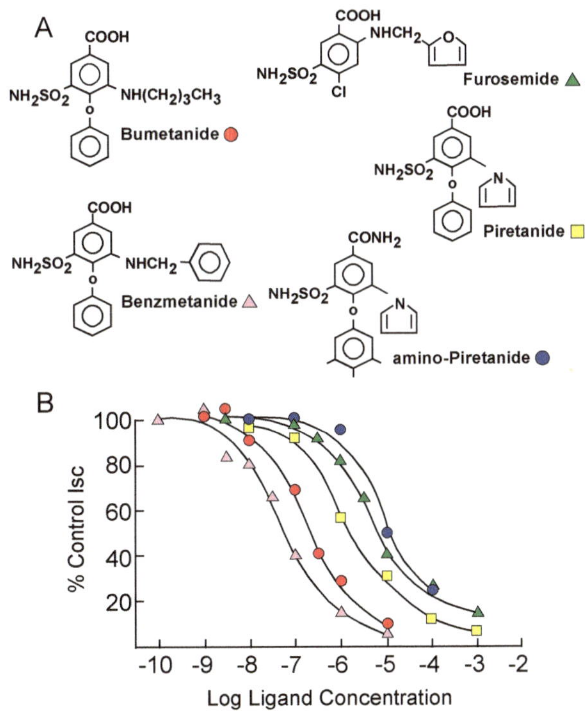

FIGURE 15.11 (A) Structures of several 5-sulfamoylbenzoic acid derivatives that inhibit Na^+-K^+-$2Cl^-$ cotransport mediated by NKCC1 in flounder intestine. (B) Dose-response curves show the effect of these derivatives on the epithelial short-circuit current (Isc). Order of potency is benzmetanide (5×10^{-8} M; $n = 4$), bumetanide (3×10^{-7} M; $n = 7$), piretanide (3×10^{-6} M; $n = 4$), furosemide (7×10^{-6} M; $n = 4$), amino-piretanide (1×10^{-5} M; $n = 4$). *Modified from O'Grady et al. (1987b).*

mention the most salient. However, when considering neuropsychiatric disorders, a contentious issue is whether bumetanide reaches therapeutic concentrations in brain tissues (Löscher & Kaila, 2022).

Besides their clinical importance, loop diuretics, particularly bumetanide, have played a crucial role in the identification of the NKCC proteins and in their subsequent cloning and functional characterization. Bumetanide is by far the most commonly used loop diuretic for research on NKCCs. However, as mentioned above, bumetanide is not a specific blocker of NKCC1; it also blocks NKCC2 at similar concentrations under physiological conditions in endogenous/native cotransporters. Bumetanide inhibits *basal* NKCC2 activity in medullary TALH tubules isolated from rat kidneys with a half inhibitory constant (IC_{50}) of 0.33 μM and, with the same IC_{50}, it inhibits *activated* NKCC1 from either rat erythrocytes or rat thymocytes (Hannaert et al., 2002). In these experiments, NKCC1 was activated by exposing the cells to hypertonic solutions. An increase in NKCC2/NKCC1 selectivity was found when comparing the diuretic sensitivity of *basal* NKCC2 versus *basal* NKCC1. For the specific case of bumetanide, the IC_{50} for *basal* NKCC1 goes up from 0.33 μM (activated state) to 2.5 μM in rat erythrocytes and 2.2 μM in rat thymocytes. In other mammalian cells expressing native/endogenous NKCC1, bumetanide inhibits the cotransporter with an IC_{50} ranging from 0.1 μM in human fibroblasts (HSWP) to 8.7 μM in Ehrlich ascites tumor cells, the cells in which NKCC was discovered (Hoffmann et al., 1986). The bumetanide IC_{50} of *activated* NKCC1 in rat DRG neurons is 5.7 μM, slightly lower than that reported for Ehrlich ascites tumor cells (Rocha-Gonzalez et al., 2008). In DRG neurons, NKCC1 was activated by cell Cl^- depletion, which also led to isosmotic cell shrinkage, and the effect of bumetanide was tested on the Na^+ dependent re-accumulation of Cl^-. Bumetanide sensitivity has also been measured in NKCC1 expressed in HEK-293 cells. Bumetanide inhibits human NKCC1 expressed in HEK-293 cells with an IC_{50} of 0.16 ± 0.03 μM (Payne et al., 1995), whereas it inhibits rat NKCC1 expressed also in HEK-293 cells with an IC_{50} of 2.4 ± 0.7 μM. Interestingly, mutating alanine 483 with cysteine (A483C) of rat NKCC1 shifts the bumetanide dose-response curve to the left, i.e., this point mutation renders NKCC1 more sensitive to bumetanide, shifting the IC_{50} from 2.4 ± 0.7 μM (wild type) to 0.37 ± 0.02 μM for the A483C mutant (Dehaye et al., 2003). Clearly, at least for mammalian NKCC1, bumetanide sensitivity (IC_{50}) varies almost two orders of magnitude; it appears to be dependent on animal species, cell type, activation state of NKCC1, and whether measurements are done in native/endogenous NKCC1 or in heterologous expression systems. *Based on bumetanide sensitivity, it is impossible to discriminate between NKCC1 and NKCC2.* However, in many publications, it is often mistakenly stated that bumetanide is a "specific NKCC1 blocker."

The sensitivity to bumetanide of the three full-length spliced variants of NKCC2 has been studied in the *Xenopus* oocyte heterologous expression system. When mouse NKCC2s are expressed in Xenopus oocytes, the IC_{50} values for bumetanide are 3.4 μM for NKCC2F, 2.0 μM for NKCC2A, and 0.6 μM for NKCC2B (Plata et al., 2002). The IC_{50} values for bumetanide inhibition of the human NKCC2 full-length splice variants expressed in *Xenopus* oocytes are 0.16 μM for NKCC2F, 0.54 μM for NKCC2A, and 0.22 μM for NKCC2B (Carota et al., 2010). As mentioned above, another study on human NKCC2A also expressed in *Xenopus* oocytes found an IC_{50} of 4.0 μM (Lykke et al., 2015). Thus, the human NKCC2A variant appears to be the least sensitive to bumetanide, whereas in mice, the least sensitive variant is NKCC2F. The reasons for these discrepancies between mouse and human NKCC2 variants remain to be elucidated. However, we can conclude that *it is impossible to discriminate between NKCC2 splice variants based on bumetanide sensitivity.*

The overall conclusion is that NKCCs' sensitivity to bumetanide varies over two orders of magnitude; it depends on whether it is assessed under basal conditions or upon activation of the cotransporters (Hannaert et al., 2002), on degree of glycosylation (Paredes et al., 2006), on animal species, on cell type, and on whether the proteins are in their native cell (endogenous) or expressed in heterologous systems like *Xenopus* oocytes or HEK-293 cells. Moreover, used alone and/or in relatively high concentrations (e.g.,> 50 μM), not even bumetanide can provide positive proof that the NKCCs mediate a given function. This is because bumetanide also blocks KCCs, although at significantly higher concentrations than those needed to block NKCCs. As discussed below, bumetanide blocks KCCs with IC_{50} values of ≈60–180 μM for KCC1 (Gillen et al., 1996; Mercado et al., 2000); 55–80 μM for KCC2 (Payne, 1997; Song et al., 2002), although values as high as 655 μM have been reported for KCC2 measuring net fluxes with a fluorescence-based assay (Delpire et al., 2009); 100 μM for KCC3 (Mount et al., 1999; Race et al., 1999), and 900 μM for KCC4 (Mercado et al., 2000). Clearly, bumetanide is a significantly less effective inhibitor of the KCCs than of the NKCCs. *Thus, bumetanide sensitivity might be used to differentiate between NKCCs and KCCs activity in functional assays.*

It is necessary to keep in mind that 5-sulfamoylbenzoic acid derivatives are nonselective inhibitors; they can inhibit channels or other carriers, although at considerably higher concentrations than those with which they inhibit NKCCs or KCCs. For instance, bumetanide blocks native CFTR Cl^- channels in sweat ducts with an IC_{50} of 300 μM (Reddy & Quinton, 1999). Furosemide blocks NMDA-induced currents in cultured spinal neurons but with an IC_{50} of 1.2 mM. Likewise, piretanide (5 mM) and bumetanide (2 mM) partially block NMDA-induced currents at millimolar concentrations (Lerma & Martin del Rio, 1992). Bumetanide and furosemide, also at very high concentrations (400 μM and 1 mM, respectively), partially block Ca^{2+}-activated Cl^- channels in rat lacrimal glands (Evans et al., 1986). Bumetanide and furosemide block vesicular glutamate uptake with an IC_{50} of 230 and 250 μM, respectively (Roseth et al., 1995). In addition, it has been shown that bumetanide reduces aquaporin 4 osmotic water flow at concentrations well above 100 μM (Migliati et al., 2009). However, in this study, a possible direct effect of bumetanide on endogenous NKCC1 of Xenopus oocytes was not ruled out. This is important given the evidence supporting the hypothesis that NKCC1 cotransports water and ions (Hamann et al., 2005, 2010; Zeuthen & Macaulay, 2012). These observations are also significant from a "translational research" perspective because local (Yan et al., 2003) as well as intravenous administration of bumetanide (O'Donnell et al., 2004) attenuate brain edema associated with transient as well as permanent experimental focal cerebral ischemia. Another point to consider when using bumetanide is that it can weakly inhibit some carbonic anhydrases. This effect could obscure experimental observations when NKCC1 and AEs coexist in the same cells, as is the case in choroid plexus epithelial cells (Carta & Supuran, 2013; Vogh & Langham, 1981). Based on the data discussed above, the conclusion is that *bumetanide at or below* 10 μM *concentrations can be considered as a* "selective" *inhibitor of NKCCs.*

Loop diuretics are anionic molecules at the physiological pH range, with very high lipid solubility, and hence can cross cell membranes (Alvarez-Leefmans et al., 1990). This raises the question of their side and site of action on NKCCs. Abundant experimental evidence suggests that bumetanide binds exclusively to the external face of NKCCs (Brandt et al., 2018; Haas, 1994). Work from Forbush, Isenring, and colleagues suggests that the central hydrophobic domain of NKCC1 determines its apparent affinities for Na^+, K^+, and Cl^- as well as for bumetanide and that the cytosolic N- and C-termini have little effect on these properties and are mainly involved in the regulation of transport activity (Isenring & Forbush, 2001). Recent single-particle cryo-EM structures supported by functional studies reveal an outward-facing conformation of human NKCC1 (hNKCC1), showing "bumetanide wedge into a pocket in the extracellular ion translocation pathway" (Zhao et al., 2022a). The authors show that bumetanide arrests NKCC1 in an outward-open conformation. This extracellular ion translocation path constitutes a set of conserved residues in human NKCC1 and NKCC2, explaining why bumetanide (and other loop diuretics) do not discriminate between these two closely related paralogs. By binding to the extracellular vestibule of NKCC1 in the open state, bumetanide inhibits by selectively plugging the extracellular ion entry pathway and sterically hindering the transporter from isomerization into other transport states.

Another cryo-EM structural study of hNKCC1 in the absence and the presence of bumetanide (or furosemide) revealed two drug inhibitor-binding sites localized at the transmembrane (TMD) and cytosolic carboxyl-terminal domains (CTD), respectively (Moseng et al., 2022). According to these studies, drug inhibition involves a coupled movement between the CTD and the TMDs. Furosemide and bumetanide binding sites in the CTD overlapped. The cover of this book (top left panel) shows the furosemide (Fu) and bumetanide (Bu) CTD binding sites based on the human NKCC1 cryo-EM-structures of hNKCC1-Fu and hNKCC1-Bu, colored cyan and orange, respectively. The inset shows an enlargement of the CTD binding sites. The CTD binding site appears to be essential for the sensitivity of the cotransporters to the inhibitors, as revealed in mutagenesis experiments. The CTD binding site is intriguing from the pharmacological perspective, given the evidence that bumetanide blocks NKCC1 from the extracellular side, not from the intracellular side (Brandt et al., 2018). Accordingly, the injection of bumetanide into *Xenopus* oocytes transfected with either human NKCC1a or NKCC1b did not inhibit NKCC1-mediated Rb^+ uptake (use as a surrogate of K^+). In contrast, when applied extracellularly, bumetanide blocks NKCC1-mediated fluxes carried by either variant with roughly the same potency (IC_{50} for hNKCC1a was 0.18 μM, and that for NKCC1b was 0.20 μM). It could be argued that the lack of effect of intracellular bumetanide in oocytes is due to intracellular diffusion barriers, a point that needs to be addressed experimentally. However, results that are complementary to those in *Xenopus* oocytes have been obtained by testing bumetanide on CSF secretion by CPECs, in which NKCC1 is located in the apical membranes facing the ventricle. Although the mechanisms by which bumetanide inhibits CSF secretion are the subject of debate (Alvarez-Leefmans, 2020a, 2020b; MacAulay & Rose, 2020), the fact is that the diuretic partially inhibits CSF secretion only when injected into the brain ventricles whose fluid content is in direct contact with the apical membrane of CPECs where NKCC1 is localized (Javaheri & Wagner, 1993; Karimy et al., 2017; Steffensen et al., 2018). In contrast, when applied systemically, bumetanide does not have an effect on CSF secretion, even though its access to the basolateral surface of the epithelium is not expected to be diffusion-limited, given the presence of fenestrated capillaries (Brightman & Reese, 1969; Keep et al., 1994; Vogh & Langham, 1981). Still, it could be that bumetanide does not reach significant intracellular concentrations because it might be extruded from the cells by organic anion transporters, many of which are expressed in the basolateral membrane of CPECs (Roth et al., 2012; Strazielle & Ghersi-Egea, 2015; Zhang et al., 2018). In conclusion, future experiments directly targeting the intracellular bumetanide binding site in the CTD are needed to clarify the issues discussed (Flygaard et al., 2023; Moseng et al., 2023).

Finally, the second bumetanide/furosemide site in the cryo-EM studies of Moseng et al. was found in the transmembrane domain (TMD) located in a cavity formed among transmembrane segments (TMs) 1,3,6,8, 9, and 10, facing the extracellular space in the hNKCC1 protomer. These observations agree with earlier mutagenesis data suggesting that the inhibitors bind within the TMD, specifically TM3, the transmembrane helix 3 (Somasekharan et al., 2012) and with the Cryo-EM observations of Zhao et al. (2022a).

5.3 Molecular structure and distribution of Na^+-K^+-Cl^- cotransport proteins

5.3.1 The Na^+-K^+-2Cl^- cotransporter 1

As mentioned in 5.1, two distinct Na^+-K^+-Cl^- cotransporter isoforms encoded by the SLC12A2 and the SLC12A1 genes have been identified by cDNA cloning and expression, NKCC1 and NKCC2, respectively. NKCC1 was originally referred to as BSC2 (bumetanide-sensitive cotransporter 2) or as "the basolateral bumetanide-sensitive Na−K−2Cl cotransporter" or as "the secretory isoform." Fortunately, these confusing and inappropriate nomenclatures have been superseded; NKCC1 is not the only cotransporter sensitive to bumetanide, neither is it exclusively located in the basolateral membrane of epithelial cells, nor is it expressed exclusively in epithelial cells, nor is it always "secretory." Thus, the current nomenclature used by the GenBank (National Center for Biotechnology Information) is recommended, and it has been universally adopted.

A cDNA encoding NKCC1 was first cloned from the shark rectal gland (Xu et al., 1994), a well-characterized secretory epithelium that has been used as a model to study Cl^- secretion in epithelial cells. More important, the shark rectal gland has a high density of Na^+-K^+-Cl^- cotransporters located in the basolateral membrane of these epithelial cells (Silva et al., 1996). Subsequently, cDNAs encoding NKCC1 were cloned from other secretory epithelia, including T84 cells, a human colonic epithelial cell line (Payne et al., 1995), and cultured inner medullary collecting duct cells (mIMCD-3) from mouse kidney (Delpire et al., 1994). Both of these mammalian NKCC1 proteins share a high degree of homology with the shark NKCC1 (71%−74%) and even higher identity with each other (91%). Later on, a cDNA encoding for the bovine aortic endothelial cells NKCC1 was also cloned, and it was found that the predicted protein exhibits ~95% amino acid identity with the T84 cell cotransporter amino acid sequence (Yerby et al., 1997).

The human SLC12A2 gene is located in chromosome five and consists of 27 exons encoding for a full-length protein referred to as NKCC1a to distinguish it from a splice variant lacking exon 21 that encodes a shorter protein (Delpire & Gagnon, 2018a; Virtanen et al., 2020). This shorter NKCC1 variant is referred to as NKCC1b and was first detected by partial cloning in mouse brain tissue (Randall et al., 1997). Exon 21 encodes 16 amino acid residues containing a dileucine motif in the carboxyl terminus of NKCC1a, the full-length NKCC1 protein (Fig. 15.12). The functional significance of the two splice variants is unclear. It has been proposed that the dileucine motif in NKCC1a determines the basolateral sorting of this protein in secretory epithelia (Carmosino et al., 2008). Experiments in cultured Madin−Darby canine kidney cells (MDCK cells) transfected with NKCC1 and NKCC2 chimeric constructs show that transfection with chimeras lacking the dileucine motif are targeted to the apical membrane. Furthermore, NKCC2 (SLC12A1), which lacks this sorting signal sequence in the carboxyl terminus domain, is targeted to the apical membrane of the thick ascending limb cells in the

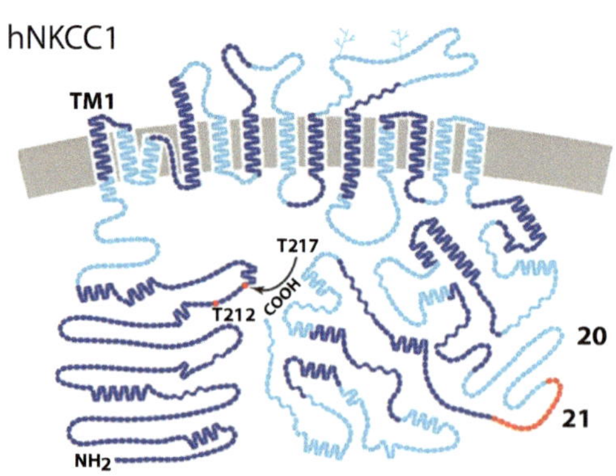

FIGURE 15.12 Structural model of the human Na^+-K^+-2Cl^- cotransporter protein NKCC1 (hNKCC1) showing the boundaries between amino acid sequences encoded by each exon (highlighted in light and dark blue, except for exon 21 highlighted in red). Each link symbolizes a single amino acid residue. "NH_2" and "COOH" indicate amino- and carboxy-termini, respectively. TM1 indicates the position of transmembrane segment 1. The branched thin lines in the long extracellular loop formed between transmembrane segments TM7 and TM8 represent N-linked glycosylation sites. Biochemically identified phosphorylated threonine (T) residues (positions T212 and T217) are indicated by red dots. The secondary structure of this protein with 12 TM helices and large intracellular amino- and carboxy-terminal domains is shared by other members of the cation-chloride co-transporter (CCC) family. The residues encoded by exons 20 and 21 in the carboxyl terminus are also illustrated. Cryo-EM observations confirm this predicted structure and add numerous significant details discussed in the text. *Adapted from Carmosino et al. (2008).*

mammalian kidney. This hypothesis of the "dileucine motif" has been tested in choroid plexus epithelial cells (CPECs), which, as mentioned, are unique among Cl^--secreting epithelial cells because NKCC1 is localized to their apical membranes (Gregoriades et al., 2019; Plotkin et al., 1997). The question was approached using an RT-PCR strategy that allowed co-amplification and quantification of the relative abundance of NKCC1a and NKCC1b transcripts, both of which were found in CPECs. The rationale was that if the sorting mechanism involving the dileucine motif in exon 21 determines basolateral polarity, CPECs should predominantly express NKCC1b. However, contrary to what was expected based on the dileucine motif hypothesis, NKCC1b was barely detectable; the relative abundance of NKCC1a/NKCC1b was $\sim$15 (Gregoriades et al., 2019). The relative abundance of NKCC1a/NKCC1b in choroid plexus epithelia was significantly higher than in brain, DRG, and spinal cord tissues but not significantly different from that in the kidney, where NKCC1 is localized to the basolateral membrane of the terminal inner medullary collecting duct (Kaplan et al., 1996). Therefore, the predominant expression of NKCC1a over NKCC1b in CPECs cannot explain the apical sorting of the cotransporter in terms of the dileucine motif hypothesis.

The absence of exon 21 predicts the loss of a protein kinase A (PKA) consensus site of the cotransporter protein. This suggests that the short NKCC1 (NKCC1b) could be subjected to different regulation than the full-length NKCC1 (NKCC1a). However, the absence of exon 21 does not appear to affect NKCC1 transport function, as suggested by experiments in MDCK epithelial cells transfected with constructs of the short NKCC1 variant (Carmosino et al., 2008). Further, NKCC1 phosphorylation related to cotransport activation is carried out by the kinases SPAK (STE20/SPS1-related proline/alanine-rich kinase) and OSR1 (oxidative stress-responsive kinase 1) at specific threonine residues located in the N-terminus of the cotransporter (Delpire & Gagnon, 2008; Gamba et al., 2009; Kahle et al., 2010), as discussed below. Little is known about whether PKA phosphorylation plays a direct role in NKCC1 activation. Clearly, more research is needed to elucidate the functional significance of the two NKCC1 splice variants. Further, the cellular specificity of NKCC1a and NKCC1b isoforms is a critical issue; these two isoforms may have different subcellular localizations and post-translational modifications that may result in different functional properties related to membrane trafficking and ion translocation (Löscher & Kaila, 2022; Virtanen et al., 2020).

Full-length NKCC1 (i.e., NKCC1a) comprises $\approx$1200 amino acids and has a molecular weight of 130−132 kDa. The predicted two-dimensional topology of NKCC1 from the shark rectal gland shows 12 α-helical transmembrane (TM) domains and hydrophilic amino- and carboxyl-terminal regions (NH_2 and COOH) flanking a central hydrophobic domain. Whereas the amino (N) terminus displays considerable variability, the large carboxyl (C) terminus is well conserved among the different NKCC proteins. This two-dimensional topology appears to be preserved in mammals and is similar to that predicted for NKCC2 and NCC (Gamba, 2009a) and the KCCs (see below). The topology of NKCC1 is supported by experimental data (Gerelsaikhan & Turner, 2000a, 2000b; Gerelsaikhan et al., 2006) and structural cryo-EM studies (Ruiz Munevar et al., 2024). Fig. 15.12 shows the predicted two-dimensional topology of human NKCC1.

NKCCs are glycoproteins and, therefore, possess consensus sites for N-linked glycosylation (*sequons*) within the large hydrophilic (extracellular) loop between TM segments 7 and 8, as shown in Fig. 15.12 (Haas & Forbush, 1998). A *sequon* is a sequence of three consecutive amino acids in a protein that can serve as the attachment site to a polysaccharide (sugar) called an N-linked glycan. N-linked protein glycosylation occurs at asparagines (N) of the NXS/T *sequon*, where X is any amino acid except proline, S is serine, and T is threonine (Ben-Dor et al., 2004). There are two conserved extracellular N-glycosylation *sequons* that correspond to asparagines 547 and 556 (N^{547} and N^{556}) in mouse NKCC1 and N^{544} and N^{553} in rat NKCC1 (Gerelsaikhan et al., 2006). Evidence that NKCC1s are N-glycosylated in vivo comes from Western blot analyses of various tissues from animals of various species including shark, frog, duck, rat, cat, rabbit and human. These analyses showed proteins with molecular weight higher than predicted for the core protein ($\approx$160 Kd). Upon enzymatic deglycosylation, the molecular weight decreases to $\approx$130 kDa, the core molecular weight [reviewed in (Di Fulvio & Alvarez-Leefmans, 2009)]. The impact of N-glycosylation on NKCC1 function has not been studied in detail. The protein in its physiological plasma membrane environment is glycosylated. It is believed that glycosylation enhances cell-surface expression of these proteins. However, experiments in COS7 cells show that NKCC1 is delivered to the plasma membrane independently of its N-glycan state (Singh et al., 2015).

NKCC1 has several potential phosphorylation sites within the predicted amino- and carboxy-terminal domains. As already mentioned, phosphorylation of NKCC1 plays a critical role in regulating its transport activity. Activation of NKCC1 by intracellular Cl^- depletion is associated with increased phosphorylation of the cotransporter at conserved threonine (T) residues located within the N-terminal domain. The pioneering studies of Lytle and Forbush in the shark rectal gland epithelium yielded the first evidence suggesting that activation of NKCC1 was associated with phosphorylation of a threonine (T) residue (Lytle & Forbush, 1992). Later studies demonstrated that the phosphoacceptor amino acid sites within the cotransporter N-terminal domain corresponded to the threonines located at positions 184, 189, and 202 of shark NKCC1. Of these residues, T189 was found to be absolutely necessary for NKCC1 to be functional. Forbush and colleagues carried out studies in vivo using an antiphospho-NKCC1 antibody raised against a synthetic peptide of the amino-terminal domain containing T212 and T217 of human NKCC1, which correspond to T184 and T189 of shark NKCC1 (Darman & Forbush, 2002; Flemmer et al., 2002) and demonstrated that activation of NKCC1 was associated with phosphorylation of these threonines (Fig. 15.12). The kinases responsible for NKCC1 activation were identified by Eric Delpire and coworkers as SPAK and OSR1, two members of the Ste20 kinases (Delpire & Gagnon, 2008, 2018a).

Immunofluorescence studies using anticotransporter antibodies showed that in secretory epithelia, NKCC1 protein is localized to the basolateral membrane (Chou et al., 2008; Evans et al., 2000; Lytle et al., 1995; McDaniel et al., 2005; Nejsum et al., 2005; Shillingford et al., 2002). As mentioned above, the only exception documented by functional and molecular methods is the choroid plexus (CP), where NKCC1 is located in the apical membrane (Gregoriades et al., 2019; Plotkin et al., 1997; Wu et al., 1998). We have already mentioned that another possible exception is the cells of the retinal pigment epithelium (RPE), where there is functional evidence for the apical location of an NKCC mechanism together with the Na^+/K^+ ATPase. Still, the precise location and molecular identity of this NKCC in RPE have not been studied with immunolabeling and molecular methods (Reichhart & Strauss, 2020).

The NKCC1 cotransporters that are localized to the basolateral membrane of epithelial cells mediate net salt influx into the cells and, in doing so, act in concert with apical Cl^- channels and basolateral K^+ channels to produce net salt and fluid secretion in these epithelia. The location of NKCC1 in the apical membrane of CP argues against its direct involvement in ion secretion into cerebrospinal fluid. There is evidence showing that the CP cotransporter is continuously active and that it functions to reabsorb K^+ from cerebrospinal fluid to blood (Alvarez-Leefmans, 2020a; Fame et al., 2023; Gregoriades et al., 2019; Wu et al., 1998; Xu et al., 2021). The location of NKCC1 in the paranodal region of the Schwann cells in peripheral nerves suggests a similar role in K^+ uptake from the peri-axonal space, thereby preventing K^+ accumulation in this critical space, which could produce abnormal changes in axonal excitability (Alvarez-Leefmans et al., 2001). Still, functional studies are needed to prove this hypothesis, which is derived mainly from immunolocalization studies.

5.3.2 The Na^+-K^+-$2Cl^-$ cotransporter 2 (NKCC2)

NKCC2 (SLC12A1) is characteristically expressed in the kidney (Gamba, 2009a; Gamba & Friedman, 2009), where it is located in the apical membrane of the TALH epithelial cells. For this reason, NKCC2 was often referred to as "the apical bumetanide-sensitive Na^+-K^+-$2Cl^-$ cotransporter," as "BSC1" (bumetanide-sensitive cotransporter 1), or as "the absorptive isoform." As with NKCC1, these nomenclatures are not recommended, among other reasons, because NKCC1 can also be located in the apical membrane of some epithelial cells, because NKCC2 is sensitive to other loop diuretics besides bumetanide, and because the expression of NKCC2 can occur in tissues other than kidney. In fact, NKCC2 is expressed in extrarenal tissues, such as pancreatic β-cells and the gastrointestinal tract (Marcoux et al., 2021; Singh et al., 2016), although the functional relevance of NKCC2 in these extrarenal tissues is largely unknown. Interestingly, mice lacking NKCC2A, a splice variant of NKCC2, show impaired glucose tolerance, glucagon, and insulin responses (Kelly et al., 2019). The NKCC2 cotransporter is also expressed in the macula densa segment of the TALH, where it functions as a NaCl sensor.

NKCC2 shares only ≈60% amino acid sequence identity with NKCC1. In fact, these two proteins are the products of independent genes; SLC12A1, located on human chromosome 15, encodes NKCC2, whereas SLC12A2, located on human chromosome 5, encodes NKCC1 (Gamba & Friedman, 2009). As mentioned, the gene encoding for NKCC2 gives rise to different NKCC2 transcripts derived from alternative splicing events. These splicing events yield three full-length NKCC2 variants: NKCC2A, NKCC2B, and NKKC2F, all of which are differentially distributed in different regions of the TALH, have different transport kinetics characteristics, and have been found in various species, including humans (Carota et al., 2010). In addition to these three full-length variants, there are shorter variants with truncated carboxyl-terminal ends. The meaning of differential splicing is that it contributes to the diversity of proteins encoded by a limited number of genes in the genome. Discussion about the possible function of each of these variants is beyond the scope of the present chapter. The interested reader should consult the comprehensive reviews by the groups of Gamba, Castrop, and Isenring (Carota et al., 2010; Castrop & Schießl, 2014; Castrop & Schnermann, 2008; Gamba, 2009a; Gamba & Friedman, 2009; Marcoux et al., 2021).

Full-length NKCC2 proteins are shorter than NKCC1, with a deduced amino acid sequence of ≈1095–99 residues and a core molecular mass of ≈110–120 kDa. Like NKCC1, NKCC2 is N-glycosylated at two sites located in the long extracellular loop between TM7 and TM8, consistent with the size of ≈150–160 kDa of the glycosylated cotransporter. N-glycosylation has been shown to impact NKCC2F function when expressed in oocytes (Gamba, 2005). Protein N-glycosylation is a co- and post-translational modification crucial for the folding, trafficking, insertion, and function of many proteins [for references, see (Ares et al., 2011; Di Fulvio & Alvarez-Leefmans, 2009)]. Prevention of N-glycosylation in rat NKCC2F by site directed mutagenesis produces a protein with reduced transport activity and decreased bumetanide sensitivity (Paredes et al., 2006). The predicted topology of NKCC2 is very similar to that of NKCC1, featuring a central hydrophobic domain comprising 12 putative transmembrane (TM) spanning regions. The difference between the two isoforms can be accounted for almost entirely by the additional amino acids at the amino terminus of NKCC1. In fact, the amino terminus is the most divergent region of the entire protein, both between the two isoforms and when comparing the same isoform between species (Haas & Forbush, 1998).

5.4 Kinetic models of Na^+-K^+-2 Cl^- cotransport

Normal operation of NKCCs requires the binding to the cotransporter of all the ions involved (i.e., 1Na^+, 1K^+, and 2Cl^-) on the same side of the membrane before ion translocation occurs. The binding order of the cotransported ions does not seem to happen randomly. Evidence shows a preferred order of ion binding resulting from an allosteric effect such that the binding of one ion increases the apparent affinity of a binding site for the next ion. However, there is no consensus about ion binding and release order. One of the best-tested models is still the one based on a systematic study of ion dependencies of self-exchange fluxes mediated by the cotransporter in duck and human red blood cells, which we now know is NKCC1 (Lytle, 2003; Lytle et al., 1998). Two partial reactions of the cotransporter cycle have been revealed: K^+-K^+ exchange in normal high-K^+ cells and Na^+-Na^+ exchange in high-Na^+ cells. The internal and external ion requirements of each mode can be explained by a reaction cycle based on ordered binding and glide symmetry. The latter means that the first ion to bind on one side of the "carrier" will be the first ion to be released on the other.

Fig. 15.13A shows the cotransport model based on ordered ion binding and glide symmetry. In this model, net cotransport influx would require that the ions bind to the cotransporter in the order shown. In the initial loading of the empty carrier at the outer face of the membrane, binding of a Na^+ in the pocket (reaction 2) induces the formation of a Cl^- site, and binding of that Cl^- (reaction 3) creates a site that can bind K^+, and so forth. All four ions are occluded momentarily in a transitional state (E_4) without access to either side of the membrane. Reaction five represents a conformational change of the NKCC such that the binding sites are now accessible from the intracellular compartment, and reactions six to nine represent the ordered release of the transported ions. Reaction ten leads to a conformational change that involves the reorientation of the totally unloaded cotransporter to an outwardly facing conformation. The model has been quantitatively tested using simulations based on flux data from duck red blood cells, HeLa cells, and epithelial cells (Benjamin & Johnson, 1997). Russell (2000) and later Delpire and Gagnon (2011) have offered stimulating critiques of this model, showing its weaknesses and strengths. Further, Delpire and Gagnon proposed an alternative model, briefly dealt with below. In addition, a mathematical model of NKCC2 has been proposed (Marcano et al., 2009). This model basically adopts the kinetic scheme proposed by Lytle and colleagues for NKCC1 but uses kinetic data reported for each splice variant of NKCC2. The authors conclude that the three long splice variants of NKCC2 differ only in ion binding affinities, a result that is consistent with published mutagenesis analysis and differences in $^{86}Rb^+$ uptake among the A, B, and F variants. NKCC2 cotransporter models will facilitate the development of larger-scale models of ion transport by TALH cells. One problem in studying the partial reactions of the transport cycle of NKCCs is that they are electroneutral transporters. Thus, unlike electrogenic transporters [e.g., (Garcia-Celma et al., 2010)], it is not possible to measure their partial reactions by electrophysiological means. Nevertheless, recent cryo-EM studies have opened new paths to tackle this problem, as discussed below.

As mentioned above, Delpire and Gagnon (2011) proposed an alternative to Lytle and colleagues' model. We refer to it as the "steady state model" to distinguish it from Lytle's "glide-symmetry model." The primary difference is that in the steady-state model, the outward-open state first binds Cl^- then Na^+, followed by a second Cl^- and then K^+, and the ions are released in reverse order to the cytoplasmic side, i.e., K^+ first, followed by Cl^-, Na^+, and Cl^- [reviewed in (Markadieu & Delpire, 2014)]. However, a recent atomistic model based on the cryo-EM structure of human NKCC1 and molecular dynamics simulations supports a substrate release order in which Cl^- is the first substrate to be released, followed by Na^+, Cl^-, and K^+ in a cooperative manner (Neumann et al., 2022). Neumann and coworkers conclude that their results do not fit into either of the models discussed, although they fit better to the "glide symmetry" one.

The NKCC1 structures recently obtained from cryo-EM have provided further insights into the workings of NKCC1 at an unprecedented resolution. These studies have confirmed that NKCC1 operates by transitioning from conformations where the ion binding sites are exposed to either the extracellular or the intracellular side of the membrane (Ruiz Munevar et al., 2024). Specifically, NKCC1 transits dynamically through three conformational states: an outward open (OO) state, in which the transporter binds ions and water from the outside of the cell; a transient occluded state (OCC); and an inward open (IO) state, in which the ions and water are released into the inside of the cell (Fig. 15.13B and C). Using a combination of molecular dynamics simulations, cryo-EM structures, and mutagenesis, these researchers conclude that NKCC1

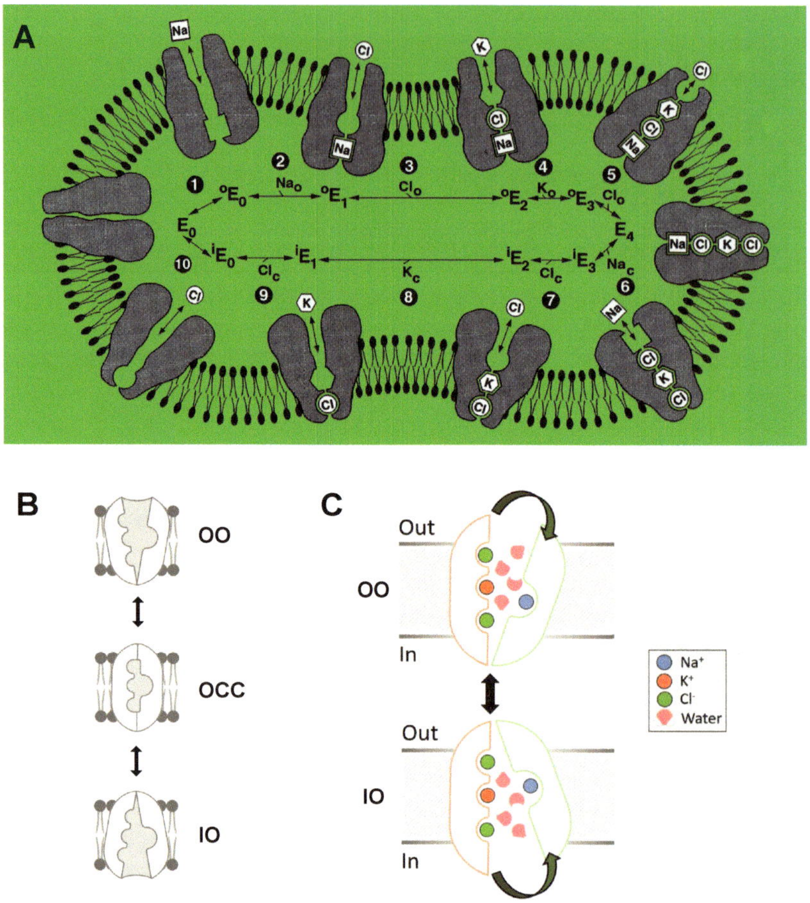

FIGURE 15.13 (A) Model of transport cycle of Na^+-K^+-$2Cl^-$ cotransport by ordered ion binding with glide symmetry. Ten intermediate states of the transporter are depicted. The model assumes that cotransport is entirely reversible and that only fully loaded (E_4) and completely unloaded (E_0) forms of cotransporter protein are capable of changing orientation from inward- to outward-facing (and vice versa). The idea is that when all sites are occupied (E_4), a major conformational change occurs, allowing bound ions to unload on the opposite side of the membrane in the order they were bound (first on, first off). When the carrier is wholly unloaded (E_0), the empty form can also undergo a major conformational change, allowing it to begin reloading ions at the opposite side of the membrane, thus facilitating net cotransport. Numbers refer to the order of binding, reorientation, and release for one complete influx cycle. (B) Representation of NKCC1 transition states along the outward open (OO)⇄occluded (OCC)⇄inward open (IO) based on cryo-EM data and molecular dynamics (MD) analysis. (C) Alternating access conformational changes in NKCC1 based on cryo-EM and MD analysis showing the "rocking bundle" mechanism where the mobile domain (light green profile) carries out an angular motion that alternates binding site accessibility to either side of the membrane. Note that fully loaded NKCC1's outward open conformation (OO) is permeable to water which, together with the three ions, is released into the cytosol in the inward open (IO) state. *(A) Modified from Lytle et al. (1998). (B and C) Modified from Ruiz Munevar et al. (2024).*

conformational transitions IO ⇄ OO can be explained through a "rocking-bundle" mechanism (Fig. 15.13C). Ruiz-Munevar and coworkers also found evidence that in addition to ions, NKCC1 transports water, consistent with experimental observations (Hamann et al., 2005, 2010; Zeuthen & Macaulay, 2012). These findings open new possibilities for studying the molecular mechanisms of NKCC1 ion and water translocation and modulation and modes of drug action on this crucial protein at unprecedented resolutions.

5.5 Functions of Na^+-K^+-$2Cl^-$ cotransport

The coupled electroneutral cotransport of Na^+, K^+, and Cl^- mediated by NKCCs serves multiple functions depending on the isoforms and cell type expressing these proteins. Some examples of NKCC functions have already been mentioned (Section 5.1) and are further discussed below in this section. The functions can be summarized as follows:

1. NKCCs maintain $[Cl^-]_i$ at levels above those predicted for electrochemical equilibrium. The resulting outward Cl^- gradient is used by epithelial cells to promote transepithelial ion and water transport. In neurons, the outward Cl^- gradient generated and maintained by NKCC1 provides the driving force for the depolarizing efflux of Cl^- through

ligand-gated Cl^- channels activated by neurotransmitters like GABA or glycine, or second messengers like intracellular Ca^{2+}. The ensuing depolarizations play essential roles in central and peripheral neurons.

2. Another major function of NKCCs is their participation in cell volume regulation, not only in cell volume recovery in response to cell shrinkage induced by hypertonic solutions but also in maintaining cell volume in euvolemic (isotonic) conditions.
3. NKCCs contribute to transcellular K^+ reabsorption in the kidney (NKCC2 in TAHL) and extracellular K^+ buffering in various cell types (e.g., apical NKCC1 in choroid plexus epithelial cells)
4. NKCC1 has been implicated in the cell cycle and proliferation.
5. NKCC1 participates in the formation of the K^+-rich endolymph in the inner ear.
6. NKCC2 is the primary salt reabsorption transport pathway in the apical membranes of the TAHL in the kidney. Loop diuretics decrease systemic blood pressure by inhibiting NKCC2 in the TAHL.

5.5.1 NKCCs are key players in transepithelial Cl^--transport

In most cases, Cl^- transporting epithelia are *secretory* (e.g., exocrine glands such as salivary, pancreatic, eccrine, lacrimal, airway epithelium, intestinal crypts, and corneal epithelium), and in a few instances, they are *absorptive* (e.g., TALH). In cells of Cl^- absorptive and Cl^--secretory epithelia, NKCCs serve as the major Cl^- entry pathway and function in concert with three other transport proteins, namely Cl^- and K^+ channels and the Na^+/K^+ pump, to carry out net transepithelial movements of salt (Fig. 15.14). The impact of the Na^+/K^+ pump in the operation of NKCCs in all cell types, whether epithelial or not, cannot be overstated; it is the primary active transport mechanism that generates and maintains the inward Na^+ gradient that, in turn, provides a substantial part of the driving force that energizes the NKCCs (see Section 5.2.3 above and Chapter 17 by Halm in this volume). Thus, the net Cl^- flux in epithelia depends on the Na^+ chemical gradient developed and maintained by primary active transport. In fact, as expected, active transepithelial Cl^- transport is abolished by Na^+ removal or by dissipating the Na^+ gradient between the extracellular fluid and the cell interior (Silva et al., 1977).

In *secretory epithelia* (Fig. 15.14A), the essential components for active transepithelial Cl^- transport are a Cl^- channel in the apical membrane and a three-component mechanism located in the basolateral membrane causing active Cl^- influx across this membrane. The three components that operate in concert to cause net Cl^- uptake across the basolateral membrane are NKCC1, the Na^+/K^+ pump, and a K^+ channel. NKCC1 maintains $[Cl^-]_i$ above electrochemical equilibrium, and transepithelial secretion is produced by the activation of apical Cl^- channels. The Na^+ that enters through NKCC1 is balanced by efflux through the Na^+/K^+-ATPase. The K^+ influx mediated by NKCC1 and the Na^+/K^+ pump is balanced by efflux via K^+ channels. The end result is that transcellular transport involves only Cl^- since both Na^+ and K^+ recycle across the basolateral membrane via the Na^+/K^+ pump and the K^+ channels, respectively (Frizzell & Hanrahan, 2012; Reuss, 1997). In Cl^- secretory epithelia, there is an apical negative transepithelial voltage, which acts as a driving force for paracellular Na^+ transport toward the lumen. The net result is NaCl secretion into the luminal compartment, which generates an osmotic driving force for water flow, resulting in net fluid secretion. Halm discusses the mechanisms and pathways for epithelial solute transport in detail in Chapter 17 of this volume. There is evidence for water being transported through both transcellular (i.e., through the cells) and paracellular (i.e., intercellular, through tight junctions in series with lateral intercellular spaces) pathways (Ayala-Torres et al., 2019; Fischbarg, 2010; Horowitz et al., 2023; Reuss, 1997; Tradtrantip et al., 2009). Paracellular mechanisms of water transport are discussed in detail in Chapter 7 by Nagarajan, Piontek, and Günzel.

Current understanding of the mechanisms and molecular pathways of transcellular water transport is incomplete and often controversial, and it is beyond the scope of the present chapter. Briefly, there are three pathways through which water crosses plasma membranes and potentially contributes to transcellular water passage in epithelia, namely: (1) the lipid bilayer (Finkelstein, 1987, p. 228); (2) aquaporins (Tradtrantip et al., 2009), and (3) cotransporters (Hamann et al., 2010; MacAulay et al., 2009; Zeuthen, 1994, 2010). Lipid membrane bilayers have relatively small permeability to water. Thus, transcellular water transport is thought to occur primarily via aquaporins and cotransporters. There is evidence showing that cotransporters, including NKCC1 and KCCs, cotransport water and ions, as mentioned above, and thus, they may contribute to transcellular fluid transport (Hamann et al., 2010; MacAulay et al., 2009). The recently resolved cryo-EM structures of NKCC1 are consistent with the hypothesis that NKCC1 transports water, as previously mentioned in Section 5.4.

The transport mechanisms in Cl^- *absorptive epithelia*, like the one lining the thick ascending limb of the loop of Henle (TALH), are illustrated in Fig. 15.14B. In this case, the Cl^- channels (ClC-Kb) are located in the basolateral membrane, and the NKCC (NKCC2) is characteristically located in the apical membrane together with K^+ channels (ROMK2). However, the Na^+/K^+ pump is located at the basolateral membrane, so that Na^+ entering the cell via NKCC2 is not

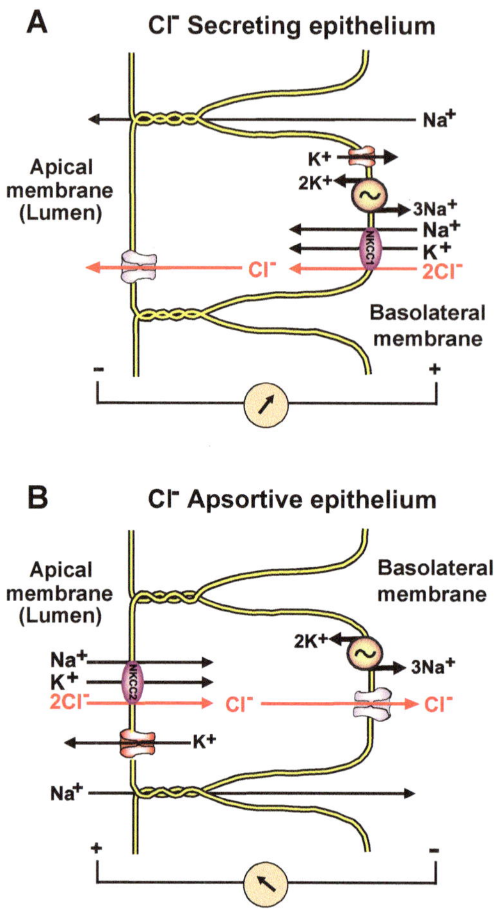

FIGURE 15.14 Basic mechanisms for active Cl^- secretion and absorption in Cl^- transporting epithelia. (A) Transport model for exocrine gland Cl^- secreting epithelium (e.g., salivary glands or tracheal epithelium). Intracellular $[Cl^-]$ is maintained above electrochemical equilibrium by the basolateral Na^+, K^+, $2Cl^-$ cotransporter NKCC1. The Na^+ influx mediated by the cotransporter is balanced by efflux via Na^+/K^+-ATPase; the K^+ influx mediated by the cotransporter is balanced by efflux via K^+ channels. Transepithelial secretion is induced by the activation of apical-membrane Cl^- channels. There is an apical negative transepithelial voltage, which acts as a driving force for paracellular Na^+ transport toward the lumen. This results in NaCl secretion into the luminal compartment, which generates an osmotic driving force for water flow, resulting in net fluid secretion. (B) Model of Cl^- transport in the thick ascending limb of Henle's loop (TAHL), a Cl^--absorptive epithelium. Cl^- entry across the apical membrane is mediated by NKCC2. Most basolateral Na^+ and Cl^- efflux proceeds through the Na^+/K^+-ATPase and Cl^- channels, respectively.

recycled but instead undergoes transcellular transport. The same occurs with Cl^-; it enters through the apical NKCC2 and exits through basolateral Cl^- channels. Thus, NKCC2 contributes to net Na^+ and Cl^- reabsorption. Most of the K^+ taken up by NKCC2 recycles across the apical membrane via inward rectifier K^+ channels ROMK2. In the TALH, Cl^- absorption is entirely transcellular, whereas Na^+ absorption is partly transcellular and partly paracellular (Gamba & Friedman, 2009). The paracellular Na^+ flux is driven by the transepithelial voltage, which, in contrast with Cl^--secreting epithelia, is lumen positive. Unlike NKCC1, NKCC2 does not appear to cotransport water (Hamann et al., 2005; Zeuthen and Macaulay, 2012).

Primary sensory neurons (PSNs) convey virtually all somatic and visceral information to the spinal cord and brainstem, including nociceptive signals, i.e., incoming signals arising from nociceptors. The latter are sensory receptors capable of transducing and encoding noxious stimuli. NKCC1 maintains $[Cl^-]_i$ of PSNs above electrochemical equilibrium throughout adulthood (Mao et al., 2012). In this case, the Na^+ and K^+ entering the cell via NKCC1 recycle through the Na^+/K^+ ATPase and K^+ channels, respectively (Fig. 15.15). Thus, $[Na^+]_i$ and $[K^+]_i$ are maintained at a constant level by the Na^+/K^+ ATPase. The outward Cl^- gradient generated by NKCC1 is maintained throughout the entire cell surface of sensory neurons, including the soma (located in DRG ganglia) and the central and peripheral processes. The expression of NKCC1 along the entire surface of sensory neurons has been ascertained with immunolabeling methods in intraspinal terminals and the cell bodies (Alvarez-Leefmans et al., 2001; Javdani et al., 2020), although in peripheral nociceptors, the

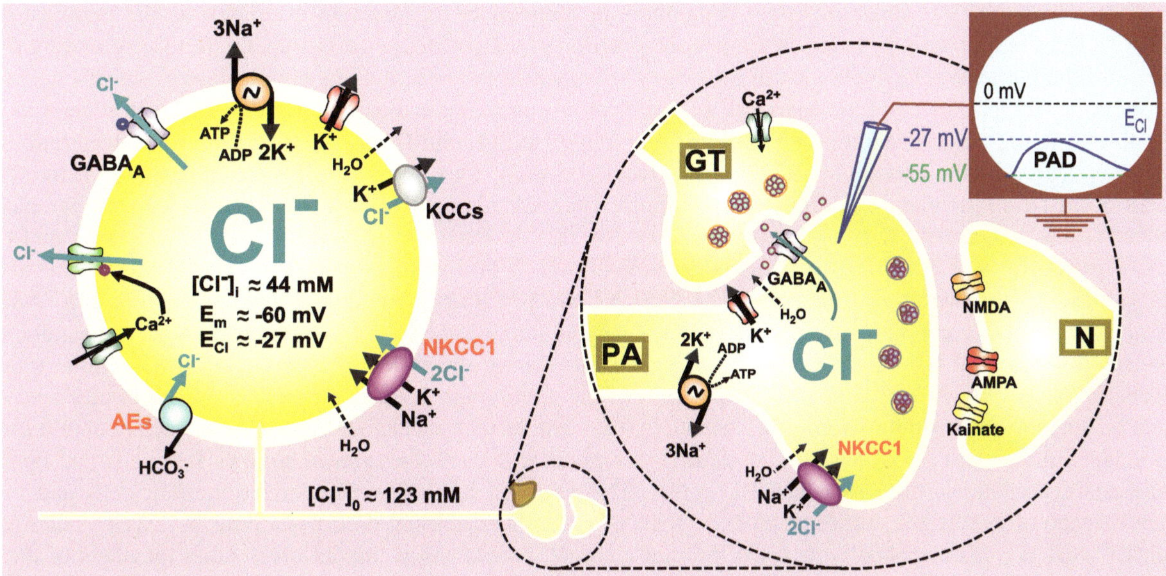

FIGURE 15.15 Schematic diagram depicting the transporters and channels involved in intracellular Cl^- regulation in primary sensory neurons and the prevailing model of generation of primary afferent depolarization (PAD) and presynaptic inhibition in the intraspinal terminals of these neurons. The left circle represents a mammalian DRG neuron showing measured average values of intracellular Cl^- concentration, $[Cl^-]_i$; membrane potential (E_m) and Cl^- equilibrium potential, E_{Cl}. The Na^+/K^+ ATPase maintains the Na^+ and K^+ gradients across the membrane. AEs, anion exchangers (AE2 and AE3); NKCC1, Na^+, K^+, $2Cl^-$ cotransporter 1; KCCs, are K^+-Cl^- cotransporters 1,3, or 4. With the possible exception of KCC3, the relative contribution of KCCs to intracellular Cl^- regulation in these neurons is not fully understood. Unlike CNS nerve cells, KCC2 is not expressed in primary sensory neurons. The diagram on the right, encircled by a dashed line, represents a magnified intraspinal primary afferent terminal from which an intracellular micropipette connected to an oscilloscope monitors E_m. PA is a glutamatergic primary afferent; GT is a GABAergic terminal from an interneuron; N is a postsynaptic terminal with glutamate receptor channels. PAD (blue trace in the "oscilloscope" screen) is generated by the activation of presynaptic GABA receptor channels, and it is associated with inhibition of neurotransmitter release from the PA terminal. *Modified from Alvarez-Leefmans (2012).*

evidence comes from pharmacological and physiological experiments (Granados-Soto et al., 2005). This Cl^- gradient is of high functional significance; at the peripheral endings, contributes to excitation of nociceptors produced by chemical mediators of itch and pain released upon tissue damage (Kim et al., 2024; Takayama & Tominaga, 2024), whereas at the central terminals, it makes possible the GABA-mediated depolarizing Cl^- efflux underlying presynaptic inhibition, a key mechanism in gating the flow of somatosensory information of various sensory modalities to the spinal cord and brain stem (Alvarez-Leefmans, 2009; Comitato & Bardoni, 2021; Fernandes et al., 2020; Guo & Hu, 2014; Kaila et al., 2014; Witschi et al., 2011). The mechanisms that make possible presynaptic inhibition in the spinal cord are depicted in Fig. 15.15. The evidence supporting this model has been gathered for over 80 years of research, although it has not remained unchallenged (Hari et al., 2022; Hochman et al., 2010; Mahrous et al., 2023). The prevailing notion is that GABA released from dorsal horn interneurons depolarizes primary afferent fibers via axo-axonic synapses. This primary afferent depolarization (PAD) inactivates Na^+ channels, thereby reducing the amplitude of the action potential as it invades the primary afferent terminals, resulting in a reduction in neurotransmitter release from these terminals (Comitato & Bardoni, 2021; Rudomin & Schmidt, 1999; Willis, 2006). The depolarizing inward current results from the opening of $GABA_A$-gated Cl^- channels expressed in intraspinal terminals of primary afferent fibers. Opening of these GABA-gated channels results in efflux of Cl^- which transiently drives the transmembrane potential toward the E_{Cl}. This depolarizing action of GABA is possible because the $[Cl^-]$ inside the nerve terminals of primary sensory neurons is maintained at a level higher than predicted from a passive distribution, thereby providing the necessary driving force for electrodiffusional (i.e., channel-mediated) depolarizing Cl^- efflux. An example of how this mechanism can impact nociceptive signaling is the antihyperalgesic action of diazepam, a $GABA_A$ receptor modulator that enhances PAD by increasing the open probability of Cl^- channels activated by GABA. By potentiating PAD and presynaptic inhibition on nociceptive primary afferents, diazepam has been found to have an antihyperalgesic action against inflammatory hyperalgesia in vivo (Witschi et al., 2011). Hyperalgesia is defined as increased pain sensitivity. Another example is the effect of NKCC1 pharmacological or molecular inactivation of NKCC1 on pain thresholds (Granados-Soto et al., 2005; Sung et al., 2000).

NKCC1 is involved in sensory transduction. Two well-documented examples are the *inner ear* and the *olfactory sensory neurons*. The mechanisms of acoustic transduction have been recently reviewed (Marcus, 2020). Here, we briefly examine the role of NKCC1 as an important component of the sensory transduction machinery in the inner ear (Watabe et al., 2017). Hearing critically depends on the ion composition of the cochlear endolymph, the fluid surrounding the upper surface of mechanosensitive hair cells. These cells carry out mechanotransduction, the conversion of a mechanical stimulus (sound-induced vibration of hair cells cilia) into an electrical response that is then processed by the central nervous system. Endolymph has a high K^+ concentration (157 mM) and a low Na^+ (1.3 mM) and Ca^{2+} ($\approx$20 μM) concentrations. Thus, the ion composition of the endolymph resembles intracellular fluid. This ionic composition depends on active transport processes in the stria vascularis, a specialized epithelium at the lateral cochlear wall. NKCC1 is located at the basolateral membrane of the strial marginal cells, together with the Na^+/K^+ ATPase and Cl^- channels (ClC-Ka/barttin and ClC-Kb/barttin), whereas the apical membrane, facing the "lumen" which in this case is the endolymph-filled scala media, expresses outwardly conducting K^+ channels (KCNQ1/KCNE1). Clearly, strial marginal cells form a *K^+- secreting epithelium*; Na^+ and Cl^- entering the cells via NKCC1 recycle through the Na^+/K^+ pump and the Cl^- channels, whereas K^+ entering through NKCC1 and the Na^+/K^+ pump exits through the apical K^+ channels. Thus, transcellular transport in these cells involves only K^+. The unique composition of the endolymph is needed because the apical mechanosensitive channels of sensory hair cells function as K^+ channels (they are, in fact, mechanosensitive nonselective cation channels). Sound-induced movement of the basilar and tectorial membranes deflects the cilia of sensory hair cells and opens these mechanosensitive channels through which K^+ enters. The resulting K^+ influx depolarizes the hair cells and triggers a complex cascade of events that culminate in exocytosis of synaptic vesicles at their basal pole. K^+ leaves outer hair cells through KCNQ4 K^+ channels and is recycled through a complex mechanism that involves other members of the SLC12 family of cation-coupled cotransporters (Hubner & Rust, 2007; Mistrik & Ashmore, 2009; Zdebik et al., 2009). The important role played by SLC12 members for inner ear ion homeostasis and endolymph secretion may explain the well-known ototoxic effects of loop diuretics (Ikeda et al., 1997) and the deafness exhibited by gene disruption of some members of the SLC12A family in knockout mouse models and humans exhibiting mutations in NKCC1 (Delpire & Koumangoye, 2023).

In *olfactory sensory neurons*, the transduction strategy differs from that of inner hair cells in the cochlea, but the mechanism also depends on NKCC1 function (Frings, 2009). Mammalian olfactory receptor neurons (ORNs) present to the air a tuft of sensory cilia expressing odorant receptors. Upon contact with odorants, these receptors trigger a cascade of events that culminate with the firing of action potentials. The cascade of events begins with the activation of odorant receptors, resulting in a rise of the second messengers cAMP and Ca^{2+} inside the cilia, a process that involves cAMP-gated-Ca^{2+}- channels. The resulting Ca^{2+} influx opens Cl^- channels TMEM16B, also known as anoctamin 2 or ANO 2 Ca^{2+}-activated channels. ORNs accumulate Cl^-, partly through NKCC1, and maintain a higher than equilibrium intracellular $[Cl^-]$ concentration, thereby setting an outward Cl^- gradient. Thus, the opening of Ca^{2+} activated Cl^- channels (TMEM16B) leads to a depolarizing Cl^- efflux from the cilia, a mechanism that amplifies the receptor potential approximately 10-fold, thus helping to excite the neuron even when stimulation is weak (Li et al., 2018). Accordingly, disruption of the NKCC1 gene and the pharmacologic suppression of Cl^- accumulation or blocking the Cl^- efflux strongly inhibit the sensory response of ORNs. Experiments with NKCC1 knockout mice have shown that NKCC1 is not the only Cl^- transporter active in ORNs. NKCC1 knockout mice retain the ability to smell, a finding suggesting an additional Cl^--uptake mechanism exists. The latter turned out to be the AE1, i.e., SLC4A1 (Hengl et al., 2010).

The role of NKCC1 in *cell volume regulation* and maintenance cannot be overstated. In many cells, NKCC1 is involved in regulatory volume increase (RVI) in response to hypertonic cell shrinkage by promoting net gain of K^+, Cl^-, and osmotically obliged water and possibly water cotransport. For detailed accounts on this subject, the reader should consult the following reviews (Hoffmann et al., 2009; Jentsch, 2016; Russell, 2000), and in this volume, Chapter 22 by Feher and Baumgarten. It is important to note that RVI is a response of cells when experimentally exposed to hyperosmotic solutions. Although the osmolarity of these experimental solutions is seldom found under physiological conditions, the study of RVI responses has yielded invaluable insights into the molecular mechanisms of cell volume control in euvolemic (physiological) conditions (O'Neill, 1999). NKCC1 is not only involved in RVI but also in *isosmotic volume control*, a mechanism by which cells keep their volume constant under isotonic and isosmotic physiologic conditions, as discussed below.

In single-cell bodies of primary sensory neurons (i.e., DRG neurons), known to express NKCC1 abundantly, it has been possible to measure changes in cell water volume resulting from NKCC1 activation in isosmotic solutions. These studies show a tight link between NKCC1 activity, water fluxes, Cl^- fluxes, and steady-state $[Cl^-]_i$ (Rocha-Gonzalez et al., 2008). *Isosmotic* removal of external Cl^- results in intracellular Cl^- depletion and concomitant cell shrinkage. The extent of the cell shrinkage is accounted for in terms of measured Cl^- depletion. Water influx occurs in isosmotic media upon readmission of the isosmotic control solution. This water influx requires the presence of external Na^+ and K^+ and is abolished by 10 μM bumetanide. Thus, net water flux is tightly linked to NKCC1 transport activity. In DRG neurons, NKCC1 determines not only the higher than passive $[Cl^-]_i$ characteristic of these cells throughout adulthood but also the *cell volume set point* through a negative feedback system in which intracellular Cl^- regulates water and Cl^- influx, thereby maintaining intracellular Cl^- concentration and cell volume constant (Alvarez-Leefmans, 2009). It is likely that due to their small volume/surface ratio, intraspinal presynaptic nerve terminals of primary sensory neurons are expected to be partially depleted of solutes (Cl^- and K^+) and water upon repetitive GABA action via $GABA_A$ receptor channels (Momose-Sato et al., 1998). It has been proposed that NKCC1 may restore ionic gradients and osmotic balance altered as a consequence of GABA action (Alvarez-Leefmans et al., 1998; Schomberg et al., 2003).

One of the most compelling experimental proofs that NKCC1 maintains cell water volume in isosmotic conditions comes from experiments in single choroid plexus epithelial cells in vitro and in situ (Gregoriades et al., 2019). In these experiments, cell water volume was measured in real time in calcein-loaded cells using a well-established method (Alvarez-Leefmans et al., 1995). The experiments show that blocking NKCC1 with bumetanide (10 μM) produces cell shrinkage (~16% decrease in relative cell water volume). However, as expected, such an effect is absent in cells from NKCC1 knockout mice ($NKCC1^{-/-}$). One of these experiments is shown on the cover of this book (right panel). Further, live-cell-imaging analysis of the cross-sectional areas of freshly dissociated choroid plexus epithelial cells from NKCC1 wild type ($NKCC1^{+/+}$) versus cells of $NKCC1^{-/-}$ shows that the latter are significantly smaller than those of wild type, an observation that was corroborated in choroid plexus epithelial cells in situ, using electron microscopy. These findings show that apical NKCC1 is continuously active under basal conditions and works in the net inward flux mode, maintaining $[Cl^-]_i$ and CWV, which are expected to be needed for CSF secretion.

As mentioned above, *NKCC1 exhibits some of the features required for an efficient extracellular K^+ buffer*; i.e., it has a substantial net inwardly directed driving force and a high affinity for external K^+ (Payne & Forbush, 1995). There is evidence for its role as a K^+-uptake mechanism that contributes to buffering and regulating extracellular K^+ in some tissues (Section 5.1). One good example is apical NKCC1 in choroid plexus epithelial cells, which, in concert with apical Na^+/K^+ ATPase, may play an essential role in the reabsorption of K^+ from the cerebrospinal fluid to the blood, thereby contributing to the buffering and regulation of CSF $[K^+]$ (Alvarez-Leefmans, 2020a; Fame et al., 2023; Gregoriades et al., 2019; Wu et al., 1998; Xu et al., 2021) (Fig. 15.17). As mentioned above, NKCC1 is also prominently located in the paranodal region of Schwann cells of myelinated axons, an area prone to extracellular K^+ accumulation resulting from axonal impulse activity (Ransom et al., 2000). This location and abundance of NKCC1 suggest its possible involvement in K^+ uptake from the periaxonal space, thereby regulating the external $[K^+]$ in this space and maintaining axonal excitability and osmotic balance (Alvarez-Leefmans, 2009; Alvarez-Leefmans et al., 2001). Consistent with this hypothesis, NKCC1b mutants show severe and progressive pathology in the peripheral nervous system, which is characterized by dysmyelination and swelling of the periaxonal space at the axon—myelin interface (Marshall-Phelps et al., 2020).

6. Electroneutral K^+-Cl^- cotransporters

The K^+-Cl^- cotransporters (KCCs) belong to the branch of the sodium-independent CCCs gene/protein family comprising four members: KCC1 to KCC4 (SLC12A4 to SLC12A7). The K^+-Cl^- cotransport mechanism was discovered more than 4 decades ago. It represents the seminal work of Peter Lauf (Lauf & Theg, 1980), Philip Dunham, Clive Ellory (Dunham et al., 1980), and their colleagues. The KCCs are carrier proteins that mediate the coupled, electroneutral movement of K^+ and Cl^- across the plasma membrane. This coupled transport is obligatory as it requires the simultaneous presence of both K^+ and Cl^- on the same side of the membrane for net transport to proceed. As with other electroneutral secondary active transport mechanisms, the direction of the net movement of K^+ and Cl^- by KCCs is determined solely by the sum of the chemical potential gradients of the two cotransported ions. Under normal physiological conditions, the outwardly directed K^+ chemical potential gradient maintained by the Na^+/K^+ pump drives the uphill movement of Cl^- against its chemical potential gradient. Hence, KCCs are efflux pathways for K^+ and Cl^- under normal conditions. However, the cotransporters are bidirectional and can mediate net ion efflux or influx, depending upon the prevailing K^+- and Cl^--chemical potential gradients. In neurons, this is crucial because KCC2, the isoform predominantly expressed in these cells, might work near its thermodynamic equilibrium and, therefore, it has the potential to mediate Cl^- fluxes in both directions (Payne, 2009).

The stoichiometry of KCC is $1K^+:1Cl^-$, as determined in rabbit red blood cells (Jennings & Adame, 2001). This means the transport process involves the one-for-one movement of K^+ with Cl^-. Thus, the overall transport process of K^+ and Cl^- via the cotransporter is not driven by the transmembrane voltage, nor does the transport process directly generate a membrane current that may change the transmembrane voltage (Kaji, 1993). The structures and activation mechanisms of human KCCs have been elucidated using cryo-EM at 2.9 to 3.6-Å resolution (Chi et al., 2021; Liu et al., 2019; Xie et al., 2020). The details of these structures and mechanisms have been recently reviewed (Portioli et al., 2021; Trejo et al., 2023).

KCCs have been implicated in regulatory volume decrease in response to hyposmotic cell swelling by promoting net efflux of K^+ and Cl^- and osmotically obliged water (Delpire & Gagnon, 2018b; Hoffmann et al., 2009). Significant hypotonic challenges can activate all four main isoforms of the cotransporter. However, the fact that they are volume-sensitive does not necessarily mean that the physiological function of these cotransporters is to regulate cell volume. As lucidly stated by Payne (2009), "… a transporter should be considered volume regulatory only after it has been demonstrated that the transport rate is a graded function of the stimulus and that the net ion transport is large enough to restore cell volume to normal. In the absence of such measurements, it is best to refer to the transporter as volume sensitive until more extensive experimentation is conducted." Indeed, experimental evidence shows that KCCs, in conjunction with volume-sensitive ion channels, mediate anisosmotic cell volume regulation (Jentsch, 2016). What is less clear is whether they play a role in isosmotic cell volume maintenance. In this respect, an exception among the four isoforms of KCCs is KCC3, which acts as a cell volume maintenance mechanism in red blood cells (Rust et al., 2007) and peripheral nerve fibers (Flores et al., 2019). Besides their function in cell volume regulation, K^+-Cl^- cotransporters play essential roles in the vectorial movement of salt and water across some epithelia and in the maintenance of $[Cl^-]_i$ (Bazúa-Valenti et al., 2016; Gamba, 2005; Reuss, 1997; Velazquez & Silva, 2003).

One of the most studied isoforms of the K^+-Cl^- cotransporters is KCC2 (encoded by SLC12A5), which is expressed predominantly in central neurons, where it carries out the uphill extrusion of Cl^- keeping E_{Cl} at more negative values than E_m, making possible the hyperpolarizing action of inhibitory neurotransmitters like GABA and glycine in the adult central nervous system (Kaila et al., 2014; Payne, 2009; Payne et al., 2003). A brief account of what is known about the function, distribution, and pathology of KCCs is presented below.

6.1 Basic features of the K^+-Cl^- cotransporters

The basic functional characteristics of KCCs have been summarized by their discoverers (Gibson et al., 2009). A defining feature of KCCs is that the transport process is Na^+ independent. Another key feature of KCCs is that the thiol-alkylating reagent N-ethylmaleimide (NEM) stimulates transport (Lauf & Theg, 1980). The cotransporters can also be activated by osmotic cell swelling (Ellory & Dunham, 1980). Reducing $[Mg^{2+}]_i$ stimulates the cotransporter, presumably via effects on regulatory kinases (Delpire & Lauf, 1991). Likewise, KCC is stimulated by the protein kinase inhibitor staurosporine (Bize & Dunham, 1994). Unlike the NKCCs or the NCC, KCCs can be inhibited by disulfonic acid stilbenes such as DIDS (Delpire & Lauf, 1992). However, this effect appears to be species-specific, given that it has been observed in sheep but not in human RBCs (Delpire et al., 2009). KCCs are also inhibited by loop diuretics, but, in general, the sensitivity for furosemide is significantly higher than for bumetanide, as discussed above in Section 5.2.4. They can also be inhibited by dihydroindenyloxyalkanoic acid (DIOA), which is often erroneously used as if it were a "specific" KCC inhibitor. A series of KCC2 inhibitors, such as N-(4-methylthiazol-2-yl)-2-(6-phenylpyridazin-3-ylthio) acetamide, have been reported (Delpire et al., 2009). The latter inhibitors are more potent than furosemide and DIOA. Loss of activity of KCC2 has emerged as a critical mechanism underlying several neurological and psychiatric disorders, including various types of epilepsy, motor spasticity, stress, anxiety, schizophrenia, morphine-induced hyperalgesia, and chronic pain. This has motivated the search for enhancers and/or modulators of KCC2 function (Gagnon et al., 2013; Jarvis et al., 2023; McMoneagle et al., 2024). The potential therapeutic impact of these drugs is high. The knowledge obtained from cryo-EM structural information on KCCs and NKCCs has opened new avenues for the structure-based design of selective modulators of these cotransporters.

6.2 Molecular structure, distribution, and functions of K^+-Cl^- cotransport proteins

Four distinct isoforms and several splice variants of these members of the cation-coupled Cl^- -cotransporter family (SLC12) have been identified. The isoforms are KCC1, KCC2, KCC3, and KCC4. They are encoded by the genes SLC12A4, SLC12A5, SLC12A6, and SLC12A7, respectively. All of these genes in vertebrates code for proteins over 1000 amino acids with core molecular weights of 120–128 KDa (Payne, 2009). The 4 KCCs exhibit alternative exon splicing and the use of alternative promoters and first exons, leading to even greater protein diversity. The four KCC proteins are 69%–75% identical when comparing their full-length sequences. The most variable segment of these proteins is the N-terminal domain, defined as the portion that extends from the first methionine to the beginning of TM1. The identity of the N-terminal domain sequence among KCCs varies between 38% and 50% (Gamba, 2005). It is important to mention that in the original cloning report of Mount and coworkers (1999), the isoform KCC3 (Hiki et al., 1999) was first named KCC4, and the latter was named KCC3. This was clarified in subsequent work. In the present chapter, the corrected nomenclature is followed.

NKCC1, NKCC2, NCC, and the 4 KCCs share a basic membrane topology consisting of a central core of 12 TM domains flanked by hydrophilic amino- and carboxy-terminal domains with a cytoplasmic orientation. The carboxy-termini of KCCs make up nearly half the size of these proteins (Fig. 15.16). When compared over their entire length, KCC proteins are 25%–30% identical to NCC and NKCCs. A significant structural difference between the KCCs and the NKCCs is the position and size of a large glycosylated extracellular loop, occurring between TM5 and TM6 in the KCCs and between TM7 and TM8 in the NKCCs and the NCC (Mount et al., 1999). This can be appreciated by comparing Fig. 15.12 with Fig. 15.16. In the KCCs, the large extracellular loop linking TM5 and TM6 comprises 120 amino acids. In NCC and NKCCs, an extracellular loop linking TM5 and TM6 exists, but it is significantly shorter, comprising 30 amino acids (Delpire & Guo, 2020).

KCC1 was the first isoform of the KCCs to be cloned and functionally characterized by heterologous expression in HEK 293 cells (Gillen et al., 1996). The gene for human KCC1 is at chromosomal location 16q22.1. KCC1 has a deglycosylated molecular mass of ~120 kDa, and it is widely expressed in mammalian tissues (brain, kidney, heart, lung, liver, stomach, colon, spleen, and placenta). This ubiquitous tissue distribution suggested that KCC1 may represent a "housekeeping" isoform responsible for cell volume regulation and maintenance and for regulating $[Cl^-]_i$. However, this "housekeeping" role is difficult to reconcile with the fact that disruption of the KCC1 gene does not produce any obvious phenotype, even in vertebrate red blood cells where KCC1 was thought to be the key player in cell volume control. In fact, KCC1 knockouts are indistinguishable from wild-type animals (Hubner & Rust, 2007; Rust et al., 2007). It is now known that red blood cells express KCC1, KCC3, and KCC4, but that the volume-sensitive K^+-Cl^- cotransporter activity in mouse and human red blood cells is mediated mainly by KCC3 (Hubner & Rust, 2007; Pan et al., 2011). Heterologous expression of KCC1 in HEK-293 cells showed that the cotransporter is little active in isosmotic media. Still, its activity can

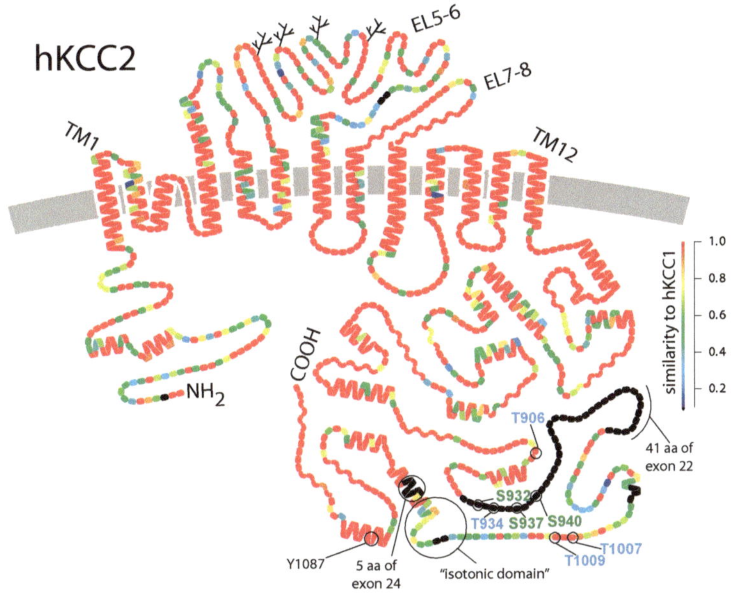

FIGURE 15.16 Structural model of the K^+-Cl^- cotransporter 2. Human KCC2 (hKCC2b) and human KCC1 (hKCC1) are compared to one another by colors that indicate the degree of similarity on a per residue basis: Red residues are identical, and black residues are absent from hKCC1. The branched thin lines in black in the large extracellular loop (EL5-6) between transmembrane segments (TM) five and six are N-linked glycosylation sites. The second extracellular loop between TM7 and TM8 (EL7-8) is shorter and highly conserved in KCSs, and it is involved in extracellular gating by salt bridge formation. Secondary structural elements are shown as helices for α-helix and wavy lines for β-sheet. The 15 amino acids region required for constitutive "isotonic" activity of KCC2 is encircled (isotonic domain). KCC2 has a large region in the carboxy terminus (COOH), exhibiting little identity with the other KCCs. This unique region of KCC2 is coded by two extra exons (exons 22 and 24), which are absent from the other KCCs. Exon 22 of KCC2 codes for 41 amino acids, whereas exon 24 codes for five amino acids. Exon 23 of KCC2 is conserved among the KCCs. However, the identity of the amino acids in exon 23 of KCC2 is low when compared to other KCCs. There are several conserved phosphorylation sites in KCC2. Protein kinase C directly phosphorylates serine 940 (S940) of KCC2. S940 phosphorylation is required for activation of KCC2 by protein kinase C. Besides S940, phosphorylation of S932, T934, and S937 enhances KCC2 activity (Radulovic et al., 2023). All of these sites are encoded by exon 22, which is only present in all vertebrate KCC2 and nontherian KCC4 isoforms. There are phosphorylation sites that enhance KCC2 activity upon dephosphorylation. These are threonines T906, T1007, and T1009. They are highly conserved among all vertebrate KCC isoforms. A consensus tyrosine-protein kinase phosphorylation site is also shown (Y1087). Mutation of this tyrosine residue (Y1087D) inhibits constitutive "isotonic" activity, as well as swelling-induced activity in heterologous, expressed KCC2. John Payne and Bliss Forbush kindly provided the backbone of the figure generated with DNAPLOT. *Modified from Alvarez-Leefmans (2012).*

be stimulated by cell swelling (~twofold) and N-ethylmaleimide and inhibited by furosemide (K_i of ~40 μM) and by bumetanide (K_i of ~60 μM). When KCC1 is expressed in *Xenopus* oocytes, it also shows minimal transport activity in isosmotic media. However, when exposed to hypoosmotic solutions, its activity increases by 20-fold (Mercado et al., 2000). The cotransporter in the *Xenopus* expression system is inhibited by bumetanide and furosemide with a K_i of 180 μM. Thus, the sensitivity to loop diuretics of KCC1 varies between expression systems but, in all cases, is significantly different than that of NKCCs. The latter are consistently more sensitive to bumetanide than furosemide; the K_i for bumetanide is at least an order of magnitude lower. Even though the expression of human KCC1 has been found in every cell and tissue tested so far, its specific physiological function remains unclear. The fact that KCl cotransport activity in red blood cells is undiminished in KCC1-knockout mice suggests that this protein has other hitherto unknown functions. KCC1 appears to be involved in red blood cell maturation and sickling. In addition, there is evidence suggesting that it might be involved in cancer development and bone resorption (Garneau et al., 2019). Although the function of KCC1 is not clear, its molecular structure using cryo-EM was first reported in 2019 for the human isoform (hKCC1) by Liu and collaborators (Liu et al., 2019), and more recently by Zhao and associates (Zhao et al., 2022b). A detailed comparative analysis of the cryo-EM structure of KCC1 with other CCSs has been recently reviewed (Delpire & Guo, 2020; Trejo et al., 2023). The cryo-EM structures have allowed a better understanding of CCCs general architecture and many functional details of the transport mechanisms.

KCC2 is often called the "neuron-specific isoform" of the K^+-Cl^- cotransporters; however, given that it is also expressed in avian cardiomyocytes (Antrobus et al., 2012) and the pancreas (Kursan et al., 2017), it is safer to say that KCC2 is "predominantly expressed" in the brain. In the pancreas, KCC2 is localized to glucagon-secreting α-cells and to insulin-secreting β-cells where it regulates insulin secretion (Kursan et al., 2017; Shimizu-Okabe et al., 2022). KCC2 was first cloned and characterized by John Payne and colleagues (Payne, 1997; Payne et al., 1996). The cryo-EM structures of KCC2 have been elucidated (Chi et al., 2021; Xie et al., 2020; Zhang et al., 2021). This cotransporter is present in the plasma membrane of cell bodies and dendrites of various types of CNS and spinal neurons but appears to be absent in axons, glial cells, and dorsal root ganglion neurons (Akhter et al., 2019; Mao et al., 2012; Stil et al., 2009; Williams et al., 1999); reviewed in Blaesse et al. (2009), Kaila et al. (2014), Payne (2009), Viemari et al. (2011). As first proposed by Payne and colleagues, KCC2 is the isoform that actively extrudes Cl^- from nerve cells, rendering E_{Cl} more negative than E_m and thus setting an inwardly directed electrochemical gradient for Cl^- making possible hyperpolarizing inhibition mediated by $GABA_A$ or glycine-gated Cl^- channels in mature neurons. Although it had long been suspected that the mechanism of active Cl^- extrusion in CNS neurons was a K^+-Cl^- cotransporter [reviewed in (Alvarez-Leefmans, 1990)], its molecular identity and its functional and kinetic features remained elusive until the pioneering work of John Payne, Kai Kaila and their colleagues demonstrating the causal role of KCC2 upregulation in the development of hyperpolarizing inhibition (Kaila et al., 2014; Rivera et al., 1999). The cloning and molecular characterization of KCC2 has been a highly significant discovery in the field of Cl^- -transport because dysfunction of this cotransporter underlies several neurological and psychiatric diseases, including epilepsy, spinal cord injury, Rett syndrome, and autism spectrum disorder (McMoneagle et al., 2024; Virtanen et al., 2021). Interestingly, KCC2 might not be the only KCC underlying hyperpolarizing inhibition; in a few neurons, KCC3 also appears to have a significant role in setting the inward Cl^- gradient needed for hyperpolarizing inhibition (Boettger et al., 2003). However, KCC3 is expressed at significantly lower levels than KCC2 in the human brain after birth (Sedmak et al., 2016). Further, unlike KCC2, KCC3 expression in the CNS does not increase with CNS development. Indeed, one of the significant discoveries in this field in the last few decades is that CCCs are developmentally regulated in the CNS. KCC2 undergoes robust developmental upregulation, NKCC1 shows moderate postnatal upregulation, and KCC3 and NKCC1 are expressed at high levels throughout life while expression of KCC1 and KCC4 is low (Kaila et al., 2014). The fact that KCC2 expression in the CNS increases with development and becomes functionally dominant over NKCC1 explains why GABA and glycine responses, which are depolarizing in embryonic or early postnatal nerve cells, become hyperpolarizing and inhibitory in the adult brain (Kaila et al., 2014; Rivera et al., 1999). Complementing these observations is the demonstration that matured spinal sensory neurons, which express NKCC1 and lack KCC2, maintain a higher than equilibrium $[Cl^-]_i$ throughout adult life and are depolarized by GABA (Mao et al., 2012).

The mammalian KCC2 gene (SLC12A5) generates two KCC2 variants named KCC2a and KCC2b, with different N termini (Uvarov et al., 2007). This N-termini heterogeneity is due to the use of alternative promoters and alternative first exons. Exon1a encodes 40 N-terminal amino acid residues, giving rise to KCC2a, and exon 1b encodes 17 amino acids, giving rise to KCC2b, the variant originally described by Payne et al. (1996). KCC2b is the predominant isoform in the mature brain. Little is known about the function of KCC2a (Virtanen et al., 2021). There is evidence that this variant might play a role in the development of the central respiratory command (Dubois et al., 2018). As mentioned above, KCC2 is assembled as a dimer, and both the cytoplasmic C-terminal and transmembrane domains are involved in dimerization. It has been speculated that KCC2a heterodimerization with KCC2b in vivo establishes a tight interaction of the respective C-termini, allowing SPAK/OSR1 to phosphorylate the C-terminal threonines of both KCC2 variants in the heterodimer (Virtanen et al., 2021).

Both KCC2 variants are prominently expressed in CNS neurons but differ in timing, level, and expression site. During postnatal development, KCC2b is steeply upregulated, whereas KCC2a is not. KCC2b is more abundant in the cortex, whereas KCC2a appears more abundant in the brainstem and spinal cord. Full KCC2 knockout mice die immediately after birth due to respiratory failure and other motor disturbances (Hubner et al., 2001). In contrast, mice lacking KCC2b, retaining normal levels of KCC2a, survive for up to 2 weeks, although they exhibit abnormal posture and gait and generalized seizures (Woo et al., 2002). Expression of either KCC2a or KCC2b in HEK 293 cells demonstrated that both variants exhibit comparable cotransport activity. Furthermore, both variants exhibit constitutive activity in isosmotic media, a functional characteristic that makes KCC2 suitable for controlling $[Cl^-]_i$ (Payne, 2009). Since most published data reflects the expression of both KCC2a and KCC2b, "KCC2" will be used here to refer to both splice variants unless otherwise stated.

Human KCC2 has a core molecular mass of 123.5 kDa (Payne, 2009), close to that of mouse and rat KCC2 (123.6 KDa). Western blots of membranes isolated from rat brains show that the glycosylated molecular mass of KCC2 is $= \approx$ 140 kDa. Upon deglycosylation, the protein migrates to near $\approx$125 kDa, close to the predicted molecular mass (Williams et al., 1999). Similarly, human KCC2 has a glycosylated molecular mass of $\approx$140 KDa, and upon deglycosylation, the protein migrates to a broad band centered around 120 KDa (Munoz et al., 2007). KCC2 expressed in HEK 293 is also stimulated by NEM and inhibited by furosemide (K_i of $\approx$25 μM) and bumetanide (K_i of $\approx$55 μM). DIDS and DIOA also significantly inhibited KCC2, but inhibition by either drug at 100 μM was not complete compared to furosemide (Payne, 1997). The main functional difference from other KCC isoforms is that KCC2 is not stimulated by osmotic swelling when expressed in HEK 293 cells. However, when expressed in Xenopus oocytes, its activity increases 3.5–20-fold by hypoosmotic swelling (Gamba, 2005; Payne, 2009). These and other discrepancies (e.g., bumetanide sensitivity in HEK cells vs. *Xenopus*) highlight the limitations of studying transporters in heterologous expression systems. A key feature of KCC2 is that it exhibits significant basal constitutive activity, i.e., it is active in isotonic media, a functional property shared with KCC3a (see below). From the functional perspective, this constitutive activity makes these KCCs ideal for establishing the steady state $[Cl^-]_i$ of adult CNS neurons. The segment of the KCC2 protein conferring constitutive "isotonic" activity, known as the "isotonic domain," comprises 15 amino acids coded by exon 23 (Fig. 15.16).

Like NKCC1, KCC2 exhibits all the features required for an efficient $[K^+]_o$ buffer: (1) it is reversible, and the direction of net transport is sensitive to subtle changes in $[K^+]_o$; (2) the affinity of the cotransporter for external K^+ is compatible with the range of the observed $[K^+]_o$ (3–10 mM), allowing a high rate of cotransport activity. These requirements seem to be a feature of both KCC2 ($K_m \approx$ 5.2 mM) and KCC3 ($K_m \approx$ 10 mM). Hence, it has been postulated that KCC2 could also act as a $[K^+]_o$ buffer (Payne, 1997). However, external $[K^+]$ in CNS interstitial fluid is tightly regulated and, under normal physiological conditions, never exceeds 3 mM. Only during extreme seizure activity $[K^+]_o$ exceeds 3 mM and it can reach up to 12 mM, the so-called *ceiling* level (Somjen, 2004). Thus, most likely, KCC2 works typically in the efflux mode.

Eric Delpire and collaborators conducted a thorough screening of small molecules in search of selective KCC2 inhibitors (Delpire et al., 2009), which are essential tools to explore the physiology and pathology of KCC2. They used a fluorescence-based thallium-flux assay for high-throughput screening and screened a library of 234,000 compounds. The method uses fluorescent dyes (BTC or fluoZin-2), which are sensitive to thallium. When thallium enters the cells, it binds to the fluorescent probe, causing a measurable change in fluorescence (Weaver et al., 2004). Cation-chloride cotransporters transport thallium, and thus, this cation can be used as a surrogate for K^+ influx mediated by KCC2 in HEK cells overexpressing KCC2 using as control untransfected HEK. In these experiments, KCC2 transport was stimulated by NEM, thereby abrogating NKCC1 function. Of particular interest was compound D4 (PubChem SID# 56435272) because of its submicromolar potency (IC_{50} = 560 nM) to selectively inhibit KCC2 while minimally affecting NKCC1. For comparison, in this study, the IC_{50} for furosemide was found to be 455 μM. In addition, they showed that D4 also inhibits KCC3 in the micromolar range, indicating that the inhibitor does not distinguish between KCC2 and KCC3. In subsequent work, Delpire and colleagues described an improved novel antagonist of KCC2 (VU046327) with an IC_{50} of 61 nM and >100-fold selectivity versus NKCC1 (Delpire et al., 2012). As expected, the selective KCC2 inhibitor VU0463271 caused epileptiform activity in hippocampal slices in vivo and has become a valuable tool to explore the role of KCC2 in seizure activity (Sivakumaran et al., 2015).

KCC3 was first identified by three groups simultaneously (Hiki et al., 1999; Mount et al., 1999; Race et al., 1999). The KCC3 gene (SL12A6) gives rise to two major splice variants, KCC3a and KCC3b. The KCC3 cDNA cloned by Hiki et al. (1999) predicted a protein of 1099 amino acids, whereas the cDNA cloned by Race et al. (1999) and Mount et al. (1999)

predicted a protein of 1150 amino acids. These two KCC3 variants are termed KCC3a (1150 aa) and KCC3b (1099 aa). They are identical except for their N-termini. Both variants come from the same gene, which is located in chromosome 15 (15q13) in humans. KCC3 proteins have a deglycosylated molecular mass of 120−128 kDa. Subsequent studies showed that the mammalian KCC3 gene has the potential to produce more than two alternative splice variants, giving rise to proteins with considerable N-termini heterogeneity (Mercado et al., 2005). Six different KCC3 transcripts coding for five different KCC3 variants exist. They are termed KCC3a, KCC3a-x2M (KCC3a minus exon 2), KCC3a-S (encodes a protein 60 amino acids shorter than KCC3a), KCC3b and KCC3b-x2M (KCC3b minus exon 2). Exon 2 encodes 15 amino acid residues. When expressed in *Xenopus* oocytes, all five variants function as K^+-Cl^- cotransporters (Mercado et al., 2005). It is not clear how the N-termini heterogeneity impacts KCC3 function. When expressed in HEK 293 cells, human KCC3a is not active in isotonic media, but it is activated by cell swelling, staurosporine, and, prominently, by NEM. Evidence shows that the activity of KCC3a in isosmotic media depends upon the phosphorylation state of the protein, suggesting that dephosphorylation increases KCC3's intrinsic transport activity (Rinehart et al., 2009). Two C-terminal Threonine residues in human KCC3, Thr991, and Thr1048, are robustly phosphorylated in isotonic media in which the transporter is quiescent but undergoes rapid dephosphorylation in response to cell swelling, which activates the transporter (Adragna et al., 2015). Genetic or pharmacologic inhibition of WNK3-SPAK prevents cell swelling in response to osmotic challenges and improves postischemic brain swelling through simultaneous inhibition of NKCC1-mediated Cl^- uptake and stimulation of KCC3-mediated Cl^- extrusion. It has been proposed that WNK3-SPAK is an integral component of the "Cl^-/volume-sensitive kinase" of the cation-Cl^-cotransporters and functions as a "molecular rheostat of cell volume in the mammalian brain" (Zhang et al., 2016). Human KCC3a expressed in HEK cells is inhibited by bumetanide ($K_i \approx 100\ \mu M$), furosemide ($K_i \approx 103\ \mu M$), and DIOA ($K_i \approx 9.3\ \mu M$). Therefore, DIOA appeared to be a better inhibitor of KCC3a than furosemide (Race et al., 1999). However, DIOA at concentrations that inhibit KCC3 can also block the KCNN4/KCa3.1 "Gardos channel" in normal and sickle red cells (Rivera et al., 2022). The most potent KCC3 inhibitor known today is compound D4 (Delpire et al., 2009). However, as mentioned above, it also inhibits KCC2. When expressed in HEK 293 cells, human KCC3b is active in isotonic media, but its activity does not increase upon hypotonic swelling. KCC3b is inhibited by bumetanide ($K_i \approx 40\ \mu M$) and furosemide ($K_i \approx 10\ \mu M$).

KCC3s exhibit different distribution patterns; KCC3a is expressed in various brain regions and the basolateral membrane of choroid plexus epithelial cells, in heart and skeletal muscle, lung, liver, pancreas, and kidney (Ferdaus et al., 2023; Pearson et al., 2001). In choroid plexus epithelium, it has been proposed that basolateral KCC3a could represent a pathway for K^+ transport from the cells to the blood (Gregoriades et al., 2019; Karadsheh et al., 2004). However, another study found minimal KCC3 transcript expression and an undetectable level of KCC3 protein in the choroid plexus (Steffensen et al., 2018). These authors provided some evidence for KCC1 being the isoform expressed in the basolateral membrane of these cells. This is an important issue that needs further investigation, given the possible role of KCC3 (or KCC1) in vectorial transport in the choroid plexus, an organ that appears to have a dual function as a K^+ absorbing and as a Na^+, K^+, Cl^-, HCO_3^-, and water-secreting-epithelium as mentioned above (Fig. 15.17A).

DRG neurons express KCC3a, KCC3a−2, and KCC3b−2 transcripts in neonatal (P0) and young adult (P17) rats. The cellular elements in which these splice variants are expressed remain to be elucidated. Developmental changes in the functional expression of KCC3 in DRG could contribute to the slight postnatal decrease in $[Cl^-]_i$ and its increased variability in DRG cells (Lucas et al., 2012; Mao et al., 2012). A study using in situ hybridization and immunohistochemical techniques in brain tissue demonstrated that KCC3a is mainly expressed in neurons and, to a much lesser extent, in some glial cells, whereas KCC3b is not expressed in brain tissue (Shekarabi et al., 2011). In human red blood cells, KCC3a is the functionally dominant isoform (Pan et al., 2011), whereas KCC3b is abundantly expressed in the kidney, localized in the basolateral membranes of proximal tubule epithelial cells where it may contribute to proximal tubular cell volume regulation (Ferdaus & Delpire, 2023; Lang et al., 2007).

Mutations in the human gene SLC12A6, which encodes KCC3s, lead to a rare neurological illness, Andermann syndrome (Howard et al., 2002). This syndrome is characterized by progressive sensorimotor neuropathy associated with complete or partial absence of corpus callosum, mental retardation, progressive physical disability, and an average life expectancy of about 33 years. Death is usually caused by respiratory complications (Dupre et al., 2003). Total KCC3 knockout (KCC3−/−) mice models not only display peripheral neuropathy but also reduced seizure threshold and deafness. The latter is due to inner ear defects in K^+ transport in the cochlea, where some of the cells involved in K^+ recirculation slowly degenerate, as do sensory hair cells (Boettger et al., 2003).

KCC3-knockout mice (KCC3−/−) also show hypertension of neurogenic origin due to elevated sympathetic tone (Rust et al., 2006) and vascular effects (Adragna et al., 2004; Meor Azlan & Zhang, 2020). Other studies show that KCC3−/− mice have higher diastolic blood pressure, lower pulse pressure, decreased aortic reactivity to adrenergic stimulation, enlarged left ventricular mass, and isosmotic polyuria but normal levels of circulating catecholamines. These findings

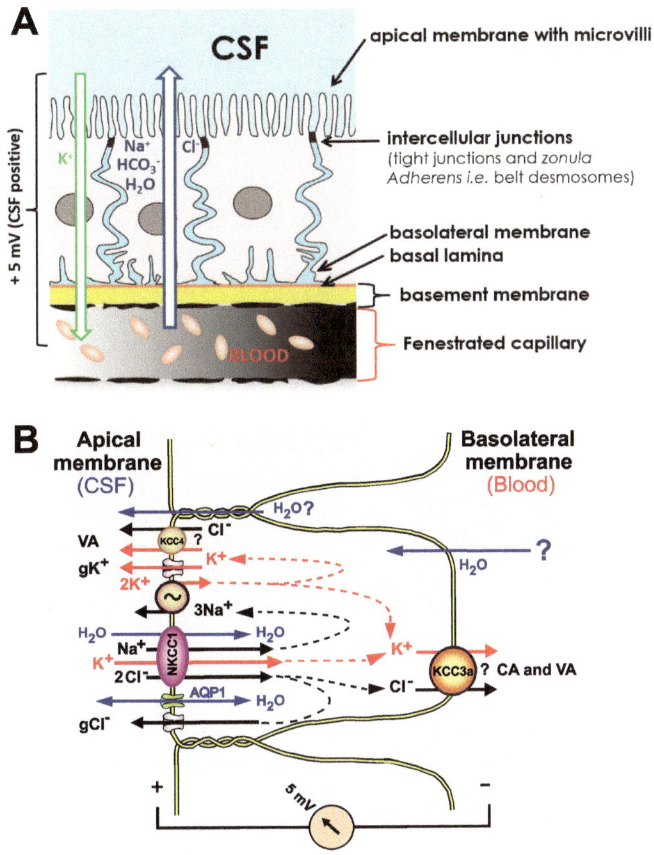

FIGURE 15.17 Net ion fluxes across the choroid plexus epithelium and apical Na^+-K^+-$2Cl^-$ cotransporter 1(NKCC1) function in choroid plexus epithelial cells (CPECs). (A) Forming the blood–cerebrospinal fluid barrier, each choroid plexus (CP) is composed of a monolayer of cuboidal epithelial cells held together by intercellular junctional complexes (tight junctions and zonula adherens) at their apical ends (luminal side facing CSF). The apical surface of CPECs exhibits abundant microvilli where NKCC1 is localized. Their basal end is anchored to the basal lamina of the basement membrane surrounding a core of fenestrated capillaries. The choroid plexus epithelium (CPE) mediates the net secretion of Na^+, Cl^-, HCO_3^-, and water (toward the lumen) and the net transcellular absorption of K^+ from CSF to blood. (B) The Na^+-K^+-$2Cl^-$ cotransporter NKCC1 plays a central role in the absorption of K^+ from CSF to the blood, in buffering and regulating CSF K^+ concentration, as well as maintaining CPECs water volume needed for CSF secretion. CSF secretion mechanisms are incompletely understood and debatable. Unlike other secreting epithelia, NKCC1 and the Na^+/K^+-ATPase (NKA) are localized to the apical membrane of the CPECs. In this configuration, Na^+ taken up via NKCC1 is recycled by the apical NKA, thereby keeping low $[Na^+]_i$ (≈ 10 mM) and thus a substantial Na^+ inward gradient needed to energize secondary active transporters. The K^+ that is taken up by NKCC1 and the NKA is recycled via apical K^+ channels (gK^+) and transported out of the cell through continuously active (CA) and volume-sensitive (VA) basolateral KCC3a. Secretion of CSF is carried out by transporters and channels other than NKCC1. The apical NKA carries out Na^+ secretion. Cl^- secretion occurs via apical KCC4 and Cl^- channels (gCl^-). Water transport in CPE is little understood. Some water might be inwardly co-transported via NKCC1. Part of the water efflux involved in secretion goes through apical aquaporin 1 (AQP1). Possible paracellular and basolateral water transport remains elusive. Bicarbonate secretion is carried out by several transporters in the apical and basolateral membranes (not shown).

suggest the involvement of vasculogenic mechanisms in the development of hemodynamic disorders due to KCC3 inactivation and a potential role for K^+-Cl^- cotransport in cardiac growth (Garneau et al., 2016; Meor Azlan & Zhang, 2020).

More recently, it has been suggested that both KCC3b and KCC3a are needed for maintaining cell volume during enhanced inward cotransport of Na-glucose in the proximal tubule and Na–HCO_3 in intercalated cells (Ferdaus & Delpire, 2023). Further, apical KCC3a in intercalated cells might couple to pendrin function to recycle Cl^-, particularly in conditions of low salt diet and, therefore, low Cl^- delivery to the distal tubule. Pendrin (SLC26A4) is an anion exchanger protein that mediates HCO_3^- exchange for Cl^- and is critical for maintaining pH and salt homeostasis in the kidney, lung, and cochlea. Coupling to Pendrin function is important in alkalemia, and KCC3a activity in the pendrin-expressing cells may contribute to the K^+ loss observed in alkalemia (Ferdaus et al., 2023).

Full KCC3-knockout mice exhibit axonal and Schwann cell swelling of sciatic nerves, ultimately resulting in neurodegeneration (Byun & Delpire, 2007). This suggests that KCC3 may play a role in cell water volume maintenance in sensorimotor axons and that this mechanism is critical for peripheral nerve integrity and function (Flores et al., 2019; Kahle et al., 2016). These observations further suggest that KCC3 might be constantly active, extruding ions and osmotically or co-transported water.

In sum, KCC3s have significant roles in transepithelial transport and regulate cell volume and $[Cl^-]_i$. These cotransporters are implicated in the pathogenesis of Andermann Syndrome and peripheral neuropathies. While KCC2 may be the key regulator of neuronal $[Cl^-]_i$, data suggest that KCC3a may have a similar role in some neurons. In the kidney, KCC3b and KCC3a are vital for maintaining cell volume during enhanced inward cotransport of Na-glucose in the proximal tubule and $Na-HCO_3$ in intercalated cells. In addition, apical KCC3a might couple to pendrin function to recycle Cl^-, particularly in conditions of a low salt diet and, therefore, low Cl^- delivery to the distal tubule (Ferdaus & Delpire, 2023). KCC3a function in the pendrin-expressing cells may contribute to the K^+ loss observed in alkalemia.

KCC4 is encoded by the SLC12A7 gene (Marcoux et al., 2017). Its monomeric structure has been resolved using cryo-EM (Reid et al., 2020). A prominent feature of KCC4 is a large extracellular domain, extending ≈35 Å above the membrane. KCC4 was first cloned and functionally characterized by Mount and coworkers (1999). The KCC4 gene is located on chromosome 5, at 5p15, and encodes for a protein product of 1083 amino acids that shares a high identity (≈75%) with KCC2. Unlike KCC2 (and KCC3), KCC4 mediates minimal K^+-Cl^- cotransport activity under isotonic conditions, but cotransport is significantly activated by osmotic cell swelling. In experiments done with *Xenopus* oocytes expressing KCC4, cotransporter activity increases 200-fold when exposed to hypotonic solution (Mercado et al., 2000). In this heterologous expression system, it was possible to establish functional comparisons between KCC1 and KCC4. Although cell swelling activates both cotransporters, KCC4 is one order of magnitude more sensitive to hypotonic activation than KCC1. Another distinctive feature of KCC4 is its relatively poor sensitivity to both furosemide ($K_i \approx 900\ \mu M$) and bumetanide ($K_i \approx 900\ \mu M$) and its sensitivity to barium and high concentrations of thiazides. Among the four KCC isoforms, KCC4 is the least sensitive to furosemide, as corroborated using the HEK 293 expression system, in which KCC2 properties were compared with KCC4 (Hartmann et al., 2010). KCC2 was more sensitive to furosemide (Ki ≈ 185 μM) than KCC4 (Ki ≈ 550 μM).

KCC4 is prominently expressed in several epithelia, including the apical membrane of choroid plexus epithelial cells (Karadsheh et al., 2004), the basolateral membrane of α-intercalated cells of the cortical collecting duct in the kidney (Boettger et al., 2002), and the apical membrane of gastric parietal cells (Fujii et al., 2009). In the apical membrane of the choroid plexus epithelial cells (CPECs), KCC4 has been proposed to be involved in K^+ recycling, working in parallel with the Na^+/K^+ pump and NKCC1 (Karadsheh et al., 2004). Given its exquisite sensitivity to changes in cell volume compared with other KCC isoforms and its apical location, relatively small cell water volume increases could result in significant KCC4 activation. Based on these arguments, it has been proposed that apical KCC4 in CPECs might work in parallel with apical NKCC1, thereby secreting K^+ and Cl^- toward the lumen, i.e., the CSF side (Fig. 15.17). Thus, NKCC1 working in coordination with KCCs, maintains the cell water volume of CPECs at a set point necessary to keep both transport mechanisms active under basal conditions, thereby maintaining cell solutes and water constant during transepithelial transport, thus making possible CSF secretion and K^+ absorption (Gregoriades et al., 2019). Consistent with the immunolabeling studies of Karadsheh and coworkers (2004) is the abundant KCC4 mRNA expression found in embryonic brain (Li et al., 2002). Moreover, in situ hybridization studies also found KCC4 mRNA in choroid plexus epithelial cells (Le Rouzic et al., 2006). Nevertheless, Steffensen, and coworkers (2018) found undetectable levels of KCC4 mRNA in mouse choroid plexus (Steffensen et al., 2018). Further research with validated antibodies and other molecular probes is needed to test for the expression of apical KCC4 and elucidate its functional role in CPECs.

In the basolateral membrane of α-intercalated cells in cortical collecting ducts in the kidney, KCC4 regulates acid secretion, working in parallel with the basolateral AE1 and the apical proton pump. The apical proton pump secretes H^+ into the lumen of the collecting duct. At the basolateral membrane, acid equivalents are transported by the AE1. Basolateral HCO_3^- efflux is coupled to Cl^- uptake, which is recycled by KCC4. Interestingly, KCC4 knockout mice exhibit renal tubular acidosis that probably results from impaired Cl^- recycling across the basolateral membrane of acid-secreting α-intercalated cells (Boettger et al., 2002). In the apical membrane of gastric parietal cells, the K^+/H^+-ATPase (i.e., the proton pump) working in parallel with KCC4 is the primary mechanism for basal HCl secretion in the apical canalicular membrane of parietal cells (Fujii et al., 2009). KCC4 is also expressed in the Deiters' cells of the inner ear, where it participates in the K^+ recycling mechanism (Boettger et al., 2002). KCC4-knockout mice are deaf due to hair cell degeneration after the beginning of hearing (Boettger et al., 2002). KCC4 has also been found to be expressed in some areas of the nervous system (the suprachiasmatic nucleus of the hypothalamus, cranial nerves, spinal cord, peripheral nerves, and DRG neurons), but its function in nervous tissue is unknown (Karadsheh et al., 2004; Le Rouzic et al., 2006;

Mao et al., 2012). KCC4 is also expressed in red blood cells, but its functional role in these cells is also unknown. More exciting is the emerging role of KCC4 in cancer cell invasiveness and tumor metastasis. It has been found that epidermal growth factor (EGF) and IGF-1 stimulate the membrane recruitment of KCC4 at lamellipodia through the myosin Va-actin trafficking route. KCC4 functions as a membrane scaffold for the assembly of signal complexes via the association with the actin-binding protein ezrin. Furthermore, molecular studies of surgical specimens suggest that the expression of KCC3, KCC4, and their stimulators, EGF or IGF-1, is closely associated with the clinical outcome of cancer patients. It has been proposed that KCC4 may be used as a biomarker to predict cancer patient outcomes (Chen et al., 2010; Marcoux et al., 2017).

6.3 Thermodynamics of K^+-Cl^- cotransport

Let us now focus on the thermodynamic aspects of K^+-Cl^- cotransport. Consider the example shown in Fig. 15.18, which depicts a model of a neuron expressing KCC, Na^+/K^+ pump, a voltage-gated Cl^- leak channel, and a GABA-gated Cl^- channel. Assuming that the K^+-Cl^- cotransport is the sole system transporting Cl^- (channels are closed) and that $[K^+]_i$ is held constant by the Na^+-K^+ pump, the driving force for the cotransport $\Delta\mu_{K,Cl}$ is the sum of the chemical potential differences of K^+ and Cl^-, as expressed in the following equation:

$$\Delta\mu_{K,Cl} = \Delta\mu_K + \Delta\mu_{Cl} \tag{15.16}$$

It follows that

$$\Delta G = \Delta\mu_{K,Cl} = RT\ln\frac{[K^+]_i}{[K^+]_o} + RT\ln\frac{[Cl^-]_i}{[Cl^-]_o} \tag{15.17}$$

This system attains thermodynamic equilibrium when

$$\Delta G = \Delta\mu_{K,Cl} = 0 \tag{15.18}$$

and therefore, for a 1:1 stoichiometry,

$$\frac{[K^+]_o}{[K^+]_i} = \frac{[Cl^-]_i}{[Cl^-]_o} \tag{15.19}$$

and hence, the $[Cl^-]_i$ when the cotransporter reaches thermodynamic equilibrium, will be

$$[Cl^-]_i = \frac{[K^+]_o[Cl^-]_o}{[K^+]_i} \tag{15.20}$$

From Eq. (15.20), one can readily calculate the $[K^+]_o$ at which the flux reverses from outward to inward or vice versa (flux reversal point or FRP):

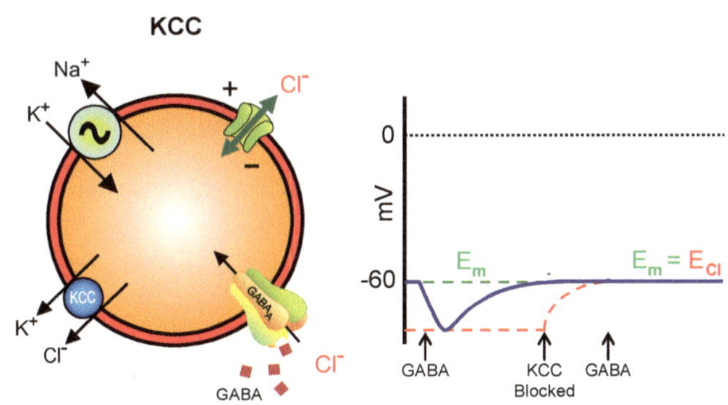

FIGURE 15.18 Model of a neuron expressing only K^+-Cl^- cotransporters (KCC), Na^+/K^+ ATPase, voltage-gated anion channels, and GABA-gated anion channels. The KCCs actively extrude Cl^- from the cell, shifting E_{Cl} (red dashed line) to a more negative value than E_m (green dashed line). Right: A puff of GABA produces a hyperpolarizing IPSP (blue trace). When KCCs are blocked, $[Cl^-]_i$ raises until reaching passive equilibrium ($E_m = E_{Cl}$). As expected, a second GABA application under these conditions does not produce a change in E_m but an increase in membrane conductance which constitutes the basis for "shunting synaptic inhibition." *Reproduced from Alvarez-Leefmans (2012).*

$$\text{FRP} = [K^+]_o = \frac{[K^+]_i[Cl^-]_i}{[Cl^-]_o} \tag{15.21}$$

The FRP is a crucial parameter (Lauf et al., 1994). At the FRP, the driving force for net K^+-Cl^- cotransport is zero ($\Delta\mu_{K,Cl} = 0$). For instance, in a cortical neuron with typical intracellular K^+ and Cl^- concentrations of 120 and 10 mM, respectively, and $[Cl^-]_o = 113$ mM, the KCC2 mediated flux will reverse from efflux to influx when $[K^+]_o = 10.6$ mM. With typical values of extra- and intracellular K^+ and Cl^- concentrations, KCCs work in the net efflux mode.

7. Electroneutral Na^+-Cl^- cotransporter

The Na^+-Cl^- cotransporter (NCC) is the principal mechanism for salt reabsorption in the apical membrane of the distal convoluted tubule (DCT) in mammals and plays a critical role in determining blood pressure. Functional characterization of this cotransporter was first made in the urinary bladder of the winter flounder (*Pseudopleuronectes americanus*), where it was demonstrated that the Na^+ and Cl^- fluxes were interdependent, suggesting that the two ions were co-transported (Renfro, 1977), the fluxes were not affected by transmembrane voltage thereby showing that the cotransport mechanisms was electroneutral (Renfro, 1975), the fluxes of the two co-transported ions were nearly identical, suggesting $1Na^+$: $1Cl^-$ stoichiometry, and were selectively blocked by thiazides (Stokes, 1984).

The NCC was the first member of the SLC12 gene family to be identified at the molecular level. It was initially cloned from epithelial cells of the urinary bladder of the winter flounder by Steve Hebert, Gerardo Gamba, and colleagues (Gamba et al., 1993). NCC is also known as NCCT, TSC (thiazide-sensitive sodium chloride cotransporter), but the preferred terminology is NCC or SLC12A3. Its mammalian homolog was isolated from rat renal cortex (Gamba, 2009b). NCC is expressed predominantly in the kidney but is also expressed in osteoblasts and intestines. In mammalian kidneys, NCC is located in the apical membrane of the epithelial cells of the distal convoluted tubule (DCT), where it functions as the principal salt reabsorption pathway. The DCT mediates reabsorption of 5%−15% of glomerular filtrate. The Na^+ gradient that drives transport from the lumen to the interstitium is generated and maintained by the Na^+/K^+ ATPase located in the basolateral membrane of DCT epithelial cells. A distinctive feature of NCC is its sensitivity to thiazides. The natriuretic effect of thiazide diuretics underlies their therapeutic effect in edema states and hypertension. In addition, luminal thiazide in the DCT has a hypocalciuric effect, implicating NCC in the fine-tuning of Ca^{2+} excretion. Mutations of NCC lead to Gitelman syndrome, a salt-wasting tubule nephropathy characterized by metabolic alkalosis, hypotension, hyperkalemia, hypocalciuria, and hypomagnesemia (Trejo et al., 2023).

Dietary potassium deficiency causes stimulation of sodium reabsorption via NCC, leading to an increased risk of hypertension. In the last few years, the mechanisms by which DCT regulates renal Na^+ absorption and modulates renal K^+ secretion have been elucidated (Ferdaus et al., 2024; Mukherjee et al., 2021; Wang et al., 2018; Wang & Lin, 2022). Changes in plasma K^+ levels following dietary K^+ intake affect the activity of inward rectifier K^+ channels Kir4.1 and Kir5.1 in the basolateral membrane of DCT, altering $[Cl^-]_i$ in these epithelial cells, which in turn affects the activity of NCC. The Kir4.1/Kir5.1 activity is an essential component of the renal K^+ sensing mechanism. When plasma K^+ levels *decrease*, Kir4.1/Kir5.1 channels become more active, leading to increased K^+ efflux, which hyperpolarizes the DCT cells. The hyperpolarization drives Cl^- out via ClC-Kb Cl^- channels, thus decreasing $[Cl^-]$i. The decrease in $[Cl^-]_i$ relieves the inhibitory effect of Cl^- on WNK4, leading to subsequent changes in SPAK activity and, ultimately, the phosphorylation and activation of NCC, thus impacting Na^+ reabsorption and K^+ balance. An *increase* in plasma K^+ levels following high K^+ intake decreases Kir4.1/Kir5 activity, thereby decreasing the Cl^- conductance, thus depolarizing the DCT cells, resulting in an increase in $[Cl^-]_i$, which in turn decreases WNK4-SPAK mediated phosphorylation of NCC ultimately decreasing NCC activity.

In humans, SLC12A3 is a 55-kb gene containing 26 exons. NCC is a membrane protein of 1002−1028 amino acid residues with a predicted topology typical of all the members of the SLC12A family. It exhibits a central hydrophobic domain comprising 12 putative transmembrane (TM) spanning regions. A hydrophilic loop connects TMs seven and eight and contains two glycosylation sites. The hydrophobic domain is flanked by a short amino-terminal domain and a long carboxy-terminal domain, both located inside the cell. Functional expression of this cotransporter in *Xenopus laevis* oocytes demonstrated the known characteristics of the cotransporter in native tissue: sensitivity to thiazide diuretics (metolazone > cyclothiazide > hydrochlorothiazide > chlorothiazide), insensitivity to amiloride and loop diuretics, K^+ independence, a lower affinity for Cl^- than for Na^+, and $1Na^+$:$1Cl^-$ stoichiometry; hence confirming that it is an electroneutral transporter. The molecular mass of the core (deglycosylated) protein is ≈ 110 kDa. It has ≈45% amino acid identity with NKCC1, 45%−48% with NKCC2, and ≈25% with KCC1 and KCC2 (Gamba, 2005, 2009b).

The structure of human NCC has been recently revealed at 2.9 Å resolution using cryo-EM (Nan et al., 2022). This study confirmed that the structural unit of NCC is a homodimer (de Jong et al., 2003). Each subunit contains 12 transmembrane segments (TMs) with both the N- and C-terminal ends on the intracellular side. Like other CCCs studied with cryo-EM, the observed NCC is trapped in an inward-open conformation occluded with substrate ions. With the extracellular gate closed, the ion translocation pathway extends from the center of the transmembrane region to the cytoplasmic side. The structure showed two Na^+ and two Cl^- binding sites. In addition, the cryo-EM structures revealed water molecules in the ion transport pathway, which suggests that NCC might also transport water. Nan et al. (2022) compared the structure of ion binding sites with those of other CCS, discussing similarities and differences, and studied the effect of mutations described in Gitelman syndrome.

More recently, the structures of human NCC alone and in complex with a commonly used thiazide diuretic (polythiazide) were also resolved at 2.8 Å resolution using cryo-EM (Fan et al., 2023). The polythiazide-bound NCC structure captured the NCC in an outward-open conformation. The structure revealed a wide-open vestibule formed by TM1, TM3, TM6, TM8, and TM10, which extends from the extracellular surface halfway into the TMD. The polythiazide binding residues appear to be distributed along the TMs mentioned above. The authors conclude that the thiazide binding site overlaps with the Cl^--binding site, locking the NCC in an outward-facing conformation. The fine structural details of NCC revealed in these cryo-EM studies provide a further understanding of the transport mechanisms of ion translocation in NCC and a framework needed for developing more selective inhibitors.

Acknowledgments

The author thanks professors Eric Delpire for enriching discussions and suggestions on the manuscript, Shmuel Muallem for providing the original data used in Fig. 15.4, and John Payne and Bliss Forbush for providing the backbone of Fig. 15.16. The work done by the author, which appears in this chapter, was supported over the years by The National Institute of Neurological Disorders, Dayton Children's Hospital, and Wright State University Boonshoft School of Medicine Research Challenge and Seed Grant Programs.

References

Achilles, K., Okabe, A., Ikeda, M., Shimizu-Okabe, C., Yamada, J., Fukuda, A., Luhmann, H. J., & Kilb, W. (2007). Kinetic properties of Cl uptake mediated by Na^+-dependent K^+-$2Cl^-$ cotransport in immature rat neocortical neurons. *Journal of Neuroscience, 27*, 8616–8627.

Adragna, N. C., Chen, Y., Delpire, E., Lauf, P. K., & Morris, M. (2004). Hypertension in K-Cl cotransporter-3 knockout mice. *Advances in Experimental Medicine & Biology, 559*, 379–385.

Adragna, N. C., Ravilla, N. B., Lauf, P. K., Begum, G., Khanna, A. R., Sun, D., & Kahle, K. T. (2015). Regulated phosphorylation of the K-Cl cotransporter KCC3 is a molecular switch of intracellular potassium content and cell volume homeostasis. *Frontiers in Cellular Neuroscience, 9*, 255.

Adrian, R. H. (1961). Internal chloride concentration and chloride efflux of frog muscle. *The Journal of Physiology, 156*, 623–632.

Aickin, C. C. (1990). Chloride transport across the sarcolemma of vertebrate smooth and skeletal. In F. J. Alvarez-Leefmans, & J. M. Russell (Eds.), *Chloride channels and carriers in nerve, muscle, and glial cells* (pp. 209–249). New York: Plenum Press.

Akhter, E. T., Griffith, R. W., English, A. W., & Alvarez, F. J. (2019). Removal of the potassium chloride co-transporter from the somatodendritic membrane of axotomized motoneurons is independent of BDNF/TrkB signaling but is controlled by neuromuscular innervation. *eNeuro, 6*.

Alper, S. L., Chernova, M. N., & Stewart, A. K. (2002). How pH regulates a pH regulator: A regulatory hot spot in the N-terminal cytoplasmic domain of the AE2 anion exchanger. *Cell Biochemistry and Biophysics, 36*, 123–136.

Alper, S. L., Vandorpe, D. H., Peters, L. L., & Brugnara, C. (2008). Reduced DIDS-sensitive chloride conductance in Ae1-/- mouse erythrocytes. *Blood Cells Molecules and Diseases, 41*, 22–34.

Alper, S. L. (2009). Molecular physiology and genetics of Na+-independent SLC4 anion exchangers. *Journal of Experimental Biology, 212*, 1672–1683.

Alvarez, B. V., Kieller, D. M., Quon, A. L., Markovich, D., & Casey, J. R. (2004). Slc26a6: A cardiac chloride-hydroxyl exchanger and predominant chloride-bicarbonate exchanger of the mouse heart. *The Journal of Physiology, 561*, 721–734.

Alvarez-Leefmans, F. J., Altamirano, J., & Crowe, W. E. (1995). Use of ion-selective microelectrodes and fluorescent probes to measure cell volume. In J. Kraicer, & S. J. Dixon (Eds.), *Methods in neurosciences* (pp. 361–391). Academic Press.

Alvarez-Leefmans, F. J., & Delpire, E. (2009a). *Physiology and pathology of chloride transporters and channels in the nervous system: From molecules to diseases*. Amsterdam; Boston; London: Elsevier/Academic.

Alvarez-Leefmans, F. J., & Delpire, E. (2009b). Thermodynamics and kinetics of chloride transport in neurons: An outline. In F. J. Alvarez-Leefmans, & E. Delpire (Eds.), *Physiology and pathology of chloride transporters and channels in the nervous system* (pp. 81–108). Amsterdam: Academic Press-Elsevier.

Alvarez-Leefmans, F. J., Gamino, S. M., Giraldez, F., & Nogueron, I. (1988). Intracellular chloride regulation in amphibian dorsal root ganglion neurones studied with ion-selective microelectrodes. *The Journal of Physiology, 406*, 225–246.

Alvarez-Leefmans, F. J., Giraldez, F., & Russell, J. M. (1990). Methods for measuring chloride transport across nerve, muscle, and glial cells. In F. J. Alvarez-Leefmans, & J. M. Russell (Eds.), *Chloride channels and carriers in nerve, muscle, and glial cells* (pp. 3–66). New York: Plenum Press.

Alvarez-Leefmans, F. J., Leon-Olea, M., Mendoza-Sotelo, J., Alvarez, F. J., Anton, B., & Garduno, R. (2001). Immunolocalization of the Na^+-K^+-$2Cl^-$ cotransporter in peripheral nervous tissue of vertebrates. *Neuroscience, 104*, 569–582.

Alvarez-Leefmans, F. J., Nani, A., & Márquez, S. (1998). Chloride transport, osmotic balance, and presynaptic inhibition. In P. Rudomin, R. Romo, & L. M. Mendell (Eds.), *Presynaptic inhibition and neural control* (pp. 50–79). New York: Oxford University Press.

Alvarez-Leefmans, F. J., & Russell, J. M. (1990). In *Chloride channels and carriers in nerve, muscle, and glial cells*. New York: Plenum Press.

Alvarez-Leefmans, F. J. (2009). Chloride transporters in presynaptic inhibition, pain and neurogenic inflammation. In F. J. Alvarez-Leefmans, & E. Delpire (Eds.), *Physiology and pathology of chloride transporters and channels in the nervous system* (pp. 439–470). Amsterdam: Academic Press-Elsevier.

Alvarez-Leefmans, F. J. (2012). Chapter 15 - Intracellular chloride regulation. In N. Sperelakis (Ed.), *Cell physiology source book* (4th ed., pp. 221–259). San Diego: Academic Press.

Alvarez-Leefmans, F. J. (2020a). CrossTalk proposal: Apical NKCC1 of choroid plexus epithelial cells works in the net inward flux mode under basal conditions, maintaining intracellular Cl^- and cell volume. *The Journal of Physiology, 598*, 4733–4736.

Alvarez-Leefmans, F. J. (1990). Intracellular Cl^- regulation and synaptic inhibition in vertebrate and invertebrate neurons. In F. J. Alvarez-Leefmans, & J. M. Russell (Eds.), *Chloride channels and carriers in nerve, muscle, and glial cells* (pp. 109–158). New York: Plenum Press.

Alvarez-Leefmans, F. J. (2020b). Rebuttal from Francisco J. Alvarez-leefmans. *The Journal of Physiology, 598*, 4741–4742.

Antrobus, S. P., Lytle, C., & Payne, J. A. (2012). K^+-Cl^- cotransporter-2 KCC2 in chicken cardiomyocytes. *American Journal of Physiology - Cell Physiology, 303*, C1180–C1191.

Ares, G. R., Caceres, P. S., & Ortiz, P. A. (2011). Molecular regulation of NKCC2 in the thick ascending limb. *American Journal of Physiology - Renal Physiology, 301*, F1143–F1159.

Arroyo, J. P., Kahle, K. T., & Gamba, G. (2013). The SLC12 family of electroneutral cation-coupled chloride cotransporters. *Molecular Aspects of Medicine, 34*, 288–298.

Ayala-Torres, C., Krug, S. M., Schulzke, J. D., Rosenthal, R., & Fromm, M. (2019). Tricellulin effect on paracellular water transport. *International Journal of Molecular Sciences, 20*.

Baasch Christensen, I., Cheng, L., Brewer, J. R., Bartsch, U., Fenton, R. A., Helle, H., & Praetorius, J. (2021). Multiple Na, K-ATPase subunits colocalize in the brush border of mouse choroid plexus epithelial cells. *International Journal of Molecular Sciences, 22*.

Bazúa-Valenti, S., Castañeda-Bueno, M., & Gamba, G. (2016). Physiological role of SLC12 family members in the kidney. *American Journal of Physiology - Renal Physiology, 311*, F131–F144.

Becker, M., Nothwang, H. G., & Friauf, E. (2003). Differential expression pattern of chloride transporters NCC, NKCC2, KCC1, KCC3, KCC4, and AE3 in the developing rat auditory brainstem. *Cell and Tissue Research, 312*, 155–165.

Bell, P. D., Komlosi, P., & Zhang, Z. R. (2009). ATP as a mediator of macula densa cell signaling. *Purinergic Signalling, 5*, 461–471.

Bellemer, A., Hirata, T., Romero, M. F., & Koelle, M. R. (2011). Two types of chloride transporters are required for GABA(A) receptor-mediated inhibition in *C. elegans*. *The EMBO Journal, 30*, 1852–1863.

Ben-Ari, Y., & Cherubini, E. (2022). The GABA polarity shift and bumetanide treatment: Making sense requires unbiased and undogmatic analysis. *Cells, 11*.

Ben-Dor, S., Esterman, N., Rubin, E., & Sharon, N. (2004). Biases and complex patterns in the residues flanking protein N-glycosylation sites. *Glycobiology, 14*, 95–101.

Benjamin, B. A., & Johnson, E. A. (1997). A quantitative description of the Na-K-2Cl cotransporter and its conformity to experimental data. *American Journal of Physiology, 273*, F473–F482.

Bevensee, M. O., & Boron, W. F. (1998). Thermodynamics and physiology of cellular pH regulation. In K. Kaila, & B. R. Ransom (Eds.), *pH and brain function* (pp. 173–194). New York: Wiley-Liss.

Bize, I., & Dunham, P. B. (1994). Staurosporine, a protein kinase inhibitor, activates K-Cl cotransport in LK sheep erythrocytes. *American Journal of Physiology, 266*, C759–C770.

Blaesse, P., Airaksinen, M. S., Rivera, C., & Kaila, K. (2009). Cation-chloride cotransporters and neuronal function. *Neuron, 61*, 820–838.

Boettger, T., Hubner, C. A., Maier, H., Rust, M. B., Beck, F. X., & Jentsch, T. J. (2002). Deafness and renal tubular acidosis in mice lacking the K-Cl cotransporter Kcc4. *Nature, 416*, 874–878.

Boettger, T., Rust, M. B., Maier, H., Seidenbecher, T., Schweizer, M., Keating, D. J., Faulhaber, J., Ehmke, H., Pfeffer, C., Scheel, O., Lemcke, B., Horst, J., Leuwer, R., Pape, H. C., Volkl, H., Hubner, C. A., & Jentsch, T. J. (2003). Loss of K-Cl co-transporter KCC3 causes deafness, neurodegeneration and reduced seizure threshold. *The EMBO Journal, 22*, 5422–5434.

Brandt, C., Seja, P., Töllner, K., Römermann, K., Hampel, P., Kalesse, M., Kipper, A., Feit, P. W., Lykke, K., Toft-Bertelsen, T. L., Paavilainen, P., Spoljaric, I., Puskarjov, M., MacAulay, N., Kaila, K., & Löscher, W. (2018). Bumepamine, a brain-permeant benzylamine derivative of bumetanide, does not inhibit NKCC1 but is more potent to enhance phenobarbital's anti-seizure efficacy. *Neuropharmacology, 143*, 186–204.

Brightman, M. W., & Reese, T. S. (1969). Junctions between intimately apposed cell membranes in the vertebrate brain. *The Journal of Cell Biology, 40*, 648–677.

Brzezinski, P. (2017). *The development of cryo-electron microscopy*. Stockholm, Sweden: The Royal Swedish Academy of Sciences.

Byun, N., & Delpire, E. (2007). Axonal and periaxonal swelling precede peripheral neurodegeneration in KCC3 knockout mice. *Neurobiology of Disease, 28*, 39–51.

Caceres, P. S., & Ortiz, P. A. (2019). Molecular regulation of NKCC2 in blood pressure control and hypertension. *Current Opinion in Nephrology and Hypertension, 28*, 474–480.

Carlton, S. M. (2014). Nociceptive primary afferents: They have a mind of their own. *The Journal of Physiology, 592*, 3403–3411.

Carmosino, M., Gimenez, I., Caplan, M., & Forbush, B. (2008). Exon loss accounts for differential sorting of Na-K-Cl cotransporters in polarized epithelial cells. *Molecular Biology of the Cell, 19*, 4341–4351.

Carota, I., Theilig, F., Oppermann, M., Kongsuphol, P., Rosenauer, A., Schreiber, R., Jensen, B. L., Walter, S., Kunzelmann, K., & Castrop, H. (2010). Localization and functional characterization of the human NKCC2 isoforms. *Acta Physiologica, 199*, 327–338.

Carta, F., & Supuran, C. T. (2013). Diuretics with carbonic anhydrase inhibitory action: A patent and literature review (2005 – 2013). *Expert Opinion on Therapeutic Patents, 23*, 681–691.

Castrop, H., & Schießl, I. M. (2014). Physiology and pathophysiology of the renal Na-K-2Cl cotransporter (NKCC2). *American Journal of Physiology - Renal Physiology, 307*, F991–f1002.

Castrop, H., & Schnermann, J. (2008). Isoforms of renal Na-K-2Cl cotransporter NKCC2: Expression and functional significance. *American Journal of Physiology - Renal Physiology, 295*, F859–F866.

Chari, A., & Stark, H. (2023). Prospects and limitations of high-resolution single-particle cryo-electron microscopy. *Annual Review of Biophysics, 52*, 391–411.

Chen, L. M., Kelly, M. L., Parker, M. D., Bouyer, P., Gill, H. S., Felie, J. M., Davis, B. A., & Boron, W. F. (2008). Expression and localization of Na-driven $Cl^{-}HCO_3^{-}$ exchanger (SLC4A8) in rodent CNS. *Neuroscience, 153*, 162–174.

Chen, Y. F., Chou, C. Y., Ellory, J. C., & Shen, M. R. (2010). The emerging role of KCl cotransport in tumor biology. *American Journal of Translational Research, 2*, 345–355.

Chew, T. A., Orlando, B. J., Zhang, J., Latorraca, N. R., Wang, A., Hollingsworth, S. A., Chen, D. H., Dror, R. O., Liao, M., & Feng, L. (2019). Structure and mechanism of the cation-chloride cotransporter NKCC1. *Nature, 572*, 488–492.

Chi, X., Li, X., Chen, Y., Zhang, Y., Su, Q., & Zhou, Q. (2021). Cryo-EM structures of the full-length human KCC2 and KCC3 cation-chloride cotransporters. *Cell Research, 31*, 482–484.

Chou, C. L., Yu, M. J., Kassai, E. M., Morris, R. G., Hoffert, J. D., Wall, S. M., & Knepper, M. A. (2008). Roles of basolateral solute uptake via NKCC1 and of myosin II in vasopressin-induced cell swelling in inner medullary collecting duct. *American Journal of Physiology - Renal Physiology, 295*, F192–F201.

Coles, J. A. (1995). Measurement of cytosolic chloride activity by ion-selective microelectrodes. In J. Kraicer, & S. J. Dixon (Eds.), *Methods in neurosciences* (pp. 340–360). Academic Press.

Comitato, A., & Bardoni, R. (2021). Presynaptic inhibition of pain and touch in the spinal cord: From receptors to circuits. *International Journal of Molecular Sciences, 22*.

Cunha, T. D. S., & Heilberg, I. P. (2018). Bartter syndrome: Causes, diagnosis, and treatment. *International Journal of Nephrology and Renovascular Disease, 11*, 291–301.

Darman, R. B., & Forbush, B. (2002). A regulatory locus of phosphorylation in the N terminus of the Na-K-Cl cotransporter, NKCC1. *Journal of Biological Chemistry, 277*, 37542–37550.

de Jong, J. C., Willems, P. H., Mooren, F. J., van den Heuvel, L. P., Knoers, N. V., & Bindels, R. J. (2003). The structural unit of the thiazide-sensitive NaCl cotransporter is a homodimer. *Journal of Biological Chemistry, 278*, 24302–24307.

Dehaye, J. P., Nagy, A., Premkumar, A., & Turner, R. J. (2003). Identification of a functionally important conformation-sensitive region of the secretory Na+-K+-2Cl- cotransporter (NKCC1). *Journal of Biological Chemistry, 278*, 11811–11817.

Delpire, E., Baranczak, A., Waterson, A. G., Kim, K., Kett, N., Morrison, R. D., Daniels, J. S., Weaver, C. D., & Lindsley, C. W. (2012). Further optimization of the K-Cl cotransporter KCC2 antagonist ML077: Development of a highly selective and more potent in vitro probe. *Bioorganic & Medicinal Chemistry Letters, 22*, 4532–4535.

Delpire, E., Days, E., Lewis, L. M., Mi, D., Kim, K., Lindsley, C. W., & Weaver, C. D. (2009). Small-molecule screen identifies inhibitors of the neuronal K-Cl cotransporter KCC2. *Proceedings of the National Academy of Sciences of the U S A, 106*, 5383–5388.

Delpire, E., & Gagnon, K. B. (2011). Kinetics of hyperosmotically stimulated Na-K-2Cl cotransporter in *Xenopus laevis* oocytes. *American Journal of Physiology. Cell Physiology, 301*, C1074–1085.

Delpire, E., & Gagnon, K. B. (2018a). Na(+) -K(+) -2Cl(-) cotransporter (NKCC) physiological function in nonpolarized cells and transporting epithelia. *Comprehensive Physiology, 8*, 871–901.

Delpire, E., & Gagnon, K. B. (2008). SPAK and OSR1: STE20 kinases involved in the regulation of ion homoeostasis and volume control in mammalian cells. *Biochemical Journal, 409*, 321–331.

Delpire, E., & Gagnon, K. B. (2018b). Water homeostasis and cell volume maintenance and regulation. *Current Topics in Membranes, 81*, 3–52.

Delpire, E., & Guo, J. (2020). Cryo-EM structures of DrNKCC1 and hKCC1: A new milestone in the physiology of cation-chloride cotransporters. *American Journal of Physiology - Cell Physiology, 318*, C225–c237.

Delpire, E., & Koumangoye, R. (2023). NKCC1 in human diseases: Is the SLC12A2 gene haploinsufficient? *American Journal of Physiology - Cell Physiology, 325*, C385–C390.

Delpire, E., & Lauf, P. K. (1992). Kinetics of DIDS inhibition of swelling-activated K-Cl cotransport in low K sheep erythrocytes. *Journal of Membrane Biology, 126*, 89–96.

Delpire, E., & Lauf, P. K. (1991). Magnesium and ATP dependence of K-Cl co-transport in low K^{+} sheep red blood cells. *The Journal of Physiology, 441*, 219–231.

Delpire, E., Lu, J., England, R., Dull, C., & Thorne, T. (1999). Deafness and imbalance associated with inactivation of the secretory Na-K-2Cl co-transporter. *Nature Genetics, 22*, 192–195.

Delpire, E., Rauchman, M. I., Beier, D. R., Hebert, S. C., & Gullans, S. R. (1994). Molecular cloning and chromosome localization of a putative basolateral Na(+)-K(+)-2Cl- cotransporter from mouse inner medullary collecting duct (mIMCD-3) cells. *Journal of Biological Chemistry, 269*, 25677–25683.

Di Fulvio, M., & Alvarez-Leefmans, F. J. (2009). The NKCC and NCC genes: An *in silico* view. In F. J. Alvarez-Leefmans, & E. Delpire (Eds.), *Physiology and pathology of chloride transporters and channels in the nervous system: From molecules to diseases* (1st ed., pp. 169–208). Amsterdam: Academic Press - Elsevier.

Duan, D. D. (2013). Phenomics of cardiac chloride channels. *Comprehensive Physiology, 3*, 667–692.

Duan, D. (2009). Phenomics of cardiac chloride channels: The systematic study of chloride channel function in the heart. *The Journal of Physiology, 587*, 2163–2177.

Dubois, C. J., Cardoit, L., Schwarz, V., Markkanen, M., Airaksinen, M. S., Uvarov, P., Simmers, J., & Thoby-Brisson, M. (2018). Role of the K^+-Cl^- cotransporter KCC2a isoform in mammalian respiration at birth. *eNeuro, 5*.

Dunham, P. B., Stewart, G. W., & Ellory, J. C. (1980). Chloride-activated passive potassium transport in human erythrocytes. *Proceedings of the National Academy of Sciences of the U S A, 77*, 1711–1715.

Dupre, N., Howard, H. C., Mathieu, J., Karpati, G., Vanasse, M., Bouchard, J. P., Carpenter, S., & Rouleau, G. A. (2003). Hereditary motor and sensory neuropathy with agenesis of the corpus callosum. *Annals of Neurology, 54*, 9–18.

Dzhala, V. I., Kuchibhotla, K. V., Glykys, J. C., Kahle, K. T., Swiercz, W. B., Feng, G., Kuner, T., Augustine, G. J., Bacskai, B. J., & Staley, K. J. (2010). Progressive NKCC1-dependent neuronal chloride accumulation during neonatal seizures. *Journal of Neuroscience, 30*, 11745–11761.

Ellory, J. C., & Dunham, P. B. (1980). Volume-dependent passive potassium transport in LK LK sheep red cells. In U. V. Lassen, H. H. Ussing, & J. O. Wieth (Eds.), *Membrane transport in erythrocytes, alfred benzon symposium* (pp. 409–427). Munksgaard.

Evans, M. G., Marty, A., Tan, Y. P., & Trautmann, A. (1986). Blockage of Ca-activated Cl conductance by furosemide in rat lacrimal glands. *Pflügers Archiv, 406*, 65–68.

Evans, R. L., Park, K., Turner, R. J., Watson, G. E., Nguyen, H. V., Dennett, M. R., Hand, A. R., Flagella, M., Shull, G. E., & Melvin, J. E. (2000). Severe impairment of salivation in Na+/K+/2Cl- cotransporter (NKCC1)-deficient mice. *Journal of Biological Chemistry, 275*, 26720–26726.

Fame, R. M., Xu, H., Pragana, A., & Lehtinen, M. (2023). Age-appropriate potassium clearance from perinatal cerebrospinal fluid depends on choroid plexus NKCC1. *Fluids and Barriers of the CNS, 20*, 45.

Fan, M., Zhang, J., Lee, C. L., Zhang, J., & Feng, L. (2023). Structure and thiazide inhibition mechanism of the human Na-Cl cotransporter. *Nature, 614*, 788–793.

Ferdaus, M. Z., & Delpire, E. (2023). The K-Cl cotransporter-3 in the mammalian kidney. *Current Opinion in Nephrology and Hypertension, 32*, 482–489.

Ferdaus, M. Z., Terker, A. S., Koumangoye, R., Wall, S. M., & Delpire, E. (2023). Bicarbonate is the primary inducer of KCC3a expression in renal cortical B-type intercalated cells. *American Journal of Physiology - Cell Physiology, 324*, C1171–c1178.

Ferdaus, M. Z., Terker, A. S., Koumangoye, R. B., Al-Qusairi, L., Welling, P. A., & Delpire, E. (2024). Deletion of KS-WNK1 promotes NCC activation by increasing WNK1/4 abundance. *American Journal of Physiology - Renal Physiology, 327*, F373–f385.

Fernandes, E. C., Pechincha, C., Luz, L. L., Kokai, E., Szucs, P., & Safronov, B. V. (2020). Primary afferent-driven presynaptic inhibition of C-fiber inputs to spinal lamina I neurons. *Progress in Neurobiology, 188*, Article 101786.

Finkelstein, A. (1987). *Water movement through lipid bilayers, pores, and plasma membranes: Theory and reality*. New York: John Wiley & Sons.

Fischbarg, J. (2010). Fluid transport across leaky epithelia: Central role of the tight junction and supporting role of aquaporins. *Physiological Reviews, 90*, 1271–1290.

Flagella, M., Clarke, L. L., Miller, M. L., Erway, L. C., Giannella, R. A., Andringa, A., Gawenis, L. R., Kramer, J., Duffy, J. J., Doetschman, T., Lorenz, J. N., Yamoah, E. N., Cardell, E. L., & Shull, G. E. (1999). Mice lacking the basolateral Na-K-2Cl cotransporter have impaired epithelial chloride secretion and are profoundly deaf. *Journal of Biological Chemistry, 274*, 26946–26955.

Flatman, P. W. (2004). Regulation of Na-K-2Cl cotransport in red cells. *Advances in Experimental Medicine & Biology, 559*, 77–88.

Flemmer, A. W., Gimenez, I., Dowd, B. F., Darman, R. B., & Forbush, B. (2002). Activation of the Na-K-Cl cotransporter NKCC1 detected with a phospho-specific antibody. *Journal of Biological Chemistry, 277*, 37551–37558.

Flores, B., Schornak, C. C., & Delpire, E. (2019). A role for KCC3 in maintaining cell volume of peripheral nerve fibers. *Neurochemistry International, 123*, 114–124.

Flygaard, R. K., Habeck, M., & Nissen, P. (2023). Technical comment on "Inhibition mechanism of NKCC1 involves the carboxyl terminus and long-range conformational coupling". *Science Advances, 9*, Article eadh0821.

Frings, S. (2009). Chloride-based signal amplification in olfactory sensory neurons. In F. J. Alvarez-Leefmans, & E. Delpire (Eds.), *Physiology and pathology of chloride transporters and channels in the nervous system* (pp. 413–424). Amsterdam: Academic Press-Elsevier.

Frizzell, R. A., & Hanrahan, J. W. (2012). Physiology of epithelial chloride and fluid secretion. *Cold Spring Harbor Perspectives in Medicine, 2*, Article a009563.

Fujii, T., Takahashi, Y., Ikari, A., Morii, M., Tabuchi, Y., Tsukada, K., Takeguchi, N., & Sakai, H. (2009). Functional association between K^+-Cl^- cotransporter-4 and H^+,K^+-ATPase in the apical canalicular membrane of gastric parietal cells. *Journal of Biological Chemistry, 284*, 619–629.

Gagnon, M., Bergeron, M. J., Lavertu, G., Castonguay, A., Tripathy, S., Bonin, R. P., Perez-Sanchez, J., Boudreau, D., Wang, B., Dumas, L., Valade, I., Bachand, K., Jacob-Wagner, M., Tardif, C., Kianicka, I., Isenring, P., Attardo, G., Coull, J. A., & De Koninck, Y. (2013). Chloride extrusion enhancers as novel therapeutics for neurological diseases. *Nature Medicine, 19*, 1524–1528.

Gamba, G., & Friedman, P. A. (2009). Thick ascending limb: The Na^+:K^+:$2Cl^-$ co-transporter, NKCC2, and the calcium-sensing receptor, CaSR. *Pflügers Archiv, 458*, 61–76.

Gamba, G., Garbarini, N., & Delpire, E. (2009). Regulation of cation-chloride cotransporters. In F. J. Alvarez-Leefmans, & E. Delpire (Eds.), *Physiology and pathology of chloride transporters and channels in the nervous system: From molecules to diseases* (1st ed., pp. 357–381). Amsterdam: Academic Press - Elsevier.

Gamba, G., Miyanoshita, A., Lombardi, M., Lytton, J., Lee, W. S., Hediger, M. A., & Hebert, S. C. (1994). Molecular cloning, primary structure, and characterization of two members of the mammalian electroneutral sodium-(potassium)-chloride cotransporter family expressed in kidney. *Journal of Biological Chemistry, 269*, 17713–17722.

Gamba, G., Saltzberg, S. N., Lombardi, M., Miyanoshita, A., Lytton, J., Hediger, M. A., Brenner, B. M., & Hebert, S. C. (1993). Primary structure and functional expression of a cDNA encoding the thiazide-sensitive, electroneutral sodium-chloride cotransporter. *Proceedings of the National Academy of Sciences of the U S A, 90*, 2749–2753.

Gamba, G. (2005). Molecular physiology and pathophysiology of electroneutral cation-chloride cotransporters. *Physiological Reviews, 85*, 423–493.

Gamba, G. (2013). *Physiology and pathophysiology of the NaCl Co-transporters in the kidney.*

Gamba, G. (2009a). The sodium-dependent chloride cotransporters. In F. J. Alvarez-Leefmans, & E. Delpire (Eds.), *Physiology and pathology of chloride transporters and channels in the nervous system: From molecules to diseases* (1st ed., pp. 307–331). Amsterdam: Academic Press - Elsevier.

Gamba, G. (2009b). The thiazide-sensitive Na+-Cl- cotransporter: Molecular biology, functional properties, and regulation by WNKs. *American Journal of Physiology - Renal Physiology, 297*, F838–F848.

Garbarini, N., & Delpire, E. (2009). Kockout models of cation-chloride cotransporters. In F. J. Alvarez-Leefmans, & E. Delpire (Eds.), *Physiology and pathology of chloride transporters and channels in the nervous system: From molecules to diseases* (1st ed., pp. 159–166). Amsterdam: Academic Press - Elsevier.

Garcia-Celma, J. J., Ploch, J., Smirnova, I., Kaback, H. R., & Fendler, K. (2010). Delineating electrogenic reactions during lactose/H+ symport. *Biochemistry, 49*, 6115–6121.

Garneau, A. P., Marcoux, A. A., Noël, M., Frenette-Cotton, R., Drolet, M. C., Couet, J., Larivière, R., & Isenring, P. (2016). Ablation of potassium-chloride cotransporter type 3 (Kcc3) in mouse causes multiple cardiovascular defects and isosmotic polyuria. *PLoS One, 11*, Article e0154398.

Garneau, A. P., Slimani, S., Fiola, M. J., Tremblay, L. E., & Isenring, P. (2020). Multiple facets and roles of Na(+)-K(+)-Cl(-) cotransport: Mechanisms and therapeutic implications. *Physiology, 35*, 415–429.

Garneau, A. P., Slimani, S., Tremblay, L. E., Fiola, M. J., Marcoux, A. A., & Isenring, P. (2019). K^+-Cl^- cotransporter 1 (KCC1): A housekeeping membrane protein that plays key supplemental roles in hematopoietic and cancer cells. *Journal of Hematology & Oncology, 12*, 74.

Geck, P., & Heinz, E. (1986). The Na-K-2Cl cotransport system. *Journal of Membrane Biology, 91*, 97–105.

Geck, P., Pietrzyk, C., Burckhardt, B. C., Pfeiffer, B., & Heinz, E. (1980). Electrically silent cotransport of Na^+, K^+ and Cl^- in Ehrlich cells. *Biochimica et Biophysica Acta, 600*, 432–447.

Gerelsaikhan, T., Parvin, M. N., & Turner, R. J. (2006). Biogenesis and topology of the secretory Na^+-K^+-$2Cl^-$ cotransporter (NKCC1) studied in intact mammalian cells. *Biochemistry, 45*, 12060–12067.

Gerelsaikhan, T., & Turner, R. J. (2000a). Membrane topology and function of the secretory Na+-K+-2Cl- cotransporter (NKCC1). *Journal of Korean Medical Science, 15*(Suppl. 1), S3–S4.

Gerelsaikhan, T., & Turner, R. J. (2000b). Transmembrane topology of the secretory Na+-K+-2Cl- cotransporter NKCC1 studied by in vitro translation. *Journal of Biological Chemistry, 275*, 40471–40477.

Gibson, J. S., Ellory, J. C., Adragna, N. C., & Lauf, P. K. (2009). Pathophysiology of the K^+-Cl^- cotransporters: Paths to discovery and overview. In F. J. Alvarez-Leefmans, & E. Delpire (Eds.), *Physiology and pathology of chloride transporters and channels in the nervous system* (pp. 27–42). Amsterdam: Academic Press-Elsevier.

Gillen, C. M., Brill, S., Payne, J. A., & Forbush, B. (1996). Molecular cloning and functional expression of the K-Cl cotransporter from rabbit, rat, and human. A new member of the cation-chloride cotransporter family. *Journal of Biological Chemistry, 271*, 16237–16244.

Gillen, C. M., & Forbush, B., 3rd (1999). Functional interaction of the K-Cl cotransporter (KCC1) with the Na-K-Cl cotransporter in HEK-293 cells. *American Journal of Physiology, 276*, C328–C336.

Goldsmith, E. J., & Rodan, A. R. (2023). Intracellular ion control of WNK signaling. *Annual Review of Physiology, 85*, 383–406.

Gonzalez-Islas, C., Chub, N., & Wenner, P. (2009). NKCC1 and AE3 appear to accumulate chloride in embryonic motoneurons. *Journal of Neurophysiology, 101*, 507–518.

Good, D. W. (1994). Ammonium transport by the thick ascending limb of Henle's loop. *Annual Review of Physiology, 56*, 623–647.

Goodspeed, K., Pérez-Palma, E., Iqbal, S., Cooper, D., Scimemi, A., Johannesen, K. M., Stefanski, A., Demarest, S., Helbig, K. L., Kang, J., Shaffo, F. C., Prentice, B., Brownstein, C. A., Lim, B., Helbig, I., De Los Reyes, E., McKnight, D., Crunelli, V., Campbell, A. J., … Lal, D. (2020). Current knowledge of SLC6A1-related neurodevelopmental disorders. *Brain Communications*, fcaa170.

Gosmanov, A. R., Schneider, E. G., & Thomason, D. B. (2003). NKCC activity restores muscle water during hyperosmotic challenge independent of insulin, ERK, and p38 MAPK. *American Journal of Physiology - Regulatory, Integrative and Comparative Physiology, 284*, R655–R665.

Granados-Soto, V., Arguelles, C. F., & Alvarez-Leefmans, F. J. (2005). Peripheral and central antinociceptive action of Na^{+}-K^{+}-$2Cl^{-}$ cotransporter blockers on formalin-induced nociception in rats. *Pain, 114*, 231–238.

Gregoriades, J. M. C., Madaris, A., Alvarez, F. J., & Alvarez-Leefmans, F. J. (2019). Genetic and pharmacological inactivation of apical Na^{+}-K^{+}-$2Cl^{-}$ cotransporter 1 in choroid plexus epithelial cells reveals the physiological function of the cotransporter. *American Journal of Physiology - Cell Physiology, 316*, C525–c544.

Grichtchenko, I. I., Choi, I., Zhong, X., Bray-Ward, P., Russell, J. M., & Boron, W. F. (2001). Cloning, characterization, and chromosomal mapping of a human electroneutral Na^{+}-driven Cl^{-} HCO_3 exchanger. *Journal of Biological Chemistry, 276*, 8358–8363.

Gundersen, D., Orlowski, J., & Rodriguez-Boulan, E. (1991). Apical polarity of Na,K-ATPase in retinal pigment epithelium is linked to a reversal of the ankyrin-fodrin submembrane cytoskeleton. *The Journal of Cell Biology, 112*, 863–872.

Guo, D., & Hu, J. (2014). Spinal presynaptic inhibition in pain control. *Neuroscience, 283*, 95–106.

Haas, B. R., & Sontheimer, H. (2010). Inhibition of the sodium-potassium-chloride cotransporter isoform-1 reduces glioma invasion. *Cancer Research, 70*, 5597–5606.

Haas, M., & Forbush, B., 3rd (2000). The Na-K-Cl cotransporter of secretory epithelia. *Annual Review of Physiology, 62*, 515–534.

Haas, M., & Forbush, B., 3rd (1998). The Na-K-Cl cotransporters. *Journal of Bioenergetics and Biomembranes, 30*, 161–172.

Haas, M., Schmidt, W. F., 3rd, & McManus, T. J. (1982). Catecholamine-stimulated ion transport in duck red cells. Gradient effects in electrically neutral [Na + K + 2Cl] Co-transport. *The Journal of General Physiology, 80*, 125–147.

Haas, M. (1994). The Na-K-Cl cotransporters. *American Journal of Physiology, 267*, C869–C885.

Hall, A. C., & Ellory, J. C. (1985). Measurement and stoichiometry of bumetanide-sensitive (2Na:1K:3Cl) cotransport in ferret red cells. *Journal of Membrane Biology, 85*, 205–213.

Hamann, S., Herrera-Perez, J. J., Bundgaard, M., Alvarez-Leefmans, F. J., & Zeuthen, T. (2005). Water permeability of Na^{+}-K^{+}-$2Cl^{-}$ cotransporters in mammalian epithelial cells. *The Journal of Physiology, 568*, 123–135.

Hamann, S., Herrera-Perez, J. J., Zeuthen, T., & Alvarez-Leefmans, F. J. (2010). Cotransport of water by the Na^{+}-K^{+}-$2Cl^{-}$ cotransporter NKCC1 in mammalian epithelial cells. *The Journal of Physiology, 588*, 4089–4101.

Hannaert, P., Alvarez-Guerra, M., Pirot, D., Nazaret, C., & Garay, R. P. (2002). Rat NKCC2/NKCC1 cotransporter selectivity for loop diuretic drugs. *Naunyn-Schmiedeberg's Archives of Pharmacology, 365*, 193–199.

Hari, K., Lucas-Osma, A. M., Metz, K., Lin, S., Pardell, N., Roszko, D. A., Black, S., Minarik, A., Singla, R., Stephens, M. J., Pearce, R. A., Fouad, K., Jones, K. E., Gorassini, M. A., Fenrich, K. K., Li, Y., & Bennett, D. J. (2022). GABA facilitates spike propagation through branch points of sensory axons in the spinal cord. *Nature Neuroscience, 25*, 1288–1299.

Harris, G. L., & Betz, W. J. (1987). Evidence for active chloride accumulation in normal and denervated rat lumbrical muscle. *The Journal of General Physiology, 90*, 127–144.

Hartmann, A. M., Wenz, M., Mercado, A., Storger, C., Mount, D. B., Friauf, E., & Nothwang, H. G. (2010). Differences in the large extracellular loop between the K^{+}-Cl^{-} cotransporters KCC2 and KCC4. *Journal of Biological Chemistry, 285*, 23994–24002.

Hartzell, H. C. (2009). Chloride channels: An historical perspective. In F. J. Alvarez-Leefmans, & E. Delipre (Eds.), *Physiology and pathology of chloride transporters and channels in the nervous system from molecules to diseases* (1st ed., pp. 3–15). Amsterdam: Academic Press-Elsevier.

Hengl, T., Kaneko, H., Dauner, K., Vocke, K., Frings, S., & Möhrlen, F. (2010). Molecular components of signal amplification in olfactory sensory cilia. *Proceedings of the National Academy of Sciences of the U S A, 107*, 6052–6057.

Hiki, K., D'Andrea, R. J., Furze, J., Crawford, J., Woollatt, E., Sutherland, G. R., Vadas, M. A., & Gamble, J. R. (1999). Cloning, characterization, and chromosomal location of a novel human K^{+}-Cl^{-} cotransporter. *Journal of Biological Chemistry, 274*, 10661–10667.

Hochman, S., Shreckengost, J., Kimura, H., & Quevedo, J. (2010). Presynaptic inhibition of primary afferents by depolarization: Observations supporting nontraditional mechanisms. *Annals of the New York Academy of Sciences, 1198*, 140–152.

Hodgkin, A. L., & Horowicz, P. (1959). The influence of potassium and chloride ions on the membrane potential of single muscle fibres. *The Journal of Physiology, 148*, 127–160.

Hoffmann, E. K., Lambert, I. H., & Pedersen, S. F. (2009). Physiology of cell volume regulation in vertebrates. *Physiological Reviews, 89*, 193–277.

Hoffmann, E. K., Schiodt, M., & Dunham, P. (1986). The number of chloride-cation cotransport sites on Ehrlich ascites cells measured with [3H] bumetanide. *American Journal of Physiology, 250*, C688–C693.

Hoffmann, E. K. (2001). The pump and leak steady-state concept with a variety of regulated leak pathways. *Journal of Membrane Biology, 184*, 321–330.

Horowitz, A., Chanez-Paredes, S. D., Haest, X., & Turner, J. R. (2023). Paracellular permeability and tight junction regulation in gut health and disease. *Nature Reviews Gastroenterology & Hepatology, 20*, 417–432.

Howard, H. C., Mount, D. B., Rochefort, D., Byun, N., Dupre, N., Lu, J., Fan, X., Song, L., Riviere, J. B., Prevost, C., Horst, J., Simonati, A., Lemcke, B., Welch, R., England, R., Zhan, F. Q., Mercado, A., Siesser, W. B., George, A. L., Jr., … Rouleau, G. A. (2002). The K-Cl cotransporter KCC3 is mutant in a severe peripheral neuropathy associated with agenesis of the corpus callosum. *Nature Genetics, 32*, 384–392.

Hubner, C. A., & Rust, M. B. (2007). Physiology of cation-chloride cotransporters. In M. Pusch (Ed.), *Chloride movements across cellular membranes* (pp. 241–277). Amsterdam: Elsevier.

Hubner, C. A., Stein, V., Hermans-Borgmeyer, I., Meyer, T., Ballanyi, K., & Jentsch, T. J. (2001). Disruption of KCC2 reveals an essential role of K-Cl cotransport already in early synaptic inhibition. *Neuron, 30*, 515–524.

Hutter, O. F., & Noble, D. (1960). The chloride conductance of frog skeletal muscle. *The Journal of Physiology, 151*, 89–102.

Ikeda, K., Oshima, T., Hidaka, H., & Takasaka, T. (1997). Molecular and clinical implications of loop diuretic ototoxicity. *Hearing Research, 107*, 1–8.

Isenring, P., & Forbush, B. (2001). Ion transport and ligand binding by the Na-K-Cl cotransporter, structure-function studies. *Comparative Biochemistry and Physiology Part A: Molecular & Integrative Physiology, 130*, 487–497.

Jarvis, R., Josephine Ng, S. F., Nathanson, A. J., Cardarelli, R. A., Abiraman, K., Wade, F., Evans-Strong, A., Fernandez-Campa, M. P., Deeb, T. Z., Smalley, J. L., Jamier, T., Gurrell, I. K., McWilliams, L., Kawatkar, A., Conway, L. C., Wang, Q., Burli, R. W., Brandon, N. J., Chessell, I. P., … Moss, S. J. (2023). Direct activation of KCC2 arrests benzodiazepine refractory status epilepticus and limits the subsequent neuronal injury in mice. *Cell Reports Medicine, 4*, Article 100957.

Javaheri, S., & Wagner, K. R. (1993). Bumetanide decreases canine cerebrospinal fluid production. In vivo evidence for NaCl cotransport in the central nervous system. *Journal of Clinical Investigation, 92*, 2257–2261.

Javdani, F., Hegedűs, K., Miranda, C. O., Hegyi, Z., Holló, K., & Antal, M. (2020). Differential expression of $Na^+/K^+/Cl^-$) cotransporter 1 in neurons and glial cells within the superficial spinal dorsal horn of rodents. *Scientific Reports, 10*, Article 11715.

Jennings, M. L., & Adame, M. F. (2001). Direct estimate of 1:1 stoichiometry of K^+-Cl^- cotransport in rabbit erythrocytes. *American Journal of Physiology - Cell Physiology, 281*, C825–C832.

Jentsch, T. J. (2008). CLC chloride channels and transporters: From genes to protein structure, pathology and physiology. *Critical Reviews in Biochemistry and Molecular Biology, 43*, 3–36.

Jentsch, T. J. (2016). VRACs and other ion channels and transporters in the regulation of cell volume and beyond. *Nature Reviews Molecular Cell Biology, 17*, 293–307.

Jurkat-Rott, K., Fauler, M., & Lehmann-Horn, F. (2006). Ion channels and ion transporters of the transverse tubular system of skeletal muscle. *Journal of Muscle Research & Cell Motility, 27*, 275–290.

Kahle, K. T., Flores, B., Bharucha-Goebel, D., Zhang, J., Donkervoort, S., Hegde, M., Hussain, G., Duran, D., Liang, B., Sun, D., Bönnemann, C. G., & Delpire, E. (2016). Peripheral motor neuropathy is associated with defective kinase regulation of the KCC3 cotransporter. *Science Signaling, 9*, Article ra77.

Kahle, K. T., Rinehart, J., & Lifton, R. P. (2010). Phosphoregulation of the Na-K-2Cl and K-Cl cotransporters by the WNK kinases. *Biochimica et Biophysica Acta, 1802*, 1150–1158.

Kahle, K. T., Simard, J. M., Staley, K. J., Nahed, B. V., Jones, P. S., & Sun, D. (2009). Molecular mechanisms of ischemic cerebral edema: Role of electroneutral ion transport. *Physiology, 24*, 257–265.

Kahle, K. T., & Staley, K. J. (2008). The bumetanide-sensitive Na-K-2Cl cotransporter NKCC1 as a potential target of a novel mechanism-based treatment strategy for neonatal seizures. *Neurosurgical Focus, 25*, E22.

Kaila, K., Price, T. J., Payne, J. A., Puskarjov, M., & Voipio, J. (2014). Cation-chloride cotransporters in neuronal development, plasticity and disease. *Nature Reviews Neuroscience, 15*, 637–654.

Kaji, D. M. (1993). Effect of membrane potential on K-Cl transport in human erythrocytes. *American Journal of Physiology, 264*, C376–C382.

Kaplan, M. R., Plotkin, M. D., Brown, D., Hebert, S. C., & Delpire, E. (1996). Expression of the mouse Na-K-2Cl cotransporter, mBSC2, in the terminal inner medullary collecting duct, the glomerular and extraglomerular mesangium, and the glomerular afferent arteriole. *Journal of Clinical Investigation, 98*, 723–730.

Karadsheh, M. F., Byun, N., Mount, D. B., & Delpire, E. (2004). Localization of the KCC4 potassium-chloride cotransporter in the nervous system. *Neuroscience, 123*, 381–391.

Karimy, J. K., Zhang, J., Kurland, D. B., Theriault, B. C., Duran, D., Stokum, J. A., Furey, C. G., Zhou, X., Mansuri, M. S., Montejo, J., Vera, A., DiLuna, M. L., Delpire, E., Alper, S. L., Gunel, M., Gerzanich, V., Medzhitov, R., Simard, J. M., & Kahle, K. T. (2017). Inflammation-dependent cerebrospinal fluid hypersecretion by the choroid plexus epithelium in posthemorrhagic hydrocephalus. *Nature Medicine, 23*, 997–1003.

Keep, R. F., Xiang, J., & Betz, A. L. (1994). Potassium cotransport at the rat choroid plexus. *American Journal of Physiology, 267*, C1616–C1622.

Kelly, L., Almutairi, M. M., Kursan, S., Pacheco, R., Dias-Junior, E., Castrop, H., & Di Fulvio, M. (2019). Impaired glucose tolerance, glucagon, and insulin responses in mice lacking the loop diuretic-sensitive Nkcc2a transporter. *American Journal of Physiology - Cell Physiology, 317*, C843–c856.

Keynes, R. D. (1963). Chloride in the squid giant axon. *The Journal of Physiology, 169*, 690–705.

Keynes, R. D. (1982). Introductory remarks. *Philosophical Transactions of the Royal Society of London - B, 299*, 367–368.

Kim, H., Shim, W. S., & Oh, U. (2024). Anoctamin 1, a multi-modal player in pain and itch. *Cell Calcium, 123*, Article 102924.

Kim, Y., & Trussell, L. O. (2009). Negative shift in the glycine reversal potential mediated by a Ca^{2+}- and pH-dependent mechanism in interneurons. *Journal of Neuroscience, 29*, 11495–11510.

Kristensen, M., & Juel, C. (2010). Potassium-transporting proteins in skeletal muscle: Cellular location and fibre-type differences. *Acta Physiologica, 198*, 105–123.

Kumar, M., & Pattnaik, B. R. (2014). Focus on Kir7.1: Physiology and channelopathy. *Channels, 8*, 488–495.

Kursan, S., McMillen, T. S., Beesetty, P., Dias-Junior, E., Almutairi, M. M., Sajib, A. A., Kozak, J. A., Aguilar-Bryan, L., & Di Fulvio, M. (2017). The neuronal K^+-Cl^- co-transporter 2 (Slc12a5) modulates insulin secretion. *Scientific Reports, 7*, 1732.

Kurschat, C. E., Shmukler, B. E., Jiang, L., Wilhelm, S., Kim, E. H., Chernova, M. N., Kinne, R. K., Stewart, A. K., & Alper, S. L. (2006). Alkaline-shifted pHo sensitivity of AE2c1-mediated anion exchange reveals novel regulatory determinants in the AE2 N-terminal cytoplasmic domain. *Journal of Biological Chemistry, 281*, 1885–1896.

Lang, F., Vallon, V., Knipper, M., & Wangemann, P. (2007). Functional significance of channels and transporters expressed in the inner ear and kidney. *American Journal of Physiology - Cell Physiology, 293*, C1187–C1208.

Lauf, P. K., Erdmann, A., & Adragna, N. C. (1994). K-Cl cotransport, pH, and role of Mg in volume-clamped low-K sheep erythrocytes: Three equilibrium states. *American Journal of Physiology, 266*, C95–C103.

Lauf, P. K., & Theg, B. E. (1980). A chloride dependent K^+ flux induced by N-ethylmaleimide in genetically low K^+ sheep and goat erythrocytes. *Biochemical and Biophysical Research Communications, 92*, 1422–1428.

Le Rouzic, P., Ivanov, T. R., Stanley, P. J., Baudoin, F. M., Chan, F., Pinteaux, E., Brown, P. D., & Luckman, S. M. (2006). KCC3 and KCC4 expression in rat adult forebrain. *Brain Research, 1110*, 39–45.

Lee, R. J., Limberis, M. P., Hennessy, M. F., Wilson, J. M., & Foskett, J. K. (2007). Optical imaging of Ca^{2+}-evoked fluid secretion by murine nasal submucosal gland serous acinar cells. *The Journal of Physiology, 582*, 1099–1124.

Lerma, J., & Martin del Rio, R. (1992). Chloride transport blockers prevent N-methyl-D-aspartate receptor-channel complex activation. *Molecular Pharmacology, 41*, 217–222.

Li, H., Tornberg, J., Kaila, K., Airaksinen, M. S., & Rivera, C. (2002). Patterns of cation-chloride cotransporter expression during embryonic rodent CNS development. *European Journal of Neuroscience, 16*, 2358–2370.

Li, R.-C., Lin, C.-C., Ren, X., Wu, J. S., Molday, L. L., Molday, R. S., & Yau, K.-W. (2018). Ca^{2+}-activated Cl current predominates in threshold response of mouse olfactory receptor neurons. *Proceedings of the National Academy of Sciences, 115*, 5570–5575.

Liu, S., Chang, S., Han, B., Xu, L., Zhang, M., Zhao, C., Yang, W., Wang, F., Li, J., Delpire, E., Ye, S., Bai, X. C., & Guo, J. (2019). Cryo-EM structures of the human cation-chloride cotransporter KCC1. *Science, 366*, 505–508.

Lobato-Álvarez, J. A., Roldán, M. L., López-Murillo, T. D., González-Ramírez, R., Bonilla-Delgado, J., & Shoshani, L. (2016). The apical localization of Na^+, K^+-ATPase in cultured human retinal pigment epithelial cells depends on expression of the β(2) subunit. *Frontiers in Physiology, 7*, 450.

Lodovichi, C., Ratto, G. M., Trevelyan, A. J., & Arosio, D. (2022). Genetically encoded sensors for chloride concentration. *Journal of Neuroscience Methods, 368*, Article 109455.

Löscher, W., & Kaila, K. (2022). CNS pharmacology of NKCC1 inhibitors. *Neuropharmacology, 205*, Article 108910.

Lucas, O., Hilaire, C., Delpire, E., & Scamps, F. (2012). KCC3-dependent chloride extrusion in adult sensory neurons. *Molecular and Cellular Neurosciences, 50*, 211–220.

Lüscher, B. P., Vachel, L., Ohana, E., & Muallem, S. (2020). Cl^- as a bona fide signaling ion. *American Journal of Physiology - Cell Physiology, 318*, C125–c136.

Lykke, K., Töllner, K., Feit, P. W., Erker, T., MacAulay, N., & Löscher, W. (2016). The search for NKCC1-selective drugs for the treatment of epilepsy: Structure-function relationship of bumetanide and various bumetanide derivatives in inhibiting the human cation-chloride cotransporter NKCC1A. *Epilepsy and Behavior E&B, 59*, 42–49.

Lykke, K., Töllner, K., Römermann, K., Feit, P. W., Erker, T., MacAulay, N., & Löscher, W. (2015). Structure-activity relationships of bumetanide derivatives: Correlation between diuretic activity in dogs and inhibition of the human NKCC2A transporter. *British Journal of Pharmacology, 172*, 4469–4480.

Lytle, C., & Forbush, B., 3rd (1992). The Na-K-Cl cotransport protein of shark rectal gland. II. Regulation by direct phosphorylation. *Journal of Biological Chemistry, 267*, 25438–25443.

Lytle, C., McManus, T. J., & Haas, M. (1998). A model of Na-K-2Cl cotransport based on ordered ion binding and glide symmetry. *American Journal of Physiology, 274*, C299–C309.

Lytle, C., Xu, J. C., Biemesderfer, D., & Forbush, B., 3rd (1995). Distribution and diversity of Na-K-Cl cotransport proteins: A study with monoclonal antibodies. *American Journal of Physiology, 269*, C1496–C1505.

Lytle, C. (2003). Na+-K+-2Cl- cotransport. In I. Bernhardt, & J. C. Ellory (Eds.), *Red cell membrane transport in health and disease* (pp. 173–195). New York: Springer, Berlin.

MacAulay, N., Hamann, S., & Zeuthen, T. (2009). Chloride transporters as water pumps: Elements in a new model of epithelial water transport. In F. J. Alvarez-Leefmans, & E. Delpire (Eds.), *Physiology and pathology of chloride transporters and channels in the nervous system* (pp. 547–568). Amsterdam: Academic Press-Elsevier.

MacAulay, N., & Rose, C. R. (2020). CrossTalk opposing view: NKCC1 in the luminal membrane of choroid plexus is outwardly directed under basal conditions and contributes directly to cerebrospinal fluid secretion. *The Journal of Physiology, 598*, 4737–4739.

Mahrous, A. A., Liang, L., Balaguer, J. M., Ho, J. C., Hari, K., Grigsby, E. M., Karapetyan, V., Damiani, A., Fields, D. P., Gonzalez-Martinez, J. A., Gerszten, P. C., Bennett, D. J., Heckman, C. J., Pirondini, E., & Capogrosso, M. (2023). GABA increases sensory transmission in monkeys. *bioRxiv: The Preprint Server for Biology.* https://doi.org/10.1101/2023.12.28.573467. PMID: 38234767; PMCID: PMC10793394.

Mandai, S., Furukawa, S., Kodaka, M., Hata, Y., Mori, T., Nomura, N., Ando, F., Mori, Y., Takahashi, D., Yoshizaki, Y., Kasagi, Y., Arai, Y., Sasaki, E., Yoshida, S., Furuichi, Y., Fujii, N. L., Sohara, E., Rai, T., & Uchida, S. (2017). Loop diuretics affect skeletal myoblast differentiation and exercise-induced muscle hypertrophy. *Scientific Reports, 7*, Article 46369.

Mao, S., Garzon-Muvdi, T., Di Fulvio, M., Chen, Y., Delpire, E., Alvarez, F. J., & Alvarez-Leefmans, F. J. (2012). Molecular and functional expression of cation-chloride cotransporters in dorsal root ganglion neurons during postnatal maturation. *Journal of Neurophysiology, 108*, 834–852.

Marcano, M., Yang, H. M., Nieves-Gonzalez, A., Clausen, C., & Moore, L. C. (2009). Parameter estimation for mathematical models of NKCC2 cotransporter isoforms. *American Journal of Physiology - Renal Physiology, 296*, F369–F381.

Marcoux, A. A., Garneau, A. P., Frenette-Cotton, R., Slimani, S., Mac-Way, F., & Isenring, P. (2017). Molecular features and physiological roles of K^+-Cl^- cotransporter 4 (KCC4). *Biochimica et Biophysica Acta (BBA) - General Subjects, 1861*, 3154–3166.

Marcoux, A. A., Tremblay, L. E., Slimani, S., Fiola, M. J., Mac-Way, F., Garneau, A. P., & Isenring, P. (2021). Molecular characteristics and physiological roles of Na^+-K^+-Cl^- cotransporter 2. *Journal of Cellular Physiology, 236*, 1712–1729.

Marcus, D. C. (2020). Ion transport across inner ear epithelia. In K. L. Hamilton, & D. C. Devor (Eds.), *Ion transport across epithelial tissues and disease: Ion channels and transporters of epithelia in health and disease* - (Vol 2, pp. 279—305). Cham: Springer International Publishing.

Markadieu, N., & Delpire, E. (2014). Physiology and pathophysiology of SLC12A1/2 transporters. *Pflügers Archiv, 466*, 91—105.

Marshall-Phelps, K. L. H., Kegel, L., Baraban, M., Ruhwedel, T., Almeida, R. G., Rubio-Brotons, M., Klingseisen, A., Benito-Kwiecinski, S. K., Early, J. J., Bin, J. M., Suminaite, D., Livesey, M. R., Möbius, W., Poole, R. J., & Lyons, D. A. (2020). Neuronal activity disrupts myelinated axon integrity in the absence of NKCC1b. *Journal of Cell Biology, 219*.

McDaniel, N., Pace, A. J., Spiegel, S., Engelhardt, R., Koller, B. H., Seidler, U., & Lytle, C. (2005). Role of Na-K-2Cl cotransporter-1 in gastric secretion of nonacidic fluid and pepsinogen. *American Journal of Physiology - Gastrointestinal and Liver Physiology, 289*, G550—G560.

McMoneagle, E., Zhou, J., Zhang, S., Huang, W., Josiah, S. S., Ding, K., Wang, Y., & Zhang, J. (2024). Neuronal K^+-Cl^- cotransporter KCC2 as a promising drug target for epilepsy treatment. *Acta Pharmacologica Sinica, 45*, 1—22.

McRoberts, J. A., Erlinger, S., Rindler, M. J., & Saier, M. H., Jr. (1982). Furosemide-sensitive salt transport in the Madin-Darby canine kidney cell line. Evidence for the cotransport of Na+, K+, and Cl. *Journal of Biological Chemistry, 257*, 2260—2266.

Meor Azlan, N. F., & Zhang, J. (2020). Role of the cation-chloride-cotransporters in cardiovascular disease. *Cells, 9*.

Mercado, A., Song, L., Vazquez, N., Mount, D. B., & Gamba, G. (2000). Functional comparison of the K^+-Cl^- cotransporters KCC1 and KCC4. *Journal of Biological Chemistry, 275*, 30326—30334.

Mercado, A., Vazquez, N., Song, L., Cortes, R., Enck, A. H., Welch, R., Delpire, E., Gamba, G., & Mount, D. B. (2005). NH2-terminal heterogeneity in the KCC3 K^+-Cl^- cotransporter. *American Journal of Physiology - Renal Physiology, 289*, F1246—F1261.

Migliati, E., Meurice, N., DuBois, P., Fang, J. S., Somasekharan, S., Beckett, E., Flynn, G., & Yool, A. J. (2009). Inhibition of aquaporin-1 and aquaporin-4 water permeability by a derivative of the loop diuretic bumetanide acting at an internal pore-occluding binding site. *Molecular Pharmacology, 76*, 105—112.

Mistrik, P., & Ashmore, J. (2009). The role of potassium recirculation in cochlear amplification. *Current Opinion in Otolaryngology & Head and Neck Surgery, 17*, 394—399.

Momose-Sato, Y., Sato, K., Hirota, A., & Kamino, K. (1998). GABA-induced intrinsic light-scattering changes associated with voltage-sensitive dye signals in embryonic brain stem slices: Coupling of depolarization and cell shrinkage. *Journal of Neurophysiology, 79*, 2208—2217.

Moseng, M. A., Su, C. C., Klenotic, P. A., Delpire, E., & Yu, E. W. (2023). Response to comment on "Inhibition mechanism of NKCC1 involves the carboxyl terminus and long-range conformational coupling". *Science Advances, 9*, Article eadi5716.

Moseng, M. A., Su, C. C., Rios, K., Cui, M., Lyu, M., Glaza, P., Klenotic, P. A., Delpire, E., & Yu, E. W. (2022). Inhibition mechanism of NKCC1 involves the carboxyl terminus and long-range conformational coupling. *Science Advances, 8*, Article eabq0952.

Mount, D. B., Mercado, A., Song, L., Xu, J., George, A. L., Jr., Delpire, E., & Gamba, G. (1999). Cloning and characterization of KCC3 and KCC4, new members of the cation-chloride cotransporter gene family. *Journal of Biological Chemistry, 274*, 16355—16362.

Mukherjee, A., Yang, C. L., McCormick, J. A., Martz, K., Sharma, A., & Ellison, D. H. (2021). Roles of WNK4 and SPAK in K^+-mediated dephosphorylation of the NaCl cotransporter. *American Journal of Physiology - Renal Physiology, 320*, F719—f733.

Munoz, A., Mendez, P., DeFelipe, J., & Alvarez-Leefmans, F. J. (2007). Cation-chloride cotransporters and GABA-ergic innervation in the human epileptic hippocampus. *Epilepsia, 48*, 663—673.

Murillo-de-Ozores, A. R., Chávez-Canales, M., de Los Heros, P., Gamba, G., & Castañeda-Bueno, M. (2020). Physiological processes modulated by the chloride-sensitive WNK-SPAK/OSR1 kinase signaling pathway and the cation-coupled chloride cotransporters. *Frontiers in Physiology, 11*, Article 585907.

Nan, J., Yuan, Y., Yang, X., Shan, Z., Liu, H., Wei, F., Zhang, W., & Zhang, Y. (2022). Cryo-EM structure of the human sodium-chloride cotransporter NCC. *Science Advances, 8*, Article eadd7176.

Nayak, S. R., Joseph, D., Höfner, G., Dakua, A., Athreya, A., Wanner, K. T., Kanner, B. I., & Penmatsa, A. (2023). Cryo-EM structure of GABA transporter 1 reveals substrate recognition and transport mechanism. *Nature Structural & Molecular Biology, 30*, 1023—1032.

Nejsum, L. N., Praetorius, J., & Nielsen, S. (2005). NKCC1 and NHE1 are abundantly expressed in the basolateral plasma membrane of secretory coil cells in rat, mouse, and human sweat glands. *American Journal of Physiology - Cell Physiology, 289*, C333—C340.

Neumann, C., Rosenbaek, L. L., Flygaard, R. K., Habeck, M., Karlsen, J. L., Wang, Y., Lindorff-Larsen, K., Gad, H. H., Hartmann, R., Lyons, J. A., Fenton, R. A., & Nissen, P. (2022). Cryo-EM structure of the human NKCC1 transporter reveals mechanisms of ion coupling and specificity. *The EMBO Journal, 41*, Article e110169.

Nielsen, S., Maunsbach, A. B., Ecelbarger, C. A., & Knepper, M. A. (1998). Ultrastructural localization of Na-K-2Cl cotransporter in thick ascending limb and macula densa of rat kidney. *American Journal of Physiology, 275*, F885—F893.

O'Donnell, M. E., Tran, L., Lam, T. I., Liu, X. B., & Anderson, S. E. (2004). Bumetanide inhibition of the blood-brain barrier Na-K-Cl cotransporter reduces edema formation in the rat middle cerebral artery occlusion model of stroke. *Journal of Cerebral Blood Flow and Metabolism, 24*, 1046—1056.

O'Grady, S. M., Palfrey, H. C., & Field, M. (1987a). Characteristics and functions of Na-K-Cl cotransport in epithelial tissues. *American Journal of Physiology, 253*, C177—C192.

O'Grady, S. M., Palfrey, H. C., & Field, M. (1987b). Na-K-2Cl cotransport in winter flounder intestine and bovine kidney outer medulla: [3H] bumetanide binding and effects of furosemide analogues. *Journal of Membrane Biology, 96*, 11—18.

O'Neill, W. C. (1999). Physiological significance of volume-regulatory transporters. *American Journal of Physiology, 276*, C995—c1011.

Oppermann, M., Mizel, D., Huang, G., Li, C., Deng, C., Theilig, F., Bachmann, S., Briggs, J., Schnermann, J., & Castrop, H. (2006). Macula densa control of renin secretion and preglomerular resistance in mice with selective deletion of the B isoform of the Na, K, 2Cl co-transporter. *Journal of the American Society of Nephrology, 17*, 2143–2152.

Oppermann, M., Mizel, D., Kim, S. M., Chen, L., Faulhaber-Walter, R., Huang, Y., Li, C., Deng, C., Briggs, J., Schnermann, J., & Castrop, H. (2007). Renal function in mice with targeted disruption of the A isoform of the Na-K-2Cl co-transporter. *Journal of the American Society of Nephrology, 18*, 440–448.

Pan, D., Kalfa, T. A., Wang, D., Risinger, M., Crable, S., Ottlinger, A., Chandra, S., Mount, D. B., Hubner, C. A., Franco, R. S., & Joiner, C. H. (2011). KCL cotransporter gene expression during human and murine erythroid differentiation. *Journal of Biological Chemistry, 286*, 30492–30503.

Panet, R., Marcus, M., & Atlan, H. (2000). Overexpression of the $Na^+/K^+/Cl^-$ cotransporter gene induces cell proliferation and phenotypic transformation in mouse fibroblasts. *Journal of Cellular Physiology, 182*, 109–118.

Paredes, A., Plata, C., Rivera, M., Moreno, E., Vazquez, N., Munoz-Clares, R., Hebert, S. C., & Gamba, G. (2006). Activity of the renal Na^+-K^+-$2Cl^-$ cotransporter is reduced by mutagenesis of N-glycosylation sites: Role for protein surface charge in Cl^- transport. *American Journal of Physiology - Renal Physiology, 290*, F1094–F1102.

Payne, J. A., & Forbush, B. (1994). Alternatively spliced isoforms of the putative renal Na-K-Cl cotransporter are differentially distributed within the rabbit kidney. *Proceedings of the National Academy of Sciences of the U S A, 91*, 4544–4548.

Payne, J. A., & Forbush, B. (1995). Molecular characterization of the epithelial Na-K-Cl cotransporter isoforms. *Current Opinion in Cell Biology, 7*, 493–503.

Payne, J. A., Rivera, C., Voipio, J., & Kaila, K. (2003). Cation-chloride co-transporters in neuronal communication, development and trauma. *Trends in Neurosciences, 26*, 199–206.

Payne, J. A., Stevenson, T. J., & Donaldson, L. F. (1996). Molecular characterization of a putative K-Cl cotransporter in rat brain. A neuronal-specific isoform. *Journal of Biological Chemistry, 271*, 16245–16252.

Payne, J. A., Xu, J. C., Haas, M., Lytle, C. Y., Ward, D., & Forbush, B., 3rd (1995). Primary structure, functional expression, and chromosomal localization of the bumetanide-sensitive Na-K-Cl cotransporter in human colon. *Journal of Biological Chemistry, 270*, 17977–17985.

Payne, J. A. (1997). Functional characterization of the neuronal-specific K-Cl cotransporter: Implications for [K+]o regulation. *American Journal of Physiology, 273*, C1516–C1525.

Payne, J. A. (2009). The potasium-chloride cotransporters: From cloning to structure and function. In F. J. Alvarez-Leefmans, & E. Delpire (Eds.), *Physiology and pathology of chloride transporters and channels in the nervous system: From molecules to diseases* (1st ed., pp. 333–356). Amsterdam: Academic Press - Elsevier.

Pearson, M. M., Lu, J., Mount, D. B., & Delpire, E. (2001). Localization of the K^+-Cl^- cotransporter, KCC3, in the central and peripheral nervous systems: Expression in the choroid plexus, large neurons and white matter tracts. *Neuroscience, 103*, 481–491.

Peña-Münzenmayer, G., Catalán, M. A., Kondo, Y., Jaramillo, Y., Liu, F., Shull, G. E., & Melvin, J. E. (2015). Ae4 (Slc4a9) anion exchanger drives Cl^- uptake-dependent fluid secretion by mouse submandibular gland acinar cells *. *Journal of Biological Chemistry, 290*, 10677–10688.

Peña-Münzenmayer, G., George, A. T., Shull, G. E., Melvin, J. E., & Catalán, M. A. (2016). Ae4 (Slc4a9) is an electroneutral monovalent cation-dependent Cl^-/HCO_3^- exchanger. *The Journal of General Physiology, 147*, 423–436.

Peti-Peterdi, J., & Harris, R. C. (2010). Macula densa sensing and signaling mechanisms of renin release. *Journal of the American Society of Nephrology, 21*, 1093–1096.

Piala, A. T., Moon, T. M., Akella, R., He, H., Cobb, M. H., & Goldsmith, E. J. (2014). Chloride sensing by WNK1 involves inhibition of autophosphorylation. *Science Signaling, 7*, Article ra41.

Piechotta, K., Lu, J., & Delpire, E. (2002). Cation chloride cotransporters interact with the stress-related kinases Ste20-related proline-alanine-rich kinase (SPAK) and oxidative stress response 1 (OSR1). *Journal of Biological Chemistry, 277*, 50812–50819.

Pizzagalli, M. D., Bensimon, A., & Superti-Furga, G. (2021). A guide to plasma membrane solute carrier proteins. *FEBS Journal, 288*, 2784–2835.

Plata, C., Meade, P., Vazquez, N., Hebert, S. C., & Gamba, G. (2002). Functional properties of the apical Na^+-K^+-$2Cl^-$ cotransporter isoforms. *Journal of Biological Chemistry, 277*, 11004–11012.

Plotkin, M. D., Kaplan, M. R., Peterson, L. N., Gullans, S. R., Hebert, S. C., & Delpire, E. (1997). Expression of the Na^+-K^+-$2Cl^-$ cotransporter BSC2 in the nervous system. *American Journal of Physiology, 272*, C173–C183.

Portioli, C., Ruiz Munevar, M. J., De Vivo, M., & Cancedda, L. (2021). Cation-coupled chloride cotransporters: Chemical insights and disease implications. *Trends in Chemistry, 3*, 832–849.

Prasad, V., Bodi, I., Meyer, J. W., Wang, Y., Ashraf, M., Engle, S. J., Doetschman, T., Sisco, K., Nieman, M. L., Miller, M. L., Lorenz, J. N., & Shull, G. E. (2008). Impaired cardiac contractility in mice lacking both the AE3 Cl^-/HCO_3^- exchanger and the NKCC1 Na^+-K^+-$2Cl^-$ cotransporter: Effects on Ca^{2+} handling and protein phosphatases. *Journal of Biological Chemistry, 283*, 31303–31314.

Quinton, P. M., Wright, E. M., & Tormey, J. M. (1973). Localization of sodium pumps in the choroid plexus epithelium. *The Journal of Cell Biology, 58*, 724–730.

Race, J. E., Makhlouf, F. N., Logue, P. J., Wilson, F. H., Dunham, P. B., & Holtzman, E. J. (1999). Molecular cloning and functional characterization of KCC3, a new K-Cl cotransporter. *American Journal of Physiology, 277*, C1210–C1219.

Radulovic, T., Rajaram, E., Ebbers, L., Pagella, S., Winklhofer, M., Kopp-Scheinpflug, C., Nothwang, H. G., Milenkovic, I., & Hartmann, A. M. (2023). Serine 937 phosphorylation enhances KCC2 activity and strengthens synaptic inhibition. *Scientific Reports, 13*, 21660.

Randall, J., Thorne, T., & Delpire, E. (1997). Partial cloning and characterization of Slc12a2: The gene encoding the secretory Na^+-K^+-$2Cl^-$ cotransporter. *American Journal of Physiology, 273*, C1267–C1277.

Ransom, C. B., Ransom, B. R., & Sontheimer, H. (2000). Activity-dependent extracellular K+ accumulation in rat optic nerve: The role of glial and axonal Na+ pumps. *The Journal of Physiology, 522*(Pt 3), 427–442.

Reddy, M. M., & Quinton, P. M. (1999). Bumetanide blocks CFTR GCl in the native sweat duct. *American Journal of Physiology, 276*, C231–C237.

Reichhart, N., & Strauss, O. (2014). Ion channels and transporters of the retinal pigment epithelium. *Experimental Eye Research, 126*, 27–37.

Reichhart, N., & Strauss, O. (2020). Regulation of ion transport through retinal pigment epithelium: Impact in retinal degeneration. In K. L. Hamilton, & D. C. Devor (Eds.), *Ion transport across epithelial tissues and disease* (2nd ed., pp. 307–332). Switzerland: Springer-American Physiological Society.

Reichhart, N., & Strauß, O. (2020). Regulation of ion transport through retinal pigment epithelium: Impact in retinal degeneration. In K. L. Hamilton, & D. C. Devor (Eds.), *Ion transport across epithelial tissues and disease: Ion channels and transporters of epithelia in health and disease* (Vol. 2, pp. 307–331). Cham: Springer International Publishing.

Reid, M. S., Kern, D. M., & Brohawn, S. G. (2020). Cryo-EM structure of the potassium-chloride cotransporter KCC4 in lipid nanodiscs. *Elife, 9*, Article e52505.

Ren, D. (2011). Sodium leak channels in neuronal excitability and rhythmic behaviors. *Neuron, 72*, 899–911.

Renfro, J. L. (1977). Interdependence of Active Na+ and Cl- transport by the isolated urinary bladder of the teleost, *Pseudopleuronectes americanus. Journal of Experimental Zoology, 199*, 383–390.

Renfro, J. L. (1975). Water and ion transport by the urinary bladder of the teleost *Pseudopleuronectes americanus. American Journal of Physiology, 228*, 52–61.

Reuss, L. (1997). Epithelial transport. In J. P. Hoffman, & J. D. Jamieson (Eds.), *Handbook of physiology section 14: Cell physiology* (pp. 309–388). New York: Oxford University Press.

Reynolds, A., Brustein, E., Liao, M., Mercado, A., Babilonia, E., Mount, D. B., & Drapeau, P. (2008). Neurogenic role of the depolarizing chloride gradient revealed by global overexpression of KCC2 from the onset of development. *Journal of Neuroscience, 28*, 1588–1597.

Rinehart, J., Maksimova, Y. D., Tanis, J. E., Stone, K. L., Hodson, C. A., Zhang, J., Risinger, M., Pan, W., Wu, D., Colangelo, C. M., Forbush, B., Joiner, C. H., Gulcicek, E. E., Gallagher, P. G., & Lifton, R. P. (2009). Sites of regulated phosphorylation that control K-Cl cotransporter activity. *Cell, 138*, 525–536.

Rivera, A., Nasburg, J. A., Shim, H., Shmukler, B. E., Kitten, J., Wohlgemuth, J. G., Dlott, J. S., Snyder, L. M., Brugnara, C., Wulff, H., & Alper, S. L. (2022). The erythroid K-Cl cotransport inhibitor [(dihydroindenyl)oxy]acetic acid blocks erythroid Ca(2+)-activated K(+) channel KCNN4. *American Journal of Physiology - Cell Physiology, 323*, C694–c705.

Rivera, C., Voipio, J., Payne, J. A., Ruusuvuori, E., Lahtinen, H., Lamsa, K., Pirvola, U., Saarma, M., & Kaila, K. (1999). The K+/Cl- co-transporter KCC2 renders GABA hyperpolarizing during neuronal maturation. *Nature, 397*, 251–255.

Rocha-Gonzalez, H. I., Mao, S., & Alvarez-Leefmans, F. J. (2008). Na^+,K^+,2Cl^- cotransport and intracellular chloride regulation in rat primary sensory neurons: Thermodynamic and kinetic aspects. *Journal of Neurophysiology, 100*, 169–184.

Romero, M. F., Chang, M.-H., & Mount, D. (2009). From cloning to structure, function, and regulation of chloride-dependent and independent bicarbonate transporters. In F. J. Alvarez-Leefmans, & E. Delipre (Eds.), *Physiology and pathology of chloride transporters and channels in the nervous system from molecules to diseases* (1st ed., pp. 43–79). Amsterdam: Academic Press-Elsevier.

Romero, M. F., Chen, A. P., Parker, M. D., & Boron, W. F. (2013). The SLC4 family of bicarbonate (HCO_3^-) transporters. *Molecular Aspects of Medicine, 34*, 159–182.

Romero, M. F., Fulton, C. M., & Boron, W. F. (2004). The SLC4 family of HCO_3^- transporters. *Pflügers Archiv, 447*, 495–509.

Roseth, S., Fykse, E. M., & Fonnum, F. (1995). Uptake of L-glutamate into rat brain synaptic vesicles: Effect of inhibitors that bind specifically to the glutamate transporter. *Journal of Neurochemistry, 65*, 96–103.

Roth, M., Obaidat, A., & Hagenbuch, B. (2012). OATPs, OATs and OCTs: The organic anion and cation transporters of the SLCO and SLC22A gene superfamilies. *British Journal of Pharmacology, 165*, 1260–1287.

Rudomin, P., & Schmidt, R. F. (1999). Presynaptic inhibition in the vertebrate spinal cord revisited. *Experimental Brain Research, 129*, 1–37.

Ruiz Munevar, M. J., Rizzi, V., Portioli, C., Vidossich, P., Cao, E., Parrinello, M., Cancedda, L., & De Vivo, M. (2024). Cation chloride cotransporter NKCC1 operates through a rocking-bundle mechanism. *Journal of the American Chemical Society, 146*, 552–566.

Russell, J. M. (1984). Chloride in the squid giant axon. *Current Topics in Membranes and Transport, 22*, 177–193.

Russell, J. M. (2009). Sodium-coupled chloride cotransporters: Discovery and newly emerging concepts. In F. J. Alvarez-Leefmans, & E. Delpire (Eds.), *Physiology and pathology of chloride transporters and channels in the nervous system* (pp. 17–26). Amsterdam: Academic Press-Elsevier.

Russell, J. M. (2000). Sodium-potassium-chloride cotransport. *Physiological Reviews, 80*, 211–276.

Rust, M. B., Alper, S. L., Rudhard, Y., Shmukler, B. E., Vicente, R., Brugnara, C., Trudel, M., Jentsch, T. J., & Hübner, C. A. (2007). Disruption of erythroid K-Cl cotransporters alters erythrocyte volume and partially rescues erythrocyte dehydration in SAD mice. *Journal of Clinical Investigation, 117*, 1708–1717.

Rust, M. B., Faulhaber, J., Budack, M. K., Pfeffer, C., Maritzen, T., Didie, M., Beck, F. X., Boettger, T., Schubert, R., Ehmke, H., Jentsch, T. J., & Hubner, C. A. (2006). Neurogenic mechanisms contribute to hypertension in mice with disruption of the K-Cl cotransporter KCC3. *Circulation Research, 98*, 549–556.

Schlatter, E., Greger, R., & Weidtke, C. (1983). Effect of "high ceiling" diuretics on active salt transport in the cortical thick ascending limb of Henle's loop of rabbit kidney. Correlation of chemical structure and inhibitory potency. *Pflügers Archiv, 396*, 210–217.

Schomberg, S. L., Bauer, J., Kintner, D. B., Su, G., Flemmer, A., Forbush, B., & Sun, D. (2003). Cross talk between the GABA(A) receptor and the Na-K-Cl cotransporter is mediated by intracellular Cl. *Journal of Neurophysiology, 89*, 159–167.

Sedmak, G., Jovanov-Milošević, N., Puskarjov, M., Ulamec, M., Krušlin, B., Kaila, K., & Judaš, M. (2016). Developmental expression patterns of KCC2 and functionally associated molecules in the human brain. *Cerebral Cortex, 26*, 4574–4589.

Shankar, S. S., & Brater, D. C. (2003). Loop diuretics: From the Na-K-2Cl transporter to clinical use. *American Journal of Physiology - Renal Physiology, 284*, F11–F21.

Shcheynikov, N., Wang, Y., Park, M., Ko, S. B., Dorwart, M., Naruse, S., Thomas, P. J., & Muallem, S. (2006). Coupling modes and stoichiometry of Cl^-/HCO_3^- exchange by slc26a3 and slc26a6. *The Journal of General Physiology, 127*, 511–524.

Shekarabi, M., Salin-Cantegrel, A., Laganiere, J., Gaudet, R., Dion, P., & Rouleau, G. A. (2011). Cellular expression of the K^+-Cl^- cotransporter KCC3 in the central nervous system of mouse. *Brain Research, 1374*, 15–26.

Shekarabi, M., Zhang, J., Khanna, A. R., Ellison, D. H., Delpire, E., & Kahle, K. T. (2017). WNK kinase signaling in ion homeostasis and human disease. *Cell Metabolism, 25*, 285–299.

Shillingford, J. M., Miyoshi, K., Flagella, M., Shull, G. E., & Hennighausen, L. (2002). Mouse mammary epithelial cells express the Na-K-Cl cotransporter, NKCC1: Characterization, localization, and involvement in ductal development and morphogenesis. *Molecular Endocrinology, 16*, 1309–1321.

Shimizu-Okabe, C., Okada, S., Okamoto, S., Masuzaki, H., & Takayama, C. (2022). Specific expression of KCC2 in the α cells of normal and type 1 diabetes model mouse pancreatic islets. *Acta Histochemica et Cytochemica, 55*, 47–56.

Silva, P., Solomon, R. J., & Epstein, F. H. (1996). The rectal gland of *Squalus acanthias*: A model for the transport of chloride. *Kidney International, 49*, 1552–1556.

Silva, P., Stoff, J., Field, M., Fine, L., Forrest, J. N., & Epstein, F. H. (1977). Mechanism of active chloride secretion by shark rectal gland: Role of Na-K-ATPase in chloride transport. *American Journal of Physiology, 233*, F298–F306.

Singh, R., Almutairi, M. M., Pacheco-Andrade, R., Almiahuob, M. Y., & Di Fulvio, M. (2015). Impact of hybrid and complex N-glycans on cell surface targeting of the endogenous chloride cotransporter Slc12a2. *International Journal of Cell Biology, 2015*, Article 505294.

Singh, R., Kursan, S., Almiahoub, M. Y., Almutairi, M. M., Garzón-Muvdi, T., Alvarez-Leefmans, F. J., & Di Fulvio, M. (2016). Plasma membrane targeting of endogenous NKCC2 in COS7 cells bypasses functional golgi cisternae and complex N-glycosylation. *Frontiers in Cell and Developmental Biology, 4*, 150.

Sivakumaran, S., Cardarelli, R. A., Maguire, J., Kelley, M. R., Silayeva, L., Morrow, D. H., Mukherjee, J., Moore, Y. E., Mather, R. J., Duggan, M. E., Brandon, N. J., Dunlop, J., Zicha, S., Moss, S. J., & Deeb, T. Z. (2015). Selective inhibition of KCC2 leads to hyperexcitability and epileptiform discharges in hippocampal slices and in vivo. *Journal of Neuroscience, 35*, 8291–8296.

Somasekharan, S., Tanis, J., & Forbush, B. (2012). Loop diuretic and ion-binding residues revealed by scanning mutagenesis of transmembrane helix 3 (TM3) of Na-K-Cl cotransporter (NKCC1). *Journal of Biological Chemistry, 287*, 17308–17317.

Somjen, G. S. (2004). *Ions in the brain*. New York: Oxford University Press.

Sonawane, N. D., Thiagarajah, J. R., & Verkman, A. S. (2002). Chloride concentration in endosomes measured using a ratioable fluorescent Cl– indicator. *Journal of Biological Chemistry, 277*, 5506–5513.

Song, L., Mercado, A., Vazquez, N., Xie, Q., Desai, R., George, A. L., Jr., Gamba, G., & Mount, D. B. (2002). Molecular, functional, and genomic characterization of human KCC2, the neuronal K-Cl cotransporter. *Molecular Brain Research, 103*, 91–105.

Steffensen, A. B., Oernbo, E. K., Stoica, A., Gerkau, N. J., Barbuskaite, D., Tritsaris, K., Rose, C. R., & MacAulay, N. (2018). Cotransporter-mediated water transport underlying cerebrospinal fluid formation. *Nature Communications, 9*, 2167.

Steinbach, H. B. (1941). Chloride in the giant axons of the squid. *Journal of Cellular and Comparative Physiology, 17*, 57–64.

Stil, A., Liabeuf, S., Jean-Xavier, C., Brocard, C., Viemari, J. C., & Vinay, L. (2009). Developmental up-regulation of the potassium-chloride cotransporter type 2 in the rat lumbar spinal cord. *Neuroscience, 164*, 809–821.

Stokes, J. B. (1984). Sodium chloride absorption by the urinary bladder of the winter flounder. A thiazide-sensitive, electrically neutral transport system. *Journal of Clinical Investigation, 74*, 7–16.

Strazielle, N., & Ghersi-Egea, J. F. (2015). Efflux transporters in blood-brain interfaces of the developing brain. *Frontiers in Neuroscience, 9*, 21.

Sulis Sato, S., Artoni, P., Landi, S., Cozzolino, O., Parra, R., Pracucci, E., Trovato, F., Szczurkowska, J., Luin, S., Arosio, D., Beltram, F., Cancedda, L., Kaila, K., & Ratto, G. M. (2017). Simultaneous two-photon imaging of intracellular chloride concentration and pH in mouse pyramidal neurons in vivo. *Proceedings of the National Academy of Sciences of the U S A, 114*, E8770–e8779.

Sung, K. W., Kirby, M., McDonald, M. P., Lovinger, D. M., & Delpire, E. (2000). Abnormal GABAA receptor-mediated currents in dorsal root ganglion neurons isolated from Na-K-2Cl cotransporter null mice. *Journal of Neuroscience, 20*, 7531–7538.

Takahashi, N., Chernavvsky, D. R., Gomez, R. A., Igarashi, P., Gitelman, H. J., & Smithies, O. (2000). Uncompensated polyuria in a mouse model of Bartter's syndrome. *Proceedings of the National Academy of Sciences of the U S A, 97*, 5434–5439.

Takayama, Y., & Tominaga, M. (2024). Interaction between TRP channels and anoctamins. *Cell Calcium, 121*, Article 102912.

Thomas, R. C. (1984). Experimental displacement of intracellular pH and the mechanism of its subsequent recovery. *The Journal of Physiology, 354*, 3P–22P.

Tóth, K., Lénárt, N., Berki, P., Fekete, R., Szabadits, E., Pósfai, B., Cserép, C., Alatshan, A., Benkő, S., Kiss, D., Hübner, C. A., Gulyás, A., Kaila, K., Környei, Z., & Dénes, Á. (2022). The NKCC1 ion transporter modulates microglial phenotype and inflammatory response to brain injury in a cell-autonomous manner. *PLoS Biology, 20*, Article e3001526.

Tradtrantip, L., Tajima, M., Li, L., & Verkman, A. S. (2009). Aquaporin water channels in transepithelial fluid transport. *Journal of Medical Investigation, 56*(Suppl. 1), 179–184.

Trejo, F., Elizalde, S., Mercado, A., Gamba, G., & de losHeros, P. (2023). SLC12A cryo-EM: Analysis of relevant ion binding sites, structural domains, and amino acids. *American Journal of Physiology - Cell Physiology, 325*, C921—c939.

Ussing, H. H. (1982). Volume regulation of frog skin epithelium. *Acta Physiologica Scandinavica, 114*, 363—369.

Uvarov, P., Ludwig, A., Markkanen, M., Pruunsild, P., Kaila, K., Delpire, E., Timmusk, T., Rivera, C., & Airaksinen, M. S. (2007). A novel N-terminal isoform of the neuron-specific K-Cl cotransporter KCC2. *Journal of Biological Chemistry, 282*, 30570—30576.

Valdivieso, G., & Santa-Coloma, T. A. (2019). The chloride anion as a signalling effector. *Biological Reviews of the Cambridge Philosophical Society, 94*, 1839—1856.

Vaughan-Jones, R. D. (1982). Chloride activity and its control in skeletal and cardiac muscle. *Philosophical Transactions of the Royal Society of London B Biological Sciences, 299*, 537—548.

Velazquez, H., & Silva, T. (2003). Cloning and localization of KCC4 in rabbit kidney: Expression in distal convoluted tubule. *American Journal of Physiology - Renal Physiology, 285*, F49—F58.

Verkman, A. S. (2009). Chemical and GFP-based fluorescent chloride indicators. In F. J. Alvarez-Leefmans, & E. Delipre (Eds.), *Physiology and pathology of chloride transporters and channels in the nervous system from molecules to diseases* (1st ed., pp. 111—123). Amsterdam: Academic Press-Elsevier.

Verkman, A. S. (1990). Development and biological applications of chloride-sensitive fluorescent indicators. *American Journal of Physiology - Cell Physiology, 259*, C375—C388.

Viemari, J. C., Bos, R., Boulenguez, P., Brocard, C., Brocard, F., Bras, H., Coulon, P., Liabeuf, S., Pearlstein, E., Sadlaoud, K., Stil, A., Tazerart, S., & Vinay, L. (2011). Importance of chloride homeostasis in the operation of rhythmic motor networks. *Progress in Brain Research, 188*, 3—14.

Virtanen, M. A., Uvarov, P., Hübner, C. A., & Kaila, K. (2020). NKCC1, an elusive molecular target in brain development: Making sense of the existing data. *Cells, 9*.

Virtanen, M. A., Uvarov, P., Mavrovic, M., Poncer, J. C., & Kaila, K. (2021). The multifaceted roles of KCC2 in cortical development. *Trends in Neurosciences, 44*, 378—392.

Vogh, B. P., & Langham, M. R., Jr. (1981). The effect of furosemide and bumetanide on cerebrospinal fluid formation. *Brain Research, 221*, 171—183.

Wang, D. D., & Kriegstein, A. R. (2009). Defining the role of GABA in cortical development. *The Journal of Physiology, 587*, 1873—1879.

Wang, H. S., Chen, Y., Vairamani, K., & Shull, G. E. (2014). Critical role of bicarbonate and bicarbonate transporters in cardiac function. *World Journal of Biological Chemistry, 5*, 334—345.

Wang, M. X., Cuevas, C. A., Su, X. T., Wu, P., Gao, Z. X., Lin, D. H., McCormick, J. A., Yang, C. L., Wang, W. H., & Ellison, D. H. (2018). Potassium intake modulates the thiazide-sensitive sodium-chloride cotransporter (NCC) activity via the Kir4.1 potassium channel. *Kidney International, 93*, 893—902.

Wang, W., Tsirulnikov, K., Zhekova, H. R., Kayık, G., Khan, H. M., Azimov, R., Abuladze, N., Kao, L., Newman, D., Noskov, S. Y., Zhou, Z. H., Pushkin, A., & Kurtz, I. (2021). Cryo-EM structure of the sodium-driven chloride/bicarbonate exchanger NDCBE. *Nature Communications, 12*, 5690.

Wang, W. H., & Lin, D. H. (2022). Inwardly rectifying K^+ channels 4.1 and 5.1 (Kir4.1/Kir5.1) in the renal distal nephron. *American Journal of Physiology - Cell Physiology, 323*, C277—c288.

Watabe, T., Xu, M., Watanabe, M., Nabekura, J., Higuchi, T., Hori, K., Sato, M. P., Nin, F., Hibino, H., Ogawa, K., Masuda, M., & Tanaka, K. F. (2017). Time-controllable Nkcc1 knockdown replicates reversible hearing loss in postnatal mice. *Scientific Reports, 7*, Article 13605.

Weaver, C. D., Harden, D., Dworetzky, S. I., Robertson, B., & Knox, R. J. (2004). A thallium-sensitive, fluorescence-based assay for detecting and characterizing potassium channel modulators in mammalian cells. *Journal of Biomolecular Screening, 9*, 671—677.

Weiner, I. D., & Verlander, J. W. (2011). Role of NH3 and NH4+ transporters in renal acid-base transport. *American Journal of Physiology - Renal Physiology, 300*, F11—F23.

Williams, J. R., Sharp, J. W., Kumari, V. G., Wilson, M., & Payne, J. A. (1999). The neuron-specific K-Cl cotransporter, KCC2. Antibody development and initial characterization of the protein. *Journal of Biological Chemistry, 274*, 12656—12664.

Willis, W. D., Jr. (1999). Dorsal root potentials and dorsal root reflexes: A double-edged sword. *Experimental Brain Research, 124*, 395—421.

Willis, W. D. (2006). John Eccles' studies of spinal cord presynaptic inhibition. *Progress in Neurobiology, 78*, 189—214.

Witschi, R., Punnakkal, P., Paul, J., Walczak, J. S., Cervero, F., Fritschy, J. M., Kuner, R., Keist, R., Rudolph, U., & Zeilhofer, H. U. (2011). Presynaptic alpha2-$GABA_A$ receptors in primary afferent depolarization and spinal pain control. *Journal of Neuroscience, 31*, 8134—8142.

Wong, J. A., Gosmanov, A. R., Schneider, E. G., & Thomason, D. B. (2001). Insulin-independent, MAPK-dependent stimulation of NKCC activity in skeletal muscle. *American Journal of Physiology - Regulatory, Integrative and Comparative Physiology, 281*, R561—R571.

Woo, N. S., Lu, J., England, R., McClellan, R., Dufour, S., Mount, D. B., Deutch, A. Y., Lovinger, D. M., & Delpire, E. (2002). Hyperexcitability and epilepsy associated with disruption of the mouse neuronal-specific K-Cl cotransporter gene. *Hippocampus, 12*, 258—268.

Wu, Q., Delpire, E., Hebert, S. C., & Strange, K. (1998). Functional demonstration of Na^+-K^+-$2Cl^-$ cotransporter activity in isolated, polarized choroid plexus cells. *American Journal of Physiology, 275*, C1565—C1572.

Wu, Y., Wang, W., Diez-Sampedro, A., & Richerson, G. B. (2007). Nonvesicular inhibitory neurotransmission via reversal of the GABA transporter GAT-1. *Neuron, 56*, 851—865.

Xie, Y., Chang, S., Zhao, C., Wang, F., Liu, S., Wang, J., Delpire, E., Ye, S., & Guo, J. (2020). Structures and an activation mechanism of human potassium-chloride cotransporters. *Science Advances, 6*.

Xu, H., Fame, R. M., Sadegh, C., Sutin, J., Naranjo, C., Della, S., Cui, J., Shipley, F. B., Vernon, A., Gao, F., Zhang, Y., Holtzman, M. J., Heiman, M., Warf, B. C., Lin, P. Y., & Lehtinen, M. K. (2021). Choroid plexus NKCC1 mediates cerebrospinal fluid clearance during mouse early postnatal development. *Nature Communications, 12*, 447.

Xu, J. C., Lytle, C., Zhu, T. T., Payne, J. A., Benz, E., Jr., & Forbush, B. (1994). Molecular cloning and functional expression of the bumetanide-sensitive Na-K-Cl cotransporter. *Proceedings of the National Academy of Sciences of the U S A, 91*, 2201–2205.

Yan, Y., Dempsey, R. J., Flemmer, A., Forbush, B., & Sun, D. (2003). Inhibition of Na^+-K^+-Cl^- cotransporter during focal cerebral ischemia decreases edema and neuronal damage. *Brain Research, 961*, 22–31.

Yerby, T. R., Vibat, C. R., Sun, D., Payne, J. A., & O'Donnell, M. E. (1997). Molecular characterization of the Na-K-Cl cotransporter of bovine aortic endothelial cells. *American Journal of Physiology, 273*, C188–C197.

Zajac, M., Chakraborty, K., Saha, S., Mahadevan, V., Infield, D. T., Accardi, A., Qiu, Z., & Krishnan, Y. (2020). What biologists want from their chloride reporters - a conversation between chemists and biologists. *Journal of Cell Science, 133*.

Zdebik, A. A., Wangemann, P., & Jentsch, T. J. (2009). Potassium ion movement in the inner ear: Insights from genetic disease and mouse models. *Physiology, 24*, 307–316.

Zeilhofer, H. U., Wildner, H., & Yévenes, G. E. (2012). Fast synaptic inhibition in spinal sensory processing and pain control. *Physiological Reviews, 92*, 193–235.

Zeuthen, T., & Macaulay, N. (2012). Cotransport of water by Na^+-K^+-$2Cl^-$ cotransporters expressed in Xenopus oocytes: NKCC1 versus NKCC2. *The Journal of Physiology, 590*, 1139–1154.

Zeuthen, T. (1994). Cotransport of K^+, Cl^- and H_2O by membrane proteins from choroid plexus epithelium of *Necturus maculosus*. *The Journal of Physiology, 478*(Pt 2), 203–219.

Zeuthen, T. (2010). Water-transporting proteins. *Journal of Membrane Biology, 234*, 57–73.

Zhang, J., Gao, G., Begum, G., Wang, J., Khanna, A. R., Shmukler, B. E., Daubner, G. M., de Los Heros, P., Davies, P., Varghese, J., Bhuiyan, M. I., Duan, J., Zhang, J., Duran, D., Alper, S. L., Sun, D., Elledge, S. J., Alessi, D. R., & Kahle, K. T. (2016). Functional kinomics establishes a critical node of volume-sensitive cation-Cl^- cotransporter regulation in the mammalian brain. *Scientific Reports, 6*, Article 35986.

Zhang, S., Zhou, J., Zhang, Y., Liu, T., Friedel, P., Zhuo, W., Somasekharan, S., Roy, K., Zhang, L., Liu, Y., Meng, X., Deng, H., Zeng, W., Li, G., Forbush, B., & Yang, M. (2021). The structural basis of function and regulation of neuronal cotransporters NKCC1 and KCC2. *Communications Biology, 4*, 226.

Zhang, Z., Tachikawa, M., Uchida, Y., & Terasaki, T. (2018). Drug clearance from cerebrospinal fluid mediated by organic anion transporters 1 (Slc22a6) and 3 (Slc22a8) at arachnoid membrane of rats. *Molecular Pharmaceutics, 15*, 911–922.

Zhao, Y., Roy, K., Vidossich, P., Cancedda, L., De Vivo, M., Forbush, B., & Cao, E. (2022a). Structural basis for inhibition of the Cation-chloride cotransporter NKCC1 by the diuretic drug bumetanide. *Nature Communications, 13*, 2747.

Zhao, Y., Shen, J., Wang, Q., Ruiz Munevar, M. J., Vidossich, P., De Vivo, M., Zhou, M., & Cao, E. (2022b). Structure of the human cation-chloride cotransport KCC1 in an outward-open state. *Proceedings of the National Academy of Sciences of the U S A, 119*, Article e2109083119.

Chapter 16

Intracellular pH regulation

Mark D. Parker
The Jacobs School of Medicine and Biomedical Sciences, The State University of New York, University at Buffalo, Buffalo, NY, United States

1. Summary

Intracellular pH (pH_i) must be maintained within a narrow range in order to preserve the activities of the proton-sensitive processes that occur within a cell. In this chapter, we consider the importance of maintaining appropriate and discrete values of pH in all intracellular compartments, including the mitochondrial matrix and other organellar lumens. Dysfunction of the proteins that regulate pH_i results in a variety of neurological, renal, gastrointestinal, and skeletal pathologies. This chapter provides an overview of (1) acid-base chemistry, (2) the key concepts of pH_i defense such as buffering and acid extrusion, (3) the proteins and sensory mechanisms that facilitate these processes in mammalian cells, and (4) the influence of pH_i on key cellular processes such as metabolism and membrane excitability. A final section in this chapter considers the mechanisms of pH_i regulation in bacteria and fungi, which are of broad interest in relation to the antimicrobial action of acids.

2. Introduction

The physiological importance of extracellular acid-base balance and its disturbance in disease has been appreciated for centuries. Early key observations include the necessity of gastric fluid acidity for digestion (17th century), the excretion of dietary and metabolic acids in urine (19th century), and the pathological loss of blood alkali with diarrhea in individuals with cholera (19th century). In fact, H^+ is the most tightly regulated ions in blood: the physiologically permissive plasma pH range of 7.35–7.45 represents a window of just 9 n moles of ionized H^+ per liter. The maintenance of extracellular pH (pH_e) within this window is particularly impressive when one considers that the typical daily acid-load for one person (the sum of dietary acid intake, metabolic acid production, and base loss in feces) is around 1 mmole H^+ per kg. The acidifying effects of this load are minimized in the short term by the action of buffers such as bicarbonate (HCO_3^-) that neutralize freely ionized H^+ to produce CO_2 that is exhaled by the lungs (Fig. 16.1A). In the long term, pH_e balance is restored by the actions of the kidneys that excrete H^+ and generate new HCO_3^- to replace that lost by buffering (see Section 5.1). In the face of an alkaline load, the rate of CO_2 exhalation can be decreased and the kidneys can excrete HCO_3^-.

Although it seems obvious to us now that cells must also actively defend their interior acid-base status in order to preserve their internal processes, prevailing thought in the early 20th century imagined a passive process that was solely dependent on chemical buffering of H^+ and the electrochemical equilibration of H^+ across the plasma membrane. It was not until the second half of the 20th century that technology allowed for the paradigm-shifting observations that form the basis of our current understanding of the mechanisms of pH_i regulation; namely, that in addition to buffers, cells employ a variety of energy-requiring acid-base transporting membrane proteins to balance intracellular acids (i.e., H^+) and bases (e.g., HCO_3^-) in order to counter challenges to pH_i, mirroring the whole-body response to disturbances in pH_e (Fig. 16.1B). The fact that these cellular processes are mediated by a diverse array of proteins has a number of important implications: (1) Each cell can tailor its response (pH set-point, H^+ sensitivity) to its own needs by the expression of a different complement or abundance of these proteins; (2) The abundance and/or activity of each transport protein can be regulated to match the cell's requirements in the face of an acid-base disturbance; (3) pH_i regulation can be disrupted by genetic mutation; and (4) pH_i regulation can be amenable to pharmacological modification.

The reason why most cellular processes are sensitive to changes in pH_i is because most biomolecules such as proteins are weak acids and bases that can be titrated by protons. For example, the structure, stability, functionality, and solubility of

Cell Physiology Sourcebook. https://doi.org/10.1016/B978-0-12-811114-7.00023-2

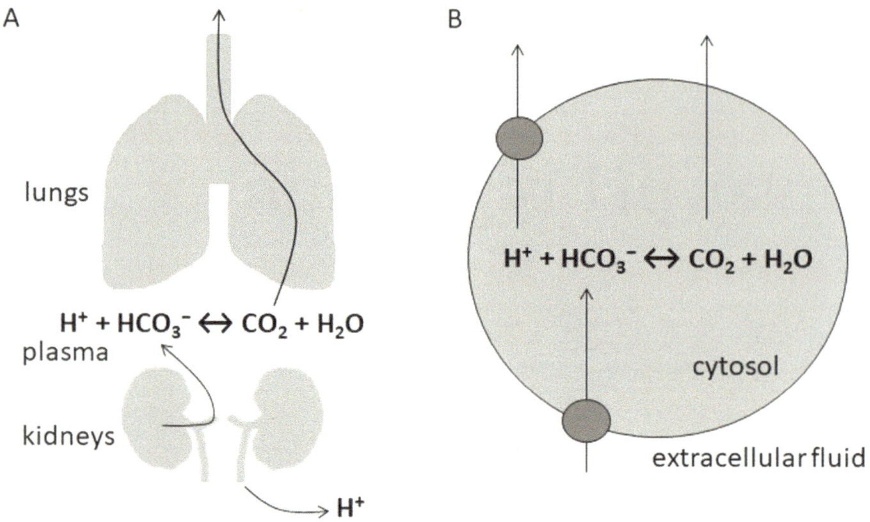

FIGURE 16.1 The mechanisms of extracellular and intracellular pH defense share many common features. (A) The pH of blood plasma and interstitial fluid is defended by buffers such as bicarbonate (HCO_3^-) that neutralize H^+ to form CO_2 that is exhaled. The excretion of H+ and generation of new HCO_3^- by the kidneys restores the status quo. (B) A parallel process occurs in service of intracellular pH defense. Cells use membrane transporters to extrude H^+ and import acid neutralizing HCO_3^-, transferring/returning their acid load to the extracellular space.

a protein depends upon the protonation state of its titratable amino-acid side-chains. Protonation state determines the protein's surface charge and the ability of key residues to form salt bridges, hydrogen bonds, contribute towards catalysis, or interact with other molecules. The protonation state of each side chain is determined by its identity and its chemical environment. For example, the imidazole group of histidine is predominantly nonprotonated and uncharged at neutral pH but becomes increasingly protonated and positively charged as pH falls. The combined influence of these numerous titratable groups on the net charge of a protein is enumerated by its isoelectric point, pI: the pH at which the net charge on the protein is zero. A positive correlation has been observed between the pH of the subcellular compartment and the pI values of the proteins expressed within, suggesting that pH_i is a factor that can drive evolutionary change. An example of the influence of pH on protein structure is the decrease in activity of chymotrypsin at pH values more alkaline than its optimum of pH 8. Besides the impact of pH on protein structure, a second key facet of pH dependence comes into play if an enzyme generates, consumes, or transfers a proton as its rate-limiting step. More than 50% of enzymatic mechanisms, including those mediated by catalytic RNAs such as ribozymes, involve proton transfer. It is the detrimental effect of lowering pH on proton transfer catalysis that explains the decreasing activity of chymotrypsin at pH values more acidic than its optimum.

These considerations underline the value of robust pH_i regulation, as a cell must preserve the pH of its compartments within the narrow, prescribed ranges within which physiological processes are optimized. However, it is worth noting that pH_i regulation has its limitations. While regulation may prevent pH_i disturbance from small insults, larger or sustained insults can temporarily overwhelm the system in which case pH_i will be perturbed until the "recovery" of pH_i into the permissive rage by pH_i regulatory mechanisms is complete. Thus, cells have developed mechanisms to sense a change in pH_i and harness it as cue to upregulate pH_i defense mechanisms (see Section 5.5). Some cells have even developed deliberately H^+-sensitive processes, essentially using H^+ as a signaling molecule (see Section 6).

3. Protons, pH, and buffering power

The concept of pH ("power of hydrogen") was introduced in 1909 by the Danish scientist Søren Sørensen as a convenient way of reporting the relative acidity of a solution. Technically, pH represents the negative of the common logarithm of H^+ activity (a_{H+}). However, in infinitely dilute solution, a_{H+} has the same value as $[H^+]$. Because the intracellular milieu typically has a very low H^+ concentration, pH can be functionally defined per Eq. (16.1):

$$pH = -\log_{10}[H^+] \qquad 16.1$$

However, Eq. (16.1) is inadequate to predict the magnitude of pH change that would be brought about in a biological fluid by a change in $[H^+]$. This is because biological fluids contain diverse titratable acids and bases that act as buffers. The concept of buffering was introduced by clinical chemist Donald Van Slyke in 1922. Van Slyke defined buffers as substances which, by their presence in solution, increase the amount of strong acid or base that must be added to cause a unit change in pH. This ability of a solution to resist pH change upon the addition of acid or base is described by its buffering power (β), per Eq. (16.2):

$$\beta = \frac{d[\text{Base}]}{d\text{pH}} \text{ or } \frac{-d[\text{Acid}]}{d\text{pH}} \text{ mM/pH unit or 'slykes'} \quad 16.2$$

The magnitude of β for a solution containing a single monovalent buffer depends on the identity of the buffer, the concentration of the buffer, as well as the pH of the solution in which it is dissolved.[1] We can illustrate this concept by considering the example of a solution that contains a single, monovalent buffer consisting of weak acid (HA) in equilibrium with its conjugate base (A^-) via an equilibrium constant, K_a', per Eqs. (16.3) and (16.4):[2]

$$HA \underset{k_2}{\overset{k_1}{\rightleftarrows}} H^+ + A^- \quad 16.3$$

$$K_a' = [H^+]\frac{[A^-]}{[HA]} \quad 16.4$$

Eq. (16.4) is Henderson's equation, which gives rise to two valuable derivatives. The first is the Henderson–Hasselbalch equation Eq. (16.5) that describes how the pH of a solution is determined by the properties of the buffer system.[3]

$$\text{pH} = \text{p}K_a' + \log\frac{[A^-]}{[HA]} \quad 16.5$$

Here pKa' is a property intrinsic to the buffer and is numerically the same as the pH of the solution at which $[A^-] = [HA]$. The second valuable derivative from Eq. (16.4) is produced by differentiating Henderson's equation to generate an expression that permits the calculation of β, per Eq. (16.6):

$$\beta = 2.3\frac{K_a'[H^+]}{(K_a' + [H^+])^2}[\text{Buffer}] \quad 16.6$$

Fig. 16.2A shows a plot of Eq. (16.6) for a hypothetical buffer for which $pK_a' = 7.2$. As can be seen, at constant [Buffer], β varies with pH and buffering power is greatest (β_{max}) when pH = pK_a'. For a solution such as this that contains a single monovalent buffer, Eq. (16.6) can restated to report the value of β_{max} (i.e., β when $K_a' = [H^+]$), per Eq. (16.7):

$$\beta_{max} = 0.58[\text{Buffer}] \quad 16.7$$

In solutions containing multiple buffers or a single polyvalent buffer that has multiple titratable groups, each with their own discrete pK_a', each buffer (or titratable group) acts independently such that the total buffering power (β_{total}) is the sum of β for the individual components, per Eq. (16.8):[4]

$$\beta_{total} = \beta_{\text{Buffer }1} + \beta_{\text{Buffer}2} + \ldots \quad 16.8$$

So far in this section we have considered only what are referred to as closed or intrinsic buffer systems, which is to say those for which HA and A^- cannot escape the system as is the case with phosphate or proteins. However, an additional and often more powerful buffering component in biological systems is the open buffer system. In an open buffer system, the buffer components are not confined to the system. The chief physiological example of this is the CO_2/HCO_3^- system that is produced by the rate-limiting equilibration of protons and HCO_3^- (the equivalent of A^-) with carbonic acid (H_2CO_3), and the subsequent, rapid dissociation of H_2CO_3 into carbon dioxide and water, per Eq. (16.9):

---

1. Solvent H_2O has no significant buffering power between pH 3 and 11 and can be ignored for calculations in the physiological range.
2. K_a is the quotient of the forward and reverse rate constants (k_1/k_2). Note that the prime designation is used to denote that K_a' bears on the concentrations rather than the activities of the buffer components.
3. The equilibrium between a weak base B and its conjugate acid BH^+ with a dissociation constant K_b' can be considered in the same way.
4. βmax cannot be calculated using Eq. (16.7) for a solution that contains multiple titratable groups because $[H^+]$ cannot simultaneously equal the K_a' of two or more buffers.

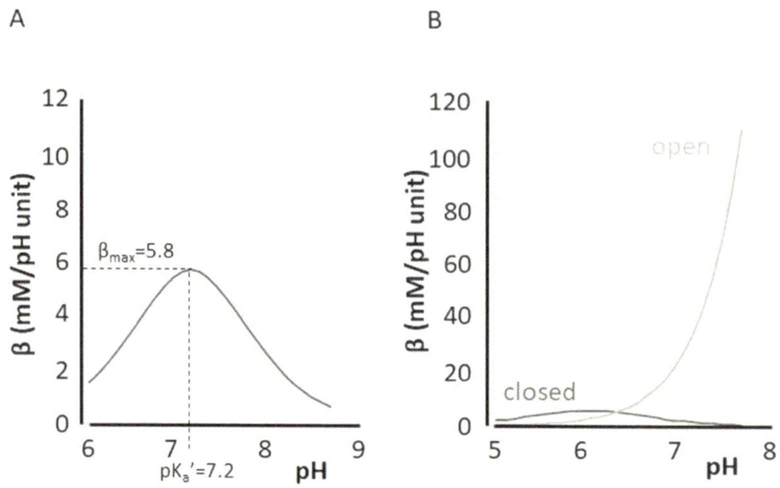

FIGURE 16.2 The power of closed and open buffer systems to resist pH change. (A) The buffering power of a 10 mM solution of a closed buffer with a pK'_a of 7.2, such as phosphate. (B) The buffering power of a 10 mM solution of a closed buffer with a pK'_a of 6.1, compared to a 10 mM solution of an open buffer (HCO_3^-) in equilibrium with a fixed CO_2 partial pressure of 40 mmHg. The ability of CO_2 to leave the open system rather than accumulate makes open buffers extremely powerful, even when their pK'_a is far removed from physiological pH.

$$H^+ + HCO_3^- \overset{K'_a}{\leftrightarrow} H_2CO_3 \leftrightarrow H_2O + CO_2 \quad 16.9$$

The weak acid component of the equilibrium, HA (per Eq. 16.3), is represented by the sum of H_2CO_3 and CO_2. Thus, the equilibrium is described by the Henderson–Hasselbalch equation, per Eq. (16.10):

$$pH = pK'_a + \log\frac{[HCO_3^-]}{[H_2CO_3] + [CO_2]} \quad 16.10$$

For the $HCO_3^-/(H_2CO_3 + CO_2)$ equilibrium, pK'_a is 6.1. At equilibrium, CO_2 is many orders of magnitude more abundant than H_2CO_3, so it is convenient to replace the term "$[H_2CO_3]+[CO_2]$" with $[CO_2]$ alone, making $[CO_2]$ a proxy for [HA]. Thus we can restate Eq. (16.10), per Eq. (16.11):

$$pH = 6.1 + \log\frac{[HCO_3^-]}{[CO_2]} \quad 16.11$$

CO_2 is a dissolved gas, so it can leave the system and does not accumulate and acidify the system like a nonvolatile "HA" buffer would, thus the system is described as open. The difference between the power of a closed and open buffer system is evident in Eq. (16.12) that describes β_{open} in which "HA" is constant:

$$\beta_{open} = 2.3[A^-] = 2.3[HCO_3^-] \quad 16.12$$

The plot of Eq. (16.12) in Fig. 16.2B shows how β is no longer maximal at pK'_a for an open buffer system (there is no theoretical β_{max}) and that an open buffer system can be far more powerful than a closed system, even at values far removed from its pK'_a.

Biological fluids, in which buffering is a sum of all the closed buffering power plus all the open system buffering capacity.

$$\beta_{total} = \sum \beta_{closed} + \sum \beta_{open} \quad 16.13$$

While Eq. (16.13) describes the magnitude of the resistance to pH change, it does not account for the rate at which buffering occurs, which can be critically important for cells. There are several other factors that enhance the ability of the cell to resist pH_i change that are discussed in Section 5, such as the catalysis of CO_2/HCO_3^- buffering by carbonic anhydrase enzymes and the rebalancing of intracellular H^+ and HCO_3^- by membrane transport proteins. Together, these systems allow cells to maintain a permissive pH_i in the face of transient as well as chronic acid/base disturbances.

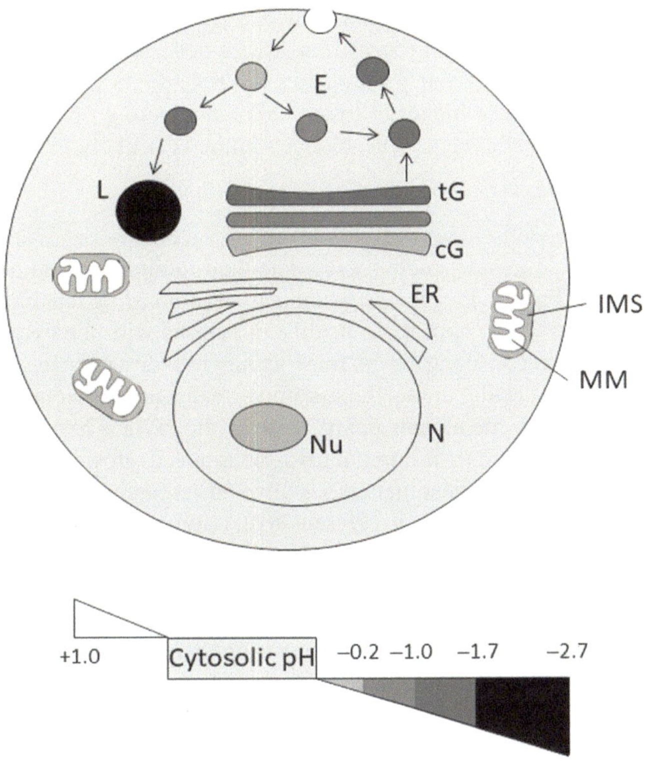

FIGURE 16.3 The pH of intracellular compartments. Most compartments actively maintain an internal pH that is different to that of the cytosol. The Golgi apparatus (trans Golgi: tG; cis Golgi: cG), endosomes (E), and lysosomes (L) are substantially more acidic. The mitochondrial matrix (MM) is substantially more alkaline. The nucleolus (Nu) and inner mitochondrial space (IMS) are slightly more acidic. The pH of the endoplasmic reticulum (ER) and nucleoplasm (N) are typically similar to cytosolic pH, but nucleoplasm:cytoplasm pH gradients have been observed in some cells.

4. Steady-state pH_i values

In this section, we will review typical values for the pH of intracellular compartments. In most literature, pH_i is synonymous with cytosolic pH but subcellular/organellar compartments maintain their own deliberately discrete luminal pH values that are often quite different from that of the cytosol (Fig. 16.3). Some pH_i measurement technologies such as H^+-selective microelectrodes and genetically encoded reporters (e.g., pHluorin) directly report the pH of the cytosol or the intracellular compartment to which they are confined. Other pH_i measurement techniques that are reliant upon partitioning of membrane permeant molecules (e.g., DMO: 5,5-dimethyl-2,4-oxazolidinedione), report a mean intracellular pH that includes contributions from the cytosol and all other cellular compartments into which the probe has permeated. However, while there may be theoretically disparities between cytosolic pH and mean intracellular pH, these are evidently small and do not practically impact the present discussion in which we consider the pH of each compartment separately. The determinants of steady-state pH, the balanced action of acid loading and acid extruding processes, is considered in Section 5.

4.1 Cytosolic pH

The mean whole-body intracellular pH for humans, as estimated by DMO distribution, is ~6.9. Average cytosolic pH values for individual mammalian cells vary according to type, but most fall within the range of 6.9–7.4. For example, in skeletal muscle pH_i is ~6.9, but in red blood cells pH_i is ~7.2. The fact that the mean whole-body intracellular pH value is on the lower end of this range likely represents the outsized contribution of muscle mass to the measured intracellular space. At 37°C, the pH of neutrality is 6.8 so all of these pH_i values are alkaline by definition. In fact, cytosolic pH is substantially more alkaline than can be accounted for by passive transmembrane distribution of H^+ according to electrochemical equilibrium. The Nernst equation predicts that cytosolic pH would be ~1 unit more acidic than plasma (i.e., pH 6.4) in most cells if that was the case.[5] The fact that cytosolic pH is held away from equilibrium demonstrates that baseline cytosolic pH regulation is an energetic process that favors net acid extrusion. Higher cytosolic pH values are described in cells from cold-blooded animals, but these values are not technically more alkaline than those from warm blooded animals; the pH of neutrality is also higher at lower temperatures (e.g., pH 7.0 at room temperature). Thus, the protonation states of biomolecules in warm versus cold-blooded animals is likely to be preserved.

4.2 Organellar pH

Most subcellular compartments are able to independently regulate their luminal pH and maintain the pH of their lumens at a value distinct to that of the cytosol (Fig. 16.3). The exception is the endoplasmic reticulum which has a luminal pH that is very similar to that of the cytosol and which appears to acidify in parallel with the cytosol. In contrast, the lumen of the Golgi apparatus is acidified from 6.7 to 6.0 along its cis-trans axis, while downstream compartments such as endosomes may be acidified even further. The most acidic intracellular compartments are lysosomes, which have a luminal pH range of 4.5−5.5, those closest to the plasma membrane being least acidic. When exocytic vesicles fuse with the plasma membrane, they release their acidic cargo and their luminal pH is neutralized. However, the lumens of recycled vesicles are gradually reacidified upon endocytosis. This phenomenon has allowed researchers to visualize the fusion and recycling of synaptic vesicles, by labeling their interiors with a pH sensitive fluorophore and following the changes in vesicular fluorescence.

The mitochondrial matrix is the only intracellular compartment that is consistently more alkaline than the cytosol. Mitochondria maintain a substantial pH gradient across their inner mitochondrial membrane (IMM) due to the action of the complexes of the electron transport chain that extruding protons from the matrix into the inner mitochondrial space (IMS). This gradient contributes to the proton motive force that drives ATP synthesis (see Section 6.1). The pH of the IMS is slightly more acidic than the surrounding cytosol, but the matrix can be more than 1 pH unit more alkaline than the IMS and is consequently more alkaline than the cytosol. The pH of the mitochondrial matrix changes in parallel with the pH of the cytosol, an effect communicated across the IMM by mitochondrial Na^+/H^+ exchangers.

A less consistent picture of pH regulation has emerged from studies of nucleoplasmic pH. It has been assumed that the presence of nuclear pores would place the pH of nucleoplasm under the secondary control of the mechanisms that control cytosolic pH. However, in recent years, nuclear to cytosolic pH gradients of varying direction and magnitude have been reported. For example, nucleoplasm has been measured as 0.5 pH units more alkaline than the cytosol in a macrophage cell line but 0.05 more acidic in a prostate cancer cell line. The pH of the nucleolus is slightly more acidic than that of the nucleoplasm. The relevance of these pH gradients, and how they are maintained, is currently unclear.

5. Maintenance of steady-state pH_i

In order to protect their intracellular processes, cells must defend pH_i from extrinsic as well as intrinsic acid-base disturbances. There are four categories of simple disturbance: metabolic acidosis due to a rise in H^+ or deficit in HCO_3^-, metabolic alkalosis due to a deficit in H^+ or an excess of HCO_3^-, respiratory acidosis due to an excess of CO_2, and respiratory alkalosis due to a deficit in CO_2. In Section 5.1, we discuss the first line of pH_i defense: the body's defense of pH_e. Per Fig. 16.1, there are many conceptual similarities between regulation of pH_e by the body and pH_i by individual cells. In Sections 5.2 and 5.4, we consider the components of the mechanisms that defend pH_i such as buffers and the balanced action of acid-loading and acid-extruding membrane proteins. Finally, in Section 5.5, we consider how the cells sense and respond to changes in pH_i.

5.1 Defense of extracellular pH

pH_i regulation cannot be considered without also considering the acid-base status of the extracellular fluid that surrounds cells, which is interstitial fluid for most cells and plasma for circulating cells like red blood cells. Interstitial fluid and plasma are in chemical equilibrium with each other across capillary walls and are considered to have the same pH and $[HCO_3^-]$. There are two ways in which extracellular acid-base status can influence pH_i. Firstly, extracellular fluid is the source of the HCO_3^- that cells import, and the sink for the acids and CO_2 that they extrude, in defense of pH_i. Secondly, defense of pH_e protects cells from extrinsic pH disturbance. The extent to which a disturbance in pH_e disturbs pH_i varies

5. Assuming an extracellular pH (pH_e) of 7.4 and a membrane potential (V_m) of −60 mV, and restating the Nernst equation to report the equilibrium potential for H^+ (E_H) as $E_H=(RT/F)*(pH_i-pH_e)$.

among cells. In a study by Salameh et al., a drop in pH_e of 0.2 units had, within 5 min, resulted in a drop in pH_i of >0.2 units in some cell types but <0.1 units in others, reflecting differences in the robustness of pH_i defense mechanisms among cell types.

The pH of plasma (permissive range 7.35–7.45) is defended by closed and open system buffers. The closed buffer systems include plasma proteins and phosphate. Although phosphate is a multivalent buffer with three titratable groups, it only has one pK within physiological range ($pK'_a = 7.2$). The open buffer system due to CO_2/HCO_3^- is catalyzed by carbonic anhydrases within, and attached to the surface of, red blood cells (see Section 5.3). The power of the open system comes from ventilation of CO_2 into the air spaces in the lungs which maintain a partial pressure of CO_2 of ~40 mmHg, fixing dissolved plasma $[CO_2]$ at 1.2 mM.[6] Together, these systems account for the total buffering power of plasma, which is 80 mM/pH unit: 25 mM/pH unit from closed non-HCO_3^- buffers and the majority 55 mM/pH unit from the open system (per Eq. 16.12). However, buffering alone doesn't fully account for whole-body pH balance; in addition, the kidneys must excrete H^+ and generate new HCO_3^- to replace that lost to buffering, maintaining a permissive plasma $[HCO_3^-]$ of ~24 mM.[7] The liver also makes important contributions to acid-base status, being involved in multiple metabolic processes that, in net, consume acids and bases. Ultimately, in this way, the daily acid load is neutralized and all bodily pH disturbances are redressed. This is not to say that defending pH_e is entirely in service of pH_i. pH_e is an independently critical parameter that directly influences processes such as immune function and neuronal excitability. In fact, during whole-body acid-stress cells can lend their intracellular buffering power to sequester extracellular H^+ in defense of pH_e.

5.2 Intracellular buffers

Buffering is the first line of defense against pH_i disturbance. The total buffering power of the cytosol is the sum of β_{closed} and β_{open}. β_{closed} includes with contributions from fixed buffers like proteins and phosphate (a multivalent buffer with three titratable groups, one of which has a pK in the physiological range, pK = 7.2), as well as small mobile buffers like homocarnosine (pK = 6.9). β_{open} is provided by the CO_2/HCO_3^- equilibrium, which is catalyzed by carbonic anhydrase (see Section 5.3). Because intracellular $[HCO_3^-]$ is lower than plasma $[HCO_3^-]$, intracellular β_{total} is also lower than extracellular β_{total} and is estimated to be 40–70 mM/pH unit. As with pH_e regulation, intracellular buffering is only part of the story as pH_i disturbances are ultimately redressed by acid loading and extruding processes (see Section 5.4).

5.3 Carbonic anhydrase

The spontaneous hydration of CO_2 represented in Eq. (16.9) is relatively slow, with a turnover of ~0.1 s^{-1}, and could be outpaced by rapid changes in H^+. In vivo, this open buffering system is optimized by carbonic anhydrase (CA) enzymes. CAs reversibly catalyze the hydration of CO_2, some at a rate a million times faster than the spontaneous process. The rate of catalysis depends on the isoform and abundance of CA, while the apparent rate is influenced by the dampening influence of closed system buffers. For example, in red blood cells CA enhances the CO_2 hydration rate to 170,000 s^{-1}, while in ventricular myocytes the rate of hydration is merely doubled by CA action. CA activity is widespread among organisms, but evolved independently several times. Animals have α-type CAs, but many other classes exist in lower eukaryotes and prokaryotes. The mechanism requires a histidine-coordinated Zn^{2+}, which lowers the pK'_a for H_2O dissociation into the physiological pH range, producing OH^- that acts as a nucleophile to attack a CO_2 molecule that is held in an adjacent hydrophobic pocket. The two-step mechanism, with the first being rate-limiting, is summarized in Eqs. (16.14) and (16.15):

$$H_2O \overset{K'_a}{\leftrightarrow} H^+ + OH^- \quad 16.14$$

$$OH^- + CO_2 \leftrightarrow HCO_3^- \quad 16.15$$

Thus, different from the spontaneous reaction, H_2CO_3, is not a reaction intermediate of the CA-catalyzed reaction. The net reaction is shown in Eq. (16.16):

$$H_2O + CO_2 \leftrightarrow H^+ + HCO_3^- \quad 16.16$$

This catalysis becomes important when the rate of dH^+/dt is faster than the spontaneous reaction of H^+ with HCO_3^-, such as might be encountered in the vicinity of a H^+ transport protein. As well as promoting H^+ neutralization, cytosolic

6. pCO_2 and dissolved $[CO_2]$ are related by a solubility coefficient, $s = 0.03$ mM/mmHg.
7. Accounting for the typical plasma pH = 7.4, per Eq. (16.11).

CA is invaluable to support the secretion of acidic and basic fluids, rapidly replenishing intracellular H^+ (e.g., osteoclasts and parietal cells) or HCO_3^- (e.g., gastrointestinal epithelia) as needed (see Section 6.5). The usefulness of CAs is underscored by their near ubiquitous expression. In mammals, there are eight cytosolic CAs, CAII being the archetypal cytosolic form that is often reported to be associated with the cytoplasmic regions of H^+- and HCO_3^--transporting membrane proteins. Only four of the other seven cytosolic isoforms have CA activity: CAI, CAIII, CAVII, and CAXIII. CAVA and CAVB are expressed in mitochondria. Other isoforms catalyze open buffering in the extracellular space and are either secreted (CAVI), tethered to the extracellular surface of the cell by a lipid anchor (CAIV and the nonhuman CAXV) or a membrane spanning helix (e.g., CAIX, CAXII, and CAXIV). Some of these extracellular CAs are reported to be associated with extracellular regions of acid-base transport proteins, facilitating buffering at the cell surface. A receptor protein tyrosine phosphatase, RPTPγ, has an extracellular carbonic anhydrase-like domain without CA activity that is involved in extracellular CO_2/HCO_3^- sensing.

5.4 Acid extruders and acid loaders

If buffer systems should become overwhelmed in the face of a substantial pH disturbance and pH_i deviates out of permissive range, restoration of pH_i becomes the goal. Most cells express some combination of acid-loading and acid-extruding proteins of diverse abundance, activity, and type: often with much apparent redundancy (Fig. 16.4). Acid extruders in the plasma membrane tend to be activated by acidic shifts in cytosolic pH and are energized by the inwardly directed Na^+-gradient either to extrude H^+ (in the case of Na^+/H^+ exchangers, see Section 5.4.1) or import HCO_3^- (in the case of Na^+-coupled HCO_3^- transporters, see Section 5.4.2) against their electrochemical gradients to redress the cytosolic pH imbalance (Fig. 16.4A). Acid loaders in the plasma membrane, Cl^-/HCO_3^- exchangers of the SLC4 and SLC26 family of proteins (see Section 5.4.3), tend to be activated by alkaline shifts in cytosolic pH and extrude HCO_3^- along its electrochemical gradient in exchange for Cl^- (Fig. 16.4B). The balance of acid-extruding and acid-loading processes determines steady-state pH_i.

Other transport processes such as H^+ pumps, H^+ channels, and $2Cl^-/H^+$ exchangers are especially valuable for control of organellar pH and are described in Sections 5.4.4 and 5.4.5. The extent to which each organellar lumen becomes acidified is determined by the balance of action between an acidifying H^+-ATPase and a tempering H^+ leak (Fig. 16.5). The leak is typically mediated by one of the intracellular cation (Na^+, K^+)/H^+ exchangers NHE6–NHE9 (see Section 5.4.1). In endosomes and lysosomes, a $2Cl^-/H^+$ exchanger, such as ClC-5 (see Section 5.4.4), also mediates H^+ efflux, yet paradoxically promotes net luminal acidification because it charge-balances the action of the H^+-ATPase. Alkalinization of mitochondrial matrix is performed by the complexes of the electron transport chain. In a microcosm of the sequestration of extracellular H^+ by intracellular buffers in times of whole-body acid stress, so too can subcellular compartments such as the Golgi apparatus sequester cytosolic H^+ in defense of cytosolic pH.

Many other H^+-coupled solute transporters such as H^+-Lactate$^-$ cotransporters likely make small contributions to steady-state pH_i, but are not the focus of this section as they are typically taking advantage of the electrochemical H^+ gradient to support solute transport rather than working against the gradient in support of pH_i regulation. The isoforms and variants of proteins referred to in this section are mammalian, in accordance with the models in which these systems are best characterized. However, it may be safe to assume that the major protein identities are transferable among vertebrates, and that the general activities and principles discussed here are transferable among all animals. Separate consideration is given to microbial H^+ transport proteins in Section 7.

5.4.1 Na^+/H^+ exchangers (NHEs)

The SLC9A family of Na^+/H^+ exchangers (NHEs) includes nine multispanning membrane proteins, each of which has a large, carboxy-terminal, cytosolic, regulatory domain. NHEs are expressed as homodimers and differ from each other in their cellular and subcellular distributions, regulation, and sensitivity to blockers (amiloride and its derivatives). NHE1-NHE5 (SLC9A1-SLC9A5) are expressed to the plasma membrane where they mediate H^+ efflux. The most widely expressed of these is NHE1, which is typically expressed in the basolateral membranes of epithelial cells and is also abundant in neurons and myocytes. The activity of NHE1 is governed by cytosolic pH, being activated by acid shifts in pH; but an obligate binding partner, CHP1 (calcineurin homologous protein 1), is necessary to shift the pK of the relationship between cytosolic pH and NHE activity into physiological range. The Na^+ influx that accompanies acid extrusion is also of consequence and can contribute to volume regulation (see Section 6.5). NHE1-null mice exhibit early mortality, ataxia, and seizures. In humans, NHE1 mutation is associated with the neurological disorder Lichtenstein–Knorr syndrome. The role of pH balance in neuronal excitability is discussed in Section 6.3. Among the other NHE isoforms, NHE3

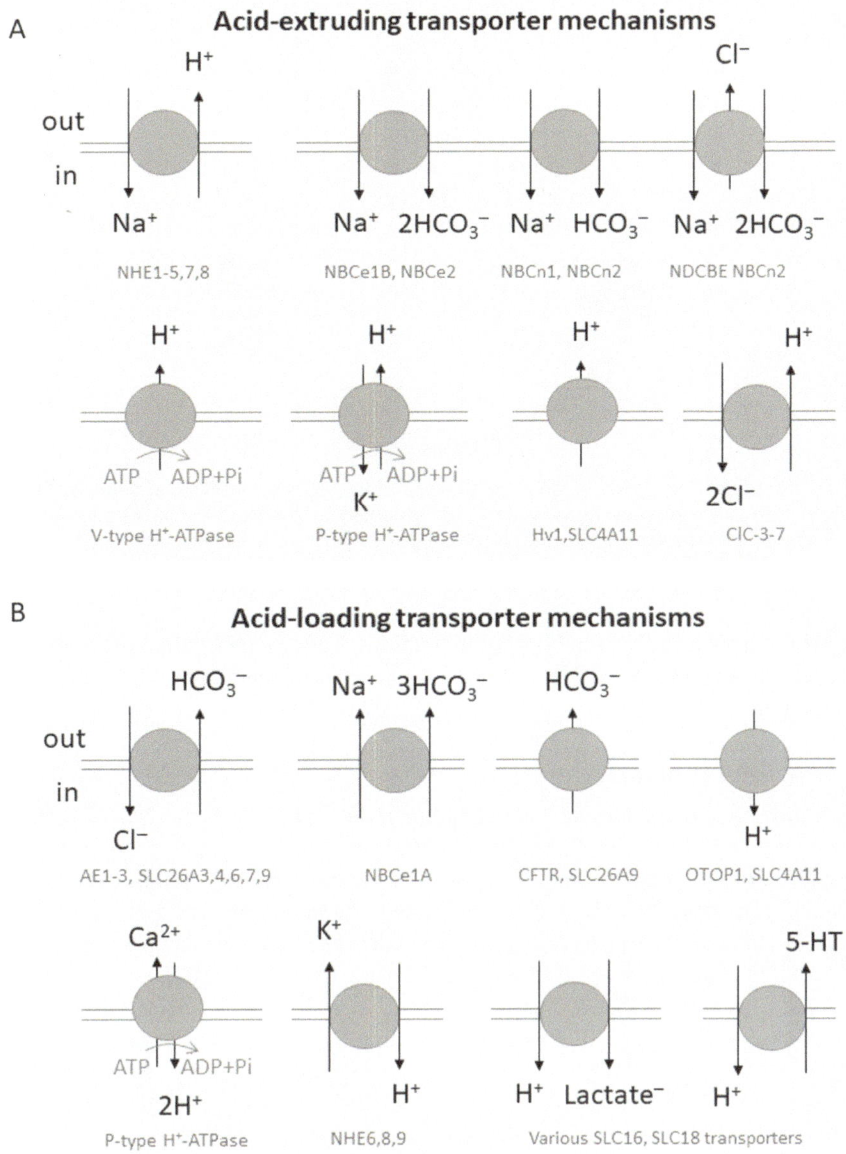

FIGURE 16.4 Membrane transport proteins that influence steady-state pH_i. (A) The actions of acid-extruders such as NHEs and NBCs counter cellular acidification. The V-type and P-type H^+-ATPases, voltage-gated H^+-channels, and $2Cl^-/H^+$ exchangers mediate H^+ extrusion (often into a subcellular lumen) and cause an alkaline load on the cell that requires compensation by acid-loaders. (B) The actions of acid-loaders such as CBEs counters cellular alkalinization. Reverse-mode-NBCs, HCO_3^- channels, H^+ leak channels, and various H^+-coupled transport processes such as H^+-coupled lactate transporters (SLC16 family) and vesicular neurotransmitter loading transporters (SLC18 family) impose an acid load on the cell that requires compensation by acid-extruders.

is notable for its role in the apical membranes of renal proximal tubule epithelia. Here it mediates the excretion of H^+ and NH_4^+ into the tubular fluid that becomes urine, contributing to pH_e balance. The widely expressed isoforms NHE6-NHE9 (SLC9A6-SLC9A9) differ from NHE1-NHE5 because they are predominantly expressed in subcellular compartments and can perform both Na^+/H^+ and K^+/H^+ exchange. NHE6 and NHE9 are thought to predominantly perform K^+/H^+ exchange in endosomes, tempering the vesicular acidification that is mediated by the V-type H^+-ATPase (see Fig. 16.5 and Section 5.4.5). NHE7 and NHE8 are expressed in the Golgi apparatus. Their role is less well understood, but NHE7 has been proposed to contribute toward endosomal and Golgi acidification and to help sequester cytosolic H^+ during periods of cytosolic acid-stress, consistent with a Na^+-coupled action. NHE8 overexpression causes Golgi alkalization, yet in some cell types such as proximal tubule epithelia, NHE8 is located in the plasma membrane where its action is reversed to mediate acid extrusion. This could be explained if NHE8 switches allegiance from K^+ to Na^+ coupling in the plasma

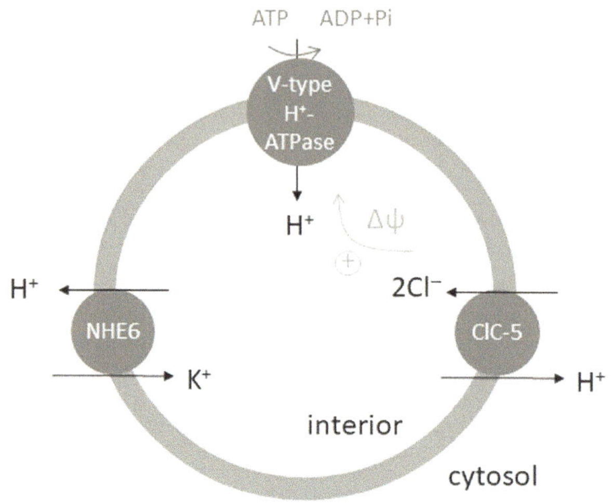

FIGURE 16.5 Organellar pH regulation. The lumens of Golgi stacks, endosomes, and lysosomes are acidified by the action of a V-type ATPase. Acidification is tempered by a H^+ leak, such as that mediated by the K^+/H^+ exchanger NHE6. Acidification is paradoxically enhanced by ClC-mediated acid extrusion from the lumen, because ClC action charge-balances H^+-loading by the ATPase. Acidic luminal pH is critical for appropriate cargo loading, sorting, and targeting.

membrane. Organellar pH regulation is discussed further in Sections 4.2 (typical luminal pH values) and 6.7 (physiological importance). Other, related gene products SLC9B1, SLC9B2, and sperm-specific SLC9C1 are not known to play a significant role in pH_i regulation.

5.4.2 Na^+-coupled HCO_3^- cotransporters

Five of the 10 SLC4 family genes encode Na^+-coupled bicarbonate transporters (NCBTs). Each is a multispanning membrane protein with a large, cytosolic, amino-terminal domain that exerts regulatory control over the transporter to the extent that transport cannot proceed without it. All five NCBTs are expressed as homodimers and most are inhibited by stilbene disulfonates. NCBTs are typically expressed in the basolateral membranes of a variety of epithelial cells and many are also abundant in neurons, glia, and myocytes. Despite a great degree of structural commonality among the NCBTs, they do not all share the same molecular actions; NBCe1 and NBCe2 (SLC4A4, SLC4A5) are electrogenic $Na^+/2HCO_3^-$ cotransporters, NBCn1 and NBCn2 (SLC4A7, SLC4A10) are electroneutral $Na^+/1HCO_3^-$ cotransporters, and NDCBE (SLC4A8) is an electroneutral Na^+-driven $Cl^-/2HCO_3^-$ exchanger.[8] Given typical transmembrane electrochemical gradients, all proteins are predicted to import HCO_3^- and therefore serve as acid extruders. However, NBCe1 undoubtedly acts as an acid loader in renal proximal tubule cells and provides a critical pathway by which plasma $[HCO_3^-]$ is repleted to regulate pH_e. A common yet still controversial hypothesis states that NBCe1 must operate with a $Na^+/3HCO_3^-$ stoichiometry in those cells in order to mediate efflux. Ultimately, our understanding of the system is incomplete.

Of the two electrogenic NCBTs, NBCe1 is the most broadly expressed and physiologically consequential. The SLC4A4 gene expresses two major products, NBCe1A and NBCe1B, from alternative promoters. The NBCe1A sequence includes a cytosolic autostimulatory domain which is replaced in NBCe1B by an autoinhibitory domain (AID). NBCe1B is activated by the AID-binding signaling molecule IRBIT (IP3 receptor binding protein released with inositol 1,4,5-trisphosphate), which presumably neutralizes the influence of the AID, placing NBCe1B action under hormonal control. This is useful, for example, in gastrointestinal epithelia in which NBCe1B must be activated to replenish intracellular HCO_3^- during secretagogue-stimulated secretion. NBCe1A is the form that acts as a renal acid loader, but the mechanism by which NBCe1 can act as a loader versus and extruder is not explained by sequence differences between NBCe1A and NBCe1B. Mutation of NBCe1 leads to proximal renal tubular acidosis, the signs of which include acidemia due to loss of NBCe1A, as well as ocular and dental defects related to loss of pH_i regulation in cells that express NBCe1B (see Section 6.5). The three electroneutral NCBTs NBCn1, NDCBE, and NBCn2 (SLC4A7, SLC4A8, and SLC4A10) are predominantly expressed in neurons and all share an NBCe1B-like AID and consequent dependence on IRBIT. Loss of any of these acid extruders in mice causes neurological defects (see Section 6.3).

8. At least for NBCe1 and NDCBE, the substrate is not $2HCO_3^-$ but the thermodynamically equivalent $1CO_3^{2-}$.

Electrogenic and electroneutral NCBTs are often expressed together in the same membrane of the same cells. The value of their differences in transport mechanism is not fully appreciated, although electrogenic NCBTs have the advantage over electroneutral NCBTs and NHEs of imposing a smaller Na^+ load on the cell while redressing pH imbalance (0.5 rather than 1 Na^+ per extruded acid-equivalent). Indeed, in NBCe1-cardiomyocytes, the increased reliance on NBCn1 and NHE1 for acid extrusion places a sufficiently greater Na^+ load on the cells that the driving force for Ca^{2+} extrusion by the cardiac $3Na^+/Ca^{2+}$ exchanger is diminished, leading to hypertrophy.

5.4.3 Cl^-/HCO_3^- exchangers

Given typical electrochemical gradients across the plasma membrane, Cl^-/HCO_3^- exchangers (CBEs) predominantly act as acid loaders that extrude HCO_3^- from cells. There are eight Cl^-/HCO_3^- exchangers in total encoded by genes from two families: SLC4 and SLC26. The SLC4 family, that also includes the NCBTs described above, encodes three electroneutral Cl^-/HCO_3^- anion exchangers AE1-3 (SLC4A1-3). The SLC26 family encodes 11 multifunctional anion transporters, 5 of which—SLC26A3, 4, 6, 7, and 9—have been described to be capable of electroneutral Cl^-/HCO_3^- exchange activity.[9] SLC4 and SLC26 proteins have similar overall structure but occupy distinct physiological niches. In broad terms, SLC4 proteins tend to be defenders of pH_i, extruding excess intracellular HCO_3^-, whereas SLC26 proteins tend to challenge pH_i as they export HCO_3^- into secreted fluids. AE1 and AE2 also have important functions that require them to operate in an acid-extruding mode (see Sections 6.5 and 6.6), but these actions are as much in service to pH_e as to pH_i regulation.

AE1-AE3 share the structure and stilbene sensitivity of their NCBT paralogs, to which they are ~40% identical at the amino acid level. AE1, commonly known as band 3, is the most abundant membrane protein in the red blood cell and promotes gas exchange by translating CO_2 changes into pH changes to influence hemoglobin O_2 affinity (the Bohr effect, see Section 6.7). The cytosolic amino-terminal domain of AE1 forms a critical anchor between the cytoskeleton and the plasma membrane via interactions with ankyrin and spectrin, conferring mechanical strength to the membrane. An amino-terminally truncated version, kAE1, is shorter by 65 amino acids and is expressed in α-intercalated epithelial cells of the kidney collecting duct. kAE1 mediates basolateral HCO_3^- extrusion into interstitial fluid, maintaining pH_i by balancing H^+ extrusion mediated by the apical H^+-ATPase in service of urinary acidification (see Section 6.4). In this way, kAE1 also contributes to pH_e balance. Loss of AE1 is associated with hemolytic anemia and distal renal tubular acidosis. AE2 and AE3, unlike AE1, are both acutely sensitive to cytosolic pH, being activated by alkaline shifts and inactivated by acidic shifts, a feature that enhances their usefulness for pH_i regulation. AE2 is expressed in a variety of cells and, in much the same way that kAE1 balances pH_i in support of renal H^+ excretion, AE2 balances cytosolic pH in gastric parietal cells and osteoclasts in support of gastric acid secretion[10] and acidification of bone-resorbing lacunas (see Section 6.5). Accordingly, AE2 loss is associated with achlorhydria and osteopetrosis. In some cells, AE2 is functionally coupled to NHE1 to mediate pH neutral NaCl uptake (see Section 6.6). AE3 is expressed in cardiomyocytes and neurons where its acid loading action is a crucial determinant of pH_i. Gain-of-function mutations in AE3 cause elevations of pH_i in cardiomyocytes and neurons, resulting in arrhythmias and epilepsy.

SLC26 proteins, like SLC4 proteins, are multispanning membrane proteins that operate as homodimers. A unique structural feature of SLC26 proteins is a cytosolic, carboxy-terminal "STAS" domain that facilitates interactions with other membrane proteins, such as CFTR, and which is necessary for plasma membrane expression of SLC26 proteins. SLC26A3 (DRA) and SLC26A6 (CFEX) are predominantly expressed to the apical membranes of intestinal epithelia, sometimes both in the same cell. However, SLC26A3 is typically more abundant in cells at the absorptive villus tip while SLC26A6 is more abundant in cells of the secretory crypts. At baseline, SLC26A3 is functionally coupled to NHE3 to mediate net pH-neutral NaCl absorption (see Section 6.5); mutations in SLC26A3 cause congenital chloride diarrhea and blockers relieve constipation. In locations such as the pancreatic duct, SLC26A6 is physically and functionally coupled to CFTR, exchanging CFTR-secreted Cl^- for HCO_3^- in order to produce HCO_3^--rich pancreatic juice. However, the main phenotype associated with SLC26A6 loss is calcium stones due to its role in mediating renal Cl^-/oxalate$^-$ exchange. SLC26A9 plays a similar role to SLC26A6 in support of CFTR activity, although perhaps this is as much due to its intrinsic anion channel activity as its Cl^-/HCO_3^- exchange activity. SLC26A9 is a modifier gene for the severity of several cystic fibrosis-related secretory defects. SLC26A4 (pendrin) is expressed in the inner ear and mediates HCO_3^- secretion into the endolymph. Loss of SLC26A4 results in endolymph acidification and hearing loss. Studies of SLC26A4-null mice reveal a role in apical Cl^-/HCO_3^- exchange activity in renal β-intercalated cells. This is evidenced by defective NaCl reabsorption and excessive

9. Electrogenic Cl^-/HCO_3^- has also been described to be a function of SLC26 proteins in some species and expression systems, but their physiological action is most consistent with electroneutral exchange.

10. It is the AE2-mediated efflux of HCO_3^- into interstitial fluid and thence plasma that results in the postprandial alkaline tide.

urine acidification under appropriately challenged conditions. Finally, among the set, SLC26A7 is unusual in being basolaterally expressed in epithelia. It is expressed alongside AE1 in renal α-intercalated cells and alongside AE2 in parietal cells. The value of this apparent redundancy is unclear.

5.4.4 Cl^-/H^+ exchangers

Four of the nine ClC proteins are Cl^- channels, while the other five mediate $2Cl^-/H^+$ exchange in endosomes (ClC-3, -4, -5, and -6) and lysosomes (ClC-7). This outwardly rectifying, electrogenic transport process favors the movement of both Cl^- and H^+ against their respective electrochemical gradients and thus ClCs act as acid loaders (Fig. 16.5). In internal compartments, ClC action charge balances luminal acid loading by V-type H^+-ATPases. Thus, although ClC exchangers permit H^+ leak from vesicles, they actually promote luminal acidification. ClC exchanger dysfunction is associated with a variety of pathologies related to endosomal and lysosomal dysfunction (see Section 6.7).

5.4.5 H^+ pumps and channels

The V-type H^+-ATPase is homologous to the vacuolar H^+-ATPase in yeast (see Section 7.2). It is a large protein composed of 13 subunits. In mammals, the V-type H^+-ATPase is expressed in the apical plasma membrane of epithelial cells and in the membranes of intracellular compartments where it transports H^+ against their electrochemical gradient in order to acidify lumenal fluid such as the renal tubular fluid that becomes urine and luminal fluid such as the contents of endosomes. The V-type H^+-ATPase plays no deliberate role in defending cytosolic pH, but is crucial for the appropriate acidification of organelles such as the Golgi apparatus and endosomes. Also crucial for the acidification of intracellular compartments is the voltage-gated H^+ channel, Hv1 which acidifies phagosomes (see Section 6.5). SLC4A11 is an unusual SLC4 family member that performs alkali-activated H^+ conduction. It is mainly expressed in corneal endothelial cells and fibroblasts in the inner ear. Its absence is associated with vision loss and progressive deafness. The allosteric activation of SLC4A11 by alkaline shift in pH is consistent with an ability to act as an acid loader to correct disturbances in pH_i, although its exact role is unknown.

There are also several P-type H^+-ATPases, which are heterodimeric proteins related to the Na^+/K^+-ATPase. None of these are known to play a major role in maintaining pH_i balance. For example, the H^+/K^+-ATPase is expressed in the luminal membrane of parietal cells where it contributes towards acidification of gastric juice. This action disturbs pH_i, requiring the corrective action of the basolateral acid loader AE2. Another P-type ATPase, the plasma membrane Ca^{2+}/H^+-ATPase (PMCA) extrudes Ca^{2+} causing a fall in neuronal pH_i during neuronal activity, invoking the corrective influence of neuronal acid extruders. An acid activated H^+-channel, Otop1, encourages rather than defends against intracellular acidification of sour-taste cells (see Section 6.3).

5.5 Sensing and responding to changes in pH_i

A cell's response to a pH_i disturbance includes four components, depending on the cell type as well as the magnitude and persistence of the disturbance. The first component is a passive response of acid loading and acid extruding proteins to the change in the transmembrane driving forces in order to promote a corrective flux of H^+ and/or HCO_3^-. The second component is the enhancement of these fluxes by allosteric regulation of these transport pathways by H^+. The third component is the sensation of altered pH and/or CO_2/HCO_3^- by signaling proteins, the action of which ultimately enhances corrective action by altering the plasma membrane abundance and/or activity of these transporters in the appropriate direction. This third mechanism takes longer to implement but is vital when the insult is sustained and the standing cohort of transporters is unable to restore pH_i. As chronic pH disturbances often arise from pH_e disturbance, many of the sensory mechanisms are involved in the sensation and correction of pH_e. These extracellular mechanisms (e.g., H^+ sensing G-coupled protein receptors) are well reviewed in articles by Buck and Levin and by Skelton et al. A fourth component of the response, not further discussed in this section, does not seek to redress the pH disturbance, but instead serves to protect the cell from the effects of the disturbance (e.g., negation of pH-sensitivity of PFK-1 in active muscle, see Section 6.2).

The allosteric activation of acid extruders and acid loaders by pH_i explains the nonlinearity of their responses to pH_i and how they balance each other's action to define a steady-state pH_i in the physiologically permissive range (Fig. 16.6). Acid extruders NHE1-3 are allosterically stimulated by cytosolic $[H^+]$. The activity of the NHE1-CHP1 complex, for example, is governed by a pK_i of ~6.8 with acid extrusion increasing from baseline activity at pH_i ~7.2 (i.e., close to typical cytosolic pH) to its maximal activity, which is an order of magnitude greater than baseline activity, as pH_i falls below 6.0. This ensures that the magnitude of acid efflux is appropriate to the magnitude of the acid disturbance. A major

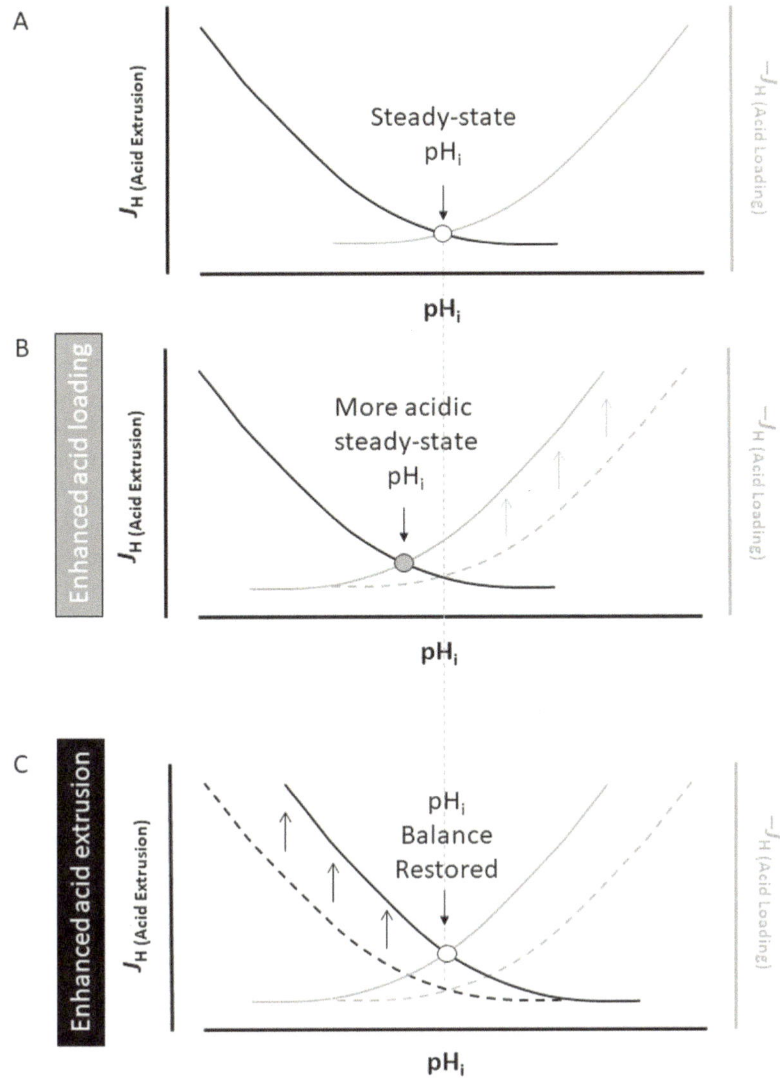

FIGURE 16.6 The value of steady-state pH$_i$ is determined by the balanced action of acid-extruding and acid-loading fluxes, which hold each other away from their individual equilibrium states. (A) The pH dependence of each flux means that each has a relatively low activity at steady-state. (B) An increased rate of acid load causes an acidic shift in pH$_i$. (C) In the face of an acid load, acid extrusion must be increased in order to preserve pH$_i$ within the permissive range.

determinant of the set point of this pH$_i$ sensitivity resides in the cytosolic regulatory domain of NHEs and sequence variations among isoforms causes each to exhibit slight differences in pH$_i$-dependence. This pH dependence is critical: if NHE action was allowed to reach equilibrium, cytosolic pH would be ~8.5. In this way, NHE activity is sufficiently small within the physiologically permissive pH range that its action is balanced by the baseline activity of acid-loaders. Of the three acid-loading CBEs from the SLC4 family, AE2 is the most exquisitely sensitive to pH$_i$, mediating a greater acid loading flux as cytosolic pH increases. Again, this pH dependence, and the balanced action of acid extruders in the permissive pH range, prevents CBE activity from reaching equilibrium. AE2 is also acutely sensitive to pH$_e$, decreasing in activity from pH 9 to pH 6 with a pK_e ~ 6.9. This feature is perhaps useful to keep the protein quiescent during biosynthesis (the lumen of the secretory pathway is acidic), or during recovery from an acid insult (ongoing acid extrusion would acidify cell surface pH, dampening baseline AE2 activity to promote the return to normal cytosolic pH). The H^+ conductor SLC4A11 exhibits the same trends of allosteric activation at alkaline pH$_i$ and pH$_e$.

Most studies of H^+- and/or CO_2/HCO_3^--sensitive signaling mechanisms have been focused on cellular responses to pH$_e$ disturbance, especially in the context of the excretion of urinary acid by renal epithelia when plasma pH and/or HCO_3^- is low. However, these mechanisms often involve the independent sensation of secondary changes in pH$_i$ caused by the extracellular disturbance, and ultimately serve to correct both pH_i and pH_e. In this sense, the role of these mechanisms in sensing and correcting pH_i per se is often underappreciated. There are two cytosolic proteins that are capable of coordinating responses to intracellular acid-base balance, even if their primary physiological function is in service of pH_e defense. The first is the cytosolic, nonreceptor-tyrosine-kinase Pyk2. Pyk2 is a direct sensor of cytosolic pH. In proximal tubule cells, the fall in pH_i that follows extracellular acidification results in the autophosphorylation of Pyk2 and the consequent activation of c-Src. This initiates a signaling cascade that results in increased plasma membrane abundance and activity of the luminal acid extruder NHE3. The main physiological purpose of this system is to promote tubular acid secretion to restore pH_e, but the sensed disturbance in pH_i is also corrected in this way. The second cytosolic protein of interest is soluble adenylyl cyclase (sAC), which is allosterically activated by rises in intracellular [HCO_3^-]. HCO_3^- binding to sAC increases the rate at which sAC generates the second messenger cAMP from ATP. sAC is expressed in many cell types but the prototypical example of its operation is in the clear cells that line the lumen of the epididymis. These cells are responsible for acidifying the luminal fluid to maintain sperm in a quiescent state. Sperm motility is triggered when the tubule is flushed with alkaline prostatic fluid. The resulting increase in lumenal [HCO_3^-] causes an increase in cytosolic [HCO_3^-] in the clear cells. This rise in intracellular [HCO_3^-] is sensed by sAC, resulting in the elevation of cAMP, promoting the cAMP-dependent translocation of V-type H^+-ATPase into the apical membrane and the subsequent reacidification of the lumen. In this case, sAC appears to be triggering a paradoxical rise in pH_i in response to elevated cytosolic [HCO_3^-], but restoration of pH_e would eliminate the original threat to pH_i and restore the status quo. Most other roles of sAC are focused on the regulation of HCO_3^--dependent processes rather than the regulation of pH_i.

For other described responses to pH disturbance, the signaling intermediates are obscure. For example, several acid-base transporters such as kAE1 and NCBTs are not intrinsically pH sensitive but their abundance increases (starting at the level of transcription) over the period of hours and days in response to chronic extracellular acidosis. In the kidney, the overarching purpose of upregulating kAE1 and NBCn1 is the preservation of pH_e by enhancing excretion of acids, but the upregulation of NCBTs in astrocytes and neurons is best explained by the requirement to preserve pH_i. Whether these responses to altered pH_e involve sensation of secondary changes in pH_i is unknown.

Finally, there are signaling pathways that upregulate acid-base transporters but that are not known to be specifically triggered by pH. The increased abundance of acid-extruders such as NHE1 and NBCn1 is a feature of cancer cells, which require robust acid extrusion processes to match their increased rate of metabolic acid production. Interestingly, this is not entirely a response to acidic pH_i: the upregulation of NBCn1 is stimulated by a hyperactive ErbB2 signaling cascade. Another example is the activation of NCBTs by IRBIT. IRBIT is part of the signaling pathway initiated by secretagogues to coordinate epithelial fluid secretion. The indirect role of IRBIT in pH_i regulation of secretory cells is that it potentiates the ability of basolateral NCBTs like NBCe1B to counter intracellular acidification.

6. Cellular processes that are sensitive to pH_i

The mechanisms that defend pH_i can be overwhelmed or disabled in pathology. Because most cellular processes, including the action of the universal Na^+/K^+-ATPase, are H^+-sensitive, the consequences of pH_i disturbance are numerous and widespread (Fig. 16.7). Some challenges to pH_i are initiated by a metabolic or respiratory disturbance to whole body acid-base status. For example, a change in pH or [HCO_3^-] in the extracellular space will alter the driving forces acting on cell transporters requiring a sustained compensatory response to preserve pH_i. Similarly, a change in CO_2 in the extracellular space will influence pH_i both directly, as a membrane-permeable modifier of the intracellular CO_2/HCO_3^- equilibrium, and also indirectly due to influence on cells ability to ventilate their own CO_2. Some challenges to pH_i are intrinsic. For example, the change in cellular metabolism of cancer cells increases the generation of H^+ and CO_2 requiring enhanced acid extruding mechanisms in those cells, and subsequently in the surrounding cells whose environment is acidified by this acid extrusion. In practice, cancer cells have a greater propensity to upregulate acid extruding transporters, giving them a selective advantage over their neighbors, whose acid extrusion abilities are ultimately overwhelmed by the local acidification of pH_e. Mutations in acid-extruders, acid-loaders, and carbonic anhydrases are all associated with a reduced ability to regulate pH_i, and potentiate the ability of acid-base disturbances to overwhelm the pH_i regulatory machinery.[11] Despite apparent functional redundancy in these systems in many cells, loss of some ubiquitously expressed proteins such as NHE1 and NBCe1 causes enhanced mortality in mice and severe disease in humans. In the following sections, we will consider the pathological consequences of pH_i disturbance as well as those processes that are deliberately sensitive to H^+ to the benefit of the cell.

11. Thus blockers of CAs and NCBTs have potential as anticancer agents.

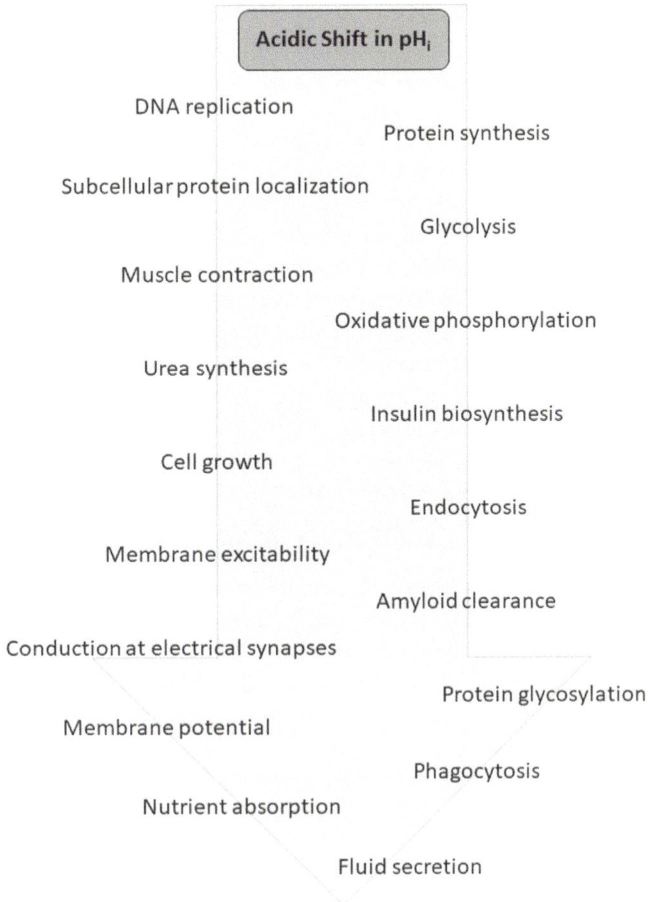

FIGURE 16.7 A selection of the diverse cellular processes that are rendered suboptimal by an acidic shift in pH_i.

6.1 DNA replication and protein synthesis

pH_i is a critical factor that affects the ability of a cell to thrive, with rising pH_i over the range 6.9–7.4 increasing the rates of DNA and protein synthesis. In the case of DNA synthesis, this observation is partly explained by the pH optima for DNA polymerases, which are typically more alkaline than pH_i. Indeed, fertilization of sea urchin and frog oocytes is associated with a cellular alkalinization, although this does not seem to be a general feature as this pH_i change has not been observed during activation of rodent oocytes.

6.2 Metabolism

It is likely that most metabolic pathways are to some extent influenced by pH_i given the pH-dependence of enzyme activity and the fact that many metabolites are themselves acids and bases. However, in this section, we will focus our attention on a couple of notable examples of pH dependent steps in major metabolic pathways and highlight their interplay with the pH-regulatory mechanisms that protect them.

Our first example is phosphofructokinase (PFK-1), which catalyzes a key rate limiting step in glycolysis. PFK-1 uses the energy of ATP hydrolysis to phosphorylate fructose-6-phosphate, generating fructose-1,6-bisphosphate. PFK-1 is allosterically inhibited by ATP, enabling downregulation of glycolysis when ATP is abundant. The binding of ATP to its allosteric site on PFK-1 is enhanced at acidic pH_i, rendering PFK-1 acutely H^+-sensitive. This is potentially an issue in myocytes in which pH_i falls during prolonged exercise as ATP could become inappropriately inhibitory to PFK-1 even at relatively low levels. However, in active muscles, gluconeogenesis produces fructose-2,6-bisphosphate, an allosteric activator of PFK-1 that eliminates the pH-dependent inhibition of PFK-1 by ATP, allowing glycolysis to proceed unhindered by acidity. Interestingly, some glycolytic enzymes such as PFK-1 and aldolase are tethered to the cytosolic tail of V-type H^+-ATPase subunits, which would presumably place them a relatively alkaline pH microenvironment. Studies in yeast suggest that these enzymes guide the assembly of V-type H^+-ATPase subunits, coupling pump activity to energy availability via the sensation of glycolytic flux. In humans, mutations in one of the pump's subunits, which disrupts the interaction with PFK1, are associated with distal renal tubular acidosis: a defect in renal acid excretion due to disabled H^+-ATPase activity.

Nitrogen metabolism is also pH_i sensitive. In the mitochondria of hepatocytes, carbamoyl phosphate synthetase 1 catalyzes the first stage of urea synthesis by harnessing the energy of ATP hydrolysis combining HCO_3^- and ammonia to form carbamoyl phosphate. This process removes toxic ammonia from circulation. However, the consumption of intracellular HCO_3^- generates an acid load that must be replenished by CA action and balanced by acid extrusion. Inhibition of Na^+/H^+ exchange in these cells limits acid extrusion and results in a fall in cytosolic pH and $[HCO_3^-]$ that greatly inhibits urea production. Mutations in mitochondrial CAVA result in hyperammonemia.

Oxidative phosphorylation is another process that is critically dependent on pH. Mitochondria maintain a matrix pH that is ~1 unit more alkaline than the inner mitochondrial space (IMS). This pH gradient across the inner mitochondrial membrane ($\Delta pH = pH_{IMS}$-pH_{matrix}) contributes to the proton motive force (Δp) that sustains ATP synthesis. Δp is the driving force for the reentry of H^+ into the mitochondrial matrix,[12] which can be calculated as the potential difference between the electrical potential across the inner mitochondrial membrane ($\Delta\psi_m = E_{IMS}$-E_{matrix}) plus the Nernstian equilibrium potential for H^+ across that membrane (−62ΔpH), per Eq. (16.17):

$$\Delta p = \Delta\psi_m - 62\Delta pH \quad (16.17)$$

For a typical $\Delta\psi_m$ of 160 mV, this equation describes a driving force of ~220 mV for H^+ reentry into the matrix, of which ΔpH accounts for ~25%. ΔpH is also important for the uptake of respiratory substrates into the matrix. For example, the phosphate, pyruvate, and glutamate carriers in the inner mitochondrial membrane are all coupled to proton transport. Disturbances in cytosolic and/or IMS pH are paralleled by changes in the pH of the matrix, which helps to maintain ΔpH and thus Δp. However, the influence of chronic pH insults upon mitochondrial action is not simple to predict as the responses do not simply reflect changes in ΔpH, but also depend on the extent of the disturbance, the status of other ambient parameters such as Ca^{2+} and O_2, and more complex considerations such as alterations in mitophagy and mitochondrial remodeling.

6.3 Membrane excitability

Neuronal activity is exquisitely sensitive to changes in both pH_i and pH_e due to the influence of pH on (1) the membrane transport processes that govern the excitability of their membranes and (2) the release/reuptake of neurotransmitters. Typically, neuronal activity is associated with a drop in pH_i and a concomitant rise in pH_e at the synaptic cleft. This is caused by a combination of metabolic acid production and the action of acid loading proteins such as the plasma membrane Ca^{2+}/H^+-ATPase or the efflux of HCO_3^- through $GABA_A$ receptors. In general, an acidic shift in pH_i is associated with a dampening of excitability. Such an acidification may help dampen excessive firing, while the concordant extracellular alkalinization would neutralize the acidic vesicular fluid that is released into synaptic clefts with neurotransmitter release. This is a necessity as the activity of postsynaptic NMDA and AMPA receptors is inhibited by extracellular acidity. Studies in mice suggest a complex link between disturbed pH_i regulation and network excitability: loss of the acid-extruding protein NDCBE increases seizure threshold, while loss of either an acid-loading (AE3) or an acid-extruding (NHE1, NBCn2) protein can decrease seizure threshold. This paradox could reflect differential expression of these transporters in excitatory versus inhibitory neurons or, for NDCBE whose action also influences intracellular $[Cl^-]$, an effect on the plasticity of GABAergic synapses.

Some sensory cells use changes in $[H^+]_i$ as an intracellular signal. In sour taste cells, extracellular acidity opens the Otop1 protein channel, resulting in drop in cytosolic pH that inhibits a H^+-sensitive K^+ channel Kir2.1. The resulting membrane depolarization triggers the firing of an action potential that releases neurotransmitters onto afferent nerve fibers.

6.4 Muscle contraction

Intracellular acidosis is associated with muscle fatigue during exercise, a phenomenon that is at least in part caused by a reduced ability of myosin to generate force at low pH. Excitation-contraction coupling is further compromised at low pH because protons displace Ca^{2+} from critical Ca^{2+}-binding proteins such as troponin. Action potential conduction at

12. In accordance with the conventions for such driving force calculations, the matrix is the equivalent of the extracellular compartment. Thus, a positive driving force represents the exit of H^+ across the inner mitochondrial membrane into the matrix.

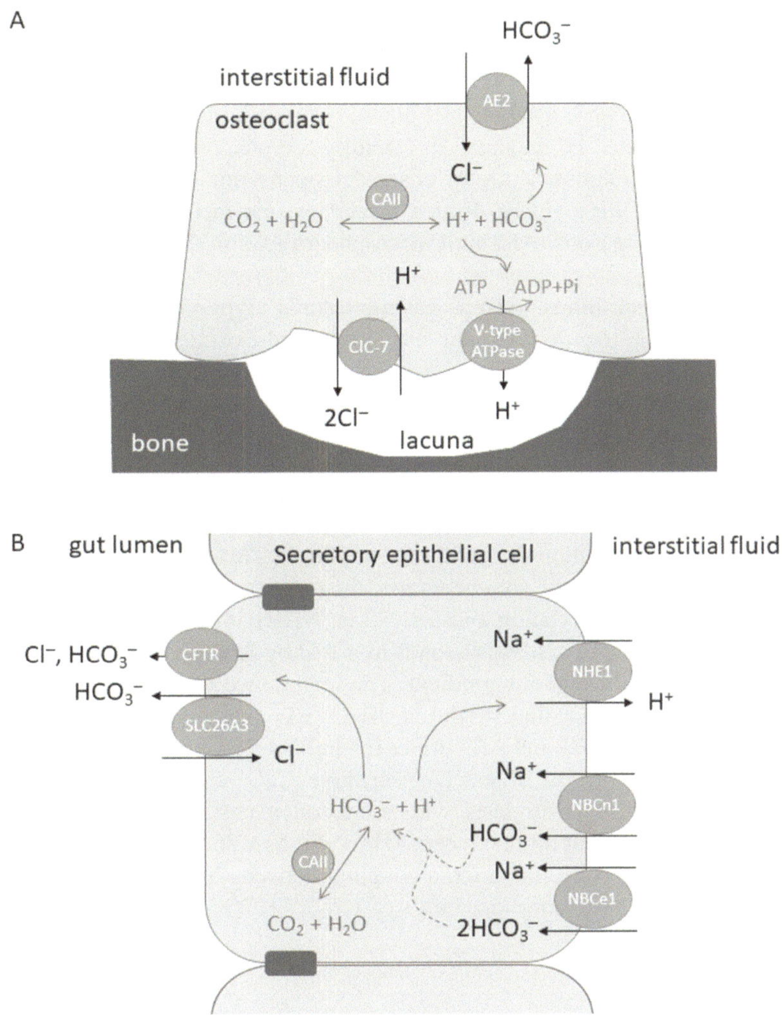

FIGURE 16.8 Balancing alkaline and acidic loads in secretory cells. (A) Osteoclasts secrete acidic bone resorbing fluid. This process is supported, without compromising cytosolic pH, by the actions of carbonic anhydrase II (CAII) and anion exchanger 2 (AE2). These proteins also support H^+ secretion by gastric parietal cells. (B) Secretory crypt epithelia in the small intestine produce a HCO_3^- containing fluid. This process is supported, without compromising cytosolic pH, by the actions of CAII, Na^+-coupled HCO_3^- transporters such as NBCn1 and NBCe1, and the Na^+/H^+ exchanger NHE1.

electrical synapses is pH_i-dependent due to the H^+-sensitivity of gap junctions, which are gated closed by short exposure to acidity and irreversibly inactivated by longer exposures to acidity. This has been suggested to a useful mechanism to prevent propagation of signals from acidic (i.e., unhealthy) cells.

6.5 Fluid secretion and nutrient absorption

Cells that secrete acidic or basic fluids require robust defense of pH_i to sustain secretion without compromising their own acid-base status. For example, bone-resorbing osteoclasts actively secrete HCl-rich fluid across their lacunar membrane in order to reclaim minerals from bone. This loss of acid would tend to cause intracellular alkalosis were it not for the combined actions of CA, which replenishes intracellular H^+, and the alkali-stimulated Cl^-/HCO_3^- exchanger AE2 in the contralacunar membrane, that disposes of the HCO_3^- generated by CA action (Fig. 16.8A). Genetic defects in CA and AE2 are associated with osteopetrosis: the formation of abnormally dense bones. In this same way, CA and AE2 support H^+ secretion by gastric parietal cells. The acute pH_i dependence of AE2 presumably allows a flexible response to variations in the robustness of acid secretion caused by neural and hormonal stimuli. In renal epithelia, pH_i is defended during H^+ secretion by CA and a variety of different basolateral acid loaders. For example, in renal proximal tubule epithelia, apical H^+ secretion into the tubular fluid that becomes urine is balanced by the action of basolateral NBCe1 operating in acid-loading mode. In collecting duct α-intercalated cells, the basolateral acid loader supporting H^+ secretion is kAE1. kAE1 is not acutely sensitive to pH_i, but its abundance is increased during metabolic acidosis. This suits the purpose of the cells which chronically upregulate H^+ secretion in response to chronic acid stress. AE2 would not be suited to this role because it is inhibited by acidosis.

Acid loaders are also valuable to balance cytosolic pH during the acidification of subcellular compartments. The actions of V-type H^+-ATPase and voltage-gated H^+ channel Hv1 acidify the phagosomes of neutrophils and macrophages. The H^+ flux mediated by Hv1 plays an additional role of charge compensation to promote the loading of reactive oxygen species into the phagosome by NADPH oxidase. Both CA and plasma membrane AE2 are required to maintain acid loading of the phagosome. For these reasons, AE2-null macrophages exhibit defective phagocytosis and pathogen killing ability.

Cells that secrete HCO_3^--containing fluids, such as gastrointestinal crypt epithelia, cells of the enamel organ, corneal endothelia, and ciliary body epithelia, require CAs and basolateral acid-extruders such as NHE1, NBCe1, and NBCn1 in order to balance pH_i and maintain a sufficiency of intracellular HCO_3^- to support anion secretion (Fig. 16.8B). NBCe1 is also valuable to balance pH_i in support of H^+-coupled nutrient uptake in absorptive epithelia. NBCe1-null mice exhibit defects in small intestinal fluid secretion and nutrient absorption.

6.6 Volume regulation, migration, and salt absorption

When cells are placed into hypertonic solution they will tend to shrink. This tendency is countered by a process known as regulatory volume increase. The restoration of volume comes about due to the activation by cell shrinkage of NHE1 via a signaling pathway that causes its phosphorylation and activation. NHE1 is functionally coupled to the alkaline-activated acid-loader AE2, such that NHE1-mediated acid extrusion is matched by AE2-mediated acid loading. The net result is a pH neutral influx of NaCl that dissipates the osmotic gradient. The importance of pH_i change to the mediation of regulatory volume increase is highlighted by the fact that pH-insensitive AE1 cannot substitute for AE2 in this mechanism. In migrating cells, this coupling between NHE1 and AE2 drives the influx of fluid at the leading edge, with the initial stimulus for NHE1 activation signaled by the action of mechanosensitive Ca^{2+} channels when actin filaments are extended. Elevated NHE1 (and NBCe1) abundance correlates with increased metastatic potential in breast cancer cell lines. In epithelial cells of the small intestine, a coupling between NHE3 and SLC26A3 mediates pH-neutral NaCl uptake for the purpose of salt absorption. In the renal collecting duct, a coupling between NDBCE and two cycles of SLC26A4 activity has also been suggested to facilitate salt reabsorption (influx of $1Na^+$, $2HCO_3^-$, and $2Cl^-$ balanced by efflux of $2HCO_3^-$ and $1Cl^-$ = net influx of $1Na^+$ and $1Cl^-$).

6.7 Erythrocyte function

The affinity of hemoglobin for O_2 is lowered by acidic shifts in pH_i, a phenomenon that underpins the Bohr effect (Fig. 16.9). In systemic capillaries, CO_2 enters the cell and is hydrated to H^+ and HCO_3^- by CA. AE1 in the red cell membrane extrudes the HCO_3^-, resulting in a drop in pH_i and a release of O_2 from hemoglobin. The opposite process occurs in lung capillaries in which the exit of CO_2 from red cells promotes the AE1-mediated influx of HCO_3^- and the increase affinity of Hb for O_2. AE1 is also a site of anchoring for soluble proteins such as glycolytic enzymes and deoxyhemoglobin. However, the major consequence of AE1 loss from red cells is hemolytic anemia due to its role in linking the red cell membrane to the underlying cytoskeleton.

6.8 Vesicular traffic and protein processing

The function of the secretory pathway is critically dependent on the luminal acidity of its compartments, as described in Section 4.2. For example, the relative acidity of the lumen of the Golgi apparatus compared the lumen of the endoplasmic reticulum provides the trigger for the retrieval of chaperone proteins with a "KDEL" motif: acidity promotes the association of these proteins with a retrieval receptor that targets them for retrograde vesicle-mediated transport back to the endoplasmic reticulum. Loss of V-type H^+-ATPase subunit ATP6V0A2 causes cutis laxa, or wrinkly-skin syndrome, the cellular manifestations of which include impaired retrograde trafficking of vesicles from the Golgi to the endoplasmic reticulum and impairment of serum protein glycosylation. The maintenance of an appropriately acidified internal milieu is also important in secretory vesicles and recycling endosomes because it provides the necessary microenvironment for the effective processing and sorting of cargo. For example, the loading of synaptic vesicles with neurotransmitters is dependent on the proton motive force established across the vesicular membrane by the V-type H^+-ATPase. Furthermore, in pancreatic beta cells, glucose metabolism triggers the acidification of maturing secretory granules, which activates the

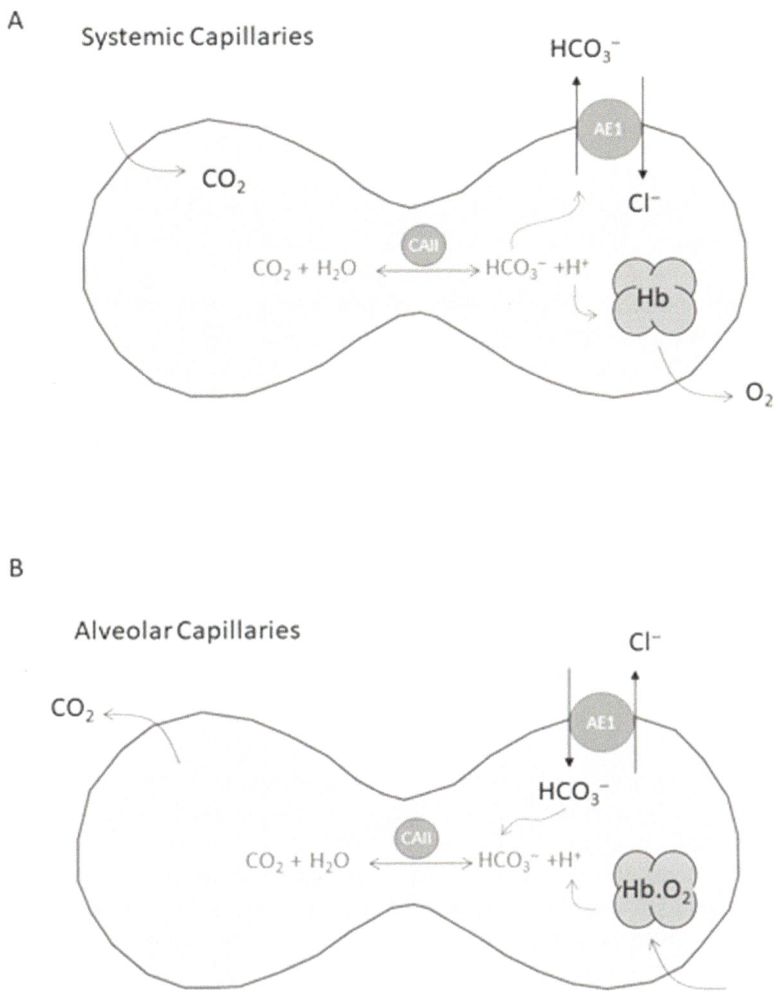

FIGURE 16.9 The Bohr effect in red cells describes the pH-dependence of the oxygen affinity of hemoglobin (Hb). (A) In systemic capillaries, CAII generates HCO_3^- and H^+ from CO_2. AE1 exports the HCO_3^-, causing an intracellular acidification that promotes O_2 release from Hb. (B) In alveolar capillaries, CO_2 leaves the cells causing CAII to consume intracellular HCO_3^- and H^+. Intracellular HCO_3^- is replenished by AE1, causing an intracellular alkalinization that increases the affinity of Hb for O_2.

proteolytic processing of proinsulin into insulin and promotes the formation of zinc-bound insulin crystals for storage prior to their release. Even the mildly acidic lumen of recycling endosomes is helpful to promote the dissociation of ligands from their receptors so that unoccupied receptors can be redeposited in the plasma membrane.

The importance of endosomal pH regulation is demonstrated by the numerous pathologies that are linked to the loss of endosomal and lysosomal H^+ transporters. For example, loss of NHE6 is associated with autism, Alzheimer's disease, and Christianson syndrome. Clues to the pathological basis of this association are provided by cell studies, which show that the excessive endosomal acidification that results from loss of the NHE6-mediated H^+-leak causes a deficit in amyloid beta clearance and an increase in amyloid deposition.[13] Loss of ClC-5 from renal epithelial cells results in impaired ability to form and acidify the early endosomes that perform the apical step of protein reabsorption. Individuals with mutations in ClC-5 present with Dent disease: proteinuria and kidney stones. Loss of ClC-7 causes osteopetrosis due to an inability of osteoclasts to form a resorption lacuna. Loss of the neuronally expressed ClC-6 results in lysosomal storage disease. However, the extent to which these phenotypes are related to pH_i disturbance alone is not clear.

Cytosolic pH is just as important: cytosolic acidity inhibits the budding of vesicles from the Golgi apparatus and may, in addition, cause it to fragment. At the level of endosomes, cytosolic acidity causes a translocation of endosomes along

13. The role of NHE6 is likely more complex as its loss is neuroprotective in an ApoE4 model of Alzheimer's disease.

microtubules toward the cell surface, disrupts the sorting of cargo into apical and basolateral transport vesicles in polarized epithelial cells, and also inhibits the retrieval of clathrin-coated vesicles from the cell surface.

7. Microbial pH$_i$ regulation

Acids are powerful antimicrobial agents. In the microbial realm, pH$_i$ regulation is based on closed buffering systems and the transport of H^+. Presumably, the CO_2/HCO_3^- system is less valuable to a cell in environments where there is either very little CO_2/HCO_3^- available or a lack of ability to control intracellular and environmental [CO_2] by regulated ventilation. Indeed, in microbes, the homologs of metazoan HCO_3^- transporters have other roles. As far as is known, Slc4-like genes are rarely present in bacterial genomes, while Slc4-like proteins in fungi are boron transporters (e.g., Bor1p in *Saccharomyces cerevisiae*). The only known microbial HCO_3^--transporter is the Slc26-like BicA, which is part of a cyanobacterial CO_2 fixation mechanism. In this section, we will summarize the proton-centric pH$_i$ regulatory mechanisms of well-characterized microbes.

7.1 Bacteria

pH$_i$ regulation is paramount in bacteria and they maintain a transmembrane, alkaline-inside, pH gradient. This transmembrane pH gradient is a substantial contributor to the proton motive force for ATP generation by H^+ synthase (as detailed for mitochondria in Section 6.1) and for bacterial mobility, as flagellar rotation is powered by the activity of proton pumps. Weak, membrane-permeant acids like acetic acid that can acidify the cytosol have potent antimicrobial effect and inhibit flagellar rotation. Neutralophilic bacteria that thrive at close to neutral pH defend their pH from both acidic and alkaline shifts in pH$_e$. At pH$_e$ = 7.0, the cytoplasmic pH of *Escherichia coli* is ~7.5. Their defense from acidity is robust and they can maintain an alkaline-inside gradient of ~2 units even when pH$_e$ is 5.0. This promotes cell survival, although they cannot grow under these conditions. On the other hand, the neutralophilic gastrointestinal ulcer causing *Helicobacter pylori* is able to maintain a neutral cytosolic pH even in stomach acid (pH$_e$ ~ 2) by virtue of the action of a H^+ gated urea transporter coupled to a cytosolic urease enzyme, which generates acid-neutralizing NH_3.

The cytosolic pH of acidophilic and alkaliphilic bacteria varies according to case but both types strive to maintain a cytosolic pH that is closer to neutral than their environment. General strategies for defense from acidic threat vary among organisms but include phosphate buffering, cell membranes that are composed to be less H^+ permeable, enhanced H^+-ATPase activity to extrude H^+, shifting metabolic pathways to favor H^+ consumption, and the expression of proteins that are either intrinsically stable at low pH, or which can be stabilized by acid-induced chaperones. Mechanisms by which bacteria defend pH$_i$ against alkaline threats include a variety of cation-coupled H^+ exchangers, such as the $Na^+/2H^+$ exchanger NhaA that imports H^+, and upregulation of the H^+ synthase that mediates a H^+ leak into the cytosol.

7.2 Fungi

As typified by the yeast *S. cerevisiae*, cytosolic pH in fungi is close to neutral and is a key determinant of growth rate. In yeast cells, cytosolic pH is mainly defended by the acid extruding actions of a P-type H^+-ATPase (Pma1) in the plasma membrane and V-type H^+-ATPase (Vma1) in the vacuolar membrane. Pma1 is activated by cytosolic acidification and cells with disabled Pma1 struggle to grow at low pH. In alkaline stress, when the proton motive force that drives growth-sustaining transport processes is weakened, a Na^+-ATPase is expressed to generate a sodium motive force and H^+-driven phosphate transporters are replaced by Na^+-driven phosphate transporters.

Vacuolar pH is more acidic than the cytosol, typically ~6.0 to promote the pH-dependent activation of the proteinases within. In addition, the vacuole is an important sink for cytosolic H^+; a fall in cytosolic pH is accompanied by a fall in vacuolar pH, likely due to Vma1 action. Both Pma1 and Vma1 are inhibited by glucose deprivation, perhaps as a mechanism to preserve cytosolic ATP. The sensing of energy status by Vma1 is mediated by its physical association with PFK, which promotes the assembly of its constituent subunits (see Section 6.1). The assembly of Vma1 activates Ras and TORC signaling pathways, promoting cell growth. The plasma membrane cation-proton exchanger Nha1 is chiefly concerned with Na^+ extrusion and does not play a dominant role in pH$_i$ regulation, although its counterpart Nhx1 balances H^+ loading in endosomes, similar to the role of NHEs 6 and 9 in mammalian endosomes (see Section 6.7).

References

Alper, S. L., & Sharma, A. K. (2013). The SLC26 gene family of anion transporters and channels. *Molecular Aspects of Medicine, 34*(2–3), 494–515. https://doi.org/10.1016/j.mam.2012.07.009

Astrup, P., & Severinghaus, J. W. (1986). *The history of blood gases, acids and bases* (1st ed.) Munksgaard.

Bates, R. G. (1964). *Determination of pH: Theory and practice*. John Wiley and Sons Inc.

Becker, H. M., Klier, M., & Deitmer, J. W. (2014). Carbonic anhydrases and their interplay with acid/base-coupled membrane transporters. *Sub-Cellular Biochemistry, 75*, 105–134. https://doi.org/10.1007/978-94-007-7359-2_7

Blakely, R. D., & Edwards, R. H. (2012). Vesicular and plasma membrane transporters for neurotransmitters. *Cold Spring Harbor Perspectives in Biology, 4*(2), Article a005595. https://doi.org/10.1101/cshperspect.a005595

Bose, S., He, H., & Stauber, T. (2021). Neurodegeneration upon dysfunction of endosomal/lysosomal CLC chloride transporters. *Frontiers in Cell and Developmental Biology, 9*, Article 639231. https://doi.org/10.3389/fcell.2021.639231

Bräuer, P., Parker, J. L., Gerondopoulos, A., Zimmermann, I., Seeger, M. A., Barr, F. A., & Newstead, S. (2019). Structural basis for pH-dependent retrieval of ER proteins from the Golgi by the KDEL receptor. *Science, 363*(6431), 1103–1107. https://doi.org/10.1126/science.aaw2859

Brady Trexler, E., Bukauskas, F. F., Bennett, M. V. L., Bargiello, T. A., & Verselis, V. K. (1999). Rapid and direct effects of pH on connexins revealed by the Connexin46 hemichannel preparation. *The Journal of General Physiology, 113*(5), 721–742.

Brett, C. L., Donowitz, M., & Rao, R. (2005). Evolutionary origins of eukaryotic sodium/proton exchangers. *American Journal of Physiology - Cell Physiology, 288*(2), C223–C239. https://doi.org/10.1152/ajpcell.00360.2004

Bublitz, M., Morth, J. P., & Nissen, P. (2011). P-type ATPases at a glance. *Journal of Cell Science, 124*(Pt 15), 2515–2519. https://doi.org/10.1242/jcs.088716

Cornelius, F., Tsunekawa, N., & Toyoshima, C. (2018). Distinct pH dependencies of Na^+/K^+ selectivity at the two faces of Na,K-ATPase. *Journal of Biological Chemistry, 293*(6), 2195–2205. https://doi.org/10.1074/jbc.RA117.000700

Davenport, H. W. (1974). *The ABC of acid-base chemistry* (6th ed.). The University of Chicago Press.

Debold, E. P., Beck, S. E., & Warshaw, D. M. (2008). Effect of low pH on single skeletal muscle myosin mechanics and kinetics. *American Journal of Physiology - Cell Physiology, 295*(1), C173–C179. https://doi.org/10.1152/ajpcell.00172.2008

Dechant, R., & Peter, M. (2014). Cytosolic pH: A conserved regulator of cell growth? *Molecular & Cellular Oncology, 1*(4), Article e969643. https://doi.org/10.4161/23723548.2014.969643

Diez-Fernandez, C., Rüfenacht, V., Santra, S., Lund, A. M., Santer, R., Lindner, M., Tangeraas, T., Unsinn, C., de Lonlay, P., Burlina, A., van Karnebeek, C. D. M., & Häberle, J. (2016). Defective hepatic bicarbonate production due to carbonic anhydrase VA deficiency leads to early-onset life-threatening metabolic crisis. *Genetics in Medicine: Official Journal of the American College of Medical Genetics, 18*(10), 991–1000. https://doi.org/10.1038/gim.2015.201

Dobson, G. P., Yamamoto, E., & Hochachka, P. W. (1986). Phosphofructokinase control in muscle: Nature and reversal of pH-dependent ATP inhibition. *American Journal of Physiology, 250*(1 Pt 2), R71–R76. https://doi.org/10.1152/ajpregu.1986.250.1.R71

DuBose, T. D., Jr. (2008). Disorders of acid-base balance. In*Brenner and rector's the kidney* (8th ed., Vol. 1, pp. 505–546). Elsevier Saunders.

Frost, S. C., & McKenna, R. (2013). *Carbonic anhydrase: Mechanism, regulation, links to disease, and industrial applications*. Springer Science & Business Media.

Fuster, D. G., & Alexander, R. T. (2014). Traditional and emerging roles for the SLC9 Na^+/H^+ exchangers. *Pfluegers Archiv European Journal of Physiology, 466*(1), 61–76. https://doi.org/10.1007/s00424-013-1408-8

Glagolev, A. N., & Skulachev, V. P. (1978). The proton pump is a molecular engine of motile bacteria. *Nature, 272*(5650), 280–282. https://doi.org/10.1038/272280a0

Gorbatenko, A., Olesen, C. W., Mørup, N., Thiel, G., Kallunki, T., Valen, E., & Pedersen, S. F. (2014). ErbB2 upregulates the Na^+,HCO_3^--cotransporter NBCn1/SLC4A7 in human breast cancer cells via Akt, ERK, Src, and Kruppel-like factor 4. *Federation of American Societies for Experimental Biology Journal: Official Publication of the Federation of American Societies for Experimental Biology, 28*(1), 350–363. https://doi.org/10.1096/fj.13-233288

Guyenet, P. G., Stornetta, R. L., Souza, G. M. P. R., Abbott, S. B. G., Shi, Y., & Bayliss, D. A. (2019). The retrotrapezoid nucleus: Central chemoreceptor and regulator of breathing automaticity. *Trends in Neurosciences, 42*(11), 807–824. https://doi.org/10.1016/j.tins.2019.09.002

Häussinger, D., & Gerok, W. (1985). Hepatic urea synthesis and pH regulation. Role of CO_2, HCO_3^-, pH and the activity of carbonic anhydrase. *European Journal of Biochemistry, 152*(2), 381–386. https://doi.org/10.1111/j.1432-1033.1985.tb09208.x

Harrington, J. T., & Lemann, J. (1970). The metabolic production and disposal of acid and alkali. *Medical Clinics of North America, 54*(6), 1543–1554.

Hayek, S. R., Rane, H. S., & Parra, K. J. (2019). Reciprocal regulation of V-ATPase and glycolytic pathway elements in health and disease. *Frontiers in Physiology, 10*, 127. https://doi.org/10.3389/fphys.2019.00127

Issaian, A., Hay, A., Dzieciatkowska, M., Roberti, D., Perrotta, S., Darula, Z., Redzic, J., Busch, M. P., Page, G. P., Rogers, S. C., Doctor, A., Hansen, K. C., Eisenmesser, E. Z., Zimring, J. C., & D'Alessandro, A. (2021). The interactome of the N-terminus of band 3 regulates red blood cell metabolism and storage quality. *Haematologica, 106*(11), 2971–2985. https://doi.org/10.3324/haematol.2020.278252

Jennings, M. L. (2021). Cell physiology and molecular mechanism of anion transport by erythrocyte band 3/AE1. *American Journal of Physiology - Cell Physiology, 321*(6), C1028–C1059. https://doi.org/10.1152/ajpcell.00275.2021

Jentsch, T. J., & Pusch, M. (2018). CLC chloride channels and transporters: Structure, function, physiology, and disease. *Physiological Reviews, 98*(3), 1493–1590. https://doi.org/10.1152/physrev.00047.2017

Johnson, D. E., Ostrowski, P., Jaumouillé, V., & Grinstein, S. (2016). The position of lysosomes within the cell determines their luminal pH. *Journal of Cell Biology, 212*(6), 677–692. https://doi.org/10.1083/jcb.201507112

Kane, P. M. (2016). Proton transport and pH control in fungi. *Advances in Experimental Medicine and Biology, 892*, 33–68. https://doi.org/10.1007/978-3-319-25304-6_3

Kath-Schorr, S., Wilson, T. J., Li, N.-S., Lu, J., Piccirilli, J. A., & Lilley, D. M. J. (2012). General acid-base catalysis mediated by nucleobases in the hairpin ribozyme. *Journal of the American Chemical Society, 134*(40), 16717–16724. https://doi.org/10.1021/ja3067429

Kellokumpu, S. (2019). Golgi pH, ion and redox homeostasis: How much do they really matter? *Frontiers in Cell and Developmental Biology, 7*, 93. https://doi.org/10.3389/fcell.2019.00093

Kornak, U., Reynders, E., Dimopoulou, A., van Reeuwijk, J., Fischer, B., Rajab, A., Budde, B., Nürnberg, P., Foulquier, F., ARCL Debré-type Study Group, Lefeber, D., Urban, Z., Gruenewald, S., Annaert, W., Brunner, H. G., van Bokhoven, H., Wevers, R., Morava, E., Matthijs, G., … Mundlos, S. (2008). Impaired glycosylation and cutis laxa caused by mutations in the vesicular H^+-ATPase subunit ATP6V0A2. *Nature Genetics, 40*(1), 32–34. https://doi.org/10.1038/ng.2007.45

Lee, S.-K., Boron, W. F., & Parker, M. D. (2013). Monitoring ion activities in and around cells using ion-selective liquid-membrane microelectrodes. *Sensors, 13*(1), 984–1003. https://doi.org/10.3390/s130100984

Leem, C.-H., & Vaughan-Jones, R. D. (1998). Out-of-equilibrium pH transients in the Guinea-pig ventricular myocyte. *The Journal of Physiology, 509*(2), 471–485. https://doi.org/10.1111/j.1469-7793.1998.471bn.x

Leem, C. H., Lagadic-Gossmann, D., & Vaughan-Jones, R. D. (1999). Characterization of intracellular pH regulation in the Guinea-pig ventricular myocyte. *The Journal of Physiology, 517*(Pt 1), 159–180.

Levin, L. R., & Buck, J. (2015). Physiological roles of acid-base sensors. *Annual Review of Physiology, 77*, 347–362. https://doi.org/10.1146/annurev-physiol-021014-071821

Liu, X., Li, T., & Tuo, B. (2018). Physiological and pathophysiological relevance of the anion transporter Slc26a9 in multiple organs. *Frontiers in Physiology, 9*, 1197. https://doi.org/10.3389/fphys.2018.01197

Loiselle, F. B., & Casey, J. R. (2010). Measurement of intracellular pH. *Methods in Molecular Biology, 637*, 311–331. https://doi.org/10.1007/978-1-60761-700-6_17

Lu, M., Ammar, D., Ives, H., Albrecht, F., & Gluck, S. L. (2007). Physical interaction between aldolase and vacuolar H^+-ATPase is essential for the assembly and activity of the proton pump. *Journal of Biological Chemistry, 282*(34), 24495–24503. https://doi.org/10.1074/jbc.M702598200

Minamino, T., Imae, Y., Oosawa, F., Kobayashi, Y., & Oosawa, K. (2003). Effect of intracellular pH on rotational speed of bacterial flagellar motors. *Journal of Bacteriology, 185*(4), 1190–1194. https://doi.org/10.1128/JB.185.4.1190-1194.2003

Morimoto, S., Harada, K., & Ohtsuki, I. (1999). Roles of troponin isoforms in pH dependence of contraction in rabbit fast and slow skeletal and cardiac Muscles1. *The Journal of Biochemistry, 126*(1), 121–129. https://doi.org/10.1093/oxfordjournals.jbchem.a022412

Myers, E. J., Marshall, A., Jennings, M. L., & Parker, M. D. (2016). Mouse Slc4a11 expressed in *Xenopus* oocytes is an ideally selective H^+/OH^- conductance pathway that is stimulated by rises in intracellular and extracellular pH. *American Journal of Physiology - Cell Physiology, 311*(6), C945–C959. https://doi.org/10.1152/ajpcell.00259.2016

Nakamura, N., Tanaka, S., Teko, Y., Mitsui, K., & Kanazawa, H. (2005). Four Na+/H+ exchanger isoforms are distributed to Golgi and post-golgi compartments and are involved in organelle pH regulation. *Journal of Biological Chemistry, 280*(2), 1561–1572. https://doi.org/10.1074/jbc.M410041200

Nuccitelli, R., & Deamer, D. W. (1982). *Intracellular pH: Its measurement, regulation, and utilization in cellular functions: Proceedings of a conference held at the kroc foundation, Santa Ynez valley, California, july 20-24, 1981* (Vol. 15). Alan R. Liss.

Parker, M. D., & Boron, W. F. (2013). The divergence, actions, roles, and relatives of sodium-coupled bicarbonate transporters. *Physiological Reviews, 93*(2), 803–959. https://doi.org/10.1152/physrev.00023.2012

Paroutis, P., Touret, N., & Grinstein, S. (2004). The pH of the secretory pathway: Measurement, determinants, and regulation. *Physiology, 19*, 207–215. https://doi.org/10.1152/physiol.00005.2004

Poburko, D., & Demaurex, N. (2012). Regulation of the mitochondrial proton gradient by cytosolic Ca^{2+} signals. *Pfluegers Archiv European Journal of Physiology, 464*(1), 19–26. https://doi.org/10.1007/s00424-012-1106-y

Prasad, H., & Rao, R. (2020). Endosomal acid-base homeostasis in neurodegenerative diseases. *Reviews of Physiology, Biochemistry & Pharmacology*. https://doi.org/10.1007/112_2020_25

Preisig, P. A. (2007). The acid-activated signaling pathway: Starting with Pyk2 and ending with increased NHE3 activity. *Kidney International, 72*(11), 1324–1329. https://doi.org/10.1038/sj.ki.5002543

Roos, A., & Boron, W. F. (1981). Intracellular pH. *Physiological Reviews, 61*(2), 296–434.

Ruffin, V. A., Salameh, A. I., Boron, W. F., & Parker, M. D. (2014). Intracellular pH regulation by acid-base transporters in mammalian neurons. *Frontiers in Physiology, 5*, 43. https://doi.org/10.3389/fphys.2014.00043

Santo-Domingo, J., & Demaurex, N. (2012). Perspectives on: SGP symposium on mitochondrial physiology and medicine: The renaissance of mitochondrial pH. *The Journal of General Physiology, 139*(6), 415–423. https://doi.org/10.1085/jgp.201110767

Santos, J. M., Martínez-Zaguilán, R., Facanha, A. R., Hussain, F., & Sennoune, S. R. (2016). Vacuolar H^+-ATPase in the nuclear membranes regulates nucleo-cytosolic proton gradients. *American Journal of Physiology - Cell Physiology, 311*(4), C547–C558. https://doi.org/10.1152/ajpcell.00019.2016

Scheiner, B., Lindner, G., Reiberger, T., Schneeweiss, B., Trauner, M., Zauner, C., & Funk, G.-C. (2017). Acid-base disorders in liver disease. *Journal of Hepatology, 67*(5), 1062–1073. https://doi.org/10.1016/j.jhep.2017.06.023

Scott, D. R., Marcus, E. A., Wen, Y., Singh, S., Feng, J., & Sachs, G. (2010). Cytoplasmic histidine kinase (HP0244)-regulated assembly of urease with UreI, a channel for urea and its metabolites, CO_2, NH_3, and NH_4^+, is necessary for acid survival of *Helicobacter pylori*. *Journal of Bacteriology, 192*(1), 94–103. https://doi.org/10.1128/JB.00848-09

Silverstein, T. P. (2021). The proton in biochemistry: Impacts on bioenergetics, biophysical chemistry, and bioorganic chemistry. *Frontiers in Molecular Biosciences, 8*, Article 764099. https://doi.org/10.3389/fmolb.2021.764099

Skelton, L. A., Boron, W. F., & Zhou, Y. (2010). Acid-base transport by the renal proximal tubule. *Journal of Nephrology, 23*(Suppl. 16), S4–S18.

Slonczewski, J. L., Fujisawa, M., Dopson, M., & Krulwich, T. A. (2009). Cytoplasmic pH measurement and homeostasis in bacteria and archaea. *Advances in Microbial Physiology, 55*(1–79), 317. https://doi.org/10.1016/S0065-2911(09)05501-5

Stewart, A. K., Chernova, M. N., Kunes, Y. Z., & Alper, S. L. (2001). Regulation of AE2 anion exchanger by intracellular pH: Critical regions of the NH2-terminal cytoplasmic domain. *American Journal of Physiology - Cell Physiology, 281*(4), C1344–C1354. https://doi.org/10.1152/ajpcell.2001.281.4.C1344

Stiernet, P., Guiot, Y., Gilon, P., & Henquin, J.-C. (2006). Glucose acutely decreases pH of secretory granules in mouse pancreatic islets: Mechanisms and influence in insulin secretion. *Journal of Biological Chemistry, 281*(31), 22142–22151. https://doi.org/10.1074/jbc.M513224200

Stock, C., & Schwab, A. (2006). Role of the Na^+/H^+ exchanger NHE1 in cell migration. *Acta Physiologica, 187*(1–2), 149–157. https://doi.org/10.1111/j.1748-1716.2006.01543.x

Teng, B., Wilson, C. E., Tu, Y.-H., Joshi, N. R., Kinnamon, S. C., & Liman, E. R. (2019). Cellular and neural responses to sour stimuli require the proton channel Otop1. *Current Biology: CB, 29*(21), 3647–3656. https://doi.org/10.1016/j.cub.2019.08.077

Van Slyke, D. D. (1922). On the measurement of buffer values and on the relationship of buffer value to the dissociation constant of the buffer and the concentration and reaction of the buffer solution. *Journal of Biological Chemistry, 52*(2), 525–570. https://doi.org/10.1016/S0021-9258(18)85845-8

Waddell, W. J., & Bates, R. G. (1969). Intracellular pH. *Physiological Reviews, 49*(2), 285–329. https://doi.org/10.1152/physrev.1969.49.2.285

Wakabayashi, S., Ikeda, T., Iwamoto, T., Pouysségur, J., & Shigekawa, M. (1997). Calmodulin-binding autoinhibitory domain controls "pH-sensing" in the Na^+/H^+ exchanger NHE1 through sequence-specific interaction. *Biochemistry, 36*(42), 12854–12861. https://doi.org/10.1021/bi9715472

Walker, W. D., Goodwin, F. J., & Cohen, R. D. (1969). Mean intracellular hydrogen ion activity in the whole body liver, heart and skeletal muscle of the rat. *Clinical Science, 36*(3), 409–417.

Weiner, I. D., & Verlander, J. W. (2013). Renal ammonia metabolism and transport. *Comprehensive Physiology, 3*(1), 201–220. https://doi.org/10.1002/cphy.c120010

Wesson, D. E. (2021). The continuum of acid stress. *Clinical Journal of the American Society of Nephrology, 16*(8), 1292–1299. https://doi.org/10.2215/CJN.17541120

Yoshida, T., Kamiya, T., Imanaka-Yoshida, K., & Sakakura, T. (1999). Low cytoplasmic pH causes fragmentation and dispersal of the Golgi apparatus in human hepatoma cells. *International Journal of Experimental Pathology, 80*(1), 51–57. https://doi.org/10.1046/j.1365-2613.1999.00097.x

Yu, Q., Liu, X., Liu, Y., Riederer, B., Li, T., Tian, D.-A., Tuo, B., Shull, G., & Seidler, U. (2016). Defective small intestinal anion secretion, dipeptide absorption, and intestinal failure in suckling NBCe1-deficient mice. *Pfluegers Archiv European Journal of Physiology, 468*(8), 1419–1432. https://doi.org/10.1007/s00424-016-1836-3

Zange, J., Gronczewski, J., & Jans, A. W. (1993). NH_4^+ metabolism and the intracellular pH in isolated perfused rat liver. *Biochemical Journal, 293*(Pt 3), 667–673. https://doi.org/10.1042/bj2930667

Chapter 17

Transepithelial transport mechanisms

Dan R. Halm

Department of Neuroscience, Cell Biology, and Physiology, Wright State University Boonshoft School of Medicine, Dayton, OH, United States

1. Introduction—Physiological roles

Epithelia form as a sheet of connected cells separating two spaces, often the inside of the body from the external environment (Cereijido et al., 2004; Le Bivic, 2013; Palmer & Andersen, 2008; Rodriguez-Boulan & Macara, 2014; Shashikanth et al., 2017). This epithelial sealing of the body compartment liberates internal cells from the potentially disruptive environmental changes that may occur and permits maintenance of a constant internal condition, *milieu interior,* as discussed in Chapter 1. Naturally then, transepithelial flows constitute the entry and exit events for all substances an organism needs for continued metabolism and for the disposal of wastes. An abundance of epithelial structures evolved, such as glands and villi, as organisms adapted to environmental challenges using ever-more intricate metabolic pathways. Just as eukaryotic cells contain internal membrane compartments such as lysosomes to isolate particular metabolic activities, adaptations of larger organisms also include internal compartments, bounded by epithelial barriers. We immediately recognize examples of these internal epithelial barriers as the endothelium lining blood vessels (Jackson, 2017; Levick & Michel, 2010) and the mesothelia lining other body cavities, including the peritoneal space, pericardial space, and pleural spaces (Flessner, 2005; Ji & Nie, 2008; Markov & Amasheh, 2014). For each epithelium, transepithelial flows depend on cellular mechanisms employing a wide array of transport proteins embedded in cell membranes as well as the proteins connecting epithelial cells together into a continuous barrier. These transepithelial pathways govern flow, allowing epithelial cells to regulate the rate and duration of flow.

A key responsibility for governing transepithelial flow involves setting a direction, with valuable substances such as nutrients entering (transepithelial absorption) into and substances in excess of metabolic requirements exiting (transepithelial secretion) from the body. Since cell membrane components operate at nanometer dimensions where thermal motions of molecules dominate, setting a direction for flow must contend with chemical and electrical forces acting on molecules and inorganic ions. The transport protein nanomachines resident in cell membranes not only provide a route for flow but also harness electrical and chemical gradients thereby assigning a direction for flow that supports cellular homeostasis. Ultimately, cellular metabolism provides the energy necessary to maintain these advantageous flow directions in the face of continual dissipation of the established gradients (Williams & Dawson, 2006). The long-term success of body activities then depends on the intricate arrangement of membrane transport nanomachines and cellular structures implemented by epithelia to guide continued nutrient acquisition for metabolic activity.

2. Transepithelial flow pathways

Intrinsic to the barrier structure of any epithelial sheet of cells (Fig. 17.1A), flow occurs either between the cells (paracellular) or through the cells (transcellular). Close apposition of the epithelial cells creates a narrow restricted paracellular route maintained by adhesion proteins connecting the cells (Frömter & Diamond, 1972; Rodriguez-Boulan & Macara, 2014). Tight-junction proteins (Chapter 7) along the apical margin further restrict the paracellular pathway and provide the primary selectivity site for passage of molecules and ions (Anderson & Van Itallie, 2009; DiBona, 1985; Furuse, 2010; Günzel & Fromm, 2012; Günzel & Yu, 2013; Günzel, 2017; Lane et al., 1992; Lee et al., 2008; Lingaraju et al., 2015; Nagarajan et al., 2023; Pinto da Silva & Kachar, 1982; Tsukita et al., 2019; Zhao et al., 2018). Following passage through the tight-junction complex, flow along the paracellular route continues via the lateral space between the cell apex and base. Closure of this paracellular route by tight junctions accentuates the intrinsic polarity of the barrier structure, separating the

Cell Physiology Sourcebook. https://doi.org/10.1016/B978-0-12-811114-7.00040-2

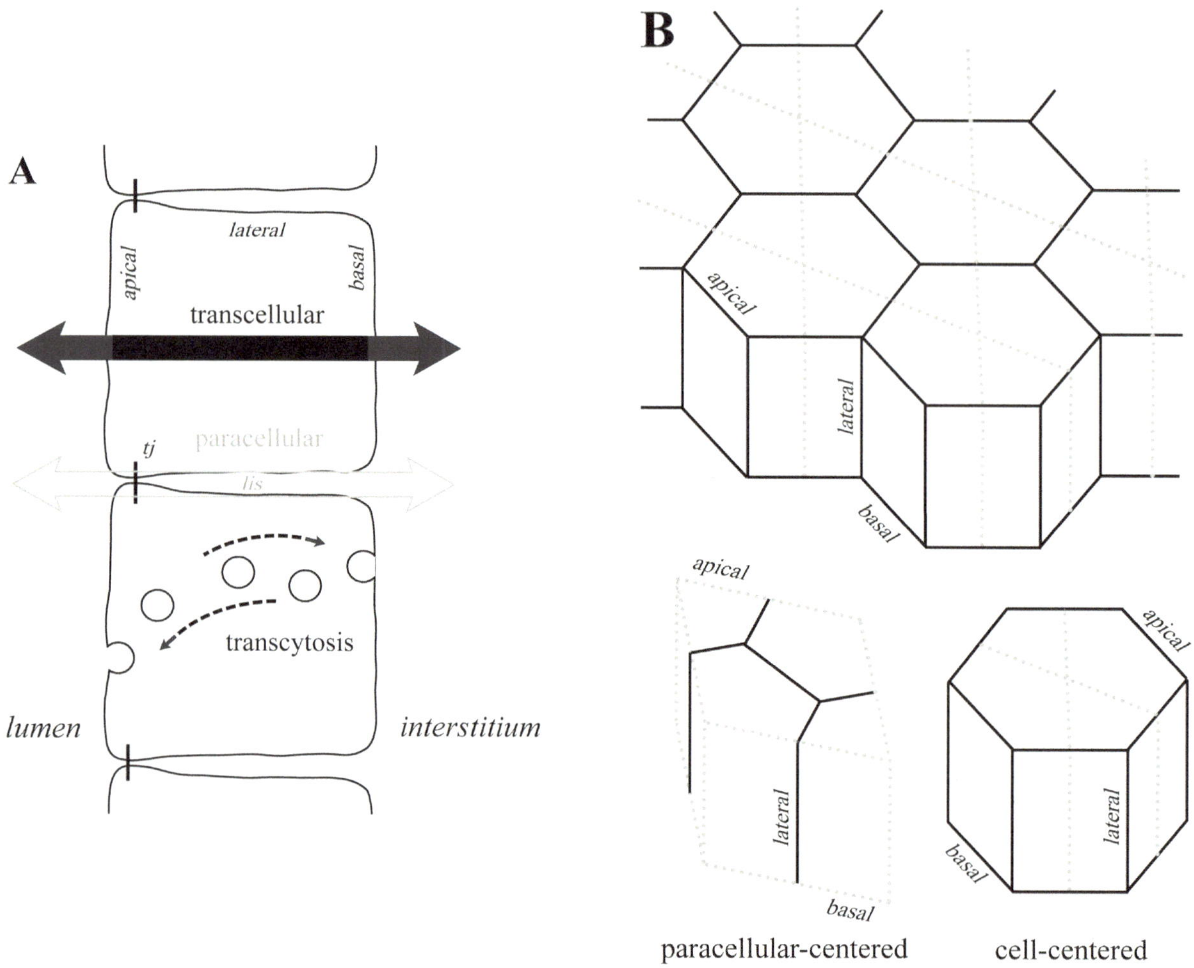

FIGURE 17.1 **Transepithelial routes for flow.** Movement across an epithelial sheet of cells occurs either through the cells (transcellular route), or between the cells (paracellular route.) (A) A transepithelial cross-section shows the transcellular route requiring sequential crossing of the apical membrane and then the basal or lateral membrane (basolateral), or in the reverse order. For small solutes, these events involve membrane proteins, including coupled transporters, channels, and pumps. Larger molecules transit through the cell by transcytosis, beginning with binding to the endocytosing membrane or engulfed in the fluid of the forming vesicle. Movement along cytoskeletal trackways carries vesicular cargo to the opposing membrane for exocytotic release. Paracellular movement involves passing through the tight-junctional proteins and along the lateral intercellular space (lis), or in the reverse order. (B) From a vantage point above an epithelium, the surface area for the sheet of cells consists of the combined apical extent of the connected cells, often arranged in a roughly hexagonal array (cell edges shown as solid lines). This apical surface area of each cell encompasses the locations of transcellular flow. Viewed from above, the lateral extent of the tight-junction along the cell edges (solid lines) encompasses the locations of paracellular flow. Fracturing the epithelial cell sheet exposes the lateral surface of the cells, showing the length of the paracellular route for flow. Each cell shares its tight junctions with an adjacent cell such that its apical surface area pairs with a paracellular route equivalent to half of its perimeter. An array of rhombuses (dotted lines) illustrates this unit contribution of the paracellular pathway, diagrammatically constructed by connecting cell centers (apical area extent equal to the cell apical extent), also shown as a paracellular-centered unit volume and cell-centered unit volume. Each effective paracellular unit includes half the value of the cell perimeter along with two tripartite vertices (a third of the six hexagon vertices).

epithelial cell surface facing the external environment (or body cavity) as apical, and the surface facing the interstitial space of the body as basal and lateral. For the transcellular route, the lipid bilayer of the cell membranes establishes a barrier to hydrophilic solutes, such that their passage occurs only if membrane embedded transport proteins permit flow (Chapters 4 and 10). The character and relative amount of flow through the paracellular and transcellular routes vary among epithelia and support the physiological role of each body compartment.

The paracellular and transcellular routes through an epithelium exemplify the nature of the size-selective barrier maintaining differences in composition of the fluids between the two sides (Bergemann, 2020; Caceres et al., 2017; Rodriguez-Boulan & Macara, 2014; Wei & Wang, 2021). The tight junctions of the paracellular route exclude passage of larger macromolecules as well as cells, such as the resident bacteria of the microbiota. Passage through the transcellular route includes two size classes of molecules, nanometer sized solutes and macromolecules. Small solutes proceed by flowing serially through nanometer-dimensioned tunnels within membrane transport proteins located in apical and basolateral membranes. Macromolecules achieve transcellular flow via endocytosis, either initially attached to the membrane or within the engulfed fluid, followed by movement of the newly formed vesicle along cytoskeletal trackways to the opposing cell membrane and subsequent release by vesicle exocytosis (Fig. 17.1A) Overall, an epithelium presents a barrier against intrusion by large objects while permitting an opportunity for high flow of selected small molecules and ions.

Histological nomenclature for epithelial cells provides useful and succinct descriptions of arrangements constituting epithelial cell layers, whether squamous, cuboidal, or columnar, simple or stratified (Lemke & Nelson, 2021; Rodriguez-Boulan & Macara, 2014). For a simple epithelial cell layer, a set of regular hexagons, squares, or triangles could cover a flat surface without gaps, but observed cell arrays exhibit a generally hexagonal cell shape with 3 cells meeting at the vertices of cell edges. Idealized epithelial cell building blocks then consist of geometrical prisms with hexagonal apical faces (Fig. 17.1B). Assembling an epithelium from these idealized cellular building blocks (Gibson & Gibson, 2009; Guillot & Lecuit, 2013; Higashi & Chiba, 2020) defines the lateral extent of the paracellular pathway as the adjacent sides of the cells, with tight junctional closures along the apical end of the lateral spaces between cells. Each cell shares its edges with an adjoining cell, so that a cell's effective complement of the epithelial paracellular route is one-half of the cell's perimeter. The apical extent of a cell (ignoring for the moment surface amplifications such as microvilli) defines how many of these cell units are required to cover a surface. Similarly, the unit of paracellular extent consists of a grid centered on the junctional vertices, with three cells meeting at the vertices in a hexagonal array. Rolling an epithelial sheet into a tube with a lumen at the center or into projections such as intestinal villi distorts these cell arrays (Andrew & Ewald, 2010; Lemke & Nelson, 2021), particularly as tight curvature restricts cell shape and volume.

Maintaining transepithelial flow requires sufficient surface area to house all of the necessary transport proteins, whether in the cell membranes or along the tight junctions. Expanding the number of epithelial cells satisfies this need by increasing both the cellular and paracellular routes in proportion. This increased number of epithelial cells, by necessity, requires an organism to change its body conformation to include the increased surface area. Selectively increasing the cellular route without increasing cell number occurs via amplifications of apical surface area seen as membrane ridges and projections such as microvilli. The microvilli of intestinal absorptive cells and renal proximal tubule cells warrant distinction as a brush-border to emphasize the high density of this surface amplification. Since tight junctions functionally isolate lateral cell membranes from the apical fluid, basal and lateral membranes together (basolateral) determine flow between the cell interior and the interstitial space. Basolateral membrane surface amplification needed to balance increases in apical flow often occurs as infolding of the basal membrane and interdigitating pleats of lateral membranes (Welling & Welling, 1976, 1988; Welling et al., 1978a, 1978b, 1981, 1983, 1987, 1996). Similarly, the extent of the paracellular pathway along the perimeter of cells can amplify by sinuosity of the cell border at the locus of the tight junction (Lynn et al., 2020). Together, these structural features combine to amplify the routes for transepithelial flow within an overall compact epithelial structure.

3. Transepithelial flow energetics

Transepithelial flow occurs through the cellular and paracellular routes driven by chemical and electrical energy gradients, as well as hydrostatic pressure differences for barriers such as capillary endothelium (Chapter 10). For ease of comparison between energy gradients, the Nernst potential formalism converts the transmembrane concentration differences representing the chemical potential of ions such as Na^+, K^+, and Cl^- into electrical units (mV), and the van't Hoff osmotic pressure relation ($\Delta\pi$) infers and converts the chemical potential of water into hydrostatic pressure units (mmHg). The total electrochemical driving force ($\Delta\mu_i/F$) for a solute (Eq. 17.1) combines the transmembrane chemical potential difference (for ions, the Nernst potential, E_i) with the transmembrane electrical potential difference (ψ_m). Similarly, the hydrostatic-osmotic driving force combines hydrostatic pressure differences (ΔP) with osmotic differences ($\Delta\pi$), $\Delta P - \Delta\pi$. For solutes without a net electrical charge, the chemical energy potential expressed in electrical units ($\Delta\mu/F = (RT/F)\ln(C_2/C_1)$) allows ready comparison with the total energy potential of electrically charged solutes. These driving forces (potentials) pair with permeability measures such as conductance or hydraulic conductivity to quantify solute flow or volume flow.

$$\Delta^{21}\mu_j = \left(\mu_{j2} - \mu_{j1}\right) = RT\ln\left(\frac{C_{j2}}{C_{j1}}\right) + z_jF(\psi_2 - \psi_1) \tag{17.1a}$$

$$\begin{aligned} \frac{\Delta^{21}\mu_j}{F} &= \frac{RT}{F}\ln\left(\frac{C_{j2}}{C_{j1}}\right) + z_j(\psi_2 - \psi_1) = z_j(\psi_m - E_j) \\ E_j &= -\frac{RT}{z_jF}\ln\left(\frac{C_{j2}}{C_{j1}}\right) \end{aligned} \qquad (17.1b)$$

Modeling epithelial flow pathways with an electrical equivalent circuit or hydraulic analog provides an aid for quantifying the contribution by each component, whether a membrane resident transport protein or a tight-junctional protein (Lewis et al., 1996; Schultz et al., 1981; Thompson, 1986). A typical starting point for a model involves a minimal set of components sufficient to account for experimental observations. For an epithelium, a minimal equivalent electrical circuit includes three conductors and associated batteries (electromotive forces, EMF) each representing either the apical membrane, basolateral membrane, or tight-junction barrier (Fig. 17.2). At each of these locations, the conductor comprises the parallel combination of all the transport proteins that allow passage of electrolytes and the battery represents the parallel combination of the Nernst potentials in proportion to the relative conductance for each electrogenic transporter.

The common distinction between tight and leaky epithelia becomes apparent from the equivalent circuit (Fig. 17.2), with an electrically tight epithelium exhibiting a conductance for the paracellular route (G_{pr}) much lower than the conductance for the transcellular route (G_{cr}), and an electrically leaky epithelium exhibiting high G_{pr} compared with G_{cr}. The total transepithelial conductance (G_{te}) simply follows then as the sum of these parallel routes for flow ($G_{te} = G_{pr} + G_{cr}$). (A frequently used alternative formulation employs transepithelial resistance, $R_{te} = R_{pr}R_{cr}/(R_{pr} + R_{cr}) = 1/G_{te}$, that had a historical advantage because of ease during equation derivation, but suffers a distinct disadvantage during physiological interpretation because R varies inversely with the concept of interest, flow.) Comparing electrically tight and leaky epithelia illustrates directly how the relative conductance of the transcellular and paracellular routes leads to dominance of the transepithelial electrical potential difference (ψ_{te}) by the EMF of the more conductive pathway (Eq. 17.2). The combined EMF of the paracellular pathway (E_{pr}) accounts for the ion concentration changes occurring on each side of the epithelium due to transepithelial flow of electrolytes and water. The transcellular route for flow includes two energetic steps, across the apical membrane (E_a) and basolateral membrane (E_b), driving flow in serial progression.

$$\begin{aligned} \psi_{te} &= \left(\frac{G_{cr}}{G_{cr} + G_{pr}}\right)(-E_a + E_b) + \left(\frac{G_{pr}}{G_{cr} + G_{pr}}\right)E_{pr} \\ G_{cr} &= G_a\left(1 + \frac{G_a}{G_b}\right)^{-1} \end{aligned} \qquad (17.2)$$

Condensing each transepithelial route into a single conductor and battery simplifies presentation but obscures details of flow needed to appreciate fully the transepithelial flow mechanisms. Expanding the cellular equivalent circuit to include all components for the apical and basolateral membranes allows analysis of electrochemical driving forces occurring at each membrane.

Transepithelial flow through the cellular route involves sequential passage across the apical and basolateral membranes, with flow from the luminal (external) space to the interstitial space producing transepithelial absorption, or in the reverse orientation producing transepithelial secretion. The electrical potential differences at the apical (ψ_a; Eq. (17.3a); Halm & Frizzell, 1991) and basolateral (ψ_b) membranes contribute to the driving forces for electrolytes during their passage through the epithelium. Two limiting cases illustrate how the combined network of epithelial conductance influences the cell membrane potential. For a very tight epithelium ($G_{pr} \approx 0$), ψ_a is dominated by features of the apical membrane

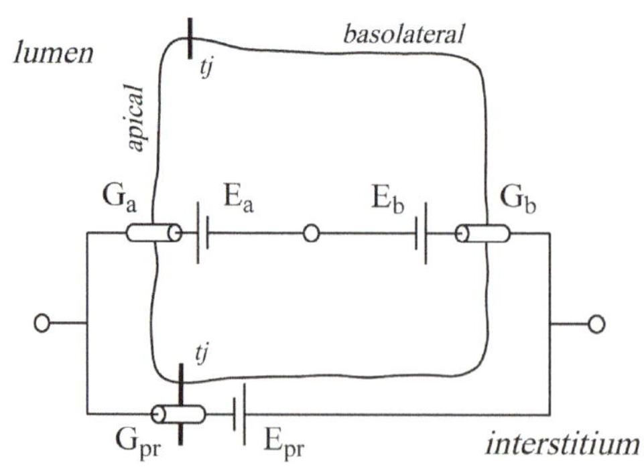

FIGURE 17.2 **Epithelial electrical equivalent circuit.** Conductors (*G*) in the apical (a) and basolateral (b) membranes as well as in the paracellular route (pr) provide pathways for ion flow. The apical and basolateral conductors include channels and other electrogenic transporters, with the conductor representing the paracellular route combining tight junction proteins (tj) and lateral space fluid conductivity. The distributed nature of the lateral membrane conductance along the lateral space (Clausen et al., 1979; Sackin & Palmer, 2013) is lumped together as part of this simplified G_{pr}. Batteries represent the *EMF's* (E_j) of the conductive ions, as in Eqs. (17.2) and (17.3).

(Eq. 17.3b). For a very leaky epithelium (high G_{pr}), where compositional differences between luminal and interstitial fluid become minimal ($E_{pr} \approx 0$), ψ_a becomes equally influenced by apical and basolateral features and thereby indistinguishable from ψ_b (Eq. 17.3c). Epithelia range between these extremes of electrical tightness so that as G_{pr} increases in relation to G_{cr}, basolateral membrane and paracellular pathway components have greater influence on ψ_a. Similarly, the relative influences of apical and paracellular components on ψ_b increase as G_{pr} increases.

$$\psi_a = \left(\frac{G_{pr}}{G_{cr} + G_{pr}}\right)\left\{\left(\frac{G_a}{G_a + G_b} + \frac{G_{cr}}{G_{pr}}\right)E_a + \left(\frac{G_b}{G_a + G_b}\right)(E_b - E_{pr})\right\} \tag{17.3a}$$

$$G_{pr} \ll G_{cr} \rightarrow \psi_a \approx E_a = \sum\left(\frac{G_j^a}{G_a}\right)E_j^a \tag{17.3b}$$

$$G_{pr} \gg G_{cr} \rightarrow \psi_a \approx \left(\frac{G_a}{G_a + G_b}\right)E_a + \left(\frac{G_b}{G_a + G_b}\right)E_b = \sum\left(\frac{G_j}{G_{cell}}\right)E_j \qquad G_{cell} = \sum G_j \tag{17.3c}$$

Looking at a simple case with the apical membrane comprised of a conductance that allows only a single ion type to flow; then, E_a equals the Nernst potential for that ion, E_j. Further, for a very tight epithelium (Eq. 17.3b), ψ_a equals E_j and the electrochemical driving force through that pathway goes to zero (Eq. 17.3b). With typical cell volume and membrane capacitance, only relatively few ions must move before attaining this electrochemical equilibrium, illustrating the general property of macroscopic electroneutrality in physiological solutions. Avoidance of this near instantaneous no flow (zero driving force) situation across the apical membrane only requires inclusion of additional apical conductive pathways with different ion selectivity so that ψ_a equals a weighted sum of the corresponding E_j's with each conductive pathway then exhibiting a nonzero driving force. In a paracellularly leaky epithelium, the EMF's of the basolateral membrane and paracellular route likely suffice to shift ψ_a away from E_a and create nonzero driving forces through apical pathways. In this way, the conductances contribute to determining electrical driving forces as well as establishing flow pathways. Each epithelium exhibits inventive strategies implemented to circumvent this electroneutrality stricture by fashioning a set of complementary flow pathways (apical, basolateral, and paracellular) that shift ψ_m away from the individual E_j values, thereby maintaining nonzero driving forces through each pathway.

Similarly, adjustments of circuit conductances permit flow through basolateral pathways to continue by setting up nonzero driving forces. The lateral space compartment between the epithelial cells creates an added issue at the basolateral membrane. For lateral intercellular spaces of narrow width, the resulting restricted flow allows the solute composition of fluid in contact with the lateral membrane to differ from the interstitial fluid in contact with the basal membrane. These steady-state standing gradients likely alter flow along the lateral spaces even if small in magnitude. Since the cell's lateral membrane conductance is distributed along this space between the cells, paracellular ion flow also alters the voltage profile along the space, resulting in differences between membrane driving forces at lateral and basal locations. A more detailed equivalent circuit, including this distributed conductance network, permits evaluation of flow through lateral and basal membranes (Clausen et al., 1979; Sackin & Palmer, 2013).

Energizing flow through these epithelial pathways involves harnessing cell metabolism to establish advantageous directions for the driving forces at each transport protein. The electrochemical gradients driving electrolyte flow consist of membrane potentials as well as the transmembrane concentration gradients (Eq. 17.1). Moreover, these cell membrane potentials, ψ_a and ψ_b, together with the transepithelial potential, ψ_{te}, result directly from those transmembrane ion concentration gradients, which correspondingly depend on the activity of cell membrane transport ATPases, such as the Na^+/K^+-pump. These ATPases create concentration gradients by coupling the favorable reaction of ATP hydrolysis into ADP with the ATPase nanomachine structure, which can bias solute flow toward the higher concentration side of the membrane. Thus, the differentiation program of epithelial cells expresses a set of membrane transport proteins that collaborate to establish steady states capable of managing cellular fluid and electrolyte balance while generating transepithelial flows that similarly support this balance on an organismal level.

4. Mechanisms for transepithelial transport

The transcellular component of transepithelial transport mechanisms fundamentally incorporates the cellular *Pump-Leak Concept* of Daniel Tosteson and Joseph Hoffman (Tosteson & Hoffman, 1960; Kay & Blaustein, 2019) in which the pumping of Na^+ and K^+ establishes an osmotic balance that maintains cell volume (Baumgarten & Feher, 2023a, 2023b; Larsen & Hoffman, 2020). The Na^+/K^+-ATPase constitutes the *Pump* (Kaplan, 2002; Larre et al., 2020) and the *Leaks* occur as flows of Na^+ and K^+ through membrane transporters driven by the requisite electrochemical gradients

(Eq. 17.1b). The resulting low intracellular Na^+ concentration, $[Na^+]_{cell}$, and high intracellular K^+ concentration, $[K^+]_{cell}$, generated by this Na^+/K^+-pump leads to E_{Na} near +70 mV and E_K near −90 mV (cell inside with respect to outside), using typical extracellular and intracellular concentrations. High relative G_K as observed for many cells leads to a ψ_m near E_K (Eq. 17.3c) contributing to anion exit from the cell, especially Cl^- ($\Delta\mu_{Cl}/F = z_{Cl}\ (\psi_m - E_{Cl}) > 0$). By providing a driving force for exit of the major extracellular anion, Cl^-, a cellular osmotic balance occurs with low $[Cl^-]_{cell}$ along with the intracellular anion concentrations contributed by impermeable anions such as metabolic intermediates and DNA (Chapter 16). In this way, the *Pump–Leak* mechanism generates a cell negative ψ_m, often referred to as the resting membrane potential (Chapter 12), which averts a perpetual osmotic crisis, but only at the expense of continual metabolic energy use (depletion of the phosphate transfer potential for ATP) by the Na^+/K^+-pump. Together these Na^+/K^+-pumps and the key *Leak* pathway established by K^+ channels serve to drive Na^+ and Cl^- from the cell.

The *Pump* as the central feature of cellular homeostasis provides the defining core for the concept of *active* transport. This realization that transport ATPases transduce the phosphate transfer potential for ATP ($\Delta G_{ATP} = \Delta G^o{}_{ATP} + RT \ln([ADP][P_i]/[ATP])$) into ionic electrochemical gradients illustrates how cells employ a wide array of transport proteins (*Leaks*) to couple solute flows to these *Pump* maintained gradients.

Transepithelial transport mechanisms deploy their *Pump–Leak* components by segregating some transport proteins to the apical membrane and others to the basolateral membrane (Weisz, 2020), such that vectorial flow occurs in either an absorptive or a secretory direction. The *Pump–Leak* mechanism also allows cells to increase or decrease cell volume by increasing or decreasing the amount of intercellular osmoles (Chapters 11 and 23). This cellular capacity for controlling cell fluid volume by adjusting intracellular osmoles permits epithelia to guide transepithelial fluid flows without unduly perturbing their epithelial cell volume. Cellular competence in avoiding osmotic crises also confers the operational flexibility to change cell volume for optimizing cell shape during transepithelial flow. In particular, adjustments in cell volume during transepithelial flow can adjust the width of the lateral intercellular spaces with impacts on paracellular water and solute flow as well as electrical coupling between apical and basolateral membranes.

Numerous signaling cascades act to coordinate the conductance/permeability of these apical, basolateral, tight junctional, and lateral space components (Hofer, 2023; Weisz, 2020). Regulation of transcellular solute flow interacts with needs for cell volume stabilization, which influences paracellular flow. This cross-talk between the apical and basolateral membrane components assures that conductances at each membrane match the requirements needed to maintain a pathway as well as a suitable driving force. Most of the specific signaling cascades have yet to be elucidated fully and many likely wait to be identified. Simulations using models for the transepithelial mechanisms (Nickerson et al., 2020) provide a means to test how the epithelial components interact and guide assessment of the influence by specific signaling connections.

Electrogenic Na^+ absorption. By elucidating a transcellular model for Na^+ absorption through frog skin, Valborg Koefoed-Johnsen and Hans Ussing introduced the archetypical cellular mechanism for generating transepithelial solute flow (Koefoed-Johnsen & Ussing, 1958; Palmer & Anderson, 2008; Ussing & Windhager, 1964). Three transport components suffice to account for Na^+ absorption, an apical membrane Na^+ conductance along with a basolateral membrane Na^+/K^+-pump, and a K^+ conductance (Fig. 17.3A) (see also Chapter 16). A ψ_{te} with the external (luminal) solution electrically negative compared with the interstitial solution results from the net transcellular cation flow, all without perturbing the net charge balance of the cell interior. Epithelial cell volume also remains stable due to balance of net osmole entry and exit. The driving forces for Na^+ entry across the apical membrane ($\Delta\mu_{Na}/F = z_{Na}\ (\psi_a - E^a{}_{Na}) < 0$) and K^+ exit across the basolateral membrane ($\Delta\mu_K/F = z_K\ (\psi_b - E^b{}_K) > 0$) depend on the concentration gradients maintained by Na^+/K^+-pumps ($E^a{}_{Na}$, $E^b{}_K$) along with the electrical potential profile across the epithelium (ψ_a, ψ_b), which in turn depends on the electrical circuit combination of these concentration gradients (Eq. 17.3). A minimal-sufficient mechanism for sustained Na^+ absorption also requires an electrical shunt pathway (Ussing & Windhager, 1964). Paracellular conductance provides this electrical connectivity such that apical Na^+ entry and basolateral K^+ exit continue because of the resulting nonzero driving forces. A small apical Na^+ driving force occurs for tight epithelia (low G_{pr}) since ψ_a remains near $E^a{}_{Na}$. In leakier epithelia (higher G_{pr}), a larger apical Na^+ driving force results as the influence of $E^b{}_K$ and E_{pr} moves ψ_a further away from $E^a{}_{Na}$ (Fig. 17.3B). This cellular mechanism generates steady-state transcellular Na^+ absorption, proceeding with a stable ψ_{te} as current flows through the circuit, inward via the cellular route and outward via the paracellular route.

Physiological consequences of an epithelium implementing the electrogenic Na^+ absorption mechanism depend on the circumstances surrounding that epithelial location. Epithelial domains where electrogenic Na^+ absorption occurs include renal distal tubules and collecting ducts, sweat gland ducts, and distal colon. For situations where extracellular body fluid bathes the epithelium on both sides, paracellular Cl^- absorption likely proceeds since $[Cl^-]_e$ is high, similar to $[Na^+]_e$, resulting in net Na^+ and Cl^- absorption via the combined transepithelial pathways. The driving force for paracellular Cl^- absorption ($\Delta\mu_{Cl}/F = z_{Cl}\ (\psi_{te} - E^p{}_{Cl}) > 0$) depends on the ψ_{te} generated by the transcellular Na^+ absorption mechanism,

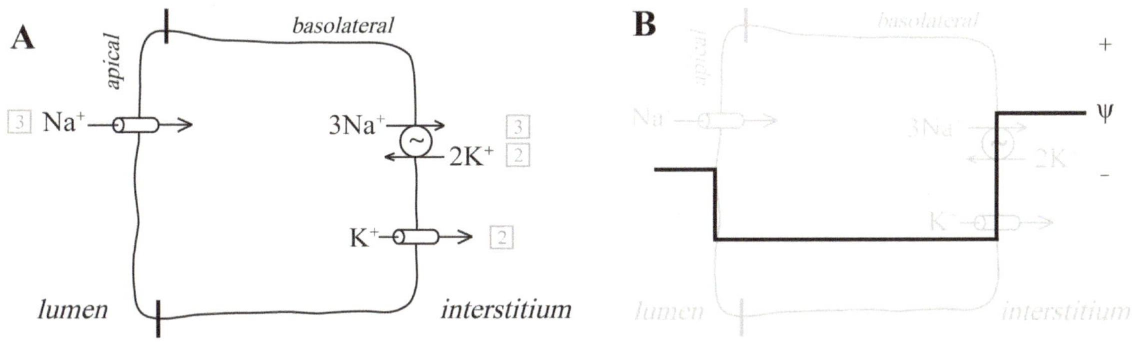

FIGURE 17.3 **Electrogenic Na^+ absorption.** (A) The Koefoed-Johnsen and Ussing model for transepithelial Na^+ absorption (Koefoed-Johnsen & Ussing, 1958) includes apical membrane Na^+ channels together with Na^+/K^+-pumps and K^+ channels in the basolateral membrane that constitute a minimal set of transporters required for transcellular absorption. During steady-state transepithelial flow, a single turnover of the Na^+/K^+-pump produces relative flows via each transporter (gray boxes), such that expenditure of one ATP results in 3 Na^+ absorbed cellularly. (B) The E_j's of the conductive ions determine the electrical potential profile across this epithelium (ψ, referenced to the interstitial space). A well-type profile occurs (solid line) with G_{pr} electrically coupling apical and basolateral membranes (Eq. 17.3), such that E_K^b hyperpolarizes ψ_a away from E_{Na}^a as well as E_{Na}^a depolarizing ψ_b away from E_K^b. The epithelial Na^+ channel protein (ENaC, *SCNN1*) comprises G^a_{Na} (Kleyman et al., 2018; Noreng et al., 2018; Song et al., 2020). The Na^+/K^+-pump constitutes the ATPase (*ATP1A*, EC3.6.3.9) generating ion gradients (Kaplan, 2002). A variety of K^+ channel types comprise G^b_K in renal principal cells (Hamilton & Devor, 2012; Wang & Lin, 2020). Resources for coupled transporters and channels, IUPHAR/BPS Guide: https://www.guidetopharmacology.org/GRAC/ReceptorFamiliesForward?type=TRANSPORTER. https://www.guidetopharmacology.org/GRAC/ReceptorFamiliesForward?type=IC.

limited by any drop in luminal Cl^- concentration due to absorption that increases the size of the opposing transepithelial concentration gradient. Paracellular Na^+ secretion also would occur, driven by ψ_{te} and any drop in luminal Na^+ concentration ($\Delta\mu_{Na}/F = z_{Na}\ (\psi_{te} - E^p_{Na}) < 0$), blunting the amount of net Na^+Cl^- absorption. Tight-junction proteins, claudins, govern both the conductance (G_{pr}) and ion selectivity, thereby constraining ion flow via the paracellular route (Günzel, 2017; Gong & Hou, 2017; Krug et al., 2014). Claudin protein isoforms (*CLDN*) range in ion selectivity (Amasheh et al., 2002; Günzel et al., 2009; Hou et al., 2010; Krug et al., 2012; Milatz et al., 2017; Tamura et al., 2011; Van Itallie et al., 2003) from cation preferring (*CLDN2, CLDN10b, CLDN15*) to anion preferring (*CLDN4, CLDN10a, CLDN11, CLDN17*). Both the renal collecting duct and distal colon surface epithelium express *CLDN4* (Günzel & Yu, 2013; Holmes et al., 2006; Rahner et al., 2001; Yu, 2015), supporting the concept of pairing a Cl^- selective paracellular pathway with transcellular electrogenic Na^+ absorption. Limited Na^+ backflow through the paracellular route in these epithelia promotes more efficient coupling of net Na^+Cl^- absorption.

The paracellular route not only provides a transepithelial flow pathway but also contributes to maintaining electrical driving forces through adjacent transcellular routes. An alternative strategy to establish a transepithelial route for Cl^- absorption adopted by frog skin and urinary bladders of some species uses mitochondrial-rich cells with Cl^- selective conductance in both the apical and basolateral membranes (Larsen, 2011). Other epithelia include similar minority cell types (Fan et al., 2012; Jakab et al., 2013; Montoro et al., 2018; Plasschaert et al., 2018). The mitochondrial-rich cell maintains its transcellular potential profile such that Cl^- enters conductively across the apical membrane ($\Delta\mu_{Cl}/F = z_{Cl}\ (\psi_a - E^a_{Cl}) < 0$) and exits conductively across the basolateral membrane ($\Delta\mu_{Cl}/F = z_{Cl}\ (\psi_b - E^b_{Cl}) > 0$). Electrically, these cellular Cl^- conductances in series ($G^a_{Cl}G^b_{Cl}/(G^a_{Cl} + G^b_{Cl})$) act like G_{pr} (Eq. 17.3) to shift ψ_a and ψ_b of the neighboring Na^+ absorbing cells such that nonzero driving forces persist for their apical Na^+ entry and basolateral K^+ exit.

Continued Na^+Cl^- absorption via transcellular electrogenic Na^+ absorption together with Cl^- absorption through parallel shunt pathways, either paracellular or transcellular, leads to decreased Na^+ and Cl^- concentrations in the luminal (external) solution. The resulting osmotic difference due to increased $[H_2O]_{lumen}$ drives H_2O flow down its concentration gradient such that H_2O absorption accompanies the Na^+Cl^- absorption. Even relatively small changes in Na^+ and Cl^- concentrations produce a substantial driving force, with a 1 mM transepithelial difference equivalent to 19 mmHg ($\Delta\pi = RT\Delta[S]$). Epithelia accentuate transcellular H_2O flow by including H_2O channels, aquaporins (*AQP*; see Chapter 18, by Nejsum et al), in the apical and basolateral membranes. In the case of principal cells found within the renal distal tubule and collecting duct, which use the electrogenic Na^+ absorption mechanism, antidiuretic hormone regulates insertion of *AQP2* into the apical membrane, such that varied rates of H_2O absorption occur relative to Na^+ absorption (Kortenoeven et al., 2020).

Epithelial transport mechanisms act by transducing metabolic energy into flows of solute and H_2O. Mechanistic insights into these physiological events require simulations that incorporate epithelial structure and electrochemical energetics of solute flow, generally accomplished with computer models. Balancing the influx and efflux of solutes within a model serves as a starting point for designing simulation strategies and provides important insights. The three-component mechanism for electrogenic Na^+ absorption (Fig. 17.3) reveals its steady-state balance by beginning with the 3:2 exchange stoichiometry of the Na^+/K^+-pump. With every turnover of the pump 3 Na^+ exit, so that steady-state Na^+ balance occurs with entry of 3 Na^+ through the apical Na^+ conductance. Balance for K^+ results from 2 K^+ exiting through the basolateral K^+ conductance, matching the 2 K^+ entering via the Na^+/K^+-pump. These steady-state balance numbers for Na^+ and K^+ represent the relative current flows through the circuit components. Instead of viewing the closed loop of K^+ flow across the basolateral membrane as a futile cycle, rather this flow underpins the electrical coupling between E_K and the driving force for apical Na^+ entry by maintaining the cell interior negative with respect to the exterior (Eq. 17.3). Experimental measurement of $[Na^+]_{cell}$ and $[K^+]_{cell}$ as well as ψ_a and ψ_b establish how the ATP used by the Na^+/K^+-pump transforms into the electrochemical gradients. Simulation using these measured values provides estimates for G^a_{Na} and G^b_K consistent with the equivalent electrical circuit of these absorbing cells (Larsen & Møbjerg, 2006).

Electrogenic glucose absorption. Essentially revising the cellular model for electrogenic Na^+ absorption, Robert Crane's concept for coupling glucose and Na^+ flow extends the *Pump–Leak* mechanism to encompass transepithelial glucose absorption (Crane et al., 1961; Crane, 1965). Instead of an apical Na^+ conductance, Na^+ enters stoichiometrically with glucose via Na^+-glucose-cotransporters, primarily in the small intestine and renal proximal tubule (Loo & Wright, 2020; Wright et al., 2011). The nanomachine structure within these cotransport proteins couples the movement of Na^+ and glucose such that the inwardly directed Na^+ electrochemical gradient drives glucose uptake across the apical membrane (Fig. 17.4). Glucose exit across the basolateral membrane occurs via glucose transporters driven by the glucose concentration gradient alone. This net transepithelial glucose absorption is emblematic of secondary active transport, a characteristic of any cellular transport mechanism capable of creating solute gradients by linking the flows of a transport ATPase, such as the Na^+/K^+-pump, with flow through a distinct transport protein, such as a Na^+-glucose-cotransporter.

Adding glucose to the transport mechanism illustrates how the flow of each solute contributes to cellular balance. The Na^+-glucose-cotransporter operating in the apical membrane of small intestinal epithelial cells couples the movement of Na^+ and glucose with a stoichiometry of 2 Na^+ for each glucose (*SGLT1*). Steady-state balance occurs with influx equaling efflux for each solute (Fig. 17.4), so that two turnovers of the basolateral membrane Na^+/K^+-pump extrudes 6 Na^+ necessitating apical entry of 6 Na^+ via the cotransporter, which through coupling also brings three glucose into the cell (3 turnovers of the cotransporter yields integer values for relative flow). Basolateral exit of three glucose molecules completes transepithelial glucose absorption. As with electrogenic Na^+ absorption (Fig. 17.3), basolateral membrane cycling of K^+ via Na^+/K^+-pump uptake and K^+ channel exit aids in maintaining a cell negative ψ_a that drives apical Na^+ entry. The nonelectrolyte glucose also influences ψ_a by providing the cosubstrate for the influx of Na^+ via the cotransporter. This simple accounting of influx and efflux results in the relative flows necessary to satisfy osmotic and electrical balance, but cellular regulatory schemes must set the precise number of transporters needed to complement the driving forces that, together, achieve these relative flows.

Transepithelial glucose absorption depletes glucose (osmoles) from the luminal (external) fluid and deposits it in the interstitial space. The cellular context for this flow rests on whole body metabolic balance that maintains interstitial glucose concentration, $[glucose]_{ECF}$, in a narrow range. The glucose concentration gradient driving glucose exit across the basolateral membrane then depends primarily on the rate of apical glucose entry that increases $[glucose]_{cell}$. Maintaining a

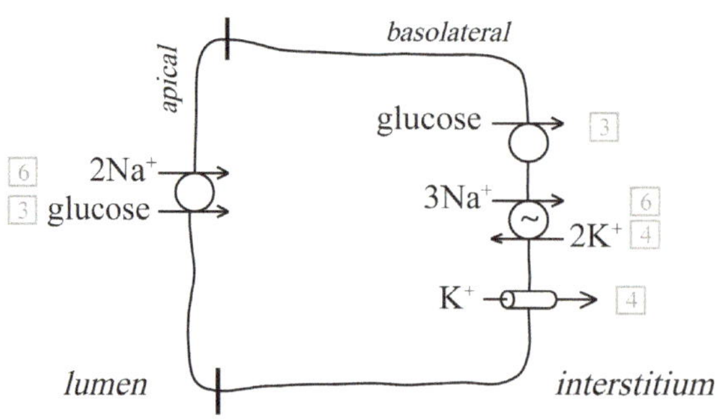

FIGURE 17.4 **Electrogenic glucose absorption.** The cellular model for intestinal glucose absorption includes apical membrane Na^+-glucose-cotransporters together with basolateral membrane Na^+/K^+-pumps, K^+ channels, and monosaccharide transporters, as a minimal set sufficient to support absorption. During steady-state transepithelial flow, two turnovers of the Na^+/K^+-pump produce relative flows via each transporter (gray boxes), such that expenditure of two ATP results in three glucose molecules absorbed cellularly together with 6 Na^+ absorbed. Na^+glucose-cotransporter proteins (SGLT1, *SLC5A1*) constitute the uptake route, and monosaccharide transporters (GLUT5, *SLC2A*) comprise the exit route (Loo & Wright, 2020).

modest elevation of $[\text{glucose}]_{\text{cell}}$ minimizes cell volume disturbance, but necessitates a large number of basolateral glucose transporters to sustain basolateral flow equal to apical glucose uptake. In addition, a modest $[\text{glucose}]_{\text{cell}}$ allows apical Na^+/glucose-cotransporters to deplete luminal glucose dramatically, until the glucose concentration gradient favoring glucose exit equals the Na^+ electrochemical gradient favoring entry (the reversal point of the Na^+/glucose-cotransporter, $\Delta\mu_{SGLT}/F = 0$). Typical values of E_{Na} and ψ_a allow a ~5000-fold lower $[\text{glucose}]_{\text{lumen}}$ compared with $[\text{glucose}]_{\text{cell}}$. Concurrently, the absorption of glucose generates a transepithelial osmotic gradient that drives H_2O absorption. The simple coupling between glucose and Na^+ entry envisioned by Crane allows the ATP expended by the Na^+/K^+-pump to assure apical membrane glucose uptake via the resulting $[Na^+]$ gradient augmented by the electrical consequence of the $[K^+]$ gradient, leading to the transepithelial osmotic gradient driving the accompanying H_2O absorption.

Electrogenic Cl^- secretion. The cellular mechanism for transepithelial Cl^- secretion illustrates another secondarily active system relying on the *Pump–Leak Concept* (Fig. 17.5), with the Na^+/K^+-pump in the basolateral membrane (DiBona & Mills, 1979; Ernst & Mills, 1977; Frizzell & Hanrahan, 2012). Four transport components account for Cl^- secretion, including an apical membrane Cl^- conductance along with a basolateral membrane transporter coupling the flow of Cl^-, Na^+, and K^+ (Degnan et al., 1977; Field, 1993; Frizzell et al., 1979; Heintze et al., 1983; Karnaky et al., 1977; Shorofsky et al., 1983, 1984; Silva et al., 1977; Smith & Frizzell, 1984; Welsh et al., 1982; Welsh, Smith et al., 1983). The *Leaks* in this Cl^- secretory mechanism consist of transport components generally used for regulatory volume adjustments, the Na^+-K^+-$2Cl^-$-cotransporter (NKCC) favoring volume increase together with Cl^- channels and K^+ channels favoring volume decrease. Positioning the Cl^- channels and K^+ channels in opposing membranes results in the observed lumen negative ψ_{te}, such that electrical circuit connectivity via the paracellular pathway (Eq. 17.3) assures the nonzero driving forces for apical membrane Cl^- exit ($\Delta\mu_{Cl}/F = z_{Cl}\ (\psi_a - E^a{}_{Cl}) > 0$) and basolateral membrane K^+ exit ($\Delta\mu_K/F = z_K\ (\psi_b - E^b{}_K) > 0$). Compensatory solute entry that maintains cell volume occurs via the electroneutral Na^+-K^+-$2Cl^-$-cotransporter with a driving force ($(\Delta\mu_{NKCC}/F) = -E^b{}_{Na} - E^b{}_K + 2E^b{}_{Cl} < 0$), establishing basolateral entry (see Chapter 16). During steady-state secretion, cellular signaling cascades by necessity coordinate these channel conductances with cotransporter turnover to balance influx with efflux to achieve stable cell volume.

Electrogenic Cl^- secretion occurs in acinar epithelia of lacrimal and pancreatic glands, secretory coils of sweat glands, as well as intestinal crypts of Lieberkühn, tracheal epithelium, and numerous other epithelial locations. The lumen negative ψ_{te} establishes an electrical driving force for paracellular ion flow with the claudins in the tight junctions determining the ion selectivity. Colonic crypt epithelial cells express *CLDN2*, *CLDN15*, and likely *CLDN10b* (Fujita et al., 2006; Gumber et al., 2014; Holmes et al., 2006; Rahner et al., 2001; Tamura et al., 2011), supporting the pairing of cation selective claudins with transcellular Cl^- secretion, which limits backflow of Cl^- via the paracellular route driven by the lumen negative ψ_{te}. Thus, primarily Na^+ secretion occurs via the paracellular route driven by a combination of the ψ_{te} generated by transcellular Cl^- secretion together with the opposing transepithelial $[Na^+]$ gradient ($\Delta\mu_{Na}/F = z_{Na}\ (\psi_{te} - E^P{}_{Na}) < 0$). The resulting transepithelial Na^+Cl^- secretion deposits solutes in the lumen creating an osmotic gradient driving H_2O flow into the lumen with a secreted fluid nearly isoosmolar with extracellular fluid.

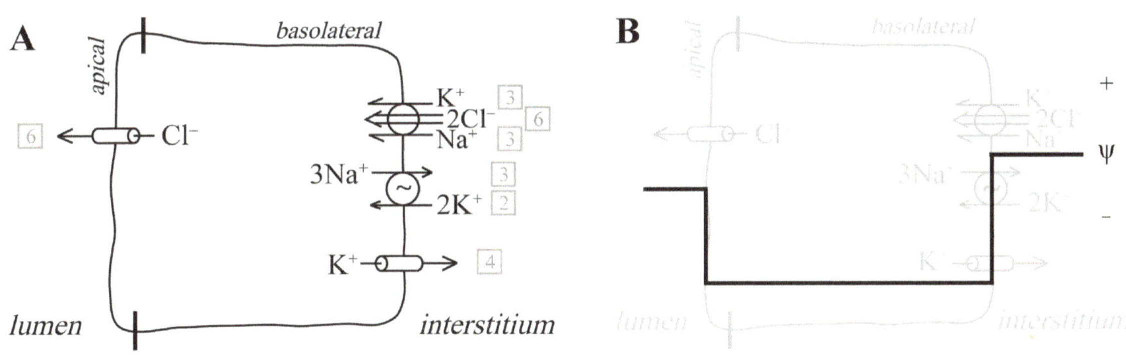

FIGURE 17.5 Electrogenic Cl^- secretion. (A) The cellular model for Cl^- secretion includes apical membrane Cl^- channels together with Na^+/K^+-pumps, Na^+-K^+-$2Cl^-$cotransporters, and K^+ channels in the basolateral membrane, as a minimal set. During steady-state transepithelial flow, a single turnover of the Na^+/K^+-pump produces relative flows via each transporter (gray boxes), such that expenditure of one ATP results in $6Cl^-$ secreted cellularly. (B) The E_j's of the conductive ions determine the electrical potential profile across this epithelium (ψ, referenced to the interstitial space). A well-type profile occurs (solid line) with G_{pr} electrically coupling apical and basolateral membranes (Eq. 17.3), such that E_K^b hyperpolarizes ψ_a away from E_{Cl}^a as well as E_{Cl}^a depolarizing ψ_b away from E_K^b. The CFTR Cl^- channel protein (*ABCC7*) constitutes $G^a{}_{Cl}$ (Al Salmani et al., 2020; Huang et al., 2012; Woodward & Guggino, 2013) with contributions from TMEM16A (*ANO1*) and other Cl^- channels (Baranovski et al., 2020; Scudieri & Galietta, 2020). Na^+-K^+-$2Cl^-$cotransporter 1 (NKCC1, *SLC12A2*) constitute the Cl^- uptake route (Delpire & Gagnon, 2018), as discussed in Chapter 16. $K_V7.1$ channels (*KCNQ1*) comprise $G^a{}_K$ with contributions from $K_{2P}5.1$ (KCNK5) and other types (Julio-Kalajzić et al., 2018).

The Cl^- secretory mechanism illustrates how segregating groups of transport proteins exclusively to either the apical or basolateral membrane harnesses the processes of cell volume control (Larsen & Hoffmann, 2020) to produce transepithelial flow of ions and H_2O. The steady-state balance of 3 Na^+ exiting via the Na^+/K^+-pump and 3 Na^+ entering via turnover of the Na^+-K^+-$2Cl^-$cotransporter drives the entry of 6 Cl^-, due to the coupling stoichiometry (Fig. 17.5). Exit of 6 Cl^- through apical Cl^- channels completes the steady-state Cl^- balance, and together with the 5 K^+ exiting via basolateral K^+ channels (also including one of the Na^+ extruded by the Na^+/K^+-pump) maintains cell volume balance during ongoing secretion. Use of the Na^+-K^+-$2Cl^-$cotransporter provides energetic efficiency for this secretory mechanism, since hydrolysis of a single ATP by the Na^+/K^+-pump secretes 6 Cl^- transcellularly. In addition, the coupling of K^+ includes an outwardly directed component to the driving force for the Na^+-K^+-$2Cl^-$cotransporter that limits excessive Na^+ entry, which could disturb cell volume control.

5. Linked cellular transport mechanisms

The three cellular transport models described above, two absorptive and one secretory, encompass the basic features of transepithelial flow, but many epithelia incorporate multiple transepithelial transport mechanisms within a single epithelial cell type. Generally, these transport mechanisms use the same core *Pump—Leak* process such that the additions operate as modules interlinked with this core energy expenditure. Inclusion of each additional mechanism involves another transport protein in the apical membrane allowing flow of a distinct solute into or out of the external environment (or luminal compartment). In some cases, this additional component augments the driving force for flow through the other apical transport proteins. In all cases, the basolateral membrane houses a set of transporters that provide the passageways between the interstitial space and the cell interior, which also contribute to establishing driving forces for flow that sustain all the transepithelial mechanisms. The increased intricacy of the combined transport mechanisms requires a more detailed simulation model to assess the range of possible physiological outcomes. The modular character of the multiple mechanisms allows cellular regulation to change the relative rates to satisfy cellular, organ, and body requirements.

Small intestinal absorption. The secondarily active cellular mechanism for glucose absorption (Fig. 17.4) allows the modular addition of other Na^+-coupled nutrient cotransporters in the apical membrane to assist in the absorption of amino acids (AA) released by protein digestion in the small intestinal lumen (Camargo et al., 2020). Together with AA-transporters in the apical and basolateral membranes, the core *Pump—Leak* mechanism accommodates AA absorption simultaneously with glucose absorption, with the only requirement that sufficient Na^+/K^+-pump turnover occurs to maintain basolateral Na^+ exit and the electrochemical driving force for apical Na^+ entry. A large number of AA-transporters operate in the apical membrane, with some Na^+-coupled AA-cotransporters and others H^+-coupled cotransporters (Camargo et al., 2020). For the H^+-AA-cotransporters in the apical membrane, the presence of apical Na^+/H^+-exchangers indirectly couples AA uptake to the $[Na^+]$ gradient generated by the Na^+/K^+-pump (Fig. 17.6A). Extrusion of H^+ by the Na^+/H^+-exchangers creates an acidic microclimate in the apical surface fluid, providing H^+ for cotransport and contributing to a driving force for uptake. Similarly, H^+-coupled cotransport of dipeptides and tripeptides across the apical membrane links to the $[H^+]$ gradient generated by Na^+/H^+-exchangers (Spanier & Rohm, 2018). After apical uptake, peptidases cleave many of these peptides into AA that then exit across the basolateral membrane. The distinct set of AA-transporters in the basolateral membrane drives the ultimate AA exit from the cell using the concentration gradients of the accumulated AA types.

At more distal locations along the small intestine (ileum), nutrient absorption occurs along with further fluid absorption driven by Na^+ and Cl^- absorption (Rao, 2019). The apical Na^+/H^+-exchangers contribute to an acidic microclimate in the exoepithelial space at the apical membrane surface, also joined by apical Cl^-/HCO_3^--exchangers (Fig. 17.6B). The microclimate pH links the driving forces of these two transporter types ($\Delta\mu_{NHE}/F = -E^a_{Na} + E^a_H$ and $\Delta\mu_{CBE}/F = E^a_{Cl} - E^a_{HCO3}$) stabilizing Na^+ and Cl^- entry across the apical membrane. Carbonic anhydrase (CA) tethered to the luminal surface of the apical membrane speeds buffering of HCO_3^- and H^+ into CO_2 and H_2O, thereby generating concentration gradients driving CO_2 and H_2O entry. Water flows inward via AQP, and CO_2 flows either through the lipid bilayer or via CO_2 permeable membrane proteins (Michenkova et al., 2021; Missner & Pohl, 2009). Intracellular depletion of HCO_3^- and H^+ by efflux via the apical exchangers drives conversion of entering CO_2 and H_2O into HCO_3^- and H^+ by intracellular CA. The combined operation of these two apical exchangers increases $[Na^+]_{cell}$ and $[Cl^-]_{cell,}$ driving H_2O entry via AQP. Balancing this drive for volume increase, basolateral K^+ channels and Cl^- channels produce solute exit driving H_2O out across the basolateral membrane and completing fluid absorption. Together with the high $[K^+]_{cell}$ established by the Na^+/K^+-pump the presence of both channel types assures that the driving forces for K^+ exit ($\Delta\mu_K/F = z_K (\psi_b - E^b_K) > 0$) and Cl^- exit ($\Delta\mu_{Cl}/F = z_{Cl} (\psi_b - E^b_{Cl}) > 0$) continue without a mandatory requirement for adjusting G_{pr} (Eq. 17.3). Although the apical uptake mechanisms proceed electroneutrally, the low $[Cl^-]_{cell}$ established by

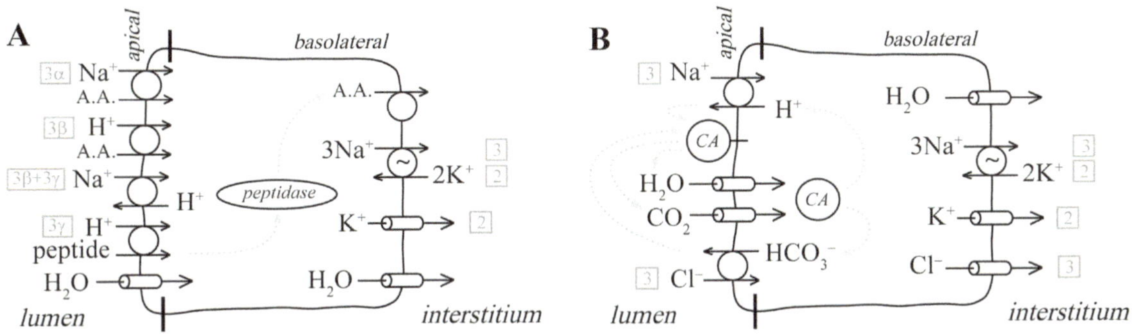

FIGURE 17.6 **Intestinal solute absorption.** (A) Cellular models for amino acid (AA) absorption in the intestine include apical membrane Na^+-coupled-AA-cotransporters, H^+-coupled-AA-cotransporters, H^+-coupled peptide cotransporters, and Na^+/H^+-exchangers together with basolateral membrane Na^+/K^+-pumps, K^+ channels, and AA-transporters. Amino acid transporters not coupled to Na^+ or H^+ also reside in the apical membrane (not shown for simplicity) and contribute to absorption. Intracellular peptidases cleave dipeptides and tripeptides, taken up across the apical membrane, into amino acids. During steady-state transepithelial flow, a single turnover of the Na^+/K^+-pump produces relative flows via each transporter (gray boxes), such that expenditure of one ATP results in 3 Na^+ absorbed cellularly together with a proportional number of amino acids (3–9). The proportions for the AA and peptide transporters turnover (α, β, γ) sum to provide the total activity ($\alpha + \beta + \gamma = 1$). Apical and basolateral membrane aquaporins support H_2O absorption driven by the osmotic gradient developed by solute absorption. Amino acid transporters contribute to the uptake and exit routes (Camargo et al., 2020). Na^+/H^+-exchangers (NHE3, *SLC9A3*) contribute to driving uptake (Kiela et al., 2020). (B) Cellular models for Na^+Cl^- absorption in the intestine interconnect modularly with the amino acid transport processes by using apical membrane Na^+/H^+-exchangers, along with apical membrane Cl^-/HCO_3^--exchangers together with basolateral membrane Cl^- channels. The Cl^-/HCO_3^--exchanger shown exhibits a one-to-one stoichiometry for Cl^- exchange with HCO_3^-. Apical membrane tethered carbonic anhydrase (CA) speeds conversion of HCO_3^- to CO_2 while intracellular CA speeds conversion of CO_2 to HCO_3^-. During steady-state transepithelial flow, a single turnover of the Na^+/K^+-pump produces relative flows via each transporter (gray boxes), such that expenditure of one ATP results in 3 Na^+ and 3 Cl^- absorbed cellularly, together with osmotically driven H_2O absorption. Cl^-/HCO_3^--exchangers (DRA, *SLC26A3*; PAT1, *SLC26A6*) constitute the Cl^- uptake route (Baranovski et al., 2020); if DRA operates with a $2Cl^-/HCO_3^-$ stoichiometry or PAT1 with a $Cl^-/2HCO_3^-$ stoichiometry, the transcellular mechanism becomes electrogenic, likely requiring additional regulatory adjustments.

the basolateral conductive Cl^- exit contributes to the driving force for Cl^- entry via Cl^-/HCO_3^--exchangers. Relying on the Na^+/K^+-pump allows absorptive cells to link transport of nutrients and electrolytes with mutually complementary driving forces that sustain absorption.

Proximal tubule bicarbonate absorption. Flow through small bore tubes, such as renal tubules, occurs with a high surface to volume ratio, allowing the surrounding epithelial cells to modify luminal composition rapidly. In this confined exoepithelial space on the apical membrane surface, alterations of luminal solute concentrations contribute to the driving forces that link flow of multiple solutes (Yu, 2017). In renal proximal tubules, bicarbonate absorption (Fig. 17.7) coexists efficiently with electrogenic glucose absorption (Fig. 17.4) because both operate using mechanisms based on the *Pump–Leak* system with Na^+/K^+-pumps and K^+ channels located in the basolateral membrane. Inclusion of apical Na^+/H^+-exchangers and basolateral Na^+-$3HCO_3^-$cotransporters (NBCe1, also known as SLC4A4), completes a minimal

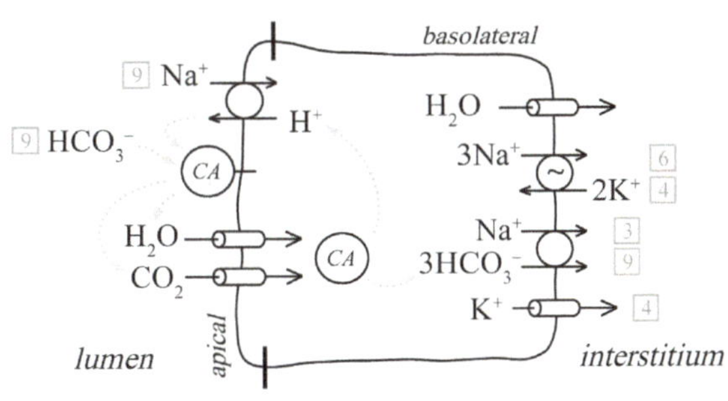

FIGURE 17.7 **Proximal tubule bicarbonate absorption.** The cellular model for bicarbonate absorption in the proximal tubule includes apical membrane Na^+/H^+-exchangers together with basolateral membrane Na^+/K^+-pumps, K^+ channels, and Na^+-$3HCO_3^-$-cotransporters. Flow of H_2O and CO_2 across the apical membrane occurs through the lipid bilayer and membrane proteins with permeability for H_2O (aquaporins) or CO_2 (Michenkova et al., 2021; Missner & Pohl, 2009). Apical membrane tethered carbonic anhydrase (CA) speeds conversion of HCO_3^- to CO_2, while intracellular CA speeds conversion of CO_2 to HCO_3^-. During steady-state transepithelial flow, turnover of the Na^+/K^+-pump produces relative flows via each transporter (gray boxes), such that expenditure of two ATP results in 9 HCO_3^- and 9 Na^+ absorbed cellularly. Electrogenic Na^+-$3HCO_3^-$-cotransporters (NBCe1-A, *SLC4A4*) constitute the HCO_3^- exit route (Brady et al., 2020; Silva & Nilufar Mohebbi, 2022). A variety of K^+ channel types comprises $G^b{}_K$ (Hamilton & Devor, 2012). Carbonic anhydrase (Purkerson & Schwatz, 2007; Silva & Nilufar Mohebbi, 2022) occurs at the apical membrane (CA-IV) and intracellularly (CA-II).

mechanism for bicarbonate absorption (Silva & Nilufar Mohebbi, 2022). Similar to the small intestinal system, apical Na^+/H^+-exchangers acidify the luminal fluid (($\Delta\mu_{NHE}/F$) = $-E^a{}_{Na} + E^a{}_H$), favoring conversion of HCO_3^- to CO_2 and H_2O, with the assistance of apically tethered carbonic anhydrase (CA; Purkerson & Schwartz, 2007). The increased $[CO_2]_{lumen}$ and $[H_2O]_{lumen}$ drive apical flow into the cells, where intracellular CA aids conversion of the elevated CO_2 and H_2O back into HCO_3^- and H^+. Electrogenic Na^+-3HCO_3^--cotransporters in the basolateral membrane establish an outwardly directed driving force ($\Delta\mu_{NBCe}/F = -2\psi_b - E^b{}_{Na} + 3E^b{}_{HCO3} > 0$) due to its coupling stoichiometry of three HCO_3^- to one Na^+. Following the path of the luminal HCO_3^- shows that its carbon atom enters across the apical membrane as CO_2, driven by the concentration gradient, followed by reconstruction as HCO_3^- via carbonic anhydrase and driven out across the basolateral membrane by the combined electrochemical gradients of Na^+ and HCO_3^-, as defined by the electrogenic Na^+-3HCO_3^--cotransporter stoichiometry.

The secondarily active operation of the cellular mechanism for proximal tubule HCO_3^- absorptions relies on the apical Na^+/H^+-exchangers to transform the [Na^+] gradient established by the Na^+/K^+-pump into luminal acidification that supports the carbonic anhydrase conversion of HCO_3^- into CO_2 (Fig. 17.7). Equally, the intracellular alkalinization produced by the apical Na^+/H^+-exchangers drives increased $[HCO_3^-]_{cell}$ via CA, which favors basolateral exit through electrogenic Na^+-HCO_3^--cotransporters. Ultimately, apical Na^+/H^+-exchangers facilitate a weak acid uptake process for HCO_3^- as its carbon atom diffuses into the cell as CO_2. The physiological outward component of the driving force for the Na^+-3HCO_3^--cotransporter derives from the [K^+] gradient generated by the Na^+/K^+-pump, expressed as ψ_b by the influence of basolateral K^+ channels. In this way, a third of the basolateral Na^+ extrusion during HCO_3^- absorption occurs via the Na^+-3HCO_3^-cotransporter and the other two thirds via the Na^+/K^+-pump (Fig. 17.7). Simultaneously with HCO_3^- absorption, proximal tubule cells absorb glucose (Fig. 17.4) and amino acids (similar to Fig. 17.6A) in a compatible manner dependent on Na^+/K^+-pumps and K^+ channels. Optimizing these flows requires a cell to produce a steady-state that balances needs for volume control (and cell shape), pH regulation (Parker, 2023), along with adjusting driving forces for all the transport proteins involved. In particular, setting $[Na^+]_{cell}$ involves tradeoffs for apical and basolateral transporters. Lower $[Na^+]_{cell}$ increases the driving force for apical entry via Na^+-coupled transporters, whereas a higher $[Na^+]_{cell}$ increases the driving force for basolateral exit of HCO_3^- via electrogenic Na^+-3HCO_3^-cotransporters. A more negative cell membrane potential, ψ_a and ψ_b, increases the driving force for both the Na^+-coupled apical glucose uptake and basolateral HCO_3^- exit, made possible by sufficient K^+ conductance. Cellular steady-state adjustments lead to an intermediate value for $[Na^+]_{cell}$, which then determines the transporter number/activity necessary to maintain flow at a physiologically required level.

Electrogenic K^+ secretion. For both electrogenic Na^+ absorption (Fig. 17.3) and electrogenic Cl^- secretion (Fig. 17.5), use of basolateral membrane Na^+/K^+-pumps establishes a cellular context for driving transcellular K^+ secretion (Halm, 2020). Simply including K^+ channels in the apical membrane permits K^+ exit into the lumen (environment) driven by the *Pump–Leak* generated apical membrane driving force for K^+ ($\Delta\mu_K/F = z_K (\psi_a - E^a{}_K) > 0$). Also, the influence of $E^a{}_K$ on ψ_a aids in assuring a driving force (Eq. 17.3) for apical Na^+ entry during electrogenic Na^+ absorption (Fig. 17.8A), as well as for apical Cl^- exit during electrogenic Cl^- secretion (Fig. 17.8B). The ability to adjust apical driving forces using apically resident K^+ channels relaxes the need to use basolateral and paracellular EMFs for this purpose, and allows adjustments of $G^b{}_K$ and G_{pr} instead to serve primarily flow through the basolateral and paracellular routes.

Transepithelial Na^+ absorption and K^+ secretion (Fig. 17.8A) occurs in the principal cells of renal distal tubules and collecting ducts as well as the surface epithelial cells of the distal colon (Berliner, 1961; Frindt & Palmer, 2009; Giebisch et al., 1967; Halm, 2020; Halm & Dawson, 1984; Malnic et al., 2013). The presence of specific claudins (*CLDN4* and *CLDN8*) in these epithelial regions confers anion selectivity to flow through the tight junctions (Amasheh et al., 2009; Fujita et al., 2006; Günzel & Yu, 2013; Holmes et al., 2006; Hou et al., 2010, 2013; Rahner et al., 2001; Yu, 2015). In situations with low $G^a{}_K$, transepithelial flow consists primarily of electrogenic Na^+ absorption via the cellular route and Cl^- absorption via the paracellular route driven by the lumen negative ψ_{te} generated by Na^+ absorption. As $G^a{}_K$ increases, K^+ secretion increases and Cl^- absorption decreases due to a less negative ψ_{te}. In this cellular mechanism, adjusting the ratio of $G^a{}_K$ and $G^b{}_K$ shifts the proportion of Na^+Cl^- absorption compared to exchange of luminal Na^+ for K^+, with the implication that Na^+Cl^- absorption creates an osmotic gradient for H_2O absorption and Na^+ exchange for K^+ increases $[K^+]_{lumen}$ without driving H_2O flow.

Electrogenic K^+ secretion accompanying Cl^- secretion (Fig. 17.8B) occurs in salivary glands, sweat glands, tracheal epithelium, and distal colon (Bijman & Quinton, 1984; Clarke et al., 1997; Frizzell et al., 1984; Halm and Frizzell, 1986; Mangos et al., 1973; Nakamoto et al., 2008; Sato & Sato, 1990; Welsh, Karp et al., 1983). Even modest $G^a{}_K$ allows an apical control scheme to adjust the driving force for apical Cl^- exit into the lumen ($\Delta\mu_{Cl}/F = z_{Cl} (\psi_a - E^a{}_{Cl}) > 0$) without requiring large changes to G_b or G_{pr} (Cook & Young, 1989; Palk et al., 2010). Similar to the Na^+ absorbing cell, adjusting the ratio of $G^a{}_K$ to $G^b{}_K$ sets the proportion of K^+ exiting into the lumen (Li & Halm, 2002; Liao et al., 2005). Increasing $G^a{}_K$ shifts ψ_a

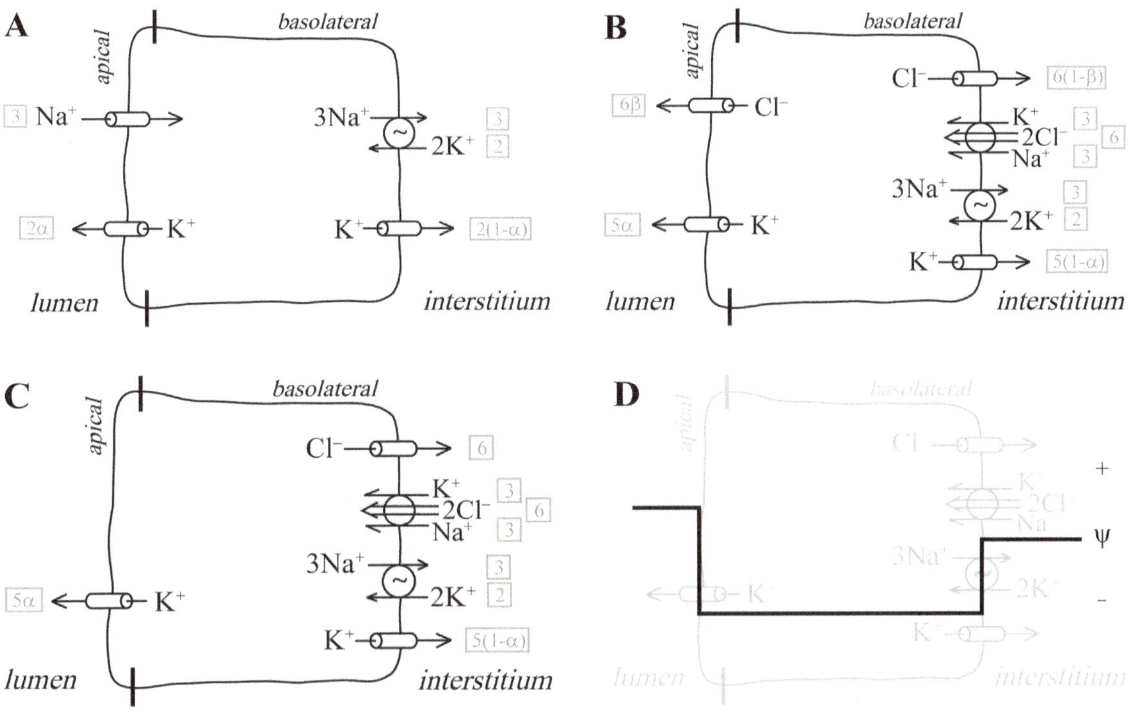

FIGURE 17.8 Electrogenic K^+ secretion. (A) Cellular models for K^+ secretion interlink effectively with electrogenic Na^+ absorption (Fig. 17.3) by including apical membrane K^+ channels. During steady-state transepithelial flow, a single turnover of the Na^+/K^+-pump produces relative flows via each transporter (gray boxes), such that expenditure of one ATP results in $3Na^+$ absorbed cellularly along with as many as $2K^+$ secreted. The proportion of conductively exiting K^+ that occurs across the apical membrane (α) determines the ratio of cellular K^+ secretion to Na^+ absorption ($2\alpha/3$). $K_{ir}1.1$ channels (*KCNJ1*) comprise G^a_K of principal cells (Berrout et al., 2014; Hebert et al., 2005; Welling, 2016) with contributions from $K_{Ca}1.1$ (*KCNMA1*) and $K_{Ca}2.3$ (*KCNN3*). $K_{Ca}1.1$ channels (*KCNMA1*) comprise G^a_K of colonic surface cells (Barmeyer et al., 2010; Flores et al., 2007; Furness et al., 2003; Hay-Schmidt et al., 2003; Sausbier et al., 2006; Zhang et al., 2012) with contributions from $K_{Ca}3.1$ (*KCNN4*). (B) Cellular models for K^+ secretion also interlink effectively with electrogenic Cl^- secretion (Fig. 17.5) by including apical membrane K^+ channels. During steady-state transepithelial flow, a single turnover of the Na^+/K^+-pump produces relative flows via each transporter (gray boxes), such that expenditure of one ATP results in as many as 6 Cl^- and 5 K^+ secreted cellularly. The apical proportion of conductively exiting K^+ (α) together with the apical proportion of conductively exiting Cl^- (β) determines the ratio of cellular K^+ secretion to Cl^- secretion ($5\alpha/6\beta$). $K_{Ca}1.1$ channels (*KCNMA1*) comprise G^a_K of colonic crypt cells (Barmeyer et al., 2010; Furness et al., 2003; Grunnet et al., 2003; Halm et al., 2006; Sandle et al., 2007; Simon et al., 2008; Sausbier et al., 2006; Zhang et al., 2012) with contributions from $K_{Ca}3.1$ (*KCNN4*) and $K_V1.3$ (*KCNA3*). (C) A cellular model for K^+ secretion alone occurs when $\beta = 0$ and depends on basolateral membrane Na^+K^+-$2Cl^-$cotransporters (NKCC1, *SLC12A2*) to provide Na^+ uptake for continued turnover of the Na^+/K^+-pumps and basolateral membrane Cl^- channels to provide a Cl^- exit pathway. During steady-state transepithelial flow, a single turnover of the Na^+/K^+-pump produces relative flows via each transporter (gray boxes), such that expenditure of one ATP results in K^+ secretion, with the apical proportion (α) determining the number of K^+ secreted cellularly. (D) The E_j's of the conductive ions determine the electrical potential profile across this epithelium (ψ, referenced to the interstitial space) transcellularly secreting K^+ alone. A well-type profile occurs (solid line) with G_{pr} electrically coupling apical and basolateral membranes (Eq. 17.3), such that E^b_{Cl} depolarizes ψ_a away from E^a_K as well as E^b_K and E^a_K hyperpolarizing ψ_b away from E^b_{Cl}.

toward E^a_K, aiding in maintaining outward Cl^- flow through G^a_{Cl}. Cation selective claudins in the tight-junction support paracellular Na^+ secretion driven by the lumen negative ψ_{te} (Fujita et al., 2006; Furukawa et al., 2017; Gumber et al., 2014; Holmes et al., 2006; Rahner et al., 2001; Tamura et al., 2011), with the resulting osmotic gradient from increased Na^+ and Cl^- driving H_2O secretion. Higher relative G^a_K also shifts the luminal fluid composition from Na^+ and Cl^- toward K^+ and Cl^-. Regulated increases in G^b_{Cl} add a further feature for adapting basolateral driving forces that support the cellular transport mechanism (Duta et al., 2006; Fischer et al., 2007; Halm, 2004; He et al., 2011; Li et al., 2003). Just as the presence of Cl^- channels and K^+ channels in the apical membrane move ψ_a to a point between E^a_K and E^a_{Cl} that sets the K^+ and Cl^- driving forces for exit, including basolateral Cl^- channels together with K^+ channels assures a driving force oriented for basolateral K^+ and Cl^- exit without a need for a large G_{pr}. Specific neural and hormonal stimulation in the distal colon eliminates G^a_{Cl}, producing a secretory mode with only K^+ secretion (Fig. 17.8C), resulting in a lumen positive ψ_{te} (Carew & Thorn, 2000; Halm & Halm, 1994, 2001; Hosoda et al., 2002; Matos et al., 2005; Rechkemmer et al., 1996; Traynor & O'Grady, 1996; Zhang et al., 2009). The inner ear secretes K^+ using this mode exclusively (Hibino & Kurachi, 2006; Marcus, 2020). A consequence of this K^+ secretion alone mode with only G^a_K requires sufficient G_{pr} so that E^b_{Cl} shifts ψ_a away from E^a_K so that K^+ exits across the apical membrane into the lumen (Fig. 17.8D).

Switching from absorption to secretion. The tracheal epithelium absorbs or secretes fluid into the thin exoepithelial space on the apical surface of the airway to promote mucociliary clearance (Hill et al., 2022; Hoegger et al., 2014; Quinton, 2017; Xie et al., 2018). Excessive absorption reduces the fluid layer height, impeding ciliary movement, whereas excessive secretion lifts the airway mucus out of contact with ciliary propulsion. Tracheal epithelial cells absorb fluid driven by electrogenic Na^+ absorption (Fig. 17.3) and secrete fluid driven by electrogenic Cl^- secretion (Fig. 17.5), controlled so that the fluid layer thickness remains optimal for propelling mucus out of the airway. Adrenergic receptors stimulate a conversion from Na^+ absorption to Cl^- secretion (Shorofsky et al., 1984; Smith & Frizzel, 1984; Welsh et al., 1982; Welsh, Smith et al., 1983), observed as an increase in G_{te} and transepithelial ion flow, I_{sc} (Fig. 17.9A). A biphasic change in ψ_a results as the cell adapts to promote increased Cl^- secretion. Just prior to onset, the positive E_a had combined with ψ_a to establish a large electrochemical driving force consistent with apical Na^+ entry (Fig. 17.9B). Together with the small G_a (Fig. 17.9C), this large driving force resulted in the initial current flow, electrogenic Na^+ absorption. During the time course of Cl^- secretory activation (Fig. 17.8), E_a rapidly decreases to a level consistent with a major influence by $E^a{}_{Cl}$, while E_b remains consistent with a dominant influence by $E^b{}_K$. A delayed activation of G_b compared with that for G_a (Fig. 17.9) produces an increase in the ratio G_a/G_b that guides ψ_a (Eq. 17.3a) closer to E_a before eventually shifting closer to E_b (Fig. 17.9B). Regulatory processes act to control apical channels for Na^+ and Cl^- such that one type dominates, leading to either fluid absorption or fluid secretion as the primary flow direction. Activation of Cl^- secretion generally initiates as opening of $G^a{}_{Cl}$ followed by opening of $G^b{}_K$ that exhibits electrically as a shift of ψ_a toward $E^a{}_{Cl}$ and then a shift back toward $E^b{}_K$ together with cell shrinkage. Steady-state Cl^- secretion occurs with increased turnover of basolateral Na^+-K^+-$2Cl^-$cotransporters (NKCC1) seen as a return of cell volume after the initial shrinkage due to channel opening (Foskett, 1990; Robertson & Foskett, 1994).

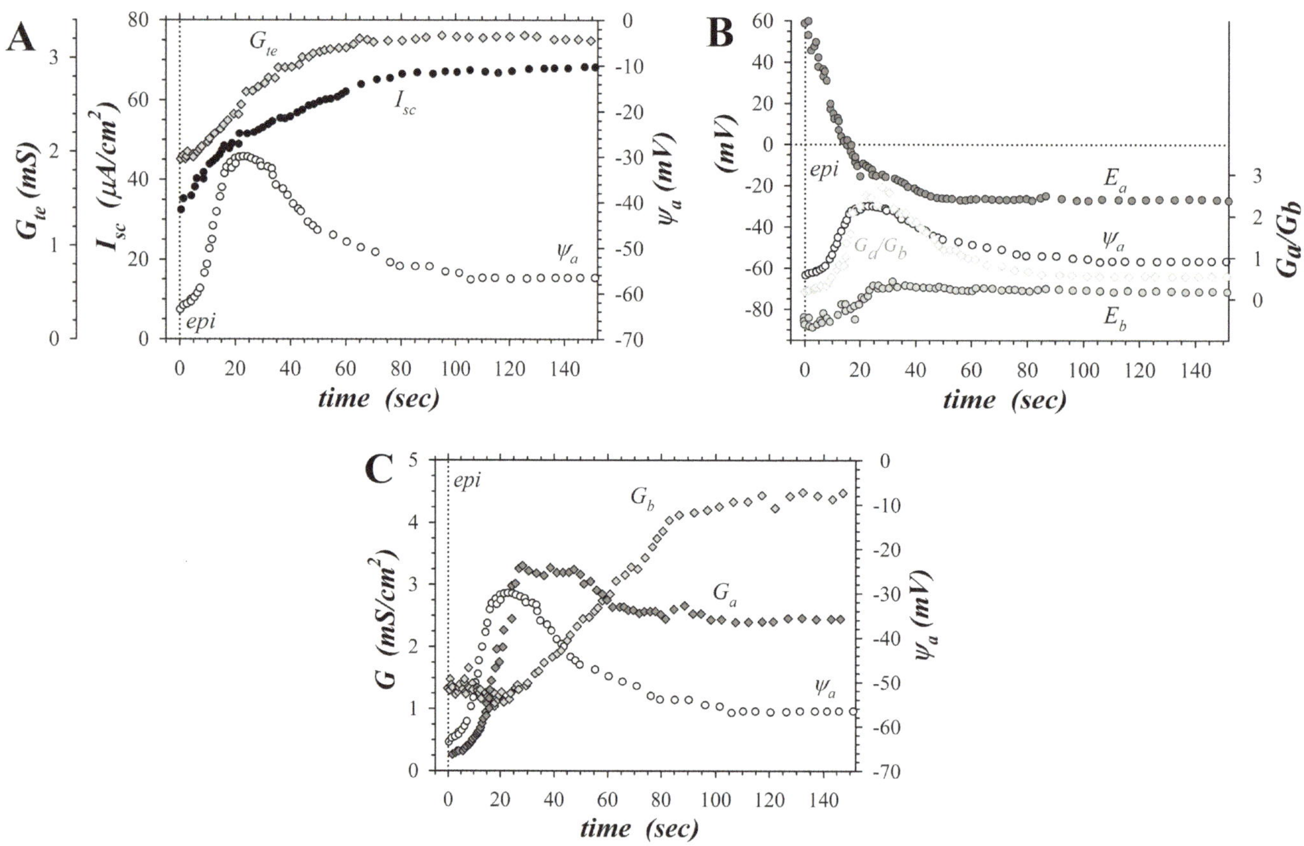

FIGURE 17.9 **Cl^- secretory activation time course.** The canine tracheal epithelium (Welsh, Smith et al., 1983) secretes Cl^- using the electrogenic mechanism (Fig. 17.5), and also absorbs Na^+ using an electrogenic mechanism (Fig. 17.3). (A) Addition of epinephrine stimulates Cl^- secretion, as monitored by the short-circuit current (, I_{sc}) and transepithelial conductance (, G_{te}). Concurrently, the apical membrane electrical potential difference (, ψ_a) changes biphasically, initially depolarizing (peak at ~25 s) and then repolarizing during the continued increase in I_{sc}. (B) The EMF for the apical membrane (, E_a) decreases, and that for the basolateral membrane (, E_b) increases over the first ~40 s of stimulation, while ψ_a remains intermediate in magnitude. The ratio of apical to basolateral conductance (, G_a/G_b) follows the same time course as ψ_a, first increasing and then decreasing (starting at 0.18, peaking at 2.5, steady-state at 0.55). (C) The increase in apical membrane conductance (, G_a) occurs during the initial 30 s of stimulation, while the increase in basolateral membrane conductance (, G_b) exhibits a delay of ~25 s and reaches maximal levels concurrent with attainment of steady-state I_{sc}.

6. Neighboring cell transport mechanisms

Many epithelia exhibit a mosaic of cells with a predominant type interspersed with another distinct type, together forming a continuous epithelial sheet with dispersed activities. Goblet cells of intestinal epithelia and the upper airways secrete mucus that covers the apical surface of neighboring columnar cells, providing a diffusive and mechanical exoepithelial barrier (Allaire et al., 2018; Davis &Dickey, 2008; Hill et al., 2022). In some cases, immunoglobulins transcytose into this macromolecular secretory pathway, adding to the protective quality of the barrier (Wei & Wang, 2021). The olfactory epithelium consists of the columnar sustentacular (support) cells making tight junctions with the olfactory bulbs of the intermingled olfactory neurons (Grubb et al., 2009; Hegg et al., 2009; Lucero, 2013; Reuter et al., 1998; Tirindelli et al., 2009; Vogalis et al., 2005). These bulbs form the base of the sensory dendrites that project into the nasal cavity. Modification of the ion composition in the apical fluid layer by columnar cells alters the Ej's of the ion channels in the sensory dendrite and the ψ_{te} generated by the columnar cells alters the ψ_m of the dendrite compared with its soma, thereby influencing the sensitivity of olfaction. The distal portions of the renal tubule and collecting duct system consist of principal cells and intercalated cells each with distinct transport characteristics that contribute to refining the final composition of urine (Kriz & Kaissling, 2013; Malnic et al., 2013). In these cases, adjacent cells specialize to facilitate distinct cellular events but exist within the same epithelial layer such that physiological transepithelial outcomes result from the combined activity of both cell types.

Renal epithelial intercalated cells. Along the renal connecting tubule and collecting duct system, two cell types comprise the epithelium, principal cells (PC) and intercalated cells (IC) (Kriz & Kaissling, 2013). Principal cells absorb Na^+ and secrete K^+ using the modified electrogenic Na^+ absorption mechanism (Fig. 17.8A) as well as absorbing H_2O transcellularly employing apical and basolateral membrane aquaporins (Kortenoeven et al., 2020). Adjacent intercalated cells energize transport using the electrogenic vacuolar-type H^+-ATPase in the plasma membrane, such that all *Pump–Leak* coupling relies on the resulting ψ_m and E_{HCO3} generated by CO_2/HCO_3^- buffering (Breton & Brown, 2013; Chambrey et al., 2013; Palmer & Frindt, 2007; Rao et al., 2019; Sabolić et al., 1999; Wieczorek et al., 1999). Membrane location of the pump distinguishes subtypes of intercalated cells with the V-type H^+-pumps in the apical membrane of A-type IC (Fig. 17.10A) and in the basolateral membrane of B-type IC (Fig. 17.10B), which orients transepithelial flow of H^+. Activity of the H^+-pump serves to maintain cell volume directly by generating a cell negative ψ_m that drives Cl^- out of the cell, and associated *Leak* pathways allow control of intracellular pH and intracellular concentrations of other solutes. Acid secretion occurs in type-A IC (Fig. 17.10A) via H^+ extrusion across the apical membrane using the H^+-pump and HCO_3^- exit across the basolateral membrane through Cl^-/HCO_3^--exchangers together with Cl^- exit/recycling through basolateral membrane channels, $G^b{}_{Cl}$ (Liu et al., 2011). The secreted H^+ titrate luminal anions (for e.g., HCO_3^-, $H_2PO_4^-$, and HSO_4^-) and ammonia (NH_3), thereby aiding excretion of excess acid produced by whole body metabolism. Secretion of HCO_3^- occurs in type-B IC (Fig. 17.10B) from rearranging the location of transporters with HCO_3^- exit across the apical membrane via Cl^-/HCO_3^--exchangers and H^+ extrusion across the basolateral membrane via H^+-pumps together with the basolateral membrane Cl^- channels. Electrical coupling via the paracellular pathway and the lumen negative influence of adjacent principal cells on ψ_{te} allows continued H^+ extrusion by the H^+-pump along with a cell negative ψ_b sufficient to establish an outward electrochemical driving force for Cl^- exit through the basolateral channels. Similar to HCO_3^- absorption in the proximal tubule, CO_2 generation by carbonic anhydrase, primarily on the interstitial side of these renal cells, produces a CO_2 concentration gradient for entry that permits intracellular carbonic anhydrase to regenerate H^+ and HCO_3^- for continued transepithelial flow.

The presence of both principal cells and intercalated cells in the same epithelial vicinity leads to a steady state that couples the transepithelial flows of both types. Principal cell Na^+ absorption influences ψ_{te} by making the lumen more negative (Fig. 17.8A) and H^+ secretion from type-A intercalated cells influences ψ_{te} by making the lumen more positive (Fig. 17.10A), such that the electrical equivalent circuit, Eq. (17.2), includes parallel terms for each cell type together with the paracellular pathway. The resulting intermediate ψ_{te} assists in maintaining nonzero driving forces across the apical and basolateral membranes of both cell types. From the opposing orientation of these two transepithelial flows, ψ_a and ψ_b shift away from the E_j's of the channels and the reversal potentials (E_{rev}) of the electrogenic transporters, such as the H^+-pump. Consequently, G_{pr} need not increase to facilitate electrical coupling and instead can adjust to optimize paracellular ion flow. For principal cells, H^+ secretion from adjacent IC_A hyperpolarizes PC ψ_a, enhancing the large driving force for apical Na^+ entry without the necessity of opening apical K^+ channels and the attendant K^+ secretion. Similarly, PC Na^+ absorption hyperpolarizes IC_A ψ_b, favoring an outward driving force for Cl^- exit through basolateral channels (Fig. 17.10A). Cellular regulatory signals controlling channel and transporter activity accommodate the required variations in these absorptive and secretory rates (Verschuren et al., 2020; Wall et al., 2020).

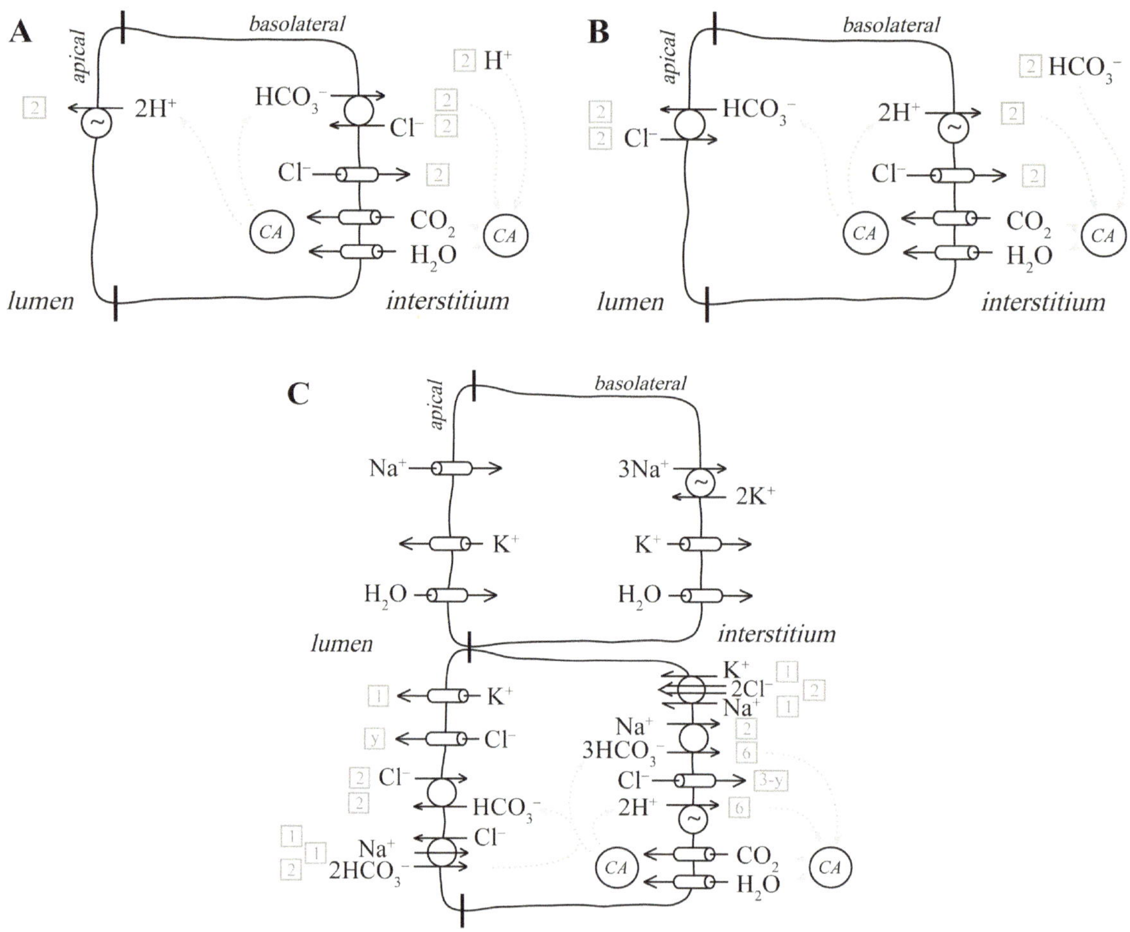

FIGURE 17.10 Renal intercalated cell acid base transport. (A) The cellular model for type-A intercalated cell H^+ secretion includes apical membrane H^+-pumps together with Cl^-/HCO_3^--exchangers and Cl^- channels in the basolateral membrane, as a minimal set. Flow of H_2O and CO_2 across the basolateral membrane occurs through the lipid bilayer and membrane proteins with permeability for H_2O (aquaporins) or CO_2 (Michenkova et al., 2021; Missner & Pohl, 2009). Intracellular and interstitial carbonic anhydrase (CA) speeds conversion of CO_2 to HCO_3^-. Basolateral exit of HCO_3^- established by the Cl^-/HCO_3^--exchanger driving force aids in maintaining a diffusive driving force for basolateral entry of CO_2. Transepithelially secreted H^+ flows across the basolateral membrane as part of an H_2O molecule. During steady-state transepithelial flow, a single turnover of the H^+-pump produces relative flows via each transporter (gray boxes), such that expenditure of one ATP results in 2 H^+ secreted cellularly. The vacuolar H^+-ATPase extrudes H^+ (Kettner et al., 2003; Wagner et al., 2004). Cl^- entry occurs via Cl^-/HCO_3^--exchangers (AE1, *SLC4A1*) and exit via CLC-Kb channels (*CLCNKB*) (Gamba et al., 2013; Stauber et al., 2012). (B) The cellular model for type-B intercalated cell HCO_3^- secretion includes apical membrane Cl^-/HCO_3^--exchangers together with H^+-pumps and Cl^- channels in the basolateral membrane, as a minimal set. Apical exit of HCO_3^- established by the Cl^-/HCO_3^--exchanger driving force aids in maintaining a diffusive driving force for basolateral entry of CO_2. Transepithelially secreted HCO_3^- flow across the basolateral membrane as a CO_2 molecule, before conversion by CA back into HCO_3^-. During steady-state transepithelial flow, a single turnover of the H^+-pump produces relative flows via each transporter (gray boxes), such that expenditure of one ATP results in 2 HCO_3^- secreted and 2 Cl^- absorbed cellularly. Apical Cl^-/HCO_3^--exchangers comprise the HCO_3^- exit route (pendrin, *SLC26A4*). (C) Intercalated cell (*lower cell*) transepithelial flow occurs adjacent to principal cell (*upper cell*) Na^+ absorption, K^+ secretion, and H_2O absorption. Cellular models for type-B intercalated cell HCO_3^- secretion occurring together with Na^+Cl^- absorption and K^+ secretion require additional modules of transporters within the intercalated cell. A contingent Na^+Cl^- absorption module includes apical membrane Na^+-driven-Cl^-/HCO_3^--exchangers together with Na^+-$3HCO_3^-$-cotransporters in the basolateral membrane, as a minimal set. A contingent K^+ secretion module includes apical membrane K^+ channels together with Na^+-K^+-$2Cl^-$-cotransporters (Ginns et al., 1996) and Na^+-$3HCO_3^-$-cotransporters (Purkerson et al., 2014) in the basolateral membrane, as a minimal set. These two transport modules interconnect via the turnover of the basolateral Na^+-$3HCO_3^-$-cotransporters. Relative flows via each transporter (gray boxes) illustrate a regulatory mode maintaining Na^+Cl^- absorption without net transepithelial HCO_3^- flow. Cl^- secretion proceeds semiindependent of H^+-pump turnover, at a rate (y) less than or equal to the net rate of Cl^- influx via the other Cl^- transporters. During steady-state transepithelial flow, turnover of the H^+-pump produces these relative flows (gray boxes), such that expenditure of 3 ATP results in 1 Na^+ absorbed, (1-y) Cl^- absorbed, and 1 K^+ secreted cellularly. $K_{Ca}1.1$ channels (*KCNMA1*) comprise G^a_K (Carrisoza-Gaytán et al., 2010, 2016; Rodan et al., 2011; Hebert et al., 2005; Welling, 2016) with contributions from $K_V1.3$ (*KCNA3*). Na^+-driven-Cl^-/HCO_3^--exchangers (NDCBE, *SLC4A8*) contribute an entry route for Na^+ and HCO_3^-, and Na^+-$3HCO_3^-$-cotransporters (AE4, *SLC4A9*) constitute a route for HCO_3^- exit (Parker & Boron, 2013).

During systemic alkalinizing conditions, such as a dietary alkaline load, the activity of type-B intercalated cells adjacent to principal cells creates distinct influences. The HCO_3^- secretion produced by the type-B IC minimal transport model (Fig. 17.10B) also generates Cl^- absorption. The resulting Na^+Cl^- absorption via separate cellular routes relieves the need for a large G_{pr} as a route for Cl^- flow. However, the low G_{cr} of IC_B would require electrical shunting via G_{pr} or PC apical K^+ channels to maintain an adequate driving force for PC apical Na^+ entry.

In most conditions, type-B intercalated cells likely include several additional transport proteins in their apical membranes together with the Cl^-/HCO_3^--exchangers (Eladari et al., 2014; Carrisoza-Gaytán et al., 2016; Edwards & Crambert, 2017; Wall et al., 2020). Inclusion of transport components into a cellular system requires compatibility with the dominant *Pump–Leak* process used in order to maintain steady-state operation, which for intercalated cells involves the H^+-pump and Cl^- channels. Each additional apical transporter requires complementary basolateral pathways to balance influx and efflux. Modular insertion/activation contributes an additional transport flow with attendant constraints on operation, but also increases the flexibility of possible transport steady-states achievable.

Both Na^+ absorption and K^+ secretion occur in type-B intercalated cells (Fig. 17.10C). A contingent Na^+ absorptive module includes apical Na^+-driven-Cl^-/HCO_3^--exchangers (NDCBE; *SLC4A8*) working in concert with the apical Cl^-/HCO_3^--exchangers (pendrin; *SLC26A4*) together with basolateral Na^+-3HCO_3^--cotransporters (AE4; *SLC4A9*) (Eladari et al., 2014). The apparently mandatory insertion of the CFTR Cl^- channel along with the pendrin-Cl^-/HCO_3^--exchanger establishes the possibility of Cl^- secretion from type-B intercalated cells (Berg et al., 2021; Wall et al., 2020). H^+-pumps power Na^+ extrusion by generating a cell negative ψ_b that contributes to the outward driving force for Na^+-3HCO_3^--cotransporters. Combined operation of apical Cl^-/HCO_3^--exchangers and Na^+-driven-Cl^-/HCO_3^--exchangers ranges from transepithelial Na^+Cl^- absorption to $Na^+HCO_3^-$ absorption depending on the relative activity of the two transporters (Chambrey et al., 2013; Edwards & Crambert, 2017; Wall et al., 2020). HCO_3^- secretion occurs whenever Cl^-/HCO_3^--exchanger turnover exceeds twice the rate of Na^+-driven-Cl^-/HCO_3^--exchangers. The contingent K^+ secretion module consists of apical K^+ channels and basolateral Na^+-K^+-2Cl^--cotransporters. Coupling of this K^+ secretory module to the H^+-pump occurs indirectly via the Na^+-3HCO_3^--cotransporters that extrude Na^+ entering via Na^+-K^+-2Cl^--cotransporters, thus generating basolateral K^+ entry (Carrisoza-Gaytán et al., 2016; Liu et al., 2011; Sabolić et al., 1999). This expanded transport model for type-B intercalated cells (Fig. 17.10C) allows simultaneous operation of the canonical HCO_3^- secretory *Pump–Leak* mode (Fig. 17.10B) with either or both the Na^+ absorptive module and the K^+ secretory module.

Type-B intercalated cells operating in the K^+ secretory mode influence ψ_{te} toward more lumen positive values (Fig. 17.10C), which enhances PC apical Na^+ entry, similar to the situation with type-A intercalated cell H^+ secretion. K^+ secretion also likely occurs in type-A intercalated cells, but the transport mechanism for Na^+ extrusion remains elusive (Carrisoza-Gaytán et al., 2016). The influence of principal cell Na^+ absorption to make ψ_{te} more lumen negative also enhances the driving force for IC apical K^+ exit. Under conditions with a more primordial diet (low in Na^+ and high in K^+, as well as alkaline) luminal $[Na^+]$ likely drops to low values due to avid renal tubular Na^+ absorption, such that the driving force for Na^+-driven-Cl^-/HCO_3^--exchangers reverses, producing Na^+ extrusion across the apical membrane (Cornelius et al., 2016). At higher rates, this apical Na^+ exit balances basolateral Na^+ entry via Na^+-K^+-2Cl^--cotransporters supporting the basolateral K^+ entry necessary for K^+ secretion, possibly eliminating a requirement for basolateral Na^+-3HCO_3^--cotransporter activity. Neighboring principal cells absorb the Na^+ secreted by these intercalated cells, such that the net epithelial result allows secretion of K^+ and HCO_3^- together with Na^+ absorption. A third minority subtype of intercalated cells, designated as non-A-non-B, expresses H^+-pumps and Cl^-/HCO_3^--exchangers in the apical membrane, but the physiological influences remain to be elucidated (Wall et al., 2020).

7. Adjacent epithelial domain transport mechanisms

Many epithelia form small-bore tubules such that fluid flows progressively toward the environment, allowing for a sequential alteration of the fluid composition along the length by epithelial domains with distinct transport capacities. Glandular epithelia generally secrete a primary fluid at the closed end using the electrogenic Cl^- secretory mechanism (Fig. 17.5) followed by ductal modification. Salivary glands and exocrine pancreas, in particular, add digestive enzymes to the primary secretion by exocytotic release from vesicles stored in the apical pole of the secretory cells (Novak & Praetorius, 2020; Thévenod, 2002). The associated ductal modifications consist primarily of increasing $[HCO_3^-]_{lumen}$ that serves to neutralize acidic fluid in the oral cavity and in the fluid arriving in the duodenum during gastric emptying. The renal tubule/duct system comprises the most elaborate example of discrete epithelial domains acting progressively on the small fluid volume in contact at each moment, allowing a sequential modification of the plasma ultrafiltrate into the final urine (Christensen et al., 2012). Specialization within these adjacent domains allows each set of epithelial cells to optimize transport for a particular task.

Sweat gland secretion and absorption. The eccrine sweat gland secretory coil secretes fluid driven by Cl^- secretion that then flows into the sweat duct (Quinton et al., 1999; Reddy, 2020; Sato, 2001). Ductal fluid modification proceeds as electrogenic Na^+ absorption that lowers sweat $[Na^+]$ and $[Cl^-]$ (Fig. 17.11). Absorbed Cl^- flows via a cellular route serially through apical Cl^- conductance and basolateral Cl^- conductance. Cellular regulation adjusts G^a_{Na}, G^b_K, G^a_{Cl}, and G^b_{Cl} such that the electrochemical driving forces for Cl^- produce entry through the apical membrane ($\Delta\mu_{Cl}/F = z_{Cl}$ $(\psi_a - E^a_{Cl}) < 0$) and exit through the basolateral membrane ($\Delta\mu_{Cl}/F = z_{Cl}$ $(\psi_b - E^b_{Cl}) > 0$). These adjustments assure that E^a_{Na} depolarizes ψ_a above E^a_{Cl} and E^b_K hyperpolarizes ψ_b below E^b_{Cl}. The small duct diameter allows the resulting large surface to luminal volume ratio to permit rapid Na^+ and Cl^- removal with limited H_2O flow, due to low ductal H_2O permeability. In a heat acclimated condition, the duct decreases $[Na^+]$ and $[Cl^-]$ of ~ 100 mM in the primary secretion to a skin surface sweat level of <20 mM. Amphibian skin establishes a similar transport loop with glands secreting fluid onto the skin surface as a thin film, followed by electrogenic Na^+ absorption together with Cl^- absorption through mitochondrial rich cells (Larsen, 2011; Larsen & Ramløv, 2013; Watlington & Huf, 1971). Positioning these oppositely directed cellular transport mechanisms as adjacent epithelial domains results in a circuit of Na^+ and Cl^- flow that generates surface water useful for evaporative cooling while minimizing salt loss.

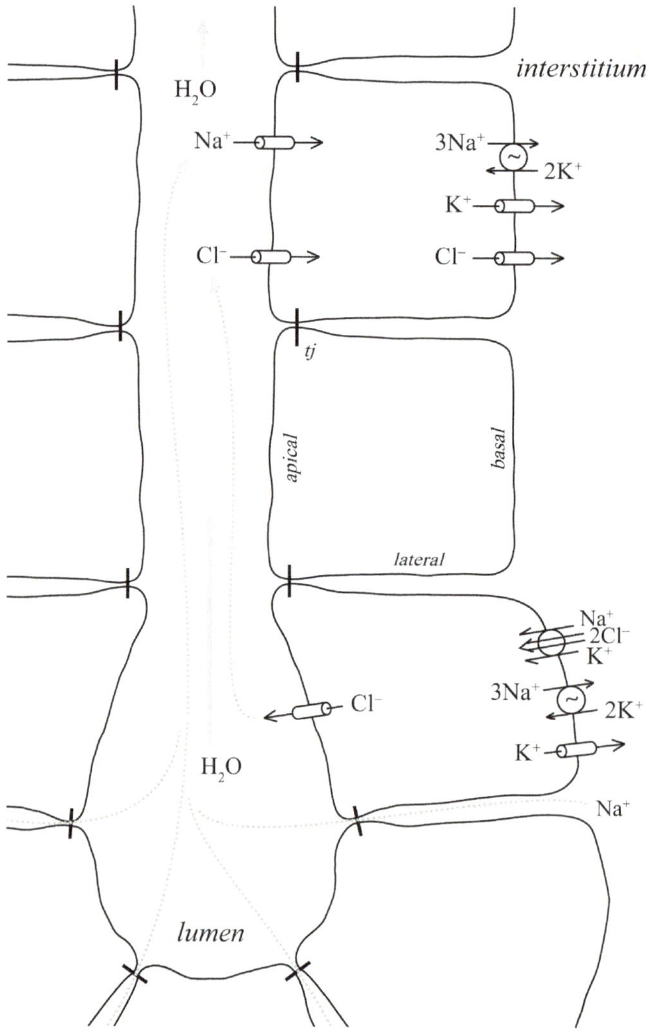

FIGURE 17.11 **Fluid production by glands.** A schematic sweat gland illustrates the primary secretion of fluid driven by electrogenic Cl^- secretion (Fig. 17.5) and the ductal modification of fluid composition by electrogenic Na^+ absorption (Fig. 17.3) together with a transcellular route for Cl^- absorption via apical and basolateral membrane Cl^- channels. Structurally, sweat glands consist of a secretory coil that increases the number of cells participating in fluid secretion, and a duct consisting of a stratified epithelium with two cell layers connected by gap junctions along lateral and basal membranes (Quinton et al., 1999; Reddy, 2020; Sato, 2001). Together with the relative impermeability of the ductal apical membrane to H_2O, absorption of a portion of the Na^+ and Cl^- in the secreted fluid leads to a reduction in the osmolarity of the final fluid exiting the gland. The Na^+ and Cl^- absorbed from the duct return to the interstitium, where capillary exchange redistributes these solutes within the body.

Proximal tubule glucose absorption. Glucose arrives in the lumen of the renal proximal tubule by glomerular filtration of plasma, where apical membrane Na^+-glucose-cotransporters provide a pathway for coupled entry as illustrated in Fig. 17.4 for the case of intestinal glucose absorption. The *Pump/Leak* mechanism determines the inward orientation of the Na^+ electrochemical gradient by lowering $[Na^+]_{cell}$ through the action of the Na^+/K^+-pump and establishing a cell negative ψ_a by the influence of basolateral K^+ channels and high $[K^+]_{cell}$. For those proximal tubule cells first receiving filtrate from the glomerulus, in the S1 and S2 segments, the apical membrane contains Na^+-glucose cotransporters exhibiting a 1:1 stoichiometry for Na^+ and glucose (SGLT2), thus permitting an apical membrane [glucose] gradient of as much as ~70-fold (Vallon & Nakagawa, 2021). The downstream S3 segment uses apical membrane $2Na^+$-glucose cotransporters 2:1 stoichiometry (SGLT1), capable of producing an apical membrane [glucose] gradient of ~5000-fold. This sequential arrangement of transport coupling plays out with ~95% of filtered glucose absorbed in the S1 and S2 segments, requiring one turnover of the Na^+/K^+-pump to drive uptake of three glucose molecules while hydrolyzing just one ATP molecule. The S3 segment absorbs the majority of the remaining glucose, but hydrolyzes 2 ATP molecules to drive uptake of three glucose molecules due to the higher coupling stoichiometry of SGLT1 (Fig. 17.4). By positioning the lower coupling stoichiometry SGLT2 first, most glucose absorption occurs more economically. Only as the luminal [glucose] reaches submillimolar levels, the S3 segment uses SGLT1 with its higher coupling stoichiometry to lower $[glucose]_{lumen}$ to extract as many glucose molecules as possible using the Na^+ electrochemical gradient of this secondary active transport system.

8. Transport mechanisms at apposition sites of multiple epithelium bounded compartments

Epithelia form the barriers surrounding and thereby defining numerous body compartments and establish the means for controlling the composition of the fluid inside them using transepithelial mechanisms similar to those already described. Of the internal compartments bounded by epithelia, the blood plasma fluid within the cardiovascular endothelium plays a central role in maintaining a uniform composition throughout the interstitial fluid space by rapidly circulating so that capillary endothelial exchange supplies nutrients and removes wastes, allowing cells to maintain their individual steady-states. For example, close apposition of capillary endothelium with specific epithelial domains of other compartments allows rapid retrieval of absorbed fluid containing nutrients (Figs. 17.4 and 17.6) and a ready supply of fluid for secretion (Fig. 17.5). In this way, the capillary endothelium redistributes these solutes and water throughout the body. Some of the highest rates of transepithelial absorption and secretion occur across domains of the epithelium that form the body's barrier with the external environment. Although a histological term is presently lacking that distinguishes the body surface epithelium from the other epithelial subsets (endothelium and mesothelia), *ectothelium* may suffice to accentuate the location of this epithelial barrier between the interstitial fluid space (extracellular fluid) and the external environment, thereby encasing the entire body. The dense capillary networks found closely opposed to the basal aspects of respiratory, renal, and intestinal domains of the ectothelium put the blood plasma fluid in these capillary lumens at the shortest manageable separation from luminal fluid flowing within ectothelial tubes either toward or away from the external environment. These ectothelial/endothelial apposition zones comprise the major sites for molecular exchange with the external environment (Caceres et al., 2017). The structure at each of these exchange zones supports the surface area needed to promote rapid flow by the transport mechanisms governed by the endothelial and ectothelial cells.

Alveolar gas exchange. Cellular respiration generates ATP from nutrients to power the many activities of cells. Within mitochondria, the conversion of nutrients to ATP uses the citric acid cycle and the electron transport chain, consuming O_2 and releasing CO_2 (Chapter 13, by Cramer and Bhaduri). These two gases then constitute the most urgent input and waste for continued cellular operation, and the alveolar exchange zone provides the site for acquiring more O_2 and relieving the CO_2 burden. Pulmonary alveolar capillaries course through the septa between lung alveoli, bringing plasma close to the external environment (Fronius et al., 2012; Hsia et al., 2016). Producing the shortest manageable distance for flow between plasma fluid and the atmosphere, squamous capillary endothelial cells meet squamous alveolar ectothelial cells at their basal aspects with a mutual thin fused basal lamina between. The transport mechanism for O_2 and CO_2 consists of diffusive flow through the four cell membranes (apical, basal, basal, and apical) in series with the interposed fluid layers of the capillary lumen, endothelial cell cytoplasm, septal interstitium, alveolar cell cytoplasm, and alveolar surface. At the air/water interface within the alveoli, a lipidic surfactant layer imposes an additional thin barrier (Knudsen & Ochs, 2018; Olmeda et al., 2017; Scarpelli, 2003). The orientation of flow follows the concentration gradients from high to low, producing O_2 uptake and CO_2 exit. The route through these cell membranes occurs between bilayer lipid molecules influenced by the particular lipid content, and in some cases augmented by membrane proteins such as aquaporins with passageways for CO_2 or O_2 (Michenkova et al., 2021; Missner & Pohl, 2009). Red blood cells also contribute to this flow by shuttling O_2 out of the lungs and CO_2 into the lungs. Ultimately, the large exchange surface area of the lungs coupled with the total cardiac output combine to balance the requirements of cellular respiration in the body.

Capillary fluid exchange accompanies the high flows of O_2 and CO_2 through the alveolar zone, potentially overwhelming the thin interstitial space in the septa between adjacent alveoli. Additionally, the thin alveolar surface fluid layer that aids rapid gas diffusion must remain substantial enough to maintain the viability of the cells in the alveolar domain of the ectothelium. A combination of routes and driving forces for fluid flow provides the means to keep the structural dimensions of the alveolar septa conducive to gas exchange during a range of physiological respiratory demands. Fluid entering the septal interstitium as filtration through the capillary endothelium exits via pulmonary lymphatic vessels located primarily in extraalveolar interstitial spaces (Beretta et al., 2021; Hsia et al., 2016; Taylor & Parker, 1985). Alveolar ectothelial cells add or subtract from this interstitial fluid by absorption or secretion, which serves to adjust the thickness of the alveolar apical surface fluid layer (Hastings et al., 2004).

Transendothelial fluid flow in capillaries follows gradients defined by the Starling forces (Beretta et al., 2021; Levick & Michel, 2010; Taylor & Parker, 1985), hydrostatic pressure difference (capillary pressure, P_c; and interstitial pressure, P_i), and osmotic pressure difference (capillary osmotic pressure, π_c; and interstitial osmotic pressure, π_i). By maintaining a high conductance for blood flow, pulmonary circulation occurs with a low P_c. Lymphatic pumping, aided by the elastic properties of the septal interstitium, acts to generate a subatmospheric value of P_i, which, together with P_c establishes a transendothelial hydrostatic driving force (ΔP_{endo}) favoring filtration. Plasma albumin constitutes the primary source of the transendothelial difference in osmotic pressure ($\Delta\pi_{endo}$), with only a limited amount of albumin entering the interstitium by endothelial transcytosis (Komarova & Malik, 2010; Simmons et al., 2019) so that $\Delta\pi_{endo}$ favors fluid reabsorption into the capillary. Fluid flows transendothelially via the paracellular route, augmented by some transcellular flow through aquaporins in the endothelial cell membranes (Verkman, 2007). The glycocalyx on the luminal surface of the endothelial cells modifies the development of the Starling forces for these continuous capillaries (Curry, 2018). Negative charges on the glycocalyx limit access of the negatively charged albumin molecules to the paracellular route, which consists of tight junctions and a restricted lateral space. Paracellular filtration fluid flow sweeps albumin from the lateral spaces, while also increasing the hydrostatic pressure in the lateral space above P_i, such that the driving force favoring filtration decreases. Fluid absorption from the abutting alveolar ectothelium also decreases the net filtration force by increasing P_i and diluting interstitial albumin, lowering π_i that favors capillary reabsorption. Since initial lymphatics appear to array at the edges of the thin portions of the alveolar septa (Schraufnagel, 2010; Sozio et al., 2012; Weber et al., 2018), the net filtration pressure calculated from the measured Starling forces best represents the quasi-annular zone surrounding the primary gas exchange sites occurring with the thinnest interstitium. Rather than a uniform transendothelial driving force for fluid flow throughout the alveolar domain, the close apposition of the capillary endothelium and the alveolar ectothelium leads to interstitial fluid flow away from the thin exchange zones out to the initial lymphatics that provide a route for fluid return to the systemic blood circulation.

Fluid absorption across the alveolar domains of the ectothelium serves to maintain the thinness of the alveolar lining fluid conducive for rapid gas exchange, and during conditions that dry this fluid layer, fluid secretion replenishes the thickness to aid the integrity of these ectothelial cells (Matthay, 2014; Verkman, 2007). Just as with the Starling forces across the endothelium, the hydrostatic-osmotic forces across the alveolar ectothelium govern fluid flow. A bias favoring absorption occurs due to the low P_i, which achieves subatmospheric values at the initial lymphatics, and the low concentration of plasma albumin in the interstitium. Transcellular Na^+Cl^- absorption increases the interstitial osmolarity, accentuating the driving force for fluid absorption. Na^+ absorption operates using the electrogenic mechanism (Fig. 17.3) with Na^+-selective channels and a portion of nonselective cation channels, which support Na^+ absorption and K^+ secretion (Koval, 2013; Lee, Maniak et al., 2003; Lee, Rhodes et al., 2003; Nielson, 1986; O'Grady & Lee, 2003; Saumon & Basset, 1993; Schlingmann et al., 2015). Combined with the Na^+ absorption, transcellular Cl^- flow occurs via apical Cl^- channels and basolateral K^+-Cl^--cotransporters (Chapter 15). The influence of the cell positive E^a_{Na} assures a driving force for conductive apical entry of Cl^- ($\Delta\mu_{Cl}/F = z_{Cl}\,(\psi_a - E^a_{Cl}) < 0$) when G_{pr} remains low enough to moderate the influence of E^b_K (Eq. 17.3a). High $[K]_{cell}$ resulting from Na^+/K^+-pump activity establishes the outward driving force for basolateral membrane K^+-Cl^--cotransporters ($(\Delta\mu/F)_{KCC} = -E^b_K + E^b_{Cl} > 0$). Regulatory demands requiring an increase in alveolar lining fluid reverse the orientation of Cl^- flow to electrogenic secretion (Fig. 17.5) by activating basolateral membrane Na^+-K^+-$2Cl^-$cotransporters for Cl^- uptake and by increasing the influence of E^b_K to produce apical Cl^- exit ($\Delta\mu_{Cl}/F = z_{Cl}$ $(\psi_a - E^a_{Cl}) > 0$) either by decreasing G^a_{Na} and/or increasing G_{pr}. Additionally, the apical membrane contains voltage-sensitive K^+ channels (K_V), which confer hypoxia sensitivity to ψ_a for adjusting fluid thickness (Bardou et al., 2009; O'Grady & Lee, 2005). Hypothetically, during conditions with a thin lining fluid, high O_2 plausibly activates these K_V channels, increasing G^a_K and shifting ψ_a so that Cl^- exits through apical channels ($z_{Cl}\,(\psi_a - E^a_{Cl}) > 0$), producing fluid secretion driven by electrogenic Cl^- secretion. Excessive alveolar fluid thickness in comparison increases the O_2 diffusion distance closing K_V channels, shifting ψ_a toward E^a_{Na} so that Cl^- enters conductively (z_{Cl} ($\psi_a - E^a_{Cl}$) < 0), establishing Na^+Cl^- absorption and thus reducing fluid layer thickness.

Glomerular filtration. Renal blood flow brings a portion of the cardiac output to endothelial/ectothelial exchange sites within the kidney for elimination of substances in excess of current body needs, wastes. Similar to other exocrine glands, the primary fluid arrives in the lumen at the blind-end of the gland (Bowman's space) followed by flow of this exoepithelial fluid through an ectothelial tube to the body surface where excretion finally occurs as the fluid discharges from the body. Starling forces at the renal glomeruli set the direction for flow through the capillary endothelium abutted to the podocyte domain of the ectothelium, delivering fluid with a composition essentially similar to interstitial fluid into Bowman's space. Now the nephron/duct domain of the ectothelium selectively absorbs all molecules *not* in excess of body needs, along with selective secretion of some in excess. This reclamation process allows the possibility of leaving an adaptive amount of water in the exoepithelial/tubular fluid when in excess. The selective absorption and secretion mechanisms occur in the renal tubular ectothelial cells matched with the abutting peritubular capillaries to carry away the absorbate and deliver molecules for secretion. This extended ectothelial/endothelial exchange zone operates in a sequential manner as exoepithelial fluid flows progressively over ectothelial cells with distinct transport mechanisms (Figs. 17.4, 17.7 and 17.10).

Fluid flow out through the glomerular exchange zone occurs entirely via the paracellular route at both the endothelium and ectothelium (podocytes). High glomerular capillary hydrostatic pressure ($P^g{}_c$) contributes to the net Starling force, favoring this filtration across the fenestrated endothelial cells in the glomerulus. The thin basal lamina (~300 nm) creates a short interstitial flow path such that the capillary endothelium and podocyte ectothelium operate as a functional syncytium, guiding this filtered fluid toward paracellular secretion through the filtration slits between adjacent podocytes and on into Bowman's space. Fenestra dimensions (60–80 nm diameter) restrain blood cells to remain in the capillary lumen, and although these glomerular endothelial cells lack diaphragms in the fenestra, the extensive negatively charged surface glycosylation restricts passage of albumin through the fenestra (Dane et al., 2015; Satchell & Braet, 2009). Negatively charged components of the basal lamina continue the restrictions for albumin passage. At the filtration slit formed by cellular interdigitations of adjacent podocytes, a continuous slit diaphragm fills the flow pathway (25–40 nm wide) with lattices of the proteins NEPHRIN (*NPHS1*) and NEPH-1 (*KIRREL1*) that include negatively charged glycosylation (Burghardt et al., 2015; Conti et al., 2017; Grahammer et al., 2016). The combined paracellular route through the endothelium and podocyte ectothelium allows ready passage of water and small solutes while restricting albumin flow (Brinkkoetter et al., 2013; Inoue & Ishibe, 2015). Proximal tubule ectothelial cells downstream absorb the small amount of albumin leaking through the barrier (Molitoris et al., 2022). The high filtration of fluid from the glomerular capillaries concentrates the restrained albumin such that the blood plasma entering the peritubular capillaries exhibits an increased osmotic pressure ($\pi^{pt}{}_c$). Combined with the lower $P^{pt}{}_c$, the peritubular Starling forces favor capillary uptake of fluid arriving from absorption via the nephron/tubule transport mechanisms. Overall, the renal waste removal system spatially separates transport mechanisms along the tubule, which permits distinct driving force orientations to guide complementary set of flows (Yu, 2017).

9. Regulated balance

Epithelial cells configure sets of membrane transport proteins in the apical and basolateral membranes that establish pathways for flow in absorptive and secretory directions, as shown in the examples above. Actuating these cellular transport mechanisms requires a coupling of solute and water flow to cellular metabolic energy sources. The key link generally occurs via Na^+/K^+-pumps and V-type H^+-pumps, such that the energy of the phosphate transfer potential for ATP (ΔG_{ATP}) essentially flows through these transport ATPases into the chemical energy of transmembrane concentration gradients for Na^+, K^+, and H^+, with further transformation into ψ_m via opening of ion channels and other electrogenic transporters. The myriad of available solute transport proteins (cotransporters, exchangers, and channels) allows this energy flow to generate additional transmembrane concentration gradients to support the flow of solutes and water in advantageous directions, according to the driving forces for those transporters. Examining the steady-state balance for minimal-sufficient transepithelial transport mechanisms provides an initial step for assessing where regulatory controls could act to adjust transport rates and direction of flow. Each of these steady states conceivably could be maintained with numerous combinations of intracellular composition and transporter turnover rates, subject to coordination maintained by intracellular signaling pathways. For this reason, the study of cellular signal transduction pathways (Chapter 24) and how they alter the activity of membrane transport proteins continues as a major focus of many epithelial research projects. Use of mathematical models to simulate the operation of these transepithelial transport mechanisms (Nickerson et al., 2020) then allows prediction of testable outcomes to guide further research. Simulations also provide a means to examine the transitions between steady states as hormonal and neural signaling elicit higher or lower transepithelial transport rates with attendant changes to cellular composition and electrical potentials. Numerous such simulations (Fong et al., 2017; Larsen & Møbjerg, 2006; Mazzuca et al., 2016; Mioni et al., 2017; Morla et al., 2016; Thorsen et al., 2014; Wang et al., 2007; Weinstein, 2015) illustrate how these transepithelial mechanisms operate and induce cellular changes; and increased use of this tool will be essential for fully appreciating how all the transporter modules interconnect within epithelial cells to produce observed physiological responses.

References

Al Salmani, M. K., Sondo, E., Balut, C., Sheppard, D. N., Singh, A. K., & Pedemonte, N. (2020). Molecular physiology and pharmacology of the cystic fibrosis transmembrane conductance regulator. In K. L. Hamilton, & D. C. Devor (Eds.) (2nd ed., *Ion channels and transporters of epithelia in health and disease: Vol 3. Studies of epithelial transporters and ion channels* (pp. 605–670). Springer.

Allaire, J. M., Morampudi, V., Crowley, S. M., Stahl, M., Yu, H., Bhullar, K., Knodler, L. A., Bressler, B., Jacobson, K., & Vallance, B. A. (2018). Frontline defenders: Goblet cell mediators dictate host-microbe interactions in the intestinal tract during health and disease. *American Journal of Physiology - Gastrointestinal and Liver Physiology, 314*, G360–G377.

Amasheh, S., Meiri, N., Gitter, A. H., Schöneberg, T., Mankertz, J., Schulzke, J. D., & Fromm, M. (2002). Claudin-2 expression induces cation-selective channels in tight junctions of epithelial cells. *Journal of Cell Science, 115*, 4969–4976.

Amasheh, S., Milatz, S., Krug, S. M., Bergs, M., Amasheh, M., Schulzke, J. D., & Fromm, M. (2009). Na^+ absorption defends from paracellular back-leakage by claudin-8 upregulation. *Biochemical and Biophysical Research Communications, 378*, 45–50.

Anderson, J. M., & Van Itallie, C. M. (2009). Physiology and function of the tight junction. *Cold Spring Harbor Perspectives in Biology, 1*, Article a002584.

Andrew, D. J., & Ewald, A. J. (2010). Morphogenesis of epithelial tubes: Insights into tube formation, elongation, and elaboration. *Developmental Biology, 341*, 34–55.

Baranovski, B. M., Fremder, M., & Ohana, E. (2020). Properties, structure, and function of the Solute Carrier 26 family of anion transporters. In K. L. Hamilton, & D. C. Devor (Eds.) (2nd ed.,*Ion channels and transporters of epithelia in health and disease: Vol 3. Studies of epithelial transporters and ion channels* (pp. 467–493). Springer.

Bardou, O., Trinh, N. T., & Brochiero, E. (2009). Molecular diversity and function of K^+ channels in airway and alveolar epithelial cells. *American Journal of Physiology - Lung Cellular and Molecular Physiology, 296*, L145–L155.

Barmeyer, C., Rahner, C., Yang, Y., Sigworth, F. J., Binder, H. J., & Rajendran, V. M. (2010). Cloning and identification of tissue-specific expression of KCNN4 splice variants in rat colon. *American Journal of Physiology: Cell Physiology, 299*, C251–C263.

Baumgarten, C. M., & Feher, J. J. (2023a). Osmotic properties of cells. In F. J. Alvarez-Leefmans, E. Delpire, & E. Kaneshiro (Eds.), *Cell physiology sourcebook: Essentials of membrane biophysics* (5th ed.). Elsevier.

Baumgarten, C. M., & Feher, J. J. (2023b). Cell volume control. In F. J. Alvarez-Leefmans, E. Delpire, & E. Kaneshiro (Eds.), *Cell physiology sourcebook: Essentials of membrane biophysics* (5th ed.). Elsevier.

Beretta, E., Romanò, F., Sancini, G., Grotberg, J. B., Nieman, G. F., & Miserocchi, G. (2021). Pulmonary interstitial matrix and lung fluid balance from normal to the acutely injured lung. *Frontiers in Physiology, 12*, Article 781874.

Berg, P., Svendsen, S. L., Sorensen, M. V., Schreiber, R., Kunzelmann, K., & Leipziger, J. (2021). The molecular mechanism of CFTR- and secretin-dependent renal bicarbonate excretion. *The Journal of Physiology, 599*, 3003–3011.

Bergemann, A. D. (2020). Establishment and maintenance of epithelial polarization. In K. L. Hamilton, & D. C. Devor (Eds.) (2nd ed.,*Ion channels and transporters of epithelia in health and disease: Vol 1. Basic epithelial ion transport principles and function* (pp. 83–114). Springer.

Berliner, R. W. (1961). Renal mechanisms for potassium excretion. In *The harvey lectures, series 55* (pp. 141–171). Academic Press.

Berrout, J., Mamenko, M., Zaika, O. L., Chen, L., Zhang, W., Pochynyuk, O., & O'Neil, R. G. (2014). Emerging role of the calcium-activated, small conductance, SK3 K^+ channel in distal tubule function: Regulation by TRPV4. *PLoS One, 9*, Article e95149.

Bijman, J., & Quinton, P. M. (1984). Influence of abnormal Cl^- impermeability on sweating in cystic fibrosis. *American Journal of Physiology: Cell Physiology, 247*, C3–C9.

Brady, C. T., Dugandžić, A., Parker, M. D., & Romero, M. F. (2020). NBCe1: An electrogenic Na^+-bicarbonate-cotransporter, in epithelia. In K. L. Hamilton, & D. C. Devor (Eds.) (2nd ed.,*Ion channels and transporters of epithelia in health and disease: Vol 3. Studies of epithelial transporters and ion channels* (pp. 93–123). Springer.

Breton, S., & Brown, D. (2013). Regulation of luminal acidification by the V-ATPase. *Physiology, 28*, 318–329.

Brinkkoetter, P. T., Ising, C., & Benzing, T. (2013). The role of the podocyte in albumin filtration. *Nature Reviews Nephrology, 9*, 328–336.

Burghardt, T., Hochapfel, F., Salecker, B., Meese, C., Gröne, H. J., Rachel, R., Wanner, G., Krahn, M. P., & Witzgall, R. (2015). Advanced electron microscopic techniques provide a deeper insight into the peculiar features of podocytes. *American Journal of Physiology - Renal Physiology, 309*, F1082–F1089.

Caceres, P. S., Benedicto, I., Lehmann, G. L., & Rodriguez-Boulan, E. J. (2017). Directional fluid transport across organ-blood barriers: Physiology and cell biology. *Cold Spring Harbor Perspectives in Biology, 9*, Article a027847.

Camargo, S. M., Poncet, N., & Verrey, F. (2020). Amino acid transporters of epithelia. In K. L. Hamilton, & D. C. Devor (Eds.) (2nd ed.,*Ion channels and transporters of epithelia in health and disease: Vol 3. Studies of epithelial transporters and ion channels* (pp. 255–323). Springer.

Carew, M. A., & Thorn, P. (2000). Carbachol-stimulated Cl^- secretion in mouse colon: Evidence of a role for autocrine prostaglandin-E_2 release. *Experimental Physiology, 85*, 67–72.

Carrisoza-Gaytán, R., Salvador, C., Satlin, L. M., Liu, W., Zavilowitz, B., Bobadilla, N. A., Trujillo, J., & Escobar, L. I. (2010). Potassium secretion by voltage-gated potassium channel Kv1.3 in the rat kidney. *American Journal of Physiology - Renal Physiology, 299*, F255–F264.

Carrisoza-Gaytán, R., Carattino, M. D., Kleyman, T. R., & Satlin, L. M. (2016). An unexpected journey: Conceptual evolution of mechanoregulated potassium transport in the distal nephron. *American Journal of Physiology: Cell Physiology, 310*, C243–C259.

Cereijido, M., Contreras, R. G., & Shoshani, L. (2004). Cell adhesion, polarity, and epithelia in the dawn of metazoans. *Physiological Reviews, 84*, 1229–1262.

Chambrey, R., Ingo Kurth, I., Peti-Peterdi, J., Houillier, P., Purkerson, J. M., Leviel, F., Hentschke, M., Zdebik, A. A., Schwartz, G. J., Hübner, C. A., & Eladari, D. (2013). Renal intercalated cells are rather energized by a proton than a sodium pump. *Proceedings of the National Academy of Sciences of the United States of America, 110*, 7928–7933.

Christensen, E. I., Wagner, C. A., & Kaissling, B. (2012). Uriniferous tubule: Structural and functional organization. *Comprehensive Physiology, 2*, 805–861.

Clarke, L. L., Chinet, T., & Boucher, R. C. (1997). Extracellular ATP stimulates K^+ secretion across cultured human airway epithelium. *American Journal of Physiology - Lung Cellular and Molecular Physiology, 272*, L1084–L1091.

Clausen, C., Lewis, S. A., & Diamond, J. M. (1979). Impedance analysis of a tight epithelium using a distributed resistance model. *Biophysical Journal, 26*, 291–317.

Conti, S., Perico, L., Grahammer, F., & Huber, T. B. (2017). The long journey through renal filtration: New pieces in the puzzle of slit diaphragm architecture. *Current Opinion in Nephrology and Hypertension, 26*, 148–153.

Cook, D. I., & Young, J. A. (1989). Effect of K^+ channels in the apical plasma membrane on epithelial secretion based on secondary active Cl^- transport. *Journal of Membrane Biology, 110*, 139–146.

Cornelius, R. J., Wang, B., Wang-France, J., & Sansom, S. C. (2016). Maintaining K^+ balance on the low-Na^+, high-K^+ diet. *American Journal of Physiology - Renal Physiology, 310*, F581–F595.

Crane, R. K., Miller, D., & Bihler, I. (1961). The restrictions on possible mechanisms of intestinal active transport of sugars. In A. Kleinzeller, & A. Kotyk (Eds.), *Membrane transport and metabolism, symposia Čsav, Svazek 1, August 1960, Prague* (pp. 439–449). Publishing House Czechoslovak Academy of Sciences.

Crane, R. K. (1965). Na^+-dependent transport in the intestine and other animal tissues. *Federation Proceedings, 24*, 1000–1006.

Curry, F. E. (2018). The molecular structure of the endothelial glycocalyx layer (EGL) and surface layers (ESL) modulation of transvascular exchange. In B. M. Fu, & N. T. Wright (Eds.), *Advances in experimental medicine and biologyMolecular, cellular, and tissue engineering of the vascular system* (pp. 29–49). Springer Nature.

Dane, M. J. C., van den Berg, B. M., Lee, D. H., Boels, M. G. S., Tiemeier, G. L., Avramut, M. C., van Zonneveld, A. J., van der Vlag, J., Vink, H., & Rabelink, T. J. (2015). A microscopic view on the renal endothelial glycocalyx. *American Journal of Physiology - Renal Physiology, 308*, F956–F966.

Davis, C. W., & Dickey, B. F. (2008). Regulated airway goblet cell mucin secretion. *Annual Review of Physiology, 70*, 487–512.

Degnan, K. J., Karnaky, K. J., Jr., & Zadunaisky, J. A. (1977). Active chloride transport in the in vitro opercular skin of a teleost (*Fundulus heteroclitus*), a gill-like epithelium rich in chloride cells. *The Journal of Physiology, 271*, 155–191.

Delpire, E., & Gagnon, K. B. (2018). Na^+-K^+-2 Cl^--cotransporter (NKCC) Physiological function in nonpolarized cells and transporting epithelia. *Comprehensive Physiology, 8*, 871–901.

DiBona, D. R., & Mills, J. W. (1979). Distribution of Na^+-pump sites in transporting epithelia. *Federation Proceedings, 38*, 134–143.

DiBona, D. R. (1985). Functional analysis of tight junction organization. *Pflügers Archiv, 405*(Suppl. 1), S59–S66.

Duta, V., Duta, F., Puttagunta, L., Befus, A. D., & Duszyk, M. (2006). Regulation of basolateral Cl^- channels in airway epithelial cells: The role of nitric oxide. *Journal of Membrane Biology, 213*, 165–174.

Edwards, A., & Crambert, G. (2017). Versatility of NaCl transport mechanisms in the cortical collecting duct. *American Journal of Physiology - Renal Physiology, 313*, F1254–F1263.

Eladari, D., Chambrey, R., Picard, N., & Hadchouel, J. (2014). Electroneutral absorption of NaCl by the aldosterone-sensitive distal nephron: Implication for normal electrolytes homeostasis and blood pressure regulation. *Cellular and Molecular Life Sciences, 71*, 2879–2895.

Ernst, S. A., & Mills, J. W. (1977). Basolateral plasma membrane localization of ouabain-sensitive sodium transport sites in the secretory epithelium of the avian salt gland. *The Journal of Cell Biology, 75*, 74–94.

Fan, S., Harfoot, N., Bartolo, R. C., & Butt, A. G. (2012). CFTR is restricted to a small population of high expresser cells that provide a forskolin-sensitive transepithelial Cl^- conductance in the proximal colon of the possum, *Trichosurus vulpecula. Journal of Experimental Biology, 215*, 1218–1230.

Field, M. (1993). Intestinal electrolyte secretion. History of a paradigm. *Archives of Surgery, 128*, 273–278.

Fischer, H., Illek, B., Finkbeiner, W. E., & Widdicombe, J. H. (2007). Basolateral Cl^- channels in primary airway epithelial cultures. *American Journal of Physiology - Lung Cellular and Molecular Physiology, 292*, L1432–L1443.

Flessner, M. F. (2005). The transport barrier in intraperitoneal therapy. *American Journal of Physiology - Renal Physiology, 288*, F433–F442.

Flores, C. A., Melvin, J. E., Figueroa, C. D., & Sepúlveda, F. V. (2007). Abolition of Ca^{2+}-mediated intestinal anion secretion and increased stool dehydration in mice lacking the intermediate conductance Ca^{2+}-dependent K^+ channel Kcnn4. *The Journal of Physiology, 583*, 705–717.

Fong, S., Chiorini, J. A., Sneyd, J., & Suresh, V. (2017). Computational modeling of epithelial fluid and ion transport in the parotid duct after transfection of human aquaporin-1. *American Journal of Physiology - Gastrointestinal and Liver Physiology, 312*, G153–G163.

Foskett, J. K. (1990). $[Ca^{2+}]_i$ modulation of Cl^- content controls cell volume in single salivary acinar cells during fluid secretion. *American Journal of Physiology: Cell Physiology, 259*, C998–C1004.

Frindt, G., & Palmer, L. G. (2009). K^+ secretion in the rat kidney: Na^+ channel-dependent and -independent mechanisms. *American Journal of Physiology - Renal Physiology, 297*, F389–F396.

Frizzell, R. A., & Hanrahan, J. W. (2012). Physiology of epithelial chloride and fluid secretion. *Cold Spring Harbor Perspectives in Medicine, 2*, Article a009563.

Frizzell, R. A., Field, M., & Schultz, S. G. (1979). Sodium-coupled chloride transport by epithelial tissues. *American Journal of Physiology - Renal Physiology, 236*, F1–F8.

Frizzell, R. A., Halm, D. R., & Krasny, E. J., Jr. (1984). Active secretion of chloride and potassium across the large intestine. In E. Skadhauge, & K. Heintze (Eds.), *Intestinal absorption and secretion, falk symposium 36, June 1983, Titisee, Germany* (pp. 313–324). MTP Press.

Frömter, E., & Diamond, J. (1972). Route of passive ion permeation in epithelia. *Nature New Biology, 235*, 9–13.

Fronius, M., Clauss, W. G., & Althaus, M. (2012). Why do we have to move fluid to be able to breathe? *Frontiers in Physiology, 3*, 146.

Fujita, H., Chiba, H., Yokozaki, H., Sakai, N., Sugimoto, K., Wada, T., Kojima, T., Yamashita, T., & Sawada, N. (2006). Differential expression and subcellular localization of claudin-7, -8, -12, -13, and -15 along the mouse intestine. *Journal of Histochemistry and Cytochemistry, 54*, 933–944.

Furness, J. B., Robbins, H. L., Selmer, I. S., Hunne, B., Chen, M. X., Hicks, G. A., Moore, S., & Neylon, C. B. (2003). Expression of intermediate conductance potassium channel immunoreactivity in neurons and epithelial cells of the rat gastrointestinal tract. *Cell and Tissue Research, 314*, 179–189.

Furukawa, C., Ishizuka, N., Hayashi, H., Fujii, N., Manabe, A., Tabuchi, Y., Matsunaga, T., Endo, S., & Ikari, A. (2017). Up-regulation of claudin-2 expression by aldosterone in colonic epithelial cells of mice fed with NaCl-depleted diets. *Scientific Reports, 7*, Article 12223.

Furuse, M. (2010). Claudins, tight junctions, and the paracellular barrier. In A. S. L. Yu (Ed.), *Current topics in membranes: Vol 65. Claudins* (pp. 1–19). Elsevier.

Gamba, G., Wang, W., & Schild, L. (2013). Sodium chloride transport in the loop of Henle, distal convoluted tubule, and collecting duct. In R. J. Alpern, O. W. Moe, & M. Caplan (Eds.), *Seldin and Giebisch's the kidney: Physiology and pathophysiology* (5th ed., pp. 1143–1179). Elsevier.

Gibson, W. T., & Gibson, M. C. (2009). Cell topology, geometry, and morphogenesis in proliferating epithelia. *Current Topics in Developmental Biology, 89*, 87–114.

Giebisch, G., Klose, R. M., & Malnic, G. (1967). Renal tubular potassium transport. *Bulletin der Schweizerischen Akademie der Medizinischen Wissenschaften, 23*, 287–312.

Ginns, S. M., Knepper, M. A., Ecelbarger, C. A., Terris, J., He, X., Coleman, R. A., & Wade, J. B. (1996). Immunolocalization of the secretory isoform of Na-K-Cl cotransporter in rat renal intercalated cells. *Journal of the American Society of Nephrology, 7*, 2533–2542.

Gong, Y., & Hou, J. (2017). Claudins in barrier and transport function-the kidney. *Pflügers Archiv, 469*, 105–113. https://doi.org/10.1007/s00424-016-1906-6. In this issue.

Grahammer, F., Wigge, C., Schell, C., Kretz, O., Patrakka, J., Schneider, S., Klose, M., Kind, J., Arnold, S. J., Habermann, A., Bräuniger, R., Rinschen, M. M., Völker, L., Bregenzer, A., Rubbenstroth, D., Boerries, M., Kerjaschki, D., Miner, J. H., Walz, G., ... Huber, T. B. (2016). A flexible, multilayered protein scaffold maintains the slit in between glomerular podocytes. *JCI Insight, 1*, Article e86177.

Grubb, B. R., Rogers, T. D., Boucher, R. C., & Ostrowski, L. E. (2009). Ion transport across CF and normal murine olfactory and ciliated epithelium. *American Journal of Physiology: Cell Physiology, 296*, C1301–C1309.

Grunnet, M., Rasmussen, H. B., Hay-Schmidt, A., & Klærke, D. A. (2003). The voltage-gated potassium channel subunit, $K_V1.3$, is expressed in epithelia. *Biochimica et Biophysica Acta, 1616*, 85–94.

Guillot, C., & Lecuit, T. (2013). Mechanics of epithelial tissue homeostasis and morphogenesis. *Science, 340*, 1185–1189.

Gumber, S., Nusrat, A., & Villinger, F. (2014). Immunohistological characterization of intercellular junction proteins in rhesus macaque intestine. *Experimental & Toxicologic Pathology, 66*, 437–444.

Günzel, D., & Fromm, M. (2012). Claudins and other tight junction proteins. *Comprehensive Physiology, 2*, 1819–1852.

Günzel, D., & Yu, A. S. (2013). Claudins and the modulation of tight junction permeability. *Physiological Reviews, 93*, 525–569.

Günzel, D., Stuiver, M., Kausalya, P. J., Haisch, L., Krug, S. M., Rosenthal, R., Meij, I. C., Hunziker, W., Fromm, M., & Müller, D. (2009). Claudin-10 exists in six alternatively spliced isoforms that exhibit distinct localization and function. *Journal of Cell Science, 122*, 1507–1517.

Günzel, D. (2017). Claudins: Vital partners in transcellular and paracellular transport coupling. *Pflügers Archiv, 469*, 35–44.

Halm, D. R., & Dawson, D. C. (1984). Control of potassium transport by turtle colon: Role of membrane potential. *American Journal of Physiology: Cell Physiology, 247*, C26–C32.

Halm, D. R. (2004). Secretory control of basolateral membrane K^+ and Cl^- channels in colonic crypt cells. In P. K. Lauf, & N. C. Adragna (Eds.), *Advances in experimental medicine and biology: Vol 559. Cell volume and signaling* (pp. 119–129). Springer.

Halm, D. R. (2020). Physiologic influences of transepithelial K^+ secretion. In K. L. Hamilton, & D. C. Devor (Eds.) (2nd ed.,*Ion channels and transporters of epithelia in health and disease: Vol 1. Basic epithelial ion transport principles and function* (pp. 337–393). Springer.

Halm, D. R., & Frizzell, R. A. (1986). Active K^+ transport across rabbit distal colon: Relation to Na^+ absorption and Cl^- secretion. *American Journal of Physiology: Cell Physiology, 251*, C252–C267.

Halm, D. R., & Frizzell, R. A. (1991). Ion transport across the large intestine. In *Supplement 19, handbook of physiology, the gastrointestinal system, intestinal absorption and secretion* (pp. 257–273). https://doi.org/10.1002/cphy.cp060408

Halm, D. R., & Halm, S. T. (1994). Aldosterone stimulates K^+ secretion prior to onset of Na^+ absorption in Guinea pig distal colon. *American Journal of Physiology: Cell Physiology, 266*, C552–C558.

Halm, D. R., & Halm, S. T. (2001). Prostanoids stimulate K^+ secretion and Cl^- secretion in Guinea pig distal colon via distinct pathways. *American Journal of Physiology - Gastrointestinal and Liver Physiology, 281*, G984–G996.

Halm, S. T., Liao, T., & Halm, D. R. (2006). Distinct K^+ conductive pathways are required for Cl^- and K^+ secretion across distal colonic epithelium. *American Journal of Physiology: Cell Physiology, 291*, C636–C648.

Hamilton, K. L., & Devor, D. C. (2012). Basolateral membrane K^+ channels in renal epithelial cells. *American Journal of Physiology - Renal Physiology, 302*, F1069–F1081.

Hastings, R. H., Folkesson, H. G., & Matthay, M. A. (2004). Mechanisms of alveolar protein clearance in the intact lung. *American Journal of Physiology - Lung Cellular and Molecular Physiology, 286*, L679–L689.

Hay-Schmidt, A., Grunnet, M., Abrahamse, S. L., Knaus, H. G., & Klærke, D. A. (2003). Localization of Ca^{2+} -activated big-conductance K^+ channels in rabbit distal colon. *Pflügers Archiv, 446*, 61–68.

He, Q., Halm, S. T., Zhang, J., & Halm, D. R. (2011). Activation of the basolateral membrane Cl^- conductance essential for electrogenic K^+ secretion suppresses electrogenic Cl^- secretion. *Experimental Physiology, 96*, 305–316.

Hebert, S. C., Desir, G., Giebisch, G., & Wang, W. (2005). Molecular diversity and regulation of renal potassium channels. *Physiological Reviews, 85*, 319–371.

Hegg, C. C., Irwin, M., & Lucero, M. T. (2009). Calcium store-mediated signaling in sustentacular cells of the mouse olfactory epithelium. *Glia, 57*, 634–644.

Heintze, K., Stewart, C. P., & Frizzell, R. A. (1983). Sodium-dependent chloride secretion across rabbit descending colon. *American Journal of Physiology - Gastrointestinal and Liver Physiology, 244*, G357–G365.

Hibino, H., & Kurachi, Y. (2006). Molecular and physiological bases of the K^+ circulation in the mammalian inner ear. *Physiology, 21*, 336–345.

Higashi, T., & Chiba, H. (2020). Molecular organization, regulation and function of tricellular junctions. *Biochimica et Biophysica Acta (BBA) - Biomembranes, 1862*, Article 183143.

Hill, D. B., Button, B., Rubinstein, M., & Boucher, R. C. (2022). Physiology and pathophysiology of human airway mucus. *Physiological Reviews, 102*, 1757–1836.

Hoegger, M. J., Fischer, A. J., McMenimen, J. D., Ostedgaard, L. S., Tucker, A. J., Awadalla, M. A., Moninger, T. O., Michalski, A. S., Hoffman, E. A., Zabner, J., Stoltz, D. A., & Welsh, M. J. (2014). Impaired mucus detachment disrupts mucociliary transport in a piglet model of cystic fibrosis. *Science, 345*, 818–822.

Hofer, A. (2023). Signal transduction. In F. J. Alvarez-Leefmans, E. Delpire, & E. Kaneshiro (Eds.), *Cell physiology sourcebook: Essentials of membrane biophysics* (5th ed.). Elsevier.

Holmes, J. L., Van Itallie, C. M., Rasmussen, J. E., & Anderson, J. M. (2006). Claudin profiling in the mouse during postnatal intestinal development and along the gastrointestinal tract reveals complex expression patterns. *Gene Expression Patterns, 6*, 581–588.

Hosoda, Y., Karaki, S. I., Shimoda, Y., & Kuwahara, A. (2002). Substance P-evoked Cl^- secretion in Guinea pig distal colonic epithelia: Interaction with PGE_2. *American Journal of Physiology - Gastrointestinal and Liver Physiology, 283*, G347–G356.

Hou, J., Renigunta, A., Yang, J., & Waldegger, S. (2010). Claudin-4 forms paracellular chloride channel in the kidney and requires claudin-8 for tight junction localization. *Proceedings of the National Academy of Sciences of the United States of America, 107*, 18010–18015.

Hou, J., Rajagopal, M., & Yu, A. S. (2013). Claudins and the kidney. *Annual Review of Physiology, 75*, 479–501.

Hsia, C. C., Hyde, D. M., & Weibel, E. R. (2016). Lung structure and the intrinsic challenges of gas exchange. *Comprehensive Physiology, 6*, 827–895.

Huang, F., Wong, X., & Jan, L. Y. (2012). International union of basic and clinical pharmacology. LXXXV: Calcium-activated chloride channels. *Pharmacological Reviews, 64*, 1–15.

Inoue, K., & Ishibe, S. (2015). Podocyte endocytosis in the regulation of the glomerular filtration barrier. *American Journal of Physiology - Renal Physiology, 309*, F398–F405.

Jackson, W. F. (2017). Boosting the signal: Endothelial inward rectifier K^+ channels. *Microcirculation, 24*. https://doi.org/10.1111/micc.12319

Jakab, R. L., Collaco, A. M., & Ameen, N. A. (2013). Characterization of CFTR high expresser cells in the intestine. *American Journal of Physiology - Gastrointestinal and Liver Physiology, 305*, G453–G465.

Ji, H. L., & Nie, H. G. (2008). Electrolyte and fluid transport in mesothelial cells. *Journal of Epithelial Biology & Pharmacology, 1*, 1–7.

Julio-Kalajzić, F., Villanueva, S., Burgos, J., Ojeda, M., Cid, L. P., Jentsch, T. J., & Sepúlveda, F. V. (2018). K2P TASK-2 and KCNQ1-KCNE3 K^+ channels are major players contributing to intestinal anion and fluid secretion. *The Journal of Physiology, 596*, 393–407.

Kaplan, J. H. (2002). Biochemistry of Na,K-ATPase. *Annual Review of Biochemistry, 71*, 511–535.

Karnaky, K. J., Jr., Degnan, K. J., & Zadunaisky, J. A. (1977). Chloride transport across isolated opercular epithelium of killifish: A membrane rich in chloride cells. *Science, 195*, 203–205.

Kay, A. R., & Blaustein, M. P. (2019). Evolution of our understanding of cell volume regulation by the pump-leak mechanism. *The Journal of General Physiology, 151*, 407–416.

Kettner, C., Bertl, A., Obermeyer, G., Slayman, C., & Bihler, H. (2003). Electrophysiological analysis of the yeast V-type proton pump: Variable coupling ratio and proton shunt. *Biophysical Journal, 85*, 3730–3738.

Kiela, P. R., Xu, H., & Ghishan, F. K. (2020). Na^+/H^+-exchangers in epithelia. In K. L. Hamilton, & D. C. Devor (Eds.) (2nd ed.,*Ion channels and transporters of epithelia in health and disease: Vol 3. Studies of epithelial transporters and ion channels* (pp. 125–209). Springer.

Kleyman, T. R., Kashlan, O. B., & Hughey, R. P. (2018). Epithelial Na^+ channel regulation by extracellular and intracellular factors. *Annual Review of Physiology, 80*, 263–281.

Knudsen, L., & Ochs, M. (2018). The micromechanics of lung alveoli: Structure and function of surfactant and tissue components. *Histochemistry and Cell Biology, 150*, 661–676.

Koefoed-Johnsen, V., & Ussing, H. H. (1958). The nature of the frog skin potential. *Acta Physiologica Scandinavica, 42*, 298–308.

Komarova, Y., & Malik, A. B. (2010). Regulation of endothelial permeability via paracellular and transcellular transport pathways. *Annual Review of Physiology, 72*, 463–493.

Kortenoeven, M. L. A., Olesen, E. T. B., & Fenton, R. A. (2020). Renal aquaporins in health and disease. In K. L. Hamilton, & D. C. Devor (Eds.) (2nd ed.,*Ion channels and transporters of epithelia in health and disease: Vol 3. Studies of epithelial transporters and ion channels* (pp. 1187–1244). Springer.

Koval, M. (2013). Claudin heterogeneity and control of lung tight junctions. *Annual Review of Physiology, 75*, 551–567.

Kriz, W., & Kaissling, B. (2013). Structural organization of the mammalian kidney. In R. J. Alpern, O. W. Moe, & M. Caplan (Eds.), *Seldin and Giebisch's the kidney: Physiology and pathophysiology* (5th ed., pp. 595–691). Elsevier.

Krug, S. M., Günzel, D., Conrad, M. P., Rosenthal, R., Fromm, A., Amasheh, S., Schulzke, J. D., & Fromm, M. (2012). Claudin-17 forms tight junction channels with distinct anion selectivity. *Cellular and Molecular Life Sciences, 69*, 2765–2778.

Krug, S. M., Schulzke, J. D., & Fromm, M. (2014). Tight junction, selective permeability, and related diseases. *Seminars in Cell & Developmental Biology, 36*, 166–176.

Lane, N. J., Reese, T. S., & Kachar, B. (1992). Structural domains of the tight junctional intramembrane fibrils. *Tissue and Cell, 24*, 291–300.

Larre, I., Cereijido, M., Paez, O., Shoshani, L., & Ponce, A. (2020). Na^+/K^+-ATPase drives most asymmetric transports and modulates the phenotype of epithelial cells. In K. L. Hamilton, & D. C. Devor (Eds.) (2nd ed.,*Ion channels and transporters of epithelia in health and disease: Vol 3. Studies of epithelial transporters and ion channels* (pp. 1–24). Springer.

Larsen, E. H., & Hoffmann, E. K. (2020). Volume regulation in epithelia. In K. L. Hamilton, & D. C. Devor (Eds.) (2nd ed.,*Ion channels and transporters of epithelia in health and disease: Vol 1. Basic epithelial ion transport principles and function* (pp. 395–460). Springer.

Larsen, E. H., & Møbjerg, N. (2006). Na^+ recirculation and isosmotic transport. *Journal of Membrane Biology, 212*, 1–15.

Larsen, E. H., & Ramløv, H. (2013). Role of cutaneous surface fluid in frog osmoregulation. *Comparative Biochemistry and Physiology Part A: Molecular & Integrative Physiology, 165*, 365–370.

Larsen, E. H. (2011). Reconciling the Krogh and Ussing interpretations of epithelial chloride transport - presenting a novel hypothesis for the physiological significance of the passive cellular chloride uptake. *Acta Physiologica, 202*, 435–464.

Le Bivic, A. (2013). Evolution and cell physiology. 4. Why invent another protein complex to build junctions in epithelial cells? *American Journal of Physiology-Cell Physiology, 305*, C1193–C1201. https://doi.org/10.1152/ajpcell.00272.2013. In this issue.

Lee, D. B., Jamgotchian, N., Allen, S. G., Abeles, M. B., & Ward, H. J. (2008). A lipid-protein hybrid model for tight junction. *American Journal of Physiology - Renal Physiology, 295*, F1601–F1612.

Lee, S. Y., Maniak, P. J., Ingbar, D. H., & O'Grady, S. M. (2003). Adult alveolar epithelial cells express multiple subtypes of voltage-gated K^+ channels that are located in apical membrane. *American Journal of Physiology: Cell Physiology, 284*, C1614–C1624.

Lee, S. Y., Maniak, P. J., Rhodes, R., Ingbar, D. H., & O'Grady, S. M. (2003). Basolateral Cl- transport is stimulated by terbutaline in adult rat alveolar epithelial cells. *Journal of Membrane Biology, 191*, 133–139.

Lemke, S. B., & Nelson, C. M. (2021). Dynamic changes in epithelial cell packing during tissue morphogenesis. *Current Biology, 31*, R1098–R1110.

Levick, J. R., & Michel, C. C. (2010). Microvascular fluid exchange and the revised Starling principle. *Cardiovascular Research, 87*, 198–210.

Lewis, S. A., Clausen, C., & Wills, N. K. (1996). Impedance analysis of epithelia. In N. K. Wills, L. Reuss, & S. A. Lewis (Eds.), *Epithelial transport: A guide to methods and experimental analysis* (pp. 118–145). Chapman & Hall.

Li, Y., & Halm, D. R. (2002). Secretory modulation of basolateral membrane inwardly rectified K^+ channel in Guinea pig distal colonic crypts. *American Journal of Physiology: Cell Physiology, 282*, C719–C735.

Li, Y., Halm, S. T., & Halm, D. R. (2003). Secretory activation of basolateral membrane Cl^- channels in Guinea pig distal colonic crypts. *American Journal of Physiology: Cell Physiology, 284*, C918–C933.

Liao, T., Wang, L., Halm, S. T., Lu, L., Fyffe, R. E. W., & Halm, D. R. (2005). The K^+ channel K_VLQT (Kcnq1) located in the basolateral membrane of distal colonic epithelium is not essential for activating Cl^- secretion. *American Journal of Physiology: Cell Physiology, 289*, C564–C575.

Lingaraju, A., Long, T. M., Wang, Y., Austin, J. R., & Turner, J. R. (2015). Conceptual barriers to understanding physical barriers. *Seminars in Cell & Developmental Biology, 42*, 13–21.

Liu, W., Schreck, C., Coleman, R. A., Wade, J. B., Hernandez, Y., Zavilowitz, B., Warth, R., Kleyman, T. R., & Satlin, L. M. (2011). Role of NKCC in BK channel-mediated net K^+ secretion in the CCD. *American Journal of Physiology - Renal Physiology, 301*, F1088–F1097.

Loo, D. D. F., & Wright, E. M. (2020). Sugar transport across epithelia. In K. L. Hamilton, & D. C. Devor (Eds.) (2nd ed.,*Ion channels and transporters of epithelia in health and disease: Vol 3. Studies of epithelial transporters and ion channels* (pp. 211–254). Springer.

Lucero, M. T. (2013). Peripheral modulation of smell: Fact or fiction? *Seminars in Cell & Developmental Biology, 24*, 58–70.

Lynn, K. S., Peterson, R. J., & Koval, M. (2020). Ruffles and spikes: Control of tight junction morphology and permeability by claudins. *Biochimica et Biophysica Acta (BBA) - Biomembranes, 1862*, Article 183339.

Malnic, G., Giebisch, G., Muto, S., Wang, W. H., Bailey, M. A., & Satlin, L. M. (2013). Regulation of K^+ excretion. In R. J. Alpern, O. W. Moe, & M. Caplan (Eds.), *Seldin and Giebisch's the kidney: Physiology and pathophysiology* (5th ed., pp. 1659–1715). Elsevier.

Mangos, J. A., McSherry, N. R., Nousia-Arvanitakis, S., & Irwin, K. (1973). Secretion and transductal fluxes of ions in exocrine glands of the mouse. *American Journal of Physiology, 225*, 18–24.

Marcus, D. C. (2020). Ion transport across inner ear epithelia. In K. L. Hamilton, & D. C. Devor (Eds.) (2nd ed.,*Ion channels and transporters of epithelia in health and disease: Vol 2. Ion transport across epithelial tissues and disease* (pp. 279–305). Springer.

Markov, A. G., & Amasheh, S. (2014). Tight junction physiology of pleural mesothelium. *Frontiers in Physiology, 5*, 221.

Matos, J. E., Robaye, B., Boeynaems, J. M., Beauwens, R., & Leipziger, J. (2005). K^+ secretion activated by luminal $P2Y_2$ and $P2Y_4$ receptors in mouse colon. *The Journal of Physiology, 564*, 269–279.

Matthay, M. A. (2014). Resolution of pulmonary edema. Thirty years of progress. *American Journal of Respiratory and Critical Care Medicine, 189*, 1301–1308.

Mazzuca, E., Aliverti, A., & Miserocchi, G. (2016). Computational micro-scale model of control of extravascular water and capillary perfusion in the air blood barrier. *Journal of Theoretical Biology, 400*, 42–51.

Michenkova, M., Taki, S., Blosser, M. C., Hwang, H. J., Kowatz, T., Moss, F. J., Occhipinti, R., Qin, X., Sen, S., Shinn, E., Wang, D., Zeise, B. S., Zhao, P., Malmstadt, N., Vahedi-Faridi, A., Tajkhorshid, E., & Boron, W. F. (2021). Carbon dioxide transport across membranes. *Interface Focus, 11*, Article 20200090.

Milatz, S., Himmerkus, N., Wulfmeyer, V. C., Drewell, H., Mutig, K., Hou, J., Breiderhoff, T., Müller, D., Fromm, M., Bleich, M., & Günzel, D. (2017). Mosaic expression of claudins in thick ascending limbs of Henle results in spatial separation of paracellular Na^+ and Mg^{2+} transport. *Proceedings of the National Academy of Sciences of the United States of America, 114*, E219–E227.

Mioni, R., Marega, A., Romano, G., & Montanaro, D. (2017). Mathematical rationalization for the renal tubular transport: Revised concepts. *Scandinavian Journal of Clinical & Laboratory Investigation, 77*, 358–372.

Missner, A., & Pohl, P. (2009). 110 years of the meyer-overton rule: Predicting membrane permeability of gases and other small compounds. *ChemPhysChem, 10*, 1405–1414.

Molitoris, B. A., Sandoval, R. M., Yadav, S. P. S., & Wagner, M. C. (2022). Albumin uptake and processing by the proximal tubule: Physiological, pathological, and therapeutic implications. *Physiological Reviews, 102*, 1625–1667.

Montoro, D. T., Haber, A. L., Biton, M., Vinarsky, V., Lin, B., Birket, S. E., Yuan, F., Chen, S., Leung, H. M., Villoria, J., Rogel, N., Burgin, G., Tsankov, A. M., Waghray, A., Slyper, M., Waldman, J., Nguyen, L., Dionne, D., Rozenblatt-Rosen, O., … Rajagopal, J. (2018). A revised airway epithelial hierarchy includes CFTR-expressing ionocytes. *Nature, 560*, 319–324.

Morla, L., Doucet, A., Lamouroux, C., Crambert, G., & Edwards, A. (2016). The renal cortical collecting duct: A secreting epithelium? *The Journal of Physiology, 594*, 5991–6008.

Nagarajan, S. K., Piontek, J., & Günzel, D. (2023). Structural and biophysical properties of tight junctions. In F. J. Alvarez-Leefmans, E. Delpire, & E. Kaneshiro (Eds.), *Cell physiology sourcebook: Essentials of membrane biophysics* (5th ed.). Elsevier.

Nakamoto, T., Romanenko, V. G., Takahashi, A., Begenisich, T., & Melvin, J. E. (2008). Apical maxi-K ($K_{Ca}1.1$) channels mediate K^+ secretion by the mouse submandibular exocrine gland. *American Journal of Physiology: Cell Physiology, 294*, C810–C819.

Nickerson, D. P., Noroozbabaee, L., Sarwar, D. M., Hamilton, K. L., & Hunter, P. J. (2020). Mathematical modeling of epithelial ion transport. In K. L. Hamilton, & D. C. Devor (Eds.) (2nd ed.,*Ion channels and transporters of epithelia in health and disease: Vol 1. Basic epithelial ion transport principles and function* (pp. 115–133). Springer.

Nielson, D. W. (1986). Electrolyte composition of pulmonary alveolar subphase in anesthetized rabbits. *Journal of Applied Physiology, 60*, 972–979.

Noreng, S., Bharadwaj, A., Posert, R., Yoshioka, C., & Baconguis, I. (2018). Structure of the human epithelial sodium channel by cryo-electron microscopy. *Elife, 7*, Article e39340.

Novak, I., & Praetorius, J. (2020). Fundamentals of bicarbonate secretion in epithelia. In K. L. Hamilton, & D. C. Devor (Eds.) (2nd ed.,*Ion channels and transporters of epithelia in health and disease: Vol 1. Basic epithelial ion transport principles and function* (pp. 461–541). Springer.

O'Grady, S. M., & Lee, S. Y. (2003). Chloride and potassium channel function in alveolar epithelial cells. *American Journal of Physiology - Lung Cellular and Molecular Physiology, 284*, L689–L700.

O'Grady, S. M., & Lee, S. Y. (2005). Molecular diversity and function of voltage-gated (Kv) potassium channels in epithelial cells. *International Journal of Biochemistry & Cell Biology, 37*, 1578–1594.

Olmeda, B., Martínez-Calle, M., & Pérez-Gil, J. (2017). Pulmonary surfactant metabolism in the alveolar airspace: Biogenesis, extracellular conversions, recycling. *Annals of Anatomy, 209*, 78–92.

Palk, L., Sneyd, J., Shuttleworth, T. J., Yule, D. I., & Crampin, E. J. (2010). A dynamic model of saliva secretion. *Journal of Theoretical Biology, 266*, 625–640.

Palmer, L. G., & Andersen, O. S. (2008). The two-membrane model of epithelial transport: Koefoed-Johnsen and Ussing (1958). *The Journal of General Physiology, 132*, 607–612.

Palmer, L. G., & Frindt, G. (2007). High-conductance K channels in intercalated cells of the rat distal nephron. *American Journal of Physiology - Renal Physiology, 292*, F966–F973.

Parker, M. D., & Boron, W. F. (2013). The divergence, actions, roles, and relatives of sodium-coupled bicarbonate transporters. *Physiological Reviews, 93*, 803–959.

Parker, M. (2023). Intracellular pH regulation. In F. J. Alvarez-Leefmans, E. Delpire, & E. Kaneshiro (Eds.), *Cell physiology source book: Essentials of membrane biophysics* (5th ed.). Elsevier.

Pinto da Silva, P., & Kachar, B. (1982). On tight-junction structure. *Cell, 28*, 441–450.

Plasschaert, L. W., Žilionis, R., Choo-Wing, R., Savova, V., Knehr, J., Roma, G., Klein, A. M., & Jaffe, A. B. (2018). A single-cell atlas of the airway epithelium reveals the CFTR-rich pulmonary ionocyte. *Nature, 560*, 377–381.

Purkerson, J. M., & Schwartz, G. J. (2007). The role of carbonic anhydrases in renal physiology. *Kidney International, 71*, 103–115.

Purkerson, J. M., Heintz, E. V., Nakamori, A., & Schwartz, G. J. (2014). Insights into acidosis-induced regulation of SLC26A4 (pendrin) and SLC4A9 (AE4) transporters using three-dimensional morphometric analysis of β-intercalated cells. *American Journal of Physiology - Renal Physiology, 307*, F601–F611.

Quinton, P. M., Elder, H. Y., Jenkinson, D. M., & Bovell, D. L. (1999). Structure and function of human sweat glands. In K. Laden (Ed.), *Antiperspirants and deodorants* (2nd ed., pp. 17–57). Marcel Dekker.

Quinton, P. M. (2017). Both ways at once: Keeping small airways clean. *Physiology, 32*, 380–390.

Rahner, C., Mitic, L. L., & Anderson, J. M. (2001). Heterogeneity in expression and subcellular localization of claudins 2, 3, 4, and 5 in the rat liver, pancreas, and gut. *Gastroenterology, 120*, 411–422.

Rao, R., Bhalla, V., & Pastor-Soler, N. M. (2019). Intercalated cells of the kidney collecting duct in Kidney Physiology. *Seminars in Nephrology, 39*, 353–367.

Rao, M. C. (2019). Physiology of electrolyte transport in the gut: Implications for disease. *Comprehensive Physiology, 9*, 947–1023.

Rechkemmer, G., Frizzell, R. A., & Halm, D. R. (1996). Active K^+ transport across Guinea pig distal colon: Action of secretagogues. *The Journal of Physiology, 493*, 485–502.

Reddy, M. M. (2020). Fundamentals of ion transport across human sweat gland in health and disease. In K. L. Hamilton, & D. C. Devor (Eds.) (2nd ed.,*Ion channels and transporters of epithelia in health and disease: Vol 2. Ion transport across epithelial tissues and disease* (pp. 143–175). Springer.

Reuter, D., Zierold, K., Schröder, W. H., & Frings, S. (1998). A depolarizing chloride current contributes to chemoelectrical transduction in olfactory sensory neurons in situ. *Journal of Neuroscience, 18*, 6623–6630.

Robertson, M. A., & Foskett, J. K. (1994). Na^+ transport pathways in secretory acinar cells: Membrane cross talk mediated by $[Cl^-]_i$. *American Journal of Physiology: Cell Physiology, 267*, C146–C156.

Rodan, A. R., Cheng, C. J., & Huang, C. L. (2011). Recent advances in distal tubular potassium handling. *American Journal of Physiology - Renal Physiology, 300*, F821–F827.

Rodriguez-Boulan, E., & Macara, I. G. (2014). Organization and execution of the epithelial polarity programme. *Nature Reviews Molecular Cell Biology, 15*, 225–242.

Sabolić, I., Herak-Kramberger, C. M., Breton, S., & Brown, D. (1999). Na^+/K^+-ATPase in intercalated cells along the rat nephron revealed by antigen retrieval. *Journal of the American Society of Nephrology, 10*, 913–922.

Sackin, H., & Palmer, L. G. (2013). Electrophysiological analysis of transepithelial transport. In R. J. Alpern, O. W. Moe, & M. Caplan (Eds.), *Seldin and Giebisch's the kidney: Physiology and pathophysiology* (5th ed., pp. 177–216). Elsevier.

Sandle, G. I., Perry, M. D., Mathialahan, T., Linley, J. E., Robinson, P., Hunter, M., & MacLennan, K. A. (2007). Altered cryptal expression of luminal potassium (BK) channels in ulcerative colitis. *The Journal of Pathology, 212*, 66–73.

Satchell, S. C., & Braet, F. (2009). Glomerular endothelial cell fenestrations: An integral component of the glomerular filtration barrier. *American Journal of Physiology - Renal Physiology, 296*, F947–F956.

Sato, K., & Sato, F. (1990). Na^+, K^+, H^+, Cl^-, and Ca^{2+} concentrations in cystic fibrosis eccrine sweat in vivo and in vitro. *The Journal of Laboratory and Clinical Medicine, 115*, 504–511.

Sato, K. (2001). The mechanism of eccrine sweat secretion. In C. V. Gisolfi, D. R. Lamb, & E. R. Nadel (Eds.), *Perspectives in exercise science and sports medicine: Vol 6. Exercise, heat, and thermoregulation* (pp. 85–117). Cooper Publishing Group.

Saumon, G., & Basset, G. (1993). Electrolyte and fluid transport across the mature alveolar epithelium. *Journal of Applied Physiology, 74*, 1–15.

Sausbier, M., Matos, J. E., Sausbier, U., Beranek, G., Arntz, C., Neuhuber, W., Ruth, P., & Leipziger, J. (2006). Distal colonic K^+ secretion occurs via BK channels. *Journal of the American Society of Nephrology, 17*, 1275–1282.

Scarpelli, E. M. (2003). Physiology of the alveolar surface network. *Comparative Biochemistry and Physiology Part A: Molecular & Integrative Physiology, 135*, 39–104.

Schlingmann, B., Molina, S. A., & Koval, M. (2015). Claudins: Gatekeepers of lung epithelial function. *Seminars in Cell & Developmental Biology, 42*, 47–57.

Schraufnagel, D. E. (2010). Lung lymphatic anatomy and correlates. *Pathophysiology, 17*, 337–343.

Schultz, S. G., Thompson, S. M., & Suzuki, Y. (1981). Equivalent electrical circuit models and the study of Na^+ transport across epithelia: Nonsteady-state current-voltage relations. *Federation Proceedings, 40*, 2443–2449.

Scudieri, P., & Galietta, L. J. V. (2020). TMEM16 proteins (Anoctamins) in epithelia. In K. L. Hamilton, & D. C. Devor (Eds.) (2nd ed.,*Ion channels and transporters of epithelia in health and disease: Vol 3. Studies of epithelial transporters and ion channels* (pp. 671–696). Springer.

Shashikanth, N., Yeruva, S., Ong, M. L. D. M., Odenwald, M. A., Pavlyuk, R., & Turner, J. R. (2017). Epithelial organization: The gut and beyond. *Comprehensive Physiology, 7*, 1497–1518.

Shorofsky, S. R., Field, M., & Fozzard, H. A. (1983). Electrophysiology of Cl secretion in canine trachea. *Journal of Membrane Biology, 72*, 105–115.

Shorofsky, S. R., Field, M., & Fozzard, H. A. (1984). Mechanism of Cl secretion in canine trachea: Changes in intracellular chloride activity with secretion. *Journal of Membrane Biology, 81*, 1–8.

Silva, P., Stoff, J., Field, M., Fine, L., Forrest, J. N., & Epstein, F. H. (1977). Mechanism of active chloride secretion by shark rectal gland: Role of Na^+-K^+-ATPase in chloride transport. *American Journal of Physiology - Renal Physiology, 233*, F298–F306.

Silva, P. H. I., & Nilufar Mohebbi, N. (2022). Kidney metabolism and acid-base control: Back to the basics. *Pflügers Archiv, 474*, 919–934.

Simmons, S., Erfinanda, L., Bartz, C., & Kuebler, W. M. (2019). Novel mechanisms regulating endothelial barrier function in the pulmonary microcirculation. *The Journal of Physiology, 597*, 997–1021.

Simon, M., Duong, J. P., Mallet, V., Jian, R., MacLennan, K. A., Sandle, G. I., & Marteau, P. (2008). Over-expression of colonic K^+ channels associated with severe potassium secretory diarrhoea after haemorrhagic shock. *Nephrology Dialysis Transplantation, 23*, 3350–3352.

Smith, P. L., & Frizzell, R. A. (1984). Chloride secretion by canine tracheal epithelium: IV. Basolateral membrane K permeability parallels secretion rate. *Journal of Membrane Biology, 77*, 187–199.

Song, C., Ma, H. P., & Eaton, D. C. (2020). Epithelial sodium channels (ENaC). In K. L. Hamilton, & D. C. Devor (Eds.) (2nd ed.,*Ion channels and transporters of epithelia in health and disease: Vol 3. Studies of epithelial transporters and ion channels* (pp. 697–803). Springer.

Sozio, F., Rossi, A., Weber, E., Abraham, D. J., Nicholson, A. G., Wells, A. U., Renzoni, E. A., & Sestini, P. (2012). Morphometric analysis of intralobular, interlobular and pleural lymphatics in normal human lung. *Journal of Anatomy, 220*, 396–404.

Spanier, B., & Rohm, F. (2018). Proton coupled oligopeptide transporter 1 (PepT1) function, regulation, and influence on the intestinal homeostasis. *Comprehensive Physiology, 8*, 843–869.

Stauber, T., Weinert, S., & Jentsch, T. J. (2012). Cell biology and physiology of CLC chloride channels and transporters. *Comprehensive Physiology, 2*, 1701–1744.

Tamura, A., Hayashi, H., Imasato, M., Yamazaki, Y., Hagiwara, A., Wada, M., Noda, T., Watanabe, M., Suzuki, Y., & Tsukita, S. (2011). Loss of claudin-15, but not claudin-2, causes Na^+ deficiency and glucose malabsorption in mouse small intestine. *Gastroenterology, 140*, 913–923.

Taylor, A. E., & Parker, J. C. (1985). Pulmonary interstitial spaces and lymphatics. In *Supplement 10, handbook of physiology, the respiratory system, circulation and nonrespiratory functions* (pp. 167–230). https://doi.org/10.1002/cphy.cp030104

Thévenod, F. (2002). Ion channels in secretory granules of the pancreas and their role in exocytosis and release of secretory proteins. *American Journal of Physiology: Cell Physiology, 283*, C651–C672.

Thompson, S. M. (1986). Relations between chord and slope conductances and equivalent electromotive forces. *American Journal of Physiology: Cell Physiology, 250*, C333–C339.

Thorsen, K., Drengstig, T., & Ruoff, P. (2014). Transepithelial glucose transport and Na^+/K^+ homeostasis in enterocytes: An integrative model. *American Journal of Physiology: Cell Physiology, 307*, C320–C337.

Tirindelli, R., Dibattista, M., Pifferi, S., & Menini, A. (2009). From pheromones to behavior. *Physiological Reviews, 89*, 921–956.

Tosteson, D. C., & Hoffman, J. F. (1960). Regulation of cell volume by active cation transport in high and low potassium sheep red cells. *The Journal of General Physiology, 44*, 169–194.

Traynor, T. R., & O'Grady, S. M. (1996). Regulation of colonic ion transport by GRP. I. GRP stimulates transepithelial K^+ and Na^+ secretion. *American Journal of Physiology: Cell Physiology, 270*, C848–C858.

Tsukita, S., Tanaka, H., & Tamura, A. (2019). The claudins: From tight junctions to biological systems. *Trends in Biochemical Sciences, 44*, 141–152.

Ussing, H. H., & Windhager, E. E. (1964). Nature of shunt path and active sodium transport path through frog skin epithelium. *Acta Physiologica Scandinavica, 61*, 484–504.

Vallon, V., & Nakagawa, T. (2021). Renal tubular handling of glucose and fructose in health and disease. *Comprehensive Physiology, 12*, 2995–3044.

Van Itallie, C. M., Fanning, A. S., & Anderson, J. M. (2003). Reversal of charge selectivity in cation or anion-selective epithelial lines by expression of different claudins. *American Journal of Physiology - Renal Physiology, 285*, F1078–F1084.

Verkman, A. S. (2007). Role of aquaporins in lung liquid physiology. *Respiratory Physiology & Neurobiology, 159*, 324–330.

Verschuren, E. H. J., Castenmiller, C., Peters, D. J. M., Arjona, F. J., Bindels, R. J. M., & Hoenderop, J. G. J. (2020). Sensing of tubular flow and renal electrolyte transport. *Nature Reviews Nephrology, 16*, 337–351.

Vogalis, F., Hegg, C. C., & Lucero, M. T. (2005). Electrical coupling in sustentacular cells of the mouse olfactory epithelium. *Journal of Neurophysiology, 94*, 1001–1012.

Wagner, C. A., Finberg, K. E., Breton, S., Marshansky, V., Brown, D., & Geibel, J. P. (2004). Renal vacuolar H^+-ATPase. *Physiological Reviews, 84*, 1263–1314.

Wall, S. M., Verlander, J. W., & Romero, C. A. (2020). The renal physiology of pendrin-positive intercalated cells. *Physiological Reviews, 100*, 1119–1147.

Wang, W. H., & Lin, D. H. (2020). Inwardly rectifying K^+ channel 4.1 regulates renal K^+ excretion in the aldosterone-sensitive distal nephron. In K. L. Hamilton, & D. C. Devor (Eds.) (2nd ed.,*Ion channels and transporters of epithelia in health and disease: Vol 3. Studies of epithelial transporters and ion channels* (pp. 823–857). Springer.

Wang, Y., Cohen, J., Boron, W. F., Schulten, K., & Tajkhorshid, E. (2007). Exploring gas permeability of cellular membranes and membrane channels with molecular dynamics. *Journal of Structural Biology, 157*, 534–544.

Watlington, C. O., & Huf, E. G. (1971). β-Adrenergic stimulation of frog skin mucous glands: Non-specific inhibition by adrenergic blocking agents. *Comparative and General Pharmacology, 2*, 295–305.

Weber, E., Sozio, F., Borghini, A., Sestini, P., & Renzoni, E. (2018). Pulmonary lymphatic vessel morphology: A review. *Annals of Anatomy, 218*, 110–117.

Wei, H., & Wang, J. Y. (2021). Role of polymeric immunoglobulin receptor in IgA and IgM transcytosis. *International Journal of Molecular Sciences, 22*, 2284.

Weinstein, A. M. (2015). A mathematical model of the rat nephron: Glucose transport. *American Journal of Physiology - Renal Physiology, 308*, F1098–F1118.

Weisz, O. A. (2020). Molecular mechanisms of apical and basolateral sorting in polarized epithelial cells. In K. L. Hamilton, & D. C. Devor (Eds.) (2nd ed.,*Ion channels and transporters of epithelia in health and disease: Vol 1. Basic epithelial ion transport principles and function* (pp. 135–158). Springer.

Welling, L. W., & Welling, D. J. (1976). Shape of epithelial cells and intercellular channels in the rabbit proximal nephron. *Kidney International, 9*, 385–394.

Welling, L. W., & Welling, D. J. (1988). Relationship between structure and function in renal proximal tubule. *Journal of Electron Microscopy Technique, 9*, 171–185.

Welling, D. J., Welling, L. W., & Hill, J. J. (1978a). Phenomenological model relating cell shape to water reabsorption in proximal nephron. *American Journal of Physiology - Renal Physiology, 234*, F308–F317.

Welling, L. W., Welling, D. J., & Hill, J. J. (1978b). Shape of cells and intercellular channels in rabbit thick ascending limb of Henle. *Kidney International, 13*, 144–151.

Welling, L. W., Evan, A. P., & Welling, D. J. (1981). Shape of cells and extracellular channels in rabbit cortical collecting ducts. *Kidney International, 20*, 211–222.

Welling, L. W., Evan, A. P., Welling, D. J., & Gattone, V. H. (1983). Morphometric comparison of rabbit cortical connecting tubules and collecting ducts. *Kidney International, 23*, 358–367.

Welling, L. W., Welling, D. J., Holsapple, J. W., & Evan, A. P. (1987). Morphometric analysis of distinct microanatomy near the base of proximal tubule cells. *American Journal of Physiology - Renal Physiology, 253*, F126–F140.

Welling, D., Urani, J., Welling, L., & Wagner, E. (1996). Fractal analysis and imaging of the proximal nephron cell. *American Journal of Physiology: Cell Physiology, 270*, C953–C963.

Welling, P. A. (2016). Roles and regulation of renal K channels. *Annual Review of Physiology, 78*, 415–435.

Welsh, M. J., Smith, P. L., & Frizzell, R. A. (1982). Chloride secretion by canine tracheal epithelium: II. The cellular electrical potential profile. *Journal of Membrane Biology, 70*, 227–238.

Welsh, M. J., Karp, P., & Ruppert, T. R. (1983). Evidence for basolateral membrane potassium conductance in canine tracheal epithelium. *American Journal of Physiology: Cell Physiology, 244*, C377–C384.

Welsh, M. J., Smith, P. L., & Frizzell, R. A. (1983). Chloride secretion by canine tracheal epithelium: III. Membrane resistances and electromotive forces. *Journal of Membrane Biology, 71*, 209–218.

Wieczorek, H., Brown, D., Grinstein, S., Ehrenfeld, J., & Harvey, W. R. (1999). Animal plasma membrane energization by proton-motive V-ATPases. *BioEssays, 21*, 637–648.

Williams, J. A., & Dawson, D. C. (2006). Cell structure and function. In M. W. Mulholland, K. D. Lillemoe, G. M. Doherty, R. V. Maier, & G. R. Upchurch (Eds.), *Greenfield's surgery: Scientific principles and practice* (4th ed., pp. 2–42). Lippincott Williams & Wilkins.

Woodward, O. M., & Guggino, W. B. (2013). Anion channels. In R. J. Alpern, O. W. Moe, & M. Caplan (Eds.), *Seldin and Giebisch's the kidney: Physiology and pathophysiology* (5th ed., pp. 1019–1045). Elsevier.

Wright, E. M., Loo, D. D., & Hirayama, B. A. (2011). Biology of human sodium glucose transporters. *Physiological Reviews, 91*, 733–794.

Xie, Y., Ostedgaard, L., Abou Alaiwa, M. H., Lu, L., Fischer, A. J., & Stoltz, D. A. (2018). Mucociliary transport in healthy and cystic fibrosis pig airways. *Annals of the American Thoracic Society, 15*(Suppl. 3), S171–S176.

Yu, A. S. (2015). Claudins and the kidney. *Journal of the American Society of Nephrology, 26*, 11–19.

Yu, A. S. L. (2017). Paracellular transport and energy utilization in the renal tubule. *Current Opinion in Nephrology and Hypertension, 26*, 398–404.

Zhang, J., Halm, S. T., & Halm, D. R. (2009). Adrenergic activation of electrogenic K^+ secretion in Guinea pig distal colonic epithelium: Desensitization via the Y2-neuropeptide receptor. *American Journal of Physiology - Gastrointestinal and Liver Physiology, 297*, G278–G291.

Zhang, J., Halm, S. T., & Halm, D. R. (2012). Role of the BK channel ($K_{Ca}1.1$) during activation of electrogenic K^+ secretion in Guinea pig distal colon. *American Journal of Physiology - Gastrointestinal and Liver Physiology, 303*, G1322–G1334.

Zhao, J., Krystofiak, E. S., Ballesteros, A., Cui, R., Van Itallie, C. M., Anderson, J. M., Fenollar-Ferrer, C., & Kachar, B. (2018). Multiple claudin-claudin cis interfaces are required for tight junction strand formation and inherent flexibility. *Communications Biology, 1*, 50.

Chapter 18

Aquaporin water channels

Michal K. Oklinski[a], Frédéric H. Login[a] and Lene N. Nejsum
Department of Clinical Medicine, Aarhus University, Aarhus, Denmark

1. Introduction

Normal epithelia are composed of tightly packed, polarized cells, which constitute barriers between biological compartments, like the inside and the outside of an organism (Nelson, 2003). A simple epithelium has three distinct plasma membrane domains: the apical plasma membrane which faces a lumen or the surface, the lateral membrane, where the cells form cell-cell adhesion and the basal membrane, where the cells adhere to the extracellular matrix. The tight junctions, positioned at the apex of the cells, form a seal between the apical and basolateral plasma membrane domains, which prevents intermixing of proteins and lipids.

Transepithelial water transport requires movement of water across epithelial plasma membranes and is essential in multiple physiological processes including regulation of body water homeostasis and fluid secretion (Knepper et al., 2015; Nejsum, 2005). While water will diffuse over a lipid bilayer, the diffusion rate is low and the bilayer thus has a low capacity for water transport. Thus, a simple diffusion model could not explain how some cells and epithelia, like red blood cells as well as renal proximal tubules and secretory glands, exhibited high water permeability. Thus, it was hypothesized that transcellular water transport in these tissues is not mediated by simple diffusion but rather via selective transport proteins. This hypothesis was further supported by the finding that mercury compounds reversibly blocked water transport in red blood cells (Macey, 1984).

In the late 1980s and early 1990s, Professor Peter Agre, a hematologist by training, studied the rhesus blood group. When purifying the 32 kDa Rh polypeptide from red blood cells, an extra band migrating around 28 kDa appeared on the gel and was first considered a contaminant or a proteolytic product (Saboori et al., 1989). Antibodies were raised against the purified polypeptide, however, it turned out to only recognize the 28 kDa protein. Thus, the band was not a proteolytic fragment but a protein of 28 kDa. The predicted structure suggested that the protein was an integral membrane protein with six transmembrane domains with the N- and C-terminal domains in the cytoplasm. The protein was named CHIP28, for Channel-like Integral membrane Protein of 28 kDa (Preston & Agre, 1991). The protein displayed high homology to MIP26, the major integral protein of lens, thought to be a channel protein, as well as to proteins located in the root tips of plants. The CHIP28 cDNA was cloned from human fetal liver and to test if indeed CHIP28 was a functional water channel, the in vitro transcribed RNA was injected into *Xenopus laevis* oocytes. *Xenopus laevis* oocytes have very low osmotic activity and can be transferred between isosmotic buffer and fresh water without apparent changes in shape. However, upon shifting the CHIP28 cDNA injected oocytes from isosmotic buffer to distilled water, the oocytes swelled and burst indicating water movement along the osmotic gradient into the oocytes. This proved that CHIP28 was a functional water channel (Harris et al., 1992; Preston et al., 1992). In 2003, Professor Agre was awarded the Nobel Prize in Chemistry for this groundbreaking discovery (Agre, 2004). CHIP28 was later renamed aquaporin-1 (AQP1) and since, a number of different AQP isoforms has been identified in all living organisms including mammals, bacteria, plants, and fungi, where permeation of water through lipid membranes is critical for physiological function. This chapter focuses mainly on mammalian AQP isoforms. Select examples of AQP localization and function in normal physiology and pathophysiology are provided.

---

[a] These authors contributed equally.

Cell Physiology Sourcebook. https://doi.org/10.1016/B978-0-12-811114-7.00016-5

2. AQP diversity

Thirteen mammalian AQPs (AQP0-12) with specific tissue and cellular expressions have been identified. Based on their permeability and the level of primary sequence identity, mammalian AQPs have been divided into three groups. The classical or orthodox AQPs, AQP0, AQP1, AQP2 AQP4, AQP5, AQP6, and AQP8 are permeable to water. Besides water, aquaglyceroporins, AQP3, AQP7, AQP9, and AQP10, are also permeable to glycerol. The last group, named superaquaporins, encompasses only two members, AQP11 and AQP12. Superaquaporins display the lowest amino acid sequence identity (around 20%) with the remaining AQPs and due to their intracellular localization, they are also referred to as subcellular AQPs (Calvanese et al., 2013; Morishita et al., 2004). Indeed, ectopic expression showed that AQP12 localizes in the cytoplasm (Itoh et al., 2005) and that AQP11 colocalizes with the endoplasmic reticulum (Morishita et al., 2005). Experiments with purified proteins inserted into liposomes showed that AQP11 and AQP1 have similar water permeability, which strongly suggests that intracellular AQPs might contribute to organelle homeostasis upon osmotic changes (Yakata et al., 2007).

Prokaryotes express fewer AQP isoforms. In *Escherichia coli* two isoforms were identified namely the water permeable channel AqpZ (Calamita et al., 1995) and the aquaglyceroporin GlpF (Sweet et al., 1990). Genomic analysis showed that orthologues of *aqpZ* and/or *glpF* are found in most bacteria and archaea (Tong et al., 2019). However, no orthologue was detected in the genome of a subset of prokaryotes suggesting that these microorganisms either express unidentified AQPs with very low sequence homology to *aqpZ* and *glpF* or transport water through AQP-independent mechanisms.

Unlike prokaryotes, plants express a relatively high number of AQP isoforms. For instance, 35 AQPs were identified in *Arabidopsis*, 34 in rice, and the largest number to date was found in soybean with 72 different isoforms (Deshmukh et al., 2015). Based on sequence homology and cellular localization, plant AQPs are classified in seven main subfamilies with versatile functions in plant physiology (Chaumont & Tyerman, 2014). Plasma membrane intrinsic proteins (PIPs) is the largest and most conserved subfamily, nodulin 26-like proteins (NIPs), and small basic intrinsic proteins (SIPs) are found in plant-cell plasma membranes. NIPs are also expressed in the symbiosome of legume root nodules and the endoplasmic reticulum of some plants. Tonoplast intrinsic proteins (TIPs) are found in the membranes of vacuoles. PIPs, TIPs, NIPs, and SIPs are widespread among terrestrial plants while X intrinsic proteins (XIPs) are mainly found in lower plants and a few terrestrial plants. Two additional aquaporin subfamilies, hybrid intrinsic proteins (HIPs) and GlpF-like intrinsic proteins (GIPs), are solely found in lower nonvascular plants. The latter subfamily is closely related to the bacterial aquaglyceroporin GlpF (Gustavsson et al., 2005).

3. AQP structure

3.1 Topology of AQPs

AQPs are small integral membrane proteins with the molecular weight of the monomers averaging 30 kDa. The AQPs all form tetramers with each monomer containing a water selective pore (Fig. 18.1). In the center of the tetramers, a fifth hydrophobic pore is formed and in plant AQPs, this central pore is believed to support diffusion of gases, in particular CO_2 (Fig. 18.1) (Kaldenhoff et al., 2014). In the vast majority of cases, AQPs form homotetramers but some plant AQPs belonging to the PIP subfamily are able to form specific heterotetramers. It was shown that different ratios of two PIP isoforms influence the cellular localization of the resulting tetramer as well as the intrinsic permeability of each monomer (Bienert et al., 2018; Yaneff et al., 2014). To date, no heterotetramers between the 13 mammalian isoforms have been described. However, it was shown that AQP4 full length (M1) and its splicing variant lacking the 22 first residues (M23) can form heterotetramers (Neely et al., 1999). Freeze-fracture electron microscopy observations of different tissues including brain, kidney, and liver have shown that AQP4 assembles in the membrane in organized supramolecular structures called orthogonal arrays of particles (OAPs) (Yang et al., 1996). While M1 homotetramers cannot form OAPs, M23 homotetramers form abundant and large OAPs. M1/M23 heterotetramers also form OAPs but of smaller sizes (Furman et al., 2003). It was shown that M23 monomers within OAPs have higher water permeability than M1 monomers, suggesting that M1/M23 expression ratio may regulate overall membrane organization and permeability of AQP4 (Silberstein et al., 2004).

All AQP monomers consist of six transmembrane helices that are connected by five loops (loop A to E) and both the N- and C-termini are located within the cytoplasm (Fig. 18.1). The cytoplasmic N- and C-termini display the lowest sequence similarity across AQPs with variation in primary sequence but also in length, which probably confer specific features to some AQPs. The hallmark of AQPs is the two conserved Asparagine-Proline-Alanine (NPA) motifs within loops B and E. These two loops form two half-helices that fold back into the membrane from opposite sides to form a seventh

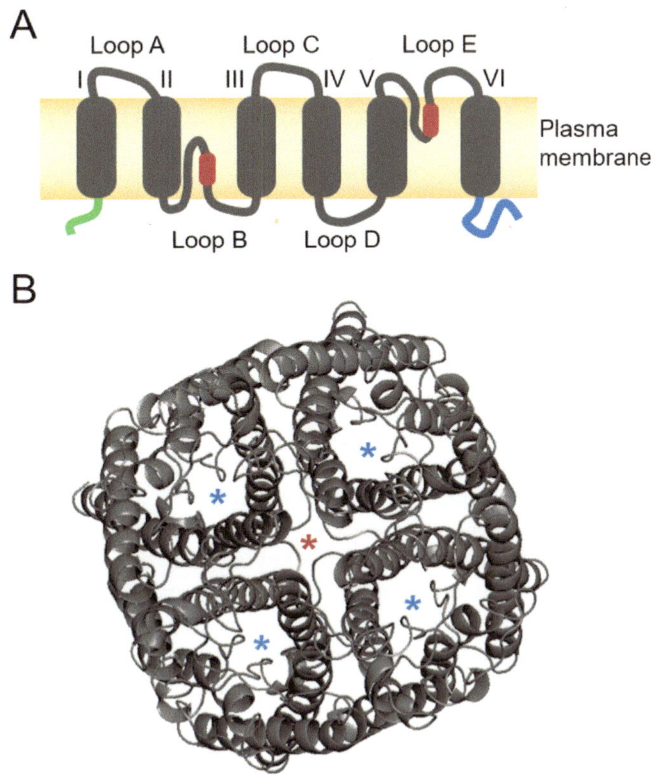

FIGURE 18.1 **Aquaporin topology.** (A) Schematic depicting the general topology of AQPs. The six transmembrane domains (I to VI) are connected by five loops (A to E). Both N- (green) and C-terminal (blue) domains are located in the cytoplasm. The two half-helices within loop B and E containing the NPA motifs (red) fold back into the membrane to form the water selective pore. (B) Top view of the 3D structure of the human AQP1 tetramer visualized with PyMOL (uniprot entry: P29972 and PDB entry: 4CSK). Red and blue asterisks respectively indicate the tetramer central pore and the water selective pore of each monomer.

pseudotransmembrane domain that generates the water selective pore (Fig. 18.1). In mammalian AQP7 and super-aquaporins as well as in several plant AQPs, the alanine residue in at least one of the NPA motifs is substituted. In mammalian AQP11 and AQP12 alanine in their loop B motifs is substituted for cysteine and threonine residues, respectively, making them NPC and NPT motifs (Ishibashi et al., 2000; Itoh et al., 2005). In fact, mammalian AQP7 has one asparagine–alanine–alanine (NAA) motif and one asparagine–proline–serine (NPS) motif. In some plants, amino acids such as cysteine, threonine, serine, valine, and leucine occupy the third residue position in one of the NPA motifs (Sonah et al., 2017). Although the main function of the NPA motifs is to form the water selective pore, site-directed mutagenesis indicated that NPA motifs might be elusively involved in proper targeting and oligomerization of some AQPs. For example, an AQP11 mutant in which the cysteine 101 in the NPC motif was substituted with an alanine, showed lower water permeability, decreased oligomerization in vitro, but identical plasma membrane localization (Ikeda et al., 2011). In contrast, substitution of alanine by threonine in the first NPA motif of AQP4 prevented plasma membrane targeting while a cysteine substitution had no effect (Guan et al., 2010).

3.2 Structural determinants of AQP selectivity

Measurements of osmotic water permeability coefficients (P_f) of AQPs expressed in *Xenopus* oocytes showed a wide range of water permeability among mammalian AQPs (Yang & Verkman, 1997). The P_f of the water selective channel AQP1 is about 6×10^{-14} cm^3/s while the aquaglyceroporin AQP3 has a P_f averaging about 2.1×10^{-14} cm^3/s. In contrast, glycerol uptake was increased by 4.5–8 folds in oocytes expressing AQP3, while no significant changes were detected for oocytes expressing AQP1 or AQP4. AQP4 displays the highest water permeability with a P_f of 24×10^{-14} cm^3/s. Similar observations were made for plant AQPs (Maurel, 1997). This disparity in water permeability relies on subtle structural differences. While AQPs support passive diffusion of water in response to osmotic gradients, they are impermeable to protons (H^+) and cations, thus preventing interference in the homeostasis of the membrane potential. Proton impermeability has been the focus of several studies since the intriguing discovery of an absence of Grotthuss H^+ transport through AQP channels. Because of its dipolar nature, H_2O molecules can be acceptor or donor of hydrogen-bonds (H-bonds) to connect with surrounding H_2O molecules but also with other H-bond donors/acceptors such as hydrophilic biological molecules. When H_2O molecules react with an H^+, hydronium (H_3O^+) ions are formed. Grotthuss H^+ transport corresponds to covalent and H-bond rearrangements that allow proton transfer from H_3O^+ to neighboring H_2O molecules or hydrophilic molecules such as proteins (Fig. 18.2) (de Grotthuss, 2006; Swanson et al., 2007). Thus, instead of conventional diffusion, in Grotthuss transport, H^+ rapidly shuttles from one molecule to the next. This mechanism is key in various biological systems including ATP synthesis by oxidative phosphorylation in mitochondria and photoexcitation of green fluorescent protein (GFP) (Kaur et al., 2021).

Structural studies using cryoelectron microscopy or X-ray crystallography on AQPs from various organisms revealed the hourglass geometry of water channels (Eriksson et al., 2013; Jung et al., 1994; Murata et al., 2000; Tajkhorshid et al., 2002; Tornroth-Horsefield et al., 2006; Walz et al., 1997). High-resolution 3D structures combined with computational modeling helped to decipher the mechanisms behind proton exclusion and motion of H_2O molecules. The selectivity of AQPs toward H_2O molecules and uncharged small solutes rely on conserved structural features within two regions along the water channel: the dual NPA motif and the aromatic arginine region (ar/R-region). With the exception of specific sites within the ar/R-region and NPA motifs, the inside of the pore is rather hydrophobic, lined with backbone carbonyl oxygens that are H-bond acceptors. The ar/R-region acts as the selectivity filter (SF) and corresponds to the narrowest point of the channel, located toward the extracellular side. The SF is formed by a highly conserved arginine that is positioned after the alanine residue of the second NPA motif in the primary protein sequence (Fig. 18.3). Three additional amino acids constitute the ar/R-region and structural data indicate that the nature of these residues dictates the constriction level as well as the hydrophobicity of the SF. Among orthodox AQPs, a conserved histidine residue also contributes to the SF and it was proposed that positive charges around the side chains of the conserved arginine and histidine establish a repulsive barrier for H_3O^+ at the channel entrance (Eriksson et al., 2013). In aquaglyceroporins, the histidine residue is replaced by a smaller nonpolar amino acid such as a glycine for AQP3, AQP7, and AQP10. The latest model indicates that H_2O molecules move pair-wise through the SF by forming H-bonds with the side chain of ar/R-region residues, weakening the connectivity of the "pair" with the H_2O molecules outside the SF thus facilitating the motion through the channel. Nitrogen of the conserved arginine and histidine side chains act as H-bond donor and acceptor respectively (Fig. 18.4). Following the SF, H_2O molecules accept H-bond from asparagine residues of the NPA motifs with each asparagine interacting with one water molecule (Eriksson et al., 2013). These protein—water H-bonds involving SF and NPA-motif residues constrain

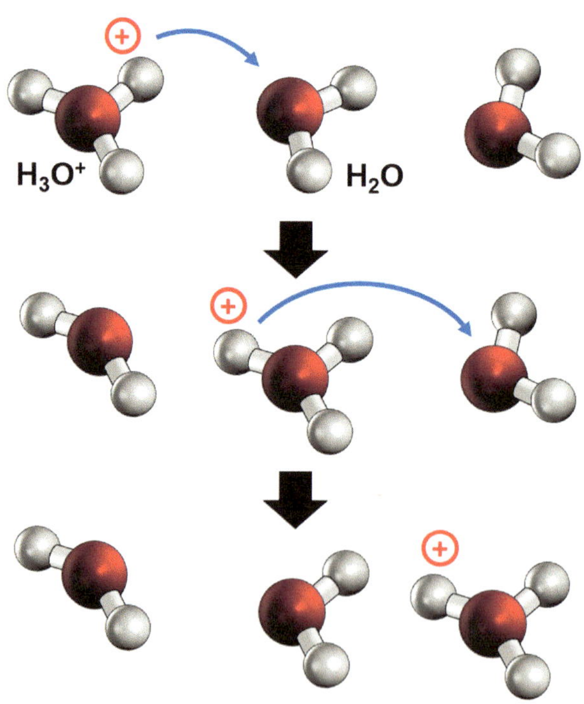

FIGURE 18.2 **The Grotthuss mechanism.** Mechanism of proton transfer from hydronium (H3O+) ions to neighboring water molecule via covalent and H-bond rearrangement.

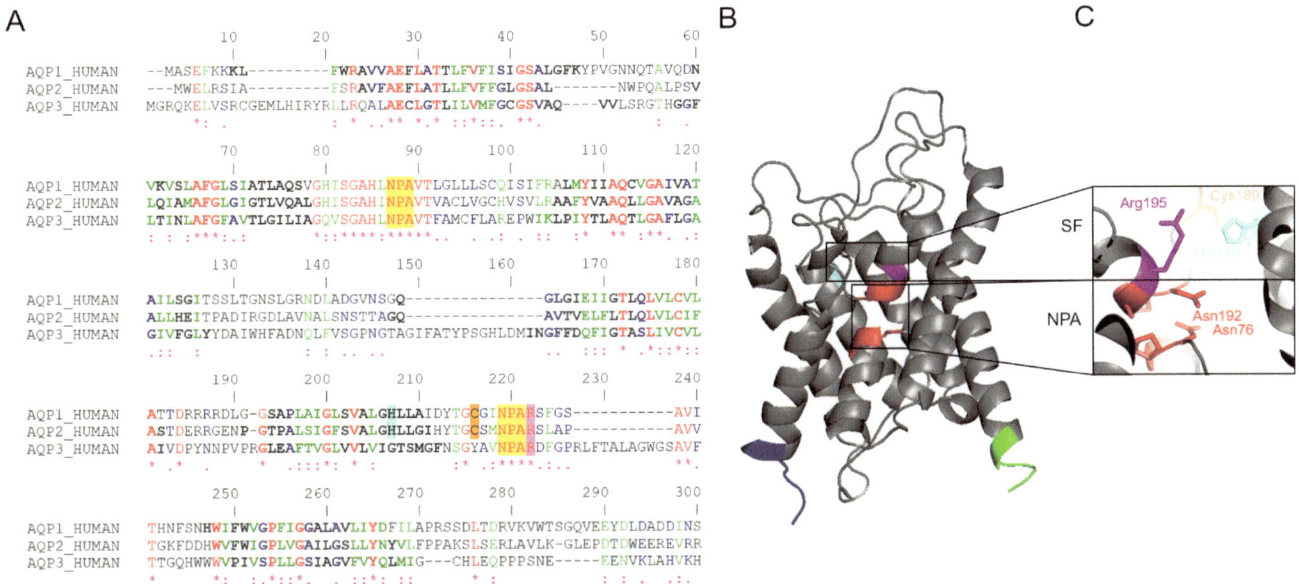

FIGURE 18.3 **Structural determinants of aquaporin selectivity.** (A) ClustalW alignment of primary sequences of human AQP1 (uniprot entry: P29972), AQP2 (uniprot entry: P41181), and AQP3 (uniprot entry: Q92482). The residues of the different transmembrane domains are in bold. The NPA motifs and the conserved arginine of the ar/R-region are highlighted in yellow and pink, respectively. The SF conserved histidine among water selective aquaporins are highlighted in blue; a glycine occupies this position in the aquaglyceroporin AQP3. Mercury-sensitive cysteine residues in AQP1 and AQP2 are highlighted in orange. (B) Side view of the 3D structure of the human AQP1 monomer visualized with PyMOL (uniprot entry: P29972 and PDB entry: 4CSK). N- and C-terminal domains are indicated in green and blue respectively. Residues critical for water selectivity are colored differentially: NPA motifs (red), Arg195 (pink), and His180 (light blue). (C) Enlarge view of the dual NPA motif and SF after 180 degrees horizontal rotation. The side chains of Asn 76 and 192 (red), Arg195 (pink), and His180 (light blue) and mercury sensitive Cys189 (light orange) are indicated.

a specific orientation to H_2O molecules that significantly perturb the connectivity between H_2O molecules preventing Grotthuss H^+ transport. In the cytoplasmic half of the channel, H_2O molecules adopt a single-file arrangement and water–water H-bonds promote the correlated motion of H_2O molecules. Thus, the water channel selectivity relies on the channel geometry, well-defined water–protein and water–water H-bonds as well as an electrostatic barrier. Electron crystallography structure showed that the AQP1 pore at the SF has a 3 Å diameter, which is slightly larger than a 2.8 Å water molecule (Murata et al., 2000). In comparison to aquaglyceroporins, the diameter of the SF in orthodox AQPs is about 1 Å narrower and is more hydrophilic which contribute to glycerol impermeability (Wang et al., 2005). Mercurial compounds such as $HgCl_2$ or p-chloromercurybenzene sulphonate inhibit water permeation by binding to accessible cysteine residues of AQPs (Macey & Farmer, 1970; Preston et al., 1993; Savage & Stroud, 2007). This inhibition can be reverse by addition of reducing agents. Site-directed mutagenesis of cysteine residues of AQP1 showed that cysteine 189, located in the SF vicinity is the mercury-sensitive residue (Preston et al., 1993). Comparison of mercury blocked and unbound structures of the bacterial AqpZ (homologue of AQP1) showed that mercury occludes the pore without provoking conformational changes (Savage & Stroud, 2007).

Several studies demonstrated that besides water and glycerol, some AQPs are capable of transporting other solutes including urea, NH_3, H_2O_2, CO_2, and O_2 (Bienert et al., 2007; Geyer et al., 2013; Kaldenhoff et al., 2014; Li et al., 2014; Wu & Beitz, 2007; Zwiazek et al., 2017). Even though NH_3 is a dipolar molecule of similar size as water, not all AQPs are permeable to NH_3. While mammalian AQP0, AQP2, and AQP5 lack NH_3 permeability, AQP1, AQP4, AQP9, and the wheat plant AQP TIP2;1 are permeable to NH_3 (Geyer et al., 2013). The physiological relevance of AQP-mediated NH_3 transport is puzzling since specialized NH_3 transporters are known to be responsible for NH_3 homeostasis. Human aquaglyceroporins AQP7 and AQP9, as well as prokaryote GlpF facilitate uptake of arsenic, which indicate one of the possible ways that toxic metalloids, such as arsenic, penetrate cells and enter the food chain (Liu et al., 2002). Recently, it was reported that human aquaglyceroporins, AQP3, AQP7, AQP9, and AQP10, could act as transporters of orthosilicic acid (H_4SiO_4), which is the dietary form of silicon (Garneau et al., 2015). In *Arabidopsis* roots, the aquaporin NIP1;2 acts as a plasma membrane transporter of malate-chelated aluminum ions from the cell wall into the cytoplasm to be further sequestrated into leaf vacuoles, thereby increasing the plant tolerance to aluminum toxicity (Wang et al., 2017). Thus, permeation of solutes other than glycerol and water implicate many AQPs in important functions such as resistance to environmental toxicity or acquisition of essential chemical elements.

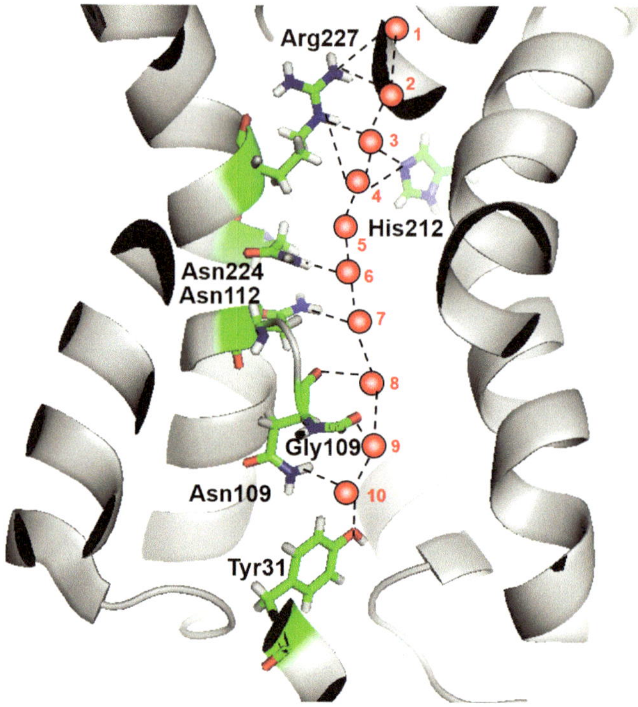

FIGURE 18.4 **Atomic model of water permeation.** Enlarge view of the pore of the sub-Ångstrom X-ray structure of Aqy1 from yeast *Pichia pastoris* (PDB entry: 3ZOJ) visualized with PyMOL. The side chain of residues implicated in water–protein interactions are colored by element (carbon atoms in green, hydrogen in gray, nitrogen in blue, and oxygen in red). The black-dashed lines illustrate H-bonds implicating residue side chains and H_2O molecules (red spheres) passing through the pore. H_2O molecules move pairwise into the SF by interacting with H-bond donor residues Arg227 and His212 (H_2O molecules one–4). Following the SF, single file H_2O molecules accept H-bonds from Asn224 and Asn112 (H_2O molecules 6 and 7). In the cytoplasmic part of the channel, H_2O molecule 8 donates an H-bond to H_2O molecule 9, which donates an H-bond to molecule 10. H_2O molecule 10 donates an H-bond to Tyr31.

4. Functional regulation of AQPs

Land-living organisms are subjected to constant changes in environmental water availability that require physiological adaptation to regulate intracellular water levels. Besides transcriptional regulation, AQP gating and cellular trafficking of AQPs are the main mechanisms used to regulate water motion efficiency through AQPs and cellular membranes, respectively.

4.1 AQP gating

All PIPs are putatively gated making regulation via gating a more common occurrence in plant AQPs than in their mammalian counterparts. Experiments with PIP bearing vesicles isolated from plasma membranes of beetroot showed a drastic decrease in water permeability upon intravesicular acidification or increased concentration of intravesicular calcium (Ca^{2+}) (Alleva et al., 2006). Site-directed mutagenesis towards two conserved serine phosphorylation sites of spinach AQP SoPIP2;1 (formerly named PM28A) promoted a significant decrease in water permeability (Johansson et al., 1998). Altogether, these observations demonstrated that cytoplasmic pH and Ca^{2+} concentration as well as phosphorylation are regulatory factors in PIP gating. Structural mechanisms underlying PIP gating were identified by comparing X-ray structures of SoPIP2;1 in closed and open states (Tornroth-Horsefield et al., 2006). In the closed state, the cytoplasmic loop D folds underneath the channel creating a hydrophobic plug restricting passage of water. The loop D conformation in the closed state is stabilized by a network of ionic and H-bond interactions between the loop D and the N-terminus. Ca^{2+} binding to a specific site or protonation of a histidine residue within the loop D due to cytoplasmic acidification, both promotes and stabilizes the closed state. In contrast, phosphorylation on serine 115 in the loop B and serine 274 in the C-terminal disturb the H-bond network leading to pore opening.

Similar regulatory mechanisms were described for mammalian AQP0. Indeed, an increase in water permeability was measured for bovine AQP0 expressed in *Xenopus* oocytes when subjected to acidic pH and low Ca^{2+} (Nemeth-Cahalan &

Hall, 2000). Also, experiments on sheep lenses showed that AQP0 phosphorylation on the C-terminal serine 235 by protein kinase A (PKA) increases water permeability (Gold et al., 2012). Regulation of water permeability by phosphorylation and Ca^{2+} has been described for AQP0 from various species, including human, and it is known that the regulation of AQP0 permeability is critical for lens transparency homeostasis (described below). Interestingly, unlike PIP gating that relies on direct intrinsic conformational rearrangements, AQP0 gating involves interaction with calmodulin (CaM). CaM is a Ca^{2+}-binding protein secondary messenger of several Ca^{2+}-signaling pathways. Upon Ca^{2+} binding, conformational changes in CaM structure allow CaM to interact with a plethora of partners to modulate their function. It was shown that in its Ca^{2+}-bound form, CaM binds to the C-terminal domain of AQP0 in a 2:1 AQP0:CaM ratio, which inhibited water motion (Reichow & Gonen, 2008). Phosphorylation of serine 235 in the C-terminal domain of AQP0 abolishes CaM interaction, thus allowing water permeation (Rose et al., 2008). Structural data showed that CaM does not operate as a plug per se but through an allosteric mechanism, CaM stabilizes cytoplasmic residues of AQP0 that maintain the channel in its closed form (Reichow et al., 2013). Expression of rat AQP5 (Rodrigues et al., 2016), human AQP3 (Zeuthen & Klaerke, 1999), AQP4 (Kaptan et al., 2015), and AQP7 (Mosca et al., 2018) in ectopic systems such as *Xenopus* oocytes and yeast also showed a pH regulation of permeability for these AQPs, thus strongly suggesting a pH-dependent gating. However, unlike AQP0, it is unclear whether pH-regulation/gating of these AQPs is relevant under normal physiological conditions.

4.2 AQP trafficking

In mammals, regulation of cellular water balance mostly relies on protein trafficking rather than gating. In response to environmental or hormonal changes, some AQPs undergo a tight regulation of total levels via regulation of transcription and/or protein stability as well as regulation of the distribution between intracellular and plasma membrane pools. Hormonal regulation of some AQPs has been described in several organs namely the gastrointestinal tract, reproductive system, and the urinary tract (described below). For example, it was shown that upon stimulation with the gastrointestinal hormone secretin, cytoplasmic AQP1 is inserted into the apical plasma membrane of epithelial rat bile duct cells, which is critical for bile secretion (Marinelli et al., 1999). During luteal phase or pregnancy, there is a significant increase in luminal fluid secretion in ovaries and uterus, which is important for the zygote movement to its eventual implantation. In rat, ovarian AQP9 expression is regulated by progesterone (Branes et al., 2005). Likewise, uterine fluctuations in expression levels and tissue localization of both AQP1 and AQP5 were observed after estrogen and progesterone treatment of mice (Richard et al., 2003). These observations highlight the importance of hormonal regulation of AQPs in fluid secretion dynamics within the female reproductive organs.

For some mammalian AQPs, posttranslational modifications (PTMs) including phosphorylation, glycosylation and ubiquitination orchestrate AQP shuttling between intracellular storage sites and the plasma membrane and regulate degradation. Phosphorylation-dependent shuttling of AQP2 in kidneys exemplifies perfectly how fine-tuned regulation of water balance is achieved by AQP trafficking (AQP2 shuttling mechanisms are detailed below). Phosphorylation-dependent trafficking has also been demonstrated for other AQPs. For instance, in astrocytes from rat cortex, concomitant Ca^{2+} influx and AQP1 phosphorylation by protein kinase C (PKC) induced rapid insertion of cytoplasmic AQP1 into the plasma membrane (Conner et al., 2012). Alanine substitution of the phosphorylation sites threonine 157 and 239 abolished membrane insertion. In mouse lung epithelial cells, AQP5 phosphorylation by PKA induces internalization of AQP5 in a short-term response while long-term phosphorylation increases plasma membrane localization (Sidhaye et al., 2005). It is unclear how phosphorylation is important for AQP5 trafficking since ectopic expression in normal epithelial human embryonic kidney (HEK293) cells (Woo, Chae, et al., 2008) and Madin−Darby canine kidney (MDCK) cells (Koffman et al., 2015) showed no effect of phosphorylation on protein localization. Regulation of AQP trafficking also occurs in plants. This can be achieved either by formation of AQP heterotetramers (described above, in the AQPs topology section) or by PTMs. It was shown that *Mc*TIP1;2 from ice plant (*Mesembryanthemum crystallinum*) relocalizes from vacuole membranes to prevacuolar compartments after osmotic stress in a glycosylation-dependent manner (Vera-Estrella et al., 2004).

Trafficking of membrane proteins to reach their proper cellular location relies on interactions with proteins of the trafficking machinery and AQPs are no exception to this. Besides proteins from the trafficking machinery, it was shown that the cellular localization of some AQPs involves interactions with specific partners. Interactions between prolactin-inducible protein (PIP) and the C-terminal of AQP5 regulate the apical localization of AQP5 in salivary and lacrimal glands (Chivasso et al., 2021; Ohashi et al., 2008).

Whether it involves gating, protein trafficking, or interactions with specific partners, regulation of water balance depends on intricate mechanisms that still require investigation to be fully unraveled.

5. AQPs in the kidneys

5.1 Localization and function of renal AQPs

Regulation of body water homeostasis is tightly controlled via balancing water intake and renal urinary water excretion. In the typical adult, approximately 180 L of plasma is filtered in the renal glomerular tuft per day and typical urine productions range between 0.5 and 2 L. The filtrate first enters the proximal tubules and then descending thin limbs, which are highly water permeable and responsible for reabsorption of the vast majority of filtered water. This reabsorption is constitutive and facilitated via AQP1 (Ma et al., 1998), which serves as both cellular entry and exit for reabsorbed water, as it localizes to both the apical and basolateral plasma membranes of proximal tubules and descending thin limbs (Nielsen, Smith, et al., 1993) (Fig. 18.5 depicts AQP localization along the renal nephron and collecting ducts). AQP1-null mice have close to normal daily urine production. However, when the animals are challenged by thirst, they are unable to concentrate their urine indicating that AQP1 is important in the urinary concentration mechanism (Ma et al., 1998). In humans, the AQP1 gene was located to the short arm of chromosome 7; the same location as the Colton blood group antigens. It turned out that indeed the Colton blood group (Co^a, Co^b and Co^{ab}) was an alanine–valine polymorphism in AQP1 on residue 45 in the first extracellular loop (Smith et al., 1994). Only a small number of Colton null individuals were

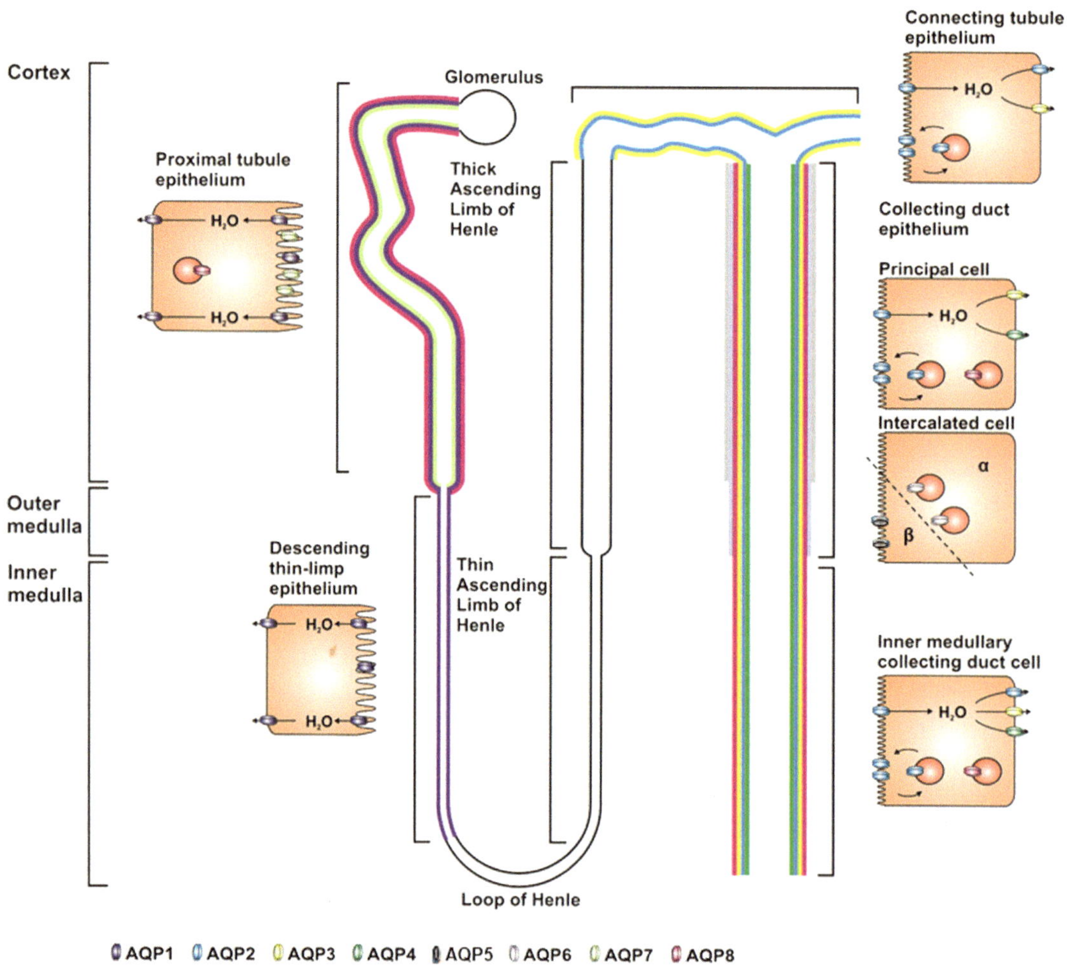

FIGURE 18.5 **Aquaporins along the renal nephron and collecting ducts.** Schematic showing the distribution of AQPs along the renal nephron and collecting ducts. AQP1 localizes to both apical and basolateral plasma membranes of the proximal tubules and thin descending limbs of Henle. AQP2 localizes to the apical plasma membrane and subapical intracellular vesicles of collecting duct principal cells. Moreover, AQP2 localizes to the basolateral membranes of connecting tubules and inner medullary collecting ducts. AQP3 localizes to the basolateral plasma membrane of connecting tubules and in collecting duct principal cells. AQP4 localizes in the basolateral plasma membrane of collecting duct principal cells. AQP5 localizes in the apical membrane of β-intercalated cells and AQP6 localizes in intracellular vesicles of acid-secreting α-intercalated cells in collecting ducts. AQP7 localizes in the apical plasma membrane of proximal tubule and AQP8 localizes in intracellular vesicles of the proximal tubule sand collecting ducts. *Modified from Nejsum (2005).*

identified. As for the AQP1-null mice, there was no obvious phenotype of the AQP1 null individuals. However, when challenged by thirsting, the individuals were unable to concentrate their urine as were the AQP1-null mice (King et al., 2001). Thus, this again indicates that AQP1 is important for the urinary concentration mechanism. Besides AQP1, the aquaglyceroporin AQP7 is localized to the apical membranes of the straight segments of proximal tubules (Nejsum et al., 2000). While both AQP1-null and AQP7-null mice have normal daily urine productions, a mouse model lacking both AQP1 and AQP7 produced a more dilute urine (Sohara et al., 2005). This indicates a compensatory relationship between AQP1 and AQP7 and a role of AQP7 in proximal tubule water handling. While the AQP7-null mice had normal daily urine productions, the mice suffered from glyceroluria (increased glycerol in the urine), indicating that AQP7 is involved in glycerol uptake in the proximal tubules (Sohara et al., 2005).

Fine-tuning of water reabsorption to balance urine secretion with the physiological requirement for water occurs in renal collecting ducts in response to hormone stimulation. In collecting duct principal cells, water entry is mediated via apical AQP2 (Nielsen, DiGiovanni, et al., 1993) and water exit is mediated via basolaterally localized AQP3 (Ecelbarger et al., 1995) and AQP4 (Terris et al., 1995) and in connecting tubules and inner medulla, also via basolaterally localized AQP2 (Christensen et al., 2003).

The signal for urine concentration is the antidiuretic hormone arginine vasopressin (AVP), which is secreted from the posterior pituitary gland into the blood stream in response to increased requirement of body water and thus, water reabsorption from the urinary filtrate. AVP stimulation of isolated perfused collecting ducts revealed that AVP stimulation lead to a rapid increase in water permeability (Nielsen et al., 1995). Upon AVP wash-out, water permeability was quickly restored to baseline levels. Immunoelectron microscopy revealed that AVP mediated a change in subcellular localization of AQP2. Before AVP stimulation, AQP2 was localized in both the apical plasma membrane and in subapical intracellular vesicles, however, upon AVP stimulation, AQP2 was mainly localized in the apical plasma membrane, which explained the increase in collecting duct water permeability. Moreover, upon AVP washout, AQP2 relocalized to intracellular vesicles as well as larger multivesicular bodies indicating that AVP wash-out induce AQP2 endocytosis (Nielsen et al., 1995). This short-term regulation of AQP2 subcellular localization in response to circulating levels of AVP results in daily urine productions ranging between 0.5 and 2L per day.

Besides the short-term regulatory effect of AVP on AQP2 shuttling, long-term changes in circulating AVP levels as well as in hydration affect AQP2 transcription levels and thus, AQP2 protein expression; high AVP and thirst increases and excessive water intake decreases expression (Nielsen, DiGiovanni, et al., 1993). As for AQP2, sustained increased levels of circulating AVP also increases AQP3 expression (Ecelbarger et al., 1995) but in contrast, AQP4 expression is unaltered (Terris et al., 1996).

5.2 AQP2 regulation

AVP binds to the basolateral AVP receptor 2 (AVPR2) of collecting duct principal cells. This leads to activation of adenylate cyclase through heterotrimeric guanosine triphosphate (GTP)-binding proteins, Gs. This increases intracellular cyclic adenosine monophosphate (cAMP) levels leading to activation of PKA. Subsequently, vesicular AQP2 is phosphorylated on serine 256 (S256) in the C-terminal tail. This leads to exocytosis of AQP2 containing intracellular vesicles, which then fuse with the apical plasma membrane (Fig. 18.6). This increases AQP2 protein levels in the apical membrane thereby instantly enhancing water reabsorption and thus, urine concentration (Fushimi et al., 1997; Katsura et al., 1997). Upon AVP removal, AQP2 is endocytosed thus decreasing collecting duct water reabsorption.

In vitro studies using phosphomimicking mutants suggested that S256 phosphorylation on three out of the four monomers was a prerequisite for AQP2 plasma membrane targeting (Christensen et al., 2000; Fushimi et al., 1997; Kamsteeg et al., 2000; Katsura et al., 1997). However, blocking global endocytosis resulted in accumulation of AQP2 in the plasma membrane despite S256 being mutated to a phosphomimic of constitutively nonphosphorylated AQP2 in serine 256 (serine to alanine) (Lu et al., 2004). This suggested that AQP2 constitutively cycles between intracellular vesicles and the plasma membrane and that subcellular localization may be regulated via modulation of the rates of both endo- and exocytosis.

Besides S256, AQP2 harbors three other confirmed phosphorylation sites in the C-terminal tail, namely S261, S264, and S269 (Bansal et al., 2010; Xie et al., 2010). Using phosphomimetics, it was shown that S256 was dominant over the other three sites in respect to AQP2 plasma membrane targeting (Van Balkom et al., 2002). With phospho-specific antibodies, it was shown that the S261 site is phosphorylated on AQP2 in intracellular vesicles. S269 is phosphorylated upon stimulation with the AVP homolog deamino D-arginine vasopressin (dDAVP) and is exclusively found in the apical plasma membrane (Moeller et al., 2009). The S269 phosphorylation was shown to be dominant to ubiquitination on lysine 270, thereby preventing AQP2 degradation (Moeller et al., 2014). The S264 site is also phosphorylated upon dDAVP treatment

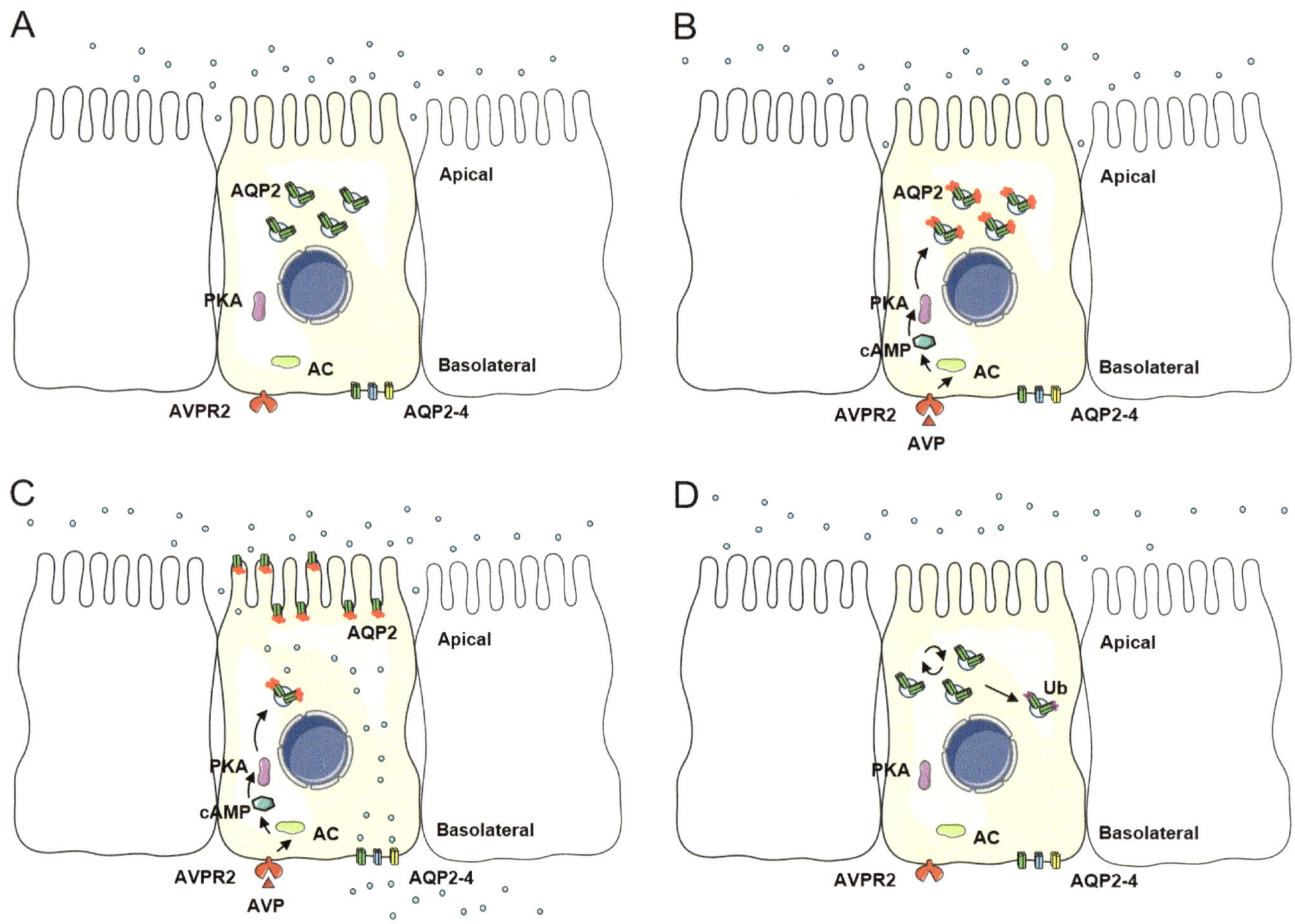

FIGURE 18.6 **Hormonal regulation of AQP2 trafficking in kidney principal cells.** (A) AQP2 tetramers predominantly localize in intracellular vesicles within collecting duct principal cells. (B) Circulating arginine vasopressin (AVP) binds the basolateral AVP receptor 2 (AVPR2) leading to activation of adenylate cyclase (AC), which increase levels of cyclic AMP (cAMP). Increased levels of cAMP activate protein kinase A (PKA) leading to AQP2 phosphorylation (red spheres). (C) Phosphorylated AQP2-bearing vesicles traffic to the apical plasma membrane and phosphorylated AQP2 is inserted in the membrane upon vesicles fusion with the plasma membrane, which increases water influx and thus urine concentration. Basolateral water exit is mediated via AQP2-4. (D) Decreased levels of AVP reduce AQP2 phosphorylation leading to AQP2 endocytosis and decreased water reabsorption. AQP2 vesicles are either recycled or targeted to degradation by ubiquitination (Ub).

and the phosphorylation is found in both apical and basolateral plasma membranes as well as in intracellular vesicles (Fenton et al., 2008). The role of S264 phosphorylation in AVP mediated AQP2 shuttling is still not clear.

Another major player in AQP2 plasma membrane targeting is filamentous actin (F-actin). In collecting duct principal cells, an F-actin network is localized subapically in close proximity to the membrane. Interestingly, in cell culture experiments, global depolymerization of F-actin resulted in AQP2 targeting to the plasma membrane (Jang et al., 2011; Klussmann et al., 2001; Tamma et al., 2001). Based on these studies, it has been hypothesized that F-actin constitutes a physical barrier between AQP2 vesicles and the plasma membrane and/or facilitates AQP2 vesicle delivery to the plasma membrane upon AQP2 S256 phosphorylation (Nedvetsky et al., 2007). Due to the small size of the AQP2 transport vesicles (below the diffraction limit of light) direct visualization of the AQP2 transport vesicles is only possible via immunoelectron microscopy but not via conventional fluorescence microscopy. Using a cell culture model combined with 3D superresolution microscopy, the AQP2 vesicular network was visualized in cell cultures alongside F-actin. Subsequent analysis revealed that a large fraction of the AQP2 vesicles were positioned in between the F-actin network and the plasma membrane, whereas the remainder localized with the F-actin network (Holst et al., 2021). This suggests that F-actin depolymerization may not be required for mobilization of all AQP2 vesicles.

Besides AVP, there are multiple AVP independent pathways, like calcium, prostaglandin E2, and the adenosine monophosphate-activated protein kinase (AMPK), mentioned below, that can orchestrate changes in AQP2 plasma membrane targeting (for recent reviews of regulation of AQP2 shuttling (Fenton et al., 2020; Olesen & Fenton, 2017)).

AQP5 is localized to apical membranes of β-intercalated cells (Procino et al., 2011); the function is still unknown. AQP6 is localized to intracellular vesicles in acid-secreting collecting duct α-intercalated cells, thus AQP6 is thought to be involved in regulation of body acid-base balance (Yasui et al., 1999a, 1999b). AQP8 localizes in intracellular vesicles in proximal tubules and to a lesser extend in collecting ducts (Elkjaer et al., 2001). AQP8-null mice produced a normal urine as did mice lacking both AQP1 and AQP8, indicating that AQP8 does not contribute to the urinary concentration mechanism (Yang et al., 2005).

5.3 Examples of water balance disorders

Dysregulation of AQP2 shuttling and/or expression is key in multiple diseases associated with water balance disorders, including central and nephrogenic diabetes insipidus (CDI and NDI, respectively) (DiGiovanni et al., 1994; Marples et al., 1995), untreated diabetes mellitus type I (Nejsum et al., 2001), congestive heart failure (CHF) (Xu et al., 1997), and chronic kidney disease (CKD) (Kwon et al., 1998). Moreover, inner medullary AQP2 is dysregulated in a model of acute pyelonephritis (Ernstsen et al., 2022).

Diabetes Insipidus (DI) is characterized by excessive thirst and excretion of large amounts of severely dilute urine where reduction of fluid intake has no effect on urine concentration. Acquired CDI can be caused by disruption or degeneration of hypothalamic neurons, due to for example a tumor, which results in inadequate levels of secreted AVP to meet the requirement for renal water uptake thus causing the patients to produce large amounts of urine.

NDI can be hereditary or acquired. In a Canadian study of males, it was estimated that hereditary DI affected 8.8 people out of 1.000.000 (Arthus et al., 2000). Of these, 90% was X-linked and due to mutations in the AVPR2 receptor rendering it unresponsive to AVP (Bichet, 1998; Bichet et al., 1998; Deen & Knoers, 1998; Nejsum et al., 2011). Mutations in the gene for AQP2 was responsible for the remainder 10% where 9% had an autosomal recessive presentation and the remainder 1% an autosomal dominant presentation (Deen et al., 1994). Mutations in the gene encoding AQP2 can result in nonfunctional proteins that are misfolded and thus degraded or mutations can be present in the C-terminal tail close to S256, which disrupts the PKA phosphorylation site. Thus, although AQP2 levels may be high, the lack of phosphorylation of S256 results in constitutive intracellular localization of AQP2 (Deen et al., 1995). Ultimately, inherited DI can lead to daily urine productions of up to 20L per day. Besides mutations that render the AVPR2 receptor unresponsive to AVP, there are also gain-of-function mutations that render the receptor constitutively active, resulting in the syndrome of inappropriate antidiuresis where the collecting ducts reabsorbs vast amounts of water despite low levels of AVP and therefore, the patients are unable to dilute their urine.

While CDI can be treated with the AVP analog desmopressin, there is currently no cure for NDI. Data from experimental animals do however show that metformin, stimulating the AMPK pathway, may be a promising intervention strategy to bypass the AVPR2 and facilitate AQP2 plasma membrane targeting in patients suffering from NDI caused by mutations in the AVPR2 (Efe et al., 2016; Klein et al., 2016).

5.4 Examples of acquired water balance disorders

Acquired NDI is much more common than inherited NDI and can be caused by for example lithium, often administered as a treatment for bipolar disorders. In experimental animals, lithium treatment results in significant downregulation of AQP2 as well as impaired plasma membrane targeting thereby rendering the animals unable to concentrate their urine (Marples et al., 1995). CKD affects more that 10% of the adult population and numbers are rising with the increase in obesity, hypertension, diabetes, and longer life-span (Chen et al., 2019; Collaboration, 2020). CKD is also associated with NDI and downregulation of renal AQPs, with subsequent polyuria.

Excess water retention is common in conditions like chronic heart failure and preeclampsia, where increased circulating AVP increases AQP2 expression and plasma membrane targeting thus leading to excess water reabsorption and body water accumulation.

6. AQPs in the nervous system

6.1 Expression of AQPs in spinal cord and brain

Approximately 80% of brain's wet weight is water and thus, osmotic homeostasis is critical for proper functionality of the central nervous system (CNS) (Badaut et al., 2014; Zador et al., 2007). The brain contains four distinct water-containing compartments namely cerebrospinal fluid (CSF), interstitial fluid (ISF), intracellular fluid (ICF), and blood (Zador et al.,

2007). Here, AQP-facilitated water transport contributes to the regulation of the extracellular space, cell volume control, and CSF production and drainage (Kimelberg, 2004; Sykova, 2004). mRNA expressions of AQP1, AQP3-9, and AQP11 have been detected in the CNS but so far, only AQP1 and AQP4 have been found to be pivotal players in the CNS and are therefore the main focus of this chapter.

6.2 AQP1 localization and function in the CNS

The choroid plexus is enclosed in the lateral, third, and fourth brain ventricle system and is responsible for producing CSF, which is essential for the removal of waste products, the transport of signaling molecules, and for the maintenance of appropriate CNS electrolyte balance. AQP1 is expressed in the apical membranes of choroid plexus epithelial cells and is a key factor in CSF production (Fig. 18.7) (Brinker et al., 2014). The CSF secretion is constant, and the total volume of approximately 150 mL is renewed four to five times within 24 h (Sakka et al., 2011). Compared to control mice, AQP1-null mice had a 25% reduction in CSF production and in a model of focal brain injury, the AQP1-null mice had a 56% reduction in intracranial pressure and increased survival rate demonstrating the importance of AQP1 in CSF production (Oshio et al., 2005; Sveinsdottir et al., 2014).

AQP1 is sparsely expressed in small capillary blood vessels in the brain and spinal cord parenchyma (Gao et al., 2006; Shields et al., 2007) and is absent in the cerebrovascular endothelium and the blood–brain barrier (Oklinski et al., 2014; Saadoun et al., 2002; Wilson et al., 2010). AQP1 is also expressed in the spinal cord dorsal horns and dorsal root ganglia, in small-diameter sensory neurons, and in the sciatic nerve and its projections as well as in healthy spinal cord and brain astrocytes within glia limitans and perivascular white matter astrocytes (Fig. 18.8) (Satoh et al., 2007).

AQP1 in the spinal cord dorsal horns of laminae I and II and dorsal root ganglion may be involved in acute pain signaling, nociception, and the development of allodynia, since AQP1-null mice have impaired firing of action potentials in small dorsal root ganglion neurons, and their allodynic response to chronic compression of the dorsal root ganglia was significantly reduced (Oshio et al., 2006; Wei et al., 2020; Zhang & Verkman, 2015).

AQP1 may also be involved in embryonic neuronal development, since AQP1 expression in neural crest cells (which will differentiate into neurons and glia of the peripheral nervous system, melanocytes, and cranial cartilage, bone, and adipose tissue) is involved in cell migration (see below) (McLennan et al., 2020).

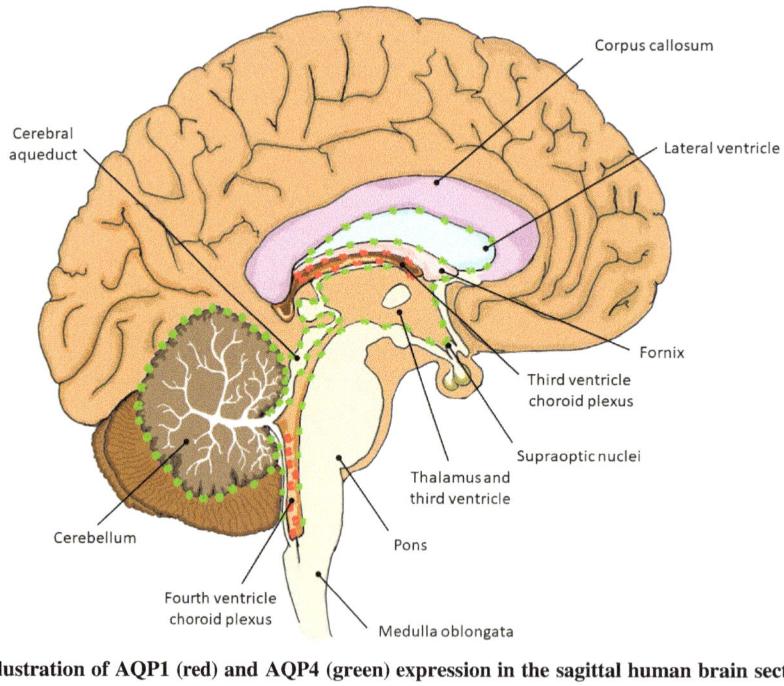

FIGURE 18.7 **Schematic illustration of AQP1 (red) and AQP4 (green) expression in the sagittal human brain section.** AQP1 is expressed in the apical membranes of choroid plexus epithelial cells enclosed in the lateral, third, and fourth brain ventricle. AQP1 is absent in the cerebrovascular endothelium and the blood–brain barrier (BBB) and is only sparsely expressed in small capillary blood vessels in the brain. The brain region with the highest AQP4 expression is the cerebellum, followed by osmosensory regions like the supraoptic nucleus and subfornical organ. The brain ventricles ependymal cells express AQP4 in their basolateral membranes similarly to astrocyte end-feet in the BBB. In the cerebellum, higher AQP4 expression is found in the granule cell layer compared to the molecular layer. In the neocortex AQP4 expression increases from deep to superficial layers with the exception of astrocytes end-feet membranes.

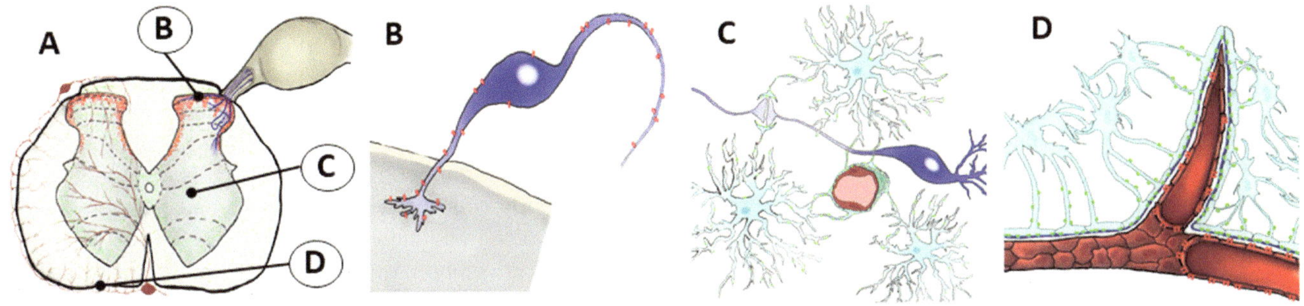

FIGURE 18.8 **AQP1 and AQP4 expression in the spinal cord.** (A) AQP1 (red) is strongly expressed at laminae I and II of the dorsal horn with decreasing signal intensity at the medial edges of dorsal horns up to lamina V. AQP4 (green) is abundantly expressed across the whole spinal cord with markedly higher intensity in the superficial lamina of dorsal horns, verges of ventral horns, glia limitans, and surroundings of blood vessels. (B) A fraction of the AQP1 signal belongs to unmyelinated neuronal cells protruding from dorsal root ganglion to the superficial laminae of dorsal horns where axons build up peripheral sensory fibers. (C) In the gray matter, protoplasmatic astrocytes AQP4 expression represents a polarized pattern with the highest intensity in the astrocytes foot processes encircling capillaries and weaker labeling in processes belonging to the tripartite synapse. (D) AQP1 labeling in the white matter is rather infrequent and found in small arterioles furcating from the arterial vasocorona. In the white matter, AQP4 is highly expressed in astrocytes foot-processes building up the glia limitans and encircling the blood vessels. *Reprinted with permission Oklinski et al. (2016).*

6.3 AQP4 localization in the CNS

The most abundantly expressed AQP in the CNS is AQP4, mostly localized in the white and gray matter of the brain and spinal cord. The highest level of AQP4 expression is found in the cerebellum, followed by osmosensory regions like the supraoptic nucleus and subfornical organ (Fig. 18.7) (Hubbard et al., 2015; Nielsen et al., 1997). AQP4 is predominantly expressed in astrocytes, where it is heterogeneously localized in different patterns in various CNS structures. AQP4 is present in astrocyte end-feet and processes that tightly ensheath the blood vessels and contribute to tripartite synapses. Here, the synaptic cleft between pre- and postsynaptic neurons is enclosed by AQP4-expressing astrocyte endings, particularly in glutamatergic synapses (Fig. 18.9) but also in subependymal astrocytes of the meninges adjacent to the pia

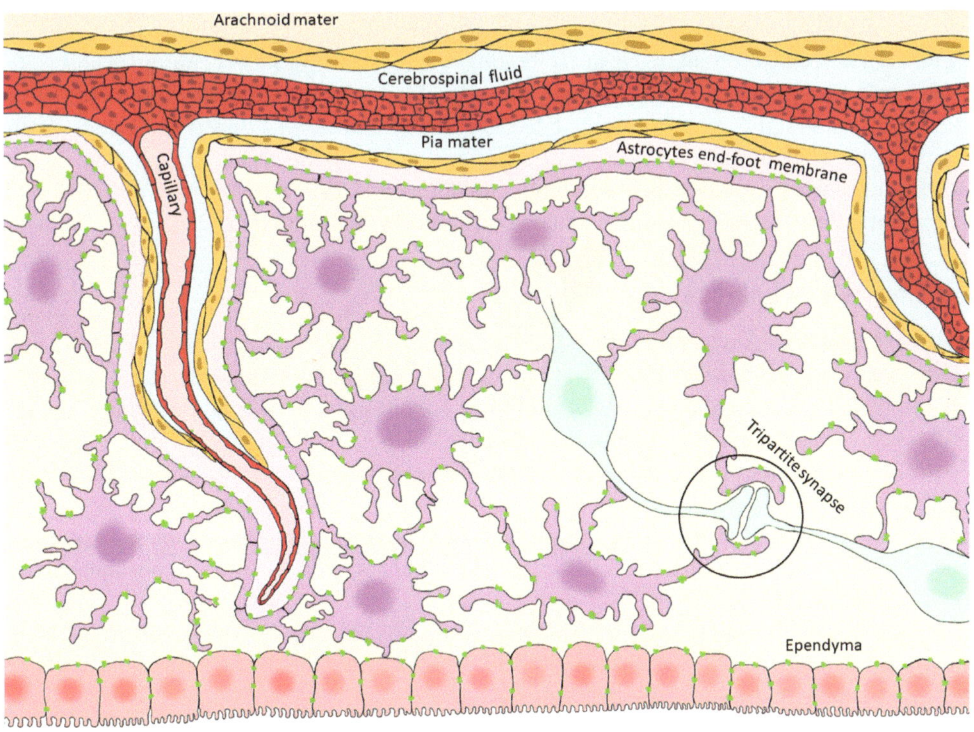

FIGURE 18.9 **Distribution of AQP4 in the brain.** AQP4 (green) is expressed in the pericapillary astrocyte processes building up the glia limitans and in ependymal cells. In the tripartite synapses, the synaptic cleft between pre- and postsynaptic neurons is enclosed by AQP4 expressing astrocyte protrusions.

mater (Bloch & Manley, 2007; Nagelhus et al., 2004; Nielsen et al., 1997; Oklinski et al., 2014). Also, brain ventricles are covered with ependymal cells where AQP4 is expressed in basolateral membranes similar to astrocyte end-feet (Bloch & Manley, 2007; Trillo-Contreras et al., 2019).

In the spinal cord, a laminar-specific AQP4 expression pattern is observed in gray matter, with highest expression in the dorsal horns laminae I and II, lower expression in lamina X around the spinal cord central canal, and lowest expression in other laminae. As in the brain, AQP4 localizes to astrocyte end-feet closely encircling endothelial cells of blood vessels. AQP4 expression in gray matter astrocytes is more polarized and primarily concentrated in astrocyte end-feet compared to white matter astrocytes where AQP4 is more dispersed along astrocyte processes and cell bodies (Fig. 18.8). AQP4 is also expressed in astrocytes around blood vessels and in astrocyte processes enveloping myelinated neuronal fibers, especially in longitudinal spinal cord white matter sections, where prominent AQP4 expression is characteristic for the astrocytes foot processes building up the spinal cord glia limitans (Oklinski et al., 2014; Smith & Verkman, 2015).

AQP4 often colocalizes with the glutamate transporter-1 (GLT-1) and the potassium channel 4.1 (Kir4.1), particularly at thin astrocyte processes in the immediate vicinity to synaptic clefts. This may indicate that AQP4 contributes to astrocyte clearing and balancing of K^+ and glutamate during synaptic activity (Butt & Kalsi, 2006; Vitellaro-Zuccarello et al., 2005).

AQP4 in the CNS has two isoforms, the M23 and the M1 isoform. While both types are highly permeable to water and can form heterotetrameric structures, only the M23 isoform forms the complex formations described as OAPs. These higher-order structures can form continuous sheets that are mostly localized in astrocytic end-feet in the glia limitans and around blood vessels (Mathiisen et al., 2010; Rash et al., 1998).

6.4 AQP4 contribution to CSF production

Approximately 80% of the CSF volume is secreted by the choroid plexus into the ventricular system, while the remaining 20% is produced by structures like the brain parenchyma. CSF flows through the ventricles, the cisterns, and subarachnoid space and is then absorbed into the venous blood at the level of the arachnoid villi (Brinker et al., 2014). However, this classical circulatory nature of CSF flow has been challenged by recent studies, and several features, like the rate and site of CSF secretion and absorption, are now being further investigated. The AQP4 expression pattern at the borders between the brain and CSF compartments directly suggests a role in CSF homeostasis and facilitation of water flow into and out of the brain. Thus, AQP4 found in the pericapillary astrocyte processes building up the glia limitans and in ependymal cells is involved in absorption and exchange of CSF and interstitial fluid (ISF). However, how AQP4 contributes to the CSF/ISF-related water movement in physiological conditions is still a matter of ongoing research.

Both AQP1 and AQP4 seem to be involved in facilitating water influx into the ventricular space. In both AQP1-null and AQP4-null mice, a decrease in ventricular volume and interventricular pressure was observed. Moreover, when mice were lacking both AQP1 and AQP4, an even larger effect on CSF drainage was observed (Trillo-Contreras et al., 2019). These findings support the hypothesis by "Bulat–Klarica–Oreskovic" where the continual CSF secretion and dissemination takes place through water exchange between brain capillaries and ISF (Oreskovic et al., 2017). The roles of AQP1 and AQP4 in CSF homeostasis are still not fully understood (Hladky & Barrand, 2022).

6.5 AQP4 and CNS water balance

Rapid and bidirectional transport of water through AQPs is a critical part of the regulation of brain and spinal cord water balance. The highly polarized expression of AQP4 at blood–CSF and blood–brain interfaces is essential in maintaining fluid homeostasis in physiological conditions. AQP4-null mice have normal appearance although, in the brains and spinal cords, a modest increase in baseline water content was observed (Bloch et al., 2006; Wu et al., 2014; Zhou et al., 2008). This suggests that AQP4 may function as a route for water efflux into the brain vessels, subarachnoid space, and ventricles from the brain parenchyma.

In the developing brain, AQP4 seems to be involved in maintaining brain water homeostasis in early development where it contributes to brain water content and AQP4 expression appears to be involved in postnatal BBB differentiation in the cerebellum (Nico et al., 2004).

Besides the function in CSF production and CNS water balance, AQP4 may also be involved in extracellular potassium buffering, important for synaptic activity, calcium signaling, and bidirectional communication between astrocytes and neurons as well as neurotransmission and synaptic plasticity.

6.6 Example of AQPs in neuronal diseases

6.6.1 *AQPs in brain edema*

Ischemic stroke and ischemia-induced cerebral tissue alterations, like cerebral edema, can result in structural damages to the brain with fatal outcome. Interestingly, AQP4-null mice had reduced brain edema following water intoxication and ischemic stroke, indicating a role of AQP4 in brain edema. In a model of mild, transient cerebral ischemic stroke, AQP4-null mice displayed lower mortality and improved motoric recovery in a 14-day period after stroke. Moreover, behavioral recovery of AQP4-null mice was connected with decreased neuroinflammation, lower neuronal cell death, and a lesser volume of brain lesions (Hirt et al., 2017).

6.6.2 *AQPs in neurodegenerative diseases*

The hallmark of Alzheimer's disease (AD) is deposits of a polymorphous amyloid-beta (Aβ) protein called senile plaques. In patients affected by AD, expressional changes of both AQP1 and AQP4 were observed. Interestingly, astrocytes in AD cortex were found to express AQP1 at low levels in the immediate vicinity of senile plaques, whereas AQP1 expression was increased in the astrocytes adjacent to plaques with low Aβ immunoreactivity (Hoshi et al., 2012). In contrast, the expression of AQP4 in astrocytes of AD brains was characterized by increased expression in the proximity to Aβ plaques (Hoshi et al., 2012). It was proposed that changes in AQP4 expression, especially in astrocytic end-feet could affect the clearance of Aβ from the perivascular space since AQP4-null mice were characterized by improved clinical scores and lower tissue inflammation (Li et al., 2009; Zago et al., 2013). Moreover, cell culture studies suggested a lack of AQP4 expression to be linked to lower secretion of proinflammatory cytokines. However, the involvement of AQP1 and AQP4 in AD pathology is still not fully understood.

6.6.3 *AQPs in traumatic brain injury*

Common occurrence of traumatic brain injury (TBI) is the development of astrogliosis, which has been shown to be associated with AQP4 upregulation and microglia activation. AQP4-null mice subjected to TBI showed increased microglial reaction and attenuated astrogliosis in comparison to controls (Fukuda et al., 2013). Similarly, siRNA-induced AQP4 knock-down demonstrated enhanced microglial activation and decreased astrogliosis on the third day after TBI. Interestingly, the AQP4-null mice were also more prone to develop epileptic seizures than WT mice after 1 month post-TBI (Lu et al., 2021).

7. AQPs in the eye

Multiple AQPs are expressed in the eye, namely AQP0, AQP1, AQP3, AQP4, AQP5, AQP7, and AQP11. AQP0 is almost exclusively expressed in lens fiber cells and is of vital importance for lens development and function. In comparison to other human AQPs, AQP0 displays low water permeability, which may be compensated by high expression and specific organization in lens fiber cells (Kushmerick et al., 1995; Zampighi et al., 2002). Moreover, lens AQP0 permeability to water depends on bilayer lipid composition and varies significantly between expression in the lens cortex and nucleus (Tong et al., 2013). In lens fiber cells, AQP0 is involved in volume regulation, nutrient supply, formation of membrane junctions, and waste excretion. The expression of AQP0 in fiber cells enhances lens transparency and structural strength (Fig. 18.10). Cell culture studies revealed that AQP0 enhanced cell–cell adhesion and AQP0-null mice lacked the characteristic dense arrangement of fiber cells (Varadaraj et al., 2010).

AQP1 is expressed in photoreceptors, retinal pigmented epithelia, ciliary epithelial cells, and the trabecular meshwork (Schey et al., 2014; Zhang et al., 2002). AQP1 in the capillary epithelium is involved in the production of aqueous humor that fills the anterior and posterior chambers of the eye. This humor is important for intraocular pressure and maintenance of the spherical shape of the eyeball (Verkman et al., 2008). In the corneal endothelium and stroma, AQP1 facilitates partial dehydration of the corneal stroma thus contributing to cornea transparency (Fig. 18.10). In lens epithelial and fiber cells, AQP1 and AQP0 contribute to the nourishing of the avascular lens and maintain lens transparency by ensuring appropriate fluid circulation (Verkman et al., 2008).

AQP3 is expressed in corneal and conjunctival epithelia. Experiments in AQP3-null mice suggest a role of AQP3 in corneal epithelial cell proliferation and glycerol transport.

AQP4 is abundantly expressed in the ciliary epithelium and Müller retina cells where it is predominantly involved in maintaining retinal fluid homeostasis. Studies employing AQP4-null mice suggested a role in eye fluid dynamics and regulation of water inflow into the vitreous body through the systemic circulation (Fig. 18.11).

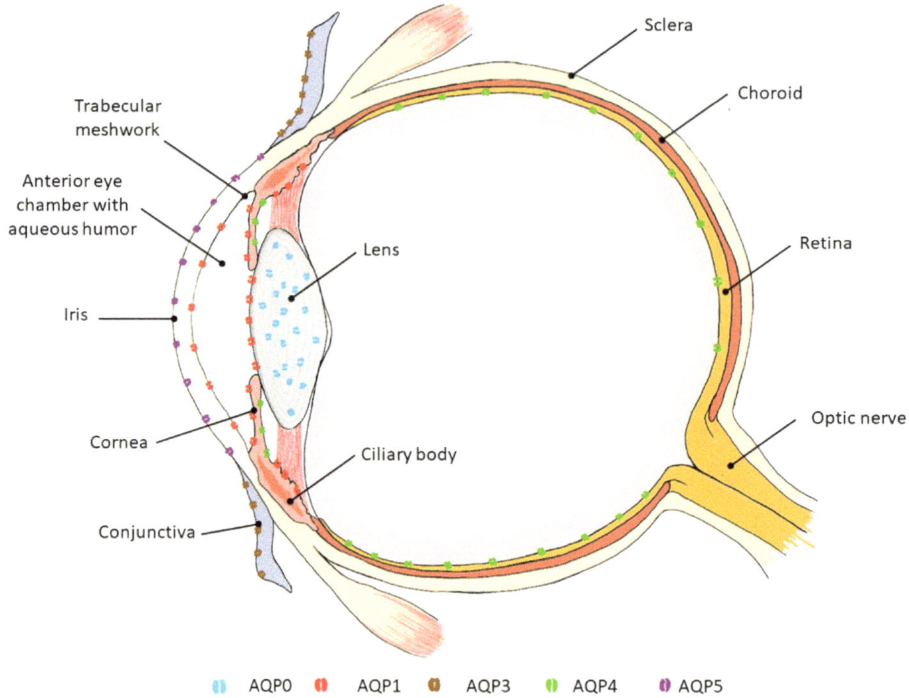

FIGURE 18.10 **Aquaporins in the eye.** AQP0 is exclusively expressed in lens fiber cells. AQP1 is expressed in photoreceptors, retinal pigmented epithelia, ciliary epithelial cells, and the trabecular meshwork. In lens epithelial and fiber cells, AQP1 and AQP0 contribute to the nourishing of the avascular lens and maintains lens transparency. AQP3 expression is localized in corneal epithelial cells. AQP4 is abundantly expressed in the ciliary epithelium and Müller retina cells. AQP5 is expressed in the outer stratified epithelium of the anterior corneal epithelium.

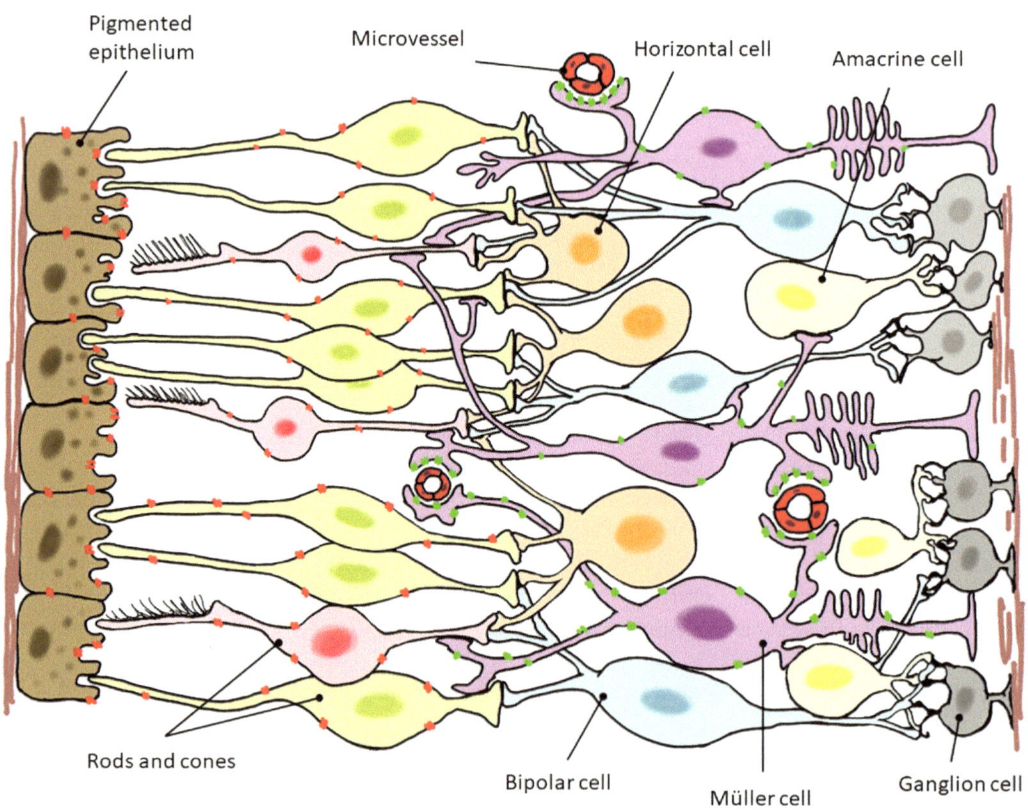

FIGURE 18.11 **Aquaporins expression in the retina** In the retina, AQP4 (green) is expressed in Müller cells, particularly in their perivascular processes. Expression of AQP1 (red) was found in photoreceptors cells and retinal pigmented epithelia.

AQP5 is expressed in the outer stratified epithelium of the anterior corneal epithelium, where AQP3 is also expressed in the epithelium and AQP1 in the endothelium, where they are involved in water fluxes across the cornea (Ren et al., 2017; Thiagarajah & Verkman, 2002). Due to direct exposure to the environment, the corneal epithelium is susceptible to injuries and recent studies indicate that AQP5 is involved in cell proliferation, migration, and acceleration of corneal wound healing (Kumari et al., 2018).

Research to investigate the roles of AQP7 and AQP11 in the eye is still ongoing.

Multiple pathologies of the eye are associated with AQP dysregulation, for example a form of congenital cataracts, which is associated with mutations in AQP0 (Berry et al., 2000).

8. AQPs in cancer biology

According to the World Health Organization, cancer is the leading cause of death worldwide with nearly 10 million casualties yearly. Countless studies found that AQP dys-, up- or downregulation are associated with multiple types of carcinoma including lung, colorectal, cervical, breast, and liver cancer. For instance, it was shown that AQP1, AQP3, and AQP5 are upregulated in colorectal (Moon et al., 2003; Yoshida et al., 2013) and breast cancer (Kang et al., 2015; Lee et al., 2014; Qin et al., 2016) and the increased expression correlated with tumor grade and poor prognosis. How AQP expression is involved in cancer development and spread has been the focus of many investigations with the ultimate goal to improve diagnosis and develop new therapeutic options.

Still, the mechanisms behind AQPs function (s) in tumor biology remain elusive. Knock-down/knock-out and ectopic expression of AQPs in different cell and animal systems have highlighted unexpected involvement of AQPs in various important cellular processes associated with cancer development and spread including cell migration, adhesion, and proliferation. Discrepancies regarding phenotypes associated with AQP expression in different cell/animal systems and cancers indicate that the phenotypic outcome differs with the cellular background and the AQP involved. Below are selected examples of AQP-associated phenotypes.

8.1 AQPs and cell migration

Both single and collective cell migration play critical roles in morphogenesis during development, in adult tissue repair, angiogenesis (de novo blood vessel formation), and the immune response. AQP1 is strongly expressed in proliferating microvessels in both healthy and tumor tissues indicating involvement of AQP1 in angiogenesis. Tumor growth as well as microvasculature density following subcutaneous implantation of melanoma cells were significantly reduced in AQP1-null mice but remained unaffected in AQP3-null mice (Saadoun, Papadopoulos et al., 2005). It was proposed that impaired angiogenesis in AQP1-null mice originates from reduced migration of AQP1-deficient endothelial cells. By using a wound healing assay where collective cell migration is monitored by microscopy, it was shown that AQP1-deficient cells migrate at a speed nearly four times slower than cells expressing AQP1. Similar observations for AQP3 (Satooka & Hara-Chikuma, 2016) and AQP5 (Jung et al., 2011) were made in the breast cancer cell lines MCF7 and MDA-MB-231 in which AQP5 and AQP3 were knocked down respectively.

To migrate, cells adopt a front-rear polarity. At the front (leading edge), cytoplasmic membrane protrusions (lamellipodia and filopodia) promoted by actin reorganization provide new adhesion points for forward motion. Conversely, at the rear, adhesion points are disassembled allowing retraction. These events at both cellular poles are tightly regulated and coordinated by different signaling pathways. Interestingly, it was shown that AQP1 localized to lamellipodia of migrating cells (Saadoun, Papadopoulos et al., 2005) similar to many ion transporters (Schwab, 2001), and AQP4 also localized at the leading edge of migrating astrocytes (Saadoun, Watanabe, et al., 2005). It was proposed that ion uptake at the leading edge creates an osmotic gradient driving water influx through AQPs, which promotes membrane protrusions and subsequently enhance cell migration. However, immunostaining of AQP3 and AQP4 in migrating inner medullary collecting duct (IMCD) cells showed no leading edge localization (Rose et al., 2021). In fact, AQP3 was not detected in the first 3 cell rows of the migrating cell sheet while a strong staining was detected in the cells behind. Also, AQP4 localized at the rear of migrating cells. Adding to the complexity, AQP1 expression in noncancerous Madin–Darby Canine Kidney (MDCK) epithelial cells had no significant effect on collective cell migration while cells expressing AQP3, AQP4 or AQP5 displayed a decrease in collective cell migration (Login et al., 2019). Thus, AQPs may influence cell migration through various mechanisms depending on the AQP(s) and the cellular context.

8.2 AQPs and cell−cell adhesion

Tissues morphogenesis and cohesion rely on cell−cell adhesion. Neighboring epithelial cells adhere to each other via tight junctions, adherens junctions, and desmosomes. Asymmetrical distribution of these cell junctional complexes is associated with specialized polarity proteins that establish an apical-basal cell polarity that is essential for epithelia homeostasis. Epithelial-mesenchymal transition (EMT) is a cellular process in which cells lose or alter their epithelial features (cell−cell junctions, apical-basal polarity, and epithelial markers) and acquire mesenchymal markers and increased cell motility (Ribatti et al., 2020). EMT and its reverse process MET are highly regulated and play a fundamental role in development and wound healing. Dysregulation of junctional protein levels and/or localization is associated with various pathologies such as cancer (Brooke et al., 2012; Cereijido et al., 2007; Vasioukhin and Harris, 2012). Malignant cells are able to escape intercellular adhesion through EMT to gain migratory and invasive characteristics allowing metastasis, which drastically increases fatal outcome probability. For example, levels of the tight junction protein ZO-1 are decreased and increased in pancreatic cancer (Doi et al., 2012) and melanoma cells (Smalley et al., 2005), respectively. Interestingly, in adenocarcinoma tissue, decreased expression of the adherens junction protein E-cadherin correlated with high levels of AQP1 (Wu et al., 2015). Alteration of junctional protein levels correlated to AQP3, AQP4, AQP5, AQP8, and AQP9 expression was reported in various cancer tissues/cells. Thus, depending on the cancer cell considered, one given AQP can either promote up- or down junctional proteins (Edamana et al., 2021). In many cases, alteration of junctional protein levels (epithelial markers) was concomitant to altered levels of mesenchymal markers like vimentin and N-cadherin. Altogether, these observations strongly suggested that in cancer cells, AQPs may contribute to EMT induction, which subsequently promotes migration, invasion, and metastasis. Ectopic expression of AQP1, AQP4, and AQP5 in noncancerous MDCK and HEK293 cells promoted lower levels of ZO-1 and the adherens junction proteins E-cadherin, β-catenin, and γ-catenin at cell-cell junctions (Jensen et al., 2018; Login et al., 2019). While AQP3 expression in MDCK cells promoted increased levels of β-catenin at cell-cell junctions, no significant changes of ZO-1 and γ-catenin levels were observed (Login et al., 2019). Intriguingly, AQP1, AQP4, and AQP5 expression also induced lower vimentin levels in MDCK cells, implying absence of EMT. How AQPs regulate junctional proteins is unknown but it was recently shown that AQP5 interacts with junctional proteins, including β-catenin and ZO-1 (Login et al., 2021). Furthermore, chimeras between AQP3 and AQP5 indicated that both the cytoplasmic C-terminal tail and loop D might be part of the mechanism. Indeed, no increase in β-catenin levels was promoted by AQP3 chimeras bearing the loop D or C-terminal tail of AQP5 (Login et al., 2019). It is thus plausible that AQP-dependent regulation of junctional proteins relies on interactions involving AQPs.

8.3 AQPs and cell proliferation

Decreased proliferation was observed in AQP3-deficient epidermal cells (Hara-Chikuma & Verkman, 2008). These cells displayed lower ATP levels due to reduced glycerol uptake, which made AQP3-null mice resistant to skin tumor development. Thus, AQP3 implication in skin tumorigenesis might relate to glycerol metabolism. Similarly, silencing of AQP3 in nonsmall cell lung cancer cells (NSCLC) decreased proliferation and intracellular glycerol levels (Xia et al., 2014). More strikingly, increased levels of p53 and cleaved caspase-3 were detected indicating that AQP3 knockdown induced cell cycle arrest and apoptosis. Thus, AQP3 may influence cell proliferation in different cell types via several mechanisms.

Within a tissue or at high cell density/confluency, cells interacting with one another inhibit proliferation and division to initiate differentiation. This phenomenon named contact inhibition of proliferation allows control of organ development and size. Contact inhibition of proliferation is tightly regulated by the Hippo signaling pathway (Rausch & Hansen, 2020). In brief, the Hippo pathway consists of kinases LATS1/LATS2 that through phosphorylation control the cellular localization and protein stability of transcriptional coactivators YAP/TAZ. When the Hippo signaling pathway is activated, LATS1/LATS2 phosphorylate YAP/TAZ and phosphorylated YAP/TAZ are either sequestrated in the cytoplasm or degraded. Inversely, in inactive conditions, nonphosphorylated YAP/TAZ are translocated to the nucleus and in complex with a transcription factor, regulate genes controlling cell proliferation and migration. Dysregulation of the Hippo pathway leads to uncontrolled proliferation. In several cancers, increased YAP/TAZ activity is associated with cancer progression and poor prognosis (Han, 2019). To date, no link between the Hippo pathway and AQP expression has been described. However, catenins (adherens junctions) and ZO proteins (tight junctions) can modulate YAP/TAZ activity through direct interaction and in a phosphorylation-independent manner. It is then plausible that AQPs influence the Hippo pathway indirectly through modulation of junctional protein levels (described above).

Most unanticipated, some AQPs can activate specific signaling pathways leading to phenotypic changes. Expression of AQP5 in the mouse fibroblast cell line NIH/3T3 activates the Ras pathway and thus proliferation (Woo, Lee, et al., 2008). The AQP5 phosphomutant in which serine 156 in the loop D was substituted to an alanine (S156A; mimics constitutively nonphosphorylated AQP5 in S156) cannot activate the Ras signaling pathway. Similarly, overexpression of AQP5 in colon cancer cells correlates with increased proliferation and activation of the Ras/extracellular receptor kinase 1/2 (ERK1/2) signaling pathway (Kang et al., 2008). Interestingly, AQP5 expression in MDCK cells promoted cell detachment from the migrating cell sheet during a wound healing assay (Login et al., 2019). This phenotype was dependent on the Ras signaling pathway activation through AQP5-S156 phosphorylation. In lung cancer cells, it was shown that AQP5 activates the epidermal growth factor receptor (EGFR)/ERK/p38 mitogen-activated protein kinase (MAPK) signaling pathway, resulting in increased proliferation (Zhang et al., 2010). In both colon and lung cancer cells, AQP1 and AQP3 expression did not activate these signaling pathways demonstrating specific features of AQP5 in these cells. Thus, AQPs seem to be at the crossroad of signaling pathways that regulate cell adhesion, migration, and proliferation. It is thus plausible that inadequate expression of AQPs in cancer cells affects the balance between these cellular processes.

References

Agre, P. (2004). Aquaporin water channels (nobel lecture). *Angewandte Chemie International Edition in English, 43*, 4278–4290.

Alleva, K., Niemietz, C. M., Sutka, M., Maurel, C., Parisi, M., Tyerman, S. D., & Amodeo, G. (2006). Plasma membrane of Beta vulgaris storage root shows high water channel activity regulated by cytoplasmic pH and a dual range of calcium concentrations. *Journal of Experimental Botany, 57*, 609–621.

Arthus, M. F., Lonergan, M., Crumley, M. J., Naumova, A. K., Morin, D., De Marco, L. A., Kaplan, B. S., Robertson, G. L., Sasaki, S., Morgan, K., Bichet, D. G., & Fujiwara, T. M. (2000). Report of 33 novel AVPR2 mutations and analysis of 117 families with X-linked nephrogenic diabetes insipidus. *Journal of the American Society of Nephrology, 11*, 1044–1054.

Badaut, J., Fukuda, A. M., Jullienne, A., & Petry, K. G. (2014). Aquaporin and brain diseases. *Biochimica et Biophysica Acta, 1840*, 1554–1565.

Bansal, A. D., Hoffert, J. D., Pisitkun, T., Hwang, S., Chou, C. L., Boja, E. S., Wang, G., & Knepper, M. A. (2010). Phosphoproteomic profiling reveals vasopressin-regulated phosphorylation sites in collecting duct. *Journal of the American Society of Nephrology, 21*, 303–315.

Berry, V., Francis, P., Kaushal, S., Moore, A., & Bhattacharya, S. (2000). Missense mutations in MIP underlie autosomal dominant 'polymorphic' and lamellar cataracts linked to 12q. *Nature Genetics, 25*, 15–17.

Bichet, D. G. (1998). Nephrogenic diabetes insipidus. *Americas Journal of Medicine, 105*, 431–442.

Bichet, D. G., Turner, M., & Morin, D. (1998). Vasopressin receptor mutations causing nephrogenic diabetes insipidus. *Proceedings of the Association of American Physicians, 110*, 387–394.

Bienert, G. P., Moller, A. L., Kristiansen, K. A., Schulz, A., Moller, I. M., Schjoerring, J. K., & Jahn, T. P. (2007). Specific aquaporins facilitate the diffusion of hydrogen peroxide across membranes. *Journal of Biological Chemistry, 282*, 1183–1192.

Bienert, M. D., Diehn, T. A., Richet, N., Chaumont, F., & Bienert, G. P. (2018). Heterotetramerization of plant PIP1 and PIP2 aquaporins is an evolutionary ancient feature to guide PIP1 plasma membrane localization and function. *Frontiers of Plant Science, 9*, 382.

Bloch, O., Auguste, K. I., Manley, G. T., & Verkman, A. S. (2006). Accelerated progression of kaolin-induced hydrocephalus in aquaporin-4-deficient mice. *Journal of Cerebral Blood Flow and Metabolism, 26*, 1527–1537.

Bloch, O., & Manley, G. T. (2007). The role of aquaporin-4 in cerebral water transport and edema. *Neurosurgical Focus, 22*, E3.

Branes, M. C., Morales, B., Rios, M., & Villalon, M. J. (2005). Regulation of the immunoexpression of aquaporin 9 by ovarian hormones in the rat oviductal epithelium. *American Journal of Physiology: Cell Physiology, 288*, C1048–C1057.

Brinker, T., Stopa, E., Morrison, J., & Klinge, P. (2014). A new look at cerebrospinal fluid circulation. *Fluids and Barriers of the CNS, 11*, 10.

Brooke, M. A., Nitoiu, D., & Kelsell, D. P. (2012). Cell-cell connectivity: Desmosomes and disease. *The Journal of Pathology, 226*, 158–171.

Butt, A. M., & Kalsi, A. (2006). Inwardly rectifying potassium channels (Kir) in central nervous system glia: A special role for Kir4.1 in glial functions. *Journal of Cellular and Molecular Medicine, 10*, 33–44.

Calamita, G., Bishai, W. R., Preston, G. M., Guggino, W. B., & Agre, P. (1995). Molecular cloning and characterization of AqpZ, a water channel from *Escherichia coli*. *Journal of Biological Chemistry, 270*, 29063–29066.

Calvanese, L., Pellegrini-Calace, M., & Oliva, R. (2013). In silico study of human aquaporin AQP11 and AQP12 channels. *Protein Science, 22*, 455–466.

Cereijido, M., Contreras, R. G., Flores-Benitez, D., Flores-Maldonado, C., Larre, I., Ruiz, A., & Shoshani, L. (2007). New diseases derived or associated with the tight junction. *Archives of Medical Research, 38*, 465–478.

Chaumont, F., & Tyerman, S. D. (2014). Aquaporins: Highly regulated channels controlling plant water relations. *Plant Physiology, 164*, 1600–1618.

Chen, T. K., Knicely, D. H., & Grams, M. E. (2019). Chronic kidney disease diagnosis and management: A review. *JAMA, 322*, 1294–1304.

Chivasso, C., Nesverova, V., Jarva, M., Blanchard, A., Rose, K. L., Oberg, F. K., Wang, Z., Martin, M., Lhotellerie, F., Zindy, E., Junqueira, B., Leroy, K., Vanhollebeke, B., Delforge, V., Bolaky, N., Perret, J., Soyfoo, M. S., Moscato, S., Baldini, C., … Delporte, C. (2021). Unraveling human AQP5-PIP molecular interaction and effect on AQP5 salivary glands localization in SS patients. *Cells, 10*.

Christensen, B. M., Wang, W., Frokiaer, J., & Nielsen, S. (2003). Axial heterogeneity in basolateral AQP2 localization in rat kidney: Effect of vasopressin. *American Journal of Physiology - Renal Physiology, 284*, F701–F717.

Christensen, B. M., Zelenina, M., Aperia, A., & Nielsen, S. (2000). Localization and regulation of PKA-phosphorylated AQP2 in response to V(2)-receptor agonist/antagonist treatment. *American Journal of Physiology - Renal Physiology, 278*, F29–F42.

Collaboration, G. B. D. C. K. D. (2020). Global, regional, and national burden of chronic kidney disease, 1990–2017: A systematic analysis for the global burden of disease study 2017. *Lancet, 395*, 709–733.

Conner, M. T., Conner, A. C., Bland, C. E., Taylor, L. H., Brown, J. E., Parri, H. R., & Bill, R. M. (2012). Rapid aquaporin translocation regulates cellular water flow: Mechanism of hypotonicity-induced subcellular localization of aquaporin 1 water channel. *Journal of Biological Chemistry, 287*, 11516–11525.

de Grotthuss, C. J. (2006). Memoir on the decomposition of water and of the bodies that it holds in solution by means of galvanic electricity. 1805. *Biochimica et Biophysica Acta, 1757*, 871–875.

Deen, P. M., Croes, H., van, A. R. A., Ginsel, L. A., & van, O. C. H. (1995). Water channels encoded by mutant aquaporin-2 genes in nephrogenic diabetes insipidus are impaired in their cellular routing. *Journal of Clinical Investigation, 95*, 2291–2296.

Deen, P. M., & Knoers, N. V. (1998). Vasopressin type-2 receptor and aquaporin-2 water channel mutants in nephrogenic diabetes insipidus. *The American Journal of the Medical Sciences, 316*, 300–309.

Deen, P. M., Verdijk, M. A., Knoers, N. V., Wieringa, B., Monnens, L. A., van-Os, C. H., & van-Oost, B. A. (1994). Requirement of human renal water channel aquaporin-2 for vasopressin-dependent concentration of urine. *Science, 264*, 92–95.

Deshmukh, R. K., Vivancos, J., Ramakrishnan, G., Guerin, V., Carpentier, G., Sonah, H., Labbe, C., Isenring, P., Belzile, F. J., & Belanger, R. R. (2015). A precise spacing between the NPA domains of aquaporins is essential for silicon permeability in plants. *The Plant Journal, 83*, 489–500.

DiGiovanni, S. R., Nielsen, S., Christensen, E. I., & Knepper, M. A. (1994). Regulation of collecting duct water channel expression by vasopressin in Brattleboro rat. *Proceedings of the National Academy of Sciences of the United States of America, 91*, 8984–8988.

Doi, Y., Yashiro, M., Yamada, N., Amano, R., Noda, S., & Hirakawa, K. (2012). VEGF-A/VEGFR-2 signaling plays an important role for the motility of pancreas cancer cells. *Annals of Surgical Oncology, 19*, 2733–2743.

Ecelbarger, C. A., Terris, J., Frindt, G., Echevarria, M., Marples, D., Nielsen, S., & Knepper, M. A. (1995). Aquaporin-3 water channel localization and regulation in rat kidney. *American Journal of Physiology, 269*, F663–F672.

Edamana, S., Login, F. H., Yamada, S., Kwon, T. H., & Nejsum, L. N. (2021). Aquaporin water channels as regulators of cell-cell adhesion proteins. *American Journal of Physiology: Cell Physiology, 320*, C771–C777.

Efe, O., Klein, J. D., LaRocque, L. M., Ren, H., & Sands, J. M. (2016). Metformin improves urine concentration in rodents with nephrogenic diabetes insipidus. *JCI Insight, 1*.

Elkjaer, M. L., Nejsum, L. N., Gresz, V., Kwon, T. H., Jensen, U. B., Frokiaer, J., & Nielsen, S. (2001). Immunolocalization of aquaporin-8 in rat kidney, gastrointestinal tract, testis, and airways. *American Journal of Physiology - Renal Physiology, 281*, F1047–F1057.

Eriksson, U. K., Fischer, G., Friemann, R., Enkavi, G., Tajkhorshid, E., & Neutze, R. (2013). Subangstrom resolution X-ray structure details aquaporin-water interactions. *Science, 340*, 1346–1349.

Ernstsen, C. V., Login, F. H., Schelde, A. B., Therkildesen, J. R., Moller-Jensen, J., Norregaard, R., Praetorius, H., & Nejsum, L. N. (2022). Acute pyelonephritis: Increased plasma membrane targeting of renal aquaporin-2. *Acta Physiologica*, Article e13760.

Fenton, R. A., Moeller, H. B., Hoffert, J. D., Yu, M. J., Nielsen, S., & Knepper, M. A. (2008). Acute regulation of aquaporin-2 phosphorylation at Ser-264 by vasopressin. *Proceedings of the National Academy of Sciences of the U S A, 105*, 3134–3139.

Fenton, R. A., Murali, S. K., & Moeller, H. B. (2020). Advances in Aquaporin-2 trafficking mechanisms and their implications for treatment of water balance disorders. *American Journal of Physiology: Cell Physiology.*

Fukuda, A. M., Adami, A., Pop, V., Bellone, J. A., Coats, J. S., Hartman, R. E., Ashwal, S., Obenaus, A., & Badaut, J. (2013). Posttraumatic reduction of edema with aquaporin-4 RNA interference improves acute and chronic functional recovery. *Journal of Cerebral Blood Flow and Metabolism, 33*, 1621–1632.

Furman, C. S., Gorelick-Feldman, D. A., Davidson, K. G., Yasumura, T., Neely, J. D., Agre, P., & Rash, J. E. (2003). Aquaporin-4 square array assembly: Opposing actions of M1 and M23 isoforms. *Proceedings of the National Academy of Sciences of the U S A, 100*, 13609–13614.

Fushimi, K., Sasaki, S., & Marumo, F. (1997). Phosphorylation of serine 256 is required for cAMP-dependent regulatory exocytosis of the aquaporin-2 water channel. *Journal of Biological Chemistry, 272*, 14800–14804.

Gao, H., He, C., Fang, X., Hou, X., Feng, X., Yang, H., Zhao, X., & Ma, T. (2006). Localization of aquaporin-1 water channel in glial cells of the human peripheral nervous system. *Glia, 53*, 783–787.

Garneau, A. P., Carpentier, G. A., Marcoux, A. A., Frenette-Cotton, R., Simard, C. F., Remus-Borel, W., Caron, L., Jacob-Wagner, M., Noel, M., Powell, J. J., Belanger, R., Cote, F., & Isenring, P. (2015). Aquaporins mediate silicon transport in humans. *PLoS One, 10*, Article e0136149.

Geyer, R. R., Musa-Aziz, R., Qin, X., & Boron, W. F. (2013). Relative CO(2)/NH(3) selectivities of mammalian aquaporins 0-9. *American Journal of Physiology: Cell Physiology, 304*, C985–C994.

Gold, M. G., Reichow, S. L., O'Neill, S. E., Weisbrod, C. R., Langeberg, L. K., Bruce, J. E., Gonen, T., & Scott, J. D. (2012). AKAP2 anchors PKA with aquaporin-0 to support ocular lens transparency. *EMBO Molecular Medicine, 4*, 15–26.

Guan, X. G., Su, W. H., Yi, F., Zhang, D., Hao, F., Zhang, H. G., Liu, Y. J., Feng, X. C., & Ma, T. H. (2010). NPA motifs play a key role in plasma membrane targeting of aquaporin-4. *IUBMB Life, 62*, 222–226.

Gustavsson, S., Lebrun, A. S., Norden, K., Chaumont, F., & Johanson, U. (2005). A novel plant major intrinsic protein in Physcomitrella patens most similar to bacterial glycerol channels. *Plant Physiology, 139*, 287–295.

Han, Y. (2019). Analysis of the role of the Hippo pathway in cancer. *Journal of Translational Medicine, 17*, 116.

Hara-Chikuma, M., & Verkman, A. S. (2008). Prevention of skin tumorigenesis and impairment of epidermal cell proliferation by targeted aquaporin-3 gene disruption. *Molecular and Cellular Biology, 28*, 326–332.

Harris, H. W., Jr., Hosselet, C., Guay-Woodford, L., & Zeidel, M. L. (1992). Purification and partial characterization of candidate antidiuretic hormone water channel proteins of M(r) 55,000 and 53,000 from toad urinary bladder. *Journal of Biological Chemistry, 267*, 22115–22121.

Hirt, L., Fukuda, A. M., Ambadipudi, K., Rashid, F., Binder, D., Verkman, A., Ashwal, S., Obenaus, A., & Badaut, J. (2017). Improved long-term outcome after transient cerebral ischemia in aquaporin-4 knockout mice. *Journal of Cerebral Blood Flow and Metabolism, 37*, 277–290.

Hladky, S. B., & Barrand, M. A. (2022). The glymphatic hypothesis: The theory and the evidence. *Fluids and Barriers of the CNS, 19*(9).

Holst, M. R., Jensen, L. G., Aaron, J., Login, F. H., Rajkumar, S., Hahn, U., & Nejsum, L. N. (2021). Regulated exocytosis: Renal aquaporin-2 3D vesicular network organization and association with F-actin. *American Journal of Physiology: Cell Physiology, 321*, C1060–C1069.

Hoshi, A., Yamamoto, T., Shimizu, K., Ugawa, Y., Nishizawa, M., Takahashi, H., & Kakita, A. (2012). Characteristics of aquaporin expression surrounding senile plaques and cerebral amyloid angiopathy in Alzheimer disease. *Journal of Neuropathology & Experimental Neurology, 71*, 750–759.

Hubbard, J. A., Hsu, M. S., Seldin, M. M., & Binder, D. K. (2015). Expression of the astrocyte water channel aquaporin-4 in the mouse brain. *ASN Neuro, 7*.

Ikeda, M., Andoo, A., Shimono, M., Takamatsu, N., Taki, A., Muta, K., Matsushita, W., Uechi, T., Matsuzaki, T., Kenmochi, N., Takata, K., Sasaki, S., Ito, K., & Ishibashi, K. (2011). The NPC motif of aquaporin-11, unlike the NPA motif of known aquaporins, is essential for full expression of molecular function. *Journal of Biological Chemistry, 286*, 3342–3350.

Ishibashi, K., Kuwahara, M., Kageyama, Y., Sasaki, S., Suzuki, M., & Imai, M. (2000). Molecular cloning of a new aquaporin superfamily in mammals. In S. Hohmann, & S. Nielsen (Eds.), *Molecular biology and physiology of water and solute transport* (pp. 123–126). Springer US.

Itoh, T., Rai, T., Kuwahara, M., Ko, S. B., Uchida, S., Sasaki, S., & Ishibashi, K. (2005). Identification of a novel aquaporin, AQP12, expressed in pancreatic acinar cells. *Biochemical and Biophysical Research Communications, 330*, 832–838.

Jang, K. J., Cho, H. S., Kang, D. H., Bae, W. G., Kwon, T. H., & Suh, K. Y. (2011). Fluid-shear-stress-induced translocation of aquaporin-2 and reorganization of actin cytoskeleton in renal tubular epithelial cells. *Integrative Biology, 3*, 134–141.

Jensen, H. H., Holst, M. R., Login, F. H., Morgen, J. J., & Nejsum, L. N. (2018). Ectopic expression of aquaporin-5 in noncancerous epithelial MDCK cells changes cellular morphology and actin fiber formation without inducing epithelial-to-mesenchymal transition. *American Journal of Physiology: Cell Physiology, 314*, C654–C661.

Johansson, I., Karlsson, M., Shukla, V. K., Chrispeels, M. J., Larsson, C., & Kjellbom, P. (1998). Water transport activity of the plasma membrane aquaporin PM28A is regulated by phosphorylation. *The Plant Cell, 10*, 451–459.

Jung, H. J., Park, J. Y., Jeon, H. S., & Kwon, T. H. (2011). Aquaporin-5: A marker protein for proliferation and migration of human breast cancer cells. *PLoS One, 6*, Article e28492.

Jung, J. S., Preston, G. M., Smith, B. L., Guggino, W. B., & Agre, P. (1994). Molecular structure of the water channel through aquaporin CHIP. The hourglass model. *Journal of Biological Chemistry, 269*, 14648–14654.

Kaldenhoff, R., Kai, L., & Uehlein, N. (2014). Aquaporins and membrane diffusion of CO2 in living organisms. *Biochimica et Biophysica Acta, 1840*, 1592–1595.

Kamsteeg, E. J., Heijnen, I., van Os, C. H., & Deen, P. M. (2000). The subcellular localization of an aquaporin-2 tetramer depends on the stoichiometry of phosphorylated and nonphosphorylated monomers. *The Journal of Cell Biology, 151*, 919–930.

Kang, S., Chae, Y. S., Lee, S. J., Kang, B. W., Kim, J. G., Kim, W. W., Jung, J. H., Park, H. Y., Jeong, J.-H., Jeong, J. Y., & Park, J.-Y. (2015). Aquaporin 3 expression predicts survival in patients with HER2-positive early. *Breast Cancer, 35*, 2775–2782.

Kang, S. K., Chae, Y. K., Woo, J., Kim, M. S., Park, J. C., Lee, J., Soria, J. C., Jang, S. J., Sidransky, D., & Moon, C. (2008). Role of human aquaporin 5 in colorectal carcinogenesis. *American Journal of Pathology, 173*, 518–525.

Kaptan, S., Assentoft, M., Schneider, H. P., Fenton, R. A., Deitmer, J. W., MacAulay, N., & de Groot, B. L. (2015). H95 is a pH-dependent gate in aquaporin 4. *Structure, 23*, 2309–2318.

Katsura, T., Gustafson, C. E., Ausiello, D. A., & Brown, D. (1997). Protein kinase A phosphorylation is involved in regulated exocytosis of aquaporin-2 in transfected LLC-PK1 cells. *American Journal of Physiology, 272*, F817–F822.

Kaur, D., Khaniya, U., Zhang, Y., & Gunner, M. R. (2021). Protein motifs for proton transfers that build the transmembrane proton gradient. *Frontiers in Chemistry, 9*, Article 660954.

Kimelberg, H. K. (2004). Water homeostasis in the brain: Basic concepts. *Neuroscience, 129*, 851–860.

King, L. S., Choi, M., Fernandez, P. C., Cartron, J. P., & Agre, P. (2001). Defective urinary-concentrating ability due to a complete deficiency of aquaporin-1. *New England Journal of Medicine, 345*, 175–179.

Klein, J. D., Wang, Y., Blount, M. A., Molina, P. A., LaRocque, L. M., Ruiz, J. A., & Sands, J. M. (2016). Metformin, an AMPK activator, stimulates the phosphorylation of aquaporin 2 and urea transporter A1 in inner medullary collecting ducts. *American Journal of Physiology - Renal Physiology, 310*, F1008–F1012.

Klussmann, E., Tamma, G., Lorenz, D., Wiesner, B., Maric, K., Hofmann, F., Aktories, K., Valenti, G., & Rosenthal, W. (2001). An inhibitory role of Rho in the vasopressin-mediated translocation of aquaporin-2 into cell membranes of renal principal cells. *Journal of Biological Chemistry, 276*, 20451–20457.

Knepper, M. A., Kwon, T. H., & Nielsen, S. (2015). Molecular physiology of water balance. *New England Journal of Medicine, 372*, 1349–1358.

Koffman, J. S., Arnspang, E. C., Marlar, S., & Nejsum, L. N. (2015). Opposing effects of cAMP and T259 phosphorylation on plasma membrane diffusion of the water channel aquaporin-5 in madin-Darby canine kidney cells. *PLoS One, 10*, Article e0133324.

Kumari, S. S., Varadaraj, M., Menon, A. G., & Varadaraj, K. (2018). Aquaporin 5 promotes corneal wound healing. *Experimental Eye Research, 172*, 152–158.

Kushmerick, C., Rice, S. J., Baldo, G. J., Haspel, H. C., & Mathias, R. T. (1995). Ion, water and neutral solute transport in Xenopus oocytes expressing frog lens MIP. *Experimental Eye Research, 61*, 351–362.

Kwon, T. H., Frokiaer, J., Knepper, M. A., & Nielsen, S. (1998). Reduced AQP1, -2, and -3 levels in kidneys of rats with CRF induced by surgical reduction in renal mass. *American Journal of Physiology, 275*, F724–F741.

Lee, S. J., Chae, Y. S., Kim, J. G., Kim, W. W., Jung, J. H., Park, H. Y., Jeong, J. Y., Park, J. Y., Jung, H. J., & Kwon, T. H. (2014). AQP5 expression predicts survival in patients with early breast cancer. *Annals of Surgical Oncology, 21*, 375–383.

Li, C., & Wang, W. (2014). Urea transport mediated by aquaporin water channel proteins. In B. Yang, & J. M. Sands (Eds.), *Urea transporters* (pp. 227–265). Springer Netherlands.

Li, L., Zhang, H., & Verkman, A. S. (2009). Greatly attenuated experimental autoimmune encephalomyelitis in aquaporin-4 knockout mice. *BMC Neuroscience, 10*, 94.

Liu, Z., Shen, J., Carbrey, J. M., Mukhopadhyay, R., Agre, P., & Rosen, B. P. (2002). Arsenite transport by mammalian aquaglyceroporins AQP7 and AQP9. *Proceedings of the National Academy of Sciences of the U S A, 99*, 6053–6058.

Login, F. H., Jensen, H. H., Pedersen, G. A., Koffman, J. S., Kwon, T. H., Parsons, M., & Nejsum, L. N. (2019). Aquaporins differentially regulate cell-cell adhesion in MDCK cells. *The FASEB Journal, 33*, 6980–6994.

Login, F. H., Palmfeldt, J., Cheah, J. S., Yamada, S., & Nejsum, L. N. (2021). Aquaporin-5 regulation of cell-cell adhesion proteins: An elusive "tail" story. *American Journal of Physiology: Cell Physiology, 320*, C282–C292.

Lu, D. C., Zador, Z., Yao, J., Fazlollahi, F., & Manley, G. T. (2021). Aquaporin-4 reduces post-traumatic seizure susceptibility by promoting astrocytic glial scar formation in mice. *Journal of Neurotrauma, 38*, 1193–1201.

Lu, H., Sun, T. X., Bouley, R., Blackburn, K., McLaughlin, M., & Brown, D. (2004). Inhibition of endocytosis causes phosphorylation (S256)-independent plasma membrane accumulation of AQP2. *American Journal of Physiology - Renal Physiology, 286*, F233–F243.

Ma, T., Yang, B., Gillespie, A., Carlson, E. J., Epstein, C. J., & Verkman, A. S. (1998). Severely impaired urinary concentrating ability in transgenic mice lacking aquaporin-1 water channels. *Journal of Biological Chemistry, 273*, 4296–4299.

Macey, R. I., & Farmer, R. E. (1970). Inhibition of water and solute permeability in human red cells. *Biochimica et Biophysica Acta, 211*, 104–106.

Macey, R. I. (1984). Transport of water and urea in red blood cells. *American Journal of Physiology, 246*, C195–C203.

Marinelli, R. A., Tietz, P. S., Pham, L. D., Rueckert, L., Agre, P., & LaRusso, N. F. (1999). Secretin induces the apical insertion of aquaporin-1 water channels in rat cholangiocytes. *American Journal of Physiology, 276*, G280–G286.

Marples, D., Christensen, S., Christensen, E. I., Ottosen, P. D., & Nielsen, S. (1995). Lithium-induced downregulation of aquaporin-2 water channel expression in rat kidney medulla. *Journal of Clinical Investigation, 95*, 1838–1845.

Mathiisen, T. M., Lehre, K. P., Danbolt, N. C., & Ottersen, O. P. (2010). The perivascular astroglial sheath provides a complete covering of the brain microvessels: An electron microscopic 3D reconstruction. *Glia, 58*, 1094–1103.

Maurel, C. (1997). Aquaporins and water permeability of plant membranes. *Annual Review of Plant Physiology and Plant Molecular Biology, 48*, 399–429.

McLennan, R., McKinney, M. C., Teddy, J. M., Morrison, J. A., Kasemeier-Kulesa, J. C., Ridenour, D. A., Manthe, C. A., Giniunaite, R., Robinson, M., Baker, R. E., Maini, P. K., & Kulesa, P. M. (2020). Neural crest cells bulldoze through the microenvironment using Aquaporin 1 to stabilize filopodia. *Development, 147*.

Moeller, H. B., Aroankins, T. S., Slengerik-Hansen, J., Pisitkun, T., & Fenton, R. A. (2014). Phosphorylation and ubiquitylation are opposing processes that regulate endocytosis of the water channel aquaporin-2. *Journal of Cell Science, 127*, 3174–3183.

Moeller, H. B., Knepper, M. A., & Fenton, R. A. (2009). Serine 269 phosphorylated aquaporin-2 is targeted to the apical membrane of collecting duct principal cells. *Kidney International, 75*, 295–303.

Moon, C., Soria, J. C., Jang, S. J., Lee, J., Obaidul Hoque, M., Sibony, M., Trink, B., Chang, Y. S., Sidransky, D., & Mao, L. (2003). Involvement of aquaporins in colorectal carcinogenesis. *Oncogene, 22*, 6699–6703.

Morishita, Y., Matsuzaki, T., Hara-chikuma, M., Andoo, A., Shimono, M., Matsuki, A., Kobayashi, K., Ikeda, M., Yamamoto, T., Verkman, A., Kusano, E., Ookawara, S., Takata, K., Sasaki, S., & Ishibashi, K. (2005). Disruption of aquaporin-11 produces polycystic kidneys following vacuolization of the proximal tubule. *Molecular and Cellular Biology, 25*, 7770–7779.

Morishita, Y., Sakube, Y., Sasaki, S., & Ishibashi, K. (2004). Molecular mechanisms and drug development in aquaporin water channel diseases: Aquaporin superfamily (superaquaporins): Expansion of aquaporins restricted to multicellular organisms. *Journal of Pharmacological Sciences, 96*, 276–279.

Mosca, A. F., de Almeida, A., Wragg, D., Martins, A. P., Sabir, F., Leoni, S., Moura, T. F., Prista, C., Casini, A., & Soveral, G. (2018). Molecular basis of aquaporin-7 permeability regulation by pH. *Cells, 7*.

Murata, K., Mitsuoka, K., Hirai, T., Walz, T., Agre, P., Heymann, J. B., Engel, A., & Fujiyoshi, Y. (2000). Structural determinants of water permeation through aquaporin-1. *Nature, 407*, 599–605.

Nagelhus, E. A., Mathiisen, T. M., & Ottersen, O. P. (2004). Aquaporin-4 in the central nervous system: Cellular and subcellular distribution and coexpression with KIR4.1. *Neuroscience, 129*, 905–913.

Nedvetsky, P. I., Stefan, E., Frische, S., Santamaria, K., Wiesner, B., Valenti, G., Hammer, J. A., 3rd, Nielsen, S., Goldenring, J. R., Rosenthal, W., & Klussmann, E. (2007). A Role of myosin Vb and Rab11-FIP2 in the aquaporin-2 shuttle. *Traffic, 8*, 110–123.

Neely, J. D., Christensen, B. M., Nielsen, S., & Agre, P. (1999). Heterotetrameric composition of aquaporin-4 water channels. *Biochemistry, 38*, 11156–11163.

Nejsum, L. N., Christensen, T. M., Robben, J. H., Milligan, G., Deen, P. M., Bichet, D. G., & Levin, K. (2011). Novel mutation in the AVPR2 gene in a Danish male with nephrogenic diabetes insipidus caused by ER retention and subsequent lysosomal degradation of the mutant receptor. *NDT Plus, 4*, 158–163.

Nejsum, L. N., Elkjaer, M., Hager, H., Frokiaer, J., Kwon, T. H., & Nielsen, S. (2000). Localization of aquaporin-7 in rat and mouse kidney using RT-PCR, immunoblotting, and immunocytochemistry. *Biochemical and Biophysical Research Communications, 277*, 164–170.

Nejsum, L. N., Kwon, T. H., Marples, D., Flyvbjerg, A., Knepper, M. A., Frokiaer, J., & Nielsen, S. (2001). Compensatory increase in AQP2, p-AQP2, and AQP3 expression in rats with diabetes mellitus. *American Journal of Physiology - Renal Physiology, 280*, F715–F726.

Nejsum, L. N. (2005). The renal plumbing system: Aquaporin water channels. *Cellular and Molecular Life Sciences, 62*, 1692–1706.

Nelson, W. J. (2003). Adaptation of core mechanisms to generate cell polarity. *Nature, 422*, 766–774.

Nemeth-Cahalan, K. L., & Hall, J. E. (2000). pH and calcium regulate the water permeability of aquaporin 0. *Journal of Biological Chemistry, 275*, 6777–6782.

Nico, B., Paola Nicchia, G., Frigeri, A., Corsi, P., Mangieri, D., Ribatti, D., Svelto, M., & Roncali, L. (2004). Altered blood-brain barrier development in dystrophic MDX mice. *Neuroscience, 125*, 921–935.

Nielsen, S., Chou, C. L., Marples, D., Christensen, E. I., Kishore, B. K., & Knepper, M. A. (1995). Vasopressin increases water permeability of kidney collecting duct by inducing translocation of aquaporin-CD water channels to plasma membrane. *Proceedings of the National Academy of Sciences of the United States of America, 92*, 1013–1017.

Nielsen, S., DiGiovanni, S. R., Christensen, E. I., Knepper, M. A., & Harris, H. W. (1993). Cellular and subcellular immunolocalization of vasopressin-regulated water channel in rat kidney. *Proceedings of the National Academy of Sciences of the United States of America, 90*, 11663–11667.

Nielsen, S., Nagelhus, E. A., Amiry-Moghaddam, M., Bourque, C., Agre, P., & Ottersen, O. P. (1997). Specialized membrane domains for water transport in glial cells: High-resolution immunogold cytochemistry of aquaporin-4 in rat brain. *Journal of Neuroscience, 17*, 171–180.

Nielsen, S., Smith, B., Christensen, E. I., Knepper, M. A., & Agre, P. (1993). CHIP28 water channels are localized in constitutively water-permeable segments of the nephron. *The Journal of Cell Biology, 120*, 371–383.

Ohashi, Y., Tsuzaka, K., Takeuchi, T., Sasaki, Y., & Tsubota, K. (2008). Altered distribution of aquaporin 5 and its C-terminal binding protein in the lacrimal glands of a mouse model for Sjogren's syndrome. *Current Eye Research, 33*, 621–629.

Oklinski, M. K., Lim, J. S., Choi, H. J., Oklinska, P., Skowronski, M. T., & Kwon, T. H. (2014). Immunolocalization of water channel proteins AQP1 and AQP4 in rat spinal cord. *Journal of Histochemistry and Cytochemistry, 62*, 598–611.

Olesen, E. T., & Fenton, R. A. (2017). Aquaporin-2 membrane targeting: Still a conundrum. *American Journal of Physiology - Renal Physiology, 312*, F744–F747.

Oreskovic, D., Rados, M., & Klarica, M. (2017). Role of choroid plexus in cerebrospinal fluid hydrodynamics. *Neuroscience, 354*, 69–87.

Oshio, K., Watanabe, H., Song, Y., Verkman, A. S., & Manley, G. T. (2005). Reduced cerebrospinal fluid production and intracranial pressure in mice lacking choroid plexus water channel Aquaporin-1. *The FASEB Journal, 19*, 76–78.

Oshio, K., Watanabe, H., Yan, D., Verkman, A. S., & Manley, G. T. (2006). Impaired pain sensation in mice lacking Aquaporin-1 water channels. *Biochemical and Biophysical Research Communications, 341*, 1022–1028.

Preston, G. M., & Agre, P. (1991). Isolation of the cDNA for erythrocyte integral membrane protein of 28 kilodaltons: Member of an ancient channel family. *Proceedings of the National Academy of Sciences of the United States of America, 88*, 11110–11114.

Preston, G. M., Carroll, T. P., Guggino, W. B., & Agre, P. (1992). Appearance of water channels in Xenopus oocytes expressing red cell CHIP28 protein. *Science, 256*, 385–387.

Preston, G. M., Jung, J. S., Guggino, W. B., & Agre, P. (1993). The mercury-sensitive residue at cysteine 189 in the CHIP28 water channel. *Journal of Biological Chemistry, 268*, 17–20.

Procino, G., Mastrofrancesco, L., Sallustio, F., Costantino, V., Barbieri, C., Pisani, F., Schena, F. P., Svelto, M., & Valenti, G. (2011). AQP5 is expressed in type-B intercalated cells in the collecting duct system of the rat, mouse and human kidney. *Cellular Physiology and Biochemistry, 28*, 683–692.

Qin, F., Zhang, H., Shao, Y., Liu, X., Yang, L., Huang, Y., Fu, L., Gu, F., & Ma, Y. (2016). Expression of aquaporin1, a water channel protein, in cytoplasm is negatively correlated with prognosis of breast cancer patients. *Oncotarget, 7*, 8143–8154.

Rash, J. E., Yasumura, T., Hudson, C. S., Agre, P., & Nielsen, S. (1998). Direct immunogold labeling of aquaporin-4 in square arrays of astrocyte and ependymocyte plasma membranes in rat brain and spinal cord. *Proceedings of the National Academy of Sciences of the U S A, 95*, 11981–11986.

Rausch, V., & Hansen, C. G. (2020). The Hippo pathway, YAP/TAZ, and the plasma membrane. *Trends in Cell Biology, 30*, 32–48.

Reichow, S. L., Clemens, D. M., Freites, J. A., Nemeth-Cahalan, K. L., Heyden, M., Tobias, D. J., Hall, J. E., & Gonen, T. (2013). Allosteric mechanism of water-channel gating by Ca2+-calmodulin. *Nature Structural & Molecular Biology, 20*, 1085–1092.

Reichow, S. L., & Gonen, T. (2008). Noncanonical binding of calmodulin to aquaporin-0: Implications for channel regulation. *Structure, 16*, 1389–1398.

Ren, Y., Lu, H., Reinach, P. S., Zheng, Q., Li, J., Tan, Q., Zhu, H., & Chen, W. (2017). Hyperosmolarity-induced AQP5 upregulation promotes inflammation and cell death via JNK1/2 Activation in human corneal epithelial cells. *Scientific Reports, 7*, 4727.

Ribatti, D., Tamma, R., & Annese, T. (2020). Epithelial-mesenchymal transition in cancer: A historical overview. *Translational Oncology, 13*, Article 100773.

Richard, C., Gao, J., Brown, N., & Reese, J. (2003). Aquaporin water channel genes are differentially expressed and regulated by ovarian steroids during the periimplantation period in the mouse. *Endocrinology, 144*, 1533–1541.

Rodrigues, C., Mosca, A. F., Martins, A. P., Nobre, T., Prista, C., Antunes, F., Cipak Gasparovic, A., & Soveral, G. (2016). Rat aquaporin-5 is pH-gated induced by phosphorylation and is implicated in oxidative stress. *International Journal of Molecular Sciences, 17*.

Rose, K. M., Wang, Z., Magrath, G. N., Hazard, E. S., Hildebrandt, J. D., & Schey, K. L. (2008). Aquaporin 0-calmodulin interaction and the effect of aquaporin 0 phosphorylation. *Biochemistry, 47*, 339–347.

Rose, R., Kemper, B., Schwab, A., Schlatter, E., & Edemir, B. (2021). Unexpected localization of AQP3 and AQP4 induced by migration of primary cultured IMCD cells. *Scientific Reports, 11*, Article 11930.

Saadoun, S., Papadopoulos, M. C., Davies, D. C., Bell, B. A., & Krishna, S. (2002). Increased aquaporin 1 water channel expression in human brain tumours. *British Journal of Cancer, 87*, 621–623.

Saadoun, S., Papadopoulos, M. C., Hara-Chikuma, M., & Verkman, A. S. (2005). Impairment of angiogenesis and cell migration by targeted aquaporin-1 gene disruption. *Nature, 434*, 786–792.

Saadoun, S., Papadopoulos, M. C., Watanabe, H., Yan, D., Manley, G. T., & Verkman, A. S. (2005). Involvement of aquaporin-4 in astroglial cell migration and glial scar formation. *Journal of Cell Science, 118*, 5691–5698.

Saboori, A. M., Denker, B. M., & Agre, P. (1989). Isolation of proteins related to the Rh polypeptides from nonhuman erythrocytes. *Journal of Clinical Investigation, 83*, 187–191.

Sakka, L., Coll, G., & Chazal, J. (2011). Anatomy and physiology of cerebrospinal fluid. *European Annals of Otorhinolaryngology, Head and Neck, 128*, 309–316.

Satoh, J., Tabunoki, H., Yamamura, T., Arima, K., & Konno, H. (2007). Human astrocytes express aquaporin-1 and aquaporin-4 in vitro and in vivo. *Neuropathology, 27*, 245–256.

Satooka, H., & Hara-Chikuma, M. (2016). Aquaporin-3 controls breast cancer cell migration by regulating hydrogen peroxide transport and its downstream cell signaling. *Molecular and Cellular Biology, 36*, 1206–1218.

Savage, D. F., & Stroud, R. M. (2007). Structural basis of aquaporin inhibition by mercury. *Journal of Molecular Biology, 368*, 607–617.

Schey, K. L., Wang, Z., Wenkie, J. L., & Qi, Y. (2014). Aquaporins in the eye: Expression, function, and roles in ocular disease. *Biochimica et Biophysica Acta, 1840*, 1513–1523.

Schwab, A. (2001). Function and spatial distribution of ion channels and transporters in cell migration. *American Journal of Physiology - Renal Physiology, 280*, F739–F747.

Shields, S. D., Mazario, J., Skinner, K., & Basbaum, A. I. (2007). Anatomical and functional analysis of aquaporin 1, a water channel in primary afferent neurons. *Pain, 131*, 8–20.

Sidhaye, V., Hoffert, J. D., & King, L. S. (2005). cAMP has distinct acute and chronic effects on aquaporin-5 in lung epithelial cells. *Journal of Biological Chemistry, 280*, 3590–3596.

Silberstein, C., Bouley, R., Huang, Y., Fang, P., Pastor-Soler, N., Brown, D., & Van Hoek, A. N. (2004). Membrane organization and function of M1 and M23 isoforms of aquaporin-4 in epithelial cells. *American Journal of Physiology - Renal Physiology, 287*, F501–F511.

Smalley, K. S., Brafford, P., Haass, N. K., Brandner, J. M., Brown, E., & Herlyn, M. (2005). Up-regulated expression of zonula occludens protein-1 in human melanoma associates with N-cadherin and contributes to invasion and adhesion. *American Journal of Pathology, 166*, 1541–1554.

Smith, A. J., & Verkman, A. S. (2015). Superresolution imaging of aquaporin-4 cluster size in antibody-stained paraffin brain sections. *Biophysical Journal, 109*, 2511–2522.

Smith, B. L., Preston, G. M., Spring, F. A., Anstee, D. J., & Agre, P. (1994). Human red cell aquaporin CHIP. I. Molecular characterization of ABH and Colton blood group antigens. *Journal of Clinical Investigation, 94*, 1043–1049.

Sohara, E., Rai, T., Miyazaki, J., Verkman, A. S., Sasaki, S., & Uchida, S. (2005). Defective water and glycerol transport in the proximal tubules of AQP7 knockout mice. *American Journal of Physiology - Renal Physiology, 289*, F1195–F1200.

Sonah, H., Deshmukh, R. K., Labbe, C., & Belanger, R. R. (2017). Analysis of aquaporins in Brassicaceae species reveals high-level of conservation and dynamic role against biotic and abiotic stress in canola. *Scientific Reports, 7*, 2771.

Sveinsdottir, S., Gram, M., Cinthio, M., Sveinsdottir, K., Morgelin, M., & Ley, D. (2014). Altered expression of aquaporin 1 and 5 in the choroid plexus following preterm intraventricular hemorrhage. *Developmental Neuroscience, 36*, 542–551.

Swanson, J. M., Maupin, C. M., Chen, H., Petersen, M. K., Xu, J., Wu, Y., & Voth, G. A. (2007). Proton solvation and transport in aqueous and biomolecular systems: Insights from computer simulations. *Journal of Physical Chemistry B, 111*, 4300–4314.

Sweet, G., Gandor, C., Voegele, R., Wittekindt, N., Beuerle, J., Truniger, V., Lin, E. C., & Boos, W. (1990). Glycerol facilitator of *Escherichia coli*: Cloning of glpF and identification of the glpF product. *Journal of Bacteriology, 172*, 424–430.

Sykova, E. (2004). Extrasynaptic volume transmission and diffusion parameters of the extracellular space. *Neuroscience, 129*, 861–876.

Tajkhorshid, E., Nollert, P., Jensen, M. O., Miercke, L. J., O'Connell, J., Stroud, R. M., & Schulten, K. (2002). Control of the selectivity of the aquaporin water channel family by global orientational tuning. *Science, 296*, 525–530.

Tamma, G., Klussmann, E., Maric, K., Aktories, K., Svelto, M., Rosenthal, W., & Valenti, G. (2001). Rho inhibits cAMP-induced translocation of aquaporin-2 into the apical membrane of renal cells. *American Journal of Physiology - Renal Physiology, 281*, F1092–F1101.

Terris, J., Ecelbarger, C. A., Marples, D., Knepper, M. A., & Nielsen, S. (1995). Distribution of aquaporin-4 water channel expression within rat kidney. *American Journal of Physiology, 269*, F775–F785.

Terris, J., Ecelbarger, C. A., Nielsen, S., & Knepper, M. A. (1996). Long-term regulation of four renal aquaporins in rats. *American Journal of Physiology, 271*, F414–F422.

Thiagarajah, J. R., & Verkman, A. S. (2002). Aquaporin deletion in mice reduces corneal water permeability and delays restoration of transparency after swelling. *Journal of Biological Chemistry, 277*, 19139–19144.

Tong, H., Hu, Q., Zhu, L., & Dong, X. (2019). Prokaryotic aquaporins. *Cells, 8*.

Tong, J., Canty, J. T., Briggs, M. M., & McIntosh, T. J. (2013). The water permeability of lens aquaporin-0 depends on its lipid bilayer environment. *Experimental Eye Research, 113*, 32–40.

Tornroth-Horsefield, S., Wang, Y., Hedfalk, K., Johanson, U., Karlsson, M., Tajkhorshid, E., Neutze, R., & Kjellbom, P. (2006). Structural mechanism of plant aquaporin gating. *Nature, 439*, 688–694.

Trillo-Contreras, J. L., Toledo-Aral, J. J., Echevarria, M., & Villadiego, J. (2019). AQP1 and AQP4 contribution to cerebrospinal fluid homeostasis. *Cells, 8*.

Van Balkom, B. W., Savelkoul, P. J., Markovich, D., Hofman, E., Nielsen, S., van der, S. P., & Deen, P. M. (2002). The role of putative phosphorylation sites in the targeting and shuttling of the aquaporin-2 water channel. *Journal of Biological Chemistry, 277*, 41473–41479.

Varadaraj, K., Kumari, S. S., & Mathias, R. T. (2010). Transgenic expression of AQP1 in the fiber cells of AQP0 knockout mouse: Effects on lens transparency. *Experimental Eye Research, 91*, 393–404.

Vasioukhin, V. (2012). Adherens junctions and cancer. In T. Harris (Ed.), *Adherens junctions: From molecular mechanisms to tissue development and disease* (pp. 379–414). Springer.

Vera-Estrella, R., Barkla, B. J., Bohnert, H. J., & Pantoja, O. (2004). Novel regulation of aquaporins during osmotic stress. *Plant Physiology, 135*, 2318–2329.

Verkman, A. S., Ruiz-Ederra, J., & Levin, M. H. (2008). Functions of aquaporins in the eye. *Progress in Retinal and Eye Research, 27*, 420–433.

Vitellaro-Zuccarello, L., Mazzetti, S., Bosisio, P., Monti, C., & De Biasi, S. (2005). Distribution of aquaporin 4 in rodent spinal cord: Relationship with astrocyte markers and chondroitin sulfate proteoglycans. *Glia, 51*, 148–159.

Walz, T., Hirai, T., Murata, K., Heymann, J. B., Mitsuoka, K., Fujiyoshi, Y., Smith, B. L., Agre, P., & Engel, A. (1997). The three-dimensional structure of aquaporin-1. *Nature, 387*, 624–627.

Wang, Y., Li, R., Li, D., Jia, X., Zhou, D., Li, J., Lyi, S. M., Hou, S., Huang, Y., Kochian, L. V., & Liu, J. (2017). NIP1;2 is a plasma membrane-localized transporter mediating aluminum uptake, translocation, and tolerance in Arabidopsis. *Proceedings of the National Academy of Sciences of the U S A, 114*, 5047–5052.

Wang, Y., Schulten, K., & Tajkhorshid, E. (2005). What makes an aquaporin a glycerol channel? A comparative study of AqpZ and GlpF. *Structure, 13*, 1107–1118.

Wei, H., Gao, W. S., Qi, L., Jia, L., Qu, Y. J., Yue, S. W., & Zhang, Y. (2020). Effect of cGMP-activated aquaporin 1 on TRPV4 in rats with allodynia induced by chronic compression of the dorsal root ganglion. *Neuroscience Letters, 716*, Article 134630.

Wilson, A. J., Carati, C. J., Gannon, B. J., Haberberger, R., & Chataway, T. K. (2010). Aquaporin-1 in blood vessels of rat circumventricular organs. *Cell and Tissue Research, 340*, 159–168.

Woo, J., Chae, Y. K., Jang, S. J., Kim, M. S., Baek, J. H., Park, J. C., Trink, B., Ratovitski, E., Lee, T., Park, B., Park, M., Kang, J. H., Soria, J. C., Lee, J., Califano, J., Sidransky, D., & Moon, C. (2008). Membrane trafficking of AQP5 and cAMP dependent phosphorylation in bronchial epithelium. *Biochemical and Biophysical Research Communications, 366*, 321–327.

Woo, J., Lee, J., Kim, M. S., Jang, S. J., Sidransky, D., & Moon, C. (2008). The effect of aquaporin 5 overexpression on the Ras signaling pathway. *Biochemical and Biophysical Research Communications, 367*, 291–298.

Wu, B., & Beitz, E. (2007). Aquaporins with selectivity for unconventional permeants. *Cellular and Molecular Life Sciences, 64*, 2413–2421.

Wu, Q., Zhang, Y. J., Gao, J. Y., Li, X. M., Kong, H., Zhang, Y. P., Xiao, M., Shields, C. B., & Hu, G. (2014). Aquaporin-4 mitigates retrograde degeneration of rubrospinal neurons by facilitating edema clearance and glial scar formation after spinal cord injury in mice. *Molecular Neurobiology, 49*, 1327–1337.

Wu, Z., Li, S., Liu, J., Shi, Y., Wang, J., Chen, D., Luo, L., Qian, Y., Huang, X., & Wang, H. (2015). RNAi-mediated silencing of AQP1 expression inhibited the proliferation, invasion and tumorigenesis of osteosarcoma cells. *Cancer Biology & Therapy, 16*, 1332–1340.

Xia, H., Ma, Y. F., Yu, C. H., Li, Y. J., Tang, J., Li, J. B., Zhao, Y. N., & Liu, Y. (2014). Aquaporin 3 knockdown suppresses tumour growth and angiogenesis in experimental non-small cell lung cancer. *Experimental Physiology, 99*, 974–984.

Xie, L., Hoffert, J. D., Chou, C. L., Yu, M. J., Pisitkun, T., Knepper, M. A., & Fenton, R. A. (2010). Quantitative analysis of aquaporin-2 phosphorylation. *American Journal of Physiology - Renal Physiology, 298*, F1018–F1023.

Xu, D. L., Martin, P. Y., Ohara, M., StJohn, J., Pattison, T., Meng, X., Morris, K., Kim, J. K., & Schrier, R. W. (1997). Upregulation of aquaporin-2 water channel expression in chronic heart failure rat. *Journal of Clinical Investigation, 99*, 1500–1505.

Yakata, K., Hiroaki, Y., Ishibashi, K., Sohara, E., Sasaki, S., Mitsuoka, K., & Fujiyoshi, Y. (2007). Aquaporin-11 containing a divergent NPA motif has normal water channel activity. *Biochimica et Biophysica Acta, 1768*, 688–693.

Yaneff, A., Sigaut, L., Marquez, M., Alleva, K., Pietrasanta, L. I., & Amodeo, G. (2014). Heteromerization of PIP aquaporins affects their intrinsic permeability. *Proceedings of the National Academy of Sciences of the U S A, 111*, 231–236.

Yang, B., Brown, D., & Verkman, A. S. (1996). The mercurial insensitive water channel (AQP-4) forms orthogonal arrays in stably transfected Chinese hamster ovary cells. *Journal of Biological Chemistry, 271*, 4577–4580.

Yang, B., Song, Y., Zhao, D., & Verkman, A. S. (2005). Phenotype analysis of aquaporin-8 null mice. *American Journal of Physiology: Cell Physiology, 288*, C1161–C1170.

Yang, B., & Verkman, A. S. (1997). Water and glycerol permeabilities of aquaporins 1-5 and MIP determined quantitatively by expression of epitope-tagged constructs in Xenopus oocytes. *Journal of Biological Chemistry, 272*, 16140–16146.

Yasui, M., Hazama, A., Kwon, T. H., Nielsen, S., Guggino, W. B., & Agre, P. (1999). Rapid gating and anion permeability of an intracellular aquaporin. *Nature, 402*, 184–187.

Yasui, M., Kwon, T. H., Knepper, M. A., Nielsen, S., & Agre, P. (1999). Aquaporin-6: An intracellular vesicle water channel protein in renal epithelia. *Proceedings of the National Academy of Sciences of the United States of America, 96*, 5808–5813.

Yoshida, T., Hojo, S., Sekine, S., Sawada, S., Okumura, T., Nagata, T., Shimada, Y., & Tsukada, K. (2013). Expression of aquaporin-1 is a poor prognostic factor for stage II and III colon cancer. *Mol Clin Oncol, 1*, 953–958.

Zador, Z., Bloch, O., Yao, X., & Manley, G. T. (2007). Aquaporins: Role in cerebral edema and brain water balance. *Progress in Brain Research, 161*, 185–194.

Zago, W., Schroeter, S., Guido, T., Khan, K., Seubert, P., Yednock, T., Schenk, D., Gregg, K. M., Games, D., Bard, F., & Kinney, G. G. (2013). Vascular alterations in PDAPP mice after anti-Abeta immunotherapy: Implications for amyloid-related imaging abnormalities. *Alzheimers Dementia, 9*, S105–S115.

Zampighi, G. A., Eskandari, S., Hall, J. E., Zampighi, L., & Kreman, M. (2002). Micro-domains of AQP0 in lens equatorial fibers. *Experimental Eye Research, 75*, 505–519.

Zeuthen, T., & Klaerke, D. A. (1999). Transport of water and glycerol in aquaporin 3 is gated by H(+). *Journal of Biological Chemistry, 274*, 21631–21636.

Zhang, D., Vetrivel, L., & Verkman, A. S. (2002). Aquaporin deletion in mice reduces intraocular pressure and aqueous fluid production. *Journal of General Physiology, 119*, 561–569.

Zhang, H., & Verkman, A. S. (2015). Aquaporin-1 water permeability as a novel determinant of axonal regeneration in dorsal root ganglion neurons. *Experimental Neurology, 265*, 152–159.

Zhang, Z., Chen, Z., Song, Y., Zhang, P., Hu, J., & Bai, C. (2010). Expression of aquaporin 5 increases proliferation and metastasis potential of lung cancer. *The Journal of Pathology, 221*, 210–220.

Zhou, J., Kong, H., Hua, X., Xiao, M., Ding, J., & Hu, G. (2008). Altered blood-brain barrier integrity in adult aquaporin-4 knockout mice. *NeuroReport, 19*, 1–5.

Zwiazek, J. J., Xu, H., Tan, X., Navarro-Rodenas, A., & Morte, A. (2017). Significance of oxygen transport through aquaporins. *Scientific Reports, 7*, Article 40411.

Chapter 19

Generation and conduction of electrical signals

Nicholas Sperelakis[†] and Simon Rock Levinson

Department of Physiology and Biophysics, University of Colorado School of Medicine, Aurora, CO, United States

1. Summary

Membrane excitability is a fundamental property of nerve and muscle cells (skeletal, cardiac, and smooth), as well as certain other cell types, such as some endocrine cells. An excitable cell is one that, in response to certain environmental stimuli (electrical, chemical, or mechanical), generates an all-or-none electrical signal or action potential (AP). The AP is sometimes called an "impulse," e.g., a nerve impulse. The AP is triggered by a decrease (i.e., "depolarization") in the resting membrane voltage membrane, which is produced by the applied stimulus. The depolarization initiates an increase in the membrane permeability to Na^+ ions via sodium channels, which then flow into the cell, causing a transient reversal in the membrane potential (i.e., the spike). The spike is terminated by the spontaneous inactivation of the Na^+ channels and the membrane is repolarized by a slower increase in the permeability of the membrane to K^+ ions, resulting in repolarization of the membrane.

The AP has several characteristics that are vital to its role as an electrical signal. Most importantly, it propagates along cell membranes, and it does so in a binary manner as an ungraded "all-or-none" (i.e., present or absent) voltage signal. Furthermore, much of the information transmitted is encoded by varying the frequency of AP generation and conduction, e.g., a higher frequency representing a stronger sensory stimulus or a command to contract muscle fibers more strongly.

Some cells display the property of automaticity (e.g., cardiac pacemaker cells, certain neurons); i.e., they produce APs spontaneously without any externally applied stimulus. Also, in some excitable cell types calcium ions (Ca^{2+}) play a role in action potential generation. Thus in cardiac cells, slow inward Ca^{2+} current results in a long plateau of the AP, while in others (e.g., cardiac pacemaker cells or smooth muscle), the Ca^{2+} current itself generates the upstroke.

The membrane currents that contribute to the AP can be studied by the voltage-clamp method, which allows isolation and characterization of each membrane current as a function of membrane potential and time. The pattern of minute currents that flow through single cells or individual voltage-dependent ion channel macromolecules can be studied using the patch-clamp method. Such measurements also provide important clues regarding the molecular mechanisms that underlie voltage-sensitive gating and ion selectivity by ion channels and are described in detail in Chapter 21.

2. Introduction

Excitability is an intrinsic membrane property that allows a cell to generate an electrical signal, the action potential (AP), in response to stimuli of sufficient magnitude. The elongated nerve axon serves to transmit information in the form of APs over long distances. However, the *biological cable* (i.e., nerve fiber or skeletal muscle fiber) is a relatively poor cable with a short *length constant* and relatively long *time constant*. Therefore, for faithful and rapid signal transmission over long distances, energy must be put into the system at each point along the way. The system evolved is that of action potential (AP) generation, which are *all-or none signals* of constant amplitude and constant propagation velocity, in addition to

[†] Deceased.

Cell Physiology Sourcebook. https://doi.org/10.1016/B978-0-12-811114-7.00038-4

having refractory periods and sharp thresholds. This is a *frequency-modulated* system, in which increasing strength of sensation or motor response is signaled by an increase in frequency of the AP signals.

The energy source for the generation of the AP is stored in the excitable cell itself. An initial depolarization is produced by a stimulus and triggers the intrinsic AP mechanism. The immediate source of energy for the AP comes from the transmembrane ionic gradients for K^+ and Na^+, which act like opposing batteries. The K^+ ion concentration gradient is mainly responsible for the generation of the rest (Chapter 14). Upon depolarization to threshold, the Na^+ ion electrochemical driving force, which is directed inward, causes a large and rapid inward Na^+ current that generates the AP upstroke. Over a longer time frame, the Na^+-K^+ pump is responsible for the generation of the Na^+ and K^+ ionic gradients and for their maintenance and restoration after repetitive AP activity. The Na^+—K^+ pump derives chemical energy from the hydrolysis of ATP. The bases of the RP and active ion transport have been discussed in earlier chapters.

3. Action potential characteristics

The first intracellular recordings of an action potential and its conduction were accomplished using the squid giant axon (Hodgkin & Huxley, 1939). This axon is large enough to allow the insertion of a fine glass micropipette to measure the transmembrane voltage directly. A typical squid giant axon AP and its method of observation are shown diagramatically in Fig. 19.1, in which the transmembrane potential is recorded as the difference between an intracellular and an extracellular electrode. Subsequent to these pioneering studies of squid axon excitability, similar characteristics have been observed and characterized in numerous other excitable cells among all animal species.

Thus all APs have the properties of being **all-or-none**, and they have a **sharp threshold** and a **refractory period** that determines their initiation. Once initiated, AP's **propagate along the axon** with a more or less constant shape, amplitude, and duration. The all-or-none property means that the single signal is not graded in amplitude (i.e., it is not an amplitude-modulated system) as it propagates and that the signal is either zero ("off" or "no") or maximum ("on" or "yes"). As such,

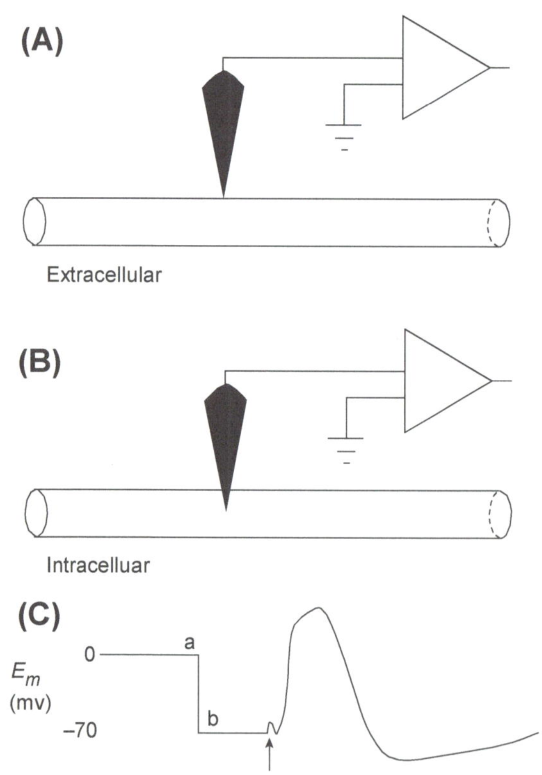

FIGURE 19.1 Recording the transmembrane potential of a squid giant nerve axon. The transmembrane potential is measured as the potential difference between the intracellular and extracellular electrodes. (A) The microelectrode is outside the axon and measures 0 mV (segment a in panel C). (B) As the electrode is advanced and crosses the membrane, the RP of −70 mV is measured (segment b in C). The cell membrane makes a tight seal around the glass electrode. (C) The membrane potential (E_m) is depicted. Stimulation (at arrow) elicits an AP.

APs constitute a *digital signaling system* in which strength of the signal to be conveyed is done by varying the frequency of the impulses. These characteristics of the AP are discussed in the next sections.

3.1 Threshold and all-or-none property

According to convention, the *direction of electrical current* is defined as the *direction of movement of positive charges*. For the active membrane, inward currents are depolarizing (bringing positive charge inside the cell) and outward currents are hyperpolarizing. A subthreshold depolarization is defined as one that does not reach threshold to elicit an AP. It is proportional to the applied stimulus and it decrements with distance along the nerve axon cable. However, if a stimulus current is of sufficient magnitude, then the resulting membrane depolarization reaches a critical value, called the *threshold potential*, at which an AP is initiated (see Fig. 19.2).

Nerve membrane responses near the site of application of brief current pulses vary depending on the magnitude and direction of the pulses (Fig. 19.2). Hyperpolarizing and subthreshold depolarizing responses are graded in magnitude according to the stimulus current. However, as can be seen, a somewhat higher intensity outward current produces a depolarizing response with a different waveform and a longer duration. This condition is referred to as the *local excitatory state* (Fig. 19.2A and B). It occurs when a small area of the membrane near the stimulus electrode comes close to threshold, but it does not generate an AP. This local membrane activity is not propagated and decays with distance along the axon. A slightly stronger stimulus is, however, sufficient to bring the membrane potential of a large enough membrane area to the *threshold potential*, thereby initiating an all-or-none AP. At the threshold potential, a large inward (depolarizing) current is initiated and the membrane continues to depolarize.

The AP parameters, including overshoot (i.e., spike height >0 mV), duration, and rate of rise, are characteristic for each type of excitable cell. For example, the duration of the AP of the squid giant axon is about 1 ms, whereas the cardiac AP

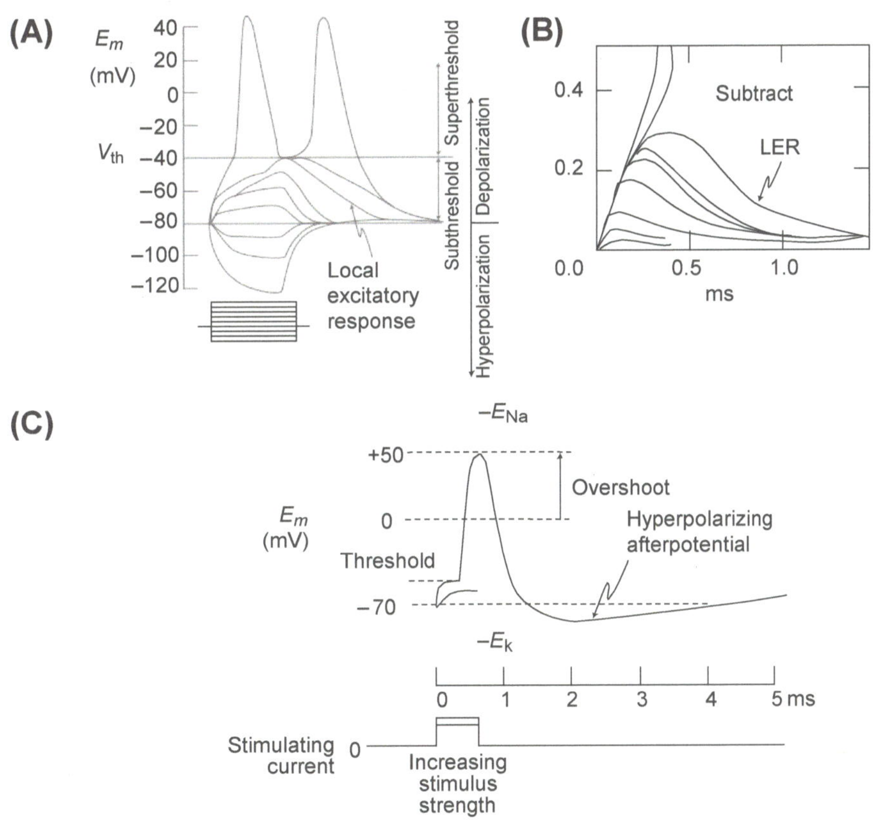

FIGURE 19.2 Diagrammatic sketches depicting initiation of the nerve impulse by membrane depolarization. (A) Membrane responses to depolarizing and hyperpolarizing current pulses are shown. The nonlinear local excitatory response (LER) occurs just below the threshold for the all-or-none AP. (B) Illustration of the LER obtained by subtraction of the membrane responses to hyperpolarizing current pulses (passive only) from the responses to depolarizing pulses (passive plus active). (C) A nerve AP is depicted, illustrating the sharp threshold, overshoot, and hyperpolarizing afterpotential.

lasts over a 100 milliseconds. These differences in the APs subserve the functions performed by the different excitable tissues. The overshoot and a hyperpolarizing afterpotential (following the spike) are illustrated in Fig. 19.2C.

3.2 Refractoriness

Once an AP is initiated, a finite and characteristic time must elapse before a second AP can be generated. This time interval is called the *refractory period* and its value depends on the type of excitable cell. Cells with long-duration APs (e.g., myocardial cells) have long refractory periods; cells with brief APs (e.g., neurons) have short refractory periods. That is, the refractory periods are related to the AP duration.

Two types of refractory periods are usually defined: an *absolute refractory period* and a *relative refractory period* (Fig. 19.3). The absolute refractory period denotes the interval during which a second AP cannot be elicited, regardless of the intensity of the applied stimulus. During the relative refractory period, a second AP may be elicited, provided that a greater than normal stimulus is applied. The second AP often is subnormal in amplitude and rate of rise. Therefore, the physiologically important refractory period is the *functional (effective) refractory period*, which is defined by the highest frequency of APs that the excitable cell can propagate. For example, if a myelinated nerve axon can propagate impulses up to 1000/s, then the functional refractory period is 1.0 ms.

Membrane excitability is greatly altered during the refractory periods. Excitability is zero during the absolute refractory period and is depressed during the relative refractory period, becoming less and less depressed as the membrane repolarizes back toward the RP. The *afterpotentials* (potentials arising after the generation of an AP spike) that many cells exhibit also affect membrane excitability: hyperpolarizing afterpotentials depress excitability (greater critical depolarization required to reach threshold) and depolarizing afterpotentials enhance excitability. A depolarizing afterpotential produces a *supernormal period of excitability*. In contrast, a hyperpolarizing afterpotential blends into and extends the relative refractory period. Thus, for example, since neurons usually exhibit hyperpolarizing afterpotentials, the functional refractory period actually includes part of the afterpotential. The significance of afterpotentials is discussed in more detail below.

3.3 AP propagation

It is imperative that the body be able to transmit a signal from one point to another very rapidly. The only way that this can be efficiently accomplished is by an electrical mechanism. Blood flow, diffusion, and signaling molecules are much too slow to allow rapid signaling. Therefore, the body makes use of electricity via AP propagation for rapid signaling in the nervous system, skeletal muscle, heart, and smooth muscles.

The continuous conduction of an AP down an axon, such as seen in the squid giant axon, is a typical form of AP propagation. In this modality, the AP can be observed as invariant in amplitude and shape when recorded at any point along the axon (Fig. 19.4). However, in contrast to the conduction of electrons in a copper wire at the speed of light, the speed of AP propagation is usually only of the order of meters per second (or miles per hour). Thus, propagation velocity is about 120 m/s in our fastest nerve fibers, about 6 m/s in skeletal muscle, about 0.5 m/s in heart, and about 0.05 m/s in smooth muscle (Table 19.1). While these AP velocities are slow relative to light speed, they are sufficient for many

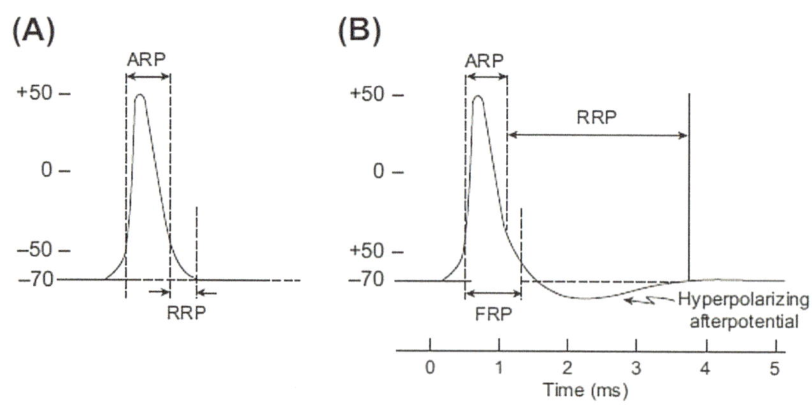

FIGURE 19.3 Refractory periods of nerve APs, both without (A) and with (B) a hyperpolarizing afterpotential. The absolute (ARP), relative (RRP), and functional (FRP) refractory periods are labeled.

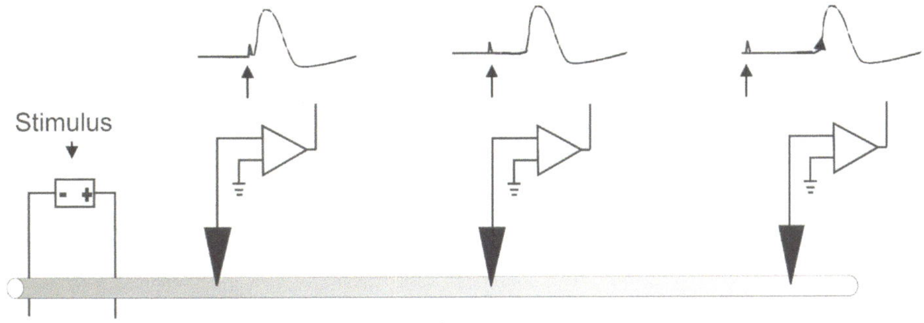

FIGURE 19.4 Continuous conduction of the AP along the axon. In the diagram above, a suprathreshold stimulus is applied to the axon, generating an AP. At increasing distances from the point of stimulation, the action potential is recorded by an intracellular electrode, as seen in the traces above the recording apparatus. Note that the amplitude and shape of the AP remain constant as the AP propagates along the axon. The trace arrows show the timing of the stimulus relative to the recording location. Thus as the AP propagates, there is an increasing delay between the time of stimulus application and the time of arrival of the AP it evokes.

TABLE 19.1 Conduction velocity as a function of fiber diameter in nerve axons and muscle fibers.

Fiber type	Fiber diameter (μm)	Propagation velocity (m/s)	Velocity/Diameter (m/s/μm)
Myelinated axons	20	120	6.0
	12	70	5.8
	5	30	6.0
Nonmyelinated axons	1.5	2.0	1.3
	1.0	1.3	1.3
Squid giant axons (20 °C)	500	25	0.05
Skeletal muscle fibers	50	6	0.12
Cardiac muscle fibers	15	0.5	0.03
Smooth muscle fibers	5	0.05	0.01

functions of the nervous system and other excitable cells, even for such complex coordinated actions as walking or avoiding a rapidly approaching threat.

Passive Factors that affect the Speed of AP propagation. For continuous conduction of an AP along a bare axon as described above, propagation velocity is faster when (1) the diameter of the axon is larger, (2) the electrical length constant is longer, and (3) the electrical time constant and capacitance are lower. These are all passive properties of the axon that strongly influence the efficacy of the active AP generating mechanism that will be discussed below. For the present, we discuss how such passive axonal characteristics will affect the spread of active electrical generation by considering the properties of the axon as a passive electrical conductor, or cable.

3.4 Biological fiber/axon as a cable

An *electrical cable* consists of two parallel conductors separated by insulation material, e.g., two copper wires separated by rubber. Usually one of the conductors is arranged as a tubular sleeve surrounding a central solid rod (wire), as depicted in Fig. 19.5A. The equivalent electrical circuit for a cable is shown in Fig. 19.5B: two parallel conductors (wires) separated by a transverse resistance (R) shown distributed along the length of the cable. The resistance of the conductors is so small, compared with the transverse insulation resistance, that it is assumed to be zero. In the case of the *biological cable* (a long narrow nerve fiber or skeletal muscle fiber), one parallel conductor is the inside fluid (cytoplasm) and the second parallel conductor is the outside fluid surrounding (bathing) the cell (the interstitial fluid). Because the conductivity of biological fluid is much less (i.e., much higher resistance) than that of copper wire and because the cross-sectional area of the cell is so small, the inside longitudinal resistance is high and cannot be ignored (Fig. 19.5C). The outside longitudinal resistance is

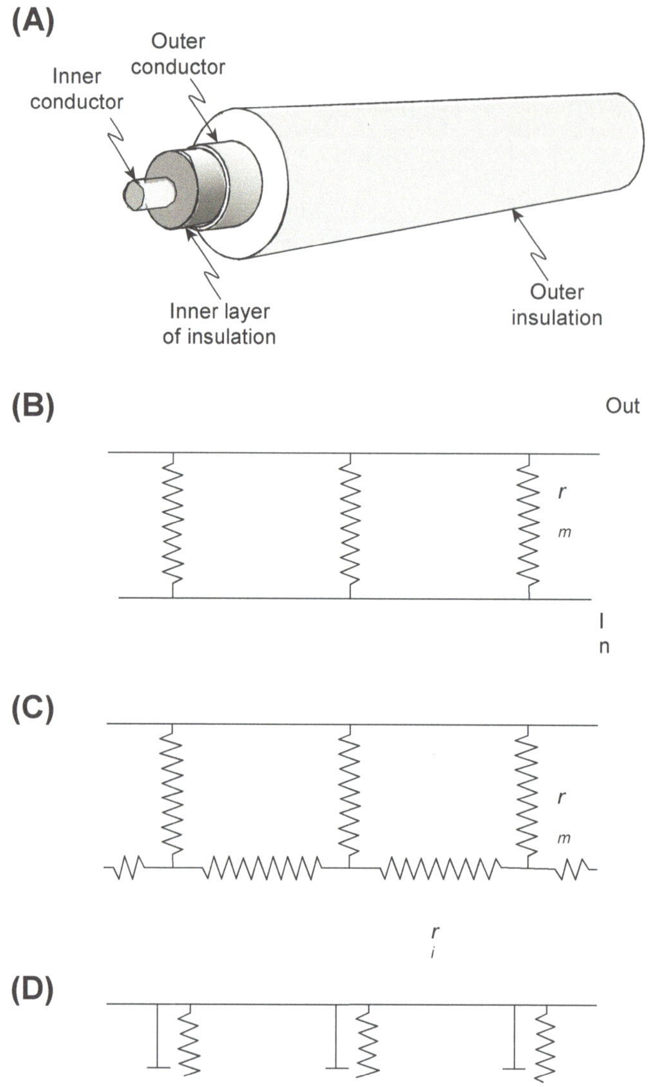

FIGURE 19.5 A coaxial cable and associated electrical equivalent circuit (A, B) and the electrical equivalent circuit for a biological cable (C, D). (A) The coaxial cable consists of an inner conductor (e.g., copper wire) and an outer concentric conductor separated by a layer of insulation (e.g., rubber). (B) The equivalent circuit for a coaxial cable. The inner and outer conductors are depicted as having nearly zero resistances and the transmembrane insulation resistance is shown distributed along the length of the cable. (C) In the biological cable, the inner conductor is the cytoplasm (axoplasm or myoplasm), which is not of negligible resistivity and therefore is depicted as r_i distributed along the length of the fiber. The transverse insulation resistance is the cell membrane resistance (r_m). (D) addition of the capacitance elements to the biological cable (c_m), which arise due to the lipid bilayer matrix of the cell membrane.

relatively small, as compared with the inside, because of the larger volume (cross-sectional area) of fluid available to carry the outside current, and therefore is assumed to be negligible (Taylor and Nastuk, 1963).

In addition, there is a *stray capacitance* distributed along the length of the cable (see Fig. 19.5D), because capacitance occurs when two parallel conductors ("plates") are separated by a high-resistance dielectric material. The dielectric constant of materials is related to vacuum, which is assigned a value of 1.0000; air has a value very close to vacuum, oils have a value of three to six and that of pure water is 81. The biological membrane, which has a matrix of phospholipid molecules, has a dielectric constant of about 5, typical of oils. The higher the dielectric constant, the higher the capacitance. The closer the parallel plates, the higher the capacitance. Because the biological membrane is so thin (approx. 70 Å or 7 nm), its capacitance is relatively high. All cell membranes have a membrane capacitance (C_m) of about 0.7–1.0 mF/cm^2 (where F stands for farads (Cole, 1968).

3.5 Length constant

In the electric cable depicted in Fig. 19.5A and B, a voltage (or signal) applied at one end would be transmitted to the distant end with little or no *decrement* (diminution or attenuation) and the so-called *length constant* would be very long or nearly infinite. In the biological cable (see Fig. 19.5C and D), however, a signal applied at one end rapidly falls off (decays) in amplitude as a function of distance, with a relatively short length constant (λ). This decay in voltage is exponential (Fig. 19.6A). An exponential process gives a straight line on a semilogarithmic plot (log V versus distance) (Fig. 19.6B). In a cable, the relationship between the voltage at any distance (x) from the applied voltage (V_0) is

$$V_x = V_o e^{-x/\lambda} \quad (19.1)$$

Thus, when $x = \lambda$

$$V_\lambda = V_0 \frac{1}{e} = .368\ V_0 \quad (19.2)$$

Hence, the distance at which the voltage decays to 36.8% of the initial value gives the length constant, λ. In nerve fibers and skeletal muscle fibers, λ has a value of *only about 1–3 mm.*

Therefore, the relatively short λ, compared with the length of the neuron (e.g., over 1.0 m for a lumbar anterior horn cell motor neuron) or skeletal muscle fiber, means that a signal applied at one end (or midpoint) would fall off very quickly with distance along the fiber (Fig. 19.7). If the length constant were 1.0 mm, then at 4 mm, the signal would become negligibly small. Hence, the electrical signal cannot be conducted passively in the biological cable, because it would decrement and disappear over relatively short distances.

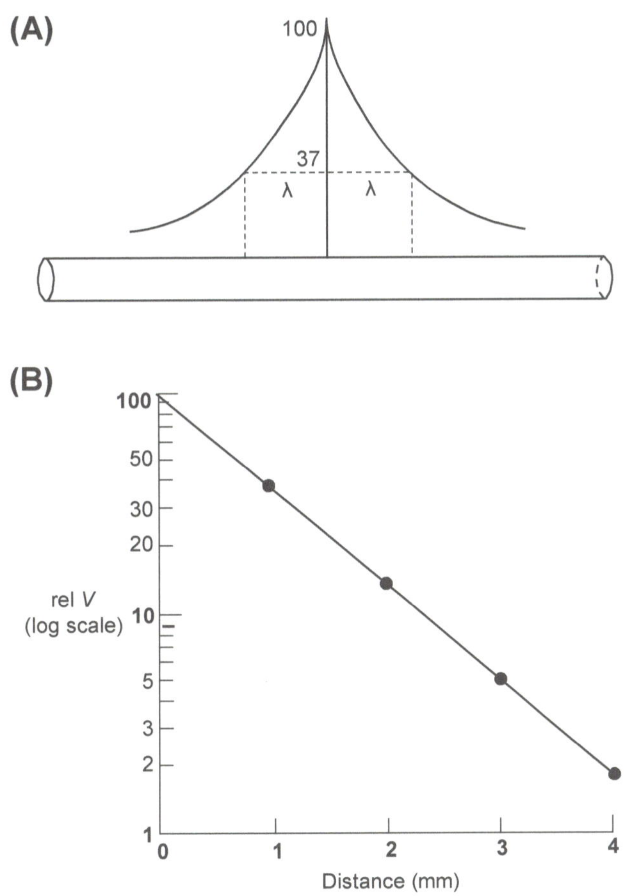

FIGURE 19.6 Length constant of the biological cable (fiber). (A) The exponential decay of voltage on both sides of an applied current (voltage) as a function of distance is diagrammed. The distance at which the voltage falls to 1/e or 36.8% of initial voltage (at x ¼ 0) gives the l value. (B) When the voltage is plotted on a log scale against distance, a straight line is obtained.

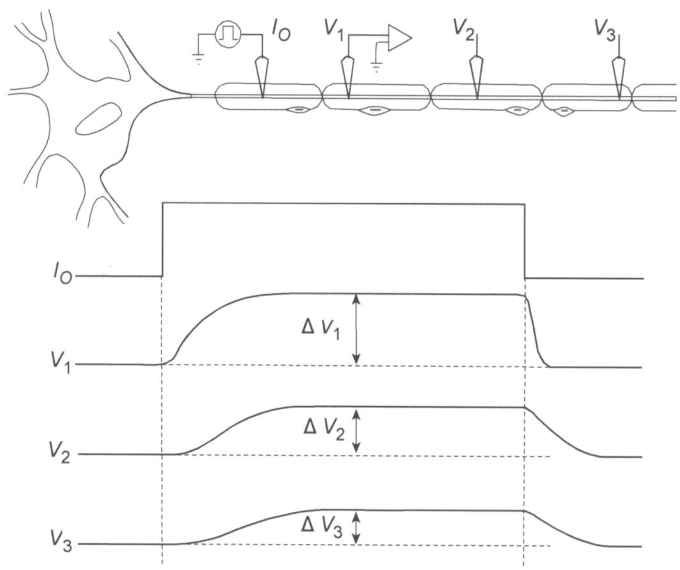

FIGURE 19.7 Diagram of a motor axon and how the voltage signal would decay with distance if the neuron were only a passive cable (nonexcitable). The example used is that of a myelinated nerve fiber, but the same considerations apply for unmyelinated axons. The voltage traces illustrate the voltage signals that would be simultaneously recorded at three different points along the axon (V_1, V_2, and V_3) from the site of injection of a rectangular current pulse (I_o). As depicted, the amplitude of the steady-state voltage pulse rapidly falls off with distance, because the length constant (L) of the biological cable is short (e.g., 1 mm) compared with the length of the axon (e.g., 1000 mm). Therefore, an active response of the cell membrane at each point is required faithfully to propagate the signal. The voltage recorded at point V_1 (D V_1) also illustrates that the membrane potential changes in an exponential manner, both at the beginning of an applied rectangular current pulse and at the end. This exponential change results from the capacitance of the cell membrane, the time constant (s) being a product of the resistance and capacitance (R_mC_m). The time it takes for the voltage to decay to 1/e (36.8%) of the initial (maximal) value at the end of the pulse, or to build up to 63.2% (11/e) of the maximal value at the beginning of the pulse, gives the s value.

Thus the AP (signal) is amplified to a constant value at each point (or each node) in the membrane, as discussed below. That is, conduction is active, not passive, with energy being put into the signal at each point to prevent any decay of the signal.

The parameters that determine the length constant of a cable are the square root of the ratio of the transverse resistance (r_m) to the sum of the inside (r_i) and outside (r_o) longitudinal resistances:

$$\lambda = \sqrt{\frac{r_m}{r_i + r_o}} \tag{19.3}$$

where r_m, r_i, and r_o are the resistances normalized for a unit length (1.0 cm) of fiber.

For surface fibers of a nerve or a muscle bundle bathed in a large volume conductor, r_o is negligibly small and Eq. (19.3) reduces to:

$$\lambda = \sqrt{\frac{r_m}{r_i}} \tag{19.4}$$

Thus, the greater the membrane resistance and the smaller the internal longitudinal resistance (larger nerve fiber diameter), the greater the λ value. We will see later that the propagation velocity is a function of λ; i.e., larger diameter fibers propagate faster. We will also see that myelination increases the effective membrane resistance (R_m) and lowers the effective capacitance, thereby increasing propagation velocity.

3.6 Time constant

Because of the high capacitance of the cell membrane (C_m), the membrane potential (E_m) cannot change instantaneously upon application of a step current pulse. Instead, E_m changes in an exponential manner (see Fig. 19.7) on both the charge and the discharge. The *membrane time constant* (τ_m) is a product of the resistance (R_m) and capacitance of the membrane and can be expressed as:

$$\tau_m = R_m C_m \tag{19.5}$$

where R_m is the resistance in ohms (Ω-cm^2) and C_m is the capacitance in farads (F/cm^2).

The discharge of the membrane (parallel RC network) is given by:

$$V_t = V_{max} e^{-t/\tau} \tag{19.6}$$

where V_t is the voltage at any time t (at the site of current injection) and V_{max} is the final maximum voltage attained during the pulse. When $t = \tau$

$$V_\tau = V_0 \frac{1}{e} = .368\ V_0 \tag{19.7}$$

Hence, the time at which the voltage decays to 36.8% of the initial (maximal) value gives the *time constant*, τ (see Fig. 19.7). In nerve fibers and skeletal muscle fibers, τ_m has a value of about 1.0 ms.

When the membrane is charging, there is a similar *exponential process*, with the identical time constant (see Fig. 19.7). The corresponding relationship is given by:

$$V_t = V_{max}\left(1 - e^{t/\tau}\right) \tag{19.8}$$

The time it takes for the voltage to build up to 63.2% of its final value gives the time constant. When $t = \tau_m$,

$$V_{\tau m} = V_{max}\left(1 - e^{-1}\right) = 0.632\ V_{max} \tag{19.9}$$

Thus, the time constant can be measured on the build-up of the pulse (time to reach 63.2% of the final voltage) or on the decay (time to decay to 36.8% of the initial voltage).

3.7 Propagation velocity determinants

The factors that determine active velocity of propagation (q) include: (1) fiber diameter, (2) length constant (λ), (3) time constant (τ_m), (4) local-circuit current intensity, (5) threshold potential, and (6) temperature. Some of these factors are interrelated, such as fiber diameter and length constant (since λ is proportional to the square root of the radius) and length constant and time constant (since both have a dependence on R_m). Propagation velocity is directly proportional to length constant and inversely proportional to time constant as:

$$\theta \propto \lambda / \tau_m \tag{19.10}$$

By substituting Eq. (19.4) for λ and Eq. (19.5) for τ_m, we have:

$$\theta \propto \frac{1}{C_m}\sqrt{\frac{a}{R_m R_i}} \tag{19.11}$$

Thus, propagation velocity is directly proportional to the square root of fiber diameter or radius (a) and inversely proportional to membrane capacitance (C_m). The larger the fiber diameter, the lower the absolute longitudinal resistance of the intracellular cytoplasm (principle of resistors in parallel) and therefore the greater the amount of local circuit current flowing longitudinally and the greater the length constant. For example, it is well known that the larger the diameter of nerve fibers, the faster they propagate (see Table 19.1). Eq. (19.11) shows that if C_m can be reduced (e.g., by myelination), then θ should increase in proportion. This is discussed later.

In addition, θ depends on the intensity of the local circuit current, and hence on the rate of rise of the AP. The greater the AP rate of rise (max dV/dt), the greater the longitudinal current and the transmembrane capacitive current (I_c). Therefore, all other factors being constant, faster-rising APs propagate faster. The AP rate of rise depends on the density of the fast Na^+ channels that carry inward current, on C_m and on temperature. Max dV/dt decreases with increased C_m, with cooling and with partial depolarization (due to the h_∞ versus E_m relationship discussed below). Cooling slows the rate of all chemical reactions, particularly those with a high Q_{10} (activation energy), such as the ion conductance changes in activated membrane.

Finally, the threshold potential (V_{th}) affects propagation velocity. If the threshold were to be shifted to a more positive voltage (more depolarized), then it would take longer for a given point in the membrane to reach threshold (and "explode") during propagation of an AP from upstream. A greater critical depolarization (difference between RP and V_{th}) would be required to bring the membrane to threshold. Therefore, the propagation velocity would be slowed. As stated earlier, some of these factors are interrelated and some actually exert opposing effects.

The foregoing discussion applies to nonmyelinated nerve axons and skeletal muscle fibers. In myelinated nerve fibers, propagation velocity is greatly increased by the myelin sheath, as discussed in the following section.

3.8 Saltatory conduction

The nerve cable has been vastly improved by the evolutionary development of myelination in vertebrates (Fig. 19.8). An electron micrograph of a myelinated nerve fiber is shown in Fig. 19.9. The myelin sheath improves the cable by increasing the effective R_m by about 100-fold and decreasing the effective C_m by about 100-fold. This increases the length constant, λ. Although τ_m tends to increase with the increase in effective R_m (due to the myelin sheath), the decrease in C_m counteracts this effect, thus acting to hold τ_m almost constant. Thus, Eqs. (19.4) and (19.5) predict that propagation velocity should increase with myelination.

One consequence of myelination, therefore, is that propagation velocity is greatly increased (see Table 19.1). Another consequence is that the energy cost of signaling is greatly decreased because passive ion leaks are limited and active current losses are restricted to the small nodes of Ranvier, which are spaced relatively far apart. An electron micrograph of a node of Ranvier is shown in Fig. 19.10. At each node, the length of exposed (naked) cell membrane is very short. The node forms an annulus around the entire perimeter of the fiber. The internodal distance is about 0.5–2.0 mm (depending on fiber diameter) and the width (length) of each node is only about 0.5–3 mm. Therefore, the degree of energy-requiring active ion transport (Na^+, K^+ and Ca^{2+}) required to maintain the steady-state ion distributions and to hold the system in a state of high potential energy is greatly reduced and to restore the amount of Na^+ gained and K^+ lost in AP generation. The low rate of oxidative metabolism in myelinated fibers reflects this lowered energy requirement.

The myelin sheath is produced by the wrapping of the Schwann cell repeatedly around the nerve fiber in a spiral, forming 20–200 wrappings, depending on axon diameter. That is, the larger axons have a thicker myelin sheath. For the purpose of our discussion, we will assume an average of 100 wrappings. The myelin sheath covers the nerve axon as a coat sleeve and is interrupted at each node. The cytoplasm of the Schwann cell in the region of the myelin sheath is nearly completely extruded during its formation, so that the sheath consists essentially of 100 cell membranes stacked in series. Because resistors in series are added to calculate the total resistance, the effective transmembrane resistance is increased 100-fold, but the effective capacitance is reduced 100-fold (because the total capacitance of capacitors in series is calculated like resistors in parallel). Since λ is directly proportional to $\sqrt{R_m}$, λ is increased accordingly. Overall, the greater λ and lowered C_m increase θ.

Myelinated nerves usually have an optimal amount of myelin, which is an amount such that the ratio of the diameter of the axon cylinder (naked axon) to the total fiber (including myelin sheath) is about 0.6–0.8. Assuming a maximal total diameter feasible (the body must pack many circuits within a limited space, e.g., sciatic nerve bundle), a greater fraction of myelin, by infringing on the diameter of the axis cylinder, would raise r_i too high, causing θ to decrease. Thus, there are two opposing factors in determining the degree of optimal thickness of the myelin sheath. The more the myelin, the greater the decrease in C_m and increase in effective R_m, but then the smaller diameter of the axis cylinder causes a higher r_i.

In saltatory conduction (from Latin *saltare, to jump*), the impulse jumps from one node to the next. The internodal membrane does not fire an AP. This is due to two reasons: (1) the internodal membrane is much less excitable (e.g., much fewer fast Na^+ channels) and (2) the depolarization of the neuron cell membrane at the internodal region is only about 1/100th of that at the node. The latter occurs because the IR voltage drop across the internodal cell membrane is only 1/100th of that across the entire series resistance network (neuron cell membrane plus 100 layers of Schwann cell membrane). Even though the internal potential in the internodal region swings positive (e.g., to +30 mV) when the adjacent nodes fire, the potential at the outer surface of the internodal membrane also swings nearly as positive (e.g., to +29 mV). Therefore, the depolarization of the neuronal membrane at the internode is only about 1 mV, which is

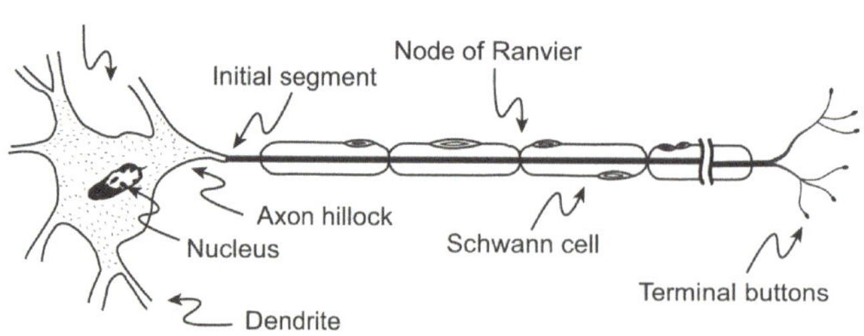

FIGURE 19.8 Diagram of a motor neuron with a myelinated axon. The major structural elements are diagrammed, including the cell body, initial segment, Schwann cell, and node of Ranvier.

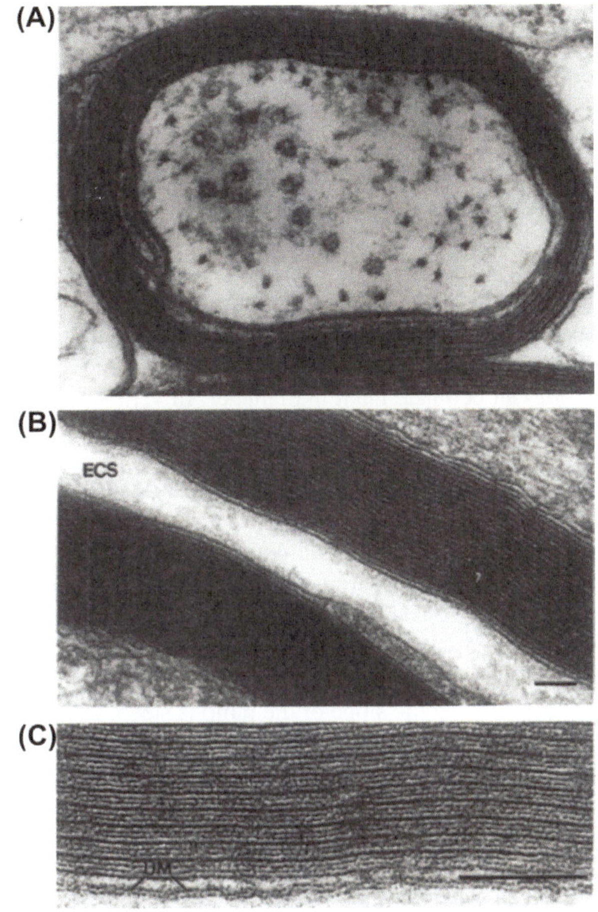

FIGURE 19.9 Myelin sheath of myelinated nerve fibers. (A) Myelinated nerve fiber in spinal cord of rat cut in cross-section. There are about seven wrappings of the Schwann cell around the axon, thus giving about 14 membranes in series with the cell membrane of the axon (axolemma). The internal mesaxon is visible at the left side (lower) of the axon and represents the beginning of the spiraling of the Schwann cell membranes (cytoplasm squeezed out) to form the myelin sheath surrounding the axon. On the upper left side, a process of an oligodendrocyte abuts on the myelin sheath. The axon shown is closely plastered up against the myelin sheath of a neighboring axon (bottom of figure). Microtubules and neurofilaments are visible in the cytoplasm of the axon. Magnification of 168 000. (B) Higher magnification of two adjacent nerves fibers (rat phrenic nerve) cut in longitudinal section to illustrate an axon with a thicker myelin sheath (approx. 23 wrappings ¼ 46 membranes). Note that between the thicker (dense) lines are two thinner lines, which can be seen more clearly in (C). The extracellular space (ECS) between the two axons is labeled. The calibration bar at the lower right equals 0.1 mm. (C) Still higher magnification of a portion of one of the myelin sheaths shown in the middle panel better illustrates the periodicity of the myelin membranes. The dense lines alternate with a pair of thin lines. The space in between the pair of thin lines is the extracellular space. The dense lines represent the intracellular (cytoplasmic) region of the Schwann cell process, with the cytoplasm extruded, thus allowing the inner (cytoplasmic face) leaflets of the cell membrane of the Schwann cell to come into close contact. The outer (extracellular face) leaflets of the Schwann cell membrane are seen as the thin lines. Thus, each cell membrane has the appearance of a double line, representing the hydrophilic surfaces of the cell membrane; the hydrophobic region of the membrane is the clear region between the dense line and the contiguous thin line on either side. UM; unit cell membrane. The scale bar on the lower right represents 0.1 mm. *(A). From Rhodin (1975), (C). Micrographs (B) and (C) courtesy of Dr. Mike Forbes, University of Virginia.*

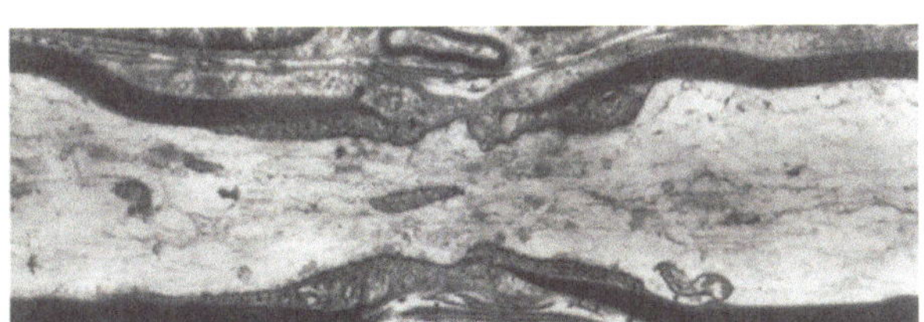

FIGURE 19.10 Electron micrograph of one node of Ranvier in a single myelinated nerve axon of rat sciatic nerve cut in longitudinal section. Dark myelin bordering the axon on each side is interrupted near the middle of the micrograph, leaving the neuronal cell membrane nude of myelin at the node region. However, cytoplasmic processes of the Schwann cells cover the nodal membrane. Collagenous fibrils of the endoneurium are visible at the region of the node. The myelin tapers and thins as it approaches the node and there is a frayed appearance caused by the successive laminae of the myelin sheath terminating as cytoplasmic swellings. The axoplasm contains neurofilaments and microtubules. Magnification of 9000×. *From Rhodin (1975).*

well below threshold, and so the internodal membrane would not fire even if sufficient ion channels to generate an AP existed in this intermodal region. The potential that controls the membrane conductances (activates the voltage dependent ion channels) is the pd (potential difference) directly across the membrane and not the absolute potential on either side.

As stated previously, the internodal membrane has only a few fast Na^+ channels, whereas the nodal membrane has a very high density. Since the cell membrane is fluid and proteins can diffuse (float) laterally in the lipid bilayer matrix, what keeps the fast Na^+ channel proteins confined (at high density) in the nodal region? There are special anchoring proteins (e.g., ankyrin) that anchor the ion channel proteins to the cytoskeletal framework, thus preventing their lateral movement into the internodal membrane.

The effect of myelin and saltatory propagation is to make propagation much faster. For example, a 20 μm diameter myelinated nerve fiber conducts even faster than a 700-μm (0.7 mm) diameter nonmyelinated nerve fiber (e.g., the giant axon in squid, lobster, earthworm): 120 m/s versus 25–50 m/s. Thus, invertebrates, to achieve fast conduction in some essential circuits, must resort to giant neurons, resulting in a lower r_i and hence fast conduction. Because of space/size limitations, only a few critical neurons can be made giant in diameter. In vertebrates, on the other hand, a large fraction of the nerve fibers in the peripheral nerves are myelinated for fast propagation.

3.9 Accommodation

Accommodation, in electrophysiology, refers to the loss of sensitivity of a cell to an applied stimulus. Sensory organs exhibit the property of accommodation, as do many neurons and other excitable membranes, such as skeletal muscle fibers. When a rectangular (square-wave) current pulse is used to depolarize a quiescent motor neuron (step depolarization) from the RP (e.g., 70 mV) to the threshold potential (V_{th}) or beyond, the neuron quickly responds with an all-or-none AP. However, if the applied pulse is ramp-shaped (triangular), the neuron may or may not respond, even if the normal V_{th} is exceeded, depending on the slope of the ramp. If the slope of the ramp is steep, the neuron will respond, but at a higher V_{th} level (more critical depolarization is required). If the slope is shallow, the neuron will fail to fire an AP, regardless of the level to which it is depolarized. This is accommodation. That is, when the membrane is depolarized gradually, the stimulus is ineffective in producing an AP response (Fig. 19.11, Top Panel). The mechanisms behind this phenomenon are discussed later in this chapter.

An exception to accommodation is found in so-called pacemaker cells (see Fig. 19.11, Bottom Panel). Automatic (spontaneously discharging) cells (e.g., nodal pacemaker cells of the heart and some sensory neurons) do not exhibit accommodation. Such cells will discharge an AP no matter how gradually the membrane is brought to the V_{th} point. In

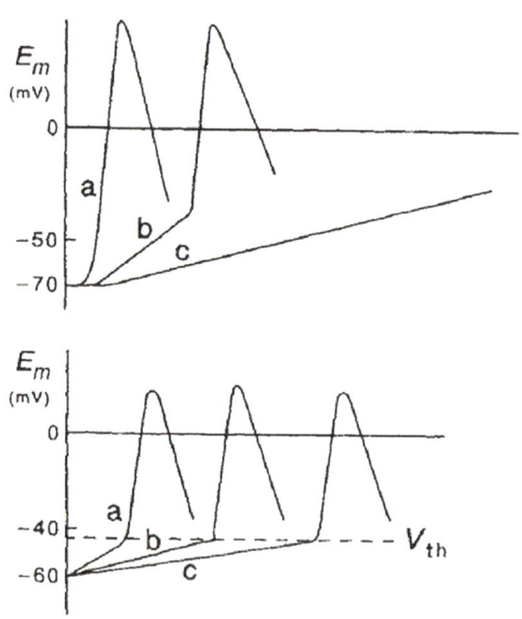

FIGURE 19.11 Sketches of accommodation and nonaccommodation. Top: Accommodation in response to a ramp stimulus in a motor neuron and Bottom: Lack of accommodation in a nodal pacemaker cell from the heart. In Top, as the slope of the ramp stimulus is decreased (from A to C), the AP response is delayed (B) and then fails completely (C).

fact, the pacemaker potential in cardiac nodal cells is of the ramp type, producing depolarization to V_{th} over a period of about 200–800 ms.

3.10 Frequency-modulated signals

Because propagating all-or-none APs are all very similar to each other (in shape, duration, amplitude, rate of rise, and propagation velocity), to make the signal stronger or weaker, the body increases or decreases, respectively, the frequency of the APs. That is, the body uses a *frequency-modulated (FM) system*, rather than an amplitude-modulated (AM) system. It is a digital system composed of on-off identical signals. At each synapse, the signal becomes graded in amplitude rather than all-or-none: the greater the amplitude and duration of the local *postsynaptic potential*, the higher the frequency of APs triggered. The same is true of the local graded *receptor potential* generated at some sensory organs/receptors. Stronger signals translate into a higher frequency of impulses, and weaker signals correspond to a lower frequency of impulses.

4. Electrogenesis of action potentials

The resting potential (RP) of cells enables the electrogenesis of *action potentials* and excitability. Early experimentation in electrophysiology focused on determining the mechanism for the generation of the AP. One important finding by Cole & Curtis (1939) was that during the AP, the membrane resistance (but not the capacitance) changed dramatically. The large reduction in membrane resistance during the AP supported the hypothesis that the AP resulted from a large increase in the ionic permeability of the membrane.

To determine which ionic species might be involved in generating the AP, subsequent experimentation was directed toward varying the concentrations of the different ions bathing the axon. When the concentration of Na^+ ions bathing the squid axon was lowered, it was found that the overshoot and the rate of rise of the AP were reduced (Hodgkin et al., 1949). This result was the first indirect demonstration that the AP resulted from an increase in the membrane permeability to Na^+ ions. Several years later, this hypothesis was confirmed directly by the voltage-clamp method.

4.1 Voltage-clamp method

The membrane current (I_M) that generates the AP is composed of ionic current (I_i) and capacitive current (I_C) according to the relationship:

$$I_M = I_i + I_C \tag{19.12}$$

The flow of ionic currents across their respective resistive membrane pathways (or channels) causes a change in the membrane potential (from Ohm's law: V=IR). The change in membrane voltage causes an I_C also to flow as:

$$I_C = C_m \, dV/dt \tag{19.13}$$

where dV/dt is the rate of change of the AP. Substituting Eq. (19.3) into Eq. (19.2) yields:

$$I_M = I_i + C_m \, dV/dt \tag{19.14}$$

Since the membrane potential during an AP is constantly changing, it is difficult to separate the contributions of these ionic and capacitive components. In addition, the total ionic current is composed of multiple individual currents carried by specific ions.

To analyze and separate the membrane currents into their capacitive and ionic components, a revolutionary method, called voltage clamping, was introduced in the early 1950s by Cole and then by Hodgkin and Huxley (1952a, 1952b, 1952c, 1952d). A diagram of the voltage-clamp method is shown in Fig. 19.12. During a voltage-clamp experiment, the membrane potential is held constant (clamped) by a negative-feedback amplifier and the amount of current that is necessary to perform this task is recorded. Since the membrane potential (V_m) is held constant, the capacitive current is equal to zero. However, at the very beginning of the depolarizing clamp pulse, there is a large transient capacitive current that occurs and again at the "off" of the pulse; these can be electronically subtracted and thus removed by a special procedure. Since V_m is constant during the clamp pulse,

$$dV / dt = 0$$

and

$$I_c = 0$$

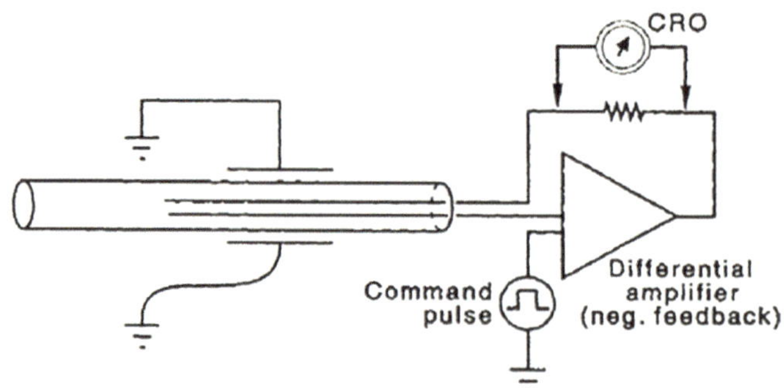

FIGURE 19.12 The voltage-clamp method used in a squid giant axon. The two wires inserted into the axon are used to measure membrane potential (V) and to pass current (I). The high-gain negative-feedback amplifier compares the command pulse with the membrane potential and outputs the amount of current necessary to hold the membrane potential constant (clamped). The magnitude of the feedback current can be measured as the IR voltage drop across a resistor and displayed on a cathode-ray oscilloscope (CRO).

Therefore,

$$I_M = I_i$$

The voltage-clamp experiment gives the magnitude and time course of the ionic currents at a given clamp potential. By clamping the membrane to many different potentials, information about the flow of ionic currents and the underlying conductance changes during the AP is obtained.

4.2 Voltage-clamp analysis

In the voltage-clamp experiments, the various ion currents (such as Na^+, Ca^{2+}, or K^+ currents) can be isolated from the total ionic current and analyzed individually. For example, in the squid axon experiments, the total ionic current consists of an early inward current followed by a delayed outward current (Fig. 19.13). By varying $[Na^+]_o$, the early inward current can be shown to be carried by Na^+ ions. Similarly, by changing $[K^+]_o$, the delayed outward current can be shown to be carried by K^+ ions. The Na^+ and K^+ currents can also be separated by blocking their pathways through the membrane. Na^+ channels can be blocked with *tetrodotoxin* (TTX), derived from the ovaries of Japanese puffer fish. K^+ channels can be blocked by several inorganic ions and organic compounds, including tetraethylammonium ions (TEA^+) and 4-aminopyridine (4-AP). The current remaining can then be subtracted from the total ionic current to reveal the time course for the current that was blocked.

The *current–voltage relationship* is obtained from measurements of the peak inward Na^+ current and peak outward K^+ current during a series of voltage clamp steps (Fig. 19.14). Depolarizing voltage steps to just above the RP first produce a

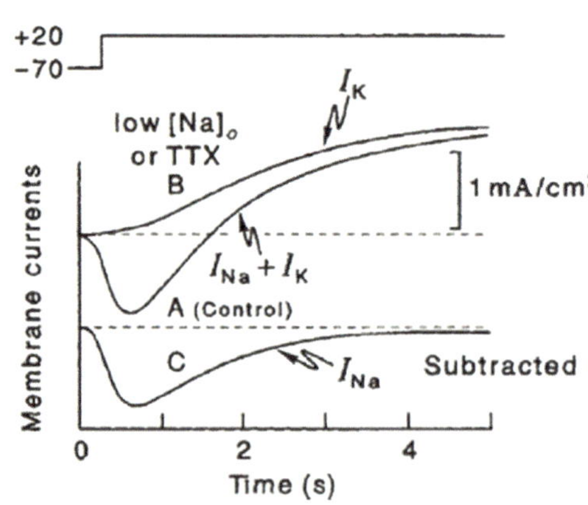

FIGURE 19.13 The ionic currents that flow when a squid giant axon is clamped from its RP (−70 mV) to a transmembrane potential of +20 mV. Trace A shows the net inward Na^+ current (I_{Na}) and outward K^+ current (I_K) in normal medium. Trace B shows the net ionic current when the axon is placed in artificial seawater with most of the Na^+ replaced by choline$^+$ (an impermeant cation), so that the intracellular and extracellular Na^+ concentrations are now equal. This current is due to K^+ only. TTX also can be used to block I_{Na}. Trace C shows the difference between curves A and B, which represents I_{Na}. *Redrawn from Hodgkin and Huxley (1952a).*

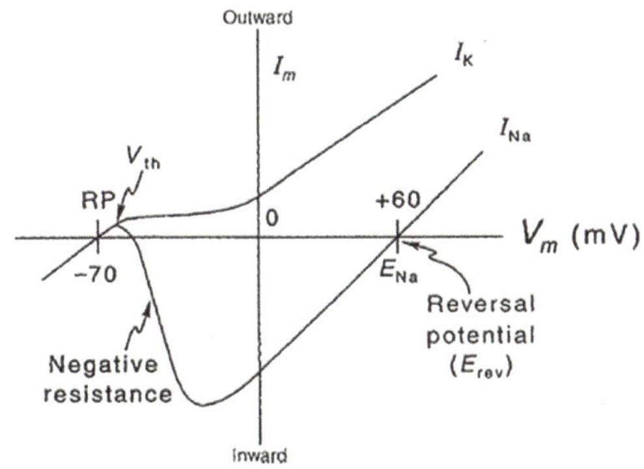

FIGURE 19.14 Current−voltage relationship for the peak early inward current and delayed outward current obtained from a squid axon under voltage-clamp. The inward current is carried by Na^+ (I_{Na}), the outward current by K^+ (I_K). The reversal potential for I_{Na} is the voltage at which the current changes from inward to outward. The region of the I_{Na} curve that has a negative slope is known as negative resistance. RP, resting potential; V_{th}, threshold; E_{Na}, Na^+ equilibrium potential; I_m, total membrane current; V_m, membrane potential. *Redrawn from Hodgkin (1964).*

small outward current. In this voltage region, the membrane behaves in an ohmic fashion. With greater depolarization, the inward I_{Na} is activated and the I−V relationship displays a negative slope or *negative resistance* region. At potentials above the peak current of the I−V curve, a positive slope is seen and the current magnitude decreases as E_{Na} is approached (and actually becomes outward at voltages above E_{Na}). The voltage at which the current reverses in direction is the *reversal potential.* The reason that the current diminishes and reverses at potentials approaching E_{Na} and beyond is that the net electrochemical driving force for Na^+ ions first becomes smaller and smaller and then outwardly directed, whereas the conductance for Na^+ ions remains constant and high over this entire voltage range, as indicated by:

$$I_{Na} = g_{Na}\left(E_m - E_{Na}\right) \tag{19.15}$$

The outward K^+ current activates above about 40 mV (in squid axon) and increases with further depolarization as:

$$I_K = g_K\left(E_m - E_K\right) \tag{19.16}$$

The voltage-clamp experiments have revealed the most fundamental property of the ionic conductances of excitable membrane: namely, that they are both voltage dependent and time dependent (Fig. 19.15). Both g_{Na} and g_K activate with depolarization, but with different time courses (Figs. 19.15 and 19.16). With time, g_{Na} spontaneously turns off or inactivates. That is, the I_{Na} current shuts off within 1−2 ms.

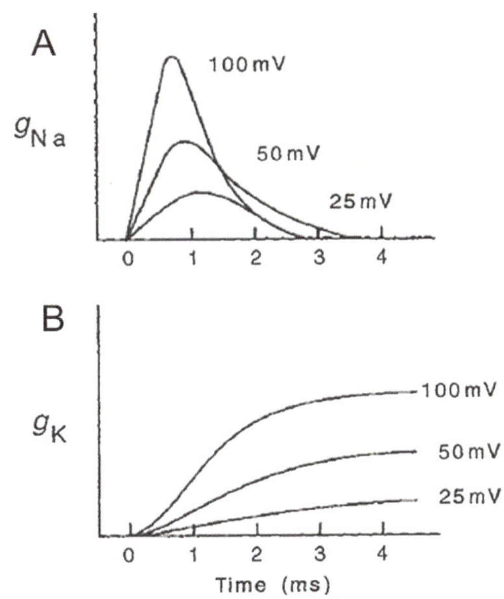

FIGURE 19.15 Voltage dependence and time dependence of the changes in Na^+ conductance (g_{Na}) and K^+ conductance (g_K) during voltage-clamp of the squid giant axon. The numbers refer to the magnitude of depolarization (in mV) from the RP. (A) g_{Na} turns on rapidly and then spontaneously declines over a brief time period. (B) g_K turns on more slowly and is sustained in amplitude during the entire clamp pulse. *Redrawn from Hodgkin (1964).*

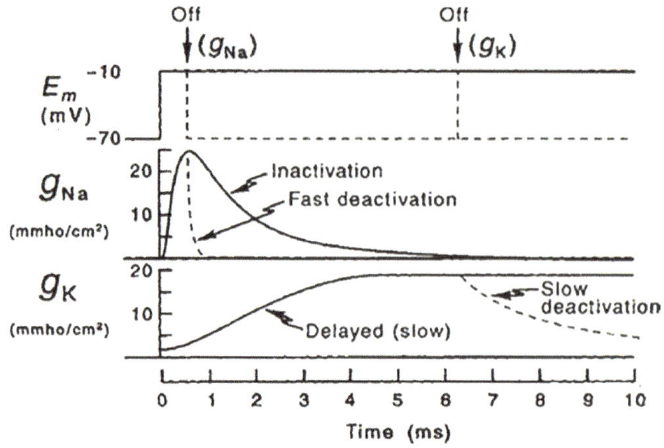

FIGURE 19.16 Na^+ conductance (g_{Na}) and K^+ conductance (g_K) changes are shown in response to voltage-clamp of the squid giant axon from the RP to a membrane potential (E_m) of -10 mV. g_{Na} inactivates after a short time, even when the voltage-clamp is maintained. In contrast, g_K remains elevated until the clamp is released. When the clamp pulse is terminated early, the g_{Na} increase is quickly turned off (deactivated); the turn-off of g_K is considerably slower. *Redrawn from Hodgkin (1958).*

A number of biological toxins that act on specific ion channels have been discovered. For example, TTX, mentioned previously, has a very high affinity for the fast Na^+ channel of nerve and some types of muscle cells. It binds to the fast Na^+ channel and blocks the passage of Na^+ ions through the channel. Several different types of toxins, including *batrachotoxin* (BTX), inhibit the inactivation process of the Na^+ channel, so that the Na^+ currents are greatly prolonged once activated. Such toxins have proven to be valuable tools in analyzing voltage-clamp currents and understanding ion channel function.

4.3 Overview of action potential generation

The increase in g_{Na} and the resulting increase in inward I_{Na} cause the regenerative depolarization of the AP. The depolarization is limited by the approach of the membrane potential toward E_{Na} and by the Na inactivation process. As the membrane is depolarized, both g_K and the driving force for I_K increase and the outward I_K repolarizes the membrane. The increase in g_K is *self-limiting*; i.e., the increase in g_K produces repolarization that, in turn, shuts off the increase in g_K.

The slow kinetics of the turn-off of g_K result in a transient hyperpolarization, the *hyperpolarizing afterpotential*. During the hyperpolarizing afterpotential, the membrane potential is brought closer to E_K than at rest.

The membrane conductance changes that occur during the nerve AP are shown in Fig. 19.17 and the time course for the resultant ionic currents is shown in Fig. 19.18. The total ionic current (I_i) is separated into its two major components, I_{Na} and I_K. Since I_K is slower to activate than I_{Na}, the inward I_{Na} predominates initially, giving rise to the fast upstroke of the

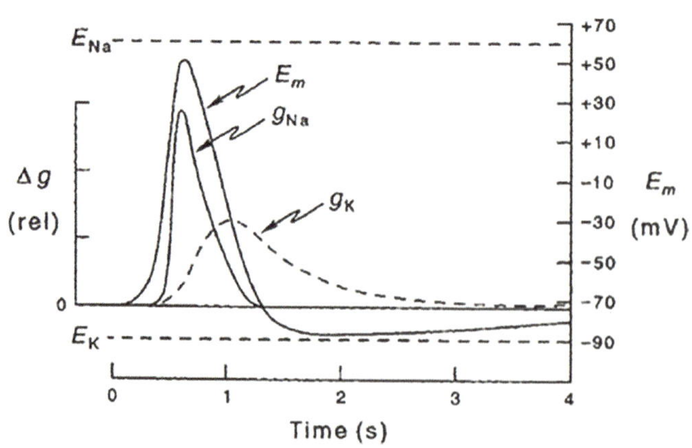

FIGURE 19.17 The relative conductance for Na^+ (g_{Na}) and K^+ (g_K) during an AP in a nerve fiber. The rising phase of the AP is caused by an increase in g_{Na}. The falling phase is due to the rise of g_K (delayed rectification) and to the decrease in g_{Na} (Na^+ inactivation). The hyperpolarizing afterpotential is explained by g_K remaining elevated for a short time following repolarization, tending to hold the membrane potential (E_m) near E_K. Δg, change in conductance. *Redrawn from Hodgkin and Huxley (1952d).*

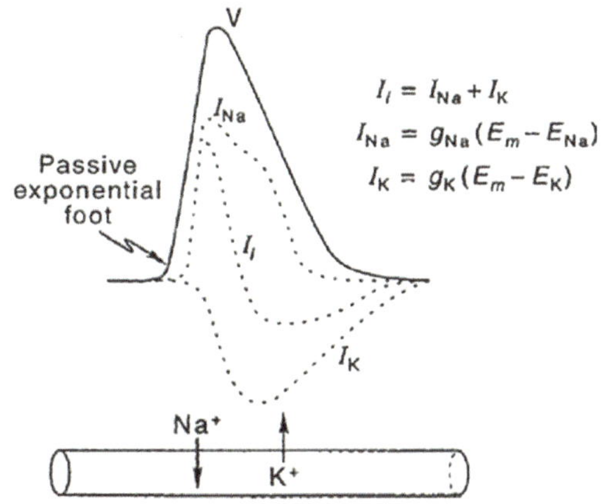

FIGURE 19.18 Ionic currents that flow during the nerve AP. The total current (I_i) is separated into an inward Na^+ current (I_{Na}) and an outward K^+ current (I_K). I_i is the algebraic sum of I_{Na} and I_K. The appropriate equations for I_i, I_{Na} and I_K are given. Also depicted is the fact that a net inward Na^+ flux occurs during the rising phase of the AP and a net K^+ efflux occurs during the repolarizing phase. *Redrawn from Hodgkin and Huxley (1952a).*

AP. Later, I_K dominates, causing a net outward current that repolarizes the membrane, i.e., it is partly responsible for the downstroke of the AP.

As stated in Eqs. (19.15) and (19.16), the specific ionic currents are a product of the membrane conductance for the ionic species and the electrochemical driving force exerted on the ion. Thus, the driving forces on Na^+ and K^+ ions continually change during the time course of the AP, as diagrammed in Fig. 19.19. At the RP, there is a large driving force for Na^+ to flow into the cell, since (E_m-E_{Na}) is large. Conversely, at the peak of the AP (E_m-E_{Na}) is at its lowest value and the driving force for Na^+ entry is small. In contrast, the driving force for K^+ efflux is largest at the peak of the AP, when (E_m-E_k) is maximal. *It is important to remember that ionic current flow depends on both conductance and driving force.* There is no net current if only one, but not the other, is present.

4.4 Fast Na^+ channel activation

During an AP, the increase in g_{Na} is related to the membrane potential E_m in a *positive-feedback* fashion, or "vicious cycle" (Fig. 19.20A). That is, a small depolarization leads to an increase in g_{Na}, which allows a larger inward I_{Na}, which causes further depolarization. This greater depolarization produces a greater increase in g_{Na}. This positive feedback process is "explosive," with a sharp trigger point (threshold) resulting from the exponential relationship (initially positive) between g_{Na} and E_m (Fig. 19.20B). It is this positive-feedback relationship that accounts for the *negative resistance* (slope) in the current—voltage curve (see Fig. 19.12). As Fig. 19.20 shows, g_{Na} reaches a maximum (saturates) at positive potentials that gives maximal activation of the population of fast Na^+ channels.

The fast Na^+ channels (and the slow Ca^{2+} channels) have a double-gating mechanism: an inactivation gate (I-gate) and an activation gate (A-gate) (Fig. 19.21). For a channel to be conducting, both the A-gate and the I-gate must be open; if either one is closed, the channel is nonconducting. The A-gate is located somewhere near the middle of the channel; it is

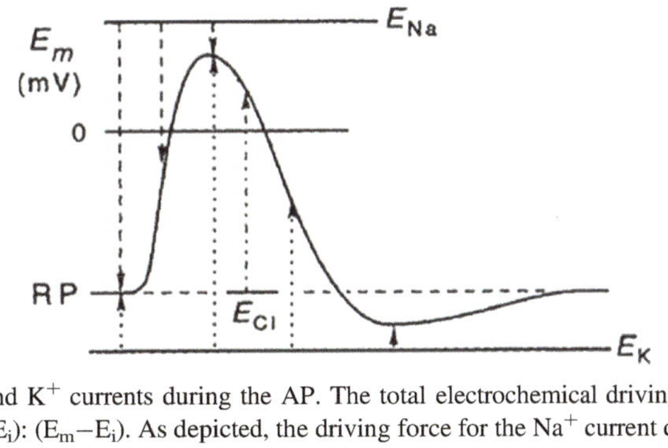

FIGURE 19.19 Driving forces for Na^+ and K^+ currents during the AP. The total electrochemical driving force is equal to membrane potential (E_m) minus the equilibrium potential for the ion (E_i): (E_m—E_i). As depicted, the driving force for the Na^+ current decreases during the AP, whereas that for K^+ increases. Even if Cl^- is passively distributed, a net driving force for Cl^- influx (outward Cl^- current) occurs during the AP.

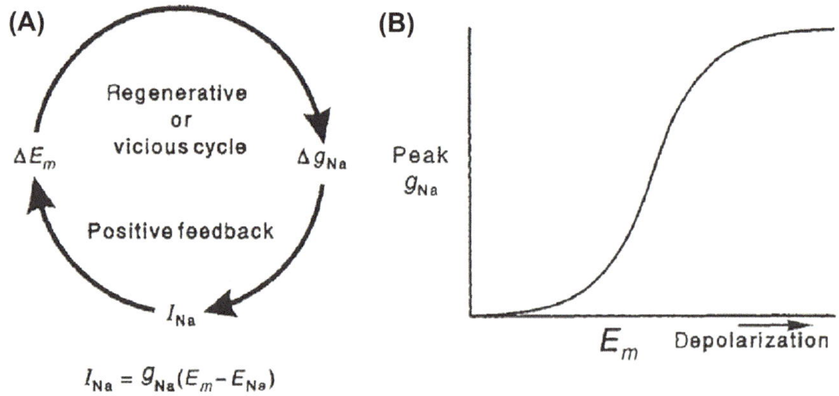

FIGURE 19.20 The positive-feedback relationship between Na^+ conductance (g_{Na}) and membrane potential E_m, leading to the all-or-none AP. (A) The increase in g_{Na} allows an increase in the inward Na^+ current, $I_{Na} = g_{Na}$ (E_mE_{Na}), which is depolarizing and so triggers a further increase in g_{Na}. This explosive feedback cycle is caused by the voltage dependency of the gated fast Na^+ channels. (B) Plot of g_{Na} versus depolarization, showing the initial exponential (positive) increase in g_{Na} as a function of voltage, followed by saturation at greater depolarization, thus giving a sigmoidal relationship. *Adapted from Hodgkin (1964).*

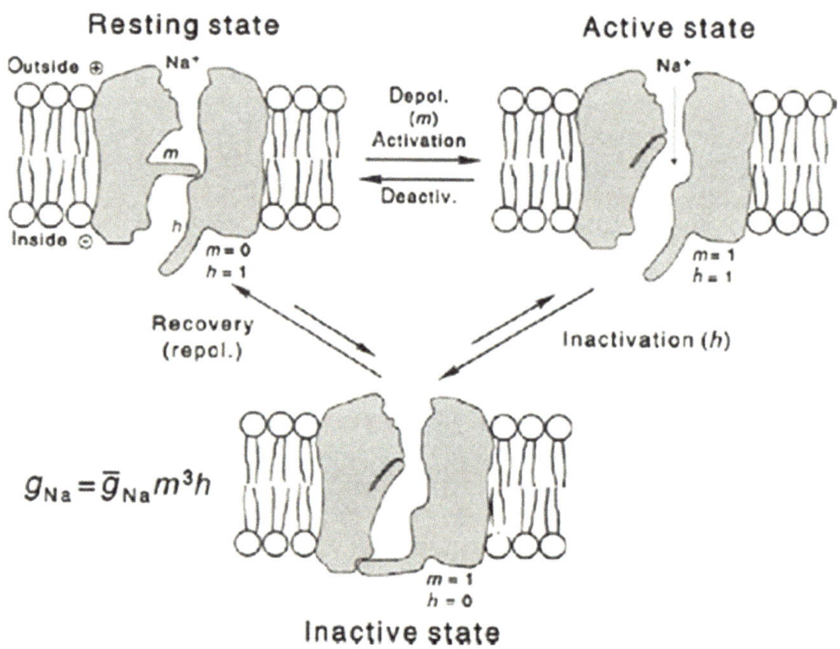

FIGURE 19.21 The three hypothetical states of the fast Na^+ channel, based on the Hodgkin–Huxley model. In the resting state, the activation gate (A or m) is closed and the inactivation gate (I or h) is open: m = 0, h = 1. Depolarization to the threshold or beyond activates the channel to the active state, the A-gate opening rapidly, and the I-gate still being open: m = 1, h = 1. The activated channel spontaneously inactivates to the inactive state because of delayed closure of the I-gate: m = 1, h = 0. The recovery process on repolarization returns the channel from the inactive state to the resting state, thus making the channel again available for reactivation. Na^+ is depicted as being bound to the outer mouth of the channel and poised for entry down its electrochemical gradient when both gates are open (active state of channel). The reaction between the resting state and the active states is reversible, whereas the other reactions may be less reversible. The Ca^{2+} slow channels pass through similar states.

not at the outer surface because even TTX does not prevent the movement of this gate, and it is not at the inner surface because proteases perfused internally do not affect it. The A-gate is closed at the resting E_m and opens rapidly on depolarization; in contrast, the I-gate is open at the resting E_m and closes slowly on depolarization.

In the Hodgkin–Huxley analysis (Hodgkin & Huxley, 1952d), the opening of the A-gate requires simultaneous occupation of three negatively charged sites by three positively charged (m^+) particles. The activation variable, m, is the probability of one site being occupied and m^3 is the probability that all three sites are occupied; therefore,

$$g_{Na} = g_{Na}^{max}\ m^3 h \tag{19.17}$$

where h is the inactivation variable and g_{Na}^{max} is the maximum conductance. Hodgkin and Huxley found that, in squid giant axon, there was an e-fold increase in g_{Na} per 4–6 mV of depolarization. To convert to ln from $\log_{10}$, the 2.303 RT/Zf factor of 58 mV (at 20C) becomes 58 mV/2.303, or 25.2 mV. Therefore, they concluded that the voltage-sensitive activation gate must have about four–six unitary charges (25/6 to 25/4). The currents arising from movement of these sites during activation ("gating currents") have been observed and are described further below, while the molecular structures involved are described in Chapter 20.

4.5 Na^+ channel inactivation

The fast I_{Na} lasts only for 1–2 ms because of the spontaneous inactivation of the fast Na^+ channels. That is, the fast Na^+ channels inactivate quickly, even if the membrane were to remain depolarized (In contrast, the slow Ca^{2+} channels inactivate slowly). Inactivation is produced in the fast Na^+ channels (and in the slow Ca^{2+} channels) by the voltage-sensitive slow closing of the inactivation gate (I-gate) (see Fig. 19.21). The I-gate is located near the inner surface of the membrane, as evidenced by the fact that addition of proteolytic enzyme to the inside of a perfused giant axon chops off the I-gate and eliminates inactivation, whereas protease added outside does not have this effect (Armstrong et al., 1973). The I-gate is presumably charged to allow it to move with changes in the membrane potential. During depolarization, the inside of the membrane becomes less negative or more positive and this causes the I-gate to close. At the normal RP, the I-gate is open (but the A-gate is closed).

The voltage dependency of inactivation is given by the h_∞ versus E_m curve (Fig. 19.22). The inactivation variable h (probability function) varies between 0 and 1.0, perhaps reflecting occupation of a negatively charged site by a positively charged h particle; h_∞ is the value of h at infinite time (>10 ms) or at steady state. When h = 1.0, the I-gates of all of the fast Na^+ channels are in the open configuration; conversely, when h = 0, all the I-gates are closed. Since g_{Na} at any time is equal to the maximal value ($\bar{g}_{Na}$) times m^3h (see Eq. 1.17), when h = 0, $g_{Na} = 0$. When h = 1.0, $g_{Na} = \bar{g}_{Na}$ (if m = 1.0). At the normal RP, h_∞ is nearly 1.0 and it diminishes with depolarization, becoming zero at about 30 mV in squid axon (Fig. 19.22A). The maximal rate of rise of the AP (max dV/dt) is directly proportional to the net inward I_{Na}, which is directly proportional to g_{Na}. The decrease in h_∞ is the cause of the decrease in g_{Na}. Therefore, depolarization by any means (e.g., elevated $[K^+]_o$ or applied depolarizing current) decreases g_{Na} and excitability disappears at about −50 mV (Fig. 19.22C).

A plot of the inactivation curve along with the activation curve is given in Fig. 19.22B to illustrate the overlap in these curves, giving rise to a so-called *window current* (or steady-state inward current) over a certain voltage region. The

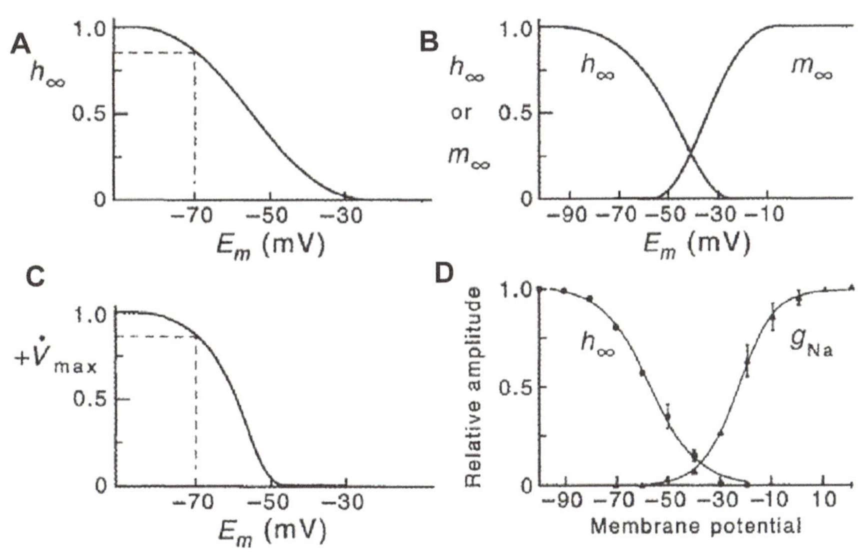

FIGURE 19.22 Voltage inactivation of the fast Na^+ channels as a function of the membrane potential (E_m). (A) h_∞ plotted against E_m, where h is the inactivation factor of Hodgkin–Huxley. The h_∞ represents h at infinite time or steady state. This graph illustrates that fast Na^+ channels begin to inactivate at about −75 mV and nearly complete inactivation occurs at about −30 mV. (B) Maximal rate of rise of the AP (max dV/dt) as a function of resting E_m. Max dV/dt is a measure of the inward current intensity, which is dependent on the number of channels available for activation. Therefore, max dV/dt decreases as h_∞ decreases. (C) Plot of both steady-state inactivation (h_∞) and activation (m_∞) against E_m to illustrate the overlap of the two curves, depicting the window current region. (D) steady-state activation and inactivation curves for the fast Na^+ current ($I_{Na(f)}$) recorded from uterine smooth muscle cells isolated from late-pregnant (18-day) rats. Note the overlap between the activation and inactivation curves, resulting in a steady-state window current between about −50 and −30 mV. *Data from Y. Inoue and N. Sperelakis, unpublished observations.*

presence of the window current means that, over the voltage range of about −50 to −30 mV, there is a steady inward Na^+ current passing through about 10%–20% of the fast Na^+ channels that are open (on a rotating population basis).

The slow Ca^{2+} channels behave much the same way as the fast Na^+ channels with respect to activation and inactivation, with one main difference being the voltage range over which the slow channels operate. For example, inactivation occurs between 50 and 0 mV for the slow channels, compared with between −100 and −30 mV for the fast Na^+ channels. Another major difference is that slow channels inactivate much more slowly than the fast channels; i.e., they have a long inactivation time constant (τ_{inact}). In myocardial cells, the h variable for the slow channel is referred to as the f variable, and the m variable as the d variable and the exponent is 2.0:

$$g_{Ca} = g_{Ca}^{max}\, d^2 f \tag{19.18}$$

4.6 Recovery from inactivation

Any Na^+ or Ca^{2+} channel that has been activated and then spontaneously inactivated must go through a recovery process before it can return to the resting state from which it can be reactivated (see Fig. 19.21). The recovery process is dependent on voltage and time. The membrane must be repolarized beyond about −50 mV before the recovery process can begin. At any given E_m, time is necessary for the recovery process to occur, namely, the time required for the charged A-gates and I-gates to move back to their resting configuration (A-gate closed, I-gate open) with the electric field. The recovery process is rapid for fast Na^+ channels (e.g., 1–10 ms) and less rapid for the slow Ca^{2+} channels. The recovery process of the slow channels is slowed by organic calcium antagonist drugs.

4.7 K^+ channel activation

The K^+ channel (outward-going delayed rectifier) is generally believed to have only an A-gate, because it does not inactivate. This gate is thought to be located near the inner surface of the membrane, because TEA^+ blocks the K^+ channel more readily from the inner surface. The block is *use dependent* or *frequency dependent*: as the A-gate opens, the TEA^+ molecule can bind in the channel behind the gate. The A-gate is believed to be positively charged and depolarization (inside going positive) opens the gate. In the Hodgkin–Huxley analysis of squid giant axon, the A-gate opens when four positively charged n^+ particles simultaneously occupy four negatively charged sites. If n is the probability that one site is occupied, then n^4 is the probability that all four sites are occupied; therefore,

$$g_K = g_K^{max}\, n^4 \tag{19.19}$$

The molecular mechanisms responsible for this gating behavior are discussed in a later chapter.

4.8 Types of K^+ channels

With the advent of powerful molecular-genetic techniques, it has been discovered that up to 100 genes for potassium channels exist in various organisms. Further, other posttranscriptional mechanisms can generate additional K^+ channel types ("isoforms) from these genes, resulting in a large family of such channels (see Chapter 12 by Sánchez).

For example, the cell membrane of some cells (e.g., myocardial cells) has at least five separate K^+ channels. As discussed previously, one type of voltage-dependent K^+ channel is the usual K^+ channel found in many types of excitable membranes. This channel slowly opens (increasing total g_K) on depolarization and is the so-called delayed rectifier [I_K or $I_{K(del)}$]. This channel allows K^+ to pass readily outward down the usual electrochemical gradient for K^+ and so is also known as the outward-going rectifier. This delayed rectifier channel in myocardial cells turns on much more slowly than in nerve, skeletal muscle, or smooth muscle and, therefore, accounts for the long duration of the AP. The activation of this channel produces the increase in total g_K that terminates the cardiac AP plateau (so-called phase 3 repolarization) (see chapter on cardiac AP).

A second type, known as the I_{K1} channel, allows K^+ ions to pass more readily inward than outward, the so-called inward-going rectifier or anomalous rectification.[1] This kinetically fast channel is responsible for the rapid decrease in K^+ conductance on depolarization (and increase in conductance with repolarization), helps to set the RP and helps bring about the terminal repolarization of the cardiac AP (phase 3). The I_{K1} channel was found to appear at a certain stage of heart development.

---

1. It is anomalous in the sense that this channel turns off with depolarization and turns on with hyperpolarization.

A third type of K^+ channel is activated by elevation of $[Ca^{2+}]_i$ and is therefore known as the Ca^{2+}-activated K^+ channel or $I_{K(Ca)}$. With Ca^{2+} influx and internal release of Ca^{2+} during the AP and contraction, these channels are activated and help the $I_{K(del)}$ channels to repolarize the AP, i.e., to bring the membrane back to the RP. The presence of this type of K^+ channel has been reported for many types of excitable cells. In actuality, two subtypes of $I_K(_{Ca})$ channel have been found: one that has a high conductance of about 400 pS (big or maxi-K channel) and one that has a lower conductance of about 120 pS (small or mini-K channel).

A fourth type of K^+ channel present in some types of cells, such as myocardial cells, is kinetically fast (compared to $I_{K(del)}$) and provides a rapid outward K^+ current that produces a small amount of initial repolarization, known as phase 1 repolarization in cardiac cells. This occurs immediately following the rapidly rising spike portion of the AP and is known as the transient outward current (I_{to}) in myocardial cells and I_A in neurons. There is some evidence that Cl^- currents may contribute to I_{to} (Cl^- influx provides an outward I_{Cl}, which is repolarizing).

A fifth type of K^+ channel is sensitive to ATP, the K_{ATP} channel and provides a current known as $I_{K(ATP)}$. This channel is regulated by ATP such that, in normal myocardial cells, this K^+ channel is inhibited (masked or silent). However, in ischemic or hypoxic conditions, when the ATP level is lowered, the $I_{K(ATP)}$ channels become unmasked and provide a large outward I_K that prematurely shortens the cardiac AP. This channel provides a protection mechanism for the heart; namely, the ischemic region of the heart develops very-abbreviated APs and hence contraction is greatly depressed. This effect acts to conserve ATP in the affected cells, enabling full recovery when the blood flow returns to normal after a short time period.

Newly discovered families of K^+ channels occur with some frequency, attesting to the vital role K^+ channels play in the generation and regulation of electrical signals.

4.9 Mechanisms of repolarization

The AP is terminated primarily by the turn-on of g_K, the *delayed rectification*. It is called the delayed rectifier because its turn-on is slower and delayed with respect to the turn-on of the fast Na^+ channels. The increase in g_K acts to bring E_m toward E_K (about 90 mV), since the membrane potential at any time is determined mainly by the ratio of g_{Na}/g_K (as discussed in a previous chapter). This type of g_K channel is activated by depolarization and turned off by repolarization. Therefore, this g_K channel is *self-limiting*, in that it turns itself off as the membrane is repolarized by its action.

In addition to the g_K turn-on, there is also some turn-off of g_{Na}, which contributes to repolarization. Two reasons for g_{Na} turnoff are (1) spontaneous inactivation of fast Na^+ channels that had been activated, i.e., closing of their I-gate (inactivation τ of 1–3 ms) and (2) a reversible shifting of activated channels directly back to the resting state because of the rapid repolarization occurring due to the g_K mechanism. Theoretically, it would be possible to have an AP that would repolarize (but more slowly) even if there were no g_K mechanism, because the g_{Na} channels would spontaneously inactivate and so E_m would slowly be restored to their original resting values. Turn-on of g_K acts to sharpen repolarization of the AP and allows higher frequency of impulses.

5. The effect of membrane potential changes on action potential generation

5.1 Accommodation mechanisms

The explanation for the phenomenon of accommodation (see above) is as follows. As the membrane is slowly depolarized toward V_{th}, the positive feedback cycle between E_m, Na^+ conductance (g_{Na}) and I_{Na} begins to operate (beginning at about 80% of the critical depolarization). Therefore, some of the fast Na^+ channels are turned on (activated), only to inactivate spontaneously (I-gates close) within 1–2 ms. If a critical number (critical mass) of fast Na^+ channels are not activated simultaneously, then the positive feedback cycle does not become explosive and a regenerative AP is not produced. That is, the slow depolarization does not allow a critical number of fast Na^+ channels to be simultaneously in the open conducting state. The channels that opened and spontaneously inactivated cannot be reactivated until they return to the resting state, which requires repolarization to near the RP. Hence, they are lost from the pool of available channels. Added to this is the fact that K^+ channels (delayed rectifier type) will open during the slow depolarization, thus increasing K^+ conductance (g_K). Whenever g_K is increased, excitability becomes depressed, because it tends to repolarize and keep the membrane from depolarizing (to produce an AP). The g_K increase lowers R_m and so lowers the effectiveness of the depolarizing current.

Therefore, accommodation to low-slope stimuli occurs for two reasons: (1) spontaneous inactivation of fast Na^+ channels that have been activated, and therefore lack of a simultaneously open critical mass and (2) increase in g_K, which depresses excitability.

The lack of accommodation in automatic cells (e.g., SA nodal cells of the heart) then may be due to (1) less spontaneous ion channel inactivation and (2) less g_K increase during the applied ramp stimulus or natural ramp pacemaker potential. Both of these conditions apparently apply to cardiac nodal cells. The inward current responsible for the rising phase of the AP is not a fast Na^+ current, but rather a slow Ca^{2+} current, which inactivates very slowly and the kinetics of the turn-on of the delayed-rectifier K^+ current is also very slow.

As stated previously, accommodation also occurs in sensory organs. For example, some stretch receptors accommodate to a sustained stretch. When the stretch is first applied, there is a burst of APs. But the bursting frequency of discharge gradually slows down and then stops, even though the stretch is maintained.

5.2 Effect of resting potential on action potential

Any agent that affects the RP has important repercussions on the AP. Depolarization reduces the rate of rise of the AP and thereby also slows its velocity of propagation. A slow spread of excitation throughout the nerve or muscle will interfere with its ability to act efficiently. This effect is progressive as a function of the degree of depolarization. If nerve, skeletal muscle fibers, or cardiac muscle cells are depolarized to about 50 mV by any means, then all excitability is lost.

Hyperpolarization usually produces only a small increase in the rate of rise. Larger hyperpolarization may actually slow the velocity of propagation, because the critical depolarization required to bring the membrane to its threshold potential is increased and it can cause propagation block.

The explanation for the effect of resting E_m (or take-off potential) on maximum rate of rise (max dV/dt) of the AP is based on the sigmoidal h_∞ versus E_m curve (see Fig. 19.22A). The I (h)-gates are open in a resting membrane and close with depolarization. As discussed previously, h, the inactivation variable for the fast Na^+ conductance is a probability factor that deals with the open ($h = 1.0$) versus closed ($h = 0$) positions of the inactivation gate of the channel (see Fig. 19.21). At the RP of 80 mV, h_∞ is 0.9−1.0 and diminishes with depolarization, becoming nearly zero at about 30 mV.

The RP also affects the duration of the AP. With use of polarizing current, depolarization lengthens the AP, whereas hyperpolarization shortens it. This occurs because of anomalous rectification[1] (i.e., g_k decreases with depolarization and increases with hyperpolarization). In contrast, when elevated $[K^+]_o$ levels are used to depolarize the cells, the AP is shortened. One important determinant of the AP duration is g_K. Agents or conditions that increase g_K, such as elevation of $[K^+]_o$, tend to shorten the duration. In contrast, agents that decrease g_K or slow its activation, such as Ba^{2+} ion or TEA^+, tend to lengthen the AP duration. Other factors are also important in determining the AP duration. For example, agents that slow the closing of the I-gates of the fast Na^+ channels, such as veratridine, prolong the AP.

5.3 Electrogenesis of afterpotentials

The APs of nerve and muscle cells usually consist of two components: an initial spike followed by an *early afterpotential* (Fig. 19.23A and B). The early afterpotentials may be of two types: *depolarizing* or *hyperpolarizing*. In addition, *late afterpotentials*, both depolarizing or hyperpolarizing, may appear following a brief train of spikes (Fig. 19.23C and D). The electrogenesis of the early and late afterpotentials is different. The early afterpotentials are due to a conductance change, whereas the late afterpotentials may be due to K^+ accumulation or depletion in restricted diffusion spaces and to electrogenic pump stimulation.

5.4 Early depolarizing afterpotentials

The AP spike in skeletal muscle fibers is immediately followed by a prominent depolarizing afterpotential (also called a negative afterpotential, based on the terminology used in external recording) (see Fig. 19.23A). The early depolarizing afterpotential of frog skeletal fibers is about 25 mV (immediately after the spike component) and gradually decays to the RP within 10−20 ms. It results from the fact that the delayed rectifier K^+ channel that opens during depolarization to terminate the spike is less selective for K^+ (ca. 30:1, K^+:Na^+) than is the K^+ channel in the resting membrane (ca. 100:1, K^+:Na^+). Therefore, from the constant-field equation (described in Chapter 12), one can predict that the membrane should be partly depolarized when the membrane is dominated by this delayed rectifier K^+ conductance. Thus, the early depolarizing afterpotential is due to the persistence of, and slow decay of, this less selective K^+ conductance.

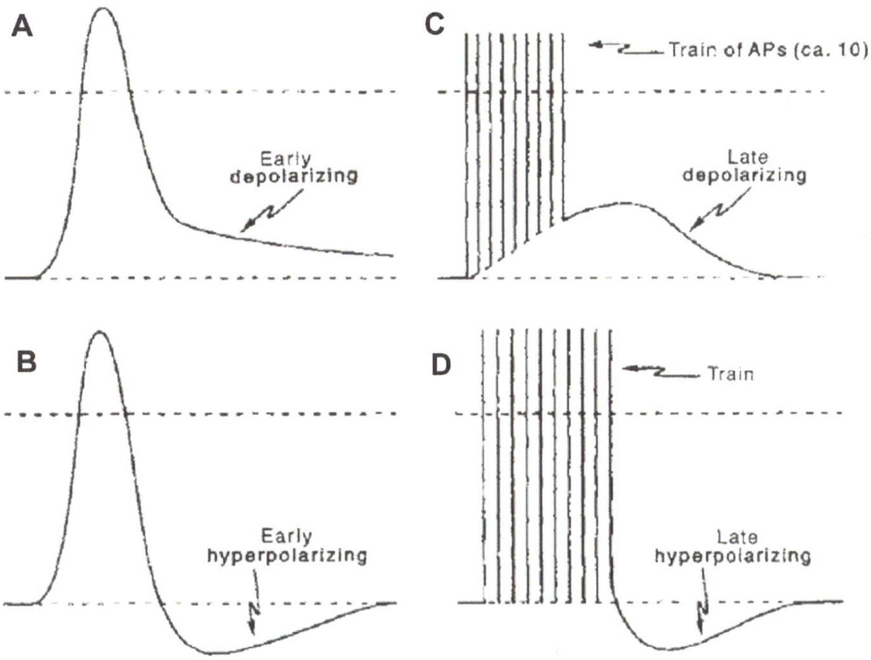

FIGURE 19.23 Examples of the different types of afterpotentials. (A) Early depolarizing (negative) afterpotential recorded after a single AP in a skeletal muscle fiber. (B) Early hyperpolarizing (positive) afterpotential recorded after a single AP in a nerve fiber. (C) Late depolarizing afterpotential recorded after a train (e.g., 10) of APs in a skeletal muscle fiber. (D) late hyperpolarizing afterpotential recorded following a train of APs in a nerve terminal.

5.5 Early hyperpolarizing afterpotentials

Neurons, heart pacemaker cells, and vascular smooth muscle cells often exhibit early hyperpolarizing (positive) afterpotentials (see Fig. 19.23B). These are due to the delayed rectifier K^+ conductance increase (which terminates the spike) persisting after the spike, thereby bringing E_m closer to E_K. The maximum amplitude possible for the afterpotential is the difference between E_K and the normal RP. The time course of this afterpotential is determined by the decay of the K^+ conductance increase.

5.6 Late depolarizing afterpotentials

The *late depolarizing afterpotential* of skeletal muscle fibers results from *accumulation of K^+ ions in the transverse (T) tubules* (see Fig. 19.23C). During the AP depolarization and turn-on of the g_K (delayed rectifier), there is a large driving force for K^+ efflux from the myoplasm coupled with a large K^+ conductance, resulting in a large outward K^+ current ($I_K = g_K (E_m E_K)$) across all surfaces of the fiber, namely, the surface sarcolemma and T-tubule walls. The K^+ efflux at the fiber surface membrane can rapidly diffuse away and mix with the relatively large interstitial fluid (ISF) volume, whereas the K^+ efflux into the T-tubules is trapped in this *restricted diffusion space*. The resulting high $[K^+]_{TT}$ decreases E_K across the T-tubule membrane and thereby depolarizes this membrane. Because of cable properties, part of this depolarization is transmitted to the surface sarcolemma and is recorded by an intracellular microelectrode. The K^+ accumulation in the T-tubules can only be dissipated relatively slowly by diffusion out of the mouth of the T-tubules and by active pumping back into the myoplasm across the T-tubule wall. Thus, the decay of the late afterpotential will be a function of these two processes.

The amplitude and duration of the late depolarizing afterpotential of frog skeletal fibers are functions of the number of spikes in the train and their frequency. That is, the greater the spike activity, the greater the amplitude and duration of the late depolarizing afterpotential. If the train consists of 20 spikes at a frequency of 50/s, a typical value for the amplitude of the afterpotential is about 20 mV. When the diameter of the T-tubules is increased by placing the fibers in hypertonic solutions, the amplitude of the afterpotential decreases, as expected, because of the greater dilution of the K^+ ions accumulating in the T-tubule lumen. When the T-tubular system is disrupted and disconnected from the surface membrane by the glycerol osmotic shock method, the late depolarizing afterpotential disappears (whereas the early depolarizing afterpotential persists).

5.7 Late hyperpolarizing afterpotentials

Some cells, neurons in particular, exhibit *late hyperpolarizing afterpotentials* following a train of spikes (see Fig. 19.23D). These hyperpolarizing afterpotentials are due to the Na^+-K^+ pump, because inhibition of the pump by any means (such as ouabain) abolishes them. Two mechanisms have been proposed for the hyperpolarization. (1) Hyperpolarization occurs because there is an increased *electrogenic* Na^+ *pump potential* (V_p), stimulated both by an increase in the $[Na^+]_i$ (since these neurons are small in diameter, and hence have a large surface area/volume ratio) and by an increase in $[K^+]_o$ (since these axons are surrounded by Schwann cells and hence have a narrow intercellular cleft and restricted diffusion space). (2) Hyperpolarization occurs due to an increased E_K caused by K^+ depletion in the intercellular cleft because of the stimulated Na^+-K^+ pump overpumping the K^+ back in. It is generally believed that the first mechanism is the most probable.

5.8 Importance of afterpotentials

All afterpotentials have physiological importance because they alter the excitability and propagation velocity of the cell. A depolarizing afterpotential enhances excitability (lower threshold) and a hyperpolarizing afterpotential depresses excitability to a subsequent AP. This is because the critical depolarization required to reach the threshold potential would be decreased or increased, respectively. A large late depolarizing afterpotential, such as that due to K^+ accumulation in the T-tubules, can trigger repetitive APs under certain pathological conditions.

The effect of afterpotentials on velocity of propagation is complex because there are two opposing factors: (1) the change in critical depolarization required and (2) the change in maximal rate of rise of the AP, which is a function of the takeoff potential (h_∞ versus E_m curve). For example, during a depolarizing afterpotential in skeletal muscle fibers, the critical depolarization required is decreased, but the maximal rate of rise of the AP is also decreased. Therefore, these two factors exert opposing effects.

6. Molecular mechanisms of AP generation

6.1 Whole-cell voltage-clamp

A newer electrophysiological method, the whole-cell voltage-clamp technique, has enabled researchers to examine the basis of excitability at the single cell level, e.g., in single myocardial cells or smooth muscle cells. The method allows the recording of the macroscopic currents that flow through the assembly of ion channels in the cell membrane. To record

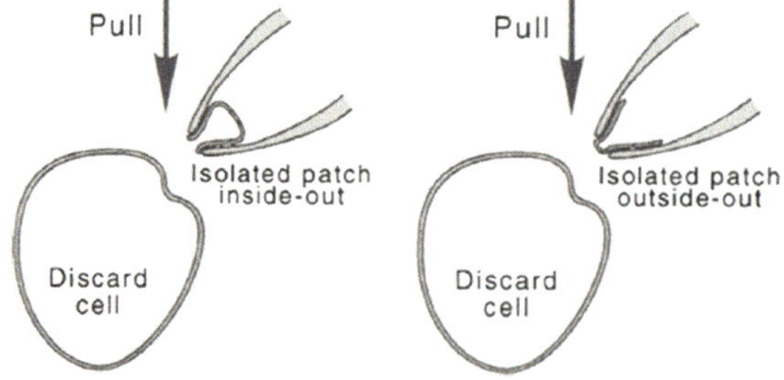

FIGURE 19.24 The whole-cell voltage-clamp technique in isolated single cells to record the macroscopic current (e.g., I_{Na}, I_{Ca}, or I_K) from the entire cell membrane. (Upper left) cell-attached patch mode, produced by applying light suction to the patch pipette to produce a gigaseal ($10^9 \Omega$). This mode is used to record the microscopic currents from only one channel or from a few channels. (Upper right) whole-cell clamp mode produced by applying strong suction to the patch pipette to blow out the membrane patch and allowing the lumen of the pipette to be continuous with the lumen of the cell. This mode is used to record the macroscopic currents from the entire complement of ion channels in the cell membrane. (Bottom) Preparation of isolated membrane patches. *Reprinted with permission from Hamill and Sakmann (1981).*

the whole-cell current, a small-tipped glass pipette, known as a *patch pipette*, is pressed against the cell membrane and negative pressure is applied to the interior of the pipette to draw a small patch of membrane into its tip. A *high-resistance seal* (e.g., $10^{10}\Omega$) spontaneously forms (Fig. 19.24). Then, strong suction is applied to the patch pipette to blow out the membrane patch and allow the lumen of the pipette to be continuous with the lumen of the cell. This technique allows recording of the whole-cell current and a complete voltage-clamp analysis can be done for small cells (e.g., Fig. 19.25). This technique also allows some control over the intracellular content of the cell; e.g., a substance can be introduced into the cell by diffusion from the patch pipette solution.

6.2 Hints at underlying molecular mechanisms

A subsequent chapter in this book (Chapter 20) describes in detail what is known about the molecular structures in channel proteins that mediate gating, ion selectivity, and pore formation, and how such knowledge was obtained. However, sophisticated electrophysiological experiments have provided clues regarding some of the mechanisms by which these structures operate.

6.3 Gating current

Since the gating of the ionic channels is voltage dependent, a part of the channel protein contains a charged group or dipole that can sense the electric field across the membrane and move in response to a change in transmembrane voltage. When the gating region moves, it causes a shift in the overall conformation of the channel, which allows it to conduct ions. A *gating current* (I_g) that corresponds to the movement of the charged m^+ particles (or rotation of an equivalent dipole) has been measured. The gating current is very small in intensity and is measured by subtracting the linear capacitive current (from a hyperpolarizing clamp step) from the total capacitive current (linear plus nonlinear) that occurs with a depolarizing clamp step beyond threshold. The outward I_g precedes the inward I_{Na}. Tetrodotoxin does not block I_g, although it does block I_{Na}. Thus, the gating current is a nonlinear outward capacitive current (not an ionic current) obtained during depolarizing clamps that reflects movement of the A-gates from the closed to the open configuration, i.e., the charge movement associated with the A-gates. The linear capacitive current results mainly from the lipid bilayer matrix.

Specific charged residues are located along the primary sequence of amino acids that comprise the channel polypeptide and they can "sense" the transmembrane electric field and move in response to changes in it. The movement of these voltage-sensing or gating residues initiates a conformational change in the channel structure, which then permits ions to flow through the central pore region of the alpha subunit of the channel (see chapter on ion channel structure and mechanisms). Gating currents have been recorded from several types of ion channels (Bezanilla, 2018).

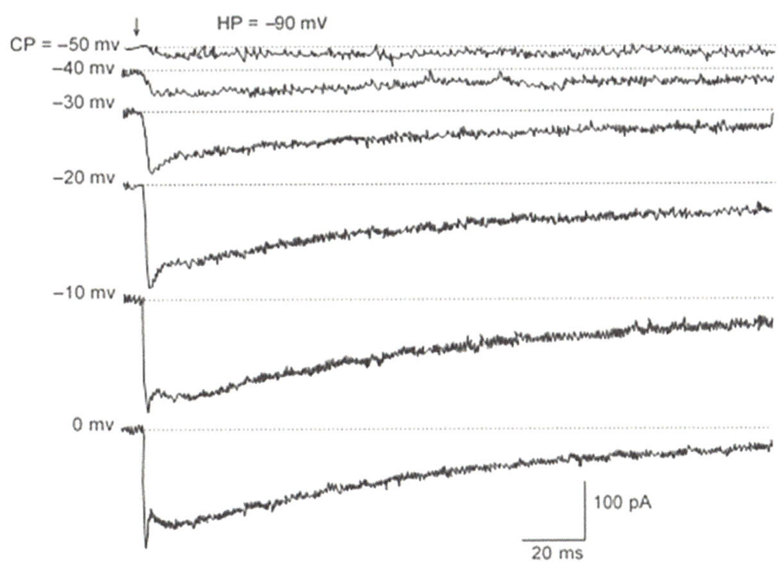

FIGURE 19.25 Two types of inward current recorded from isolated single myometrial cells from pregnant rat using whole cell voltage-clamp with a patch-pipette technique. Depolarizing potential steps (50—0 mV) were applied from a holding potential (HP) of 90 mV. The arrow indicates the beginning of a voltage step that continues to the end of the trace. The bath contained K^+-free solution with 150 mM Na^+ and 2 mM Ca^{2+}. The pipette solution contained high Cs^+. Cell capacitance was 80 pF. CP, command potential. *Reproduced with permission from Ohya and Sperelakis (1989).*

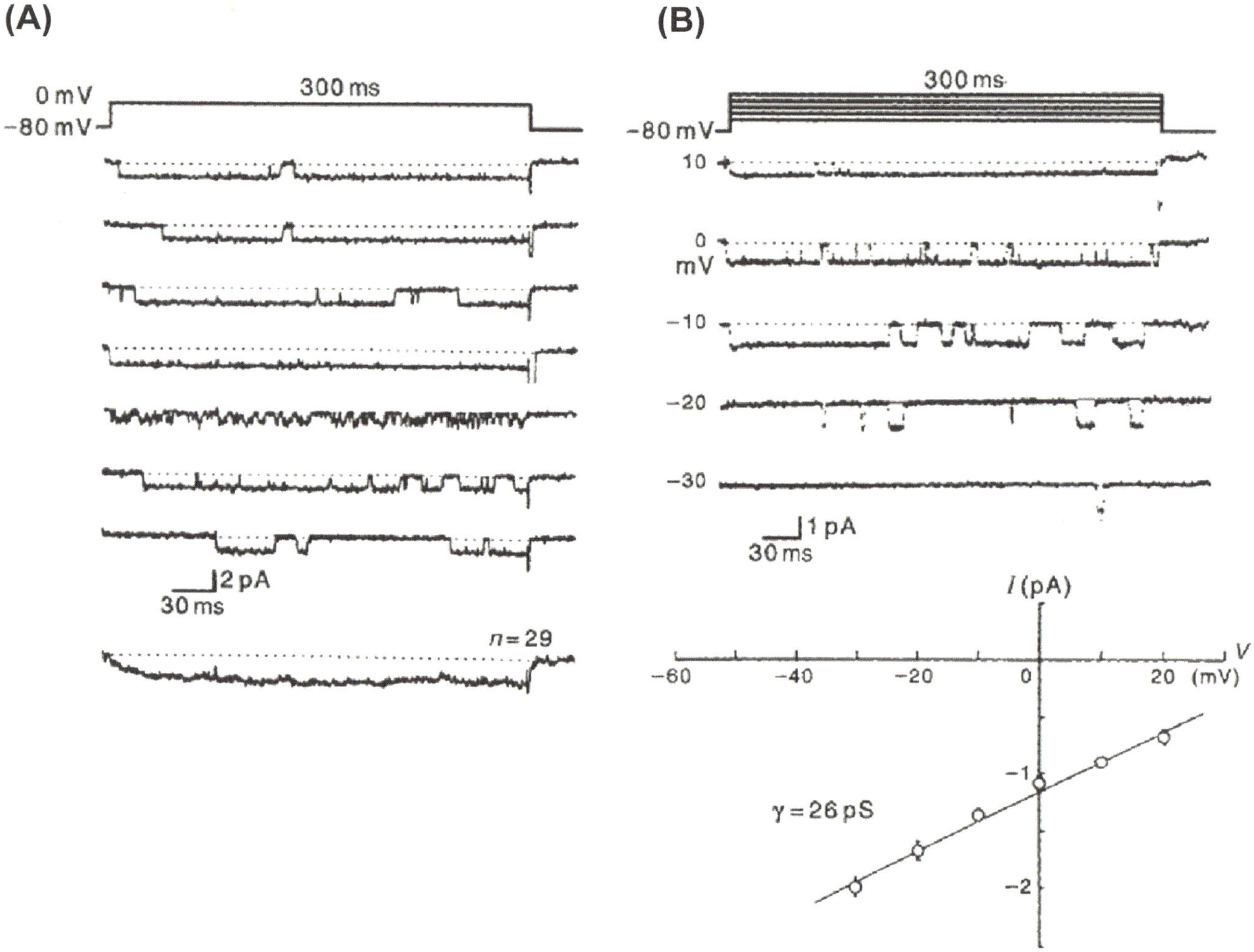

FIGURE 19.26 Single-channel currents recorded from a slow (L-type) Ca^{2+} channel, using the cell-attached patch clamp technique, in a single cardiomyocyte isolated from a young (3-day old) embryonic chick heart. These Ca^{2+} channels in young embryonic/fetal chick and rat heart cells exhibit a high incidence of long openings. Recordings in a cell illustrating the inward currents recorded at five different command step potentials (−30, −20, −10, 0 and + 10 mV) from an HP of −80mV. Note the presence of numerous long openings. The amplitude of the unitary current became smaller at the higher (more positive) command potentials. The current amplitudes are plotted below in the current-voltage curve (each point being the mean SE of 5−10 experiments). The data points were fitted by a straight line giving a slope conductance of 26 pS for the single-channel conductance. *Modified from Tohse and Sperelakis (1990).*

6.4 Single channel observations

The function of single channel macromolecules can be observed using the patch-clamp variation of the whole-cell voltage clamp technique described above (Fig. 19.24). In this method the initial formation of the patch under the pipette is maintained, and if the patch area is small enough it will contain only a few ion channels whose openings and closings can be observed individually (Fig. 19.26). Such single channel records reveal that most voltage-dependent channels have a two-state gating behavior in which they are either closed or fully open. Thus the single channel currents are seen as square steps, and the magnitude of these steps shows that the opening of a single channel allows several million ions/sec to flow. This high rate of transport strongly suggests that the transmembrane pathway is an aqueous pore through the lipid bilayer, rather than a carrier or pump transport mechanism. This sophisticated technique, which earned the Nobel Prize in Physiology and Medicine in 1991 for E. Neher and B. Sakmann, allows the biophysical study of the behavior of a single protein molecule and thus represents electrophysiology at the molecular level (Neher, 1991; Sakmann, 1991).

References

Armstrong, C. M., Bezanilla, F., & Rojas, E. (1973). Destruction of sodium conductance inactivation in squid axons perfused with pronase. *Journal of General Physiology, 62*, 375–391.

Bezanilla, F. (2018). Gating currents. *Journal of General Physiology, 150*(7), 911–932.

Cole, K. S. (1968). *Membranes, ions and impulses: A chapter of classical biophysics*. Berkeley: University of California Press.

Cole, K. S., & Curtis, H. J. (1939). Electric impedance of the squid giant axon during activity. *Journal of General Physiology, 22*, 649–687.

Hamill, O. P., & Sakmann, B. (1981). Multiple conductance states of single acetylcholine receptor channels in embryonic muscle cells. *Nature, 294*, 462–464.

Hodgkin, A. L., & Katz, B. (1949). The effect of sodium ions on the electrical activity of the giant axon of the squid. *The Journal of Physiology, 108*, 37–77.

Hodgkin, A. L, & Huxley, A. F. (1939). Action potentials from inside a nerve fibre. *Nature, 144*(3651), 710–711.

Hodgkin, A. L., & Huxley, A. F. (1952d). A quantitative description of membrane current and its application to conduction and excitation in nerve. *The Journal of Physiology, 117*, 500–544.

Hodgkin, A. L., & Huxley, A. F. (1952a). Currents carried by sodium and potassium ions through the membrane of the giant axon of Loligo. *The Journal of Physiology, 116*, 449–472.

Hodgkin, A. L., & Huxley, A. F. (1952b). The components of membrane conductance in the giant axon of Loligo. *The Journal of Physiology, 116*, 473–496.

Hodgkin, A. L., & Huxley, A. F. (1952c). The dual effect of membrane potential on sodium conductance in the giant axon of Loligo. *The Journal of Physiology, 116*, 497–506.

Hodgkin, A. L. (1958). Ionic movements and electrical activity in giant nerve fibres. *Proc Roy Soc B, 148*, 1.

Hodgkin, A. L. (1964). *The conduction of the nervous impulse*. Liverpool: Liverpool University Press.

Neher, E. (1991). Ion channels for communication between and within cells. *Nobel Lecture*, 10–25. https://www.nobelprize.org/uploads/2018/06/neher-lecture.pdf.

Ohya, Y., & Sperelakis, N. (1989). Fast $Na^{þ}$ and slow $Ca^{2þ}$ channels in single uterine muscle cells from pregnant rat. *Am J Physiol Cell, 257*, C408–C412.

Rhodin, J. A. G. (1975). *An atlas of histology*. New York: Oxford University Press.

Sakmann, B. (1991). Elementary steps in synaptic transmission revealed by currents through single channels. *Nobel Lecture*, 31–59. https://www.nobelprize.org/uploads/2018/06/sakmann-lecture.pdf.

Taylor, R. E. (1963). Cable theory. In W. L. Nastuk (Ed.), *Physical techniques in biological research* (Vol. 6, pp. 219–262).

Tohse, N., & Sperelakis, N. (1990). Long-lasting openings of single slow (L-type) $Ca^{2þ}$ channels in chick embryonic heart cells. *American Journal of Physiology, 259*, H639–H642.

Recommended for further study

Hille, B. (2001). *Ion channels of excitable membrane*. Sunderland, MA: Sinauer Associates.

Chapter 20

Structure and mechanism of voltage-gated ion channels

Simon Rock Levinson[1,2] and William A. Sather[1,2]

[1]Department of Pharmacology, University of Colorado School of Medicine, Aurora, CO, United States; [2]Department of Physiology and Biophysics, University of Colorado School of Medicine, Aurora, CO, United States

1. Summary

The structure, mechanism, and expression of ion channels are currently intense areas of research interest. The first advances in these areas were made by the application of biochemical techniques to the problems of channel purification. Such studies have told us the size of intact channels and their subunit composition and have yielded clues regarding functionally important domains (such as pores, gates, modulation sites, and nonprotein modifications). For Na^+ and Ca^{2+} channels, attention has been focused on a large polypeptide (the so-called α peptide) found in each purified preparation. These α proteins apparently have all the molecular apparatus required for channel operation, while smaller associated subunits may play other roles in channel modulation, synthesis, or cellular localization. Purified material has also been important in the construction of probes to identify clones of channel-encoding DNA fragments.

Recent work has applied recombinant DNA technology to learn the primary amino acid sequence of channels through "cloning." These sequences have revealed widespread sequence homology among the superfamily of voltage-gated ion channels, for example between Na^+ and Ca^{2+} channel α polypeptides. Furthermore, the large α proteins have four domains of homology within their sequences, suggesting that pore formation occurs via a common "staves of a barrel" architecture. K^+ channel sequences, determined via a genetic approach, are relatively shorter, and they partially correspond to a single internal repeat of the larger alpha peptides. Hence, potassium channels are thought to consist of tetramers with each subunit contributing to the pore wall. Other analyses of amino acid sequences have given clues regarding the mechanistically and structurally important domains of ion channels. One analysis was based on thermodynamic (hydropathy) considerations to identify possible membrane-spanning domains. In addition, the sequences were subjected to an empirical analysis that predicts regions of secondary structure, such as helixes and sheets. From such predictions detailed models of channel structure and function have been developed. These models have been tested using immunological and mutagenesis approaches, and candidate domains for ion selectivity, pore wall formation, gating, and modulation have been identified. More recently, sophisticated techniques to determine the high-resolution structure of voltage-gated channels have resulted in new insights into the molecular mechanisms of pore-formation, ion selectivity, and voltage-sensitive gating.

Finally, molecular studies have revealed an astonishing diversity of ion channels at all levels of organization. In particular, the genomes of most organisms are capable of expressing multiple channel isoforms, sometimes coexisting within the same living cell. There remains much to be learned about channel function and the role of channel isoforms in the biology of both simple and complex organisms.

2. Introduction: How is ion channel structure studied?

Previous chapters have described how the flow of ions across cell membranes is the basis for electrical excitability and signaling in the nervous system. These flows are controlled by a special class of macromolecules known as **ion channels** that form gated pores in the cell membrane.

What are the questions that should be addressed in this chapter? Most fundamentally, we wish to know how these macromolecular structures give rise to the important functional properties of ion channels, namely the formation of

Cell Physiology Sourcebook. https://doi.org/10.1016/B978-0-12-811114-7.00035-9

transmembrane pores, the selective transport of specific ions, and the ability to open and close the pore. That is, how do ion channels function as molecular mechanisms? As we will see, these considerations will also lead into a brief consideration of the structural diversity among ion channels, and the origin and possible purposes of such diversities.

Cellular ion channels are basically proteins. In the earliest structural studies, biochemical methods were employed in order to purify channel molecules from excitable tissues. This approach has been essential to the current state of knowledge of ion channel structure. However, ion channels have certain physicochemical properties that have limited the amount of information obtained solely through biochemical characterization. This "bottleneck" was broken through the application of recombinant DNA methodologies, which have uncovered the primary structures of numerous ion channel types, while providing a powerful means to study higher order structure and function. However, this approach has its own limitations, and ultimately we will need to "see" more directly the actual structure of the channels in order to understand more fully their molecular mechanisms of action. In this regard, it is encouraging that significant progress has been made recently in elucidating the structure of purified channels using X-ray crystallographic and cryoelectron microscopy techniques.

3. Biochemistry of ion channels: Purification and characterization of voltage-gated channels

Before the advent of recombinant DNA methods, ion channels were extensively studied by first purifying channel proteins from an excitable tissue and then characterizing the purified molecules by a combination of chemical and physical techniques. While this biochemical approach fell somewhat out of use in favor of DNA cloning techniques, it seems certain to reemerge as a necessary adjunct to structural studies that are now feasible (see below). Thus a brief introduction to channel biochemistry seems appropriate here. The first voltage-gated channel to be purified was the Na^+ channel from the electric eel *Electrophorus electricus*, and the account of how this was achieved will serve to illustrate the basic rationale for the biochemical approach (see Miller et al., 1983).

The basic procedure for purification of membrane-associated proteins is as follows (Fig. 20.1). First, one usually prepares **an enriched fraction of membranes** by disrupting the source tissue and its cells mechanically, and subsequently separating the insoluble fraction containing membranes and connective tissue components from the soluble fraction consisting of cytoplasmic protein. Although relatively nonspecific, this step also often gives substantial enrichment of ion channel preparations. Next, the **membranes are solubilized** using detergents. This step disperses the membranes into minute droplets of lipid, protein, and detergent known as **micelles**. This is often the most problematic step of a purification, since the micellar fraction must be dispersed well enough so that no more than one protein molecule is present in each

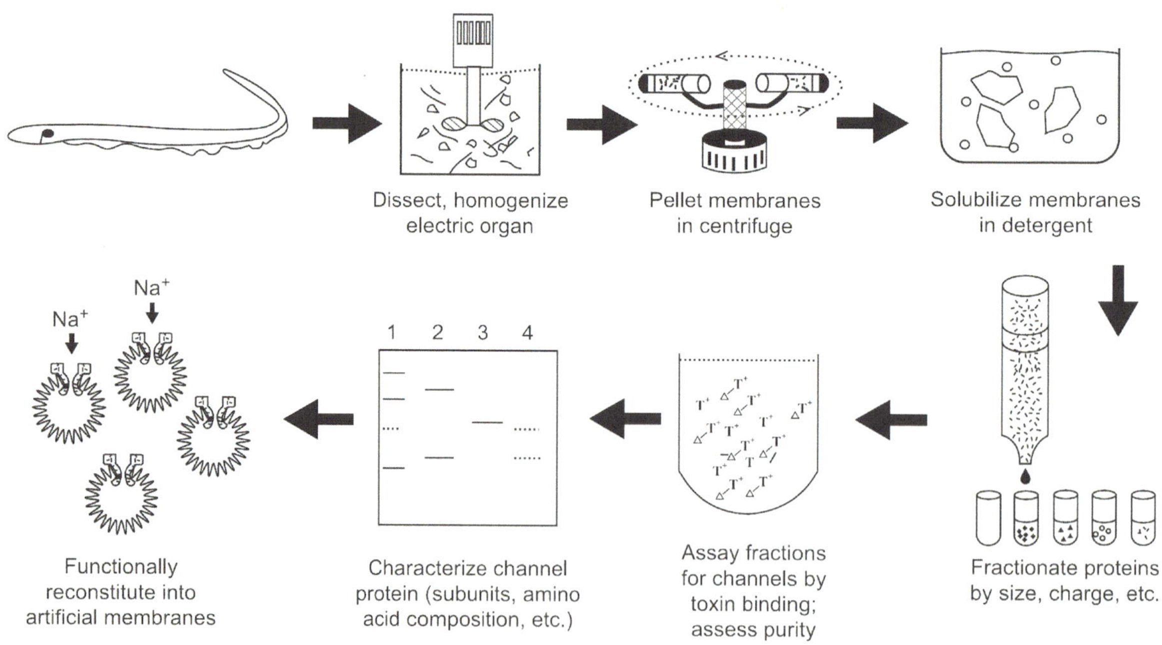

FIGURE 20.1 Scheme used for the biochemical isolation of Na^+ channels from electric tissue of the eel *Electrophorus electricus*. Purification protocols devised for other ion channels utilize similar general principles but differ in details.

micelle. Too much detergent often results in irreversible denaturation of channels, hence preventing their identification during the purification process. Thus, the solubilization process must be optimized, and often many different detergents at various concentrations are tested under a plethora of conditions before the right combination is found. Finally, the dispersed micelles are **fractionated** based on their physical properties or affinities for certain compounds until a pure sample of channel protein is isolated.

3.1 Electric fish are a rich source of ion channels

The first requirement for purification of any protein is to have a rich source. Actually, this can be a serious difficulty, since ion channels are present in very small amounts in most excitable tissues. The rarity of channels might at first seem paradoxical in view of the ubiquity of excitability phenomena in nearly all cell membranes. However, the reader should recall from previous chapters e.g. Chapter 12 that only minute numbers of channels are needed to mediate the flow of ions required to cause the changes in transmembrane voltage during electrical excitation (see Table 20.1).

Fortunately, there exist some highly specialized "freaks" of nature that utilize large ionic currents for purposes other than signaling and stimulus transduction. These are the **strongly electric fish**, which are capable of generating powerful electrical discharges to stun prey and defend themselves. The **electric organs** of these animals contain large numbers of ion channels that are present at high density on the surface of the excitable cells of the electric organ. As a result, the first two ion channels to be purified were the acetylcholine receptor channel of the electric ray *Torpedo* and the Na^+ channel of the electric eel *Electrophorus*.

3.2 Toxins and drugs as markers for ion channels during purification

Having a rich source of channels does not guarantee a successful isolation of channels. One must also have a very specific and sensitive assay for the presence of solubilized channels in order to follow the process of fractionation. Unfortunately, unlike enzymes or other chemically active proteins, the activity of ion channels is to transport ions across membranes. Since this barrier is disrupted during the first stage of purification, the compartments essential to channel function no longer exist, and hence the electrically detectable flow of ions through channels no longer occurs. Instead, one may use the binding of powerful neurotoxins to channels as a means to assess the degree of purity attained during fractionation.

For Na^+ channels, there exist highly selective inhibitors of sodium transport known as **guanidinium toxins**. The most commonly used of these are **tetrodotoxin (TTX)**, obtained from certain species of toxic puffer fish, and **saxitoxin (STX)**, a product of certain dinoflagellates that "bloom" during oceanic "red tides." Either toxin may be radioactively labeled, and these substances very specifically interact with Na^+ channels from a wide variety of sources with high affinity. To assess purity, one may thus compare the number of toxin binding sites in a fraction with the amount of protein present.

TABLE 20.1 Sodium channel densities in selected excitable tissues.

Tissue	Surface density (Channels/μm^2)	Tissue density tissue (μg channel/g tissue)
Mammalian		
Vagus nerve (unmyelinated)	110	28
Node of Ranvier	2100	–
Skeletal muscle (various)	266–557	6–14
Other animals		
Squid giant axon	166–533	~1
Frog sartorius muscle	280	6
Electric eel electroplax (excitable surface)	550	38
Garfish olfactory nerve	35	96
Lobster walking leg nerve	90	24

3.3 Isolation of channel molecules based on fractionation procedures

With a good source of channels, efficient solubilization procedures, and a specific and sensitive assay, one may proceed to purification attempts. For Na^+ channels, fractionation based on charge, followed by size fractionation of the mixed channel/lipid/detergent micelle, resulted in highly purified preparations of Na^+ channels. Alternatively, in some protocols channel proteins have been specifically purified using affinity chromatography, in which channels are selectively retained by a column to which channel-binding substances, such as drugs, toxins, or antibodies, are attached.

Skeletal muscle Ca^{2+} channels have also been extensively purified and characterized using much the same approaches as described for Na^+ channels. In this case, it has been the L-type channels that were isolated, using radiolabeled drugs of the dihydropyridine class of drugs to tag channels irreversibly in the intact muscle membrane. This allowed the investigators to isolate the channel protein by following the fractionation of the bound radiolabeled drug (Sharp et al., 1987).

3.4 Information about channel structure from channel purification?

There is much to be learned about channel structure from the physicochemical characteristics of the purified molecule. Some of this information forms the basis for separating the molecule from others in membrane extracts (such as size or charge differences), as described above, while other measurements may be done of the purified material itself. Thus, for Na^+ channels, it was found that the channel is very negatively charged. Furthermore, in detergent solution, the channel appears quite large; in micellar form, its size is of the order of several million molecular weight. Both characteristics were important clues regarding the chemical composition of Na^+ channels, which in addition to protein often contains extensive carbohydrate and lipid domains.

The subunit composition of a channel is an especially important characteristic to be determined from the purified material. For a number of enzymes, separate functions have been attributed to different subunits, hence breaking down the structural entities represents an important basis for further structure/function correlations. In brief, one usually determines the subunit composition of the purified material by analyzing it using denaturing electrophoresis, in which the intersubunit bonds are broken and each subunit separated according to size in an electric field. The most commonly used technique is sodium dodecyl sulfate-polyacrylamide gel electrophoresis (SDS-PAGE), where each polypeptide species appears as a separate band on staining the polyacrylamide sieve used to separate polypeptides. Unfortunately, this is often more difficult than it appears, since a) one is often not sure that different bands on a gel are contaminants rather than true subunits and b) the stoichiometry of subunits is difficult to establish due to variations in the intrinsic staining intensity among various polypeptides. As a result, considerable effort is made by various researchers to determine (and argue about!) subunit composition using a wide variety of approaches. For the voltage-gated Na^+ channel, it appears that a single large polypeptide accounts for pore formation, ion selectivity, and gating. This protein appears on SDS-PAGE as broad band at high molecular weight.

Given the large size of the eel electric organ, enough protein can be purified to allow study of its chemical composition. Not too surprisingly, Na^+ channels have an elevated proportion of hydrophobic amino acids compared to soluble globular proteins, likely reflecting domains of the channel that insert into the lipid membrane. However, the channel molecule appeared to be far more hydrophobic than could be accounted for from its amino acid composition. Further chemical analysis found that a large number of lipid molecules were bound to the eel protein. Finally, it has been found that Na^+ channels are modified by the presence of extensive domains of carbohydrate. This **glycosylation** makes up about 30% of the molecule (by weight).

Na^+ channels have also been purified from mammalian brain and muscle tissue. Like the eel channel, they consist primarily of heavily glycosylated large polypeptides. However, unlike the electric organ channel, they are found to be in association with one or several smaller polypeptides that have been thought to be channel subunits (Table 20.2). The large polypeptide common to Na^+ channels from all sources is referred to as **α** while smaller subunits have been designated as **β** (muscle) or **$\beta 1$, $\beta 2$, $\beta 3$, $\beta 4$** (brain). Evidence suggests that these smaller subunits aid in channel expression and modulate channel gating (Namadurai et al., 2015).

For the skeletal muscle L-type Ca^{2+} channel, it was found that while a single large polypeptide ($\alpha 1$) was affinity labeled by the dihydropyridine (DHP) marker, four other polypeptides ($\alpha 2$, β, γ, δ) copurified with the DHP label. As with the Na^+ channel, $\alpha 1$ seems to have the mechanisms required for ion channel operation. However, there is evidence that the other subunits are important accessory proteins that may be involved in the key role of this channel as the voltage-sensor in excitation–contraction coupling that is described in a later chapter.

Also shown in Table 20.2 are data for voltage-gated K^+ channels. The characterization of K^+ channel subunits was achieved primarily through molecular genetic means, as will be discussed below.

TABLE 20.2 Subunit composition of some representative voltage-gated ion channels.

Channel/tissue	Subunit	Molecular weight[a]
Na^+ channel		
Eel electric organ	α	260 (208)
Rat brain	α	260 (220)
	β1 or β3	36 (23)
	β2 or β4	33
Rat skeletal muscle	α	260 (209)
	β	36
Chick heart	α	235
Ca^{2+} channel		
L-type, rabbit skeletal muscle	α1	170 (212)
	α2	143 (125)[b]
	β	54 (58)
	γ	30 (25)
	δ	24-27 (27)[b]
K^+ channel		
Drosophila shaker A	—	65-85 (70)
Rat brain	α (4 copies)	130 (95)
	β1 or β2 (4 copies)	(45)

[a]*First set of values is the apparent molecular weight determined biochemically (SDS-PAGE); weights in parentheses were obtained from cloned DNA sequences.*
[b]*The biochemically observed calcium channel α2 and δ subunits are derived from proteolysis of a full-length α2 translation product during biosynthesis.*

3.5 Reconstitution of purified proteins confirms their identity as channels

An important criterion for purification was the reconstitution of the isolated material into artificial membranes and demonstrating that voltage-gated, sodium-selective, toxin-inhibitable channels were formed (Morera et al., 2007). Thus, these experiments showed that purified toxin-binding material could be reinserted into small artificial lipid vesicles (**liposomes**) or planar lipid films (**lipid bilayers**) where they retained their ability to transport sodium selectively in response to voltage changes. These reconstituted channels had the same ability to respond to different pharmacological compounds, such as toxins and anesthetics as the channels in natural tissue (see Section 5). Similar experiments have demonstrated that purified DHP receptors are capable of forming functional Ca^{2+} channels in bilayers (Catterall, 1988).

3.6 Limitations of biochemical characterization of channels

The biochemical approach to channel purification thus gave important information about the size, chemical composition, and polypeptide makeup of Na^+ and L-type Ca^{2+} channels. Classically, the next step would be to sequence the purified material and attempt to get high-resolution structure from crystallographic approaches. However, while these techniques have worked well for a relatively small number of soluble globular proteins, they were until recently relatively unsuccessful for membrane proteins in general and voltage-gated channels in particular. In the first place, sequencing a large polypeptide requires that it be enzymatically or chemically cleaved into overlapping smaller fragments amenable to Edman-degradation techniques (which can only sequence from 25 to 50 residues at a time). In such an approach, each small fragment has to be separately purified from the fragmented preparation. For the large polypeptides of Na^+ and Ca^{2+} channels, the number of fragments generated by such cleavages is just too great for all of them to be purified separately. Furthermore, many of the interesting membrane-spanning segments do not fragment or isolate well due their high degree of **hydrophobicity** ("fears water"). Finally, the "amphipathic" (i.e., polarized hydrophilic/hydrophobic) nature of membrane proteins inhibits their crystallization; instead, in the absence of lipids or detergents, such purified preparations usually form amorphous aggregates or precipitates with little intrinsic order. Hence, until recently, the elaboration of detailed structure of biochemically purified channels was hampered by the unfavorable physical properties of membrane proteins.

4. Channel structure investigation through manipulation of DNA sequences encoding channel polypeptides

Fortunately, the limitations of classic biochemistry may be partly overcome by the application of techniques that allow one to identify and sequence DNA that encodes channel peptides. The next sections describe how the powerful methods of molecular biology are being applied to questions of interest to the cell physiologist. These approaches have given important insights into the structure of channels and how they work. Molecular biology has also shown us that voltage-gated channels exist in a rich diversity of forms even within a single organism. First, a brief description of how recombinant DNA methods are used in such studies is provided for the novice reader (Fig. 20.2).

4.1 Primary structure of ion channels determined using recombinant DNA technology

One starts with the same tissues shown to be enriched in channel protein because these are also likely to be enriched **in messenger RNAs (mRNAs)** encoding these channels. Using the retroviral enzyme **reverse transcriptase**, one may then make **complementary DNA** copies (**cDNAs**) of the mRNAs. Using enzymatic "scissors" (called **restriction enzymes**), the cDNAs obtained may each be inserted separately into a **circular DNA plasmid**. These plasmids have the ability to be replicated along with other genetic material in bacteria, and also they encode an enzyme that will destroy certain antibiotics that would otherwise kill their bacterial hosts. The recombinant plasmids are next introduced into a **bacterial host** such that each bacterium will contain no more than a single plasmid with its unique cDNA insert. The bacteria are then diluted and spread over a plate of agar containing an antibiotic. Thus, bacteria without a plasmid are killed, while plasmid-containing

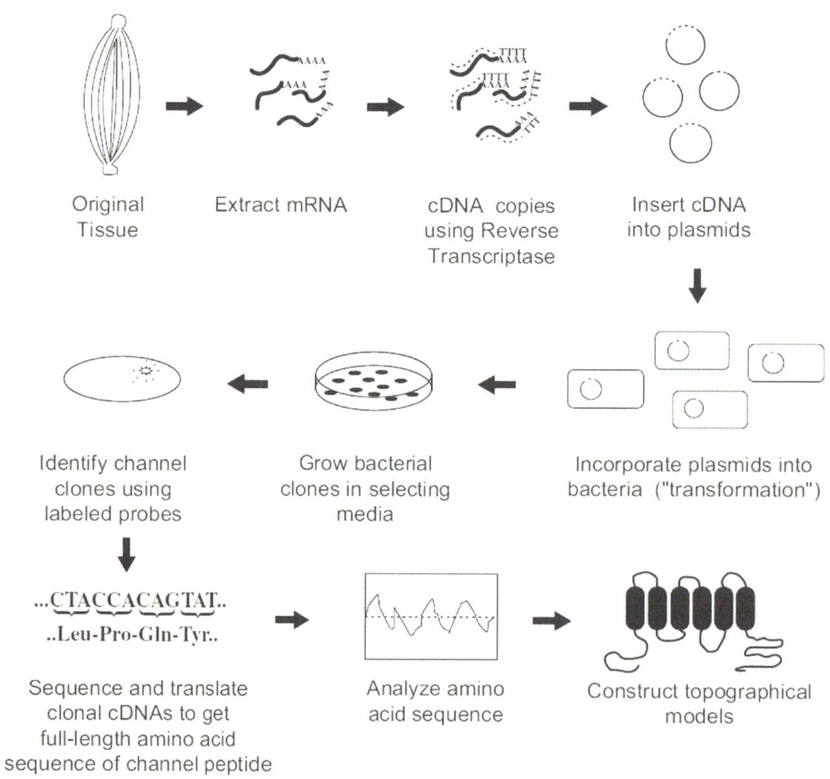

FIGURE 20.2 General strategy used to determine ion channel structure using recombinant DNA approaches.

bacteria can survive and replicate to form visible colony plaques. Since each such colony arises from a single original plasmid-transformed cell, they are known as "clones." Each agar plate may contain many thousands of such clones, each clone having many cells with plasmids containing the identical cDNA insert.

The trick is to identify which of the many colonies has a particular cDNA that encodes for the channel protein of interest, among all of the cloned cDNAs that encode the myriad proteins of the original tissue. The severity of this problem can be appreciated from the fact that even highly enriched tissues such as eel electric organ may have at most only 1 message out of 10,000 or so that encodes a Na^+ channel. Hence, there must be a rapid method for screening the colonies to identify the channel cDNA clones. This is where information from biochemical purification proved to be vital.

Despite the near impossibility of totally sequencing large Na^+ or Ca^{2+} channel polypeptides, from a proteolytic digest one can readily purify a homogeneous preparation of a particular fragment. This fragment may then be partially sequenced using standard Edman degradation techniques. In practice, only one needs a few (perhaps six or so) residues of sequence to be reasonably sure that a unique segment of the protein is encoded. From this short amino acid segment, one can then synthesize (by automated means) a short DNA fragment, called an **oligonucleotide**, that encodes the segment. Further, this synthetic DNA may be radiolabeled, and thus becomes a binding probe to identify cDNA inserts. Identification is done by placing a piece of special paper briefly on top of the agar plate, and lifting off a few bacteria from each colony plaque onto the paper. The paper is then placed in a special solution that lyses the bacteria but binds the released DNA immediately on the surface. The radiolabeled synthetic probe may then be washed over the paper; it will bind by complementary interactions to the **antisense strand** of clonal plasmids encoding the channel protein containing the original sequence. After washing away unbound probe, the paper is overlaid in the dark with X-ray film. Channel clones are thus identified as dark spots (usually only a few) on the film and may be traced back to the plaque on the original agar plate. This plaque is then lifted from the plate with a toothpick and grown in a nutrient broth to produce many bacteria. This provides enough cells from which the plasmids are purified by simple chemical means, and the cDNA insert may be readily sequenced. Once the cDNA sequence is known, the genetic code is applied to derive the primary amino acid sequence of the channel protein.

The above description of clonal sequencing leaves out many details. For example, the length of DNA that one may clone or sequence at one time is limited, often requiring complex strategies to sequence the entire length of long channel polypeptides. In addition, ion channels that are heteromeric complexes will require that each subunit polypeptide be separately sequenced. Despite these apparent complexities, with group endeavors and automation, in a relatively short time hundreds of thousands of clones may be screened and thousands of bases of interesting cDNA inserts sequenced.

4.2 Analysis of the primary structure of large polypeptide voltage-gated channels

The electroplax Na^+ channel was the first voltage-gated channel to be sequenced (Noda et al., 1984). As expected from the biochemistry, it was found to be a long polypeptide of 1820 amino acids in length (Fig. 20.3). Subsequent cloning of mammalian channels showed them to be similar in size and sequence (see Section 8). What can one learn about channel structure from such primary sequence data? Typically, one subjects the primary sequence data to several forms of analysis to derive a **model** of what the higher order structure may look like. The first striking characteristic revealed in the Na^+ channel sequence was the presence of four repeats (of about 200 amino acids each) that were highly (but not completely) homologous to one another. Furthermore, these internal repeats were found to have great potential for forming membrane-spanning domains of the channel. From such analyses, the model shown in Fig. 20.6 was obtained.

How are such analyses performed? Basically, the analyses occur in two or three separate steps. First, one analyzes the sequence of the peptide for regions of high hydrophobicity, on the expectation that membrane spanning regions should have a high proportion of hydrophobic amino acids. To do this, **hydropathy values** are assigned to each amino acid residue that reflect its relative hydrophobicity. The assignment of such values is largely based on indirect physical measurements of the hydrophobicity, such as the ability of a given residue to partition from an aqueous solution to an oil phase. Such values may or may not reflect the chemical behavior of a residue in a long polypeptide chain, where chemical properties may be further influenced by neighboring residues. As a result, a number of such hydropathy tables have been developed that assign hydrophobicity values to the various amino acid residues. One of the most popular is that of Kyte & Doolittle (1982), as shown in Fig. 20.4. In most cases, the assignments are qualitatively obvious, especially for expected hydrophobic groups (for example, the hydrocarbon chains of leucine or valine) or hydrophilic groups(e.g., the charged residues such as glutamate or lysine).

Fig. 20.5 shows hydropathy plots for the Na^+ channel sequences given in Fig. 20.3. The plot is obtained by averaging the hydropathy values of a fixed segment (perhaps 7–15 residues). Starting with the first residues at the N-terminal end, this average is computed and the value plotted versus the number of the residue in the middle of the segment. This averaging "frame" is then shifted by one residue toward the carboxyl end, the new hydropathy average recomputed (i.e.,

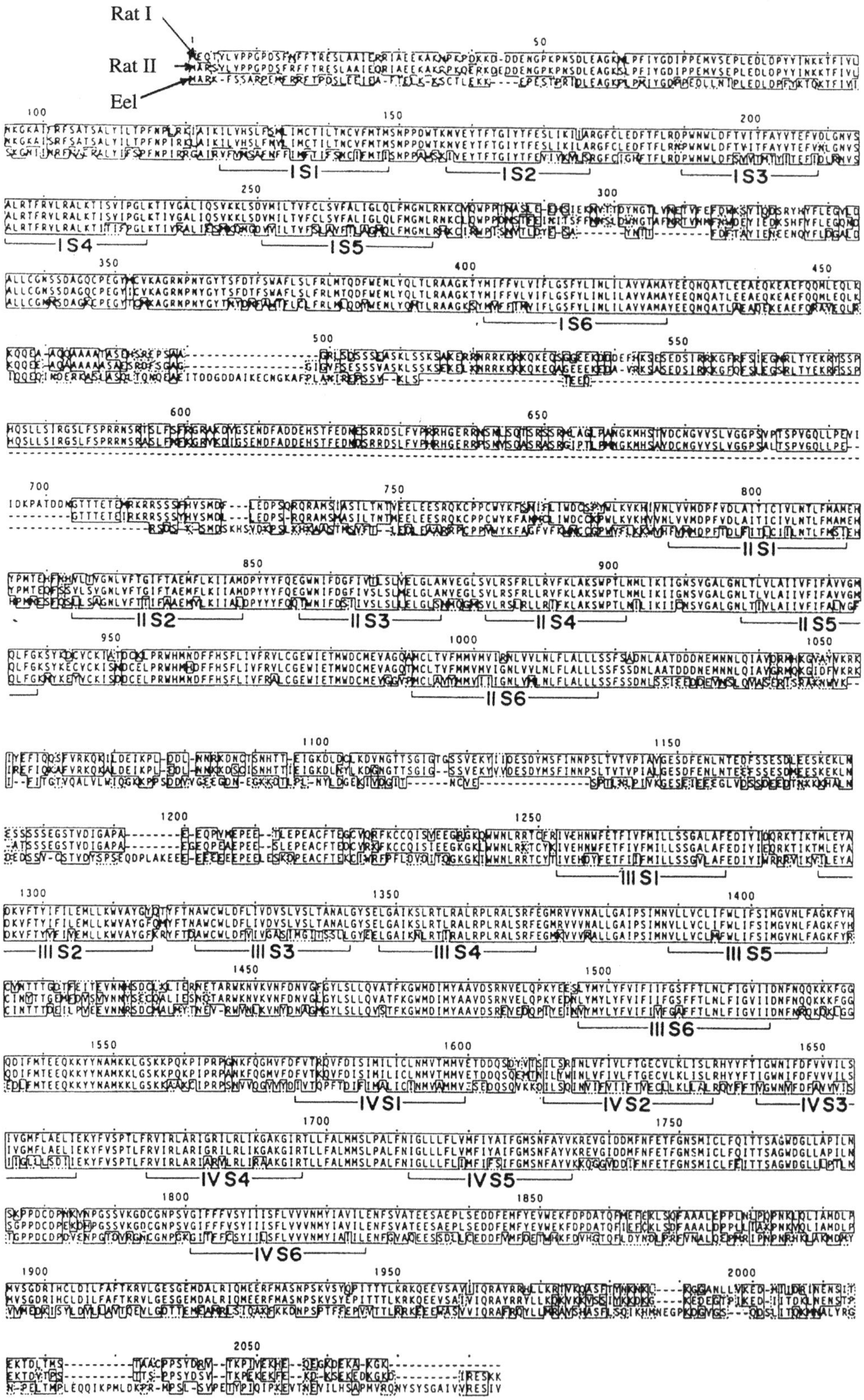

FIGURE 20.3 Amino acid sequences of three Na^+ channels obtained by cDNA cloning techniques. Rat I and II were obtained from rat brain cDNA libraries, while the eel Na^+ channel is that expressed in the electric organ of *Electrophorus electricus*. Boxed residues show homology among the three channels, while dashed lines designate absent segments. Labeled brackets underneath the sequences refer to parts of the highly conserved putative transmembrane repeats (see text and Figs. 20.6 and 20.7 for details). *From Numa and Noda (1986).*

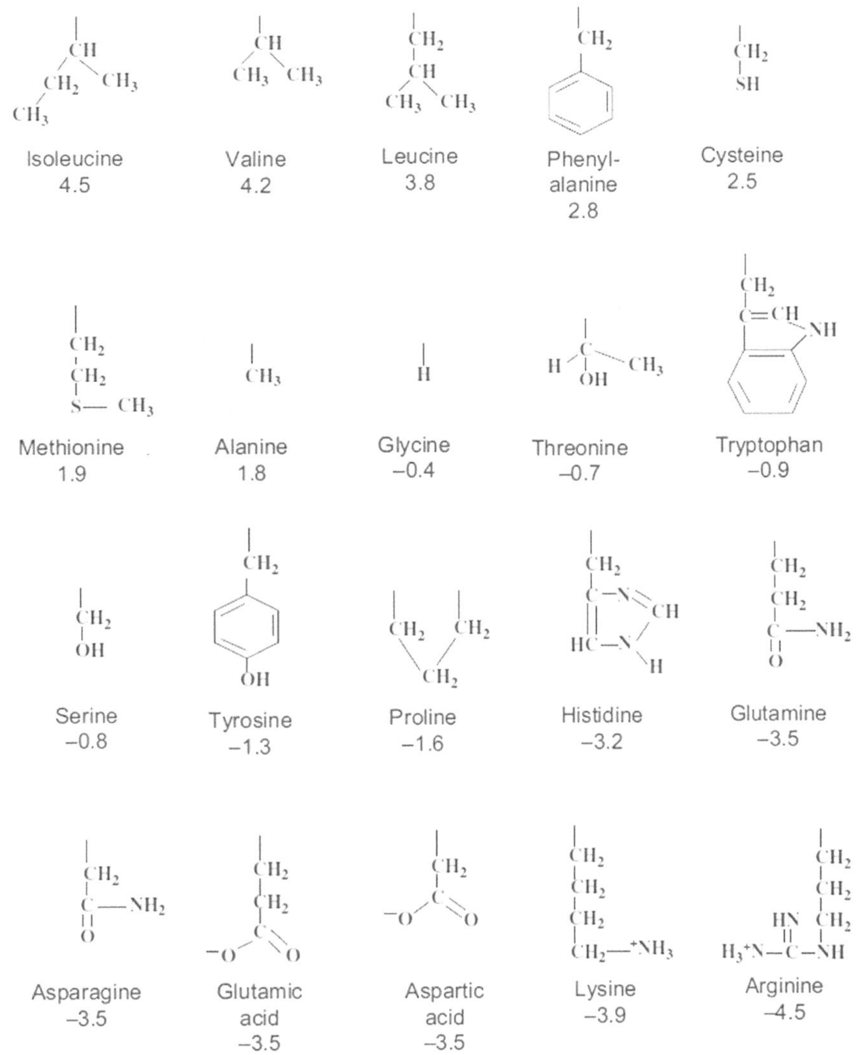

FIGURE 20.4 Structure and hydropathy values of amino acid residues. Arranged from most hydrophobic to most hydrophilic according to relative values assigned by Kyte and Doolittle (Kyte & Doolittle, 1982). Only side groups are shown with hydropathy values underneath.

the change due to the loss of the N-terminal side residue and the gain of the next C-terminal side residue in the sequence), and the value plotted by incrementing the residue number by one. This "frame-shift" average is thus continued until the end of the protein is reached. The purpose of using the averaging frame is to "smooth out" sharp variations in hydropathy, so that a pattern may be more readily seen.

As can be seen in Fig. 20.5, there is a repeated pattern of hydropathy for Na^+ channels that reflects the conserved homology of the four internal repeat domains I—IV. Within each domain, there are five or six peaks of hydrophobicity that are each about 20 amino acids long. This is significant because such a length is considered optimal to span the hydrophobic interior of the lipid bilayer. However, hydrophobicity alone is not sufficient to predict a membrane spanning region; instead, the residues must assume a favorable **secondary structure**, most commonly assumed to be an **alpha-helix**. In the helical conformation, the hydrophobic side chains project radially away from the axis of the helix and presumably into the surrounding lipid. Such a configuration prevents the very hydrophilic peptide backbone of the protein from exposure to the hydrophobic lipid environment, and thus represents a favorable low-free energy arrangement of the protein/lipid interface.

How are such secondary structures predicted from primary sequence? This too is a highly empirical process (Chou & Fasman, 1978). Basically, investigators have analyzed in detail a number of proteins for which there are high resolution X-ray structures available at the amino acid level. By cataloging the frequency which a given amino acid or short combinations of residues are found in helices, **beta sheets**, or turns, a set of empirical rules have been developed to predict what

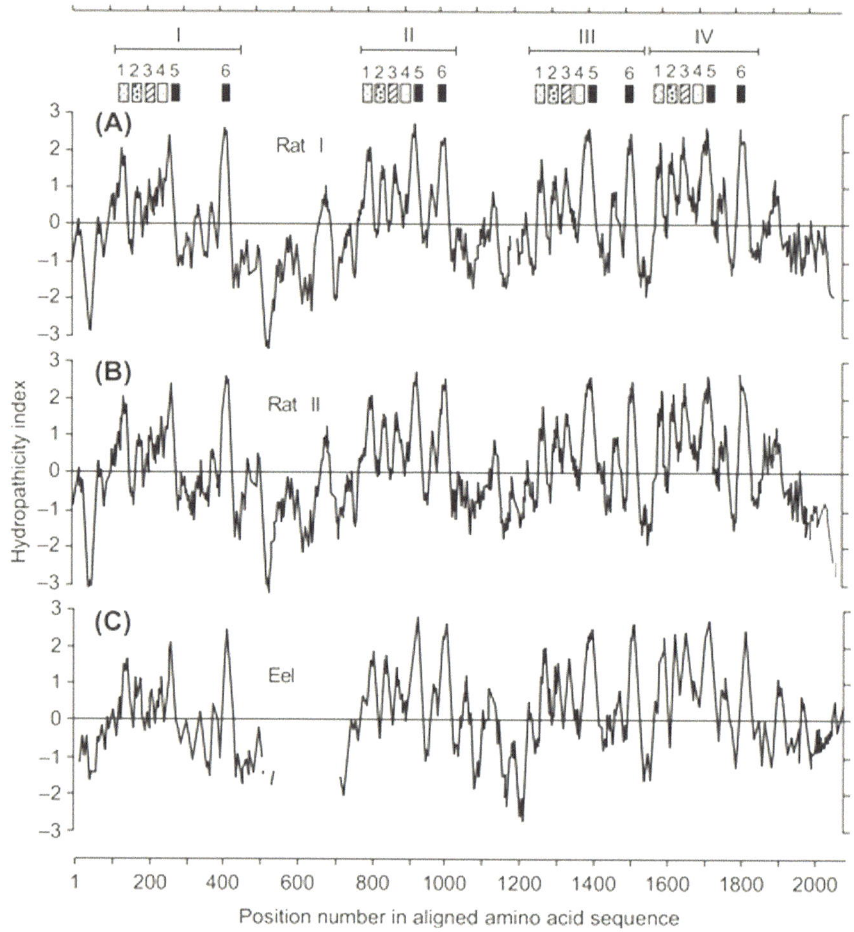

FIGURE 20.5 Hydropathy plots for the three Na^+ channel sequences of Fig. 20.4. Brackets and numbered patterns at top refer to location of homologous repeats and transmembrane segments (see text, Fig. 20.7). *From Numa and Noda (1986).*

secondary structures a given sequence of amino acids are most likely to form (see Fig. 20.5). Although the general accuracy of such predictions has been debated, when combined with hydropathy information, it appears that the method is rather accurate for the prediction of membrane-spanning domains in membrane proteins. This is probably because of the severe thermodynamic constraints faced by polypeptides needing to cross a boundary of high hydrophobicity, yet to interface with a highly hydrophilic aqueous phase on either side of the membrane and in the aqueous channel pore.

4.3 How is channel structure formed?

In any case, such primary sequence analyses suggest that each internal domain has six alpha-helical membrane spanning segments (see Section 8). The question to consider next is how such an arrangement might form a transmembrane pore for the passage of ions. The answer is that the four transmembrane regions probably contribute equally to the pore walls in a "staves of a barrel" configuration (Fig. 20.6; see Guy & Seetharamulu, 1986). In fact, such a model was comforting to the original investigators, because a similar arrangement had previously been shown to exist among the separate subunits of the synaptic acetylcholine-gated channel, the first ion channel to be purified and sequenced via cDNA cloning.

4.4 Sequence homology among the transmembrane domains of ion channels

Analysis of the DHP-sensitive Ca^{2+} channel alpha1 peptide revealed it also to have four internal homologous repeats (Fig. 20.7; Tanabe et al., 1987). Furthermore, these repeats (but not the intervening loops) were found to be highly homologous to the Na^+ channel repeats. The "staves of a barrel" configuration in which either homologous internal repeats

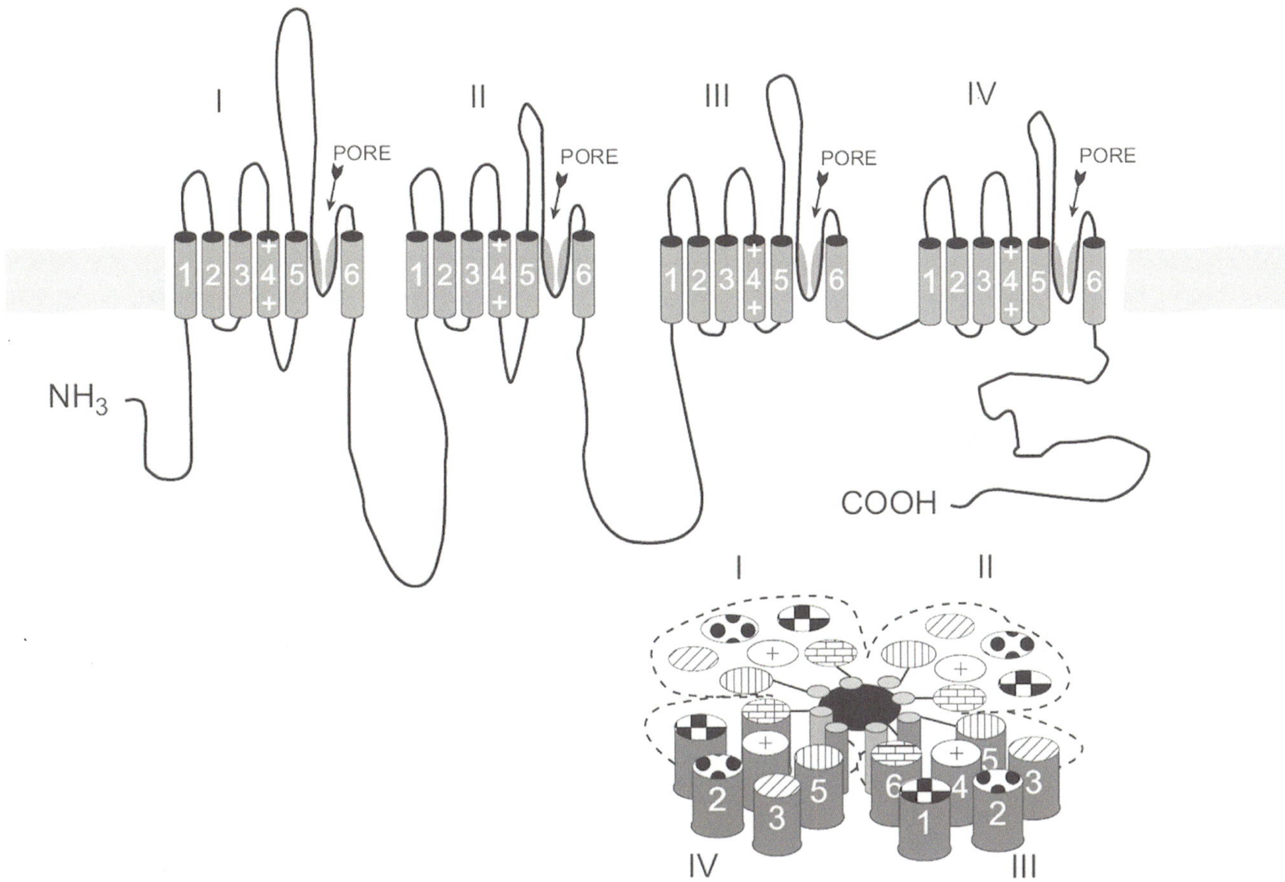

FIGURE 20.6 Model of Na^+ channel structure based on analysis of primary sequence data. Top drawing shows linear placement of four homologous domains, loops and termini (outer surface of membrane is at the top). Possible pore and selectivity regions are labeled and shown as the angled segments between transmembrane segments five and six. Drawing on bottom right shows hypothetical arrangement of homology repeats in the "staves of a barrel" configuration to form a pore (placement of transmembrane segments one to six is somewhat arbitrary).

(for Na^+ and L-type Ca^{2+} channels) or highly homologous subunits (as in the case of transmitter-gated and K^+ channels, see below) form a pore seems to be a widespread architecture among ion channels in general.

4.5 How are topographical predictions for channel structure tested?

The modeling approach based on primary structure is highly uncertain and needs to be empirically confirmed for each channel. One of the most popular ways this has been done is through the use of "site-directed" antibodies raised to synthetic peptides encoding a short stretch (i.e., 10–20 residues) of the channel that one wishes to know the topographical orientation of. These antibodies may be applied to channels in their native membranes or intact cells, and the relative "sidedness" of the epitope determined from the side of the membrane to which the antibody is bound. The approach has been used for Na^+ channels in several laboratories, and the topographical model shown in Fig. 20.6 has largely been confirmed (Catterall, 1992). Similar approaches may be used in which impermeant group-reactive chemical reagents are applied to either intact cells or liposomes with oriented channels, followed by some method to identify the site of interaction on the protein (e.g., sequencing of proteolytic fragments). Alternatively, a synthetic, but highly antigenic, short epitope may be recombinantly inserted into the channel sequence and the mutant expressed in cultured cells. In this case, the same antibody may be used in all experiments to determine on which side of the membrane the "epitope tag" is located as a function of its position in the native channel sequence.

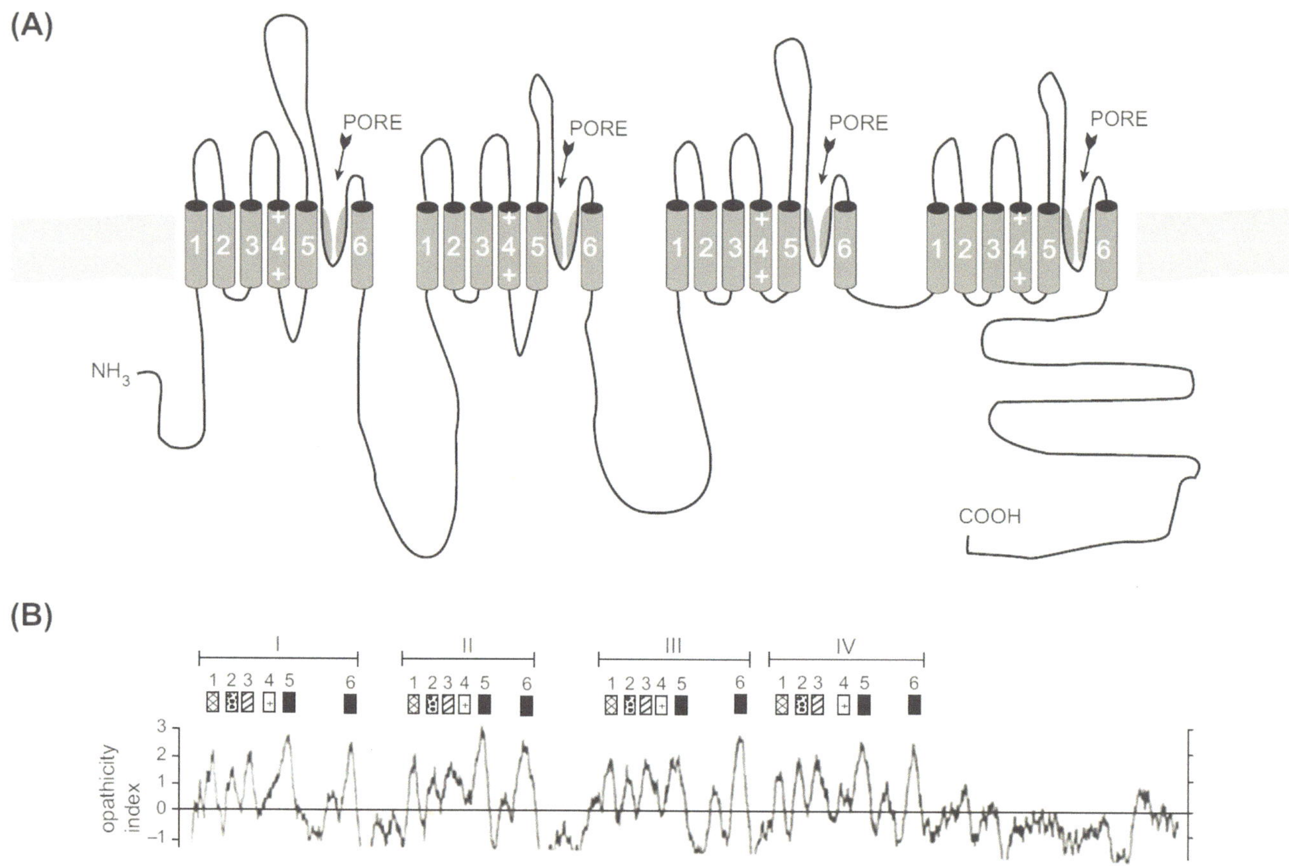

FIGURE 20.7 Model (A) and hydropathy plot (B) of L-type skeletal muscle Ca^{2+} channel alpha-1 subunit. Compare to Figs. 20.6 and 20.7.

4.6 The genetic approach used to identify nucleic acid clones coding for K^+ channels

K^+ channels are an extremely important channel type in excitation and transduction phenomena. Partly because of the way these channels were cloned, and partly because their structure is a simplified version of the long Na^+ and Ca^{2+} channels, they have been very valuable systems with which to study the general structures and mechanisms of the voltage-gated ion channel class. Such channels were not originally cloned via the usual biochemical purification-oligonucleotide probe approach described above because there were few good chemical "labels" like tetrodotoxin or DHP.

A highly creative, yet complex, approach was taken independently by several laboratories that utilized behavioral mutants of the fruit fly **Drosophila** (e.g., Papazian et al., 1987). In brief, mutant strains were identified that had uncontrolled shaking when exposed to ether. Electrophysiology then established that the muscles and certain neurons of such flies had missing or altered K^+ currents of a specific type known as **A-type** currents that, like sodium currents, inactivate upon depolarization. The locus of the genetic defect causing this behavior could be seen as an actual physical defect (translocation) in the polytene salivary chromosomes of these so-called ***Shaker* mutants**. Since extensive libraries of cloned identified chromosomal DNA fragments of Drosophila have long been available (called "genomic" clones), the investigators were able to obtain the DNA fragment of normal flies where the mutant defect occurred. This allowed them to sequence such fragments, and by comparison to *Shaker* mutant sequences and by identification of homologous cDNA fragments derived (as described above), the sequence of the A-type K^+ channel could be inferred. Since the *Shaker* mutation allowed a completely genetic elucidation of the channel primary amino acid sequence, the A-current it manifests has become known as the *Shaker*-type K^+ channel.

The recent advent of advanced technologies has allowed the rapid sequencing of entire genomes from a wide variety of organisms, as well as the mRNA sequences from the entire "expressome" of expressed genes. Thus, these sequences can rapidly searched for homology to sequences from previously sequenced channels, allowing the ready identification of ion channel sequences from most organisms.

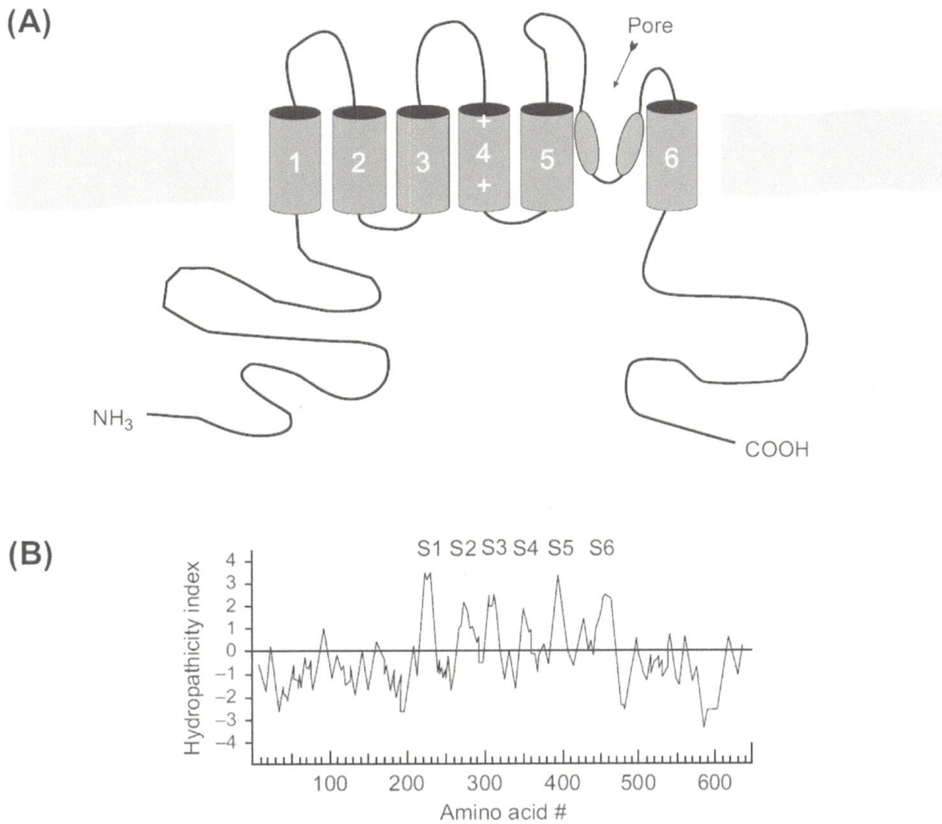

FIGURE 20.8 Model (A) and hydropathy plot (B) of *Shaker* B-type K^+ channel. Compare to Figs. 20.6–20.8.

4.7 K^+ channels are homologous to a single internal repeat of Na^+ channels

An initially surprising finding when the *Shaker* A-type channel sequence was analyzed was that the length of the sequence was much shorter than the other voltage-gated channels, but contained a single domain that was reminiscent of (but strictly speaking not very homologous to) the internal repeat domains of the Na^+ or Ca^{2+} channels (Fig. 20.8; see Pongs, 1992). An obvious hypothesis for channel structure for the *Shaker* channel was that four molecules of this sequence (i.e., subunits) combined to form the "staves of a barrel" structure postulated for the long-channel polypeptides (MacKinnon, 1991). This tetrameric assembly model has ultimately been confirmed by crystallographic studies (see below).

5. Molecular mechanisms of channel function: How does one investigate them?

We now turn to the methods one may use to determine which parts of an elucidated channel sequence are involved in the basic channel properties of pore formation, ion selectivity, and voltage-dependent gating. Here too the methods of recombinant DNA technology have proven to be of great utility in that they allow the investigator to change the nucleic acid sequence encoding a channel in the manner desired. Using such **mutagenesis** techniques, the basic idea is to alter the amino acid sequence via deletion, addition, or change in the primary sequence in specific locations, and then to test how such changes affect the function of the mutant channels. The hope is that by obtaining a number of such "structure/function correlations," one might not only infer the part of the sequence that underlies, for example, channel activation, but also how this structure works as a mechanism.

5.1 Choosing interesting sites and segments as targets for mutagenesis

How does one decide which of the many amino acids in a channel polypeptide to change? This is where the modeling approach described above is so important, for it provides clues or testable hypotheses that focus mutagenesis attempts on reasonable guesses. For example, one might assume that a highly conserved sequence in channels from phylogenetically distant animals might reflect a functionally essential part of the molecule that cannot withstand evolutionary alteration

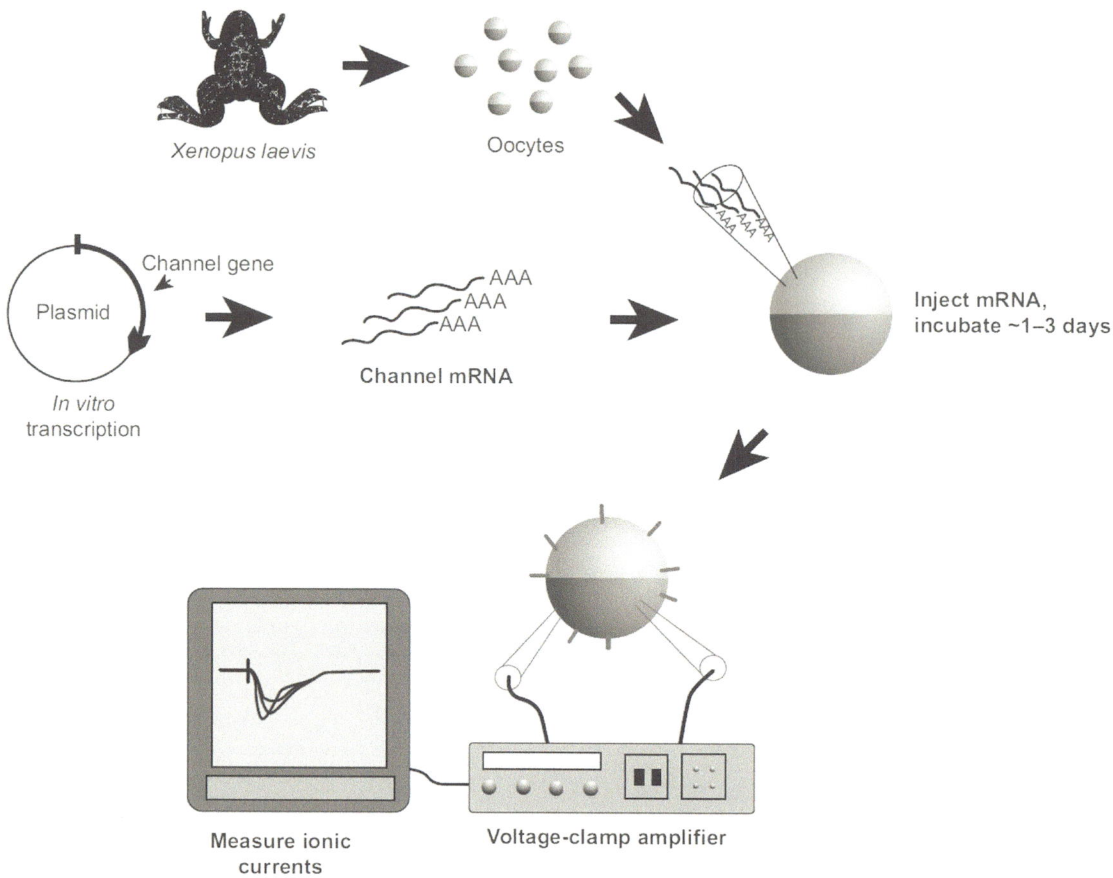

FIGURE 20.9 *Xenopus* oocyte expression system. For details, see text.

without being deleterious. Another strategy might be to compare the regions of **divergence** in the sequence of the various voltage-gated channels in the hopes of finding candidate regions involved in different functions among channels. For example, the difference in ion selectivity among Na^+, K^+, and Ca^{2+} channels must lie somewhere in the variations in their amino acid sequence. Both approaches have had a certain degree of success, although "homology" is a relative term and in reality large areas of channels are highly variable, even among isoforms of channels from the same organism (see below). This makes it difficult to guess which of these many differences relate to specific functions.

Yet another approach is to take advantage of certain genetic disorders in which mutations in channel structure occur naturally with functional consequences. There are a number of human diseases and animal models shown to be caused by channel defects. The sequence of such channels can give important clues into structure/function relationships in channels, while directing artificial mutagenesis experiments (see Catterall, 2012).

5.2 Use of expression systems in mutagenesis studies

Another essential part of mutagenesis experimentation is the ability to introduce artificially altered genetic material into a cell that is capable of translating it into a protein molecule, processing the molecule appropriately (i.e., folding the polypeptide, assembling subunits, adding sugars, or lipids) and inserting the protein into the cell surface. Then the protein may be functionally characterized using the biophysical methods described other chapters. In earlier studies, the most popular such "expression" system was the oocyte of the South African frog *Xenopus laevis* (Fig. 20.9; Leonard et al., 1991). In this system, one prepares mRNA from clones of the mutant channel and injects this message into the large (1–2 mm diameter) oocytes with a micropipette. After waiting a few days for channel synthesis and membrane insertion to occur in sufficient numbers, the expressed ionic currents are observed via voltage-clamp methods using large micro-electrodes to impale the cell. Alternatively, one may also use a very large diameter electrode to patch clamp a large area of membrane; such "macropatch" methods give better voltage control than the impaled electrode technique while allowing macroscopic current characterization. Finally, conventional single-channel analysis may be done in all the usual modes using standard patch pipettes. Thus, the oocyte allowed the investigator to compare both the macroscopic and single channel properties of the mutants with the normal "wild-type" channel.

However, other expression systems have been developed in which the altered genetic material may be introduced into cells transiently via viruses or integrated permanently into the host genome (**stable transfection**). These expression systems use cell lines that may be indefinitely cultured, providing a constant source of channel-expressing cells for study. In addition, the culture of stably-transfected cell lines may scaled up to provide a source of channel protein for structural studies.

Due to its great potential in elucidating channel mechanisms, mutagenesis is currently used by a large number of investigators. As a result, both new and altered concepts of channel function appear frequently in the scientific literature. In the brief account that follows, we summarize some of the basics of gating, pore formation, and ion selectivity as determined by mutagenesis.

5.3 Domains involved in voltage-dependent activation

The ability to respond to changes in transmembrane potential is a hallmark of voltage-gated channels. Thus investigators looked for common motifs in channel sequences suggestive of voltage-sensor domains that might target mutagenesis among the large number of residues in cloned voltage-gated channels. In comparing Na^+, DHP-sensitive Ca^{2+}, and *Shaker*-type K^+ channels, the fourth predicted helix in the repeat/subunit motif was seen to display a highly conserved sequence pattern (Figs. 20.6–20.8). This segment, known as "S4," consists of a repeated triad of a positively charged residue (arginine or lysine), followed by two highly hydrophobic residues (such as valine, leucine, isoleucine, etc., see Fig. 20.10). While the number of such triads in the various S4 segments range from 3 to 8, its structure of both charged and hydrophobic residues suggests an element capable of responding to transmembrane voltage changes (see Catterall, 1992). Accordingly, a number of experiments have now been done in which both the charged and hydrophobic elements have been changed. In brief, S4 mutagenesis usually causes significant changes in the steady-state activation behavior of the affected channel (Fig. 20.11; Lopez et al., 1991; Papazian et al., 1991). Thus shifts of the steady-state activation curve along the voltage axis occur with most changes to either charged or hydrophobic residues. In addition, changes in activation slope (indicative of the number of gating charges involved) are frequently seen.

5.4 Limitations to mutagenesis in the study of channel mechanisms

However, the mutagenesis approach is inherently limited in its ability to distinguish between alterations that **directly** affect a mechanistically important part of the molecule or those that indirectly cause functional effects through structural changes that propagate from mechanistically unrelated domain. This problem arises because protein structures are notoriously sensitive to long range effects of such alterations, generically known as allosteric (or "other site") effects.

The naive reader can understand the problem from a crude analogy in which a many of us as children tried to understand the workings of a mechanical watch by a similar approach using simple tools. The child understands the function of the mechanism only by the movement of the hands. However, as we all know, a mechanical watch is quite a delicate machine. Thus even the act of carelessly opening the case may cause the mechanism to cease working. Are we to conclude that the case is therefore part of the mechanism? Given access to the mechanism itself, the situation is highly complex,

	+	+	+	+	+	+	+	+
Electric eel NaCh IV-S4	LFR	VIR	LAR	IAR	VLR	LIR	AAK	GIR
Rat brain NaCh IV-S4	LFR	VIR	LAR	IGR	ILR	LIK	GAK	GIR
Rat skeletal muscle NaCh IV-S4	LFR	VIR	LAR	IGR	VLR	LIR	GAK	GIR
Drosophila NaCh S4	LLR	VVR	VFR	IGR	ILR	LIK	AAK	GIR
Rabbit skel. muscle CaCh IV-S4	SSA	FFR	LFR	VMR	LIK	LLS	RAE	GVR
Mouse brain KCh S4	ILR	VIR	LVR	VFR	IFK	LSR	HSK	
Drosophila Shaker S4	ILR	VIR	LVR	VFR	IFK	LSR	HSK	

FIGURE 20.10 Comparison of S4 segments from different voltage-gated channels. Note the repeating triadic motif of positive charge (R = arginine, K = lysine) followed by two hydrophobic residues. Hatched areas show conservation of charge among channels, all amino acids designated by single letter codes.

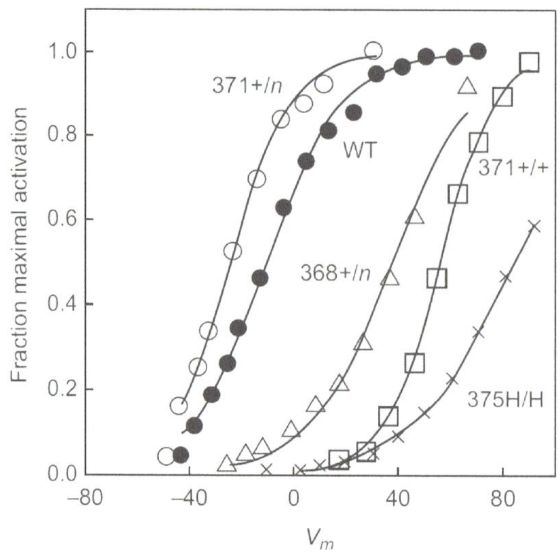

FIGURE 20.11 Effect of S4 mutations on *Shaker* K^+ channel activation expressed in oocytes. Shown are steady-state activation curves for wild type (closed circles) and various single amino acid changes. "+/n" = change from positively charged residue to a neutral one; "+/+" = change of positively charged residue for other charged residue; "H/H" change of hydrophobic residue for another hydrophobic residue. *Data redrawn from Lopez et al. (1991), Papazian et al. (1987).*

since the movement of many parts is required for the ultimate movement of the hands. For example, the result of stopping the second hand by interfering with a given gear tells one very little of the role of that gear in how the watch works.

For quite some time, it has been known that the interactions of chemical substances or structural alterations with a site on one side of a protein can dramatically effect the function of another site at some distance, say on the other side of the protein. This concern and the potentially complex nature of channel gating mechanisms should motivate the responsible practitioner of mutagenesis to display caution in interpreting his results in mechanistic terms.

How, then, in the face of such limitations does one make progress in elucidating the mechanisms of channel function? In fact, the approach is philosophically identical to the way one attempts to prove any hypothesis, namely to establish via as many independent tests as possible a circumstantial case. In the example of S4 and mutagenesis, for example, one might wish to demonstrate that mutations to any other part of the molecule have no effect on voltage-dependent activation as a sufficient (but not necessary) test. Unfortunately, mutations to other parts of channel sequence can have equally profound effects as S4 alterations (for example, to the pore forming region described above, see Yool & Schwarz, 1991). Thus, while these other mutations do not **disprove** a role for S4 as a sensor (or any other role), they do suggest, as in the mechanical watch, that a number of segments of the sequence are either directly involved in a complex mechanism of voltage-gating or that they indirectly affect a gating process through allosteric/structural influences.

5.5 The mechanism of channel inactivation

Genetic engineering studies of the inactivation process currently provide a better example of what one may learn mechanistically from the mutagenesis approach. Some of the most important such studies have involved the *Shaker* A-type channel. Here early workers found important clues to the segment involved in the inactivation process by comparing naturally occurring variants of the channel ("isoforms," see below). Thus Fig. 20.12 shows the pattern of four such isoforms, called A, B, C, and D (Timpe et al., 1988). ***Shaker* B, C,** and **D** were identified from further testing of cDNA clones using probes constructed from the original ***Shaker* A** sequenced as described above. When expressed in oocytes, these isoforms displayed very different kinetics of inactivation. Analysis of the regions of homology and difference among the isoforms shown in Fig. 20.12 indicates that fast inactivation properties correlate most strongly with the nature of the N-terminal segment (thought to be cytoplasmic, see Fig. 20.8).

Further evidence for the role of this segment in inactivation was obtained by constructing channel "chimeras" in which the ***Shaker* A** clone was modified by replacing its N- terminal segment with those from other isoforms. In each case the mutant channel assumed the inactivation properties of the isoform donating the N-terminal segment, despite the fact that the rest of the molecule had the sequence of the fast-inactivating Shaker isoform (Aldrich et al., 1990).

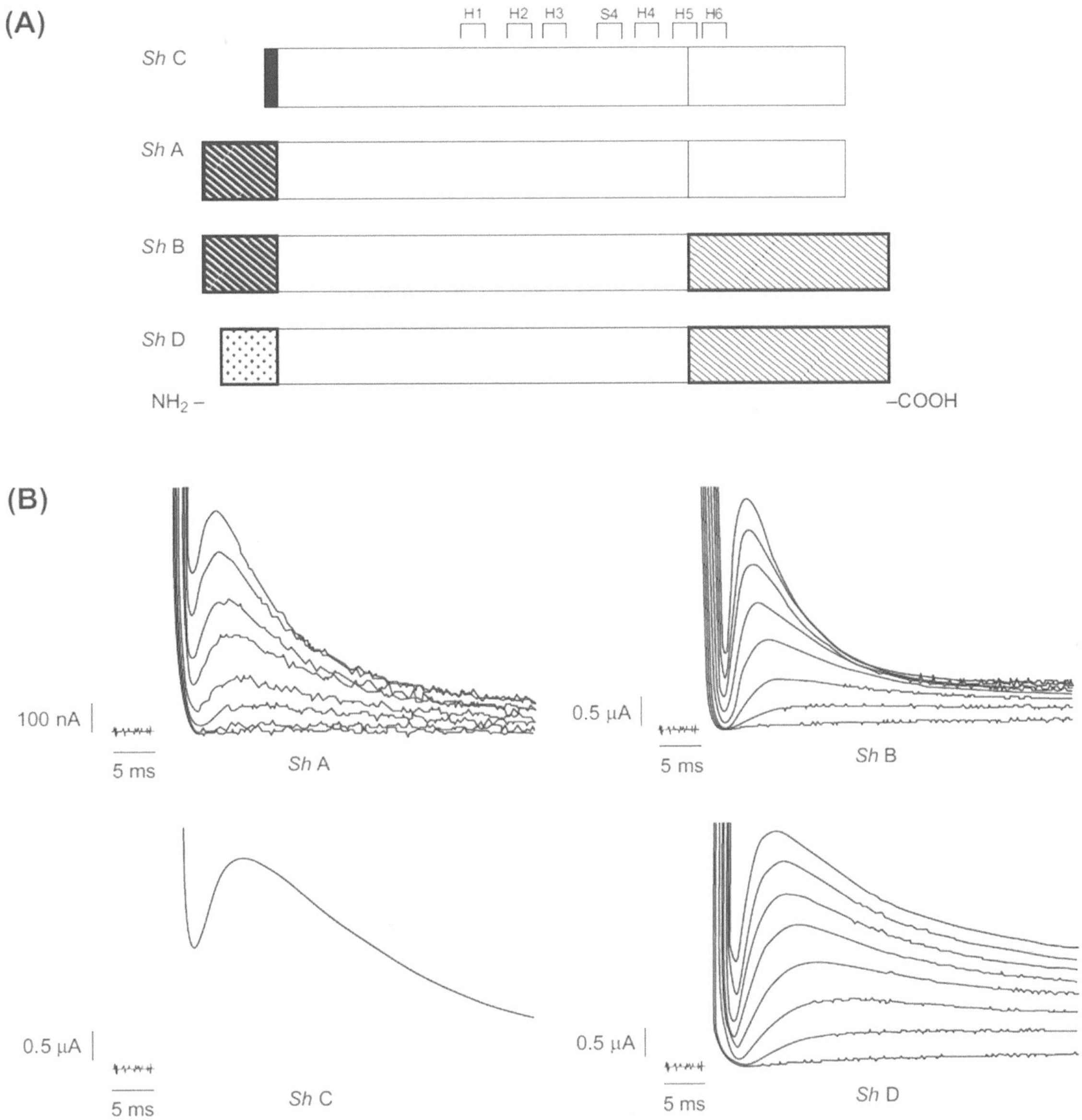

FIGURE 20.12 Gating kinetics of alternatively spliced *Shaker* isoforms expressed in oocytes. Top diagram compares sequences, with patterned regions representing alternatively spliced domains that are different among isoforms (other unpatterned regions are identical in sequence). Both current traces obtained from oocyte voltage-clamp show currents resulting from each isoform. *From Timpe et al. (1988).*

Finally, investigators mutagenized specific regions of the fast inactivating *Shaker* B segment to localize those residues involved in the inactivation process (Hoshi et al., 1990). Both by deleting small segments and by replacing others it was found that the crucial part of the molecule was a stretch of residues from positions 6–83 (i.e., the N-terminal tail of the molecule seen in Fig. 20.8). In this segment, there were distinct effects of the alterations. Thus, mutations to the residues in the 23–83 residue region tended to produce changes in inactivation kinetics, while changes to those in the N-terminal side (Aldrich et al., 1990; Armstrong & Bezanilla, 1977; Catterall, 1988; Chou & Fasman, 1978; Guy & Seetharamulu, 1986; Kyte & Doolittle, 1982; Leonard et al., 1991; Lopez et al., 1991; MacKinnon, 1991; Noda et al., 1984; Numa & Noda, 1986; Papazian et al., 1987, 1991; Pongs, 1992; Tanabe et al., 1987; Timpe et al., 1988; Yool & Schwarz, 1991) of the segment tended to destroy inactivation completely (see Fig. 20.14).

What might be the mechanism of *Shaker* inactivation? Much earlier it had been reported by Armstrong and coworkers that internal perfusion of proteases in the squid destroyed the ability of Na^+ channels to inactivate. On the basis of this and other experiments, Armstrong proposed that the inactivation mechanism is a cytoplasmic ball attached to a peptide tether that swings into an open channel to block it from the inside (Fig. 20.13; Armstrong & Bezanilla, 1977). In this hypothesis,

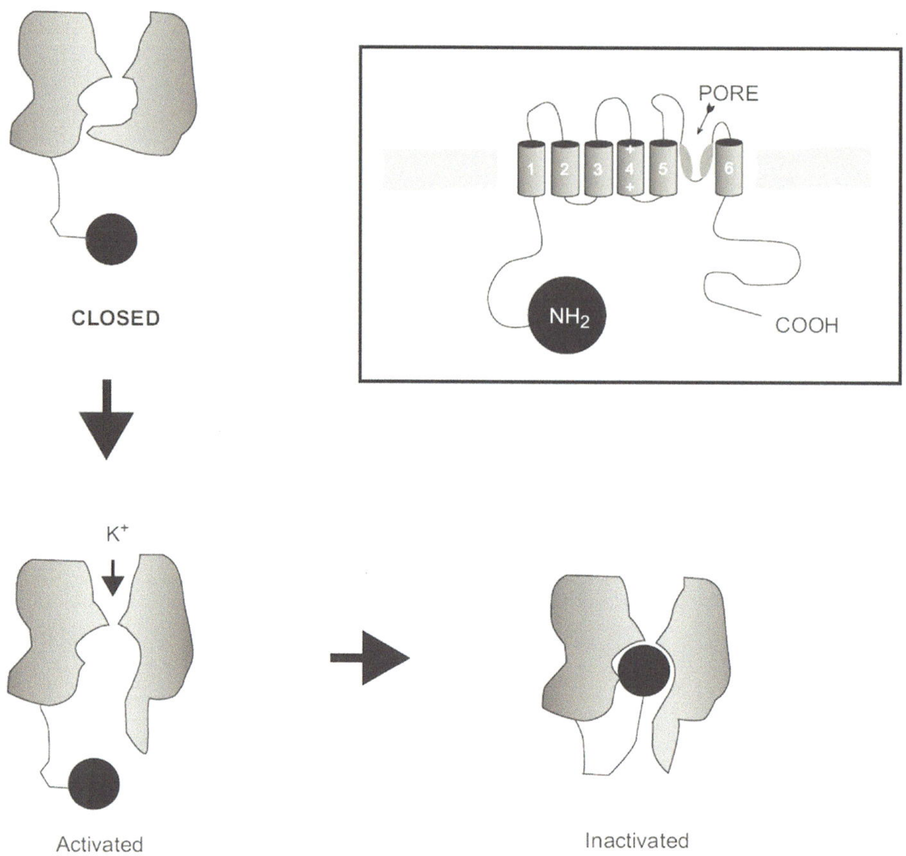

FIGURE 20.13 "Ball and chain" model of inactivation. Not shown are the three other "balls" that the other subunits of the channel will contribute.

the ball blocks the channel by binding to a receptor in the inside of the pore that is accessible only when the channel has opened (activated). The reader may recall the evidence that inactivation is coupled to activation, but may itself be intrinsically voltage-independent. Thus, the "ball and chain" model also explains these properties since the cytoplasmic ball would experience little of the transmembrane field but could only swing into the channel after it had activated and revealed its receptor (hence the coupling properties).

In the case of the *Shaker* channel, Aldrich and colleagues proposed that the N-terminal segment that they identified was just such a mechanism. Thus changes to the length of the 23−83 segment altered inactivation kinetics (as might be expected from changing the length of the "chain"), while changes to the N-terminal segment (such as alterations of charge) that eliminated inactivation were postulated to prevent the "ball" from binding to its receptor in the activated channel.

How might such a model be further tested? Here the investigators performed a particularly elegant experiment (Fig. 20.14; Zagotta et al., 1990). First they constructed a mutant *Shaker* channel in which inactivation had been eliminated by a deletion of its own N-terminal segment. Next, a synthetic peptide was prepared that had the exact sequence of the deleted region. The mutant channel was then expressed in oocytes, where both excised macropatch and single channel observations could be made. The investigators found that when the peptide was applied to the solution bathing the inside surface of the membrane, inactivation was restored to these mutant channels. Furthermore, the kinetics of the restored inactivation depended on the concentration of the peptide in the bath, and the inactivation could be reversed by washing away the peptide-containing solution. These observations are thus consistent with a direct blocking interaction of the peptide with a site in the open channel and constitute strong evidence for the "ball and chain" mechanism of inactivation in A-type K^+ channels.

5.6 Other gating-related domains in voltage-sensitive channels

Upon analysis of the homology of the various cloned Na^+ channels, it was found that the longest segment of conservation lay in the postulated internal loop between repeats III and IV, even among animals of some phylogenetic distance

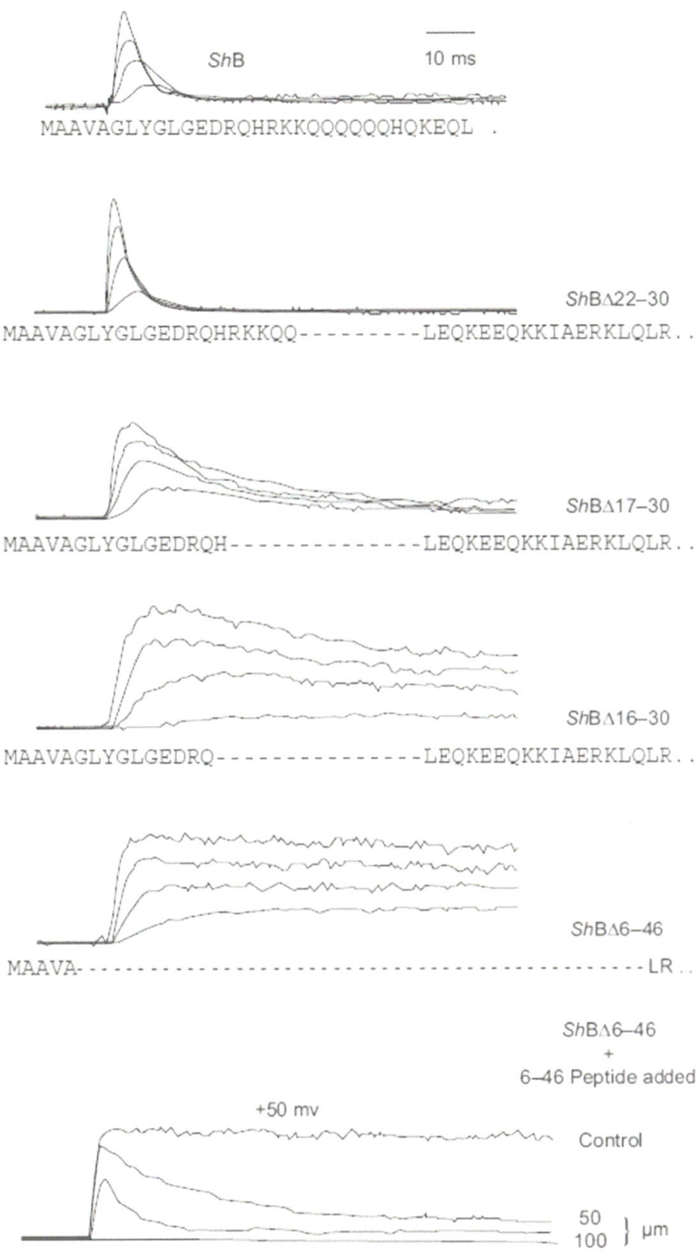

FIGURE 20.14 Identification of a putative "ball and chain" in *Shaker* B channels expressed in oocytes. Panels two to five from the top show the effect of deleting increasing lengths of a segment near the N terminus of the *Shaker* protein. Note that the deletion of residues 6–46 results in a noninactivating K^+ current. The bottommost panel shows that inactivation properties of this latter mutant may be restored by application of the synthetic 6–46 peptide to the inner channel surface in a dose-dependent fashion. *Reorganized from Hoshi et al. (1990), Zagotta et al. (1990).*

(Fig. 20.3). This high degree of conservation suggested a functionally important role for the segment. This idea was initially tested by constructing mutant Na^+ channels in which the channel was expressed in oocytes in two pieces, with the genetic "cut" within this segment (Stuhmer et al., 1989). Interestingly, such artificial dimeric channels were assembled and expressed in the oocyte and displayed relatively normal activation and conductance. However, such channels did not inactivate. Further, site-directed antibodies raised to a synthetic peptide encoding this segment were found to slow or eliminate inactivation when they were applied to the cytoplasmic surface of cells expressing normally inactivating channels (Vassilev et al., 1989). Finally, mutagenesis directed to this segment produces channels with altered inactivation (Catterall, 1992).

However, the sequence of this segment has little resemblance to the N-terminal "ball and chain" Shaker domain, and it is additionally tethered at both ends. Nonetheless, several lines of evidence now suggest that this domain operates analogously to the Shaker "ball and chain" in producing sodium channel inactivation (Armstrong & Hollingworth, 2018). Most notably, via mutagenesis to the III—IV loop one can produce "inactivationless" sodium channels; in such mutants inactivation can be restored with the internal application of the same synthetic Shaker peptide that restored inactivation to noninactivating potassium channels in the experiments described above.

For DHP-sensitive Ca^{2+} channels, an isoform to the originally cloned skeletal muscle channel has been cloned from the heart. In the parent tissues, these channels share some characteristics, but differ greatly in the speed by which their calcium currents activate (heart channel currents being much more rapid in responding to voltage). Although the isoforms share considerable homology, there is still enough difference in the sequences to make it difficult to guess which residues might be responsible for activation kinetics. This problem was elegantly addressed by Beam, Tanabe, Numa, and coworkers who made chimeras of the two channels (Fig. 20.15; Tanabe et al., 1991). To focus the mutagenesis it was reasoned that only the predicted transmembrane repeats should be involved in voltage-gated activation. Thus the chimeras swapped the homologous repeats among constructs. As shown in Fig. 20.15, it was found that the kinetics of activation correlated almost completely with the kinetics of the donor of the first (I) repeat. This finding suggests that despite their approximate homology, the repeats in a given channel contribute differently to gating properties. This is perhaps expected from the observation that the S4 regions among these domains have somewhat different structures and numbers of charges. Having identified the domain responsible, these investigators were able to narrow their search to shorter segments and ultimately identified the residues involved in this activation phenomena (Nakai et al., 1994).

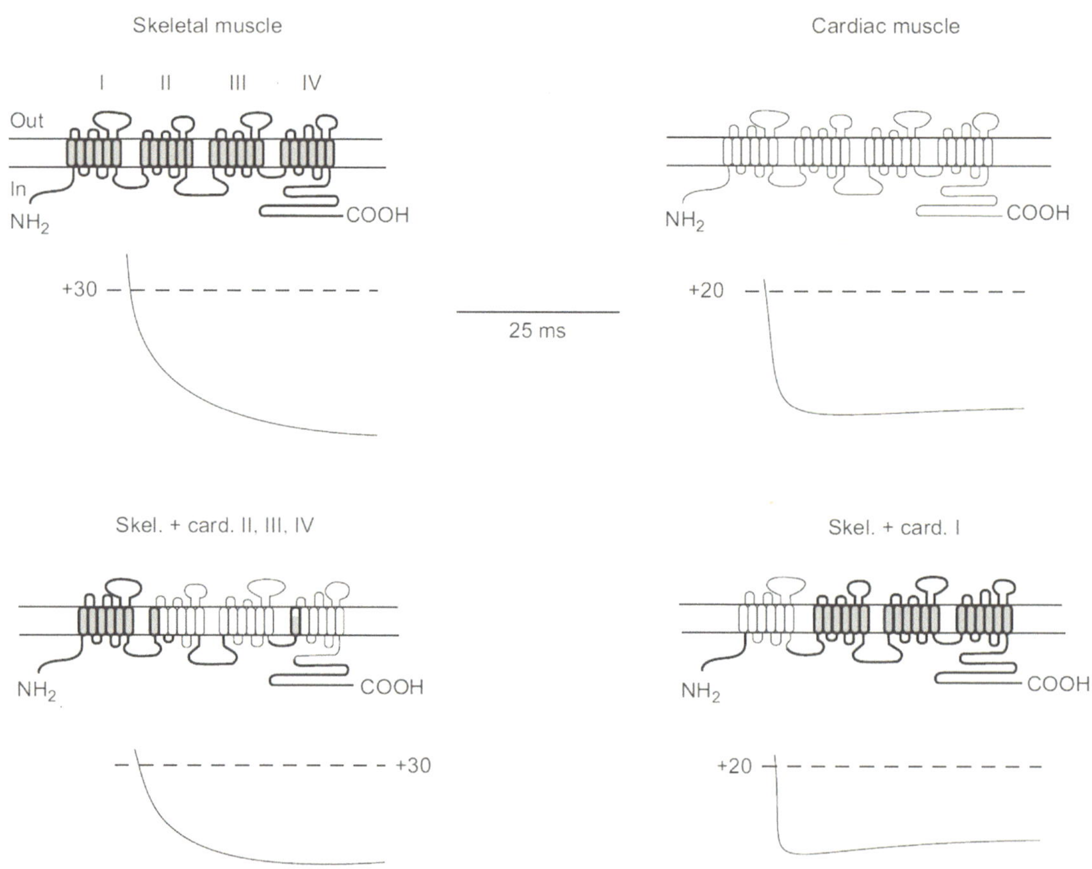

FIGURE 20.15 Identification of regions determining activation kinetics in L-type Ca^{2+} channels from skeletal muscle and heart. In these experiments, hybrids ("chimeras") of skeletal muscle and cardiac channels were made in various combinations to localize the domains responsible for the differences in Ca^{2+} current activation kinetics in heart and skeletal muscle. cDNA encoding each construct was injected in cultured muscle cells (myotubes) from mutant mice unable to synthesize active L-type channels. Thus, these mutant cells would express only currents resulting from the genetically engineered DNA introduced into them. The top two panels show the current kinetics resulting from DNA encoding the wild-type skeletal muscle and cardiac channels. The bottom panels show two chimeras in which current kinetics correlate with the donor of repeat domain I. Most of the other possible combination channels were also made, with the same correlation of kinetics with domain I. *Redrawn from Nakai et al. (1994).*

5.7 Pore formation and ion selectivity

Pores are the defining structures of all ion channels. As such, they provide both an aqueous pathway for ions to traverse the membrane and a means for discriminating among ions for access to this path. Studies of the mechanisms of pores provide an especially vivid example of the integration of theoretical modeling, mutagenesis, and molecular structural approaches.

Even before the advent of structural studies, Hille hypothesized that ion selectivity and pore formation were performed by the same parts of the channel structure (2001). Experimental evidence also indicated that charged or polar oxygen groups lined the narrowest region of the pore, where cation recognition occurs (see Nakai et al., 1994). This latter finding was a sensible one considering the chemistry of cations in solution. In aqueous solution, metal cations such K^+, Na^+, and Ca^{2+} are stabilized by electrostatic interaction with the electronegative oxygen atoms of water molecules. In the narrowest region of the pore, water molecules must be removed in order for the channel to recognize the preferred cation; lining this narrow region with oxygen-bearing groups provides a chemically favorable "selectivity filter."

Based on this early work, investigators sought short protein segments that might line the aqueous pore with oxygen atoms. A candidate sequence was initially identified using sophisticated analysis of primary sequence information (Guy & Seetharamulu, 1986). This sequence is 20–25 amino acids long and forms part of the loop connecting the fifth and sixth predicted transmembrane helices in the voltage-gated cation channels (see Figs. 20.6–20.8). Tests of this putative pore-lining loop's ("P-loop") role in pore formation were soon forthcoming.

One type of test was to determine whether site-directed mutations in the P-loop sequence affected the action of known pore-blocking agents on heterologously expressed mutant channels. Among the agents tested in this way were the K^+ channel blockers tetraethylammonium and charybdotoxin and the Na^+ channel blocker tetrodotoxin (MacKinnon et al., 1990; MacKinnon & Yellen, 1990; Terlau et al., 1991). Block by the externally effective agents tetrodotoxin or charybdotoxin was specifically affected by mutations at the extracellular end of the P-loop. Tetraethylammonium block of K^+ channels, which occurs by blocker binding at either end of the narrow pore, was altered by mutations at both ends of the P-loop. Results such as these confirmed that the P-loop does indeed form at least part of the pore in voltage-gated ion channels. Does the P-loop also form the ion selectivity filter?

The answer to this question also comes from site-directed mutagenesis studies. In *Shaker*-type K^+ channels, point mutations introduced into the P-loop profoundly altered the channel's preference for K^+ over Rb^+ or NH_4^+, as shown in Fig. 20.16 (Yool & Schwarz, 1991). However, nonconducting channels were produced whenever mutations were introduced into one stretch of eight amino acids in K^+ channel P-loops. This so-called K^+ channel signature sequence is highly conserved in K^+ channels found in bacteria to humans, and it was therefore suspected to form the narrow selectivity filter in K^+ channels. Site-directed mutations in the P-loops of Na^+ channels and Ca^{2+} channels also affect ion selectivity (Heinemann et al., 1992; Yang et al., 1993). In Na^+ channels, a locus of four residues has been identified that confers selectivity for Na^+ over Ca^{2+} (Fig.20.17). Each of the four P-loops of a Na^+ channel contributes one residue to the locus, which includes two carboxylate-bearing residues (aspartate, glutamate) and two noncarboxylate residues (lysine, alanine). Substitution of glutamate residues for the lysine and alanine residues dramatically reduces Na^+ selectivity and allows

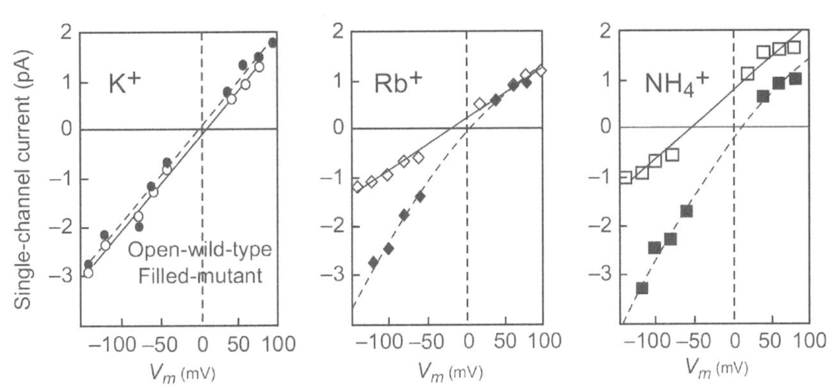

FIGURE 20.16 Alteration of *Shaker* B K^+ channel ion selectivity caused by a mutation in the putative pore-forming region. In these experiments, the phenylalanine at position 433 was mutated to serine. The oocyte system to express both wild-type and mutant channels as described in the text. Single-channel currents were measured using patch-clamp methods in which the ion indicated in each panel was applied to the external surface of the channel with K^+ on the inside in all cases. In these biionic conditions, inward (negative) currents will be carried mainly by the test ion, while outward (positive) currents will be mostly K^+ ion fluxes. As can be seen from the increased inward currents, permeability of the channel to rubidium and ammonium ions was greatly enhanced in the mutants, while K^+ was unaffected. *Data redrawn from Yool and Schwarz (1991).*

	Motif I					Motif II					Motif III					Motif IV				
	−2	−1	0	1	2	−2	−1	0	1	2	−2	−1	0	1	2	−2	−1	0	1	2
CaCh	T	M	**E**	G	W	T	G	**E**	D	W	T	f	**E**	G	W	T	G	**E**	A	W
NaCh	T	Q	**D**	X	W	C	G	**E**	-	W	T	F	**K**	G	W	T	S	**A**	G	W

FIGURE 20.17 Alignment of Ca^{2+} and Na^{+} channel pore sequences. Position 0 in the alignments (shaded) marks the critical residues in the selectivity filter (EEEE locus in Ca^{2+} channels, DEKA locus in Na^{+} channels). High-voltage activated calcium channels are identical to one another in these pore-sequences, except at the −1 position in motif III: The "f" indicates that this position is occupied by either phenylalanine (F) or glycine (G). In sodium channels, the residue at the +1 position in motif I (x) varies depending on channel isoform. Na^{+} channel sensitivity to tetrodotoxin depends partly on the residue present at this position.

significant amounts of Ca^{2+} to permeate this mutant Na^{+} channel. The analogous locus in Ca^{2+} channels comprises four glutamate residues. Functional analysis of the effects of mutations in this locus has shown that these four glutamate residues form the core of the selectivity filter in Ca^{2+} channels (Ellinor et al., 1995).

As described above, oxygen atoms have long been thought to line the selectivity filter in voltage-gated, cation-selective channels. The presence of multiple carboxylate groups in the selectivity filters of Ca^{2+} and Na^{+} channels is consonant with the earlier work, although in principle, the carboxylates may either project into the pore or away from it. In fact, Ca^{2+} channel carboxylate groups apparently project into the aqueous pore, and thereby provide the electronegative oxygen atoms needed to stabilize Ca^{2+} there. Evidence supporting this picture of selectivity filter structure derives from the fact that protonation of these carboxylate groups blocks current through Ca^{2+} channels; mutational replacement of the carboxylates by nonprotonatable groups predictably alters proton block. Orientation of selectivity filter side-chains has also been investigated using a technique called the "substituted cysteine accessibility method." In this method, a sulfhydryl-containing cysteine residue is substituted for the wild-type residue, and a sulfhydryl-modifying agent is bath-applied to the mutant channel. If the side-chain of the substituted cysteine projects into the pore, where it may be accessible to the sulfhydryl-modifying agent, then the modifying agent will attach itself to the cysteine sulfhydryl and obstruct ion flow. Interruption of channel current thus reports that a particular side-chain is oriented into the aqueous pore, whereas absence of block indicates that the side-chain faces away from the pore. Application of this method to Na^{+} and Ca^{2+} channels indicates that the side-chains in the carboxylate-containing loci of these channels (Chiamvimonvat et al., 1996) all project into the pore. The extreme narrowness of K^{+} channel pores has prevented this technique from being effectively employed with these channels. However, the selectivity filter of K^{+} channels may be structurally distinct from that of Ca^{2+} channels and perhaps Na^{+} channels.

5.8 Crystal structure of a bacterial K^{+} channel pore region

Results of these site-directed mutagenesis studies have been strikingly confirmed by the first X-ray crystallographic structure obtained for a member of the family of voltage-gated ion channels (Doyle et al., 1998). This structure was solved for a bacterial K^{+} channel, which though not in fact voltage-gated, is nevertheless a member of the family of voltage-gated ion channels: most significantly, it possesses the hallmark of all K^{+} selective channels, the K^{+} channel signature sequence. The crystal structure reveals that the selectivity filter is formed by part of the P-loop, with two K^{+} ions stabilized therein, probably by main chain carbonyl oxygen atoms (Fig. 20.18). That uncharged carbonyl oxygens apparently line the selectivity filter in the bacterial K^{+} channel actually agrees with predictions, based on mutagenesis studies, for other K^{+} channels.

Why might K^{+} channels have evolved a selectivity structure that employs neutral, main-chain carbonyl oxygen atoms, whereas their evolutionary descendants, Ca^{2+} channels, apparently line their selectivity filter with negatively charged, side-chain carboxylate oxygens? Perhaps the negative charge of carboxylate oxygens is necessary in selecting for Ca^{2+}, a charge-dense divalent cation, whereas uncharged carbonyl oxygen atoms are ideal in selecting for certain monovalent cations, such as K^{+}. How then to account for Na^{+} channel selectivity structure? Na^{+} channels have evolved from Ca^{2+} channels, and the Na^{+} channel selectivity filter apparently includes two side-chain carboxylate oxygens. K^{+} and Na^{+} have the same ionic charge but Na^{+} has a higher charge density owing to its smaller ionic radius. As in the case for Ca^{2+}, the

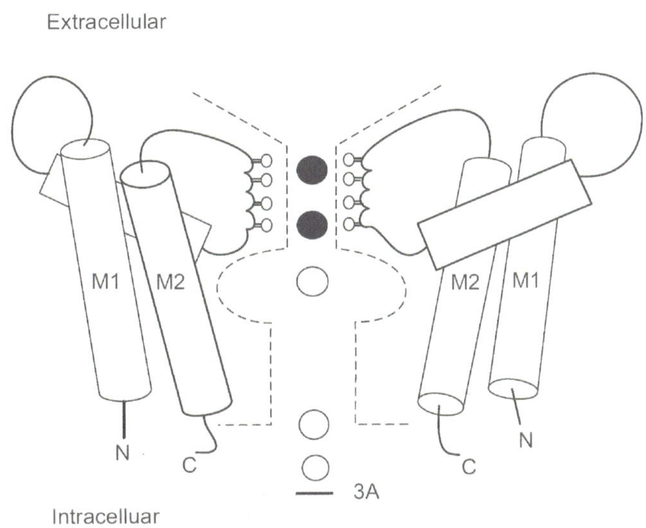

FIGURE 20.18 Cartoon illustration of the bacterial K^+ channel structure. Only two of the four subunits of the channel are shown. Each subunit has a P-loop and two helices (M1, M2), similar to the structure of inwardly rectifying K^+ channels. The segment of the P-loop that lines the ion permeation pathway projects carbonyl oxygen atoms into the pore. A pair of K^+ ions are shown stabilized in the selectivity filter by the rings of carbonyl oxygens. *Adapted with permission from Choe, S., & Robinson, R. (1998). An ingenious filter: The structural basis for ion channel selectivity. Neuron 20, 821–823. Copyright 1998 Cell Press.*

relatively high charge density of Na^+ may require carboxylate oxygens for selectivity, though only two rather than the four present in the Ca^{2+} channel selectivity filter.

More recently, techniques have been developed to deal with the problems of crystallization of channel protein hydropathy domains (see above). This has allowed determination of high-resolution structures of mammalian potassium channels and the development of models of voltage-sensitive gating (Chen et al., 2010).

5.9 High-resolution structure of a eukaryotic sodium channel

Recent application of "cryo-electron microscopy" (CRM) methods has also yielded high-resolution structures for ion channels (see (Milne et al., 2013) for a review of CRM for general readers). In brief, this method starts by spreading a small amount of highly purified channel protein on a small grid and then rapidly freezing it to very low temperatures (e.g., to that of liquid nitrogen). In this state, the native protein structure is highly preserved and is more resistant to destruction by the ionizing effect of the electron beam used to image it. Molecular structures of the channel are then obtained by averaging high-resolution digital images of the spread protein obtained from the electron microscope. Such analysis uses sophisticated computational methods to combine thousands of images of single channel particles in a variety of different orientations to obtain a high-resolution image of an average channel structure.

Nonetheless, the resultant structures obtained by the CRM approach rarely have the resolution of crystallographic methods and so rely on constraints obtained by other methods (e.g., mutagenesis, pharmacological effects, as discussed above) to assign primary and secondary amino acid structures to obtain amino acid residue-level resolution. In this way, researchers have been able to obtain detailed structures for a mammalian skeletal muscle calcium channel and a cockroach sodium channel (Shen, Zhou, et al., 2017; Wu et al., 2016; Fig. 20.19). Such structures clearly show the channel pores and paddle-shaped voltage-sensitive domains containing the S4 helix. A unique aspect of this CRM approach was its success in freezing individual channel proteins in various gating conformational transitions. In addition, the investigators were able to determine the sodium channel binding site of certain gating toxins with known functional effects. Thus, this has enabled the elucidation of highly detailed molecular mechanisms for the electromechanical coupling involved in channel opening, closing, and inactivation. Clearly, CRM is a powerful approach, it might be expected that it will be applied to the study of molecular mechanisms for nearly all other membrane-associated transporters in the near future.

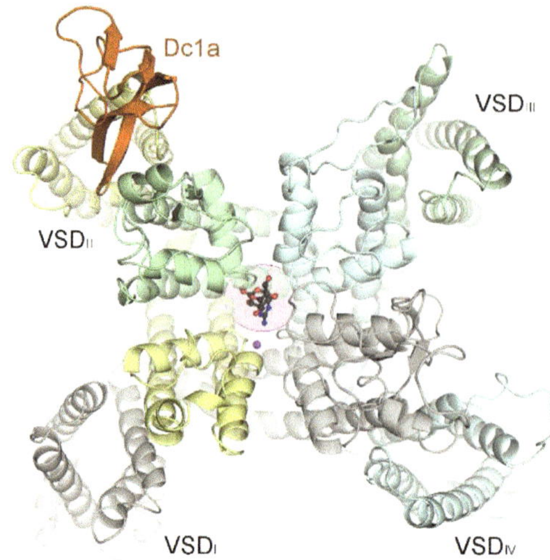

FIGURE 20.19 High-resolution structure of a eukaryotic voltage-gated sodium channel. This structure was obtained with two neurotoxins bound to the channel, namely the pore blocker saxitoxin (center of structure) and an activation gate spider toxin Dc1a. Looking at the outside surface of the channel, the pseudotetrameric structure is shown in different colors with the transmembrane pore in the center and the S4-containing voltage-sensitive domains ("VSDs") as paddle-like structures on the periphery. Thus, these toxin molecules identify structures involved in pore formation and ion selectivity (saxitoxin) as well as voltage-dependent activation (Dc1a). The location of the spider toxin site is especially illuminating, because this toxin causes a persistent activation of the channel, freezing the channel in its open, conducting state. Thus the binding site appears to identify part of the electromechanical linkage between S4 movement and channel opening. *Reprinted with permission from Shen, Li, et al. (2017).*

6. Isoforms of voltage-gated channels as part of a large superfamily

One of the most striking results of recombinant DNA analysis of channel clones has been the discovery of numerous channel "isoforms" expressed within the same organism and even within the same cell. Naturally, some degree of homology among channels of different organisms was expected, and electrophysiological recordings showed that channels could be functionally diverse at the cellular level. Even a structural similarity among Na^+, Ca^{2+}, and K^+ channels had been predicted by some investigators. However, the number and diversity of channels within a single organism revealed by molecular biology was rather unexpected. Channels of a functionally similar class (usually based on ion selectivity) are now called "isotypes" or "isoforms," by analogy to the phenomenon of isoenzymes.

We have previously discussed the *Shaker* isoforms. Shaker-derived sequences have also been used to construct probes that have discovered other families of K^+ channel in flies, mammals, and in other species (Salkoff et al., 1992). These families are so extensive that they challenge investigators to develop appropriate nomenclature for their classification (see Table 20.3; Moulton et al., 2003). In any case, although they are all to some degree homologous with the original *Shaker* family, they have much higher degrees of sequence homology among members of their class. Functionally, they run the gamut from inactivating A-type channels to noninactivating channels of the classic delayed rectifier sort. Interestingly, there is usually a much higher degree of sequence homology among channels of the same class from different animals than there is between channels of separate classes in the same animal. For example, *Shaker* channels from *Drosophila* are much more homologous to *Shaker* channels of mouse brain than they are to the **Drosophila** *Shab*-type channels.

TABLE 20.3 Gene subfamilies of voltage-gated ion channels.[a]

Voltage-gated gene family	# Subfamilies	Total # genes	Subfamily nomenclature
Potassium channels	12	40	$K_vX.y$, X = 1–12, y = subfamily isoform #
Calcium channels (α1-subunit)	3	10	$Ca_vX.y$, X = 1–3, y = subfamily isoform #
Sodium channels	1	9	$Na_v1.x$, x = 1-9

[a]*Only diversity in main pore-forming α-subunits are considered here.*

Recently many other subfamilies of potassium channels have been discovered in mammalian and invertebrate genomes (see Guide to Receptors and Channels, 2011). Two of these, formerly known as the *erg* and KQT subfamilies, are important in mammalian cardiac function. Mutants of these channels have been associated with serious inherited heart and nervous system disorders in humans, in particular the long QT (LQT) syndrome. In any case, in mammals the number of potassium channel isoforms is already quite large (see Table 20.3), with more being discovered at frequent intervals.

Considerable diversity is also found among Na^+ and Ca^{2+} channels (Table 20.3; Catterall et al., 2005; Hofmann et al., 1994). For the main channel-forming α-subunits (Figs. 20.6 and 20.7), there are thought to be nine Na^+ channel and 10 Ca^{2+} channel genes in the mammalian genome. Some of these isoforms are expressed very selectively in certain tissues, such as brain, or in only certain cells types or at very specific stages of development.

The term "isoform" is rather nonspecific in that it does not convey a sense of relative relatedness among channels. An evolving system of classification refers to the set of voltage-gated channels as a "superfamily," while the selectivity types (i.e., Na^+, Ca^{2+}, or K^+ channels) form families. Due to the impressive degree of diversity found for K^+ channels investigators have developed a uniform system of nomenclature, as well as separate nomenclatures for isoforms of Ca^{2+} channel and Na^+ channel subunits have also been developed (Guide to Receptors and Channels, 2011). Further consideration of channel classification will undoubtedly be useful in understanding the process of channel evolution.

6.1 How do isoforms arise?

In general, there are two separate mechanisms that have been identified that create such diversity. First, there is the expected situation in which distinctly separate genes encode different channels in the same organism. These typically originate from gene duplication during evolution. Such as largely the situation for mammalian Na^+ channel and Ca^{2+} channel isoforms. However, a second mechanism has been found called "alternative splicing," in which a given segment of a channel polypeptide may be encoded in several different stretches of genomic DNA known as "exons." This is the case for *Shaker* in *Drosophila*; thus in Fig. 20.12 above, the various *Shaker* isoforms are generated by mixing and matching the transcripts of different exons. During transcription some sort of regulatory mechanism exists that decides which one of the alternative coding RNAs for a given segment will be incorporated in the final mRNA. The other undesired homologous segments are then excised from the transcript and the desired RNA segments spliced together. Currently, there is evidence for the operation of both mechanisms in vertebrates and invertebrates. However, most mammalian systems studied to date appear to rely more heavily on separate genes than alternative splicing, while in the fruit fly both mechanisms appear common.

More recently, a phenomenon known as **RNA editing** has been implicated in the generation of functionally distinct isoforms. Here, specialized enzymes may chemically change a discrete nucleotide base in the mRNA encoding a channel, thus leading to an amino acid change in sequence that results in a functional change. Such a phenomenon has been shown to occur in an octopus potassium channel as part of a temperature-adaptation process (Garrett & Rosenthal, 2012).

Finally, for all multisubunit channels, there is the possibility of diversification through various combinations of each set of subunit isoforms. The best studied examples are the heteromeric formation of channels by different K^+ channel subunit isoforms (Salkoff et al., 1992). For example, if the same cell synthesized *Shaker* A and *Shaker* B channel subunits, one might find heteromeric complexes of A and B types as well as the homomeric types. In fact, experiments with the oocyte expression system into which both types of mRNA have been injected show that such heteromeric assembly can occur, yielding functional channels with characteristics different from homomultimers of either channel type. Thus, it is theoretically possible that distinct channel properties may be generated by heteromeric assembly of K^+ channel subunits. Whether this occurs in nature is currently being investigated.

On the other hand, oocyte experiments show that heteromeric association among subunits from different families (for example, *Shaker* and *Shal*) does not occur. Apparently different families have unique structural domains that allow only homomeric assembly among member of their family. This discovery in turn led to a set of chimeric experiments in which domains among a fly *Shaker* channel and a distantly related mammalian delayed rectifier channel (drk1) were exchanged in an attempt to identify the segments specifying assembly specificity (Li et al., 1992). Such a segment was found on the N-terminal part of the peptide. Thus substitution of the *Shaker* segment into the drk1 cDNA produced a channel drk1 subunit that was able to coassemble with native *Shaker*.

6.2 Why are channels so diverse in a given organism?

One obvious reason for the presence of so many channel isoforms in the same animal is that each isoform has different functional properties that are appropriate to the function of its host cell. For example, high-frequency, repetitive firing requires short duration action potentials that could be created by rapidly gating Na^+ channels and fast inactivating A-type K^+ channels. In addition, isoforms may be differentially sensitive to intracellular modulators that allow cellular mechanisms to alter the properties of specific channels selectively in response to physiological effectors. On the other hand, many of the discovered isoforms have no significant functional differences among them. In this case, it is possible that it is not the differences in the channels themselves that are important, but rather the way the isoforms are expressed. Thus in a number of cases it has been shown that expression of one isoform over another occurs developmentally or in response to a physiological stimulus (Ribera et al., 1992). Such specific expression is thought be controlled by genetic elements outside the coding region for the channel, known as regulatory elements. These stretches of DNA respond specifically to one of a number of possible soluble factor by increasing (in the case of promotors and enhancers) or inhibiting (silencers) transcription of the channel gene (Maue et al., 1990). Separate genes, then, potentially allow the cell to express selectively a given channel in response to a given stimulus or at a specific time during development. Finally, there is the possibility that a given isoform sequence encodes information used by the cellular machinery to localize or cluster it in very discrete locations, such as the nodes of Ranvier in myelinated nerve fibers (Dugandzija-Novakovic et al., 1995), neuronal cell bodies, or dendrites (Maletic-Savatic et al., 1994). The diversity of channels and their biological purpose will undoubtedly continue to be a highly active area of cell biology.

6.3 The role of ion channels in disease

Although beyond the scope of this chapter, we make a brief mention of the association of voltage-gated ion channels with human disease. First, these channels have long been primary targets for various drugs used to treat a number of conditions. Primary examples include local anesthetics for pain management (act on sodium channels), antiarrhythmics (act on sodium, calcium, and/or potassium channels), antiseizure drugs (act on sodium channels), and certain drugs used to manage psychiatric disorders (act on sodium channels). Further, genetic defects in these ion channels have been shown to cause a growing number of diseases, especially forms of epilepsy, certain muscular diseases, and a wide variety of other neurological diseases such as pain syndromes, movement disorders, and cognitive dysfunction. Due to their importance, such diseases are now known as "channelopathies" and have become a major interest to clinicians. The reader is referred to Catterall (2012), for a discussion of this field and further references.

References

Aldrich, R. W., Hoshi, T., & Zagotta, W. N. (1990). Differences in gating among amino- terminal variants of *Shaker* potassium channels. *Cold Spring Harbor Symposia on Quantitative Biology, 55*, 19–27.

Armstrong, C. M., & Bezanilla, F. (1977). Inactivation of the sodium channel. II. Gating current experiments. *Journal of General Physiology, 70*, 567–590.

Armstrong, C. M., & Hollingworth, S. (2018). A perspective on Na and K channel inactivation. *The Journal of General Physiology, 150*(1), 7–18.

Catterall, W. A., Goldin, A. L., & Waxman, S. G. (2005). International Union of Pharmacology. XLVII. Nomenclature and structure-function relationships of voltage-gated sodium channels. *Pharmacological Reviews, 57*, 397–409.

Catterall, W. A. (1992). Cellular and molecular biology of voltage-gated sodium channels. *Physiological Reviews, 72*(Suppl. 1), S15–S48.

Catterall, W. A. (1988). Molecular properties of dihydropyridine sensitive calcium channels in skeletal muscle. *Journal of Biological Chemistry, 263*, 3535–3538.

Catterall, W. A. (2012). Voltage-gated sodium channels at 60: Structure, function and pathophysiology. *Journal of Physiology, 590*, 2577–2589.

Chen, X., Wang, Q., Ni, F., & Ma, J. (2010). Structure of the full-length Shaker potassium channel Kv1.2 by normal-mode-based X-ray crystallographic refinement. *Proceedings of the National Academy of Sciences, 107*, 11352–11357.

Chiamvimonvat, N., Perez-Garcia, M. T., Ranjan, R., Marban, E., & Tomaselli, G. F. (1996). Depth asymmetries of the pore-lining segments of the Na^+ channel revealed by cysteine mutagenesis. *Neuron, 16*, 1037–1047.

Chou, P. Y., & Fasman, G. D. (1978). Empirical predictions of protein conformation. *Annual Review of Biochemistry, 47*, 251–276.

Doyle, D. A., Cabral, J. M., Pfuetzner, R. A., Kuo, A., Gulbis, J. M., Cohen, S. L., Chair, B. T., & MacKinnon, R. (1998). The structure of the potassium channel: Molecular basis of K^+ conduction and selectivity. *Science, 280*, 69–77.

Dugandzija-Novakovic, S., Koszowski, A. G., Levinson, S. R., & Shrager, P. (1995). Clustering of Na^+ channels and node of Ranvier formation in remyelinating axons. *Journal of Neuroscience, 15*, 492–503.

Ellinor, P. T., Yang, J., Sather, W. A., Zhang, J.-F., & Tsien, R. W. (1995). Ca^{2+} channel selectivity at a single locus for high affinity Ca^{2+} interactions. *Neuron, 15*, 1121–1132.

Garrett, S., & Rosenthal, J. J. C. (2012). RNA editing underlies temperature adaptation in K^+ channels from polar octopuses. *Science, 335*, 848–851.

Guide to receptors and channels (GRAC), 5th edition. *British Journal of Pharmacology, 164*(Suppl. 1), (2011), S1–S2. https://doi.org/10.1111/j.1476-5381.2011.01649_1.x

Guy, H. R., & Seetharamulu, P. (1986). Molecular model of the action potential sodium channel. *Proceedings of the National Academy of Sciences of the United States of America, 83*, 508–512.

Heinemann, S. H., Terlau, H., Stuhmer, W., Imoto, K., & Numa, S. (1992). Calcium channel characteristics conferred on the sodium channel by single mutations. *Nature, 356*, 441–443.

Hille, B. (2001). *Ion channels of excitable membranes* (3rd ed.). Sunderland, Mass: Sinauer Associates.

Hofmann, F., Biel, M., & Flockerzi, V. (1994). Molecular basis for Ca^{2+} channel diversity. *Annual Review of Neuroscience, 17*, 399–418.

Hoshi, T., Zagotta, W. N., & Aldrich, R. W. (1990). Biophysical and molecular mechanisms of Shaker potassium channel inactivation. *Science, 250*, 533–538.

Kyte, J., & Doolittle, R. F. (1982). A simple method for displaying the hydropathic character of a protein. *Journal of Molecular Biology, 157*, 105–132.

Leonard, J., & Snutch, T. P. (1991). The expression of neurotransmitter receptors and ion channels in *Xenopus* oocytes. In D. M. Glover, & B. D. Hanes (Eds.), *Molecular neurobiology: A practical approach* (pp. 161–182). New York: Oxford University Press.

Li, M., Jan, Y. N., & Jan, L. Y. (1992). Specification of subunit assembly by the hydrophilic amino-terminal domain of the shaker potassium channel. *Science, 257*, 1225–1230.

Lopez, G. A., Jan, Y. N., & Jan, L. Y. (1991). Hydrophobic substitution mutations in the S4 sequence alter voltage-dependent gating in Shaker K+ channels. *Neuron, 7*, 327–336.

MacKinnon, R. (1991). Determination of the subunit stoichiometry of a voltage-activated potassium channel. *Nature (London), 350*, 232–235.

MacKinnon, R., Heginbotham, L., & Abramson, T. (1990). Mapping the receptor site for charybdotoxin, a pore-blocking potassium channel inhibitor. *Neuron, 5*, 767–771.

MacKinnon, R., & Yellen, G. (1990). Mutations affecting TEA blockade and ion permeation in voltage-activated K^+ channels. *Science, 250*, 276–279.

Maletic-Savatic, M., Lenn, N. J., & Trimmer, J. S. (1994). Differential spatiotemporal expression of K^+ channel polypeptides in rat hippocampal neurons developing *in situ* and *in vitro*. *Journal of Neuroscience, 15*, 3840–3851.

Maue, R. A., Kraner, S. D., Goodman, R. H., & Mandel, G. (1990). Neuron-specific expression of the rat brain type II sodium channel gene is directed by upstream regulatory elements. *Neuron, 4*, 223–231.

Miller, J. A., Agnew, W. S., & Levinson, S. R. (1983). Principal glycopeptide of the tetrodotoxin/saxitoxin binding protein from Electrophorus electricus: Isolation and partial physical and chemical characterization. *Biochemistry, 22*, 462–470.

Milne, J. L., Borgnia, M. J., Bartesaghi, A., Tran, E. E., Earl, L. A., Schauder, D. M., Lengyel, J., Pierson, J., Patwardhan, A., & Subramaniam, S. (2013). Cryo-electron microscopy – a primer for the non-microscopist. *FEBS Journal, 280*, 28–45.

Morera, F. J., Vargas, G., González, C., Rosenmann, E., & Latorre, R. (2007). Ion-Channel reconstitution. In A. M. Dopico (Ed.), *Methods in molecular biology: Vol 400. Methods in membrane lipids*. Humana Press.

Moulton, G., Attwood, T. K., Parry-Smith, D. J., & Packer, J. C. (2003). Phylogenomic analysis and evolution of the potassium channel gene family. *Receptors and Channels, 9*, 363–377.

Nakai, J., Adams, B. A., Imoto, K., & Beam, K. G. (1994). Critical roles of the S3-segment and S3-S4 linker of repeat I in activation of L-type calcium channels. *Proceedings of the National Academy of Sciences, 91*, 1014–1018.

Namadurai, S., Yereddi, N. R., Cusdin, F. S., Huang, C. L.-H., Chirgadze, D. Y., & Jackson, A. P. (2015). A new look at sodium channel β subunits. *Open Biology, 5*, Article 140192.

Noda, M., Shimizu, S., Tanabe, T., Takai, T., Kayano, T., Ikeda, T., Takahashi, H., Nakayama, H., Kanaoka, Y., Minamino, N., et al. (1984). Primary structure of electrophorus electricus sodium channel deduced from cDNA sequence. *Nature, 312*, 121–127.

Numa, S., & Noda, M. (1986). Molecular structure of sodium channels. *Annals of the New York Academy of Sciences, 479*, 338–355.

Papazian, D. M., Timpe, L. C., Jan, Y. N., & Jan, L. Y. (1991). Alteration of voltage- dependence of Shaker potassium channel by mutations in the S4 sequence. *Nature, 349*, 305–310.

Papazian, D. M., Timpe, L. C., Jan, Y. N., & Jan, L. Y. (1987). Cloning of genomic and complementary DNA from *Shaker*, a putative potassium channel gene from *Drosophila. Science, 237*, 749–753.

Pongs, O. (1992). Molecular biology of voltage-dependent potassium channels. *Physiological Reviews, 72*(Suppl. 1), S69–S88.

Ribera, A. B., & Spitzer, N. C. (1992). Developmental regulation of potassium channels and the impact on neuronal differentiation. In T. Narahashi (Ed.), *Ion channels* (Vol. 3). New York: Plenum Press.

Salkoff, L., Baker, K., Butler, A., Covarrubias, M., Pak, M. D., & Wei, A. (1992). An essential set of K^+ channels conserved in flies, mice, and humans. *Trends in Neurosciences, 15*, 161–166.

Sharp, A. H., Imagawa, T., Leung, A. T., & Campbell, K. P. (1987). Identification and characterization of the dihydropyridine-binding subunit of the skeletal muscle dihydropyridine receptor. *Journal of Biological Chemistry, 262*, 12309–12315.

Shen, H., Zhou, Q., Pan, X., Li, Z., Wu, J., & Yan, N. (2017). Structure of a eukaryotic voltage-gated sodium channel at near-atomic resolution. *Science, 355*(6328).

Shen, H., Li, Z., Jiang, Y., Pan, X., Wu, J., Cristofori-Armstrong, B., Smith, J. J., Chin, Y. K. Y., Lei, J., Zhou, Q., King, G. F., & Yan, N. (2017). Structural basis for the modulation of voltage-gated sodium channels by animal toxins. *Science*. https://doi.org/10.1126/science.aau2596

Stuhmer, W., Conti, F., Suzuki, H., Wang, X., Noda, M., Yahagi, N., Kubo, H., & Numa, S. (1989). Structural parts involved in the activation and inactivation of sodium channels. *Nature, 339*, 597–603.

Tanabe, T., Adams, B. A., Numa, S., & Beam, K. G. (1991). Repeat I of the dihydropyridine receptor is critical in determining sodium channel activation kinetics. *Nature, 352*, 800–803.

Tanabe, T., Takeshima, H., Mikami, A., Flockerzi, V., Takahashi, H., Kangawa, K., Kojima, M., Matsuo, H., Hirose, T., & Numa, S. (1987). Primary structure of the receptor for calcium channel blockers from skeletal muscle. *Nature, 328*, 313–318.

Terlau, H., Heinemann, S. H., Stuhmer, W., Pusch, M., Conti, F., Imoto, K., & Numa, S. (1991). Mapping the site of block by tetrodotoxin and saxitoxin of sodium channel II. *FEBS Letters, 293*, 93–96.

Timpe, L. C., Jan, Y. N., & Jan, L. Y. (1988). Four cDNA clones from the Shaker locus of Drosophila induce kinetically distinct A-type potassium currents in Xenopus oocytes. *Neuron, 1*, 659–667.

Vassilev, P., Scheuer, T., & Catterall, W. A. (1989). Inhibition of inactivation of single sodium channels by a site-directed antibody. *Proceedings of the National Academy of Sciences of the United States of America, 86*, 8147–8151.

Wu, J., Yan, Z., Li, Z., Qian, X., Lu, S., Dong, M., Zhou, Q., & Yan, N. (2016). Structure of the voltage-gated calcium channel Cav1.1 at 3.6 A °resolution. *Nature, 537*, 191–196.

Yang, J., Ellinor, P. T., Sather, W. A., Zhang, J.-F., & Tsien, R. W. (1993). Molecular determinants of Ca^{2+} selectivity and ion permeation in L-type Ca^{2+} channels. *Nature, 366*, 158–161.

Yool, A. J., & Schwarz, T. L. (1991). Alteration of ionic selectivity of a K^{+} channel by mutation of the H5 region. *Nature, 349*, 700–704.

Zagotta, W. N., Hoshi, T., & Aldrich, R. W. (1990). Restoration of inactivation in mutants of Shaker potassium channels by a peptide derived from ShB. *Science, 250*, 568–571.7

Chapter 21

Mechanosensitive channels: What are they and why are they important

Boris Martinac and Charles D. Cox
Mechanosensory Biophysics Laboratory, Victor Chang Cardiac Research Institute & St Vincent's Clinical School, Darlinghurst, NSW, Australia

1. Introduction

Mechanical force is omnipresent. It manifests itself in many forms, including gravity, osmotic forces, hydrostatic pressure, as well as stretching, compressing, bending, or breaking a material. With the exception of gravity, which acts on every object in the universe through the gravitational force field, mechanical force requires physical contact for its action. As objects experiencing force in the form of mechanical stimuli such as touch, sound waves, or osmotic pressure, living cells have throughout evolution learned to cope with and respond to various mechanical stimuli by possessing a range of proteins associated with their cellular envelope, including cell wall, cytoskeleton (CSK), extracellular matrix (ECM), and membrane proteins. Among them, mechanosensitive (MS) ion channels are force-sensing integral membrane proteins, whose function is tightly linked to the lipid bilayer of cell membranes. MS channels are usually found at the very origin of cellular signaling pathways involved in mechanosensory transduction processes. Given the omnipresence of mechanical stimuli acting on the cell membrane, it is important to note that the function of integral membrane proteins other than MS channels, such as G protein coupled receptors (GPCRs) or membrane associated phospholipases, can also be modulated by the application of mechanical force (Lehtonen & Kinnunen, 1995; Storch et al., 2012; Xu et al., 2018). MS ion channels, however, present an excellent example of coupling membrane proteins and their structural dynamics to the mechanics of the cell membrane. Over the last 30 years, a large number of MS channels have been identified at the molecular level in organisms encompassing all types of life forms from bacteria to humans. The 3D structure of many of these channels has now been determined using X-ray crystallography or, more recently, by cryo-electron microscopy (cryo-EM) (Bass et al., 2002; Brohawn et al., 2012; Chang et al., 1998; Dong et al., 2015; Ge, 2015; Han et al., 2024). Their function has been studied in a great variety of cells and tissues using a range of experimental and computational approaches, as described in several reviews, some of which are listed here (Chalfie, 2009; Gillespie & Walker, 2001; Hamill & Martinac, 2001; Kocer, 2015; Kung, 2005; Martinac, 2011; Sachs, 2010). However, with the exception of bacterial MS channels MscL and MscS, we are still in the infancy of understanding how MS channels detect mechanical stimuli and convert them into the structural transitions between the closed and open channel structures, as well as which cellular components are directly involved in these structural changes. In this chapter, we provide a brief overview of what has been learnt about MS channels since their discovery over 30 years ago and focus on the most recent developments in this exciting area of mechanobiology research.

1.1 What are mechanosensitive channels and why are they important?

MS ion channels are pore-forming membrane proteins that gate in response to mechanical stimuli exerted on the cell membrane. By switching between the closed and open conformations, these channels allow ions and other solutes to flow across the cellular membranes (Hamill & Martinac, 2001). They have been shown to play a key role in many physiological processes associated with mechanosensory transduction, including osmoregulation in plants and bacteria as well as hearing, touch, proprioception, and blood flow regulation in mammalian cells (Honore et al., 2015; Martinac, 2004; Ranade et al., 2015). MS channels are specialized membrane proteins that act as molecular transducers of mechanical stimuli exerted on cell membranes. They play key roles in sensing and transducing mechanical stimuli into electrical and/or

Cell Physiology Sourcebook. https://doi.org/10.1016/B978-0-12-811114-7.00039-6

chemical signals by passing millions of ions and solutes through their pores when they are opened (Martinac, 2013). Good examples are the bacterial MscL (mechanosensitive channel of large conductance) and MscS (mechanosensitive channel of small conductance) which open upon activation by mechanical forces quickly releasing osmolytes from bacterial cells, when they swell due to a hypoosmotic environment leading to an increase in turgor pressure. By responding to membrane tension resulting from the increase in turgor, MscL and MscS protect the cells from bursting when they become exposed to hypoosmotic environments (Cox et al., 2018; Hamill & Martinac, 2001; Martinac, 2011).

To accurately define an MS ion channel as a truly mechanically activated channel, its activity must be controlled across its entire dynamic range, from the fully closed state to the a fully open state, solely by mechanical force, regardless of the channel gating mechanism (Fig. 21.1). The forces to which these channels are exposed in living organisms are equivalent to membrane tensions of 10−15 pN/nm, which roughly corresponds to the average lytic tension of the lipid bilayer of cellular membranes (Bavi et al., 2014; Nomura et al., 2012). Worth noting is that besides the true MS channels, there are ion channels that can be modulated by mechanical stimuli, but their activity is primarily dependent on factors other than mechanical force, such as membrane potential or ligand concentration (Martinac & Cox, 2017).

Given that a large proportion of our knowledge on MS channels comes from "non-specialized" cells, the necessary and sufficient criteria that have generally been accepted as proof of the true physiologically relevant mechanosensitivity of an ion channel are (Ridone et al., 2019):

1. The ion channel is required for cellular mechanotransduction and its removal abolishes the mechanosensory response.
2. Structural mutations introduced into the channel alter its functional properties and consequently affect the mechanosensory cellular response.
3. Heterologous expression of the channel gives rise to robust mechanosensory responses.

It is, nevertheless, important to remember that although very useful for experimental characterization of MS channels, the above basic criteria cannot account for the biological complexities arising from the structural plasticity of many ion channel families and functional redundancy characteristic of many physiological systems, which can make it difficult to unequivocally determine the molecular origins of mechanosensitive currents encountered in a large variety of living cells.

The role in various physiological processes of many MS channels is well-established across all kingdoms of life. More recently, the medical importance of MS channels has been recognized given the role that abnormal MS channel activity plays in the pathophysiology of many diseases, which can collectively be called *mechanochannelopathies*, which will be dealt with in more detail in Section 4.

1.2 Ancient origins of mechanosensory transduction

For most of its 4.5 billion years, the Earth has been inhabited by a vast number of life forms, of which microbes have formed the largest group to date (Martinac & Kloda, 2003). Given the importance of water for all life forms, including microbes, coping with osmotic forces has been of utmost importance for survival of microbial cells. To prevent damage

FIGURE 21.1 Examples of inherently mechanically activated ion channels. MscL, Piezo1, and TREK-1 are activated and controlled by mechanical force over their whole dynamic range from the fully closed to the fully open state (Ridone et al., 2019).

resulting from sudden changes in environmental osmolarity microbes have, in most cases, been protected by a cell wall that could sustain the increase in turgor pressure of up to 30 atm, which is about 10 times larger than the pressure a car tire has to resist. Since the cell wall can sustain such large pressures only to some extent, microbes are also equipped with different types of MS channels, of which *E. coli*, for example, has seven of them belonging to the prokaryotic families of MscL- and MscS-like channels. It is worth noting that a difference in osmolarity of only a few milliosmols (e.g., 5 mOsm equals 0.13 atm) across a bacterial membrane is sufficient to fully activate both MscL and MscS channels (Martinac & Jackson, 1993). Consequently, from very early on, MS channels might have served as cellular osmoregulators that responded to changes in turgor pressure generated by small variations in environmental osmolarity as the concentration of solutes and water across microbial membranes changed. With the appearance of multicellular eukaryotic organisms, MS channels had to be either reinvented or redesigned and diversified given the great differences in their molecular structure. The existence of the currently known large variety of mechanical force sensors playing central roles in specialized forms of mechanosensory transduction, such as pollen germination in plants (Hamilton et al., 2015), proprioception in mammals (Proske & Gandevia, 2012; Woo et al., 2015), or mechanoelectrical feedback in the heart (Bode et al., 2001; Friedrich et al., 2012; Gottlieb et al., 2004), has very likely been driven by the evolutionary convergence toward the same or similar function in mechanotransduction processes.

2. A quest for the general mechanism of gating mechanosensitive channels by mechanical force

The exact mechanism(s) of MS channels gating are currently largely unknown due to the relatively young age of 30 odd years of this research area in mechanobiology/physiology. Despite the inherent complexities in mechanical force transduction in mammalian systems, there are two main gating paradigms that have generally been accepted to apply to gating of MS channels in both prokaryotic and eukaryotic cells. One of the paradigms is known as the *force-from-lipids* (Kung, 2005; Martinac et al., 1990; Teng et al., 2015), and the other one as the *force-from-filament* (Fig. 21.2) (Chalfie, 2009; Katta et al., 2015).

2.1 Force-from-lipids or force-from filament?

The force-from-lipids (FFL) gating paradigm implies that the mechanical force that activates the MS channels, is transmitted exclusively through the lipid bilayer, with no requirement for other cellular components (Cox et al., 2019). More specifically, according to this model, it is the lateral stretch force from the lipid membrane that ultimately opens MS channels, not the external tether nor the internal cytoskeleton. This model was proposed over 35 years ago in conjunction with studies on bacterial MS channels (Martinac et al., 1990). The historical development of the FFL model and the experimental evidence backing it have been recently reviewed (Martinac & Kung, 2022). Unlike bacterial cells, animal and human cells do not have a rigid cell wall, but they possess instead membrane invaginations in the form of ruffles, folds, and microvilli. These invaginations contribute excess membrane area, which, by unfolding upon stress, protects the cell membrane from excessive strain. In addition, the membrane of mammalian cells is supported by the extracellular matrix (ECM) as external scaffolding, whereas the cytoskeleton (CSK) strengthens the membrane from the intracellular side and its contraction and expansion locally regulate the mechanosensitivity of MS channels in animal and human cells. In sum, in the force-from-filament (FFF) paradigm, force transmission requires the ECM and the CSK (Cox et al., 2019).

For a long time, the prevailing belief was that only prokaryotic (bacterial and archaeal) MS channels were gated according to the FFL paradigm (Martinac, 2004; Sachs, 1988, 1991). This notion stemmed from the fact that bacteria possess

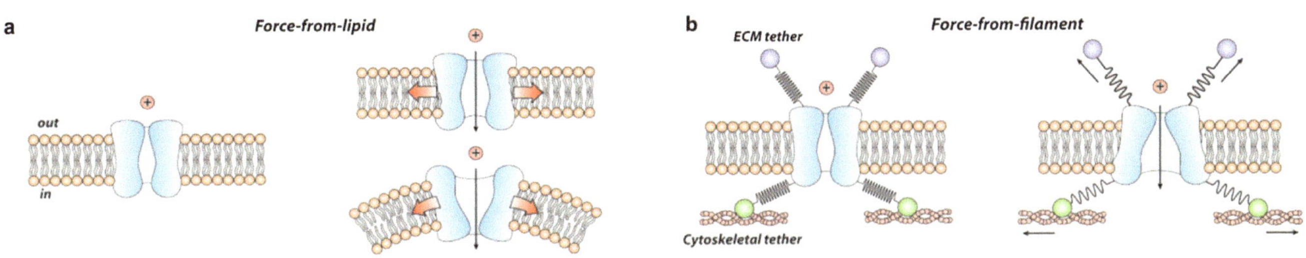

FIGURE 21.2 **Gating paradigms of MS channels.** The two main generally accepted gating paradigms of MS channels are defined according to the cell membrane components lipid bilayer (a) or ECM/CSK (b) transmitting the activation force directly to MS channels.

a rudimentary CSK compared to eukaryotes, from which the more complex mammalian CSK might have evolved from this simpler form. This further suggested that bacteria might rely more directly on membrane mechanics for mechanosensation (Barry & Gitai, 2011). Further, bacteria largely rely on the cell wall to protect their fragile cytoplasmic membrane against various stresses, including osmotic pressure. However, it has recently been shown that mammalian MS channels, notably K2P-type TREK-1, TREK-2, and TRAAK (Berrier et al., 2013; Brohawn, Campbell, & MacKinnon, 2014; Brohawn, Su, & MacKinnon, 2014) ion channels as well as Piezo1 channels (Cox, Bae, et al., 2016; Cox, Bavi, & Martinac, 2016; Syeda et al., 2016), are also gated through the FFL mechanism. This is in addition to eukaryotic members of the OSCA/TMEM63 family of ion channels (Murthy et al., 2018). In contrast, NOMPC channels, belonging to the TRPN family, which are involved in a large array of mechanosensitive responses in *Drosophila* (Yan et al., 2013; Zhang et al., 2013), have been shown to be activated by mechanical force pulling on microtubules via the ankyrin tether. Specifically, N-terminal ankyrin repeats act as a tether to gate the channel (Zhang et al., 2015). This result provides strong support for the FFF mechanism, given that the pulling force acting on ankyrin tethers is transferred directly onto the channel. However, this does not exclude the role that membrane lipids, as a complementary mechanism to the FFF gating mechanisms, may also play in TRPN mechanosensitivity (Cox et al., 2019).

2.2 Role of extracellular matrix and cytoskeleton in mechanosensitive channel gating

Together with MS channels, the CSK is another firmly established cellular mechanosensor. Although several mammalian ion channels including K2P-type (Berrier et al., 2013; Brohawn, Campbell, & MacKinnon, 2014; Brohawn, Su, & MacKinnon, 2014), Piezo1 (Cox, Bae, et al., 2016; Syeda et al., 2016), and NMDA (Kloda, Lua, et al., 2007) ion channels can be activated by tension developed in the lipid bilayer alone, other types of MS channels, including MEC-4/MEC-10 and TRP-type channels may, in addition to the lipid bilayer, require cytoskeletal and/or ECM linkages for their proper function, as has been shown for the TRPN ion channels (Zhang et al., 2015). This indicates that mechanotransduction in mammalian cells may involve different mechanosensory mechanisms relying on the CSK and the ECM as alternative or additional force-transducing cellular components (Hudspeth, 1989). Given that both ECM and CSK may be tethered to MS channels and may contribute to the channel activation by mechanical force, this model of MS channel gating has been named the "dual-tether model" and was first proposed for the hair cell transduction in vertebrates (Chalfie, 2009; Martinac, 2013). In the context of the role that CSK and ECM may play in the mechanism of the MS channel activation by mechanical force two structural features intrinsic to the mammalian cell membranes, historically referred to as "*lipid rafts*" and *cellular force foci*, such as focal adhesions, illustrate the complexity of the ways living cells have learned to cope with various mechanical stimuli acting on them throughout evolution.

"Lipid rafts" are historically referred to as membrane microdomains composed of combinations of cholesterol, glycosphingolipids, and protein receptors. Compared with the surrounding lipid bilayer, they are more ordered and tightly packed (Pralle et al., 2000; Simons & Toomre, 2000). Such bilayer platforms could serve not only as the gating linker in hair cells but also to play a role of the "spring" transmitting the mechanical force to MS channels in general (Anishkin & Kung, 2013). Interestingly, cholesterol-enriched platforms are universally present in cellular force foci, which can be defined as points of force reception characteristic of specialized forms of mechanotransduction, including hearing and touch, as well as of more general cadherin foci and integrin foci (Anishkin & Kung, 2013). Integrins, for example, form adhesions on the surface of mammalian cell membranes allowing mechanical stimuli to be focused on different CSK components, which directly or indirectly transmit these stimuli to membrane proteins (Wang et al., 1993). To fulfill this role, CSK is organized as "tensegrity" structures prestressing the cells to keep the cell shape stabilized by the network of opposing tension and compression components (Ingber, 2006) (see Chapter 33 by Sackman). Thus, lipid rafts could provide a hybrid mechanism between the FFL and FFF mechanisms of MS channel gating by mechanical force, which seems to make the "either-or-distinction" between the FFL and FFF paradigms superfluous. As demonstrated for Piezo1 channels, which are gated according to the FFL paradigm (Syeda et al., 2016), the important function of the CSK and ECM is, therefore, to mechanically protect the lipid bilayer by absorbing mechanical stresses (Cox, Bae, et al., 2016) and possibly also to alter the forces that reach the channel locality (Hamill & Martinac, 2001; Young et al., 2023).

2.3 Transbilayer pressure profile and gating of MS channels

Given that the distinction between the FFL and FFF mechanism of gating MS channels by mechanical force seems to be a false dichotomy, confusing rather than clarifying the molecular principles of mechanotransduction, the question about a general mechanism that would reduce the two paradigms to a single basic principle of MS channel mechanosensitivity remains to be addressed and answered. We suggest the *transbilayer pressure profile* (Fig. 21.3) as the possible common

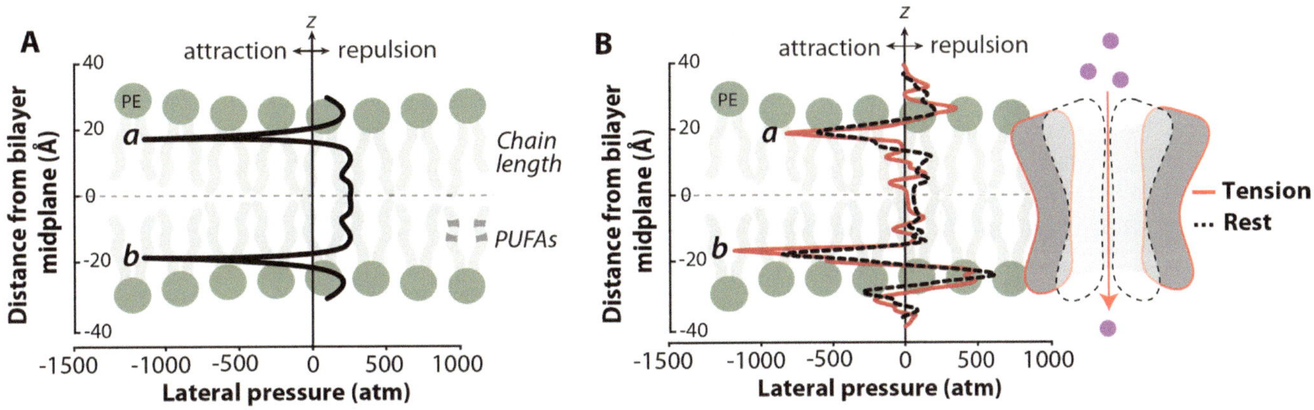

FIGURE 21.3 **Transbilayer pressure profile.** (A) The transbilayer pressure profile is largely inhomogeneous across the bilayer thickness, which can be modified by lipid properties. An idealized symmetrical transbilayer pressure profile is shown with characteristic negative peaks at the water–lipid interface and repulsive positive peaks in the headgroup and tail region. z indicates the bilayer thickness direction. (B) In the presence of a membrane protein, the pressure profile in a symmetrical lipid bilayer become noticeably asymmetric. Peak a and peak b represent the rise in the pressure profile at the lipid–solvent interface (for more details see (Bavi et al., 2016; Cox, Bavi, & Martinac, 2016; Ridone et al., 2019)).

denominator of the two gating mechanisms to reduce them to the FFL paradigm only, which thus would present the basic principle underlying mechanosensory transduction processes mediated by MS channels.

Transbilayer pressure profile (Fig. 21.3A) refers to a strong anisotropic internal stress in the lipid bilayer resulting from the amphipathic nature of the lipids, which drives the bilayer self-assembly in aqueous environment by minimizing the exposure of the phospholipid aliphatic chains to water (Cantor, 1999). The pressure inside the lipid bilayer is about 300 atm, whereas around the head groups, it can be up to 1000 atm (Gullingsrud & Schulten, 2004; Ridone et al., 2019). These stress/pressure distributions within the bilayer have been mostly estimated using a variety of computational approaches. Recently, however, the transbilayer pressure profile of lipid bilayers made of mono- and polyunsaturated phospholipids was experimentally determined using NMR spectroscopy, which largely confirmed the results from the computational studies. Concomitantly with patch-clamp experiments, the NMR results demonstrated that changes in transbilayer pressure profile were directly related to modulation of MS channel gating (Ridone et al., 2018) in agreement with previous findings showing that the stress heterogeneity along the bilayer thickness alters the conformation of membrane proteins (Cantor, 1998). Reciprocally, membrane proteins, including MS channels, can redistribute the transbilayer pressure profile causing a noticeable asymmetry in the pressure profile characterized by a reduction of the pressure peaks in the tails and significant asymmetry between the peaks at the lipid–solvent interface (Fig. 21.3B) (Cantor, 1999; Lundbaek et al., 2010). Thus, when an MS channel becomes displaced from the bilayer by pulling on a tether attached to it, such as TRPN, for example (Zhang et al., 2015), it is likely that the change in the pressure profile asymmetry resulting from the channel movement relative to the bilayer can lead to the channel opening (Kung, 2005). A similar conclusion has been reached for the TREK-2 channel, a mammalian two pore domain (K2P) K^+ channel important for mechanical nociception in sensory neurons (Clausen et al., 2017). In addition, transbilayer pressure profile is highly susceptible to physical and chemical stimuli (Cantor, 1999), which seems to explain the effect of the insertion of amphipathic compounds into a single leaflet of the lipid bilayer, including conical lipids such as lysophosphatidylcholine (LPC) or phosphatidic acid (PA), on the activity of various MS channels, including Piezo1 (Maingret et al., 2000; Perozo, Kloda, et al., 2002; Ridone et al., 2019; Syeda et al., 2016). Together, both the movement of an MS channel caused by pulling a tether and insertion of amphipaths into one leaflet of the lipid bilayer can activate MS channels via changes of the transbilayer pressure profile asymmetry. Thus, the evidence supports the FFL paradigm of MS channel gating by mechanical force as a general evolutionary conserved physicochemical principle underlying the MS channel-mediated mechanosensory transduction in cells of all living organisms. Potentially, the FFL principle may apply to other mechanosensitive membrane proteins, including GPCRs (Storch et al., 2012) and phospholipases (Lehtonen & Kinnunen, 1995). To delve deeper into the FFL paradigm, the transbilayer profile, and the mechanisms of MS gating, the reader is encouraged to explore various sources that offer in-depth reviews and analyses (Bavi et al., 2023; Martinac et al., 2018; Martinac & Kung, 2022; Ridone et al., 2019).

3. Diversity of mechanosensitive channels

MS channels exist in cells of organisms from all three kingdoms of life on Earth: Bacteria, Archaea, and Eukarya (Martinac & Kloda, 2003). Based on their evolutionary provenance and ubiquity, MS channels can largely be divided into two major groups: prokaryotic and eukaryotic ion channels, respectively (Martinac, 2004).

3.1 MS channels of prokaryotes

Bacteria and Archaea represent the two evolutionary domains of prokaryotic cells (Pace, 1997; Woese, 1994). In the plasma membranes of these cells only two types of MS channels belonging to the MscL- and MscS-like families have been found (Fig. 21.4). The phylogenetic distribution analysis of the two families of prokaryotic MS channels indicates that the *mscL* and *mscS* genes have followed separate evolutionary pathways. Whereas MscL-like channels are mainly found in bacteria and archaea, MscS-like channels are present in all types of cell-walled organisms, including bacteria, archaea, plants, and fungi (Martinac et al., 2013; Pivetti et al., 2003).

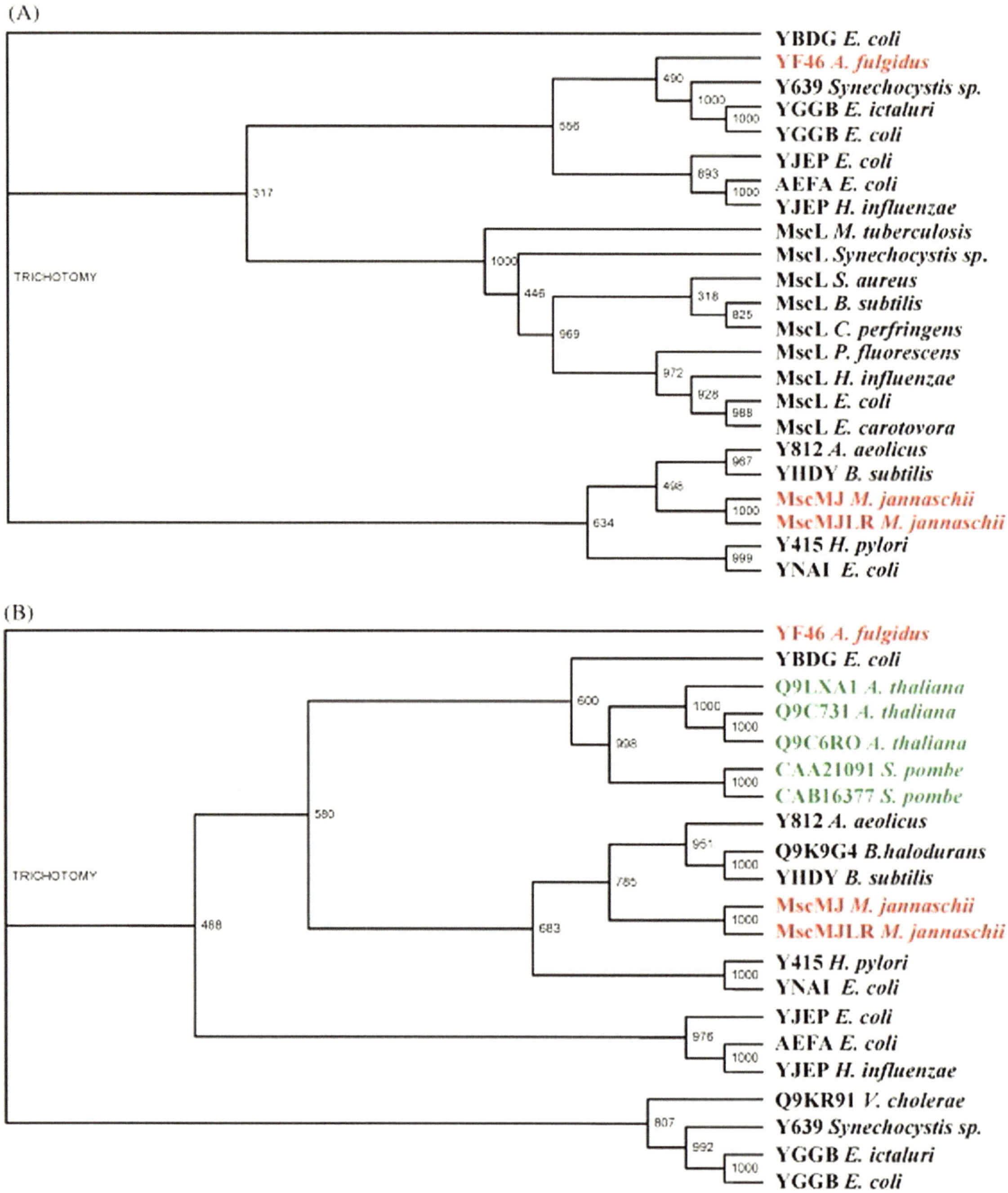

FIGURE 21.4 **Families of prokaryotic MS channels.** (A) Family of MscL-like bacterial channels. (B) Family of MscS-like channel homologues. Bacterial MS channels are shown in black, archaeal channels are shown in red, and homologues found in fungi and plants are shown in green. *Reproduced with permission from Martinac and Kloda (2003).*

EcMscL, the MscL channel of *E. coli*, was the first MS channel to be cloned and identified at the molecular level (Sukharev et al., 1994). It is also the first MS channel, whose 3D structure in the closed state of TbMscL, a homologue of *M. tuberculosis*, was solved using X-ray crystallography (Chang et al., 1998). The 3D structure revealed that MscL forms a homopentameric channel made of monomers consisting of two transmembrane domains, TM1 and TM2, cytoplasmically located N- and C-terminal domains, and a loop facing the periplasm (Steinbacher et al., 2007). The open state of EcMscL was solved by electron paramagnetic resonance spectroscopy, EPR (Perozo, Cortes et al., 2002) and FRET (Corry et al., 2010; Wang et al., 2014) spectroscopy, revealing an open pore of ~28 Å in diameter corresponding to the large channel conductance of 3 nS (Cruickshank et al., 1997; Häse et al., 1995; Sukharev et al., 1994). MscS of *E. coli* (EcMscS), was identified at the molecular level a few years later (Levina et al., 1999). Its 3D crystal structure, solved shortly after its cloning, revealed a homoheptameric channel in an intermediate nonconducting conformation. A single monomer of EcMscS consists of three transmembrane helices, TM1, TM2, and TM3, the N-terminus located at the periplasmic side, and a large C-terminal domain located in the cytoplasm (Bass et al., 2002). The structure of the A106V mutant channel is currently the only 3D structure representing a partly open conformation (Pliotas et al., 2012), which conducts ions below the standard unitary conductance of EcMscS of ~1 nS (Martinac et al., 1987; Sukharev et al., 1993).

Both MscL and MscS function as osmosensors of membrane tension in prokaryotic cells, which protect the cells from sudden changes in environmental osmolarity, preventing membrane rupture upon exposure to hypo-osmotic conditions (Levina et al., 1999; Wood, 2015). They do it gradually given that MscS is more sensitive to membrane tension than MscL corresponding to their half activation (open probability $P_o = 0.5$) by membrane tension of ~6 pN/nm and ~12 pN/nm, respectively (Nomura et al., 2012). MscS and MscL can compensate for each other's absence since both of them have to be knocked out to make bacteria sensitive to osmotic down-shock. In addition to MscL and MscS, other paralogues of MscS, including *ybdG*, *ybiO*, *ynaI*, and *yjeP* (Booth & Blount, 2012; Edwards et al., 2012), seem to contribute to bacterial osmoregulation in a more subtle way depending on the magnitude and rate of the hypoosmotic shock (Boer et al., 2011). As the simplest model for mechanosensory transduction MscL and MscS are gated by mechanical force transmitted through the lipid bilayer and present the best example of channels following the FFL paradigm of MS channel gating.

3.2 MS channels of eukaryotes

A great diversity of MS ion channels is found in eukaryotic cells. Touch, hearing, and proprioception, as well as control of cellular turgor and blood flow regulation, are examples of the role that MS channels may play in mechanosensory transduction in eukaryotic organisms (Hamill & Martinac, 2001). There is quite a large literature available on electrophysiology and functional characterization of eukaryotic MS ion channels, however, only recently the 3D structure of several members of this diverse class of ion channels has been obtained, including TREK-2 (Dong et al., 2015) and TRAAK (Brohawn et al., 2012) of the two-pore-domain (2P-domain) family of weakly inward-rectifying K^+ channels, and Piezo1, a member of the recently discovered distinct family of MS channels (Coste et al., 2010; Guo & MacKinnon, 2017; Saotome et al., 2018; Zhao et al., 2018). The two other major types of MS channels in eukaryotes are either members of the superfamily of TRP-like ion channels (Christensen & Corey, 2007) that are relatively nonselectively permeable to cations, including Na^+, Ca^{2+}, and Mg^{2+}, or the DEG/ENaC family of amiloride sensitive, Na^+-selective ion channels (Chalfie, 2009). The trimeric 3D structure of the acid-sensing ion channel ASIC, a nonmechanosensitive member of the DEG/ENaC family, is the only information available on their molecular architecture (Jasti et al., 2007). Advances in cryoelectron microscopy (cryo-EM) have resulted in structural information related to TRPV1 (Liao et al., 2013), TRPV2 (Huynh et al., 2016; Zubcevic et al., 2016), and TRPA1 (Paulsen et al., 2015). These channels are heat sensitive and have also been reported to function as mechanosensors, although the evidence is not as strong as for other MS channels (Zheng, 2013). Moreover, mammalian TRP-like channels cannot be activated by membrane stretch (Cox et al., 2024; Nikolaev et al., 2019). In addition to these four major groups of eukaryotic MS channels, there are a few other ion channels worth mentioning here briefly.

TMC1 and TMC2 are transmembrane channel-like proteins 1 and 2 belonging to a family of eight membrane proteins. They are required for normal auditory function in mammalian species (Kim & Fettiplace, 2013; Pan et al., 2013; Pan et al., 2018). Given the current information about the biophysics of the auditory MS channel, including the large conductance of 80–320 pS and a large conformational change of 4 nm the channel is undergoing during its opening (Fettiplace & Kim, 2014), both TMC1 and TMC2 satisfy these requirements. However, given that no mechanosensitive currents could be recorded from heterologously expressed mammalian TMCs, it is at present unclear that they represent the long sought-after auditory MS channel. However, a recent study provides evidence that human TMC1/2 are indeed pore-forming MS channels, supporting their roles as mechanotransduction channels in hair cells (Fu et al., 2025).

The NMDA glutamate receptor is another mechanosensitive channel, which can be qualified as an ion channel modulated by mechanical force (Belin et al., 2022; Johnson et al., 2019; Kloda, Lua, et al., 2007, Kloda, Martinac, & Adams, 2007; Parnas et al., 2009), although it is primarily a ligand-gated channel (Bliss & Collingridge, 1993) gated by glutamate and glycine, as discussed in detail in Chapter 23 by Tovar and Westbrook in this volume. NMDA receptor/channels have long been implicated in learning and memory processes (Dupuis et al., 2023; Johnson et al., 2019; Luscher & Malenka, 2012). NMDA receptor channels form heterotetramers of NR1 subunit, which is a product of a single gene encoding eight splice variants, and four distinct NR2 subunits (NR2A–NR2D) or, in some cases of two NR3 subunits encoded by discrete genes (Dingledine et al., 1999; Paoletti & Neyton, 2007). The crystal structure of the heterotetrameric GluNR1/GluNR2B NMDA receptor channel has been solved at a resolution of 4 Å (Karakas & Furukawa, 2014).

4. Mechanosensitive channel function/dysfunction in physiology and pathophysiology of mechanotransduction

The importance the malfunctioning MS ion channels as contributors to the pathology of a variety of hereditary diseases of previously unknown etiology has increasingly caught the attention in medical reserach. This includes a variety of mechanochannelopathies where point mutations in an MS channel can be traced as the causative entity, such as hereditary Xerocytosis (also known as dehydrated stomatocytosis), a familial anemia (Alper and Gottlieb, 2017, Chapter 4; Bae et al., 2013; Zarychanski et al., 2012), arthrogryposis, a congenital joint contractures in different areas of the body (Alper and Gottlieb, 2017, Chapter 4; Coste et al., 2013) and autosomal-dominant polycystic kidney disease (ADPKD), a prevalent genetic disorder that occurs in approximately 0.1%–0.2% of the population and may be caused by abnormal Ca^{2+} signaling by polycystins acting as Ca^{2+}-regulated/Ca^{2+}-permeable MS ion channels (Chen et al., 1999). Furthermore, there are multifactorial conditions where MS channel dysfunction is a contributing factor, for example, in pain (Ferrari et al., 2015) and cardiac hypertrophy (Seo et al., 2014; Yu et al., 2022). Cardiac hypertrophy and atrial fibrillation, for example, may, among other causes, also originate from malfunctioning TRPC6 and/or TRPC3 channels (Seo et al., 2014) or TRPM4 channels (Constantine et al., 2016; Demion et al., 2014; Guo et al., 2021), respectively. Another multifactorial disease in which MS channels play a role is muscular dystrophy, which is a collective name for a group of muscular degenerations that cause progressive weakness and loss of muscle mass resulting from abnormal dystrophin genes (Franco & Lansman, 1990). Although here it is important to note that it is unklikely that most TRP channels represent primary mechanosensors. As the research on MS ion channels continues to progress and the new knowledge of MS channel structure and function continues to grow, new MS channel proteins may be discovered, and new links between malfunctioning MS channels and related mechanopathophysiology will be found. As our awareness of diseases linked to malfunctioning MS channels increases MS channels will become targets for the development of new drugs to treat a variety of mechanochannelopathies.

5. Pharmacology of mechanosensitive channels

At present, there are only a few agents and compounds generally used for applications in MS channel research. These are the lanthanides Gd^{3+} and La^{3+}, the aminoglycoside antibiotics such as streptomycin and gentamicin, and the GsMTx-4 peptide isolated from the toxin of the spider *Grammostola spatulata*, which are known to block most of the known MS channels (Hamill & McBride, 1996). Another group of compounds, such as amphipaths chlorpromazine and LPC, have been shown to activate both prokaryotic and eukaryotic MS channels (Hamill & McBride, 1996). In addition to these "*typical*" MS channel blockers and activators, there are more specific compounds affecting only a certain type of MS channels. For example, mibefradil and fluoxetine block and general anesthetics activate TREK-1 K2P channels (Heurteaux et al., 2004; Honore et al., 2006), whereas ruthenium red and Yoda-1 block and activate Piezo1, respectively (Coste et al., 2012; Syeda et al., 2015). A comprehensive list of blocking and activating agents of various types of MS channels can be found in Martinac and Cox (2017).

6. Conclusions

In this chapter, we attempted to illustrate what makes MS channels unique and why it is important to study them and to understand their diversity and mode(s) of operation. From bacterial osmoregulation to the senses of touch and hearing, these channels are the molecules that provide information about the outside world through the direct contact cells have with their surroundings. Sometimes this information may not be exactly transduced into proper intracellular signaling due to aberrant MS channels, as is apparently the case in hereditary mechanopathologies we briefly reported here. Our knowledge of the structure and function of MS channels comes to the rescue by helping us understand the correct etiology of these pathologies. This emphasizes the need for more knowledge and a better understanding of the MS channel diversity and their equally diverse structure and function, which can only be achieved through continuous discoveries and painstaking research. Future developments that may open up new avenues for the treatment of a whole host of mechanochannelopathies, which should enrich our knowledge and improve human health conditions.

References

Alper, S. L. (2017). Piezo channels. In P. A. Gottlieb (Ed.), *Current topics in membranes* (Vol. 79). Academic Press, Elsevier Inc.

Anishkin, A., & Kung, C. (2013). Stiffened lipid platforms at molecular force foci. *Proceedings of the National Academy of Sciences of the United States of America, 110*, 4886–4892. https://doi.org/10.1073/pnas.1302018110

Bae, C., Gnanasambandam, R., Nicolai, C., Sachs, F., & Gottlieb, P. A. (2013). Xerocytosis is caused by mutations that alter the kinetics of the mechanosensitive channel PIEZO1. *Proceedings of the National Academy of Sciences of the United States of America, 110*, E1162–E1168. https://doi.org/10.1073/pnas.1219777110

Barry, R. M., & Gitai, Z. (2011). Self-assembling enzymes and the origins of the cytoskeleton. *Current Opinion in Microbiology, 14*, 704–711. https://doi.org/10.1016/j.mib.2011.09.015

Bass, R. B., Strop, P., Barclay, M., & Rees, D. C. (2002). Crystal structure of *Escherichia coli* MscS, a voltage-modulated and mechanosensitive channel. *Science, 298*, 1582–1587. https://doi.org/10.1126/science.1077945

Bavi, N., et al. (2014). Biophysical implications of lipid bilayer rheometry for mechanosensitive channels. *Proceedings of the National Academy of Sciences of the United States of America, 111*, 13864–13869. https://doi.org/10.1073/pnas.1409011111

Bavi, N., et al. (2016). The role of MscL amphipathic N terminus indicates a blueprint for bilayer-mediated gating of mechanosensitive channels. *Nature Communications, 7*. https://doi.org/10.1038/ncomms11984

Bavi, N., Cox, C. D., Nikolaev, Y. A., & Martinac, B. (2023). Molecular insights into the force-from-lipids gating of mechanosensitive channels. *Current Opinion in Physiology, 36*, 100706. https://doi.org/10.1016/j.cophys.2023.100706

Belin, S., Maki, B. A., Catlin, J., Rein, B. A., & Popescu, G. K. (2022). Membrane stretch gates NMDA receptors. *The Journal of Neuroscience, 42*(29), 5672–5680.

Berrier, C., et al. (2013). The purified mechanosensitive channel TREK-1 is directly sensitive to membrane tension. *Journal of Biological Chemistry, 288*, 27307–27314. https://doi.org/10.1074/jbc.M113.478321

Bliss, T. V., & Collingridge, G. L. (1993). A synaptic model of memory: Long-term potentiation in the hippocampus. *Nature, 361*, 31–39. https://doi.org/10.1038/361031a0

Bode, F., Sachs, F., & Franz, M. R. (2001). Tarantula peptide inhibits atrial fibrillation. *Nature, 409*, 35–36. https://doi.org/10.1038/35051165

Boer, M., Anishkin, A., & Sukharev, S. (2011). Adaptive MscS gating in the osmotic permeability response in *E. coli*: The question of time. *Biochemistry, 50*, 4087–4096. https://doi.org/10.1021/bi1019435

Booth, I. R., & Blount, P. (2012). The MscS and MscL families of mechanosensitive channels act as microbial emergency release valves. *Journal of Bacteriology, 194*, 4802–4809. https://doi.org/10.1128/JB.00576-12

Brohawn, S. G., Campbell, E. B., & MacKinnon, R. (2014). Physical mechanism for gating and mechanosensitivity of the human TRAAK K+ channel. *Nature, 516*, 126–130. https://doi.org/10.1038/nature14013

Brohawn, S. G., del Marmol, J., & MacKinnon, R. (2012). Crystal structure of the human K2P TRAAK, a lipid- and mechano-sensitive K+ ion channel. *Science, 335*, 436–441. https://doi.org/10.1126/science.1213808

Brohawn, S. G., Su, Z., & MacKinnon, R. (2014). Mechanosensitivity is mediated directly by the lipid membrane in TRAAK and TREK1 K+ channels. *Proceedings of the National Academy of Sciences of the United States of America, 111*, 3614–3619. https://doi.org/10.1073/pnas.1320768111

Cantor, R. S. (1999). Lipid composition and the lateral pressure profile in bilayers. *Biophysical Journal, 76*, 2625–2639. https://doi.org/10.1016/S0006-3495(99)77415-1

Cantor, R. S. (1998). The lateral pressure profile in membranes: A physical mechanism of general anesthesia. *Toxicology Letters, 100–101*, 451–458.

Chalfie, M. (2009). Neurosensory mechanotransduction. *Nature Reviews Molecular Cell Biology, 10*, 44–52. https://doi.org/10.1038/nrm2595

Chang, G., Spencer, R. H., Lee, A. T., Barclay, M. T., & Rees, D. C. (1998). Structure of the MscL homolog from *Mycobacterium tuberculosis*: A gated mechanosensitive ion channel. *Science, 282*, 2220–2226.

Chen, X. Z., et al. (1999). Polycystin-L is a calcium-regulated cation channel permeable to calcium ions. *Nature, 401*, 383–386. https://doi.org/10.1038/43907

Christensen, A. P., & Corey, D. P. (2007). TRP channels in mechanosensation: Direct or indirect activation? *Nature Reviews Neuroscience, 8*, 510–521. https://doi.org/10.1038/nrn2149

Clausen, M. V., Jarerattanachat, V., Carpenter, E. P., Sansom, M. S. P., & Tucker, S. J. (2017). Asymmetric mechanosensitivity in a eukaryotic ion channel. *Proceedings of the National Academy of Sciences of the United States of America, 114*(40), E8343–E8351.

Constantine, M., et al. (2016). Heterologously-expressed and liposome-reconstituted human transient receptor potential melastatin 4 channel (TRPM4) is a functional tetramer. *Scientific Reports, 6*, Article 19352. https://doi.org/10.1038/srep19352

Corry, B., et al. (2010). An improved open-channel structure of MscL determined from FRET confocal microscopy and simulation. *The Journal of General Physiology, 136*, 483–494. https://doi.org/10.1085/jgp.200910376

Coste, B., et al. (2013). Gain-of-function mutations in the mechanically activated ion channel PIEZO2 cause a subtype of Distal Arthrogryposis. *Proceedings of the National Academy of Sciences of the United States of America, 110*, 4667–4672. https://doi.org/10.1073/pnas.1221400110

Coste, B., et al. (2012). Piezo proteins are pore-forming subunits of mechanically activated channels. *Nature, 483*, 176–181. https://doi.org/10.1038/nature10812

Coste, B., et al. (2010). Piezo1 and Piezo2 are essential components of distinct mechanically activated cation channels. *Science, 330*, 55–60. https://doi.org/10.1126/science.1193270

Cox, C. D., Bavi, N., & Martinac, B. (2016). In *Current topics in membranes*. Academic Press.

Cox, C. D., et al. (2016). Removal of the mechanoprotective influence of the cytoskeleton reveals PIEZO1 is gated by bilayer tension. *Nature Communications, 7*, Article 10366. https://doi.org/10.1038/ncomms10366

Cox, C. D., Bavi, N., & Martinac, B. (2018). Bacterial mechanosensors. *Annual Review of Physiology, 80*, 71–93.

Cox, C. D., Bavi, N., & Martinac, B. (2019). Biophysical principles of ion-channel-mediated mechanosensory transduction. *Cell Reports, 29*(1), 1–12. https://doi.org/10.1016/j.celrep.2019.08.075

Cox, C. D., Poole, K., & Martinac, B. (2024). Re-evaluating TRP channel mechanosensitivity. *Trends in Biochemical Sciences*. PMID: 38851904.

Cruickshank, C. C., Minchin, R. F., Le Dain, A. C., & Martinac, B. (1997). Estimation of the pore size of the large-conductance mechanosensitive ion channel of *Escherichia coli*. *Biophysical Journal, 73*, 1925–1931. https://doi.org/10.1016/S0006-3495(97)78223-7

Demion, M., et al. (2014). Trpm4 gene invalidation leads to cardiac hypertrophy and electrophysiological alterations. *PLoS One, 9*, Article e115256. https://doi.org/10.1371/journal.pone.0115256

Dingledine, R., Borges, K., Bowie, D., & Traynelis, S. F. (1999). The glutamate receptor ion channels. *Pharmacological Reviews, 51*, 7–61.

Dong, Y. Y., et al. (2015). K2P channel gating mechanisms revealed by structures of TREK-2 and a complex with Prozac. *Science, 347*, 1256–1259. https://doi.org/10.1126/science.1261512

Dupuis, J. P., Nicole, O., & Groc, L. (2023). NMDA receptor functions in health and disease: Old actor, new dimensions. *Neuron, 111*(15), 2312–2328. https://doi.org/10.1016/j.neuron.2023.05.002

Edwards, M. D., et al. (2012). Characterization of three novel mechanosensitive channel activities in *Escherichia coli*. *Channels, 6*, 272–281. https://doi.org/10.4161/chan.20998

Ferrari, L. F., Bogen, O., Green, P., & Levine, J. D. (2015). Contribution of Piezo2 to endothelium-dependent pain. *Molecular Pain, 11*. https://doi.org/10.1186/s12990-015-0068-4

Fettiplace, R., & Kim, K. X. (2014). The physiology of mechanoelectrical transduction channels in hearing. *Physiological Reviews, 94*, 951–986. https://doi.org/10.1152/physrev.00038.2013

Franco, A., Jr., & Lansman, J. B. (1990). Calcium entry through stretch-inactivated ion channels in mdx myotubes. *Nature, 344*, 670–673. https://doi.org/10.1038/344670a0

Friedrich, O., Wagner, S., Battle, A. R., Schurmann, S., & Martinac, B. (2012). Mechano-regulation of the beating heart at the cellular level–mechanosensitive channels in normal and diseased heart. *Progress in Biophysics and Molecular Biology, 110*, 226–238. https://doi.org/10.1016/j.pbiomolbio.2012.08.009

Fu, S., Pan, S., Lu, M, Dong, J., & Yan, Z. (2025). Human TMC1 and TMC2 are mechanically gated ion channels. *Neuron, 113*(3), 411–425. https://doi.org/10.1016/j.neuron.2024.11.009

Ge, J., et al. (2015). Architecture of the mammalian mechanosensitive Piezo1 channel. *Nature, 527*, 64–69. https://doi.org/10.1038/nature15247

Gillespie, P. G., & Walker, R. G. (2001). Molecular basis of mechanosensory transduction. *Nature, 413*, 194–202. https://doi.org/10.1038/35093011

Gottlieb, P. A., Suchyna, T. M., Ostrow, L. W., & Sachs, F. (2004). Mechanosensitive ion channels as drug targets. *Current Drug Targets: CNS Neurological Disorders, 3*, 287–295.

Gullingsrud, J., & Schulten, K. (2004). Lipid bilayer pressure profiles and mechanosensitive channel gating. *Biophysical Journal, 86*, 3496–3509. https://doi.org/10.1529/biophysj.103.034322

Guo, Y. R., & MacKinnon, R. (2017). Structure-based membrane dome mechanism for Piezo mechanosensitivity. *Elife, 6*. https://doi.org/10.7554/eLife.33660

Han, Y., Zhou, Z., Jin, R., Dai, F., Ge, Y., Ju, X., Ma, X., He, S., Yuan, L., Wang, Y., Yang, W., Yue, X., Chen, Z., Sun, Y., Corry, B., Cox, C. D., & Zhang, Y. (2024). Mechanical activation opens a lipid-lined pore in OSCA ion channels. *Nature, 628*(8009), 910–918. https://doi.org/10.1038/s41586-024-07256-9. PMID: 38570680.

Häse, C. C., Le Dain, A. C., & Martinac, B. (1995). Purification and functional reconstitution of the recombinant large mechanosensitive ion channel (MscL) of *Escherichia coli*. *Journal of Biological Chemistry, 270*, 18329–18334.

Guo, Y., Yu, Z.-Y., Wu, J., Gong, H., Kesteven, S., Iismaa, S. E., Chan, A. Y., Holman, S., Pinto, S., Pironet, A., Cox, C. D., Graham, R. M., Vennekens, R., Feneley, M. P., & Martinac, B. (2021). The Ca^{2+}-activated cation channel TRPM4 is a positive regulator of pressure overload-induced cardiac hypertrophy. *eLife, 10*, Article e66582. https://doi.org/10.7554/eLife.66582

Hamill, O. P., & Martinac, B. (2001). Molecular basis of mechanotransduction in living cells. *Physiological Reviews, 81*, 685–740.

Hamill, O. P., & McBride, D. W., Jr. (1996). The pharmacology of mechanogated membrane ion channels. *Pharmacological Reviews, 48*, 231–252.

Hamilton, E. S., et al. (2015). Mechanosensitive channel MSL8 regulates osmotic forces during pollen hydration and germination. *Science, 350*, 438–441. https://doi.org/10.1126/science.aac6014

Heurteaux, C., et al. (2004). TREK-1, a K+ channel involved in neuroprotection and general anesthesia. *The EMBO Journal, 23*, 2684–2695. https://doi.org/10.1038/sj.emboj.7600234

Honore, E., Martins, J. R., Penton, D., Patel, A., & Demolombe, S. (2015). The Piezo mechanosensitive ion channels: May the force Be with you. *Reviews of Physiology, Biochemistry & Pharmacology, 169*, 25–41. https://doi.org/10.1007/112_2015_26

Honore, E., Patel, A. J., Chemin, J., Suchyna, T., & Sachs, F. (2006). Desensitization of mechano-gated K2P channels. *Proceedings of the National Academy of Sciences of the United States of America, 103*, 6859–6864. https://doi.org/10.1073/pnas.0600463103

Hudspeth, A. J. (1989). How the ear's works work. *Nature, 341*, 397–404. https://doi.org/10.1038/341397a0

Huynh, K. W., et al. (2016). Structure of the full-length TRPV2 channel by cryo-EM. *Nature Communications, 7*, Article 11130. https://doi.org/10.1038/ncomms11130

Ingber, D. E. (2006). Cellular mechanotransduction: Putting all the pieces together again. *The FASEB Journal: Official Publication of the Federation of American Societies for Experimental Biology, 20*, 811–827. https://doi.org/10.1096/fj.05-5424rev

Jasti, J., Furukawa, H., Gonzales, E. B., & Gouaux, E. (2007). Structure of acid-sensing ion channel 1 at 1.9 A resolution and low pH. *Nature, 449*, 316–323. https://doi.org/10.1038/nature06163

Johnson, L. R., Battle, A. R., & Martinac, B. (2019). Remembering mechanosensitivity of NMDA receptors. *Frontiers in Cellular Neuroscience, 13*(533). https://doi.org/10.3389/fncel.2019.00533

Karakas, E., & Furukawa, H. (2014). Crystal structure of a heterotetrameric NMDA receptor ion channel. *Science, 344*, 992–997. https://doi.org/10.1126/science.1251915

Katta, S., Krieg, M., & Goodman, M. B. (2015). Feeling force: Physical and physiological principles enabling sensory mechanotransduction. *Annual Review of Cell and Developmental Biology, 31*, 347–371. https://doi.org/10.1146/annurev-cellbio-100913-013426

Kim, K. X., & Fettiplace, R. (2013). Developmental changes in the cochlear hair cell mechanotransducer channel and their regulation by transmembrane channel-like proteins. *The Journal of General Physiology, 141*, 141–148. https://doi.org/10.1085/jgp.201210913

Kloda, A., Lua, L., Hall, R., Adams, D. J., & Martinac, B. (2007). Liposome reconstitution and modulation of recombinant N-methyl-D-aspartate receptor channels by membrane stretch. *Proceedings of the National Academy of Sciences of the United States of America, 104*, 1540–1545. https://doi.org/10.1073/pnas.0609649104

Kloda, A., Martinac, B., & Adams, D. J. (2007). Polymodal regulation of NMDA receptor channels. *Channels, 1*, 334–343.

Kocer, A. (2015). Mechanisms of mechanosensing - mechanosensitive channels, function and re-engineering. *Current Opinion in Chemical Biology, 29*, 120–127. https://doi.org/10.1016/j.cbpa.2015.10.006

Kung, C. (2005). A possible unifying principle for mechanosensation. *Nature, 436*, 647–654. https://doi.org/10.1038/nature03896

Lehtonen, J. Y., & Kinnunen, P. K. (1995). Phospholipase A2 as a mechanosensor. *Biophysical Journal, 68*, 1888–1894. https://doi.org/10.1016/S0006-3495(95)80366-8

Levina, N., et al. (1999). Protection of *Escherichia coli* cells against extreme turgor by activation of MscS and MscL mechanosensitive channels: Identification of genes required for MscS activity. *The EMBO Journal, 18*, 1730–1737. https://doi.org/10.1093/emboj/18.7.1730

Liao, M., Cao, E., Julius, D., & Cheng, Y. (2013). Structure of the TRPV1 ion channel determined by electron cryo-microscopy. *Nature, 504*, 107–112. https://doi.org/10.1038/nature12822

Lundbaek, J. A., Collingwood, S. A., Ingolfsson, H. I., Kapoor, R., & Andersen, O. S. (2010). Lipid bilayer regulation of membrane protein function: Gramicidin channels as molecular force probes. *Journal of the Royal Society, Interface/the Royal Society, 7*, 373–395. https://doi.org/10.1098/rsif.2009.0443

Luscher, C., & Malenka, R. C. (2012). NMDA receptor-dependent long-term potentiation and long-term depression (LTP/LTD). *Cold Spring Harbor Perspectives in Biology, 4*(6), a005710. https://doi.org/10.1101/cshperspect.a005710

Maingret, F., Patel, A. J., Lesage, F., Lazdunski, M., & Honore, E. (2000). Lysophospholipids open the two-pore domain mechano-gated K(+) channels TREK-1 and TRAAK. *Journal of Biological Chemistry, 275*, 10128–10133.

Martinac, B. (1993). In M. B. Jackson (Ed.), *CRC thermodynamics of membrane receptors and channels* (pp. 327–352). CRC Press.

Martinac, B., Adler, J., & Kung, C. (1990). Mechanosensitive ion channels of *E. coli* activated by amphipaths. *Nature, 348*, 261–263. https://doi.org/10.1038/348261a0

Martinac, B. (2011). Bacterial mechanosensitive channels as a paradigm for mechanosensory transduction. *Cellular Physiology and Biochemistry, 28*, 1051–1060.

Martinac, B., Bavi, N., Ridone, P., Nikolaev, Y. A., Martinac, A. D., Nakayama, Y., Rhode, P. R., & Bavi, O. (2018). Tuning ion channel mechanosensitivity by asymmetry of the transbilayer pressure profile. *Biophysical Reviews, 10*, 1377–1384.

Martinac, B., Buechner, M., Delcour, A. H., Adler, J., & Kung, C. (1987). Pressure-sensitive ion channel in *Escherichia coli*. *Proceedings of the National Academy of Sciences of the United States of America, 84*, 2297–2301.

Martinac, B., et al. (2013). Bacterial mechanosensitive channels: Models for studying mechanosensory transduction. *Antioxidants and Redox Signaling*. https://doi.org/10.1089/ars.2013.5471

Martinac, B., & Cox, C. D. (2017). Mechanosensory transduction: Focus on ion channels. *Reference module in life sciences*. Amsterdam: Elsevier. https://www.sciencedirect.com/science/article/pii/B9780128096338080948).

Martinac, B., & Kloda, A. (2003). Evolutionary origins of mechanosensitive ion channels. *Progress in Biophysics and Molecular Biology, 82*, 11–24.

Martinac, B. (2004). Mechanosensitive ion channels: Molecules of mechanotransduction. *Journal of Cell Science, 117*, 2449–2460. https://doi.org/10.1242/jcs.01232

Martinac, B. (2013). The ion channels to cytoskeleton connection as potential mechanism of mechanosensitivity. *Biochimica et Biophysica Acta*. https://doi.org/10.1016/j.bbamem.2013.07.015

Martinac, B., & Kung, C. (2022). The force-from-lipid principle and its origin, a 'what is true for *E. coli* is true for the elephant' refrain. *Journal of Neurogenetics, 36*(2-3), 44–54. https://doi.org/10.1080/01677063.2022.2097674

Murthy, S. E, Dubin, A. E., Whitwam, T., Jojoa-Cruz, S., Cahalan, S. M., Mousavi, S. A. R., & Patapoutian, A. (2018). OSCA/TMEM63 are an evolutionarily conserved family of mechanically activated ion channels. *eLife, 1*(7), e41844. https://doi.org/10.7554/eLife.41844. PMID: 30382938.

Nikolaev, Y. A., Cox, C. D., Ridone, P., Rohde, P. R., Cordero-Morales, J. F., Vásquez, V., ... Martinac, B. (2019). Mammalian TRP ion channels are insensitive to membrane stretch. *Journal of Cell Science, 132*(23), jcs238360. https://doi.org/10.1242/jcs.238360

Nomura, T., et al. (2012). Differential effects of lipids and lyso-lipids on the mechanosensitivity of the mechanosensitive channels MscL and MscS. *Proceedings of the National Academy of Sciences of the United States of America, 109*, 8770–8775. https://doi.org/10.1073/pnas.1200051109

Pace, N. R. (1997). A molecular view of microbial diversity and the biosphere. *Science, 276*, 734–740.

Pan, B., Akyuz, N., Liu, X. P., Asai, Y., Nist-Lund, C., Kurima, K., Derfler, B. H., György, B., Limapichat, W., Walujkar, S., & Wimalasena, L. N. (2018). TMC1 forms the pore of mechanosensory transduction channels in vertebrate inner ear hair cells. *Neuron, 9*(4), 736–753. PMID: 30138589.

Pan, B., et al. (2013). TMC1 and TMC2 are components of the mechanotransduction channel in hair cells of the mammalian inner ear. *Neuron, 79*, 504–515. https://doi.org/10.1016/j.neuron.2013.06.019

Paoletti, P., & Neyton, J. (2007). NMDA receptor subunits: Function and pharmacology. *Current Opinion in Pharmacology, 7*, 39–47. https://doi.org/10.1016/j.coph.2006.08.011

Parnas, M., et al. (2009). Membrane lipid modulations remove divalent open channel block from TRP-like and NMDA channels. *Journal of Neuroscience: The Official Journal of the Society for Neuroscience, 29*, 2371–2383. https://doi.org/10.1523/JNEUROSCI.4280-08.2009

Paulsen, C. E., Armache, J. P., Gao, Y., Cheng, Y., & Julius, D. (2015). Structure of the TRPA1 ion channel suggests regulatory mechanisms. *Nature, 520*, 511–517. https://doi.org/10.1038/nature14367

Perozo, E., Cortes, D. M., Sompornpisut, P., Kloda, A., & Martinac, B. (2002). Open channel structure of MscL and the gating mechanism of mechanosensitive channels. *Nature, 418*, 942–948. https://doi.org/10.1038/nature00992

Perozo, E., Kloda, A., Cortes, D. M., & Martinac, B. (2002). Physical principles underlying the transduction of bilayer deformation forces during mechanosensitive channel gating. *Nature Structural Biology, 9*, 696–703. https://doi.org/10.1038/nsb827

Pivetti, C. D., et al. (2003). Two families of mechanosensitive channel proteins. *Microbiology and Molecular Biology Reviews: Microbiology and Molecular Biology Reviews, 67*, 66–85.

Pliotas, C., et al. (2012). Conformational state of the MscS mechanosensitive channel in solution revealed by pulsed electron-electron double resonance (PELDOR) spectroscopy. *Proceedings of the National Academy of Sciences of the United States of America, 109*, E2675–E2682. https://doi.org/10.1073/pnas.1202286109

Pralle, A., Keller, P., Florin, E. L., Simons, K., & Horber, J. K. (2000). Sphingolipid-cholesterol rafts diffuse as small entities in the plasma membrane of mammalian cells. *The Journal of Cell Biology, 148*, 997–1008.

Proske, U., & Gandevia, S. C. (2012). The proprioceptive senses: Their roles in signaling body shape, body position and movement, and muscle force. *Physiological Reviews, 92*, 1651–1697. https://doi.org/10.1152/physrev.00048.2011

Ranade, S. S., Syeda, R., & Patapoutian, A. (2015). Mechanically activated ion channels. *Neuron, 87*, 1162–1179. https://doi.org/10.1016/j.neuron.2015.08.032

Sachs, F. (1991). Mechanical transduction by membrane ion channels: A mini review. *Molecular and Cellular Biochemistry, 104*, 57–60.

Ridone, P., Grage, S. L., Patkunarajah, A., Battle, A. R., Ulrich, A. S., Martinac, B., et al. (2018). "Force-from-lipids" gating of mechanosensitive channels modulated by PUFAs. *Journal of the Mechanical Behavior of Biomedical Materials, 79*, 158–167. https://doi.org/10.1016/j.jmbbm.2017.12.026

Ridone, P., Vassalli, M., & Martinac, B. (2019). Piezo1 mechanosensitive channels: What are they and why are they important. *Biophysical Reviews, 11*, 795–805.

Sachs, F. (1988). Mechanical transduction in biological systems. *Critical Reviews in Biomedical Engineering, 16*, 141–169.

Sachs, F. (2010). Stretch-activated ion channels: What are they? *Physiology, 25*, 50–56. https://doi.org/10.1152/physiol.00042.2009

Saotome, K., et al. (2018). Structure of the mechanically activated ion channel Piezo1. *Nature, 554*, 481–486. https://doi.org/10.1038/nature25453

Seo, K., et al. (2014). Combined TRPC3 and TRPC6 blockade by selective small-molecule or genetic deletion inhibits pathological cardiac hypertrophy. *Proceedings of the National Academy of Sciences of the United States of America, 111*, 1551–1556. https://doi.org/10.1073/pnas.1308963111

Simons, K., & Toomre, D. (2000). Lipid rafts and signal transduction. *Nature Reviews Molecular Cell Biology, 1*, 31–39. https://doi.org/10.1038/35036052

Steinbacher, S., Bass, R., Strop, P., & Rees, D. C. (2007). *Mechanosensitive ion channels*. In O. P. Hamill (Ed.) (pp. 1–24). Elsevier Academic Press, Inc.

Storch, U., Mederos y Schnitzler, M., & Gudermann, T. (2012). G protein-mediated stretch reception. *American Journal of Physiology - Heart and Circulatory Physiology, 302*, H1241–H1249. https://doi.org/10.1152/ajpheart.00818.2011

Sukharev, S. I., Blount, P., Martinac, B., Blattner, F. R., & Kung, C. (1994). A large-conductance mechanosensitive channel in *E. coli* encoded by mscL alone. *Nature, 368*, 265–268. https://doi.org/10.1038/368265a0

Sukharev, S. I., Martinac, B., Arshavsky, V. Y., & Kung, C. (1993). Two types of mechanosensitive channels in the *Escherichia coli* cell envelope: Solubilization and functional reconstitution. *Biophysical Journal, 65*, 177–183. https://doi.org/10.1016/S0006-3495(93)81044-0

Syeda, R., et al. (2015). Chemical activation of the mechanotransduction channel Piezo1. *Elife, 4*. https://doi.org/10.7554/eLife.07369

Syeda, R., et al. (2016). Piezo1 channels are inherently mechanosensitive. *Cell Reports, 17*, 1739–1746. https://doi.org/10.1016/j.celrep.2016.10.033

Teng, J., Loukin, S., Anishkin, A., & Kung, C. (2015). The force-from-lipid (FFL) principle of mechanosensitivity, at large and in elements. *Pflügers Archiv*. https://doi.org/10.1007/s00424-014-1530-2

Wang, N., Butler, J. P., & Ingber, D. E. (1993). Mechanotransduction across the cell surface and through the cytoskeleton. *Science, 260*, 1124–1127.

Wang, Y., et al. (2014). Single molecule FRET reveals pore size and opening mechanism of a mechano-sensitive ion channel. *Elife, 3*, Article e01834. https://doi.org/10.7554/eLife.01834

Woese, C. R. (1994). There must be a prokaryote somewhere: microbiology's search for itself. *Microbiological Reviews, 58*, 1–9.

Woo, S. H., et al. (2015). Piezo2 is the principal mechanotransduction channel for proprioception. *Nature Neuroscience, 18*, 1756–1762. https://doi.org/10.1038/nn.4162

Wood, J. M. (2015). Bacterial responses to osmotic challenges. *The Journal of General Physiology, 145*, 381–388. https://doi.org/10.1085/jgp.201411296

Xu, J., et al. (2018). GPR68 senses flow and is essential for vascular physiology. *Cell, 173*, 762–775. https://doi.org/10.1016/j.cell.2018.03.076

Yan, Z., et al. (2013). Drosophila NOMPC is a mechanotransduction channel subunit for gentle-touch sensation. *Nature, 493*, 221–225. https://doi.org/10.1038/nature11685

Young, M. N., Sindoni, M. J., Lewis, A. H., Zauscher, S., & Grandl, J. M. (2023). The energetics of rapid cellular mechanotransduction. *Biophysical Journal, 122*(3), 89a–90a. PMID: 36795747.

Yu, Z. Y., Gong, H., Kesteven, S., et al. (2022). Piezo1 is the cardiac mechanosensor that initiates the cardiomyocyte hypertrophic response to pressure overload in adult mice. *Nature Cardiovascular Research, 1*, 577–591. https://doi.org/10.1038/s44161-022-00082-0

Zarychanski, R., et al. (2012). Mutations in the mechanotransduction protein PIEZO1 are associated with hereditary xerocytosis. *Blood, 120*, 1908–1915. https://doi.org/10.1182/blood-2012-04-422253

Zhang, W., et al. (2015). Ankyrin repeats convey force to gate the NOMPC mechanotransduction channel. *Cell, 162*, 1391–1403. https://doi.org/10.1016/j.cell.2015.08.024

Zhang, W., Yan, Z., Jan, L. Y., & Jan, Y. N. (2013). Sound response mediated by the TRP channels NOMPC, NANCHUNG, and INACTIVE in chordotonal organs of Drosophila larvae. *Proceedings of the National Academy of Sciences of the United States of America, 110*, 13612–13617. https://doi.org/10.1073/pnas.1312477110

Zhao, Q., et al. (2018). Structure and mechanogating mechanism of the Piezo1 channel. *Nature, 554*, 487–492. https://doi.org/10.1038/nature25743

Zheng, J. (2013). Molecular mechanism of TRP channels. *Comprehensive Physiology, 3*, 221–242. https://doi.org/10.1002/cphy.c120001

Zubcevic, L., et al. (2016). Cryo-electron microscopy structure of the TRPV2 ion channel. *Nature Structural & Molecular Biology, 23*, 180–186. https://doi.org/10.1038/nsmb.3159

Chapter 22

Cell volume regulation

Clive M. Baumgarten and Joseph J. Feher

Department of Physiology and Biophysics, Virginia Commonwealth University School of Medicine, Richmond, VA, United States

1. Summary

The initial response to osmotic swelling or shrinking produces lines on a Boyle–van't Hoff plot that do not pass through the origin. This is interpreted to indicate an "osmotically inactive" volume that comprises a significant fraction of cell volume. The double Donnan or pump-leak hypothesis remains a theoretical backbone of isosmotic cell volume regulation that highlights the important role of the Na^+-K^+-ATPase. However, other transporters and pumps are also important. In many cells, regulatory volume increase (RVI) occurs upon cell shrinking resulting in activation of transporters to increase cellular osmolyte concentrations and cell volume. The signal pathways and transporters involved vary with the type of cell, but the most important transporters for RVIs are NKCC1, NHE1, AE2, and organic osmolyte transporters such as TauT. Regulatory volume decrease (RVD) occurs upon volume expansion in many cells in which transporters extrude solutes from the cell to reduce cell volume. Although the transporters and signaling pathways vary with tissues, the main transporters for RVD appear to be the volume-regulated anion channel (volume-regulated anion channel [VRAC], volume-sensitive organic anion channel [VSOAC], and $I_{Cl,swell}$), Ca^{2+}-activated K^+ channels of big, intermediate, and small conductances; KCC1-4, and TMEM16, or anoctamin. The molecular basis for VRAC, VSOAC, and $I_{Cl,swell}$ has been identified as heterogenous hexameric assemblies of LRRC8A-E subunits, although other channel families also can produce volume-sensitive Cl^- currents. It appears that there is not a single sensor for volume but various volume regulatory responses may use molecular crowding and its consequent liquid–liquid phase separation of signaling molecules, or changes in ionic strength, Cl^- or Ca^{2+} concentration, or signals originating upon changes in membrane tension or cytoskeletal elements or in response to other factors.

2. Cell osmometry

In human whole blood, red blood cells are suspended in the plasma, the fluid part of blood, with about 5×10^6 red blood cells per mm^3 of whole blood. These cells lack a nucleus and are shaped like biconcave disks about 7 μm in diameter and about 2 μm thick. The normal hematocrit, or fraction of blood volume occupied by the red blood cells, is about 0.45. From the hematocrit and the number of cells per unit volume, we can calculate that the average volume of the red blood cells is about 90 fL. We can determine the effect of the composition of the plasma on the volume of the red blood cells by replacing known volumes of plasma with equal volumes of solutions containing various concentrations of NaCl, glucose, or urea, as described in the legend to Fig. 22.1. The relative volume of the cells can be determined by the hematocrit. Fig. 22.2 shows the results of such an experiment using rabbit blood. When plasma is replaced by 0.9% NaCl (w/v), the red blood cells retain their shape and volume. When plasma is replaced with higher concentrations of salt, the red blood cells shrink (the hematocrit becomes less). Under a microscope, the cells appear as spheres with spikes all over their surfaces. These are described as being *crenated*. When plasma is replaced with lower concentrations of salt, the cells swell and become spherical. Solutions that cause neither shrinking nor swelling are called **isotonic** solutions, whereas those that cause the cells to shrink are called **hypertonic** solutions, and those that cause cells to swell are **hypotonic** solutions. Thus, 0.9% NaCl or 5% glucose are both isotonic solutions. The concentration of isotonic saline, 0.9%, is 0.9 g per 100 mL of water, or 0.154 M NaCl; the concentration of 5% glucose, 5 g per 100 mL of water, is 0.278 M glucose. If plasma is replaced with 1.8% urea (0.300 M urea), the red blood cells swell so much that they burst, and the observed hematocrit is zero. Part of the goal of this chapter is to explain these elementary phenomena: Why do cells swell when placed in solution

Cell Physiology Sourcebook. https://doi.org/10.1016/B978-0-12-811114-7.00011-6

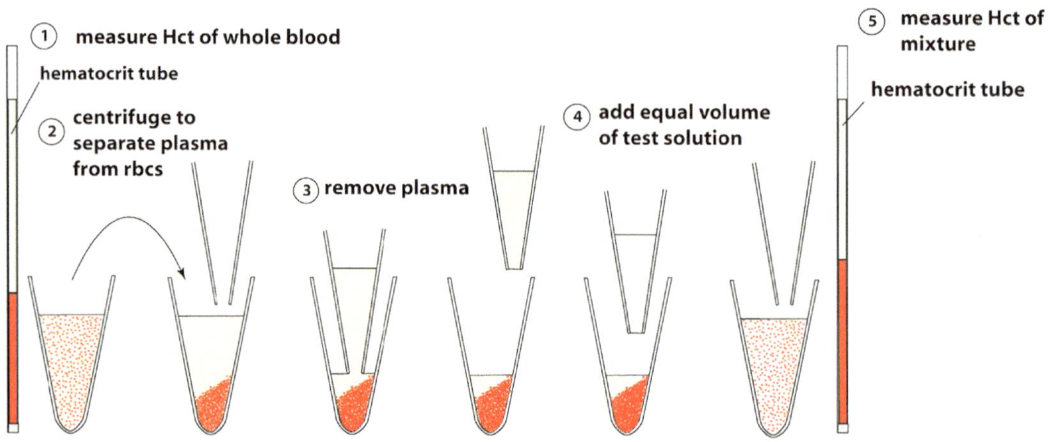

FIGURE 22.1 Procedure for testing the tonicity of solutions. First, the hematocrit (Hct) of whole blood is obtained to determine the volume of the red blood cells under isotonic conditions. This is done by centrifuging an aliquot of the blood in a narrow cylindrical tube and taking the ratio of the height of the packed cells to the total height of the blood column (1). A known volume of blood is centrifuged to separate the plasma from the red blood cells (2). Most of the volume of plasma is removed (3) and then replaced with an equal volume of test solution of varying composition (4). The suspension is then mixed and equilibrated, and the hematocrit is measured again (5). An increase in Hct reflects red blood cell swelling and a decrease reflects cell shrinkage.

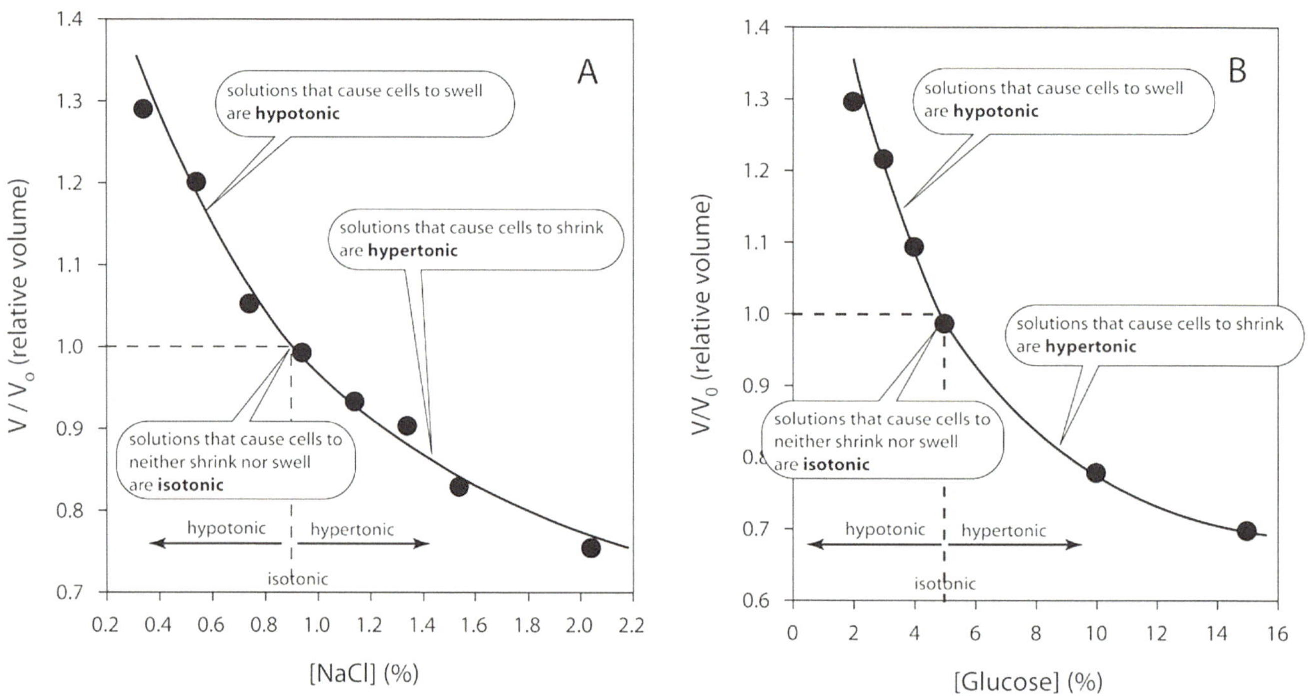

FIGURE 22.2 Swelling or shrinking of rabbit erythrocytes when an aliquot of plasma was replaced with solutions containing various concentrations of NaCl (panel A) or glucose (panel B), as shown in Fig. 22.1. When plasma was replaced with 0.9% NaCl or 5% glucose, the cells retained the volume they had in whole blood. These solutions are described as isotonic. When [NaCl] exceeded 0.9%, or [glucose] exceeded 5%, the cells shrank. These solutions are described as hypertonic. When the [NaCl] was less than 0.9%, or [glucose] was less than 5%, the erythrocytes swelled, and these solutions of NaCl or glucose are said to be hypotonic. *From Feher (2017).*

with low concentrations of NaCl or glucose? Why do they shrink when exposed to higher concentrations of NaCl or glucose? Why do they burst regardless of the urea concentration?

Nearly all animal or plant cells will change their volume when faced with anisotonic conditions. Once swollen or shrunk, some cells will maintain the new volume, but other cells can engage in volume regulatory responses. A wide variety of cells can sense their increased or decreased cell volume and trigger mechanisms to return the volume towards normal. Another goal of this chapter is to explain these regulatory responses: How do the cells detect deviations from the isotonic condition? What mechanisms are used to either shrink the swollen cell, or swell the shrunken one? What signaling pathways are involved?

The answers to these questions center on the basic concepts of **osmosis**, **osmolytes**, **channels**, and **transporters**. *Osmosis* refers to water movement in response to differences in *solute* concentrations across a membrane. Dissolution of solutes in water produces an **osmotic pressure**, defined as the pressure needed to be applied to a solution to stop fluid movement when the solution is separated from pure water by a semipermeable membrane, one that is permeable to water but not to the solutes. For this reason, these solutes are referred to as **osmolytes**. The water movement across biological membranes has several routes but mostly occurs through specific channels called **aquaporins** (discussed in Chapter 18), whereas the osmolytes move via ion channels, transporters, and pumps. If aquaporins are not present or are blocked, water will pass through the lipid bilayer itself, although much more slowly than via aquaporins (Suleymanian & Baumgarten, 1996). The water permeability of the membrane is discussed in Chapter 11 of this volume.

Equilibration of the osmotic pressure across red blood cell membranes occurs rapidly compared with any solute movement, reaching osmotic equilibrium within 0.6 s (Zhurova et al., 2014). If we assume no movement of solute during osmotic equilibration, then the total content of osmolytes of the cell should be the same before and after equilibration, although the volumes are different. Thus, we may write

$$\begin{aligned} V_c\,\Sigma C_c &= V_0\,\Sigma C_0 \\ V_c\,\Sigma RTC_c &= V_0\,\Sigma RTC_0 \\ V_c\,\pi_c &= V_0\,\pi_0 \end{aligned} \tag{22.1}$$

where V_C is the volume of the cell under anisotonic conditions, V_0 is its volume under isotonic conditions, and π_C and π_0 are the effective osmotic concentrations, approximately proportional to the osmotic pressures, under the two conditions. This equation, however, does not reflect the fact that some part of the cell volume is occupied by molecules that do not participate in the osmotic equilibrium, the **osmotically inactive volume**. Red blood cells lack intracellular organelles that might form part of the osmotically inactive volume, but they are filled with hemoglobin, comprising 33.4 g/100 mL of packed cell volume. The specific volume of hemoglobin is about 0.749 cm^3/g (historical values reviewed by DeMoll et al., 2007). Thus, the volume of hemoglobin within the red blood cell is 0.334 g Hb/cm^3 red cell volume × 0.749 cm^3 Hb volume/g Hb = 0.25 cm^3 Hb volume/cm^3 red cell volume. That is to say, hemoglobin occupies 25% of red cell volume. We might expect other constituents to further increase this volume. We can correct Eq. (22.1) for this osmotically inactive volume by writing

$$(V_c - V_b)\pi_c = (V_0 - V_b)\pi_0 \tag{22.2}$$

where V_b is the "osmotically inactive volume." This equation may be rearranged to give

$$\frac{V_c}{V_0} = \left(1 - \frac{V_b}{V_0}\right)\frac{\pi_0}{\pi_c} + \frac{V_b}{V_0} \tag{22.3}$$

(Dick, 1966). Recall that V_0 and π_0 refer to the volume of the cell and the osmotic pressure under isotonic conditions. Thus, a plot of V_c/V_0 against π_0/π_c should give a straight line with a y-intercept of V_b/V_0. This is commonly referred to as the Boyle–van't Hoff plot, and the results are shown in Fig. 22.3 for rabbit red blood cells diluted into varying concentrations of NaCl (data from Fig. 22.2). From these results in the rabbit, $V_b/V_0 = 0.379$, in reasonable agreement with previous results with human red blood cells ($V_b/V_0 = 0.392$; Lefevre, 1964). Correction of the cell volume for the "osmotically inactive volume" ($V_0 - V_b$), however, does not give the water content of the cell, W, obtained by gravimetric analysis after desiccation and, typically, $W > V_0 - V_b$. The ratio $(V_0 - V_b)/W$, denoted **R**, is called Ponder's **R**. This variable is generally less than 1.0 and was first interpreted as the fraction of water that is "free," with the rest bound to proteins or other structures in the cell. However, this "bound water" can be considered part of the nonideal behavior of the erythrocyte contents, and modification of the Boyle–van't Hoff equation, Eq. (22.3), to include nonideality largely eliminates the difference between W and $(V_0 - V_b)$ determined from the graphs (Prickett et al., 2008). Values of V_b/V_0 obtained from Boyle–van't Hoff plots vary with the cell type. For example, in ventricular myocytes, V_b/V_0 is 0.338 (Drewnowska & Baumgarten, 1991), and in lymphocytes it is 0.32 (Grinstein et al., 1984).

The linearity of a Boyle–van't Hoff plot is based on the flux of water alone and depends critically on the lack of movement of ions during osmotic equilibration. Movement of ions in response to an osmotic challenge is generally part of RVIs or RVDs and causes non-osmometric behavior and curvature of the plot (Grinstein et al., 1984).

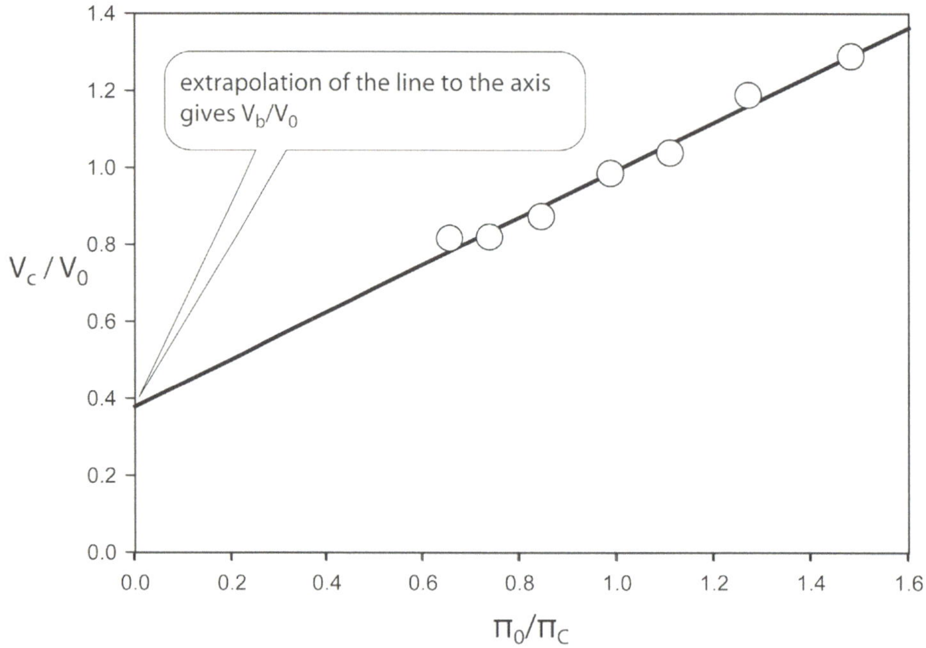

FIGURE 22.3 Boyle–van't Hoff plot for rabbit red blood cells. Volumes of red blood cells were determined from hematocrits after cells were equilibrated with solutions of different osmolarity, using NaCl. The graph refers to Eq. (22.3) in which the relative volume (V_c/V_0) is plotted against relative osmolarity (π_0/π_c). Extrapolation of the line to infinite osmolarity (y-intercept) gives a measure of the relative "osmotically inactive" volume (V_b/V_0).

3. Regulation of cell volume under isosmotic conditions

3.1 Cells require metabolism to maintain volume

In large multicellular organisms, cells lie adjacent to the interstitial fluid whose composition and volume is maintained fairly constant by cooperation of multiple systems—the intestines for absorption of water and osmolytes and loss of these through the feces; the lungs and skin that generally lose fluid; the heart that circulates the blood and provides pressure for the kidneys that conserve or waste water or electrolytes as needed. As long as homeostasis of the extracellular fluid is maintained and cells are provided with nutrients and gas exchange, cells will maintain their volume. However, cells deprived of energy, such as in anaerobic settings, will swell even under isoosmolar conditions. Thus, even at rest, cells must work to maintain their volume. T.H. Wilson (1954) first proposed that " … the tendency of salts and water to enter cells owing to the intracellular colloid osmotic pressure is, under normal conditions of oxygenation … exactly balanced by the active transport of ions and passive movement of water out of the cell". Frederick George Donnan (1911) was the first to show that impermeant anions would attract smaller, permeant cations and repel the small permeant anions, leading to an electrical potential, now called the Donnan potential, and this effect would also cause an osmotic disequilibrium that would eventually lead to cell rupture. Leaf (1956) and Tosteson and Hoffman (1960) later showed that this catastrophe could be avoided by a "double Donnan" or "pump-leak" system in which the effect of impermeant anions within the cell is balanced by the effective impermeability of Na^+ ions outside of the cell due to its energy-dependent (i.e., active) extrusion.

3.2 The Gibbs–Donnan equilibrium and membrane potential

To see how a Gibbs-Donnan equilibrium develops—often simply called a Donnan equilibrium, consider the situation shown in Fig. 22.4. A cell initially contains 150 mM K^+ and 130 mM of impermeant anions, A^-, and a balance of 20 mM Cl^-. The impermeant anions consist of polyvalent proteins, which are generally negatively charged at intracellular pH, and organic phosphates (ADP, ATP, creatine phosphate, etc.). These concentrations meet the condition of **macroscopic electroneutrality**:

$$\sum_i C_i^+ = \sum_i C_i^- \tag{22.4}$$

which says that the total concentration of positive charges in any volume must be equal to the total concentration of negative charges. This equation is an approximate one. Separation of charge occurs across cell membranes and is responsible

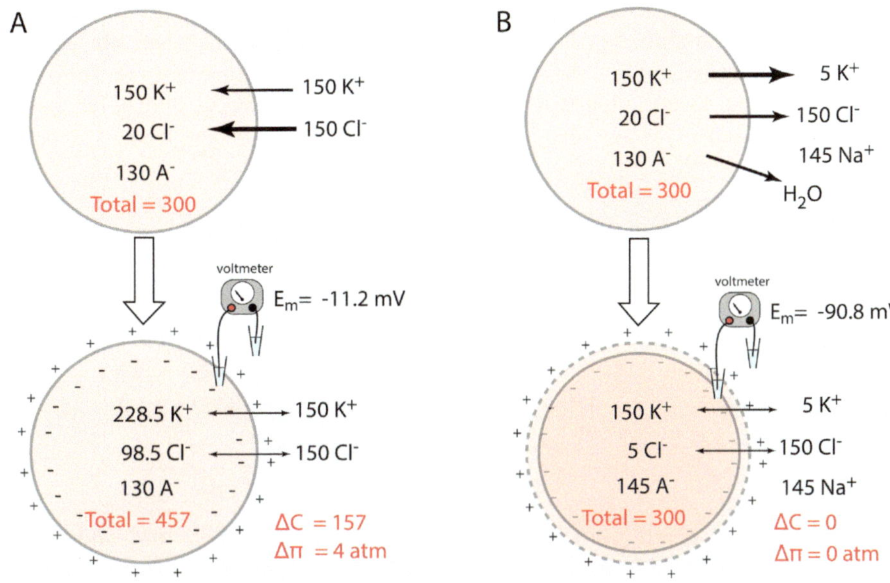

FIGURE 22.4 Donnan equilibrium (panel A) and double Donnan equilibrium (panel B). Panel A: A cell bathed in 150 mM KCl is filled with 150 mM K^+, 20 mM Cl^-, and 130 mM of impermeant anions, A^-. Initially, there is no potential and no osmotic gradient. Cl^- enters the cell down its electrochemical gradient, creating a separation of charge and movement of K^+ follows until $[K^+]_{out}\cdot[Cl^-]_{out} = [K^+]_{in}\cdot[Cl^-]_{in}$. The ionic fluxes that establish the Donnan equilibrium potential also create an osmotic gradient. Equilibrium is reached only if cell volume cannot change, either because the membrane is rigid or it disallows water movement. With a biological membrane that is neither rigid nor impermeant to water, water would enter the cell, lowering $[K^+]_{in}\cdot[Cl^-]_{in}$ and causing further entry of solute and water until the cell bursts. Panel B: Part of the extracellular K^+ is replaced by Na^+, which is assumed to be impermeant. With impermeant ions on both sides of the membrane (A^- inside and Na^+ outside), equilibrium is attained without an osmotic gradient. Here K^+, Cl^-, and water leave the cell until, at equilibrium, $[K^+]_{out}\cdot[Cl^-]_{out} = [K^+]_{in}\cdot[Cl^-]_{in}$. The amount of A^- inside the cell is fixed, and so, the decrease in cell volume increases $[A^-]_{in}$.

for the electrical potential difference across membranes. However, the number of separated charges that produce the membrane potential is negligible compared with the total number of charges in a cell, and so, Eq. (22.4) is generally valid even for volumes as small as those of a single cell (Baumgarten & Feher, 2012). Initially, we consider that the cell is impermeable to water and the anions A^- but is freely permeable to the small electrolytes, K^+ and Cl^-. Suppose we place the cell in a solution containing 150 mM K^+ and 150 mM Cl^- as shown in Fig. 22.4A. Initially, Cl^- will move into the cell because the concentration gradient favors that movement. The movement of Cl^- into the cell creates a separation of charge that produces a membrane potential, negative inside the cell, which attracts K^+, causing K^+ to move against its concentration gradient but down its electrochemical gradient. Eventually an equilibrium is established that can be analyzed using the thermodynamic principle that the free energy change at equilibrium is zero. For ideal solutions, we write this as

$$\begin{aligned}
\Delta\mu_K &= \mu_{K,final} - \mu_{K,initial}\\
\Delta\mu_K &= \mu_K^0 + RT\ \ln\left[K^+\right]_{in} + z_K\Im\psi_{in} - \mu_K^0 - RT\ \ln\left[K^+\right]_{out} - z_K\Im\psi_{out}\\
\Delta\mu_{Cl} &= \mu_{Cl,final} - \mu_{Cl,initial}\\
\Delta\mu_{Cl} &= \mu_{Cl}^0 + RT\ \ln\left[Cl^-\right]_{in} + z_{Cl}\Im\psi_{in} - \mu_{Cl}^0 - RT\ \ln\left[Cl^-\right]_{out} - z_{Cl}\Im\psi_{out}
\end{aligned} \tag{22.5}$$

where μ^0 is the chemical potential of the standard state, z_i is the valence of species i, $\Im$ is the Faraday (96,485 coulombs/mol) that converts the charge to coulombs, ψ is the electrical potential (volts), and R and T have their usual meanings. At equilibrium, the change in free energy for each permeant species to cross the membrane is zero, and Eq. (22.5) can be rearranged to

$$\begin{aligned}
E_m &= \psi_{in} - \psi_{out} = \frac{RT}{z_K\Im}\ln\frac{\left[K^+\right]_{out}}{\left[K^+\right]_{in}}\\
E_m &= \psi_{in} - \psi_{out} = \frac{RT}{z_{Cl}\Im}\ln\frac{\left[Cl^-\right]_{out}}{\left[Cl^-\right]_{in}}
\end{aligned} \tag{22.6}$$

where the membrane potential, E_m, is taken as $\psi_{in} - \psi_{out}$. The potentials given by Eq. (22.6) are identical to the Nernst equilibrium potentials for the ions, E_K and E_{Cl}. Thus, in a Donnan system where the membrane is permeable to both ions, the ions distribute themselves so that the single potential, E_m, is simultaneously equal to both E_K and E_{Cl}. Inserting $z_K = +1$ and $z_{Cl} = -1$, and equating the two expressions in Eq. (22.6) because they are both equal to E_m, we find

$$\frac{[K^+]_{out}}{[K^+]_{in}} = \frac{[Cl^-]_{in}}{[Cl^-]_{out}} \tag{22.7}$$

which is often expressed as

$$[K^+]_{out} \cdot [Cl^-]_{out} = [K^+]_{in} \cdot [Cl^-]_{in} \tag{22.8}$$

Using this relation, we can calculate the equilibrium ion distribution in the situation of Fig. 22.4. Here we assume that the outside volume is sufficiently large that movement of K^+ and Cl^- do not appreciably change the outside concentrations. If Δm are the concentrations of K^+ and Cl^- that move, we solve for Δm as

$$\frac{[K^+]_{out}}{[K^+]_{in}} = \frac{150\ mM}{150\ mM + \Delta m} = \frac{20\ mM + \Delta m}{150\ mM} = \frac{[Cl^-]_{in}}{[Cl^-]_{out}} \tag{22.9}$$

and find $\Delta m = 78.5$ mM. Thus, at equilibrium $[K^+]_{in} = 228.5$ mM, $[Cl^-]_{in} = 98.5$ mM and $E_m = (8.314\ \text{J/mol} \cdot \text{K} \times 310\ \text{K}/96{,}485\ \text{C/mol}) \times \ln(150\ \text{mM}/228.5\ \text{mM}) = -0.0112$ V. Now Δm cannot be precisely the same for K^+ and Cl^-. A very small imbalance of anion and cation fluxes leads to the separation of charge that produces the potential difference across the membrane. The discrepancy required is "microscopic" and does not alter ionic concentrations by a physiologically relevant amount for a typical cell.

3.3 Gibbs–Donnan equilibrium and osmotic equilibrium

Initially, we supposed that the cell in Fig. 22.4 was permeable to K^+ and Cl^- but not permeable to A^- and water. What if the membrane were permeable to water? The inside concentration of osmolytes after Donnan equilibration is 130 mM A^- + 228.5 mM K^+ + 98.5 mM Cl^- = 457 mM, whereas the concentration outside is 300 mM. The osmotic pressure difference between the inside and outside would be approximately given by the van't Hoff equation: $\Delta\pi = RT\Delta C = 0.082\ \text{L} \cdot \text{atm/mol} \cdot \text{K} \times 310\ \text{K} \times 0.157\ \text{mol/L} = 3.99$ atm. This is an enormous pressure. Entrance of K^+ and Cl^- as a consequence of the Donnan equilibrium would increase π_{in} and, if the cell is permeable to water, would cause an influx of water. This would dilute the internal solution, and more K^+ and Cl^- would then enter to reestablish the Donnan equilibrium. And then more water would enter. This would continue until the cell bursts. Thus, the Donnan equilibrium describes the equilibrium of electrolytes, but not water, and the whole system does not come to an equilibrium with an intact cell unless the water permeability is zero.

The Donnan equilibrium is not merely a theoretical exercise. It predicts that increasing extracellular $[K^+]$ in physiological solutions will cause a cell obeying Donnan equilibrium to take up K^+ and Cl^- and swell. Some tissues appear to follow this prediction. Boyle and Conway (1941) immersed frog sartorius muscles in solutions with varying $[K^+]_{out}$ and carefully determined cell water, $[K^+]_{in}$ and $[Cl^-]_{in}$. For $[K^+]_{out}$ greater than 6 mM, the ratio $[K^+]_{out} \cdot [Cl^-]_{out} / [K^+]_{in} \cdot [Cl^-]_{in}$ was very nearly 1.0, in agreement with Eq. (22.8). The deviations at low $[K^+]_{out}$ occur because Na^+ influx causes E_m to deviate from E_K. The accompanying changes in cell water in these experiments also agreed with theory. Donnan equilibrium behavior also may be relevant in certain clinical settings. The volume of isolated cardiac ventricular myocytes was consistent with the Donnan $[K^+] \cdot [Cl^-]$ ratio under experimental conditions that mimicked cold cardioplegia used to arrest the heart during surgical procedures (Drewnowska et al., 1991). Similar considerations are likely to apply to the cold transport of other organs for transplantation.

3.4 The double Donnan or pump-leak hypothesis

In principle, there are several ways to stabilize the Donnan system shown in Fig. 22.4. One is to have membranes that are impermeable to water. Most cells do not fit this description. A second way is to increase the hydrostatic pressure within the cell to arrest the inward transmembrane water flow. In view of the enormous pressures required, this is not a feasible solution for animal cells. Transmembrane hydrostatic pressures differences are very low, and the membrane is insufficiently strong to withstand much pressure difference. Third, an equilibrium could also be established by restricting an ionic species in the extracellular compartment to match the ones restricted in the intracellular compartment. This arrangement is called a double Donnan (Leaf, 1956; Macknight & Leaf, 1977) or a pump-leak system (Tosteson & Hoffman, 1960). To illustrate this, we consider again the cell in Fig. 22.4. The cell is filled with 150 mM K^+ and 150 mM Cl^-, but this time it is immersed in a solution of 5 mM K^+, 145 mM Na^+, and 150 mM Cl^- (Fig. 22.4B). This is identical to the case of Fig. 22.4A except we have replaced most of the extracellular permeable K^+ with impermeant Na^+. In this case, $[K^+]_{in} \cdot [Cl^-]_{in} > [K^+]_{out} \cdot [Cl^-]_{out}$ and KCl must leave the cell to establish a Donnan equilibrium. When it does, it lowers the intracellular osmotic pressure, and water follows the KCl into the external solution. The extracellular concentrations do not change because their volume is assumed to be much larger than the intracellular volume. The outward movement of water concentrates the intracellular impermeant anions until such time that the osmotic pressure equilibrates, and the system reaches a global equilibrium of K^+, Cl^-, and water. At this point, $E_m = E_K = E_{Cl} = -90.8$ mV, and the osmotic pressure of the impermeant anions and their counterions (commonly referred to as the **colloid osmotic pressure**) is balanced by the osmotic pressure of ions and their counterions that are restricted to the extracellular fluid. As a result, there is no net osmotic pressure across the membrane.

It is important to realize that an ion does not need to be impermeant to be *effectively* restricted to the extracellular space and thus counterbalance the osmotic pressure of the intracellular macromolecules. Na^+ is permeant, but it adequately plays this role anyway because the leak of Na^+ down its electrochemical gradient into the cell is matched by its export back out. As long as this continues, cell volume will be stable. To explain cell volume regulation, Leaf (1956) postulated the existence of a pump that actively extrudes Na^+ against its concentration gradient. Subsequently, Skou and others identified the Na^+-K^+-ATPase, which transports three Na^+ ions out of the cell and two K^+ into the cell with the hydrolysis of each ATP (reviewed by Skou, 1990). Because energy is continuously consumed to extrude Na^+, the cell is in a steady state rather than in an equilibrium. Inhibition of the pump or of metabolism that feeds ATP into the pump will result in cell swelling.

3.5 Conditions of steady-state volume regulation

The electroneutrality of cellular contents allows us to write

$$[K^+]_{in} + [Na^+]_{in} = [Cl^-]_{in} + |z|[A^-]_{in} \tag{22.10}$$

because intracellular K^+ and Na^+ are essentially the only intracellular cations, and A^- is taken to mean all of the intracellular anions that are not Cl^-. Mostly, these are proteins and organic phosphates. Here, $|z|$ is the absolute value of the mean valence (a negative number) of the impermeant intracellular anions. Simultaneous osmotic equilibrium requires

$$[K^+]_{in} + [Na^+]_{in} + [Cl^-]_{in} + [X]_{in} + [A^-]_{in} = \frac{\pi_{out}}{RT} = \Sigma C_{out} \tag{22.11}$$

where $[X]$ denotes uncharged osmolytes within the cell and ΣC_{out} is the sum of the osmotically active solutes outside of the cell. This represents the extracellular concentration, which is maintained fairly constant by organismal homeostasis. Inserting Eq. (22.10) for $[K^+]_{in} + [Na^+]_{in}$ and multiplying by the cell volume, we find

$$V = \frac{2M_{in}^{Cl} + M_{in}^{X} + (1 + |z|)M_{in}^{A}}{\Sigma C_{out}} \tag{22.12}$$

where V is the volume of the cell and M denotes the *amount* of each substance in the cell, in moles. Thus, if the external osmolarity decreases, but the total cellular osmolytes do not, the cell swells. Conversely, if the number of cellular osmolytes decreases, but extracellular osmolarity does not change, the cells shrink. This equation clarifies that the cell volume is determined by the intracellular content of Cl^-, X and the net charge on the intracellular impermeant macromolecules (Hoffmann et al., 2009). Thus, regulation of cell volume under isosmotic conditions depends on a continuous balance of all influx and efflux pathways for osmolytes.

3.6 Isosmotic volume regulation: The ventricular myocyte

Cells differ in many important regards, including their intracellular composition and the mechanisms by which they maintain their volume. Here, we use the particular example of the ventricular myocyte. As predicted from the pump-leak model, cell swelling occurs in many cells in which the Na^+-K^+-ATPase is inhibited (Hoffmann et al., 2009). However, this does not predict how fast such swelling might occur. In rabbit cardiac myocytes, inhibition of the Na^+-K^+-ATPase with ouabain did not affect cell volume for 20 min. These cells nevertheless rapidly swelled in hypoosmotic solutions or shrank in hyperosmotic solutions, without a RVD or RVI for 20 min (Drewnowska & Baumgarten, 1991). Ouabain treatment increased swelling in hypoosmotic media, and increased shrinkage in hyperosmotic media, indicating that the Na^+-K^+-

ATPase acted to limit volume changes under anisosmotic conditions. Inhibition of the Na^+-K^+-$2Cl^-$ cotransporter (NKCC1, gene SLC12A2) by bumetanide decreased cell volume under isosmotic conditions and enhanced cell swelling and decreased cell shrinking under anisosmotic conditions. These results indicate a role for NKCC1 in isosmotic regulation of cell volume. In a later study using a combination of simulations and experiments in guinea pig ventricular myocytes, the observation that ouabain did not cause swelling in isosmotic media was confirmed, and inhibition of Na^+-Ca^{2+} exchanger (NCX, gene SLC8A1) in the presence of ouabain resulted in cell swelling (Takeuchi et al., 2006). In this model, the plasma membrane Ca^{2+}-ATPase (PMCA1, gene ATP2B1) pumps Ca^{2+} out of the cell and NCX exchanges three Na^+ out for one Ca^{2+} in, and the Ca^{2+} is pumped back out again, forming a futile cycle for Ca^{2+} that powers Na^+ efflux. In this way, the combined PMCA and NCX help adjust cell volume when intracellular $[Na^+]$ rises. On the other hand, when Na^+ is normal (the Na^+-K^+-ATPase is not inhibited), the NCX may be going in the opposite direction, extruding Ca^{2+} and taking up Na^+. Regulation of cell volume without the Na^+-K^+ pump is not unique to cardiac myocytes. Mature erythrocytes from carnivores lack a Na^+-K^+ pump, have similar intracellular and extracellular Na^+ and K^+ concentrations, and also regulate volume by the PMCA working in parallel with NCX (for references, see Baumgarten, 2006). The relative rates of Na^+ transport in cardiac cells at rest have been estimated by Bers et al. (2003), as shown in Table 22.1. Rabbit, rat, guinea pig, and human are likely to have significant differences in the relative importance of these pathways even for the same type of cell. The Kyoto model of cardiac ion fluxes used by Takeuchi et al. (2006) (see Fig. 22.5) includes many more osmolyte pathways. Despite this complexity, the Kyoto model omits some transport pathways and otherwise does not completely model the volume dependence of the osmolyte pathways (Baumgarten, 2006). What these studies show is that the entire set of transporters, with their regulation, must be taken into account to fully understand volume regulation. In addition, there is no general answer to the question about which transporter contributes how much to volume regulation. Cells vary in the isoforms of transporters, in the level of their expression, and how they are regulated. As an example, NKCC1 in the heart has been shown to be regulated by ANP (atrial natriuretic peptide) via cGMP (Clemo et al., 1992) produced by ANP binding to membrane receptor guanylate cyclase A (GC-A) (Kuhn, 2015). Cardiomyocytes do not show volume regulatory increases following cell shrinkage, but cells that do often use NKCC1 to mediate RVI (Pederson et al., 2006).

4. Regulation of cell volume under anisosmotic conditions

When exposed to anisotonic media, most cells initially shrink or swell in accordance with the Boyle—van't Hoff relation (Eq. 22.3) because their membranes are more permeable to water than to solutes. In many cells, the initial swelling activates compensatory processes that lead to an efflux of osmolytes and reduction in cell volume. This is called a RVD for regulatory volume decrease. On the other hand, in many cells, shrinking activates an influx of osmolytes that leads to a compensatory swelling, referred to as a RVI (regulatory volume increase).

There are three aspects to these compensatory volume responses:

- sensing the volume change
- signaling between volume sensor and transporters
- activation of osmolyte transporters or levels of organic osmolytes

TABLE 22.1 Contribution of various transporters to Na^+ transport in cardiac myocytes.

Source	Transport	Resting flux (mM/min)
Na^+-K^+-ATPase	3 Na^+ out: 2 K^+ in	0.77
Na^+-Ca^{2+} exchanger (NCX)	3 Na^+: 1 Ca^{2+} (reversible)	0.28
Na^+ channel (I_{Na})	Uncoupled	0.14
Na^+-H^+ exchanger (NHE)	1 Na^+: 1 H^+	0.12
Na^+-HCO_3^- cotransporter (NBC)	1 Na^+: 1 HCO_3^-	0.12
Na^+-K^+-$2Cl^-$ cotransporter (NKCC)	1 Na^+: 1 K^+: 2 Cl^-	0.15
Na^+-Mg^{2+} exchanger (NaMgX)	2 Na^+: 1 Mg^{2+}	0.08

Values from Bers et al. (2003).

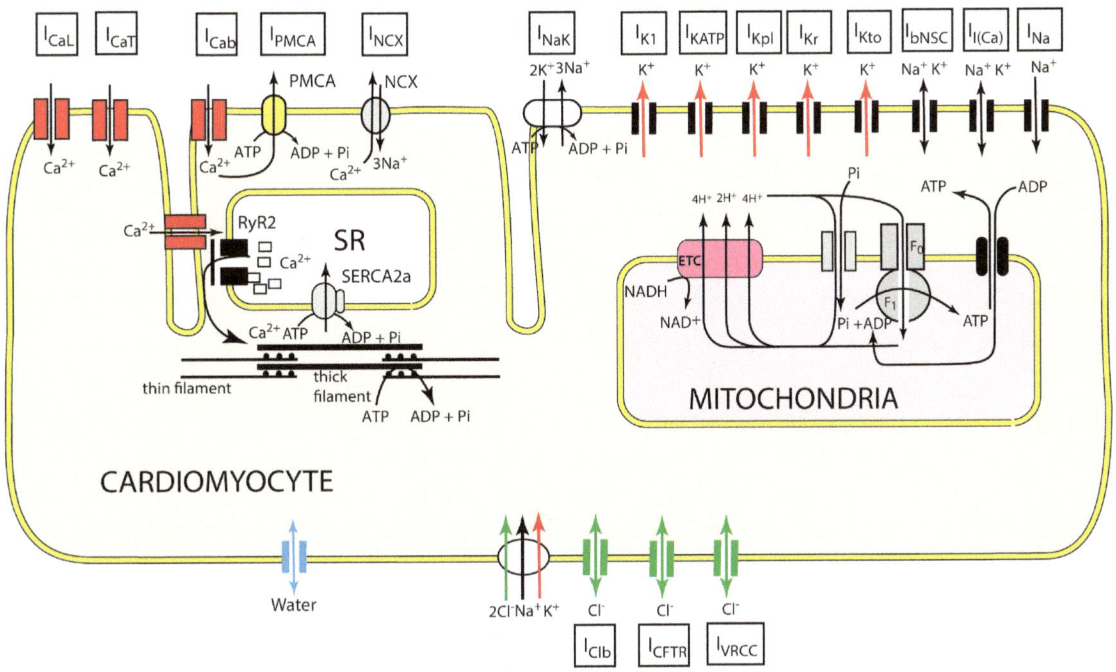

FIGURE 22.5 Kyoto model of ion fluxes in cardiac myocytes illustrates that multiple ion channels, transporters, and pumps play a role in volume regulation, and their contributions depend on conditions. For example, cardiac myocytes resist cell swelling when the Na^+-K^+ pump is inhibited. Under these conditions, cell volume is maintained by the operation of the Na^+-Ca^{2+} exchanger (NCX) and Ca^{2+}-ATPase (PMCA1) working in parallel to extrude Na^+ while hydrolyzing ATP; Ca^{2+} influx via NCX is matched by Ca^{2+} efflux via PMCA1. *Adapted from Takeuchi et al. (2006).*

These three aspects of compensatory volume responses form a causative cascade—they generally occur in the order given. However, our understanding of the system typically has evolved in the reverse order. We often experimentally identify osmolyte transporters first, identify the signals second, and figure out how they are initiated last. The presentation of the entire cascade is best understood as a recapitulation once all the parts are known.

A summary of the major transporters involved in regulatory volume responses is shown in Fig. 22.6 (Jentsch, 2016). In isosmotic conditions, water exchanges freely across the membrane, but there is no net driving force for net water movement. Similarly, ions are transported in both directions, but the net transport is zero. Cells exposed to hypertonic media shrink due to water movement out of the cell, concentrating cellular contents until osmotic equilibrium is reached. These cells undergo RVIs by accumulating osmolytes and thereby drawing water into the cell by osmosis. Different cells may use different sets of transporters, and the transporters that are used to effect RVIs are not exclusively, or even mainly, used for volume regulation. The principal mechanisms of RVI involve the Na^+-K^+-$2Cl^-$ transporter (primarily NKCC1, gene SLC12A2), discussed in Chapter 14. Na^+ influx is increased by activation of the Na^+-H^+ exchanger (NHE, mainly NHE1, gene SLC9A1) and taurine transporter (TauT, SLC6A6). Often the Cl^--HCO_3^- anion exchanger (AE, mostly AE2, SLC4A2) is increased in parallel with NHE. In some cells, betaine transporter (BGT1, SLC6A12) is used to accumulate betaine, and others accumulate myo-inositol using SMIT1 (SLC5A3). In other cells, intracellular synthesis of organic osmolytes such as myo-inositol and sorbitol is increased.

Cell swelling initiates an RVD by reducing the osmolytes within the cell. Swelling activates efflux pathways for K^+ and Cl^- through an electroneutral K^+-Cl^- cotransporter (mainly KCC1, KCC3, and KCC4, genes SLC12A4, 6, and 7, respectively) or through separate K^+ and Cl^- channels. Some of these are regulated by intracellular [Ca^{2+}], such as TMEM16A (also known as anoctamin) or BK_{Ca}, for big K^+ conductance activated by Ca^{2+}. The volume-regulated anion channel (VRAC) and $I_{Cl,swell}$ are used to describe channels that carry currents with similar properties. Nevertheless, they appear to be part of heterogenous groups of anion channels with certain phenotypic differences including channels that also transport organic osmolytes such as taurine (volume-sensitive organic anion channels, VSOACs). Expression of these distinct channel proteins varies between cells, and multiple types can be expressed in the same cell.

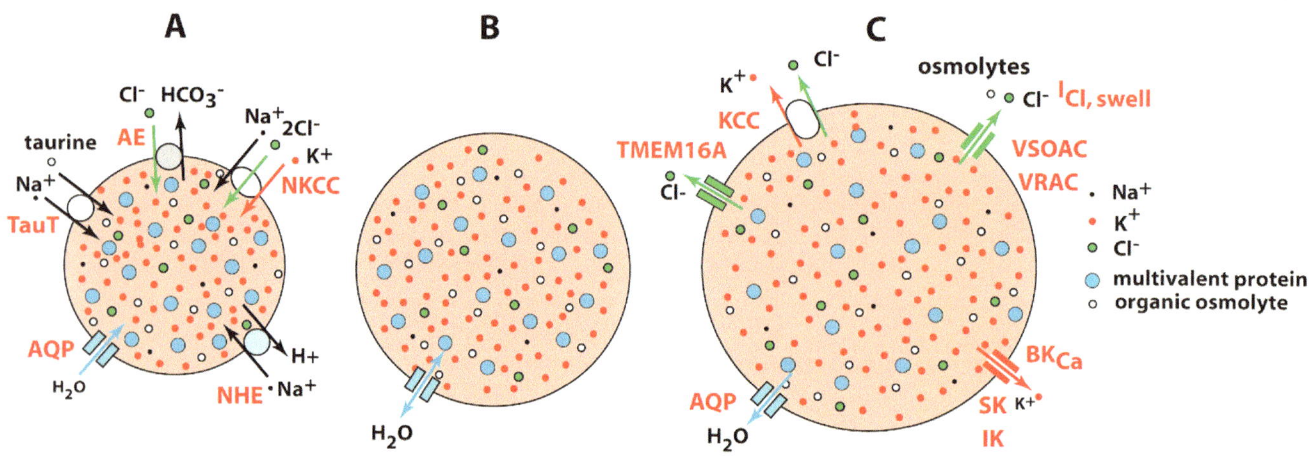

FIGURE 22.6 Principal transporters involved in regulatory volume increases (RVIs) and regulatory volume decreases (RVDs). Panel B: The normal isosmotic situation in which K^+ ions constitute the main intracellular cation and multivalent proteins or phosphates constitute the main intracellular anions. Panel A: Cell shrinkage activates transporters to increase cellular content of solutes, generally by increasing Na^+ influx coupled to Cl^- either directly, as in the Na^+-K^+-$2Cl^-$ transporter (primarily NCKK1 = SLC12A2) or indirectly through Cl^- channels (not shown). Na^+ influx is increased by activation of NHE (mainly NHE1 = SLC9A1) and TauT (SLC6A6), a Na^+- and Cl^--dependent taurine transporter. In some cells, betaine transporters (BGT1 = SLC6A12) are used to accumulate betaine, and others accumulate myo-inositol (SMIT1 = SLC5A3). In other cells, intracellular synthesis of organic osmolytes such as myo-inositol and sorbitol is increased. Cl^--HCO_3^- exchange (AE, mostly AE2 = SLC4A2) is also increased in many cells during RVI in parallel with NHE. Panel C: Cell swelling activates efflux pathways for K^+ and Cl^- through an electroneutral K^+-Cl^- cotransporter (mainly KCC1, KCC3, and KCC4 = SLC12A4, 6, and 7) or through separate K^+ and Cl^- channels. Some of these are regulated by intracellular Ca^{2+}, such as TMEM16A (also known as anoctamin), or BK_{Ca}, for big K^+ conductance activated by Ca^{2+}. The volume-regulated anion channel (VRAC) and $I_{Cl,swell}$ appear to be part of a heterogenous group of channels that also transport organic osmolytes such as taurine (VSOAC). In both RVI and RVD, aquaporins equilibrate the osmotic pressure and the Na^+-K^+-ATPase exchanges Na^+ influx (for RVIs) or K^+ efflux (for RVDs) with its opposing cation. *Adapted from Jentsch (2016).*

5. RVI mechanisms after shrinking

5.1 The Na^+-H^+ exchanger

The Na^+-H^+ exchanger (NHE) is a family of exchange proteins of the SLC9 gene family that has at least 11 unique genes. The primary structures of the proteins vary from 48 to 99 kDa and can be divided into three principal clusters on a phylogenetic basis (Orlowski & Grinstein, 2011). The first cluster contains SLC9A1-5, corresponding to proteins NHE1-5. These are the plasma membrane type of NHE, although this description is misleading as NHE3 and NHE5 are in a dynamic equilibrium with endosomal pools and are not restricted to the cell surface. The second cluster contains NHE6-9 (genes SLC9A6-9) that are localized mainly in intracellular compartments in the secretory or endocytic pathways. Plasma membrane NHEs export H^+ out of the cell in exchange for Na^+ entry, driven by the free energy (electrochemical potential) for Na^+ entry provided by the negative intracellular potential and the much higher Na^+ concentration in extracellular fluid than inside the cell. This is the "forward" operation of the NHEs, but they can also work in the "reverse" mode. In most cells, the function of NHE is to regulate the intracellular pH. Cells produce H^+ continuously as a consequence of their metabolism, and the H^+ must be continually exported. The kidney epithelium expresses NHE2 and NHE3 on the apical membrane that enables a large fraction of reabsorption of filtered NaCl. These NHE isoforms are also involved in HCO_3^- reabsorption and NH_4^+ excretion and so they contribute importantly to organismal pH regulation.

In many cells, NHE1 is important for volume regulation. Cala (1980) provided some of the earliest data in support of NHE1 involvement in RVI when he showed that the RVI after osmotic shrinking of red blood cells from *Amphiuma* could be blocked by amiloride, was stimulated by extracellular HCO_3^-, resulted in medium acidification, and was electrically neutral. Cala postulated that the RVI was accomplished by coupling NHE with Cl^--HCO_3^- exchange so that the coupled processes resulted in accumulation of NaCl with water following osmotically. The net efflux of both H^+ and HCO_3^- by the coupled processes means that the pH of the cell is unchanged by the two working together. Mutant Chinese hamster ovary (CHO) cells that lack NHE do not exhibit an RVI upon cell shrinking. The wild-type CHO cells express NHE and exhibit an RVI upon shrinking that is blocked by amiloride (Rotin & Grinstein, 1989). Lymphocytes that are shrunk osmotically do not show an RVI. However, if these cells are swollen by exposure to hypotonic media, they show an RVD. If these swollen cells that have undergone an RVD are then returned to an isotonic media, they shrink and then show an RVI that is blocked by amiloride (Grinstein et al., 1983). NHE1-deficient fibroblasts lack an RVI response to hypertonic cell shrinkage, but transfection with NHE1 restores the RVI response (Maeno et al., 2006).

How the NHE1 is activated to produce an RVI is unclear. NHE1 is an integral membrane phosphoglycoprotein with 815 amino acids and an approximate molecular weight of 110 kDa. It has 12 transmembrane α-helices and a 300 amino acid cytoplasmic tail. The cytoplasmic tail contains serine residues that are constitutively phosphorylated in quiescent cells and that have increased phosphorylation in response to growth factors and other signals (Putney et al., 2002). However, hypertonic activation of NHE1 occurs without detectable changes in phosphorylation (Grinstein et al., 1992; McSwine et al., 1996). One hypothesis is that the cytoplasmic domain has allosteric regulatory sites for H^+ that alter its sensitivity to intracellular $[H^+]$ so that acidification greatly activates the NHE1 (Aronson et al., 1982).

5.2 The Cl^--HCO_3^- exchanger

The anion exchangers move bicarbonate into or out of cells in exchange for chloride (Chapter 15). Moving HCO_3^- into the cell alkalinizes the cell, and moving HCO_3^- out of the cell acidifies it. In addition to maintaining cell pH, these transporters also help regulate cell volume and acid/base secretion. Anion exchangers are members of two distinct gene families, SLC4A and SLC26A (Bonar & Casey, 2008). The SLC4A family consists of the Na^+-independent anion exchangers AE1, AE2, and AE3 encoded by genes SLC4A1, SLC4A2, and SLC4A3, respectively; Na^+-dependent Cl^--HCO_3^- transporters (NDCBE, SLC4A8); and Na^+-dependent HCO_3^- transporters (NBCe1, NBCe2, NBCn1, and NBCn2) (Romero et al., 2013). The SLC26A genes encode a family of anion exchangers that transport Cl^-, HCO_3^-, sulfate, oxalate, I^-, and formate (Alper & Sharma, 2013). Pendrin (SLC26A4) has important functions for I^- release into the thyroid colloid (Yoshida et al., 2002) and in the pH regulation of the endolymph (Hibino et al., 2010), and prestin (SLC26A5) is important for the electromotility of outer hair cells in the cochlea (Liberman et al., 2002; Ludwig et al., 2001). For our purposes here, AE2 is the most important of these anion exchangers because it has a specific role in volume regulation when coupled with NHE1, as described before. Hypertonic stress activates Cl^--HCO_3^- exchange in a variety of cell types. AE2 expressed in *Xenopus* oocytes has been shown to be activated by hypertonic stress, whereas AE1 lacks this response (Humphreys et al., 1995).

5.3 The Na^+-K^+-$2Cl^-$ and K^+-Cl^- cotransporters

The SLC12A gene family has nine members, seven of which (SLC12A1-7) have been shown to be electroneutral **c**ation-coupled **c**hloride **c**otransporters (Arroyo et al., 2013), or CCCs. CCCs come in a variety of types. There are two types of Na^+-K^+-$2Cl^-$ transporters, NKCC1 (SLC12A2) and NKCC2 (SLC12A1), and the remaining members of the family are the Na^+-Cl^- cotransporter (NCC) and K^+-Cl^- cotransporters (KCC1–4). These CCCs contribute to multiple functions of cells, including regulation of intracellular $[Cl^-]$, transepithelial transport of ions and water, and regulation of cell volume as discussed in Chapter 14. The location and function of the CCCs are shown in Table 22.2. Most cells express NKCC1 on their surface membrane, but the expression of NKCC2 is restricted to the apical membranes of cells of the thick ascending limb of the loop of Henle (reviewed in Delpire & Gagnon, 2018). NKCC2 plays a crucial role in salt reabsorption and the organismal regulation of salt and water. NKCC1 is the main cotransporter involved in RVI.

TABLE 22.2 The SLC12 family of cation-chloride cotransporters.

Gene/Protein	Location	Physiological function
SLC12A1/NKCC2	Thick ascending limb of loop of henle	NaCl reabsorption; regulate organismal osmolarity
SLC12/A2/NKCC1	Ubiquitous	Cell volume regulation: RVI
SLC12A3/NCC	Distal convoluted tubule of kidney tubule	NaCl reabsorption in DCT; regulate Ca^{2+} and K^+ excretion
SLC12A4/KCC1	Ubiquitous	Cell volume regulation: RVD
SLC12A5/KCC2	CNS	Regulation neuronal $[Cl^-]$
SLC12A6/KCC3	Lung, heart, kidney, brain	Volume regulation in the brain; K^+ recycling in the kidney
SLC12A7/KCC4	Lung, heart, kidney, brain	Acid secretion in α intercalated cells of collecting duct

From Arroyo et al. (2013) and de los Heros et al. (2018).

The common characteristic of the CCC family is that (1) they all couple transport of one cation, either Na^+ or K^+, for each anion and that anion is always Cl^-. Therefore, these cotransporters are all electroneutral; (2) they are all modulated by cell volume and by the intracellular Cl^- concentration indirectly through protein phosphorylation/dephosphorylation. The NKCC1, NKCC2, and NCC all move Cl^- into the cell, while KCCs move Cl^- out of the cell. Thus, NKCCs and NCC increase the cellular osmolytes and are involved in RVI, whereas the KCCs decrease the cellular osmolytes and are involved in RVD. The NKCCs and KCCs are reciprocally regulated during RVIs and RVDs. Thus, the transporters that promote RVDs are turned off when the transporters that promote RVIs are turned on (de los Heros et al., 2018).

NKCC1 has multiple regulatory threonine phosphorylation sites on the N-terminus, whereas the KCCs have multiple regulatory phosphorylation sites on the C-terminus. Phosphorylation activates NKCC1 and NCC and inhibits KCCs (Delpire & Gagnon, 2018). The regulation of the phosphorylation of the CCCs is complicated, involving regulation of kinases and phosphatases that set the phosphorylation state of these cotransporters. Kinases called SPAK (Ste10-related proline—alanine-rich kinase) and OSR1 (oxidative stress—responsive kinase-1) directly phosphorylate the CCCS (Dowd & Forbush, 2003; de los Heros et al., 2014). SPAK/OSR1 kinases themselves are activated by phosphorylation by the regulatory kinase WNK (with no lysine kinase), a family of serine—threonine kinases encoded by genes WNK1-4 (de los Heros et al., 2018). WNKs are activated by autophosphorylation consequent to osmotic stress. WNK1 and WNK4 respond directly to low intracellular [Cl^-], whereas WNK3 does not appear to be regulated by [Cl^-] (Murillo de Ozores et al., 2020). Chloride binds to the WNK1 catalytic domain and inhibits its autophosphorylation and kinase activity (Piala et al., 2014). The WNKs are also regulated by a mitogen-activated protein kinase cascade initiated by a member of the MAP3K (mitogen-activated protein kinase kinase kinase) family, called ASK3 (apoptosis signal-regulating kinase 3) (Zhou et al., 2016). In the MAPK pathway, the most upstream group are the MAPK kinase kinases or MAP3Ks, such as ASK3, that phosphorylate serine or threonine residues in MAPK kinases (MAP2K or MAPKKs). In turn, MAP2Ks phosphorylate MAPKs. Examples of MAPKs that are involved in volume regulation include extracellular signal-regulated kinase (ERK1/2), c-jun N terminal kinase (JNK), and p38MAPK (Zhou et al., 2016). WNK4 is phosphorylated through the p38MAPK pathway but only for hypotonic stress. Hypotonic stress activation of WNK4 requires ASK3, but hypertonic stress activation of WNK4 does not require ASK3. Thus, the upstream modulators may depend on the type of osmotic stimulus (Maruyama et al., 2016). Activation of WNK1, on the other hand, is reduced by ASK3 activation by hypotonic stress (Naguro et al., 2012). There is evidence of yet further upstream regulation of this pathway, as ASK3 itself is reversibly phosphorylated and is dephosphorylated within minutes after hyperosmotic stress by PP6 (protein phosphatase 6) (Watanabe et al., 2018). Fig. 22.7 illustrates the signaling cascade involved in the RVI response of the cation-coupled chloride cotransporters.

5.4 Osmolyte transporters

Organic, electroneutral osmolytes including taurine, betaine, myo-inositol, sorbitol, and glycerophosphocholine can significantly affect cell volume. Taurine is particularly interesting in that it is synthesized in liver, brain, lungs, and skeletal muscle, but there are no degradative pathways, and it is excreted either in urine or as conjugated bile acids in feces. Cells take up taurine by using two transporters, TauT and the β-amino acid transporter, PAT1. Activity of TauT is inhibited by osmotic cell swelling, PKC, and ROS (reactive oxygen species) and is activated by PKA (Lambert et al., 2015). Long-term modulation of TauT activity involves changing the number of transporters in the membrane through increased transcription of the gene. Downregulation follows long-term exposure to hypotonic solutions or high extracellular [taurine] and involves activation of the tumor suppressor protein p53. Up-regulation involves activation of NFAT5 (nuclear factor of activated T cells), also known as TonEBP or OREBP (tonicity response element binding protein; osmotic response element binding protein), apparently by increased stability of the TonEBP/NFAT5 mRNA by high [NaCl] (Burg et al., 2007). TonEBP/NFAT5 binds to a variety of genes to increase their expression and to increase the amount of the final proteins produced from the set of mRNA (Burg et al., 2007). TonEBP is a transcription factor for TauT, aldose reductase (AR) that synthesizes sorbitol from glucose, betaine/GABA transporter (BGT1) that couples betaine import to that of Cl^- and Na^+, SMIT (sodium myo-inositol transporter) that couples import of inositol to 2 Na^+, and NTE (neuropathy target esterase) that produces glycerophosphocholine (GPC) from phosphatidylcholine (Burg et al., 2007).

Taurine exits cells by two pathways, a volume-sensitive pathway (VSOAC) that responds to cell swelling and a volume-insensitive pathway. In lung epithelial cell line A549 in culture, swelling-induced taurine efflux was suppressed by inhibitors of 5-lipoxygenase or the cysteinyl leukotriene 1 receptor ($CysLT_1$). Hypoosomotic swelling in these cells produces reactive oxygen species (ROS) through NADPH oxidase (NOX). Thus, swelling-induced taurine release is suppressed by scavenging ROS and potentiated by H_2O_2 (Holm et al., 2013). These ROS-dependent effects are similar to those observed in cardiac cells with activation of $I_{Cl,swell}$ as a final effect, as discusseed in Section 7.2. Thus, this pathway

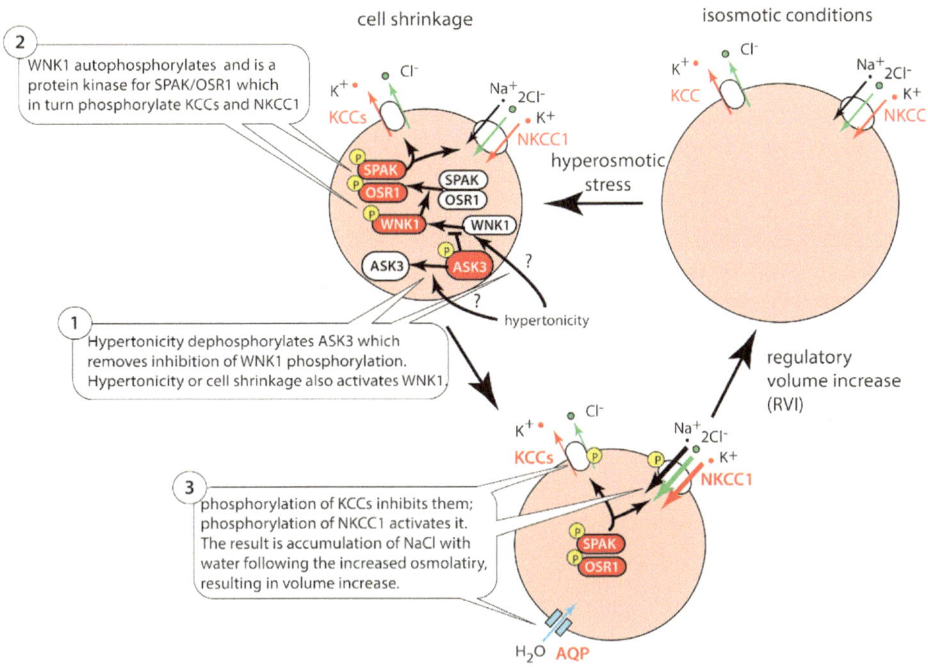

FIGURE 22.7 Signaling pathways for regulation of CCCs (cation-coupled chloride cotransporters) during regulatory volume increase (RVI). Exposure of cells to a hyperosmotic medium results in fast cell shrinkage due to water exit. This causes dephosphorylation of ASK3 (apoptosis signal-regulating kinase 3), a member of the mitogen-activated protein kinase (MAPK) cascade. It is a MAPK3 or MAP kinase kinase kinase. Through its MAPK cascade, phosphorylated ASK3 inhibits phosphorylation of WNK1 (with no lysine kinase 1). Dephosphorylation of ASK3 promotes phosphorylation of WNK1. WNK1 phosphorylation is also caused directly by cell shrinkage by unknown mechanisms. Phosphorylated WNK1 is an active protein kinase that phosphorylates SPAK or OSR1 (Ste10-related proline–alanine-rich kinase or oxidative stress–responsive kinase-1) that then phosphorylates KCC (K^+-Cl^- cotransporter) and NKCC1 (Na^+-K^+-$2Cl^-$ cotransporter). Phosphorylation inhibits KCCs and activates NKCC1. The result is a reduction in efflux of K^+ and Cl^- over the KCCs and increase in influx of Na^+, K^+, and Cl^- over the NKCC1. The increased osmolyte concentration draws water into the cell over AQP (aquaporin), leading to an increase in cell volume, or RVI. The figure does not explain everything about RVI, only the signaling pathways for KCCs and NKCC1. Other pathways, shown in Fig. 22.6, also participate in RVIs.

appears to involve swelling-induced activation of phospholipase A_2 and release of arachidonic acid from membrane lipids, oxidation of the arachidonic acid to eicosanoids by 5-lipoxygenase, and binding to $CysLT_1$. The full molecular identity of VSOAC is not yet definitive, but it involves the LRRC8 (**l**eucine-**r**ich **r**epeat **c**ontaining protein **8**) heteromers that also are thought to comprise the channel underlying VRAC and $I_{Cl,swell}$, except the taurine carrier has a different composition of LRRC8A-E subunits (Schober et al., 2017).

6. RVD mechanisms after cell swelling

6.1 Volume-regulated anion channel

In most cells, swelling activates a chloride current, $I_{Cl,swell}$, that is mediated by a channel variously called volume-regulated anion channel (VRAC) or volume-sensitive organic anion channel (VSOAC) (Pedersen et al., 2015). In bovine endothelial cells, $I_{Cl,swell}$ is activated by reducing the ionic strength, Γ, at a constant osmolarity and [Cl^-] (Nilius et al., 1998). In these cells, $I_{Cl,swell}$ is activated through a protein tyrosine kinase. Inhibitors of protein tyrosine kinase blocked the activation of $I_{Cl,swell}$ by low Γ, suggesting that low Γ elicits $I_{Cl,swell}$ by activating the protein tyrosine kinase rather than by a direct effect on VRAC. In CHO cells, VSOAC was shown to become activated specifically by low ionic strength (Cannon et al., 1998). However, local Ca^{2+} signaling is required for proper activation of VRAC (Benedetto et al., 2016). The molecular identity of VRAC remained elusive for many years and was finally identified as being a heteromer of LRRC8 (**l**eucine-**r**ich **r**epeat **c**ontaining protein **8**A) (Qui et al., 2014; Voss et al., 2014). The LRRC8 family has five known members, LRRC8A-E. The monomer has a M_r of about 95 kDa and associates in the membrane as a homo- or heterohexamer (Bertelli et al., 2021). The LRRC8A forms an essential part of the VRAC. Different heteromers appear to have different conductivities and specificities, with some transporting organic osmolytes such as taurine (Voss et al., 2014). Moreover, LRRC8A/E heteromers are activated >10-fold by cysteine oxidation of 8E at C424 and C448 forming a disulfide bond, whereas 8A/C and 8A/D heterodimers are inhibited by oxidation (Bertelli et al., 2022; Gradogna et al., 2017). Insertion of the purified proteins into lipid bilayers showed that the LRRC8 heteromers could become activated by lowering Γ in the absence of any other regulatory proteins (Syeda et al., 2016). These authors suggested that VRAC itself has a sensor for ionic strength. However, the changes in ionic strength that were used in these studies are quite large and unphysiological. It is possible that local ionic strengths due to water fluxes could reach these levels, but alternative routes of activation are available (Bertelli et al., 2021). LRRC8A is colocalized with caveolin (Benedetto et al., 2016) and caveolin-1β modulates the activity of VRAC (Trouet et al., 1999), suggesting that distortion of caveolae upon swelling is an important part of activation of VRAC.

6.2 TMEM16A/anoctamin

Swelling activated Cl^- currents and the consequent RVD can still be observed in cells that have no LRRC8A, pointing to the importance of Cl^- channels other than the LRRC8A channel in RVDs (Sirianant et al., 2016). One of these channels is a member of the TMEM16 family of proteins. TMEM16, also known as the anoctamins, forms a family of 10 proteins that can be subdivided into two major subgroups. TMEM16A and TMEM16B (ANO1 and ANO2) are Ca^{2+}-activated Cl^- channels, whereas the other TEMEM members appear to act primarily as Ca^{2+}-activated lipid scramblases (reviewed in Kunzelmann, 2015). However, TMEM16F (ANO6) also produces a Ca^{2+}-activated Cl^- channel. TMEM16A (ANO1) is activated by local increases in $[Ca^{2+}]$ caused by release of Ca^{2+} from ER stores by the inositol trisphosphate receptor (IP_3R) (Thorn et al., 1993) and aided by tethering of ANO1 to the IP_3R (Jin et al., 2016). TMEM16A and LRRC8A/C interact by their shared dependence on the local $[Ca^{2+}]$. TMEM16A augments activation of LRRC8A/C by enhancing the swelling-activated local intracellular $[Ca^{2+}]$, presumably by virtue of the interaction of TMEM16A with the IP_3R. Thus, TMEM16A contributes to VRAC by its own Cl^- conduction and its enhancement of the swelling-induced increase in $[Ca^{2+}]$ augments current carried by LRRC8A/C (Centeio et al., 2020; Jin et al., 2016; Kunzelmann, 2015).

6.3 Bestrophins

A third type of Ca^{2+}-activated Cl^- channel are the bestrophins, a family of four members named Best1, Best2, Best3, and Best4. Best1 is mainly expressed in the basolateral membrane of cells in the human retinal pigment epithelium (RPE), and mutations in this protein produce Best vitelliform macular dystrophy (Best disease or BVMD) (Owji et al., 2021). Best1 is crucial for volume regulation in RPE cells, whereas LRRC8 is not (Milenkovic et al., 2015). These authors noted that Best1 is most highly expressed in mouse spermatozoa and that lack of Best1 in these cells renders them unable to cope with changes in osmotic pressure (Milenkovic et al., 2015). Like ANO1, Best1 has been postulated to be partially localized in the ER and to modulate Ca^{2+} handling by the ER (Neussert et al., 2010).

Although VRAC shares certain common characteristics in almost all cells, it now is clear that a single channel is not responsible. Rather, a diversity of at least three distinct families of Cl^- channels contribute (LRRC8, TEMEM16, and Bestrophin) with distinct phenotypes and functional and regulatory properties that are tailored to the specific requirements of the various cell types in which they are found.

6.4 Ca^+-activated K^+ channels

Ca^{2+}-activated K^+ channels (K_{Ca}) are widely expressed in excitable and nonexcitable tissues. They constitute a heterogenous family of channels whose open probability is increased by elevated cytosolic $[Ca^{2+}]$. Along with the KCC (K^+-Cl^- cotransporter), they provide a major efflux pathway for K^+ during the RVD response to cell swelling.

The K_{Ca} family are divided into three groups based on their single-channel conductance and pharmacological sensitivities. The big conductance (BK) group (100−300 pS) includes the K_{Ca} 1.1 channel; the intermediate conductance (IK) group (25−100 pS) consists of the K_{Ca} 3.1 channel; the small conductance (SK) group (2−25 pS) includes K_{Ca} 2.1, K_{Ca} 2.2, and K_{Ca} 2.3 channels. All of these are gated by intracellular $[Ca^{2+}]$, but BK_{Ca} is also activated by depolarization (Vergara et al., 1998). BK_{Ca} consists of a tetramer of pore-forming α-subunits, which may associate with four distinct β subunits, β1−4, or four distinct γ subunits, γ1−4 (Sancho & Kyle, 2021). BK_{Ca} channels are specifically inhibited by the exogenous poisons charybdotoxin and iberiotoxin and are regulated by a variety of physiologically important signals such as PKA, PKG, PKC, membrane lipids (phosphatidyl inositol bisphosphate, PIP_2, cholesterol, and omega-3 fatty acids) and redox derivatives (DNTB, NAD, glutathione, peroxinitrate) (Sancho & Kyle, 2021). Upstream signals regulating BK_{Ca} include NO, angiotensin II, endothelin-1, 5-hydroxytryptamine, CGRP, and prostaglandins (Guntur et al., 2021). Most reviews of BK_{Ca} focus on its function in cell excitability or K^+ handling, without specifically mentioning its importance in volume regulation (Guntur et al., 2021; Vetri et al., 2014). However, inhibition of BK_{Ca} by the specific inhibitor, iberiotoxin, abolished the RVD in eel intestinal cells following hypotonic swelling (Lionetto et al., 2008).

7. Volume sensors

In general, changes in the volume of cells impacts the cell in four separate ways (Hoffmann et al., 2009). These are

- macromolecular crowding
- ionic strength or concentrations of specific ions
- mechanical or chemical changes in the bilayer
- stress on the cytoskeleton

Importantly, each of these may modulate transport proteins and the signaling pathways that regulate them.

7.1 Macromolecular crowding

"Macromolecular crowding" refers to the reduction of available water when the cytoplasm becomes crowded with proteins. It is related to the "excluded volume" that gives rise to the nonideality of osmotic pressure caused by proteins (see Chapter 11 of this volume). Molecular crowding can manifest itself in increased osmotic pressure, conformational changes in proteins, and increased viscosity. If we were to add a test molecule to the cytoplasm, the volume available to the test molecule depends on its size, and so does the free energy change for the transfer. If the test molecule is small, most of the volume not occupied by other molecules would be available to the test molecule, as shown in Fig. 22.8A. If the test molecule were a sphere of radius r, then the center of the test molecule could come no closer to the surfaces of the other

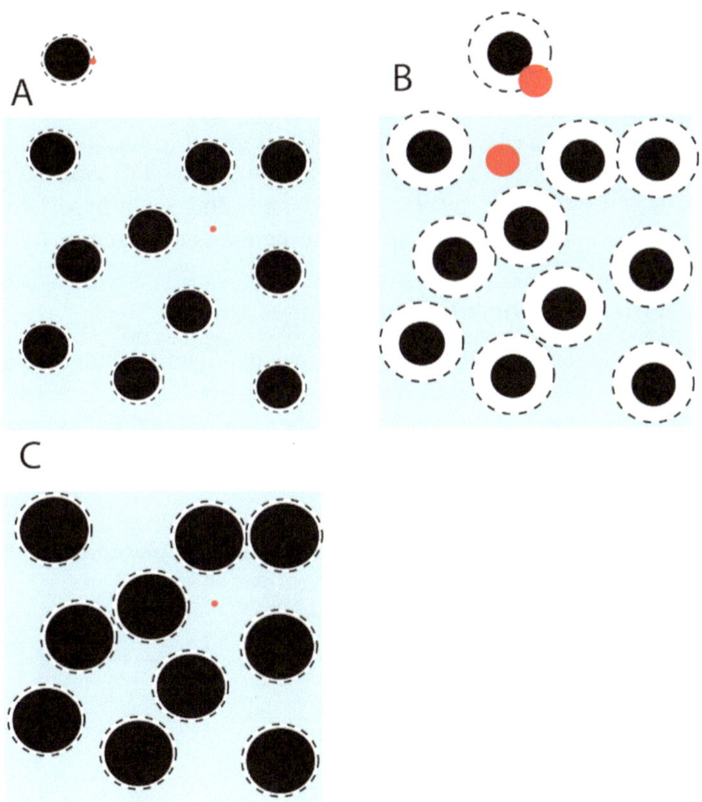

FIGURE 22.8 The principle of macromolecular crowding. A test molecule, shown as small solid circles, is added to a solution of macromolecules. The volume of water available to the test molecule is shown as the open space between molecules. The center of the test molecule cannot come nearer to any surface than the radius, r, of the test molecule. Therefore, any volume closer than r to the surfaces of other molecules is unavailable or excluded for the test molecule (see insets above panels A and B). In panel A, the test molecule is small. In this case, each macromolecule excludes only slightly more than the volume it occupies. The test molecule occupies a volume smaller than the total volume, and so its apparent activity is higher than that calculated on the basis of total volume. If water is removed, the activity increases proportionately more than the fraction of water removed. In panel B, the test molecule is larger than in panel A. Therefore, the volume accessible to the center of the test molecule is much smaller, and the effect of the macromolecules on the activity of the test molecule is much larger. In panel C, the macromolecules are larger and occupy a larger fraction of the volume for the same concentration of macromolecules. In this case, the region excluded at the surface of each macromolecule is greater. Thus, macromolecular crowding increases with the molecular weight of the macromolecule. *Adapted from Minton (2001).*

molecules than its radius, *r*. The available volume can be visualized by drawing a shell of thickness *r* around all of the remaining molecules, as shown in Fig. 22.8B. Because there is less available volume, the effective concentration of solutes, formally, their activity, is altered by the presence of macromolecules. The ratio of the effective concentration to the actual concentration is *defined* as the activity coefficient. The dependence of the activity coefficient of proteins on their concentration is highly nonlinear. The activity coefficient of hemoglobin, for example, can be deduced directly from the osmotic coefficient and is large, ~3–5, over the normal range of hemoglobin concentrations in red blood cells (Minton, 2001). In contrast, the activity coefficient for the major inorganic anions and cations is <1. Changes in cell volume change the activity of solutes, which in turn alters the equilibria between free solutes and binding sites or alters equilibria of protein conformational states (Speer et al., 2022). Summers et al. (1997) tested this hypothesis in barnacle muscle cells by isotonic replacement of large macromolecules with smaller solutes (sucrose) that leave a larger available volume and found that macromolecular crowding could explain the activation of a verapamil-sensitive Ca^{2+} influx channel upon swelling. Ca^{2+} influx is necessary for the RVD in barnacle muscle because the RVD is abolished in the absence of extracellular Ca^{2+} or by blockade of the Ca^{2+} influx channel with verapamil. A possible proposal for the mechanism is shown in Fig. 22.9. Macromolecular crowding can also affect proteins in biological membranes. Macromolecular crowding reduces the lateral diffusion within the plane of the membrane, can potentially alter membrane shape (Lowe et al., 2020), and can affect catalysis by membrane enzymes such as phospholipase (Wei et al., 2019).

A corollary of macromolecular crowding may be the formation of liquid–liquid phase separations (Andre & Spruijt, 2020). Liquid–liquid phase separations occur when a uniformly dispersed molecule spontaneously separates into a highly concentrated, condensed phase and a dilute phase, forming structures variously called membraneless organelles, liquid droplets, granules, phase-separated bodies, biomolecular condensates, and assemblies (Zhang et al., 2021). These structures undergo dynamic and rapid phase transitions. There is a long list of biomolecules that undergo liquid–liquid phase separations as a consequence of cell volume perturbation (Ishihara et al., 2021). One of these is ASK3 (apoptosis signal-regulating kinase 3, a member of the MAP kinase kinase kinase family), which bidirectionally responds to osmotic stress and contributes to cell volume responses. Macromolecular crowding during hypertonic stress induces liquid phase condensates of ASK3 that inactivates it (Watanabe et al., 2021). As discussed in Section 5.3, ASK3 inactivation releases its inhibition of WNK1 so that the Na^+-K^+-2Cl^- cotransporters can be activated to load the cell with Na^+, K^+, and Cl^- to help form the RVI. Thus, at least in this case, macromolecular crowding upon hypertonic stress appears to form condensates of ASK3, and the formation of condensates appears to be the volume sensor for the cell.

7.2 Ionic strength or specific ion concentration changes

Cell swelling or cell shrinking by osmotic equilibration entails the addition or subtraction of water from the cell. At least temporarily, the activities of ions will be reduced, in the case of cell swelling, or increased in the case of cell shrinking. Theoretically, this could change the membrane potential and alter ion fluxes through alteration of their electrochemical driving force and through voltage-dependent conductances. Cells are quite sensitive to volume changes, responding to

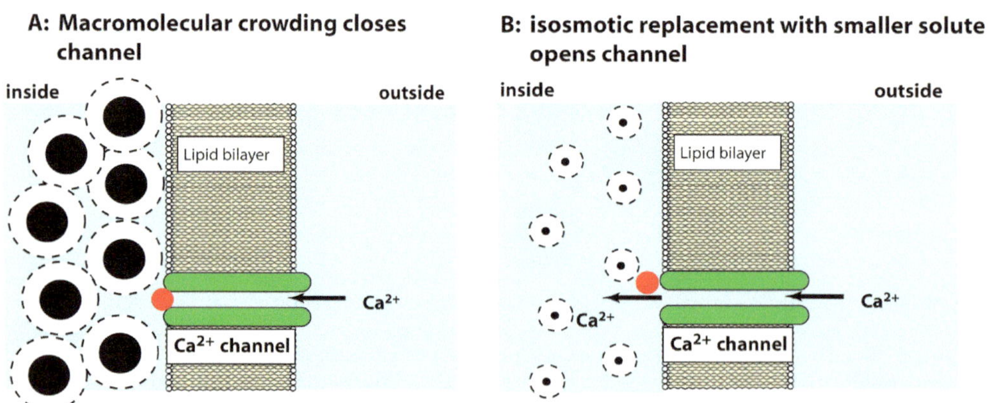

FIGURE 22.9 Model for control of a channel by macromolecular crowding. Panel A: Macromolecules restrict the water volume near a channel, increasing the activity of the channel regulator (small solid circle) and causing inactivation of the channel. Dashed circles in white indicate the volume of fluid around the macromolecules that is inaccessible to the center of the regulator. Panel B: Isotonic replacement of the large macromolecules with smaller ones increases the volume that is accessible to the regulator, reducing its activity, resulting in relief of channel inhibition. *Adapted from Summers et al. (1997).*

changes of less than 5% of volume. In a cell that obeys the Nernst equation, such a change would alter the equilibrium potentials, such as E_K, by only about 2 mV, which is too small to have much effect. Thus, changes in membrane potential are unlikely to be the mechanisms by which cells sense their volume. The activation of $I_{Cl,swell}$ by the reduction in ionic strength, Γ, was discussed earlier (Section 6.1), but $I_{Cl,swell}$ also can be activated without an apparent change in Γ as discussed in the next section.

7.3 Mechanical or chemical changes in the bilayer

As mentioned before, cell swelling activates a chloride current called $I_{Cl,swell}$. How is this current activated? In addition to the possibilities of lower macromolecular crowding and reduced ionic strength accompanying swelling, investigators have proposed a role for tension in the bilayer mediated either through membrane proteins attached to the extracellular matrix and the cytoskeleton or release of signaling molecules from the bilayer lipids.

In cardiomyocytes, Browe and Baumgarten (2003) showed that pulling on β_1 integrins can activate a current that had the characteristics of $I_{Cl,swell}$ under isosmotic conditions, and this effect was blocked by inhibitors of tyrosine protein kinase FAK or Src, which are known to interact with complexes of proteins in the costameres of striated muscle. Integrins are heterodimers that bind to both extracellular matrix proteins and intracellular cytoskeletal proteins, thereby making a mechanical link between forces inside and outside of the muscle cell. In vertebrates, the integrin family has 18 α-subunit members and 8 different β-subunits, and these assemble into only 24 known heterodimers (Takada et al., 2007). Cardiomyocytes primarily express $\alpha_1\beta_1$, $\alpha_5\beta_1$, and $\alpha_7\beta_1$ integrin heterodimers, which predominately bind extracellular collagen (Col), fibronectin (FN), and laminin (LN), respectively (Israeli-Rosenberg et al., 2014). On the cytoplasmic side, integrins bind to a complex of proteins at costameres. These proteins link the integrins to the cytoskeleton (e.g., talin and vinculin) or provide a scaffold for binding of other proteins (e.g., PINCH [particularly interesting new cysteine—histidine-rich protein], parvin and paxillin) or provide a signal function (e.g., FAK and Src). Further work on the stretch-activated chloride current in cardiomyocytes showed that angiotensin II (AngII) also activated the current and that blockade of the angiotensin II receptors (AT_1) with losartan blocked the stretch-activated Cl^- current, suggesting that stretch signaled downstream through AT_1 receptors (Browe & Baumgarten, 2004). They deduced that the signaling cascade involved NADPH oxidase and generation of H_2O_2—catalase and inhibitors of NADPH oxidase blocked the response to stretch, whereas adding H_2O_2 elicited the current. Cardiac myocytes express all of the components of a phagocyte-like NADPH oxidase: a core heterodimer consisting of a large $gp91^{phox}$ (NOX2) and a smaller $p22^{phox}$ subunit, and four cytosolic regulatory subunits: $p40^{phox}$, $p47^{phox}$, $p67^{phox}$, and rac1, a small GTP-binding protein (Li et al., 2002). Translocation of the cytosolic subunits and rac1 to the membrane and their assembly with $gp91^{phox}$ and $p22^{phox}$ involves PKC, Src, and phosphatidylinositol-3-kinase (PI-3K) (Seshiah et al., 2002). The activated NADPH oxidases uses cytosolic NADPH to reduce molecular oxygen to superoxide (O_2^-), which is then dismutated by extracellular superoxide dismutase to form membrane permeant H_2O_2.

Stretch, integrins and AngII transactivate epidermal growth factor receptor (EGFR) kinase, which in turn activates PI-3K. Inhibitors of EGFR kinase block the stretch-activated Cl^- current, suggesting that EGFR kinase is downstream of the stretch signal, known now to be mediated by the AT_1 receptor. EGF exposure also activates a Cl^- current, and inhibitors of PI-3K blocked the activation by EGFR. The stretch and EGF-induced currents are also volume-sensitive, suggesting they are carried by the same channels as $I_{Cl,swell}$ (Browe & Baumgarten, 2006). In addition, ET-1 also activated $I_{Cl,swell}$ via ET_A receptors (Deng, Baki, & Baumgarten, 2010). Because ET-1-induced current was abolished by inhibiting EGFR kinase or PI-3K, these kinases must be downstream of ET-1. Activation of $I_{Cl,swell}$ by osmotic swelling or AngII was suppressed by ET_A blockade, but AT_1 receptor blockade did not suppress ET-1-induced Cl^- current. Thus, ET-1 receptor is downstream of the AT_1 receptor. These results established the order of activation: first AT_1 receptor, then ET-1 receptor, and then EGFR, followed by PI-3K, activation of NOX2 and production of O_2^- leading to H_2O_2. The H_2O_2 then activates the Cl^- current. Suppression of O_2^- production by NOX2 with inhibitors, including highy selective gp91ds-tat, or supression of O_2^- production by cardiac mitochondria with rotenone, a complex I inhibitor, completely blocked the ET-1-induced Cl^- current as well as the swelling-induced $I_{Cl,swell}$. Importantly, activation of mitochondrial ROS production at complex III with antimycin A or diazoxide evoked $I_{Cl,swell}$ after block of NOX2, and antimycin A-induced $I_{Cl,swell}$ was unaffected by osmotic shrinkage. This suggests that these sources of ROS act in series rather than by independent pathways and that mitochondrial ROS production is downstream to that by NOX2 and downstream to signalling triggered by osmotic shrinkage (Deng, Baki, & Baumgarten, 2010). The central role of mitochondrial ROS production in cardiac $I_{Cl,swell}$ activation was further confirmed in studies on antiviral HIV protease inhibitors (Deng, Baki, & Yin et al., 2010). Under isosmotic conditions, ritonavir and lopinavir elicited $I_{Cl,swell}$ that was fully supressed by the $I_{Cl,swell}$ inhibitor DCPIB and the mitochondrial e^- transport blocker rotenone but was insensitive to NOX2 inhibitors. These antivirals also

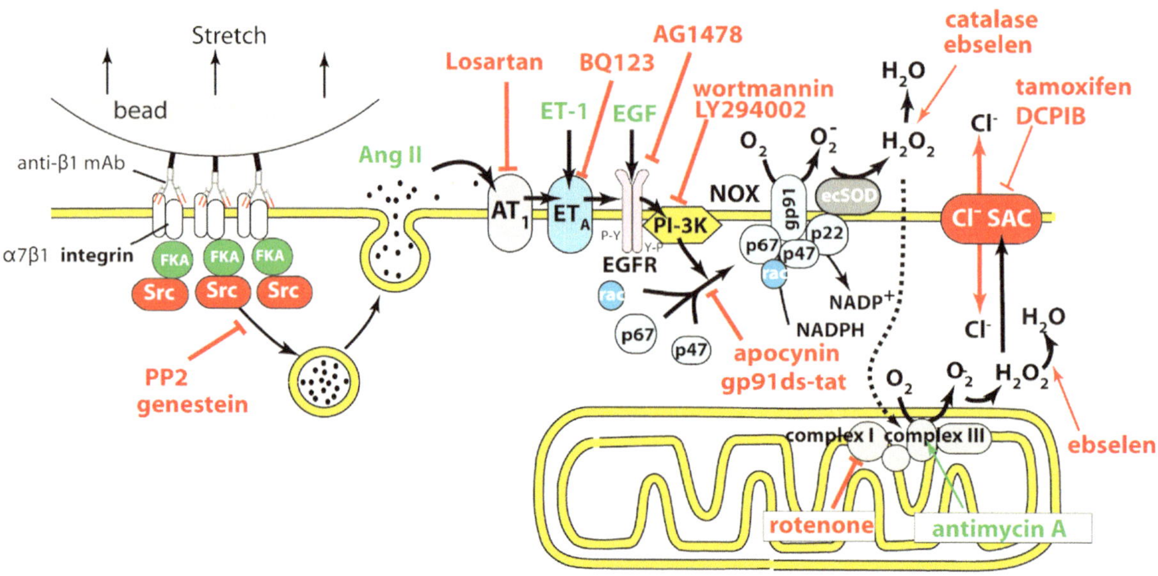

FIGURE 22.10 Proposed scheme for the regulation of $I_{Cl,swell}$ in cardiac myocytes. Osmotic swelling or stretch activates $I_{Cl,swell}$. Stretch was produced by binding a paramagnetic bead to surface integrins through monoclonal antibodies attached to the bead, and then subjecting the bead to a magnetic force. The signal cascade for activation of $I_{Cl,swell}$ was deduced by the judicious use of both inhibitors, shown in red (bar), and agonists, shown in green (arrow). Agonists applied exogenously activated $I_{Cl,swell}$ under isosmotic conditions. FAK and Src are protein tyrosine kinases associated with the complex of proteins binding integrins to the cytoskeleton. Ang II is angiotensin II, which is released from these cells upon stretch and activates AT_1 receptors in an autocrine and paracrine manner. ET_A, endothelin receptor type A; EGF, epidermal growth factor; PI-3K, phosphatidylinositol-3-kinase; gp91 and p22 are membrane-bound subunits of NADPH oxidase 2 (NOX); p67, p47, and rac are cytoplasmic subunits that assemble to form active NOX2 with the help of PI-3K; ecSOD, extracellular superoxide dismutase, which converts superoxide (O_2^-) that is produced by NOX2 to membrane permeant hydrogen peroxide (H_2O_2). $I_{Cl,swell}$ also is activated by exogenous application of H_2O_2.

depolarized mitochondrial membrane potential, $\Delta\Psi_m$, and augmented intracellular ROS as measured with fluorescent probes. Further, antiviral-induced ROS production was prevented by pretreatment with rotenone but not by pretreatment with a highly selective NOX2 inhibitor. Fig. 22.10 summarizes the current model of control of $I_{Cl,swell}$ in cardiac myocytes mediated through the ROS mechanism. Importantly, $I_{Cl,swell}$ can be activated under isosmotic conditions by mechanical stretch, signaling mediators and ROS without an apparent change in Γ.

Membrane lipids may also help control $I_{Cl,swell}$. Exogenous C_2-ceramide, a membrane permeant ceramide, activates $I_{Cl,swell}$, and endogenous ceramides produced by exogenous sphingomyelinase could also activate the current in cardiac myocytes (Raucci et al., 2010). The signaling mechanism appears to involve some of the same pathways as shown in Fig. 22.10, and activation of $I_{Cl,swell}$ by exogenous sphinogomyelinase was blocked by ebselen, a cell-permeant glutathione peroxidase mimetic.

References

Alper, S. L., & Sharma, A. K. (2013). The SLC26 gene family of anion transporters and channels. *Molecular Aspects of Medicine, 34*, 494–515.

Andre, A. A. M., & Spruijt, E. (2020). Liquid-liquid phase separations in crowded environments. *International Journal of Molecular Sciences, 21*, 5908.

Aronson, P. S., Nee, J., & Suhm, M. A. (1982). Modifier role of internal H^+ in activating the Na^+-H^+ exchanger in renal microvillus membrane vesicles. *Nature, 299*, 161–163.

Arroyo, J. P., Kahle, K. T., & Gamba, G. (2013). The SLC12 family of electoneutral cation-coupled chloride cotransporters. *Molecular Aspects of Medicine, 34*, 288–298.

Baumgarten, C. M. (2006). Cell volume regulation in cardiac myocytes: A leaky boat gets a new bilge pump. *Journal of General Physiology, 128*, 487–489.

Baumgarten, C. M., & Feher, J. J. (2012). Osmosis and regulation of cell volume. In N. Sperelakis (Ed.), *Cell Physiology: Source book* (4th ed.). Waltham MA: Academic Press.

Benedetto, R., Siriant, L., Pankonien, I., Wanitchakool, P., Ousingsawat, J., Cabrita, I., Schreiber, R., Amaral, M., & Kunselmann, K. (2016). Relationship between TMEM16A/anoctamin 1 and LRRC8A. *Pflügers Archiv, 468*, 1751–1763.

Bers, D. M., Barry, W. H., & Despa, S. (2003). Intracellular Na^+ regulation in cardiac myocytes. *Cardiovascular Research, 57*, 897–912.

Bertelli, S., Remigante, A., Zuccolini, P., Barbieri, R., Ferrera, L., Picco, C., Gavazzo, P., & Pusch, M. (2021). Mechanisms of activation of LRRC8 volume regulated anion channels. *Cellular Physiology and Biochemistry, 55*, 41–56.

Bertelli, S., Zuccolini, P., Gavazzo, P., & Pusch, M. (2022). Molecular determinants underlying volume-regulated anion channel subunit-dependent oxidation sensitivity. *Journal of Physiology, 600*, 3965–3982.

Bonar, P. T., & Casey, J. R. (2008). Plasma membrane Cl^-/HCO_3^- exchangers. *Channels, 2*, 337–345.

Boyle, P. J., & Conway, E. J. (1941). Potassium accumulation in muscle and associated changes. *Journal of Physiology, 100*, 1–63.

Browe, D. M., & Baumgarten, C. M. (2003). Stretch of β1 integrin activates an outwardly rectifying chloride current via FAK and Src in rabbit ventricular myocytes. *Journal of General Physiology, 122*, 689–702.

Browe, D. M., & Baumgarten, C. M. (2004). Angiotensin II (AT1) receptors and NADPH oxidase regulate Cl^- current elicited by β1 integrin stretch in rabbit ventricular myocytes. *Journal of General Physiology, 124*, 273–287.

Browe, D. M., & Baumgarten, C. M. (2006). EGFR kinase regulates volume-sensitive chloride current elicited by integrin stretch via PI-3K and NADPH oxidase in ventricular myocytes. *Journal of General Physiology, 127*, 237–251.

Burg, M. B., Ferraris, J. D., & Dmitrieva, N. I. (2007). Cellular response to hyperosmotic stress. *Physiological Reviews, 87*, 1441–1474.

Cala, P. M. (1980). Volume regulation by amphiuma red blood cells. *Journal of General Physiology, 76*, 683–708.

Cannon, C. L., Basavappa, S., & Strange, K. (1998). Intracellular ionic strength regulates the volume sensitivity of a swelling-activated anion channel. *American Journal of Physiology, 275*, C416–C422.

Centeio, R., Ousingsawat, J., Schreiber, R., & Kunzelmann, K. (2020). Ca^{2+} dependence of volume-regulated VRAC/LRRC8 and TMEM16A Cl^- channels. *Frontiers in Cell and Developmental Biology, 8*, Article 596879.

Clemo, H. F., Feher, J. J., & Baumgarten, C. M. (1992). Modulation of rabbit ventricular cell volume and $Na^+/K^+/2Cl^-$ cotransport by cGMP and atrial natriuretic factor. *Journal of General Physiology, 100*, 89–114.

de los Heros, P., Alessi, D. R., Gourlay, R., Campbell, D. G., Deak, M., MacCartney, T. J., Kahle, K. T., & Zhang, J. (2014). The WNK-regulated SPAK/OSR1 kinases directly phosphorylate and inhibit the K^+-Cl^- co-transporters. *Biochemical Journal, 458*, 559–573.

de los Heros, P., Pacheo-Alvarez, D., & Gamba, G. (2018). Role of WNK kinases in the modulation of cell volume. *Current Topics in Membranes, 81*, 207–235.

Delpire, E., & Gagnon, K. B. (2018). Water homeostasis and cell volume maintenance and regulation. *Current Topics in Membranes, 81*, 3–52.

DeMoll, E., Cox, D. J., Daniel, E., & Riggs, A. F. (2007). Apparent specific volume of human hemoglobin: Effect of ligand state and contribution of heme. *Analytical Biochemistry, 363*, 196–203.

Deng, W., Baki, L., & Baumgarten, C. M. (2010). Endothelin signaling regulates volume-sensitive Cl^- current via NADPH oxidase and mitochondrial reactive oxygen species. *Cardiovascular Research, 88*, 93–100.

Deng, W, Baki, L, Yin, J, Zhou, H, & Baumgarten, CM (2010). HIV protease inhibitors elicit volume-sensitive Cl^- current in cardiac myocytes via mitochondrial ROS. *Journal of Molecular and Cellular Cardiology, 49*, 746–752.

Dick, D. A. T. (1966). *Cell water*. Washington: Butterworths.

Donnan, F. G. (1911). Theorie der membranegleichgewichte und membranpotentiale bie verhandensein von nicht dialysierenden elektrolyten. Ein beitrag zur physikalish-chemischen physiologie. *Zeitshrift fur Elektrochemie und angewandte Chemie, 17*, 572–581 [translated and reprinted as: Theory of membrane equilibria and membrane potentials in the presence of non-dialyzing electrolytes. J. Membr. Sci. 100: 45-55 (1995)].

Dowd, B. F. K., & Forbush, B. (2003). PASK (proline-alanine-rich STE20-related kinase), a regulatory kinase of the Na-K-Cl cotransporter (NKCC1). *Journal of Biological Chemistry, 278*, 27347–27353.

Drewnowska, K., & Baumgarten, C. M. (1991). Regulation of cellular volume in rabbit ventricular myoctyes: Bumetanide, chlorothiazide and ouabain. *American Journal of Physiology, 260*, C122–C131.

Drewnowska, K., Clemo, H. F., & Baumgarten, C. M. (1991). Prevention of myocardial intracelllar edema induced by St. Thomas' Hospital cardioplegic solution. *Journal of Molecular and Cellular Cardiology, 23*, 1215–1221.

Feher, J. J. (2017). *Quantitative human physiology*. Academic Press.

Gradogna, A., Gavazzo, P., Boccaccio, A., & Pusch, M. (2017). Subunit-dependent oxidative stress sensitivity of LRRC8 volume-regulated anion channels. *Journal of Physiology, 595*, 6719–6733.

Grinstein, S., Clarke, C. A., & Rothstein, A. (1983). Activation of Na^+/H^+ exchange in lymphocytes by osmotically induced volume changes and cytoplasmic acidification. *Journal of General Physiology, 82*, 619–638.

Grinstein, S., Rothstein, A., Sarkadi, B., & Gelfand, E. W. (1984). Responses of lymphocytes to anisotonic media: Volume regulating behavior. *American Journal of Physiology, 246*, C204–C215.

Grinstein, S., Woodside, M., Sardet, C., Pouyssegue, J., & Rotin, D. (1992). Activation of the Na^+/H^+ antiporter during cell volume regulation. *Journal of Biological Chemistry, 267*, 23823–23828.

Guntur, D., Olshewski, H., Enedi, P., Csaki, R., Olsehewski, A., & Nagaraj, C. (2021). Revisiting the large-conductance calcium-activated potassium (BK_{Ca}) channels in the pulmonary circulation. *Biomolecules, 11*, 1629.

Hibino, H., Nin, F., Tsuzuki, C., & Kurachi, Y. (2010). How is the highly positive endocochlear potential formed? The specific architecture of the stria vascularis and the roles of the ion-transport apparatus. *Pflügers Archiv, 459*, 521–533.

Hoffmann, E. K., Lambert, I. H., & Pedersen, S. F. (2009). Physiology of cell volume regulation in vertebrates. *Physiological Reviews, 89*, 193–277.

Holm, J. B., Grygorczyk, R., & Lambert, I. H. (2013). Volume-sensitive release of organic osmolytes in the human lung epithelial cell line A549: Role of the 5-lipoxygenase. *American Journal of Physiology, 305*, C48–C60.

Humphreys, B. D., Jiang, L., Chernova, M. N., & Alper, S. L. (1995). Hypertonic activation of AE2 exchanger in Xenopus oocytes via NHE-mediated intracellular alkalinization. *American Journal of Physiology, 268*, C201–C209.

Ishihara, S., Ichijo, H., & Watanabe, K. (2021). A novel lens for cell volume regulation: Liquid-liquid phase separations. *Cellular Physiology and Biochemistry, 55*(S1), 135–160.

Israeli-Rosenberg, S., Manso, A. M., Okada, H., & Ross, R. S. (2014). Integrins and integrin-associated proteins in the cardiac myocyte. *Circulation Research, 114*, 572–586.

Jentsch, T. J. (2016). VRACs and other ion channels and transporters in the regulation of cell volume and beyond. *Nature Reviews Molecular Cell Biology, 17*, 293–307.

Jin, X., Shah, S., Du, X., Zhang, H., & Gamper, N. (2016). Activation of Ca^{2+}-activated Cl^- channel and ANO1 by localized Ca^{2+} signals. *Journal of Physiology, 594*, 19–30.

Kuhn, M. (2015). Cardiac actions of atrial natriuretic peptide. *Circulation Research, 116*, 1278–1280.

Kunzelmann, K. (2015). TMEM16, LRRC8A, bestrophin: Chloride channels controlled by Ca^{2+} and cell volume. *Trends in Biochemical Sciences, 40*, 535–543.

Lambert, I. H., Kristensen, D. M., Holm, J. B., & Mortensen, O. H. (2015). Physiological role of taurine- from organism to organelle. *Acta Physiologica, 213*, 191–212.

Leaf, A. (1956). On the mechanism of fluid exchange of tissues in vitro. *Biochemistry, 62*, 241–248.

Lefevre, P. G. (1964). The osmotically functional water content of the human erythrocyte. *Journal of General Physiology, 47*, 585–603.

Li, J.-M., Gall, N. P., Grieve, D. J., Chen, M., & Shah, A. M. (2002). Activation of NADPH oxidase during progression of cardiac hypertrophy to failure. *Hypertension, 40*, 477–484.

Liberman, M. C., Gao, J., He, D. Z. Z., Wu, X., Jia, S., & Zuo, J. (2002). Prestin is required for electomotility of the outer hair cell and for the cochlear amplifier. *Nature, 419*, 300–304.

Lionetto, M. G., Rizzello, A., Giordano, M. E., Maffia, M., deNuccio, F., Nicolardi, G., Hoffmann, E. K., & Schettino, T. (2008). Molecular and functional expression of high conductance Ca^{2+} activated K^+ channels in the eel intestinal epithelium. *Cellular Physiology and Biochemistry, 21*, 373–384.

Lowe, M., Kalacheva, M., Boersma, A. J., & Kedrov, A. (2020). The more the merrier: Effects of macromolecular crowding on the structure and dynamics of biological membranes. *FEBS Journal, 287*, 5039–5067.

Ludwig, J., Oliver, D., Frank, G., Klocker, N., Gummer, A. W., & Fakler, B. (2001). Reciprocal electromechanical propertiesof rat prestin: The moto molecule from rat outer hair cells. *Proceedings of the National Academy of Sciences, 98*, 4178–4183.

Macknight, A. D. C., & Leaf, A. (1977). Regulation of cellular volume. *Physiological Reviews, 57*, 510–573.

Maeno, E., Takahashi, N., & Okada, Y. (2006). Dysfunction of regulatory volume increase is a key component of apoptosis. *FEBS Letters, 580*, 6513–6517.

Maruyama, J., Kobayashi, Y., Umeda, T., Vandewalle, A., Takeda, K., Ichijo, H., & Naguro, I. (2016). Osmotic stress induces the phosphorylation of WNK4 Ser575 via the p38MAPK-MK pathway. *Scientific Reports, 6*, 18710.

McSwine, R. L., Li, J., & Villereal, M. L. (1996). Examination of the role for Ca^{2+} in regulation and phosphorylation of the Na^+/H^+ antiporter NHE1 via mitogen and hypertonic stimulation. *Journal of Cellular Physiology, 168*, 8–17.

Milenkovic, A., Brandl, C., Milenkovic, V. M., Jendryke, T., Sirianant, L., Wanitchakool, P., Zimmermann, S., Reiff, C. M., Horling, F., Schrewe, H., Schreiber, R., Kunzelmann, K., Wetzel, H., & Weber, B. H. F. (2015). Bestrophin 1 is indispensable for volume regulation in human retinal pigment epithelium cells. *Proceedings of the National Academy of Sciences of the United States of America, 112*, E2630–E2639.

Minton, A. P. (2001). The influence of macromolecular crowding and macromolecular confinement on biochemical reactions in physiological media. *Journal of Biological Chemistry, 276*, 10577–10580.

Murillo de Ozores, A. R., Chávez-Canales, M., de Los Heros, P., Gamba, G., & Castañeda-Bueno, M. (2020). Physiological processes modulated by the chloride-sensitive WNK-SPAK/OSR1 kinase signaling pathway and the cation-coupled chloride cotransporters. *Frontiers in Physiolgy, 11*, 585907.

Naguro, I., Umeda, T., Kobayashi, Y., Maruyama, J., Hattori, K., Shimizu, Y., Kataoka, K., Kim-Mitsuyama, S., Uchida, S., Vandewalle, A., Noguchi, T., Nishitoh, H., Matsuzawa, A, Takeda, K., & Ichijo, H. (2012). ASK3 responds to osmotic stress and regulates blood pressure by suppressing WNK1-SPAK/OSR1 signaling in the kidney. *Nature Communications, 3*, 1285.

Neussert, R., Muller, C., Milenkovic, V. M., & Strauss, O. (2010). The presence of bestrophin-1 modulates the Ca^{2+} recruitment from Ca^{2+} stores in the ER. *Pflügers Archiv, 460*, 163–175.

Nilius, B., Prene, J., Voets, T., Eggermount, J., & Droogmans, G. (1998). Activation of volume-regulated chloride currents by reduction of intracellular ionic strength in boving endothelial cells. *Journal of Physiology, 506*, 353–361.

Orlowski, J., & Grinstein, S. (2011). Na^+/H^+ exchangers. *Comprehensive Physiology, 1*, 2083–2100.

Owji, A. P., Kittredge, A., Zhang, Y., & Yang, T. (2021). Structure and function of the bestrophin family of calcium-activated chloride channels. *Channels, 15*, 604–623.

Pedersen, S. F., Klausen, T. K., & Nilius, B. (2015). The identification of a volume-regulated anion channel: An amazing odyssey. *Acta Physiologica, 213*, 868–881.

Pederson, S. F., O'Donnell, M. E., Anderson, S. E., & Cala, P. M. (2006). Physiology and pathophysiology of Na^+/H^+ exchange and Na^+-K^+-$2Cl^-$ cotransport in the heart, brain and blood. *American Journal of Physiology, 291*, R1–R25.

Piala, A. T., Moon, T. M., Akella, R., He, H., Cobb, M. H., & Goldsmith, E. J. (2014). Chloride sensing by WNK1 involves inhibition of autophosphorylation. *Science Signaling, 7*, 324.

Prickett, R. C., Elliott, J. A. W., Hakda, S., & McGann, L. E. (2008). A non-ideal replacement for the Boyle van't Hoff equation. *Cryobiology, 57*, 130–136.

Putney, L. K., Denker, S. P., & Barber, D. L. (2002). The changing face of the Na^+/H^+ exchanger, NHE1: Structure, regulation, and cellular actions. *Annual Review of Pharmacology and Toxicology, 42*, 527–552.

Qui, Z., Dubin, A. E., Mathur, J., Tu, B., Reddy, K., Miraglia, L. J., Reinhardt, J., Orth, A. P., & Patapoutian, A. (2014). SWELL1, a plasma membrane protein, is an essential component of volume-regulated anion channel. *Cell, 157*, 447–458.

Raucci, F. J., Wijesinghe, D. S., Chalfant, C. E., & Baumgarten, C. M. (2010). Exogenous and endogenous ceramides elicit volume-sensitive chloride current in ventricular myocytes. *Cardiovascular Research, 86*, 55–62.

Romero, M. F., Chen, A.-P., Parker, M. D., & Boron, W. F. (2013). The SLC4 family of bicarbonate (HCO_3^-) transporters. *Molecular Aspects of Medicine, 34*, 159–182.

Rotin, D., & Grinstein, S. (1989). Impaired cell volume regulation in Na^+-H^+ exchange-deficient mutants. *American Journal of Physiology, 257*. C1158-C115.

Sancho, M., & Kyle, B. D. (2021). The large-conductance, calcium-activated potassium channel: A big key regulator of cell physiology. *Frontiers in Physiology, 12*, Article 750615.

Schober, A. L., Wilson, C. S., & Mongin, A. A. (2017). Molecular composition and heterogeneity of the LRRC8-containing swelling-activated osmolyte channels in primary rat astrocytes. *Journal of Physiology, 595*, 6939–6951.

Seshiah, P. N., Weber, D. S., Rocic, P., Valppu, L., Taniyama, Y., & Griendling, K. K. (2002). Angiotensin II stimulation of NAD(P)H oxidase activity. *Circulation Research, 91*, 406–413.

Sirianant, L., Wanitchakool, P., Ousingsawat, J., Benedetto, R., Zormpa, A., Cabrita, I., Schreiber, R., & Kunzelmann, K. (2016). Non-essential contribution of LRRC8A to volume regulation. *Pflügers Archiv, 468*, 805–816.

Skou, J. C. (1990). The energy coupled exchange of Na^+ for K^+ across the cell membrane. The Na^+,K^+-pump. *FEBS Letters, 268*, 314–324.

Speer, S. L., Stewart, C. J., Sapir, L., Harries, D., & Pielak, G. J. (2022). Macromolecular crowding is more than hard-core repulsions. *Annual Review of Biophysics, 51*, 267–300.

Suleymanian, M. A., & Baumgarten, C. M. (1996). Osmotic gradient-induced water permeation across the sarcolemma of rabbit ventricular myocytes. *Journal of General Physiology, 107*, 503–514.

Summers, J. C., Trais, L., Lajvardi, R., Hergan, D., Buechler, R., Chang, H., Pena-Rasgado, C., & Rasgado-Flores, H. (1997). Role of concentration and size of intracellular macromolecules in cell volume regulation. *American Journal of Physiology, 273*, C360–C370.

Syeda, R., Qiu, Z., Dubin, A. E., Murthy, S. E., Florendo, M. N., Mason, D. E., Mathur, J., Cahalan, S. M., Peters, E. C., Montal, M., & Patapoutian, A. (2016). LLRC8 proteins form volume-regulated anion channels that sense low ionic strength. *Cell, 164*, 499–511.

Takada, Y., Ye, X., & Simon, S. (2007). The integrins. *Genome Biology, 8*, 215.

Takeuchi, A., Tatsumi, S., Sarai, N., Terasima, K., Matsuoka, S., & Noma, A. (2006). Ionic mechanisms of cardiac cell swelling induced by blocking Na^+/K^+ pump as reveealed by experiements and simulation. *Journal of General Physiology, 128*, 495–507.

Thorn, P., Lawrie, A. M., Smith, P. M., Gallacher, D. V., & Petersen, O. H. (1993). Local and global cytosolic Ca^{2+} oscillations in exocrine glands evoked by agonists and inositol trisphosphate. *Cell, 74*, 661–668.

Tosteson, D. C., & Hoffman, J. F. (1960). Regulation of cell volume by active cation transport in high and low potassium sheep red cells. *Journal of General Physiology, 44*, 169–194.

Trouet, D., Nilius, B., Jacobs, A., Remache, C., Droogmans, G., & Eggermont, J. (1999). Caveolin-1 modulates the activity of the volume-regulated chloride channel. *Journal of Physiology, 520*, 113–119.

Vergara, C., Latorre, R., Marrion, N. V., & Adelman, J. P. (1998). Calcium-activated potassium channels. *Current Opinion in Neurobiology, 8*, 321–329.

Vetri, F., Roy, M. S., Pelligrino, D. A., & Sundivakkam, P. (2014). BK_{Ca} channels as physiological regulators: A focused review. *Journal of Receptor, Ligand and Channel Research, 7*, 3–13.

Voss, F. K., Ullrich, F., Munch, J., Lazarow, K., Lutter, D., Mah, N., Andrade-Navarro, M. A., von Kries, J. P., Stauber, T., & Jentsch, T. J. (2014). Identification of LRRC8 heteromers as an essential component of the volume-regulated anion channel VRAC. *Science, 344*, 634–638.

Watanabe, K., Morishita, K., Zhou, X., Shiizaki, S., Uchiyama, Y., Koike, M., Naguro, I., & Ichijo, H. (2021). Cells recognize osmotic stress through liquid-liquid phase separation lubricated with poly (ADP-ribose). *Nature Communications, 12*, 1353.

Watanabe, K., Umeda, T., Niwa, K., Naguro, I., & Ichijo, H. (2018). A PP6-ASK3 module coordinates the bidirectional cell volume regulation under osmotic stress. *Cell Reports, 22*, 2809–2817.

Wei, Y., Mayoral-Delgado, I., Stewart, N. A., & Dymond, M. K. (2019). Macromolecular crowding and membrane binding proteins: The case of phospholipase A_1. *Chemistry and Physics of Lipids, 218*, 91–102.

Wilson, T. H. (1954). Ionic permeability and osmotic swelling of cells. *Science, 120*, 104–105.

Yoshida, A., Taniguchi, S., Hisatome, I., Royauz, I. E., Green, E. D., Kohn, L. D., & Suzuki, K. (2002). Pendrin is an iodide-specific apical porter responsible for iodide efflux from thyroid cells. *Journal of Clinical Endocrinology & Metabolism, 87*, 3356–3361.

Zhang, J. Z., Mehta, S., & Zhang, J. (2021). Liquid-liquid phase separation: A principal organizer of the cell's biochemical activity architecture. *Trends in Pharmacological Sciences, 42*, 845–856.

Zhou, X., Naguro, I., Ichijo, H., & Watanabe, K. (2016). Mitogen-activated protein kinases as key players in osmotic stress signaling. *Biochimica et Biophysica Acta, 1860*, 2037–3052.

Zhurova, M., McGann, L.e., & Acker, J. P. (2014). Osmotic parameters of red blood cells from umbilical cord blood. *Cryobiology, 68*, 379.

Chapter 23

Ligand-gated ion channels

Kenneth R. Tovar[1] and Gary L. Westbrook[2]

[1]*University of California Santa Barbara, Santa Barbara, CA, United States;* [2]*Vollum Institute, Oregon Health & Science University, Portland, OR, United States*

1. Introduction

Ion channels are fundamental signaling molecules in virtually all cells. Studies of ion channel in the plasmalemma of excitable cells have provided the foundation for our knowledge of these important molecules. In the nervous system, voltage-gated ion channels mediate action potentials (APs) and trigger the release of neurotransmitters , whereas ligand-gated channels are responsible for chemical signaling mediated by classical fast-acting transmitters. Neurotransmitters also trigger slower synaptic responses that are mediated by G protein-coupled receptors (Chapter 24 by Hofer).

Until the 1980s, the molecular properties of ion channel proteins were largely inferred from physiological and biophysical studies (Hille, 2001). In retrospect, many of the predictions from early biophysical studies were confirmed and extended following the elucidation of the amino acid sequence of voltage- and ligand-gated ion channels by molecular cloning. However, the increasing knowledge of the three-dimensional molecular structure of ion channels and their homologues in lower organisms now permits sophisticated correlations of protein structure with function. The structures of ligand-gated channels have been determined by X-ray crystallography over the past 20 years, but increasingly, the use of single-particle cryo-electron microscopy (cryoEM) methods promises to take our knowledge to yet another level, including interactions of ligand-gated channels as components of multiprotein complexes at synapses. These advances have dramatically increased our understanding of the molecular operation of these prototypic membrane proteins. Ion channels have captured the interest of many outstanding scientists whose studies of ion channel function and structure have led to several Nobel Prizes (MacKinnon, 2004; Neher, 1992; Sakmann, 1992). Neurological disorders can also be caused by mutations in ion channels, known as "channelopathies" (Kullmann, 2010), and by autoimmune mechanisms that target ligand-gated ion channel proteins (Dalmau et al., 2017).

A biophysical approach to ion channels was pioneered by studies of voltage-gated ion channels underlying the action potentials in squid axon. These studies established that a conductance change was associated with the action potential, and that this conductance change could be attributed to selective increases in the membrane permeability to Na^+ and K^+ ions, with the energy for the process derived from the transmembrane ion gradients created by the Na^+, K^+-ATPase (Chapters 19 and 20). These findings suggested that there were discrete pores in the membrane that accounted for the Na^+ and K^+ conductance. However, it was not until the 1970s that the existence of ion channels as discrete membrane proteins was confirmed, first by measurements of the statistical properties of currents through populations of ion channels, a technique called **fluctuation** or **noise analysis** (Katz & Miledi, 1972). Later, this conclusion was refined by the introduction of **patch-clamp recording,** which allowed measurement of the current through single ion channels (Neher, 1992; Sakmann, 1992). Patch-clamp recording revolutionized the functional study of ion channels because it allowed detailed biophysical studies of the electrical activity of single molecules. Protein purification and molecular cloning revealed that ion channels are a large and diverse group of membrane proteins that are activated by voltage (Chapter 20), mechanical stimuli (Chapter 21), as well as the ligand-gated channels discussed in this chapter. Although ligand-gated ion channels vary in structural details, they share a common basic structure in that they are composed of multiple subunits arranged around a central water-filled pore. Each subunit is a polypeptide encoded by a separate gene. Generally, there are a large number of possible subunit combinations; thus, the particular subunits expressed in a certain class of neurons can result in channels with distinct functional characteristics. Ligand-gated channels are not limited to neurons, but are also expressed in glial cells in the brain (Lin & Bergles, 2004).

Cell Physiology Sourcebook. https://doi.org/10.1016/B978-0-12-811114-7.00022-0

This chapter is divided into three topic areas. In the first, the categories of ligand-gated ion channels are briefly reviewed. The basic physiological properties and general molecular structure of ligand-gated channels are then discussed using the muscle nicotinic acetylcholine receptor (AChR) and the N-methyl-D-aspartate (NMDA)-type glutamate receptor as the prototypes. In the third section, the features of neurotransmitter-gated channels that are responsible for fast synaptic transmission are discussed with an emphasis on their distinctive characteristics. Channels gated by acetylcholine (ACh), γ-aminobutyric acid (GABA), glycine, and glutamate are included as representative examples. Related topics on ion channels are discussed in other chapters of this book (Chapters 12, 19–21). The reader is also referred to the excellent monographs and reviews listed in the references for further details.

2. Classes of ligand-gated ion channels

Ligands that activate ion channels in nerve cell membranes can be roughly divided into two major categories, neurotransmitters and intracellular ligands (Table 23.1). The neurotransmitter-gated ion channels involved in fast chemical synaptic transmission have been extensively studied. These include acetylcholine (ACh), which is the transmitter at the vertebrate neuromuscular junction and in autonomic ganglia, and the amino acids, L-glutamate and GABA, which mediate the majority of fast excitatory and inhibitory synaptic transmission, respectively, in the vertebrate central nervous system. These ligands also activate receptors that are coupled to second messengers through GTP-binding (G) proteins. G protein-coupled receptors generally mediate slower and neuromodulatory transmembrane signaling. Other neurotransmitters that activate ligand-gated ion channels in various cell types include serotonin, glycine, histamine, and adenosine triphosphate (ATP). For example, the $5HT_3$ receptor activated by serotonin is a ligand-gated ion channel (Julius, 1991). ATP is released from nerve terminals in several pathways and activates the P2X family of channels (Browne et al., 2010; Burnstock, 2015). The Transient Receptor Potential (TRP) channels form a particularly large and diverse group that are activated not by classical neurotransmitters, but rather by diverse chemical and physical stimuli (Wu et al., 2010; Zhang et al., 2023).

In addition to neurotransmitter-gated ion channels, increasing attention has focused on ligand-gated ion channels that are activated by intracellular ligands, also known as second messengers, such as Ca^{2+}, cyclic nucleotides, ATP, and inositol trisphosphate (IP_3). Intracellular ligands can increase (Ca^{2+}, cAMP, IP_3) or decrease (cGMP, ATP) activity of the associated channel. These channels play important roles in cell function. For example, Ca^{2+}-dependent potassium, chloride, and nonspecific cation channels modify the excitability of the nerve cell membrane and thus can alter the AP and release of transmitter from nerve terminals. The cyclic nucleotide (cAMP, cGMP)-gated channels mediate sensory transduction in the visual and olfactory systems. Light results in the closing of cGMP channels in vertebrate photoreceptors, whereas odors trigger the opening of cAMP-gated channels in olfactory receptor neurons. See Chapters 28 (McDonald and Porter) and Chapter 29 (Bigiani) for more discussion of channels gated by cyclic nucleotides.

TABLE 23.1 Ligand-gated ion channels.

Ligand	Receptor	Ion selectivity[a]
Neurotransmitters		
Acetylcholine	Muscle, neuronal AChRs	NS
Glutamate	AMPA, kainate, NMDA	NS
GABA	$GABA_A$	Cl
Glycine	GlyR	Cl
Serotonin	$5HT_3$ receptor	NS
ATP	P2X receptor	NS
Intracellular ligands calcium	Calcium-dependent channels	K, Cl, NS
Cyclic nucleotides	cGMP and cAMP receptors	NS
ATP	ATP-dependent channel	K
IP_3	Calcium release channel	Ca

[a]*Abbreviations for ion selectivity are NS (nonselective cation permeability, some with calcium permeability); Cl (permeable to chloride); K (permeable to potassium); and Ca (permeable to calcium).*

3. Basic physiological features

Patch-clamp techniques, developed in the early 1980s, allow scientists to record from one or a few channels with a glass micropipette "patched" to the cell membrane (single channel recording), or to break that membrane patch with pipette suction, and record from all channels in the cell (whole-cell recording). Using these techniques, investigators have defined in great detail the three fundamental properties of any ion channel—**conductance, selective permeability,** and **gating**—in molecular terms. In addition to these fundamental properties, **modulation, channel block,** and **desensitization** are processes that alter the activity of ligand-gated ion channels, and are important in shaping the impact of channel activity on membrane excitability. Modern analysis of these properties involves a combination of functional studies complemented by structural analysis by X-ray crystallography and single particle cryoEM. The biophysical basis of these properties will be discussed first, and then the structural features of ligand-gated channels that control these properties will be considered.

Conductance is a measure of the flux of charged ions through the channel, as measured in picoSiemens (pS). Most ligand-gated ion channels have a conductance of 5–50 pS, which corresponds to the movement of millions of ions per second through the channel pore. These high flux rates initially suggested that ligand-gated channels must contain a water-filled pore, because this rate is much higher than that predicted for other membrane transport mechanisms, such as pumps or exchangers. The size of the single-channel conductance (γ) is characteristic for a given channel. Several openings of a glutamate-activated channel in a hippocampal neuron are shown in Fig. 23.1. The conductance of the open channel in this case is near 50 pS, but drops occasionally to subconductance states of 8, 35, 40, and 45 pS. The conductance of a channel increases as the permeant ion concentration is raised, but eventually saturates at concentrations well above physiological levels. This flux of ions through the channel pore can be described by Michaelis–Menten kinetics, suggesting that ions bind to sites within the pore rather than simply obeying the laws of free diffusion. Most permeant ions have a low K_m of approximately 100 mM (K_m is the concentration at which the conductance is half-maximal, as defined by the Michaelis–Menten equation); thus the ions bind for only a microsecond or so before continuing through the pore.

The current carried by the opening of a single channel is measured in picoamperes (pA), with the single-channel current given by

$$i = \gamma(V - V_{eq})$$

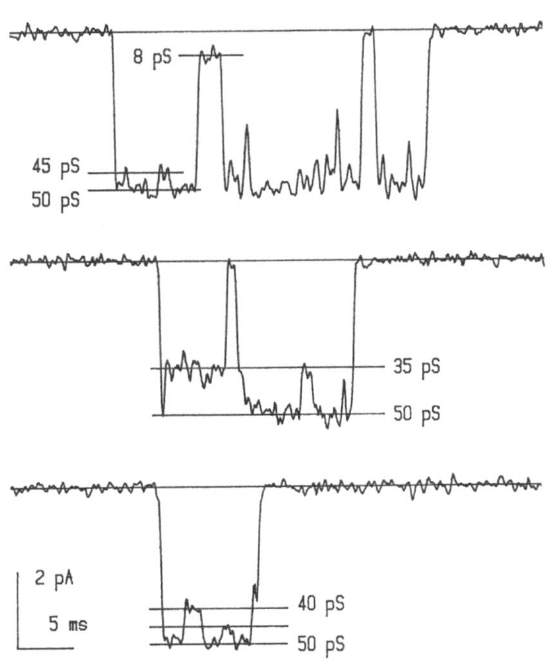

FIGURE 23.1 Activity of single ligand-gated channels as measured by patch clamp recording. Examples of NMDA-type glutamate channel recording in an outside-out membrane patch obtained from a cultured hippocampal neuron. Downward deflections indicate opening of the channel in the presence of ligand (in this case, 20 μM NMDA). The channel usually opens to the 50 pS level, but may switch to lower levels of 8, 35, 40 or 45 pS before closing to the baseline level. *Modified with permission from Jahr and Stevens (1987).*

where γ is the single channel conductance, V is the membrane potential, and V_{eq} is the equilibrium potential at which no net current is measured. This equation is simply Ohm's law, where $(V - V_{eq})$ is the electrochemical driving force for ions that can pass through the channel. For example, if the extracellular and cytoplasmic concentrations of permeant ions are equal, then V_{eq} is 0 mV. Thus, as the membrane potential changes, the size of the single-channel current also changes; a plot of this relationship is called a **current-voltage** or ***I*—*V* plot.** For channels such as muscle AChRs that are permeable to Na^+ and K^+, the reversal potential is 0 mV, and the I—V curve is nearly linear. However, some channels deviate from linear behavior and show inward or outward rectification (that is, the channel passes current in one direction better than in the other), analogous to the behavior of channels in which channel opening and closing (i.e., gating) is itself dependent on the membrane voltage.

Channels are not equally permeable to all ions, that is, they exhibit **selective permeability.** As listed in Table 23.1, most ligand-gated ion channels are permeable to either monovalent cations or anions. The selective permeability of channels is partly due to the physical size of the pore. For example, the muscle AChR is permeable to cations with diameters up to about 6.5 Å. In general, cationic ligand-gated ion channels are less selective than voltage-gated ion channels, which are highly selective for Na^+, K^+, or Ca^{2+}, whereas some channels have very large pores such as gap junction channels (See Chapter 8 by Veenstra), which are permeable to small molecular weight molecules such as cyclic nucleotides as well as to ions. Because anions and cations are approximately the same size, it is obvious that the pore dimensions cannot totally account for selective permeability. As discussed later, positively and negatively charged amino acid residues at the entrances to the channel and within the pore provide a means for this discrimination. Selective permeability is extremely important in considering the function of an ion channel. For example, channels that are permeable to Na^+ and K^+ do not significantly alter the ion concentrations on either side of the membrane, but instead provide an electrical signal that depolarizes the neuron and brings the membrane potential closer to the threshold for action potential generation. However, channels with significant calcium permeability can transiently increase the cytoplasmic Ca^{2+} concentration, and thus act as a biochemical signal. The relative Ca^{2+} permeability of ligand-gated channels such as the NMDA receptor and some nicotinic receptors is an important aspect of their role in neuronal function. The reader is referred to the superb monograph by Hille (2001) for a full discussion of channel permeability and its measurement.

Channel **gating** refers to the conformational change in the ion channel protein that is triggered by ligand binding. The conformational change results in a rapid switch between conducting (open) and nonconducting (closed) states of the channel. The gating behavior of channels has many parallels with the allosteric behavior of enzymes, with the binding of ligand providing the free energy necessary to maintain the channel in the open conformation. Gating between the open and closed configuration is extremely rapid (approximately 10 μs), and thus is beyond the resolution of standard electrophysiological recording methods. Each channel opening usually lasts a few milliseconds, but can, on some occasions, last up to several hundred milliseconds. Binding of more than one agonist molecule is usually necessary to open a ligand-gated channel. The binding of multiple agonist molecules creates a sigmoidal dose-response relationship, characteristic of the cooperative binding of substrates to enzymes. The steep activation created by sigmoidal kinetics has the advantage of preventing the channel from opening in the presence of a low concentration of agonist, which can be quite important in preventing desensitization of synaptic receptors. In general, the higher the binding affinity of the agonist, the longer the channel can remain open until the agonist dissociates.

The gating of ligand-gated ion channels can be described by multistate kinetic diagrams as shown in Fig. 23.2A. The minimal kinetic model for a ligand-gated channel usually has at least four states, including closed and unbound, closed and bound, open and bound, and nonconducting (desensitized). Such state diagrams have been developed for many ligand-gated ion channels, based on the analysis of open and closed time distributions obtained in single channel recording. For example, the presence of two exponentials in an open time histogram implies the existence of two open states. For the case shown in Fig. 23.2A and B, note that the channel opens only when two agonist molecules are bound. For fast-acting neurotransmitters, the concentration of transmitter in the synaptic cleft is typically very high for a brief period. At high agonist concentrations, receptors quickly reach the fully liganded (but closed) state because transition to this state is determined by the agonist concentration as well as the binding rate, r_b. Thus the rising and falling phase of the synaptic response largely reflects the unbinding rate r_u, the channel opening and closing rates (β and α) and, in some cases, the rates in and out of the desensitized state, R_d. Classical pharmacological terms can be interpreted in terms of these kinetic schemes (Colquhoun, 1998). Agonists open the channel, whereas antagonists bind, but do not cause a sufficient conformational change in the channel protein to open the channel. The fraction of time the bound channel spends in the open state is called the **open probability,** and may differ between agonists—an agonist that activates the channel with higher probability therefore has a higher efficacy in the nomenclature of classical receptor pharmacology. As discussed below, the atomic structure of a number of ligand-gated channels captured in closed, open, and desensitized states provide a view of the underlying conformational movements.

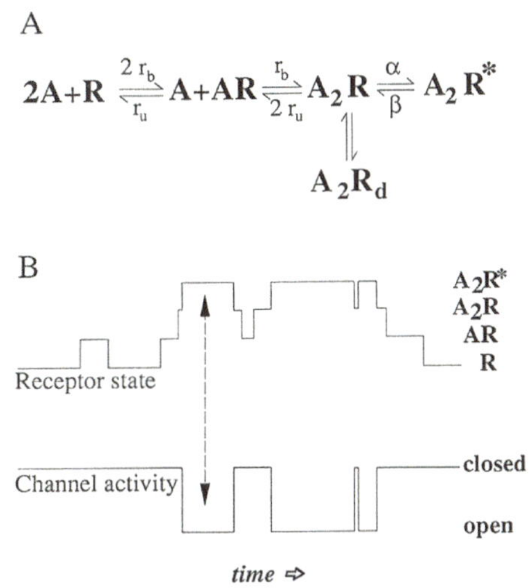

FIGURE 23.2 Kinetics of ligand-gated channels. (A) Example of a four-state kinetic scheme used to describe the behavior of ligand-gated ion channels. In this example, two agonist molecules (A) must bind before the channel can open (A_2R^*) or enter a nonconducting desensitized state (A_2R_d). The rate constants for agonist binding and unbinding are designated r_b and r_u, and the rate constants for channel opening and closing are designated as β and α. (B) Relationship of the state of the receptor to the open and closing of the channel. The channel opens only when the receptor is in the A_2R^* state as indicated by the arrows.

Earlier studies of the activity of ligand-gated channels were generally limited to equilibrium measurements because of the limits in the speed of application of agonists to cells or cell-free patches. Measurements in the continuous presence of a neurotransmitter are difficult to interpret in terms of state diagrams such as that shown in Fig. 23.2. In addition, the duration of fast-acting neurotransmitters such as glutamate is generally brief, suggesting that the synapse itself operates under nonequilibrium conditions. However, rapid solution exchange techniques allow agonists to be applied within several hundred microseconds to membrane patches, usually in the "outside-out" configuration, to allow an agonist to access the extracellular face of the receptor. Such approaches more closely resemble conditions at synapses and are particularly useful in evaluating the role of channel modulation, channel block, and desensitization on synaptic responses. Ligand-gated channels have relatively low affinity for agonist, compared with typical G-protein coupled receptors or hormone receptors. Several mechanisms, such as neurotransmitter catalysis, active transport, or neurotransmitter out of the extracellular space, keep the neurotransmitter concentration low. Thus, activation of ligand-gated channels may primarily occur through nonequilibrium agonist presentation at synapses. An example of the response of a membrane patch from a hippocampal neuron to a brief application of glutamate is shown in Fig. 23.3.

In general, the conductance and selective permeability of ligand-gated channels are not subject to regulation. However, allosteric control mechanisms can profoundly affect gating (Changeux & Edelstein, 1998). These mechanisms may manifest as **desensitization, channel block,** or **allosteric modulation. Desensitization** refers to the loss of the response during the continued presence of an agonist, and was first noted in studies of the muscle AChR. At the single-channel level, desensitization reflects a ligand-bound, but nonconducting state of the channel. This phenomenon occurs in most ligand-gated channels, and can be described by including a ligand-bound, but nonconducting, state in the kinetic scheme (see Fig. 23.2A). Depending on the channel, the nonconducting state may be accessible from either the closed state (A_2R) or the open state (A_2R^*). Desensitization represents a distinct conformation of the receptor as has been demonstrated by structural studies of several ligand-gated channels. Very rapid desensitization may play a role in terminating the neuronal response to neurotransmitters at some synapses, whereas slower desensitization may actually prolong synaptic responses as channels re-enter open states after transiting through desensitized states (Jones & Westbrook, 1996). Although we refer here to desensitization of channels that remain in the membrane, ligands in some cases can trigger internalization of receptors, including ligand-gated channels, resulting in loss of responsiveness.

Ligand-gated channels can be plugged by ions or drugs, a phenomenon referred to as **channel block.** If the blocking particle is charged, such as a large ion, the block will be influenced by the voltage across the membrane. Thus, the reduction in channel activity will be more pronounced at some membrane potentials. The voltage-dependent block of NMDA-type glutamate channels by extracellular Mg^{2+} is an example of this phenomenon (Ascher & Nowak, 1987). At the single channel level, channel block is usually seen as rapid interruptions ("flickers") of the open state as illustrated in Fig. 23.4A and B. A characteristic feature of most forms of channel block is that the blocker is only effective when the

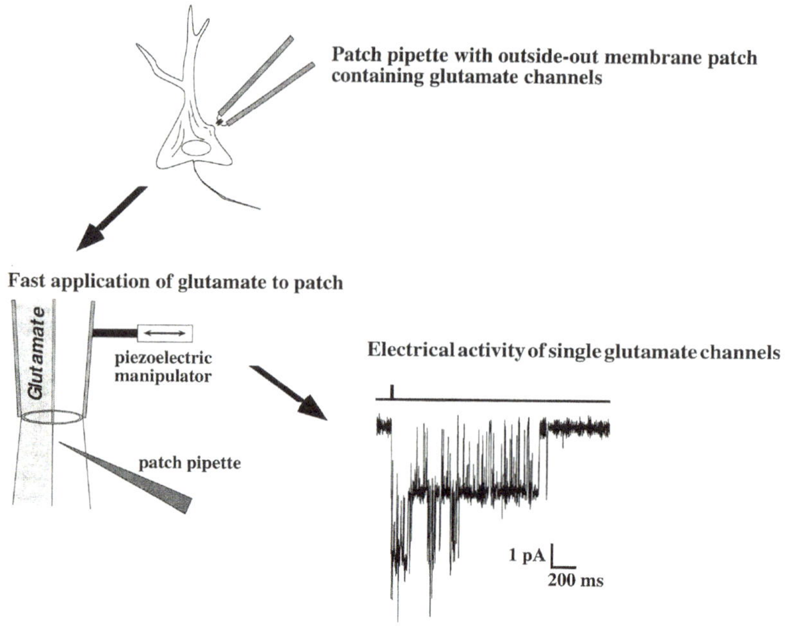

FIGURE 23.3 Use of rapid application methods to study the kinetics of ligand-gated channels. The patch-clamp technique allows small membrane patches to be removed from a cell, such as a hippocampal neuron, with the extracellular side of the membrane facing the bath (outside-out configuration). The tip of the pipette is then placed near the interface of two rapidly moving streams of solution. A piezoelectric manipulator moves the solutions to change the solution flow over the patch in less than 1 ms., which mimics agonist presentation at glutamate release sites. The response, recorded as the opening of glutamate channels of the NMDA type, greatly outlasts the brief application of the agonist. The bar above the channel activity indicates the duration of the glutamate application.

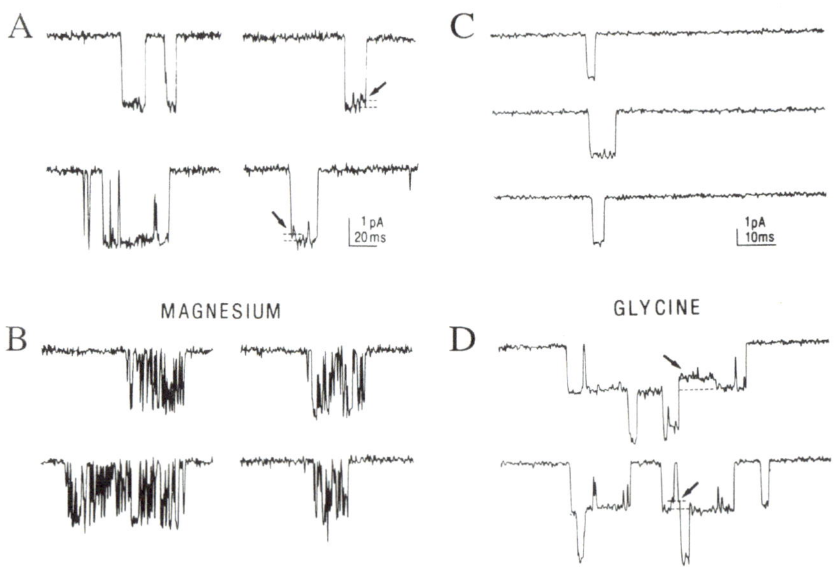

FIGURE 23.4 NMDA-type glutamate receptors illustrate two common mechanisms of channel regulation: channel block and allosteric modulation. (A and B) Channels activated by NMDA in a membrane patch from a hippocampal neuron in the absence of extracellular Mg^{2+} remain open (downward deflections) for several milliseconds before closing (A). However, in the presence of Mg^{2+} (B), the openings are interrupted by brief closures due to plugging of the pore by Mg^{2+} ions. (C and D) The frequency of NMDA channel opening is increased in the presence of extracellular glycine (D) compared with glycine-free solutions (C). The stepwise increase in current in part D indicates that at least two channels were present in the membrane patch. Because glycine is required for NMDA channel opening, the occasional openings in part C probably reflect contamination of the solution with low levels of glycine. *Modified with permission from Ascher and Nowak (1987).*

channel is open. This phenomenon indicates that the binding site for the blocker resides within the pore of the channel, or at least at a site that is only available in the open conformation of the channel. Channel block is the mechanism of action for a number of drugs such as the local anesthetic lidocaine, the dissociative anesthetic ketamine, and psychoactive compounds such as phencyclidine (PCP).

Channel gating is also subject to **modulation** by either covalent modification such as phosphorylation (Swope et al., 1999), or through the noncovalent binding of modulators to the channel. Most ligand-gated channels undergo phosphorylation of intracellular portions of the channel protein; however, the resulting effect on channel function is dependent on the specific channel and the type of kinase involved. Phosphate groups are highly charged and thus can modify interactions in local regions. For example, phosphorylation of the AChR enhances desensitization and is also important in clustering of AChR molecules at sites of innervation. The list of noncovalent modulators is long and includes Ca^{2+}, toxins, drugs, and some endogenous substances such as protons and steroid hormones. These modulators usually act in a noncompetitive manner, that is, they do not directly interfere with ligand binding. Most of these reagents affect gating by altering the rate of channel opening or the time spent in the open state. For example, glycine binds to the NMDA channel and results in an increased probability of opening (Fig. 23.4C and D). This is a dramatic example of allosteric regulation because the channel will not open unless both glycine and the transmitter (glutamate) are bound. A less profound, but equally important, example is the up-regulation of $GABA_A$ receptor activity by benzodiazepines (Olsen & Sieghart, 2009).

4. Molecular structure

Ligand-gated ion channels can be divided into three main structural classes: the Cys-loop receptors, the glutamate receptors, and the P2X receptors. Channels in these three families are formed as combinations of 5, 4, and 3 subunits, respectively (Fig. 23.5). We discuss here structural information on the pentameric Cys-loop receptor superfamily that includes the nicotinic, $GABA_A$, glycine, and 5-HT3 receptors, and the tetrameric glutamate receptor family that includes the AMPA, kainate and NMDA receptors. For the Cys-loop and glutamate receptor families, structural studies continue to provide a more and more detailed understanding of the main functional features of ligand-gated ion channels: binding of ligand, the gating and ion selectivity of the channel, and the coupling of ligand binding to channel gating. For information on the unique structure and functional roles of the trimeric P2X receptors activated by ATP, the reader is referred to these reviews (Browne et al., 2010; Burnstock, 2015).

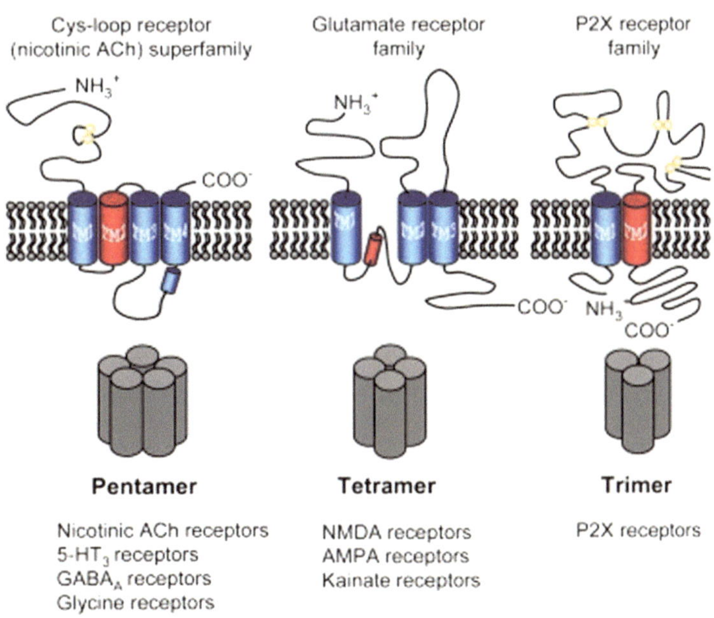

FIGURE 23.5 The topology of the 3 families of ligand-gated ion channel subunits. The general topologies of ligand-gated receptor subunits for each of the main receptor families are shown in the upper row. The stoichiometry of subunits in an assembled receptor are shown in the lower row. The Cys-loop family of receptors are pentameric whereas the glutamate receptors are tetrameric and P2X receptors are trimeric. The three protein families also differ in the general structure of extracellular, intramembrane and intracellular domains. The red cylinders indicate the intramembrane pore region through which ions flow, M2 (TM2) in the case of Cys-loop and P2X receptors and the P loop for glutamate receptors. *Modified with permission from Collingridge et al. (2009).*

The AChR led the way in revealing the molecular structure of ligand-gated channels, beginning with biochemical studies of the muscle-type receptor (Karlin, 1991). This analysis took advantage of the high density of AChRs in the electric organ of the ray *(Torpedo californica),* and the high-affinity snake toxin, α-bungarotoxin, that binds to muscle-type AChRs. Protein purification revealed an approximately 250-kDa complex with two α-bungarotoxin binding sites per complex, and separation on denaturing gels revealed polypeptides of 40, 50, 60, and 65 kDa designated as the α, β, γ, and δ subunits. Because of the molecular weight of the complex and the binding of toxin molecules, a pentameric structure with two α subunits and single copies of β, γ, and δ was proposed. Using the partial amino acid sequence of the purified subunits, the cDNAs encoding AChRs in the *Torpedo* electric organ were cloned and sequenced in the early 1980s. Incorporation of the purified protein complex into lipid bilayers or expression of its mRNA in *Xenopus* oocytes confirmed the identity of the protein as a ligand-gated channel. Consistent with the proposed combination of subunits in intact receptors, the expression of AChRs in oocytes was much greater when all four subunit mRNAs were included. The α subunit was essential in order to obtain responses to acetylcholine.

Based on the hydrophobicity of amino acid residues in the primary sequence, the proposed topology included four putative transmembrane domains (M1-M4), a large N-terminal domain and a short C-terminal extracellular domain, as well as two cytoplasmic loops—a short one between M1 and M2 and a longer loop containing phosphorylation sites between M3 and M4. The first evidence that the M2 domain was the pore came from amino acid substitutions within that domain. In addition, a 23-amino-acid peptide fragment containing the M2 region of the *Torpedo* δ AChR subunit formed a cation channel when incorporated into lipid bilayers. The analysis of the primary amino acid sequence provides limited information about three-dimensional protein structure. Subsequently, 2-dimensional crystals of AChRs from the *Torpedo* electric organ were analyzed with the electron microscope (Miyazawa et al., 1999), providing a first estimate of the overall shape and structural features of the AChR.

Yet it was the identification and high-resolution crystal structure of an acetylcholine-binding protein (AChBP) from the freshwater snail (*Lymnaea stagnalis*) that provided the next level of resolution (Brejc et al., 2001; Smit et al., 2001). The AChBP is a pentamer of identical subunits that is homologous to the extracellular ACh-binding portion of the AChR, and thus provided improved molecular resolution of the extracellular ligand-binding domain. Combining information from the AChBP with two-dimensional arrays of *Torpedo* channels began to reveal the relationship of the extracellular binding domains and the structural movements that lead to opening of the intramembrane channel (Unwin, 2005). Crystal structures of two full-length homomeric channels related to the Cys-loop family then allowed the first glimpse of the structure of an entire protein of this class. Interestingly, these structures were not those of well-studied members of the Cys-loop family, but rather a pentameric ligand-gated channel from bacteria (Hilf & Dutzler, 2009) and a glutamate-gated chloride channel from *C.elegans* (Hibbs & Gouaux, 2011).

Each of the two binding pockets in the Cys-loop receptors consists of hydrophobic and aromatic amino acid residues contributed by discontinuous segments of an alpha and a nonalpha subunit. The Cys-loop receptors get their name from an N-terminal 13 amino acid sequence flanked by cysteines that form a disulfide bond and make a closed loop in the extracellular domain. The binding pocket is capped by the "C-loop" contributed by the alpha subunit, which must be open for agonist access and is locked in place when the agonist is bound and thus fully enveloped. The structure also suggested a mechanism by which agonist binding can result in channel opening through linkers that connect to the first transmembrane domain. The most important advance in understanding of Cys-loop receptors in the last few years has been a series of reports of X-ray and cryoEM structures of Cys-loop family members including the $\alpha 4\beta 2$ nNAChR (Morales-Perez et al., 2016; Walsh et al., 2018), 5-HT3 (Hassaine et al., 2014), glycine (Du et al., 2015), and $GABA_A$ receptors (Laverty et al., 2019; Miller & Aricescu, 2014). The new Cys-loop receptor structures and particularly the ability of single particle cryoEM to detect different conformational variants have provided a ever increasing picture of the receptor movements that underlie the open, closed, and desensitized states as well as the molecular states associated with binding of agonists, antagonists and allosteric modifiers such as benzodiazepines on the $GABA_A$ receptor (Phulera et al., 2018; Zhu et al., 2018).

As one might expect for a multisubunit molecule, the assembly of Cys-loop receptors as homomers, heteromers and even triheteromers is a complex cell biological problem, and is carried out by molecular chaperones and associated proteins in the endoplasmic reticulum for subsequent transport to the cell membrane. In some cases, as occurs for glutamate receptor-associated proteins, these associated proteins may be retained in the membrane and affect channel gating. The details of receptor biogenesis and trafficking are not explored further in this chapter.

As with the Cys-loop receptors, the proposed quaternary structure of glutamate channels involves multisubunit complexes surrounding a central pore (Fig. 23.6). However, the cloning of glutamate subunits in the 1990s revealed that they differed substantially from the Cys-loop receptors in their primary amino acid sequence (Hollmann & Heinemann, 1994; Nakanishi, 1992). Notable differences include the much larger extracellular N-terminal domain (ca. 500 residues). In

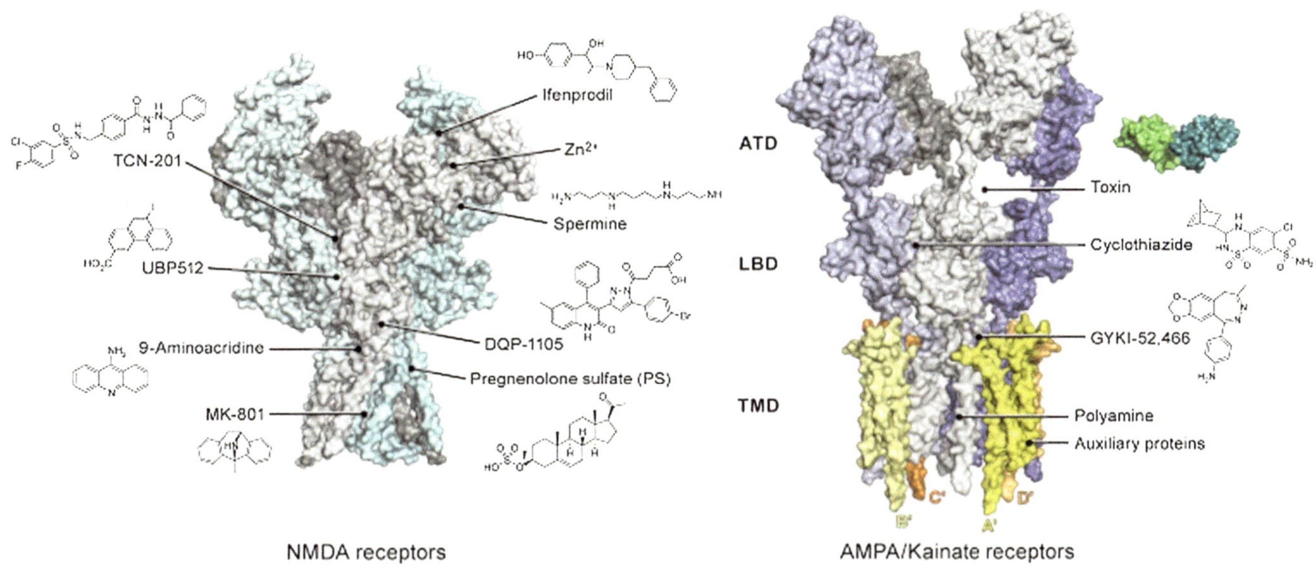

FIGURE 23.6 The domain structure of glutamate receptors is shown in a side view for the tetrameric NMDA receptors (left) and AMPA/Kainate receptors (right). The four subunits form a central pore for ion movement through the TMD. Note that the receptors show overall similarity in the extracellular ATD and LBD domains as well as the intramembrane TMD. The ligand, glutamate (and glycine for NMDA receptors), binds within the LBD, whereas allosteric modulators (e.g., ifenprodil), channel blocking drugs (e.g., MK-801), and auxiliary proteins such as TARPS bind at sites in each of the three main domains. *Modified with permission from Zhu and Gouaux (2017).*

addition, the second hydrophobic domain of glutamate subunits is a reentrant loop that enters and exits the membrane from the cytoplasmic side without fully traversing the membrane. This "P" loop shows homology to the P loop of voltage-gated channels that line the channel pore, and controls ion selectivity and conductance. Interestingly, studies of a prokaryotic glutamate receptor suggest an evolutionary link between voltage-gated potassium channels and glutamate receptors (Chen et al., 1999). Thirdly, the region between the third and fourth hydrophobic domains is extracellular in glutamate subunits, whereas it is cytoplasmic in the AChR subunits. Finally, the C-terminus of the glutamate subunits is cytoplasmic. These unique features provided clues to the location of the transmitter binding sites, the behavior of the channel pore, and sites of interaction with proteins that anchor and/or localize glutamate receptor subunits at synapses.

The extracellular component of glutamate receptors constitutes 2 separate domains (Fig. 23.6). The amino terminal domain (ATD) of approximately 380 amino acids is involved in subunit assembly and is a site of binding of allosteric modulators. The agonist/ligand-binding domain (LBD) constitutes the other extracellular domain. The LBD is discontinuous, being formed from the first extracellular region (D1) and the extracellular loop between the second and third transmembrane regions (D2). The ATD and LBD are homologous to bacterial proteins that bind amino acids. This homology was recognized by several groups soon after the cloning of glutamate receptors and led to a proposed model for glutamate binding based on the crystal structure of bacterial periplasmic binding proteins. This idea was subsequently confirmed based on chimeric receptors with domains from two glutamate subunits with distinct pharmacology (Stern-Bach et al., 1994).

Unlike AMPA or kainate receptors, which can assemble as homomeric channels, functional NMDA channels are heteromeric, requiring two GluN1 and two GluN2 subunits. In the GluN1 subunits of the NMDA receptor, glycine occupies the agonist-binding domain whereas glutamate binds to the GluN2 subunits. Selectivity for glycine over glutamate is conferred by the volume of the agonist-binding cavity resulting from steric constraints from residues present only in GluN1 (Furukawa & Gouaux, 2003). The general tetrameric organization of glutamate receptors was confirmed with the solving of a homomeric AMPA receptor structure (Sobolevsky et al., 2009). The overall structure of glutamate receptors is arranged in three layers with the amino-terminal domain (ATD), the ligand-binding domain (LBD), and the transmembrane domain (TM). An important insight from the initial tetrameric structure was that the 4 subunits are arranged in a dimer of dimer configuration. Furthermore, A/B and C/D dimer pairs in the ATD domain swap position to become A/D and B/C pairs in the LBD domain. This structure of a homomeric AMPA receptor was followed by structures for the GluN1/GluN2B NMDA receptor (Karakas & Furukawa, 2014; Lee et al., 2014), as well as structures for AMPA, kainate and NMDA receptors in a variety of conformations, i.e., with or without bound agonists or toxins (for review, see Zhu & Gouaux, 2017). Unlike Cys-loop receptors, agonist binding in glutamate receptors occurs in a subunit-autonomous manner, requiring no intersubunit interactions. Within a subunit, ligand binding occurs at the D1/D2 interface, which results in the closure of the LBD domain in a clamshell fashion. These domain movements are associated with channel gating behavior, specifically, opening, closing and desensitization.

5. Neuronal acetylcholine receptor channels

Studies of acetylcholine receptors began with the skeletal muscle isoforms (Steinbach, 1989), but AChRs are also widely expressed in neurons (McGehee & Role, 1995). The embryonic muscle AChR consists of an $\alpha_2\beta\gamma\delta$ pentamer, but later in development the γ subunit is replaced by an ε subunit. A much greater heterogeneity of receptor subunits is involved in the expression of AChRs at central synapses (Dani, 2015). Neuronal AChRs (nAChRs) mediate excitatory synaptic transmission in autonomic ganglia and at the recurrent collateral synapse of spinal motoneurons onto Renshaw cells (inhibitory interneurons). The functional role of nAChRs in the central nervous system was initially difficult to establish despite the well-known action of nicotine (through tobacco use) on human behavior. However, it is now clear that these receptors are widely expressed in brain and innervation by cholinergic neurons is pervasive in virtually all brain regions. nAChRs differ in several respects from muscle AChRs. First, α-bungarotoxin blocks muscle AChRs, but not most nAChRs. However, α-bungarotoxin binding sites in the CNS represent a distinct subtype of nAChR that has faster kinetics and a higher calcium permeability than other nAChR subtypes. The Ca^{2+} permeability of some nAChRs, as well as changes in channel properties during synaptogenesis, suggest that nAChR subtypes play a role in synaptic plasticity and development. The activity of nAChRs is also modulated by phosphorylation. The kinase activation can be due to the actions of neuropeptides such as vasoactive intestinal peptide and substance P that are coreleased with acetylcholine at some synapses.

Perhaps even more than for other ligand-gated channels, the cloning and characterization of subunit genes greatly accelerated studies of the structure and function of nAChRs (Dani, 2015; Role, 1992). Unlike the muscle AChRs, the nAChR pentamer can consist of only one or two classes of subunits, aptly named α and β. Subunits with cysteines at residues 192 and 193 are designated as α; subunits lacking these cysteines are called β. The genes for 9 α (α_{2-10}) and 3 β (β_{2-4}) subunits were cloned from rat and chick brain cDNA libraries. The stoichiometry of each assembled channel can be either homo-oligomeric (five α subunits) or heteromeric (usually with 2 α and 3 β subunits). Expression of single subunits and combination of subunits in *Xenopus* oocytes narrowed the possible combinations of subunits in functional channels. For example, α_{2-6} and β_{2-4} form heteromeric channels, whereas α_7, α_8, and α_9 can form as homodimers. Combinations of subunits differ in their single-channel properties, ligand affinity, and sensitivity to antagonists such as α-bungarotoxin. The $\alpha_4\beta_2$ receptor is a common heteromer in the CNS. It has a relatively high affinity for nicotinic agonists, and thus is a principal target of the nicotine in tobacco smoke, which delivers 20–100 nM nicotine that in turn activates and desensitizes these receptors. Homo-oligomeric α_7 channels have a relatively low affinity for ACh and rapidly desensitize. This subunit has the closest homology to the snail AChBP and has been used to probe the structural substrates of channel properties such as desensitization, ion flux, and selectivity. Homo-oligomeric α_7, α_8, and α_9 channels are sensitive to α-bungarotoxin, but α_7 is the predominant subunit of these three in brain. Presynaptic homomeric α_7 receptors with their high calcium permeability can enhance the release of glutamate and other transmitters. The α_9 subunit is unique due to its limited expression in the outer hair cells of the cochlea as well as its mixed muscarinic/nicotinic pharmacology.

Many central neurons have responses to exogenous application of nicotinic agonists, and there is a wide distribution of nAChR subunit genes as revealed by in situ hybridization. As we would expect, there is a reasonable correlation between pathways with functional nAChRs (e.g., the cholinergic input to the hippocampus and medial habenula from the medial septum and diagonal band), the presence of nicotinic binding sites, and the expression of nAChR subunits. Studies of nAChRs in different neuronal cell types have revealed a spectrum of channel properties and pharmacological characteristics, suggesting that individual neurons express several combinations of nAChR subunits. For example, in situ hybridization studies of neurons in the medial habenular nucleus have revealed two α and three β subunit genes. As for many classes of ligand-gated channels, the role of specific nAChR subunits has been investigated using genetically-engineered knockout and knockin mice to investigate subunit-specific functions. Interest in the nAChRs has been further fueled by studies indicating that mutations in α_4 or β_2 subunits cause a genetic form of human epilepsy.

6. γ-Aminobutyric acid and glycine receptor channels

Synaptic inhibition in the central nervous system (CNS) is mediated largely by $GABA_A$ and glycine receptors. These ligand-gated receptor channels are selectively permeable to anions, principally Cl^- under physiological conditions. GABA-gated Cl^- channels are designated $GABA_A$ receptors to distinguish them from the G-protein-coupled $GABA_B$ receptor (Padgett & Slesinger, 2010). $GABA_A$ and glycine receptors are members of the Cys-loop receptor family. Unlike other mammalian Cys-loop receptors that are nonselective cation channels, $GABA_A$ and glycine channels are selectively permeable to anions.

Virtually all CNS neurons have $GABA_A$ receptors, whereas the anatomical distribution of glycine receptors is generally restricted to the brainstem and spinal cord. $GABA_A$ receptors are often localized on proximal dendrites of central neurons, but are also expressed on the cell bodies, axon initial segments, and distal dendrites. Because the Cl^- equilibrium potential in mature neurons is more negative than the resting potential, the opening of $GABA_A$ or glycine channels hyperpolarizes the cell membrane potential thereby reducing neuronal excitability. In addition to hyperpolarizing the membrane potential, the opening of large numbers of these channels lowers the membrane electrical resistance. Thus, opening of $GABA_A$ channels, for example, at proximal dendrites, effectively "shunts" excitation traveling down the dendrite from excitatory synapses on more distal dendritic branches. In most central neurons, particularly during embryonic and early postnatal development, the Cl^- equilibrium is more positive than the resting potential, resulting in depolarizing $GABA_A$ or glycine responses. The exeptions are dorsal root ganglion and sympathetic ganglion neurons, which continue to be depolarized by GABA throughout adulthood. The polarity of the $GABA_A$-induced responses is determined by the intracellular Cl^- concentration, which in turn is mainly determined by the relative functional expression of electroneutral Cl^- cotransporters such as KCC2 and NKCC1, as discussed in depth in Chapter 15 by Alvarez-Leefmans.. Finally, some inhibitory synapses in the spinal cord and brainstem contain both $GABA_A$ and glycine receptors. Analysis of unitary release events at these sites indicates that single synaptic vesicles contain both GABA and glycine, and that a subpopulation of postsynaptic sites contains both receptor types (Jonas et al., 1998). Analogous to the scaffolding proteins such as PSD-95 at excitatory synapses, molecular studies have revealed anchoring and regulatory proteins at inhibitory synapses that interact with glycine and $GABA_A$ receptors. These molecules include gephyrin (Fritschy et al., 2008) and GABA receptor-associated protein (GABARAP; Mohrluder et al., 2009) as well as other molecules including collybistin and neuroligin 2. Gephyrin was identified as a cytoplasmic protein that interacts directly with glycine receptors, but also interacts with $GABA_A$ receptors. Gephyrin interacts with tubulin and the actin-binding protein profilin and thus acts as a bridge between glycine receptors and the cytoskeleton. GABARAP interacts with many $GABA_A$ receptor subtypes, as well as binding to gephyrin and tubulin. Interaction with these cytoplasmic factors can alter the localization and trafficking of $GABA_A$ and glycine receptors as well as create zones of localized signal transduction.

The behavior of single $GABA_A$ and glycine channels can be described by a kinetic scheme similar to that of the nAChR, with the binding of two agonist molecules required for channel opening (Macdonald & Twyman, 1992). Analysis of the openings and closings of single $GABA_A$ channels suggests that the channel may open briefly following the binding of a single $GABA_A$ molecule, and into two longer-lived open states from the doubly liganded configuration. Comparison of the total open duration of singly- and doubly-liganded receptors demonstrates that occupancy of both agonist sites results in many more channel openings. Channels may close and re-enter longer-lived open states before the agonist dissociates from the receptor. These so-called **bursts** are composed of short closings interrupting a series of openings, and they can last tens of milliseconds. Desensitization of $GABA_A$ channels results in long closed intervals that are grouped with bursts into **clusters** lasting up to several hundred milliseconds. These clusters are important in determining the duration of inhibitory postsynaptic potentials at some synapses (Jones & Westbrook, 1996).

The drugs that act on $GABA_A$ and glycine receptor/channels comprise a fascinatingly rich assortment of clinically important compounds (Gallagher et al., 2022; Olsen et al., 1991; Sieghart & Savić, 2018). Because these channels are so important for synaptic inhibition in the CNS, enhancement or reduction in their activity can lead to profound changes in brain function, including amnesia (increased $GABA_A$ activity) or seizures (decreased $GABA_A$ activity). Antagonists for these receptors include strychnine, which inhibits glycine receptors; bicuculline, which inhibits $GABA_A$ receptors; and picrotoxin, which inhibits both receptor types. The $GABA_A$ receptor is also the target of sedative-hypnotic drugs, such as the benzodiazepines and barbiturates. Benzodiazepines (BDZ) increase the probability of channel opening, whereas barbiturates appear to act by prolonging long channel openings (bursts). The pharmacology of benzodiazepine modulation of the $GABA_A$ receptor is particularly interesting because compounds either can enhance channel opening (BDZ agonists), reduce channel opening (BDZ inverse agonists), or block the effects of BDZ agonists (BDZ antagonists). $GABA_A$ receptor activity is also modulated by alcohol, volatile anesthetics such as isoflurane, and some steroid anesthetics (or their endogenous equivalents, the neurosteroids).

Using benzodiazepines and strychnine as selective ligands, $GABA_A$ and glycine receptors were purified as multimeric protein complexes, each with molecular weights of approximately 50–60 kDa. The solubilized receptor complex had a molecular weight of approximately 250 kDa, suggesting that, as for the AChR, five subunits constitute a receptor. Subsequent molecular cloning identified a series of receptor subunits for both receptors. The glycine subunits include the strychnine-binding subunit (α), of which four have been cloned, and a single β subunit, with a stoichiometry of $(\alpha)_2(\beta)_3$ for receptors from mature animals. Interestingly, the immature form of the glycine receptor contains only α subunits. Gephyrin binds to the β subunit, thus the interaction between gephyrin and glycine receptors is limited to the adult form. Nineteen $GABA_A$ subunits have been identified and grouped according to sequence similarity. These include six α, three β, three γ, three ρ subunits and single δ, $\in$, π, and Θ subtypes (Olsen & Sieghart, 2009; Wisden & Seeburg, 1992). In heterologous systems, expression of single $GABA_A$ or glycine receptor subunits can result in functional homomeric receptors. However, given the broad coexpression patterns of many $GABA_A$ and glycine receptor subunits and the functional heterogeneity of native receptors, homomeric receptors probably occur rarely. The large number of $GABA_A$ receptor subunits provides a formidable challenge in determining which combinations form functional receptors in neurons. $GABA_A$ and glycine receptor subunit expression also varies during development and with neuronal cell type. Based on pharmacology, expression, biochemistry and subcellular localization, at least 26 different types of native $GABA_A$ receptors have been identified in CNS neurons (Olsen & Sieghart, 2009).

Subunit composition can have a strong influence on the biophysical and pharmacological properties of $GABA_A$ and glycine receptors. The GABA and benzodiazepine binding sites reside at the interface between an α subunit and a β or γ subunit (usually γ_2), respectively. The γ_2 subunit is broadly and highly expressed in the CNS and genetic deletion greatly reduces BDZ binding sites in the brain. Interestingly, the α_6 subunit has a low affinity for BDZ agonists, but still can bind BDZ inverse agonists or antagonists, which may explain benzodiazepine-insensitive $GABA_A$ receptors in some neurons. Homomeric receptors composed of the $GABA_A$ receptor ρ subunit are bicuculline-insensitive, weakly antagonized by picrotoxin and insensitive to BDZs, barbiturates and neurosteroids. These channels also show distinct gating properties and conductances compared with other $GABA_A$ receptors. They were initially referred to as $GABA_C$ receptors. However, due to their sequence similarity and proposed structure, they are currently considered as a subtype of $GABA_A$ receptors. The three ρ subunits ($\rho1$, $\rho2$ and $\rho3$) are expressed throughout the CNS but expression is predominantly in several cell types in the retina.

7. Glutamate receptor channels

Although it was known from the early 1950s that L-glutamate was a neuroexcitant, it was not until the 1980s that the role of glutamate-gated ion channels in central synaptic transmission was widely accepted (Collingridge & Lester, 1989; Westbrook & Jahr, 1989). It soon became clear that glutamate receptor channels mediate a substantial portion of fast excitatory transmission in the vertebrate brain and spinal cord through the simultaneous activation of two types of ion channels colocalized at excitatory synapses. Of note, related proteins are expressed at the neuromuscular junction in invertebrates thus serving the role that acetylcholine receptors play in vertebrates. Characterization of glutamate receptor channels was initially based on selective activation by the exogenous amino acid ligands: NMDA, kainate, and quisqualate. AMPA (α-amino-3-hydroxy-5-methyl-4-isoxazole proprionic acid) replaced quisqualate as one of the prototypic ligands because quisqualate also activates a G protein-coupled receptor. Use of specific antagonists made it clear that NMDA receptors were distinct from kainate and AMPA receptors, but for some time it was debated whether kainate and AMPA (i.e., non-NMDA) responses arose from the same or different receptors. Molecular cloning revealed two subunit families (GluA1-4 and GluK1-5) that combine as tetramers in AMPA and kainate receptors, respectively (Hansen et al., 2021).

Initial studies of native AMPA/kainate receptor channels in neurons demonstrated many of the same features as AChRs. In hippocampal pyramidal neurons, AMPA receptor activation leads to brief openings (1−5 ms) of monovalent cation channels that show little or no voltage dependence. However, in some neurons, particularly interneurons, AMPA receptors have a higher calcium permeability and are inwardly rectifying. AMPA receptors rapidly desensitize when gated by glutamate such that both agonist unbinding and desensitization can contribute to the decay of the synaptic current. The behavior of AMPA channels seems well suited to the rapid relay of information that characterizes the fast component of excitatory postsynaptic potentials in the brain.

The NMDA gated channel provides a particularly interesting example of tight coupling between fundamental properties of ion channels and their effect on electrical activity of single cells and complex behavioral phenomena (Bekkers & Stevens, 1990; McBain & Mayer, 1994). The fundamental features of NMDA channels include: (1) voltage-dependent block of the open channel by extracellular Mg^{2+}; (2) a relatively high permeability to Ca^{2+}; and (3) slow channel kinetics resulting in long lasting channel activity. Because Mg^{2+} ions are positively charged and thus sense the membrane electric field they are drawn into the open NMDA receptor channel at negative membrane potentials. Mg^{2+} binds in the channel pore and impedes the flow of permeant ions, even though the agonists (glutamate and glycine) remain bound. However, during synaptic activity as the cell membrane depolarizes, Mg^{2+} falls out of the channel and permeant cations flow through the channel. Thus ion flux through the channel is voltage-dependent. This mechanism differs from other voltage-dependent ion channels such as the sodium or calcium channel, where voltage-dependence results from an intrinsic conformational change in the channel protein. However, the end result is the same; current flows through the NMDA receptor channel more effectively with depolarization, and thus acts as a positive feedback on synaptic activation.

Unlike many AMPA channels, Ca^{2+} contributes a substantial fraction of the flux through NMDA channels. However, the rate of flux of Ca^{2+} through the NMDA channel is slower than Na^+ because Ca^{2+} transiently binds with higher affinity to amino acid sidechains within the pore of the channel. In biophysical terms, these ion binding sites are called "energy wells." Thus most of the current through open NMDA channels is due to Na^+, which is present in 50-fold excess in the extracellular space compared with calcium. Nonetheless, the 5%−10% of the current that is carried by Ca^{2+} is sufficient to act as a biochemical signal for such processes as the induction of long-term potentiation in the hippocampus, an experimental model of associative learning. NMDA receptor activity is also involved in structural plasticity of neural circuits during development and in the adult. Excessive activation of glutamate receptors may also cause neuronal damage, presumably through elevations of intracellular Ca^{2+} concentration and subsequent activation of proteases and free radical formation. This mechanism, called **excitotoxicity,** has been implicated in brain damage due to prolonged seizures, following strokes, and may also contribute to the loss of neurons in neurological diseases.

Because no high-affinity ligands were available for purification of glutamate receptors, molecular biologists were forced to use the somewhat laborious task of expression cloning to isolate cDNA clones for glutamate receptor subunits. The first AMPA receptor subunit (GluA1) was isolated by this technique in 1989, and the first NMDA receptor clone (GluN1) was also isolated by expression cloning in 1991, followed by homology cloning of related subunits (for review see Nakanishi, 1992; Seeburg, 1993; Hollmann & Heinemann, 1994). Using these initial AMPA and NMDA clones to screen for homologous sequences, 16 related glutamate receptor subunits were isolated from rat and mouse cDNA libraries. On the basis of glutamate receptor sequence homology and the characteristics of the expressed receptors, the subunits can be grouped into three categories: four AMPA subunits (GluA1-4), five high-affinity kainate subunits (GluK1-5) and seven NMDA subunits (GluN1, GluN2A-D, GluN3A-B). GluD1 and GluD2 have significant homology to glutamate receptor subunits yet they do not bind glutamate agonists. They are referred to as orphan subunits and their function remains unclear. The GluD2 subunit is particularly intriguing because a mutation in the gene encoding this protein is responsible for the neurological phenotype of the *Lurcher* mouse. Relatively less is known about GluN3 subunits and it is unclear to what degree neurons express functional NMDA receptors that contain NR3 subunits. Interestingly, when GluN1 (the glycine binding subunit in conventional NMDA receptors) and GluN3A or GluN3B are coexpressed in heterologous cells, they form excitatory channels that are gated by glycine rather than glutamate.

Some AMPA receptor subunits can combine to form functional homo-oligomeric receptors, and this has revealed some interesting and curious phenomena. For example, the genes for GluA2, GluK1, and GluK2 code for a glutamine in the P loop region. However, the mRNA undergoes editing such that the expressed protein has an arginine residue at this site (Higuchi et al., 1993). This switch has a profound effect on the behavior of the channel. The current evoked by homo-oligomeric expression of unedited subunits shows marked inward rectification and an increased permeability to Ca^{2+}. The edited versions have a linear current-voltage relationship and are permeable only to monovalent cations (Na^+ and K^+). Because the synaptic response mediated by AMPA receptors in hippocampal pyramidal cells has a linear current-voltage relationship, most AMPA receptors in these neurons contain one or more edited copies of GluA2. However, in interneurons in cortex and hippocampus as well as in dorsal horn neurons of the spinal cord, AMPA receptors lacking edited GluA2 subunits show inward rectification and increased permeability to Ca^{2+}.

Although studies of homomeric glutamate channels have been an important experimental tool, nearly all native receptors consist of at least two different subunits in the tetramer. In addition, there are studies suggesting that NMDA receptors are composed of 3 subunits (triheteromeric, Rauner & Köhr, 2011; Tovar et al., 2013; Lü et al., 2017). In the case of hippocampal pyramidal neurons, this arrangement includes 2 GluN1, 1 GluN2A and 1 GluN2B subunits. Given that these subunits differ in their functional properties and allosteric modulation sites, this allows for a wide range of effects on receptor function.

Glutamate receptors are regulated by a wide variety of allosteric mechanisms and by phosphorylation. Some of the reagents that can bind to glutamate receptors are illustrated in Fig. 23.6. The most dramatic are the actions of glycine as a coagonist that is required for the opening of NMDA channels, and extracellular Mg^{2+} that blocks NMDA channels. Binding of two molecules of glutamate and two molecules of glycine is required to activate an NMDA channel. Because glutamate receptor ion channels are tetramers, it is thought that each subunit contributes to agonist binding. The glycine concentration in cerebrospinal fluid exceeds the EC_{50} for the glycine site, thus the glycine site may be tonically saturated. Endogenous D-serine can also bind to the glycine site, with approximately the same affinity and efficacy as glycine, and so may substitute for glycine in some brain regions. Although AMPA receptor channels do not require a coagonist, given their homology with NMDA receptor channels, four molecules of glutamate are probably necessary to activate AMPA channels (Rosenmund et al., 1998). Other regulatory factors that affect NMDA channel activity include extracellular protons, Zn^{2+}, polyamines, redox potential, and intracellular Ca^{2+}. The ability of these factors to affect NMDA receptors can depend on subunit composition. For example, GluN1/GluN2A receptors are almost 100-fold more sensitive to Zn^{2+} than GluN1/GluN2B receptors. There are numerous phosphorylation sites on the C-terminal domains of AMPA and NMDA receptor subunits that can be regulated by protein kinase A and C, tyrosine kinases and Ca^{2+}/calmodulin-dependent kinase (Chen & Roche, 2007). Kinase activation can alter any of a number of channel properties including gating and membrane trafficking. Phosphorylation of AMPA and NMDA receptors can also affect receptor trafficking that can then govern the ability of synapses to respond to stimuli that trigger longterm changes in synaptic efficacy. Activation of the NMDA receptor itself and subsequent Ca^{2+} influx can result in PKC activation, which can then alter AMPA receptor trafficking. These regulatory mechanisms have important an impact on synaptic function (Lee, 2006).

As with other neurotransmitter receptors, glutamate receptors are clustered in the postsynaptic membrane at synaptic sites. A number of proteins are involved in the clustering and localization of glutamate receptor channels. Glutamate channels are imbedded in the postsynaptic density (PSD) along with regulatory proteins, and cytoskeletal proteins. Biochemical studies suggest that calmodulin and cytoskeletal proteins can interact with C-terminal domains of several NMDA receptor subunits at domains that are involved in the regulation of NMDA receptor desensitization, suggesting an intriguing link between dynamic regulation of channel activity (e.g., by compartmentalized regulatory proteins) and structural features such as channel anchoring or localization at synapses. A family of proteins with a PDZ domain are involved in interactions with AMPA and NMDA subunits (Ehlers et al., 1996; Sheng & Pak, 2002). TARPs (transmembrane AMPA receptor regulatory proteins) are a family of proteins that share homology with auxiliary subunits of voltage-gated Ca^{2+} channels. TARPs alter the gating and pharmacological sensitivity of AMPA receptors (Milstein & Nicoll, 2008) and thus are considered auxiliary AMPA receptor subunits (Mayer, 2005). The use of cryoEM to examine receptor-TARP molecular complexes has revealed that TARPs bind underneath the LBD domain with extensive interactions with the transmembrane domain and also with D2 lobe of LBD to affect channel gating (Twomey et al., 2016; Zhao et al., 2016). Other auxiliary proteins such as CKAMP44 may have similar regulatory actions. The interactions between glutamate receptors and their associated proteins is important in receptor lateral mobility and trafficking, synapse development and the dynamic regulation of activity of ligand-gated channels in a broad range of brain functions including learning and memory.

8. Summary

Ligand-gated ion channels are membrane proteins that are fundamental signaling molecules in neurons. These molecules are localized in the plasmalemma and on intracellular organelles and can be gated by extracellular and intracellular ligands. The neurotransmitter-gated ion channels discussed in this chapter mediate fast excitation and inhibition in the nervous system, and have been well characterized by physiological and molecular studies. Studies using the technique of voltage and patch-clamp recording have examined the three basic features of an ion channel: gating, conductance, and selective permeability. Ligand-gated ion channels can be described by kinetic models involving the binding of two or more ligand molecules that induce a conformational change in the protein. As a result, a central, water-filled pore opens and conducts ions at rates of up to 10^7 ions per second. Channel activity is terminated when the channel closes or when it enters a nonconducting (desensitized) state. In vertebrates, acetylcholine and glutamate receptors are cation channels, whereas GABA and glycine receptors are anion channels. The ACh, GABA and glycine receptors constitute the Cys-loop receptor **superfamily** that presumably evolved from a common ancestral gene. Glutamate receptors comprise a separate protein family. The ligand-binding domain of glutamate channels shares homology with bacterial amino acid binding proteins, whereas the pore shares homology with the pore of voltage-dependent ion channels.

Each of the neurotransmitter-gated channels is comprised of a hetero-oligomeric complex of closely related subunits surrounding a central ion pore. For each receptor type, there are a number of possible subunit combinations that can form functional channels. AChR subunits have four transmembrane domains with the second transmembrane domain lining the pore. Charged amino acid residues near the mouth of the pore are important in determining the selectivity and conductance of the individual channel types. The glutamate subunits have three transmembrane domains and a reentrant loop domain that forms a portion of the pore. Structural studies of ligand-gated channels using X-ray crystallography and single particle cryo electron microscopy have provided atomic level resolution of the conformational movements involved in binding and gating of these channels. Continued structural studies should soon reveal how these channels interact with other proteins to determine synaptic function.

References

Ascher, P., & Nowak, L. (1987). Electrophysiological studies of NMDA receptors. *Trends in Neurosciences, 10*, 284–288.

Bekkers, J. M., & Stevens, C. F. (1990). Computational implications of NMDA receptor channels. *Cold Spring Harbor Symposia on Quantitative Biology, 55*, 131–135.

Brejc, K., van Dijk, W. J., Klaassen, R. V., Schuurmans, M., van Der Oost, J., Smit, A. B., & Sixma, T. K. (2001). Crystal structure of Ach-binding protein reveals the ligand- binding domain of nicotinic receptors. *Nature, 411*, 269–276.

Browne, L. E., Jiang, L. H., & North, R. A. (2010). New structure enlivens interest in P2X receptors. *Trends in Pharmacological Sciences, 31*, 229–237.

Burnstock, G. (2015). Physiopathological roles of P2X receptors in the central nervous system. *Current Medicinal Chemistry, 22*, 819–844.

Changeux, J.-P., & Edelstein, S. J. (1998). Allosteric receptors after 30 years. *Neuron, 21*, 959–980.

Chen, B. S., & Roche, K. W. (2007). Regulation of NMDA receptors by phosphorylation. *Neuropharmacology, 53*, 362–368.

Chen, G. Q., Cui, C., Mayer, M. L., & Gouaux, E. (1999). Functional characterization of a potassium-selective prokaryotic glutamate receptor. *Nature, 402*, 817–821.

Collingridge, G. L., & Lester, R. A. (1989). Excitatory amino acid receptors in the vertebrate central nervous system. *Pharmacological Reviews, 41*, 143–210.

Collingridge, G. L., Olsen, R. W., Peters, J., & Spedding, M. (2009). A nomenclature for ligand- gated ion channels. *Neuropharmacology, 56*, 2–5.

Colquhoun, D. (1998). Binding, gating, affinity and efficacy: The interpretation of structure-activity relationships for agonists and of the effects of mutating receptors. *British Journal of Pharmacology, 125*, 923–947.

Dalmau, J., Geis, C., & Graus, F. (2017). Autoantibodies to synaptic receptors and neuronal cell surface proteins in autoimmune diseases of the central nervous system. *Physiological Reviews, 97*, 839–887.

Dani, J. A. (2015). Neuronal nicotinic acetylcholine receptor structure and function and response to nicotine. *International Review of Neurobiology, 124*, 3–19.

Du, J., Lu, W., Cheng, Y., & Gouaux, E. (2015). Glycine receptor mechanism elucidated by electron cryo-microscopy. *Nature, 526*, 224–229.

Ehlers, M. D., Mammen, A. L., Lau, L. F., & Huganir, R. L. (1996). Synaptic targeting of glutamate receptors. *Current Opinion in Cell Biology, 8*, 484–489.

Fritschy, J. M., Harvey, R. J., & Schwarz, G. (2008). Gephyrin: Where do we stand, where do we go? *Trends in Neurosciences, 31*, 257–264.

Furukawa, H., & Gouaux, E. (2003). Mechanisms of activation, inhibition and specificity: Crystal structures of the NMDA receptor NR1 ligand-binding core. *EMBO Journal, 22*, 2873–2885.

Gallagher, C. I., Ha, D. A., Harvey, R. J., & Vandenberg, R. J. (2022). Positive allosteric modulators of glycine receptors and their potential use in pain therapies. *Pharmacological Reviews, 74*, 933–961.

Hansen, K. B., Wollmuth, L. P., Bowie, D., Furukawa, H., et al. (2021). Structure, function, and pharmacology of glutamate receptor ion channels. *Pharmacological Reviews, 73*(4), 298–487.

Hassaine, G., Deluz, C., Grasso, L., Wyss, R., Tol, M. B., Hovius, R., Graff, A., Stahlberg, H., Tomizaki, T., Desmyter, A., Moreau, C., Li, X. D., Poltevin, F., Vogel, H., & Nury, H. (2014). X-ray structure of the mouse serotonin 5-HT3 receptor. *Nature, 512*, 276–281.

Hibbs, R. E., & Gouaux, E. (2011). Principles of activation and permeation in an anion-selective Cys-loop receptor. *Nature, 474*, 54–60.

Higuchi, M., Single, F. N., Kohler, M., Sommer, B., Sprengel, R., & Seeburg, P. H. (1993). RNA editing of AMPA receptor subunti GluR-B: A base-paired intron-exon structure determines position and efficiency. *Cell, 75*, 1361–1370.

Hilf, R. J., & Dutzler, R. (2009). A procaryotic perspective on pentameric ligand-gated ion channel structure. *Current Opinion in Structural Biology, 19*, 418–424.

Hille, B. (2001). *Ionic channels of excitable membranes* (3rd ed.). Sunderland, MA: Sinauer Associates Inc.

Hollmann, M., & Heinemann, S. (1994). Cloned glutamate receptors. *Annual Review of Neuroscience, 17*, 31–108.

Jahr, C. E., & Stevens, C. F. S. (1987). Glutamate activates multiple single channel conductances in hippocampal neurons. *Nature, 325*, 522–525.

Jonas, P., Bischofberger, J., & Sandkühler, J. (1998). Corelease of two fast neurotransmitter at a central synapse. *Science, 281*, 419–424.

Jones, M. V., & Westbrook, G. L. (1996). The impact of receptor desensitization on fast synaptic transmission. *Trends in Neurosciences, 19*, 96–101.

Julius, D. (1991). Molecular biology of serotonin receptors. *Annual Review of Neuroscience, 14*, 335–360.

Karakas, E., & Furukawa, H. (2014). Crystal structure of a heterotetrameric NMDA receptor ion channel. *Science, 344*, 992–997.

Karlin, A. (1991). Explorations of the nicotinic acetylcholine receptor. *The Harvey Lectures, 85*, 71–107.

Katz, B., Miledi, R., et al. (1972). The statistical nature of the acetycholine potential and its molecular components. *Journal of Physiology, 224*, 665–669.

Kullmann, D. M. (2010). Neurological channelopathies. *Annual Review of Neuroscience, 33*, 151–172.

Lu, W., Du, J., Goehring, A., & Gouaux, E. (2017). Cryo-EM structures of the triheteromeric NMDA receptor and its allosteric modulatioin. *Science, 355*, Article eaal3729.

Lee, C.-H., Lu, W., Michel, J. C., Goehring, A., Du, J., Song, X., & Gouaux, E. (2014). NMDA receptor structures reveal subunit arrangement and pore architecture. *Nature, 511*, 191–197.

Laverty, D., Desai, R., Uchański, T., Masiulis, S., Stec, W. J., Malinauskas, T., Zivanov, J., Pardon, L., Steyaert, J., Miller, K. W., & Aricescu, R. (2019). Cryo-EM structure of the human $\alpha 1\beta 3\gamma 2$ $GABA_A$ receptor in a lipid bilayer. *Nature, 565*(7740), 516–520.

Lee, H.-K. (2006). Synaptic plasticity and phosphorylation. *Pharmacology & Therapeutics, 112*, 810–832.

Lin, S. C., & Bergles, D. E. (2004). Synaptic signaling between neurons and glia. *Glia, 47*, 290–298.

Macdonald, R. L., & Twyman, R. E. (1992). Kinetic properties and regulation of $GABA_A$ receptor channels. *Ion Channels, 3*, 315–343.

MacKinnon, R. (2004). Nobel Lecture. Potassium channels and the atomic basis of selective ion conduction. *Bioscience Reports, 24*, 75–100.

Mayer, M. L. (2005). Glutamate receptor ion channels. *Current Opinion in Neurobiology, 15*, 282–288.

McBain, C., & Mayer, M. (1994). N-methyl-D-aspartic acid receptor structure and function. *Physiological Reviews, 74*, 723–759.

McGehee, D. S., & Role, L. W. (1995). Physiological diversity of nicotinic acetylcholine receptors expressed by vertebrate neurons. *Annual Review of Physiology, 57*, 521–546.

Miller, P. S., & Aricescu, A. R. (2014). Crystal structure of a human $GABA_A$ receptor. *Nature, 512*, 270–275.

Milstein, A. D., & Nicoll, R. A. (2008). Regulation of AMPA receptor gating and pharmacology by TARP auxiliary subunits. *Trends in Pharmacological Sciences, 29*, 333–339.

Miyazawa, A., Fujiyoshi, Y., Stowell, M., & Unwin, N. (1999). Nicotinic acetylcholine receptor at 4.6 A resolution: Transverse tunnels in the channel wall. *Journal of Molecular Biology, 288*, 765–786.

Mohrluder, J., Schwarten, M., & Willbold, D. (2009). Structure and potential function of gamma-aminobutyrate type A receptor-associated protein. *FEBS Journal, 279*, 4989–5005.

Morales-Perez, C. L., Noviello, C. M., & Hibbs, R. E. (2016). X-ray structure of the human a4b2 nicotinic receptor. *Nature, 538*, 411–415.

Nakanishi, S. (1992). Molecular diversity of glutamate receptors and implications for brain function. *Science, 258*, 597–603.

Neher, E. (1992). Ion channels for communication between and within cells. *Science, 256*, 498–502.

Olsen, R. W., Sapp, D. M., Bureau, M. H., Turner, D. M., & Kokka, N. (1991). Allosteric actions of central nervous system depressants including anesthetics on subtypes of the inhibitory gamma-aminobutyric $acid_A$ receptor-chloride channel complex. *Ann. NY Acad. Sci., 625*, 145–154.

Olsen, R. W., & Sieghart, W. (2009). $GABA_A$ receptors: Subtypes provide diversity of function and pharmacology. *Neuropharmacology, 56*, 141–148.

Padgett, C. L., & Slesinger, P. A. (2010). $GABA_B$ receptor coupling to G-proteins and ion channels. *Advances in Pharmacology, 58*, 123–147.

Phulera, S., Zhu, H., Yu, J., Claxton, D. P., Yoder, N., Yoshioka, C., & Gouaux, E. (2018). Cryo-EM structure of the benzodiazepine-sensitive a1b1g2S tri-heteromeric GABAA receptor in complex with GABA. *eLife pii*, Article e39383.

Rauner, C., & Köhr, G. (2011). Triheteromeric NR1/NR2A/NR2B receptors constitute the major N-methyl-D-aspartate receptor population in adult hippocampal synapses. *Journal of Biological Chemistry, 286*, 7558–7566.

Role, L. W. (1992). Diversity in primary structure and function of neuronal nicotinic acetylcholine receptor channels. *Current Opinion in Neurobiology, 2*, 254–262.

Rosenmund, C., Stern-Bach, Y., & Stevens, C. F. (1998). The tetrameric structure of a glutamate receptor channel. *Science, 280*, 1596–1599.

Sakmann, B. (1992). Nobel Lecture. Elementary steps in synaptic transmission revealed by currents through single ion channels. *Neuron, 8*, 613–629.

Seeburg, P. H. (1993). The molecular biology of mammalian glutamate receptor channels. *Trends in Neurosciences, 16*, 359–365.

Sheng, M., & Pak, D. T. (2002). Ligand-gated ion channel interactions with cytoskeletal and signaling proteins. *Annual Review of Physiology, 62*, 755–778.

Sieghart, W., & Savić, M. M. (2018). International union of basic and clinical pharmacology. CVI: $GABA_A$ receptor subtype- and function-selective ligands: Key issues in translation to humans. *Pharmacological Reviews, 70*(4), 836–878. https://doi.org/10.1124/pr.117.014449

Smit, A. B., Syed, L. I., Schaap, D., van Minnen, J., Klumperman, J., Kits, K. S., Lodder, H., van der Schors, R. C., van Elk, R., Sorgedrager, B., Brejc, K., Sixma, T. K., & Geraerts, W. P. (2001). A glial-derived acetylcholine-binding-protein that modulates synaptic transmission. *Nature, 411*, 261–268.

Sobolevsky, A. I., Rosconi, M. P., & Gouaux, E. (2009). X-ray structure, symmetry and mechanism of an AMPA-subtype glutamate receptor. *Nature, 462*, 745–756.

Steinbach, J. H. (1989). Structural and functional diversity in vertebrate skeletal muscle nicotinic acetylcholine receptors. *Annual Review of Physiology, 51*, 353–365.

Stern-Bach, Y., Bettler, B., Hartley, M., Sheppard, P. O., O'Hara, P. J., & Heineman, S. F. (1994). Agonist selectivity of glutamate receptors is specified by two domains structurally related to bacterial amino acid-binding proteins. *Neuron, 13*, 1345–1357.

Swope, S. L., Moss, S. J., Raymond, L. A., & Huganir, R. L. (1999). Regulation of ligand-gated ion channels by protein phosphorylation. *Advances in Second Messenger and Phosphoprotein Research, 33*, 49–78.

Tovar, K. R., McGinley, M. J., & Westbrook, G. L. (2013). Triheteromeric NMDA receptors at hippocampal synapses. *Journal of Neuroscience, 22*, 9150–9160.

Twomey, E. C., Yelshanskaya, M. V., Grassucci, R. A., Frank, J., & Sobolevsky, A. I. (2016). Elucidation of AMPA receptor-stargazin complexes by cryo-electron microscopy. *Science, 353*, 83–86.

Unwin, N. (2005). Refined structure of the nicotinic acetylcholine receptor at 4 A resolution. *Journal of Molecular Biology, 346*, 967–989.

Walsh, R. M., Jr., Roh, S. H., Gharpure, A., Morales-Perez, C. L., Teng, J., & Hibbs, R. E. (2018). Structural principles of distinct assemblies of the human a4b2 nicotinic receptor. *Nature, 557*, 261–265.

Westbrook, G. L., & Jahr, C. E. (1989). Glutamate receptors in excitatory neurotransmission. *Seminars in Neuroscience, 1*, 103–114.

Wisden, W., & Seeburg, P. H. (1992). $GABA_A$ receptor channels: From subunits to functional entities. *Current Opinion in Neurobiology, 2*, 263–269.

Wu, L. J., Sweet, T. B., & Clapham, D. E. (2010). International Union of Basic and Clinical Pharmacology. LXXVI. Current progress in the mammalian TRP ion channel family. *Pharmacological Reviews, 262*, 381–404.

Zhang, M., Ma, Y., Ye, X., et al. (2023). TRP (transient receptor potential) ion channel family: structures, biological functions and therapeutic interventions for diseases. *Signal Transduction and Targeted Therapy, 8*(261). https://doi.org/10.1038/s41392-023-01464-x

Zhao, Y., Chen, S., Yoshioka, C., Baconguis, I., & Gouaux, E. (2016). *Architecture of fully occupied GluA2 AMPA receptor-TARP complex elucidated by cryo-EM.*

Zhu, S., & Gouaux, E. (2017). Structure and symmetry inform gating principles of ionotropic glutamate receptors. *Neuropharmacology, 112*, 11–15.

Zhu, S., Noviello, C. M., Teng, J., Walsh, R. M., Jr., Kim, J. J., & Hibbs, R. E. (2018). Structure of a human synaptic $GABA_A$ receptor. *Nature, 559*, 67–72.

Further reading

Mayer, M. L. (2006). Glutamate receptors at atomic resolution. *Nature, 440*, 456–462.

Chapter 24

Signal transduction and second messengers

Aldebaran M. Hofer
VA Boston Healthcare System and the Department of Surgery, Brigham and Women's Hospital and Harvard Medical School, West Roxbury, MA, United States

1. What is signal transduction?

Primeval unicellular organisms undoubtedly possessed rudimentary detection systems to sense nutrient and temperature gradients, to react to variations in light, and to avoid noxious stimuli in their extracellular environments. With the advent of multicellularity, eukaryotic organisms required progressively more sophisticated means to respond to an ever larger array of external cues and to coordinate activities between different cells. Increasingly complex signal transduction networks evolved in higher organisms to control intricate physiological processes such as cellular differentiation, embryonic development, memory, and learning.

Classically defined, signal transduction was originally regarded as a process of information transfer mediated by soluble extracellular chemical stimuli (for example, hormones, neurotransmitters, or cytokines) that bind to transmembrane receptors at the cell surface to initiate signaling events inside the cell. These processes typically involve the rapid and transient production of intracellular "second messengers", small molecule chemical mediators that include calcium ions, cAMP, cGMP, diacylglycerol (DAG) and inositol trisphosphate ($InsP_3$). Second messengers then activated or inhibited a target (for example causing phosphorylation by a kinase), culminating in a change in a physiological process.

From this narrow definition many additional modes of intracellular communication are now recognized. Signal transduction has therefore become more broadly defined to encompass any type of signaling circuit used by cells to monitor external and internal states in order to effect a change in cellular activity. Most of these pathways do not require the generation of soluble second messenger molecules, and they do not even necessarily involve the detection of a soluble extracellular ligand. Moreover, many seemingly specialized signaling pathways, for example those formerly of interest principally to developmental biologists such as Wnt, hedgehog, and Notch pathways, are now receiving attention from physiologists as well, as these highly conserved signaling mechanisms continue to exert homeostatic effects on the organism throughout its lifetime. Knowledge about the details of signaling circuits and their elaborate interrelationships is continually evolving. In this chapter we will survey this complex topic in order to provide a general framework for understanding signaling as it relates to the study of physiology.

2. General principles

The human genome consists of about ~23–25,000 protein-coding genes, but there are actually far more biological processes than unique proteins (including splice variants), indicating that many proteins are "recycled" for multiple purposes. Proteins dedicated to signal transduction are highly represented in the human genome. It has been estimated that there are 1543 genes for receptors, 518 protein kinases and 150 different protein phosphatases. Signals are used to rapidly control cellular functions such as ion channel activity, secretion or motility, but most signaling cascades ultimately converge on transcription factors (of which there are at least 1850) to execute programs of gene expression that result in long-term changes in cellular and organismal physiology (Papin et al., 2005). Differentiated cells (of which there are about 200 different types in humans) express a particular repertoire of signaling components that allow them to respond appropriately to some types of stimuli while ignoring others.

Cell Physiology Sourcebook. https://doi.org/10.1016/B978-0-12-811114-7.00009-8

A number of factors contribute to the enormous diversification that is a hallmark of signaling circuits. First, nearly all types of signaling proteins exist as a multiplicity of distinct **isoforms.** For example, there are 10 isoforms of the adenylyl cyclases, the enzymes that catalyze the production of cAMP, and 12 protein kinase C (PKC) family members. While at first glance this may seem redundant, the presence of specific isoforms for receptors, effectors (the proteins mediating the signal) and final target proteins that have varying affinities for their ligands actually permits further "tuning" of responses, as it allows the pathway to discriminate between small- and large-amplitude stimulus strength. Moreover, individual isoforms of signaling proteins receive **customized inputs** from other signaling circuits to differentially regulate their activity.

Second, while there is great diversity in the individual constituents of signaling cascades, there is also some degree of **redundancy** coming from the fact that multiple signaling pathways frequently converge on the same target to give a similar physiological outcome. An example of this is in transcription factors, which are generally the target of multiple parallel signaling inputs.

A third factor that generates both diversity and specificity among signaling circuits is their **spatial organization**. Scaffolding molecules increase the efficiency of the process by helping to assemble **signaling platforms**. These large multiprotein complexes provide physically coupled activation–deactivation cycles of kinases, their phosphatases, upstream activators and downstream effectors. It is important to remember that the enzymatic reactions taking place in these large, immobilized aggregates may not necessarily obey simple models of enzymes kinetics. A recent variation on this theme has come from the recognition that signaling proteins can undergo a biophysical process of **liquid-liquid phase separation** to yield discrete signaling compartments (droplets) that are not confined by a membrane (Zhang et al., 2020). Messages can also be **temporally encoded**, i.e., some effectors are tuned to respond to short lived signals, while other processes require sustained signaling to exert an effect. Thus the "**spatiotemporal**" features of a message, even highly diffusible second messengers such as Ca^{2+} and cAMP, can dictate the physiological response to that signal.

Subtle **remodeling** of signaling properties takes place throughout the lifetime of a cell due to changes in expression levels of elements within a cascade. This can give rise to a certain degree of heterogeneity even among cells of the same type. Remodeling has been considered to be a key feature in the progression of certain disease states (e.g., cancer, cardiovascular disease).

Finally, there is an incredible degree of **interconnectivity** amongst the various signaling pathways, which are also subject to complex feedback and feed-forward regulation. The signaling properties of any given cell would therefore be more realistically represented as a **nonlinear network of interacting circuits**, although generally depicted (as they are in this chapter) as simplified linear sequences of biochemical events.

3. General types of signal transduction cascades and their components

More than a decade ago, Sir Michael Berridge categorized the major intracellular signal transduction cascades into ~19 different groups (Berridge, 2009), but can now be updated to include two new ones. Most of these pathways (Fig. 24.1) are highly conserved across animal and even plant species, and many are initiated following engagement of two general types of cell surface receptors: G-protein coupled receptors (**GPCR**s; see also Chapter 9) or receptor tyrosine kinases (**RTK**s). Signals are often transduced and amplified through cycles of reversible covalent modification of target proteins, catalyzed by an activator and deactivator. For example, protein kinases phosphorylate protein targets, and their activity is terminated through the action of phosphatases. Another general class of transducer is the small GTP-binding proteins, whose action is accelerated by guanine-nucleotide exchange factors (**GEF**s) and terminated by GTPase-activating proteins (**GAP**s). Transcription factors are frequently the ultimate target for many of these inputs (Brivanlou & Darnell, 2002). These can be resident in the nucleus, shuttle between nucleus and cytosol, or require activation in the cytosol in order to gain access to the nucleus (see Table 24.1 for examples). We will begin our discussion by briefly introducing some of the common players in signal transduction, and then address each pathway separately.

3.1 GPCRs and classical second messenger signaling

GPCRs comprise an enormous family represented by over 800 genes in the human genome. They have a common structure characterized by seven transmembrane-spanning domains. While a significant percentage of these genes encode receptors for odorants, there are ~360 distinct GPCRs that recognize a remarkably diverse assortment of known nonodorant ligands. They serve as detectors for neurotransmitters, hormones, peptides, photons, bile acids, amino acids, Ca^{2+} ions and tastants (including bitter, sweet, and umami taste). There are also ~120 orphan receptors for which the natural ligand remains unidentified. As of 2009, eight Nobel prizes have been awarded for contributions in the field of signal transduction through

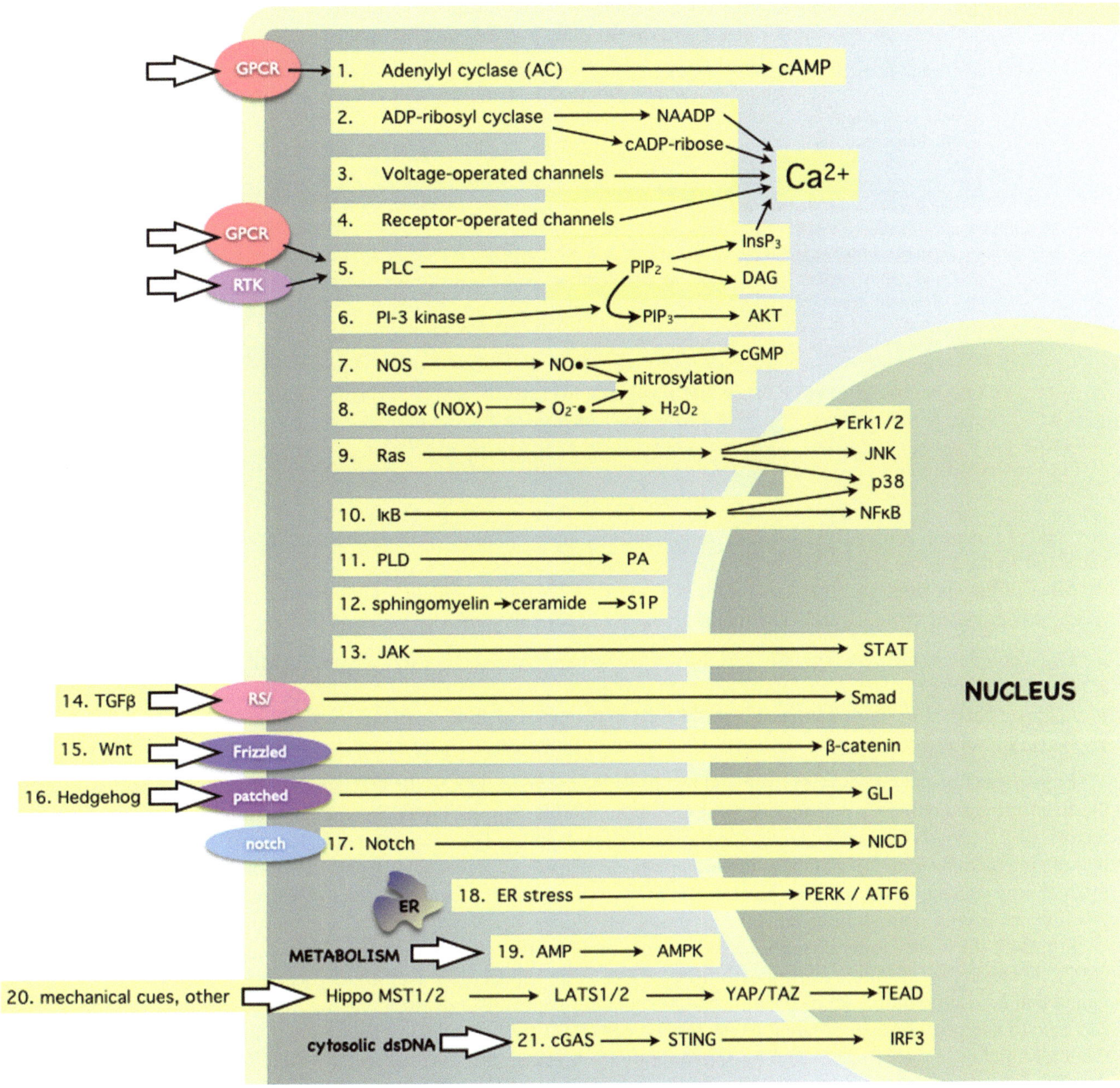

FIGURE 24.1 **Inventory of major intracellular signaling pathways.** (1) Cyclic AMP signaling pathway, (2) Cyclic ADP-ribose and nicotinic acid-adenine dinucleotide phosphate (NAADP), (3) Voltage-operated channels (VOCs), (4) Receptor-operated channels (ROCs), (5) Phospholipase C-dependent hydrolysis of PtdIns4,5P$_2$ (PIP$_2$), (6) PtdIns 3-kinase signaling (phosphorylation of PIP$_2$ to form PIP$_3$), (7) Nitric oxide (NO)/cyclic GMP signaling pathway, (8) Redox signaling: NADPH oxidase (NOX)→ reactive oxygen species (ROS), (9) Mitogen-activated protein kinase (MAPK) signaling, (10) Nuclear factor κB (NF-κB) signaling pathway, (11) Phospholipase D pathway: phosphatidylcholine→phosphatidic acid (PA), (12) Sphingomyelin signaling pathway: Formation of ceramide and sphingosine 1-phosphate (S1P)., (13) Janus kinase (JAK)/signal transducer and activator of transcription (STATs)., (14) Smad signaling pathway: TGFβ→ Smad transcription factors., (15) Wnt signaling pathway, (16) Hedgehog signaling: Generation of the transcription factor GLI, (17) Notch signaling: Generation of the transcription factor NICD (notch intracellular domain), (18) Endoplasmic reticulum stress signaling: Information transfer from the ER to the nucleus, (19) AMP signaling: Metabolic sensor, (20) Hippo pathway, (21) cGAS-STING cytosolic DNA sensing. *Adapted from Berridge (2009). See http://www.cellsignallingbiology.org for further details.*

G proteins and second messengers. GPCRs are also targets (directly or indirectly) for *at least* 30% of all currently prescribed drugs, further attesting to the physiological and medical importance of this class of proteins (Pierce et al., 2002).

G-protein coupled receptors are so-named because they interact with a class of proteins known as **heterotrimeric G-proteins** (Oldham & Hamm, 2008), consisting of a $\mathbf{G_\alpha}$ subunit (represented by 16 genes) and fused $\mathbf{G_\beta}$ and $\mathbf{G_\gamma}$ subunits

TABLE 24.1 Examples of SIGNAL-dependent transcription factors.

Most transcription factors are regulated by multiple signaling inputs; a classic example is CREB (cAMP response element binding), which is phosphorylated by both PKA and Ca^{2+}/calmodulin-dependent kinase IV (CaMKIV).

Nuclear receptors
(e.g., steroid hormone receptors)
GR, ER, PR, TR, RARs, RXRs, PPARs
Regulated by internal states
(e.g., ER stress, DNA damage, hypoxia)
SREBP, ATF6, p53, HIF
Regulated by signals generated at the cell surface:
Resident in nucleus
CREB, E2F1, ETS, ATMs, SRF, FOS-JUN, MEF2, Myc, MeCP2
Shuttle between nucleus and cytosol
DREAM, FOXO
Become activated in the cytosol, translocate to nucleus
STATs, SMADs, NFκB, β-catenin, NFAT, Tubby, GLI, MITF

From Brivanlou and Darnell (2002).

(six and 12 genes, respectively). **Lipid modifications** (e.g., palmitoylation, prenylation) keep the various subunits anchored to the plasma membrane where they interact with GPCRs.

Most of the specificity of the signal derives from the G_α subunit, which is allowed to couple to an effector when the receptor is engaged by its ligand. The major types of G_α subunits are:

- $\mathbf{G_{\alpha s}}$ - couples to adenylyl cyclases (AC1-AC9) to generate the second messenger, cAMP.
- $\mathbf{G_{\alpha i/o}}$ — inhibits the activity of most adenylyl cyclases.
- $\mathbf{G_{\alpha q/11}}$ - couples to phospholipase C and the Ca^{2+}-phopshoinositide pathway.
- $\mathbf{G_{\alpha 12/13}}$ — ultimately engages the small GTP-binding protein Rho A, to alter the cytoskeleton.

Engagement of the GPCR with its ligand results in exchange of a GDP for GTP on the G_α subunit, and dissociation of the heterotrimeric complex into free G_α and $G_{\beta\gamma}$ dimers. This permits the GTP-bound G_α subunit to couple with its effectors (e.g., adenylyl cyclases or phospholipase C; see below). The $G_{\beta\gamma}$ subunits also have biological activities that include interactions with ion channels, monomeric G-proteins (e.g., Rac, Cdc42), and PI 3-kinase. GPCRs can additionally interact (through heterotrimeric G proteins) with ion channels, for example the **GIRK** channels (G protein-coupled inwardly rectifying potassium channels).

Fig. 24.2 illustrates some of the more common stimuli and biological outputs of this signaling system (Neves et al., 2002). There are a few things worth pointing out. First, because of GPCR isoform diversity, a single type of extracellular ligand can activate multiple kinds of intracellular signaling pathways. For example, the neurotransmitter serotonin (5-HT) can bind to any one of seven different receptor subtypes (5-HT1-7), with additional diversity from splice variants, bringing the total number of GPCRs activated by serotonin to 15. It is also the case that certain GPCRs are promiscuous, i.e., it is possible for particular GPCRs to couple to more than one type of G_α subunit in order to simultaneously activate multiple pathways (Hofer & Brown, 2003). It is noteworthy that communication through the vast majority of GPCR ligands ultimately converges on just 2 s messenger systems: the **Ca^{2+}/phosphoinositide** (through $G_{\alpha q/11}$) and **cAMP** signaling systems (through $G_{\alpha s}$ and $G_{\alpha i/o}$). How these simple messengers manage to control so many biological functions remains an exciting area of continuing investigation.

3.2 Receptor tyrosine kinases and receptor serine/threonine kinases

Numerous growth factors, survival factors, and hormones such as insulin exert their actions through receptor tyrosine kinases (**RTKs**). These are single-membrane spanning proteins that are represented by ~20 different subfamilies. They generally form **dimers** upon engagement with their ligand (an exception is the insulin receptor, which is already in a dimeric configuration in the absence of insulin), causing tyrosine residues in a cytoplasmic domain of the receptor to become autophosphorylated. The RTK then assumes an active conformation that permits it to interface with several types of downstream signaling transducers. RTKs are also linked to scaffolding proteins that facilitate these processes. Major intracellular pathways that are turned on by RTK stimulation include:

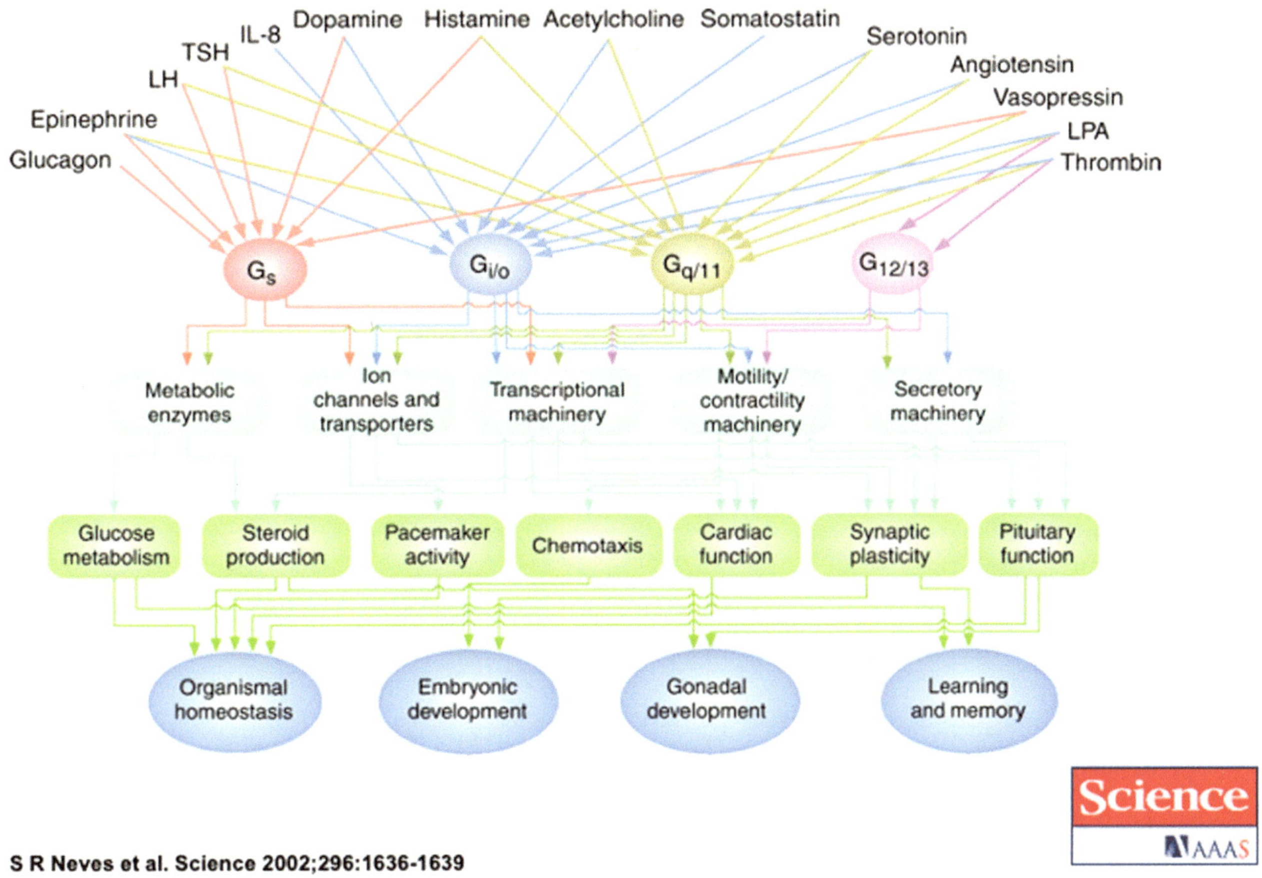

FIGURE 24.2 **Regulation of biological functions through heterotrimeric G-protein pathways.** Examples of regulatory networks that control functions through the four major classes of heterotrimeric G-proteins are depicted. See text for further details. *Taken from Neves et al. (2002) AAAS publishers (reference #6; permissions in progress).*

1. **Ras**- The activated RTK is linked through adaptors proteins (**Grb2** and **SOS**) to the small G protein ras, which in turn activates another monomeric G-protein known as **raf.** Raf is important in turning on the MAP kinase cascade (e.g., ERK1/2; see below).
2. **PLCγ**- Activation of phospholipase Cγ generates $InsP_3/Ca^{2+}$ and DAG (see Ca^{2+}-phosphoinositide pathway below).
3. **PI-3Kinase**- Generates the lipid PIP_3 to activate Akt (Protein Kinase B; see below).
4. **Src family kinases**- These are part of a large family of nonreceptor tyrosine kinases. They work by both helping to assemble signaling complexes, and to phosphorylate tyrosine residues on targets within the complex.

The **receptor serine/threonine kinases** (**RS/TKs**) of the transforming growth factor β (TGFβ) superfamily are also single-membrane spanning receptors. RS/TKs are detectors for certain growth stimuli, including bone morphogenetic protein (BMP) and TGFβ (Moustakas et al., 2001). These receptors generally form **tetramers** and couple to the Smad signaling pathway (discussed below).

3.3 Small GTP-binding proteins

As the name suggests, these are small monomeric proteins that bind GTP, which results in the G protein assuming an "on" state. This reaction is generally catalyzed by a large assortment of guanine nucleotide exchange factors (GEFs). When GTP is hydrolyzed to GDP, the G proteins are in the "off" state. GTPase-activating proteins (GAPs) accelerate this process (Bos et al., 2007).

Monomeric G proteins (~150 different ones) are important intermediaries in numerous types of signal transduction systems and can be subdivided into five general families: **Ras, Rho, Rab, Ran and Arf. Ras** proteins (36 genes) are control points for multiple types of signaling pathways (e.g., MAP kinase and PI-3 kinase pathways; see below). Many of the other G proteins (e.g., **Rho** family members) have key roles in the control of the cytoskeleton, while others (e.g., **Rab**, **Arf**) are involved in membrane trafficking. **Ran** is involved in nuclear transport.

4. Phosphorylation by kinases and other posttranslational modifications

The most common motif for altering the activity of a target is through reversible **phosphorylation** of specific amino acid residues by a protein kinase (Pearce et al., 2010). These are usually serine or threonine residues, or in the case of tyrosine kinase family members, a tyrosine residue. Protein kinases have regulatory roles in all aspects of eukaryotic cell function and provide tremendous **signal amplification**. The domains of some of the more common kinases with similarity to the classical protein kinases A, C and G (**PKA**, **PKC**, **PKG**) are illustrated in Fig. 24.3. **Phosphatases** accelerate the removal of phosphate groups on phosphorylated proteins to terminate the signal.

Other types of reversible modifications to amino acid residues are also used to control the activity of intermediates and final targets in signaling cascades. These include:

- **oxidation** of cysteine residues by the redox signaling pathway (see NADPH oxidase pathway below)
- **nitrosylation** (from nitric oxide pathway; see below)
- **SUMOylation**
- **acetylation**
- **glycosylation**
- **lipidation** (myristoylation, prenlylation, palmitoylation)
- **protein methylation**
- **ubiquitination**

For every type of enzymatic process that generates a posttranslational modification, there is a corresponding enzyme to reverse the modification (e.g., deacetylases, deubiquinating enzymes), and are thus critical for terminating the signal.

5. Intracellular signal transduction pathways

5.1 Cyclic AMP pathway

The foundations of the second messenger concept were established nearly half a century ago when Sutherland and Rall identified a small heat stable factor that mediated the intracellular actions of the hormones glucagon and epinephrine on glycogen metabolism in the liver. The mysterious factor turned out to be **cAMP**, a discovery that earned Sutherland the 1971 Nobel Prize in Physiology or Medicine (Rehmann et al., 2007).

This small highly diffusible second messenger is produced mostly through the activation of GPCRs that are **coupled** through **Gα_s to adenylyl cyclases** (**AC**s; see Fig. 24.4). The conventional ACs are a family of nine membrane spanning proteins (**AC1-AC9**) that are regulated differentially by other signaling pathways, (e.g., Ca^{2+}) (Hofer, 2012). A given cell type may express several different isoforms of AC. Apart from the conventional ACs there is also another route to cAMP production through a so-called **"soluble" adenylyl cyclase** (**sAC** or **AC10**). This AC is not a transmembrane protein, but it can nevertheless be localized to discrete regions of the cell (e.g., organelles) to provide localized cAMP signals. Because sAC is activated mainly by HCO_3^-/CO_2 (it is also regulated by Ca^{2+}) it has been considered to function as a metabolic sensor.

Once cAMP is generated, it can bind to and alter the activity of three general types of effector proteins:

- **Protein Kinase A** (**PKA**)
- **Cyclic nucleotide-gated channels**
- **Epac** (exchange protein activated by cAMP; see below)

PKA is without question the most important and well-characterized target of the cAMP signal. At rest, PKA exists as a four-protein complex known as a heterotetramer because it consists of two catalytic subunits and two regulatory subunits. When cAMP binds the regulatory subunits, it releases the catalytic subunits, which are then free to phosphorylate other proteins within the cell. Important PKA targets include ion channels (e.g., CFTR, AMPA receptors), metabolic enzymes (e.g., lipase) and cAMP-dependent transcription factors such as the prototypical **CREB** (cAMP response element binding) protein. Inputs from other signaling cascades, including Ca^{2+} and MAP kinase pathways, also converge on CREB to regulate its function.

Epac ("exchange protein activated by cAMP") is a more recently identified effector of the cAMP signal for which there are two isoforms, Epac1 and Epac2. The best understood action of Epac is to enhance the GTP-ase activity of the small G proteins of the Ras superfamily, Rap1 and Rap2.

The cAMP signal is terminated by phosphodiesterases (**PDE**s), a complex family of enzymes with more than 20 members. Some of these isoforms of PDE also preferentially metabolize cyclic GMP (see below). PDEs are of major

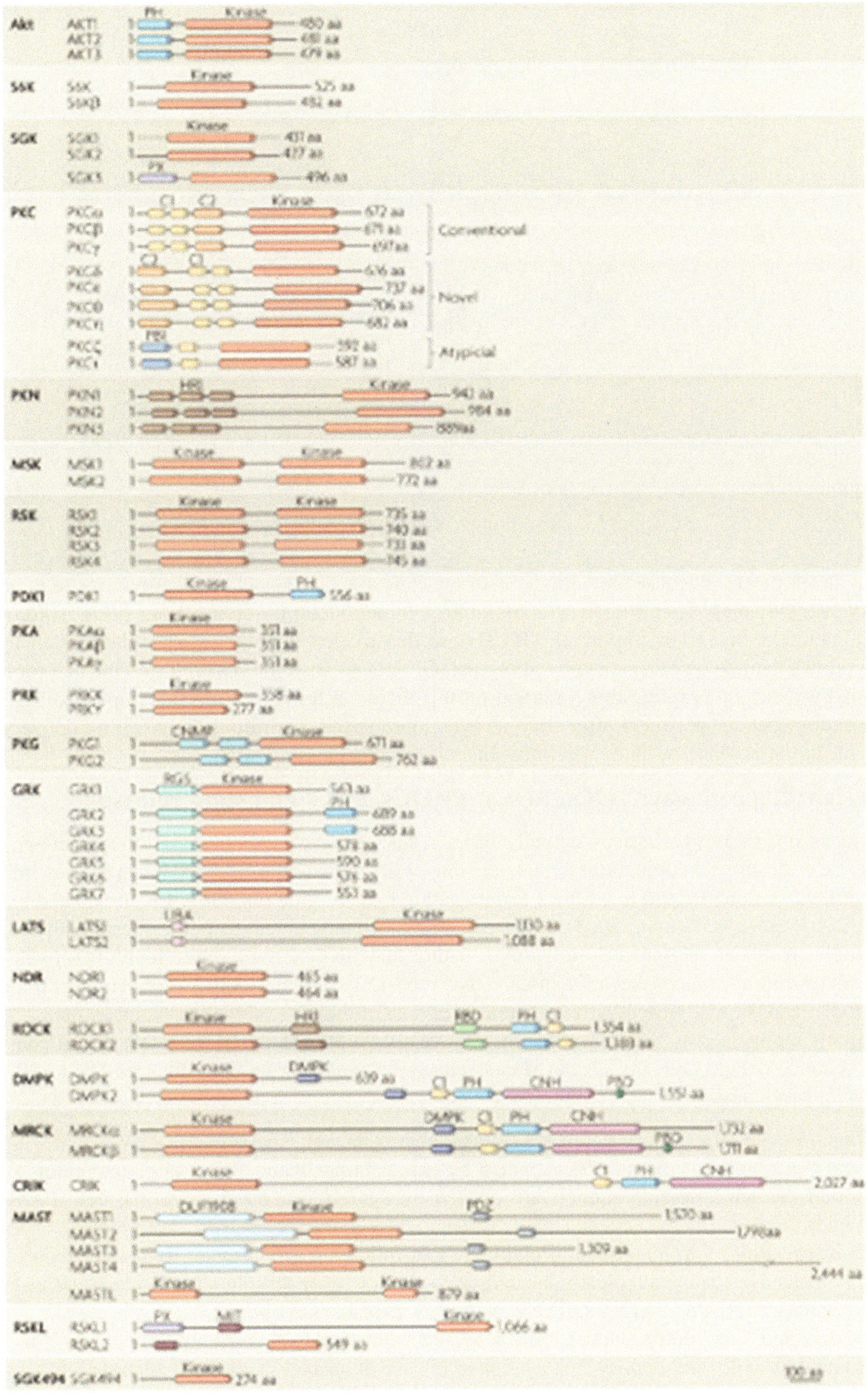

Nature Reviews | Molecular Cell Biology

FIGURE 24.3 Examples of kinases with structural similarities to protein kinase A (PKA), protein kinase G (PKG), and protein kinase C (PKC). *From Pearce et al. (2010).*

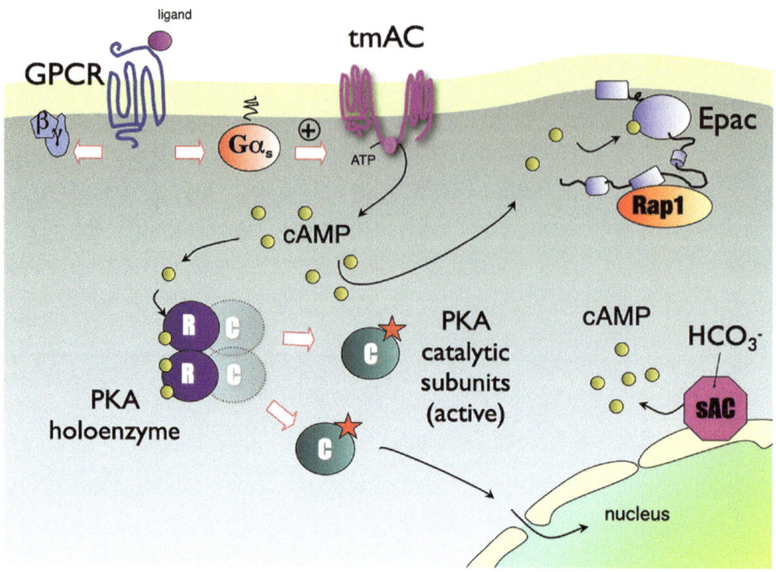

FIGURE 24.4 Basic features of the cyclic AMP signaling cascade.

interest for drug development. Because the various isoforms can have tissue-specific distributions (e.g., in airway and cardiovascular systems), they may provide an avenue for controlling particular tissue functions using isoform-selective drugs.

Prompt termination of signals is important. GPCRs are deactivated directly by phosphorylation by **GRKs** (GPCR kinases), which allows them to be recognized by **arrestins**. GPCRs are also internalized into the cell for recycling through the endolysosomal system. In an interesting variation on this theme, a few types of $G_{\alpha s}$-coupled GPCRs can continue to signal from the endosomal compartment **after** they've been internalized, permitting long-lasting delivery of cAMP to the interior of the cell in the absence of an extracellular ligand (Calebiro et al., 2010).

5.2 Calcium signaling pathway: VOCs, ROCs, SMOCs, SOCs and store release

Calcium is a major second messenger used by virtually all cells. It is at the same time the simplest messenger, since no enzymes are needed to create or destroy this ubiquitous signal, and one of the most versatile, since it can convey information in many different compartments within the cell (*e.g.*, local events within cytosol, mitochondria, ER lumen and extracellular space) (Berridge et al., 2003; Hofer & Brown, 2003). Moreover, signaling targets can respond differentially to specific types of signals. For example, many Ca^{2+} signals are oscillatory in nature, and this pattern may preferentially activate one kind of target (e.g., an ion channel) while a sustained elevation in Ca^{2+} may ultimately activate another (e.g., a transcriptional cascade). This is a very broad topic, and we will present only a general overview of the Ca^{2+} signaling pathway here.

Ca^{2+} is normally maintained at very low levels in the cytosol (**~50 nM**-this is very low if you consider that distilled water contains about 5–10 μM free Ca^{2+}!) but it can rapidly become elevated as much as 10- to 20-fold through any number of different mechanisms:

1. **VOCs**: Voltage-operated Ca^{2+} entry pathways (in excitable cells; see Chapter 25).
2. **ROCs**: Receptor operated Ca^{2+} channels, activated by extracellular ligands such as glutamate or ATP.
3. **SMOCs**: Second messenger-gated channels are opened by ligands generated inside the cell (cAMP, cGMP, DAG, arachidonic acid).
4. Release from **intracellular Ca^{2+} stores**, especially the **ER**, following activation of intracellular Ca^{2+} release channels. These are Ca^{2+} selective channels typically located on the ER or SR membrane, and include the $InsP_3R$, release channels activated through cADPR, and RyR. These pathways are discussed below.
5. **SOCs**: Store-operated Ca^{2+} entry channels are activated when the free concentration of Ca^{2+} in the ER lumen is reduced, for example following agonist/$InsP_3$-induced release described in the next section.

5.3 The Ca^{2+}-phosphoinositide pathway

GPCRs and RTKs elicit Ca^{2+} signals through the Ca^{2+}-phosphoinositide pathway, a ubiquitous mechanism found in nearly all cell types (including excitable cells). Receptor engagement results in increased turnover by phospholipases (**PLCβ** for GPCRS and **PLCγ** for RTKs) that hydrolyze the plasma membrane lipid **PIP_2** to produce **DAG** (which remains associated with the bilayer) and a small diffusible hydrophilic molecule, **InsP$_3$**. While DAG works to stimulate certain **PKC isoforms**, InsP$_3$ binds to a Ca^{2+} release channel on the ER membrane known as the InsP$_3$ receptor. The ER continually sequesters Ca^{2+} through a pump known as the **SERCA** (sarco-endoplasmic Ca^{2+}-ATPase) so that the free resting [Ca^{2+}] is ~500 μM in the ER lumen. Once the InsP$_3$ receptor is opened, the cation floods out of the ER into the cytoplasm, generating a spike in cytosolic Ca^{2+} that subsides as the stores become emptied.

There is then a second and very interesting phase of the signal. The reduction in **free [Ca^{2+}] within the ER lumen** is sensed by the luminal domain of an ER transmembrane protein known as **STIM1**. The decreased [Ca^{2+}] causes STIM1 to aggregate in discrete zones of ER just under the plasma membrane and this leads to the activation of Ca^{2+} selective channels (mostly of the family of channels known as **Orai** channels) in the plasma membrane through a physical linkage of STIM and Orai proteins. Entry of Ca^{2+} through these channels typically generates a sustained phase of the Ca^{2+} signal that serves both as a source of Ca^{2+} for refilling the ER stores, and as a signal for cellular targets that require a more persistent elevation of Ca^{2+} for their activation. When the initiating stimulus is removed and InsP$_3$ production ceases, the stores refill through the action of the SERCAs, thereby terminating the signal.

5.4 Cyclic ADP-ribose and NAADP pathways

Cyclic ADP-ribose (**cADPR**) and nicotinic acid-adenine dinucleotide phosphate (**NAADP**) are related Ca^{2+}-mobilizing messengers that are synthesized by a single enzyme (ADP ribosyl cyclase) from NAD^+ and NADP, respectively. Curiously, the same synthetic enzyme also degrades cADPR. While cADPR is established to release Ca^{2+} from the ER, the physiological role of cADPR as a messenger has been somewhat controversial. The stimuli leading up to its production and the precise mechanisms by which it releases Ca^{2+} are, however, an active area of investigation.

NAADP is an extremely potent releaser of intracellular Ca^{2+} stores. The "**two-pore channels**" (TPC1 and TPC2) that reside in acidic endo-lysosomal compartments, are considered to be the major targets of NAADP (Zhu et al., 2010). Since other intracellular Ca^{2+} release channels on the ER/SR (InsP$_3$R and RyR) increase their open probability with increasing levels of cytosolic [Ca^{2+}], a small spurt of Ca^{2+} from these acidic compartments may serve to "prime" the InsP$_3$R and/or RyR, resulting in the generation of global Ca^{2+} signals.

5.5 PtdIns 3-kinase signaling

The membrane lipid constituent **PIP$_2$** has many biological functions that include regulation of ion channels and exocytosis. It is also an important precursor for other biologically active lipids (Wymann & Schneiter, 2008). As noted above, stimulation of GPCRs and RTKs can lead to activation of phospholipases (PLCβ or PLCγ, respectively) that metabolize PIP$_2$ into InsP$_3$ and DAG. Following receptor engagement, another parallel arm of these reactions can also be set in motion that involves **phosphorylation of PIP$_2$** by the enzyme **phosphatidyl inositol 3-kinase** (**PI3-kinase**). A major target of the newly generated lipid species, **PIP$_3$**, is another kinase, protein kinase B (**PKB**), more commonly known as **AKT**.

AKT is in effect recruited to the inner leaflet of the plasma membrane by the freshly generated PIP$_3$ because it contains a domain known as the **Pleckstrin Homology (PH)** domain, a common lipid-binding motif found in many lipid-dependent signaling proteins. Another player in this cascade, that also has a PH-domain, is **PDK1** (phosphoinositide-dependent protein kinase 1). When AKT and PDK1 are both recruited to the plasma membrane at the same time, PDK1 can phosphorylate and turn on AKT. Once activated, AKT phosphorylates or interacts with a wide range of targets (for example, the kinase mTOR) to govern processes such as growth, protein, lipid, and glycogen synthesis, and apoptosis and proliferation.

PI3K/AKT signaling is arrested when PIP$_3$ in the membrane becomes metabolized. This occurs through a specific enzyme known as **PTEN.** PTEN is a phosphoinositide phosphatase, i.e., an enzyme capable of removing phosphate groups from PIP$_3$. The timely termination of the signal by PTEN is crucial. For example, during growth factor stimulation, the absence of PTEN leads to overstimulation of this pathway and inappropriate entry into the cell cycle even after the initiating stimulus has been removed. In fact, PTEN is defined as a tumor suppressor gene and PTEN mutations and deficiencies are prevalent in many types of human cancers. PTEN and other elements of the PI3K pathway are actively being investigated as targets for anticancer drugs.

5.6 Nitric oxide (NO)/cyclic GMP signaling pathway

Nitric oxide (**NO**) is a highly reactive, membrane permeable gaseous transmitter that is produced from the amino acid L-arginine by three different nitric oxide synthase (NOS) enzymes: **eNOS** (endothelial NOS), **nNOS** (neuronal NOS), and **iNOS** (inducible NOS) (Pacher et al., 2007). The nNOS and eNOS isoforms were named for the tissues in which they were originally identified, and iNOS was so named because its expression could be induced in macrophages during inflammation, however, all three enzymes are actually widely distributed across different cell types (Lundberg & Weitzberg, 2022).

NO is typically generated by eNOS and nNOS following an elevation in intracellular Ca^{2+}. The iNOS enzyme, on the other hand, appears to be constitutively active (although regulated by other signaling molecules, particularly calmodulin) and can produce large amounts of NO for prolonged periods.

Once synthesized, the NO molecule has two major jobs. The first is to activate the so-called "soluble" **guanylyl cyclase (sGC)**, which catalyzes the formation of **cGMP** (Rehmann et al., 2007) (which is far more stable than NO) from GTP. Cyclic GMP in turn activates a cGMP-dependent protein kinase (**PKG**) that can phosphorylate numerous downstream targets. The cyclic nucleotide can also open cyclic nucleotide-gated (CNG) cation channels in the plasma membrane of certain cells, such as retinal photoreceptors, olfactory neurons, and spermatozoa.

Note that there is also a nitric oxide-*independent* route to cGMP production that occurs when certain peptide hormones such as **atrial natriuretic factor (ANP)** and **guanylin** bind to the transmembrane receptor known as **particulate guanylyl cyclase** (pGC). In addition to stimulating the formation of cGMP, the other major action of NO is to cause **S-nitrosylation** of the thiol-side chains of cysteine residues within protein targets, thereby altering their activity.

Ferid Murad, Louis Ignarro, and Robert Furchgott were awarded the 1998 Nobel prize for their discovery of NO, underscoring the physiological and clinical importance of this messenger. NO (originally known as **EDRF** or endothelial-derived relaxing factor) regulates smooth muscle relaxation and hence controls vasodilatation, bladder function, penile erection, and gastrointestinal motility, to name but a few of its functions. Nitric oxide is also used by fireflies to elicit flashes of light, and by plants to evade pathogens. Other endogenous "**gasotransmitters**" have recently been described, including **carbon monoxide (CO)** and **hydrogen sulfide (H_2S)** gas.

5.7 Redox signaling through NADPH oxidase (Nox)

It has long been appreciated that immune cells such as neutrophils take advantage of the regulated production of reactive oxygen species (ROS) through **NADPH oxidases (Nox)** in order to neutralize invading pathogens. It is now known that nearly all cell types utilize Nox (of which there are five isoforms Nox1-Nox5) and the related **dual-oxidase enzymes Duox** (represented by two isoforms, Duox1 and Duox2) to generate reactive oxygen species and peroxide (H_2O_2) metabolites that, in effect, serve as short-lived second messengers to regulate a wide variety of cell functions.

One route to the activation of NADPH oxidases occurs at the plasma membrane after stimulation of receptors for hormones, growth factors or cytokines linked to PI3−K. This enzyme converts PIP_2 into PIP_3 (see above), and the resulting PIP_3 stimulates Nox, producing superoxides that can inactivate many protein targets (e.g., transcription factors, tyrosine phosphatases, PTEN) in the localized microdomain surrounding the Nox complex. It should be noted that substantial amounts of ROS can also be generated as a by-product of mitochondrial respiration.

5.8 Mitogen-activated protein kinase (MAPK) signaling

The **MAPK pathway** is a major multikinase signaling circuit that is often thought of in the context of growth factor stimulation (e.g., gene expression leading to growth and proliferation). In fact, it controls a multitude of cellular functions including metabolism, motility, apoptosis, and long-term memory, to name a few. It receives inputs from multiple signaling systems, including GPCRs, RTKs, Toll receptors, and integrins, and can also transduce the effects of environmental stresses (e.g., osmotic stress, UV radiation).

There are three major types of MAPK cascades named after their major transducers:

1. **Extracellular-signal-related kinase (ERK)**
2. **c-Jun N-terminal kinase (JNK)**
3. **p38**

As shown in Fig. 24.5, the general scheme of this pathway involves the initial activation of a **MAPKKK**, which phosphorylates a **MAPKK**, which in turn phosphorylates ERK, JNK, or p38 (the **MAPK**). Although counterintuitive, it is thought that this arrangement of having three kinases in row may actually conduct the signal faster than systems employing fewer numbers of kinases. Once phosphorylated, ERK, JNK, and p38 frequently translocate to the nucleus to regulate gene expression.

5.9 Nuclear factor κB (NF-κB) signaling pathway

The **NF-κB/Rel** family (NF-κB1, NF-κB2, NF-κB3, RelB, and cRel) consists of a group of **master transcription factors** that remain latent in the cytoplasm until activated (Hayden et al., 2006; Perkins, 2007). It is commonly regarded as a "**first responder**" to a large, diverse, panel of stress-induced stimuli that include cytokines (e.g., tumor necrosis factor, TNFα),

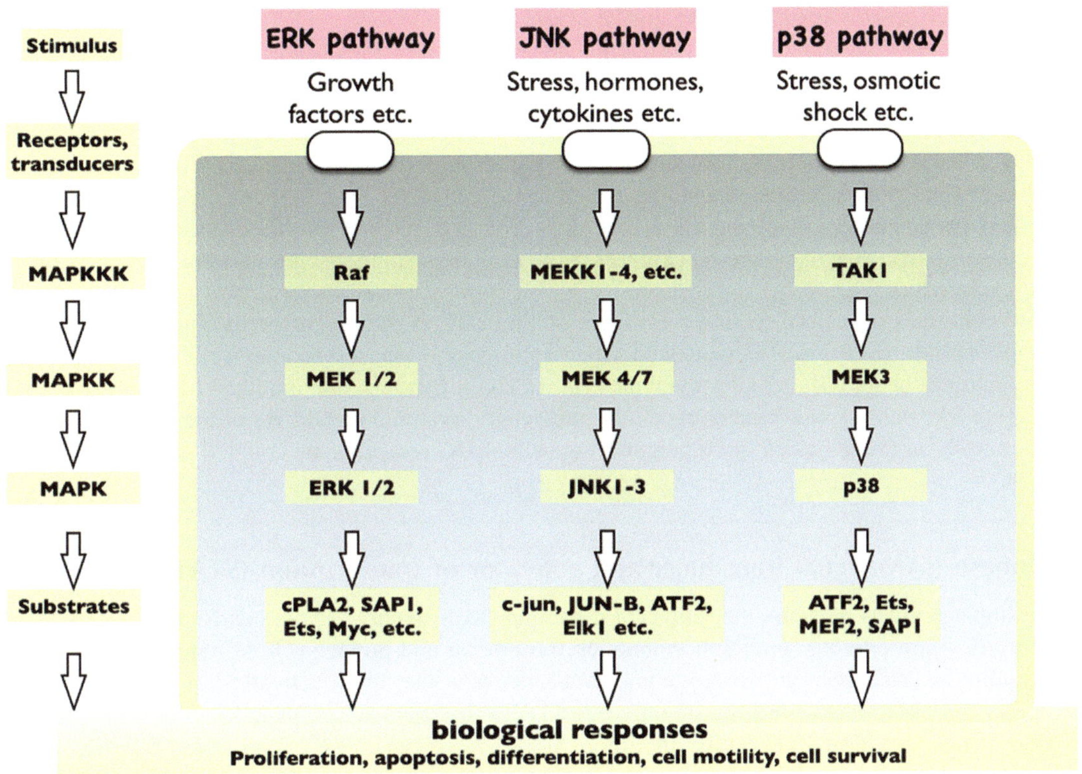

FIGURE 24.5 MAP kinase cascades.

lymphokines, mitogens, hormones, carcinogens, inflammatory agents, tumor promoters, drugs, UV light, oxidative stress, and certain growth factors. There are at least 150 types of extracellular stimuli known to activate this signaling system, which is present in nearly all cells. It also is involved in the control of osteoclastogenesis through the RANKL receptor, and activation of receptors known as Toll-like receptors, whose job it is to recognize pathogens and inflammatory cytokines. Each of these types of inputs requires a complex cast of players (not detailed here) that ultimately converge on the NF-κB pathway according to the following general scheme:

At rest, NF-κB is retained in the cytoplasm (and is therefore kept inactive) by **"IκB"** (inhibitor of NF-κB). IκB is a cytosolic resident that belongs to a family of six proteins responsible for masking the nuclear localization signal of NF-κB. However, in response to diverse stimuli, a kinase known as **"IKK"** (inhibitor of NF-κB kinase) becomes activated. There are three forms of IKK (**IKKα, IKKβ, and "NEMO",** a regulatory protein also known as **IKK**γ), which phosphorylate IκB, thereby tagging it for degradation by the **proteasome**. Liberated from IκB, NF-κB is now free to enter the nucleus, where it can bind to DNA and turn on expression of large gene sets, e.g., those involved in innate immune and stress responses. The NF-κB pathway also receives inputs from numerous other signaling systems to modulate its activity and is an important object of drug discovery efforts.

5.10 Phospholipase D pathway

There are more than 1000 types of biological lipids. Phospholipase D enzymes (two isoforms: **PLD1** and **PLD2**) hydrolyze **phoshatidylcholine**, a ubiquitous lipid constituent of the bilayer, to generate the bioactive lipid, **phosphatidic acid (PA)**. PLD1 is located principally on intracellular membranes (the ER, Golgi, and endosomal membranes) and in caveolae of the plasma membrane. PLD1 becomes activated by elements downstream of GPCR or RTK stimulation (e.g., through PKC). PLD2 appears to *not* be activated by these stimuli and is mainly localized to the plasma membrane. Once synthesized, PA acts as a second messenger to regulate a number of enzymes, including protein phosphatases, mTOR, and other targets that control budding and fusion, secretion, changes in cytoskeleton, and proliferation (Cazzolli et al., 2006).

5.11 Sphingomyelin/ceramide signaling pathway

Ceramides are bioactive lipids that can be generated enzymatically in response to stressful stimuli that include TNFα, radiation, chemotherapy drugs and through the activation of "death receptors" by a transmembrane protein known as Fas ligand on a neighboring cell. There are two routes to its formation: (1) from the breakdown of sphingomyelin (a constituent of the lipid bilayer, particularly the plasma membrane); or (2) through de novo synthesis (Hannun & Obeid, 2008).

Ceramide has several protein targets, including the protease cathepsin D, a particular isoform of PKC (PKCξ), and ceramide-activated phosphatases. However, its action does not stop there. Ceramide is further converted to the lipid messenger **sphingosine**, which activates certain kinases. Like ceramide, sphingosine can drive apoptosis (programmed cell death) and cell cycle arrest.

This cascade continues even further, however, since sphingosine is the precursor for another potent bioactive lipid, **sphingosine 1-phosphate (S1P)**. S1P is generated when sphingosine kinase becomes activated by diverse stimuli that include growth factors (PDGF, IGF, VEGF), cytokines (IL-1, TNF), hypoxia, and oxidized LDL to name a few (Wymann & Schneiter, 2008). S1P can act as a local extracellular messenger to stimulate GPCRs of the "**EDG**" family (specifically EDG1, 3, 5, six and 8) on the same cell or on neighboring cells. EDG receptors are coupled to the Ca^{2+}-phosphoinositide signaling pathways.

5.12 Janus kinase (JAK)/signal transducer and activator of transcription (STATs)

This relatively simple pathway mediates the actions of more than 30 different hormones and cytokines, including numerous interleukins (e.g., IL-2), interferons, growth hormone, erythropoietin and prolactin. It is frequently referred to as a **"fast-track"** for signaling because there are so few components between the initial activation of the receptor (RTK, GPCR, cytokine receptor etc.) and the translocation of activated STATs to the nucleus, where they bind the promoter region of responsive genes. There are **four JAK family members** (JAK1-JAK3 and TYK1) and **seven STATs**.

Typically, JAKs "prime" the ligand-bound cytokine receptor for its interaction with the STATs by phosphorylating specific tyrosine residues on the dimerized receptor. STATs dock onto the phosphorylated receptor, and then themselves become phosphorylated by JAKs (which remain associated with the receptor). At this point the activated STATs disengage from the receptor and form dimers, rendering them competent to enter into the nucleus to complete their work as transcription factors.

5.13 Smad signaling pathway

Instructions from the **TGF-β super-family** of growth factors, including Bone Morphogenetic Proteins (**BMP**s), are funneled through their serine-threonine kinase receptors on the cell surface into an intracellular system of transcription factors and regulators known as **Smads** (Moustakas et al., 2001). Smads exist as three different types: **R-Smads, co-Smad,** and **I-Smads.** The R-Smads are receptor-regulated and become directly phosphorylated when TGF-β binds its receptor (through a multistep process), activating the R-Smads, Smad2 or Smad3. Alternatively, agonists of the BMP-type bind to their cognate receptors to activate Smad1, Smad5, or Smad8. Smad4 is known as a co-Smad, which serves to facilitate signaling by forming dimers with R-Smads that then translocate to the nucleus to elicit expression of target genes. I-Smads (Smad6, 7) compete with Smad4 to inhibit this process.

In addition to controlling developmental processes, these systems maintain important physiological functions in the adult. For example, BMP is involved in skeletal remodeling and controlling proliferation of cells within the intestinal stem cell niche. TGF-β also directs differentiation of maturing cells within the intestinal villus, a structure characterized by extremely rapid cell turnover.

5.14 Wnt signaling pathway

Wnt, Notch, and **Hedgehog** (along with the TGF-β and receptor tyrosine kinase systems described above) are the major signaling pathways controlling early development throughout the animal kingdom. These pathways (Fig. 24.6) also exert important homeostatic and remodeling functions in adult organisms and are being increasingly indicted in the genesis of disease states, including tumorigenesis.

Wnts are extracellular lipoprotein growth factors that are encoded by 19 different genes in humans (Gordon & Nusse, 2006). Wnt binds to a receptor known as "**Frizzled**" (of which there are 10 types) that resembles the seven membrane-spanning receptors of the GPCR family. In the "canonical" (i.e., orthodox) **Wnt/ß-catenin pathway**, Wnt also binds to

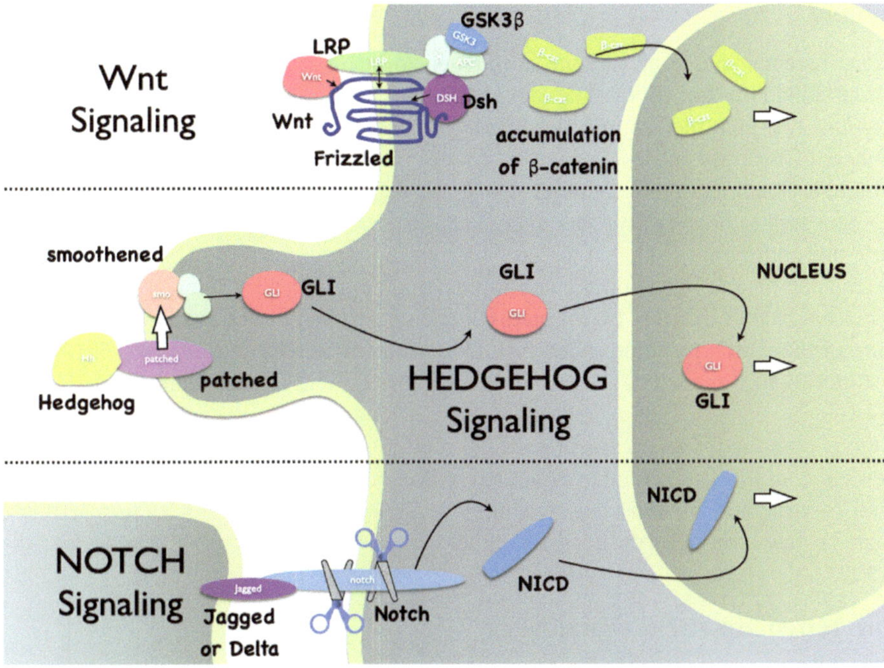

FIGURE 24.6 Wnt, hedgehog, and notch pathways.

a coreceptor of the **LRP** family and these proteins then form a complex with an intracellular protein known as **disheveled**. The Wnt pathway is somewhat unusual in that ligand binding leads to downstream steps that turn off constitutive degradation of the transcription factor **ß-catenin** by a **destruction complex** formed by **Axin**, adenomatous polyposis coli (**APC**) protein and glycogen synthase kinase 3β (**GSK-3**). This allows β-catenin to accumulate to high levels, and act as a transcription factor in the nucleus. There also exist ß-catenin-independent modes of Wnt/Frizzled signaling. Among its many functions, the Wnt pathway is known to turn on proliferation of intestinal stem cells. It is noteworthy that mutations in members of this pathway are frequently observed in colon cancer cells.

5.15 Hedgehog signaling

The hedgehog gene was so named following the discovery of a *Drosophila* mutant possessing anterior-posterior patterning defects. This animal also presented a spiky hedgehog-like appearance due to the condensed solid lawn of denticles (cuticular extensions) on the body surface. The hedgehog protein (of which there are three in vertebrates, **Indian hedgehog**, **desert hedgehog** and **sonic hedgehog**) is a secreted **morphogen**. This pathway is active in adult hematopoietic stem cells, and hair follicles, where it signals the transition to the growth phase. Components of Hedgehog signaling are frequently mutated in cancer cells and are targets of drug development. Of note, several of these anticancer compounds have been approved for the treatment of basal cell carcinoma.

Hedgehog (often abbreviated **Hh**) is secreted from cells by an unusual mechanism (Chen et al., 2007). The cytosolic protein first acquires certain unique lipid modifications that allow it to be recognized and exported from the cell by a transmembrane enzyme known as **dispatched**. Carrier proteins in the extracellular milieu facilitate the delivery of Hh to target cells, where it binds a receptor called **patched**. At rest, the function of patched is to repress another transmembrane protein, **smoothened**. The job of smoothened is to keep transcription factors of the GLI family (three in mammals: **GLI1-GLI3**) in a dormant state. Binding of Hh to patched therefore effectively allows for the activation of GLI. An interesting feature of the preceding steps is that they take place almost entirely on the **primary cilium** (plural "cilia") of the target cell, a specialized surface projection present in virtually every vertebrate cell type. Once active GLI is generated, it translocates to the nucleus where it controls transcription of genes involved in proliferation and patterning.

Of note, primary cilia also harbor a very specific subset of GPCRs that are generally coupled to the cAMP pathway through $\mathbf{G_{\alpha s}}$ (to stimulate cAMP production) or $\mathbf{G_{\alpha i}}$ (to inhibit cAMP production) (Jiang et al., 2019). One important reason for the existence of this ciliary cAMP signaling circuit may be related to the fact that PKA strongly regulates the activity of the Gli transcription factors (Mukhopadhyay & Rohatgi, 2014).

5.16 Notch signaling

Notch is a cell surface receptor that is an integral component of a relatively simple signaling pathway used extensively throughout the animal kingdom for controlling development (Bray, 2006). It takes its name from a notched wing phenotype in *Drosophila* first identified in genetic studies performed nearly a century ago. Flies with the notch mutation have increased numbers of neurons but less epidermal tissue.

There are four human Notch receptors (**Notch1-4**). Notch is unusual because most of its ligands are transmembrane proteins, therefore it becomes activated only when there is contact between receptor and ligand on a neighboring cell. This pathway is also very straightforward in that it signals without any amplification. The cascade starts when the Notch receptor encounters a transmembrane ligand on the communicating cell. One such ligand is known as **Jagged** and another is known as **Delta**. This interaction initiates two proteolytic cleavage events (not detailed here) that ultimately liberate an intracellular fragment of the Notch receptor called **NICD** (**Notch intracellular domain**). NICD works as a transcription factor that then translocates to the nucleus to alter gene expression. The signal is terminated through proteolytic degradation of NICD.

It is possible to envision how this arrangement permits a cell to influence gene expression in another, and thereby allow an organism to assemble complex multicellular structures with similar or dissimilar phenotypes. In adult organisms Notch signaling is involved in structural homeostasis, response to injury and repair. For example, this can take the form of influencing of the fate of intestinal cells destined for either secretory or absorptive lineages, and during angiogenesis.

5.17 Endoplasmic reticulum stress signaling

This pathway provides a route for transmitting information from the ER to the nucleus (Ron & Walter, 2007). The ER stress response starts when polypeptides that are being processed in the ER lumen lose their tertiary structure, or there is an overload of ER proteins. This can occur physiologically, for example if there is a sudden up-regulation of protein production so that the protein-folding machinery in the ER becomes overwhelmed, and during viral infections. It also occurs when there is a persistent loss of Ca^{2+} from the ER lumen, causing Ca^{2+}-dependent chaperones to allow unfolding of the proteins they are escorting (note that this underscores the overall importance of reliable ER Ca^{2+} homeostasis, apart from its signaling function).

The presence of unfolded polypeptides in the lumen is sensed by several different systems (Lai et al., 2007). For example, this condition elicits processing and release of an ER-resident transcription factor **ATF6**, which ultimately travels to the nucleus, where it induces the expression of more chaperone proteins. Another attempt to correct the problem is mediated by the ER kinase **PERK**, which ultimately turns off protein synthesis. Persistent unfolded protein responses result in apoptosis through other ER stress sensors.

5.18 AMPK signaling: Metabolic sensors

Cells need a way to communicate and adapt to changes in energetic status. When ATP levels are low, AMP is increased, and this is detected by a fuel-sensing kinase, **AMPK** (**AMP-activated protein kinase**) (Steinberg & Kemp, 2009). AMPK directs processes that allow the cell to correct its energy deficit by turning on ATP-generating processes (glycolysis, fatty acid oxidation and mitochondrial biogenesis) while inhibiting activities that use up ATP (gluconeogenesis, glycogen, fatty acid and protein synthesis). It also directly influences appetite-regulation in the hypothalamus.

AMPK is interesting in that it not only senses local AMP concentrations, but it also integrates the actions of hormones involved in energy balance such as leptin, ghrelin, adiponectin, glucocorticoids, insulin, as well as cannabinoids. It has generated recent excitement due to its apparent role in mediating the longevity-producing effects of calorie restriction.

5.19 Hippo signaling

During the formation and remodeling of tissues, there must be a mechanism to tell the developing organism to stop growing. Hippo is a kinase cascade that was uncovered following observations of chunky over-sized Drosophila mutants in which there was abnormal differentiation and overgrowth of tissues and organs, hence the name "hippo" (Misra & Irvine, 2018).

Cells respond to mechanical cues, changes in cell adhesion and alterations in cell energy state through hippo signaling intermediates. While the identity of a specific extracellular ligand or transmembrane receptor mediating hippo signaling is not certain, it is known that G-protein-coupled receptors can regulate the pathway. Cell-cell contacts, cell polarity and cytoskeletal tension are also sensed by hippo, owing to the localization of these components at cell-cell junctions.

These various cues converge on the first major element of the kinase cascade, **MST1/2** (the two mammalian orthologs of the drosophila hippo protein), which phosphorylate a second key kinase, **LATS1/2**. The LATS kinases phosphorylate the closely related transcriptional coregulators **YAP** and **TAZ**. The overall effect of these phosphorylation events is to repress transcriptional activity by preventing the interaction of YAP/TAZ with transcription factors such as **TEAD**, thereby suppressing programs of cell growth. Multiple other proteins participate in this highly complex cascade, and as is generally the case, there is much crosstalk with other signaling pathways. The hippo pathway is a promising target in the development of anticancer drugs and in tissue engineering.

5.20 Cytosolic DNA sensing: cGAS-STING

Recognition of foreign DNA by cells (for example following bacterial or viral invasion) is an essential aspect of the innate immune system. Recently the pathway that permits cytosolic DNA surveillance, **cGAS-STING,** has been elucidated (Decout et al., 2021). The process, which involves the generation of a novel soluble di-cyclic nucleotide, cyclic GMP-AMP (**cGAMP**), starts when the **cGAS** enzyme (cGAMP synthase) encounters double-stranded DNA in the cytosol. Recall DNA should normally be confined only to the nucleus or the mitochondrion, but cell damage may also lead to inappropriate localization of self-DNA. Binding of foreign or out-of-context self-DNA causes cGAS to dimerize and catalyze the production of cGAMP from GTP and ATP. The soluble messenger then binds to **STING**, a transmembrane receptor that is localized to the endoplasmic reticulum. Through a series of steps involving ER budding and vesicular transport through the Golgi, STING ultimately triggers the activation of a protein known as **IRF3** (interferon regulatory factor 3). IRF3 in turn enters the nucleus to regulate programs of gene expression involved in inflammation, such as interferon production. Like all signaling pathways, the cGAS-STING signaling system is being actively pursued for pharmaceutical development. Promising cGAS and STING inhibitors for the treatment of autoimmune disease and many other disease conditions are on the horizon.

6. Conclusions

The preceding provides an abbreviated and highly simplified snapshot of the major signaling pathways involved in the control of physiological processes. It highlights their beautiful diversity but also hints at their overwhelming intricacy. For the physiologist, the interconnected nature of these signaling networks means that perturbation of any one component may result in unforeseen alterations in other aspects of a cell's function. Systems biology approaches and new methods for monitoring signaling molecules with high spatiotemporal resolution (e.g., fluorescence- and GFP-based sensors) are among the tools that are currently being used to help unravel this complexity. Optical approaches for manipulating signaling pathways with light ("optogenetics") are also providing a wealth of new information. The good news for those of us studying the regulation of physiological function is that each cell type and organ system will present us with exciting challenges for a long time to come.

References

Berridge, M. (2009). In *Cell signalling pathways*. Portland Press Limited. wwwcellsignallingbiologyorg.

Berridge, M. J., Bootman, M. D., & Roderick, H. L. (2003). Calcium signalling: Dynamics, homeostasis and remodelling. *Nature Reviews Molecular Cell Biology, 4*, 517–529.

Bos, J. L., Rehmann, H., & Wittinghofer, A. (2007). GEFs and GAPs: Critical elements in the control of small G proteins. *Cell, 129*, 865–877.

Bray, S. J. (2006). Notch signalling: A simple pathway becomes complex. *Nature Reviews Molecular Cell Biology, 7*, 678–689.

Brivanlou, A. H., & Darnell, J. E., Jr. (2002). Signal transduction and the control of gene expression. *Science, 295*, 813–818.

Calebiro, D., Nikolaev, V. O., Persani, L., & Lohse, M. J. (2010). Signaling by internalized G-protein-coupled receptors. *Trends in Pharmacological Sciences, 31*, 221–228.

Cazzolli, R., Shemon, A. N., Fang, M. Q., & Hughes, W. E. (2006). Phospholipid signalling through phospholipase D and phosphatidic acid. *IUBMB Life, 58*, 457–461.

Chen, M. H., Wilson, C. W., & Chuang, P. T. (2007). SnapShot: Hedgehog signaling pathway. *Cell, 130*, 386.

Decout, A., Katz, J. D., Venkatraman, S., & Ablasser, A. (2021). The cGAS-STING pathway as a therapeutic target in inflammatory diseases. *Nature Reviews Immunology, 21*, 548–569.

Gordon, M. D., & Nusse, R. (2006). Wnt signaling: Multiple pathways, multiple receptors, and multiple transcription factors. *Journal of Biological Chemistry, 281*, 22429–22433.

Hannun, Y. A., & Obeid, L. M. (2008). Principles of bioactive lipid signalling: Lessons from sphingolipids. *Nature Reviews Molecular Cell Biology, 9*, 139–150.

Hayden, M. S., West, A. P., & Ghosh, S. (2006). SnapShot: NF-kappaB signaling pathways. *Cell, 127*, 1286–1287.

Hofer, A. M. (2012). Interactions between calcium and cAMP signaling. *Current Medicinal Chemistry, 19*, 5768–5773.

Hofer, A. M., & Brown, E. M. (2003). Extracellular calcium sensing and signalling. *Nature Reviews Molecular Cell Biology, 4*, 530–538.

Jiang, J. Y., Falcone, J. L., Curci, S., & Hofer, A. M. (2019). Direct visualization of cAMP signaling in primary cilia reveals up-regulation of ciliary GPCR activity following Hedgehog activation. *Proceedings of the National Academy of Sciences of the U S A, 116*, 12066–12071.

Lai, E., Teodoro, T., & Volchuk, A. (2007). Endoplasmic reticulum stress: Signaling the unfolded protein response. *Physiology, 22*, 193–201.

Lundberg, J. O., & Weitzberg, E. (2022). Nitric oxide signaling in health and disease. *Cell, 185*, 2853–2878.

Misra, J. R., & Irvine, K. D. (2018). The hippo signaling network and its biological functions. *Annual Review of Genetics, 52*, 65–87.

Moustakas, A., Souchelnytskyi, S., & Heldin, C. H. (2001). Smad regulation in TGF-beta signal transduction. *Journal of Cell Science, 114*, 4359–4369.

Mukhopadhyay, S., & Rohatgi, R. (2014). G-protein-coupled receptors, Hedgehog signaling and primary cilia. *Seminars in Cell & Developmental Biology, 33*, 63–72.

Neves, S. R., Ram, P. T., & Iyengar, R. (2002). G protein pathways. *Science, 296*, 1636–1639.

Oldham, W. M., & Hamm, H. E. (2008). Heterotrimeric G protein activation by G-protein-coupled receptors. *Nature Reviews Molecular Cell Biology, 9*, 60–71.

Pacher, P., Beckman, J. S., & Liaudet, L. (2007). Nitric oxide and peroxynitrite in health and disease. *Physiological Reviews, 87*, 315–424.

Papin, J. A., Hunter, T., Palsson, B. O., & Subramaniam, S. (2005). Reconstruction of cellular signalling networks and analysis of their properties. *Nature Reviews Molecular Cell Biology, 6*, 99–111.

Pearce, L. R., Komander, D., & Alessi, D. R. (2010). The nuts and bolts of AGC protein kinases. *Nature Reviews Molecular Cell Biology, 11*, 9–22.

Perkins, N. D. (2007). Integrating cell-signalling pathways with NF-kappaB and IKK function. *Nature Reviews Molecular Cell Biology, 8*, 49–62.

Pierce, K. L., Premont, R. T., & Lefkowitz, R. J. (2002). Seven-transmembrane receptors. *Nature Reviews Molecular Cell Biology, 3*, 639–650.

Rehmann, H., Wittinghofer, A., & Bos, J. L. (2007). Capturing cyclic nucleotides in action: Snapshots from crystallographic studies. *Nature Reviews Molecular Cell Biology, 8*, 63–73.

Ron, D., & Walter, P. (2007). Signal integration in the endoplasmic reticulum unfolded protein response. *Nature Reviews Molecular Cell Biology, 8*, 519–529.

Steinberg, G. R., & Kemp, B. E. (2009). AMPK in health and disease. *Physiological Reviews, 89*, 1025–1078.

Wymann, M. P., & Schneiter, R. (2008). Lipid signalling in disease. *Nature Reviews Molecular Cell Biology, 9*, 162–176.

Zhang, J. Z., Lu, T. W., Stolerman, L. M., Tenner, B., Yang, J. R., Zhang, J. F., Falcke, M., Rangamani, P., Taylor, S. S., Mehta, S., & Zhang, J. (2020). Phase separation of a PKA regulatory subunit controls cAMP compartmentation and oncogenic signaling. *Cell, 182*.

Zhu, M. X., Ma, J., Parrington, J., Galione, A., & Evans, A. M. (2010). TPCs: Endolysosomal channels for Ca^{2+} mobilization from acidic organelles triggered by NAADP. *FEBS Letters, 584*, 1966–1974.

Chapter 25

Synaptic transmission

Matthew R. Bowers[1,2,a], Mallory C. Shields[1,2,a] and Noreen E. Reist[1,2]

[1]*Colorado State University, Department of Biomedical Sciences, Fort Collins, CO, United States;* [2]*Colorado State University, Molecular, Cellular and Integrative Neurosciences Program, Fort Collins, CO, United States*

1. Introduction

Nervous system function relies upon the transmission of signals between neurons. Neurons transmit information across long distances through the nervous system by sending an electrical impulse, known as an action potential, down its axon (see Chapter 10). However, these electrical signals generally do not propagate between spatially distinct cells. To cross the gap between nerve cells, the majority of cell-to-cell signaling relies on small diffusible chemicals known as neurotransmitters. These neurotransmitters are released from active neurons in a calcium dependent manner at specialized cell-to-cell contact sites called synapses. It is important to note that direct electrical communication can occur between nerve cells through specialized structures called gap junctions (see Chapter 8). Chemical synaptic transmission is primarily unidirectional. Thus, the transmission of information required for canonical neuronal processes, such as sensation, movement, and thought, is primarily from the presynaptic cell to the postsynaptic cell. Retrograde signaling from the postsynaptic cell back to the presynaptic cell occurs, but will not be addressed in this chapter. Slower and less-directed release of chemical transmitters occurs in the autonomic nervous system and neuroendocrine pathways, including release of proteinaceous signals from large dense core vesicles. These mechanisms will not be addressed here. This chapter is focused on the Ca^{2+}-dependent, release of classical chemical neurotransmitters from an activated presynaptic neuron and their effects on their postsynaptic target.

Pioneering studies by Bernard Katz and colleagues in the 1950s first identified that the release of neurotransmitters from nerve cells at the frog neuromuscular junction occurrs through unitary packets (Fatt & Katz, 1951, 1952; Del Castillo et al., 1954). These studies were foundational to the modern understanding of chemical synaptic transmission (Section 3.2). In the 1960s and 1970s, experiments in rabbits and *Aplasia* demonstrated that synaptic transmission was highly plastic in response to activity by documenting changes in synaptic strength underlying changes in behavior (Bliss & Lomo, 1973; Lomo, 1966; Pinsker et al., 1970) (Section 7). In the 1970s, extracellular molecules located in the synaptic cleft were discovered to direct the differentiation of synaptic contacts (Burden et al., 1979; Sanes et al., 1978). Throughout the 1970−1980s, many postsynaptic receptor proteins were identified (Bradley et al., 1986; Cartaud et al., 1973; Changeux et al., 1970; Honore et al., 1982; Nickel & Potter, 1973; Pfeiffer et al., 1982; Takeuchi & Onodera, 1972). Essential players in the synaptic cleft began to be identified throughout the 1980s and 1990s. These include extracellular molecules such as agrin, the neural cadherins, and others (Bartley et al., 1994; Beckmann et al., 1994; Cheng & Flanagan, 1994; Hatta et al., 1985; Hirai et al., 1987; Ichtchenko et al., 1995; Magill et al., 1987; Nitkin et al., 1987; Ushkaryov et al., 1992). Also in the 1980s and 1990s, the molecular identity of the presynaptic neurotransmitter release machinery was identified using a combination of purification and cloning techniques in conjunction with clostridial neurotoxin studies (Block et al., 1988; Clary et al., 1990; Clary & Rothman 1990; Schiavo et al., 1993; Sollner et al., 1993b). Since then, these mechanisms have been found to be conserved at the majority of synapses.

2. Structure of the chemical synapse

While synaptic structure varies between cell types, many basic features are relatively conserved. This chapter will focus on generalized neuron-to-neuron synapses in the central nervous system (Fig. 25.1A and B) and neuron-to-muscle synapses

[a] Equally contributing.

Cell Physiology Sourcebook. https://doi.org/10.1016/B978-0-12-811114-7.00021-9

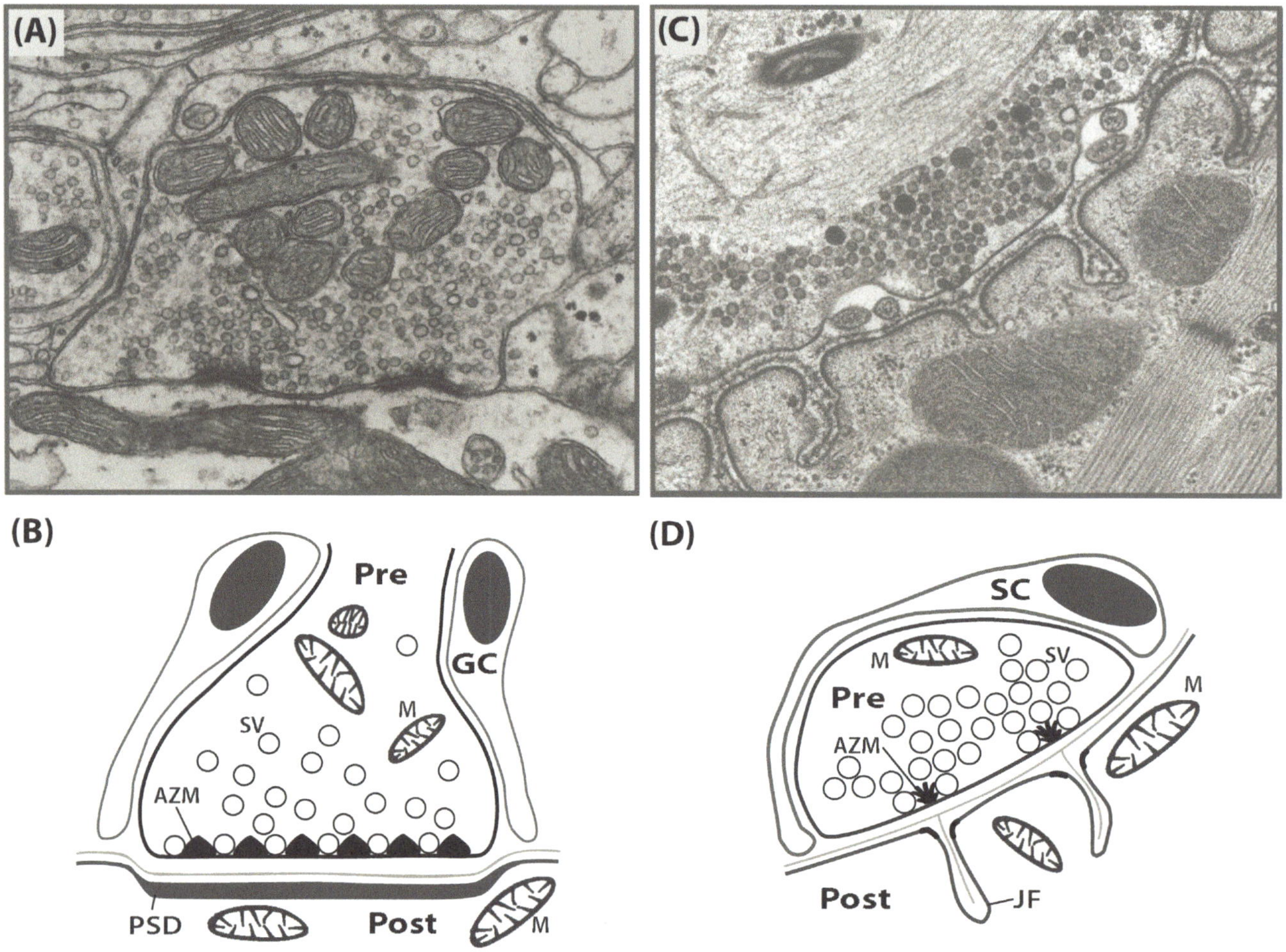

FIGURE 25.1 Synaptic structure. (A) Electron micrograph of a synapse in the central nervous system (CNS). (B) The presynaptic (PRE) and postsynaptic (POST) neurons of a CNS synapse mediating fast synaptic transmission showing mitochondria (M) and a population of small, clear synaptic vesicles (SV) containing neurotransmitter. Some vesicles are docked on the presynaptic membrane adjacent to dense material containing the protein machinery for vesicle fusion, active zone material (AZM). Opposite the docked vesicles and AZM, the membrane of the postsynaptic neuron appears thickened due to the presence of neurotransmitter receptors and scaffolding proteins, referred to as the postsynaptic density (PSD). Glial cells (GC) envelop and insulate the synapse. (C) Electron micrograph of a neuromuscular junction. (D) The presynaptic neuron (PRE) and postsynaptic striated muscle fiber (POST) of a neuromuscular junction showing similar mitochondria and neurotransmitter filled SVs with a subset docked adjacent to AZM. Invaginations of the postsynaptic membrane, junctional folds (JF), are found immediately opposite the AZM. Receptors for neurotransmitter and scaffolding proteins are especially concentrated near the tops of the JFs. The insulating glial cell is a Schwann cell (SC). *Micrographs courtesy of John Heuser.*

peripherally (Fig. 25.1C and D). Chemical synapses possess a pool of small, clear synaptic vesicles clustered around electron dense synaptic machinery in the presynaptic cell. The pre- and postsynaptic cells are separated by a narrow space, the synaptic cleft, and are insulated by glial cells. Finally, the postsynaptic cells have defined, electron-dense domains containing receptors positioned opposite release sites.

2.1 Presynaptic structure

The presynaptic terminal is generally a distinct bouton or varicosity filled with synaptic vesicles containing small, fast-acting chemical neurotransmitters (Section 3.1). Synaptic vesicles must fuse with the presynaptic membrane to release their neurotransmitters into the synaptic cleft and initiate synaptic transmission. Thus, presynaptic terminals contain the machinery required to: (1) synthesize neurotransmitter, (2) package it into vesicles, (3) couple incoming electrical signals with an increase in cytosolic [Ca^{2+}], (4) fuse the vesicles with the presynaptic membrane in response to this Ca^{2+} influx, and (5) recycle the vesicle membrane and associated proteins back into the presynaptic terminal to make new synaptic vesicles.

Within presynaptic terminals, vesicles are clustered at specialized fusion sites along the presynaptic membrane, called active zones (Couteaux & Pecot-Dechavassinein, 1970). Active zones consist of specialized scaffolding proteins that organize the presynaptic machinery required for the Ca^{2+}-dependent, vesicular fusion events that release neurotransmitter. Electron microscopy has shed light onto the precise structural organization at the active zone (Section 4.1). This organization provides precise alignment of synaptic vesicles with the fusion machinery complex and Ca^{2+} channels. This highly ordered structure is critical for proper triggering of neurotransmitter release in response to action potentials.

2.2 Synaptic cleft structure

The synaptic cleft contains structural proteins that control the surface area of the synaptic interface and the proximity between the pre- and postsynaptic membranes. Synaptic cleft width is related to the type of transmission at the synapse. Small synapses, with closer apposition of the pre- and postsynaptic cells (20–50 nm), are fast acting with localized effects. Larger synapses with wide synaptic clefts can have slower time scales and a wider range of postsynaptic targets (Hall & Sanes, 1993). Cell-adhesion molecules and scaffolding proteins control cleft width (Missler et al., 2012). Regulatory proteins align domains on the pre- and postsynaptic membranes and direct the organization of the pre- and postsynaptic specializations required for efficient synaptic communications (Magill et al., 1987; Specht & Triller, 2008).

Other synaptic cleft molecules provide clearance of neurotransmitter. Following release, neurotransmitters must be efficiently removed to terminate the signal from the pre-to the postsynaptic cell. Neurotransmitter may be enzymatically cleaved within the cleft or removed by uptake machinery on the surface of the surrounding glial cells, the presynaptic terminal, or the postsynaptic cell (Section 5).

2.3 Postsynaptic structure

The postsynaptic cell must transduce the chemical signal of the neurotransmitter back into an electrical signal. Required specializations within the postsynaptic membrane include a high concentration of neurotransmitter receptors, regulatory molecules, and ion channels to produce the electrical signal (reviewed in (Sheng & Kim, 2011)). The postsynaptic response may be a fast, specific response or a slower, more generalized response. It can be excitatory (depolarizing) or inhibitory (hyperpolarizing). The postsynaptic response depends on the specific neurotransmitter receptors present in the postsynaptic membrane, which are highly variable between cell populations, but generally fall into two main categories: ionotropic and metabotropic (Section 6.1). Ionotropic receptors are ligand-gated ion channels; the binding of neurotransmitter opens the channel allowing specific ions to cross the membrane. The selective permeability for specific ions determines whether the cell is depolarized or hyperpolarized. Metabotropic receptors are integral membrane proteins that bind G-proteins, called G-protein coupled receptors (GPCRs). Neurotransmitter binding to GPCRs results in activation of the coupled G-protein, triggering second messenger cascades that cause downstream secondary effects. These include opening of ion channels, posttranslational modification of proteins, or transcriptional changes.

3. Neurotransmitters

The presynaptic element is specialized to synthesize and release neurotransmitters. The type of neurotransmitter released varies between neuron types, with many neurons releasing more than one kind of neurotransmitter. During synaptic transmission, neurotransmitter-filled synaptic vesicles undergo a stereotypical cycle: docking at active zones, priming, Ca^{2+}-triggered fusion, vesicle reformation, refilling with neurotransmitter, and trafficking back to the active zone (Fig. 25.2). This continuous cycle provides a constantly replenished supply of neurotransmitter ready for release.

3.1 Neurotransmitter synthesis, loading and clearance

To be classified as a neurotransmitter, a chemical must meet several criteria. There must be enzymatic machinery located at the synapse to synthesize the neurotransmitter. It must be released from the terminal in response to an electrical signal. Finally, there must be machinery to degrade or clear the neurotransmitter from the synaptic cleft.

The small, fast-acting neurotransmitters, shown in Table 25.1, can be grouped into a few general classes: cholinergic, amino acidergic, monoaminergic, and purinergic. While each has its own specific synthesis, loading, and clearance machinery (reviewed in (McMahon & Nicholls, 1991)), they share many common characteristics.

Vesicle loading for these neurotransmitters relies on a proton (H^+) gradient driving force. A proton pump, or vesicular-ATPase, hydrolyzes ATP to transport H^+ ions into the vesicles. Transporters for specific neurotransmitters rely on the resultant proton gradient to move neurotransmitter into the vesicle in exchange for allowing protons to exit (reviewed in (Blakely & Edwards, 2012; Marshansky & Futai, 2008)).

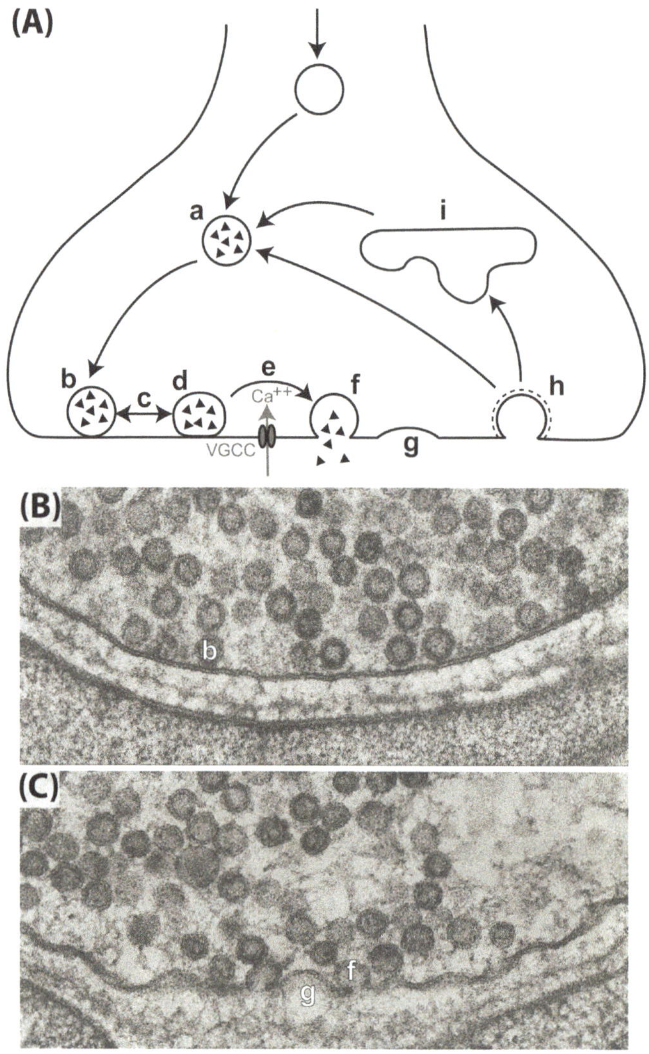

FIGURE 25.2 The synaptic vesicle cycle. (A) Schematic of the synaptic vesicle cycle depicting vesicle loading with neurotransmitter (a), vesicle docking (b), variable vesicle priming (c), maximally-primed, docked vesicle (d), depolarization-triggered Ca^{2+} entry (e), vesicle fusion with presynaptic membrane (f), vesicle membrane collapse (g), endocytosis (h), and recycling endosome (i). (B) Electron micrograph showing docked vesicles (b) at a neuromuscular junction. (C) Electron micrograph showing fusing vesicles with omega morphology (f) and vesicle membrane collapse (g). *Micrographs courtesy of John Heuser.*

TABLE 25.1 Common neurotransmitters and their receptors.

Category	Neurotransmitter	Ionotropic receptors	Metabotropic receptors
Cholinergic	Acetylcholine	nAChR	mAChR
Amino acidergic	Glutamate	Kainate, AMPA, NMDA	mGluR
	Glycine	Gly-R	
	GABA	GABAa, GABAc	GABAb
Monoaminergic	Dopamine		D_1, D_2
	Norepinepherine		α_1, α_2, β_1, β_2
	Serotonin	$5\text{-}HT_3$	$5\text{-}HT_1$, $5\text{-}HT_2$
Purinergic	ATP, adenosine	P_{2x}	P_1, P_{2y}

3.1.1 Cholinergic neurons

Cholinergic neurons synthesize and release the neurotransmitter acetylcholine (ACh). In mammals, ACh is present in the central, peripheral and autonomic nervous systems. ACh is the neurotransmitter in motor neurons that innervate skeletal muscle, making it essential for movement. It is one of two predominant transmitters in the parasympathetic division of the autonomic nervous system and is increasingly thought to play an important role in cognition. ACh is synthesized by choline acetyltransferase, which attaches choline to acetyl coenzyme A. Acetyl coenzyme A and choline acetyltransferase are synthesized in cholinergic neurons, but choline must be transported into these neurons by a Na^{+}-dependent carrier. Following synthesis of ACh, the neurotransmitter is loaded into vesicles by the vesicular acetylcholine transporter using the proton gradient mentioned above. Once released, ACh is cleared from the synaptic cleft by acetylcholinesterases that hydrolyze ACh. This produces acetate and choline, much of which is recaptured by the presynaptic terminal for synthesis of new ACh (reviewed in (Sarter & Parikh, 2005; Soreq, 2015)).

3.1.2 Amino acidergic neurons

Amino acidergic neurons release glycine, glutamate, or gamma-aminobutyric acid (GABA), which are common neurotransmitters released broadly throughout the CNS. Neurons synthesize glycine from serine using the enzyme serine hydroxymethyltransferase. Glycine can have excitatory or inhibitory effects on the postsynaptic cell depending on the receptor types present. Following release, specific transporters sequester it into glia or back into nerve terminals. Neurons are unable to synthesize glutamate or GABA de novo from glucose. They rely on glutamine production in surrounding glia as sources of the precursors for neurotransmitter production and reuptake of released neurotransmitter. In glia, alpha-ketoglutaric acid of the tricarboxylic acid cycle is converted to glutamate by the enzyme transaminase. This glutamate is then converted to glutamine, which is transported to the nerve terminal. Once there, the enzyme glutaminase catalyzes the conversion of glutamine back to glutamate. GABA is generated from glutamate by the enzyme glutamic acid dehydrogenase. Once synthesized, each is loaded into synaptic vesicles, again utilizing the proton gradient. Following release, they are transported from the cleft back into the nerve terminal for reuse or into surrounding glial cells. In glial cells, the neurotransmitter is converted back to glutamine for transport back to the presynaptic cell. This glutamine-glutamate cycle between the neuron and glial cell provides the synapse with the necessary pool of neurotransmitter (reviewed in (Bak et al., 2006; Schousboe et al., 2014)).

3.1.3 Monoaminergic neurons

Monoaminergic neurons release a wide array of neurotransmitters (reviewed in (Carlsson, 1987; Ng et al., 2015)). One class, known as catecholaminergic neurons, release dopamine, norepinephrine, or epinephrine. The catecholamines are synthesized from tyrosine through multiple intermediaries. Tyrosine is initially converted to L-dihydroxyphenylalanine (L-DOPA) by tyrosine hydroxylase, the rate-limiting step in catecholamine production. L-DOPA is decarboxylated to dopamine by L-amino acid decarboxylase. In dopaminergic neurons, dopamine is loaded into vesicles through its specific monoamine transporter. In neurons that release norepinephrine, dopamine is loaded into vesicles containing the enzyme dopamine-β-hydroxylase which converts the dopamine into norepinephrine. In epinephrine-releasing neurons, norepinephrine leaks out of vesicles, where it is converted to epinephrine by the cytosolic enzyme phenolethanolamide-N-methyltransferase. This epinepherine is then loaded into its own vesicle population. Following release, dopamine, norepinephrine and epinephrine are cleared from the cleft by Na^{+}- and Cl^{-}-dependent monoamine transporters that move them back into nerve terminals. In the terminal, these monoamines are either broken down by mitochondrial monoamine oxidase (MAO) or reloaded into new synaptic vesicles.

Other monoaminergic neurons release the indoleamine, serotonin. In serotonergic terminals, tryptophan-hydroxylase converts the amino acid tryptophan into 5-hydroxytryptophan, which is decarboxylated by 5-hydroxytryptophan decarboxylase to serotonin (5-hydroxytryptamine or 5-HT). Serotonin is loaded into vesicles through vesicular monoamine transporters. Following release, serotonin is cleared from the cleft and degraded in the cytosol of the presynaptic terminal by MAO or reloaded into synaptic vesicles.

3.1.4 Purinergic neurons

Purinergic neurons synthesize and release the nucleotides ATP and adenosine (reviewed in (Abbracchio et al., 2009; Burnstock, 2014)). ATP and adenosine are important for numerous cellular processes, and are therefore sequestered into different populations. At the synapse, they are loaded into vesicles to prevent other cellular processes from depleting the supply necessary for release. This class of neurotransmitter is often coreleased with other neurotransmitters throughout the peripheral and central nervous systems. Once released, ATP and adenosine are degraded in the cleft by cell surface enzymes known as ectonucleotidases.

3.2 Quantal release hypothesis

Identification of synaptic vesicles as the source of neurotransmitter release was a groundbreaking step in our understanding of synaptic transmission. Bernard Katz and his colleagues proposed the quantal release hypothesis in the early 1950s (Del Castillo et al., 1954; Fatt & Katz, 1951, 1952). In pioneering electrophysiological studies at the frog neuromuscular junction, they identified spontaneous, small postsynaptic changes (Fig. 25.3A) that occurred in the absence of neural activity. These depolarizations resembled end plate potentials (EPPs) following an action potential (Fig. 25.3B), albeit on a much smaller scale. These small, spontaneous events are referred to as miniature end-plate potentials (mEPPs). Since mEPPs appeared to be random and similarly-sized (between 0.3 and 0.5 mV), this suggested neurotransmitter was being released in discrete packets, or quanta. When they evoked EPPs with presynaptic action potentials in the presence of excess Mg^{2+}, which decreases Ca^{2+} entry, the amplitude of the EPP was decreased and highly variable (Fig. 25.3C). Importantly, they noticed that the EPP amplitudes fell in roughly even steps, with the step height being multiples of the spontaneous mEPP amplitude. These findings led to the hypothesis that neurotransmitter was being released in multimolecular quantized packets. Furthermore, an action potential causes the synchronous release of a large number of these packets, generating a large EPP (Fig. 25.3B) in the postsynaptic cell. Later studies at the mammalian neuromuscular junction clearly demonstrated the step-wise distribution of EPP amplitudes under similar conditions (Fig. 25.3D).

The concept that synaptic transmission could be explained by the quantized release of neurotransmitter was proposed around the same time that small membranous vesicles within nerve terminals were first identified by electron microscopy (De Robertis & Bennett, 1955; Palay, 1956). Together, these findings led to the vesicular hypothesis for neurotransmitter release, which states that each synaptic vesicle contains a similar amount of neurotransmitter (1 quantum), and that the transmitter release following an action potential results from a discrete number of vesicles fusing synchronously with the plasma membrane.

This hypothesis allows for neurotransmitter release at the synapse to be understood in terms of the relatively simple equation $m = np$, where m is the average number of quanta released following a single nerve impulse (referred to as quantal content). The parameter m depends on the number of vesicles immediately available for release (n) and the probability (p) that any given vesicle will fuse with the membrane after a single impulse. This probability depends on multiple factors, including how "ready" the vesicle is to fuse, which we will refer to as its degree of priming, and the amount of Ca^{2+} influx into the presynaptic terminal (see Sections 4.1 and 4.2). Statistical analysis of neurotransmitter release extends beyond this foundational concept (reviewed in (Bennett & Kearns, 2000)).

4. Presynaptic vesicle cycle

As introduced above, synaptic vesicles must undergo a distinct, tightly regulated series of steps. They must be loaded with neurotransmitter (Fig. 25.2A(a)), transported to the active zone, and docked to the presynaptic membrane by way of specific molecular interactions (Fig. 25.2A(b) and B(b)). The population of docked vesicles at a synapse is thought to constitute a readily releasable pool of neurotransmitter. However, only a subset of docked vesicles are fusion competent at any given time, namely, those that are maximally primed (Fig. 25.2A(d)). In a resting terminal, individual maximally-primed vesicles can spontaneously fuse with the presynaptic membrane, resulting in the release of a single quantum of neurotransmitter. In an activated neuron, the influx of Ca^{2+} through voltage-gated Ca^{2+} channels (Fig. 25.2A(e)) triggers the fusion of multiple maximally-primed vesicles with the presynaptic membrane and results in a larger release of neurotransmitter (Fig. 25.2A(f,g) and C(f,g)). This is accomplished by Ca^{2+}-sensing proteins activating the fusion machinery. Neurotransmitter release occurs in two temporally distinct Ca^{2+}-dependent phases: synchronous and asynchronous. After neurotransmitter release, the fusion machinery is disassembled and recycled for subsequent use. Concurrently, the synaptic vesicle membrane and associated proteins are retrieved through endocytosis (Fig. 25.2A(h)) to form new synaptic vesicles. These vesicles are reloaded with neurotransmitter and transported back to active zones, completing the cycle. The proteins covered in this and subsequent sections can be found in Table 25.2.

4.1 The active zone

Functionally, the role of the active zone is to transduce an electrical nerve terminal depolarization into neurotransmitter release. Couteaux and Pecot-Dechavassinein coined the term "active zone" in 1970 when they noticed synaptic vesicles at the frog neuromuscular junction docked to the presynaptic membrane adjacent to electron dense material in electron micrographs (Couteaux & Pecot-Dechavassinein, 1970). Subsequent ultrastructural studies revealed similar morphologies across organisms. The structure of this active zone material (AZM) varies among different types of synapses, but must include the requisite protein machinery for vesicle docking, priming, and fusion. The major constituents of active zones include: active zone core proteins, essential fusion machinery, Sec1/Munc18-like (SM) proteins, Ca^{2+} channels, and Ca^{2+}

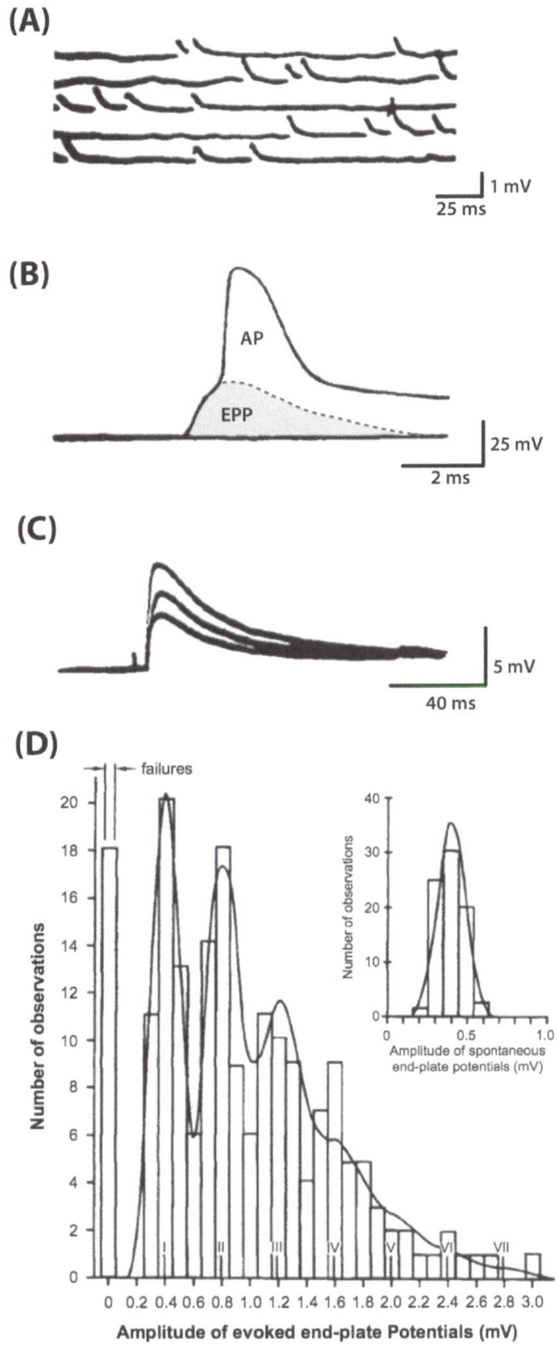

FIGURE 25.3 Intracellular recordings at the neuromuscular junction. (A) Spontaneous, mEPPs recorded from a resting (not stimulated) frog neuromuscular junction. Note that these small depolarizing potentials are less than 1 mV in amplitude and occur randomly. (B) Postsynaptic response to a presynaptic action potential. The initial hump on the recorded waveform is the EPP elicited by ACh released from the presynaptic terminal. Note that the EPP is a large depolarization (>40 mV), sufficient to bring the end-plate to threshold and trigger a muscle action potential (AP). (C) Fluctuations in EPPs when neurotransmitter output has been reduced by adding 10 mM Mg^{2+} to the bathing medium. (D) Distribution of EPP amplitudes recorded from a mammalian end-plate under conditions of reduced transmitter release in high (12.5 mM) Mg^{2+}. The inset shows a histogram of spontaneous mEPPs recorded from a resting junction. Note that EPP amplitudes group around multiples of mean mEPP amplitude. The number of experimentally observed failures (0 quanta released) and single, double, triple, or more quantal responses, fit the theoretical distribution (solid curve) calculated from the Poisson equation. *(A andB) Adapted with permission from (Fatt et al., 1952), ©1952 The Physiological Society. (C) Adapted with permission from (Del Castillo et al., 1954) ©1954 The Physiological Society. (D) Adapted with permission from (Boyd & Martin, 1956) ©1956 The Physiological Society.*

TABLE 25.2 Synaptic proteins.

Compartment	Protein	Function
Presynaptic	RIM	Active zone material
	Munc13	Active zone material
	RIM-BP	Active zone material
	Alpha-liprin	Active zone material
	ELKS/CAST	Active zone material
	synaptobrevin2	Fusion machinery
	Syntaxin	Fusion machinery
	SNAP-25	Fusion machinery
	Munc18	SM protein
	Synaptotagmin 1/2	Ca^{2+} sensor, synchronous
	Synaptotagmin 7	Ca^{2+} sensor, asynchronous
	Doc2	Ca^{2+} sensor, asynchronous
	NSF	SNARE disassembly
	Clathrin	Endocytosis
	Dynamin	Endocytosis
Synaptic cleft	Agrin	Synaptogenesis
	MuSK	Synaptogenesis
	LRP4	Synaptogenesis
	Neuroligins	Synaptogenesis/plasticity
	Neurexins	Synaptogenesis/plasticity
	Cadherins	Scaffolding/plasticity
	Eph tyrosine kinases	Scaffolding/plasticity
	Epherins	Scaffolding/plasticity
Postsynaptic	Ionotropic receptors	Changes in postsynaptic potential
	Metabotropic receptors	Changes in postsynaptic potential/activation of second messenger cascades

sensors. The core proteins covered in this chapter are Rab3-interacting molecules (RIM), Munc13, RIM binding protein (RIM-BP), alpha-liprin, and ELKS/CAST (Südhof, 2012). The essential fusion machinery consists of the soluble N-ethlymaleimide sensitive factor adaptor protein receptors, or SNAREs. The Ca^{2+}-sensing proteins covered in this chapter that trigger evoked neurotransmitter release are synaptotagmin 1, 2, 7, and double C_2 containing protein (Doc2). Numerous additional regulatory and adaptor molecules are present at or near active zones, but are not covered.

The AZM holds a cluster of synaptic vesicles near release sites, including those docked against the presynaptic membrane. The precise structure of the AZM can vary in size and shape, depending on synaptic type and/or organism. In certain visual and auditory neurons, ribbon-like structures have been observed across several species. In contrast, neuromuscular junction ultrasturcture varies across organisms. In *Drosophila,* the neuromuscular AZM has a "T-bar" shape surrounded by synaptic vesicles. At mammalian neuromuscular junctions, the AZM contains short bars of electron-dense material located on either side of two adjacent docked vesicles with additional associated vesicles above. Frog neuromuscular junctions exhibit a long midline AZM with two rows of docked vesicles on either side (see Fig. 25.6) surrounded by additional synaptic vesicles. Despite their differences, *Drosophila*, frog, and mammalian neuromuscular junctions share common synaptic machinery. Recent electron tomography studies at the frog and mouse neuromuscular junctions reveal similar ultrastructural AZM components (Jung et al., 2016; Nagwaney et al., 2009). The molecular identity of the individual proteins comprising these structures remains an active area of synaptic research.

4.1.1 Active zone core proteins

Some of the proteins that are located in the core of the active zone include RIM, Munc13, RIM-BP, alpha-liprin, and ELKS/CAST. Together with additional adaptor proteins, they recruit synaptic vesicles, fusion machinery, and Ca^{2+} channels to the active zone. RIM and Munc13 work together to link synaptic vesicles to active zones, as discussed below. In addition, RIM binds Ca^{2+} channels, which are concurrently bound by RIM-BP and ELKS/CAST, creating a trimeric complex essential for recruiting Ca^{2+} channels to active zones. Munc13 is additionally involved in enhancing vesicle fusion competence (see Section 4.2). Alpha-liprins interact with multiple proteins that regulate active zone formation (Zhen & Jin, 2004). They are scaffold proteins that may act as anchors to recruit and stabilize molecules at the active zone, as they are critical for recruiting RIM (Schoch et al., 2002; Spangler et al., 2013) and trafficking synaptic vesicles (Miller et al., 2005). Studies in invertebrates suggest a role for alpha liprin in active zone morphogenesis as mutants exhibit larger, less dense active zones (Kaufmann et al., 2002). The invertebrate homolog of ELKS/CAST is implicated in active zone cytoskeletal formation (Wagh et al., 2006) and Ca^{2+} channel recruitment (Kittel et al., 2006).

4.1.2 Fusion machinery

The minimum machinery required to fuse a vesicle with its target membrane is the SNARE fusion complex, which is comprised of a vesicle-associated SNARE protein (vSNARE) and target-membrane associated SNARE proteins (tSNAREs). At the synapse, the vSNARE is synaptobrevin-2, also known as vesicle-associated membrane protein 2 (VAMP2) and the tSNAREs are syntaxin-1 and synaptosomal-associated protein of 25 kDa (SNAP-25) (Fig. 25.4).

Each SNARE protein contains at least one SNARE motif of ~65 residues with a propensity to form coiled coils (Lupas, 1996). Synaptotobrevin and syntaxin each contain one SNARE motif, and SNAP-25 contains two (Sutton et al., 1998). Prior to vesicle docking, syntaxin is found in a stable, closed conformation, with its SNARE motif bound to an internal three-helix-bundle regulatory domain, making it inaccessible for interactions with other SNARE proteins (Dulubova et al., 1999). Syntaxin undergoes a conformational change triggered by Munc13 (see Section 4.2), which allows its SNARE motif to interact with the SNARE motifs of synaptobrevin and SNAP-25. This creates a *trans*-SNARE complex which is associated with both the vesicular membrane and the presynaptic plasma membrane. On any given vesicle, tight coiling of multiple SNARE complexes (Yersin et al., 2003) provides the energy required to drive fusion of the vesicle and target membranes (Fig. 25.4) (Li et al., 2007; Sollner et al., 1993a). Following membrane fusion, all members of the SNARE complex are located in the target membrane, and the complex is referred to as a *cis*-SNARE complex.

SNARE complexes mediate both constitutive and regulated vesicle fusion events throughout cells. At the synapse, multiple regulatory elements suppress constitutive SNARE-mediated fusion events and facilitate the fast, synchronous fusion of multiple synaptic vesicles upon Ca^{2+} influx. Synchronous fusion of multiple vesicles is essential for neuronal signaling and relies on the docking of vesicles at the active zone and their priming to a fusion ready state (Section 4.2).

The basic concept of the minimal machinery necessary to fuse a vesicular with its target membrane is as follows: SNARE proteins on both the vesicular and target membranes associate to form tight coiled-coil configurations. When

FIGURE 25.4 Minimal fusion machinery. Schematic of a vesicle just prior to fusion showing two (of multiple) maximally-coiled, *trans*-SNARE complexes pulling the vesicle membrane flat against the presynaptic membrane. *Reprinted by permission from (Sutton et al., 1998), Springer Nature: Springer, Nature, Crystal structure of an SNARE complex involved in synaptic exocytosis at 2.4 Å resolution. R Brian Sutton, Dirk Fasshauer, Reinhard Jahn, Axel T. Brunger, ©1998.*

maximally coiled, *trans*-SNARE complexes force the two membranes together, thereby destabilizing the membranes' natural curvature (Fig. 25.4). The coiling of *trans*-SNARE complexes all the way into their transmembrane domains provides the energy to drive fusion of the membranes (Castillo, 2012; Kameyama et al., 1998; Sollner et al., 1993a). Following fusion, the vSNAREs and tSNARES are now in a *cis*-SNARE complex, and the vesicular contents are released.

4.1.3 SM proteins

SM proteins function during virtually all SNARE-dependent fusion reactions. These soluble proteins fold into general "clasp" shapes (Misura et al., 2000) with a propensity to bind four helix bundles. At least one SM protein is necessary for any SNARE-mediated fusion event to occur. SM proteins are believed to be essential to stabilize SNARE complexes spatially and support SNARE complex assembly. Munc18, a member of the SM protein family, binds the N-terminal region of syntaxin in its closed conformation (Dulubova et al., 1999; Misura et al., 2000; Yang et al., 2000). This SM/SNARE interaction prevents SNARE complex assembly, thereby inhibiting fusion events. Once syntaxin undergoes a conformational change to its open configuration (see Section 4.2), Munc18 remains bound but alters its binding to interact directly with the four helix bundle of the assembling *trans*-SNARE complex, aiding in fusion (Shen et al., 2007; Dulubova et al., 2007). In this way, SM proteins are both negative and positive regulators of SNARE-mediated fusion.

4.1.4 Ca^{2+} channels

The presence of voltage-gated Ca^{2+} channels at active zones is necessary to couple neuronal depolarization to vesicle fusion events. Not only do these channels need to be present at active zones, their precise localization within an active zone can have immense effects on the efficacy of synaptic transmission. Ca^{2+} channels are located within 50 nm from docked vesicles (Eggermann et al., 2011; Harlow et al., 2001; Heuser et al., 1979). When the cell is depolarized, the Ca^{2+} channels briefly open, creating a nanodomain of particularly high $[Ca^{2+}]$ immediately adjacent to each channel. This $[Ca^{2+}]$ reaches hundreds of μM at the mouth of the channel (Llinas et al., 1992) and drops off rapidly with distance (Fig. 25.5) (Chad & Eckert, 1984; Fogelson & Zucker, 1985; Naraghi & Neher, 1997; Simon & Llinas, 1985). As the availability of Ca^{2+} is limited temporally and spatially, the closer the channel is located to the Ca^{2+} sensor, the greater the odds of saturating sensor binding and triggering vesicle fusion. Since the sensor for fast, synchronous neurotransmitter release is a low affinity sensor (see below), even a 5 nm change in the distance between the channel and the sensor has profound effects on vesicle release probability (Shahrezaei & Delaney, 2004).

4.1.5 Ca^{2+} sensors

Synaptotagmins 1, 2, and 7, as well as Doc2 regulate two distinct forms of Ca^{2+}-dependent neurotransmitter release: fast, synchronous release and a longer lasting, asynchronous release (Section 4.3). While other Ca^{2+} sensors exist, only the previously mentioned proteins will be covered as they are currently the best understood.

Synaptotagmins are a large family of integral membrane proteins that contain two Ca^{2+}-binding motifs called C_2 domains. Synaptotagmin one and two are located on synaptic vesicles and play functionally homologous roles in different parts of the nervous system. In mammals, synaptotagmin one is expressed predominantly in the cerebral hemispheres, while synaptotagmin two is expressed predominantly in the brainstem, spinal cord, and at the neuromuscular junction

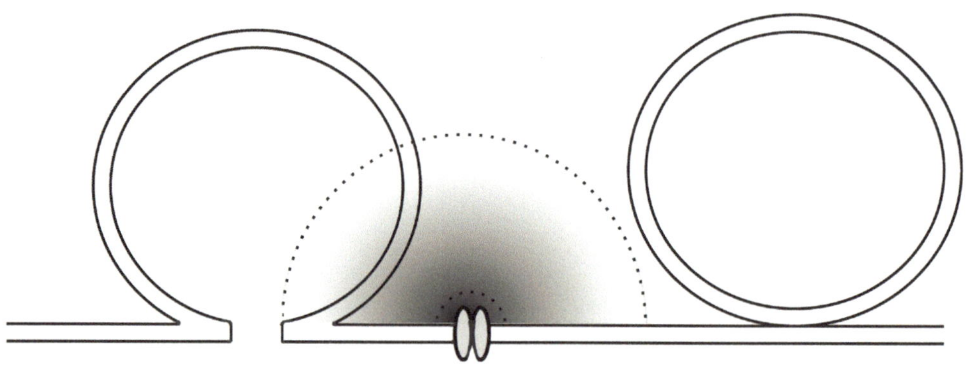

FIGURE 25.5 Ca^{2+} nanodomains. Schematic depicting the rapid decay of $[Ca^{2+}]$ with distance from the Ca^{2+} channel. The darker the gradient, the higher the $[Ca^{2+}]$. A Ca^{2+} sensor on the left vesicle would experience a very high $[Ca^{2+}]$, which would maximize the probability of fusion. A Ca^{2+} sensor on the right vesicle would be exposed to a much lower $[Ca^{2+}]$ and have a correspondingly lower probability of fusion.

(Marqueze et al., 1995; Ullrich et al., 1994). Synaptotagmins 1/2 are low-affinity Ca^{2+} sensors, meaning they only bind Ca^{2+} when intracellular [Ca^{2+}] is very high, as found in the Ca^{2+} nanodomains discussed above. In the Ca^{2+}-bound state, the C_2 domains of synaptotagmin interact with both SNARE proteins (Chapman et al., 1995; Lynch et al., 2007; Schupp et al., 2016; Vrljic et al., 2010; Wang et al., 2016; Zhou et al., 2015) and negatively-charged phospholipid membranes in vitro (Bai et al., 2002; Hui et al., 2009; Li et al., 2006). Upon Ca^{2+} binding in vivo, synaptotagmins 1/2 trigger the fast, synchronous phase of neurotransmitter release (Mackler et al., 2002) primarily through membrane interactions (Paddock et al., 2008, 2011), although Ca^{2+}-dependent SNARE interactions likely contribute as well.

Synaptotagmin seven is another member of the synaptotagmin family enriched at synapses. It is an integral membrane protein localized in the presynaptic membrane (Sugita et al., 2001) that binds Ca^{2+} with a higher affinity than synaptotagmins 1/2. Doc2 is also enriched at synapses and contains two Ca^{2+}-binding C_2 domains with a high affinity for Ca^{2+} (Yao et al., 2011). Unlike the synaptotagmins, Doc2 is a cytosolic protein. Both synaptotagmin seven and Doc2 are attractive candidates for triggering asynchronous release (Section 4.3) due to their high Ca^{2+} affinities and slow Ca^{2+}-interaction kinetics (Bacaj et al., 2013).

4.2 Docking and priming at the active zone

To become fusion competent, synaptic vesicles must be precisely targeted to the active zone and held in direct contact with the presynaptic membrane. First, vesicles are tethered near the presynaptic membrane at active zones. When the vesicle membrane touches the presynaptic membrane, the vesicle is considered docked at the release site. Once docked, additional interactions, referred to as priming, increase the probability of vesicle fusion.

Adaptor proteins piccolo/bassoon are thought to guide synaptic vesicles toward active zones in vertebrates (Hallermann et al., 2010; Mukherjee et al., 2010). Once there, RIM and Munc13 contribute to vesicle docking (Weimer et al., 2006). RIM binds to Munc13, creating a RIM/Munc13 heterodimer. This heterodimer binds an accessory protein, either Rab3 or Rab27, and this complex provides the scaffold needed to precisely target and hold synaptic vesicles to active zones (Fig. 25.2A(b)) (Dulubova et al., 2005). These vesicles are now considered docked.

Additional intermolecular interactions are required before vesicles are maximally primed and fully fusion competent. Munc13, an essential active zone protein, is thought to orchestrate the first step in fusion machinery assembly (note: Munc13 should not to be confused with the SM protein, Munc18, discussed in Section 4.1). Munc13 binds to the closed conformation of syntaxin and facilitates a conformational change in syntaxin from its "closed" to its "open" state (Dulubova et al., 1999; Ma et al., 2011). Only then can syntaxin's SNARE motif interact with the SNARE motifs of SNAP-25 and synaptobrevin to form a *trans*-SNARE complex. Upon syntaxin's conformational change, the Munc18 interaction adjusts to facilitate the assembly of the *trans*-SNARE complex (Shen et al., 2007; Dulubova et al., 2007), thereby contributing to the fusion-competency of the vesicle.

Recent work in the McMahan lab suggests that vesicle priming is in a state of dynamic equilibrium (Fig. 25.6A, arrows), such that only docked vesicles with the largest area of contact with the presynaptic membrane are maximally primed and ready to fuse upon Ca^{2+} influx (Jung et al., 2016). One mechanism of accomplishing variable priming would be a balance between the repulsive forces of the vesicular and presynaptic membranes that result in SNARE complex uncoiling to favor minimal priming (Fig. 25.6A, left) and SNARE complex coiling to favor maximal priming (Fig. 25.6A, right) (Jung et al., 2016). At any given moment, only a subpopulation of docked synaptic vesicles have a sufficient number of maximally-coiled *trans*-SNARE complexes to produce maximal SV-PM contact area. These synaptic vesicles constitute the maximally-primed population that is most likely to fuse upon Ca^{2+} influx. Once triggered by Ca^{2+}, these maximally-primed synaptic vesicles complete the SNARE coiling process all the way into the transmembrane domains of syntaxin and synaptobrevin (Stein et al., 2009). The synaptic vesicle membrane merges with the presynaptic membrane, resulting in *cis*-SNARE complex formation. Thus, the coiling of SNARE complexes provides the energy needed to induce membrane fusion.

4.3 Vesicle fusion

Fusion events can occur spontaneously or can be evoked. When an action potential depolarizes the presynaptic terminal, voltage-gated Ca^{2+} channels open. The subsequent influx of Ca^{2+} triggers a series of intermolecular interactions that results in the fusion of multiple maximally-primed vesicles with the presynaptic membrane (Figs. 25.2A,C and 25.7).

4.3.1 Ca^{2+}-independent neurotransmitter release

Ca^{2+}-independent neurotransmitter release is referred to as spontaneous release. This form of neurotransmitter release occurs in resting nerve terminals when a single vesicle fuses with the presynaptic membrane in the absence of presynaptic

Ca^{2+} influx (Fig. 25.3A). Classically, spontaneous release is considered to be random fusion events caused by low-probability conformational changes that complete coiling of the SNARE proteins and result in fusion. Since these spontaneously fusing vesicles would be in a maximally-primed, fusion-competent state, little energy is required to overcome the fusion barrier.

Spontaneous and evoked fusion events can occur at distinct active zones (Melom et al., 2013; Peled et al., 2014) from separate recycling pools (Chung et al., 2010; Fawley et al., 2014). Thus, location may partially determine the probability of

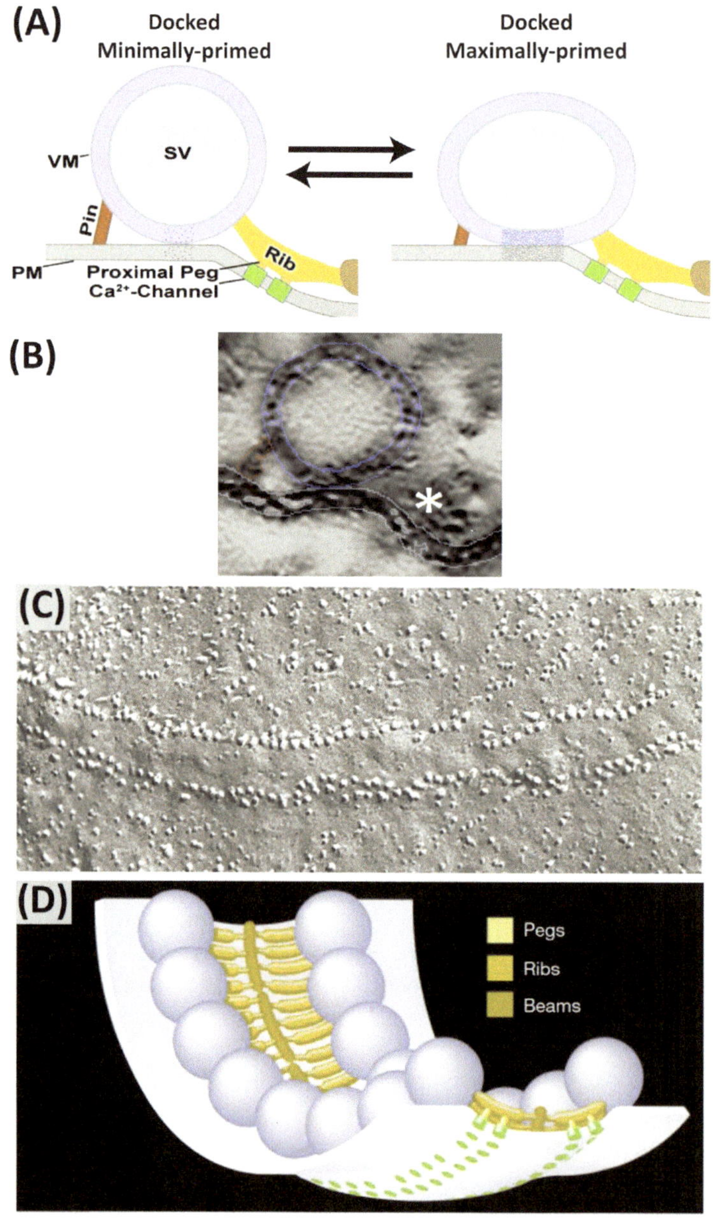

FIGURE 25.6 Docking and variable priming of synaptic vesicles. (A) Left, schematic depicting a docked, minimally-primed vesicle at the presynaptic membrane, defined by the small area of contact between the vesicle and plasma membranes. Right, the fusion machinery located in the AZM (ribs, pegs) and in pins pulls the vesicle and presynaptic membranes together, increasing the area of contact, resulting in a maximally-primed vesicle that has the highest fusion probability. The degree of priming is in a state of dynamic equilibrium (*arrows*). Those vesicles that happen to be maximally primed when Ca^{2+} influx occurs are thought to be the vesicles that fuse. (B) A docked synaptic vesicle viewed by electron tomography. White * represents AZM. (C) Freeze-fracture electron micrograph at the frog neuromuscular junction showing the characteristic parallel double-rows of particles that comprise the pegs of the AZM and include the voltage-gated Ca^{2+} channels required for triggering fusion. (D) Schematic showing the relationship of docked synaptic vesicles, the AZM, and the double row of particles that include Ca^{2+} channels at the frog neuromuscular junction. *(A) Adapted from (Jung et al., 2016). (B) Adapted from (Szule et al., 2012). (C). Adapted from (Heuser et al., 1979) ©1979 Journal of Cell Biology. 81:275–300. https://doi.org/10.1083/jcb.81.2.275. (D). Reprinted with permission from (Harlow et al., 2001), Springer Nature: Springer, Nature, The Architecture of active zone material at the frog's neuromuscular junction. Mark L Harlow, David Ress, Arne Stoschek, Robert M Marshall, Uel J McMahan, ©2001.*

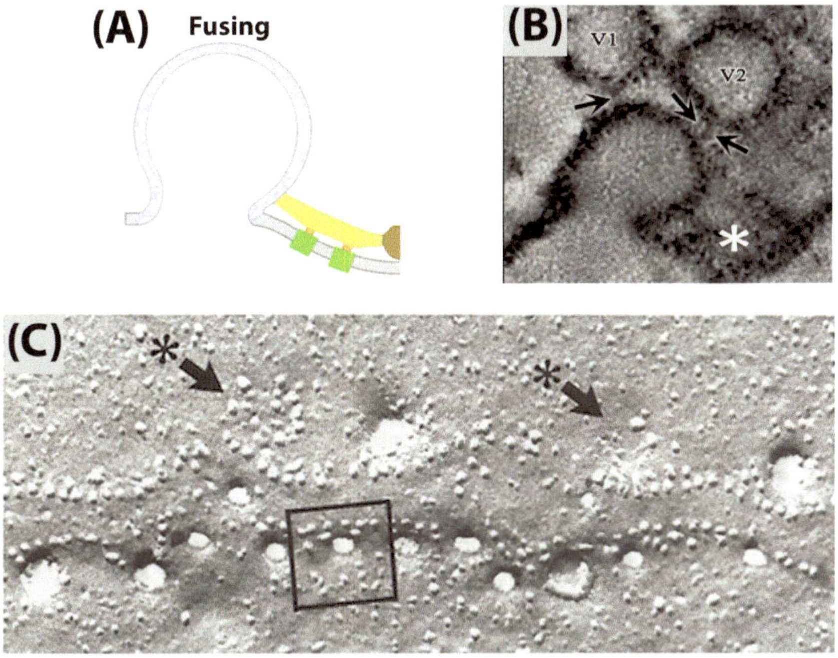

FIGURE 25.7 Vesicle fusion. (A) Schematic depicting a vesicle that has just fused with the presynaptic membrane such that the vesicle membrane is now incorporated into the plasma membrane, forming an omega figure. (B) Electron tomograph of a fusing vesicle. V1 and V2 represent additional synaptic vesicles located near the active zone that are linked to the fusing vesicle (*arrows*). White * represents AZM. (C) Freeze-fracture electron micrograph showing the openings of fusing vesicles (*box*) just lateral to the double-rows of particles found at active zones. Cluster of particles characteristic of fully-collapsed vesicles (*black * bold arrows*) are also found lateral to the double-row of particles. *(A and B) Adapted from (Szule et al., 2012). (C) Adapted from (Heuser et al., 1979) ©1979 Journal of Cell Biology. 81:275–300. https://doi.org/10.1083/jcb.81.2.275*

a vesicle participating in spontaneous versus evoked neurotransmitter release. These recent studies, together with the classic spontaneous release hypothesis, suggest multiple mechanisms occurring during spontaneous fusion, with a subpopulation originating from maximally-primed vesicles reacting to transient changes in their environment, and another subpopulation occurring at specialized, Ca^{2+}-independent fusion sites.

4.3.2 Ca^{2+}-dependent neurotransmitter release

Ca^{2+}-dependent neurotransmitter release is differentiated temporally into two phases: fast, synchronous release and a slower, asynchronous release (Fig. 25.8) (Atluri & Regehr, 1998; Goda & Stevens, 1994). The fast, synchronous phase is the large burst of neurotransmitter release that occurs within milliseconds of the arrival of the action potential and constitutes the classical neurotransmitter release pathway typically described in textbooks (Fig. 25.8, upper trace). However, there is an additional slower, or asynchronous, phase of elevated release that is still Ca^{2+}-dependent and lasts for tens to

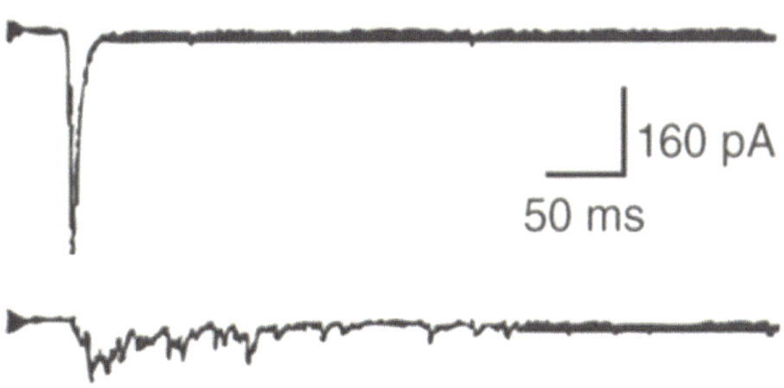

FIGURE 25.8 Ca^{2+}-dependent neurotransmitter release at cultured hippocampal synapses. Upper, voltage trace depicts the fast, synchronous neurotransmitter release in cultures from wild type mice. Lower, voltage trace depicts the slower, asynchronous neurotransmitter release in cultures from synaptotagmin one knock out mice. *Reprinted from (Geppert et al., 1994), Cell, 79/4, Martin Geppert, Yukiko Goda, Robert E Hammer, Cai Li, Thomas W Rosahl, Charles F Stevens, Thomas C Sudhof, Synaptotagmin I: a major Ca^{2+} sensor for transmitter release at a central synapse, p. 717–27. ©1994, with permission from Elsevier.*

hundreds of milliseconds (Atluri & Regehr, 1998; Goda & Stevens, 1994). This phase of release is most notable following periods of high frequency stimulation, at specialized synapses, and at synaptotagmin one knock out synapses (Fig. 25.8, lower trace) (Atluri & Regehr, 1998; Best & Regehr, 2009; Daw et al., 2009; Hefft & Jonas, 2005; Jensen et al., 2000).

4.3.3 Synchronous release

Fast, synchronous release of neurotransmitter occurs when multiple synaptic vesicles fuse within milliseconds of depolarization (Fig. 25.8, upper trace). The structure of the active zone provides optimal conditions to accomplish this task. Synaptotagmin 1/2 is a low-affinity Ca^{2+} sensor with fast binding/unbinding kinetics. Thus, saturated Ca^{2+} binding requires the exceptionally high $[Ca^{2+}]$ that only occurs in nanodomains around open Ca^{2+} channels. Since this high $[Ca^{2+}]$ drops off rapidly in time and space by diffusion (Chad & Eckert, 1984; Fogelson & Zucker, 1985; Naraghi & Neher, 1997; Simon & Llinas, 1985), only synaptotagmin 1/2 located within such a nanodomain would be expected to trigger fusion; the rapid decrease in high $[Ca^{2+}]$ ensures that the signal to fuse is terminated. Therefore, synaptotagmin's precise location relative to the presynaptic Ca^{2+} channels has profound effects on the release probability of the vesicle. As the SNAREs coil and pull the vesicle and plasma membranes together, not only does the membrane contact area increase, the distance between the Ca^{2+} channel and synaptotagmin is predicted to decrease (Jung et al., 2016). According to this hypothesis, only the subpopulation of docked synaptic vesicles that are maximally primed would situate synaptotagmin at its closest to Ca^{2+} channels. The closer synaptatogamin 1/2 is to the Ca^{2+} channel at the time of Ca^{2+} influx, the higher the $[Ca^{2+}]$ that reaches its C_2 domains, favoring fusion.

Upon Ca^{2+} binding, the net charge of the Ca^{2+} binding pockets of synaptotagmin changes from negative to positive. This enhances interactions with the negatively charged presynaptic membrane and potentially with *trans*-SNARE complexes (Bai et al., 2002; Chapman et al., 1995; Chapman & Davis, 1998; Zhang et al., 2002). Thus, Ca^{2+} binding allows synaptotagmin 1/2 to act as an "electrostatic switch" (Davletov et al., 1998; Murray & Honig, 2002; Striegel et al., 2012; Ubach et al., 1998). Switching from electrostatic repulsion to electrostatic attraction permits hydrophobic residues in the C_2 domains of synaptotagmin 1/2 to escape the hydrophilic cytosol by penetrating into the hydrophobic core of the presynaptic membrane. Therefore, synaptotagmin 1/2 can be thought of as providing the final push, much like popping a tightly-inflated balloon, to trigger fusion. This hypothesis can account for both the spatial and temporal restriction observed during fast, synchronous neurotransmitter release.

4.3.4 Asynchronous release

Asynchronous release continues for several 100's of milliseconds up to a second (Fig. 25.8, lower trace). Asynchronous neurotransmitter release is observed in many neuronal types during and following extended high frequency activity (Jensen et al., 2000). Under these circumstances, the $[Ca^{2+}]$ builds up since Ca^{2+} clearance is slower than Ca^{2+} entry. During the prolonged Ca^{2+} clearance phase, the residual $[Ca^{2+}]$ is thought to drop below saturating binding levels for synaptotagmin 1/2 (Goda & Stevens, 1994), but it remains high enough to activate a high-affinity Ca^{2+} sensor. While most synapses exhibit little to no asynchronous release following a single action potential, asynchronous release is predominant in some specialized synapses (Best & Regehr, 2009; Daw et al., 2009; Hefft & Jonas, 2005). Asynchronous release has been implicated in modulating postsynaptic responses (Iremonger & Bains, 2007).

The identity of the Ca^{2+} sensor(s) for asynchronous release is still debated. Like the fast, synchronous Ca^{2+} sensor, the asynchronous Ca^{2+} sensor needs be present at synapses and enriched at fusion sites. However, it must have a higher affinity for Ca^{2+} so that it can bind at lower $[Ca^{2+}]$, permitting this protein to trigger release over a longer time course. Synaptotagmin seven and Doc2 are current candidates for the Ca^{2+} sensor for triggering asynchronous neurotransmitter release, as they exhibit these characteristics (Bacaj et al., 2013; Hui et al., 2005; Yao et al., 2011).

4.4 SNARE and vesicle recycling

After a synaptic vesicle fuses, the SNARE complex is in a *cis* conformation since the vesicle membrane is incorporated into the presynaptic membrane. Cytosolic adaptor proteins specifically recognize and bind to *cis*-SNARE complexes and recruit N-ethylmaleimide-sensitive factor (NSF). NSF is an ATPase responsible for disassembling the SNARE complex (Sollner et al., 1993a). This disassembly is required before the vSNAREs can be endocytosed for incorporation into new vesicles and before the tSNAREs can be reused in subsequent rounds of fusion events.

The synaptic vesicle membrane and all associated vesicle proteins must be rapidly recycled within nerve terminals to maintain the pool of vesicles available for fusion. There are multiple proposed mechanisms of synaptic vesicle recycling, but clathrin-mediated endocytosis (CME) is currently the best understood. CME was first described over 50 years ago (Heuser &

Reese, 1973) and occurs in many cell types (reviewed in (Kaksonen & Roux, 2018)). It was first implicated in synaptic vesicle recycling in 1973 when it was shown to occur away from active zones following full collapse of the synaptic vesicle membrane into the presynaptic membrane (Fig. 25.2A and H) (Heuser & Reese, 1973). During CME, a plethora of adaptor and accessory proteins facilitate appropriate membrane invagination and recruitment of clathrin to form spherical clathrin-coated membrane pits that project into the presynaptic cytoplasm (Reider & Wendland, 2011). The GTPase, dynamin, is then recruited to mediate a scission event that separates the clathrin-coated pit from the presynaptic membrane. The new vesicle is uncoated, targeted to endosomes or synapses, reloaded with neurotransmitter, and transported back to active zones. This completes the synaptic vesicle cycle, and can occur as fast as ~20–40 s (Loerke et al., 2009).

Recently, a clathrin-independent endocytic mechanism was reported in both invertebrate (Watanabe et al., 2013a) and vertebrate (Watanabe et al., 2013b) neurons. This mechanism, called ultrafast endocytosis, occurs 200 times faster than CME. However, it must be stressed that this pathway still requires the action of clathrin for reformation of synaptic vesicles. The clathrin-independent bulk internalization of membrane produces large vesicles that form endosomes. The reformation of synaptic vesicles requires clathrin-mediated vesicle budding from these endosomes (Watanabe et al., 2014). Another mechanism postulated to mediate vesicle recycling is known as "kiss and run", where vesicles undergo a transient connection to the plasma membrane through a fusion pore, but the vesicle never collapses (Ceccarelli et al., 1972). While this mechanism occurs in endocrine cells (Neher & Marty, 1982), its role in nerve terminals is questionable (reviewed in (He & Wu, 2007)).

5. Synaptic cleft

The synaptic cleft contains proteins and glycoproteins that support cell adhesion and regulate synaptic development and stability. Cleft proteins also clear neurotransmitter. The vast majority of proteins in the synaptic cleft provide structural support to the pre- and postsynaptic membranes from development through maturity. Many of these proteins are transmembrane cell adhesion molecules (reviewed in (Missler et al., 2012)). Their cytosolic domains generally interact with intracellular synaptic proteins while their extracellular domains bind to transmembrane synaptic proteins of the opposing cell. Some extracellular and transmembrane synaptic proteins direct synaptogenesis, while others are important for neurotransmitter clearance, including enzymes that break them down and transporters that sequester them.

The protein machinery and organization varies by synapse type and location. A CNS synapse connects two neurons, and in some locations, it is functionally important for the strength of CNS synapses to be easily modified. At the neuromuscular junction, the postsynaptic cell is a muscle fiber where there is less need for rapid changes in synaptic strength. Nevertheless, many general processes are conserved across synapses. Machinery for synaptogenesis, scaffolding, and neurotransmitter clearance are universally required, but the individual molecular interactions will be synapse specific.

5.1 Synaptogenesis

A specific role for an extracellular matrix molecule in regulating the formation and maintenance of synaptic connections has been best elucidated at the frog neuromuscular junction. Dr. McMahan and colleagues discovered that the synaptic basal lamina contains molecules that direct the formation of synaptic specializations in both regenerating nerve terminals (Sanes et al., 1978) and regenerating muscle fibers (Burden et al., 1979). They purified the basal lamina molecule responsible for directing postsynaptic specializations and named it agrin (Nitkin et al., 1987).

Agrin is a critical signal for synaptogenesis and synaptic maintenance (McMahan, 1990; Tintignac et al., 2015). It is secreted from presynaptic nerve terminals and is stably incorporated into the synaptic basal lamina. It binds to and activates a receptor complex in the plasma membrane of the muscle fiber, which then induces postsynaptic specializations (McMahan, 1990). This receptor complex is composed of the low-density lipoprotein receptor-related protein 4 (LRP4), which binds agrin, and a muscle-specific kinase (MuSK), which triggers a postsynaptic signaling cascade (Kim et al., 2008; Weatherbee et al., 2006; Zhang et al., 2008). This cascade recruits acetylcholine receptors, acetylcholinesterase (AChE), and numerous other components of the postsynaptic apparatus to postsynaptic sites immediately opposite the nerve terminal (Okada et al., 2006). Precise alignment is critical for the formation, organization, maintenance, and function of the neuromuscular junction.

CNS synaptogenesis is not well understood. However, the neuroligin and neurexin families of proteins may play a role. The neuroligins are postsynaptic cell adhesion molecules (Lise & El-Husseini, 2006) that bind to the presynaptic cell adhesion molecules, the neurexins (Missler & Sudhof, 1998). This heterotypic interaction across the cleft aggregates the machinery (both pre- and postsynaptic) necessary for a functional synapse (Graf et al., 2004; Scheiffele et al., 2000). Neuroligin-1 is present exclusively in excitatory synapses (Song et al., 1999), while neuroligin-2 is present exclusively in inhibitory synapses (Varoqueaux et al., 2004). Knockdown of neuroligins (Chih et al., 2005) or interference with neuroligin-neurexin signaling (Levinson et al., 2005) diminishes the number of fully developed synapses. This research has led to the hypothesis that in addition to offering the structural stability of cell adhesion molecules at developing synapses, the postsynaptic neuroligins and presynaptic neurexins are critical for aggregating synaptic components and assigning the valence of the synapse.

5.2 Scaffolding

The stability of synaptic connections is tightly regulated based on the functional requirements of the particular synapse. Some synapses, such as neuromuscular junctions, are relatively stable and provide reliable transmission for proper locomotion over long time periods. While there is minor synaptic remodeling, the function at this synapse remains relatively constant in healthy individuals. Other synapses are short lived or exhibit synaptic strengths that vary widely within seconds. In the cerebral cortex, synaptic plasticity may strengthen some synapses in response to heavy activity while pruning others that are less active. Such alterations in synaptic strength are foundational to the mechanisms of learning and memory first posited by Hebb (1949). Scaffolding molecules spanning the synaptic cleft can either maintain and support a synapse or promote retraction and weakening.

The cadherins are a family of transmembrane cell adhesion molecules (reviewed in (Pokutta & Weis, 2007)) present at most synapses. They are thought to provide critical support for synapse stability, particularly in dendritic spines (Mendez et al., 2010; Saglietti et al., 2007). They are present both pre- and postsynaptically and bridge the synaptic cleft through homotypic binding interactions. Cadherins are generally concentrated around the edges of synapses (Uchida et al., 1996). This localization may form a diffusion barrier that helps sequester receptors and maintain the structural integrity of the synapse (Renner et al., 2012).

Other classes of transmembrane proteins have more varied affects. The Eph receptor tyrosine kinases and their binding partners, epherins, interact across the cleft through heterotypic interactions. Signals in both antero- and retrograde directions are postulated to promote remodeling of synaptic structure (Arvanitis & Davy, 2008; Lai & Ip, 2009). Different subtypes of Eph receptors and epherins have alternate affects. Knockdown of EphB2 leads to decreased synapse number and loss of synaptic superstructure (Henkemeyer et al., 2003). Therefore, EphB2 signaling encourages the stability of synapses (Kayser et al., 2006). Ephrin-A3 and EphA4 signaling, on the other hand, results in decreased synapse number and retraction of synaptic ultrastructure (Murai et al., 2003). Thus, these scaffolding molecules are critical for both maintenance and alteration of existing synapses across time, and their trans-synaptic signaling is necessary during many neurological processes.

5.3 Neurotransmitter clearance

Removal and/or breakdown of neurotransmitter by cleft proteins is pivotal for regulating the temporal aspects of the signal strength between neurons. The dopamine transporter, a well-defined example of this role (Vaughan & Foster, 2013), is the primary regulator of dopamine concentration in the synaptic cleft in a key reward center in the brain. Inhibition of this transporter by cocaine prevents the removal of dopamine from the presynaptic cell into the cleft (Huang et al., 2009). The resulting build up causes increased signaling to the postsynaptic cell and consequent over-activity of the reward circuitry, leading to addiction.

Enzymatic breakdown is the other way to clear neurotransmitter after release. A well-characterized example of this clearance mechanism is the enzyme AChE (Soreq & Seidman, 2001), which hydrolyzes ACh. The action of AChE is so fast that the rate-limiting step in the breakdown of ACh is diffusion of the transmitter. This rapid breakdown is critical for proper synaptic function as it regulates ACh concentration and thus the postsynaptic response. AChE inhibitors provide effective treatment for diseases involving deficits in ACh signaling. In myasthenia gravis, the immune system attacks ACh receptors such that normal ACh release from motor neurons cannot provide sufficient muscle stimulation. By slowing down the clearance of ACh, AChE inhibitors counteract the muscle's decreased ability to respond (Pohanka, 2012). Presynaptic defects are similarly amenable to treatment. According to the cholinergic hypothesis of Alzheimer's disease (Francis et al., 1999), loss of cholinergic neurons may underlie some cognitive deficits. AChE inhibitors are again used to slow down clearance, resulting in an accumulation of the ACh released by the remaining cholinergic neurons. Thus, it is essential that proteins specific for clearance of each type of neurotransmitter(s) released are present in the synaptic cleft.

6. Postsynaptic function

The postsynaptic membrane is uniquely specialized to detect neurotransmitters and transduce this chemical signal back into an electrical signal. Binding of neurotransmitters to receptors can elicit either a fast, transient or slower, longer-lasting response, which may activate or inhibit electrical signaling in the postsynaptic cell. The specific response depends on the neurotransmitter released as well as the postsynaptic receptor subtype(s) present. Ionotropic receptors are ion channels that open upon neurotransmitter binding. Metabotropic receptors are GPCRs that activate G-protein second messenger systems, leading to changes in ion channel opening. Activating either type of receptor can result in postsynaptic depolarization or hyperpolarization depending on the ion selectivity of the activated channels.

6.1 Postsynaptic receptors

6.1.1 Ionotropic receptors

Ionotropic receptors are ligand-gated ion channels that form pores in the membrane allowing the passage of ions down their electrochemical gradients (Smart & Paoletti, 2012). Ion channels are generally closed at rest. Binding of their specific neurotransmitter causes a conformational change, opening the pore and allowing ion influx. Early studies at the frog neuromuscular junction measured ACh-induced changes in ion conductances in muscle fibers demonstrating channel opening in response to a ligand (Anderson & Stevens, 1973). This process is extremely rapid, allowing immediate ion influx that abruptly ceases when the ligand unbinds, and the channel closes. In addition, some channels inactivate, stopping ionic conductance even while a ligand remains bound. In this way, ionotropic channels generate fast and transient electrical signals in response to the presence of neurotransmitter.

6.1.2 Metabotropic receptors

Metabotropic receptors have extracellular binding sites for specific neurotransmitters and bind to intracellular G-proteins (Huang & Thathiah, 2015). Binding of neurotransmitter to these GPCRs causes activation of the intracellular G-protein, which dissociates from the receptor and may activate intracellular kinase cascades, phosphatases, and/or other second messengers, which then act on various downstream effectors, including opening of ion channels. The necessity of multiple protein interactions before the opening of an ion channel means that the time scale of metabotropic receptor effects will be much slower than those of ionotropic receptors. Second messenger cascades can also amplify the original signal. One neurotransmitter molecule can activate a G-protein, which can activate multiple kinases, which can activate multiple ion channels. In addition, the downstream second messengers may remain activated after the ligand comes unbound from the GPCR. Thus, the postsynaptic response generated by metabotropic receptors, though slower, tend to be larger and more persistent.

Second messenger systems can have additional wide-ranging affects independent of electrical changes in the postsynaptic cell. Some second messengers act as transcription factors, altering gene expression (Kvachnina et al., 2005). Others cause release of internal Ca^{2+} stores and lead to changes in synaptic strength (Miyata et al., 2000). The potential downstream effects of multiple, simultaneously activated signaling pathways can vary widely depending on the downstream machinery present in the postsynaptic cell.

6.2 Generation of postsynaptic potentials

The postsynaptic membrane's ability to transduce the chemical signal into an electrical signal is dependent on the opening of ion channels (either ionotropic receptors, or downstream effectors of GPCRs) that allow ions to travel down their electrochemical gradients. Thus, the ion selectivity of the activated channels determines the net potential generated in the postsynaptic cell. At a typical excitatory synapse, the open ion channels allow positively-charged ions, such as Na^+ and Ca^{2+}, to flow down their electrochemical gradients into the postsynaptic cell. The influx of positive charge depolarizes the postsynaptic cell, otherwise known as an excitatory postsynaptic potential (EPSP). If the activated channels are permeable to either Cl^- or K^+, the postsynaptic membrane will be hyperpolarized. The concentration gradient for each of these ions is great enough to drive them against the electrical gradient, so Cl^- will enter or K^+ will leave the cell, which drives the membrane potential to a more negative value. Such a hyperpolarization is called an inhibitory postsynaptic potential (IPSP).

The net change in membrane potential depends on the summation of all synaptic inputs. Mammalian neuromuscular junctions are exclusively excitatory. At these synapses, sufficient ACh is released to ensure that the postsynaptic muscle fiber is depolarized past threshold, resulting in muscle fiber contraction (Fig. 25.3B). In the CNS, all synaptic inputs to the postsynaptic neuron, which may include both EPSPs and IPSPs, are summated. If the postsynaptic neuron is sufficiently depolarized at its axon initial segment, its firing rate will increase. If the net effect at the axon initial segment is a hyperpolarization, its firing rate will decrease. Thus, it is the net activity that determines the overall response in the postsynaptic cell (reviewed in (Blomfield, 1974)).

7. Modulation of the chemical synapse

The response of a given postsynaptic cell to the same level of activity in a presynaptic cell can vary depending on the efficacy of the synapse. Synaptic efficacy is based on many factors, including the quantal content (number of neurotransmitter-filled synaptic vesicles that fuse in response to a given stimulus, discussed in Section 5.2) and the type and number of postsynaptic receptors. Importantly, these pre- and postsynaptic parameters can be regulated by activity patterns. The capacity to modulate synaptic efficacy is an important mechanism underlying learning and memory. These activity-dependent changes are known as synaptic plasticity.

Experimentally, synaptic plasticity is measured both pre- and postsynaptically. Presynaptically, the amount of neurotransmitter released can be decreased or increased, leading to depressed or facilitated responses, respectively. Specifically, the quantal content of a synapse (m) depends on the number of vesicles available to fuse (n), and the release probability of each vesicle (p) (Section 5.2). The release probability is dependent on the magnitude and duration of Ca^{2+} influx, the speed of Ca^{2+} clearance, and the proximity of the Ca^{2+} sensor to the channel (Section 4.1). Higher quantal content causes larger responses in the postsynaptic cell, barring any postsynaptic adaptations. Synapses with a high quantal content result from some combination of a large readily releasable pool of vesicles, high vesicle release probability, high Ca^{2+} influx, and/or slow Ca^{2+} clearance. Alternatively, low quantal content synapses have the opposite characteristics and result in smaller postsynaptic responses.

Postsynaptically, changes in synaptic activity can lead to changes in the morphological makeup of the postsynaptic membrane through insertion or removal of receptors. Any change in receptor density results in a change in response strength to a given signal. Therefore, activity-dependent alteration of the postsynaptic machinery allows for flexible modulation of postsynaptic responses to neurotransmitter release.

7.1 Depression

Synaptic depression is defined as a reduction in the amplitude of the postsynaptic response following a given presynaptic stimulus. This can be observed at high quantal content synapses during paired-pulse experiments, in which two closely spaced stimuli are administered to a cell (Debanne et al., 1996; Thomson et al., 1993). The first response, or EPSP, is robust, while the second response is reduced. At a high quantal content synapse, the first stimulation is sufficient to release a large fraction of the readily releasable vesicles. Accordingly, there are fewer vesicles available to fuse during the second stimulus, which results in a reduced EPSP. Another example of synaptic depression is seen during a high-frequency stimulus train to the presynaptic cell (Parker, 1995; Parnas, 1972). Since exocytosis is more rapid than endocytosis, such stimulation depletes the readily releasable pool of synaptic vesicles, even at low quantal content synapses, and quantal content is decreased.

7.2 Facilitation

Facilitation is an incremental increase in the postsynaptic response to a given presynaptic stimulation observed in paired-pulse experiments at low quantal content synapses. Since the first stimulus at such a synapse results in the fusion of only a small fraction of the readily releasable pool, the number of vesicles available for fusion following the second impulse is not significantly reduced. If the delay between the two stimuli is within tens of milliseconds, not all the Ca^{2+} that entered during the first pulse has been removed by the time of the second pulse. Therefore, the $[Ca^{2+}]$ experienced by vesicles during the second pulse is higher and triggers a greater number of synaptic vesicles to fuse. If the delay between stimuli is long enough to allow the Ca^{2+} from the first pulse to be removed, the second response is not facilitated. This is known as the residual Ca^{2+} hypothesis (Dodge & Rahamimoff, 1967; Katz & Miledi, 1965, 1967), and this form of presynaptic facilitation is very short-lived.

7.3 Long-term potentiation

Long-term potentiation (LTP) is characterized by a long-lasting facilitated response in the postsynaptic cell following a brief tetanic stimulation of a strong input (Fig. 25.9A) or simultaneous stimulation of multiple weaker inputs. Since its discovery in the rabbit hippocampus in 1966 (Lomo, 1966), the study of LTP has grown into an immense field. In the intervening 50 years, multiple forms of LTP have been detected in many different cells and brain regions. These changes can last from hours to days and rely on presynaptic and postsynaptic mechanisms (reviewed in (Castillo, 2012; Malenka &

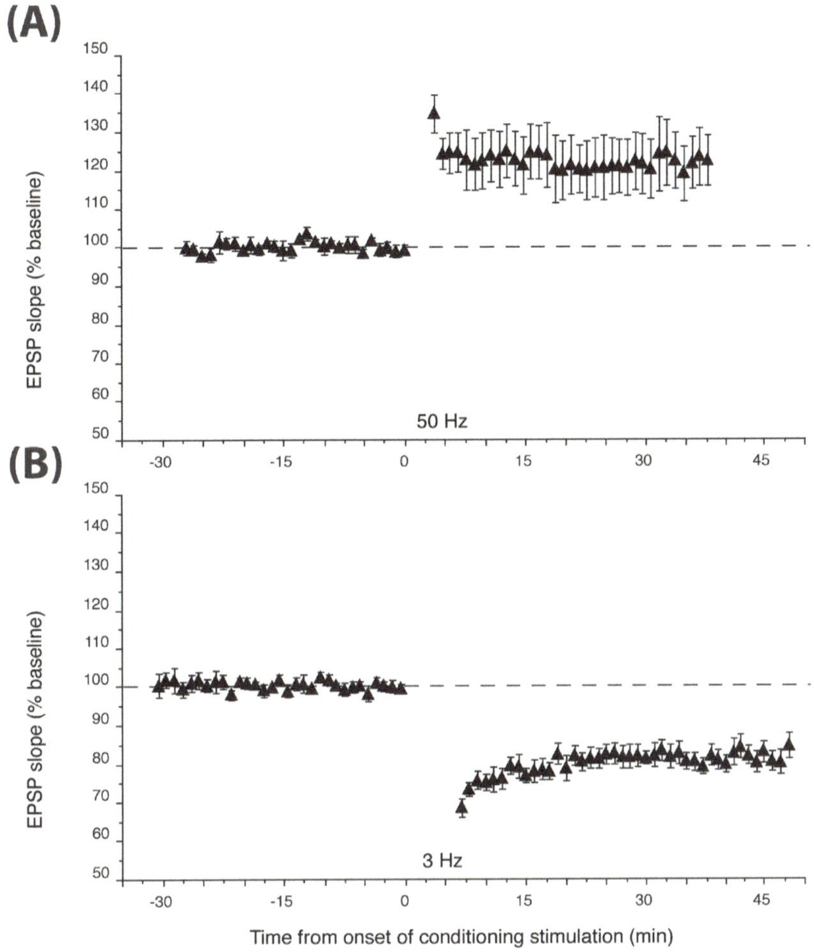

FIGURE 25.9 Long-term potentiation and long-term depression. (A) Example of long-term potentiation induced by a brief high frequency stimulation (50 Hz for 900 pulses). (B) Example of long-term depression induced by a longer low frequency stimulation (3 Hz for 900 pulses). *Adapted from (Dudek & Bear, 1992).*

Bear, 2004)). Here we describe a prototypic form of postsynaptic LTP, namely NMDA receptor-dependent LTP at excitatory synapses in the hippocampus (reviewed in (Herring & Nicoll, 2016)).

In this form of LTP, the postsynaptic neuron contains two classes of ionotropic glutamate receptors, AMPA receptors and NMDA receptors. NMDA receptors have extracellular Mg^{2+} bound in their pores at the resting membrane potential. Following a single stimulus, glutamate binds to both types of receptors. The AMPA receptors open, which allows Na^+ influx and results in an EPSP. However, little to no current passes through the NMDA receptors due to the voltage-dependent Mg^{2+} block (Mayer et al., 1984; Nowak et al., 1984).

During tetantic stimulation, the postsynaptic cell is sufficiently depolarized to remove the Mg^{2+} block from the NMDA receptor pore. With the block removed, glutamate binding results in a robust influx of Ca^{2+} through the NMDA receptor (Dingledine, 1983). This is the essential step in the induction of LTP (Collingridge et al., 1983; Morris et al., 1986; Wigstrom & Gustafsson, 1984). The increase in intracellular Ca^{2+} activates Ca^{2+}/calmodulin-dependent kinase II (CaMKII), resulting in the addition of more AMPA receptors in the postsynaptic membrane (Durand et al., 1996; Liao et al., 1995; Shi et al., 1999). This is an essential step in the expression of LTP (Lledo et al., 1995; Malenka et al., 1989; Malinow et al., 1989; Pettit et al., 1994; Silva et al., 1992), and AMPA receptor phosphorylation is likely involved (Barria et al., 1997; Barria, Derkach, & Soderling, 1997; McGlade-McCulloh et al., 1993; Tan et al., 1994). The additional AMPA receptors may be locally inserted from intracellular pools and/or captured by the postsynaptic density from extrasynaptic receptors already in the membrane. While the precise mechanisms mediating the increase in AMPA receptors remain an active area of research, the presence of additional AMPA receptors in the postsynaptic membrane causes a given amount of glutamate release to elicit a potentiated response in the postsynaptic cell.

7.4 Long-term depression

Much like LTP, long-term depression (LTD) consists of multiple forms found in many different cell types and brain regions (reviewed in (Bellone et al., 2008; Gladding et al., 2009; Heifets & Castillo, 2009)). This chapter covers NMDA receptor-dependent LTD in the hippocampus. LTD is a long-lasting, activity-dependent change in the efficacy of a synapse that results in a depressed postsynaptic response to a given presynaptic stimulus. It can be induced by prolonged low-frequency stimulation (Fig. 25.9B) (Dudek & Bear, 1992). As in LTP, this form of LTD requires both AMPA and NMDA receptors and is dependent on Ca^{2+} entry through unblocked NMDA receptors (Christie & Abraham, 1992; Dudek & Bear, 1992; Kandler et al., 1998; Lee et al., 1998; Mulkey & Malenka, 1992; Yang et al., 1999). Long duration, low-frequency synaptic stimulation results in a consistent, long-lasting, but smaller Ca^{2+} influx than the large, short-lived Ca^{2+} influx that causes LTP. This lower $[Ca^{2+}]$ does not activate CaMKII (Lee et al., 1998). Instead, it activates the phosphatases calcineurin and protein phosphatase 1 (Kameyama et al., 1998; Kirkwood & Bear, 1994; Lee et al., 1998; Mulkey et al., 1993, 1994). These phosphatases modify the phosphorylation state of AMPA receptors (Barria, Derkach, & Soderling, 1997; Lee et al., 1998, 2000; Roche et al., 1996) and tag them for internalization by clathrin and dynamin-dependent endocytosis (Carroll et al., 1999; Heynen et al., 2000; Man et al., 2000). The decrease in the number of AMPA receptors causes a depressed postsynaptic response. The structural changes in the number of postsynaptic AMPA receptors following either LTP or LTD induction allow the change in synaptic strength to last for a prolonged period of time.

References

Abbracchio, M. P., et al. (2009). Purinergic signalling in the nervous system: An overview. *Trends in Neurosciences, 32*(1), 19–29.

Anderson, C. R., & Stevens, C. F. (1973). Voltage clamp analysis of acetylcholine produced end-plate current fluctuations at frog neuromuscular junction. *Journal of Physiology, 235*(3), 655–691.

Arvanitis, D., & Davy, A. (2008). Eph/ephrin signaling: Networks. *Genes & Development, 22*(4), 416–429.

Atluri, P. P., & Regehr, W. G. (1998). Delayed release of neurotransmitter from cerebellar granule cells. *Journal of Neuroscience, 18*(20), 8214–8227.

Bacaj, T., et al. (2013). Synaptotagmin-1 and synaptotagmin-7 trigger synchronous and asynchronous phases of neurotransmitter release. *Neuron, 80*(4), 947–959.

Bai, J., Wang, P., & Chapman, E. R. (2002). C2A activates a cryptic Ca(2+)-triggered membrane penetration activity within the C2B domain of synaptotagmin I. *Proceedings of the National Academy of Sciences of the U S A, 99*(3), 1665–1670.

Bak, L. K., Schousboe, A., & Waagepetersen, H. S. (2006). The glutamate/GABA-glutamine cycle: Aspects of transport, neurotransmitter homeostasis and ammonia transfer. *Journal of Neurochemistry, 98*(3), 641–653.

Barria, A., et al. (1997). Regulatory phosphorylation of AMPA-type glutamate receptors by CaM-KII during long-term potentiation. *Science, 276*(5321), 2042–2045.

Barria, A., Derkach, V., & Soderling, T. (1997). Identification of the Ca2+/calmodulin-dependent protein kinase II regulatory phosphorylation site in the alpha-amino-3-hydroxyl-5-methyl-4-isoxazole-propionate-type glutamate receptor. *Journal of Biological Chemistry, 272*(52), 32727–32730.

Bartley, T. D., et al. (1994). B61 is a ligand for the ECK receptor protein-tyrosine kinase. *Nature, 368*(6471), 558–560.

Beckmann, M. P., et al. (1994). Molecular characterization of a family of ligands for eph-related tyrosine kinase receptors. *EMBO Journal, 13*(16), 3757–3762.

Bellone, C., Luscher, C., & Mameli, M. (2008). Mechanisms of synaptic depression triggered by metabotropic glutamate receptors. *Cellular and Molecular Life Sciences, 65*(18), 2913–2923.

Bennett, M. R., & Kearns, J. L. (2000). Statistics of transmitter release at nerve terminals. *Progress in Neurobiology, 60*(6), 545–606.

Best, A. R., & Regehr, W. G. (2009). Inhibitory regulation of electrically coupled neurons in the inferior olive is mediated by asynchronous release of GABA. *Neuron, 62*(4), 555–565.

Blakely, R. D., & Edwards, R. H. (2012). Vesicular and plasma membrane transporters for neurotransmitters. *Cold Spring Harbor Perspectives in Biology, 4*(2).

Bliss, T. V., & Lomo, T. (1973). Long-lasting potentiation of synaptic transmission in the dentate area of the anaesthetized rabbit following stimulation of the perforant path. *Journal of Physiology, 232*(2), 331–356.

Block, M. R., et al. (1988). Purification of an N-ethylmaleimide-sensitive protein catalyzing vesicular transport. *Proceedings of the National Academy of Sciences of the U S A, 85*(21), 7852–7856.

Blomfield, S. (1974). Arithmetical operations performed by nerve cells. *Brain Research, 69*(1), 115–124.

Boyd, I. A., & Martin, A. R. (1956). The end-plate potential in mammalian muscle. *Journal of Physiology, 132*(1), 74–91.

Bradley, P. B., Humphrey, P. P., & Williams, R. H. (1986). Evidence for the existence of 5-hydroxytryptamine receptors, which are not of the 5-HT2 type, mediating contraction of rabbit isolated basilar artery. *British Journal of Pharmacology, 87*(1), 3–4.

Burden, S. J., Sargent, P. B., & McMahan, U. J. (1979). Acetylcholine receptors in regenerating muscle accumulate at original synaptic sites in the absence of the nerve. *The Journal of Cell Biology, 82*(2), 412–425.

Burnstock, G. (2014). Purinergic signalling: From discovery to current developments. *Experimental Physiology, 99*(1), 16–34.

Carlsson, A. (1987). Perspectives on the discovery of central monoaminergic neurotransmission. *Annual Review of Neuroscience, 10*, 19–40.

Carroll, R. C., et al. (1999). Rapid redistribution of glutamate receptors contributes to long-term depression in hippocampal cultures. *Nature Neuroscience, 2*(5), 454–460.

Cartaud, J., et al. (1973). Presence of a lattice structure in membrane fragments rich in nicotinic receptor protein from the electric organ of Torpedo marmorata. *FEBS Letters, 33*(1), 109–113.

Castillo, P. E. (2012). Presynaptic LTP and LTD of excitatory and inhibitory synapses. *Cold Spring Harbor Perspectives in Biology, 4*(2).

Ceccarelli, B., Hurlbut, W. P., & Mauro, A. (1972). Depletion of vesicles from frog neuromuscular junctions by prolonged tetanic stimulation. *The Journal of Cell Biology, 54*(1), 30–38.

Chad, J. E., & Eckert, R. (1984). Calcium domains associated with individual channels can account for anomalous voltage relations of CA-dependent responses. *Biophysical Journal, 45*(5), 993–999.

Changeux, J. P., et al. (1970). Extraction from electric tissue of gymnotus of a protein presenting several typical properties characteristic of the physiological receptor of acetylcholine. *Comptes rendus hebdomadaires des séances de l'Académie des sciences. Série D, 270*(23), 2864–2867.

Chapman, E. R., & Davis, A. F. (1998). Direct interaction of a Ca2+-binding loop of synaptotagmin with lipid bilayers. *Journal of Biological Chemistry, 273*(22), 13995–14001.

Chapman, E. R., et al. (1995). Ca2+ regulates the interaction between synaptotagmin and syntaxin 1. *Journal of Biological Chemistry, 270*(40), 23667–23671.

Cheng, H. J., & Flanagan, J. G. (1994). Identification and cloning of ELF-1, a developmentally expressed ligand for the Mek4 and Sek receptor tyrosine kinases. *Cell, 79*(1), 157–168.

Chih, B., Engelman, H., & Scheiffele, P. (2005). Control of excitatory and inhibitory synapse formation by neuroligins. *Science, 307*(5713), 1324–1328.

Christie, B. R., & Abraham, W. C. (1992). NMDA-dependent heterosynaptic long-term depression in the dentate gyrus of anaesthetized rats. *Synapse, 10*(1), 1–6.

Chung, C., et al. (2010). Acute dynamin inhibition dissects synaptic vesicle recycling pathways that drive spontaneous and evoked neurotransmission. *Journal of Neuroscience, 30*(4), 1363–1376.

Clary, D. O., Griff, I. C., & Rothman, J. E. (1990). SNAPs, a family of NSF attachment proteins involved in intracellular membrane fusion in animals and yeast. *Cell, 61*(4), 709–721.

Clary, D. O., & Rothman, J. E. (1990). Purification of three related peripheral membrane proteins needed for vesicular transport. *Journal of Biological Chemistry, 265*(17), 10109–10117.

Collingridge, G. L., Kehl, S. J., & McLennan, H. (1983). Excitatory amino acids in synaptic transmission in the Schaffer collateral-commissural pathway of the rat hippocampus. *Journal of Physiology, 334*, 33–46.

Couteaux, R., & Pecot-Dechavassine, M. (1970). Synaptic vesicles and pouches at the level of "active zones" of the neuromuscular junction. *Comptes Rendus Hebdomadaires des Seances de l'Academie des Sciences, D271*, 2346–2349.

Davletov, B., Perisic, O., & Williams, R. L. (1998). Calcium-dependent membrane penetration is a hallmark of the C2 domain of cytosolic phospholipase A2 whereas the C2A domain of synaptotagmin binds membranes electrostatically. *Journal of Biological Chemistry, 273*(30), 19093–19096.

Daw, M. I., et al. (2009). Asynchronous transmitter release from cholecystokinin-containing inhibitory interneurons is widespread and target-cell independent. *Journal of Neuroscience, 29*(36), 11112–11122.

De Robertis, E. D., & Bennett, H. S. (1955). Some features of the submicroscopic morphology of synapses in frog and earthworm. *The Journal of Biophysical and Biochemical Cytology, 1*(1), 47–58.

Debanne, D., et al. (1996). Paired-pulse facilitation and depression at unitary synapses in rat hippocampus: Quantal fluctuation affects subsequent release. *Journal of Physiology, 491*(Pt 1), 163–176.

Del Castillo, J., & Katz, B. (1954). Quantal components of the end-plate potential. *Journal of Physiology, 124*(3), 560–573.

Dingledine, R. (1983). N-methyl aspartate activates voltage-dependent calcium conductance in rat hippocampal pyramidal cells. *Journal of Physiology, 343*, 385–405.

Dodge, F. A., Jr., & Rahamimoff, R. (1967). Co-operative action a calcium ions in transmitter release at the neuromuscular junction. *Journal of Physiology, 193*(2), 419–432.

Dudek, S. M., & Bear, M. F. (1992). Homosynaptic long-term depression in area CA1 of hippocampus and effects of N-methyl-D-aspartate receptor blockade. *Proceedings of the National Academy of Sciences of the U S A, 89*(10), 4363–4367.

Dulubova, I., et al. (1999). A conformational switch in syntaxin during exocytosis: Role of munc18. *EMBO Journal, 18*(16), 4372–4382.

Dulubova, I., et al. (2005). A Munc13/RIM/Rab3 tripartite complex: From priming to plasticity? *The EMBO Journal, 24*(16), 2839–2850.

Dulubova, I., et al. (2007). Munc18-1 binds directly to the neuronal SNARE complex. *Proceedings of the National Academy of Sciences, 104*(8), 2697–2702.

Durand, G. M., Kovalchuk, Y., & Konnerth, A. (1996). Long-term potentiation and functional synapse induction in developing hippocampus. *Nature, 381*(6577), 71–75.

Eggermann, E., et al. (2011). Nanodomain coupling between Ca(2)(+) channels and sensors of exocytosis at fast mammalian synapses. *Nature Reviews Neuroscience, 13*(1), 7–21.

Fatt, P., & Katz, B. (1951). An analysis of the end-plate potential recorded with an intra-cellular electrode. *Journal of Physiology, 115*(3), 320–370.

Fatt, P., & Katz, B. (1952). Spontaneous subthreshold activity at motor nerve endings. *Journal of Physiology, 117*(1), 109–128.

Fawley, J. A., Hofmann, M. E., & Andresen, M. C. (2014). Cannabinoid 1 and transient receptor potential vanilloid 1 receptors discretely modulate evoked glutamate separately from spontaneous glutamate transmission. *Journal of Neuroscience, 34*(24), 8324–8332.

Fogelson, A. L., & Zucker, R. S. (1985). Presynaptic calcium diffusion from various arrays of single channels. Implications for transmitter release and synaptic facilitation. *Biophysical Journal, 48*(6), 1003–1017.

Francis, P. T., et al. (1999). The cholinergic hypothesis of Alzheimer's disease: A review of progress. *Journal of Neurology, Neurosurgery & Psychiatry, 66*(2), 137–147.

Geppert, M., et al. (1994). Synaptotagmin I: A major Ca2+ sensor for transmitter release at a central synapse. *Cell, 79*(4), 717–727.

Gladding, C. M., Fitzjohn, S. M., & Molnar, E. (2009). Metabotropic glutamate receptor-mediated long-term depression: Molecular mechanisms. *Pharmacological Reviews, 61*(4), 395–412.

Goda, Y., & Stevens, C. F. (1994). Two components of transmitter release at a central synapse. *Proceedings of the National Academy of Sciences of the U S A, 91*(26), 12942–12946.

Graf, E. R., et al. (2004). Neurexins induce differentiation of GABA and glutamate postsynaptic specializations via neuroligins. *Cell, 119*(7), 1013–1026.

Hall, Z. W., & Sanes, J. R. (1993). Synaptic structure and development: The neuromuscular junction. *Cell, 72*(Suppl. 1), 99–121.

Hallermann, S., et al. (2010). Bassoon speeds vesicle reloading at a central excitatory synapse. *Neuron, 68*(4), 710–723.

Harlow, M. L., et al. (2001). The architecture of active zone material at the frog's neuromuscular junction. *Nature, 409*(6819), 479–484.

Hatta, K., Okada, T. S., & Takeichi, M. (1985). A monoclonal antibody disrupting calcium-dependent cell-cell adhesion of brain tissues: Possible role of its target antigen in animal pattern formation. *Proceedings of the National Academy of Sciences of the U S A, 82*(9), 2789–2793.

He, L., & Wu, L. G. (2007). The debate on the kiss-and-run fusion at synapses. *Trends in Neurosciences, 30*(9), 447–455.

Hebb, D. O. (1949). *The organization of behavior.* New York: John Wiley and Sons, Inc.

Hefft, S., & Jonas, P. (2005). Asynchronous GABA release generates long-lasting inhibition at a hippocampal interneuron-principal neuron synapse. *Nature Neuroscience, 8*(10), 1319–1328.

Heifets, B. D., & Castillo, P. E. (2009). Endocannabinoid signaling and long-term synaptic plasticity. *Annual Review of Physiology, 71*, 283–306.

Henkemeyer, M., et al. (2003). Multiple EphB receptor tyrosine kinases shape dendritic spines in the hippocampus. *The Journal of Cell Biology, 163*(6), 1313–1326.

Herring, B. E., & Nicoll, R. A. (2016). Long-term potentiation: From CaMKII to AMPA receptor trafficking. *Annual Review of Physiology, 78*, 351–365.

Heuser, J. E., et al. (1979). Synaptic vesicle exocytosis captured by quick freezing and correlated with quantal transmitter release. *The Journal of Cell Biology, 81*(2), 275–300.

Heuser, J. E., & Reese, T. S. (1973). Evidence for recycling of synaptic vesicle membrane during transmitter release at the frog neuromuscular junction. *The Journal of Cell Biology, 57*(2), 315–344.

Heynen, A. J., et al. (2000). Bidirectional, activity-dependent regulation of glutamate receptors in the adult hippocampus in vivo. *Neuron, 28*(2), 527–536.

Hirai, H., et al. (1987). A novel putative tyrosine kinase receptor encoded by the eph gene. *Science, 238*(4834), 1717–1720.

Honore, T., Lauridsen, J., & Krogsgaard-Larsen, P. (1982). The binding of [3H]AMPA, a structural analogue of glutamic acid, to rat brain membranes. *Journal of Neurochemistry, 38*(1), 173–178.

Huang, X., Gu, H. H., & Zhan, C. G. (2009). Mechanism for cocaine blocking the transport of dopamine: Insights from molecular modeling and dynamics simulations. *Journal of Physical Chemistry B, 113*(45), 15057–15066.

Huang, Y., & Thathiah, A. (2015). Regulation of neuronal communication by G protein-coupled receptors. *FEBS Letters, 589*(14), 1607–1619.

Hui, E., et al. (2009). Synaptotagmin-mediated bending of the target membrane is a critical step in Ca(2+)-regulated fusion. *Cell, 138*(4), 709–721.

Hui, E., et al. (2005). Three distinct kinetic groupings of the synaptotagmin family: Candidate sensors for rapid and delayed exocytosis. *Proceedings of the National Academy of Sciences of the U S A, 102*(14), 5210–5214.

Ichtchenko, K., et al. (1995). Neuroligin 1: A splice site-specific ligand for beta-neurexins. *Cell, 81*(3), 435–443.

Iremonger, K. J., & Bains, J. S. (2007). Integration of asynchronously released quanta prolongs the postsynaptic spike window. *Journal of Neuroscience, 27*(25), 6684–6691.

Jensen, K., Lambert, J. D., & Jensen, M. S. (2000). Tetanus-induced asynchronous GABA release in cultured hippocampal neurons. *Brain Research, 880*(1–2), 198–201.

Jung, J. H., et al. (2016). Variable priming of a docked synaptic vesicle. *Proceedings of the National Academy of Sciences of the U S A, 113*(8), E1098–E1107.

Kaksonen, M., & Roux, A. (2018). Mechanisms of clathrin-mediated endocytosis. *Nature Reviews Molecular Cell Biology, 19*(5), 313–326.

Kameyama, K., et al. (1998). Involvement of a postsynaptic protein kinase A substrate in the expression of homosynaptic long-term depression. *Neuron, 21*(5), 1163–1175.

Kandler, K., Katz, L. C., & Kauer, J. A. (1998). Focal photolysis of caged glutamate produces long-term depression of hippocampal glutamate receptors. *Nature Neuroscience, 1*(2), 119–123.

Katz, B., & Miledi, R. (1965). The effect of calcium on acetylcholine release from motor nerve terminals. *Proceedings of the Royal Society of London B Biological Sciences, 161*, 496–503.

Katz, B., & Miledi, R. (1967). The release of acetylcholine from nerve endings by graded electric pulses. *Proceedings of the Royal Society of London B Biological Sciences, 167*(1006), 23–38.

Kaufmann, N., et al. (2002). Drosophila liprin-α and the receptor phosphatase dlar control synapse morphogenesis. *Neuron, 34*, 27–38.

Kayser, M. S., et al. (2006). Intracellular and trans-synaptic regulation of glutamatergic synaptogenesis by EphB receptors. *Journal of Neuroscience, 26*(47), 12152–12164.

Kim, N., et al. (2008). Lrp4 is a receptor for Agrin and forms a complex with MuSK. *Cell, 135*(2), 334–342.

Kirkwood, A., & Bear, M. F. (1994). Homosynaptic long-term depression in the visual cortex. *Journal of Neuroscience, 14*(5 Pt 2), 3404–3412.

Kittel, R. J., et al. (2006). Bruchpilot promotes active zone assembly, Ca2+ channel clustering, and vesicle release. *Science, 312*(5776), 1051–1054.

Kvachnina, E., et al. (2005). 5-HT7 receptor is coupled to G alpha subunits of heterotrimeric G12-protein to regulate gene transcription and neuronal morphology. *Journal of Neuroscience, 25*(34), 7821–7830.

Lai, K. O., & Ip, N. Y. (2009). Synapse development and plasticity: Roles of ephrin/eph receptor signaling. *Current Opinion in Neurobiology, 19*(3), 275–283.

Lee, H. K., et al. (1998). NMDA induces long-term synaptic depression and dephosphorylation of the GluR1 subunit of AMPA receptors in hippocampus. *Neuron, 21*(5), 1151–1162.

Lee, H. K., et al. (2000). Regulation of distinct AMPA receptor phosphorylation sites during bidirectional synaptic plasticity. *Nature, 405*(6789), 955–959.

Levinson, J. N., et al. (2005). Neuroligins mediate excitatory and inhibitory synapse formation: Involvement of PSD-95 and neurexin-1beta in neuroligin-induced synaptic specificity. *Journal of Biological Chemistry, 280*(17), 17312–17319.

Li, F., et al. (2007). Energetics and dynamics of SNAREpin folding across lipid bilayers. *Nature Structural & Molecular Biology, 14*(10), 890–896.

Li, L., et al. (2006). Phosphatidylinositol phosphates as co-activators of Ca2+ binding to C2 domains of synaptotagmin 1. *Journal of Biological Chemistry, 281*(23), 15845–15852.

Liao, D., Hessler, N. A., & Malinow, R. (1995). Activation of postsynaptically silent synapses during pairing-induced LTP in CA1 region of hippocampal slice. *Nature, 375*(6530), 400–404.

Lise, M. F., & El-Husseini, A. (2006). The neuroligin and neurexin families: From structure to function at the synapse. *Cellular and Molecular Life Sciences, 63*(16), 1833–1849.

Lledo, P. M., et al. (1995). Calcium/calmodulin-dependent kinase II and long-term potentiation enhance synaptic transmission by the same mechanism. *Proceedings of the National Academy of Sciences of the U S A, 92*(24), 11175–11179.

Llinas, R., Sugimori, M., & Silver, R. (1992). Microdomains of high calcium concentration in a presynaptic terminal. *Science, 256*(5057), 677–679.

Loerke, D., et al. (2009). Cargo and dynamin regulate clathrin-coated pit maturation. *PLoS Biology, 7*(3), e57.

Lomo, T. (1966). Frequency potentiation of excitatory synaptic activity in the dentate area of the hippocampal formation. *Acta Physiologica Scandinavia, 277*, 128.

Lupas, A. (1996). Prediction and analysis of coiled-coil structures. *Methods in Enzymology, 266*, 513–525.

Lynch, K. L., et al. (2007). Synaptotagmin C2A loop 2 mediates Ca2+-dependent SNARE interactions essential for Ca2+-triggered vesicle exocytosis. *Molecular Biology of the Cell, 18*(12), 4957–4968.

Ma, C., et al. (2011). Munc13 mediates the transition from the closed syntaxin–Munc18 complex to the SNARE complex. *Nature Structural & Molecular Biology, 18*, 542.

Mackler, J. M., et al. (2002). The C(2)B Ca(2+)-binding motif of synaptotagmin is required for synaptic transmission in vivo. *Nature, 418*(6895), 340–344.

Magill, C., et al. (1987). Agrin. *Progress in Brain Research, 71*, 391–396.

Malenka, R. C., et al. (1989). An essential role for postsynaptic calmodulin and protein kinase activity in long-term potentiation. *Nature, 340*(6234), 554–557.

Malenka, R. C., & Bear, M. F. (2004). LTP and LTD: An embarrassment of riches. *Neuron, 44*(1), 5–21.

Malinow, R., Schulman, H., & Tsien, R. W. (1989). Inhibition of postsynaptic PKC or CaMKII blocks induction but not expression of LTP. *Science, 245*(4920), 862–866.

Man, H. Y., et al. (2000). Regulation of AMPA receptor-mediated synaptic transmission by clathrin-dependent receptor internalization. *Neuron, 25*(3), 649–662.

Marqueze, B., et al. (1995). Cellular localization of synaptotagmin I, II, and III mRNAs in the central nervous system and pituitary and adrenal glands of the rat. *Journal of Neuroscience, 15*(7 Pt 1), 4906–4917.

Marshansky, V., & Futai, M. (2008). The V-type H+-ATPase in vesicular trafficking: Targeting, regulation and function. *Current Opinion in Cell Biology, 20*(4), 415–426.

Mayer, M. L., Westbrook, G. L., & Guthrie, P. B. (1984). Voltage-dependent block by Mg2+ of NMDA responses in spinal cord neurones. *Nature, 309*(5965), 261–263.

McGlade-McCulloh, E., et al. (1993). Phosphorylation and regulation of glutamate receptors by calcium/calmodulin-dependent protein kinase II. *Nature, 362*(6421), 640–642.

McMahan, U. J. (1990). The agrin hypothesis. *Cold Spring Harbor Symposia on Quantitative Biology, 55*, 407–418.

McMahon, H. T., & Nicholls, D. G. (1991). The bioenergetics of neurotransmitter release. *Biochimica et Biophysica Acta, 1059*(3), 243–264.

Melom, J. E., et al. (2013). Spontaneous and evoked release are independently regulated at individual active zones. *Journal of Neuroscience, 33*(44), 17253–17263.

Mendez, P., et al. (2010). N-cadherin mediates plasticity-induced long-term spine stabilization. *The Journal of Cell Biology, 189*(3), 589–600.

Miller, K. E., et al. (2005). Direct observation demonstrates that liprin-α is required for trafficking of synaptic vesicles. *Current Biology, 15*, 684–689.

Missler, M., & Sudhof, T. C. (1998). Neurexins: Three genes and 1001 products. *Trends in Genetics, 14*(1), 20–26.

Missler, M., Sudhof, T. C., & Biederer, T. (2012). Synaptic cell adhesion. *Cold Spring Harbor Perspectives in Biology, 4*(4). a005694.

Misura, K. M. S., Scheller, R. H., & Weiss, W. I. (2000). Three-dimensional structure of the neuronal Sec1-syntaxin 1a complex. *Nature, 404*(6776), 355–362.

Miyata, M., et al. (2000). Local calcium release in dendritic spines required for long-term synaptic depression. *Neuron, 28*(1), 233–244.

Morris, R. G., et al. (1986). Selective impairment of learning and blockade of long-term potentiation by an N-methyl-D-aspartate receptor antagonist, AP5. *Nature, 319*(6056), 774–776.

Mukherjee, K., et al. (2010). Piccolo and bassoon maintain synaptic vesicle clustering without directly participating in vesicle exocytosis. *Proceedings of the National Academy of Sciences of the U S A, 107*(14), 6504–6509.

Mulkey, R. M., Herron, C. E., & Malenka, R. C. (1993). An essential role for protein phosphatases in hippocampal long-term depression. *Science, 261*(5124), 1051–1055.

Mulkey, R. M., et al. (1994). Involvement of a calcineurin/inhibitor-1 phosphatase cascade in hippocampal long-term depression. *Nature, 369*(6480), 486–488.

Mulkey, R. M., & Malenka, R. C. (1992). Mechanisms underlying induction of homosynaptic long-term depression in area CA1 of the hippocampus. *Neuron, 9*(5), 967–975.

Murai, K. K., et al. (2003). Control of hippocampal dendritic spine morphology through ephrin-A3/EphA4 signaling. *Nature Neuroscience, 6*(2), 153–160.

Murray, D., & Honig, B. (2002). Electrostatic control of the membrane targeting of C2 domains. *Molecular Cell, 9*(1), 145–154.

Nagwaney, S., et al. (2009). Macromolecular connections of active zone material to docked synaptic vesicles and presynaptic membrane at neuromuscular junctions of mouse. *Journal of Comparative Neurology, 513*(5), 457–468.

Naraghi, M., & Neher, E. (1997). Linearized buffered Ca2+ diffusion in microdomains and its implications for calculation of [Ca2+] at the mouth of a calcium channel. *Journal of Neuroscience, 17*(18), 6961–6973.

Neher, E., & Marty, A. (1982). Discrete changes of cell membrane capacitance observed under conditions of enhanced secretion in bovine adrenal chromaffin cells. *Proceedings of the National Academy of Sciences of the U S A, 79*(21), 6712–6716.

Ng, J., et al. (2015). Monoamine neurotransmitter disorders–Clinical advances and future perspectives. *Nature Reviews Neurology, 11*(10), 567–584.

Nickel, E., & Potter, L. T. (1973). Ultrastructure of isolated membranes of Torpedo electric tissue. *Brain Research, 57*(2), 508–517.

Nitkin, R. M., et al. (1987). Identification of agrin, a synaptic organizing protein from Torpedo electric organ. *The Journal of Cell Biology, 105*(6 Pt 1), 2471–2478.

Nowak, L., et al. (1984). Magnesium gates glutamate-activated channels in mouse central neurones. *Nature, 307*(5950), 462–465.

Okada, K., et al. (2006). The muscle protein Dok-7 is essential for neuromuscular synaptogenesis. *Science, 312*(5781), 1802–1805.

Paddock, B. E., et al. (2008). Ca2+-dependent, phospholipid-binding residues of synaptotagmin are critical for excitation-secretion coupling in vivo. *Journal of Neuroscience, 28*(30), 7458–7466.

Paddock, B. E., et al. (2011). Membrane penetration by synaptotagmin is required for coupling calcium binding to vesicle fusion in vivo. *Journal of Neuroscience, 31*(6), 2248–2257.

Palay, S. L. (1956). Synapses in the central nervous system. *The Journal of Biophysical and Biochemical Cytology, 2*(4 Suppl. l), 193–202.

Parker, D. (1995). Depression of synaptic connections between identified motor neurons in the locust. *Journal of Neurophysiology, 74*(2), 529–538.

Parnas, I. (1972). Differential block at high frequency of branches of a single axon innervating two muscles. *Journal of Neurophysiology, 35*(6), 903–914.

Peled, E. S., Newman, Z. L., & Isacoff, E. Y. (2014). Evoked and spontaneous transmission favored by distinct sets of synapses. *Current Biology, 24*(5), 484–493.

Pettit, D. L., Perlman, S., & Malinow, R. (1994). Potentiated transmission and prevention of further LTP by increased CaMKII activity in postsynaptic hippocampal slice neurons. *Science, 266*(5192), 1881–1885.

Pfeiffer, F., Graham, D., & Betz, H. (1982). Purification by affinity chromatography of the glycine receptor of rat spinal cord. *Journal of Biological Chemistry, 257*(16), 9389–9393.

Pinsker, H., et al. (1970). Habituation and dishabituation of the gill-withdrawal reflex in Aplysia. *Science, 167*(3926), 1740–1742.

Pohanka, M. (2012). Acetylcholinesterase inhibitors: A patent review (2008 - present). *Expert Opinion on Therapeutic Patents, 22*(8), 871–886.

Pokutta, S., & Weis, W. I. (2007). Structure and mechanism of cadherins and catenins in cell-cell contacts. *Annual Review of Cell and Developmental Biology, 23*, 237–261.

Reider, A., & Wendland, B. (2011). Endocytic adaptors–Social networking at the plasma membrane. *Journal of Cell Science, 124*(Pt 10), 1613–1622.

Renner, M., et al. (2012). Diffusion barriers constrain receptors at synapses. *PLoS One, 7*(8), Article e43032.

Roche, K. W., et al. (1996). Characterization of multiple phosphorylation sites on the AMPA receptor GluR1 subunit. *Neuron, 16*(6), 1179–1188.

Südhof, T. C. (2012). The presynaptic active zone. *Neuron, 75*, 11–25.

Saglietti, L., et al. (2007). Extracellular interactions between GluR2 and N-cadherin in spine regulation. *Neuron, 54*(3), 461–477.

Sanes, J. R., Marshall, L. M., & McMahan, U. J. (1978). Reinnervation of muscle fiber basal lamina after removal of myofibers. Differentiation of regenerating axons at original synaptic sites. *The Journal of Cell Biology, 78*(1), 176–198.

Sarter, M., & Parikh, V. (2005). Choline transporters, cholinergic transmission and cognition. *Nature Reviews Neuroscience, 6*(1), 48–56.

Scheiffele, P., et al. (2000). Neuroligin expressed in nonneuronal cells triggers presynaptic development in contacting axons. *Cell, 101*(6), 657–669.

Schiavo, G., et al. (1993). Identification of the nerve terminal targets of botulinum neurotoxin serotypes A, D, and E. *Journal of Biological Chemistry, 268*(32), 23784–23787.

Schoch, S., et al. (2002). RIM1α forms a protein scaffold for regulating neurotransmitter release at the active zone. *Nature, 415*, 321.

Schousboe, A., et al. (2014). Glutamate metabolism in the brain focusing on astrocytes. *Advances in Neurobiology, 11*, 13–30.

Schupp, M., et al. (2016). Interactions between SNAP-25 and synaptotagmin-1 are involved in vesicle priming, clamping spontaneous and stimulating evoked neurotransmission. *Journal of Neuroscience, 36*(47), 11865–11880.

Shahrezaei, V., & Delaney, K. R. (2004). Consequences of molecular-level Ca2+ channel and synaptic vesicle colocalization for the Ca2+ microdomain and neurotransmitter exocytosis: A Monte Carlo study. *Biophysical Journal, 87*(4), 2352–2364.

Shen, J., et al. (2007). Selective activation of cognate SNAREpins by sec1/munc18 proteins. *Cell, 128*(1), 183–195.

Sheng, M., & Kim, E. (2011). The postsynaptic organization of synapses. *Cold Spring Harbor Perspectives in Biology, 3*(12).

Shi, S. H., et al. (1999). Rapid spine delivery and redistribution of AMPA receptors after synaptic NMDA receptor activation. *Science, 284*(5421), 1811–1816.

Silva, A. J., et al. (1992). Alpha calcium/calmodulin kinase II mutant mice: Deficient long-term potentiation and impaired spatial learning. *Cold Spring Harbor Symposia on Quantitative Biology, 57*, 527–539.

Simon, S. M., & Llinas, R. R. (1985). Compartmentalization of the submembrane calcium activity during calcium influx and its significance in transmitter release. *Biophysical Journal, 48*(3), 485–498.

Smart, T. G., & Paoletti, P. (2012). Synaptic neurotransmitter-gated receptors. *Cold Spring Harbor Perspectives in Biology, 4*(3).

Sollner, T., et al. (1993a). A protein assembly-disassembly pathway in vitro that may correspond to sequential steps of synaptic vesicle docking, activation, and fusion. *Cell, 75*(3), 409–418.

Sollner, T., et al. (1993b). SNAP receptors implicated in vesicle targeting and fusion. *Nature, 362*(6418), 318–324.

Song, J. Y., et al. (1999). Neuroligin 1 is a postsynaptic cell-adhesion molecule of excitatory synapses. *Proceedings of the National Academy of Sciences of the U S A, 96*(3), 1100–1105.

Soreq, H. (2015). Checks and balances on cholinergic signaling in brain and body function. *Trends in Neurosciences, 38*(7), 448–458.

Soreq, H., & Seidman, S. (2001). Acetylcholinesterase–new roles for an old actor. *Nature Reviews Neuroscience, 2*(4), 294–302.

Spangler, S. A., et al. (2013). Liprin-alpha2 promotes the presynaptic recruitment and turnover of RIM1/CASK to facilitate synaptic transmission. *The Journal of Cell Biology, 201*(6), 915–928.

Specht, C. G., & Triller, A. (2008). The dynamics of synaptic scaffolds. *BioEssays, 30*(11–12), 1062–1074.

Stein, A., et al. (2009). Helical extension of the neuronal SNARE complex into the membrane. *Nature, 460*(7254), 525–528.

Striegel, A. R., et al. (2012). Calcium binding by synaptotagmin's C2A domain is an essential element of the electrostatic switch that triggers synchronous synaptic transmission. *Journal of Neuroscience, 32*(4), 1253–1260.

Sugita, S., et al. (2001). Synaptotagmin VII as a plasma membrane Ca(2+) sensor in exocytosis. *Neuron, 30*(2), 459–473.

Sutton, R. B., et al. (1998). Crystal structure of a SNARE complex involved in synaptic exocytosis at 2.4 A resolution. *Nature, 395*(6700), 347–353.

Szule, J. A., et al. (2012). Regulation of synaptic vesicle docking by different classes of macromolecules in active zone material. *PLoS One, 7*(3), Article e33333.

Takeuchi, A., & Onodera, K. (1972). Effect of bicuculline on the GABA receptor of the crayfish neuromuscular junction. *Nature New Biology, 236*(63), 55–56.

Tan, S. E., Wenthold, R. J., & Soderling, T. R. (1994). Phosphorylation of AMPA-type glutamate receptors by calcium/calmodulin-dependent protein kinase II and protein kinase C in cultured hippocampal neurons. *Journal of Neuroscience, 14*(3 Pt 1), 1123–1129.

Thomson, A. M., Deuchars, J., & West, D. C. (1993). Large, deep layer pyramid-pyramid single axon EPSPs in slices of rat motor cortex display paired pulse and frequency-dependent depression, mediated presynaptically and self-facilitation, mediated postsynaptically. *Journal of Neurophysiology, 70*(6), 2354–2369.

Tintignac, L. A., Brenner, H. R., & Ruegg, M. A. (2015). Mechanisms regulating neuromuscular junction development and function and causes of muscle wasting. *Physiological Reviews, 95*(3), 809–852.

Ubach, J., et al. (1998). Ca2+ binding to synaptotagmin: How many Ca2+ ions bind to the tip of a C2-domain? *EMBO Journal, 17*(14), 3921–3930.

Uchida, N., et al. (1996). The catenin/cadherin adhesion system is localized in synaptic junctions bordering transmitter release zones. *The Journal of Cell Biology, 135*(3), 767–779.

Ullrich, B., et al. (1994). Functional properties of multiple synaptotagmins in brain. *Neuron, 13*(6), 1281–1291.

Ushkaryov, Y. A., et al. (1992). Neurexins: Synaptic cell surface proteins related to the alpha-latrotoxin receptor and laminin. *Science, 257*(5066), 50–56.

Varoqueaux, F., Jamain, S., & Brose, N. (2004). Neuroligin 2 is exclusively localized to inhibitory synapses. *European Journal of Cell Biology, 83*(9), 449–456.

Vaughan, R. A., & Foster, J. D. (2013). Mechanisms of dopamine transporter regulation in normal and disease states. *Trends in Pharmacological Sciences, 34*(9), 489–496.

Vrljic, M., et al. (2010). Molecular mechanism of the synaptotagmin-SNARE interaction in Ca2+-triggered vesicle fusion. *Nature Structural & Molecular Biology, 17*(3), 325–331.

Wagh, D. A., et al. (2006). Bruchpilot, a protein with homology to ELKS/CAST, is required for structural integrity and function of synaptic active zones in Drosophila. *Neuron, 49*(6), 833–844.

Wang, S., Li, Y., & Ma, C. (2016). Synaptotagmin-1 C2B domain interacts simultaneously with SNAREs and membranes to promote membrane fusion. *Elife, 5*.

Watanabe, S., et al. (2014). Clathrin regenerates synaptic vesicles from endosomes. *Nature, 515*(7526), 228–233.

Watanabe, S., et al. (2013a). Ultrafast endocytosis at Caenorhabditis elegans neuromuscular junctions. *Elife, 2*, Article e00723.

Watanabe, S., et al. (2013b). Ultrafast endocytosis at mouse hippocampal synapses. *Nature, 504*(7479), 242–247.

Weatherbee, S. D., Anderson, K. V., & Niswander, L. A. (2006). LDL-receptor-related protein 4 is crucial for formation of the neuromuscular junction. *Development, 133*(24), 4993–5000.

Weimer, R. M., et al. (2006). UNC-13 and UNC-10/Rim localize synaptic vesicles to specific membrane domains. *Journal of Neuroscience, 26*(31), 8040–8047.

Wigstrom, H., & Gustafsson, B. (1984). A possible correlate of the postsynaptic condition for long-lasting potentiation in the Guinea pig hippocampus in vitro. *Neuroscience Letters, 44*(3), 327–332.
Yang, B., et al. (2000). nSec1 binds a closed conformation of syntaxin1A. *The Journal of Cell Biology, 148*(2), 247–252.
Yang, S. N., Tang, Y. G., & Zucker, R. S. (1999). Selective induction of LTP and LTD by postsynaptic [Ca2+]i elevation. *Journal of Neurophysiology, 81*(2), 781–787.
Yao, J., et al. (2011). Uncoupling the roles of synaptotagmin I during endo- and exocytosis of synaptic vesicles. *Nature Neuroscience, 15*(2), 243–249.
Yersin, A., et al. (2003). Interactions between synaptic vesicle fusion proteins explored by atomic force microscopy. *Proceedings of the National Academy of Sciences of the U S A, 100*(15), 8736–8741.
Zhang, B., et al. (2008). LRP4 serves as a coreceptor of agrin. *Neuron, 60*(2), 285–297.
Zhang, X., et al. (2002). Ca2+-dependent synaptotagmin binding to SNAP-25 is essential for Ca2+-triggered exocytosis. *Neuron, 34*(4), 599–611.
Zhen, M., & Jin, Y. (2004). Presynaptic terminal differentiation: Transport and assembly. *Current Opinion in Neurobiology, 14*(3), 280–287.
Zhou, Q., et al. (2015). Architecture of the synaptotagmin-SNARE machinery for neuronal exocytosis. *Nature, 525*(7567), 62–67.

Biography: Research in the Reist lab has focused on the functional roles of the protein machinery mediating synaptic transmission. Specifically, they have used genetic, electrophysiological, and ultrastructural techniques to elucidate the molecular mechanisms of the synaptic vesicle cycle.

Chapter 26

Pancreas and insulin

Juan V. Sanchez-Andres[1], Yukari Takeda[2] and Colin A. Leech[3]

[1]*Department of Medicine, University Jaume I-UJI, Castellon, Spain;* [2]*Department of Bioinformatics, Ritsumeikan University, Shiga, Japan;* [3]*Department of Medicine, State University of New York (SUNY) Upstate Medical University, Syracuse, NY, United States*

1. General physiology of the endocrine pancreas

Food intake typically takes place discontinuously in the daily life of animals. This irregular meal consumption would drive into abrupt oscillations in glycemia (blood glucose concentration) in the absence of a systemic buffering mechanism. The endocrine pancreas provides such a buffer system, and thus glycemia is maintained within a narrow range. The tight regulation of plasma glucose level is commonly referred to as glucose homeostasis. Maintained high glucose levels (hyperglycemia), a hallmark of diabetes mellitus, may cause neuropathy, retinopathy, and nephropathy (Allaman et al., 2015). On the other hand, decreases in glucose levels (hypoglycemia) can be deleterious by producing a metabolic deficit that impedes brain function because of the strict metabolic dependence of neurons on glucose levels in the extracellular fluid (Mergenthaler et al., 2013). Control of glycemia is therefore essential for the healthy physiological function of the body.

Glucose homeostasis is primarily regulated by pancreatic hormones secreted from endocrine cells located within the islets of Langerhans (see Section 2). The islets of Langerhans contain α-, β-, δ-, γ (PP)-, and ε-cells secreting glucagon, insulin, somatostatin, pancreatic polypeptide, and ghrelin, respectively. The pivotal pancreatic hormones involved in glucose homeostasis are insulin and glucagon, which exhibit opposing roles.

Insulin, discovered in 1921 by Branting and Best, is the only hormone able to lower blood glucose concentration, and has been clinically used as a key remedy for diabetic patients. Insulin is released by β-cells within the islets of Langerhans when glucose levels rise after food intake. Insulin allows peripheral cells to uptake glucose and subsequently metabolize it to energy forms (e.g., ATP). Excess glucose may be stored as glycogen, mainly in skeletal muscle and the liver, or fats (triglycerides), and in adipose cells, when glycogen stores are filled. Note that liver may also store triglycerides. Healthy individuals remove blood glucose rapidly when glucose is in excess, whereas insulin-stimulated glucose disposal is impaired in insulin-resistant and type 2 diabetic subjects (see Section 8).

Conversely, glucagon, the principal plasma glucose elevating hormone, is released by the α-cells of the islets of Langerhans when the blood sugar drops between meals. Glucagon stimulates liver cells to degrade glycogen to glucose and release it into the blood to restore euglycemia. This insulin-glucagon system is a paradigmatic example of feedback regulation in physiology that is strictly regulated by neurohormonal/autocrine/paracrine systems to maintain glucose homeostasis (see Sections 5 and 6 and Aronoff et al., 2004). Any perturbation in the system may thus cause pathophysiological conditions.

2. Microscopic anatomy of the islets of Langerhans

The Islets of Langerhans are the endocrine pancreas functional units. They act as microorgans responsible for maintaining glucose homeostasis. Islets contain five different endocrine cell types, which reacts to changes in plasma nutrient levels with the release of islet hormones into the portal vein. α- and β-cells were both described first in 1907 by Lane, while Bloom first recognized δ cells in 1931. Both PP and ε cells were discovered later (In't Veld and Smeets, 2015).

The development of pancreatic islets begins in humans around the embryonic stage of 12 weeks of gestation. At 13–16 weeks of gestation, primitive endocrine precursor cells budding from the wall of the duct constitutes aggregates of cells, resulting in an initial primordium of islets (Fig. 26.1). At 17–20 weeks of gestation, the direct connection between

Cell Physiology Sourcebook. https://doi.org/10.1016/B978-0-12-811114-7.00006-2

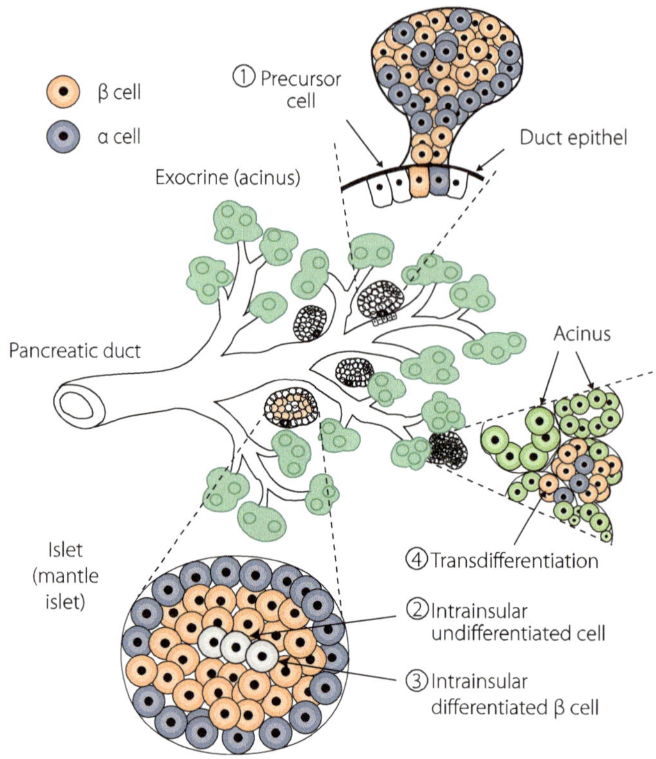

FIGURE 26.1 Development of new islets (islet neogenesis). (1) New islets derive mostly from primitive precursor cells located in the ductal wall that form budding of aggregated endocrine cells as a primordium of islets. (2) The islets grow in size and detach from the duct wall to become isolated. Within the islet, undifferentiated cells develop to differentiate into specific endocrine cell types, such as β-cells or α-cells, or others. (3) Islet β-cells can also replicate themselves to provide new β cells. (4) Under metabolic stress or environmental stimuli, acinar cells can transdifferentiate into islet endocrine cells. *From Yagihashi et al. (2016).*

the islet and duct wall is lost, and the newly-formed islets become isolated. At 21–26 weeks of gestation, β-cells distribute mostly into the central area of the islets. During maturation, islets grow in size due to replicating β-cells. Occupancy of β-cells in the pancreatic parenchyma reaches the adult level at the age of 5–10 years. In the third decade of life, the weight of the pancreas becomes stable, remaining almost being constant until 60 years of age. After that, the pancreatic weight slowly decreases. The replication rate of β-cells is preserved in adult life while rather low, at a rate of 0.01%–0.1% in subjects either with or without diabetes. The lifespan of human β-cells has been estimated to be more than 20 years. Human β-cells retain the potential for replication along its lifespan (Yagihashi et al., 2016). This potential was early considered a way to treat or manage diabetes (Ham & Cormarck, 1984).

β-cells constitute the largest population of cells in the islets, estimated between 50% and 80% of the organ. β-cells secrete insulin, a 51-amino-acid peptide with hypoglycemic action. Insulin, similar to other peptide hormones, is derived from an inactive precursor, proinsulin, which is made up of three parts. The A and B chains remain connected by two sulfur bridges forming the biologically active insulin molecule, and a C chain (Connecting peptide). α-cells secrete glucagon, a 29-amino-acid peptide with hyperglycemic activity. The peptide is derived from proglucagon (180 amino acids) by proteolytic cleavage. The percentage of α cells is estimated at 15%–20% of the total cells in the islets. δ-cells release the hormone somatostatin, a peptide that was formerly identified in the hypothalamus that counteracts growth hormone action. In the islets, somatostatin is a potent inhibitor of insulin and glucagon.

PP cells produce the hormone Pancreatic Polypeptide. Information on the role of PP is somewhat limited. However, it would be expected to be significant as the human ventral pancreas PP cell mass is comprised of up to 80% of these cells. Finally, ε cells secrete ghrelin. As in the case of the PP cells, its role in the pancreas is not well established, but recent research shows potentially high relevance, as ghrelin expression defines a multipotent progenitor lineage giving rise to α, PP and β cells (Arnes et al., 2012).

A traditional view of the islets of Langerhans shows them as static clusters with some regional specificities, more apparent in rodents than in humans, where the endocrine cells are distributed through the islets without apparent organization. As shown in Fig. 26.1, such view seems to be far from reality as the islets change throughout life. Care should be taken as a more accurate description of islets has to be dynamic rather than static. Changes occur in composition from progenitor cells and transdifferentiation (metaplasia), under the pressure of physiological and pathological demands. Aging (Fig. 26.2) drives differentiated β-cell responsiveness. Molecular labeling studies following normal aging, show increasing demands (obesity) or pathological disruption (diabetes). The result is islets with an adapted constitution, and even exhausted functionality allowing disease emergence (Fig. 26.3). A significant challenge in the years to come will be to understand the islets of Langerhans not as a fixed scenario but as a dynamic scenario with adaptations to pressure and, sometimes, giving up permitting pathology emergence.

3. Insulin secretion from pancreatic β-cells

Pancreatic β-cells secrete insulin by converting metabolic signals to electrical membrane activity in response to elevated blood glucose (Malaisse, 2014). More specifically, β-cells generate "bursts," which are characteristic patterns of oscillatory electrical activity, consisting of depolarized plateau potential, on which action potentials are superimposed (Fig. 26.4). Oscillations of $[Ca^{2+}]i$ accompany this rhythmic burst activity. As glucose levels elevate, burst duration increases causing more Ca^{2+} influx and thus insulin secretion. The glucose-dependent response generates a classical negative feedback loop where insulin, through the mechanisms described above, induces glucose uptake by peripheral tissues keeping blood glucose within the normal range (4.4−6.1 mmol/L-79.2−110 mg/dL).

The cell membrane separates an intracellular environment from the extracellular milieu and provides key characteristics to individual cell types. These include different physiological variables such as ionic concentrations that determine cellular excitability. Excitability requires precise interchanges between intra- and extracellular milieu, which is controlled by ion channels. These channels open and close in response to appropriate stimuli. Stimuli allow ions to cross the membrane, thereby modifying the membrane potential since these ions carry electric charges. Electrophysiological techniques are useful in monitoring these ion movements. The driving force that moves ions across the membrane is the concentration gradient of specific ions; if there is no concentration gradient for that ion or if channels for that ion are close, no movement occurs. When channels open, the movement is allowed, and the specific ion tries to reach its equilibrium going in or out of the cell. This movement of charges modifies the membrane potential. At rest, cells exhibit a relative intracellular negative potential. Then, when a channel carrying positive charges opens, the membrane potential becomes more positive (depolarization) provided the ion get inside the cell. Conversely, when a channel for a positively charged ion leaves the cell due to its gradient, this will reduce positive charges inside making the inside more negative (repolarization). Changes in membrane potential are key mechanisms of insulin release by beta cells.

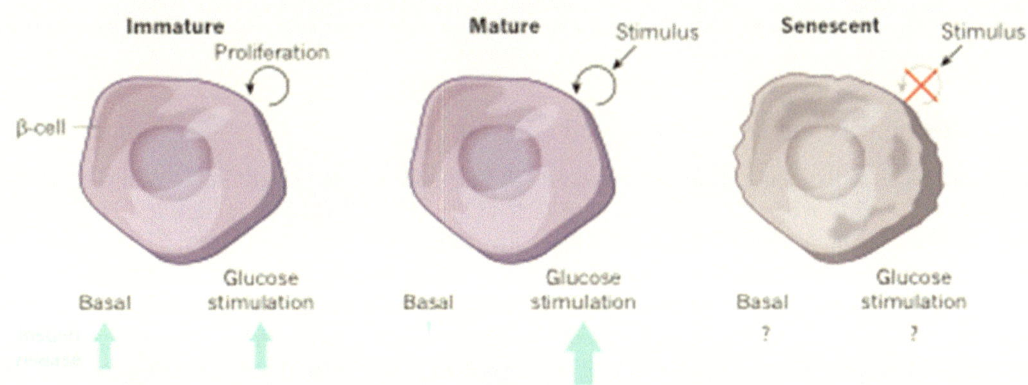

FIGURE 26.2 β-cell subtypes heterogeneity. The insulin-secreting β-cells of the pancreas are thought to renew slowly, so, at any one time, cells of different ages will be present in the tissue. Immature cells are highly proliferative and do not respond robustly to glucose stimulation, but secrete some insulin even at basal (*low*) blood-glucose levels. Mature cells proliferate in response to external stimuli and release insulin robustly, secreting little or no insulin under basal conditions. Old cells cease to proliferate (they become senescent) and may be dysfunctional. *From Bonne-Weir and Aguayo-Mazzucato (2016).*

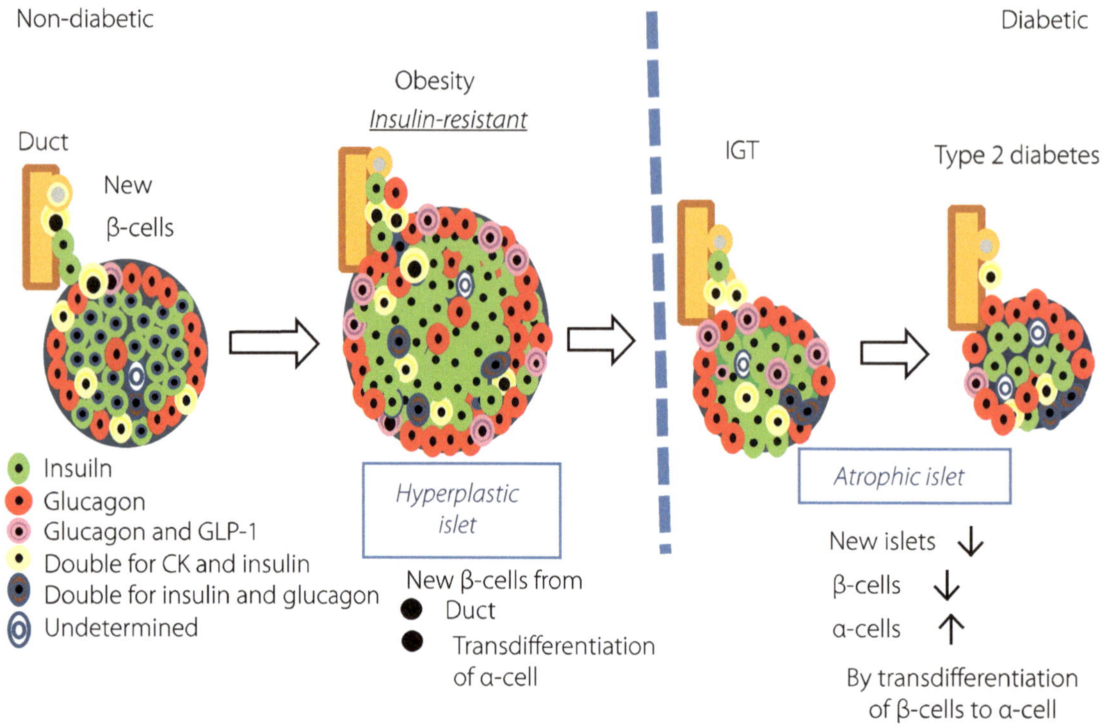

FIGURE 26.3 Emerging concept of islet endocrine cell transdifferentiation (metaplasia). There is remodeling of islet endocrine cells in obese or diabetic subjects. In healthy control subjects, islet β-cells undergo compensatory hyperplastic changes in response to obesity. For this background, it is proposed that increased β-cells derive from α-cells by transdifferentiation in obese insulin-resistant subjects. In contrast, it has been hypothesized that continuous metabolic stress results in promoting transdifferentiation of α-cells to β-cells. Such transdifferentiation of β-cells to α-cells are supposed to contribute to decreased β-cell volume density in subjects with impaired glucose tolerance (IGT) and type 2 diabetes. *CK*, cytokeratin; *GLP-1*, glucagon-like peptide-1. *From Yagihashi et al. (2016).*

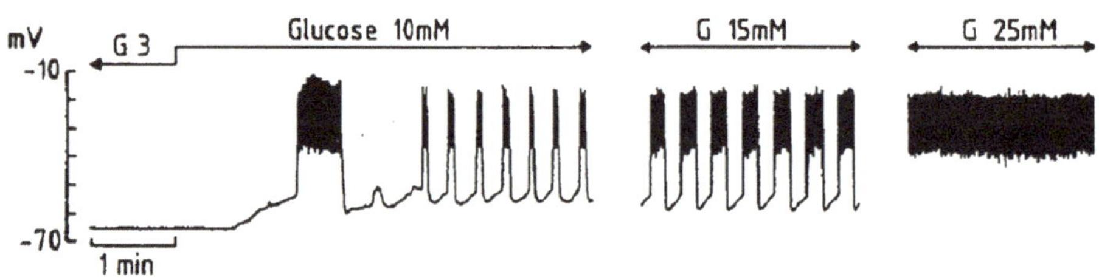

FIGURE 26.4 Intracellular recording of a pancreatic β-cell showing the voltage dependence on glucose concentrations and the glucose induced depolarization that includes oscillations of the membrane potential with associated calcium action potentials on top of the depolarizations (bursts).

3.1 Glucose sensing mechanism of the β-cell

The first step in the β-cell activation is glucose entry through the glucose transporter Glut2. While Glut2 has a low affinity for glucose ($K_m = \sim 17$ mM), it is a very efficient carrier. Glut2 is expressed at a very high levels in pancreatic β-cells, as well as the basolateral membranes of intestinal and kidney epithelial cells and hepatocytes. The presence of Glut2 ensures fast equilibration of glucose between the extracellular space and the cytosol. In pancreatic beta cells, the rate of glucose metabolism is controlled at the glucose phosphorylation step, catalyzed by the glucokinase (GCK) enzyme, that is rate limiting for insulin secretion. Typically modulation of Glut2 surface expression does not regulate metabolism unless its reduction is sufficient to limit glucose access to the kinases, as in the case of diabetic conditions in β-cells (Thorens & Mueckler, 2010). Glucose goes through glycolysis and is converted to pyruvate (the end product of glycolysis) which enters mitochondrial metabolism. Mitochondrial oxidative metabolism is essential for glucose-induced insulin secretion in human islets, and the application of mitochondrial uncouplers promptly inhibits glucose-induced insulin secretion (Henquin et al., 2006).

A key mechanism by which the energy status modulates electrical excitability is the ATP-sensitive K^+ (K_{ATP}) channel. The conductance of the K_{ATP} is regulated by concentrations of ATP and also by ADP, a hydrolyzed form (Hopkins et al., 1992). It is commonly said that the "ATP/ADP" ratio determines the activity of K_{ATP}. However, it is not as simple as the ratio term. The β-cell K_{ATP} channel is a complex octameric unit of two different proteins: an inward rectifier, Kir6.2, and the sulfonylurea receptor, SUR-1. Kir6.2 is the K^+ conducting pore and is regulated by ATP. Increase in ATP concentration decreases K_{ATP} conductance. On the other hand, the SUR-1 belongs to a superfamily of ATP-binding cassette proteins, and it seems to be positively and negatively regulated by ADP. The K_{ATP} conductance increases as ADP concentration increases to ~100 μM, whereas it starts to decrease when the ADP concentration further rises. A simulation study from Magnus and Keizer (1998) showed how the ATP and ADP regulate the K_{ATP} activity. At nonstimulatory glucose concentration, the low ambient level of intracellular ATP keeps the K_{ATP} channel conductance high enough to prevent bursting activities and subsequent insulin secretion. When the glucose supply increases after a meal, an elevation in ATP, accompanied by a reduction in ADP, deactivates (closes) K_{ATP} channel, leading to membrane depolarization and Ca^{2+} influx through voltage-gated Ca^{2+} channels, causing exocytosis of insulin vesicles (Ashcroft et al., 1988). Sulfonylurea drugs bind to the SUR-1 and deactivate the K_{ATP} channels, which stimulate electrical activity and, in turn, increases insulin release. These drugs, therefore, have been used as the therapeutics of type-2 diabetes for 60 years; however, they may cause hypoglycemia because its action mechanisms are independent of glucose concentrations. Ideally, the target floor to be reached for hypoglycemic drugs in diabetes is euglycemia, while sulfonylureas are very capable of breaking the floor and driving to hypoglycemia with physio-pathological consequences.

3.2 Triggering pathway of glucose-dependent insulin secretion

Experimental and theoretical studies have proposed fundamental ionic mechanisms underlying periodic bursting activity following glucose-dependent deactivation (closure) of K_{ATP} channel (see review by Rorsman et al., 2011). Briefly, depolarization induced by the reduction of K_{ATP} conductance triggers regenerative activation of VGCC (voltage-gated Ca^{2+} channels, primarily L-type Ca^{2+} channels for rodent whereas T- and P/Q-types for humans), which is responsible for the upstroke of the action potentials. Subsequent activation of large-conductance (BK) Ca^{2+}-activated K^+ channels and voltage-gated (delayed rectifier) K^+ channels cause action potential repolarization. During the interval between action potentials, voltage-gated Ca^{2+} channels recover from inactivation allowing new action potentials production. Increase in $[Ca^{2+}]_i$ during a series of action potentials results in ATP consumption for intracellular Ca^{2+} clearance through activation of Ca^{2+}-ATPases on the endoplasmic reticulum (SERCA) and plasma membrane (PMCA). Reduced ATP/ADP ratio causes sufficient activation of the K_{ATP} channels leading to the termination of bursting and eventually brings the plateau potential down to interburst potential. Among the numerous theoretical studies published, Cha et al. (2011) were successful in applying mathematical analysis to model the burst properties observed in beta cells, and they suggested a role for inactivation of voltage-gated Ca^{2+} channels and deactivation of Na^+-Ca^{2+} exchanger in bursting termination. Note that glycolysis has the characteristic of intrinsic oscillation, most likely due to positive feedback mediated by the phosphofructokinase-1. Although, Ca^{2+} is not a prerequisite for the metabolic oscillation, changes in Ca^{2+} links metabolic, and subsequent ATP/ADP oscillations to electrical oscillations (Merrins et al., 2016).

As glucose level elevates, burst duration increases (Fig. 26.4), causing more Ca^{2+} influx through voltage-gated Ca^{2+} channels and thus insulin secretion. Simultaneously, interburst interval also decreases with increasing glucose concentrations and eventually culminates in a continuous burst. Interestingly, β-cells are more sensitive to glucose in vivo than when recorded in ex vivo microdissected islets, suggesting a role for other (e.g., neural or hormonal) factors in determining the islet sensitivity to glucose (Sanchez-Andres et al., 1995).

The increase in intracellular Ca^{2+} during the electrical oscillations is the primary trigger for exocytosis of insulin-containing secretory vesicles through activation of the exocytotic machinery responsible for docking, priming, and fusion of large dense-core insulin-containing vesicles. These vesicles have diameters ranging from 200 to 500 nm (Fig. 26.5). Pancreatic β-cells express the full complement of proteins identified as playing a role in neurotransmitter release, including synaptotagmin (Ca^{2+} sensor) the SNARE (syntaxin-1, SNAP-25, and VAMP2), the SNARE-interacting proteins (Munc 18−1, Munc 13−1), complexin 1; and the GTPase Rab3A. In addition to these proteins, insulin secretion requires the action of non neuronal proteins, such as granulophilin, Noc2, and Rab27A (Kasai et al., 2015). When increases in $[Ca^{2+}]_i$ are detected by synaptotagmin, the protein binds to the SNARE complex triggering the transport of insulin granules to the plasma membrane, their fusion to the membrane, and extrusion of insulin.

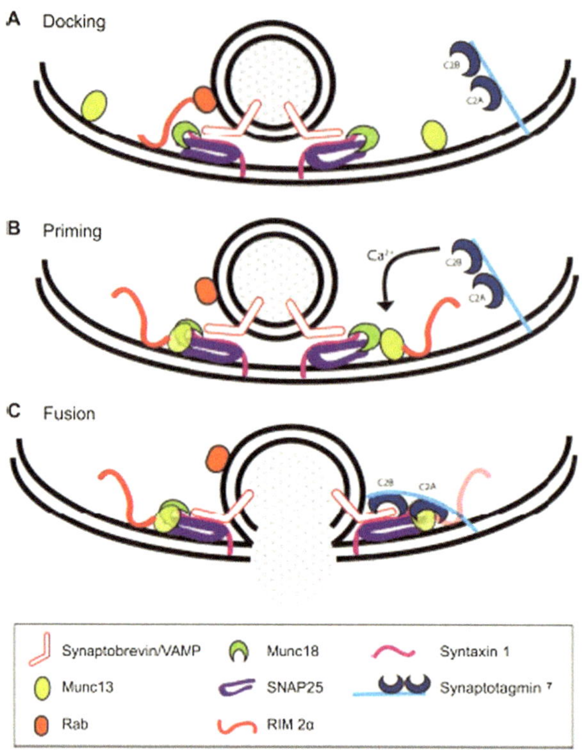

FIGURE 26.5 Model of regulated exocytosis. (A) During granule docking, the SNARE complex assembles on the plasma membrane, and Rim2α binds to Rab. (B) Munc13 association with syntaxin one and Rim2α primes the granule for exocytosis. Upon stimulation, an increase in intracellular calcium levels mobilizes synaptotagmin 7. (C) Synaptotagmin seven interacts with the SNARE complex to elicit exocytosis. *From Röder et al. (2016).*

3.3 Amplifying pathways of glucose-dependent insulin secretion

Although elevation in Ca^{2+} underlying the glucose-dependent electrical oscillations is the primary trigger for insulin exocytosis (triggering pathway), this mechanism is enhanced by metabolic amplification pathways as well as by neurohormonal/autocrine/paracrine pathways. The metabolic amplification pathways involve augmentation of insulin secretion by increasing the efficacy of insulin granules exocytosis without affecting $[Ca^{2+}]_{i,}$ although the detailed mechanisms have not been thoroughly investigated. Nevertheless, it has been shown that ligand-dependent activation of intracellular cAMP-PKA/Epac and IP_3/diacylglycerol (DAG)/PKC signaling cascades are the predominant mechanisms for the neurohormonal/autocrine/paracrine pathways that tightly regulate the Ca^{2+}-dependent exocytosis (Zawalich & Rasmussen, 1990).

The mechanism underlying the enhancement of the glucose-dependent insulin secretion through cAMP-PKA/Epac signaling pathway has been most intensively investigated. Activation of the cAMP-PKA/Epac signaling modifies the activity of ion channels at the plasma membrane (e.g., K_{ATP} channels, nonselective cation channels, and voltage-gated Ca^{2+} channels) as well as at ER membrane (e.g., IP_3R, RYR, and SERCA) to increase membrane excitability and Ca^{2+} -movement. The cAMP-PKA/Epac signaling also directly regulates the exocytotic machinery (Song et al., 2011), so it becomes more sensitive to Ca^{2+} (Fig. 26.6). It is worthwhile to emphasize that the glucose-ATP-Ca^{2+}-dependent triggering pathway and the intracellular signaling-based amplifying pathways exhibit *a complex interplay so that increase in insulin secretion through activation of the cAMP-PKA/Epac signaling is glucose dependent.* An example of a downstream target molecule which activity is dually regulated by glucose and cAMP-PKA/Epac signaling system is IP3R since it is regulated by ATP, Ca^{2+}, and phosphorylation (Takeda et al., 2016). Furthermore, the cAMP-PKA/Epac signaling also enhances insulin transcription and cell proliferation and prevents cell death. The enhancement of glucose-dependent insulin secretion through *IP3/DAG/PKC* signaling is attributed to very similar mechanisms to those involved in cAMP-PKA/Epac signaling, since downstream target effectors are in part common between both pathways. Further details on ligands involved in neurohormonal/autocrine/paracrine pathways that regulate *cAMP-PKA/Epac* and *IP3/DAG/PKC* signaling pathways will be introduced in Section 6.2.

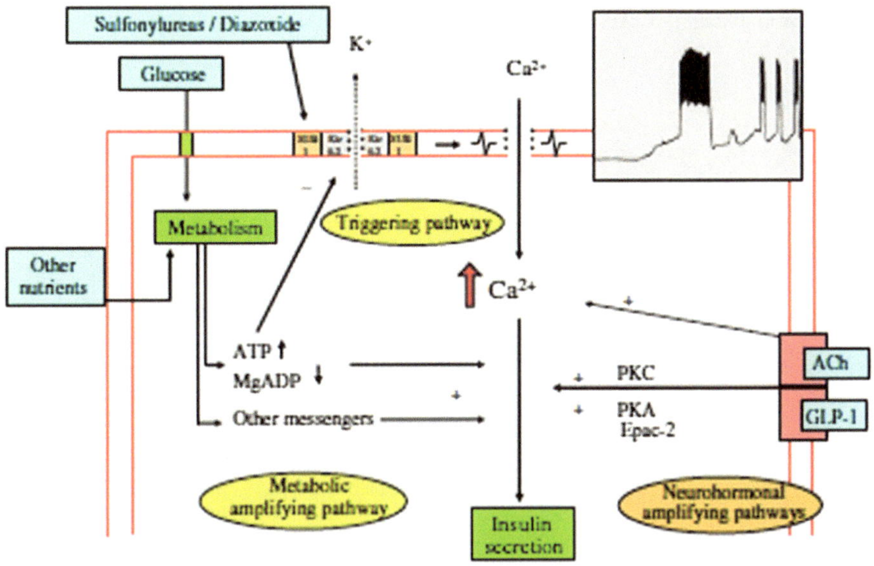

FIGURE 26.6 Schematic representation of the triggering and metabolic amplifying pathways mediating the stimulation of insulin secretion by glucose, and of the neurohormonal amplifying pathways mediating the effects of neurotransmitters and hormones. The KATP channel made up of SUR1, and Kir6.2 subunits is the transducer of metabolic changes into membrane potential changes. The inset illustrates the depolarization and the electrical activity induced by glucose in 1 mouse beta cell within an intact islet. The dotted line indicates a decreased flow. +, stimulation; −, inhibition; *ACh*, acetylcholine; *Epac-2*, Exchange protein directly activated by cAMP 2; *GLP-1*, glucagon-like peptide 1; *Kir*, inwardly rectifying K channel; *PKA*, protein kinase A; *PKC*, protein kinase C; *SUR1*, sulfonylurea receptor 1. *From Henquin (2009).*

4. Glucagon secretion from pancreatic α-cells

Unlike β-cells, α-cells produce action potentials in the absence of glucose, which are associated with the Ca^{2+} influx and glucagon secretion. That said, glucagon exocytosis is maximally inhibited by small increases in glucose concentration, even at levels that have a low stimulatory effect on insulin secretion.

Similarly, to β-cells (Fig. 26.7A), it has been proposed that K_{ATP} channel are also the principal regulator for stimulus-secretion coupling of glucagon secretion in pancreatic α-cells. K_{ATP} in α-cells, however, seem to have higher sensitivity to ATP, so that α-cells are more depolarized at low glucose concentration (~1 mM) to fire large-amplitude action potentials (Fig. 26.7B, *a*). These action potentials are associated with repetitive activation of voltage-activated Na^+ channels as well as T- and P/Q-type VGCCs, which trigger Ca^{2+} entry and the exocytosis of glucagon-containing granules. As shown in Fig. 26.7B, *b*, an increase of extracellular glucose above 4–5 mM closes the remaining K_{ATP} channels, and further depolarizes the α-cells into a voltage range of −40 to −35 mV, where Na^+ channels and T-type VGCCs are largely inactivated. Subsequently, the action potential amplitude decreases due to a decrease in the probability of P/Q-type Ca^{2+} channel to be open, hence associated with a reduction in Ca^{2+} entry and thus, reduced glucagon secretion (Ashcroft & Rorsman, 2013).

Although glucose sensitivity for exocytosis of secretory granules is entirely different in β-cells and α-cells, the mechanisms underlying stimulus-secretion coupling are to some extent similar (Goepel et al., 2000). However, whether glucagon secretion is primarily controlled by glucose-dependent membrane excitability or by neurohormonal/autocrine/paracrine remains controversial. Indeed, it has been proposed that cAMP signaling plays a more critical stimulatory role, in comparison with Ca^{2+}, since glucagon secretion occurs independently of the rise in $[Ca^{2+}]_i$ in α-cells (Li et al., 2015). Furthermore, in response to increasing glucose levels, cAMP is degraded in α-cells whereas it is generated in β-cells (Tengholm & Gylfe, 2017). Therefore, the effects of cAMP need to be considered not only for neurohormonal/autocrine/paracrine-dependent amplification of on glucagon/insulin secretion, but also for glucose-dependent secretions.

5. Local and humoral regulation of the islets of Langerhans

Although glucose is the primary stimulus for the triggering pathway of insulin secretion, as described in the previous section, some other fuel secretagogues (initiators), such as leucine, glyceraldehyde and ketoisocaproate, are able to stimulate exocytosis of insulin granules by, at least in part, increasing ATP via metabolism (Fig. 26.8). In physiology,

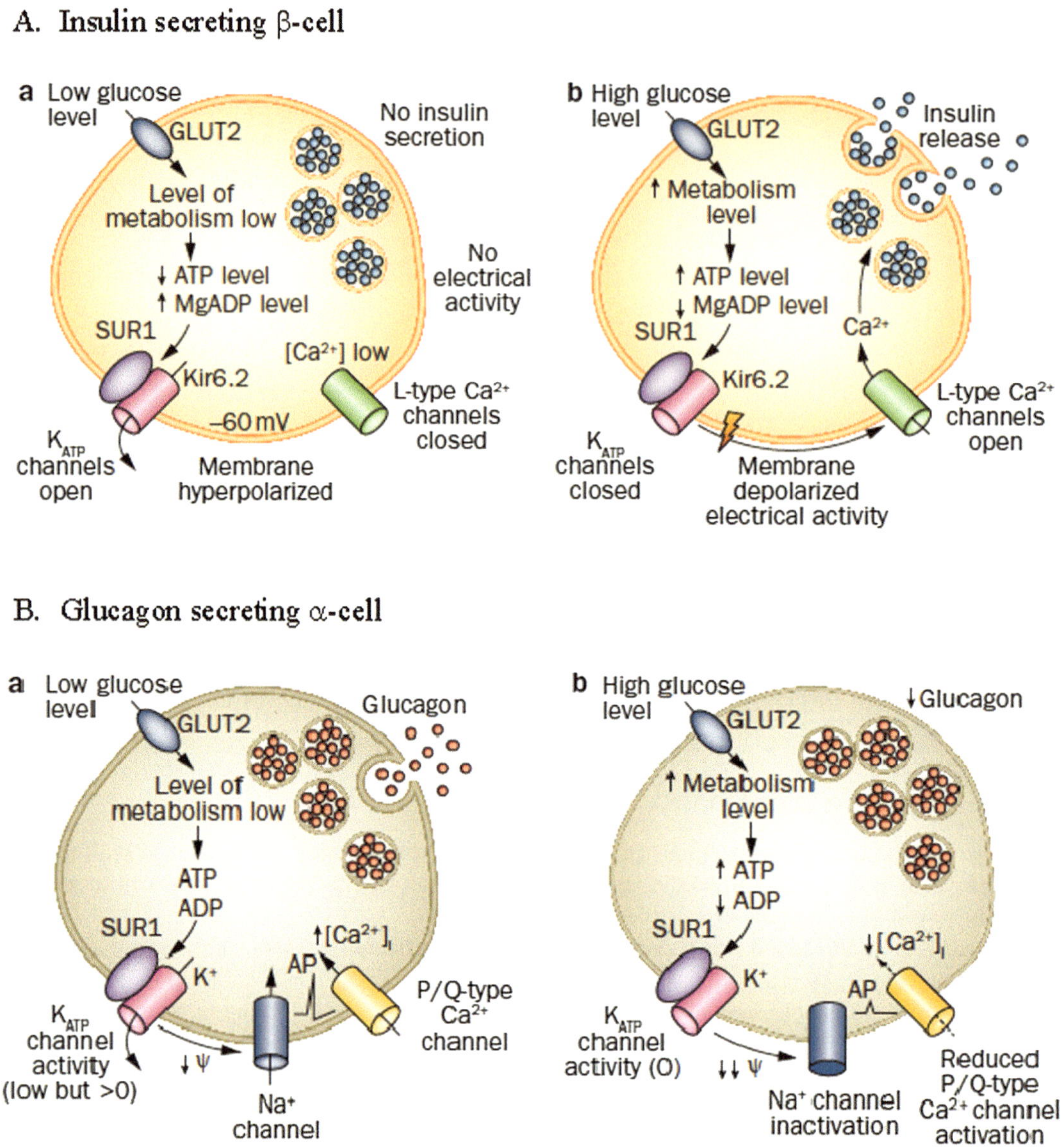

FIGURE 26.7 (A) and (B): SSC (stimulus-secretion coupling) in pancreatic islet β-cells (A) and α-cells (B), both prominent metabolic sensors cells. Both cells use closure of ATP-sensitive K^+ (KATP) channels as a means of coupling a rise in cytosolic [ATP]/[ADP] to cell depolarization and regulation of voltage-dependent Ca^{2+} entry resulting in secretory granule exocytosis. In the case of β-cells, quiescent at resting levels of glucose, closure of KATP channels provides the depolarization to open high-voltage-activated (HVA) Ca^{2+} channels. In contrast, in the case of α-cells, already depolarized and exhibiting Ca^{2+} entry-dependent exocytosis at resting levels of glucose, further depolarization inactivates Na+ and T-type Ca^{2+} channels, reducing the amplitude of AP underlain by the latter and curtailing exocytosis dependent on Ca^{2+} entry through HVA Ca^{2+} channels. *From Ashcroft and Rorsman (2013).*

exocytotic events through glucose-dependent triggering pathway are regulated by intracellular signaling, only at permissive glucose concentrations, and under the control of neurohormonal/autocrine/paracrine systems. Generally, "potentiators" activate G_s or G_q protein-coupled receptors and subsequent cAMP-PKA/Epac or IP_3/DAG/PKC signal transduction pathways, respectively, and enhance glucose-dependent insulin secretion. In contrast, "inhibitors" activate G_i protein-coupled receptors, thus inhibiting cAMP-PKA/Epac signalings, and exocytotic events. These potentiators and inhibitors may also regulate the secretion of other endocrine hormones in the islet of Langerhans. The neurohormonal/autocrine/paracrine control for insulin and glucagon secretion is summarized in Table 26.1. The mechanism of action in insulin secretion is through direct binding to the SUR-1 receptor and blocking the K_{ATP} channels (see Section 4), not by regulation of G protein-coupled receptors. In this section, the paracrine/autocrine (VI.1) and endocrine (VI.2) controls of the islet of Langerhans are discussed. Neuronal regulation will be introduced in the later section.

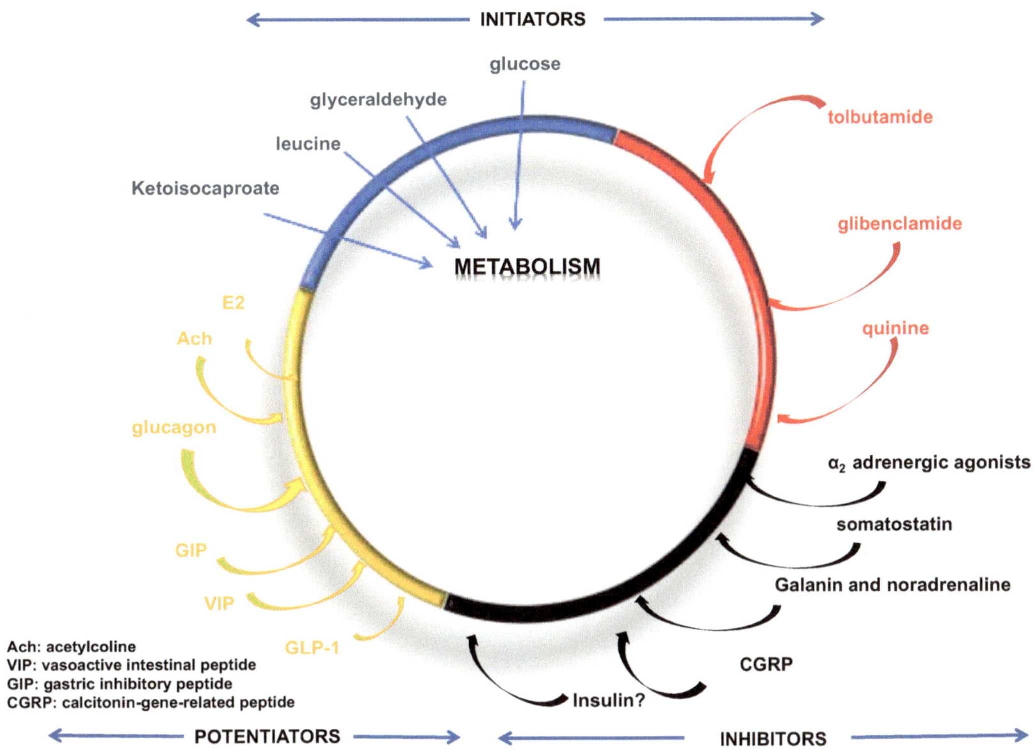

FIGURE 26.8 Islet regulators. Initiators (*red*, *blue*) stimulate insulin secretion without requiring any other molecule while potentiators (*yellow*) increase the initiator's effect but are unable to stimulate by themselves. Inhibitors (*black*) reduce and/or cancel the stimulator's effect. *Modified from Lenzen and Panten (1988).*

5.1 Local intraislet (paracrine/autocrine) regulation

As described in Section 2, the islets of Langerhans are composed by at least five secretory cell types: α-, β-, δ-, PP-, and ε-cell, secreting glucagon, insulin, somatostatin, pancreatic polypeptide, and ghrelin, respectively. Each of these cells regulates the secretory function of neighboring cells in a paracrine manner and even its own function in an autocrine manner (see Table 26.1).

While the glucagon secreted by α-cells enhances insulin secretion from β-cells, the insulin effects on α- and β-cells are more controversial, certainly small if any. That said, it has been demonstrated that β-cells co-release ATP and GABA alongside insulin. ATP seems to further stimulate insulin secretion (autocrine regulation) through activating $P2Y_1$-purinoceptors, while GABA, through $GABA_A$ receptors, seems to inhibit the secretory function of α-cells (paracrine regulation). Although the GABA effect is attributed to hyperpolarization of α-cells through activation of Cl^- channels (as in the nervous system), the mechanism remains controversial.

It is generally agreed that somatostatin, secreted from δ-cells, inhibits the secretion of both insulin and glucagon, and this is mediated by G_i-coupled inhibition of AC activity. It is interesting that glucose stimulation alone decreases cAMP levels in α-cells; however, the effect is independent of somatostatin (paracrine) signaling (Tengholm, 2007). Ghrelin, secreted from ε-cells are reported to increase somatostatin secretion and indirectly inhibit insulin and glucagon secretion from β- and α-cells, respectively. However, direct inhibitory effects on β- and α-cells are yet to be established.

Finally, the secretion of Pancreatic Polypeptide is stimulated by eating, exercising, and fasting. It may inhibit pancreatic exocrine secretion, but its role in glucose homeostasis is uncertain.

5.2 Endocrine islet regulation

The question of a role for regional regulation of insulin secretion was raised by the classical observation that a bolus of glucose is more effective and produces an early peak of insulin when administered orally then intravenously. The explanation was that gut is able to produce β−cells stimulating messengers, even before nutrients were digested and glucose was incorporated into the bloodstream. This early insulin peak was considered a preparatory condition for the

TABLE 26.1 Neurohormonal/autocrine/paracrine control of the endocrine pancreas.

Regulator	Effector	Target/Receptor	Function	References
α-cells	glucagon	β-cells (glucagon receptor/G_s)	insulin secretion ↗↗	Samols et al. (1965)
		α-cells (glucagon receptor/G_s)	glucagon secretion ↗↗	Ma et al. (2005)
β-cells	insulin	β-cells	controversial	Rhodes et al. (2013)
		α-cells	controversial	Li et al. (2015)
	ATP	β-cells ($P2Y_1/G_q$)	insulin secretion ↗↗	Tengholm (2014)
	GABA	α-cells ($GABA_A = Cl^-$ channel)	glucagon secretion ↘↘	Rorsman et al. (1989)
δ-cells	somatostatin	α-cells ($SSTR_2/G_i$)	glucagon secretion ↘↘	Li et al. (2015)
		β-cells ($SSTR_{1,\ 2,\ 5}/G_i$)	insulin secretion ↘↘	Li et al. (2015), Tengholm and Gylfe (2017)
PP-cells (or γ-cells)	pancreatic polypeptide	unknown	unknown	N/A
ε-cells	ghrelin	δ-cells (Ghsr)	Somatostatin secretion ↗↗	Adriaenssens et al. (2016)
Gastrointestinal L-cells	GLP-1	β-cells (GLP-1 receptor/G_s)	insulin secretion ↗↗	Tengholm and Gylfe (2017)
		α-cells (unknown mechanism)	glucagon secretion ↘↘	Tengholm and Gylfe (2017)
Gastrointestinal K-cells	GIP	β-cells (GIP receptor/G_s)	insulin secretion ↗↗	Tengholm and Gylfe (2017)
		α-cells (GIP receptor/G_s)	glucagon secretion ↗↗	Tengholm and Gylfe (2017)
Gastrointestinal I-cells	Cholecystokinin (CCK)	β-cells (CCKBR or CCK/G_q)	insulin secretion ↗↗	Rehfeld and Stadil (1973)
Gastrointestinal G-cells	gastrin			Karlsson and Ahrén (1992)
Gastrointestinal S-cells	Secretin	β-cells (SCTR receptor/G_s)	insulin secretion ↗↗	Kraegen et al. (1970)
Released from intestinal tract and vagus nerves	vasoactive intestinal polypeptide (VIP)	β-cells (VPAC2 receptor/G_s)	insulin secretion ↗↗	Ahrén (2000)
Sympathetic	Adrenaline/noradrenaline	α-cells (β2/G_s)	glucagon secretion ↗↗	Schuit and Pipeleers (1986)
		β-cells (α2/G_i)	insulin secretion ↘↘	Schuit and Pipeleers (1986)
Parasympathetic	Muscarinic	α-cells (M3/G_q)	glucagon secretion ↗↗	Duttaroy et al. (2004)
		β-cells (M3/G_q)	insulin secretion ↗↗	Verspohl et al. (1990)

coming nutrients and will be further explained in the next section when we will review parasympathetic neural islet regulation. This phenomenon was explained after the finding that enhanced insulin secretion resulted from the release of metabolic hormones from the GI (gastrointestinal) tract following food intake. These "metabolic hormones" are now collectively named incretins, and GIP (glucose-dependent insulinotropic polypeptide, formerly called gastric inhibitory peptide) and GLP-1 (glucagon-like peptide-1) are recognized as incretins up to date. GIP is released from K crypt cells in the proximal duodenum and jejunum after meals containing glucose or long-chain fatty acids. GLP-1, secreted from L-cells in the small intestine, is a proglucagon-derived polypeptide, a cleavage product of glucagon, and further processed into glucagon-like peptide-1.

Similarly to the paracrine signaling of glucagon within the islets of Langerhans, both GIP and GLP-1 act on Gs-protein-coupled receptors and activate cAMP-PKA/Epac signal transduction systems, potentiating insulin secretion through a mechanism described in Section 3.3 (see Fig. 26.9, Holz & Habener, 1992). For patients with T2DM, GLP-1 is recognized to be more effective in increasing insulin, in comparison with GIP, and GLP-1 based therapies have been one of the most widely applied treatments for type-2 diabetes.

5.3 Autonomic neural regulation of the islets of Langerhans

Islets have sympathetic, parasympathetic, and sensory innervation. The nerve fibers contain noradrenaline, acetylcholine, and several neuropeptides. Neural regulation of insulin secretion was observed early. By 1935, Juan Planelles, a Ivan Pavlov disciple, following the case of the salivary glands classical conditioning described a parasympathetically-mediated equivalent in insulin secretion called *conditioned hypoglycemia* (Planelles and Luwisch, 1935). Further research has been devoted to obtaining a better understanding of this phenomenon (Woods, 2013), the consequence of a vagal discharge centrally originated from food perception. The result is a preabsortive increase in insulin release, constituting the cephalic phase of insulin secretion. In this phenomenon, regional components such as incretins are involved (see previous section).

Evidence that autonomic nerves influence pancreatic hormone secretion was established early with the rule that sympathetic nerves inhibit insulin and stimulate glucagon secretion and that parasympathetic nerves markedly stimulate insulin and modestly stimulate glucagon secretion (Ahrén, 2000). Noradrenaline (the sympathetic neurotransmitter) binds to α_2 adrenergic receptors, hyperpolarizing the β—cells, reducing Ca^{2+} entry, and inhibiting insulin release. Acetylcholine (the parasympathetic neurotransmitter) binds to M_3 muscarinic receptor depolarizing the β-cells, which allows for Ca^{2+} entry to potentiate glucose-induced insulin release. Noradrenaline and adrenaline stimulations of β-cells participate in stress-induced hyperglycemia by inhibiting insulin secretion.

Islet autonomic nerve terminals contain and co-release several neuropeptides active in modifying insulin secretion. Vasoactive intestinal polypeptide (VIP), pituitary adenylate cyclase activating polypeptide (PACAP), and gastrin-releasing peptide (GRP) are all found in parasympathetic nerves and they stimulate insulin secretion, whereas galanin and neuropeptide Y (NPY) are found in the sympathetic nerves, and they both inhibit insulin secretion. There is a convergence of actions between the classical autonomic neurotransmitters and neuropeptides that suggests a modulatory role for the latter.

6. Action mechanisms of insulin and glucagon

6.1 Action mechanisms of insulin

6.1.1 The insulin receptor

Freychet et al. (1971) discovered that insulin binds to cell surface receptors in the liver. Later research showed that the insulin receptor (IR) is broadly localized in tissues, including the endocrine pancreas. There are described two types, IR-A and IR-B that belong to the family of tyrosine kinase receptors. Tyrosine kinase receptors have the property of phosphorylating tyrosine residues in target proteins to change their activity. Insulin receptors are complex polymers formed by two extracellular α subunits anchored to the plasma membrane and bound to the transmembrane β subunits. Therefore, the α subunits contain the domains that bind to the circulating insulin while the β subunits contain the catalytic tyrosine kinase capability that only becomes active when insulin binds to the α subunits. In the absence of insulin, the complex stays inactive, but when two insulin molecules bind to the α subunits a conformational change happens such that kinase activates phosphorylation of tyrosine residues in target proteins, which then act as second messengers. Every cellular type utilizes a set of second messengers to fullfill its physiological role (for a review see Ullrich, 2015, Chapter 23). One of a critical action of insulin is to increase glucose uptake by adipose tissues and striated (skeletal and cardiac) muscles. Another important function of insulin is to inhibit gluconeogenesis in liver and stimulate fatty acid synthesis in adipocytes.

6.1.2 The glucose transporter

It is worth to emphasize that glucose is unable to cross membranes, and glucose transporters (GLUT), integral membrane proteins, are required to transport glucose across the plasma membrane. Up to 14 members of the GLUT family have been described, and GLUT1–4 are the better characterized glucose transporters. They are expressed in a tissue- and cell-specific manner, exhibiting distinct kinetic and regulatory properties that reflect their specific functional roles. GLUT4, expressed predominantly in adipose tissues and striated muscles, is the only subtype regulated by insulin. Under low insulin concentration, most GLUT4 is found in intracellular vesicles. Upon elevation of insulin, activation of the insulin receptor signaling cascade leads to translocation of GLUT4 from these vesicles to the plasma membrane and increases the uptake of glucose by facilitated diffusion, with no energy consumption. The GLUT transporter may equally allow the glucose movement from the cytoplasm to the blood when intracellular glucose concentration is relatively higher (Bell et al., 1990).

GLUT1–3 are insulin-insensitive glucose transport proteins. GLUT1 is a widely expressed isoform that provides many cells with their basal glucose requirements. GLUT2 is a high-K_m isoform expressed in hepatocytes, pancreatic β-cells, and the basolateral membranes of intestinal and renal epithelial cells. GLUT3 is a low-K_m isoform responsible for glucose uptake into neurons. Other GLUTs are out of the scope of this chapter. For this section, the relevant isoform is GLUT4 (Fig. 26.9).

6.1.3 Second messengers and final effects of insulin

The primary outcome of insulin action is allowing an increase in intracellular glucose concentration. Using a simplified view, peripheral tissues can be classified into two groups: 1. Tissues that use glucose for their own consumption, and 2. Tissues that apart from self-sustenance, act as metabolic stores releasing the sugar later when, between meals, plasma glucose tends to get lower as it is consumed by tissues of group 1. Mechanisms for self-sustenance are common for both groups, but mechanisms for storage are only expressed in tissues of group 2.

6.1.3.1 Direct metabolic effects of insulin

Increased intracellular glucose levels are phosphorylated by the glucokinase enzyme, getting into the glycolysis sequence to produce ATP and finally pyruvate which enters the Krebs cycle to produce additional ATP. This process is described in detail in any Biochemistry textbook.

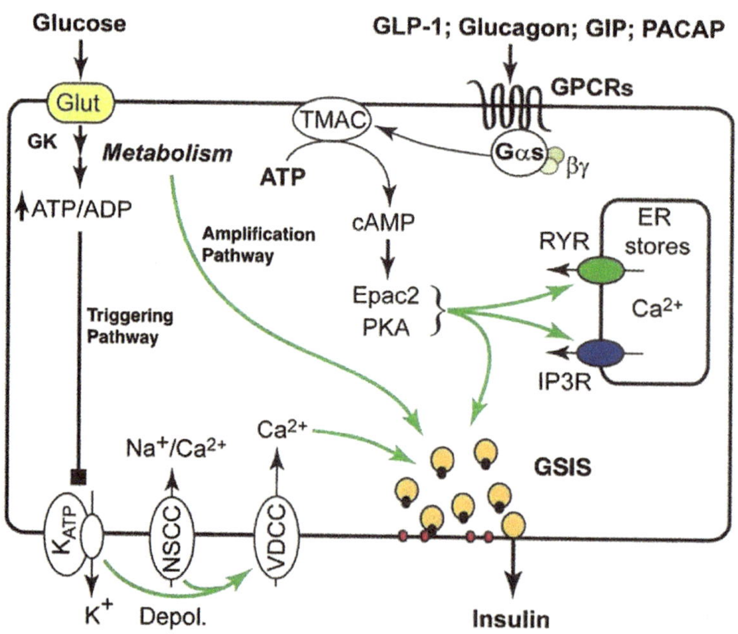

FIGURE 26.9 Incretin action on the pancreatic β-cells. GPCRs (G protein-coupled receptors) are activated by incretins producing cAMP that activates Epac2 and PKA to potentiate glucose stimulated insulin secretion (GSIS). The triggering pathway (see also Fig. 26.6) results in insulin exocytosis that is enhanced by PKA generated in the amplification pathway. Note that Ca^{2+} entry generates the Ca^{2+} signal critical to exocytosis through voltage-dependent Ca^{2+} channels located in the plasma membrane or by the mobilization of Ca^{2+} from intracellular Ca^{2+} stores. Intracellular Ca^{2+} release channels (IP3R, RYR) located on the endoplasmic reticulum (ER) are targets of Epac2 and PKA, thereby allowing cAMP to facilitate the glucose-dependent release of Ca^{2+} from the ER. *From Holz et al. (2015).*

6.1.3.2 Insulin effects in energy storage

Multiple tissues are capable of energy storage: liver, muscle, and fat. Because insulin promotes the creation of stores, it is considered a primary anabolic hormone. In the muscles and liver, insulin promotes the conversion of glucose-6-phosphate to glucose-1-phosphate, a glycogen precursor. It has been demonstrated that insulin activates a phosphatase that dephosphorylates the active form of the glycogen synthase 3enzyme. Then, increased intracellular glucose concentrations, together with activated synthase, cause the accumulation of glycogen. The body's capability to store glycogen is however limited (500 g, as an average). Further energy storage has to take place as fat in liver and adipose tissue. In hepatocytes, high glucose concentrations activate the transcription factor known as carbohydrate response element-binding protein (ChREBP), a key inducer of glycolysis and lipogenesis (Dentin et al., 2006). ChREBP controls approximately 50% of de novo lipogenesis in the liver (Lizuka, 2017). On the other hand, insulin activates the lipoprotein lipase (LPL) facilitating the mobilization of free fatty acids (FFA) from the liver towards the adipose tissue, where they are stored as triglycerides.

6.2 Action mechanisms of glucagon

6.2.1 The glucagon receptor

The glucagon receptor is a 62 kDa protein that is activated by glucagon. It is a member of class B G-protein coupled family of receptors (GPCR) (Brubaker & Drucker, 2002). Stimulation of the receptor results in the activation of adenylate cyclase leading to increased levels of intracellular cAMP. In humans, the glucagon receptor is encoded by the GCGR gene. Glucagon receptors are mainly expressed in liver and kidney with lesser amounts found in heart, adipose tissue, spleen, thymus, adrenal glands, pancreas, cerebral cortex, and gastrointestinal tract.

6.2.2 Mechanism of action

Glucagon is secreted continuously in the interprandial state to maintain hepatic glucose production. The composition of the meal dictates the pattern of subsequent glucagon secretion in nondiabetic subjects. It is classically said that glucagon is a hormone typically catabolic, working in the opposite direction of insulin. While insulin is responsible for the storage of energetic molecules ingested in meals, glucagon is responsible for the gradual release of these stores, according to the body's demands.

Glucagon receptor activation stimulates the production of cAMP which activates Protein Kinase A (PKA). PKA phosphorylates several enzymes, particularly in the liver and in the adipose tissue. Again, the liver plays a central role in energetic metabolism while the role of the adipose tissue is that of a reservoir. In the liver, glucose-1-P is produced from glycogen (glycogenolysis) after the inactivation of the glycogen synthase. At the same time, gluconeogenesis is activated. In summary: in the liver, the anabolic actions of insulin are canceled while glucose is being synthesized. In the adipose tissue, several glucagon sensitive lipases are activated by cAMP, they hydrolyze fat, leading to the release of free fatty acids and glycerol into the blood. These substrates circulate and when they arrive at the liver, they are incorporated into gluconeogenesis. As a consequence, the hepatocytes have higher intracellular glucose concentrations than the plasma, favoring the return of glucose to the bloodstream through facilitated diffusion.

At this point, a question can be easily raised: how would glucose transporters that are induced by insulin be ready to extrude glucose out of hepatocytes upon glucagon stimulation? At a first glance, there seems to be a contradiction: hyperglycemia activates insulin release and glucose transporters. Then, when hypoglycemia activates glucagon secretion, it is expected that the newly produced glucose be trapped inside the cells, but this is not the case. The answer to this apparent enigma is that in any case, under physiological conditions, never glycemia goes so low as to prevent insulin secretion. Under fasting conditions, the concentration of insulin in the portal vein of humans is 440 $\pm$ 25 pmol/L (Song et al., 2000), and islets of Langerhans secrete not only insulin but also glucagon, which protects against hypoglycemia (Shahidul Islam, 2015). Therefore, it is expected that some glucose transporters are active and ready to move glucose in one or other direction, depending on gradient. Following this rationale, one concludes that insulin and glucagon are not antagonistic to each other but rather, complementary. Both hormones work simultaneously in tilting the balance, one way or the other, depending on the body metabolic state.

7. Basic physiopathology of diabetes

Diabetes is a disease known from the time of the ancient Greeks that reported a deleterious illness characterized by patient's sweet urine that attracted flies. Clinically, diabetes patients show a syndromic triade of polyuria, polydipsia and severe slimming. These signs are the consequence of the tissue inability to incorporate the glucose ingested with the meals. This defect causes the two other capital signs of the disease: i. when glucose stays elevated (*hyperglycemia*) in the bloodstream and arrives at the kidney where a glucose tubular reabsorption mechanism recovers the filtrated glucose; and ii. when the transporter threshold is overtaken by hyperglycemia glucose escapes and appears in urine (*glycosuria*).

A complete physiopathological understanding was not possible until the discovery of insulin by Banting et al. (1921), and in 1922 they applied the hormone successfully for the first time to a human patient (Banting et al., 1922). The circle was closed: diabetes was the hormone in charge of saving glucose after meals. When insulin is absent, glucose stays higher in the blood (hyperglycemia) without being incorporated into the tissues that have to use other energy sources for sustenance. The lack of competence to assimilate glucose drives to severe slimming. The tubular kidney transporter is exceeded, and glucose gets lost in urine (glycosuria). The glucose containing urine has a high osmolarity that only can be eliminated carrying water that is lost in excess (polyuria), and the patient gets thirsty (polydipsia). After this milestone, over 600,000 papers were published on the subject providing a more accurate view of the disease, but also showing a puzzle of increasing complexity.

The classical physiopathological scheme corresponds more closely to the first type of described diabetes (juvenile or type 1 diabetes, T1D) characterized by early onset in life where autoimmune insulitis conducts to a loss of β−cells function that fires the physiopathologic cascade. It is a diabetes type originated in the islet (central pole) that a good extent can be treated through replacement therapy with exogenous insulin. Nevertheless, it is worth to keep in mind that T1D associates with secondary effects that frequently are not well prevented by insulin therapy.

Later, it was described the second type as late onset diabetes (adult diabetes, type 2 diabetes, T2D) that was initially thought to be based in the peripheral pole where tissues become resistant to insulin. Here the hyperglycemia is not derived from an insulin deficit but by the incompetence of peripheral tissues to activate the insulin derived mechanisms to incorporate glucose (increased insulin resistance). It is now frequent to associate T2D with a Western life style characterized by sedentarism, unhealthy food, and overweight (increased in body mass index or BMI). Insulin replacement therapy plays a minor role in T2D, and research on therapies is an extremely active field in the search for solutions to a disease that is behaving like a global epidemy. While the first descriptions of T2D considered it a disease of the peripheral pole (tissues), later it has been observed that the central pole (islets) shows changes in volume, composition, and functionality: as BMI increases the islets experience hyperplasic, apparently adaptive changes, to respond to the increasing demand. On the other hand, studies in Eastern cohorts have shown that in contrast to the typical western-world, obese, type-2 diabetis, Japanese type-2 diabetics are lean. Furthermore, Japanese subjects with obesity or impaired glucose tolerance do not show hyperplasic compensatory responses or hyperinsulinemia, which makes them more susceptible to the overt disease.

The focus on islet progression along the physio-pathologic process is summarized in Fig. 26.3, where it is shown that increased peripheral resistance to insulin determines hyperplasic/metaplasic adaptations consisting in increased islet volume and cellularity, at least in part derived from transdifferentiation (metaplasia) of α-cells to β-cells. This response lasts until the islet fails, i.e., when hyperplastic islets become unable to maintain euglycemia and symptomatic diabetes arises while the islets become atrophic and transdifferentiation occurs in the inverse direction, from β-cells to α-cells (for a review, see Yagihashi et al., 2016). In short, the physio-pathologic process should then be considered as highly dynamic and progressive process, in absence of therapeutic intervention.

References

Adriaenssens, A. E., Svendsen, B., Lam, B. Y., Yeo, G. S., Holst, J. J., Reimann, F., & Gribble, F. M. (2016). Transcriptomic profiling of pancreatic alpha, beta and delta cell populations identifies delta cells as a principal target for ghrelin in mouse islets. *Diabetologia, 59*(10), 2156−2165.

Ahrén, B. (2000). Autonomic regulation of islet hormone secretion-implications for health and disease. *Diabetologia, 43*, 393−410.

Allaman, I., Bélanger, M., & Magistretti, P. J. (2015). Methylglyoxal, the dark side of glycolysis. *Frontiers in Neuroscience, 9*(23), 1−12.

Arnes, L., Hill, J. T., Gross, S., et al. (2012). Ghrelin expression in the mouse pancreas defines a unique multipotent progenitor population. *PLoS One, 7*, Article e52026.

Aronoff, S. L., Berkowitz, K., Shreiner, B., & Want, L. (2004). Glucose metabolism and regulation: Beyond insulin and glucagon diab. *Spect, 17*, 183−190.

Ashcroft, F. M., Ashcroft, S. J., & Harrison, D. E. (1988). Properties of single potassium channels modulated by glucose in rat pancreatic β-cells. *The Journal of Physiology, 400*, 501−527.

Ashcroft, F. M., & Rorsman, P. (2013). K(ATP) channels and islet hormone secretion: New insights and controversies. *Nature Reviews Endocrinology, 9*, 660−669.

Banting, F. G., Best, C. H., Collip, J. B., Campbell, W. R., & Fletcher, A. A. (1922). Pancreatic extracts in the treatment of diabetes mellitus. *Canadian Medical Association Journal, 12*(3), 141−146.

Bell, G. I., Kayano, T., Buse, J. B., Burant, C. F., Takeda, J., Lin, D., Fukumoto, H., & Seino, S. (1990). Molecular biology of mammalian glucose transporters. *Diabetes Care, 13*, 198–208.

Bonne-Weir, S., & Aguayo-Mazzucato, C. (2016). Pancreatic β-cell heterogeneity revisited. *Nature, 535*, 365–366.

Brubaker, P. L., & Drucker, D. J. (2002). Structure-function of the glucagon receptor family of G protein-coupled receptors: The glucagon, GIP, GLP-1, and GLP-2 receptors. *Receptors and Channels, 8*, 179–188.

Cha, C. Y., Santos, E., Amano, A., Shimayoshi, T., & Noma, A. (2011). Time-dependent changes in membrane excitability during glucose-induced bursting activity in pancreatic β cells. *The Journal of General Physiology, 138*(1), 39–47.

Dentin, R., Denechaud, P. D., Benhamed, F., Girard, J., & Postic, C. (2006). Hepatic gene regulation by glucose and polyunsaturated fatty acids: A role for ChREBP. *Journal of Nutrition, 136*, 1145–1149.

Duttaroy, A., Zimliki, C. L., Gautam, D., Cui, Y., Mears, D., & Wess, J. (2004). Muscarinic stimulation of pancreatic insulin and glucagon release is abolished in m3 muscarinic acetylcholine receptor-deficient mice. *Diabetes, 53*(7), 1714–1720.

Freychet, P., Roth, J., & Neville, D. M. J. (1971). Insulin receptors in the liver: Specific binding of ^{125}I insulin to the plasma membrane and its relation to insulin bioactivity. *Proceedings of the National Academy of Sciences, 68*, 1833–1837.

Goepel, S. O., Kanno, T., Barg, S., Weng, X. G., Gromada, J., & Rorsman, P. (2000). Regulation of glucagon release in mouse α cells by K_{ATP} channels and inactivation of TTX-sensitive Na channels. *The Journal of Physiology, 528*, 509–520.

Ham, A. W., & Cormarck, D. H. (1984). Hams's histology. In *Interamericana*.

Henquin, J. C. (2009). Regulation of insulin secretion: A matter of phase control and amplitude modulation. *Diabetologia, 52*, 739–751.

Henquin, J. C., Dufrane, D., & Nenquin, M. (2006). Nutrient control of insulin secretion in isolated normal human islets. *Diabetes, 55*, 73470–73477.

Holz, G., Chepurny, O. G., Leech, C. A., Song, W.-J., & Hussain, M. A. (2015). Molecular basis of cAMP signaling in pancreatic β-cells. In M. Shahidul Islam (Ed.), *Islets of Langerhans* (2nd ed., pp. 565–603). Dordrecht: Springer (chapter 20).

Holz, G., & Habener, J. F. (1992). Signal transduction crosstalk in the endocrine system: Pancreatic β-cells and the glucose competence concept. *Trends in Biochemical Sciences, 17*, 388–393.

Hopkins, W. F., Fatherazi, S., Peter-Riesch, B., Corkey, B. E., & Cook, D. L. (1992). Two sites for adenine-nucleotide regulation of ATP-sensitive potassium channels in mouse pancreatic beta-cells and HIT cells. *Journal of Membrane Biology, 129*(3), 287–295.

In't Veld, P., & Smeets, S. (2015). Microscopic anatomy of the human islet of Langerhans. In M. Shahidul Islam (Ed.), *Islets of Langerhans* (2nd ed., pp. 19–38). Dordrecht: Springer (chapter 2).

Karlsson, S., & Ahrén, B. (1992). Cholecystokinin and the regulation of insulin secretion. *Scandinavian Journal of Gastroenterology, 27*(3), 161–165.

Kasai, H., Hatakeyama, H., Ohno, M., & Takahashi, N. (2015). Exocytosis in islet β-cells. In M. Shahidul Islam (Ed.), *Islets of Langerhans* (2nd ed., pp. 475–510). Dordrecht: Springer (chapter 16).

Kraegen, E. W., Chisholm, D. J., Young, J. D., & Lazarus, L. (1970). The gastrointestinal stimulus to insulin release. II. A dual action of secretin. *Journal of Clinical Investigation, 49*(3), 524–529.

Lenzen, S., & Panten, U. (1988). Signal recognition by pancreatic Acells. *Biochemical Pharmacology, 37*, 371–378.

Li, J., Yu, Q., Ahooghalandari, P., Gribble, F. M., Reimann, F., Tengholm, A., & Gylfe, E. (2015). Submembrane ATP and Ca^{2+} kinetics in α-cells: Unexpected signaling for glucagon secretion. *The FASEB Journal, 29*(8), 3379–3388.

Lizuka, K. (2017). The role of carbohydrate response element binding protein in intestinal and hepatic fructose metabolism. *Nutrients, 22*. https://doi.org/10.3390/nu9020181

Ma, X., Zhang, Y., Gromada, J., Sewing, S., Berggren, P. O., Buschard, K., Salehi, A., Vikman, J., Rorsman, P., & Eliasson, L. (January 2005). Glucagon stimulates exocytosis in mouse and rat pancreatic alpha-cells by binding to glucagon receptors. *Molecular Endocrinology, 19*(1), 198–212.

Magnus, G., & Keizer, J. (1998). Model of beta-cell mitochondrial calcium handling and electrical activity. I. Cytoplasmic variables. *American Journal of Physiology, 274*(4 Pt 1), C1158–C1173.

Malaisse, W. J. (2014). Insulin release: The receptor hypothesis. *Diabetologia, 57*, 1287–1290.

Mergenthaler, P., Lindauer, U., Dienel, G. A., & Meisel, A. (2013). Sugar for the brain: The role of glucose in physiological and pathological brain function. *Trends in Neurosciences, 36*(10), 587–597.

Merrins, M. J., Poudel, C., McKenna, J. P., Ha, J., Sherman, A., Bertram, R., & Satin, L. S. (2016). Phase analysis of metabolic oscillations and membrane potential in pancreatic islet βCells. *Biophysical Journal, 110*, 691–699.

Planelles, J., & Luwisch, D. (1935). La acción hipoglucemiante del apetito, reflejo condicionado (English translation: Apetyte hypoglycemic action, conditioned reflex). *Archivos de Neurobiologia, XVI*, 383–386.

Rehfeld, J. F., & Stadil, F. (1973). The effect of gastrin on basal- and glucose-stimulated insulin secretion in man. *Journal of Clinical Investigation, 52*(6), 1415–1426.

Rhodes, C. J., White, M. F., Leahy, J. L., & Kahn, S. E. (2013). Direct autocrine action of insulin on β-cells: Does it make physiological sense? *Diabetes, 62*(7), 2157–2163.

Röder, P. V., Wong, X., Hong, W., & Han, W. (2016). Molecular regulation of insulin granule biogénesis and exocytosis. *Biochemical Journal, 473*, 2737–2756.

Rorsman, P., Berggren, P. O., Bokvist, K., Ericson, H., Möhler, H., Ostenson, C. G., & Smith, P. A. (1989). Glucose-inhibition of glucagon secretion involves activation of GABAA-receptor chloride channels. *Nature, 341*(6239), 233–236.

Rorsman, P., Eliasson, L., Kanno, T., Zhang, Q., & Gopel, S. (2011). Electrophysiology of pancreatic βcells in intact mouse islets of Langerhans. *Progress in Biophysics and Molecular Biology, 107*, 224–235.

Samols, E., Marri, G., & Marks, V. (1965). Promotion of insulin secretion by glucagon. *Lancet, 2*(7409), 415–416.

Sanchez-Andres, J. V., Gomis, A., & Valdeolmillos, M. (1995). The electrical activity of mouse pancreatic β-cells recorded *in vivo* shows glucose-dependent oscillations. *Journal of Physiology (London), 486*(1), 223–228.

Schuit, F. C., & Pipeleers, D. G. (1986). Differences in adrenergic recognition by pancreatic A and B cells. *Science, 232*(4752), 875–877.

Shahidul Islam, M. (2015). Calcium signaling in the islets. In M. Shahidul Islam (Ed.), *Islets of Langerhans* (2nd ed., pp. 605–632). Dordrecht: Springer (chapter 21).

Song, S. H., McIntyre, S. S., Shah, H., Veldhuis, J. D., Hayes, P. C., & Butler, P. C. (2000). Direct measurement of pulsatile insulin secretion from the portal vein in human subjects. *Journal of Clinical Endocrinology & Metabolism, 85*, 4491–4499.

Song, W. J., Seshadri, M., Ashraf, U., Mdluli, T., Mondal, P., Keil, M., Azevedo, M., Kirschner, L. S., Stratakis, C. A., & Hussain, M. A. (2011). Snapin mediates incretin action and augments glucose-dependent insulin secretion. *Cell Metabolism, 13*, 308–319.

Takeda, Y., Shimayoshi, T., Holz, G. G., & Noma, A. (2016). Modeling analysis of inositol 1,4,5-trisphosphate receptor-mediated Ca^{2+}mobilization under the control of glucagon-like peptide-1 in mouse pancreatic β-cells. *American Journal of Physiology - Cell Physiology, 310*, C337–C347.

Tengholm, A. (2007). Cyclic AMP: Swing that message! *Cellular and Molecular Life Sciences, 64*(4), 382–385.

Tengholm, A. (2014). Purinergic P2Y1 receptors take centre stage in autocrine stimulation of human beta cells. *Diabetologia, 57*(12), 2436–2439.

Tengholm, A., & Gylfe, E. (2017). cAMP signalling in insulin and glucagon secretion. *Diabetes, Obesity and Metabolism, 19*(Suppl 1), 42–53. https://doi.org/10.1111/dom.12993

Thorens, B., & Mueckler, M. (2010). Glucose transporters in the 21st century. *American Journal of Physiology. Endocrinology and Metabolism, 298*, E141–E145.

Ullrich, S. (2015). IGF-1 and insulin-receptor signalling in insulin-secreting cells: From function to survival. In M. Shahidul Islam (Ed.), *Islets of Langerhans* (2nd ed., pp. 659–685). Dordrecht: Springer (chapter 23).

Verspohl, E. J., Tacke, R., Mutschler, E., & Lambrecht, G. (1990). Muscarinic receptor subtype in rat pancreatic islets: Binding and functional studies. *European Journal of Pharmacology, 178*, 303–311.

Woods, S. C. (2013). From conditioned hypoglycemia to obesity: Following the data. *Physiology & Behavior, 121*, 19–24.

Yagihashi, S., Inaba, W., & Mizukami, H. (2016). Dynamic pathology of islet endocrine cells in type 2 diabetes: βCell growth, death, regeneration and their clinical implications. *Journal of Diabetes Invest, 7*, 155–165.

Zawalich, W. S., & Rasmussen, H. (1990). Control of insulin secretion: A model involving Ca^{2+}, cAMP and diacylglycerol. *Molecular and Cellular Endocrinology, 70*, 119–137.

Further reading

Mueckler, M. (1994). Facilitative glucose transporters. *European Journal of Biochemistry, 219*, 713–725.

Santulli, G., Lombardi, A., Sorriento, D., Anastasio, A., Del Giudice, C., Formisano, P., Béguinot, F., Trimarco, B., Miele, C., & Iaccarino, G. (2012). Age-related impairment in insulin release: The essential role of β(2)-adrenergic receptor. *Diabetes, 61*(3), 692–701.

Seino, S., & Shibasaki, T. (2005). PKA-dependent and PKA-independent pathways for cAMP-regulated exocytosis. *Physiological Reviews, 85*, 1303–1342.

Simonsson, E., Karlsson, S., & Ahrén, B. (2000). The cyclic AMP-protein kinase A pathway restrains islet phospholipase A(2) activation. *Biochemical and Biophysical Research Communications, 269*(1), 242–246.

Chapter 27

Bioluminescence: Cell physiology, diversity, and new evolutionary insights

Tom Iwanicki[1,2,3], **Tamara Frank**[4] **and Megan L. Porter**[3]

[1]*The Earth Commons Institute, Georgetown University, Washington, DC, United States;* [2]*Smithsonian Institution, National Museum of Natural History, Department of Invertebrate Zoology, Washington, DC, United States;* [3]*Department of Biology, University of Hawai'i at Mānoa, Honolulu, HI, United States;* [4]*Halmos College of Natural Sciences and Oceanography, Nova Southeastern University, Dania Beach, FL, United States*

1. Introduction

Bioluminescence, or the production of light by a living organism, has intrigued and inspired humans for centuries. It conjures imagery of glowing blue, crashing ocean waves, or the starlight-like dazzlement of fireflies in a field. Bioluminescence, however, is much more than a beautiful novelty; it plays a crucial role in many organisms' existence. It is used for a variety of functions, ranging from offense, defense, and communication. Bioluminescence is present in all habitats on earth, but it is virtually ubiquitous in the marine environment. Off the coast of Monterey Bay, California, 76% of all organisms observed from the surface to 3900 m depth were capable of bioluminescence (Martini & Haddock, 2017). Bioluminescence is estimated to have evolved independently upwards of 94 times across living taxa (Lau & Oakley, 2021), with representatives in a broad diversity of phyla (e.g., Bacteria, Dinoflagellata, Ctenophora, Cnidaria, Arthropoda, Annelida, Chordata, and Fungi). The many independent origins invariably lead to a variety of regulatory mechanisms used to control the intensity of light emission (e.g., quorum sensing, action potentials, pH, Ca^{2+} channels, and oxygen regulation). Furthermore, recent investigations into organisms previously known to possess bioluminescence reveal novel functions. Finally, lineages utilizing bioluminescence for communication and courtship are associated with greater species richness compared with nonluminescent sister clades and lineages using luminescence for camouflage only (Davis et al., 2016; Ellis & Oakley, 2016).

It is difficult to underscore the impact discoveries in bioluminescent systems have had on medical diagnostics, molecular imaging, and other technologies. The use of bioluminescent assays is common for pathology screening, environmental monitoring, and in the future may be used to efficiently illuminate roads and urban areas. The list of bioluminescent applications is exhaustive and beyond the scope of this review. Suffice it to say fundamental research into bioluminescence has yielded a wealth of applications.

This review will briefly describe the biochemistry and regulation of bioluminescent pathways from a diversity of species. Some bioluminescent systems are well described, while work is still preliminary in other systems. This review aims to distill the basic luciferin-luciferase biochemistry and regulation of each well-described system and emphasize the breadth of independently evolved bioluminescent systems. The possibility of phototransduction associated with bioluminescent organs across a variety of animal taxa and novel evolutionary insights utilizing bioluminescent taxa are briefly summarized at the end of this chapter.

2. Diversity, physiology, and chemical mechanisms for bioluminescence

Bioluminescence is the result of a substrate (luciferin) becoming oxidized resulting in a highly unstable molecule (oxyluciferin) that emits light as it returns to its ground state; the reaction is catalyzed by a protein (luciferase). Although this is an oversimplification, all bioluminescent reactions involve these basic components (Fig. 27.1).

Cell Physiology Sourcebook. https://doi.org/10.1016/B978-0-12-811114-7.00014-1

$$\text{Luciferin} + O_2 \xrightarrow{\text{Luciferase}} \text{Oxyluciferin}^* \longrightarrow \text{Oxyluciferin} + \text{light}$$

* denotes excited state

FIGURE 27.1 The basic chemical reaction resulting in bioluminescence.

While the terms luciferin and luciferase are used ubiquitously to describe the molecules involved in producing bioluminescence across diverse taxa, it is important to note there are many independently evolved forms of luciferin and luciferase, with a wide variety of chemical compositions, by-products, and photoproducts. Because the terms luciferin and luciferase have the potential to mislead the reader into thinking these molecules are homologous across taxa, we will refer to luciferin and luciferase based on the taxa they originate from (e.g., dinoflagellate luciferin, bacterial luciferase) and clarify the chemical composition of the luciferin when known. There are presently nine well-characterized luciferins: bacterial luciferin (Cormier & Strehler, 1953; McElroy & Green, 1955), dinoflagellate luciferin (Nakamura et al., 1989), coelenterazine, used in a taxonomically broad group of animals, including ctenophores, cnidarians, fish, and crustaceans (Inoue et al., 1977; Rees et al., 1990; Shimomura et al., 1980), firefly luciferin (Bitler & McElroy, 1957), cypridinid (i.e., ostracod crustacean) luciferin (Shimomura et al., 1957), mollusc (*Latia*) luciferin (Shimomura & Johnson, 1968), two annelid luciferins (i.e., *Diplocardia* and *Fridericia*) (Ohtsuka et al., 1976; Petushkov et al., 2014), and fungal luciferin (Purtov et al., 2015). See Fig. 27.2 for the chemical structure of known luciferins.

2.1 Bacteria

2.1.1 Biochemistry

Bacterial luciferin is a reduced flavin mononucleotide ($FMNH_2$) and long-chain fatty aldehyde. $FMNH_2$ reacts with NADH catalyzed by FMN-reductase to form the highly unstable reduced flavin mononucleotide (McElroy & Green, 1955). $FMNH_2$ reacts with oxygen to produce peroxy flavin, which then reacts with myristyl aldehyde to form the acid and luciferase-bound hydroxyl flavin in its excited state (Strehler & Cormier, 1953; Strehler et al., 1954). Most bacterial luciferins have a peak emission of approximately 490 nm, though in vivo the peak emission can vary between blue and yellow depending on the species. Variation in color can be attributed to antenna proteins interacting with the luciferase binding pocket affecting the bioluminescent reaction (Petushkov et al., 1996). The interaction of the antenna protein can shift the wavelength of maximum absorption from 495 to 540 nm (Macheroux et al., 1987).

2.1.2 Regulation

Bacteria regulate luminescence through a mechanism called quorum sensing (see Chapter 34) and the genes associated with luminescence are regulated by the lux operon (Engebrecht et al., 1983). The utility of a single bacterium producing large numbers of photons is low. The signal to noise ratio of a single luminescent bacterium is diminished due to the filtering properties of water. Quorum sensing bacteria release chemical signals, called autoinducers, and autoinducers increase proportionally with density (Nealson, 1977). Once the optimal density of autoinducers, or "quorum", has been reached, *lux* expression is upregulated and the bacteria emit light (Nealson & Hastings, 2006).

2.1.3 Symbioses

It is a common misconception that most bioluminescence in Eukaryotes is the result of bacterial symbioses. Symbioses have been demonstrated in five clades, namely, Cephalopoda, Nematoda, Chondrichthyes, Actinopterygii, and Tunicata (Haddock et al., 2010), but most Eukaryotic bioluminescence does not involve bacteria. Some examples of bacterial symbioses include the anglerfish, who host luminescent bacterial symbionts in their esca, or "lure", used to attract prey. Bioluminescent bacterial symbioses provide excellent models for how symbiotic relationships can evolve. Symbioses can occur independently among closely related species; however, the symbiont appears to drive predictable patterns in host transcription (Pankey et al., 2014). *Xenorhabdus luminescens* is a luminous bacterium that forms a symbiotic relationship with the nematode *Heterorhabditis bacteriophora* (Akhurst, 1982). The *lux* luciferase genes of *X. luminescens*, a terrestrial luminescent bacterium, are homologous to those found in the marine luminescent bacterium, *Vibrio fischeri*, suggesting a common origin of both terrestrial and marine bioluminescence (Frackman et al., 1990). Hawaiian bobtail squid, *Euprymna scolopes*, form a symbiotic relationship with the luminescent bacteria, *V. fischeri*. This relationship is horizontally transmitted and, once developed, bobtail squid vent *V. fischeri* daily (Lee & Ruby, 1994; Miyashiro & Ruby, 2012). The interaction between cephalopods and their luminescent symbiont are developmentally linked. Morphogenesis of the light

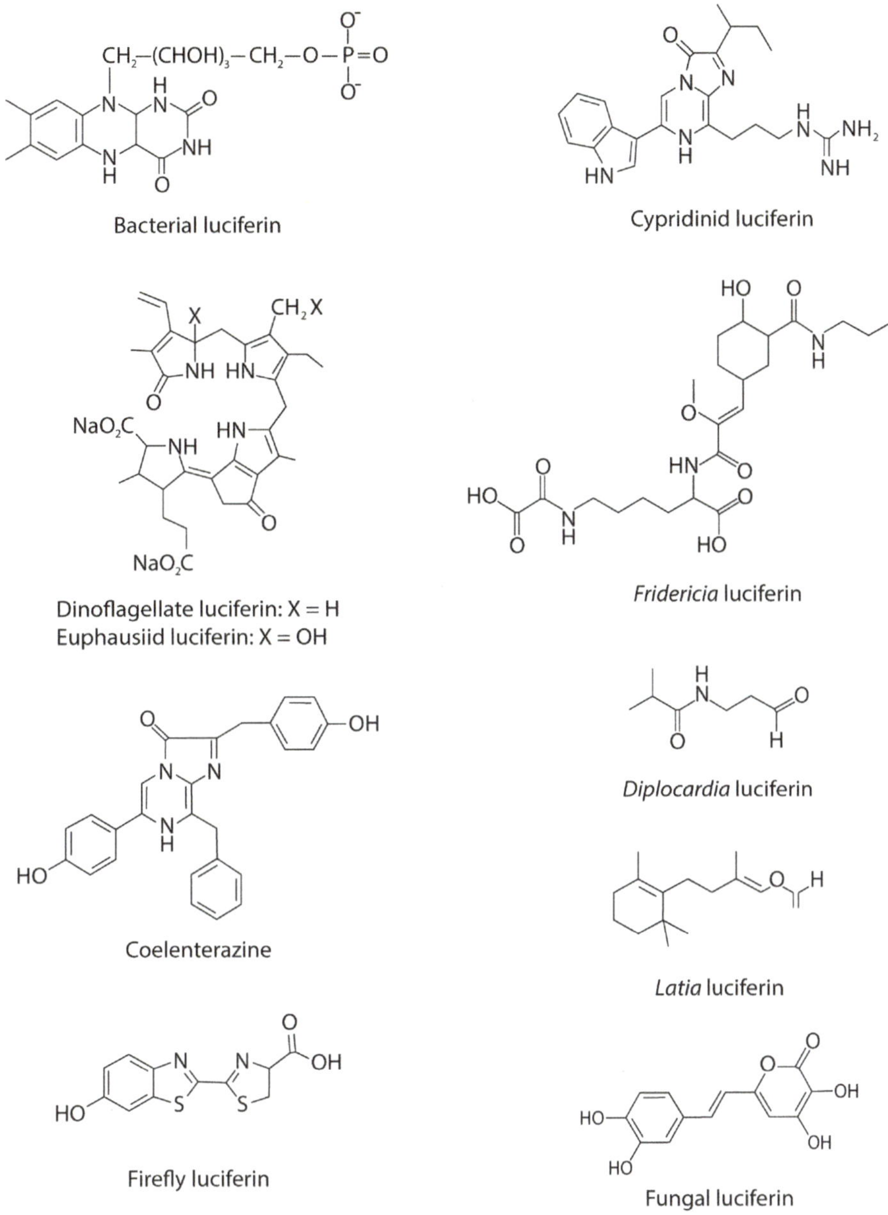

FIGURE 27.2 Chemical structures of known luciferins.

organ in juvenile *E. scolopes* is accelerated in the presence of *V. fischeri*, and either does not occur or proceeds at a reduced rate in their absence (McFall-Ngai et al., 2012).

2.2 Dinoflagellate

2.2.1 Biochemistry and cell biology

Dinoflagellates are responsible for the brilliant optical displays often seen in marine waters, which people often imagine when they think about bioluminescence (Fig. 27.3). Dinoflagellate luciferin was first described in *Pyrocystis lunula* and is a tetrapyrrole similar to luciferins found in euphausiids (Nakamura et al., 1989). Several different dinoflagellate luciferases have been characterized and they all appear to cross-react with the *P. lunula* luciferin (Schmitter et al., 1976). Dinoflagellate luciferin is oxidized using a similar molecular mechanism to most luciferin-luciferase pathways. Interestingly, the oxyluciferin may not be the source of luminescence, and early investigations suggest that the luciferin or an analog is in fact luminous (Wang and Liu, 2016). Chlorophyll is also a tetrapyrrole, and the synthesis of dinoflagellate luciferin is

FIGURE 27.3 The brilliant blue glow produced by dinoflagellates when disturbed by the wake of a boat in transit. It is dinoflagellate bioluminescence people often evoke when thinking about bioluminescence. *Photo credit: Danté Fenolio.*

thought to be derived from chlorophyll catabolism (Topalov & Kishi, 2001). However, many bioluminescent dinoflagellates originate from lineages without photosynthesis. Recent phylotranscriptomic analyses discovered a plastid tetrapyrrole pathway in *Noctiluca* that provides a mechanism for dinoflagellate luciferin biosynthesis in nonphotosynthetic species (Janouskovec et al., 2017).

2.2.2 Regulation

Regulation of dinoflagellate luminescence is controlled by pH and circadian clocks. Light emission is triggered by vibrations of the cytoplasmic membrane, causing an action potential that triggers a proton flux from the vacuole into the luminous organelles called scintillons (approx. 0.5 μm). The initiation of the bioluminescent reaction appears to be initiated by stretch-activated and ion-gated pathway channels at the plasma membrane (Jin et al., 2013). The reaction does not proceed at pH 8. The reaction is signaled when pH drops to 6. This is due to the protein structure of dinoflagellate luciferase, which has a barrel structure that "opens" to the binding of luciferin at low pH. Furthermore, dinoflagellate circadian rhythms control the abundance of scintillons; they are abundant at night and scarce during the day. This involves daily synthesis and destruction of proteins to regulate luminescence.

2.3 Ctenophores and cnidarians

2.3.1 Biochemistry and cell biology

Cnidarian and ctenophore luciferin is coelenterazine (Inoue et al., 1977) and the bioluminescent reaction among these groups produces wavelengths ranging from 440 to 506 nm (Haddock & Case, 1999). Bioluminescence plays an important ecological role with 91.8% of ctenophores and more than 98% of cnidarians capable producing light (Martini & Haddock, 2017). Coelenterazine is a common bioluminescent substrate among a taxonomically broad range of species, ranging from cnidarians and ctenophores (Fig. 27.4) to fish and crustaceans (Shimomura et al., 1980). In cnidarians and ctenophores, coelenterazine is oxidized with the luciferase to yield a very stable peroxide, aequorin (Shimomura, 2005). The presence of Ca^{2+} and aequorin yields coelenteramide in its excited state, which emits light as it returns to the ground state. Aequorin was discovered in the hydromedusan, *Aequorea aequorea*, alongside green fluorescent protein (GFP) (Shimomura et al., 1962). Coelenteramide is capable of initiating a chain reaction with its luminescence exciting GFP causing the emission of green light from the fluorescent protein. The luciferin-luciferase reaction alone emits blue light, and in vivo emission of green light is chiefly due to the energy transfer from coelenteramide to GFP (Lorenz et al., 1991).

2.3.2 Regulation

Bioluminescence in cnidarians and ctenophores is regulated by Ca^{2+} channeling. The stability of aequorin facilitates long term storage within photocytes, epithelial cells that compartmentalize the aequorin away from Ca^{2+} (see Charbonneau et al., 1985; Shimomura, 2006). Action potentials open calcium channels, allowing the passage of Ca^{2+} into the photocytes to serve as cofactors required for the bioluminescent reaction. Aequorin is comprised of a protein bound to its ligand (coelenterazine); when Ca^{2+} is introduced, the protein conformation changes, exposing the coelenterazine-peroxide group.

FIGURE 27.4 Clockwise from top left: ctenophore, siphonophore (Physophora hydrostatica), and cnidarian (Periphyla sp.). *Photo credit: Danté Fenolio.*

This exposed group breaks down, releasing CO_2 and an excited coelenteramide, which emits light as it returns to its ground state (Head et al., 2000). Coelenteramide, apoaequorin, and molecular oxygen can react using a thiol reducing agent in the absence of Ca^{2+} to regenerate aequorin.

2.4 Fireflies (Insecta: Coleoptera: Lampyridae)

2.4.1 Biochemistry and cell biology

The firefly luciferin-luciferase reaction is one of the most well studied bioluminescent reactions (Fig. 27.5). The firefly light organ (or lantern) is composed of photocytes clustered around trachea (Smith, 1963). Firefly luciferin is a benzothiazoyl-thiazole that reacts with oxygen, adenosine triphosphate (ATP), magnesium ions (Mg^{2+}), and luciferase forming a luciferyl adenylate intermediate (Fraga et al., 2004; McElroy & DeLuca, 1978). Luciferyl adenylate is oxidized through a single electron-transfer pathway with a peroxide intermediate, which eventually results in oxyluciferin in its excited state (Branchini et al., 2015).

2.4.2 Regulation

Regulation of firefly bioluminescence is generally considered to be mediated by the availability of free oxygen; however, the precise mechanism controlling oxygen availability is unknown. The peripheral cytoplasm of photocytes in the lantern is packed with mitochondria and their activity may limit oxygen availability. Nitric oxide can stimulate bioluminescence in fireflies, and it potentially blocks respiration of the mitochondria, thereby increasing available oxygen that can facilitate luminescence. Furthermore, the discovery that nitric oxide synthase (*nos*) is highly expressed in the cells interposed

FIGURE 27.5 A female firefly emits light from its posterior end of the abdomen to communicate with mates. Some female fireflies from the genus Photuris mimic receptive females of the genus Photinus to lure and capture males. *Photo credit: Danté Fenolio.*

between nerve endings and photocytes supports nitric oxide as a plausible mechanism for control of bioluminescent displays (Trimmer et al., 2001). Another plausible mechanism involves modulating the amount of fluid in the tracheoles supplying the photocytes. Oxygen diffusion is much faster in air than water. When the tracheoles are filled with fluid, the luminescence is OFF, and when the tracheoles are filled with air, the luminescence is ON (Timmins et al., 2001). However, Tsai et al. (2014) found that oxygen consumption by mitochondria exceeds the maximum rate of oxygen diffusion through the tracheal system to the photocytes, and an additional mechanism is required (e.g., nitric oxide).

2.5 Crustaceans

2.5.1 Biochemistry and cell biology

Bioluminescence in crustaceans is remarkably diverse. Oplophorid shrimp use coelenterazine as luciferin. All species from the families Oplophoridae and Acanthephyridae emit bioluminescence as a secretion (Fig. 27.6). When startled, a luminescent secretion is discharged from the mouth, which emits light at approximately 460 nm (Latz et al., 1988; Shimomura et al., 1978). Some species of Oplophoridae also possess cuticular photophores, which emit light at approximately 475 nm (Herring, 1983; Latz et al., 1988). cDNA from the catalytic domain of secretion luciferase shows no homology to known luciferases that catalyze the oxidation of coelenterazine and must be unique to oplophorids (Inouye et al., 2000).

Cypridinid ostracods have a unique luciferin, a substituted imidazopyrazine nucleus (similar to coelenterazine) (Kishi et al., 1966; Shimomura et al., 1957). Cypridinid luciferin reacts with oxygen forming a cyclic peroxide, which breaks down into CO_2 and an excited carbonyl (Shimomura & Johnson, 1971).

Euphausiid shrimp luciferin is a tetrapyrrole, strikingly similar to dinoflagellate luciferin, and in fact euphausiid and dinoflagellate luciferins and luciferases cross-react (Dunlap et al., 1980). The similarity among euphausiid and dinoflagellate luciferin suggest a dietary origin within euphausiids, but the question remains, how did the luciferases evolve? Euphausiid luciferase has not yet been sequenced, and once available, it may offer insight into convergent molecular evolution in distantly related organisms.

The dietary origin of luciferin across taxa is a prevalent phenomenon. The mysid shrimp, *Neognathophausia ingens*, is capable of a secretory bioluminescence. *N. ingens* held for 2 months in the laboratory and fed a diet restricted to tissues from nonbioluminescent organisms lost the ability to bioluminescence after 2 months (Frank et al., 1984). Furthermore, bioluminescent ability was rapidly regained after *N. ingens* consumed prey that contains the coelenterate bioluminescent system. Starvation studies on *Metridia longa* revealed bioluminescent potential persisted after 3 weeks without food, suggesting metabolic maintenance of the luciferin substrate (Buskey & Stearns, 1991). Later studies confirmed Metridinid copepods are capable of de novo synthesis of their coelenterazine luciferin and may provide the dietary basis for many higher trophic bioluminescent taxa (Oba et al., 2009; Tessler et al., 2018).

2.5.2 Regulation

Luminescence is stimulated by 5-hydroxytryptamine in the euphausiid, *Meganyctiphanes norvegica*, and the stimulation is enhanced when used together with the muscle relaxants papaverine and verapamil (Krönström et al., 2009). Krönström et al. (2009) hypothesize that sphincter structures around the euphausiid photophore control the amount of hemolymph,

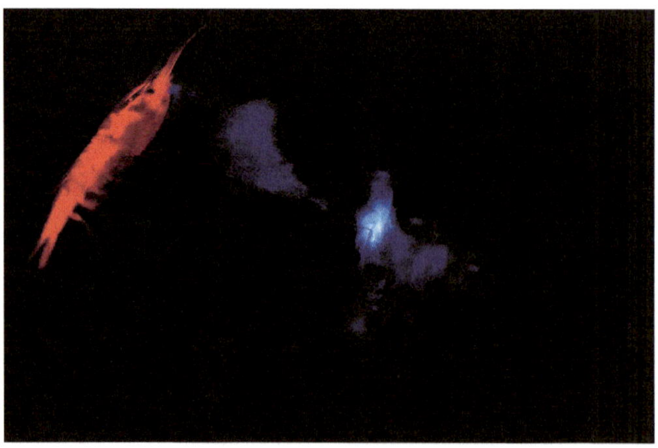

FIGURE 27.6 Crustaceans from the family Oplophoridae are capable of releasing a startling bioluminescent secretion when threatened. Some species within the family also possess cuticular photophores presumably for counter-illumination. *Photo credit: Danté Fenolio.*

and hence oxygen, bathing the photophore, providing the oxygen required for the bioluminescence reaction. Oplophorid shrimp express genes associated with phototransduction within cuticular photophores and may offer an internal mechanism for detecting and regulating light emission for counter-illumination (Bracken-Grissom et al., 2020).

2.6 Other taxa

Bioluminescence is a taxonomically broad phenomenon, and not all bioluminescent systems are thoroughly studied. The following list describes representatives from clades capable of bioluminescence but lack a complete description of the chemistry underpinning the luciferin-luciferase reaction and its regulation.

2.6.1 Fish (Chordata: Actinopterygii and Chondrichthyes)

Many midwater fishes utilize bioluminescence for counter-illumination and offense (Fig. 27.7) including the kitefin shark, the largest known vertebrate capable of bioluminescence to date (Mallefet et al., 2021). The luciferin in many ray-finned and cartilaginous fishes is coelenterazine (Mallefet & Shimomura, 1995) but there is a high degree of complexity among these two groups. Although the luciferase enzyme is present in the genome of many bioluminescent fishes, the preponderance of coelenterazine in the digestive tract of fishes suggests the luciferin is derived from diet rather than synthesized by the fish themselves (Mallefet & Shimomura, 1995; Shimomura et al., 1980). Autogenic (nonbacterial in origin) luminescence was recently described in the lanternshark, *Etmopterus spinax*, and the luminescence is under hormonal control (Claes & Mallefet, 2009; Duchatelet, Delroisse, & Mallefet, 2020). More precisely, melatonin and prolactin induce bioluminescence, alpha-melanocyte-stimulating hormone and adrenocorticotropic hormone inhibit it, and nonocular phototransduction controls the aperture of light organs through an iris-like structure (Duchatelet, Sugihara, et al., 2020). However, the identity of the luciferin in *E. spinax* has yet to be discovered. Although coelenterazine is found in the digestive tract of *E. spinax,* coelenterazine, euphausiid luciferin (tetrapyrrole), and cypridinid luciferin did not react with shark catalyst extract, indicating either an alternative storage form of an already described luciferin, or an unidentified luciferin molecule (Renwart & Mallefet, 2013). The midshipman, *Porichthys notatus*, is a bioluminescent fish distributed along the Pacific coast of North America. Nonbioluminescent populations are present in Puget Sound, and the addition of cypridinid luciferin from the marine ostracod, *Vargula*, induced the ability for bioluminescence. The bioluminescent ability induced in nonluminescent *P. notatus* involves the recycling of luciferin that persists for more than 2 years with no additional luciferin required (Thompson et al., 1988).

2.6.2 Molluscs

The firefly squid, *Watasenia scintillans*, emits blue luminescence from ventral photophores and is a classic example of counter-illuminating camouflage (Young & Roper, 1976) (Fig. 27.8). The firefly squid luciferin is coelenterazine disulfate (Inoue et al., 1975) and the reaction was predicted to be catalyzed by luciferase in the presence of molecular oxygen, ATP, and Mg^{2+} (Tsuji, 2002, 2005). The pathway proposed by Tsuji (2005) was recently supported by Ding and Liu (2017),

FIGURE 27.7 Barbeled dragonfish, Stomias affinis, with the bioluminescent barbel and eye spots clearly visible. Photophores on the ventral side are also visible, presumably used for counter-illumination. *Photo credit: Danté Fenolio.*

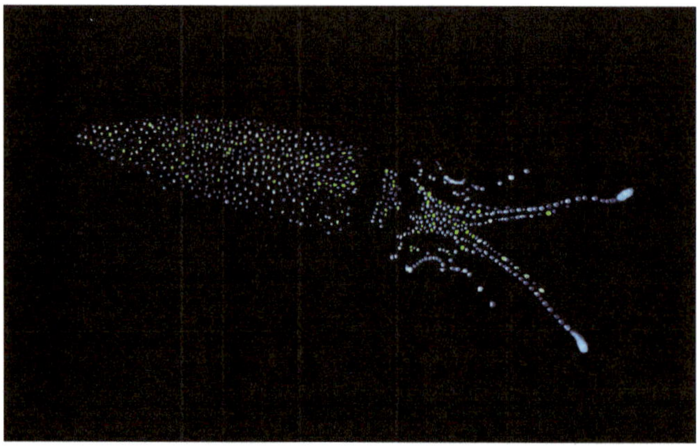

FIGURE 27.8 Firefly squid, Watasenia scintillans, with illuminated ventral photophores used for camouflage by counter-illumination. *Photo credit: Danté Fenolio.*

with the oxygenation involving a single electron-transfer mechanism and light emission produced by gradually reversible charge-transfer-induced luminescence.

The freshwater limpet, *Latia neritoides*, uses a novel luciferin-luciferase pathway. *Latia* luciferin is a formyl ester catalyzed by luciferase in the presence of molecular oxygen (Shimomura et al., 1972). The luminescence appears to involve a purple protein, but it fluoresces at wavelengths dissimilar to the in situ bioluminescence, so how it is involved is unclear.

2.6.3 Myriapods

Millipedes are capable of emitting light from their exoskeleton at approximately 500 nm (Fig. 27.9) requiring only four components: a photoprotein, ATP, Mg^{+2} and O_2 (Shimomura, 1981). The reaction is oxygen-dependent, and the addition of ATP to the luminescent extract increased overall intensity (Hastings & Davenport, 1957). The luminescent millipede, *Luminodesmus sequoiae*, is also fluorescent, using the light emitter 7,8-dihydropterin-6-carboxylic acid (Kuse et al., 2001).

2.6.4 Annelids

Annelids exhibit remarkable diversity of bioluminescent chemistry including some of the shortest wavelength emissions documented in Terebellidae (444 nm, violet) (Kanie et al., 2021), and recent studies are only now exposing just how diverse the mechanisms for luminescence in this phylum truly is. Sipunculates form a basal branch for annelid radiation, and the first bioluminescent sipunculan worm, *Nephasoma pellucidum*, was recently described producing a greenish light with a Renilla-like luciferase in its transcriptome (Oliveira et al., 2021). The structure of two annelid luciferins has been

FIGURE 27.9 Motyxia sequoia alia displaying bioluminescence (*left*) and displaying fluorescence emitted in response to exposure to UV light (*right*). *Photo credit: Danté Fenolio.*

described (i.e., *Diplocardia* and *Fridericia* luciferin). Luciferin from the annelid, *Diplocardia longa*, is a simple aldehyde, N-isovaleryl-3-aminopropanal, with an amide functional group; the luciferase, although not yet sequenced, is a copper-containing protein (Ohtsuka et al., 1976). *Diplocardia* luciferin, when mixed with hydrogen peroxide and copper ions emits light in the absence of luciferase extract (Rudie & Wampler, 1979). Ca^{2+} is a cofactor in the luminescent reaction in the annelid, *Henle* sp. (Rodionova et al., 2002). Further investigation into the role of metal ions as activators or cofactors will aid in understanding the biochemistry of N-isovaleryl-3-aminopropanal bioluminescence pathway. *Fridericia heliota* is a small luminescent annelid that emits light at approximately 478 nm. A recent study isolated the novel annelid *Fridericia* luciferin, which did not cross-react with any other known annelid luciferin or luciferase extracts. The novel luciferin is ATP-dependent and the peptide structure sets it apart from all known luciferins (Petushkov et al., 2014).

Bioluminescence was recently characterized in three deep-sea polychaetes: an unidentified species of the genus *Tomopteris*, *Poeobius meseres*, and *Flota flabelligera* (Francis et al., 2016). All three produce blue-green light, which is unusual for *Tomopteris* sp., as all other tomopterids, with the exception of one record (Gouveneaux et al., 2016), produce yellow bioluminescence (Haddock et al., 2010). *Tomopteris* oxyluciferin may be an anthraquinone, previously only found in plants (Francis et al., 2014). Bioluminescence in *Tomopteris helgolandica* is controlled through activation of nicotinic cholinergic receptors. Pharmacological experiments showed that carbachol induced flashes at low concentrations and steady glow at higher concentrations. The mechanism is also calcium dependent, involving L-type calcium channels (Gouveneaux & Mallefet, 2013).

2.6.5 Echinoderms

Bioluminescence occurs in four of five classes of echinoderms (Herring, 1995). The luciferase appears to be homologous to *Renilla* luciferase, though nonluminous echinoderms also possess homologs. It is therefore possible that bioluminescence in Cnidarians and Echinoderms independently evolved from a homologous gene (Delroisse et al., 2017). Echinoderm bioluminescence uses coelenterazine luciferin and *Amphiura filiformis* acquires coelenterazine through diet (Mallefet et al., 2020).

2.6.6 Porifera

Sponge bioluminescence has long been reported anecdotally but not confirmed until recently. A carnivorous sponge species from the group Cladorhizidae, collected at 4000 m off of Monterey, California, was recently shown to exhibit coelenterazine-based bioluminescence, the first recorded for the phylum (Martini et al., 2020). This exciting new discovery contributes to a broadening understanding of the utility and relatively lower abundance of bioluminescence in the marine benthic habitat compared with the pelagic realm (Martini et al., 2019).

2.6.7 Fungi

Fungi bioluminescence utilizes luciferin and luciferase resulting in light emission approximately 530 nm (Airth & McElroy, 1959) (Fig. 27.10). Fungal luciferin is 3-hydrohispidin, derived from hispidin, a known fungal and plant secondary metabolite (Purtov et al., 2015). Hispidin is the luciferin precursor and is enzymatically transformed in the presence of reduced nicotinamide adenine dinucleotide phosphate (NADPH). Once 3-hydrohispidin is available, luciferase and molecular oxygen are all that is required to complete the bioluminescent reaction (Purtov et al., 2015). Luminous strains of *Panellus stipticus* exhibit greater expression of extracellular manganese superoxide dismutase (MnSOD) and other proteins involved in posttranscriptional modification (Vydryakova & Bissett, 2016). Inhibition of MnSOD resulted in the loss of bioluminescence, suggesting a role in either regulation or biosynthesis of the fungal luciferin. A theoretical framework for the near-continuous emission of light was proposed involving the synergistic departure of carbon dioxide and dissociation of peroxides (Wang & Liu, 2020). The function of fungal bioluminescence has not been clearly demonstrated. Mogilnaya et al. (2016) hypothesize the role of peroxidases in the production of bioluminescence serves to neutralize peroxide compounds under stress.

3. Functions of bioluminescence

There are a wide variety of functions associated with bioluminescence in nature. They can be generalized as offensive, defensive, or communicative (Table 27.1). Empirically testing many of these functions can be difficult because most organisms utilize bioluminescence for multiple functions. What follows is a brief overview of the variety of bioluminescent functions, with an emphasis on studies that empirically demonstrate the function in situ where available.

FIGURE 27.10 Many fungi are capable of bioluminescence; however, its function has not been empirically demonstrated. Top image: Fungi under broad spectrum light; bottom image: Fungi emitting bioluminescence. *Photo credit: Danté Fenolio.*

TABLE 27.1 Common functions of bioluminescence and how they are achieved.

Function	Strategy	How it's used
Offense	Aid in vision	Illuminate prey to aid in capture
	Attract	Lure, often continuous light
Defense	Startle	Bright flashes
	Diversion, smoke screen	Secreted luminous cloud
	Predators learn to avoid	Aposematic signal
	Camouflage	Ventral emission, counter-illumination
	Burglar alarm	Luminescence attracts secondary predator
	Diversion, sacrificial lure	Autotomized luminescent body part
Communication	Congener recognition	Photophore pattern
	Mate attraction	Flash signals

3.1 Offensive bioluminescence

The majority of animals in the deep sea are only capable of detecting blue light. Functionally, one would expect deep-sea organisms to be sensitive to blue light because, due to the filtering properties of water, below 200 m depth the blue wavelengths are the only wavelengths left from sunlight with sufficient intensity to stimulate a visual response. Bioluminescence may be used by predators to actively illuminate their prey (Fig. 27.11) and in some cases that light is undetectable to their prey. Certain dragonfish, from the group called "loosejaws", have a cunning adaptation using red bioluminescence as a flashlight. Using far-red emitting photophores located beneath their eyes, the loosejaws illuminate their surroundings with red light. The production of red light alone would not serve an animal with only blue visual

FIGURE 27.11 Some organisms use bioluminescence to actively illuminate their prey. Three species of dragonfish possess suborbital photophores that emit red light, presumably through fluorescence or filtering of the blue bioluminescence (Herring & Cope, 2005). All deep-sea organisms studied to date, with the exception of the dragonfish with red photophores, are unable to detect red light, allowing dragonfish to illuminate their prey undetected.

FIGURE 27.12 Bioluminescence may be used to attract or "lure" prey within range of capture. The classic example is the anglerfish using light produced in the esca, dangling near their gaping mouth, ready to capture unsuspecting prey lured toward the light.

sensitivity well. Loosejaws from the genus *Aristostomias* also have novel red-sensitive rhodopsins that, when bound to a 3,4-dehydroretinal, are maximally sensitive to red light; their visual pigment repertoire is sensitive to 531, 548, and 588 nm (Douglas, Partridge, & Marshall, 1998). The red-sensitive chemistry in the genus *Malacosteus* is different, utilizing a chlorophyll photosensitizer, which is maximally sensitive to approximately 671 nm (Douglas, Partridge, Dulai, et al., 1998). The red visual pigments in loosejaw eyes are presumably tuned to detect the red light produced by their photophores (Kenaley et al., 2014; Partridge & Douglas, 1995).

Some bioluminescent predators also use bioluminescent lures to attract prey (Fig. 27.12). A classic example of bioluminescence used aggressively for prey capture is observed in the anglerfish. Anglerfish dangle their bioluminescent organ, the esca, near their mouth. Prey organisms are attracted to the high intensity light produced from the small esca and are drawn to within striking distance of the anglerfish. The Australian glow worm, *Arachnocampa flava* (order Diptera), also employs a "sit and wait" predator technique. The glow worm lowers sticky mucous tubes that luminesce from a posterior light organ to attract and ensnare phototactic prey. The energetic cost of bioluminescence is comparable with the resting state. Once the web is built a glow worm can attract a meal with little additional metabolic investment (Willis et al., 2011).

3.2 Defensive bioluminescence

Defensive strategies for bioluminescence include use as diversions, camouflage through counter-illumination, aposematic signaling, and burglar alarms. Bioluminescence can be used to startle or distract potential predators through the release of a bright, luminescent cloud (Fig. 27.13). This type of defense is analogous to cephalopods (e.g., squid) releasing ink to avoid predation (Wood et al., 2010). In both cases, a visual predator's sightline is disrupted facilitating the prospective prey's escape. Many species of shrimp (oplophorids, acanthephyrids, lophogastrids, pandalids) secrete a bright blue

FIGURE 27.13 Diversion using a "smoke screen" is employed by many bioluminescent shrimp. When startled, the shrimp secretes a cloud of bright blue light that impairs the predators' vision and facilitates its escape.

bioluminescent cloud from their mouth to deter predators, and the secretion is viscous and can remain luminous for many minutes (Herring, 1976). The "smoke-screen" may also deter predators through interference with multiple sensory systems (e.g., olfaction and gustation). Other groups that are able to emit luminous clouds include ctenophores, ostracods, amphipods, copepods, squid, and fish (see Widder, 1999 for review).

Another defensive strategy utilizing bioluminescence is aposematic signaling. Aposematic signaling offers a warning to potential predators, with an organism signaling the unprofitability or unpalatabillity of consuming them (Fig. 27.14). Bioluminescent aposematic signaling is found in both marine and terrestrial species. The lanternshark, *Etmopterus spinax,* uses bioluminescence to illuminate its dorsal defensive spines, which are partially translucent, as a defensive display discouraging piscivorous fishes from risking punctures (Claes et al., 2013). Millipedes in the genus *Motxyia* are nocturnal detritivores that produce bioluminescent aposematic signals. *Motxyia* naturally produce hydrogen cyanide, which is unpalatable and toxic to potential predators. Using live and clay-model nonluminescent millipedes, Marek et al. (2011) demonstrated that nonluminescent millipedes were attacked up to four times more often than luminescent *Motxyia*, supporting the aposematic hypothesis.

Using light to camouflage oneself may at first seem counter-intuitive, however, even in the deep-sea, organisms' bodies block faint down-welling light from the sun. The result is a silhouette or shadow visible to predators below. Many marine organisms, from crustaceans and cephalopods to fish, use bioluminescent organs on the ventral side of their body to produce light matching the downwelling irradiance, effectively camouflaging their silhouette (Fig. 27.15). The camouflage is achieved through one of three mechanisms: disruptive camouflage, countershading, and counter-illumination. Disruptive camouflage makes the shadow cast by an individual unrecognizable to predators below (Fig. 27.15B; pandas, orcas, and leopards offer good examples of disruptive coloration). Countershading in the marine environment involves having a dark back and a light belly. The lighter belly color blends in better with the brighter downwelling light, while the darker back is needed to blend in with the dimmer upwelling light. Counter-illumination is accomplished by precisely tuning light to

FIGURE 27.14 Bioluminescence may be used as an aposematic signal. An organism communicates to potential predators that they are either distasteful or dangerous. Recently it was discovered that lanternsharks (Etmopterus spinax) illuminate their sharp dorsal spines, which deters piscivorous fishes (Claes et al., 2013).

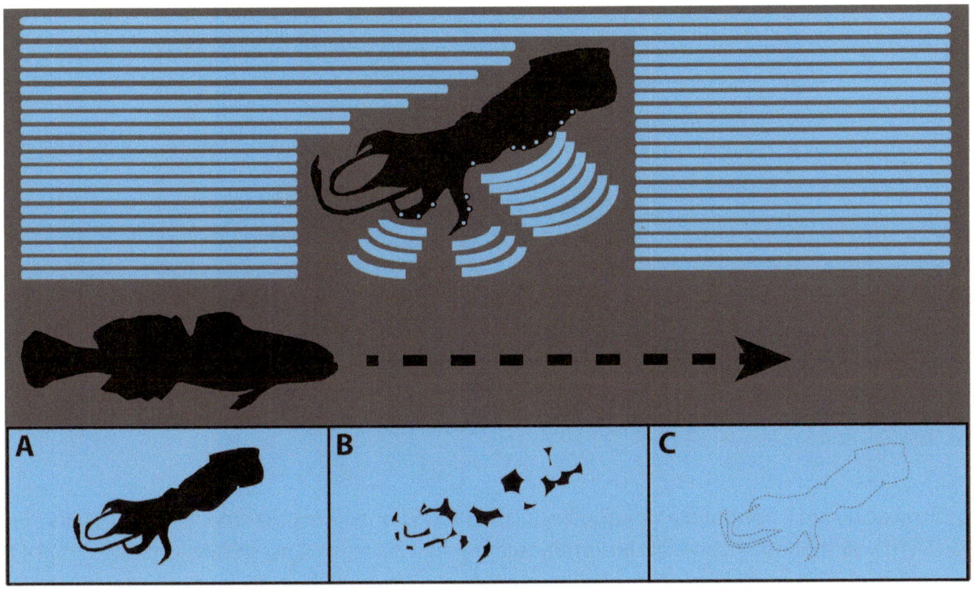

FIGURE 27.15 Counter-illumination is a common mechanism for camouflage in the deep-sea. Ventral photophores can be tuned to match faint, downwelling sunlight (horizontal blue lines) to camouflage the shadow cast by the organism's opaque body. Panels (A–C): predator's view from below prey given three scenarios; (A) no bioluminescence, conspicuous shadow; (B) bioluminescence used for disruptive camouflage, making the silhouette unrecognizable to the predator below; (C) bioluminescence used for precise background matching, with no conspicuous shadow visible to the predator.

match downwelling irradiance to replace the light blocked by the body (Fig. 27.15C). Intensity, color, and angular distribution are required for high-fidelity counter-illumination. Sergestid shrimp tune the light intensity produced by their ventral photophores to exactly match the ambient downwelling light (Warner et al., 1979). Most species cannot adjust the color of their bioluminescence, with the diel migrating squid *Abraliopsis* sp. and *Abralia trigonura* being some rare exceptions capable of adjusting the color of bioluminescence in response to changes in water temperature with depth. The temperatures and color correspond well to predicted downwelling light and depth (e.g., cold water and deeper depths during the day, warm water and shallower depths at night as they migrate up in the water column) (Young & Mencher, 1980). Lanternshark, *E. spinax,* in addition to the aforementioned aposematic signal produced on their dorsal fins, have ventral photophores that are used for counter-illumination (Claes et al., 2010). There is a paucity of empirical behavioral evidence for counter illumination except in the shrimp *Sergestes similis* (Warner et al., 1979), due to the difficulties of collecting in situ behavioral observations, or collecting live specimens in good enough condition to conduct laboratory experiments. Counter-illumination may perfectly match downwelling light making an organism invisible to predators below, or the ventral photophores may function in a disruptive manner, breaking up the silhouette into small, unrecognizable fragments. Johnsen et al. (2004) modelled vision of predators with varying spatial acuity and found that even predators with low spatial acuity could see squid and fish photophores at 1 meter; counter-illumination only provided effective camouflage at 8 meters or greater.

Bioluminescence can function like a burglar alarm. When a predator disturbs its prey, the resultant flash by prospective prey can alert larger secondary predators to eliminate the threat (Fig. 27.16). Bioluminescent dinoflagellates utilize this burglar alarm when being preyed upon by mysid shrimp or copepods. Laboratory experiments have demonstrated significantly reduced grazing by copepods on day-phase, nonluminescent dinoflagellates versus night-phase, luminescent dinoflagellates (Esaias & Curl, 1972; White, 1979). Midshipman fish, alerted to the presence of the mysid shrimp by dinoflagellate luminescence, have significantly increased predation rates on mysids when the density of dinoflagellates is higher (Mesinger & Case, 1992). The improved secondary predation rates on mysids foraging on bioluminescent dinoflagellates were also demonstrated with the cephalopods, *Sepia officinalis* and *Euprymna scolopes* (Fleisher & Case, 1995). Another example of a bioluminescent burglar alarm is found in the deep-sea sea cucumber, *Enypniastes eximia,* which is a particularly fragile benthic-grazer. Direct observations using submersibles and remotely operated vehicles indicated that, when physically disturbed by a predator, the skin of *E. eximia* sloughs off and creates a glowing cloud that acts as a burglar alarm (Robison, 1992).

In many cases, despite bioluminescent defensive strategies, a predator successfully captures its bioluminescent prey. Although empirical evidence is scant, it is predicted some prey species have evolved a strategy for this situation; they

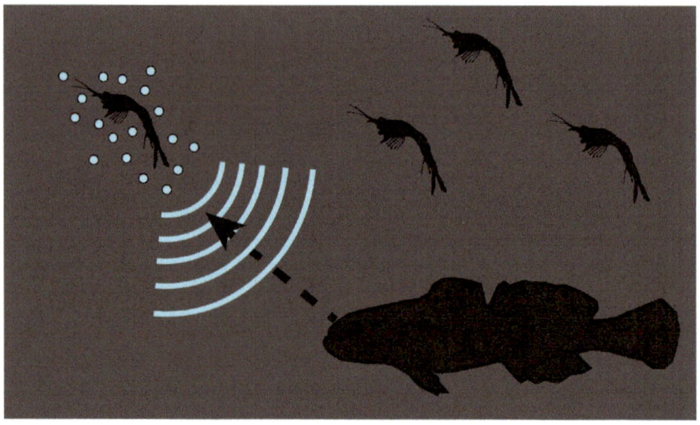

FIGURE 27.16 Small bioluminescent organisms (e.g., dinoflagellates) can set a "burglar alarm" on their primary predator (e.g., mysid shrimp) that can attract secondary predators to end the threat.

autotomize a limb or scale that continues to luminesce, distracting the predator and facilitating the escape of the prey (Figs. 27.17 and 27.18). Brittle stars produce bioluminescence that functions as an aposematic signal (Jones & Mallefet, 2013). However, Basch (1988) postulates that when captured by a crab, brittle stars autotomize the captured limb, which then produces bright bioluminescent flashes (Herring, 1995). The polychaete worm, *Swima bombiviridis* (Acrocirridae), can release autotomized glowing "bombs" and scaleworms (Polynoidae) release autotomized glowing scales (Osborn et al., 2009). Deep-sea scyphozoans (*Atolla parva* and *Periphylla periphylla*) release secretory bioluminescent displays, but it is not known whether they are autotomized lures or used to startle visual predators (Herring & Widder, 2004).

3.3 Communication and sexual selection

In some species, bioluminescence is used for congener recognition and courtship displays. As discussed above, the most studied system is the firefly bioluminescent mating display. Firefly bioluminescence varies from blue-green to yellow-green and the color is species specific. The visual sensitivity of fireflies corresponds well with the spectra of bioluminescence they produce (Lall et al., 1980). The wavelength and periodicity of flashes are used by females to attract conspecific males for mating. These signals, however, can also be exploited for aggressive means. *Photuris* females mimic the bioluminescent signals of female *Photinus* and capture and consume male *Photinus* who arrive to mate (Lloyd, 1965).

The scyphozoan, *Periphylla periphylla*, possesses a large repertoire of bioluminescent displays. Recent behavioral observations found that *P. periphylla* form ephemeral social groups. Individuals swim laterally and once directly below a conspecific, change trajectory and swim vertically at 5–10 times normal speed toward the conspecific (Kaartvedt et al., 2015). The swim pattern of an approaching individual did not follow the path of the conspecific, suggesting the approaching individuals are not following a scent trail. Instead, Kaartvedt et al. (2015) postulate *P. periphylla* are using visual cues, including bioluminescent signals, which is remarkable given the simple visual system many jellyfish possess.

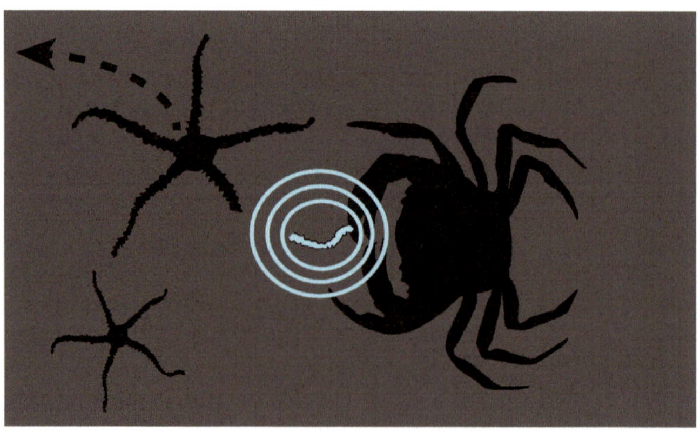

FIGURE 27.17 Some animals, such as the brittle star, upon capture, sacrifice a limb to distract predators in order to facilitate escape.

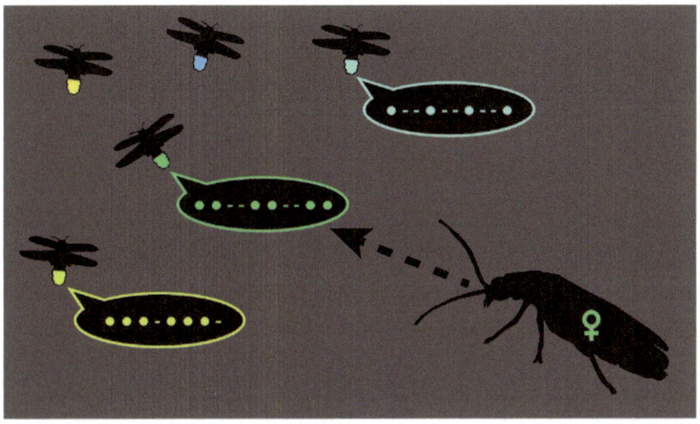

FIGURE 27.18 Fireflies are capable of emitting light ranging from blue-green to yellow-green, depending on the species. Males will display their color and flash pattern to females during courtship displays and, if the female responds with the correct flash pattern, will land near the females to mate. Hypothetical flash patterns are depicted in this cartoon as Morse code: dots = bioluminescent flashes; dashes = pause or no flash.

Due to logistical constraints, there is a paucity of direct observations of marine organisms using bioluminescence for congener recognition or sexual selection. One of the best studied systems is that used by shallow water ostracods. Myodocopid ostracod males (*Vargula annecohenae*) exhibit intricate bioluminescent displays that are characterized by distinct phases. The initial phase involves three bright pulses as the male swims at intermediate speeds (~5 cm/s) in a loop, likely used to draw females' attention. The second phase involves the male rapidly swimming up (~10 cm/s) in a helical pattern, releasing about a dozen evenly placed luminescent secretions that serve as a luminescent pathway for approaching females (Rivers & Morin, 2008). Other taxa have well described bioluminescent systems but lack direct observational evidence of their luminescent displays. All deep-sea shrimp from the family Oplophoridae possess a secretory bioluminescence (mentioned briefly above). Several species within the family also possess light-emitting photophores that produce different spectra than the secretion (Herring, 1983). All oplophorids that have photophores have a unique visual system, with sensitivity to both blue and UV light (Frank & Case, 1988), due to the presence of a long wavelength and short wavelength opsin (Wong et al., 2015). It is predicted that the additional UV sensitivity peak provides a "dual-sensitivity system" permitting the discrimination between the two forms of bioluminescence (Cronin & Frank, 1996).

4. Phototransduction-mediated bioluminescence

Recent transcriptome evidence from a taxonomically broad range of species suggests both autogenic and symbiotic bioluminescence may involve phototransduction in the regulation of light output. Hawaiian bobtail squid (*Euprymna scolopes*) bioluminescent organs, which contain bioluminescent symbiotic bacteria, may be capable of light detection. The organs contain proteins for phototransduction, including rhodopsin, G-proteins expressed in eyes, rhodopsin kinase, and visual arrestin (Tong et al., 2009). Four genes critical for eye development, pax6, eya, six, and dac, are expressed in both the eye and light organ of *E. scolopes* (Peyer et al., 2014). In the light organ, gene expression depends on cues provided by bacterial symbionts, *Vibrio fischeri*. Regulation of pax6 and eya transcripts was affected by *V. fischeri* cues. Down regulation of the six transcript only occurred in response to wild type *V. fischeri*, but not nonluminescent *V. fischeri*, suggesting that the six transcript responded to bacterial luminescence. The ctenophore, *Mnemiopsis leidyi*, has opsin and calcium regulated photoprotein genes in its bioluminescent photocytes, suggesting that photocytes might have both light sensing and light producing functions (Schnitzler et al., 2012). Lanternshark, *Etmopterus spinax*, photophores express genes involved in photoemission as well as phototransduction. Encephalopsin, an extraocular opsin, was identified in the lanternshark photogenic ventral tissue transcriptome (Duchatelet, Sugihara, et al., 2020). Deep-sea shrimp from the family Sergestidae, briefly mentioned above, can match the intensity of down-welling sunlight remarkably well (Warner et al., 1979). Recent evidence suggests that the photophores of several species in the family Oplophoridae are photosensitive (Bracken-Grissom et al., 2020), suggesting that photosensitivity within the photophores of luminescent organisms may be the long sought-after mechanism for how this finely tuned counter-illumination is achieved.

Phototransduction in nonimage forming organs is not limited to bioluminescent organisms, and can be involved in similar camouflage strategies. Opsin transcripts and proteins were found in the eyes and fins and ventral skin samples of squid and cuttlefish, which actively use visual displays and camouflage using pigmented chromatophores (Kingston et al., 2015; Mäthger et al., 2010). Ramirez and Oakley (2015) demonstrated that isolated octopus skin responded to light at wavelengths corresponding to their opsin maximum sensitivity. Furthermore, opsins present in the transcriptome of flatfish skin tissue may mediate hormonal changes that regulate metamorphosis (Shao et al., 2017). Phototransduction-mediated regulation of bioluminescence may have coopted other nonvisual hormone-mediating phototransduction pathways.

5. Broader evolutionary implications

Exciting new research has revealed that some bioluminescent organisms can acquire luciferase through their diet by way of a "kleptoprotein" process (Bessho-Uehara et al., 2020). *Parapriacanthus ransonneti*, a bioluminescent fish named the golden sweeper, acquires its luciferase from its ostracod prey, and this kleptoprotein process offers a novel pathway for the acquisition of complex functional traits without encoding the protein genetically. Bioluminescence may also contribute to the rapid rates of evolutionary novelty. Clades that use luminescence for communication are significantly more species-rich than lineages that do not use luminescence for communication. Bioluminescent teleost fish lineages, particularly those inhabiting the deep-sea that use luminescence for communication, exhibit remarkable species richness relative to clade age (Davis et al., 2016). The relationship between bioluminescence and species richness is not limited to fishes. Ellis and Oakley (2016) found the phenomenon is taxonomically broad (e.g., Mollusca, Crustacea, Arthropoda, Annelida, Chordata) (Fig. 27.19). Furthermore, lineages using bioluminescence for communication and courtship have higher speciation rates than sister lineages using luminescence for camouflage only (Ellis & Oakley, 2016). Although the substrate need not be bioluminescence, e.g., changes in ornamentation (luminescent or otherwise) are associated with increased species richness (Gomes et al., 2016), bioluminescence appears to provide an ideal substrate from which sexual selection can act upon slight deviations. This chapter describes the breadth and diversity of functions, taxa, and molecules involved in bioluminescence. Based on current studies, bioluminescence has independently evolved more than 94 times, demonstrating that biological light production is a common evolutionary solution to the selection pressures of life in the dark. Furthermore, bioluminescence is a prevalent ecological phenomenon, particularly in the open ocean in which more than 75% of all organisms are capable of producing light (Martini & Haddock, 2017). The physiological basis of bioluminescence is labile and natural selection may capitalize on slight modifications in the pathways producing luciferins and luciferases and accelerating speciation.

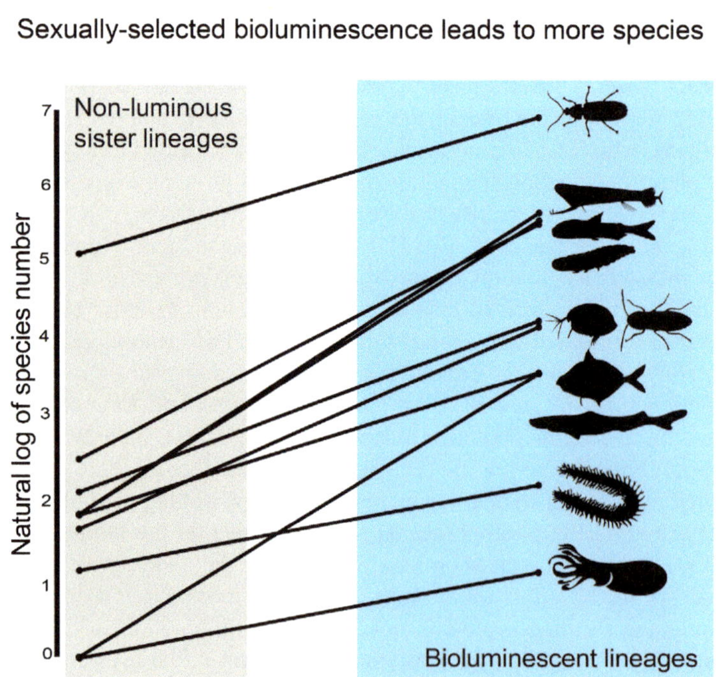

FIGURE 27.19 Lineages with bioluminescent courtship are associated with high species richness relative to nonluminous sister lineages and clades with bioluminescence for camouflage only. *Figure reproduced with permission from Ellis and Oakley (2016).*

References

Airth, R. L., & McElroy, W. D. (1959). Light emission from extracts of luminous fungi. *Journal of Bacteriology, 77*(2), 249.

Akhurst, R. J. (1982). Antibiotic activity of Xenorhabdus spp., bacteria symbiotically associated with insect pathogenic nematodes of the families Heterorhabditidae and Steinernematidae. *Microbiology, 128*(12), 3061–3065.

Basch, L. V. (1988). Bioluminescent anti-predator defense in a subtidal ophiuroid. In R. D. Burke, P. V. Mladenov, P. Lambert, & R. L. Parsley (Eds.), *Echinoderm biology* (pp. 565–573). Rotterdam: A.A. Balkema.

Bessho-Uehara, M., Yamamoto, N., Shigenobu, S., Mori, H., Kuwata, K., & Oba, Y. (2020). Kleptoprotein bioluminescence: Parapriacanthus fish obtain luciferase from ostracod prey. *Science Advances, 6*(2). https://doi.org/10.1126/sciadv.aax4942

Bitler, B., & McElroy, W. D. (1957). The preparation and properties of crystalline firefly luciferin. *Archives of Biochemistry and Biophysics, 72*(2), 358–368.

Bracken-Grissom, H. D., Deleo, D. M., Porter, M. L., Iwanicki, T., Sickles, J., & Frank, T. M. (2020). Light organ photosensitivity in deep-sea shrimp may suggest a novel role in counterillumination. *Scientific Reports*, 1–10. https://doi.org/10.1038/s41598-020-61284-9

Branchini, B. R., Behney, C. E., Southworth, T. L., Fontaine, D. M., Gulick, A. M., Vinyard, D. J., & Brudvig, G. W. (2015). Experimental support for a single electron-transfer oxidation mechanism in firefly bioluminescence. *Journal of the American Chemical Society, 137*(24), 7592–7595.

Buskey, E. J., & Stearns, D. E. (1991). The effects of starvation on bioluminescence potential and egg release of the copepod Metridia longa. *Journal of Plankton Research, 13*(4), 885–893. https://doi.org/10.1093/plankt/13.4.885

Charbonneau, H., Walsh, K. A., McCann, R. O., Prendergast, F. G., Cormier, M. J., & Vanaman, T. C. (1985). Amino acid sequence of the calcium-dependent photoprotein aequorin. *Biochemistry, 24*(24), 6762–6771.

Claes, J. M., Aksnes, D. L., & Mallefet, J. (2010). Phantom hunter of the fjords: Camouflage by counterillumination in a shark (Etmopterus spinax). *Journal of Experimental Marine Biology and Ecology, 388*(1), 28–32.

Claes, J. M., Dean, M. N., Nilsson, D. E., Hart, N. S., & Mallefet, J. (2013). A deepwater fish with 'lightsabers'—Dorsal spine-associated luminescence in a counterilluminating lanternshark. *Scientific Reports, 3*, 1308.

Claes, J. M., & Mallefet, J. (2009). Hormonal control of luminescence from lantern shark (Etmopterus spinax) photophores. *Journal of Experimental Biology, 212*(22), 3684–3692.

Cormier, M. J., & Strehler, B. L. (1953). The identification of KCF: Requirement of long-chain aldehydes for bacterial extract luminescence. *Journal of the American Chemical Society, 75*(19), 4864–4865.

Cronin, T. W., & Frank, T. M. (1996). A short-wavelength photoreceptor class in a deep-sea shrimp. *Proceedings of the Royal Society of London B Biological Sciences, 263*(1372), 861–865.

Davis, M. P., Sparks, J. S., & Smith, W. L. (2016). Repeated and widespread evolution of bioluminescence in marine fishes. Edited by E.V. Thuesen. *PLOS One, 11*(6), e0155154.

Delroisse, J., Ullrich-Lüter, E., Blaue, S., Ortega-Martinez, O., Eeckhaut, I., Flammang, P., & Mallefet, J. (2017). A puzzling homology: A brittle star using a putative cnidarian-type luciferase for bioluminescence. *Open Biology, 7*(4). https://doi.org/10.1098/rsob.160300

Ding, B. W., & Liu, Y. J. (2017). Bioluminescence of firefly squid via mechanism of single electron-transfer oxygenation and charge-transfer-induced luminescence. *Journal of the American Chemical Society, 139*(3), 1106–1119.

Douglas, R. H., Partridge, J. C., Dulai, K., Hunt, D., Mullineaux, C. W., Tauber, A. Y., & Hynninen, P. H. (1998). Dragon fish see using chlorophyll. *Nature, 393*(6684), 423–424.

Douglas, R. H., Partridge, J. C., & Marshall, N. J. (1998). The eyes of deep-sea fish I: Lens pigmentation, tapeta and visual pigments. *Progress in Retinal and Eye Research, 17*(4), 597–636.

Duchatelet, L., Delroisse, J., & Mallefet, J. (2020). Bioluminescence in lanternsharks: Insight from hormone receptor localization. *General and Comparative Endocrinology, 294*, Article 113488. https://doi.org/10.1016/j.ygcen.2020.113488

Duchatelet, L., Sugihara, T., Delroisse, J., Koyanagi, M., Rezsohazy, R., Terakita, A., & Mallefet, J. (2020). From extraocular photoreception to pigment movement regulation: A new control mechanism of the lanternshark luminescence. *Scientific Reports, 10*(1), 10195. https://doi.org/10.1038/s41598-020-67287-w

Dunlap, J. C., Hastings, J. W., & Shimomura, O. (1980). Crossreactivity between the light-emitting systems of distantly related organisms: Novel type of light-emitting compound. *Proceedings of the National Academy of Sciences, 77*(3), 1394–1397.

Ellis, E. A., & Oakley, T. H. (2016). High rates of species accumulation in animals with bioluminescent courtship displays. *Current Biology, 26*(14), 1916–1921.

Engebrecht, J., Nealson, K., & Silverman, M. (1983). Bacterial bioluminescence: Isolation and genetic analysis of functions from Vibrio fischeri. *Cell, 32*(3), 773–781.

Esaias, W. E., & Curl, H. C. (1972). Effect of dinoflagellate bioluminescence on copepod ingestion rates. *Limnology & Oceanography, 17*(6), 901–906.

Fleisher, K. J., & Case, J. F. (1995). Cephalopod predation facilitated by dinoflagellate luminescence. *The Biological Bulletin, 189*(3), 263–271.

Frackman, S., Anhalt, M., & Nealson, K. H. (1990). Cloning, organization, and expression of the bioluminescence genes of Xenorhabdus luminescens. *Journal of Bacteriology, 172*(10), 5767–5773.

Fraga, H., Esteves da Silva, J. C., & Fontes, R. (2004). Identification of luciferyl adenylate and luciferyl coenzyme a synthesized by firefly luciferase. *ChemBioChem, 5*(1), 110–115.

Francis, W. R., Powers, M. L., & Haddock, S. H. (2016). Bioluminescence spectra from three deep-sea polychaete worms. *Marine Biology, 163*(12), 255.

Francis, W. R., Powers, M. L., & Haddock, S. H. (2014). Characterization of an anthraquinone fluor from the bioluminescent, pelagic polychaete Tomopteris. *Luminescence, 29*(8), 1135–1140.

Frank, T. M., & Case, J. F. (1988). Visual spectral sensitivities of bioluminescent deep-sea crustaceans. *The Biological Bulletin, 175*(2), 261–273.

Frank, T. M., Widder, E. A., Latz, M. I., & Case, J. F. (1984). Dietary maintenance of bioluminescence in a deep-sea mysid. *Journal of Experimental Biology, 109*(1), 385–389.

Gomes, A. C. R., Sorenson, M. D., & Cardoso, G. C. (2016). Speciation is associated with changing ornamentation rather than stronger sexual selection. *Evolution, 70*(12), 2823–2838.

Gouveneaux, A., Flood, P. R., & Mallefet, J. (2016). Unexpected diversity of bioluminescence in planktonic worms. *Luminescence, 32*(3), 394–400.

Gouveneaux, A., & Mallefet, J. (2013). Physiological control of bioluminescence in a deep-sea planktonic worm, Tomopteris helgolandica. *Journal of Experimental Biology, 216*(22), 4285–4289.

Haddock, S. H. D., & Case, J. F. (1999). Bioluminescence spectra of shallow and deep-sea gelatinous zooplankton: Ctenophores, medusae and siphonophores. *Marine Biology, 133*(3), 571–582.

Haddock, S. H., Moline, M. A., & Case, J. F. (2010). Bioluminescence in the sea. *Annual Review of Marine Science, 2*, 443–493.

Hastings, J. W., & Davenport, D. (1957). The luminescence of the millipede, Luminodesmus sequoiae. *The Biological Bulletin, 113*(1), 120–128.

Head, J. F., Inouye, S., Teranishi, K., & Shimomura, O. (2000). The crystal structure of the photoprotein aequorin at 2.3 Å resolution. *Nature, 405*(6784), 372–376.

Herring, P. J. (1976). Bioluminescence in decapod Crustacea. *Journal of the Marine Biological Association of the United Kingdom, 56*(4), 1029–1047.

Herring, P. J. (1995). Bioluminescent echinoderms: Unity of function in diversity of expression. *Echinoderm Research*, 1–17.

Herring, P. J. (1983). The spectral characteristics of luminous marine organisms. *Proceedings of the Royal Society of London B Biological Sciences, 220*(1219), 183–217.

Herring, P. J., & Cope, C. (2005). Red bioluminescence in fishes: On the suborbital photophores of Malacosteus, Pachystomias and Aristostomias. *Marine Biology, 148*, 383–394.

Herring, P. J., & Widder, E. A. (2004). Bioluminescence of deep-sea coronate medusae (Cnidaria: Scyphozoa). *Marine Biology, 146*(1), 39–51.

Inoue, S., Kakoi, H., Murata, M., Goto, T., & Shimomura, O. (1977). Complete structure of Renilla luciferin and luciferyl sulfate. *Tetrahedron Letters, 18*(31), 2685–2688.

Inoue, S., Sugiura, S., Kakoi, H., Hasizume, K., Goto, T., & Iio, H. (1975). Squid bioluminescence II. Isolation from Watasenia scintillans and synthesis of 2-(p-hydroxybenzyl)-6-(p-hydroxyphenyl)-3, 7-dihydroimidazo [1, 2-a] pyrazin-3-one. *Chemistry Letters, 4*(2), 141–144.

Inouye, S., Watanabe, K., Nakamura, H., & Shimomura, O. (2000). Secretional luciferase of the luminous shrimp Oplophorus gracilirostris: cDNA cloning of a novel imidazopyrazinone luciferase. *FEBS Letters, 481*(1), 19–25.

Janouškovec, J., Gavelis, G. S., Burki, F., Dinh, D., Bachvaroff, T. R., Gornik, S. G., Bright, K. J., Imanian, B., Strom, S. L., Delwiche, C. F., & Waller, R. F. (2017). Major transitions in dinoflagellate evolution unveiled by phylotranscriptomics. *Proceedings of the National Academy of Sciences, 114*(2), E171–E180.

Jin, K., Klima, J. C., Deane, G., Dale Stokes, M., & Latz, M. I. (2013). Pharmacological investigation of the bioluminescence signaling pathway of the dinoflagellate Lingulodinium polyedrum: Evidence for the role of stretch-activated ion channels. *Journal of Phycology, 49*(4), 733–745.

Johnsen, S., Widder, E. A., & Mobley, C. D. (2004). Propagation and perception of bioluminescence: Factors affecting counterillumination as a cryptic strategy. *The Biological Bulletin, 207*(1), 1–16.

Jones, A., & Mallefet, J. (2013). Why do brittle stars emit light? Behavioural and evolutionary approaches of bioluminescence. *Cahiers de Biologie Marine, 54*, 729–734.

Kaartvedt, S., Ugland, K. I., Klevjer, T. A., Røstad, A., Titelman, J., & Solberg, I. (2015). Social behaviour in mesopelagic jellyfish. *Scientific Reports, 5*.

Kanie, S., Miura, D., Jimi, N., Hayashi, T., Nakamura, K., Sakata, M., Ogoh, K., Ohmiya, Y., & Mitani, Y. (2021). Violet bioluminescent Polycirrus sp. (Annelida: Terebelliformia) discovered in the shallow coastal waters of the Noto Peninsula in Japan. *Scientific Reports, 11*(1), 1–9. https://doi.org/10.1038/s41598-021-98105-6

Kenaley, C. P., DeVaney, S. C., & Fjeran, T. T. (2014). The complex evolutionary history of seeing red: Molecular phylogeny and the evolution of an adaptive visual system in deep-sea dragonfishes (Stomiiformes: Stomiidae). *Evolution, 68*(4), 996–1013.

Kingston, A. C., Kuzirian, A. M., Hanlon, R. T., & Cronin, T. W. (2015). Visual phototransduction components in cephalopod chromatophores suggest dermal photoreception. *Journal of Experimental Biology, 218*(10), 1596–1602.

Kishi, T., Goto, T., Hirata, Y., Shimomura, O., & Johnson, F. H. (1966). Cypridina bioluminescence I Sructure of cypridina luciferin. *Tetrahedron Letters, 7*(29), 3427–3436.

Krönström, J., Karlsson, W., Johansson, B. R., & Holmgren, S. (2009). Involvement of contractile elements in control of bioluminescence in Northern krill, Meganyctiphanes norvegica (M. Sars). *Cell and Tissue Research, 336*(2), 299.

Kuse, M., Kanakubo, A., Suwan, S., Koga, K., Isobe, M., & Shimomura, O. (2001). 7, 8-Dihydropterin-6-carboxylic acid as light emitter of luminous millipede, Luminodesmus sequoiae. *Bioorganic & Medicinal Chemistry Letters, 11*(8), 1037–1040.

Lall, A. B., Seliger, H. H., Biggley, W. H., & Lloyd, J. E. (1980). Ecology of colors of firefly bioluminescence. *Science, 210*(4469), 560–562.

Latz, M. I., Frank, T. M., & Case, J. F. (1988). Spectral composition of bioluminescence of epipelagic organisms from the Sargasso Sea. *Marine Biology, 98*(3), 441–446.

Lau, E. S., & Oakley, T. H. (2021). Multi-level convergence of complex traits and the evolution of bioluminescence. *Biological Reviews, 96*(2), 673–691. https://doi.org/10.1111/brv.12672

Lee, K. H., & Ruby, E. G. (1994). Effect of the squid host on the abundance and distribution of symbiotic Vibrio fischeri in nature. *Applied and Environmental Microbiology, 60*(5), 1565–1571.

Lloyd, J. E. (1965). Aggressive mimicry in Photuris: Firefly femmes fatales. *Science, 149*(3684), 653–654.

Lorenz, W. W., McCann, R. O., Longiaru, M., & Cormier, M. J. (1991). Isolation and expression of a cDNA encoding Renilla reniformis luciferase. *Proceedings of the National Academy of Sciences, 88*(10), 4438–4442.

Mäthger, L. M., Roberts, S. B., & Hanlon, R. T. (2010). Evidence for distributed light sensing in the skin of cuttlefish, Sepia officinalis. *Biology Letters, 6*(5), 600–603.

Macheroux, P., Schmidt, K. U., Steinerstauch, P., Ghisla, S., Colepicolo, P., Buntic, R., & Hastings, J. W. (1987). Purification of the yellow fluorescent protein from Vibrio fischeri and identity of the flavin chromophore. *Biochemical and Biophysical Research Communications, 146*(1), 101–106.

Mallefet, J., Duchatelet, L., & Coubris, C. (2020). Bioluminescence induction in the ophiuroid Amphiura filiformis (Echinodermata). *Journal of Experimental Biology, 223*(4). https://doi.org/10.1242/jeb.218719

Mallefet, J., & Shimomura, O. (1995). Presence of coelenterazine in mesopelagic fishes from the Strait of Messina. *Marine Biology, 124*(3), 381–385.

Mallefet, J., Stevens, D. W., & Duchatelet, L. (2021). Bioluminescence of the largest luminous vertebrate, the kitefin shark, Dalatias licha: First insights and comparative aspects. *Frontiers in Marine Science, 8*(633582), 1–13. https://doi.org/10.3389/fmars.2021.633582

Marek, P., Papaj, D., Yeager, J., Molina, S., & Moore, W. (2011). Bioluminescent aposematism in millipedes. *Current Biology, 21*(18), R680–R681.

Martini, S., & Haddock, S. H. (2017). Quantification of bioluminescence from the surface to the deep sea demonstrates its predominance as an ecological trait. *Scientific Reports, 7*.

Martini, S., Kuhnz, L., Mallefet, J., & Haddock, S. H. D. (2019). *Distribution and quantification of bioluminescence as an ecological trait in the deep sea benthos* (Submitted).

Martini, S., Schultz, D. T., Lundsten, L., & Haddock, S. H. D. (2020). Bioluminescence in an undescribed species of carnivorous sponge (Cladorhizidae) from the deep sea. *Science, 7*(December), 1–10. https://doi.org/10.3389/fmars.2020.576476

McElroy, W. D., & DeLuca, M. (1978). Chemistry of firefly luminescence. *Bioluminescence in Action, 109*.

McElroy, W. D., & Green, A. A. (1955). Enzymatic properties of bacterial luciferase. *Archives of Biochemistry and Biophysics, 56*(1), 240–255.

McFall-Ngai, M., Heath-Heckman, E. A., Gillette, A. A., Peyer, S. M., & Harvie, E. A. (2012). The secret languages of coevolved symbioses: Insights from the Euprymna scolopes–Vibrio fischeri symbiosis. In *Seminars in immunology* (Vol. 24(1), pp. 3–8). Academic Press.

Mesinger, A. F., & Case, J. F. (1992). Dinoflagellate luminescence increases susceptibility of zooplankton to teleost predation. *Marine Biology, 112*(2), 207–210.

Miyashiro, T., & Ruby, E. G. (2012). Shedding light on bioluminescence regulation in Vibrio fischeri. *Molecular Microbiology, 84*(5), 795–806.

Mogilnaya, O. A., Ronzhin, N. O., & Bondar, V. S. (2016). Comparative evaluation of total peroxidase and catalase activities during light emission of luminous fungus Neonothopanus nambi. *Mycosphere, 7*(4), 499–510.

Nakamura, H., Kishi, Y., Shimomura, O., Morse, D., & Hastings, J. W. (1989). Structure of dinoflagellate luciferin and its enzymatic and nonenzymatic air-oxidation products. *Journal of the American Chemical Society, 111*(19), 7607–7611.

Nealson, K. H. (1977). Autoinduction of bacterial luciferase. *Archives of Microbiology, 112*(1), 73–79.

Nealson, K. H., & Hastings, J. W. (2006). Quorum sensing on a global scale: Massive numbers of bioluminescent bacteria make milky seas. *Applied and Environmental Microbiology, 72*(4), 2295–2297.

Oba, Y., Kato, S. ichi, Ojika, M., & Inouye, S. (2009). Biosynthesis of coelenterazine in the deep-sea copepod, Metridia pacifica. *Biochemical and Biophysical Research Communications, 390*(3), 684–688. https://doi.org/10.1016/j.bbrc.2009.10.028

Ohtsuka, H., Rudie, N. G., & Wampler, J. E. (1976). Structural identification and synthesis of luciferin from the bioluminescent earthworm, Diplocardia longa. *Biochemistry, 15*(5), 1001–1004.

Oliveira, A. G., Amaral, D. T., Hannon, M. C., & Schulze, A. (2021). First record of bioluminescence in a Sipunculan worm. *Frontiers in Marine Science, 8*, Article 762706. https://doi.org/10.3389/fmars.2021.762706

Osborn, K. J., Haddock, S. H., Pleijel, F., Madin, L. P., & Rouse, G. W. (2009). Deep-sea, swimming worms with luminescent "Bombs". *Science, 325*(5943), 964–964.

Pankey, M. S., Minin, V. N., Imholte, G. C., Suchard, M. A., & Oakley, T. H. (2014). Predictable transcriptome evolution in the convergent and complex bioluminescent organs of squid. *Proceedings of the National Academy of Sciences, 111*(44), E4736–E4742.

Partridge, J. C., & Douglas, R. H. (1995). Far-red sensitivity of dragon fish. *Nature, 375*, 21–22.

Petushkov, V. N., Ketelaars, M., Gibson, B. G., & Lee, J. (1996). Interaction of Photobacterium leiognathi and Vibrio fischeri Y1 luciferases with fluorescent (antenna) proteins: Bioluminescence effects of the aliphatic additive. *Biochemistry, 35*(37), 12086–12093.

Petushkov, V. N., Tsarkova, A. S., Dubinnyi, M. A., Rodionova, N. S., Marques, S. M., da Silva, J. C. E., Shimomura, O., & Yampolsky, I. V. (2014). CompX, a luciferin-related tyrosine derivative from the bioluminescent earthworm Fridericia heliota. Structure elucidation and total synthesis. *Tetrahedron Letters, 55*(2), 460–462.

Peyer, S. M., Pankey, M. S., Oakley, T. H., & McFall-Ngai, M. J. (2014). Eye-specification genes in the bacterial light organ of the bobtail squid Euprymna scolopes, and their expression in response to symbiont cues. *Mechanisms of Development, 131*, 111–126.

Purtov, K. V., Petushkov, V. N., Baranov, M. S., Mineev, K. S., Rodionova, N. S., Kaskova, Z. M., Tsarkova, A. S., Petunin, A. I., Bondar, V. S., Rodicheva, E. K., & Medvedeva, S. E. (2015). The chemical basis of fungal bioluminescence. *Angewandte Chemie, 127*(28), 8242–8246.

Ramirez, M. D., & Oakley, T. H. (2015). Eye-independent, light-activated chromatophore expansion (LACE) and expression of phototransduction genes in the skin of Octopus bimaculoides. *Journal of Experimental Biology, 218*(10), 1513–1520.

Rees, J. F., Thompson, E. M., Baguet, F., & Tsuji, F. I. (1990). Detection of coelenterazine and related luciferase activity in the tissues of the luminous fish, Vinciguerria attenuata. *Comparative Biochemistry and Physiology Part A: Physiology, 96*(3), 425–430.

Renwart, M., & Mallefet, J. (2013). First study of the chemistry of the luminous system in a deep-sea shark, Etmopterus spinax Linnaeus, 1758 (Chondrichthyes: Etmopteridae). *Journal of Experimental Marine Biology and Ecology, 448*, 214–219.

Rivers, T. J., & Morin, J. G. (2008). Complex sexual courtship displays by luminescent male marine ostracods. *Journal of Experimental Biology, 211*(14), 2252–2262.

Robison, B. H. (1992). Bioluminescence in the benthopelagic holothurian Enypniastes eximia. *Journal of the Marine Biological Association of the United Kingdom, 72*(02), 463–472.

Rodionova, N. S., Bondar, V. S., & Petushkov, V. N. (September 2002). Ca2+-Activator of the luminescence system of the earthworms Henlea sp.,(Annelida: Clitellata: Oligochaeta: Enchytraeidae). In *Doklady biochemistry and biophysics* (Vol. 386(1–6), pp. 260–263). Kluwer Academic Publishers-Plenum Publishers.

Rudie, N. G., & Wampler, J. E. (1979). A unique aldehyde chemiluminescence involving peroxide and copper I in acidic DMSO. *Photochemistry and Photobiology, 29*(1), 171–174.

Schmitter, R. E., Njus, D., Sulzman, F. M., Gooch, V. D., & Hastings, J. W. (1976). Dinoflagellate bioluminescence: A comparative study of invitro components. *Journal of Cellular Physiology, 87*(1), 123–134.

Schnitzler, C. E., Pang, K., Powers, M. L., Reitzel, A. M., Ryan, J. F., Simmons, D., Tada, T., Park, M., Gupta, J., Brooks, S. Y., & Blakesley, R. W. (2012). Genomic organization, evolution, and expression of photoprotein and opsin genes in Mnemiopsis leidyi: A new view of ctenophore photocytes. *BMC Biology, 10*(1), 107.

Shao, C., Bao, B., Xie, Z., Chen, X., Li, B., Jia, X., … Hamre, K. (2017). The genome and transcriptome of Japanese flounder provide insights into flatfish asymmetry. *Nature Genetics, 49*(1), 119–124.

Shimomura, O. (1981). A new type of ATP-activated bioluminescent system in the millipede Luminodesmus sequoiae. *FEBS Letters, 128*(2), 242–244.

Shimomura, O. (2006). *Bioluminescence: Chemical principles and methods*. World Scientific.

Shimomura, O., Goto, T., & Hirata, Y. (1957). Crystalline cypridina luciferin. *Bulletin of the Chemical Society of Japan, 30*(8), 929–933.

Shimomura, O., Inoue, S., Johnson, F. H., & Haneda, Y. (1980). Widespread occurrence of coelenterazine in marine bioluminescence. *Comparative Biochemistry and Physiology Part B: Comparative Biochemistry, 65*(2), 435–437.

Shimomura, O., & Johnson, F. H. (1968). The structure of Latia luciferin. *Biochemistry, 7*(5), 1734–1738.

Shimomura, O., Johnson, F. H., & Kohama, Y. (1972). Reactions involved in bioluminescence systems of limpet (Latia neritoides) and luminous bacteria. *Proceedings of the National Academy of Sciences, 69*(8), 2086–2089.

Shimomura, O., & Johnson, F. H. (1971). Mechanism of the luminescent oxidation of Cypridina luciferin. *Biochemical and Biophysical Research Communications, 44*(2), 340–346.

Shimomura, O. (2005). The discovery of aequorin and green fluorescent protein. *Journal of Microscopy, 217*(1), 3–15.

Shimomura, O., Johnson, F. H., & Saiga, Y. (1962). Extraction, purification and properties of aequorin, a bioluminescent protein from the luminous hydromedusan, Aequorea. *Journal of Cellular and Comparative Physiology, 59*(3), 223–239.

Shimomura, O., Masugi, T., Johnson, F. H., & Haneda, Y. (1978). Properties and reaction mechanism of the bioluminescence system of the deep-sea shrimp Oplophorus gracilorostris. *Biochemistry, 17*(6), 994–998.

Smith, D. S. (1963). The organization and innervation of the luminescent organ in a firefly, Photuris pennsylvanica (Coleoptera). *Journal of Cell Biology, 16*(2), 323.

Strehler, B. L., & Cormier, M. J. (1953). Factors affecting the luminescence of cell-free extracts of the luminous bacterium, Achromobacter fischeri. *Archives of Biochemistry and Biophysics, 47*(1), 16–33.

Strehler, B. L., Harvey, E. N., Chang, J. J., & Cormier, M. J. (1954). The luminescent oxidation of reduced riboflavin or reduced riboflavin phosphate in the bacterial luciferin-luciferase reaction. *Proceedings of the National Academy of Sciences, 40*(1), 10–12.

Tessler, M., Gaffney, J. P., Crawford, J. M., Trautman, E., Gujarati, N. A., Alatalo, P., Pieribone, V. A., & Gruber, D. F. (2018). Luciferin production and luciferase transcription in the bioluminescent copepod Metridia lucens. *PeerJ, 6*(e5506), 1–14. https://doi.org/10.7717/peerj.5506

Thompson, E. M., Toya, Y., Nafpaktitis, B. G., Goto, T., & Tsuji, F. I. (1988). Induction of bioluminescence capability in the marine fish, Porichthys notatus, by Vargula (crustacean)[14C] luciferin and unlabelled analogues. *Journal of Experimental Biology, 137*(1), 39–51.

Timmins, G. S., Robb, F. J., Wilmot, C. M., Jackson, S. K., & Swartz, H. M. (2001). Firefly flashing is controlled by gating oxygen to light-emitting cells. *Journal of Experimental Biology, 204*(16), 2795–2801.

Tong, D., Rozas, N. S., Oakley, T. H., Mitchell, J., Colley, N. J., & McFall-Ngai, M. J. (2009). Evidence for light perception in a bioluminescent organ. *Proceedings of the National Academy of Sciences, 106*(24), 9836–9841.

Topalov, G., & Kishi, Y. (2001). Chlorophyll catabolism leading to the skeleton of dinoflagellate and krill luciferins: Hypothesis and model studies. *Angewandte Chemie International Edition, 40*(20), 3892–3894.

Trimmer, B. A., Aprille, J. R., Dudzinski, D. M., Lagace, C. J., Lewis, S. M., Michel, T., … Zayas, R. M. (2001). Nitric oxide and the control of firefly flashing. *Science, 292*(5526), 2486–2488.

Tsai, Y. L., Li, C. W., Hong, T. M., Ho, J. Z., Yang, E. C., Wu, W. Y., Margaritondo, G., Hsu, S. T., Ong, E. B., & Hwu, Y. (2014). Firefly light flashing: Oxygen supply mechanism. *Physical Review Letters, 113*(25), 258103.

Tsuji, F. I. (2002). Bioluminescence reaction catalyzed by membrane-bound luciferase in the "firefly squid," Watasenia scintillans. *Biochimica et Biophysica Acta (BBA)-Biomembranes, 1564*(1), 189–197.

Tsuji, F. I. (2005). Role of molecular oxygen in the bioluminescence of the firefly squid, Watasenia scintillans. *Biochemical and Biophysical Research Communications, 338*(1), 250–253.

Vydryakova, G. A., & Bissett, J. (2016). Differential regulation of proteins and a possible role for manganese superoxide dismutase in bioluminescence of Panellus stipticus revealed by suppression subtractive hybridization. *Advances in Microbiology, 6*(9), 613–626.

Wang, M. Y., & Liu, Y. J. (2016). Theoretical study of dinoflagellate bioluminescence. *Photochemistry and Photobiology, 93*(2), 511–518.

Wang, M., & Liu, Y. (2020). *Chemistry in fungal bioluminescence: A theoretical study from luciferin to light emission*. https://doi.org/10.1021/acs.joc.0c02788

Warner, J. A., Latz, M. I., & Case, J. F. (1979). Cryptic bioluminescence in a midwater shrimp. *Science, 203*(4385), 1109–1110.

White, H. H. (1979). Effects of dinoflagellate bioluminescence on the ingestion rates of herbivorous zooplankton. *Journal of Experimental Marine Biology and Ecology, 36*(3), 217–224.

Widder, E. A. (1999). Bioluminescence. In *Adaptive mechanisms in the ecology of vision* (pp. 555–581). Netherlands: Springer.

Willis, R. E., White, C. R., & Merritt, D. J. (2011). Using light as a lure is an efficient predatory strategy in Arachnocampa flava, an Australian glowworm. *Journal of Comparative Physiology B, 181*(4), 477–486.

Wong, J. M., Pérez-Moreno, J. L., Chan, T. Y., Frank, T. M., & Bracken-Grissom, H. D. (2015). Phylogenetic and transcriptomic analyses reveal the evolution of bioluminescence and light detection in marine deep-sea shrimps of the family Oplophoridae (Crustacea: Decapoda). *Molecular Phylogenetics and Evolution, 83*, 278–292.

Wood, J. B., Maynard, A. E., Lawlor, A. G., Sawyer, E. K., Simmons, D. M., Pennoyer, K. E., & Derby, C. D. (2010). Caribbean reef squid, Sepioteuthis sepioidea, use ink as a defense against predatory French grunts, Haemulon flavolineatum. *Journal of Experimental Marine Biology and Ecology, 388*(1), 20–27.

Young, R. E., & Mencher, F. M. (1980). Bioluminescence in mesopelagic squid: Diel color change during counterillumination. *Science, 208*(4449), 1286–1288.

Young, R. E., & Roper, C. F. (1976). Bioluminescent countershading in midwater animals: Evidence from living squid. *Science, 191*(4231), 1046–1048.

Chapter 28

Evolution and physiology of animal visual systems

Marisa McDonald and Megan L. Porter
School of Life Science, University of Hawai'i at Mānoa, Honolulu, HI, United States

1. Overview

Modern visual systems likely arose at the time of the Cambrian explosion, around 515–520 million years ago, when the range and diversity of life forms erupted over a short geological timeframe. It was during this short period of evolutionary time that the majority of extant phyla formed, with vision being crucial for the expansion of life as we know it (Nilsson, 2004). In this chapter, we will review the evolution and physiology of visual systems across extant animal taxa, with a particular focus on photoreceptor cells. In Section 3 we will describe the basic structures of the eye. The simplest definition of an eye is a photoreceptor combined with a pigment cell that allows for directional sensing of light. Natural selection has molded that basic theme into a wide variety of visual systems. This chapter will cover the main eye structures found in animals today. In Section 4 we will discuss the structure within the eye responsible for light detection and image formation, the photoreceptor. We will focus on rhabdomeric and ciliary photoreceptors, the main types of photoreceptors found in protostome and vertebrate visual system, respectively. Section 5 will discuss the function and evolution of the visual pigment, an opsin protein bound to its chromophore, the molecular unit responsible for absorbing light in all visual systems. Additionally, Section 5 will explore the phototransduction cascade following light absorption, which results in the opening or closing of ion channels creating an electrical signal, which is then transmitted to the brain. Finally, Section 6 will discuss the evolution of visual systems, including the ongoing debate for and against the "monophyly of eyes". The debate refers to the hypothesis that all animal eyes, despite their diversity, share a common evolutionary origin from a single ancestral eye (monophyly). This contrasts with the idea that eyes evolved independently multiple times (polyphyly).

2. Introduction

Light is an important environmental cue for the majority of life forms. Mechanisms to sense and utilize light are therefore evolutionarily significant, as evidenced by the multitude of light detection systems across animal diversity. While the diversity of photodetection systems includes light sensitive pigments involved in circadian rhythms and biological clocks located in a variety of tissues (Porter, 2016), it is through the sensory structures known as eyes that light is detected and formed into an image. Although eyes are found throughout the animal kingdom, there is an enormous diversity in structure and types among extant animals, with differences in morphology, photoreceptor type, and lens composition (Gehring, 2005). It is believed that the modern eye evolved from a simple light sensor, or photoreceptor, that measured circadian and seasonal rhythms (Lamb, 2011). Photoreceptors, sensory neurons that house the biochemical phototransduction machinery and generate an electrical response when stimulated by light, are the cellular structures that initiate vision in animals (Pearring et al., 2013). At the most basic level, eyes are defined as light sensitive structures composed of a photoreceptor shielded on one side by pigment allowing for the directional detection of light (Hayakawa et al., 2015). In this chapter, we will discuss the current understanding of photoreceptor physiology and their evolution within animal eyes.

Cell Physiology Sourcebook. https://doi.org/10.1016/B978-0-12-811114-7.00008-6

3. Basic eye structure

When discussing visual systems, describing variations in the structure and function of extant eyes is crucial for understanding the evolution of vision. There are several general types of eyes structurally ranging from very simple to complex that result in differences in visual system function. It is also important to consider the evolutionary context and how such complex structures evolved across so many taxa. Although early estimates proposed that the eye has evolved independently somewhere between 40 and 65 times, this is now believed to be an overestimation (Gehring, 2005). Many now propose that all eyes arose from a single ancestry, which is supported by recent genetic work; this hypothesis is still being actively investigated, and will be discussed in more detail later in the chapter (Gehring, 2011).

Before delving into the components that make up an eye, it is important to note the differences in the way similar components are arranged in different eye types. Eyes, and specifically photoreceptor cells, can be classified structurally in a number of ways, including overall body position, photoreceptor cell origins, and orientation. Although eyes are typically thought of as cephalic, or related to the head, there are many animals with ectopic eyes on different areas of the body, including the pallial eyes on the mantle bivalves or on the trunk and crown of annelids (Purschke et al., 2006). Eyes can superficially look the same but have different developmental origins. For example, vertebrates, including human camera-type eyes, have photoreceptor cells that differentiate from the central nervous system, while invertebrate photoreceptor cells, such as those in cephalopod camera type-eyes, differentiate from the epidermis (Nilsson, 1996).

Eyes are a structure that can detect photons of light and their spatial orientation, and in the most basic forms require a photoreceptor cell housing a photosensitive visual pigment and a nerve center to interpret the chemical signal produced in response to the absorption of light energy (Serb, 2008). In a basic eye the visual pigment is concentrated on one side with the other open to light, allowing the photoreceptor to perceive direction. In Bilateria, eyes are found to be either everse or inverse, with everse eyes containing photoreceptors projecting away from the pigment cup towards the light, and inverse eyes projecting towards the pigment cup away from light (Arendt et al., 2002). Considering all these variations in developmental origins and structure, animal eyes can be classified into several categories, the most common of which will be described in detail below.

3.1 Simple eye and ocelli

The ancestral prototype eye likely consisted of a single photoreceptor cell and a single pigment cell. This eye type can be found in trochophore larvae and planarians. While still a relatively simple form, the next level of complexity in eye types is ocelli, which have the structure of a simple eye with the addition of support cells that provide directionality. Ocelli resemble a pigment cup where the photoreceptors are embedded on a cup shaped layer of pigment, allowing the light to only enter at the "cup" opening leading to directional sensing (Fig. 28.1A) (Arendt, 2003). Several photoreceptors can be present as well as the pigment, but it remains a basic structure designed to sense direction and little more. Typically, no lens is involved. This type of eye is most commonly found in planktonic larvae across many different phyla, as well as in frontal eyes of crustaceans, pigment cup eyes of flatworms, and in planula jellyfish larvae (Nilsson, 2013). Cases where ocellar eyespots develop additional structures are known as ocelloids. Ocelloids are found in dinoflagellates and contain structures such as a cornea, lens, and a pigment cup that make them similar to a camera type eye (Hayakawa et al., 2015).

3.2 Pinhole eye

Pinhole eyes (Fig. 28.1B) are found only in extant Nautilidae, a family of cephalopods known as the pearly nautiluses, and are believed to represent an ancient eye type (Serb & Eernisse, 2008). Nautiluses lack a lens or cornea but contain an adjustable pupil and have potentially very strong spatial resolution (Serb, 2008). The interior of the eye is in direct contact with the surrounding sea through the pinhole, or pupil. Light reflecting off of an object passes through the pupil to form an inverted image on the back of the retina, in the posterior of the eye (Serb & Eernisse, 2008.) Pinhole eyes are optically poor due to tradeoffs required between spatial resolution and light sensitivity. In order for an increase in sensitivity and the formation of a bright image on the retina, the pupil must expand. However, for increased resolution and a sharp image, the pupil must contract, thereby decreasing the amount of light reaching the retina. Because of these two contrasting features, pinhole eyes cannot achieve both sensitivity and resolution at the same time, and must choose between a bright defocused image and a dim but sharp image (Colicchia, 2006).

3.3 Camera eyes

The camera-type complex eye is a common structural arrangement found in vertebrates, cephalopods, and some cnidarians. The characteristic camera-type eye has a single lens that focuses light onto the retina. This type of eye consists of an iris,

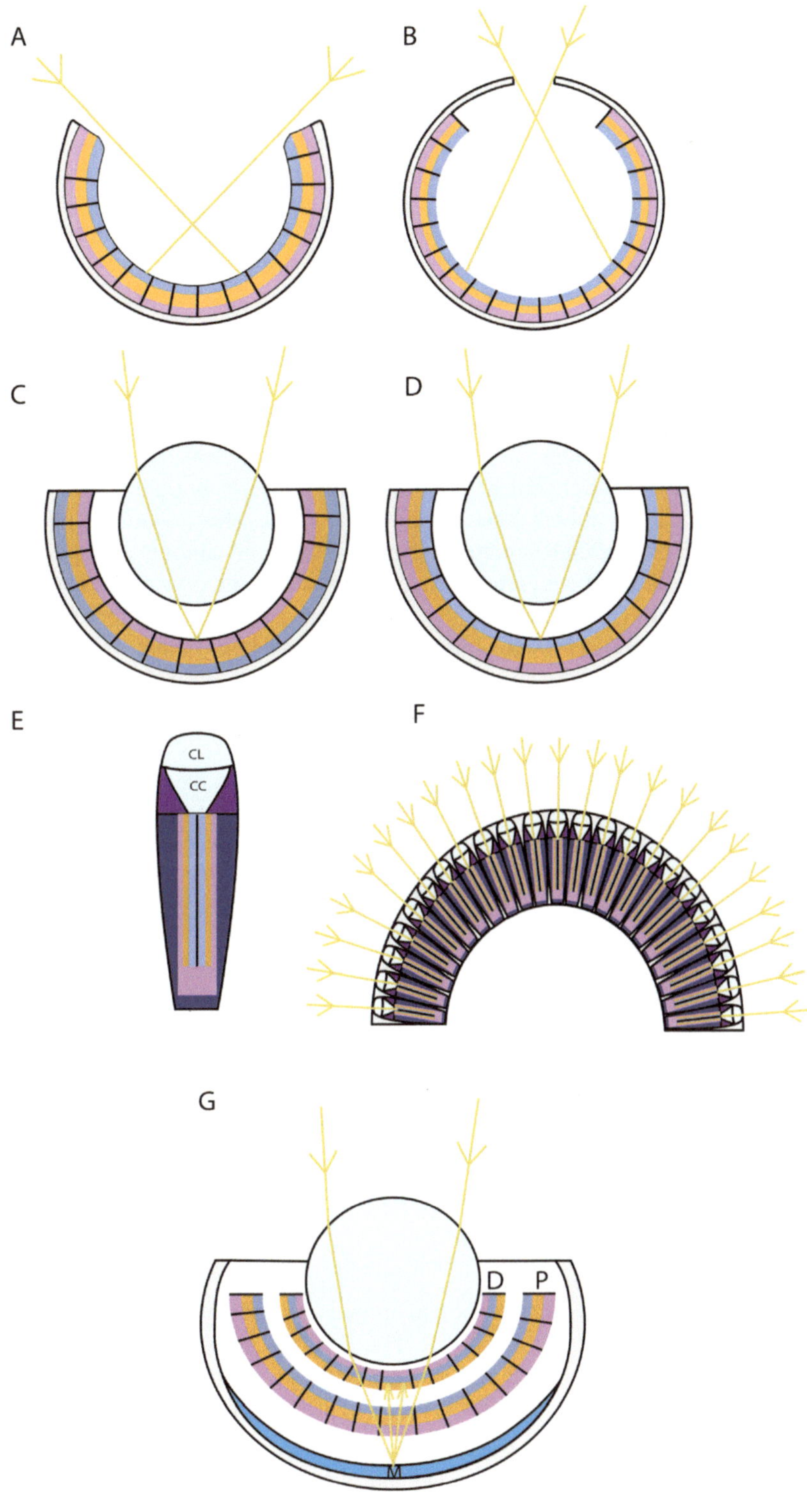

FIGURE 28.1 Representations of the most common animal eye structures. (A) Pigment cup eye; (B) pinhole eye; (C) vertebrate camera eye; (D) cephalopod camera eye; (E) longitudinal-section of single arthropod ommatidium, containing both a focusing lens (CL) and a crystalline cone (CC); (F) apposition compound eye composed of ommatidium with lenses facing the light source; (G) a scallop concave mirror eye with a distal (D) ciliary row of photoreceptors and a proximal (P) rhabdomeric row of photoreceptors, both in front of a concave mirror (M) structure at the back of the eye. Scallop eye structure for figure developed with reference to Barber et al. (1967). Ciliary photoreceptors appear to have an unusual orientation, with the synaptic region extending from the bottom and wrapping around to the top of the photoreceptor where the light first hits, followed by the outer segment and inner segment (Barber et al., 1967). In each figure, the blue, orange, and purple segments represent the photoreceptor cells, with blue representing the outer segment, orange the inner segment containing the mitochondria, and purple the synaptic and nuclear region; gray represents the visual pigments, light blue circles represent lenses, and arrows represent the direction of light.

lens, and retina; in addition, vertebrates also have a cornea (Fernald, 2000). Vertebrate eyes develop from the neural ectoderm, which induces the lateral head ectoderm to form a lens (Fig. 28.1C). While being structurally similar to vertebrates, cephalopod eyes are formed through a series of invaginations of the lateral head ectoderm (Nilsson, 2004). Cephalopods possess an everted retina with the photoreceptive cells membranes containing the visual pigments located directly behind the lens (Fig. 28.1D). This means that they are directed towards the incoming light with connection to the optic nerve behind the retina. Vertebrate camera eyes utilize inverted photoreceptor cells, where the neural cells are behind the lens and the retinal orientation of the photoreceptor cells is reversed (Serb & Eernisse, 2008). These camera-type eyes are a prominent example of convergent evolution (Serb, 2008).

3.4 Compound eyes

Compound eyes are made up of individual units known as ommatidia. They are one of the most common eye types in the animal kingdom, present in the Bilateria, particularly within the Arthropoda and Polychaeta (Nilsson, 2005). Structurally resembling a pigment cup, each ommatidium consists of a set of photoreceptor cells containing visual pigments with a lens of its own (Fig. 28.1E). Typically, ommatidia are composed of a corneal lens, two primary pigment cells, and eight rhabdomeric photoreceptor cells, but this arrangement can vary. Although it is not uncommon for the numbers of receptor cells, cone cells, or pigment cells to vary among species, within a species these numbers remain constant (Henze et al., 2012). While the organization of ommatidial structures varies within and among animal phyla, all have these basic components. Arthropods have an additional structure in the eye, known as the crystalline cone, which creates a refractive index gradient and behaves as a sort of telescope, typically arranged in a set of 4 cells that focus light onto the underlying photoreceptor cells (Nilsson, 2005). To form a single compound eye, large populations of ommatidia are arranged in arrays (Fig. 28.1F). These arrays range in complexity from just a few ommatidia to hundreds or thousands of ommatidia. By modifications of the basic ommatidial structures, particularly in regionalized subsets of ommatidia, compound eyes in different animals have been optimized for different functions, such as color vision, heightened spatial resolution or light sensitivity, and polarized vision (Marshall et al., 2007).

3.5 Concave mirror eye

Scallops have a concave mirror eye, composed of a cornea, a lens that sits in front of two retinas (distal and proximal), followed by a fluid filled cavity before a concave mirror layer (Fig. 28.1G) (Speiser & Johnsen, 2008). These eyes are found along the mantle of the animal. The eye is believed to function by using a lens to focus the light onto the distal retina and a concave mirror to reflect light onto the proximal retina, so the proximal retina is used for light intensity purposes and the distal for movement detection (Barber et al., 1967). The lens size and shape vary among species, as well as the cavity size between the mirror and the retinas (Speiser & Johnsen, 2008). The cavity size is believed to have a large effect on the eye function, and when the cavity is small the light will be focused on the distal retina and when the cavity is large the light is focused onto the proximal retina. It is hypothesized that the distal retina is used for important tasks such as predator detection, while the proximal retina is used more by swimming species for tasks such as visual habitat detection (Speiser & Johnsen, 2008).

4. Photoreceptors

Photoreceptor cells are sensory neurons that initiate vision in animals by generating an electrical response when stimulated by light through photon absorption (Pearring et al., 2013). Here we discuss photoreceptors within the context of high-resolution spatial vision mediated by eyes, but photoreceptors are not exclusively found within visual systems. Light sensitive photoreceptor cells are widespread in nearly all forms of life and tissue types, including animals, plants, fungi, unicellular eukaryotes, and prokaryotes (Nilsson, 2004; Porter, 2016).

Photoreceptors are critical for vision, as they are responsible for the conversion of photons of light into neuronal electrical signals (Nie et al., 2012). Photoreceptor cells are the first component of the visual system, housing the biochemical machinery responsible for phototransduction, which will be discussed in further detail later in this chapter. All photoreceptor cells have an expansion of the apical membrane that is enlarged for the storage of photopigments, the light sensitive molecules responsible for absorbing photons of light (Fig. 28.2). As well, all photoreceptors consist of four distinct compartments, the inner segment, outer segment, nuclear region, and synaptic region (Pearring et al., 2013). Each photoreceptor type employs similar yet distinct activation mechanisms, with light absorption by photopigments initiating a G-protein cascade that elicits an electrical response in the cell (Shichida & Matsuyama, 2009).

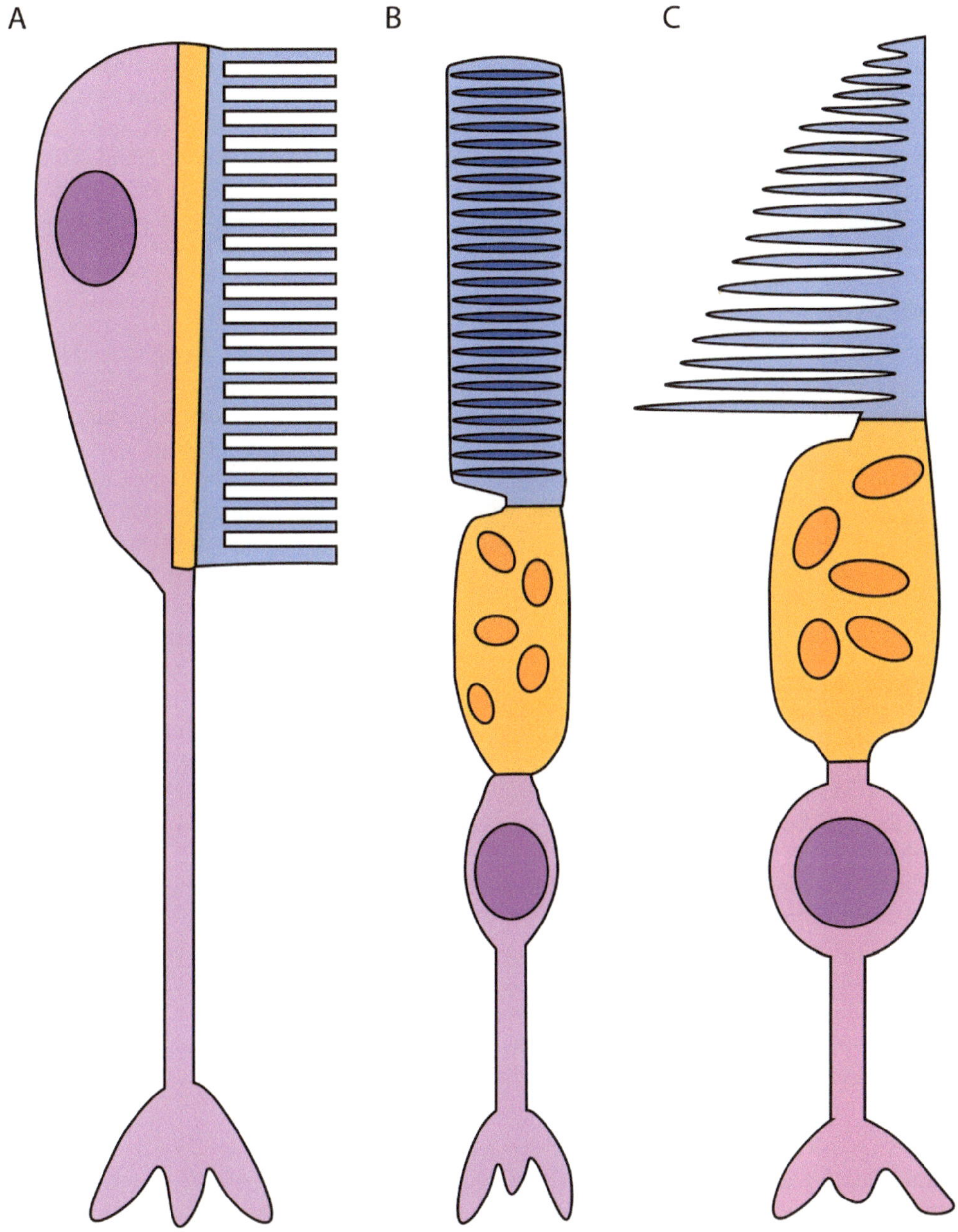

FIGURE 28.2 Photoreceptor cell types. (A) Representation of a rhabdomeric photoreceptor cell, commonly found in protostomes. (B) Representation of a rod photoreceptor, the dominant vertebrate photoreceptor responsible for vision in dim light. It is composed of highly organized stacks flattened membranous sacs in the outer segment. (C) Representation of the cone photoreceptor, responsible for color vision and vision in bright light in vertebrates. It is known for its tapered shape containing the stacked membranous discs. In each panel, the regions of the photoreceptor cells are defined by color: Blue is the outer segment, orange the inner segment, and purple the synaptic region. Mitochondria appear in the inner segment as darker organelles, and the nucleus is enclosed in the synaptic region.

In visual systems, there are two general classes of photoreceptor cells depending on the type of membrane expansion: the ciliary or c-type and the rhabdomeric or r-type. Ciliary photoreceptors typically form vertebrate rod and cone receptors, while rhabdomeric photoreceptors are mainly found in protostomes (Lamb, 2013). Although these two classes are an oversimplification of photoreceptor diversity, the anatomical division represents fundamental differences in underlying function (Fain et al., 2010). However, it is now recognized that there are several groups, including amphioxus, polychaetes, scallops, and possibly others, where it is not possible to categorize the photoreceptors into these two strict classes (Lamb et al., 2008). We will discuss the two main photoreceptor cell types, well-characterized exceptions to this classification scheme, and the major taxonomic groups using each type for vision in the following sections.

4.1 Rhabdomeric photoreceptors

Rhabdomeric photoreceptors, typically found in protostomes such as crustaceans, insects, and annelids, are formed through the eversion of microvilli, which then group together to form structures called "rhabdomeres" (Mahato et al., 2014). The microvilli are extended from the apical plasma membrane in finger-like projections that are membranous, actin rich, and arranged in densely packed, parallel bundles (Fig. 28.2A) (Porter et al., 2011; Purschke et al., 2006). These extensions work to increase the surface area of the cell exposed to light, allowing the photons to interact more effectively with visual pigments (Plachetzki et al., 2005). Upon activation, rhabdomeric photoreceptors employ a distinct phototransduction cascade that simulates a cellular response through depolarization; the specifics of this cascade will be discussed in detail later in the chapter.

4.1.1 Arthropods

Rhabdomeric photoreceptors are the dominant type found in the phylum Arthropoda and play a vital role in their vision. The most common eye type in this phylum is the compound eye, but there is wide diversity within the phylum, particularly among the subphylum Crustacea and Hexapoda (Cronin & Porter, 2008). The compound eye can be formed in many different ways depending on ommatidia arrangement. For example, crustaceans often have eight photoreceptor cells arranged into two photoreceptive units. A single short wavelength sensitive cell known as the R8 sits on top of the remaining 7 cells R1-7, which form a light sensitive tube known as a rhabdom, with their microvilli pointing inwards and overlapping, and serve as a single middle wavelength sensitive receptor (Cronin & Porter, 2008). The tiered arrangement of the R8 receptor on top of the R1-7 receptor cells allows for different portions of the visible spectrum to be absorbed separately for distinct functions.

The animal with the most complex compound eyes yet characterized is arguably the stomatopod, commonly known as the mantis shrimp. Stomatopods, and their eyes, diverged from other crustaceans hundreds of millions of years ago, and while they have ommatidia built from the standard crustacean arrangement, some regionalized sets of ommatidia exhibit distinct anatomical specializations (Porter et al., 2010). Stomatopods have 16 spectrally and anatomically different rhabdomeric photoreceptor types, the majority of which lie in an area of enlarged ommatidia that circle the equator of the eye known as the midband. There is variation among species, with species having anywhere from zero to six parallel rows of ommatidia in the midband that sample different areas of planar space (Cronin & Porter, 2008). The rhabdomeric photoreceptors in rows one through four are arranged differently compared with the average rhabdom as well. Instead of R1-7 fused into a single receptor, cells 1,4, and 5 are fused into a receptor that sits on top of a second receptor formed from cells 2,3,6, and 7, with the R8 cell remaining at the top of the rhabdom. This creates a three-tiered photoreceptor system that allows for added complexity and functional specialization (Cronin & Porter, 2008; Marshall et al., 2007). Through these specialized ommatidial structural modifications, stomatopods are able to achieve specific visual capabilities such as color vision, heightened spatial resolution, light sensitivity, and polarization vision (Marshall et al., 2007). However, because all the photoreceptor specializations are packed into the narrow equatorial midband region, the eye must constantly move in order to integrate the information across a visual scene. This is one area of specialization that continues to fascinate researchers and continuing studies are investigating how such complex forms arose (Cronin et al., 2022; Cronin & Porter, 2008).

4.1.2 Cephalopod

Rhabdomeric photoreceptors are also present in the Cephalopoda, a type of mollusk that includes octopuses, squids, cuttlefish, and nautiluses. Despite superficial similarities in eye structure with vertebrates, cephalopod rhabdomeric photoreceptors have an everse orientation in the retina. Nautiloids (i.e., *Nautilus*) have a relatively unsophisticated, basic pinhole eye containing only a retina. However, because the nautilus eye is large, it has similar resolving power to an insect eye with a densely packed retina of rhabdomeric photoreceptors (Serb, 2008). Though the eye is simple, with ocular motor reflexes *Nautilus* is able to detect motion and form images.

The coleoid cephalopods, which include octopuses, cuttlefish, and squid, have a camera-style eye superficially similar to vertebrates (Yoshida & Ogura, 2011). In particular, coleoid eyes appear to resemble elasmobranch eyes in terms of morphology (Serb, 2008). However, unlike the ciliary-based photoreceptors in the vertebrate eye, the photoreceptors in coleoid camera eyes are rhabdomeric, forming from microvilli. Often, the cephalopod camera eye is seen as a notable example of convergent evolution (Serb, 2008).

4.2 Ciliary photoreceptors

Cilium based photoreceptors are typically found in vertebrate animals, including humans. In ciliary photoreceptors, the cilium is folded into stacked flattened membranous disks. An enlarged cellular compartment is utilized to maximize surface area and house large quantities of visual pigment, increasing photon capture (Arendt et al., 2004). The modified cilium is characterized by a 9 + 0 microtubule organization (Lamb et al., 2008). The ciliary photoreceptors generally utilize different phototransduction mechanisms than rhabdomeric cells, and upon absorption of a photon, the cells hyperpolarize. Details of the ciliary phototransduction cascade will be discussed later in the chapter.

4.2.1 Vertebrates

The main difference between vertebrate ciliary photoreceptor cell types lies in the morphology of the membrane expansion, which varies based on evolutionary history. The simplest vertebrates contain photoreceptors arranged longitudinally similar, while hagfish have membranes splayed laterally from the cilium, and lamprey have neatly stacked membranes most similar to jawed vertebrates (Lamb et al., 2008). The jawed vertebrates contain photoreceptors with ciliary membrane expansions arranged in neatly stacked discs.

In jawed vertebrates there are generally two types of ciliary photoreceptors, often referred to as "rods" (Fig. 28.2B) and "cones" (Fig. 28.2C) (Porter et al., 2011). These two photoreceptor cell types exhibit both morphological and functional differences. Cones are responsible for vision in bright light and mediate color vision (Sung & Chuang, 2010), while rods are responsible for vision in dim light and are the dominant receptor in the majority of vertebrate retinas. There are a few exceptions, including turtles, lizards, some fish, and the ground squirrel, all of which have cone dominated retinas (Pearring et al., 2013). While the rod and cone photoreceptors are closely associated with specific networks of phototransduction proteins, they contain different isoforms of all of the major proteins. These differences account for the lower sensitivity and faster response of cones compared with rods, while rods contain more visual pigments to increase photon capture and sensitivity in comparison with cones.

Rod cells are remarkably sensitive; they can detect and respond to a single photon of light, which triggers a cellular response (Reingruber et al., 2015; Rieke & Baylor, 1998). Morphologically, rods contain around 2000 flattened membranous sacs forming disks that are surrounded by the plasma membrane in the outer segment of the cell (Fig. 28.2B). The disks house the visual pigments, which are responsible for absorbing light and interacting with a network of proteins in the plasma membrane and cytosol that converts the chemical changes in the cell to electric signals. The outer segment rises from the cilium base that connects to the inner segment of the cell. The inner segment contains the mitochondria and the nucleus of the cell and ends in a synaptic terminal that relays the electrical signal produced upon light absorption (Stryer, 1986). Cones are similar, but the outer segment is tapered, while rod outer segments are cylindrical (Fig. 28.2C) (Hárosi & Novales Flamarique, 2012). The tapering of the cone is variable, but it follows an ellipsoid shape that allows for the concentration of light onto the outer segment, which contains the stacked disks. This structure is a trade-off that allows for metabolic savings and accelerated regeneration of visual pigment, but in doing so, it is less sensitive than the rod (Hárosi & Novales Flamarique, 2012).

4.2.2 Cnidarians

Cnidarians are believed to contain the most primitive example of extant eyes, and as such, it has been hypothesized that Cnidaria, or an earlier common ancestor, are the common origin of the photoreceptor (Martin, 2002). On an evolutionary timescale, Cnidaria arose earlier than the Bilateria. While the majority of eye types evolved in the bilaterian lineage, Cnidaria also exhibit a range of eye types from ocelli to complex lens camera-type eyes (Martin, 2002). Cnidarian camera-like eyes have been described from the Cubozoa (box jellyfish) and contain ciliary photoreceptors with membrane stacks similar to vertebrate rods (Kozmik et al., 2008, Martin, 2002). In cnidarians, photoreceptors in the camera eye are believed to be connected to the nerve-net of the jellyfish, which acts as a central nervous system and is sophisticated enough to accomplish spectral discrimination (e.g., discrimination among different wavelengths of light).

4.3 Animals with unexpected photoreceptor types

4.3.1 Scallops and clams

Despite the early view of eyes using either rhabdomeric or ciliary photoreceptors, several animals have eyes that contain both types of cells. One example is the mirror eyes of scallops. Unlike the majority of mollusks with visual systems dominated by rhabdomeric photoreceptors, scallops utilize both types of photoreceptors in their eyes. Scallops have pallial eyes that are clustered thickly along the mantle and that contain a double retina, comprised of the proximal and the distal retinas (Fig. 28.1G). The distal retina is composed of ciliary photoreceptors and is responsible for image formation. The proximal retina is composed of rhabdomeric photoreceptors and is responsible for detecting changes in light intensity. These two photoreceptor types, segregated into distinct retinal layers, have distinct phototransduction cascades as well as opposite physiological responses. The distal retina depolarizes and the proximal retina hyperpolarizes when activated by light, as expected based on other similar structures (Serb, 2008). Therefore, while containing both photoreceptor types, the two types function as expected based on current knowledge of these structures. Arc clams also have pallial eyes across the mantle. Unlike the scallop mirror eye, however, arc clam species have compound eyes superficially similar in structure to those of arthropods, but composed of ciliated receptor cells instead of microvilli photoreceptors (Nilsson, 2005).

4.3.2 Ciliary and rhabdomeric photoreceptors beyond visual systems

While the main focus of this chapter is on photoreceptor types within animal visual systems, it is important to note that in many species there is evidence of both photoreceptor types in different areas of the body. Because of this, it has been hypothesized that the last ancestor of both invertebrates and vertebrates in the Urbilateria contained both photoreceptor cell types (Purschke et al., 2006). The marine rag worm *Platynereis* utilizes a rhabdomeric photoreceptor in the eyes and a ciliary photoreceptor in the brain (Arendt et al., 2004). The ciliary photoreceptor expresses a c-opsin, the functional visual pigment for ciliary photoreceptors in vertebrates, while the rhabdomeric photoreceptor expresses an r-opsin found in protostomes. This is still the only example of a c-opsin found in a polychaete at this time (Ramirez et al., 2016). This implies that polychaete and vertebrate photoreceptors are homologous, and that the photoreceptor types may have evolved prior to the last common ancestor of the Bilaterians.

5. Phototransduction

Phototransduction is the physiological process by which light is converted into an electrical signal within photoreceptor cells in the retina. This process begins when a light-sensitive visual pigment absorbs a photon of light. Photon absorption is the first step of phototransduction, and begins with a G-protein coupled receptor (GPCR), comprised of a protein, the opsin, covalently bound to a vitamin A derivative chromophore, most commonly 11-*cis* retinal. Upon photon absorption by the chromophore, the opsin undergoes a conformational change. Second, a paired heterotrimeric G-protein is activated (Yarfitz & Hurley, 1994). G-proteins, or guanine nucleotide binding proteins, are the major modulators of cellular response. In visual systems, G proteins mediate and relay the signal from an activated visual pigment. Each class of G-protein activates a different effector enzyme, or ion channel (Terakita et al., 2008) and the coupling of the opsin GPCR and its complementary G-protein is relatively strict. Upon activation, a cascade of interacting proteins is initiated, ultimately resulting in an electrical response in the cell. The two main photoreceptor types,rhabdomeric and ciliary, exhibit opposite polarity responses to light; rhabdomeric photoreceptors depolarize in response to light, whereas ciliary photoreceptors hyperpolarize. This is not the only difference; while the mechanisms of photon absorption are similar, differences in phototransduction among these photoreceptor types are found in the opsin protein utilized and continue in the downstream pathways and interactions. In this section, we will first examine the best characterized first step of the phototransduction cascade, the activation of visual pigments, as well as the basics of physiological processes and the differences among photoreceptor types.

5.1 Light sensing molecules: The opsins

The best understood part of phototransduction is the first step, the activation of a visual pigment, composed of an opsin protein covalently bound to a vitamin A chromophore (Shichida & Matsuyama, 2009). After activation, the visual pigment initiates a G-protein mediated cascade and causes an electrical response in the cell (Kojima et al., 1997). Opsin molecules react to light through conformational change upon photon absorption by the chromophore. This results in a change in the molecular structure, initiating the G-protein cascading response. Opsins are a member of a large superfamily of proteins, the G-Protein Coupled Receptors (GPCRs). GPCRs activate signal pathways in response to extracellular stimuli including Ca^{2+}, amines, hormones, peptides, as well as the light response (Palczewski et al., 2000). While there are many types of GPCRs, all have a similar conserved structure, which was first crystallized and described in detail by Palczewski et al. (2000).

Structurally opsins are seven transmembrane α-helix receptors covalently bound to a retinal chromophore (Palczewski et al., 2000). Opsins have a binding pocket formed by the transmembrane α-helices in which the chromophore sits (Briscoe & Chittka, 2001). The retinal chromophore can take many forms of isomers but typically is a vitamin A based, 11-*cis*-retinal chromophore (Shichida & Matsuyama, 2009). The chromophore is bound to the opsin through a Schiff base linkage at a conserved lysine in the seventh transmembrane helix (Yokoyama, 2008).

Opsins are essential for animals' ability to use light for diverse biological functions and are the major light sensitive molecule throughout the animal kingdom. The number and type of opsins possessed by animals are strongly associated with their environment (Yokoyama, 2008). Opsins, while initiating vision, are also responsible for different types of vision.

Visual pigments are highly diverse and allow for differences in functionality and spectral tuning (Terakita, 2005), which will be discussed in detail further in the chapter. Much of the early work on opsins was limited to those expressed in vertebrates and arthropods, but more recent studies have expanded on this work across further phyla. Current research describes nine opsin paralogs in the bilaterian ancestor (Fig. 28.3) (Ramirez et al., 2016). While most studies focus on opsins within visual systems, opsins have also been found in many tissues outside of the eye, including the brain, ganglion cells, and epithelial cells, where they detect light for nonvisual functions (Kingston et al., 2015).

There have been many subgroups and classification systems throughout the history of opsin studies. Historically, it was believed that there were two basic classes, the "r-type" and the "c-type", reflecting the photoreceptor cell in which the protein was housed ; the "r-type" was expressed in rhabdomeric photoreceptors, while the "c-type" was found in ciliary photoreceptors (Porter et al., 2011). However, current work suggests that there are further classes of opsin beyond these two main groups. It has been suggested that all opsin lineages originated from a common ancestor in the eumetazoan, and gene duplications allowed for the various types that are seen today (Feuda et al., 2012). The most recent study of opsin diversity described four eye opsin paralogs, the canonical c-opsin, the canonical r-opsin, the xenopsins, and the Go-opsins (Ramirez et al., 2016). The other five opsin paralogs described at this time, while important, function in areas outside of the

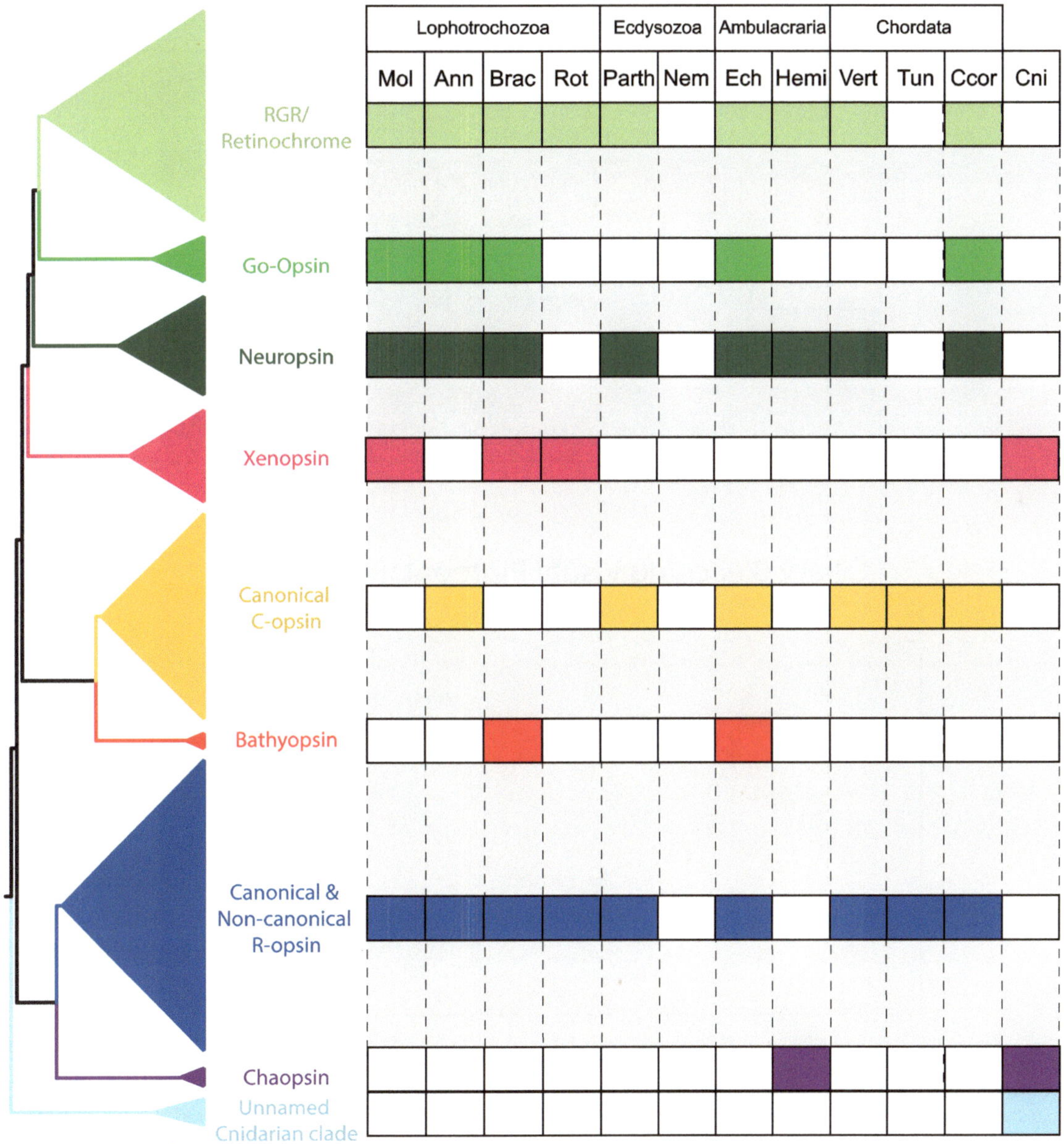

FIGURE 28.3 Phylogeny of the opsin dataset from Ramirez et al. (2016), reconstructed as in Battelle et al. (2016), and collapsed into clades and colored to highlight the major opsin lineages. The taxonomic composition of sequences in each clade is displayed to the right of the phylogeny, with colored boxes representing the presence of at least one sequence from that taxonomic group, and white boxes indicating the absence of sequences from that group. Taxonomic abbreviations are as follows: Mol - Mollusca; Ann - Annelida; Brach - Brachiopoda; Rot - Rotifera; Parth - Panarthropoda (e.g., insects, crustaceans, tardigrades, and onychophorans); Nem - Nematoda; Ech -Echinodermata; Hemi - Hemichordata; Vert - Vertebrata; Tun - Tunicata; Ccor -Cephalochordata; Cni - Cnidaria.

eye and therefore will not be explored in depth here. The four opsin paralogs that are found in eyes will be explored further in this section; however, it is important to note that many taxonomic groups and species remain understudied and so there is still much to discover about the true extent of opsin diversity (Porter et al., 2011).

5.1.1 Canonical r-opsin

Canonical r-opsins are G_q-coupled receptors generally expressed in rhabdomeric photoreceptors. However, they have also been found to be expressed in neurons, gonads, and chromatophores (Tsukamoto & Terakita, 2010). Opsin subfamilies in this group are found in the eyes of lophotrochozoans, platyhelminthes, and arthropods, as well as vertebrate nonvisual pigments, of which melanopsins are the most widespread (Koyanagi & Terakita, 2008; Ramirez et al., 2016). G_q mediated systems activate phospholipase C through the G_q protein, which initiates a cascade that results in depolarization of the cells (discussed further below). R-opsins are bistable, meaning they retain the chromophore in dark and light states throughout the activation process (Fig. 28.4A) (Porter et al., 2011).

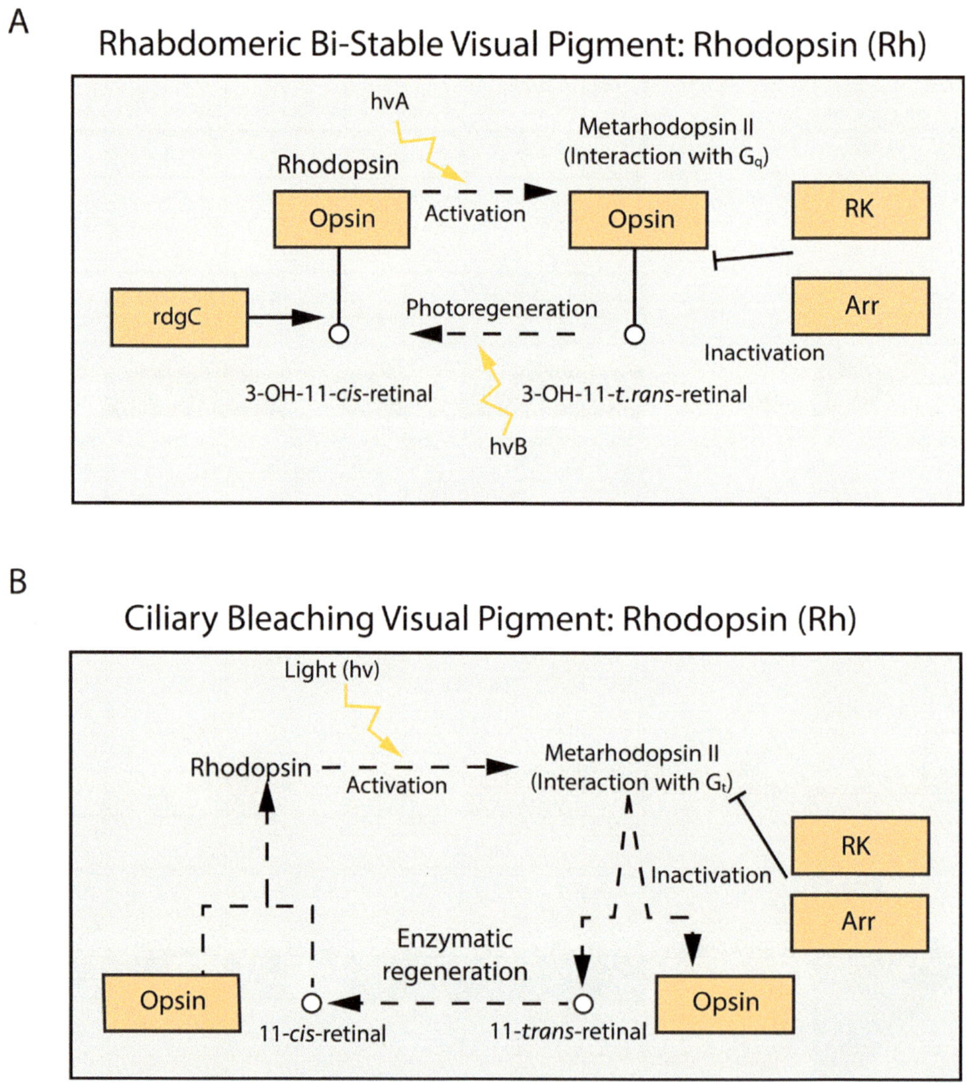

FIGURE 28.4 Visual pigment regeneration after light activation for ciliary (A) and rhabdomeric (B) photoreceptor types, modified from Kyoto encyclopedia of genes and genomes, KEGG (Kanehisa et al., 2017; Kanehisa et al., 2016, Kanehisa & Goto, 2000). (A) Representation of a bleaching ciliary visual pigment. The light (hv) activates the rhodopsin pigment, isomerizing the 11-*cis*-retinal chromophore to an all-*trans*-retinal configuration. When activated, rhodopsin forms metarhodopsin. This reaction leads to the activation of the cascade. This is then inhibited by RK (G protein-coupled receptor kinase 7) and Arr (arrestin), which allows for enzymatic regeneration and the chromophore is converted back to its base state. (B) Representation of a bi-stable rhabdomeric visual pigment. Rhodopsin (Rh) is activated by light, resulting in a conformational change of the chromophore, forming metarhodopsin, which initiates the G-protein cascade. Arr (arrestin) and RK (G protein-coupled receptor kinase 1) inhibit the metarhodopsin, and with a second light source the cell undergoes photoregeneration reverting the chromophore back to the base state. Retinal degeneration C (rdgC) interacts with the chromophore to assist in reforming Rh. In each panel, solid arrows represent molecular interactions., Dashed arrows represent indirect effects. A barred arrow represents inhibition. Straight lines represent bound proteins. Open circles represent other molecular chemical compounds.

G_q coupled visual pigments have been historically difficult to study due to difficulties of purification and expression in culture on a large scale (Terakita et al., 2008). In the G_q coupled opsin group, subgroups form based on their sensitivity to different wavelengths. Vertebrates and arthropods are known to have color vision, which is accomplished through the expression of a diversity of opsins in their photoreceptors. These differences in absorption are caused either by changes in the amino acid sequences of the opsin protein or changes in the chromophores that they bind to (Briscoe & Chittka, 2001). While in some cases multiple chromophore types are used, typically the same chromophore is used, meaning that it is variation in the opsin sequence that results in the differences in maximal wavelength absorption among different visual pigments (Porter et al., 2006).

While the spectral range of invertebrates can be very wide depending on the spectral photoreceptor types, typically there are conservative patterns of wavelength positioning. For arthropods, there are at least four phylogenetic clades of opsins that roughly correspond to wavelength sensitivity of the visual pigments formed; long wavelength (LW) green-yellow sensitive (>500 nm), middle wavelength (MW) blue-green sensitive (400−500 nm), short wavelength (SW) blue sensitive (377−400 nm) and ultraviolet (UV) (<377 nm) sensitive opsins (Briscoe & Chittka, 2001; Gärtner & Towner, 1995.) For a recent review see (Yilmaz et al., 2022). Cephalopods use similar G_q-coupled opsins in their visual systems, but typically express only a single, blue-green sensitive opsin and therefore appear to lack the ability to perceive color. The lack of color vision begs the question of how cephalopods achieve their remarkable camouflage capabilities; some suggest this is accomplished through polarized light (Chung & Marshall, 2016). There does appear to be a possibility that a small group of mid-water squid is expanding their visual range through the use of different chromophores (Chung & Marshall, 2016). These extraocular photoreceptors have been found to express rhodopsin, retinochrome, and G_q transcripts; however, the G_q transcript in the dermal chromatophores appear to be different than those found in the retina of the same animal and therefore it is currently unknown if the same signaling pathway is used in the chromatophores and the retinal photoreceptors (Kingston et al., 2015).

The canonical r-opsin lineage also contains the melanopsins, a significant group of nonimage forming opsins (Provencio et al., 1998). These are involved in nonimaging forming visual tasks across all vertebrate classes and several invertebrate classes (Bellingham et al., 2006). While the melanopsins are not the only nonvisual opsin lineage in the canonical r-opsin group, it appears to be the most widespread across taxa. These melanopsins are believed to be responsible for assisting in the regulation of light responses such as circadian rhythm, seasonal timing of reproduction, and melatonin production (Provencio et al., 2000). These responses exist in the absence of eyes, and therefore are regulated by extraocular photoreceptors.

5.1.2 Canonical c-opsin

The canonical c-opsins are found expressed in the ciliary photoreceptors of vertebrate eyes, as well as neural tissues of vertebrates, arthropods, and the annelid *Playtneris dumerilii.* In vertebrate visual systems, canonical c-opsins couple with a G_t protein, also known as transducin. C-opsins activate a cGMP-specific phosphodiesterase (PDE) through transducin, leading to hyperpolarization of the cell. C-type opsins bleach upon photoactivation, i.e., when the 11-*cis*-retinal is isomerized to all-*trans*-retinal upon light absorption, the all-*trans*-retinal chromophore is released from the opsin binding pocket, which is referred to as "bleached" (Fig. 28.4B) (Tsukamoto & Terakita, 2010).

The c-type opsin used in vertebrate vision can be described based on the cell type they are associated with; rhodopsin is found in rod photoreceptors, while cone opsins are found in cone photoreceptors (Terakita, 2005). There is only one class of rhodopsins, and they are used in rod receptors that are responsible for vision in dim light. The cone opsins can then be further subdivided into four evolutionary classes based on the absorption properties of the visual pigments they form. The main opsin types expressed in cones are long wavelength sensitive (LWS), green middle wavelength sensitive (RH2), short wavelength sensitive 1 (SWS1) with UV/violet sensitivity, and short wavelength sensitive 2 (SWS2) with blue sensitivity. LWS and RH2 are one evolutionary clade, while SWS1 and SWS2 are distinct clades. Not all vertebrates have each of these four types (Terakita, 2005). For example, mammals in general only use two of these opsin clades, lacking the SWS2 and RH2 genes. Humans, however, have regained green sensitivity through a duplication of the LWS gene, with the duplicated copies diverging in absorption properties to red and green, respectively (Nathans et al., 1986).

5.1.3 G_{o}-opsins

The eumetazoan opsin class known as the 'Tetraopsins' are coupled with G_o-proteins, and contain opsin lineages such as the G_o-coupled opsins, retinochromes, RGR opsins, and neuropsins. In this section we focus only on the G_o-coupled opsins, as they are the only Tetraopsin lineages found within visual systems (Gühmann et al., 2015). The G_o-coupled opsins are a functionally diverse set of genes with representatives in chordates, mollusks, echinoderms, and cephalochordates (Ramirez et al., 2016) and are expressed in ciliary photoreceptors as well as neuronal, gonad, retinal epithelial, and skin cells (Porter et al., 2011).

The G_o coupled opsins are an independent evolutionary lineage, separate from the G_t and G_q related opsin classes, and they are considered an evolutionarily ancient opsin (Gühmann et al., 2015). Scallops, as discussed previously, have both rhabdomeric and ciliary photoreceptor types. The rhabdomeric photoreceptors contain the expected G_q-type opsins that mediated a phototransduction cascade leading to a depolarizing response. In scallop ciliary photoreceptors, however, the expressed opsin was found to be neither canonical r-opsins nor canonical c-opsins, which led to the conclusion that a third type of opsin, a G_o-coupled opsin was responsible for the observed hyperpolarizing response in ciliary scallop photoreceptors, with different molecular mechanisms than vertebrate ciliary photoreceptors (Kojima et al., 1997). The G_o-type opsin have also been described in marine annelids, where it was coexpressed with r-opsins in depolarizing cells of the larvae that contain both photoreceptor types. In this system, the G_o-type opsin is responsible for absorption in the blue-light spectrum, with a peak at 480 nm (Gühmann et al., 2015). Unlike hyperpolarizing G_o-opsin expressing ciliary cells in scallop retinas, the annelid larvae G_o-coupled receptors elicited a depolarizing response.

5.1.4 Xenopsins

Xenopsins are a major evolutionary clade of opsins that appear to be the main group of photopigments found in cnidarians, paired with a G_s type G-protein. Cnidaria are thought to make up the earliest branching animal phylum with a well-developed visual system, and are thought to be the sister group of Bilateria (Kozmik et al., 2008). Xenopsins are expressed in the retinal ciliary photoreceptors, CNS, and gonad cells of cnidarians (Porter et al., 2011). Xenopsins also have representatives from a variety of lophotrochozoan protostomes such as molluscs, rotifers, and brachiopods (Ramirez et al., 2016). Similar to vertebrates and cephalopods, some Cnidaria, such as the Cubozoa "box jellyfish", have camera-type eyes with a cornea, lens, and retina, which contain ciliary photoreceptors (Kozmik et al., 2008). However, while they contain similar photoreceptors to vertebrates, box jellyfish eyes contain xenopsins rather than canonical c-opsins (Bielecki et al., 2014). While previous research suggested that the xenopsins are closely related to the c-opsins, clustering with vertebrate c-type opsins in a phylogenetic analysis, the most recent analysis provides strong evidence for a monophyletic clade (Ramirez et al., 2016). This protein also triggers a different cascade. Although housed in ciliary photoreceptors in cnidarian eyes, the xenopsins trigger a different phototransduction cascade than found in vertebrate or scallop ciliary receptors; based on current studies, xenopsin visual pigments interact with a G_s-type heterotrimeric G-protein, causing an increase in cAMP levels through a G_s-adenylyl cyclase when exposed to light (Koyanagi et al., 2008).

5.2 Depolarizing response to light

In the dark, the photoreceptor is in a resting state (Fig. 28.5A). Upon light absorption rhabdomeric photoreceptors in the visual systems of arthropods, annelids, and cephalopods activate a G_q protein cascade resulting in cell depolarization (Fig. 28.5B). Upon activation, G_q-coupled receptors initiate a phospholipase C (PLC) cascade. In arthropods and cephalopods, the main taxonomic groups utilizing rhabdomeric photoreceptors, the amino acid residue tyrosine in the seventh transmembrane helix stabilizes the positive charge of the Schiff base during photoactivation (Porter et al., 2006). These proteins are known to be bistable, meaning they retain the chromophore in light and dark states (Fig. 28.4B) (Porter et al., 2011).

Activation of phototransduction begins with light converting the chromophore from the *cis* to all-*trans* configuration, generating metarhodopsin and activating the G_q protein. The G_q protein then releases its α subunit, activating phospholipase C (PLCβ4 encoded by the *norpA* gene), ultimately resulting in the opening of two Ca^{2+} permeable channels known as TRP and TRPL (Fig. 28.5B) (Hardie et al., 2002). While PLC is the effector enzyme, exactly how these channels are opened has long been controversial, but Ca^{2+} is known to play an essential role in the opening and closing of channels. During activation, there is a rapid rise in the concentration of Ca^2, followed by rapid termination and fall in concentration (Hardie et al., 2002). Ultimately, the TRP and TRPL channels are opened leading to depolarization of the cell. The transduction cascade is then terminated through the binding of arrestin, which halts the GTPase activity of the G_q protein, closing the channels, and returning the cell to the resting state (Hardie & Raghu, 2001). After activation, the thermostable metarhodopsin will be directly re-isomerized back to rhodopsin (Fig. 28.4A) (Hardie & Raghu, 2001).

5.3 Hyperpolarizing response to light

In the dark, vertebrate ciliary photoreceptors maintain a resting depolarized state (Fig. 28.6A). Upon light absorption and activation of the G_t protein cascade, the cell hyperpolarizes (Fig. 28.6B). Phototransduction takes place in the ciliary outer

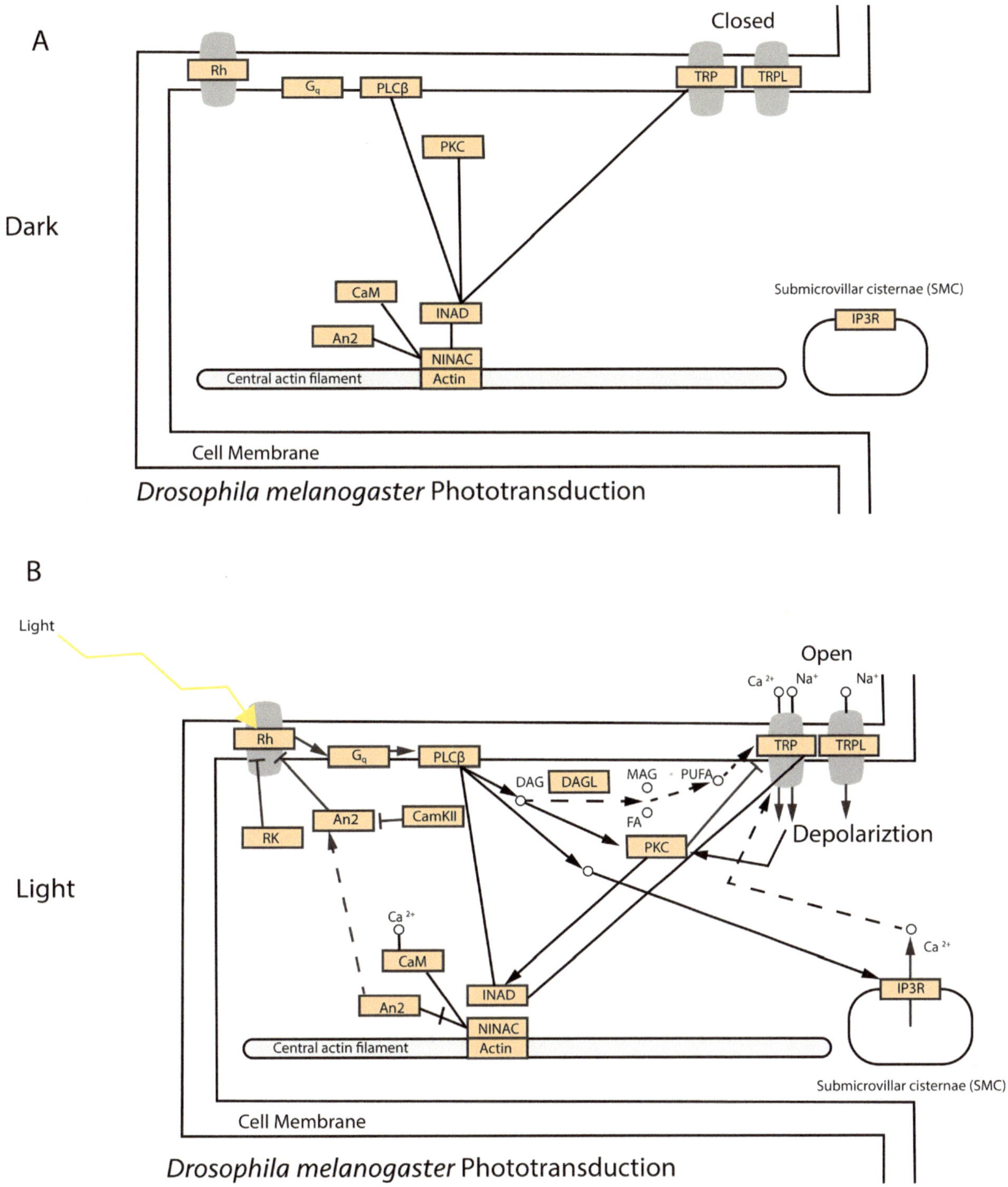

FIGURE 28.5 Phototransduction cascade of *Drosophila melanogaster*, modified from Kyoto encyclopedia of genes and genomes, KEGG (Kanehisa et al., 2017; Kanehisa et al., 2016, Kanehisa & Goto, 2000). (A) An inactivated resting state (dark) of the *Drosophila melanogaster* phototransduction cascade, where rhodopsin (RH) and two Ca^{2+} permeable cation channels (TRP and TRPL) are embedded in the cell membrane of the photoreceptor. As well, the G protein (G_q) and the effector of the G-protein, the phospholipase Cβ (PLCβ), are on the inner membrane of the cell. The TRP and TRPL channels are closed in the dark resting state. The PLCβ, PKC (classical protein kinase C alpha type), and TRP channels are bound to an inactivation no afterpotential D protein (INAD), which is in turn bound to NINAC (myosin III). This is bound to calmodulin (CaM) and Arr2 (arrestin). Inositol triphosphate receptor (IP3R) is found on the submicrovillar cisternae in the cell. (B) A light-activated phototransduction cascade of a *Drosophila melanogaster* photoreceptor. When light hits, RH activates the G_q protein through therelease of the α subunit, leading to the activation of the PLC generating IP3R (inositol) and DAGL (snl-specific diacylglycerol lipase). The reaction leads to the opening of the TRP and TRPL channels, allowing Ca^{2+} and Na^{+} to enter the cell, resulting in membrane depolarization. In each panel, orange boxes correspond to the coding protein. Solid arrows are molecular interactions. Dashed arrows represent indirect effects. A barred arrow represents inhibition. Straight lines represent bound proteins. Open circles represent other molecular chemical compounds.

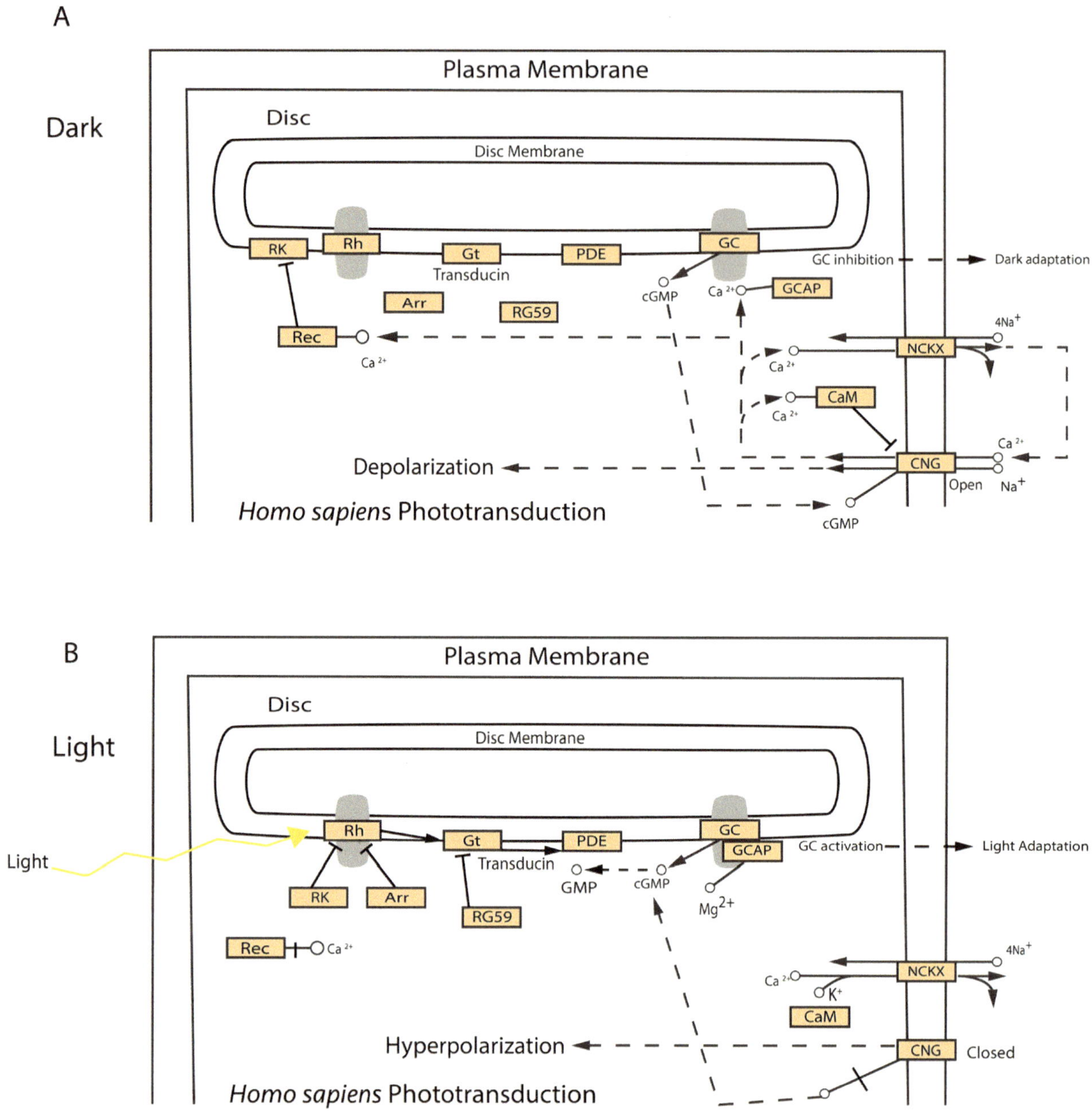

FIGURE 28.6 Diagram of the phototransduction cascades of human (*Homo sapiens*) ciliary photoreceptors. The diagram is modified from Kyoto encyclopedia of genes and genomes, KEGG (Kanehisa et al., 2017; Kanehisa et al., 2016, Kanehisa & Goto, 2000). (A) resting state (dark) rod cell phototransduction cascade. Rhodopsin (Rh) and guanine cyclase (GC) are embedded in the membrane of the disc. GC is inhibited in the dark. Rh consists of a chromophore bound to an opsin. Also found on the disc membrane are rhodopsin kinase (RK), the G protein transducin (G_t), and transducin phosphodiesterase (PDE). When in the dark, arrestin (Arr), recoverin (REC), G-protein regulator (RGs9), and guanylate cyclase activator (GCAP) are found within the cell, but are inhibited. The cyclic nucleotide-gated (CNG) cation channels and the K^+-dependent Na^+/Ca^{2+}-exchanger (NCKX) work in concert to regulate ion fluxes. In the dark, a high concentration cGMP keeps the CNG channels open, allowing a continuous inward current of Na^+ and Ca^{2+} ions, which is known as the "dark current". This dark current depolarizes the photoreceptor cell membrane. When light hits the retina, the level of cGMP inside the photoreceptors decreases, causing the CNG channels to close, leading to membrane hyperpolariztion. Simultaneously, the (NCKX), also located on the plasma membrane, exchanges 4 Na^+ ions for 1 K^+ ion and 1 Ca^{2+} ion. The exchanger uses the energy stored in the Na^+ gradient to move Ca^{2+} out of the cell and K^+ into the cell thereby keeping a low intracellular Ca^{2+} concentration. (B) A light-activated phototransduction cascade of a human photoreceptor. The visual pigment chromophore is isomerized in rhodopsin (Fig. 28.4B), triggering the hydrolysis of cGMP through the activation of a PDE cascade, resulting in the closure of the CNG channels leading to hyperpolarization. RK initiates recovery before being capped by Arr. Transducin is deactivated through RGS9. In each panel, orange boxes correspond to the coding protein. Solid arrows are molecular interactions. Dashed arrows represent indirect effects. A barred arrow represents inhibition. Straight lines represent bound proteins. Open circles represent other molecular chemical compounds.

segment, composed of membrane discs that house the visual pigments (Cheng et al., 2009). It only takes one absorbed photon to activate numerous G proteins, resulting in signal amplification and triggering of the cascade (Cheng et al., 2009). Upon absorbing light, the visual pigments become unstable and a bleaching photoproduct known as metarhodopsin II is produced (Fig. 28.4B) (Fain et al., 2001). In contrast to invertebrate photopigments, lowering the pH of the system results in decreased absorbance due to deprotonation of the retinylidene Schiff base. This then activates the G_t protein transducin, stimulating the cyclic nucleotide PDE that hydrolyzes 3′,5′-cyclic monophosphate (cGMP). In darkness, the cGMP-gated cation channels remain in an open state due to high amounts of cGMP in the system, maintaining a steady inward "dark current" that keeps the cell in a depolarized state (Fig. 28.6A). The cell remains at a membrane potential of −30 mV, which sustains glutamate release. When exposed to light,the cascade begins, and these gated channels are closed. This stops glutamate release and hyperpolarizes the cell (Cheng et al., 2009). After this occurs, it is essential for a termination of the signal. The opsins are not bistable as in invertebrate cascades. Instead the metarhodopsin is phosphorylated by rhodopsin kinase, followed by termination of G protein activity through arrestin (Fig. 28.4B). In rods a specific GTPase-activating protein is also required, and finally the channels are reopened through synthesis of cGMP in the cell.

An important process for these photoreceptor cells is light adaptation, which allows the cascade to remain effective at varying light levels (Fain et al., 2010). There is evidence that the modulation of the cascade for adaptation is occurring through the second messenger Ca^{2+}, which is believed to affect the transduction mechanism by altering the gain of the transduction cascade and regulating the affinity of cGMP for the channels of the outer segment (Fain et al., 2001).

The scallop ciliary photoreceptors hyperpolarize through a different cascade, and a different G protein, the G_o protein. While typically photoreceptors utilize Ca^{2+} for their transduction cascade, scallop G_o channels appear to utilize a light induced increases in a K^+ conductance leading to cell hyperpolarization (Gomez & Nasi, 2000). It appears that the cascade results in the opening of cGMP sensitive potassium channels, as opposed to the closing of cGMP cationic channels in other invertebrate visual systems. To accomplish this, the G_0-type opsin couples with the effector enzyme guanylyl-cyclase, elevating the cytosolic cGMP concentration to open the K^+ channels and hyperpolarize the membrane (Kojima et al., 1997). This is a similar mechanims to the G_o-cascade documented in the mammalian nervous system. Arrestin is also used by these photoreceptors to terminate the light response (Gomez & Nasi, 2000).

6. Evolution of the visual system

There are clear differences in visual systems among taxa. Because of this, throughout the history of the study of visual systems there has been much debate over how something as complex as eyes was able to evolve. The debate began with Charles Darwin (1809−1882), who argued that the eye evolved through natural selection, illustrated by functional variation in eyes between a light-sensitive patch of cells and a full complex eye (Darwin, 1859). However, scientists later argued that the visual system must have evolved separately multiple times due to morphological differences only possible through convergent evolution (Gehring, 2005). The debate continues today, and viewpoints vary depending on the the visual system being studied.

There is still a question of how many components of the visual system described above in this chapter, including the photoreceptor lineages, opsin diversity, and phototransduction cascades, evolved independently. Early studies found clear-cut evolutionary differences between protostomes and vertebrate eyes, with separate lineages exhibiting distinct photoreceptor cascades and cellular responses. However, there are many taxa, such as the polychaetes and scallops, and honeybees, which contain one photoreceptor type in their eyes and another in their brain (Purschke et al., 2006; Serb, 2008; Velarde et al., 2005). This leads to the theory that both photoreceptor types were present in an ancient lineage, but were subsequently specialized for different tasks as the species diverged. It has been hypothesized that vertebrate retinal neurons are ancestrally related to rhabdomeric photoreceptors, meaning that in vertebrate lineages the ciliary photoreceptors were specialized in image formation, while the rhabdomeric photoreceptors were ancestrally related to the retinal ganglion cells (Erclik et al., 2009).

The discovery that the homeobox gene *Pax6* acts as a "master control gene" for eye morphogenesis in flies and vertebrates led to the proposal that complex animal eyes are monophyletic (from a single common ancestor) in origin (Erclik et al., 2009). Paired box domain (*Pax*) transcription factors are highly conserved and involved in a wide range of developmental pathways, the major genes being *Pax6, Pax6(5a),* and *Pax2,* all of which are involved in neural development (Skottheim Honn et al., 2016). The *Pax6* protein in particular is involved in the formation of eyes during development and is essential to lens induction or retinal neural differentiation (Carbe et al., 2013). The *Pax6* gene's ability to induce ectopic eyes in diverse species is a key piece of evidence supporting the idea that animal visual development has a monophyletic origin. Specifically, the observation that the vertebrate *Pax6* transcription factor can trigger eye development when introduced into *Drosophila* (and other organisms) suggests a shared, conserved genetic pathway for eye formation. This conserved function, despite differences in eye structure across various groups, suggests a common ancestral origin of the *Pax6* gene and the developmental pathways it regulates (Chow et al., 1999).

The importance of *Pax6* is evident through the consequences of interrupted expression. When *Pax6* is disrupted, it results in the absence of eyes and often further disruption to the craniofacial and central nervous system of animals, including *Drosophila*, humans, and mice (Skottheim Honn et al., 2016). When humans are heterozygous for *Pax6,* they often develop blindness or other eye problems. Mice that are null for *Pax6* fail to form a mature eye structure (Carbe et al., 2013). Mutations of the gene have resulted in the absence of eyes or reduced eyes in both mammals and flies (Loosli et al., 1996). It is believed that the gene predates the development of visual systems; homologs of the gene are also found in the nervous system of animals beyond those with complex visual systems, such as sponges, which have been found to contain a homologue of the vertebrate *Pax-2/5/8* gene (Hoshiyama et al., 1998). In contrast, Cnidaria, which have lens-containing eyes, lack a *Pax6* gene homolog, relying instead on a different class of *Pax* genes in their visual development (Kozmik, 2008). This means that *Pax6* is likely ancestral in bilaterian genes, and postdates the divergence of cnidarians and bilaterians (Kozmik, 2008). The evolution of the *Pax* genes can be traced back to a single *Pax2* gene in sponges; the emergence of the visual system in bilateria coincides with the separation of the *Pax2* and *Pax6* genes (Carbe et al., 2013).

There are still some discrepancies in the theory of *Pax6* as the "master control gene" of eye development, as there are examples of bilaterian animals that lack the *Pax6* control of eye development. *Platynereis,* a polychaete, and several other species rely on different genes for eye development and regeneration (Arendt & Wittbrodt, 2001). It is unclear why this is if the argument for a monophyletic eye origin with the *Pax* gene as the ancestral control is true. Another issue is that there have been very few species with extensive studies on the *Pax6* development, instead relying on a generally good correlation between *Pax6* expression and the presence of eyes; a wider and more in-depth study on a range of animals would therefore be necessary to better understand the *Pax6* gene control of visual systems. In sum, there has been no consensus at this point about whether the visual system arose monophyletically , paraphyletically (shared ancestry but excluding some descendants) or polyphyletically (convergent evolution); what is known is that visual systems are extremely complex and important to continue to study and unravel to better understand evolution in the animal kingdom.

7. Summary

There is a remarkable diversity in animal eyes. In most cases, regardless of anatomical complexity and physiological sophistication eyes require a fundamental set of building blocks. While bilaterians have extremely diverse eye structures, the majority of animal visual systems use either a rhabdomeric or ciliary photoreceptor to house a wide variety of opsin visual pigments. Despite the ongoing debate for a monophyletic or paraphyletic origin of these structures, the eye remains an evolutionary success that has contributed to the remarkable speciation that has occurred since the Cambrian explosion.

References

Arendt, D., Tessmar, K., de Campos-Baptista, M. M., Dorresteijn, A., & Wittbrodt, J. (2002). Development of pigment- cup eyes in the polychaete platynereis dumerilii and evolutionary conservation of larval eyes in bilateria. *Development, 129*(5), 1143.

Arendt, D., Tessmar-Raible, K., Snyman, H., Dorresteijn, A. W., & Wittbrodt, J. (2004). Ciliary photoreceptors with a vertebrate- type opsin in an invertebrate brain. *Science, 306*(5697), 869.

Arendt, D., & Wittbrodt, J. (2001). Reconstructing the eyes of urbilateria. *Philosophical Transactions: Biological Sciences, 356*(1414), 1545–1563.

Arendt, D. (2003). Evolution of eyes and photoreceptor cell types. *International Journal of Developmental Biology, 47*(7), 563–571.

Barber, V., Evans, E., & Land, M. (1967). The fine structure of the eye of the mollusc pecten maximus. *Zeitschrift Für Zellforschung, 76*(3), 295–312.

Battelle, B., Ryan, J. F., Kempler, K. E., Saraf, S. R., Marten, C. E., Warren, W. C., Minx, P. J., Montague, M. J., Green, P. J., Schmidt, S. A., Fulton, L., Patel, N. H., Protas, M. E., Wilson, R. K., & Porter, M. L. (2016). Opsin repertoire and expression patterns in horseshoe crabs: Evidence from the genome of limulus polyphemus (arthropoda: Chelicerata). *Genome Biology and Evolution, 8*(5), 1571–1589.

Bellingham, J., Chaurasia, S. S., Melyan, Z., Liu, C., Cameron, M. A., Tarttelin, E. E., Iuvone, P. M., Hankins, M. W., Tosini, G., & Lucas, R. J. (2006). Evolution of melanopsin photoreceptors: Discovery and characterization of a new melanopsin in nonmammalian vertebrates (shedding light on melanopsin evolution). *PLoS Biology, 4*(8), Article e254.

Bielecki, J., Zaharoff, A. K., Leung, N. Y., Garm, A., Oakley, T. H., & Warrant, E. J. (2014). Ocular and extraocular expression of opsins in the rhopalium of tripedalia cystophora (cnidaria: Cubozoa). *PLoS One, 9*(6).

Briscoe, A. D., & Chittka, L. (2001). The evolution of color vision in insects. *Annual Review of Entomology, 46*, 471.

Carbe, C., Garg, A., Cai, Z., Li, H., Powers, A., & Zhang, X. (2013). An allelic series at the paired box gene 6 (Pax6) locus reveals the functional specificity of pax genes. *Journal of Biological Chemistry, 288*(17), Article 12130.

Cheng, N., Tsunenari, T., & King-Wai, Y. (2009). Intrinsic light response of retinal horizontal cells of teleosts. *Nature, 460*(7257), 899.

Chow, R. L., Altmann, C. R., Lang, R. A., & Hemmati-Brivanlou, A. (1999). Pax6 induces ectopic eyes in a vertebrate. *Development, 126*(19), 4213–4222.

Chung, W., & Marshall, N. J. (2016). Comparative visual ecology of cephalopods from different habitats. *Proceedings of the Royal Society B: Biological Sciences, 283*(1838).

Colicchia, G. (2006). Ancient cephalopod scavenges successfully with its pinhole eye. *Physics Education, 41*(1), 15–17.

Cronin, T., & Porter, M. (2008). Exceptional variation on a common theme: The evolution of crustacean compound eyes. *Evolution: Education and Outreach, 1*(4), 463–475.

Cronin, T. W., Porter, M. L., Bok, M. J., Caldwell, R. L., & Marshall, J. (2022). Colour vision in stomatopod crustacean. *Philosophical Transactions of the Royal Society B, 3777*, 20210278. https://doi.org/10.1098/rstb.2021.0278

Darwin, C. (1859). *On the origin of species by means of natural selection* (1st). London: John Murray.

Erclik, T., Hartenstein, V., Mcinnes, R. R., & Lipshitz, H. D. (2009). Eye evolution at high resolution: The neuron as a unit of homology. *Developmental Biology, 332*(1), 70–79.

Fain, G. L., Hardie, R., & Laughlin, S. B. (2010). Phototransduction and the evolution of photoreceptors. *Current Biology, 20*(3), R114–R124.

Fain, G. L., Matthews, H., Cornwall, M., & Koutalos, Y. (2001). Adaptation in vertebrate photoreceptors. *Physiological Reviews, 81*(1), 117.

Fernald, R. D. (2000). Evolution of eyes. *Current Opinion in Neurobiology, 10*(4), 444–450.

Feuda, R., Hamilton, S. C., Mcinerney, J. O., & Pisani, D. (2012). Metazoan opsin evolution reveals a simple route to animal vision. *Proceedings of the National Academy of Sciences of the U S A, 109*(46), Article 18868.

Gärtner, W., & Towner, P. (1995). Invertebrate visual pigments. *Photochemistry and Photobiology, 62*(1), 1.

Gühmann, M., Jia, H., Randel, N., Verasztó, C., Bezares-Calderón, L., Michiels, N., Yokoyama, S., & Jékely, G. (2015). Spectral tuning of phototaxis by a go- opsin in the rhabdomeric eyes of *Platynereis. Current Biology, 25*(17), 2265–2271.

Gehring, W. J. (2011). Chance and necessity in eye evolution. *Genome Biology and Evolution, 3*, 1053–1066.

Gehring, W. J. (2005). New perspectives on eye development and the evolution of eyes and photoreceptors. *Journal of Heredity, 96*(3), 171–184.

Gomez, M. P., & Nasi, E. (2000). Light transduction in invertebrate hyperpolarizing photoreceptors: Possible involvement of a go- regulated guanylate cyclase. *Journal of Neuroscience, 20*(14), 5254.

Hárosi, F. I., & Novales Flamarique, I. (2012). Functional significance of the taper of vertebrate cone photoreceptors. *The Journal of General Physiology, 139*(2), 159–187.

Hardie, R. C., Martin, F., Cochrane, G. W., Juusola, M., Georgiev, P., & Raghu, P. (2002). Molecular basis of amplification in drosophila phototransduction: Roles for G protein, phospholipase C, and diacylglycerol kinase. *Neuron, 36*(4), 689–701.

Hardie, R. C., & Raghu, P. (2001). Visual transduction in drosophila. *Nature, 413*(6852), 186.

Hayakawa, S., Takaku, Y., Hwang, J. S., Horiguchi, T., Suga, H., Gehring, W., Ikeo, K., & Gojobori, T. (2015). Function and evolutionary origin of unicellular camera-type eye structure. *PLoS One, 10*(3), Article e0118415.

Henze, M., Dannenhauer, K., Kohler, M., Labhart, T., & Gesemann, M. (2012). Opsin evolution and expression in arthropod compound eyes and ocelli: Insights from the cricket *Gryllus bimaculatus. BMC Evolutionary Biology, 12.*

Hoshiyama, D., Suga, H., Iwabe, N., Koyanagi, M., Nikoh, N., Kuma, K., Matsuda, F., Honjo, T., & Miyata, T. (1998). Sponge pax cDNA related to pax-2/5/8 and ancient gene duplications in the pax family. *Journal of Molecular Evolution, 47*(6), 640–648.

Kanehisa, M., Furumichi, M., Tanabe, M., Sato, Y., & Morishima, K. (2017). Kegg: New perspectives on genomes, pathways, diseases and drugs. *Nucleic Acids Research, 45*, D353–D361.

Kanehisa, M., & Goto, S. (2000). KEGG: Kyoto encyclopedia of genes and genomes. *Nucleic Acids Research, 28*(1), 27.

Kanehisa, M., Sato, Y., Kawashima, M., Furumichi, M., & Tanabe, M. (2016). KEGG as a reference resource for gene and protein annotation. *Nucleic Acids Research, 44*, D457.

Kingston, A. C. N., Kuzirian, A. M., Hanlon, R. T., & Cronin, T. W. (2015). Visual phototransduction components in cephalopod chromatophores suggest dermal photoreception. *Journal of Experimental Biology, 218*(10), 1596.

Kojima, D., Terakita, A., Ishikawa, T., Tsukahara, Y., Maeda, A., & Shichida, Y. (1997). A novel go- mediated phototransduction cascade in scallop visual cells. *Journal of Biological Chemistry, 272*(37), Article 22979.

Koyanagi, M., Takano, K., Tsukamoto, H., Ohtsu, K., Tokunaga, F., & Terakita, A. (2008). Jellyfish vision starts with cAMP signaling mediated by opsin-G_s cascade. *Proceedings of the National Academy of Sciences, 105*(40), Article 15576.

Koyanagi, M., & Terakita, A. (2008). Gq- coupled rhodopsin subfamily composed of invertebrate visual pigment and melanopsin. *Photochemistry and Photobiology, 84*(4).

Kozmik, Z., Ruzickova, J., Jonasova, K., Matsumoto, Y., Vopalensky, P., Kozmikova, I., Strnad, H., Kawamura, S., Piatigorsky, J., Paces, V., et al. (2008). Assembly of the cnidarian camera-type eye from vertebrate-like components. *Proceedings of the National Academy of Sciences of the U S A, 105*(26), 8989.

Kozmik, Z. (2008). The role of pax genes in eye evolution. *Brain Research Bulletin, 75*(2–4), 335–339.

Lamb, T., Pugh, E., & Collin, S. (2008). The origin of the vertebrate eye. *Evolution: Education and Outreach, 1*(4), 415–426.

Lamb, T. D. (2013). Evolution of phototransduction, vertebrate photoreceptors and retina. *Progress in Retinal and Eye Research, 36*, 52–119.

Lamb, T. D. (2011). Evolution of the eye. *Scientific American, 305*(1), 64–69.

Loosli, F., Kmita-Cunisse, M., & Gehring, W. J. (1996). Isolation of a pax-6 homolog from the ribbonworm lineus sanguineus. *Proceedings of the National Academy of Sciences of the U S A, 93*(7), 2658.

Mahato, S., Morita, S., Tucker, A. E., Liang, X., Jackowska, M., Friedrich, M., Shiga, Y., & Zelhof, A. C. (2014). Common transcriptional mechanisms for visual photoreceptor cell differentiation among pancrustaceans (rhabdomeric photoreceptor cell differentiation). *PLoS Genetics, 10*(7), Article e1004484.

Marshall, J., Cronin, T. W., & Kleinlogel, S. (2007). Stomatopod eye structure and function: A review. *Arthropod Structure & Development, 36*(4), 420–448.

Martin, V. J. (2002). Photoreceptors of cnidarians. *Canadian Journal of Zoology, 80*(10), 1703–1722.

Nathans, J., Thomas, D., & Hogness, D. S. (1986). Molecular genetics of human color vision: The genes encoding blue, green, and red pigments. *Science, 232*(4747), 193–202.

Nie, J., Mahato, S., Mustill, W., Tipping, C., Bhattacharya, S. S., & Zelhof, A. C. (2012). Cross species analysis of prominin reveals a conserved cellular role in invertebrate and vertebrate photoreceptor cells. *Developmental Biology, 371*(2), 312–320.

Nilsson, D. (1996). Eye ancestry: Old genes for new eyes. *Current Biology, 6*(1), 39–42.

Nilsson, D. (2013). Eye evolution and its functional basis. *Visual Neuroscience, 30*(1–2), 5–20.

Nilsson, D. (2004). Eye evolution: A question of genetic promiscuity. *Current Opinion in Neurobiology, 14*(4), 407–414.

Nilsson, D. (2005). Photoreceptor evolution: Ancient siblings serve different tasks. *Current Biology, 15*(3), R94–R96.

Palczewski, K., Kumasaka, T., Hori, T., Behnke, C., Motoshima, H., Fox, G., Trong, I., Teller, D., OkadaT, S. R., et al. (2000). Crystal structure of rhodopsin: A G protein- coupled receptor. *Science, 289*(5480), 739–745.

Pearring, J. N., Salinas, R. Y., Baker, S. A., & Arshavsky, V. Y. (2013). Protein sorting, targeting and trafficking in photoreceptor cells. *Progress in Retinal and Eye Research, 36*, 24–51.

Plachetzki, D. C., Serb, J. M., & Oakley, T. H. (2005). New insights into the evolutionary history of photoreceptor cells. *Trends in Ecology & Evolution, 20*(9), 465–467.

Porter, M. L. (2016). Beyond the eye: Molecular evolution of extraocular photoreception. *Integrative and Comparative Biology, 56*(5), 842–852.

Porter, M. L., Blasic, J. R., Bok, M. J., Cameron, E. G., Pringle, T., Cronin, T. W., & Robinson, P. R. (2011). Shedding new light on opsin evolution. *Proceedings of the Royal Society B: Biological Sciences, 279*(1726), 3.

Porter, M. L., Cronin, T. W., McClellan, D. A., & Crandall, K. A. (2006). Molecular characterization of crustacean visual pigments and the evolution of pancrustacean opsins. *Molecular Biology and Evolution, 24*(1), 253–268.

Porter, M. L., Zhang, Y., Desai, S., Caldwell, R. L., & Cronin, T. W. (2010). Evolution of anatomical and physiological specialization in the compound eyes of stomatopod crustaceans. *Journal of Experimental Biology, 213*, 3473.

Provencio, I., Jiang, G., De Grip, W. J., Hayes, W. P., & Rollag, M. D. (1998). Melanopsin: An opsin in melanophores, brain, and eye. *Proceedings of the National Academy of Sciences of the U S A, 95*(1), 340–345.

Provencio, I., Rodriguez, I. R., Jiang, G., Hayes, W. P., Moreira, E. F., & Rollag, M. D. (2000). A novel human opsin in the inner retina. *Journal of Neuroscience, 20*(2), 600.

Purschke, G., Arendt, D., Hausen, H., & Müller, M. C. M. (2006). Photoreceptor cells and eyes in annelida. *Arthropod Structure & Development, 35*(4), 211–230.

Ramirez, M. D., Pairett, A. N., Pankey, M. S., Serb, J. M., Speiser, D. I., Swafford, A. J., & Oakley, T. H. (2016). The last common ancestor of most bilaterian animals possessed at least 9 opsins. *Genome Biology and Evolution, 8*(12), 3640–3652.

Serb, J., & Eernisse, D. (2008). Charting Evolution's trajectory: Using molluscan eye diversity to understand parallel and convergent evolution. *Evolution: Education and Outreach, 1*(4), 439–447.

Reingruber, J., Holcman, D., & Fain, G. L. (2015). How rods respond to single photons: Key adaptations of a G-protein cascade that enable vision at the physical limit of perception. *Bioessays, 37*(11), 1243–1252. https://doi.org/10.1002/bies.201500081

Rieke, F., & Baylor, D. A. (1998). Single-photon detection by rod cells of the retina. *Reviews of Modern Physics, 70*(3), 1027–1036. https://doi.org/10.1103/RevModPhys.70.1027

Serb, J. M. (2008). Toward developing models to study the disease, ecology, and evolution of the eye in Mollusca. *American Malacological Bulletin, 26*(1), 3–18.

Shichida, Y., & Matsuyama, T. (2009). Evolution of opsins and phototransduction. *Philosophical Transactions of the Royal Society of London - Series B: Biological Sciences, 364*(1531), 2881.

Skottheim Honn, J., Johansson, L., & Rasmuson Lestander, Å. (2016). Regulation of twin of eyeless during drosophila development. *Gene Expression Patterns, 20*(2), 120–129.

Speiser, D. I., & Johnsen, S. (2008). Comparative morphology of the concave mirror eyes of scallops (Pectinoide). *American Malacological Bulletin, 26*(1), 27–33.

Stryer, L. (1986). Cyclic GMP cascade of vision. *Annual Review of Neuroscience, 9*, 87.

Sung, C., & Chuang, J. (2010). The cell biology of vision. *The Journal of Cell Biology, 190*(6), 953.

Terakita, A., Tsukamoto, H., Koyanagi, M., Sugahara, M., Yamashita, T., & Shichida, Y. (2008). Expression and comparative characterization of Gq-coupled invertebrate visual pigments and melanopsin. *Journal of Neurochemistry, 105*(3), 883–890.

Terakita, A. (2005). The opsins. *Genome Biology, 6*(3), 213.

Tsukamoto, H., & Terakita, A. (2010). Diversity and functional properties of bistable pigments. *Photochemical and Photobiological Sciences, 9*(11), 1435–1443.

Velarde, R. A., Sauer, C. D., Walden Kk, O., Fahrbach, S. E., & Robertson, H. M. (2005). Pteropsin: A vertebrate-like non-visual opsin expressed in the honey bee brain. *Insect Biochemistry and Molecular Biology, 35*(12), 1367–1377.

Yarfitz, S., & Hurley, J. B. (1994). Transduction mechanisms of vertebrate and invertebrate photoreceptors. *Journal of Biological Chemistry, 269*(20), Article 14329.

Yilmaz, A., Hempel de Ibarra, N., & Kelber, A. (2022). High diversity of arthropod colour vision: from genes to ecology. *Philosophical Transactions of the Royal Society B, 377*(1862). https://doi.org/10.1098/rstb.2021.0273

Yokoyama, S. (2008). Evolution of dim- light and color vision pigments. *Annual Review of Genomics and Human Genetics, 9*, 259–282.

Yoshida, M., & Ogura, A. (2011). Genetic mechanisms involved in the evolution of the cephalopod camera eye revealed by transcriptomic and developmental studies. *BMC Evolutionary Biology, 11*, 180.

Chapter 29

Taste and smell chemoreceptors

Albertino Bigiani
Dipartimento di Scienze Biomediche, Metaboliche e Neuroscienze, Università di Modena e Reggio Emilia, Modena, Italy

1. Summary

Taste and smell are sensory activities used by vertebrates to survey their chemical environment. Chemosensory information relayed to the brain is critical for the regulation of vital processes such as nutrition, individual recognition, and reproduction. The ability to detect exogenous chemicals and convert them into the only language understood by central neurons, namely action potentials, is due to specific sensory cells known as taste and smell chemoreceptors, which are strategically located at the beginning of the digestive and respiratory systems, respectively. Taste chemoreceptors are specialized epithelial cells found in the oral cavity, whereas smell chemoreceptors are true neurons located in the nasal cavity. Regardless, the logic of their operation is similar: they have molecular sensors that detect chemicals and rely on intracellular signaling pathways that lead to the activation of afferent nerve fibers. Central processing of incoming gustatory and olfactory electrical impulses directs multiple behavioral outcomes based on body demands.

2. Introduction

The ability to detect environmental chemicals is widespread among living organisms. In bacteria and protozoans, the cell itself monitors the external medium and translates the molecular signals into cellular responses. In metazoans, complex cellular systems have evolved for recognizing and interpreting exogenous substances. Such systems mediate the so-called *chemical senses*, taste and smell (Bartoshuk & Beauchamp, 1994), which are critical for individual and species survival (Breslin, 2013; Ihara et al., 2013). Typically, taste provides information on the chemical composition of food, also preventing the ingestion of toxic substances, whereas smell allows probing the external world in search of food or mating partners. In humans, taste and smell also play a key role in hedonics, that is, the appreciation (pleasantness or unpleasantness) of food, beverages, and fragrances. Table 29.1 summarizes the anatomical and functional characteristics of the gustatory (taste) and olfactory (smell) systems in mammals.

Taste and smell systems rely on peripheral cellular elements known as *chemoreceptors* to probe chemicals. Taste chemoreceptors are specialized epithelial cells found in the oral cavity that detect the presence of substances in amounts compatible with bodily metabolic and homeostatic needs while also warning of potentially harmful molecules. Smell chemoreceptors are true neurons found in the mucosa of the nasal cavity that allow the brain to recognize trace amounts of chemicals that guide social-sexual behavior or aid in the discovery of food sources. *Transduction* is the process by which taste and smell chemoreceptors convert exogenous chemicals into cellular signals, such as a membrane depolarization, a change in cytosolic Ca^{2+} concentration, or both. This process includes the interaction of chemical stimuli with specific receptor proteins, as well as the intracellular molecular signaling triggered by that interaction. The cellular responses elicited during transduction are required for the generation of action potentials in afferent nerve fibers.

In this chapter, I will look at the molecular and cellular events that underpin the chemical detection mechanisms in taste and smell chemoreceptors. The majority of data comes from studies using laboratory rodents (mainly mice and rats) as animal models. As a result, I will discuss current theories about how taste and smell chemoreceptors work in these species. It is important to note that different sensory mechanisms may operate in other vertebrates or in invertebrates depending on feeding habits and social behaviors (Kaupp, 2010; Liman et al., 2014).

Cell Physiology Sourcebook. https://doi.org/10.1016/B978-0-12-811114-7.00034-7

TABLE 29.1 Taste and smell in mammals: Anatomical and functional characteristics.

		Smell	
	Taste	**Olfactory mucosa**	**Vomeronasal organ**
Adequate stimulus	Water-dissolved chemicals	Airborne chemicals (Odorants and pheromones)	Volatile and fluid-phase chemicals (Pheromones)
Localization of chemoreceptors	Mouth	Nose	Nose
Nature of chemoreceptors	Epithelial cells	Neurons	Neurons
Peripheral synapses	Yes	No	No
Chemoreceptor turnover	Yes	Yes	Yes
Molecular sensors	Ion channels GPCRs	GPCRs	GPCRs
Cranial nerves	VII − IX − X	I	Vomeronasal nerve
First central neurons	Nucleus of the solitary tract (Medulla)	Main olfactory bulb (Forebrain)	Accessory olfactory bulb (Forebrain)
Functional roles (examples)	Nutrition Defense against toxics	Nutrition Reproduction	Socio-sexual behaviors
Sensory qualities in humans	Salty, sour, sweet, bitter, umami	Thousands	a

GPCR, G-protein-coupled receptor; *I*, olfactory nerve; *IX*, glossopharyngeal nerve; *VII*, facial nerve; *X*, vagus nerve.
[a]*VNO is absent in humans.*

3. Taste chemoreceptors

3.1 General aspects of the gustatory function

3.1.1 Sensory structures

The existence of specialized peripheral structures, the *taste buds*, strategically located in the rostral region of the gastroenteric tract, is required for gustatory sensitivity in mammals. This allows for a quick assessment of the nutritional composition of food as well as the presence of potentially harmful substances prior to ingestion (Northcutt, 2004). Taste buds are mostly confined to small mucosa prominences of the tongue termed *fungiform*, *foliate*, and *circumvallate papillae* (Breslin & Huang, 2006; Witt, 2019). Isolated taste buds can also be found throughout the palate and epiglottis mucosa (Breslin & Huang, 2006).

Taste buds are onion-shaped clusters of 50−100 specialized epithelial cells, called *taste cells* (Royer & Kinnamon, 1994; Witt, 2019). Each taste bud communicates with the oral cavity through a tiny hole known as the *taste pore*, through which chemicals dissolved in saliva (taste stimuli) can interact with taste cells (Yang et al., 2020). Taste cells activate *afferent nerve endings* through synaptic contacts (Kinnamon et al., 1985, 1988). Sensory information in the form of action potentials can thus be transmitted to the central nervous system. Taste cells reaching the taste pore are *elongate*, *spindle-shaped elements* that typically extend from the base to the top of taste buds (Royer & Kinnamon, 1994; Yang et al., 2020). *Tight junctions* are located just beneath the taste pore and divide the cell membrane into two distinct regions: the *apical membrane*, which protrudes into the taste pore and is bathed by saliva; and the *basolateral membrane*, which is bathed by interstitial fluid and is located below the tight junctions. The apical membrane recognizes chemicals, while the basolateral membrane is specialized for cell-to-cell communication. Thus, elongate taste cells are *functionally polarized sensory cells*. Taste cells, such as the epithelial cells of the surrounding mucosa, undergo *continuous turnover*. Cell renewal occurs in laboratory rodents in 2−3 weeks (Barlow, 2015; Perea-Martinez et al., 2013). At the base of the taste buds, round or ovoid cells coexist with elongate cells. These *basal cells* are postmitotic precursors of all elongate cells and derive from specific epithelial progenitors located around the taste buds (Barlow, 2015; Finger & Barlow, 2021). Taste cells arise from local epithelium rather than neurogenic ectoderm during embryonic development (Stone et al., 1995).

3.1.2 Taste stimuli

Taste stimuli are chemical substances recognized by taste buds. In terrestrial vertebrates, they are essentially *water-soluble substances*. Taste stimuli have a diverse chemical composition, ranging from simple inorganic ions, such as Na^+, to complex organic molecules, such as sugars (sucrose, fructose) and alkaloids (strychnine, quinine, caffeine), as well as proteins (monellin, thaumatin) (Cagan, 1973). Despite this chemical diversity, humans can only recognize a few taste sensations (Bartoshuk & Wixted, 2018; Schiffman & Erickson, 1971), which represent the *basic* or *primary tastes*: *salty*, *sour*, *sweet*, *bitter*, and *umami*. With animals, this terminology is not appropriate; instead, behavioral responses in terms of preference (acceptance), avoidance (rejection), or indifference are preferred.

Salty is the sensation evoked by the table salt, NaCl. The presence of Na^+ is the main cause of saltiness. The biological significance of salty taste is linked to electrolyte homeostasis, since it drives mineral ingestion in terrestrial animals to compensate for sodium loss through sweat and urine (Bigiani, 2021; Bigiani, 2020a). A *sour* taste is elicited by H^+ in food products. Because H^+ is a reactive ion, a very sour taste is thought to be a warning indication of spoiled food, where bacterial growth is associated with the production of organic acids. *Sweet* taste signals the presence of energy-rich substances in foods and beverages (Hladik et al., 2002). Common table sugar, sucrose, is an example of a sweet chemical. Natural sugars provide glucose, which is an energy substrate required for cellular metabolism, particularly in neurons and red blood cells. Poisons and, in general, all toxic chemicals for our bodies, including many drugs, have a *bitter* taste (Wooding et al., 2021). Alkaloids such as strychnine and caffeine induce this taste sensation. Bitter taste serves as a warning signal, so taste cells must be extremely sensitive to potentially harmful compounds (Hladik et al., 2002). *Umami*, a Japanese term that means "delicious," describes the sensation elicited by certain chemicals, such as free L-glutamate and some nucleotides that cannot be explained using traditional tastes (Lugaz et al., 2002). Umami is a flavor found in aged cheeses (such as parmesan cheese) and beef broth, but it has also been linked to several vegetables (such as, tomatoes and mushrooms). As the umami flavor refers to the detection of some amino acids (mainly monosodium glutamate for humans), it may indicate the presence of protein-rich foods (i.e., meat). In cookery, umami flavor is used to improve the palatability of a variety of processed meals (Diepeveen et al., 2022).

Usually, taste stimuli such as Na^+ and sugars must be present in the saliva at concentrations of millimolars or more to activate taste buds (Heath et al., 2006; Zhao et al., 2003). On the other hand, poisonous compounds can stimulate taste buds when present in extremely small amounts, even micromolars or less (Bufe et al., 2002; Heath et al., 2006; Meyerhof, 2005). These differences in detection thresholds are most likely due to the biological importance of taste cues. Sugars, for example, are needed in large quantities to sustain energy metabolism, so identifying them in minute amounts is pointless. Natural poisons (e.g., strychnine), on the other hand, must be detected even in trace amounts to avoid ingestion of life-threatening chemicals.

3.2 Type II and type III taste cells

Elongate, spindle-shaped taste cells are rather heterogeneous as to their morphological and structural features and can be subdivided into three groups: *Type I*, *Type II*, and *Type III cells* (Finger & Barlow, 2021). Major discoveries in understanding the functional roles of these morphotypes have been made in recent years utilizing a mix of approaches (functional testing, immunostaining, and single-cell RT-PCR) (Bigiani & Prandi, 2011; Chaudhari & Roper, 2010; Kinnamon & Finger, 2019; Roper & Chaudhari, 2017). Some of the molecular and biophysical properties of elongate taste cells, as well as their functional properties, are summarized in Table 29.2. *Type I cells* are assumed to be *glial-like cells* given the presence of enzymes responsible for neurotransmitter clearance and of potassium channels involved in extracellular K^+ buffering. *Type II cells* have the molecular machinery to detect organic compounds. Finally, *Type III cells* are able to sense sour stimuli. Taste cells detecting Na^+, named *sodium cells* (Nomura et al., 2020), likely represent a subset of Type II cells, as suggested by recent studies (Nomura et al., 2020; Ohmoto et al., 2020). In short, *mammalian taste buds contain three main groups of chemoreceptors*: *Type II cells* for sweet, bitter, and umami compounds, *Type III cells* for acids, and *sodium cells* for sodium ions. These cells, also called *taste receptor cells*, communicate with afferent nerve endings through the release of specific neurotransmitters. Interestingly, Type II cells have short microvilli in their apical membrane, whereas Type III cells have a thick apical microvillus (Yang et al., 2020). Furthermore, Type II cells do not have typical synaptic contacts with nerve endings because they lack synaptic vesicles, whereas Type III cells do (Finger & Barlow, 2021; Taruno et al., 2021).

TABLE 29.2 **Molecular and functional properties of elongate taste cells in mammalian taste buds.**

Cell types	*Receptors, ion channels and other proteins*	*Physiology*
Type I cells (Glial-like cells)	[1] Ecto-ATPase [2] GLAST [3] $K_{IR}1.1$ [4] P2Y	[1] Extracellular ATP hydrolysis [2] Glutamate uptake [3] Extracellular K^+ redistribution [4] Paracrine signaling (ATP receptor)
Type II cells (Chemoreceptors that use ATP as neurotransmitter)	[1] T1Rs, T2Rs [2] Gustducin [3] PLCβ2 + IP_3R_3 + TRPM5 [4] $K_{IR}2.1$ [5] $Na_V1.7$, $Na_V1.3$ [6] Voltage-gated K^+ channels [7] CALHM1/CALHM3 [8] 5-HT1A	[1,2,3] Sensory transduction of sweet, bitter, umami compounds [4] Resting potential [5,6] Action potential firing [7] ATP release [8] Paracrine signaling (serotonin receptor)
Sodium cells (Chemoreceptors that use ATP as neurotransmitter)	[1] ENaC [2] Voltage-gated Na^+ channels [3] Voltage-gated K^+ channels [4] CALHM1/CALHM3	[1] Sensory transduction of Na^+ (salty) [2,3] Action potential firing [4] ATP release
Type III cells (Chemoreceptors that use serotonin as neurotransmitter)	[1] OTOP1 [2] $K_{IR}2.1$ [3] $Na_V1.2$ [4] Voltage-gated K^+ channels [5] $Ca_V2.1$, $Ca_V1.2$ [6] SNAP25	[1,2] Sensory transduction of H^+ (sour) [2] Resting potential [3,4] Action potential firing [5,6] Calcium-dependent exocytosis [5,6] Serotonin release

5-HT1A, serotonin receptor type 1A; *CALHM*, Calcium homeostasis modulator channel; *Ca_V*, voltage-gated Ca^{2+} channel; *ENaC*, epithelial sodium channel; *GLAST*, Glutamate aspartate transporter; *IP_3R_3*, Inositol 1,4,5-trisphosphate receptor type 3; *K_{IR}*, Inwardly rectifying K^+ channel; *Na_V*, voltage-gated Na^+ channel; *OTOP1*, Otopetrin1 proton channel; *P2Y*, metabotropic purinergic receptor; *PLCβ2*, Phospholipase C β2; *SNAP25*, Synaptosomal-associated protein 25; *T1R*, *T2R*, Taste receptors; *TRPM5*, Transient receptor potential M5 channel.

3.2.1 Electrophysiology

Taste chemoreceptors were thought to be "passive" electrical elements with a membrane behaving like an ohmic resistor until the 1980s, when Stephen D. Roper's groundbreaking work revolutionized our understanding of taste cell physiology. His seminal microelectrode recordings with the large taste receptor cells of the amphibian *Necturus maculosus* revealed that these cells are indeed *electrically excitable*, that is, capable of firing action potentials (Fig. 29.1A) (Roper, 1983). With the development of patch-clamp techniques, it was possible to definitively show that mammalian taste chemoreceptors can generate action potentials (Fig. 29.1B) and that their membrane electrical activity differs dramatically from linearity (Béhé et al., 1990; Chen et al., 1996; Kimura et al., 2014; Ma et al., 2017). Indeed, these cells express a variety of *voltage-gated ion channels* similar to those seen in other excitable cells like neurons (Furue & Yoshii, 1997; Gao et al., 2009; Herness & Sun, 1995; Liu et al., 2005; Medler et al., 2003). For many years, the relevance of action potentials in cells without fiber-like processes has been a mystery. Other sensory receptors, such as those in the cochlea and retina, cannot generate action potentials despite having many different types of ion channels. However, some epithelial-like cells, such as pancreatic β-cells, are able to fire action potentials during insulin release. As described later, the substantial membrane depolarization associated with the action potential is currently thought to be essential to activate the molecular machinery involved in neurotransmitter release (Taruno, Matsumoto, et al., 2013, 2021; Vandenbeuch & Kinnamon, 2009).

3.2.2 Cell-to-cell communication

Taste chemoreceptors communicate not only with nerve endings but also with each other. *Dye-coupling* with Lucifer yellow in *Necturus* taste buds (Yang & Roper, 1987) provided the first indirect evidence of *electrical communication*, demonstrating the presence of gap junction between neighboring cells. The patch-clamp technique has allowed researchers to study the properties of these electrical connections (Bigiani & Roper, 1994, 1995). Taste cells in mice have also been shown to be electrically coupled (Yoshii, 2005). Nonetheless, the role of cell-to-cell electrical transmission in taste bud physiology is currently unknown (Roper, 2021). On the other hand, a wealth of knowledge has been obtained about the

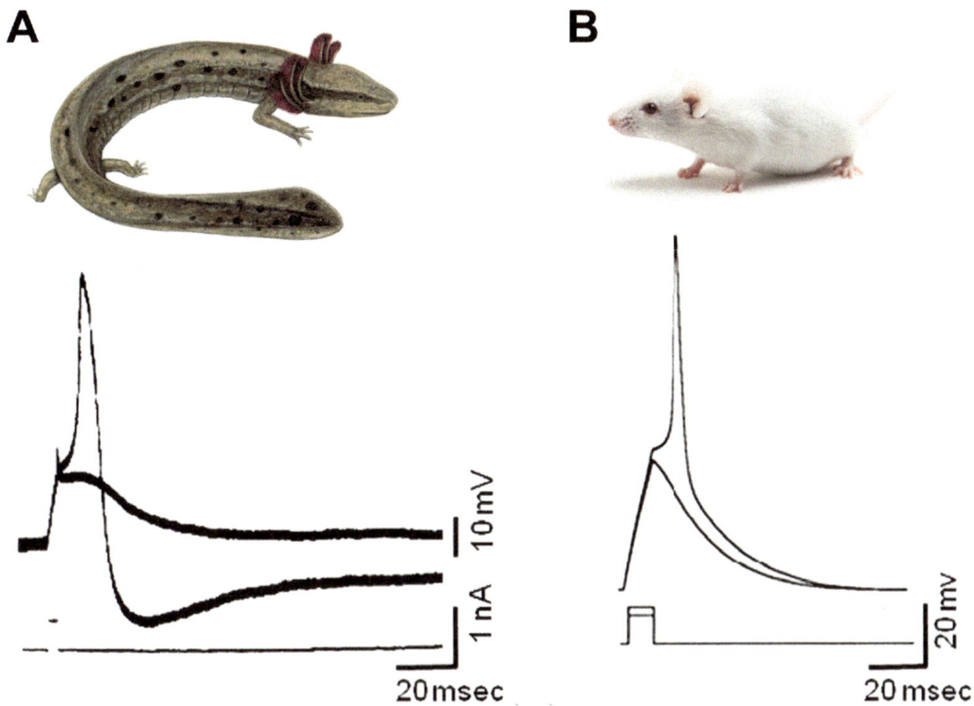

FIGURE 29.1 **Vertebrate taste cells are electrically excitable.** (A) Intracellular recording from a *Necturus* taste cell demonstrates an overshooting action potential induced by a depolarizing current injected through the recording microelectrode. (B) Patch clamp recording (current clamp mode) from a mouse taste cell demonstrates an action potential triggered by a depolarizing current pulse injected through the patch electrode. The similarity of the waveforms between the two animals indicates that the ionic mechanisms underlying these electrical signals do not differ significantly between species. *(A) Modified from Roper (1983), (B) modified from Bigiani et al. (2002).*

chemical communication between taste chemoreceptors in laboratory rodents. According to molecular biology, immunocytochemistry, and functional studies (Cao et al., 2009; Dando & Roper, 2009, 2012; Dvoryanchikov et al., 2011; Herness et al., 2005; Huang et al., 2009, Huang, Pereira, et al., 2011, Huang, Stone, et al., 2011; Kaya et al., 2004), *paracrine* and *autocrine signaling* exist inside taste buds. When taste cells are stimulated with specific substances, complex feedback systems based on lateral interactions are activated, and this is what some authors propose as the *signal processing inside the taste buds* (Chaudhari, 2014; Roper, 2013, 2021). The main molecules involved in cell-to-cell communications are *ATP* and *acetylcholine* released by Type II cells, and *serotonin* and *GABA* released by Type III cells (Roper, 2021). Recent research has suggested that Type I "glial-like" cells may also play a role in the intercellular communication inside taste buds (Baumer-Harrison et al., 2020; Dvoryanchikov et al., 2011; Guarascio et al., 2021; Rodriguez et al., 2021).

3.3 Sensory transduction

3.3.1 Taste receptors

The interaction of taste stimuli with specific proteins (*taste receptors*) in the apical membrane characterizes the initial events of the transduction process. There are two main categories of taste receptors: *ion channels* and *G protein-coupled receptors* (*GPCRs*) (Töle et al., 2019). It is not surprising that ion channels can function as receptors because the flow of ions requires the channel protein to recognize the permeating ion. In mammals, two electrolytes, Na^+ (salty) and H^+ (sour), are detected through ion channels. Apical GPCRs, on the other hand, allow the detection of organic compounds that are responsible for sweet, bitter, and umami tastes.

Na^+ receptors – The *epithelial sodium channel* (*ENaC*) is the most well-studied sodium receptor (Bigiani, 2016). The effect of the diuretic drug, *amiloride*, on the nerve response to tongue stimulation with NaCl solutions was the first to show the presence of ENaC in taste cells (Heck et al., 1984). ENaC is an ion channel found in many adsorbing epithelia, including the renal tubule and colon lining (Garty & Palmer, 1997). Several patch-clamp studies evaluating the so-called

response to amiloride (Fig. 29.2B), a membrane current change induced by amiloride application (Bigiani, 2016), have then provided information on the biophysical and pharmacological properties of taste ENaC. This channel is characterized by selectivity for Na^+ over K^+, is constitutively open, is not voltage-gated, and is inhibited at submicromolar concentrations by amiloride (Bigiani, 2016). According to molecular biology and immunocytochemistry studies, taste ENaC in laboratory rodents is likely formed of the same three subunits (α, β, and γ) that characterize members of the degenerin/epithelial Na^+ channel (DEG/ENaC) superfamily (Fig. 29.2A) (Bianchi, 2021; Kellenberger & Schild, 2002). The precise stoichiometry of ENaC subunit composition in taste cells, however, remains unknown (Lossow et al., 2020). Nonetheless, the apical membrane of "sodium cells" contains functional ENaCs (Yoshida et al., 2009). In humans, an additional subunit, δ-ENaC, is expressed in taste tissues. This subunit, which is missing in rodents, is homologous to the α-ENaC (Bigiani, 2020b).

The administration of amiloride to the tongue only partially abolishes the response to Na^+ recorded from taste nerves, indicating that sodium receptors other than ENaC must exist in taste buds (Bachmanov et al., 2014; Bigiani, 2020a; Roper,

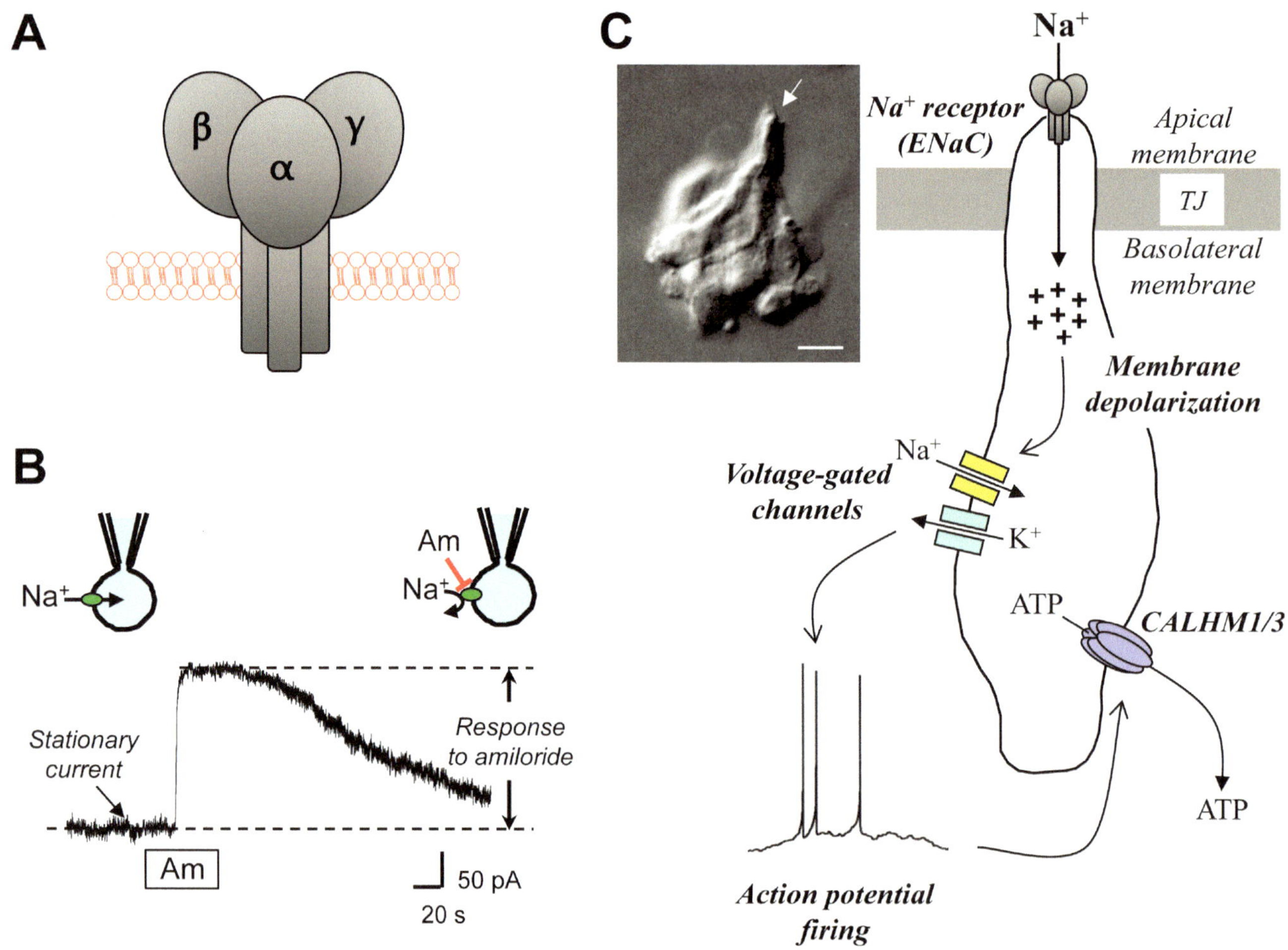

FIGURE 29.2 **Taste transduction pathway for Na + detection in sodium cells.** In this model, the epithelial sodium channel (ENaC) at the apical membrane represents the sodium receptor. (A) ENaC membrane topology and subunit stoichiometry. In a typical ENaC, the channel pore is formed by all of the subunits (α, β, γ). (B). A response to amiloride was recorded with the patch-clamp technique (whole-cell, voltage clamp mode) in a rat fungiform taste cell. At a membrane potential of −80 mV and with the cell immersed in physiological saline containing 150 mM Na^+, a stationary current directed toward the interior of the cell and caused by sodium ions entering through open ENaCs can be recorded. Bath application of amiloride (1 μM; Am, box under record) results in a decrease in stationary current (response to amiloride) due to amiloride blockage of open channels. (C) Na^+ detection and transduction working hypothesis. Increases in the concentration of Na^+ in the saliva cause an influx of Na^+ driven by its electrochemical gradient through the apical ENaCs. The apical inward Na^+ current (transduction current) causes membrane depolarization (receptor potential), which activates voltage-gated ion channels and causes action potential firing. This electrical excitation is required to open the CALHM1/3 ion channel complex, which is responsible for the nonvesicular release of neurotransmitter (ATP), which affects nerve endings or other adjacent taste cells. (**Inset**) A taste bud isolated from a rat fungiform papilla observed with a differential interference contrast microscope. The arrow points to the apical end of taste cells. Scale bar: 5 μm. *TJ,* tight junction. *(C) modified from Bigiani (2016).*

2015). The relative contribution of amiloride-sensitive (ENaC-mediated) and amiloride-insensitive components to sodium sensing differs substantially between mammalian species. In general, species with a low sodium intake, i.e., herbivores, and some omnivores, such as rodents, utilize ENaC as a specific sodium receptor (Chandrashekar et al., 2010), whereas carnivores may utilize a more nonselective ion channel (Bachmanov et al., 2014; Lindemann, 1996). Both amiloride-sensitive and amiloride-insensitive components have been described for sodium taste in humans (Feldman et al., 2009; Schiffman et al., 1983), although the role of ENaC is not well understood (Bigiani, 2020b). Multiple mechanisms may underpin the amiloride-insensitive component of the salt response, according to studies with mouse circumvallate taste cells. Chemical-physical properties of salt solutions, such as ionic strength (Oka et al., 2013), osmotic pressure (Lewandowski et al., 2016), anion size (Lewandowski et al., 2016; Roebber et al., 2019), and the presence of chloride ions (Roebber et al., 2019) can all alter taste cell activity.

H^+ receptors – Although it was obvious from the early observations that taste chemoreceptors use an apical channel to monitor the H^+ content (sour) in the saliva (Bigiani et al., 2003), it took several studies to develop a comprehensive model of sour transduction. The great chemical reactivity of protons, which can affect nearly any protein, has been the principal obstacle in exploring H^+ detection. Furthermore, there appears to be a lot of variation in the detecting mechanisms among species. Hamster taste cells seem to use the apical ENaC to mediate a proton influx when the saliva concentration of H^+ increases (Gilbertson et al., 1992). Indeed, typical ENaC shows a considerable permeability to H^+ (Garty & Palmer, 1997). In the rat, molecular evidence suggests that an *acid-sensitive ion channel* (*ASIC*) mediates proton detection (Shimada et al., 2006). Together with ENaC, this channel belongs to the DEG/ENaC family of ion channels, which are typically Na^+-permeable pores (Bianchi, 2021; Kellenberger & Schild, 2015). Rat taste cells express ASIC2a and ASIC2b, which form a heteromeric complex to function as presumptive sour receptor (Ugawa et al., 2003). ASIC is a *proton-gated channel* that opens transiently when the extracellular pH drops, enabling Na^+ to flow in. The lack of understanding about ASIC's location on the apical membrane of taste cells, on the other hand, raises concerns about its role in acid detection.

In recent years, studies on mice have provided compelling evidence for the role of a proton channel as a molecular detector of acidic stimuli (Chang et al., 2010). This channel has distinct characteristics: it is permeable to H^+, it is not voltage-dependent, and it is inhibited by extracellular Zn^{2+} but not by amiloride. Transcriptome analysis of mouse taste cells has led to the identification of *otopetrin1* (*OTOP1*) as the proton channel in acid detecting cells (Tu et al., 2018). Remarkably, the OTOP1 channel is a protein required for the development of otoconia, bio-crystals found in the vestibular system for sensing gravity (Lundberg et al., 2015). Further studies have definitely demonstrated that the OTOP1 proton channel functions as an acid receptor in mouse Type III cells (Teng et al., 2019; Zhang et al., 2019). Although Zn^{2+} is known to also affect the activity of ASICs (Kellenberger & Schild, 2015), ASIC2 is not expressed in mouse taste cells, and, consequently, ASIC2 knock-out mice respond normally to sour stimuli (Richter et al., 2004). It was possible to demonstrate the apical localization of the proton channel in single mouse taste cells by combining the suction electron recording approach with a controlled focal increase in intracellular H^+ concentration based on UV flash photolysis of caged protons (Chang et al., 2010). This is significant because the apical membrane is immediately exposed to the majority of taste stimuli.

G Protein-Coupled Receptors - By using molecular biology and genetics methodologies, it has been feasible to uncover two families of genes and related proteins that function as taste receptors: *T1Rs* or *TAS1Rs* (Nelson et al., 2001, 2002; Sainz et al., 2001), and *T2Rs* or *TAS2Rs* (Adler et al., 2000; Chandrashekar et al., 2000). All of these receptors have seven transmembrane helices that are characteristic of G protein-coupled receptors. T1Rs and T2Rs are coupled to a heterotrimeric G-protein composed of a specific α subunit, *gustducin* (Wong et al., 1996) and the $\beta3$ and $\gamma13$ subunits (Huang et al., 1999). Gustducin is related to transducin found in photoreceptors (McLaughlin et al., 1992).

T1Rs are classified into three subtypes in humans and many other mammals: T1R1, T1R2, and T1R3 (Nei et al., 2008). T1Rs heterodimerize to form the operating taste receptor. In particular, the heterodimer *T1R2/T1R3* is the molecular detector of sweet compounds (Nelson et al., 2001), whereas *T1R1/T1R3* is required to detect umami substances (Nelson et al., 2002). The human sweet receptor T1R2/T1R3 responds to many natural sugars (e.g., sucrose, fructose, lactose) and synthetic sweeteners (e.g., acesulfame K, saccharin, aspartame, cyclamate), as well as to sweet-tasting proteins, such as monellin and thaumatin (Li et al., 2002). T1R2/T1R3 has multiple ligand binding sites to interact with such a variety of sweet chemicals (Cui et al., 2006). Nevertheless, the signaling cascade activated by T1R2/T1R3 is common to all ligands. The human heterodimer T1R1/T1R3 responds to umami substances, such as L-glutamate, and activates the same signaling cascade controlled by the sweet receptor.

T2Rs comprise several subtypes (25 in humans and 35 in mice) (Nei et al., 2008), which interact with bitter substances (e.g., alkaloids). T2Rs, unlike T1Rs, do not form dimers but instead function as a single protein molecule responding to a wide range of bitter compounds (Behrens & Meyerhof, 2006). T2Rs, surprisingly, activate the same signaling pathway as sweet and umami receptors (Zhang et al., 2003). However, despite the fact that T1Rs and T2Rs are both found in Type II cells, each cell expresses only one type of receptor, either a sweet receptor (T1R2/T1R3), an umami receptor (T1R1/T1R3), or bitter receptors (T2Rs). Thus, there are dedicated Type II cells for each of the primary tastes: sweet, umami, and bitter (Chandrashekar et al., 2006).

3.3.2 Cellular responses

The interaction of chemical stimuli with their corresponding taste receptors results in a cellular response that is characterized by the appearance of specific biological signals: *membrane depolarization, intracellular Ca^{2+} change*, or both. The production of these signals is an important step in the transduction process.

Response to Na^+- Increases in the sodium ion concentration in the oral cavity are thought to be transduced by a *direct inflow of Na^+ into taste cells* through open ENaCs at the apical membrane. In turn, the inflow of Na^+ causes a *membrane depolarization* (*receptor potential*), which is followed by action potential firing (Fig. 29.2C). Thus, the taste stimulus (Na^+) enters into taste cells and produces an electrical signal (Nomura et al., 2020): in other words, there is a *transduction current* carried by the stimulus itself. The excess of Na^+ is then pumped out at the basolateral membrane through the Na^+/K^+-ATPase (Lindemann, 1996), and the current loop is closed through the paracellular pathway through the tight junctions (Bigiani, 2020a).

Na^+ concentrations in saliva typically range from 2−3 to 60−100 mM, depending on the rate of salivary production (Matsuo et al., 2015). With an estimated resting potential of about −70/-80 mV (Béhé et al., 1990) and a low concentration of Na^+ in the cytosol, there must be a steady-state flow of salivary Na^+ into taste cells down its electrochemical gradient. This sodium current represents a background chemical signal on top of which taste cells read sodium concentration changes. Thus, salivary sodium content is a physiological *adapting solution* for sodium taste detection. The sodium concentration in the oral cavity can rise significantly when consuming sodium-containing foods. A popular canned French onion soup can have 300 mM NaCl or more (Bigiani, 2020a). The extracellular sodium concentration at the apical membrane of taste cells will rise because of the high sodium content. Na^+ influx will increase significantly, causing membrane depolarization despite the background, adapting solution.

Response to H^+- Taste chemoreceptors detect saliva acidification (pH < 7) and provide a conscious feeling in humans known as "sour taste." The mechanism of activation of taste cells by protons seems to vary considerably among species (see above). In short, it is likely that a single mechanism for H^+ detection common to all mammals does not exist. Since most of the recent work has been done on mice, I will present the current hypothesis for sour transduction in this rodent.

H^+ from strong acids (e.g., HCl) permeates the apical proton channel (OTOP1), producing an inward *transduction current* that depolarizes Type III cells (Fig. 29.3). Intracellular acidification (Fig. 29.3, *Inset B_1*), in turn, inhibits a specific potassium channel, $K_{ir}2.1$ (Ye et al., 2016), with a potentiation of the initial depolarization (*signal amplification*). $K_{IR}2.1$ is an inwardly rectifying K^+ channel that contributes to setting the membrane potential to negative values at rest. The resulting receptor potential triggers impulse firing with consequent opening of voltage-gated Ca^{2+} channels, an increase in intracellular Ca^{2+} concentration (Fig. 29.3, *Inset B_2*), and calcium-dependent exocytosis of the neurotransmitter, serotonin, stored in synaptic vesicles (Vandenbeuch et al., 2010). This model also explains how weak acids such as acetic acid (found in vinegar) and citric acid (found in lemon and lime) activate taste cells. These acids are membrane permeable in their undissociated, neutral form, and can diffuse into the cytoplasm (Richter et al., 2003). Here, due to the cell's resting pH of about 7, they dissociate, releasing protons that block the $K_{IR}2.1$ channel (Ye et al., 2016) with consequent depolarization and release of the neurotransmitter (Huang et al., 2008). The involvement of the $K_{IR}2.1$ channels minimizes the need for substantial proton currents to produce membrane depolarization (Ye et al., 2016). This mechanism explains why weak acids are effective as taste stimuli at higher pHs and why they taste more sour than strong acids at equal pHs (Beatty & Cragg, 1935; Richards, 1898).

Response to organic molecules - Transduction of complex organic molecules, such as sugars and alkaloids, is achieved through the activation of T1Rs and T2Rs in the apical membrane of Type II taste cells. These receptors control a common intracellular signaling pathway involving the enzyme *phospholipase C β2* (PLCβ2) and the *transient receptor potential melastatin 5* (*TRPM5*) *channel*, a taste-specific cation channel (Pérez et al., 2002; Zhang et al., 2003). However, as described earlier, receptors for sugars (T1R2/T1R3), for umami substances (T1R1/T1R3), and for bitter compounds (T2Rs) segregate in separate cells; that is, they are not expressed in the same cell. When the taste stimulus binds to the apical GPCR, the G protein is activated, releasing the $\beta_3\gamma_{13}$ complex. This activates PLCβ2 with the consequent formation of diacylglycerol (DAG) and inositol-1,4,5-triphosphate (IP_3) from the membrane lipid phosphatidyl inositol bisphosphate (PIP_2). In this model, IP_3, a highly water-soluble molecule, diffuses into the cytosol and binds to its Type III receptor (IP_3R_3) in the endoplasmic reticulum membrane, leading to a transient release of Ca^{2+} (Fig. 29.4A). Thus, the first effect of taste stimulation is the mobilization of Ca^{2+}. The *increase in cytosolic Ca^{2+} concentration* gates the TRPM5 channel,

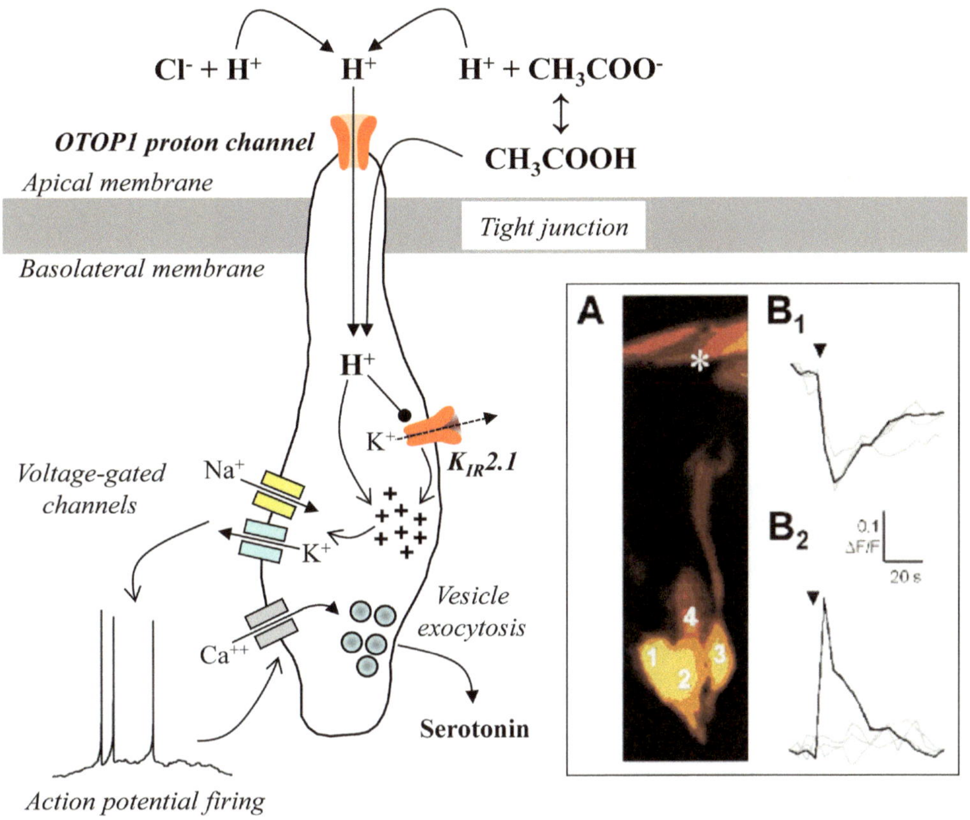

FIGURE 29.3 **Taste transduction pathway for H^+ detection in mouse type III taste cells.** In this model, the OTOP1 proton channel in the apical membrane mediates an inflow of H^+ when the saliva pH decreases. As a result, the membrane depolarizes (receptor potential) (+). This electrical signal is further potentiated by the blocking effect of cytosolic acidification on the resting K^+ conductance ($K_{IR}2.1$). Action potentials are generated when the firing threshold is reached. This electrical excitation opens voltage-gated Ca^{2+} channels, resulting in an increase in cytosolic Ca^{2+}, which is required to initiate synaptic vesicle exocytosis. Serotonin, the neurotransmitter used by sour-detecting taste cells, is finally released onto nerve endings. Weak acids may cause cytosolic acidification through a different mechanism. These acids, in their undissociated form, can cross the cell membrane and then dissociate in the cytosol, lowering intracellular pH without requiring an apical proton current. This additional mechanism supplements the one produced by the influx of free H^+. (**Inset**) Calcium imaging and pH imaging of taste cells with scanning laser confocal microscopy. Four taste cells were loaded with a Ca^{2+} indicator (CaO, red) and a pH indicator (BCECF-G, green) to simultaneously measure changes in intracellular $[Ca^{2+}]$ and pH (*A*). Overlapping red and green is visible as *yellow*. Focal application (*arrowhead*) of citric acid (100 mM, pH ~3) to the taste pore (*asterisk*) produces a reduction in the intracellular pH in all four labeled cells (*B_1*). On the contrary, the same acid stimulus induces an increase in intracellular $[Ca^{2+}]$ only in 1 cell (*B_2, black* trace). This implies that only this latter cell could convert the sour stimulus into a biological signal, despite the fact that the intracellular pH is reduced in all 4 cells (*gray* traces). *(Inset) Modified from Richter et al. (2003).*

which is selective for monovalent cations (Huang & Roper, 2010; Liu & Liman, 2003). As a result, Na^+ enters the cells and causes a *membrane depolarization* (*receptor potential*), which triggers the action potential firing required for neurotransmitter release (Fig. 29.4).

TRPM5 is a member of a broad superfamily of transient receptor potential channels that includes roughly 30 channels divided into six subfamilies in mammals (Wu et al., 2010). Most TRP channels are permeable to both monovalent (e.g., Na^+) and divalent (e.g., Ca^{2+}) cations, with the exception of TRPM4 and TRPM5, which are the only monovalent-selective TRP channels (Liu & Liman, 2003; Wu et al., 2010). TRPM5 is directly gated by intracellular Ca^{2+} (Fig. 29.4B), with a half-maximal concentration of 21 μM in the dose-response curve (Liu & Liman, 2003). Studies on TRPM5-KO mice indicate that the TRPM4 channel, which is activated by increased intracellular Ca^{2+} like TRPM5, contributes to stimulus transduction in taste cells (Dutta et al., 2018). Because TRPM5 is more calcium sensitive than TRPM4, TRPM5 and TRPM4 are thought to act in sequence when the intracellular Ca^{2+} concentration rises (Dutta et al., 2018).

The fate of the Gα subunit, gustducin, is not well understood. Gustducin knock-out mice have impaired sensitivity to bitter, sweet, and umami stimuli, implying that it plays a key role in taste transduction (He et al., 2004; Wong et al., 1996). Like transducin in photoreceptors, gustducin activates phosphodiesterase (PDE) in vitro, which lowers cytosolic cAMP

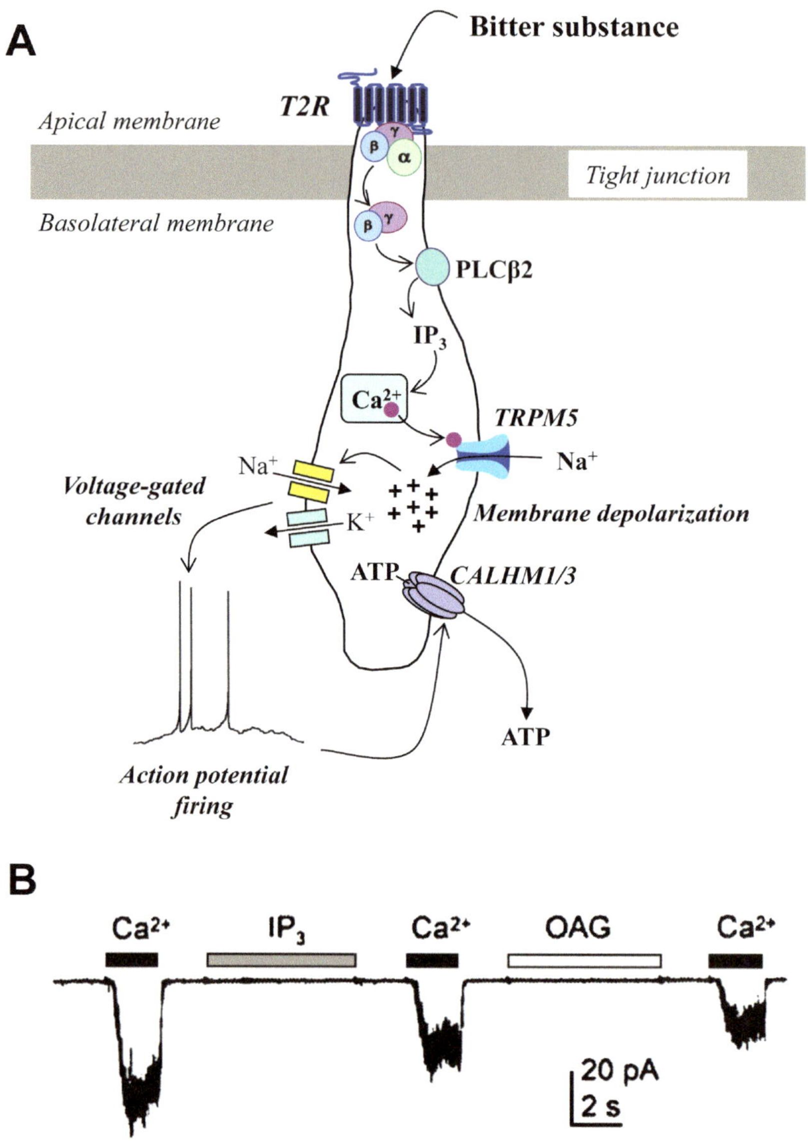

FIGURE 29.4 **Taste transduction pathway for the detection of organic molecules in mouse type II taste cells.** This model depicts the response to bitter compounds. (A) The taste stimulus binds to a G protein-coupled receptor (T2R) in the apical membrane. The βγ complex activates phospholipase C β2 (PLC β2), a membrane enzyme that hydrolyses a membrane lipid to form inositol-1,4,5-triphosphate (IP_3), a second messenger. IP_3 binds to its receptor on the membrane of calcium stores, causing Ca^{2+} to be released. Increased cytosolic Ca^{2+} opens a transient receptor potential M5 (TRPM5) channel, allowing Na^+ to enter taste cells and to depolarize the membrane. Action potentials are generated when the firing threshold is reached. This electrical excitation activates the CALHM1/3 channel complex, a large-conductance channel that allows ATP to exit (nonvesicular release). ATP directly affects nerve endings, but it may also influence the activity of other nearby taste cells. (B) Ca^{2+} sensitivity of TRPM5 current in a patch excised (inside-out configuration) from a TRPM5-transfected CHO−K1 cell. The membrane patch was held at −80 mV and was exposed to three putative second messengers: 40 μM Ca^{2+}, 10 μM IP_3, and 100 μM OAG (1-oleoyl-2-acetyl-*sn*-glycerol, an analog of DAG). Only Ca^{2+} could induce an inward current in the membrane patch. *Modified from Liu & Liman (2003); Copyright (2003) National Academy of Sciences, U.S.A.*

(McLaughlin et al., 1994). It has been proposed that gustducin ensures proper sensitivity of taste cells by maintaining a low cAMP level, as cAMP inhibits Ca^{2+} signaling (Clapp et al., 2008).

3.4 Activation of nerve endings

Once stimuli have been translated into biological signals, taste cells need to communicate with nerve endings to inform the brain: that is, taste cell activation must be converted into patterns of action potentials in afferent nerve fibers. This process is only partially understood and is a hotly debated topic (e.g., (Barretto et al., 2015; Chen et al., 2011; Tomchik et al., 2007;

Wu et al., 2015)). Despite this, a substantial amount of information has been gathered on the first step of sensory information transfer, that is, the release of neurotransmitters from taste chemoreceptors.

Immunohistochemical observations provided the first concrete evidence that taste cells may use chemical substances akin to known neurotransmitters (e.g., (Kim & Roper, 1995)). However, the role of these chemicals in synaptic transmission between taste chemoreceptors and nerve endings has only recently been proven with certainty. The use of molecular biology techniques and genetic manipulations (e.g., (Finger et al., 2005)), as well as the availability of a functional bioassay (biosensor technology) to detect the release of substances from taste cells during chemostimulation in real time (e.g., (Huang et al., 2005)), made this possible. It is now well established, at least in mice, that *ATP* represents the primary neurotransmitter involved in signaling between Type II taste cells and nerve endings (Finger & Kinnamon, 2021; Kinnamon and Finger, 2013, 2021). However, other substances, such as *serotonin* released by Type III cells, may play a role in transferring the taste information to nerve endings (Jaber et al., 2014; Larson et al., 2015).

How do taste chemoreceptors release neurotransmitters? In taste buds, two distinct mechanisms are at work: one for ATP release by Type II cells (and sodium cells), and another for serotonin release by Type III cells. In Type II cells, which lack synaptic vesicles, all the available evidence points to a *nonvesicular release of ATP* through large, nonspecific channels opened by membrane depolarization during action potential firing (Finger & Barlow, 2021; Taruno et al., 2021). For several years, the molecular identification of these channels has been a point of contention. Early studies suggested either pannexin 1 or connexin hemichannels might be involved (Huang et al., 2007; Romanov et al., 2007). Recent experiments, however, indicate that *calcium homeostasis modulator 1* (*CALHM1*) *channels* are essential for ATP release from Type II taste cells and sodium cells (Nomura et al., 2020; Taruno, Vingtdeux, et al., 2013). CALHM1 channels are associated with unique mitochondria, which appear to be the immediate source of releasable ATP and are found near the plasma membrane at the contact points with nerve endings (Romanov et al., 2018). Importantly, the biophysical properties of the native ATP releasing channel are identical to those of the *hetero-hexameric CALHM1/CALHM3 channel*, which therefore represents the *ATP-permeable channel complex* in Type II cells and in sodium cells (Ma et al., 2018; Nomura et al., 2020; Taruno et al., 2021). It remains to establish the role of both pannexin and connexin hemichannels, given their expression in taste cells (Huang et al., 2007; Romanov et al., 2007). In rat sodium cells, pannexin hemichannels might modulate the activity of CALHM1 channels by functional coupling (Bigiani, 2017). Serotonin release, on the contrary, takes place according to the conventional calcium-dependent exocytosis of synaptic vesicles (Finger & Barlow, 2021; Vandenbeuch et al., 2010). Action potential firing is required for the release of both ATP and serotonin.

Released ATP acts on purinergic receptors (P2X2/P2X3) in taste nerve endings (Bo et al., 1999; Ishida et al., 2009) as genetic elimination of these receptors abolishes taste responses in gustatory fibers (Finger et al., 2005). Surprisingly, P2X2/P2X3 double knockout mice do not respond to citric acid, which is detected by Type III cells, and NaCl, which is detected by sodium cells, in addition to sweet, bitter, and umami compounds detected by Type II cells (Finger et al., 2005). Thus, *ATP appears to be the primary taste neurotransmitter for all taste stimuli* (Finger & Kinnamon, 2021; Vandenbeuch et al., 2015). These results raise questions about the role of serotonin released by Type III cells during sour stimulation (Huang et al., 2005, 2008). The 5-HT1A serotonin receptor is found in Type II cells, whereas the 5-HT3A subtype is expressed in nerve endings (Herness & Chen, 2000; Jaber et al., 2014; Kaya et al., 2004; Larson et al., 2015; Wang et al., 2002). According to the most recent findings, serotonin acting on 5-HT3A receptors plays a minor role in the transmission of taste information to the central nervous system (Larson et al., 2015). It has also been suggested that serotonin might play a role in "shaping the afferent taste signals prior to their transmission across gustatory nerves" (Jaber et al., 2014). In conclusion, ATP is the key neurotransmitter for sour taste reception (Finger & Kinnamon, 2021), although its source remains a mystery. Type II cells do not produce the ATP required for the transmission of sour information, as mice lacking these taste cells respond normally to sour stimuli (Larson et al., 2020).

3.5 Termination of taste signals

The cellular response triggered by the interaction of taste stimuli with the apical channels/receptors is switched off upon the removal (dilution) of stimuli from the chemosensitive ends of taste chemoreceptors. Extracellular recordings of taste cell action potentials have clearly shown the existence of a second mechanism that reduces the response during prolonged, continuing stimulation (*sensory adaptation*) (Avenet & Lindemann, 1991; Gilbertson et al., 1992). The cellular events responsible for taste cell adaptation, however, are still unknown.

4. Smell chemoreceptors

4.1 General aspects of the olfactory function

4.1.1 Sensory structures

The *olfactory mucosa* in the nose detects airborne substances. Chemical cues (*odorants*) from other individuals, food sources, and the environment can, therefore, be tracked as soon as they enter the airways through the nostrils. Furthermore, substances generated during food chewing might travel through the posterior nasal openings (internal nostrils or choanae) and reach the olfactory mucosa. This sensory information is critical for developing food flavor perception in specific parts of the cortex, such as the orbitofrontal cortex, where taste and olfactory information are combined (Rolls, 2015). In addition to the olfactory mucosa, many mammals have a second olfactory structure in their noses: the *vomeronasal organ* (*VNO*) or Jacobson's organ, which plays a key role for sensing compounds secreted or excreted by conspecifics (*pheromones*) and that are necessary for reproductive and social behaviors (Døving & Trotier, 1998). VNO is a bony-encased structure that is found bilaterally at the base of nasal septum and that it opens anteriorly into the nasal cavities through the vomeronasal duct (Vaccarezza et al., 1981). Unlike the olfactory mucosa, which is exclusively stimulated by airborne odorants, the VNO can also be activated by nonvolatile compounds, such as proteins present in rodent urine or bodily fluids such as tears or vaginal secretions (Ihara et al., 2013; Liberles, 2014; Niimura et al., 2021). Interestingly, also hemoglobin plays a role as a chemosensory signal, eliciting behavioral responses in mice in a state-dependent fashion (Osakada et al., 2022). It is worth noting that the olfactory mucosa can sense some pheromones as well (Ihara et al., 2013; Liberles, 2014; Sanchez-Andrade & Kendrick, 2009). Indeed, parallel processing of olfactory cues detected by the two olfactory systems does occur in mammals (Spehr et al., 2006). Given the presence of this additional olfactory structure, the sensory tissue in the olfactory mucosa is commonly referred to as *main olfactory epithelium* (*MOE*). MOE and VNO project to different nervous centers: the main olfactory bulb and the accessory olfactory bulb, respectively (Mori et al., 2000; Spehr et al., 2006). During embryonic development, these olfactory structures derive from neurogenic placodes (Bhattacharyya & Bronner-Fraser, 2008; Katreddi & Forni, 2021).

Main olfactory epithelium - The olfactory mucosa is made up of a 100-μm-thick pseudo-stratified epithelium (*main olfactory epithelium, MOE*) that contains three main types of cells: olfactory receptor neurons, supporting cells, and basal progenitor cells (Morrison & Costanzo, 1992). Because it contains neurons, this epithelium is also known as *neuroepithelium*.

The *olfactory receptor neurons* (*ORNs*) are *chemoreceptors*. They are *bipolar neurons* with an apical dendrite that reaches the mucosal surface and an unmyelinated, unbranched axon that projects directly to the main olfactory bulb in the central nervous system (Mori et al., 2000). The distal end of the dendrite is the chemosensitive region. Here, the dendrite is a bit enlarged (*dendritic knob*, a few micrometers in diameter) and houses 10–30 *olfactory cilia* (Fig. 29.5A). These cilia have a diameter of 200–300 nm and penetrate the 30 μm-thick mucus film covering the mucosa surface. *The olfactory cilia are the structures that detect odorant molecules*. It is important to emphasize that all airborne compounds must dissolve into the thin mucus layer in order to reach the sensitive olfactory cilia. *Bowman's glands* in the underlying lamina propria produce the majority of the mucus that covers the olfactory mucosa's surface, but supporting cells also contribute (Morrison & Costanzo, 1992). The mucus layer protects the neuroepithelium from drying and thermal shocks. Through tight junctions, ORN distal dendrites are mechanically hooked onto nearby cells, particularly supporting cells. The unmyelinated axon allows ORNs to send action potentials directly to the central nervous system. Unlike taste buds, no peripheral synaptic contacts are present in the sensory neuroepithelium.

Supporting cells of the olfactory mucosa, such as astrocytes in the central nervous system, are thought to provide metabolic support to ORNs and regulate the ionic composition of the interstitial fluid, which is constantly modified by ORN activity. Supporting cells are found throughout the neuroepithelium, from the basal membrane to the mucosa surface, and are intermingled with ORNs (Liang, 2020).

Basal cells are globular or flattened elements found near the neuroepithelial basal membrane. They are undifferentiated stem cells that ensure normal olfactory receptor neuron turnover. Rodent ORNs do, in fact, have a short lifespan (from 2–3 weeks to 2 months) and are replaced by new cells derived from the basal cells (Yu & Wu, 2017). Due to their unique position (olfactory mucosa), these sensory neurons are exposed to numerous microorganisms (bacteria and viruses) hosted in the nasal cavities as well as inhaled dust particles and toxic pollutants. Cell turnover ensures the presence of a stable population of sensory neurons despite the continuous environmental stressors that cause deterioration, rapid aging, and eventually the death of ORNs.

Vomeronasal organ - In rodents, the VNO is a blind-ended tubular organ of a few millimeters in length that contains a pseudostratified neuroepithelium (about 100 μm thick) with cell types similar to those found in the MOE, namely sensory neurons, supporting cells, and basal cells (Vaccarezza et al., 1981). *Vomeronasal sensory neurons* (*VSNs*) are

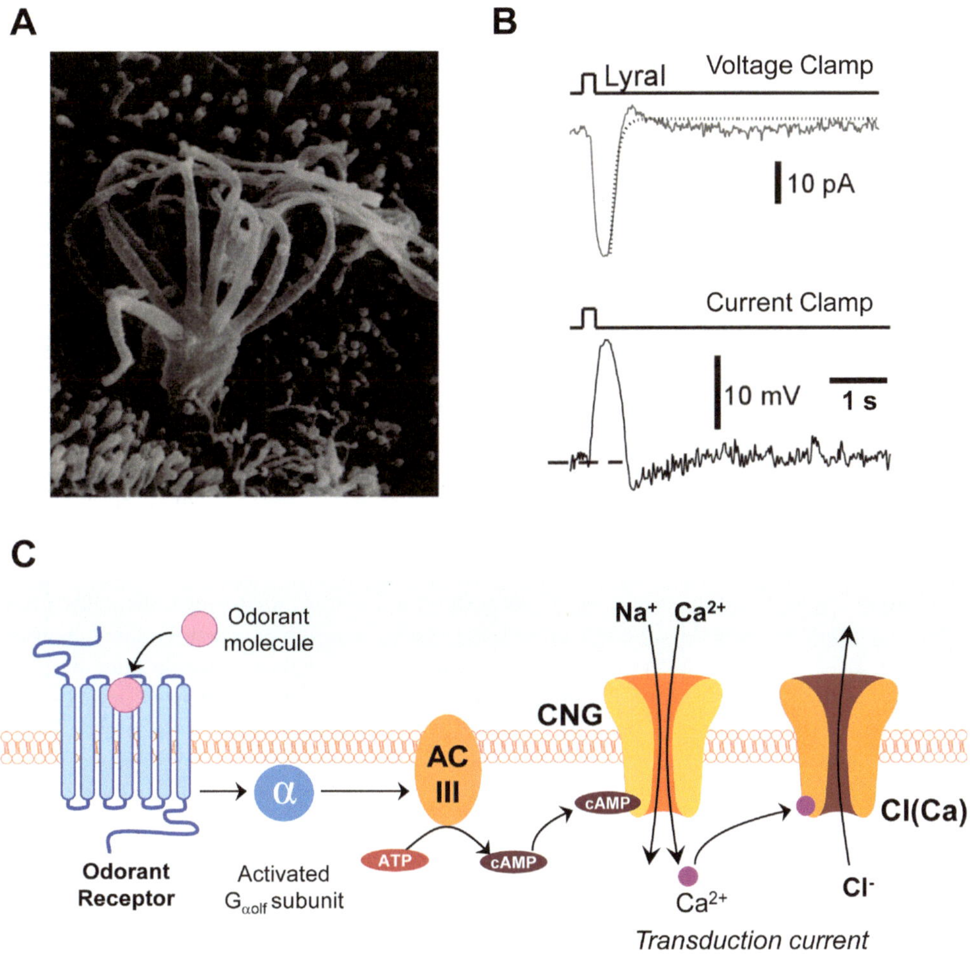

FIGURE 29.5 **Sensory transduction in olfactory receptor neurons.** (A) The molecular events that underpin odorant detection and lead to receptor potential generation occur in the olfactory cilia. In this picture, a human dendritic knob is observed with scanning electron microscopy. Note the numerous cilia extending over the epithelial surface. x18,525. (B). Whole-cell patch clamp recording from the dendritic knob of a mouse olfactory receptor neuron. Response to the odorant lyral (300 μM) applied to olfactory cilia by pressure ejection from a multibarrel pipette and recorded in voltage clamp mode (upper record: Inward transduction current) and in current clamp mode (lower record: Depolarizing receptor potential). (C). Signaling cascade during olfactory transduction. A specific odorant receptor (OR), which is a G protein-coupled receptor, recognizes the odorant molecule. The G protein is composed of three subunits (α, β, and γ): odorant binding activates the α subunit ($G_{\alpha olf}$), which then activates the membrane-bound enzyme type III adenylyl cyclase (ACIII). ACIII converts ATP to cyclic AMP (cAMP), which directly gates the CNG channels: Na^+ and Ca^{2+} from mucus flow into the cilia, causing membrane depolarization. Furthermore, Ca^{2+} activates a calcium-dependent chloride channel, $Cl_{(Ca)}$, resulting in Cl^- efflux and, as a result, membrane depolarization. Because of the similar concentrations of Cl^- in mucus and ciliary cytosol, the Nernst equilibrium potential for this anion is close to 0 mV, whereas the resting membrane potential of olfactory receptor neurons is around −70 mV. Thus, an influx of Na^+ and Ca^{2+}, as well as an efflux of Cl^-, carries the overall depolarizing transduction current. The chloride current can be thought of as an amplifying mechanism that increases the olfactory cilium's odorant sensitivity. *(A) Modified from Morrison & Costanzo (1990); (B) modified from Grosmaitre et al. (2006).*

chemoreceptors. There are two distinct layers of VSNs: *apical* and *basal neurons*. This distinction is based on the dendrite length, cell body position, and expression of specific G protein subunits and receptors, as described later. Numerous *microvilli*, a few micrometers long, protrude into the VNO lumen at the dendritic end. VSNs do not have olfactory cilia and the *apical microvilli are structures that detect pheromones* (Dibattista et al., 2012; Liman et al., 1999; Menco et al., 2001; Takigami et al., 1999). Each neuron sends a single unmyelinated axon to the accessory olfactory bulb, a central structure dorsal to the main olfactory bulb (Tirindelli et al., 2009). Sensory neurons in VNO have a cell turnover rate of about 2−3 months (Døving & Trotier, 1998). Supporting cells are intermingled with sensory neurons and are mechanically attached to them through membrane junctions. Supporting cells reach the epithelium surface, where they show microvilli. Adult humans lack a structured VNO (Bigiani et al., 2005; Tirindelli et al., 2009). However, typical vomeronasal receptors may exist in the human olfactory mucosa (Rodriguez et al., 2000).

4.1.2 Olfactory stimuli

Odorants - The odorants are the chemical substances that activate ORNs (Hudson, 2000). These *olfactory stimuli* are typically *small volatile organic molecules*. Odorants communicate information about many aspects of the external environment, such as food, partners, and habitat. As a result, olfactory cues are extremely important in animal life, not only for food search but also for reproduction and social communication (Ihara et al., 2013). Indeed, some pheromones, such as urinary volatiles and steroids, can stimulate ORNs (Ihara et al., 2013; Liberles, 2014; Spehr et al., 2006). In general, olfactory stimuli are detected by ORNs at much lower concentrations than water-soluble chemicals required to stimulate taste cells (Sarrafchi et al., 2013).

Smell enables humans to distinguish between a wide variety of odors and perfumes (about 10.000). It is debatable whether there are *basic* or *primary odors*, as there are in taste (Auffarth, 2013; Triller et al., 2008). It is important to note that the term "odor" refers to the conscious perception of an "odorant," which is the molecule detected by ORNs (Hudson, 2000). Until the 1990s, there was a generally held belief that it was possible to interpret all types of olfactory stimuli using only a few primary odors, such as "ethereal" (acetone, ethyl ether), "camphoraceous" (camphor, eucalyptol), and so on. This hypothesis was based on the observation that some people were insensitive to certain odors, a condition known as *specific anosmia* (Amoore, 1967), which is the olfactory equivalent of colorblindness in vision. For two main reasons, it is currently thought that such a classification is a rather coarse approximation. First, more than 70 distinct anosmias are now recognized, and second, many presumed olfactory sensations are actually caused by activation of trigeminal nerve endings, which are distributed throughout the nasal mucosa (Hummel & Frasnelli, 2019). This is true, for example, of the "pungent" odors of ammonia and chlorine. It is thus preferable to refer to generic *odor categories* rather than primary odors (Fráter et al., 1998), keeping in mind that they do not cover completely the wide range of olfactory substances capable of eliciting a conscious perception. It is also important to highlight that what we perceive as a single odor is often the result of the integration of sensory information from a mixture of substances, each with its own odor (Price, 1987; Thomas-Danguin et al., 2014).

Pheromones - Karlson and Lüscher (Karlson & Luscher, 1959) coined the term "pheromone" to describe chemicals released by an individual that can affect the behavior or physiological status of another individual of the same species (conspecifics). MOE and VNO are both involved in pheromone reception (Liberles, 2014). Pheromones have a striking structural diversity, ranging from small volatile molecules such as aliphatic ketones to excreted proteins, and are typically found in bodily fluids such as urine, sweat, saliva, and tears (Cavaliere et al., 2020; Ihara et al., 2013; Liberles, 2014; Tirindelli et al., 2009). Pheromone detection causes sexual-social responses such as sexual attraction, male aggressiveness, and individual recognition, to name a few (Cavaliere et al., 2020; Ihara et al., 2013; Liberles, 2014; Tirindelli et al., 2009). Some pheromones, such as mouse farnesene and 2-heptanone, have extremely low detection thresholds (around 10^{-11} M), making pheromonal communication extremely sensitive (Leinders-Zufall et al., 2000). For a long time, scientists have debated the existence of human pheromones. Despite evidence that pheromonal communication occurs between individuals, a VNO is thought to be absent or vestigial in humans (Bigiani et al., 2005; Tirindelli et al., 2009). The main olfactory epithelium may then be involved in pheromone reception in humans.

4.2 Olfactory receptor neurons

In comparison with taste chemoreceptors, olfactory receptor neurons are rather homogeneous as far as their structural and molecular features are concerned. All these neurons are chemosensitive and project to the olfactory bulb. The initial events of olfactory transduction take place in the cilia, which contain the molecular machinery to transform the presence of odorants into cellular signals.

4.2.1 Olfactory cilia

Olfactory cilia are designed to collect odorants by expanding the membrane surface of the dendritic knob, hence increasing the chances of odorant contact. Cilia are narrow tubes (length: $\sim 10-20$ μm; diameter: ~ 0.2 μm) with a very small volume (McEwen et al., 2008). This physical factor amplifies the concentration changes of molecules produced inside the cilium during the transduction process. In particular, even minor variations in the actual number of a given molecule result in large concentration changes. An olfactory cilium works as a self-contained, functional unit (Leinders-Zufall et al., 1998). Many cilia converge onto the same dendritic knob, and this suggests spatial integration of the responses to multiple odorant molecules (Kleene, 2008).

Because olfactory cilia are embedded in the mucous layer, the extracellular ionic environment that bathes their membrane likely reflects the ionic conditions found in mucus. This is significant for understanding the transduction

TABLE 29.3 Resting ionic conditions at the apical end (dendritic knob and olfactory cilia) of rat olfactory receptor neurons.

				Transduction current at resting potential (−70 mV)	
Ion	*[Ion]$_{mucus}$* (mM)	*[Ion]$_{cytosol}$* (mM)	*E_{ion}* (mV)	*CNG*	*Cl(Ca)*
Na^+	55	53	+1	Inward	
Free Ca^{2+}	4.8	40 nM[a]	+156	Inward	
Cl^-	55	54	0		Inward (Cl^- flowing out)
K^+	69	172	−24	Inward	

The concentrations reported (in approximate average values) are the best estimates available (reviewed in (Kleene, 2008)). These values represent total ionic concentrations rather than free ionic concentrations, which are still unknown. None of the major ions present in the cilia are at the diffusion equilibrium predicted by the Nernst equation, assuming a resting potential of about −70 mV.
$Cl_{(Ca)}$, calcium-activated chloride channel; *CNG*, cyclic-nucleotide gated channel; E_{ion}, Nernst equilibrium potential.
[a]*Mammalian ORNs have not been determined for this value. In the tiger salamander* (Ambystoma tigrinum)*, the intracellular free Ca^{2+} concentration is about 40 nM in the cilia (Leinders-Zufall et al., 1997).*

mechanism because the mucus ion composition differs from that of interstitial fluid: Na^+ and Cl^- are less concentrated, whereas K^+ is more abundant (Reuter et al., 1998). By using ion-sensitive microelectrodes, ion-sensitive fluorescent dyes, and energy-dispersive X-ray microanalysis (e.g., (Kaneko et al., 2001; 2004; Reuter et al., 1998)), it has been possible to estimate the ion concentrations in the dendritic knob (likely reflecting those inside cilia). Table 29.3 compares mucosal ionic conditions with those found in the intracellular fluid of rat ORNs (Kleene, 2008). The dendritic knob actively accumulates Cl^- through the Na^+-K^+-$2Cl^-$ cotransporter (NKCCl) to levels comparable with those found in mucus (Kaneko et al., 2004). Cl^- accumulation in the cilia may also be facilitated by a Cl^-/HCO^{3-} exchanger (Hengl et al., 2010). As a result, Cl^- is evenly distributed across the cilium membrane, and its Nernst equilibrium potential is close to 0 mV. This has a significant impact on the amplification of transduction current, as explained later.

4.2.2 Electrophysiology

Resting potentials of ORNs typically range from −70 to −90 mV (Schild & Restrepo, 1998) (see Discussion in (Lagostena & Menini, 2003)). Because the resting membrane resistance is quite high (up to 40 GΩ), these neurons are extremely sensitive to even the smallest current (1 pA) flowing through a single open channel (Lynch & Barry, 1989). The olfactory cilia and the cell body should have the same resting potential because they are electrically neighboring membrane compartments. Ionic flow from mucus to cytoplasm is caused by the application of odorant molecules to the cilia. Under voltage-clamp conditions, this flow appears as an inward current that depolarizes the native neuron (Fig. 29.5B) (Grosmaitre et al., 2006). This depolarization represents the *receptor potential.* The voltage-gated Na^+ and K^+ channels that underpin action potential generation are expressed in the ORN cell body (Lagostena & Menini, 2003; Narusuye et al., 2003). These channels are activated by odorant-induced depolarization that spreads electrotonically from the olfactory cilia to the cell body. The greater the depolarization, the higher the frequency of action potentials is (Liman & Corey, 1996).

Recent studies have shown that the ORNs discharge action potentials even in the absence of odorant stimulation. Interestingly, the rate of basal activity is dependent on the type of odorant receptor expressed by OSN and most likely reflects spontaneous (i.e., without the presence of an odorant molecule) receptor activation (Connelly et al., 2013). The significance of this spontaneous activity is being investigated. Firing may play a role in regulating the growth of renewing neurons, which must connect to specific olfactory bulb locations (Pietra et al., 2016).

4.3 Vomeronasal sensory neurons

Unlike ORNs, which are directly stimulated by airborne chemicals while breathing, VSNs are stimulated by pheromones through a mechanism called "vomeronasal pump". The VNO is a tubular structure with a blind end, and pheromone-containing fluids are thus conveyed to its lumen from the nasal cavity through an active vascular pumping mechanism controlled by the autonomic nervous system (Døving & Trotier, 1998; Meredith, 1998).

4.3.1 Apical microvilli

Unlike olfactory receptor neurons, VSNs in rodents lack cilia in their apical membrane. They do, however, have typical microvilli that protrude into the VNO lumen (Vaccarezza et al., 1981). The reason for this difference is unknown. However, there is evidence for the presence of olfactory cells with apical microvilli rather than cilia also in MOE, though their role in olfactory transduction is mysterious (Elsaesser & Paysan, 2007). It is unknown what the ionic composition of the mucus bathing microvilli is. Given the activity of the vomeronasal pump, which draws mucus and dissolved substances in (Døving & Trotier, 1998; Meredith, 1998), the ionic content of the fluid in contact with the microvilli is likely to reflect, at least in part, that of the main nasal mucosa. Because of their high resting Cl^- permeability (Ghiaroni et al., 2003), supporting cells may play a role in controlling the ionic conditions of fluid bathing the microvillar layer.

4.3.2 Electrophysiology

Patch-clamp recordings have yielded a wealth of data on the electrical properties of rodent VSNs. These neurons fire action potentials in response to current injections (Liman & Corey, 1996; Shimazaki et al., 2006; Ukhanov et al., 2007) or to pheromone application (Holy et al., 2000; Inamura et al., 1999; Leinders-Zufall et al., 2000). The interplay of voltage-gated ion channels, including TTX-sensitive Na^+ channels, L-type and T-type Ca^{2+} channels, and several types of K^+ channels (Fieni et al., 2003; Inamura et al., 1997; Liman & Corey, 1996; Shimazaki et al., 2006; Ukhanov et al., 2007), determines firing activity. VSNs have a resting potential greater than -70 mV (Ukhanov et al., 2007). VSNs, such as ORNs, are highly excitable, as evidenced by their sensitivity to injected currents (Liman & Corey, 1996). This excitability most likely reflects the high membrane resistance in resting conditions (Liman & Corey, 1996; Shimazaki et al., 2006). When not stimulated by pheromones, VSNs exhibit spontaneous activity (Shimazaki et al., 2006; Ukhanov et al., 2007). The significance of spontaneous firing is unknown. It could play a role in establishing the right projection to specific areas of the accessory olfactory bulb during cell turnover, just as it does in ORNs.

4.4 Sensory transduction

Different transduction mechanisms exist in taste chemoreceptors depending on the type of chemical stimulus (inorganic ion vs. organic molecule). Sensory transduction in ORNs, on the other hand, is based on a single, shared mechanism that is characterized by a series of molecular events that lead to membrane depolarization. Different odorants are distinguished by a molecular and neural code that involves the expression of specific odorant receptors in ORNs and the convergence of related nerve fibers to the same spots in the olfactory bulb (Mombaerts et al., 1996). The transduction process causes membrane depolarization in VSNs as well. Unlike ORNs, coding is dependent on the spatial separation of apical neurons from basal neurons, which are specifically tuned to detect different types of pheromones (Tirindelli, 2021).

G protein-coupled receptors (GPCRs) are the molecular receptors for odorants and pheromones. The MOE contains two major GPCR families: odorant receptors (ORs) and trace amine-associated receptors (TAARs). These receptors are housed in the olfactory cilia. The VNO contains other GPCR families, the most important of which are the vomeronasal receptors (VRs). These receptors are found in VSN apical microvilli.

4.4.1 Odorant receptors

Odorant receptors (ORs) form the largest GPCR family (Buck & Axel, 1991) comprising about 1000 genes in mice, and about 350 genes in humans (Nei et al., 2008). However, each ORN expresses only one type of receptor (the "one neuron-one receptor" rule) (Serizawa et al., 2004). The binding region for the odorant molecules is most likely found in transmembrane domains III, IV, and V. A single OR binds several types of odorants (broad selectivity), and a single odorant can interact with multiple ORs at the same time. Thus, coding of sensory information in the olfactory epithelium is based on the seemingly random activation of different ORNs, but their nerve fibers converge on specific glomeruli in the olfactory bulb (DeMaria & Ngai, 2010). OR ligands are volatile molecules, and binding is typically characterized by a low affinity interaction. Odorant receptors are linked to a heterotrimeric G protein (G_{olf}) that contains an ORN-specific α-subunit, $G\alpha_{olf}$ (Jones & Reed, 1989), in addition to the β and γ subunits. The very short (<1 ms) odorant dwell-time is an interesting property of olfactory transduction, making the probability of activating the G_{olf} very low (Bhandawat et al., 2005). This means that the very first step of signal transduction (activation of G protein) lacks any significant amplification, as seen in photoreceptors, where photoisomerized rhodopsin lives long enough to activate several molecules of the G protein, transducin (Genovese et al., 2021). In olfactory transduction, amplification occurs at later stages of the signaling cascade and involves interaction between two separate transduction channels (Fig. 29.5C). Typically, ORNs respond to odorant concentrations in the $1-100\ \mu M$ range (Grosmaitre et al., 2006), which is greater than the minimal pheromone concentration detected by vomeronasal neurons (10 pM or less) (Leinders-Zufall et al., 2000, 2004). The reduced sensitivity could be attributed to the broad specificity of ORs, which bind a variety of ligands (Breer, 2003). As a result, the binding affinity should be low (Kleene, 2008).

4.4.2 Trace amine-associated receptors

Trace amine-associated receptors (TAARs) are a smaller class of olfactory receptors found in MOE (Liberles & Buck, 2006). TAARs share similarities with biogenic amine receptors but are not related to ORs (Liberles, 2015). TAARs, such as ORs, are G protein-coupled receptors. However, unlike ORs, which respond to a wide range of odorants, TAARs are specifically tuned to detect volatile amines, which signal social cues and/or the presence of predators (Ihara et al., 2013; Liberles, 2015). TAARs have been found to be evolutionary conserved among vertebrates, with 15 members in mice and six in humans (Gainetdinov et al., 2018; Nei et al., 2008). TAARs, such as ORs, are linked to $G\alpha_{olf}$ (Liberles & Buck, 2006). Surprisingly, TAAR-expressing neurons are highly sensitive to amines, with responses to concentrations as low as 10^{-12} M (Zhang et al., 2013). In this regard, TAAR-expressing neurons are similar to vomeronasal neurons, which have a high sensitivity to pheromones (Leinders-Zufall et al., 2000, 2004).

4.4.3 Vomeronasal receptors

VSNs express two major families of GPCRs (Dulac & Axel, 1995; Herrada & Dulac, 1997; Matsunami & Buck, 1997; Ryba & Tirindelli, 1997): *vomeronasal receptors type 1* (*V1Rs*) and *type 2* (*V2Rs*). In mice, around 180 V1Rs and 120 V2Rs have been identified (Liberles, 2014; Nei et al., 2008). V1Rs are structurally similar to ORs and topologically related to taste receptors for bitter substances. On the contrary, V2Rs have a long N-terminal extracellular domain that neither V1Rs nor ORs have, and which is thought to contain the binding site for waterborne chemicals, such as peptides or proteins (Silva & Antunes, 2017; Tirindelli, 2021). In rodent VNO, sensory neurons expressing VRs are segregated into different zones of the neuroepithelium, apical and basal. *Apical neurons* express V1Rs in addition to the G protein subunit $G\alpha_{i2}$, whereas *basal neurons* express V2Rs in addition to the G protein subunit $G\alpha_o$ (Bigiani et al., 2005; Tirindelli, 2021; Tirindelli et al., 2009). The $G\gamma$ subunit, such as the $G\alpha$ subunit, exhibits remarkable layer-specific segregation in VNO, with apical neurons expressing preferentially $G\gamma_2$ and basal neurons $G\gamma_8$, despite the fact that both types of neurons share the same $G\beta_2$ subunit (Rünnenburger et al., 2002). In rodents, V1Rs are believed to detect small volatile chemicals, e.g., 2-heptanone promoting estrous, whereas V2Rs interact with nonvolatile compounds, e.g., urinary proteins promoting sexual attraction or aggression toward intruders (Ihara et al., 2013; Liberles, 2014; Tirindelli et al., 2009). Unlike ORs, V1Rs are not broadly tuned to various pheromone molecules (Tirindelli, 2021; Tirindelli et al., 2009). Humans have five intact V1R genes but no functional V2R genes (Nei et al., 2008). The mRNA of a V1R gene is expressed in the human olfactory mucosa (Rodriguez et al., 2000).

4.4.4 Cellular responses

Response to odorants - Specific second messengers (cAMP and Ca^{2+}) characterize the transduction cascade that occurs in the olfactory cilia. These chemical signals control various aspects of the molecular processes during odorant detection (Fig. 29.5C). Odorant binding to ORs causes the exchange of GTP for GDP in $G\alpha_{olf}$, which then dissociates from the other two subunits (β and γ) and activates a membrane enzyme, *type III adenylate cyclase* (*ACIII*). As a result, ACIII catalyzes the conversion of adenosine triphosphate (ATP) to *3′,5′-cyclic AMP* (*cAMP*), the first intracellular signal produced during the transduction process. According to estimates (Takeuchi & Kurahashi, 2005), the concentration of cAMP may rise to more than 100 μM. The molecular target of cAMP is a cationic channel gated by this nucleotide, known as *cyclic nucleotide-gated (CNG) channel*, similar to the one found in retinal photoreceptors (Genovese et al., 2021). In ORNs, this channel is closed in the absence of olfactory stimuli, but it opens in response to cAMP during odorant stimulation. The open channel allows Na^+ and Ca^{2+} to enter the ORN (*transduction current*), resulting in membrane depolarization (*receptor potential*). This latter is the biological signal produced by the transduction process (Fig. 29.5B).

The presence of an ionic conductance regulated by cAMP in the olfactory cilia was first observed in excised membrane patches of the toad olfactory epithelium (Nakamura & Gold, 1987). Subsequent research confirmed that a cyclic nucleotide-gated channel is an important molecular component in the olfactory transduction pathway (Kleene, 2008; Pifferi, Boccaccio, & Menini, 2006). The olfactory CNG channel is a tetrameric protein composed of three types of subunits (CNGA2, CNGA4, and CNGB1b) with a 2:1:1 stoichiometry (Zheng & Zagotta, 2004). Each CNG channel has four binding sites for cAMP, one for each subunit. Channel activation occurs through subunit cooperativity, as indicated by a dose-response curve with a Hill coefficient of about 2 (Dhallan et al., 1990). The CNG channel is a cationic channel that is permeable to Na^+, K^+, and Ca^{2+}, does not inactivate, and has slight outward rectification (Kleene, 2008). Intraciliary Ca^{2+} influences CNG currents through *calmodulin*, which binds to the CNG channels (Bradley et al., 2004) and shifts the dose-response relationship to higher cAMP concentrations, that is, it makes the channel less sensitive to cAMP (Balasubramanian et al., 1996).

Ca^{2+} entry through the CNG channel is critical in the physiology of the olfactory cilium. The electric charge carried by calcium depolarizes the membrane, and thus Ca^{2+} contributes to the generation of receptor potential on its own. Furthermore, Ca^{2+} plays a role in the activation of a *calcium-activated chloride current* [$Cl_{(Ca)}$] in the ciliary membrane. A Hill function with a half-maximal concentration of about 2–5 μM can be used to fit the relationship between intracellular Ca^{2+} concentration and chloride currents (Pifferi, Pascarella, et al., 2006; Reisert et al., 2003). During olfactory transduction, the Ca^{2+} concentration in the cilia has been estimated to be as high as 100 μM (Delgado & Bacigalupo, 2004). As a result, $Cl_{(Ca)}$ channels open and collaborate with CNG channels to generate receptor current (Fig. 29.5C). This is made possible by the chloride ion concentration gradient (approximately 55 mM in both mucus and cilium cytosol), which sets its Nernst equilibrium potential near 0 mV (Table 29.3). As a result, the neuron's negative resting potential drives Cl^- out of the cilium: electrically, this outflow of negative charges corresponds to an inflow of positive charges, causing membrane depolarization. Thus, *anionic current* through *$Cl_{(Ca)}$ channels amplifies the effect of primary cationic current* through *CNG channels*. Given the cilium's high surface-to-volume ratio, coupling between these two ion channels is expected to be very effective (Kleene, 2008). Interestingly, this amplification mechanism is similar to the one recently described for sour taste transduction (see Fig. 29.3). It is estimated that the chloride current accounts for up to 90% of the total transduction current in mouse ORNs (Boccaccio & Menini, 2007). Recent modeling has emphasized the importance of activating a chloride current as the primary amplification step to ensure consistent responses to odorants (Reisert & Reingruber, 2019). However, the role of these channels in smelling remains debatable. Mice with the gene for $Cl_{(Ca)}$ channels knocked out apparently do not exhibit any significant changes in odor discrimination or odor sensitivity (Billig et al., 2011). Nonetheless, the same animal model demonstrated poor performance in odor-guided food finding (Pietra et al., 2016) and tracking unfamiliar, weak odors (Neureither et al., 2017). Interestingly, $Cl_{(Ca)}$ channel activity appears to be important for regulating the firing activity of olfactory sensory neurons, ensuring proper fiber targeting to the correct areas in the olfactory bulb (Pietra et al., 2016). $Cl_{(Ca)}$ channels are TMEM (Transmembrane member)16 proteins (anoctamins), a large family of proteins involved in a variety of cellular processes (Pedemonte & Galietta, 2014). ORNs specifically express *TMEM16b/ANO2*, which mediates calcium-activated chloride currents recorded from olfactory cilia (Pifferi et al., 2012; Stephan et al., 2009).

Response to Pheromones - Although apical and basal vomeronasal neurons express different G proteins, they share downstream signaling molecules such as phospholipase C (PLC), diacylglycerol (DAG), and the *transient receptor potential canonical 2* (*TRPC2*) ion channel (Liman et al., 1999). The following is the sequence of molecular activation of VSNs (Fig. 29.6A) (Bigiani et al., 2005; Chamero et al., 2012; Mohrhardt et al., 2018; Tirindelli et al., 2009). Pheromone binding to VRs in microvilli (Menco et al., 2001) activates PLC through the *$G\beta\gamma$ complex*, which catalyzes the production of 2 s messengers, DAG and inositol-1,4,5-triphosphate (IP_3), from the membrane lipid phosphatidyl inositol bisphosphate (PIP_2). *DAG opens the TRPC2 channel* (Lucas et al., 2003) located in the apical microvilli (Fig. 29.6B), causing an influx of Na^+ and Ca^{2+} (*transduction current*) (Fig. 29.6C). The mechanism of its activation by DAG is still unknown. TRPC2 is an important component of the pheromone transduction mechanism (Zufall et al., 2005) and is rarely expressed outside of the VNO. However, recent studies have shown that it exists in some neurons of the mouse MOE (Omura & Mombaerts, 2014). The $G\alpha$ subunit does not appear to be involved in the initial signaling events that lead to the transduction current. However, it is required for a functional VNO in terms of pheromone-mediated behavior, as demonstrated by conditional, cell-specific $G\alpha_o$ or $G\alpha_{i2}$ ablation (Chamero et al., 2011; Oboti et al., 2014; Trouillet et al., 2019). The function of IP_3 in the signaling cascade is not completely understood (Tirindelli et al., 2009). According to a recent study, the Type III IP_3 receptor, which mediates the IP3-induced signaling cascade, is not essential for TRPC2 channel gating (Chamero et al., 2017).

Like in ORNs, the *Ca^{2+}-activated Cl^- channel* from the TMEM16/anoctamin family is an additional component of the transduction cascade also in VSNs (Amjad et al., 2015; Dibattista et al., 2012). Given the high intracellular Cl^-concentration (~80 mM) (Kim et al., 2015), this channel likely mediates the outflow of Cl^- from microvilli (depolarizing current) during pheromone stimulation (Kim et al., 2011). Thus, this channel may be as important in vomeronasal transduction as it is in ORNs (Fig. 29.6A). Vomeronasal neurons express two isoforms of the Ca^{2+}-activated Cl^- channel, *TMEM16a/ANO1* and *TMEM16b/ANO2*. TMEM16a/ANO1 appears to play an important role in the generation of Ca^{2+}-activated Cl^- currents and in modulating the spontaneous activity of VNs (Hernandez-Clavijo et al., 2021; Münch et al., 2018). Surprisingly, when stimulated with diluted urine, both isoforms affect VSN firing activity (Hernandez-Clavijo et al., 2021). The role of these channels in pheromone-mediated behaviors, however, is unclear (Münch et al., 2018).

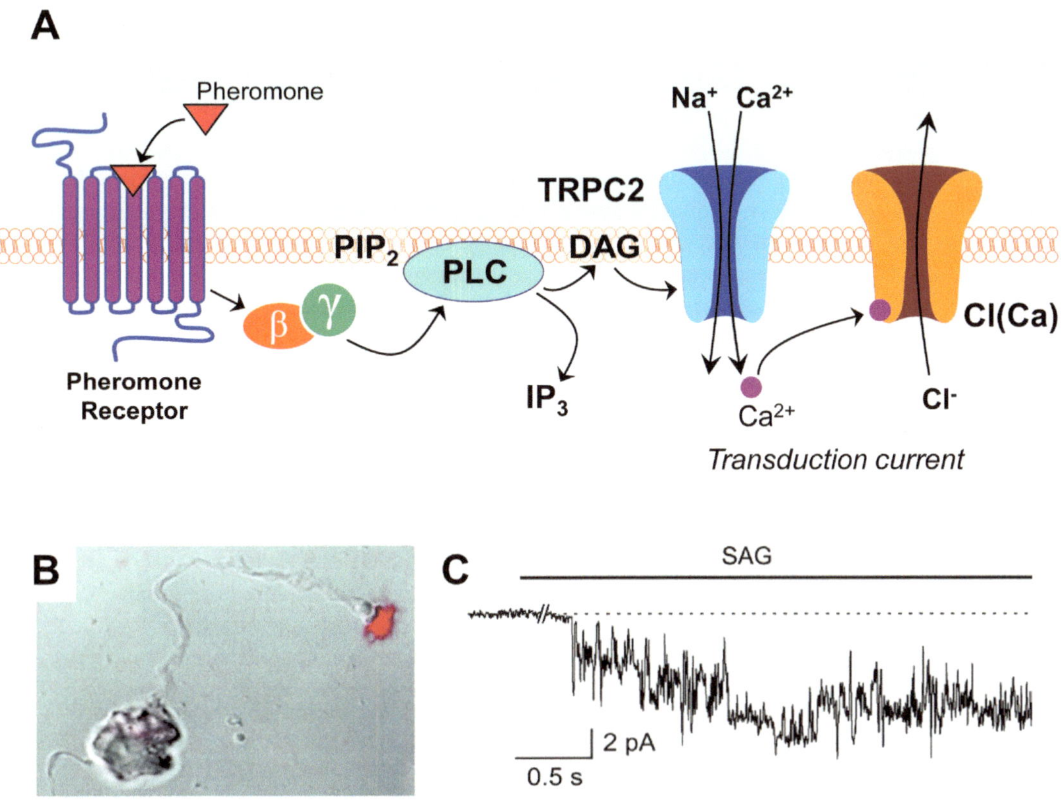

FIGURE 29.6 **Sensory transduction in vomeronasal sensory neurons.** (A) The molecular events underlying pheromone detection and leading to receptor potential generation occur in the apical microvilli of vomeronasal sensory neurons. Vomeronasal receptors (VRs), which are G protein-coupled receptors (GPCRs), recognize pheromones. The G protein is made up of three subunits (α, β, and γ): pheromone binding to VR activates the G protein, resulting in βγ complex detachment and activation of the membrane-bound enzyme phospholipase C (PLC). PLC hydrolyzes the membrane lipid phosphatidyl inositol bisphosphate (PIP_2) into diacylglycerol (DAG) and inositol-1,4,5-triphosphate (IP_3). DAG opens the transient receptor potential canonical 2 (TRPC2) channel: Na^+ and Ca^{2+} flow into the cell, producing a depolarizing transduction current. An increase in intracellular Ca^{2+} concentration may affect a calcium-dependent chloride channel [$Cl_{(Ca)}$], causing Cl^- efflux and thus potentiating the transduction current. This latter mechanism is similar to that found in olfactory transduction. (B) Immunocytochemistry of TRP2 protein in an isolated vomeronasal neuron dissociated from rat vomeronasal organ. Immunolabeling (*red*) is found at the apical end. (C) Patch-clamp recording from an inside-out patch excised from the dendritic tip of a mouse vomeronasal neuron (voltage-clamp mode, holding potential: −80 mV). Bath application (i.e., to the cytoplasmic side of the membrane patch) of 100 μM SAG (1-stearoyl-2-arachidonoyl-*sn*-glycerol, a DAG analog) elicited sustained channel activity in the membrane patch. *(B) Modified from Liman et al. (1999); (C) modified from Lucas et al. (2003).*

4.5 Transfer of olfactory signals to the brain

Receptor potentials generated in the olfactory cilia spread passively towards the cell body and the beginning of the axon, where voltage-gated Na^+ and K^+ channels are found. Action potentials are generated at this level. Odorant-induced graded depolarizations (analogic signals) in the cilia are therefore encoded into a train of all-or-none impulses (digital signals) required to transmit sensory information to the olfactory bulb. In addition to coding olfactory information, spiking activity appears to play an important role in establishing the correct connection with glomerular structures (Pietra et al., 2016). Vomeronasal sensory neurons have a similar mechanism of electrical excitation and transmission.

4.6 Termination of olfactory signals

Short applications of odorant to the olfactory cilia typically result in a transient inward current that decays over time (Kleene, 2008). The termination of ORN excitation is caused by a number of different mechanisms (Kleene, 2008). The removal of odorant molecules from mucus (through dispersion in breathed air) is, of course, the first step in terminating olfactory stimulation. The hydrolysis of cAMP to AMP by the enzyme phosphodiesterase (PDE), which removes the ligand (cAMP) required to open the CNG channel (Fig. 29.5C), is a second important mechanism. A third mechanism involves Ca^{2+} extrusion into the external space through the Ca^{2+}-ATPase (calcium pump) and the Na^{+}/Ca^{2+} exchanger. When intraciliary Ca^{2+} is removed, $Cl_{(Ca)}$ channels close.

Even when olfactory stimuli are persistent, the transduction current amplitude decreases, a phenomenon known as *sensory adaptation* (also described as *desensitization*) (Kleene, 2008). The significance of this reduced response could be twofold. On the one hand, it may aid in the elimination of background chemical signals that are useless to the central nervous system. On the other hand, it may be useful to adjust olfactory sensitivity to a new range of odorant concentrations (Kurahashi & Menini, 1997). Olfactory adaptation appears to be based on a number of mechanisms, all of which share an increase in ciliary Ca^{2+} concentration. Ca^{2+} entry into olfactory cilia during odorant stimulation has both an electrical (membrane depolarization) and a chemical (an increase in Ca^{2+} concentration) effect. As demonstrated by removing Ca^{2+} from the extracellular solution, the latter represents a second biological signal that is responsible for the adaptation (Kleene, 2008). As a result, adaptation occurs following the opening of CNG channels (Kurahashi & Menini, 1997). Ca^{2+} reduces the sensitivity of the CNG channel to cAMP by binding to the protein *calmodulin* (*CaM*). This negative feedback on the transduction channel causes the neuron to rapidly adapt; that is, the neuron's response tends to decrease in the continuous presence of the stimulus. This Ca^{2+}-dependent regulatory mechanism is thought to be important for evaluating changes in environmental odors in the presence of background stimuli. Furthermore, adaptation shifts and broadens the *dynamic range* (the range of stimulus concentrations that produce a response) of the neuron response (Boccaccio et al., 2006), implying that it is a mechanism for resetting ORN sensitivity without saturating response capability (Kleene, 2008). Ca^{2+}-CaM inhibits the CNG channel through interactions with the CNGA4 and CNGB1b subunits.

Sensory adaptation occurs in vomeronasal neurons as well, thanks to a Ca^{2+}-calmodulin feedback mechanism: TRPC2 channel activation is inhibited by Ca^{2+}-CaM (Spehr et al., 2009). Even in the adapted state, VSNs continue to generate action potentials, albeit at a lower frequency. This could explain why sensory adaptation in these neurons was not observed in earlier studies (e.g., (Holy et al., 2000)).

4.7 Odor masking

The use of pleasant smells to mask unpleasant ones is common in everyday life. Many aspects of our olfactory behavior are guided by odor masking, as evidenced by spray deodorants and environmental fragrances (Thomas-Danguin et al., 2014). It is now well established that odor suppression occurs at the very beginning of the sensory pathway, in the olfactory receptor neurons. Some odorants have the ability to act as agonists or antagonists on olfactory receptors, suppressing the response to other odorants (Oka et al., 2004). Furthermore, many types of odorant molecules can *directly* reduce current through the CNG channel (Takeuchi et al., 2009). Thus, the CNG channel is involved in both sensory adaptation (through Ca^{2+}-CaM feedback inhibition) and masking (through a direct block of the channel by some odorants). The importance of competitive antagonism between odorants at the level of olfactory receptors has been highlighted by modeling the processing of odor mixtures by ORNs (Reddy et al., 2018). Several recent studies have also demonstrated the importance of antagonistic interactions in combinatorial receptor coding (Kurian et al., 2021). Obviously, the final perception of odor mixtures necessitates sensory information processing at higher levels of the olfactory pathway (Thomas-Danguin et al., 2014).

References

Adler, E., Hoon, M. A., Mueller, K. L., Chandrashekar, J., Ryba, N. J., & Zuker, C. S. (2000). A novel family of mammalian taste receptors. *Cell, 100*(6), 693–702.

Amjad, A., Hernandez-Clavijo, A., Pifferi, S., Maurya, D. K., Boccaccio, A., Franzot, J., et al. (2015). Conditional knockout of TMEM16A/anoctamin1 abolishes the calcium-activated chloride current in mouse vomeronasal sensory neurons. *The Journal of General Physiology, 145*(4), 285–301.

Amoore, J. E. (1967). Specific anosmia: A clue to the olfactory code. *Nature, 214*(5093), 1095–1098.

Auffarth, B. (2013). Understanding smell–the olfactory stimulus problem. *Neuroscience and Biobehavioral Reviews, 37*(8), 1667–1679.

Avenet, P., & Lindemann, B. (1991). Noninvasive recording of receptor cell action potentials and sustained currents from single taste buds maintained in the tongue: The response to mucosal NaCl and amiloride. *Journal of Membrane Biology, 124*(1), 33–41.

Béhé, P., DeSimone, J. A., Avenet, P., & Lindemann, B. (1990). Membrane currents in taste cells of the rat fungiform papilla. Evidence for two types of Ca currents and inhibition of K currents by saccharin. *The Journal of General Physiology, 96*(5), 1061–1084.

Bachmanov, A. A., Bosak, N. P., Lin, C., Matsumoto, I., Ohmoto, M., Reed, D. R., et al. (2014). Genetics of taste receptors. *Current Pharmaceutical Design, 20*(16), 2669–2683.

Balasubramanian, S., Lynch, J. W., & Barry, P. H. (1996). Calcium-dependent modulation of the agonist affinity of the mammalian olfactory cyclic nucleotide-gated channel by calmodulin and a novel endogenous factor. *Journal of Membrane Biology, 152*(1), 13–23.

Barlow, L. A. (2015). Progress and renewal in gustation: New insights into taste bud development. *Development (Cambridge, England), 142*(21), 3620–3629.

Barretto, R. P., Gillis-Smith, S., Chandrashekar, J., Yarmolinsky, D. A., Schnitzer, M. J., Ryba, N. J., et al. (2015). The neural representation of taste quality at the periphery. *Nature, 517*(7534), 373–376.

Bartoshuk, L. M., & Beauchamp, G. K. (1994). Chemical senses. *Annual Review of Psychology, 45*, 419–449.

Bartoshuk, L. M. (2018). Taste. In J. T. Wixted (Ed.), *Stevens' handbook of experimental psychology and cognitive neuroscience* (4th ed., pp. 1–33). John Wiley & Sons.

Baumer-Harrison, C., Raymond, M. A., Myers, T. A., Sussman, K. M., Rynberg, S. T., Ugartechea, A. P., et al. (2020). Optogenetic stimulation of type I GAD65(+) cells in taste buds activates gustatory neurons and drives appetitive licking behavior in sodium-depleted mice. *Journal of Neuroscience: The Official Journal of the Society for Neuroscience, 40*(41), 7795–7810.

Beatty, R. M., & Cragg, L. H. (1935). The sourness of acids. *Journal of the American Chemical Society, 57*, 2347–2351.

Behrens, M., & Meyerhof, W. (2006). Bitter taste receptors and human bitter taste perception. *Cellular and Molecular Life Sciences: CMLS, 63*(13), 1501–1509.

Bhandawat, V., Reisert, J., & Yau, K. W. (2005). Elementary response of olfactory receptor neurons to odorants. *Science (New York, NY), 308*(5730), 1931–1934.

Bhattacharyya, S., & Bronner-Fraser, M. (2008). Competence, specification and commitment to an olfactory placode fate. *Development (Cambridge, England), 135*(24), 4165–4177.

Bianchi, L. (2021). DEG/ENaC ion channels in the function of the nervous system: From worm to man. *Advances in Experimental Medicine and Biology, 1349*, 165–192.

Bigiani, A. (2016). Electrophysiology of sodium receptors in taste cells. *Journal of Biomedical Science and Engineering, 9*, 367–383.

Bigiani, A. (2017). Calcium homeostasis modulator 1-like currents in rat fungiform taste cells expressing amiloride-sensitive sodium currents. *Chemical Senses, 42*(4), 343–359.

Bigiani, A. (2020a). Salt taste. In B. Fritzsch, & W. Meyerhof (Eds.), *The senses: A comprehensive reference* (2nd ed., pp. 247–263). Elsevier, Academic Press.

Bigiani, A. (2020b). Does ENaC work as sodium taste receptor in humans? *Nutrients, 12*(4).

Bigiani, A. (2021). The origin of saltiness: Oral detection of NaCl. *Current Opinion in Physiology, 19*, 156–161.

Bigiani, A., Cristiani, R., Fieni, F., Ghiaroni, V., Bagnoli, P., & Pietra, P. (2002). Postnatal development of membrane excitability in taste cells of the mouse vallate papilla. *Journal of Neuroscience: The Official Journal of the Society for Neuroscience, 22*(2), 493–504.

Bigiani, A., Ghiaroni, V., & Fieni, F. (2003). Channels as taste receptors in vertebrates. *Progress in Biophysics and Molecular Biology, 83*(3), 193–225.

Bigiani, A., Mucignat-Caretta, C., Montani, G., & Tirindelli, R. (2005). Pheromone reception in mammals. Reviews of physiology. *Biochemistry and Pharmacology, 154*, 1–35.

Bigiani, A., & Prandi, S. (2011). Functional diversity of taste cells. A review. *Flavour and Fragrance Journal, 26*, 214–217.

Bigiani, A., & Roper, S. D. (1995). Estimation of the junctional resistance between electrically coupled receptor cells in Necturus taste buds. *The Journal of General Physiology, 106*(4), 705–725.

Bigiani, A., & Roper, S. D. (1994). Reduction of electrical coupling between Necturus taste receptor cells, a possible role in acid taste. *Neuroscience Letters, 176*(2), 212–216.

Billig, G. M., Pál, B., Fidzinski, P., & Jentsch, T. J. (2011). Ca2+-activated Cl− currents are dispensable for olfaction. *Nature Neuroscience, 14*(6), 763–769.

Bo, X., Alavi, A., Xiang, Z., Oglesby, I., Ford, A., & Burnstock, G. (1999). Localization of ATP-gated P2X2 and P2X3 receptor immunoreactive nerves in rat taste buds. *NeuroReport, 10*(5), 1107–1111.

Boccaccio, A., Lagostena, L., Hagen, V., & Menini, A. (2006). Fast adaptation in mouse olfactory sensory neurons does not require the activity of phosphodiesterase. *The Journal of General Physiology, 128*(2), 171–184.

Boccaccio, A., & Menini, A. (2007). Temporal development of cyclic nucleotide-gated and Ca2+ -activated Cl- currents in isolated mouse olfactory sensory neurons. *Journal of Neurophysiology, 98*(1), 153–160.

Bradley, J., Bönigk, W., Yau, K. W., & Frings, S. (2004). Calmodulin permanently associates with rat olfactory CNG channels under native conditions. *Nature Neuroscience, 7*(7), 705–710.

Breer, H. (2003). Olfactory receptors: Molecular basis for recognition and discrimination of odors. *Analytical and Bioanalytical Chemistry, 377*(3), 427–433.

Breslin, P. A. (2013). An evolutionary perspective on food and human taste. *Current Biology: CB, 23*(9), R409–R418.

Breslin, P. A. S., & Huang, L. (2006). Human taste: Peripheral anatomy, taste transduction, and coding. *Advances in Oto-Rhino-Laryngology, 63*, 152–190.

Buck, L., & Axel, R. (1991). A novel multigene family may encode odorant receptors: A molecular basis for odor recognition. *Cell, 65*(1), 175–187.

Bufe, B., Hofmann, T., Krautwurst, D., Raguse, J. D., & Meyerhof, W. (2002). The human TAS2R16 receptor mediates bitter taste in response to beta-glucopyranosides. *Nature Genetics, 32*(3), 397–401.

Cagan, R. H. (1973). Chemostimulatory protein: A new type of taste stimulus. *Science (New York, NY), 181*(4094), 32–35.

Cao, Y., Zhao, F. L., Kolli, T., Hivley, R., & Herness, S. (2009). GABA expression in the mammalian taste bud functions as a route of inhibitory cell-to-cell communication. *Proceedings of the National Academy of Sciences of the United States of America, 106*(10), 4006–4011.

Cavaliere, R. M., Silvotti, L., Percudani, R., & Tirindelli, R. (2020). Female mouse tears contain an anti-aggression pheromone. *Scientific Reports, 10*(1), 2510.

Chamero, P., Katsoulidou, V., Hendrix, P., Bufe, B., Roberts, R., Matsunami, H., et al. (2011). G protein G(alpha)o is essential for vomeronasal function and aggressive behavior in mice. *Proceedings of the National Academy of Sciences of the United States of America, 108*(31), 12898–12903.

Chamero, P., Leinders-Zufall, T., & Zufall, F. (2012). From genes to social communication: Molecular sensing by the vomeronasal organ. *Trends in Neurosciences, 35*(10), 597–606.

Chamero, P., Weiss, J., Alonso, M. T., Rodríguez-Prados, M., Hisatsune, C., Mikoshiba, K., et al. (2017). Type 3 inositol 1,4,5-trisphosphate receptor is dispensable for sensory activation of the mammalian vomeronasal organ. *Scientific Reports, 7*(1), Article 10260.

Chandrashekar, J., Hoon, M. A., Ryba, N. J., & Zuker, C. S. (2006). The receptors and cells for mammalian taste. *Nature, 444*(7117), 288–294.

Chandrashekar, J., Kuhn, C., Oka, Y., Yarmolinsky, D. A., Hummler, E., Ryba, N. J., et al. (2010). The cells and peripheral representation of sodium taste in mice. *Nature, 464*(7286), 297–301.

Chandrashekar, J., Mueller, K. L., Hoon, M. A., Adler, E., Feng, L., Guo, W., et al. (2000). T2Rs function as bitter taste receptors. *Cell, 100*(6), 703–711.

Chang, R. B., Waters, H., & Liman, E. R. (2010). A proton current drives action potentials in genetically identified sour taste cells. *Proceedings of the National Academy of Sciences of the United States of America, 107*(51), 22320–22325.

Chaudhari, N., & Roper, S. D. (2010). The cell biology of taste. *The Journal of Cell Biology, 190*(3), 285–296.

Chaudhari, N. (2014). Synaptic communication and signal processing among sensory cells in taste buds. *The Journal of Physiology, 592*(16), 3387–3392.

Chen, X., Gabitto, M., Peng, Y., Ryba, N. J., & Zuker, C. S. (2011). A gustotopic map of taste qualities in the mammalian brain. *Science (New York, NY), 333*(6047), 1262–1266.

Chen, Y., Sun, X. D., & Herness, S. (1996). Characteristics of action potentials and their underlying outward currents in rat taste receptor cells. *Journal of Neurophysiology, 75*(2), 820–831.

Clapp, T. R., Trubey, K. R., Vandenbeuch, A., Stone, L. M., Margolskee, R. F., Chaudhari, N., et al. (2008). Tonic activity of Galpha-gustducin regulates taste cell responsivity. *FEBS Letters, 582*(27), 3783–3787.

Connelly, T., Savigner, A., & Ma, M. (2013). Spontaneous and sensory-evoked activity in mouse olfactory sensory neurons with defined odorant receptors. *Journal of Neurophysiology, 110*(1), 55–62.

Cui, M., Jiang, P., Maillet, E., Max, M., Margolskee, R. F., & Osman, R. (2006). The heterodimeric sweet taste receptor has multiple potential ligand binding sites. *Current Pharmaceutical Design, 12*(35), 4591–4600.

Døving, K. B., & Trotier, D. (1998). Structure and function of the vomeronasal organ. *Journal of Experimental Biology, 201*(Pt 21), 2913–2925.

Dando, R., & Roper, S. D. (2012). Acetylcholine is released from taste cells, enhancing taste signalling. *The Journal of Physiology, 590*(13), 3009–3017.

Dando, R., & Roper, S. D. (2009). Cell-to-cell communication in intact taste buds through ATP signalling from pannexin 1 gap junction hemichannels. *The Journal of Physiology, 587*(Pt 24), 5899–5906.

Delgado, R., & Bacigalupo, J. (2004). Cilium-attached and excised patch-clamp recordings of odourant-activated Ca-dependent K channels from chemosensory cilia of olfactory receptor neurons. *European Journal of Neuroscience, 20*(11), 2975–2980.

DeMaria, S., & Ngai, J. (2010). The cell biology of smell. *The Journal of Cell Biology, 191*(3), 443–452.

Dhallan, R. S., Yau, K. W., Schrader, K. A., & Reed, R. R. (1990). Primary structure and functional expression of a cyclic nucleotide-activated channel from olfactory neurons. *Nature, 347*(6289), 184–187.

Dibattista, M., Amjad, A., Maurya, D. K., Sagheddu, C., Montani, G., Tirindelli, R., et al. (2012). Calcium-activated chloride channels in the apical region of mouse vomeronasal sensory neurons. *The Journal of General Physiology, 140*(1), 3–15.

Diepeveen, J., Moerdijk-Poortvliet, T. C. W., & van der Leij, F. R. (2022). Molecular insights into human taste perception and umami tastants: A review. *Journal of Food Science, 87*(4), 1449–1465.

Dulac, C., & Axel, R. (1995). A novel family of genes encoding putative pheromone receptors in mammals. *Cell, 83*(2), 195–206.

Dutta, B. D., Martin, L. E., Freichel, M., Torregrossa, A. M., & Medler, K. F. (2018). TRPM4 and TRPM5 are both required for normal signaling in taste receptor cells. *Proceedings of the National Academy of Sciences of the United States of America, 115*(4).

Dvoryanchikov, G., Huang, Y. A., Barro-Soria, R., Chaudhari, N., & Roper, S. D. (2011). GABA, its receptors, and GABAergic inhibition in mouse taste buds. *Journal of Neuroscience: The Official Journal of the Society for Neuroscience, 31*(15), 5782–5791.

Elsaesser, R., & Paysan, J. (2007). The sense of smell, its signalling pathways, and the dichotomy of cilia and microvilli in olfactory sensory cells. *BMC Neuroscience, 8*(Suppl. 3), S1.

Feldman, G. M., Heck, G. L., & Smith, N. L. (2009). Human salt taste and the lingual surface potential correlate. *Chemical Senses, 34*(5), 373–382.

Fieni, F., Ghiaroni, V., Tirindelli, R., Pietra, P., & Bigiani, A. (2003). Apical and basal neurones isolated from the mouse vomeronasal organ differ for voltage-dependent currents. *The Journal of Physiology, 552*(Pt 2), 425–436.

Finger, T., & Kinnamon, S. (2021). Purinergic neurotransmission in the gustatory system. *Autonomic Neuroscience: Basic and Clinical, 236*, Article 102874.

Finger, T. E., & Barlow, L. A. (2021). Cellular diversity and regeneration in taste buds. *Current Opinion in Physiology, 20*, 146–153.

Finger, T. E., Danilova, V., Barrows, J., Bartel, D. L., Vigers, A. J., Stone, L., et al. (2005). ATP signaling is crucial for communication from taste buds to gustatory nerves. *Science (New York, NY), 310*(5753), 1495–1499.

Fráter, G., Bajgrowicz, J. A., & Kraft, P. (1998). Fragrance chemistry. *Tetrahedron, 54*, 7633–7703.

Furue, H., & Yoshii, K. (1997). In situ tight-seal recordings of taste substance-elicited action currents and voltage-gated Ba currents from single taste bud cells in the peeled epithelium of mouse tongue. *Brain Research, 776*(1–2), 133–139.

Gainetdinov, R. R., Hoener, M. C., & Berry, M. D. (2018). Trace amines and their receptors. *Pharmacological Reviews, 70*(3), 549–620.

Gao, N., Lu, M., Echeverri, F., Laita, B., Kalabat, D., Williams, M. E., et al. (2009). Voltage-gated sodium channels in taste bud cells. *BMC Neuroscience, 10*, 20.

Garty, H., & Palmer, L. G. (1997). Epithelial sodium channels: Function, structure, and regulation. *Physiological Reviews, 77*(2), 359–396.

Genovese, F., Reisert, J., & Kefalov, V. J. (2021). Sensory transduction in photoreceptors and olfactory sensory neurons: Common features and distinct characteristics. *Frontiers in Cellular Neuroscience, 15*, Article 761416.

Ghiaroni, V., Fieni, F., Tirindelli, R., Pietra, P., & Bigiani, A. (2003). Ion conductances in supporting cells isolated from the mouse vomeronasal organ. *Journal of Neurophysiology, 89*(1), 118–127.

Gilbertson, T. A., Avenet, P., Kinnamon, S. C., & Roper, S. D. (1992). Proton currents through amiloride-sensitive Na channels in hamster taste cells. Role in acid transduction. *The Journal of General Physiology, 100*(5), 803–824.

Grosmaitre, X., Vassalli, A., Mombaerts, P., Shepherd, G. M., & Ma, M. (2006). Odorant responses of olfactory sensory neurons expressing the odorant receptor MOR23: A patch clamp analysis in gene-targeted mice. *Proceedings of the National Academy of Sciences of the United States of America, 103*(6), 1970–1975.

Guarascio, D. M., Gonzalez-Velandia, K. Y., Hernandez-Clavijo, A., Menini, A., & Pifferi, S. (2021). Functional expression of TMEM16A in taste bud cells. *The Journal of Physiology, 599*(15), 3697–3714.

He, W., Yasumatsu, K., Varadarajan, V., Yamada, A., Lem, J., Ninomiya, Y., et al. (2004). Umami taste responses are mediated by alpha-transducin and alpha-gustducin. *Journal of Neuroscience: The Official Journal of the Society for Neuroscience, 24*(35), 7674–7680.

Heath, T. P., Melichar, J. K., Nutt, D. J., & Donaldson, L. F. (2006). Human taste thresholds are modulated by serotonin and noradrenaline. *Journal of Neuroscience: The Official Journal of the Society for Neuroscience, 26*(49), 12664–12671.

Heck, G. L., Mierson, S., & DeSimone, J. A. (1984). Salt taste transduction occurs through an amiloride-sensitive sodium transport pathway. *Science (New York, NY), 223*(4634), 403–405.

Hengl, T., Kaneko, H., Dauner, K., Vocke, K., Frings, S., & Möhrlen, F. (2010). Molecular components of signal amplification in olfactory sensory cilia. *Proceedings of the National Academy of Sciences of the United States of America, 107*(13), 6052–6057.

Hernandez-Clavijo, A., Sarno, N., Gonzalez-Velandia, K. Y., Degen, R., Fleck, D., Rock, J. R., et al. (2021). TMEM16A and TMEM16B modulate pheromone-evoked action potential firing in mouse vomeronasal sensory neurons. *eNeuro, 8*(5).

Herness, M. S., & Chen, Y. (2000). Serotonergic agonists inhibit calcium-activated potassium and voltage-dependent sodium currents in rat taste receptor cells. *Journal of Membrane Biology, 173*(2), 127–138.

Herness, M. S., & Sun, X. D. (1995). Voltage-dependent sodium currents recorded from dissociated rat taste cells. *Journal of Membrane Biology, 146*(1), 73–84.

Herness, S., Zhao, F. L., Kaya, N., Shen, T., Lu, S. G., & Cao, Y. (2005). Communication routes within the taste bud by neurotransmitters and neuropeptides. *Chemical Senses, 30*(Suppl. 1), i37–i38.

Herrada, G., & Dulac, C. (1997). A novel family of putative pheromone receptors in mammals with a topographically organized and sexually dimorphic distribution. *Cell., 90*(4), 763–773.

Hladik, C. M., Pasquet, P., & Simmen, B. (2002). New perspectives on taste and primate evolution: The dichotomy in gustatory coding for perception of beneficent versus noxious substances as supported by correlations among human thresholds. *American Journal of Physical Anthropology, 117*(4), 342–348.

Holy, T. E., Dulac, C., & Meister, M. (2000). Responses of vomeronasal neurons to natural stimuli. *Science (New York, NY), 289*(5484), 1569–1572.

Huang, L., Shanker, Y. G., Dubauskaite, J., Zheng, J. Z., Yan, W., Rosenzweig, S., et al. (1999). Ggamma13 colocalizes with gustducin in taste receptor cells and mediates IP3 responses to bitter denatonium. *Nature Neuroscience, 2*(12), 1055–1062.

Huang, Y. A., Dando, R., & Roper, S. D. (2009). Autocrine and paracrine roles for ATP and serotonin in mouse taste buds. *Journal of Neuroscience: The Official Journal of the Society for Neuroscience, 29*(44), 13909–13918.

Huang, Y. A., Maruyama, Y., Stimac, R., & Roper, S. D. (2008). Presynaptic (Type III) cells in mouse taste buds sense sour (acid) taste. *The Journal of Physiology, 586*(12), 2903–2912.

Huang, Y. A., Pereira, E., & Roper, S. D. (2011). Acid stimulation (sour taste) elicits GABA and serotonin release from mouse taste cells. *PLoS One, 6*(10), Article e25471.

Huang, Y. A., & Roper, S. D. (2010). Intracellular Ca(2+) and TRPM5-mediated membrane depolarization produce ATP secretion from taste receptor cells. *The Journal of Physiology, 588*(Pt 13), 2343–2350.

Huang, Y. A., Stone, L. M., Pereira, E., Yang, R., Kinnamon, J. C., Dvoryanchikov, G., et al. (2011). Knocking out P2X receptors reduces transmitter secretion in taste buds. *Journal of Neuroscience: The Official Journal of the Society for Neuroscience, 31*(38), 13654–13661.

Huang, Y. J., Maruyama, Y., Dvoryanchikov, G., Pereira, E., Chaudhari, N., & Roper, S. D. (2007). The role of pannexin 1 hemichannels in ATP release and cell-cell communication in mouse taste buds. *Proceedings of the National Academy of Sciences of the United States of America, 104*(15), 6436–6441.

Huang, Y. J., Maruyama, Y., Lu, K. S., Pereira, E., Plonsky, I., Baur, J. E., et al. (2005). Mouse taste buds use serotonin as a neurotransmitter. *Journal of Neuroscience: The Official Journal of the Society for Neuroscience, 25*(4), 843–847.

Hudson, R. (2000). Odor and odorant: A terminological clarification. *Chemical Senses, 25*(6), 693.

Hummel, T., & Frasnelli, J. (2019). The intranasal trigeminal system. *Handbook of Clinical Neurology, 164*, 119–134.

Ihara, S., Yoshikawa, K., & Touhara, K. (2013). Chemosensory signals and their receptors in the olfactory neural system. *Neuroscience, 254*, 45–60.

Inamura, K., Kashiwayanagi, M., & Kurihara, K. (1997). Inositol-1,4,5-trisphosphate induces responses in receptor neurons in rat vomeronasal sensory slices. *Chemical Senses, 22*(1), 93–103.

Inamura, K., Matsumoto, Y., Kashiwayanagi, M., & Kurihara, K. (1999). Laminar distribution of pheromone-receptive neurons in rat vomeronasal epithelium. *The Journal of Physiology, 517*(Pt 3), 731–739.

Ishida, Y., Ugawa, S., Ueda, T., Yamada, T., Shibata, Y., Hondoh, A., et al. (2009). P2X(2)- and P2X(3)-positive fibers in fungiform papillae originate from the chorda tympani but not the trigeminal nerve in rats and mice. *The Journal of Comparative Neurology, 514*(2), 131–144.

Jaber, L., Zhao, F. L., Kolli, T., & Herness, S. (2014). A physiologic role for serotonergic transmission in adult rat taste buds. *PLoS One, 9*(11), Article e112152.

Jones, D. T., & Reed, R. R. (1989). Golf: An olfactory neuron specific-G protein involved in odorant signal transduction. *Science (New York, NY), 244*(4906), 790–795.

Kaneko, H., Nakamura, T., & Lindemann, B. (2001). Noninvasive measurement of chloride concentration in rat olfactory receptor cells with use of a fluorescent dye. *American Journal of Physiology - Cell Physiology, 280*(6), C1387–C1393.

Kaneko, H., Putzier, I., Frings, S., Kaupp, U. B., & Gensch, T. (2004). Chloride accumulation in mammalian olfactory sensory neurons. *Journal of Neuroscience: The Official Journal of the Society for Neuroscience, 24*(36), 7931–7938.

Karlson, P., & Luscher, M. (1959). Pheromones': A new term for a class of biologically active substances. *Nature, 183*(4653), 55–56.

Katreddi, R. R., & Forni, P. E. (2021). Mechanisms underlying pre- and postnatal development of the vomeronasal organ. *Cellular and Molecular Life Sciences: CMLS, 78*(12), 5069–5082.

Kaupp, U. B. (2010). Olfactory signalling in vertebrates and insects: Differences and commonalities. *Nature Reviews Neuroscience, 11*(3), 188–200.

Kaya, N., Shen, T., Lu, S. G., Zhao, F. L., & Herness, S. (2004). A paracrine signaling role for serotonin in rat taste buds: Expression and localization of serotonin receptor subtypes. *American Journal of Physiology - Regulatory, Integrative and Comparative Physiology, 286*(4), R649–R658.

Kellenberger, S., & Schild, L. (2002). Epithelial sodium channel/degenerin family of ion channels: A variety of functions for a shared structure. *Physiological Reviews, 82*(3), 735–767.

Kellenberger, S., & Schild, L. (2015). International Union of Basic and Clinical Pharmacology. XCI. structure, function, and pharmacology of acid-sensing ion channels and the epithelial Na+ channel. *Pharmacological Reviews, 67*(1), 1–35.

Kim, D. J., & Roper, S. D. (1995). Localization of serotonin in taste buds: A comparative study in four vertebrates. *The Journal of Comparative Neurology, 353*(3), 364–370.

Kim, S., Ma, L., Unruh, J., McKinney, S., & Yu, C. R. (2015). Intracellular chloride concentration of the mouse vomeronasal neuron. *BMC Neuroscience, 16*, 90.

Kim, S., Ma, L., & Yu, C. R. (2011). Requirement of calcium-activated chloride channels in the activation of mouse vomeronasal neurons. *Nature Communications, 2*, 365.

Kimura, K., Ohtubo, Y., Tateno, K., Takeuchi, K., Kumazawa, T., & Yoshii, K. (2014). Cell-type-dependent action potentials and voltage-gated currents in mouse fungiform taste buds. *European Journal of Neuroscience, 39*(1), 24–34.

Kinnamon, J. C., Sherman, T. A., & Roper, S. D. (1988). Ultrastructure of mouse vallate taste buds: III. Patterns of synaptic connectivity. *The Journal of Comparative Neurology, 270*(1), 1-10–56-57.

Kinnamon, J. C., Taylor, B. J., Delay, R. J., & Roper, S. D. (1985). Ultrastructure of mouse vallate taste buds. I. Taste cells and their associated synapses. *The Journal of Comparative Neurology, 235*(1), 48–60.

Kinnamon, S., & Finger, T. (2021). The role of ATP and purinergic receptors in taste signaling. *Handbook of Experimental Pharmacology, 275.*

Kinnamon, S. C., & Finger, T. E. (2013). A taste for ATP: Neurotransmission in taste buds. *Frontiers in Cellular Neuroscience, 7*, 264.

Kinnamon, S. C., & Finger, T. E. (2019). Recent advances in taste transduction and signaling. *F1000Research, 8.*

Kleene, S. J. (2008). The electrochemical basis of odor transduction in vertebrate olfactory cilia. *Chemical Senses, 33*(9), 839–859.

Kurahashi, T., & Menini, A. (1997). Mechanism of odorant adaptation in the olfactory receptor cell. *Nature, 385*(6618), 725–729.

Kurian, S. M., Naressi, R. G., Manoel, D., Barwich, A. S., Malnic, B., & Saraiva, L. R. (2021). Odor coding in the mammalian olfactory epithelium. *Cell and Tissue Research, 383*(1), 445–456.

Lagostena, L., & Menini, A. (2003). Whole-cell recordings and photolysis of caged compounds in olfactory sensory neurons isolated from the mouse. *Chemical Senses, 28*(8), 705–716.

Larson, E. D., Vandenbeuch, A., Anderson, C. B., & Kinnamon, S. C. (2020). Function, innervation, and neurotransmitter signaling in mice lacking type-II taste cells. *eNeuro, 7*(1).

Larson, E. D., Vandenbeuch, A., Voigt, A., Meyerhof, W., Kinnamon, S. C., & Finger, T. E. (2015). The role of 5-HT3 receptors in signaling from taste buds to nerves. *Journal of Neuroscience: The Official Journal of the Society for Neuroscience, 35*(48), 15984–15995.

Leinders-Zufall, T., Brennan, P., Widmayer, P., Prashanth, C. P., Maul-Pavicic, A., Jäger, M., et al. (2004). MHC class I peptides as chemosensory signals in the vomeronasal organ. *Science (New York, NY), 306*(5698), 1033–1037.

Leinders-Zufall, T., Greer, C. A., Shepherd, G. M., & Zufall, F. (1998). Imaging odor-induced calcium transients in single olfactory cilia: Specificity of activation and role in transduction. *Journal of Neuroscience: The Official Journal of the Society for Neuroscience, 18*(15), 5630–5639.

Leinders-Zufall, T., Lane, A. P., Puche, A. C., Ma, W., Novotny, M. V., Shipley, M. T., et al. (2000). Ultrasensitive pheromone detection by mammalian vomeronasal neurons. *Nature, 405*(6788), 792–796.

Leinders-Zufall, T., Rand, M. N., Shepherd, G. M., Greer, C. A., & Zufall, F. (1997). Calcium entry through cyclic nucleotide-gated channels in individual cilia of olfactory receptor cells: Spatiotemporal dynamics. *Journal of Neuroscience: The Official Journal of the Society for Neuroscience, 17*(11), 4136–4148.

Lewandowski, B. C., Sukumaran, S. K., Margolskee, R. F., & Bachmanov, A. A. (2016). Amiloride-insensitive salt taste is mediated by two populations of type III taste cells with distinct transduction mechanisms. *Journal of Neuroscience: The Official Journal of the Society for Neuroscience, 36*(6), 1942–1953.

Li, X., Staszewski, L., Xu, H., Durick, K., Zoller, M., & Adler, E. (2002). Human receptors for sweet and umami taste. *Proceedings of the National Academy of Sciences of the United States of America, 99*(7), 4692–4696.

Liang, F. (2020). Sustentacular cell enwrapment of olfactory receptor neuronal dendrites: An update. *Genes, 11*(5).

Liberles, S. D., & Buck, L. B. (2006). A second class of chemosensory receptors in the olfactory epithelium. *Nature, 442*(7103), 645–650.

Liberles, S. D. (2014). Mammalian pheromones. *Annual Review of Physiology, 76*, 151–175.

Liberles, S. D. (2015). Trace amine-associated receptors: Ligands, neural circuits, and behaviors. *Current Opinion in Neurobiology, 34*, 1–7.

Liman, E. R., Corey, D. P., & Dulac, C. (1999). TRP2: A candidate transduction channel for mammalian pheromone sensory signaling. *Proceedings of the National Academy of Sciences of the United States of America, 96*(10), 5791–5796.

Liman, E. R., & Corey, D. P. (1996). Electrophysiological characterization of chemosensory neurons from the mouse vomeronasal organ. *Journal of Neuroscience: The Official Journal of the Society for Neuroscience, 16*(15), 4625–4637.

Liman, E. R., Zhang, Y. V., & Montell, C. (2014). Peripheral coding of taste. *Neuron, 81*(5), 984–1000.

Lindemann, B. (1996). Taste reception. *Physiological Reviews, 76*(3), 719–766.

Liu, D., & Liman, E. R. (2003). Intracellular Ca2+ and the phospholipid PIP2 regulate the taste transduction ion channel TRPM5. *Proceedings of the National Academy of Sciences of the United States of America, 100*(25), 15160–15165.

Liu, L., Hansen, D. R., Kim, I., & Gilbertson, T. A. (2005). Expression and characterization of delayed rectifying K+ channels in anterior rat taste buds. *American Journal of Physiology - Cell Physiology, 289*(4), C868–C880.

Lossow, K., Hermans-Borgmeyer, I., Meyerhof, W., & Behrens, M. (2020). Segregated expression of ENaC subunits in taste cells. *Chemical Senses, 45*(4), 235–248.

Lucas, P., Ukhanov, K., Leinders-Zufall, T., & Zufall, F. (2003). A diacylglycerol-gated cation channel in vomeronasal neuron dendrites is impaired in TRPC2 mutant mice: Mechanism of pheromone transduction. *Neuron, 40*(3), 551–561.

Lugaz, O., Pillias, A. M., & Faurion, A. (2002). A new specific ageusia: Some humans cannot taste L-glutamate. *Chemical Senses, 27*(2), 105–115.

Lundberg, Y. W., Xu, Y., Thiessen, K. D., & Kramer, K. L. (2015). Mechanisms of otoconia and otolith development. *Developmental Dynamics: An Official Publication of the American Association of Anatomists, 244*(3), 239–253.

Lynch, J. W., & Barry, P. H. (1989). Action potentials initiated by single channels opening in a small neuron (rat olfactory receptor). *Biophysical Journal, 55*(4), 755–768.

Münch, J., Billig, G., Hübner, C. A., Leinders-Zufall, T., Zufall, F., & Jentsch, T. J. (2018). Ca(2+)-activated Cl(-) currents in the murine vomeronasal organ enhance neuronal spiking but are dispensable for male-male aggression. *Journal of Biological Chemistry, 293*(26), 10392–10403.

Ma, Z., Saung, W. T., & Foskett, J. K. (2017). Action potentials and ion conductances in wild-type and CALHM1-knockout type II taste cells. *Journal of Neurophysiology, 117*(5), 1865–1876.

Ma, Z., Taruno, A., Ohmoto, M., Jyotaki, M., Lim, J. C., Miyazaki, H., et al. (2018). CALHM3 is essential for rapid ion channel-mediated purinergic neurotransmission of GPCR-mediated tastes. *Neuron, 98*(3).

Matsunami, H., & Buck, L. B. (1997). A multigene family encoding a diverse array of putative pheromone receptors in mammals. *Cell., 90*(4), 775–784.

Matsuo, R., & Carpenter, G. H. (2015). The role of saliva in taste transduction. In R. L. Doty (Ed.), *Handbook of olfaction and gustation* (3rd ed., pp. 625–636). Hoboken, New Jersey: Wiley.

McEwen, D. P., Jenkins, P. M., & Martens, J. R. (2008). Olfactory cilia: Our direct neuronal connection to the external world. *Current Topics in Developmental Biology, 85*, 333–370.

McLaughlin, S. K., McKinnon, P. J., & Margolskee, R. F. (1992). Gustducin is a taste-cell-specific G protein closely related to the transducins. *Nature, 357*(6379), 563–569.

McLaughlin, S. K., McKinnon, P. J., Spickofsky, N., Danho, W., & Margolskee, R. F. (1994). Molecular cloning of G proteins and phosphodiesterases from rat taste cells. *Physiology and Behavior, 56*(6), 1157–1164.

Medler, K. F., Margolskee, R. F., & Kinnamon, S. C. (2003). Electrophysiological characterization of voltage-gated currents in defined taste cell types of mice. *Journal of Neuroscience: The Official Journal of the Society for Neuroscience, 23*(7), 2608–2617.

Menco, B. P., Carr, V. M., Ezeh, P. I., Liman, E. R., & Yankova, M. P. (2001). Ultrastructural localization of G-proteins and the channel protein TRP2 to microvilli of rat vomeronasal receptor cells. *The Journal of Comparative Neurology, 438*(4), 468–489.

Meredith, M. (1998). Vomeronasal function. *Chemical Senses, 23*(4), 463–466.

Meyerhof, W. (2005). Elucidation of mammalian bitter taste. Reviews of physiology. *Biochemistry and Pharmacology, 154*, 37–72.

Mohrhardt, J., Nagel, M., Fleck, D., Ben-Shaul, Y., & Spehr, M. (2018). Signal detection and coding in the accessory olfactory system. *Chemical Senses, 43*(9), 667–695.

Mombaerts, P., Wang, F., Dulac, C., Chao, S. K., Nemes, A., Mendelsohn, M., et al. (1996). Visualizing an olfactory sensory map. *Cell, 87*(4), 675–686.

Mori, K., von Campenhause, H., & Yoshihara, Y. (2000). Zonal organization of the mammalian main and accessory olfactory systems. *Philosophical Transactions of the Royal Society of London - Series B: Biological Sciences, 355*(1404), 1801–1812.

Morrison, E. E., & Costanzo, R. M. (1992). Morphology of olfactory epithelium in humans and other vertebrates. *Microscopy Research and Technique, 23*(1), 49–61.

Morrison, E. E., & Costanzo, R. M. (1990). Morphology of the human olfactory epithelium. *The Journal of Comparative Neurology, 297*(1), 1–13.

Nakamura, T., & Gold, G. H. (1987). A cyclic nucleotide-gated conductance in olfactory receptor cilia. *Nature, 325*(6103), 442–444.

Narusuye, K., Kawai, F., & Miyachi, E. (2003). Spike encoding of olfactory receptor cells. *Neuroscience Research, 46*(4), 407–413.
Nei, M., Niimura, Y., & Nozawa, M. (2008). The evolution of animal chemosensory receptor gene repertoires: Roles of chance and necessity. *Nature Reviews Genetics, 9*(12), 951–963.
Nelson, G., Chandrashekar, J., Hoon, M. A., Feng, L., Zhao, G., Ryba, N. J., et al. (2002). An amino-acid taste receptor. *Nature, 416*(6877), 199–202.
Nelson, G., Hoon, M. A., Chandrashekar, J., Zhang, Y., Ryba, N. J., & Zuker, C. S. (2001). Mammalian sweet taste receptors. *Cell, 106*(3), 381–390.
Neureither, F., Stowasser, N., Frings, S., & Möhrlen, F. (2017). Tracking of unfamiliar odors is facilitated by signal amplification through anoctamin 2 chloride channels in mouse olfactory receptor neurons. *Physiological Reports, 5*(15).
Niimura, Y., Tsunoda, M., Kato, S., Murata, K., Yanagawa, T., Suzuki, S., et al. (2021). Origin and evolution of the gene family of proteinaceous pheromones, the exocrine gland-secreting peptides, in rodents. *Molecular Biology and Evolution, 38*(2), 634–649.
Nomura, K., Nakanishi, M., Ishidate, F., Iwata, K., & Taruno, A. (2020). All-electrical Ca(2+)-independent signal transduction mediates attractive sodium taste in taste buds. *Neuron, 106*(5).
Northcutt, R. G. (2004). Taste buds: Development and evolution. *Brain, Behavior and Evolution, 64*(3), 198–206.
Oboti, L., Pérez-Gómez, A., Keller, M., Jacobi, E., Birnbaumer, L., Leinders-Zufall, T., et al. (2014). A wide range of pheromone-stimulated sexual and reproductive behaviors in female mice depend on G protein Gαo. *BMC Biology, 12*, 31.
Ohmoto, M., Jyotaki, M., Foskett, J. K., & Matsumoto, I. (2020). Sodium-taste cells require skn-1a for generation and share molecular features with sweet, umami, and bitter taste cells. *eNeuro, 7*(6).
Oka, Y., Butnaru, M., von Buchholtz, L., Ryba, N. J., & Zuker, C. S. (2013). High salt recruits aversive taste pathways. *Nature, 494*(7438), 472–475.
Oka, Y., Omura, M., Kataoka, H., & Touhara, K. (2004). Olfactory receptor antagonism between odorants. *The EMBO Journal, 23*(1), 120–126.
Omura, M., & Mombaerts, P. (2014). Trpc2-expressing sensory neurons in the main olfactory epithelium of the mouse. *Cell Reports, 8*(2), 583–595.
Osakada, T., Abe, T., Itakura, T., Mori, H., Ishii, K. K., Eguchi, R., et al. (2022). Hemoglobin in the blood acts as a chemosensory signal via the mouse vomeronasal system. *Nature Communications, 13*(1), 556.
Pérez, C. A., Huang, L., Rong, M., Kozak, J. A., Preuss, A. K., Zhang, H., et al. (2002). A transient receptor potential channel expressed in taste receptor cells. *Nature Neuroscience, 5*(11), 1169–1176.
Pedemonte, N., & Galietta, L. J. (2014). Structure and function of TMEM16 proteins (anoctamins). *Physiological Reviews, 94*(2), 419–459.
Perea-Martinez, I., Nagai, T., & Chaudhari, N. (2013). Functional cell types in taste buds have distinct longevities. *PLoS One, 8*(1), Article e53399.
Pietra, G., Dibattista, M., Menini, A., Reisert, J., & Boccaccio, A. (2016). The Ca2+-activated Cl- channel TMEM16B regulates action potential firing and axonal targeting in olfactory sensory neurons. *The Journal of General Physiology, 148*(4), 293–311.
Pifferi, S., Boccaccio, A., & Menini, A. (2006). Cyclic nucleotide-gated ion channels in sensory transduction. *FEBS Letters, 580*(12), 2853–2859.
Pifferi, S., Cenedese, V., & Menini, A. (2012). Anoctamin 2/TMEM16B: A calcium-activated chloride channel in olfactory transduction. *Experimental Physiology, 97*(2), 193–199.
Pifferi, S., Pascarella, G., Boccaccio, A., Mazzatenta, A., Gustincich, S., Menini, A., et al. (2006). Bestrophin-2 is a candidate calcium-activated chloride channel involved in olfactory transduction. *Proceedings of the National Academy of Sciences of the United States of America, 103*(34), 12929–12934.
Price, S. (1987). Effects of odorant mixtures on olfactory receptor cells. *Annals of the New York Academy of Sciences, 510*, 55–60.
Rünnenburger, K., Breer, H., & Boekhoff, I. (2002). Selective G protein beta gamma-subunit compositions mediate phospholipase C activation in the vomeronasal organ. *European Journal of Cell Biology, 81*(10), 539–547.
Reddy, G., Zak, J. D., Vergassola, M., & Murthy, V. N. (2018). Antagonism in olfactory receptor neurons and its implications for the perception of odor mixtures. *Elife, 7*.
Reisert, J., Bauer, P. J., Yau, K. W., & Frings, S. (2003). The Ca-activated Cl channel and its control in rat olfactory receptor neurons. *The Journal of General Physiology, 122*(3), 349–363.
Reisert, J., & Reingruber, J. (2019). Ca(2+)-activated Cl(-) current ensures robust and reliable signal amplification in vertebrate olfactory receptor neurons. *Proceedings of the National Academy of Sciences of the United States of America, 116*(3), 1053–1058.
Reuter, D., Zierold, K., Schröder, W. H., & Frings, S. (1998). A depolarizing chloride current contributes to chemoelectrical transduction in olfactory sensory neurons in situ. *Journal of Neuroscience: The Official Journal of the Society for Neuroscience, 18*(17), 6623–6630.
Richards, T. W. (1898). The relation of the taste of acids to their degree of dissociation. *American Chemical Journal, 20*, 121–126.
Richter, T. A., Caicedo, A., & Roper, S. D. (2003). Sour taste stimuli evoke Ca2+ and pH responses in mouse taste cells. *The Journal of physiology, 547*(Pt 2), 475–483.
Richter, T. A., Dvoryanchikov, G. A., Roper, S. D., & Chaudhari, N. (2004). Acid-sensing ion channel-2 is not necessary for sour taste in mice. *Journal of Neuroscience: The Official Journal of the Society for Neuroscience, 24*(16), 4088–4091.
Rodriguez, I., Greer, C. A., Mok, M. Y., & Mombaerts, P. (2000). A putative pheromone receptor gene expressed in human olfactory mucosa. *Nature Genetics, 26*(1), 18–19.
Rodriguez, Y. A., Roebber, J. K., Dvoryanchikov, G., Makhoul, V., Roper, S. D., & Chaudhari, N. (2021). Tripartite synapses" in taste buds: A role for type I glial-like taste cells. *Journal of Neuroscience: The Official Journal of the Society for Neuroscience, 41*(48), 9860–9871.
Roebber, J. K., Roper, S. D., & Chaudhari, N. (2019). The role of the anion in salt (NaCl) detection by mouse taste buds. *Journal of Neuroscience: The Official Journal of the Society for Neuroscience, 39*(32), 6224–6232.
Rolls, E. T. (2015). Taste, olfactory, and food reward value processing in the brain. *Progress in Neurobiology, 127–128*, 64–90.
Romanov, R. A., Lasher, R. S., High, B., Savidge, L. E., Lawson, A., Rogachevskaja, O. A., et al. (2018). Chemical synapses without synaptic vesicles: Purinergic neurotransmission through a CALHM1 channel-mitochondrial signaling complex. *Science Signaling, 11*(529).

Romanov, R. A., Rogachevskaja, O. A., Bystrova, M. F., Jiang, P., Margolskee, R. F., & Kolesnikov, S. S. (2007). Afferent neurotransmission mediated by hemichannels in mammalian taste cells. *The EMBO Journal, 26*(3), 657–667.

Roper, S. (1983). Regenerative impulses in taste cells. *Science (New York, NY), 220*(4603), 1311–1312.

Roper, S. D., & Chaudhari, N. (2017). Taste buds: Cells, signals and synapses. *Nature Reviews Neuroscience, 18*(8), 485–497.

Roper, S. D. (2021). Chemical and electrical synaptic interactions among taste bud cells. *Current Opinion in Physiology, 20*, 118–125.

Roper, S. D. (2013). Taste buds as peripheral chemosensory processors. *Seminars in Cell and Developmental Biology, 24*(1), 71–79.

Roper, S. D. (2015). The taste of table salt. *Pfluegers Archiv European Journal of Physiology, 467*(3), 457–463.

Royer, S. M., & Kinnamon, J. C. (1994). Application of serial sectioning and three-dimensional reconstruction to the study of taste bud ultrastructure and organization. *Microscopy Research and Technique, 29*(5), 381–407.

Ryba, N. J., & Tirindelli, R. (1997). A new multigene family of putative pheromone receptors. *Neuron, 19*(2), 371–379.

Sainz, E., Korley, J. N., Battey, J. F., & Sullivan, S. L. (2001). Identification of a novel member of the T1R family of putative taste receptors. *Journal of Neurochemistry, 77*(3), 896–903.

Sanchez-Andrade, G., & Kendrick, K. M. (2009). The main olfactory system and social learning in mammals. *Behavioural Brain Research, 200*(2), 323–335.

Sarrafchi, A., Odhammer, A. M., Hernandez Salazar, L. T., & Laska, M. (2013). Olfactory sensitivity for six predator odorants in CD-1 mice, human subjects, and spider monkeys. *PLoS One, 8*(11), Article e80621.

Schiffman, S. S., & Erickson, R. P. (1971). A psychophysical model for gustatory quality. *Physiology and Behavior, 7*(4), 617–633.

Schiffman, S. S., Lockhead, E., & Maes, F. W. (1983). Amiloride reduces the taste intensity of Na+ and Li+ salts and sweeteners. *Proceedings of the National Academy of Sciences of the United States of America, 80*(19), 6136–6140.

Schild, D., & Restrepo, D. (1998). Transduction mechanisms in vertebrate olfactory receptor cells. *Physiological Reviews, 78*(2), 429–466.

Serizawa, S., Miyamichi, K., & Sakano, H. (2004). One neuron-one receptor rule in the mouse olfactory system. *Trends in Genetics: Trends in Genetics, 20*(12), 648–653.

Shimada, S., Ueda, T., Ishida, Y., Yamamoto, T., & Ugawa, S. (2006). Acid-sensing ion channels in taste buds. *Archives of Histology and Cytology, 69*(4), 227–231.

Shimazaki, R., Boccaccio, A., Mazzatenta, A., Pinato, G., Migliore, M., & Menini, A. (2006). Electrophysiological properties and modeling of murine vomeronasal sensory neurons in acute slice preparations. *Chemical Senses, 31*(5), 425–435.

Silva, L., & Antunes, A. (2017). Vomeronasal receptors in vertebrates and the evolution of pheromone detection. *Annual Review of Animal Biosciences, 5*, 353–370.

Spehr, J., Hagendorf, S., Weiss, J., Spehr, M., Leinders-Zufall, T., & Zufall, F. (2009). Ca2+ -calmodulin feedback mediates sensory adaptation and inhibits pheromone-sensitive ion channels in the vomeronasal organ. *Journal of Neuroscience: The Official Journal of the Society for Neuroscience., 29*(7), 2125–2135.

Spehr, M., Spehr, J., Ukhanov, K., Kelliher, K. R., Leinders-Zufall, T., & Zufall, F. (2006). Parallel processing of social signals by the mammalian main and accessory olfactory systems. *Cellular and Molecular Life Sciences: CMLS, 63*(13), 1476–1484.

Stephan, A. B., Shum, E. Y., Hirsh, S., Cygnar, K. D., Reisert, J., & Zhao, H. (2009). ANO2 is the cilial calcium-activated chloride channel that may mediate olfactory amplification. *Proceedings of the National Academy of Sciences of the United States of America, 106*(28), 11776–11781.

Stone, L. M., Finger, T. E., Tam, P. P., & Tan, S. S. (1995). Taste receptor cells arise from local epithelium, not neurogenic ectoderm. *Proceedings of the National Academy of Sciences of the United States of America, 92*(6), 1916–1920.

Töle, J. C., Behrens, M., & Meyerhof, W. (2019). Taste receptor function. *Handbook of Clinical Neurology, 164*, 173–185.

Takeuchi, H., Ishida, H., Hikichi, S., & Kurahashi, T. (2009). Mechanism of olfactory masking in the sensory cilia. *The Journal of General Physiology, 133*(6), 583–601.

Takeuchi, H., & Kurahashi, T. (2005). Mechanism of signal amplification in the olfactory sensory cilia. *Journal of Neuroscience: The Official Journal of the Society for Neuroscience, 25*(48), 11084–11091.

Takigami, S., Osada, T., Yoshida-Matsuoka, J., Matsuoka, M., Mori, Y., & Ichikawa, M. (1999). The expressed localization of rat putative pheromone receptors. *Neuroscience Letters, 272*(2), 115–118.

Taruno, A., Matsumoto, I., Ma, Z., Marambaud, P., & Foskett, J. K. (2013). How do taste cells lacking synapses mediate neurotransmission? CALHM1, a voltage-gated ATP channel. *BioEssays: News and Reviews in Molecular, Cellular and Developmental Biology, 35*(12), 1111–1118.

Taruno, A., Nomura, K., Kusakizako, T., Ma, Z., Nureki, O., & Foskett, J. K. (2021). Taste transduction and channel synapses in taste buds. *Pfluegers Archiv European Journal of Physiology, 473*(1), 3–13.

Taruno, A., Vingtdeux, V., Ohmoto, M., Ma, Z., Dvoryanchikov, G., Li, A., et al. (2013). CALHM1 ion channel mediates purinergic neurotransmission of sweet, bitter and umami tastes. *Nature, 495*(7440), 223–226.

Teng, B., Wilson, C. E., Tu, Y. H., Joshi, N. R., Kinnamon, S. C., & Liman, E. R. (2019). Cellular and neural responses to sour stimuli require the proton channel Otop1. *Current Biology: CB, 29*(21).

Thomas-Danguin, T., Sinding, C., Romagny, S., El Mountassir, F., Atanasova, B., Le Berre, E., et al. (2014). The perception of odor objects in everyday life: A review on the processing of odor mixtures. *Frontiers in Psychology, 5*, 504.

Tirindelli, R. (2021). Coding of pheromones by vomeronasal receptors. *Cell and Tissue Research, 383*(1), 367–386.

Tirindelli, R., Dibattista, M., Pifferi, S., & Menini, A. (2009). From pheromones to behavior. *Physiological Reviews, 89*(3), 921–956.

Tomchik, S. M., Berg, S., Kim, J. W., Chaudhari, N., & Roper, S. D. (2007). Breadth of tuning and taste coding in mammalian taste buds. *Journal of Neuroscience: The Official Journal of the Society for Neuroscience, 27*(40), 10840–10848.

Triller, A., Boulden, E. A., Churchill, A., Hatt, H., Englund, J., Spehr, M., et al. (2008). Odorant-receptor interactions and odor percept: A chemical perspective. *Chemistry and Biodiversity, 5*(6), 862–886.

Trouillet, A. C., Keller, M., Weiss, J., Leinders-Zufall, T., Birnbaumer, L., Zufall, F., et al. (2019). Central role of G protein Gαi2 and Gαi2(+) vomeronasal neurons in balancing territorial and infant-directed aggression of male mice. *Proceedings of the National Academy of Sciences of the United States of America, 116*(11), 5135–5143.

Tu, Y. H., Cooper, A. J., Teng, B., Chang, R. B., Artiga, D. J., Turner, H. N., et al. (2018). An evolutionarily conserved gene family encodes proton-selective ion channels. *Science (New York, NY), 359*(6379), 1047–1050.

Ugawa, S., Yamamoto, T., Ueda, T., Ishida, Y., Inagaki, A., Nishigaki, M., et al. (2003). Amiloride-insensitive currents of the acid-sensing ion channel-2a (ASIC2a)/ASIC2b heteromeric sour-taste receptor channel. *Journal of Neuroscience: The Official Journal of the Society for Neuroscience, 23*(9), 3616–3622.

Ukhanov, K., Leinders-Zufall, T., & Zufall, F. (2007). Patch-clamp analysis of gene-targeted vomeronasal neurons expressing a defined V1r or V2r receptor: Ionic mechanisms underlying persistent firing. *Journal of Neurophysiology, 98*(4), 2357–2369.

Vaccarezza, O. L., Sepich, L. N., & Tramezzani, J. H. (1981). The vomeronasal organ of the rat. *Journal of Anatomy, 132*(Pt 2), 167–185.

Vandenbeuch, A., & Kinnamon, S. C. (2009). Why do taste cells generate action potentials? *Journal of Biology, 8*(4), 42.

Vandenbeuch, A., Larson, E. D., Anderson, C. B., Smith, S. A., Ford, A. P., Finger, T. E., et al. (2015). Postsynaptic P2X3-containing receptors in gustatory nerve fibres mediate responses to all taste qualities in mice. *The Journal of Physiology, 593*(5), 1113–1125.

Vandenbeuch, A., Zorec, R., & Kinnamon, S. C. (2010). Capacitance measurements of regulated exocytosis in mouse taste cells. *Journal of Neuroscience: The Official Journal of the Society for Neuroscience, 30*(44), 14695–14701.

Wang, Z. Y., Keith, I. M., Olson, E. B., Jr., Vidruk, E. H., & Bisgard, G. E. (2002). Expression of 5-HT3 receptors in primary sensory neurons of the petrosal ganglion of adult rats. *Autonomic Neuroscience: Basic and Clinical, 95*(1–2), 121–124.

Witt, M. (2019). Anatomy and development of the human taste system. *Handbook of Clinical Neurology, 164*, 147–171.

Wong, G. T., Gannon, K. S., & Margolskee, R. F. (1996). Transduction of bitter and sweet taste by gustducin. *Nature, 381*(6585), 796–800.

Wooding, S. P., Ramirez, V. A., & Behrens, M. (2021). Bitter taste receptors: Genes, evolution and health. *Evolution, Medicine, and Public Health, 9*(1), 431–447.

Wu, A., Dvoryanchikov, G., Pereira, E., Chaudhari, N., & Roper, S. D. (2015). Breadth of tuning in taste afferent neurons varies with stimulus strength. *Nature Communications, 6*, 8171.

Wu, L. J., Sweet, T. B., & Clapham, D. E. (2010). International Union of Basic and Clinical Pharmacology. LXXVI. Current progress in the mammalian TRP ion channel family. *Pharmacological Reviews, 62*(3), 381–404.

Yang, J., & Roper, S. D. (1987). Dye-coupling in taste buds in the mudpuppy, *Necturus maculosus. Journal of Neuroscience: The Official Journal of the Society for Neuroscience, 7*(11), 3561–3565.

Yang, R., Dzowo, Y. K., Wilson, C. E., Russell, R. L., Kidd, G. J., Salcedo, E., et al. (2020). Three-dimensional reconstructions of mouse circumvallate taste buds using serial blockface scanning electron microscopy: I. Cell types and the apical region of the taste bud. *The Journal of Comparative Neurology, 528*(5), 756–771.

Ye, W., Chang, R. B., Bushman, J. D., Tu, Y. H., Mulhall, E. M., Wilson, C. E., et al. (2016). The K+ channel KIR2.1 functions in tandem with proton influx to mediate sour taste transduction. *Proceedings of the National Academy of Sciences of the United States of America, 113*(2), E229–E238.

Yoshida, R., Horio, N., Murata, Y., Yasumatsu, K., Shigemura, N., & Ninomiya, Y. (2009). NaCl responsive taste cells in the mouse fungiform taste buds. *Neuroscience, 159*(2), 795–803.

Yoshii, K. (2005). Gap junctions among taste bud cells in mouse fungiform papillae. *Chemical Senses, 30*(Suppl. 1), i35–i36.

Yu, C. R., & Wu, Y. (2017). Regeneration and rewiring of rodent olfactory sensory neurons. *Experimental Neurology, 287*(Pt 3), 395–408.

Zhang, J., Jin, H., Zhang, W., Ding, C., O'Keeffe, S., Ye, M., et al. (2019). Sour sensing from the tongue to the brain. *Cell, 179*(2).

Zhang, J., Pacifico, R., Cawley, D., Feinstein, P., & Bozza, T. (2013). Ultrasensitive detection of amines by a trace amine-associated receptor. *Journal of Neuroscience: The Official Journal of the Society for Neuroscience, 33*(7), 3228–3239.

Zhang, Y., Hoon, M. A., Chandrashekar, J., Mueller, K. L., Cook, B., Wu, D., et al. (2003). Coding of sweet, bitter, and umami tastes: Different receptor cells sharing similar signaling pathways. *Cell, 112*(3), 293–301.

Zhao, G. Q., Zhang, Y., Hoon, M. A., Chandrashekar, J., Erlenbach, I., Ryba, N. J., et al. (2003). The receptors for mammalian sweet and umami taste. *Cell, 115*(3), 255–266.

Zheng, J., & Zagotta, W. N. (2004). Stoichiometry and assembly of olfactory cyclic nucleotide-gated channels. *Neuron, 42*(3), 411–421.

Zufall, F., Ukhanov, K., Lucas, P., Liman, E. R., & Leinders-Zufall, T. (2005). Neurobiology of TRPC2: From gene to behavior. *Pfluegers Archiv European Journal of Physiology, 451*(1), 61–71.

Chapter 30

Microtubules, dynein and eukaryotic cilia

David R. Mitchell[1], Taviare L. Hawkins[2] and Kenneth W. Foster[3]

[1]*Cell and Developmental Biology, SUNY Upstate Medical University, Syracuse, NY, United States;* [2]*Department of Physics, Wagner College, Staten Island, NY, United States;* [3]*Physics Department, Syracuse University, Syracuse, NY, United States*

1. Introduction

1.1 Brief overview of what microtubule-based motility does for a cell and an organism

Microtubules (Fig. 30.1) are essential cytoskeletal elements in eukaryotic cells, but are not observed in prokaryotes. Eukaryotic cells are, overall, much larger than prokaryotic cells and benefit from mechanisms that overcome the limitations of simple diffusion for the transport and distribution of macromolecular assemblies. Evolution of microtubules was a key component in making this possible. Microtubules, along with actin filaments, form an essential scaffold that governs distribution of typical eukaryotic organelles, including the nucleus, endoplasmic reticulum, Golgi apparatus and mitochondria. Molecular motor proteins move along these cytoskeletal tracks expanding cytoskeletal function, enabling eukaryotic cells to control the distribution of cellular constituents and dynamically change cell shape. In this chapter, we describe the regulated assembly of the microtubule cytoskeleton, and how microtubules function as tracks for the movement of two classes of motors. We will also look at the underlying ability of these motor proteins to do work utilizing the mechanochemical coupling between enzymatic hydrolysis of ATP and changes in protein conformation.

All these functions are made possible by the interesting ways in which the microtubule building block proteins interact and by the varied mechanisms that have evolved to control microtubule assembly and disassembly. Initially we look at basic physics of the cellular environment and then delve into the microtubule building blocks and their interactions. Their physical properties set limits on the ways microtubules can be used.

Many prokaryotic genomes encode proteins related to actin and sometimes these prokaryotic actin-like proteins polymerize into filaments. In the bacterium *Mixococcus xanthus*, the actin-like MreB protein forms filaments that spiral around the cell just under the membrane, and a motor complex (unrelated to eukaryotic myosins) travels along these spiral paths. Coupling of this motor to transmembrane adhesion sites results in bacterial gliding as the cell is pulled, along with the MreB filaments, past stationary motor complexes. Thus, actin-like proteins were probably used in some forms of motility even in the prokaryotic predecessors of eukaryotic cells. In contrast, present evidence suggests that both the ability to form microtubules and the evolution of motors that can travel along them are strictly a eukaryotic invention. Bacterial FtsZ, a distant homolog of tubulins, can polymerize into curved filaments whose assembly directs cell wall formation during bacterial cell division, but these filaments do not form tubes and are not known to work in conjunction with motor proteins.

A benefit of regulated assembly of the cytoskeleton, combined with motility along cytoskeletal filaments, enabled dynamic cellular shape changes and movement across surfaces by amoeboid motion or gliding (Chapter 34 by Erich Sackman). Further, it enabled aquatic swimming with beating cilia made from doublet microtubules (Fig. 30.2). The four-hundred-fold increase in the power of a typical cilium versus a bacterial flagellum of similar length enabled the motility of much larger organisms (Chapter 32, by Van Houten and Valentine). This increase in power was made possible because a cilium is completely intracellular and motors can be distributed all along its length rather than only at one end.

To avoid perpetuating naming confusion, both eukaryotic flagella and cilia are here called cilia, since they are structurally and biochemically distinct from the evolutionarily unrelated rotatory bacterial flagella, which are now called simply flagella, and the unrelated rotatory archaeal flagella now called archaella (see Chapters 5 and 31).

Cell Physiology Sourcebook. https://doi.org/10.1016/B978-0-12-811114-7.00030-X

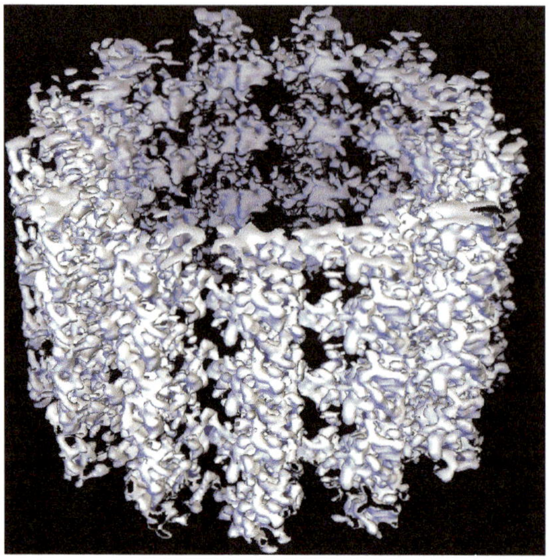

FIGURE 30.1 Overall view of the microtubule structure at 8 Å resolution with the plus end of the microtubule toward the top. *Modified from Li et al. (2002).*

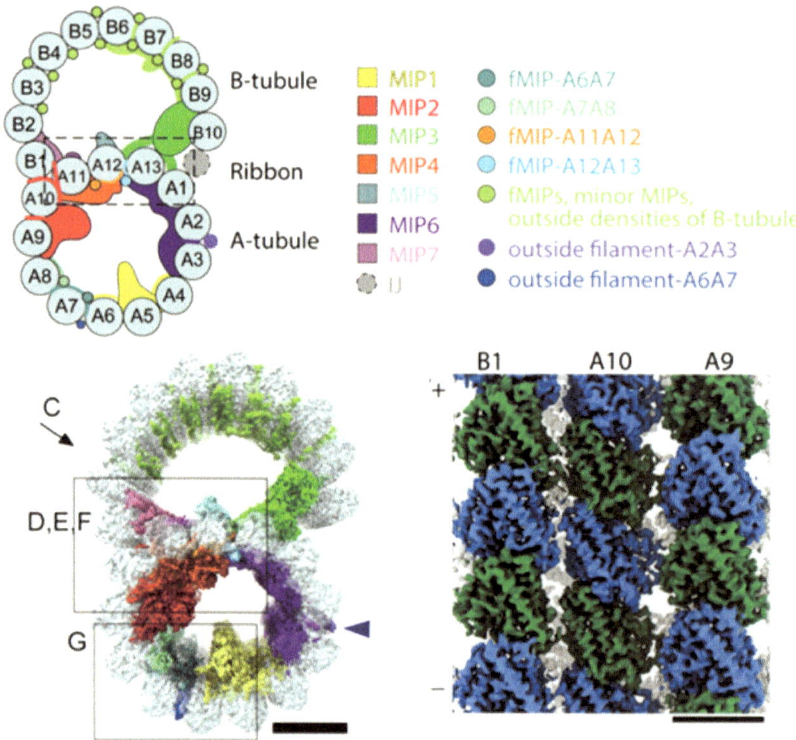

FIGURE 30.2 Networks of the MIPs are woven into the tubulin lattice. (A) Schematic cartoon of the doublet from *Tetrahymena* viewed from the tip of the cilia. PF numbers are shown, and MIPs are colored as on the Right. The PF ribbon region is indicated as the dashed box. The inner junction (IJ), not presented in our structure, is known to bridge the B- and A-tubules (60). (B) Surface rendering of the 48-nm unit of the doublet is colored according to A. Blue arrowhead indicates the outside filament-A2A3/the outside molecular ruler (19). (Scale bar, 10 nm) Other views C−G indicated in B can be found in the reference Ichikawa et al. (2019). Lower right. The complex tubulin lattice within the 48-nm repeating unit of the doublet showing the exterior view of surface rendering of the doublet at PFs A9, A10, and B1. α- and β-tubulins are colored green and blue. The seam is between A9 and A10. *From Ichikawa et al. (2019).*

The superior stiffness of microtubules, compared with other cytoskeletal filaments, led to their adoption as the stiffening elements of many cellular projections in addition to cilia, most notably neuronal axons and dendrites in metazoa. Plants use membrane-associated microtubule arrays as linear templates for depositing oriented strands of their cellulosic cell walls. Among unicellular eukaryotes, in addition to providing anchoring structures (microtubule roots) for ciliary basal bodies, membrane-associated microtubule arrays such as those in *Euglena* maintain and modify elaborate cell surface shapes. They also support feeding structures, such as the ventral groove of Excavates, suctorian tentacles of some Ciliates, haptonemes of Haptophytes, axopodia of Heliozoa and filopodial nets of foraminiferan Rhizaria.

The transition from prokaryotic to eukaryotic organization resulted in a tremendous increase in genome size, enabling larger and more complex cells. In this process, new mechanisms evolved to ensure the genome was properly replicated and distributed to new cells in each generation. This need for accurate genome distribution was one selective force driving the evolution of a microtubule-based mitotic apparatus, which links directly to chromosomes and generates the required forces for their precise separation into two identical new nuclei. We end looking at the important roles of microtubule assembly and disassembly processes, as well as the function of microtubule-associated motor proteins, in mitosis.

1.2 Environment for microtubules and cilia

1.2.1 Viscosity and the Low Reynolds number regime

Critical to understanding motors and cell motility is appreciating the viscous Low Reynolds number regime they inhabit. Viscosity is due to the attractive forces between molecules in a liquid. If one volume element (Fig. 30.3A) is in motion relative to another, the contact forces, which oppose this motion, are called frictional forces. The frictional or viscous drag force F_v is proportional to the relative velocity between the elements dv and the contact area of the surface elements dA, inversely proportional to the distance between the elements dx, and the constant between the elements and the force is the dynamic viscosity η ($F_v = \eta(dv/dx)dA$) (see Chapter 9, by Alvarez-Leefmans).

For a particle moving under an applied force F in free space, where $F_{net} = ma$. m is the mass and a is the acceleration, where $a = dv/dt$. The particle's velocity would continuously increase if it were not for the opposing viscous drag force. At low velocities, the viscous force is proportional to the product of the velocity and the coefficient of friction. The frictional coefficient for a spherical particle of radius r is $f = 6\,\pi\,\eta\,r$. The equation of motion will then be $m \cdot (dv/dt) = F - fv$.

If the velocity is constant, the cell's thrust force necessary to overcome the viscosity is

$F_v = fv = \eta\left(\frac{dv}{dx}\right)dA = \eta\left(\frac{v}{L}\right)L^2 = \eta vL$ where L is the characteristic dimension (length) of the object, typically modeled as the diameter of a sphere.

Mass can be written as the product of density and volume $m = \rho_m L^3$, hence the inertial force F_i ($m\,(dv/dt)$) may be written as $\rho_m L^3(v/t)$ and since $L/t = v$, as $\rho_m L^2 v^2$. The ratio of the inertial force F_i relative to the viscous force, F_v, is the Reynolds number $R_e = F_i/F_v = \rho_m L^2 v^2/(\eta vL) = vL\rho_m/\eta$, where v is the velocity of the fluid with respect to the object, L is the characteristic dimension (length) of the object, ρ_m is the density of the fluid medium, and η is the dynamic viscosity. When R_e is small, friction dominates, stirring produces laminar flow, and if an external force stops fluid flow it

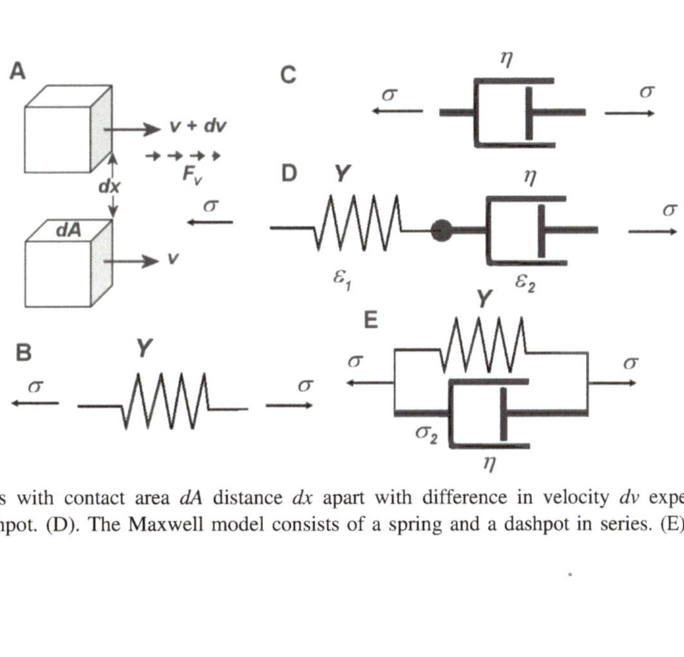

FIGURE 30.3 (A). Two elements with contact area dA distance dx apart with difference in velocity dv experience a viscous drag force $F_v = \eta(dv/dx)dA$. (B). Spring. (C). Dashpot. (D). The Maxwell model consists of a spring and a dashpot in series. (E). Kelvin-Voigt model of spring and dashpot in parallel.

stops almost immediately. For example, a eukaryotic swimming cell, such as *Chlamydomonas* (120 μm/s and $L = 10$ μm), has a R_e of 1.3×10^{-3}.

If a swimming cell stops applying force F or thrust, it comes to a stop exponentially, as illustrated using our equation of motion given above with $f = 6\ \pi\ \eta\ r$.

$m(dv/dt) = -6\pi\eta r v$, hence $dv/v = -6\pi\eta r/m\, dt$ and hence $v(t) = v(0)\, e^{-t/\tau}$, where $\tau = m/6\pi\eta r$. Since mass equals volume times density, $m = (4/3\pi r^3)\rho_c$ then $\tau = \frac{2}{9}\frac{r^2\rho_c}{\eta}$. If $r = 5$ μm, $v(0) = 120\ \mu$m/s, density of the cell $\rho_c = 1000\ kg/m^3$, and $\eta = 0.001$ Pa s. The time constant of the exponential for coasting is $\tau \sim L^2\rho_c/18\eta$, with ($L = 2\ r$) then $\tau \approx 5.6$ μs and distance to stop $d = \int_0^\infty v(t)dt = \tau\, v(0) = 0.67$ nm. Such a cell is the microorganism *Chlamydomonas*, which has a thrust $F = 11.3$ pN (10^{-12}N) and power $P = Fv$ of 1.4 fW (10^{-15}W).

Organisms such as *Volvox* (1-mm diameter, 1 mm/s velocity) occupy the borderline between low and modest Reynolds numbers with a Reynolds number R_e of ~ 1, a τ of 200 ms and a $d_{coast} = 0.2$ mm. This border can also be crossed by somewhat smaller organisms under special circumstances. *Paramecium* (typically 0.2 mm, 1 mm/s) has a R_e of ~ 0.2 but can transiently go as fast as 10 mm/s with a R_e of ~ 2 and a d_{coast} of ~ 20 μm (Hamel et al., 2011). It might jump most of its length in this burst of motion, but even with a R_e of ~ 2 coasting is still a minimal part of it. The Low Reynolds number regime is characterized by the uselessness of reciprocal motions as there is nothing to be gained by inertia and hence propulsive systems involving wave propagation down appendages and crawling are necessary, as we will discuss (Nelson, 2008). Rotating appendages and nonmicrotubule nonactin-related crawling are discussed in Chapter 31.

1.2.2 A biological environment is not however just viscous, but viscoelastic

In addition to microtubules and cilia being viscoelastic the media of microtubules and cilia exhibits both viscous and elastic behavior, somewhere between a solid and a fluid. Linear viscoelasticity may be modeled by combinations of linear elastic springs and linear viscous dashpots. Stress σ is force per unit area (pressure) and is typically expressed in units of Pascals (Pa or N/m^2). Due to stress σ along an axis, an object is deformed along that axis; this is called strain ε (a dimensionless ratio of deformation such as stretching or compressing over initial length). If in the stress limit for Hooke's law, the strain $\varepsilon = \sigma/Y$, where Y is Young's or elastic modulus (Fig. 30.3B). Hooke's law is that the extension is linearly related to the applied force, namely $F = k\,x$, where F is the force, k is the spring constant, and x is the extension. Hence, Y is a measure of the stiffness of a linear elastic solid material. If just the elastic force is present, such an object is called a spring. When pulled it will undergo an instantaneous elastic strain upon loading, hold it as long as loaded, and then if unloaded it will instantaneous de-strain. The shear modulus G is the ratio of shear stress to shear strain $G = \frac{F/A}{\Delta x/l}$, where F/A is the shear stress and Δx/l is the transverse displacement. The shear modulus is related to Young's elastic modulus by the Poisson ratio, $G = Y/2(1+v)$ where v is the Poisson ratio, which for an object under uniaxial stress is the contraction in the transverse direction relative to the extension in the direction of the applied stress.

A viscous dashpot is modeled as a fluid filled cylinder with a leaky piston with a defined relationship, namely with its strain rate $\dot{\varepsilon}$ (or $d\varepsilon/dt$) related to stress by $\dot{\varepsilon} = \sigma/\eta$, where η is the viscosity. Then, if a stress in the form of a load is applied, the strain will linearly increase without bound and when the load is removed the strain will be permanent (Fig. 30.3C). This is seen by integration $\varepsilon = \sigma_0 t/\eta$, where σ_0 is the initial stress.

If the spring and dashpot are connected in series, they form the Maxwell model (Fig. 30.3D). Since stress must be in the same direction in both the dashpot and spring, this leads to the strain rate $\dot{\varepsilon} = \sigma/\eta + \dot{\sigma}/Y$. If pulled the spring will stretch immediately and the dashpot will leave a permanent strain when the pulling stops.

If the object is subjected to a constant strain ε_0 at $t = 0$, then

$\dot{\varepsilon} = 0 = \sigma/\eta + \dot{\sigma}/Y$ *and for* $(t > t_0)\ \sigma(t) = -\lambda\dot{\sigma}$, where $\lambda = \eta/Y$ is the ratio of viscosity to Young's modulus and the stress $\sigma(t)$ will relax slowly. λ is the *relaxation time* of the stress $\sigma(t - t_0) = Y\lambda_0 \bullet e^{-(t-t0)/\lambda}$, where $Y\lambda_0 = \sigma_0$ is the stress at time t_0.

On the other hand if the spring and dashpot are in parallel, the strain experienced by the spring is the same as that experienced by the dashpot, so that $\sigma = Y\varepsilon + \eta\dot{\varepsilon}$, the Kelvin-Voigt model (Fig. 30.3D). With the initial condition $\varepsilon(0) = 0$, $\varepsilon(t) = \sigma_0/Y(1 - e^{-t/\lambda})$, where λ is now called the retardation time, the time for the strain to accumulate. If unloaded with zero stress then $Y\varepsilon + \eta\dot{\varepsilon} = 0$ and stress $\varepsilon(t) \propto e^{-t/\lambda}$, decays exponentially. Neither model by itself is realistic, but more complex models may be made with more components.

To describe viscoelastic solutions that microtubules or cilia may be moving in, the Deborah *De* and Weissenberg *Wi* numbers are frequently used. The Deborah number is the ratio of the relaxation time λ of a material as already discussed to the observation or experimental time. $De = \lambda/t_{\text{observation}}$, where the observation time might be, for example, the duration of a ciliary beat. The viscoelastic response is then proportional to De, since $\sigma\ (t - t_0) = Y\ \lambda_0 \bullet e^{-(t-t0)/\lambda} \propto e^{-1/De}$, the more solid-like the medium, the higher *De* and more liquid-like the lower. The *Weissenberg* number *Wi* is the ratio of viscous forces to elastic forces of a material, which is equal to the relaxation time λ times the strain rate $\dot{\varepsilon}$.

1.2.3 Environment is altered by swimming in free solution or near surfaces (Hunt et al., 1994)

1.2.3.1 Near a surface

The drag coefficients $c_{\parallel}$, $c_{\perp}$, and c_v, denote the drag force per unit velocity per unit length of an infinite cylinder near a surface. They apply to creeping flow (i.e., vanishingly low Reynolds number). The cylinder radius is r, the height of the cylinder axis above the surface is h, and the dynamical viscosity of the fluid is η.

For motion parallel to the cylinder axis,

$$c_{\parallel} = 2\pi\eta \,/\, cosh^{-1}(h\,/\,r) = 2\pi\eta \,/\, ln\left[(h\,/\,r) + \left[(h/r)^2 - 1\right]^{1/2}\right]$$

For motion perpendicular to the cylinder axis and parallel to the wall

$$c_{\perp} = 4\pi\eta/cosh^{-1}(h\,/\,r) = 2\,c_{\parallel}$$

For vertical motion, meaning perpendicular to both the cylinder axis and the wall

$$c_v = 1\,/\,\left[c_{\perp}^{-1}\,c_{*}^{-1}\right]$$

c^* is the rotational drag coefficient, defined as follows: for rotation of the cylinder about its axis with angular velocity ω the drag torque per unit length is $c^*\omega r^2$, with coefficient

$$c_* = 4\pi\eta\,/\,\left[1 - (r/h)^2\right]^{1/2}$$

To complete the set of drag coefficients, we note that from $c_{\parallel}$ one can directly estimate the drag torque T of a long but finite cylinder ($L\,/h \gg 1$), rotating about a vertical axis located at a distance L_1 from one end and L_2 from the other, with $L_1 + L_2 = L$. For an angular velocity ω

$$T \cong c_{\perp}\omega\left(L_1^3 + L_2^3\right)/\,3.$$

Unlike the preceding results, however, this corresponds to no exact fluid solution.

1.2.3.2 Distant from a surface

We use a Γ to denote total drag coefficients for a particle, in contrast to the per-length drag coefficients given above for the case of being near a surface. For a cylinder of length L in unbounded fluid, with ratio $p = L/2r$, where r is the radius.

$$\Gamma_{\parallel} = (2\pi\eta L)\,/\,\left(\ln p + \gamma_{\parallel}\right), \Gamma_{\perp} = (4\pi\eta L)\,/\,(\ln p + \gamma_{\perp}), \Gamma_r = \left(1/3\pi\eta L^3\right)/\,(\ln p + \gamma_r)$$

$\Gamma_{\parallel}$ and $\Gamma_{\perp}$ are the total translational drag coefficients of the cylinder for motion along and perpendicular to the cylinder axis, respectively. For a rotation with angular velocity ω about an axis perpendicular to the cylinder axis, the drag torque is $\Gamma_r\,\omega$.

The end-correction terms $\gamma_{\parallel}$, $\gamma_{\perp}$, and γ_r, are not constant but, in fact, depend on the axial ratio $p = L/2r$.

Consult Hunt et al. (1994) for more details on this and for non-Newtonian (e.g., viscoelastic) media.

1.2.4 Brownian motion

What is sometimes not appreciated because of the static nature of drawings and X-ray diffraction is that motors and molecules are enormously buffeted about by Brownian motion due to collisions with water and other molecules (see Chapter 9). For a motor to walk on a microtubule, it is like a person trying to walk along a path in a hurricane. It is no wonder they pop off if they let go for a second.

1.2.5 Diffusion and transport

Diffusion is the dominant form of material transport at small scales like within a cell (see Chapter 9). Thermal energy k_BT (~4.1 pN nm or 25 meV [milli-electron volt] at room-body temperatures, k_B Boltzmann constant, T absolute temperature in Kelvin) is not small. For linear diffusion, $fD = k_BT$ (Einstein relation) where D is the diffusion constant, and for a sphere of radius, r, the frictional coefficient $f = 6\pi\,\eta\,r$. For rotational diffusion, $\zeta\theta = k_BT$, where θ is the rotational diffusion constant with frictional coefficient $\zeta = 8\pi\,\eta\,r^3$ for a sphere. Note that the radius is raised to the third power, such that the rate of change of orientation of a cell, particle or molecule changes radically with r. Particularly useful to describe molecular and cell motion is the rotational relaxation time, which is given in terms of the average $\overline{\cos\phi}$, where ϕ is the angle relative to the initial direction. The relaxation time τ is the time during which $1/e$ of the change in $\overline{\cos\phi}$ from the initial orientation to another has occurred. For a sphere $\tau = \zeta/(2k_BT)$ or $1/2\theta$. For the ellipsoidal *Chlamydomonas* gamete ($7 \times 7 \times 10$ μm) τ is about 2 s.

One further fact regarding the scales involved here is the relationship between diffusion and the need for molecular transport. Without aid by transport along cytoskeletal elements such as microtubules, the distance molecules could travel is limited by diffusion. For example, given $< x^2 > = 2Dt$ for diffusion in one direction, where x is distance and t is time, a small amino-acid with a diffusion constant D of 600 $\mu m^2/s$ would diffuse in one second the root-mean-squared distance $\sqrt{< x^2 >} \sim 35$ μm. For ATP in cytoplasm with a diffusion constant of 150 $\mu m^2/s$, the anticipated distance in one second would be ~17 μm. The time to reach a distance corresponding to the 42-μm length of a *Ciona* sperm would be 8.3 s. The corresponding time to reach the end for the longest cilium would be 1.1×10^7 s or 131 days under the unrealistic assumption of no ATP usage along the cilium. Nature found a better way in the adenylate kinase shuttle, which can move ATP outward with an approximately constant velocity. Without some means of transport of energy to motors along a cilium, the wave would only go about 15 μm (Machin, 1958; Tombes et al., 1987), the distance ATP would diffuse before being used up by the dynein motors. Due to the line like geometry of cilia, most of the biochemistry has to be done with the aid of shuttles once one gets beyond the first few micrometers from the base, as further discussed in Section 3.6.

1.3 Structures

1.3.1 Mechanical properties of structures

Flexural rigidity κ is the bending resistance of a filament and may be defined as Young's modulus Y times the second moment of area or the area moment of inertia. The second moment of inertia about the x-axis is $I_x = \iint A\, y^2 dxdy$. For a solid cylindrical rod of radius $\boldsymbol{r}$, $\kappa = \pi\, Y\boldsymbol{r}^4/4$.

Persistence length ξ_p is κ/k_BT. If the persistence length is significantly greater than its contour length L, then the object will appear to be rod-like. However, if the persistence length is much less than L, it will appear as squiggly or rope-like. For an axial load, the axial stiffness is YA, where A is the cross-sectional area. Note that the flexural rigidity may be altered by compression on the ends of a structure. Where s is the distance along an arc from the base end, the tangent correlation function $\langle \mathbf{t}(0)^* \mathbf{t}(s) \rangle = e^{(-s/\xi p)}$.

Microtubules (Fig. 30.1) made up of tubulin with 13 protofilaments have a linear density about 160 kDa/nm (Da Dalton), about 10 times that of actin filaments, and they maintain their stiffness over the length scale of a μm. Their flexural rigidity is about ~12 pN μm^2, hence a persistence length of 2–6 mm. For DMTs (doublet microtubules) (Fig. 30.2) with 23 protofilaments (10 of B-lattice connected to 13 of A-lattice), Ishijima and Hiramoto (1994) measured 14 pN μm^2, but with a potential for bond changes between them when subjected to shear their flexural rigidity could be altered. This is remarkably similar in value to the singlet microtubule.

Interestingly, cilia, which have nine doublet microtubules attached together and might be expected to be stiffer, are not. The static flexural rigidity of a cilium is about 21 pN•μm^2, but dynamically it increases to 168 pN•μm^2 when beating at 40 Hz for the central part of a cilium and 86 pN•μm^2 for the more flexible basal part of cilia (Foster et al., 2017). This suggests that the flexural rigidities are strongly tuned to meet functional requirements.

1.3.2 Springs on cilia

Linear springs obey Hooke's law, namely $F = k\,x$, where F is the force, k is the spring constant and x is the extension. For a cilium, the spring constant has been measured as 2 nN/μm per μm of cilium length (Minoura et al., 1999), further confirmed by Foster et al. (2017).

1.3.3 An axoneme as a composite beam

An axoneme's flexural rigidity is not simple for a cilium with its nine doublet microtubules and a central pair connected by several protein complexes that restrict their motion. Their flexural rigidity is beating frequency dependent as seen in

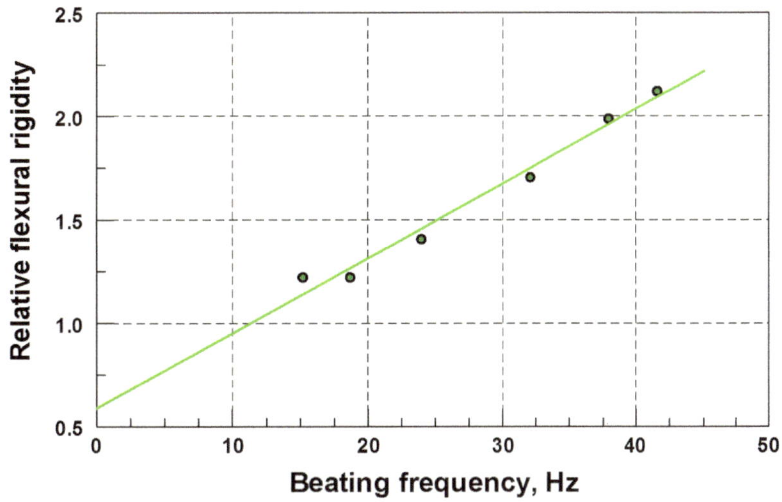

FIGURE 30.4 Beating frequency dependence of the flexural rigidity of the central region of a *Ciona* cilium. *Modified from Foster et al. (2017).*

Fig. 30.4. From the figure, $\kappa_c = \kappa_0 + \kappa_{cv} \cdot \nu = 0.588 + 0.0362 \cdot \nu$ μm$\cdot E_m$(ν in Hz), where $E_m \approx 80$ pNμm (data from Brokaw (1996), see Foster et al., 2017 for more details). Since there are connections between the DMTs, bending the axoneme requires bending all the DMTs at once, thus significantly increasing the flexural rigidity of the axoneme relative to a DMT.

1.4 Description of shape of structure or path

The convenient nomenclature of the Frenet-Serret basis (Fig. 30.5) is useful for describing the cilium-bending plane, which is the plane defined by **n** and **t,** and is useful when thinking about the path of swimming organisms. It is an essential piece of knowledge when working with the geometry of paths or structures. Every point of a curve can be defined through the Frenet-Serret basis set of vectors. The tangent vector **t** is the derivative of the position vector with respect to the arc length *s* and points outward along the cilium or swimming path and of course is tangent to that path. The curvature $\kappa(s)$ is defined by $d\mathbf{t}/ds = \mathbf{t}'(s) = \kappa(s)\,\mathbf{n}(s)$. Let ψ be the angle in a plane between the tangent of the curve and a fixed axis. In order to be consistent with $\kappa(s) = d\psi/ds$ for curves in a plane, it is convenient that $\kappa(s)$ have the same sign as ψ. This is to avoid **n**(s) switching direction every time one passes through an inflection point. The normal **n**(s) points toward the local center of curvature if ψ is positive and in the opposite direction if ψ negative (e.g., the ciliary beating plane is defined by the **n** and **t** vectors). In three-dimensional space, it means taking **n**(s) in the same or opposite direction as $\boldsymbol{t}'(s)$ and then maintaining that direction, avoiding sudden reversals by changing the sign of $\kappa(s)$. The binormal $\mathbf{b}(s) = \boldsymbol{t}(s) \times \mathbf{n}(s)$ is orthogonal to both. Torsion $\tau(s)$ measures how a space curve twists as a point moves along it as defined by $d\mathbf{b}(s)/ds = \mathbf{b}'(s) = -\tau(s)\boldsymbol{n}(s)$. Any curved path or cilium centerline can be defined by its local curvature $\kappa(s)$ and local torsion $\tau(s)$, further $\mathbf{n}' = \tau(s)\,\mathbf{b}(s) - \kappa(s)\,\mathbf{t}(s)$. A plane curve does not have torsion; however, a cilium even in a plane may twist. If one is concerned with the direction of helical swimming path, which is common to many

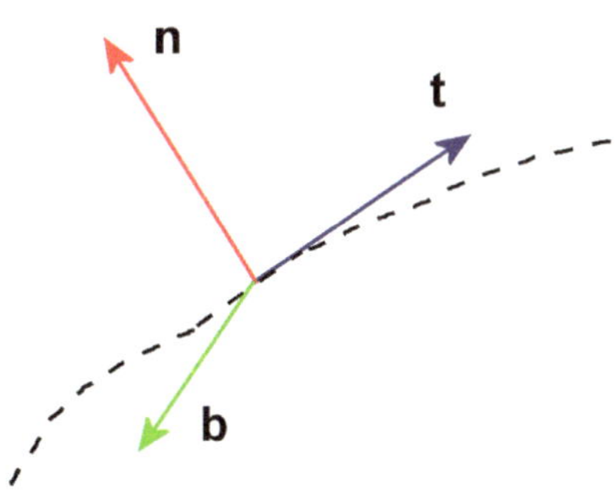

FIGURE 30.5 The Frenet-Serret basis: for a cilium the bending plane is defined by the normal **n** and the tangent **t** pointing outward from the base of the cilium, and the binormal **b** perpendicular to the bending plane.

microorganism such as *Chlamydomonas,* then the direction $\vec{\Omega}$ of the helix (Darboux vector) is relevant given by $\vec{\Omega}(s) = \tau(s)\vec{T}(s) + \kappa(s)\vec{B}(s)$ and its spatial derivative $\frac{d\vec{\Omega}}{ds} = \frac{d\tau}{ds}\vec{T} + \frac{d\kappa}{ds}\vec{B}$.

1.5 Compliant mechanisms

A compliant mechanism is one that transforms motion, force, or energy, in a way that stores energy as the strain energy of flexible elastic members or beams for future use. For industrial purposes, compliant mechanisms are claimed to have two advantages (a) cost efficiency (a dramatic reduction in number of parts, reduced assembly time, simplified manufacturing) and (b) increased performance (increased precision and reliability and reduced wear, weight, and maintenance) (Howell, 2001). These are all highly valued attributes in nature so compliant mechanisms for mechanical systems are a major theme as will be illustrated here and in Chapter 31.

Energy can be stored or transformed to be released later or in a different manner. A familiar example is the bow and arrow. Energy is stored in the limbs of the bow as the archer draws the bow, the elastic strain energy of the bent bow is then transformed into the kinetic energy of the arrow. A further advantage is that these mechanisms may easily be miniaturized. Consequently, a recurring theme is also elasticity, which provides a common means for energy storage. To load the energy into these elastic storehouses requires an energy source provided by motors in cilia. Additionally, for real world materials, it is important to remember that biological materials are not purely elastic, but viscoelasticity as already discussed.

1.6 Mechanical energy

1.6.1 In a Newtonian fluid

In the case of cilia in a Newtonian fluid, such as water, the total mechanical energy **M** of a compliant mechanical system is the sum of the positive strain energy **E** and the potential energy **U** of external forces. The strain energy is formed by the motors bending the cilium (Fig. 30.6B) and pulling the longitudinal interdoublet springs except near the base in the wave growing region where they are not attached (the negative of the work done by the internal forces). The potential energy **U** of external forces comes from the axial compression, which shortens the tip to base distance, due to the hydrodynamics of moving the whole system in an immersing fluid (Jones, 2006) (negative of work done by external forces). **M = E + U**.

Das et al. (2008) have considered a linear elastic theory for a microtubule and the surrounding medium. For non-Newtonian fluids they include the leading nonlinearity terms for the elasticity of the surrounding matrix. For their

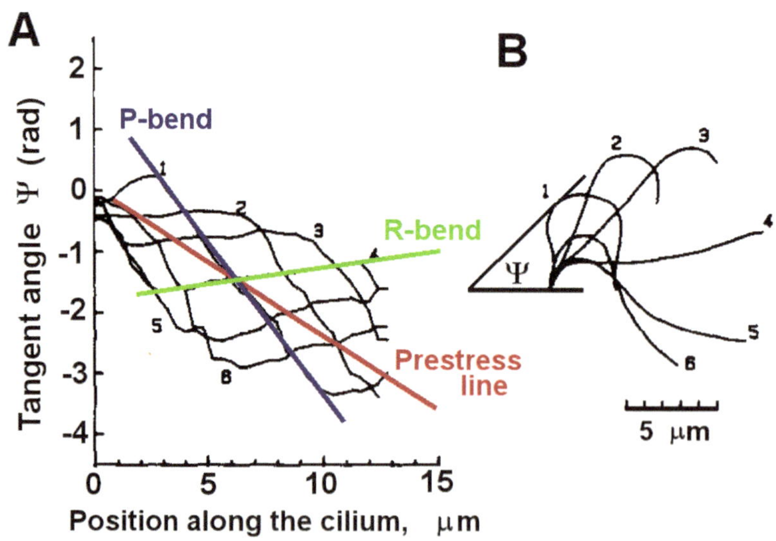

FIGURE 30.6 Based on averaging the tangent angles of six consecutive images of a *Chlamydomonas* cilium. The tangent angle ψ, as shown in B, is the angle in radians of a tangent to the actual curve with respect to a reference axis, CCW positive. The tangent angle ψ is conveniently a single-valued function of s, the contour length of the cilium from the base as shown in A. The slope of a $\psi - s$ plot, $d\psi/ds$, is the curvature, C, of the cilium, dC/ds is the shear. The P-bend (principle) has negative curvature, and the R-bend (recovery) positive curvature. The prestress line (called "static curvature" by Brokaw) corresponds to the fairly constant radius of curvature of the cilium that the traveling wave follows. Plotted, as it is here in (A), on a $\psi - s$ plot it is a straight line. (B) The diagram of a full cilium was calculated by integrating the data at the left. *Modified from Brokaw et al. (1982).*

theory they assume the microtubule is subjected to a compressive load f at one end denoted by $x = 0$. The resulting elastic deformation energy can be written in terms of its transverse and longitudinal displacements, $u(x)$ and $v(x)$:

If the surrounding medium is simply viscous such as pure water and assuming the structure is relatively very long then the mechanical energy for this purely viscous system is

$\mathbf{M}_{pv} = \mathbf{E} + \mathbf{U} = \int_0^\infty \left\{ \frac{\kappa}{2} u''(x)^2 \right] dx - f_{pv}\, v(0)$, where $v(x) = \int_x^\infty \frac{1}{2} u'(y)^2\, dy$ as is required for an incompressible rod. In terms of θ, for instance, $v(s) = \int_x^\infty (1 - \cos\theta) ds$ while the curvature is $d\theta/ds$.

The first term corresponds to the bending energy of the rod; κ is the bending rigidity. The second term represents the axial compression energy released by bending. Classical Euler buckling, in which a long-wavelength bend of a rod of length L occurs at a threshold force of $\pi^2\kappa/L^2$, results from these two terms. This applies to a microtubule as well as to a cilium. For a cilium one must add the additional energy stored by the pulling of the internal springs that are connected along most of the length.

Hence, if one may assume constant curvature for a rod (a reasonable approximation for a cilium), the strain energy per length is $E/L = \frac{1}{2}\, \kappa\, C^2$, where C is the curvature (1/(the radius of curvature)). However, if the cilium is curved in the unstrained condition, then the strain energy stored is dependent instead on the curvature relative to the static curvature condition. The static unstrained curvature has been termed the prestress due to the equilibrium of DMT prestress and the stretching of springs, etc. (see Section 4.6.1). The relevant curvature that stores strain energy is then $C = 4\psi_{Cmax}/\lambda$, where ψ_{Cmax} is ψ_{max} (the maximum tangent angle Fig. 30.6B) minus the net prestress. Hence, for the case of a cilium, for a bend of a quarter of a wavelength $\lambda/4$, $E \approx 2\kappa_c(\nu)\, \psi_{Cmax}{}^2/\lambda$ where $\kappa_c(\nu)$ is the flexural rigidity that is dependent on the beating frequency ν (Fig. 30.7). In addition to bending of the elastic cilium, the mechanical strain energy E in a ciliary half-wave arises from stretching the longitudinal interdoublet springs due to sliding of the DMTs by the dynein motors (see Section 3.3). The energy of the springs can be approximated by $E_s \approx \frac{1}{2} k_s \Delta^2$ or, since the DMT sliding $\Delta \approx a\psi(s)$ where a is the diameter of the axoneme, $E_s \approx (1/2) k_s\, a^2 \psi_{max}{}^2$. An approximation for $\mathbf{E}$ in the bend of the central region of a cilium where there are springs is $\boldsymbol{E} \approx E_b + E_s$.Here, $\boldsymbol{a}$ is the axenome diameter, k_s is the spring constant of the longitudinal springs. Importantly, the κ here is that of the axoneme bending *only*, as if the axoneme had no springs or sliding friction.

The axial compression releases potential energy **U** due to buckling, because **dU** is negative during wave formation as buckling increases. The axial compression comes from the flexible axoneme pushing forward on the cell body and backward against the fluid reducing the total mechanical energy. The motors work to shape the geometry of the cilium and contribute to its dynamic flexural rigidity κ. The strain energy in the cilium is incorporated by the action of the motors to bend the axoneme and extend the longitudinal springs. All the energy of the motors becomes dissipated in moving the immersing fluid and changing the shape of the cilium.

To gain further insight into the buckling as it applies to a cilium, consider a constant force on the cell body during the beat cycle. Further, consider the net compressive force as F_{axial} on the initiating wave (directed toward the cell body in the longitudinal direction [Timoshenko, 1936; see Fig. 30.7]).

This leads to the buckle wave equation for a quarter wave, $\kappa_b(\nu)\frac{\partial^2\phi}{\partial s^2} + F_{axial}\sin\phi = 0$, where $\kappa_b(\nu)$ *is* the ν -dependent flexural rigidity κ_b as for example in the basal region of a cilium. This equation form is similar to that of a simple undamped pendulum where arc length s has been substituted for time t. ϕ changes along the cilium in the same way as the angle of the pendulum changes from the vertical. Assuming small ϕ, the wavelength λ_B in the basal region would give for the forming $\lambda_B/4$ wave: $\lambda_B \approx 2\pi\sqrt{\frac{\kappa_b(\nu)}{F_{axial}}}$ (an exact calculation in the most extreme case of high curvature of the

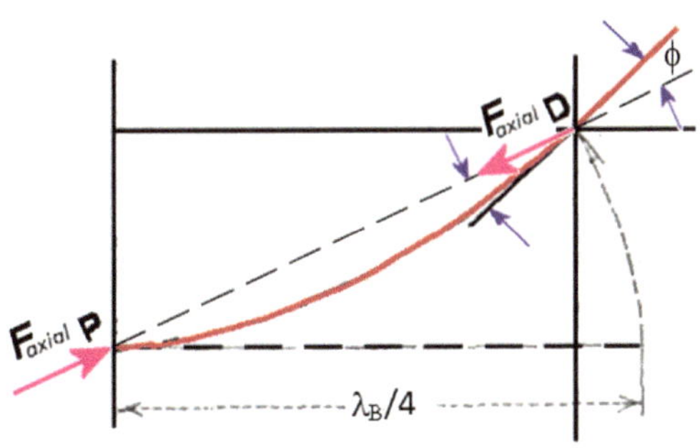

FIGURE 30.7 Compliant curve of a bent axoneme showing some of the compressive forces that buckle the axoneme. Note in the figure the direction of the force vectors shown corresponds to the approximation Foster et al. (2017) assumed when using the buckle equation. Two of these are the F_{axial} 's on the ends of the quarter-wave next to the cell body.

cilium would yield an increase of λ_B of at most 12%). As we will see, buckling turns out to play a crucial role in initiating traveling waves and controlling the beating frequency.

1.6.2 In a non-Newtonian fluid such as in a cell

In the case of cilia in a non-Newtonian viscoelastic fluid such as mucus or microtubules inside of a cell, then the mechanical energy must take into account its interaction with the fluid. Three additional sources of strain energy must be added due to the linear elastic transverse and longitudinal mechanical couplings between the microtubule or cilium and the surrounding viscoelastic media and the leading term of the nonlinear transverse coupling to the medium.

Following Das et al. (2008), then the mechanical energy for a viscoelastic system

$\mathbf{M}_{ve}\int_0^\infty \left\{\frac{\kappa}{2}u''(x)^2\right]dx - f_{ve}\,v(0) + \int_0^\infty \frac{\alpha_\perp}{2}u(x)^2dx + \int_0^\infty\left[\frac{\alpha_\parallel}{2}\,v(x)^2 + \frac{\beta}{4}\,u(x)^4\right]dx$, where f_{ve} is not the same as f_{pv} as for example, in the case of pushing a cell the compressive force experienced pushing a viscoelastic fluid is markedly more than pushing a purely viscous fluid of the same apparent viscosity.

The elastic constant $\alpha_\perp$ corresponds to the linear elastic transverse coupling between microtubules and the surrounding intracellular network or the cilium and the extracellular viscoelastic matrix. The transverse coupling is related to the shear modulus G of the surrounding medium: $\alpha_\perp = 4\pi G/\ln(\lambda/a)$, where λ is the characteristic wavelength of lateral displacement and a is a small wavelength cut-off. This third term by itself leads to buckling on a short wavelength $\lambda = 2\pi l_0$, where

$l_0 = (\kappa/\alpha_\perp)^{1/4}$, and at a force threshold $f_c = 8\pi^2\kappa/\lambda^2 = 2(\kappa\alpha_\perp)^{1/2}$, which is larger than for Euler buckling.

The elastic constant $\alpha_\parallel$ corresponds to the longitudinal mechanical coupling with the surrounding intracellular network or extracellular viscoelastic matrix. It is not directly related to the elasticity of the surrounding medium, but depends on the degree of attachment between the microtubule or cilium and the surrounding network. It falls in the range $0 \le \alpha_\parallel \le 2\pi G/\ln(\lambda/a)$, where the lower limit corresponds to an uncoupled microtubule that slides freely along its axis and the upper limit corresponds to strong coupling with the medium along its length. The term proportional to $\beta > 0$ represents the leading nonlinear transverse coupling to the medium.

2. Microtubules and the cytoskeleton

2.1 Overview of the cytoskeleton

For proper eukaryotic cell function, they must be spatially organized and able to mechanically interact with each other, as well as with their environments. The inner workings of cells are in constant motion. Cytoskeletal filaments and their motors proteins, through the processes of regulated assembly and disassembly, and intracellular transport, govern much of this motion. Cytoskeletal filaments are also responsible for mitosis and developmental changes in cellular architecture. The three major structural proteins of the cytoskeleton are actin, microtubules, and intermediate filaments. Each filament has distinct biological, dynamic, and mechanical roles it must fulfill to ensure proper cellular function. In addition to acting independently, they must also act collectively to cooperatively promote normal cellular shape, strength, and motility.

Microtubules are the thickest and stiffest of the cytoskeletal filaments. They give cells shape and are responsible for cell motility in fluid environments, intracellular transport, and generating necessary force for chromosome segregation during mitosis and meiosis. Microtubules accomplish their functions with the assistance of motor proteins, which transport cargo, arrange, and organize the cell's interior. Actin is the thinnest and most flexible of the cytoskeletal filaments and is responsible for membrane support, cell shape, muscle contraction and crawling-type cell motility (Chapter 31 and 33). The intermediate filaments are a diverse family of filament types, most of which are restricted to the metazoan branch of eukaryotic organisms. They are the least studied of the three cytoskeletal filaments. Intermediate filaments appear to be responsible for supplying the cell with mechanical strength to resist physical stresses it may encounter (Fletcher & Mullins, 2010; Young & Kothary, 2011). Even some prokaryotic cells have cytoskeletal analogs. Originally, biologists thought that bacteria and archaea did not require the sophisticated filament system of the eukaryotic cytoskeleton, however, bacteria and archaea contain homologs of actin and tubulin and sometimes these proteins even form cytoskeletal filaments (Herrmann & Strelkov, 2011), although their functions may be different from those of their eukaryotic counterparts.

Each filament can grow to various lengths, and length regulation alone frequently governs the organization and proper function of the cytoskeleton. Short microtubules dominate in the mitotic spindle (Ribbeck & Mitchison, 2006), whereas long microtubules serve as highways in the axon for vesicular transport. Assembly of short actin filaments generates force to push the cell cortex during cell motility in crawling epithelial cells (Ananthakrishnan & Ehrlicher, 2007), whereas long, rope-like actin cables are essential for intracellular transport in budding yeast cells (Fagarasanu et al., 2006) and actin filaments in skeletal muscle maintain a fixed length during muscle contraction. The actin and microtubule cytoskeleton systems both support the transport of intracellular organelles. Typically, microtubule tracks and their associated motors manage long-range bidirectional transport between the cell center and the periphery. At the cell periphery, actin and its motor protein take over and manage short-range movements. This coordination between the microtubule and actin cytoskeleton systems is critical for proper intracellular compartmental positioning.

2.2 Tubulin

Microtubules are composed of two globular building block proteins of approximately 55 kDa called α- and β-tubulin, which form an obligate dimer shortly after their synthesis through strong, noncovalent interactions. Although α- and β-tubulins are structurally similar, they only share about 40% of their amino-acid sequence (Nogales et al., 1998). In contrast, each type of monomer has been highly conserved across wide evolutionary spans, with yeast and human α-tubulins about 75% identical in amino-acid sequence. In mammals, there are at least six α-tubulin family members and another six β-tubulins, each encoded by a different gene. Each tubulin monomer has a nucleotide-binding site for guanosine triphosphate (GTP). In αβ dimers, the α-tubulin GTP binding site is trapped at the interface between the two subunits; it cannot exchange nor is this GTP ever hydrolyzed (Menéndez et al., 1998). The β-tubulin GTP binding pocket remains open and will allow nucleotide exchange until the dimer is incorporated into a growing microtubule lattice (Nogales, 1999), at which time the end-to-end interaction of one dimer with another dimer closes this binding pocket and can stimulate hydrolysis of the trapped β-tubulin GTP. While the resulting phosphate can escape immediately, GDP remains bound to β-tubulin until the dimer is released during microtubule disassembly, at which time GDP can exchange for a new GTP.

2.3 Microtubule structure and assembly

Tubulin dimers interact in three ways: head-to-tail interactions, which can lead to long chains (protofilaments) of uniformly oriented dimers but are too weak to maintain more than very short polymers on their own, side-to-side interactions in which α-tubulins interact with β-tubulins (A-lattice interactions), and side-to-side interactions in which α-tubulins interact with other α-tubulins and β-tubulins interact with β-tubulins (B-lattice interactions). Side-to-side interactions of either type generate a slight angle between the dimers as well as a slight translational offset but are also relatively weak. Therefore, under ideal conditions of a high concentration of dimers and in the presence of GTP, Mg^{2+} and a favorable environment (pH, temperature), dimers will spontaneously self-assemble into microtubules, but there can be a considerable initial lag until a stable seed is formed for addition of new dimers. Such a seed has enough side-to-side interactions to close into a short cylinder, most frequently with 13 dimers around the circumference, but flexibility in side-to-side bond angles can result in cylinders with anywhere from 8 to 17 protofilaments. In addition, variability in the degree of offset of adjacent dimers around the circumference can lead to differences in helical pitch of the microtubule lattice that result when the cylinder closes. A 13 protofilament microtubule with a helical pitch offset by 2.5 dimers has protofilaments that are straight (Fig. 30.8), whereas lesser or greater numbers of protofilaments result in protofilaments that gradually twist around the surface of the microtubule. Most microtubules in vivo have 13 protofilaments (Tilney et al., 1973) formed with B-lattice interactions (Ichikawa et al., 2017; McIntosh et al., 2009), although some exceptions exist (Chaaban & Brouhard, 2017). One of the interesting consequences of this configuration is that although most of the protofilaments have B-lattice interactions with their neighbors, one A-lattice interaction is needed to complete the structure (Fig. 30.9), and this A-lattice seam alters the interactions of proteins with the microtubule wall and reduces microtubule stability (Katsuki et al., 2014). In vitro, microtubules with different numbers of protofilaments and different helical pitches can form, some of which have a uniform surface with no seam (Fig. 30.10) (Zhang & Nogales, 2015). The typical 13-protofilament microtubule has an outer diameter of 25 nm and an inner diameter of 17 nm (Fig. 30.8). The wide, cylindrical structure of a microtubule makes it stiff and able to resist compression. The bending stiffness of a microtubule has an average measured flexural rigidity of 2.5 $pN\mu m^2$ or a persistence length of 0.6 mm (Hawkins et al., 2013), with a torsional rigidity of 0.927 $pN\mu m^2$ (Chełminiak et al., 2010). Note that values depend significantly on their specific conditions. Depending on the cell type, microtubule lengths can vary from 1 μm to 1 mm (Milo & Phillips, 2015).

Within the lattice of tubulin dimers, the interconnected dimers are held together longitudinally and laterally by weak noncovalent interactions. Although a single longitudinal bond is stronger than a single lateral bond, the sum of the lateral bonds between neighboring protofilaments is stronger than longitudinal αβ-tubulin bonds because there are more lateral bonds than longitudinal bonds. The only exception to this bond counting occurs at the ends of the microtubule, where there

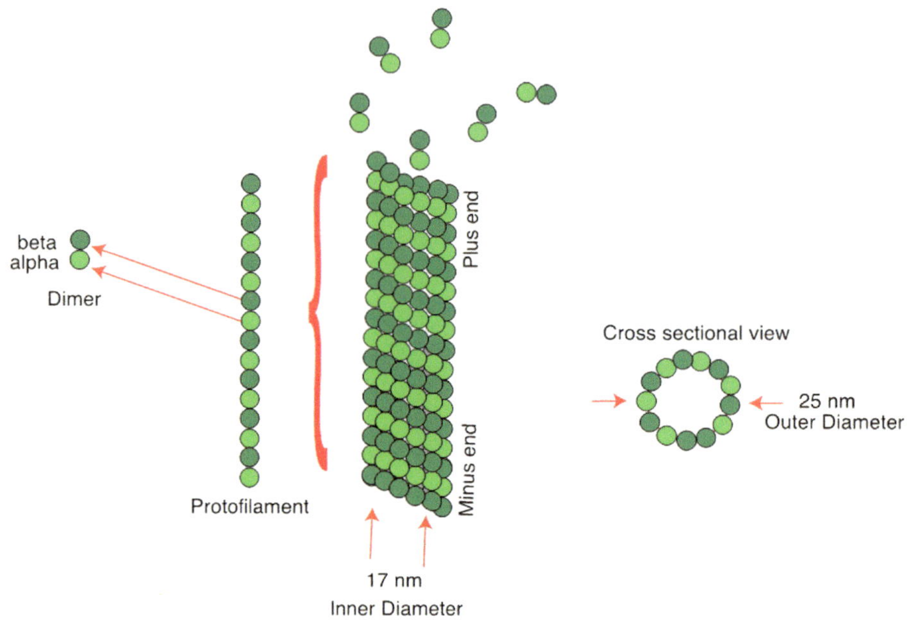

FIGURE 30.8 Microtubule structure and dynamics. Microtubules are polymer filaments made from α-β tubulin dimers. The microtubule is a cylinder with an outer diameter of 25 nm and an inner diameter of 17 nm.

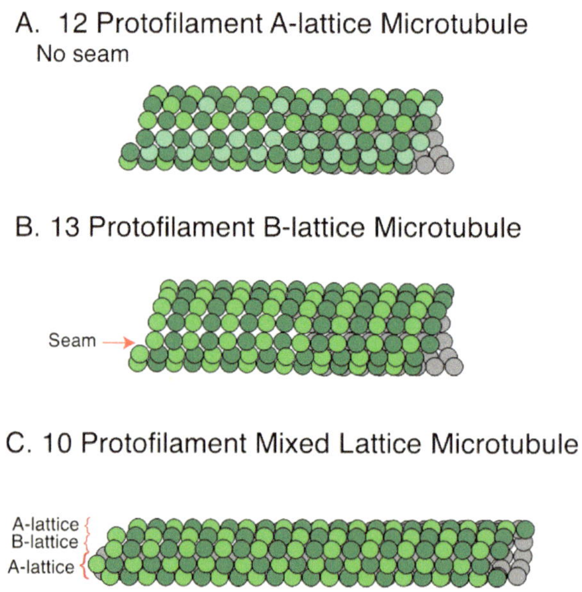

FIGURE 30.9 Microtubules made of A, B, and A-B Mixed lattice types. (A) 12-protofilament A-lattice type. (B) 13-Protofilament B-lattice type. Note there is a discontinuity in the tubulin dimer pattern between protofilaments, this is called a seam. (C) 10-protofilament Mixed lattice type.

are fewer bonds, and it is easier for dimers to add and subtract. This allows for rapid assembly (polymerization) and disassembly (depolymerization) of dimers from the ends. Microtubules will undergo net elongation if tubulin dimers are in excess, and will shrink if the dimer concentration is too low. The rate of addition to the two ends of the microtubule is different, with addition to the end with exposed β-tubulin occurring faster. The entire microtubule thus has a polarity, determined by the uniform orientation of all its dimers, and describe as **plus** for the fast-growing end, and **minus** for the slow-growing end. As a microtubule grows, its ends have newly added dimers with GTP β-tubulin (a GTP cap), but as growth rates slow due to depletion of free dimers, the rate of hydrolysis can exceed the rate of new subunit addition. Even at steady-state, with no overall change in average microtubule length, polymerization-depolymerization switches stochastically in a process known as dynamic instability. This process is regulated in part by changes in dimer conformation

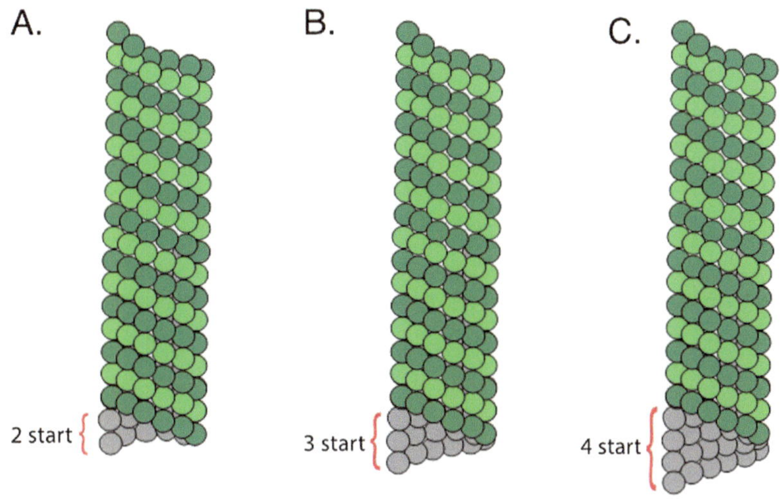

FIGURE 30.10 Helical structure of microtubules. All three microtubules shown are made from 13 protofilaments (A) 2+ start. (B) 3+ start. (C) 4+ start.

that are linked to hydrolysis of GTP by β-tubulin, which only occurs within the assembled microtubule. GTP hydrolysis and release of P_i changes longitudinal contacts between adjacent dimers in a protofilament, creating strain as the longitudinal dimer-dimer bonds attempt to change angle, but does not alter lateral interaction strength (Alushin et al., 2014). GTP tubulin is straight or gently curved up to 5 degree bend per αβ-tubulin (Brouhard & Sept, 2012; Chrétien et al., 1995; Mandelkow et al., 1991), whereas GDP tubulin has a natural 12 degree bend per αβ–tubulin dimer. The GTP cap physically stabilizes the microtubule lattice structure by locking GDP dimers into a straight configuration. When microtubules lose their GTP cap, they rapidly disassemble (termed a **catastrophe**) with GDP tubulin protofilament bonds breaking and dimers falling off the microtubule lattice. The addition of new dimers to the shrinking end can stop the disassembly process and reinitiate growth (termed a **rescue**). Amazingly, proteins that associate with the shrinking plus end of a microtubule, and used to do work in cells such as moving chromosomes (Driver et al., 2017), can capture the release of strain energy into curving protofilaments. The chemotherapeutic drug Taxol (Table 30.1) stabilizes microtubules by blocking the change in bend angle that follows GTP hydrolysis (Alushin et al., 2014).

2.4 In vivo nucleation and stability

Microtubule dynamics is different in cells than in solutions of pure tubulin. In cells, microtubules exhibit a much higher polymerization rate, greater catastrophe frequency, and extended pauses in growth. Some microtubules turn over within seconds to minutes, while others remain stable for extended periods of time, on the order of days to weeks for ciliary axonemal microtubules. Rapid turnover of most cytoplasmic microtubules in motile cells such as fibroblasts allows for continuous remodeling of the cytoplasm.

Although microtubules can assemble from a population of pure tubulin dimers in vitro, spontaneous assembly is rarely observed in vivo, due to a microtubule-nucleating complex that is found almost universally in eukaryotic organisms. Three contributing factors selected for its evolution. First is the energetically unfavorable association of several dimers into a stable seed that can close into a tube and elongate. Without nucleation, there is always a long lag even when all other conditions (tubulin concentration, temperature, presence of GTP, etc.) are favorable. An ability to assemble microtubules without a lag provides a great advantage in cellular environments, which are highly dynamic and must respond quickly to changing conditions. Second, by maintaining the tubulin concentration at a level close to the critical concentration for assembly (Al-Bassam, 2017), cells can use the activation and localization of nucleating sites as a way to control where and when a microtubule is assembled. Finally, since microtubules assembled in vitro can have varied lattice geometries, a nucleating site can force all typical cytoplasmic microtubules to have precisely 13 protofilaments, which form straight tracks for motor proteins, with a B-lattice wall and an A-lattice seam conferring specific protein binding sites and stability control.

TABLE 30.1 **Effects of drugs on microtubule stability. Microtubule inhibitors are drugs that make microtubules more or less stable during polymerization. They are used to block mitosis, treat cancers, and work as pesticides for plant based microtubules.**

Drug name	Application	Used in the treatment of:	What does it do:
Paclitaxel (Taxol)	Chemotherapeutic	Ovarian and breast cancers	Prevents mitosis. Taxol binds to microtubules and makes them more stable by preventing tubulin dimers from dissociating.
Vinblastine		Cancer	Prevents polymerization
Demecolcine			Depolymerizes microtubules
Colchicine		Gout	Prevents polymerization, causing microtubules to disappear. This blocks the migration of white blood cells responsible for inflammation in the disease.
Zoxamide	Pesticide	Plant fungus	Binds specifically to fungal tubulins and prevents fungal growth in potatoes.
Rotenone			
Nocodazole			Prevents polymerization

The most common and universal microtubule nucleator in cells is based on gamma tubulin, a paralog of alpha and beta tubulin. In order to act as an efficient nucleator, gamma tubulin must associate with other proteins to form a gamma tubulin ring complex (γTuRC), in which 13 gamma tubulin molecules are held in a ring with a helical twist that exactly matches the helical pitch of a 13-protofilament MT (Fig. 30.11B). The exposed face of each gamma tubulin in the γTuRC complex acts as a binding site for alpha tubulin in a dimer. The active γTuRC looks much like the (+) end of an assembled MT. Structural studies of γTuRC activation in budding yeast show that most gamma tubulin in the cytoplasm is found in a smaller complex of two gamma tubulins and two helper proteins, the gamma tubulin small complex (γTuSC; Fig. 30.11A). Seven of these tetramers then interact with an activating protein (Spc110) to form the γTuRC, which actually has 14 gamma tubulin molecules. Because of the helical pitch of this complex, the first gamma tubulin is hidden beneath the last one, so only 13 gamma tubulins are exposed to act as assembly templates (Fig. 30.11C).

In vivo, active γTuRCs only form at microtubule-required sites, which, in the case of budding yeast, is limited to a specialized site on the nuclear membrane called a spindle pole body (SPB). During interphase, the yeast SPB supports assembly of microtubules on the inside of the nucleus that maintain an attachment to each chromosome, while on the outside of the nucleus, microtubule assembly provides a cytoplasmic microtubule array. The yeast SPB is a simple form of microtubule organizing center, or MTOC. Most cells have a single prominent MTOC that, like the yeast SPB, supports assembly of MTs for both a cytoplasmic array and for interaction with chromosomes during mitosis. However, additional sites of MT nucleation occur in many cells and can be quite variable from 1 cell type to another and at different times during the cell cycle. We will first trace the evolutionary origin and mechanisms of biochemical regulation of the single, primary MTOC in typical animal cells, the **centrosome**, and then briefly discuss alternative patterns of microtubule organization and assembly regulation.

All of the many branches of eukaryotic organisms have species with the same sophisticated mode of motility, the cilium or eukaryotic flagellum. Not only single celled swimming organisms such as *Paramecium caudatum* and *Euglena gracilis*, but also the male gametes of even such diverse multicellular organisms as some higher plants (*Ginko biloba*), giant kelp (*Macrocystis pyrifera*), thallus-forming fungi (*Allomyces macrogynus*), slime mold amoebae (*Physarum polycephalum*) and animals have motile cilia. In the vast majority of eukaryotic organisms, all the microtubules in a cell are organized around pairs of basal bodies that anchor these cilia, and even in cell types that don't assemble a cilium, or that must disassemble their cilium to go through cell division, the paired basal bodies still retain their status as the primary MTOC (Yubuki & Leander, 2013). Most eukaryotic organisms with cilia retain traces of an ancestral set of four microtubule bundles (roots R1-R4) that project away from the basal bodies in different directions, anchoring cilia against the torque generated by their beating motility and providing a cytoskeleton that organizes cellular contents. In the branch of life that includes metazoa, and therefore in most mammalian cells, these microtubule bundles are missing, and the paired basal bodies are surrounded by a multilayered halo of proteins that regulate and anchor γTuRC complexes, from which single microtubules radiate out in all directions. This combination of basal bodies and surrounding material is typically located near the nucleus and, likely from forces generated by microtubule assembly, remains near the center of the cell. Its central

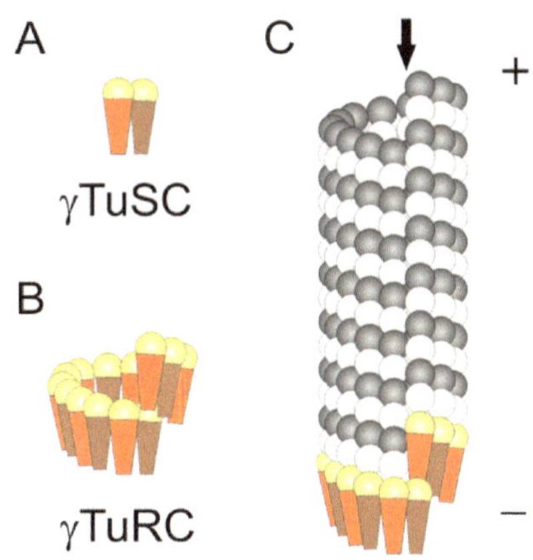

FIGURE 30.11 Gamma tubulin complexes support microtubule assembly. (A) Most gamma tubulin in yeast cells is present as a dimer of two gamma tubulins and two accessory proteins, the gamma tubulin small complex or γTuSC. Activation involves assembly of 14 γTuSCs into a helical ring, the gamma tubulin ring complex or γTuRC (B), which can then nucleate the addition of tubulin dimers through interaction of alpha tubulin with gamma tubulin (C). The resulting 13 protofilament MTs have a seam (arrow), where dimers are shifted in adjacent protofilaments so that alpha tubulin is next to beta tubulin.

location and its clear involvement in organizing other cellular organelles gave this MTOC the name centrosome, and the basal bodies contained within it **centrioles**.

Centrosomes in mammalian cells form through assembly of several proteins that can anchor and/or activate γTuRC complexes, including ninein, pericentrin, Cep215, AKAP450, NEDD1, Cep192 and CDK5RAP2. Pericentrin and AKAP450 are distant homologs of yeast Spc110, which activates the yeast γTuRC and anchors it to the yeast SPB. Regulation of the nucleating activity of centrosomes has been linked to both the amount of these centrosomal proteins and to their phosphorylation by protein kinases. Organization of cellular organelles by the centrosome then follows through their distribution along the polarized array of single microtubules that assemble onto centrosomal γTuRCs. These microtubules all have their (−) ends on the centrosome and (+) ends radiating out through the cytoplasm, and serve as tracks for motor proteins dynein and kinesin (discussed below) that attach to organelles and distribute them along this microtubule scaffold.

Not all cell types limit microtubule assembly to centrosomes. In some differentiated cell types, which are no longer undergoing cell division, centrosome activity is reduced or lost and γTuRCs associate with new MTOCs. Examples include neurons, where microtubules play important roles as motility tracks in long neurites (axons and dendrites), and columnar epithelial cells, where most microtubule (−) ends are associated with cortical regions at the apical end of the cell. In both cases, microtubule polarity governs the direction of vesicle transport by motor proteins. Axons have many other microtubule-associated proteins, which generate and maintain these cytoskeletal elements. As axons grow, both existing microtubules and unpolymerized tubulin move into the elongating tube of cytoplasm. To increase the number of microtubules, and especially to begin new axon branches, existing microtubule are cut by members of a distinct family of microtubule severing AAA ATPases that includes katanins, fidgetins and spastins. Once cut, the exposed (−) ends are protected from disassembly by (−) end binding CAMSAP proteins and the exposed (+) ends can continue to elongate. In addition, new axonal MTs can be nucleated from the side of existing MTs by the anchoring and activation of γTuRCs. Such branched microtubules require a special γTuRC-anchoring complex, augmin, which also supports formation of branched microtubules in the mitotic apparatus. In axons, the newly assembled branched microtubules are then released by katanin or spastin severing activity.

In many polarized cells, especially cells that migrate through extracellular matrices in response to cytokine signals, directed migration requires a polarized microtubule array to support vesicular transport, which brings vesicles toward the expanding leading edge of the cell. Many such cells have, in addition to centrosomal microtubules, a second microtubule array that assembles from the surface of Golgi vesicles. GM130, a membrane protein found in cis-Golgi lamellae, recruits AKAP450 and CDK5RAP2, which in turn anchor γTuRCs. Successful assembly of Golgi-derived MTs also requires CLASPs, proteins that bind near microtubule (+) ends, promote assembly, and protect against catastrophes. CLASPs are only one of a growing number of proteins known to associate specifically with the growing (+) ends of microtubules. These proteins, collectively called **(+) TIPs**, interact either with the GTP cap, or with tubulin dimers that have the unique curved conformation seen only during (+) end assembly. The best studied of these (+) TIP proteins is end-binding 1 (EB1), which is widely conserved. Because EB1 only recognizes tubulin in the GTP cap, and microtubule bound EB1 rapidly exchanges with a diffuse cytoplasmic pool, a fluorescently tagged EB1 will continuously track growing ends of cytoplasmic microtubules.

An additional level of regulation of microtubule function beyond dynamic instability and the binding of microtubule-associated proteins involves direct posttranslational modification of tubulin itself. Common modifications fall into two categories, the addition or removal of amino acids at or near the C-terminus (glutamylation, glycylation or tyrosination), and the addition or removal of an acetate group to a specific lysine residue (acetylation). Together these modifications act as a code, much like the histone code, that modifies microtubule properties (Roll-Mecak, 2020). C-terminal modifications primarily work through changes in the number of negatively charged side chains exposed on the microtubule surface, altering the strength of electrostatic interactions with other proteins. Acetylation alters a residue that is not exposed on the outside of an assembled microtubule, and this modification is largely associated with stable, long-lasting microtubules including those in cilia. Surprisingly, acetylation works by reducing the strength of lateral contacts between protofilaments, allowing some sliding between adjacent protofilaments when microtubules are bent and thus protecting against mechanical breakage (Xu et al., 2017), especially important in cilia that may bend up to 80 times per second.

3. Motor proteins

3.1 Cytoplasmic motor proteins for MT-based movement

The first microtubule motor proteins characterized were **dyneins** (derived from the Greek word "dyne," meaning force), discovered in the 1960s as ATPases that power the beating of cilia. A second microtubule motor family, **kinesins** (kinesis = movement), was discovered 20 years later as the ATPases that power vesicle movement in nerve axons. These ATPases drive stepwise movement along microtubules. Dyneins always step toward microtubule (−) ends and conventional kinesins step toward (+) ends. Together with various cargo adaptors, these motors enable the regulated distribution of cellular constituents along microtubule tracks.

To analyze motor activity, purified motor proteins and microtubules have been assembled in vitro. Microtubules can be settled onto microscope coverslips coated with motor proteins. The microtubules, after the addition of MgATP, may then glide over the surface. By adding a marker for the (+) or (−) end of the microtubules, these gliding assays determine the direction and velocity of motor walking. By tethering motor proteins to fluorescent beads or coupling them directly to small fluorescent tags, the motion of single motor molecules along microtubules can be followed with high spatial and temporal resolution using total internal reflection fluorescent (TIRF) microscopy. By tracking single motors it is possible to measure step size and how long a motor can walk along a microtubule before it diffuses away, a property called processivity. In addition, by trapping a bead in an optical trap, the force generated by motors attached to the bead can be measured (Belyy & Yildiz, 2017). A single kinesin can take 8 nm steps, move at 1 μm/sec and generate a force of 5 pN, while a single cytoplasmic dynein can take 8, 16, or even 24 nm steps, move at up to 1 μm/sec and generate ~1−3 pN of force. These parameters are in an ideal range for the movement of cellular cargoes at useful velocities over the distances typically encountered in eukaryotic cells.

3.2 Kinesin

Kinesins are a diverse family of motor proteins, all with globular "head" domains that bind to microtubules and change conformation during a cycle of ATP hydrolysis (Fig. 30.12). In many kinesin family members these globular heads are associated with more extended alpha helical coiled-coil stalk domains that dimerize, generating a two-headed motor. In these two-headed kinesins, a cycle of hydrolysis by one kinesin head couples to movement of the motor (and any attached cargo) toward the plus end of a microtubule, while the other head remains associated with the microtubule. By alternating the activity of the two heads, the motor "walks" along a microtubule surface, and by having one head always attached to the microtubule track, a single motor can translocate up to 1 μm before dissociating (Isojima et al., 2016; Mori et al., 2007).

The enzymatic cycle of a walking kinesin begins with hydrolysis of substrate (ATP) bound in the catalytic pocket of one microtubule-attached head (the lagging head, nearest the microtubule (−) end, labeled 1 in Fig. 30.12B, step 1). Upon ATP hydrolysis and release of inorganic phosphate (P_i), this head changes conformation enough, markedly reducing its

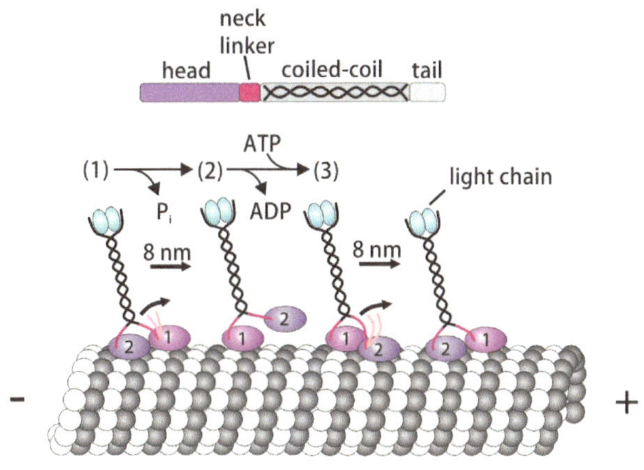

FIGURE 30.12 (A) linear diagram of structural domains in a typical kinesin heavy chain. An N-terminal globular head domain contains both the catalytic site and the microtubule-binding site. The neck linker connects the head to an alpha helical region that can dimerize through coiled-coil interactions. (B) The mechanochemical cycle of a dimeric kinesin walking along a microtubule. Both heads are bound to the microtubule at (1), but release of Pi shifts the lagging head to a low-affinity state and allows the neck linker of the leading head to swing, bringing the lagging head and the cargo-binding light chains toward the microtubule (+) end (2). Release of ADP and binding of a fresh ATP returns the first (now leading) head to a high-affinity state (3). Each step moves the molecule and associated cargo by 8 nm toward the (+) end.

affinity for tubulin. In this low-affinity conformation, the lagging head is free to release from the microtubule surface and assume a new position relative to its partner by swinging two tubulin dimers toward the microtubule plus end (Fig. 30.12B, step 2). Re-binding in this new position is coupled to the release of ADP and binding of a new ATP, which in turn allows this head to return to its original conformation (Fig. 30.12B, step 3). This conformational change in an attached head creates a spring tension force that pulls on the remaining second head, and any cargo attached to the dimerized tail domains, toward the microtubule plus end. The second head, which has now become the lagging head, can, in turn, hydrolyze ATP and release it from the microtubule. Once this second head swings forward, the first head becomes the lagging head and is in position to begin a new cycle of ATP binding and hydrolysis. Coupling between the two heads in a dimer ensures that each, in turn, biases the other to rebind only toward the plus end of the microtubule and reduces the likelihood that the motors or their cargo will diffuse away from the microtubule track.

To convert the relatively small conformational changes associated with substrate binding, catalysis, and product release at the catalytic site into the large steps involved in walking along a microtubule, these kinesin motors have a lever-arm or neck linker between the catalytic and coiled-coil domains. A small change in angle at one end of this neck linker is translated into a much larger movement at the other end. Structural analysis of kinesin motor catalytic pockets and lever arms has shown that they are nearly superimposable on those of myosin motors. Despite the almost complete lack of sequence homology between myosins and kinesins, they likely evolved from a common ancestor capable of coupling ATP hydrolysis to conformational change. Somewhat surprisingly, some kinesin family members can walk (albeit slowly) in the "wrong" direction, toward microtubule minus ends. In these motors, the orientation of the lever arm has changed so that the direction of movement coupled to catalysis is reversed.

Most organisms express a great variety of kinesins, each adapted to carry one or more specific types of cargo. Kinesins usually interact with their cargoes through smaller subunits (light chains) or through cargo-specific adaptor proteins that bind to the kinesin tail domain. Kinesin motors play important roles in vesicle trafficking, as exemplified by axonal transport of vesicles from the nerve cell body toward the synaptic terminal, and in the elongation of the mitotic apparatus during chromosome segregation. In cells of multicellular animals (metazoans), which usually have microtubule arrays organized by centrosomes, one type of kinesin associates with endoplasmic reticulum membranes and generates a network of membrane tubules that extend from the outer nuclear membrane (near the centrosome) toward microtubule plus ends. Another kinesin isoform is specific to cilia, where it is important for plus-end trafficking of a membrane-associated complex (the intraciliary (or flagellar) transport or ICT (IFT) complex) and for use as a gliding motor. Not all kinesins are motors, however. The kinesin-13 family instead binds at the plus ends of microtubules and acts as a regulated microtubule depolymerizer, with roles in regulating the distribution and length of both cytoplasmic and ciliary microtubules.

3.3 Dynein

Dyneins, like kinesins, have globular head domains that hydrolyze ATP and bind to microtubules, and extended tails that interact with cargoes. However, these two motor families differ greatly in overall size, domain architecture, and motor mechanism. Most AAA superfamily ATPases have only a single AAA domain. Such AAA ATPases are widespread in both prokaryotes and eukaryotes, where many of them function as hexameric chaperone complexes that use ATP binding and hydrolysis to aid in the assembly or disassembly of macromolecular complexes. In contrast, dyneins are ~500 kDa AAA ATPase motors in which a noncatalytic amino-terminal tail domain is joined by a linker to six carboxy-terminal AAA domains that fold together to form a hexameric ring (Fig. 30.13). Only the first four AAA domains retain nucleotide-binding sites, and only AAA1 has been characterized as an ATPase whose activity is directly linked to motility. Inserted between dynein AAA domains 4 and 5 is an antiparallel α-helical coiled-coil stalk tipped by a small loop that acts as a microtubule-binding domain (MTBD).

Dyneins act exclusively as minus-end directed motors in both the cytoplasm, where they are invariably two-headed homodimers, and in cilia, where isoforms with one, two, or three heads are found. Unlike the highly diverse kinesin genes expressed in most cell types, cytoplasmic dyneins are encoded by only two genes, one for general cytoplasmic minus-end motility, such as vesicular transport and mitosis, and another limited to minus-end trafficking of ICT complexes in cilia. The mechanochemical coupling between ATPase enzymatic activity and microtubule motor activity differs considerably between kinesins and dyneins (Cianfrocco et al., 2015). The overall mechanism of force generation is similar for cytoplasmic dyneins and axonemal dyneins (Ishibashi et al., 2020). Unlike kinesins, in dyneins, the ATP-bound form has a low affinity for microtubules (Fig. 30.13B, step 2). This low affinity is generated by a shift in the register of coiled-coil interactions along the MTBD stalk, which in turn disrupts interactions of the terminal loop with a site on the tubulin lattice (Gibbons et al., 2005). ATP hydrolysis in the unbound form is linked to a rotation of the large AAA ring relative to

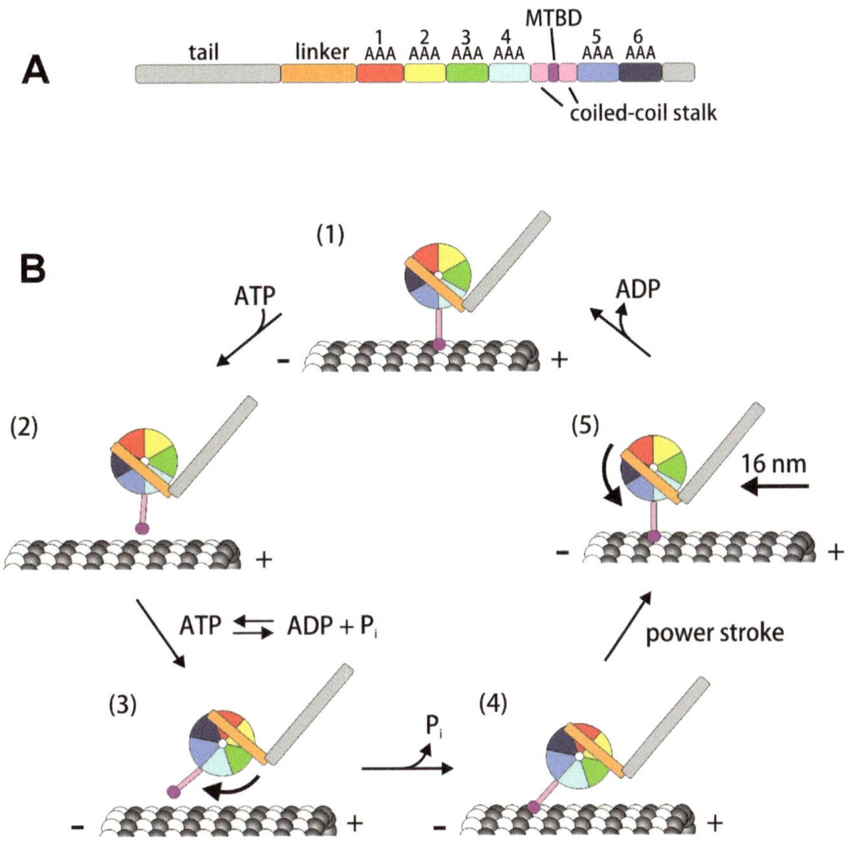

FIGURE 30.13 (A) Linear diagram of a dynein catalytic heavy chain. A cargo-binding tail and a linker domain are followed by six AAA domains. AAA1 has ATPase activity that is linked to motility. The microtubule-binding domain (MTBD) is on a loop flanked by two alpha helical regions that interact in an antiparallel coiled-coil stalk. The six AAA domains interact to form a hexameric ring, as shown in B. (B) Steps in the dynein mechanochemical cycle. Binding of ATP to an attached motor (1) leads to release of the MTBD (2) and rotation of the hexameric head (3). Release of Pi allows rebinding of the MTBD (4), followed by the power stroke (5), which moves dynein and associated cargo by as much as 24 nm (16 nm in this diagram) toward the MT (−) end. Loss of ADP allows return to the initial state.

the N-terminal tail domain, which moves the stalk and MTBD to a new binding site toward the (−) end on the same tubulin protofilament (Fig. 30.13B, step 3). The release of Pi restores the MTBD to a high affinity state, promoting binding to the microtubule (step 4). This, in turn, triggers the relaxation of the rotated AAA ring, pulling the motor toward the microtubule minus end (step 5). The second product, ADP, must be released before the cycle can begin again (step 1).

The motor activity of cytoplasmic dynein has been linked to many forms of vesicular transport, including retrograde transport of vesicles from the endoplasmic reticulum to the Golgi, endosomal trafficking, vesicular and mitochondrial trafficking in axons and dendrites, and tethering of both the nucleus and the Golgi apparatus to the centrosome. This motor also participates in the focusing of microtubule minus ends at spindle poles and movement of chromosomes during early stages of mitosis, and in spindle orientation and elongation at later stages, as described in more detail in Section 5 below. Because of their high importance to the function and maintenance of neurons, defects in dynein- and kinesin-dependent transport have been linked to several neurodegenerative diseases (Millecamps & Julien, 2013) (see Box on Neurodegenerative Diseases and MT-based Transport Defects).

Box on neurodegenerative diseases and MT-based transport defects

Microtubules play an especially important role in the formation and function of neurons. Axons, the longest processes of any cell type, depend on microtubules for their formation and maintenance, and dendrites, while generally shorter, are equally dependent on a microtubule cytoskeleton. It should come as no surprise that defects in microtubule-based transport have been linked to neurodegenerative diseases, as listed below. Perhaps more importantly, neurodegenerative diseases that have not been directly linked to microtubule transport defects may yet show indirect connections to the essential roles that microtubules play in these cells.

Alzheimer's Disease (AD) – Common features of (neurofibrillary tangles) of the microtubule-associated protein Tau, whose normal role is to bind to and stabilize axonal MT. These tangles, like the extracellular deposits of amyloid protein also common in this disease, are more likely symptoms of the disease process rather than direct causes of neurodegeneration. Known genetic defects, such as mutations in the β-amyloid precursor protein, result in changes in proteolytic cleavage of this vesicle protein that may alter vesicular transport. Early changes in axonal transport, which may over time lead to neural degeneration, have also been observed in **Huntington's Disease** and **Amyotrophic Lateral Sclerosis (ALS)**.

Charcot-Marie-Tooth disease – A missense mutation (H306R) in the tail of cytoplasmic dynein heavy chain has been identified as causal for CMT-type 20 neuropathy, whereas a mutation in kinesin Kif1B is linked to CMT-type 2A.

Hereditary Spastic Paraplegia – Although over 70 loci have been linked to genetic predisposition to this syndrome, one of the best-characterized is the locus encoding a kinesin, Kif5A, which works as a vesicle motor in neurons.

Lissencephaly – Mutations in dynein regulatory protein Lis1 disrupt neuronal precursor cell migration and prevent normal formation of neuronal layers in the brain, resulting in disrupted brain morphology and function.

Peripheral Spinal Muscular Atrophy – Mutations in p150, a subunit of the dynactin complex, or in BicD2, a protein that associates with the dynein/dynactin complex, can cause this rare autosomal dominant disease, in which distal limb (hand, foot) motor neurons degenerate.

Cytoplasmic dyneins have multiple smaller subunits (intermediate chains, light-intermediate chains and light chains) associated with their tail domains that are thought to aid catalytic heavy chain dimerization and to facilitate interaction with cargoes (Fig. 30.14). In addition, most interactions of dynein with cytoplasmic cargoes require a separate adaptor complex called **dynactin**, which includes a microtubule-binding protein (p150) dimer and multiple copies of an actin-related protein (Arp1). Additional cargo adapters, such as BICD2, generate specificity for different cargoes. This complex seems to play two separate roles, acting both as an interface between cargo and motor, and also as a tether to maintain motor contact with the microtubule surface during multiple rounds of ATP hydrolysis. This tethering function provides greatly increased processivity to cytoplasmic dynein. Without dynactin, dynein motors only move for short distances along a microtubule before dissociating and diffusing away, along with their cargoes. However, when dynein and its cargo are trapped in a thin space that does not permit the motor to diffuse away from its track, tethering by dynactin is not needed. This situation occurs both for ICT dynein, which is trapped between ciliary microtubules and the ciliary membrane, and for the dyneins that directly power ciliary beating, which are trapped between pairs of doublet microtubules.

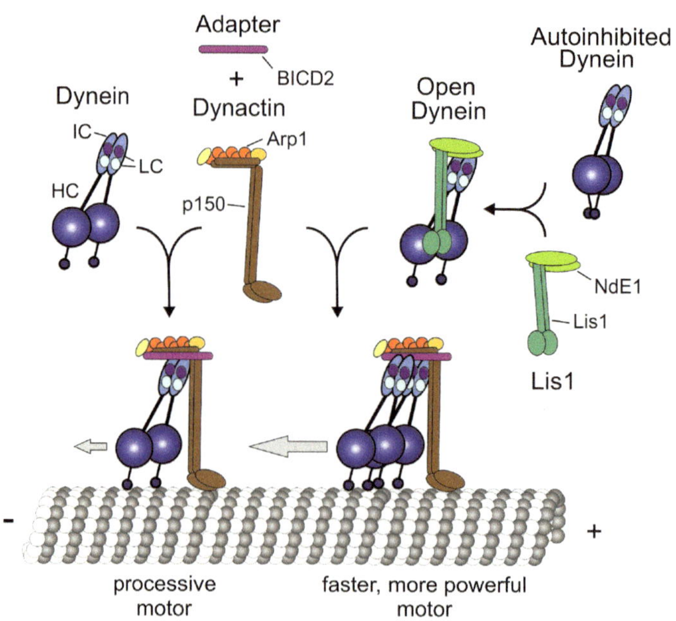

FIGURE 30.14 Complexes that interact to form (−) end-directed cytoplasmic dynein motors. Cytoplasmic dynein heavy chains dimerize through interactions with dimeric intermediate chains (IC) and light chains (LC). These dimeric dynein complexes can then interact with dynactin complexes (left) and cargo adapters such as BICD2 to form processive motors. The multisubunit dynactin complex includes a short actin-like filament of Arp1 subunits and a long dimer of p150 subunits. Interaction with dynein allows the dimeric p150 heads to tether dynein to the MT surface, increasing processivity, while the Arp1 side-arm and adapters provide an interface with cargoes such as vesicles. The heterotetrameric Lis1 complex is built from dimers of NudE and Lis1 and can bind to both the catalytic heads and the IC complex of dynein, shifting dynein from a closed, autoinhibited conformation to an open conformation. Two open conformation dimers can bind simultaneously to one dynactin/adapter complex, creating a more powerful motor.

Dynein activity can also be modulated by interaction with Lis1 (Fig. 30.14), a protein initially discovered through the analysis of mutations that affect brain development in humans (resulting in a condition called *lissencephaly*). This human condition results from an inability to generate the proper layers of neurons during brain development. Characterization of nuclear distribution or *Nud* mutants, which disrupt the normal distribution of nuclei in syncytial hyphae of the fungus *Aspergillus nidulans*, led to the identification of a fungal Lis1 homolog, NudF, and a NudF interacting protein, NudE (human orthologs NdE1 and NdEL1). What do these two mutant phenotypes have in common? Both result from improper movement of nuclei within cells, and evidence supports a role for the Lis1 complex in modifying cytoplasmic dynein from a motor that generates low force (perfect for transport of small vesicles) to one that generates high force (ideal for movement of bulky nuclei through the molecularly crowded cytoplasm). Dynein can interact with either dynactin or Lis1, but not with both at the same time (McKenney et al., 2011). Binding of the Lis1 complex shifts cytoplasmic dynein from an autoinhibited state to an open state and increases its affinity for the Dynactin/adapter complex, leading to formation of dynactin complexes that have two dimeric dyneins instead of just one (Fig. 30.14; reviewed in Markus et al., 2020). In addition to their interactions with cytoplasmic dynein, Lis1 homologs have also been found associated with ciliary dyneins, but the functional significance of this observation is not yet known.

4. Cilia as specialized MT-based motility organelles

4.1 Advantages of actively bending cilia compared with thin rotary propellers

Machin (1958) found that due to the energy dissipation of moving a thin beam in a fluid, any wave initiated at the base would die in about 1.5 wavelengths unless energy is applied all along its length. Since bacterial and archaeal flagella are too narrow, 20 nm and 10–14 nm respectively, to supply energy to motors along their length, they must be rigid and rotate to move cells or fluid (see Chapter 31). However, the much larger 200-nm-diameter eukaryotic cilia can incorporate ATP-driven motors down their entire length to maintain a traveling wave of bending down a length as long as $30,000\ \mu m$! Relative to prokaryotic flagella, the power can be increased one thousand-fold per length, making it possible for eukaryotic microorganisms to go more places a lot faster with bigger and more complex organisms or larvae in tow. Of course, there are problems with this solution, such as how do you get ATP and/or other energy molecules and controlling signals down the entire length beyond the few μm that diffusion can reach.

4.2 The doublet and triplet MT skeleton of cilia

In addition to their function as tracks for the distribution and motility of cellular organelles, microtubules play essential roles in two specialized structures, the mitotic apparatus and the motile cilium. Here we summarize the basic properties of cilia and describe their assembly and function as both motile and sensory organelles.

Cilia are long, narrow protrusions of the cell supported by a cylinder of nine doublet microtubules (Fig. 30.15). Each doublet consists of a 13-protofilament A-tubule, like singlet microtubules seen in the cytoplasm, but in addition has a second, 11-protofilament B-tubule attached along one side (Fig. 30.15A). In vitro, modified buffer properties can induce tubulin dimers to assemble partial B-tubule-like sheets of variable numbers of protofilaments on the side of existing microtubules, without the need for additional proteins. Several sheets, which appear as hooks in cross section, may assemble onto a single microtubule; however, these sheets never close into B-tubules. In contrast, ciliary doublet B-tubules have precisely 10 protofilaments made from tubulin dimers, and are closed back onto the A-tubule at the inner junction by an 11th non-tubulin protofilament (Yanagisawa et al., 2014). Doublets likely evolved to provide superior mechanical properties to cilia and to generate unique A- and B-tubule surfaces for attachment of motor proteins.

In a poorly understood process, yet a third partial microtubule can form on the B-tubule of doublets to form short cylinders of triplet microtubules (Fig. 30.15A, right). These triplets provide the 9-fold framework for basal bodies, structures that can dock onto the plasma membrane, nucleate the assembly of ciliary doublets, and act as cytoplasmic anchors that resist the forces exerted on doublets by ciliary beating (Fig. 30.15B). The 9-fold symmetry of basal bodies is provided by a cartwheel with nine spokes that self-assembles from two proteins, Sas6 and Cep135, which in turn interact with nucleation sites that support growth of triplet microtubules (Gönczy, 2012; Guichard et al., 2017). Basal bodies are anchored to the membrane through distal appendages that project from each triplet microtubule to as yet poorly characterized membrane specializations. In many multiciliated cells two additional appendages, the striated rootlet and basal foot, project from opposite sides of each basal body into the adjacent cytoplasm in an orientation aligned with the planar ciliary bend direction (Zhang & Mitchell, 2015). They help to anchor basal bodies and distribute the bending moment generated by each cilium throughout the rest of the cell.

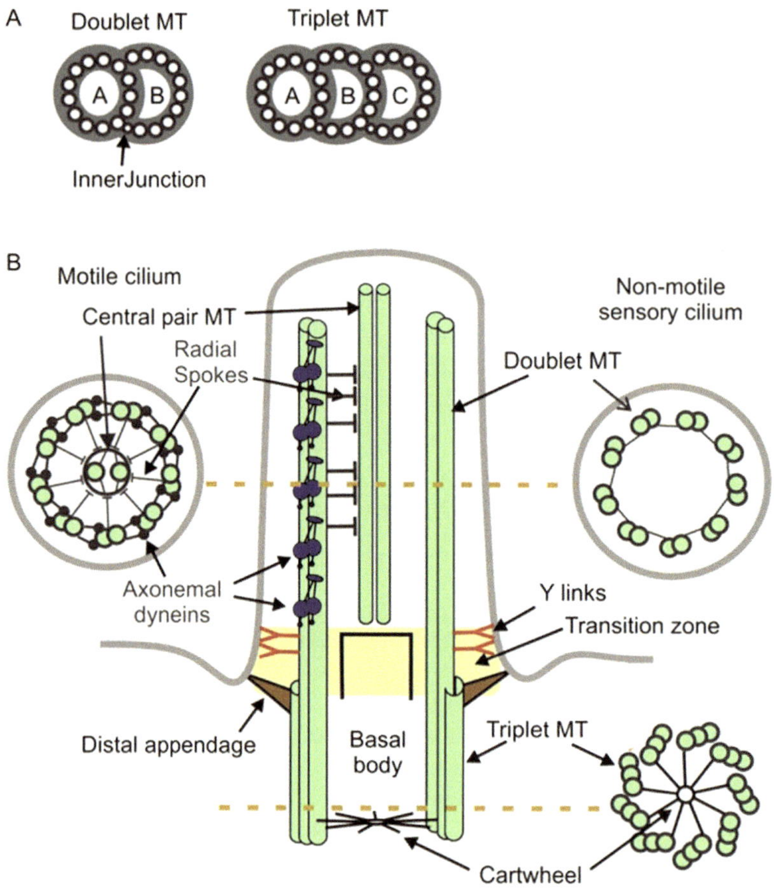

FIGURE 30.15 Cilia are built from singlet, doublet and triplet microtubules. (A) Diagrams of cross sections through doublet and triplet microtubules, composed of 13 tubulin protofilament A-tubules and 10 tubulin protofilament B- and C-tubules. The B- and C-tubules close back onto the A-tubules through a nontubulin protofilament at the inner junction. (B) A cartwheel structure templates assembly of 9 triplet MT to form a basal body. Distal appendages link the basal body to the cell membrane, and doublet MT assemble from basal body A- and B-tubules to form the ciliary axoneme. Cross sections of nonmotile cilia (*right*) display a 9 + 0 structure, whereas motile cilia (*left*) have two singlet central pair MT, generating a 9 + 2 structure. Doublets are linked to the membrane by Y-links in a transition zone just above the basal body. Doublets form a cytoskeleton for attachment of axonemal dyneins, radial spokes, and central pair proteins in motile cilia.

During differentiation, multiciliated cells such as those in airway epithelia assemble hundreds of basal bodies in the cytoplasm and dock them onto the apical cell surface, with the (+) ends of each triplet MT facing the membrane. Doublet microtubules then assemble by addition of dimers onto the A- and B-tubules of each triplet, pushing out the membrane to form a cilium. Additional links between doublet MT and the membrane define a short transition zone just above the basal body (Fig. 30.15B). In motile cilia, two additional singlet (13-protofilament) microtubules, the central pair, assemble in the center of the cylinder of doublets beginning above the transition zone (9 + 2 pattern). Protein complexes that localize to transition zones play important roles in regulating communication between the ciliary compartment and the rest of the cell, thereby restricting free diffusion of both soluble cytoplasmic proteins and proteins associated with ciliary versus nonciliary plasma membranes in both motile and nonmotile cilia. One such transition zone complex forms Y links, which form attachments between doublet MT and the membrane.

4.3 Centrioles

In most organisms, basal bodies can perform a dual role. In addition to acting as anchoring sites for cilia, they are organizing sites for assembly of cytoplasmic microtubules and hence organization of many cellular constituents. In typical mammalian cells, a pair of basal bodies and surrounding material form a single MT organizing center or centrosome (Fig. 30.16, Cell Cycle). Only one of the paired basal bodies, or centrioles, docks to the plasma membrane through distal appendages and supports assembly of a cilium. Many cells in metazoan organisms display a single, nonmotile cilium

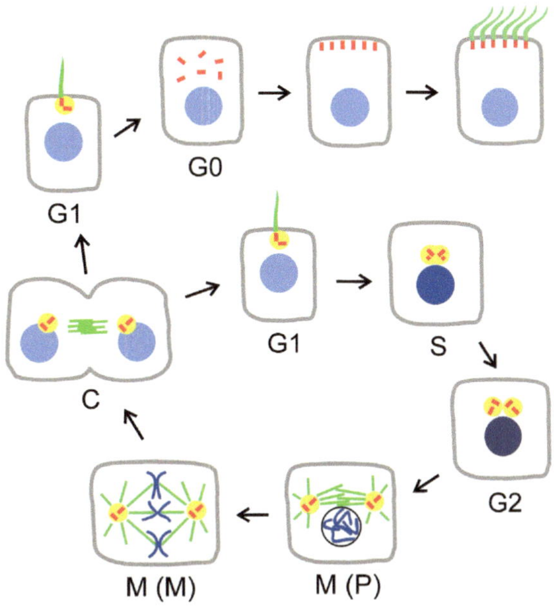

FIGURE 30.16 Duplication of basal bodies and centrosomes in cycling monociliated and in terminally differentiated multiciliated cells. Upper pathway (terminal differentiation) shows the transition from a cell in G1 of the cell cycle, with one centrosome (*yellow*) supporting cytoplasmic microtubules and one mature centriole (basal body) supporting a primary cilium, to a multiciliated cell. Lower circular pathway (cell cycle) shows that duplication of centrioles occurs during S phase, but separation of the two-centriole pairs into two centrosomes does not occur until prophase of mitosis, M (P). Only the older of the two centrioles in a centrosome can serve as a basal body for primary cilium assembly in G1 phase, and this cilium is disassembled at the start of S phase.

(primary cilium) specialized for signal transduction. These cilia completely lack central pair microtubules and all the other proteins specifically needed to generate and regulate ciliary bending (9 + 0 pattern). The centriole or basal body in each pair that supports ciliary assembly is called the mother centriole and is always at least 1 cell cycle older than the nonciliated (daughter) centriole. When cells enter the cell cycle and begin DNA replication (S-phase), the cilium disassembles, and a new centriole begins to assemble next to each existing centriole (Fig. 30.16, S). At mitosis, the two-centriole pairs separate to create two microtubule-organizing centers that can then function as ends of the mitotic spindle. After mitosis and cell division, the older centriole of each pair can once again form distal appendages and support cilium assembly (Fig. 30.16, G1). Because of this tight cell cycle coupling between basal body/centriole duplication and the number of microtubule organizing centers, cells that are destined to become multiciliated must exit the cell cycle before beginning the unusual process of building hundreds of basal bodies in the cytoplasm (Fig. 30.16, terminal differentiation pathway). Individual basal bodies (not pairs) then dock to the apical plasma membrane and support the assembly of motile cilia. Some organisms, such as higher plants and many fungi, evolved life cycles that do not depend on cilia at all. No doublet or triplet microtubules appear in these organisms and genes encoding delta, epsilon and zeta tubulin (only needed for centriole assembly), along with several hundred additional cilia-specific genes, have been lost from their genomes.

4.4 Axonemes

In 9 + 2 cilia, hundreds of proteins in addition to tubulin associate with outer doublet and central pair microtubules in precise arrays to generate a sophisticated machine for generating bends, the motile ciliary axoneme. All of these proteins repeat along the axoneme at periodicities related to the underlying 8 nm size of the tubulin dimers in each protofilament (Fig. 30.17). In most organisms, two rows of axonemal dyneins decorate the A-tubules of each doublet, an outer row (ODAs) composed of a single dynein isoform that attaches every 24 nm, and a more complex inner row (IDAs) with multiple different isoforms attached in a 96 nm repeat pattern. This 96 nm repeat, specified by a pair of molecular ruler proteins, forms the basis for attachment of many other proteins that associate with doublet microtubules, including a dynein regulatory complex that coordinates dynein activity and links adjacent doublets once every 96 nm, and groups of radial spokes that project in toward the central pair microtubules (Lin et al., 2014; Ma et al., 2019). Each spoke ends in a broad head that appears to rest against a cylindrical surface generated by the projections of the central pair microtubule complex or central apparatus (CA) (Fig. 30.15B). Interactions between radial spoke heads and the CA, and between the base of

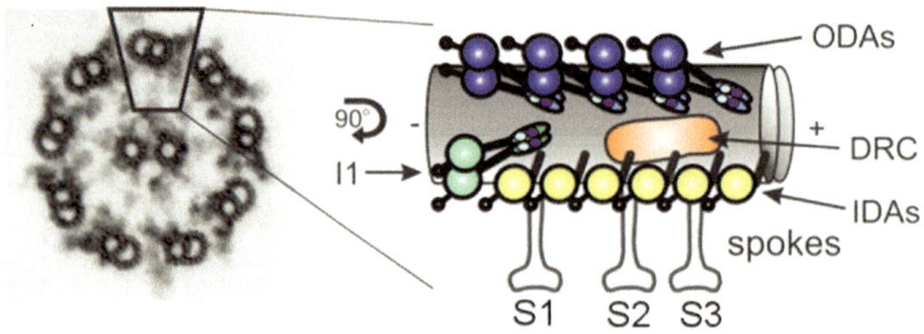

FIGURE 30.17 Details of the arrangement of dyneins and regulatory complexes in motile cilia. Left, electron micrograph of a cross section through a ciliary axoneme. Right, diagram of one 96 nm repeat module of an outer doublet microtubule. Each repeat contains four dimeric outer dynein arms (ODAs), one dimeric inner dynein arm (I1), and six monomeric inner dynein arms (IDAs). The dynein regulatory complex (DRC) links adjacent doublets and interacts with several dyneins. Radial spokes also interact with several dyneins as well as with the central pair complex.

radial spokes and doublet-associated dynein regulatory complexes, provide a regulatory network that aids in the coordination of ciliary dyneins to generate specific beat parameters. The interaction of radial spokes and the CA may also provide structural support to prevent collapse of the axoneme due to compressive forces generated by bend propagation in dissipative media.

Motility generated by axonemal dyneins is based on a simple geometric mechanism. Each dynein attaches to an A-tubule by its tail, so that the A-tubule acts as a transported cargo and to an adjacent B-tubule by its head, so that the B-tubule serves as a track (Fig. 30.18, left). Since dyneins only move toward microtubule (−) ends, dynein activation on one A-tubule will result in sliding (shear) forces in one direction between each pair of doublets. Because doublets are usually anchored at their (−) ends to basal body triplets, but not at their (+) ends, sliding can only occur if the doublet pair bends. Each doublet pair therefore can generate a bending force in the direction of an arrow pointing from the A-tubule of doublet N toward the B-tubule of adjacent doublet N + 1. Circular arcs (Fig. 30.18, right) can often approximate the resulting bend. Sliding displacement Δl accumulates throughout a bend due to the difference in lengths of the arcs for doublets N and $N+1$, which in turn is directly related to the difference between radii, $r1$ and $r2$, and the size of the bend angle in radians, θ. If dyneins on all nine doublets were active at once, these bending forces would cancel and the axoneme could not bend. Activity progressing sequentially from one adjacent doublet pair to the next consistently in either direction, with each bend activating dyneins on the next doublet, would create helical waves. Helical bending waves are in fact seen in some simplified cilia and sperm tails that lack radial spokes and central pair microtubules. Nutation seen in some primary cilia is simply helical bending waves on a short cilium. The cilium does not rotate, but the tip wobbles on its axis, moving around in a circle without significant torsion of the structure. In most cilia and sperm flagella, the additional level of regulation and resistance to compression provided by central pair-radial spoke interactions, together with links that limit sliding between one of the nine pairs of doublets (typically 1−2(5−6) see Fig. 30.25), allows generation of powerful planar bending waves, as described in greater detail below.

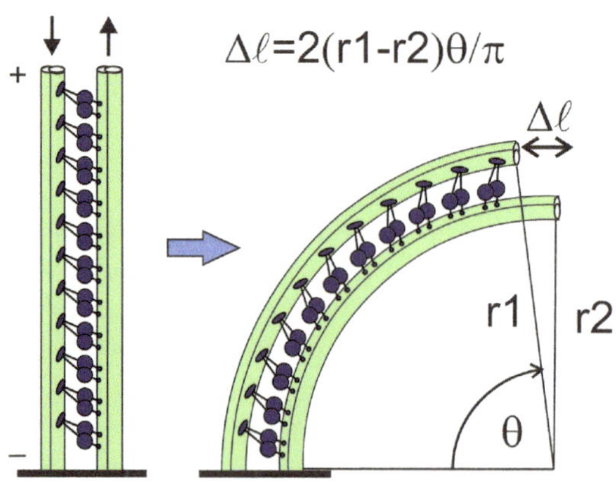

FIGURE 30.18 The basic geometry of ciliary bend formation when there is no sliding at the base and without strong compression forces at the open end. Left, two adjacent doublets with a row of dimeric dyneins attached to the A-tubule of doublet N. Dynein MTBDs interact with the B-tubule of doublet N+1, and generate shear (small arrows) through their (−) end motor activity. Right, shear leads to sliding between the doublets only if a bend forms. Differences in the bend radius (r) for doublets N ($r1$) and $N+1$ ($r2$) are related to the sliding displacement (Δl) and the bend angle (θ) by the formula $\Delta l = (r1 - r2)\,\theta$.

4.5 Intraciliary transport ICT (Synonym: Intraflagellar transport IFT)

Cilia do not contain ribosomes; therefore, all proteins needed for their assembly must be synthesized on free or ER-associated ribosomes in the cytoplasm, moved through the transition-zone diffusion barrier, and carried to their final destination (usually the growing ciliary tip). Specialized motility complexes of an ICT system aid this movement (Fig. 30.19). These complexes evolved from similar complexes that form an interface between cytoplasmic microtubule motors (kinesin and dynein) and membrane-bound vesicles, such as COPI and COPII vesicle coat proteins (Jékely & Arendt, 2006). Since the closed structure of cilia prevents vesicles from fusing directly with ciliary membranes, this ICT mechanism likely evolved when a tight connection formed between basal bodies and the plasma membrane, to allow continued trafficking of membrane proteins to this specialized ciliary membrane domain. At the same time, protein subunits of the cilium itself, including axonemal dyneins and tubulin, although not directly associated with membranes, need to use the same ICT transport system to gain access to their assembly sites at the tips of growing cilia (Lechtreck, 2015). ICT cargoes transport through association with proteins of two large interacting complexes, IFTA and IFTB, which associate with two motor proteins: a kinesin specialized for transporting these complexes toward ciliary tips (anterograde), and a cytoplasmic dynein isoform used only for transport back toward basal bodies (retrograde). These ICT complexes concentrate around basal bodies in a staging area for cargo binding, and then group together into trains of several particles before traveling up in the space between doublet microtubules and the ciliary membrane. Transport along cilia by ICT complexes is remarkably fast. ICT kinesin moves ICT trains anterograde at up to 2.5 μm/s along B-tubules and, after a brief pause and rearrangement, ICT dynein returns these complexes at about 4 μm/s along A-tubules. Mutations in a mouse gene encoding an ICT protein (IFT88), which disrupted normal ciliary assembly in kidney tubules, provided the first link between cilia and a wide range of human syndromes with strange and apparently unrelated symptoms, now known collectively as ciliopathies (Mitchison & Valente, 2017) (see diseases in Box on Ciliopathies).

Box on Ciliopathies

Several human syndromes characterized by strange and difficult to explain combinations of symptoms have in common defects in some aspect of ciliary function. These can largely be divided into defects in ciliary motility (termed primary ciliary dyskinesia or PCD) and defects in cilia-dependent signal transduction.

Motile cilia defects (PCD) – Mutations in axonemal proteins essential for motility can affect all motile cilia and sperm flagella, or selective tissue specific ciliary types, depending on the pattern of expression of the affected gene. Mutations in nonaxonemal proteins needed for assembly of axonemal dyneins (chaperones, transport factors) can have similar effects. If nodal cilia are defective, patients may have problems due to improper development of left-right asymmetry, including congenital heart defects. Airway ciliary dyskinesia leads to continuous upper airway infections and lung dysfunction, defects in sperm flagella are linked to infertility in men, and dysmotility of oviduct cilia leads to reduced fertility in women.

Ciliary signaling defects - The variable and pleiotropic symptoms of many confusing conditions stem from mutations in large protein complexes localized to the transition zone (also known as the ciliary pore or ciliary gate), a region responsible for controlling entry and exit of ciliary membrane receptors. Most of these diseases have one or more of the following symptoms: blindness, anosmia, cystic kidneys, liver fibrosis, obesity, polydactyly, congenital heart disease, and mental disabilities.

Bardet-Biedl syndrome (**BBS**) – most proteins encoded by BBS genes associate into a complex that functions as an adapter to regulate trafficking of transmembrane receptor proteins from trans-Golgi membranes to cilia. The BBS complex also associates with IFT complexes and may function as an IFT adaptor for the transport of receptors into the ciliary compartment.

Nephronophthisis (**NPHP**), Meckel-Gruber syndrome (**MKS**), Oral-facial-digital syndrome Type 1 (**OFD1**), and Joubert syndrome (**JBST**) have all been linked to defects in protein complexes that localize specifically to ciliary transition zones. These complexes may function as part of the ciliary gate or pore, or may help in loading receptor cargoes onto the BBS/IFT transport machinery.

Autosomal dominant polycystic kidney disease (**AD-PKD**) is caused by defects in *PKD1* or *PKD2*, genes encoding two interacting ciliary membrane proteins that together form a regulated cation channel. Similar symptoms arise from defects in the channel proteins and from recessive mutations that disrupt the trafficking of these proteins to ciliary membranes. Although the most typical symptoms arise from loss of kidney function, AD-PKD also leads to liver, spleen and pancreatic cysts.

Alstrom syndrome – *ALMS1* mutations can cause blindness, hearing loss, obesity, diabetes, cardiomyopathy, and multiple organ failure. The protein localizes to centrioles/basal bodies, and mutations may affect endosomal trafficking, including trafficking of essential ciliary signaling proteins.

In addition to working as escalators to bring construction material to the tips of cilia during their initial assembly, ICT complexes play ongoing roles in cilia maintenance and in the regulated movement of signaling proteins into and out of cilia

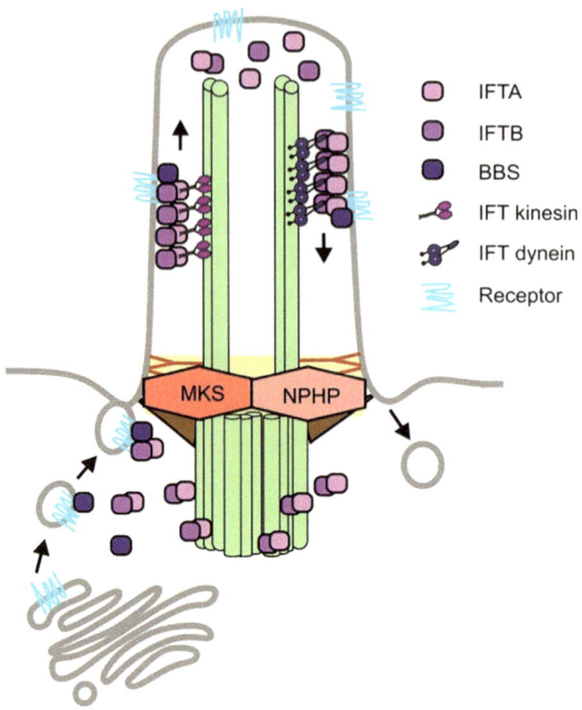

FIGURE 30.19 The intraciliary/intraflagellar transport (ICT/IFT) complex supports ciliary assembly and display of ciliary membrane receptors. Short trains of IFTA and IFTB complexes are transported toward the ciliary tip (anterograde) by IFT kinesin, and toward the basal body (retrograde) by IFT dynein. MKS and NPHP complexes form a selective barrier at the transition zone, preventing free diffusion of membrane receptors into the ciliary compartment. Membrane vesicles carrying ciliary proteins from the trans-Golgi must deliver their cargo to the cell membrane, where the BBS complex interacts with IFT trains and aids in recognition and transport of receptors.

(Mourão et al., 2016). One of the best studied of these signaling pathways, Hedgehog, regulates widespread developmental patterning in vertebrates such as control of digit (finger and toe) formation; Hedgehog defects link to polydactyly. Processing of signals downstream from the binding of Hedgehog to its membrane receptor (Patched) requires transport of another membrane protein (Smoothened) into cilia, where it triggers modification of an intermediate in the pathway (Gli) in the ciliary compartment. The modified Gli can then exit the cilium, enter the nucleus, and regulate gene transcription. Mutations that disrupt this pathway have been traced not only to subunits of ICT complexes, but also to a related complex (BBS or Bardet-Biedl Syndrome) that is not essential for ciliary assembly, but rather appears to be specialized for the regulated transport of cilia-associated receptors. The BBS complex, like ICT complexes, travels in the space between doublet microtubules and the ciliary membrane but relies on an interaction with the ICT complexes for its job and does not appear able to interact directly with the kinesin and dynein that power ICT traffic (Fig. 30.19). BBS proteins have also been implicated in trafficking of post-Golgi vesicles carrying ciliary membrane proteins to the periciliary plasma membrane, where they can fuse and deliver their cargo to a staging area for ICT transport.

Membrane proteins transported into the ciliary compartment release from ICT and BBS complexes and can diffuse along the ciliary membrane but cannot readily diffuse out onto the rest of the cell surface. Similarly, membrane proteins common to the rest of the cell membrane cannot diffuse into the ciliary compartment. Two interacting protein complexes that localize to ciliary transition zones, the Nephronophthisis (NPHP) and Meckle Gruber Syndrome (MKS) complexes appear to play important roles in this selective barrier (Fig. 30.19). Mutations that disrupt these complexes have been linked to defects in many organ systems that depend on the sensory functions of ciliary membrane proteins (see Box on Ciliopathies).

4.6 Crawling or gliding motility

Many single-celled aquatic eukaryotes take advantage of either anterograde or retrograde ICT movement as an alternative propulsive force for cell motility. Most of these organisms have only one or two long cilia (usually called eukaryotic flagella). Ciliary transmembrane adhesion proteins are used by these cells to stick to surfaces in their environment and coupling these adhesion receptors to ICT motors generates ciliary gliding motility.

For example, *Peranema* glides at 33 $\mu m/s$ (Saito et al., 2003) on the "hard" air-water interface under the control of rhodopsin photoreceptors (Saranak & Foster, 2005). The rate of gliding is proportional to the anterior ciliary length plus about 9 $\mu m/s$, which is probably due to the posteriorly directed cilium and works just as well without any other cell attachment to the substrate. Beads attached to the ciliary surface move as fast. In *Chlamydomonas* the motion of the ciliary surface moves from tip to base dragging the cell body at about 1.6 $\mu m/s$. FMG-1B is the membrane protein in contact with the substrate and it is powered by dynein1, which also drives ICT trains toward the base (Bloodgood, 2009). Membrane viscosity slows the motion of FMG-1B to less than the speed of ICT trains. A [Ca^{2+}] dependent signaling pathway (Shih et al., 2013) connects the two systems.

4.7 Energy supplied to cilia

ATP in the presence of Mg^{2+} and a simple buffered salt solution can reactivate the motility of isolated cilia in vitro, and the resulting beating closely mimics normal in vivo motility patterns. Under these controlled conditions, beating frequency correlates precisely with variation in MgATP concentration, and the relationship between beating frequency and concentration can be fit to Michaelis−Menten kinetics (Fig. 30.20).

To maintain the beat frequencies observed in vivo, the ciliary compartment must maintain an ATP concentration of about 1 mM. To get this [ATP] inside the ciliary compartment, ATP generated by oxidative phosphorylation in mitochondria, or by glycolysis in the cytoplasm, must diffuse into cilia through the transition zone. Although the transition zone appears highly permeable to small metabolites such as ATP, diffusion through a long, narrow cylinder such as a cilium likely limits substrate concentration available to axonemal dyneins at the distal end as energy use may exceed the diffusion rate. To alleviate this problem, mammalian spermatozoa have mitochondria wrapped around the proximal half of each sperm axoneme for continuous ATP generation. They can also take up glucose from seminal fluid, which is then metabolized by glycolytic enzymes distributed along doublet-associated dense fibers to generate ATP. While cilia in most other cell types lack both mitochondria and glycolysis and depend entirely on ATP diffusion and its transport from the bulk cytoplasm, mechanisms have evolved to generate additional ATP in the ciliary compartment by direct phosphorylation of ADP by enzymatic shuttles within the ciliary compartment, based on phosphotransferase reactions. Shuttle systems act as chains to form a constant-velocity transport circuit that can reach much farther and faster than is possible by diffusion. The key for the shuttles is that these enzymes operate close to equilibrium. The net effect is to establish a gradient of disequilibrium of the kinase reactions (XP + ADP ↔ ATP + X) along the cilium so that their concerted action can sustain the ATP concentration in the distal regions of the cilium without the need for diffusion over long distances (Fig. 30.21). For example, the sperm of invertebrates such as sea urchins lack both mitochondria and a source of sugars for glycolysis. Instead, they store phosphate bond energy in the form of phosphocreatine and use creatine phosphokinase to re-

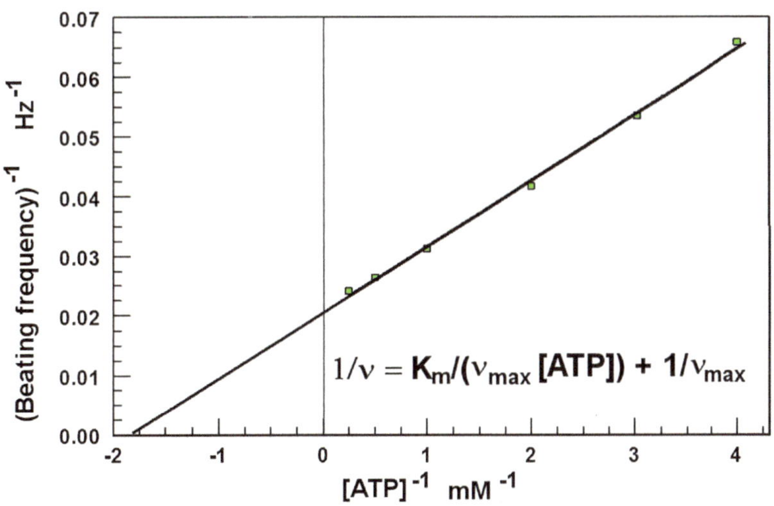

FIGURE 30.20 Double reciprocal plot of beating frequency and [ATP], graph of ν^{-1} versus $[ATP]^{-1}$ at 1.35 mPa s viscosity. Underlying the beating frequency is motor stepping, which likely has a Michaelis−Menten relationship with [ATP], namely the stepping rate $V = V_{max}[ATP]/(K_m + [ATP])$. Since the stepping rate V is the number of steps N in one beating cycle to form the basal wave times the beating frequency ν, $V = N\nu$, then the reciprocal of beating frequency can be written: $\nu^{-1} = K_m/(\nu_{max})[ATP]^{-1} + \nu_{max}^{-1}$. It predicts a linear relation between ν^{-1} and $[ATP]^{-1}$, supported by the graph, with $K_m = 0.53$ mM and $\nu_{max} = 45$ Hz and that N is a constant. Error bars on frequency are all $\leq 1\%$. *Data from Brokaw (1996).*

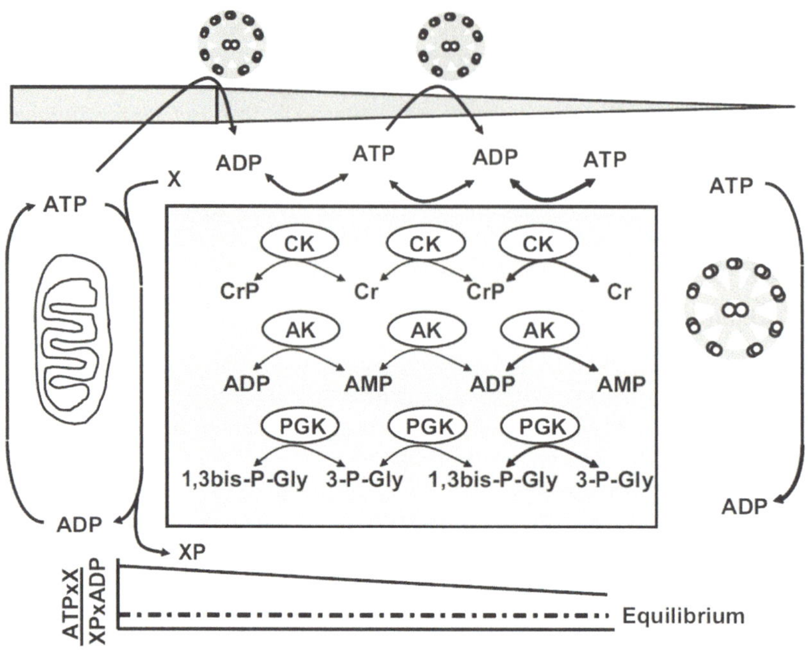

FIGURE 30.21 ATP transfer shuttles seen in some in cilia. Along the cilium ATP is hydrolyzed to ADP by the ATPase motors to enable sliding of adjacent DMTs. ADP is rephosphorylated by phosphate transfer from creatine phosphate (CrP), ADP or 1,3-bisphosphoglycerate (1,3bis-P-Gly), catalyzed by creatine kinase (CK), adenylate kinase (AK) or 3-phosphoglycerate (3-P-Gly) kinase (PGK) respectively. These reactions establish a gradient of kinase reactions (XP+ADP ↔ ATP+X) along the cilium to sustain [ATP] in more distal regions of the cilium over distances not sustainable by diffusion alone. Note that other shuttles are also required to make the system work. *From Ford (2006).*

phosphorylate ADP, making use of a phosphocreatine and an adenylate kinase shuttle. Even organisms such as *Chlamydomonas,* which only has two short 12.8 μm cilia, supplement ATP diffusion with a shuttle made up of adenylate kinases, which can catalyze the reversible transfer of a phosphate from one ADP to another ADP, generating AMP and ATP and thus boosting local ATP concentrations. *Chlamydomonas* cilia also have a shuttle for a glycolytic intermediate, 3-phosphoglycerate, which is metabolized within the ciliary compartment to generate additional ATP. Ciliates such as *Paramecium* use another high-energy phosphate source, phosphoarginine, that is generated through intermediary metabolism in the cytoplasm and diffuses into cilia.

4.8 Bending motility – free swimming with attached cilia

The mechanism of propulsion is by a traveling wave of lateral displacement due to bending and sliding of DMTs down the cilium's length. If the cilium's radius of curvature is sufficiently small than the cilium will appear to beat as in a breaststroke (ciliary beat), but if the curvature is relatively straight then the motion will be whip like (flagellar beat). Using a bendable cantilever to make an effective motile structure in an aqueous environment faces several issues. First is the geometry, for if the end does not slide at the cell body then as the wave grows at the base the concave side has to move more outward than the convex side (Fig. 30.18). However, for the traveling wave (Fig. 30.26), the outside or convex side has to move outward more relative to the concave side. To make this seamless the growing wave sets a curvature and when complete is replaced along its whole length by the traveling wave whose motor activity is set by that curvature that then pushes the convex side outward (Fig. 30.28). The dynein motors directed toward the outside of bends have more room and freedom to move than those directed toward the inside of curves (Cibert & Ludu, 2010).

Bending against the aqueous environment and reshaping the cilium dissipates energy. Hence, motors sliding two filaments relative to each other (Fig. 30.26) must do so all along the length. To avoid unnecessary controls, the beating would be best self-organized with a stable limit cycle and therefore be driven by a nonlinear oscillator dependent only on local physical information. In fact, because, as was shown by Foster et al. (2017), the initiation of beating is independent of time in that a wave attains precisely the same energy no matter the speed of its creation, only purely physical models of beating initiation are likely. This is not a surprise because the high frequency of the beating, most typically 40–80 Hz, already makes chemical control improbable relative to mechanical control. Chemical controls exist at <10 Hz.

4.8.1 How free-swimming sperm, larvae and microorganisms swim at low Reynolds number

Figs. 30.6 and 30.22 show respectively the cilia of *Chlamydomonas* and *Ciona* sperm with definition of their ciliary parameters. Typically, as for *Ciona* spermatozoa to move, a bend forms at the base and then travels outward (other organisms may have inward waves or go in either direction) (Fig. 30.23). Traveling waves travel down an arc of constant curvature (the axoneme prestress) (Eshel & Brokaw, 1987), which apparently is significantly due to the central pair. The rightward traveling waves are readily apparent for flagellar mode beating. The term prestress emphasizes the analogy to the prestress of a rod or beam, which results in its static curvature.

In the organism *Chlamydomonas*, changing the prestress with [Ca^{2+}] changes its beating from a ciliary (breaststroke, $-0.22\ \mu m^{-1}$ curvature) to flagellar (whip-like, $-0.02\ \mu m^{-1}$ curvature) beat. *Chlamydomonas* can switch rapidly between ciliary and flagellar mode beating with their different prestresses and yet subtracting the prestress makes the waves similar. The mode has significant consequences for different parameters of the motion. Differences in prestress will for the same wavelength significantly change the degree of sliding $\mathbf{\Delta}$, which will strongly influence the stretching of the presently unknown connections between the DMTs that give the cilium its longitudinal elasticity and its transverse forces. Cilia must, hence, be controlled by parameters that are insensitive to the prestress.

The forces that drive a cilium and its accompanying cell in its highly dissipative medium are shown in Fig. 30.24. The well-known Scallop theorem (Purcell) in Newtonian media shows that you cannot go anywhere by reciprocal motion, hence you either rotate like flagella and archaella (Chapter 31) or send a traveling bending wave down a cilium as discussed here. However, in non-Newtonian media, which are present in many biological environments, it would be possible to generate motility using reciprocal motion (Qiu et al., 2014).

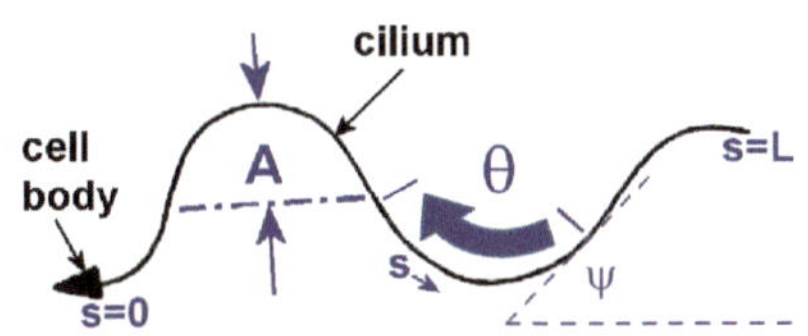

FIGURE 30.22 The geometry of cilia: A *Ciona*-sperm planar cilium; *A* is amplitude, ψ is the tangent angle (positive in the counterclockwise direction) and θ is twice ψ_{max}. The wavelength λ is the distance along the arc s of the cilium between repetitions of the same waveform and not the shorter distance in real space. *L* is the length of the cilium along its arc.

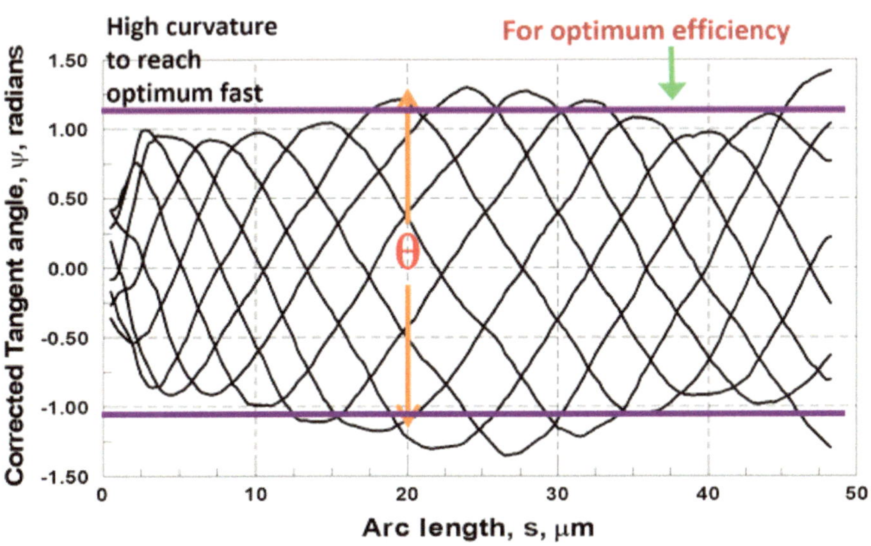

FIGURE 30.23 Shows the tangent angle for a single ciliary beating cycle of a *Ciona* sperm cilium, plotted as successive instances of the bending waves going rightward, in 1.35 mPa s viscosity and 2 mM ATP, calculated from Fig. 1D of Brokaw (1996). The prestress curvature (linear function of **s**) of $-0.014\ \mu m^{-1}$ has been subtracted from ψ defined in Fig. 30.22 and then the mean (a space average) of the remainder has been subtracted to show more clearly the initiating and traveling wave at 3.33 ms intervals. Note the high curvature (slope of lines) near the base to quickly reach the most efficient amplitude indicated by the purple lines.

4.8.2 To generate propulsive force DMT sliding produces bends that travel down the cilium

Consider the cross-section of a cilium and its nine DMTs (Fig. 30.25). Note there are two independent DMT motor sliding systems to accommodate the fact that dynein motors can only actively walk on the adjacent DMT toward the base (Fig. 30.18). Dynein motors provide the shear forces to produce the relative sliding of the DMTs, which, through transverse forces, bend the cilium. To slide DMT 1(5) outward relative to 6(1) requires the motors of 6(1), 7(2), 8(3), and 9(4) (the − side or the principal bending side). To slide 6(1) outward relative to 1(5) and 2(6) requires the motors of 2(6), 3(7), 4(8) and 5(9) (the + side or reverse or recovery bending side). The wave travels rightward toward the tip (Figs. 30.23 and 30.24). The motors either slide outward the top or bottom DMTs. Note the right-hand rule of *all* cilia, with thumb pointing outward along a cilium, the motors from A to B are *always* in the direction of the fingers (Fig. 30.25).

For conceptual simplicity of the planar traveling wave, the 9 + 2 structure of the axoneme may be replaced by a pair of doublets separated by a distance a of 180 nm (Fig. 30.26). Since DMT filaments are nonextensile, all sliding produces transverse forces that cause bending and potentially torsion or twisting of the DMTs and the absence of bending implies no relative sliding. For there to be motion of a traveling wave, the outer half-wave must move more rightward than the inner half-wave (Fig. 30.28). The sliding displacement $\Delta(s, t)$ is the relative displacement of the upper group relative to the lower group. Hence $\Delta(s,t) = \Delta_B(t) + \boldsymbol{a}(\psi(s,t) - \psi(0,t))$, where $\Delta_B(t)$ is the sliding at the base and $\boldsymbol{a}$ is the distance between the two sides.

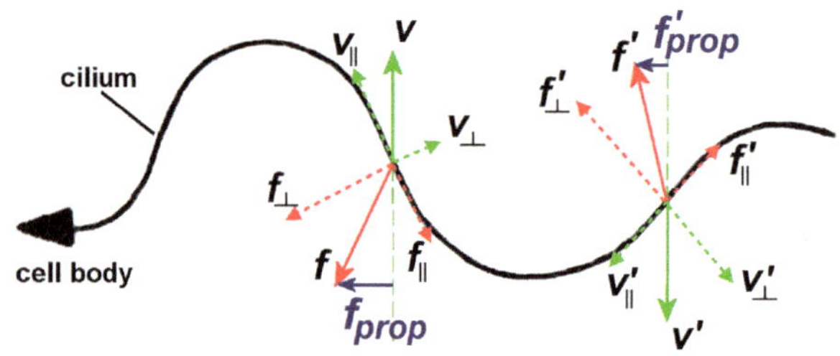

FIGURE 30.24 Shows normal $f_\perp$ and tangential f_2 forces on a planar cilium of a swimming organism (in the opposite direction of the forces of the cilium on the fluid which are proportional to the indicated local velocities) since the normal drag is stronger than the tangential, the organism as depicted in the drawing would go to the left as indicated by the f_{prop} vectors. On the other hand if there were stiff mastigonemes or hairs extending normal to the cilium, making the tangential drag stronger than the normal drag, then contrary to this figure the direction of swimming would be rightward, as f_{prop} would point rightward.

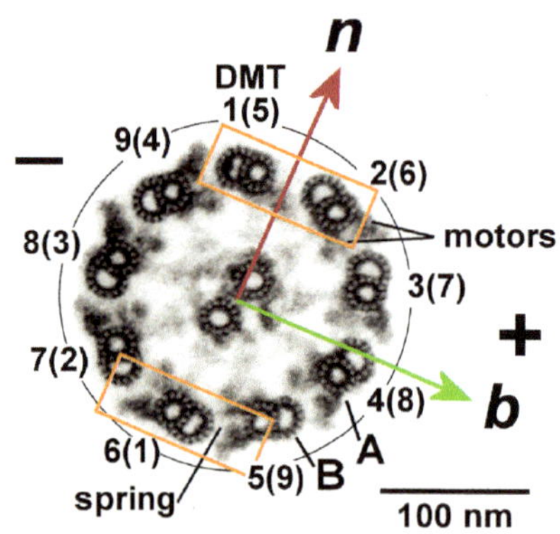

FIGURE 30.25 Basal view (Gibbons, 1961) of *Chlamydomonas* axoneme cross-section with the vertical aligned with the body axis. The two "arms" coming out from each doublet microtubule DMT with numbering *Chlamydomonas* (*Ciona* or mammalian) (Lin et al., 2012) are the dynein motors that walk on the adjacent DMT. In walking the "A" microtubule feels force acting toward the base. Springs are attached from A to B. The tangent vector is directed away from the base and the normal vector is defined in the bending plane as in the direction 90° CCW from the tangent. The tangent vector $\mathbf{t} = \mathbf{n} \times \mathbf{b}$. The motors on the binormal + **b** side marked + promote positive curvatures ($d\psi/ds$ positive) in propagating waves such as depicted in Figs. 30.23 and 30.24 with normal **n** pointing upward. Motors on the -**b** side marked - promote negative curvatures in propagating waves. The central pair is oriented with C1 microtubule at the top in the **n** direction and C1b is alongside DMTs 6(1) to 9(4). C1b is the site of the glycolytic enzyme enolase which yields phosphoenolpyruvate that maybe converted rapidly by the cytosolic enzyme pyruvate kinase with large negative Gibbs free energy to pyruvate and ATP. The immediacy of this reaction to produce ATP maybe sufficient for *Chlamydomonas* to take advantage of the central pair having rotated to be adjacent the motors that drive the negative curvature of the cilium (Mitchell, 2000; Zhang & Mitchell, 2004). *Modified from Foster et al. (2017).*

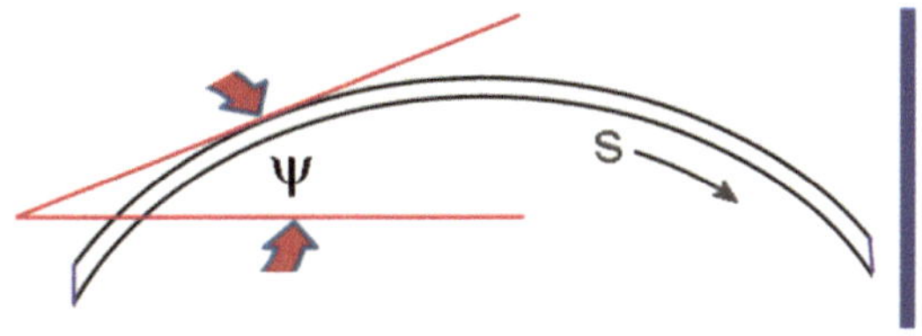

FIGURE 30.26 For the traveling wave, depicting DMTs 1(5) and 2(6) by the single top curve and DMT 6(1) by the equal-length bottom curve, the 9 + 2 structure of the axoneme is replaced by a pair of doublets for the purpose of analyzing planar beats (Foster et al., 2017). Note as the wave passes the bar on the right, the upper curve that is initially behind the lower curve becomes in front by sliding faster to the right.

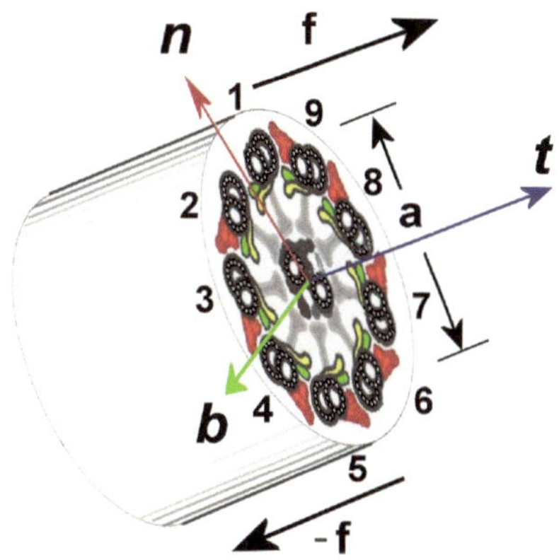

FIGURE 30.27 Distal view of cross-section. The bending plane contains the tangent vector ***t***, pointing away from the base and 90° CCW from **t**, the normal direction **n**. **b** is the binormal direction. The complex system may be simplified to a force per length **f** acting on the upper DMTs and −**f** acting on the lower DMTs, located a distance **a** apart.

4.8.3 The traveling wave

When the motors of DMT 9(4) walk on DMT 1(5) in the direction of the base, there is counterforce per length $f(s)$ due to the motors and springs acting on DMT 1(5) directed toward the tip (Fig. 30.27). This walking in turn, as discussed, produces motion rightward in Fig. 30.26 of the upper (convex side) curve (towards the tip). In addition, the motors on DMT 6(1) walk toward the base on DMT 7(2). This corresponds to a shearing tangential force per unit length $-f(s)$ on DMT 6(1) directed toward the base, again consistent with the relative motion of the upper curve towards the tip. It is important to realize that axonemal dyneins are "rowers" that only transiently bind DMTs rather than "porters" that may progressively hold onto a microtubule. To achieve fast sliding, the arms of dyneins must bind DMTs only briefly so that multiple arms in tandem on a DMT do not interfere with each other. Lin and Nicastro (2018) refer to these active inner and outer dyneins in the traveling wave as "intermediate-conformation dyneins." With cryoelectron tomography, they see these dyneins having blurred stalk density, implying positional heterogeneity because of their transient binding to the DMTs. This is expected since the dynein motors immediately initiate mechanical power strokes if they bind resulting in steps of sliding that, in the presence of ATP, cause an almost immediate dissociation from the microtubule (Roberts et al., 2012). As anticipated, this motor activity is only found within the bends on the side that would promote the convex outer curve to move outward (see Fig. 30.29 for the outer dyneins). As might be expected in *Chlamydomonas* cilia the ATP-synthesis side of the central-pair complex rotates to face and provide extra fuel to the motors that drive the convex side of bends (Mitchell, 2003; Zhang & Mitchell, 2004) (Fig. 30.25). Further details could be obtained by measuring the effect of motor binding, namely by measuring the local viscoelasticity along the cilium as the traveling wave passes by, for example, by use of a pair of optical traps (Memet et al., 2018). This would be valuable since flexural rigidity is typically proportional to the degree of binding between DMTs.

Since the cilium is thin and the distance between the doublets is small, even a short walk can produce a significant bend in the cilium (e.g., 100 nm relative sliding of a doublet may induce a bend as large as 50 degrees). For ciliary traveling waves, the motors that drive the convex side are active. Once initiated, the traveling wave travels all along the cilium and beyond about a wavelength from the base, does not depend on the length of the cilium or on the details of its initiation. Travelling waves carry energy outward from the initiating disturbance. Barring disruptions, the energy stored in the forming wave is transported outward without loss. For a cilium, this energy dissipates as it moves the surrounding fluid and

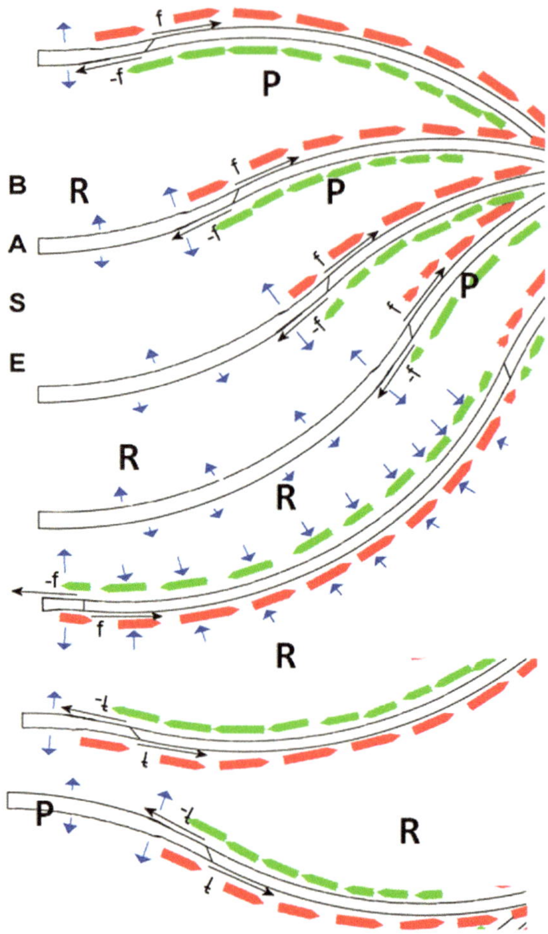

FIGURE 30.28 If the DMTs are fixed at the base and the top and bottom lines are of equal length the inner curve (concave side) would extend beyond the outer (convex side). For this to occur the inner curve must be pulled outward by the motors of the distal curved cilium of opposite curvature (shown by the arrowheads), as shown in four stages of pulling out the concave side (Foster et al., 2017). Since DMT filaments are nearly nonextensile, if there is no bending there is no motor driven longitudinal shear. Hence, since in the already formed part of the forming basal wave there is no change of shape then there must be no motor activity there. Going downward in the figure are shown the cilium near the base at successive times. The arrows on the side of each depiction of the cilium show the location of the region and direction of sliding caused by the motors in the case of the traveling wave. In the earliest times of the drawings the sliding outward is greatest on the inside or concave side of the forming bend, i.e., is more rightward than on the convex side, as it is pulled by the activity of motors driving the outer side of the more distal adjacent bend of opposite curvature. However, once the wave starts to propagate as a traveling wave the sliding is rightward on the outside or convex side of the bend. Inward and outward arrows show the formation and closing of the gap between the two sides of the cilium that existed during the forming wave process.

changes the shape of the cilium, generating thrust and enabling the attached cell to move, or fluid to move past the cell. The dissipated energy is perfectly balanced by the dynein motors as long as sufficient bending torque *af* (Izawa & Shingyoji, 2020) (see Fig. 30.30) is available. Although the wavelength λ of the propagated wave varies along the cilium, the same energy from node to node in the wave is maintained. This is because the density of motors and [ATP] is uniform along most of the ciliary length and (a) the time these motors spend in supporting the wave is inversely proportional to the propagation velocity v_p, which is proportional to the wavelength λ while (b) the number of motors contributing is directly proportional to the wavelength λ.

4.8.4 The nonlinear basal-wave oscillator that initiates the traveling waves

An oscillator with a limit cycle, which necessarily limits the wave's amplitude by some nonlinear process, must initiate all traveling waves. The question for years has been what limits the basal ciliary wave.

If the DMTs are *not* fixed at the base, then right at the base, the convex side must be pulled outward and the concave side pushed back by the neighboring distal wave of opposite curvature that is beginning to propagate outward. This

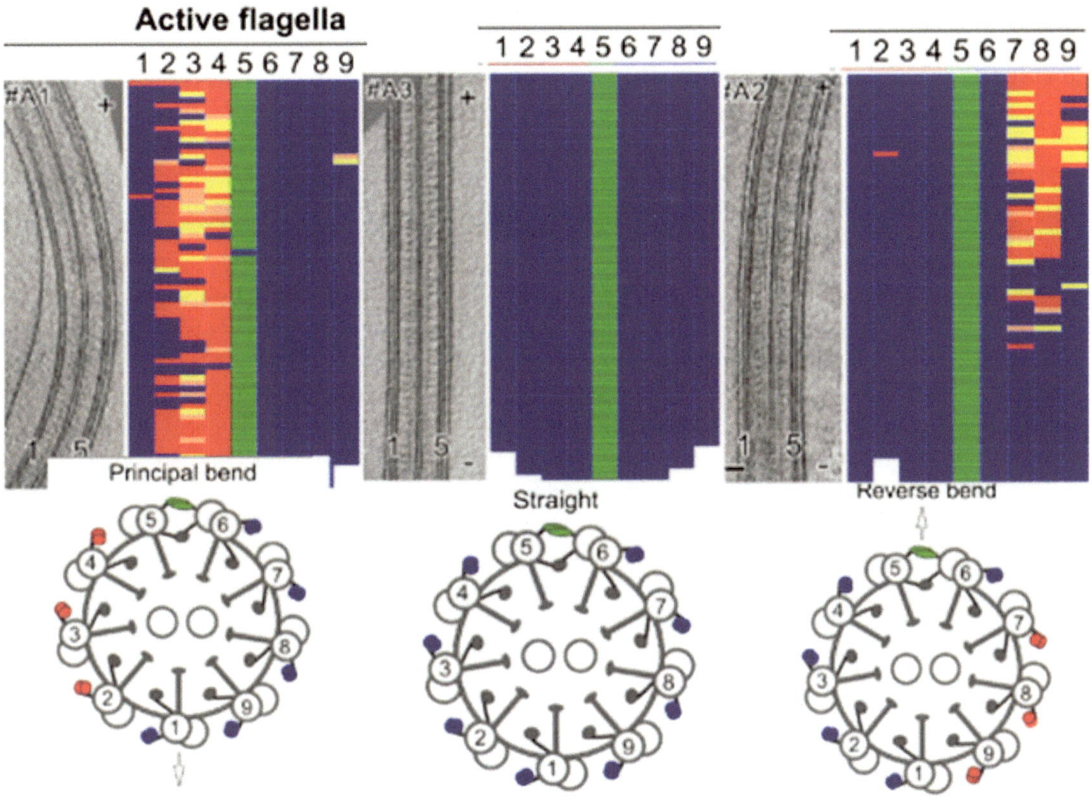

FIGURE 30.29 Shows the activity of outer dyneins that bend the convex side of cilia during traveling waves; (A) the principle bend primarily of dyneins labeled as 2–4 and (B) the recovery or reverse bend of dyneins 7–9. This is for beating in a relatively low viscosity so that more motors can be anticipated to be active in higher viscosities. Green is the fixed connection, red (IM1), orange (IM3) and yellow (IM2). Inner dyneins, a, g, d and I1 are similarly distributed. *From Fig. 3C of Lin and Nicastro (2018).*

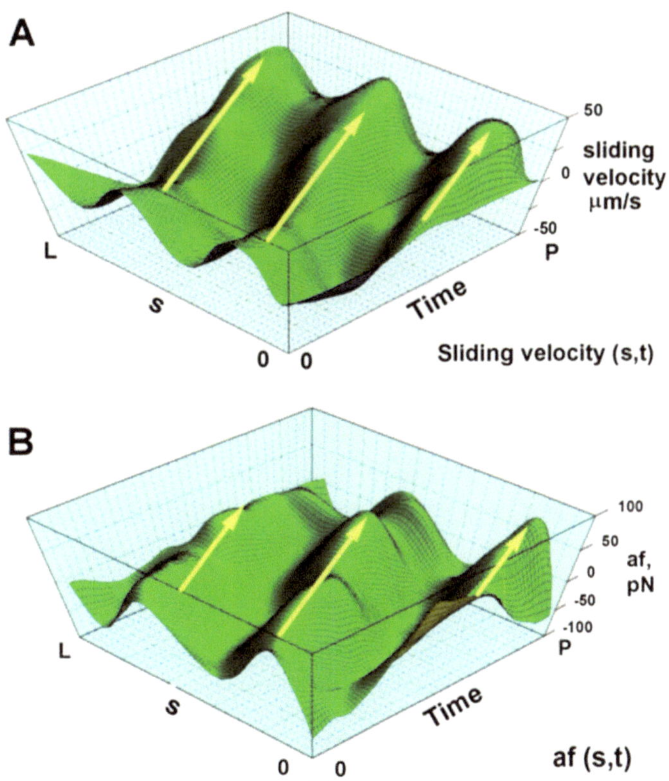

FIGURE 30.30 The sliding velocity and shear force or torque per length ***af*** during a beating cycle. (A) Sliding velocity as a function of time during a beating cycle and position s along the cilium, with the attachment of the cilium to the cell at zero. (B) The ***af*** (normal) shear force or torque per length as a function of time during a beating cycle and the positon *s* along the cilium. It is approximately in phase with the sliding velocity. *From Fig. 12A and B, Foster et al. (2017).*

advance of the concave side makes it possible for the convex side's motion to dominate just as it does for the traveling wave. On the other hand, as in most cases, the DMTs *are not motile* at the base; therefore the basal wave (Fig. 30.28) that initiates the traveling wave must be different from the traveling wave. Geometry requires the *concave* DMTs to move more rightward than the *convex* ones as one travels away from the base, as shown in Fig. 30.18. The distal travelling wave puts tension on the inside (concave) and compression on the outside (convex) so that in this case the inner and outer sides separate, as shown by the arrows (Fig. 30.28), forming a gap between the sides due to the differential in their buckling or bowing. This suggests that a gap likely forms, preventing *both* the + and − motor sets from attaching to their binding sites on the adjacent DMTs, resulting in no local bending. As shown in Fig. 30.28, the basal bend only grows at its distal end. This phenomenon was first reported in *Chlamydomonas* (Brokaw et al., 1982) and is also observed in *Ciona* as shown in Fig. 30.31 and in sea urchin spermatozoa as well. A relative displacement or gap of the inner and outer doublets, as per Fig. 30.28, is seen by the basal region's diameter ***a*** of the cilium larger by 16% in *Chlamydomonas* (Lindemann & Mitchell, 2007) relative to distal bends as diagrammed in Fig. 30.28. Also, as expected, while the basal wave is growing, the dynein motors are not attached, the average flexural rigidity of the basal region is several fold less than in the central region during beating (Foster et al., 2017). Due to internal friction, pulling by the adjacent distal wave can only effectively pull a small section of DMTs, which is the leading edge of the forming initiating wave (Fig. 30.28) as suggested by Sale (1986) who also found that in this basal region the inner and outer sides of a demembranated and reactivated cilium would spontaneously split (Fig. 30.32). At some length, the wave is observed to be complete. The pulling stops and the tension on the concave side stops, resulting in the two sides rejoining as shown at the bottom of Fig. 30.28.

The growing wave region proximal to the base is specialized with no attached springs, as Foster et al. (2017) report, and Bui et al. (2012) found the b dynein to be missing. This permits much higher curvatures as observed without jamming and much higher energies to be stored in a short section, enabling reaching optimum efficient tangent angles rapidly (Fig. 30.23). In the proximal region, the inner side squishes by 4% compared with 1% in the central region.

After thus forming the basal wave, which is now one-quarter wavelength long, with its propagation (the startup of the traveling wave), at any particular ***s*** that point is moving backward along the already formed bend so the sliding direction must then reverse. Hence, once the newly formed wave is moving outward, as a traveling wave, rightward sliding occurs on the *convex* side of the bend all the way to the next newly forming wave. In doing so, the motors simply maintain the existing curvature of the formed bend, with motors simultaneously attaching and propagating the existing wave outward. The preformed sign of the bend has conveniently set the side of the motors that is active, with the convex side always being moved, requiring no extra control. A new wave at the base is then initiated (Fig. 30.28). By this self-organizing mechanism, an initiating wave is created first in one direction and then again in the opposite direction, sending down the cilium a square wave of motor activity (driven by alternating plus side and minus side motors) (Fig. 30.33).

The sliding velocity is effectively an integral of the motor activity and hence a triangular wave (Fig. 30.33). Since no springs appear to be involved in this basal region (Foster et al., 2017), the process is quite simple resulting in high curvatures and high amplitude waves. Further out from the base this initiating wave is turned into a traveling wave with the aid

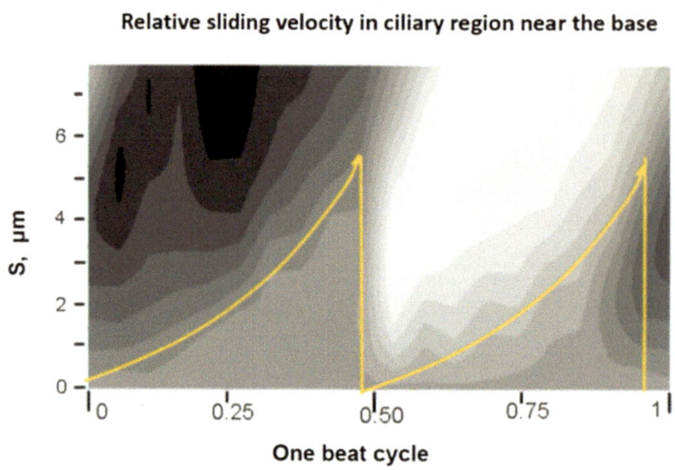

FIGURE 30.31 The sliding velocity of the basal region in a contour plot, from maximum positive (*black*) to negative (*white*) velocity, from analysis of Fig. 2A of Brokaw (1996) with 0.1 mM MgATP. After bending has been restored all the way to the base, there is zero sliding velocity during the basal wave formation with only sliding at its leading edge (below the quadratic curves), and then there is sudden resumption of sliding of the outer side of the bend along all of its length. *Modified from Fig. 7A Foster et al. (2017).*

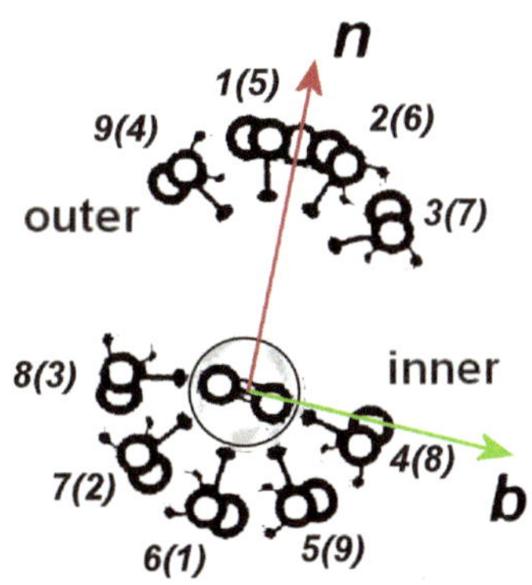

FIGURE 30.32 The observed splitting of the demembranated and reactivated axonemes in the basal region of sea urchin sperms (Sale, 1986). Numbering and orientation is the same as in Fig. 23–25.

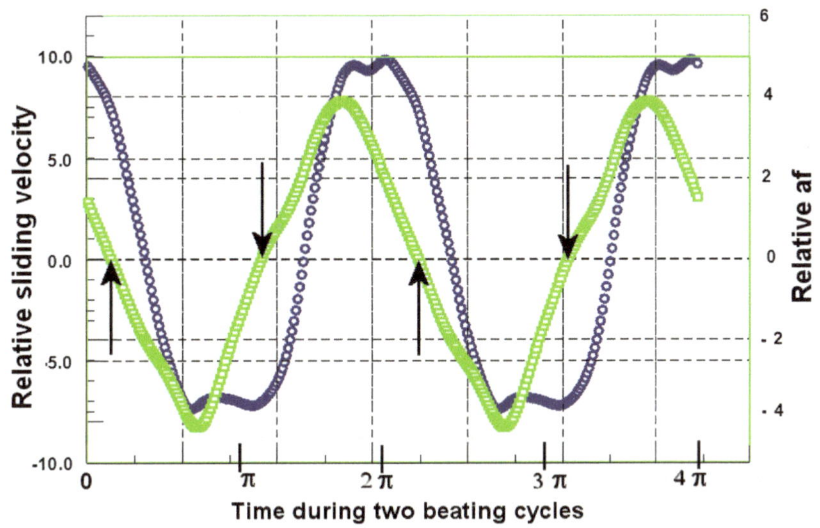

FIGURE 30.33 Plot of ***af*** (○) and sliding velocity (□) as a function of time within two beating cycles outside of but near the basal wave forming region (Foster et al., 2017). One may note the relatively square wave of ***af*** as one expects from the mechanism of formation of the initiating wave and the triangular shape of the sliding velocity due to its apparent integration of the square wave of motor activity. The curves are relatively simple due to the absence of springs in the basal region (Foster et al., 2017). The relative phase of the two waves is probably due to the relative timing of the two waves on the cilium at this particular location along the cilium. The down arrows indicate the beginning of negative curvature and the up arrows the beginning of positive curvature of the cilium. Maximum power occurs in the middle of each half-wave.

of springs that change the phase of the relationship between the drivers, motors plus springs, and the sliding such that the driving bending torque becomes approximately in phase with the sliding velocity. This optimizes the thrust produced by the cilium. The effect of the springs is clearly seen in plots of the relative velocity versus *af* in the basal region relative to the central region (Fig. 30.34). In the second and fourth quadrants, the springs contract to assist sliding from zero speed. In the first and third quadrants, motors stretch the springs substantially slowing sliding from 7.5 to 1.8 and −8.5 to −2.5 in relative units.

Cilia with DMT sliding at the base might not require a gap. The pulling of the outer side distal wave on the initiating basal wave could pull out the outer side right from the beginning of the wave, after which the motors driving the outside could be continuously active as the wave propagates.

4.8.5 The constant energy of the traveling wave and its implications

As already discussed (Fig. 30.20), with the viscosity held constant, the beating frequency ν, determined in the near-base region, is simply related to [ATP] (Brokaw, 1967). Since each ATP molecule causes a motor to make a single step, the stepping rate should increase with [ATP] according to the first-order kinetics of Michaelis-Menten:

stepping rate $= \frac{V_{\max}[ATP]}{K_M + [ATP]}$. Here, V_{max} is the maximum stepping rate and K_M is the Michaelis constant (the value of [ATP] for which the stepping rate is half maximum). Fig. 30.20 implies that the ciliary beating frequency ν is directly proportional to the stepping rate, and the number of steps to form the wave proximal to the base is constant and independent of [ATP]. The constancy of the number of steps would be expected automatically if each motor stepped once per beat cycle; however, we know that more steps per cycle are needed since the bend amplitude of the wave is too large. We also know the wavelength λ varies, implying the number of involved motors also varies. In any case, given a constant number of steps, if each step or motor's powerstroke contributed the same energy, then the total energy put into making the wave by the motors E_{mot} would be the same. Independent of that Foster et al. (2017) found in the case of *Ciona* spermatozoa that the energy of a wave is constant no matter the [ATP] or even for a 4-fold variation in the viscosity.

That the initiating basal bend expands to a constant energy independent of the frequency of the ciliary beating ν means *its control is insensitive to time,* suggesting it is mechanically and not chemically regulated. A wave of a slow beat, where the shear dC/dt is small, has the same energy as when shear is large. Foster et al. (2017) hypothesize that the basal bend stops growing when it reaches maximum mechanical energy. The total mechanical energy **M** of a compliant mechanical system is the sum of the internal strain energy **E** and the potential energy of external forces **U**. The positive strain energy **E** is formed by bending the elastic cilium (the negative of the work done by the internal forces). The bending is done by motors sliding DMTs and axial compression on the ends of the wave as the flexible axoneme pushes forward on the cell body and backward against the fluid. The outward pulling activity of motors immediately distal to the basal wave is proposed to do the internal work **E** on the inner concave side of the forming wave (see Fig. 30.28). The lack of basal interdoublet springs makes this easier as seen by the higher velocities reached near the base (Fig. 30.34 at $s = 6$ μm relative to s = 16 μm). The axial compression also shortens the tip to base distance, reducing the wave's mechanical potential energy **U**. This is the potential energy due to external forces (Jones, 2006) (negative of work done by external forces). Since the axial compression releases potential energy of external forces **U**, during wave formation dU/dt is negative. The wave reaches its maximum size and mechanical energy when $dM/dt = dE/dt + dU/dt = 0$. The DMTs of the growing basal wave will buckle when the axial compression force N_C equals $\pi^2\kappa_b/L^2$ (Das et al., 2008; Wierzbicki, 2013), where κ_b is the flexural rigidity of the cilium's basal bend and L is the growing length of the basal bend. See Wierzbicki (2013) for its derivation from the two components of its strain energy, namely due to the increasing bending storing more energy **E,** and the increased shortening of the cilium due to the axial load that reduces the strain energy **U**. In other words, with the inverse quadratic dependence of N_C it becomes easier for it to buckle quadratically with its increasing length. Since the axial compression force N_C is proportional to the flexural rigidity κ_b, one may write $N_C = g\kappa_b$, where g is the proportionality constant. Then $g = \pi^2/L^2$ for buckling, and defining $\beta = \pi^2/g$ enables the simple result $L^2 = \beta$. Then, β is the constant that sets the length at which buckling occurs and the lack of dependence of β on κ_b explains the insensitivity of buckling to variations in flexural rigidity. Forming the basal bend is a relatively easy process since there are no active springs to push or pull against. Therefore, high curvatures can quickly build up the wave amplitude to the optimum diameter for efficient swimming.

The velocity of the leading edge of the basal wave increases linearly with time, possibly due to increased availability of ATP, which could be proportional to the length of the existing basal wave (Fig. 30.31) as produced ATP is not used over the already formed portion, but only at the leading edge. Hence, its arc length increases quadratically with time, $L(t) = \boldsymbol{a}\, t^2$ (in Fig. 30.31, the area under the quadratic curve shows where there is zero sliding) and hence $\beta = \boldsymbol{a}^2\, t^4$. With the fourth power of time, buckling will be reached at a well-defined basal wavelength and time. This narrow time window explains the high Q or quality factor of >80 (beating frequency divided by its standard deviation) found for the cilia of *Chlamydomonas*.

With buckling, the relative DMT-sliding velocity at the leading edge of the basal wave also goes to zero, abruptly stopping wave growth. The relative tension on the concave side and compression of the convex side stop causing the gap to close and launch the traveling wave all along the length of the formerly growing wave. Consequently, it is not surprising that cilia do not beat until completion of the growth of the specialized basal region (see figure of Bottier et al., 2019) and there is "an abrupt transition" to beating once the basal curvature reaches a threshold (Geyer et al., 2016). The observed sudden reaching of maximum M, namely the buckling, is the nonlinearity we sought stabilizing the limit and beating cycle of cilia, as mentioned earlier. It sets the wave frequency and the wavelength without the need of signaling. As this is a buckling process, the basal wavelength will obey the equation:

$$\lambda_B \approx 2\pi\sqrt{\frac{\kappa_b(\nu)}{F_{axial}}}$$

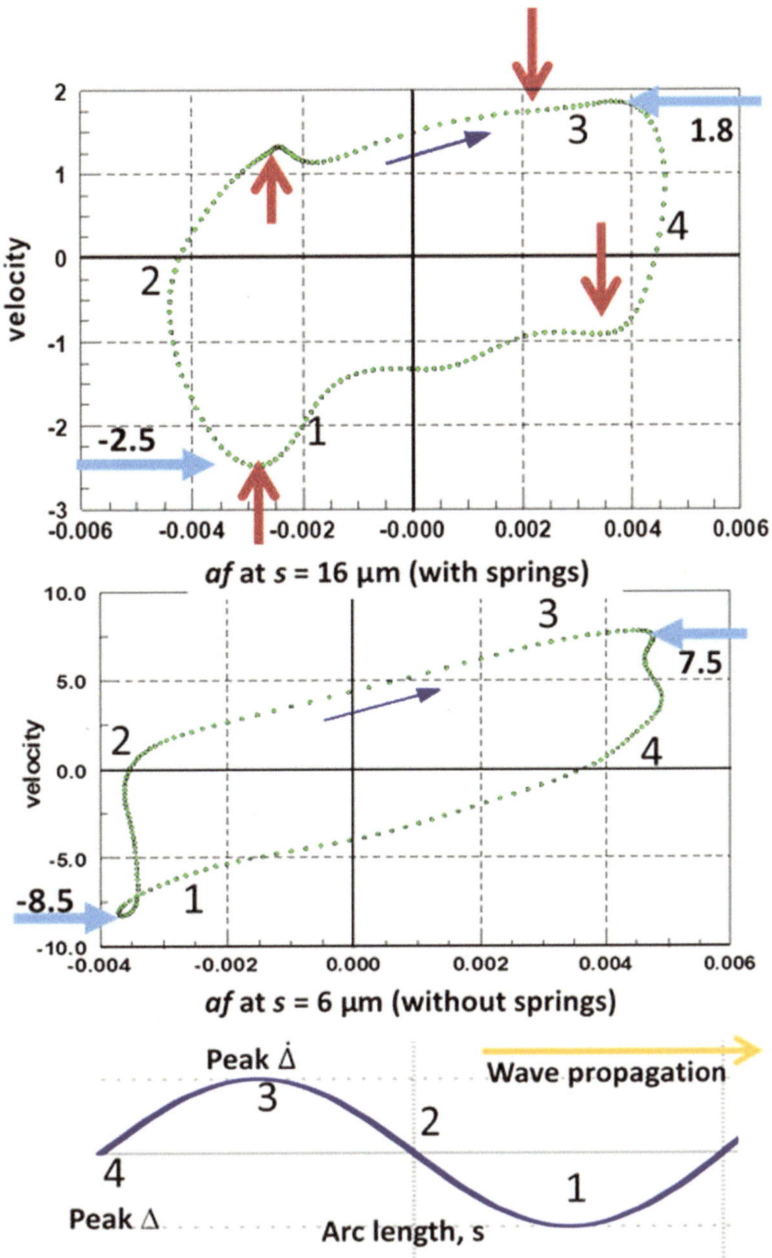

FIGURE 30.34 Shows the sliding velocity and shear force or torque per length *af* during a beating cycle. (*Top*) Sliding velocity as a function of time during a beating cycle at position s = 16 μm along the cilium from the base. (*Middle*) The *af* (normal) shear force or torque per length as a function of time during a beating cycle at the position $s = 6$ μm (end of wave growing region) along the cilium. It is approximately in phase with the sliding velocity. (*Bottom*) Key to the numbering above. *Middle from Fig. 8B, Foster et al. (2017).*

4.8.6 The dynamic stiffness − the relationship between force and sliding

In addition to recently gained greater knowledge of structure, significant advances in understanding the physiology of cilia are coming from the analysis of the forces and energies associated with cilia, using hydrodynamic calculations. From a minimum of eight images of cilia per beat cycle, with proper interpolation, one can obtain sufficiently accurate local velocities along their length to obtain local forces. Since the Reynolds number based on the length of the cilium and its stroke velocity is about 10^{-3} or smaller, the flow induced by the cilia beating can be described by the simple Stokes equations of motion. Furthermore, since the diameter of a cilium (~ 200 nm) is much smaller than its length (12−14 μm, as with *Chlamydomonas* or much more), slender body theory may be used to determine the force exerted by the fluid along the cilia. In free solution, a microorganism with its cilia would, if it were not experiencing linear or angular acceleration, have no force or torque (moment) being applied to it; hence, to evaluate its motion, one may form equations that effectively enforce both force balance and torque balance. One obtains the net (sum of both sides of the cilium) relationship between the motor forces and doublet sliding along the ciliary length with time during the beating cycle (for details see Sangani et al., 2016).

4.8.6.1 For example

The hydrodynamic force $F_j(s,t)$, together with the cilium position $r_j(s.t)$, s being the distance along the cilium measured from its base, and t the time, may be used to determine the motor forces along the cilium. Since inertial effects are negligible, the elastic forces arising from the bending of the doublets and the internal forces consisting of active forces due to motor activity and passive restraining forces due to nexins, dyneins, and other DMT connections balance the hydrodynamic force. The connections between adjacent doublets provide an elastic resistance to longitudinal sliding with a complex dependence on stretch (estimated to be about 2 nN/μm per μm of cilium (Minoura et al., 1999).

DMT are inextensible, but flexible with a single microtubule being anisotropic with a shear modulus of ~1.4 MPa and a Young's modulus of ~100 MPa (Kis et al., 2002). The sum of the internal forces at any cross-section of an axoneme must vanish even though some doublets are under tension and others under compression.

To approximate planar beats conveniently the 9 + 2 structure of axonemes can be equivalently replaced by two elastic filaments (groups of doublets) separated by distance a of 180 nm (Fig. 30.26) and bending flexural rigidity κ, both of which can be estimated from the known geometry and mechanical properties of the axoneme. Let f be the internal shear force acting along the filament on the outer side of the bend, and -f the shear force on the inner one.

Since the sliding velocity is the time derivative of $\Delta(s,t) = \Delta_0(t) + \boldsymbol{a}(\psi(s,t) - \psi(0,t))$, the tangent angles $\psi(s,t)$ (CCW positive), the DMT-sliding velocity is obtained. Now the shearing force per length $f(s,t)$ can be obtained from fits to the following equation:

$$af(s,t) = -\frac{\partial\kappa_r(s,\nu)C(s,t)}{\partial s} - \frac{\partial^2\kappa_i C(s,t)/\omega}{\partial t\partial s} - \vec{n}\cdot\int_s^L \vec{f}_h(s,t)ds$$

where C is the curvature $\partial\psi/\partial s$, $\kappa(s,\nu)$ is the dynamic κ of the restoring torque due to the axonemal viscoelasticity (the real part κ_r associated with elastic bending of the axoneme and the imaginary part κ_i associated with internal viscous sliding during bending). ω is the angular beating frequency, $\vec{n}$ is the unit normal to the cilium in the plane of the beating, and L is the cilium length. f_h is the hydrodynamic force exerted on the fluid by the cilium per unit length counterclockwise with respect to the cilium tangent. It is approximately proportional to the external medium viscosity η. $af(s,t)$ is the internal torque/length due to springs and motors. In addition, $f(s)$ is the internal shear force per length toward the right on the upper DMTs relative to the bottom and a the distance between the two sides. Since all the quantities except for $f(s)$ can be determined from the tangent-angle ψ data and hydrodynamic analysis, f(s) can be determined along the cilium. $af(s)$ is important as a threshold af is required to maintain sliding and wave propagation in a beating cilium (Izawa & Shingyoji, 2020) or otherwise the wave will stop. These authors report that there is a threshold wave propagation velocity v_p for the wave to travel. This seems reasonable since the bending torque $af(s)$, which one would anticipate to be the parameter necessary to drive the wave propagation, is proportional to the sliding velocity, which in turn is proportional to the wave propagation velocity v_p times the relatively constant curvature.)

Since the ciliary beating is periodic in time it is convenient to use the Fourier components of the sliding $\tilde{\Delta}$ and the tangential force per length $\tilde{f}$. Let $\Delta(s,t) = \tilde{\Delta}(s)\exp(i\omega t) + \tilde{\Delta}^*(s)\exp(-i\omega t)$ with $\omega = 2\pi\nu$, and use identical notation for the sliding velocity $\dot{\Delta}(s,t)$ and the shearing force per length $f(s,t)$. The relationship between forces due to motors aided by springs and the resulting DMT sliding displacement Δ and velocity in the central region of the cilium (Fig. 30.35) can then be expressed by the complex response function $\tilde{f} = \chi\tilde{\Delta}$, where the complex χ is the dynamic stiffness as a function of location in the cilium (Fig. 30.36).

By this criterion, there are three regions of the cilium (Fig. 30.36), a wave formation region near the base, a wave propagation central region along most of its length and a tail region, which presumably has less motors and DMTs. As discussed, the basal region is notable for growing an oscillatory wave. The central region is notable for providing most of the thrust of the cell with its traveling waves. Since χ_R is about zero in the central region the force/length or torque is linear and in phase with DMT sliding velocity $\dot{\Delta}$; as expected for a traveling wave The response function may then be more

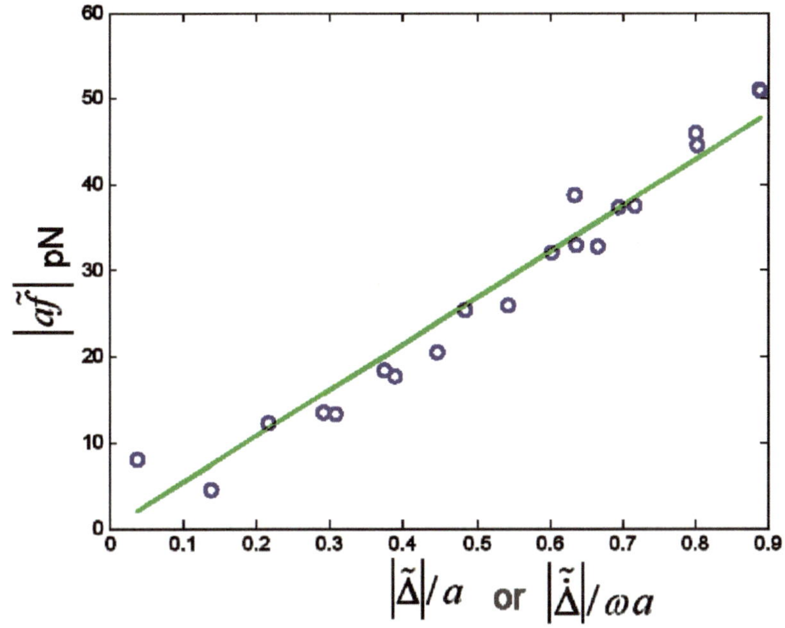

FIGURE 30.35 Absolute value of bending torque versus the sliding amplitude or sliding velocity amplitude for cilia beating in methylcellulose [MC] giving an apparent $\eta = 4.7$ mPa•s and 2 mM ATP. The effective cilium diameter **a** was assumed to be 180 nm. The force per length ***f*** works out to about 25 pN/μm or 0.7 pN/motor. The sliding velocity $\dot{\Delta} \sim a\nu$ is about 3 μm/s. The shear force ***af*** is about 50 pN.

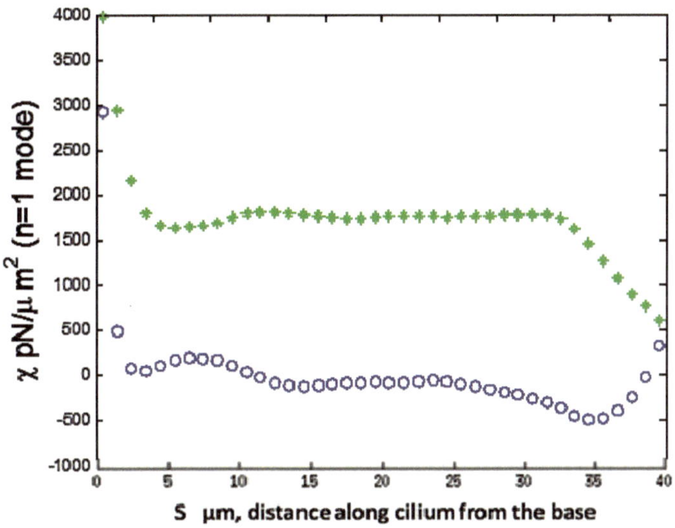

FIGURE 30.36 Dynamic stiffness. Analysis of $\psi(s,t)$ and $f(s,t)$ curves (Foster et al., 2017). Let $\Delta(s,t) = \tilde{\Delta}(s)\exp(i\omega t) + \tilde{\Delta}^*(s)\exp(-i\omega t)$ with $\omega = 2\pi\nu$, and use identical notation for the sliding velocity $\dot{\Delta}(s,t)$ and the shearing force per length $f(s,t)$. This assumes that the fundamental ν-mode dominates ($n = 1$), so that a single ν describes the vibrations. The complex-stiffness response function χ is defined by $\tilde{f} = \chi\tilde{\Delta}$. The figure shows for one cycle the real $(\chi_{R,}\text{O})$ and imaginery $(\chi_I, *)$ parts of response coefficient χ as a function of s, from the base (0) to the tip (L) ($\kappa_R = 23$ pN $\cdot\mu$m^2 and $\kappa_I = 156$ pN $\cdot\mu$m^2, but χ is not sensitive to the chosen values). Ciliary motions relative to the cell body can be described by $\psi(s)$. Taking advantage of the low Reynolds number (< 1) and slender-body hydrodynamic theory $\psi(s,t)$ data from Brokaw (1996) Fig. 1C (4.7 mPa•s, 2 mM ATP) was used to calculate the ciliary motion relative to the fluid and the hydrodynamic forces. The active and passive internal forces in a cilium are balanced by the restoring torque of axonemal elasticity and by hydrodynamic forces.

simply written as, $\tilde{f}^{(n)} = \chi_{nI}\tilde{\Delta}^{(n)}/n\omega$, where n is the Fourier mode. The phase is nearly independent of the external viscosity and [ATP].

The dynamic stiffness depends on the external viscosity η plus the internal viscosity or friction ο ("in") Fig. 30.37. In terms of viscosity units ο is equivalent to about 2.4 mPa s compared with ~1 mPa s for water, and hence potentially a significant dampener of motion that stabilizes the beating. Noteworthy is that the capacity for force generation for swimming in high viscosity can be increased two to three times, and power is observed to increase almost twice with viscosity. Hence, it is not surprising that the number of dyneins involved at "water" viscosity is less than those possible (Lin & Nicastro, 2018, Fig. 30.29).

The beating frequency depends on how long it takes the basal region initiating wave to reach maximum mechanical energy, which depends on the ratio of the flexural rigidity to the axial compression force. Consequently, if in a purely viscous environment, ν will decrease with increasing η (Brokaw, 1996), because while the strain energy is thought to not change much, the extra work (as shown by the increase dynamic stiffness with viscosity as shown in Fig. 30.37) of the motors required to increase the strain energy at high η takes more time as shown by the slower growth of the length of the wave.

That two independent motor bending systems are needed to produce planar waves was understood many years ago (Wais-Steider & Satir, 1979), the "switch point hypothesis" that different sets of motors are switched between sides twice a cycle, the question that remained was how. It is now apparent that motor activity arises from the self-organized dynamics. Further, the cilium's properties suggest the basal region is optimized to initiate stable oscillation and the central part is optimized for the successful propagation of a traveling wave.

4.8.7 Beating cycle refinements that optimize the effectiveness of ciliary beating

a) The tapering of mammalian sperm is an adaptation for high viscosity. Since flexural rigidity κ is proportional to the fourth power of the diameter for an isotropic solid, the tapering of mammalian sperm from wide at the base to thin at the tip makes it possible for it to be rigid near the base and much more flexible at the tip. Since $E_{wave} \propto \kappa C^2$, the curvature C increases along the cilium by having E_{wave}/κ increase. The stiffening of the basal region further suppresses excessive buckling due to the motion against high viscosity (Gadêlha & Gaffney, 2019). The tapered mammalian sperm cilium is then well adapted for high viscosity, enabling a sperm to swim through 100 times as viscous an environment as water.
b) The cell controls ciliary length for optimum function of ciliary beating, since a cilium must be a specific length (either too short or too long will result in a less efficient beating pattern). The question has been how a cell does it with so low a standard deviation, but while genes affecting length have been identified, the mechanism is still not completely clear.
c) For flow measurement by mechanosensitive cilia, length controls optimization for measurement of different flow rates.
d) The central pair rotates in some cilia in order that ATP synthesis (See Fig. 30.25) is brought near the most active motors or that ciliary controls are positioned at a constant phase with respect to the bending.
e) There is regulation of the cilium's prestress or static curvature to switch between different modes of beating. For example, while the ciliary beating mode (breaststroke) is not as efficient as the flagellar beating mode for traveling

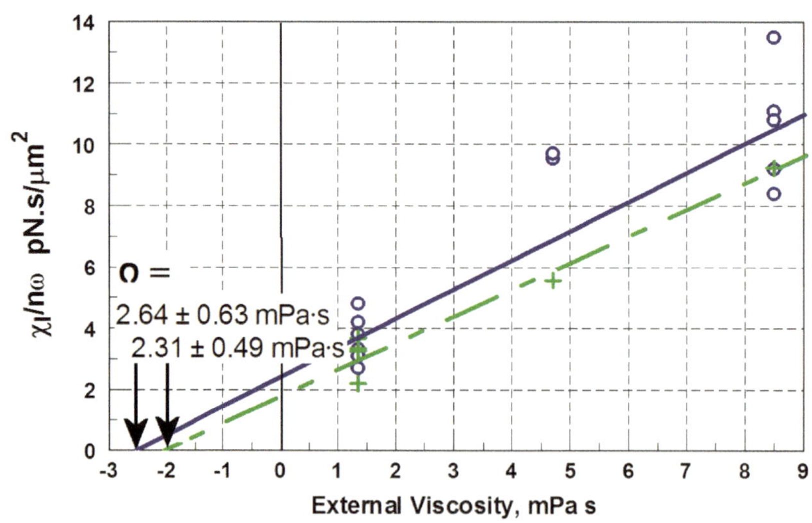

FIGURE 30.37 The stiffness χ divided by the mode and the angular beating frequency (the ordinate axis) is proportional to the internal plus external viscosity, $\chi_{nI}/n\omega \propto (o + \eta)$ as observed here. For the central (near-base) region [solid, O (dashed, +)] the proportionality is 0.94 ± 0.09 (0.84 ±0.09), ο is 2.64 ± 0.63 (2.31 ±0.49) mPa•s and with nine (six) available data sets the probability (P-value) that $\chi_{nI}/n\omega$ is not correlated with $o + \eta$ is 0.02% (0.07%).

from A to B, it maximizes the chemical sampling of the environment by bringing the medium from a wide distance for the cell body past the cell, hence, it is desired if that is advantageous for the organism. Further switching between ciliary beating and flagellar beating provides an easy route to reversing direction rapidly. In many metazoans, ciliary beating aids transport of mucus over epithelia, through the interaction of ciliary tips with a layer of high viscosity mucins, while the cilia themselves beat in an environment of mucins that have a lower viscosity.

4.8.8 Behavioral controls – the temporal modification of the physical parameters of beating

The behavioral functions of cilia have been well studied, but should be explored in a greater variety of organisms. It is likely that they have evolved independent pathways for their control. A few examples from work on the unicellular green alga *Chlamydomonas* are provided as suggestions for topics and techniques that have general potential application.

4.8.8.1 Caltractin

Chlamydomonas, a unicellular green alga, utilizes its cilia for locomotion and to direct its movement towards or away from a light source, a phenomenon known as phototaxis, as discussed in Section 4.8.10 below. As observed by Josef et al. (2005), the angle between the cilia of *Chlamydomonas* is linearly proportional to the intensity of light and is stable, not adapting. Calcium ions are well-suited for signaling roles within cells due to their ability to be sequestered or extruded from the cell, preventing their precipitation within the cellular environment. This means that responses can be transient, as a temporary increase in intracellular [Ca^{2+}] can be terminated. A distal striated fiber containing caltractin (also known as centrin) runs between the basal bodies of *Chlamydomonas*. It turns out that light raises the [Ca^{2+}] linearly and contraction of the caltractin reduces the angle between the cilia. The mechanism by which this benefits the cell is not yet known, although a sufficient increase in [Ca^{2+}] will result in a substantial change associated with the switch from ciliary to flagellar mode beating.

4.8.8.2 Ion channel controls

Light controls the opening of ion channels in *Chlamydomonas*, which , in turn, influences various aspects of the physical properties. The most obvious is the effect of the resulting increase in [Ca^{2+}] , in that it determines the prestress or static curvature along the entire length of the cilium. This, in turn, controls whether one has a ciliary mode or flagellar mode beat. This is a proportional effect. There are also separate transient responses to [Ca^{2+}].

4.8.8.3 ATP

ATP concentration determines beating frequency as it controls the rate of motor stepping. It can hence also transiently control the beating frequency. Like Ca^{2+}, it is an ideal transient signal because rapid feedback loops exist to return the level immediately to a specific [ATP] (Foster, 2009).

4.8.8.4 cAMP- braking to mate

In Chlamydomonas, gamete activation and fusion are triggered by a rapid increase in cyclic AMP (cAMP), specifically a 15-fold rise. This surge in cAMP is achieved through the activation of adenylyl cyclase(s) following ciliary adhesion and the interaction of sex-specific adhesion molecules. The increased cAMP levels then initiate a cascade of cellular events that prepare the gametes for fusion. Studies suggest that various second messengers, including cAMP, can affect the assembly state of flagella. For instance, alterations in cAMP levels can induce flagellar elongation or retraction. While the precise conditions that would lead to complete ciliary arrest (stoppage of beating) mediated by cAMP alone are still being investigated, significant disruptions in the ciliary beat pattern or frequency, potentially triggered by altered cAMP signaling, could effectively slow or halt the organism's movement. Of unknown generality, but interestingly, cAMP appears to control ciliary beating that slows or stops swimming, facilitating mating. There is a downside in that the same pathway controls the balance between ballistic and diffusive processes. By raising cAMP, the system is then biased to greater diffusive swimming which further restricts the exploration of the environment away from the optimum for finding places with more nutrients. Despite the growing understanding of ciliary biology, much remains unknown about the specific roles of many kinases and phosphatases within the cilium proteome.

4.8.9 Cell rotation

The cilia in *Chlamydomonas* have a specific orientation with respect to the cell body, resulting in the body traveling in a left-handed spiral (Fig. 30.38). Looking down from above (Fig. 30.39), near the base, the #1 doublet microtubule faces the

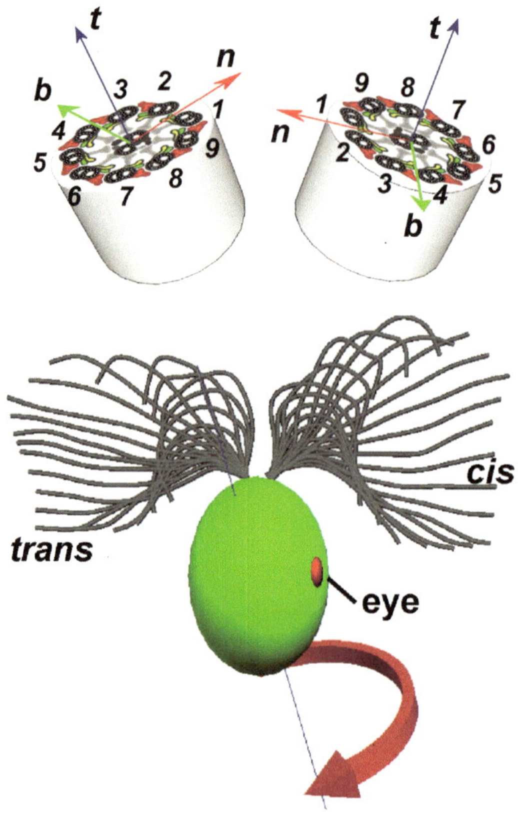

FIGURE 30.38 Shows attachment of the cilia to the cell body with C2 symmetry (from animation of Jyothish Vidyadharan and the axoneme cross-section image of Fig. 23–27). Note the left-hand rotation of the cell, the ciliary beating patterns and the eye (Foster, 2009).

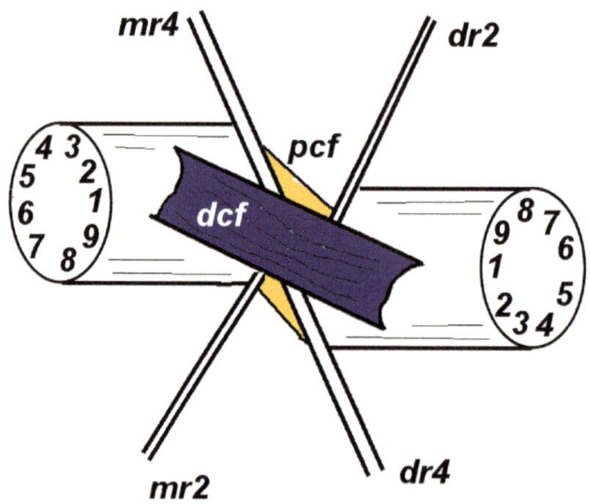

FIGURE 30.39 Schematic diagram of the basal body apparatus of *C. reinhardtii* as seen from the exterior of the cell. Numbers identify the relative positions of triplet microtubules that form the walls of the basal bodies. dr2, dr4 represent two- and four-membered daughter microtubule rootlets, mr2, mr4 represent two- and four-membered mother microtubule rootlets, respectively. Distal connected or striated fiber dcf; proximal connected fibers pcf.

other cilium, in two-fold rotation or C2 symmetry, with the cilia relatively offset from the centerline, which guarantees the consistent left-handed cell rotation. This left-handed cell rotation makes possible reproducible scanning of the environment with its single "eye" located beside daughter rootlet dr4, creating a consistent temporal signal that can be used for phototaxis. The position of the "eye" is strategically phase advanced with respect to the plane of the cilia to compensate for the inevitable delay of any responding system and thus optimizing the phototactic response. Böhm and Kreimer (2020) have extensively reviewed the "eye", a specialized light-sensing organelle also known as the eyespot .

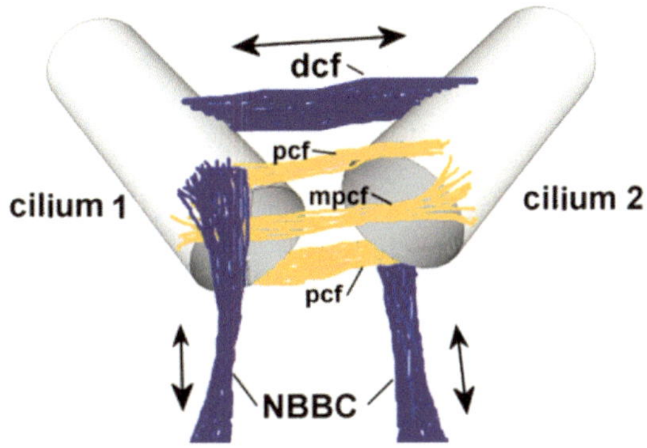

FIGURE 30.40 An angled view through the basal bodies and the base of the cilia of *C. reinhardtii.* dcf, distal connecting striated fiber joining the two basal bodies; pcf, proximal connecting fiber joining the two basal bodies; medial proximal connecting fibers joining the two basal bodies; NBBC, nucleus-basal body connectors (Geimer & Melkonian, 2004, 2005; Silflow & Lefebvre, 2001). The NBBC and pcf are contractile. Note the offset of the ends of the two cilia at the base.

The problem with travel in 3D is the noncommutative result of successive motions; a left turn followed by a right turn is not, in general, the same as a right turn followed by a left. Fishes and birds can take advantage of their sense of gravity to convert their 3D space to 2D providing a stable reference to relatively roll, yaw and pitch. If they wish, they can stay in a plane from which they may commutatively yaw or pitch.

For very small organisms, the force due to gravity is insignificant, so a small aquatic organism spirals along a helix. Consequently, the precise shallow figure eight beating of the cilia of *Chlamydomonas* is refined to be out of plane such that after each cycle, the cell maintains its same orientation with respect to the axis of its spiral motion. This design simplifies the motion in 3D space to small angle changes with respect to the helical axis of its swimming. This may be either as a single or as a double helix (see Fig. 30.41 respectively, the left and the middle drawing) resulting in relatively stable pathways. This rotation also makes it possible to modulate the light that provides the signal for phototaxis.

Further, cells may control their beating by changing the properties of their basal connection. In the case of *Chlamydomonas* the two cilia (Fig. 30.39) are connected to each other through proximal fibers (pf and mpcf) at the base plate and distal fibers (df) at about 250 nm from the base (Fig. 30.40). As a result, the dynamics of the two cilia are connected, and the cell may use this connection to synchronize or desynchronize the beating pattern. In addition to the distal fibers linking the dynamics of the two cilia, the nuclear-basal body connectors (NBBC) in the basal body region may play a significant role in cellular control of the beating. The NBBC and dcf contain centrin (caltractin), which exhibits calcium-sensitive contractile or elastic behavior. Centrin-based flagellar roots are contractile under conditions of elevated intraflagellar calcium in a variety of eukaryotes, including *Chlamydomonas* and *Tetraselmis*. In addition, a fine filament runs between the centers of the proximal ends of each basal body that would be very sensitive to the relative motion of either cilium.

The distal striated fiber dcf (Fig. 30.40) of nominal length 280 nm can contract to as little as 220 nm resulting in the decreased angle between the two cilia from about 65 degrees to 55 degrees. Contraction of the NBBC would also aid this movement and pull the base end inward so that the cilia can more easily exit the holes in the cell wall. This change will induce an additional force, in addition to the force due to sliding caused by dyneins. Now, the sum of the forces due to base sliding and connection must be balanced by the component of the total hydrodynamic force parallel to the cilium at the base.

4.8.10 Phototaxis

Overlaid on top of the self-organized beating and its optimizations are cellular controls to achieve a rich variety of functions. These include control of cell steering associated with phototaxis and chemotaxis, which is hypothesized to involve structures near or at a different radial spoke than the control of mating and of the balance between ballistic and diffusive motion. The temporary turning off of phototaxis when an obstacle is encountered involves calmodulin/[Ca^{2+}] and not cAMP. Even though the radial spokes are thought of as primarily solid-state effectors, local chemical perturbations of the environment can be independently perceived. This is true in the short term, even though the radial spoke sites such as the S1 site and S2 sites (Fig. 30.17) are only 48 nm apart, there is enough separation for these controls to be relatively independent in the 18–20 ms of a beat cycle.

By way of an example, consider some of the temporal characteristics of phototaxis control as measured in *Chlamydomonas* (Foster, 2009) (Fig. 30.41). The correlation of the ciliary responses with stimuli has been measured with a quad-

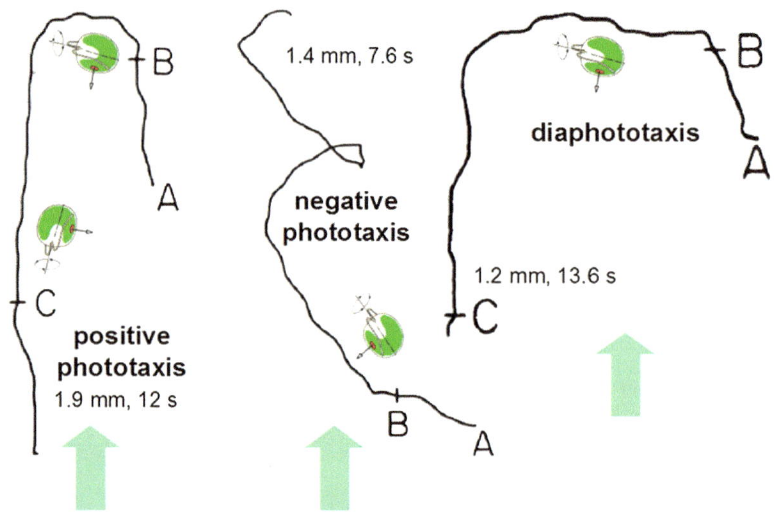

FIGURE 30.41 Three of the seven types of phototaxis: helical positive, superhelical negative and diaphototaxis. *Modified from Foster (2009).*

photodiode ciliary monitor that monitored the two cilia independently. Most classical analysis is done in the frequency domain because of the simplicity of its interpretation and its ability to distinguish simply between the beating cycle at 50–60 Hz and response control in the 2–6 Hz range. Bode plots showing the ciliary beating frequency CBF (Fig. 30.42A and B) and stroke-velocity SV responses (Fig. 30.42 D and E) represent this approach. The output response was correlated with a 196-s long input light stimulus. The slopes of the gain plot and the phase changes provide directly the nature of the

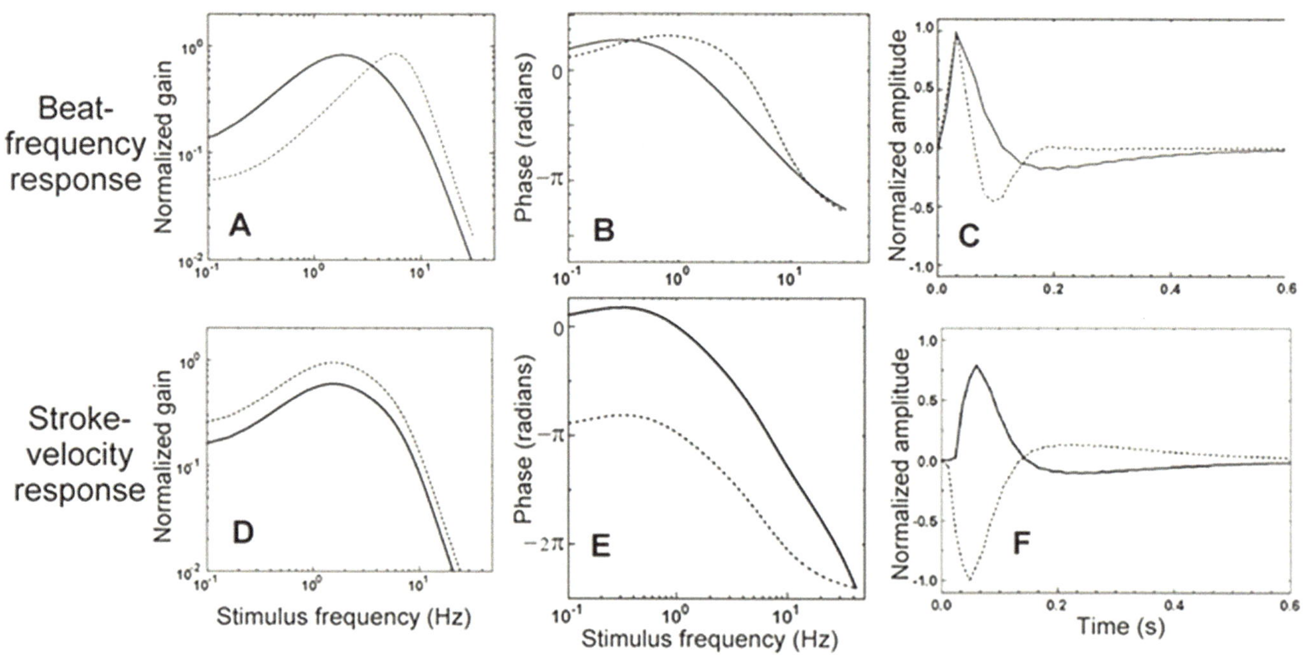

FIGURE 30.42 Typical ciliary beating frequency (CBF) of the negatively phototactic strain 806 in response to green light (543 nm) modulated by a pseudorandom Gaussian-white-noise (GWN) stimulus. The cis- and trans-CBF for each record were separately cross-correlated with log10 (stimulus intensities) to produce respective impulse responses (C). The *cis* and *trans* responses were then separately averaged (solid lines for *cis* and dotted lines for *trans*). The Fourier transform of the averaged impulse responses yielded the gains (A) and phases (B). Typical ciliary stroke-velocity response to green light modulated by a GWN stimulus. Stroke velocities for the *cis* and *trans* cilia for each record were independently cross-correlated with log 10 (stimulus intensities) to produce impulse responses (F) that were then separately averaged for *cis* cilium (solid line) and *trans* cilium (dotted line). The Fourier transform of the averaged impulse responses yielded the gains (D) and phases (E). It is anticipated that future work with the many mutants available it will possible to extend into molecular detail the processes involved both in ciliary beating and its behavioral control of the mechanical properties of cilia beating. *From Foster (2009).*

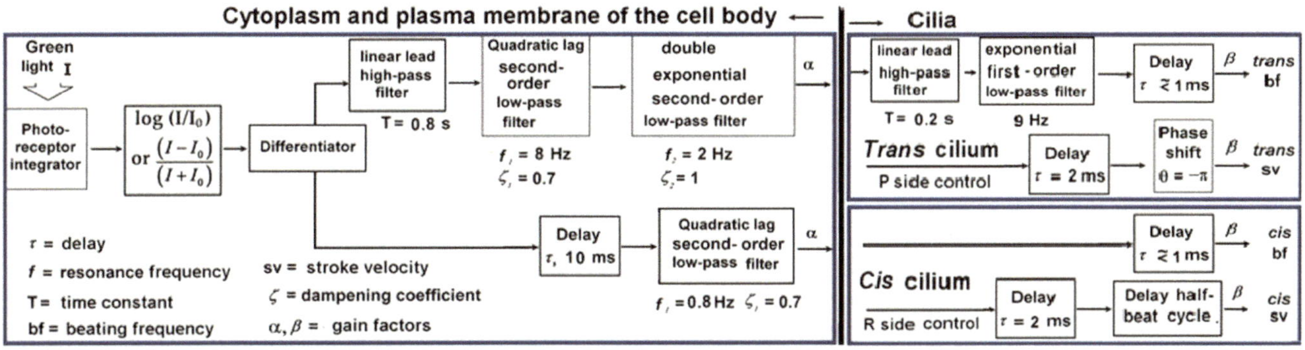

FIGURE 30.43 Foster's current model for the green light controlled negative phototaxis network based on the responses of ciliary beating and cell currents to light stimuli. With the limited bandwidth of signals to the cilia, the three subsystems (the big boxes in the figure) are treated as having their own inputs and outputs. The cell cytoplasm and plasma-membrane compartment decides by weighing inputs from the environment and other compartments (cilia, mitochondria, chloroplasts, etc.) and principally uses its ion channels, receptors, and messengers to receive and send signals. At present, chemical signals are known to control the angle between the cilia (Josef et al., 2006) and phototaxis direction. There is also the electric field signal carried by the plasma membrane that mediates the temporal course of the beating frequency and the stroke velocity subject to the state set by the chemically diffused signals. Increased pH and ATP levels by photosynthesis increase the beating frequency with a several hundred-millisecond delay, so they could be diffused signals from the cell body. Parameters that change the cell direction also change the stroke-velocity responses.

functional processes. For example, the slope at high frequency gives the number of processes involved in the cascade of the processing making functional modeling relatively easy. The temporal responses (Fig. 30.42C and F) are mathematically equivalent and with this representation it is easier to see how the cell would respond as it rotates, but functional modeling is less transparent. The *trans* and *cis* cilia are different. They have their own unique as well as common responses. For example, the CBF and SV responses differ significantly. On the other hand, the response to changes in availability of ATP (not shown here) appears to be similar. Note in particular that the *trans* cilium has a much higher frequency response for CBF than the *cis*. Further, that the phase of the *trans* cilium is about π different from the *cis* and that the delay is markedly, actually reproducibly, longer than for the *trans* cilium.

Such results provide insight into what the signal processing system does to optimize the cell's phototaxis. In spite of the noisiness of individual ciliary responses, high temporal resolution is achieved by averaging over 196 s. Note that the observed delays to response are short relative to a single beating cycle of ~18 ms. Furthermore, these measured responses show significant *cis-trans* differences, which contribute to how the cell steers. Note there is an initial delay of the *cis*-cilium stroke velocity in Fig. 30.42F and further the *cis* cilium has a slower CBF response than the *trans* in Fig. 30.42A. The peak times for *cis* and *trans* CBF are approximately 40 and 30 ms, respectively, whereas the peak times for the stroke velocity are 80 and 50 ms. Why the system has *cis* slower than *trans* for both responses is not known. Fig. 30.42F also shows that steering for phototaxis involves briefly altering the relative effectiveness of the stroke velocity of the *cis* relative to the *trans* cilium. The changes are as if the shape of the *trans* beat exclusively involves the outer dynein arms, and the *cis* beat shape is exclusively determined by the inner arms, but with the frequency and power contributed by the outer arms. A simple linear model, asfully described in Foster (2009), captures 80% of the observed variance. Fig. 30.43 shows the network of signal processing that occurs. There is now a beginning to look at what happens correspondingly in 3D (Cortese & Wan, 2021; Mojiri et al., 2021), beyond the well known figure-eight motion seen by Rüffer and Nultsch (1991), which, as discussed, stabilizes orientation in every cycle of beating along its helical path.

4.8.11 Attractors and repellers — a useful and relatively universal way to think of behavior

To make the best use of analysis options in the literature, the control parameters of behavior can be thought of as a space of variables. These variables modulate the physical parameters associated with ciliary beating, such as the frequency of beating or the amplitude of the waves. A stimulus causes the current point in this multidimensional space of these variables to jump to a new location. A response is falling from this nonequilibrium position to the steady-state equilibrium position of these variables, namely the attractor while avoiding regions of the space, which are the repellers (Blair et al., 2011).

4.9 Platform for sensory functions

Cilia play varied roles in sensory loops: as sensory receptor platforms for signals that are then carried to other parts of the cell (and on to other cells), as responders to sensory input from receptors elsewhere on the cell surface, and at times as part of a complete, closed organelle-autonomous sensory-response transduction pathway. The types of signals transduced and the role of cilia in signal responses depend in part on whether the cilia retain motility or have been reduced to simple nonmotile signaling platforms and in part on which signal transduction pathways are expressed in the cell.

Motile cilia typically respond at the end of a sensory loop by changing their beating frequency or waveform. For example, many single-cell organisms, as well as some simple metazoans, show phototactic changes in ciliary beating, allowing the organism to swim toward or away from a light source (Jékely, 2009). In many cases, the photosensitive receptor is localized elsewhere on the cell, not present in the cilia themselves, but in some unicellular organisms (in Euglenoids and in Stramenopiles such as the Chrysophyceae, Xanthophyceae, Phaeophyceae and Eustigmatophyceae) the photoreceptor seems to be part of the cilium. Since phototactic responses evolved independently in many different branches of eukaryotes, the photopigments, signal transduction pathways, and mechanisms of ciliary motility regulation vary tremendously. Light receptors typically trigger a change in membrane potential that then spreads throughout the cell. If the photosensitive cell is ciliated, these potential changes can open voltage-gated calcium channels in ciliary membranes, and changes in ciliary calcium concentration can then control ciliary orientation or beat parameters. In metazoa such as the larvae of some marine invertebrates, phototaxis starts with light receptors on distant neurons that then communicate, through synapses, with ciliated cells to regulate ciliary beating patterns and steer the organism.

The range of receptor types found on cilia is broad, and includes receptor tyrosine kinases (RTKs), G-protein coupled receptors (GPCRs), transient receptor potential (TRP) ion channels, and voltage-gated ion channels. When such sensory loops begin with receptors that localize directly to ciliary membranes, an entire signal transduction loop can exist within one organelle. Touch sensation through mechanosensitive ion channels on *Paramecium* cilia allow these single-celled ciliates to quickly change swimming direction when an object is encountered. In some biflagellate brown algal gametes, phototaxis is mediated by flavoproteins localized on the posterior flagellum, which steers the cell by bending in response to light. In mammals, signal transduction loops in ciliated cells of bronchial and nasal epithelia provide a rapid defense against airway bacterial infections. Compounds from gram-negative bacteria can bind directly to T2R-type GPCRs (bitter taste receptors) in airway ciliary membranes, triggering a signaling cascade that results both in increased ciliary beat frequency, to aid in clearing bacteria from the airway, and in direct release of antibacterial compounds from the cell. In many organisms, sperm depend on membrane receptors to detect chemotactic compounds released by eggs and respond by changes in their swimming patterns that bring them closer to the egg and their goal of fertilization.

Nonmotile 9 + 0 cilia can only function as the start of a sensory loop, but such cilia play important roles in animal development and physiology (Falk et al., 2015). Olfactory cilia project from dendritic extensions of nerve cells into the nasal cavity, and display GPCRs on the ciliary membrane that respond to odorants (Fig. 30.44A see Chapter 28). and cilia on retinal photoreceptors are highly modified to respond to light (Fig. 30.44C; see Chapter 27). The role of nonmotile cilia in transduction of Hedgehog signals, which play important roles in many developmental processes, has already been described above. Other nonmotile cilia with well-characterized receptor pathways include cilia that protrude into fluid-transporting tubes or recesses, such as kidney tubules, bile ducts, and an embryonic structure (the node) essential to generation of proper left-right asymmetry during development (Babu & Roy, 2013). Primary cilia in these 3 cell types function as lever arms. All display Pkd2 TRPV membrane proteins, which may act as mechanoreceptor channels to allow calcium influx into these cilia in response to fluid flow (Figs. 30.44B and 30.45).

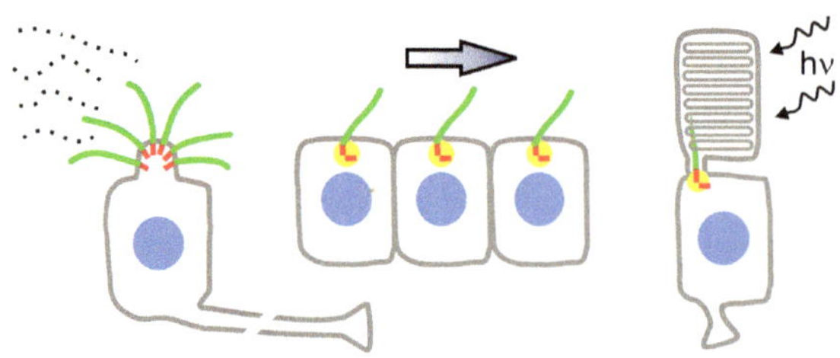

FIGURE 30.44 Three types of sensory cilia. (*Left*) A sensory neuron of the olfactory epithelium has multiple nonmotile sensory cilia on a short dendritic process. Olfactory receptors are displayed on the ciliary membrane. An axon (truncated in this diagram) sends signals to the central nervous system. (*Middle*) Mechanosensory cilia that display Pkd1/2 receptors sense fluid flow on kidney tubule epithelial cells. (*Right*) A retinal rod photoreceptor cell displays opsins on membranes of a modified primary cilium. Light absorbed by the receptors induces signaling through an axon at the opposite end of the cell.

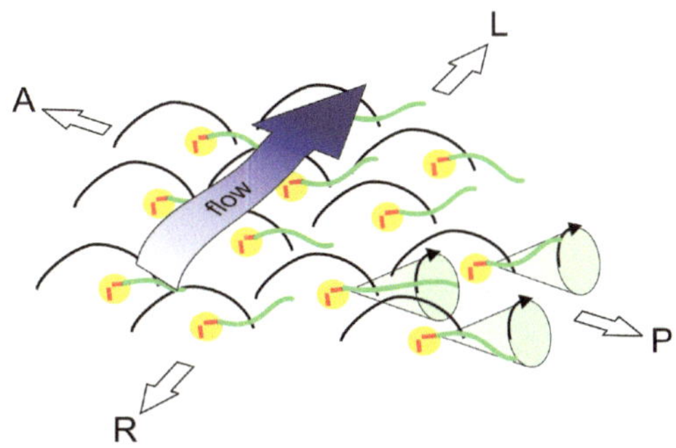

FIGURE 30.45 The embryonic node of a typical mammal has cells with motile cilia that all tilt toward the posterior end of the embryo. A circular beating pattern induces extracellular fluid flow toward the left side of the embryo, where flow interacts with nonmotile mechanosensory cilia. This mechanism transforms information from an existing anterior-posterior axis to generate a left-right axis. *A*, anterior; *L*, left; *P*, posterior; *R*, right.

4.10 Determinants and indicators of cell orientation and polarity

Many cellular functions are governed by asymmetric interactions of a cell with its environment. One simple form of asymmetry is polarity, as seen in the difference between leading and trailing edges of an amoeboid cell, or the apical and basolateral sides of an intestinal absorptive epithelial cell. For many cell types, polarity is regulated through changes in the relative position of the nucleus and centrosome and, in cells that assemble cilia, in the surface on which a cilium assembles. This link is evident in single celled flagellates, in which the orientation of the cell in its environment is determined directly by ciliary motility, but a similar relationship also occurs in many cells in multicellular organisms. Migrating fibroblasts typically have a primary (nonmotile) cilium and respond to cytokines, such as PDGFα, through receptors that are displayed on the ciliary membrane. When these cells migrate toward increasing cytokine concentrations (chemotaxis), such as during wound healing, the centrosome moves to the leading edge of the nucleus and the cilium points toward the leading edge of the cell. In other cases, cell polarization results from cell-cell interactions, such as junctions between neighboring cells in epithelial sheets, which may precede migration of the centrosome to the apical surface and assembly of a cilium. Correct positioning of a primary cilium (kinocilium) in differentiating hair cells of the inner ear provides an essential early cue for the later assembly of actin-based stereocilia bundles that respond directly to sound vibrations. In a similar fashion, when epithelial cells at the center of the mammalian embryonic node assemble single motile primary cilia, polarity cues that generate the anterior-posterior axis of the embryo also direct basal body migration to one side of each domed apical cell surface, so that all nodal cilia tilt toward the posterior (Fig. 30.45). These nodal cilia retain axonemal dyneins but lack radial spokes and central pair microtubules and, therefore, beat in a circular pattern. Their circular motion generates unidirectional leftward fluid flow that leads, through activation of a mechanosensitive Pkd2-dependent signaling pathway on nonmotile cilia, to correct left-right asymmetry in the embryo (Fig. 30.46).

5. Microtubules and motors in mitosis

5.1 Overview

Cell division is a perilous time when any error in segregation of genetic material can lead to disaster. This dilemma is more easily handled in prokaryotic cells, where most genetic information is typically contained in a single, circular chromosome and simple mechanisms are enough to ensure accurate assignment of one new chromosome copy to each new cell. Eukaryotes, with multiple linear chromosomes, face a tougher challenge that is largely met by the careful deployment of a unique microtubule-based mitotic apparatus. Although the following discussion focuses solely on events involving microtubules and their associated motors, readers should be aware that tremendous advances in understanding many fascinating details of the regulatory cascades that govern progress through mitosis (and the rest of the cell cycle) are known but have been omitted here for brevity. Following DNA replication (S phase) and the simultaneous duplication of centrioles (Fig. 30.16), cells pass through a brief interval (G2 phase) during which replication errors are corrected and proteins needed for mitosis are synthesized. Mitosis itself (M phase of the cell cycle) begins, by definition, with condensation of chromosomes and separation of the duplicated centriole pairs to form two microtubule organizing centers (Fig. 30.46).

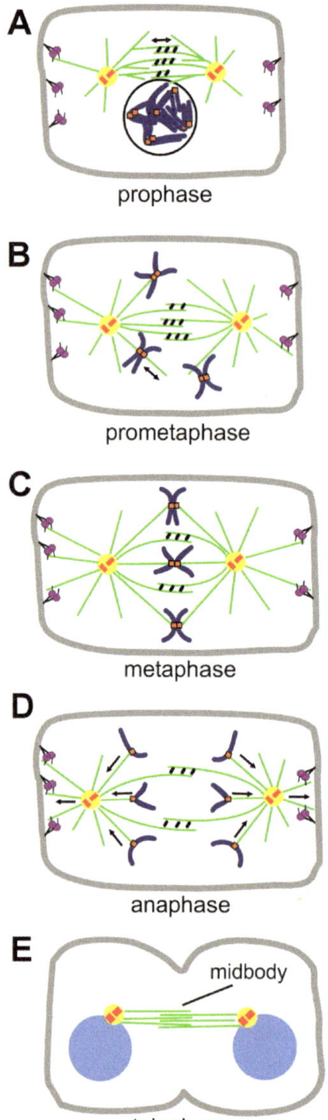

FIGURE 30.46 The role of microtubules and MT-associated motors during different stages of mitosis. (A) Prophase with condensing chromosomes, branched microtubules, and kinesins that power centrosome separation. Kinetochores (orange squares) assemble on centromeric chromatin. (B) Prometaphase. The nuclear membrane breaks down, kinetochores capture microtubules, and the spindle continues to elongate. Dyneins at the cell cortex engage astral microtubules. (C) Metaphase occurs when balanced forces on all kinetochores align chromosomes in the center of the spindle. (D) Separation of chromatids in anaphase. Arrows show movement of chromatids toward each spindle pole, kinesin-dependent sliding of overlapping central spindle microtubules, and dynein-mediated pulling on astral microtubules. (E) Chromosomes decondense and the nucleus reforms in telophase. Cytokinesis is positioned by the mitotic spindle, and the MT overlap zone becomes a midbody, which contributes to the regulation of the final stages of cytokinesis.

5.2 Prophase

During the first stage of mitosis, **prophase**, a change in microtubule stability leads most cytoplasmic microtubules to disassemble. At the same time, the centrosome separates into two organizing centers, and the number of nucleation sites on each centrosome also increases (Fig. 30.46A). To supplement microtubules that assemble from nucleating sites on centrosomes, additional microtubules can assemble from complexes that bind to the sides of existing microtubules, generating branched microtubule arrays. The fixed orientation of these branching nucleators assures that the microtubules originating in each half of the forming spindle all have the same orientation, with (+) ends pointed away from each centrosome. Newly assembled microtubules that overlap between the two centrosomes (spindle poles) attract homotetrameric kinesins such as kinesin-5 family members that actively push the poles apart to form a bipolar spindle, while microtubules that radiate away from the central spindle form astral (star-like) arrays that can interact with dynein motors anchored at the cell cortex (Heald & Khodjakov, 2015).

5.3 Prometaphase, metaphase and kinetochore function

Prometaphase (Fig. 30.46B) begins when the nuclear membrane breaks down into vesicles, allowing microtubules that assemble from spindle poles to reach the condensed chromosomes. On each duplicated chromosome copy, centromeric DNA sequences define an attachment site for proteins that form an interface between DNA and microtubules called the **kinetochore**. These complex structures include microtubule-binding proteins, microtubule motor proteins (kinesins and cytoplasmic dynein), and regulatory complexes that can respond to tension. However, kinetochore composition has been difficult to define because it changes rapidly during progression through mitosis (Stukenberg & Foltz, 2010). As an example, the inner centromere complex provides a protective interface between chromosomal DNA and the pulling forces exerted on kinetochores during mitosis. Many of its constituent proteins, including those of the Chromosome Passenger Complex (CPC), which assembles through phase separation into a liquid-like organelle (Trivedi et al., 2019). The CPC remains associated with centromeres during early phases of mitosis, but then relocates to midzone microtubules during anaphase. Kinetochore proteins capture microtubules that assemble near kinetochores during prometaphase, initially by lateral interactions. While only one of the two kinetochores of a duplicated chromosome is attached, the entire chromosome may move bidirectionally along this kinetochore microtubule through the action of cytoplasmic dynein and kinesin-7. Stable attachment only follows capture of microtubules from the opposite pole by the second kinetochore of each chromosome pair, leading to a balance of forces on the two kinetochores and to attachment of each kinetochore microtubule by its (+) end. In a typical mammalian cell, each kinetochore may gather dozens of microtubules, whereas in baker's yeast (*S. cerevisiae*), each small kinetochore remarkably attaches to only a single microtubule.

When a chromosome comes under tension from interaction of both of its kinetochores with microtubules from opposite spindle poles (bipolar attachment), it aligns midway between poles due to the balance of forces exerted on the kinetochore microtubules (**metaphase**, Fig. 30.46C). This force is in large part maintained by kinetochore complexes that simultaneously hold onto the microtubule (+) ends and attempt to disassemble those same (+) ends. Chromosome arms are further pushed toward the middle of the mitotic apparatus by kinesins that associate with chromatin through DNA binding domains (chromokinesins) that walk toward the (+) ends of central spindle MT and generate a polar ejection force measured at 0.5−1 pN (Almeida & Maiato, 2018). Alignment of all chromosomes in the spindle midzone and equal tension on all kinetochores triggers a transition to the next phase of mitosis, **anaphase**, when the two halves of each duplicated chromosome (chromatids) finally separate and move toward opposite spindle poles (Fig. 30.46D).

5.4 Anaphase movement and cytokinesis

Anaphase movements in most cell types take the form of two relatively independent activities. The first of these, termed Anaphase A, results in the separation of each chromatid of a pair into a region close to one of the spindle poles. Chromatid movement toward each pole is accomplished by shortening of kinetochore microtubules, primarily through kinetochore-regulated disassembly at the MT (+) ends, but also in some cells through disassembly of kinetochore MT at the poles and continued activity of (−) end motors including cytoplasmic dynein (Asbury, 2017). The forces generated are typically very small, but under imposed resistance, they may reach as high as 700 pN per chromosome (~12 pN/MT). Simultaneously, chromatids are further separated from each other by elongation of the spindle itself, termed Anaphase B. This increase in distance between spindle poles is achieved by the dual action of dynein motors associated with the cell cortex, which exert a pulling force by walking toward the (−) ends of astral microtubules, and of kinesin motors that bridge overlapping central spindle microtubules, which push the poles apart by walking toward MT (+) ends. The effectiveness of these central spindle kinesins depends on the continued overlap of microtubules from opposite poles, which must necessarily grow longer to accommodate the observed spindle elongation.

Telophase (Fig. 30.46E) is defined as the time after anaphase movement is complete when chromosomes decondense and a new nuclear membrane forms. Most cells simultaneously assemble actin-based machinery to divide the mother cell into two daughters (**cytokinesis**) on a plane through the former plane of chromosome alignment, the metaphase plate, through positional cues generated by the midzone microtubules. In metazoan cells this overlap zone performs another essential role later during cytokinesis. Protein complexes assembled at the midzone play a role in the final separation of one cell into two daughter cells, with the final scission site usually located just on one side of the midzone, which has by this time condensed into a structure called a midbody. The cell with the older mother centriole, which usually retains this midbody structure, often assembles a primary cilium more rapidly than the other daughter cell, and in many tissues, the two cells go on to follow unique differentiation fates.

References

Al-Bassam, J. (2017). Revisiting the tubulin cofactors and Arl2 in the regulation of soluble αβ-tubulin pools and their effect on microtubule dynamics. *Molecular Biology of the Cell, 28*, 359–363. https://doi.org/10.1091/mbc.E15-10-0694

Almeida, A. C., & Maiato, H. (2018). Chromokinesins. *Current Biology, 28*, R1131–R1135.

Alushin, G. M., Lander, G. C., Kellogg, E. H., Zhang, R., Baker, D., & Nogales, E. (2014). High-Resolution microtubule structures reveal the structural transitions in αβ-tubulin upon GTP hydrolysis. *Cell, 157*, 1117–1129. https://doi.org/10.1016/j.cell.2014.03.053

Ananthakrishnan, R., & Ehrlicher, A. (2007). The forces behind cell movement. *International Journal of Biological Sciences, 3*(5), 303–317.

Asbury, C. L. (2017). Anaphase A: Disassembling microtubules move chromosomes toward spindle Poles. *Biology, 6*, 15.

Böhm, M., & Kreimer, G. (2020). Orient in the world with a single eye: The green algal eyespot and phototaxis. In F. M. Cánovas, U. Lüttge, M. C. Risueño, & H. Pretzsch (Eds.), *Progress in botany* (Vol 82). Cham: Springer. https://doi.org/10.1007/124_2020_38

Babu, D., & Roy, S. (2013). Left–right asymmetry: Cilia stir up new surprises in the node. *Open Biol, 3*.

Belyy, V., & Yildiz, A. (2017). Studying the mechanochemistry of processive cytoskeletal motors with an optical trap. *Methods in Enzymology, 582*, 31–54. https://doi.org/10.1016/bs.mie.2016.09.037

Blair, H. A., Saranak, J., & Foster, K. W. (2011). Reverse engineering cellular decisions for hybrid reconfigurable network modeling. In I. X. Nanoengineering, H. Szu, & L. Dai (Eds.), *Independent component analyses, wavelets, neural networks, biosystems* (Vol 8058). Proc. of SPIE, 80581L-1 - 80581L-14.

Bloodgood, R. A. (2009). From central to rudimentary to primary: The history of an underappreciated organelle whose time has come. The primary cilium. *Methods in Cell Biology, 94*, 3–52.

Bottier, M., Thomas, K. A., Dutcher, S. K., & Bayly, P. V. (2019). How does cilium length affect beating? *Biophysical Journal, 116*, 1292–1304.

Brokaw, C. J. (1967). Adenosine triphosphate usage by flagella. *Science, 156*, 76–78.

Brokaw, C. J., Luck, D. J. L., & Huang, B. (1982). Analysis of the movement of *Chlamydomonas* flagella: The function of the radial-spoke system is revealed by comparison of wild-type and mutant flagella. *The Journal of Cell Biology, 92*, 722–732.

Brokaw, C. J. (1996). Microtubule sliding, bend initiation, and bend propagation parameters of Ciona sperm flagella altered by viscous load. *Cell Motility and the Cytoskeleton, 33*, 6–21.

Brouhard, G., & Sept, D. (2012). Microtubules: Sizing up the GTP cap. *Current Biology: CB, 22*(18), R802–R803. https://doi.org/10.1016/j.cub.2012.07.050

Bui, K. H., Yagi, T., Yamamoto, R., Kamiya, R., & Ishikawa, T. (2012). Polarity and asymmetry in the arrangement of dynein and related structures in the *Chlamydomonas* axoneme. *The Journal of Cell Biology, 198*, 913–925.

Chaaban, S., & Brouhard, G. J. A. (2017). Microtubule bestiary: structural diversity in tubulin polymers. *Molecular Biology of the Cell, 28*, 2924–2931. https://doi.org/10.1091/mbc.E16-05-0271

Chełminiak, P., Dixon, J. M., & Tuszyński, J. A. (2010). Torsional elastic deformations of microtubules within continuous sheet model. *European Physical Journal E: Soft Matter, 31*, 215–227. https://doi.org/10.1140/epje/i2010-10562-x

Chrétien, D., Fuller, S. D., & Karsenti, E. (1995). Structure of growing microtubule ends: Two-dimensional sheets close into tubes at variable rates. *The Journal of Cell Biology, 129*(5), 1311–1328.

Cianfrocco, M. A., DeSantis, M. E., Leschziner, A. E., & Reck-Peterson, S. L. (2015). Mechanism and regulation of cytoplasmic dynein. *Annual Review of Cell and Developmental Biology, 31*, 83–108. https://doi.org/10.1146/annurev-cellbio-100814-125438

Cibert, C., & Ludu, A. (2010). Is the curvature of the flagellum involved in the apparent cooperativity of the dynein arms along the "9+2" axoneme? *Journal of Theoretical Biology, 265*, 95–103. https://doi.org/10.1016/j.jtbi.2010.04.004

Cortese, D., & Wan, K. Y. (2021). Control of helical navigation by three-dimensional flagellar beating. *Physical Review Letters, 126*, Article 088003.

Das, M., Alex, J., Levine, J. A., & MacKintosh, F. C. (2008). Buckling and force propagation along intracellular microtubules. *EPL, 84*, Article 18003. https://doi.org/10.1209/0295-5075/84/1800

Driver, J. W., Geyer, E. A., Bailey, M. E., Rice, L. M., & Asbury, C. L. (2017). Direct measurement of conformational strain energy in protofilaments curling outward from disassembling microtubule tips. *Elife, 6*, Article e28433. https://doi.org/10.7554/eLife.28433

Eshel, D., & Brokaw, C. J. (1987). New evidence for a "biased baseline" mechanism for calcium-regulated asymmetry of flagellar bending. *Cell Motility and the Cytoskeleton, 7*, 160–168.

Fagarasanu, M., Fagarasanu, A., & Rachubinski, R. A. (2006). Sharing the wealth: Peroxisome inheritance in budding yeast. *Biochimica et Biophysica Acta, 1763*, 1669–1677. https://doi.org/10.1016/j.bbamcr.2006.08.01

Falk, N., Lösl, M., Schröder, N., & Gießl, A. (2015). Specialized cilia in mammalian sensory systems. *Cells, 4*, 500–519. https://doi.org/10.3390/cells4030500

Fletcher, D. A., & Mullins, R. D. (2010). Cell mechanics and the cytoskeleton. *Nature, 463*(7280), 485–492. https://doi.org/10.1038/nature08908

Ford, W. C. L. (2006). Glycolysis and sperm motility does a spoonful of sugar help the flagellum go around? *Human Reproduction Update, 12*, 269–274.

Foster, K. W. (2009). Analysis of the ciliary/flagellar beating of *Chlamydomonas*. In S. M. King, & G. J. Pazour (Eds.), *Cilia: Structure and motility. Methods in cell biology* (Vol 91, pp. 173–239).

Foster, K. W., Vidyadharan, J., & Sangani, A. (2017). Evidence for a self-organized complaint mechanism for the spontaneous steady beating of cilia. *Cytoskeleton, 74*, 260–280.

Gönczy, P. (2012). Towards a molecular architecture of centriole assembly. *Nature Reviews Molecular Cell Biology, 13*, 425–435. https://doi.org/10.1038/nrm3373

Gadêlha, H., & Gaffney, E. A. (2019). Flagellar ultrastructure suppresses buckling instabilities and enables mammalian sperm navigation in high-viscosity media. *Journal of The Royal Society Interface, 16*, Article 20180668.

Geimer, S., & Melkonian, M. (2004). The ultrastructure of the *Chlamydomonas reinhardtii* basal apparatus: Identification of an early marker of radial asymmetry inherent in the basal body. *Journal of Cell Science, 117*, 2663–2674. https://doi.org/10.1242/jcs.01120

Geimer, S., & Melkonian, M. (2005). Centrin scaffold in *Chlamydomonas reinhardtii* revealed by immunoelectron microscopy. *Eukaryotic Cell, 4*, 1253–1263. https://doi.org/10.1128/EC.4.7.1253–1263.2005

Geyer, V. F., Sartori, P., Friedrich, B. M., Jülicher, F., & Howard, J. (2016). Independent control of the static and dynamic components of the Chlamydomonas flagellar beat. *Current Biology, 26*, 1098–1103. https://doi.org/10.1016/j.cub.2016.02.053

Gibbons, I. R. (1961). Structural asymmetry in cilia and flagella. *Nature, 190*, 1128–1129.

Gibbons, I. R., Garbarino, J. E., Tan, C. E., Reck-Peterson, S. L., Vale, R. D., & Carter, A. P. (2005). The affinity of the dynein microtubule-binding domain is modulated by the conformation of its coiled-coil stalk. *Journal of Biological Chemistry, 280*, 23960–23965.

Guichard, P., Hamel, V., Le Guennec, M., Banterle, N., Iacovache, I., Nemčíková, V., Flückiger, I., Goldie, K. N., Stahlberg, H., Lévy, D., Zuber, B., & Gönczy, P. (2017). Cell-free reconstitution reveals centriole cartwheel assembly mechanisms. *Nature Communications, 8*, Article 14813. https://doi.org/10.1038/ncomms14813

Hamel, A., Fisch, C., Combettes, L., Dupuis-Williams, P., & Baroud, C. N. (2011). Transitions between three swimming gaits in *Paramecium* escape. *Proceedings of the National Academy of Sciences of the United States of America, 108*, 7290–7295.

Hawkins, T. L., Sept, D., Mogessie, B., Straube, A., & Ross, J. L. (2013). Mechanical properties of doubly stabilized microtubule filaments. *Biophysical Journal, 104*(7), 1517–1528. https://doi.org/10.1016/j.bpj.2013.02.026

Heald, R., & Khodjakov, A. (2015). Thirty years of search and capture: The complex simplicity of mitotic spindle assembly. *The Journal of Cell Biology, 211*, 1103–1111. https://doi.org/10.1083/jcb.201510015

Herrmann, H., & Strelkov, S. V. (2011). History and phylogeny of intermediate filaments: Now in insects. *BMC Biology, 9*, 16. https://doi.org/10.1186/1741-7007-9-16

Howell, L. L. (2001). *Compliant mechanisms* (p. 459). New York: John Wiley.

Hunt, A. J., Gittes, F., & Howard, J. (1994). The force exerted by a single kinesin molecule against a viscous load. *Biophysical Journal, 67*, 766–781.

Ichikawa, M., Khalifa, A. A. Z., Kubo, S., Dai, D., Basuc, K., Maghrebia, M. A. F., Vargasa, J., & Bui, K. H. (2019). Tubulin lattice in cilia is in a stressed form regulated by microtubule inner proteins. *Proceedings of the National Academy of Sciences of the United States of America, 116*, 19930–19938. https://doi.org/10.1073/pnas.1911119116

Ichikawa, M., Liu, D., Kastritis, P. J., Basu, K., Hsu, T. C., Yang, S., & Bui, K. H. (2017). Subnanometre-resolution structure of the doublet microtubule reveals new classes of microtubule associated proteins. *Nature Communications, 8*, Article 15035. https://doi.org/10.1038/ncomms15035

Ishibashi, K., Sakakibara, H., & Oiwa, K. (2020). Force-generating mechanism of axonemal dynein in solo and ensemble. *International Journal of Molecular Sciences, 21*(2843), 17.

Ishijima, J., & Hiramoto, Y. (1994). Flexural rigidity of echinoderm sperm flagella. *Struct. Func., 19*, 349–362.

Isojima, H., Io, R., Niitani, Y., Noji, H., & Tomishige, M. (2016). Direct observation of intermediate states during the stepping motion of kinesin-1. *Nature Chemical Biology, 12*, 290–297. https://doi.org/10.1038/nchembio.2028

Izawa, Y., & Shingyoji, C. (2020). Mechanical induction of oscillatory movement in demembranated, immotile flagella of sea urchin sperm at very low ATP concentrations. *Journal of Experimental Biology, 223*, 1–15. https://doi.org/10.1242/jeb.225797

Jékely, G., & Arendt, D. (2006). Evolution of intraflagellar transport from coated vesicles and autogenous origin of the eukaryotic cilium. *BioEssays, 28*, 191–198. https://doi.org/10.1002/bies.20369

Jékely, G. (2009). Evolution of phototaxis. *Philosophical Transactions of the Royal Society B, 364*, 2795–2808. https://doi.org/10.1098/rstb.2009.0072

Jones, R. M. (2006). *Buckling of bars, plates, and shells* (p. 824). Blacksburg, Virginia: Bull Ridge Publishing.

Josef, K., Saranak, J., & Foster, K. W. (2005). An electro-optic monitor of the behavior of *Chlamydomonas reinhardtii* cilia. *Cell Motility and the Cytoskeleton, 61*, 83–96.

Josef, K., Saranak, J., & Foster, K. W. (2006). Linear systems analysis of the ciliary steering behavior associated with negative-phototaxis in *Chlamydomonas reinhardtii. Cell Motility and the Cytoskeleton, 63*, 758–777.

Katsuki, M., Drummond, D. R., & Cross, R. A. (2014). Ectopic A-lattice seams destabilize microtubules. *Nature Communications, 5*, 3094. https://doi.org/10.1038/ncomms4094

Kis, A., Kasas, S., Babić, B., Kulik, J., Benoît, W., Briggs, G. A. D., Schönenberger, C., Catsicas, S., & Forró, L. (2002). Nanomechanics of microtubules. *Physical Review Letters, 89*, 248101. https://doi.org/10.1103/PhysRevLett.89.248101

Lechtreck, K. F. (2015). IFT-cargo interactions and protein transport in cilia. *Trends in Biochemical Sciences, 40*, 765–778. https://doi.org/10.1016/j.tibs.2015.09.003

Li, H., DeRosier, D. J., Nicholson, W. V., Nogales, E., & Downing, K. H. (2002). Microtubule structure at 8 Å resolution. *Structure, 10*, 1317–1328.

Lin, J., Heuser, T., Song, K., Fu, X., & Nicastro, D. (2012). One of the nine doublet microtubules of eukaryotic flagella exhibits unique and partially conserved structures. *PLoS One, 7*, Article e46494. https://doi.org/10.1371/journal.pone

Lin, J., & Nicastro, D. (2018). Asymmetric distribution and spatial switching of dynein activity generates ciliary motility. *Science, 360*, Article eaar1968. https://doi.org/10.1126/science.aar1968

Lin, J., Okada, K., Raytchev, M., Smith, M. C., & Nicastro, D. (2014). Structural mechanism of the dynein power stroke. *Nature Cell Biology, 16*, 479–485. https://doi.org/10.1038/ncb2939

Lindemann, C. B., & Mitchell, D. R. (2007). Evidence for axonemal distortion during the flagellar beat of *Chlamydomonas*. *Cell Motility and the Cytoskeleton, 64*, 580–589.

Ma, M., Stoyanova, M., Rademacher, G., Dutcher, S. K., Brown, A., & Zhang, R. (2019). Structure of the decorated ciliary doublet microtubule. *Cell, 179*, 909–922. https://doi.org/10.1016/j.cell.2019.09.030

Machin, K. E. (1958). Wave propagation along flagella. *Journal of Experimental Biology, 35*, 796–806.

Mandelkow, E. M., Mandelkow, E., & Milligan, R. A. (1991). Microtubule dynamics and microtubule caps: A time-resolved cryo-electron microscopy study. *The Journal of Cell Biology, 114*(5), 977–991. https://doi.org/10.1083/jcb.114.5.977

Markus, S. M., Marzo, M. G., & McKenney, R. J. (2020). New insights into the mechanism of dynein motor regulation by lissencephaly-1. *Elife, 9*, Article e59737. https://doi.org/10.7554/eLife.59737

McIntosh, J. R., Morphew, M. K., Grissom, P. M., Gilbert, S. P., & Hoenger, A. (2009). Lattice structure of cytoplasmic microtubules in a cultured mammalian cell. *Journal of Molecular Biology, 394*(2), 177–182. https://doi.org/10.1016/j.jmb.2009.09.033

McKenney, R. J., Weil, S. J., Scherer, J., & Vallee, R. B. (2011). Mutually exclusive cytoplasmic dynein regulation by nude-lis1 and dynactin. *Journal of Biological Chemistry, 286*, 39615–39622. https://doi.org/10.1074/jbc.M111.289017

Memet, E., Hilitski, F., Morris, M. A., Schwenger, W. J., Dogic, Z., & Mahadevan, L. (2018). Microtubules soften due to cross-sectional flattening. *Elife, 7*, Article e34695. https://doi.org/10.7554/eLife.34695

Menéndez, M., Rivas, G., Diaz, J. F., & Andreu, J. M. (1998). Control of the structural stability of the tubulin dimer by one high affinity bound magnesium ion at nucleotide n-site. *Journal of Biological Chemistry, 273*(1), 167–176. https://doi.org/10.1074/jbc.273.1.167

Millecamps, S., & Julien, J.-P. (2013). Axonal transport deficits and neurodegenerative diseases. *Nature Reviews Neuroscience, 14*, 161–176. https://doi.org/10.1038/nrn3380

Milo, R., & Phillips, R. (2015). *Cell biology by the numbers* (p. 358). New York: Garland Science.

Minoura, I., Yagi, T., & Kamiya, R. (1999). Direct measurement of inter-doublet elasticity in flagellar axonemes. *Cell Structure and Function, 24*, 27–33.

Mitchell, D. R. (2000). *Chlamydomonas* flagella. *Journal of Phycology, 36*, 261–273.

Mitchell, D. R. (2003). Orientation of the central pair complex during flagellar bend formation in *Chlamydomonas*. *Cell Motility and the Cytoskeleton, 56*, 120–129.

Mitchison, H. M., & Valente, E. M. (2017). Motile and non-motile cilia in human pathology: From function to phenotypes. *The Journal of Pathology, 241*, 294–309. https://doi.org/10.1002/path.4843

Mojiri, S., Isbaner, S., Mühle, S., Jang, H., Bae, A. J., Gregor, I., Gholami, A., & Enderlein, J. (2021). Rapid multi-plane phase-contrast microscopy reveals torsional dynamics in flagellar motion. *Biomedical Optics Express, 12*, 3169–3180. https://doi.org/10.1364/BOE.419099

Mori, T., Vale, R. D., & Tomishige, M. (2007). How kinesin waits between steps. *Nature, 450*, 750–754. https://doi.org/10.1038/nature06346

Mourão, A., Christensen, S. T., & Lorentzen, E. (2016). The intraflagellar transport machinery in ciliary signaling. *Current Opinion in Structural Biology, 41*, 98–108. https://doi.org/10.1016/j.sbi.2016.06.009

Nelson, P. (2008). *Biological physics: Energy, information, life*. New York: W.H. Freeman.

Nogales, E. (1999). A structural view of microtubule dynamics. *Cellular and Molecular Life Sciences CMLS, 56*(1–2), 133–142. https://doi.org/10.1007/s000180050012

Nogales, E., Wolf, S. G., & Downing, K. H. (1998). Structure of the αβ tubulin dimer by electron crystallography. *Nature, 391*(6663), 199. https://doi.org/10.1038/34465

Qiu, T., Lee, T.-C., Mark, A. G., Morozov, K. I., Münster, R., Mierka, O., Turek, S., Alexander, M., Leshansky, A. M., & Fischer, P. (2014). Swimming by reciprocal motion at low Reynolds Number. *Nature Communications, 5*, 5119. https://doi.org/10.1038/ncomms6119

Rüffer, U., & Nultsch, W. (1991). Flagellar photoresponses of *Chlamydomonas* cells held on micropipettes: II. Change in flagellar beat pattern. *Cell Motility and the Cytoskeleton, 18*, 269–278.

Ribbeck, K., & Mitchison, T. J. (2006). Meiotic spindle: Sculpted by severing. *Current Biology: CB, 16*(21), R923–R925. https://doi.org/10.1016/j.cub.2006.09.048

Roberts, A. J., Malkova, B., Walker, M. L., Sakakibara, H., Numata, N., Kon, T., Ohkura, R., Edwards, T. A., Knight, P. J., Sutoh, K., et al. (2012). ATP-Driven remodeling of the linker domain in the dynein motor. *Cell, 20*, 1670–1680.

Roll-Mecak, A. (2020). The tubulin code in microtubule dynamics and information coding. *Developmental Cell, 54*, 7–20.

Saito, A., Suetomo, Y., Arikawa, M., Omura, G., Khan, S. M. M. K., Kakuta, S., Etsuko Suzaki, E., Kataoka, K., & Suzaki, T. (2003). Gliding movement in *Peranema trichophorum* Is powered by flagellar surface motility. *Cell Motility and the Cytoskeleton, 55*, 244–253.

Sale, W. S. (1986). The axonemal axis and Ca^{2+}-induced asymmetry of active microtubule sliding in sea urchin sperm tails. *The Journal of Cell Biology, 102*, 2042–2052.

Sangani, A. S., Vidyadharan, J., & Foster, K. W. (2016). Analysis of the forces acting on beating cilia. *Journal of Physics D Applied Physics, 49*, Article 255401.

Saranak, J., & Foster, K. W. (2005). The photoreceptor for curling behavior in *Peranema trichophorum* and the evolution of eukaryotic rhodopsin. *Eukaryotic Cell, 4*, 1605–1612. https://doi.org/10.1128/EC.4.10.1605–1612.2005

Shih, S. M., Engel, B. D., Kocabas, F., Bilyard, T., Gennerich, A., Marshall, W. F., & Yildiz, A. (2013). Intraflagellar transport drives flagellar surface motility. *Elife, 2*, Article e00744. https://doi.org/10.7554/eLife.00744

Silflow, C. D., & Lefebvre, P. A. (2001). Assembly and motility of eukaryotic cilia and flagella. Lessons from *Chlamydomonas reinhardtii*. *Plant Physiology, 127*, 1500–1507.

Stukenberg, P. T., & Foltz, D. R. (2010). Kinetochores: Orchestrating the chromosomal minuet. *Current Biology, 20*, R522–R525. https://doi.org/10.1016/j.cub.2010.05.017

Tilney, L. G., Bryan, J., Bush, D. J., Fujiwara, K., Mooseker, M. S., Murphy, D. B., & Snyder, D. H. (1973). Microtubules: Evidence for 13 protofilaments. *The Journal of Cell Biology, 59*, 267–275. https://doi.org/10.1083/jcb.59.2.267

Timoshenko, S. P. (1936). *Theory of elastic stability* (1st ed.). McGraw-Hill Book Company.

Tombes, R. M., Brokaw, C. J., & Shapiro, B. M. (1987). Creatine kinase-dependent energy transport in sea urchin spermatozoa. Flagellar wave attenuation and theoretical analysis of high energy phosphate diffusion. *Biophysical Journal, 52*, 75–86. https://doi.org/10.1016/S0006-3495(87)83190-9

Trivedi, P., Palomba, F., Niedzialkowska, E., Digman, M. A., Gratton, E., & Stukenberg, P. T. (2019). The inner centromere is a biomolecular condensate scaffolded by the chromosomal passenger complex. *Nature Cell Biology, 21*, 1127–1137. https://doi.org/10.1038/s41556-019-0376-4

Wais-Steider, J., & Satir, P. (1979). Effect of vanadate on gill cilia: Switching mechanism in ciliary beat. *J. Supramolecular Structure, 11*, 339–347.

Wierzbicki, T. (2013). *Structural mechanics. MIT course number 2.080J/1.573J.*

Xu, Z., Schaedel, L., Portran, D., Aguilar, A., Gaillard, J., Marinkovich, M. P., Théry, M., & Nachury, M. V. (2017). Microtubules acquire resistance from mechanical breakage through intralumenal acetylation. *Science, 356*, 328–332. https://doi.org/10.1126/science.aai8764

Yanagisawa, H. A., Mathis, G., Oda, T., Hirono, M., Richey, E. A., Ishikawa, H., Marshall, W. F., Kikkawa, M., & Qin, H. (2014). FAP20 is an inner junction protein of doublet microtubules essential for both the planar asymmetrical waveform and stability of flagella in *Chlamydomonas. Molecular Biology of the Cell, 25*, 1472–1483. https://doi.org/10.1091/mbc.E13-08-0464

Young, K. G., & Kothary, R. (2011). Intermediate filament interactions in neurons. In R. A. Nixon, & A. Yuan (Eds.), *Advances in neurobiology 3Cytoskeleton of the nervous system* (pp. 379–410). New York: Springer. https://doi.org/10.1007/978-1-4419-6787-9_17

Yubuki, N., & Leander, B. S. (2013). Evolution of microtubule organizing centers across the tree of eukaryotes. *The Plant Journal, 75*, 230–244. https://doi.org/10.1111/tpj.12145

Zhang, H., & Mitchell, D. R. (2004). Cpc1, a *Chlamydomonas* central pair protein with an adenylate kinase domain. *Journal of Cell Science, 117*, 4179–4188.

Zhang, R., & Nogales, E. (2015). A new protocol to accurately determine microtubule lattice seam location. *Journal of Structural Biology, 192*, 245–254. https://doi.org/10.1016/j.jsb.2015.09.015

Zhang, S., & Mitchell, B. J. (2015). Basal bodies in *Xenopus*. *Cilia, 5*, 2. https://doi.org/10.1186/s13630-016-0024-6

Further reading

Alberts, B., Johnson, A., Lewis, J., Raff, M., Roberts, K., & Walter, P. (2002). *Molecular biology of the cell* (4th ed., p. 1463). New York: Garland Science.

Amos, L. A., & Klug, A. (1974). Arrangement of subunits in flagellar microtubules. *Journal of Cell Science, 14*, 523–549.

Bayly, P. V., & Dutcher, S. K. (2016). Steady dynein forces induce flutter instability and propagating waves in mathematical models of flagella. *Journal of The Royal Society Interface, 13*, Article 20160523. https://doi.org/10.1098/rsif.2016.0523

Bouissou, A., Vérollet, C., de Forges, H., Haren, L., Bellaïche, Y., Perez, F., Merdes, A., & Raynaud-Messina, B. (2014). γ-tubulin ring complexes and EB1 play antagonistic roles in microtubule dynamics and spindle positioning. *The EMBO Journal, 33*, 114–128. https://doi.org/10.1002/embj.201385967

Boyd, J. S., Gray, M. M., Thompson, M. D., Horst, C. J., & Dieckmann, C. (2011). The daughter four-membered microtubule rootlet determines anterior-posterior positioning of the eyespot in *Chlamydomonas reinhardtii. Cytoskeleton (Hoboken), 68*, 459–469.

Brouhard, G. J. (2015). Dynamic instability 30 years later: Complexities in microtubule growth and catastrophe. *Molecular Biology of the Cell, 26*, 1207–1210. https://doi.org/10.1091/mbc.E13-10-0594

Dutcher, S. K. (2019). Asymmetries in the cilia of *Chlamydomonas. Philosophical Transactions of the Royal Society B, 375*, Article 20190153. https://doi.org/10.1098/rstb.2019.0153

Geyer, V. F., 1 Pablo Sartori, P., Friedrich, B. M., Jülicher, F., & Howard, J. (2020). Independent control of the static and dynamic components of the *Chlamydomonas* flagellar beat. *Current Biology, 26*, 1098–1103.

Maheshwari, A., Obbineni, J. M., Bui, K. H., Shibata, K., Toyoshima, Y. Y., & Ishikawa, T. (2015). α- and β-tubulin lattice of the axonemal microtubule doublet and binding proteins revealed by single particle cryo-electron microscopy and tomography. *Structure, 23*, 1584–1595. https://doi.org/10.1016/j.str.2015.06.017

Mitchell, B. F., Pedersen, L. B., Feely, M., Rosenbaum, J. L., & Mitchell, D. R. (2005). ATP production in *Chlamydomonas reinhardtii* flagella by glycolytic enzymes. *Molecular Biology of the Cell, 16*, 4509–4518.

Wan, K. Y. (2018). Coordination of eukaryotic cilia and flagella. *Essays in Biochemistry, 62*(8), 829–883.

Wan, K. Y., & Goldstein, R. E. (2016). Coordinated beating of algal flagella is mediated by basal coupling. *PNAS, 113*, E2784–E2793. https://doi.org/10.1073/pnas.1518527113

Wan, K. Y., & Jekely, G. (2021). Origins of eukaryotic excitability. *Philosophical Transactions of the Royal Society B, 376*, Article 20190758.

Chapter 31

Motion by rotatory flagella and archaella, twitching, gliding, springs, and catapults

Kenneth W. Foster
Physics Department, Syracuse University, Syracuse, NY, United States

1. Introduction

A major theme here is the ubiquitous use of compliant mechanisms that transforms motion, force, or energy in a way that stores energy in the strain energy of flexible members for future use. For commercial applications in industry, compliant mechanisms have two advantages (a) cost efficiency (a dramatic reduction in number of parts, reduced assembly time, simplified manufacturing) and (b) increased performance (increased precision and reliability and reduced wear, weight, and maintenance). These are all attributes highly valued by nature as illustrated here and in Chapter 30. "Compliant mechanisms offer… better solutions to many mechanical-design problems," particularly "to transfer or transform motion, force or energy." (Howell, 2001, p. 459).

Additional themes are the evolutionary conservatism of nature making multiple uses of components that can be repurposed in different machines such as for propulsion and molecular transport (Celia et al., 2019). A further theme is that the physics of the situation results in evolutionary distant lineages evolving analogous devices that achieve locomotion by sending traveling waves outward by bending (Chapter 30 Fig. 31.24) or by rotating helical appendages as discussed here. Nanomotors in general are reviewed by Guo et al. (2016).

2. Rotatory motor-driven slender appendages

Although the bacterial and archaeal flagellum are both very slender rotating cylinders that enable a cell to move, since they are completely unrelated structures of separate evolutionary origin, we have adopted their names respectively as *flagellum* and *archaellum* as suggested by Jarrell and Albers (2012). To avoid confusion with the completely unrelated eukaryotic flagella, both eukaryotic flagella and cilia are now trending to be exclusively called *cilia* and are discussed in Chapter 31. A recent comprehensive review can be found in Beeby et al. (2020).

The flagella and archaella of respectively bacteria and archaea rotate for a simple physical reason. As pointed out by Machin (1958), you cannot drive a travelling bending wave down a slender motile structure immersed in a dissipative fluid without providing energy along its length. If you are so small that your motility device is too narrow, as flagella (20 nm) and archaella (10−14 nm) are to house motors along their length, to produce thrust like cilia, then their only option is to rotate as a rigid helical propeller.

A further limitation is the rate of signal transport provided by diffusion in these organisms. For example, with bacteria if you make an *E. coli* too long it can no longer function properly for chemotaxis because the peritrichously distributed flagella do not switch at similar enough times, due to differing diffuse delays in the signal to switch direction. Nevertheless, bacteria and archaea have been enormously successful in colonizing earth with their greater relative simplicity and ability to survive under extreme conditions. A typical flagellum uses the information from about 50 genes in sharp contrast to the estimated 650 genes required for eukaryotic cilia.

2.1 Viscosity and the low Reynolds number regime

Critical to understanding motors, cell motility, and application of molecular devices is appreciating the viscous low Reynolds number regime they inhabit (Berg, 1983). All liquids have viscosity since there are attractive forces between

Cell Physiology Sourcebook. https://doi.org/10.1016/B978-0-12-811114-7.00032-3

different portions of a liquid. If a volume element is in motion relative to another, the local forces, which oppose the motion, are called frictional forces. The frictional force F will be proportional to the relative velocity between elements dv and the contact area between the elements dA and inversely proportional to the distance between the elements dx with the constant between the elements and the force being the dynamic viscosity η ($F = \eta(dv/dx)dA$).

For a particle moving under an applied force F in free space where $F = ma$, where m is the mass, and acceleration $a = dv/dt$, it would continuously increase in velocity if it were not for the opposing viscous force. At low velocities, the viscous force will be proportional to the velocity; hence fv, where f is the frictional coefficient, which for a spherical particle of radius r is $f = 6\pi\eta r$.

The equation of motion will then be $m(dv/dt) = F - fv$.

If the acceleration is zero, then the cell thrust force to overcome the viscosity is $F = fv$ or $\eta(dv/dx)dA$ or $\eta(v/L)L^2$.

The ratio of inertial forces (m (dv/dt)) or $\rho_m L^3(v/t)$ relative to viscous forces is the Reynolds number $R = \rho_m L^3(v/t)/[\eta(v/L)L^2] = vL\rho_m/\eta$, where *v* is the velocity of the fluid with respect to the object, L is the characteristic dimension (length) of the object, typically the diameter for a sphere, ρ_m is the density of the fluid medium, and η is the dynamic viscosity. Even when Re is up to a few hundred with inertial forces dominating, the viscous forces play a significant role.

If a swimming cell stops applying force F or thrust, it comes exponentially to a stop, as illustrated using our equation of motion given above with $f = 6\pi\eta r$,

$m(dv/dt) = -6\pi\eta r v$, hence $dv/v = -6\pi\eta r/m\, dt$ and hence $v(t) = v(0)\exp(-t/\tau)$, where $\tau = m/6\pi\eta r$. Since $m = 4/3\pi r^3 \rho_c$ then $\tau = 2/9\, r^2\rho_c/\eta$.

For example, consider a 2-μm long saugage or bacillus shaped bacterium of 0.5-μm diameter that swims about 30 μm/s (v_0) with a Reynolds number R of 3×10^{-5}. If it were to stop rotation of its flagella, it would deaccelerate exponentially with a time constant τ of 0.2 μs and it would coast a distance ($\tau \cdot v_0$) of $d_{coast} \sim 6$ pm or picometers or 10^{-12} m (distance between two carbon atoms in a molecule is about 100 pm). With these parameters, an *E. coli* is capable of a thrust *F* of about $6\pi\eta r v$ where r is 0.5 μm, v = 30 μm/s; hence, F equals 0.47 pN and has a power P of about 14 aW (attowatt 10^{-18} W) about 1/400th of a cilium of similar length.

How motion from a rotatory helical-shaped appendage or propeller occurs without inertia in this low Reynolds number regime is shown in Fig. 31.1.

2.2 Flagella (synonym bacterial flagella)

Flagellated bacteria, such as *Escherichia coli*, swim by rotating thin helical filaments, each driven at their base by a reversible 50-nm-diameter nanopowertrain powered by ion flux (Fig. 31.2). These powertrains develop maximum torque at stall but can rotate at several hundred Hz. A sensory system controls its rotation direction making it possible for cells to accumulate in favored regions. A great deal is known about the nanopowertains's structure, genetics, assembly, and function, and how it changes rotation direction but perhaps not precisely as desired the generation of torque. Much remains to be discovered on how it senses and regulates its response to load and details of all its dynamic processes. Interestingly,

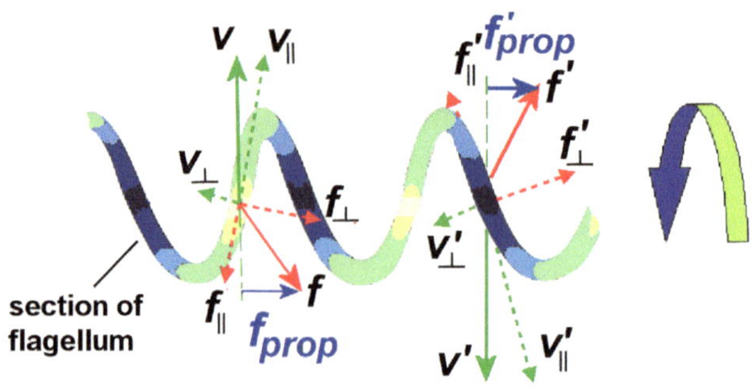

FIGURE 31.1 Shows normal and tangential **forces on a helical flagellum** or archaellum of a swimming organism (in the opposite direction of the **forces** of the flagellum or archaellum **on the fluid**, which are proportional to the indicated local velocities of the flagellum). Note the lighter colors are toward the reader and the darker colors are away. Since the normal drag is stronger than the tangential, the flagellum will push an organism to the right as indicated by the f_{prop} vectors. This particular case is for an organism like *E. coli* during a run in which the flagellum is left-handed and rotating CCW as viewed from outside the cell (right-handed rotation as directed outward from the cell). Rotating CW forces the helix to become right handed, but doing so causes the bundle of flagella to fly apart. Note the similar generation of thrust to a planar beating cilium and its outward traveling wave. The archaella of *Halobacterium salinarium* are on the other hand right handed with forward pushing motion when rotated CW as expected, but will back up if rotated CCW as the archaellum maintains its handedness.

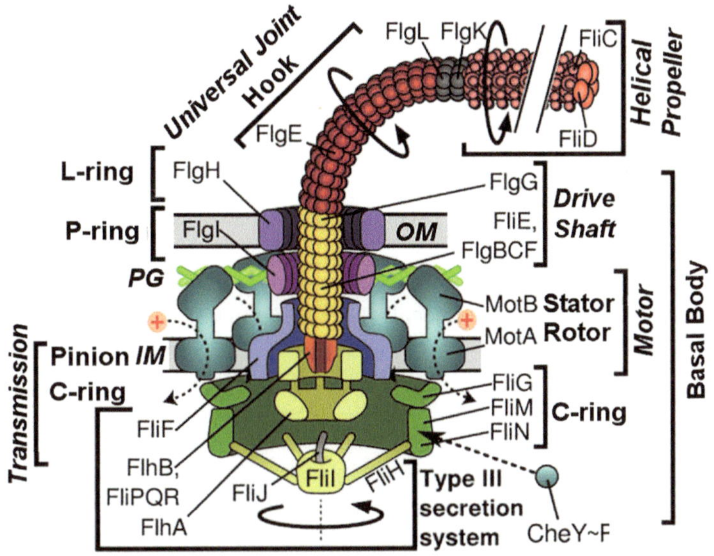

FIGURE 31.2 Schematic of the flagellar powertrain including the motor or engine, transmission and associated other parts from model enteric bacteria such as *Salmonella* or *E. coli*. IM: inner membrane, PG: peptidoglycan, OM: outer membrane. The rigid PG anchors a PMF motor with its stationary stator motB. Revolving around it is a rotor motA, which is also a 5-tooth pinion gear that meshes with a 34- tooth gear (C-ring) forming the transmission that increases the torque. The axle of the larger gear is a drive shaft that turns a universal joint (hook) that couples power to the helical propeller, the flagellar filament that protrudes from the cell. *Modified from Fig. 31.3D, Beeby et al. (2020).*

the word flagellum etymologically implies a whip like motion, but in reality, they are functionally rigid helical propellers and not whips, only cilia (from Latin eyelash) have a whip like motion.

The next sections on the structure, energy supplies, and mechanism of motion follow the reviews by Sowa and Berry (2008), Beeby et al. (2020), Biquet-Bisquert et al. (2021), Deme et al. (2020), Chang et al. (2020, 2021), Santiveri et al., 2020, and Hu et al. (2021). However, it is also further informed by the reviews of Berg (2003), Zhao et al. (2014), Rossez et al. (2015) and Minamino and Imada (2015), and Khan & Scholey (2018). A brief discussion of the associated behavior follows that.

2.2.1 Cell attachment configurations

Bacteria with a single flagellum are called monotrichous, those with multiple flagella grouped together are called lophotrichous, those with a single flagellum at opposite ends are called amphitrichous (with only one flagellum operating at a time enabling cell reversal by switching which flagellum is active), and those with flagella projecting in all directions are called peritrichous. Some bacteria, such as most spirochetes, have two or more specialized flagella (endoflagella), which rotate within the periplasmic space making possible motion through extremely viscous environments like within connective tissue. Their rotation causes the entire bacterium to move forward in a corkscrew-like manner. An *E. coli* bacterium has an average of four flagella with each attached by a flexible hook to the basal body of a motor.

2.2.2 Structure of the flagellar nanopowertrain

The flagellum is a ~20-nm-thin ~10-μm-long rigid helical filament, which is part of a molecular powertrain that rotates it like a propeller (schematic Fig. 31.2). About 13 different proteins and a further ~25 proteins are required for its expression and assembly to make it. The flagellar powertrain is analogous to a microboat engine with its motor, transmission, drive shaft, and driven propeller. The components include a transmembrane ion-motive-force IMF driven motor that converts chemical energy into mechanical work, a stator to anchor it, a transmission, a drive shaft (rod), a universal joint (hook), and a helical propeller. The stator(s) anchors it to the peptidoglycan layer, which is the rigid cell wall that gives the bacterium its shape. The transmission is somewhat analogous to a spur gear with 5 teeth meshed with a somewhat analogous spur gear with 34 or 46 teeth that increase the torque 34/5 or 46/5 times and can reverse the direction. The universal joint couples the drive shaft to the propeller. The rigid propeller's rotation produces thrust as shown in Fig. 31.1.

To measure flagellar rotation, the current preferred method is to attach submicron polystyrene beads to truncated flagellar filaments of immobilized cells and to record their rotation with either back-focal-plane interferometry or high-speed fluorescence microscopy. The viscous drag coefficient of a 0.5-μm diameter bead on a truncated filament is similar to that of a flagellar filament. The smaller the bead, the lower the load and the higher the speeds that can be recorded.

2.2.3 Drive shaft, universal joint, and propeller

The hollow hook and filament are single-protein thin tubular polymers consisting of 11 helical protofilaments. Pentameric cap complexes regulate monomer incorporation. Each polymer has a long and short form, which creates the helical structures of the hook and filament. During steady rotation, the filament is a rigid propeller. Motor switching in *E. coli* causes a torsionally induced transformation between the alternative filament forms changing the handedness of the filament helix. The transformed filament breaks up the bundle of filaments currently propelling the cell and thus leads to cell reorientation. To see this, flagellar filaments have been fluorescent-labeled (Fig. 31.3) and swimming cells have been recorded with stroboscopic laser illumination and high-speed video microscopy. Response to external forces has been studied with optical tweezers. The hook is much more flexible than the filament and works as a universal joint to allow several filaments from motors all over the cell to rotate together in a bundle in peritrichiously flagellated species. The hook is so flexible that early attempts to see any stepping were frustrated. The hook twists when an external torque is applied so that even when flagellar hook is minimally driven at ~150 pN-nm, it twists the hook 0.35 radians. In a subsequent step, the motion of an attached bead will be damped with a relaxation time equal to the viscous drag coefficient divided by the spring constant of the hook. To detect steps, the time between steps must be greater than this relaxation time. This hook property smooths out any roughness in the turning of the gears.

2.2.4 The motor and transmission

Transmissions use gear trains to provide speed, torque conversion, and reversibility from a rotator to another device. In this case, the motor is incorporated into the first or pinion gear, the smaller of a meshed gear system, in the transmission. The best studied flagellar transmissions Figs. 31.2 and 31.4 are those of the peritrichously flagellated bacterium *Escherichia*

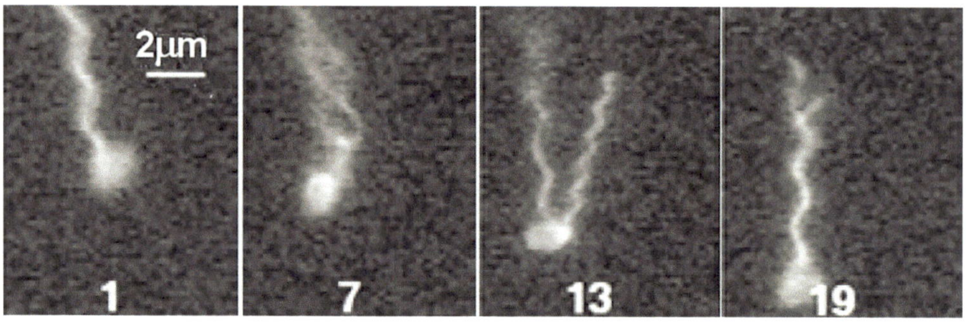

FIGURE 31.3 Stroboscopic images of a fluorescently labeled swimming *E. coli* cell. The numbers are frame numbers, at 1/60 s intervals. Between frames 7 and 13, a single filament leaves the bundle and undergoes polymorphic transitions that cause a change in swimming direction known as a "tumble" (Sowa & Berry, 2008).

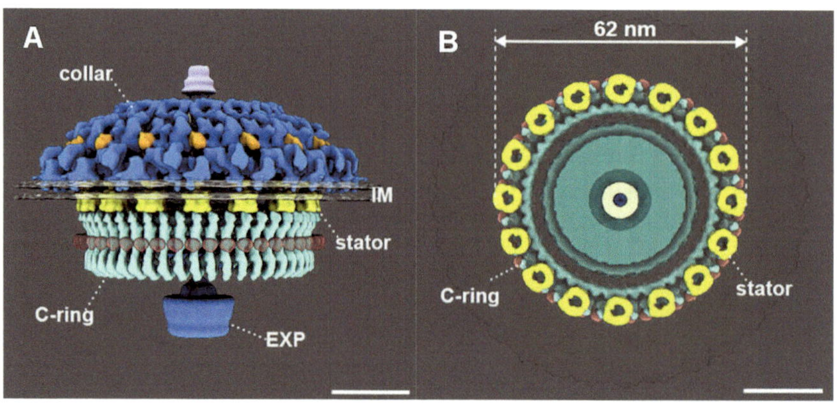

FIGURE 31.4 (A) The spirochete *Borrelia burgdorferi*'s flagella machine's motor structure in 3D with the side view of its unique collar structure, the PMF pinion in yellow, the C-ring in cyan, IM: inner membrane, (B) Top view of the stator ring on the top of the C-ring. *(A) Modified from Fig. 31.1I, Chang et al. (2020). (B) Modified from Fig. 31.1H, Chang et al. (2020).*

coli and the related *Salmonella enterica* Sv *typhimurium* (*S. typhimurium*), and others such as *Vibrio alginolyticus*, the spirochete *Borrelia burgdorferi, Campylobacter jejuni*, but many others have been studied (Minamino & Imada, 2015). As seen in Fig. 31.4 the "stator ring" containing the driving pinion-gear motor sits above the C-ring, a larger gear that rotates the drive shaft that connects to the hook and propeller. The ratio of diameters to teeth numbers is consistent with analogous real gears. The ratio of the gear sizes does the torque conversion, increasing the torque to the propeller in proportion to the ratio of their number of "teeth." This transmission can switch the output drive shaft between rotating counterclockwise (CCW, viewed from the outside or right handed with thumb pointing outward) with formation of a bundle of flagella that produces smooth swimming, but if rotated clockwise (CW, left handed) causes tumbles or rapid rotational diffusion. That bacterial flagella rotate was first observed by the rotation of the cell body tethered to a surface by its flagellar hook. These tethered cells rotated up to ~20 Hz. They spin faster with lower loads with the maximum-recorded speed so far being 1700 Hz in the Na^+-driven motor of the marine *Vibro alginolyticus* at 37°C.

The bacterial flagellar powertrain is an integration of components of different evolutionary origin (Beeby et al., 2020). Combining information from homologous powertrains from different species has aided research. Single particle image reconstruction from cryo-electron microscopy (cryo-EM, for a concise explanation, see Zhao et al., 2014) has determined the structure of the flagellar powertrain, and their components have been identified by genetic and biochemical studies. With typical biological efficiency, the torque producing driving motor's rotor motA or pomA (~8 nm) is also a pinion of the transmission, Fig. 31.5 (ExbB or the TonB motor of *Serratia marcescens*). Again, with typical biological efficiency, the twistable, but not net rotating-axle of the rotor motB or pomB, is fixed to the peptidoglycan layer acting as a stator and provides a pathway for the ions to pass. Pinion rotation under normal conditions is always CCW or right handed, and in turn, it meshes with and rotates the larger 50-nm (*E. coli*) or 62 nm (*Borrelia burgdorferi*) diameter main gear of the transmission (Fig. 31.6).

Fig. 31.4B shows in yellow the pinions sitting above and meshing with the C-ring. Hence, this main gear spins relative to the cell rotating the helical filament or propeller that drives the cell. The geometrical configuration of the gears is not conventional since the main driven gear top has its teeth pointing toward the outside of the cell, and the interacting mini-motors bottom is on top with its teeth pointing toward the inside of the cell. The transmission exists in two important regulated functional states of different diameters, for example, from 55 to 62 nm (Fig. 31.6). Switching between the two functional states of the gear constitutes a directional reversal mechanism for the transmission in that when regulator protein cheY-P is attached or not to the gear, the pinion effectively rotates it in one direction or the other. If the pinion rotates against the inside (proximal to the axis) of the main gear when cheY-P is attached (Fig. 31.6B), the main gear turns in the same CW (left-handed) direction (Fig. 31.7B), whereas if cheY-P is not attached the pinion rotates against the outside of the gear (Fig. 31.6A) rotating it in the CCW direction (Fig. 31.7A). Taking advantage of the inherent biological advantage of compliant components, multiple pinions or motors can simultaneously increase the load capability of the driven main gear (up to 11 in the *E.coli,* 16 in spirochetes *Borrelia burgdorferi*).

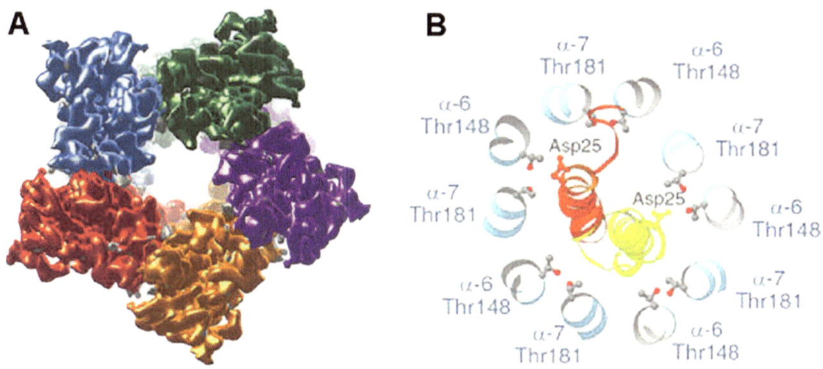

FIGURE 31.5 (A) Representation of the pentameric structure of ExbB of *Serratia marcescens,* the rotor of its PMF motor in its TonB dependent transporter, solved by cryo-EM view from the cytoplasmic side. (B) Arrangement of ExbD α-helices in the ExbB-ExbD subcomplex. Colored in orange and yellow the TM domains of ExbD in a ribbon representation are viewed from the cytoplasm. Ball and sticks represent the essential Asp25 residues and the ring of conserved threonines 148 on ExbB TM α6 and 181 on TM α7. *(A) Modified from Fig. 31.2B, Biou et al. (2021). (B) Modified from Fig. 31.2A of Celia et al. (2019).*

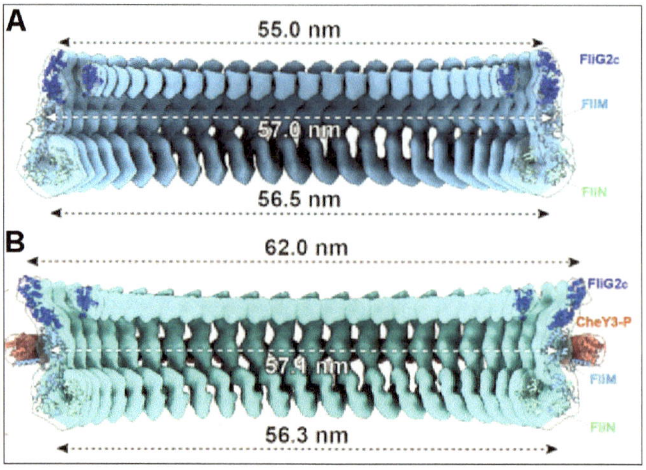

FIGURE 31.6 Comparison of the C-ring structures in CCW and CW rotation of spirochete *Borrelia burgdorferi*. (A) Diameters of the FliG2C, FliM and FliN rings in the C-ring with CCW rotation with 55 nm being relevant to the transmission gearing. (B) Diameters of the FliG2, FliM and FliN rings in the C-ring with CW rotation. With CheY3-P attached leading to 62 nm being relevant to the transmission gearing. *Modified from extended data set Fig. 31.8B and D, Chang et al. (2019).*

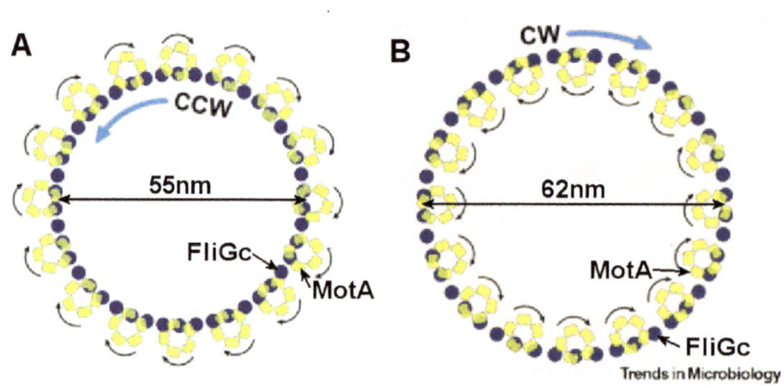

FIGURE 31.7 Model for the mechanism of rotational switching. (A) Interactions of the five toothed pinion with FliG2 in the C-ring during CCW rotation. When there is no bound CheY3-P, the FliG2 proteins interact with the part of the pinion (colored in yellow) toward the axle of the main gear. When the pinion, rotor of the driving motor, rotates CW, its cytoplasmic side rotates the C-ring (blue main gear) CCW. (B) When CheY3-P is bound the C-ring is enlarged so that the FliG2 proteins interact with the part of the pinion (colored in yellow) furthest from the axle. Consequently CW rotation of the pinion drives the C-ring CW. *Modified from Fig. 31.4 F and I, Chang et al. (2021).*

2.2.5 The pinion motor

The motA pinion is homologous to the ubiquitous TonB motor complex ExbD and to the TolQR complex (Fig. 31.8) with obvious similarly to a spur gear (see Deme et al., 2020). The TonB-ExbB-ExbD molecular motor harnesses the proton motive force PMF across the bacterial inner membrane to couple energy to transporters at the outer membrane. These facilitate uptake of essential nutrients such as iron and cobalamine (Celia et al., 2019). TonB physically interacts with the nutrient-loaded transporter to exert a force that opens an import pathway across the outer membrane. It belongs to the larger family of motors involved in nutrient import across the outer membrane of Gram-negative bacteria (ExbBD) (Biou et al., 2021). ExbB alone behaves as a stable pentamer, and the complex displays the $ExbB_5$-$ExbD_2$ stoichiometry. Ton B-dependent transporters are reviewed by Noinaj et al. (2010).

For Gram-negative bacteria, the core of the main gear or rotor is called the basal body and consists of a set of rings that span three layers of the cell envelope, significantly borrowed and assembled by structures homologous to the Type III secretion system (Beeby et al., 2020). The L (lipopolysaccharide) and P (peptidoglycan) rings are thought to be embedded in the outer lipopolysaccharide membrane and peptidoglycan cell wall and act as a bushing between the rotor and the outer parts of the cell envelope. The drive shaft (FlgB, C, F, and G) connects the universal joint or hook to the MS (membrane/supramembrane) ring located at the cytoplasmic membrane. The S, M, and C (cytoplasmic) rings of the basal body make up the main gear that turns the flagellum. FliG proteins (34 copies in some, 46 in others) are placed around the periphery of the C ring specifically individually interacting sterically and/or electrostatically with loops or cleavages between the five subunits of MotA of the torque-generating pinion with MotB as its stator connecting to the peptidoglycan layer, so it does

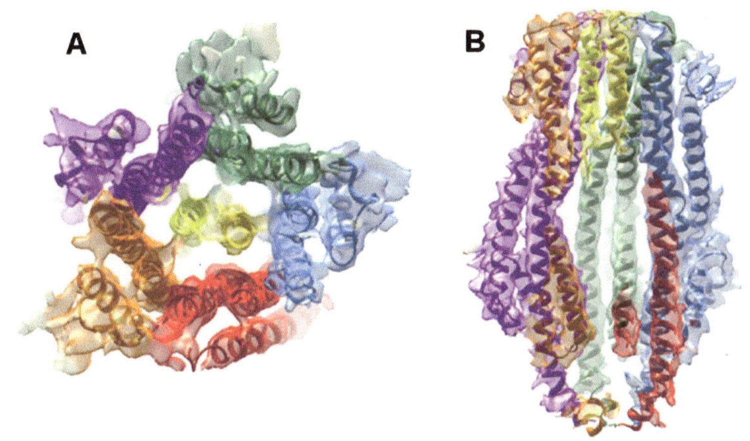

FIGURE 31.8 Structure of *Serratia marcescens* ExbBD based on cryo-EM. (A) View from the periplasmic space with five ExbB monomer in distinct shades of blue and two ExbD in red and magenta. (B) Clipped side view, with the helix of each ExbD monomer in red and magenta. *From Fig. 31.3B and C of Biou et al. (2021).*

not rotate with the main-gear rotor. Depending on the load, a powertrain can have between 1 and 11 engaged pinion motors (traditionally confusingly called stators in the literature while only the motB part is a stator). In some, the MS ring consists of 34 copies of a single protein, FliF. After the MS ring is assembled, the rest of the powertrain is built from it. The cytoplasmic face of the MS ring is attached to the C-ring with 34-fold symmetry in *E. coli,* which contains the proteins FliG, FliM, and FliN, which by being turned by the pinion results in torque generation of the main gear. The motors are a complex of two proteins: five MotAs and two MotBs in H + -driven flagellar motors such as those of *E. coli*. The filament grows from its distal end supplied by a channel down its center. To do this, an export apparatus inside the C ring pumps proteins down the center to make the hook and extend the filament.

2.2.6 Energy transduction

The pinions or motors consume the ion motive force (IMF), the electrochemical gradient across the inner membrane that results from cellular respiration, converting the electrochemical energy of translocated ions into mechanical energy.

Molecular motors convert electro-chemical energy to mechanical work and operate close to thermal energy, k_BT (k_B Boltzmann constant, T absolute temperature). In the flagellar motor, the free energy input from a single ion passing through the cytoplasmic membrane is a single electric charge times the ion-motive force (IMF; either proton-motive force [PMF] or sodium-motive force [SMF] depending on the driving ion). The IMF is the electro-chemical potential, $IMF = V_m + k_BT/q \ln (C_i/C_o)$, where V_m is the transmembrane voltage (inside minus outside) and q, C_i, and C_o are the charge, inside, and outside concentrations of the coupling ion, respectively. With a typical IMF of around -150 mV, the free energy of a single ion transit is $\sim 6\ k_BT$. Torque is defined as the product of force, and the perpendicular distance to an axis of rotation and therefore has dimensions of Newton-meters or energy. Because the Reynolds number for a spinning flagellar motor is $<<1$, inertia is negligible and torque can be calculated as $M = f\ \omega$, where f is the rotational drag coefficient and ω is the angular velocity.

"The force exerted by each force-generating unit is substantial but not large on an absolute scale. If we take an estimate for the stall torque of 4000 pN·nm and assume that force-generating units act at the periphery of the rotor at a radius of about 20 nm, then 200 pN is applied. If there are eight independent force generating units, then each contributes 25 pN. This force is about equal in magnitude to that between two electrons 4.8 Å apart in a medium of dielectric constant 40 (midway between water, 80, and lipid, about 2). So, almost any kind of chemistry will do" (Berg, 2003).

2.2.7 Ion selectivity

Bacterial flagellar motors are driven by ions and not ATP hydrolysis like muscle or ciliary motors or archaella. Flagella are driven by the membrane potential and/or by the hydronium ion or [Na^+] gradients.

2.2.8 Motor dependence upon IMF

The IMFs drive flagellar motors oriented in the membrane, and ions travel through the motor in a particular direction. Ion drives permit the high pinion-motor turnover speeds and high power output. The free energy of $\sim 6\ k_BT$, corresponding to one ion crossing the membrane, is smaller than the free energy of hydrolysis of ATP ($\sim$22 k_BT). This low amount of energy per ion transported creates an interesting and difficult design problem to make a flagellum rotate with useful propulsive power, particularly in viscous media. In addition, the enthalpic contribution is proportional to the continuously variable and reversible membrane voltage.

By voltage clamping the membrane voltage at the flagellar motor and monitoring the powertrain motion by video microscopy of markers attached to the hooks of powertrains, their rotation speed has been found to be proportional to the applied voltage up to −150 mV (Fig. 31.9). By varying the concentration of driving ions and diffusion potential, the electrical and chemical components of the IMF have been found to be equivalent at high load in H^+-driven flagellar motors (Fig. 31.10). Subsequently the speed of the fast powertrain and hence the motors that drive it at low load has been found also proportional to IMF (Fig. 31.11).

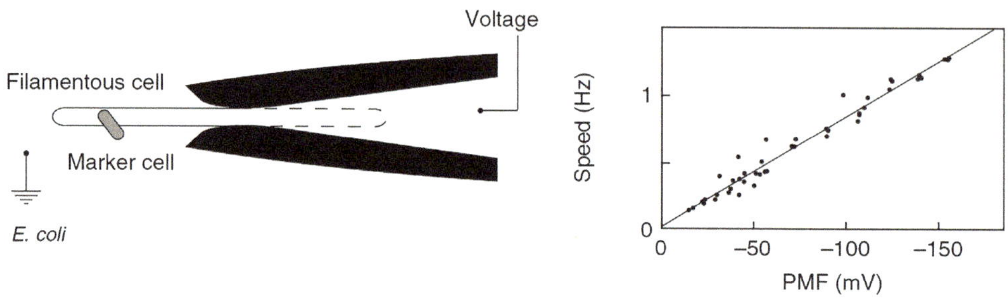

FIGURE 31.9 Torque versus IMF. Left: Schematic of a voltage-clamp method using *E. coli* cells held in micropipettes. The part of the membrane inside the pipette (indicated by the dashed line) is made permeable using the ionophore gramicidin S. Powertrain speed from observation of hook markers attached to the external hook of the powertrain (Fig. 31.2). Right: Powertrain speed found proportional to membrane voltage (PMF) between 0 and −150 mV (Sowa & Berry, 2008).

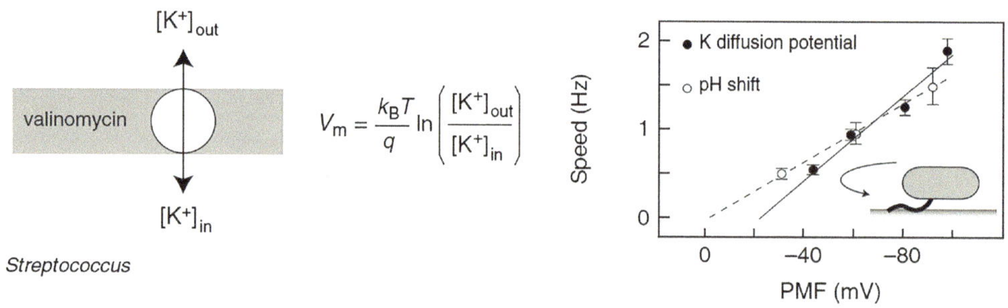

FIGURE 31.10 Left: Membrane voltages in *Streptococcus* controlled by a K^+ diffusion potential in the presence of valinomycin. Right: The speed of tethered *Streptococcus* cells found proportional to PMF and membrane voltage found equivalent to pH gradient (Sowa & Berry, 2008).

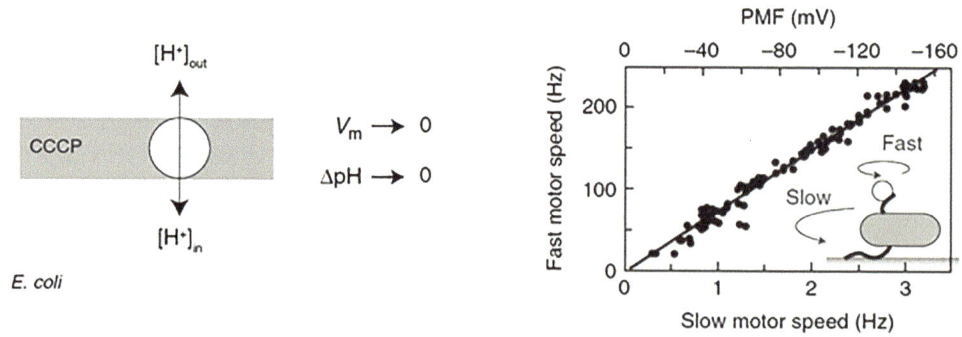

FIGURE 31.11 Left: PMF can be varied slowly from −150 mV down to 0 by adding small concentrations of carbonyl cyanide m-chlorophenylhydrazone (CCCP) or sodium azide. Right: The speed of a tethered *E. coli* powertrain (x-lower axis) was used as a proxy for PMF (x-upper axis, absolute value shown). The speed of a second powertrain on the same cell, attached to a sub-micron bead, was found to be proportional to PMF (Sowa & Berry, 2008).

2.2.9 Torque versus speed—A constant torque motor

The torque–speed relationship of the flagellar powertrain (but frequently confusingly just called the "motor" for short in the literature not distinguishing it from the actual motor component) has been measured to clarify the mechanochemical cycle. The speed can be measured over a range of viscous load or externally applied torque. With the latter method, the torque of the powertrain can be measured when forced to rotate backward or forward faster than its zero-load speed. The electrorotation method was particularly effective; a rotating electric field (megahertz) polarizes the cell body and torque is exerted on the cell due to a phase lag between the field and induced dipole moment. Fig. 31.13 summarizes the data from spinning the cell body in both directions at speeds up to ~1 kHz. There is a plateau of nearly constant torque up from

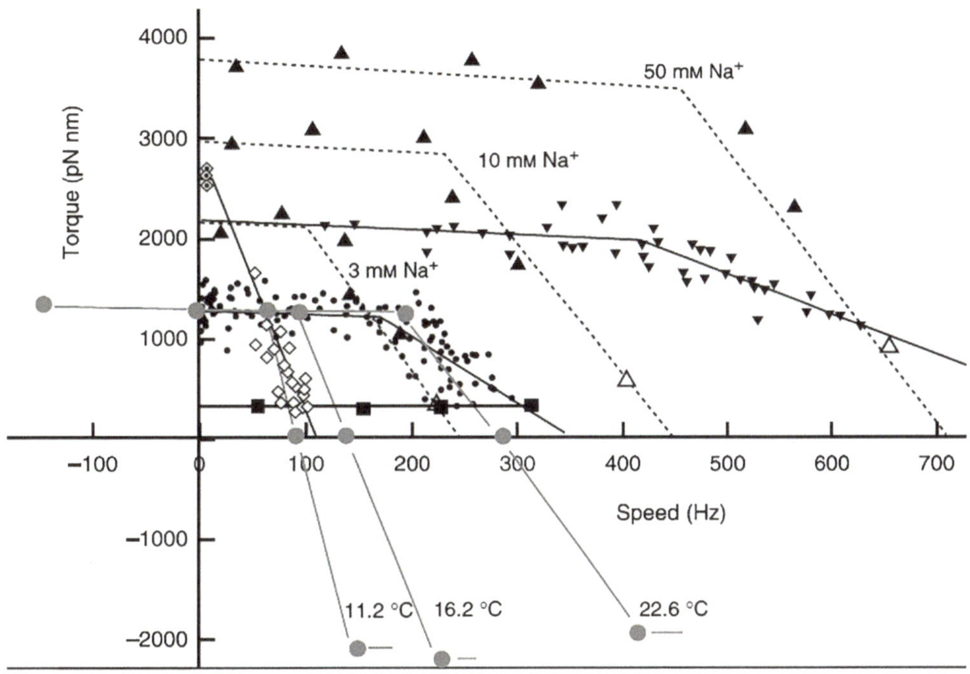

FIGURE 31.12 Torque-speed relationships for flagellar powertrains of diverse species measured using various methods. All measurements were made at room temperature except where indicated. The *E. coli* experiments using electrorotation (grey circles) and beads (black circles) did not result in absolute torques, so these curves are scaled to a stall torque of 1260 pN nm (Sowa & Berry, 2008, where more details may be sought).

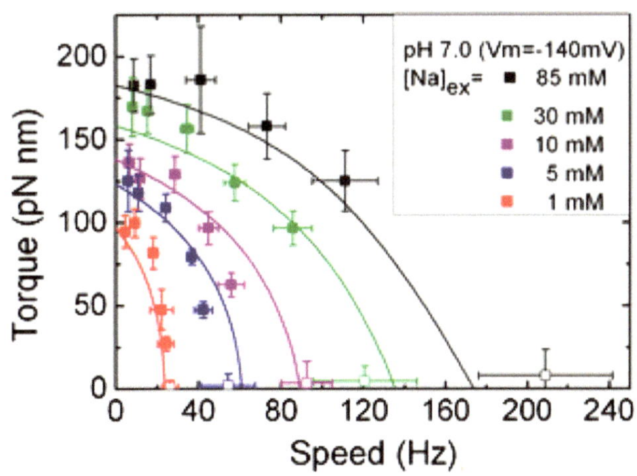

FIGURE 31.13 Torque–speed relationships of flagellar powertrains with only a single active pinion/stator or motor unit. All data points were measured using polystyrene beads attached to sticky filaments (solid symbols) except for those at the lowest load, which were measured by gold particles attached to the hook (open symbols). Lines are calculated torque–speed curves from the best fitting parameter set of a reduced three-state kinetic model (Lo et al., 2013, see reference for more conditions).

backwards speeds of about ~100 Hz to a "knee" forward speed of several 100 Hz. Then, the torque falls linearly towards the zero-torque speed at ~400 Hz (Fig. 31.12, circles, triangles), at which point, the powertrain resists rotation with a torque of similar magnitude to the plateau torque (Fig. 31.13, large grey circles). The absolute magnitude of torque generated by the *E. coli* flagellar powertrain was determined to be 1260 ± 190 pN nm using measurements with polystyrene beads of 1 μm diameter. This means the torque generated by its motor is ~185 pN nm (corrected from the powertrain measurement by the ratio of teeth of pinion gear (5) to teeth of main gear (34) x 1260 pN nm) The length of the plateau depends significantly on the different numbers of motors providing torque with more utilized when the load is higher. Torque-speed curves of main gear for a single motor are shown in Fig. 31.13.

In the torque plateau, transitions linked to ion flux are not rate limiting so speed is only mechanically limited by the load on the motor and depends only on the IMF, a thermodynamic quantity. The continuity of torque on either side of stall indicates that there is no irreversible step in the mechanochemical cycle. Faster than the "knee," transitions in the motor are rate-limiting, possibly due to a requirement that ions must be gated across the structure. In this case, motor speed depends on absolute temperature, the hydrogen isotope, and which component of the SMF is dominant. High rotation speed decreases the efficiency of ions going across the ion gate through the motor. For more details read Xing et al. (2006), Biquet-Bisquert et al. (2021).

2.2.10 Multiple pinion motors engaged with a single transmission

Tellingly when functional Mot proteins were introduced into a mot-minus mutant strain, the powertrain speed increased in discrete increments, implying independent multiple active motors provide the torque. At medium to high load up to 11 speed increments were seen, consistent with the EM images and similarly for the Na^+-driven chimera in *E. coli.* There are also transient speed changes in normally expressed motors, indicating that the attachment of motors are not permanent, but in constant turnover. An average of 22 GFP-MotB molecules per cell has been counted consistent with 11 stator complexes each with the A_5B_2 stoichiometry predicted biochemically. Furthermore, a mobile pool of ~200 GFP-MotB molecules was seen in the cell membrane exchanging with GFP-MotB in the motor on a timescale of minutes, confirming that stator units are in dynamic flux. How they are recruited is not clear. The number of MotAB motors driving the main gear depends on the type of bacteria.

As mentioned, the low amount of energy available from a single transport of an ion represents an interesting design problem to make an effective rotating flagellum. As mentioned, the solution appears to be many (16 "stators") motor rotating pinions with five steps ("teeth") Fig. 31.5 driving a larger 62 nm meshed gear (C-ring) (Fig. 31.6) with 46 "teeth" FliG in the case of spirochete *Borrelia burgdorferi. Vibrio alginolyticus* gear has 34 teeth and up to 11 minimotors. These several motors with motB/pomB stators attached to the peptidoglycan layer generate torque with the motA/pomA rotor making use of the previously mentioned ion flow, hydronium ion/H^+ for motA, and Na^+ for pomA from the periplasm to the cytoplasm (Fig. 31.14). Specific regions of motAs interact with the "torque helix" of FlaG of the main gear that turns the drive shaft or rod, the hook, and the flagellum, much as the meshing of a pinion and its gear.

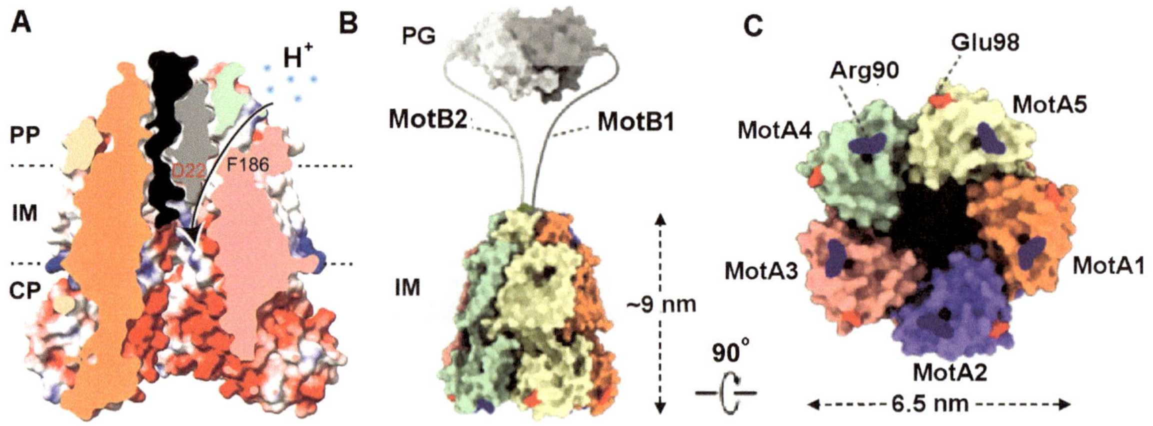

FIGURE 31.14 (A) View from within the membrane of the *Campylobacter jejuni* (Cj)MotAB pinion motor showing the potential proton channel of the CjMotAB PMF motor in the on-state conformation. (B) Shows the complete two MotB and five MotA proteins connecting to the peptidoglycan, based on the crystal structure of *Salmonella* MotB periplasmic domain (5Y40) on top of the cryo-EM structure of the *Campylobacter jejuni* cytoplasmic domain (6KYP). (C) Bottom view of the MotA pentamer of *C. jejuni* showing conserved charged residues Arg90 (blue) and Glu98 (red) (using *E. coli* numbering). *(A) From Fig. 31.3D, Hu et al. (2021). (B and C) From Fig. 31.3E and F, Chang et al. (2021).*

When the gears of the transmission are not connected, the pinion-motors are in an inactive state with its channel plugged (Fig. 31.15). When the pinion contacts the larger main gear, the interaction between MotA and FliG triggers rearrangement of the C-terminal domain of MotB such that the PG-binding domain of MotB extends to the peptidoglycan PG layer, enabling activation of its ion channel and generation of torque. As with other biological motors, they work only in one direction, namely, the motA rotating in a left-handed way around the motB stator acts as pinion on the main gear by electrostatic and probably steric interactions rotating the main gear attached to the flagellum via the individual FlaG proteins.

2.2.11 Mechanism of torque generation: Interactions between rotor and stator of pinion

Hu et al. (2021) have a proposed provisional model for MotA rotation. A universally conserved feature of the pinion or MotAB motor is its "plug motif" or ion-channel off-state in which it is not fixed to the peptidoglycan layer (Fig. 31.15). On the other hand, the on-state structure reveals a potential proton channel that links the periplasmic space to the conserved acidic residue D22 on TM helix of MotB chain 2 (Fig. 31.14). To the inside of MotA cytoplasmic domain, there are many negatively charged residues found to be critical for ion transport. Shielding this channel is the conserved hydrophobic residue F186 of TM4 of MotA, the side chain of which adopts two different conformations, likely controlling the ion flux, ensuring efficient IMF utilization.

D22 on the TM of the Mot B Gram-negative bacterium *Campylobacter jejuni* CjMotB chain one is buried in a hydrophobic environment; therefore, it is more likely to accommodate a protonated (or hydronium interacting) D22 compared with the MotB chain 2 (Fig. 31.16 left, compare to Fig. 31.5B for a different homologous motor). No major conformational differences are seen between the on and off states. Instead, rotation of MotA around MotB is assumed to be the cause of torque generation.

MotA rotation mechanism: In the on-state of the active pinion motor, the PGB domains of the two MotB chains are dimerized and anchored to the peptidoglycan layer, and only MotA can rotate around MotB. Based on this analysis, MotB chain 1 D22 is assumed bound to a proton or hydronium ion and engaged to MotA. It is ready to move, but rotation is blocked because the negatively charged MotB chain 2 D22 cannot move across the hydrophobic surface of the inside face of the rotor MotA. Once MotB chain 2 D22 receives a neutralizing proton or a hydronium ion from the periplasmic side, so both charge sites are neutralized MotA can rotate (Fig. 31.16 middle). Then, MotB chain 1 D22 can be deprotonated to release the bound proton or hydronium into the cytoplasm. When it deprotonates, the D22 flips upward from pointing down toward the cytoplasmic interface to pointing to the periplasm in the deprotonated state kicking MotA 1/10 revolution in a clockwise direction. After MotB chain 1 D22 has released its proton, MotB chains one and two have now functionally switched roles, and the cycle can start again. After 10 transported protons, motA will have gone all the way around with its five teeth (Fig. 31.16 right).

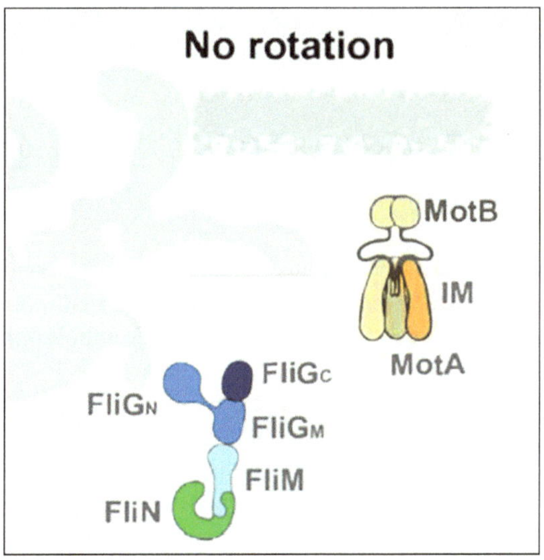

FIGURE 31.15 Pinion-PMF motors motA and stator motB float along the inner membrane (IM) when they are not associated with the transmission represented by the Fli proteins Gc, Gn, Gm, M and N of the C-ring. *From Fig. 31.4A of Chang et al. (2021).*

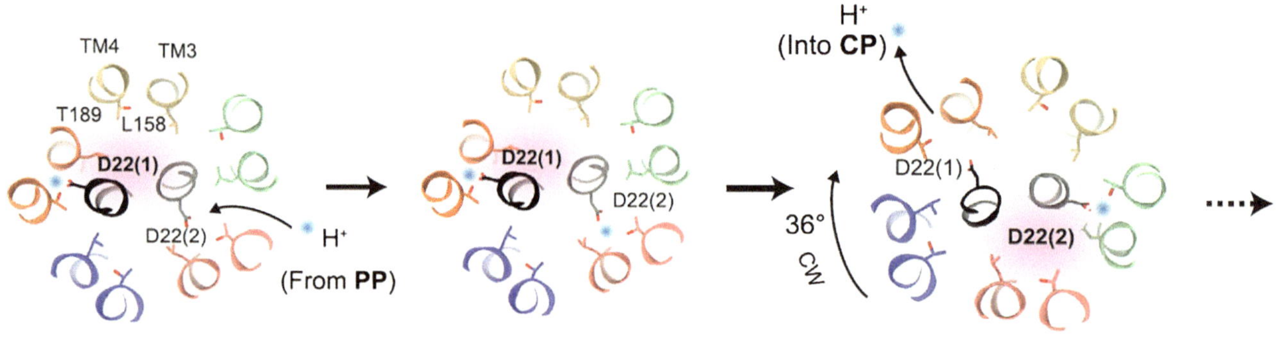

FIGURE 31.16 Proposed model for PMF rotation of MotA around MotB. D22 from MotB engaged with MotA is highlighted with pink halos Left: MotB D22 of chain 1 [D22(1)] is coupled with a proton or hydronium and MotB D22 of chain 2 [D22(2)] receives a proton or hydronium from the periplasmic (PP) side. Middle: Proton or hydronium binding neutralizes MotB D22(2) charge. Right: MotA rotates 36 degrees clockwise (CW) by a power stroke. MotB D22(1) releases the proton or hydronium into the cytoplasm (CP) and MotB D22(2) (charge is neutralized) gets in a position similar to the MotB D22(1) in the left panel. *From Fig. 31.3E, Hu et al. (2021).*

2.2.12 The transmission pinion-main gear interaction

The C-ring is a major part of the transmission. In *E. coli* and *Salmonella*, the C-ring is a complex of three proteins: FliG, FliM, and FliN (Fig. 31.2) with variations in other bacterial species. Because of basic homology, studies on many species have been relevant. For example, the crystal structure of the cytoplasmic domain of FliF (FliFC) in complex with FliGN from *Helicobacter pylori* revealed the assembling interface between the C-ring and MS ring,

FliG comprises three domains: an N-terminal domain (FliGN), responsible for binding to the MS-ring; a C-terminal domain (FliGC), containing that torque helix that meshes with the motA pinion gear; and a middle domain (FliGM), providing the binding site for FliM. This interaction allows attachment of the C-ring close to the inner membrane. Cryo-EM analysis of the MS-ring from *Salmonella typhimurium* reveals that the MS-ring adopts a 34-fold symmetry. The C-ring also contains 34 FliG protomers. FliM also has three domains. The N-terminal domain of FliM (FliMN) contains the binding site for the phosphorylated chemotaxis signaling protein CheY (CheY-P). The middle domain of FliM (FliMM) binds to FliGM. The C-terminal domain (FliMC) dimerizes with FliN located at the bottom of the C-ring.

Conserved charged residues located in the large cytoplasmic domain of MotA are proposed to be its interaction interface with FliGN. Electrostatic interaction contributes to torque generation and the proper assembly of pinion around the rotor for both proton motive force driven motors and those using a sodium gradient. In CjMotAB, the FliG torque helix-binding site is most likely located at the cleft between two MotA subunits, involving the positively charged residues R89 and R90 from one MotA subunit and the negatively charged residues D97 and D117 from the adjacent MotA subunit (Fig. 31.17, compare to similar Fig. 31.14C). During MotA rotation around MotB, at least one of these binding sites incorporates the FliG torque helix through interactions by reversed charged residues.

2.2.13 Torques of various flagella

To adapt to differently viscous environments, powertrains have evolved with a range of achievable torques as shown in Fig. 31.18, which illustrates this correlation.

2.2.14 Transmission reversal of drive shaft

The direction of myosin and kinesin motions is thought to be biased by thermal motions of their head domains, determined by the conformational bias of their neck regions so F-actin-myosins, microtubule-kinesins, and microtubule dyneins are all unidirectional. However, when the bias is produced by proton motive force PMF, which can be forced to be in either direction, the motor can be reversed. Under natural conditions, with an IMF of around −150 mV, motors spontaneously switch direction stochastically every second or so, but this could be due instead to instability of the transmission forward-reverse state and not the IMF. Nonswitching mutants rotate in the opposite direction when the PMF is reversed. This was achieved using a K^+ diffusion potential in *Streptococcus* and a voltage clamp in *E. coli*. In both cases, only a fraction of motors rotated when the PMF was reversed. In the *E. coli* experiment, removal of the normal PMF appeared to cause the detachment of stators after a few revolutions. These results indicate that the mechanochemical cycle of the flagellar motor is essentially reversible.

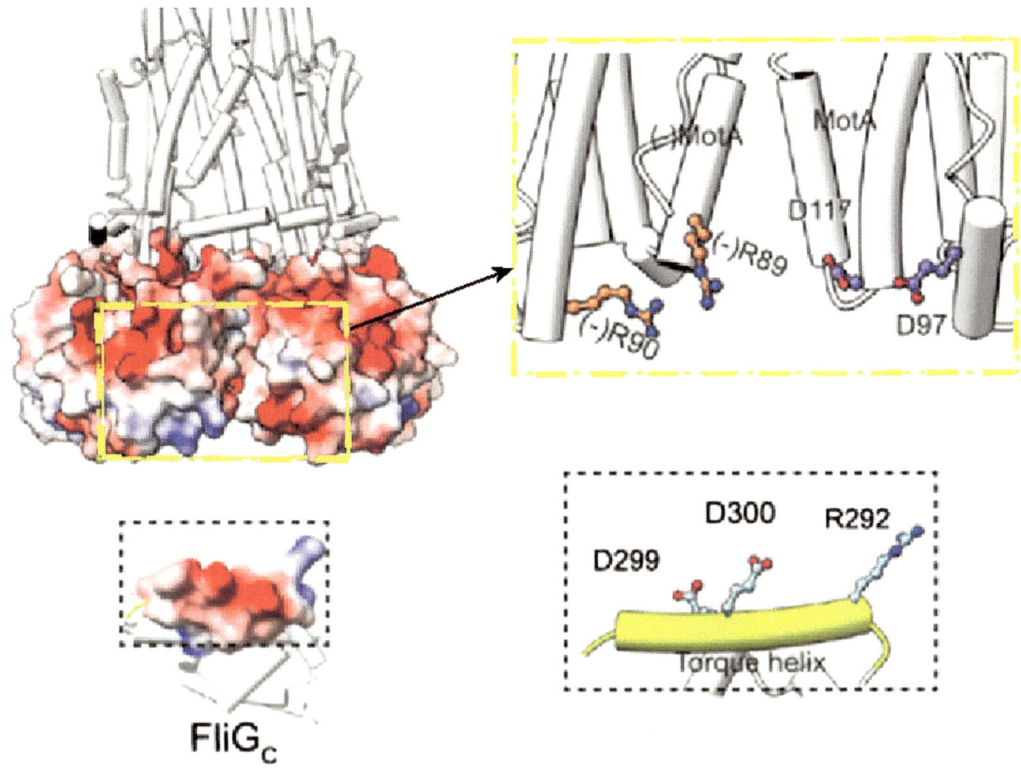

FIGURE 31.17 Left: The calculated electrostatic potential (negatively charged, red; positively charged, blue) of the *Campylobacter jejuni* (Cj)MotA cytoplasmic and the CjFliG C-terminal domain that contains the torque-generating helix. Right: Charged residues, shown in stick representation, on the potential interaction interface between the cytoplasmic domain of CjMotA and the torque helix of CjFliG.

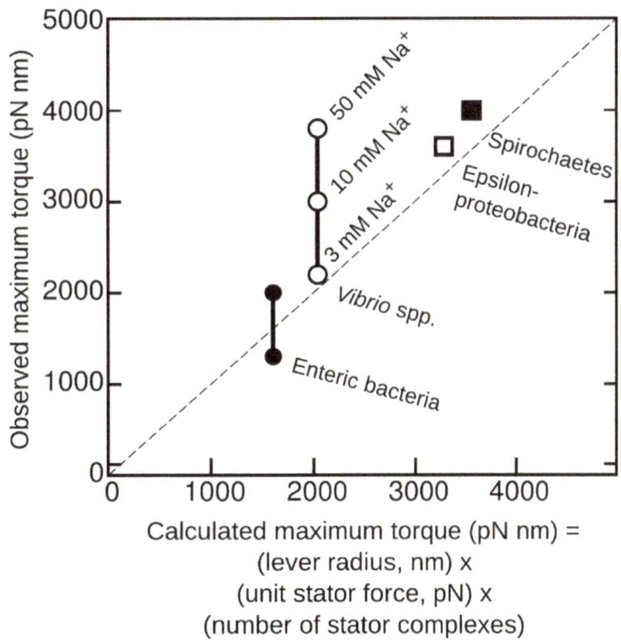

FIGURE 31.18 Comparison of observed and predicted torques of flagellar powertrains from various species. Filled circles: Enteric bacteria; open circles: *Vibrio* spp.; open squares: e-proteobacteria; and filled squares: spirochetes. Where there were no direct measurements of torque, those of *C. jejuni* and *B. burgdorferi* were inferred from closely related species with similar powertrains namely *H. pylori* and *Leptospira*, respectively. Relative torque strengths were validated by their swimming speeds and their ability to swim through viscous fluids. The dotted line represents a perfect prediction (Beeby et al., 2016).

The flagella of peritrichously flagellated bacteria like *E. coli* rotate in both CW and CCW directions. They swim forward when they rotate CCW and tumble if CW. On the other hand, *V. alginolyticus* has a single polar flagellum that pushes with CCW rotation or pulls with CW rotation. Spirochetes, which are spiral-shaped, contain periplasmic flagella lying in between the inner and outer membranes. If the flagella at their poles rotate in opposite directions, the cell moves forward; if rotate in the same direction, they flex in place. In *Sinorhizobium meliloti* and *Rhodobacter sphaeroides,* the flagella rotate only CW and reorient by Brownian motion when stopped. It is worthwhile to consider how those that can switch do so.

As discussed the FliG torque helix of the C-ring interacts with the torquing pinion. Binding of the active phosphorylated form of the chemotaxis response regulator CheY (CheY-P) to FliM on the C-ring (Fig. 31.19) expands the C-ring through a conformational change of the FliGs around the ring. In the CCW rotational mode, FliG adopts a closed, compact form, whereas, in the CW rotational mode, it adopts an extended and open form changing the interaction interface between the MotA pinion and rotor. The gear becomes larger 55–62 nm in the spirochete (Fig. 31.6). In this process, the FliG multimers have undergone a dramatic conformational change during the flagellar motor switching between the CCW and CW rotational directions.

Note the interaction is top of FliG to bottom of MotA and the orientation of the bottom of MotA is flipped 180 degrees between rotating the FliG on the inside or outside (Fig. 31.7). Subsequently, it may be necessary to change the orientation of the FliG torque helix direction depending on whether the side of the MotA is far or near the axis of the C-ring (see Figs. 31.19 and 31.20 to see that orientation has changed, but not clear specifically). This is a concerted conformational

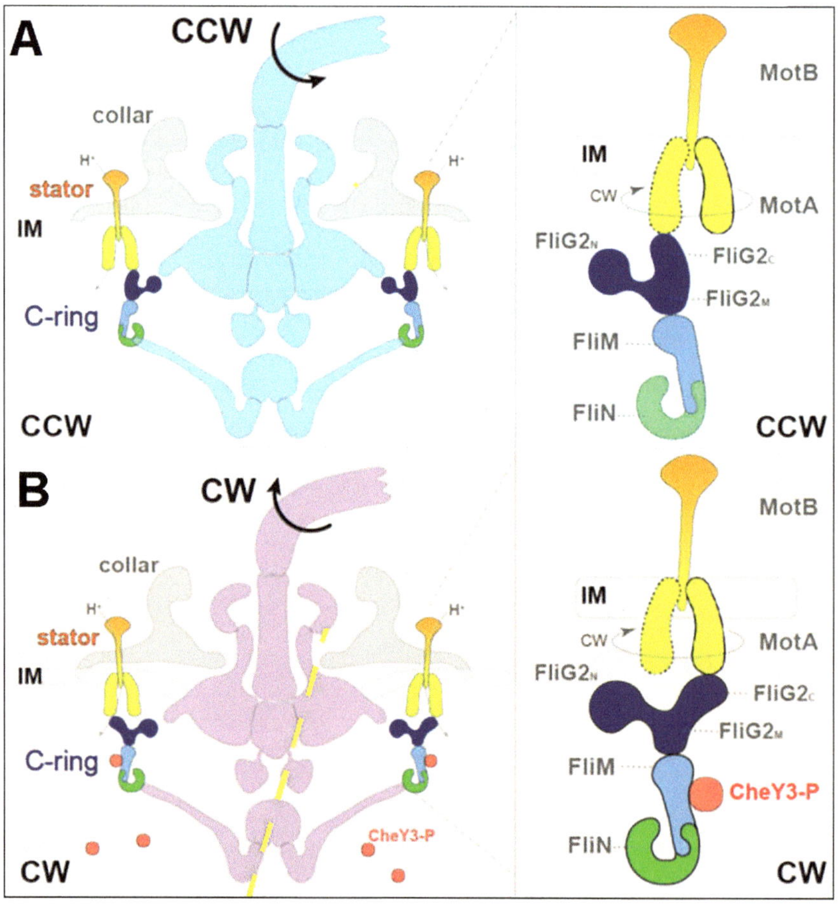

FIGURE 31.19 Model for the mechanism of rotational switching of the transmission. (A) Interactions of MotA with FliG2 in the C-ring during CCW rotation. When there is no bound CheY3-P, the FliG2 proteins interact with MotA rotor or pinion (colored in yellow) closest to C-ring axle. With the pinion spins CW, the C-ring (blue) spins CCW. At right: Detailed view of the interaction between the C-ring and the MotA pinion. (B) CheY3-P induced conformational changes in the C-ring result in altered interactions between the MotA and C-ring causing it to spin CW. At right: When CheY3-P binds to FliM on the exterior surface of the C-ring, its binding triggers the shift so that FliG2C interacts with the pinion distal to C-ring axle. *From Fig. 31.5, Chang et al. (2020).*

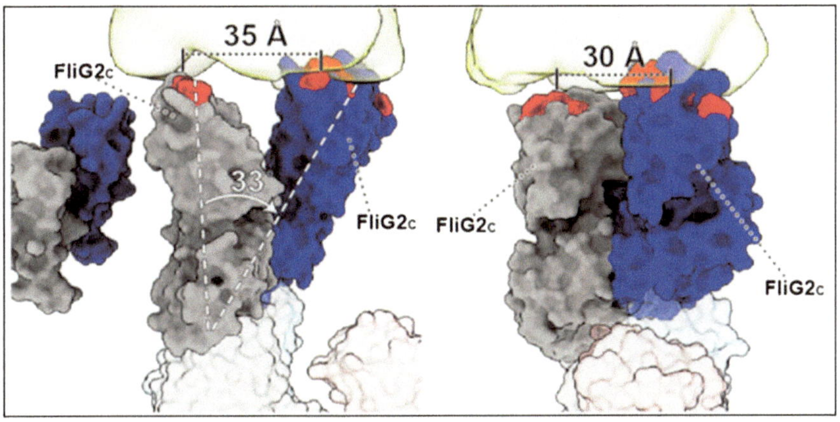

FIGURE 31.20 Comparison between C-ring models before (grey) and after (colored) CheY3-P binding. FliG2 undergoes a large tilt altering the interactions between FliG2 and MotA. The charged residues (Lys275, Arg292, Glu299, and Asp300) in the C-terminal domain of FliG2 (FliG2C) are colored in red. Left view orthogonal to the right view. Unfortunately, the orientation of the putative torque helix is not shown.

change of the C-ring, where prior to binding is grey and after is blue. By concerted is meant the steep dependence of motor bias (probability of CW rotation) upon CheY-P binding, with a Hill coefficient >16 (Yuan & Berg, 2013) (Fig. 31.21).

Sowa and Berry (2008) suggest that the flagellar motor and its powertrain is the first ion-driven molecular machine that can be studied at this level of mechanical detail. The chimeric sodium-driven motor in *E. coli* and the newly developed fluorescent methods to measure both components of the SMF offer exciting new possibilities. Because the enthalpic (membrane voltage) and entropic (ion concentration gradient) components of the IMF are of similar magnitude, unlike ATP hydrolyis where the dominant free-energy change is usually enthalpic, this work will offer insight into the fundamental question of how molecular motors convert the different components of free energy into mechanical work (recent review Biquet-Bisquert et al., 2021).

2.2.15 Chemotaxis

A natural question is what are these amazing flagellar powertrains utilized for given they exhibit such elaborate optimized engineering. In *E. coli,* they enable directed motion in spite of a passive orientation relaxation time of about 1 s due to the bacterium's rotational diffusion (Fig. 31.22). This directed motion aids search for food, biofilm formation, quorum sensing, bacterial pathogenesis, finding an optimum temperature (Yoney & Salman, 2015), and host infection (reviewed in Karmakar, 2021 and Matilla & Krell, 2019). Perhaps then, it is not a surprise that their control for chemotaxis is also highly optimized often in surprising ways (Parkinson et al., 2015). From a physiological point of view understanding how the system works is perhaps best described from a simple linear prospective as represented by a Bode plot and then adding on the additional nonlinear modifications that have given it enormously increased dynamic range from what you might anticipate, temperature robustness, and extraordinary sensitivity.

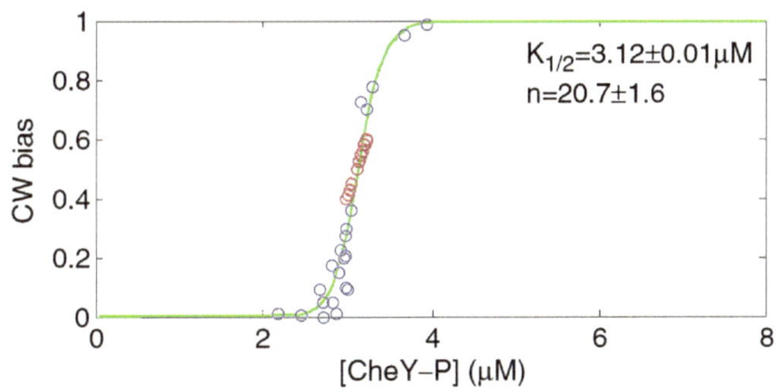

FIGURE 31.21 CW bias of the motor as a function of the CheY-P signal input. The blue circles below the red circles were obtained from attractant addition, while the blue circles above the red circles were obtained by attractant removal. The red circles were the initial values for the adapted motors. The green curve is the corresponding Hill-function fit. The error values are from the fit. Input−output relationship is for motors with an adapted CW bias of 0.5 ± 0.1 (Yuan & Berg, 2013).

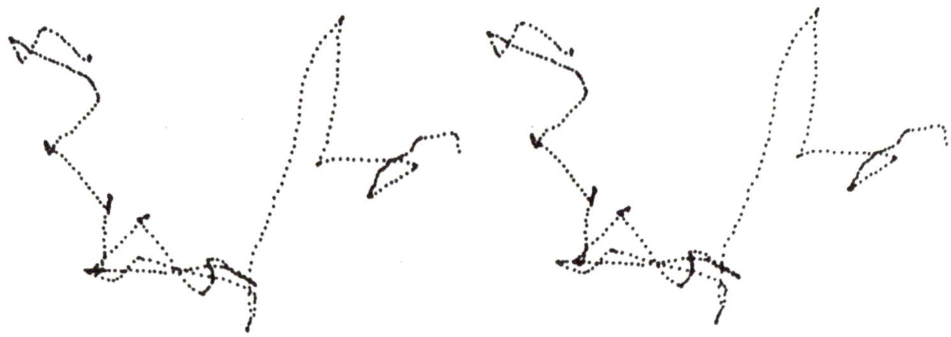

FIGURE 31.22 Swimming track of *E. coli* (Berg & Brown, 1972). This stereo plot shows about 30 s of an *E. coli* K-12 bacterium in an isotropic homogenous medium. The track range is about 0.1 mm and shows 26 runs and tumbles with the longest run being 3.6 s. The mean speed is about 21 μm/s. For 3D effect focus your left eye on left image and your right eye on right image.

The gain-frequency Bode plot (Fig. 31.23) shows that functionally the linear description is, in no particular order, a differentiator of the input signal and two quadratic-lag second-order low pass filters (~0.2 and ~1.6 Hz), each responsible for a change of minus two in the slope (going from left to right in the figure). The differentiator is usually most associated with comparison of receptor occupancy and methylation level. Fig. 31.24 shows the equivalent response in the time domain. When fitting Bode plots, it is worthwhile to consider what the curve should look like. As in this case, one should expect slopes of integer value and quadratic lags of minus two slope change as well as differentiators of plus one slope are very common. Each of these functions will likely correspond to single specific reactions so more effort should be made to identify the components that cause these dynamics. Since there is such an extraordinary vast literature on measurements of *E. coli* chemotaxis behavior, for the moment, it is left to the reader to assign components and the reactions that are likely responsible. Further optimization is obtained by a so-called trimers-of-dimers-architecture cheA—cheW complex. It would appear that for every receptor molecule activated, 36 kinases are activated, a considerable amplification.

It is worth noting here that the exact plus one slope of the Bode plot at low frequencies implies perfect differentiation of the input signal. The transfer function between the input, a pulse of L-aspartate, to the bias, the probability the flagellum spins CCW (viewed from outside the cell, right handed) is not changed. Some people have called the differentiation transfer function adaptation, but most engineers would *not* call that adaptation, reserving the term for when the transfer function is actually adapted to a change in environment such as temperature or food sources. Differentiation of an input signal is common in biology and in industry as it gives a transient response to a signal or a change in some signal level and often used to predict what is to come. Of course, there are also signaling responses that are simply proportional to signaling inputs.

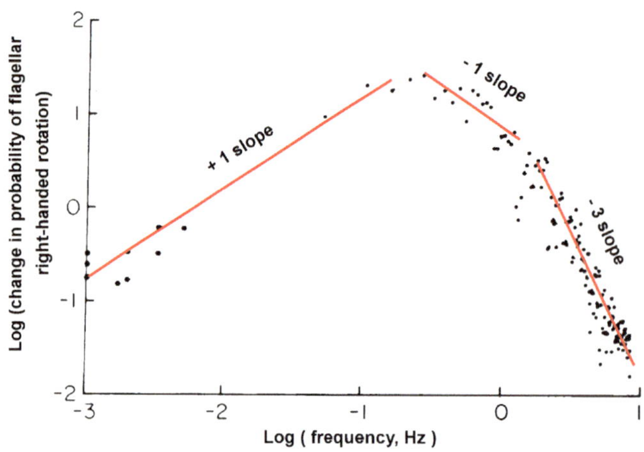

FIGURE 31.23 Amplitude-frequency domain Bode plot of the change in the CW-CCW bias in flagellar rotation of *E. coli* resulting from modulation of attractant, plotted as log(change in bias) versus log(frequency) of temporal modulation of the concentration of a-methylaspartate. The low-frequency data set comes from exponentiated sine-wave stimuli, while the high-frequency data comes from Fourier transforming the impulse response. The red curves show the probable dependences associated with the biochemical pathways. *Modified from Foster (2009).*

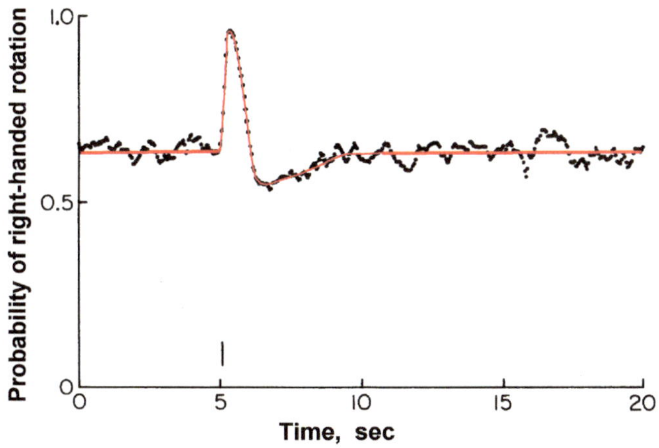

FIGURE 31.24 Impulse response to attractant in wild-type cells. The dotted curve is the probability determined from repetitive stimulation that tethered cells of strain AW405 spin CCW or right handed when exposed to pulses of L-aspartate or a-methyl-DL-aspartate beginning at 5.06 s (vertical bar). The smooth curve is a fit to a sum of exponentials. The curve was constructed from 378 records comprising 7566 reversals of 17 cells. Points were determined every 0.05 s (Segall et al., 1986).

A further interesting characteristic of the *E. coli* response system is the extraordinary sensitivity of response to the concentration of CheY-P the regulator of switching the rotation from CCW to CW (Fig. 31.21). Normally, the system adjusts to varying concentrations of regulator by varying the number of FliM subunits on the switching complex. With a fixed number of subunits, the cooperativity of those subunits is remarkable with a Hill coefficient of about 21. The conformational spread model of motor switching explains this ultrasensitivity by the large number of protomers (~ 34) and the bistability of the protomer structure due to the low energy difference between the CW and CCW states of each protomer compared to the strong coupling between neighboring protomers. This ultrasensitivity is crucial for coordination of multiple motors in a cell subject to fluctuations or bursts in the concentration of CheY-P (Yuan & Berg, 2013). For an informative review on the issue of ultrasensitivity, see Khan (2020).

In addition to this signaling system, there are many other more newly discovered second messengers systems in bacteria reviewed by Yoon and Waters (2021).

2.2.16 Fluctuation analysis

One might anticipate that since a responding system cannot distinguish between noise input and signal; then, it must respond at low levels of signaling molecules to noise input in the same way as small real signals. There is discussion of how the response network is tuned to simultaneously maximize both the random spread of cells in the absence of attractants and the cellular response to gradients of attractant. Emonet and Cluzel (2008) pursue this advanced topic profitably. Comparison of the other bacterial chemotactic systems to that of *E. coli* has shown that most other systems are more complex, even involving cascaded feedback loops, which bring greater stability to response. This also is an advanced topic worth reading about (Hamadeh et al., 2011).

2.2.17 The chemotaxis pathway of E. coli

Combating a slow rotational diffusion *E. coli* (Fig. 31.25) swims with a bundle of four to six left-handed flagella CW, powered by right-handed CCW rotating pinion motors. To increase the rate of rotational diffusion, one or more of the flagellar machines switch from right (CCW) to left-handed rotation (CW), causing the flagellar bundle to break apart temporarily and the cell to swim off in a new direction. During the CCW rotation runs, all flagellar motors are strongly correlated, but during the CW tumble, individual motors are uncorrelated over a long time (Dominick & Wu, 2021). Chemotaxis is enabled by modulation of the frequency of these tumbling events in response to stimuli. A pathway comprising transmembrane receptors (methyl-accepting chemotaxis proteins; MCPs), a histidine protein kinase (CheA), an adaptor protein (CheW), a response regulator (CheY), and two "adaptation," really differentiation, proteins, CheB and CheR, controls the switching of flagellar rotation.

Cells respond to temporal changes in attractant and repellent levels, which are monitored by their clustered chemoreceptors (Jones & Armitage, 2015), by suppressing signals to reverse flagellar transmissions during runs that carry them in favorable directions. *E. coli* swims toward sugars (glucose, galactose, ribose, and maltose), amino acids (aspartic acid and serine), dipeptides, pyrimidines, and electron acceptors (oxygen, nitrate, and *fumarate*) and away from fatty acids,

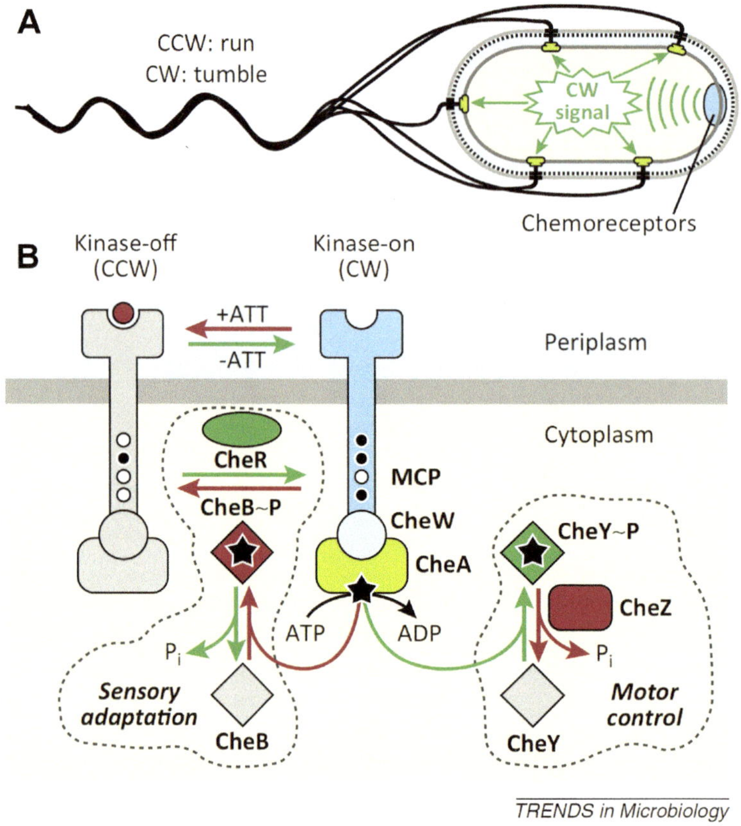

FIGURE 31.25 The chemotaxis control pathway of *E. coli*. (A) Chemoreceptor clusters send out a signal to the flagella that rotate CCW as viewed from the outside by default. Those signals (green) trigger transient CW rotation. Cytoplasm, yellow; inner membrane, gray; periplasmic space, light blue; peptidoglycan, black and broken; outer membrane, light gray. (B) A two-state model of receptor signaling. Gray components are inactive. Green components and their reactions promote [phospho-CheY] CW signals; red components and their reactions reduce [phospho-CheY]. Small circles on receptor molecules (MCP) indicate unmodified (white) and methylated (black) sensory differentiation (adaptation) sites. Stars indicate signaling phosphoryl groups (Parkinson et al., 2015).

alcohols, and some divalent cations, which might be harmful. *E. coli* is also attracted to a quorum-sensing signal AI-2 and to a norepinephrine metabolite. Ternary signaling complexes are formed between chemoreceptors, CheA, a histidine autokinase, and CheW. These signaling complexes have kinase-on and kinase-off states. The probability of reversing motor rotation depends on the proportions of receptor signaling complexes in the ON and OFF states. Attractants shift receptors toward the OFF state, slowing the flux of CheA phosphoryl groups to two response regulators, CheB and CheY. Phospho-CheY interacts with the flagellar basal body to promote CW rotation, transiently because of rapid dephosphorylation by phosphatase CheZ. The short half-life of phospho-CheY allows the cell to repeatedly trigger rapid motor responses to chemotactic stimuli. Phospho-CheB is part of a signal-differentiation feedback circuit that resets the ON−OFF ratio to its prestimulus level. With the noted signal differentiation (Fig. 31.23), the cell is able to detect and respond to changes over a wide range of concentrations of stimuli. It controls receptor signaling by covalent modification of several glutamyl residues in the cytoplasmic portion of the methyl-accepting chemotaxis protein (MCP) molecule. A methyltransferase CheR interacts with the OFF-state receptors and catalyzes a glutamyl methylation reaction that pushes the receptor toward the ON state. A dedicated methylesterase CheB interacts with ON-state MCPs and hydrolyzes glutamyl methyl groups to glutamic acid so that receptors move toward the OFF state. The OFF-state substrate binding of CheR and the ON-state binding of CheB account for the negative feedback of the sensory differentiation circuit. Phosphorylation of CheB enhances its catalytic activity controlling the rate of the sensory differentiation process.

2.2.18 An alternative switching mechanism in uniflagellated bacteria

Controlled switching to new directions is crucial for motile bacteria to enable them to move relative to environmental gradients they sense. While switching to a new direction in many bacteria as discussed is the result of changing the rotation direction of their several flagella, this is not true for many marine bacteria that have a single flagellum.

These uni-flagellated bacteria (Son et al., 2013) exploit the compliance of their flagellar hook to achieve direction switching. In forward swimming, these hooks experience compression and that extra load causes the hook to buckle in a controlled mechanical failure. If the swimming speed is reduced below the buckling threshold, reorientations are suppressed. The compression in forward swimming comes from the drag of the cell body and the thrust of the flagellum. On the other hand, if the bacterium is swimming flagellum first, then thrust and drag are oppositely directed with tension on the hook and no buckling occurs (Figs. 31.26 and 31.27).

2.2.19 Cell–cell communication

Cell–cell communication among bacteria coordinates their movement, growth, and biochemical activities. It can significantly enhance the chemotactic migration of bacterial populations. *E. coli* cells respond to the chemoattractant gradient not only by biasing their own random-walk swimming pattern through the well-known intracellular chemotaxis signaling but also by actively secreting a chemical signal into the extracellular medium. This extracellular signaling molecule is a strong chemoattractant that attracts distant cells to the food source. The observed behavior may represent a common evolved solution to accelerate the function of biochemical networks of interacting cells (Long et al., 2017) (Fig. 31.28).

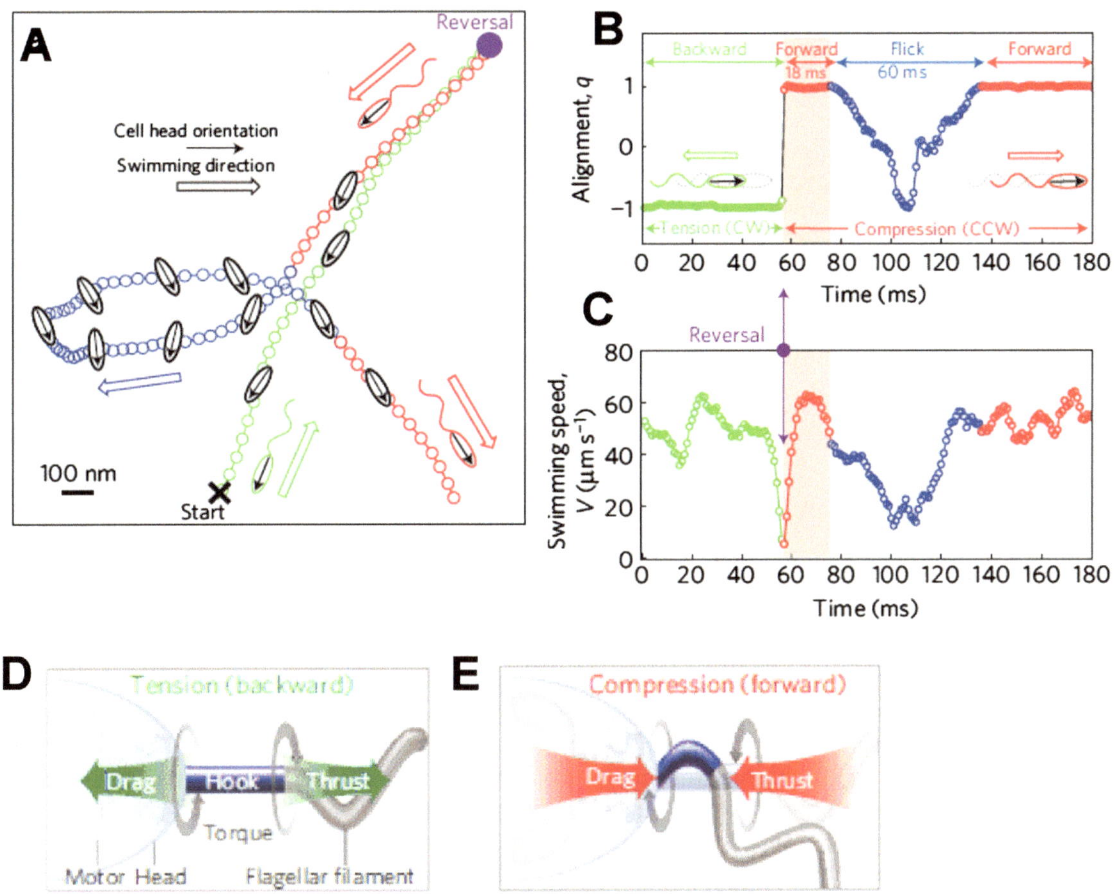

FIGURE 31.26 (A) Cell swimming trajectory of *Vibrio alginolyticus* at 1 ms intervals containing a flagellar-hook buckling event ("flick") after the beginning of forward swimming. A not-to-scale schematic of the head orientation at selected times is overlaid. (B) The alignment, q (directional cosine between the cell head and the swimming direction) for the trajectory in A. During the swimming cycle of *V. alginolyticus*, there is a 18 ms forward swimming segment (long; red) before the 60-m buckling event (blue). (C) Swimming speed during a flick (same color scheme as in B. (D and E) Schematics (not to scale) of the flagellar filament, hook and rotary motor during backward swimming (D), when the hook is in tension, and during forward swimming (E), when the hook is in compression (Son et al., 2013).

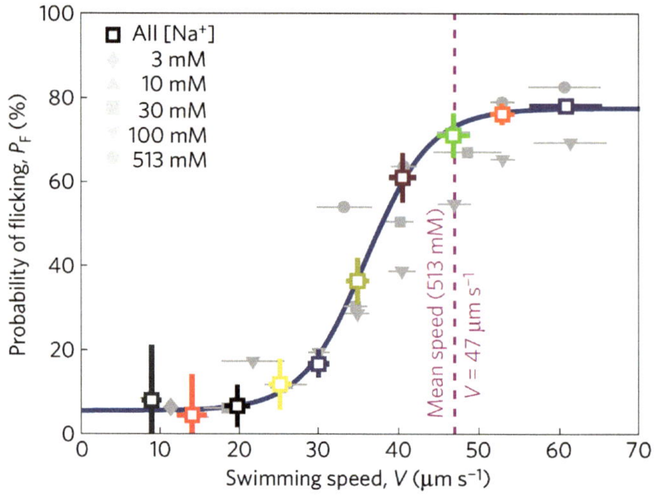

FIGURE 31.27 The probability P_F of flagellar-hook buckling ("flick") sharply increases with increasing load on the hook due to increased swimming speed V (open squares).The mean swimming speed, V = 47 μm/s, for cells swimming at $[Na^+] = 513$ mM (typical of natural marine environments) is indicated by the dashed magenta line. The blue curve is a logistic fit. The horizontal error bars are the standard deviations in cell speed. The vertical error bars are the errors in classifying reorientations. The dependence of P_F on V was confirmed by measuring P_F for each $[Na^+]$ tested and binning each data set by swimming speed (filled grey symbols, >300 trajectories each). The variation in cell speed was obtained by that naturally arising within the population. This rules out physiological effects associated with changes in $[Na^+]$ as a possible cause of the suppression of buckling events (Son et al., 2013).

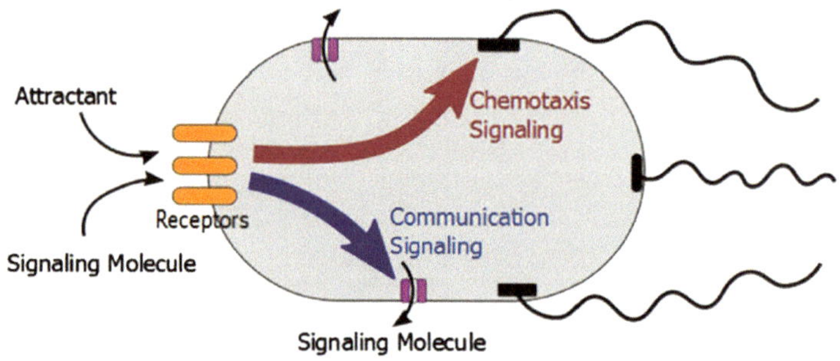

FIGURE 31.28 Model of chemotaxis and population communication signaling in *E. coli*. The red arrow represents the well-understood chemotaxis signaling pathway. The attractant gradients are first sensed by chemoreceptors and further transduced via a set of cytoplasmic signaling proteins to the flagellar transmission. The blue arrow stands for the hypothesized communication signal-transduction pathway, by which a population of bacterial cells actively secrete signaling molecules into the surrounding medium. This aids population response to sensed steep gradients of preferred metabolic substrates (Long et al., 2017).

2.3 Archaella (synonym archaeal flagella)

Another rotating external appendage, filament, or rigid helical propeller is the archaellum (Fig. 31.29) of the Archaea domain (reviewed by Jarrell et al., 2021). It has no homology with the flagellum of bacteria. However, just as for bacterial flagella since the archaella are very narrow, they cannot support motors down their length and hence must rotate to produce thrust giving them a superficial similarity to the flagella of bacteria. Swimming about 3 μm/s, they swim about one order slower than *E. coli*. The movement of archaella, bacterial type IV pili (T4P), and type 2 and type 4 secretion (T2S and T4S) systems are driven by superfamily ATPases. Their rotation rate is inversely proportional to load (Fig. 31.30).

The helices are always right handed, no matter the rotation direction unlike bacterial flagella that have multiple conformations, which means that when archaellum reverse their motors they backup (see, Fig. 31.1). Tethered cells have been observed to step in either 60 degrees or 36 degrees steps, six or 10 per revolution, which may hint at the mechanism of rotation. Very fast swimming speeds have been observed by Herzog and Wirth (2012) from 3 to 380 μm/s for different Archaea. Unlike *E. coli* Archaea do not swim with smooth paths followed by tumbles to alter swimming direction, but rather they rapidly swim a long distance taking a relatively straight path but once reaching near a surface make zigzag motions at much lower speed.

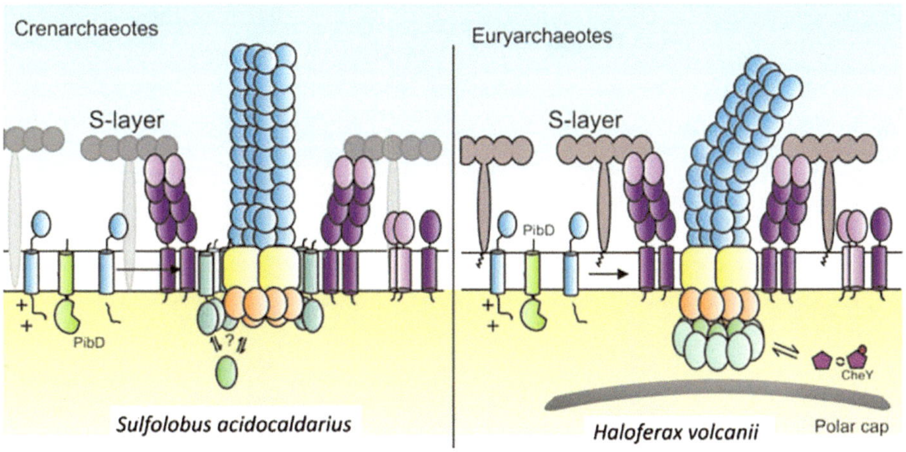

FIGURE 31.29 Models for two examples of archaella in Crenarchaeotes and Euryarchaeotes. *From Jarrell et al. (2021), for details see reference.*

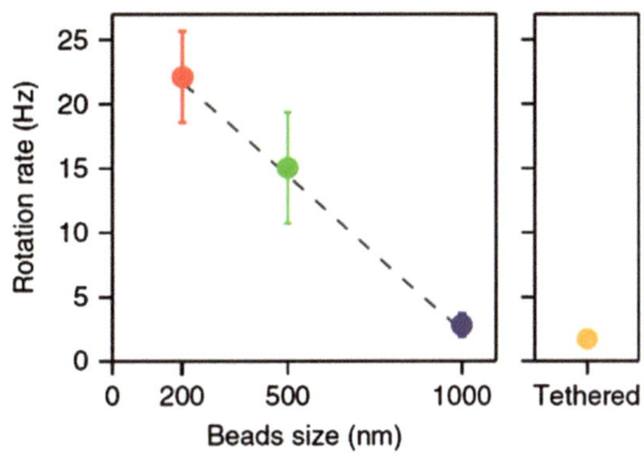

FIGURE 31.30 Archaella rotation rate is inversely proportional to load. Rotation rates of markers of different sizes with their standard deviations. Dotted line shows a linear fit, $V = A/\phi + B$, where V is rotation rate; ϕ diameter of bead; B = 27 Hz, an intercept; and $A = -0.024\ \text{Hz nm}^{-1}$, a proportional coefficient (Fig. 31.2E, Iwata et al., 2019).

See Reindl et al. (2013) for details of the hexamer structure (6 FlaIs arranged in a circle) and potential conformation changes. However, it does seem the rotation mechanism is not yet well understood. A more recent proposal to explain Iwata et al.'s estimated torque of 160 pN nm, is based on the concern that a single rotation requires work of 2π times the torque or ~1000 pN nm. This would require the hydrolysis of 12 ATP molecules (two per FlaI subunit) given that ΔG ATP is 80 pN nm in cells (Iwata et al., 2019, Fig. 31.31).

3. Bacterial gliding

The bacterium *Flavobacterium johnsoniae* colonizes and moves over surfaces (Shrivastava et al., 2015, 2016). It makes a mobile and adhesive external tread (in yellow Fig. 31.32) that is stuck to the substrate by adhesins. This tread moves on an internal right-handed track (in red Fig. 31.32) in the direction of the arrow forcing the cell forward in the opposite direction namely that of the green arrow in a spiral fashion. It is a self-propelled screw with external adhesion. The adhesins (SprB) move along their external threads at about 2 μm/s. While this bacterium uses PMF to driven its motor, but unlike the constant torque motor that drives flagella, its motor appears to be a constant speed motor, agl.

The motors for gliding motility in *Myxococcus xanthus*, also known as adventurous (A-motility), are proton channels formed by three proteins AglR, AglQ, and AglS. This proton channel/motor complex is homologous to the *Escherichia coli* flagella pinion-stator complex MotAB (AglR is a MotA homologue, while AglQ and AglS are MotB homologs). The gliding motors move rapidly along helical tracks in the inner membrane. They slow down and assemble into multicomponent gliding complexes at the putative focal adhesion sites and propagate PMF onto the cell surface. *M. xanthus* coordinates with an additional motility system, the S-mobility system, which uses a version of the typeIV Pilus system with

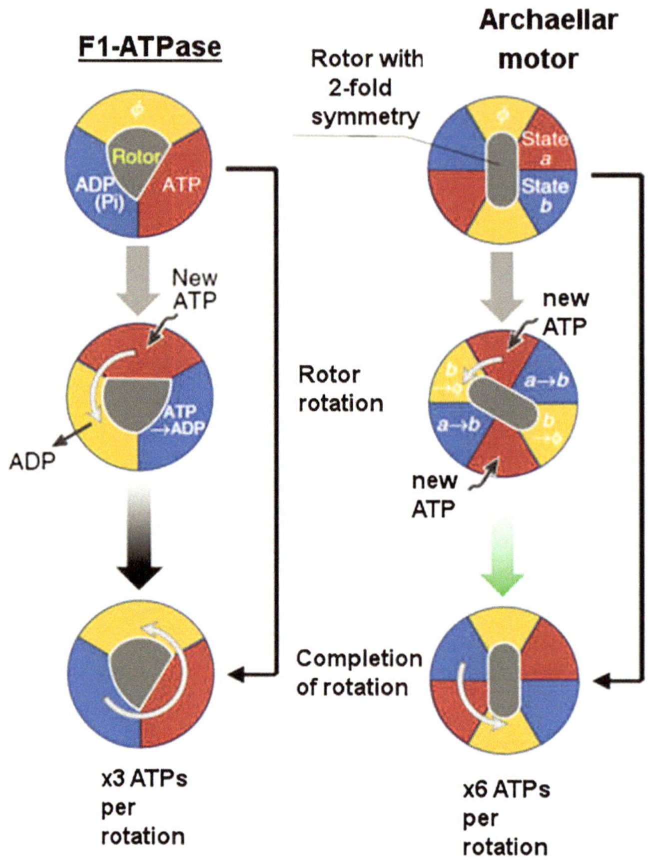

FIGURE 31.31 To explain the estimated torque of 160 pN nm of Iwata et al. (2019) being larger than the available work from hydrolysis of six ATP molecules, one per FlaI, they suggest more are needed per FlaI. The analogous F1-ATPase has three catalytic subunits in which three ATPs are hydrolyzed per turn of the rotor (left side of figure) and the expected six in archaella (right side of figure). The hypothesis is more than one ATP can be utilized per FlaI. Two per FlaI would be enough to meet the need of at least 12 ATP hydrolyzes per rotation. *Modified from Iwata et al. (2019).*

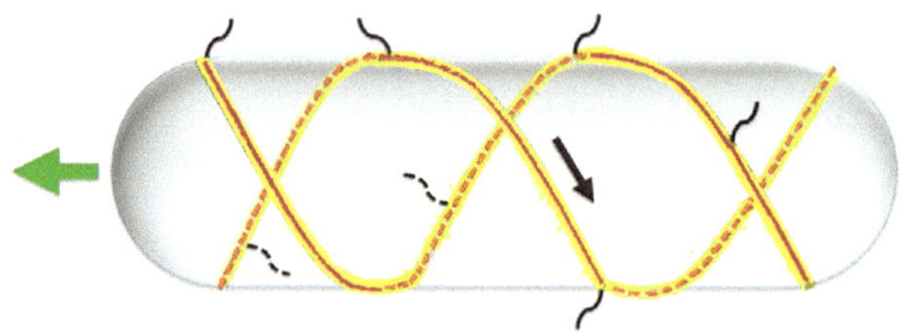

FIGURE 31.32 A screw-like mechanism for bacterial gliding. A track (red) is fixed on the cell surface. SprB (black) and a tread (yellow) move relative to the cell surface over the fixed track. The arrow shows the right-handed motion of the SprB and the tread. Attachment of SprB with an external substratum results in forward motion and right-handed rotation of the cell body (Shrivastava et al., 2016).

the usual extenstion and retraction, to move unidirectionally along the long cell axis. To achieve optimal motility efficiency, cells reverse their moving direction regularly and control the reversal frequency in a narrow range (Nan & Zusman, 2016; Pogue et al., 2018).

There would appear to be many potentially fascinating gliding mechanisms out there among bacteria that remain to be studied.

4. Type IVa pilus machine and its twitching motion

Many bacteria have several micrometer long, 5–7 nm diameter appendages called type IVa pili (T4P) Fig. 31.33 (Chang et al., 2016; Craig et al., 2019). On the surface of bacterial cells T4P are "grappling hooks" that extend, attach to surfaces,

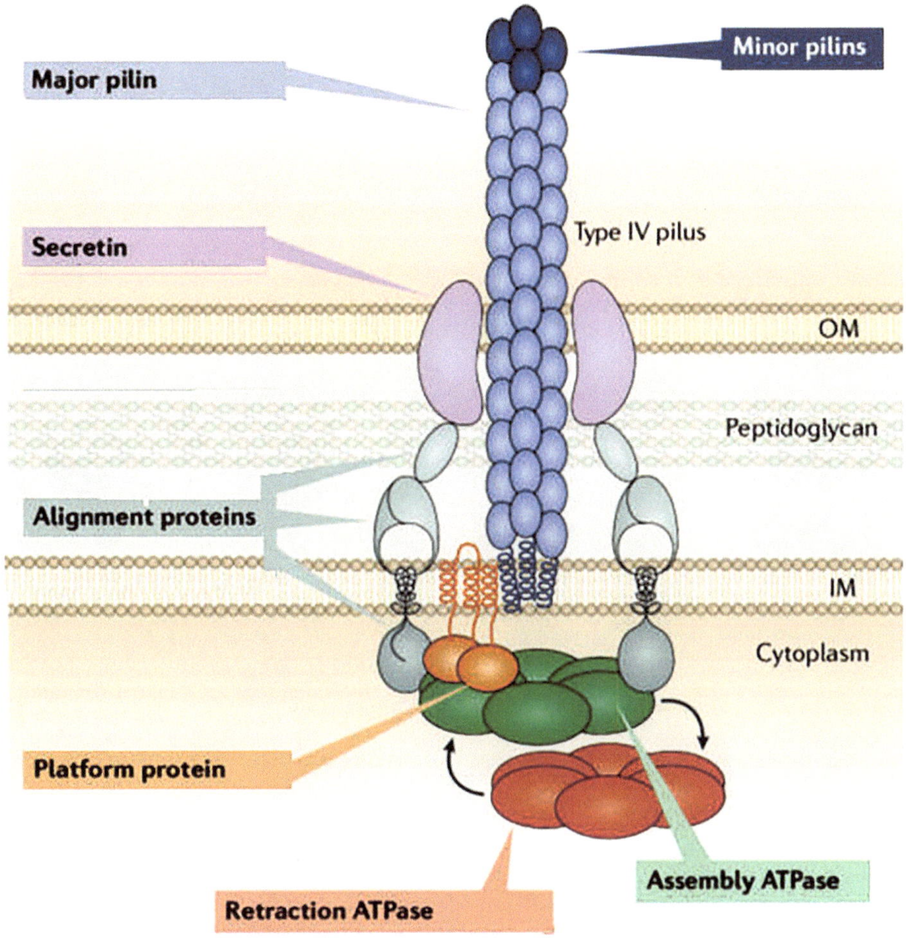

FIGURE 31.33 The type IV pilus (T4P) machine utilizes a cytoplasmic assembly ATPase (shown in green) to drive polymerization and pilus growth from the inner membrane (IM), across the periplasm and through the outer membrane (OM) secretin channel (purple). Most T4P have a second retraction ATPase (red) to drive pilus depolymerization. Both ATPases likely interact with the cytoplasmic domains of the IM platform protein (orange) and are stabilized by cytoplasmic domains of an alignment protein (grey). ATP binding and hydrolysis-induced conformational changes incrementally extrude the pilus upon each pilin subunit addition during assembly ATPase and release pilins back into their membrane reservoir during retraction. *Modified from Fig. 31.1 Craig et al. (2019).*

and then retract to pull the bacteria to the point of attachment. This is referred to as twitching motility, involves very rapid retraction, and forces in excess of 100 pN (compares to ~ 2 pN for a single dynein motor). Because of their homology to the type IVb pilus system, the type II secretion system, archaella, and bacterial competence systems, they are important to understand. While T4P extend and retract instead of rotate, twitch instead of swim, and have no anchor they have many similarities to archaella. Their archaellins are similar to type iv pilins; they utilize ATP hydrolysis and grow from their base, have similar N-terminal sequences on the pilins/archaellins, have class iii signal peptide homology, and may have glycoslyation of structural proteins (Jarrell et al., 2021).

Most interesting are the assembly and retraction motors. These ATPases of the T4P systems belong to the family of secretion ATPases including the motor that drives type II secretion. Following the argument of Craig et al. (2019), all secretion ATPases possess an N- terminal PAS-like domain and a conserved C-terminal RecA-like motor domain (Fig. 31.34A). Assembly ATPases also possess additional N-terminal GSPII domains. These motors are hexameric rings (Fig. 31.34B) with subunits having stable interactions between the C-terminal domain (CTD) and the N-terminal domain (NTD) of adjacent polypeptides N and N + 1, respectively (Fig. 31.34B and C). ATP is thought to bind to subunits in the cleft between NTDN and the conserved nucleotide-binding loop in the CTDN. The favorable interaction of positively charged and invariant NTD arginine side chains with the negatively charged γ-phosphate of ATP pulls NTDN toward CTDN to clamp ATP in place, exerts force on the adjacent subunit, N + 1. This conformational change likely leads to ATP hydrolysis and/or to cleft opening and the release of bound ADP in subunit N + 1 (Fig. 31.34C and D) occurring in a

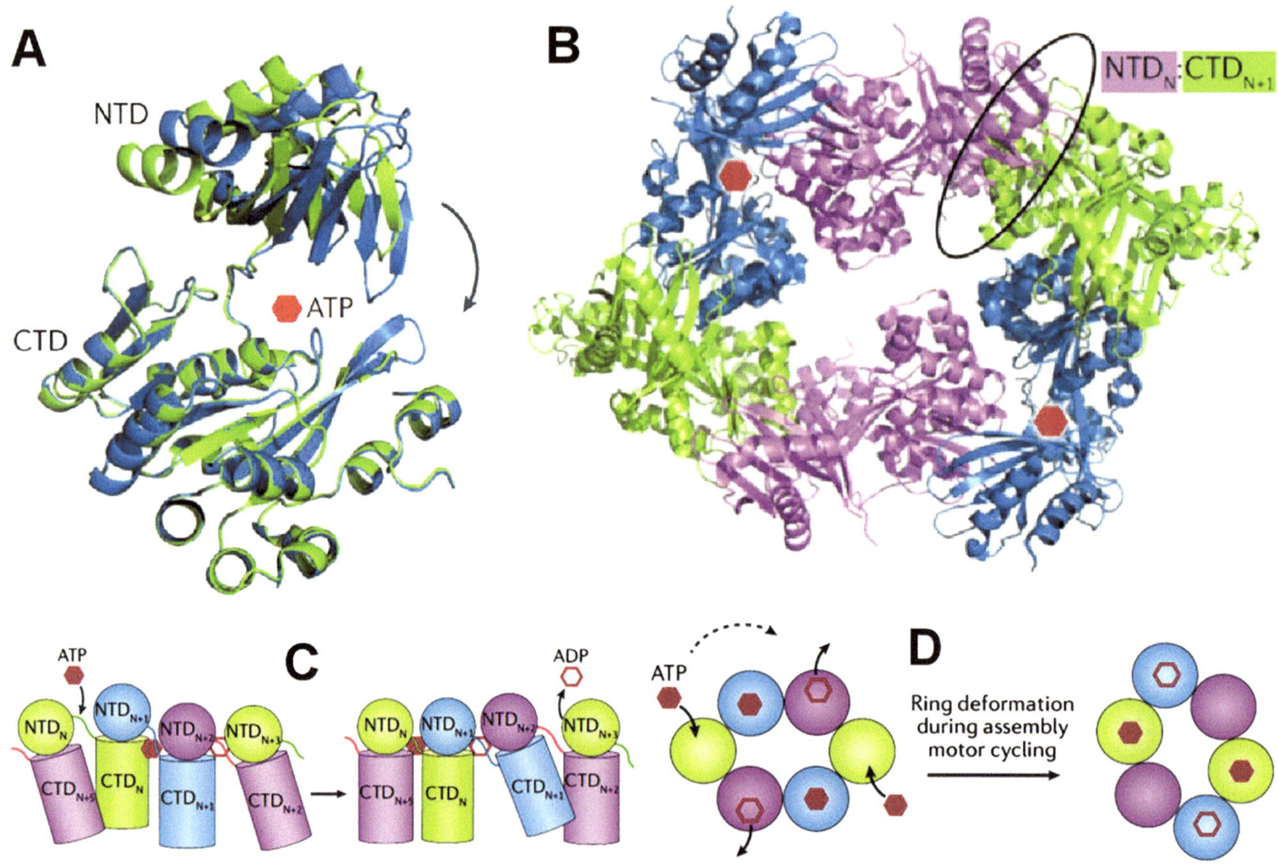

FIGURE 31.34 This figure depicts a model (Craig et al., 2019) for ATP-induced conformational changes in assembly and retraction of ATPases. (A) Monomers of the assembly and retraction ATPases have a bilobed structure with an amino-terminal domain (NTD) and a carboxyterminal domain (CTD) (*Pseudomonas aeruginosa* PilT). ATP enters an open site between NTDN and CTDN and causes closing of the nucleotide cleft. (B) Assembly and retraction ATPases form oblong hexamers in which opposing subunits adopt the same conformation (top view of *Geobacter sulfurreducens* assembly ATPase). Numerous interactions between NTDN and CTDN+1stabilize the ring. (C) The four subunits shown schematically form a hexamer. As ATP binds in the cleft of subunit N, closure exerts force on subunit N + 1. This may lead to ATP hydrolysis, ADP release and opening or closing of subsequent subunits in the hexamer in a cooperative manner. opposing subunit pairs in equivalent subunits have the same colors. Motions are in 3D affecting dynamic positioning to the platform protein lying above the ring. (D) Sequential cycles of ATP binding, ATP hydrolysis and ADP + Pi release cause structural deformations that moves around the motor hexamer. The open versus closed state of subsequent subunits differs for the assembly and retraction so they have substantially different pore deformations with implications for motions of the platform proteins. *Modified from Fig. 31.3, Craig et al. (2019).*

cooperative and directional manner. The structures are oblong two-fold symmetric (C2) hexamers, in which each pair of opposing subunits adopts a distinct nucleotide-dependent state with relatively more open or closed arrangements (Fig. 31.34B). A possible scenario for assembly and retraction is suggested in Fig. 31.34 caption.

In any case, major pilins diffuse from their inner-membrane IM reservoir into the base of the growing pilus, attracted in part by charge complementarity between their Glu5 and the positively charged amine in the terminal pilin in the growing pilus, which neutralizes these charges in the hydrophobic acyl phase of IM. Subunits add to the pilus along a right-handed helix, where they are related by a ~100 degrees rotation and an 8−10 Å rise. Thus, subunits enter the pilus base from different directions separated by 100 degrees around its axis, and the filament is extruded 8−10 Å perpendicular to the membrane with each addition, creating a gap for the next subunit to dock (Craig et al., 2019).

5. Biological springs

5.1 Contractile springs

The centrin-based family of springs typically contract, sometimes slowly, but occasionally dramatically. These calcium-binding proteins are associated with centrioles and basal bodies in many eukaryotes including humans. Centrin (also called caltractin) assembles into fibers that can contract when the calcium concentration is increased and, in some organisms, the contraction of centrin fibers are used to steer cell motility by changing the angle at which cilia emerge from the cell surface of *Chlamydomonas* (Foster, 2009). The distal fibers connecting the two cilia and the nuclear-basal body connectors respond in a monotonic fashion to the internal calcium concentration. They set the base impedance for the attachment of a cell's cilia to the cell body. Their contraction is much faster than their extension. While the exact purposes of the settings are not understood, this response does facilitate the angle changes needed for flagellar versus ciliary beating.

On the other hand, spasmin, which also belongs to this family, is used (40%–60% of the spasmoneme dry mass) in the contractile spring spasmoneme. They are an example of a molecular spring designed to release a large amount of stored energy almost all at once and then slowly recover (Marshall, 2008). Calcium release triggers further release of calcium in positive feedback. Contraction of isolated spasmonemes is driven by increasing the calcium concentration from 10^{-8} M to 10^{-6} M. Multiple cycles of contraction and extension require multiple switches of calcium concentration requiring energy. The spasmoneme is triggered when a *Vorticella* cell feels threatened, contracting its tail at a rate of 60–100 mm/s with a tensile force up to 500 nN (Figs. 31.35 and 31.36) (Upadhyaya et al., 2008). The speed works out to about 200 lengths per second (L/s), faster than the fastest muscles, which contract at around 20 L/s. Since the spasmoneme winds helically inside the stalk, the contraction collapses the straight stalk into a helix. The reverse process (recovery) is powered by dissociation of Ca^{2+} ions from spasmin and active sequestration back into the calcium storage sites. Of interest are what molecular mechanisms drive the contraction and how is the contraction coordinated along the length of the entire structure. Using high-speed video microscopy, the rate of contraction has been measured as a function of the viscosity of the surrounding media (see Fig. 31.37) (Misra et al., 2010). From the scaling relation between maximum speed and viscosity, one can conclude that the speed is limited by the power dissipated by dragging the top of the *Vorticella* through the surrounding viscous media. A question arises as to how the contraction-triggering signal can be transmitted over the whole length of the spasmoneme within a few milliseconds. The movement of beads attached to the *Vorticella* stalk show that contraction initiates near the body of the *Vorticella* and propagates like a wave down the stalk (see Fig. 31.37). Since contraction is driven by calcium binding, this suggests a calcium wave mediated by calcium-triggered calcium release from the endoplasmic reticulum. In vivo experiments show that a Ca^{2+} concentration as small as 10^{-7} M is sufficient to trigger the release of stored calcium. However, it is unclear whether such a wave can be as fast as the observed 100 mm/s rate. Perhaps, there is an electrical signal or a propagating wave of protein conformational change within the spasmoneme. Presumably, the spasmoneme evolved from other centrin containing structures under selective pressure to contract at high speeds. Detailed molecular comparisons of centrin and spasmin might shed light on the adaptations that allow spasmin to contract so fast.

5.2 Expanding springs

The acrosomal bundle (cross-linked actin fibers) of the horseshoe crab (*Limulus*) sperm cell is a molecular spring (see Fig. 31.38) (Tam et al., 2009). The actin bundle is stored at the base of the sperm cell as a twisted coil. Calcium binding to the actin cross-linking protein, scruin, changes its conformation. With further calcium binding along the bundle along its

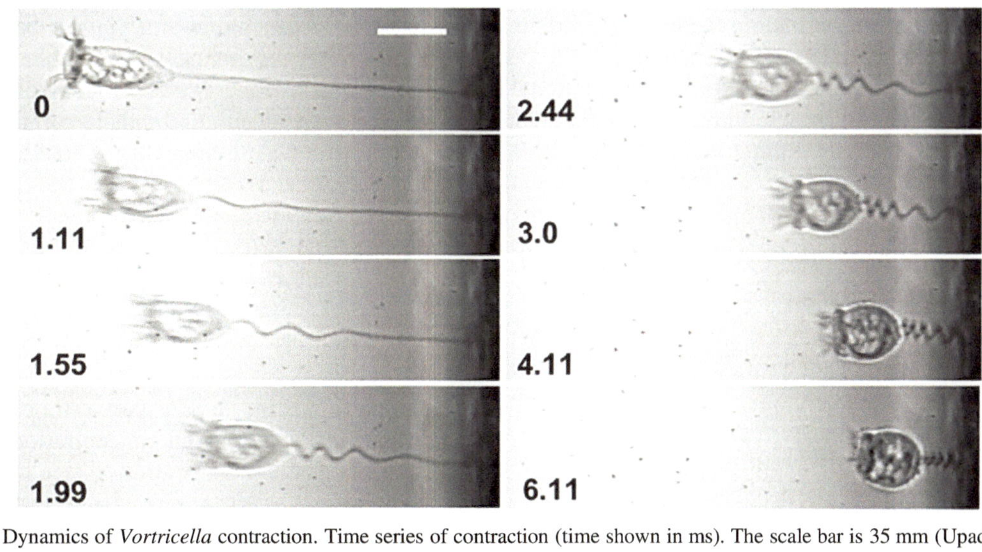

FIGURE 31.35 Dynamics of *Vortricella* contraction. Time series of contraction (time shown in ms). The scale bar is 35 mm (Upadhyaya et al., 2008).

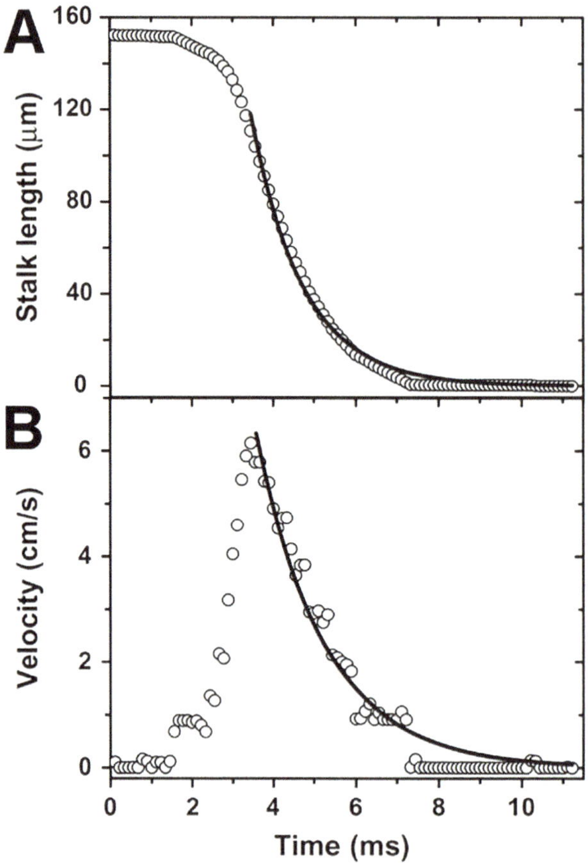

FIGURE 31.36 (A) *Vorticella* stalk length as a function of time during a contraction. The solid curve is an exponential fit (Upadhyaya et al., 2008). (B) Instantaneous velocity of the cell body as a function of time. The solid curve is an exponential fit to the decaying part of the velocity (Upadhyaya et al., 2008).

length, the acrosome extends from the sperm cell in order to penetrate and fertilize the horseshoe crab egg. Provided external calcium is continuously available, the 60-μm long bundle extends in 15 s, fast relative to either polymerization or motor protein drive movements. This device illustrates how conformation energy in the form of a twisted compliant bundle can be rapidly released by calcium binding to generate significant cellular movements. During the acrosome reaction, a 60-μm long coiled and twisted bundle of actin filaments straightens and extends from a sperm cell, penetrating the vitelline layer surrounding the egg. A subtle over twist of 0.23 degrees/subunit (5 nm in size) relative to the natural chiral structure of an actin filament as a straight rod underlies the mechanochemical basis for the extension of this actin spring. Upon calcium activation, this conformational strain energy is converted to mechanical work, generating the force required to extend the bundle through the vitelline layer. The calcium sequentially binds to calmodulin molecules decorating the actin filaments leading to a collective wave of untwisting of the actin filaments. About a million subunits of actin-scruin-calmodulin achieve a constant expansion of 10 μm/s (3 μm/s at 10°C, 37 μm/s at 30°C (see Fig. 31.39) with a maximum of force of 1.9 nN and a puncturing pressure of 1.6 MPa (about 15 atm). This force of extension is three times larger than that needed to puncture the vitelline layer. Mahadevan et al. (2011) proposes the following model for this mechanochemical engine: (1) Ca^{2+} binds to scruin; (2) the scruin—actin interactions subtly change; (3) a small region of the coil begins to untwist mediated by the scruin which decorates neighboring actin filaments; (4) driven by the potential energy difference between the coil and true states of the bundle, this untwisting deforms the bundle of actin without completely uncross-linking it in the binding region; and (5) later Ca^{2+} binds to calmodulin releasing the twisted state, leading to the propagation of a front of untwisting that travels along the bundle converting the twisted state irreversibly to the extended state. The volumetric rate of expansion is constant at a given concentration of calcium, and the volumetric extension rate is proportional to the continuously available calcium concentration (see Fig. 31.40). These observations (Tam et al., 2009) make sense if the binding of Ca^{2+} to the scruin—CaM complex is according to the simple reaction mechanism:

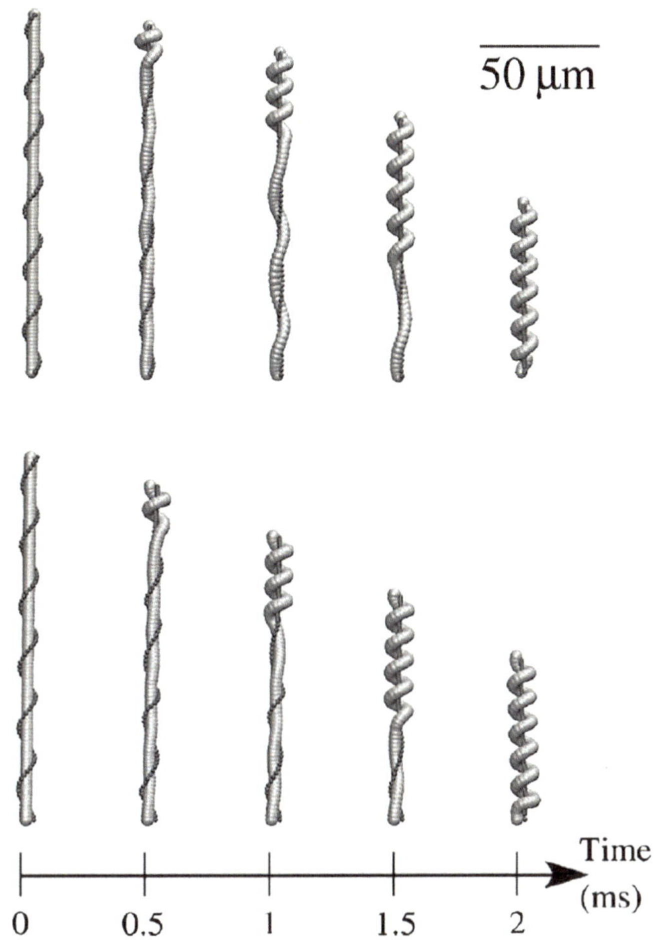

FIGURE 31.37 Time-lapse images of the initial phase of *Vorticella* contraction driven by a propagating calcium signal (for clarity, the head is not shown). (Top) Young's modulus Y = 1 kPa. (Bottom) Y = 4 kPa (Misra et al., 2010).

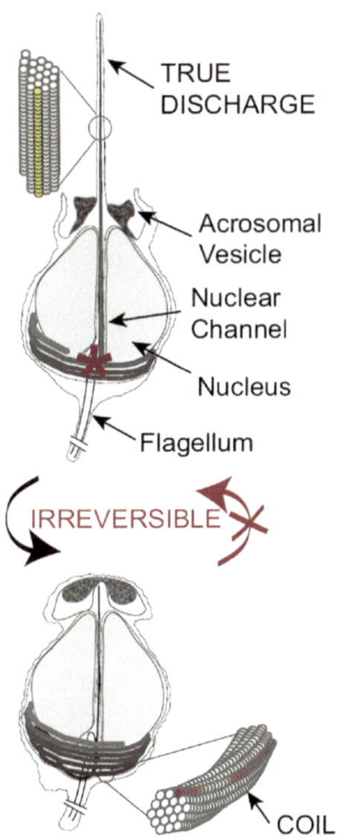

FIGURE 31.38 Expansion of actin spring. The geometry and dynamics of the acrosomal reaction. Geometry shows that, in the presence of Ca^{2+}, the bundle switches irreversibly from coil (bottom) to true discharge (top) (Mahadevan et al., 2011).

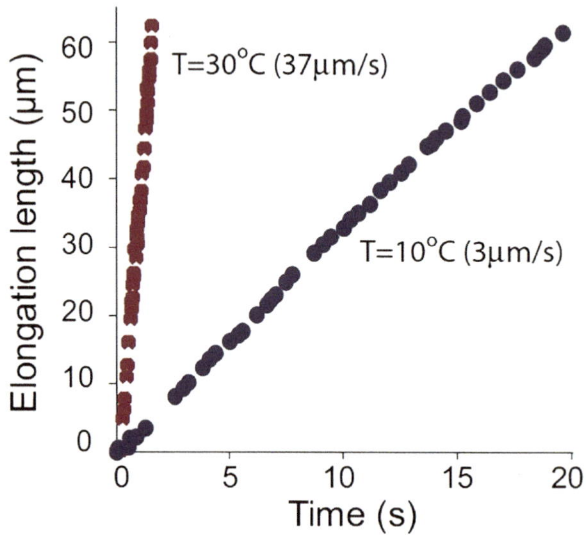

FIGURE 31.39 Extrusion occurs at constant speed. The two traces correspond to extrusion at 10°C (3 mm/s) and 30°C (37 mm/s) (Mahadevan et al., 2011).

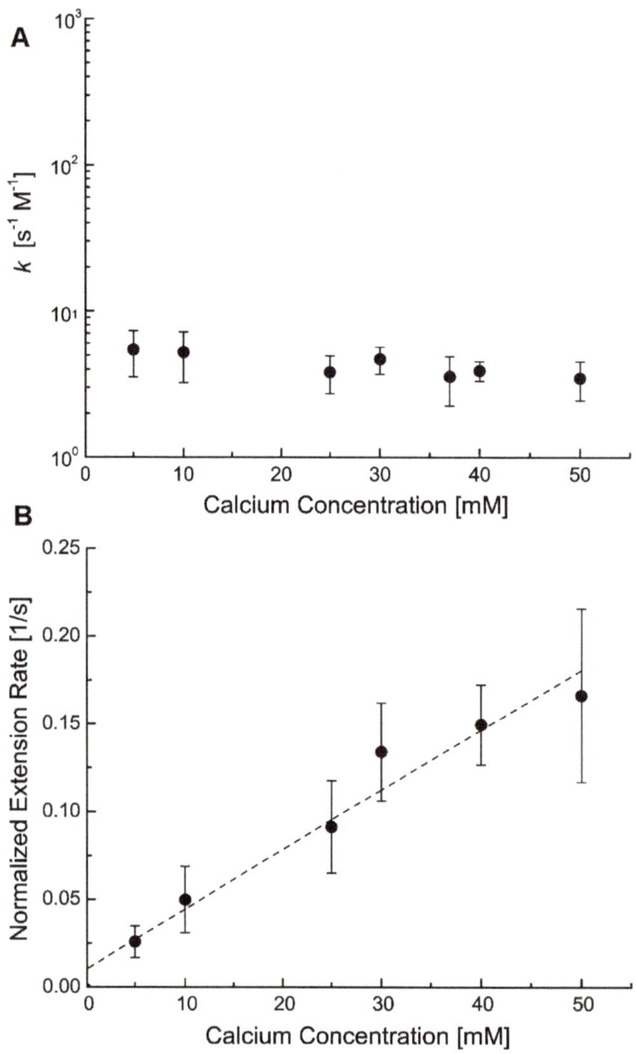

FIGURE 31.40 (A) Rate constant k calculated using a non-cooperative, constant rate CaM-Ca^{2+} binding mechanism. The slight lack of constancy of [Ca^{2+}] suggests perhaps an overestimation of [CaM] participating in the reaction. (B) Normalized volume extension rates. Values were calculated from least-squares fitting of volumetric extension profiles as functions of time. The linear behavior (dashed line) suggests a rate mechanism involving calmodulin with one calcium-binding site (Tam et al., 2009).

$CaM + Ca^{2+} \xrightarrow{k} CaM{\cdot}Ca^{2+}$, where k is an effective rate constant with units of $[M^{-1}\ s^{-1}]$, which governs the binding of Ca^{2+} to the scruin–CaM complex associated with the twisted filaments. How the strain energy is incorporated does not appear to be understood.

6. Cannons, catapults, and other shooters

When one is discussing shooting projectiles like spore cases, one has to be aware this is no longer the low Reynolds number regime. The velocities are very large and the viscosities low. The situation is much more complicated. A useful approximate equation is $F_{drag} = ½\ \rho A v^2\ C_d$, with ρ = density of the fluid $[kg{\cdot}m^3]$, A = cross-sectional area (πr^2 for a sphere) $[m^2]$, v = velocity of the projectile [m/s], and Cd = drag coefficient (i.e., a dimensionless quantity that quantifies what is not accounted for by the explicit variables).

Catapults were ancient siege engines that used stored energy to hurl a projectile without the need for explosives. The energy was stored by the potential energy of strained elastic compliant components, such as wood or rope. When this stored potential energy was suddenly released, it launched the projectile. These ancients were mimicking nature. Here we mostly follow the review of Sakes et al. (2016), which is recommended reading for the full range of the catapult mechanisms in fungi, plants, and animals. Further reviews are Forterre (2013), Guo et al. (2015), and Poppinga et al. (2015). Water maybe condensed on the outer surface or absorbed into the cells or evaporated from the cells of the shooting mechanism.

Consider the case of fungi shooting their spores or sporangia. They experience a large drag, which is the rate of removal of momentum from a moving fluid, as they fly. Once the sporangium is fired like a cannon ball, it is no longer self-propelled and comes rapidly to a stop due to viscous drag and of course sticking to objects.

6.1 Fired by fluid pressure

One method is to accentuate the turgor pressure Δp and store the energy in stretched spore cases (Vogel, 2005) as elastic potential energy. The absorption of water into the cell results in this high turgid pressure against the rigid cell wall. Alternatively, another way to store potential energy is from the evaporation of water causing a negative pressure (relative to ambient) inside the cell that pulls away from the cell wall, resulting in a plasmolyzed cell.

At the desired precise time and place (usually at the tip), a weakness is created that fires the spores and fluid out within the case at incredible velocities. The purpose of these "cannons" or catapults is to distribute seeds or, for example, spores from coprophilous fungi, like the zygomycete *Pilobolus* (Fig. 31.41), *the* squirt-gun fungus, (range about 0.8 m, launch velocity 20 m/s), away from the dung they are living on out onto forage plants where they may more probably be eaten. "Obulus" means throw. *Pilobolus* has a stalk (sterigma) Fig. 31.41, a balloon-like vesicle (Fig. 31.41) and a sporangium (containing 30,000–90,000 spores) is formed at the tip (Fig. 31.41) (Yafetto et al., 2008). The vesicle swells due to osmotic absorption of water and the hydrostatic pressure in it increases. Water diffuses through a semi-permeable membrane due to a concentration gradient of ions, sugars, or alcohols between two adjacent solutions. These osmolytes are transported actively through the cell membrane up a concentration gradient by transmembrane ATPase protein pumps or by coupled transport pumps that use the potential energy of an existing electrochemical gradient. At about 5.5 atm (0.55 MPa), the sporangium breaks free in 10–30 μs and with a jet of cell sap is propelled with an acceleration of up to ~21,000 g and a launch velocity of 16 m/s (mean 9 m/s), resulting in a launch distance up to 2.5 m. The firing of the ejected spores and collapse of the stalk of the sporangiophore is powered by the release of elastic energy from the contracting wall of the vesicle. To first order, the distance will be proportional to the diameter squared times the firing velocity (given that the other variables such as density of the projectile and the viscosity of the medium are more or less the same). The energy in a spore case or ascus that can be stored is proportional to the pressure squared and the radius cubed. Given that the biomaterial is basically the same in these organisms, the limit of wall tension is likely to be similar limiting the range of design. The wall tension is due to the pressure difference Δp (inside minus outside) times the radius, so this product is fairly constant among organisms to prevent premature blowout. This means that large asci will have a lower pressure. Hence, a large ascus might have a turgor pressure of 3 atm while a small one up to 15 atm. The energy stored in the stretching of the chamber, which starts out crinkled up, presumably is what leads to its rapid contraction maintaining the pressure during the emptying of the chamber resulting in a complete release of the contents. The simple consideration that the work, ΔpV, where V is volume available due to turgor pressure, should be equal to the consequent kinetic energy of the projectile, $\rho V v^2/2$, where ρ is the particle density and v is the launch velocity leads to:

$$\text{the launch velocity: } v = \sqrt{\frac{2\Delta p}{\rho}}$$

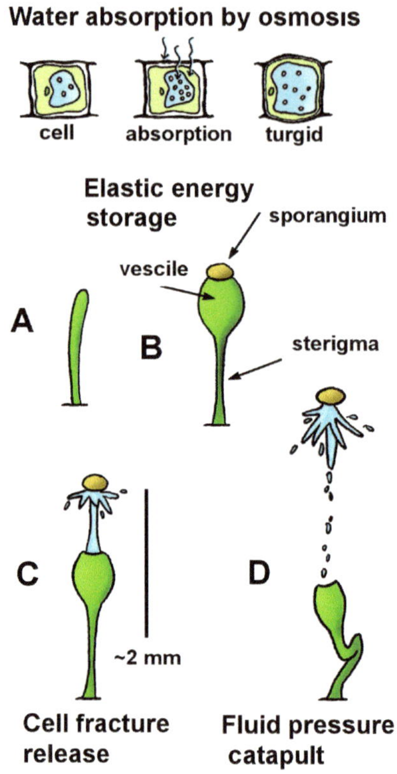

FIGURE 31.41 Fluid-pressure catapult mechanism of the zygomycete *Pilobolus kleinii*. (A) Early stage of sporangiophore growth. (B) Sporangiophore with its sterigma (stalk), balloon like vesicle, and the sporangium at the tip. (C) When a critical pressure ($\sim$0.55 MPa relative to ambient) is reached; the sporangium and cell sap is squirted by the contracting vesicle wall. (D) The sporangiophore collapses. Scale bar 2 mm (Sakes et al., 2016).

However, the energy storage capability in the larger chamber more than makes up for the lower pressure. The osmolytes include sugars, but probably primarily come from ions pumped into the chamber. Mechanisms for triggering the breaking of the tip, firing the cannon, do not seem to be described. There is a helical orientation of wall fibers as is typical for pressurized systems in nature. Also note that the hoop stress is double the longitudinal stress. A further sort of amazing aspect is the incredibly large accelerations seen. This is of course possible due to the small mass of the projectiles, $a = F/m$ (acceleration = force/mass). For example, the *Gibberella zeae* sporangium exhibits an acceleration of 8.5×10^6 m/s^2, while the relatively large 200 μm diameter *Pilobolus* (Fig. 31.41) projectile *only* has an acceleration of 5×10^5 m/s^2 (50,000 times the acceleration of earth's gravity). Finally, these systems are interesting from the hydrodynamic point of view as they have an aerodynamic drag problem. The *Pilobolus* sporangium loses 98% of its potential range due to drag (if it could fire into a vacuum it would be really amazing and maybe have a 40 m range). Perhaps not surprisingly spore shapes seem to be optimized to minimize their drag (Roper et al., 2008).

Basidiobolus ranarum has a spore-producing structure similar to that of *Pilobolus*, except that it discharges only a single spore (conidium). The wall of the spore-bearing structure ruptures around its circumference with an estimated internal pressure of 0.01–0.72 MPa relative to ambient. The spore's peak launch acceleration is 150,000 g, its peak launch velocity is 9 m/s (mean 4 m/s), and peak launch distance is 20 mm. The reduced distance between *Pilolobus* and *Basidiobolus* is because of the smaller size of the conidium versus the spongium. This leads to a relatively greater effect of viscous drag.

In the above, the cannons fire projectiles, however, for every action, there is a reaction, and sometimes the reaction is quite important, likely getting oneself out of the way of a predator. *Paramecium* fires trichocysts from its body enabling a transient speed of 10 mm/s, approximately 10 times the speed it achieves with its synchronized waves of ciliary beating (see Fig. 31.42) (Hamel et al., 2011).

6.2 Fired by water condensation

Energy need not be stored elsewhere in the cell wall. In the Basidiomycota, two water droplets coalesce (the so-called Buller's drop and adaxial drop) on the spore surface generating the energy to discharge the spores by momentum

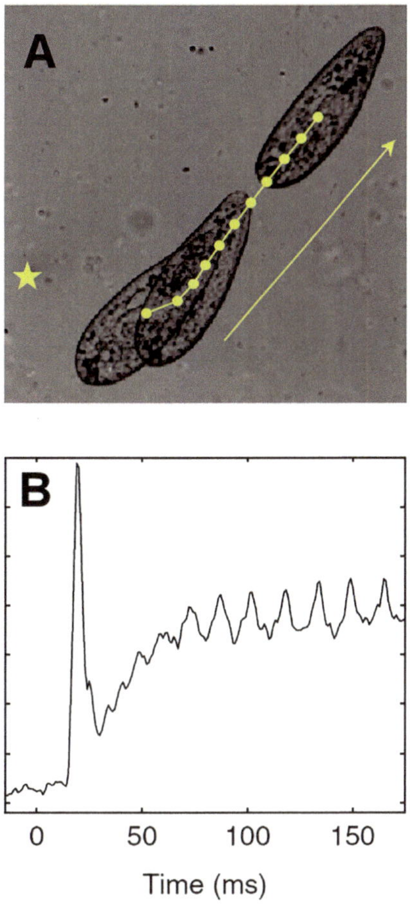

FIGURE 31.42 (A) Images showing *Paramecium* at different times following a nearby laser pulse. (B) Radial velocity of the center of mass away from the laser position for the experiments shown above. Time t = 0 corresponds to the moment the laser is switched on (Hamel et al., 2011).

transfer. The nearly spherical Buller's drop and the hemispherical adaxial drop grow due to the secretion of osmolytes onto the spore surface, lowering the center of mass of the spore (Fig. 31.43A and B). When Buller's drop and the adaxial drop reach a critical size, contact each other, and begin to coalesce, they generate a compression force on the stalk and a counterforce on the spore (Fig. 31.43B and C). With further coalescence, the momentum transfer from the merged drop to the spore and the braking of the drop at the spore's tip break the hilum releasing the spore (Fig. 31.43D). Launch accelerations vary from 3300 to 25,500g, velocities from 0.1 to 1.8 m/s, and distances from of a few μm to a few mm.

6.3 Fired by cavitation

Cavitation based spore discharge occurs in the Ascomycota *Deightoniella torulosa* (incertae sedis). This fungus grows its spores on stalks called conidiophores. When dehydration causes the cell membranes to shrink, the cell walls begin to cave inward increasing the tension within these walls deforming the structure. When negative pressure exceeds the tensile strength of the cytoplasm, the cytoplasm fractures, resulting in the sudden formation of a gas or cavitation bubble that releases the wall tension. The walls rapidly release their absorbed strain energy returning to their original shape firing the spores out of the conidiophore. A similar cavitation coiling catapult mechanism is also observed in *Zygophiala jamaicensism*, which has a conidiophore with two sporogenous cells and conidia at the apex of the conidiophore (Fig. 31.44A). Drying of its conidiophore deforms it into an S-shape (Fig. 31.40B). At the sudden appearance of a gas bubble, the conidiophore springs back to its former shape, firing the conidia from the sporogenous cells (Fig. 31.44C and D).

6.4 Fired by osmotic-powered eversion or snap buckling

In coprophilous *Sphaerobolus*, the cannonball fungus, the stored elastic energy is released by the eversion of a membrane. The fungus fires a spore-filled package called a gleba with a diameter of 1 mm. The immature base of *Sphaerobolus* is a closed sphere (Fig. 31.45A). At maturity, this sphere splits radially from its apex, forming a star-shaped cup that envelops the

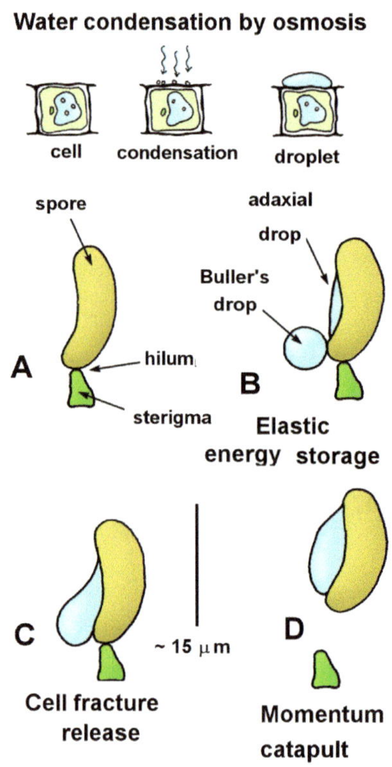

FIGURE 31.43 The momentum-firing mechanism of an *Auricularia auricular* (Basdiomycota). (A) The sporogenous cell before discharge with the spore attached to the sterigma at the hilum. (B) By secreting osmolytes, Buller's drop and the adaxial drop grow on the surface of the spore. (C) When Buller's drop reaches a critical size the two drops coalesce. The loss in surface area releases some surface tension energy generating a compression force on the sterigma which is converted into the kinetic energy to launch the spore. (D) Further the rapid shift of the joint center of mass of the spore and the coalesced drops puts more tension on the hilum and when a critical tensile stress is reached, the hilum breaks and the spore (together with the coalesced drops) is discharged. Scale bar 15 μm [μm] (Sakes et al., 2016).

gleaba (Fig. 31.45B). The cup consists of a firm outer case and an elastic inner membrane (peridium). By solubilization of glycogen and subsequent absorption of water, the turgor pressure within the radially oriented cells of the elastic peridium increases. As the peridium has a concave form, the exposed ends of the radially orientated cells are more compressed than their bases storing strain energy within the cell walls. The osmotic-powered sudden eversion of the peridium surrounding the package, similar to the kids toy "jumping poppers," relieves the strain (Fig. 31.46). The spore package is fired (approximately vertically) with an estimated peak launch velocity of 10 m/s, reaching up to 6 m from its base (Fig. 31.45C and D). The large launch distance (partly) compensates for a low efficacy of wind dispersal, as the gleba is too heavy to be swept away by the wind. A similar event occurs with the snap-buckling of the Venus flytrap (*Dionaea muscipula*) (Fig. 31.47).

6.5 Animal cannons

The contraction of muscle fibers near or in the projectile in animals can generate the energy for firing "cannons." Examples include tongue and other appendage protrusion to stun and capture prey and shooting of love darts in mating. The muscle sarcomeres contain actin and myosin filaments, which form cross-bridges. ATP-driven mechanical power strokes, which attach and reattach myosin along the actin track, shorten the muscles. The elastic energy is stored in collagen or other fibrous structures such as resilin. These elastic compliant tissue structures can recoil much faster than muscles can shorten. This stored energy on brain commend can be transformed into the kinetic energy of the projectile. In fact, their peak mass-specific power outputs can be 500 times that of muscles and are less temperature sensitive. The stored elastic energy for shooting maybe released by eversion of the projectile itself as in cnidarians, relaxation of collagen fibers in the ballistic tongues in chameleons and lungless salamanders, or release of a latch in stomatopods.

7. A few lessons learned

1. Motion is commonly a result of conformation change. Whether it is flagellar motors or dynein, kinesin, or myosin or centrin contraction, always, it has been found that the key is protein conformation change. This is unlike the mechanical

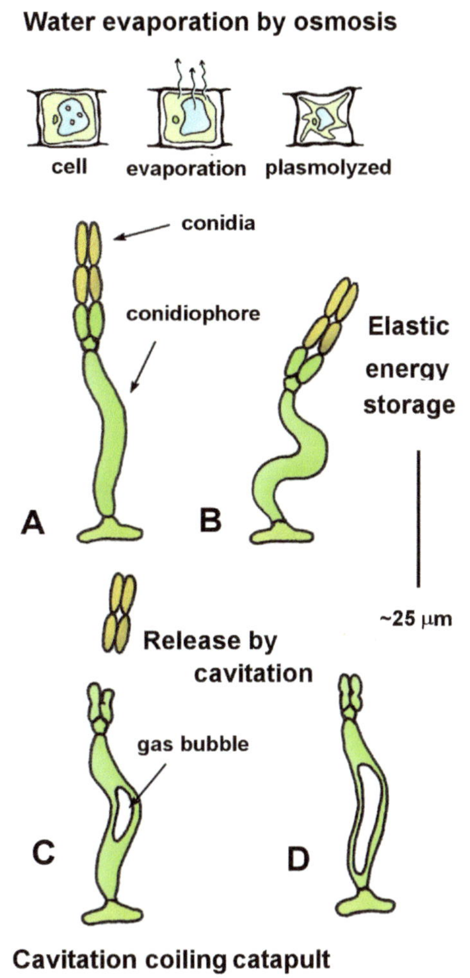

FIGURE 31.44 The cavitation-firing mechanism in *Zygophalia jamaicensis*. (A) A conidiophore with two divergent conidia at its apex. (B) Drying causes an S-shaped compression of the conidiophore. (C) When a critical negative pressure (compared to ambient) is reached a gas bubble suddenly forms releasing the stored elastic energy in the compliant cell walls and discharges the conidia from the sporogenous cells. (D) Conidiophore after discharge. Scale bar 25 μm (Sakes et al., 2016).

devices that humans have designed where the shape of stators, pinions, and rotors do not change. The components simply move relative to each other under electromagnetic forces. There are of course biomimetic motors that do depend on repeated conformation change as, for example, ultrasonic motors. An advantage of ultrasonic motors and all the biological motors described here is that they do not have to be energized to hold their place, which is an enormous economy for portable devices like biological cells.

2. Note continuity of torque or force on each side of stall in each case implying the reversibility of actions, "there is no irreversible step in the mechanochemical cycle" (Sowa & Berry, 2008).
3. Elastic storage mechanisms abound making compliance mechanisms important for many organisms.
4. Component compliance is also taken advantage of to control swimming direction as in the buckling in the universal joint or "hook" of flagella in marine bacteria.

8. Final comments

With the enlargement of the web, a search engine will place before you animations and real movies of much of what is described here. Be careful in your work to be accurate, for example, never say something rotates clockwise without saying from what direction one is looking; otherwise the statement is meaningless.

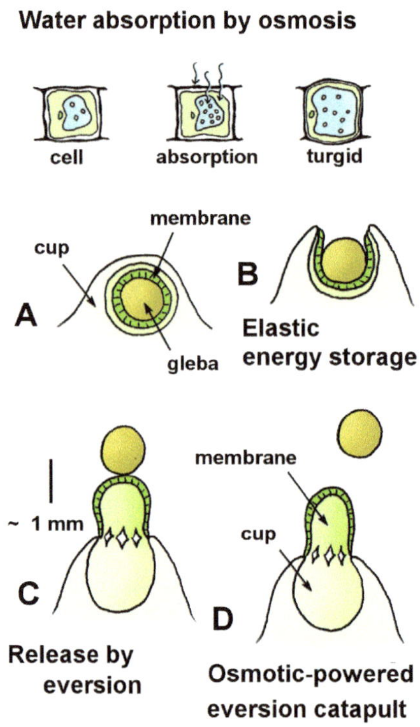

FIGURE 31.45 The osmotic-powered eversion firing mechanism of Sphaerobolus. (A) Immature fruiting body. (B) The developed fruiting body with the exposed gleba (yellow, 1 mm diameter, containing the spores) that is supported by an elastic membrane (peridium) within a firm outer case of the cup. (C and D) Once a critical pressure is reached inside the cells the elastic compliant peridium rapidly everts discharging the gleba from the cup. Scale bar 1 mm (Sakes et al., 2016).

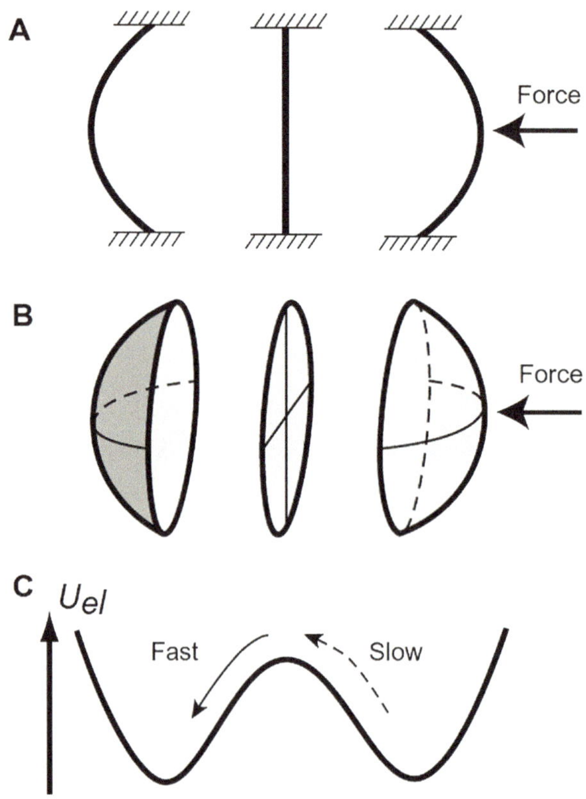

FIGURE 31.46 Principle of mechanical instability brought about by the geometrical constraint of a "jumping popper". (A) Snap-buckling of an elastic rod attached to two rigid walls separated by a distance shorter than the rod length. It is energetically difficult to get from the right to the left state because of the energy barrier of the intermediate state. (B) Snap-buckling instability of an elastic shell. If it has enough energy to bust through the intermediate state it will explosively take advantage of the lower energy of the state at left. (C) Sketch of the elastic energy path in both cases. For a jumping popper and bioanalogs, the left energy well is likely far deeper than the right well making the action dramatic. When a lateral force is applied to the system, the system first slowly stores elastic energy until the barrier is crossed and the elastic energy is rapidly released (Forterre, 2013) just as jumping poppers jump to the ceiling if properly released.

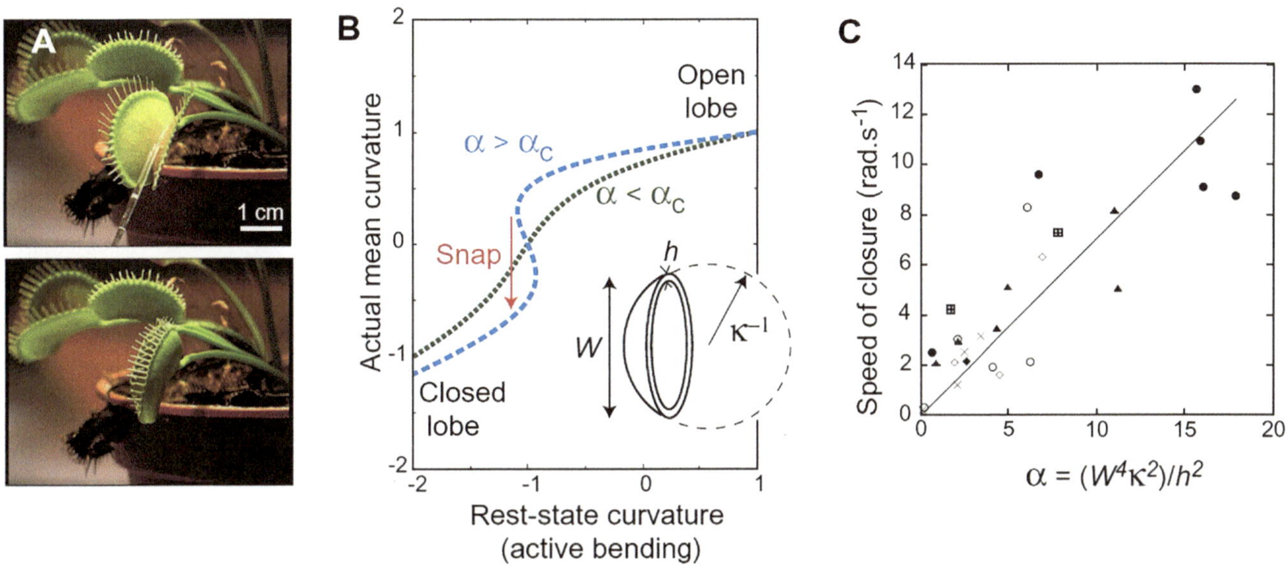

FIGURE 31.47 Snap-buckling of the Venus flytrap (*Dionaea muscipula*). (A) Venus flytrap in the open (upper panel) and closed (lower panel) state. (B) Predicted smooth-snapping transition by an elastic shell model as a function of the geometrical parameter $\alpha = W^4\kappa^2/h^2$, where W is the lobe size or width, κ is the curvature of the lobe in the open state, and h is the lobe thickness. For $\alpha < \alpha_c$, no instability and a smooth transition is predicted between the open and closed state. On the other hand, for $\alpha > \alpha_c$, there is a region of multi-stability leading to a rapid 'snapping' transition. (C) Average speed of trap closure trap closure as a function of the parameter α (Forterre, 2013).

References

Beeby, M., Ferreira, J. L., Tripp, P., Albers, S.-V., & Mitchell, D. R. (2020). Propulsive nanomachines: The convergent evolution of archaella, flagella and cilia. *FEMS Microbiology Reviews, 44*, 253–304. https://doi.org/10.1093/femsre/fuaa006

Beeby, M., Ribardo, D. A., Brennan, C. A., Ruby, E. G., Jensen, G. J., & Hendrixson, D. R. (2016). Diverse high-torque bacterial flagellar motors assemble wider stator rings using a conserved protein scaffold. *Proceedings of the National Academy of Sciences of the United States of America, 113*, E1917–E1926.

Berg, H. C., & Brown, D. A. (1972). Chemotaxis in *Escherichia coli* analyzed by three-dimensional tracking. *Nature, 239*, 500–504.

Berg, H. C. (1983). *Random walks in biology* (p. 142). Princeton: Princeton University Press.

Berg, H. C. (2003). The rotary motor of bacterial flagella. *Annual Review of Biochemistry, 7*, 19–54.

Biou, V., Adaixo, R. J. D., Chami, M., Coureus, P.-D., Laurent, B., Ntsogo, Y., de Amorim, G. C., Izadi-Pruneyres, N., Malosse, C., Chamot-Rooke, J., Stahlberg, H., & Delepelaire, P. (2021). Functional and structural characterization of *Serratia marcescens* ExbB: Determinants of the interaction with HasB/TonB. *bioRxiv Preprint*. https://doi.org/10.1101/2021.04.21.440815

Biquet-Bisquert, A., Labesse, G., Pedaci, F., & Nord, A. L. (2021). The dynamic ion motive force powering the bacterial flagellar motor. *Frontiers in Microbiology, 12*, Article 659464. https://doi.org/10.3389/fmicb.2021.659464

Celia, H., Botos, I., Ni, X., Fox, T., De Val, N., Lioubes, R., Jiang, J., & Buchanan, S. K. (2019). Cryo-EM structure of the bacterial Ton motor subcomplex ExbB–ExbD provides information on structure and stoichiometry. *Communications Biology, 2*, 358. https://doi.org/10.1038/s42003-019-0604-2

Chang, Y.-W., Rettberg, L. A., Treuner-Lange, A., Iwasa, J., Søgaard-Andersen, L., & Jensen, G. J. (2016). Architecture of the type IVa pilus machine. *Science, 351*. https://doi.org/10.1126/science.aad2001

Chang, Y., Carroll, B. L., & Liu, J. (2021). Structural basis of bacterial flagellar motor rotation and switching. *Trends in Microbiology, 29*, 1024–1033. https://doi.org/10.1016/j.tim.2021.03.009

Chang, Y., Zhang, K., Carroll, B. L., Zhao, X., Charon, N. W., Norris, S. J., Motaleb, M. A., Li, C., & Liu, J. (2020). Molecular mechanism for rotational switching of the bacterial flagellar motor. *Nature Structural & Molecular Biology, 27*, 1041–1047.

Craig, L., Katrina, T., Forest, K. T., & Berenike Maier, B. (2019). Type IV pili: Dynamics, biophysics and functional consequences. *Nature Reviews Microbiology, 17*, 429–440. https://doi.org/10.1038/s41579-019-0195-4

Deme, J. C., Johnson, S., Vickery, O., Aron, A., Monkhouse, H., Griffiths, T., Rory Hennell James, R. H., Berks, B. C., Coulton, J. W., Phillip, J., Stansfeld, P. J., & Lea, S. M. (2020). Structures of the stator complex that drives rotation of the bacterial flagellum. *Nature Microbiology, 5*, 1553–1564. https://doi.org/10.1038/s41564-020-0788-8

Dominick, C. N., & Wu, X.-L. (2021). Bacterial cell-body rotation driven by a single flagellar motor and by a bundle. *Biophysical Journal, 120*, 2454–2460.

Emonet, T., & Cluzel, P. (2008). Relationship between cellular response and behavioral variability in bacterial chemotaxis. *Proceedings of the National Academy of Sciences of the United States of America, 105*, 3304–3309.

Forterre, Y. (2013). Slow, fast and furious: Understanding the physics of plant movements. *Journal of Experimental Botany, 64*, 4745–4760. https://doi.org/10.1093/jxb/ert230

Foster, K. W. (2009). Analysis of the ciliary/flagellar beating of *Chlamydomonas*. In S. M. King, & G. J. Pazour (Eds.), *Methods in cell biology: Vol. 91. Cilia: Structure and motility* (pp. 173–239). San Diego: Academic Press.

Guo, P., Noji, H., Yengo, C. M., Zhao, Z., & Grainge, I. (2016). Biological nanomotors with a revolution, linear, or rotation motion mechanism. *Microbiology and Molecular Biology Reviews, 80*, 161–186. https://doi.org/10.1128/MMBR.00056-15

Guo, Q., Dai, E., Han, X., Xie, S., Chao, E., & Chen, Z. (2015). Fast nastic motion of plants and bioinspired structures. *Journal of The Royal Society Interface, 12*, Article 20150598. https://doi.org/10.1098/rsif.2015.0598

Hamadeh, A., Roberts, M. A. J., August, E., McSharry, P. E., Maini, P. K., Armitage, J. P., & Papachristdoulou, A. (2011). Feedback control architecture and the bacterial chemotaxis network. *PLoS Computational Biology, 7*, Article e1001130. https://doi.org/10.1371/journal.pcbi.1001130

Hamel, A., Fisch, C., Combettes, L., Dupuis-Williams, P., & Baroud, C. N. (2011). Transitions between three swimming gaits in *Paramecium* escape. *Proceedings of the National Academy of Sciences of the United States of America, 108*, 7290–7295.

Herzog, B., & Wirth, R. (2012). Swimming behavior of selected species of *archaea*. *Applied and Environmental Microbiology, 78*, 1670–1674. https://doi.org/10.1128/AEM.06723-11

Howell, L. L. (2001). *Compliant mechanisms*. New York: John Wiley.

Hu, H., Santiveri, M., Wadhwa, N., Berg, H. C., Erhardt, M., & Taylor, M. M. I. (2021). Structural basis of torque generation in the bi-directional bacterial flagellar motor. In *Trends biochemical sciences*. https://doi.org/10.1016/j.tibs.2021.06.005

Iwata, S., Kinosita, Y., Uchida, N., Nakane, D., & Nishizaka, T. (2019). Motor torque measurement of *Halobacterium salinarum* archaellar suggests a general model for ATP-driven rotary motor. *Communications Biology, 2*, 199. https://doi.org/10.1038/s42003-019-0422-6

Jarrell, K. F., Albers, S.-V., & Machado, J. N.deS. (2021). A comprehensive history of motility and Archaellation in Archae. *FEMS Microbes, 2*, Article xtab002. https://doi.org/10.1093/femsmc/xtab002

Jarrell, K. F., & Albers, S.-V. (2012). The archaellum: An old motility structure with a new name. *Trends in Microbiology, 20*, 307–312. https://doi.org/10.1016/j.tim.2012.04.007

Jones, C. W., & Armitage, J. P. (2015). Positioning of bacterial chemoreceptors. *Trends in Microbiology, 23*, 247–256. https://doi.org/10.1016/j.tim.2015.03.004

Karmakar, R. (2021). State of the art of bacterial chemotaxis. *Journal of Basic Microbiology, 61*, 366–379. https://doi.org/10.1002/jobm.202000661

Khan, S. (2020). The architectural dynamics of the bacterial flagellar motor switch. *Biomolecules, 10*, 33. https://doi.org/10.3390/biom10060833

Khan, S., & Scholey, J. M. (2018). Assembly, functions and evolution of archaella, flagella, and cilia. *Current Biology, 28*, R278–R282.

Lo, C.-J., Sowa, Y., Pilizota, T., & Berry, R. M. (2013). Mechanism and kinetics of a sodium-driven bacterial flagellar motor. *Proceedings of the National Academy of Sciences, 110*, E2544–E2551. https://doi.org/10.1073/pnas.1301664110

Long, Z., Quaife, B., Salman, H., & Oltvail, A. N. (2017). Cell-cell communication enhances bacterial chemotaxis toward external attractants. *Scientific Reports, 5*, Article 12855. https://doi.org/10.1038/s41598-017-13183-9

Machin, K. E. (1958). Wave propagation along flagella. *Journal of Experimental Biology, 35*, 796–806.

Mahadevan, L., Riera, C. S., & Shin, J. H. (2011). Structural dynamics of an actin spring. *Biophysical Journal, 100*, 839–844.

Marshall, W. F. (2008). Don't blink: Observing the ultra-fast contraction of spasmonemes. *Biophysical Journal, 94*, 4–5.

Matilla, M. A., & Krell, T. (2019). The effect of bacterial chemotaxis on host infection and pathogenicity. *Communications Biology, 2*, 199. https://doi.org/10.1038/s42003-019-0422-6

Minamino, T., & Imada, K. (2015). The bacterial flagellar motor and its structural diversity. *Trends in Microbiology, 23*, 267–274. https://doi.org/10.1016/j.tim.2014.12.011

Misra, G., Dickinson, R. B., & Ladd, A. J. C. (2010). Mechanics of *Vorticella* contraction. *Biophysical Journal, 98*, 2923–2932.

Nan, B., & Zusman, D. R. (2016). Novel mechanisms power bacterial gliding motility. *Molecular Microbiology, 101*, 186–193.

Noinaj, N., Guillier, M., Barnard, T. J., & Buchanan, S. K. (2010). TonB-dependent transporters: Regulation, structure, and function. *Annual Review of Microbiology, 64*, 43–60.

Parkinson, J. S., Hazelbauer, G. I., & Falke, J. J. (2015). Signaling and sensory adaptation in *Escherichia coli* chemoreceptors: 2015 update. *Trends in Microbiology, 23*, 257–266. https://doi.org/10.1016/j.tim.2015.03.003

Pogue, C. B., Zhou, T., & Nan, B. (2018). PlpA, a PilZ-like protein, regulates directed motility of the bacterium *Myxococcus xanthus*. *Molecular Microbiology, 107*, 214–228.

Poppinga, S., Weisskopf, C., Westermeier, A. S., Masselter, T., & Speck, T. (2015). Fastest predators in the plant kingdom: Functional morphology and biomechanics of suction traps found in the largest genus of carnivorous plants. *AoB PLANTS, 8*, Article plv140. https://doi.org/10.1093/aobpla/plv140

Reindl, S., Abhrajyoti Ghosh, A., Williams, G. J., Kerstin Lassak, K., Tomasz Neiner, T., A Henche, A.-L., Albers, S.-V., & Tainer, J. A. (2013). Insights into FlaI functions in archaeal motor assembly and motility from structures, conformations, and genetics. *Molecular Cell, 49*, 1069–1082.

Roper, M., Squires, T. M., & Brenner, M. P. (2008). Symmetry unbreaking in the shapes of perfect projectiles. *Physics of Fluids, 20*, Article 093606.

Rossez, Y., Wolfson, E. B., Holmes, A., Gally, D. L., & Holden, N. J. (2015). Bacterial flagella: Twist and stick, or dodge across the kingdoms. *PLoS Pathogens, 11*, Article e1004483. https://doi.org/10.1371/journal.ppat.1004483

Sakes, A., van der Wiel, M., Henselmans, P. W. J., van Leeuwen, J. L., Dodou, D., & Breedveld, P. (2016). Shooting mechanisms in nature: A systematic review. *PLoS One, 11*, Article e0158277. https://doi.org/10.1371/journal.pone.0158277

Santiveri, M., Roa-Eguiara, A., Kuhne, C., Wadhwa, N., Hu, H., Berg, H. C., Erhardt, M., & Taylor, N. M. I. (2020). Structure and function of stator units of the bacterial flagellar motor. *Cell, 183*, 244–257. https://doi.org/10.1016/j.cell.2020.08.016

Segall, J. E., Block, S. M., & Berg, H. C. (1986). Temporal comparisons in bacterial chemotaxis. *Proceedings of the National Academy of Sciences of the United States of America, 83*, 8987–8991.

Shrivastava, A., Lele, P. P., & Berg, H. C. (2015). A rotary motor drives *Flavobacterium* gliding. *Current Biology, 25*, 338–341. https://doi.org/10.1016/j.cub.2014.11.045

Shrivastava, A., Roland, T., & Berg, H. C. (2016). The screw-like movement of a gliding bacterium is powered by spiral motion of cell-surface adhesins. *Biophysical Journal, 111*, 1008–1013. https://doi.org/10.1016/j.bpj.2016.07.043

Son, K., Guasto, J. S., & Stocker, R. (2013). Bacteria can exploit a flagellar buckling instability to change direction. *Nature Physics, 9*, 494–498. https://doi.org/10.1038/nphys2676

Sowa, Y., & Berry, R. M. (2008). Bacterial flagellar motor. *Quarterly Reviews of Biophysics, 41*, 103–132.

Tam, B. K., Shin, J. H., Pfeiffer, E., Matsudaira, P., & Mahadevan, L. (2009). Calcium regulation of an actin spring. *Biophysical Journal, 97*, 1125–1129.

Upadhyaya, A., Baraban, M., Wong, J., Matsudaira, P., van Oudenaarden, A., & Mahadevan, L. (2008). Power-limited contraction dynamics of *Vorticella convallaria*: An ultrafast biological spring. *Biophysical Journal, 94*, 265–272.

Vogel, S. (2005). Living in a physical world II. The bio-ballistics of small projectiles. *Journal of Bio-Science, 30*, 167–175.

Xing, J., Bai, F., Berry, R., & Oster, G. (2006). Torque–speed relationship of the bacterial flagellar motor. *Proceedings of the National Academy of Sciences of the United States of America, 103*, 1260–1265. https://doi.org/10.1073/pnas.0507959103

Yafetto, L., Carroll, L., Cui, Y., Davis, D. J., Fischer, M. W. F., et al. (2008). The fastest flights in nature: High-speed spore discharge mechanisms among fungi. *PLoS One, 3*(9), Article e3237. https://doi.org/10.1371/journal.pone.0003237

Yoney, A., & Salman, H. (2015). Precision and variability in bacterial temperature sensing. *Biophysical Journal, 108*, 2427–2436.

Yoon, S.h., & Waters, C. M. (2021). The ever-expanding world of bacterial cyclic oligonucleotide second messengers. *Current Opinion in Microbiology, 6*, 96–103.

Yuan, J., & Berg, H. C. (2013). Ultrasensitivity of an adaptive bacterial motor. *Journal of Molecular Biology, 425*, 1760–1764. https://doi.org/10.1016/j.jmb.2013.02.016

Zhao, X., Norris, S. J., & Liu, J. (2014). Molecular architecture of the bacterial flagellar motor in cells. *Biochemistry, 53*, 4323–4333. https://doi.org/10.1021/bi500059y

Further reading

Nakamura, S., Kami-ike, N., Yokota, J. P., Minamino, T., & Namba, K. (2010). Evidence for symmetry in the elementary process of bidirectional torque generation by the bacterial flagellar motor. *Proceedings of the National Academy of Sciences of the United States of America, 107*, 17616–17620.

Nelson, P. (2008). *Biological physics: Energy, information, life*. New York: W.H. Freeman.

Chapter 32

Cell migration, taxis, kinesis, quorum sensing, and tropism

Judith Van Houten[1] and Megan Valentine[2]

[1]*Department of Biology, University of Vermont, Burlington, VT, United States;* [2]*State University of New York at Plattsburgh, Department of Biological Sciences, Plattsburgh, NY, United States*

1. Summary

Migration, taxis, kinesis, quorum sensing, and tropism all are behaviors of eukaryotic cells that are driven by sensory systems designed to detect environmental cues including chemicals (signifying food, toxic conditions, oxygen levels, pH), light, temperature, touch, electric fields. For the most part, we address free living cells in this chapter, but general principles apply to cells from multicellular organisms, such as leukocytes and multiciliated cells of metazoa. We begin with some definitions and examples.

2. Introduction

2.1 Migration

Eukaryotic cells readily move in their environment. They may use flagella and cilia to swim through an aqueous milieu or they may crawl on substrate using their cilia or cytoskeleton to alter shape and gain traction to move. We describe free swimming ciliates such as *Euplotes* and *Paramecium* and for, amoeboid movement, we primarily describe *Dictyostelium discoideum* and leukocytes as they migrate.

2.2 Taxis and kinesis

Both cells propelled by cilia or crawling on a substrate move by random locomotion, that is, in a nonguided way (Dunn et al., 1981). When these cells move in response to a stimulus, they employ either of two locomotory strategies: taxis or kinesis (Fraenkel and Gunn, 1961). When performing a taxis, the organism orients and moves directly toward or away from the stimulus. Amoebae, such as *D. discoideum*, orient and move toward the source of cyclic AMP (chemotaxis) as part of the developmental program of forming a multicellular organism, as discussed in detail in Chapter 33 by Sackman. Cells can also follow a stiffness gradient of the substrate they travel on through a process of *durotaxis.* As you will see, the way in which the cell responds to and moves up or down a gradient of stimulus requires an orchestration of receptors, intracellular signaling, and cytoskeletal components.

If performing a kinesis, the cell does not orient relative to the source of the stimulus. Instead, it changes its random walk by biasing its speed or frequency of turning as it approaches or moves away from the stimulus. Biasing speed is called orthokinesis; biasing turning is called klinokinesis. Using these biased random walks, organisms can efficiently move up and down gradients. Some organisms, such as *Paramecium tetraurelia*, even modulate both speed and turning frequency to accumulate or disperse by a biased random walk (Van Houten, 1978, 1979). Other organisms, such as motile bacteria, are attracted or dispersed by klinokinesis, as Howard Berg (1983) reviews in his book Random Walks in Biology. As we review amoeboid and ciliary motility, we will revisit these terms about how cells move around their environment.

There is some confusion about the distinction between taxis and kinesis because taxis, in its broadest use, is the net directional response of a cell in response to a stimulus (Gooday, 1981). Historically, chemotaxis has been used for bacterial

Cell Physiology Sourcebook. https://doi.org/10.1016/B978-0-12-811114-7.00025-6

response when it clearly qualifies as a klinokinesis (Gooday, 1981). Here, we will try to respect the fields' preference for terms but make clear the mechanism underlying the motility response of the organism.

Quorum sensing, as the name implies, allows cells to take a measure of their numbers. The outcome of this census influences the migration behavior of individual cells such that they aggregate (*Dictoystelium*) to develop into a multicellular form or undergo division. Quorum sensing is carried out by eukaryotes and prokaryotes, with some important signals conserved. Among prokaryotes, perhaps fluorescence in fish due to *Vibrio* bacteria might come to mind. We address cell migration and quorum sensing together.

Tropism refers to growth toward or away from a stimulus. Whereas *chemotaxis* (e.g., migratiomn of *Dictyostelium* amoebae toward a cyclic AMP gradient), involves orientation of the cells and movement up or down the gradient of stimulus, *chemotropism* is the growth of a cell toward or away from the stimulus. For example, a yeast cell responding to mating pheromone changes its cytoskeleton and does not move but rather grows toward the stimulus. Signal transduction pathways control the movement of *Dictyostelium* and other amoeboid cells, such as leukocytes, as they move up and down gradients. Likewise, yeast cells have active signal transduction pathways guiding their growth toward a stimulus. There are a surprising number of elements in common between the signaling pathways for chemotaxis and chemotropism as reviewed by Arkowitz (1999).

2.3 Cell migration, taxis, and amoeboid movement

Chemotaxis, the migration of cells directed by chemical cues, underlies many processes that result in tissue formation, wound healing, immune responses, and disease states of asthma, arthritis, atherosclerosis, and cancers (see Willard and Devreotes, 2006 and Roussos et al., 2011 for a discussion). The model system of the slime mold *D. discoideum* offers numerous insights into the coordinated and dynamic events underlying cell migration and cell motility, which can be extrapolated to other motile cells such as leukocytes (Artemenko et al., 2014). We will focus first on this slime mold system and return to leukocytes further in this chapter. We will not discuss the mesenchymal migration of cells such as fibroblasts, which require strong attachment to and proteolysis of the substrate.

The slime mold cells migrate to aggregate into a multicellular fruiting body whose cysts enable their survival in starvation conditions and future regrowth as amoebae (Fig. 32.1). This is called *sorocarpic multicellularity* and is found not only in *Dictyostelia* but also in myxobacteria (Du et al., 2015). Schaap (2016) has compared the cell–cell communication strategies of *Dictyostelia*, the largest group of eukaryotes that aggregate to achieve multicellularity. We highlight here the signaling and resultant cellular mechanisms that allow *D. discoideum* free moving cells to orient, move, and come together during their complex developmental plan to form a temporary multicellular entity (Fig. 32.1). Reduction in the food (bacteria) for the motile amoebae and resulting starvation is a major trigger for the aggregation to begin. The loss of food accompanied by the prestarvation factor (PSF) levels determines when chemotaxis and migration will commence. PSF serves as a quorum sensing molecule that is secreted constantly by the vegetative amoebae grazing for food on their substratum. When food declines and PSF is high (i.e., population density is high), the pathway in Fig. 32.1 is initiated with a few founder cells pulsing secretions of cyclic AMP (cAMP). Waves of cAMP propagate through the cell populations, and the stimulated cells move by chemotaxis toward the source. A second quorum sensing pathway, conditioned medium factor (CMF), comes into play after the combination of PSF and high cell numbers stimulate chemotaxis (Fig. 32.1). Sufficient CMF is required for the cyclic AMP surface receptor (CarA) mediated signal transduction to proceed.

Neighboring cells entrain each other to synchronously secrete cAMP, resulting in waves of cAMP 6–8 min apart (Loomis, 2014). The extracellular phosphodiesterase PdsA controls the height and shape of the waves of secreted cAMP radiating outward from the center of the aggregate. Thus the starving and primed cells follow the trails of cAMP toward a founder cell source and also relay the cAMP pulses to more outlying cells. Consequently, cells move inward in trails in a spiral to meet up for the formation of a mound, and subsequently a slug and fruiting body. The key players, receptor CarA, adenylyl cyclase AcaA, and phosphodiesterase PdsA coordinate the pulses of cAMP and the orderly procession of cells to gather together in a mound.

The CarA receptor for cAMP is a G protein–coupled receptor. Upon cAMP ligand binding, its bound tripartite G protein is released into the cell as free G_α and $G_{\beta\lambda}$ subunits. You might expect that the G_α subunit directly stimulates the adenylyl cyclase AcaA, but the dissociated G protein subunits activate GTP exchange factors for Ras that in its GTP bound form leads indirectly to a strong induction of the membrane associated AcaA (see Loomis for review by Loomis, 2014).

Remarkably, cAMP plays roles through these later developmental steps as well as in chemotaxis (see yellow highlight in Fig. 32.1). Following induction of chemotaxis, the cAMP pulses upregulate genes for aggregation and development (Du et al., 2015). During development after aggregation, cAMP continues to be a critical signal, but a different adenylyl cyclase (AcgA) is now in play. Yet other signals for differentiation include one that is necessary for stalk cell differentiation, cyclic di-GMP (Schaap, 2014). The Cyclic di-nucleotide monophosphates (c-di-GMP and c-di-AMP) were thought to be used only in the prokaryotic world for signaling, but now it is clear that eukaryotes avail themselves of these signals as well. We will come across them again in our discussion of interkingdom interactions. We will also find that the sensor histidine kinase two-component signaling is critical to the development of *Dictyostelium* (Fig. 32.1) just as it is in prokaryotic quorum sensing and chemotaxis (Schaap, 2014).

3. Basal cell motility and chemotaxis

Devreotes and others (Devreotes et al., 2017) argue that almost all living cells move spontaneously, and also with directionality due to detection and response to extracellular cues. The latter requires cell motility, cell polarization, polarized sensitivity, and directional sensing through receptors that are part of a signal transduction excitable network (STEN). We describe here the amoeboid movement of cells that migrate by differentiating their front and rear with polymerizing actin in the leading edge protrusions (pseudopodia or lamellipodia) and myosin II at the sides and rear (Artemenko et al., 2014; Devreotes et al., 2017). We will use the term pseudopodia for the leading edge membrane protrusions of *Dictyostelium* and leukocyte cells (see also Chapter 33 by Sackman).

Dictyostelium cells move randomly to graze on bacteria on the substratum. Fig. 32.2 shows the multiple pseudopodia that are characterized by the relatively high levels of PIP3 (Willard and Devreotes, 2006). Cells rely on the rearrangement of actin filaments to form and retract pseudopodia, which have only weak interactions with the surface but are sufficient to pull the cell forward (Lammermann et al., 2008). A stronger interaction could slow cell migration, as observed in fibroblasts migrating along collagen threads, which migrate at a rate of 0.75−1.25 mm per day (Cornwell et al., 2004). Clearly, it is the dynamic cytoskeleton that makes it possible for the cell to move. *Dictyostelium* cells, even those without

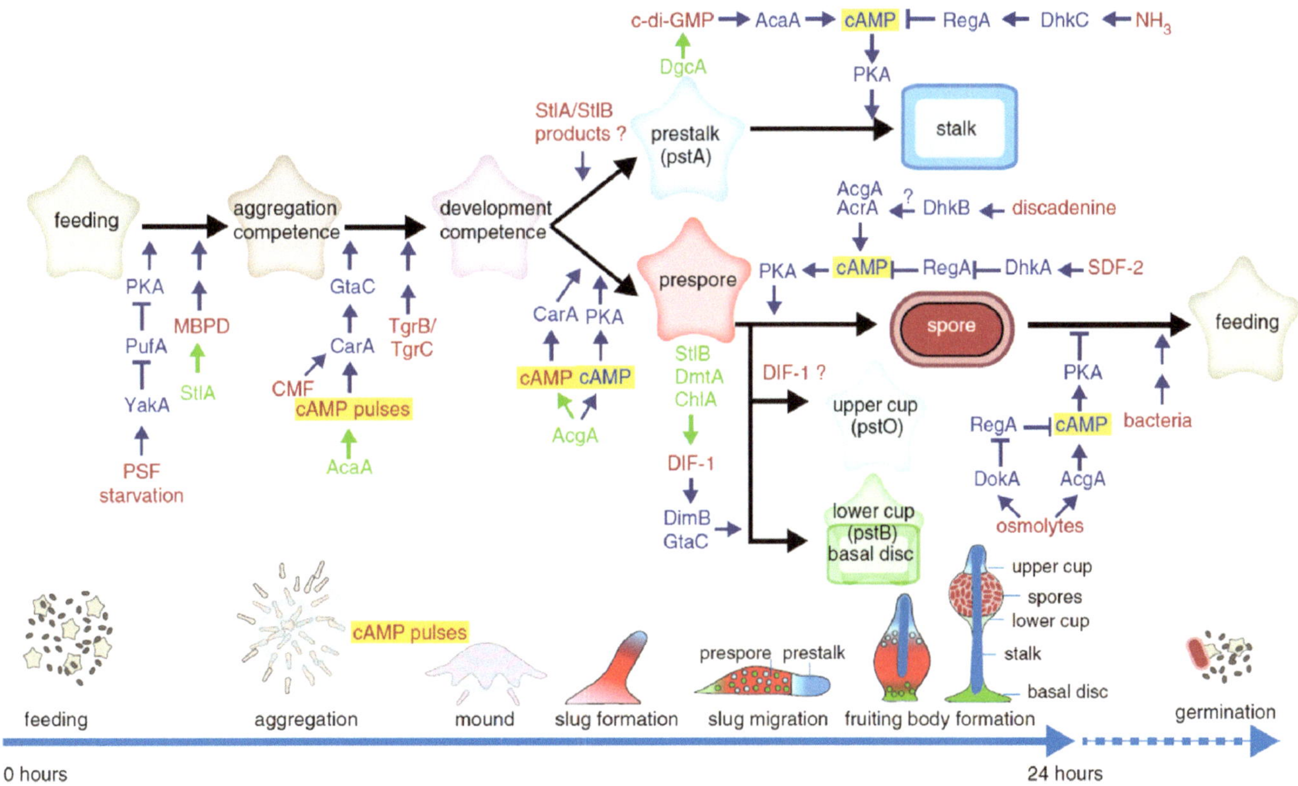

Current Opinion in Genetics & Development

FIGURE 32.1 Developmental signaling during the *Dictyostelium discoideum* 24 h life cycle. Starving amoebae aggregate by secreting and replaying cAMP pulses to form mounds. The mound continues to emit cAMP pulses, attracting cells from underneath, which push the tip upward to form a slug. In the slug, the cells differentiate into precursors of spore and stalk, basal disc, and upper and lower cup cells. Note that the initial stages of the pathway have the beginnings of chemotaxis. Cyclic AMP plays multiple roles as an extracellular and intracellular signal (Schaap, 2016). *Car*, cAMP receptor1; *CMF*, conditioned medium factor; *PKA*, protein kinase A called YakA; *PSF*, prestarvation factor.

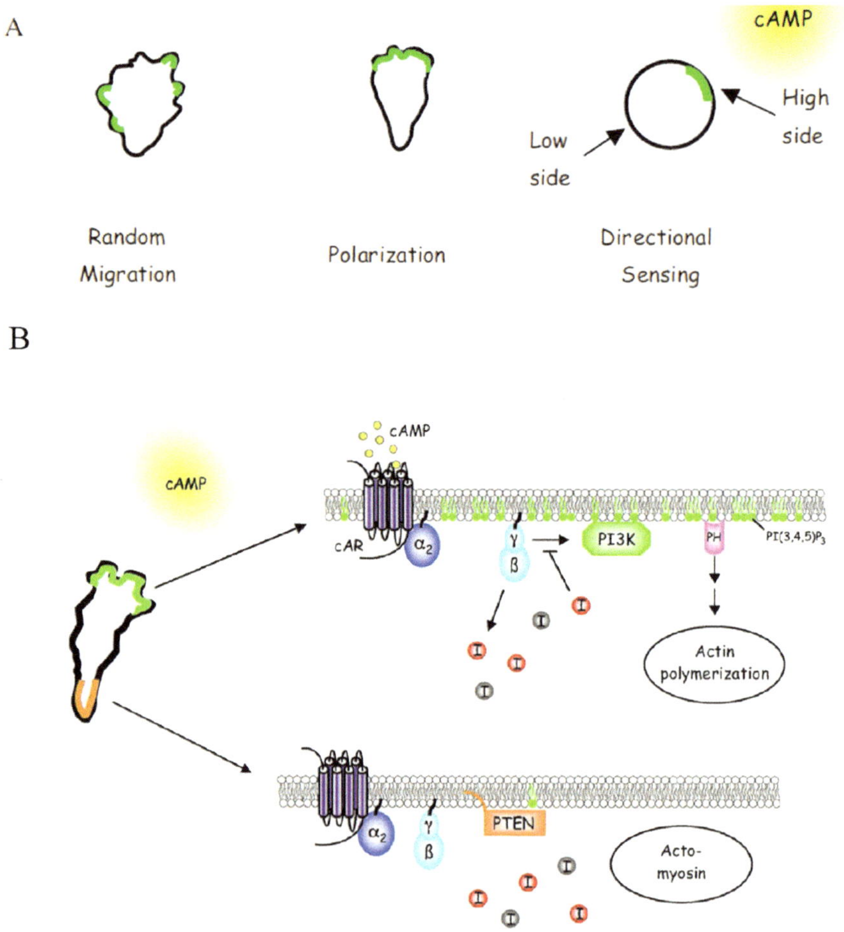

FIGURE 32.2 Cellular responses and signaling in *Dictyostelium discoideum* chemotaxis. (A). Three typical cellular responses during which PIP3 mark specific areas of the plasma membrane. First, during random migration, which occurs regardless of nutritional status, PIP3 is a marker for pseudopod formation. Second, starved amoebae exhibit polarity during chemotaxis toward cAMP; PIP3 is a marker for the side of the cell experiencing the highest concentration of cAMP. Third, in round immobile cells exposed to a gradient of chemoattractant, PIP3 labels the high side of the cell. (B). Major signaling events at the high and low sides of the cell during chemotaxis up a concentration gradient (Willard and Devreotes, 2006).

chemoreceptors, form pseudopods through the polymerization of actin (VanHaastert et al., 2004). There can be one to two pseudopodia that form about every 30–60 s and pull the cell in random directions.

This basal motility is "harnessed" for oriented motility, i.e., chemotaxis (Kay et al., 2008; Willard and Devreotes, 2006). In general, for chemotaxis, there must be an integration of motility, directional sensing and polarization of the cell (Artemenko et al., 2016). Fig. 32.2A shows a cartoon of the polarization and directional sensing needed for chemotaxis. Polarization is the elongation of the cell and the distribution of specific proteins at either pole of the cell. Multiple signal transduction pathways (Fig. 32.2B) work in parallel to bias actin polymerization to create the single leading edge pseudopod protrusion in the direction of the gradient and myosin II assembly at the sides to suppress pseudopod formation there and at the back end uropod for retraction (Kimmel and Parent, 2003; VanHaastert et al., 2004). We will start with the front end of the cell, the pseudopod, and return to the uropod function later (Fig. 32.2).

To polarize a cell and produce directional sensing such that the cell's single pseudopod would lead the cell up the gradient, one might expect that the receptors for the attractant in the gradient would concentrate at the leading edge of the cell. This is not the case. The cAMP receptors for *Dictyostelium* (Xiao et al., 1997) remain evenly distributed across the cell surface. Instead, *Dictyostelium* establishes polarity through an intracellular gradient of a second messenger phosphatidylinositol (3,4,5)-trisphosphate (PIP3) across the cell. However, receptor occupancy is not even around the cell.

As a gradient of extracellular cAMP forms, there will be a side of the cell with the highest cAMP receptor occupancy where the second messenger PIP3 is highest (Fig. 32.2) (Franca-Koh et al., 2006). The released G proteins of the occupied cAMP receptors activate Ras, which recruits PI3K, the enzyme that synthesizes PIP3 from PIP2, from the cytoplasm to the membrane at the front of the cell (Artemenko et al., 2014; Loomis, 2014; Sasaki et al., 2004). PTEN, the enzyme that degrades PIP3, comes off the membrane at the front, but remains bound at the membrane in the back of the cell (Funamoto et al., 2002; Iijima and Devreotes, 2002). Thus, the cell's reorganization at its membrane reinforces its new orientation and highest PIP3 concentrations toward the highest levels of stimulus. The PIP3 at the leading edge recruits an array of pleckstrin homology (PH) domain proteins (see Artemenko et al., 2014; Loomis, 2014; Parent, 2004; VanHaastert et al., 2004 for more on the PH domain proteins.) The PIP3 at the front of the cell promotes the actin polymerization that propagates into the leading edge of the cell and causes protrusion of the cell into a broad pseudopod in *Dictyostelium* (Inagaki and Katsuno, 2016). PIP3 is enriched in actin waves, which are made of filamentous actin and associated proteins. PIP3 may even be sufficient, with G protein Rac1, to initiate these waves. The actin waves generate the force needed to move the cell forward by the coupling the growing actin arrays to the adhesive substrate through "clutch" molecules on the membrane.

The transition of actin monomers into actin filaments (F-actin), sometimes containing small branches for the actin arrays and waves, is a process that is highly conserved among organisms (reviewed in Devreotes and Horwitz, 2015). Formins (mDia1 and mDia2) help to regulate F-actin by weakly attaching to the barbed ends of the growing filaments while adding more actin monomers (Goode and Eck, 2007). Where the barbs branch from the linear filaments, a complex of proteins helps to stabilize the branches, including the actin-related proteins 2 and 3 (Arp2/3), WASP (Wiskott–Aldrich syndrome protein), and SCAR/WAVE (suppressor of cyclic AMP receptor mutation/WASP related verprolin homologs protein) (Pollitt and Insall, 2009). These complexes are under the control of the Rho GTPases and help to drive the extension of the lamellipodia (reviewed in Insall and Machesky, 2009) (see also Chapter 33 for more details).

At the back end of the cell, the uropod is also established in polarization. Here the actin is depolymerized and the uropod contracts by myosin-II motor proteins (Bugyi and Carlier, 2010), which bundle into filaments. The myosin filaments contract using ATP hydrolysis to pull on the actin cytoskeleton (Laevsky and Knecht, 2003) and to tug the cell forward. The myosin II filaments, which are composed of two myosin heavy chains (MHC), two essential light chains (ELC) and two regulatory light chains (RLC), are concentrated at the posterior and along the periphery of the cell (reviewed in Bosgraaf and Van Haastert, 2006). *Dictyostelium* myosin II mutants have been very helpful in deciphering the contributions of myosin II light and heavy chains in the chemotaxis of cells. For example, the myosin heavy chain mutant *mhcA*$^-$ is inefficient at chemotaxis and slow compared with wild type cells (Wessels et al., 1988). Myosin activity has been extensively studied in *Dictyostelium* where it is better understood than in leukocytes. Myosin II activity is regulated by phosphorylation; the phosphorylation of the RLC is important for regulating myosin motor activity (Griffith et al., 1987), while phosphorylation of the tail end of MHC regulates the formation of myosin filaments (Egelhoff et al., 1993).

Unique to *Dictyostelium*, the extracellular stimulation by cAMP or folate (a signal of the presence of bacteria food that does not initiate development or aggregation) causes a brief increase in the intracellular concentration of cGMP through the activation of two guanylyl cyclases as well as inhibition of phosphodiesterases (PDEs) (Parent, 2004; VanHaastert et al., 2004) (Fig. 32.3). The intracellular cGMP increase ultimately results in the relocation of myosin in the rear of the cell that is necessary for chemotaxis (Veltman and van Haastert, 2008). The main target of intracellular cGMP is cGMP binding protein C (GbpC), which performs two functions: formation of myosinII filaments that inhibit pseudopod formation in the rear of the cell and increases the phosphorylation of the myosin regulatory light chain (RLC) that is necessary for motility (Bosgraaf and Van Haastert, 2006). This cGMP signaling process is vital to the aggregation of *Dictyostelium* cells, and when it is not fully functional, the cell-to-cell contacts are weakened and the directional migration is impaired (Bosgraaf and Van Haastert, 2002; Franca-Koh et al., 2006; VanHaastert et al., 2004; Veltman and van Haastert, 2008). A recent history of research on *Dictyostelium* cGMP in motility and chemotaxis provides an excellent overview (Van Haastert et al., 2021).

In discussing white blood cell migration, we use the terms leukocyte and the subset neutrophil interchangeably. Neutrophils migrate from the blood stream to sites of infection and inflammation signaled by formylated peptides that are bacterial products (e.g., f-met-leu-phe), platelet-activating factor PAF, attractant/activation protein-1 (NAP-1) (Leonard et al., 1990), numerous chemokines (Baggiolini, 1998) (reviewed in Moser et al., 2004), and complement component 5 cleavage product (C5a) (Sha'afi and Molski, 1988).

The mechanisms of neutrophil migration show such strong evolutionary conservation with chemotaxis of *Dictyostelium* amoebae that slime mold and neutrophil motility have been contrasted in many articles (Artemenko et al., 2014; Devreotes and Parent, 1989; Parent, 2004; VanHaastert et al., 2004). Both *Dictyostelium* amoebae and neutrophils detect and respond to gradients through pseudopod extension, directional sensing and polarization (Fig. 32.3) (Devreotes and Parent, 1989; Parent, 2004; VanHaastert et al., 2004). Leukocyte extracellular signals for chemotaxis are numerous, as are their receptors, compared to the two signals (cAMP and folate) for *Dictyostelium* chemotaxis (Artemenko et al., 2014).

In both cell types, slime mold and neutrophils, receptors are uniformly distributed across the cells (Parent, 2004), while signals from the occupied receptors at the upgradient side of the cell set into motion the polarization of the cells. Signal transduction modules (e.g., Ras module, PI3K module, TOR/PKB module, Rho module) that are connected into networks

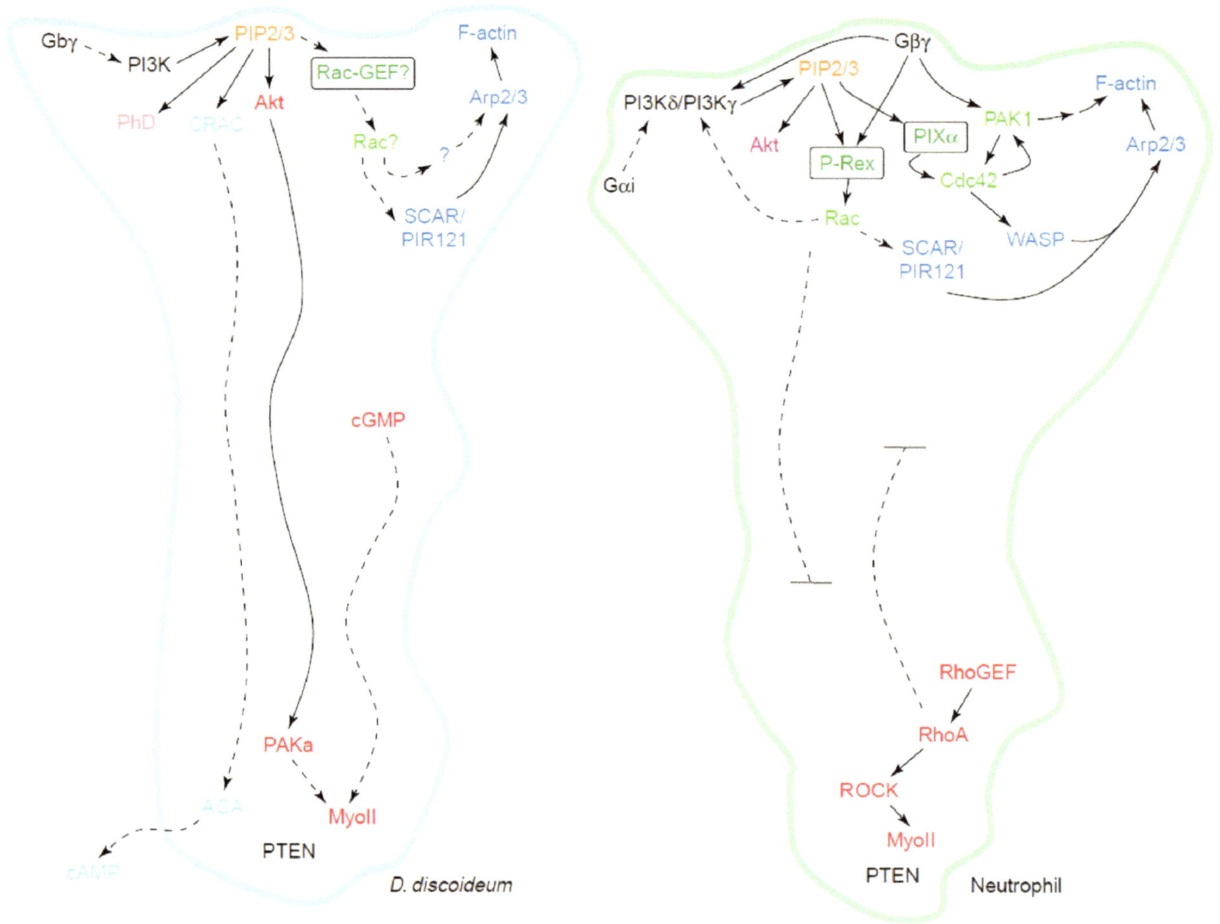

Current Opinion in Cell Biology

FIGURE 32.3 Schematic representation of polarized signaling events in *Dictyostelium discoideum* cells (*left*) and neutrophils (*right*). The cartoon depicts cells as they are moving in response to a gradient of chemoattractant; the leading edge of the cells is at the top of the figures. Although both chemoattractant receptors (not shown) and heterotrimeric G proteins are mostly uniformly distributed, the majority of the downstream effectors become segregated as the cells polarize. Sustained levels of PIP3 at the leading edge are caused by the reciprocal cellular distribution of PI3K at the front end and PTEN at the back of both cell types. For both cells, PIP3 polarized distribution leads to the leading edge recruitment of actin filaments, PH domain containing proteins (PhD and CRAC) for *Dictyostelium* and Akt for both cell types. Akt in *Dictyostelium* activates PAKa that participates in myosin II assembly, as does cGMP. In contrast in neutrophils, myosin II assembly requires activation of RhoA and downstream effector ROCK. ACA is highly enriched at the backs of *Dictyostelium* cells, but not neutrophils, for cAMP signal relay (Parent, 2004).

orchestrate actin polymerization at the anterior of the cell and cortical myosin formation at the rear, both of which are essential for movement up attractant gradients.

In both cell types, PIP3 concentrations that are critical for actin polymerization and pseudopod formation are highest where receptor occupancy is highest. PIP3 is implicated in actin waves at the pseudopod in both cell types (Inagaki and Katsuno, 2016). Also, in both cell types, the PIP3 levels rise at the anterior leading edge because PI3K is activated there through the G protein cascade from the activated receptors. Reciprocally PTEN levels are low at the leading edge and rise at the rear of the cell.

Differences between slime mold and neutrophils manifest themselves in the specific intracellular signaling molecules of the modules and underlying networks. For example, Fig. 32.3 shows that the G protein subunits that signal from the occupied receptor differ between the two cell types yet converge on the same outcome of increased PIP3 at the leading edge of the cell (Parent, 2004). Rac is a key actor in the slime mold leading edge actin polymerization and forward movement, while Cdc42 plays that role in leukocytes (Artemenko et al., 2014; Parent, 2004) (Fig. 32.3).

Cyclic GMP is the major regulator of myosinII function to suppress pseudopods in the lateral side of the cell and retract the uropod for movement in *Dictyostelium* cells, while in neutrophils the myosinII performs similar functions but is regulated by Rho kinases (Fig. 32.3) (Parent, 2004; VanHaastert et al., 2004). Another departure is that leukocytes' PTEN affects motility speed in leukocytes, but not the directionality of the cells (Lacalle et al., 2004). A systematic comparison of the chemotaxis components and especially the intracellular signaling network modules can be found in Artemenko et al. (2014) and Van Haastert and Devreotes (2004).

Discussed before, actin waves are associated with cell polarization and forward movement in *Dictyostelium*. These waves are also found in neutrophil migration and polarization. For example, neutrophils in response to *N*-formyl-methionyl-leucyl-phenylalanine peptide (a bacterial product and sign of infection) generate actin waves followed by membrane protrusion at the leading edge (Brunetti et al., 2021; Inagaki and Katsuno, 2016).

The directed collective migration of cells is a characteristic of far more than *Dictyostelium* and neutrophil cells. Shellard and Mayor (2020) review the importance of directed migration for immune cells surveillance, cancer metastasis, neural crest formation, and wound healing. Importantly, the analysis of directed migration in *Dictyostelium* to aggregate has relevance for all of these behaviors (Wang et al., 2014). *Dictyostelium* cells are capable of migrating in streams with cells moving in single file, touching head to tail. This collective migration involves cell—cell and cell—substrate adhesions for the coordination of the physically touching cells (Qin et al., 2021; Wang et al., 2014), which respond differently if they are a leader or follower cell. The migrating cells are guided by the chemical signals for this chemotaxis that we discussed before. In addition, there are mechanical signals for guidance of this cell migration in the stiffness of the substratum, and the resulting behavior is known as durotaxis (Sunyer and Trepat, 2020). Durotaxis cues act on the same set of cellular components in common with chemotaxis to cooperatively control migration of cells (Shellard and Mayor, 2021). They certainly converge at the actin network in the lamellipodia, which are mechanoresponsive and sensitive to the substratum (Machesky and Insall, 2021; Mehidi et al., 2021). Another mechanical stimulus for *Dictyostelium* and leukocytes is shear flow, which induces migration, also by activating the same signal transduction network of chemoattractants (Cole et al., 2021).

In *Dictyostelium*, a connection between mechanosensation and cell migration could be the mechanosensitive, nonspecific cation channel, Piezo1 (Srivastava et al., 2020) (see Section 7 in Chapter 33, by Sackman, in this volume). Calcium entering through this channel controls the form of migration through blebs or pseudopodia at the leading edge, but its role in durotaxis is not yet clear.

Migrating cells such as *Dictyostelium*, neutrophils, and many other types are not likely to encounter one stimulus in isolation. They likely have to respond concurrently to chemical cues, substratum characteristics, and even guidance cues bound into a gradient (haptotaxis). Therefore, the concept of "mixotaxis" has been proposed to account for the sharing of sensor and transducer functions by both biochemical and biophysical signals (Barriga and Theveneau, 2020; Espina et al., 2021). Insights into the molecular mechanisms of integration of two or more taxis processes and the nature of the behavioral outcomes are needed.

The most fascinating differences between *Dictyostelium* and neutrophils are in their uropods (Fais and Malorni, 2003; Hind et al., 2016). The neutrophil uropod is the site of myosinII accumulation and contraction of the cell that contributes to forward movement. However, the neutrophil uropod is a much larger and more enduring structure than the slime mold counterpart. This is most evident when cells encounter a change in the direction of a stimulus and should execute a turn or restructure to repolarize themselves so that the leading edge and pseudopod are now in a different location on the cell. Neutrophils will make a U turn and retain the uropod in the same location, while *Dictyostelium* cells will relocate their pseudopod and uropod to align themselves with the new gradient. In addition, the uropod of the neutrophil serves key sensory functions in the immune response. Neutrophils circulate in blood as resting cells and leave the blood stream to locomote to an area of inflammation. Here, cells take in information about their environment and participate in cell-to-cell interaction through touching of uropods (Fais et al., 2003) (see Hind et al., 2016 for a more thorough discussion, and Chapter 33).

4. Cilia-based motility

Cilia are slim projections from the eukaryotic cell surface that have the internal apparatus to beat and cause a cell to swim or fluid to move past the cell. We will use the term cilium, but intend that the description of a cilium also applies to the term flagellum, as you will find used with *Chlamydomonas* and sperm. We will focus on ciliary motility that powers the motility and behavior of protozoa and sperm.

Eukaryotic cilia generally are motile with a highly conserved physical structure of an axoneme made of nine doublets of microtubules with their inner and outer dynein arms and radial spokes that attach the doublets to the inner doublet (Fig. 32.4) (Lindemann and Leisch, 2010; Satir et al., 2014). This is the basic apparatus of each cilium that beats by the hydrolysis of ATP by the dynein arms. This in turn causes the translocation of doublets relative to each other and the bending of the cilium during the act of swimming (Satir et al., 2014).

Cilia can be classified as motile or immotile, usually called primary cilia (Hilgendorf et al., 2024). Loss of cilia or critical components for function, whether in motile or primary cilia, results in severe sensory and developmental problems

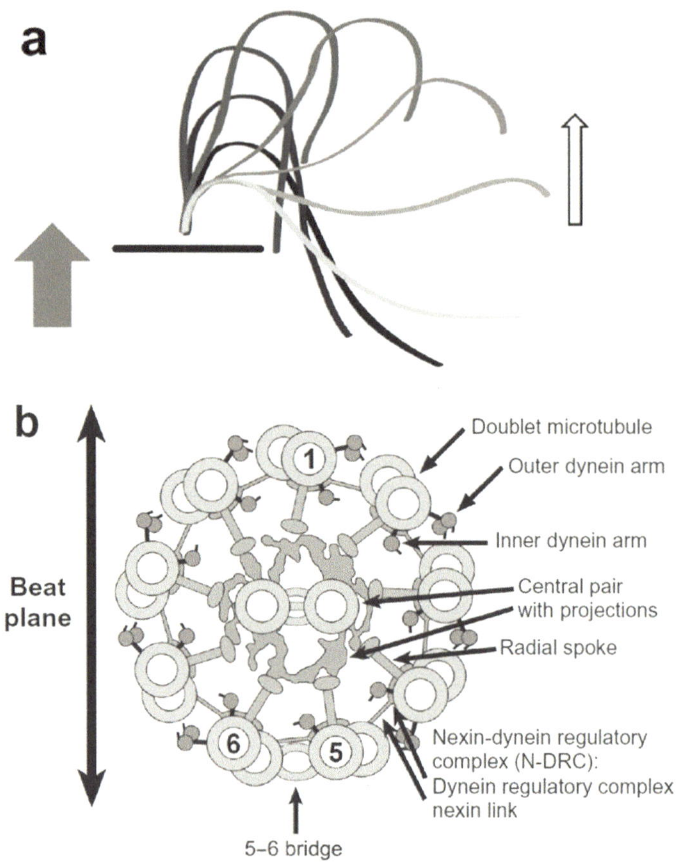

FIGURE 32.4 A. Bending patterns of cilia. Black bar is 5 μm; the gray arrow indicates the direction of cell movement; open arrow indicates direction of the forward ciliary band; open arrow indicates direction of forward ciliary bend. B. Transverse section of motile cilium illustrating the main features of the 9 + 2 axoneme from motile metazoan cilia, including the outer doublet, central pair of microtubules, outer and inner dynein arms, the central pair projections, the 5–6 bridge, the radial spokes, and the dynein regulatory complex with nexin links (Satir et al., 2014).

in humans (Reiter and Leroux, 2017). Loss of cilia function in multiciliated epithelia is associated with loss of nodal flow that compromises left–right asymmetry in development and flow of fluids in the body, e.g., cerebrospinal fluid flow, respiratory airway clearance, or fallopian tube flow for ovum transport (Spassky and Meunier, 2017).

At the base of the cilium is a basal body made of microtubule triplets above which is a set of structures that gate what enters and what leaves the cilium. The gate resides in the transition zone, which is an area of intense scrutiny as many of its components are associated with ciliopathies (diseases caused by dysfunction of cilia) (Reiter et al., 2012; Reiter and Leroux, 2017). Both motile and immotile primary cilia have this transition zone, but the ciliopathies are mostly associated with the loss of function in the primary cilium (Reiter and Leroux, 2017) (for reviews of ciliopathies, see Braun and Hildebrandt, 2017 and Reiter and Leroux, 2017 and, in *Paramecium*, see Valentine and Van Houten, 2021).

The cilium is covered with a membrane that is contiguous with the plasma membrane of the cell body. It is this membrane that has receptors, ion channels, and other sensory molecules (Kleene and Van Houten, 2014) that make the cilium an antennae for detecting and responding to environmental cues (Malicki and Johnson, 2017). The antennae aspects of cilia are often described in primary nonmotile cilia of metazoan tissues, but the principles are shared with motile cilia, e.g., of metazoans, protists and sperm (Bloodgood, 2010). The antennae sensory functions of motile cilia allow the cells they propel to change their swimming behavior and respond to their environment.

Modulation of the ciliary bending and power stroke of the cilia of a protist like *Paramecium* (Fig. 32.5) allows this free swimming cell to speed up, turn, and reverse direction in response to attractants, repellents, mechanical stimuli that probably mean the touch of a predator, gravity, and temperature among other stimuli. The cilia beat with a strong power stroke toward the posterior of the cell and a weaker recovery stroke to bring the cilia back to the beginning of the stroke. This beating that is due to the organized movement of the axonemal doublets relative to each other drives the cell forward through the pond water (Satir et al., 2014). When the cell encounters a stimulus (e.g., bumps into an object (Fig. 32.6) or enters an area of high ionic strength), the cell membrane depolarizes (Fig. 32.7). If the depolarization rises above the

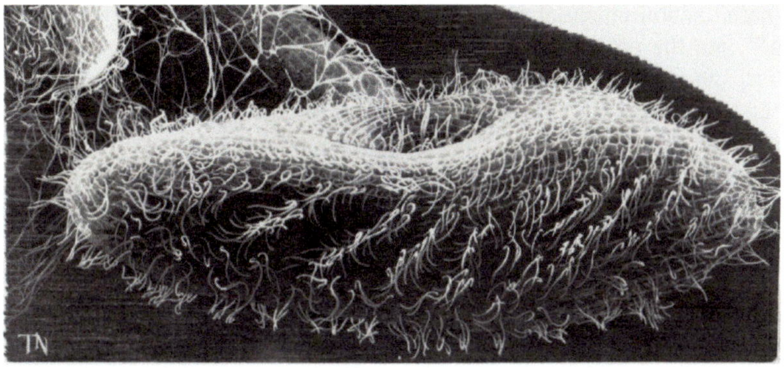

FIGURE 32.5 Scanning electron micrograph of *Paramecium tetraurelia* cell showing the metachronal waves of cilia beating with the power stroke toward the posterior (left end of cell). J. Van Houten, Grass Calendar 1985.

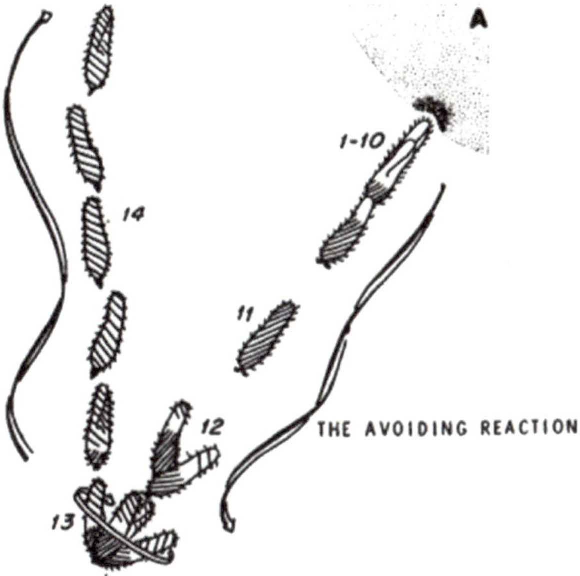

FIGURE 32.6 Schematic of a *Paramecium* cell encountering a solid object while swimming. The cell transiently reverses ciliary beat, backs up, rotates in place, and moves off in a new direction (Eckert, 1972; Jennings, 1906).

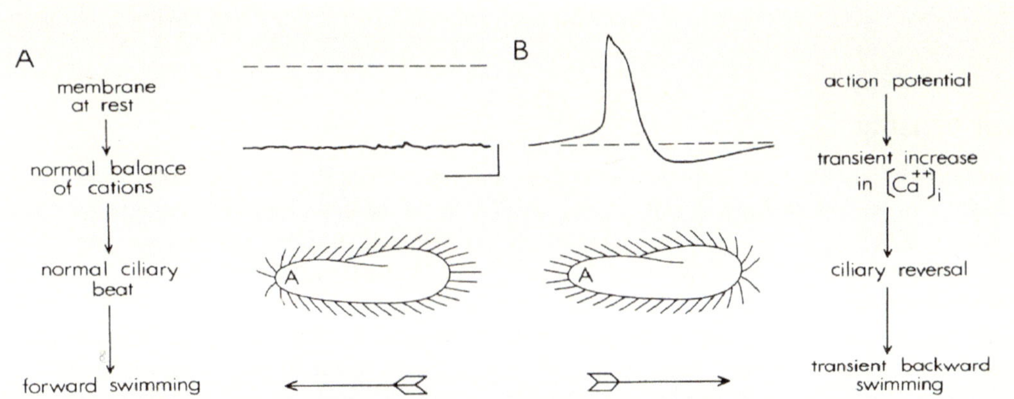

FIGURE 32.7 Relationship of membrane potential and locomotor behavior of *Paramecium*. (A) Top to bottom: Membrane potential at resting level. This level varies from 25 mv in the standard K−Ca solution to nearly zero in certain Ba solutions (4 mM KCl, 1 mM $CaCl_2$ with 1 mM Tris, pH 7.2). A normal balance of cations is maintained, especially a very low internal Ca^{2+} concentration, $[CaCl_2]_i$. This state correlates with the normal ciliary beat in which the power stroke beats away from the anterior of the cell. The metachronal beat of thousands of cilia propels the paramecium forward (to the left). (B) Action potential in a stimulating $BaCl_2$ solution. The excited membrane allows entry of Ca^{2+} into the cilia. The increased internal concentration of Ca^{2+}causes the cilia to beat in a reversed direction, and hence the cell backs up. The situation quickly reverses when the active depolarization is over and excess internal Ca^{2+} is removed by active pumping. For details, see Eckert 1972 and Kung et al. (1975).

threshold for a calcium-based action potential, the voltage-gated calcium channels of the cilia open transiently (Eckert, 1972). The inrush of Ca^{2+} into the cilia changes the axonemal timing, and the strong power stroke is now toward the anterior, causing the cell to swim backward. This does not last for long as the Ca^{2+} in the cilium is brought back to its previous nM basal level and forward swimming resumes. In addition, Ca^{2+} inactivates the channel. If the cell encounters an attractant or lower ionic strength, the resulting hyperpolarization of the membrane will cause the cilia to beat faster and the swimming to speed up (Machemer, 1974, 1989). Therefore, the speed of swimming and frequency of turning are dependent upon the membrane electrical properties of the cell.

These modulations of ciliary activity are most easily envisioned in the examples of chemoresponse. Paramecia encountering attractants hyperpolarize and speed up. Intracellular and intraciliary cAMP are known to be produced by hyperpolarization in low K^+ solutions and for at least one attractant, glutamate (Preston and Van Houten, 1987; Van Houten, 1998; Yang et al., 1997). The intraciliary cAMP has an effect upon the axoneme by cAMP-dependent phosphorylation of an outer arm protein p29 (Satir et al., 1993). Repellents cause slower swimming and frequent turns due to frequent action potentials. These behaviors can be attributed to raised Ca^{2+} in the cilia due to opening of the voltage-gated calcium channels, which are expressed exclusively in the cilia (Dunlap, 1977; Lodh et al., 2016). A ciliary response to Ca^{2+} is not unique to ciliates; it manifests itself as an arrest of ciliary beat in mussel gill cilia and other multiciliated cells.

As seen in Fig. 32.5, the cilia over the *Paramecium* cell beat in metachronal waves and do not tangle. Similarly, the cilia of multiciliated epithelial cells (e.g., respiratory system, reproductive tracts, and even brain) beat in metachronal waves that are hydrodynamically coupled (Spassky and Meunier, 2017). The spacing of the multiple cilia on epithelial cells is accomplished with beautiful arrays of actin filaments and microtubules, and anchoring through striated rootlets (Werner et al., 2011; Werner and Mitchell, 2012). The spacing of cilia on the *Paramecium* surface is constrained by the regimented surface and subsurface cytoskeletal organization such that there are one or two cilia arising from each hexagonal cortical unit (Fig. 32.5).

Sperm swim by movement of their single flagellum. They modify their motility upon sensing chemical and physical cues that signal the presence of their target, the egg (Kaupp and Strunker, 2017). In sea urchin sperm, the receptor is the membrane bound guanylyl cyclase that binds peptides released from the egg to make a gradient for the sperm to follow (Alvarez et al., 2014). This chemoreceptor is exceptional with very high affinity, high density in the membrane, rapid inactivation, and low signal amplification that enables the sperm to remain sensitive to chemoattractant while it continues to probe for the target (Pichlo et al., 2014). This is in contrast to the *Paramecium* chemoreceptors with μM to mM affinities that signify food or toxic conditions: seven-transmembrane receptor for cAMP, glycosylphosphatidyl inositol anchored protein for folate, N-methyl-D-aspartate like-receptor for glutamate (Valentine et al., 2008).

As in *Paramecium*, a calcium channel on the ciliary membrane is key to the change in ciliary (flagellar) beat and waveform that results in sperm chemoresponse (Kaupp and Strunker, 2017). While the array of other channels and transporters in the flagellar membrane of sperm differ among sea urchin, human, mouse, and zebra fish, the common element is the calcium channel and influx of Ca^{2+} that has a direct impact on the motile apparatus. A comparison of sea urchin and mammalian sperm chemotaxis swimming strategies and stimuli can be found in Eisenbach and Giojalas' review (Eisenbach and Giojalas, 2006). In addition to chemotaxis, mammalian sperm must also navigate by thermotaxis to locate the egg.

5. Quorum sensing

Bacteria are social; they communicate among themselves with molecules that they release called autoinducers, which signal that populations are low or high in cell density and control gene expression in response (Schauder and Bassler, 2001). This process of taking a census, i.e., a measure of cell density, is called quorum sensing. Quorum sensing in prokaryotes has a rich literature describing in detail the many signals that bacteria use to communicate and influence each other within and between species (Gross, 2017; Schauder and Bassler, 2001; Whiteley et al., 2017). Gram-negative bacteria use acyl-homoserine lactones as signaling molecules, while gram-positive bacteria use oligopeptides to secrete into the local environment as a measure of cell density. At sufficiently high concentration (proxy for cell density), transcription of specific sets of genes is set in motion. In the most famous example, light is produced from the light organ of a squid infected with *Vibrio fischeri* that have reached a sufficient density measured by concentration of the *N*-acyl-homoserine lactone. Other quorum sensing systems regulate biofilm development, antibiotic resistance, and the production and release of toxins (reviewed earlier in this book and in Jayaraman and Wood, 2008 and Papenfort and Bassler, 2016). The bacteria have very active communications with each other and with those of different species.

Density matters in infection, virulence, biofilm formation, and other bacterial behaviors. Schauder and Bassler review the chemical languages of bacteria (Schauder and Bassler, 2001), while others review how the hosts translate and respond to these bacterial signals (Hartmann and Schikora, 2012; Hughes and Sperandio, 2008a; Pacheco and Sperandio, 2009; Teplitski et al., 2011). Mammalian cells can fight back against bacterial pathogens by recognizing and destroying the secreted HSLs. If the host also mounts an immune response, some bacteria sense the signals of the activated innate immune system and modulate their virulence gene expression. Plants likewise have a defense against pathogenic bacteria by secreting molecules that mimic the signals and confuse the bacterial signaling system.

Here we describe some quorum sensing pathways and prokaryotic signals coopted by eukaryotes and follow this with a discussion of interkingdom signaling where some shared signals exist between cells belonging to different organisms.

5.1 Quorum sensing (cell density signaling) in *Dictyostelium*

The aforementioned discussion of slime mold cell motility and development into a multicellular structure addressed quorum sensing in the amoebae as they chemotax and aggregate. As cell numbers increase, the motile amoebae run out of food and begin to starve. At the same time, the cells are counting their numbers with the proxy prestarvation factor (PSF). PSF serves as a quorum sensing molecule that is secreted constantly by the vegetative amoebae grazing for food on their substratum (Loomis, 2014). When food declines and PSF is high (i.e., population density is high), the pathway in Fig. 32.1 is initiated (e.g., gene protein kinase A YakA transcribed) with a few founder cells pulsing secretions of cyclic AMP (cAMP).

After attraction by chemotaxis to cAMP, the dense collection of cells in the aggregate risk running out of food. The indirect quorum sensing maneuver of repulsion allows other cells to avoid and move away from the aggregate, through a suppression of pseudopods on the side of the cell nearest the aggregate where the autocrine proliferation repressor protein (AprA) is highest (Kirolos and Gomer, 2021). The PI3K/Akt/protein kinase B and GC pathways of the chemoattraction pathways are not involved in this directed movement. Instead, AprA inhibits local Ras activation, thereby inhibiting pseudopod formation on the side of cells closest to the aggregate and enabling cells to move away from the AprA source.

A second quorum sensing pathway, conditioned medium factor (CMF), comes into play when development is induced by starvation, after the combination of PSF and high cell numbers stimulate chemotaxis (Fig. 32.1) (Loomis, 2014). This delay is accomplished by the receptor for CMF, which is not located on the cell surface until after the onset of starvation. Sufficient CMF is required for the cyclic AMP surface receptor (CarA)-mediated signal transduction to proceed. The cells need a short exposure to CMF before the G proteins liberated from the cAMP receptor can stimulate the adenylyl cyclase ACA for the production of the extracellular signaling cAMP for aggregation.

Yet another factor signals to the aggregating cells so that they do not exceed the optimal number of about 2×10^4 (Tang et al., 2002) by decreasing adhesion and increasing motility. This has the effect of breaking up the streams of cells spiraling in toward the founder cell to aggregate into a mound and further development. The size of the cAMP pulses is also decreased by the counting factor (simply called CF), which is a complex of five proteins (Tang et al., 2002).

5.2 Quorum sensing in yeast

Quorum sensing has recently been described in some fungal species. Quorum sensing examples discussed here all involve the transition from a filamentous, or mycelial, fungal form to a yeast form. In *Histoplasma capsulatum*, a human pathogen, the cells exist in a filamentous form, but at higher densities, transition to yeast form. At high densities, the yeast emit a glucan which they use in their cell walls, increasing the virulence (Kugler et al., 2000). The causative fungus in Dutch Elm disease, *Ceratocystis ulmi*, also exists in a filamentous form at low densities. At higher densities, the cells emit a substance, which acts similarly to autoinducers in bacteria, causing the cells to switch to a yeast form. This substance that is released has yet to be identified, but studies of transferring the filamentous cells at very low densities into a culture that previously contained high concentrations of the yeast, or conditioned medium, cause the fungal cells to rapidly switch to the yeast form, even though their densities were not high enough to induce the switch (Hornby et al., 2004).

A quorum sensing molecule quorum sensing–like peptide 1 (Qsp1) has been identified in the human pathogenic fungus *Cryptococcus neoformans*. This small peptide is synthesized and secreted by the fungus as a propeptide. The propeptide is cleaved by a cell-associated protease aptly named proQsp1 protease 1 (Pqp1). The generated Qsp1 peptide is transported back into the cell by the oligopeptide transporter, Opt1. Once inside the cell, Qsp1 induces changes that increase the virulence of the cell such as growth, changing the cell surface properties (cell wall), and increasing secreted proteases (Homer et al., 2016). The implications of the quorum sensing identified here suggest new targets for treatments and drug therapies, which could help in the fight against dangerous and costly infections.

6. Prokaryotic pathways found in eukaryotic signaling

6.1 Kinase signaling pathways

Both prokaryotes and eukaryotes utilize the phosphorylation of proteins in their signal transduction pathways, but the method of phosphate transfer can be very different. Prokaryotic signaling famously includes two component histidine kinase systems (Schaller et al., 2011). Upon stimulus acting upon a sensor subunit (e.g., chemoreceptor), there is a transfer of a phosphate group from a histidine on the kinase to an aspartate on a response regulator (e.g., transcription factor or enzyme). That is, the phosphate is physically relayed from one molecule to another. Some stimuli may also reverse this process by transfer of the phosphoryl group from the response regulator to the histidine kinase (Schaller et al., 2011).

In contrast, eukaryotic signal transduction pathways were thought not to treat phosphorylation as a replay, but rather to catalyze the addition of a phosphate to a substrate by kinase hydrolysis of ATP or removal of a phosphate by phosphatase hydrolysis. The phosphate is not handed off from one pathway component to another.

Our understanding of eukaryotic signaling is expanding to include eukaryotic two component systems in plants, animals, and protists including *Dictyostelium* (Schaller et al., 2011) (see Fig. 32.1 and Schaap, 2014; Schaap, 2016). Starvation of *Dictyostelium* amoebae initiates extracellular signaling of pulses of cyclic AMP. Amoebae follow the gradient of cyclic AMP to migrate together in anticipation of forming a slug. As described more thoroughly elsewhere in this chapter, cells that aggregate into a slug will go on to form stalk and spore cells, once again in response to cyclic AMP but with a different set of receptors. While starving, the amoebae also release NH_3 from proteolysis. This NH_3 keeps cyclic AMP-activated protein kinase A (PKA) in check and thus differentiation in check through a two-component system. NH_3 activated histidine kinase Dhkc, which phosphorylates phosphodiesterase RegA through a relay of a phosphoryl group from kinase DhcK to RegA. Once the slug tip projects upward, the NH_3 is lost by diffusion and the two-component pathway inhibition of PKA is finished. The subsequent development of cells into prestalk and then prespore cells depends on the histidine kinases Dhkc and DhKA. In the latter case, the P group is relayed from RegA to Hpt to DhkA, whereas in prestalk from Dhkc sensor to reg to RegA. Cyclic AMP then initiates prespore formation.

6.2 Bacterial signal cyclic di-GMP can Be synthesized in eukaryotes

Cyclic di-GMP (c-di-GMP) was thought to be a messenger in prokaryotes alone until 2012 when Schaap and workers identified it as a signal in *D. discoideum* development. C-di-GMP triggers stalk cell differentiation and serves as the morphogen for stalk cell phenotype (Fig. 32.1) (Chen and Schaap, 2012; Du et al., 2015). While c-di-GMP has now been found in other Dictyostelids, this messenger has not been identified widely in eukaryotes. When found in mammalian cells, it can originate from the bacteria responsible for an infection (Jenal et al., 2017). (See below for more details on cyclic dinucleotide messengers.)

7. Interkingdom signaling

7.1 Bacterial quorum sensing signals sensed by eukaryotes

Plant and human pathogenic bacteria cluster their virulence genes under quorum sensing control (reviewed in Hartmann and Schikora, 2012). Therefore, it would be advantageous for the host to detect and respond to the presence of quorum sensing signals such as homoserine lactones (HSLs) to mount a defense. There is evidence that the plant and animal hosts can distinguish among the HSL molecule side chains and respond differently to them with resistance or immune responses (reviewed in Hartmann and Schikora, 2012; Teplitski et al., 2011).

The commensal microbes of the host gut are important for nutrient assimilation and development of the innate immune system (Hughes and Sperandio, 2008b). However, pathogenic microbes can "highjack" the signals used for the commensal relationship for activation of virulence genes (Hughes and Sperandio, 2008b). The bacteria use signals to develop virulence and build the biofilms that are dangerous to the host. In turn, the host immune system can use these bacterial signals to respond to infections. One kind of host response for defense against biofilms is to degrade the signal HSL (Hastings, 2004; Pacheco and Sperandio, 2009). Quorum sensing molecules seem to talk with the host to induce inflammation (as in the cystic fibrosis lung) and apoptosis.

The mammalian host communicates information to the bacteria through its cytokines, hormones and neurotransmitters (Atkinson and Williams, 2009; Teplitski et al., 2011). Some pathogenic bacteria recognize and use these host signals to activate virulence factors. (See Hughes and Sperandio, 2008a; Pacheco and Sperandio, 2009 and Hughes and Sperandio for discussion of catecholamines and endogenous opioids on bacterial signaling (Atkinson and Williams, 2009) and Boukerb et al. for a review on interkingdom signaling (Boukerb et al., 2021).) Using a different approach to confuse the signaling of pathogens, plants have "mastered the art of AHL mimicry" (Hughes and Sperandio, 2008b).

While the examples of bacterial quorum sensing signals on host immune responses are established, a more unusual example of behavioral change of a eukaryote in response to *N*-acyl-homoserine lactone (AHL) is the algae *Ulva intestinalis* and biofilm bacteria. The motile zoospores of the marine algae settle on substrates but will detach if they are not optimal. The zoospores preferentially settle on biofilms that produce AHL, apparently by slowing down at the location of appropriate biofilms, i.e., by a chemokinesis behavior (Joint et al., 2002, 2007). Settling on AHL-producing biofilms has trade-offs, as the AHL will also negatively impact germination and growth (Twigg et al., 2014).

7.2 Bacterial cyclic di-nucleotide signals sensed by eukaryotes

Other prokaryotic signaling molecules that are sensed by eukaryotes are cyclic-di-GMP (Jenal et al., 2017; Rö et al., 2013) and cyclic-di-AMP (Corrigan and Grundling, 2013). These molecules are synthesized and utilized by many species of bacteria. C-di-GMP and c-di-AMP are uniquely bacterial in origin and signal about infection to mammalian cells. We will address in the following another cyclic di-nucleotide c-GMP-AMP that is synthesized in bacteria, and probably arose as an antiviral defense in bacteria (Morehouse et al., 2020) but importantly also in eukaryotic cells in response to infection (Cohen et al., 2019; Kato et al., 2017; Rö et al., 2013). All of these signals appear to have the same receptor in immune cells to set in motion signaling pathways culminating in type I interferon and proinflammatory cytokines.

Bacterial c-di-GMP is widely recognized as important second messenger for a large number of both gram-negative and gram-positive bacteria. C-di-GMP participates in the regulation of bacterial motility, virulence, differentiation, and biofilm formation among other functions; high and low concentrations of c-di-GMP control an astonishing number of phenotypes (Hengge, 2021; Hengge et al., 2016; Jenal et al., 2017; Patterson et al., 2021; Rö et al., 2013; Quantity Meets Quality at mTOR). See (Hengge, 2021) for Hengge's review of the conundrum of local and global signaling pathways all using the same second messenger c-di-GMP.

This signaling molecule profoundly affects eukaryotes in two ways: by control of the pathogenicity features of bacteria (such as virulence, adherence, biofilm formation, cytotoxicity, among others) and by modulation and stimulation of the innate immune system (Fig. 32.8). C-di-GMP administered to mice triggers the release of multiple cytokines and chemokines. At least part of the mechanism seems to be through binding of c-di-GMP to membrane protein STING (stimulator of interferon genes), and release of type I interferons (Burdette et al., 2011). C-di-GMP, which is not found in mammalian cells, is recognized by the host as a signal of bacterial infection by the mammalian immune system. It is this interkingdom interaction and recognition that is of interest for this chapter.

Another player in the ancient signaling systems of bacteria is cyclic di-AMP, which is associated with bacterial resistance, sporulation, lipid and cell wall homeostasis, DNA integrity, and other phenotypes (Corrigan and Grundling, 2013). This cyclic di-nucleotide is detected by the mammalian immune system through the STING pathway as a signal of DNA that does not belong in this host cell (Corrigan and Grundling, 2013) (Fig. 32.8). For example, the intracellular pathogen *Listeria monocytogenes* enters a host cell and at first resides in a vacuole. When the bacteria leave the vacuole and enter the cytoplasm, they secrete c-di-GMP and c-dix-AMP into the host cell. Once these signaling molecules are in the host cytoplasm, they converge on activation of the immune response to infection (Corrigan and Grundling, 2013) by binding c-di-GMP directly to STING or c-di-AMP with a coreceptor (helicase DDX41) to STING to induce interferon synthesis.

As with the other nucleotide signals (c-di-GMP and c-di-AMP), c-GMP-AMP activates the innate immune system that is the first line of defense against nonself during bacterial and viral infections. Note in Fig. 32.8 that double-stranded DNA from infection or cell damage activates cyclic guanosine monophosphate—adenosine monophosphate (cGAMP) synthase (cGAS), which is the enzyme that produces c-GMP-AMP. The c-GMP-AMP signals to the host to contribute to the production of interferons and cytokines by binding to its receptor STING (Gao et al., 2013; Kato et al., 2017; Li et al., 2013).

The predominant role of cGAS is a positive one of sensing viral DNA during infection (Ablasser and Chen, 2019). Unfortunately, some autoimmune diseases can also be linked back to pGAS sensing of self-DNA (reviewed in Ablasser and Chen, 2019). Cyclic GMP-AMP can be spread among neighbors by intercellular transfer through gap junction channels to amplify the immune response (Ablasser et al., 2013). This system of immunity is highly conserved and not limited to mammals. For example, it is also utilized by Drosophila to respond to viral infections and RNA (Holleufer et al., 2021; Slavik et al., 2021).

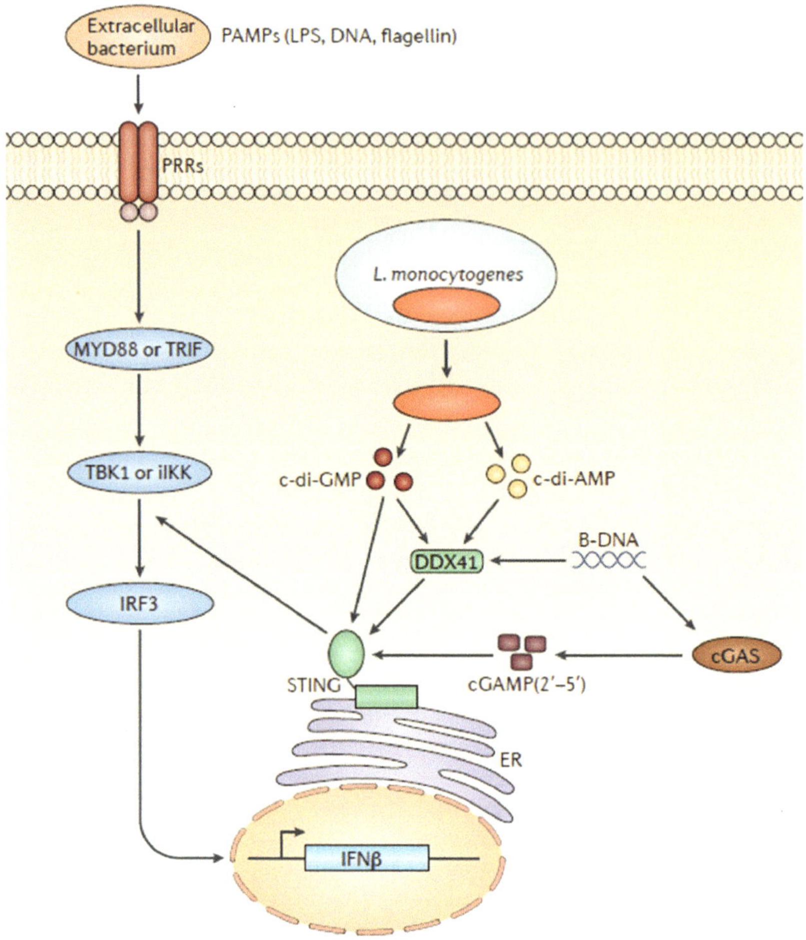

FIGURE 32.8 Induction of type I interferon response by bacteria. Recognition of extracellular bacteria occurs through pathogen-associated molecular patterns (PAMPs) such as lipopolysaccharide, DNA, and flagellin, leading to activation of kinases or the TRIF pathway (left side of the figure). By contrast, the intracellular bacterial pathogen *Listeria monocytogenes* enters eukaryotic host cells and is initially contained within a vacuole where it avoids detection. However, the bacteria can escape the vacuole and enter the cytosol, where some of their pools of cyclic di-GMP and cyclic di-AMP can be secreted. There is evidence that c-di-AMP binds to the cell's helicase DDX41, which in turn interacts with and activates coreceptor STING, or alternatively binds directly to STING. STING activation leads to activation of TBK1, subsequent phosphorylation and translocation of IRF3 to the nucleus for activation of IFNβ production. STING is also the receptor for c-di-GMP. STING can also be activated by c-GAMP which is produced in response to the double-stranded DNA (evidence of infection) that activates cGAMP synthase (cGAS). The c-GAMP produced by cGAS further activates STING (Corrigan and Grundling, 2013).

An ongoing challenge is to determine how cGAS distinguishes double stranded DNA that signals damage, stress, or infection from self-DNA, in particular from nuclear chromatin (Wischnewski and Ablasser, 2021). How can cGAS, as a nuclear protein, interact with nuclear chromatin and not wreak havoc on the cell? What is the role of liquid–liquid phase separation with chromatin in the activation of cGAS (Wischnewski and Ablasser, 2021; Zhao et al., 2022)? What are the nonimmune roles for cGAS and similar enzymes to respond as sensors to metabolites such as folate (Kranzusch, 2019)?

8. Chemotropism in yeast

As we have seen, the hallmarks of chemotaxis in *Dictyostelium* are motility, directional sensing, and polarization (Van-Haastert et al., 2004). Yeast are not motile, but they have the capability of directional sensing and polarization in chemotropism, i.e., asymmetric growth in a gradient. Arkowitz, 1999, 2009 in his comparison of *Dictyostelium* with yeast emphasizes the common aspects of seven-transmembrane receptors, coupled G proteins for G_β downstream signaling, stimulus in a gradient, and actin polymerization (Arkowitz, 2009; Ismael et al., 2016) in the pole with the highest concentration of stimulus.

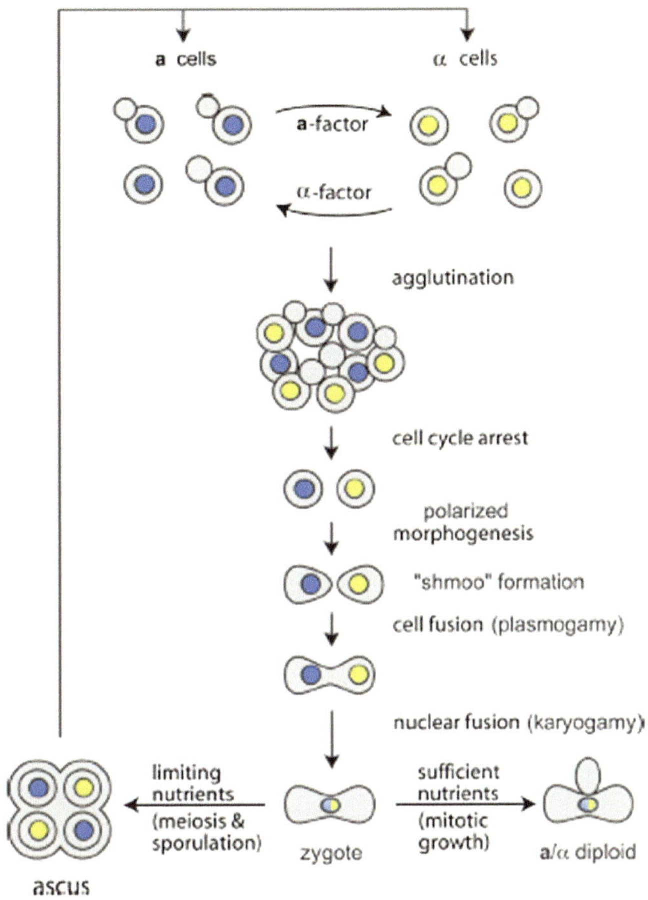

FIGURE 32.9 Overview of events that occur in pheromone-induced conjugation (mating) of the two haploid mating types of budding yeast *Saccharomyces cerevisiae* (Alvaro and Throner, 2016).

Yeast *Saccharomyces cerevisiae* have two cell types, a and α (Fig. 32.9) (Alvaro and Thorner, 2016; Arkowitz, 2009). Each secretes a specific mating pheromone, a-factor from a cells and α-factor from α cells. These cell types express receptors specific for a or α mating pheromones, which are 12- and 13-amino-acid peptides, respectively, with a lipid modification on the a-factor. This system allows cells to determine whether there are appropriate cells for mating in the vicinity. Surrounded by a field of yeast cells, a cell will respond to the complementary mating type by arresting the cell cycle and polarizing with the protruding end toward the cell emitting the complementary pheromone. The resulting shmoo shaped (pear shaped) cell (Fig. 32.9) has set into action the complex signaling seen in Fig. 32.10 as the prelude to mating of cells of two different mating types. This signaling allows the shmoo cells to grow toward each other and not toward others for fusion.

Changes in shape of yeast cells occur in contexts other than mating. The bud for cell division forms opposite the previous bud, thus polarizing the cell. However, it is the response of a cell to the steepest gradient of mating pheromone emanating from a complementary cell that underlies directional growth (Arkowitz, 2009).

We have focused on yeast, but sex pheromone receptor—based chemotropism has been identified in other fungi. For filamentous fungi, *Neurospora* serves as a model system for directed growth of hyphae toward a mating partner (Kronstad and Staben, 1997). Common features are mating-type receptors and mitogen-activated protein kinase (MAPK) and cascades, while differences abound. Pathogenic fungi also use hyphal growth, but in this case toward chemical signals from the host plant. The plant pathogen *Fusarium oxysporum* grows toward plant roots (Turrà et al., 2015). The underlying signaling pathways utilize the MAPK cascade and receptors similar to the mating-type seven-tansmembrane receptors of yeast. The work of Turrà and others (Turrà et al., 2015) points to the mobilization of pheromone machinery for colonization of plant roots by the fungi.

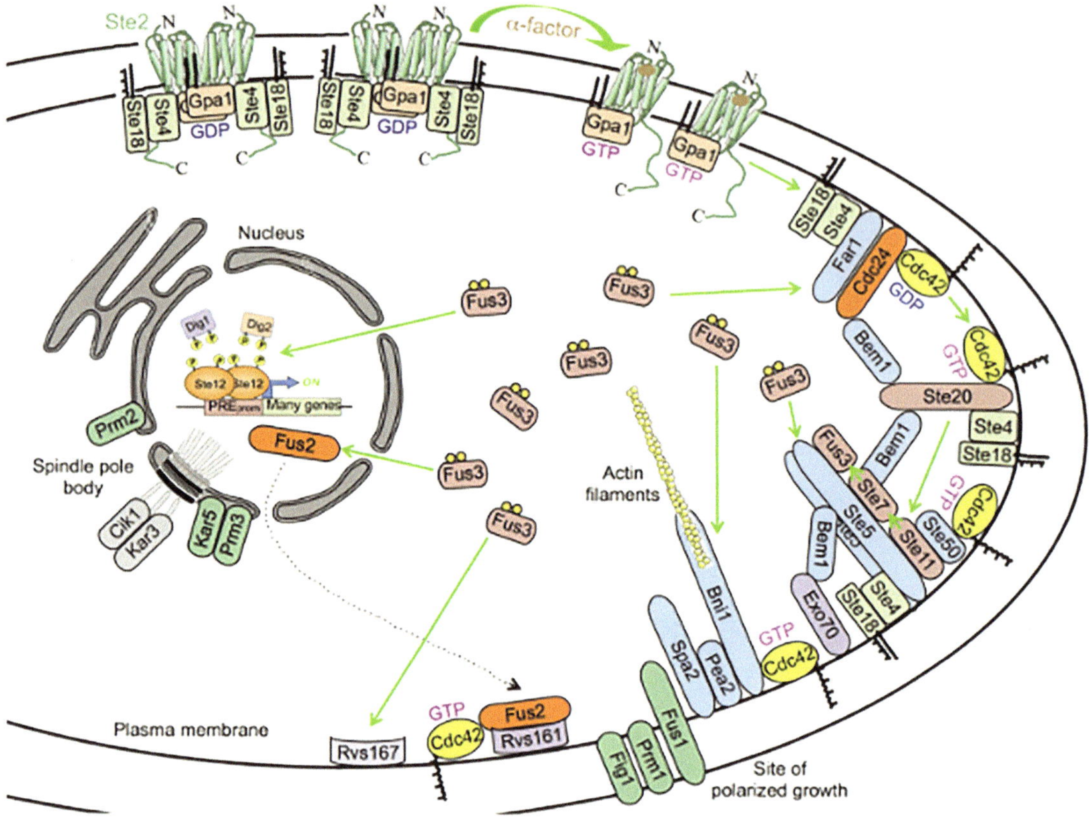

FIGURE 32.10 Schematic representation of the GPCR-initiated biochemical processes required for the execution of the mating pheromone response program (Alvaro and Thorner, 2016).

References

Ablasser, A., et al. (2013). Cell intrinsic immunity spreads ty bystander cells via the intercellular transfer or cGAMP. *Nature, 503*, 510–534.

Ablasser, A., & Chen, Z. J. (2019). cGAS in action: Expanding roles in immunity and inflammation. *Science, 363*(6431), Article eaat8657.

Alvarez, L., et al. (2014). The computational sperm cell. *Trends in Cell Biology, 24*(3), 198–207.

Alvaro, C. G., & Thorner, J. (2016). Heterotrimeric G protein-coupled receptor signaling in yeast mating pheromone response. *Journal of Biological Chemistry, 291*(15), 7788–7795.

Arkowitz, R. A. (1999). Responding to attraction: Chemotaxis and chemotropism in Dictyostelium and yeast. *Trends in Cell Biology, 9*, 20–27.

Arkowitz, R. (2009). Chemical gradients and chemotropism in yeast. *Cold Spring Harbor Perspectives in Biology, 1*, 1–20.

Artemenko, Y., et al. (2016). Chemical and mechanical stimuli act on common signal transduction and cytoskeletal networks. In *Proceedings of the national academy of sciences*.

Artemenko, Y., Lampert, T. J., & Devreotes, P. N. (2014). Moving towards a paradigm: Common mechanisms of chemotactic signaling in Dictyostelium and mammalian leukocytes. *Cellular and Molecular Life Sciences, 71*(19), 3711–3747.

Atkinson, S., & Williams, P. (2009). Quorum sensing and social networkingin the microbial world. *Journal of The Royal Society Interface*.

Baggiolini, M. (1998). Chemokines and leukocyte traffic. *Nature, 392*(6676), 565–568.

Barriga, E. H., & Theveneau, E. (2020). In vivo neural crest cell migration is controlled by "mixotaxis". *Frontiers in Physiology, 11*(1537).

Berg, H. C. (1983). *Random walks in Biology*. Princeton, NJ: Princeton University Press.

Bloodgood, R. (2010). Sensory reception is an attribute of both primary cilia and motile cilia. *Journal of Cell Science, 123*, 505–509.

Bosgraaf, L., & Van Haastert, P. J. M. (2002). A model for cGMP signal transduction in Dictyostelium in perspective of 25 years of cGMP research. *Journal of Muscle Research & Cell Motility, 23*(7–8), 781–791.

Bosgraaf, L., & van Haastert, P. J. (2006). The regulation of myosin II in Dictyostelium. *European Journal of Cell Biology, 85*(9–10), 969–979.

Boukerb, A. M., et al. (2021). Inter-kingdom signaling of stress hormones: Sensing, transport and modulation of bacterial physiology. *Frontiers in Microbiology, 12*, Article 690942.

Braun, D., & Hildebrandt, F. (2017). Ciliopathies. *Cold Spring Harbor Perspectives in Biology, 9*, Article a028191.

Brunetti, R. M., et al. (2021). WASP integrates substrate topology and cell polarity to guide neutrophil migration. *Journal of Cell Biology, 221*(2), Article e202104046.

Bugyi, B., & Carlier, M. F. (2010). Control of actin filament treadmilling in cell motility. *Annual Review of Biophysics, 39*, 449–470.

Burdette, D. L., et al. (2011). STING is a direct innate immune sensor of cyclic di-GMP. *Nature*.

Chen, Z.-H., & Schaap, P. (2012). The prokaryote messenger c-di-GMP triggers stalk cell differentiation in Dictyostelium. *Nature, 488*(7413), 680–683.

Cohen, D., et al. (2019). Cyclic GMP–AMP signalling protects bacteria against viral infection. *Nature, 574*(7780), 691–695.

Cole, A., et al. (2021). Differential roles of actin crosslinking proteins filamin and α-actinin in shear flow-induced migration of *Dictyostelium discoideum. Frontiers in Cell and Developmental Biology, 9*(2276).

Cornwell, K. G., Downing, B. R., & Pins, G. D. (2004). Characterizing fibroblast migration on discrete collagen threads for applications in tissue regeneration. *Journal of Biomedical Materials Research Part A, 71*(1), 55–62.

Corrigan, R. M., & Grundling, A. (2013). Cyclic di-AMP: Another second messenger enters the fray. *Nature Reviews Microbiology, 11*(8), 513–524.

Devreotes, P., & Horwitz, A. R. (2015). Signaling networks that regulate cell migration. *Cold Spring Harbor Perspectives in Biology, 7*(8), Article a005959.

Devreotes, P., & Parent, C. A. (1989). Chemotaxis inEukaryotic cells: A focus on leukocytes and Dictyostelium. *Annual Review of Cell Biology, 4*, 649–686.

Devreotes, P. N., et al. (2017). Excitable signal transduction networks in directed cell migration. *Annual Review of Cell and Developmental Biology, 33*(1), 103–125.

Du, Q., Kawabe, Y., Schilde, C., Chen, Z-h, & Schaap, P. (2015). The evolution of aggregative multicellularity and cell-cell communication in Dictyostelia. *Journal of Molecular Biology, 427*, 3711–3733.

Dunlap, K. (1977). Localization of calcium channels in Paramecium caudatum. *The Journal of Physiology, 271*(1), 119–133.

Dunn, G. (1981). Biology of the chemotactic response. In J. M. Lackie, & P. C. Wilkinson (Eds.), *Society for experimental Biology seminar series 12*. London: Cambridge University Press.

Eckert, R. (1972). Bioelectric control of ciliary activity. *Science, 176*(4034), 473–481.

Egelhoff, T. T., Lee, R. J., & Spudich, J. A. (1993). Dictyostelium myosin heavy chain phosphorylation sites regulate myosin filament assembly and localization in vivo. *Cell, 75*(2), 363–371.

Eisenbach, M., & Giojalas, L. (2006). Sperm guidance in mammals - an unpaved road to the egg. *Nature Reviews Molecular Cell Biology, 7*, 276–285.

Espina, J., Marchant, C., & Barriga, E. (2021). Durotaxis: The mechanical control of directed cell migration. *FEBS Journal*, 1–19.

Fais, S., & Malorni, W. (2003). Leukoocyte uropod formation and membrane/cytoskeleton linkage in immune interactions. *Journal of Leukocyte Biology, 73*, 556–563.

Fraenkel, G. S., & Gunn, D. L. (1961). *The Orientation of animals*. New York: Dover Publications.

Franca-Koh, J., Kamimura, Y., & Devreotes, P. (2006). .Navigating signaling networks: Chemotaxis in *Dictyostelium discoideum. Current Opinion in Genetics & Development, 16*(4), 333–338.

Funamoto, S., et al. (2002). Spatial and temporal regulation of 3-phosphoinositides by PI 3-kinase and PTEN mediates chemotaxis. *Cell, 109*(5), 611–623.

Gao, P., et al. (2013). Cyclic [G(22,52)pA(32,52)p] is the metazoan second messenger produced by DNA-activated cyclic GMP-AMP synthase. *Cell, 153*(5), 1094–1107.

Gooday, G. (1981). Chemotaxis in eukaryotic microbes. In J. Lackie, & P. Wilkinson (Eds.), *Society for experimental Biology seminar series 12* (Vol 12, pp. 115–138). London: Cambridge University Press.

Goode, B. L., & Eck, M. J. (2007). Mechanism and function of formins in the control of actin assembly. *Annual Review of Biochemistry, 76*, 593–627.

Griffith, L. M., Downs, S. M., & Spudich, J. A. (1987). Myosin light chain kinase and myosin light chain phosphatase from Dictyostelium: Effects of reversible phosphorylation on myosin structure and function. *The Journal of Cell Biology, 104*(5), 1309–1323.

Gross, M. (2017). Shining new light on quorum sensing. *Current Biology, 27*(24), R1293–R1296.

Hartmann, A. C., & Schikora, A. (2012). Quorum sensing of bacteria and trans-kingdom interactions of N-acyl homoserine lactones with eukaryotes. *Journal of Chemical Ecology*.

Hastings, J. (2004). Bacterial quorum-sensing signals are inactivated by mammalian cells. *Proceedings of the National Academy of Sciences of the United States of America, 101*, 3993–3994.

Hengge, R., et al. (2016). Bacterial signal transduction by cyclic di-GMP and other nucleotide second messengers. *Journal of Bacteriology, 198*(1), 15–26.

Hengge, R. (2021). High-specificity local and global c-di-GMP signaling. *Trends in Microbiology, 29*(11), 993–1003.

Hilgendorf, K., Myers, B., & Reiter, J. (2024). Emerging mechanistic understanding of cilia function in cellular signalling. *Nature Reviews Molecular and Cell Biology, 25*(7), 555–573. https://doi.org/10.1038/s41580-023-00698-5

Hind, L., Vincent, W., & Huttenlocher, A. (2016). Leading from the back: The role of the uropod in neutrophil polarization and migration. *Developmental Cell, 38*, 161–169.

Holleufer, A., et al. (2021). Two cGAS-like receptors induce antiviral immunity in Drosophila. *Nature*.

Homer, C. M., et al. (2016). Intracellular action of a secreted peptide required for fungal virulence. *Cell Host & Microbe, 19*(6), 849–864.

Hornby, J. M., et al. (2004). Inoculum size effect in dimorphic fungi: Extracellular control of yeast-mycelium dimorphism in Ceratocystis ulmi. *Applied and Environmental Microbiology, 70*(3), 1356–1359.

Hughes, D. T., & Sperandio, V. (2008a). Inter-kingdom signalling: Communication between bacteria and their hosts. *Nature Reviews Microbiology, 6*.

Hughes, D. T., & Sperandio, V. (2008b). Inter-kingdom signalling: Communication between bacteria and their hosts. *Nature Reviews Microbiology, 6*(2), 111–120.

Iijima, M., & Devreotes, P. (2002). Tumor suppressor PTEN mediates sensing of chemoattractant gradients. *Cell, 109*(5), 599–610.

Inagaki, N., & Katsuno, H. (2016). Actin waves: Origin of cell polarization and migration? *Trends in Cell Biology, 27*(7), 515–526.

Insall, R. H., & Machesky, L. M. (2009). Actin dynamics at the leading edge: From simple machinery to complex networks. *Developmental Cell, 17*(3), 310–322.

Ismael, A., et al. (2016). Gβ promotes pheromone receptor polarization and yeast chemotropism by inhibiting receptor phosphorylation. *Science Signaling, 9*(423), Article ra38-ra38.

Jayaraman, A., & Wood, T. K. (2008). Bacterial quorum sensing: Signals, circuits, and implications for biofilms and disease. *Annual Review of Biomedical Engineering, 10*, 145–167.

Jenal, U., Reinders, A., & Lori, C. (2017). Cyclic di-GMP: Second messenger extraordinaire. *Nature Reviews Microbiology, 15*(5), 271–284.

Jennings, H. (1906). *Behavior of the lower organisms* (1962 Edition). Bloomington, Indiana: Indiana University Press.

Joint, I., et al. (2002). Cell-to-cell communication across the prokaryote-eukaryote boundary. *Science, 298*, 1207.

Joint, I., Tait, K., & Wheeler, G. (2007). Cross-kingdom signalling: Exploitation of bacterial quorum sensing molecules by the green seaweed Ulva. *Philosophical Transactions of the Royal Society B: Biological Sciences, 362*(1483), 1223–1233.

Kato, K., et al. (2017). Cyclic GMP–AMP as an endogenous second messenger in innate immune signaling by cytosolic DNA. *Annual Review of Biochemistry, 86*(1), 541–566.

Kaupp, U. B., & Strunker, T. (2017). Signaling in sperm: More different than similar. *Trends in Cell Biology, 27*, 101–109.

Kay, R., et al. (2008). Changing directions in the study of chemotaxis. *Nature Reviews Molecular Cell Biology, 9*, 455–463.

Kimmel, A. R., & Parent, C. A. (2003). The signal to move: D. discoideum go orienteering. *Science, 300*, 1525–1529.

Kirolos, S. A., & Gomer, R. H. (2021). A chemorepellent inhibits local Ras activation to inhibit pseudopod formation to bias cell movement away from the chemorepellent. *Molecular Biology of the Cell, 33*(1), ar9.

Kleene, S., & Van Houten, J. (2014). Electrical signaling in motile and primary cilia. *BioScience, 64*, 1092–1102.

Kranzusch, P. J. (2019). cGAS and CD-NTase enzymes: structure, mechanism, and evolution. *Current Opinion in Structural Biology, 59*, 178–187.

Kronstad, J. W., & Staben, C. (1997). Mating type in filamentous fungi. *Annual Review of Genetics, 31*(1), 245–276.

Kugler, S., et al. (2000). Phenotypic variation and intracellular parasitism by histoplasma Capsulatum. *Proceedings of the National Academy of Sciences of the United States of America, 97*(16), 8794–8798.

Kung, C., et al. (1975). Genetic dissection of behavior in Paramecium. *Science, 188*, 898–904.

Lacalle, R. A., et al. (2004). PTEN regulates motility but not directionality during leukocyte chemotaxis. *Journal of Cell Science, 117*(Pt 25), 6207–6215.

Laevsky, G., & Knecht, D. A. (2003). Cross-linking of actin filaments by myosin II is a major contributor to cortical integrity and cell motility in restrictive environments. *Journal of Cell Science, 116*(Pt 18), 3761–3770.

Lammermann, T., et al. (2008). Rapid leukocyte migration by integrin-independent flowing and squeezing. *Nature, 453*(7191), 51.

Leonard, E. J., et al. (1990). Leukocyte specificity and binding of human neutrophil attractant activation protein-1. *Journal of Immunology, 144*(4), 1323–1330.

Li, X., et al. (2013). Pirotal roles of cGAS-cGAMP signaling in antiviral defense and immune adjuvant effects. *Science, 341*, 1390–1394.

Lindemann, C., & Leisch, K. (2010). Flagellar and ciliary beating: The proven and the possible. *Journal of Cell Science, 123*, 519–528.

Lodh, S., et al. (2016). Voltage-gated calcium channels of Paramecium cilia. *Journal of Experimental Biology, 219*(19), 3028–3038.

Loomis, W. (2014). Cell signaling during development of Dictyostelium. *Developmental Biology, 391*, 1–16.

Machemer, H. (1989). Cellular behavior modulated by ions: Electrophysiological implications. *Journal of Protozoology, 36*, 463–487.

Machemer, H. (1974). Frequency and directional responses of cilia to membrane potential changes in Paramecium. *Journal of Comparative Physiology, 92*, 293–316.

Machesky, L. M., & Insall, R. H. (2021). WAVE complex regulation by force. *Nature Cell Biology, 23*(11), 1111–1112.

Malicki, J. J., & Johnson, C. A. (2017). The cilium: Cellular antenna and central processing unit. *Trends in Cell Biology, 27*(2), 126–140.

Mehidi, A., et al. (2021). Forces generated by lamellipodial actin filament elongation regulate the WAVE complex during cell migration. *Nature Cell Biology, 23*(11), 1148–1162.

Morehouse, B. R., et al. (2020). STING cyclic dinucleotide sensing originated in bacteria. *Nature.*

Moser, B., et al. (2004). Chemokines: Multiple levels of leukocyte migration control. *Trends in Immunology, 25*(2), 75–84.

Pacheco, A., & Sperandio, V. (2009). Ubter-kingdom signaling: Chemical language between bacteria and host. *Current Opinion in Microbiology, 12*, 192–198.

Papenfort, K., & Bassler, B. L. (2016). Quorum sensing signal-response systems in Gram-negative bacteria. *Nature Reviews Microbiology, 14*(9), 576–588.

Parent, C. A. (2004). Making all the right moves: Chemotaxis in neutrophils and Dictyostelium. *Current Opinion in Cell Biology, 16*(1), 4–13.

Patterson, D. C., et al. (2021). Differential ligand-selective control of opposing enzymatic activities within a bifunctional c-di-GMP enzyme. *Proceedings of the National Academy of Sciences, 118*(36), Article e2100657118.

Pichlo, M., et al. (2014). High density and ligand affinity confer ultrasensitive signal detection by a guanylyl cyclase chemoreceptor. *The Journal of Cell Biology, 206*(4), 541–557.

Pollitt, A. Y., & Insall, R. H. (2009). WASP and SCAR/WAVE proteins: The drivers of actin assembly. *Journal of Cell Science, 122*(Pt 15), 2575–2578.

Preston, R. R., & Van Houten, J. L. (1987). Chemoreception in paramecium tetraurelia: Acetate and folate-induced membrane hyperpolarization. *Journal of Comparative Physiology, 160*(4), 525–535.

Qin, L., et al. (2021). Roles of leader and follower cells in collective cell migration. *Molecular Biology of the Cell, 32*(14), 1267–1272.

Quantity meets quality at mTOR Junctionâ™¦. *Journal of Biological Chemistry, 285*(35), e99963.

Römling, U., Galperin, M. Y., & Gomelsky, M. (2013). Cyclic di-GMP: The first 25 Years of a universal bacterial second messenger. *Microbiology and Molecular Biology Reviews: Microbiology and Molecular Biology Reviews, 77*(1), 1–52.

Reiter, J. F., & Leroux, M. R. (2017). Genes and molecular pathways underpinning ciliopathies. *Nature Reviews Molecular Cell Biology, 18*(9), 533–547.

Reiter, J. F., Blacque, O. E., & Leroux, M. R. (2012). The base of the cilium: Roles for transition fibres and the transition zone in ciliary formation, maintenance and compartmentalization. *EMBO Reports, 13*(7), 608–618.

Roussos, E. T., Condeelis, J. S., & Patsialou, A. (2011). Chemotaxis in cancer. *Nature Reviews Cancer, 11*, 573.

Sasaki, A. T., et al. (2004). Localized Ras signaling at the leading edge regulates P13K, cell polarity, and directional cell movement. *Journal of Cell Biology, 167*(3), 505–518.

Satir, P., Barkalow, K., & Hamasaki, T. (1993). The control of ciliary beat frequency. *Trends in Cell Biology, 3*, 409–412.

Satir, P., Heuser, T., & Sale, W. (2014). A structural basis for how motile cilia beat. *BioScience, 64*, 1073–1083.

Schaap, P. (2016). Evolution of developmental signaling in Dictyostelid social amoebas. *Genetics & Development, 39*, 29–34.

Schaap, P. (2014). Prokaryote-like signaling controls eukaryote development. *Microbe Magazine, 9*, 198–203.

Schaller, G. E., Shiu, S.-H., & Armitage, J. P. (2011). Two-component systems and their Co-option for eukaryotic signal transduction. *Current Biology: CB, 21*(9), R320–R330.

Schauder, S., & Bassler, B. (2001). The language of bacteria. *Genes & Development, 15*, 1468–1488.

Sha'afi, R. I., & Molski, T. F. (1988). Activation of the neutrophil. *Progress in Allergy, 42*, 1–64.

Shellard, A., & Mayor, R. (2020). All roads lead to directional cell migration. *Trends in Cell Biology, 30*(11), 852–868.

Shellard, A., & Mayor, R. (2021). Collective durotaxis along a self-generated stiffness gradient in vivo. *Nature.*

Slavik, K. M., et al. (2021). cGAS-like receptors sense RNA and control 3′2′-cGAMP signaling in Drosophila. *Nature.*

Spassky, N., & Meunier, A. (2017). The development and functions of multiciliated epithelia. *Nature Reviews Molecular Cell Biology, 18*(7), 423–436.

Srivastava, N., et al. (2020). Pressure sensing through Piezo channels controls whether cells migrate with blebs or pseudopods. *Proceedings of the National Academy of Sciences, 117*(5), 2506–2512.

Sunyer, R., & Trepat, X. (2020). Durotaxis. *Current Biology, 30*(9), R383–R387.

Tang, L., et al. (2002). A cell number-counting factor regulates the cytoskeleton and cell motility in Dictyostelium. *Proceedings of the National Academy of Sciences, 99*(3), 1371–1376.

Teplitski, M., Mathesius, U., & Rumbaugh, K. (2011). Perception and degradation of N-acyl homoserine lactone quorum sensing signals by mammalian and plant cells. *Chemical Reviews, 111*, 100–116.

Turrà, D., et al. (2015). Fungal pathogen uses sex pheromone receptor for chemotropic sensing of host plant signals. *Nature, 527*, 521.

Twigg, M., et al. (2014). Interference with germination and growth of Ulva zoospores by quorum-sensing molecules from Ulva-associated epiphytic bacteria. *Environmental Microbiology, 16*, 445–453.

Valentine, M., & Van Houten, J. (2021). Using paramecium as a model for ciliopathies. *Genes, 12*, 1493.

Valentine, M., Yano, J., & Van Houten, J. (2008). Chemosensory transduction in paramecium. *Japanese Journal of Protozoology, 41*, 1–7.

Van Haastert, P., et al. (2021). Firty-five years of cGMP research in Dictyostelium: Understanding the regulation and function of the cGMP pathway for cell movement and chemotaxis. *Molecular Biology of the Cell, 32*, 1–14.

Van Houten, J. L. (1998). Chemosensory transduction in paramecium. *European Journal of Protistology, 34*, 301–307.

Van Houten, J. L. (1979). Membrane potential changes during chemokinesis in Paramecium. *Science, 204*, 1100–1103.

Van Houten, J. L. (1978). Two mechanisms of chemotaxis in Paramecium. *Journal of Comparative Physiology, 127*, 167–174.

Van Haastert, P. J. M., & Devreotes, P. N. (2004). Chemotaxis: Signalling the way forward. *Nature Reviews Molecular Cell Biology, 5*(8), 626–634.

Veltman, D. M., & van Haastert, P. J. M. (2008). The role of cGMP and the rear of the cell in Dictyostelium chemotaxis and cell streaming. *Journal of Cell Science, 121*(1), 120–127.

Wang, C., et al. (2014). The interplay of cell—cell and cell—substrate adhesion in collective cell migration. *Journal of the Royal Society Interface, 11*(100), Article 20140684.

Werner, M. E., et al. (2011). Actin and microtubules drive differential aspects of planar cell polarity in multiciliated cells. *The Journal of Cell Biology.*

Werner, M. E., & Mitchell, B. J. (2012). Planar cell polarity: Microtubules make the connection with cilia. *Current Biology, 22*(23), R1001–R1004.

Wessels, D., et al. (1988). Cell motility and chemotaxis in Dictyostelium amebae lacking myosin heavy chain. *Developmental Biology, 128*(1), 164–177.

Whiteley, M., Diggle, S. P., & Greenberg, E. P. (2017). Progress in and promise of bacterial quorum sensing research. *Nature, 551*, 313.

Willard, S. S., & Devreotes, P. N. (2006). Signaling pathways mediating chemotaxis in the social amoeba, *Dictyostelium discoideum. European Journal of Cell Biology, 85*(9–10), 897–904.

Wischnewski, M., & Ablasser, A. (2021). Interplay of cGAS with chromatin. *Trends in Biochemical Sciences, 46*(10), 822–831.

Xiao, Z., et al. (1997). Dynamic distribution of chemoattractant receptors in living cells during chemotaxis and persistent stimulation. *Journal of Cell Biology, 139*(2), 365–374.

Yang, W., et al. (1997). Cyclic nucletotides in gluamate chemosensory signal transduction of Paramecium. *Journal of Cell Science, 110*, 1567–1572.

Zhao, M., et al. (2022). The stress granule protein G3BP1 promotes pre-condensation of cGAS to allow rapid responses to DNA. *EMBO Reports, 23*(1), Article e53166.

Chapter 33

Cell locomotion driven by pulsed solitary actin gelation waves and actin-microtubule crosstalk

Erich Sackmann

Physics Department, Technical University Munich, Garching, Germany

Abbreviations

AD Adhesion domains
Cas Crk-associated substrate. It forms a forced-dependent adaptor cascade with Crk
Crk Adaptor protein recruiting cytoplasmic proteins close to Src Kinases
DAG Diacylglycerol
FAK Focal adhesion kinase
GAP GTPase-activating protein which deactivates GTPases*
GAP Guanin hydrolyzing protein accelerating the deactivation of GTPase
GEF Guanine exchange factor; activates GTPases by GDP ⇔ GTP exchange
GEF Guanine exchange factor accelerating the activation of GTPases by replacement of GDP by GTP
MT Microtubule
PCK Protein kinase C
PI-3.4.5-P3 Phosphatidylinositol-3,4,5-phosphate
PI-3K Phosphatidylinositol-3-kinase acts as PIP2 → PIP3 transformer
PI-4,5-P2 Phosphatidylinositol-4,5-phosphate
PM Plasma membrane
PSL-γ Phospholipase gamma
Src Tyrosine kinase which can associate with FAK

1. Introduction

Polarized cell locomotion is triggered by exogenous and endogenous signals, such as chemoattractants and endogenous electric fields, respectively. The cell movement is driven by ongoing biochemical remodeling of the composite cell envelope and the associated actin network. Thereby, the cellular shape changes are guided by crosstalking between the actin heterogel and the aster-shaped microtubule skeleton. Cell locomotion is instrumental in embryonic development, immune responses, and wound repair. Moreover, cell shape changes and cell adhesion are closely related. Quantitative studies of cell locomotion can provide valuable insights into the dysfunction of the interplay between cell signaling and material properties of the cells caused by diseases, mutations, and genetic defects, such as cancer metastasis and inflammatory disorders.

Over the last decades, insight into the correlation between cell adhesion and locomotion has been gained by studying biomimetic systems containing the basic ingredients of cell adhesion and amoeboid cells (such as Dictyostelia cells) migrating on biofunctionalized surfaces (see Schindl et al., 1995; Tanaka & Sackmann, 2005). Microinterferometric studies showed that the contact zone between cells and tissues is associated with the decay into microdomains of strong adhesion (with interfacial distances of h $\sim$ 15 nm) separated by nonadhering zones (exhibiting distances of $\sim$50 nm). The transition from the free to the adhering (bound) state of soft shells can be described in terms of the Cahn theory of first-order wetting (Bruinsma et al., 2000; Sackmann, 2015).

Cell Physiology Sourcebook. https://doi.org/10.1016/B978-0-12-811114-7.00036-0

In the first part of this chapter (Section 33), I briefly summarize the physical basis of cell adhesion as learned from biomimetic studies. I introduce basic concepts of cell adhesion and locomotion by considering, as an example, the penetration of leukocytes across the endothelial cell layer that forms the walls of blood vessels, a process also known as diapedesis, or transendothelial migration.

In Sections 3–6, I describe the phenomenology of locomotion in terms of the treadmilling model (Le Clainche & Carlier, 2008). I first discuss how the cell front is pushed forward in a rhythmic fashion by sequential actin polymerization and gelation pulses emanating from nascent adhesion domains. These are formed by clustering of cell adhesion molecules (CAMs), such as integrins, that serve as biochemical reaction centers, which transiently activate specific actin polymerization promoters, such as WASP and VASP, through the antagonistic interplay of the GTPases Rac-1 and RhoA. These reaction centers (also called hot spots) are switched on by recruitment and activation of the Rac-1 activating guanine exchange factor (GEF) to the nascent ADs by superaffine (strong binding interaction) lipid anchors PI-3,4,5-P3. Then I discuss how pseudopods and filopodia protrude in parallel by the "lateral inhibition" of GTPases Rac-1 and Rho A. Lateral inhibition in the context of GTPases like Rac-1 and Rho A refers to the mutually inhibitory feedback mechanisms where the activation of one promotes the inactivation of the other.

Section 5 deals with the major mechanisms mediating the retraction of the uropod (the rear of the cell, involved in cell adhesion and retraction). We discuss the evidence showing that it is primarily determined by the RhoA switch and its associated kinase ROCK . In Section 7, I discuss the major role played by intracellular Ca^{2+} pulses that are generated by the opening of Ca^{2+}-channels, mostly resulting from the opening of mechanosensitive channels. The second messenger Ca^{2+} initiates the disassembly of the adhesion domains by activating the protease calpain. In parallel, the ROCK kinase switches on retraction forces by phosphorylating and thereby activating the myosin II motor.

In cell migration, polarization refers to the asymmetric distribution of cellular components, creating a distinct front (leading edge) and rear end (trailing edge), which is crucial for directed movement. This front-rear polarity allows cells to crawl forward by coordinating the extension of the leading edge with the retraction of the highly contractile rear end. The front is characterized by protrusions, such as lamellipodia, while the rear is characterized by high actomyosin and low adhesion, enabling the retraction of the uropod. In Section 7, I address the issue of the biochemically and force-induced global polarization of migrating cells. I show that the Ca^{2+}-sensitive actin-microtubule linking protein IQGAP, forming a complex with calmodulin, plays a key role. By recruiting the actin-binding protein Filamin A and the Rac-1* breaker RacGAP, it serves as a strain and Ca^{2+}-sensitive control system. I discuss how the IQGAP–RacGAP complex can enforce the directed motion by suppressing lateral extensions (that are triggered by forces exerted by the tissue) through inhibition of the actin–gelatin machines. Filamin A can further adapt the adhesion and the pushing force strength at the front to match extrinsic forces exerted by the tissue. This adaptation is achieved by exposing integrin binding domains above threshold forces of about 10 pN (Rognoni et al., 2012) which enables the strengthening of the adhesion domain by recruiting new CAMs of the tissue. In contrast, in the force free state, Filamin A suppresses the actin gelation by suppressing Rac-1 activation by recruitment of the GTP hydrolyzing protein Rac-GAP.

2. The shell rope model of cells

As mentioned above, migrating cells exhibit global polarization; they establish a distinct "front" or leading edge, characterized by the extension of lamellipodia and filopodia, and a "rear" or trailing edge (uropod). Thus, polarization in the context of cell migration refers to the asymmetric distribution of cellular components, creating a distinct front and rear end. Polarization is fundamental for coordinated cell movement (Hind et al., 2016). The global polarization of migrating cells is regulated by the interaction between actin filaments and microtubules (MTs), through crosstalk mediated by coupling of MTs to adhesion domains at the leading and trailing edges, through passive and active actin–microtubule linking proteins. Two ubiquitous linkers are: passive linkers (called plus-end tracking proteins or +TIP (Akhmanova & Steinmetz, 2010) presented in Fig. 33.1, and the Ca^{2+}sensitive coupling protein complex IQGAP/CLIP70 that plays a key role in the control of the direction of cell migration.

Many of these chemomechanical processes can be understood on the basis of the shell-rope model presented in Box 33.1. The model is based on the observation that the flexible MTs cannot transmit bending moments (Heinrich & Sackmann, 2006). The mechanical stability is therefore mediated by anchoring individual MTs at the centrosome (microtubule organization center (MTOC)) and at the viscoelastic actin cortex. Due to the lack of bending stiffness, the tensile stresses in the microtubules must be compensated by tangential stresses in the viscoelastic cell envelope (Box 33.1).

The structure of cells from tension-transmitting microtubules coupled to an elastic actin shell shares common features with the famous construction principle of the "tensegrity structures" introduced by the American architect Richard Buckminster-Fuller. These structures consist of elastic rods connected by cables. By pre-tensioning the cables

BOX 33.1 The endothelium as an active semipermeable barrier for white blood cells

The ability of cells to sense and respond to external forces by restructuring the cell envelope and intracellular cytoskeleton can be studied particularly well on the semipermeable monolayer of endothelial cells (also called endothelium). The mechanical stability and cell permeability of the endothelium are determined by several families of cell adhesion molecules (CAMs).

- The endothelial cells form the inner wall of blood vessels, which is lined by a 20–200 nanometer-thick basement membrane. This coupling is mediated by integrins that bind to various ligands of the basement membrane, such as collagen IV and laminin.
- The lateral cell–cell adhesion is mainly realized by homophilic CAMs of the cadherin family. Strong attraction forces are exerted by the head groups of the linkers forming two-dimensional quasi-crystalline arrays (Wua et al., 2010). The lateral cell–cell adhesion is, also regulated by another homophilic CAM, namely PECAM-1 (Platelet Endothelial Cell Adhesion Molecule-1), also known as CD31. This CAM is part of a stress sensor that is involved in the adaptation of endothelial cells adhesion strength to the speed and direction of blood flow. It also plays a significant role in the control of leukocyte permeation through the endothelium (Newman and Newman, 2003). Another role of PECAM-1 is the suppression of cell death, particularly apoptosis.

Another mechanism for adjusting the stress balance in the cell envelope is the activation of the membrane-associated molecular motor myosin Ib. Under mechanical load, this motor can withstand very high lateral stresses and actively generate forces (Laakso et al., 2008). The sum of all these forces determines the dynamic changes in the shape and direction of movement of the cells. They also ensure that cells always return to a mechanical ground state even after very strong deformations. In this sense, they behave like tensegrity structures (Ingber, 2000).

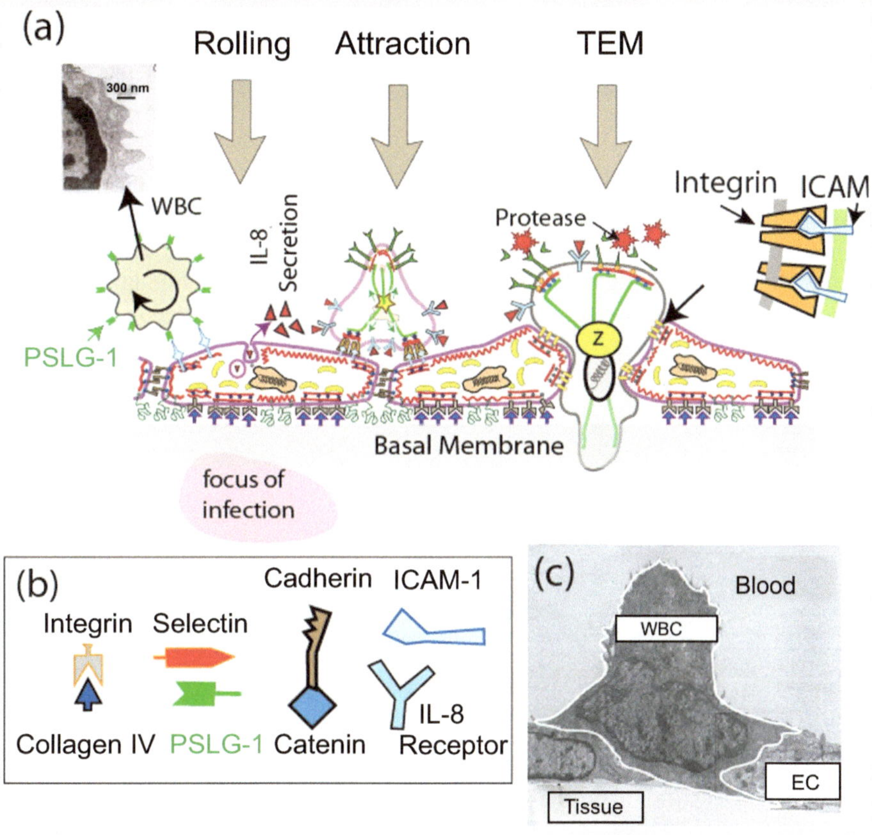

BOX FIG. 1 Transendothelial migration. Penetration of the endothelial cell layer by a lymphocyte, triggered by chemoattractants generated by a hostile intruder (e.g., bacteria) in the body. The endothelial cells (ECs) are attached to the basement membrane via adhesion domains formed by integrin clusters linked to tissue ligands, such as collagen IV, fibronectin or laminin. The lateral aggregation of ECs is predominantly mediated by clusters of cadherins. (a) *Left* side (Rolling): A white blood cell (WBC) rolling along the ECs "searching" for the site of inflammation. The WBC (leukocite) harbors about 5000 microvilli (length 0.3–0.5 μm, width 150 nm). At their tip, the microvilli expose about 10 glycoproteins of the class PSLG-1 (P-selectin glycoprotein ligand-1, represented in green), with which they bind to the selectins of the endothelial cells. The extracellular domains of both CAMs are about 40 nm long and semiflexible. Therefore, the glycocalyx does not interfere with the connection between the two CAMs, which can thus act like an entropy spring (Hammer & Apte, 1992). The central part (Attraction) shows the situation when rolling leucocytes come close to the foci of infection. The roundish cell is polarized within seconds. It docks firmly on the endothelium with one side, while with the other it squeezes through the cell layer and the basement membrane to enter the tissue, a process known as transendothelial migration (TEM) or diapedesis. To this end, the WBC degrades actin filaments on the uropod and can thus squeeze in between two endothelial cells (ECs). The cell adhesion molecules ICAM-1 (Intercellular Adhesion Molecule 1) and integrins ensure thorough contact between these two cell types (insert in panel a) (b) Symbols of some essential cell adhesion molecules (CAM) mediating the cell-tissue and cell-cell adhesion. (c) Electron microscopy image of a thin section through a penetrating WBC and ca surrounding endothelial cell. *Modified from Cinamon et al. (2001)*

Continued

BOX 33.1 The endothelium as an active semipermeable barrier for white blood cells—cont'd

When the rolling leukocytes approach the foci of infection, a dramatic change in their behavior takes place. The round cell/ is polarized in seconds, docks firmly at the endothelial cell (EC) with one side, and squeezes through the cell layer and the basement membrane to enter the tissue (Box Fig. 1). This dramatic change in cell behavior is triggered by enhancing the affinity of the integrins for coreceptors of the ICAM-1 family. This increase in adhesion is induced by interleukin-8 (IL-8), which acts as a chemoattractant. It is released by the endothelial cells and binds to the IL-8-specific receptors located on the cell surface. Interestingly, the receptor functions like other hormone or photon amplifiers. It binds to and thereby activates the heterogeneous GTPases ($G_{\alpha\beta\chi}$), which generate second-messenger molecules (see below and Chapter 24). In this way, several hundred IL-8 molecules are enough to mediate the change in WBC behavior.

The adhesion of WBCs to the endothelium is mediated by the formation of microdomains generated by the tight coupling of a specific integrin to ICAM-I, also called **I**ntercellular **A**dhesion **M**olecule. The integrin exposes β2 chains. The glycoproteins exposing large head groups, such as CD43, CD44, or PSLG-1, are moved toward the cell rear end (the uropod). A driving force behind this global separation of integrins and glycoproteins is steric repulsion, which can be quantified in terms of the Dolan-Edwards potential (Bruinsma et al., 2000). Another reason for the polarization of the cell is the different properties of the actin membrane couplers talin and ezrin/moesin. Talin binds (often together with kindlin) preferentially to the integrin beta chains, while ezrin exhibits an additional microtubule attachment sequence.

In sum, the change in WBC behavior is initiated by enhancing the affinity of the integrins for the coreceptors of the ICAM-1 family. This dramatic increase in adhesion is triggered by the chemoattractant interleukin-8 (IL-8), which is released by endothelial cells (ECs) and binds to IL-8-specific receptors. The IL-8 receptor functions like an olfactory signal amplifier based on the heterotrimeric GTPase $G_{\alpha\beta\chi}$. Signals are amplified by the generation of second messengers (Baggiolini & Clark-Lewis, 1992). In this way, some 100 molecules suffice to mediate the change in WBC behavior.

appropriately, these tensegrity elements stabilize themselves. Such structures have the advantage of being extremely deformable under external loads without breaking. After the load is removed, they always return to their original state. For cells survival in a tissue this cell structure has many advantages; the cells become mechanically robust, despite the fluid nature of the cytoplasmic space. This fluidity allows rapid intracellular transport while utilizing the mechanical stabilizers of actin and MT as transport rails (Ingber et al., 2014).

The MT-actin coupling is very robust. If, for example, magnetic tweezers are applied to an MT perpendicular to the cell longitudinal axis, the MT plus end (+) and the centrosome move in opposite directions in about 0.1 sec (arrows in Fig. 33.1). Thereafter, the cell returns to the ground state with relaxation times of ~10 sec (Heinrich & Sackmann, 2006).

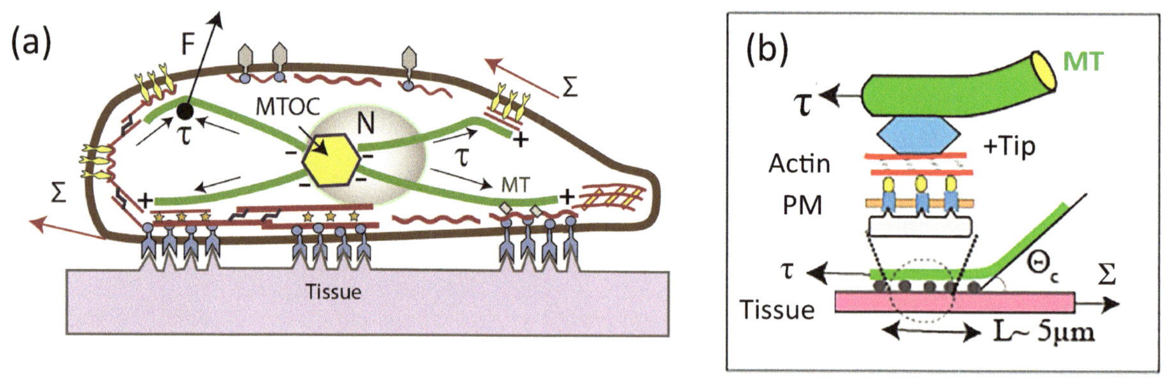

FIGURE 33.1 Shell rope tensegrity model of cells. The mechanical stabilization of cells with fluid cytoplasmic space is achieved by coupling of the aster shaped microtubule (MT) assembly (centrosome) to actin gel clusters. The minus ends (−) of all MTs (*green lines*) are mechanically coupled together via the centrosome ("Microtubule Organization Center": MTOC) and the cell shell (Sackmann, 2015). (a) A force F applied on an MT by means of magnetic tweezers, causes a traction τ, which is compensated by a tangential stress Σ in the cell envelope and the sum of tensile forces arising at the other MTs coupled to the MTOC (centrosome). The latter must also be compensated by the tangential stresses in the composite membrane. Thus, the shape of the cell is determined by the balance of the traction forces and the membrane tensions. Traction forces (τ) in the (20 nm thick) MTs are balanced by tangential stresses in the viscoelastic actin cortex (Σ), which is generally isotropic. (b) The MTs are coupled to actin gel patches by passive linkers denoted as + tip, and actively by the complex between IQGAP and the actin binding protein CLIP 170 (not shown). IQGAP is an MT binding protein harboring a Ca^{2+}-binding calmodulin domain. Passive coupling of a MT to the actin cortex by MT plus-ends (+) binding proteins (plus-end tracking protein: + Tip), such as CLIP-170. This stabilizes the MT plus ends. The energy gained from the MT-actin coupling induces a one-dimensional spreading pressure in the MTs.

This design principle allows cells to deform drastically without breaking. Most importantly for their survival, the cells return to their steady-state rest position after each deformation.

The microtubule plus ends (+ in Fig. 33.1a) are coupled to the actin cortex, most likely to actin gel patches, stabilizing the adhesion domains by passive and active linkers. Passive tensile forces are generated by binding of MT plus ends to F-actin via MT plus-end binding proteins (+ Tip in Fig. 33.1b). One can understand the emergence of a tensile force as a one-dimensional spreading pressure. Its strength is determined by the energy w gained by the MT–actin coupling. The passive traction τ is then given by.

$$w = \tau(1 - \cos\theta_c) \tag{33.1}$$

where θ_c is the contact angle and w the binding energy per unit length.

2.1 Tensile forces induced membrane stresses

The tensile forces generated by internal or external mechanisms in an MT induce lateral stresses in the cell shell (i.e., the composite membrane also referred to as cell envelope), which is determined by the length L_i of the actin gel clusters as follows,

$$\Sigma_i = \tau_i L_i^{-1} \cos\theta_c \tag{33.2}$$

Here, θ_c is the angle between the longitudinal axis of the microtubules and the plasma membrane surface. Starting from an equilibrium state and applying a force F to the MT, the centrosome moves in the cytoplasmic space until the sum of all tensile forces is zero. Since the cell envelope is "plastic", the contact points of the MT move laterally. Another mechanism for adjusting the tensional balance in the cell envelope is the activation of the membrane-associated myosin Ib. In the force free state, this motor behaves in a nonprocessive fashion. Under a load, it can withstand very high lateral stresses and actively generate forces (Laakso et al., 2008). The interaction of all these forces regulates the dynamic changes in the shape and direction of movement of cells. This force balance also ensures that cells always strive for a stationary mechanical ground state after each deformation. In this sense, they behave like tensegrity structures.

In summary, if we start from an equilibrium state and apply a force F onto a MT, the centrosome moves in the cytoplasmic space until the sum of all tensile forces is zero. Since the cell envelope is viscoelastic, the Actin-MT contacts move in different lateral directions until the cell returns to the stationary equilibrium state in which the lateral tension is isotropic (Heinrich & Sackmann, 2011; Ingber, 2000).

3. Phenomenology of cell locomotion

Two intensively studied examples of migrating cells are lymphocytes searching our body for harmful intruders and the foraging of amoebae *Dictyostelium discoideum.* In both cases, the creep process can be understood as a repetitive sequence of two steps. In the first step, a cell protrusion forms at the front, known as pseudopodium. This is followed by the withdrawal of the cell's back end (the uropod). This cell behaviour is reminiscent of the way caterpillars move. The pseudopodium moves across the surface at a constant speed and then stops abruptly. After a stabilization phase, the cell drags its rear end behind it, forming a kind of cat's hump back (Fig. 33.2 left panel). This periodically changes the adhesion surface of the cell, which it uses to maintain contact with the surface. Under normal conditions, pseudopodia form randomly in all directions and often simultaneously at several locations on the cell. However, this cell behaviour changes drastically when a direction-dependent stimulus is present in the environment, typically a chemotactic gradient of a signaling molecule. Then, the cells preferentially move in the direction in which the chemotactic agent is located (Meier et al., 2011).

The characteristic behavior of a vegetative *Dictyostelium* cell moving on hard surfaces, has been studied by Reflection Interference Contrast Microscopy (RICM). The front of the pseudopod consists of a thin, about 300 nm thick lobe, which pushes over the surface at a constant speed v_p and abruptly stops after a time τ_P (typically $t_P \sim 6$ s). During this step, the cell front is pushed forward by a distance ΔL (typically $\Delta L \approx 7$nm). After a period of stabilization, the cell retracts its back end (uropod), forming the cat back hump mentioned above. As a result, the adhesion area A_c of the cell changes periodically (Schindl et al., 1995) (see Fig. 33.2 right panel). Numerous experiments have yielded the following insights into the phenomenology of cell movement.

1. The variation of the contact area depends on the strength of the adhesion. On albumin-coated glass, where the cells adhere strongly, the contact area varies by a factor of 1.5, while it fluctuates by a factor of four on freshly split

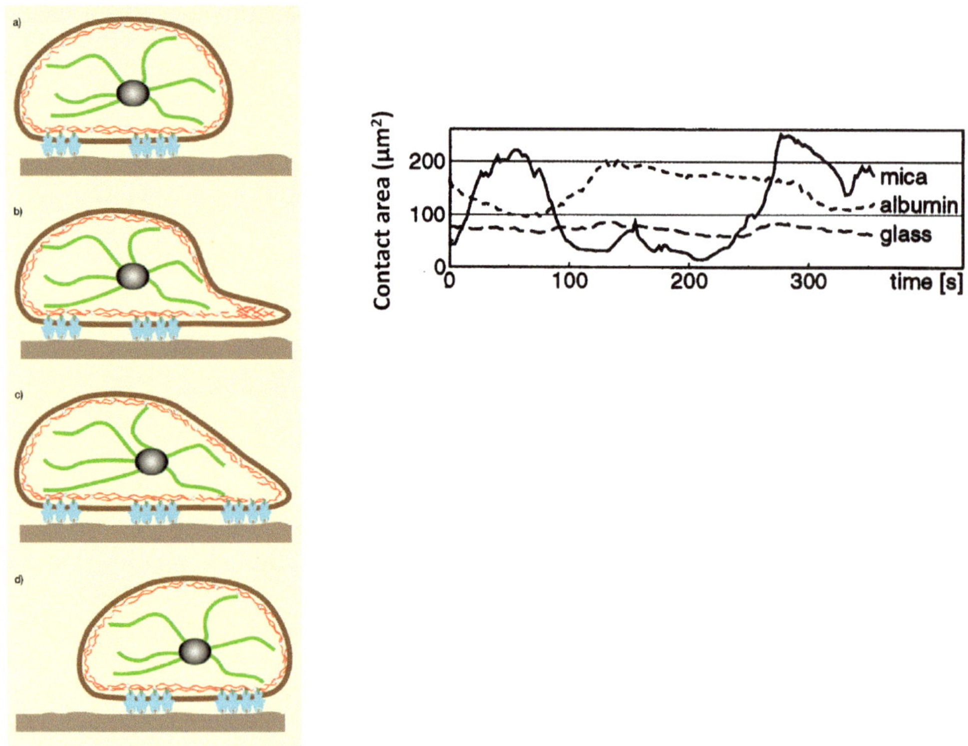

FIGURE 33.2 **Rhythmic progression-retraction cycle of migrating eukaryotic cells.** (a) A diagram illustrating typical crawling cycle movement of a eukaryotic is shown on the left (a—d). This movement cycle lasts about five minutes on a mica substrate. (a) starting point (initial state), (b) a membrane flap spreads across the surface, (c) then the cell draws its back end, and in doing so, it rounds off and forms a "cat's back" state. Due to these changes in shape, the contact area between the cell and the substrate also changes cyclically, (d) After the trailing ends, the cell retracts (Heinrich et al., 2018). *Right*: Oscillation of the area of contact of a *Dictyostelium* cell migrating on freshly cleaved mica and albumin-coated glass (Schindl et al., 1995).

mica, on which the cells adhere weakly due to van der Waals forces. This result suggests that the cell movement on solids is essentially determined by nonspecific forces.

2. In the absence of a chemotactic gradient, pseudopodia form randomly in all directions and often simultaneously at several sites of the cell periphery. The center of gravity of the cell moves randomly on large scales. In contrast, in the presence of a chemotactic gradient, the cells move preferentially in the direction of the gradient. For instance, in gradients of the chemotactic signal molecule cAMP, the cells preferentially move in the direction of increasing concentrations, with the direction of migration fluctuating statistically (Li et al., 2008). The reason for the directed movement is a global polarization of the cell, which will be discussed below.

The sequential formation of pseudopodia is driven by a pulse-like sequence of actin polymerization, which pushes the cell front forward. The actin polymerization pulse can be observed using fluorescence microscopy in cell lines derived from *Dictyostelium discoideum* coexpressing the mRFP-tagged LimEΔ construct and GFP-tagged actin, which allows visualization of freshly polymerized F-actin (Etzrodt et al., 2006). The solitary waves propagate at a rate of about 0.2 μm/s, and thus their velocity is of the same order of magnitude as the spreading speed of the pseudopods. Solitary actin polymerization waves have also been shown to drive the uptake of bacteria by phagocytosis.

4. Treadmill model of the sequential advance of the pseudopod

The treadmill model provides a phenomenological description of the progression of pseudopods and the filopodia sprouting from them . The model entails the dynamic assembly and disassembly of actin filaments; and it is based on the idea that the transient actin polymerization is coupled to the protrusion of the cell's leading edge where acting is organized forming a dendritic array of branched filaments. Thus, actin polymerization plays a central role in cell motility and migration by pushing the plasma membrane forward (Le Clainche & Carlier, 2008). In brief, in the treadmill model, actin monomers are added to the plus end (barbed end) of the filaments, resulting in growth at the leading edge of the cell protrusion, which pushes the membrane forward and enables cell movement. Simultaneously, at the base of the pseudopod, actin monomers are trimmed at their minus ends, resulting in shortening of the filaments. The combined effect of growth at one end and shortening at the other keeps the filament lenght unchanged but allows the newly added actin monomers to move forward with the filaments. . Actin exists in two forms: monomeric G-actin (globular actin), which is the building block that polymerizes to form filamentous actin (F-actin). Thus, individual G-actin monomers combine to form F-actin. Actin undergoes cyclic transitions between the monomeric G-form and the filamentous F-form, driving cell motility. G-actin binds ATP, forming ATP-G-actin. The latter polymerizes into F-actin. As discussed above, actin filament growth is determined by the balance between the addition and removal of actin monomers. During the polymerization process, the ATP bound to G-actin is hydrolyzed to ADP. Actin ATP hydrolysis energetically drives the cyclic assembly and disassembly process. The process of F-actin polymerization (assembly) and disassembly is regulated by the interplay of various actin binding proteins (See Fig. 33.3). These proteins regulate the availability of free actin monomers (G-actin) for polymerization, and include Profilin, β-Thimosin, and cofilin. Profilin catalyzes the exchange of ADP for ATP in actin monomers, creating a pool of profilin-bound ATP-actin monomers that become available for polymerization. The profilin-actin complexes bind to the barbed end of actin filaments, promoting filament growth and elongation (Fig. 33.3b). β-Thimosin sequesters actin monomers, acting as a buffer or reservoir of G-actin that can be utilized for polymerization (left part of Fig. 33.3b). β-Thimosin can inhibit treadmilling by sequestering actin monomers, thereby reducing the pool available for polymerization at the barbed end. Cofilin severs actin filaments, thereby promoting the rapid depolymerization of actin filaments. In brief, profilin facilitates monomer addition, while cofilin enhances filament turnover by severing filaments and accelerating depolymerization. In addition, there is a series of filament assembly

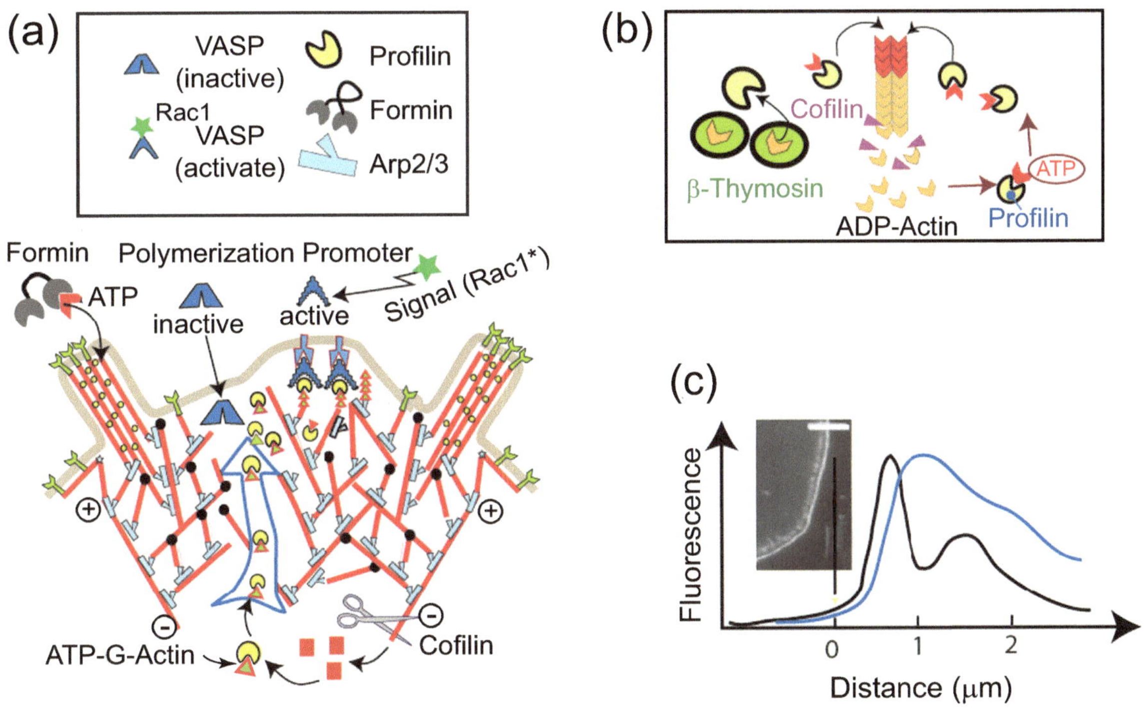

FIGURE 33.3 The treadmilling model of actin gelation. (a) Sequential advancement of the cell front (leading edge) pushed by solitary actin polymerization pulses emanating from adhesion domains. Theese polymerization pulses arise by coordinated growth of the filaments at the front and degradation of inactive monomers (GDP-G-actin) at the back of the actin cortex. At the front, the monomers are activated by ATP binding mediated by the actin-binding protein profilin (b). This effector is accumulated at the surface of the plasma membrane. Profilin interacts with monomeric G-actin, promoting its nucleotide exchange, thus making it available for polymerization. The gelator Arp2/3 complex is activated by members of the WASP/VASP family of growth promoters (acting as scaffold proteins). These are in turn activated by biochemical switches of the Ra-1 family and recruited to the membrane, thus making the binding site accessible to Arp2/3. In parallel with pseudopod formation, filipodia may sprout out of the front. Their outgrowth is driven by a tandem dimer of formin that can alternately couple a GTP-G-actin monomer to the single strand. (b) Two sources of G-actin: left side: ADP-actin captured by β-thymosin is decoupled. On the right side: cleavage of G-actin at the minus ends of the actin filament by cofilin and its acquisition and activation by profilin. A second source of GDP-G-actin (not shown here) is provided by the pool bound to β-thymosin .(c) Distribution of the framework protein VASP (GFP-labeled) and the actin (marked with phalloidin) at the front edge of the pseudopods. VASP is enriched at the interface between the plasma membrane and the actin cortex and binds to the plasma membrane after activation by the GTPase. The fluorescence image (insert) is reproduced from Kaverina et al. (2002).

proteins such as Formin, Arp2/3, and the WASP family, which initiate the formation of new actin filaments or promote their elongation.

It is worth noting that profilin, that regulates actin polymerization and filament turnover, binds to both PI(4,5)P2 and PI(3,4,5)P3. It is therefore enriched along the inner leaflet of the plasma membrane (see Glossary "Profilin") and can be delivered to the site of action by two-dimensional diffusion along the membrane.

5. The thrust generating actin gelation machines. Adhesion domains acting as biochemical reaction and force transmission centers

Cell protrusions can be actively driven by two mechanisms, which are often activated in parallel and by different molecular switches. The growth of F-actin bundles, which drives the finger-like outgrowths of the cell envelope (filopodia), is activated by the molecular switch RhoA. The broad lamellipodia generating the thrust of cell migration are pushed forward by the growth of branched actin gels . They are triggered by the molecular switches Rac-1 or cdc42, both of which are GTPases of the Rho family. Often, both processes are activated in parallel, in such a way that fingerlike protrusions sprout out of the lamellipodia, at distances of a few microns. This parallel growth of branched gels and bundles on the cell front is observed especially in axon growth cones during the search for target cells (Gomez & Letourneau, 2014).

The growth of actin bundles pushing the filopodia is mediated by the actin polymerizer formin (Fig. 33.3). It consists of a homodimer exposing two binding sites for actin that are activated alternately by the GTPase RhoA. They displace the very strongly binding capping proteins (like ZCAP and CAP) from the actin plus-end to dock onto a GTP-binding G-actin. The thrust of the actin bundles is generated by the formin-induced ATP cleavage. The free energy of 30 kJ/M available for mechanical work could generate a maximum force of around 10 pN (Kovar & Pollard, 2004; Mogilner & Oster, 1996).

The formation of branched actin gels is initiated by activation of the crosslinker Arp2/3 complex. It binds to the longitudinal side of existing filaments and drives the growth of new filaments, which form an angle of about 60° with the mother filament. In this way, Bethe-like meshworks of F-actin are formed. Since the basal activity of Arp2/3 is very small, these actin gelators must be activated by polymerization promoters such as VASP, WAVE or CARMIL (Fig. 33.3).

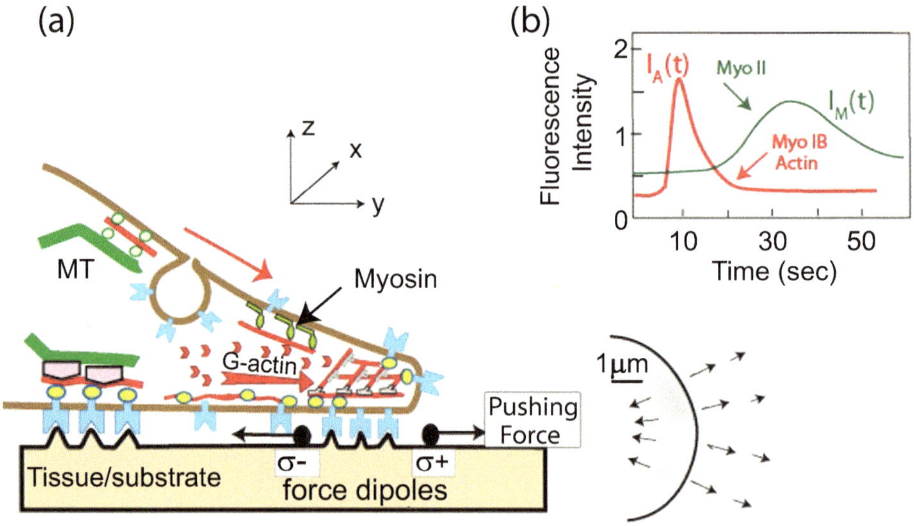

FIGURE 33.4 **Pushing force generation between advancing pseudopodia and a substrate or a tissue.** (a) Schematic side view of a pseudopod forming focal adhesion domains (ADs) by clustering of integrins, which are a crucial type of cell adhesion molecule (CAMs). Force transmission between cells and the surface is mediated by lock and key forces between cell surface receptors (e.g., integrins) and the tissue or substrate surface. The model is based on force microscopy measurements using magnetic beads as force transducers and non-magnetic beads as optical markers to measure local shear substrate deformations caused by the cells. From the shear deformation, it is possible to determine the tension generated by the advancing cell front (for details see Ladam et al., 2005). (b) Observation of the solitary pulse (wave) (I_A (t)) of actin polymerization at the front of a newly formed pseudopodium and the delayed activation of the myosin II motor (I_M (t)). The wave was triggered by a local cAMP gradient applied at time $t = 0$ (after (Etzrodt et al., 2006)). The image was taken by quantitative fluorescence microscopy. Myosin IB shows a similar time course as F-actin. However, its activation is delayed by about 1 min. Note: The increase and decrease of the F-actin concentration corresponding to a solitary wave passing a resting observer. The curve I_M (t) shows that the enrichment of the motor protein myosin II associated with the polarization of the cell occurs at the back of the cells with a delay of approximately 1–2 min. *Image modified from Etzrodt et al. (2006).*

Summarizing, pulse-like actin polymerization waves originating from integrin clusters in adhesion domains provide the thrust for the pseudopodia advancement. At the same time, filopodia grow out of the broad protuberances like fingers of a hand. The polymerization bursts are generated at the site of the integrin clusters of adhesion domains acting as biochemical reaction centers. The broad lamellipodia are driven by branched actin networks, whose growth is activated by the protein Arp2/3. VASP/WASP cooperate with Arp2/3 to facilitate lamellipodial growth. In cooperation with VASP/WASP. The transiently formed filopodia are everted by formin-mediated formation of actin bundles. This suggests a division of labor between Arp2/3 and formins in cell motility. The evidence comes from experiments showing that knockdown or inhibition of formin or Arp2/3 impairs lamellipodia formation and associated cell movements in various cell types, including *Dictyostelium* (Swaney & Li, 2016).

The sites of freshly generated actin gel patches can be visualized by GFP tagged LIME, The LIME protein acts as linker between F-actin and microtubules. It is therefore assumed to enable the visualization of sites of MT-actin coupling. (see Glossary "LIME").

The activation of formin occurs through the active GTPase RhoA*, while the gelator Arp2/3 is activated by the GTPase Rac-1 (via the promoters VASP/WAVE). Mutual antagonistic interaction between Rac-1 and RhoA, and the crosstalk between microtubules and the actin cortex act as organizers of the coordinated movement of the front (leading edge) and the rear end of the cells (Fig. 33.7). Rac-1 is predominantly active at the leading edge of the cells, promoting the formation of actin-rich structures like lamellipodia, which are essential for cell protrusion. RhoA is primarily involved in the rear retraction of the cell by promoting actomyosin contractility and stress fiber formation (Ridley, 2015).

The actin bundle formation requires very strong binding of the growth activators to displace capping proteins (such as CP and Z-CAP) from the plus ends of actin filaments ((Edwards et al., 2014; Jung et al., 2001); Uruno et al., 2006). These include CARMIL (Capping Protein Regulator and Myosin 1 Linker) and Z-CAP. CARMILs are proteins that bind directly to CP (caping protein) and modulate its actin-capping activity. Specifically, CARMIL binding induces a conformational change in CP that reduces its affinity for the + end of actin filaments, allowing for renewed actin polymerization. CARMIL exhibits several specific features. First, it harbors a domain (PPP) for the binding of the motor protein myosin I. Second, it triggers the growth of F-actin by displacing the cap proteins (such as Z-CAP). The intracellular concentration of CARMIL in *Dictyostelia* is $\approx 2 \times 10^{-6}$ M, and the dissociation constant of the CARMIL-CP complex is $K_d = 4 \times 10^{-7}$ M. Thus, CP can be released when about 20% of the promoters are activated (Uruno et al., 2006). Interestingly, CARMIL can both drive the F-actin growth and generate high membrane tensions through Myosin Ib. It can thus push filopodia outgrowths forward against high tensions of the cell envelope.

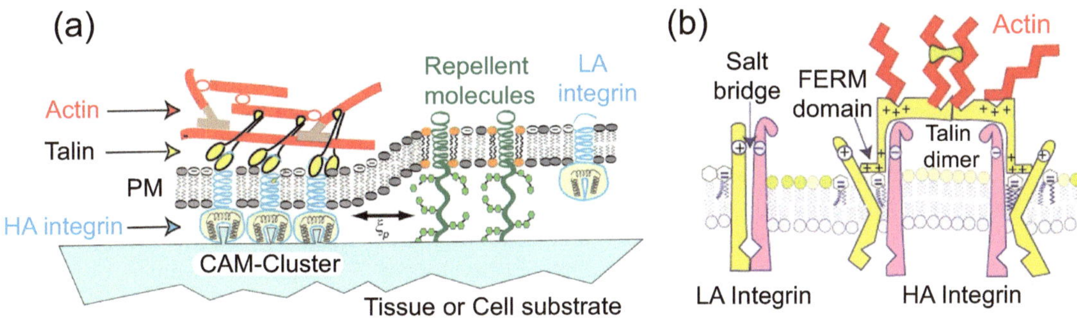

FIGURE 33.5 **Mechanisms of adhesion domains formation.** (a) Coupling of actin gel patches to a CAM cluster of the integrin family by Talin. The latter binds to the cytoplasmic tails of integrin beta subunits (blue coils), triggering conformational changes that increase the integrin's affinity for ligands (extracellular matrix components like fibronectin or collagen) (Sackmann & Smith, 2014). The adhesion strength is reinforced by increasing the integrin affinity through binding to talin (or kindlin). The talin-stabilized adhesion domains can be directly visualized by fluorescence labeling (Tsujioka et al., 2004). *HA*, high affinity, and *LA*, low affinity. Repellent molecules (green coils) are cell surface proteins characterized by long extracellular tails or a high negative charge that generate repulsion forces that help to prevent cells from forming strong or non-specific adhesions. Selectins and CD43 are examples of repellent molecules. The breadth of the interface between the two phases is determined by the persistence length of membrane deformations exhibiting typical values of $\zeta_p \sim 50$nm. (b) Molecular model of integrin activation by binding of the talin FERM domain to its β chain, resulting in the opening of the integrin binding pocket. Modified from Wegener et al. (2007). The *left* side shows the low-affinity state of integrin stabilized by a salt bridge. The *right* side shows the high-affinity state triggered by binding of a talin dimer both to the integrin β chain (breaking up the salt bridge) and to the membrane through electrostatic—hydrophobic forces. Note that talin exposes several binding sites for F-actin. In escence, talin acts as a molecular switch, initiating a cascade of events that transforms integrins from a low-affinity to a high-affinity state for binding to their extracellular ligands.

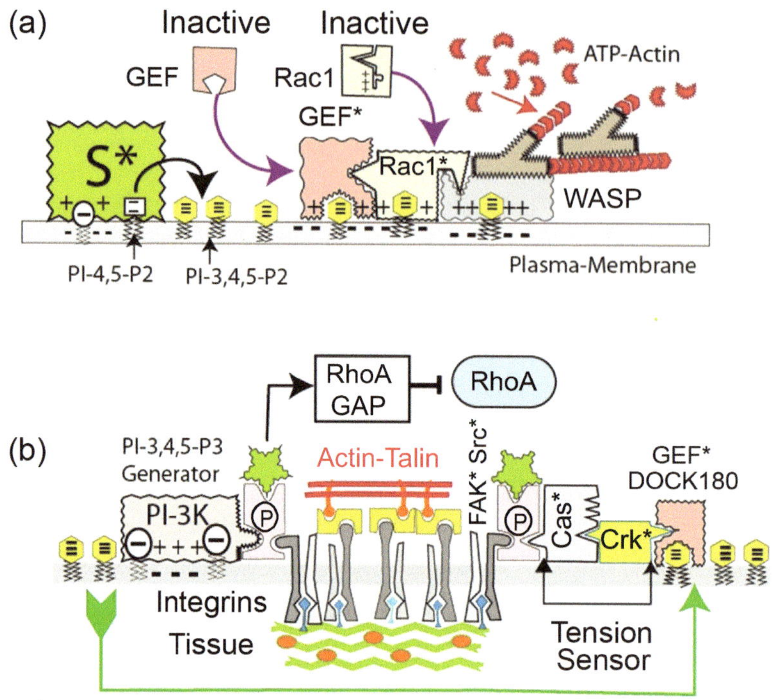

FIGURE 33.6 **Protrusion force generation by forward directed actin filament network formation/gelation at integrin clusters of nascent adhesion domains.** (A) Molecular mechanisms of actin filament network formation. Activation of GEF (guanin nucleotide exchange factor) facilitates the exchange of GDP for GTP on Rac1, leading to its activation. Activated Rac1 binds to WASP proteins, which in turn activate the Arp2/3 complex (brown). The latter nucleates the formation of new, branched actin filaments, driving lamellipodia formation. The double kinase FAK/src. FAK* activates the PI-3K kinase of the switch S. S produces the superaffine lipid anchor PI-3,4,5-P3, which recruits the Rac1-specific guanine exchange protein (DOCK180) to the cell membrane (Côté & Vuori, 2007), in turn DOCK180* recruited and activated Rac1, which turns-on Arp2/3 via the polymerization promoter (VASP). (b) DOCK180 is attracted to the newly formed integrin clusters, through the adapter tandem Cas and Crk, which serves simultaneously as strain-sensitive adaptor. The strain-sensitive adapter tandem Cas and Crk recruits the GEF-factor DOCK180 to the adhesion domain. Another role of the double kinase FAK/src is the inhibition of the activity of RhoA, by locally activating the GTP hydrolyzing protein (GAP) p120RhoGAP, by the src kinase.

6. Adhesion domains serve as biochemical relay stations and centers of cell advancement

The phenomenological treadmilling model teaches us that swarms of actin monomers (G-actin) are repeatedly transferred to the front of the cell, where they are primed by ATP coupling through the action of profilin (Fig. 33.3). This actin-activator is accumulated at the cell envelope by binding to the PI-4,5-P2 lipid anchor. The pool of GTP-G-actin can serve the formation of branched gels or actin bundles, depending upon whether the molecular switches Rac-1 (switching on the polymerization promoter VASP) or RhoA (switching on the bundle generator formin) are activated. Since the basal activity of the actin complex Arp2/3 is low, it must be activated by polymerization promoters (such as VASP/WASP).

Let us now address the question of how the centers of actin gelation are activated. I will first discuss that newly formed adhesion domains simultaneously act as universal biochemical activators of GTPases and as centers of force transmission. Then, I will consider the control of local and global polarization of migrating cells by the dynamic antagonistic interaction of the GTPases RhoA and Rac-1. We will see that the constant advancement of the cell front by ΔL can also be determined by the dynamic interaction between Rac-1 and RhoA.

In Box 33.2 on adhesion domain formation, I briefly discuss that strong cell tissue adhesion is ensured by clusters of integrin linked to ligands of the tissue by lock-an-key forces. Also, they are mechanically stabilized by linking actin gel clusters to the cytoplasmic domain of these CAMs. The integrin clusters of the adhesion domains function as biochemical reaction centers, through sequential recruitment of molecular GTPase switches and activators of actin gelation. The key biochemical switch of these processes is the "Focal Adhesion Kinase" (FAK), in combination with the tyrosine kinase src (Fig. 33.6).

---

* Stars indicate activated states of molecular switches (such as GTPases) and activators (such as Cas and Crk).

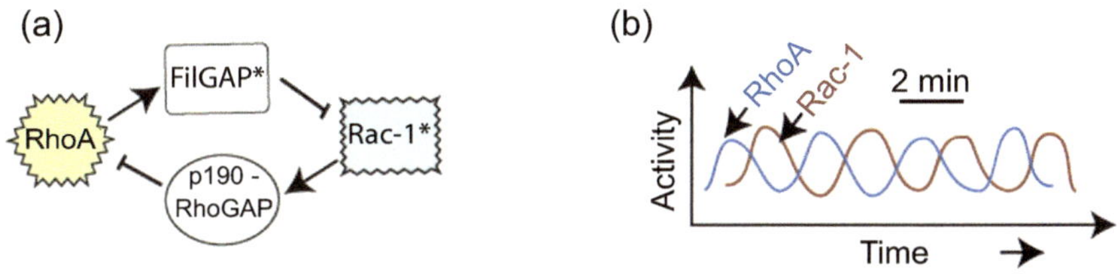

FIGURE 33.7 **Antagonistic interplay between Rac1 and RhoA.** (a) Antagonistic interaction of Rac-1 and RhoA by GAP-mediated hydrolysis of GTPase switches. Rac-1* (or RhoA*) deactivates RhoA* (or Rac-1*) via the specific GTP-hydrolyzing protein p190RhoGAP (or FilGAP). (b) Oscillation of the activity of the two GTPases revealed by the experiments of Machacek and coworkers (Machacek et al., 2009).

BOX 33.2 Adhesion domain formation. A short overview of what we have learned from biomimetic studies (Sackmann & Smith, 2014)

Cell adhesion is controlled by a competitive interplay of specific and generic interfacial forces. These include attractive short range (~15 nm) lock-and-key forces between cell surface receptors and ligands exposed by tissues and medium range repulsion forces (~30 nm) mediated by glycoproteins. For that reason, the adhesion is inevitably accompanied by lateral phase separation, leading to the formation of adhesion domains tightly coupled by the two linkers with intermembrane distances of 15 nm. These are separated by slightly coupled contact zones (exhibiting a typical distance of 40–50 nm). Taken together, biomimetic studies show that cell adhesion can be understood as a first-order wetting transition that results in the formation of adhesion plaques. These can form huge sites of focal adhesion by the recruitment of actin crosslinking proteins, such as vinculin (for a recent review see Katsuta et al. (2024)). Most importantly, this strategy allows cells to form robust sites of contact with a relatively small number of receptors.

An important generic regulator of adhesion strength is the membrane bending elasticity, which is sensitively determined by the actin network coupled to the integrin clusters of the ADs. The nonspecific Van der Waals adhesion of many stroma cells (such as Leukocytes and Macrophages) is counteracted by long-range repulsive forces generated by thermally or actively generated membrane bending undulations (Auth et al., 2007; Raedler and Sackmann, 1992; Zidovska & Sackmann, 2011). The entropic forces play two important roles. First, they suppress cell adhesion by Van der Waals forces and accelerate the nucleation of CAM clusters by generating pushing forces toward the target cell in the tissue (reference see Bruinsma et al., 2000).

We first discuss some fundamental features of the FAK/Src kinase. FAK (Focal Adhesion Kinase) is a critical regulator of cell adhesion, migration, proliferation, and survival. It acts as an adapter protein, exposing several (about 5) tyrosine residues that can be phosphorylated, leading to activation of other signalling molecules. FAK harbors a FERM (four-point-one, ezrin, radixin, moesin) domain which plays a crucial role in its ability to bind to integrins. FAK's FERM domain is crucial for its role in both regulating its own activity and interacting with integrin β chains after interaction with activated src kinase (src*) (Schaller, 2011). Since FAK does not have its own receptor function, it works primarily through interaction with other kinases, such as the nonreceptor tyrosine kinase src. By coupling the FAK/src double kinase with integrin clusters, biochemical reaction centers with potentiated kinase activity are formed.

The interaction between the tyrosine kinases (src) and FAK is a key step in signaling processes triggered by adhesion domain formation. FAK* recruits, above all, adapter proteins with SH2 domains. These include phosphoinositol-3-kinase (PI−3K). This key switch triggers the membrane recruitment of many functional proteins by transiently increasing the concentration of the seven-fold negatively charged lipid anchor PI(3,4,5)P3.

A target protein of central importance is the GTP exchange protein (GEF) DOCK 180, which activates actin polymerization by recruiting and activating the switch Rac-1 (Fig. 33.6) (Côté & Vuori, 2007). Another important protein activated by binding to PI(3,4,5)P3 is protein kinase B (often referred to as Akt), which plays a central role in preventing apoptosis. The superaffine PI(3,4,5)P3 is rapidly transformed back into PI-4,5-P2 by the PI-3 phosphatase PTEN (phosphatase and tensin homologue). By converting PI(3,4,5)P3 back to PI(4,5)P2, PTEN effectively dampens the PI3K-Akt signaling pathway. This crucial negative feedback mechanism helps to prevent the "overreaction" or excessive activation of cells that could otherwise lead to uncontrolled growth and potentially tumorigenesis.

6.1 The thrust generating actin gelation machine

Each step of sequential actin gelation at the cell front by Rac-1-mediated activation of VASP or other Arp2/3 activators is initiated by the formation of new integrin clusters and activation of the FAK*/src* tandem kinase. This complex triggers the general switch S (shown in Fig. 33.6), which recruits and thereby activates two modules serving the optimal Rac-1 activation: first, the PI-3,4,5-generator PI3K and second, the strain-sensitive adapter cascade made up of Cas and Crk. In principle, the kinase PI3K could activate the exchange factor DOCK-180 alone and trigger a solitary impulse of actin gelation. However, to adapt the thrust of the F-actin formed by an integrin cluster to exogenous forces, DOCK180 is coupled to the cascade of strain-sensitive Cas and Crk adapters.

All three modules are coupled to the plasma membrane by electrostatic-hydrophobic forces, resulting in their activation. Most importantly, the PI-3, 4, 5-P3 lipid anchors produced by PI3K* enhance the local membrane coupling of DOCK-180 (and thus of Rac-1). This is because in contrast to other Rac-specific exchange factors, DOCK180 does not bind to PIP3 lipids through conventional pleckstrin domains but via a polybasic C2-domains (Côté & Vuori, 2007).[1] This unique coupling mechanism provides a positive feedback between Rac-1 activation and PI-3K kinase activity.

6.1.1 *Spatiotemporal switching of the cell front behavior by Rac-1 and RhoA antagonism*

How is the stepwise advancement of the cell front regulated? How can pseudopodia and filopodia arise in parallel at the cell front? How do cells regulate the global polarization and the coordinated advance of the front and the retreat of the uropod?

Numerous studies show that the coordinated migration of cells is regulated by the antagonistic interplay of the GTPases RhoA and Rac1 (Machacek et al., 2009; Nguyen et al., 2016; Sackmann, 2015). This a mutual antagonistic interaction is a key factor in controlling the balance between cell protrusion (driven by Rac1) and retraction (driven by RhoA), which is necessary for efficient cell migration. As shown in Fig. 33.7a the antagonistic interaction of RhoA and Rac-1 is based on the turning down of the switches by the mutual activation of GTP-hydrolyzing proteins of the GAP superfamily. By recruiting these molecular GAP switches to activated membrane domains their effect can be controlled spatiotemporally in a logistic manner, following the scheme of Fig. 33.7a.

The antagonistic interplay of the GTPases Rac-1 and RhoA mediated by the GTP hydrolyzing DAP-proteins is illustrated in Fig. 33.7. The RhoA activation in the region of the adhesion domains is suppressed by recruitment of the GAP protein p190RhoGAP to avoid the impediment of the forward directed growth of the actin gels controlled by Rac-1*. In turn, RhoA* disables the Rac-1 switch at the force free backside of the pseudopods by recruiting the GTP-hydrolyzing protein FilGAP to the stress-free integrin chains. However, a necessary condition for the suppression of Rac-1 activation is the phosphorylation of FilGAP by the kinase ROCK (Ohta et al., 2006). This kinase activated by RhoA is mainly active in the rear of the cell which behaviour is dominated by RhoA. Here, it is also important to note that under stress free conditions FilGAP remains bound to filamin A. It thus suppresses the actin gelation in the cell-substrate contact zone at the rear of the advancing front, with the exception of the lateral boundaries of the polarized cell where exogenous forces may pull at the cell envelope.

Since the GTPases and helper proteins GEF and GAP can rapidly diffuse in the cell, the range of Rac—Rho antagonism (typically $<5\ \mu m$) is much smaller than the cell diameter. For this reason, adherent cells form protrusions simultaneously at several sites, unless they are globally polarized. Before addressing the issue of global polarization, we must first discuss the question of how the antagonistic interplay is controlled in space and time, and second, whether it is involved in the control of the step length ΔL of the pseudopod. Elegant experiments by Machacek, and coworkers (Machacek et al., 2009) show that at the front of the migrating cells Rac-1 and RhoA oscillate in time. The period of oscillation may determine the timing of new adhesion domain formation and the duration of actin gelation. Evidence for this role is provided by the observation that the period of the oscillation, about 3 min, coincides with the period of advancement of *Dictyostelia* cells, which is about 4 min. In sum, the work of Machacek and colleagues, indicates that the leading edge of migrating cells exhibits a dynamic interplay between Rac1 and RhoA activities, involving a timed oscillation that orchestrates the cyclical events of protrusion and retraction during cell migration.

Taken together, the observations described above suggest the following scenario for the cyclic advancement of the fixed-pitch cell front. The newly formed adhesion domains activate actin gelation, (growing actin filament network) by activating Rac-1 and thus generate a constant thrust for a fixed time. During this first step, the activation of integrin by talin

1. Similar to zinkfinger proteins, C2 homology domains penetrate deeply into the hydrophobic area of the membrane with two hydrophobic chains. They are often additionally coupled by coupling to negative lipids via a divalent ion (such as Ca^{++} and Zn^{++}) forming salt bridge with phosphatidylserine.

plays an important role, and RhoA must remain inactive in the area of the adhesion domains during this phase. The force generating step is terminated by switching-off Rac-1* by the GAP protein FilGAP. However, since FilGAP is activated only after its phosphorylation by ROCK, this is only possible after deactivation of Rac1* by RhoA (i.e., after about 1 min).

The question arises as to whether there are thresholds that govern directed cell motion in the context of chemotaxis (movement in response to chemical gradients). The issue has been experimentally addressed in *Dictyostelium* cells by exposing them to linear gradients of cyclic AMP using microfluidic technologies (Song et al., 2006) (see also Muljadi et al. (2022)). Song and coworkers found that when exposed to fluids without cAMP or in cAMP gradients of less than 10^{-3} nM/μm, the cells exhibited random motion with no directional preference. In other words, the threshold for the chemotactic response was 10^{-3} nM/μM. Above this threshold, the cells exhibited directed motion toward the chemoattractant, which was governed by the absolute steepness of the chemoattractant gradient. Thus, the motion was independent of the local chemoattractant concentration. The chemotactic speed and motility of the cells above threshold increased with increasing steepness of the gradient until 10^{-1} nM/μm. No further increase in the chemotactic response was observed in steeper gradients up to 1 nM/μm. They also found that at gradients above 3.3 nM/μm, the cell behaviour returned to that observed at subthreshold values. The demonstration of the directed motion threshold using the microfluidic technology opened the path to address other important questions, such as directional sensing in alternating gradient fields. In elegant studies, Meir and colleagues (Meier et al., 2011) investigated the chemotactic responses to modifications in the shape and position of gradients as a function of switching frequencies. This allowed them to determine the limits of cell migration switching direction when exposed to alternating gradients. To this end, they used a double T-junction chamber equipped with three inlets, one central and two on each side (SF1 and SF2). The central one contained no chemoattractant, while the other two (SF1 and SF2) contained cAMP as a chemotactic agent. The introduction of the chemoattractant on the SF1 side induced cell migration in that direction, whereas the introduction of the chemoattractant on the SF2 side triggered a switch in polarization, causing the cells to migrate in the opposite direction. They found that at a gradient switching rate of 0.02 Hz (corresponding to a time period of 50 s), the cells migrated in the direction of the new chemotactic stimulus. However, at a faster switching rate of 0.1 Hz (time period 10 s), the cells were stalled in a what they called a "chemotactically trapped cell state", a kind of "refractory period", suggesting that their intracellular feedback signaling systems are incapable of responding to rapid changes in external stimuli in directed migration, due to slower internal repolarization times.

6.1.1.1 The G-actin delivery and activation model of the step length ΔL of pseudopod protrusions

Actin monomers are provided by two sources: first, by depolymerization of mature actin network (that is, after ATP hydrolysis of the actin monomers) behind the front mediated by cofilin[2] and second by decoupling of actin from the pool sequestered by β-thymosin. The actin depolymerizing factor cofilin is inactivated by phosphorylation at Ser-3 by LIM kinase (LIMK) and reactivated by dephosphorylation by the cofilin-phosphatase Slingshot (SLP). Most importantly, the slingshot phosphatase is activated by the PI-3K and is thus active as long as the PI(3,4,5)P3 generator. Consequently, one could expect that the lifetime of the active state of the PI-3K determines the time during which G-actin can be provided by cofilin. However, the LIM kinase is activated by the ROCK kinase (ROCK*). This means that in regions dominated by RhoA (at a given time), cofilin is inhibited and cannot depolymerize actin to increase the reservoir of G-actin. Therefore, the rhythm of actin polymerization is most likely determined by the oscillation of the Rac-1 and RhoA activation (Figs. 33.7 and 33.8).

6.1.1.2 Interim recapping

The first step in forming an actin gelation pulse at the cell front is the assembly of new integrin clusters through the nucleation and growth of new adhesion domains (ADs). This is usually triggered by talin-mediated coupling of F-actin to the cytoplasmic β-chains, as illustrated in Fig. 33.5. In a second step, the dual kinase FAK/src complex is switched-on. This complex attracts the strain-sensitive adaptor Cas/CrK, which recruits one or several guanine exchange factors (GEF) DOCK 180 and thus initiates the activation of Rac-1. Simultaneously, the FAK*/src* complex triggers the RhoA-deactivating GAP-protein p12RhoGAP, which suppresses active RhoA* in the area of active adhesion domains.

As a consequence of the antagonistic interaction between Rac1 and RhoA, a broad lamellipodium and several filopodia can be formed simultaneously on the cell front, since RhoA is only suppressed by p190RhoGAP in the region of the adhesion domains (Fig. 33.3). This is a classical situation of pattern formation by lateral inhibition.

---

2. To generate F-actin by spontaneous growth or by polymerization promoters (such as Arp2/3), the monomer has to be primed by replacement of ADP by ATP.

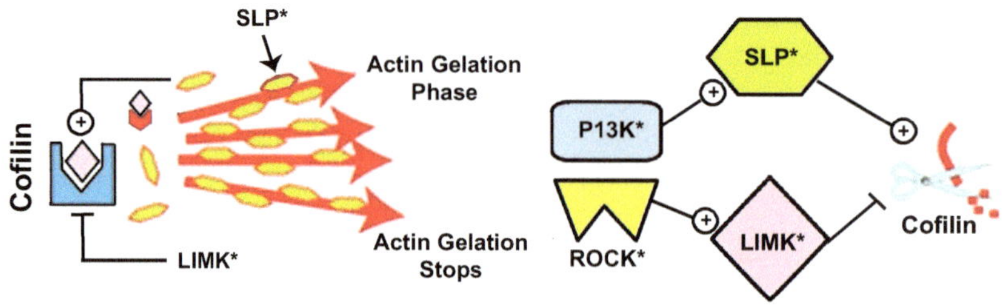

FIGURE 33.8 Control of step length ΔL of pseudopod advancement. This is a crucial aspect of cell migration, and it is achieved through the regulation of the time of activation of cofilin through the interplay of the slingshot phosphatases (SLP) and the LIM kinase (LIMK). The period of G-actin delivery is determined by the time during which LIMK* is downregulated. When RhoA dominates the behavior, the active kinase is switched on in periods determined by the RhoA activity. The *right* side illustrates that the PI-3,4,5-generator PI-3K provides the key for actin polymerization by activation both the high affinity lipid anchor for the Rac-1 switch, and by switching on the cofilin-based G-actin delivery machine slingshot phosphatase (SLP). The Rho-associated coiled-coil-containing protein kinase (ROCK) is a serine/threonine kinase that also plays a significant role in regulating the actin cytoskeleton and cell migration, including pseudopod advancement. ROCK can indirectly influence cofilin activity through its regulation of LIMK. Specifically, ROCK phosphorylates and activates LIMK, leading to cofilin phosphorylation and inactivation. Studies have shown that blocking ROCK activity can increase pseudopod length and reduce pseudopod retraction. This suggests that ROCK, through its influence on cofilin and actomyosin contractility, plays a role in regulating the dynamic process of pseudopod extension and retraction, and thus controls the "step length" of pseudopod advancement.

6.1.2 Mechanical control of polarity of migration by strain-sensitive adapters (Cas and Filamin)—How can Filamin act as a thrust enhancer?

Migrating cells require mechanisms to probe and sense the mechanical properties of their environment such as tissue stiffness. Moreover, they need strategies to adjust their protruding force to the mechanical impedance of the tissue (Mechanical impedance refers to a mechanical system's opposition to motion when subjected to a force) (Sackman, 2019; Sackmann et al., 2010; van Helvert et al., 2018). Stromal cells are constantly subjected to exogenous forces exerted by the tissue. To survive, it is essential that they establish an ongoing stationary mechanical equilibrium to avoid apoptosis. This interactive adaptation of the strain of the cell to exogenous forces exerted by the tissue is called "tensional homeostasis" (Brown et al., 1998). One essential control factor for survival is the enzyme protein kinase B (Akt). Akt suppresses tension-induced apoptosis (Kim et al., 2012) . Growth factors signaling through the phosphoinositide 3-kinase/Akt pathway (PI3K/ Akt) promote cell survival especially under stress-free conditions.

The mechanical control of cell movement is accomplished by two strain-sensitive regulators: the adapter protein Cas and the actin-binding protein Filamin A (FLNA), both of which act as mechanosensors (Braniš et al., 2017; Razinia et al., 2012). There is evidence highlighting the role of Cas (Crk-associated substrate) proteins, particularly p130Cas, as mechanosensors, transducing force into mechanical extension and priming phosphorylation and downstream signalling (Sawada et al., 2006). Cas respond to mechanical signals from the surrounding environment, such as the rigidity, tension, and stretching of the extracellular matrix. Mechanical force or tension leads to the mechanical extension of a specific domain within Cas proteins, like the substrate domain. This extension makes certain sites on the protein more accessible to

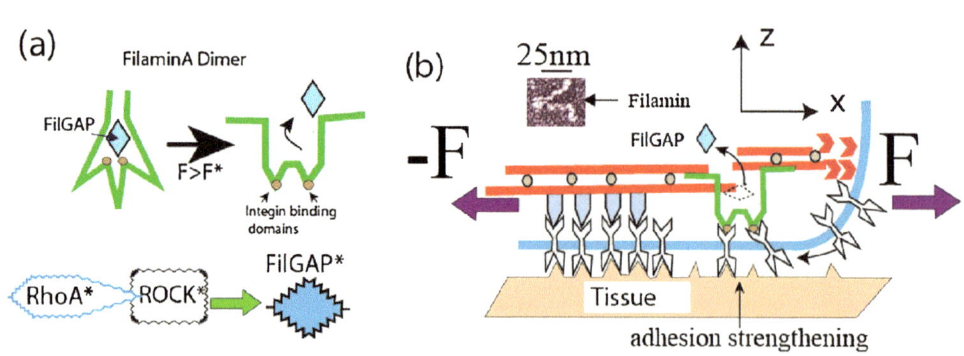

FIGURE 33.9 **Tension-induced control of cell adhesion strength by filamin.** (a) Control of inhibition of Rac1 activation by the guanine-hydrolyzing protein FilGAP. The GAP protein is sequestered by filamin-A dimers in the force-free state. Under the action of local shear stress (Rognonia et al., 2012) FilGAP decouples and escapes into the cytoplasm. A necessary condition for the activation of FilGAP is, however, phosphorylation by active ROCK kinase. Its function is thus restricted to the rear area of the cell that is dominated by RhoA. (b) Model of local strengthening of cell adhesion at sites of strong forces by filamin-A mediated recruitment of integrins to existing integrin clusters. The diagram shows the situation of b induced by increasing the thrust at the cell front.

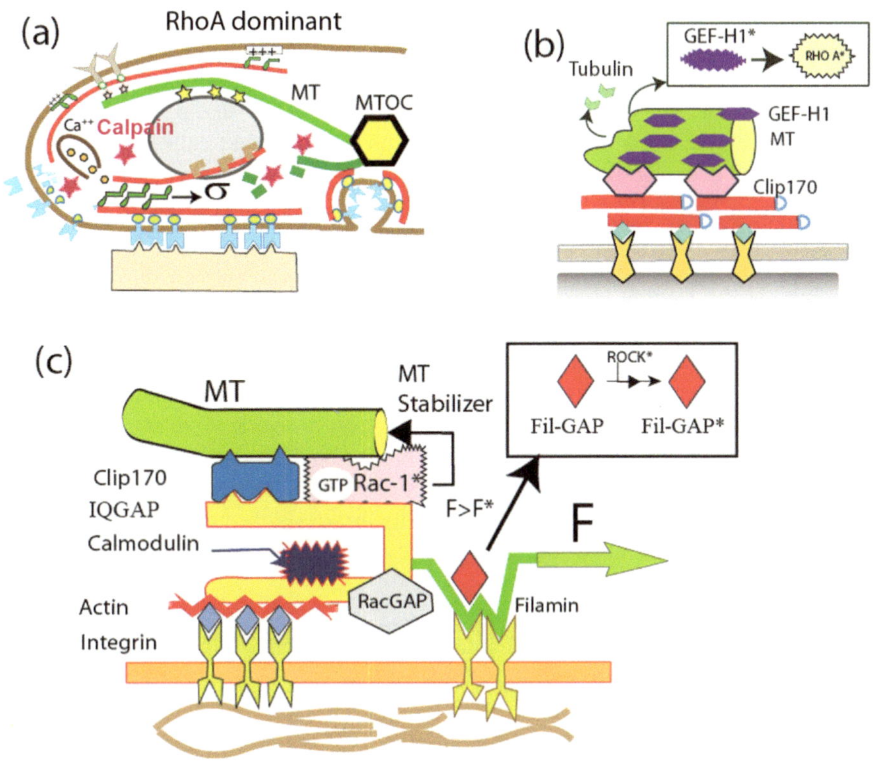

FIGURE 33.10 **Mechanism regulating the uropod retraction.** (a) Polarization and coordinated withdrawal of cell ends. At the cell end (uropod), (*left*), old adhesion domains are dismantled by calcium-induced local activation of the protease calpain. Parallel activation of actin-myosin-II stress fibers by Ca^{2+} and Rho A* releases weakened ADs. Parts of the adhesion domains can also be disposed of by endocytosis. The advancement of the front edge can be facilitated by fusion of vesicles with the plasma membrane, and second, that the resting voltage of the cell envelope is significantly determined by myosin I motors. The extreme *left* side shows the degradation of the adhesion domains by the Ca^{2+}-dependent protease calpain, that mediates the degradation of focal ADs, facilitating the retraction of the uropod. (b) *Left*: Activation of the RhoA switches by degradation of the MT plus ends at the uropod and the concomitant release of the Rho A specific exchange factor GEF-H1. Its activation is generally suppressed by binding to MT and the GEF can only contribute to the RhoA activation after the degradation of the MT-plus ends. (c) Control of global polarization via actin MT coupling by the supramolecular complex CLIP-170/IQGAP, which performs several functions described in the text. In addition, IQGAP binds calmodulin, which in turn binds calcium thereby suppressing the action of this messenger on the cell front.

kinases, facilitating their phosphorylation. Phosphorylation of the Cas substrate domain, particularly the tyrosine-rich regions, is critical for cell migration and regulation of the actin cytoskeleton. Cas protein phosphorylation determines the rate of Rac1 activation (Chodniewicz & Klemke, 2004). As mentioned above, Rac1 plays a key role in cell migration, leading edge extension, and cell spreading. One way Cas achieves this is by facilitating the interaction between Crk and DOCK180, which then leads to Rac1 activation. p130Cas knockout can negatively impact cell migration by reducing the speed of actin polymerization, which affects cell spreading and the overall movement process (Meenderink et al., 2010).

Cas can achieve its function in two ways. One is based on increasing the affinity of its phosphotyrosine-binding sites for Crk by forces in the pN range (Sawada et al., 2006). Since Cas couples via talin (or paxilin) to the integrin chains, its activity is regulated by the shear stress in the actin cortex. A second pathway of control is based on the fact that Cas has several binding sites (~4) for the adapter Crk. This allows several Rac1 switches to be activated in parallel allowing cells to drive the growth of the actin on a broad front.

Experiments by the group of Matthias Rief (Rognoni et al., 2012) showed that filamin A play a key role in adapting the structure of the actin cortex and cell adhesion to intracellular (endogenous) or external (exogenous) forces. By forming V-shaped dimers, they generate a cryptic binding site for integrin β-chains and a trap for Rac1* hydrolyzing the protein FilGAP. Crucial is the ability of dimers to increase the affinity of their cryptic binding sites for integrins in a force-dependent manner by two orders of magnitude. Since FLNA binds to F-actin with both arms and can bridge two filaments over distances of around 150 nm, the integrin binding sites are subjected to the shear forces exerted by the force dipole at the moving cell front. This has two consequences, first, filamin can trap integrins diffusing in the membrane and thus increase their adhesiveness by enlargement of the adhesion domains. Second, inhibition of Rac-1 activation is abolished, and the actin cortex can be stiffened by Rac-1* -induced compaction of the actin gel patches such as by decreasing its mesh size. This effect could play an important role in controlling the stress homeostasis. In particular, it prevents the mechanical overstretching of cells and the tissue (such as the tissue connected to the Z-band of skeletal muscles). The importance of this effect is underpinned by the observation that cells are rapidly eliminated by apoptosis after switching off the FILGAP gene (Shifrin et al., 2009)

6.2 How filamin can act as a thrust enhancer

Filamin-A is often enriched at the leading edge of migrating cells. Measurements of the shear forces provide strong evidence that at this position they function to facilitate the thrust of the cell front overcoming the impedance of the tissue. *Dictyostelium* cells migrating on a soft polymer produce shear forces of up to 1 nN, corresponding to lateral shear stresses of $\Sigma \sim 3 \times 10^{-4}\ \mathrm{N\,m^{-1}}$ (Ladam et al., 2005). These induced shear stresses are around 50 times larger than the lateral tension of weakly adherent cells at rest. The mesh size of the actin network is about 100 nm and therefore each actin filament is subjected to a tensile force of about $\tau * \approx 10\ pN$. This is above the threshold Σ_t at which the filamin fork opens and exposes the cryptic binding sites for integrins (Rognoni et al., 2012). Therefore, filamins bound to the actin cortex at the cell front can capture new integrins diffusing in the cell membranes in order to drive the growth of the actin against high exogenous forces or to stabilize the adhesion strength.

7. Control of global cell polarization by microtubules and Ca^{2+}gradients

The basic steps of mesenchymal cell migration, e.g, leukocytes, can be summarized as follows: cells are attached to the extracellular matrix (ECM) via integrins and focal adhesions (FA). The FAs act as crucial anchor points, physically connecting the F-actin cytoskeleton to the ECM. This enables the cell to generate the forces necessary to move across the ECM. Polymerization of actin at the leading edge of the cell extends filamentous actin (F-actin) protrusions, establishing front-rear polarization. Formation of new FAs attaches the F-actin protrusions to the ECM followed by a backwards movement of F-actin, known as actin retrograde flow. Disassembly of the rear FA and myosin II contraction at the back of cell (uropod) generates a pushing force crucial to move the cell forward. Thus, uropod retraction is a crucial step in cell migration; it is coupled with the cell's extension at the leading edge, creating a coordinated "push–pull" mechanism. Uropode retraction is a complex process that involves the disassembly of focal adhesion proteins, which link the cell's cytoskeleton to the extracellular matrix at their rear end, allowing the cell to detach and move forward. Several mechanisms controlling the retraction of the uropod have been established, which are summarized below.

- First, there is strong evidence that uropode retraction is initiated by local elevations of intracellular Ca^{2+} concentration at the cell's rear end (Brundage et al., 1991; Wheal et al., 2014)$^{2+}$ at the uropod begins prior to retraction. Loading the cells with the intracellular Ca^{2+} chelator BAPTA abolishes both the Ca^{2+} transients and the retraction . Another Ca^{2+} imaging study using migrating vascular smooth muscle cells demonstrated the presence of intracellular Ca^{2+} waves during migration, which travel from the uropod toward the leading edge. Abolishing these waves with BAPTA reduces retraction, thus supporting a causal role for Ca^{2+} in the initiation of retraction (Espinosa-Tanguma et al., 2011). The source and mechanism of intracellular Ca^{2+} elevation have been the subject of intense research over the last decades. The increase in intracellular Ca^{2+} during cell migration is caused by the opening of mechanosensitive (MS) ion channels, also known as stretch-activated channels (Canales Coutiño & Mayor, 2021). As their name implies, these channels sense mechanical forces or deformations of the plasma membrane. As the cell moves forward, the uropod experiences tension and mechanical stress due to the stretching of the cell membrane. MS ion channels are permeable to cations, allowing the influx of Ca^{2+}, which triggers the activity of biochemical signalling pathways involved in regulating cell motility. Two families of MS ion channels are known to regulate cell migration: the transient receptor potential (TRP) and the Piezo channels (Piezo1 and Piezo2). The most extensively studied in terms of cell migration are Piezo1 channels; they are exclusively activated by mechanical stimuli. Piezo1 (originally named as Fam38A for the gene and protein now known as Piezo1). Piezo channels are only activated by mechanical stimuli. The increase in Ca^{2+} has two main consequences leading to retraction: First, the protease calpain is activated, which weakens the adhesion strength of the ADs at the cell end by degradation of the talin–integrin connection. Second, similar to the kinase ROCK, Ca^{2+} stimulates the biochemical myosin II motors by activating the MLC kinase thereby triggering the retraction force of the stress fibers. The activation of Calpain by Ca^{2+} is suppressed at the cell front due to its capture by the IQGAP/

calmodulin complex. As a result, this second messenger contributes significantly to the global polarization of cell movement (Franco et al., 2004).

- The suppression of Rac1* activity at the force-free adhesion domains near the uropod, accelerates the retraction of the uropod for another reason. Rac1* constantly deactivates the MT dismantling protein stathmin. Consequently, the plus ends of the microtubules are rapidly degraded, resulting in the induced release of the RhoA-specific exchange factor GEF-H1 (see Fig. 33.10b and Krendel et al., 2002). This RhoA-specific GEF factor activates RhoA* and enhances the activity of myosin motors by the kinase ROCK.

The actin/IQGAP/MT super complex exerts three functions. First, IQGAP recruits the activated switch Rac-1. Second, owing to the binding of the Ca^{2+} by calmodulin, IQGAP suppresses the activity of this second messenger at concentrations below 10 μM. Thirdly, apparently counteracting the first function, it recruits the Rac-1* deactivator Rac-1-GAP to the adhesion domains.

The first function suppresses the instability of MT by inhibiting the activity of the MT-degrading factor stathmin (see text and references in (Sackmann and Smith, 2014). The second function plays a key role for the global polarization of the migrating cells as described below (see also (Sackmann, 2015)). The third function suppresses the formation of protrusions (such as secondary pseudopodia) from the lateral rim of the migrating cells oriented perpendicular to the direction of motion.

7.1 On the suppression of lateral pseudopodia by MT/IQGAP/actin complex

The suppression of Rac1 activity by binding of FilGAP to adhesion domains, as illustrated in Fig. 33.10a, works in the force-free regions of the cells and in the presence of ROCK*. However, the long edges of cells, moving in the tissue are constantly subjected to exogenic forces or chemotactic signals exerted by the tissue. Therefore, new integrin clusters are constantly being formed, which disturb the directed movement and try to induce lateral protuberances. Due to the centripetal tensile stresses on the cell directed toward the center of the cell, it is expected that filamin will also form new integrin clusters and activate Rac1 to form outgrowths perpendicular to the polarization or to stiffen the membrane. In fact, one constantly observes the formation of transient protuberances that are quickly withdrawn. Two groundbreaking experiments showed how cells solve the problem by suppressing the local Rac-1 activity.

Fluorescence Resonance Energy Transfer (FRET) has been applied to determine the state of activation of integrins within the adhesion zone of the cells (Askari et al., 2010). These studies found that only the integrins in the active junctions are activated by the binding of talin, whereas the receptors in immature integrins remain in a low affinity state. Nevertheless, these domains could activate Rac1 to form new zones of actin gelation.

Jacquemet and collaborators showed in elegant biochemical experiments that this activation is inhibited by a second Ra-1 (or Cdc42) specific GTP hydrolyzing protein Rac1GAP as follows (Jacquemet et al., 2013). RacGAP1 forms a trimeric complex with the scaffolding protein IQGAP and Filamin, which binds to newly formed integrin clusters (for instance, at the rim of the cell) and suppresses the activation of Rac1. This follows from the observation that the removal of IQGAP and Filamin results in erratic cell protrusions. In this context, it is important to note that the whole cytoplasmic surface of migrating cells is covered by nearly parallel assemblies of microtubules (which are separated by about ~ 2 μm). The plus ends are coupled to actin gel patches of adhesion domains by the linking protein Clip170 (Fig. 33.10). An interesting protein involved in the actin MT coupling is LIME. Taken together, these observations show that the global polarization of migrating cells is controlled by microtubule-actin crosstalk. Further work provides additional evidence for the central role of MTs a regulator of global cell polarization in combination with molecular motors and the complex of IQGAP with the calcium-activated protein calmodulin. The IQGAP/Calmodulin complex is often constitutively coupled to the adhesion domains where it can suppress RAC-1* (Fukata et al., 2002).

At the protruding front of the cell, the IQGAP/CLIP170/MT-complex coupled to actin gel patches fulfills three important functions:

(1) It can attract the activated GTPase Rac-1 and promote the actin polymerization at the front and thus contribute to the generation of pushing forces.
(2) The activated GTPase Rac1 (Rac1*) suppresses the activity of the MT-Destabilizer Stathmin by catalyzing its phosphorylation through a specific kinase.
(3) IQGAP binds to the Ca^{2+}-binding protein calmodulin, which sequesters Ca^{2+} up to a concentration of ~1 μM. IQGAP thus prevents the action of the protease (such as calpain) at the cell front.

7.1.1 Summary

Cell locomotion is a vital biological process. Quantitative studies of cellular shape change and protrusion forces can provide valuable insights into the remodeling of the composite cell envelope by transmembrane signal transduction (mediated by chemoattractants such as IL8 or cAMP) or by cell tissue adhesion. Parallel measurements of the cell adhesion strength and the viscoelastic impedances of the cell envelope can quantify dysfunctions caused by mutations of actin binding proteins (Simson et al., 1998). Dysfunctions of migration induced by specific mutations yield valuable insights into the assembly of functional domains in multicomponent membranes achieved through logistically controlled delivery of functional modules to specific sites on the cell envelope, such as integrin clusters of nascent adhesion domains. A key takeaway from the studies summarized here is that the proteins of the membrane-associated machines generally swim in the cytoplasm in a self-inhibited state. They are recruited to the membrane (and thereby activated) by electrostatic hydrophobic forces. The site of assembly is generally determined by the generation of swarms of the high-affinity anchors PI(3,4,5)P3. We refer to this pathway of functional membrane domain formation as logistically controlled self-organization. Ultimately, we can understand how the antagonistic interplay of molecular switches regulates the spatiotemporal formation of competing processes in cells, such as the parallel protrusion of pseudopods and filopodia.

Glossary

DdLimE is a LIM protein that plays a role in cell motility and cytokinesis in Dictyostelium. It is suggested that the cortical actin network be linked to the microtubule system to regulate its dynamics. Studies using DdLimE-null mutants reveal defects in cytokinesis and a delayed disassembly of the mitotic spindle, suggesting a connection between DdLimE, the actin network, and microtubule function during cell division. It also affects the length of interphase microtubules and the depolymerization during prophase. Thus, DdLimE is a LIM protein that acts as a linker between the cortical actin network and the microtubule system, regulating microtubule dynamics and playing a crucial role in cytokinesis and cell motility.

Profilin as Regulator of G-actin activation by ATP binding: Profilin binds to G-actin, preventing it from spontaneously polymerizing and also enhances the exchange of ADP for ATP in G-actin, which is necessary for polymerization. The total concentration of profilin (50 μM). A large fraction of profilin is located at the plasma membrane and the organelles, often together with the phosphoinositide scavenger protein MARCKS, comprising 10 μM. Taken together, it is likely that the lipid anchor phosphoinositols are covered by profilin and MARCKS and are thus protected from the unwanted activation of cells by the phospholipase C. This key enzyme generates two second-messenger molecules, IP3 and the lipid anchor diacylglycerol (DAG), which could induce unwanted overexcitations of cells.

References

Akhmanova, A., & Steinmetz, M. O. (2010). Microtubule +TIPs at a glance. *Journal of Cell Science, 123*, 3415–3419.

Askari, J. A., Tynan, C. J., Webb, S. E. D., Martin-Fernandez, M. L., Ballestrem, C., & Humphries, M. J. (2010). Focal adhesions are sites of integrin extension. *Journal of Cell Biology, 188*(6), 891–903.

Auth, T., Safran, S., & Gov, N. (2007). Fluctuations of coupled fluid and solid membranes with application to red blood cells. *Physical Review E—Statistical, Nonlinear, and Soft Matter Physics, 76*, Article 051910.

Baggiolini, M., & Clark-Lewis, I. (1992). Interleukin-8, a chemotactic and inflammatory cytokine. *FEBS Letters, 307*, 97–101.

Braniš, J., Pataki, C., Spörrer, M., Gerum, R. C., Mainka, A., Cermak, V., Goldmann, W. H., Fabry, B., Brabek, J., & Rosel, D. (2017). The role of focal adhesion anchoring domains of CAS in mechanotransduction. *Scientifi Reports, 13*(7), 46233. https://doi.org/10.1038/srep46233

Brown, R., Prajapati, R., McGrouther, D. A., Yannas, I., & Eastwood, M. J. (1998). Tensional homeostasis in dermal fibroblasts: Mechanical responses to mechanical loading in three-dimensional substrates. *Journal of Cellular Physiology, 175*, 323–332.

Bruinsma, R., Behrisch, A., & Sackmann, E. (2000). Adhesive switching of membranes: Experiment and theory. *Physics Reviews E, 61*, 4253–4267.

Brundage, R. A., Fogarty, K. E., Tuft, R. A., & Fay, F. S. (1991). Calcium gradients underlying polarization and chemotaxis of eosinophils. *Science, 254*(5032), 703–706.

Canales Coutiño, B., & Mayor, R. (2021). Mechanosensitive ion channels in cell migration. *Cells Development, 166*, 203683. https://doi.org/10.1016/j.cdev.2021.203683

Côté, J. F., & Vuori, K. (2007). GEF what? Dock180 and related proteins help rac to polarize cells in new ways. *Trends in Cell Biology, 17*, 383–393.

Chodniewicz, D., & Klemke, R. L. (2004). Regulation of integrin-mediated cellular responses through assembly of a CAS/Crk scaffold. *Biochimica et Biophysica Acta, 1692*, 63–76.

Cinamon, G., et al. (2001). Shear forces promote lymphocyte migration across vascular endothelium bearing apical chemokines. *Nature Immunology, 6*, 515–521.

Edwards, M., Zwolak, A., Schafer, D., Sept, D., Dominguez, R., & Cooper, J. (2014). Capping protein regulators fine-tune actin assembly dynamics. *Nature Reviews Molecular Cell Biology, 15*(10), 677–689. https://doi.org/10.1038/nrm3869

Espinosa-Tanguma, R., O'Neil, C., Chrones, T., Pickering, J. G., & Sims, S. M. (2011). Essential role for calcium waves in migration of human vascular smooth muscle cells. *American Journal of Physiology-Heart and Circulatory Physiology, 3o1*, 315–323.

Etzrodt, M., Ishikawa, H., Dalous, J., Mueller-Taubenberger, A., Bretschneider, T., & Gerisch, G. (2006). Time-resolved responses to chemoattractant, characteristic of the front and tail of Dictyostelium cells. *FEBS Letters, 580*, 6707–6713.

Franco, S. J., Rodgers, M. A., Perrin, B. J., Han, J., Bennin, D. A., Critchley, et al. (2004). Calpain-mediated proteolysis of talin regulates adhesion dynamics. *Nature Cell Biology, 6*, 977–983.

Fukata, M., Watanabe, T., Noritake, J., Nakagawa, M., Yamaga, M., et al. (2002). Rac1 and Cdc42 capture microtubules through IQGAP1 and CLIP-170. *Cell, 109*, 873–885.

Gomez, T. M., & Letourneau, P. C. (2014). Actin dynamics in growth cone motility and navigation. *Journal of Neurochemistry, 129*(2), 221–234. https://doi.org/10.1111/jnc.12506

Hammer, D. A., & Apte, S. (1992). Simulation of cell rolling and adhesion on surfaces in shear flow: General results and analysis of selectin-mediated neutrophil adhesion biophys. *D-J Series, 83*, 35–57.

Heinrich, D., Götz, M., & Sackmann, E. (2018). Die Zelle – erstaunlich physikalisch. Zellmechanik und -dynamik. *Physik in unserer Zeit, 49*(2), 64–70. https://doi.org/10.1002/piuz.201801455

Heinrich, D., & Sackmann, E. (2006). Ative mechanical stabilisation of the viscoplastic intracellular space of Dictyostelia cells by microtubule actin cross talk. *Acta Biomaterialia, 2*, 619–631.

van Helvert, S., Storm, C., & Friedl, P. (2018). Mechanoreciprocity in cell migration. *Nature Cell Biology, 21*(1), 8–20. https://doi.org/10.1038/s41556-017-0012-0

Hind, L. E., Vincent, W. J. B., & Huttenlocher, A. (2016). Leading from the back: The role of the uropod in neutrophil polarization and migration. *Developmental Cell, 38*, 161–169.

Ingber, D. E. (2000). Opposing views on tensegrity as a structural framework for understanding cell mechanics. *Journal of Applied Physiology, 89*, 1663–1670.

Ingber, D. E., Wang, N., & Stamenović, D. (2014). Tensegrity, cellular biophysics, and the mechanics of living systems. *Reports on Progress in Physics, 77*(4). https://doi.org/10.1088/0034-4885/77/4/046603

Jacquemet, G., Morgan1, M., Byron, A., Humphries, J., et al. (2013). Rac1 is deactivated at integrin activation sites through an IQGAP1–filaminA–Rac GAP1 pathway. *Journal of Cell Science, 126*, 4121–4135.

Jung, G., Remmert, K., Wu, X., Volosky, J. M., & Hammer, J. A. (2001). CARMIL is a potent capping protein antagonist: Identification of a conserved CARMIL domain that inhibits the activity of capping protein and uncaps capped actin filaments. *Journal of Biological Chemistry, 281*, 10635–10650.

Katsuta, H., Sokabe, M., & Hirata, H. (2024). From stress fiber to focal adhesion: A role of actin crosslinkers in force transmission. *Frontiers in Cell and Developmental Biology, 12*. https://doi.org/10.3389/fcell.2024.1444827

Kaverina, I., Krylyshkina, O., & Small, J. V. (2002). Regulation of substrate adhesion dynamic during cell motility. *The International Journal of Biochemistry & Cell Biology, 34*, 746–761.

Kim, C. H., Hao, J., Ahn, H. Y., & Kim, S. W. (2012). Activation of Akt/protein kinase B mediates the protective effects of mechanical stretching against myocardial ischemia-reperfusion injury. *Journal of Veterinary Science, 13*(3), 235–244. https://doi.org/10.4142/jvs.2012.13.3.235

Kovar, D. R., & Pollard, T. D. (2004). Insertional assembly of actin filament barbed ends in association with s produces piconewton forces. *Proceedings of the National Academy of Sciences of the United States of America, 101*, 14725–14730.

Krendel, M., Zenke, F., & Gary, B. G. (2002). Nucleotide exchange factor GEF-H1 mediates cross-talk between microtubules and the actin cytoskeleton. *Nature Cell Biology, 4*, 204–301.

Laakso, J., Lewis, J., Henry Shuman, H., & Ostap, E. M. (2008). Myosin Ib myosin-I can act as a molecular force sensor. *Science, 32*, 1133–1136.

Ladam, G., Vonna, L., & Sackmann, E. (2005). Protrusion force transmission of amoeboid cells crawling on soft biological tissue. *Acta Biomaterialia, 1*, 485–497.

Le Clainche, C., & Carlier, M. F. (2008). Regulation of actin assembly associated with protrusion and adhesion in cell migration. *Physiological Reviews, 88*, 489–513.

Li, L., Nørrelykke, E., & Cox, C. (2008). Zigzag persistent cell motion in the absence of external signals: A search strategy for eukaryotic cells. *PLoS One, 3*, Article e2093.

Machacek, M., Hodgson, L., Welch, C., Elliott, H., et al. (2009). Coordination of Rho GTPase activities during cell protrusion. *Nature, 461*, 99–103.

Meenderink, L. M., Ryzhova, L., Donato, D. M., Gochberg, D. F., Kaverina, I., & Hanks, S. K. (2010). P130Cas Src-binding and substrate domains have distinct roles in sustaining focal adhesion disassembly and promoting cell migration. *PLos One, 5*(10). https://doi.org/10.1371/journal.pone.0013412

Meier, B., Zielinski, A., Weber, C., Arcizet, D., Franosch, T., Rädler, J. O., & Heinrich, D. (2011). Chemotactic cell trapping in controlled alternating gradient fields. *PNAS, 108*, 11417–11422.

Mogilner, A., & Oster, G. (1996). The physics of lamellipodial protrusion. *European Biophysics Journal, 25*, 47–53.

Muljadi, M., Fu, Y.-C., & Cheng, C.-M. (2022). Understanding the cell's response to chemical signals: Utilisation of microfluidic technology in studies of cellular and *Dictyostelium discoideum* chemotaxis. *Micromachines, 13*(1737). https://doi.org/10.3390/mi13101737

Newman, P., & Newman, D. (2003). Signal transduction pathways mediated by PECAM-1. *Arteriosclerosis, Thrombosis, and Vascular Biology, 23*, 953–964.

Nguyen, L. K., Kholodenko, B. N., & von Kriegsheim, A. (2016). Rac1 and RhoA: Networks, loops and bistability. *Small GTPases, 9*(4), 316–321. https://doi.org/10.1080/21541248.2016.1224399

Ohta, Y., Hartwig, J. H., & Stossel, T. (2006). FilGAP, a Rho- and ROCK regulated GAP for Rac binds filaminA to control actin remodelling. *Nature Cell Biology, 8*, 803−814.

Raedler, J., & Sackmann, E. (1992). On the measurement of weak repulsive and frictional colloidal forces by reflection interference contrast microscopy. *Langmuir, 8*, 848−853.

Razinia, Z., Mäkelä, T., Ylänne, J., & Calderwood, D. A. (2012). Filamins in mechanosensing and signaling. *Annual Review of Biophysics, 41*, 227−246. https://doi.org/10.1146/annurev-biophys-050511-102252

Ridley, A. (2015). Rho GTPase signalling in cell migration. *Current Opinion in Cell Biology, 36*, 103−112.

Rognoni, L., Stigler, J., Pelz, B., Ylänne, J., & Rief, M. (2012). Dynamic force sensing of filamin revealed in single-molecule experiments. *PNAS, 109*(48), 19679−19684. https://doi.org/10.1073/pnas.1211274109

Sackmann, E. (2015). How actin/myosin crosstalks guides the adhesion, locomotion and polarization of cells. *Biochimica et Biophysica Acta, 1853*, 3132−3142.

Sackman, E. (2019). Viscoelasticity of single cells-from subcellular to cellular level. *Seminars in Cell & Developmental Biology, 93*, 2−15.

Sackmann, E., Keber, F., & Heinrich, D. (2010). Physics of cellular movements. *Annual Review of Condensed Matter Physics, 1*, 257−276.

Sackmann, E., & Smith, A. (2014). Physics of cell adhesion: Some lessons from cell mimetic systems. *Soft Matter, 10*, 1644−1659.

Sawada, Y., Tamada, M., Dubin-Thaler, B., Cherniavskaya, O., Ryuichi, S., Tanaka, S., & Sheetz, M. (2006). Force sensing by mechanical extension of the src family kinase substrate p130Cas. *Cell, 127*, 1015−1026.

Schaller, M. D. (July 25, 2001). Biochemical signals and biological responses elicited by the focal adhesion kinase. *Biochimica et Biophysica Acta, 1540*(1), 1−21.

Schindl, M., Wallraff, E., Deubzer, B., Witke, W., Gerisch, G., & Sackmann, E. (1995). Cell-substrate interactions and locomotion of Dictyostelium wild-type and mutants defective in three cytoskeletal proteins. *Biophysical Journal, 68*, 1177−1190.

Shifrin, Y., Arora, P., Ohta, Y., & Calderwood, D. (2009). The role of FilGAP-filamin A interactions in mechanoprotection. *Molecular Biology of the Cell, 20*, 1269−1279.

Simson, R., Wallraff, E., Faix, J., Niewöhner, J., Gerisch, G., & Sackmann, E. (1998). Membrane bending modulus and adhesion energy of wild-type and mutant cells of Dictyostelium lacking talin or cortexillins. *Biophysical Journal, 74*, 514−522.

Song, L., Nadkarni, S. M., Bodeker, H. U., Beta, C., Bae, A., Franck, C., Rappel, W.-J., Loomi, W. F., & Bodenschatz, E. (2006). *Dictyostelium discoideum* chemotaxis: Threshold for directed motion. *European Journal of Cell Biology, 85*, 981−989.

Swaney, K. F., & Li, R. (2016). Function and regulation of the Arp2/3 complex during cell migration in diverse environments. *Current Opinion in Cell Biology, 42*, 63−72.

Tanaka, M., & Sackmann, E. (2005). Polymer-supported membranes as models of the cell surface. *Nature, 437*. https://doi.org/10.1038/nature04164

Tsujioka, M., Yoshida, K., & Inouye, K. (2004). Talin B is required for force transmission in morphogenesis of Dictyostelium. *EMBO Journal, 23*, 2216−2225.

Uruno, T., Remmert, K., & Hammer, J. A. (2006). CARMIL is a potent capping protein. *Journal of Biological Chemistry, 28*, 10635−10650.

Wegener, K., Partridge, A., Han, J., Pickford, A. R., Liddington, R., Ginsberg, M. H., & Campbell, I. D. (2007). Structural basis of integrin activation by talin. *Cell, 128*, 171−182.

Wheal, B. D., Beach, R. J., Tanabe, N., Dixon, J., & Sims, S. M. (2014). Subcellular elevation of cytosolic free calcium is required for osteoclast migration. *Journal of Bone and Mineral Research, 29*(3), 725−734.

Wua, Y., Jina, X., Harrison, O., Shapiro, L., Honigaand, B. H., & Ben-Shaul, A. (2010). Cooperativity between trans and cis interactionsin cadherin-mediated junction formation. *PNAS, 107*, 17592−17597.

Zidovska, A., & Sackmann, E. (2011). On the mechanical stabilization of filopodia. *Biophysical Journal, 100*, 1−10.

Further reading

Chang, F., Lemmon, C., Park, D., & Romer, L. (2007). FAK potentiates Rac1 activation and localization to matrix adhesion sites: A role for βPIX. *Molecular Biology of the Cell, 18*, 253−264.

Clark, A., & Petty, H. (2008). Observation of calcium microdomains at the uropod of living morphologically polarized human neutrophils using flashlamp-based fluorescence microscopy. *Cytometry A, 73*, 673−678.

Ehrlicher, A. J., Nakamura, F., Hartwig, J. H., Weitz, D. A., & Stossel, T. (2011). Mechanical strain in actin networks regulates FilGAP and integrin binding to filamin A. *Nature, 478*, 260−263.

Landgraf, K., Pilling, C., & Falke, J. (2011). The GRP1 PH domain, like the AKT1 PH domain, possesses a sentry glutamate residue essential for specific targeting to plasma membrane PI(3,4,5)P3. *Biochemistry, 50*, 9845−9856.

Maekawa, M., Ishizaki1, T., Boku, S., et al. (1999). Signaling from Rho to the actin cytoskeleton through protein kinases ROCK and LIM-kinase. *Science, 285*, 895−898.

Mateer, S., Wang, N., & Bloom, G. S. (2003). IQGAPs integrators of the cytoskeleton, cell adhesion machinery, and signaling networks. *Cell Motility and the Cytoskeleton, 55*, 147−155.

McHugh, B., Buttery, W., Lad, Y., Banks, S., & Sethi, H. C. (2009). T. Integrin activation by Fam38A uses a novel mechanism of R-Ras targeting to the endoplasmic reticulum. *Journal of Cell Science, 123*, 51−60.

Moehl, C., Kirchgessner, N., Schaefer, C., Hoffmann, B., & Rudolf Merkel, R. (2011). Quantitative mapping of averaged focal adhesion dynamics in migrating cells by shape normalization. *Journal of Cell Science, 125*, 155−165.

Noritake, J., Watanabe, T., Sato, K., Wang, S., & Kaibuchi, K. (2005). IQGAP1: A key regulator of adhesion and migration. *Journal of Cell Science, 118*, 2085–2092.

Pierres, A., Benoliel, A., Touchard, D., & Bongrand, P. (2008). How cells tiptoe on adhesive surfaces before sticking. *Biophysical Journal, 94*, 4114–4122.

Remmert, K., Olszewski, T. E., Bowers, M. B., Dimitrova, M., Ginsburg, A., & Hammer, J. A. (2004). CARMIL is a bona fide capping protein interactant. *Journal of Biological Chemistry, 279*, 3068–3077.

Rognonia Small, J. V., Geiger, B., Kaverina, I., & Bershadsky, A. (2002). How do microtubules guide migrating cells? *Nature Reviews Molecular Cell Biology, 3*.

Sackmann, E. (2011). Quantal concept of T-cell activation: Adhesion domains as immunological synapses. *New Journal of Physics, 13*, Article 065013.

Sackmann, E., & Zidovska, A. (2011). Brownian motion of nucleated cell envelopes impedes adhesion. *Biophysical Journal, 100*, 1–10.

Schneider, N., Weber, I., Faix, J., Prassler, J., Müller-Taubenberger, A., Köhler, J., Burghardt, E., Gerisch, G., & Marriott, G. (2003). A Lim protein involved in the progression of cytokinesis and regulation of the mitotic spindle. *Cell Motility and Cytoskeleton, 56*, 130–139.

Weiner, O., Marganski, W. A., Altschuler, S., & Kirschner, M. W. (2011). An actin-based wave generator organizes cell motility. PLoS biology 5 (2007) e221cells by shape normalization. *Journal of Cell Science, 125*, 155–165.

Chapter 34

Chronobiology of single-cell organisms

David Lloyd
Microbiology, School of Biosciences, Cardiff University, Wales, United Kingdom

1. Introduction: Oscillations, rhythms and clocks

The integrated and coherent characteristics of living systems remain as a central unsolved problem of biology. Lower eukaryotes, and especially single-cell organisms, have proved extremely useful in the unraveling of molecular mechanisms involved in chronobiology and have set a firm foundation for work on the cultured cells, tissues and whole organism studies on higher organisms from mice to humans. Investigations of sleep disorders, some forms of depression, jet lag, and the health problems of shift workers has drawn heavily on chronobiology research. Although rather slow to spread to clinic and hospital, the principle messages of the discipline, only within the last decade is filtering into pharmacology and general medical practice. Biological timekeeping and the deciphering of synchrony and coherence in the model protists, as described in this Chapter, is continuing to provide detailed information on the core processes and events typical of ancestral characteristics basic to more highly evolved and elaborated animal and plants. The most recent attempts to achieve a synthesis bring several powerful new techniques to bear on an age-old enigma. In a universe where increasing disorder rules, driven by inescapable thermodynamic forces, islands of self-organization may seem only remotely possible. Yet, provided that conditions are held within a set of defined boundaries, the development and maintenance of the living state proceeds apace and in a truly robust manner. A basic feature of its open system spatiotemporal dynamic complexity is both oscillatory and rhythmic. Functional explanations require more than biochemical and biophysical details of the molecular structures of components, their synthesis, inter-conversion and control, and the kinetics of these processes. Thus we need to be able to probe integrated physiological functions at both descriptive and mechanistic levels.

Successive attempts to come to terms with this problem at a theoretical level have included systems theory, cybernetics, and now "systems biology", actually not a new concept, but one made ever more feasible by the latest techniques. Matching the ideas emanating from control engineering with biochemical information on metabolic control has yielded exciting and valuable results and insights. Even before the power of digital computing was unleashed upon theoretical biology, realistic simulations of glycolytic control were obtained using analogue methods. The discoveries of intracellular oscillations in the concentrations of coenzymes and metabolites during the utilization of glucose by yeast and mammalian cells gave impetus to these activities. Thus one of the most characteristic properties of living organisms is their inherent rhythmicity − their autodynamic nature. As objects of the highest complexity, their constituent functions operate over a wide range of time scales (Fig. 34.1), and on each of these time scales, coordination, regulation and control is paramount. The system has also to be coordinated with external perturbations and environmental change. The day-night cycle is a central reference providing circadian control, and the physiology of higher organisms as well as single cell eukaryotes are entrained to the earth's rotation. In the infradian and ultradian domains we see events and processes occurring on time scales longer and shorter than a day respectively.

2. High frequency periodic phenomena

From attoseconds (in the trapping of photons in the photosynthetic light absorption events), through picoseconds (ligation of CO to the haem of hemoglobin), nanoseconds (in fluorescence energy transfer phenomena), microseconds (in conformational state changes in membranes), milliseconds (for electron and proton transport steps and electrical oscillations in neurons and in brain and heart tissues), to seconds (for metabolic inter-conversions) are considered to be high frequency periodic phenomena (Fig. 34.2).

Cell Physiology Sourcebook. https://doi.org/10.1016/B978-0-12-811114-7.00017-7

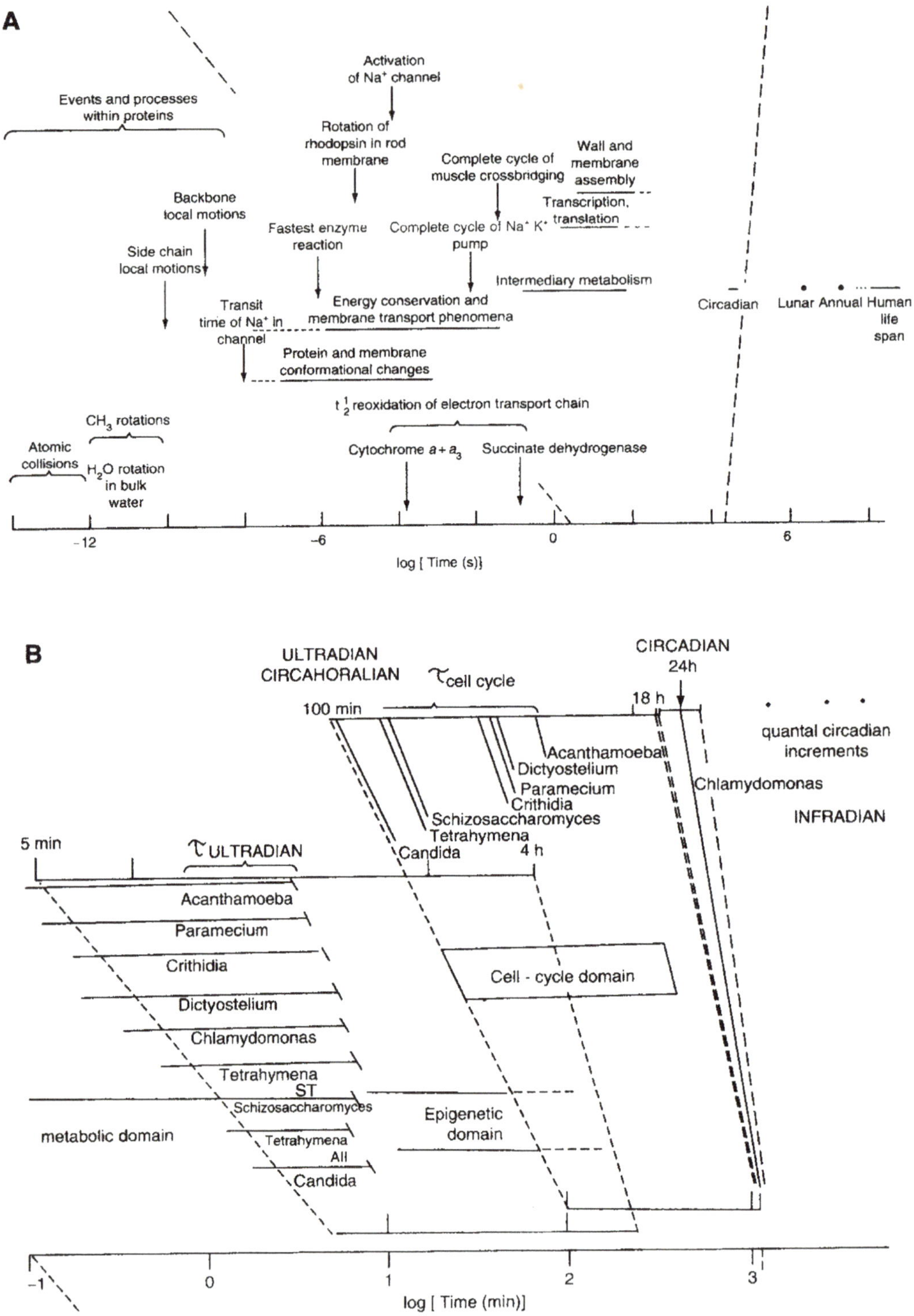

FIGURE 34.1 The time domains of living systems. (A) Approximate relaxation times (to a return of one/e of a small perturbative effect are indicated. (B) Ultradian clock and cell division times for some protists and a yeast.

Transcription, translation, and the final stages of assembly of membranes, cell walls, and organelles may take many seconds or minutes, and segregation and cell divisions many minutes or hours (Fig. 34.1). On and between all these levels of the temporal hierarchy there is a necessity for timekeeping, so that convergence of parallel processing is ensured and that the whole system is coherent. The analogy with a computer is apt but only partial. The magnificent machine also creates itself as it performs, creation requires continuing and simultaneous destruction; complexity *par excellence*!

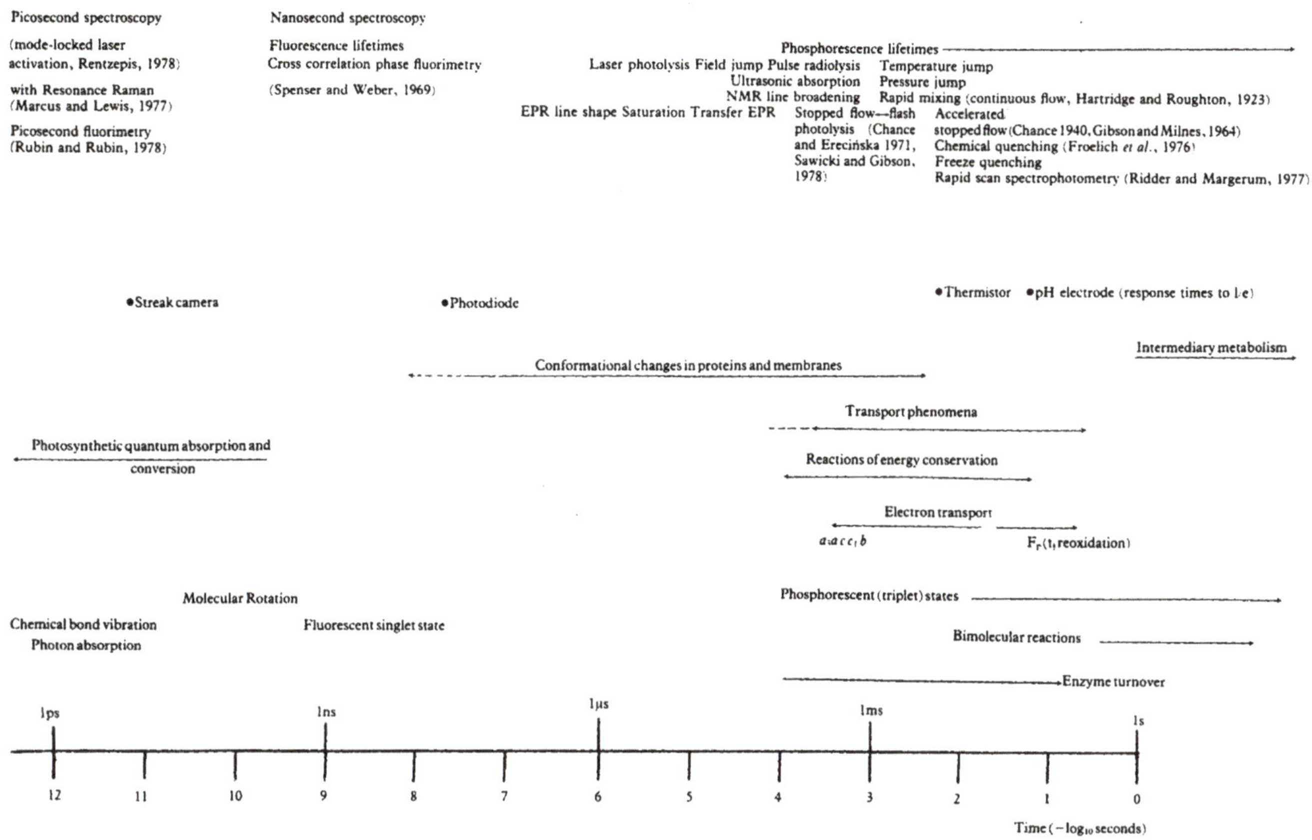

FIGURE 34.2 Fast processes and events occurring within a second in living systems, together with some of the techniques devised for their study.

3. Glycolytic and respiratory oscillations

Although studied for almost 60 years, and still yielding new information on the control mechanisms governing the principle pathway of glucose utilization, a clear function for oscillations in the intracellular concentrations of the intermediates and cofactors of glycolysis is yet to be uncovered. Their periods are often of the order of minutes, are temperature sensitive, highly damped, and are usually triggered (e.g., by a sudden transition to anaerobic conditions). Potential relevance to the maintenance of cellular synchronization in yeast using acetaldehyde as a key coupling agent does not fully explain the kinetics of fast synchronization dynamics provoked by glucose only when the substrate transporter is not fully saturated. Disappearance of oscillations in all the population together with the synchrony at low cell counts suggests a quorum-sensing mechanism of chemical communication, and candidates proposed include acetaldehyde as a small, highly diffusible and volatile molecule. Techniques for studying glycolytic oscillations in single ischemic cardiomyocytes and in isolated yeast organisms held in "laser tweezers" provide new insights into metabolic control mechanisms as well as possible functions. The recent observation that the aqueous intracellular phase shows oscillatory relaxation properties (as revealed by spectral shifts in the fluorescent emission from Acdan, Prodan, or Laurdan probes (Fig. 34.3), that are tightly coupled with glycolytic oscillations opens up new vistas for investigation.

Spontaneous respiratory oscillations (period of about 100 s) are also observed in populations and in single yeast cells: mitochondrial and cytosolic contributions to NAD(P)H can be separated using 2-photon fluorescence excitation at 740 nm (Fig. 34.4).

4. Ultradian rhythms and clocks

A day, or even a minute, is relatively a long time in the life of the cell, so as well as becoming entrained and matched to the daily external environmental changes, there is another timekeeping requirement for life. This is the internal time base, upon which all the diverse biochemical events, sequences and processes must be coordinated. Time scales in this case are much less than a day, hence these higher frequency processes "Ultradian" (or more precisely "Circahoralian") are highly

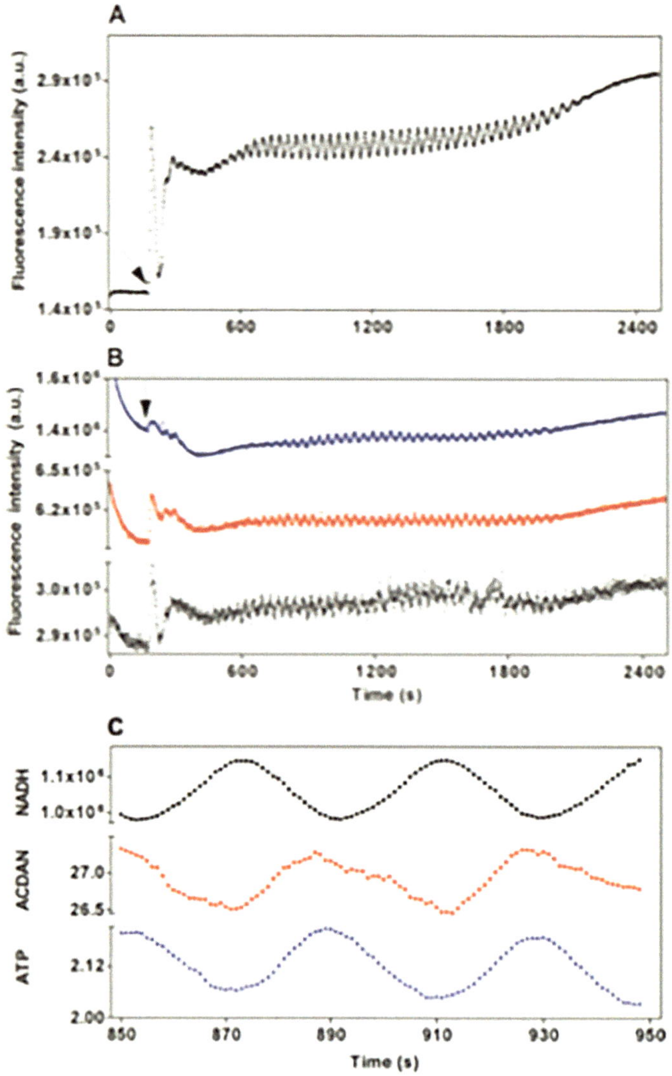

FIGURE 34.3 Oscillations of glycolysis in a yeast suspension, as measured by fluorescence: In A that of NADH, and in B, that of Acdan, Prodan and Laurdan (in descending order) after an addition of KCN. (C) indicates that intracellular ATP also oscillates in phase with Acdan, but out of phase with NADH. *Adapted with permission from Henrik Thoke, and Drs Lars Olsen and Luis Bagatolli.*

conserved. The most detailed elucidation of an ultradian clock comes from the yeast, *Saccharomyces cerevisiae*, in which a 40 min periodicity has been revealed. But it should be stressed that this phenomenon is not a special case, not a mere curiosity, but a general phenomenon. The strict rhythmic self-organization evident here is evident in another yeast, *Schizosaccharomyces pombe*, in mammalian cells (*e.g.*, hepatocytes), and probably in every living organism: its pervasive ubiquity emphasizes its conserved central importance. In other yeasts its period is different, as it is in protozoa and unicellular algae, in plants and in humans; the synchrony that the ultradian clock provides is a necessity for the coherence of all life processes.

The synchronized yeast system has many advantages; its most important property is the availability of continuous readout methods, so that the clock output may be followed second upon second, minute upon minute, hour upon hour, day upon day, up to several months. In a continuous culture vessel, temperature, pH, stirring rate, aeration rate and nutrient supply are strictly and automatically controlled. Residual dissolved O_2 is monitored continuously, and as the supply rate does not vary, any changes in O_2 consumption are reflected in a reciprocal change in dissolved O_2 levels, *i.e.*, it is the measured respiratory activity of the organisms. Initially, cells are grown in a batch culture where near the onset of nutrient and stored-energy reserve (trehalose) depletion, the dissolved O_2 begins to oscillate. Continuous growth is then initiated after all the carbon sources have been exhausted (by starting the flow of culture medium). The immersed O_2 electrode

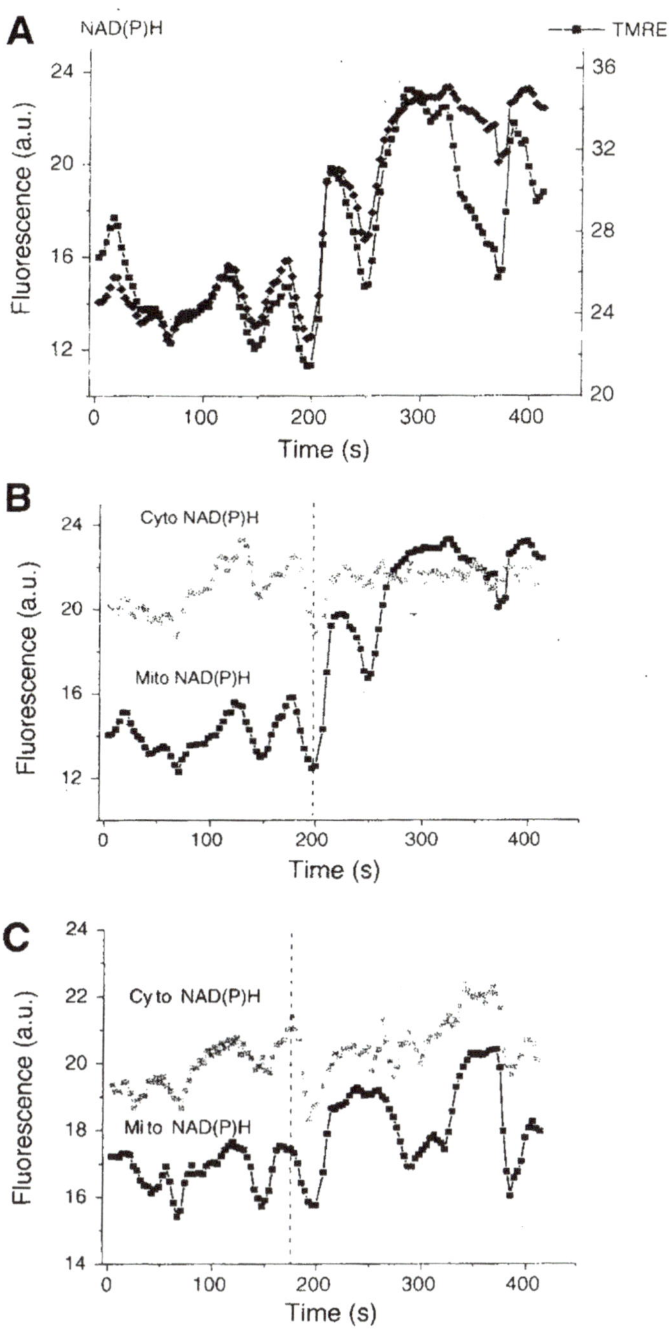

FIGURE 34.4 Spontaneous respiratory oscillations in mitochondrial and cytosolic NAD(P)H in a single organism of an aerobic glucose-fed cell suspension. Autofluorescence (measured with 2-photon excitation (740-420 nm). *Data of Dr Katie Lemar and Dr Miguel A Aon.*

provides a stable device for continuous measurement. The output is not sinusoidal, but rather has a distinct wave-form; its trough-to-peak amplitude is about half that of the highest levels of dissolved O_2 attained, and its period about 40 min (Fig. 34.5).

Experimental changes of temperature through a series of discrete steps reveal the clock-like characteristic of the compensation of period length. Other characteristics suggest that well-known signaling pathways are involved; these include dose-dependent prolongation of period length by Li^+, and perturbation of waveform and period by A-type monoamine oxidase inhibitors such as phenelzine. Many hundreds of measured variables (metabolite and cofactor concentrations, transcription rates, enzyme protein levels and activities, organelle activities, biosynthetic processes, *etc.*) are

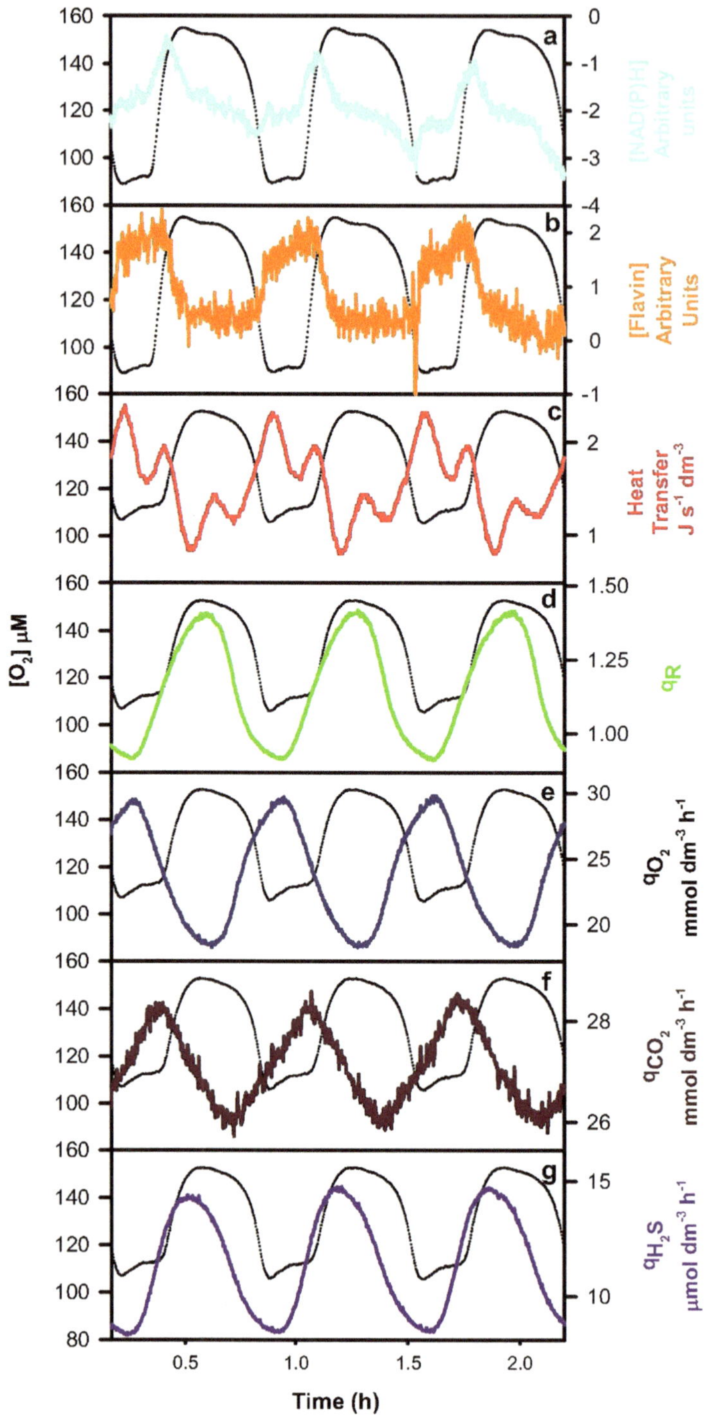

FIGURE 34.5 The ultradian clock in an aerobic self-synchronized continuous culture of yeast (*Saccharomyces cerevisiae*). The dissolved O_2 rises and falls in 40 min-long cycles, as does the intracellular redox state of the population (indicated by fluorescence of NAD(P)H) and oxidized flavins, as do calorimetric measurements, dissolved CO_2, and dissolved H_2S. The cell-budding cycle takes 8 h under the conditions employed here, but may be varied by adjustment of medium flow rate. *Data of Dr Douglas B Murray.*

coupled in a network of phase relationships that provides ordered sequences of cellular composition and function. During one 40 min ultradian respiratory cycle, cellular redox state alternates so as to give an oxidative and a reductive phase and a resulting optimized segregation of energy production and utilization stages.

5. Cell division cycle timekeeping

Oscillations in continuous cultures of yeast are commonplace. These open systems, especially when of large volume (e.g., on pilot plant or full industrial scales), even under the most strictly controlled conditions, operate in pseudo-steady states, and the literature is vast. Many examples are typified by reactions to triggering events (transient anaerobiosis, pulsatile pH control by acid or alkaline titration, drop-wise nutrient addition etc). In other cases, synchronized cell growth and division (budding) as characterized by oscillatory O_2 consumption, partly as a consequence of changing trehalose utilization, leads to oscillatory states. It should be emphasized that unlike the ultradian clock-linked sequence of coupled events described above, the overall cell cycle as well as many of its component processes are highly temperature-dependent, and hence the frequently employed term "cell cycle clock", is in fact a misnomer (Fig. 34.6).

However, even before the characterization of the oscillating cyclin control system components, many investigators treat the cell cycle as an oscillator-driven process ever since the pioneering concepts of Sel'kov, Klevecz, and Gilbert in the

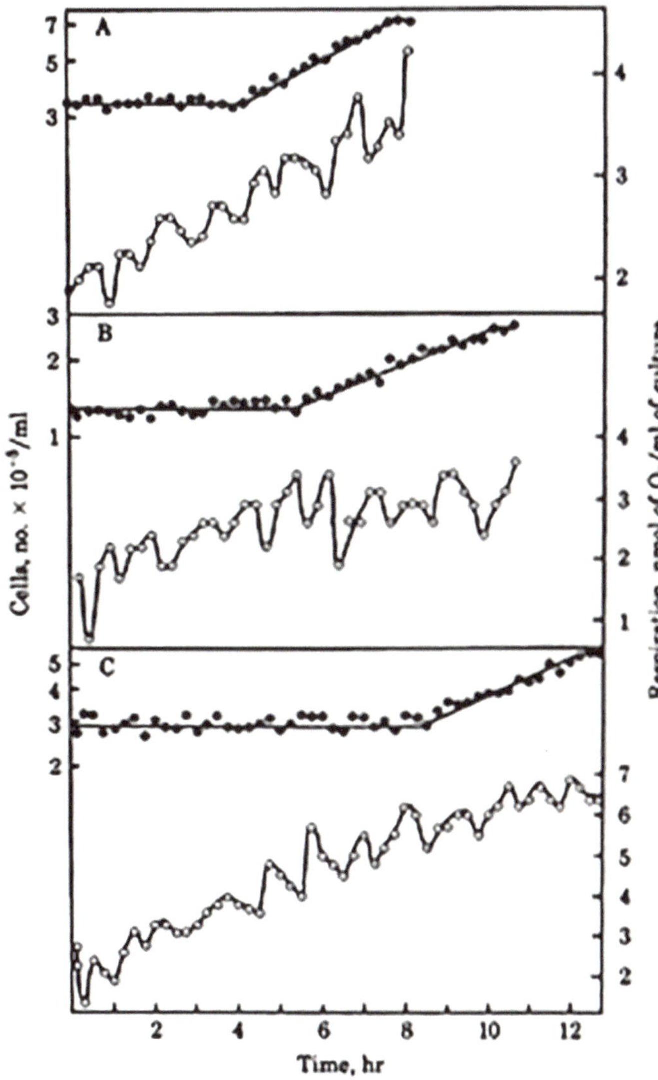

FIGURE 34.6 The temperature-sensitive cell division cycle of *Acanthamoeba castellanii* displays an integrally-increasing number of ultradian clock subcycles as growth temperature is decreased (i.e., quantized organization). *Reproduced by permission of Professor Steven W. Edwards.*

early 1970s. Models have suggested candidates for the molecular components of cell division control machinery and the "quantized" ultradian clock mechanisms that time cell division cycles. Now that molecular genetics has provided many details, key control points have been revealed, and many recently-discovered oscillators described. Modeling proceeds apace and, provided that a full and realistic consideration of physiologically scenarios are fulfilled, new understanding of the variables important for biological phenomena (duration of cell cycle stages, identity of checkpoints, transcriptional noise) may emerge.

"Gating" of the cell cycle by the circadian clock has been studied extensively, and in the natural environment this phenomenon is extremely important and widespread, from planktonic dinoflagellates, cyanobacteria, the green alga *Chlamydomonas reinhardtii* (Fig. 34.7) and other protists, to the proliferative cells of humans in culture.

A mechanism involving cyclic AMP signaling has been established in *C. reinhardtii*. Nuclear regulation of the chloroplast circadian expression has been confirmed.

6. Circadian cycles

We are born into regularly periodic circumstances; the capabilities of living organisms to use the separation of environmental light from dark have yielded an extraordinary repertoire of means for its exploitation and for the fitness of our species. Thereby properties of the organism have become closely enmeshed with the planet's geophysical parameters by the operation of a "Biological Clock". This biochemical device, the Circadian clock (period about 24 h), is composed of an intricate web of interacting components (genes, messenger substances, proteins and small molecules), but can be likened to a mechanical or electronic timekeeper. It has to be composed of an energy source, input pathways, an oscillator, output pathways, and a read-out mechanism, as well as an internal temperature—compensated biological mechanism, which is both re-settable (at dawn and at dusk, Fig. 34.6) and entrainable by environmental (*e.g.*, feeding), or social stimuli or cues. In humans and most other mammals the master pacemaker is located in the brain's suprachiasmatic nucleus (SCN), and peripheral slave oscillators in virtually all body cells. A general model of circadian rhythm consists of its genetically encoded protein components and transcriptional and/or translational feed-forwards and feed backwards control systems. The complex networks are different in different organisms, but in all organisms strong coupling to core components results in daily cycling of many hundreds of gene products.

Circadian clocks occur in all organisms from cyanobacteria to humans; their presence in some other bacteria has been claimed, but these claims are still a matter of dispute. Several examples of eukaryotic unicellular organisms (*Chlamydomonas*, *Tetrahymena*, *Paramecium*, *Gonyaulax* (renamed *Lingulodinium*), and yeasts) have provided excellent systems for study, as has also the mycelial fungus, *Neurospora*.

Many of the fundamental advances in understanding have come from studies of these organisms and have resulted in an enormous literature. The mechanism of temperature compensation, a key characteristic of circadian periodicity, has been intensively studied in *Lingulodinium*, and modeled specifically in *Neurospora*, as well as in more general terms. Transcriptomics and functional proteomics enable a rich diversity of components involved in expression of circadian systems and reveal evolutionary exploitation of very many distinct scenarios. Interaction with environmental cues, not only through photic pathways, but also by nutrient input (especially researched in animal systems) can act on chromatin remodeling. This increases rhythmic complexity and disturbs circadian regularity, sometimes leading to pathophysiological states.

7. Functions of periodic phenomena

Table 34.1 summarizes the possible uses to which rhythmic behavior has been assigned. Autodynamic properties are necessary for the performance of cellular and molecular functions on each and every level of the spatiotemporal hierarchy. Steady-state operation is hardly possible in the dynamic whirlpool of life; even "structural" elements (cytoskeletal proteins) are being turned-over at rates that have been consistently underestimated. Homeodynamics ensures an apparent constancy of the structure-function continuum within limits of amplitude.

8. Integration of cell physiology

In man-made timepieces, typically there is a single ultra-fast oscillator (e.g., that of a quartz crystal), which is geared down several times to generate second, minute, hour and day chronometers. What of biological clocks? It is known theoretically that fast oscillators can be coupled to generate slower rhythms where period is determined in part by coupling topology (Figs. 34.8 and 34.9).

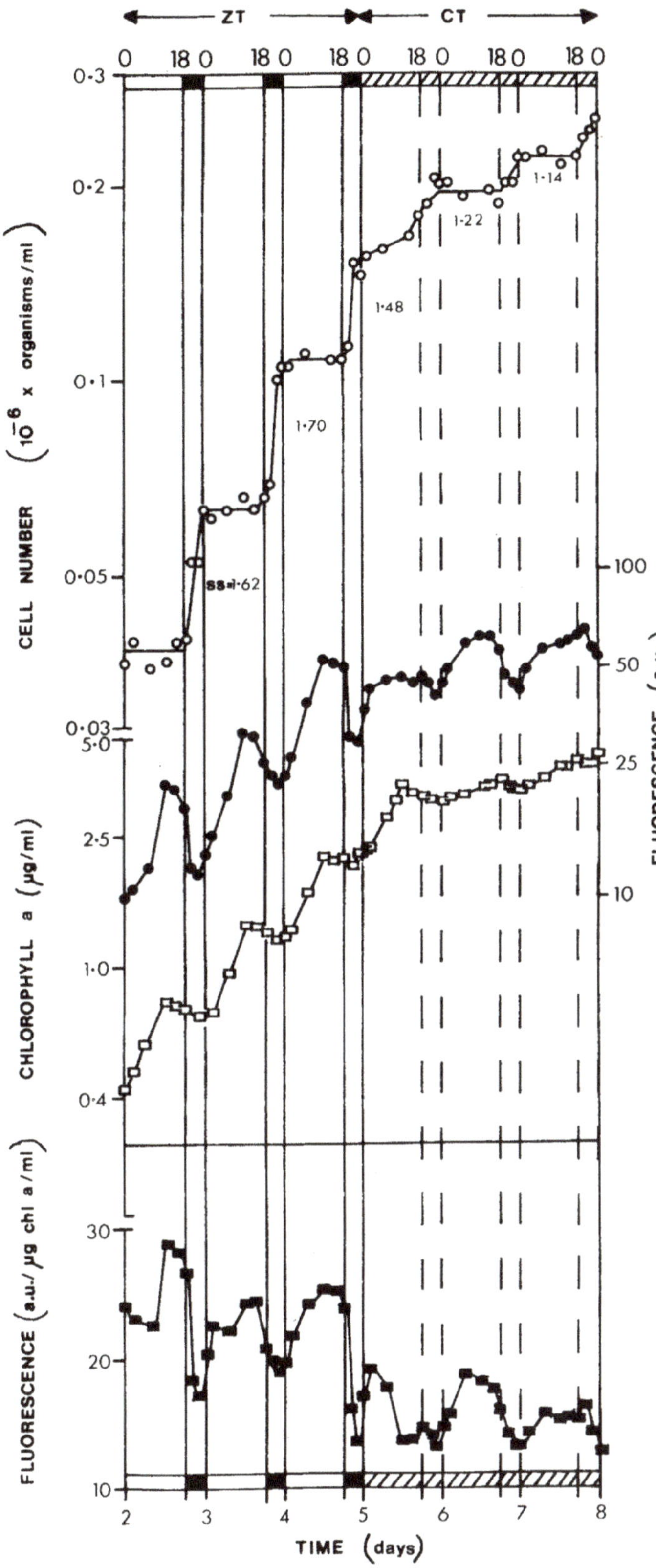

FIGURE 34.7 The cell division cycle of *Chlamydomonas reinhardtii* is gated by the circadian cycle as monitored by (°) cell counts, (•) in vivo, (□) extracted, and (■) in vivo chlorophyll fluorescence (418⇒ 685 nm) per ml culture. *Data of Dr Helen Jenkins.*

TABLE 34.1 Functions of oscillations, rhythms, and clocks.

1. Increased energy efficiency? e.g., in glycolysis
2. Signaling e.g., by Ca^{2+}, superoxide radical ions, by amplitude modulation or frequency modulation
3. Spatial organization e.g., during development in insects; segmentation, by the somite "clock".
4. Temporal organization
   a. Matching intracellular/interorganism events with the environmental cycles e.g., circadian rhythms, seasonal rhythm.
   b. Separating incompatible processes, e.g., photosynthetic production of oxygen from oxygen sensitive nitrogen fixation
   c. Coordinating events in different compartments of the cell e.g., mitochondrion/nucleus/cytosol, or organism e.g., organ/organ
   d. Convergence of pathways after parallel processing
   e. Synchrony of populations and in tissues by cell-cell communication
5. Entrainment by a. phase, or b. frequency
6. Prediction e.g., by dawn or dusk by circadian rhythms
7. Responsiveness to stimuli
8. Biological "fitness": vitality (energetic advantage) or viability (reproductive advantage)

There are also lines of evidence suggesting that the circahoralian and circadian oscillators are closely related in some way, and development of the longer period clock from a higher frequency one has been extensively studied (e.g., in maturing crayfish). Circadian timekeeping itself involves more than one clock in some organisms: multiple timepieces exist to satisfy various needs. Data acquisition systems usually have built-in clocks, and simple bench-top chronometers time routine tasks.

Scale-invariant (self-similar, fractal) features of temporal organization are indicated by several independent analytical procedures (power spectral, relative dispersion, and wavelet analyses), in yeast, in the locomotor activity of quails, and in cardiac function. Thus coherence and synchronization of temporal architecture can be confirmed over time scales of days, hours, and minutes. Reactions and processes occurring in different time domains are interconnected not in a hierarchy, but in a heterarchy, with strong interactions between different time domains. In yeast cultures, and in single cells, where fast

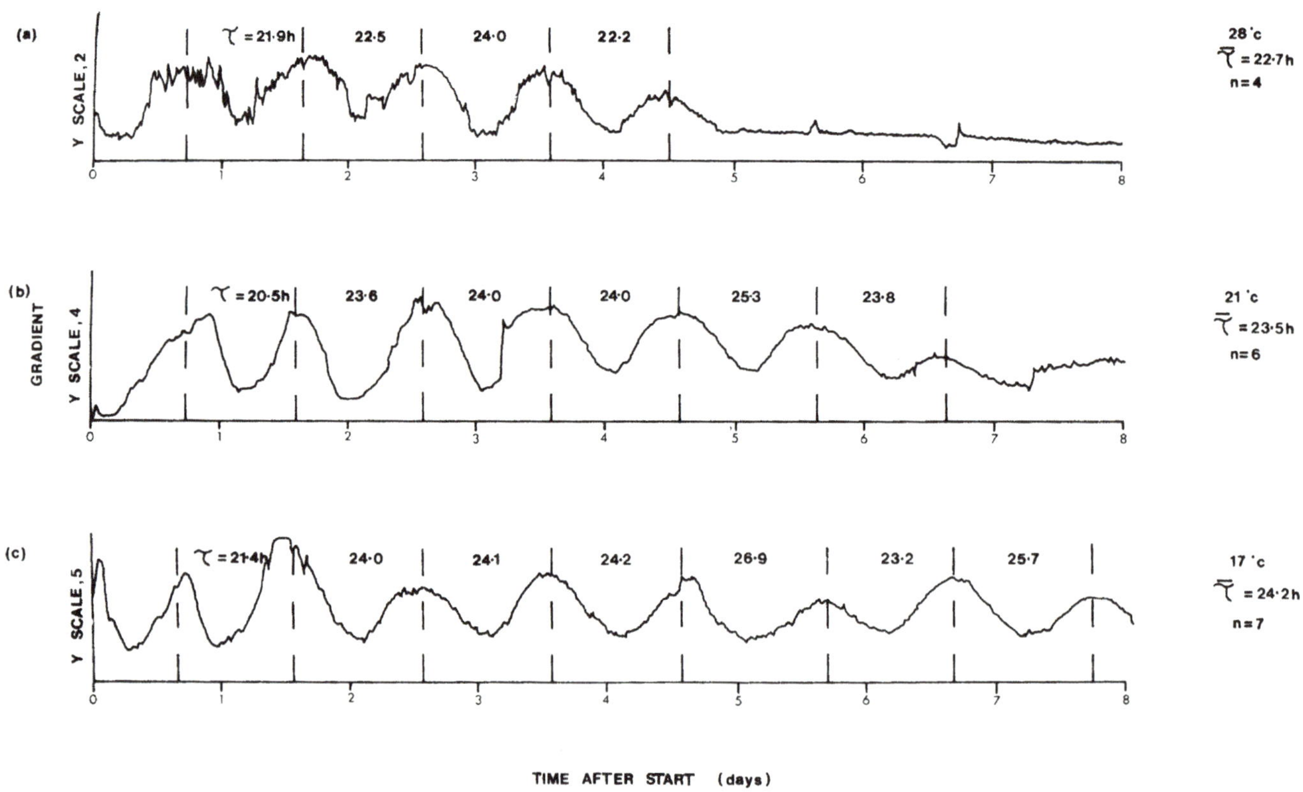

FIGURE 34.8 The circadian cycle of *Chlamydomonas reinhardtii* phototaxis is temperature compensated. *Data of Dr Helen Jenkins.*

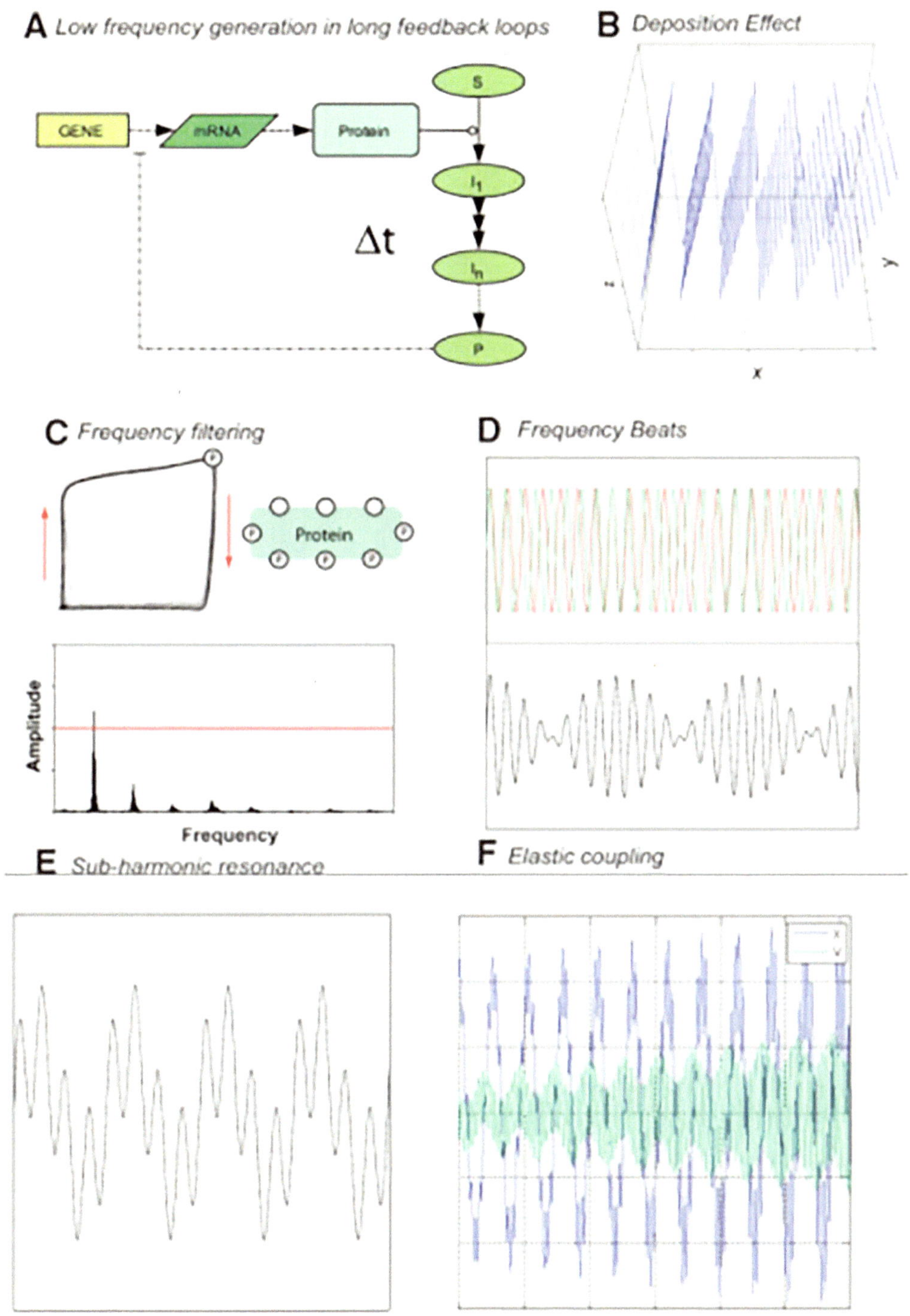

FIGURE 34.9 Many distinct mechanisms are suggested for the construction of long period oscillations, rhythms and clocks from shorter ones.

sampling and hence statistical analysis is more easily achieved: ever-faster temporal relationships in have been described over seconds and subsecond scales using minimally-perturbing methods. Thus coherence and synchronization of temporal architecture can be confirmed over time scales of days, hours, and minutes (Fig. 34.10).

Reactions and processes occurring in different time domains are interconnected not in a hierarchy, but in a heterarchy. In self-synchronized continuously-cultured yeast, where fast sampling and hence statistical analysis is more easily achieved: ever-faster temporal relationships have been described over seconds and subsecond scales. Deterministic chaos has

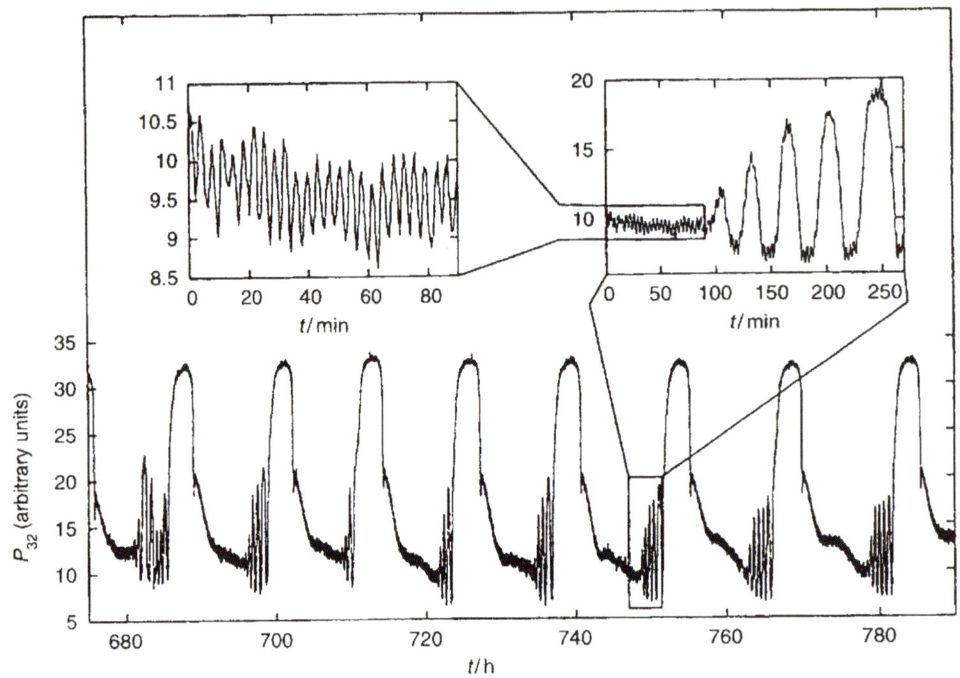

FIGURE 34.10 Membrane inlet mass spectrometry monitoring of an aerobic spontaneously synchronous yeast culture for dissolved O_2 (m/z = 32 relative to Ar at 40). Sampling was in a 12 s cycle time. *Data of Professor Marc Roussel.*

been demonstrated in such cultures by extended monitoring of dissolved gas concentrations (O_2, CO_2, and H_2S), over periods of months at sampling intervals of 12 s.

9. The cell in its environment

Wherever possible experiments with organisms or cell cultures should as far as possible mimic their natural environments and care should be taken to minimize physical disturbances (e.g., exposure to harsh centrifugation, and bright light). Changes in nutrient concentrations, pH values, ionic strength, CO_2 and osmotic balancing are all factors that may perturb outputs. Most neglected is O_2 concentration, especially for organisms that inhabit low O_2 niches within tissues or within cells and therefore normally experiencing near anaerobic conditions.

10. Cell-cell communication

Synchronization of cell populations depend on intercellular signaling by physical, chemical and/or electrical means. Experiments with synchronous cultures indicate that oscillating extracellular concentrations of low molecular mass, volatile, rapidly diffusing and oxidizable molecules (e.g., acetaldehyde, H_2S) serve this function in yeast. In this system, spontaneous oscillations in redox states initiate and spread through the mitochondria and initiate cell-cell interactions locally, before further dissemination to the entire cellular population (Fig. 34.1). This communal propensity is presumably involved in biofilm and higher levels of organization such as tissue formation (Fig. 34.11).

11. Evolution of rhythmic behavior

Self-sustaining oscillations have been studied in continuously fed chemical systems and entrainment by temperature or light can be demonstrated. It is therefore quite conceivable that such systems may have developed on the prebiotic earth. In bacteria, control systems are ubiquitous and oscillatory states commonplace.

We may conjecture that systems that may have originated as somewhat "sloppy" control circuitry must have been exploited for all the known functions of biological oscillators (Table 34.1), and that a more precise and robust performance gradually evolved. Oscillations became rhythms, and the development of temperature compensation permitted time-keeping. This would have occurred on and between every timescale in the heterarchy of spatiotemporal organization, so

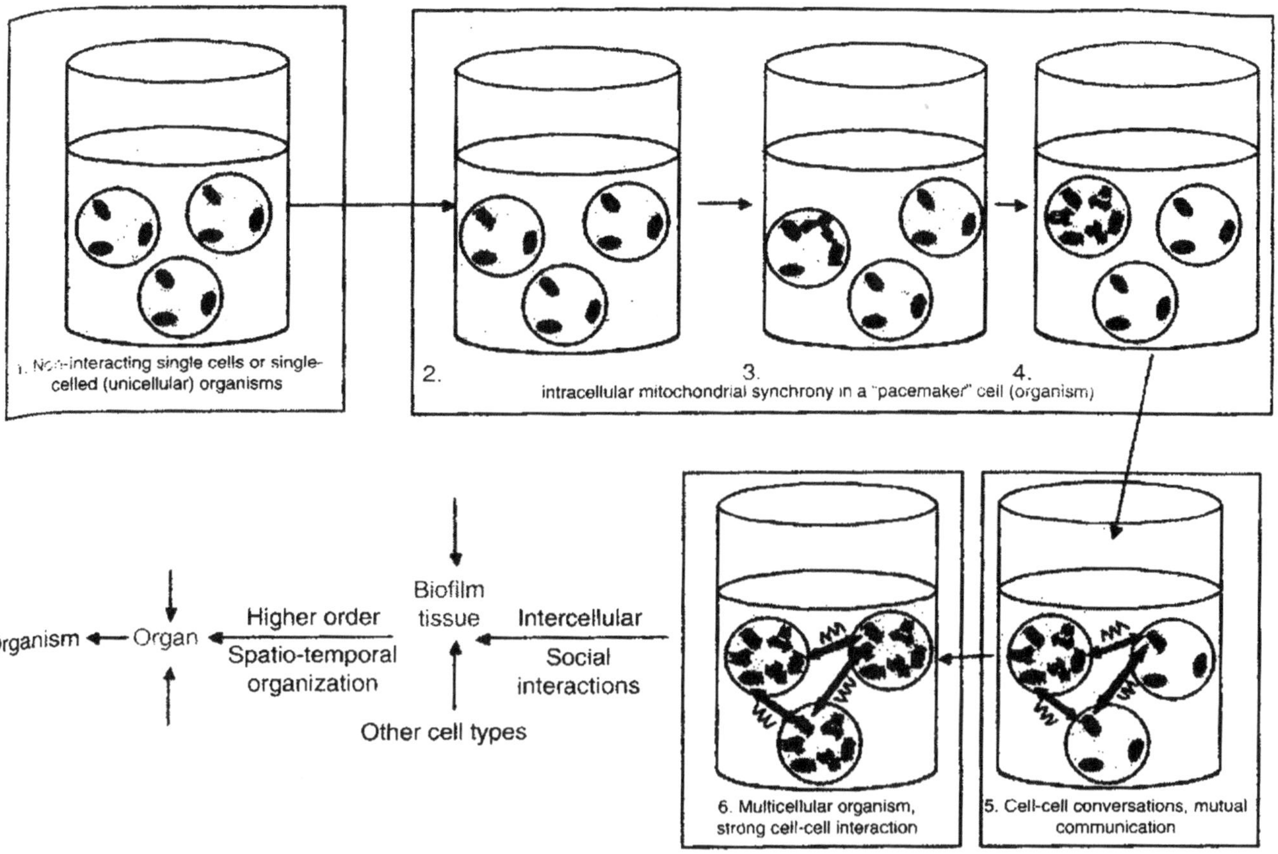

FIGURE 34.11 The evolution of communities of single cells and the development of multicellular organisms has utilized ultradian rhythms for both intracellular coherence and intercellular signaling and cell interactions. *Illustration by Dr Victoria Gray.*

that the massively interconnected network self-optimized to meet the demands of both external matching with geophysical cycles and internal coherence.

The most dramatic step in evolution, the origin of eukaryotes has been the subject of much speculation. One model that now finds growing support is the acquisition of a sulfide-oxidizing α-proteobacterial symbiont by an Archaeal host that produced H_2S. The ultradian organization of the clock in yeast suggests that accommodation to this inter-relationship involved separation in time of the reductive and oxidative stages thereby required, also indicating that the ultradian clock preceded the circadian clock and the evolutionary development of the latter as a more recent phenomenon is echoed today in all higher organisms. This relationship becomes most evident in those mutant organisms where disruption of circadian rhythms uncovers the underlying ultradians. One of the most pressing problems is to discover how the c.24 h rhythm was first constructed during evolution (and presumably is still being constructed) from higher frequency ones: hypotheses abound for counters, couplers and rhythm generators, but good experiments are overdue. Those with crayfish and with plant, animal cell cultures indicate that alterations of the ultradian clock period produce commensurate alterations in circadian time. Ablation of circadian rhythmicity (by mutation or low temperature) reveals the underlying ultradian infrastructure. The circadian "clock genes" may be regarded as an integral part of a global oscillating system that involves many more control circuits than those currently studied in flies, molds, mice and humans. Implications of these concepts for the differentiated organization of higher plants showing infradian (seasonal and annual) rhythms have recently been elucidated (e.g., in the timing of flowering). Thus the multioscillator systems of present-day organisms consist not just of the outputs of a small number of "canonical circadian genes" but of the entire metabolic circuitry as well as its translational and transcriptional networks.

> *We need to be extremely careful in not imposing our constructions on what exists. This becomes evident when we attempt to deal with multiple parallel processes going on in the same space … The coherence of the system which may be impossible to define at the global level, is assuredly generated by the properties of the elements, because the system exists and has survived the test of natural selection.*
>
> Brenner (1999).

Acknowledgments

The author thanks colleagues, postgraduate and postdoctoral coworkers for their insights and excellent experimental achievements that made this synthesis possible.

Bibliography

Anserinsky, E., & Kleitman, N. (1953). Regularly occurring periods of eye motility, and concomitant phenomena during sleep. *Science, 118*, 273–274.

Aon, M. A., & Cortassa, S. (1997). *Dynamic biological organization: Fundamentals as applied to living systems*. London: Chapman and Hall.

Aon, M. A., Cortassa, S., Lemar, K. M., Hayes, A. J., & Lloyd, D. (2006). Single and cell population respiratory oscillations in yeast: A 2-photon scanning laser microscopy study. *FEBS Letters, 581*, 8–14.

Aon, M. A., Roussel, M. R., Cortassa, S., O'Rourke, B., Murray, D. B., Beckmann, M., & Lloyd, D. (2008). The scale-free dynamics of eukaryotic cells. *PLoS One, 3*, Article e3624.

Barrio, R. S., Zhang, L., & Maini, P. K. (1997). Hierarchically coupled ultradian oscillations generating robust circadian rhythms. *Bulletin of Mathematical Biology, 59*, 517–532.

Bashford, C.,L., Chance, B., Lloyd, D., & Poole, R. K. (1980). Oscillations of redox states in synchronized cultures of Acanthamoeba castellanii and Schizosaccharomyces pombe. *J Biophys, 29*, 1–11.

Berrill, N. J., & Goodwin, B. C. (2010). The life of form: Emergent patterns of morphological development. *Rivista di Biologia, 103*, 273–288.

Bertalanffy, L. Von (1952). *Problems of life*. New York: Harper.

Boyd, C. A. R., & Noble, D. (1993). *The logic of life: The challenge of integrative physiology*. Oxford: Oxford University Press.

Brodsky, V. J. (2006). Direct cell-cell communication. A new approach due to recent data on the nature and self organization of ultradian cirahoralian intracellular rhythms. *Biological Reviews, 81*, 143–162.

Brodsky, V. J. (1975). Protein synthesis rhythm. *Journal of Theoretical Biology, 55*, 167–200.

Bunning, E. (1964). *The physiological clock*. Berlin: Springer-Verlag.

Carré, I. A., & Edmunds, L.,N., Jr. (1993). Oscillator control of cell division in euglena: Cyclic AMP oscillations mediate the phasing of the cell division cycle by the circadian clock. *Journal of Cell Science, 104*, 1163–1173.

Chance, B., Estabrook, R. W., & Ghosh, A. K. (1964). Damped sinusoidal oscillations of cytoplasmic reduced pyridine nucleotide in cells. *Proceedings of the National Academy of Sciences of the United States of America, 51*, 1244–1251.

Chance, B., Pye, E. K., Ghosh, A. K., & Hess, B. (Eds.). (1973). *Biological and biochemical oscillations*. New York: Academic Press.

Chandrashekaran, M. K. (2005). *Time in the living world*. Hyderabad: Universities Press (India).

Chin, S. L., Marcus, I. M., Klevecz, R. R., & Li, C. M. (2012). Dynamics of oscillatory phenotypes in *Saccharomyces cerevisiae* reveal a network of genome-wide oscillators. *FEBS Journal, 279*, 1119–1130.

Conrad, M. (1986). What is the use of chaos? In A. V. Holden (Ed.), *Chaos* (pp. 3–14). Manchester: Univ Press.

Edmunds, L. N., Jr. (1984). *Cell cycle clocks*. New York: Marcell Dekker.

Edmunds, L. N., Jr. (1988). *Cellular and molecular bases of biological clocks*. New York: Springer.

Engelmann, W. (2015). Photoperiodism: The calendar of plants. In S. Mancuso, & S. Shabala (Eds.), *Rhythms in plants: Dynamic responses in a dynamic environment* (pp. 191–230). Cham, Switzerland: Springer Science + Business Media.

Fell, D. A. (1997). *Understanding the control of metabolism*. London: Portland Press.

Garfinkel, D. (1971). Simulation of the Krebs cycle and closely related metabolites in perfused rat liver. *Computers and Biomedical Research, 4*, 18–42.

Gilbert, D. A. (1984). The nature of tumor cell proliferation. *Nature, 311*, 610.

Goodwin, B. C. (1964). A statistical mechanics of temporal organization in cells. *Symposia of the Society for Experimental Biology, 18*, 301–326.

Goodwin, B. C., & Cohen, M. H. (1969). A phase shift model for the spatial and temporal organisation of developing systems. *Journal of Theoretical Biology, 25*, 49–107.

Grassberger, P., & Procaccia, I. (1983). Measuring the strangeness of strange attractors. *Physica, 9D*, 189–208.

Halberg, F., Cornelissen, G., Faraone, P., Poeggeler, B., Hardeland, R., Katinas, G., Schwartzkopff, O., Otsuka, K., & Bakken, E. E. (2005). Prokaryote and eukaryote unicellular chronomics. *Biomedicine and Pharmacotherapy, 59*(Suppl. 1), S192–S202.

Hastings, J. W. (2013). Circadian rhythms in dinoflagellates: What is the purpose of synthesis and destruction of proteins? *Microorganisms, 18*, 26–32.

Hastings, J. W., & Sweeney, B. M. (1957). On the mechanism of temperature independence in a biological clock. *Proceedings of the National Academy of Sciences of the United States of America, 43*, 804–811.

Hastings, J. W. (2007). The Gonyaulax clock at 50: Translational control of circadian expression. *Cold Spring Harbor Symposia on Quantitative Biology, 72*, 141–144.

Jenkins, H., Griffiths, A. J., & Lloyd, D. (1990). Selection synchronized Chlamydomonas reinhardii display ultradian but not circadian rhythms. *Journal of Interdisciplinary Cycle Research, 21*, 75–80.

Jenkins, H., Griffiths, A. J., & Lloyd, D. (1989). Simltaneous operation of ultradian and circadian rhythms in Chlamydomonas reinhardii. *Journal of Interdisciplinary Cycle Research, 20*, 257–264.

Kacser, H., & Burns, J. (1973). In D. D. Davies (Ed.), *Rate control of biological processes* (pp. 65–104). Cambridge: Cambridge Univ Press.

Klevecz, R. R. (1992). A precise circadian clock from chaotic cell cycle oscillations. In D. Lloyd, & E. L. Rossi (Eds.), *Ultradian rhythms in life processes* (pp. 41–70). London: Springer-Verlag.

Klevecz, R. R. (1976). Quantized generation times in mammalian cells as an expression of the cellular clock. *Proceedings of the National Academy of Sciences of the United States of America, 73*, 4012–4016.

Kucho, K.–I., Okamoto, K., Isuchiya, Y., Nomura, S., Nango, M., Kanehisa, M., & Ishiura, M. (2005). Global analysis of circadian expression in the cyanobacterium Synechocystis sp Strain PCC 6803. *Journal of Bacteriology, 187*, 2190–2199.

Kurtz, F. K., Kembro, J. M., Flesis, A. G., Armoundas, A. A., Cortassa, S., Aon, M. A., & Lloyd, D. (2016). Analytical tools of complex systems in metabolic, organelle, and cellular networks: From experiments to to models and back. *Wiley Interdisciplinary Reviews: Systems Biology and Medicine, 7*. https://doi.org/10.1002/wsbm.1352

Lloyd, A. L., & Lloyd, D. (1995). Chaos: Its significance and detection in biology. *Biological Rhythm Research, 26*, 233–252.

Lloyd, A. L., & Lloyd, D. (1993). Hypothesis: The central oscillator of the circadian clock is controlled chaotic oscillator. *Biosystems, 29*, 77–85.

Lloyd, D., & Gilbert, D. A. (1998). Temporal organization of the cell division cycle of eukaryote microbes. *Symposium of the Society for General Microbiology, 56*, 251–278.

Lloyd, D., & Murray, D. B. (2007). Redox rhythmicity: Clocks at the core of temporal coherence. *BioEssays, 29*, 465–474.

Lloyd, D., Aon, M. A., & Cortassa, S. (2001). Why homeodynamics, not homeostasis? *Scientific World, 1*, 133–145.

Lloyd, D. (2016). Biological timekeeping: The business of a blind watchmaker. *Science in Progress, 99*, 1–21.

Lloyd, D. (1998). Circadian and ultradian clock controlled rhythms in unicellular microorganisms. *Advances in Microbial Physiology, 39*, 292–339.

Lloyd, D., & Edwards, S. W. (1984). Epigenetic oscillations during the cell cycles of lower eukaryotes are coupled to a clock: life's slow dance to the music of time. In L. Edmunds (Ed.), *Cell clocks and cell cycles* (pp. 26–27). New York: Plenum Press.

Lloyd, D., Edwards, S. W., & Fry, J. C. (1982). Temperature-compensated oscillations in respiration and cellular protein content in synchronous cultures of Acanthamoeba castellanii. *Proceedings of the National Academy of Sciences of the United States of America, 79*, 3786–3788.

Lloyd, D. (2003). Effects of uncoupling of mitochondrial energy conservation on the ultradian clock-driven oscillations in *Saccharomyces cerevisiae* continuous culture. *Mitochondrion, 3*, 139–146.

Lloyd, D., & Kippert, F. (1987). A temperature compensated ultradian clock explains temperature-dependent quantal cell cycle times. In *Temperature and animal cells, bowler K and fuller BJ* (pp. 135–155). Cambridge: Cambridge University Press.

Lloyd, D., Lemar, K. M., Salgado, L. E. J., Gould, T. M., & Murray, D. B. (2003). Respiratory oscillations in yeast: Mitochondrial reactive oxygen species, apoptosis and time; an hypothesis. *FEMS Yeast Research, 3*, 333–339.

Lloyd, D., Lloyd, A. L., & Olsen, L. F. (1992). The cell division cycle: A physiologically plausible dynamic model can exhibit chaotic solutions. *Biosystems, 27*, 17–24.

Lloyd, D., & Murray, D. B. (2000). Redox cycling of intracellular thiols: State variables for ultradian, cell division cycle and circadian cycles? In T. Vanden Driessche, et al. (Eds.), *The redox state and circadian rhythms* (pp. 85–94). Amsterdam: Kluwer.

Lloyd, D., & Murray, D. B. (2006). The temporal architecture of eukaryotic ultradian rhythms in life processes. *FEBS Letters, 580*, 2830–2835.

Lloyd, D., & Murray, D. B. (2005). Ultradian metronome: Timekeeper for orchestration of cellular coherence. *Trends in Biochemical Sciences, 30*, 373–377.

Lloyd, D., Poole, R. K., & Edwards, S. W. (1982). *The cell division cycle: Temporal organization and control of cellular growth and reproduction.* London: Academic Press.

Lloyd, D., & Rossi, E. R. (1992). *Ultradian rhythms in life processes.* London: Springer.

Lloyd, D., Salgado, E. J., Turner, M. P., & Murray, D. B. (2002). Respiratory oscillations in yeast: Clock-driven mitochondrial cycles if energization. *FEBS Letters, 519*, 41–44.

Lloyd, D., Salgado, E. J., Turner, M. P., Suller, M. T. E., & Murray, D. B. (2002). Cycles of mitochondrial energization driven by the ultradian clock in a continuous culture of *Saccharomyces cerevisiae*. *Microbiology, 148*, 3715–3724.

Lloyd, D., & Stupfel, M. (1991). The occurrence and functions of ultradian rhythms. *Biological Reviews, 66*, 275–299.

Lloyd, D. (2006). Ultradian rhythms and clocks in plants and yeast. *Biological Rhythm Research, 37*, 281–296.

Lloyd, D., & Volkov, E. I. (1990). Quantized cell cycle times: Interaction between a relaxation oscillator and ultradian clock pulses. *Biosystems, 23*, 305–310.

Lloyd, D., & Volkov, E. I. (1991). The ultradian clock: Timekeeping for intracellular dynamics. In E. Mosekilde, & L. Mosekilde (Eds.), *Complexity, chaos and biological evolution* (pp. 51–60). New York: Plenum Press.

Lloyd, D., & Williams, C. F. (2015). Avoid excessive oxygen levels in experiments with organisms, cells, and tissues. *Advances in Microbial Physiology, 67*.

Mancuso, S., & Shabala, S. (Eds.). (2006, 2015). *Rhythms in plants: Dynamic responses in a dynamic environment.* Cham, Switzerland: Springer.

Maroto, M., & Monk, N. A. M. (Eds.). (2008). *Cellular oscillatory mechanisms.* New York: Springer Science + Business Media LLC & Landes Bioscience, and Adv Exp Med 641.

Mittag, M., & Wagner, V. (2003). The circadian clock of the unicellular eukaryotic model organism Chlamydomonas reinhardii. *Journal of Biological Chemistry, 384*, 689–695.

Morré, D. J., Church, P. J., Pletcher, T., Tang, X., Wu, L. Y., & Morré, D. M. (2002b). Biochemical basis for the biological clock. *Biochem, 41*, 11941–11945.

Murray, D.,B., Beckmann, M., & Kitano, H. (2007). Regulation of yeast oscillatory dynamics. *Proceedings of the National Academy of Sciences of the United States of America, 104*, 2241–2246.

Murray, D. B., Amariei, C., Sasidharan, K., Machne, R., et al. (2014). In M. A. Aon, V. Saks, & U. Schattner (Eds.), *Systems biology of metabolic and signaling networks: Energy, mass and information transfer* (pp. 323–349). New York: Springer.

Murray, D. B., Englen, F. A. A., Keulers, M., Kuriyama, H., & Lloyd, D. (1998). NO^+, but not NO inhibits respiratory oscillations in ethanol-grown continuous cultures of *Saccharomyces cerevisiae. FEBS Letters, 431*, 297—299.

Murray, D. B., Keulers, M., Engelen, F., Lloyd, D., & Kuriyama, H. (1999). Involvement of glutathione in the regulation of respiratory oscillations during continuous culture of *S. cerevisiae. Microbiology, 145*, 2739—2745.

Murray, D. B., Klevecz, R. R., & Lloyd, D. (2003). Generation and maintenance of synchrony in *Saccharomyces cerevisiae* continuous culture. *Experimental Cell Research, 287*, 10—15.

Murray, D. B. (2004). On the temporal organisation of *Saccharomyces cerevisiae. Current Genomics, 5*, 665—671.

Murray, D. B., Roller, S., Kuriyama, H., & Lloyd, D. (2001). Clock control of ultradian respiratory oscillation found during yeast continuous culture. *Journal of Bacteriology, 183*, 7253—7259.

Nakajima, M., Imai, K., Ito, H., Mishawaka, T., Mahayana, Y., Iwasaki, H., Osama, T., & Kondo, T. (2005). Reconstitution of circadian oscillation of cyanobacterial Kai C phosphorylation in vitro. *Science, 308*, 414—415.

Noble, D. (2002). Modeling the heart — from genes to cells to the whole organ. *Science, 295*, 1678—1682.

Olsen, L. F., & Degn, H. (1977). Chaos in an enzymatic reaction. *Nature, 267*, 177—178.

Ott, E., Grebogi, C., & Yorke, J. A. (1990). Controlling chaos. *Physical Review Letters, 64*, 1196—1199.

Paetkau, V., Edwards, R., & Illner, R. (2006). A model for generating circadian rhythm by coupling ultradian oscillators. *Theoretical Biology and Medical Modelling, 3*, 12.

Pirt, S. J. (1975). *Principles of microbe and cell cultivation.* Oxford: Blackwell Scientific.

Poole, R. K., Lloyd, D., & Kemp, R. B. (1973). Respiratory oscillations and heat production in *Schizosaccharomyces pombe* 972 h-. *Journal of General Microbiology, 77*, 209—220.

Pyragas, K. (2002). Analytical properties and optimization of time-delayed feedback control. *Physical Review E - Statistical, Nonlinear and Soft Matter Physics, 66*, Article 026207.

Pyragas, K. (1992). Continuous control of chaos by self-controlling feed-back. *Physics Letters, 170*, 421—428.

Pyragiene, T., & Pyragas, K. (2005). Delayed feedback control of forced self-sustained oscillations. *Physical Review E - Statistical, Nonlinear and Soft Matter Physics, 72*, Article 026203.

Salgado, E., Murray, D. B., & Lloyd, D. (2002). Some antidepressant agents (Li^+, monoamine oxidase type A inhibitors) perturb the ultradian clock in *S. cerevisiae. Biological Rhythm Research, 33*, 351—361.

Satroutdinov, A. D., Kuriyama, H., & Kobayashi, H. (1992). Oscillatory metabolism of *Saccharomyces cerevisiae* in continuous culture. *FEMS Microbiology Letters, 77*, 261—267.

Schibler, U., & Naef, F. (2005). Cellular oscillators: Rhythmic gene expression and metabolism. *Current Opinion in Cell Biology, 17*, 223—229.

Schneider, E.,D., & Sagan, D. (2005). Into the cool. In *Energy flow thermodynamics and life*. Chicago: Univ Press.

Searcy, D. G. (2003). Metabolic integration during the evolutionary origin of mitochondria. *Cell Research, 4*, 229—234.

Sel'kov, E. E. (1970). Two alternative self-oscillating stationary states in thiol metabolism - two alternative types of cell division normal and malignant ones. *Biophysika, 15*, 1065—1073.

Sohn, H.-Y., & Kuriyama, H. (2001). Ultradian metabolic oscillation of *Saccharomyces cerevisiae* during aerobic continuous culture: H_2S, a population synchronizer is produced by sulphite reductase population synchrony. *Yeast, 18*, 125—135.

Sohn, H.-Y., Murray, D. B., & Kuriyama, H. (2000). Ultradian oscillation of *Saccharomyces cerevisiae* during aerobic continuous culture: Hydrogen sulphide mediates population synchrony. *Yeast, 16*, 1185—1190.

Strässle, C., Sonnleitner, B., & Fiechter, A. (1988). A predictive model for the spontaneous synchronization of yeast (*Saccharomyces cerevisiae*) grown in continuous culture. *Journal of Biotechnology, 7*, 299—318.

Sweeney, B. M. (1982). Interaction of the circadian cycle with the cell cycle in Pyrocystis fusiformis. *Plant Physiology, 70*, 272—276.

Thoke, H. S., Tobiesen, A., Brewer, J., Hansen, P. L., et al. (2015). Tight coupling of metabolic oscillations and intracellular water dynamics in *Saccharomyces cerevisiae. PLoS One, 10*, Article 0117308.

Tyson, J. J., Chen, K. C., & Novak, B. (2003). Sniffers, buzzers, toggles and blinkers: Dynamics of regulatory and signaling pathways in the cell. *Current Opinion in Cell Biology, 15*, 221—231.

Weber, G. (1990). Whither biophysics? *Annual Review of Biophysical Chemistry, 19*, 1—6.

Wicken, J. (1987). *Evolution, thermodynamics and information: Extending the darwin program.* New York: Oxford University Press.

Wiener, N. (1961). *Cybernetics* (2nd ed.). Cambridge: MIT Press.

Wolf, A., Swift, J. B., Swinney, H. L., & Vastano, J. A. (1985). Determining Lyapunov exponents from a times series. *Physica, 16D*, 285—318.

Yang, T. H., Hunt, B. R., & Ott, E. (2000). Optimal periodic orbits of continuous time chaotic systems. *Physical Review E Statistical Physics Plasmas Fluids Related Interdisciplinary Topics Topics, 62*, 1950—1959.

Yates, F. E. (1992). Fractal applications in biology. Scaling in biochemical networks. *Methods in Enzymology, 210*, 636—675.

Yates, F. E. (1992). Outline of a physical theory of physiological systems. *Journal of Physiology and Pharmacology, 60*, 217—248.

Chapter 35

Smooth muscle excitability

Amanda J. Stolarz[1,2], Neil D. Detweiler[3], Chinwendu Amazu[4], Sarah K. England[4] and Nancy J. Rusch[1,2]
[1]*Department of Pharmacology and Toxicology, College of Medicine, University of Arkansas for Medical Sciences, Little Rock, AR, United States;* [2]*Department of Pharmaceutical Sciences, College of Pharmacy, University of Arkansas for Medical Sciences, Little Rock, AR, United States;* [3]*Department of Biology, Goshen College, Goshen, IN, United States;* [4]*Department of Obstetrics and Gynecology, School of Medicine, Washington University, Saint Louis, MO, United States*

1. Summary

The excitability profiles of smooth muscle cells (SMCs) in different organs of the body are highly diverse. Accordingly, they determine organ-specific patterns of contraction and relaxation in the vasculature, airways, gastrointestinal tract, uterus, bladder and other SMC–containing tissues. Unique populations of ion channels have evolved in the SMCs of these different organ systems, which underlie the diverse excitability profiles in, and support the distinct physiological functions of, each organ. This chapter initially focuses on the contribution of K^+ and Ca^{2+}-conducting channels to SMC excitation-contraction coupling using vascular smooth muscle as the central model. However, subsequent sections summarize the electrical profiles of SMCs in other organ systems, which express their own unique blueprints of ion channels. These ion channels are modulated by biological influences including neurotransmitters, autocoids, and other specialized input to determine the final level of smooth muscle tone.

2. Introduction

Smooth muscle cells (SMCs) compose many different tissues including blood vessels, airways, gastrointestinal tract, bladder, uterus, and lymphatics. These organ systems have highly diverse functions, and so not surprisingly, their SMCs exhibit distinctive patterns of electrical excitability that have evolved to support their unique physiological roles. For example, the SMCs in large arteries generally lack spiking action potentials that would drive rhythmic contraction and relaxation, and thereby cause widespread and rapid changes in blood pressure and blood flow. Instead more stable resting membrane potentials (E_m) and slower depolarizing patterns are characteristic of these SMCs. This incremental pattern of excitation enables Ca^{2+} influx through voltage-gated Ca^{2+} channels to incrementally regulate vessel diameter. In contrast, arterioles of the microcirculation may show low-amplitude action potentials that mediate rhythmic vasomotion, a phenomenon that may help to finely match local blood flow to the metabolic demands of distal tissues. Similarly, under normal conditions, the SMCs of airways and bladder lack coordinated electrical spiking that could trigger tonic bronchial constriction or bladder contraction, respectively. However, coordinated action potentials driven by neurotransmitters or environmental cues can mediate sustained contraction. In contrast, gastrointestinal (GI) smooth muscle regularly shows spontaneous electrical activity that drives rhythmic contractions and relaxations to enable digestion and peristalsis. Similarly, the lymphatic system shows rhythmic action potentials and contractions that enable it to pump extracellular fluid from tissues to the venous circulation. Finally, the pattern of action potentials and contractions of uterine smooth muscle are modulated by gestational stage. Clearly, patterns of SMC excitability are highly heterogeneous across, and even within, organ systems to enable their diverse functions (Fig. 35.1; Table 35.1). Further diversity in patterns of SMC excitation is generated by a wide array of physiological factors including neurotransmitters, autocoids, circulating factors, and mechanical forces that can quickly modify the level of SMC excitability.

At least several key differences exist between SMCs and the striated muscle cells that compose skeletal and cardiac tissues. By definition, SMCs lack striations (aligned dark and light bands at the light microscope level) and instead have a less strictly organized arrangement of contractile fibers that mediate graded tone. Another important difference is the very

Cell Physiology Sourcebook. https://doi.org/10.1016/B978-0-12-811114-7.00013-X

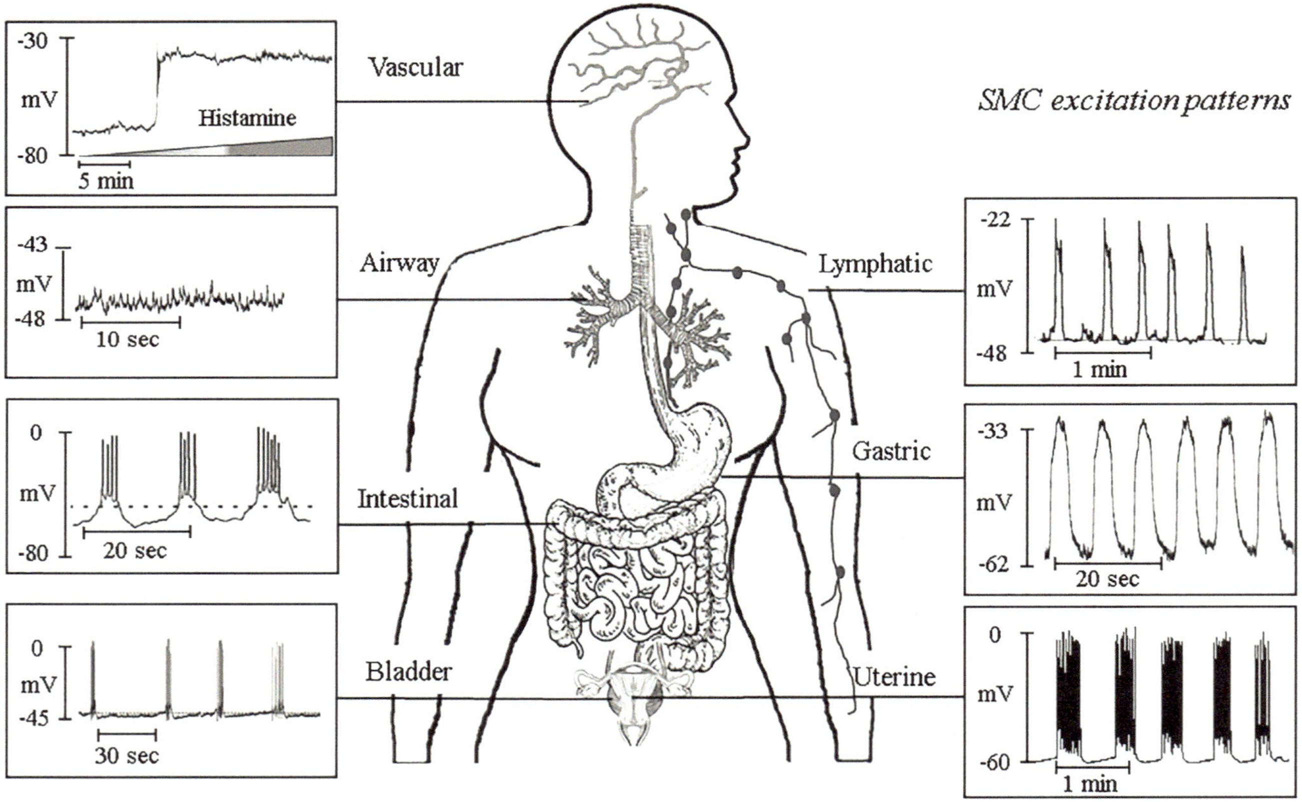

FIGURE 35.1 There are many organs in the body composed of smooth muscle, including the vasculature, airways, gastrointestinal tract, uterus, bladder, and lymphatic system. The patterns of excitation are highly heterogeneous, to support the unique functions of each of the organs. *Reproduced with permission from (A) Gokina and Bevan (2000). (B) Bramich (2000). (C) Sanders et al. (2006). (D) Hashitani et al. (2004). (E) von der Weid et al. (2008). (F) Forrest et al. (2009). (G) Courtesy England S.K.*

TABLE 35.1 Representative electrical profiles of different types of smooth muscle.

Preparation	Species	Resting E_m (mV)	SW or AP amplitude (mV) and frequency (/min)	References
Vascular				
Saphenous artery	Rat	−39	n/a	Berczi et al. (1992)
Hepatic artery	Rat	−61	n/a	Edwards et al. (1998)
Mesenteric artery	Rat	−54	n/a	Lindeman et al. (2018)
Cerebral artery	Rat	−55	n/a	Albarwani et al. (2003)
Cerebral artery	Rabbit	−37	n/a	Brayden and Nelson (1992)
Subcutaneous artery	Human	−55	n/a	Coleman et al. (2001)
Internal mammary artery	Human	−61	n/a	Ma et al. (2015)
Airway				
Trachea	Ferret	−60	n/a	Lee et al. (1990)
Bronchus	Ferret	−70	n/a	Lee et al. (1990)
Trachea	Dog	−60	n/a	Kamei et al. (1994)
Bronchus	Dog	−70	n/a	Kamei et al. (1994)
Bronchus	Human	−52	n/a	Morin et al. (2007)
Gastrointestinal				
Gastric antrum	Mouse	−65	SW (25 mV, 8/min)	Forrest et al. (2009)
Small intestine	Dog	−55	SW (25 mV, 8/min)	Sanders et al. (2006)
Colon	Dog	−75	SW (55 mV, 4/min)	Sanders et al. (2006)
Small intestine	Human	−55	SW spiking (15 mV, 8/min)	Sanders et al. (2006)
Colon	Human	−50	SW bursts (20 mV, 4/min)	Sanders et al. (2006)
Uterine (gestation)				
Myometrium[b](0 days)	Rat	−56	AP bursts (55 mV, 1/min)	Kuriyama and Suzuki (1976)
Myometrium[a] (17–18 days)	Rat	−43	AP bursts (29 mV, 4/min)	Anderson et al. (1981)
Myometrium[a] (21 days)	Rat	−38	AP bursts (30 mV, 1–2/min)	Anderson et al. (1981)
Myometrium[a] (25–55 days)	Sheep	−50	AP bursts (23–67 mV, 1–2/min)	Parkington (1985)
Myometrium[a] (100 days)	Sheep	−50	AP bursts (56 mV, 4/min)	Parkington (1985)
Myometrium (37–40 weeks)	Human	−59	AP bursts (40 mV, 0.1/min)	Parkington et al. (1999)

Continued

TABLE 35.1 Representative electrical profiles of different types of smooth muscle.—cont'd

Preparation	Species	Resting E_m (mV)	SW or AP amplitude (mV) and frequency (/min)	References
Bladder				
Detrusor	Mouse	−42	AP (47 mV, 6/min)	Hayase et al. (2009)
Detrusor	Guinea pig	−46	AP (61 mV, 15/min)	Hashitani et al. (2004)
Detrusor	Pig	−43	AP bursts (40 mV, 1−2/min)	Hashitani and Brading (2003)
Detrusor	Human	−41	AP bursts (40 mV, 8/min)	Hashitani and Brading (2003)
Lymphatic				
Mesenteric vessel	Rat	−48	AP (25 mV, 8/min)	von der Weid et al. (2014)
Mesenteric vessel	Guinea pig	−66	AP (40 mV, 6/min)	van Helden (1993)
Mesenteric vessel	Guinea pig	−50	AP (20 mV, 8/min)	Fox and von der Weid (2002)
Mesenteric vessel	Guinea pig	−55	AP (33 mV, 6/min)	Imtiaz et al. (2007)
Mesenteric vessel	Human	−45	AP (31 mV, 4/min)	Telinius et al. (2015)

Values illustrate the adaptation of smooth muscle to the specific needs of each organ. Gastrointestinal smooth muscle exhibits rhythmic and regular depolarizations known as slow waves (**SW**), which mediate peristalsis. The level of E_m in airway is relatively negative to ensure airway relaxation and air flow. The E_m of vascular smooth muscle also is variable; E_m is more depolarized in SMCs of pressurized arteries that show higher vascular tone. Lymphatic and bladder (detrusor) SMCs exhibit spontaneous action potentials (**AP**), which permit rhythmic contractions.

[a]*circular muscle.*

[b]*longitudinal muscle. Averaged values were obtained from text and tables of corresponding references, or estimated from microelectrode recordings.*

small size of SMCs compared with skeletal or cardiac muscle cells, which requires their dependence on multicellular interactions to coordinate smooth muscle contraction and relaxation even at the local level. Some SMCs exhibit electrical or chemical coupling, whereby a change in the level of resting membrane potential (E_m) or altered concentrations of intracellular signaling molecules in individual cells can be propagated to adjacent cells to produce coordinated contraction or relaxation (Figuero & Duling, 2009; Haddock & Hill, 2005; Hashitani et al., 2001; Lammers, 2015; Sanders, 2008; Young, 2016). Finally, SMCs appear to have a mechanism for sustaining contractions for long periods of time with relatively low energy cost, a concept referred to as the "latch-state". This phenomenon refers to the ability of SMCs to sustain cross-bridge attachments and contraction in the face of falling intracellular Ca^{2+} concentrations, cross-bridge phosphorylation, and ATP consumption. The mechanisms that enable the "latch-state" are reviewed elsewhere (Murphy & Rembold, 2005), but one key concept is that the graded excitation patterns in many types of SMCs combined with their unique ability to sustain contraction at low energy cost permits *smooth muscle tone*, a term that refers to graded, sustained levels of smooth muscle contraction.

Importantly, some organ systems regarded as predominantly smooth muscle also may exhibit a subset of "hybrid" muscle cells that coexpress proteins associated with striated muscle. For example, the muscle cells of the lymphatic vasculature exhibit SMC-like characteristics; they lack striations, are small in size, and their ion channel populations are more similar to SMCs than cardiac or skeletal myocytes. However, subsets of lymphatic "smooth muscle cells" also express striated muscle markers including β-myosin heavy chain and cardiac and skeletal isoforms of α-actin (Breslin, 2014; Dougherty et al., 2008; Muthuchamy et al., 2003; Scallan et al., 2016). These findings emphasize the challenge of precisely defining the electrical properties of SMCs even within a single tissue, which may contain subsets of muscle cells that express unique genes involved in excitation-contraction coupling.

3. Determination of resting membrane potential in SMCs

The origin of the resting membrane potential (E_m) is detailed in Chapter 12. However a short discussion is warranted here, since some determinants of E_m are tissue-specific and unique to SMCs. First, SMCs do not densely express voltage-gated sodium (Na^+) channels in their plasma membrane, which is a hallmark feature of skeletal and cardiac muscle cells. In these striated muscle cells, Na^+ influx enabled by the opening of voltage-gated Na^+ channels mediates the rapid depolarization or "upstroke" of the action potential. However, action potentials (if present) in SMCs are primarily mediated by Ca^{2+} influx through voltage-gated Ca^{2+} channels (see Section 4); the voltage-gated Na^+ channels are sparse or absent. Second, the resting E_m in SMCs is generally more positive (-60 mV to -40 mV) than the more negative E_m level (-90 mV to -60 mV) observed in skeletal and cardiac muscle cells (see Table 35.1). Importantly, the more positive resting E_m of the SMCs is near or even resides within the lower E_m range for opening of voltage-gated Ca^{2+} channels. Thus, even small-amplitude depolarizations will result in the opening of voltage-gated Ca^{2+} channels, resulting in Ca^{2+} influx and SMC activation. Notably, smooth muscle can contract in either a phasic (short term/rhythmic) or a tonic (long term/sustained) pattern depending on the properties of the particular SMC and its input from excitatory or inhibitory signals. When smooth muscle initiates phasic contractions such as during peristalsis of the GI tract, or the rhythmic contractions of lymph vessels, the phasic contractions usually mirror and rely on phasic depolarizations that trigger the voltage-gated Ca^{2+} influx required for SMC activation. Other smooth muscle is characterized by tonic contraction, for example the smooth muscle of sphincters or large airway or arteries, and in this case the level of E_m also is relatively constant under basal conditions. Thus, the E_m of SMCs can be finely tuned to ensure that the intensity of contraction is placed precisely on the continuum between full relaxation and maximal contraction.

The tight voltage control of SMC contraction is evident when using a current pulse to directly depolarize segments of rat basilar arteries (Navarro-Gonzales et al., 2009). The sustained depolarization of SMCs is mirrored by tonic vasoconstriction of the artery (Fig. 35.2A, a-b). Note that vessel diameter dynamically tracks changes in E_m, and small voltage oscillations super-imposed on tonic depolarization elicit similar oscillating changes in vessel diameter. In this preparation, initiation of tonic contraction was attributed to activation of T-type Ca^{2+} channels, whereas high-threshold L-type Ca^{2+} channels were implicated in oscillatory behavior. Even in the absence of external stimuli, the SMCs of the same artery exhibit low-amplitude Ca^{2+}-dependent action potentials that mediate rhythmic contractions or "vasomotion" (Fig. 35.2B, a-b) (Navarro-Gonzales et al., 2009). The action potentials and vasomotion are eliminated by nifedipine, a selective antagonist of L-type Ca^{2+} channels (Fig. 35.2B, c-d). In addition to inherent vasomotion, many physiological stimuli also rely on tight coupling between E_m, voltage-dependent Ca^{2+} influx, and contraction to modify SMC tone. For example, the SMCs of an arteriole in the hamster cheek pouch initially depolarize and constrict in response to phenylephrine (PE), an α_1-adrenergic receptor agonist (Fig. 35.2C, a-b) (Welsh et al., 1998). After return to baseline, exposure of the same arteriole to a high arterial oxygen concentration elicits sustained SMC depolarization concomitant with tonic vasoconstriction

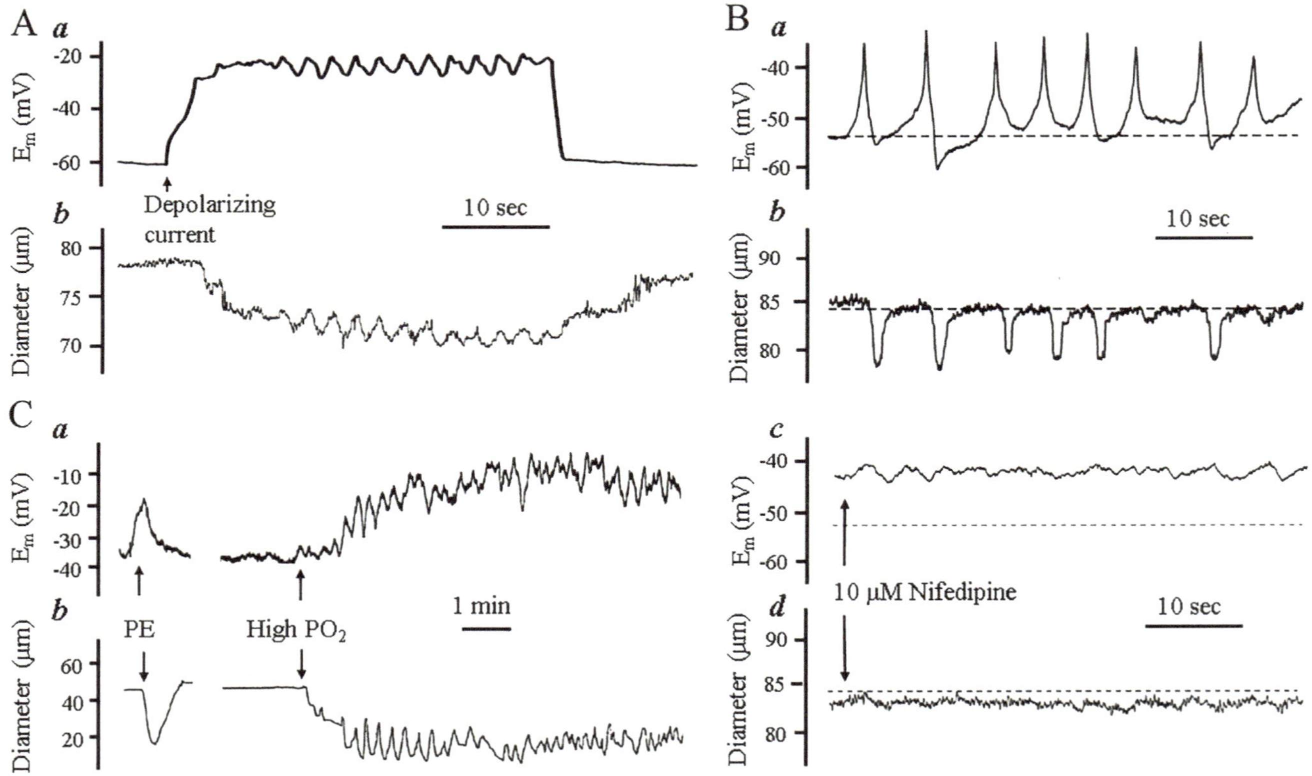

FIGURE 35.2 Role of L-type Ca^{2+} channels in vascular tone. (Aa-b) Applying a depolarizing current to activate L-type Ca^{2+} channels in an isolated rat basilar artery results in vessel constriction. (Ba-b) arteries may exhibit small oscillations in E_m that mediate small oscillations in diameter termed vasomotion. (Bc-d) addition of nifedipine (10 μM) abolishes E_m oscillations and vasomotion. (Ca-b) the vasoconstrictor response of a hamster cheek pouch arteriole to high PO_2 acts as a compensatory mechanism to normalize oxygen delivery to the distal tissues. The E_m and diameter were recorded simultaneously. The arteriole was sequentially exposed to the α-adrenergic receptor agonist, phenylephrine (PE), and then after washout of PE, exposed to high ambient PO_2. High PO_2 elicited a sustained depolarization with superimposed oscillations in E_m that were associated with a tonic, vacillating vasoconstriction. *Reproduced with permission from Navarro-Gonzales et al. (2009) and Welsh et al. (1998).*

mediated by Ca^{2+} channels. In this case of oxygen over-delivery, the arteriole appropriately reduces diameter to normalize the amount of oxygen reaching the distal tissues.

Notably, the control of resting E_m in SMCs is accomplished by a complex system of ion channels, pumps, and exchangers, which regulate the movement of ions across the lipid bilayer of the plasma membrane. Active transport systems (ATP-driven ion pumps and exchangers) maintain nonequilibrium concentration gradients of ions across the plasma membrane. Although there are multiple active transport systems in SMCs, their summed activity is responsible only for a relatively small component of resting E_m. The most prominent ATP-driven pump in SMCs is Na^+/K^+-ATPase, an electrogenic pump in the plasma membrane that generates a hyperpolarizing current to maintain resting E_m. A single cycle of the Na^+/K^+-ATPase results in the extrusion of three Na^+ ions from the cell but entry of only two K^+ ions into the cell, resulting in a net intracellular charge of −1. This hyperpolarizing influence of Na^+/K^+-ATPase may account for 5%−20% of the level of resting E_m in vascular SMCs; loss-of-function can result in a net Ca^{2+} gain of SMCs, enhanced Ca^{2+} signaling, and increased vascular tone (Blaustein et al., 2016). Additionally, because Na^+/K^+-ATPase establishes a high intracellular concentration of K^+ (135−145 mmol/L) compared with extracellular K^+ (3.5−5 mmol/L), it plays an important secondary role in establishing resting E_m. The high transmembrane K^+ gradient in SMCs combined with abundant K^+ channels results in basal K^+ efflux, and this hyperpolarizing K^+ current is the main determinant of the negative resting E_m in SMCs. Thus, pharmacological block of K^+ channels in SMCs causes a profound membrane depolarization. In contrast, the minimal ionic flux through other ion channels has less influence on resting E_m, but these ion channels may critically contribute to SMC contraction. For example, voltage-gated, L-type Ca^{2+} channels and cation−permeable transient receptor potential (TRP) channels provide activator Ca^{2+} for SMC contraction. Although chloride (Cl^-) channels are expressed in SMCs, their contribution to resting E_m is less clear and the reader is referred to detailed reviews on this topic (Leblanc et al., 2015; Picollo et al., 2015; Boedtkjer et al., 2016; Oh and Jung, 2016). For the sake of brevity, this chapter will focus only on the better characterized K^+ and Ca^{2+} -permeable ion channels in SMCs using vascular SMCs as a model (Fig. 35.3). Final sections will discuss key aspects of excitability in airway, gastrointestinal, uterine, bladder and lymphatic SMCs.

4. Potassium channels

The plasma membrane of SMCs expresses diverse populations of K^+ channels. The primary role of these K^+ channels is to mediate K^+ efflux, membrane hyperpolarization and relaxation of SMCs. Distinct types of K^+ channels work in concert to optimize SMC tone in different organs. This section briefly reviews three major K^+ channel families, which will include the voltage-gated K^+ (K_V) channels, high-conductance Ca^{2+}-sensitive K^+ (BK_{Ca}) channels, and inward-rectifier K^+ (K_{IR}) channels. Excellent reviews of other K^+ channel types including the ATP-sensitive K^+ channels and the two-pore domain K^+ channels are available elsewhere (Foster & Coetzee, 2016; Tykocki et al., 2017; Wiedmann et al., 2016). Intracellular

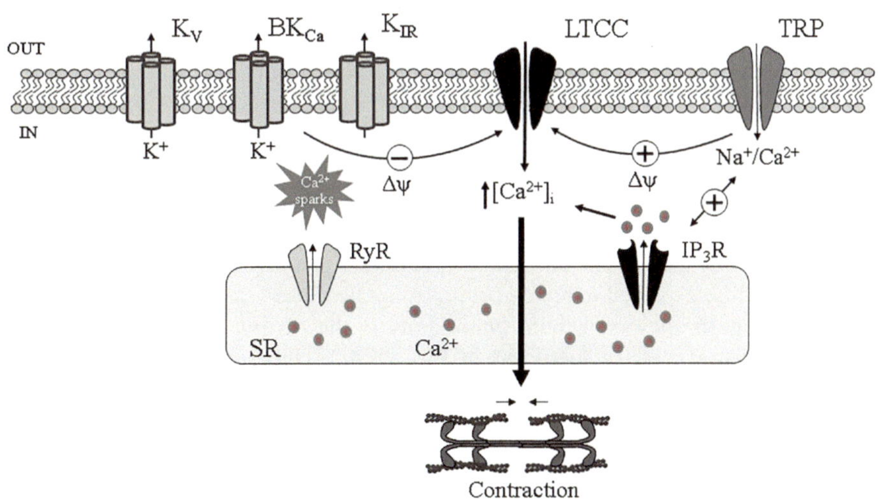

FIGURE 35.3 Some key families of ion channels in SMCs. The K^+ channels include the voltage-gated (K_v), inwardly rectifying (K_{IR}), and high-conductance, Ca^{2+}-sensitive (BK_{Ca}) K^+ channels. The interplay between transient receptor potential (TRP) channels and L-type Ca^{2+} channels (LTCC) provides activator Ca^{2+} for SMC contraction. The channels interact with the ryanodine receptors (RyR) and inositol triphosphate receptors (IP_3R) to modulate SMC excitability.

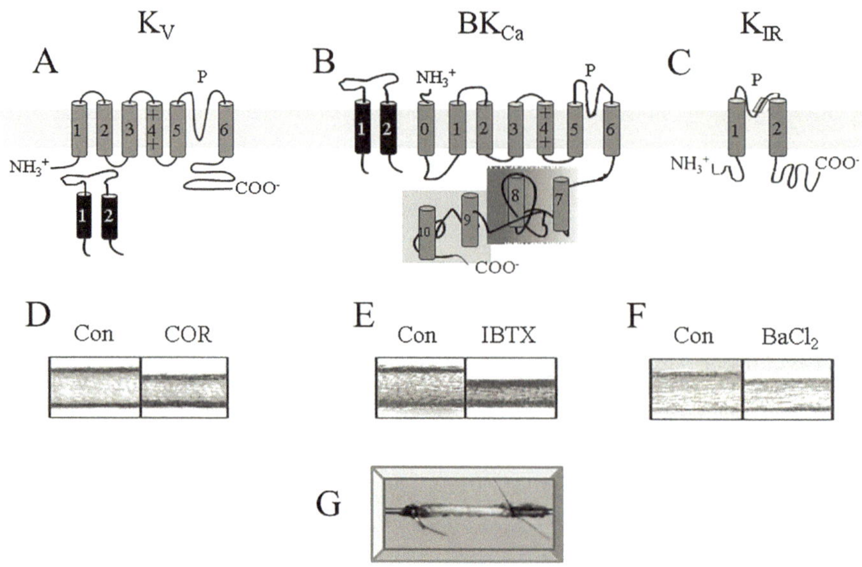

FIGURE 35.4 Proposed topology of K^+ channel α-subunits in SMCs and their vasodilator role in cerebral arteries. Details are provided in the text. (A and B) The multiprotein K_v and BK_{Ca} channels are presumed to be composed of pore-forming α-subunits and regulatory β-subunits. The β-subunit of the K_v channel is assumed to be located intracellularly, whereas the β-subunit of the BK_{Ca} channel is membrane-delineated. (C) The α-subunit of the K_{IR} channel is composed of only two transmembrane domains spanning a pore loop. All three K^+ channel types are composed of four α-subunits that coassemble to form a functional tetramer. An additional four β-subunits compose the K_v and BK_{Ca} channels. (D–F) All three types of K^+ channels contribute to the resting diameter of a rat middle cerebral artery. The artery was cannulated and perfused at an intraluminal pressure of 80 mm Hg. In each case, the left panels show the control resting diameter. The right panels show the decrease in vessel diameter after addition of: 10 μmol/L correolide (COR) to block *Shaker*-like K_v channels (D), 100 nmol/L iberiotoxin (IBTX) to block BK_{Ca} channels (E), and 10 μmol/L $BaCl_2$ to block K_{IR} channels (F). Drugs were washed out and the artery returned to resting diameter between drug responses (not shown). (G) A photograph depicting a rat cerebral artery cannulated on glass micropipettes and perfused at 80 mm Hg.

K^+ channels also have been characterized in the nucleus, mitochondria and other organelles of some cell types, and although findings in SMCs are sparse, the reader should be aware of these emerging findings (Checchetto et al., 2016; Szabo et al., 2012; Toro et al., 2014).

4.1 Voltage-gated K^+ channels

The voltage-gated K^+ (K_v) channels are a diverse superfamily of K^+ channels that share the common properties of a potassium-selective pore and voltage-dependent activation. The α-subunits are composed of six hydrophobic transmembrane segments (S1–S6) flanked by hydrophilic amino- and carboxyl termini located in the cell interior (Fig. 35.4A). An intrinsic voltage sensor conferred by positively-charged amino acids in transmembrane domain S4 confers channel gating, while the pore loop (P) confers K^+ selectivity. Four α-subunits coassemble in the endoplasmic reticulum to form the pore-forming structure of the K_v channel; they further associate with four regulatory β-subunits, which serve as chaperone proteins and modulate the properties of the multiprotein K_v channel (Bocksteins, 2016; Cox & Fromme, 2016; Nystoriak et al., 2017). Once the fully functional K_v channels are translocated to the plasma membrane, intracellular scaffolding proteins may colocalize the K_v channels with cytosolic signaling molecules that regulate their function.

Although K_v channels would be expected to reside primarily in the closed state at the negative voltages associated with resting E_m, ample evidence suggests they contribute importantly to the resting E_m of some SMCs. For example, in cannulated and pressurized rat cerebral arteries (depicted in Fig. 35.4G), pharmacological block by correolide (COR) of "*Shaker-type*" K_v channels depolarizes and contracts the SMCs, ultimately reducing arterial diameter (Fig. 35.4D). Similarly, gene deletion of K_v1 α-subunits in mouse coronary arteries results in vasoconstriction and lower levels of resting blood flow to the heart (Ohanyan et al., 2015). Recent studies reveal that the tonic phosphorylation and opening of *Shaker*-type K_v channels at resting E_m is enabled by their colocalization on intracellular scaffolds with protein kinase A (Moore et al., 2014; Brignell et al., 2015). Additional excitatory stimuli that depolarize and contract SMCs further activate voltage-dependent K_v channels. The resulting hyperpolarizing K^+ current limits Ca^{2+} influx through voltage-dependent Ca^{2+} channels to buffer SMC contraction. Thus, K_v channels have a well-recognized role in counteracting excitation in vascular and other SMC types (Albarwani et al., 2003; Bardou et al., 2009; Chen et al., 2006; Currò, 2016; Haick & Byron, 2016; Petkov, 2011; Tykocki et al., 2017). A loss of functional K_v channels in vascular SMCs is associated with diseases of anomalous vascular tone, which include systemic hypertension (Amberg & Santana, 2006; Tobin et al., 2009), pulmonary hypertension (Hayabuchi, 2017), and diabetes (Nieves-Cintron et al., 2015).

4.2 Large-conductance, Ca^{2+}-sensitive K^+ channels

The large-conductance, Ca^{2+}-sensitive K^+ channels, which are referred to as "BK_{Ca}" or "maxi-K" channels due to their large single-channel current amplitudes, are densely expressed in SMCs. The α-subunit shows partial homology with the K_v channel in six (S1−S6) of its seven (S0−S6) transmembrane domains including the S4 voltage sensor (Fig. 35.4B). Although BK_{Ca} channels are only weakly opened by depolarization in the absence of intracellular calcium ($[Ca]_i$), the voltage-sensitivity of the BK_{Ca} channel increases markedly when $[Ca]_i$ rises. The "Ca^{2+}-sensitivity" of the α-subunit is conferred by cytosolic domains S7−S10 and by its β-subunit that interacts with the N-terminal S0 domain (Toro et al., 2014). Despite their low level of opening at the resting E_m of SMCs, the densely expressed BK_{Ca} channels can mediate a significant hyperpolarizing current to sustain resting E_m and arterial relaxation in some vascular beds (Knot et al., 1998; Lohn et al., 2001; Krishnamoorthy et al., 2014; Krishnamoorthy-Natarajan & Koide, 2016). For example, iberiotoxin (IBTX)-induced block of BK_{Ca} channels in cerebral arterial SMCs results in a sustained depolarization and elevation of $[Ca]_i$ mediated by Ca^{2+} influx through L-type Ca^{2+} channels. The functional outcome is vasoconstriction (Fig. 35.4E). In contrast, the BK_{Ca} channel in the skeletal muscle circulation apparently is inactive at resting E_m. It exhibits a low Ca^{2+}-sensitivity that may be conferred by a unique α-subunit isoform, or alternatively, by a scarcity of Ca^{2+}-sensitizing β-subunits (Jackson & Blair, 1998). However, in most vascular beds, it appears that BK_{Ca} channels work in concert with K_v channels during SMC excitation to provide a negative feedback mechanism to buffer excessive depolarization and SMC contraction. Notably, a loss of vascular BK_{Ca} channels has been linked to essential hypertension in humans (Cheng et al., 2016).

4.3 Inwardly rectifying K^+ channels

Inwardly rectifying K^+ (K_{IR}) channels share the unique property of "inward rectification", which means that they are activated at negative membrane potentials and are inactivated during depolarization, in direct contrast to the voltage-dependence of K_v and BK_{Ca} channels. The property of inward rectification is at least partially conferred by voltage-dependent block of the channel pore by Mg^{2+} and polyamines at the cytoplasmic side (Park et al., 2008). A second unique feature of K_{IR} channels is their activation by elevations of extracellular K^+ concentration. Like the K_v and BK_{Ca} channels, the K_{IR} channels are tetrameric structures composed of four α-subunits with $K_{IR}2$ isoforms primarily expressed in SMCs; however, no corresponding β-subunit has been identified (Fig. 35.4C). Since the K_{IR} channels are not voltage-dependent, but instead primarily open at negative voltages, they contribute to the resting E_m of many cell types including SMCs. Thus, inhibition of K_{IR} channels by barium chloride ($BaCl_2$) results in a loss of hyperpolarizing K^+ current, depolarization and Ca^{2+}-dependent vasoconstriction in cerebral arteries (Fig. 35.4F). Another well-recognized role of K_{IR} channels is mediating the increased blood flow of metabolic hyperemia in the cerebral, coronary and skeletal muscle circulations (Burns et al., 2004; Knot et al., 1996; Longden & Nelson, 2015). In this case, activation of K_{IR} channels at least partially relies on the small increases in extracellular K^+ concentrations that accompany neuronal activity, cardiac challenge, and skeletal muscle exercise (Burns et al., 2004; Knot et al., 1996; Longden & Nelson, 2015). Importantly, the K_{IR} channel −mediated hyperpolarization and relaxation of arterial SMCs embedded in metabolically challenged organs can be conducted cell-to-cell to more proximal arteries, thereby amplifying the ability of K_{IR} channels and other vasodilator influences to improve tissue perfusion (Longden & Nelson, 2015). The K_{IR} channels in vascular SMCs appear to be less abundant during cardiovascular diseases, including hypertension (Tajada et al., 2012; Seo et al., 2014), suggesting that compromised metabolic vasodilator responses may exist in the crucial coronary and cerebral vascular beds.

5. Voltage-dependent calcium channels

At least two gene families of voltage-gated Ca^{2+} channels are expressed in the plasma membrane of SMCs (Ghosh et al., 2017). The *long-lasting (L-type)* Ca^{2+} *channels* encoded by the *$Ca_V1.2$* gene are high voltage-activated channels, which can open in response to membrane depolarization within the range of E_m found in SMCs. As inferred by their designation of "long-lasting", the opening of L-type Ca^{2+} channels is sustained during depolarization, resulting in a persistent influx of Ca^{2+} current that provides activator Ca^{2+} to drive excitation-contraction coupling in SMCs. In contrast, *transient (T-type)*

Ca^{2+} channels encoded by the *Ca_V3* gene activate at more negative E_m and then rapidly close to provide a transient "burst" of Ca^{2+} influx during SMC depolarization. The T-type Ca^{2+} channels have been described in subsets of vascular and nonvascular SMCs (Cribbs, 2006; Fry et al., 2006; Ghosh et al., 2017; Kuo et al., 2011). In most types of smooth muscle, T-type Ca^{2+} channels are thought to contribute to spontaneous electrical activity and rhythmic contractions. Conversely, recent evidence suggests that Ca^{2+} influx through a subset of T-type Ca^{2+} channels may open Ca^{2+}-dependent K^+ channels in SMCs to cause relaxation, thereby acting as a contraction brake (Harraz et al., 2014, 2015). The coexpression of the T-type and L-type Ca^{2+} channels also would be predicted to broaden the voltage range over which Ca^{2+} influx occurs in SMCs to allow titration of Ca^{2+}-dependent processes at all levels of cell excitation (Kuo et al., 2011).

As inferred above, the contribution of T-type Ca^{2+} channels to excitation-contraction coupling in SMCs remains unclear. We lack a highly selective pharmacological antagonist for these channels and, additionally, three different isoforms of T-type Ca^{2+} channels ($Ca_v3.1$, $Ca_v3.2$; $Ca_v3.3$) may enable functional diversity. Another difficulty is that human SMCs may express $Ca_v3.3$ channels rather than the $Ca_v3.1$ channels found in rodent SMCs (Harraz et al., 2015), thereby confounding interspecies comparisons of T-type Ca^{2+} channel function. Regardless, studies using divalent cations or mibefradil to inhibit T-type Ca^{2+} channels, or utilizing Ca_v3 isoform-specific knockout mice, suggest that Ca^{2+} influx through $Ca_v3.1$ channels (rodent) and $Ca_v3.3$ channels (human) mediate SMC contraction. Accordingly, $Ca_v3.1$ channels are reported to contribute to the Ca^{2+}-dependent tone of rat skeletal muscle, mesenteric and cerebral arterioles, particularly at low perfusion pressures (Jensen et al., 2017). Similarly, Ca^{2+} influx mediated by $Ca_v3.3$ channels has been implicated in constriction of human cerebral arteries (Harraz et al., 2015). In contrast, Ca^{2+} influx through T-type Ca^{2+} channels composed of the $Ca_v3.2$ isoform has been linked to SMC relaxation (Harraz et al., 2014, 2015). For example, the short bursts of Ca^{2+} influx mediated by $Ca_v3.2$ channels are reported to activate BK_{Ca} channels in SMCs of human cerebral arteries, thereby mediating hyperpolarization and vasodilation (Harraz et al., 2015). Similarly, coronary arteries from $Ca_v3.2$ knockout mice exhibit defective relaxation, implying that $Ca_v3.2$ channels are essential for normal relaxation of coronary SMCs (Chen et al., 2003). However, studies using $Ca_v3.1$ and $Ca_v3.2$ knockout mice report normal blood pressures in these animals, providing no evidence that T-type Ca^{2+} channels regulate vascular tone in vivo (Chiang et al., 2009; Mangoniet al., 2006). In summary, the physiological roles of T-type Ca^{2+} channels in SMCs remain uncertain, and the design of Ca_v3 isoform- and SMC- specific knockout mice coupled to further studies in human SMCs is needed to pinpoint their contribution to SMC function.

In contrast to the T-type Ca^{2+} channels, there is over-whelming evidence that L-type Ca^{2+} channels are the primary pathways for voltage-dependent Ca^{2+} influx in most types of SMCs. These channels are multisubunit complexes composed of a large α_{1C} subunit that forms the channel pore (Ghosh et al., 2017). The central pore-forming α_{1C} subunit confers most functional properties to the channel including voltage sensing and Ca^{2+} permeability; the α_{1C} subunit also is the binding site for clinical calcium channel antagonists (Godfraind, 2017). Smaller regulatory subunits (β, $\alpha_2\delta$) may finely tune pore-protein function and promote its expression at the cell surface (Fig. 35.5A). The L-type Ca^{2+} channels are the critical link between membrane depolarization and contraction (electromechanical coupling) in SMCs (Ghosh et al., 2017). More positive E_m levels cause more frequent Ca^{2+} channel openings, more Ca^{2+} influx, and greater Ca^{2+}-dependent activation of contractile proteins. Furthermore, since L-type Ca^{2+} channels inactivate slowly during sustained depolarization, the Ca^{2+} influx permitted by a small fraction of these channels may be sufficient to mediate tonic contraction of SMCs. Accordingly, direct depolarization of vascular SMCs by application of high extracellular concentrations of potassium chloride (KCl) or activation of excitatory α-adrenergic receptors by phenylephrine (PE), which are stimuli that open L-type Ca^{2+} channels, contract small tibialis arteries of normal, wild-type mice (Fig. 35.5B). However, similar arteries of genetically modified mice, in which the pore-forming α_{1C} subunit of the L-type Ca^{2+} channel is deleted, only weakly contract to the same stimuli (Fig. 35.5C) (Mossmang et al., 2003). Thus, the L-type Ca^{2+} channels permit many tissues composed of SMCs to properly perform their contractile functions that may involve regulating blood flow (vascular SMCs) or air flow (airway SMCs), or enabling peristalsis and digestion (intestinal SMCs). The Ca^{2+}-dependent SMC tone also provides an excitatory template upon which endogenous substances may act to further modify organ function.

6. Transient receptor potential channels

Twenty-eight mammalian transient receptor potential (TRP) channels have been identified to date with many of them detected in SMCs. The TRP channels are classified into six related protein subfamilies on the basis of amino acid sequence homology (Earley & Brayden, 2015; Guibert et al., 2011; Montell, 2005; Yue et al., 2015). Three subfamilies of TRP channels that have drawn the most attention in SMCs include the "classical" or "canonical" (TRPC), vallaniod (TRPV), and melastatin (TRPM) channels. TRP channels are formed by the assembly of four polypeptide subunits as homo- or heterotetramers giving rise to variable channel properties. Each α-subunit has six transmembrane spanning domains (S1-6),

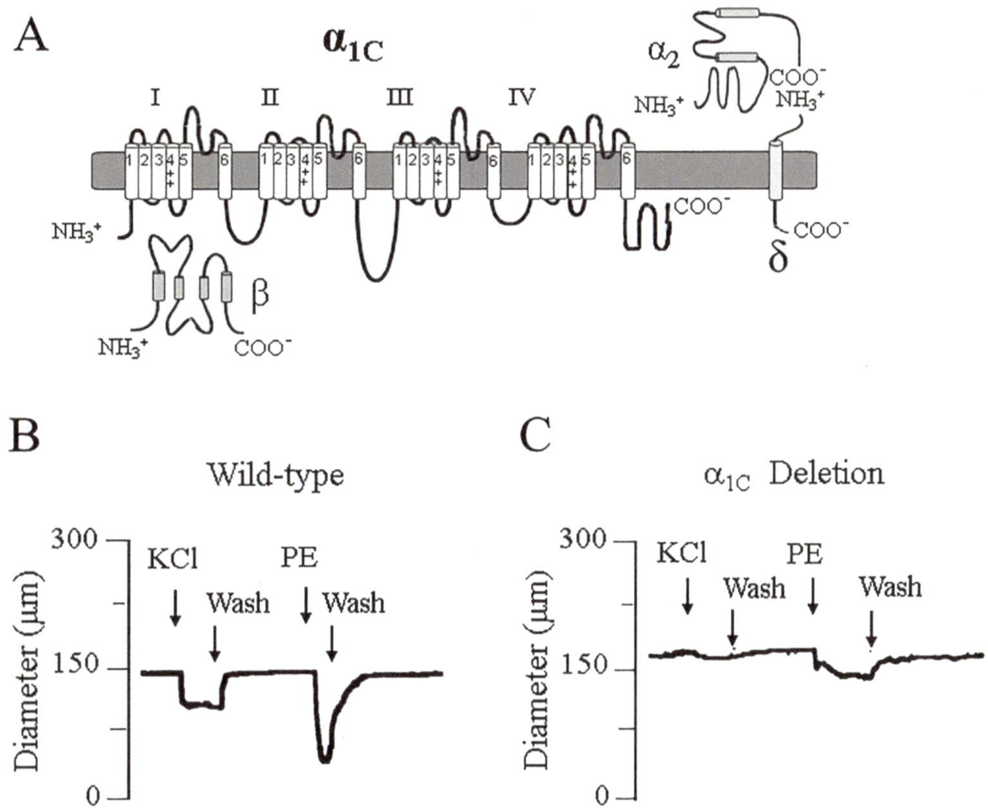

FIGURE 35.5 (A) The α_{1C} subunit of the L-type Ca^{2+} channel is critical for mediating the response to many vasoconstrictor stimuli. Diameter tracings demonstrate that contractions elicited by a depolarizing concentration of KCl or by phenylephrine (PE) are robust in tibialis arteries from wild type mice (B), but greatly attenuated in mice lacking the pore-forming α_{1C} subunit of the L-type Ca^{2+} channel (C). *Reproduced with permission from Moosmang et al. (2003).*

a pore-forming loop between S5 and S6 and intracellularly located NH_2 and COOH termini (Fig. 35.6A). The S4 segment lacks the positively-charged residues necessary for voltage sensing (Gaudet, 2008). Unlike L-type Ca^{2+} channels, the TRP channel pore-protein is not necessarily Ca^{2+}-selective and voltage-sensitive. Rather TRP channels generally are voltage-insensitive and they can selectively or nonselectively conduct a variety of cation species, particularly Na^+ and Ca^{2+} ions. For example, TRPV5 and TRPV6 are highly selective for Ca^{2+}, while TRPM4 and TRPM5 are impermeable to Ca^{2+} and selectively conduct monovalent ions, including Na^+ and K^+. Yet TRPC3 and TRPC6 channels appear to nonselectively conduct both Ca^{2+} and Na^+ (Earley & Brayden, 2015). Based on their mechanism of activation, TRP channels can be grouped as receptor-operated, store-operated or stretch-activated channels with some TRP channels exhibiting more than one property. For example, store- and receptor-operated TRP channels are activated by muscarinic (M2 and M3) receptor stimulation in visceral SMCs and by α_1-adrenergic receptors in vascular SMCs (Dietrich et al., 2006). Other excitatory native stimuli can open TRP channels through G-protein coupled receptors in the plasma membrane. A number of signaling molecules generated by receptor occupation can regulate the activity of TRP channels in SMCs (Albert & Large, 2006; Beech et al., 2004).

Despite initial research, the functional significance of specific TRP channel subtypes in SMCs is unclear. Most TRP channel subtypes lack selective pharmacological blockers, which has led to the use of blocking antibodies and gene deletion as tools to tease out their role in SMC tone. Although a number of knockout mice with specific TRP channel deletions have been generated, the impact of TRP gene deletion on SMC function is not straightforward due to the redundant expression of many TRP subtypes in a single SMC. For example, TRPC6 deletion in mouse arteries results in increased agonist-induced contractile responses instead of the loss of contraction that would be expected after deletion of a Ca^{2+}-conducting channel. This unexpected finding was explained by a compensatory upregulation of TRPC3 channels in the vascular SMCs, which mediated accentuated cation influx and contraction (Tsvilvoskyy et al., 2009). Mechanical stimuli including osmotic stress, membrane stretch and shear force also regulate SMC tone in most organs, and TRP

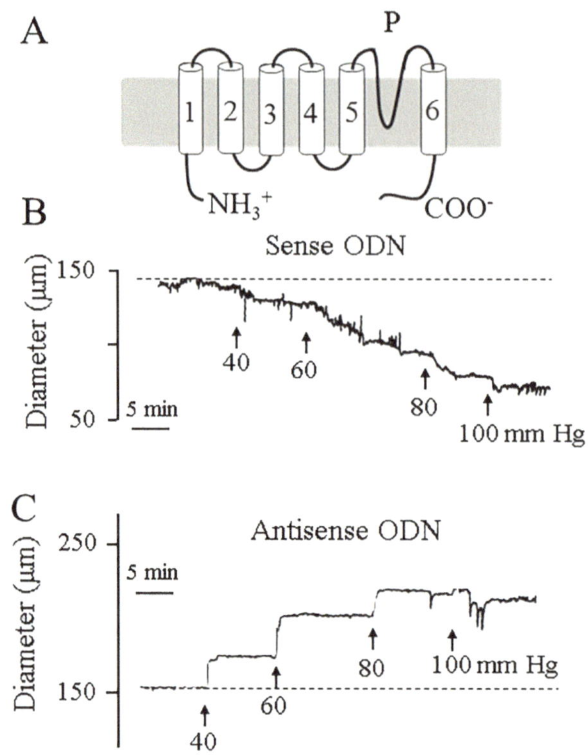

FIGURE 35.6 Structure and function of TRP channels. (A) The α-subunit contains six transmembrane spanning domains (S1 to S6) with the pore (P) formed between S5 and S6. The S4 domain lacks the positive residues that confer voltage sensitivity. (B) In control vessels treated with sense oligonucleotides (ODNs), step elevations of intraluminal pressure in isolated rat cerebral arteries result in pressure-induced constriction. (C) Arteries treated with TRPC6 antisense ODNs fail to show pressure-induced constriction and instead show passive dilation. The horizontal dotted line represents the arterial diameter before intraluminal pressure was increased. *Reproduced with permission from Welsh et al. (2002).*

channel families contain member subtypes that are mechanosensitive. But, again, the role of TRP channels in regulating the contractile responses to mechanical stimuli is controversial (Dietrich & Gudermann, 2011). For example, antisense-mediated knockdown of TRPC6 channels reportedly eliminates pressure-induced vasoconstriction in pressurized rat cerebral arteries (Fig. 35.6B and C), implying the presence of mechanosensitive TRPC6 channels in the vascular SMCs (Welsh et al., 2002). However, other investigators failed to provide evidence for mechanosensitive TRPC6 channels in studies comparing contraction between arteries of wild-type and *Trpc6* knockout mice (Dietrich et al., 2005). A more recent study has described a TRCP6/TRPM4 network dependent on phospholipase activity to illicit pressure-induced vasoconstriction in cerebral arteries, suggesting integrated signaling pathways for TRP channels (Gonzales et al., 2014). Regardless of these disparate findings, the current tenet is that the inward cation current mediated by TRP channels depolarizes SMCs leading to the opening of L-type Ca^{2+} channels, Ca^{2+} influx and contractile protein activation. Thus, one important function of TRP channels in SMCs may be to serve as a triggering mechanism for the activation of L-type Ca^{2+} channels (Gudermann et al., 2004; Dietrich et al., 2006). However, some TRP channels may also play a role in SMC relaxation. Activation of TRPV4 channels in cerebral arterial myocytes generates Ca^{2+} sparklets, activation of BK_{Ca} channels, and SMC hyperpolarization (Earley et al., 2005; Mercado et al., 2014; Yue et al., 2015). Additionally, cerebral arteries of TRPV4 knockout mice show enhanced vasoconstriction to angiotensin II, suggesting that TRPV4 channels may be involved in signaling mechanisms that oppose agonist-induced vasoconstriction (Mercado et al., 2014).

7. Excitability profiles of nonvascular smooth muscle

The SMCs of different tissues exhibit unique excitability profiles that determine their contractile phenotypes. Extremes of the excitability spectrum include the stable resting E_m of some large conduit arteries and airways compared with the spiking action potentials that drive the rhythmic contractions in lymphatic SMCs. A spectrum of excitability phenotypes dictated by distinct functional requirements for SMCs can be found even within the same physiological system. Even the same SMC exposed to different environments can show divergent excitation patterns. Thus, the action potentials that underlie contractions of uterine smooth muscle are altered at different stages of gestation. The goal of the ensuing sections is to briefly acknowledge the diverse excitation patterns of SMCs in the airways, gastrointestinal (GI) tract, uterus, bladder and lymphatic system that ultimately lead to their ability to perform their unique functions in the body. Each section identifies comprehensive reviews for readers seeking more detail.

7.1 Airway smooth muscle

The airway SMCs express a diverse assortment of K^+ channels, voltage-dependent Ca^{2+} channels and TRP channels similar to vascular SMCs. The resting E_m of airway smooth muscle ranges between −70 mV and −30 mV (Hirota et al., 2007). It varies slightly between the tracheal and bronchial SMCs with the former value slightly more negative (−50 mV) than the latter (−40 mV to −50 mV) (Honda & Tomita, 1987; Oonuma et al., 2000). Human airways possess inherent tone which is largely the result of contractile 5-lipoxygenase products (Watson et al., 1997). Although the source of contractile leukotrienes is unclear, recent studies have implicated activation of TRPV4 channels in airway SMCs as a mediator of leukotriene-dependent contractions (McAlexander et al., 2014); however, bronchial epithelium and inflammatory cells associated with the SMCs also are potential candidates. These electrical characteristics and signaling pathways of airway SMCs support the function of the respiratory system. The more positive resting E_m in the bronchial SMCs compared with tracheal SMCs may reflect the fact that the bronchi play a far more important role in determining resistance to airflow, and the more positive resting E_m may lend susceptibility to excitatory stimuli. Notably, the bronchi also are the locus of inflammation and other functional/structural changes associated with asthma and other conditions of airway hyperexcitability.

The similarities of ion channels and excitatory mechanisms between airway smooth muscle and other types of SMCs will not be re-iterated here; instead differences will be emphasized. First, the SMCs of airways are uniquely exposed to a broad spectrum of environmental factors that can regulate airway diameter and they also are exposed to autacoids released by mast cell activation. The excitatory ligands may bind to surface receptors on airway SMCs to initiate Ca^{2+} influx through Ca^{2+}-permeable ion channels and/or release of Ca^{2+} from the sarcoplasmic reticulum (SR) to elevate $[Ca]_i$. In contrast to the central importance of voltage-dependent Ca^{2+} influx in activation of vascular SMCs, the main trigger for bronchoconstriction may be the release of Ca^{2+} sequestered within the SR. Thus, the two major Ca^{2+} release channels in the SR, the inositol-1, 4, 5-trisphosphate receptor (IP_3R) and the ryanodine receptor (RyR), appear to be prominent features in airway SMC contraction. The properties of the IP_3R and RyR channels in airway SMCs are reviewed elsewhere (Bai & Sanderson, 2006; Boie et al., 2017; Croisier et al., 2015; Tazzeo et al., 2008). Second, agonist-activation of SMCs in airway often evokes recurring $[Ca^{2+}]_i$ transients that distinguish airway SMCs from other cell types. This pattern of excitation is characterized by intracellular Ca^{2+} oscillations, which initially rely on mobilization of SR Ca^{2+} stores by activation of IP_3Rs and RyRs. During longer periods of activation when the SR Ca^{2+} stores become depleted, the maintenance of Ca^{2+} oscillations depends on store-operated Ca^{2+} entry. During this phase, Ca^{2+} released from the SR activates Orai, TRPC1, and TRPV1 channels to facilitate Ca^{2+} influx across the plasma membrane and replenish SR Ca^{2+} store (Boie et al., 2017; Croisier et al., 2013; Koopmans et al., 2014; Yocum et al., 2017). Additionally, the frequency of Ca^{2+} oscillations is inversely dependent on agonist concentration. For example, activation of excitatory cholinergic receptors on bronchial SMCs by methacholine (MCh) triggers Ca^{2+} oscillations in the cell cytosol (Fig. 35.7A) (Bai & Sanderson, 2006). These Ca^{2+} oscillations are associated with SMC contraction (Fig. 35.7B) and a reduction of airway area (Fig. 35.7C). The subsequent addition of isoproterenol (ISO), a β-adrenergic receptor agonist with bronchodilator properties, transiently increases the frequency of Ca^{2+} oscillations, an event that is coupled to a temporary loss of contractile force and increased airway diameter.

7.2 Gastrointestinal smooth muscle

The excitation patterns of SMCs in the gastrointestinal (GI) tract are designed to provide the contractions required for mixing ingested food with digestive enzymes, and then propulsion of the resulting gastric contents. These two functions are accomplished by highly coordinated movements of circular and longitudinal smooth muscles, which are aligned perpendicular to each other in the wall of the GI tract. Additionally, different organs within the GI tract are required to perform different functions. For example, the stomach and small intestine enable digestion and absorption, whereas the large intestine enables the drying and compaction of waste. Clearly a complex scheme of excitation is needed to integrate the different functions of the GI tract and electrical mechanisms provide an important source of modulation to regulate organ function. For example, the resting E_m in SMCs varies between −85 mV and −40 mV along the GI tract. The SMCs in the small intestine often are more depolarized (~−55 mV) than those in stomach or colon (~−75 mV) (Sanders,

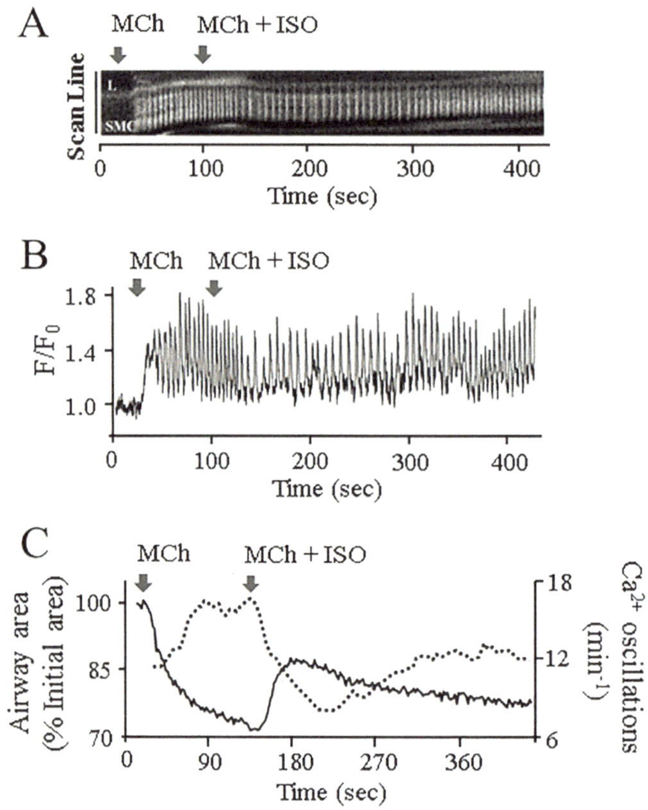

FIGURE 35.7 The effect of methacholine (MCh) and isoproterenol (ISO) on Ca^{2+} signaling of airway SMCs. (A) Line-scan plot constructed from a fluorescent image of an airway SMC loaded with Oregon Green 488 BAPTA-1 AM. Images were obtained by a two-photon microscope and the sequence of recorded images of Ca^{2+} fluorescence in response to MCh (200 nM) and ISO (1 μM) are depicted. (B) Corresponding changes in force generated by the same concentrations of MCh and ISO. (C) The correlation of the Ca^{2+} oscillation frequency (dotted line) and airway area (solid line) in a single cell during contraction induced by MCh (200 nM) and relaxation induced by ISO (10 μM). The frequency of the Ca^{2+} oscillations was inversely coupled to the contraction of the airway. *Reproduced with permission from Bai and Sanderson (2006).*

2008). This electrical heterogeneity is partially achieved by the site-specific expression of diverse populations of ion channels. The main classes of ion channels expressed in GI smooth muscle mirror those in vascular SMCs. Thus, GI smooth muscle expresses multiple gene families of K^+ channels that set the resting E_m and buffer SMC contraction; conversely, voltage-gated Ca^{2+} channels and cationic TRPC channels mediate contraction. The tight regulation of resting E_m is vital because it determines whether the SMCs can respond to depolarizing stimuli originating from the interstitial cells of Cajal (*explained later*) and cell-surface receptors (Lee et al., 2015; Lyford & Farrugia, 2003; Sanders et al., 2006).

Calcium influx through voltage-gated L-type Ca^{2+} channels is indispensable for GI motility. Slow waves depolarize the SMC membrane to open L-type Ca^{2+} channels and, thus, pharmacological block of L-type Ca^{2+} channels attenuates contraction in GI SMCs (Horowitz et al., 1999). Additionally, members of the TRPC family are gaining attention for their role in SMC excitability. As mentioned in the previous section, intestinal SMCs of TRPC4/6 knockout mice show defective muscarinic receptor -mediated cationic current and depolarization (Tsvilovskyy et al., 2009). Although the SMCs ultimately execute force development, a type of interstitial cell in the GI tract called the interstitial cells of Cajal (ICC) plays a vital role in coordinating SMC contraction. The ICC are classified based on morphology: stellate-shaped ICC are found mainly between the muscle layers of the stomach; spindle-shaped ICC largely reside between the smooth muscle fibers of the intestine (Fig. 35.8) (Hirst & Edwards, 2001; Horowitz et al., 1999; Lyford & Farrugia, 2003). They are electrically coupled to each other, the enteric nervous system (ENS) and the SMCs of the GI tract through gap junctions. Networks of ICC generate pacemaker activity and propagate "slow wave" depolarizations across and down the GI tract. Pacemaker activity is sustained by Na^+-K^+-$2Cl^-$ cotransport to maintain a Cl^- gradient and regulated by intracellular Ca^{2+} release. Intracellular Ca^{2+} release in the ICC network also is responsible for slow wave duration and amplitude (Drumm et al., 2017; Sanders et al., 2006, 2014, 2016; Zhu et al., 2015, 2016). GI motility relies on these rhythmic depolarizations that spread through the GI SMCs at a low frequency (<5/min) and activate voltage-dependent Ca^{2+}

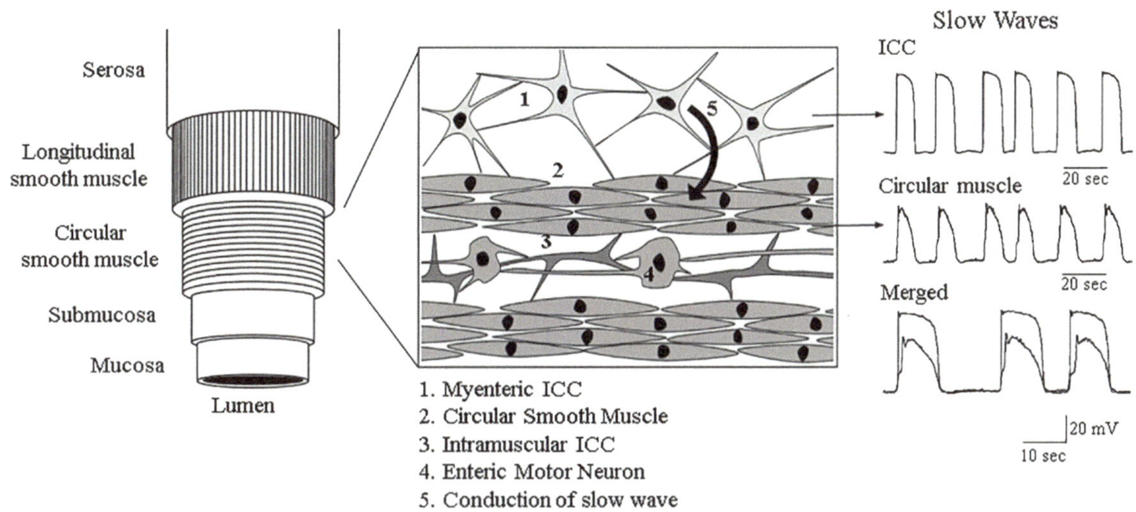

FIGURE 35.8 A cutaway view showing the different layers of the intestinal tract (left). A magnified view of the smooth muscle layers and surrounding layers of ICC (center). Membrane potential recordings of ICC and circular smooth muscle showing the propagation of slow waves from the myenteric ICC to the circular smooth muscle (right). *Reproduced with permission from Hirst and Edwards (2001) and Horowitz et al. (1999).*

channels to cause Ca^{2+}-dependent electrical spiking and SMC contraction (Fig. 35.8) (Sanders, 2008). A key feature of the slow waves is that they propagate faster circumferentially (depolarizing circular SMCs) but slow as they pass down the GI tract (depolarizing longitudinal SMCs) to allow enough time for digestion.

Although the SMCs of the vasculature and GI tract share many ion channels, there are major differences in their electrical profiles. Vascular SMCs generally show a stable resting E_m and a tonic level of contraction under resting conditions, whereas the SMCs of the GI tract exhibit tonic and phasic patterns of contractions that reflect stable or fluctuating E_m levels, respectively. The spontaneous excitability of the SMCs in the GI tract emanates from the unique influence of the ICC, which also receives signals from the enteric nervous system (ENS) and paracrine substances. Similarly, the vascular SMCs also are subject to unique stimuli not found in the GI tract. For example, the intraluminal pressure inside blood vessels can be a powerful force for depolarization and contraction of arterial SMCs, and this stimulus for cell excitation is unique to the vascular system. In addition to voltage-dependent Ca^{2+} influx mediated by L-type Ca^{2+} channels, the influx of cations through TRP channels may contribute to pressure-induced constriction. Other factors that influence vascular SMC excitability not discussed in this short chapter include neurotransmitters, circulating factors and the release of vasoactive substances from the endothelial cells that line the blood vessel lumen.

7.3 Uterine smooth muscle

The uterus is divided into three regions, the endometrium, the myometrium and the serosa (or perimetrium). The innermost layer, the endometrium, is highly permeated by the reproductive blood vessels and is the site of embryo implantation. The outermost layer is the serosa, which is comprised of superficial mesothelium and a thin layer of connective tissue. The myometrium, a type of visceral smooth muscle comprised of multiple layers of smooth muscle, lies in the middle and comprises the majority of the uterus. This layer is composed of blood, lymph vessels, immune cells and importantly for this chapter, bundles of smooth muscle anchored by connective tissue. In humans, three layers comprise the myometrium, the inner layer, the junctional or subendometrial layer, and the outer layer (Aguilar & Mitchell, 2010). In the mouse, the myometrium is comprised of the inner circular muscle and the outer longitudinal smooth muscle layer. The long, spindle-shaped myometrial SMCs (MSMCs) are the major units of uterine contraction.

The resting E_m of MSMCs is dominated by K^+ conductance similar to other SMCs. Also similar to other SMCs, myometrial smooth muscle has a slight resting permeability to Na^+ and this additional contribution from the Na^+ ion makes the resting E_m more positive than the equilibrium potential for K^+ (E_K). The wide variation in E_m observed in uterine SMCs (Table 35.1) likely arises from measurements taken at various stages of the estrous cycle, different stages of pregnancy, hormonal status, and the species (Aguilar & Mitchell, 2010). Uterine smooth muscle is spontaneously active as a result of innate electrical activity within the MSMCs (Fig. 35.9A). Phasic uterine contractions result from slow-wave action potentials reminiscent of those found in the intestine; however, unlike the intestine that contains specialized

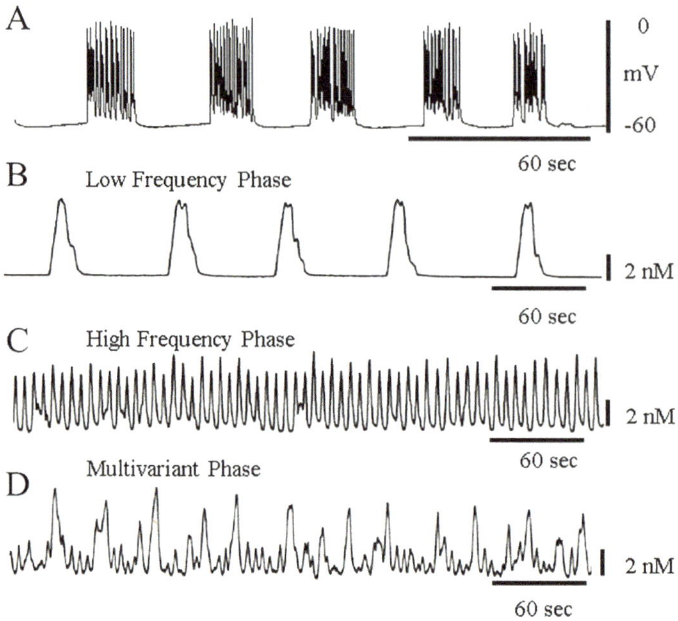

FIGURE 35.9 Types of spontaneous activity exhibited by uterine smooth muscle throughout the estrous cycle. (A) Spontaneous action potentials measured in uterine smooth muscle. Membrane potential and contractile activity vary based on phase of estrous cycle. Depicted here are three different patterns (phases) of uterine contraction, including (B) low frequency, (C) high frequency, and (D) multivariant frequency. *Reproduced with permission from Dodds et al. (2015).*

regions that initiate this activity (i.e., ICC), the initiation of electrical activity can vary and shift between sites in the uterus (Lammers et al., 2008, 2015; Marshall, 1962). The electrical activity may consist of intermittent bursts of spiking action potentials or action potentials that are more plateau-type in nature (Aguilar & Mitchell, 2010; Chamley & Parkington, 1984). The phasic depolarization and repolarization of the plasma membrane of myometrial SMCs, which contract and relax the uterus, respectively, can vary in frequency, amplitude and duration (Fig. 35.9B–D).

In order to elicit an action potential and contraction, the MSMC membrane potential must reach a threshold suitable to activate voltage-sensitive channels. This regulation is conducted, in part, by leak channels in the MSMCs which consist of nonselective, nongated, constitutively active channels that allow the flow of cations across the cell membrane with a reversible potential close to 0 mV, under physiological conditions (Reinl et al., 2015). The MSMC action potentials arise due to voltage and time-dependent changes in membrane permeability to various ions. The number of ion channels, pumps, and exchangers that modulate myometrial excitability is vast and likely underlies the complex electrical signaling required for the multiple functions of the uterus (Chan et al., 2014). The primary stimulus for initiating and maintaining spontaneous contractions is an increase in the level of intracellular level calcium (Ca^{2+}) (Marshall, 1990; Parkington et al., 1999; Wray et al., 2003). This rise occurs through multiple mechanisms (Aguilar & Mitchell, 2010; Szal et al., 1994; Tribe et al., 2000), including: (i) calcium influx through voltage-gated Ca^{2+} channels in response to membrane depolarization; (ii) agonist binding to receptor-operated channels that allow extracellular Ca^{2+} to enter the cell; (iii) agonist [e.g., oxytocin] binding to receptors that induce Ca^{2+} release from intracellular stores, including the sarcoplasmic reticulum, by activating signal transduction pathways; and (iv) store-operated Ca^{2+} entry (SOCE) in response to sarcoplasmic reticulum store depletion. The rise in $[Ca^{2+}]_i$ activates Ca^{2+}-calmodulin, myosin light chain kinase, and the actomyosin machinery, thus leading to MSMC contraction (Aguilar & Mitchell, 2010; Parkington et al., 1999). MSMC relaxation occurs by several mechanisms. First, myosin light chain kinase and myosin light chain phosphatase are inhibited by phosphorylation and dephosphorylation, respectively (Haeberle et al., 1985). Additionally, $[Ca^{2+}]_i$ is reduced by extrusion of Ca^{2+} from the cytosol by plasma membrane Ca^{2+}-ATPases and sequestration of Ca^{2+} into internal stores by sarcoplasmic reticulum Ca^{2+}-ATPases (Tribe et al., 2000). Finally, K^+ efflux through multiple types of K^+ channels that include K_v channels, ATP-gated K^+ channels, and BK_{Ca} channels repolarize the MSMC membrane, thereby closing voltage-dependent Ca^{2+} channels and returning the cell to a resting state and $[Ca^{2+}]_i$ to basal levels (Khan et al., 1997; Pierce & England, 2010).

The contractile apparatus of MSMCs contains thick, thin, and intermediate myofilaments that are arranged randomly. Therefore, MSMCs have slower development of force compared with skeletal and cardiac muscle. This contraction pattern helps serve the diverse processes of the uterus including menses, embryo implantation, quiescence throughout pregnancy, forceful and coordinated contractions during labor, and similar forceful contractions postlabor during delivery of the placenta. As such, a highly complex electrical signal is required for these various functions. The multiple ion channels, pumps, and exchangers in the MSMC plasma membrane, in addition to intercellular gap junctions, underlie the excitable nature of these cells and are necessary for female reproductive tract functions. RNAseq analysis using next-generation sequencing to reveal RNA species in human uterine tissue reveals that the myometrium expresses more than 30 different ion channels; however, functional investigations still continue to dissect their individual roles in modulating myometrial activity (Atia et al., 2016; Chan et al., 2014). The wide variety of ion channels expressed in the myometrium may provide functional redundancy, create a more robust system for contraction, and fine-tune myometrial electrical signaling across regions of the uterus and spanning the reproductive life cycle.

7.4 Bladder smooth muscle

The bladder is a urine-containing reservoir with the SMC-containing detrusor and trigone muscles providing the reservoir wall. As reviewed elsewhere (Andersson & Arner, 2004; Fowler et al., 2008), a sophisticated interplay between musculoskeletal, neurologic, and psychological functions regulates bladder function. However, the detrusor muscle is the main effector for emptying of the bladder, and its activity is coordinated with the urethral sphincter. During storage of urine, the release of norepinephrine from sympathetic nerves maintains relaxation of the bladder SMCs by binding to β-adrenergic receptors (Andersson & Arner, 2004). Conversely, voiding of urine is initiated by corelease of acetylcholine and adenosine triphosphate (ATP) from parasympathetic nerve endings; these neurotransmitters bind to muscarinic and purinergic receptors, respectively, to mediate SMC contraction (Fetscher et al., 2002; Matsui et al., 2000, 2002; Yamanishi et al., 2000). Not surprisingly, nonobstructive chronic urinary retention in humans often relates to compromised contractions of the bladder SMCs. This condition may arise from injury to nerves that innervate the bladder, or direct damage to the bladder that compromises the contractility of the SMCs (Kaplan et al., 2008).

Most knowledge of SMC excitability patterns in the bladder arises from studies performed with isolated strips of detrusor muscle (Brading, 2006; Hashitani & Brading, 2003; Hashitani et al., 2004; Hayase et al., 2009), including preparations from human detrusor (Brading, 2006; Hashitani & Brading, 2003). These studies indicate that the resting E_m in detrusor SMCs from mouse, guinea pig, pig and human bladder ranges between −50 mV and −40 mV. These values are more positive than the values for resting E_m exhibited by many other types of SMCs, but closely resemble the resting E_m level of SMCs in some arterial beds (Table 35.1). The more positive resting E_m in detrusor and vascular SMCs presumably facilitates close coupling between SMC depolarization and opening of voltage-gated L-type Ca^{2+} channels to enable graded Ca^{2+}-dependent tone in the bladder and circulatory system. Additionally, detrusor SMCs from all species exhibit spontaneous action potentials, which may appear as trains of single action potentials or bursting activity superimposed on resting E_m (Fig. 35.10). Although the mechanistic basis of the electrical pacemaker in detrusor SMCs is unclear, the initial depolarizing phase of the action potential activates L-type Ca^{2+} channels. The subsequent Ca^{2+} influx drives the upstroke of the action potential, which may consist of a phasic depolarization of 40−60 mV associated with SMC contraction. However, these contractions called "micromotions" remain localized and do not contribute to bladder emptying due to poor electrical coupling between the SMC bundles of the bladder (Biallosterski et al., 2011; Drake et al., 2017; Hammad et al., 2014; Streng et al., 2006). Instead, micromotions may evoke afferent nerve activity to the CNS during bladder filling to optimize the length-tension relationship of bladder SMCs, thereby "priming" the bladder for later emptying when localized contractions are synchronized by parasympathetic nerve activity (Heppner et al., 2015).

Despite the specialized physiological role of the urinary bladder, there is no evidence for unique populations or properties of ion channels in detrusor SMCs compared with other SMC types. Indeed, studies searching for differences have reported negative results (Malysz et al., 2004). Instead, it appears that a novel and complex environment influences the activity of already-recognized ion channels to account for the level of resting E_m and phasic action potential activity in detrusor SMCs. For example, similar to other types of SMCs, the resting E_m of detrusor SMCs appears to rely on multiple K^+ channel types including voltage-gated K^+ (K_v) channels, Ca^{2+}-activated K^+ channels (K_{Ca}), inward-rectifying K^+ (K_{IR}) channels, and two-pore-domain K^+ (K2P) channels (Petkov, 2011). Similarly, the contribution of L-type Ca^{2+} channels in detrusor SMCs to the upstroke of the action potential is reminiscent of its role in intestinal, uterine, and lymphatic SMCs. The repolarizing phase of the action potential and relaxation of phasic contractions in detrusor SMCs appear to be highly dependent on K^+ efflux through densely-expressed BK_{Ca} channels (Darblade et al., 2006; Petkov, 2014). Thus, pharmacological block of BK_{Ca} channels decreases the frequency and increases the duration of spontaneous action potentials in detrusor SMCs (Buckner et al., 2002; Heppner et al., 1997; Herrera et al., 2000). Similarly, gene deletion of the BK_{Ca} channel pore-forming α−subunit markedly reduces the frequency and accentuates the amplitude of

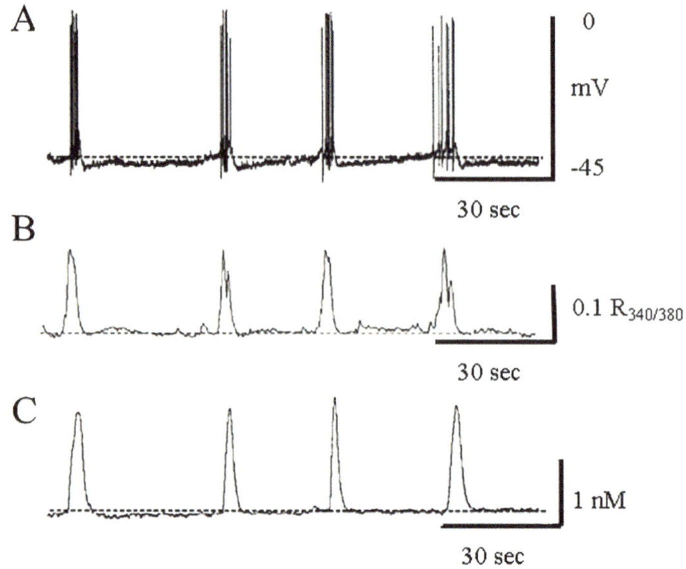

FIGURE 35.10 Correlation between electrical and mechanical activity in the detrusor smooth muscle of the bladder. Spontaneous action potentials can be generated individually or (A) in bursts. Each action potential is accompanied by (B) a transient rise in intracellular calcium followed by a (C) phasic muscle contraction. *Reproduced with permission from Hashitani et al. (2004).*

spontaneous and nerve-evoked contractions of murine bladder, abnormalities associated with high bladder pressure and incontinence (Petkov, 2014; Thorneloe et al., 2005; Werner et al., 2007). Dysfunction of the BK_{Ca} channel also is implicated in detrusor overactivity and overactive bladder in humans (Chang et al., 2010; Hristov et al., 2013; Oger et al., 2010), implying that BK_{Ca} channel openers may represent potential therapeutics to alleviate these conditions (Petkov, 2014; Soder & Petkov, 2011). Another strategy explored earlier for restoring BK_{Ca} channel function to hyper-excitable detrusor SMCs was injection of cDNA encoding the BK_{Ca} channel α-subunit in a rat model of partial bladder obstruction, which alleviated bladder hyperactivity (Christ et al., 2001). Subsequent safety outcomes from an initial Phase 1 clinical trial of BK_{Ca} gene transfer were promising in suggesting lack of toxicity (Melman et al., 2007). Thus, targeting ion channels in detrusor SMCs using gene transfer or pharmacological approaches may lead to improved medications for bladder dysfunction (Christ, 2011; Petkov, 2014).

7.5 Lymphatic muscle

The lymphatic vasculature, often referred to as “the third circulation”, is found throughout the entire body, including in the brain, heart, lungs, kidney and mesentery (El-Chemaly et al., 2008; Fanous et al., 2007; Ishikawa et al., 2006; Klotz et al., 2015; Louveau et al., 2015; McIntosh & Morris, 1971; Park & Jeong, 2013; Ratajska et al., 2014; Schmid-Schönbein, 1990; Takahashi et al., 1997). The primary “mechanical” job of the lymphatic vasculature is to propel fluid (containing protein, lymphocytes, lipids, and other cells) from the extracellular space to the central venous ducts to prevent excess fluid accumulation in tissues (Aukland & Reed, 1993; Choi et al., 2012; Gashev, 2002; Hansen et al., 2015; Nipper & Dixon, 2011; Schmid-Schönbein, 1990; Zawieja, 2009). The lymphatic vasculature begins with blind-ended lymphatic capillaries intertwined with the vascular capillary beds, which then drain into collecting lymphatic vessels composed of a single layer of endothelial cells surrounded by a single layer of specialized lymphatic muscle cells (LMCs) (Gashev, 2002; Ji, 2008; Muthuchamy et al., 2003; van Helden, 2014). Collecting lymphatics also contain bicuspid valves to prevent back-flow of lymph fluid and the valves also serve to separate lymphatic vessels into functional units termed lymphangions (Fig. 35.11A) (Davis et al., 2011; van Helden, 2014). The transport of lymph fluid relies on contributions from extrinsic and intrinsic forces. Extrinsic forces or mechanisms of passive lymph transport shared by all vascular beds include formation of new lymph fluid, arterial pulsations, and fluctuations in central venous pressure. Passive lymph transport also is driven by site-specific forces. For example, transport through lymph vessels (LVs) buried in skeletal muscle can be facilitated by muscle contraction, whereas respiration can drive lymph transport through lung lymphatics (Bridenbaugh et al., 2003; Negrini et al., 1994; Schmid-Schönbein, 1990).

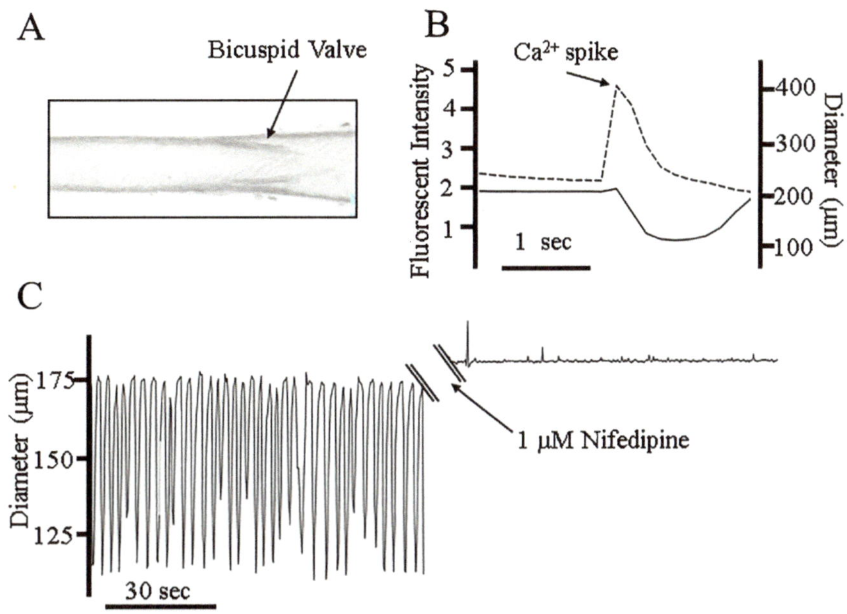

FIGURE 35.11 Calcium-dependent contractions exhibited by an isolated and cannulated rat mesenteric lymph vessel. (A) Image of lymph vessel with bicuspid valve. (B) The change in $[Ca^{2+}{}_i]$ in the mesenteric lymph vessel was measured by Fluo-4 fluorescence; peaks (Ca^{2+} spikes) precede a subsequent lymph vessel contraction. (C) Spontaneous, rhythmic contractions were measured in the lymph vessel using edge-detection software. Then the vessel was superfused with 1 μM nifedipine, a pharmacological antagonist of the L-type Ca^{2+} channel, which abolished spontaneous contractions. *Courtesy of Stolarz, S. J., & Rusch, N. J. University of Arkansas for Medical Sciences.*

Although extrinsic forces play a role in lymph transport in vivo, the majority of studies regarding lymphatic function have focused on mechanisms of active lymph transport mediated by the intrinsic lymphatic pump. This intrinsic lymphatic pump is fueled by the spontaneous contractions of LVs thought to originate in pacemaker muscle cells, which are believed to compose 1%–5% of the muscle cell population (Imtiaz et al., 2007; Toland et al., 2000; van Helden, 1993). The remaining LMCs have been described as a specialized hybrid of smooth muscle and striated muscle as evidenced by functional and molecular studies (Breslin, 2014; Dougherty et al., 2008; Muthuchamy et al., 2003; Scallan et al., 2016).

The resting E_m of LMCs in guinea pig approximates −55 mV with spontaneous depolarizations of ~30–40 mV (Imtiaz et al., 2007; van Helden, 1993). Resting E_m is primarily determined by K_v channels (Allen & McHale, 1988; Telinius et al., 2014) and Ca^{2+}-activated Cl^- (Cl_{Ca}) channels (Imtiaz et al., 2007; van Helden, 1993). The cellular events that underlie the lymphatic pacemaker and spontaneous membrane depolarizations of LMCs are poorly defined, but it's proposed that endothelin-1 (ET-1) binding to endothelin receptors on LMCs leads to inositol trisphosphate (IP_3) production, Ca^{2+} release from IP_3 receptors, and opening of Cl_{Ca} channels on the plasma membrane. The resulting Cl^- efflux depolarizes the plasma membrane, thereby enabling Ca^{2+} influx through L-type Ca^{2+} channels to initiate contraction (Beckett et al., 2007; Ferrusi et al., 2004; Imtiaz et al., 2007; Toland et al., 2000; van Helden, 1993; Zhao & van Helden, 2003). Although other ion channels similar to those that control pacemaking in the sino-atrial node of the heart also have been identified in LMCs, including Na^+ channels (Hollywood et al., 1997; Telinius et al., 2015), T-type Ca^{2+} channels (Hollywood et al., 1997; Lee et al., 2014), hyperpolarization-activated cyclic nucleotide-gated (HCN) channels (McCloskey et al., 1999), and *ether-à-go-go* related gene (ERG) channels (Gui et al., 2014), the roles of these channels in lymphatic pacemaking are unclear.

Regardless of the cellular identity of the pacemaking current in LMCs, it is well documented that spontaneous depolarization of LMCs opens L-type Ca^{2+} channels in the plasma membrane, resulting in Ca^{2+} influx and activation of the contractile proteins to support the rhythmic contractions of lymph vessels (Choi et al., 2012; Gashev, 2002; Nipper & Dixon, 2011; Schmid-Schönbein, 1990; Zawieja, 2009). Thus, the pharmacological antagonists of the L-type Ca^{2+} channel, including nifedipine, inhibit lymphatic contractions (Fig. 35.11C) (Lee et al., 2014; Telinius et al., 2014). The transient rise in $[Ca^{2+}]_i$ or "Ca^{2+} spike" (Fig. 35.11B) in the LMCs is mediated by activation of L-type Ca^{2+} channels that also may mobilize intracellular Ca^{2+} stores through activation of IP_3 receptors and RyRs on the SR, where IP_3 receptors appear to play a more predominant role (Atchison et al., 1998; Atchison & Johnston, 1997; Imtiaz et al., 2007; Zhao & van Helden, 2003). In addition, the activation of Kv channels (Telinius et al., 2014), BK_{Ca} channels (Cotton et al., 1997) and

ATP-sensitive K^+ (K_{ATP}) channels (Mizuno et al., 1999; Mathias & von der Weid, 2013) can hyperpolarize the LMC plasma membrane and inhibit spontaneous contractions.

The composition of contractile proteins in LMCs also indicates that these specialized muscle cells are a hybrid of skeletal, cardiac, and smooth muscle. Contractions of rat LMCs display a distinctly lower Ca^{2+} sensitivity compared with mesenteric arteries or veins, which is associated with reduced expression of CPI-17 (Koyama et al., 2000), a SMC-specific myosin light chain phosphatase inhibitory protein (Davis et al., 2011). However, LMCs also express β-MHC, which is a striated muscle-specific myosin heavy chain. In addition, four actin genes associated with striated and smooth muscle (vascular α-actin, cardiac α-actin, skeletal α-actin, and enteric γ-actin) are expressed in rat mesenteric lymphatic vessels, whereas only vascular α-actin is expressed in arterioles (Muthuchamy et al., 2003). These studies indicate that the genotypical and phenotypical profiles exhibited by lymphatic muscle result from a unique combination of smooth and striated muscle contractile components. Therefore, mechanisms of Ca^{2+} signaling and regulation in LMCs may reflect a combination of smooth muscle and striated muscle regulatory processes.

8. Concluding remarks

This chapter provides a succinct overview of the basis for SMC excitability with an emphasis on vascular smooth muscle, but also includes additional information on other smooth muscle types, including airway, gastrointestinal, uterine, urinary and lymphatic. The latter sections serve to emphasize the heterogeneous electrical properties of SMCs from different organ systems, which have evolved to ensure that all smooth muscle –containing systems of the body can attain normal physiological function. Additionally, we have summarized some of the key ion channels in SMCs, and due to brevity, referenced detailed reviews for missing information. These ion channels constitute a large and highly diverse population of proteins that act in concert to finely tune the levels of contraction and relaxation in the SMCs of the vasculature and other SMC –containing physiological systems. Scrutiny of different types of SMCs has revealed organ and site -specific patterns of cell excitability that are finely tuned to provide unique and localized electrical responses that collectively enable the entire organism to flourish.

Acknowledgments

The authors were supported by a PhRMA Predoctoral Fellowship (A.J.S.), NIH K12 GM088021 (N.D.D), an American Physiological Society Porter Physiology Development Fellowship (C.A.), NIH R01 HD037831 (S.K.E.), NIH R01 HD088097 (S.K.E.), March of Dimes Grant 21-FY15-147 (S.K.E.), NIH R21 1CA131164 (N. J. R.), NIH R37 CA282349-01 (A.J.S.) and the UAMS Sturgis Charitable Trust (N.J.R).

References

Aguilar, H. N., & Mitchell, B. F. (2010). Physiological pathways and molecular mechanisms regulating uterine contractility. *Human Reproduction Update, 16*, 725–744.

Albarwani, S., Nemetz, L. T., Madden, J. A., Tobin, A. A., England, S. K., Pratt, P. F., & Rusch, N. J. (2003). Voltage-gated K^+ channels in small rat cerebral arteries: Molecular identity of the functional channels. *The Journal of Physiology, 551*, 751–763.

Albert, A. P., & Large, W. A. (2006). Signal transduction pathways and gating mechanisms of native TRP-like cation channels in vascular myocytes. *The Journal of Physiology, 570*, 45–51.

Allen, J. M., & McHale, N. G. (1988). The effect of known K^+-channel blockers on the electrical activity of bovine lymphatic smooth muscle. *Pflügers Archiv, 411*, 167–172.

Amberg, G. C., & Santana, L. F. (2006). Kv2 channels oppose myogenic constriction of rat cerebral arteries. *American Journal of Physiology, 291*, C348–C356.

Anderson, G. F., Kawarabayashi, T., & Marshall, J. M. (1981). Effect of indomethacin and aspirin on uterine activity in pregnant rats: Comparison of circular and longitudinal muscle. *Biology of Reproduction, 24*, 359–372.

Andersson, K. E., & Arner, A. (2004). Urinary bladder contraction and relaxation: Physiology and pathophysiology. *Physiological Reviews, 84*, 935–986.

Atchison, D. J., & Johnston, M. G. (1997). Role of extra- and intracellular Ca^{2+} in the lymphatic myogenic response. *American Journal of Physiology, 272*, R326–R333.

Atchison, D. J., Rodela, H., & Johnston, M. G. (1998). Intracellular calcium stores modulation in lymph vessels depends on wall stretch. *Canadian Journal of Physiology and Pharmacology, 76*, 367–372.

Atia, J., McCloskey, C., Shmygol, A. S., Rand, D. A., van den Berg, H. A., & Blanks, A. M. (2016). Reconstruction of cell surface densities of ion pumps, exchangers, and channels from mRNA expression, conductance kinetics, whole-cell calcium, and current-clamp voltage recordings, with an application to human uterine smooth muscle cells. *PLoS Computational Biology, 12*, Article e1004828.

Aukland, K., & Reed, R. (1993). Interstitial-lymphatic mechanisms in the control of extracellular fluid volume. *American Journal of Physiology, 73*, 1–78.

Bai, Y., & Sanderson, M. J. (2006). Airway smooth muscle relaxation results from a reduction in the frequency of Ca^{2+} oscillations induced by a cAMP-mediated inhibition of the IP_3 receptor. *Respiratory Research, 7*, 34–53.

Bardou, O., Trinh, N. T., & Brochiero, E. (2009). Molecular diversity and function of K^+ channels in airway and alveolar epithelial cells. *American Journal of Physiology, 296*, L145–L155.

Beckett, E. A. H., Hollywood, M. A., Thornbury, K. D., & McHale, N. G. (2007). Spontaneous electrical activity in sheep mesenteric lymphatics. *Lymphatic Research and Biology, 5*, 29–43.

Beech, D. J., Muraki, K., & Flemming, R. (2004). Non-selective cationic channels of smooth muscle and the mammalian homologues of drosophila TRP. *The Journal of Physiology, 1559*, 685–706.

Berczi, V., Stekiel, W. J., Contney, S. J., & Rusch, N. J. (1992). Pressure-induced activation of membrane K^+ current in rat saphenous artery. *Hypertension, 19*, 725–729.

Biallosterski, B. T., van Koeveringe, G. A., van Kerrebroeck, P. E., Gillespie, J. I., & de Wachter, S. G. (2011). Nonvoiding activity of the Guinea pig bladder. *The Journal of Urology, 186*, 721–727.

Blaustein, M. P., Chen, L., Hamlyn, J. M., Leenen, F. H. H., Lingrel, J. B., Wier, W. G., & Zhang, J. (2016). Pivotal role of $\alpha 2$ Na^+ pumps and their high affinity ouabain binding site in cardiovascular health and disease. *The Physiologist, 594*, 6079–6103.

Bocksteins, E. (2016). Kv5, Kv6, Kv8, and Kv9 subunits: No simple silent bystanders. *The Journal of General Physiology, 147*, 105–125.

Boedtkjer, E., Matchkov, V. V., Boedtkjer, D. M., & Aalkjaer, C. (2016). Negative news: Cl^- and $HCO3^-$ in the vascular wall. *The Physiologist, 31*, 370–383.

Boie, S., Chen, J., Sanderson, M. J., & Sneyd, J. (2017). The relative contributions of store-operated and voltage-gated Ca^{2+} channels to the control of Ca^{2+} oscillations in airway smooth muscle. *The Journal of Physiology, 595*, 3129–3141.

Brading, A. F. (2006). Spontaneous activity of lower urinary tract smooth muscles: Correlation between ion channels and tissue function. *The Journal of Physiology, 570*(1), 13–22.

Bramich, N. J. (2000). Electrical behavior of Guinea pig tracheal smooth muscle. *American Journal of Physiology, 278*, L320–L328.

Brayden, J. E., & Nelson, M. T. (1992). Regulation of arterial tone by activation of calcium-dependent potassium channels. *Science, 256*(5056), 532–535.

Breslin, J. W. (2014). Mechanical forces and lymphatic transport. *Microvascular Research, 96*, 46–54.

Bridenbaugh, E. A., Gashev, A. A., & Zawieja, D. C. (2003). Lymphatic muscle: A review of contractile function. *Lymphatic Research and Biology, 1*, 147–158.

Brignell, J. L., Perry, M. D., Nelson, C. P., Willets, J. M., Challiss, R. A. J., & Davies, N. W. (2015). Steady-state modulation of voltage-gated K^+ channels in rat arterial smooth muscle by cyclic AMP-dependent protein kinase and protein phosphatase 2B. *PLoS One, 10*(3), Article e0121285.

Buckner, S. A., Milicic, I., Daza, A. V., Coghlan, M. J., & Gopalakrishnan, M. (2002). Spontaneous phasic activity of the pig urinary bladder smooth muscle: Characteristics and sensitivity to potassium channel modulators. *British Journal of Pharmacology, 135*, 639–648.

Burns, W. R., Cohen, K. D., & Jackson, W. F. (2004). K^+-induced dilation of hamster cremasteric arterioles involves both the Na^+/K^+-ATPase and inward-rectifier K^+ channels. *Microcirculation, 11*, 279–293.

Chamley, W. A., & Parkington, H. C. (1984). Relaxin inhibits the plateau component of the action potential in the circular myometrium of the rat. *The Journal of Physiology, 353*, 51–65.

Chan, Y. W., van den Berg, H. A., Moore, J. D., Quenby, S., & Blanks, A. M. (2014). Assessment of myometrial transcriptome changes associated with spontaneous human labour by high-throughput RNA-seq. *Experimental Physiology, 99*, 510–524.

Chang, S., Gomes, C. M., Hypolite, J. A., Marx, J., Alanzi, J., Zderic, S. A., Malkowicz, B., Wein, A. J., & Chacko, S. (2010). Detrusor overactivity is associated with downregulation of large-conductance calcium- and voltage-activated potassium channel protein. *American Journal of Physiology, 298*, F1416–F1423.

Checchetto, V., Teardo, E., Carraretto, L., Leanza, L., & Szabo, I. (2016). Physiology of intracellular potassium channels: A unifying role as mediators of counterion fluxes? *Biochimica et Biophysica Acta, 1857*(8), 1258–1266. https://doi.org/10.1016/j.bbabio.2016.03.011

Chen, C. C., Lamping, K. G., Nuno, D. W., Barresi, R., Prouty, S. J., Lavoie, J. L., Cribbs, L. L., England, S. K., Sigmund, C. D., Weiss, R. M., Williamson, R. A., Hill, J. A., & Campbell, K. P. (2003). Abnormal coronary function in mice deficient in α1HT-type Ca^{2+} channels. *Science, 302*, 1416–1418.

Chen, T. T., Luykenaar, K. D., Walsh, E. J., Walsh, M. P., & Cole, W. C. (2006). Key role of Kv1 channels in vasoregulation. *Circulation Research, 99*, 53–60.

Cheng, J., Mao, L., Wen, J., Li, P.-Y., Wang, N., Tan, X.-Q., Zhang, X.-D., Zeng, X.-R., Xu, L., Xia, X.-M., Xia, D., He, K., Su, S., Yao, H., & Yang, Y. (2016). Different effects of hypertension and age on the function of large conductance calcium- and voltage-activated potassium channels in human mesentery artery smooth muscle cells. *Journal of American Heart Association, 5*, Article e003913.

Chiang, C. S., Huang, C. H., Chieng, H., Chang, Y. T., Chang, D., Chen, J. J., Chen, Y. C., Chen, Y. H., Shin, H. S., Campbell, K. P., & Chen, C. C. (2009). The $Ca_v3.2$ T-type Ca^{2+} channel is required for pressure overload-induced cardiac hypertrophy in mice. *Circulation Research, 104*, 522–530.

Choi, I., Lee, S., & Hong, Y. (2012). The new era of the lymphatic system : No longer secondary to the blood vascular system. *Cold Spring Harbor Perspectives in Medicine, 2*, 1–24.

Christ, G. J., Day, N. S., Day, M., Santizo, C., Zhao, W., Sclafani, T., Zinman, J., Hsieh, K., Venkateswarlu, K., Valcic, M., & Melman, A. (2001). Bladder injection of "naked" *hSlo*/pcDNA3 ameliorates detrusor hyperactivity in obstructed rats in vivo. *American Journal of Physiology, 281*, R1699–R1709.

Christ, G. J. (2011). Potential applications of gene therapy/transfer to the treatment of lower urinary tract diseases/disorders. *Handbook of Experimental Pharmacology, 202*, 255–265.

Coleman, H. A., Tare, M., & Parkington, H. C. (2001). K^+ currents underlying the action of endothelium-derived hyperpolarizing factor in Guinea-pig, rat and human blood vessels. *The Journal of Physiology, 531*(2), 359–373.

Cotton, K. D., Hollywood, M. A., McHale, N. G., & Thornbury, K. D. (1997). Outward currents in smooth muscle cells isolated from sheep mesenteric lymphatics. *Journal of Physiology, 503*, 1–11.

Cox, R. H., & Fromme, S. (2016). Comparison of voltage gated K^+ currents in arterial myocytes with heterologously expressed Kv subunits. *Cell Biochemistry and Biophysics, 74*, 499–511.

Cribbs, L. L. (2006). T-Type Ca^{2+} channels in vascular smooth muscle: Multiple functions. *Cell Calcium, 40*, 221–230.

Croisier, H., Tan, X., Chen, J., Sneyd, J., Sanderson, M. J., & Brook, B. S. (2015). Ryanodine receptor sensitization results in abnormal calcium signaling in airway smooth muscle cells. *American Journal of Respiratory Cell and Molecular Biology, 53*, 703–711.

Croisier, H., Tan, X., Perez-Zoghbi, J. F., Sanderson, M. J., Sneyd, J., & Brook, B. S. (2013). Activation of store-operated calcium entry in airway smooth muscle cells: Insight from a mathematical model. *PLoS One, 8*, Article e69598.

Currò, D. (2016). The modulation of potassium channels in the smooth muscle as a therapeutic strategy for disorders of the gastrointestinal tract. *Advances in Protein Chemistry and Structural Biology, 104*, 263–305.

Darblade, B., Behr-Roussel, D., Oger, S., Hieble, J.-P., Lebret, T., Gorny, D., Benoit, G., Alexandre, L., & Giuliano, F. (2006). Effects of potassium channel modulators on human detrusor smooth muscle myogenic phasic contractile activity: Potential therapeutic targets for overactive bladder. *Urology, 68*, 442–448.

Davis, M. J., Rahbar, E., Gashev, A. A., Zawieja, D. C., & Moore, J. E. (2011). Determinants of valve gating in collecting lymphatic vessels from rat mesentery. *American Journal of Physiology, 301*, H48–H60.

Dietrich, A., Chubanov, V., Kalwa, H., Rost, B. R., & Gudermann, T. (2006). Cation channels of the transient receptor potential superfamily: Their role in physiological and pathophysiological processes of smooth muscle cells. *Pharmacology & Therapeutics, 112*, 744–760.

Dietrich, A., & Gudermann, T. (2011). TRP channels in the cardiopulmonary vasculature. *Advances in Experimental Medicine & Biology, 704*, 781–810.

Dietrich, A., Mederos y Schnitzler, M., Gollasch, M., Gross, V., Storch, U., Dubrovska, G., Obst, M., Yildirim, E., Salanova, B., Kalwa, H., Essin, K., Pinkenburg, O., Luft, F. C., Gudermann, T., & Birnbaumer, L. (2005). Increased vascular smooth muscle contractility in $TRPC6^{-/-}$ mice. *Molecular and Cellular Biology, 25*, 6980–6989.

Dodds, K. N., Staikopoulos, V., & Beckett, E. A. H. (2015). Uterine contractility in the nonpregnant mouse: Changes during the estrous cycle and effects of chloride channel blockade. *Biology of Reproduction, 92*, 141–156.

Dougherty, P. J., Davis, M. J., Zawieja, D. C., & Muthuchamy, M. (2008). Calcium sensitivity and cooperativity of permeabilized rat mesenteric lymphatics. *American Journal of Physiology, 294*, R1524–R1532.

Drake, M. J., Kanai, A., Bijos, D. A., Ikeda, Y., Zabbarova, I., Vahabi, B., & Fry, C. H. (2017). The potential role of unregulated autonomous bladder micromotions in urinary storage and voiding dysfunction; overactive bladder and detrusor underactivity. *BJU International, 119*, 22–29.

Drumm, B. T., Hennig, G. W., Battersby, M. J., Cunningham, E. K., Sung, T. S., Ward, S. M., Sanders, K. M., & Baker, S. A. (2017). Clustering of Ca^{2+} transients in interstitial cells of Cajal defines slow wave duration. *The Journal of General Physiology, 149*, 703–725.

Earley, S., & Brayden, J. E. (2015). Transient receptor potential channels in the vasculature. *Physiological Reviews, 95*, 645–690.

Earley, S., Heppner, T. J., Nelson, M. T., & Brayden, J. E. (2005). TRPV4 forms a novel Ca^{2+} signaling complex with ryanodine receptors and BK_{Ca} channels. *Circulation Research, 97*, 1270–1279.

Edwards, G., Dora, K. A., Gardener, M. J., Garland, C. J., & Weston, A. H. (1998). K^+ is an endothelium-derived hyperpolarizing factor in rat arteries. *Nature, 396*, 269–272.

El-Chemaly, S., Levine, S. J., & Moss, J. (2008). Lymphatics in lung disease. *Annals of the New York Academy of Sciences, 1131*, 95–202.

Fanous, M. Y. Z., Phillips, A. J., & Windsor, J. A. (2007). Mesenteric lymph: The bridge to future management of critical illness. *Journal of the Pancreas, 8*, 374–399.

Ferrusi, I., Zhao, J., van Helden, D. F., & von der Weid, P. Y. (2004). Cyclopiazonic acid decreases spontaneous transient depolarizations in Guinea pig mesenteric lymphatic vessels in endothelium-dependent and -independent manners. *American Journal of Physiology, 286*, H2287–H2295.

Fetscher, C., Fleichman, M., Schmidt, M., Krege, S., & Michel, M. C. (2002). M_3 muscarinic receptors mediate contraction of human urinary bladder. *British Journal of Pharmacology, 136*, 641–643.

Figuero, X. F., & Duling, B. R. (2009). Gap junctions in the control of vascular function. *Antioxidants & Redox Signaling, 11*, 251–256.

Forrest, A. S., Hennig, G. W., Jokela-Willis, S., Park, C. D., & Sanders, K. M. (2009). Prostaglandin regulation of gastric slow waves and peristalsis. *American Journal of Physiology, 296*(6), G1180–G1190.

Foster, M. N., & Coetzee, W. A. (2016). K_{ATP} channels in the cardiovascular system. *Physiological Reviews, 96*, 177–252.

Fowler, C. J., Griffiths, D., & de Groat, W. C. (2008). The neural control of micturition. *Nature Reviews Neuroscience, 9*, 453–466.

Fox, J. L. R., & von der Weid, P.-Y. (2002). Effects of histamine on the contractile and electrical activity in isolated lymph vessels of the Guinea-pig mesentery. *British Journal of Pharmacology, 136*, 1210–1218.

Fry, C. H., Sui, G., & Wu, C. (2006). T-type Ca^{2+} channels in non-vascular smooth muscles. *Cell Calcium, 40*, 231–239.

Gashev, A. A. (2002). Physiologic aspects of lymphatic contractile function. *Annals of the New York Academy of Sciences, 979*, 178–187.

Gaudet, R. (2008). TRP channels entering the structural era. *The Journal of Physiology, 586*, 3565–3575.

Ghosh, D., Syed, A. U., Prada, M. P., Nystoriak, M. A., Santana, L. F., Nieves-Cintrón, M., & Navedo, M. F. (2017). Calcium channels in vascular smooth muscle. *Advances in Pharmacology, 78*, 49–87.

Godfraind, T. (2017). Discovery and development of calcium channel blockers. *Frontiers in Pharmacology, 8*, 286.

Gokina, N. I., & Bevan, J. A. (2000). Histamine-induced depolarization: Ionic mechanisms and role in sustained contraction of rabbit cerebral arteries. *American Journal of Physiology, 278*, H2094–H2104.

Gonzales, A. L., Yang, Y., Sullivan, M. N., Sanders, L., Dabertrand, F., Hill-Eubanks, D. C., Nelson, M. T., & Earley, S. (2014). A PLCγ1-dependent, force-sensitive signaling network in the myogenic constriction of cerebral arteries. *Science Signaling, 7*, Article ra49.

Gudermann, T., Mederos y Schnitzler, M., & Dietrich, A. (2004). Receptor-operated cation entry–more than esoteric terminology? *Science Signaling, 243*, Article pe35.

Gui, P., Li, M., Hill, M. A., & Davis, M. J. (2014). KCNQ and ERG channels control the rate of diastolic depolarization and electrical pacemaking frequency in lymphatic muscle. *The FASEB Journal, 28*(Suppl. 1), 666.

Guibert, C., Ducret, T., & Savineau, J. P. (2011). Expression and physiological roles of TRP channels in smooth muscle cells. *Advances in Experimental Medicine & Biology, 704*, 687–706.

Haddock, R. E., & Hill, C. E. (2005). Rhythmicity in arterial smooth muscle. *The Journal of Physiology, 566*, 645–656.

Haeberle, J. R., Hathaway, D. R., & DePaoli-Roach, A. A. (1985). Dephosphorylation of myosin by the catalytic subunit of a type-2 phosphatase produces relaxation of chemically skinned uterine smooth muscle. *Journal of Biological Chemistry, 260*, 9965–9968.

Haick, J. M., & Byron, K. L. (2016). Novel treatment strategies for smooth muscle disorders: Targeting Kv7 potassium channels. *Pharmacology & Therapeutics, 165*, 14–25.

Hammad, F. T., Stephen, B., Lubbad, L., Morrison, J. F. B., & Lammers, W. J. (2014). Macroscopic electrical propagation in the Guinea pig urinary bladder. *American Journal of Physiology, 307*, F172–F182.

Hansen, K. C., D'Alessandro, A., Clement, C. C., & Santambrogio, L. (2015). Lymph formation, composition and circulation: A proteomics perspective. *International Immunology, 27*, 219–227.

Harraz, O. F., Abd El-Rahman, R. R., Bigdely-Shamloo, K., Wilson, S. M., Brett, S. E., Romero, M., Gonzales, A. L., Earley, S., Vigmond, E. J., Nygren, A., Menon, B. K., Mufti, R. E., Watson, T., Starreveld, Y., Furstenhaupt, T., Muellerleile, P. R., Kurjiaka, D. T., Kyle, B. D., Braun, A. P., & Welsh, D. G. (2014). $Ca_V3.2$ channels and the induction of negative feedback in cerebral arteries. *Circulation Research, 115*, 650–661.

Harraz, O. F., Visser, F., Brett, S. E., Goldman, D., Zechariah, A., Hashad, A. M., Menon, B. K., Watson, T., Starreveld, Y., & Welsh, D. G. (2015). $Ca_V1.2/Ca_V3.x$ channels mediate divergent vasomotor responses in human cerebral arteries. *The Journal of General Physiology, 145*, 405–418.

Hashitani, H., & Brading, A. F. (2003). Electrical properties of detrusor smooth muscles from the pig and human urinary bladder. *British Journal of Pharmacology, 140*, 146–158.

Hashitani, H., Brading, A. F., & Suzuki, H. (2004). Correlation between spontaneous electrical, calcium and mechanical activity in detrusor smooth muscle of the Guinea-pig bladder. *British Journal of Pharmacology, 141*, 183–193.

Hashitani, H., Fukuta, H., Takano, H., Klemm, M. F., & Suzuki, H. (2001). Origin and propagation of spontaneous excitation in smooth muscle of the Guinea pig urinary bladder. *The Journal of Physiology, 530*, 273–286.

Hayabuchi, Y. (2017). The action of smooth muscle cell potassium channels in the pathology of pulmonary arterial hypertension. *Pediatric Cardiology, 38*, 1–14.

Hayase, M., Hashitani, H., Kohri, K., & Suzuki, H. (2009). Role of K^+ channels in regulating spontaneous activity in detrusor smooth muscle in situ in the mouse bladder. *The Journal of Urology, 181*, 2355–2365.

Heppner, T. J., Bonev, A. D., & Nelson, M. T. (1997). Ca^{2+}-activated K^+ channels regulate action potential repolarization in urinary bladder smooth muscle. *American Journal of Physiology, 273*, C110–C117.

Heppner, T. J., Tykocki, N. R., Hill-Eubanks, D., & Nelson, M. T. (2015). Transient contractions of urinary bladder smooth muscle are drivers of afferent nerve activity during filling. *The Journal of General Physiology, 147*, 323–335.

Herrera, G. M., Heppner, T. J., & Nelson, M. T. (2000). Regulation of urinary bladder smooth muscle contractions by ryanodine receptors and BK and SK channels. *American Journal of Physiology, 279*, R60–R68.

Hirota, S., Helli, P., & Janssen, L. J. (2007). Ionic mechanisms and Ca^{2+} handling in airway smooth muscle. *European Respiratory Journal, 30*, 114–133.

Hirst, G. D., & Edwards, F. R. (2001). Generation of slow waves in the antral region of Guinea-pig stomach–a stochastic process. *The Journal of Physiology, 535*, 165–180.

Hollywood, M. A., Cotton, K. D., Thornbury, K. D., & McHale, N. G. (1997). Isolated sheep mesenteric lymphatic smooth muscle cells possess both T- and L-type calcium currents. *The Journal of Physiology, 501*, 109P–110P.

Honda, K., & Tomita, T. (1987). Electrical activity in isolated human tracheal muscle. *The Japanese Journal of Physiology, 37*, 333–336.

Horowitz, B., Ward, S. M., & Sanders, K. M. (1999). Cellular and molecular basis for electrical rhythmicity in gastrointestinal muscles. *Annual Review of Physiology, 61*, 19–43.

Hristov, K. L., Afeli, S. A. Y., Parajuli, S. P., Cheng, Q., Rovner, E. S., & Petkov, G. V. (2013). Neurogenic detrusor overactivity is associated with decreased expression and function of the large conductance voltage- and Ca^{2+}-activated K^+ channels. *PLoS One, 8*(7), Article e68052.

Imtiaz, M. S., Zhao, J., Hosaka, K., von der Weid, P. Y., Crowe, M., & van Helden, D. F. (2007). Pacemaking through Ca^{2+} stores interacting as coupled oscillators via membrane depolarization. *Biophysical Journal, 92*, 3843–3861.

Ishikawa, Y., Akasaka, Y., Kiguchi, H., Akishima-Fukasawa, Y., Hasegawa, T., Ito, K., Kimura-Matsumoto, M., Ishiguro, S., Morita, H., Sato, S., Soh, S., & Ishii, T. (2006). The human renal lymphatics under normal and pathological conditions. *Histopathology, 49*, 265–273.

Jensen, L. J., Nielsen, M. S., Salomonsson, M., & Sørensen, C. M. (2017). T-type Ca^{2+} channels and autoregulation of local blood flow. *Channels, 11*, 183–195.

Jackson, W. F., & Blair, K. L. (1998). Characterization and function of Ca^{2+}-activated K^+ channels in arteriolar muscle cells. *American Journal of Physiology, 274*, H27–H34.

Ji, R. C. (2008). Lymphatic endothelial cells, lymphedematous lymphangiogenesis, and molecular control of edema formation. *Lymphatic Research and Biology, 6*, 123–137.

Kamei, K., Yoshida, S., Imagawa, J., Nabata, H., & Kuriyama, H. (1994). Regional and species differences in glyburide-sensitive K^+ channels in airway smooth muscles as estimated from actions of KC 128 and levcromakalim. *British Journal of Pharmacology, 113*, 889–897.

Kaplan, S. A., Wein, A. J., Staskin, D. R., Roehrborn, C. G., & Steers, W. D. (2008). Urinary retention and post-void residual urine in men: Separating truth from tradition. *The Journal of Urology, 180*, 47–54.

Khan, R. N., Smith, S. K., Morrison, J. J., & Ashford, M. L. (1997). Ca^{2+} dependence and pharmacology of large-conductance K^+ channels in nonlabor and labor human uterine myocytes. *American Journal of Physiology, 273*, C1721–C1731.

Klotz, L., Norman, S., Vieira, J. M., Masters, M., Rohling, M., Dubé, K. N., Bollini, S., Matsuzaki, F., Carr, C. A., & Riley, P. R. (2015). Cardiac lymphatics are heterogeneous in origin and respond to injury. *Nature, 522*, 62–67.

Knot, H. J., Standen, N. B., & Nelson, M. T. (1998). Ryanodine receptors regulate arterial diameter and wall [Ca^{2+}] in cerebral arteries of rat via Ca^{2+}-dependent K^+ channels. *The Journal of Physiology, 508*, 211–221.

Knot, H. J., Zimmermann, P. A., & Nelson, M. T. (1996). Extracellular K^+-induced hyperpolarizations and dilatations of rat coronary and cerebral arteries involve inward rectifier K^+ channels. *The Journal of Physiology, 492*, 419–430.

Koopmans, T., Anaparti, V., Castro-Piedras, I., Yarova, P., Irechukwu, N., Nelson, C., Perez-Zoghbi, J., Tan, X., Ward, J. P., & Wright, D. B. (2014). Ca^{2+} handling and sensitivity in airway smooth muscle: Emerging concepts for mechanistic understanding and therapeutic targeting. *Pulmonary Pharmacology & Therapeutics, 29*, 108–120.

Koyama, M., Ito, M., Feng, J., Seko, T., Shiraki, K., Takase, K., Hartshorne, D. J., & Nakano, T. (2000). Phosphorylation of CPI-17, an inhibitory phosphoprotein of smooth muscle myosin phosphatase, by Rho-kinase. *FEBS Letters, 475*, 197–200.

Krishnamoorthy, G., Sonkusare, S. K., Heppner, T. J., & Nelson, M. T. (2014). Opposing roles of smooth muscle BK channels and ryanodine receptors in the regulation of nerve-evoked constriction of mesenteric resistance arteries. *American Journal of Physiology, 306*, H981–H988.

Krishnamoorthy-Natarajan, G., & Koide, M. (2016). BK channels in the vascular system. *International Review of Neurobiology, 128*, 401–438.

Kuo, I. Y.-T., Wolfle, S. E., & Hill, C. E. (2011). T-Type calcium channels and vascular function: The new kid on the block? *The Journal of Physiology, 589*(4), 783–795.

Kuriyama, H., & Suzuki, H. (1976). Change in electrical properties of rat myometrium during gestation and following hormonal treatments. *The Journal of Physiology, 260*, 315–333.

Löhn, M., Lauterbach, B., Haller, H., Pongs, O., Luft, F. C., & Gollasch, M. (2001). Beta1-subunit of BK channels regulates arterial wall [Ca^{2+}] and diameter in mouse cerebral arteries. *Journal of Applied Physiology, 91*, 1350–1354.

Lammers, W. J. E. P. (2015). Normal and abnormal electrical propagation in the small intestine. *Acta Physiologica, 213*, 349–359.

Lammers, W. J., Mirghani, H., Stephen, B., Dhanasekaran, S., Wahab, A., Al Sultan, M. A., & Abazer, F. (2008). Patterns of electrical propagation in the intact pregnant Guinea pig uterus. *American Journal of Physiology, 294*, R919–R928.

Lammers, W. J., Stephen, B., Al-Sultan, M. A., Subramanya, S. B., & Blanks, A. M. (2015). The location of pacemakers in the uteri of pregnant Guinea pigs and rats. *American Journal of Physiology, 309*, R1439–R1446.

Leblanc, N., Forrest, A. S., Ayon, R. J., Wiwchar, M., Angermann, J. E., Pritchard, H. A., Singer, C. A., Valencik, M. L., Britton, F., & Greenwood, I. A. (2015). Molecular and functional significance of Ca^{2+}-activated Cl^- channels in pulmonary arterial smooth muscle. *Pulmonary Circulation, 5*, 244–268.

Lee, H. K., Leikauf, G. D., & Sperelakis, N. (1990). Electromechanical effects of endothelin on ferret bronchial and tracheal smooth muscle. *Journal of Applied Physiology, 68*, 417–420.

Lee, J. Y., Ko, E., Ahn, K. D., Kim, S., & Rhee, P. L. (2015). The role of K^+ conductances in regulating membrane excitability in human gastric corpus smooth muscle. *American Journal of Physiology, 308*, G625–G633.

Lee, S., Roizes, S., & von der Weid, P. Y. (2014). Distinct roles of L- and T-type voltage-dependent Ca^{2+} channels in regulation of lymphatic vessel contractile activity. *The Journal of Physiology, 592*, 5409–5427.

Lindeman, J., Khammy, M. M., Lundegaard, P. R., Aalkjaer, C., & Jepps, T. A. (2018). Microtubule regulation of Kv7 channels orchestrates cAMP-mediated vasorelaxations in rat arterial smooth muscle. *Hypertension, 71*.

Longden, T. A., & Nelson, M. T. (2015). Vascular inward rectifier K^+ channels as external K^+ sensors in the control of cerebral blood flow. *Microcirculation, 22*, 183–196.

Louveau, A., Smirnov, I., Keyes, T. J., Eccles, J. D., Rouhani, S. J., Peske, J. D., Derecki, N. C., Castle, D., Mandell, J. W., Lee, K. S., Harris, T. H., & Kipnis, J. (2015). Structural and functional features of central nervous system lymphatic vessels. *Nature, 523*, 337–341.

Lyford, G. L., & Farrugia, G. (2003). Ion channels in gastrointestinal smooth muscle and interstitial cells of Cajal. *Current Opinion in Pharmacology, 3*, 583–587.

Ma, Y., Zhang, P., Li, J., Lu, J., Ge, J., Zhao, Z., Ma, X., Wan, S., Yao, X., & Shen, B. (2015). Epoxyeicosatrienoic acids act through TRPV4-TRPC1-K_{Ca}1.1 complex to induce smooth muscle membrane hyperpolarization and relaxation in human internal mammary arteries. *Biochimica et Biophysica Acta, 1852*, 552–559.

Malysz, J., Buckner, S. A., Daza, A. V., Milicic, I., Perez-Medrano, A., & Gopalakrishnan, M. (2004). Functional characterization of large conductance calcium-activated K^+ channel openers in bladder and vascular smooth muscle. *Naunyn-Schmiedeberg's Archives of Pharmacology, 369*, 481–489.

Mangoni, M. E., Traboulsie, A., Leoni, A. L., Couette, B., Marger, L., Le Quang, K., Kupfer, E., Cohen-Solal, A., Vilar, J., Shin, H. S., Escande, D., Charpentier, F., Nargeot, J., & Lory, P. (2006). Bradycardia and slowing of the atrioventricular conduction in mice lacking Cav3.1/α1G T-type calcium channels. *Circulation Research, 98*, 1422–1430.

Marshall, J. M. (1962). Regulation of activity in uterine smooth muscle. *Physiological Reviews, 5*, 213–227.

Marshall, J. M. (1990). Relation between membrane potential and spontaneous contraction of the uterus. In R. E. Garfield (Ed.), *Uterine contractility. Mechanisms of control. Norwell, MA: Sereno symposia USA* (pp. 3–7).

Mathias, R., & von der Weid, P. Y. (2013). Involvement of the NO-cGMP-K_{ATP} channel pathway in the mesenteric lymphatic pump dysfunction observed in the Guinea pig model of TNBS-induced ileitis. *American Journal of Physiology, 304*, G623–G634.

Matsui, M., Motomura, D., Fujikawa, T., Jiang, J., Takahashi, S., Manabe, T., & Taketo, M. M. (2002). Mice lacking M2 and M3 muscarinic acetylcholine receptors are devoid of cholinergic smooth muscle contractions but still viable. *Journal of Neuroscience, 22*, 10627–10632.

Matsui, M., Motomura, D., Karasawa, H., Fujikawa, T., Jiang, J., Komiya, Y., Takahashi, S., & Taketo, M. M. (2000). Multiple functional defects in peripheral autonomic organs in mice lacking muscarinic acetylcholine receptor gene for the M3 subtype. *PNAS, 97*, 9579–9584.

McAlexander, M. A., Luttmann, M. A., Hunsberger, G. E., & Undem, B. J. (2014). Transient receptor potential vanilloid 4 activation constricts the human bronchus via the release of cysteinyl leukotrienes. *Journal of Pharmacology and Experimental Therapeutics, 349*, 118–125.

McCloskey, K. D., Toland, H. M., Hollywood, M. A., Thornbury, K. D., & McHale, N. G. (1999). Hyperpolarization-activated inward current in isolated sheep mesenteric lymphatic smooth muscle. *The Journal of Physiology, 521*, 201–211.

McIntosh, G. H., & Morris, B. (1971). The lymphatics of the kidney and the formation of renal lymph. *The Journal of Physiology, 214*, 365–376.

Melman, A., Bar-Chama, N., McCullough, A., Davies, K., & Christ, G. (2007). Plasmid-based gene transfer for treatment of erectile dysfunction and overactive bladder: Results of a phase I trial. *The Israel Medical Association Journal, 9*, 143–146.

Mercado, J., Baylie, R., Navedo, M. F., Yuan, C., Scott, J. D., Nelson, M. T., Brayden, J. E., & Santana, L. F. (2014). Local control of TRPV4 channels by AKAP150-targeted PKC in arterial smooth muscle. *The Journal of General Physiology, 143*, 559–575.

Mizuno, R., Ono, N., & Ohhashi, T. (1999). Involvement of ATP-sensitive K^+ channels in spontaneous activity of isolated lymph microvessels in rats. *American Journal of Physiology, 277*, H1453–H1456.

Montell, C. (2005). The TRP superfamily of cation channels. *Science Signaling, 272*, Article re3.

Moosmang, S., Schulla, V., Welling, A., Feil, R., Feil, S., Wegener, J. W., Hofmann, F., & Klugbauer, N. (2003). Dominant role of smooth muscle L-type calcium channel Cav1.2 for blood pressure regulation. *The EMBO Journal, 22*, 6027–6034.

Morin, C., Sirois, M., Echave, V., Gomes, M. M., & Rousseau, E. (2007). Functional effects of 20-HETE on human bronchi: Hyperpolarization and relaxation due to BK_{Ca} channel activation. *American Journal of Physiology, 293*, L1037–L1044.

Moore, C. L., Nelson, P. L., Parelkar, N. K., Rusch, N. J., & Rhee, S. W. (2014). Protein kinase A–phosphorylated K_V1 channels in PSD95 signaling complex contribute to the resting membrane potential and diameter of cerebral arteries. *Circulation Research, 114*, 1258–1267.

Murphy, R. A., & Rembold, C. M. (2005). The latch-bridge hypothesis of smooth muscle contraction. *Canadian Journal of Physiology and Pharmacology, 83*, 857–864.

Muthuchamy, M., Gashev, A. A., Boswell, N., Dawson, N., & Zawieja, D. C. (2003). Molecular and functional analysis of the contractile apparatus in lymphatic muscle. *The FASEB Journal, 17*, 920–922.

Navarro-Gonzales, M. F., Grayson, T. H., Meaney, K. R., Cribbs, L. L., & Hill, C. E. (2009). Non L-type voltage dependent calcium channels control vascular tone of the rat basilar artery. *Clinical and Experimental Pharmacology and Physiology, 36*, 55–66.

Negrini, D., Ballard, S. T., & Benoit, J. N. (1994). Contribution of lymphatic myogenic activity and respiratory movements to pleural lymph flow. *Journal of Applied Physiology, 76*, 2267–2274.

Nieves-Cintrón, M., Nystoriak, M. A., Prada, M. P., Johnson, K., Fayer, W., Dell'Acqua, M. L., Scott, J. D., & Navedo, M. F. (2015). Selective down-regulation of $K_V2.1$ function contributes to enhanced arterial tone during diabetes. *Journal of Biological Chemistry, 290*, 7918–7929.

Nipper, M. E., & Dixon, J. B. (2011). Engineering the lymphatic system. *Cardiovascular Engineering and Technology, 2*, 296–308.

Nystoriak, M. A., Zhang, D., Jagatheesan, G., & Bhatnagar, A. (2017). Heteromeric complexes of aldo-keto reductase auxiliary $K_V\beta$ subunits (AKR6A) regulate sarcolemmal localization of $K_V1.5$ in coronary arterial myocytes. *Chemico-Biological Interactions, 2797*, 30317–30324.

Oger, S., Behr-Roussel, D., Gorny, D., Bernabé, J., Comperat, E., Chartier-Kastler, E., Denys, P., & Giuliano, F. (2010). Effects of potassium channel modulators on myogenic spontaneous phasic contractile activity in human detrusor from neurogenic patients. *BJU International, 108*, 604–611.

Oh, U., & Jung, J. (2016). Cellular functions of TMEM16/anoctamin. *Pflügers Archiv, 468*, 443–453.

Ohanyan, V., Yin, L., Bardakjian, R., Kolz, C., Enrick, M., Hakobyan, T., Kmetz, J., Bratz, A., Luli, J., Nagane, M., Khan, N., Hou, H., Kuppusamy, P., Graham, J., Fu, F. K., Janota, D., Oyewumi, M. O., Logan, S., Lindner, J. R., & Chilian, W. M. (2015). Requisite role of Kv1.5 channels in coronary metabolic dilation. *Circulation Research, 117*, 612–621.

Oonuma, H., Nakajima, T., Nagata, T., Iwasawa, K., Wang, Y., Hazama, H., Morita, Y., Wang, Y., Yamamoto, K., Nagai, R., & Omata, M. (2000). Endothelin-1 is a potent activator of non-selective cation currents in human bronchial smooth muscle cells. *American Journal of Respiratory Cell and Molecular Biology, 23*, 213–221.

Park, J. H., & Jeong, J. O. (2013). Therapeutic applications targeting the cardiac lymphatics in heart disease. In *The cardiac lymphatic system* (pp. 73–85). New York: Springer.

Park, W. S., Han, J., & Earm, Y. E. (2008). Physiological role of inward rectifier K^+ channels in vascular smooth muscle cells. *Pflügers Archiv, 457*, 137–147.

Parkington, H. C. (1985). Some properties of the circular myometrium of the sheep throughout pregnancy and during labour. *The Journal of Physiology, 359*, 1–15.

Parkington, H. C., Tonta, M. A., Davies, N. K., Brennecke, S. P., & Coleman, H. A. (1999). Hyperpolarization and slowing of the rate of contraction in human uterus in pregnancy by prostaglandins E2 and F2α: Involvement of the Na^+ pump. *The Journal of Physiology, 514*, 229–243.

Petkov, G. V. (2014). Central role of the BK channel in urinary bladder smooth muscle physiology and pathophysiology. *American Journal of Physiology, 307*, R571–R584.

Petkov, G. V. (2011). Role of potassium ion channels in detrusor smooth muscle function and dysfunction. *Nature Reviews Urology, 9*(1), 30–40.

Picollo, A., Malvezzi, M., & Accardi, A. (2015). TMEM16 proteins: Unknown structure and confusing functions. *Journal of Molecular Biology, 427*, 94–105.

Pierce, S. L., & England, S. K. (2010). SK3 channel expression during pregnancy is regulated through estrogen and Sp factor-mediated transcriptional control of the KCNN3 gene. *American Journal of Physiology, 299*, E640–E646.

Ratajska, A., Gula, G., Flaht-Zabost, A., Czarnowska, E., Ciszek, B., Jankowska-Steifer, E., Niderla-Bielinska, J., & Radomska-Lesniewska, D. (2014). Comparative and developmental anatomy of cardiac lymphatics. *Science World Journal, 2014*, 1–9.

Reinl, E. L., Cabeza, R., Gregory, I. A., Cahill, A. G., & England, S. K. (2015). Sodium leak channel, non-selective contributes to the leak current in human myometrial smooth muscle cells from pregnant women. *Molecular Human Reproduction, 21*, 816–824.

Sanders, K. M., Kito, Y., Hwang, S. J., & Ward, S. M. (2016). Regulation of gastrointestinal smooth muscle function by interstitial cells. *The Physiologist, 31*, 316–326.

Sanders, K. M., Koh, S. D., & Ward, S. M. (2006). Interstitial cells of Cajal as pacemakers in the gastrointestinal tract. *Annual Review of Physiology, 68*, 307–343.

Sanders, K. M. (2008). Regulation of smooth muscle excitation and contraction. *Neuro-Gastroenterology and Motility, 20*(Suppl. 1), 39–53.

Sanders, K. M., Ward, S. M., & Koh, S. D. (2014). Interstitial cells: Regulators of smooth muscle function. *Physiological Reviews, 94*, 859–907.

Scallan, J. P., Zawieja, S. D., Castorena-Gonzalez, J. A., & Davis, M. J. (2016). Lymphatic pumping: Mechanics, mechanisms and malfunction. *The Journal of Physiology, 594*, 5749–5768.

Schmid-Schönbein, G. W. (1990). Microlymphatics and lymph flow. *Physiological Reviews, 70*, 987–1028.

Seo, E. Y., Kim, H. J., Zhao, Z. H., Jang, J. H., Jin, C. Z., Yoo, H. Y., Zhang, Y.-H., & Kim, S. J. (2014). Low K^+ current in arterial myocytes with impaired K^+-vasodilation and its recovery by exercise in hypertensive rats. *Pflügers Archiv, 466*, 2101–2111.

Soder, R. P., & Petkov, G. V. (2011). Large conductance Ca^{2+}-activated K^+ channel activation with NS1619 decreases myogenic and neurogenic contractions of rat detrusor smooth muscle. *European Journal of Pharmacology, 670*(1), 252–259.

Streng, T., Hedlund, P., Talo, A., Andersson, K. E., & Gillespie, J. I. (2006). Phasic non-micturition contractions in the bladder of the anaesthetized and awake rat. *BJU International, 97*, 1094–1101.

Szabo, I., Leanza, L., Gulbins, E., & Zoratti, M. (2012). Physiology of potassium channels in the inner membrane of mitochondria. *Pflügers Archiv: European Journal of Physiology, 463*(2), 231–246. https://doi.org/10.1007/s00424-011-1058-7

Szal, S. E., Repke, J. T., Seely, E. W., Graves, S. W., Parker, C. A., & Morgan, K. G. (1994). $[Ca^{2+}]_i$ signaling in pregnant human myometrium. *American Journal of Physiology, 267*, E77–E87.

Tajada, S., Cidad, P., Moreno-Domínguez, A., Pérez-García, M. T., & López-López, J. R. (2012). High blood pressure associates with the remodelling of inward rectifier K^+ channels in mice mesenteric vascular smooth muscle cells. *The Journal of Physiology, 590*, 6075–6091.

Takahashi, T., Shibata, M., & Kamiya, A. (1997). Mechanism of macromolecule concentration in collecting lymphatics in rat mesentery. *Microvascular Research, 54*, 193–205.

Tazzeo, T., Zhang, Y., Keshavjee, S., & Janssen, L. J. (2008). Ryanodine receptors decant internal Ca^{2+} store in human and bovine airway smooth muscle. *European Respiratory Journal, 32*, 275–284.

Telinius, N., Kim, S., Pilegaard, H., Pahle, E., Nielsen, J., Hjortdal, V., Aalkjaer, C., & Boedtkjer, D. B. (2014). The contribution of K^+ channels to human thoracic duct contractility. *American Journal of Physiology, 307*, H33–H43.

Telinius, N., Majgaard, J., Kim, S., Katballe, N., Pahle, E., Nielsen, J., Hjortdal, V., Aalkjaer, C., & Boedtkjer, D. B. (2015). Voltage-gated sodium channels contribute to action potentials and spontaneous contractility in isolated human lymphatic vessels. *The Journal of Physiology, 593*, 3109–3122.

Telinius, N., Mohanakumar, S., Majgaard, J., Kim, S., Pilegaard, H., Pahle, E., Nielsen, J., de Leval, M., Aalkjaer, C., Hjortdal, V., & Boedtkjer, D. B. (2014). Human lymphatic vessel contractile activity is inhibited in vitro but not in vivo by the calcium channel blocker nifedipine. *The Journal of Physiology, 592*, 4697–4714.

Thorneloe, K. S., Meredith, A. L., Knorn, A. M., Aldrich, R. W., & Nelson, M. T. (2005). Urodynamic properties and neurotransmitter dependence of urinary bladder contractility in the BK channel deletion model of overactive bladder. *American Journal of Physiology, 289*, F604–F610.

Tobin, A. A., Joseph, B. K., Al-Kindi, H. N., Albarwani, S., Madden, J. A., Nemetz, L. T., Rusch, N. J., & Rhee, S. W. (2009). Loss of cerebrovascular *shaker*-type K^+ channels: A shared vasodilator defect of genetic and renal hypertensive rats. *American Journal of Physiology, 297*, H293–H303.

Toland, H. M., McCloskey, K. D., Thornbury, K. D., McHale, N. G., & Hollywood, M. A. (2000). Ca(2+)-activated Cl(-) current in sheep lymphatic smooth muscle. *American Journal of Physiology. Cell Physiology, 279*(5), C1327–C1335. https://doi.org/10.1152/ajpcell.2000.279.5.C1327

Toro, L., Li, M., Zhang, Z., Singh, H., Wu, Y., & Stefania, E. (2014). MaxiK channel and cell signaling. *Pflügers Archiv, 466*, 875–886.

Tribe, R. M., Moriarty, P., & Poston, L. (2000). Calcium homeostatic pathways change with gestation in human myometrium. *Biology of Reproduction, 63*, 748–755.

Tsvilovskyy, V. V., Zholos, A. V., Aberle, T., Philipp, S. E., Dietrich, A., Zhu, M. X., Birnbaumer, L., Freichel, M., & Flockerzi, V. (2009). Deletion of TRPC4 and TRPC6 in mice impairs smooth muscle contraction and intestinal motility in vivo. *The Gastroenterologist, 137*, 1415–1424.

Tykocki, N. R., Boerman, E. M., & Jackson, W. F. (2017). Smooth muscle ion channels and regulation of vascular tone in resistance arteries and arterioles. *Comprehensive Physiology, 7*, 485–581.

van Helden, D. F. (1993). Pacemaker potentials in lymphatic smooth muscle of the Guinea-pig mesentery. *The Journal of Physiology, 471*, 465–479.

van Helden, D. F. (2014). The lymphangion: A not so "primitive" heart. *The Journal of Physiology, 592*, 5353–5354.

von der Weid, P.-Y., Lee, S., Imtiaz, M. S., Zawieja, D. C., & Davis, M. J. (2014). Electrophysiological properties of rat mesenteric lymphatic vessels and their regulation by stretch. *Lymphatic Research and Biology, 12*, 66–75.

von der Weid, P. Y., Rahman, M., Imtiaz, M. S., & van Helden, D. F. (2008). Spontaneous transient depolarizations in lymphatic vessels of the Guinea pig mesentery: Pharmacology and implication for spontaneous contractility. *American Journal of Physiology, 295*, H1989–H2000.

Watson, N., Magnussen, H., & Rabe, K. F. (1997). Inherent tone of human bronchus: Role of eicosanoids and the epithelium. *British Journal of Pharmacology, 121*, 1099–1104.

Welsh, D. G., Jackson, W. F., & Segal, S. S. (1998). Oxygen induces electromechanical coupling in arteriolar smooth muscle cells: A role for L-type Ca^{2+} channels. *American Journal of Physiology, 274*, H2018–H2024.

Welsh, D. G., Morielli, A. D., Nelson, M. T., & Brayden, J. E. (2002). Transient receptor potential channels regulate myogenic tone of resistance arteries. *Circulation Research, 90*, 248–250.

Werner, M. E., Knorn, A.-M., Meredith, A. L., Aldrich, R. W., & Nelson, M. T. (2007). Frequency encoding of cholinergic- and purinergic-mediated signaling to mouse urinary bladder smooth muscle: Modulation by BK channels. *American Journal of Physiology, 292*, R616–R624.

Wiedmann, F., Schmidt, C., Lugenbiel, P., Staudacher, I., Rahm, A.-K., Seyler, C., Schweizer, P. A., Katus, H. A., & Thomas, D. (2016). Therapeutic targeting of two-pore-domain potassium (K2P) channels in the cardiovascular system. *Clinical Science, 130*, 643–650.

Wray, S., Jones, K., Kupittayanant, S., Li, Y., Matthew, A., Monir-Bishty, E., Noble, K., Pierce, S. J., Quenby, S., & Shmygol, A. V. (2003). Calcium signaling and uterine contractility. *Journal of the Society for Gynecologic Investigation, 10*, 252–264.

Yamanishi, T., Chapple, C. R., Yasuda, K., & Chess-Williams, R. (2000). The role of M_2-muscarinic receptors in mediating contraction of the pig urinary bladder in vitro. *British Journal of Pharmacology, 131*, 1482–1488.

Yocum, G. T., Chen, J., Choi, C. H., Townsend, E. A., Zhang, Y., Xu, D., Fu, X. W., Sanderson, M. J., & Emala, C. W. (2017). Role of transient receptor potential vanilloid 1 in the modulation of airway smooth muscle tone and calcium handling. *American Journal of Physiology, 312*, L812–L821.

Young, R. C. (2016). Mechanotransduction mechanisms for coordinating uterine contractions in human labor. *Reproduction, 152*, R51–R61.

Yue, Z., Xie, J., Yu, A. S., Stock, J., Du, J., & Yue, L. (2015). Role of TRP channels in the cardiovascular system. *American Journal of Physiology, 308*, H157–H182.

Zawieja, D. C. (2009). Contractile physiology of lymphatics. *Lymphatic Research and Biology, 7*, 87–96.

Zhao, J., & van Helden, D. F. (2003). ET-1-associated vasomotion and vasospasm in lymphatic vessels of the Guinea-pig mesentery. *British Journal of Pharmacology, 140*, 1399–1413.

Zhu, M. H., Sung, T. S., Kurahashi, M., O'Kane, L. E., O'Driscoll, K., Koh, S. D., & Sanders, K. M. (2016). Na^+-K^+-Cl^- cotransporter (NKCC) maintains the chloride gradient to sustain pacemaker activity in interstitial cells of Cajal. *American Journal of Physiology, 311*, G1037–G1046.

Zhu, M. H., Sung, T. S., O'Driscoll, K., Koh, S. D., & Sanders, K. M. (2015). Intracellular Ca^{2+} release from endoplasmic reticulum regulates slow wave currents and pacemaker activity of interstitial cells of Cajal. *American Journal of Physiology, 308*, C608–C620.

Chapter 36

Excitation-contraction coupling in skeletal muscle

Judith A. Heiny
Department of Pharmacology and Systems Physiology, University of Cincinnati, College of Medicine, Cincinnati, OH, United States

1. Summary

Skeletal muscle cells (fibers) are activated to contract by a sequence of fast, electrically driven events collectively termed excitation-contraction coupling (ECC). ECC occurs at specialized membrane junctions, called triads, which connect the outer plasma membrane with the intracellular sarcoplasmic reticulum. Two key proteins – a voltage-dependent calcium channel in the plasma membrane ($Ca_V1.1$) and a calcium channel (ryanodine receptor, RyR1) on the sarcoplasmic reticulum (SR) – physically interact at the triad junction to release stored Ca^{2+} from the SR. The resulting increase in cytosolic Ca^{2+} rapidly activates the contractile proteins, actin and myosin, to generate force.

2. Introduction

Skeletal muscle cells function in a bipolar manner, with long periods of inactivity punctuated by brief periods of intense contraction. Excitation-contraction coupling describes the sequence of electrical and biochemical signals that initiate contraction. In ECC, an electrical signal on the outer plasma membrane is transduced into a trigger for Ca^{2+} release from the intracellular sarcoplasmic reticulum, where Ca^{2+} is sequestered. The resulting rise in cytosolic Ca^{2+} allows actin and myosin to interact and generate force. This chapter focuses on ECC in a mammalian fast-twitch skeletal muscle, which has been extensively characterized [reviewed in (Bolaños & Calderón, 2022; Dirksen et al., 2022; Franzini-Armstrong, 2018; Hernández-Ochoa & Schneider, 2018; Shishmarev et al., 2022)].

Skeletal muscle is the fastest contracting of the major muscle types and is evolutionarily related to cardiac and smooth muscle. The same two key proteins – a voltage-dependent calcium channel (Ca_V) on the plasma membrane and a Ca^{2+} release channel (ryanodine receptor, RyR1) on the sarcoplasmic reticulum – mediate excitation-contraction coupling in cardiac, smooth, and skeletal muscle. The different types of muscle express tissue-specific subtypes of these proteins. Mammalian skeletal muscles use $Ca_V1.1$ (also termed skeletal L-type Ca channel, or dihydropyridine receptor, DHPR) and RyR1. Importantly, each muscle type has evolved specialized mechanisms of using these proteins to serve its distinct functions. Cardiac and smooth muscle primarily mediate continuous, relatively slow, contractile processes that are under autonomic and humoral control. In contrast, skeletal muscles primarily mediate rapid, willed bodily movements under control of the central nervous system. The defining features of skeletal muscle activation are speed and voluntary control, as needed for fine control of body movements.

3. Overview of excitation-contraction coupling in skeletal muscles

Muscle activation (Fig. 36.1) begins when excitatory input from a motor neuron depolarizes the postsynaptic membrane (termed muscle end plate) of the neuromuscular junction. The depolarization opens voltage-dependent Na^+ channels and triggers a propagating action potential, which spreads the depolarization to the entire plasma membrane.

Every excitatory input from the motor nerve elicits an action potential on the muscle and triggers contraction. This strict responsiveness of muscle contraction to nerve input is ensured by the high density of voltage-dependent Na^+ channels on

Cell Physiology Sourcebook. https://doi.org/10.1016/B978-0-12-811114-7.00026-8

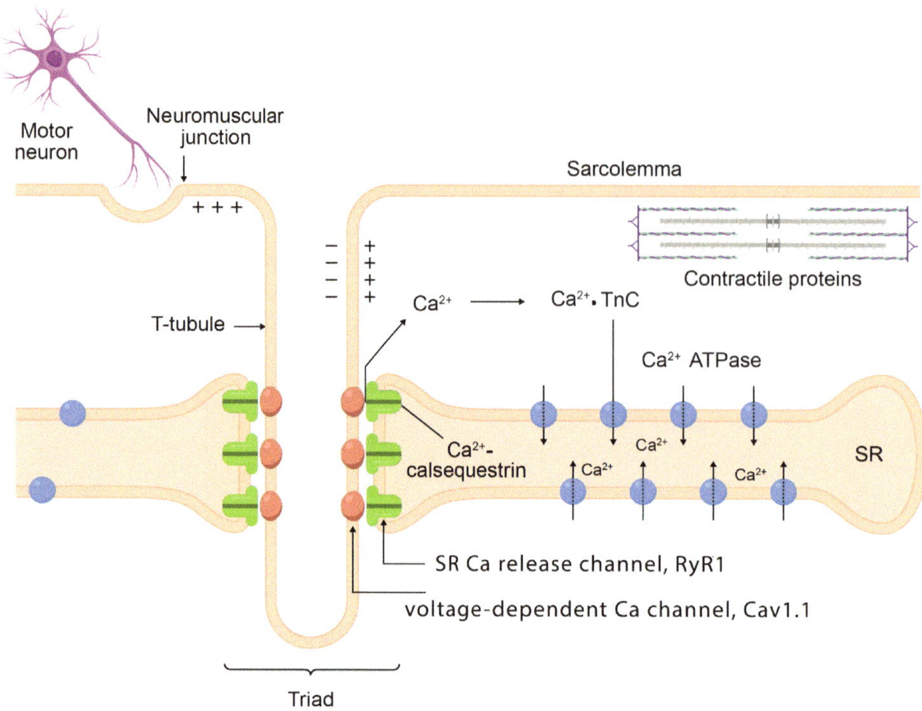

1 Motor nerve triggers an Action Potential on the post-synaptic neuromuscular junction
2 The Action Potential propagates the depolarization along the outer plasma membrane and into T-tubules
3 Depolarization activates voltage-dependent Ca channels, Cav1.1, at a subcellular junction (triad)
4 Cav1.1 physically transmits a signal to Ca release channels, RyR1, on the SR membrane
5 RyR1 channels open and release Ca^{2+} into the cytosol
6 Ca^{2+} binds to troponin C (TnC) to activate the contractile proteins, actin and myosin. Contraction proceeds.
7 Ca ATPase (SERCA) pumps Ca^{2+} back into the SR, Contraction stops.
8 Within the SR, Ca^{2+} binds to calsequestrin and is stored

FIGURE 36.1 Key steps in the activation of a vertebrate skeletal muscle fiber.

the muscle end plate and the very negative resting potential of skeletal muscles (−80 to −90 mV). This negative resting potential minimizes slow inactivation of Na^+ channels and maintains them in an available state (In contrast, neurons typically maintain resting membrane potentials of about −70 mV, near the mid-point of Na^+ channel slow inactivation, which allows Na^+ channel responsiveness to be up- or down-modulated). Because each muscle fiber is individually innervated, the force produced by a whole muscle is graded in proportion to the number of fibers recruited.

From the centrally positioned neuromuscular junction, the action potential propagates the depolarization bidirectionally along the outer plasma membrane and into the fiber interior through a system of plasma membrane invaginations termed **transverse tubules (T-tubules)** (Peachey, 1965). T-tubules are continuous with and invaginate from the plasma membrane at periodic intervals to form a planar network that spans the entire fiber cross-section. T-tubules provide the conduit for the action potential to reach the fiber interior and bring the plasma membrane near the internal sarcoplasmic reticulum at specialized junctions called **triads**. At the triad junction, the action potential depolarization activates voltage-dependent Ca channels ($Ca_V1.1$) on the plasma membrane which, in turn, transmit a physical trigger to Ca^{2+} release channels (RyR1) on the opposing SR membrane. RyR1 channels open rapidly and release Ca^{2+} into the cytosol. Ca^{2+} binds to troponin C (TnC), thereby allowing the contractile proteins to interact and generate force. Troponin C (TnC) is a molecular ON/OFF switch for the contractile proteins. At resting cytosolic Ca^{2+} concentration (0.1 μM), unbound TnC inhibits actin-myosin interaction; when cytosolic Ca^{2+} rises above μM levels, Ca-bound TnC relieves this inhibition, allowing actin and myosin to interact and produce force.

In this way, as in many other cell processes, Ca^{2+} serves as the central intracellular messenger (Fig. 36.2), as discussed in Chapter 15 in this volume. Contraction is inhibited at low cytosolic Ca^{2+} concentration (100 nM) and proceeds when

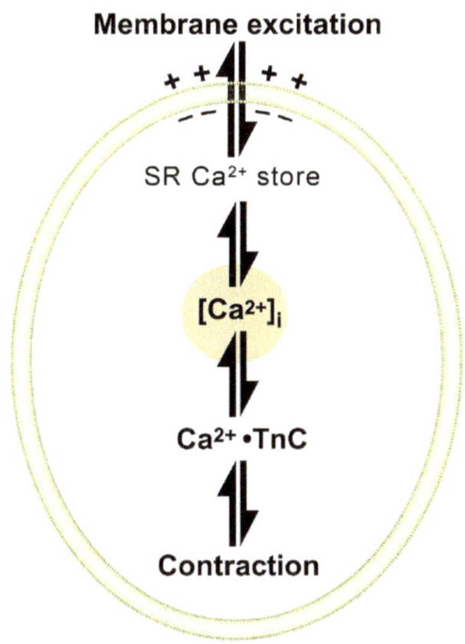

FIGURE 36.2 **Cytosolic Ca^{2+} is the central link between membrane excitation and contraction.** At rest, Ca^{2+} is sequestered in the SR, which keeps the cytosolic Ca^{2+} concentration, $[Ca^{2+}]_i$, low. The cytosolic Ca^{2+} concentration increases transiently by more than 100-fold when the SR releases Ca^{2+} in response to membrane depolarization. Ca^{2+} binding to Troponin C (TnC) allows actin and myosin to interact and produce force.

$[Ca^{2+}]_i$ rises to micromolar concentrations. The force produced by each skeletal muscle fiber is proportional to the cytosolic Ca^{2+} concentration (Fig. 36.3). Repetitive nerve stimulation is the normal physiological pattern of muscle activation. Cytosolic Ca^{2+} summates during repeated SR Ca^{2+} release events, thereby allowing force to increase in proportion to the frequency and duration of action potentials elicited by the motor nerve.

ECC operates as an ON/OFF switch under strict control of the plasma membrane voltage. Positively charged, transmembrane helices of $Ca_V1.1$ serve as the voltage sensor in this process. The action potential depolarization drives voltage-sensing helices outward and these movements, in turn, drive conformational changes that are transmitted to RyR1 to open its pore. The physical communication between $Ca_V1.1$ and RyR1 is achieved through multiple associations between the cytoplasmic regions of Cav1.1, the cytoplasmic domains of RyR1, and additional proteins within the junctional gap. In this way, voltage-driven conformational transitions of a plasma membrane Ca^{2+} channel control the opening of a separate Ca channel on the intracellular SR membrane.

The SR is evolutionarily related to the endoplasmic reticulum of eukaryotic cells. The SR of skeletal muscle, is highly specialized for controlling the cytosolic Ca^{2+} concentration, $[Ca^{2+}]_i$. A Ca^{2+} ATPase, **SERCA1**, actively pumps Ca^{2+} into the SR to maintain cytosolic $[Ca^{2+}]_i$ at submicromolar levels in noncontracting muscle and restores Ca^{2+} homeostasis after each Ca^{2+} release event. Within the SR, Ca^{2+} is stored as a complex bound to the Ca^{2+}-binding protein, **calsequestrin,**

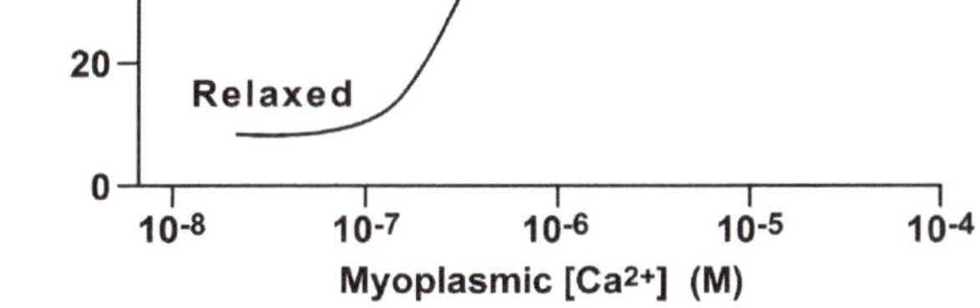

FIGURE 36.3 The force produced by a skeletal muscle fiber is proportional to the cytosolic Ca^{2+} concentration $[Ca^{2+}]_i$.

present in the lumen of terminal cisternae. Upon membrane depolarization, RyR1 rapidly releases massive amounts of Ca^{2+} from the SR, at flux rates approaching 100 μM/ms. The cytosolic Ca^{2+} concentration increases up to 30 μM, dependent on fiber type. Ca^{2+} release by RyR1 occurs at junctional regions of SR. Ca^{2+} reuptake by SERCA1 occurs on nonjunctional SR.

Ca^{2+} release is brief and terminates within milliseconds. Contraction ends when the cytosolic Ca^{2+} concentration has been restored to its basal level and unbinds from TnC. Multiple processes interact to return cytosolic Ca^{2+} to basal levels (Fig. 36.4). As the T-tubule membrane repolarizes, $Ca_V1.1$ channels return to their initial conformation, thereby physically closing the RyR1 pore. RyR1 is itself deactivated by negative feedback mechanisms that depend on cytosolic $[Ca^{2+}]_i$. SERCA1 is rapidly stimulated by contraction-related factors and pumps Ca^{2+} back to the SR. With its affinity and high capacity, SERCA1 is the primary mechanism for reducing cytosolic Ca^{2+} at early times after Ca^{2+} release (Fig. 36.1). Soluble Ca^{2+}-binding proteins with fast kinetics (parvalbumins, etc.) act as an intermediate sink that buffers Ca^{2+} while SERCA1 returns Ca^{2+} to the SR.

In addition to these early Ca^{2+} handling processes, mechanisms exist to extrude, sequester, and/or recycle Ca^{2+} transiently during repetitive muscle activation lasting tens of milliseconds to minutes. During intense contractile activity, cytoplasmic Ca^{2+} levels can exceed the capacity of SERCA1. In this condition, additional membrane processes extrude or sequester Ca^{2+} transiently. Soluble Ca^{2+}-binding proteins with fast kinetics (parvalbumins, etc.) act as an intermediate sink that buffers Ca^{2+} while SERCA1 returns Ca^{2+} to the SR. The plasma membrane Ca^{2+} ATPase (PMCA) and Na/Ca exchanger (NCX), which localize to the T-tubules, extrude Ca^{2+} to the extracellular space. NCX, despite its rapid kinetics

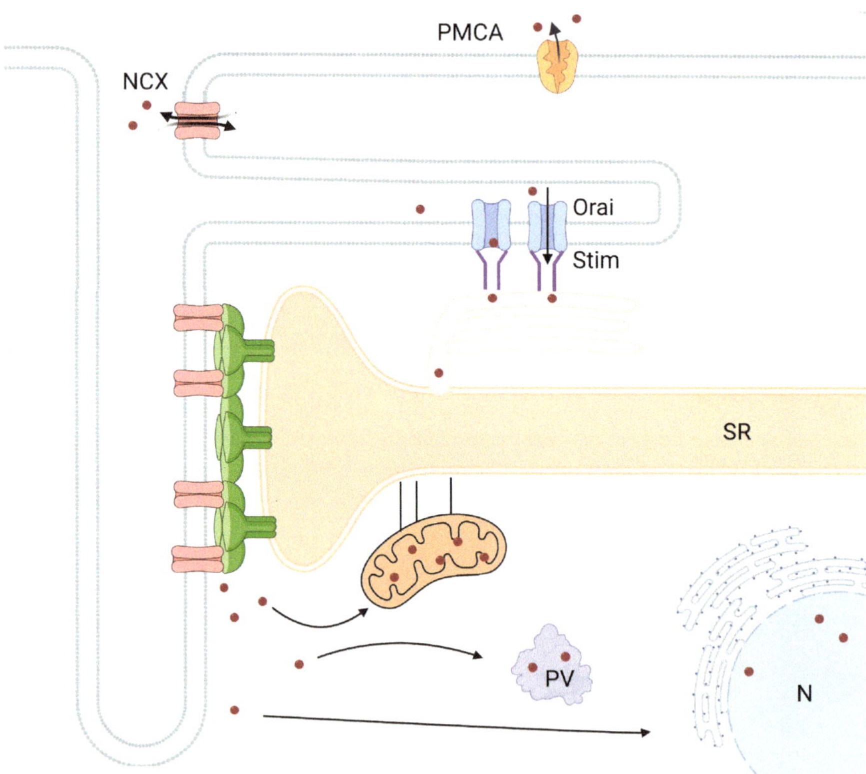

FIGURE 36.4 **Ca^{2+} handling during intense or prolonged contractile activity.** Multiple mechanisms interact to extrude or sequester Ca^{2+} (*red* spheres) transiently when the cytosolic Ca^{2+} concentration exceeds the capacity of SERCA1. Soluble Ca^{2+} buffers (PV, parvalbumin etc.) with low affinity provide an intermediate sink for holding Ca^{2+} while SERCA1 returns it to the SR. Some Ca^{2+} is extruded to the extracellular space by a Ca^{2+} ATPase (PMCA1) and the Na^+/Ca^{2+} exchanger (NCX) in the T-tubules. Mitochondria near the triad take up Ca^{2+} using the mitochondrial calcium uniporter (MCU). Some Ca^{2+} also enters nuclei. Mechanisms also exist to refill the SR during periods of prolonged contractile activity when SR Ca^{2+} stores can become depleted. A process, termed Store Operated Ca^{2+} Entry (SOCE), imports Ca^{2+} from the extracellular space to the SR using the stromal interaction molecule 1 (STIM1), a Ca^{2+} sensor of the SR, and Orai calcium release-activated calcium modulator 1 (Orai1), a Ca^{2+} channel in the T-tubules. STIM1 and Orai1 form specialized contacts (Calcium Entry Units) between longitudinally oriented branches of T-tubules and the SR membranes. Prolonged muscle contraction promotes the formation of additional interaction sites between STIM1 and Orai1. The STIM1-Orai1 pathway recycles Ca^{2+} extruded by PMCA and NCX back to the SR.

and localization near the sites of Ca^{2+} release, is rapidly saturated and does not compete measurably with TnC for released Ca^{2+} at early times. Ca^{2+} extrusion by NCX is linked to the intracellular Na concentration generated by Na^+/K^+-ATPase α2 isoform transport, which is dramatically stimulated by contraction and remains elevated for minutes following contraction. The resulting secondary increase in NCX activity contributes to muscle relaxation and Ca^{2+} homeostasis during and following contraction. Mitochondria, which comprise up to 15% of the muscle fiber volume, take up Ca^{2+} using the mitochondrial calcium uniporter (MCU). Mitochondria in skeletal muscle are positioned mostly near the triads where they form specific associations with the SR. The slower kinetics but large capacity of MCU help restore Ca^{2+} homeostasis during prolonged contractile activity.

Mechanisms also exist to refill the SR to allow contraction to continue during prolonged muscle activity, when SR Ca^{2+} stores can become depleted. Using a process termed store operated Ca^{2+} entry (SOCE), the SR imports Ca^{2+} from the extracellular space at contact sites between STIM1 (stromal interaction molecule 1), a Ca^{2+} sensor of the SR, and Orai1, a Ca^{2+} channel in the T-tubules. Prolonged muscle contraction can promote the formation of additional interaction sites between STIM1 and Orai1. These calcium entry units form where stacks of SR cisternae associate with longitudinally oriented branches of the T-tubules. In this way, the STIM-Orai1 pathway recycles Ca^{2+} extruded by PMCA and NCX back to the SR.

Altogether, this complex interplay of Ca^{2+} buffering, extrusion, and sequestration in intracellular organelles, as well as SR refilling, maintains the ability of muscles to contract for extended periods of time. Although contraction ends when cytosolic Ca^{2+} has returned to the resting concentration, the final restoration of Ca^{2+} homeostasis is not achieved until the Ca^{2+} concentrations in the cytosol, extracellular spaces, SR, mitochondria, and other organelles have all returned to basal levels.

4. Speed of skeletal muscle activation

Activation of a skeletal muscle fiber is extremely fast. In mammals, the process from nerve excitation to SR Ca^{2+} release is complete within ~5 thousandths of a second (Fig. 36.5). The resulting force peaks with a half-time of ~20 ms. This speed contrasts with the tens of milliseconds required for calcium-induced calcium release in cardiac muscle cells, which rely on diffusion of Ca^{2+} imported through plasma membrane calcium channels. This speed is especially remarkable given that skeletal muscle fibers are among the largest of mammalian cells. Mature skeletal muscle fibers form from the fusion of hundreds of single myoblasts, achieving diameters of 30–150 μm and lengths ranging from millimeters to centimeters.

To produce effective mechanical force, the contractile proteins in all regions of the fiber must shorten simultaneously. This synchronous activation is achieved by: (i) the speed of action potential propagation, which spreads the depolarization at speeds up to 120 m/s; (ii) the anatomy and abundance of triad junctions, which cover over 80% of the T-tubule membrane in mammals (Dulhunty, 1984); (iii) direct protein-protein interactions between $Ca_V1.1$, RyR1 and associated proteins (instead of diffusion); and (iv) the large conductance of open RyR1 channels, which releases Ca^{2+} into the cytosol at flux rates approaching 100 μM Ca^{2+} per millisecond.

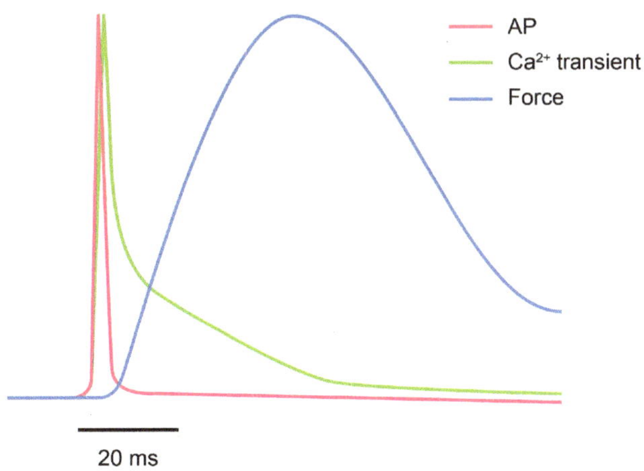

FIGURE 36.5 **Speed of excitation-contraction coupling (ECC) in a mammalian skeletal muscle fiber.** Relative time courses of the action potential (AP, *red*), cytosolic Ca^{2+} transient (*green*), and force (*blue*) in a mouse fast twitch muscle fiber. These times are for a single twitch. Prolonged and greater forces are achieved by repetitive action potential activity and summation of Ca^{2+} release events.

The time course of muscle relaxation follows the time course of Ca^{2+} removal from the cytoplasm. The half-time to relax from peak force occurs in tens of milliseconds, dependent on fiber type and the peak cytosolic Ca^{2+} concentration.

5. The triad junction

The triad junction is highly organized to provide a stable platform to rapidly communicate the plasma membrane depolarization into a physical trigger for Ca^{2+} release from the SR. Fig. 36.6 illustrates the three-dimensional architecture of the outer plasma membrane and T-tubules, triad junctions, and SR of a vertebrate skeletal muscle fiber, and their physical relationships to the contractile proteins (Peachey, 1965). The T-tubules comprise most of the plasma membrane, with a total surface area that is 3–4 times greater than the outer membrane. T-tubules invaginate from the outer membrane at periodic intervals where they branch and interconnect to form a planar, mesh-like network that spans the fiber cross-section (Fig. 36.7). This geometry ensures that all Ca^{2+} binding sites on troponin-C (TnC) are within a one-micron diffusion distance from Ca^{2+} release channels. In mammalian skeletal muscles, two planes of T-tubules form in each sarcomere, near regions where actin and myosin overlap (A-I band).

The triad junction is the structural platform where ECC occurs (Fig. 36.8). A T-tubule is flanked on each side by junctional regions of SR (**terminal cisternae**), thereby creating a tripartite appearance in cross-section (Fig. 36.8A). The lumen of a T-tubule is ~80 nm in diameter and flattens at the junction to oppose the SR across a gap of ~12 nm. The principal triad proteins are $Ca_V1.1$ and RyR1. $Ca_V1.1$ embeds in the T-tubule membrane and RyR1 embeds in the SR membrane. RyR1 has a mushroom-like shape in cross-section. Four transmembrane regions of RyR1 insert into the SR to form a Ca^{2+} conducting pore, and four cytosolic globular domains extend far into the junctional gap. These cytosolic regions contribute to the electron-opaque material seen in electron micrographs of the triad. Ca_V and RyR1 channels are positioned on the T-tubule and SR membranes, respectively, in a precise alignment (Fig. 36.8B and C). On the SR, RyR1 channels form a para crystalline lattice with corner-to-corner attachments and a center-to-center spacing of ~30 nm. One RyR1 channel on the SR interacts with four $Ca_V1.1$ channels on the T-tubule membrane (termed a **tetrad**) (Fig. 36.8D). In many mammalian skeletal muscle fibers, only every *other* RyR1 associates with a tetrad of $Ca_V1.1$ channels. This configuration of $Ca_V1.1$ tetrads and RyR1 tetramers is unique to skeletal muscle. At cardiac muscle junctions, type 2

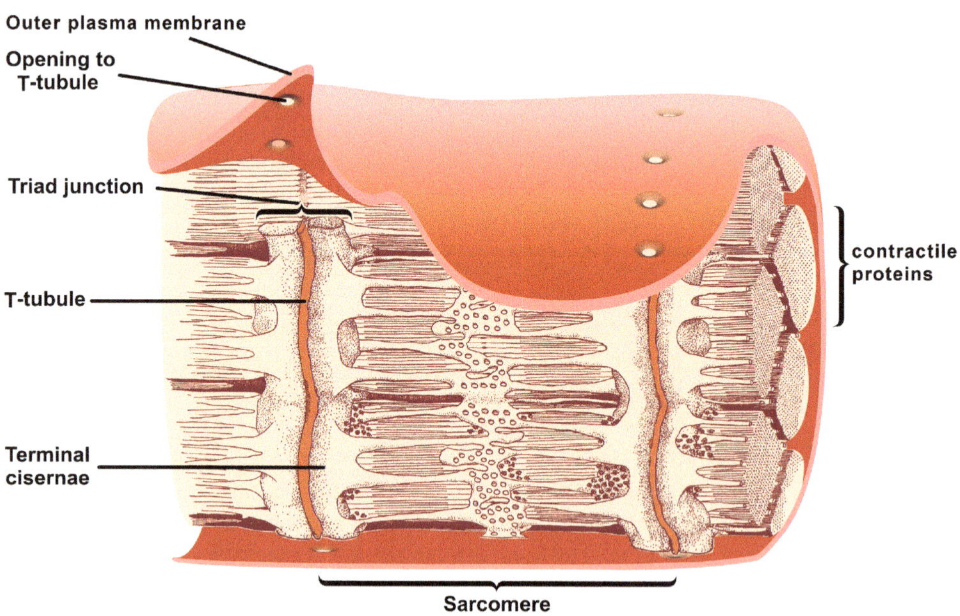

FIGURE 36.6 **Three-dimensional reconstruction of membranes involved in excitation-contraction coupling (ECC) in skeletal muscle.** The outer cell membrane (*red*) surrounds bundles of contractile proteins. T-tubules invaginate from the outer cell membrane at periodic intervals. T-tubules form triad junctions with the SR along most of their length. At the triad junctions, SR membranes dilate into sacs (terminal cisternae), which closely face the T-tubules and contain the Ca^{2+} release channels, RyR1. Nonjunctional regions of SR contain the Ca^{2+} ATPase (SERCA1) that pumps Ca^{2+} into the SR. A single force-generating unit (sarcomere) is shown. This reconstruction is from a frog skeletal muscle. In mammalian skeletal muscles, two T-tubule networks form per sarcomere. *Adapted from Peachey (1965).*

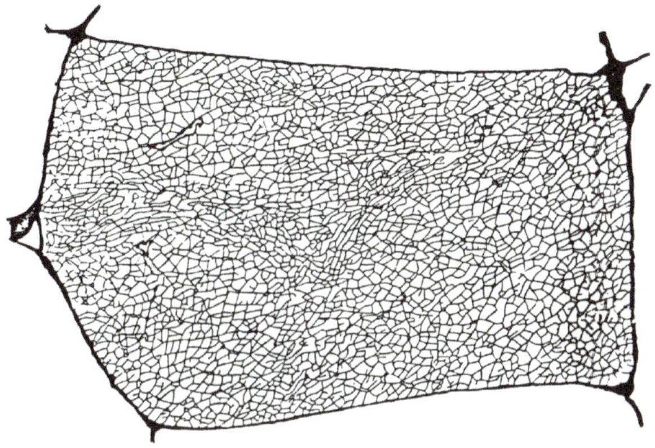

FIGURE 36.7 **Reconstruction of the T-tubule network in one cross-sectional plane of a frog twitch muscle fiber.** Tubules are 40–80 nm in diameter and branch every 80–100 nm. The reconstruction was made from electron micrographs of transverse serial slices of a muscle fiber perfused with an electron-dense, membrane-impermeant molecule that diffused into the T-tubules. *From Peachey and Eisenberg (1978).*

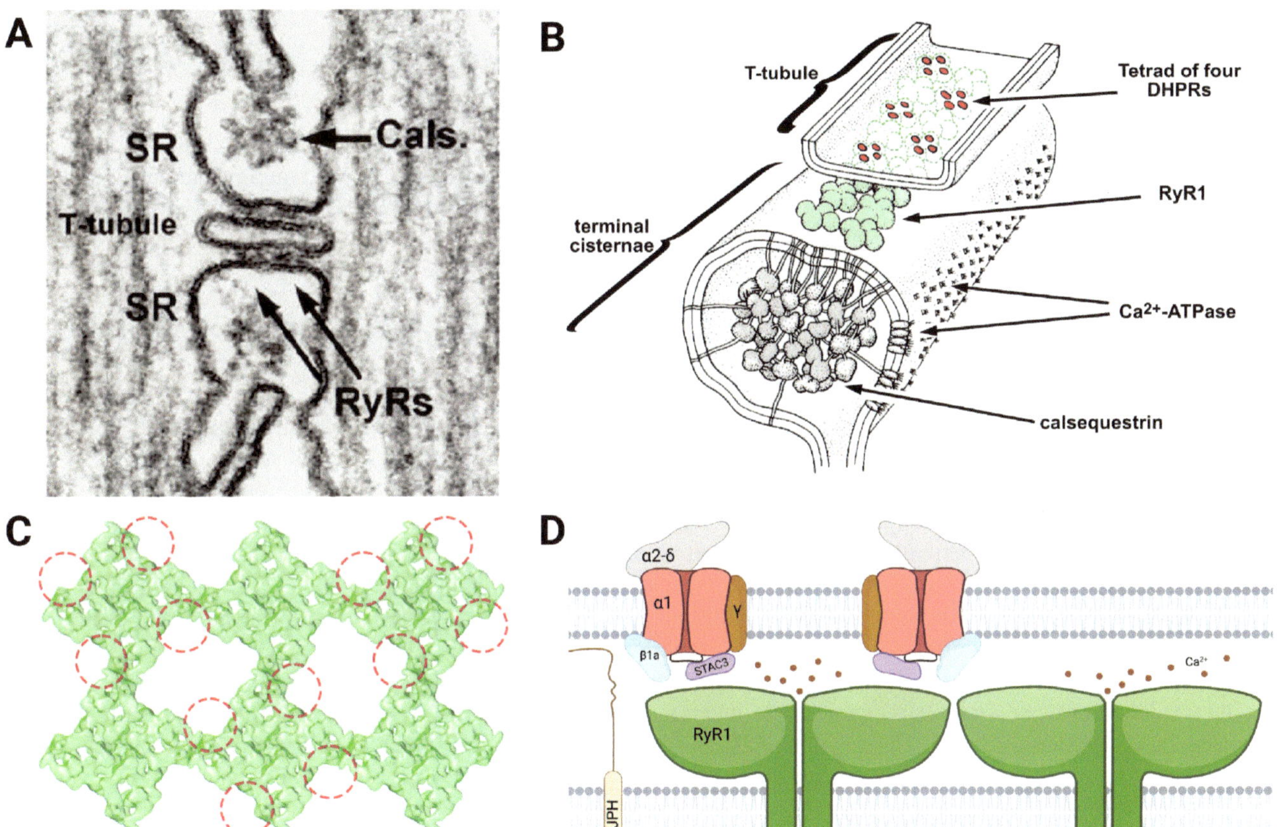

FIGURE 36.8 **The triad junction.** (A) Electron micrograph of triad junctions showing the typical tripartite architecture. A flattened T-tubule is flanked on both sides by terminal cisternae regions of SR. RyR1 channels appear as electron-dense material within the junctional gap. (B) Three-dimensional diagram of a half triad showing the alignment of $Ca_v1.1$ clusters (tetrads, *red*) in the T-tubule membrane relative to RyR1 channels (*green*) in the SR. Ca-ATPase localizes to the nonjunctional SR membrane. Within the SR, calcium is stored bound to calsequestrin, which associates with luminal domains of RyR1. (C) Alignment of $Ca_v1.1$ channels (dashed *red* outline) in the T-tubule membrane relative to RyR1 (*green, top view*) on the SR. Cytosolic domains of RyR1 associate with each other at the corners in a para-crystalline lattice. Each of four cytosolic domains of one RyR1 associates with one $Ca_v1.1$ channel. Four $Ca_v1.1$ channels (a tetrad) associate with one RyR1. (D) Schematic of key proteins of the triad junction: The T-tubule Ca_v channel ($Ca_v1.1.\alpha1s$, $\alpha2$-δ, $\beta1_a$, and $\gamma1$ subunits), RyR1, STAC3, and junctophilins (JPH). This cross-sectional view depicts two of the four cytosolic domains of each RyR. In many mammalian skeletal muscles, only every *other* RyR1 associates with a tetrad of $Ca_v1.1$ channels (two of the 4 $Ca_v1.1$ channels in a tetrad are depicted). *Red*, $Ca_v1.1$, *red*; β_{1a}, *blue*; γ_1, gold; $\alpha2$-δ, *grey*; STAC3, *purple*; RyR1, *green*; JPH, *yellow*. *(A). From Franzini-Armstrong and Peachey (1981). (B) Adapted from Sun et al. (1995). (C) From Samsó (2017).*

ryanodine receptors (RyR2) form arrays but they do not align precisely with cardiac-type Ca_V channels, which distribute more randomly on the plasma membrane.

The triad junction has strong structural integrity and maintains a constant gap of ~12 nm over a wide range of sarcomere lengths, even as the fiber sustains maximal contractile forces. Multiple protein-protein and protein-lipid interactions and a unique lipid composition provide this structural integrity. Interestingly, Ca_V1.1 and RyR1, although essential for ECC, are not required for triad formation. Triads form in muscles of mice that lack Ca_V1.1, RyR1, or both (Felder et al., 2002). Junctophilins, proteins which connect the plasma membrane and the endo/sarcoplasmic reticulum in a range of cell types (Lehnart & Wehrens, 2022), tether the T-tubule to the SR membrane (Perni, 2022) (see section V). The junctional T-tubule membrane is enriched in cholesterol and sphingomyelin, which confer a more ordered and rigid lipid environment than the outer plasma membrane (Barrientos et al., 2017). Caveolin-3, the muscle-specific member of the caveolin family of proteins, associates with cholesterol and provides a scaffold that further supports membrane order. Additional proteins that bind lipids or regulate phospholipid metabolism participate in triad formation and maintenance (Al-Qusairi & Laporte, 2011).

Fig. 36.8 is based on a fish skeletal muscle fiber. A similar configuration is seen in triads and peripheral couplings of skeletal muscles from the frog, rat, chicken, mouse, and human (Franzini-Armstrong & Jorgensen, 1994). Variations on this structure occur in skeletal muscles of other species. These include **dyads,** in which the junctional SR membrane faces a single short segment of T-tubule, and **peripheral couplings,** in which the SR forms junctions with the surface membrane (Franzini-Armstrong & Jorgensen, 1994). Triads, dyads, and peripheral couplings are structurally and functionally equivalent.

6. Key proteins of ECC

Ca_V1.1 and RyR1 form the hub of a large macromolecular complex that operates as a functional unit to control Ca^{2+} release. [*Reviewed in* (Dulhunty et al., 2018; Zalk et al., 2015):]. The complete ECC complex includes Ca_V1 and its auxiliary subunits (α2-δ, $\beta 1_a$, and γ1), JPH-1/2, STAC3, RyR1 (Fig. 36.8D) and supporting proteins. Five of these proteins - Ca_V1.1, $\beta 1_a$, STAC3, RyR1, and Junctophilin 2 – can reconstitute functional ECC when expressed in nonmuscle cells (Perni et al., 2017). Therefore, this set is considered the minimum ECC complex for achieving voltage driven Ca^{2+} release from the SR.

6.1 The voltage-dependent Ca channel

The main function of **Ca_V1.1** is to sense the action potential depolarization and transmit the trigger for RyR1 to release Ca^{2+}. In adult skeletal muscle, the original role of Ca_V1.1 – importing Ca^{2+} from the external space – has been suppressed during evolution in exchange for mutations that allow it to serve as a voltage-dependent controller of RyR1 opening (Mackrill & Shiels, 2020). Ca_V1.1 is the only alpha isoform in the Ca_V family that is capable of efficiently coupling membrane depolarization to activation of an SR ryanodine receptor using a direct, physical coupling mechanism that does not require entry of extracellular Ca^{2+} (Bannister & Beam, 2013). Although movement of voltage sensor domains of Ca_V1.1 are able to induce pore opening in some conditions, the resulting Ca^{2+} influx is insignificant during a single action potential and is not required for ECC (Ríos, 2018). In contrast to other L-type Ca channels, Ca_V1.1 has a low probability of opening and small single-channel currents. Several mechanisms combine to suppress the ion conducting function of Ca_V1.1. A developmentally regulated sequence in an extracellular loop of Ca_V1.1 reduces Ca^{2+} flux and shifts the voltage dependence of activation to more positive potentials, outside the physiological range. Additionally, interactions with the γ subunit promote channel inactivation and reduce pore conductance (El Ghaleb et al., 2022).

The skeletal muscle Ca_V channel (Fig. 36.9) is composed of four muscle-specific subunits– a primary voltage-sensing and pore-forming alpha subunit (Ca_V1.1; also named $\alpha 1_S$), a large extracellular α2-δ subunit, a cytosolic β_{1a} subunit, and a small transmembrane γ1 subunit – in a ratio of 1:1:1:1:1 (Catterall, 2011). The auxiliary subunits play roles in the expression, localization, and physiological functions of the channel. The primary Ca_V1.1 subunit (Fig. 36.9A) belongs to the superfamily of voltage-dependent Ca_V channels. Four homologous domains, each composed of six transmembrane α-helices (S1–S6), assemble to form an integral membrane protein with a central pore. Ca_V1.1 has two main functional regions. Helices S5 and S6 and an alpha helical fold between them (termed the pore helix) form the channel pore, while helices S1–S4 (blue) regulate pore opening. The S4 helix serves as the voltage sensor. It is highly charged, having a positively charged Arg or Lys residue on every third amino acid. The action potential depolarization drives S4 helices outward and their movement, interacting with helices S1–S3, controls pore opening. Ca_V1.1 has five cytosolic regions –

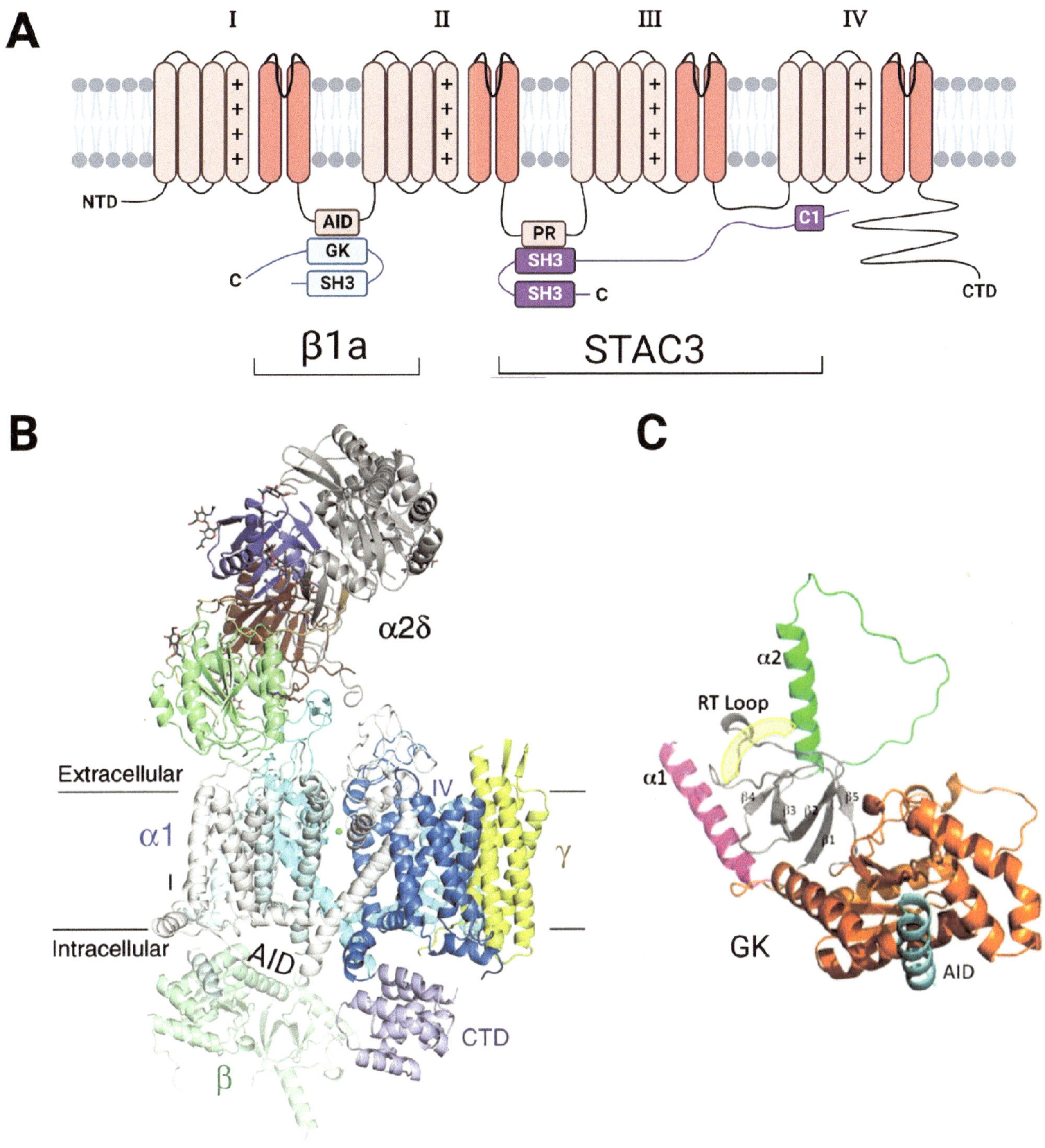

FIGURE 36.9 **The skeletal muscle Ca_v1.1 channel.** (A) The primary alpha subunit, Ca_v1.1, is composed of four homologous membrane domains, I–IV, that are connected by cytosolic linkers I-II, II-III, III-IV, and extracellular linkers. N- and C-terminal domains (NTD, CTD) are cytosolic. Each membrane domain is composed of six membrane-spanning alpha helices (S1–S6). Helices 5 and 6 (*red*) form the pore and helices 1–4 (tan) control pore activation. The β1a subunit (*blue*) and the junctional protein STAC3 (*purple*) form interactions with Ca_v1.1 that are essential for ECC. The GK domain of $β1_a$ binds the AID domain (*grey*) of the I-II linker. The first tandem SH3 domain of STAC3 binds a critical proline rich (PR) segment of the II-III linker. A C-terminal domain (C1) domain of STAC3 interacts with the CTD of Cav1.1. (B) Protein structure of the complete Ca_v channel in the membrane determined at 4.2 Å resolution using cryo-electron microscopy. (C) X-ray crystal structure of $β1_a$ (*orange*) complexed with the AID (aqua) domain of Cav1.1. A groove in the GK region of $β1_a$ binds the AID domain of Ca_v1.1 (Norris et al., 2017). *(B) From Wu et al. (2015).*

an N-terminal domain (NTD), linker sequences between transmembrane domains I—II, II—III, III—IV, and a C-terminal domain (CTD).

The cytosolic II-III linker is essential for ECC (Bannister & Beam, 2013; Tanabe et al., 1990). A central proline-rich sequence of the II-III linker associates with the first SH3 domain of STAC3, and the C1 region of STAC3 interacts with the CTD of $Ca_V1.1$. The II-III loop has a high intrinsic mobility and its structure can change upon activation of $Ca_V1.1$ during the action potential (Polster et al., 2012). For these reasons, it is proposed that the II—III loop might transmit the physical trigger to RyR1, either directly or through interactions with STAC3 or other coupling proteins. The I-II linker contains an alpha helical interaction domain (AID) that binds the $\beta 1_a$ subunit in its GK domain. The III-IV linker forms a globular domain that associates with the CTD of $Ca_V1.1$. The cytosolic CTD and NTD are not essential for ECC.

The structure of the complete skeletal muscle Ca_V channel in closed and activated states has been determined at 4.2 Å and at 3.6 Å resolution, using cryo-electron microscopy [(Wu et al., 2015, 2016) (Fig. 36.9B)]. Transmembrane helices S5 and S6 (green and grey) insert in the membrane to form a central channel; the four pore helices (red) align in a ring-like pattern near the intracellular mouth of the channel and move like an iris diaphragm to constrict or widen the pore. During an action potential, outward movements of S4 helices and their interactions with S1—S3 drive the pore helices outward to open the pore. In the case of $Ca_V1.1$, the pore does not become fully conductive. Rather, these rearrangements are transmitted to RyR1, either directly or through additional coupling proteins, to open its pore. The four S4 helices likely move in a cooperative fashion, but the exact nature of their movement remains to be determined (Banks et al., 2021). The ensemble movement of charged S4 helices can be detected electrically as a nonionic, capacitive current termed "voltage dependent charge movement" (Schneider & Chandler, 1973).

The cytosolic **$\beta 1_a$ subunit** (Fig. 36.9C**37**) belongs to the family of membrane-associated proteins that contain conserved Src homology 3 (SH3) and guanylate kinase (GK) domains (Chen et al., 2004). $\beta 1_a$ is essential for ECC (Perni et al., 2017). $\beta 1_a$ is capable of multiple interactions with $Ca_V1.1$, STAC3 and RyR1 and is the only mammalian β isoform capable of mediating skeletal muscle-type ECC. Its SH3 and a proximal C-terminal domain promote Ca_V targeting to the triad and assembly of tetrads, and work cooperatively to support voltage-sensing by $Ca_V1.1$ (Dayal et al., 2022). The GK domain is not enzymatically active, but a hydrophobic groove in this region binds with high affinity to the alpha interacting domain (AID) domain of $Ca_V1.1$ (Van Petegem et al., 2004). Short peptides in this region can activate isolated RyR1s and can immunoprecipitate the entire RyR1. $\beta 1_a$ is unique among mammalian β subunits in that it is altered significantly by the interaction with AID. Shifts in voltage sensing domains of $Ca_V1.1$ and the S6 helix of domain I are transmitted to the I-II helix in association with a displacement of $\beta 1_a$ (Wu et al., 2017), and these shifts are reversible. For these reasons, it is proposed that the $\beta 1_a$ subunit may transmit the physical trigger to RyR1. However, unambiguous identification of a corresponding binding site on RyR1 remains to be established.

The hydrophobic **γ1 subunit** stabilizes $Ca_V1.1$ in the membrane and associates with a transmembrane domain of $\alpha 1_S$. Interactions of γ1 with $\alpha 1_S$ (Melzer, 2022; Wu et al., 2017) exert inhibitory actions on the channel pore (Fig. 36.8A). These interactions enhance inactivation and reduce the maximal Ca^{2+} conductance, and may underlie the extremely small and slow Ca^{2+} flux characteristic of skeletal muscle $Ca_V1.1$. The extracellular **α2-δ1 subunit** interacts with extended extracellular loops of $Ca_V1.1$. It participates in membrane targeting and anchoring and may play a regulatory role in ECC. Neither γ1 nor α2-δ1 is essential for ECC.

6.2 RyR1

The skeletal muscle ryanodine receptor (Fig. 36.10) belongs to a superfamily of intracellular Ca^{2+} release channels, referred to as ryanodine receptors due to their ability to bind the toxic plant alkaloid ryanodine. The toxicity of ryanodine is due to its ability to inhibit ryanodine receptors. The ryanodine receptors (RyRs) mediate Ca^{2+} release from the SR/ER in a wide range of cellular processes. Structurally and functionally, RyRs are related to inositol trisphosphate receptors (IP3R), the other family of ion channels that release Ca^{2+} from intracellular stores (Woll & Van Petegem, 2022).

Mammals express three isoforms of ryanodine receptors that share about 70% homology. RyR1 is the dominant isoform in skeletal muscle. RyR2 is expressed mainly in cardiac muscle and is also present in the diaphragm, smooth muscle, brain, and other tissues. RyR3 is expressed mainly in neurons and is present on peri-junctional regions of SR in skeletal muscles of some species (Franzini-Armstrong & Protasi, 1997). All RyRs are activated by Ca^{2+}. However, RyR1 is unique in that it has acquired the ability to be activated by a physical interaction with $Ca_V1.1$, in addition to activation by Ca^{2+} (Nakai et al., 1997). Activation of RyR1 by physical coupling to $Ca_V1.1$ is the dominant mechanism of activation in mammalian skeletal muscle.

RyRs are one of the largest known ion channel proteins, with a mass of ~2.3 MDa (Wagenknecht & Samsó, 2002). RyR1 is a homotetramer formed from four identical subunits that assemble in a mushroom-like shape with a “stalk”-like

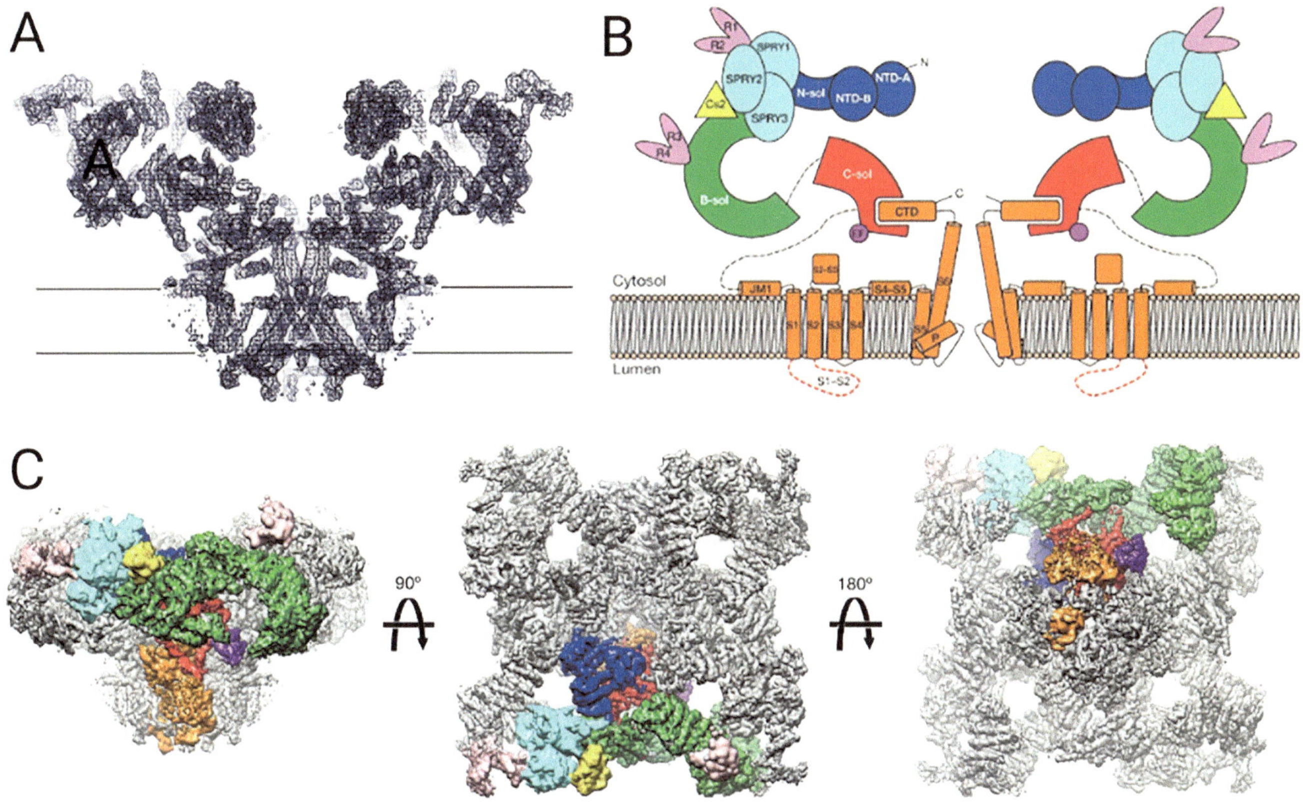

FIGURE 36.10 **Protein structure of ryanodine receptor isoform 1 (RyR1) at 4.8 Å resolution.** (A) Density map showing two of the four monomers of a mammalian RyR1. In cross-section, its shape resembles a mushroom, with a transmembrane stalk and a large cytosolic cap that extends far into the junctional gap. (B) Key domains of RyR1. Six α helices (S1−S6; *orange*) from each monomer form the ion channel. The cytosolic assembly is built on an extended α-solenoid scaffold. Its outer shell is composed of two N-terminal domains (NTD-A, NTD-B) that form a vestibule above the pore for exit of Ca^{2+} ions, an N-solenoid domain (N-sol), three SPRY domains composed of antiparallel β-sheets (SPRY1, SPRY2, and SPRY3), a Ca^{2+} binding domain, ryanodine binding domains (R1−R4), and a bridge solenoid (B-sol). The bridge solenoid connects to a core solenoid (C-sol; *red*) that contains EF-hand Ca^{2+} coordinating motifs. C-sol interacts with ion channel regions JM1, S1−S3, CTD, and S6. (C, left to right) RyR1 viewed from the side, above, and below the plane of the membrane. Domains of one monomer are colored following panel B. *From Zalk et al. (2015).*

region that inserts into the SR and forms the channel, and a large cytosolic "cap" that extends far into the junctional gap (Fig. 36.10A). The "cap" or cytosolic assembly comprises more than 90% of the protein mass. It contains the pathway for exit of Ca^{2+} ions from the SR and interaction sites with the T-tubule Ca_V channel.

RyR1 has three major functional domains (Melville et al., 2022). These include an N-terminal outer shell, a central core solenoid, and a C-terminal ion channel. The outer shell contains interaction sites with theT-tubule Ca_V channel and a pathway for exit of Ca^{2+} ions. The core solenoid transmits the signal from the outer shell to the RyR1 pore and controls pore opening. Each domain contains subdomains that are tightly linked, even though their primary sequences are far apart. This connectedness allows communication across long distances of RyR1, to support physical communication from Ca_V channel to the RyR1 pore.

The massive N-terminal cytosolic assembly is built on a scaffold of α-solenoid repeats. The outer shell contains three N-terminal domains (NTD-A, NTD-B, and N-terminal solenoid, N-sol), three SPRY domains, binding domains for Ca^{2+} and other activating ligands, ryanodine binding domains (R1- R4), and a large bridge solenoid (B-sol). The outer shell connects to a central core solenoid (C-sol), which contains two EF hand Ca^{2+} binding motifs. The core solenoid, in turn, connects with the CTD. The bridge solenoid is dynamic and can induce significant structural rearrangements in the central domain.

The channel domain belongs to the superfamily of voltage-dependent cation-selective ion channels (e.g., Fig. 36.9A) composed of four domains, each containing six transmembrane α-helices (S1−S6). Helices S5−S6 and an alpha helical pore helix form the channel, while helices S1−S4 regulate pore opening. The pore helices align in a horizontal direction near the intracellular surface and widen or constrict to open or close the pore. An extended peptide (P segment) with negative charges lines the opening to the SR lumen. The P segment is structurally analogous to, but less selective than, the selectivity filter of potassium channels. RyR1 has a large conductance for both divalent and monovalent cations, while maintaining a divalent to monovalent selectivity ratio of ~7. Two unique features of the channel domain of RyR1 are the presence of a cytosolic segment between S2–S3 that interacts closely with paired EF-hand motifs in the cytosolic assembly, and an extended S6 helix with a final C-terminal domain (CTD) that interacts with the central core.

The cytosolic assembly contains interaction sites for numerous ligands, both physiological and nonphysiological, which can modulate RyR1 (Meissner, 2017). RyR1 channels are activated by Ca, ATP, and caffeine, and inhibited by cytosolic Mg^{2+} and ryanodine. ATP complexed with Mg^{2+} modulates the activation of RyR1 by Ca^{2+}. Lumenal regions of RyR1 contain Ca^{2+} coordinating sites that negatively regulate pore opening depending on the Ca concentration within the SR. During ECC, the Ca^{2+} concentration within the SR exerts a negative feedback effect on Ca^{2+} release by promoting RyR1 channel inactivation. As expected for a physical coupling mechanism, interactions between $Ca_V1.1$ and RyR1 are bidirectional. RyR1 activity can influence Ca^{2+} currents through $Ca_V1.1$ through "retrograde" coupling actions (Dirksen, 2002). Many accessory proteins such as the FKBPs, calmodulin, S1001, triadin, junctin, and anchoring proteins for kinases and phosphatases, interact with RyR1 and influence its activity (Meissner, 2017).

6.3 STAC3

STAC3 (SH3 and cysteine-rich domain 3) is a small, soluble adaptor protein expressed primarily in skeletal muscle (Rufenach & Van Petegem, 2021). It consists of a protein kinase C-like domain (PKC C1, zinc finger) connected through a disordered region to tandem SH3 domains (Fig. 36.11) (Wong King Yuen et al., 2017). STAC3 localizes to the triad and is essential for ECC (Horstick et al., 2013; Nelson et al., 2013). STAC3 forms direct associations with both $Ca_V1.1$ and RyR1. The SH3 domains of STAC3 bind a conserved sequence within the critical II-III loop of $Ca_V1.1$ (Polster et al., 2018). STAC3 also associates with RyR1 in triads of genetically modified mice that lack $Ca_V1.1$ and β_{1a}. For these reasons, it is proposed that STAC3 may provide both structural and coupling functions that allosterically transmit voltage-driven movements of $Ca_V1.1$ to RyR1.

6.4 Junctophilins

Junctophilins (JPHs) comprise a family of structural proteins that connect the plasma membrane to intracellular organelles in a range of cell types (Lehnart & Wehrens, 2022). All junctophilin isoforms share a common structure (Fig. 36.12). A membrane occupation and recognition nexus (MORN) domain associates with the cytosolic leaflet of the T-tubule membrane, an alpha helical bridge extends to the SR, and a C-terminal transmembrane domain (tail anchor) inserts into the SR. Skeletal muscles express two isoforms, JPH1 and JPH2. Both are expressed at triads, where they tether the T-tubule membrane to the SR (Perni, 2022). Junctophilins are essential for triad formation and maintenance. In addition to their structural roles, JPHs bind and regulate proteins at the triad and act as a scaffold for assembly of the Ca^{2+} release complex. JPH1 and JPH2 form homo- and heterodimers that can interact with multiple triad proteins, including $Ca_V1.1$, $\beta1_a$, $\alpha2$-δ, and RyR1, caveolin-3, and the TRPC3 channel. JPHs are also required to support the SOCE pathway.

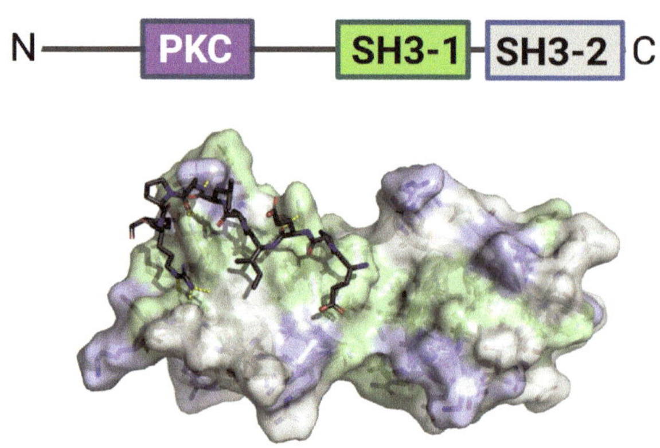

FIGURE 36.11 STAC3 structural domains and surface representation of human STAC3 aligned with the II-III loop of $Ca_V1.1$ (PDB: 6B27). The II-III loop associates mostly with conserved residues (*green*) of STAC3. Human STAC3 is colored according to conservation with human STAC1 and human STAC2. *Green*, completely conserved; *blue*, residue property conserved; *white*, not conserved. Nearly all direct interactions of the II-III loop with STAC2 are through conserved residues. *From Rufenach et al. (2020).*

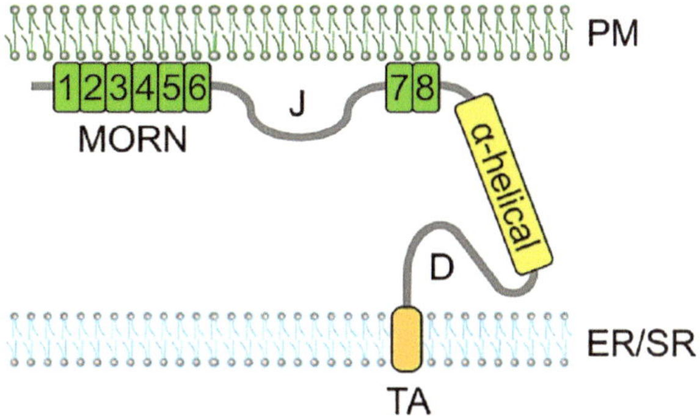

FIGURE 36.12 Predicted organization of junctophilin within a subcellar junction. Two sets of membrane occupation and recognition nexus domains (MORN1–6, and 7–8), connected by a flexible domain (J), associate with the cytosolic leaflet of the plasma membrane (PM). An α-helical region and divergent (D) region tether to the C-terminal tail that anchors into the ER/SR membrane using a transmembrane tail anchor (TA). *From Lehnart and Wehrens (2022).*

Altogether, these interactions stabilize the junctional complex. Either JPH1 or JPH2, coexpressed with $Ca_V1.1$, $\beta 1_a$, STAC3, and RYR1 in nonmuscle cells, can reconstitute voltage induced Ca^{2+} release (Perni & Beam, 2022). However, junctophilins do not appear to play a direct role in transmitting the activation signal from $Ca_V1.1$ to RyR1.

7. Mechanism of RyR1 activation

In the absence of $Ca_V1.1$, isolated RyR1s behave as ligand gated ion channels with cytosolic Ca^{2+} as the primary activator (Meissner, 2017). Cryo-EM studies of purified RyR1s in closed and open states have revealed the structural basis of activation by Ca^{2+} at near atomic resolution (des Georges et al., 2016). RyR1 activation by Ca^{2+} uses a conserved **Ca^{2+}-activation module** that includes Ca^{2+} coordinating residues in the outer shell, C-sol and EF hand motifs in the central region, and interaction domains with the transmembrane ion channel. Ca^{2+} binding to sites in the outer shell rearranges the Ca^{2+} activation module, which allosterically transmits rigid changes through the central core to the transmembrane ion channel. These Ca^{2+} coordinating residues in the outer shell and changes leading to pore opening are conserved among the RyR and IP3 families. Moreover, the binding sites for two other RyR activators, ATP, and caffeine, are located near the Ca^{2+} coordinating residues. The finding that different activating ligands induce the same allosteric transitions leading to pore opening suggests that the Ca^{2+} activation module represents a conserved network for RyR activation. This network may also be a final pathway for pore opening induced by the physical trigger from $Ca_V1.1$.

In contrast to RyR1 activation by Ca^{2+} and other ligands, RyR1 activation during ECC is initiated remotely by voltage-driven movements of $Ca_V1.1$. The physical coupling of RyR1 activation to voltage-sensing movements of $Ca_V1.1$ ensures that skeletal muscle activation remains under strict control of the action potential depolarization. A central question is: *How is the signal from Cav1.1 communicated to RyR1?* The key players and broad outlines of this communication are well established (Fig. 36.13). Because direct, high affinity interactions between $Ca_V1.1$ and RyR1 have not been found, it is thought that voltage-sensor movements of $Ca_V1.1$ are physically transmitted to RyR1 through a coupling protein(s), possibly β1a and/or STAC3, interacting with the II-III loop of $Ca_V1.1$. Multiple interactions between $Ca_V1.1$ and β1a, and STAC3 have been found. However, unambiguous evidence for direct interactions between RyR1 and β1a or STAC3 is lacking. This leaves open the possibility that other coupling proteins remain to be found. Alternatively, it is possible that weak interactions between multiple regions of RyR1 and $Ca_V1.1$, β1a and/or STAC3 may combine to control RyR1 pore opening. This possibility is suggested by the 4:1 ratio of Ca_V channels to RyR1. Weak interactions between four $Ca_V1.1$ channels and associated proteins, acting at the outer corners of one RyR1, could initiate torque-like movements that transmit through the RyR1 core to open the RyR1 pore. Complete details of the conformational movements transmitted from $Ca_V1.1$ to RyR1 will await structural information of each interaction at atomic level resolution.

A related question is: *How are RyR1 channels that are not directly associated with a tetrad of Ca_V channels opened?* The alternate arrangement of $Ca_V1.1$ tetrads opposite every *other* RyR1 (Fig. 36.7) suggests a dual mode of RyR1 activation. Those RyR1s associated with a tetrad are activated remotely by the physical coupling mechanism, while RyR1 channels not associated with a tetrad may be activated by Ca^{2+} released from neighboring RyR1s using the conserved

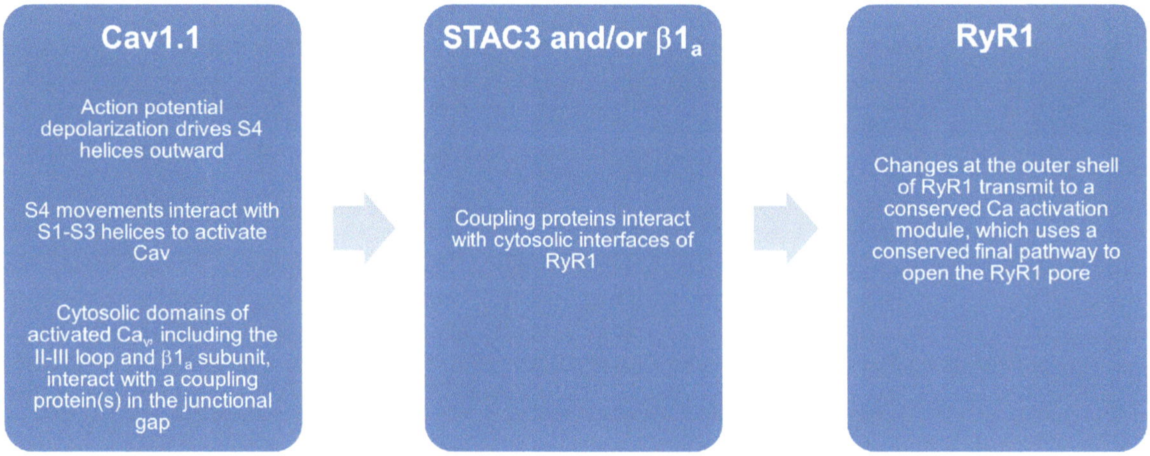

FIGURE 36.13 Depolarization dependent activation of RyR1 by $Ca_V1.1$ in skeletal muscle ECC.

Ca^{2+} activation module. An alternate hypothesis is that cooperative interactions between physically touching RyRs may open all RyR1s simultaneously – i.e., clusters of RyR1 channels, both coupled and uncoupled, may operate as a functional unit to elicit a synchronous global change in cytosolic Ca^{2+}.

8. Disorders of ECC

Hereditary or acquired defects in $Ca_V1.1$ (*CACNA1S* gene), RyR1, or their associated proteins underlie several muscle disorders of humans [(reviewed in (Rossi et al., 2022)].

Malignant hyperthermia (MH) is an inherited disease that causes uncontrolled release of Ca^{2+} from the SR when the affected person receives volatile halogenated anesthetics. The resulting muscle contractions produce an uncontrolled increase in body temperature that is fatal unless treated immediately. MH-linked mutations are distributed throughout the RyR1 coding sequence and cause RyR1 to open more readily in response to depolarization (Lawal et al., 2020). In addition to RyR1 mutations, a small percentage of MH cases have been linked to mutations in the coding sequence of the $Ca_V1.1$ gene (*CACNA1S*) (Flucher, 2020). More rare mutations in $Ca_V1.1$ functionally mimic disease-causing mutations in RyR1 itself. In one form of MH, replacement of a highly conserved arginine residue in the III-IV linker of $Ca_V1.1$ replaces a positively charged amino acid with histidine, and shifts the voltage-dependence of $Ca_V1.1$ activation to lower voltages. The increased sensitivity of $Ca_V1.1$ to activation, in turn, renders RyR1 more sensitive to opening by depolarization or caffeine.

Central core disease (CCD) is a congenital myopathy characterized by regions of muscle fibers devoid of mitochondria and oxidative enzymes that appear as "cores" in histological specimens. Over 60 mutations in RyR1 are linked to CCD and increase RyR1 channel activity, resulting in a leaky SR and Ca^{2+} dysregulation.

Hypokalemic Periodic Paralysis (HypoPP) is an autosomal dominant disease characterized by episodes of generalized muscle weakness accompanied by low serum potassium levels (<3.5 mM). HypoPP is often associated with progressive muscle weakness and degeneration. Episodes can be triggered by rest after exercise, hypokalemia following the intake of carbohydrates or insulin, and exposure to cold. Approximately 60% of HypoPP cases are caused by mutations in $Ca_V1.1$ that neutralize gating charges of the voltage sensing S4 helices. These mutations create an anomalous ion conducting pathway through the voltage-sensing domains of $Ca_V1.1$ (unrelated to the channel pore domain). Depolarizing ion currents through these anomalous conducting pathways (termed "omega currents") inactivate voltage-dependent Na^+ channels and cause temporary weakness or paralysis.

A mutation in SH3-1 of human STAC3 disrupts $Ca_V1.1$ and ECC and is the direct cause of the severe muscle disease, **STAC3 Disorder** (also known as Native American Myopathy (Horstick et al., 2013; Polster et al., 2016). In STAC3 Disorder, a reduction of functioning STAC3 protein prevents Ca^{2+} release by RyR. Affected persons suffer muscle weakness and poor muscle tone from birth.

The JPH1 gene is a modifier gene of the hereditary motor and sensory neuropathy, **Charcot-Marie-Tooth disease** (Pla-Martín et al., 2015).

References

Al-Qusairi, L., & Laporte, J. (2011). T-tubule biogenesis and triad formation in skeletal muscle and implication in human diseases. *Skelet Muscle, 1*, 26. https://doi.org/10.1186/2044-5040-1-26

Banks, Q., Bibollet, H., Contreras, M., Bennett, D. F., Bannister, R. A., Schneider, M. F., & Hernández-Ochoa, E. O. (2021). Voltage sensor movements of Ca(V)1.1 during an action potential in skeletal muscle fibers. *Proceedings of the National Academy of Sciences of the U S A, 118*. https://doi.org/10.1073/pnas.2026116118

Bannister, R. A., & Beam, K. G. (2013). Ca(V)1.1: The atypical prototypical voltage-gated Ca^{2+} channel. *Biochimica et Biophysica Acta, 1828*, 1587–1597. https://doi.org/10.1016/j.bbamem.2012.09.007

Barrientos, G., Sánchez-Aguilera, P., Jaimovich, E., Hidalgo, C., & Llanos, P. (2017). Membrane cholesterol in skeletal muscle: A novel player in excitation-contraction coupling and insulin resistance. *Journal of Diabetes Research, 2017*, Article 3941898. https://doi.org/10.1155/2017/3941898

Bolaños, P., & Calderón, J. C. (2022). Excitation-contraction coupling in mammalian skeletal muscle: Blending old and last-decade research. *Frontiers in Physiology, 13*, Article 989796. https://doi.org/10.3389/fphys.2022.989796

Catterall, W. A. (2011). Voltage-gated calcium channels. *Cold Spring Harbor Perspectives in Biology, 3*, Article a003947. https://doi.org/10.1101/cshperspect.a003947

Chen, Y. H., Li, M. H., Zhang, Y., He, L. L., Yamada, Y., Fitzmaurice, A., Shen, Y., Zhang, H., Tong, L., & Yang, J. (2004). Structural basis of the alpha1-beta subunit interaction of voltage-gated Ca^{2+} channels. *Nature, 429*, 675–680. https://doi.org/10.1038/nature02641

Dayal, A., Perni, S., Franzini-Armstrong, C., Beam, K. G., & Grabner, M. (2022). The distal C terminus of the dihydropyridine receptor β(1a) subunit is essential for tetrad formation in skeletal muscle. *Proceedings of the National Academy of Sciences of the U S A, 119*, Article e2201136119. https://doi.org/10.1073/pnas.2201136119

des Georges, A., Clarke, O. B., Zalk, R., Yuan, Q., Condon, K. J., Grassucci, R. A., Hendrickson, W. A., Marks, A. R., & Frank, J. (2016). Structural basis for gating and activation of RyR1. *Cell, 167*. https://doi.org/10.1016/j.cell.2016.08.075

Dirksen, R. T. (2002). Bi-directional coupling between dihydropyridine receptors and ryanodine receptors. *Frontiers in Bioscience, 7*, d659–d670. https://doi.org/10.2741/a802

Dirksen, R. T., Eisner, D. A., Ríos, E., & Sipido, K. R. (2022). Excitation-contraction coupling in cardiac, skeletal, and smooth muscle. *The Journal of General Physiology, 154*. https://doi.org/10.1085/jgp.202213244

Dulhunty, A. F. (1984). Heterogeneity of T-tubule geometry in vertebrate skeletal muscle fibres. *Journal of Muscle Research & Cell Motility, 5*, 333–347. https://doi.org/10.1007/bf00713111

Dulhunty, A. F., Beard, N. A., & Casarotto, M. G. (2018). Recent advances in understanding the ryanodine receptor calcium release channels and their role in calcium signalling. *F1000Res, 7*. https://doi.org/10.12688/f1000research.16434.1

El Ghaleb, Y., Ortner, N. J., Posch, W., Fernández-Quintero, M. L., Tuinte, W. E., Monteleone, S., Draheim, H. J., Liedl, K. R., Wilflingseder, D., Striessnig, J., et al. (2022). Calcium current modulation by the γ1 subunit depends on alternative splicing of CaV1.1. *The Journal of General Physiology, 154*. https://doi.org/10.1085/jgp.202113028

Felder, E., Protasi, F., Hirsch, R., Franzini-Armstrong, C., & Allen, P. D. (2002). Morphology and molecular composition of sarcoplasmic reticulum surface junctions in the absence of DHPR and RyR in mouse skeletal muscle. *Biophysical Journal, 82*, 3144–3149. https://doi.org/10.1016/s0006-3495(02)75656-7

Flucher, B. E. (2020). Skeletal muscle Ca(V)1.1 channelopathies. *Pflügers Archiv, 472*, 739–754. https://doi.org/10.1007/s00424-020-02368-3

Franzini-Armstrong, C. (2018). The relationship between form and function throughout the history of excitation-contraction coupling. *The Journal of General Physiology, 150*, 189–210. https://doi.org/10.1085/jgp.201711889

Franzini-Armstrong, C., & Jorgensen, A. O. (1994). Structure and development of E-C coupling units in skeletal muscle. *Annual Review of Physiology, 56*, 509–534. https://doi.org/10.1146/annurev.ph.56.030194.002453

Franzini-Armstrong, C., & Peachey, L. D. (1981). Striated muscle-contractile and control mechanisms. *The Journal of Cell Biology, 91*, 166s–186s. https://doi.org/10.1083/jcb.91.3.166s

Franzini-Armstrong, C., & Protasi, F. (1997). Ryanodine receptors of striated muscles: A complex channel capable of multiple interactions. *Physiological Reviews, 77*, 699–729. https://doi.org/10.1152/physrev.1997.77.3.699

Hernández-Ochoa, E. O., & Schneider, M. F. (2018). Voltage sensing mechanism in skeletal muscle excitation-contraction coupling: Coming of age or midlife crisis? *Skelet Muscle, 8*, 22. https://doi.org/10.1186/s13395-018-0167-9

Horstick, E. J., Linsley, J. W., Dowling, J. J., Hauser, M. A., McDonald, K. K., Ashley-Koch, A., Saint-Amant, L., Satish, A., Cui, W. W., Zhou, W., et al. (2013). Stac3 is a component of the excitation-contraction coupling machinery and mutated in Native American myopathy. *Nature Communications, 4*, 1952. https://doi.org/10.1038/ncomms2952

Lawal, T. A., Todd, J. J., Witherspoon, J. W., Bönnemann, C. G., Dowling, J. J., Hamilton, S. L., Meilleur, K. G., & Dirksen, R. T. (2020). Ryanodine receptor 1-related disorders: An historical perspective and proposal for a unified nomenclature. *Skelet Muscle, 10*, 32. https://doi.org/10.1186/s13395-020-00243-4

Lehnart, S. E., & Wehrens, X. H. T. (2022). The role of junctophilin proteins in cellular function. *Physiological Reviews, 102*, 1211–1261. https://doi.org/10.1152/physrev.00024.2021

Mackrill, J. J., & Shiels, H. A. (2020). Evolution of excitation-contraction coupling. *Advances in Experimental Medicine & Biology, 1131*, 281–320. https://doi.org/10.1007/978-3-030-12457-1_12

Meissner, G. (2017). The structural basis of ryanodine receptor ion channel function. *The Journal of General Physiology, 149*, 1065–1089. https://doi.org/10.1085/jgp.201711878

Melville, Z., Kim, K., Clarke, O. B., & Marks, A. R. (2022). High-resolution structure of the membrane-embedded skeletal muscle ryanodine receptor. *Structure, 30*. https://doi.org/10.1016/j.str.2021.08.001

Melzer, W. (2022). From α1s splicing to γ1 function: A new twist in subunit modulation of the skeletal muscle L-type Ca^{2+} channel. *The Journal of General Physiology, 154*. https://doi.org/10.1085/jgp.202213182

Nakai, J., Ogura, T., Protasi, F., Franzini-Armstrong, C., Allen, P. D., & Beam, K. G. (1997). Functional nonequality of the cardiac and skeletal ryanodine receptors. *Proceedings of the National Academy of Sciences of the U S A, 94*, 1019−1022. https://doi.org/10.1073/pnas.94.3.1019

Nelson, B. R., Wu, F., Liu, Y., Anderson, D. M., McAnally, J., Lin, W., Cannon, S. C., Bassel-Duby, R., & Olson, E. N. (2013). Skeletal muscle-specific T-tubule protein STAC3 mediates voltage-induced Ca^{2+} release and contractility. *Proceedings of the National Academy of Sciences of the U S A, 110*, 11881−11886. https://doi.org/10.1073/pnas.1310571110

Norris, N. C., Joseph, S., Aditya, S., Karunasekara, Y., Board, P. G., Dulhunty, A. F., Oakley, A. J., & Casarotto, M. G. (2017). Structural and biophysical analyses of the skeletal dihydropyridine receptor β subunit β(1a) reveal critical roles of domain interactions for stability. *Journal of Biological Chemistry, 292*, 8401−8411. https://doi.org/10.1074/jbc.M116.763896

Peachey, L. D. (1965). The sarcoplasmic reticulum and transverse tubules of the frog's sartorius. *The Journal of Cell Biology, 25*(Suppl. l), 209−231. https://doi.org/10.1083/jcb.25.3.209

Peachey, L. D., & Eisenberg, B. R. (1978). Helicoids in the T system and striations of frog skeletal muscle fibers seen by high voltage electron microscopy. *Biophysical Journal, 22*, 145−154. https://doi.org/10.1016/s0006-3495(78)85480-0

Perni, S. (2022). The builders of the junction: Roles of Junctophilin1 and Junctophilin2 in the assembly of the sarcoplasmic reticulum-plasma membrane junctions in striated muscle. *Biomolecules, 12*. https://doi.org/10.3390/biom12010109

Perni, S., & Beam, K. (2022). Junctophilins 1, 2, and 3 all support voltage-induced Ca^{2+} release despite considerable divergence. *The Journal of General Physiology, 154*. https://doi.org/10.1085/jgp.202113024

Perni, S., Lavorato, M., & Beam, K. G. (2017). De novo reconstitution reveals the proteins required for skeletal muscle voltage-induced $Ca(^{2+})$ release. *Proceedings of the National Academy of Sciences of the U S A, 114*, 13822−13827. https://doi.org/10.1073/pnas.1716461115

Pla-Martín, D., Calpena, E., Lupo, V., Márquez, C., Rivas, E., Sivera, R., Sevilla, T., Palau, F., & Espinós, C. (2015). Junctophilin-1 is a modifier gene of GDAP1-related Charcot-Marie-Tooth disease. *Human Molecular Genetics, 24*, 213−229. https://doi.org/10.1093/hmg/ddu440

Polster, A., Nelson, B. R., Olson, E. N., & Beam, K. G. (2016). Stac3 has a direct role in skeletal muscle-type excitation-contraction coupling that is disrupted by a myopathy-causing mutation. *Proceedings of the National Academy of Sciences of the U S A, 113*, 10986−10991. https://doi.org/10.1073/pnas.1612441113

Polster, A., Nelson, B. R., Papadopoulos, S., Olson, E. N., & Beam, K. G. (2018). Stac proteins associate with the critical domain for excitation-contraction coupling in the II-III loop of Ca(V)1.1. *The Journal of General Physiology, 150*, 613−624. https://doi.org/10.1085/jgp.201711917

Polster, A., Ohrtman, J. D., Beam, K. G., & Papadopoulos, S. (2012). Fluorescence resonance energy transfer (FRET) indicates that association with the type I ryanodine receptor (RyR1) causes reorientation of multiple cytoplasmic domains of the dihydropyridine receptor (DHPR) α(1S) subunit. *Journal of Biological Chemistry, 287*, 41560−41568. https://doi.org/10.1074/jbc.M112.404194

Ríos, E. (2018). Calcium-induced release of calcium in muscle: 50 years of work and the emerging consensus. *The Journal of General Physiology, 150*, 521−537. https://doi.org/10.1085/jgp.201711959

Rossi, D., Catallo, M. R., Pierantozzi, E., & Sorrentino, V. (2022). Mutations in proteins involved in E-C coupling and SOCE and congenital myopathies. *The Journal of General Physiology, 154*. https://doi.org/10.1085/jgp.202213115

Rufenach, B., Christy, D., Flucher, B. E., Bui, J. M., Gsponer, J., Campiglio, M., & Van Petegem, F. (2020). Multiple sequence variants in STAC3 affect interactions with Ca(V)1.1 and excitation-contraction coupling. *Structure, 28*. https://doi.org/10.1016/j.str.2020.05.005

Rufenach, B., & Van Petegem, F. (2021). Structure and function of STAC proteins: Calcium channel modulators and critical components of muscle excitation-contraction coupling. *Journal of Biological Chemistry, 297*, Article 100874. https://doi.org/10.1016/j.jbc.2021.100874

Samsó, M. (2017). A guide to the 3D structure of the ryanodine receptor type 1 by cryoEM. *Protein Science, 26*, 52−68. https://doi.org/10.1002/pro.3052

Schneider, M. F., & Chandler, W. K. (1973). Voltage dependent charge movement of skeletal muscle: A possible step in excitation-contraction coupling. *Nature, 242*, 244−246. https://doi.org/10.1038/242244a0

Shishmarev, D., Rowland, E., Aditya, S., Sundararaj, S., Oakley, A. J., Dulhunty, A. F., & Casarotto, M. G. (2022). Molecular interactions of STAC proteins with skeletal muscle dihydropyridine receptor and excitation-contraction coupling. *Protein Science, 31*, Article e4311. https://doi.org/10.1002/pro.4311

Sun, X. H., Protasi, F., Takahashi, M., Takeshima, H., Ferguson, D. G., & Franzini-Armstrong, C. (1995). Molecular architecture of membranes involved in excitation-contraction coupling of cardiac muscle. *The Journal of Cell Biology, 129*, 659−671. https://doi.org/10.1083/jcb.129.3.659

Tanabe, T., Beam, K. G., Adams, B. A., Niidome, T., & Numa, S. (1990). Regions of the skeletal muscle dihydropyridine receptor critical for excitation-contraction coupling. *Nature, 346*, 567−569. https://doi.org/10.1038/346567a0

Van Petegem, F., Clark, K. A., Chatelain, F. C., & Minor, D. L., Jr. (2004). Structure of a complex between a voltage-gated calcium channel beta-subunit and an alpha-subunit domain. *Nature, 429*, 671−675. https://doi.org/10.1038/nature02588

Wagenknecht, T., & Samsó, M. (2002). Three-dimensional reconstruction of ryanodine receptors. *Frontiers in Bioscience, 7*, d1464−d1474. https://doi.org/10.2741/a853

Woll, K. A., & Van Petegem, F. (2022). Calcium-release channels: Structure and function of IP(3) receptors and ryanodine receptors. *Physiological Reviews, 102*, 209−268. https://doi.org/10.1152/physrev.00033.2020

Wong King Yuen, S. M., Campiglio, M., Tung, C. C., Flucher, B. E., & Van Petegem, F. (2017). Structural insights into binding of STAC proteins to voltage-gated calcium channels. *Proceedings of the National Academy of Sciences of the U S A, 114*, E9520–e9528. https://doi.org/10.1073/pnas.1708852114

Wu, J., Yan, N., & Yan, Z. (2017). Structure-function relationship of the voltage-gated calcium channel Ca(v)1.1 complex. *Advances in Experimental Medicine & Biology, 981*, 23–39. https://doi.org/10.1007/978-3-319-55858-5_2

Wu, J., Yan, Z., Li, Z., Qian, X., Lu, S., Dong, M., Zhou, Q., & Yan, N. (2016). Structure of the voltage-gated calcium channel Ca(v)1.1 at 3.6 Å resolution. *Nature, 537*(7619), 191–196. https://doi.org/10.1038/nature19321

Wu, J., Yan, Z., Li, Z., Yan, C., Lu, S., Dong, M., & Yan, N. (2015). Structure of the voltage-gated calcium channel Cav1.1 complex. *Science, 350*, Article aad2395. https://doi.org/10.1126/science.aad2395

Zalk, R., Clarke, O. B., des Georges, A., Grassucci, R. A., Reiken, S., Mancia, F., Hendrickson, W. A., Frank, J., & Marks, A. R. (2015). Structure of a mammalian ryanodine receptor. *Nature, 517*, 44–49. https://doi.org/10.1038/nature13950

Index

A

D

E

F

G

H

I

J

K

L

Q

R

S